bucco- *cheek* buccolabial, pertaining to the cheek and lip

calor- *heat* calories, a measure of energy

capill- *hair* blood and lymph capillaries

caput- *head* decapitate, remove the head

carcin- *cancer* carcinogen, a cancer-causing agent

cardi, cardio- *heart* cardiotoxic, harmful to the heart

carneo- *flesh* trabeculae carneae, ridges of muscle in the ventricles of the heart

carot- *1) carrot, 2) stupor* 1) carotene, an orange pigment; 2) carotid arteries in the neck, blockage causes fainting

cata- *down* catabolism, chemical breakdown

caud- *tail* caudal (directional term)

cec- *blind* cecum of large intestine, a blind-ended pouch

cele- *abdominal* celiac artery, in the abdomen

cephal- *head* cephalometer, an instrument for measuring the head

cerebro- *brain, especially the cerebrum* cerebrospinal, pertaining to the brain and spinal cord

cervic-, cervix *neck* cervix of the uterus

chiasm- *crossing* optic chiasma, where optic nerves cross

chole- *bile* cholesterol; cholecystokinin, a bile-secreting hormone

chondr- *cartilage* chondrogenic, giving rise to cartilage

chrom- *colored* chromosome, so named because they stain darkly

cili- *small hair* ciliated epithelium

circum- *around* circumnuclear, surrounding the nucleus

clavic- *key* clavicle, a "skeleton key"

co-, con- *together* concentric, common center, together in the center

coccy- *cuckoo* coccyx, which is beak-shaped

cochlea *snail shell* the cochlea of the inner ear, which is coiled like a snail shell

coel- *hollow* coelom, the ventral body cavity

commis- *united* gray commissure of the spinal cord connects the two columns of gray matter

concha *shell* nasal conchae, coiled shelves of bone in the nasal cavity

contra- *against* contraceptive, agent preventing conception

corn-, cornu- *horn* stratum corneum, outer layer of the skin composed of (horny) cells

corona *crown* coronal suture of the skull

corp- *body* corpse, corpus luteum, hormone-secreting body in the ovary

cort- *bark* cortex, the outer layer of the brain, kidney, adrenal glands, and lymph nodes

cost- *rib* intercostal, between the ribs

crani- *skull* craniotomy, a skull operation

crypt- *hidden* cryptomenorrhea, a condition in which menstrual symptoms are experienced but no external loss of blood occurs

cusp- *pointed* bicuspid, tricuspid valves of the heart

cutic- *skin* cuticle of the nail

cyan- *blue* cyanosis, blue color of the skin due to lack of oxygen

cyst- *sac, bladder* cystitis, inflammation of the urinary bladder

cyt- *cell* cytology, the study of cells

de- *undoing, reversal, loss, removal* deactivation, becoming inactive

decid- *falling off* deciduous (milk) teeth

delta *triangular* deltoid muscle, roughly triangular in shape

den-, dent- *tooth* dentin of the tooth

dendr- *tree, branch* dendrites, telodendria, both branches of a neuron

derm- *skin* dermis, deep layer of the skin

desm- *bond* desmosome, which binds adjacent epithelial cells

di- *twice, double* dimorphism, having two forms

dia- *through, between* diaphragm, the wall through or between two areas

dialys- *separate, break apart* kidney dialysis, in which waste products are removed from the blood

diastol- *stand apart* cardiac diastole, between successive contractions of the heart

diure- *urinate* diuretic, a drug that increases urine output

dors- *the back* dorsal; dorsum; dorsiflexion

duc-, duct *lead, draw* ductus deferens which carries sperm from the epididymis into the urethra during ejaculation

dura *hard* dura mater, tough outer meninx

dys- *difficult, faulty, painful* dyspepsia, disturbed digestion

ec-, ex-, ecto- *out, outside, away from* excrete, to remove materials from the body

ectop- *displaced* ectopic pregnancy; ectopic focus for initiation of heart contraction

edem- *swelling* edema, accumulation of water in body tissues

ef- *away* efferent nerve fibers, which carry impulses away from the central nervous system

ejac- *to shoot forth* ejaculation of semen

embol- *wedge* embolus, an obstructive object traveling in the bloodstream

en-, em- *in, inside* encysted, enclosed in a cyst or capsule

enceph- *brain* encephalitis, inflammation of the brain

endo- *within, inner* endocytosis, taking particles into a cell

entero- *intestine* enterologist, one who specializes in the study of intestinal disorders

epi- *over, above* epidermis, outer layer of skin

erythr- *red* erythema, redness of the skin; erythrocyte, red blood cell

eso- *within* esophagus

eu- *well* euesthesia a normal state of the senses

excret *separate* excretory system

exo- *outside, outer layer* exophthalmos, an abnormal protrusion of the eye from the orbit

extra- *outside, beyond* extracellular, outside the body cells of an organism

extrins- *from the outside* extrinsic regulation of the heart

fasci-, fascia- *bundle, band* superficial and deep fascia

fenestr- *window* fenestrae of the inner ear; fenestrated capillaries

ferr- *iron* transferrin, ferritin, both iron-storage proteins

flagell- *whip* flagellum, the tail of a sperm cell

flat- *blow, blown* flatulence

folli- *bag, bellows* hair follicle

fontan- *fountain* fontanels of the fetal skull

foram- *opening* foramen magnum of the skull

foss- *ditch* fossa ovalis of the heart; mandibular fossa of the skull

gam-, gamet- *married, spouse* gametes, the sex cells

gangli- *swelling, or knot* dorsal root ganglia of the spinal nerves

gastr- *stomach* gastrin, a hormone that influences gastric acid secretion

gene *beginning, origin* genetics

germin- *grow* germinal epithelium of the gonads

gero-, geront- *old man* gerontology, the study of aging

gest- *carried* gestation, the period from conception to birth

glauc- *gray* glaucoma, which causes gradual blindness

glom- *ball* glomeruli, clusters of capillaries in the kidneys

glosso- *tongue* glossopathy, any disease of the tongue

gluco-, glyco- *sugar* gluconeogenesis, the production of glucose from non-carbohydrate molecules

glute- *buttock* gluteus maximus, largest muscle of the buttock

gnost- *knowing* the gnostic sense, a sense of awareness of self

gompho- *nail* gomphosis, the term applied to the joint between tooth and jaw

gon-, gono- *seed, offspring* gonads, the sex organs

gust- *taste* gustatory sense, the sense of taste

hapt- *fasten, grasp* hapten, a partial antigen

hema-, hemato-, hemo- *blood* hematocyst, a cyst containing blood

hemi- *half* hemiglossal, pertaining to one-half of the tongue

hepat- *liver* hepatitis, inflammation of the liver

hetero- *different or other* heterosexuality, sexual desire for a person of the opposite sex

hiat- *gap* the hiatus of the diaphragm, the opening through which the esophagus passes

hippo- *horse* hippocampus of the brain, shaped like a seahorse

hirsut- *hairy* hirsutism, excessive body hair

hist- *tissue* histology, the study of tissues

holo- *whole* holocrine glands, whose secretions are whole cells

hom-, homo- *same* homeoplasia, formation of tissue similar to normal tissue; homocentric, having the same center

hormon- *to excite* hormones

humor- *a fluid* humoral immunity, which involves antibodies circulating in the blood

hyal- *clear* hyaline cartilage, which has no visible fibers

hydr-, hydro- *water* dehydration, loss of body water

hyper- *excess* hypertension, excessive tension

hypno- *sleep* hypnosis, a sleeplike state

hypo- *below, deficient* hypodermic, beneath the skin; hypokalemia, deficiency of potassium

hyster-, hystero- *uterus or womb* hysterectomy, removal of the uterus; hysterodynia, pain in the womb

ile- *intestine* ileum, the last portion of the small intestine

im- *not* impermeable, not permitting passage, not permeable

inter- *between* intercellular, between the cells

intercal- *insert* intercalated discs, the end membranes between adjacent cardiac muscle cells

intra- *within, inside* intracellular, inside the cell

iso- *equal, same* isothermal, equal, or same, temperature

jugul- *throat* jugular veins, prominent vessels in the neck

juxta- *near, close to* juxtaglomerular apparatus, a cell cluster next to the glomeruli in the kidneys

karyo- *kernal nucleus* karyotype, the assemblage of the nuclear chromosomes

kera- *horn* keratin, the water-repellent protein of the skin

kilo- *thousand* kilocalories, equal to one thousand calories

kin-, kines- *move* kinetic energy, the energy of motion

labi-, labri- *lip* labial frenulum, the membrane which joins the lip to the gum

lact- *milk* lactose, milk sugar

lacun- *space, cavity, lake* lacunae, the spaces occupied by cells of cartilage and bone tissue

lamell- *small plate* concentric lamellae, rings of bone matrix in compact bone

lamina *layer, sheet* basal lamina, part of the epithelial basement membrane

lat- *wide* latissimus dorsi, a broad muscle of the back

laten- *hidden* latent period of a muscle twitch

later- *side* lateral (directional term)

leuko- *white* leukocyte, white blood cell

leva- *raise, elevate* levator labii superioris, muscle that elevates upper lip

lingua- *tongue* lingual tonsil, adjacent to the tongue

lip-, lipo- *fat, lipid* lipophage, a cell that has taken up fat in its cytoplasm

lith- *stone* cholelithiasis, gallstones

luci- *clear* stratum lucidum, clear layer of the epidermis

lumen *light* lumen, center of a hollow structure

lut- *yellow* corpus luteum, a yellow, hormone-secreting structure in the ovary

lymph *water* lymphatic circulation, return of clear fluid to the bloodstream

macro- *large* macromolecule, large molecule

macula *spot* macula lutea, yellow spot on the retina

magn- *large* foramen magnum, largest opening of the skull

mal- *bad, abnormal* malfunction, abnormal functioning of an organ

mamm- *breast* mammary gland, breast

mast- *breast* mastectomy, removal of a mammary gland

mater *mother* dura mater, pia mater, membranes that envelop the brain

meat- *passage* external auditory meatus, the ear canal

medi- *middle* medial (directional term)

medull- *marrow* medulla, the middle portion of the kidney, adrenal gland, and lymph node

mega- *large* megakaryocyte, large precursor cell of platelets

meio- *less* meiosis, nuclear division that halves the chromosome number

melan- *black* melanocytes, which secrete the black pigment melanin

men-, menstru- *month* menses, the cyclic menstrual flow

meningo- *membrane* meningitis, inflammation of the membranes of the brain

mer-, mero-, *a part* merocrine glands, the secretions of which do not include the cell

meso- *middle* mesoderm, middle germ layer

meta- *beyond, between, transition* metatarsus, the part of the foot between the tarsus and the phalanges

metro- *uterus* metroscope, instrument for examining the uterus

micro- *small* microscope, an instrument used to make small objects appear larger

mictur- *urinate* micturition, the act of voiding the bladder

mito- *thread, filament* mitochondria, small, filamentlike structures located in cells

mnem- *memory* amnesia

mono- *single* monospasm, spasm of a single limb

morpho- *form* morphology, the study of form and structure or organisms

multi- *many* multinuclear, having several nuclei

mural *wall* intramural ganglion, a nerve junction within an organ

muta- *change* mutation, change in the base sequence of DNA

myelo- *spinal cord, marrow* myeloblasts, cells of the bone marrow

myo- *muscle* myocardium, heart muscle

nano- *dwarf* nanometer, one billionth of a meter

narco- *numbness* narcotic, a drug producing stupor or numbed sensations

natri- *sodium* atrial natriuretic factor, a sodium-regulating hormone

necro- *death* necrosis, tissue death

neo- *new* neoplasm, an abnormal growth

nephro- *kidney* nephritis, inflammation of the kidney

neuro- *nerve* neurophysiology, the physiology of the nervous system

noci- *harmful* nociceptors, receptors for pain

nom- *name* innominate artery; innominate bone

noto- *back* notochord, the embryonic structure that precedes the vertebral column

nucle- *pit, kernel, little nut* nucleus

nutri- *feed, nourish* nutrition

ob- *before, against* obstruction, impeding or blocking up

oculo- *eye* monocular, pertaining to one eye

odonto- *teeth* orthodontist, one who specializes in proper positioning of the teeth in relation to each other

olfact- *smell* olfactory nerves

oligo- *few* oligodendrocytes, neuroglial cells with few branches

onco- *a mass* oncology, study of cancer

oo- *egg* ocyte, precursor of female gamete

ophthalmo- *eye* ophthalmology, the study of the eyes and related disease

orb- *circular* orbicularis oculi, muscle that encircles the eye

orchi- *testis* cryptorchidism, failure of the testes to descend into the scrotum

org- *living* organism

ortho- *straight, direct* orthopedic, correction of deformities of the musculoskeletal system

osm- *smell* anosmia, loss of sense of smell

osmo- *pushing* osmosis

osteo- *bone* osteodermia, bony formations in the skin

oto- *ear* otoscope, a device for examining the ear

ov-, ovi- *egg* ovum, oviduct

Additional resources included with your textbook purchase.

As the owner of this book, you can get the most from your financial investment and study time by using a variety of study aids that accompany the book at no additional charge.

Each new copy of the text includes the following items:

- An 18-month, prepaid subscription to **The A&P Place**
- **Study Partner CD-ROM version 2.0**⋆
- *Atlas of the Human Skeleton,* with photographs by Ralph T. Hutchings⋆

How to activate your prepaid subscription to The A&P Place:

1. Point your web browser to **www.anatomyandphysiology.com**.
2. Select Marieb, *Human Anatomy & Physiology,* **Fifth Edition**.
3. Click "Register Here".
4. Enter your pre-assigned access code exactly as it appears below in the activation ID and password fields:

Activation ID	APANST18135601
Password	TEMPLE

5. Select "Submit"
6. Complete the online registration form to create your own personal user ID and password.
7. Once your personal ID and password are confirmed via email, go back to **www.anatomyandphysiology.com** to enter the site with your new ID and password.

Your activation ID and password can be used only once to establish your subscription, which is not transferable.

If you have purchased a used copy of this book, this activation ID and password may not be valid. However, if your instructor is recommending or requiring use of **The A&P Place**, you can find more information on purchasing a subscription directly on-line at **www.anatomyandphysiology.com**. You also have the option of purchasing a subscription at your local college bookstore by placing an order for ISBN: 0-8053-4924-3.

The A&P Place contains several tools to help you prepare for exams, including 40 review questions per chapter, interactive clinical case studies, bone review anatomy quizzing, web research activities, on-line histology, a glossary with audio pronunciation guide, and A&P related news articles.

Turn the page to learn more…

⋆ These items are available only with **NEW** copies of *Human Anatomy & Physiology,* **Fifth Edition**, and are not sold separately.

How to use the Study Partner CD-ROM:

The **Study Partner CD-ROM** includes multiple-choice review questions and practice testing for each chapter, interactive case studies and art exercises, and a glossary with pronunciation guide. To help you locate relevant Study Partner review materials, each chapter summary in your text includes Study Partner references, which are signaled by an icon **SP** .

How to contact Customer Service and Technical Support:

To purchase additional study aids, including the **Student Study Guide**, **Interactive Physiology CD-ROM's**, **Student Video Series**, and coloring books, please contact your local textbook retailer, visit our website (**www.awl.com/bc**), or call Addison Wesley Longman Customer Service at 800-282-0693.

For technical support for **The A&P Place**, **Study Partner CD-ROMs**, or **Interactive Physiology CD-ROMs**, please visit our technical support website at:**www.awl.com/techsupport**. For help with CD-ROMs, you may also send an email to **media.support@pearsoned.com**. For help with **The A&P Place**, please send an e-mail to **online.support@pearsoned.com**. Please include a detailed description of your computer system and the technical problem.

How to contact us:

We invite your comments and suggestions via email at **question@awl.com**. As you complete your A&P course, please take a few minutes to fill out the postage-paid questionnaire at the end of the text.

Allied Health Student Scholarship

Each year, Benjamin Cummings offers a scholarship award to five students who are enrolled in allied health programs in the United States. For more information and application instructions, please visit our website at **www.awl.com/bc** and select "Student Resources."

In the meantime, we wish you great success in your A&P course!

HUMAN ANATOMY & PHYSIOLOGY

FIFTH EDITION

Elaine N. Marieb, R.N., Ph.D.
Holyoke Community College

Benjamin
Cummings

San Francisco Boston New York
Capetown Hong Kong London Madrid Mexico City
Montreal Munich Paris Singapore Sydney Tokyo Toronto

Publisher: Daryl Fox
Senior Project Editor: Kay Ueno
Senior Developmental Editor: Mark Wales
Managing Editor: Wendy Earl
Associate Project Editor: Tiffany Barnes
Associate Editor: Susan Teahan
Editorial Assistant: Richard Gallagher
Production Supervisor: Sharon Montooth
Art and Design Director: Bradley Burch
Art Manager: Michele Mangelli
Principal Artists/Studio for 5th. edition: Wendy Hiller Gee,
Imagineering Scientific and Technical Artworks, Inc.,
Kristin Mount, Tomo Narashima, Nadine Sokol

Photo Researcher: Stuart Kenter Associates
Art and Photo Coordinator: David Novak
Assistant Art Coordinator: Anne Hikido
Art Coordination Interns: Charlotte Ross and Olivia Ross
Chapter Opener and Cover Design: tani hasegawa
Copyeditor: Anita Wagner
Proofreader: Martha Ghent
Indexer: Karen Hollister
Manufacturing Coordinator: Stacey Weinberger
Senior Marketing Manager: Lauren Harp

Cover Photograph: Mia Hamm, © Annie Leibovitz, 1999

Text, photography, and illustration credits appear following the Appendices.

Library of Congress Cataloging-in Publication Data

Marieb, Elaine Nicpon
   Human anatomy & physiology / Elaine N. Marieb—5th ed.
     p. cm.
   Includes index.
   ISBN 0-8053-4950-2
     1. Human physiology. 2. Human anatomy. I. Title: Human anatomy and
   physiology. II. Title.

QP34.5 .M265 2000
612—dc21   00-039805

ISBN 0–8053–4989–8 (For sale copy with CD-ROM)

ISBN 0–8053–4992–8 (Professional copy with CD-ROM)

     5 6 7 8 9—VHP—04 03 02

ABOUT THE AUTHOR

By any measure, Elaine N. Marieb is an accomplished educator and author. Her academic career began at Springfield College, where she taught anatomy and physiology to physical education majors. She joined the faculty of the Biological Science Division of Holyoke Community College in 1969, with a Ph.D. in zoology from the University of Massachusetts at Amherst. Like other fine, inquisitive teachers, however, Dr. Marieb found there was more to learn—and more to understand—in order to meet the needs of her students, many of whom were pursuing degrees in nursing. To that end, Dr. Marieb, while teaching full time, continued her education in a different field, which culminated in a master of science degree with a clinical specialization in gerontology from the University of Massachusetts. As a result, she gained an understanding of the relationship between the scientific study of the human body and the clinical aspects of nursing practice. From this experience and stories from the field—former students in health careers—Dr. Marieb developed the unique perspective for which her widely used texts and laboratory manuals are known.

While actively engaged as an author, Dr. Marieb also serves as series consultant for the Benjamin Cummings—A.D.A.M.®, CD-ROM series Interactive Physiology, and she is an active member of the Human Anatomy and Physiology Society (HAPS). In 1994 Dr. Marieb received the Benefactor Award from the National Council for Resource Development, American Association of Community Colleges, which recognizes her ongoing sponsorship of student scholarships, faculty teaching awards, and other academic contributions to Holyoke Community College. In May 2000, the science building at Holyoke Community College was named in her honor. *Human Anatomy and Physiology*, Fifth Edition, is only the latest expression of Dr. Marieb's ongoing commitment to student needs in the study of A&P.

PREFACE TO THE INSTRUCTOR

As human anatomy and physiology teachers and educators, we all face the same challenge. We must provide information in a framework that fosters genuine understanding, devise new presentations to get students over conceptual hurdles, and help students to apply what they have learned to new situations. All the while, we hope to inspire in them a real love of the subject.

After many years of teaching human anatomy and physiology to nursing students, my curiosity about the clinical aspects of anatomy and physiology led me to the classroom—not as a teacher, but as a nursing student. When I sat on the other side of the desk among allied health students of all ages, it was easy to see how my explanations of anatomy and physiology could be improved. I became convinced that new approaches to many topics could excite and challenge students' natural curiosity, and I decided to write this book. I was very fortunate to have a publisher that shared my goal: to set new standards for pedagogical and visual effectiveness in an anatomy and physiology text.

SHAPING THE REVISION

As I began preparing the fifth edition of this text, I looked back to review how well my strategies—both in personal preparation and professional writing—have worked. As to my personal education, I find that I made a wise decision. My clinical nursing background has served my teaching and writing purposes well, as the left hand to work with my right hand, which is biological theory. Perhaps even more important, it has allowed me to view my presentations through my students' eyes and from the vantage points of their career interests. But now, even after 30 years of classroom experience and over 20 years as a textbook author, I realize that I'm not done yet. I'm still honing and refining my work. The fruit of my progress thus far is this edition.

So how did we begin this opus this time? As in the past, we used the feedback Benjamin Cummings had gathered from both instructors and students on the effectiveness of the book as a teaching and learning tool. This information guided us to specific areas that needed to be updated or streamlined in order to make the text an even better teaching tool. Overall, the positive nature of this feedback reassured me that my pyramid approach, building a broad base to support all that comes later, is still viable. In this edition, I continue to explain fundamental principles and unifying themes first, and then reinforce these explanations with comfortable analogies and familiar examples that focus on mechanisms and cause-and-effect relationships. Thus students can gain an integrated and continuously enhanced understanding of the workings of the human body.

Unifying Themes

The story of anatomy and physiology is not coherent or logical without unifying themes. The three integrating themes selected to organize, unify, and set the tone of the first edition of this text continue to be valid and are retained in this edition. These are:

- **Interrelationships of body organ systems:** The fact that nearly all regulatory mechanisms require interaction of several organ systems is continually emphasized. For example, Chapter 6, which describes the growth and maintenance of bone tissue, emphasizes the importance of muscle pull on bones to maintain bone strength; Chapter 26, which deals with the structure and function of the urinary system, discusses the vital importance of the kidneys not only in maintaining adequate blood volume to ensure normal blood circulation, but also in continually adjusting the chemical composition of blood so that all body cells remain healthy. The unique *Making Connections* feature is the culmination of this approach and should help students think of the body as a dynamic community of interdependent parts rather than as a number of isolated structural units.

- **Homeostasis:** The normal and most desirable condition of body functioning is homeostasis. Its loss or destruction always leads to some type of pathology—temporary or permanent. Pathological conditions are introduced and integrated with the text material as appropriate. However, clinical examples are used to clarify and illuminate normal functioning, not as an end in and of themselves. For example, Chapter 20, which deals with the structure and function of blood vessels, explains how the ability of healthy arteries to expand and recoil ensures continuous blood flow and proper circulation. Also discussed in this chapter, however, are the effects on homeostasis when arteries lose their elasticity: high blood pressure and all of its attendant problems. These homeostatic imbalances are indicated visually by a red seesaw symbol that has been tipped off balance. Whenever students see the unbalanced seesaw in text, the concept of disease as a loss of homeostasis is reinforced.

- **Complementarity of structure and function:** Students are encouraged to understand the structure of an organ, a tissue, or a cell as a prerequisite to comprehending its function. The fundamental concepts of physiology are carefully explained and related to structural characteristics that promote or allow the various functions to occur. For example, the lungs can act as a gas exchange site because the walls of its air sacs present an incredibly thin barrier between blood and air.

Pedagogical Features Retained

Pedagogical features that have proven their worth and are still with us include the following:

- Chapter outlines and student objectives preface each chapter and guide students to concepts that are most important to know

- Illustrated tables summarize complex information and serve as "one-stop shopping" study tools

- Key questions accompany selected figures and prod the student to interpret the concepts or illustrated processes, or to make predictions about "what's next." Answers can be found upside down on the same page.

- A Closer Look boxes on timely subjects such as medical technology, new discoveries in medical research, and important societal issues broaden the students' horizons

- Homeostatic icons alert the students to consequences when the body is not functioning optimally

- Two-page Connections features highlight relationships between body systems and provide an opportunity for students to apply what they've learned by reading a case study, diagnosing and evaluating a patient's problem, and suggesting appropriate actions. Answers to the Clinical Connections questions are provided in Appendix F.

- Related clinical terms help students prepare for the clinical world

- Comprehensive chapter summaries with page references provide excellent study aids

- Review questions that help students to evaluate their progress in the subject

- Critical thinking and clinical application questions identified by a stethoscope logo spur students to synthesize the material and solve mini-case studies

- The Study Partner (CD-ROM), packaged with each book, provides high quality multimedia backup to the text

New to the Fifth Edition

- **Enhanced readability.** Always aware that students will not read what they cannot understand, I spent many hours searching for synonyms to enhance the palatability of the text to the student reader without compromising the depth of the coverage with oversimplification.

- **Factual material updates.** Another major undertaking of this edition was to update the factual material in all subject areas. The breadth of literature in areas that provide subject matter for an A&P course makes this a monumental task that demands painstaking selectivity. There is never any doubt that certain areas will be updated in every edition—immunity, cell signaling, and the offerings of genetic engineering. Although information changes even as a textbook goes to press, be assured that this intent and responsibility was carried out to the best of my ability.

■ **Reorganization of the nervous system unit, Chapters 12–15.** Organization of nervous system topics is always a challenge—it seems like we never get it quite right. Somehow Chapter 15, Neural Integration, always seemed to be hanging "out there." Now, with information on sensory receptors and transduction, and a discussion of effector synapses from Chapter 11, plus information from Chapter 12 on the ascending (sensory) and descending (motor) pathways of the spinal cord, Chapter 15 is finally "one-piece" relative to sensory-motor integration and a good deal of redundant information has been deleted. I worked hard on the integration, I hope you like it.

■ **Revised art program.** Every piece of art has been touched—the subject of many of my nightmares during this revision. The color palette has been revised to provide art that has richer, more vibrant color (something our readers have asked for). Bone and muscle art (Chapters 7 and 10 respectively) has been enlarged, and more bone photos have been added. Many views—particularly reflective pieces of bone, muscle, and vascular anatomy—have been touched up to enhance their detail. Many of the physiology figures have been redrawn to reflect the more recent research findings about functional processes. As before, the art has a consistent color protocol. Every time you see an ATP starburst, it is bright yellow; oxygen is red; cellular cytoplasm is routinely beige; muscles and muscle tissues are terra cotta, and so on. This encourages automatic learning on a grand scale, a very desirable pedagogical goal.

Additionally, we have many more cadaver photos and photomicrographs in this edition, and those retained from earlier editions have been upsized in most cases. As pointed out by a conscientious reviewer that "if you can't see it clearly, then it's not worth having." Right on!

■ **Multimedia offerings enhanced.** Our new edition of the "Study Partner" CD-ROM, packaged free inside each new book, now includes 3 new case studies plus additional interactive exercises and animations. As before, this interactive study tool involves students in engrossing case studies where they develop hypotheses and sleuth out patients' diagnoses. In addition, we offer a myriad of other exciting ancillaries including the new A&P Place website and the Benjamin Cummings Digital Library described in more detail in the opposite column.

■ **New IP and SP icons.** New Interactive Physiology **IP** and Study Partner **SP** icons in chapter summaries provide students with specific references to our CD-ROM technology.

A COMPREHENSIVE TEACHING AND LEARNING PACKAGE

The supplementary material developed by Benjamin Cummings for the Fifth Edition of *Human Anatomy and Physiology* is more innovative and thorough than ever before—in print, on CD-ROM, and on the Web—for instructor and student alike.

New! The Anatomy and Physiology Place (www. anatomyandphysiology.com) From The A & P Place, students and instructors can access a Special Edition for *Human Anatomy and Physiology*, which includes numerous activities. Included are 50 practice quizzes per chapter, interactive clinical case studies with labeling and matching exercises, bone review anatomy quizzing, web research activities, on-line histology, a pronunciation glossary with an audio feature, and A&P related news articles. Instructors will find a useful Syllabus Manager to help organize the course. A free 18-month subscription to this powerful web site is included with each new student copy of the text. Instructors will receive a six month subscription, which can be renewed free of charge by adopters. A 3-day trial option is also available at the web site.

Course Management Systems, including WebCT and Blackboard Useful for on-line course management and distance learning courses. Available stand-alone or packaged with the text. Content includes selections from The A&P Place, and the entire computerized Testbank. (www.cms@awl.com)

***Benjamin Cummings Digital Library* for *Human Anatomy and Physiology*, Fifth Edition** Includes *all* illustrations—approximately 700 images—from the book (excluding photographs), which can be edited and customized for lecture presentation or testing purposes. Available for easy use in PowerPoint or on the Web in three formats: images with labels and leaders, images with leaders only, and images with no labels or leaders. Included with this edition are illustrated tables.

Study Partner CD-ROM Version 2.0 Included with every text, this version of Study Partner, which had its premiere in the fourth edition of Marieb's *Human Anatomy and Physiology*, is even better. We have added three new clinical case studies, for a total of 16. Included in each case study are relevant labeling and matching exercises. Learning Activities include two to three art exercises for every chapter, plus 15 Interactive Physiology animation exercises. Practice Quizzes in a multiple choice format give students an opportunity to assess their understanding of the material in each chapter and get immediate feedback. A random testing feature allows students to construct tests that cover multiple chapters, with the total score revealed at the end of the test.

Atlas of the Human Skeleton Inspired by Elaine N. Marieb, who chose the skeletal views and organization, Benjamin Cummings has produced this photographic atlas to provide you with an important reference for the laboratory and the classroom. You will find approximately 104 four-color images, with labels and leaders, and beautifully photographed by Ralph T. Hutchings. Packaged free with each new textbook.

Illustration Notebook Approximately 250 full-sized text illustrations, without labels, are included in this notebook. Pages are perforated and three-hole punched for easy removal and filing with lecture notes.

Transparency Acetates This expanded package includes all illustrations—approximately 700—from the book, excluding photographs. All images and labels have been brightened and enlarged for easy viewing in the classroom or lecture hall. Includes illustrated tables from the text.

AWL Tutoring Center Free tutoring is available to students who purchase a copy of *Human Anatomy and Physiology*, Fifth Edition. The AWL Tutoring Center is staffed by qualified anatomy instructors who can tutor students on all the material covered in the text, including the art and multiple choice/matching questions at the end of each chapter. Tutoring content is restricted to text material, and tutors will only discuss answers that are provided in the textbook. After an instructor authorizes student access to this service, students can contact the Tutoring Center by phone, fax, e-mail, or the internet. The AWL Tutoring Center is open Sunday through Thursday, 5:00 P.M. to 12:00 A.M. EST. Instructors should contact their local Benjamin Cummings sales representative for information about authorizing student access to the Tutoring Center.

Instructor's Resource Guide Written and revised by Barbara Stewart, Burlington County College, to accompany the Fifth Edition of *Human Anatomy and Physiology*, this edition includes a new "Resource Appendix," with suggestions for using web and media resources effectively in the classroom, plus 24 Interactive Physiology review worksheets created by Shirley Whitescarver, Lexington Community College. Based upon the popular Interactive Physiology CD-ROM modules produced by Benjamin Cum-

mings and A.D.A.M.®, these worksheets help to reinforce specific physiology concepts and can be used for homework, lab assignments, review, or quizzing. As always, instructors are provided with a variety of other useful features for each chapter in the book, including chapter synopses, suggested lecture outlines, media resources, lists of correlated transparency acetates, lecture tips, suggested readings, and answers to short essay and critical thinking questions in the text.

Test Bank Written and revised by Wayne Seifert, Brookhaven College, to accompany the Fifth Edition of *Human Anatomy and Physiology*, this testbank includes line drawings and more than 4,000 test questions in a variety of formats: multiple choice, matching, short answers, and essay. The cross-platform CD-ROM in TestGen-EQ allows instructors to generate tests via a user-friendly interface in which they can easily view, edit, sort, and add questions. QuizMaster-EQ makes testing even easier for instructors. This free and fully networkable program enables the instructor to create and save tests and quizzes using TestGen-EQ so that students can take tests or quizzes on a computer network. The program automatically grades exams and allows the instructor to view or print reports for individual students, classes, or courses.

Student Study Guide Written and revised by Elaine N. Marieb to accompany the Fifth Edition of *Human Anatomy and Physiology*, students will find a wide variety of exercise formats that address different learning styles. The three major sections: Building the Framework, Challenging Yourself, and Covering All Your Bases, help the students to build a base of knowledge using recall, reasoning, and imagination that can be applied to solving problems in both clinical and nonclinical situations.

Human Anatomy and Physiology Laboratory Manuals Elaine N. Marieb's three widely used and acclaimed laboratory manuals complement this textbook and are designed to meet the varying needs of most laboratory courses: *Human Anatomy and Physiology Lab Manual: Main Version*, Fifth Edition (1999); *Human Anatomy and Physiology Lab Manual, Cat Version*, Sixth edition (1999); and *Human Anatomy and Physiology Lab Manual: Pig Version*, Sixth edition (1999). The *Main Version* is designed for labs that use no dissection animal. Shrink-wrapped with each laboratory manual is the Physio-Ex 2.0 CD-ROM, which features seven easy-to-use computerized physiology experiments (also in the laboratory manuals as written exercises), providing convenient laboratory access for students enrolled in Internet-based distance learning courses.

Additional Supplements Available
from Benjamin Cummings

A.D.A.M.® *Interactive Physiology* CD-ROMs: Muscular System: Cardiovascular System; Respiratory System; Nervous System I; Nervous System II; Urinary System; Fluid, Electrolyte, and Acid/Base Balance.

A.D.A.M.® *Interactive Anatomy Student Lab Guide*
By Mark Lafferty and Samuel Panella

A.D.A.M.® *Interactive Anatomy Student Package*

A.D.A.M.® *Anatomy Practice*

Human Cadaver Dissection Videos
By Rose Leigh Vines, et al

Student Video Series for Human Anatomy and Physiology, Volume 1

Student Video Series for Human Anatomy and Physiology, Volume 2

Anatomy Flashcards
By Glenn Bastian

Anatomy and Physiology Coloring Workbook: A Complete Study Guide, Sixth Edition
By Elaine N. Marieb

The Physiology Coloring Book, Second Edition
By Wynn Kapit, Robert I. Macey, Esmail Meisami

The Anatomy Coloring Book, Second Edition
By Winn Kapit and Lawrence M. Elson

A Color Atlas of Histology
By Dennis Strete

Histology for the Life Sciences
By Allen Bell and Victor Eroshenko

Contact your Benjamin Cummings sales representative or your campus bookstore for more information.

PREFACE TO
THE STUDENT

This book is written for you. In a way, it is written by you (students) because it incorporates your suggestions, answers to questions most often asked, and explanations that have produced the greatest success in helping you learn about the human body. Human anatomy and physiology is more than just interesting—it is fascinating. To help get you involved in the study of this exciting subject, a number of special features are incorporated throughout the book.

The tone of this book is intentionally informal and unintimidating. There is absolutely no reason that you can't enjoy your learning task. This book is meant to be a guide to the understanding of your own body, not an encyclopedia of human anatomy and physiology. I have tried to be selective about the information included and have chosen facts that stress essential concepts. Physiological concepts are explained thoroughly; abundant analogies are used and examples are taken from familiar events whenever possible.

The illustrations and tables are designed with your learning needs in mind. The tables are summaries of important information in the text and should be valuable resources for reviewing for an exam. In all cases, the figures are referenced where their viewing would be most advantageous to help in understanding the text. Those depicting physiological mechanisms often use the flowchart format so that you can see where you've been and where you're going. Additionally, several *key figures* come equipped with questions that will help you to interpret the figures or to apply and integrate what you have learned. (Answers are provided at the bottom of the same page.) Special topic boxes, called *A Closer Look*, alert you to advances in medicine or present scientific information that can be applied to your daily life. The *Making Connections* feature includes three elements to support your studies. *Systems Connections* recaps the information that you should know when you complete each body system. *Closer Connections* broadens your horizon and enriches your understanding of the system being studied—you may or may not have covered all of its concepts. *Clinical Connections* challenges you to apply your learning to clinical situations. Please use this feature; it will help you to learn this subject.

Each chapter begins with an outline of its major topics (and their page references), with specific learning objectives listed for each topic. Important terms within the chapter text are bold-faced (in dark type), and phonetic spellings are provided for terms that are likely to be unfamiliar to you. To use these phonetic aids, you will need to remember the following rules:

1. Syllables are separated by dashes.
2. Accent marks follow stressed syllables. Primary stress is shown by ' and secondary stress by ".
3. Unless otherwise noted, assume that vowels at the ends of syllables are long. Assume that all other vowels are short unless a bar, indicating a long vowel sound, appears above them.

For example, the phonetic spelling of "thrombophlebitis" is throm"bo-fleh-bi'tis. The fourth syllable (bi′) receives the greatest stress, and the first syllable (throm″) gets the secondary stress. The "o" in the second syllable and "i" in the fourth syllable are long. Further explanation of the pronunciation system can be found at the beginning of the glossary.

Any exam causes anxiety. To help you prepare for an exam or comprehend the material you have just read, thorough summaries complete with page references are found at the ends of the chapters, as are review questions that use a combination of testing techniques (multiple choice, matching, short answer essay, critical thinking, and clinical applications). Also remember that key figure questions are sprinkled throughout the chapter.

I hope that you enjoy *Human Anatomy and Physiology* and that this book makes learning about the body's structures and functions an exciting and rewarding process. Perhaps the best bit of advice I can give you is that memory depends on understanding. Thus, if you strive to achieve understanding instead of rote memorization, your memory will not fail you very often.

I would appreciate hearing from you about your experiences with this textbook or suggestions for improvements in future editions.

Elaine N. Marieb

Elaine N. Marieb
Anatomy and Physiology
Benjamin Cummings Science
1301 Sansome Street
San Francisco, CA 94111

ACKNOWLEDGMENTS

They say that the more things change, the more they stay the same. Each edition of this book undergoes some major modifications, in both its text and art. Yet, the work involved in the process never really changes—it grinds away, the grist finer with each edition. Lucky for me, it's a labor of love.

I have not toiled alone. Many dedicated people shared the work of this revision and they deserve their proper due. Daryl Fox, who as a neophyte to the company sponsored the 4th edition, is still with me. Inundated with work himself, he still found time to listen any time I was confronted with what I perceived as an insurmountable problem, and he still puts himself on the line to underwrite the best educational product (text and multimedia) possible for the team to produce. Right on, Daryl.

Lauren Fogel, the project manager for 4e, has moved on to a new, more prestigious position as the executive producer of all multimedia for Benjamin Cummings, but she still keeps her hand on the "fine tuning buttons" to make sure her first babies (the A&P websites and the Study Partner) continue to add pedagogical value to the text presentations. My present project manager, Kay Ueno, is as different from Lauren as she can possibly be. Small and feisty, she's a buzzsaw that can do virtually any job that gets pushed her way (isn't that right, Daryl?). Charged with overseeing my text revision, she took on a developmental art editor's role in preparing the art for revision or rerendering, inherited the total management of the new Atlas (even to convincing Ralph Hutchings of England that he wanted to shoot new bone photos for us), spent a number of hours on-line searching out requested photos that the photoresearchers were having trouble finding, oversaw the supplements package, and then, at the end of the day, still found the energy to act as a sounding board for my myriad ideas for reorganization and additions to the art program. I thought the past year was hectic; I can't imagine how Kay perceived it. Mark Wales, Senior Developmental Editor, lent his editorial expertise in anatomy and physiology to the project by proofing every piece of art. Without "behind the scenes" editorial support on text and supplements, where would we be? Kudos for playing often thankless but always valuable roles go to these people: Carol Pritchard-Martinez, Tiffany Barnes, Susan Teahan, and Richard Gallager. Thanks also to Lauren Harp, the senior marketing manager who keeps *Human Anatomy and Physiology* in touch with the professors and students it serves while providing feedback that helps us to meet the needs and desires of the marketplace.

I was lucky again to have the team at Wendy Earl Productions on board for this edition—they always produce a beautiful book of the highest quality. Bradley Burch of WEP did a fine job as the art director, as did David Novak in his role as photo and supplements coordinator. Thanks also go to Kenter Associates, the main photoresearchers for this edition. Sharon Montooth, Production Supervisor, shepherded the book through production quietly and competently. Aiding and abetting the process was my eagle-eyed and conscientious copy editor Anita Wagner, Martha Ghent who proofed the work, and Karen Hollister, the indexer. Michele Mangelli, the art manager, had the thankless and delicate task of acting as the go-between for the editorial staff at

Benjamin Cummings and the author (me) on the one hand and the artists on the other—always trying to make sure the work proceeded while difficult areas were being clarified or ironed out. Michele's assistants, Anne Hikido and Charlotte Ross and Olivia Ross, kept her sane during the dark moments. The creative talents of the artists are greatly appreciated, particularly because we worked them so hard. They provided the part of the book that prompts the imagination to soar. Tomo Narashima produced all the new and incredibly beautiful reflective pieces for the Cell chapter, as well as the showcase pieces for skin and alimentary canal histology. The team of illustrators at Imagineering generated the large amount of digital computer art that graces this edition, and Kristin Mount breathed new life into the original reflective pieces of the muscle and bone art—once again they are smashing! Karl Miyajima ably updated many of the pieces from previous editions. A monumental vote of thanks and an offer to wash his feet have gone out to Ralph T. Hutchings of London, a renowned photographer of human anatomy, formerly of The College of Surgeons of England, for his beautiful bone photos that grace this edition of the book and our new bone Atlas packaged with the book. His lightning-fast (and still courteous) responses to our requests for photos will be remembered for a long, long time. Not to be forgotten is F. tani Hasegawa who came up with the chapter openers and the stunning cover (using a Mia Hamm photo shot by the creative photographer Annie Liebovitz). GTS Graphics assembled the final pages under an incredibly tight schedule and much to their credit did so with very few errors. Harvey Howell, my beloved courier, ran the manuscript to the UPS box daily without ever a complaint. I also owe a huge debt of gratitude to my reviewers and colleagues from the United States and abroad who, over the past three years, have made helpful suggestions and offered constructive criticisms to improve this work. A complete list of these reviewers follows. As with each edition, I am happy in its completion and do not look forward to the next. But the angst fades and inspiration is renewed with time. Please—if you are so inclined—let me know what you think of this edition. Your suggestions and comments provide much of the impetus for the next edition, and I would appreciate hearing from you about ways in which this book might be improved still further in future editions.

Elaine N. Marieb

Elaine N. Marieb
Anatomy and Physiology
Benjamin Cummings Science
1301 Sansome Street
San Francisco, CA 94111

Reviewers for Human Anatomy and Physiology, Fifth Edition

Joan Barber
Delaware Technical and Community College

Satish Chandran
Kennedy-King Community College

William Dunscombe
Union County College

Brent DeMars
Lakeland Community College

Chris Eckel
Salt Lake Community College

Esther Fleischman
University of Maryland, Baltimore County

Chaya Gopalan
St. Louis Community College

Carol Haspel
LaGuardia Community College

Janet L. Haynes
Long Island University, Brooklyn Campus

William H. Hightower, Jr.
Southside Virginia Community College

Dr. Carl Hirtzel
Rose State College

Linda Kollett
Massasoit Community College

Theresa Martin
College of San Mateo

Tamara L. McNutt-Scott
Augusta State University

William A. Olexik
Montgomery College

Elsa Price
Wallace Community College

Dr. Dianne Snyder
Augusta State University

Cynthia Surmacz
Bloomsburg University

Larry R. Wilmore
Lamar State College at Orange

International Reviewers

Richard Brightwell
Edith Cowan University, Australia

Leslie Fleming
University of New Brunswick

Lars Hedin
Goteborg University, Sweden

Ingibjorg H. Jonsdottir
Halmstad University, Sweden

Rudi Klein
RMIT University, Melbourne

Gary M. Lee
University of Sydney

Lidia Mayner
The Flinders University of South Australia

Adrian Verrinder
La Trobe Bendigo, Australia

Instructor Art Focus Group

Billie S. Lane
Chattanooga State Technical College

Gail O. McKenzie
Samford University

Robert J. McDonough
Georgia Perimeter College

Andrew J. Penniman
Georgia Perimeter College

Benjamin R. Wall, Jr.
Bevill State College, Walker College Campus

A&P Instructor Focus Group

Sara Brenzier
Shelton State Community College

Teresa Brandon
Dona Ana Branch Community College

Linda Falkow
Mercer County College

Chaya Golupan
St. Louis Community College

Noah Henley
Rowan Cabarrus Community College

Javanika Mody
Anne Arundel Community College

Colwell Cook
University of Evansville

Karen LaFleur
Greenville Area Technical College

James Janik
Miami University of Ohio, Middleton Campus

Art Advisory Board

Linda Adamchak
Hudson Valley Community College

Patricia Ahanotu
Georgia Perimeter College, Central

Roger Biduaka
St. Phillips College

Ed Bradel
Raymond Walters College

Kenneth Bynum
University of North Carolina, Chapel Hill

Russell Garcia
San Antonio College

Delaine Gilcrease
Mesa Community College

Edwin Gines-Candelaria
Miami Dade Community College, Wolfson

James Horwitz
Palm Beach Community College

Christine Ilitis
Salt Lake Community College

Bonnie Kallison
Mesa Community College

Alice Mills
Middle Tennessee State University

Betsy Ott
Tyler Junior College, Texas

Linda Powell
Community College of Philadelphia

Robert Seegmiller
Brigham Young University

Sharon Simpson
Broward Community College

R. W. Stevens
Old Dominion University

Pamela Vaughn
Laredo Community College

Michael Vitale
Daytona Beach Community College

Deloris Wenzel
University of Georgia

Art Zeitlin
Kingsborough Community College

Student Survey Participants

We would like to thank Clint Benjamin, Lower Columbia College; Nancy Rauch, Merritt College; Nancy Kincaid, Troy State University, Montgomery; and Elsa Price, J. Spivey, and K. Jamison, Wallace Community College, who allowed us to conduct a survey of their students during the 1998–1999 term. We asked these students to remark on the Fourth Edition of *Human Anatomy and Physiology* and on their A&P course needs in general. While there were too many student participants to include individual names here, we wish to express our gratitude for the incredibly thoughtful and thorough responses we received from these students. Their comments were invaluable in helping us develop the fifth edition.

CONTENTS

■ CHAPTER 4 ■
TISSUE: THE LIVING
FABRIC 114

UNIT TWO Covering,
Support, and Movement
of the Body

■ CHAPTER 5 ■
THE INTEGUMENTARY
SYSTEM 148

■ CHAPTER 6 ■
BONES AND SKELETAL
TISSUES 172

■ CHAPTER 7 ■
THE SKELETON 198

HUMAN ANATOMY & PHYSIOLOGY

1 THE HUMAN BODY: AN ORIENTATION

An Overview of Anatomy and Physiology (pp. 3–4)

1. Define anatomy and physiology and describe their subdivisions.

2. Explain the principle of complementarity.

Levels of Structural Organization (pp. 4–7)

3. Name (in order of increasing complexity) the different levels of structural organization that make up the human body, and explain their relationships.

4. List the 11 organ systems of the body, identify the components of each, and briefly explain the major function(s) of each system.

Maintaining Life (pp. 8–10)

5. List and define the functional characteristics necessary to maintain life in humans.

6. List the survival needs of the body.

Homeostasis (pp. 10–13)

7. Define homeostasis and explain its significance.

8. Describe how negative and positive feedback are involved in maintaining body homeostasis.

9. Describe the relationship between homeostatic imbalance and disease.

The Language of Anatomy (pp. 14–23)

10. Describe the anatomical position.

11. Use correct anatomical terms to describe body directions, regions, and body planes or sections.

12. Locate and name the major body cavities and their subdivisions, and list the major organs contained within them.

13. Name the specific serous membranes and indicate their common function.

14. Name the nine regions or four quadrants of the abdominopelvic cavity and list the organs they contain.

As you make your way through this book, you will be learning about one of the most fascinating subjects possible—your own body. Such a study is not only highly personal, but timely as well. The current information blizzard brings news of some medical advance almost daily. If you are to appreciate emerging discoveries in genetic engineering, to understand new techniques for detecting and treating disease, and to make use of published facts on how to stay healthy, it is important to learn about the workings of your body. For those of you preparing for a career in the health sciences, the study of anatomy and physiology has added rewards because it provides the strong foundation needed to support your clinical experiences.

In this chapter we define and contrast anatomy and physiology and discuss how the human body is organized. Then we review needs and functional processes common to all living organisms. Three essential concepts—the *complementarity of structure and function, hierarchy of structural organization,* and *homeostasis*—will unify and form the bedrock for your study of the human body. The final section of the chapter deals with the language of anatomy—terminology that anatomists use when they describe the body or its parts.

AN OVERVIEW OF ANATOMY AND PHYSIOLOGY

Two complementary branches of science—anatomy and physiology—provide the concepts that help us to understand the human body. **Anatomy** studies the *structure* of body parts and their relationships to one another. Anatomy has a certain appeal because it is concrete. Body structures can be seen, felt, and examined closely; it is not necessary to *imagine* what they look like. **Physiology** concerns the *function* of the body's structural machinery, in other words, how all the body parts work and carry out their life-sustaining activities. When all is said and done, physiology is explainable only in terms of the underlying anatomy.

Topics of Anatomy

Anatomy is a broad field with many subdivisions, each providing enough information to be a course in itself. **Gross,** or **macroscopic, anatomy** is the study of large body structures visible to the naked eye, such as the heart, lungs, and kidneys. Indeed, the term *anatomy* (derived from the Greek words meaning "to cut apart") relates most closely to gross anatomy because in such studies preserved animals or their organs are dissected (cut up) to be examined. Gross anatomy can be approached in different ways. In **re-**gional anatomy, all the structures (muscles, bones, blood vessels, nerves, etc.) in one particular region of the body, such as the abdomen or leg, are examined at the same time. In **systemic** (sis-tem'ik)* **anatomy,** the gross anatomy of the body is studied system by system. For example, when studying the cardiovascular system, you would examine the heart and the blood vessels of the entire body. Another subdivision of gross anatomy is **surface anatomy,** the study of internal body structures as they relate to the overlying skin surface. You use surface anatomy when you identify the bulging muscles beneath a body builder's skin, and clinicians use it to locate the appropriate blood vessels in which to feel pulses and draw blood.

Microscopic anatomy concerns structures too small to be seen with the naked eye. For most such studies, exceedingly thin slices of body tissues are stained and mounted on slides to be examined under the microscope. Subdivisions of microscopic anatomy include **cytology** (si-tol'o-je), which considers the cells of the body, and **histology** (his-tol'o-je), the study of tissues.

Developmental anatomy traces structural changes that occur in the body throughout the life span. **Embryology** (em"bre-ol'o-je), a subdivision of developmental anatomy, concerns developmental changes that occur before birth and helps to explain birth defects.

Some highly specialized branches of anatomy are used primarily for medical diagnosis and scientific research. For example, *anatomic pathology,* also called *pathological anatomy,* studies structural changes caused by disease. *Radiographic anatomy* studies internal structures as visualized by X-ray images or specialized scanning procedures. In *molecular biology,* the structure of biological molecules (chemical substances) is investigated. Molecular biology is actually a separate branch of biology, but it falls under the anatomy "umbrella" when we push anatomical studies to the subcellular level. As you can see, subjects of interest to anatomists range from easily seen structures down to the smallest molecule.

As you will quickly learn, one of the most important "tools" for studying anatomy is a mastery of anatomical terminology. Others are observation, manipulation, and, in a living person, palpation (feeling organs with your hands) and auscultation (listening to organ sounds with a stethoscope). A simple example will illustrate how some of these tools work together in an anatomical study. Let's assume that your topic is freely movable joints of the body. In the laboratory, you will be able to *observe* an

---

*For the pronunciation guide rules, see the Preface to the Student.

animal joint, noting how its parts fit together. You can work the joint (*manipulate* it) to determine its range of motion. Then, using *anatomical terminology*, you can name its parts and describe how they are related so that other students (and your instructor) will have no trouble understanding you. The list of word roots (on the inside of the book cover) and the glossary will help you with this special vocabulary.

Although most of your observations will be made with the naked eye or with the help of a microscope, medical technology has developed a number of sophisticated tools that can peer into the body without disrupting it. These exciting medical imaging techniques are discussed in *A Closer Look* on pp. 20–21.

Topics of Physiology

Like anatomy, physiology has many subdivisions, most of which consider the operation of specific organ systems. For example, **renal physiology** concerns kidney function and urine production; **neurophysiology** explains the workings of the nervous system; and **cardiovascular physiology** examines the operation of the heart and blood vessels. While anatomy provides us with a static image of the body's architecture, physiology reveals the dynamic nature of the living body.

Physiology often focuses on events at the cellular or molecular level. This is because the body's abilities depend on those of its individual cells, and cells' abilities ultimately depend on the chemical reactions that go on within them. An understanding of physiology also rests on principles of physics, which help to explain electrical currents, blood pressure, and the way muscles use bones to cause body movements among other things. Thus, basic chemical and physical principles are presented in Chapter 2 and throughout the book as needed to explain physiological topics.

Complementarity of Structure and Function

Although it is possible to study anatomy and physiology in isolation from one another, they are really inseparable because function always reflects structure. That is, what a structure can do depends on its specific form. This is called the **principle of complementarity of structure and function.** For example, bones can support and protect body organs because they contain hard mineral deposits; blood flows in one direction through the heart because the heart has valves that prevent backflow; and the lungs can serve as a site for gas exchange because the walls

of their air sacs are extremely thin. Throughout this book, a description of the anatomy of a structure is accompanied by an explanation of its function, and structural characteristics contributing to that function are emphasized.

LEVELS OF STRUCTURAL ORGANIZATION

The human body has many levels of structural organization (Figure 1.1). The simplest level of the structural hierarchy is the **chemical level,** which we study in Chapter 2. At this level, *atoms,* tiny building blocks of matter, combine to form *molecules* such as water, sugar, and proteins. Molecules, in turn, associate in specific ways to form *organelles,* basic components of the microscopic cells. *Cells* are the smallest units of living things. The **cellular level** is examined in Chapter 3. Individual cells vary widely in size and shape, reflecting their unique functions in the body. All cells have some common functions, but only certain cell types form the transparent lens of the eye, secrete mucus, or conduct nerve impulses.

The simplest living creatures are composed of single cells, but in complex organisms such as human beings, the hierarchy continues on to the **tissue level.** *Tissues* are groups of similar cells that have a common function. The four basic tissue types in the human body are epithelium, muscle, connective tissue, and nervous tissue. Each tissue type has a characteristic role in the body, which we explore in detail in Chapter 4. Briefly, epithelium covers the body surface and lines its cavities; muscle provides movement; connective tissue supports and protects body organs; and nervous tissue provides a means of rapid internal communication by transmitting electrical impulses.

An *organ* is a discrete structure composed of at least two tissue types (four is more commonplace) that performs a specific function for the body. At the **organ level,** extremely complex functions become possible. Let's take the stomach for an example. Its lining is an epithelium that produces digestive juices; the bulk of its wall is muscle which churns and mixes stomach contents (food); its connective tissue reinforces the soft muscular walls; and its nerve fibers increase digestive activity by stimulating the muscle to contract more vigorously and the glands to secrete more digestive juices. The liver, the brain, and a blood vessel are very different from the stomach, but they are organs as well. You can think of each organ of the body as a specialized functional center responsible for a necessary activity that no other organ can perform.

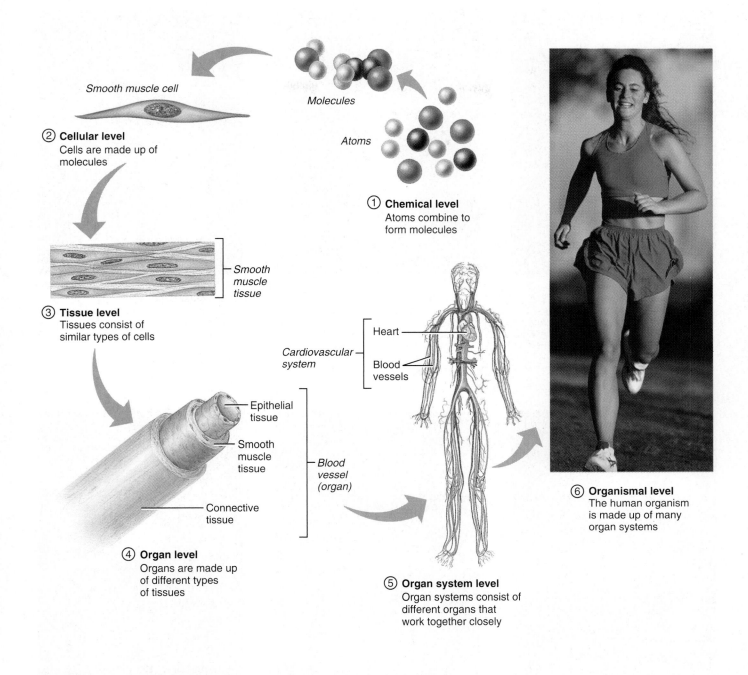

Smooth muscle cell

Molecules

② **Cellular level**
Cells are made up of
molecules

Atoms

① **Chemical level**
Atoms combine to
form molecules

Smooth
muscle
tissue

③ **Tissue level**
Tissues consist of
similar types of cells

Epithelial
tissue

Smooth
muscle
tissue

Connective
tissue

④ **Organ level**
Organs are made up
of different types
of tissues

Cardiovascular
system

Heart

Blood
vessels

Blood
vessel
(organ)

⑤ **Organ system level**
Organ systems consist of
different organs that
work together closely

⑥ **Organismal level**
The human organism
is made up of many
organ systems

FIGURE 1.1 *Levels of structural organization.* In this diagram, components of
the cardiovascular system are used to illustrate the various levels of structural
organization in a human being.

The next level of organization is the **organ system.** Organs that work closely with one another to accomplish a common purpose make up an *organ system.* For example, organs of the cardiovascular system—mainly the heart and blood vessels—see to it that blood circulates continuously to carry oxygen and nutrients to all body cells. Besides the cardiovascular system, the other organ systems of the body are the integumentary, skeletal, muscular, nervous, en-

docrine, respiratory, digestive, lymphatic, urinary, and reproductive systems. Figure 1.2 provides a brief overview of the 11 organ systems, which we study in more detail in Units 2–5.

The highest level of organization is the *organism,* the living human being. The **organismal level** represents the sum total of all structural levels working together to promote life.

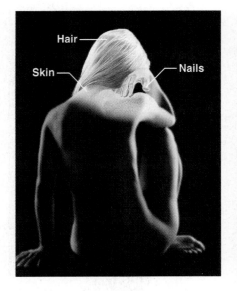

(a) Integumentary System
Forms the external body covering; protects deeper tissues from injury; synthesizes vitamin D; site of cutaneous (pain, pressure, etc.) receptors, and sweat and oil glands.

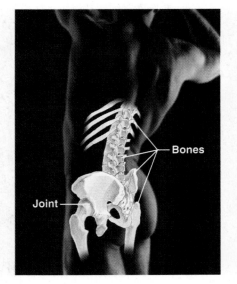

(b) Skeletal System
Protects and supports body organs; provides a framework the muscles use to cause movement; blood cells are formed within bones; stores minerals.

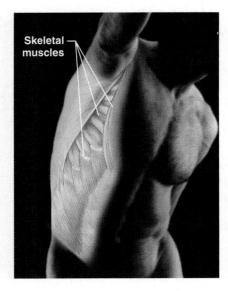

(c) Muscular System
Allows manipulation of the environment, locomotion, and facial expression; maintains posture; produces heat.

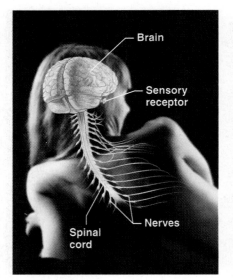

(d) Nervous System
Fast-acting control system of the body; responds to internal and external changes by activating appropriate muscles and glands.

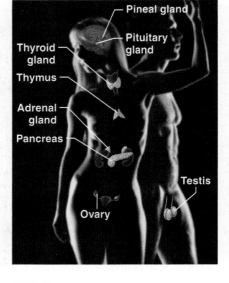

(e) Endocrine System
Glands secrete hormones that regulate processes such as growth, reproduction and nutrient use (metabolism) by body cells.

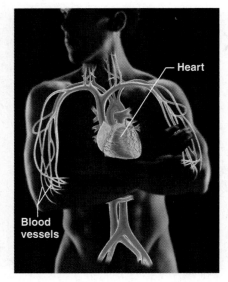

(f) Cardiovascular System
Blood vessels transport blood, which carries oxygen, carbon dioxide, nutrients, wastes, etc.; the heart pumps blood.

FIGURE 1.2 *Summary of the body's organ systems.* The structural components of each organ system are illustrated in the diagrammatic view. The major functions of the organ system are listed beneath each illustration.

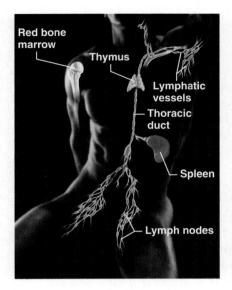

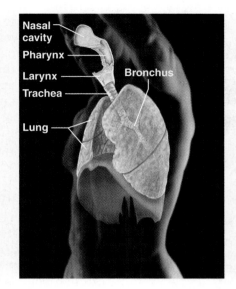

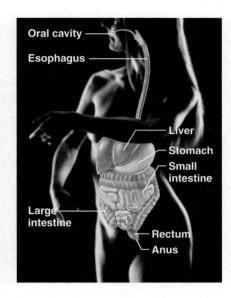

(g) Lymphatic System/Immunity
Picks up fluid leaked from blood vessels and returns it to blood; disposes of debris in the lymphatic stream; houses white blood cells (lymphocytes) involved in immunity. The immune response mounts the attack against foreign substances within the body.

(h) Respiratory System
Keeps blood constantly supplied with oxygen and removes carbon dioxide; the gaseous exchanges occur through the walls of the air sacs of the lungs.

(i) Digestive System
Breaks down food into absorbable units that enter the blood for distribution to body cells; indigestible foodstuffs are eliminated as feces.

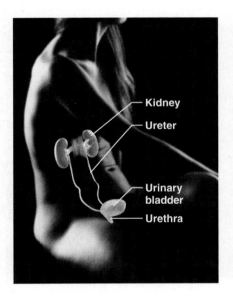

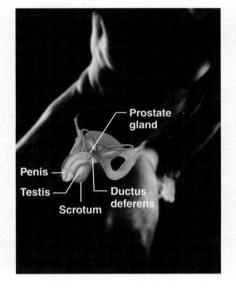

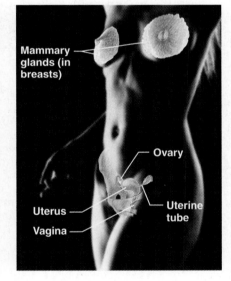

(j) Urinary System
Eliminates nitrogenous wastes from the body; regulates water, electrolyte and acid-base balance of the blood.

(k) Male Reproductive System

(l) Female Reproductive System
Overall function is production of offspring. Testes produce sperm and male sex hormone; ducts and glands aid in delivery of sperm to the female reproductive tract. Ovaries produce eggs and female sex hormones; remaining structures serve as sites for fertilization and development of the fetus. Mammary glands of female breasts produce milk to nourish the newborn.

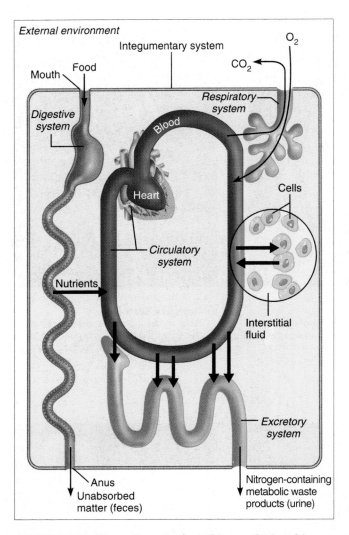

FIGURE 1.3 *Examples of selected interrelationships among body organ systems.* The integumentary system protects the body as a whole from the external environment. The digestive and respiratory systems, in contact with the external environment, take in nutrients and oxygen, respectively, which are then distributed by the blood to all body cells. Elimination from the body of metabolic wastes is accomplished by the urinary and respiratory systems.

MAINTAINING LIFE

Necessary Life Functions

Now that you know the structural levels composing the human body, the question that naturally follows is: What does this highly organized human body do? Like all complex animals, human beings maintain their boundaries, move, respond to environmental changes, take in and digest nutrients, carry out metabolism, dispose of wastes, reproduce themselves, and grow. We will discuss each of these necessary life functions briefly here and in more detail in later chapters.

It cannot be emphasized too strongly that the multicellular state and parceling out of vital body functions to several *different* organ systems result in

interdependence of all body cells. No individual's organ systems work in isolation. Instead, they work cooperatively to promote the well-being of the entire body. Because this theme is emphasized throughout this book, it is appropriate to identify some of the organ systems making the major contributions to selected functional processes (Figure 1.3). Also, as you read this section you may want to refer back to the more detailed descriptions of the organ systems in Figure 1.2.

Maintaining Boundaries

Every living organism must **maintain its boundaries** so that its internal environment (inside) remains distinct from the external environment surrounding it (outside). In single-celled organisms, the external boundary is a limiting membrane that encloses its contents and admits needed substances while restricting entry of potentially damaging or unnecessary substances. Similarly, all the cells of our body are surrounded by a selectively permeable membrane. Additionally, the body as a whole is enclosed and protected by the integumentary system or skin, which protects our internal organs from drying out (a fatal change), bacteria, the damaging effects of heat, sunlight, and an unbelievable number of chemicals in the external environment.

Movement

Movement includes all the activities promoted by the muscular system, such as propelling ourselves from one place to another by running or swimming, and manipulating the external environment with our nimble fingers. The skeletal system provides the bony framework that the muscles pull on as they work. Movement also occurs when substances such as blood, foodstuffs, and urine are propelled through internal organs of the cardiovascular, digestive, and urinary systems, respectively. On the cellular level, the muscle cell's ability to move by shortening is more precisely called **contractility.**

Responsiveness

Responsiveness, or **irritability,** is the ability to sense changes (stimuli) in the environment and then respond to them. For example, if you cut your hand on broken glass, a withdrawal reflex occurs—you involuntarily pull your hand away from the painful stimulus (the broken glass). It is not necessary to think about it—it just happens! Likewise, when carbon dioxide in your blood rises to dangerously high levels, chemical sensors respond by sending messages to brain centers controlling respiration, and your breathing rate speeds up.

Because nerve cells are highly irritable and communicate rapidly with each other via electrical impulses, the nervous system is most involved with responsiveness. However, all body cells exhibit irritability to some extent.

Digestion

Digestion is the process of breaking down ingested foodstuffs to simple molecules that can be absorbed into the blood. The nutrient-rich blood is then distributed to all body cells by the cardiovascular system. In a simple, one-celled organism such as an amoeba, the cell itself is the "digestion factory," but in the multicellular human body, the digestive system performs this function for the entire body.

Metabolism

Metabolism (mĕ-tab′o-lizm; "a state of change") is a broad term that includes all chemical reactions that occur within body cells. It includes breaking down substances into their simpler building blocks (more specifically called *catabolism*), synthesizing more complex cellular structures from simpler substances *(anabolism)*, and using nutrients and oxygen to produce (via *cellular respiration*) ATP, the energy-rich molecules that power cellular activities. Metabolism depends on the digestive and respiratory systems to make nutrients and oxygen available to the blood and on the cardiovascular system to distribute these needed substances throughout the body. Metabolism is regulated largely by hormones secreted by endocrine system glands.

Excretion

Excretion is the process of removing *excreta* (ek-skre′tah), or wastes, from the body. If the body is to operate as we expect it to, it must get rid of nonuseful substances produced during digestion and metabolism. Several organ systems participate in excretion. For example, the digestive system rids the body of indigestible food residues in feces, and the urinary system disposes of nitrogen-containing metabolic wastes, such as urea, in urine. Carbon dioxide, a byproduct of cellular respiration, is carried in the blood to the lungs, where it leaves the body in exhaled air.

Reproduction

Reproduction can occur at the cellular or organismal level. In cellular reproduction the original cell divides, producing two identical daughter cells that may then be used for body growth or repair. Reproduction of the human organism, or making a whole new person, is the major task of the reproductive system. When a sperm unites with an egg, a fertilized egg forms, which then develops into a baby within the mother's body. The reproductive system is directly responsible for producing offspring, but its function is exquisitely regulated by hormones of the endocrine system.

Because males produce sperm and females produce eggs (ova), there is a division of labor in the reproductive process, and the reproductive organs of males and females are quite different (see Figure 1.2). Additionally, the female's reproductive structures provide the site for fertilization of eggs by sperm, then protect and nurture the developing fetus until birth.

Growth

Growth is an increase in size of a body part or the organism. It is usually accomplished by increasing the number of cells. However, individual cells also increase in size when not dividing. For true growth to occur, constructive activities must occur at a faster rate than destructive ones.

Survival Needs

The ultimate goal of all body systems is to maintain life. However, life is extraordinarily fragile and requires that several factors be present. These factors, which we will call *survival needs*, include nutrients (food), oxygen, water, and appropriate temperature and atmospheric pressure.

Nutrients, taken in via the diet, contain the chemical substances used for energy and cell building. Most plant-derived foods are rich in carbohydrates, vitamins, and minerals, whereas most animal foods are richer in proteins and fats. Carbohydrates are the major energy fuel for body cells. Proteins, and to a lesser extent, fats, are essential for building cell structures. Fats also cushion body organs, form insulating layers, and provide a reserve of energy-rich fuel. Selected minerals and vitamins are required for the chemical reactions that go on in cells and for oxygen transport in the blood. For example, the mineral calcium helps to make bones hard and is required for blood clotting.

All the nutrients in the world are useless unless **oxygen** is also available. Because the chemical reactions that release energy from foods are *oxidative* reactions that require oxygen, human cells can survive for only a few minutes without oxygen. Approximately 20% of the air we breathe is oxygen. It is made available to the blood and body cells by the cooperative efforts of the respiratory and cardiovascular systems.

Water accounts for 60 to 80% of body weight and is the single most abundant chemical substance in the body. It provides the watery environment necessary for chemical reactions and the fluid base for body secretions and excretions. Water is obtained chiefly from ingested foods or liquids and is lost from the body by evaporation from the lungs and skin and in body excretions.

If chemical reactions are to continue at life-sustaining rates, **normal body temperature** must be maintained. As body temperature drops below 37°C (98°F), metabolic reactions become slower and slower, and finally stop. When body temperature is too high, chemical reactions occur at such a frantic pace that body proteins lose their characteristic shape and stop functioning. At either extreme, death occurs. Most body heat is generated by the activity of the muscular system.

Atmospheric pressure is the force that air exerts on the surface of the body. Breathing and gas exchange in the lungs depend on *appropriate* atmospheric pressure. At high altitudes, where atmospheric pressure is lower and the air is thin, gas exchange may be inadequate to support cellular metabolism.

The mere presence of these survival factors is not sufficient to sustain life. They must be present in *appropriate* amounts; excesses and deficits may be equally harmful. For example, oxygen is essential, but excessive amounts are toxic to body cells. Similarly, the food we eat must be of high quality and in proper amounts; otherwise, nutritional disease, obesity, or starvation is likely. Also, while the needs listed above are the most crucial, they do not even begin to encompass all of the body's needs. For example, we can live without gravity if we must, but the quality of life suffers.

HOMEOSTASIS

When you think about the fact that your body contains trillions of cells in nearly constant activity, and that remarkably little usually goes wrong with it, you begin to appreciate what a marvelous machine your body is. Walter Cannon, an American physiologist of the early twentieth century, spoke of the "wisdom of the body," and he coined the word **homeostasis** (ho"me-o-sta'sis) to describe its ability to maintain relatively stable internal conditions even though the outside world changes continuously. Although the literal translation of homeostasis is "unchanging," the term does not really mean a static, or unchanging, state. Rather, it indicates a *dynamic* state of equilibrium, or a balance, in which internal conditions vary, but always within relatively narrow limits. In general, the body is in homeostasis when its needs are adequately met and it is functioning smoothly.

Maintaining homeostasis is much more complicated than it appears at first glance. Virtually every organ system plays a role in maintaining the constancy of the internal environment. Adequate blood levels of vital nutrients must be continuously present, and heart activity and blood pressure must be constantly monitored and adjusted so that the blood is propelled to all body tissues. Also, wastes must not be allowed to accumulate, and body temperature must be precisely controlled. A wide variety of chemical, thermal, and neural factors act and interact in complex ways—sometimes helping and sometimes hindering the body as it works to maintain its "steady rudder."

Homeostatic Control Mechanisms

Communication within the body is essential for homeostasis. Communication is accomplished chiefly by the nervous and endocrine systems, which use electrical impulses delivered by nerves or blood-borne hormones respectively as information carriers. The details of how these two great regulating systems operate are covered in later chapters, but the basic characteristics of control systems that promote homeostasis are explained here.

Regardless of the factor or event being regulated—this is called the **variable**—all homeostatic control mechanisms have at least three interdependent components (Figure 1.4). The first component is a **receptor.** Essentially, it is some type of sensor that monitors the environment and responds to changes, called *stimuli*, by sending information (input) to the second component, the *control center*. The flow of information from the receptor to the control center occurs along the so-called *afferent pathway*. The **control center,** which determines the *set point* (the level or range) at which a variable is to be maintained, analyzes the input it receives and then determines the appropriate response or course of action.

The third component is the **effector.** The effector provides the means for the control center's response (output) to the stimulus. Information flows from the control center to the effector along the *efferent pathway*. The results of the response then *feed back* to influence the stimulus, either depressing it (negative feedback) so that the whole control mechanism is shut off or enhancing it (positive feedback) so that the reaction continues at an even faster rate.

Now that we have examined the basic flow of information in control systems, we are ready to explain how negative and positive feedback mechanisms help maintain body homeostasis.

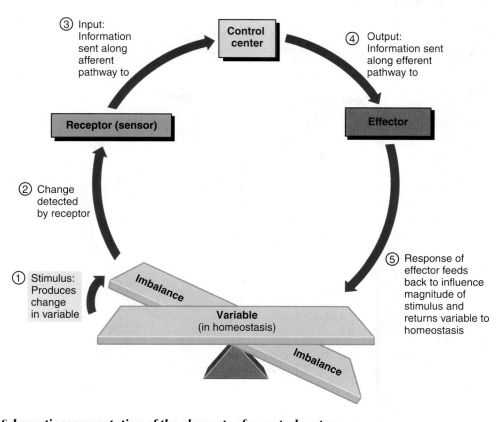

FIGURE 1.4 *Schematic representation of the elements of a control system.* Communication between the receptor, control center, and effector is essential for normal operation of the system.

Negative Feedback Mechanisms

Most homeostatic control mechanisms are **negative feedback mechanisms.** In such systems, the output of the system shuts off the original stimulus or reduces its intensity. These mechanisms cause the variable to change in a direction *opposite* to that of the initial change, returning it to its "ideal" value; thus the name "negative" feedback mechanisms.

A good example of a nonbiological negative feedback system is a home heating system connected to a temperature-sensing thermostat. The thermostat houses both the receptor and the control center. If the thermostat is set at 20°C (68°F), the heating system (effector) is triggered ON when the house temperature drops below that setting. As the furnace produces heat and warms the air, the temperature rises, and when it reaches 20°C or slightly higher, the thermostat triggers the furnace OFF. This process results in a cycling of "furnace-ON" and "furnace-OFF" so that the temperature in the house stays very near the desired temperature of 20°C. Your body "thermostat," located in a part of your brain called the hypothalamus, operates in a similar fashion.

Regulation of body temperature is only one of the many ways the nervous system maintains the constancy of the internal environment. Another type of neural control mechanism is seen in the *withdrawal reflex* referred to earlier, in which the hand is jerked away from a painful stimulus such as broken glass. The endocrine system is equally important in maintaining homeostasis, however, and a good example of a hormonal negative feedback mechanism is the control of blood glucose levels by pancreatic hormones (Figure 1.5).

To carry out normal metabolism, body cells need a continuous supply of glucose, their major fuel for producing cellular energy, or ATP. Blood sugar levels are normally maintained around 90 milligrams (mg) of glucose per 100 milliliters (ml) of blood.* Let's assume that you have lost your willpower and have just downed four jelly doughnuts. The doughnuts are quickly broken down to sugars by your digestive system, and as the sugars flood into the bloodstream, blood sugar levels spike upward, disrupting homeostasis. The rising glucose levels stimulate the insulin-producing cells of the pancreas, which respond by secreting insulin into the blood. Insulin accelerates the uptake of glucose by most body cells. It

---

*The metric system is described in Appendix A.

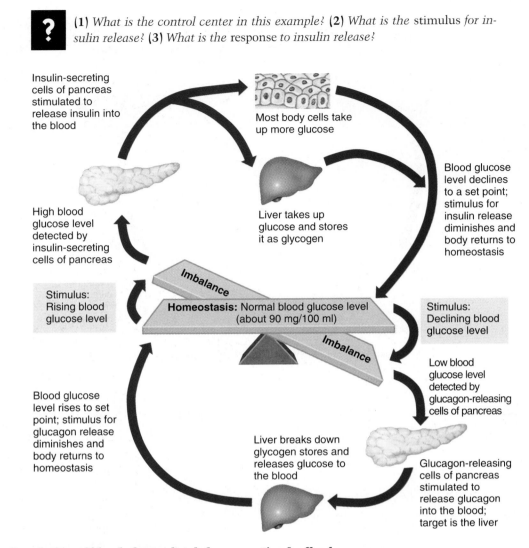

? (1) *What is the control center in this example?* (2) *What is the* stimulus *for insulin release?* (3) *What is the* response *to insulin release?*

Insulin-secreting cells of pancreas stimulated to release insulin into the blood

Most body cells take up more glucose

Blood glucose level declines to a set point; stimulus for insulin release diminishes and body returns to homeostasis

High blood glucose level detected by insulin-secreting cells of pancreas

Liver takes up glucose and stores it as glycogen

Imbalance

Stimulus: Rising blood glucose level

Homeostasis: Normal blood glucose level (about 90 mg/100 ml)

Stimulus: Declining blood glucose level

Imbalance

Blood glucose level rises to set point; stimulus for glucagon release diminishes and body returns to homeostasis

Liver breaks down glycogen stores and releases glucose to the blood

Low blood glucose level detected by glucagon-releasing cells of pancreas

Glucagon-releasing cells of pancreas stimulated to release glucagon into the blood; target is the liver

FIGURE 1.5   *Regulation of blood glucose levels by a negative feedback mechanism involving pancreatic hormones.*

also encourages storage of excess glucose as glycogen in the liver and muscles so that a reserve supply of glucose is "put into the larder," so to speak. Consequently, blood sugar levels ebb back toward the normal set point, and the stimulus for insulin release diminishes.

Glucagon, another pancreatic hormone, has the opposite effect. Its release is triggered as blood sugar levels decline below the set point. Suppose, for example, that you have skipped lunch and it is now about 2 PM. Since your blood sugar level is running low, glucagon secretion is stimulated. Glucagon targets the liver, causing it to release its glucose reserves

from glycogen into the blood. Consequently, blood sugar levels increase back into the homeostatic range.

The body's ability to regulate its internal environment is fundamental, and all negative feedback mechanisms have the same goal: preventing sudden severe changes within the body. Body temperature and blood glucose levels are only two of the variables that need to be regulated. There are hundreds! Other negative feedback mechanisms regulate heart rate, blood pressure, the rate and depth of breathing, and blood levels of oxygen, carbon dioxide, and minerals. We will discuss many of these mechanisms when we consider the different organ systems of the body. Now, however, we will look at the other type of feedback control mechanism—positive feedback.

(1) *The pancreas.* (2) *Rising blood glucose levels.* (3) *Blood glucose levels fall as body cells take it up from the blood.*

(1) *Why is this control mechanism called a "positive feedback mechanism"?* **(2)** *What event ends the cascade or chain reaction seen in this positive feedback control mechanism?*

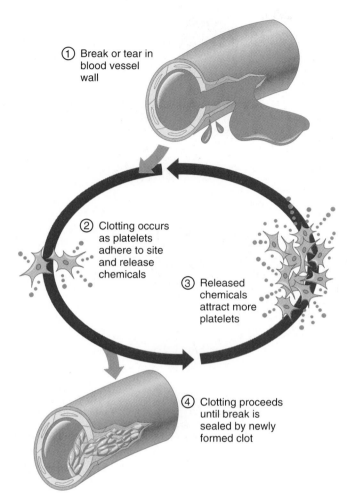

① Break or tear in blood vessel wall

② Clotting occurs as platelets adhere to site and release chemicals

③ Released chemicals attract more platelets

④ Clotting proceeds until break is sealed by newly formed clot

FIGURE 1.6 *Summary of the positive feedback mechanism regulating blood clotting.*

Positive Feedback Mechanisms

In **positive feedback mechanisms,** the result or response enhances or exaggerates the original stimulus so that the activity (output) is accelerated. This feedback mechanism is said to be "positive" because the change that occurs proceeds in the *same* direction as the initial disturbance, causing the variable to deviate further and further from its original value or range. In contrast to negative feedback controls, which maintain some physiological function or keep

blood chemicals within narrow ranges, positive feedback mechanisms usually control infrequent events that do not require continuous adjustments. Typically, they set off a series of events that may be self-perpetuating and that, once initiated, have an amplifying or waterfall effect. Positive feedback mechanisms are often referred to as *cascades* (from the Italian word meaning "to fall") because of these characteristics. Positive feedback mechanisms are likely to race out of control, and so they are rarely used to promote the moment-to-moment well-being of the body. However, there are at least two familiar examples of their use as homeostatic mechanisms—blood clotting and enhancement of labor contractions during birth.

Blood clotting is a normal response to a break in the lining of a blood vessel and is an excellent example of an important body function controlled by positive feedback. Basically, as related in Figure 1.6, once vessel damage has occurred ①, blood elements called platelets immediately begin to cling to the injured site ② and release chemicals that attract more platelets ③. This rapidly growing pileup of platelets initiates the sequence of events that finally forms a clot ④.

The positive feedback mechanism in which oxytocin, a hypothalamic hormone, intensifies labor contractions during the birth of a baby is described in Chapter 29 (see Figure 29.16). Oxytocin causes the contractions to become both more frequent and more powerful until the baby is finally born, an event that ends the stimulus for oxytocin release and shuts off the positive feedback mechanism.

Homeostatic Imbalance

Homeostasis is so important that most disease is regarded as a result of its disturbance, a condition called **homeostatic imbalance.** As we age, our body's control systems become less efficient. As a result, our internal environment becomes less and less stable. These events increase our risk for illness and produce the changes we associate with aging.

Another important source of homeostatic imbalance occurs when the usual negative feedback mechanisms are overwhelmed and destructive positive feedback mechanisms take over. Some instances of heart failure reflect this phenomenon.

Examples of homeostatic imbalance are provided throughout this book to enhance your understanding of normal physiological mechanisms. These homeostatic imbalance sections are preceded by the symbol ⚡ to alert you to the fact that an abnormal condition is being described.

---

(1) *Because the response leads to an even greater response (i.e., platelets cling to damaged vessels and release chemicals that attract more platelets which release more chemicals, etc.) rather than shutting off the stimulus.* **(2)** *When the clot seals the break in the vessel, the cascade ends.*

As you study this figure, determine precisely where you would be injured if you (1) pulled a muscle in your inguinal region, (2) cracked a bone in your olecranal region, and (3) received a cut on the fibular region.

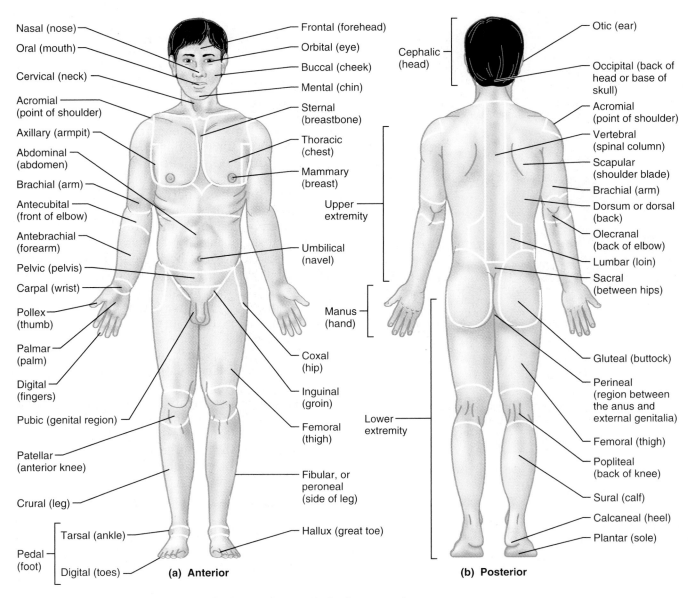

Anterior labels (left side):
- Nasal (nose)
- Oral (mouth)
- Cervical (neck)
- Acromial (point of shoulder)
- Axillary (armpit)
- Abdominal (abdomen)
- Brachial (arm)
- Antecubital (front of elbow)
- Antebrachial (forearm)
- Pelvic (pelvis)
- Carpal (wrist)
- Pollex (thumb)
- Palmar (palm)
- Digital (fingers)
- Pubic (genital region)
- Patellar (anterior knee)
- Crural (leg)
- Pedal (foot) — Tarsal (ankle), Digital (toes)

Anterior labels (right side):
- Frontal (forehead)
- Orbital (eye)
- Buccal (cheek)
- Mental (chin)
- Sternal (breastbone)
- Thoracic (chest)
- Mammary (breast)
- Upper extremity
- Umbilical (navel)
- Manus (hand)
- Coxal (hip)
- Inguinal (groin)
- Femoral (thigh)
- Fibular, or peroneal (side of leg)
- Hallux (great toe)

(a) Anterior

Posterior labels (left side):
- Cephalic (head)

Posterior labels (right side):
- Otic (ear)
- Occipital (back of head or base of skull)
- Acromial (point of shoulder)
- Vertebral (spinal column)
- Scapular (shoulder blade)
- Brachial (arm)
- Dorsum or dorsal (back)
- Olecranal (back of elbow)
- Lumbar (loin)
- Sacral (between hips)
- Gluteal (buttock)
- Perineal (region between the anus and external genitalia)
- Femoral (thigh)
- Popliteal (back of knee)
- Sural (calf)
- Calcaneal (heel)
- Plantar (sole)

Lower extremity

(b) Posterior

FIGURE 1.7   *Regional terms used to designate specific body areas.* The body is depicted in anatomical position in (**a**). In (**b**), the heels are raised slightly to show the plantar surface of the foot.

THE LANGUAGE OF ANATOMY

Although most of us are naturally curious about our bodies, our interest sometimes dwindles when we are confronted with the terminology used in the study of anatomy and physiology. Let's face it. You can't just pick up an anatomy and physiology book and read it as though it were a novel. Unfortunately, confusion is inevitable without such specialized terminology. For example, if you are looking at a ball, "above" always means the area over the top of the ball. Other directional terms can also be used consistently because the ball is a totally symmetrical object, a sphere. The human body, of course, has many bends and protrusions. Thus, the question becomes: Above what? To prevent misunderstanding, anatomists use universally accepted terms to identify body structures precisely and with a minimum

(1) Your groin. (2) Posterior aspect of the elbow. (3) The side of your leg.

of words. This language of anatomy is presented and explained next.

Anatomical Position and Directional Terms

To describe body parts and position accurately, we need an initial reference point and must indicate direction. The anatomical reference point is a standard body position called the **anatomical position.** In the anatomical position, the body is erect with feet only slightly apart. This position is easy to remember because it resembles "standing at attention," except that the palms face forward and the thumbs point away from the body. You can see the anatomical position in Figure 1.7a. It is essential to understand the anatomical position because most of the directional terms used in this book refer to the body *as if it were in this position, regardless of its actual position.* Another point to remember is that the terms "right" and "left" refer to those sides of the person or the cadaver being viewed—not those of the observer.

Directional terms allow us to explain exactly where one body structure is in relation to another. For example, we could describe the relationship between the ears and the nose informally by stating, "The ears are located on each side of the head to the right and left of the nose." Using anatomical terminology, this condenses to, "The ears are lateral to the nose." Clearly, using anatomical terms saves words and is less ambiguous. Commonly used orientation and directional terms are defined and illustrated in Table 1.1. Many of these terms are also used in everyday conversation, but keep in mind that their anatomical meanings are very precise.

Regional Terms

The two fundamental divisions of our body are its *axial* and *appendicular* (ap"en-dik'u-lar) parts. The **axial part,** which makes up the main *axis* of our body, includes the head, neck, and trunk. The **appendicular part** consists of the *appendages,* or *limbs,* which are attached to the body's axis. **Regional terms** used to designate specific areas within the major body divisions are indicated in Figure 1.7. The common term for each of these body regions is also provided (in parentheses).

Body Planes and Sections

For anatomical studies, the body is often *sectioned* (cut) along a flat surface called a plane. The most frequently used body planes are sagittal, frontal, and transverse planes, which lie at right angles to one another (Figure 1.8 on p. 17). A section is named for the plane along which it is cut. Thus, a cut along a sagittal plane produces a sagittal section.

A **sagittal plane** (saj'ĭ-tal; "arrow") is a vertical plane that divides the body into right and left parts. A sagittal plane that lies exactly in the midline is the **median plane,** or **midsagittal plane** (Figure 1.8c). All other sagittal planes, offset from the midline, are **parasagittal planes** (*para* = near).

Frontal planes, like sagittal planes, lie vertically. Frontal planes, however, divide the body into anterior and posterior parts (Figure 1.8a). A frontal plane is also called a **coronal plane** (kŏ-ro'nal; "crown").

A **transverse,** or **horizontal, plane** runs horizontally from right to left, dividing the body into superior and inferior parts (Figure 1.8b). Of course, many different transverse planes exist at every possible level from head to foot. A transverse section is also called a **cross section.** Cuts made diagonally between the horizontal and the vertical planes are called **oblique sections.** Such sections are often confusing and difficult to interpret and are not used much.

The ability to interpret sections made through the body, especially transverse sections, is increasingly important in the clinical sciences. New medical imaging devices (described on pp. 20–21) produce sectional images rather than three-dimensional images. It can be difficult to decipher an object's overall shape from sectioned material. A cross section of a banana, for example, looks like a circle and gives no indication of the whole banana's crescent shape. Likewise, sectioning the body or an organ along different planes often results in very different views. For example, a transverse section of the body trunk at the level of the kidneys would show kidney structure in cross section very nicely; a frontal section of the body trunk would show a different view of kidney anatomy; and a midsagittal section would miss the kidneys completely. With practice, you will gradually learn to relate two-dimensional sections to three-dimensional shapes.

Body Cavities and Membranes

Within the axial portion of the body are two large cavities called the dorsal and ventral body cavities. These cavities are closed to the outside and each contains internal organs.

Text continues on page 18.

TABLE 1.1	**Orientation and Directional Terms**	
Term	*Definition*	*Example*
Superior (cranial)	Toward the head end or upper part of a structure or the body; above	The head is superior to the abdomen
Inferior (caudal)	Away from the head end or toward the lower part of a structure or the body; below	The navel is inferior to the chin
Anterior (ventral)*	Toward or at the front of the body; in front of	The breastbone is anterior to the spine
Posterior (dorsal)*	Toward or at the back of the body; behind	The heart is posterior to the breastbone
Medial	Toward or at the midline of the body; on the inner side of	The heart is medial to the arm
Lateral	Away from the midline of the body; on the outer side of	The arms are lateral to the chest
Intermediate	Between a more medial and a more lateral structure	The collarbone is intermediate between the breastbone and shoulder
Proximal	Closer to the origin of the body part or the point of attachment of a limb to the body trunk	The elbow is proximal to the wrist
Distal	Farther from the origin of a body part or the point of attachment of a limb to the body trunk	The knee is distal to the thigh
Superficial (external)	Toward or at the body surface	The skin is superficial to the skeletal muscles
Deep (internal)	Away from the body surface; more internal	The lungs are deep to the skin

*Whereas the terms *ventral* and *anterior* are synonymous in humans, this is not the case in four-legged animals. *Ventral* specifically refers to the "belly" of a vertebrate animal and thus is the inferior surface of four-legged animals. Likewise, although the dorsal and posterior surfaces are the same in humans, the term *dorsal* specifically refers to an animal's back. Thus, the dorsal surface of four-legged animals is their superior surface.

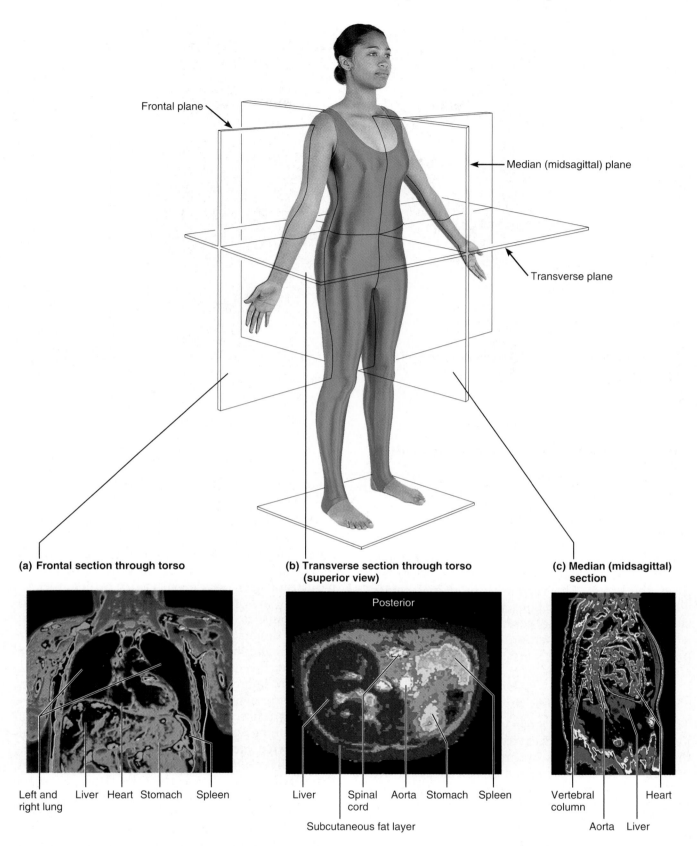

(a) Frontal section through torso

Left and right lung   Liver   Heart   Stomach   Spleen

(b) Transverse section through torso (superior view)

Posterior

Liver   Spinal cord   Aorta   Stomach   Spleen

Subcutaneous fat layer

(c) Median (midsagittal) section

Vertebral column   Heart

Aorta   Liver

Frontal plane

Median (midsagittal) plane

Transverse plane

FIGURE 1.8 *Planes of the body.* The three major planes of space (frontal, transverse, and median [midsagittal]) are superimposed on the photograph of a young woman shown in the anatomical position. Selected areas of the body, visualized using magnetic resonance imaging (MRI) scans of the body taken at corresponding planes, are illustrated below the photograph.

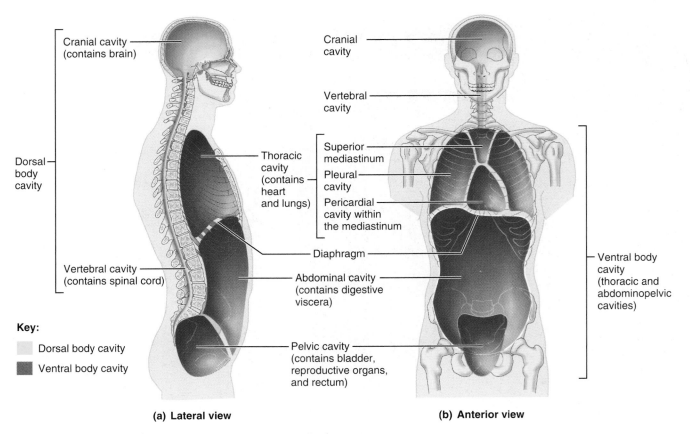

Key:
- Dorsal body cavity
- Ventral body cavity

(a) Lateral view   (b) Anterior view

FIGURE 1.9 *Dorsal and ventral body cavities and their subdivisions.*

Dorsal Body Cavity

The **dorsal body cavity,** which protects the fragile nervous system organs (Figure 1.9), has two subdivisions. The **cranial cavity,** within the skull, encases the brain. The **vertebral,** or **spinal cavity,** which runs within the bony vertebral column, encloses the delicate spinal cord. Since the spinal cord is essentially a continuation of the brain, the cranial and spinal cavities are continuous with one another.

Ventral Body Cavity

The more anterior and larger of the closed body cavities is the **ventral body cavity** (see Figure 1.9). Like the dorsal cavity, it has two major subdivisions, the *thoracic cavity* and the *abdominopelvic cavity.* The ventral body cavity houses a group of internal organs that are collectively called the **viscera** (vis′er-ah; *viscus* = an organ in a body cavity), or **visceral organs.**

The superior subdivision, the **thoracic** (tho-ras′ik) **cavity,** is surrounded by the ribs and muscles of the chest. The thoracic cavity is further subdivided into lateral **pleural** (ploo′ral) **cavities,** each housing a lung, and the medial **mediastinum** (me″de-ah-sti′num). The mediastinum contains the **pericardial** (per″ĭ-kar′de-al) **cavity,** which encloses the heart, and it also surrounds the remaining thoracic organs (esophagus, trachea, and others).

The thoracic cavity is separated from the more inferior **abdominopelvic** (ab-dom′ĭ-no-pel′vic) **cavity** by the diaphragm, a dome-shaped muscle important in breathing. The abdominopelvic cavity, as its name suggests, has two parts. However, these regions are not physically separated by a muscular or membrane wall. Its superior portion, the **abdominal cavity,** contains the stomach, intestines, spleen, liver, and other organs. The inferior part, the **pelvic cavity,** lies within the bony pelvis and contains the bladder, some reproductive organs, and the rectum. As Figure 1.9a reveals, the abdominal and pelvic cavities are not aligned with each other. Instead, the bowl-shaped pelvis tips away from the perpendicular.

Homeostatic Imbalance

When the body is subjected to physical trauma (as often happens in an automobile accident), the abdominopelvic organs are most vulnerable. This is because the walls of the abdominal cavity are formed only by trunk muscles and are not reinforced by bone. The pelvic organs receive a somewhat greater degree of protection from the bony pelvis. ∎

Membranes in the Ventral Body Cavity

The walls of the ventral body cavity and the outer surfaces of the organs it contains are covered by an exceedingly thin, double-layered membrane, the **serosa** (se-ro′sah), or **serous membrane.** The part of the membrane lining the cavity walls is called the **parietal** (pah-ri′ĕ-tal; *parie* = wall) **serosa.** It folds in on itself to form the **visceral serosa,** covering the organs in the cavity.

You can visualize the relationship between the serosal layers by pushing your fist into a limp balloon (Figure 1.10a). The part of the balloon that clings closely to your fist can be compared to the visceral serosa clinging to the organ's external surface. The outer wall of the balloon then represents the parietal serosa that lines the walls of the cavity. (However, unlike the balloon, it is never exposed but is always fused to the cavity wall.) In the body, the serous membranes are separated not by air but by a thin layer of lubricating fluid, called **serous fluid,** which is secreted by both membranes. Although there is a potential space between the two membranes, the slitlike space is filled with serous fluid.

The slippery serous fluid allows the organs to slide without friction across the cavity walls and one another as they carry out their routine functions. This freedom of movement is especially important for mobile organs such as the pumping heart and the churning stomach.

The serous membranes are named for the specific cavity and organs with which they are associated. For example, as shown in Figure 1.10b, the *parietal pericardium* lines the pericardial cavity, while the *visceral pericardium* covers the heart within that cavity. Likewise, the *parietal pleura* (ploo′rah) lines the walls of the thoracic cavity, and the *visceral pleura* covers the lungs; and the *parietal peritoneum* (per″ĭ-to-ne′um) is associated with the walls of the abdominopelvic cavity, while the *visceral peritoneum* covers most of the organs within that cavity. (The pleural and peritoneal serosae are illustrated in Figure 4.9c on p. 136.)

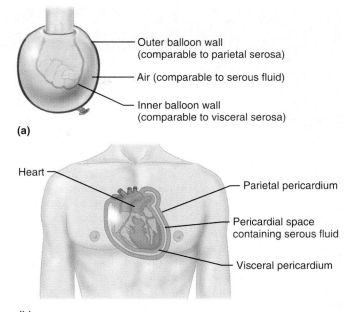

(a)

Outer balloon wall (comparable to parietal serosa)

Air (comparable to serous fluid)

Inner balloon wall (comparable to visceral serosa)

Heart

Parietal pericardium

Pericardial space containing serous fluid

Visceral pericardium

(b)

FIGURE 1.10 *Serous membrane relationships.* (a) A fist is thrust into a flaccid balloon to demonstrate the relationship between the parietal and visceral serous membrane layers. **(b)** The serosae associated with the heart. The parietal pericardium is the outer layer lining the pericardial cavity; the visceral pericardium clings to the external heart surface.

Homeostatic Imbalance

When the serous membranes are inflamed, they typically produce less of the lubricating serous fluid. This leads to excruciating pain as the organs stick together and drag across one another, as anyone who has experienced *pleurisy* (inflammation of the pleurae) or *peritonitis* (inflammation of the peritonea) knows. ∎

Other Body Cavities

In addition to the large closed body cavities, there are several smaller body cavities. Most of these are in the head and most open to the body exterior:

1. Oral and digestive cavities. The oral cavity, commonly called the mouth, contains the teeth and tongue. This cavity is part of and continuous with the cavity of the digestive organs, which opens to the exterior at the anus.

2. Nasal cavity. Located within and posterior to the nose, the nasal cavity is part of the respiratory system passageways.

Text continues on page 22.

A CLOSER LOOK Medical Imaging: Illuminating the Body

Physicians have long sought ways to examine the body's internal organs for evidence of disease without subjecting the patient to the shock and pain of exploratory surgery, but until about 30 years ago, the magical but murky X ray was the only means to extract information from within a living body. Produced by directing *X rays*, electromagnetic waves of very short wavelength, at the body, an **X ray** or **radiograph** is essentially a shadowy negative of the internal structures of the body. Dense structures absorb the X rays most and so appear as light areas on the radiograph. Hollow air-containing organs and fat, which absorb the X rays less, show up as dark areas. What X rays do best is visualize hard, bony structures and locate abnormally dense structures (tumors, tuberculosis nodules) in the lungs.

The 1950s saw the advent of nuclear medicine, which uses radioisotopes to scan the body, and ultrasound techniques. In the 1970s, CT, PET, and MRI scanning techniques were introduced. These new imaging techniques not only reveal the structure of our "insides" but also wring out information about the private, and until now secret, workings of their molecules.

The best known of the new imaging devices is **computed tomography (CT,** formerly called **computerized axial tomography, CAT)**, a refined version of X-ray equipment. As the patient is slowly moved through the doughnut-shaped CT machine, its X-ray tube rotates around the body and sends beams from all directions to a specific level of the patient's body. Because at any moment its beam is confined to a "slice" of the body about as thick as a dime, CT ends the confusion resulting from images of overlapping structures seen in conventional X rays. The device's computer translates this information into a detailed, cross-sectional picture of each of the consecutive body regions scanned.

CT scans are at the forefront of medical technology for evaluating most problems that affect the brain [see photo (a) for localization of a brain tumor] and abdomen. CT's clarity has all but eliminated exploratory surgery.

Xenon CT is a CT brain scan enhanced with xenon, a radioactive gas to quickly trace blood flow. Inhaled xenon rapidly enters the bloodstream and distributes to different body tissues in proportion to their blood flow. Absence of xenon from a part of the brain indicates that a stroke is occurring there, allowing better treatment planning.

Ultrafast CT scanners are used in a technique called **dynamic spatial reconstruction (DSR)**, which not only provides three-dimensional images of body organs from any angle but also allows their movements and changes in their internal volumes to be scrutinized at normal speed, in slow motion, and at a specific moment. Its greatest value has been to visualize the heart beating and blood flowing through blood vessels, allowing heart defects, constricted or blocked blood vessels, and the status of coronary bypass grafts to be evaluated.

Another computer-assisted X-ray technique is **digital subtraction angiography (DSA)** (*angiography* = "vessel pictures"). This technique provides an unobstructed view of small arteries. Its principle is simple: Conventional radiographs are taken before and after a contrast medium is injected into an artery. Then the computer subtracts the "before" image from the "after" image, eliminating all traces of body structures that obscure the vessel. DSA is often used to identify blockages in the arteries that supply the heart wall and the brain [photo (b)].

Just as the X ray spawned "big brothers," so too did nuclear medicine in the form of **positron emission tomography (PET)**. PET excels in observing *metabolic processes*. The patient is given an injection of radioisotopes tagged to biological molecules (such as glucose) and is then positioned in the PET scanner.

As the radioisotopes are absorbed by the most active brain cells, high-energy gamma rays are produced. The computer analyzes the gamma-ray emission and produces a live-action picture of the brain's biochemical activity in vivid colors. PET's greatest value has been its ability to provide insights into brain activity in people affected by mental illness, stroke, Alzheimer's disease, and epilepsy. One of its most exciting uses has been to determine which specific regions of the *healthy* brain are most active during certain tasks (e.g., speaking, listening to music, or figuring out a mathematical problem), thereby providing direct evidence of the functions of specific brain regions.

Sonography, or **ultrasound imaging**, has some distinct advantages over the approaches examined so far. First, the equipment is inexpensive. Second, the high-frequency sound waves (ultrasound) used as its energy source seem to be safer than ionizing forms of radiation used in nuclear medicine. The body is probed with pulses of sound waves that cause echoes when reflected and scattered to different extents by body tissues. A computer analyzes these echoes to construct somewhat blurry visual images of the outlines of body organs of interest. A single, handheld device is used both to emit the sound and to pick up the echoes. This device is easy to move around the body, so sections can be scanned from many different body planes.

Because of its safety, ultrasound is the imaging technique of choice in obstetrics for determining fetal age and position and locating the placenta [see photo (c)]. It is also used to visualize the gallbladder, and increasingly to search for atherosclerotic plaques in the arteries. Because sound waves have very low penetrating power and rapidly dissipate in air, sonography is of little value for looking at air-filled structures (the lungs) or those surrounded by bone (the brain and spinal cord).

Magnetic resonance imaging (MRI) is a technique with tremendous appeal because it produces high-

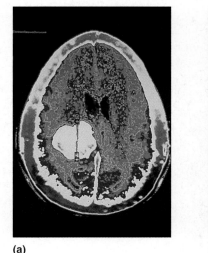
(a)

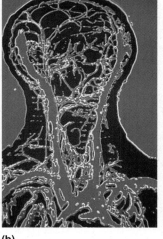
(b)

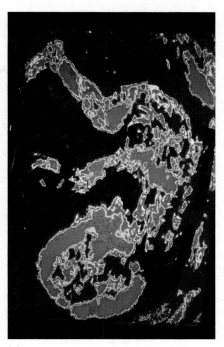

(c)

Three different methods for illuminating the body. (**a**) CT scan show-ing a brain tumor (oval yellow area on left side of the brain scan). (**b**) DSA of head and neck blood vessels. (**c**) Computer-enhanced ultrasound image of a fetus in utero; the head, trunk, and limbs are clearly visible.

contrast images of our *soft* tissues, an area in which X rays and CT scans are weak. As initially developed, MRI primarily maps the element hydrogen. In our body, most of the hydrogen is in water. The technique subjects the body to magnetic fields up to 60,000 times stronger than that of the earth to pry information from the body's molecules. The patient lies in a chamber within a huge magnet. Hydrogen molecules spin like tops in the magnetic field, and their energy is further enhanced by radio waves. When the radio waves are turned off, the energy released is translated into a visual image.

Because MRI tends to distinguish body tissues from one another on the basis of differences in water content, it can differentiate between the fatty white matter and the more watery gray matter of the brain and enable physicians to see the delicate nerve fibers in the spinal cord. Because dense structures do not show up at all in MRI, it peers easily into the skull and vertebral column. MRI is also particularly good at detecting tumors and degenerative disease of various kinds. Multiple sclerosis plaques do not show up well in CT scans, but are dazzlingly clear in MRI scans. MRI can also tune in on metabolic

reactions, such as processes that generate energy-rich ATP molecules. A newer variation of MRI, called **magnetic resonance spectroscopy (MRS),** maps the distribution of elements other than hydrogen to reveal more about how disease changes body chemistry. Furthermore, advances in computer techniques are allowing MRI scans to be displayed in three dimensions and used to guide laser surgery. In 1992 MRI technology leaped forward with the development of the **functional MRI,** which allows tracking of blood flow into the brain in real time. Until then, matching thoughts, deeds, and disease to corresponding brain activity had been the sole domain of PET. Because functional MRI does not require injections of tracer elements and can pinpoint much smaller brain areas than PET, it may provide a desirable alternative. Computer scientists are now using advanced volume rendering techniques to assemble multiple MRI scans into three-dimensional images. As more powerful computers become available, the speed of these recon-structions is increasing rapidly.

Despite its advantages, the powerful magnets of the clanging,

claustrophobia-inducing MRI present some thorny problems. For example, they can "suck" metal objects, such as implanted pacemakers and loose tooth fillings, through and from the body. Moreover, there is no convincing evidence that such strong magnetic fields can be used without risk to the body, although the procedure is currently considered safe.

Although they are stunning, the pictures produced by the new imaging devices are abstractions assembled within the "mind" of a computer. They are artificially enhanced for sharpness and artificially colored to increase contrast (all colors are "phony"). The images are not inaccurate; they are, however, several steps removed from direct visual observation.

As you can see, medical science has some remarkable new diagnostic tools that are advancing rapidly. Soon, it should be possible to generate immediate 3-D images while patients are still in the doctor's office or surgical suite, tremendously simplifying diagnoses and aiding surgical decisions. With modern telecommunications, surgeons will even be able to direct operations on patients great distances away.

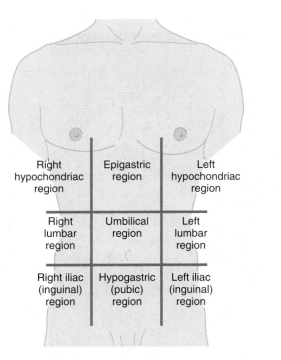

(a)

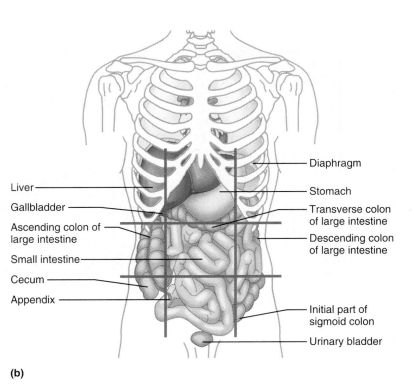

- Diaphragm
- Stomach
- Transverse colon of large intestine
- Descending colon of large intestine
- Initial part of sigmoid colon
- Urinary bladder

Liver
Gallbladder
Ascending colon of large intestine
Small intestine
Cecum
Appendix

(b)

FIGURE 1.11 *The nine abdominopelvic regions.* (**a**) Division of the abdominopelvic cavity into nine regions delineated by four planes. The superior transverse plane is just inferior to the ribs; the inferior transverse plane is just superior to the hip bones; and the parasagittal planes lie just medial to the nipples. (**b**) Anterior view of the abdominopelvic cavity showing the superficial organs.

3. Orbital cavities. The orbital cavities (orbits) in the skull house the eyes and present them in an anterior position.

4. Middle ear cavities. The middle ear cavities carved into the skull lie just medial to the eardrums. These cavities contain tiny bones that transmit sound vibrations to the organ of hearing in the inner ears.

5. Synovial (sĭ-no've-al) **cavities.** Synovial cavities are joint cavities. They are enclosed within fibrous capsules that surround freely movable joints of the body (such as the elbow and knee joints). Like the serous membranes of the ventral body cavity, membranes lining the synovial cavities secrete a lubricating fluid that reduces friction as the bones move across one another.

Abdominopelvic Regions and Quadrants

Because the abdominopelvic cavity is large and contains many organs, it is helpful to divide it into smaller areas for study. One division method, used primarily by anatomists, uses two transverse and two parasagittal planes. These planes, positioned like a tic-tac-toe grid on the abdomen, divide the cavity into nine **regions** (Figure 1.11):

- The **umbilical region** is the centermost region deep to and surrounding the umbilicus (navel).

- The **epigastric region** is located superior to the umbilical region (*epi* = upon, above; *gastri* = belly).

- The **hypogastric (pubic) region** is located inferior to the umbilical region (*hypo* = below).

■ The **right** and **left iliac,** or **inguinal** (ing'gwĭ-nal), **regions** are located lateral to the hypogastric region (*iliac* = superior part of the hip bone).

■ The **right** and **left lumbar regions** lie lateral to the umbilical region (*lumbus* = loin).

■ The **right** and **left hypochondriac regions** flank the epigastric region laterally (*chondro* = cartilage).

Medical personnel usually use a simpler scheme to localize the abdominopelvic cavity organs (Figure 1.12). In this scheme, one transverse and one median sagittal plane pass through the umbilicus at right angles. The resulting **quadrants** are named according to their positions from the subject's point of view: the **right upper quadrant (RUQ), left upper quadrant (LUQ), right lower quadrant (RLQ),** and **left lower quadrant (LLQ).**

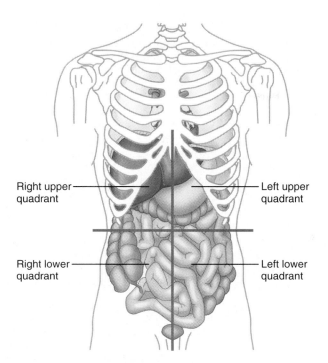

FIGURE 1.12 *The four abdominopelvic quadrants.* In this scheme, the abdominopelvic cavity is divided into four quadrants by two planes. The figure shows superficial organs within each quadrant.

CHAPTER SUMMARY

Media study tools that could provide you additional help in reviewing specific key topics of Chapter 1 are referenced below. **SP** *= Study Partner;* **IP** *= Interactive Physiology.*

An Overview of Anatomy and Physiology (pp. 3–4)

1. Anatomy is the study of body structures and their relationships. Physiology is the science of how body parts function.

Topics of Anatomy (pp. 3–4)

2. Major subdivisions of anatomy include gross anatomy, microscopic anatomy, and developmental anatomy.

Topics of Physiology (p. 4)

3. Typically, physiology concerns the functioning of specific organs or organ systems. Examples include cardiac physiology, renal physiology, and muscle physiology.

4. Physiology is explained by chemical and physical principles.

Complementarity of Structure and Function (p. 4)

5. Anatomy and physiology are inseparable: What a body can do depends on the unique architecture of its parts. This is called the complementarity of structure and function.

Levels of Structural Organization (pp. 4–7)

1. The levels of structural organization of the body, from simplest to most complex, are: chemical, cellular, tissue, organ, organ system, and organismal.

2. The 11 organ systems of the body are the integumentary, skeletal, muscular, nervous, endocrine, cardiovascular, lymphatic, respiratory, digestive, urinary, and reproductive systems. The immune system is a functional system closely associated with the lymphatic system. (For functions of these systems see pp. 6–7)

Maintaining Life (pp. 8–10)

Necessary Life Functions (pp. 8–9)

1. All living organisms carry out certain vital functional activities necessary for life, including maintenance of boundaries, movement, responsiveness, digestion, metabolism, excretion, reproduction, and growth.

Survival Needs (pp. 9–10)

2. Survival needs include nutrients, water, oxygen, appropriate body temperature, and atmospheric pressure.

Homeostasis (pp. 10–13)

1. Homeostasis is a dynamic equilibrium of the internal environment. All body systems contribute to homeostasis, but the nervous and endocrine systems are most important. Homeostasis is necessary for health.

Homeostatic Control Mechanisms (pp. 10–13)

2. Control systems of the body contain at least three elements: receptor(s), control center, and effector(s).

3. Negative feedback mechanisms reduce the original stimulus, and are essential for maintaining homeostasis. Body temperature, heart rate, breathing rate and depth, and blood levels of glucose and certain ions are regulated by negative feedback mechanisms.

4. Positive feedback mechanisms intensify the initial stimulus, leading to an enhancement of the response. Positive feedback mechanisms rarely contribute to homeostasis, but blood clotting and labor contractions are regulated by such mechanisms.

SP Exercise: Chapter 1, Dorsal and Ventral Body Cavities.

Homeostatic Imbalance   *(p. 13)*

5. With age, the efficiency of negative feedback mechanisms declines, and positive feedback mechanisms occur more frequently. These changes underlie certain disease conditions.

The Language of Anatomy   (pp. 14–23)

Anatomical Position and Directional Terms   *(pp. 15–16)*

1. In the anatomical position, the body is erect, facing forward, feet together, arms at sides with palms forward.

2. Directional terms allow body parts to be located precisely. Terms used to describe body directions and orientation include: superior/inferior; anterior/posterior; ventral/dorsal; medial/lateral; intermediate; proximal/distal; and superficial/deep.

Regional Terms   *(pp. 14–15)*

3. Regional terms are used to designate specific areas of the body (see Figure 1.7).

Body Planes and Sections   *(pp. 15, 17)*

4. The body or its organs may be cut along planes, or imaginary lines, to produce different types of sections. Frequently used planes are sagittal, frontal, and transverse.

SP Exercises: Chapter 1, Body Planes; Case Studies: Congenital Defect; Skeletal Disorder.

Body Cavities and Membranes   *(pp. 15, 18–19, 22)*

5. The body contains two major closed cavities. The dorsal cavity, subdivided into the cranial and spinal cavities, contains the brain and spinal cord. The ventral cavity is subdivided into the superior thoracic cavity, which houses the heart and lungs, and the inferior abdominopelvic cavity, which contains the liver, digestive organs, and reproductive structures.

6. The walls of the ventral cavity and the surfaces of the organs it contains are covered with thin membranes, the parietal and visceral serosae, respectively. The serosae produce a thin fluid that decreases friction during organ functioning.

7. There are several smaller body cavities. Most of these are in the head and open to the exterior.

SP Exercise: Chapter 1, Dorsal and Ventral Body Cavities; Case Studies: Congentital Defect and Skeletal Disorders.

Abdominopelvic Regions and Quadrants   *(pp. 22–23)*

8. The abdominopelvic cavity may be divided by four planes into nine abdominal regions (epigastric, umbilical, hypogastric, right and left iliac, right and left lumbar, and right and left hypochondriac), or by two planes into four quadrants. (For boundaries and organs contained see Figures 1.11 and 1.12.)

REVIEW QUESTIONS

Multiple Choice/Matching

(Some questions have more than one correct answer.)

1. The correct sequence of levels forming the structural hierarchy is (a) organ, organ system, cellular, chemical, tissue, organismal; (b) chemical, cellular, tissue, organismal, organ, organ system; (c) chemical, cellular, tissue, organ, organ system, organismal; (d) organismal, organ system, organ, tissue, cellular, chemical.

2. The structural and functional unit of life is (a) a cell, (b) an organ, (c) the organism, (d) a molecule.

3. Which of the following is a *major* functional characteristic of all organisms? (a) movement, (b) growth, (c) metabolism, (d) responsiveness, (e) all of these.

4. Two of these organ systems bear the *major* responsibility for ensuring homeostasis of the internal environment. Which two? (a) nervous system, (b) digestive system, (c) cardiovascular system, (d) endocrine system, (e) reproductive system.

5. In (a)–(e), a directional term [e.g., distal in (a)] is followed by terms indicating different body structures or locations (e.g., the elbow/the wrist). In each case, choose the structure or organ that matches the given directional term.
(a) distal: the elbow/the wrist
(b) lateral: the hip bone/the umbilicus
(c) superior: the nose/the chin
(d) anterior: the toes/the heel
(e) superficial: the scalp/the skull

6. Assume that the body has been sectioned along three planes: (1) a median sagittal plane, (2) a frontal plane, and (3) a transverse plane made at the level of each of the organs listed below. Which organs would not be visible in all three cases? (a) urinary bladder, (b) brain, (c) lungs, (d) kidneys, (e) small intestine, (f) heart.

7. Relate each of the following conditions or statements to either the dorsal body cavity or the ventral body cavity.
(a) surrounded by the bony skull and the vertebral column
(b) includes the thoracic and abdominopelvic cavities
(c) contains the brain and spinal cord
(d) contains the heart, lungs, and digestive organs

8. Which of the following relationships is *incorrect*?
(a) visceral peritoneum/outer surface of small intestine,
(b) parietal pericardium/outer surface of heart, (c) parietal pleura/wall of thoracic cavity.

9. Which ventral cavity subdivision has no bony protection? (a) thoracic cavity, (b) abdominal cavity, (c) pelvic cavity.

Short Answer Essay Questions

10. According to the principle of complementarity, how does anatomy relate to physiology?

11. Construct a table that lists the 11 systems of the body, name two organs of each system (if appropriate), and describe the overall or major function of each system.

12. List and describe briefly five external factors that must be present or provided to sustain life.

13. Define homeostasis.

14. Compare and contrast the operation of negative and positive feedback mechanisms in maintaining homeostasis. Provide two examples of variables controlled by negative feedback mechanisms and one example of a process regulated by a positive feedback mechanism.

15. Describe and assume the anatomical position. Why is an understanding of this position important? What is the importance of directional terms?

16. Define plane and section.

17. Provide the anatomical term that correctly names each of the following body regions: (a) arm, (b) thigh, (c) chest, (d) fingers and toes, (e) anterior aspect of the knee.

18. (a) Make a diagram showing the nine abdominopelvic regions, and name each region. Name two organs (or parts of organs) that could be located in each of the named regions. (b) Make a similar sketch illustrating how the abdominopelvic cavity may be divided into quadrants, and name each quadrant.

 ### Critical Thinking and Clinical Application Questions

1. John has been suffering agonizing pain with each breath and has been informed by the physician that he has pleurisy. (a) Specifically, what membranes are involved in this condition? (b) What is their usual role in the body? (c) Explain why John's condition is so painful.

2. At the clinic, Harry was told that blood would be drawn from his antecubital region. What body part was Harry asked to hold out? Later, the nurse came in and gave Harry a shot of penicillin in his deltoid region. Did Harry take off his shirt or drop his pants to receive the injection? Before Harry left, the nurse noticed that Harry had a nasty bruise on his gluteal region. What part of his body was black and blue?

3. A man is behaving abnormally, and his physician suspects that he has a brain tumor. Which of the following medical imaging devices would best localize the tumor in the man's brain (and why)? Conventional X ray, DSA, PET, sonography, MRI.

4. When we become dehydrated, we usually feel thirsty, which causes us to drink fluids. Decide whether the thirst sensation is part of a negative or positive feedback control system and defend your choice.

2 CHEMISTRY COMES ALIVE

PART 1: BASIC CHEMISTRY

Definition of Concepts: Matter and Energy (pp. 27–28)

1. Differentiate clearly between matter and energy and between potential energy and kinetic energy.
2. Describe the major energy forms.

Composition of Matter: Atoms and Elements (pp. 28–32)

3. Define chemical element and list the four elements that form the bulk of body matter.
4. Define atom. List the subatomic particles; describe their relative masses, charges, and positions in the atom.
5. Define atomic number, atomic mass, atomic weight, isotope, and radioisotope.

How Matter Is Combined: Molecules and Mixtures (pp. 32–33)

6. Distinguish between a compound and a mixture. Define molecule.
7. Compare solutions, colloids, and suspensions.

Chemical Bonds (pp. 33–38)

8. Explain the role of electrons in chemical bonding and in relation to the octet rule.
9. Differentiate between ionic and covalent bonds. Contrast these bonds with hydrogen bonds.
10. Compare and contrast polar and nonpolar compounds.

Chemical Reactions (pp. 38–41)

11. Identify three major types of chemical reactions (synthesis, decomposition, and exchange). Comment briefly on the nature of oxidation-reduction reactions and their importance.
12. Explain why chemical reactions in the body are often irreversible.
13. Describe factors that affect chemical reaction rates.

PART 2: BIOCHEMISTRY

Inorganic Compounds (pp. 41–45)

14. Explain the importance of water and salts to body homeostasis.
15. Define acid and base, and explain the concept of pH.

Organic Compounds (pp. 45–60)

16. Describe and compare the building blocks, general structures, and biological functions of carbohydrates, lipids, proteins, and nucleic acids.
17. Explain the role of dehydration synthesis and hydrolysis in the formation and breakdown of organic molecules.
18. Describe the four levels of protein structure.
19. Describe the general mechanism of enzyme activity.
20. Describe the function of molecular chaperones.
21. Compare and contrast DNA and RNA.
22. Explain the role of ATP in cell metabolism.

Why study chemistry in an anatomy and physiology course? The answer is simple. The food you eat and the medicines you take when you are ill are composed of chemical substances. Indeed, your entire body is made up of chemicals, thousands of them, continuously interacting with one another at an incredible pace. Although it is possible to study anatomy without much reference to chemistry, chemical reactions underlie all physiological processes—movement, digestion, the pumping of your heart, and even your thoughts. This chapter presents the basics of chemistry and biochemistry (the chemistry of living material), providing the background you will need to understand body functions.

■ PART 1 ■
BASIC CHEMISTRY

DEFINITION OF CONCEPTS: MATTER AND ENERGY

Matter

Matter is the "stuff" of the universe. More precisely, matter is anything that occupies space and has mass. For all practical purposes, we can consider mass to be the same as weight, and we will use these terms interchangeably. However, this usage is not quite accurate. The *mass* of an object, which is equal to the actual amount of matter in the object, remains constant wherever the object is. In contrast, weight varies with gravity. So while your mass is the same at sea level and on a mountaintop, you weigh just slightly less on that mountaintop. The science of chemistry studies the nature of matter, especially how its building blocks are put together and interact with one another.

States of Matter

Matter exists in *solid, liquid,* and *gaseous states.* Examples of each state are found in the human body. Solids, like bones and teeth, have a definite shape and volume. Liquids such as blood plasma have a definite volume, but they conform to the shape of their container. Gases have neither a definite shape nor a definite volume. The air we breathe is a gas.

Energy

Compared with matter, **energy** is less tangible. Energy has no mass and does not take up space. It can be measured only by its effects on matter. Energy is defined as the capacity to do work, or to put matter into motion. The greater the work done, the more energy is used doing it. A baseball player who has just hit the ball over the fence uses much more energy than a batter who just bunts the ball back to the pitcher.

Kinetic Versus Potential Energy

Energy exists in two forms, or work capacities, each transformable to the other. **Kinetic** (ki-net′ik) **energy** is energy in action. Kinetic energy is seen in the constant movement of the tiniest particles of matter (atoms) as well as in larger objects (a bouncing ball). It does work by moving objects, which in turn can do work by moving or pushing on other objects. For example, a push on a swinging door sets it into motion.

Potential energy is stored energy, that is, inactive energy that has the *potential,* or capability, to do work but is not presently doing so. The batteries in an unused toy have potential energy, as do your leg muscles when you sit still on the couch. When potential energy is released, it becomes kinetic energy and so is capable of doing work. For example, dammed water becomes a rushing torrent when the dam is opened, and that rushing torrent can move a turbine at a hydroelectric plant, or charge a battery.

Actually, energy is a topic of physics, but matter and energy are inseparable. Matter is the substance, and energy is the mover of the substance. All living things are composed of matter and they all require energy to grow and function. It is the release and use of energy by living systems that gives us the elusive quality called life. Thus, it is worth taking a brief detour to introduce the forms of energy used by the body as it does its work.

Forms of Energy

The body uses energy in several forms. **Chemical energy** is the form stored in the bonds of chemical substances. When chemical reactions occur that rearrange the atoms of the chemicals in a certain way, the potential energy is unleashed and becomes kinetic energy.

For example, some of the energy in the foods you eat is eventually converted into the kinetic energy of your moving arm. However, food fuels cannot be used to energize body activities directly. Instead, some of the food energy is captured temporarily in the bonds of a chemical called **adenosine triphosphate** (ah-den′o-sēn tri″fos′fāt), or **ATP.** Later, ATP's bonds are broken and the stored energy is released as needed to do cellular work. Chemical energy in the form of ATP is the most fundamental form of energy in living systems because it is used to run all functional processes. (ATP is described more fully on pp. 59–60.)

Electrical energy reflects the movement of charged particles. In your home, electrical energy is found in the flow of electrons along the household wiring. In your body, electrical currents are generated as charged particles called *ions* move across cell membranes. The nervous system uses electrical currents, called *nerve impulses,* to transmit messages from one part of the body to another. Electrical currents traveling across the heart stimulate it to contract (beat) and pump blood. (This is why a strong electrical shock, which interferes with such currents, can cause death.)

Mechanical energy is energy *directly* involved in moving matter. When you ride a bicycle, your legs provide the mechanical energy that moves the pedals.

Radiant energy, or **electromagnetic** (e-lek″tro-mag-net′ik) **energy,** is energy that travels in waves. These waves, which vary in length, are collectively called the *electromagnetic spectrum* and include visible light, infrared waves, radio waves, ultraviolet waves, and X rays (see Figure 16.13, p. 574). When light strikes the retinas of our eyes, it sets off a chain of reactions that eventually results in vision. Ultraviolet waves cause sunburn, but they also stimulate our body to make vitamin D. X rays have little to do with normal body functioning, but they are important in medical diagnosis.

With few exceptions, energy can easily be converted from one form to another. For example, chemical energy (gasoline) that powers the motor of a speedboat is converted into the mechanical energy of the whirling propeller that allows the boat to skim across the water. Energy conversions are quite inefficient; some of the initial energy supply is always "lost" to the environment as heat. (It is not really lost because energy cannot be created or destroyed, but that part given off as heat is at least partly *unusable.*) It is easy to demonstrate this principle. Electrical energy is converted into light energy in a light bulb. But if you touch a lit bulb, you will soon discover that some of the electrical energy is producing heat instead. Likewise, all energy conversions that occur in the body liberate heat. This heat makes us warm-blooded animals and contributes to our relatively high body temperature, which has an important influence on body functioning. For example, when matter is heated, the kinetic energy of its particles increases and they begin to move more quickly. The higher the temperature, the faster the body's chemical reactions occur. We will learn more about this later.

COMPOSITION OF MATTER: ATOMS AND ELEMENTS

All matter is composed of fundamental substances that are called **elements.** Elements are unique substances that cannot be broken down into simpler substances by ordinary chemical methods. Among the well-known elements are oxygen, carbon, gold, silver, copper, and iron. At present, 112 elements are known with certainty (and numbers 114, 116, and 118 are alleged). Of the known elements, 92 occur in nature. The rest are made artificially in particle accelerator devices. Four elements—carbon, oxygen, hydrogen, and nitrogen—make up about 96% of body weight; 20 others are present in the body in trace amounts. The elements contributing to body mass and their importance are given in Table 2.1. The **periodic table** in Appendix D provides a more complete listing of the known elements.

Each element is composed of more or less identical particles or building blocks, called **atoms.** The smallest atoms are less than 0.1 nanometer (nm) in diameter, and the largest only about five times as large. (1 nm = 0.0000001 [or 10^{-7}] centimeter [cm], or 40 billionths of an inch!)

Each element's atoms differ from those of all other elements and give the element its unique physical and chemical properties. *Physical properties* are those we can detect with our senses (such as color and texture) or measure (such as boiling point and freezing point). *Chemical properties* pertain to the way atoms interact with other atoms (bonding behavior) and account for the facts that iron rusts, gasoline burns in air, animals can digest their food, and so on.

Each element is designated by a one- or two-letter chemical shorthand called an **atomic symbol,** usually the first letter(s) of the element's name. For example, C stands for carbon, O for oxygen, and Ca for calcium. In a few cases, the atomic symbol is taken from the Latin name for the element. For example, sodium is indicated by Na, from the Latin word *natrium.*

Atomic Structure

The word *atom* comes from the Greek word meaning "indivisible." However, we now know that atoms are clusters of even smaller particles called protons, neutrons, and electrons and that even those subatomic particles can be subdivided with high-technology tools. Even so, the old idea of atomic indivisibility is still helpful because an atom loses the unique properties of its element when it is split into its subatomic particles.

An atom's subatomic particles differ in mass, electrical charge, and position in the atom. An atom has a central **nucleus** containing protons and neutrons tightly bound together. The nucleus, in turn, is surrounded by orbiting electrons (Figure 2.1). **Protons** (p^+) bear a positive electrical charge, while **neutrons** (n^0) are neutral. Thus, the nucleus is positively charged overall. Protons and neutrons are heavy particles and have approximately the same mass, which is arbitrarily designated as 1 **atomic mass unit** (1 amu). Since all of the heavy subatomic particles are

TABLE 2.1		Common Elements Composing the Human Body*	
Element	**Atomic symbol**	**Approx. % body mass**[†]	**Functions**

Major (96.1%)

Element	Atomic symbol	Approx. % body mass	Functions
Oxygen	O	65.0	A major component of both organic (carbon-containing) and inorganic (non-carbon-containing) molecules; as a gas, it is needed for the production of cellular energy (ATP)
Carbon	C	18.5	A primary component of all organic molecules, which include carbohydrates, lipids (fats), proteins, and nucleic acids
Hydrogen	H	9.5	A component of all organic molecules; as an ion (proton), it influences the pH of body fluids
Nitrogen	N	3.2	A component of proteins and nucleic acids (genetic material)

Lesser (3.9%)

Element	Atomic symbol	Approx. % body mass	Functions
Calcium	Ca	1.5	Found as a salt in bones and teeth; its ionic (Ca^{2+}) form is required for muscle contraction, conduction of nerve impulses, and blood clotting
Phosphorus	P	1.0	Part of calcium phosphate salts in bones and teeth; also present in nucleic acids; part of ATP
Potassium	K	0.4	Its ion (K^+) is the major positive ion (cation) in cells; necessary for conduction of nerve impulses and muscle contraction
Sulfur	S	0.3	Component of proteins, particularly muscle proteins
Sodium	Na	0.2	As an ion (Na^+), sodium is the major positive ion found in extracellular fluids (fluids outside of cells); important for water balance, conduction of nerve impulses, and muscle contraction
Chlorine	Cl	0.2	Ionic chlorine (Cl^-) is the most abundant negative ion (anion) in extracellular fluids
Magnesium	Mg	0.1	Present in bone; also an important cofactor in a number of metabolic reactions
Iodine	I	0.1	Needed to make functional thyroid hormones
Iron	Fe	0.1	Component of hemoglobin (which transports oxygen within red blood cells) and some enzymes

Trace (less than 0.01%)

Chromium (Cr); Cobalt (Co); Copper (Cu); Fluorine (F); Manganese (Mn); Molybdenum (Mo); Selenium (Se); Silicon (Si); Tin (Sn); Vanadium (V); Zinc (Zn)

These elements are referred to as *trace elements* because they are required in very minute amounts; many are found as part of enzymes or are required for enzyme activation.

*A listing of the elements by ascending order of atomic number appears in the periodic table, Appendix D.

[†]Percentage of "wet" body mass; includes water.

concentrated in the nucleus, the nucleus is fantastically dense and accounts for nearly the entire mass (99.9%) of the atom. The tiny **electrons** (e^-) bear a negative charge equal in strength to the positive charge of the proton. However, an electron has only about 1/2000 the mass of a proton, and the mass of an electron is usually designated as 0 amu.

Because all atoms are electrically neutral, the number of protons in an atom must be precisely balanced by its number of electrons (the + and − charges will then cancel the effect of each other). Thus, hydrogen has one proton and one electron,

and iron has 26 protons and 26 electrons. For any atom, the number of protons and electrons is always equal.

The **planetary model,** illustrated in Figure 2.1a, is a simplified (and now outdated) model of atomic structure. As you can see, it depicts electrons moving around the nucleus in fixed, generally circular orbits. But we can never determine the exact location of electrons at a particular time because they jump around following unknown trajectories. So, instead of speaking of discrete orbits, chemists talk about **orbitals**—regions around the nucleus in

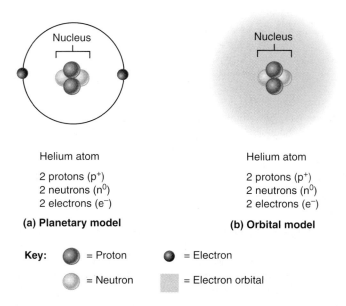

Helium atom

2 protons (p^+)
2 neutrons (n^0)
2 electrons (e^-)

(a) Planetary model

Helium atom

2 protons (p^+)
2 neutrons (n^0)
2 electrons (e^-)

(b) Orbital model

Key: ● = Proton   ● = Electron

○ = Neutron   ▨ = Electron orbital

FIGURE 2.1 *The structure of an atom.* The dense central nucleus contains the protons and neutrons. **(a)** In the planetary model of atomic structure, the electrons move around the nucleus in fixed orbits. **(b)** The orbital model recognizes that we never know exactly where electrons are; therefore electrons are shown as a cloud of negative charge (electron cloud).

which a given electron or electron pair is likely to be found most of the time. This more modern model of atomic structure, called the **orbital model,** is more useful for predicting the chemical behavior of atoms. As illustrated in Figure 2.1b, the orbital model depicts *probable* regions of greatest electron density by denser shading (this is called the *electron cloud*). However, because of the simplicity of the planetary model, most descriptions of atomic structure in this text will use that model.

Hydrogen, with just one proton and one electron, is the simplest atom. You can visualize the spatial relationships in the hydrogen atom by imagining it enlarged until its diameter equals the length of a foot-

ball field. In that case, the nucleus could be represented by a lead ball the size of a gumdrop in the exact center of the sphere and its lone electron pictured as a fly buzzing about unpredictably within the sphere. Though not completely accurate, this mental picture should serve as a reminder that most of the volume of an atom is empty space, and nearly all of its mass is concentrated in the central nucleus.

Identifying Elements

All protons are alike, regardless of the atom considered. The same is true of all neutrons and all electrons. So what determines the unique properties of each element? The answer is that atoms of different elements are composed of *different numbers* of protons, neutrons, and electrons.

The simplest and smallest atom, hydrogen, has one proton, one electron, and no neutrons (Figure 2.2). Next in size is the helium atom, with two protons, two neutrons, and two orbiting electrons. Lithium follows with three protons, four neutrons, and three electrons. If we continued this step-by-step progression, we would get a graded series of atoms containing from 1 to 112 protons, an equal number of electrons, and a slightly larger number of neutrons at each step. However, all we really need to know to identify a particular element are its atomic number, mass number, and atomic weight. Taken together, these provide a fairly complete profile of each element.

Atomic Number

The **atomic number** of any atom is equal to the number of protons in its nucleus and is written as a subscript to the left of its atomic symbol. Hydrogen, with one proton, has an atomic number of 1 ($_1$H); helium, with two protons, has an atomic number of 2 ($_2$He); and so on. The number of protons is always equal to the number of electrons in an atom; so the atomic number *indirectly* tells us

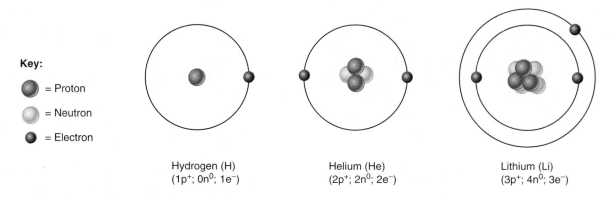

Key:

● = Proton

○ = Neutron

● = Electron

Hydrogen (H)
($1p^+$; $0n^0$; $1e^-$)

Helium (He)
($2p^+$; $2n^0$; $2e^-$)

Lithium (Li)
($3p^+$; $4n^0$; $3e^-$)

FIGURE 2.2 *Atomic structure of the three smallest atoms.*

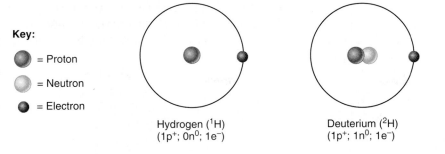
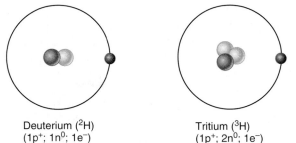

Key:

⬤ = Proton

◯ = Neutron

● = Electron

Hydrogen (^{1}H)
($1p^+$; $0n^0$; $1e^-$)

Deuterium (^{2}H)
($1p^+$; $1n^0$; $1e^-$)

Tritium (^{3}H)
($1p^+$; $2n^0$; $1e^-$)

FIGURE 2.3 *Isotopes of hydrogen.*

the number of electrons in the atom as well. As explained shortly, this is important information indeed, because electrons determine the chemical behavior of atoms.

Mass Number and Isotopes

The **mass number** of an atom is the sum of the masses of its protons and neutrons. (The mass of the electrons is so small that it is ignored.) Hydrogen has only one proton in its nucleus, so its atomic and mass numbers are the same: 1. Helium, with two protons and two neutrons, has a mass number of 4. The mass number is usually indicated by a superscript to the left of the atomic symbol. Thus, helium can be described by the notation $^4_2$He. This simple notation allows us to deduce the total number and kinds of subatomic particles in any atom because it indicates the number of protons (the atomic number), the number of electrons (equal to the atomic number), and the number of neutrons (mass number minus atomic number).

From what we have said so far, it may appear as if each element has one, and only one, type of atom representing it. This is not the case. Nearly all known elements have two or more atomic variants called **isotopes** (i'so-tōps), which have the same atomic number but vary in their mass numbers (and therefore their weights). Another way of saying this is that all the isotopes of an element have the same number of protons (and electrons), but differ in the number of neutrons they contain. Earlier, when we said that hydrogen has a mass number of 1, we were speaking of ^{1}H, its most abundant isotope. Some hydrogen atoms have a mass of 2 or 3 amu (atomic mass units), which means that they have one proton and, respectively, one or two neutrons (Figure 2.3). Carbon has several isotopic forms. The most abundant of these are ^{12}C, ^{13}C, and ^{14}C. Each of the carbon isotopes has six protons (otherwise it would not be carbon), but ^{12}C has six neutrons, ^{13}C has seven, and ^{14}C has eight. Isotopes are also written with the mass number following the symbol: C-14, for example.

Atomic Weight

It would seem that atomic weight should be the same as atomic mass, and this would be so if atomic weight referred to the weight of a single atom. However, **atomic weight** is an average of the relative weights (mass numbers) of *all* the isotopes of an element, taking into account their relative abundance in nature. As a rule, the atomic weight of an element is approximately equal to the mass number of its most abundant isotope. For example, the atomic weight of hydrogen is 1.008, which reveals that its lightest isotope (^{1}H) is present in much greater amounts in our world than its ^{2}H or ^{3}H forms.

Radioisotopes

The heavier isotopes of many elements are unstable, and their atoms decompose spontaneously into more stable forms. This process of atomic decay is called *radioactivity,* and isotopes that exhibit this behavior are called **radioisotopes** (ra"de-o-i'so-tōps). The disintegration of a radioactive nucleus may be compared to a tiny explosion. It occurs when *alpha (α) particles* (packets of 2p + 2n), *beta (β) particles* (electronlike negative particles), or *gamma (γ) rays* (electromagnetic energy) are ejected from the atomic nucleus. Why this happens is complex, and you only need to know that the dense nuclear particles are composed of even smaller particles called *quarks* that associate in one way to form protons and in another way to form neutrons. Apparently, the nuclear "glue" that holds these quarks together is less effective in the heavier isotopes.

Radioisotopes may produce one or more forms of radiation and gradually lose their radioactive behavior. The time required for a radioisotope to lose one-half of its activity is called its *half-life.* The half-lives of radioisotopes vary dramatically from hours to thousands of years.

Because radioactivity can be detected with scanners, radioisotopes are valuable tools for biological research and medicine. Most radioisotopes used in the clinical setting are used for diagnosis, that is, to localize damaged or cancerous tissues. For example, iodine-131 is used to determine the size and activity of the thyroid gland and to detect thyroid cancer. The sophisticated PET scans described in *A Closer Look* on pp. 20–21 use radioisotopes to probe the workings of molecules deep within our bodies. All types of radioactivity, regardless of the purpose for which they are used, damage living tissue.

Alpha emission has the lowest penetrating power and is least damaging to living tissue; γ emission has the greatest penetrating power. Radium-226, cobalt-60, and certain other radioisotopes are used to destroy localized cancers.

HOW MATTER IS COMBINED: MOLECULES AND MIXTURES

Molecules and Compounds

Most atoms do not exist in the free state, but instead are chemically combined with other atoms. Such a combination of two or more atoms held together by chemical bonds is called a **molecule.** (The nature of the uniting chemical bonds is described shortly.)

If two or more atoms of the *same* element combine, the resulting substance is called a *molecule of that element.* When two hydrogen atoms bond, the product is a molecule of hydrogen gas and is written as H_2. Similarly, when two oxygen atoms combine, a molecule of oxygen gas (O_2) is formed. Sulfur atoms commonly combine to form sulfur molecules containing eight sulfur atoms (S_8).

When two or more *different* kinds of atoms bind, they form molecules of a **compound.** Two hydrogen atoms combine with one oxygen atom to form the compound water (H_2O), and four hydrogen atoms combine with one carbon atom to form the compound methane (CH_4). Notice again that molecules of methane and water are compounds, but molecules of hydrogen gas are not, because compounds always contain atoms of at least two different elements.

Compounds are chemically pure, and all of their molecules are identical. Thus, just as an atom is the smallest particle of an element that still exhibits the properties of the element, a molecule is the smallest particle of a compound that still displays the specific characteristics of the compound. This is an important concept because the properties of compounds are usually very different from those of the atoms they contain. Indeed, it is next to impossible to tell what atoms are in a compound without analyzing it chemically.

Mixtures

Mixtures are substances composed of two or more components *physically intermixed.* Although most matter in nature exists in the form of mixtures, there are only three basic types: *solutions, colloids,* and *suspensions.*

Solutions

Solutions are *homogeneous* mixtures of components that may be gases, liquids, or solids. Examples include the air we breathe (a mixture of gases), seawater (a mixture of salts, which are solids, and water), and rubbing alcohol (a mixture of two liquids, alcohol and water). The substance present in the greatest amount is called the **solvent** (or dissolving medium). Solvents are usually liquids. Substances present in smaller amounts are called **solutes.**

Water is the body's chief solvent. Most solutions in the body are *true solutions* containing gases, liquids, or solids dissolved in water. True solutions are usually transparent. Examples are saline solution (table salt [NaCl] and water) and a mixture of glucose and water. The solutes of true solutions are minute, usually in the form of individual atoms and molecules. Consequently, they are not visible to the naked eye, do not settle out, and do not scatter light. If a beam of light is passed through a true solution, you will not see the path of light.

Concentration of Solutions True solutions are described in terms of their *concentration,* which may be indicated in various ways. Solutions used in a college laboratory or a hospital are often described in terms of the **percent** (parts per 100 parts) of the solute in the solution. This designation always refers to the solute percentage, and unless otherwise noted, water is assumed to be the solvent.

Another way to express the concentration of a solution is in terms of its **molarity** (mo-lar'ĭ-te), or moles per liter, indicated by *M*. This method is more complicated but much more useful. To understand molarity, you must first understand what a mole is. A **mole** of any element or compound is equal to its atomic weight or **molecular weight** (sum of the atomic weights) weighed out in grams. This concept is easier than it seems, as illustrated by the following example.

Glucose is $C_6H_{12}O_6$, which indicates that it has 6 carbon atoms, 12 hydrogen atoms, and 6 oxygen atoms. To compute the molecular weight of glucose, you would look up the atomic weight of each of its atoms in the periodic table (see Appendix D) and compute its molecular weight as follows:

Atom	Number of atoms		Atomic weight		Total atomic weight
C	6	$\times$	12.011	=	72.066
H	12	$\times$	1.008	=	12.096
O	6	$\times$	15.999	=	95.994
					180.156

Then, to make a *one-molar* solution of glucose, you would weigh out 180.156 grams (g), called a gram molecular weight, of glucose and add enough water to make 1 liter (L) of solution. Thus, a one-molar solution (abbreviated 1.0 *M*) of a chemical substance is one gram molecular weight of the substance (or one gram atomic weight in the case of elemental substances) in 1 L (1000 ml) of solution.

The beauty of using the mole as the basis of preparing solutions is its precision. One mole of any substance always contains exactly the same number of solute particles, that is, 6.02×10^{23}. This number is called **Avogadro's** (av"o-gad'rōz) **number.** So whether you weigh out 1 mole of glucose (180 g) or 1 mole of water (18 g) or 1 mole of methane (16 g), in each case you will have 6.02×10^{23} molecules of that substance.[*] This allows almost mind-boggling precision to be achieved.

Colloids

Colloids (kol'oidz), also called *emulsions,* are *heterogeneous* mixtures that often appear translucent or milky. Although the solute particles are larger than those in true solutions, they still do not settle out. However, they do scatter light, and so the path of a light beam shining through a colloidal mixture is visible.

Colloids have many unique properties, including the ability of some to undergo **sol-gel transformations,** that is, to change reversibly from a fluid (sol) state to a more solid (gel) state. Jell-O, or any gelatin product, is a familiar example of a nonliving colloid that changes from a sol to a gel when refrigerated (and that will liquefy again if placed in the sun). Cytosol, the semifluid material in living cells, is also a colloid, and its sol-gel changes underlie many important cell activities, such as cell division.

Suspensions

Suspensions are *heterogeneous* mixtures with large, often visible solutes that tend to settle out. An example of a suspension is a mixture of sand and water. So is blood, in which the living blood cells are suspended in the fluid portion of blood (blood plasma). If left to stand, the suspended cells will settle out unless some means—mixing, shaking, or in the body, circulation—is used to keep them in suspension.

---

[*]The important exception to this rule concerns molecules that ionize and break up into charged particles (ions) in water, such as salts, acids, and bases (see p. 42). For example, simple table salt (sodium chloride) breaks up into two types of charged particles; therefore, in a 1.0-*M* solution of sodium chloride, there are actually *2 moles* of solute particles in solution.

As you can see, all three types of mixtures are found in both living and nonliving systems. In fact, living material is the most complex mixture of all, since it contains all three kinds of mixtures interacting with one another.

Distinguishing Mixtures from Compounds

Now we are ready to zero in on how to distinguish mixtures and compounds from one another. Mixtures differ from compounds in several important ways:

1. The chief difference between mixtures and compounds is that no chemical bonding occurs between the components of a mixture. The properties of atoms and molecules are not changed when they become part of a mixture. Remember they are only physically intermixed.

2. Depending on the mixture, its components can be separated by physical means—straining, filtering, evaporation, and so on. Compounds, by contrast, can be separated into their constituent atoms only by chemical means (breaking bonds).

3. Some mixtures are homogeneous, whereas others are heterogeneous. To say that a substance is *homogeneous* means that a sample taken from any part of the substance has exactly the same composition (in terms of the atoms or molecules it contains) as any other sample. A bar of 100% pure (elemental) iron is homogeneous, as are all compounds. *Heterogeneous* substances vary in their makeup from place to place. For example, iron ore is a heterogeneous mixture that contains iron and many other elements.

CHEMICAL BONDS

As noted earlier, when atoms combine with other atoms, they are held together by **chemical bonds.** A chemical bond is not a physical structure like a pair of handcuffs linking two people together. Instead, it is an energy relationship between the electrons of the reacting atoms, and it is made or broken in less than a trillionth of a second.

The Role of Electrons in Chemical Bonding

Electrons forming the electron cloud around the nucleus of an atom occupy regions of space called **electron shells** that consecutively surround the atomic nucleus. The atoms known so far can have electrons in seven shells (numbered 1 to 7 from the nucleus outward), but the actual number of

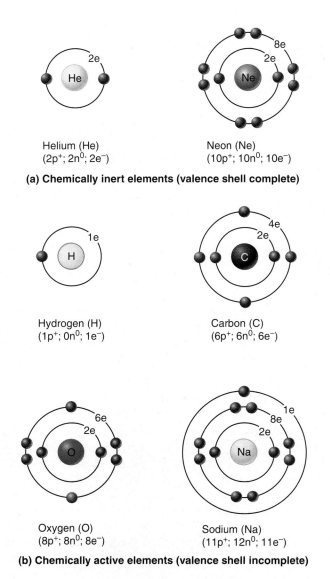

Helium (He)
(2p$^+$; 2n^0; 2e$^-$)

Neon (Ne)
(10p$^+$; 10n^0; 10e$^-$)

(a) Chemically inert elements (valence shell complete)

Hydrogen (H)
(1p$^+$; 0n^0; 1e$^-$)

Carbon (C)
(6p$^+$; 6n^0; 6e$^-$)

Oxygen (O)
(8p$^+$; 8n^0; 8e$^-$)

Sodium (Na)
(11p$^+$; 12n^0; 11e$^-$)

(b) Chemically active elements (valence shell incomplete)

FIGURE 2.4   *Chemically inert and reactive elements.*
(**a**) Helium and neon are chemically inert because in each case the outermost energy level (valence shell) is fully occupied by electrons. (**b**) Elements in which the valence shell is incomplete are chemically reactive. Such atoms tend to react with other atoms to gain, lose, or share electrons to fill their valence shells. (*Note:* To simplify the diagrams, each atomic nucleus is shown as a circle with the atom's symbol in it; individual protons and neutrons are not shown.)

electron shells occupied in a given atom depends on the number of electrons that atom possesses. Each electron shell contains one or more orbitals.

It is important to understand that each electron shell represents a different **energy level,** because this prompts you to think of electrons as particles with a certain amount of potential energy. In general, the terms *electron shell* and *energy level* are used interchangeably.

The amount of potential energy an electron has depends on the energy level it occupies, because the

attraction between the positively charged nucleus and negatively charged electrons is greatest closest to the nucleus and falls off with increasing distance. This statement explains why electrons farthest from the nucleus (1) have the greatest potential energy (it takes more energy to overcome the nuclear attraction and reach the more distant energy levels), and (2) are most likely to interact chemically with other atoms (they are the least tightly held by their own atomic nucleus and the most easily influenced by other atoms and molecules).

Each electron shell can hold a specific number of electrons. Shell 1, the shell immediately surrounding the nucleus, can accommodate only 2 electrons. Shell 2 holds a maximum of 8; shell 3 has room for 18 electrons. Subsequent shells hold larger and larger numbers of electrons. The shells tend to be filled with electrons consecutively. For example, shell 1 is completely filled before any electrons appear in shell 2.

When atoms form bonds, the important electrons are those in the atom's outermost energy level. Inner electrons usually do not take part in bonding because they are more tightly held by the atomic nucleus. When the outermost energy level of an atom is filled to capacity or contains eight electrons, the atom is stable. Such atoms are *chemically inert* (in-ert'), that is, unreactive. A group of elements called the *noble gases,* which include helium and neon, typify this condition (Figure 2.4a). On the other hand, atoms in which the outermost energy level contains fewer than eight electrons (Figure 2.4b) tend to gain, lose, or share electrons with other atoms to achieve stability.

A possible point of confusion must be addressed here. In atoms that have more than 20 electrons, the energy levels beyond shell 2 can contain *more* than eight electrons. However, the number of electrons that can participate in bonding is still limited to a total of eight. The term **valence** (va'lens) **shell** is used specifically to indicate an atom's outermost energy level *or that portion of it* containing the electrons that are chemically reactive. Hence, the key to chemical reactivity is the **octet** (ok-tet') **rule,** or **rule of eights.** Except for shell 1, which is filled with two electrons, atoms tend to interact in such a way that they have eight electrons in their valence shell.

Types of Chemical Bonds

Three major types of chemical bonds—*ionic, covalent,* and *hydrogen bonds*—result from attractive forces between atoms.

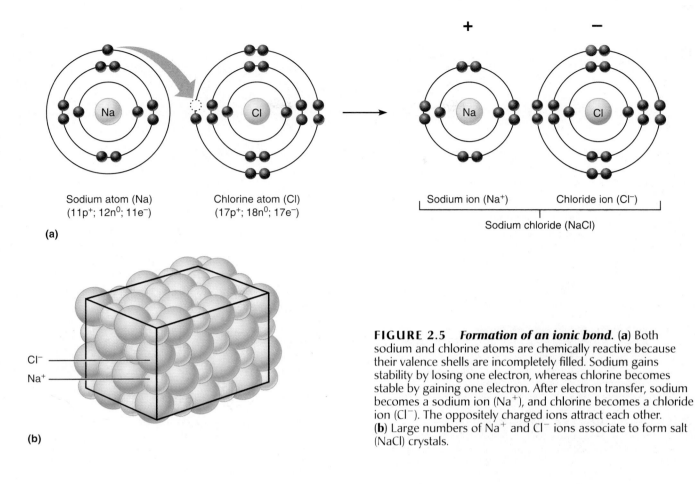

Sodium atom (Na)
($11p^+$; $12n^0$; $11e^-$)

Chlorine atom (Cl)
($17p^+$; $18n^0$; $17e^-$)

(a)

Sodium ion (Na^+)

Chloride ion (Cl^-)

Sodium chloride (NaCl)

Cl^-
Na^+

(b)

FIGURE 2.5 *Formation of an ionic bond.* **(a)** Both sodium and chlorine atoms are chemically reactive because their valence shells are incompletely filled. Sodium gains stability by losing one electron, whereas chlorine becomes stable by gaining one electron. After electron transfer, sodium becomes a sodium ion (Na^+), and chlorine becomes a chloride ion (Cl^-). The oppositely charged ions attract each other. **(b)** Large numbers of Na^+ and Cl^- ions associate to form salt (NaCl) crystals.

Ionic Bonds

Atoms are electrically neutral. However, electrons can be transferred from one atom to another, and when this happens, the precise balance of + and − charges is lost and charged particles called **ions** are formed. The atom that gains one or more electrons, the *electron acceptor*, acquires a net negative charge and is called an **anion** (an'i-on). The atom that loses electrons (the *electron donor*) acquires a net positive charge and is called a **cation** (kat'i-on). (It might help to think of the "t" in "cation" as a + sign.) Both anions and cations are formed whenever electron transfer between atoms occurs. Because opposite charges attract, these ions tend to stay close together. The result is an **ionic** (i-on'ik) **bond.**

One example of ionic bonding is the formation of sodium chloride (NaCl), by interaction of sodium and chlorine atoms (Figure 2.5). Sodium, with an atomic number of 11, has only one electron in its valence shell, and it would be very difficult to attempt to fill this shell by adding seven more. However, if this single electron is lost, shell 2 with eight electrons becomes the valence shell (outermost energy level containing electrons) and is full. Thus, by losing the lone electron in its third energy level, sodium achieves stability and becomes a

cation (Na^+). On the other hand, chlorine, atomic number 17, needs only one electron to fill its valence shell. By accepting an electron, chlorine becomes an anion and achieves stability. When these two atoms interact, this is exactly what happens. Sodium donates an electron to chlorine, and the ions created in this exchange attract each other, forming sodium chloride. Ionic bonds are commonly formed between atoms with one or two valence shell electrons (the metallic elements, such as sodium, calcium, and potassium) and atoms with seven valence shell electrons (such as chlorine, fluorine, and iodine).

Most ionic compounds fall in the chemical category called *salts*. In the dry state, ionic compounds such as sodium chloride do not exist as individual molecules. Instead, they form **crystals,** large arrays of cations and anions held together by ionic bonds (see Figure 2.5b).

Sodium chloride is an excellent example of the difference in properties between a compound and its constituent atoms. Sodium is a silvery white metal, and chlorine in its molecular state is a poisonous green gas used to make bleach. However, sodium chloride is a white crystalline solid that we sprinkle on our food.

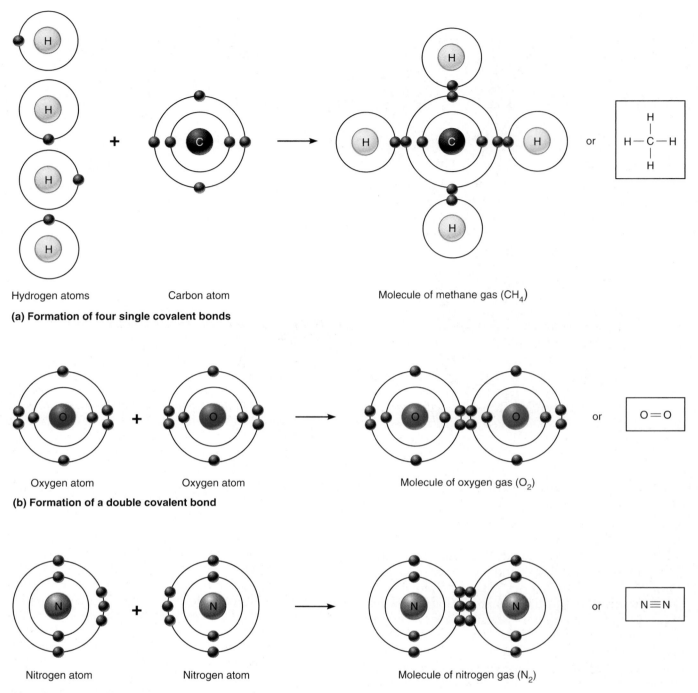

Hydrogen atoms   Carbon atom   Molecule of methane gas (CH_4)

(a) Formation of four single covalent bonds

Oxygen atom   Oxygen atom   Molecule of oxygen gas (O_2)

(b) Formation of a double covalent bond

Nitrogen atom   Nitrogen atom   Molecule of nitrogen gas (N_2)

(c) Formation of a triple covalent bond

FIGURE 2.6   *Formation of covalent bonds.* The electron–sharing patterns of interacting atoms are shown to the left. In the shaded boxes to the far right, each covalent bond is indicated by a single line between the atoms that share an electron pair. **(a)** Formation of a molecule of methane; carbon shares four electron pairs with four hydrogen atoms. **(b)** Formation of a molecule of oxygen gas; each oxygen atom shares two electron pairs with its partner, forming a double covalent bond. **(c)** When a molecule of nitrogen gas is formed, the two nitrogen atoms share three electron pairs, forming a triple covalent bond.

Covalent Bonds

Electrons do not have to be completely transferred for atoms to achieve stability. Instead, they may be *shared* so that each atom is able to fill its outer electron shell for at least part of the time. Electron sharing produces molecules in which the shared electrons occupy a single orbital common to both atoms and constitute **covalent** (ko-va'lent) **bonds.**

Hydrogen with its single electron can fill its only shell (shell 1) by sharing a pair of electrons with another atom. When it shares with another hydrogen atom, a molecule of hydrogen gas is formed. The shared electron pair orbits around the molecule as a whole, satisfying the stability needs of each atom. Hydrogen can also share an electron pair with different kinds of atoms to form a compound (Figure 2.6a). Carbon has four electrons in its outermost shell, but needs eight to achieve stability, whereas hydrogen has one electron, but needs two. When a methane molecule (CH_4) is formed, carbon shares four pairs of electrons with four hydrogen atoms (one pair with each hydrogen). Again, the shared electrons orbit and "belong to" the whole molecule, ensuring the stability of each atom. When two atoms share one pair of electrons, a *single covalent bond* is formed (indicated by a single line connecting the atoms, such as H—H). In some cases, atoms share two or three electron pairs (Figure 2.6b and c), resulting in *double* or *triple covalent bonds* (indicated by double or triple lines such as O=O or N≡N).

Polar and Nonpolar Molecules In the covalent bonds discussed thus far, the shared electrons are shared equally between the atoms of the molecule. The molecules formed are electrically balanced and are called **nonpolar molecules** (because they do not have separate + and − poles of charge). This is not always the case. When covalent bonds are formed, the resulting molecule always has a specific three-dimensional shape, with the bonds formed at definite angles. A molecule's shape helps determine what other molecules or atoms it can interact with; it may also result in unequal electron pair sharing and **polar molecules.** This is especially true in nonsymmetrical molecules that have atoms with different electron-attracting abilities. In general, *small* atoms with six or seven valence shell electrons, such as oxygen, nitrogen, and chlorine, are electron-hungry and attract electrons very strongly. This capability is called **electronegativity.** Conversely, most atoms with only one or two valence shell electrons tend to be **electropositive;** that is, their electron-attracting ability is so low that they usually lose *their* valence shell electrons to other atoms. Potassium and sodium, each with one valence shell

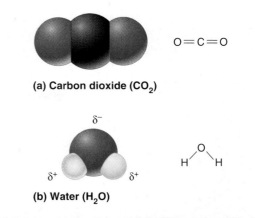

(a) Carbon dioxide (CO₂)

(b) Water (H₂O)

FIGURE 2.7 *Molecular models illustrating the three-dimensional structure of carbon dioxide and water molecules.*

electron, are good examples of electropositive atoms.

Carbon dioxide and water illustrate how molecular shape and the relative electron-attracting abilities of atoms determine whether a covalently bonded molecule is nonpolar or polar. Carbon dioxide (CO_2) forms when carbon shares four electron pairs with two oxygen atoms (two pairs are shared with each oxygen). Oxygen is very electronegative and so attracts the shared electrons much more strongly than does carbon. However, since the carbon dioxide molecule is linear (Figure 2.7a), the electron-pulling ability of one oxygen atom is offset by that of the other, like a standoff between equally strong teams in a game of tug-of-war. As a result, the shared electrons are shared equally and carbon dioxide is a nonpolar compound.

On the other hand, a water molecule (H_2O) is V-shaped (Figure 2.7b). The two hydrogen atoms are located at the same end of the molecule, and oxygen is at the opposite end. This arrangement allows oxygen to pull the shared electrons toward itself and away from the two hydrogen atoms. In this case, the electron pairs are *not* shared equally, but spend more time in the vicinity of oxygen. Because electrons are negatively charged, the oxygen end of the molecule is slightly more negative (indicated with a delta and minus as δ^-) and the hydrogen end slightly more positive (indicated by δ^+). Because water has two poles of charge, it is a *polar molecule,* or **dipole** (di'pōl). Polar molecules orient themselves toward other dipoles or toward charged particles (such as ions and some proteins), and they play essential roles in chemical reactions in body cells. The polarity of water is particularly significant, as you will see later in this chapter.

Different molecules exhibit different degrees of polarity, and we can see a gradual change from

? *Assume imaginary compound XY has a covalent polar bond. How does its charge distribution differ from that of XX molecules?*

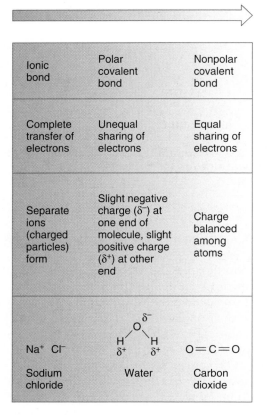

Ionic bond	Polar covalent bond	Nonpolar covalent bond
Complete transfer of electrons	Unequal sharing of electrons	Equal sharing of electrons
Separate ions (charged particles) form	Slight negative charge (δ^-) at one end of molecule, slight positive charge (δ^+) at other end	Charge balanced among atoms
Na$^+$ Cl$^-$		
Sodium chloride	Water	Carbon dioxide

FIGURE 2.8   Comparison of ionic, polar covalent, and nonpolar covalent bonds. This display compares the status of electrons involved in bonding and the electrical charge distribution in the molecules formed.

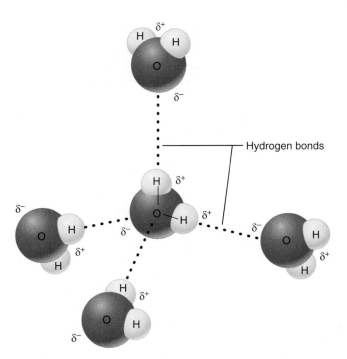

Hydrogen bonds

FIGURE 2.9   Hydrogen bonding between polar water molecules. The slightly positive ends (indicated by δ^+) of the water molecules become aligned with the slightly negative ends (indicated by δ^-) of other water molecules.

ionic to nonpolar covalent bonding as summarized in Figure 2.8. Ionic bonds (complete electron transfer) and nonpolar covalent bonds (equal electron sharing) are the extremes of a continuum, with various degrees of unequal electron sharing in between.

Hydrogen Bonds

Unlike the stronger ionic and covalent bonds, hydrogen bonds are too weak to bind atoms together to form molecules. Hydrogen bonds form when a hydrogen atom, already covalently linked to one electronegative atom (usually nitrogen or oxygen), is attracted by another electron-hungry atom, and

forms a "bridge" between them. Hydrogen bonding is common between dipoles such as water molecules, because the slightly negative oxygen atoms of one molecule attract the slightly positive hydrogens of other molecules (Figure 2.9). The tendency of water molecules to cling together and form films, referred to as *surface tension*, helps explain why water beads up into spheres when it sits on a hard surface.

Hydrogen bonds are also important as *intramolecular bonds*, which bind different parts of a single large molecule together into a specific three-dimensional shape. Some large biological molecules, such as proteins and DNA, have numerous hydrogen bonds that help maintain and stabilize their structures.

CHEMICAL REACTIONS

As noted earlier, all particles of matter are in constant motion because of their kinetic energy. Movement of atoms or molecules in a solid is usually limited to vibration because the particles are united by fairly rigid bonds. But in liquids or gases, particles dart about randomly, sometimes colliding with one another and interacting to undergo chemical reactions. A **chemical reaction** occurs whenever chemical bonds are formed, rearranged, or broken.

Because X and Y are different atoms, one is bound to be more electronegative than the other. Consequently, there will be a separation of charge in XY, which is a dipole. In XX, the two atoms are the same and there will be equal ˉ sharing; no dipole is formed.

Chemical Equations

Chemical reactions can be written in symbolic form as **chemical equations.** For example, the joining of two hydrogen atoms to form hydrogen gas is indicated as

$$\text{H} + \text{H} \rightarrow \text{H}_2 \text{ (hydrogen gas)}$$

(reactants)   (product)

and the combining of four atoms of hydrogen and one atom of carbon to form methane is written

$$4\text{H} + \text{C} \rightarrow \text{CH}_4 \text{ (methane)}$$

Notice that a number written as a *subscript* indicates that the atoms are joined by chemical bonds. But a number written as a *prefix* denotes the number of *unjoined* atoms or molecules. Hence, CH_4 reveals that four hydrogen atoms are bonded together with carbon to form the methane molecule, but 4H signifies four unjoined hydrogen atoms.

A chemical equation is like a sentence describing what happens in a reaction. It contains the following information: the number and kinds of reacting substances, or **reactants;** the chemical composition of the **product(s);** and in balanced equations, the relative proportion of each reactant and product. In the examples used above, the reactants are atoms, as indicated by their atomic symbols (H, C). The product in each case is a molecule, as represented by its **molecular formula** (H_2, CH_4). The equation for the formation of methane may be read as *either* "four atoms of hydrogen plus one atom of carbon yield one molecule of methane" *or* "four moles of hydrogen atoms plus one mole of carbon yield one mole of methane." Using moles is more practical because it is impossible to measure out one atom or one molecule of anything!

Patterns of Chemical Reactions

Most chemical reactions exhibit one of three major patterns: they are either *synthesis, decomposition,* or *exchange reactions.*

When atoms or molecules combine to form a larger, more complex molecule, the process is a **synthesis,** or **combination, reaction.** A synthesis reaction, which always involves bond formation, can be represented (using arbitrary letters) as

$$\text{A} + \text{B} \rightarrow \text{AB}$$

Synthesis reactions are the basis of constructive, or **anabolic,** activities in body cells, such as joining small molecules called amino acids into large protein molecules (Figure 2.10a). Synthesis reactions are conspicuous in rapidly growing tissues.

Which of the reaction types depicted occurs when fats are digested in your small intestine?

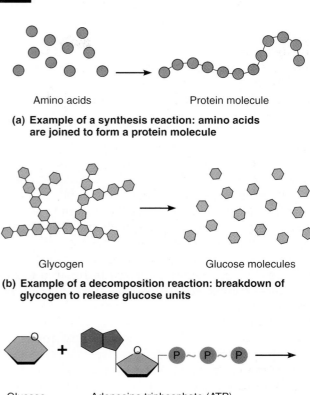

(a) Example of a synthesis reaction: amino acids are joined to form a protein molecule

Amino acids   Protein molecule

Glycogen   Glucose molecules

(b) Example of a decomposition reaction: breakdown of glycogen to release glucose units

Glucose   Adenosine triphosphate (ATP)

Glucose phosphate   Adenosine diphosphate (ADP)

(c) Example of an exchange reaction: ATP transfers its terminal phosphate group to glucose to form glucose-phosphate

FIGURE 2.10 *Patterns of chemical reactions.* **(a)** In synthesis reactions, smaller particles (atoms, ions, or molecules) are bonded together to form larger, more complex molecules. **(b)** Decomposition reactions involve bond breaking. **(c)** In exchange, or displacement, reactions, bonds are both made and broken.

A **decomposition reaction** occurs when a molecule is broken down into smaller molecules or its constituent atoms:

$$\text{AB} \rightarrow \text{A} + \text{B}$$

Essentially, decomposition reactions are reverse synthesis reactions; bonds are broken. Decomposition reactions underlie all degradative, or **catabolic,**

Digestion of fats (lipids) to their fatty acid and glycerol building blocks is an example of **(b)** *a decomposition reaction.*

processes that occur in body cells. For example, the bonds of large glycogen molecules are broken to release simpler molecules of glucose sugar (Figure 2.10b).

Exchange, or **displacement, reactions** involve both synthesis and decomposition; that is, bonds are both made and broken. In an exchange reaction, parts of the reactant molecules change partners, so to speak, producing different product molecules:

$$AB + C \rightarrow AC + B \quad \text{and} \quad AB + CD \rightarrow AD + CB$$

An exchange reaction occurs when ATP reacts with glucose and transfers its end phosphate group (indicated by a circled P in Figure 2.10c) to glucose, forming glucose-phosphate. At the same time, the ATP becomes ADP. This important reaction occurs whenever glucose enters a body cell and it effectively traps the glucose fuel molecule inside the cell.

Another group of very important chemical reactions in living systems are **oxidation-reduction reactions—redox reactions** for short. Oxidation-reduction reactions are decomposition reactions in that they are the basis of all reactions in which food fuels are catabolized for energy (that is, in which ATP is produced). They are also a special type of exchange reaction because electrons are exchanged between the reactants. The reactant losing the electrons is referred to as the *electron donor* and is said to be **oxidized;** the reactant taking up the transferred electrons is called the *electron acceptor* and is said to become **reduced.**

Redox reactions occur when ionic compounds are formed. Recall that in the formation of NaCl (see Figure 2.5), sodium loses an electron to chlorine. Consequently, sodium is oxidized and becomes a sodium ion, and chlorine is reduced and becomes a chloride ion. However, not all oxidation-reduction reactions involve *complete transfer* of electrons—some simply change the pattern of electron sharing in covalent bonds. For example, a substance is oxidized both by losing hydrogen atoms and by combining with oxygen. The common factor in these events is that electrons that formerly "belonged" to the reactant molecule are lost—either entirely (as hydrogen is removed and takes its electron with it) or relatively (the shared electrons spend more time in the vicinity of the very electronegative oxygen atom).

To understand the importance of oxidation-reduction reactions in living systems, take a look at the overall equation for *cellular respiration,* which represents the major pathway by which glucose is broken down for energy in body cells:

$$C_6H_{12}O_6 + 6O_2 \rightarrow 6CO_2 + 6H_2O + ATP$$
glucose   oxygen   carbon   water   cellular
                     dioxide   energy

As you can see, it is an oxidation-reduction reaction. Glucose is oxidized to carbon dioxide as it loses hydrogen atoms, and oxygen is reduced to water as it accepts the hydrogen atoms. This reaction is described in detail in Chapter 25, along with other topics of cellular metabolism.

Energy Flow in Chemical Reactions

Because all chemical bonds represent stored chemical energy, all chemical reactions ultimately result in net absorption or release of energy. Reactions that release energy are called **exergonic reactions.** These reactions yield products that have less energy than the initial reactants, but they also provide energy that can be harvested for other uses. With a few exceptions, catabolic and oxidative reactions are exergonic. On the other hand, the products of energy-absorbing, or **endergonic,** reactions contain more potential energy in their chemical bonds than did the reactants. Energy is required to make the bonds of most large molecules in the body, and anabolic reactions are typically energy-absorbing endergonic reactions. Essentially this is a case of "one hand washing the other"—the energy released as fuel molecules are broken down (oxidized), is captured in ATP molecules and then used to synthesize the complex biological molecules the body needs to sustain life.

Reversibility of Chemical Reactions

All chemical reactions are theoretically reversible. If chemical bonds can be made, they can be broken, and vice versa. Reversibility is indicated by a double arrow. When the arrows differ in length, the longer arrow indicates the major direction in which the reaction proceeds:

$$A + B \rightleftharpoons AB$$

In this example, the forward reaction (reaction going to the right) predominates. Over time, the product (AB) accumulates and the reactants (A and B) decrease in amount.

When the arrows are of equal length, as in

$$A + B \rightleftharpoons AB$$

neither the forward reaction nor the reverse reaction is dominant; that is, for each molecule of product (AB) formed, one product molecule breaks down, releasing the reactants A and B. Such a chemical reaction is said to be in a state of **chemical equilibrium.** Once chemical equilibrium is reached, there is no further *net change* in the amounts of reactants and products. Product molecules are still formed and broken down, but the situation already established when equilibrium is reached (such as greater numbers of product molecules) remains unchanged. This is analogous to the admission scheme used by many large museums in which tickets are sold according to time of entry. If 300 tickets are issued for the 9 AM admission, 300 people will be admitted when the doors open. Thereafter, when 6 people leave, 6 are admitted, and when another 15 people leave, 15 more are allowed in. Although there is a continual turnover, the museum contains about 300 art lovers throughout the day.

Although all chemical reactions are reversible, many show so little tendency to go in the reverse direction that they are irreversible for all practical purposes. This is true of many biological reactions. Chemical reactions that release energy when going in one direction will not go in the opposite direction unless energy is put back into the system. For example, as mentioned above, when our cells break down glucose via the reactions of cellular respiration to yield carbon dioxide and water, some of the energy released is trapped in the bonds of ATP. Because the cells need ATP's energy for various functions (and more glucose will be along with the next meal), this particular reaction is never reversed in our cells. Furthermore, if a product of a reaction is continuously removed from the reaction site, it is unavailable to take part in the reverse reaction. This situation occurs when the carbon dioxide that is released during glucose breakdown leaves the cell, enters the blood, and is eventually removed from the body by the lungs.

Factors Influencing the Rate of Chemical Reactions

For atoms and molecules to react chemically, they must *collide* with enough force to overcome the repulsion between their electrons. Interactions between valence shell electrons—the basis of bond making and breaking—cannot occur long distance. The force of collisions depends on how fast the particles are moving. Solid, forceful collisions between rapidly moving particles are much more likely to cause reactions than are those in which the particles graze each other lightly.

Temperature

Increasing the temperature of a substance increases the kinetic energy of its particles and the force of their collisions. Therefore, chemical reactions proceed quicker at higher temperatures. When the temperature drops, particle movement and reactions occur more slowly.

Particle Size

Smaller particles move faster than larger ones (at the same temperature) and therefore tend to collide more frequently and more forcefully. Hence, the smaller the reacting particles, the faster a chemical reaction goes at a given temperature and concentration.

Concentration

Chemical reactions progress most rapidly when the reacting particles are present in high concentration, because the larger the number of randomly moving particles in a given space, the greater the chance of successful collisions. As the concentration of the reactants declines, chemical equilibrium eventually occurs unless additional reactants are added or products are removed from the reaction site.

Catalysts

Although many chemical reactions in nonliving systems can be speeded up simply by heating, drastic increases in body temperature are life threatening because important biological molecules are destroyed. Still, at normal body temperatures, most chemical reactions would proceed far too slowly to maintain life were it not for the presence of catalysts. **Catalysts** (kat'ah-lists) are substances that increase the rate of chemical reactions without themselves becoming chemically changed or part of the product. Biological catalysts are called **enzymes** (en'zīmz). The mode of action of enzymes is described later in this chapter.

■ PART 2 ■ BIOCHEMISTRY

Biochemistry is the study of the chemical composition and reactions of living matter. Biological compounds fall into one of two major classes: organic or inorganic compounds. **Organic compounds** contain carbon. All organic compounds are covalently bonded molecules, and many are large.

All other chemicals in the body are considered **inorganic compounds.** These include water, salts, and many acids and bases. Organic and inorganic compounds are equally essential for life. Trying to decide which is more valuable is like trying to decide whether the ignition system or the engine is more essential to run your car!

INORGANIC COMPOUNDS

Water

Water, the most abundant and important inorganic compound in living material, makes up 60% to 80%

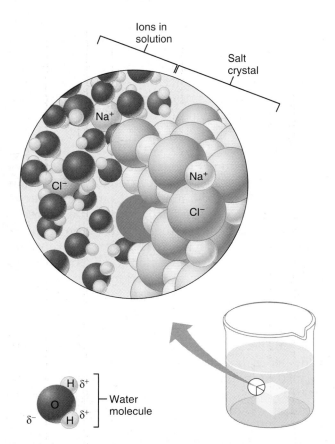

Ions in solution

Salt crystal

Na^+

Na^+

Cl^-

Cl^-

H δ^+
O
δ^- H δ^+
] Water molecule

FIGURE 2.11 Dissociation of a salt in water. The slightly negative ends of the water molecules (δ^-) are attracted to Na^+, whereas the slightly positive ends of water molecules (δ^+) orient toward Cl^-, causing the ions to be pulled off the crystal lattice.

of the volume of most living cells. The incredible versatility of this vital fluid reflects its properties:

1. High heat capacity. Water has a high heat capacity; that is, it absorbs and releases large amounts of heat before changing appreciably in temperature itself. Its abundance in the body prevents sudden changes in temperature caused by external factors, such as sun or wind exposure, or by internal conditions that release heat rapidly, such as vigorous muscle activity. As part of blood, water redistributes heat among body tissues, ensuring temperature homeostasis.

2. High heat of vaporization. When water evaporates, or vaporizes, it changes from a liquid to a gas (water vapor). This transformation requires that large amounts of heat be absorbed to break the hydrogen bonds that hold water molecules together. This property is extremely beneficial when we sweat. As perspiration (mostly water) evaporates from our skin, large amounts of heat are removed from the body and lost to the external environment, providing an efficient cooling mechanism.

3. Polar solvent properties. Water is an unparalleled solvent; indeed, it is often called the **universal solvent.** Biochemistry is "wet chemistry." Biological

molecules do not react chemically unless they are in solution, and virtually all chemical reactions occurring in the body depend on water's solvent properties.

Because of their polar nature, water molecules orient themselves with their slightly negative ends toward the positive ends of the solutes, and vice versa, first attracting them, then surrounding them. This property of water explains why ionic compounds and other small reactive molecules (such as acids and bases) *dissociate* in water, their ions separating from each other and becoming evenly scattered in the water, forming true solutions (Figure 2.11). Water also forms **hydration layers** (layers of water molecules) around large charged molecules such as proteins, shielding them from the effects of other charged substances in the vicinity and preventing them from settling out of solution. Such protein-water mixtures are *biological colloids.* Cerebrospinal fluid is an example of a colloid. So is blood plasma.

Water serves as the body's major transport medium because it is such an excellent solvent. Nutrients, respiratory gases, and metabolic wastes carried throughout the body are dissolved in blood plasma, and many metabolic wastes are excreted from the body in urine, another watery fluid.

Specialized molecules that lubricate the body also use water as their dissolving medium. Such lubricating fluids include mucus, which eases feces through the bowel, and the serous fluids that reduce friction among visceral organs.

4. Reactivity. Water is an important *reactant* in many chemical reactions. For example, foods are digested to their building blocks by adding a water molecule to each bond to be broken. Such decomposition reactions are more specifically called **hydrolysis** (hi-drol'ĭ-sis; "water splitting") **reactions.** Conversely, when large carbohydrate or protein molecules are synthesized from smaller molecules, a water molecule is removed for every bond formed, a reaction called **dehydration synthesis.**

5. Cushioning. Finally, by forming a resilient cushion around certain body organs, water helps protect them from physical trauma. The cerebrospinal fluid surrounding the brain exemplifies water's cushioning role.

Salts

A **salt** is an ionic compound containing cations other than H^+ and anions other than the hydroxyl ion (OH^-). As already noted, when salts are dissolved in water, they dissociate into their component ions (Figure 2.11). For example, sodium sulfate (Na_2SO_4) dissociates into two Na^+ ions and one SO_4^{2-} ion. This occurs easily because the ions are already formed. All that remains is for water to overcome the attraction between the oppositely charged ions. All ions are **electrolytes** (e-lek′tro-līts), substances that conduct an electrical current in solution. (Note that groups of atoms that bear an overall charge, such as sulfate, are called *polyatomic ions*.)

Salts commonly found in the body include NaCl, Ca$_2$CO$_3$ (calcium carbonate), and KCl (potassium chloride). However, the most plentiful salts are the calcium phosphates that make bones and teeth hard. In their ionized form, salts play vital roles in body function. For instance, the electrolyte properties of sodium and potassium ions are essential for nerve impulse transmission and muscle contraction. Ionic iron forms part of the hemoglobin molecules that transport oxygen within red blood cells, and zinc and copper ions are important to the activity of some enzymes. Other important functions of the elements found in body salts are summarized in Table 2.1 (p. 29).

Homeostatic Imbalance

Maintaining proper ionic balance in our body fluids is one of the most crucial homeostatic roles of the kidneys. When this balance is severely disturbed, virtually nothing in the body works. All the physiological activities listed above and thousands of others are disrupted and grind to a stop. ■

Acids and Bases

Like salts, acids and bases are electrolytes; that is, they ionize and dissociate in water and can then conduct an electrical current. However, unlike salts, acids and some bases are covalently bonded molecules. Therefore, ions have to be formed before dissociation can occur.

Acids

Acids have a sour taste, can react with many metals, and "burn" a hole in your rug. But the most useful definition of an acid for our purposes is that it is a substance that releases **hydrogen ions** (H$^+$) in detectable amounts. Since a hydrogen ion is just a hydrogen nucleus, or "naked" proton, acids are also defined as **proton donors.**

When acids dissolve in water, they release hydrogen ions (protons) and anions. It is the concentration of protons that determines the acidity of a solution. The anions have little or no effect on acidity. For example, hydrochloric acid (HCl), an acid produced by stomach cells that aids digestion, dissociates into a proton and a chloride ion:

$$HCl \rightarrow H^+ \quad Cl^-$$
$$\text{proton} \quad \text{anion}$$

Other acids found or produced in the body include acetic acid (HC$_2$H$_3$O$_2$, commonly abbreviated as HAc), the acidic portion of vinegar; and carbonic acid (H$_2$CO$_3$). The molecular formula for an acid is easy to recognize because the hydrogen is written first.

Bases

Bases have a bitter taste, feel slippery, and are **proton acceptors**—that is, they take up hydrogen ions (H$^+$) in detectable amounts. Common inorganic bases include the *hydroxides* (hi-drok′sīds), such as magnesium hydroxide (milk of magnesia) and sodium hydroxide (lye). Like acids, hydroxides dissociate when dissolved in water, but in this case **hydroxyl** (hi-drok′sil) **ions** (OH$^-$) and cations are liberated. For example, ionization of sodium hydroxide (NaOH) produces a hydroxyl ion and a sodium ion; the hydroxyl ion then binds to (accepts) a proton present in the solution. This produces water and simultaneously reduces the acidity (hydrogen ion concentration) of the solution:

$$NaOH \rightarrow Na^+ + OH^-$$
$$\text{cation} \quad \text{hydroxyl ion}$$

and then

$$OH^- + H^+ \rightarrow H_2O$$
$$\text{water}$$

Bicarbonate ion (HCO$_3$$^-$), an important base in the body, is particularly abundant in blood. **Ammonia (NH$_3$)**, a common waste product of protein breakdown in the body, is also a base. It has one pair of unshared electrons that strongly attracts protons. By accepting a proton, ammonia becomes an ammonium ion:

$$NH_3 + H^+ \rightarrow NH_4{}^+$$
$$\text{ammonium} \\ \text{ion}$$

pH: Acid-Base Concentration

The more hydrogen ions in a solution, the more acidic the solution is. Conversely, the greater the concentration of hydroxyl ions (the lower the concentration of H$^+$), the more basic, or *alkaline* (al′kuh-lin), the solution becomes. The relative concentration of hydrogen ions in various body fluids is measured in concentration units called **pH** (pe-āch′) **units.**

The idea for a pH scale was devised by a Danish biochemist and part-time beer brewer named Sören Sörensen in 1909. He was searching for a convenient means of checking the acidity of his alcoholic product to prevent its spoilage by bacterial action. (Many bacteria are inhibited by acidic conditions.) The pH scale that resulted is based on the concentration of hydrogen ions in a solution, expressed in terms of moles per

Concentration in moles/liter

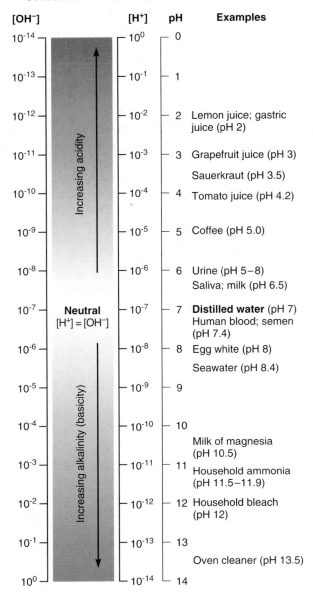

FIGURE 2.12 *The pH scale and pH values of representative substances.* The pH scale is based on the number of hydrogen ions in solution. The actual concentration of hydrogen ions ([H⁺]) (expressed in moles per liter) and the corresponding hydroxyl concentration ([OH⁻]) are indicated for each pH value noted. At a pH of 7, the concentrations of hydrogen and hydroxyl ions are equal and the solution is neutral.

liter, or molarity. The pH scale runs from 0 to 14 and is logarithmic; that is, each successive change of one pH unit represents a tenfold change in hydrogen ion concentration (Figure 2.12). The pH of a solution is thus defined as the negative logarithm of the hydrogen ion concentration [H⁺] in moles per liter or $-\log[\text{H}^+]$.

At a pH of 7 (at which [H⁺] is 10^{-7} *M*), the number of hydrogen ions exactly equals the number of hydroxyl ions, and the solution is *neutral*—neither acidic nor basic. Absolutely pure (distilled) wa-

ter has a pH of 7. Solutions with a pH below 7 are acidic; the hydrogen ions outnumber the hydroxyl ions. The lower the pH, the more acidic the solution. A solution with a pH of 6 has ten times as many hydrogen ions as a solution with a pH of 7.

Solutions with a pH higher than 7 are alkaline, and the relative concentration of hydrogen ions decreases by a factor of 10 with each higher pH unit. Thus, solutions with pH values of 8 and 12 have, respectively, 1/10 and 1/100,000 (1/10 × 1/10 × 1/10 × 1/10 × 1/10) as many hydrogen ions as a solution of pH 7. Notice that as the hydrogen ion concentration decreases, the hydroxyl ion concentration rises, and vice versa. The approximate pH of several body fluids and of a number of common substances also appears in Figure 2.12.

Neutralization

When acids and bases are mixed, they react with each other in displacement reactions to form water and a salt. For example, when hydrochloric acid and sodium hydroxide interact, sodium chloride (a salt) and water are formed.

$$\text{HCl} + \text{NaOH} \rightarrow \text{NaCl} + \text{H}_2\text{O}$$
$$\text{acid} \qquad \text{base} \qquad \text{salt} \qquad \text{water}$$

This type of reaction is called a **neutralization reaction,** because the joining of H⁺ and OH⁻ to form water neutralizes the solution. Although the salt formed in this reaction is written in molecular form (NaCl), remember that it actually exists as dissociated sodium and chloride ions when dissolved in water.

Buffers

Living cells are extraordinarily sensitive to even slight changes in the pH of the environment. In high concentrations, acids and bases are extremely damaging to living tissue. Homeostasis of acid-base balance is carefully regulated by the kidneys and lungs and by chemical systems (proteins and other types of molecules) called **buffers.** Buffers resist abrupt and large swings in the pH of body fluids by releasing hydrogen ions (acting as acids) when the pH begins to rise and by binding hydrogen ions (acting as bases) when the pH drops. Because blood comes into close contact with nearly every body cell, regulation of its pH is particularly critical. Normally, blood pH varies within a very narrow range (7.35 to 7.45). If blood pH varies from these limits by more than a few tenths of a unit, it may be fatal.

To comprehend how chemical buffer systems operate, you must thoroughly understand strong and weak acids and bases. The first important concept is that the acidity of a solution reflects *only* the free hydrogen ions, not those still bound to anions. Consequently, acids that dissociate completely and irreversibly in water are called **strong acids,** because they can dramatically change the pH of a solution. Examples are hydrochloric acid and sulfuric acid. If we could count out 100 hydrochloric acid molecules and place them in 1 ml of water, we could expect to end up with 100 H^+, 100 Cl^-, and no undissociated hydrochloric acid molecules in that solution.

Acids that do not dissociate completely, like carbonic acid (H_2CO_3) and acetic acid (HAc), are said to be **weak acids.** If we were to place 100 acetic acid molecules in 1 ml of water, the reaction would be something like this:

$$100 \; HAc \rightarrow 90 \; HAc + 10 \; H^+ + 10 \; Ac^-$$

Because undissociated acids do not affect pH, the acetic acid solution is much less acidic than the HCl solution. Weak acids dissociate in a predictable way, and molecules of the intact acid are in dynamic equilibrium with the dissociated ions. Consequently, the dissociation of acetic acid may also be written as

$$HAc \rightleftharpoons H^+ + Ac^-$$

This viewpoint allows us to see that if H^+ (released by a strong acid) is added to the acetic acid solution, the equilibrium will shift to the left and some H^+ and Ac^- will recombine to form HAc. On the other hand, if a strong base is added and the pH begins to rise, the equilibrium shifts to the right and more HAc molecules dissociate to release H^+. This characteristic of weak acids allows them to play extremely important roles in the chemical buffer systems of the body.

The concept of strong and weak bases is more easily explained. Remember that bases are proton acceptors. Thus, **strong bases** are those, like hydroxides, that dissociate easily in water and quickly tie up H^+. On the other hand, sodium bicarbonate (commonly known as baking soda) ionizes incompletely and reversibly. Because it accepts relatively few protons, its released bicarbonate ion is considered a **weak base.**

Now let's examine how one buffer system helps to maintain pH homeostasis of the blood. Although there are other chemical blood buffers, the **carbonic acid–bicarbonate system** is a very important one. Carbonic acid (H_2CO_3) dissociates reversibly, releasing bicarbonate ions (HCO_3^-) and protons (H^+):

Response to rise in pH

$$H_2CO_3 \xrightleftharpoons[\text{Response to drop in pH}]{} HCO_3^- + H^+$$

H^+ donor H^+ acceptor   proton
(weak acid)   (weak base)

The chemical equilibrium between carbonic acid (a weak acid) and bicarbonate ion (a weak base) resists changes in blood pH by shifting to the right or left as

H1 ions are added to or removed from the blood. As blood pH rises (i.e., becomes more alkaline due to the addition of a strong base), the equilibrium shifts to the right, forcing more carbonic acid to dissociate. Similarly, as blood pH begins to drop (i.e., becomes more acidic due to the addition of a strong acid), the equilibrium shifts to the left as more bicarbonate ions begin to bind with protons. As you can see, strong bases are replaced by a weak base (bicarbonate ion) and protons released by strong acids are tied up in a weak one (carbonic acid). In either case, the blood pH changes much less than it would in the absence of the buffering system. Acid-base balance and buffers are discussed in more detail in Chapter 27.

ORGANIC COMPOUNDS

Molecules unique to living systems—proteins, carbohydrates, lipids (fats), and nucleic acids—all contain carbon and hence are organic compounds. Although organic compounds are distinguished by the fact that they contain carbon and inorganic compounds are defined as compounds that lack carbon, there are a few irrational exceptions to this generalization that you should be aware of. Carbon dioxide, carbon monoxide, and carbides, for example, all contain carbon but are considered inorganic compounds.

What makes carbon so special that "living" chemistry depends on its presence? To begin with, no other *small* atom is so precisely **electroneutral.** The consequence of its electroneutrality is that carbon never loses or gains electrons; it always shares them. Furthermore, with four valence shell electrons, carbon forms four covalent bonds with other elements, as well as with other carbon atoms. As a result, carbon is found in long, chainlike molecules (common in fats), ring structures (typical of carbohydrates and steroids), and many other structures that are uniquely suited for specific roles in the body.

Carbohydrates

Carbohydrates, a group of molecules that includes sugars and starches, represent 1 to 2% of cell mass. Carbohydrates contain carbon, hydrogen, and oxygen, and generally the hydrogen and oxygen atoms occur in the same 2:1 ratio as in water. This ratio is reflected in the word *carbohydrate* (meaning "hydrated carbon").

A carbohydrate can be classified according to size and solubility as a monosaccharide ("one sugar"), disaccharide ("two sugars"), or polysaccharide ("many sugars"). Monosaccharides are the structural units, or building blocks, of the other carbohydrates. In general, the larger the carbohydrate molecule, the less soluble it is in water.

Monosaccharides

Monosaccharides (mon"o-sak'ah-rīdz), or *simple sugars*, are single-chain or single-ring structures containing from three to seven carbon atoms (Figure 2.13a). Usually, the carbon, hydrogen, and oxygen atoms occur in the ratio 1:2:1, so a general formula for a monosaccharide is $(CH_2O)_n$, where n is the number of carbons in the sugar. Glucose, for example, has six carbon atoms, and its molecular formula is $C_6H_{12}O_6$. Ribose, with five carbons, is $C_5H_{10}O_5$.

Monosaccharides are named generically according to the number of carbon atoms they contain. Most important in the body are the pentose (five-carbon) and hexose (six-carbon) sugars. For example, the pentose *deoxyribose* (de-ok"sĭ-ri'bōs) is part of DNA, and *glucose*, a hexose, is blood sugar. Two other hexoses, *galactose* and *fructose*, are **isomers** (i'so-mers) of glucose. That is, they have the same molecular formula ($C_6H_{12}O_6$), but their atoms are arranged differently, giving them different chemical properties (see Figure 2.13a).

Disaccharides

A **disaccharide** (di-sak'ah-rīd), or *double sugar*, is formed when two monosaccharides are joined by **dehydration synthesis** (Figure 2.13b). In this synthesis reaction, a water molecule is lost as the bond is made, as illustrated by the synthesis of sucrose (soo'krōs):

$$2C_6H_{12}O_6 \quad \rightarrow \quad C_{12}H_{22}O_{11} \quad + \quad H_2O$$

glucose + fructose   sucrose   water

Notice that the molecular formula for sucrose contains two hydrogen atoms and one oxygen atom less than the total number of hydrogen and oxygen atoms in glucose and fructose, because a water molecule is released during bond formation.

Important disaccharides in the diet are *sucrose* (glucose + fructose), which is cane or table sugar; *lactose* (glucose + galactose), found in milk; and *maltose* (glucose + glucose), also called malt sugar (see Figure 2.13b). Since disaccharides are too large to pass through cell membranes, they must be digested to their simple sugar units to be absorbed from the digestive tract into the blood. This decomposition process is called **hydrolysis** and is essentially the reverse of dehydration synthesis. A water molecule is added to each bond, breaking the bonds and releasing the simple sugar units (see Figure 2.13b).

Polysaccharides

Polysaccharides (pol"e-sak'ah-rīdz) are long chains of simple sugars linked together by dehydration synthesis (Figure 2.13c). Such long, chainlike molecules made of many similar units are called **polymers.** Because polysaccharides are large, fairly insoluble molecules, they are ideal storage products. Another consequence of their large size is that they lack the sweetness of the simple and double sugars. Only two polysaccharides are of major importance to the body: starch and glycogen. Both are polymers of glucose. Only their degree of branching differs.

Starch is the storage carbohydrate formed by plants. The number of glucose units composing a starch molecule is high and variable. When we eat starchy foods such as grain products and potatoes, the starch must be digested for its glucose units to be absorbed. (We are unable to digest *cellulose*, another polysaccharide found in all plant products. However, it is important in providing the *bulk* (one form of fiber) that helps move feces through the colon.)

Glycogen (gli'ko-jen), the storage carbohydrate of animal tissues, is stored primarily in skeletal muscle and liver cells. Like starch, it is highly branched and is a very large molecule (see Figure 2.13c). When blood sugar levels drop sharply, liver cells break down glycogen and release its glucose units to the blood. Since there are many branch endings from which glucose can be released simultaneously, body cells have almost instant access to glucose fuel.

Carbohydrate Functions

The major function of carbohydrates in the body is to provide a ready, easily used source of cellular fuel. Most cells can use only a few simple sugars, and glucose is at the top of the "cellular menu." As described in our earlier discussion of oxidation-reduction reactions (p. 40), glucose is broken down and oxidized within cells. During these chemical reactions, electrons are transferred. This relocation of electrons releases the bond energy stored in glucose, and this energy is used to synthesize ATP. When ATP supplies are sufficient, dietary carbohydrates are converted to glycogen or fat and stored. Those of us who have gained weight from eating too many carbohydrate-rich snacks have personal experience with this conversion process!

---

*Notice that in Figure 2.13 the carbon (C) atoms present at each angle of the carbohydrate ring structures are not illustrated. For example, the illustrations below show the full structure of glucose on the left and the shorthand structure on the right. This type of shorthand is used nearly for all organic ringlike structures illustrated in this chapter.

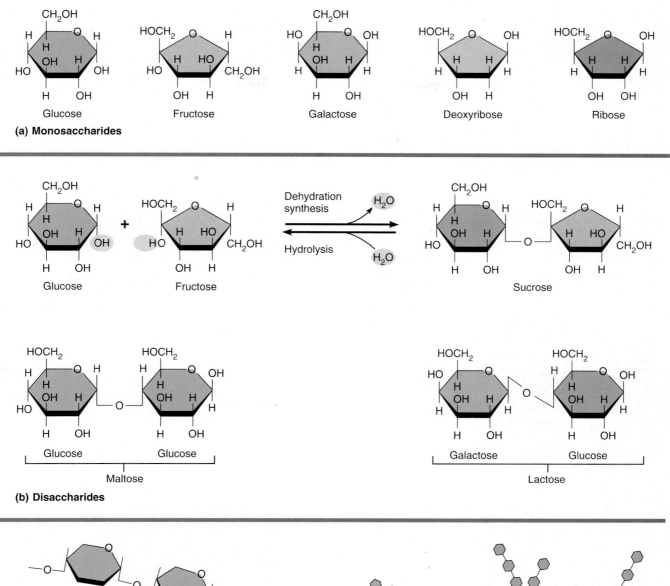

(a) Monosaccharides

Glucose   Fructose   Galactose   Deoxyribose   Ribose

Dehydration synthesis H_2O

Hydrolysis H_2O

Glucose   Fructose   Sucrose

Glucose   Glucose   Galactose   Glucose

Maltose   Lactose

(b) Disaccharides

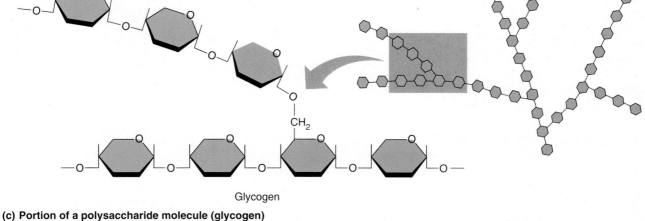

CH_2

Glycogen

(c) Portion of a polysaccharide molecule (glycogen)

FIGURE 2.13   *Carbohydrate molecules.* * **(a)** Monosaccharides important to the body. The hexose sugars glucose, fructose, and galactose are isomers: They have the same molecular formula ($C_6H_{12}O_6$), but the arrangement of their atoms differs, as shown here. Deoxyribose and ribose are pentose sugars. **(b)** Disaccharides are two monosaccharides linked together by dehydration synthesis, a process involving the removal of a water molecule at the bond site. The formation of sucrose ($C_{12}H_{22}O_{11}$) from glucose and fructose molecules is illustrated. In the reverse reaction (hydrolysis), sucrose is broken down to glucose and fructose by the addition of a water molecule to the bond. Other important disaccharides are maltose and lactose (both $C_{12}H_{22}O_{11}$). **(c)** Simplified representation of part of a glycogen molecule, a polysaccharide formed from glucose units.

*See footnote on *facing page* concerning the complete structure of these sugars.

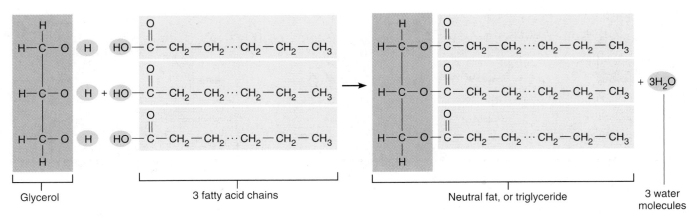

Glycerol   3 fatty acid chains   Neutral fat, or triglyceride   3 water molecules

(a) Formation of a triglyceride

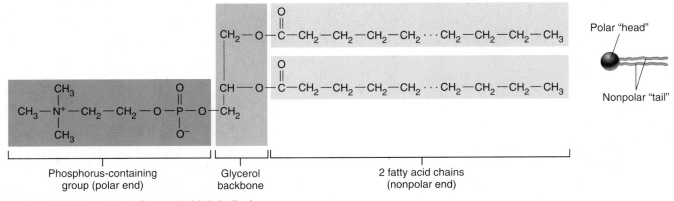

Polar "head"

Nonpolar "tail"

Phosphorus-containing group (polar end)   Glycerol backbone   2 fatty acid chains (nonpolar end)

(b) Phospholipid molecule (phosphatidyl choline)

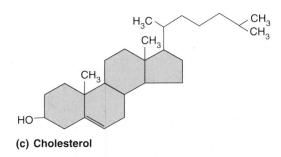

(c) Cholesterol

FIGURE 2.14   Lipids. (a) The neutral fats, or triglycerides, are synthesized by dehydration synthesis. In this process, three fatty acid chains are attached to a single glycerol molecule, and a water molecule is lost at each bond site. **(b)** Structure of a typical phospholipid molecule. Two fatty acid chains and a phosphorus-containing group are attached to the glycerol backbone. In the commonly used schematic to the right, the polar end ("head") is depicted as a sphere, and the nonpolar end ("tail") is depicted as two wavy lines. **(c)** The generalized structure of cholesterol. Cholesterol is the basis for all steroids formed in the body. The steroid nucleus is shaded.

Only small amounts of carbohydrates are used for structural purposes. For example, some sugars are found in our genes. Others are attached to the external surfaces of cells where they act as "road signs" to guide cellular interactions. When protein intake is too low, the liver converts some sugars into building blocks needed to produce proteins.

Lipids

Lipids are insoluble in water but dissolve readily in other lipids and in organic solvents such as alcohol and ether. Like carbohydrates, all lipids contain carbon, hydrogen, and oxygen, but the proportion of oxygen in lipids is much lower. In addition, phosphorus is found in some of the more complex lipids. Lipids are a diverse group that includes *neutral fats*, *phospholipids* (fos"fo-lip'idz), *steroids* (stĕ'roidz), and a number of other lipoid substances. Table 2.2 gives the locations and functions of some lipids found in the body.

Neutral Fats

The **neutral fats** are commonly known as *fats* when solid or *oils* when liquid. A neutral fat is composed of two types of building blocks, **fatty acids** and **glycerol** (glis'er-ol) (Figure 2.14a). Fatty acids are linear chains of carbon and hydrogen atoms (hydrocarbon chains) with an organic acid group (—COOH) at one end. Glycerol is a modified simple sugar (a sugar alcohol). Fat synthesis involves attaching three fatty acid chains to a single glycerol molecule by dehydration synthesis; the result is an E-shaped molecule. Because of the 3:1 fatty acid to glycerol ratio, the neutral fats are also called **triglycerides** (tri-glis'er-īdz) or **triacylglycerols** (tri-as"il-glis'er-olz; *acyl* = acid). The glycerol backbone is the same in all neutral fats, but the fatty acid chains vary, resulting in different kinds of neutral fats. Neutral fats are large molecules, often consisting of hundreds of atoms, and ingested fats and oils must be broken down to their building blocks before they can be absorbed. Neutral fats are the body's most concentrated source of usable energy fuel, and when they are oxidized they yield large amounts of energy.

The hydrocarbon chains make neutral fats nonpolar molecules. Because polar and nonpolar molecules do not interact, oil (or fats) and water do not mix. Consequently, neutral fats are well suited for storing energy fuel in the body. Deposits of neutral fats are found mainly beneath the skin, where they insulate the deeper body tissues from heat loss and protect them from mechanical trauma. Women are usually more successful English Channel swimmers than men. This is due partly to their thicker subcutaneous fatty layer, which helps insulate them from the bitterly cold water of the channel.

The length of a neutral fat's fatty acid chains and their degree of saturation determine how solid a neutral fat is at a given temperature. Fatty acid chains with only single covalent bonds between carbon atoms are referred to as **saturated.** Fatty acids that contain one or more double bonds between carbon atoms are said to be **unsaturated** (**monounsaturated** and **polyunsaturated** respectively). Neutral fats with short fatty acid chains or unsaturated fatty acids are liquid at room temperature and are typical of plant lipids. We are familiar with these unsaturated fats as oils used for cooking. Olive and peanut oils are rich in monounsaturated fats. Corn, soybean, and safflower oils contain a high percentage of polyunsaturated fatty acids. Longer fatty acid chains and more saturated fatty acids are common in animal fats such as butter fat and the fat of meats, which are solid at room temperature.

Phospholipids

Phospholipids are modified triglycerides with a phosphorus-containing group and two, rather than three, fatty acid chains (Figure 2.14b). The phosphorus-containing group gives phospholipids their distinctive chemical properties. Although the hydrocarbon portion (the "tail") of the molecule is nonpolar and interacts only with nonpolar molecules, the phosphorus-containing part (the "head") is polar and attracts other polar or charged particles, such as water or ions. Molecules that have both polar and nonpolar regions are said to be *amphipathic* (am-fe-path'ik). As you will see in Chapter 3, cells use this unique characteristic of phospholipids in building their membranes. Some biologically important phospholipids and their functions are listed in Table 2.2.

Steroids

Structurally, steroids differ quite a bit from fats. **Steroids** are basically flat molecules made of four interlocking hydrocarbon rings. Like neutral fats, steroids are fat soluble and contain little oxygen. The single most important molecule in our steroid chemistry is *cholesterol* (ko-les'ter-ol) (Figure 2.14c). Because cholesterol is actually a steroid alcohol, it is more properly called a *sterol.* We ingest cholesterol in animal products such as eggs, meat, and cheese, and our liver produces a certain amount.

Cholesterol has earned bad press because of its role in arteriosclerosis, but it is absolutely essential for human life. Cholesterol is found in cell membranes and is the raw material of vitamin D, steroid hormones, and bile salts. Although steroid hormones are present in the body in only small quantities, they are vital to homeostasis. Without sex hormones, reproduction would be impossible, and a total lack of the corticosteroids produced by the adrenal gland is fatal.

Eicosanoids

The **eicosanoids** (i-ko'sah-noyds) are diverse lipids chiefly derived from a 20-carbon fatty acid (arachidonic acid) found in all cell membranes. Most important of these are the *prostaglandins* and their relatives, which play roles in various body processes (see Table 2.2) including blood clotting, inflammation, and labor contractions.

Proteins

Protein composes 10 to 30% of cell mass and is the basic structural material of the body. However, not all proteins are construction materials; many play vital roles in cell function. Proteins, which include enzymes (biological catalysts), hemoglobin of the blood, and contractile proteins of muscle, have the most varied functions of any molecules in the body. All proteins contain carbon, oxygen, hydrogen, and nitrogen, and many contain sulfur and phosphorus as well.

TABLE 2.2	Representative Lipids Found in the Body
Lipid type	**Location/function**
Neutral fats (triglycerides)	In fat deposits (subcutaneous tissue and around organs); protects and insulates body organs; the major source of *stored* energy in the body
Phospholipids (phosphatidyl choline; cephalin; others)	Chief component of cell membranes; may participate in the transport of lipids in plasma; prevalent in nervous tissue
Steroids	
Cholesterol	The structural basis for manufacture of all body steroids
Bile salts	Breakdown products of cholesterol; released by the liver into the digestive tract, where they aid fat digestion and absorption
Vitamin D	A fat-soluble vitamin produced in the skin on exposure to UV radiation; necessary for normal bone growth and function
Sex hormones	Estrogen and progesterone (female hormones) and testosterone (a male hormone) are produced in the gonads and are necessary for normal reproductive function
Adrenal cortical hormones	Cortisol, a glucocorticoid, is a metabolic hormone necessary for maintaining normal blood glucose levels; aldosterone helps to regulate salt and water balance of the body by targeting the kidneys
Other lipoid substances	
Fat-soluble vitamins:	
A	Found in orange-pigmented vegetables and fruits; converted in the retina to retinal, a part of the photoreceptor pigment involved in vision
E	Found in plant products such as wheat germ and green leafy vegetables; claims have been made (but not proved in humans) that it promotes wound healing and contributes to fertility; may help to neutralize highly reactive particles called free radicals believed to be involved in triggering some types of cancer
K	Made available to humans largely by the action of intestinal bacteria; also prevalent in a wide variety of foods; necessary for proper clotting of blood
Eicosanoids (prostaglandins; leukotrienes; thromboxanes)	Group of molecules derived from fatty acids found in all cell membranes; the potent prostaglandins have diverse effects, including stimulation of uterine contractions, regulation of blood pressure, control of gastrointestinal tract motility, and secretory activity; both prostaglandins and leukotrienes are involved in inflammation; thromboxanes are powerful vasoconstrictors
Lipoproteins	Lipoid and protein-based substances that transport fatty acids and cholesterol in the bloodstream; major varieties are high-density lipoproteins (HDLs) and low-density lipoproteins (LDLs)

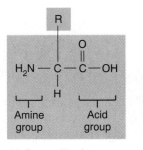

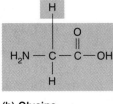

(a) **Generalized structure of all amino acids**

(b) **Glycine** (the simplest amino acid)

(c) **Aspartic acid** (an acidic amino acid)

(d) **Lysine** (a basic amino acid)

(e) **Cysteine** (a sulfur-containing amino acid)

FIGURE 2.15 *Amino acid structures.* (a) Generalized structure of amino acids. All amino acids have both an amine (—NH₂) group and an acid (—COOH) group; they differ only in the atomic makeup of their R groups (green).

(b–e) Specific structures of four amino acids. (b) The simplest (glycine) has an R group consisting of a single hydrogen atom. (c) An acid group in the R group, as in aspartic acid, makes the amino acid more acidic. (d) An amine group in

the R group, as in lysine, makes the amino acid more basic. (e) The presence of a sulfhydryl (—SH) group in the R group of cysteine suggests that this amino acid is likely to participate in intramolecular bonding.

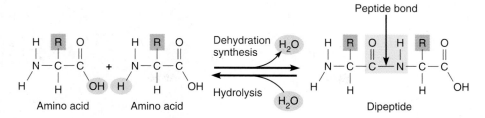

FIGURE 2.16 *Amino acids are linked together by dehydration synthesis.*
The acid group of one amino acid is bonded to the amine group of the next, with
loss of a water molecule. The bond formed is called a peptide bond. Peptide bonds
are broken when water is added to the bond (i.e., during hydrolysis).

Amino Acids and Peptide Bonds

The building blocks of proteins are molecules called
amino acids, of which there are 20 common types
(see Appendix B). All amino acids have two important functional groups: a basic group called an *amine*
(ah'mēn) *group* (—NH$_2$), and an organic *acid group*
(—COOH). An amino acid may therefore act either
as a base (proton acceptor) or an acid (proton donor).
In fact, all amino acids are identical except for a single group of atoms called their *R group.* Differences
in the numbers and arrangement of the atoms making up the R group make each amino acid unique in
its chemical behavior (Figure 2.15).

Proteins are long chains of amino acids joined
together by dehydration synthesis, with the amine
end of one amino acid linked to the acid end of the
next. The resulting bond produces a characteristic
arrangement of linked atoms and is called a **peptide
bond** (Figure 2.16). Two united amino acids form a
dipeptide, three a *tripeptide,* and ten or more a
polypeptide. Although polypeptides containing more
than 50 amino acids are called proteins, most proteins are **macromolecules,** large, complex molecules
containing from 100 to over 10,000 amino acids.

Because each type of amino acid has distinct
properties, the sequence in which they are bound together produces proteins that vary widely in both
structure and function. We can think of the 20
amino acids as a 20-letter "alphabet" used in specific
combinations to form "words" (proteins). Just as a
change in one letter can produce a word with an entirely different meaning (flour → floor) or that is
nonsensical (flour → fllur), changes in the kinds or
positions of amino acids can yield proteins with different functions or proteins that are nonfunctional.
Nevertheless, there are thousands of different proteins in the body, each with distinct functional properties, and all constructed from different combinations of the 20 amino acids.

Structural Levels of Proteins

Proteins can be described in terms of four structural
levels. The linear sequence of amino acids composing the polypeptide chain is called the *primary structure* of a protein. This structure, which resembles a
strand of amino acid "beads," is the backbone of the
protein molecule (Figure 2.17a on p. 52).

Proteins do not normally exist as simple, linear
chains of amino acids. Instead, they twist or bend
upon themselves to form a more complex *secondary
structure.* The most common type of secondary
structure is the **alpha (α) helix,** which resembles a
Slinky toy or the coils of a telephone cord (Figure
2.17b). The α helix is formed by coiling of the primary chain and is stabilized by hydrogen bonds
formed between NH and CO groups of amino acids
in the primary chain, which are approximately four
amino acids apart. Hydrogen bonds in α helices always link different parts of the *same* chain together.
In another type of secondary structure, the **beta (β)-
pleated sheet,** the primary polypeptide chains do
not coil, but are linked side by side by hydrogen
bonds to form a pleated, ribbonlike structure (Figure
2.17c) that resembles an accordion. Notice that in
this type of secondary structure, the hydrogen
bonds may link together *different polypeptide*
chains as well as *different parts* of the same chain
that has folded back on itself. A single polypeptide
chain may exhibit both types of secondary structure
at various places along its length.

Many proteins have *tertiary* (ter'she-a"re) *structure,* the next higher level of complexity, which is superimposed on secondary structure. Tertiary structure is achieved when α-helical or β-pleated regions
of the polypeptide chain fold upon one another to
produce a compact ball-like, or globular, molecule.
For example, the tertiary structure (and underlying
secondary structure) of a beta polypeptide chain of
hemoglobin, the oxygen-binding protein found in red
blood cells, is shown in Figure 2.17d. This unique
structure is maintained by both covalent and hydrogen bonds between amino acids that are often far
apart in the primary chain. When two or more
polypeptide chains are arranged, or aggregate, in a
regular manner to form a complex protein, the protein has *quaternary* (kwah'ter-na"re) *structure.* Hemoglobin (Figure 2.17e) exhibits this structural level.

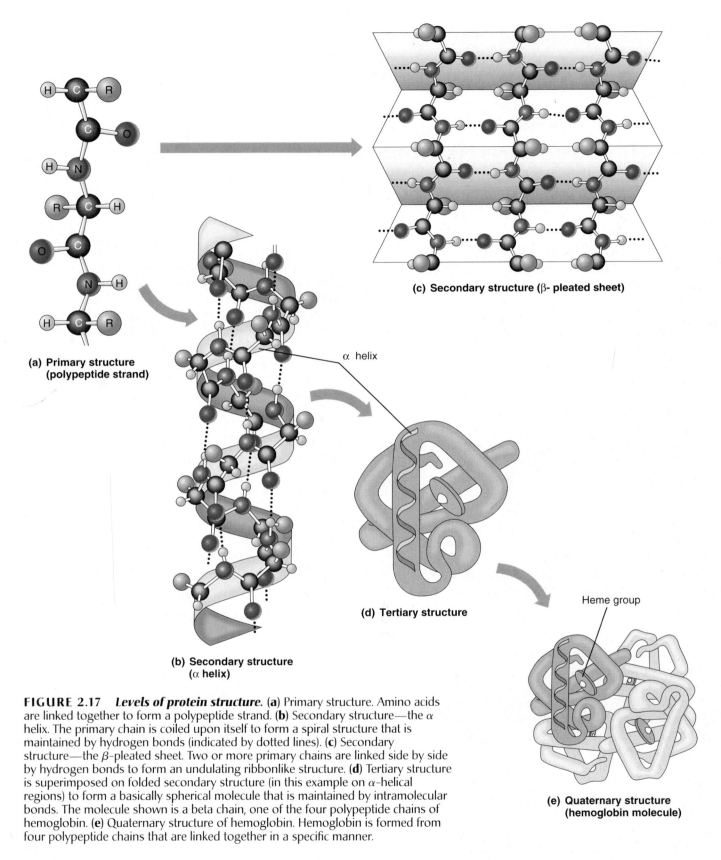

(a) Primary structure (polypeptide strand)

(b) Secondary structure (α helix)

(c) Secondary structure (β- pleated sheet)

α helix

(d) Tertiary structure

Heme group

(e) Quaternary structure (hemoglobin molecule)

FIGURE 2.17 *Levels of protein structure.* (**a**) Primary structure. Amino acids are linked together to form a polypeptide strand. (**b**) Secondary structure—the α helix. The primary chain is coiled upon itself to form a spiral structure that is maintained by hydrogen bonds (indicated by dotted lines). (**c**) Secondary structure—the β-pleated sheet. Two or more primary chains are linked side by side by hydrogen bonds to form an undulating ribbonlike structure. (**d**) Tertiary structure is superimposed on folded secondary structure (in this example on α-helical regions) to form a basically spherical molecule that is maintained by intramolecular bonds. The molecule shown is a beta chain, one of the four polypeptide chains of hemoglobin. (**e**) Quaternary structure of hemoglobin. Hemoglobin is formed from four polypeptide chains that are linked together in a specific manner.

Although a protein with tertiary or quaternary structure looks a bit like a clump of congealed pasta, the ultimate overall structure of any protein is very specific and is dictated by its primary structure. This reflects the fact that the types and relative positions of amino acids in the protein backbone determine where bonds can form to produce the complex coiled or folded structures that keep water-loving amino acids near the surface and water-fleeing amino acids buried in the protein's core.

Fibrous and Globular Proteins

The overall structure of a protein determines its biological function. For example, the ability of hemoglobin to transport oxygen is determined by its unique structure. In general, proteins are classified according to their overall appearance and shape as either fibrous or globular.

Fibrous proteins are extended and strandlike. Most exhibit only secondary structure, but some have quaternary structure as well. For example, *collagen* (kol'ah-jen) is a composite of the helical tropocollagen molecules packed together side by side to form a strong ropelike structure (Figure 2.18). Fibrous proteins are insoluble in water, and very stable—qualities ideal for providing mechanical support and tensile strength to the body's tissues. Besides collagen, which is the single most abundant protein in the body, the fibrous proteins include keratin, elastin, and certain contractile proteins of muscle (Table 2.3). Because fibrous proteins are the chief building materials of the body, they are also known as **structural proteins.**

Globular proteins are compact, spherical proteins that have at least tertiary structure; some also exhibit quaternary structure. The globular proteins are water soluble, chemically active molecules, and they play crucial roles in virtually all biological processes. Consequently, some refer to this group as **functional proteins.** Some of these proteins (antibodies) help to provide immunity, others (protein-based hormones) regulate growth and development, and still others (enzymes) are catalysts that oversee just about every chemical reaction in the body. The roles of these and selected other proteins found in the body are summarized in Table 2.3.

Protein Denaturation

Fibrous proteins are very stable, but globular proteins are quite the opposite. The activity of a protein depends on its specific three-dimensional structure, and intramolecular bonds, particularly hydrogen bonds, are important in maintaining that structure. However, hydrogen bonds are fragile and easily broken by many chemical and physical factors, such as excessive acidity or heat. Although individual pro-

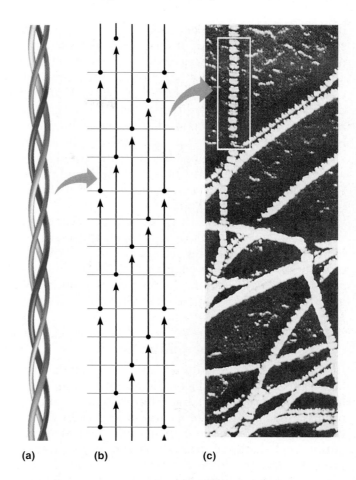

(a)   **(b)**   **(c)**

FIGURE 2.18 *The structure of collagen fibers.* The fibrous protein collagen is made up of tropocollagen molecules packed together to form fibers. The tropocollagen molecule is a triple helix. **(a)** The triple-helical secondary structure of a tropocollagen molecule. **(b)** Tropocollagen molecules align side by side in a staggered fashion to form the collagen fiber. **(c)** An electron micrograph of collagen at lower magnification shows the crisscrossing of many fibers, with the periodic pattern clearly visible in each.

teins vary in their sensitivity to environmental conditions, hydrogen bonds begin to break when the pH drops or the temperature rises above normal (physiological) levels, causing proteins to unfold and lose their specific three-dimensional shape. In this condition, a protein is said to be **denatured.** Fortunately, the disruption is reversible in most cases, and the "scrambled" protein regains its native structure when desirable conditions are restored. However, if the temperature or pH change is so extreme that protein structure is damaged beyond repair, the protein is *irreversibly denatured.* The coagulation of egg white (primarily albumin protein) that occurs when you boil or fry an egg is an example of irreversible protein denaturation. There is no way to restore the white, rubbery protein to its original translucent form.

When globular proteins are denatured, they can no longer perform their physiological roles. This is because their function depends on the presence of specific arrangements of atoms, called **active sites,** on their surfaces. The active sites are regions that fit and interact chemically with other molecules of complementary shape and charge. Because atoms contributing to an active site may actually be very far apart in the primary chain, disruption of intramolecular bonds separates them and destroys the active site (Figure 2.19). Hemoglobin becomes totally unable to bind and transport oxygen when blood pH is too acidic, because the structure needed for its function has been destroyed.

Most body proteins are described in conjunction with the organ systems or functional processes to which they are closely related. However, two groups of proteins—*molecular chaperones* and *enzymes*—are intimately involved in the normal functioning of all cells, and so these incredibly complex molecules are considered here.

Molecular Chaperones

In addition to the ubiquitous enzymes, all cells contain a class of unrelated globular proteins called **molecular chaperones,** or **chaperonins,** which, among other things, help proteins to achieve their functional three-dimensional structure. Although its amino acid sequence determines the precise way a protein folds, the folding process also requires the help of molecular chaperones to ensure that the folding is quick and accurate. Apparently, molecular chaperones have numerous protein-related roles to play. For example, specific chaperonin types

- Prevent accidental, premature, or incorrect folding of polypeptide chains or their association with other polypeptides
- Aid the desired folding and association process
- Help to translocate proteins across cell membranes
- Promote the breakdown of damaged or denatured proteins

The first such proteins discovered were called *heat shock proteins (hsp)* because they accumulate in cells exposed to abnormal temperature jumps and seem to protect the cells from the destructive effects of heat. It was later found that these proteins are also produced in response to a variety of traumatizing stimuli—for example, in the oxygen-deprived cells of a heart attack patient—and the name *stress proteins* replaced hsp for that particular group of chaperonins. It is now clear that chaperonins are vitally important to cell function regardless of the circumstances. One group of chaperonins, the *metallochaperones,* aid

 *What event, which is possible in situation (**a**), can no longer occur in situation (**b**)?*

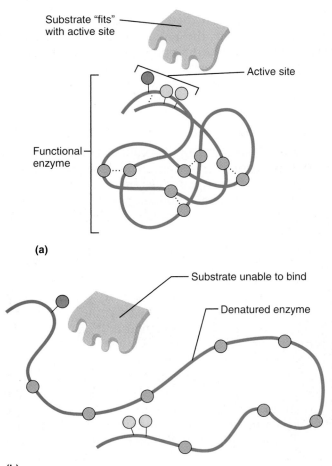

(a)

(b)

FIGURE 2.19   *Denaturation of a globular protein such as an enzyme.* (**a**) The molecule's globular structure is maintained by intramolecular bonds. Atoms composing the active site of the enzyme are shown as stalked particles. The substrate, or interacting molecule, has a corresponding binding site, and the two sites fit together precisely. (**b**) Breaking the intramolecular bonds that maintain the secondary and tertiary structure of the enzyme results in a linear molecule, with atoms of the former active site widely separated. Enzyme-substrate binding can no longer occur.

the transit of certain metal ions (copper, iron, zinc) across the plasma membrane helping to avoid excesses which might be toxic. Once in the cytosol, the metal ions are then safely delivered to the appropriate protein receptors within the cell.

*The substrate can no longer bind to the enzyme's active site in (**b**) because the atoms composing the active site are no longer in their functional relationship to one another.*

TABLE 2.3		Representative Types of Proteins in the Body

Classification according to

Overall structure	General function	Examples from the body
Fibrous	Structural framework/ mechanical support	*Collagen,* found in all connective tissues, is the single most abundant protein in the body. It is responsible for the tensile strength of bones, tendons, and ligaments.
		Keratin is the structural protein of hair and nails and a waterproofing material of skin.
		Elastin is found, along with collagen, where durability and flexibility are needed, such as in the ligaments that bind bones together.
		Spectrin internally reinforces and stabilizes the surface membrane of some cells, particularly red blood cells. *Dystrophin* reinforces and stabilizes the surface membrane of muscle cells. *Titin* helps organize the intracellular structure of muscle cells and accounts for the elasticity of skeletal muscles.
	Movement	*Actin and myosin,* contractile proteins, are found in substantial amounts in muscle cells, where they cause muscle cell shortening (contraction); they also function in cell division in all cell types. Actin is important in intracellular transport, particularly in nerve cells.
Globular	Catalysis	*Protein enzymes* are essential to virtually every biochemical reaction in the body; they increase the rates of chemical reactions by at least a millionfold. Examples include salivary amylase (in saliva), which catalyzes the breakdown of starch, and oxidase enzymes, which act to oxidize food fuels.
	Transport	*Hemoglobin* transports oxygen in blood, and *lipoproteins* transport lipids and cholesterol. Other transport proteins in the blood carry iron, hormones, or other substances.
	Regulation of pH	Many plasma proteins, such as *albumin,* function reversibly as acids or bases, thus acting as buffers to prevent wide swings in blood pH.
	Regulation of metabolism	*Peptide* and *protein hormones* help to regulate metabolic activity, growth, and development. For example, *growth hormone* is an anabolic hormone necessary for optimal growth; *insulin* helps regulate blood sugar levels.
	Body defense	*Antibodies* (immunoglobulins) are specialized proteins released by immune cells that recognize and inactivate foreign substances (bacteria, toxins, some viruses). *Complement proteins,* which circulate in blood, enhance both immune and inflammatory responses. *Molecular chaperones* aid folding of new proteins in both healthy and damaged cells and transport of metal ions into and within the cell.

Enzymes and Enzyme Activity

Characteristics of Enzymes Enzymes are globular proteins that act as biological catalysts. *Catalysts* are substances that regulate and accelerate the rate of biochemical reactions but are not used up or changed in those reactions. More specifically, enzymes can be thought of as chemical traffic cops that keep our metabolic pathways flowing. Enzymes cannot force chemical reactions to occur between molecules that would not otherwise react; they can only increase the speed of reaction. Without enzymes, biochemical reactions proceed so slowly that for practical purposes they do not occur at all. Enzymes increase reaction rates by more than a millionfold.

Some enzymes are purely protein. In other cases, the functional enzyme consists of two parts, collectively called a **holoenzyme**—an **apoenzyme** (the protein portion) and a **cofactor.** Depending on the enzyme, the cofactor may be an ion of a metal element such as copper or iron, or an organic molecule needed to assist the reaction in some particular way. Most organic cofactors are derived from vitamins (especially the B complex vitamins); this type of cofactor is more precisely called a **coenzyme.**

Each enzyme is chemically specific. Some enzymes control only a single chemical reaction. Others exhibit a broader specificity in that they can bind with similar (but not identical) molecules and thus regulate a small group of related reactions. The

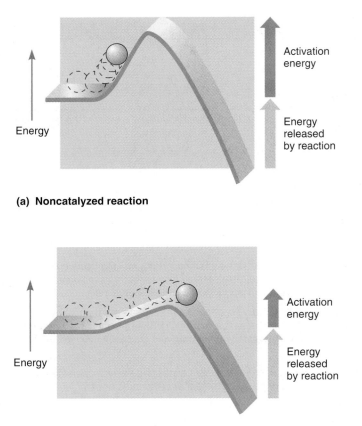

(a) Noncatalyzed reaction

(b) Enzyme-catalyzed reaction

FIGURE 2.20 *Comparison of the energy barrier in a noncatalyzed reaction and in the same reaction catalyzed by an enzyme.* In each case, the reacting particles, indicated simply by a ball, must achieve a certain energy level before they can interact. The amount of energy that must be absorbed to reach that state and overcome the energy barrier, represented by the peak, is called activation energy. In (**a**), the noncatalyzed reaction, the amount of activation energy needed is much higher than that needed in (**b**), the enzyme-catalyzed reaction.

presence of specific enzymes thus determines not only which reactions will be speeded up, but also which reactions will occur—no enzyme, no reaction. This also means that unwanted or unnecessary chemical reactions do not occur. Most enzymes are named for the type of reaction they catalyze. *Hydrolases* (hi′druh-lās-es) add water during hydrolysis reactions; *oxidases* (ok′sĭ-dās-es) add oxygen; and so on. Most enzyme names can be recognized by the suffix *-ase.*

Some enzymes are produced in an inactive form and must be activated in some way before they can function. For example, digestive enzymes produced in the pancreas are activated when they reach the small intestine, where they actually do their work. If they were produced in active form, the pancreas would digest itself. Sometimes, enzymes are inactivated immediately after they have performed their

catalytic function. This is true of enzymes that promote blood clot formation when the wall of a blood vessel is damaged. Once clotting is triggered, those enzymes are inactivated. Otherwise, you would have blood vessels full of solid blood instead of one protective blood clot.

Mechanism of Enzyme Activity How do enzymes perform their catalytic role? As we said earlier, a chemical reaction cannot occur unless the interacting molecules reach a certain energy level, which is reflected in their speed of movement. More precisely, every reaction requires that a certain amount of energy, called **activation energy,** be absorbed to prime the reaction. The activation energy pushes the reactants to an energy level where their random collisions are forceful enough to ensure interaction. This is true regardless of whether the overall reaction is ultimately energy absorbing or energy releasing.

One obvious way to increase molecular energy is to increase the temperature, but in living systems this would denature proteins. (This is why a high fever can be a serious event.) Enzymes allow reactions to occur at normal body temperature by decreasing the amount of activation energy required (Figure 2.20). Exactly how enzymes accomplish this remarkable feat is not fully understood. However, we know that they decrease the randomness of reactions by binding the reacting molecules temporarily to the enzyme surface and presenting them to each other in the proper position for chemical interaction to occur.

Three basic steps appear to be involved in the mechanism of enzyme action (Figure 2.21).

① The enzyme must first bind with the substance(s) on which it acts. These substances are called the **substrates** of the enzyme. Substrate binding occurs at the active site on the enzyme's surface. Substrate binding causes the active site to change shape so that the substrate and the active site fit together very precisely. This idea of the process by which an enzyme recognizes the proper substrate is called the *induced-fit model.*

② The enzyme-substrate complex undergoes internal rearrangements that form the product.

③ The enzyme releases the product of the reaction. This step shows the catalytic role of an enzyme: If the enzyme became part of the product, it would be a reactant and not a catalyst.

Because the unaltered enzymes can act again and again, cells need only small amounts of each enzyme. Catalysis occurs with incredible speed; most enzymes can catalyze millions of reactions per minute.

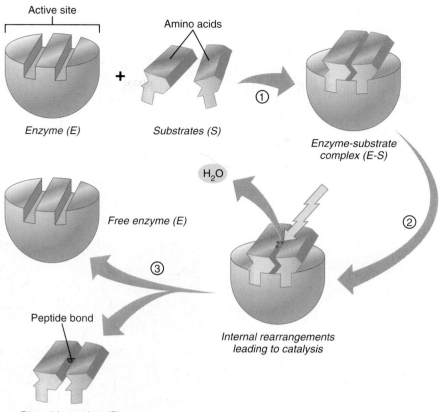

FIGURE 2.21 *Mechanism of enzyme action.* Each enzyme is highly specific in terms of the reaction(s) it can catalyze and bonds properly to only one or a few substrates. In this example, the enzyme catalyzes the formation of a dipeptide from specific amino acids.
Step 1: The enzyme–substrate complex (E–S) is formed.
Step 2: Internal rearrangements occur. In this case, energy is absorbed (indicated by the yellow arrow) as a water molecule is removed and a peptide bond is formed.
Step 3: The enzyme releases the product (P) of the reaction, the dipeptide. The free enzyme has not changed in the course of the reaction and is now available to catalyze another such reaction.
Summary: E + S → E-S → P + E

Nucleic Acids (DNA and RNA)

The **nucleic** (nu-kle′ic) **acids,** composed of carbon, oxygen, hydrogen, nitrogen, and phosphorus, are the largest molecules in the body. Their structural units, called **nucleotides,** are quite complex. Each nucleotide consists of three components (Figure 2.22a): a nitrogen-containing base, a pentose sugar, and a phosphate group. Five major varieties of nitrogen-containing bases can contribute to nucleotide structure: **adenine** (ad′ĕ-nēn), abbreviated A; **guanine** (gwan′ēn), G; **cytosine** (si′to-sēn), C; **thymine** (thi′mēn), T; and **uracil** (u′rah-sil), U. Adenine and guanine are large, two-ring bases (called purines), whereas cytosine, thymine, and uracil are smaller, single-ring bases (called pyrimidines).

The stepwise synthesis of a nucleotide involves the attachment of a base to the pentose sugar to form first a *nucleoside,* named for the nitrogenous base it contains. The nucleotide is formed when a phosphate group is bonded to the sugar of the nucleoside.

The nucleic acids include two major classes of molecules, **deoxyribonucleic** (de-ok″sĭ-ri″bo-nu-kle′ik) **acid (DNA)** and **ribonucleic acid (RNA).** Although DNA and RNA are both composed of nucleotides, they differ in many respects. Typically, DNA is found in the nucleus (control center) of the cell, where it constitutes the *genetic material,* or *genes.* DNA has two fundamental roles: It replicates (reproduces) itself before a cell divides, ensuring that the genetic information in the descendant cells is identical, and it provides instructions for building

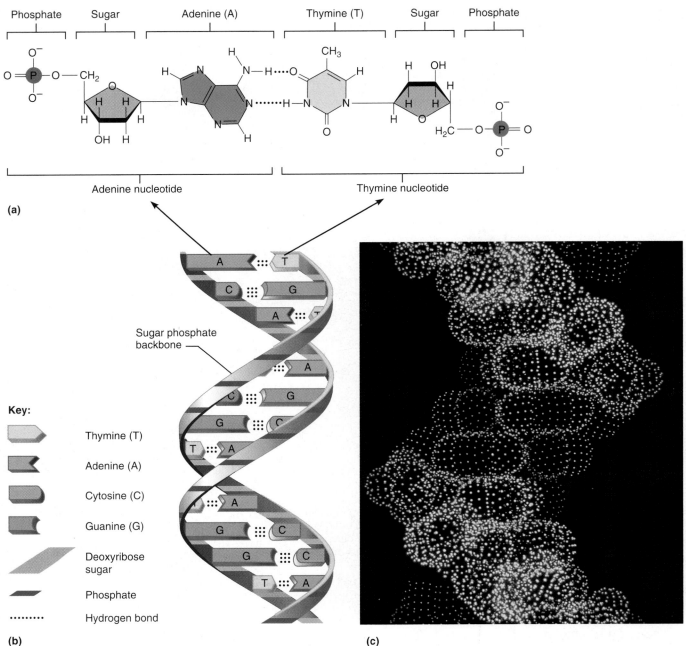

FIGURE 2.22 phosphate units. The rungs are formed

Phosphate | Sugar | Adenine (A) | Thymine (T) | Sugar | Phosphate

Adenine nucleotide

Thymine nucleotide

(a)

Sugar phosphate backbone

Key:

Thymine (T)

Adenine (A)

Cytosine (C)

Guanine (G)

Deoxyribose sugar

Phosphate

Hydrogen bond

(b)

(c)

FIGURE 2.22   *Structure of DNA.*
(a) The unit of DNA is the nucleotide, which is composed of a deoxyribose sugar molecule linked to a phosphate group, with a base attached to the sugar. Two nucleotides, linked by hydrogen bonds between their complementary bases, are illustrated. **(b)** DNA is a coiled double polymer of nucleotides (a double helix). The backbones of the ladderlike molecule are formed by alternating sugar and phosphate units. The rungs are formed by the binding together of complementary bases (A-T and G-C) by hydrogen bonds (shown by dotted lines). **(c)** Computer-generated image of a DNA molecule.

every protein in the body. By providing the information for protein synthesis, DNA determines what type of organism you will be—frog, human, oak tree—and directs your growth and development. Although we have said that enzymes govern all chemical reactions, remember that enzymes, too, are proteins formed at the direction of DNA. RNA is located chiefly outside the nucleus and can be considered a "molecular slave" of DNA. That is,

RNA carries out the orders for protein synthesis issued by DNA. (Viruses in which RNA [rather than DNA] is the genetic material are an exception to this generalization.)

DNA is a long, double-stranded polymer—a double chain of nucleotides (Figure 2.22b and c). The bases in DNA are A, G, C, and T, and its pentose sugar is *deoxyribose* (as reflected in its name). Its two nucleotide chains are held together by hy-

TABLE 2.4	Comparison of DNA and RNA	
Characteristic	**DNA**	**RNA**
Major cellular site	Nucleus	Cytoplasm (cell area outside the nucleus)
Major functions	Is the genetic material; directs protein synthesis; replicates itself before cell division	Carries out the genetic instructions for protein synthesis
Sugar	Deoxyribose	Ribose
Bases	Adenine, guanine, cytosine, thymine	Adenine, guanine, cytosine, uracil
Structure	Double strand coiled into a double helix	Single strand, straight or folded

drogen bonds between the bases, so that a ladder-like molecule is formed. Alternating sugar and phosphate molecules of each chain form the *backbones* or "uprights" of the "ladder," and the joined bases form the "rungs." The whole molecule is coiled into a spiral staircase–like structure called a **double helix.** Bonding of the bases is very specific: A always bonds to T, and G always bonds to C. A and T are therefore called **complementary bases,** as are C and G. Hence, ATGA on one nucleotide strand would necessarily be bonded to TACT (a complementary base sequence) on the other strand.

RNA molecules are single strands of nucleotides. RNA bases include A, G, C, and U (U replaces the T found in DNA), and its sugar is *ribose* instead of deoxyribose. The three varieties of RNA are distinguished by their relative size and shape, and each has a specific role to play in carrying out DNA's instructions. We discuss DNA replication and the relative roles of DNA and RNA in protein synthesis in Chapter 3. Table 2.4 compares DNA and RNA.

Adenosine Triphosphate (ATP)

Although glucose is the most important cellular fuel, none of the chemical energy contained in its bonds is used directly to power cellular work. Instead, energy released during glucose catabolism is coupled to **adenosine triphosphate (ATP)** synthesis—that is, it is captured and stored as small packets of energy in the bonds of ATP. ATP, in turn, acts as a chemical "drive shaft" to provide a form of energy that is immediately usable by all body cells.

Structurally, ATP is an adenine-containing RNA nucleotide to which two additional phosphate groups have been added (Figure 2.23). Chemically, the triphosphate tail of ATP can be compared to a tightly coiled spring ready to uncoil with tremendous energy when the catch is released. Actually, ATP is a very unstable energy-

storing molecule because its three negatively charged phosphate groups are closely packed and repel each other. When its terminal high-energy phosphate bonds are broken (hydrolyzed), the chemical "spring" relaxes and the molecule as a whole becomes more stable. Cells tap ATP's bond energy during coupled reactions by using enzymes to transfer the terminal phosphate groups from ATP to other compounds. The newly *phosphorylated* molecules are said to be "primed" and temporarily become more energetic and capable of

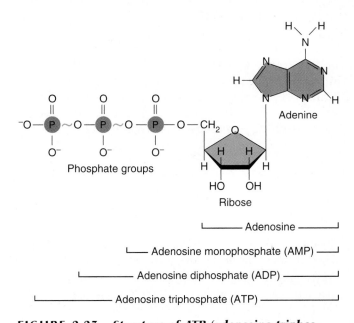

FIGURE 2.23 *Structure of ATP (adenosine triphosphate).* ATP is an adenine nucleotide to which two additional phosphate groups have been attached during breakdown of food fuels. (The phosphate bonds that are hydrolyzed to liberate free energy are indicated by the wavy lines.) When the terminal phosphate group is cleaved off, energy is released to do useful work and ADP (adenosine diphosphate) is formed. When the terminal phosphate group is cleaved off ADP, a similar amount of energy is released and AMP (adenosine monophosphate) is formed.

performing some type of cellular work. In the process of doing their work, they lose the phosphate group. The amount of energy released and transferred during ATP hydrolysis corresponds closely to that needed to drive most biochemical reactions. As a result, cells are protected from excessive energy release that might be damaging, and energy squandering is kept to a minimum.

Cleaving the terminal phosphate bond of ATP yields a molecule with two phosphate groups—*adenosine diphosphate (ADP)*—and an inorganic phosphate group, indicated by P_i, accompanied by a transfer of energy:

$$\text{ATP} \underset{\text{H}_2\text{O}}{\overset{\text{H}_2\text{O}}{\rightleftharpoons}} \text{ADP} + P_i + \text{energy}$$

As ATP is hydrolyzed to provide energy for cellular needs, ADP accumulates. Cleavage of the terminal phosphate bond of ADP liberates a similar amount of energy and produces adenosine monophosphate (AMP). ATP supplies are replenished as glucose and other fuel molecules are oxidized and their bond energy is released. The same amount of energy that is liberated when ATP's terminal phosphates are cleaved off must be captured and used to reverse the reaction to reattach phosphates and reform the energy-transferring phosphate bonds. Without ATP, molecules cannot be made or degraded, cells cannot transport substances across their membrane boundaries, muscles cannot shorten to tug on other structures, and life processes cease (Figure 2.24).

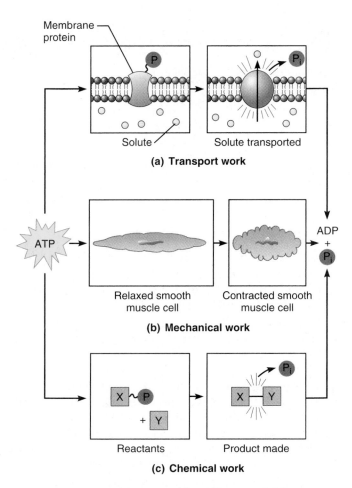

FIGURE 2.24 *Three examples of how ATP drives cellular work.* The high-energy bonds of ATP are like coiled springs that release energy for use by the cell when they are broken. **(a)** ATP drives the transport of certain solutes (amino acids, for example) across cell membranes. **(b)** ATP activates contractile proteins in muscle cells so that the cells can shorten and perform mechanical work. **(c)** ATP provides the energy to drive endergonic (energy-absorbing) chemical reactions.

RELATED CLINICAL TERMS

Acidosis (as″ĭ-do′sis; *acid* = sour, sharp) A condition of acidity or low pH (below 7.35) of the blood; high hydrogen ion concentration.

Alkalosis (al″kah-lo′sis) A condition of basicity or high pH (above 7.45) of the blood; low hydrogen ion concentration.

Heavy metals Metals with toxic effects on the body, including arsenic, mercury, and lead; iron, also included in this group, is toxic in high concentrations.

Ionizing radiation Radiation that causes atoms to ionize, e.g., radioisotope emissions and X rays.

Ketosis (ke-to′sis) A form of acidosis resulting from excessive ketones (breakdown products of fats) in the blood; common during starvation and acute attacks of diabetes mellitus.

Radiation sickness Disease resulting from exposure of the body to radioactivity; digestive system organs are most affected.

CHAPTER SUMMARY

Media study tools that could provide you additional help in reviewing specific key topics of Chapter 2 are referenced below. **SP** = *Study Partner;* **IP** = *Interactive Physiology.*

Part 1: Basic Chemistry

Definition of Concepts: Matter and Energy (pp. 27–28)

Matter (p. 27)

1. Matter is anything that takes up space and has mass. Energy is the capacity to do work or put matter into motion.

Energy (pp. 27–28)

2. Energy exists as potential energy (stored energy or energy of position) and kinetic energy (active or working energy).

3. Forms of energy involved in body functioning are chemical, electrical, radiant, and mechanical. Of these, chemical (bond) energy is most important.

4. Energy may be converted from one form to another, but some energy is always unusable (lost as heat) in such transformations.

Composition of Matter: Atoms and Elements (pp. 28–32)

1. Elements are unique substances that cannot be decomposed into simpler substances by ordinary chemical methods. Four elements (carbon, hydrogen, oxygen, and nitrogen) make up 96% of body weight.

Atomic Structure (pp. 28–30)

2. The building blocks of elements are atoms.

3. Atoms are composed of positively charged protons, negatively charged electrons, and uncharged neutrons. Protons and neutrons are located in the atomic nucleus, constituting essentially the atom's total mass; electrons are outside the nucleus in the electron shells. In any atom, the number of electrons equals the number of protons.

Identifying Elements (pp. 30–31)

4. Atoms may be identified by their atomic number (p^+) and mass number ($p^+ + n^0$). The notation ^4_2He means that helium (He) has an atomic number of 2 and a mass number of 4.

5. Isotopes of an element differ in the number of neutrons they contain. The atomic weight of any element is approximately equal to the mass number of its most abundant isotope.

Radioisotopes (pp. 31–32)

6. Many heavy isotopes are unstable (radioactive). These so-called radioisotopes decompose to more stable forms by emitting α or β particles or γ rays. Radioisotopes are useful in medical diagnosis and treatment and in biochemical research.

How Matter Is Combined: Molecules and Mixtures (pp. 32–33)

Molecules and Compounds (p. 32)

1. A molecule is the smallest unit resulting from the chemical bonding of two or more atoms. If the atoms are different, they form a molecule of a compound.

Mixtures (pp. 32–33)

2. Mixtures are physical combinations of solutes in a solvent. Mixture components retain their individual properties.

3. The types of mixtures, in order of increasing solute size, are solutions, colloids, and suspensions.

4. Solution concentrations are typically designated in terms of percent or molarity.

Distinguishing Mixtures from Compounds (p. 33)

5. Compounds are homogeneous; their elements are chemically bonded. Mixtures may be homogeneous or heterogeneous; their components are physically combined and separable.

Chemical Bonds (pp. 33–38)

The Role of Electrons in Chemical Bonding (pp. 33–34)

1. Electrons of an atom occupy areas of space called electron shells or energy levels. Electrons in the shell farthest from the nucleus (valence shell) are most energetic.

2. Chemical bonds are energy relationships between valence shell electrons of the reacting atoms. Atoms with a full valence shell or eight valence shell electrons are chemically unreactive (inert); those with an incomplete valence shell interact with other atoms to achieve stability.

Types of Chemical Bonds (pp. 34–38)

3. Ionic bonds are formed when valence shell electrons are completely transferred from one atom to another.

4. Covalent bonds are formed when atoms share electron pairs. If the electron pairs are shared equally, the molecule is nonpolar; if they are shared unequally, it is polar (a dipole).

5. Hydrogen bonds are weak bonds formed between hydrogen and nitrogen or hydrogen and oxygen. They bind together different molecules (e.g., water molecules) or different parts of the same molecule (as in protein molecules).

Chemical Reactions (pp. 38–41)

Chemical Equations (p. 39)

1. Chemical reactions involve the formation, breaking, or rearrangement of chemical bonds.

Patterns of Chemical Reactions (pp. 39–40)

2. Chemical reactions include synthesis, decomposition, and exchange reactions. Oxidation-reduction reactions may be considered a special type of exchange (or catabolic) reaction.

Energy Flow in Chemical Reactions (p. 40)

3. Bonds are energy relationships and there is a net loss or gain of energy in every chemical reaction.

4. In exergonic reactions, energy is liberated; in endergonic reactions, energy is absorbed.

Reversibility of Chemical Reactions (pp. 40–41)

5. If reaction conditions remain unchanged, all chemical reactions eventually reach a state of chemical equilibrium in which the reaction proceeds in both directions at the same rate.

6. All chemical reactions are theoretically reversible, but many biological reactions go in only one direction because of energy requirements or the removal of reaction products.

Factors Influencing the Rate of Chemical Reactions (p. 41)

7. Chemical reactions occur only when particles collide and valence shell electrons interact.

8. The smaller the reacting particles, the greater their kinetic energy and the faster the reaction rate. Higher temperature or reactant concentration, as well as the presence of catalysts, increases chemical reaction rates.

Part 2: Biochemistry

Inorganic Compounds (pp. 41–45)

1. Most inorganic compounds do not contain carbon. Those found in the body include water, salts, and inorganic acids and bases.

Water (pp. 41–42)

2. Water is the single most abundant compound in the body. It absorbs and releases heat slowly, acts as a universal solvent, participates in chemical reactions, and cushions body organs.

Salts (pp. 42–43)

3. Salts are ionic compounds that dissolve in water and act as electrolytes. Calcium and phosphorus salts contribute to the hardness of bones and teeth. Ions of salts are involved in many physiological processes.

Acids and Bases (pp. 43–45)

4. Acids are proton donors; in water, they ionize and dissociate, releasing hydrogen ions (which account for their properties) and anions.

SP Exercise: Chapter 2, The pH Scale and pH Values.

 Fluid, Electrolyte, and Acid/Base Balance CD-ROM: Introduction to Body Fluids, pages 1–8.

5. Bases are proton acceptors. The most important inorganic bases are the hydroxides; bicarbonate ion and ammonia are important bases in the body.

6. pH is a measure of hydrogen ion concentration of a solution (in moles per liter). A pH of 7 is neutral; a higher pH is alkaline, and a lower pH is acidic. Normal blood pH is 7.35–7.45. Buffers help to prevent excessive changes in the pH of body fluids.

SP Case Study: Chapter 2, Lung Disease.

IP Fluid, Electrolyte, and Acid/Base Balance CD-ROM: Acid Base Homeostasis, pages 1–12, 16, 17.

Organic Compounds (pp. 45–60)

1. Organic compounds contain carbon. Those found in the body include carbohydrates, lipids, proteins, and nucleic acids, all of which are synthesized by dehydration synthesis and digested by hydrolysis. All of these biological molecules contain C, H, and O. Proteins and nucleic acids also contain N.

Carbohydrates (pp. 45–48)

2. Carbohydrate building blocks are monosaccharides, the most important of which are hexoses (glucose, fructose, galactose) and pentoses (ribose, deoxyribose).

3. Disaccharides (sucrose, lactose, maltose) and polysaccharides (starch, glycogen) are composed of linked monosaccharide units.

4. Carbohydrates, particularly glucose, are the major energy fuel for forming ATP. Excess carbohydrates are stored as glycogen or converted to fat for storage.

SP Exercise: Chapter 25, Nutrient Use by Body Cells.

Lipids (pp. 48–49)

5. Lipids dissolve in fats or organic solvents, but not in water.

6. Neutral fats are composed of fatty acid chains and glycerol. They are found chiefly in fatty tissue where they provide insulation and reserve body fuel.

7. Phospholipids are modified phosphorus-containing neutral fats that have polar and nonpolar portions. They are found in all plasma membranes.

8. The steroid cholesterol is found in cell membranes and is the basis of steroid hormones, bile salts, and vitamin D.

Proteins (pp. 49–57)

9. The unit of proteins is the amino acid; 20 common amino acids are found in the body.

10. Many amino acids joined by peptide bonds form a polypeptide. A protein (one or more polypeptides) is distinguished by the number and sequence of amino acids in its chain(s) and by the complexity of its three-dimensional structure.

11. Fibrous proteins, such as keratin and collagen, have secondary (α helix or β pleated sheet) and perhaps quaternary structure. Fibrous proteins are used as structural materials.

12. Globular proteins achieve tertiary or quaternary structure and are generally spherical, soluble molecules. Globular proteins (e.g., enzymes, some hormones, antibodies, hemoglobin) perform special functional roles for the cell (e.g., catalysis, molecule transport).

13. Proteins are denatured by extremes of temperature or pH. Denatured globular proteins are unable to perform their usual function.

SP Exercise: Chapter 2, Mechanism of Enzyme Action; Case Study: Sickle Cell Anemia.

14. Enzymes are biological catalysts; they increase the rate of chemical reactions by decreasing the amount of activation energy needed. They do this by combining with the reactants and holding them in the proper position to interact. Many enzymes require cofactors to function.

15. Molecular chaperones assist in folding proteins into their functional 3-D shape. They are synthesized in greater amounts when cells are stressed by environmental factors and begin to accumulate denatured proteins.

Nucleic Acids (DNA and RNA) (pp. 57–59)

16. Nucleic acids include deoxyribonucleic acid (DNA) and ribonucleic acid (RNA). The structural unit of nucleic acids is the nucleotide, which consists of a nitrogenous base (adenine, guanine, cytosine, thymine, or uracil), a sugar (ribose or deoxyribose), and a phosphate group.

17. DNA is a double-stranded helix: it contains deoxyribose and the bases A, G, C, and T. DNA specifies protein structure and replicates itself exactly before cell division.

18. RNA is single-stranded; it contains ribose and the bases A, G, C, and U. RNA is involved in carrying out DNA's instructions for protein synthesis.

Adenosine Triphosphate (ATP) (pp. 59–60)

19. ATP is the universal energy compound of body cells. Some of the energy liberated by the breakdown of glucose and other food fuels is captured in the bonds of ATP molecules and transferred via coupled reactions to energy-consuming reactions.

SP Exercise: Chapter 25, Aerobic Respiration Factory.

REVIEW QUESTIONS

Multiple Choice/Matching

1. Which of the following forms of energy are in use during vision? (a) chemical, (b) electrical, (c) mechanical, (d) radiant.

2. All of the following are examples of the four major elements contributing to body mass except (a) hydrogen, (b) carbon, (c) nitrogen, (d) sodium, (e) oxygen.

3. The mass number of an atom is (a) equal to the number of protons it contains, (b) the sum of its protons and neutrons, (c) the sum of all of its subatomic particles, (d) the average of the mass numbers of all of its isotopes.

4. A deficiency in this element can be expected to reduce the hemoglobin content of blood: (a) Fe, (b) I, (c) F, (d) Ca, (e) K.

5. Which set of terms best describes a proton? (a) negative charge, massless, in the orbital; (b) positive charge, 1 amu, in the nucleus; (c) uncharged, 1 amu, in the nucleus.

6. The subatomic particles responsible for the chemical behavior of atoms are (a) electrons, (b) ions, (c) neutrons, (d) protons.

7. Which of the following are molecules of a compound? (a) N_2, (b) C, (c) $C_6H_{12}O_6$, (d) NaOH, (e) S_8.

8. Which of the following does *not* describe a mixture? (a) properties of its components are retained, (b) chemical bonds are formed, (c) components can be separated physically, (d) includes both heterogeneous and homogeneous examples.

9. A mixture that is homogeneous and transparent and does not scatter light is (a) a true solution, (b) a colloid, (c) a compound, (d) a suspension.

10. When a pair of electrons is shared between two atoms, the bond formed is called (a) a single covalent bond, (b) a double covalent bond, (c) a triple covalent bond, (d) an ionic bond.

11. Molecules formed when electrons are shared unequally are (a) salts, (b) polar molecules, (c) nonpolar molecules.

12. Which of the following covalently bonded molecules are polar?

$$\text{(a) } H—Cl \quad \text{(b) } H—\overset{\displaystyle H}{\underset{\displaystyle H}{\overset{|}{\underset{|}{C}}}}—H \quad \text{(c) } Cl—\overset{\displaystyle H}{\underset{\displaystyle Cl}{\overset{|}{\underset{|}{C}}}}—Cl \quad \text{(d) } N{=}N$$

13. Identify each reaction as (a) a synthesis reaction, (b) a decomposition reaction, or (c) an exchange reaction.

(1) $2Hg + O_2 \rightarrow 2HgO$
(2) $HCl + NaOH \rightarrow NaCl + H_2O$

14. Factors that accelerate the rate of chemical reactions include all but (a) the presence of catalysts, (b) increasing the temperature, (c) decreasing the temperature, (d) increasing the concentration of the reactants.

15. Which of the following molecules is an inorganic molecule? (a) sucrose, (b) cholesterol, (c) collagen, (d) sodium chloride.

16. Water's importance to living systems reflects (a) its polarity and solvent properties, (b) its high heat capacity, (c) its high heat of vaporization, (d) its chemical reactivity, (e) all of these.

17. Acids (a) release hydroxyl ions when dissolved in water, (b) are proton acceptors, (c) cause the pH of a solution to rise, (d) release protons when dissolved in water.

18. A chemist, during the course of an analysis, runs across a chemical composed of carbon, hydrogen, and oxygen in the proportion 1:2:1 and having a six-sided molecular shape. It is probably (a) a pentose, (b) an amino acid, (c) a fatty acid, (d) a monosaccharide, (e) a nucleic acid.

19. A neutral fat consists of (a) glycerol plus up to three fatty acids, (b) a sugar-phosphate backbone to which two amino groups are attached, (c) two to several hexoses, (d) amino acids that have been thoroughly saturated with hydrogen.

20. A chemical has an amine group and an organic acid group. It does not, however, have any peptide bonds. It is (a) a monosaccharide, (b) an amino acid, (c) a protein, (d) a fat.

21. The lipid(s) used as the basis of vitamin D, sex hormones, and bile salts is/are (a) neutral fats, (b) cholesterol, (c) phospholipids, (d) prostaglandin.

22. Enzymes are organic catalysts that (a) alter the direction in which a chemical reaction proceeds, (b) determine the nature of the products of a reaction, (c) increase the speed of a chemical reaction, (d) are essential raw materials for a chemical reaction that are converted into some of its products.

Short Answer Essay Questions

23. Define or describe energy, and explain the relationship between potential and kinetic energy.

24. Some energy is lost in every energy conversion. Explain the meaning of this statement. (Direct your response to answering the question, Is it really lost? If not, what then?)

25. Provide the atomic symbol for each of the following elements: (a) calcium, (b) carbon, (c) hydrogen, (d) iron, (e) nitrogen, (f) oxygen, (g) potassium, (h) sodium.

26. Consider the following information about three atoms:

$$^{12}_{6}C \qquad ^{13}_{6}C \qquad ^{14}_{6}C$$

(a) How are they similar to one another? (b) How do they differ from one another? (c) What are the members of such a group of atoms called? (d) Using the planetary model, draw the atomic configuration of $^{12}_{6}C$ showing the relative position and numbers of its subatomic particles.

27. How many moles of aspirin, $C_9H_8O_4$, are there in a bottle containing 450 g by weight? (*Note:* The approximate atomic weights of its atoms are C = 12, H = 1, and O = 16.)

28. Given the following types of atoms, decide which type of bonding, ionic or covalent, is most likely to occur: (a) two oxygen atoms; (b) four hydrogen atoms and one carbon atom; (c) a potassium atom $(^{39}_{19}K)$ and a fluorine atom $(^{19}_{9}F)$.

29. What are hydrogen bonds and how are they important in the body?

30. The following equation, which represents the oxidative breakdown of glucose by body cells, is a reversible reaction.

glucose + oxygen → carbon dioxide + water + ATP

(a) How can you indicate that the reaction is reversible? (b) How can you indicate that the reaction is in chemical equilibrium? (c) Define chemical equilibrium.

31. (a) Differentiate clearly between primary, secondary, and tertiary protein structure. (b) What level of structure do most fibrous proteins achieve? (c) How do globular proteins relate to the structural levels?

32. Dehydration and hydrolysis reactions are essentially opposite reactions. How are they related to the synthesis and degradation (breakdown) of biological molecules?

33. Describe the mechanism of enzyme activity. In your discussion, explain how enzymes decrease activation energy requirements.

34. Explain the importance of molecular chaperones.

Critical Thinking and Clinical Application Questions

1. As Ben jumped on his bike and headed for the freshwater lake, his mother called after him, "Don't swim if we have an electrical storm—it looks threatening." This was a valid request. Why?

2. Some antibiotics act by binding to certain essential enzymes in the target bacteria. (a) How might these antibiotics influence the chemical reactions controlled by the enzymes? (b) What is the anticipated effect on the bacteria? On the person taking the antibiotic prescription?

3. Mrs. Roberts, in a diabetic coma, has just been admitted to Noble Hospital. Her blood pH indicates that she is in severe acidosis, and measures are quickly instituted to bring her blood pH back within normal limits. (a) Define pH and note the normal pH of blood. (b) Why is severe acidosis problematic?

3 CELLS: THE LIVING UNITS

All living organisms are cellular in nature, from one-celled "generalists" like amoebas to complex multicellular organisms such as humans, dogs, or trees. Just as bricks and timbers are the structural units of a house, **cells** are the structural units of all living things. The human body has 50 to 106 trillion of these tiny building blocks.

This chapter focuses on structures and functions shared by all of our cells. Specialized cells and their unique functions are addressed in later chapters.

OVERVIEW OF THE CELLULAR BASIS OF LIFE

The English scientist Robert Hooke first observed plant cells with a crude microscope in the late 1600s. However, it was not until the 1830s that two German scientists, Matthias Schleiden and Theodor Schwann, were bold enough to insist that all living things are composed of cells. The German pathologist Rudolf Virchow extended this idea by contending that cells arise only from other cells. Virchow's proclamation was revolutionary because it openly challenged the widely accepted *theory of spontaneous generation*, which held that organisms arise spontaneously from garbage or other nonliving matter. Since the late 1800s, cell research has been exceptionally fruitful and provided us with four concepts collectively known as the **cell theory:**

1. A cell is the basic structural and functional unit of living organisms. So when you define cell properties you are in fact defining the properties of life.

2. The activity of an organism depends on both the individual and collective activities of its cells.

3. According to the **principle of complementarity,** the biochemical activities of cells are dictated by the specific subcellular structures of cells.

4. Continuity of life has a cellular basis.

All of these concepts will be expanded as we progress. For now, let us consider the idea that the cell is the smallest living unit. Whatever its form, however it behaves, the cell is the microscopic package that contains all the parts necessary to survive in an ever-changing world. It follows then that loss of cellular homeostasis underlies virtually every disease that can confront us.

A striking property of a cell is its complex organization. Chemically, cells are composed chiefly of carbon, hydrogen, nitrogen, oxygen, and trace amounts of several other elements. These substances are found in the air around us and in the ground beneath our feet, but within the cell they take on the special characteristics of life. Life stems from the way living matter is organized and carries out metabolic processes—its chemical composition is only part of the equation (see *A Closer Look* box on pp. 93–94).

The trillions of cells in the human body include some 200 different cell types that vary greatly in shape, size, and function. The spherical fat cells, disc-shaped red blood cells, branching nerve cells, and cubelike cells of kidney tubules are just a few examples of the shapes cells take. Depending on type, cells also vary greatly in length—ranging from 2 micrometers (1/12,000 of an inch) in the smallest cells to over a meter in the nerve cells that cause you to wiggle your toes. A cell's shape reflects its function. For example, the flat, tilelike epithelial cells that line the inside of your cheek fit closely together, forming a living barrier that protects underlying tissues from bacterial invasion.

Although many cell types exist, all cells have the same basic parts and some common functions. Thus, it is possible to speak of a **generalized,** or **composite, cell** to acquaint you with cell parts and components (Figure 3.1). Human cells have three main parts: the plasma membrane, the cytoplasm, and the nucleus. The *plasma membrane*, a fragile barrier, is the outer boundary of the cell. Internal to this membrane is the *cytoplasm* (si′to-plazm), which is packed with organelles, small structures that perform specific cell functions. The *nucleus* (nu′kle-us), which is surrounded by the cytoplasm, controls cellular activities and lies near the cell's center. These cell structures are summarized in Table 3.1 on pp. 66 and 67 and described beginning on page 68.

TABLE 3.1	Parts of the Cell: Structure and Function	
Cell Part	**Structure**	**Functions**
Plasma membrane (Figure 3.2)		
	Membrane made of a double layer of lipids (phospholipids, cholesterol, and so on) within which proteins are embedded; proteins may extend entirely through the lipid bilayer or protrude on only one face; externally facing proteins and some lipids have attached sugar groups	Serves as an external cell barrier; acts in transport of substances into or out of the cell; maintains a resting potential that is essential for functioning of excitable cells; externally facing proteins act as receptors (for hormones, neurotransmitters, and so on) and in cell-to-cell recognition
Cytoplasm		
	Cellular region between the nuclear and plasma membranes; consists of fluid **cytosol,** containing dissolved solutes, **inclusions** (stored nutrients, secretory products, pigment granules), and **organelles,** the metabolic machinery of the cytoplasm	
Cytoplasmic organelles		
• Mitochondria (Figure 3.15)	Rodlike, double-membrane structures; inner membrane folded into projections called cristae	Site of ATP synthesis; powerhouse of the cell
• Ribosomes (Figures 3.16, 3.17)	Dense particles consisting of two subunits, each composed of ribosomal RNA and protein; free or attached to rough ER	The sites of protein synthesis
• Rough endoplasmic reticulum (Figures 3.16, 3.17)	Membrane system enclosing a cavity, the cisterna, and coiling through the cytoplasm; externally studded with ribosomes	Sugar groups are attached to proteins within the cisternae; proteins are bound in vesicles for transport to the Golgi apparatus and other sites; external face synthesizes phospholipids and cholesterol
• Smooth endoplasmic reticulum (Figure 3.16)	Membranous system of sacs and tubules; free of ribosomes	Site of lipid and steroid synthesis, lipid metabolism, and drug detoxification
• Golgi apparatus (Figures 3.18, 3.19)	A stack of smooth membrane sacs and associated vesicles close to the nucleus	Packages, modifies, and segregates proteins for secretion from the cell, inclusion in lysosomes, and incorporation into the plasma membrane
• Lysosomes (Figure 3.20)	Membranous sacs containing acid hydrolases	Sites of intracellular digestion
• Peroxisomes (Figure 3.1)	Membranous sacs of oxidase enzymes	The enzymes detoxify a number of toxic substances; the most important enzyme, catalase, breaks down hydrogen peroxide
• Microtubules (Figures 3.22–3.25)	Cylindrical structures made of tubulin proteins	Support the cell and give it shape; involved in intracellular and cellular movements; form centrioles
• Microfilaments (Figures 3.22, 3.23)	Fine filaments of the contractile protein actin	Involved in muscle contraction and other types of intracellular movement; help form the cell's cytoskeleton
• Intermediate filaments (Figure 3.22)	Protein fibers; composition varies	The stable cytoskeletal elements; resist mechanical forces acting on the cell
• Centrioles (Figure 3.24)	Paired cylindrical bodies, each composed of nine triplets of microtubules	Organize a microtubule network during mitosis to form the spindle and asters; form the bases of cilia and flagella
• Cilia (Figure 3.25)	Short, cell surface projections; each cilium composed of nine pairs of microtubules surrounding a central pair	Move in unison, creating a unidirectional current that propels substances across cell surfaces
• Flagella	Like cilium, but longer; only example in humans is the sperm tail	Propels the cell
Nucleus (Figure 3.26)		
	Largest organelle; surrounded by the nuclear envelope; contains fluid nucleoplasm, nucleoli, and chromatin	Control center of the cell; responsible for transmitting genetic information and providing the instructions for protein synthesis

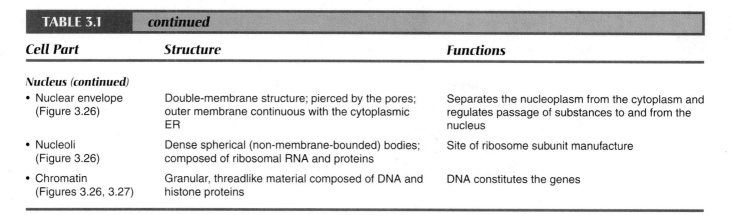

TABLE 3.1	*continued*	
Cell Part	**Structure**	**Functions**
Nucleus (continued)		
• Nuclear envelope (Figure 3.26)	Double-membrane structure; pierced by the pores; outer membrane continuous with the cytoplasmic ER	Separates the nucleoplasm from the cytoplasm and regulates passage of substances to and from the nucleus
• Nucleoli (Figure 3.26)	Dense spherical (non-membrane-bounded) bodies; composed of ribosomal RNA and proteins	Site of ribosome subunit manufacture
• Chromatin (Figures 3.26, 3.27)	Granular, threadlike material composed of DNA and histone proteins	DNA constitutes the genes

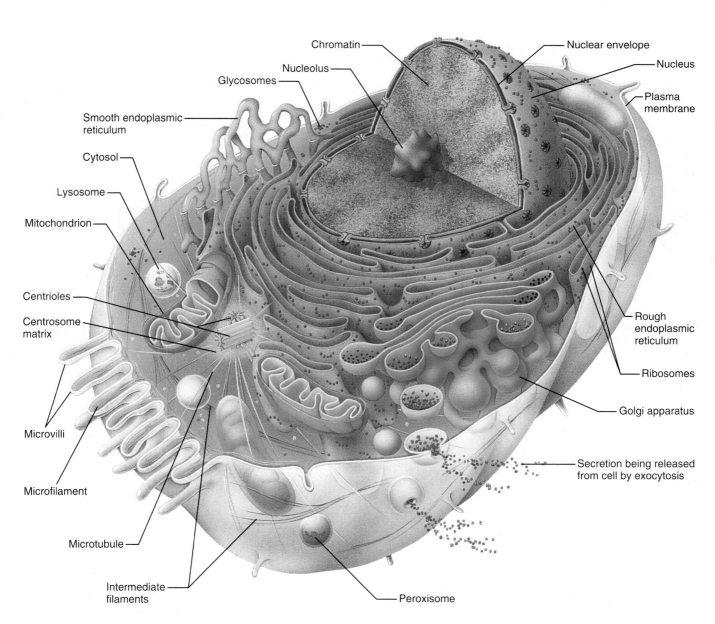

FIGURE 3.1   *Structure of the generalized cell.* No cell is exactly like this one, but this composite illustrates features common to many human cells. Note that not all of the organelles are drawn to the same scale in this illustration.

? *Some membrane proteins float freely in the lipid phase of the membrane. Others are secured in specific locations. According to this diagram, what would serve as the anchoring structures in the latter case?*

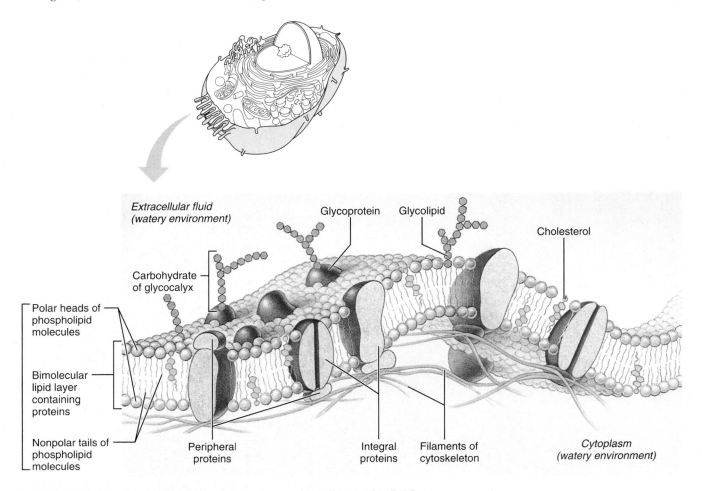

Extracellular fluid
(watery environment)

Glycoprotein   Glycolipid

Cholesterol

Carbohydrate
of glycocalyx

Polar heads of
phospholipid
molecules

Bimolecular
lipid layer
containing
proteins

Nonpolar tails of
phospholipid
molecules

Peripheral
proteins

Integral
proteins

Filaments of
cytoskeleton

Cytoplasm
(watery environment)

FIGURE 3.2   *Structure of the plasma membrane according to the fluid mosaic model.*

THE PLASMA MEMBRANE: STRUCTURE

The flexible **plasma membrane** defines the extent of the cell, thereby separating two of the body's major fluid compartments—the *intracellular* fluid within cells and the *extracellular* fluid outside and between cells. The term *cell membrane* is commonly used, but since nearly all cellular organelles are membranous, we will specifically designate the cell's surface or outer limiting membrane as the plasma membrane. The plasma membrane is much more than a passive envelope. As you will see, its unique structure allows it to play a dynamic role in many cellular activities.

The Fluid Mosaic Model

The **fluid mosaic model** of membrane structure (Figure 3.2) depicts the plasma membrane as an exceedingly thin (7–8 nm) structure composed of a double layer, or bilayer, of lipid molecules with protein molecules dispersed in it. The proteins, many of which float in the fluid lipid *bilayer*, form a constantly changing mosaic pattern; hence the name of the model.

The lipid bilayer, which forms the basic "fabric" of the membrane, is constructed largely of *phospholipids*, with smaller amounts of *cholesterol* and *glycolipids*. Each lollipop-shaped phospholipid molecule has a polar "head" that is charged and **hydrophilic** (*hydro* = water, *philic* = loving), and an uncharged, nonpolar "tail" made of two fatty acid chains that is **hydrophobic** (*phobia* = hating). The polar heads are attracted to water—the main constituent of both the cytoplasm and the fluid external to the cell—so they lie on both the inner and outer faces of the membrane. The nonpolar tails, by contrast, avoid water and line up in the center of the membrane. The result is that all biological membranes share a common sandwich-like structure: They are composed of two parallel sheets of phospholipid molecules lying tail to tail, with their polar heads exposed to water inside and outside the cell. This self-orienting property of phospholipids encourages biological membranes to self-assemble into closed, generally spherical, structures and to reseal (repair) themselves quickly when torn. It also enables scientists to suck out and replace a cell's nucleus in *cloning experiments,* a hot scientific area today.

The inner and outer layers of the membrane differ somewhat in the specific kinds of lipids they contain. About 10% of the externally facing phospholipid molecules are **glycolipids** (gli″ko-lip′idz) with attached sugar groups. The sugar groups, like the phosphate-containing groups of phospholipids, make that end of the glycolipid molecule polar, whereas the fatty acid tails are nonpolar. Some 20% of membrane lipids are cholesterol, which stabilizes the lipid membrane by wedging its platelike hydrocarbon rings between the phospholipid tails and immobilizing parts of them (Figure 3.2). However, cholesterol also makes the membrane less fluid. This characteristic of cholesterol helps to explain why atherosclerotic arteries (in which excessive cholesterol has been deposited in the membranes of cells lining the blood vessels) lose their flexibility.

There are two distinct populations of membrane proteins, integral and peripheral (see Figure 3.2). Proteins make up about half of the membrane by weight and are responsible for most of the specialized functions of the plasma membrane (Figure 3.3). **Integral proteins** are firmly inserted into the lipid bilayer. Some integral proteins protrude from one membrane face only, but most are *transmembrane proteins* that span the entire width of the membrane and protrude on both sides. Whether they extend entirely through the membrane or not, all integral proteins have both hydrophobic and hydrophilic regions. This structural feature allows them to interact with the nonpolar tails of the lipids buried within the membrane and with water inside and outside the cell. Transmembrane proteins are mainly involved in transport functions (see Figure 3.3). Some cluster together to form *channels,* or pores, through which small, water-soluble molecules or ions can move, thus bypassing the lipid part of the membrane. Others act as *carriers* that bind to a substance and move it through the membrane. Proteins that face *only* the external environment are usually receptors for hormones or other chemical messengers and act to relay a message to the cell interior (a process called signal transduction).

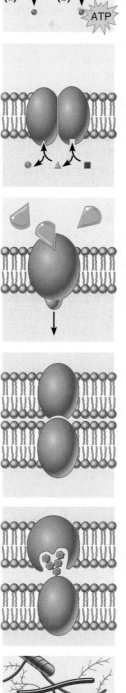

Transport
(a) A protein that spans the membrane may provide a hydrophilic channel across the membrane that is selective for a particular solute. **(b)** Some transport proteins hydrolyze ATP as an energy source to actively pump substances across the membrane.

Enzymatic activity
A protein built into the membrane may be an enzyme with its active site exposed to substances in the adjacent solution. In some cases, several enzymes in a membrane act as a team that catalyzes sequential steps of a metabolic pathway as indicated (right to left) here.

Receptors for signal transduction
A membrane protein exposed to the outside of the cell may have a binding site with a specific shape that fits the shape of a chemical messenger, such as a hormone. The external signal may cause a conformational change in the protein that initiates a chain of chemical reactions in the cell.

Intercellular joining
Membrane proteins of adjacent cells may be hooked together in various kinds of intercellular junctions. Some membrane proteins (CAMs) of this group provide temporary binding sites that guide cell migration and other cell-to-cell interactions.

Cell-cell recognition
Some glycoproteins (proteins bonded to short chains of sugars) serve as identification tags that are specifically recognized by other cells.

Attachment to the cytoskeleton and extracellular matrix (ECM)
Elements of the cytoskeleton (cell's internal supports) and the extracellular matrix (ECM) may be anchored to membrane proteins, which help maintain cell shape and fix the location of certain membrane proteins. Others play a role in cell movement or bind adjacent cells together.

FIGURE 3.3 *Some functions of membrane proteins.* A single protein may perform some combination of these tasks.

Peripheral proteins, in contrast, are not embedded in the lipid at all. Instead, they attach rather loosely to integral proteins or membrane lipids on the inner and outer faces of the membrane and are easily removed without disrupting the membrane. Peripheral proteins include a network of filaments that helps support the membrane from its cytoplasmic side. Some peripheral proteins are enzymes. Others are involved in mechanical functions, such as changing cell shape during cell division and muscle cell contraction, or linking cells together.

Many of the proteins that abut the extracellular space are glycoproteins that sport branching sugar groups. The term **glycocalyx** (gli″ko-kal′iks; "sugar covering") is used to describe the fuzzy, somewhat sticky carbohydrate-rich area at the cell surface. You can think of your cells as being "sugar-coated." The glycocalyx is enriched by glycoproteins secreted by the cell that cling to its surface, and by the glycolipids mentioned previously.

Because every cell type has a different pattern of sugars in its glycocalyx, the glycocalyx provides highly specific biological markers by which approaching cells recognize each other. For example, a sperm recognizes the ovum (egg cell) by the ovum's unique glycocalyx, and cells of the immune system identify bacteria and virus particles by binding to certain membrane glycoproteins. These matters are discussed in greater detail on pp. 82–83.

⚠ *Homeostatic Imbalance*

Definite changes in the glycocalyx occur in a cell that is transforming into a cancerous cell. In fact, a cancer cell's glycocalyx may change almost continuously, allowing it to keep ahead of immune system recognition mechanisms and avoid destruction. (Cancer is discussed in Chapter 4.) ■

The plasma membrane is a dynamic fluid structure that is in a constant state of flux. Its consistency is like that of olive oil. The lipid molecules move freely from side to side, but their polar-nonpolar interactions prevent them from flip-flopping or moving from one lipid layer to the other. Some of the proteins float freely. Others, particularly the peripheral proteins, are restricted in their movements because they are "tethered" to intercellular structures that make up the *cytoskeleton* (see Figure 3.3).

Specializations of the Plasma Membrane

Microvilli

Microvilli (mi″kro-vil′i; "little shaggy hairs") are minute, fingerlike extensions of the plasma membrane that project from a free, or exposed, cell surface (see Figure 3.4). They increase the plasma membrane surface area tremendously and are most often found on the surface of absorptive cells such as intestinal and kidney tubule cells. Microvilli have a core of actin filaments. Actin is a contractile protein, but in microvilli it appears to function as a mechanical "stiffener."

Membrane Junctions

Although certain cell types—blood cells, sperm cells, and some phagocytic cells—are "footloose" within the body, many cells, particularly epithelial cells, are knit into tight communities. Typically, three factors act to bind cells closely together:

1. Glycoproteins forming part of the glycocalyx act as a kind of adhesive.

2. The wavy contours of the membranes of adjacent cells fit together in a tongue-and-groove fashion.

3. Special membrane junctions (described next) are the most important of these factors.

Tight Junctions   In **tight junctions,** some of the protein molecules in adjacent plasma membranes fuse together, forming an *impermeable junction* that encircles the cell (Figure 3.4). Tight junctions help prevent molecules from passing through the extracellular space between adjacent cells of an epithelial membrane. For example, tight junctions between epithelial cells lining the digestive tract keep digestive enzymes and microorganisms in the intestine from seeping into the bloodstream. Although called "impermeable" junctions, some tight junctions are somewhat leaky and may allow certain types of ions to pass.

Desmosomes   Desmosomes (des′mo-sōmz; "binding bodies") are *anchoring junctions*—mechanical couplings scattered like rivets along the sides of abutting cells that prevent their separation. Desmosomes have a complicated structure (Figure 3.4). On the cytoplasmic face of each plasma membrane is a buttonlike thickening called a *plaque.* Adjacent cells do not actually touch, but are held together by thin linker proteins (cadherins) that extend from the plaques and interdigitate like the teeth of a zipper in the intercellular space. Thicker protein (intermediate) filaments (which form part of the cytoskeleton) extend from the cytoplasmic side of the plaque across the width of the cell to anchor to the plaque on the cell's opposite side. Thus, desmosomes not only bind adjacent cells together, they also contribute to a continuous internal network of strong "guy-wires," many of which extend from cell to cell. This arrangement distributes tension throughout a

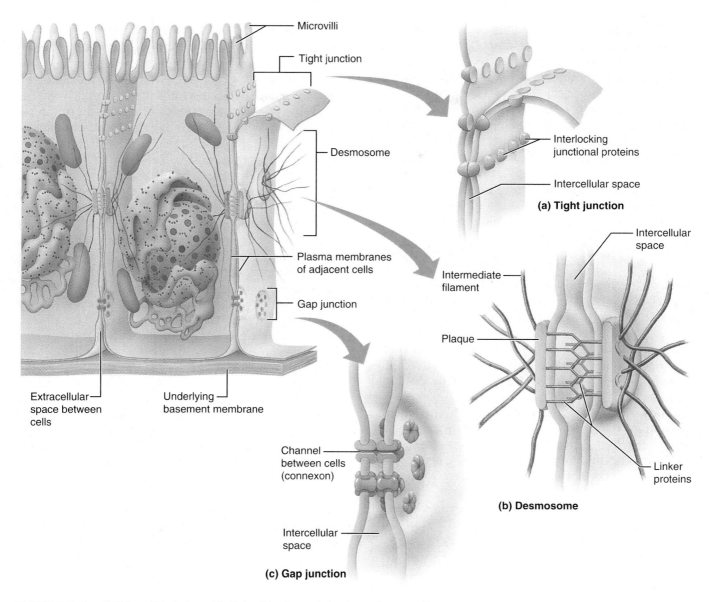

FIGURE 3.4   *Cell junctions.* An epithelial cell is shown joined to adjacent cells by three common types of cell junctions: tight junctions, desmosomes, and gap junctions.

cellular sheet and reduces its chance of tearing when it is subjected to pulling forces. Desmosomes are abundant in tissues subjected to great mechanical stress, such as skin, heart muscle, and the neck of the uterus.

Gap Junctions A **gap junction,** or *nexus* (nek'sus; "bond"), allows chemical substances to pass between adjacent cells. At gap junctions the adjacent plasma membranes are very close, and the cells are connected by hollow cylinders called *connexons* (kŏ-nek'sonz), composed of transmembrane proteins. Ions, sugars, and other small molecules pass through these water-filled channels from one cell to the next (Figure 3.4). Gap junctions between embryonic cells are vital in distributing nutrients before

the circulatory system is established. In adult tissues, gap junctions exist in electrically excitable tissues, such as the heart and smooth muscle, where ion passage from cell to cell helps synchronize their electrical activity and contraction.

THE PLASMA MEMBRANE: FUNCTIONS
Membrane Transport

Our cells are continuously bathed in an extracellular fluid called **interstitial** (in"ter-stish'al) **fluid** that is derived from the blood. Interstitial fluid is like a rich, nutritious "soup." It contains thousands of ingredients, including amino acids, sugars, fatty acids, vitamins, regulatory substances such as hormones and neurotransmitters, salts, and waste products. To remain healthy, each cell must extract from this soup the exact amounts of the substances it needs at specific times and reject the rest.

Although there is continuous traffic across the plasma membrane, it is a **selectively,** or **differentially, permeable** barrier, meaning that it allows some substances to pass while excluding others. Thus, it allows nutrients to enter the cell, but keeps many undesirable substances out. At the same time, valuable cell proteins and other substances are kept within the cell, and wastes are allowed to pass out of it.

Homeostatic Imbalance

Selective permeability is a characteristic of healthy, intact cells. When the cell (or its plasma membrane) is severely damaged, the membrane becomes permeable to virtually everything, and substances flow into and out of cells freely. This phenomenon is evident when someone has been severely burned. Precious fluids, proteins, and ions "weep" from the dead and damaged cells of the burned areas. ■

Movement of substances through the plasma membrane happens in essentially two ways—passively or actively. In **passive processes,** substances penetrate the membrane without any energy input from the cell. In **active processes,** the cell provides the metabolic energy (ATP) to move the substance across the membrane. The various transport processes that occur in cells are summarized in Table 3.2 on page 80. Let's examine how each of these types of membrane transport works.

Passive Processes: Diffusion

Diffusion (di-fu'zhun) is an important means of passive membrane transport for every cell of the body. By contrast, the other passive transport process, *filtration,* generally occurs only across capillary walls.

Diffusion is the tendency of molecules or ions to scatter evenly throughout the environment (Figure 3.5). Recall that all molecules possess kinetic energy and are in constant motion (see Chapter 2, p. 27). As molecules move about randomly at high speeds, they collide and ricochet off one another, changing direction with each collision. The overall effect of this erratic movement is that molecules move away from areas where they are in higher concentration to areas where their concentration is lower, so we say that molecules diffuse *along,* or *down,* their **concentration gradient.** The greater the difference in concentration between the two areas, the faster the net diffusion of the particles.

Because the driving force (energy source) is the kinetic energy of the molecules themselves, the speed of diffusion is influenced by the *size* of molecules (the smaller, the faster) and by *temperature* (the warmer, the faster). In a closed container, diffusion will eventually produce a uniform mixture of the different kinds of molecules. In other words, the system reaches equilibrium, with molecules moving equally in all directions (no *net* movement). Examples of diffusion are familiar to everyone. When you peel onions, you get teary-eyed because the cut onion releases volatile substances that diffuse through the air, dissolve in the fluid film covering your eyes, and form an irritating sulfuric acid.

Because of its hydrophobic interior, the plasma membrane is a physical barrier to free diffusion. However, a molecule *will* diffuse passively through

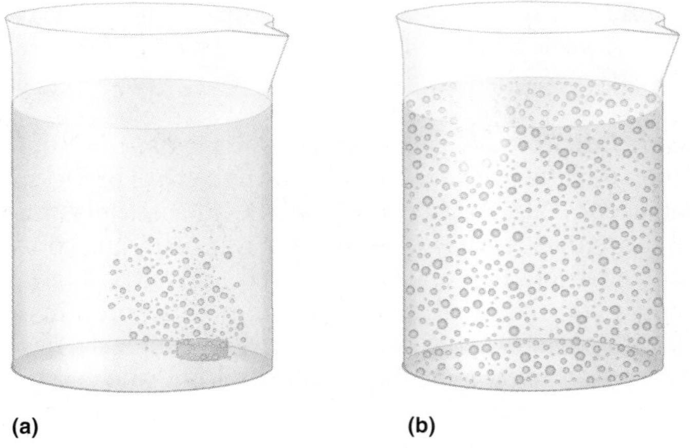

FIGURE 3.5 *Diffusion.* Molecules in solution move continuously and collide constantly with other molecules. As a result, molecules tend to move away from areas of their highest concentration and become evenly distributed. (**a**) Diagram of dye molecules diffusing from a dye pellet into the surrounding water. (**b**) Dye molecules from pellet have diffused evenly throughout the water.

(a)   (b)

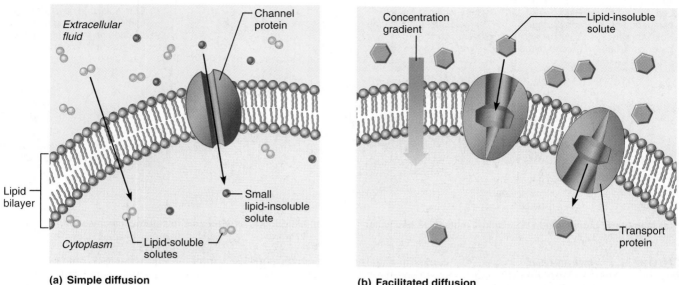

FIGURE 3.6 *Diffusion through the plasma membrane.* **(a)** Simple diffusion. As depicted on the left, fat-soluble molecules diffuse directly through the lipid bilayer of the plasma membrane, in which they can dissolve. On the right, small polar or charged particles (water molecules or small ions) are shown diffusing through membrane channels constructed by channel proteins. **(b)** Facilitated diffusion moves large, lipid-insoluble molecules (e.g., glucose) across the membrane. The substance to be transported binds to a transmembrane carrier protein.

the plasma membrane if it is (1) lipid soluble, (2) small enough to pass through membrane channels, or (3) assisted by a carrier molecule. The unassisted diffusion of lipid-soluble or very small particles is called *simple diffusion*. A special name, *osmosis* (oz-mo'sis; *osmos* = pushing), is given to the unassisted diffusion of a solvent (usually water) through a membrane. Assisted diffusion is known as *facilitated diffusion*.

Simple Diffusion Substances that are nonpolar and lipid soluble diffuse directly through the lipid bilayer (Figure 3.6a). Such substances include oxygen, carbon dioxide, fat-soluble vitamins, and alcohol. Oxygen is always in higher concentration in the blood than in tissue cells, so oxygen continuously enters the cells, whereas carbon dioxide (in higher concentration within the cells) diffuses from cells into the blood.

Most water-soluble particles are unable to diffuse through the lipid bilayer because they are repelled by its core of nonpolar fatty acid chains, but polar and charged particles can diffuse through the membrane if they are small enough to pass through the water-filled pores constructed by the channel proteins (Figure 3.6a). Pore size varies, and the channels tend to be selective in what they allow to pass. Additionally, while some pores are always open, others are *gated*; that is, they can be opened or closed in response to various chemical or electrical signals.

Facilitated Diffusion Certain molecules, notably glucose and other simple sugars, are too polar to dissolve in the lipid bilayer and too large to pass through plasma membrane channels. However, they move through the membrane very rapidly by a passive transport process called **facilitated diffusion,** in which they combine with protein *carrier* molecules in the plasma membrane and are released into the cytoplasm. Although it was initially believed that the integral proteins that act as carriers either flip-flopped or physically crossed the membrane like ferryboats, the most popular model, shown in Figure 3.6b, indicates that changes in the shape of the carrier allow it to first envelop and then release the transported substance, shielding it en route from the nonpolar regions of the membrane. Hence, the binding site is moved from one face of the membrane to the other by changes in the conformation of the carrier protein.

As in any diffusion process, glucose is transported down its concentration gradient. Glucose is normally in higher concentrations in the blood than in the cells, where it is rapidly used for ATP synthesis, so within the body glucose transport is *typically* unidirectional—into the cells. However, carrier-mediated transport is *limited* by the number of receptors present. For example, when all the glucose carriers are "engaged," they are said to be *saturated,* and glucose transport is occurring at its maximum rate.

Oxygen, water, and glucose are vitally important to cellular homeostasis. Thus, their passive transport by diffusion represents a tremendous saving of

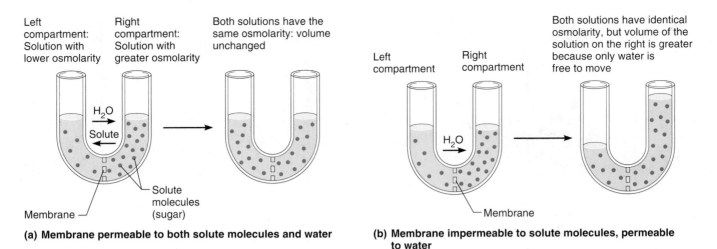

(a) **Membrane permeable to both solute molecules and water**

(b) **Membrane impermeable to solute molecules, permeable to water**

FIGURE 3.7 *Influence of membrane permeability on diffusion and osmosis.* **(a)** In this system, the membrane is permeable to both water and solute (sugar) molecules. Water moves from the solution with lower osmolarity (left compartment) to the solution with greater osmolarity (right compartment). The solute moves along its own concentration gradient in the opposite direction. When the system comes to equilibrium (right), the solutions have the same osmolarity and volume. **(b)** This system is identical to that in (a) except that the membrane is impermeable to the solute. Water moves by osmosis from the left to the right compartment, until its concentration and that of the solutions are identical. Because the solute is prevented from moving, the volume of the solution in the right compartment increases.

cellular energy. Indeed, if all these substances (and carbon dioxide) had to be transported actively, cell expenditures of ATP would increase exponentially!

Osmosis The diffusion of a solvent, such as water, through a selectively permeable membrane like the plasma membrane is called **osmosis.** Even though water is highly polar, it passes readily through the lipid bilayer. This is surprising because you'd expect it to be repelled by the hydrophobic lipid tails. Although still hypothetical, one explanation is that water is a small molecule and when the membrane lipids move randomly about, small gaps open up between their wiggling tails, allowing water to slip and slide its way through the membrane by moving from gap to gap. Water also moves through membrane pores constructed by transmembrane proteins called **aquaporins.**

Osmosis occurs whenever the water concentration differs on the two sides of the membrane. We will use some examples from nonliving systems to illustrate the process of osmosis and then describe osmosis across living membranes.

If distilled water is present on both sides of a selectively permeable membrane, no *net* osmosis occurs, even though water molecules continue to move in both directions through the membrane. If, however, the solute concentration on the two sides of the membrane differs, water concentration differs as well, because as solute concentration increases, water concentration decreases. The extent to which water's concentration is decreased by solutes depends on the *number*, not the *type*, of solute particles in a solution, because one molecule or one ion of solute

(theoretically) displaces one water molecule. The total concentration of all solute particles in a solution is referred to as the solution's **osmolarity** (oz″mo-lar′ĭ-te). When equal volumes of solutions of different osmolarity are separated by a membrane that is *permeable to all molecules* in the system, net diffusion of both solute and water occurs, each moving down its own concentration gradient (Figure 3.7a). Eventually, equilibrium is reached, and the water and solute concentrations will be the same in both compartments. If we consider the same system, but make the membrane *impermeable to the solute* molecules, we see quite a different result (Figure 3.7b). Water quickly diffuses from the left to the right compartment and continues to do so until its concentration (and that of the solution) is the same on both sides of the membrane. Notice that in this case equilibrium results from the movement of water alone (the solutes are prevented from moving). Notice also that the movement of water leads to dramatic changes in the volumes of the two compartments.

The last example considered mimics the osmotic events that occur across plasma membranes of living cells, with one major difference. In our examples, the volumes of the compartments are infinitely expandable and the effect of pressure exerted by the added weight of the higher fluid column is not considered. In living plant cells, which have rigid cell walls external to their plasma membranes, this is not the case. As water diffuses into the cell, the point is finally reached where the **hydrostatic pressure** (the back pressure exerted by water against the membrane) within the cell is equal to its **osmotic**

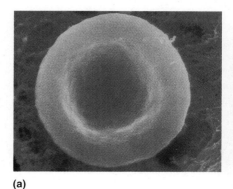
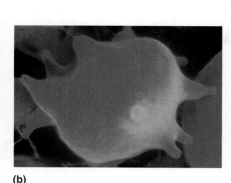
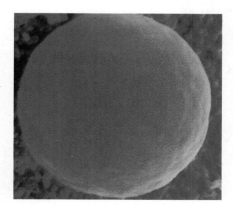

(a)   **(b)**   **(c)**

FIGURE 3.8 *The effect of solutions of varying tonicities on living red blood cells.* **(a)** In isotonic solutions (same solute/water concentrations as inside cells), cells retain their normal size and shape. **(b)** In a hypertonic solution (containing more solutes than are present inside the cells), the cells lose water and shrink (crenate). **(c)** In a hypotonic solution (containing fewer solutes than are present inside the cell), the cells take in water by osmosis until they become bloated and burst (lyse).

pressure—its tendency to resist further (net) water entry. As a rule, the higher the amount of nondiffusible (or nonpenetrating) solutes within the cell, the higher the osmotic pressure and the greater the hydrostatic pressure that must be developed to resist further net water entry.

However, such major changes in hydrostatic (and osmotic) pressures do not occur in living animal cells, which lack the rigid cell walls of plants. Osmotic imbalances cause animal cells to swell or shrink (due to net water gain or loss) until the solute concentration on both sides of the plasma membrane is the same (reaches equilibrium), or the membrane is stretched to its breaking point.

This leads us to the important concept of *tonicity* (to-nis′ĭ-te). As noted, many molecules, particularly intracellular proteins and selected ions, are prevented from diffusing through the plasma membrane. Consequently, any change in their concentration alters the water concentration on the two sides of the membrane and results in net loss or gain of water by the cell.

The ability of a solution to change the tone or shape of cells by altering their internal water volume is called **tonicity** (*tono* = tension). Solutions with the same concentrations of nonpenetrating solutes as those found in cells (0.9% saline or 5% glucose) are **isotonic** (literally, "the same tonicity"). Cells exposed to such solutions retain their normal shape, and exhibit no net loss or gain of water (Figure 3.8a). As you might expect, the body's extracellular fluids and most intravenous solutions (solutions infused into the body via a vein) are isotonic. Solutions with a higher concentration of nonpenetrating solutes than the cell (for example, a strong saline solution) are **hypertonic.** Cells immersed in hypertonic solutions lose water and shrink, or *crenate* (kre′nāt) (Figure 3.8b). Solutions that are more dilute (contain fewer nonpenetrating solutes) than cells are called **hypotonic.** Cells placed in hypotonic solutions plump up rapidly as water rushes into them (Figure 3.8c). Distilled water represents the most extreme example of hypotonicity. Because it contains *no* solutes, water continues to enter cells until they finally burst or *lyse.*

Notice that osmolarity and tonicity are not the same. The key factor in determining tonicity is the nonpenetrating solutes, whereas osmolarity is the total solute spectrum. Osmolarity is expressed as osmoles per liter (osmol/L); 1 osmol is equal to 1 mole of nonionizing molecules.* A 0.3-osmol/L solution of NaCl is isotonic because sodium ions are usually prevented from diffusing freely through the plasma membrane. But if the cell is immersed in a 0.3-osmol/L solution of a penetrating solute, the solute will enter the cell and water will follow. The cell will then swell and burst, just as if it had been placed in pure water.

Osmosis is extremely important in determining distribution of water in the various fluid-containing compartments of the body (in cells, in blood, and so

---

*Osmolarity (Osm) is determined by multiplying molarity (*M*) by the number of particles resulting from ionization. For example, since NaCl ionizes to Na$^+$ + Cl$^-$, a 1-*M* solution of NaCl is a 2-Osm solution. For substances that do not ionize (e.g., glucose), molarity and osmolarity are the same.

(1) *Which ions are being pumped against their concentration gradient?* **(2)** *Which ions are being pumped against their electrical gradient?*

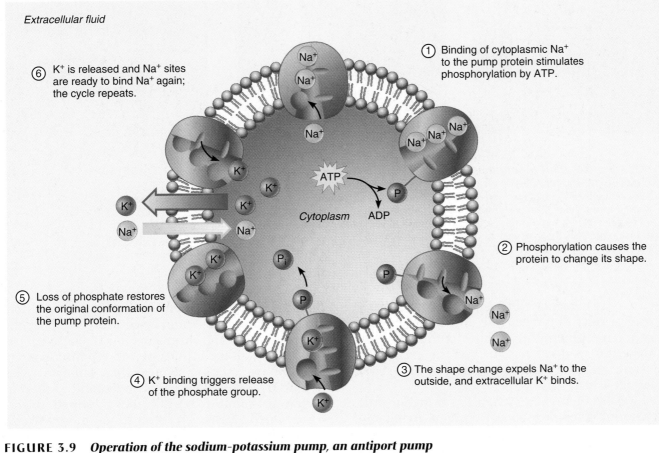

Extracellular fluid

⑥ K⁺ is released and Na⁺ sites are ready to bind Na⁺ again; the cycle repeats.

① Binding of cytoplasmic Na⁺ to the pump protein stimulates phosphorylation by ATP.

ATP

P

Cytoplasm   ADP

② Phosphorylation causes the protein to change its shape.

⑤ Loss of phosphate restores the original conformation of the pump protein.

④ K⁺ binding triggers release of the phosphate group.

③ The shape change expels Na⁺ to the outside, and extracellular K⁺ binds.

FIGURE 3.9 *Operation of the sodium-potassium pump, an antiport pump (Na⁺-K⁺ ATPase).* Hydrolysis of a molecule of ATP provides energy for a "pump" protein to move three sodium ions out of the cell and two potassium ions into the cell. Both ions are moved against their concentration gradients, indicated by colored arrows moving through the membrane (yellow arrow = Na⁺ gradient; green arrow = K⁺ gradient). Hence, this pump is an *antiport*.

on). In general, osmosis continues until pressures acting at the membrane (osmotic and hydrostatic pressures) are equal. For example, water is forced out of capillary blood by the hydrostatic pressure of the blood against the capillary wall, but the presence in the blood of solutes that are too large to cross the capillary membrane draws water back into the bloodstream. As a result, very little net loss of plasma fluid occurs.

Simple diffusion and osmosis are not very selective processes. In those processes, whether a molecule can pass through the membrane depends chiefly on its size or solubility in lipid, not on its unique

structure. Facilitated diffusion, on the other hand, *is* highly selective; the carrier for glucose combines specifically with glucose, in much the same way an enzyme binds to its specific substrate.

◭ *Homeostatic Imbalance*

Hypertonic solutions are sometimes infused intravenously into the bloodstream of edematous patients (those swollen because water is retained in their tissues) to draw the excess water out of the extracellular space and move it into the bloodstream so that it can be eliminated by the kidneys. Hypotonic solutions may be used (with care) to rehydrate the tissues of extremely dehydrated patients. In less extreme cases of dehydration, drinking hypotonic fluids, including colas, apple juice, and sports drinks, usually does the trick. ■

their electrical gradient.
tration gradient. (2) Sodium ions are being pumped against
(1) Both kinds of ions are being pumped against their concen-

Passive Processes: Filtration

Filtration is the process that forces water and solutes through a membrane or capillary wall by *fluid,* or *hydrostatic,* pressure. Like diffusion, filtration is a passive transport process and involves a gradient. However, the gradient for filtration is a **pressure gradient** that actually pushes solute-containing fluid (filtrate) from a higher-pressure area to a lower-pressure area. As already mentioned, in the body hydrostatic pressure exerted by blood forces fluid from the capillaries, and this fluid contains solutes that are vital to the tissues. Filtration also provides the fluid ultimately excreted by the kidneys as urine. Filtration is not selective; only blood cells and protein molecules too large to pass through the membrane pores are held back.

Active Processes

When a cell uses bond energy of ATP to move solutes across the membrane against their concentration gradients, the process is referred to as *active.* Substances moved across the plasma membrane by active means are usually unable to pass in the necessary direction by any of the passive diffusion processes. They may be too large to pass through the pores, incapable of dissolving in the bilipid membrane core, or unable to move with their concentration gradient. There are two major mechanisms of active membrane transport: active transport and vesicular transport.

Active Transport **Active transport** is similar to facilitated diffusion in that it requires carrier proteins that combine *specifically* and *reversibly* with the transported substances. However, facilitated diffusion always honors concentration gradients because its driving force is kinetic energy. In contrast, the active transport protein transporters or **solute pumps** move solutes, most importantly ions (such as Na^+, K^+, and Ca^{2+}), "uphill" *against* their concentration gradients. To do this work, cells must expend the energy of ATP to energize the transport process.

Active transport processes are distinguished according to the source of energy that drives the transport process. In *primary active transport,* the energy is provided directly by hydrolysis of ATP, whereas in *secondary active transport,* transport is driven indirectly by energy stored in ionic gradients created by operation of primary active transport pumps. Secondary active transport systems are all *coupled systems;* that is, they move more than one substance at a time. If the two transported substances are moved in the same direction, the system is a **symport** (*sym* = same) **system.** If the transported substances "wave to each other" as they cross the membrane in opposite directions, it is called an **antiport** (*anti* = opposite, against) **system.** Let's examine these processes more carefully.

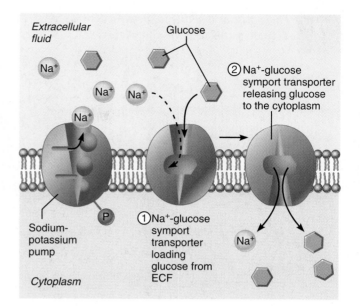

FIGURE 3.10 *Secondary active transport.* (Sequence of events moves from left to right.) In this example, the ATP-driven Na^+-K^+ pump stores energy by creating a steep concentration gradient for Na^+ entry into the cell. As Na^+ diffuses back across the membrane through a membrane cotransporter protein, it escorts glucose into the cell.

1. Primary active transport. In primary active transport, hydrolysis of ATP results in the phosphorylation of the transport protein, which in turn causes it to change its conformation in such a manner that it "pumps" the bound solute across the membrane.

The most investigated example of a primary active transport system is the operation of the **sodium-potassium (Na^+-K^+) pump,** for which the carrier is an enzyme called **Na^+-K^+ ATPase.** The concentration of K^+ inside the cell is some 10–20 times higher than that outside, and the reverse is true of Na^+. These ionic concentration differences are essential for excitable cells like muscle and nerve cells to function normally and for all body cells to maintain their normal fluid volume. Because Na^+ and K^+ leak slowly but continuously through the plasma membrane along their concentration gradients (and cross more rapidly in stimulated muscle and nerve cells), the Na^+-K^+ pump operates more or less continuously as an antiport to simultaneously drive Na^+ out of the cell against a steep gradient and pump K^+ back in. The operation of the Na^+-K^+ pump is illustrated in Figure 3.9. Other examples of primary active transport include the calcium pumps that actively sequester ionic calcium from the intracellular fluid into specific organelles or eject it from the cell.

2. Secondary active transport (Figure 3.10). A single ATP-powered pump, such as the Na^+-K^+ exchange pump that maintains the sodium gradient, can indirectly drive the transport of several other

Phagocytic cells gather in the air sacs of the lungs, especially in the lungs of smokers. What is the connection?

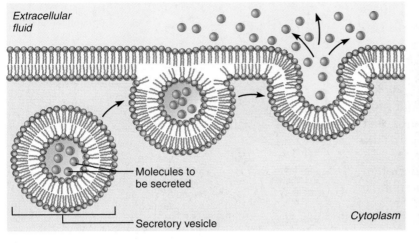

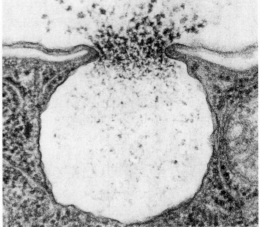

(a)   (b)

FIGURE 3.11 *Exocytosis.* **(a)** The membrane-bounded vesicle containing the substance to be secreted migrates to the plasma membrane, and the two membranes fuse. The fused site opens and releases the contents of the secretory vesicle into the intercellular space. **(b)** Photograph of a secretory vesicle releasing its contents to the cell exterior.

solutes (secondary active transport). By moving sodium across the plasma membrane against its concentration gradient, the pump stores energy (in the ion gradient). Then, just as water pumped uphill can do work (turn a turbine or water wheel for instance) as it flows back down, a substance pumped across a membrane can do work as it leaks back, propelled downhill along its concentration gradient. Thus, as sodium diffuses back into the cell with the help of a carrier protein (facilitated diffusion), other substances are "dragged along" or cotransported by the same carrier proteins. For example, some sugars, a variety of amino acids, and many ions are cotransported in this way into cells lining the small intestine. Though the cotransported substances both move passively, sodium has to be pumped back into the lumen of the intestine to maintain its diffusion gradient. Ion gradients can also be used to drive antiport systems such as those that help to regulate intracellular pH by using the sodium gradient to expel hydrogen ions (H^+).

Regardless of whether the energy is provided directly or indirectly for the active transport process, each membrane pump transports only specific substances. Hence, active transport systems provide a way for the cell to be very selective in cases where

substances cannot pass by diffusion. (No pump—no transport.)

Vesicular Transport Large particles and macromolecules are transported across plasma membranes by **vesicular transport.** Like solute pumping, this transport process is energized by ATP [or *GTP* (guanosine triphosphate, another energy-rich compound) in selected cases]. The two major kinds of vesicular transport are exocytosis and endocytosis.

Exocytosis (ek"so-si-to'sis; "out of the cell") is a mechanism that moves substances from the cell interior into the extracellular space. It accounts for hormone secretion, neurotransmitter release, mucus secretion, and in some cases, ejection of wastes. In exocytosis, the substance or cell product to be released is first enclosed in a membranous sac called a **vesicle.** The vesicle migrates to the plasma membrane, fuses with it, and then ruptures, spilling the sac contents out of the cell (Figure 3.11). The mechanism involves a "docking" process in which proteins on the vesicles recognize certain plasma membrane proteins and bind with them, bringing the two membranes close enough to fuse. As described shortly, membrane material added by exocytosis is removed by endocytosis—the reverse process.

Endocytosis (en"do-si-to'sis; "into the cell") provides a means for large particles or macromolecules to enter cells. The substance to be taken into the cell is progressively enclosed by an infolding portion of

Lungs, open to the external environment, collect dust and other airborne debris. In the lungs of smokers, carbon particles are added to this debris.

the plasma membrane. Once the membranous vesicle is formed, it pinches off from the plasma membrane and moves into the cytoplasm, where its contents are digested. Based on the nature and quantity of material taken up and the mechanism of uptake, three types of endocytosis are recognized: phagocytosis, bulk-phase endocytosis, and receptor-mediated endocytosis.

In **phagocytosis** (fag″o-si-to′sis), literally, "cell eating," cytoplasmic extensions called pseudopods (soo′do-pahdz; *pseudo* = false, *pod* = foot) protrude and flow around some relatively large or solid material, such as a clump of bacteria or cell debris, and engulf it (Figure 3.12a). The vesicle thus formed is called a **phagosome** (fag′o-sōm; "eaten body"). In most cases, the phagosome then fuses with a *lysosome* (li′so-sōm), a specialized cell structure containing digestive enzymes (see Figure 3.19, p. 87), and its contents are digested.

Only a few cell types of the human body—macrophages and certain white blood cells—are "experts" at phagocytosis. They help to police and protect the body by ingesting and disposing of bacteria, other foreign substances, and dead tissue cells. Most phagocytes move about by **amoeboid** (ah-me′boyd; "changing shape") **motion;** that is, the flowing of their cytoplasm into temporary pseudopods allows them to creep along.

In **bulk-phase endocytosis** (Figure 3.12b), a bit of infolding plasma membrane surrounds a droplet of extracellular fluid containing dissolved molecules. This droplet enters the cell in a tiny membranous vesicle. Unlike phagocytosis, bulk-phase endocytosis is a routine activity of most cells, affording them a nonselective way of sampling the extracellular fluid. It is particularly important in cells that absorb nutrients, such as cells that line the intestines.

In both phagocytosis and bulk-phase endocytosis, bits of the plasma membrane are removed when the membranous sacs are internalized. However, these membranes are recycled back to the plasma membrane by exocytosis, so the surface area of the plasma membrane remains remarkably constant.

Receptor-mediated endocytosis is the main mechanism for the specific uptake of most macromolecules by body cells, and it is exquisitely selective (Figure 3.12c). Substances taken up by receptor-mediated endocytosis include enzymes, insulin (and some other hormones), low-density lipoproteins (such as cholesterol attached to a transport protein), and iron. Flu viruses and diphtheria toxin use this route to enter and attack our cells.

The receptors for this process are plasma membrane proteins that bind only with certain substances. Both the receptors and attached molecules are internalized in a small vesicle called a *coated pit,* a term that refers to the bristlelike **clathrin**

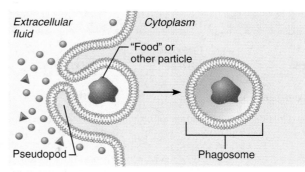

(a) Phagocytosis

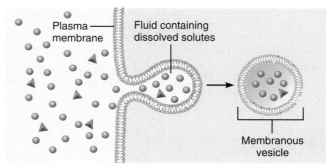

(b) Bulk-phase endocytosis

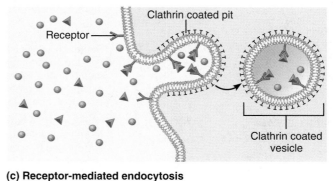

(c) Receptor-mediated endocytosis

FIGURE 3.12 *Three types of endocytosis.*

(kla′thrin; "lattice") protein coating on the cytoplasmic face of the vesicle. The clathrin coat acts both to deform the membrane to form the vesicle and in cargo selection.

Once inside, the vesicle loses its fuzzy coat and then fuses with a processing vesicle called an *endosome,* where the ingested molecules detach from the receptors. Contents of the endosome may be (1) recycled back to the plasma membrane (applies to many receptors and membrane components), (2) combined with a lysosome where the ingested substance is degraded or released (if iron or cholesterol), or (3) transported completely across the cell and released by exocytosis on its opposite side, a process called *transcytosis.* Transcytosis is common in the endothelial cells lining blood vessels because it provides a quick means to get substances from the blood to the interstitial fluid.

TABLE 3.2	Membrane Transport Processes		

Process	Energy source	Description	Examples
Passive processes			
Simple diffusion	Kinetic energy	Net movement of particles (ions, molecules, etc.) from an area of their higher concentration to an area of their lower concentration, that is, along their concentration gradient	Movement of fats, oxygen, carbon dioxide through the lipid portion of the membrane, and ions through protein channels under certain conditions
Facilitated diffusion	Kinetic energy	Same as simple diffusion, but the diffusing substance is attached to a lipid-soluble membrane carrier protein	Movement of glucose into cells
Osmosis	Kinetic energy	Simple diffusion of water through a selectively permeable membrane	Movement of water into and out of cells via membrane pores
Filtration	Hydrostatic pressure	Movement of water and solutes through a semipermeable membrane from a region of higher hydrostatic pressure to a region of lower hydrostatic pressure, that is, along a pressure gradient	Movement of water, nutrients, and gases through a capillary wall; formation of kidney filtrate
Active processes			
Active transport (solute pumping)	ATP (cellular energy)	Movement of a substance through a membrane against a concentration (or electrochemical) gradient; requires a membrane carrier protein	Movement of amino acids and most ions across the membrane
Vesicular transport			
• Exocytosis	ATP	Secretion or ejection of substances from a cell; the substance is enclosed in a membranous vesicle, which fuses with the plasma membrane and ruptures, releasing the substance to the exterior	Secretion of neurotransmitters, hormones, mucus, etc.; ejection of cell wastes
• Phagocytosis (endocytosis)	ATP	"Cell eating": A large external particle (proteins, bacteria, dead cell debris) is surrounded by a "seizing foot" and becomes enclosed in a plasma membrane sac	In the human body, occurs primarily in protective phagocytes (some white blood cells and macrophages)
• Bulk-phase endocytosis	ATP	Plasma membrane sinks beneath an external fluid droplet containing small solutes; membrane edges fuse, forming a fluid-filled vesicle	Occurs in most cells; important for taking in solutes by absorptive cells of the kidney and intestine
• Receptor-mediated endocytosis	ATP	Selective endocytosis process; external substance binds to membrane receptors, and coated pits are formed	Means of intake of some hormones, cholesterol, iron, and other molecules

⚠ *Homeostatic Imbalance*

In an inherited disease called *familial hypercholes-terolemia*, the protein receptors for cholesterol delivering LDLs are missing. As a result, cholesterol cannot enter body cells and accumulates in the blood. If the high LDL levels are untreated, atherosclerosis develops early and there is a high risk of coronary artery disease. ■

Caveolae (ka"ve-o'le; "little caves"), flask-shaped inpocketings of the plasma membrane seen in many cell types, also seem to be involved in a kind of receptor-mediated endocytosis. Like clathrin-coated pits, they capture specific molecules from the extracellular fluid in coated vesicles. However, caveolae are smaller than coated pits, and their coat is thinner and composed of a different protein. Their precise role in the cell is still being worked out.

Generating and Maintaining a Resting Membrane Potential

As you're now aware, the plasma membrane is more permeable to some molecules than to others. This differential permeability can lead to dramatic osmotic flows, but that is not its only consequence. An equally important result is the generation of a **membrane potential,** or voltage, across the membrane. A *voltage* is electrical potential energy resulting from the separation of oppositely charged particles. In cells, the oppositely charged particles are ions, and the barrier that keeps them apart is the plasma membrane.

In their resting state, all body cells exhibit a **resting membrane potential** that typically ranges from −20 to −200 millivolts (mV), depending on the cell type. Hence, all cells are said to be **polarized.** The sign before the voltage indicates that the *inside* of the cell is negative compared to its outside. This voltage (or charge separation) exists only at the membrane. If we were to add up all the negative and positive charges in the cell cytoplasm, we would find that the cell interior is electrically neutral. Likewise, the positive and negative charges in the extracellular fluid balance exactly.

So how does the resting membrane potential come about, and how is it maintained? Although many different kinds of ions are found both inside cells and in the extracellular fluid, the resting membrane potential is determined mainly by concentration gradients of two cations, Na^+ and K^+, and by the differential permeability of the plasma membrane to those ions. As mentioned before and illustrated in Figure 3.13, K^+ predominates inside body cells and the extracellular fluid contains relatively more Na^+. At rest, the plasma membrane is slightly permeable to K^+, but nearly impermeable to Na^+. Potassium diffuses out of the cell along its concentration gradient. Sodium is strongly attracted to the cell interior by its concentration gradient, but since the membrane is less permeable to sodium than to potassium, sodium influx is inadequate to balance potassium outflow. This unequal diffusion of Na^+ and K^+ across the membrane results in a relative deficit of positive ions within the cell, which establishes the resting membrane potential. Although it is tempting to believe that massive flows of ions occur to generate the resting potential, this is not the case. Surprisingly, the number of ions producing the membrane potential is so small that it does not change ion concentrations in any significant way.

In the polarized state, sodium and potassium concentrations are *not* at equilibrium. If only passive forces were at work, sodium and potassium concentrations would eventually become equal inside and outside the cell. Instead, the cell exhibits a *steady*

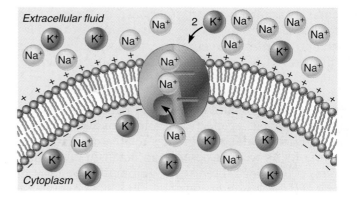

FIGURE 3.13 *Summary of forces that generate and maintain membrane potentials.* The ionic imbalances that produce the membrane potential reflect passive diffusion of ions (sodium diffuses into the cell more slowly than potassium diffuses out of the cell because of differential membrane permeability to these two ions). The net effect is that the outside membrane face becomes more electrically positive (more positive ions build up) than the inside, which is relatively more negative. The active transport of sodium and potassium ions (in a ratio of 3:2) by the Na^+-K^+ pump maintains these conditions.

state condition in which ionic imbalances due to diffusion polarize the membrane and active transport processes work to maintain that membrane potential. As we have seen, potassium and sodium ions leak passively according to their driving forces, but the rate of active transport is equal to, and depends on, the rate of diffusion of sodium ions into the cell. If more sodium enters, more is pumped out. (This is like being in a leaky boat. The more water that comes in, the faster you bail!) The Na^+-K^+ pump couples sodium and potassium transport and, on average, each "turn" of the pump ejects $3Na^+$ out of the cell and carries $2K^+$ back in (see Figure 3.13). Because the membrane is always slightly more permeable to K^+, the ATP-dependent Na^+-K^+ pump maintains both the membrane voltage and the osmotic balance. Indeed, were sodium not continuously removed from cells, so much would accumulate intracellularly that the osmotic gradient would draw water into the cells, causing them to burst.

Now that we've introduced the membrane potential, we may add a detail or two to our discussion of diffusion. Earlier we said that solutes diffuse down their concentration gradients. This is true for uncharged solutes, but only partially true for ions or other charged molecules. The negatively and positively charged faces of the plasma membrane can help or hinder diffusion of ions driven by a concentration gradient. It is more correct to say that ions diffuse according to **electrochemical gradients,** thereby recognizing the effect of both electrical and concentration (chemical) forces. Consequently, if we take a second look at the diffusion of K^+ and Na^+ across the plasma membrane, we see that although diffusion of K^+ is aided by the membrane's greater permeability to it and by its concentration gradient, it is resisted somewhat by the positive charge on the cell exterior. On the other hand, Na^+ is drawn into the cell by a steep electrochemical gradient; the limiting factor is the membrane's relative impermeability to it. As we will describe in detail in later chapters, "upsetting" the resting membrane potential by transient opening of ion (Na^+ and K^+) channels in the plasma membrane is a normal means of activating neurons and muscle cells.

Cell-Environment Interactions

Cells are biological minifactories and, like other factories, they receive and send orders from and to the outside community. But *how* does a cell interact with its environment, and what activates it to synthesize proteins or carry out its other homeostatic functions?

Cells do not always interact directly with other cells. In many cases cells respond to extracellular chemicals, such as hormones and neurotransmitters distributed in body fluids. Cells also interact with molecules in the substratum that act as signposts to guide cell migration during development and repair.

Whether cells interact directly or indirectly, the glycocalyx is always involved. The best understood of the participating glycocalyx molecules fall into two large families—adhesion molecules and membrane receptors (see Figure 3.3).

Roles of Cell Adhesion Molecules

Thousands of **cell adhesion molecules (CAMs)** are found on almost every cell in the body. CAMs play key roles in embryonic development and wound repair (situations where cell mobility is important) and in immunity. These sticky glycoproteins (*cadherins* and *integrins*) are

1. the molecular "Velcro" that cells use to anchor themselves to molecules in the extracellular space and to each other,

2. the "arms" that migrating cells use to haul themselves past one another, and

3. the SOS signals (in the form of selectin CAMs sticking out from the blood vessel lining) that rally protective white blood cells to a nearby infected or injured area.

Roles of Membrane Receptors

Membrane receptors are a huge and diverse group of integral proteins and glycoproteins that serve as binding sites. Some function in contact signaling, others in chemical signaling, and still others in electrical signaling. Let's take a look.

Contact Signaling Contact signaling, the actual coming together and touching of cells, is the means by which cells recognize other cells. It is particularly important for normal development and immunity. Some bacteria and other infectious agents use contact signaling to identify "preferred" target tissues or organs.

Electrical Signaling Certain plasma membrane receptors are channel proteins that respond to changes in membrane voltage by opening or closing the "ion gates" associated with the channel. Such voltage-regulated receptors are common in excitable tissues like neural and muscle tissues, and indispensable to their normal functioning.

Chemical Signaling Most membrane-bound receptors are involved in chemical signaling, and this group will receive the bulk of our attention. Signaling chemicals that bind specifically to membrane receptors are called **ligands.** Among these ligands are most *neurotransmitters* (nervous system signals), *hormones* (endocrine system signals), and *paracrines* (chemicals that act locally and are rapidly destroyed). Different cells respond in different ways to the same ligand. Acetylcholine, for instance, stimulates skeletal muscle cells to contract, but it inhibits heart muscle. Thus, target cell response depends on the internal machinery that the receptor is linked to, not the specific ligand that binds to it.

Though cell responses to receptor binding vary widely, there is a fundamental similarity. When a ligand binds to a receptor, the receptor's structure changes, and cell proteins are always altered in some way—muscle proteins change shape to generate force, enzymes are activated or inactivated, channel proteins open and close ion channels, and so on.

Some of these membrane receptor proteins are *catalytic proteins* that function as enzymes. Others, such as the *chemically gated channel-linked receptors* common in muscle and nerve cells, respond to ligands by transiently opening or closing the associated ion gates or channels, which in turn changes the excitability of the cell (see the discussion of membrane potentials on p. 81). Still other receptors are coupled to enzymes or ion channels by a regulatory molecule called a G protein. Since nearly every cell in the body displays at least some of these receptors, we will spend a bit more time with them.

G protein–linked receptors exert their effect indirectly through a **G protein,** which acts as a middleman or relay to activate (or inactivate) a membrane-bound enzyme or ion channel (Figure 3.14). As a result, one or more intracellular chemical signals, commonly called **second messengers,** are generated and connect plasma membrane events to the internal metabolic machinery of the cell. Two very important second messengers are **cyclic AMP** and ionic calcium, both of which typically activate protein kinase enzymes. By transferring phosphate groups from ATP to other proteins, the protein kinases can activate a whole series of enzymes (including some other kinases) that bring about the desired cellular activity. Since a single enzyme can catalyze hundreds of reactions, the amplification effect of such a chain of events is tremendous. These and other receptor systems are described in greater detail in Chapter 17.

One more signaling molecule must be mentioned, even though it doesn't fit any of the mechanisms described. *Nitric oxide (NO),* made of a single atom of nitrogen and one of oxygen, is one of nature's simplest molecules, an environmental pollutant, and the first gas known to act as a biological messenger. Because of its tiny size, it slips into and out of cells easily. Its one unpaired electron makes it a highly reactive free radical that reacts with head-spinning speed with other key molecules to spur cells into a broad array of activities. You will be hearing more about *NO* in later (neural, cardiovascular, and immune) chapters.

THE CYTOPLASM

Cytoplasm ("cell forming material") is the cellular material between the plasma membrane and the nucleus. It is the site where most cellular activities are accomplished. Although early microscopists thought that the cytoplasm was a structureless gel, the electron microscope has revealed that it consists of three major elements: the cytosol, organelles, and inclusions.

The **cytosol** (si'to-sol) is the viscous, semitransparent fluid substance in which the other cytoplasmic elements are suspended. Dissolved in the cytosol, which is largely water, are proteins, salts, sugars, and a variety of other solutes. Hence, the cytosol is a complex mixture with properties of both a colloid and a true solution.

The **cytoplasmic organelles,** described in detail shortly, are the metabolic machinery of the cell. Each type of organelle is "engineered" to carry out a specific function for the cell as a whole. Some synthesize proteins, others package those proteins, and so on.

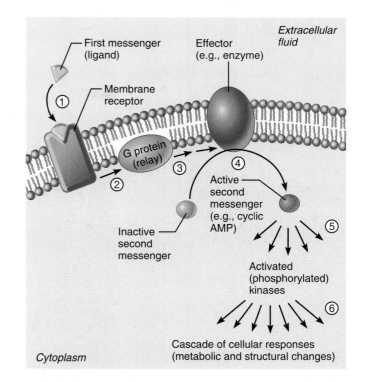

FIGURE 3.14 *Model of the operation of a G protein–linked receptor.* In this simplified diagram: ① an extracellular molecule (ligand) is a first messenger that binds to a specific receptor protein. ② The receptor activates a G protein that acts as a relay to ③ stimulate an effector protein. ④ The effector is an enzyme that produces a second messenger inside the cell. ⑤ The second messenger, cyclic AMP in this model, activates kinase enzymes. ⑥ The kinase enzymes can activate a whole series of enzymes that trigger the various responses of the cell.

Inclusions are not functional units but chemical substances that may or may not be present, depending on the specific cell type. Examples include stored nutrients, such as *glycosomes* ("sugar bodies"), glycogen granules abundant in liver and muscle cells; lipid droplets common in fat cells; pigment (melanin) granules seen in certain cells of the skin and hairs; and crystals of various types.

Cytoplasmic Organelles

The cytoplasmic organelles, literally, "little organs," are specialized cellular compartments, each performing its own job to maintain the life of the cell. Some organelles, the so-called *nonmembranous organelles,* lack membranes. Examples are the cytoskeleton, centrioles, and ribosomes. However,

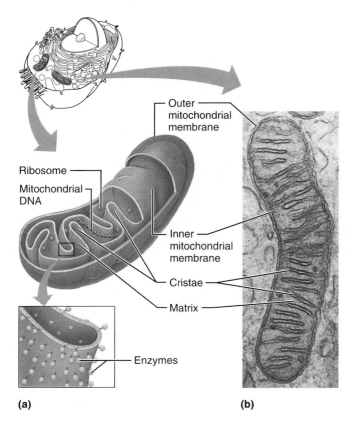

FIGURE 3.15 *Mitochondrion.* **(a)** Diagrammatic view of a longitudinally sectioned mitochondrion. **(b)** Electron micrograph of a mitochondrion (175,900×).

most organelles are bounded by a membrane similar in composition to the plasma membrane (minus the glycocalyx), which enables them to maintain an internal environment different from that of the surrounding cytosol. This compartmentalization is crucial to cell functioning. Without it, thousands of enzymes would be randomly mixed and biochemical activity would be chaotic. Besides providing "splendid isolation," an organelle's membrane unites it with the rest of an interactive intracellular system called the *endomembrane system* (see p. 89). The system's *membranous organelles* include the mitochondria, peroxisomes, lysosomes, endoplasmic reticulum, and Golgi apparatus. Now, let us consider what goes on in each of the workshops of our cellular factory.

Mitochondria

Mitochondria (mi″to-kon′dre-ah) are tiny threadlike (*mitos* = thread) or sausage-shaped organelles (Figure 3.15). In living cells they squirm, elongate, and change shape almost continuously. Mitochondria are the power plants of a cell, providing most of its ATP supply. The density of mitochondria in a particular cell reflects that cell's energy requirements, and mitochondria are generally clustered where the ac-

tion is. Busy cells like muscle and liver cells have hundreds of mitochondria, whereas cells that are relatively inactive (such as unchallenged lymphocytes) have just a few.

Mitochondria are enclosed by *two* membranes, each with the general structure of the plasma membrane. The outer membrane is smooth and featureless, but the inner membrane folds inward, forming shelflike **cristae** (krĭ′ste; "crests"). The cristae protrude into the *matrix,* the gel-like substance within the mitochondrion. Intermediate products of food fuels (glucose and others) are processed and broken down to water and carbon dioxide by teams of enzymes, some dissolved in the mitochondrial matrix and others forming part of the crista membrane.

As the metabolites are broken down and oxidized, some of the energy released is captured and used to attach phosphate groups to ADP molecules to form ATP. This multistep mitochondrial process is generally referred to as *aerobic* (a-er-o′bik) *cellular respiration* because it requires oxygen. It is described in detail in Chapter 25.

Mitochondria are complex organelles: They contain their own DNA and RNA and are able to divide to reproduce themselves and increase their cell number. Although mitochondrial genes (some 37 of them) direct the synthesis of some of the proteins required for mitochondrial function, the DNA of the nucleus encodes the remaining proteins needed to carry out cellular respiration. When cellular requirements for ATP increase, the mitochondria synthesize more cristae or simply pinch in half (a process called *fission*) to increase their number, then grow to their former size. Intriguingly, mitochondria are very similar to a specific group of bacteria (purple bacteria phylum), and mitochondrial DNA is a circular DNA molecule similar to that found in bacterial cells. It is now widely believed that mitochondria arose from bacteria that invaded the ancient ancestors of plant and animal cells.

Ribosomes

Ribosomes (ri′bo-sōmz) are small, dark-staining granules composed of proteins and a variety of RNA specifically called *ribosomal RNA.* Each ribosome has two globular subunits that fit together like the body and cap of an acorn (see Figure 3.16c). Ribosomes are sites of protein synthesis, a function we discuss in detail later in this chapter.

Some ribosomes float freely in the cytoplasm; others are attached to membranes, forming a complex called the *rough endoplasmic reticulum.* These two ribosomal populations appear to divide the chore of protein synthesis. **Free ribosomes** make soluble proteins that will function in the cytosol. **Membrane-bound ribosomes** synthesize proteins destined for incorporation into cellular membranes or

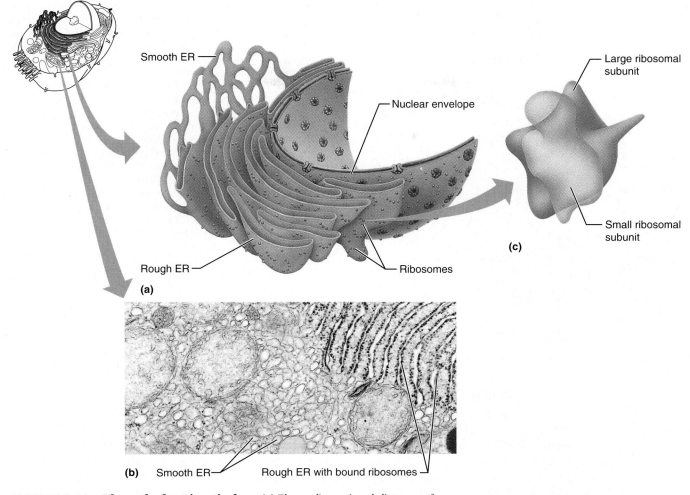

Smooth ER

Nuclear envelope

Large ribosomal subunit

Rough ER

Ribosomes

Small ribosomal subunit

(a)

(c)

(b) Smooth ER — Rough ER with bound ribosomes

FIGURE 3.16 *The endoplasmic reticulum.* (a) Three–dimensional diagram of the rough ER in a liver cell; its connections to smooth ER areas are also illustrated. (b) Electron micrograph of smooth and rough endoplasmic reticulum (approx. 26,500×). (c) Diagram of a ribosome showing the large and small subunits.

for export from the cell. Ribosomes can switch back and forth between these two functions, attaching to and detaching from the membranes of the endoplasmic reticulum, according to the type of protein they are making at a given time.

Endoplasmic Reticulum

The **endoplasmic reticulum** (en"do-plaz'mik re-tik'u-lum), or **ER,** is literally the "network within the cytoplasm." As shown in Figure 3.16a, it is an extensive system of interconnected tubes and parallel membranes enclosing fluid-filled cavities, or **cisternae** (sis-ter'ne), that coils and twists through the cytosol. The ER is continuous with the nuclear membrane and accounts for about half of the cell's membranes. There are two distinct varieties of ER: rough ER and smooth ER.

Rough Endoplasmic Reticulum The external surface of the **rough ER** is studded with ribosomes (Figure 3.16a). Proteins assembled on these ribosomes thread their way into the fluid-filled interior of the ER cisternae where various fates await them (as described shortly). The rough ER has several functions. Its ribosomes manufacture all proteins that are secreted from cells. Thus, the rough ER is particularly abundant and well developed in most secretory cells, antibody-producing plasma cells, and liver cells, which produce most blood proteins. The rough ER is also the cell's "membrane factory" because the integral proteins and phospholipids that form part of all cellular membranes are manufactured there. The enzymes needed to catalyze lipid synthesis have their active sites on the external (cytosolic) face of the ER membrane, where the needed substrates are readily available.

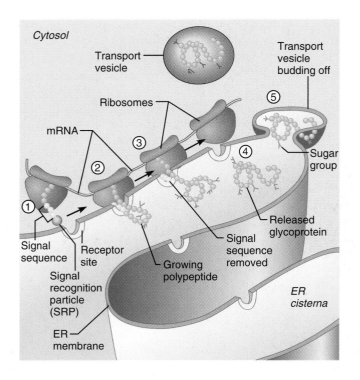

FIGURE 3.17 *The signal mechanism targets ribosomes to the ER for protein synthesis.* Enlarged view of a portion of the rough ER membrane bearing ribosomes and the ER cisterna. According to the signal mechanism of protein synthesis: ① Presence of a short signal sequence on a newly forming protein causes the mRNA–ribosome complex to be directed to the rough ER by a signal recognition particle (SRP), which binds to a receptor site that includes a pore and an enzyme to clip the signal sequence. ② Once attached to the ER receptor site, the SRP is released and the growing polypeptide snakes through the ER membrane into the cisterna. ③ The signal sequence initially remains attached to the receptor but shortly it is clipped off by an enzyme. As protein synthesis continues, sugar groups may be added to the protein. ④ In this example, the completed protein is released from the ribosome and folds into its 3-D conformation, a process aided by molecular chaperones (see p. 54). Some proteins are only partially translocated and remain embedded in the membrane. ⑤ The protein is enclosed within a membranous transport vesicle that pinches off the ER. The transport vesicles make their way to the Golgi apparatus where further processing of the proteins occurs (see Figure 3.18).

Now that we have provided an overview of the rough ER's functions, we will examine events that occur there as proteins are synthesized by its resident ribosomes. When a short "leader" peptide segment called a **signal sequence** is present in the protein being synthesized, the associated ribosome attaches to the membrane of the rough ER (Figure 3.17). This signal equence (with its attached "cargo" of the ribosome and messenger RNA) is guided to the appropriate receptor sites on the ER membrane by a **signal-recognition particle (SRP),** which cycles between the ER and the cytosol. The subsequent events occurring at the ER are detailed in Figure 3.17.

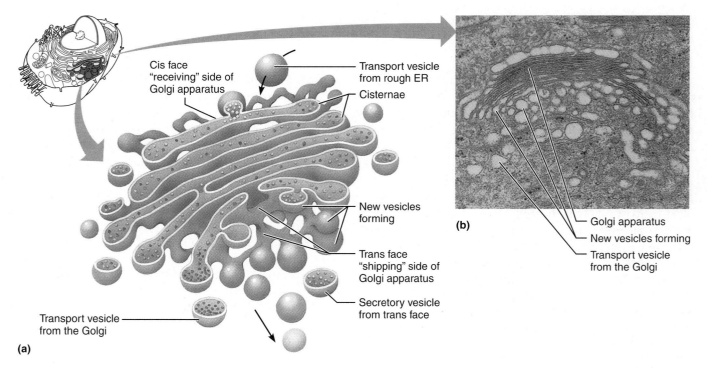

FIGURE 3.18 *Golgi apparatus.*
(a) Three-dimensional diagrammatic view of the Golgi apparatus.

(b) Electron micrograph of the Golgi apparatus (96,700×).

Notice the vesicles in the process of pinching off from the membranous Golgi apparatus.

Smooth Endoplasmic Reticulum The **smooth ER** (see Figures 3.1 and 3.16) is a continuation of the rough ER and consists of tubules arranged in a looping network. Its enzymes (all integral proteins forming part of its membranes) play no role in protein synthesis. Instead, they catalyze reactions involved with the following processes:

1. Lipid metabolism and synthesis of cholesterol and of the lipid components of lipoproteins (in liver cells)

2. Synthesis of steroid-based hormones such as sex hormones (testosterone-synthesizing cells of the testes are full of smooth ER)

3. Absorption, synthesis, and transport of fats (in intestinal cells)

4. Detoxification of drugs, certain pesticides, and carcinogens (in the liver and kidneys)

5. Breakdown of stored glycogen to form free glucose (in liver cells especially)

Additionally, skeletal and cardiac muscle cells have an elaborate smooth ER (called the sarcoplasmic reticulum) that plays an important role in calcium ion storage and release during muscle contraction. Except for the examples given above, most body cells contain little, if any, true smooth ER

Golgi Apparatus

The **Golgi** (gol'je) **apparatus** consists of stacked and flattened membranous sacs, shaped like hollow dinner plates, associated with swarms of tiny membranous vesicles (Figure 3.18). The Golgi apparatus is the principal "traffic director" for cellular proteins. Its major function is to modify, concentrate, and package the proteins and membranes made at the rough ER in specific ways, depending on their ultimate destination. The transport vesicles that bud off from the rough ER move on microtubule tracks to and fuse with the membranes at its convex *cis face,* the "receiving" side of the Golgi apparatus (Figures 3.18 and 3.19). Inside the apparatus, the protein cargos are modified: Some sugar groups are trimmed while others are added, and in some cases, phosphate groups are added. The various proteins are

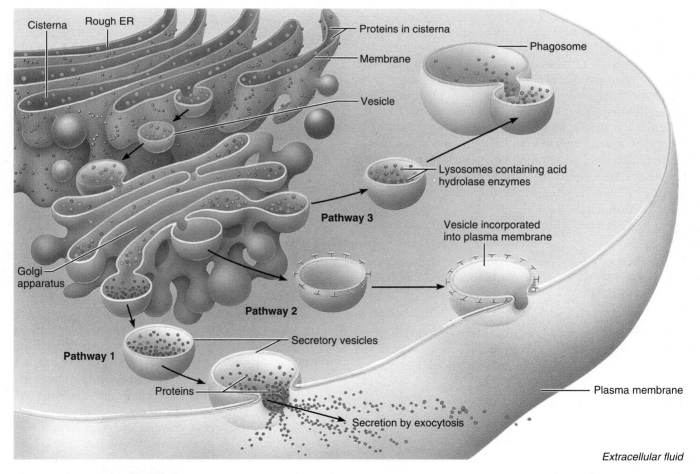

FIGURE 3.19   *Role of the Golgi apparatus in packaging proteins for cellular use and for secretion.* The sequence of events from protein synthesis on the rough ER to the final distribution of those proteins. Protein-containing vesicles pinch off the rough ER and migrate to fuse with the membranes of the Golgi apparatus. Within the Golgi compartments, the proteins are modified; they are then packaged within different Golgi vesicle types, depending on their ultimate destination (pathways 1–3 as shown).

"tagged" for delivery to a specific address, sorted, and packaged in at least three distinct types of vesicles that bud from the concave *trans face* (the "shipping" side) of the Golgi stack.

Vesicles containing proteins destined for export pinch off from the trans face as **secretory vesicles,** or **granules,** which migrate to the plasma membrane and discharge their contents from the cell by exocytosis. (See pathway 1, Figure 3.19.) Specialized secretory cells such as the enzyme-producing cells of the pancreas have a very prominent Golgi apparatus. Besides its packaging-for-release function, the Golgi apparatus pinches off vesicles containing lipids and transmembrane proteins destined for a "home" in the plasma membrane (pathway 2, Figure 3.19) or other membranous organelles. It also packages hydrolytic (digestive) enzymes into membranous sacs called *lysosomes* that remain in the cell (pathway 3 in Figure 3.19).

Lysosomes

Lysosomes ("disintegrator bodies") are spherical membranous bags containing digestive enzymes (Figure 3.20). As you might guess, lysosomes are large and abundant in phagocytes. Lysosomal enzymes can digest almost all kinds of biological molecules. They work best in acidic conditions (approximately pH 5) and thus are called *acid hydrolases* (hi"drah-la'siz). The lysosomal membrane is adapted to serve lysosomal functions in two important ways: (1) It contains hydrogen ion (proton) "pumps" that gather hydrogen ions from the surrounding cytosol to maintain the organelle's low internal pH, and (2) it retains the dangerous acid hydrolases while permitting the final products of digestion to escape so that they can be used by the cell or excreted. Hence, lysosomes provide sites where digestion can proceed *safely* within a cell.

Lysosomes function as a cell's "demolition crew" by

- Digesting particles taken in by endocytosis, particularly important in disarming ingested bacteria, viruses, and toxins

- Degrading worn-out or nonfunctional organelles

- Performing metabolic functions, such as the breakdown of stored glycogen and the release of thyroid hormone from its storage form in thyroid cells

- Breaking down nonuseful tissues, such as the webs between the fingers and toes of a developing fetus and the uterine lining during menstruation

- Breaking down bone to release calcium ions into the blood

The lysosomal membrane is ordinarily quite stable, but it becomes fragile when the cell is injured or

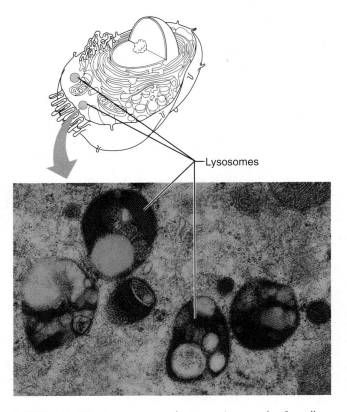

FIGURE 3.20 *Lysosomes.* Electron micrograph of a cell containing lysosomes (217,120×).

deprived of oxygen and when excessive amounts of vitamin A are present. Lysosomal rupture results in self-digestion of the cell, a process called **autolysis** (aw"tol'ĭ-sis). Autolysis is the basis for desirable destruction processes (see fourth point above), and for certain *autoimmune* diseases such as rheumatoid arthritis (see Chapter 8).

Homeostatic Imbalance

Glycogen and certain lipids in the brain are degraded at a relatively constant rate by lysosomes. In *Tay-Sachs disease,* an inherited condition seen mostly in Jews from Central Europe, the lysosomes lack an enzyme needed to break down a specific glycolipid abundant in nerve cell membranes. As a result, the lysosomes of nerve cells become swollen with the undigested lipids, which interfere with nervous system functioning. Affected infants typically have doll-like features and pink translucent skin. At 3 to 6 months of age, the first signs of disease appear (listlessness, motor weakness). These progress to mental retardation, seizures, blindness, and ultimately death within a year and a half. ■

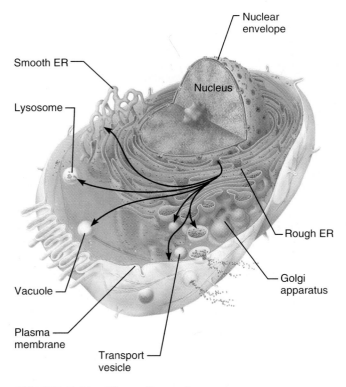

Nuclear envelope

Smooth ER

Nucleus

Lysosome

Rough ER

Vacuole

Golgi apparatus

Plasma membrane

Transport vesicle

FIGURE 3.21 *The endomembrane system.*

Summary of Interactions in the Endomembrane System

The **endomembrane system** (Figure 3.21) is a system of organelles (most described above) that work together mainly (1) to produce, store, and export biological molecules, and (2) to degrade potentially harmful substances. It includes the ER, Golgi apparatus, secretory vesicles, and lysosomes, as well as the nuclear membrane—that is, all of the membranous organelles or elements that are either continuous structurally or arise or interact via forming or fusing transport vesicles. There are continuities between the nuclear envelope (itself an extension of the rough ER) and the rough and smooth ER. The plasma membrane, though not actually an *endo* membrane, is also functionally part of this system. Besides these direct structural relationships, a wide variety of indirect interactions (also indicated by the arrows on the figure) occur among the members of the system. Some of the vesicles "born" in the ER migrate to and fuse with the Golgi apparatus or the plasma membrane, and vesicles arising from the Golgi apparatus can become part of the plasma membrane, secretory vesicles, or lysosomes.

The figure also reveals another interesting point: Cell products that initially enter the ER cisternae can be secreted by the cell or enter its nucleus without ever actually crossing a cellular membrane. Because the endomembrane system meanders through a good part of the cytosol and thus could (conceiv-

ably) hinder intracellular transport, this is an important point indeed.

Peroxisomes

Peroxisomes (pĕ-roks′ĭ-sōmz; "peroxide bodies") are membranous sacs containing a variety of powerful enzymes, the most important of which are oxidases and catalases. Oxidases use molecular oxygen (O_2) to detoxify a number of harmful or toxic substances, including alcohol and formaldehyde. However, their most important function is to neutralize dangerous free radicals, which they convert to hydrogen peroxide. Hydrogen peroxide is also reactive and dangerous but it is quickly converted to water by catalase enzymes. **Free radicals** (such as superoxide ion (O_2^-) and hydroxyl radical) are highly reactive chemicals with unpaired electrons that can scramble the structure of biological molecules. Free radicals and hydrogen peroxide are normal by-products of cellular metabolism, but they have devastating effects on cells if they are allowed to accumulate. Peroxisomes are especially numerous in liver and kidney cells, which are very active in detoxification.

Although peroxisomes look like small lysosomes (see Figure 3.1), they are self-replicating organelles formed by a simple pinching in half of preexisting peroxisomes. Unlike lysosomes, they do not arise by budding from the Golgi apparatus.

Cytoskeleton

The **cytoskeleton**, literally, "cell skeleton," is an elaborate series of rods running through the cytosol. This network acts as a cell's "bones," "muscles," and "ligaments" by supporting cellular structures and providing the machinery to generate various cell movements. The three types of rods in the cytoskeleton are *microtubules, microfilaments,* and *intermediate filaments.* None of these is membrane covered.

Microtubules (mi″kro-tu′būlz), the elements with the largest diameter, are hollow tubes made of spherical protein subunits called *tubulins* (Figure 3.22 on p. 90). All microtubules radiate from a small region of cytoplasm near the nucleus called the *centrosome* (see Figure 3.24). These stiff but bendable microtubules determine the overall shape of the cell, as well as the distribution of cellular organelles. Mitochondria, lysosomes, and secretory granules attach to the microtubules like ornaments hanging from the limbs of a Christmas tree. These attached organelles are continually pulled along the microtubules and repositioned by **motor proteins** (*kinesins, dyneins,* and others) that act like train engines on the microtubular "railroad tracks" (see Figure 3.23a). Microtubules are remarkably dynamic organelles, constantly growing out from the centrosome, disassembling, and then reassembling.

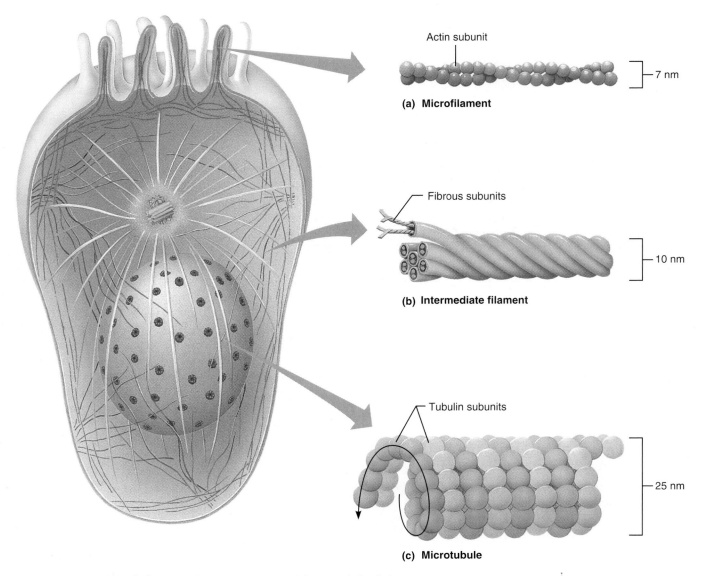

Actin subunit

(a) Microfilament   7 nm

Fibrous subunits

(b) Intermediate filament   10 nm

Tubulin subunits

(c) Microtubule   25 nm

FIGURE 3.22   Cytoskeleton. Diagrammatic views of the cytoskeletal elements. Left: General orientation of the cytoskeletal elements in the cell. Right: Detailed structure of the three types of cytoskeletal elements.

Microfilaments (mi"kro-fil'ah-ments), the thinnest elements of the cytoskeleton, are strands of the protein *actin* ("ray"). Each cell has its own arrangement of microfilaments; thus no two cells are alike. However, nearly all cells have a fairly dense crosslinked network of microfilaments (Figure 3.22) attached to the cytoplasmic side of their plasma membrane that braces and strengthens the cell surface. Most microfilaments are involved in cell motility or changes in cell shape. You could say that cells move when they get their act(in) together. For example, actin microfilaments interact with another protein, **myosin** (mi'o-sin), to generate contractile forces within a muscle cell (see Figure 3.23) and to form the cleavage furrow that pinches a cell in two during cell division. Microfilaments attached to CAMs (see Figure 3.3, p. 69) are responsible for the crawling movements of amoeboid motion, and for membrane changes that accompany endocytosis and exocytosis. Except in muscle cells, where they are highly developed and permanent, microfilaments are constantly breaking down and re-forming from smaller subunits whenever and wherever their services are needed.

Intermediate filaments are tough, insoluble protein fibers, with a diameter between those of microfilaments and microtubules (Figure 3.22). Constructed like woven ropes, intermediate filaments are the most stable and permanent of the cytoskeletal elements and have high tensile strength. They act as internal guy wires to resist pulling forces on the cell, and they help form desmosomes (the anchoring junctions described on p. 70). Because the protein composition of intermediate filaments varies

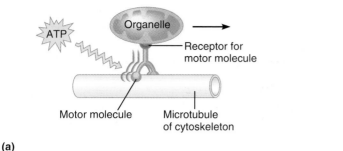

(a)

(b)

FIGURE 3.23 *Interaction of motor molecules with cytoskeletal elements.* Microtubules and microfilaments function in motility by interacting with protein complexes called motor molecules. The various types of motor molecules, all powered by ATP, work by changing their shapes, moving back and forth something like microscopic legs. With each cycle of shape changes, the motor molecule releases at its free end and grips at a site farther along the microtubule or microfilament. **(a)** Motor molecules can attach to receptors on organelles, such as mitochondria or ribosomes, and enable the organelles to "walk" along the microtubules of the cytoskeleton. **(b)** In some types of cell motility, motor molecules attached to one element of the cytoskeleton can cause it to slide over another element. For example, the sliding of one set of microfilaments past another underlies muscle contraction, and the sliding of neighboring microtubules moves cilia.

in different cell types, there are numerous names for these cytoskeletal elements. For example, they are called neurofilaments in nerve cells and keratin filaments in epithelial cells.

Centrosome and Centrioles

As mentioned, many microtubules appear to be anchored at one end in a region near the nucleus called the **centrosome.** The centrosome, which acts as a *microtubule organizing center,* has few distinguishing marks other than the fact that it has a granular–looking matrix that contains paired **centrioles,** small, barrel-shaped organelles oriented at right angles to each other (Figure 3.24). Each centriole consists of a pinwheel array of nine *triplets* of microtubules, arranged to form a hollow tube. Centrioles are best known for their role of organizing the mitotic spindle (see Figure 3.30 on pp. 100–101) in cell division. They also form the bases of cilia and flagella.

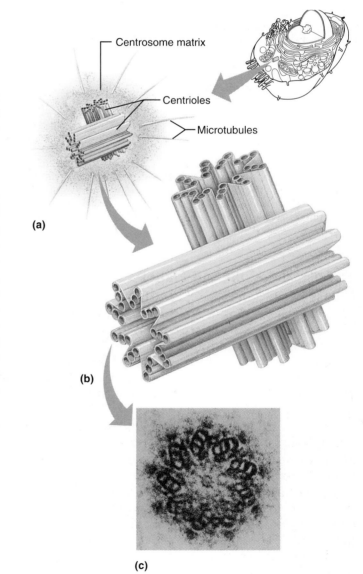

FIGURE 3.24 *Centrioles.* **(a)** Three-dimensional view of a centriole pair oriented at right angles, as they are usually seen in the cell. The centrioles are located in a nonconspicuous region to one side of the nucleus called the centrosome, or cell center. **(b)** An electron micrograph showing a cross section of a centriole (381,600✕). Notice that it is composed of nine microtubule triplets.

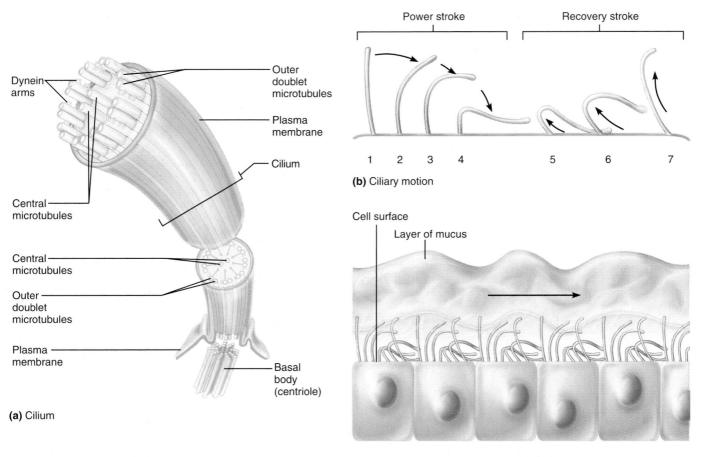

(a) Cilium

(b) Ciliary motion

Power stroke   Recovery stroke

1   2   3   4   5   6   7

(c) Movement of mucus across cell surfaces

Cell surface
Layer of mucus

FIGURE 3.25 *Cilia structure and function.* (a) Three-dimensional diagram of a cross section through a cilium, showing its nine pairs of peripheral microtubules and one central microtubule pair. (b) Diagram of the phases of ciliary motion. 1–4 are part of the power (propulsive) stroke; 5–7 are phases of the recovery (nonpropulsive) stroke when the cilium is resuming its initial position. (c) A representation of the traveling wave created by the activity of many cilia acting together to propel mucus across the cell surfaces.

Cilia and Flagella

Cilia (sil′e-ah; "eyelashes") are whiplike, motile cellular extensions that occur, typically in large numbers, on the exposed surfaces of certain cells. Ciliary action is important in moving substances in one direction across cell surfaces. For example, ciliated cells that line the respiratory tract propel mucus laden with dust particles and bacteria upward away from the lungs.

When a cell is about to form cilia, the centrioles multiply and line up beneath the plasma membrane at the free cell surface. Microtubules then begin to "sprout" from each centriolar region, forming the ciliary projections by exerting pressure on the plasma membrane. When the projections formed by centrioles are substantially longer, they are called **flagella** (flah-jel′ah). The single example of a flagellated cell in the human body is a sperm, which has one propulsive flagellum, commonly called a tail. Notice that cilia *propel other substances* across a cell's surface, whereas a flagellum *propels the cell itself.*

Centrioles forming the bases of cilia and flagella are commonly referred to as **basal** (ba′sal) **bodies** (Figure 3.25a) because they were thought to be different from the structures seen in the centrosome. We now know that centrioles and basal bodies are identical. However, the pattern of microtubules in the cilium or flagellum itself (nine *doublets*, or pairs, of microtubules encircling one central pair [Figure 3.25a]) differs slightly from that of a centriole (nine microtubule *triplets*).

Just how the activity of the cilia is coordinated is not fully understood, but microtubules are definitely involved. Extending from the microtubule doublets are arms composed of the motor protein *dynein*. Ciliary movement is generated when the dynein side arms grip adjacent doublets and start to crawl along their length, much like cats dig in their claws to climb a tree (see Figure 3.23b). The collective bending action of all the doublets causes the cilium to bend.

Text continues on page 94.

A CLOSER LOOK

Tensegrity: The Cell's Architect

Life is the ultimate complexity. All organisms are constructed from huge numbers of components, which are made of even smaller subcomponents—atoms and molecules, those highly dynamic building blocks. When these subcomponents merge in a larger functional unit, such as an organ, or even a cell, new and often unpredictable properties appear. However, identifying the parts or chemistry of a living system, though important, does not tell us how the whole system is put together and its structure maintained once assembled. After all, molecules and cells are continually removed and replaced—it is not the individual subcomponents but the maintenance of pattern and architecture that we call life. But guess what? Scientists are beginning to piece together the answer. Only recently they have discovered that living things, whether they be cells, organs, or whole organisms, exhibit two important properties in common:

1. *Self-assembly.* For example, large molecules self-assemble into cells, which self-assemble into tissues, which self-assemble into organs, and so on.

2. *Tensegrity.* An architectural system that stabilizes itself mechanically, not by the strength of its individual elements but by the way tension and compressive forces are distributed and balanced within the structure, is said to have tensegrity (tensional integrity). Perhaps the best examples of tensegrity structures in nonliving systems are the gravity-defying cable and strut sculptures of the sculptor Kenneth Snelson (photo a).

Tensegrity structures gain shape and strength by counteracting elements that resist compression with other elements that are under tension, thereby creating a *prestressed system*. The rigid compressed elements stretch (or tense) the flexible tension-bearing members, while the tension-bearing members compress the rigid struts. A critical feature of

such a structure is that the forces are transmitted across the whole structure—if one element is stressed more, all elements are. So increases in tension occur globally and are balanced by increases in compression throughout the structure. In this way, tensegrity structures offer maximum strength for a given amount of structural building material.

But what about cells? How does tensegrity apply to them? It appears that the cytoskeleton is the main actor in making the cell into a tensegrity structure (photo b on page 94): The contractile microfilaments, which extend throughout the cell, exert tension, that is, pull the plasma membrane toward the nucleus. This is opposed by two types of compression elements: (1) the extracellular matrix outside the cell and (2) the microtubules (or large bundles of microfilaments) inside the cell. The third cytoskeletal element, the intermediate filaments, are the unifiers. The intermediate filaments connect the compression-bearing struts to each other, the plasma membrane, and the nucleus, and act as guy-wires to secure the nucleus in place. This firmly wired geodesic structure not only stabilizes cell shape but enables the cell to change its shape.

An interesting consequence of the cell's tensegrity structure is the physical forces transmitted across or acting at the cell surface can stimulate changes in the cytoskeleton, in

cell flexibility, and in cell metabolism. For example, simply tugging on a specific plasma membrane receptor causes a nearly instant deformation and rearrangement of nuclear components and cells that form many adhesion junctions with the extracellular matrix tend to flatten out. Tugging on cells that lack such attachments makes them more rounded or dome shaped. It also seems that such mechanical restructuring of the cytoskeleton "tells" the cell what to do. For example, when the skin surface is broken, adjacent cells flatten out and spread. Apparently, this delivers the signal that more cells are needed to cover the underlying surface and the cells gear up to proliferate and fill the gap. On the other hand, cell rounding often indicates cell crowding, in which case apoptosis (see *Clinical Terms* p. 110) may be stimulated to prevent tumor formation. However, stress applied to cell surface receptors involved with metabolism does *not* lead to force changes within the cells; thus, cellular reactions to local mechanical stress occur along very specific pathways.

Tensegrity also applies at the molecular level. For example, when a signal-bearing molecule binds to a transmembrane receptor protein, conformational changes occur in the opposite end of the receptor molecule, which typically initiate specific intracellular metabolic changes. Thus,

(a)

A CLOSER LOOK   *continued*

the cytoskeleton also serves as a communication system that plays a role in signal transduction leading to gene activation, cell division, and so on.

Whether we are talking about molecules or your musculoskeletal system, tensegrity, which "tones" the whole system mechanically as one, is the preferred architecture. Only this building system can explain how movement of your leg makes your skin stretch, your cells distort, and their cytoskeletons feel the stretch and respond mechanically—all without breaking a thing.

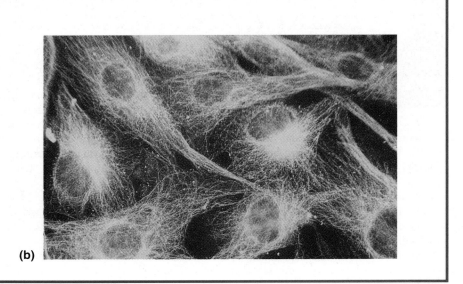

(b)

As a cilium moves, it alternates rhythmically between a propulsive *power stroke,* when it is nearly straight and moves in an arc, and a *recovery stroke,* when it bends and returns to its initial position (Figure 3.25b). With these two strokes, the cilium produces a pushing motion in a single direction. The activity of cilia in a particular region is coordinated so that the bending of one cilium is quickly followed by the bending of the next and then the next, creating a current at the cell surface that brings to mind the *traveling waves* that pass across a field of grass on a windy day (Figure 3.25c).

THE NUCLEUS

Anything that works, works best when it is controlled. For cells, the control center is the gene-containing **nucleus** (*nucle* = pit, kernel). The nucleus can be compared to a computer, design department, construction boss, and board of directors—all rolled into one. As the genetic library, it contains the instructions needed to build nearly all the body's proteins. Additionally, it dictates the kinds and amounts of proteins to be synthesized at any one time in response to signals acting on the cell.

Most cells have only one nucleus, but some, including skeletal muscle cells, bone destruction cells, and some liver cells, are **multinucleate** (mul″tĭ-nu′kle-āt); that is, they have many nuclei. The presence of more than one nucleus usually signifies that a larger-than-usual cytoplasmic mass must be regu-

lated. With one exception, all of our body cells are nucleated. The exception is mature red blood cells, whose nuclei are ejected before the cells enter the bloodstream. These **anucleate** (a-nu′kle-āt; *a* = without) cells cannot reproduce and therefore live in the bloodstream for only three to four months before they begin to deteriorate. Without a nucleus, a cell cannot make any more proteins, and when its enzymes and cell structures start to break down (as all eventually do), they cannot be replaced.

The nucleus, averaging 5 μm in diameter, is larger than any of the cytoplasmic organelles. While most often spherical or oval, its shape usually conforms to the shape of the cell. If a cell is elongated, for example, the nucleus may also be elongated. The nucleus has three distinct regions or structures: the *nuclear envelope (membrane), nucleoli,* and *chromatin* (Figure 3.26).

Nuclear Envelope

The nucleus is bounded by the **nuclear envelope,** which is a *double* membrane barrier (separated by a fluid-filled space) similar to the mitochondrial membrane. The outer nuclear membrane is continuous with the rough ER of the cytoplasm and is studded with ribosomes on its external face. The inner face of the envelope is lined by a network of protein filaments (the *nuclear lamina*) that maintains the shape of the nucleus.

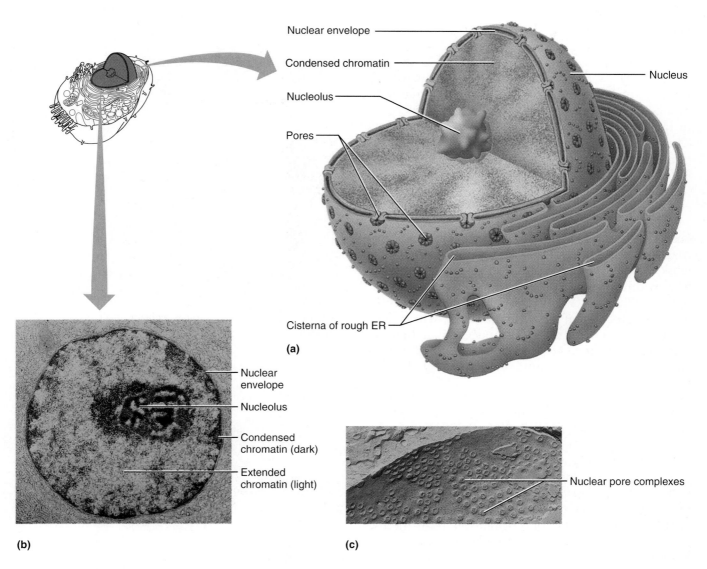

FIGURE 3.26 *The nucleus.*
(a) Three-dimensional diagrammatic view of the nucleus, showing the continuity of its double membrane with the ER.

(b) Transmission electron micrograph (45,000×) of the nucleus showing nuclear envelope, nuclear pores, a nucleolus, and condensed chromatin regions.

(c) Scanning electron micrograph (163,640×) of the external surface of the nuclear envelope. Notice the raised nuclear pore (NP) complexes.

At various points, the two layers of the nuclear envelope fuse, and **nuclear pores** penetrate these regions. An intricate complex of proteins, called a *pore complex*, lines each pore and regulates the entry and exit of large particles.

Like other cellular membranes, the nuclear envelope is selectively permeable, but the passage of substances is much freer than elsewhere. Protein molecules imported from the cytoplasm and RNA molecules exported from the nucleus pass easily through the relatively large pores.

The nuclear envelope encloses a jellylike fluid called nucleoplasm (nu'kle-o-plazm) in which other nuclear elements are suspended. Like the cytosol,

the nucleoplasm contains dissolved salts, nutrients, and other essential solutes.

Nucleoli

Nucleoli (nu-kle'o-li; "little nuclei") are the dark-staining spherical bodies found within the nucleus; they are sites of ribosome production (see Figure 3.26). They are not membrane bounded. Typically, there are one or two nucleoli per cell, but there may be more. Nucleoli are sites where ribosome subunits are assembled; consequently, they are usually very large in growing cells that are making large amounts of tissue proteins. Nucleoli are associated with *nucleolar organizer regions*, chromatin regions containing the DNA that issues genetic instructions for synthesizing ribosomal RNA (rRNA). As molecules of rRNA are synthesized, they are combined with proteins to form the two kinds of ribosomal subunits. (The proteins are manufactured on ribosomes in the cytoplasm and "imported" into the nucleus.) These subunits leave the nucleus through the nuclear pores and enter the cytoplasm, where they join to form functional ribosomes.

Chromatin

Seen through a light microscope, **chromatin** (kro′-mah-tin) appears as a fine, unevenly stained network, but special techniques reveal it as a system of bumpy threads weaving their way through the nucleoplasm (Figure 3.27a). Chromatin is composed of approximately equal amounts of **DNA**, which constitutes our genetic material, and globular **histone** (his′tōn) **proteins. Nucleosomes** (nu′kle-o-sōmz; "nuclear bodies"), the fundamental units of chromatin, consist of discus-shaped cores or clusters of eight histone proteins connected like beads on a string by a DNA molecule that winds (like a piece of velcro) around each of them and then continues on to the next cluster via *linker DNA* segments (Figure 3.27b). In addition to providing a physical means for packing the very long DNA molecules in a compact, orderly way, the histones play an important role in gene regulation. For example, changes in the shape of the histones by addition of phosphate or methyl groups, in a nondividing cell expose different DNA segments, or genes, so that they can "dictate" the specifications for protein synthesis. Such active chromatin segments, referred to as *extended chromatin*, are not usually visible under the light microscope. The generally inactive *condensed chromatin* segments are darker staining and so are more easily detected (see Figure 3.26b). Understandably, the most active body cells have much larger amounts of extended chromatin.

When a cell is preparing to divide, the chromatin threads coil and condense enormously to form short, barlike bodies called **chromosomes** ("colored bodies") (Figure 3.27b). Chromosome compactness avoids entanglement and breakage of the delicate chromatin strands during the movements that occur during cell division (pp. 100–101). The functions of DNA and the events of cell division are described in the next section.

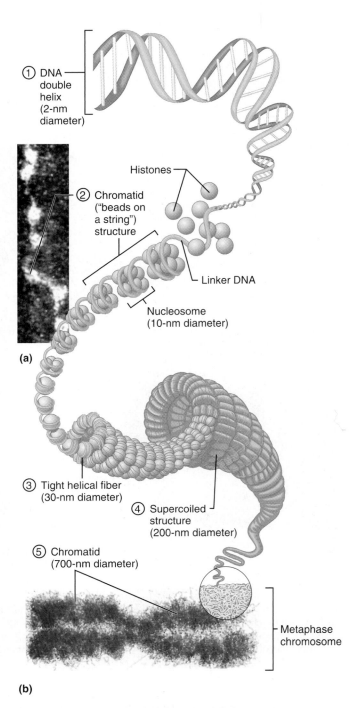

(a)

(b)

FIGURE 3.27   *Chromatin and chromosome structure.*
(**a**) Electron micrograph of chromatin fibers, which have the appearance of beads on a string (216,000×). (**b**) DNA packing in a chromosome. The levels of increasing structural complexity (coiling) from the DNA helix to the metaphase chromosome are indicated in numerical order. (Metaphase is a stage of nuclear division that occurs before the genetic material is distributed to the daughter cells.) Notice the structure of the nucleosomes, the fundamental units of chromatin that give it the "beads on a string" appearance. Each nucleosome is composed of eight histone proteins wrapped by two winds of the DNA helix.

CELL GROWTH AND REPRODUCTION

The Cell Life Cycle

The **cell life cycle** is the series of changes a cell goes through from the time it is formed until it reproduces itself. The cycle encompasses two major periods: *interphase,* in which the cell grows and carries on its usual activities, and *cell division,* or the *mitotic phase,* during which it divides into two cells (Figure 3.28).

Interphase

Interphase includes the total period from cell formation to cell division. Early cytologists, unaware of the constant molecular activity in cells and impressed by the obvious movements of cell division, called interphase the resting phase of the cell cycle. (The term *interphase* reflects this idea of a stage *between* cell divisions.) However, this image is grossly misleading because during interphase a cell is carrying out all its routine activities and is "resting" only from dividing. Perhaps a more accurate name for this phase would be *metabolic phase* or *growth phase.*

In addition to carrying on its life-sustaining reactions, an interphase cell prepares for the next cell division. Interphase is divided into G_1, S, and G_2 subphases (G = gaps before and after the S phase). During **G_1 (gap 1),** the first portion of interphase, cells are metabolically active, synthesize proteins rapidly, and grow vigorously. This is the most variable phase in terms of length. Cells with rapid division rates have a G_1 phase typically lasting several minutes to hours; in those that divide slowly, it may last for days or even years. Cells that permanently cease dividing are said to be in the **G_0 phase.**

For most of G_1, virtually no activities directly related to cell division occur. However, as G_1 ends, the centrioles start to replicate in preparation for cell division. During the next phase, the **S (synthetic) phase,** DNA replicates itself, ensuring that the two future cells will receive identical copies of the genetic material. New histones are made and assembled into chromatin. (We will describe DNA replication next.) The final phase of interphase, called **G_2 (gap 2),** is very brief. During this period, enzymes and other proteins needed for division are synthesized and moved to their proper sites. By the end of G_2, centriole replication (begun in G_1) is complete. Throughout the S and G_2 phases, the cell continues to grow and carries on with business as usual.

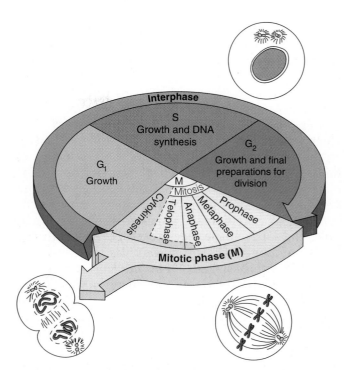

FIGURE 3.28   *The cell cycle.* During G_1, cells grow rapidly and carry out their routine functions; the centrioles begin to replicate as this phase ends. The S phase begins when DNA synthesis starts, and it ends when the DNA has replicated. In the short G_2 phase, materials needed for cell division are synthesized and growth continues. During the M phase (cell division), mitosis and cytokinesis occur, producing two daughter cells. The length of the cell cycle varies with different cell types, but the G_1 phase is the longest and most variable phase in all cells.

DNA Replication   Before a cell can divide, its DNA must be replicated exactly, so that identical copies of the cell's genes can be passed on to each of its offspring. The replication process begins simultaneously on several chromatin threads and continues until all of the DNA has been replicated.

The process has several steps.

1. The DNA helices unwind from the nucleosome.

2. A *helicase* enzyme untwists the double helix and gradually separates the DNA molecule into its two complementary nucleotide chains, exposing the nitrogenous bases (Figure 3.29). The site of separation is called a *replication bubble;* the Y-shaped region at each end of the replication bubble is called a *replication fork.*

3. Each freed nucleotide strand then serves as a *template*, or set of instructions, for building a new complementary nucleotide strand from free DNA precursors in the nucleoplasm (Figure 3.29).

Recall that nucleotide base pairing is always complementary: *adenine (A)* bonds to *thymine (T)*, and *guanine (G)* bonds to *cytosine (C)* (see p. 59). Hence the order of the nucleotides on the template strand determines the order on the strand being built. For example, a TACTGC sequence on a template strand would bond to new nucleotides with the order ATGACG. This process is catalyzed by the enzyme **DNA polymerase**, which positions and then links the DNA nucleotides together. Although the DNA building blocks are *nucleotides* the substrates for DNA synthesis are DNA *nucleoside* triphosphates, which have three phosphate groups. As in ATP, their terminal phosphates are secured by high-energy bonds. As each DNA nucleoside triphosphate joins the growing nucleotide chain, hydrolysis of its terminal two phosphates provides the energy to drive the polymerization process. A pretty neat system indeed.

DNA polymerase works only in one direction. Consequently, one strand, the *leading strand*, is synthesized continuously following the movement of the replication fork. The other strand, called the *lagging strand*, is constructed in segments in the opposite direction.

4. The short segments of DNA are then spliced together by **DNA ligase.** The end result is that two DNA molecules are formed from and are identical to the original DNA helix, and each consists of one old and one newly assembled nucleotide strand. This mechanism of DNA replication is referred to as **semiconservative replication.**

5. As soon as replication ends, histones (synthesized in the cytoplasm and imported into the nucleus) associate with the DNA, completing the formation of two new chromatin strands. The chromatin strands condense to form **chromatids** (see Figure 3.27) and are united by a centromere. The chromatids remain attached until the cell has entered the anaphase stage of mitotic cell division. They are then distributed to the daughter cells, as described next, ensuring that each has identical genetic information.

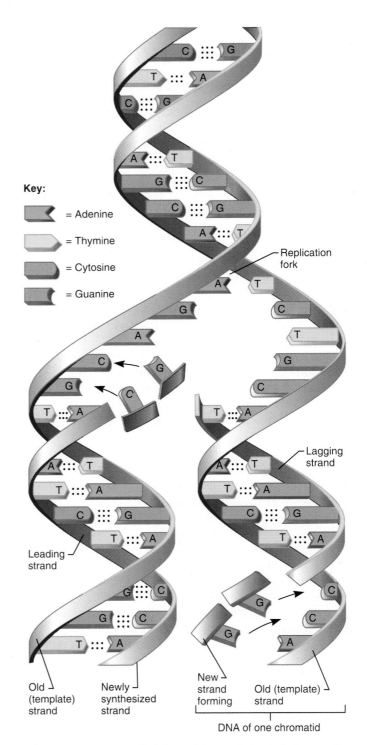

Key:

= Adenine

= Thymine

= Cytosine

= Guanine

FIGURE 3.29 *Replication of DNA.* The DNA helix uncoils, and the hydrogen bonds between its base pairs are broken. Then each nucleotide strand of the DNA acts as a template for the construction of a complementary nucleotide strand, as illustrated in the bottom portion of the diagram. Because the DNA polymerases (not illustrated) work in one direction only, the two new strands (leading and lagging) are synthesized in opposite directions. After replication is completed, two DNA molecules exist, each identical to the original DNA molecule and to one another. Each DNA molecule formed consists of one old (template) strand and one newly assembled strand and constitutes a chromatid of a chromosome.

Cell Division

Cell division is essential for body growth and tissue repair. Cells that continually wear away, such as cells of the skin and intestinal lining, reproduce themselves almost continuously. Others, such as liver cells, divide more slowly (maintaining the organ they compose at a particular size), but retain the ability to reproduce quickly if the organ is damaged. For the most part, cells of nervous tissue, skeletal muscle, and heart muscle lose their ability to divide when they are fully mature, and repairs are made with scar tissue (a fibrous type of connective tissue).

Events of Cell Division In most body cells, cell division or the **M (mitotic) phase** of the cell life cycle (see Figure 3.28) involves two distinct events: *mitosis* (mi-to′sis), or division of the nucleus, and *cytokinesis* (si″to-ki-ne′sis), or division of the cytoplasm. A somewhat different process of nuclear division called *meiosis* (mi-o′sis) produces sex cells (ova and sperm) with only half the number of genes found in other body cells. (Then when two sex cells combine in fertilization, the full complement of genetic material is restored.) The details of meiosis are discussed in Chapter 28. Here we concentrate on mitotic cell division.

Mitosis Mitosis (*mit* = thread; *osis* = process) is the series of events that parcel out the replicated DNA of the mother cell to two daughter cells. Mitosis is described in terms of four phases: **prophase, metaphase, anaphase,** and **telophase,** but it is actually a continuous process, with one phase merging smoothly into the next. Its duration varies according to cell type, but it typically lasts about an hour or less in human cells from start to finish. Mitosis is described in Figure 3.30 on pp. 100–101.

Cytokinesis Cytokinesis (*kines* = movement), or the division of the cytoplasm, begins during late anaphase and is completed after mitosis ends (see Figure 3.30). The plasma membrane over the center of the cell (the spindle equator) is drawn inward to form a **cleavage furrow** by the activity of a *contractile ring* made of actin and myosin filaments. The cleavage furrow deepens until the original cytoplasmic mass is pinched into two parts, so that at the end of cytokinesis there are two daughter cells. Each is smaller and has less cytoplasm than the mother cell, but is *genetically* identical to it. The daughter cells then enter the interphase portion of their life cycle and grow and carry out normal cell activities until it is their turn to divide.

Control of Cell Division The signals that prod cells to divide are poorly understood, but we know that *surface-volume relationships* are important. The amount of nutrients a growing cell requires is directly related to its volume. Volume increases with the cube of cell radius, whereas surface area increases with the square of the radius. Thus, a 64-fold increase in cell volume is accompanied by only a 16-fold increase in surface area. Consequently, the surface area of the plasma membrane becomes inadequate for nutrient and waste exchange when a cell reaches a certain critical size. Cell division solves this problem because the smaller daughter cells once again have a favorable surface-volume relationship. These surface-volume relationships help explain why most cells are microscopic in size.

Other mechanisms that influence when cells divide include chemical signals (growth factors, hormones, and others) released by neighboring or distant cells and the availability of space. Normal cells stop proliferating when they begin touching, a phenomenon called *contact inhibition*. Cancer cells escape many of the normal controls of cell division and divide wildly, which makes them dangerous to their host.

The cell cycle is controlled by a distinct cell-cycle control system that has been compared to an automatic washer's timer control. Like that timer, the cell cycle control system proceeds on its own, driven by a built-in clock. However, just as the washer's cycle is subject to adjustments (by regulating the faucet or by an internal water-level sensor) the cell cycle is regulated by both internal and external factors.

Although all the triggers for cell division are still being researched, we do know that a number of "switches" and checkpoints are involved. Thus far, two groups of proteins appear crucial to the ability of a cell to enter mitosis—**cyclins** (regulatory proteins whose levels rise and fall during each cycle) and **Cdks** (cyclin-dependent kinases), which are present in a constant concentration in the cell and are activated by binding to particular cyclins. In response to specific signals, a new batch of cyclins accumulates during each interphase (Figure 3.31a). Subsequent joining of specific Cdk and cyclin proteins initiates enzymatic cascades that phosphorylate histones and other proteins needed to direct or perform the tasks of the various stages of cell division. At the end of mitosis, the cyclins are abruptly destroyed.

Text continues on page 102.

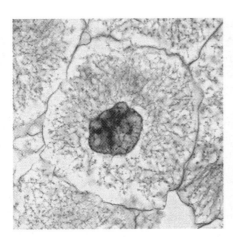

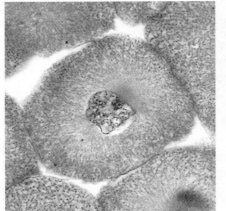

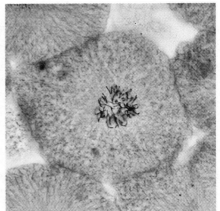

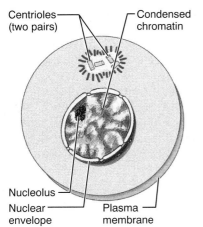

Centrioles (two pairs) — Condensed chromatin

Nucleolus

Nuclear envelope — Plasma membrane

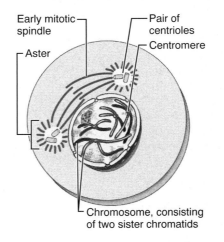

Early mitotic spindle — Pair of centrioles

Aster — Centromere

Chromosome, consisting of two sister chromatids

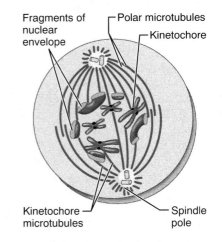

Fragments of nuclear envelope — Polar microtubules

Kinetochore

Kinetochore microtubules — Spindle pole

Interphase

Interphase is the period of a cell's life when it is carrying out its normal metabolic activities and growing. During interphase, the chromosomal material is seen in the form of extended and condensed chromatin, and the nuclear membrane and nucleolus are intact and visible. Microtubule arrays (asters) are seen extending from the centrosomes. During various periods of this phase, the centrioles begin replicating (G_1 through G_2), DNA is replicated (S), and the final preparations for mitosis are completed (G_2). The centriole pair finishes replicating into two pairs during G_2.

Early prophase

As mitosis begins, microtubule arrays called *asters* ("stars") are seen extending from the centrosome matrix around the centrioles. Early in *prophase*, the first and longest phase of mitosis, the chromatin threads coil and condense, forming barlike *chromosomes* that are visible with a light microscope. Since DNA replication has occurred during interphase, each chromosome is actually made up of two identical chromatin threads, now called *chromatids*. The chromatids of each chromosome are held together by a small, buttonlike body called a *centromere*. After the chromatids separate, each is considered a new chromosome.

As the chromosomes appear, the nucleoli disappear, and the cytoskeletal microtubules disassemble. The centriole pairs separate from one another. The centrioles act as focal points for growth of a new assembly of microtubules called the **mitotic spindle.** As these microtubules lengthen, they push the centrioles farther and farther apart, propelling them toward opposite ends (poles) of the cell.

Late prophase

While the centrioles are still moving away from each other, the nuclear membrane fragments, allowing the spindle to occupy the center of the cell and to interact with the chromosomes. Meanwhile, some of the growing spindle microtubules attach to special protein–DNA complexes, called *kinetochores* (ki-ne′to-korz), on each chromosome's centromere. Such microtubules are called *kinetochore microtubules.* The remaining spindle microtubules, which do not attach to any chromosomes, are called *polar microtubules.* The tips of the polar microtubules are linked near the center; these push against each other forcing the poles apart. The kinetochore microtubules, on the other hand, pull on each chromosome from both poles, resulting in a tug-of-war that ultimately draws the chromosomes to the middle of the cell.

FIGURE 3.30 *The stages of mitosis.* These particular cells are from an early embryo of a whitefish. (Micrographs approx. 600×.) For simplicity, only four chromosomes are drawn.

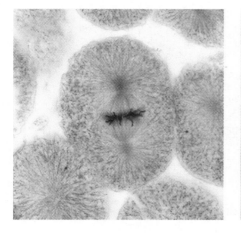

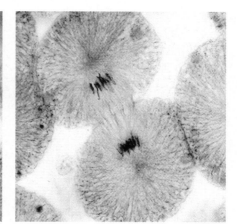

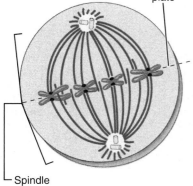

Metaphase plate

Spindle

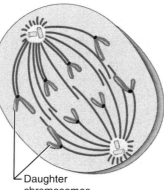

Daughter chromosomes

Nucleolus forming

Contractile ring at cleavage furrow

Nuclear envelope forming

Metaphase

Metaphase is the second phase of mitosis. The chromosomes cluster at the middle of the cell, with their centromeres precisely aligned at the exact center, or *equator,* of the spindle. This arrangement of the chromosomes along a plane midway between the poles is called the *metaphase* plate.

Anaphase

Anaphase, the third phase of mitosis, begins abruptly as the centromeres of the chromosomes split, and each chromatid now becomes a chromosome in its own right. The kinetochore fibers, moved along by motor proteins in the kinetochores, rapidly disassemble at their kinetochore ends by removing tubulin subunits, and gradually pull each chromosome toward the pole it faces. By contrast the polar microtubules slide past each other and lengthen (a process presumed to be driven by kinesin motor molecules), and push the two poles of the cell apart, causing the cell to elongate. Anaphase is easy to recognize because the moving chromosomes look V-shaped. The centromeres, which are attached to the kinetochore microtubules, lead the way, and the chromosomal "arms" dangle behind them. Anaphase is the shortest stage of mitosis; it typically lasts only a few minutes.

This process of moving and separating the chromosomes is helped by the fact that the chromosomes are short, compact bodies. Diffuse threads of extended chromatin would tangle, trail, and break, which would damage the genetic material and result in its imprecise "parceling out" to the daughter cells.

Telophase and cytokinesis

Telophase begins as soon as chromosomal movement stops. This final phase is like prophase in reverse. The identical sets of chromosomes at the opposite poles of the cell uncoil and resume their threadlike extended–chromatin form. A new nuclear membrane, derived from the rough ER, reforms around each chromatin mass. Nucleoli reappear within the nuclei, and the spindle breaks down and disappears. Mitosis is now ended. The cell, for just a brief period, is binucleate (has two nuclei) and each new nucleus is identical to the original mother nucleus.

As a rule, as mitosis draws to a close, *cytokinesis* completes the division of the cell into two daughter cells. Cytokinesis occurs as a contractile ring of peripheral microfilaments forms at the *cleavage furrow* and squeezes the cells apart. Cytokinesis actually begins during late anaphase and continues through and beyond telophase.

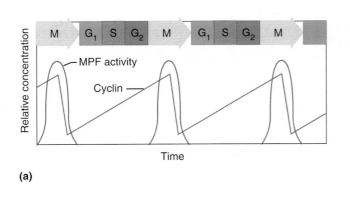

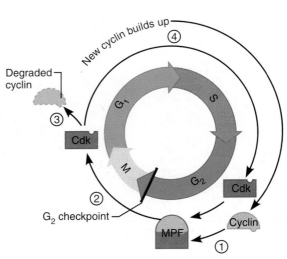

(b)

FIGURE 3.31 *Cell-cycle control at the G₂ checkpoint.* The stepwise processes of the cell cycle are timed by rhythmic fluctuations in the activity of protein kinases called cyclin-dependent kinases (Cdks), because they are active only when bound to a cyclin, a protein whose concentration varies cyclically. Here we focus on a Cdk-cyclin complex called MPF, which acts at the G_2 checkpoint to trigger mitosis. **(a)** The graph shows how MPF activity fluctuates with the level of cyclin in the cell. The cyclin level rises throughout interphase (G_1, S, and G_2 phases), then falls abruptly during mitosis (M phase). The peaks of MPF activity and cyclin concentration correspond. The Cdk itself is present at a constant level (not shown). **(b)** ① By the G_2 checkpoint (black bar), enough cyclin is available to produce many molecules of MPF. ② MPF promotes mitosis by phosphorylating various proteins, including other enzymes ③ One effect of MPF is the initiation of a sequence of events leading to the breakdown of its own cyclin. ④ The Cdk component of MPF is recycled. Its kinase activity will be restored by association with new cyclin that accumulates during interphase.

A number of crucial checkpoints occur throughout interphase. One important checkpoint, and the first to be understood, occurs late in G_2 (see Figure 3.31) when a threshold amount of a protein complex called **MPF** is required to give the okay signal to enter the mitotic (M) phase. Although called *maturation promoting factor,* MPF may be better thought of as *M-phase promoting factor* because it allows the cell to pass the G_2 checkpoint and enter mitosis. Later in M phase, MPF initiates a cascade leading to its own inactivation by causing destruction of its cyclin.

Protein Synthesis

In addition to directing its own replication, DNA serves as the master blueprint for protein synthesis. Although cells also make lipids and carbohydrates, DNA does not dictate their structure. DNA specifies *only* the structure of protein molecules, including the enzymes that catalyze the synthesis of all classes of biological molecules. Most of the metabolic machinery of the cell is concerned in some way with protein synthesis. This is not surprising, because structural proteins constitute most of the dry cell material, and functional proteins direct and underlie all cellular activities. Essentially, cells are miniature protein factories that synthesize the huge variety of proteins that determine the chemical and physical nature of cells—and therefore of the body as a whole.

Proteins, you will recall, are composed of polypeptide chains, which in turn are made up of amino acids (see Figure 2.17, p. 52). For purposes of this discussion, we can define a **gene** as a segment of a DNA molecule that carries instructions for one polypeptide chain. However, selected genes specify the structure of certain varieties of RNA as their final product.

The four nucleotide bases (A, G, T, and C) are the "letters" of the genetic dictionary, and the information of DNA is found in the sequence of these bases. Each sequence of three bases, called a **triplet,** can be thought of as a "word" that specifies a particular amino acid. For example, the triplet AAA calls for the amino acid phenylalanine, and CCT calls for glycine. The sequence of the triplets in each gene forms a "sentence" that tells exactly how a particular polypeptide is to be made; it specifies the number, kinds, and order of amino acids needed to build that protein. Variations in the arrangement of A, T, C, and G allow our cells to make all the different kinds of proteins needed. Even an unusually "small" gene has an estimated 2100 base pairs in sequence. Since the ratio between DNA bases in the gene and amino acids in the polypeptide is 3:1, the polypeptide specified by such a gene would be expected to contain 700 amino acids. Additionally, most genes of higher organisms contain **exons,** amino acid–specifying informational sequences, separated by introns. **Introns** are noncoding segments that range from 60 to 100,000 nucleotides. Since a single gene may have as many as 50 introns, most genes are much larger than you might expect.

The Role of RNA

By itself, DNA is rather like a strip of magnetic recording tape: The information it contains cannot be used without a decoding mechanism. Furthermore, polypeptides are manufactured at ribosomes in the cytoplasm, but in interphase cells DNA never leaves the nucleus. So, DNA requires not only a decoder, but a messenger as well. The decoding and messenger functions are carried out by RNA, the second type of nucleic acid.

As you learned in Chapter 2, RNA differs from DNA in being single stranded and in having the sugar ribose instead of deoxyribose and the base uracil (U) instead of thymine (T). Three forms of RNA act together to carry out DNA's instructions for polypeptide synthesis:

1. Transfer RNA (tRNA), small and roughly cloverleaf-shaped molecules

2. Ribosomal RNA (rRNA), which forms part of the ribosomes

3. Messenger RNA (mRNA), relatively long nucleotide strands resembling "half-DNA" molecules, or one of the two strands of a DNA molecule

All three types of RNA are formed on the DNA in the nucleus in much the same way as DNA replicates itself: The DNA helix separates and one of its strands serves as a template for synthesizing a complementary RNA strand. Once formed, the RNA molecule is released from the DNA template and migrates into the cytoplasm. Its job done, DNA simply recoils into its helical, inactive form. Most of the nuclear DNA codes for the synthesis of short-lived mRNA, so named because it carries the "message" containing the instructions for building a polypeptide from the gene to the ribosomes. Thus, the end product of most genes is a polypeptide. Only small amounts of the DNA code for the synthesis of rRNA (the nucleolar organizer regions mentioned previously) and tRNA, both of which are long-lived and stable types of RNA. Since rRNA and tRNA do not transport codes for synthesizing other molecules, they are the final products of the genes that code for them. Ribosomal RNA and tRNA act together to "translate" the message carried by mRNA.

The rules by which the base sequence of a DNA gene are translated into protein structure (amino acid sequence) are called the **genetic code.** Essentially, polypeptide synthesis involves two major steps: (1) *transcription,* in which DNA's information is encoded in mRNA, and (2) *translation,* in which the information carried by mRNA is decoded and used to assemble polypeptides. These steps are summarized in Figure 3.32 and described in more detail next.

Transcription

The word *transcription* often refers to one of the jobs done by a secretary—converting notes taken in shorthand into a typed or electronic copy. In other words, the same information is transformed, or transferred, from one form or format to another. In cells, **transcription** involves the transfer of information from a DNA gene's base sequence to the complementary base sequence of an mRNA molecule. The form is different, but the same information is being conveyed. Once the mRNA molecule is made, it detaches and leaves the nucleus. Only DNA and mRNA are involved in the transcription process.

Let us follow the transcription process when a particular polypeptide is to be made.

Making the mRNA Complement The process begins when a chemical messenger called a *transcription factor* binds to a special DNA site adjacent to the "start" sequence, called the promoter. The transcription factor, in turn, mediates the binding of **RNA polymerase,** the enzyme that oversees the synthesis of mRNA. Once bound, the polymerase opens up the DNA helix and the DNA segment coding for that protein is uncoiled. One DNA nucleotide strand, called the *sense strand,* serves as the template for the construction of a complementary mRNA molecule (see Figure 3.32, step ①). For example, if a particular DNA triplet sequence is AGC, the mRNA sequence synthesized at that site will be UCG. The DNA nucleotide strand not used for a template is called the *antisense strand.*

For each triplet, or three-base sequence, on DNA the corresponding three-base sequence on mRNA is called a **codon,** reflecting the fact that it is the coded information of mRNA that is actually used in protein synthesis. Since there are four kinds of RNA (or DNA) nucleotides, there are 4^3, or 64, possible codons. Three of these 64 codons are "stop signs" that call for termination of a polypeptide. All the rest code for amino acids. Because there are only about 20 amino acids, some are specified by more than one codon. This redundancy in the genetic code helps protect against problems due to transcription (and translation) errors. The genetic code and a complete codon list is found in Figure 3.33.

Editing of mRNA Although it would appear that translation can begin as soon as the complementary messenger is made, the process is a bit more complex. As mentioned before, mammalian DNA (like ours) has coding regions (exons) separated by noninformational regions (introns). Since the DNA gene is transcribed sequentially, the mRNA initially made, called *pre-mRNA* or *primary transcript,* is littered with intron "nonsense." Before the newly formed RNA can be used as a messenger, it must be edited or processed—sections corresponding to introns must be removed. The remaining exon-coded sections are spliced together in the order in which they occurred in the DNA gene to produce the functional mRNA that directs translation at the ribosome.

Translation

A translator takes a message in one language and restates it in another. In the **translation** step of protein synthesis, the language of nucleic acids (base sequence) is translated into the language of proteins (amino acid sequence). The process of translation occurs in the cytoplasm and involves all three varieties of RNA (see Figure 3.32, steps ②–⑤).

When the mRNA molecule carrying instructions for a particular protein reaches the cytoplasm, it binds to a small ribosomal subunit by base pairing to rRNA. Then transfer RNA (tRNA) comes into the picture. As its name suggests, tRNA has the job of *transferring* amino acids to the ribosome. There are approximately 20 different types of tRNA, each capable of binding with a specific amino acid. The attachment process is controlled by a synthetase enzyme in each case and is activated by ATP. Once its amino acid is loaded, the tRNA migrates to the ribosome, where it maneuvers the amino acid into the proper position, as specified by the codons on the mRNA strand. This is more complex than it seems. Not only must tRNA bring an amino acid to the protein synthesis site, but it must also "recognize" the codon calling for that amino acid.

The structure of the tiny tRNA molecule is well suited for its dual function. The amino acid is bound to one end of tRNA, at a region called the tail. At the other end, the head, is its **anticodon** (an"ti-ko'don), a specific three-base sequence complementary to the mRNA codon calling for the amino acid carried by that particular tRNA. Anticodons form hydrogen bonds with complementary codons, so the tiny tRNAs serve as the links between the languages of nucleic acids and proteins. Thus, if the mRNA codon is UUU, which specifies phenylalanine, the tRNAs carrying phenylalanine will have the anticodon AAA, which can bind to the correct codons.

The translation process begins when a special "initiator" tRNA binds to an mRNA molecule and a small ribosomal subunit. This stimulates a large ribosomal subunit to bind and results in the formation of a functional ribosome with the mRNA positioned in the "groove" between the two ribosomal subunits. Downstream on the mRNA is the "start" codon, AUG, which is recognized and bound by the anticodon (UAC) of a tRNA carrying the amino acid methionine (see Figure 3.33). This initiation process requires the help of a number of initiation factors and is energized by GTP.

The ribosome is more than just a passive attachment site for mRNA and tRNA. Besides its binding site for mRNA, it has three binding sites for tRNA (an A site for incoming tRNA, a P site for the tRNA holding the growing polypeptide chain, and an E [exit] site for outgoing tRNA). Like a vise, the ribosome holds the tRNA and mRNA close together to coordinate the coupling of codons and anticodons, and properly positions the next (incoming) amino acid for addition to the growing polypeptide chain. Once the first tRNA has been positioned on a codon, the ribosome slides the mRNA strand along, bringing the next codon into position to be "read" by another tRNA. As successive amino acids are brought to their proper positions, peptide bonds are formed between them and the polypeptide chain becomes longer and longer. When each amino acid is bonded

FIGURE 3.32 *Protein synthesis.* ① *Transcription.* The DNA segment, or gene, specifying one polypeptide uncoils, and one strand acts as a template for the synthesis of a complementary mRNA molecule. (②–⑤) *Translation.* Messenger RNA from the nucleus attaches to a small ribosomal subunit in the cytoplasm ②. Transfer RNA transports amino acids to the mRNA strand and recognizes the mRNA codon calling for its amino acid by base pairing with it (via its anticodon). The ribosome then assembles and translation begins ③. The ribosome moves along the mRNA strand as each codon is read sequentially ④. As each amino acid is bound to the next by a peptide bond, its tRNA is released ⑤. The polypeptide is released when the stop codon is read. (For simplicity, the mRNA editing that occurs in phase 1 is not shown.)

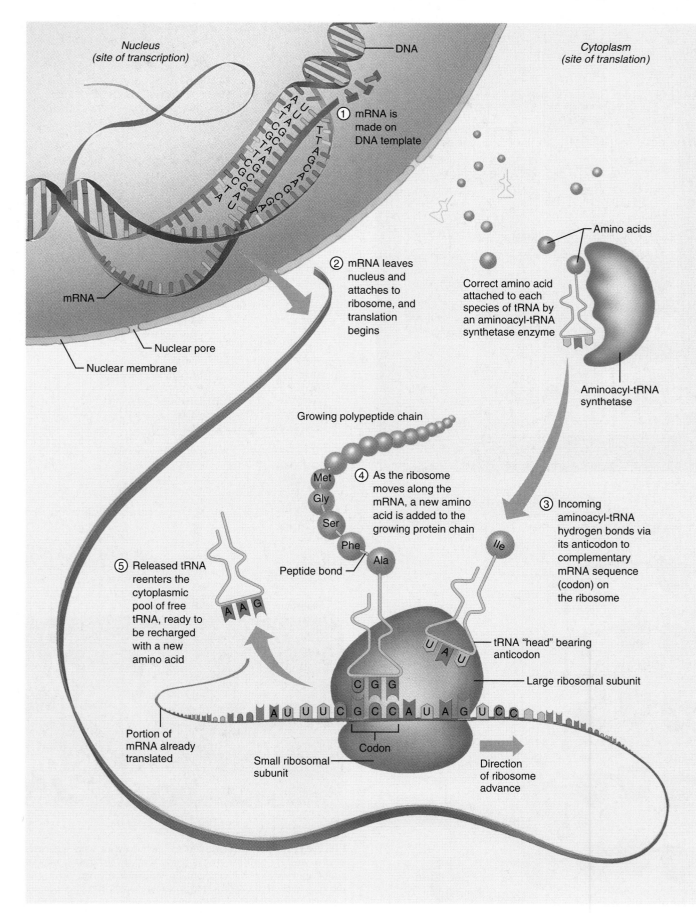

Nucleus
(site of transcription)

DNA

Cytoplasm
(site of translation)

① mRNA is made on DNA template

mRNA

Nuclear pore

Nuclear membrane

② mRNA leaves nucleus and attaches to ribosome, and translation begins

Amino acids

Correct amino acid attached to each species of tRNA by an aminoacyl-tRNA synthetase enzyme

Aminoacyl-tRNA synthetase

Growing polypeptide chain

Met

Gly

Ser

Phe

Ala

Peptide bond

④ As the ribosome moves along the mRNA, a new amino acid is added to the growing protein chain

③ Incoming aminoacyl-tRNA hydrogen bonds via its anticodon to complementary mRNA sequence (codon) on the ribosome

Ile

⑤ Released tRNA reenters the cytoplasmic pool of free tRNA, ready to be recharged with a new amino acid

A A G

tRNA "head" bearing anticodon

U A U

Large ribosomal subunit

C G G

A U U U C G C C A U A G U C C

Portion of mRNA already translated

Codon

Small ribosomal subunit

Direction of ribosome advance

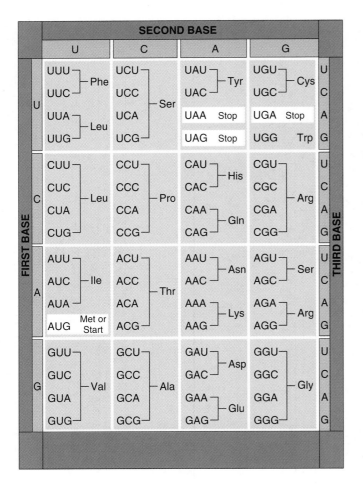

		SECOND BASE		

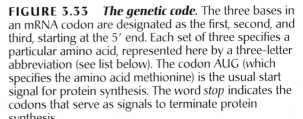

FIGURE 3.33 *The genetic code.* The three bases in an mRNA codon are designated as the first, second, and third, starting at the 5′ end. Each set of three specifies a particular amino acid, represented here by a three-letter abbreviation (see list below). The codon AUG (which specifies the amino acid methionine) is the usual start signal for protein synthesis. The word *stop* indicates the codons that serve as signals to terminate protein synthesis.

Abb.*	Amino acid
Ala	alanine
Arg	arginine
Asn	asparagine
Asp	aspartic acid
Cys	cysteine
Glu	glutamic acid
Gln	glutamine
Gly	glycine
His	histidine
Ile	isoleucine
Leu	leucine
Lys	lysine
Met	methionine
Phe	phenylalanine
Pro	proline
Ser	serine
Thr	threonine
Trp	tryptophan
Tyr	tyrosine
Val	valine

*Abb. = abbreviation for the amino acid

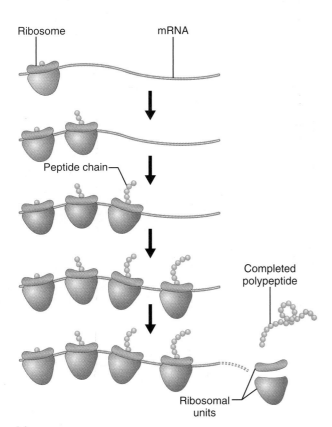

(a)

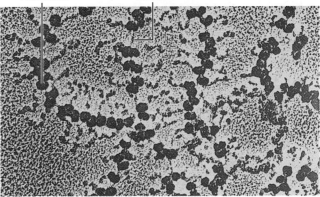

(b)

FIGURE 3.34 *Polyribosomes.* Each polyribosome consists of one strand of mRNA that is being read by several ribosomes simultaneously. In this diagram, the mRNA is moving through the ribosome to the left and the "oldest" functional ribosome is farthest to the right. Polyribosome arrays allow a single strand of mRNA to produce hundreds of the same polypeptide molecules in a short time (500,000×).

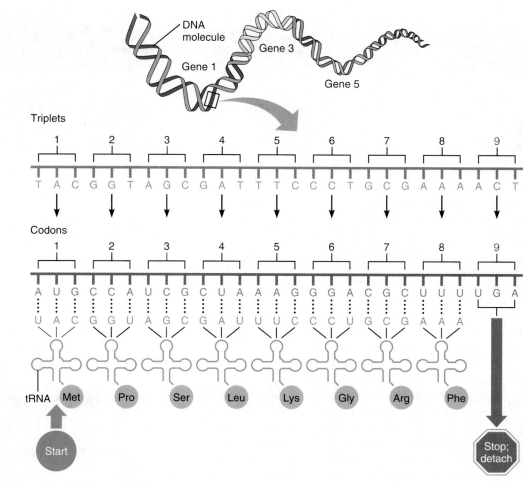

DNA base sequence (triplets) of the gene coding for the synthesis of a particular polypeptide chain

Base sequence (codons) of the transcribed mRNA

Consecutive base sequences of tRNA *anticodons* capable of recognizing the mRNA codons calling for the amino acids they transport

Amino acid sequence of the polypeptide chain

FIGURE 3.35 *Information transfer from DNA to RNA.* Information is transferred from the DNA of the gene to the complementary messenger RNA molecule, whose codons are then "read" by transfer RNA anticodons. Notice that the "reading" of the mRNA by tRNA anticodons reestablishes the base (triplet) sequence of the DNA genetic code (except that T is replaced by U).

to the next, its tRNA is released from the P site and moves to the E site and then away from the ribosome to pick up another amino acid. As mRNA is progressively read, its initial portion passes through the ribosome and may become attached successively to several other ribosomes, all reading the same message simultaneously. Such a multiple ribosome-mRNA complex is called a *polyribosome* (Figure 3.34), and it provides an efficient system for producing many copies of the same protein. The mRNA strand continues to be read sequentially until its last codon, the *stop codon* (one of UGA, UAA, or UAG) enters the ribosomal groove. The stop codon is the "period" at the end of the mRNA sentence that tells the ribosome that translation of that mRNA is finished. The completed polypeptide chain is then released from the ribosome (Figures 3.34 and 3.35).

The ribosome then separates into its subunits, if not engaged by another mRNA molecule.

The genetic information of a cell is translated into the production of proteins via a sequence of information transfer that is completely directed by complementary base pairing (Figure 3.35). Thus, the transfer of information goes from DNA base sequence (triplets) to the complementary base sequence of mRNA (codons) and back to the DNA base sequence (anticodons).

Cytosolic Protein Degradation

All cellular proteins are eventually degraded (Figure 3.36). Organelle proteins are digested within lysosomes, but lysosomal enzymes do not have access to soluble proteins in the cytosol that need to be disposed of because they are damaged, incorrectly

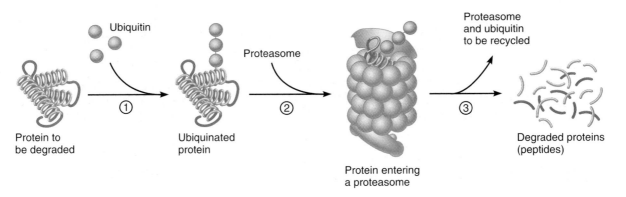

FIGURE 3.36 *Degradation of a protein by a proteasome.* ① Enzymes in the cytosol attach ubiquitin molecules to a protein. (It is not known how the protein is selected.) ② A proteasome recognizes the ubiquinated protein and takes it into its central cavity. ③ Enzymatic components of the proteasome cut the protein into small peptides, which can be further degraded by cytosolic enzymes. Steps 1 and 3 require ATP.

folded, or, like cyclin at the end of mitosis, simply no longer needed for a particular job. To prevent such proteins from accumulating while preventing wholesale destruction of virtually all soluble proteins by the cytosolic enzymes, doomed proteins are marked for attack (proteolysis) by attachment of a protein called *ubiquitin* (u-bĭ′kwĭ-tin) in an ATP-dependent reaction. The tagged proteins are hydrolyzed into small peptides by soluble enzymes or by giant complexes of protein-digesting enzymes called *proteasomes.* (Structurally the proteasome is remarkably similar to one variety of molecular chaperone [see p. 54] and may in fact be one of that protein group.)

EXTRACELLULAR MATERIALS

Extracellular materials are substances contributing to body mass that are found outside the cells. One class of extracellular materials is *body fluids,* represented by interstitial fluid, blood plasma, and cerebrospinal fluid. These fluids are important as transport and dissolving media. *Cellular secretions,* also extracellular materials, include substances that aid in digestion (intestinal and gastric fluids) and that act as lubricants (saliva, mucus, and serous fluids).

By far the most abundant of the extracellular materials is the *extracellular matrix.* Most body cells are in contact with a jellylike substance composed of proteins and polysaccharides. These molecules are secreted by the cells and self-assemble into an organized mesh in the extracellular space, where they serve as a universal "cell glue" that helps to hold body cells together. In connective tissues the extracellular matrix is particularly abundant—so much so, in some cases, that it (rather than living cells) accounts for the bulk of that tissue type. Depending on

the structure to be formed, the extracellular matrix in connective tissue may be soft, tough and ropelike, or rock-hard, as described in the next chapter.

DEVELOPMENTAL ASPECTS OF CELLS

We all begin life as a single cell, the fertilized egg, from which arise all the cells of our body. Very early in development, cells begin to specialize; some become liver cells, some nerve cells, and others the transparent lens of the eye. Since all our cells carry the same genes, how can one cell be so different from another? This is a fascinating question. Apparently, cells in various regions of the embryo are exposed to different chemical signals that channel them into specific pathways of development. When the embryo consists of just a few cells, slight differences in oxygen and carbon dioxide concentrations between the more superficial and the deeper cells may be the major signals. But as development continues, cells release chemicals that influence development of neighboring cells by triggering processes (such as adding methyl groups to specific genes) that switch some of their genes "off." Some genes are active in all cells. For example, all cells carry out protein synthesis and make ATP. However, genes for enzymes that catalyze the synthesis of specialized products such as hormones or neurotransmitters are activated only in certain cell populations—only cells of the thyroid gland can produce thyroxine. Hence, the secret of cell specialization lies in the kinds of proteins made and reflects differential gene activation in different cell types. Cell specialization leads to *structural* variation—different organelles come to predominate in different cells. For example, muscle cells make large amounts of actin and myosin, and their cytoplasm fills with microfilaments. Liver and phagocytic cells produce more lysosomes. The development of specific and distinctive features in cells is called **cell differentiation.**

During early development, cell death and destruction are normal events. Nature takes a few chances. More cells than are needed are produced and excesses are eliminated later. This is particularly true in the nervous system. Most organs are well formed and functional long before birth, but the body continues to grow and enlarge by forming new cells throughout childhood and adolescence. Once adult size has been reached, cell division is important mainly to replace short-lived cells and repair wounds.

During young adulthood, cell numbers remain fairly constant. However, local changes in the rate of cell division are fairly common. For example, when one is anemic, bone marrow undergoes **hyperplasia** (hi″per-pla′ze-ah), or accelerated growth (*hyper* = over; *plas* = grow), so that red blood cells are produced at a faster rate. If the anemia is remedied, the excessive marrow activity ceases. **Atrophy** (at′ro-fe), a decrease in size of an organ or body tissue, can result from loss of normal stimulation. Muscles that lose their nerve supply atrophy and waste away, and lack of exercise leads to thinned, brittle bones.

Cell aging occurs and it accounts for most problems associated with old age. Cell aging is a complicated process with many causes. The *wear-and-tear theory* attributes aging to little chemical insults and formation of free radicals that have cumulative effects throughout life. For example, environmental toxins such as pesticides, alcohol, and bacterial toxins may damage cell membranes, poison enzyme systems, or cause "mistakes" in DNA replication. Temporary lack of oxygen, which occurs increasingly with age as our blood vessels clog with fatty materials, leads to accelerated rates of cell death throughout the body. Most free radicals are produced in our mitochondria, because these organelles have the highest rate of metabolism. This has led to the proposal that diminished energy production by radical-damaged mitochondria weakens (and ages) our cells. X rays and other types of radiation, and some chemicals, also generate huge numbers of free radicals, which can help overwhelm the peroxisomal enzymes. Vitamins C and E act as antioxidants in the body and may help to prevent excessive free radical formation. Attack comes from within as well. With age, glucose (blood sugar) becomes party to cross linking our proteins together, which severely disrupts protein function and accelerates the course of arteriosclerosis.

Another theory attributes cell aging to progressive disorders in the immune system. According to this theory, cell damage results from (1) autoimmune responses in which the immune system turns against one's own tissues, and (2) a progressive weakening of the immune response, so that the body is less and less able to get rid of cell-damaging pathogens.

Perhaps the most widely accepted theory of cell aging is the *genetic theory*, which suggests that cessation of mitosis and cell aging are "programmed into our genes." One interesting notion here is that a *telomere clock* determines the number of times a cell can divide. **Telomeres** (*telo* = end; *mer* = piece) are strings of nucleotides that cap the end of chromosomes, protecting them from fraying or fusing with other chromosomes. In human telomeres, the base sequence TTAGGG is repeated a thousand times or more (like a DNA stutter). Though the telomeres carry no genes, they appear to be vital for chromosomal survival, because each time DNA is replicated, 50 to 100 of the end nucleotides are lost and the telomeres get a bit shorter. When telomeres reach a certain minimum length, the stop-division signal is given. The idea that cell longevity depends on telomere integrity is supported by the 1994 discovery of *telomerase,* an enzyme that protects telomeres from degrading. Pegged as the "immortality enzyme," telomerase is almost universally found in cancer cells, but is barely detectable or absent in other adult cell types.

* * *

We have described here the structure and function of the generalized cell. One of the wonders of the cell is the disparity between its minute size and its extraordinary activity, which reflects the amazing diversity of its organelles. The evidence for division of labor and functional specialization among organelles is inescapable. Only ribosomes synthesize proteins, while protein packaging is the bailiwick of the Golgi apparatus. Membranes compartmentalize most organelles, allowing them to work without hindering or being hindered by other cell activities, and the plasma membrane regulates molecular traffic across the cell's boundary. Now that you know what cells have in common, you are ready to explore how they differ in the various body tissues.

RELATED CLINICAL TERMS

Anaplasia (an"ah-pla′ze-ah; *an* = without, not; *plas* = to grow) Abnormalities in cell structure; e.g., cancer cells typically lose the appearance of the parent cells and come to resemble undifferentiated or embryonic cells.

Apoptosis (ap"o-to′sis; "falling away") Programmed cell death. This process of controlled cellular suicide eliminates cells that are stressed, unneeded, excessive, or aged. In response to damaged macromolecules within the cell or to some extracellular signal, a series of intracellular enzymes is activated. These enzymes, called capsases, destroy the cell's DNA, cytoskeleton, etc., producing a quick, neat death. The apoptotic cell shrinks without leaking its contents into the surrounding tissue, detaches from other cells, rounds up, and is immediately phagocytized by nearby cells. Cancer cells fail to undergo apoptosis, but oxygen-starved cells do so excessively (heart-muscle and brain cells during heart attacks and strokes, for example).

Dysplasia (dis-pla′ze-ah; *dys* = abnormal) A change in cell size, shape, or arrangement due to chronic irritation or inflammation (infections, etc.).

Hyperplasia (hi"per-pla′ze-ah; "excess shape") Excessive cell proliferation. It differs from cancer because hyperplastic cells retain their normal form and arrangement within tissues.

Hypertrophy (hi-per′tro-fe) Growth of an organ or tissue due to an increase in the size of its cells. Hypertrophy is a normal response of skeletal muscle cells when they are challenged to lift excessive weight; differs from hyperplasia, which is an increase in size due to an increase in cell number.

Liposomes (lip′o-sōmz) Hollow microscopic sacs formed of phospholipids that can be filled with a variety of drugs. Serve as multipurpose vehicles for drugs, genetic material, and cosmetics.

Mutation A change in DNA base sequence leading to incorporation of incorrect amino acids in particular positions in the resulting protein; the affected protein may remain unimpaired or may function abnormally or not at all, leading to disease.

Necrosis (nĕkro′sis; *necros* = death; *osis* = process) Death of a cell or group of cells due to injury or disease. Acute injury causes the cells to swell and burst, and induces the inflammatory response. (This is *uncontrolled* cell death, in contrast to apoptosis described above.)

CHAPTER SUMMARY

Media study tools that could provide you additional help in reviewing specific key topics of Chapter 3 are referenced below. **SP** = *Study Partner;* **IP** = *Interactive Physiology.*

Overview of the Cellular Basis of Life (pp. 65–67)

1. All living organisms are composed of cells—the basic structural and functional units of life. Cells vary widely in both shape and size.

2. The principle of complementarity states that the biochemical activity of cells reflects the operation of organelles.

3. The generalized cell is a concept that typifies all cells. The generalized cell has three major regions—the nucleus, cytoplasm, and plasma membrane.

The Plasma Membrane: Structure (pp. 68–71)

1. The plasma membrane encloses cell contents, mediates exchanges with the extracellular environment, and plays a role in cellular communication.

The Fluid Mosaic Model (pp. 68–70)

2. The fluid mosaic model depicts the plasma membrane as a fluid bilayer of lipids (phospholipids, cholesterol, and glycolipids) within which proteins are inserted.

3. The lipids have both hydrophilic and hydrophobic regions that organize their aggregation and self-repair. The lipids form the structural part of the plasma membrane.

4. Most proteins are integral transmembrane proteins that extend entirely through the membrane. Some, appended to the integral proteins, are peripheral proteins.

5. Proteins are responsible for most specialized membrane functions: some are enzymes, some are receptors, and others mediate membrane transport functions. Externally facing glycoproteins contribute to the glycocalyx.

Specializations of the Plasma Membrane (pp. 70–71)

6. Microvilli are extensions of the plasma membrane that increase its surface area for absorption.

7. Membrane junctions join cells together and may aid or inhibit movement of molecules between or past cells. Tight junctions are impermeable junctions; desmosomes mechanically couple cells into a functional community; gap junctions allow joined cells to communicate.

SP Exercise: Chapter 3, Structure of the Plasma Membrane.

The Plasma Membrane: Functions (pp. 71–83)

Membrane Transport (pp. 71–80)

1. The plasma membrane acts as a selectively permeable barrier. Substances move across the plasma membrane by passive processes, which depend on the kinetic energy of molecules or on pressure gradients, and by active processes, which depend on the use of cellular energy (ATP).

2. Diffusion is the movement of molecules (driven by kinetic energy) down a concentration gradient. Fat-soluble solutes can diffuse directly through the membrane by dissolving in the lipid. Charged small molecules or ions move by diffusion through the membrane if they are small enough to pass through the protein channels. Some protein channels are selective.

3. Facilitated diffusion is the passive movement of certain solutes across the membrane by their combination with a membrane carrier protein. As with other diffusion processes, it is driven by kinetic energy, but the carriers are selective.

4. Osmosis is the diffusion of a solvent, such as water, through a selectively permeable membrane. Water diffuses through membrane pores (aquaporins) or directly through

the lipid portion of the membrane from a solution of lesser osmolarity to a solution of greater osmolarity.

5. The presence of impermeable solutes leads to changes in cell tone that may cause the cell to swell or shrink. Net osmosis ceases when the solute concentration on both sides of the plasma membrane reaches equilibrium.

6. Solutions that cause a net loss of water from cells are hypertonic; those causing net water gain are hypotonic; those causing neither gain nor loss of water are isotonic.

7. Filtration occurs when a filtrate is forced across a membrane by hydrostatic pressure. It is nonselective and limited only by pore size. The pressure gradient is the driving force.

8. Active transport (solute pumping) depends on a carrier protein and ATP. Substances transported move against concentration or electrical gradients. In primary active transport, such as the Na^+-K^+ pump, ATP directly provides the energy. In secondary active transport, the energy of an ion gradient (produced by a primary active transport process) is used to transport a substance passively. Many secondary active transport systems (pumps) are coupled, that is, cotransported substances move in the same (symport) or opposite (antiport) directions across the membrane.

9. Vesicular transport also requires that ATP be provided. Exocytosis ejects substances (hormones, wastes, secretions) from the cell. Endocytosis brings substances into the cell. If the substance is particulate, the process is called phagocytosis; if the substance is dissolved molecules in general, the process is bulk-phase endocytosis. Receptor-mediated endocytosis is selective; engulfed particles attach to receptors on the membrane before endocytosis occurs.

Generating and Maintaining a Resting Membrane Potential   (pp. 81–82)

10. All cells in the resting stage exhibit a voltage across their membrane, called the resting membrane potential. Because of the membrane potential, both concentration and electrical gradients determine the ease of an ion's diffusion.

11. The membrane potential is generated by concentration gradients of and differential permeability of the plasma membrane to sodium and potassium ions. Sodium is in high extracellular–low intracellular concentration, and the membrane is poorly permeable to it. Potassium is in high concentration in the cell and low concentration in the extracellular fluid. The membrane is more permeable to potassium than to sodium.

12. The greater outward diffusion of potassium (than inward diffusion of sodium) leads to a charge separation at the membrane (inside negative). This charge separation is maintained by the operation of the sodium-potassium pump.

SP Exercise: Chapter 3, Membrane Potentials.

IP Nervous System I CD–ROM: Ion Channels, pages 3, 18, 19; Membrane Potentials, pages 1–17.

Cell-Environment Interactions   (pp. 82–83)

13. Cells interact directly and indirectly with other cells. Indirect interactions involve extracellular chemicals carried in body fluids or forming part of the extracellular matrix.

14. Molecules of the glycocalyx are intimately involved in cell-environment interactions. Most are cell adhesion molecules or membrane receptors.

15. Activated membrane receptors act as catalysts, regulate channels, or like G protein–linked receptors, act through second messengers such as cyclic AMP and Ca^{2+}. Ligand binding results in changes in protein structure or function within the targeted cell.

SP Case Study: Chapter 3, Diabetes Mellitus.

The Cytoplasm   (pp. 83–94)

1. The cytoplasm, the cellular region between the nuclear and plasma membranes, consists of the cytosol (fluid cytoplasmic environment), inclusions (nonliving nutrient stores [lipid droplets, glycosomes], pigment granules, crystals, etc.), and cytoplasmic organelles.

Cytoplasmic Organelles   (pp. 83–94)

2. The cytoplasm is the major functional area of the cell. These functions are mediated by cytoplasmic organelles.

3. Mitochondria, organelles limited by a double membrane, are sites of ATP formation. Their internal enzymes carry out the oxidative reactions of cellular respiration.

4. Ribosomes, composed of two subunits containing ribosomal RNA and proteins, are the sites of protein synthesis. They may be free or attached to membranes.

5. The rough endoplasmic reticulum is a ribosome-studded membrane system. Its cisternae act as sites for protein modification. Its external face acts in phospholipid and cholesterol synthesis. Vesicles pinched off from the ER transport the proteins to other cell sites.

6. The smooth endoplasmic reticulum synthesizes lipid and steroid molecules. It also acts in fat metabolism and in drug detoxification. In muscle cells, it is a calcium ion depot.

7. The Golgi apparatus is a membranous system close to the nucleus that packages protein secretions for export, packages enzymes into lysosomes for cellular use, and modifies proteins destined to become part of cellular membranes.

8. Lysosomes are membranous sacs of acid hydrolases packaged by the Golgi apparatus. Sites of intracellular digestion, they degrade worn-out organelles, and tissues that are no longer useful, and release ionic calcium from bone.

9. Peroxisomes are membranous sacs containing oxidase enzymes that protect the cell from the destructive effects of free radicals and other toxic substances by converting them first to hydrogen peroxide and then water.

10. The cytoskeleton includes microfilaments, intermediate filaments, and microtubules. Microfilaments, formed of contractile proteins, are important in cell motility or movement of cell parts. Microtubules organize the cytoskeleton and are important in intracellular transport. Motility functions involve motor proteins. Intermediate filaments help cells resist mechanical stress and connect other elements.

11. Centrioles form the mitotic spindle and are the basis of cilia and flagella.

The Nucleus   (pp. 94–96)

1. The nucleus is the control center of the cell. Most cells have a single nucleus. Without a nucleus, a cell cannot divide or synthesize more proteins; thus, it is destined to die.

2. The nucleus is surrounded by the nuclear envelope, a double membrane penetrated by fairly large pores.

3. Nucleoli are nuclear sites of ribosome subunit synthesis.

4. Chromatin is a complex network of slender threads containing histone proteins and DNA. The chromatin units are called nucleosomes. When a cell begins to divide, the chromatin coils and condenses, forming chromosomes.

Cell Growth and Reproduction   (pp. 97–108)

The Cell Life Cycle   (pp. 97–102)

1. The cell life cycle is the series of changes that a cell goes through from the time it is formed until it divides.

2. Interphase is the nondividing phase of the cell life cycle. Interphase consists of G_1, S, and G_2 subphases. During G_1, the cell grows and centriole replication begins; during the S phase, DNA replicates; and during G_2, the final preparations for division are made.

3. DNA replication occurs before cell division; it ensures that all daughter cells have identical genes. The DNA helix uncoils, and each DNA nucleotide strand acts as a template for the formation of a complementary strand. Base pairing provides the guide for the proper positioning of nucleotides.

4. The products of the semiconservative replication of a DNA molecule are two DNA molecules identical to the parent molecule, each formed of one "old" and one "new" strand.

SP Case Study: Chapter 3, Skeletal Disorder.

5. Cell division, essential for body growth and repair, occurs during the M phase. Cell division is stimulated by certain chemicals (including growth factors and some hormones) and increasing cell size. Lack of space and inhibitory chemicals deter cell division. Cell division is regulated by cyclin-Cdk complexes, of which one example is MPF. Cell division consists of two distinct phases: mitosis and cytokinesis.

6. Mitosis, consisting of prophase, metaphase, anaphase, and telophase, results in the parceling out of the replicated chromosomes to two daughter nuclei, each genetically identical to the mother nucleus. Cytokinesis, which begins late in mitosis, divides the cytoplasmic mass into two parts.

Protein Synthesis   (pp. 102–108)

7. A gene is defined as a DNA segment that provides the instructions for the synthesis of one polypeptide chain. Since the major structural materials of the body are proteins, and all enzymes are proteins, this amply covers the synthesis of all biological molecules.

8. The base sequence of DNA provides the information for protein structure. Each three-base sequence (triplet) calls for a particular amino acid to be built into a polypeptide chain.

9. The three varieties of RNA are synthesized on single strands of the DNA template. RNA nucleotides are joined following base-pairing rules.

SP Exercise: Chapter 3, Protein Synthesis.

10. Ribosomal RNA forms part of the protein synthesis sites; messenger RNA carries instructions for making a polypeptide chain from the DNA to the ribosomes; transfer RNA ferries amino acids to the ribosomes and recognizes codons on the mRNA strand specifying its amino acid.

11. Protein synthesis involves (1) transcription, synthesis of a complementary mRNA, and (2) translation, "reading" of the mRNA by tRNA and peptide bonding of the amino acids into the polypeptide chain. Ribosomes coordinate translation.

12. Soluble proteins that are damaged or no longer needed are targeted for destruction by attachment of ubiquitin. Such proteins are degraded by cytosolic enzymes or proteasomes.

Extracellular Materials   (p. 108)

1. Extracellular materials are substances found outside the cells. These include body fluids, cellular secretions, and extracellular matrix. Extracellular matrix is particularly abundant in connective tissues.

Developmental Aspects of Cells   (pp. 108–109)

1. The first cell of an organism is the fertilized egg. Early in development, cell specialization begins and reflects differential gene activation.

2. During adulthood, cell numbers remain fairly constant; cell division occurs primarily to replace lost cells.

3. Cellular aging may reflect chemical insults, progressive disorders of immunity, or a genetically programmed decline in the rate of cell division with age.

REVIEW QUESTIONS

Multiple Choice/Matching

1. The smallest unit capable of life by itself is (a) the organ, (b) the organelle, (c) the tissue, (d) the cell, (e) the nucleus.

2. The major types of lipid found in the plasma membranes are (choose two) (a) cholesterol, (b) neutral fats, (c) phospholipids, (d) fat-soluble vitamins.

3. Membrane junctions that allow nutrients or ions to flow from cell to cell are (a) desmosomes, (b) gap junctions, (c) tight junctions, (d) all of these.

4. A person drinks a six-pack of beer and has to make several trips to the bathroom. This increase in urination reflects an increase in what process occurring in the kidneys? (a) diffusion, (b) osmosis, (c) solute pumping, (d) filtration.

5. The term used to describe the type of solution in which cells will lose water to their environment is (a) isotonic, (b) hypertonic, (c) hypotonic, (d) catatonic.

6. Osmosis always involves (a) a differentially permeable membrane, (b) a difference in solvent concentration, (c) diffusion, (d) active transport, (e) a, b, and c.

7. A physiologist observes that the concentration of sodium inside a cell is decidedly lower than that outside the cell. Yet sodium does diffuse easily across the plasma membrane of such cells when they are dead, but *not* when they are alive. Which of the following terms applies best to this cellular function that is lacking in dead cells? (a) osmosis, (b) diffusion, (c) active transport (solute pumping), (d) dialysis.

8. The solute-pumping variety of active transport is accomplished by (a) exocytosis, (b) phagocytosis, (c) electrical forces in the cell membrane, (d) conformational-positional changes in carrier molecules in the plasma membrane.

9. The endocytotic process in which a sampling of particulate matter is engulfed and brought into the cell is called (a) phagocytosis, (b) bulk-phase endocytosis, (c) exocytosis.

10. The nuclear substance composed of histone proteins and DNA is (a) chromatin, (b) the nucleolus, (c) nuclear sap, or nucleoplasm, (d) nuclear pores.

11. The information sequence that determines the nature of a protein is the (a) nucleotide, (b) gene, (c) triplet, (d) codon.

12. Mutations may be caused by (a) X rays, (b) certain chemicals, (c) radiation from ionizing radioisotopes, (d) all of these.

13. The phase of mitosis during which centrioles reach the poles and chromosomes attach to the spindle is (a) anaphase, (b) metaphase, (c) prophase, (d) telophase.

14. Final preparations for cell division are made during the life cycle subphase called (a) G_1, (b) G_2, (c) M, (d) S.

15. The RNA synthesized on one of the DNA strands is (a) mRNA, (b) tRNA, (c) rRNA, (d) all of these.

16. The RNA species that carries the coded message, specifying the sequence of amino acids in the protein to be made, from the nucleus to the cytoplasm is (a) mRNA, (b) tRNA, (c) rRNA, (d) all of these.

17. If DNA has a sequence of AAA, then a segment of mRNA synthesized on it will have a sequence of (a) TTT, (b) UUU, (c) GGG, (d) CCC.

18. A nerve cell and a lymphocyte are presumed to differ in their (a) specialized structure, (b) suppressed genes and embryonic history, (c) genetic information, (d) a and b, (e) a and c.

Short Answer Essay Questions

19. (a) Name the organelle that is the major site of ATP synthesis. (b) Name three organelles involved in protein synthesis or modification or both. (c) Name two organelles that contain enzymes, and describe their relative functions.

20. Explain why mitosis can be thought of as cellular immortality.

21. If a cell loses or ejects its nucleus, what is its fate and why?

22. The external faces of some of the proteins in the plasma membrane have carbohydrate groups attached to them. What role do such "sugar-coated" proteins play in the life of the cell?

23. Cells are living units. However, three classes of non-living substances are found outside the cells. What are these classes, and what functional roles do they play?

24. Comment on the role of the sodium-potassium pump in maintaining a cell's resting membrane potential.

25. Differentiate clearly between primary and secondary active transport processes.

Critical Thinking and Clinical Application Questions

1. Explain why limp celery becomes crisp and the skin of your fingertips wrinkles when placed in tap water. (The principle is exactly the same.)

2. A "red-hot" bacterial infection of the intestinal tract irritates the intestinal cells and interferes with digestion. Such a condition is often accompanied by diarrhea, which causes loss of body water. On the basis of what you have learned about osmotic water flows, explain why diarrhea may occur.

3. Two examples of chemotherapeutic drugs (drugs used to treat cancer) and their cellular actions are listed below. Explain why each drug could be fatal to a cell.

• Vincristine: Damages the mitotic spindle

• Adriamycin: Binds to DNA and blocks mRNA synthesis

4. The normal function of one tumor-suppressor gene is to prevent cells with damaged chromosomes and DNA from "progressing from G_1 to S," whereas another tumor-suppressor gene prevents "passage from G_2 to M." When these tumor suppressor genes fail to work, cancer can result. Explain what the phrases in quotations mean.

5. In their anatomy lab, many students are exposed to the chemical preservatives phenol, formaldehyde, and alcohol. Our cells break down these toxins very effectively. What cellular organelle is responsible for this?

6. Dynein is missing from the cilia and flagella of individuals with a specific inherited disorder. These individuals have severe respiratory problems and, if males, are sterile. What is the structural connection between these two symptoms?

7. Explain why alcoholics are likely to have much more smooth ER than teetotalers.

4 TISSUE: THE LIVING FABRIC

Amoebas and other unicellular (one-cell) organisms are rugged individualists. Each cell alone obtains and digests its food, ejects its wastes, and carries out all the other activities necessary to keep itself alive and "buzzin' around on all cylinders." But in the multicellular human body, cells do not operate independently. Instead, they form tight cell communities that live and work together.

Individual body cells are specialized, with each type performing specific functions that help maintain homeostasis and benefit the body as a whole. Cell specialization is obvious: How muscle cells look and act differs greatly from skin cells, which in turn are easy to distinguish from brain cells. Cell specialization allows the body to function in sophisticated ways, but division of labor has certain hazards. When a particular group of cells is indispensable, its loss or injury can severely disable or even destroy the body.

Groups of cells that are similar in structure and perform a common or related function are called **tissues** (*tissu* = woven). Four primary tissue types interweave to form the "fabric" of the body. These basic tissues are epithelial (ep"i-the'le-ul), connective, muscle, and nervous tissue, and each has numerous subclasses or varieties. If we had to assign a single term to each primary tissue type that would best describe its general role, the terms would most likely be *covering* (epithelial), *support* (connective), *movement* (muscle), and *control* (nervous). However, these terms reflect only a fraction of the functions that each tissue performs.

As explained in Chapter 1, tissues are organized into organs such as the kidneys and the heart. Most organs contain all four tissue types, and their arrangement determines the organ's structure and capabilities. The study of tissues, or **histology,** complements the study of gross anatomy. Together they provide the structural basis for understanding organ physiology.

EPITHELIAL TISSUE

Epithelial tissue, or an **epithelium** (plural: epithelia), is a sheet of cells that covers a body surface or lines a body cavity (*epithe* = laid on, covering). It occurs in the body as (1) *covering and lining epithelium* and (2) *glandular epithelium.* Covering and lining epithelium forms the outer layer of the skin, dips into and lines the open cavities of the cardiovascular, digestive, and respiratory systems, and covers the walls and organs of the closed ventral body cavity. Glandular epithelium fashions the glands of the body.

Epithelia form boundaries between different environments. For example, the epidermis of the skin lies between the inside and the outside of the body, and epithelium lining the urinary bladder separates underlying cells of the bladder wall from urine. Furthermore, nearly all substances received or given off by the body must pass through an epithelium.

In its role as an interface tissue, epithelium accomplishes many functions, including (1) protection, (2) absorption, (3) filtration, (4) excretion, (5) secretion, and (6) sensory reception. Each of these functions is described in detail later, but to illustrate these functions briefly: The epithelium of the skin protects underlying tissues from mechanical and chemical injury and bacterial invasion and contains nerve endings that respond to various stimuli acting at the skin surface (pressure, heat, etc.). The epithelium lining the digestive tract is specialized to absorb substances; and that found in the kidneys performs nearly the whole functional "menu"—excretion, absorption, secretion, and filtration. Secretion is the specialty of glands.

Special Characteristics of Epithelium

Epithelial tissues have many characteristics that distinguish them from other tissue types.

1. Cellularity. Epithelial tissue is composed almost entirely of close-packed cells. Only a tiny amount of extracellular material lies in the narrow spaces between them.

2. Specialized contacts. Epithelial cells fit close together to form continuous sheets. Adjacent cells are bound together at many points by lateral contacts, including *tight junctions* and *desmosomes* (see Chapter 3, p. 70).

3. Polarity. All epithelia have an **apical surface,** a free surface exposed to the body exterior or the cavity of an internal organ, and an attached **basal surface.** All epithelia exhibit *polarity,* meaning that cell regions near the apical surface differ from those near the basal surface in both structure and function.

Although some apical surfaces are smooth and slick, most have **microvilli,** fingerlike extensions of the plasma membrane. Microvilli tremendously increase the exposed surface area, and in epithelia that absorb or secrete substances (those lining the intestine or kidney tubules, for instance), the microvilli are often so dense that the cell apices have a fuzzy appearance called a *brush border.* Some epithelia, such as that lining the trachea, have motile **cilia** that propel substances along their free surface.

Lying adjacent to the basal surface of an epithelium is a thin supporting sheet called the **basal lamina** (lam'ĭ-nah; "sheet"). This noncellular, adhesive sheet consists largely of glycoproteins secreted by the epithelial cells. The basal lamina acts as a selective filter that determines which molecules diffusing

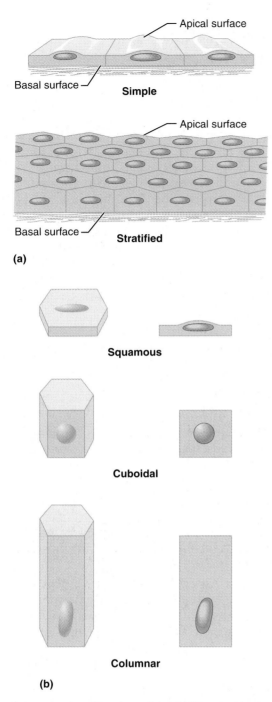

(a)

(b)

FIGURE 4.1 *Classification of epithelia.* (**a**) Classification on the basis of number of layers. (**b**) Classification on the basis of cell shape. For each category, a whole cell is shown on the left and a longitudinal section is shown on the right.

from the underlying connective tissue will be allowed to enter the epithelium. The basal lamina also acts as a scaffolding along which epithelial cells can migrate to repair a wound.

4. Supported by connective tissue. All epithelial sheets rest upon and are supported by connective tissue. Just deep to the basal lamina is the **reticular lamina,** a layer of extracellular material containing a fine network of collagen protein fibers that "belongs to" the underlying connective tissue. Together the two laminae form the **basement membrane.** The basement membrane reinforces the epithelial sheet, helping it to resist stretching and tearing forces, and defines the epithelial boundary.

5. Avascular but innervated. Although epithelium is *innervated* (supplied by nerve fibers), it is *avascular* (contains no blood vessels). Epithelial cells are nourished by substances diffusing from blood vessels in the underlying connective tissue.

6. Regeneration. Epithelium has a high regenerative capacity. Some epithelia are exposed to friction and their surface cells are removed by abrasion. Others are damaged by hostile substances in the external environment (bacteria, acids, smoke). As long as epithelial cells receive adequate nutrition, they can replace lost cells rapidly by cell division.

⚠ *Homeostatic Imbalance*

An important characteristic of cancerous epithelial cells is their failure to respect this boundary, which they penetrate to invade the tissues beneath. ∎

Classification of Epithelia

Each epithelium is given two names. The first name indicates the number of cell layers present; the second describes the shape of its cells (Figure 4.1). Based on the number of cell layers, there are simple and stratified epithelia. **Simple epithelia** are composed of a single cell layer. They are typically found where absorption and filtration occur and a thin epithelial barrier is desirable. **Stratified epithelia,** consisting of two or more cell layers stacked one on top of the other, are common in high-abrasion areas where protection is important, such as the skin surface and the lining of the mouth.

In cross section, all epithelial cells have six (somewhat irregular) sides, and an apical surface view of an epithelial sheet looks like a honeycomb. This polyhedral shape allows the cells to be closely packed. However, epithelial cells vary in height, and on that basis, there are three common shapes of epithelial cells. **Squamous** (skwa′mus) **cells** are flattened and scalelike (*squam* = scale); **cuboidal** (ku-boi′dahl) **cells** are boxlike, approximately as tall as they are wide; and **columnar** (kŏ-lum′nar) **cells** are tall and column shaped. In each case, the shape of the nucleus conforms to that of the cell. The nucleus of a squamous cell is disc shaped; that of a cuboidal cell is spherical; and a columnar cell nucleus is elongated from top to bottom and is usually located close to the cell base. Keep nuclear shape in mind when you attempt to identify epithelial types.

Simple epithelia are easy to classify by cell shape because all cells in the layer usually have the same shape. In stratified epithelia, however, the cell shapes usually differ among the different cell layers. To avoid ambiguity, stratified epithelia are named according to the shape of the cells in the *apical* layer. This naming system will become clearer as we explore the specific epithelial types.

As you read about the epithelial classes, study their illustrations in Figure 4.2. Using the photomicrographs, try to pick out the individual cells within each epithelium. This is not always easy, because the boundaries between epithelial cells often are indistinct. Furthermore, the nucleus of a particular cell may or may not be visible, depending on the precise plane of the cut made to prepare the tissue slides.

Simple Epithelia

The simple epithelia are most concerned with absorption, secretion, and filtration. Because they consist of a single cell layer and are usually very thin, protection is not one of their "specialties."

Simple Squamous Epithelium The cells of a **simple squamous epithelium** are flattened laterally, and their cytoplasm is sparse (Figure 4.2a). In a surface view, the close-fitting cells resemble a tiled floor. When the cells are cut perpendicular to their free surface, they resemble fried eggs seen from the side, with their cytoplasm wisping out from the slightly bulging nucleus. Thin and often permeable, this epithelium is found where filtration or the exchange of substances by rapid diffusion is a priority. In the kidneys, simple squamous epithelium forms part of the filtration membrane; in the lungs, it forms the walls of the air sacs across which gas exchange occurs.

Two simple squamous epithelia in the body have special names that reflect their location. **Endothelium** (en"do-the′le-um; "inner covering") provides a slick, friction-reducing lining in lymphatic vessels and in all hollow organs of the cardiovascular system—blood vessels and the heart. Capillaries consist exclusively of endothelium, where its exceptional thinness encourages the efficient exchange of nutrients and wastes between the bloodstream and surrounding tissue cells. **Mesothelium** (mez"o-the′le-um; "middle covering") is the epithelium found in serous membranes lining the ventral body cavity and covering its organs. (The serous membranes are described in more detail on p. 137.)

(a) Simple squamous epithelium

Description: Single layer of flattened cells with disc-shaped central nuclei and sparse cytoplasm; the simplest of the epithelia.

Function: Allows passage of materials by diffusion and filtration in sites where protection is not important; secretes lubricating substances in serosae.

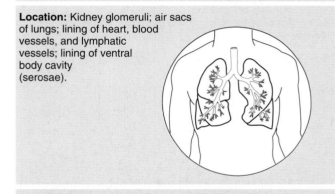

Location: Kidney glomeruli; air sacs of lungs; lining of heart, blood vessels, and lymphatic vessels; lining of ventral body cavity (serosae).

Photomicrograph: Simple squamous epithelium forming part of the alveolar (air sac) walls (240×).

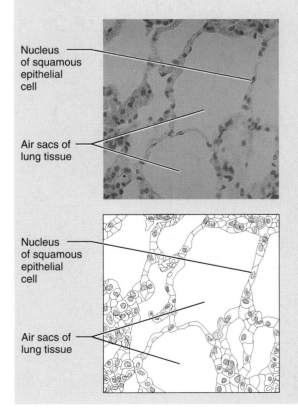

Nucleus of squamous epithelial cell

Air sacs of lung tissue

Nucleus of squamous epithelial cell

Air sacs of lung tissue

FIGURE 4.2 ***Epithelial tissues.*** Simple epithelium **(a)**.

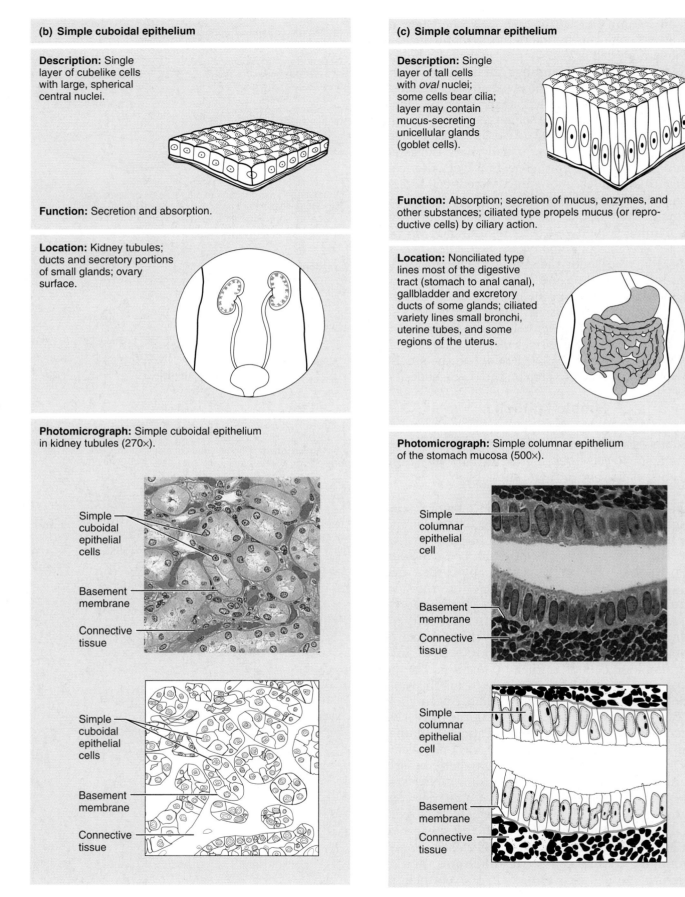

(b) Simple cuboidal epithelium

Description: Single layer of cubelike cells with large, spherical central nuclei.

Function: Secretion and absorption.

Location: Kidney tubules; ducts and secretory portions of small glands; ovary surface.

Photomicrograph: Simple cuboidal epithelium in kidney tubules (270×).

Simple cuboidal epithelial cells

Basement membrane

Connective tissue

Simple cuboidal epithelial cells

Basement membrane

Connective tissue

(c) Simple columnar epithelium

Description: Single layer of tall cells with *oval* nuclei; some cells bear cilia; layer may contain mucus-secreting unicellular glands (goblet cells).

Function: Absorption; secretion of mucus, enzymes, and other substances; ciliated type propels mucus (or reproductive cells) by ciliary action.

Location: Nonciliated type lines most of the digestive tract (stomach to anal canal), gallbladder and excretory ducts of some glands; ciliated variety lines small bronchi, uterine tubes, and some regions of the uterus.

Photomicrograph: Simple columnar epithelium of the stomach mucosa (500×).

Simple columnar epithelial cell

Basement membrane

Connective tissue

Simple columnar epithelial cell

Basement membrane

Connective tissue

FIGURE 4.2 (continued)
Simple epithelia (**b**, **c**, and **d**).

(d) Pseudostratified columnar epithelium

Description: Single layer of cells of differing heights, some not reaching the free surface; nuclei seen at different levels; may contain goblet cells and bear cilia.

Function: Secretion, particularly of mucus; propulsion of mucus by ciliary action.

Location: Nonciliated type in male's sperm–carrying ducts and ducts of large glands; ciliated variety lines the trachea, most of the upper respiratory tract.

Trachea

Photomicrograph: Pseudostratified ciliated columnar epithelium lining the human trachea (800×).

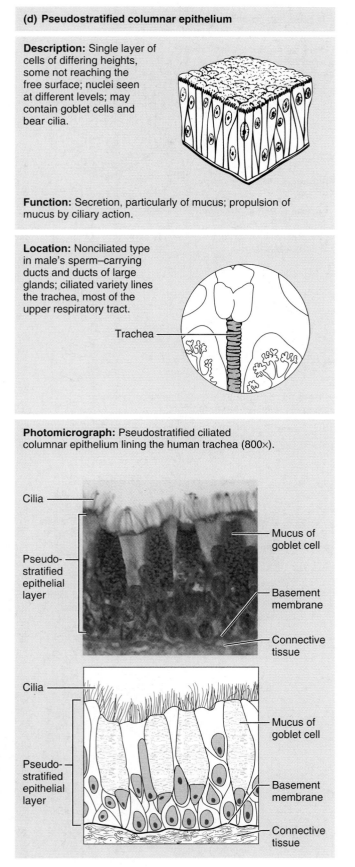

Cilia

Pseudo-stratified epithelial layer

Mucus of goblet cell

Basement membrane

Connective tissue

Cilia

Mucus of goblet cell

Pseudo-stratified epithelial layer

Basement membrane

Connective tissue

Simple Cuboidal Epithelium Simple cuboidal epithelium consists of a single layer of cells as tall as they are wide (Figure 4.2b). The spherical nuclei stain darkly, causing the layer to look like a string of beads when viewed microscopically. Important functions of simple cuboidal epithelium are secretion and absorption. A common location is kidney tubules. In glands, it forms both the secretory portions and some of the small ducts.

Simple Columnar Epithelium Simple columnar **epithelium** is seen as a single layer of tall, closely packed cells, aligned like soldiers in a row (Figure 4.2c). It lines the digestive tract from the stomach to the rectum. Columnar cells are mostly associated with absorption and secretion, and the digestive tract lining has two distinct modifications that reflect that dual function: (1) dense microvilli on the apical surface of absorptive cells and (2) **goblet cells** that secrete a protective lubricating mucus. The goblet cells are named for their goblet–shaped, "cups" of mucus that occupy most of the apical cell volume (see Figure 4.2d). Some simple columnar epithelia display cilia on their free surfaces.

Pseudostratified Columnar Epithelium The cells of **pseudostratified** (soo"do-strắ′tĭ-fīd) **columnar epithelium** vary in height (Figure 4.2d). All of its cells rest on the basement membrane, but only the tallest reach the free surface of the epithelium. The cell nuclei lie at different levels above the basement membrane, giving the false (pseudo) impression that several cell layers are present, hence "pseudostratified." This epithelium, like the simple columnar variety, secretes or absorbs substances. A ciliated version containing goblet cells lines most of the respiratory tract. Here the motile cilia propel sheets of dust-trapping mucus superiorly away from the lungs.

(e) Stratified squamous epithelium

Description: Thick membrane composed of several cell layers; basal cells are cuboidal or columnar and metabolically active; surface cells are flattened (squamous); in the keratinized type, the surface cells are full of keratin and dead; basal cells are active in mitosis and produce the cells of the more superficial layers.

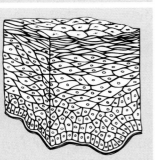

Function: Protects underlying tissues in areas subjected to abrasion.

Location: Nonkeratinized type forms the moist linings of the esophagus, mouth, and vagina; keratinized variety forms the epidermis of the skin, a dry membrane.

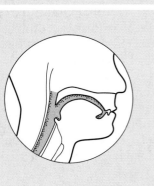

Photomicrograph: Stratified squamous epithelium lining of the esophagus (151×).

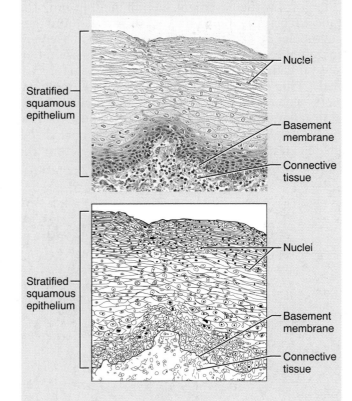

(f) Stratified columnar epithelium

Description: Several cell layers; basal cells usually cuboidal; superficial cells elongated and columnar.

Function: Protection; secretion.

Location: Rare in the body; small amounts in male urethra and in large ducts of some glands.

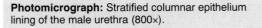

Uretha

Photomicrograph: Stratified columnar epithelium lining of the male urethra (800×).

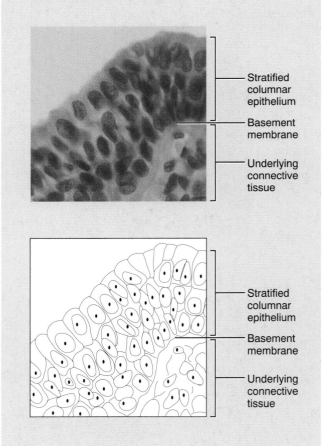

FIGURE 4.2 (continued)
Stratified epithelia (**e**, **f**, and **g**).

(g) Transitional epithelium

Description: Resembles both stratified squamous and stratified cuboidal; basal cells cuboidal or columnar; surface cells dome shaped or squamouslike, depending on degree of organ stretch.

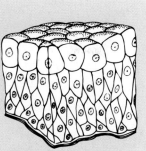

Function: Stretches readily and permits distension of urinary organ by contained urine.

Location: Lines the ureters, bladder, and part of the urethra.

Photomicrograph: Transitional epithelium lining of the bladder, relaxed state (265×); note the bulbous, or rounded, appearance of the cells at the surface; these cells flatten and become elongated when the bladder is filled with urine.

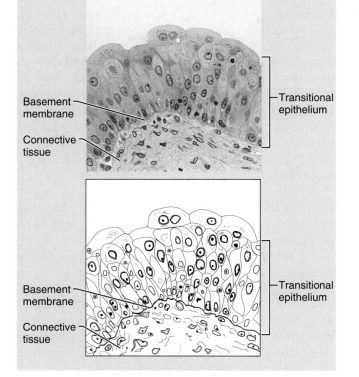

Basement membrane

Connective tissue

Transitional epithelium

Basement membrane

Connective tissue

Transitional epithelium

Stratified Epithelia

Stratified epithelia consist of two or more cell layers. They regenerate from below; that is, the basal cells divide and push apically to replace the older surface cells. Considerably more durable than the simple epithelia, the major (but not the only) role of stratified epithelia is protection.

Stratified Squamous Epithelium Stratified squamous epithelium is the most widespread of the stratified epithelia (Figure 4.2e). Composed of several layers, it is thick and well suited for its protective role in the body. Its free surface cells are squamous; cells of the deeper layers are cuboidal or columnar. This epithelium is found in areas subjected to wear and tear, and its surface cells are constantly being rubbed away and replaced by mitotic division of its basal cells. Since epithelium depends on diffusion of nutrients from a deeper connective tissue layer, the epithelial cells farther from the basement membrane are less viable and those at the apical surface are often flattened and atrophied.

To avoid memorizing all its locations simply remember that this epithelium forms the external part of the skin and extends a short distance into every body opening that is directly continuous with the skin. The outer layer, or *epidermis*, of the skin is *keratinized* (ker'ah-tin"ĭzd), meaning its surface cells contain *keratin*, a tough protective protein. (The epidermis is discussed in Chapter 5.) The other stratified squamous epithelia of the body are *nonkeratinized*.

Stratified Columnar Epithelia Stratified columnar epithelium is a rare type of tissue, forming little else than the large ducts of some glands. (Specific locations are mentioned in Figure 4.2f.)

Transitional Epithelium Transitional epithelium forms the lining of urinary organs, which are subjected to considerable stretching as they fill with urine (Figure 4.2g). Cells of its basal layer are cuboidal or columnar. The apical cells vary in appearance, depending on the degree of distension of the organ. When the organ is not stretched, the membrane is many-layered and the superficial cells are rounded and domelike. When the organ is distended with urine, the epithelium thins from about six cell layers to three, and its apical cells flatten and become squamouslike. The ability of transitional cells to change their shape allows a greater volume of urine to flow through a tubelike organ. In the bladder, it allows more urine to be stored.

Glandular Epithelia

A **gland** consists of one or more cells that make and secrete (export) a particular product. This product, called a **secretion,** is an aqueous (water-based) fluid that usually contains proteins, but there is variation—for example, some glands release a lipid- or steroid-rich secretion. Secretion is an active process; glandular cells obtain needed substances from the blood and transform them chemically into a product that is then discharged from the cell. Notice that the term *secretion* can refer to both the gland's *product* and the *process* of making and releasing that product.

Glands are classified based on

1. Site of product release. According to this criterion, there are *endocrine* ("internally secreting") *glands,* which release their product into the extracellular space, and *exocrine* ("externally secreting") *glands,* which send their product to an epithelial surface inside or outside the body.

2. Relative number of cells forming the gland: *unicellular* ("one-celled") versus *multicellular* ("many-celled"). Unicellular glands are scattered within epithelial sheets. By contrast, most multicellular epithelial glands form by invagination (inward growth) or evagination (outward growth) from an epithelial sheet and, at least initially, most have **ducts,** tubelike connections to the epithelial sheets.

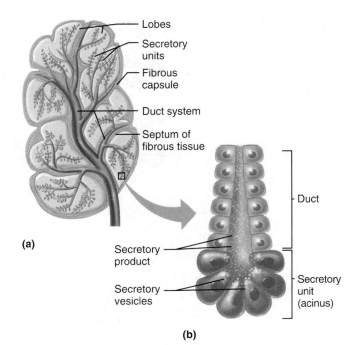

FIGURE 4.3   *General structure of a typical multicellular exocrine gland.* (a) The gland is enclosed by a fibrous capsule. Septa of fibrous tissue divide the gland into lobes. **(b)** A secretory unit and the associated portion of the duct system.

Endocrine Glands

Because **endocrine glands** eventually lose their ducts, they are often called **ductless glands.** They produce **hormones,** regulatory chemicals that they secrete by exocytosis directly into the extracellular space. From there the hormones enter the blood or lymphatic fluid and travel to specific target organs. Each hormone prompts its target organ(s) to respond in some characteristic way. For example, hormones produced by certain intestinal cells cause the pancreas to release enzymes that help digest food in the digestive tract.

Although most endocrine glands are compact multicellular organs, some individual hormone-producing cells are scattered in the digestive tract mucosa and in the brain, giving rise to their collective description as the *diffuse endocrine system.* Endocrine glands are structurally diverse, so one description does not fit all. Their secretions are also varied, ranging from modified amino acids to peptides and proteins, glycoproteins, and steroids. Since not all endocrine glands are epithelial derivatives, we defer consideration of their structure and function to Chapter 17.

Exocrine Glands

Exocrine glands are far more numerous than endocrine glands, and many of their products are familiar. All exocrine glands secrete their products onto body surfaces (skin) or into body cavities—the unicellular glands directly (by exocytosis) and the multicellular glands via an epithelial-walled duct that transports the secretion to the epithelial surface. Exocrine glands are a diverse lot. They include mucous, sweat, oil, and salivary glands, the liver (which secretes bile), the pancreas (which synthesizes digestive enzymes), and many others.

Unicellular Exocrine Glands The only important example of a **unicellular** (or one-celled) **gland** is the *goblet cell.* True to its name, a goblet cell is shaped like a goblet (a drinking glass with a stem). Goblet cells are sprinkled in the epithelial linings of the intestinal and respiratory tracts amid columnar cells with other functions (see Figure 4.2d). In humans, all such glands produce **mucin** (mu′sin), a complex glycoprotein that dissolves in water when secreted. Once dissolved, mucin forms **mucus,** a slimy coating that both protects and lubricates surfaces.

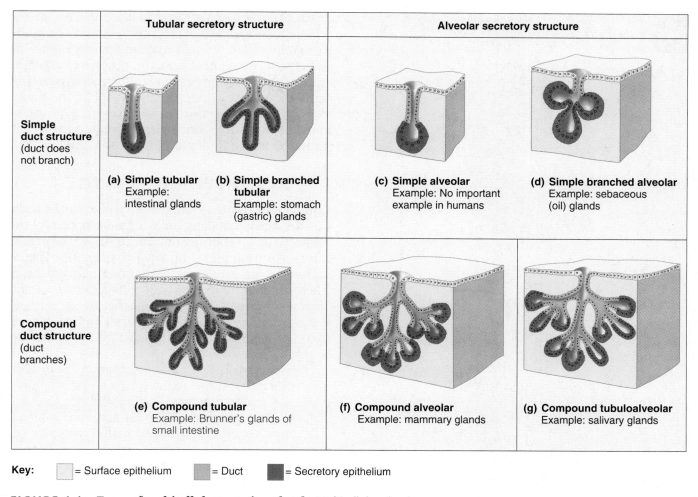

	Tubular secretory structure		Alveolar secretory structure	
Simple duct structure (duct does not branch)	**(a) Simple tubular** Example: intestinal glands	**(b) Simple branched tubular** Example: stomach (gastric) glands	**(c) Simple alveolar** Example: No important example in humans	**(d) Simple branched alveolar** Example: sebaceous (oil) glands
Compound duct structure (duct branches)	**(e) Compound tubular** Example: Brunner's glands of small intestine		**(f) Compound alveolar** Example: mammary glands	**(g) Compound tubuloalveolar** Example: salivary glands

Key: ☐ = Surface epithelium ▨ = Duct ■ = Secretory epithelium

FIGURE 4.4 *Types of multicellular exocrine glands.* Multicellular glands are classified according to duct type (simple or compound) and the structure of their secretory units (tubular, alveolar, or tubuloalveolar).

Multicellular Exocrine Glands Compared to the unicellular glands, the **multicellular exocrine glands** are structurally much more complex. They have two basic parts: an epithelium-derived *duct* and a *secretory unit* consisting of secretory cells (acini). In all but the simplest glands, *supportive connective tissue* surrounds the secretory unit and supplies it with blood vessels and nerve fibers. Often, the connective tissue forms a *fibrous capsule* that extends into the gland proper and divides the gland into *lobes* (Figure 4.3).

Structural Classification On the basis of their duct structures, multicellular exocrine glands are either simple or compound (Figure 4.4). **Simple glands** have a single unbranched duct, whereas **compound glands** have a branched duct. The glands are further categorized by their secretory units as (1) **tubular** if the secretory cells form tubes; (2) **alveolar** (al-ve′o-lar) if the secretory cells form small, flasklike sacs (*alveolus* = small hollow cavity); and (3) **tubuloalveolar** if they contain both types of secretory units. Note that the term **acinar** (as′ĭ-nar; "berrylike") is used interchangeably with alveolar.

Modes of Secretion Multicellular exocrine glands have different modes of secretion (secrete their products in different ways), so they can also be described functionally. Most are **merocrine** (mer′o-krin) **glands,** which secrete their products by exocytosis as they are produced. The secretory cells are not altered in any way. The pancreas, most sweat glands, and salivary glands belong to this class (Figure 4.5a).

Which of the gland types illustrated would have the highest rate of cellular mitosis? Why?

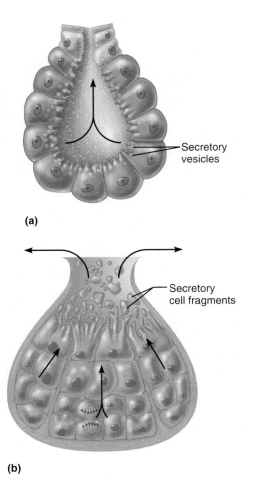

(a)

(b)

FIGURE 4.5   *Chief modes of secretion in human exocrine glands.* (a) Merocrine glands secrete their products by exocytosis. (b) In holocrine glands, the entire secretory cell ruptures, releasing secretions and dead cell fragments.

Secretory cells of **holocrine** (hol′o-krin) **glands** accumulate their products within them until they rupture. (They are replaced by the division of underlying cells.) Because holocrine gland secretions include the synthesized product plus dead cell fragments (*holos* = all), you could say that their cells "die for their cause." Sebaceous (oil) glands of the skin are the only true example of holocrine glands (Figure 4.5b).

Although *apocrine* (ap′o-krin) *glands* are definitely present in other animals, there is some controversy over whether humans have this third gland type. Like holocrine glands, apocrine glands accumulate their products, but in this case only just be-

The holocrine gland, because cells that are fragmented and secreted have to be replaced.

neath the free surface. Eventually, the apex of the cell pinches off (*apo* = from, off), releasing the secretory granules and a small amount of cytoplasm. The cell repairs its damage and the process is repeated again and again. The best possibility in humans concerns the release of free lipid droplets by the mammary glands, but most histologists classify mammary glands as merocrine glands because this is the means by which milk proteins are secreted.

CONNECTIVE TISSUE

Connective tissue is found everywhere in the body. It is the most abundant and widely distributed of the primary tissues, but its amount in particular organs varies. For example, skin consists primarily of connective tissue, while the brain contains very little. There are four main classes of connective tissue and several subclasses. The main classes are (1) *connective tissue proper* (which includes fat and the fibrous tissue of ligaments), (2) *cartilage*, (3) *bone*, and (4) *blood*.

Connective tissue does much more than *connect* body parts; it has many forms and many functions. Its major functions include (1) *binding and support*, (2) *protection*, (3) *insulation*, and as blood, (4) *transportation* of substances within the body. For example, bone and cartilage support and protect body organs by providing the hard underpinnings of the skeleton; fat cushions insulate and protect body organs and provide reserve energy fuel.

Common Characteristics of Connective Tissue

Despite their many and varied functions in the body, connective tissues have some common properties that set them apart from other primary tissues:

1. Common origin. All connective tissues arise from **mesenchyme** (an embryonic tissue) and hence have a common kinship (Figure 4.6).

2. Degrees of vascularity. Connective tissues run the entire gamut of vascularity. Cartilage is avascular; dense connective tissue is poorly vascularized; and the other types of connective tissue have a rich supply of blood vessels.

3. Extracellular matrix. Whereas all other primary tissues are composed mainly of cells, connective tissues are largely nonliving **extracellular matrix** (ma′triks; "womb"), which separates, often widely, the living cells of the tissue. Because of this matrix, connective tissue is able to bear weight, withstand great tension, and endure abuses, such as physical trauma and abrasion, that no other tissue would be able to tolerate.

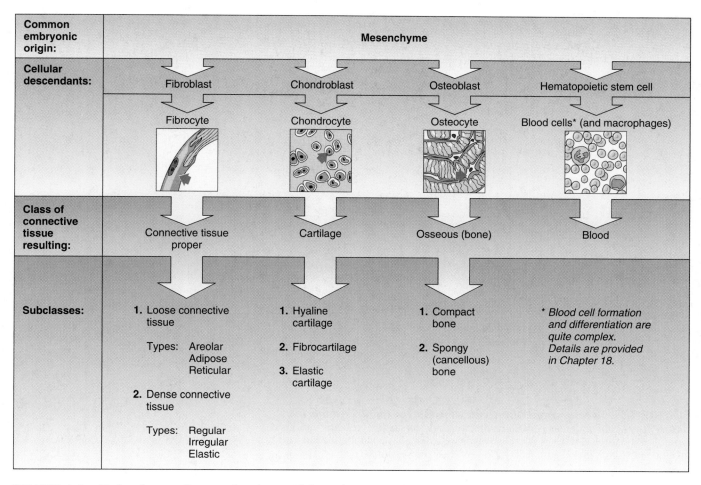

Common embryonic origin:	Mesenchyme			
Cellular descendants:	Fibroblast	Chondroblast	Osteoblast	Hematopoietic stem cell
	Fibrocyte	Chondrocyte	Osteocyte	Blood cells* (and macrophages)
Class of connective tissue resulting:	Connective tissue proper	Cartilage	Osseous (bone)	Blood
Subclasses:	1. Loose connective tissue Types: Areolar Adipose Reticular 2. Dense connective tissue Types: Regular Irregular Elastic	1. Hyaline cartilage 2. Fibrocartilage 3. Elastic cartilage	1. Compact bone 2. Spongy (cancellous) bone	* Blood cell formation and differentiation are quite complex. Details are provided in Chapter 18.

FIGURE 4.6 *Major classes of connective tissue.* All these classes arise from the same embryonic tissue type (mesenchyme).

Structural Elements of Connective Tissue

Connective tissues have three main elements: *ground substance, fibers,* and *cells.* Ground substance and fibers make up the extracellular matrix. (Note that some authors use the term *matrix* to indicate the ground substance only.)

The properties of the cells and the composition and arrangement of extracellular matrix elements vary tremendously. The result is an amazing diversity of connective tissues, each uniquely adapted to perform its specific function in the body. For example, the matrix can be delicate and fragile to form a soft "packing" around an organ, or it can form "ropes" (tendons and ligaments) of incredible strength. Nonetheless, connective tissues have a common structural plan, and we use *areolar* (ah-re'o-lar) *connective tissue* as our *prototype,* or model for this group of tissues (Figure 4.7). All other subclasses are simply variants of this common tissue type.

Ground Substance

Ground substance is the unstructured material that fills the space between the cells and contains the fibers. It is composed of *interstitial (tissue) fluid, cell adhesion proteins,* and *proteoglycans* (pro"te-o-gli'kanz). Cell adhesion proteins *(fibronectin, laminin, and others)* serve mainly as a connective tissue glue that allows connective tissue cells to attach themselves to matrix elements. The proteoglycans consist of a protein core to which *glycosaminoglycans* (gli"kos-ah-me"no-gli'kanz) (GAGs) are attached. The strandlike GAGs are large, negatively charged polysaccharides that stick out from the core protein like the fibers of a bottle brush. Important examples of GAGs in connective tissues include *chondroitin sulfates* and *hyaluronic* (hi"ah-lu-ron'ik) *acid.* The GAGs intertwine and trap water, forming a substance that varies from a fluid to a viscous gel. In general, the higher its GAG content, the stiffer the ground substance is.

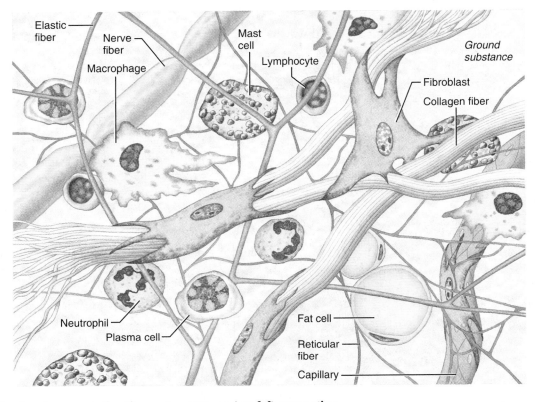

Elastic fiber
Nerve fiber
Macrophage
Mast cell
Lymphocyte
Ground substance
Fibroblast
Collagen fiber
Neutrophil
Plasma cell
Fat cell
Reticular fiber
Capillary

FIGURE 4.7 *Areolar connective tissue: A prototype (model) connective tissue.* This tissue underlies epithelia and surrounds capillaries. Note the various cell types and the three classes of fibers (collagen, reticular, elastic) embedded in the ground substance. (See Figure 4.8b for a less idealized version.)

The ground substance holds large amounts of fluid and functions as a molecular sieve, or medium, through which nutrients and other dissolved substances can diffuse between the blood capillaries and the cells. The fibers embedded in the ground substance make it less pliable and impede diffusion somewhat.

Fibers

The *fibers* of connective tissue provide support. Three types of fibers are found in connective tissue matrix: collagen, elastic, and reticular fibers. Of these, collagen fibers are by far the strongest and most abundant.

Collagen fibers are constructed primarily of the fibrous protein *collagen*. Collagen molecules are secreted into the extracellular space, where they are assembled spontaneously into cross-linked fibers, which, in turn, are bundled together into the thick collagen fibers seen with a microscope. Collagen fibers are extremely tough and provide high tensile strength (that is, the ability to resist longitudinal stress) to the matrix. Indeed, stress tests show that collagen fibers are stronger than steel fibers of the same size! When fresh, they have a glistening white appearance; they are therefore also called *white fibers*.

Elastic fibers are long, thin fibers that form branching networks in the extracellular matrix. These fibers contain a resilient rubberlike protein, *elastin*, that allows them to stretch and recoil like rubber bands. Connective tissue can stretch only so much before its thick, ropelike collagen fibers become taut. Then, when the tension lets up, elastic fibers snap the connective tissue back to its normal length and shape. Elastic fibers are found where greater elasticity is needed, for example, in the skin, lungs, and blood vessel walls. Since fresh elastic fibers appear yellow, they are sometimes called *yellow fibers*.

Reticular fibers are fine collagenous fibers (with a slightly different chemistry and form) and are continuous with collagen fibers. They branch extensively, forming delicate networks (*reticul* = network) that surround small blood vessels and support the soft tissue of organs. They are particularly abundant where connective tissue abuts other tissue types, for example, in the basement membrane of epithelial tissues, and around capillaries, where they form fuzzy "nets."

Cells

Each major class of connective tissue has a fundamental cell type that exists in immature and mature forms (see Figure 4.6). The undifferentiated cells, typically indicated by the suffix *blast* (literally, "bud," or "sprout," but meaning "forming"), are actively mitotic cells that secrete the ground substance and the fibers characteristic of their particular matrix. The primary blast cell types by connective tissue class are (1) connective tissue proper: **fibroblast;** (2) cartilage: **chondroblast** (kon′dro-blast); (3) bone: **osteoblast** (os′te-o-blast″); and (4) blood: **hematopoietic stem cell.**

Once they synthesize the matrix, the blast cells assume their less active, mature mode, indicated by the suffix *cyte* (see Figure 4.6). The mature cells maintain the health of the matrix. However, if the matrix is injured, they can easily revert to their more active state to repair and regenerate the matrix. (The blood-forming hematopoietic stem cell found in bone marrow is always actively mitotic.)

Additionally, connective tissue is home to an assortment of other cell types, such as nutrient-storing *fat cells* and mobile cells that migrate into the connective tissue matrix from the bloodstream. The latter include defensive **white blood cells** (neutrophils, eosinophils, lymphocytes) and other cell types concerned with tissue response to injury, such as *mast cells, macrophages* (mak′ro-fāj-es), and antibody-producing **plasma cells.** This wide variety of cells is particularly obvious in our prototype, areolar connective tissue (Figure 4.7).

All of these accessory cell types are described in later chapters, but mast cells and macrophages are so important to overall body defense that they deserve a brief mention here. The oval **mast cells** typically cluster along blood vessels. These cells act as sensitive sentinels to detect foreign substances (e.g., bacteria, fungi) and initiate local inflammatory responses against them. In the mast cell cytoplasm are conspicuous secretory granules containing chemicals that mediate inflammation, especially in severe allergies. These include (1) *heparin* (hep′ah-rin), an anticoagulant chemical that prevents blood clotting when free in the bloodstream (but its significance in human mast cells is uncertain); (2) *histamine* (his′tah-mēn), a substance that makes capillaries leaky; and *proteases* (protein-degrading enzymes). (The inflammatory process is discussed in Chapter 22.)

Macrophages (*macro* = large; *phago* = eat) are large, irregularly shaped cells that avidly phagocytize a broad variety of foreign materials, ranging from foreign molecules to entire bacteria to dust particles. These "big eaters" also dispose of dead tissue cells, and they are central actors in the immune system.

In connective tissues, they may be attached to connective tissue fibers (fixed) or may migrate freely through the matrix.

Macrophages are peppered throughout loose connective tissue, bone marrow, and lymphatic tissue. Those in certain sites are given specific names; for example, those in loose connective tissue are called *histiocytes* (his′te-o-sītz″). Some macrophages have selective appetites; for example, macrophages of the spleen primarily dispose of aging red blood cells, but they will not turn down other "delicacies" that come their way.

Types of Connective Tissue

As noted, all classes of connective tissue consist of living cells surrounded by a matrix. Their major differences reflect cell type and fiber type and relative amount. The connective tissues described in this section are illustrated in Figure 4.8. Mature connective tissues arise from a common embryonic tissue, which we describe here as well.

Embryonic Connective Tissue: Mesenchyme

Mesenchyme (mes′en-kīm) is the first definitive tissue formed from the mesoderm germ layer (Figure 4.13, p. 144). Mesenchyme is composed of star-shaped mesenchymal cells and a fluid ground substance containing fine fibrils (Figure 4.8a). It arises during the early weeks of embryonic development and eventually differentiates (specializes) into all other connective tissues. However, some mesenchymal cells remain and provide a source of new cells in mature connective tissues.

Mucous connective tissue is a temporary tissue derived from mesenchyme and similar to it. *Wharton's jelly,* which supports the umbilical cord of the fetus, is the best example of this scant embryonic tissue.

Connective Tissue Proper

Connective tissue proper has two subclasses: the **loose connective tissues** (areolar, adipose, and reticular) and **dense connective tissues** (dense regular, dense irregular, and elastic). Except for bone, cartilage, and blood, all mature connective tissues belong to this class.

Areolar Connective Tissue Areolar connective tissue, as you may have guessed, is pretty special. Its functions, shared by some but not all connective tissues, include (1) supporting and binding other tissues (the job of the fibers); (2) holding body fluids (the ground substance's role); (3) defending against infection (via the activity of white blood cells and

Embryonic connective tissue

(a) Mesenchyme

Description: Embryonic connective tissue; gel-like ground substance containing fibers; star-shaped mesenchymal cells.

Function: Gives rise to all other connective tissue types.

Location: Primarily in embryo.

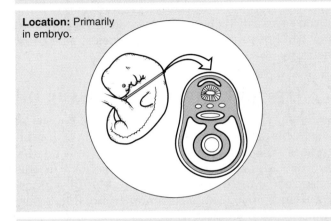

Photomicrograph: Mesenchymal tissue, an embryonic connective tissue (523×); the clear-appearing background is the fluid ground substance of the matrix; notice the fine, sparse fibers.

Mesenchymal cell

Ground substance

Fibers

Mesenchymal cell

Ground substance

Fibers

Connective tissue proper: Loose connective tissue (b to d)

(b) Areolar connective tissue

Description: Gel-like matrix with all three fiber types; cells: fibroblasts, macrophages, mast cells, and some white blood cells.

Function: Wraps and cushions organs; its macrophages phagocytize bacteria; plays important role in inflammation; holds and conveys tissue fluid.

Location: Widely distributed under epithelia of body, e.g, forms lamina propria of mucous membranes; packages organs; surrounds capillaries.

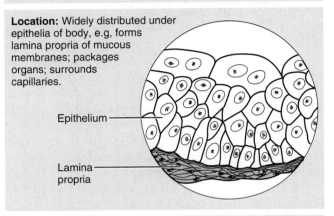

Epithelium

Lamina propria

Photomicrograph: Areolar connective tissue, a soft packaging tissue of the body (200×).

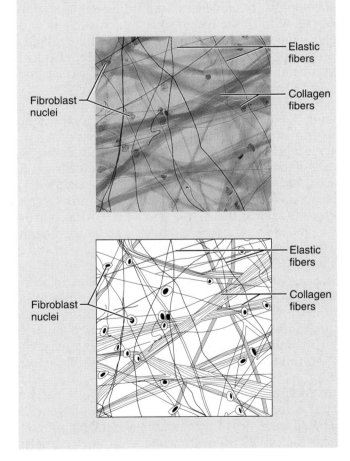

Fibroblast nuclei

Elastic fibers

Collagen fibers

Fibroblast nuclei

Elastic fibers

Collagen fibers

FIGURE 4.8   *Connective tissues.* Embryonic connective tissue (**a**) and connective tissue proper (**b** and **c**).

Loose connective tissue (continued)

(c) Adipose tissue

Description: Matrix as in areolar, but very sparse; closely packed adipocytes, or fat cells, have nucleus pushed to the side by large fat droplet.

Function: Provides reserve food fuel; insulates against heat loss; supports and protects organs.

Location: Under skin; around kidneys and eyeballs; within abdomen; in breasts.

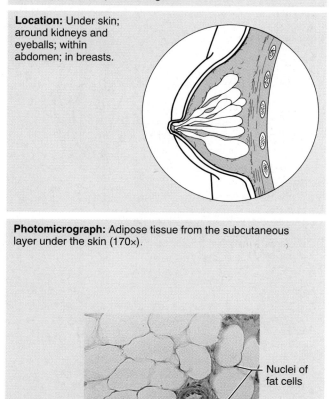

Photomicrograph: Adipose tissue from the subcutaneous layer under the skin (170×).

Nuclei of fat cells

Blood vessel

Vacuole containing fat droplet

Nuclei of fat cells

Blood vessel

Vacuole containing fat droplet

macrophages); (4) storing nutrients as fat (in fat cells). Its semifluid or gelatinous ground substance is formed primarily of hyaluronic acid in which all three fiber types are loosely dispersed (Figure 4.8b). **Fibroblasts,** flat, branching cells that appear spindle shaped in profile, predominate, but numerous macrophages are also seen and present a formidable barrier to invading microorganisms. Fat cells appear singly or in small clusters, and occasional mast cells are identified easily by the large, darkly stained cytoplasmic granules that often obscure their nuclei. Other cell types are scattered throughout.

The most obvious structural feature of this tissue is the loose *arrangement* of its fibers; hence, its classification as a *loose* connective tissue is fitting. The rest of the matrix, occupied by ground substance, appears to be empty space when viewed through the microscope; in fact, the Latin term *areola* means "a small open space." Because of its loose nature, areolar connective tissue provides a reservoir of water and salts for surrounding body tissues, always holding approximately as much fluid as there is in the entire bloodstream. Essentially all body cells obtain their nutrients from and release their wastes into this "tissue fluid." However, its high content of hyaluronic acid makes the ground substance quite viscous, like molasses, which may hinder the movement of cells through it. Some white blood cells, which protect the body from disease-causing microorganisms, secrete the enzyme hyaluronidase, to liquefy the ground substance and ease their passage. (Unhappily, some potentially harmful bacteria have the same ability.) When a body region is inflamed, the areolar tissue in the area soaks up excess fluids like a sponge, and the affected area swells and becomes puffy, a condition called **edema** (ĕ-de′mah).

Areolar connective tissue is the most widely distributed connective tissue in the body and it serves as a kind of universal packing material between other tissues. It binds body parts together while allowing them to move freely over one another; wraps small blood vessels and nerves; surrounds glands; and forms the subcutaneous tissue, which cushions and attaches the skin to underlying structures. It is present in all mucous membranes as the *lamina propria.* (As described on p. 137, mucous membranes line body cavities open to the exterior.)

Adipose (Fat) Tissue Adipose (ad′ĭ-pōs) **tissue** is similar to areolar tissue in structure and function but its nutrient-storing ability is much greater. Consequently, **adipocytes** (ad′ĭ-po-sītz), commonly called *adipose* or *fat cells,* predominate and account for 90% of this tissue's mass. The matrix is scanty and the cells are packed closely together, giving a chicken wire appearance to the tissue. A glistening oil droplet (almost pure neutral fat) occupies most of a fat cell's volume and displaces the nucleus to one side so that only a thin rim of surrounding cytoplasm is seen (Figure 4.8c). Mature adipocytes are among the largest cells in the body and cannot divide. As they take up or release fat, they become plumper or more wrinkled looking, respectively.

Adipose tissue is richly vascularized, indicating its high metabolic activity. Without the fat stores in our adipose tissue, we could not live for more than a few days without eating. Adipose tissue is certainly abundant: It constitutes 18% of an average person's body weight. Indeed, a chubby person's body can be 50% fat without being morbidly obese.

Adipose tissue may develop almost anywhere areolar tissue is plentiful, but it usually accumulates in subcutaneous tissue, where it also acts as a shock absorber and as insulation. Because fat is a poor conductor of heat, it helps prevent heat loss from the body. Other sites where fat accumulates include surrounding the kidneys, behind the eyeballs, and at genetically determined fat depots such as the abdomen and hips.

The adipose tissue just described is sometimes called *white fat*, or *white adipose tissue*, to distinguish it from **brown fat** or **brown adipose tissue.** Whereas white fat stores nutrients, brown fat consumes its nutrient stores to generate heat to warm the body. The richly vascular brown fat occurs only in babies who (as yet) lack the ability to produce body heat by shivering. Most such deposits are located between the shoulder blades, on the anterolateral neck, and on the anterior abdominal wall.

Reticular Connective Tissue

Reticular connective tissue **Reticular connective tissue** resembles areolar connective tissue, but the only fibers in its matrix are reticular fibers. The interwoven reticular fibers form a delicate network along which fibroblasts called **reticular cells** lie scattered (Figure 4.8d). Although reticular fibers are widely distributed in the body, reticular tissue is limited to certain sites. It forms a labyrinthlike **stroma,** (literally, "bed" or "mattress"), or internal framework, that can support many free blood cells (largely lymphocytes) in lymph nodes, the spleen, and bone marrow.

Dense Regular Connective Tissue

Dense regular connective tissue **Dense regular connective tissue** (Figure 4.8e) is one variety of the dense connective tissues, all of which have fibers as their predominant element. For this reason, the dense connective tissues are often referred to as **fibrous connective tissues.**

Dense regular connective tissue contains closely packed bundles of collagen fibers running in the same direction, parallel to the direction of pull. This results in white, flexible structures with great resistance to tension (pulling forces) where the tension is

---

Loose connective tissue (*continued*)

(d) Reticular connective tissue

Description: Network of reticular fibers in a typical loose ground substance; reticular cells lie on the network.

Function: Fibers form a soft internal skeleton (stroma) that supports other cell types.

Location: Lymphoid organs (lymph nodes, bone marrow, and spleen).

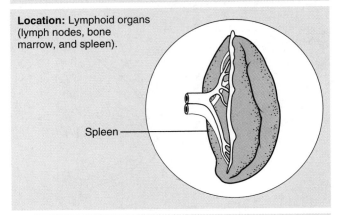

Spleen

Photomicrograph: Dark-staining network of reticular connective tissue fibers forming the internal skeleton of the spleen (650×).

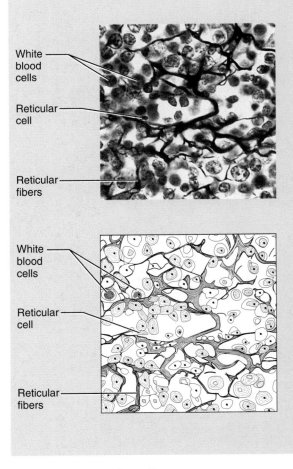

White blood cells

Reticular cell

Reticular fibers

White blood cells

Reticular cell

Reticular fibers

FIGURE 4.8 (continued)
Connective tissue proper (**d**, **e**, and **f**).

Connective tissue proper: Dense connective tissue (e and f)

(e) Dense regular connective tissue

Description: Primarily parallel collagen fibers; a few elastin fibers; major cell type is the fibroblast.

Function: Attaches muscles to bones or to muscles; attaches bones to bones; withstands great tensile stress when pulling force is applied in one direction.

Location: Tendons, most ligaments, aponeuroses.

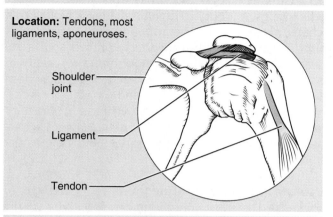

Shoulder joint

Ligament

Tendon

Photomicrograph: Dense regular connective tissue from a tendon (200×).

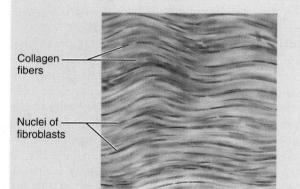

Collagen fibers

Nuclei of fibroblasts

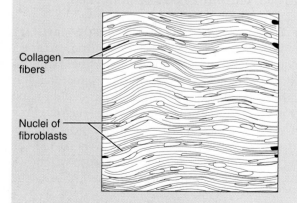

Collagen fibers

Nuclei of fibroblasts

Dense connective tissue *(continued)*

(f) Dense irregular connective tissue

Description: Primarily irregularly arranged collagen fibers; some elastic fibers; major cell type is the fibroblast.

Function: Able to withstand tension exerted in many directions; provides structural strength.

Location: Dermis of the skin; submucosa of digestive tract; fibrous capsules of organs and of joints.

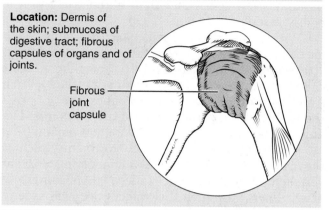
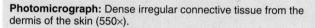

Fibrous joint capsule

Photomicrograph: Dense irregular connective tissue from the dermis of the skin (550×).

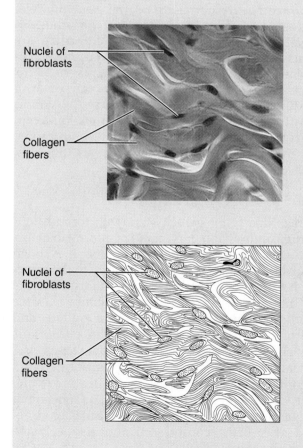

Nuclei of fibroblasts

Collagen fibers

Nuclei of fibroblasts

Collagen fibers

exerted in a single direction. Crowded between the collagen fibers are rows of fibroblasts that continuously manufacture fibers and a scant ground substance. As seen in Figure 4.8e, collagen fibers are slightly wavy. This allows the tissue to stretch a little, but once the fibers are straightened out by a pulling force, there is no further "give" to this tissue. Unlike our model (areolar) connective tissue, this tissue has few cells other than fibroblasts and it is poorly vascularized.

With its enormous tensile strength, dense regular connective tissue forms the *tendons,* cords that attach muscles to bones, and flat, sheetlike tendons called *aponeuroses* (ap"o-nu-ro'sēz) that attach muscles to other muscles or to bones. It also forms the *ligaments* that bind bones together at joints. Ligaments contain more elastic fibers than tendons and are slightly more stretchy. A few ligaments, such as the *ligamenta nuchae* and *flava* connecting adjacent vertebrae, are very elastic. Indeed, their content of elastic fibers is so high that the connective tissue in those structures is referred to as **elastic connective tissue.**

Dense Irregular Connective Tissue
Dense irregular connective tissue has the same structural elements as the regular variety. However, the bundles of collagen fibers are much thicker and they are arranged irregularly; that is, they run in more than one plane (Figure 4.8f). This type of tissue forms sheets in body areas where tension is exerted from many different directions. It is found in the skin as the leathery *dermis,* and it forms fibrous joint capsules and the fibrous coverings that surround some organs (testes, kidneys, bones, cartilages, muscles, and nerves).

Cartilage

Cartilage (kar'tĭ-lij), which stands up to both tension *and* compression, has qualities intermediate between dense connective tissue and bone. It is tough but flexible, providing a resilient rigidity to the structures it supports. Cartilage is avascular and lacks nerve fibers. Its ground substance contains large amounts of the GAGs chondroitin sulfate and hyaluronic acid. The ground substance contains firmly bound collagen fibers and in some cases elastic fibers, and is usually quite firm. Cartilage matrix also contains an exceptional amount of tissue fluid; in fact, cartilage is up to 80% water! The movement of tissue fluid in its matrix enables cartilage to rebound after being compressed and also helps to nourish the cartilage cells.

Chondroblasts, the predominant cell type in growing cartilage, produce new matrix until the skeleton stops growing at the end of adolescence.

The firm cartilage matrix prevents the cells from becoming widely separated, so **chondrocytes,** or mature cartilage cells, are typically found in small groups within cavities called **lacunae** (lah-ku'ne; "pits").

⧫ *Homeostatic Imbalance*

Because cartilage is avascular and aging cartilage cells lose their ability to divide, cartilages heal slowly when injured. This phenomenon is excruciatingly familiar to those who have experienced sports injuries. During later life, cartilages tend to calcify or even ossify (become bony). In such cases, the chondrocytes are poorly nourished and die. ■

There are three varieties of cartilage: *hyaline cartilage, elastic cartilage,* and *fibrocartilage.* Each of these is dominated by a particular fiber type as described below.

Hyaline Cartilage Hyaline (hi'ah-lī n) **cartilage,** or *gristle,* is the most abundant cartilage type in the body. Although it contains large numbers of collagen fibers, they are not apparent and the matrix appears amorphous and glassy (*hyalin* = glass) blue-white when viewed by the unaided eye (Figure 4.8g). Chondrocytes account for only 1–10% of the cartilage volume.

Hyaline cartilage provides firm support with some pliability. It covers the ends of long bones as *articular cartilage,* providing springy pads that absorb compression at joints. Hyaline cartilage also supports the tip of the nose, connects the ribs to the sternum, and forms most of the respiratory system passages (see Figure 6.1, p. 174). Most of the embryonic skeleton is formed of hyaline cartilage before bone is formed. Hyaline cartilage persists during childhood as the *epiphyseal* (e"pĭ-fis'e-ul) *plates,* actively growing regions near the end of long bones that provide for continued growth in length.

Elastic Cartilage Histologically, **elastic cartilage** (Figure 4.8h) is nearly identical to hyaline cartilage. However, there are many more elastin fibers in elastic cartilage, which gives this tissue a greater tolerance for repeated bending. Found where strength and exceptional stretchability are needed, elastic cartilage forms the "skeletons" of the external ear and the epiglottis. (The epiglottis is the flap that covers the opening to the respiratory passageway when we swallow, preventing food or fluids from entering the lungs.)

Fibrocartilage **Fibrocartilage** is often found where hyaline cartilage meets a true ligament or a tendon. Fibrocartilage is a perfect structural intermediate

Cartilage: (g to i)

(g) Hyaline cartilage

Description: Amorphous but firm matrix; collagen fibers form an imperceptible network; chondroblasts produce the matrix and when mature (chondrocytes) lie in lacunae.

Function: Supports and reinforces; has resilient cushioning properties; resists compressive stress.

Location: Forms most of the embryonic skeleton; covers the ends of long bones in joint cavities; forms costal cartilages of the ribs; cartilages of the nose, trachea, and larynx.

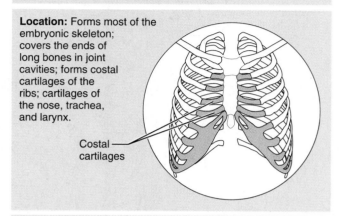

Costal cartilages

Photomicrograph: Hyaline cartilage from the trachea (375×).

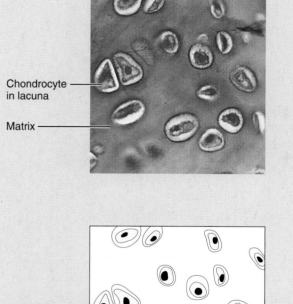

Chondrocyte in lacuna

Matrix

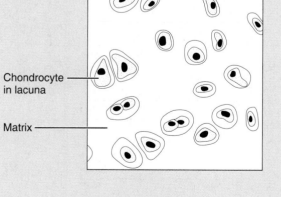

Chondrocyte in lacuna

Matrix

Cartilage (continued)

(h) Elastic cartilage

Description: Similar to hyaline cartilage, but more elastic fibers in matrix.

Function: Maintains the shape of a structure while allowing great flexibility.

Location: Supports the external ear (pinna); epiglottis.

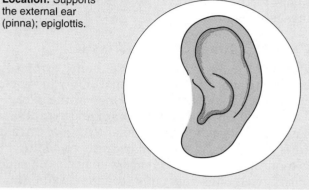

Photomicrograph: Elastic cartilage from the human ear pinna; forms the flexible skeleton of the ear (158×).

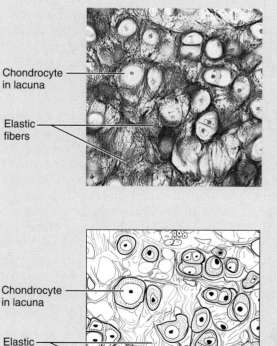

Chondrocyte in lacuna

Elastic fibers

Chondrocyte in lacuna

Elastic fibers

FIGURE 4.8 (continued)
Cartilage (**g** and **h**).

between hyaline cartilage and dense regular connective tissues. Its rows of chondrocytes (a cartilage feature) alternate with rows of thick collagen fibers (a feature of dense regular connective tissue) (Figure 4.8i). Because it is compressible and resists tension well, fibrocartilage is found where strong support and the ability to withstand heavy pressure are required. For example, the intervertebral discs (resilient cushions between the bony vertebrae) and the spongy cartilages of the knee are fibrocartilage structures (see Figure 6.1, p. 174).

Bone (Osseous Tissue)

Because of its rocklike hardness, **bone,** or **osseous** (os'e-us) **tissue,** has an exceptional ability to support and protect body structures. Bones of the skeleton also provide cavities for fat storage and synthesis of blood cells. Bone matrix is similar to that of cartilage but is harder and more rigid because, in addition to its more abundant collagen fibers, bone has an added matrix element—inorganic calcium salts (bone salts).

Osteoblasts produce the organic portion of the matrix; then bone salts are deposited on and between the fibers. Mature bone cells, or **osteocytes,** reside in the lacunae within the matrix they have made (Figure 4.8j). Unlike cartilage, the next firmest connective tissue, bone is very well supplied by invading blood vessels. We will consider the structure, growth, and metabolism of bone further in Chapter 6.

Blood

Blood, the fluid within blood vessels, is the most atypical connective tissue. It does *not* connect things or give mechanical support. It is classified as a connective tissue because it develops from mesenchyme and consists of blood *cells,* surrounded by a nonliving fluid matrix called *blood plasma* (Figure 4.8k). The "fibers" of blood are soluble protein molecules that become visible only during blood clotting. Blood functions as the transport vehicle for the cardiovascular system, carrying nutrients, wastes, respiratory gases, and many other substances throughout the body. Blood is discussed in detail in Chapter 18.

EPITHELIAL MEMBRANES: COVERINGS AND LININGS

Now that we have described both connective and epithelial tissues, we can consider the epithelial membranes that incorporate both types of tissue. Our previous classification of epithelia by cell shape and arrangement allows each individual epithelium to be described precisely, but it reveals nothing about the tissue's location in the body. Here we will describe

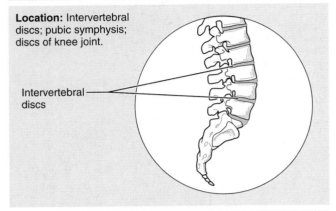

Cartilage (*continued*)

(i) Fibrocartilage

Description: Matrix similar but less firm than in hyaline cartilage; thick collagen fibers predominate.

Function: Tensile strength with the ability to absorb compressive shock.

Location: Intervertebral discs; pubic symphysis; discs of knee joint.

Intervertebral discs

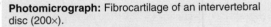

Photomicrograph: Fibrocartilage of an intervertebral disc (200×).

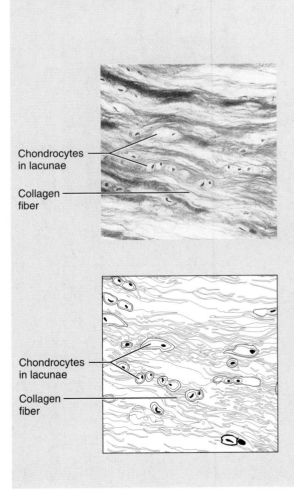

Chondrocytes in lacunae

Collagen fiber

Chondrocytes in lacunae

Collagen fiber

FIGURE 4.8 (continued)
Cartilage (**i**).

Others: (j and k)

(j) Bone (osseous tissue)

Description: Hard, calcified matrix containing many collagen fibers; osteocytes lie in lacunae. Very well vascularized.

Function: Bone supports and protects (by enclosing); provides levers for the muscles to act on; stores calcium and other minerals and fat; marrow inside bones is the site for blood cell formation (hematopoiesis).

Location: Bones

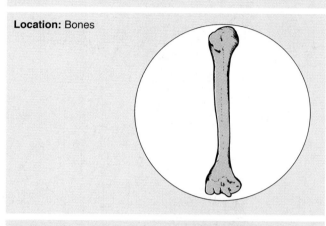

Photomicrograph: Cross-sectional view of bone (110×).

Osteocytes in lacunae

Osteocytes in lacunae

FIGURE 4.8 (continued)
Bone **(j)**.

Others (continued)

(k) Blood

Description: Red and white blood cells in a fluid matrix (plasma).

Function: Transport of respiratory gases, nutrients, wastes and other substances.

Location: Contained within blood vessels.

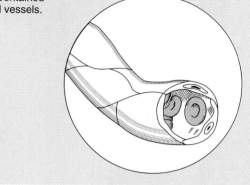

Photomicrograph: Smear of human blood (1100x); two white blood cells (neutrophil in upper left and lymphocyte in lower right) are seen surrounded by red blood cells.

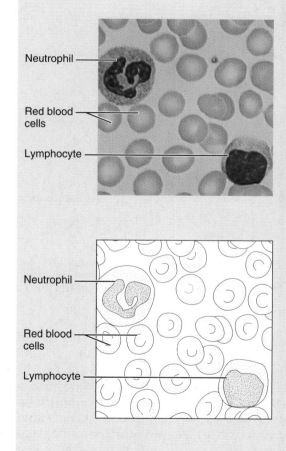

Neutrophil

Red blood cells

Lymphocyte

Neutrophil

Red blood cells

Lymphocyte

FIGURE 4.8 (continued)
Blood **(k)**.

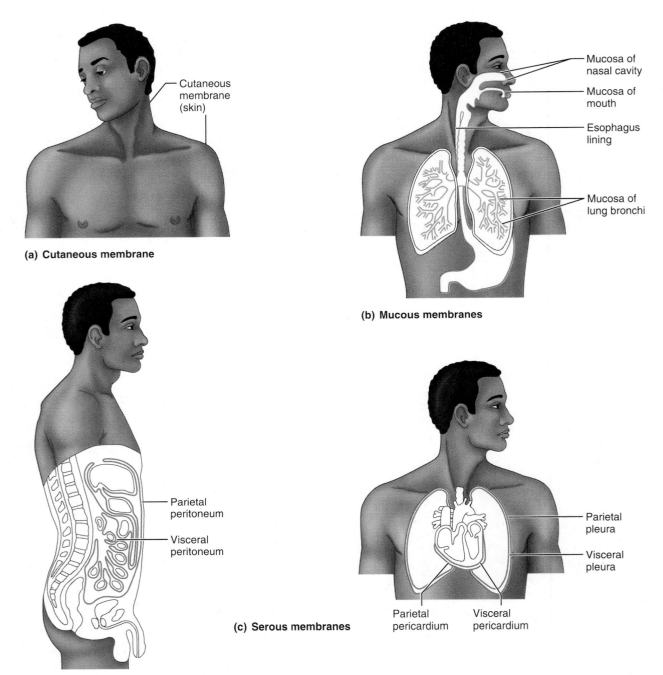

(a) Cutaneous membrane

Cutaneous membrane (skin)

Mucosa of nasal cavity
Mucosa of mouth
Esophagus lining
Mucosa of lung bronchi

(b) Mucous membranes

Parietal peritoneum
Visceral peritoneum

(c) Serous membranes

Parietal pleura
Visceral pleura
Parietal pericardium   Visceral pericardium

FIGURE 4.9   *Classes of epithelial membranes.* (a) Cutaneous membrane, or skin. **(b)** Mucous membranes line body cavities that are open to the exterior. **(c)** Serous membranes line ventral body cavities that are closed to the exterior.

the covering and lining epithelia using terms that indicate their special locations in the body and their general functional qualities.

There is tremendous variety in the way the term "epithelial membrane" is used. Here, **epithelial membrane** is defined as a continuous multicellular sheet composed of at least two primary tissue types: an epithelium bound to an underlying layer of connective tissue proper. Hence, epithelial membranes can be considered to be simple organs. These membranes are of three types: *cutaneous, mucous,* or *serous.*

Cutaneous Membrane

The **cutaneous** (ku-ta′ne-us; *cutis* = skin) **membrane** is your skin (Figure 4.9a). It is an organ system consisting of a keratinized stratified squamous epithelium (epidermis) firmly attached to a thick layer of dense irregular connective tissue (dermis). Unlike other epithelial membranes, the cutaneous membrane is exposed to the air and is a dry membrane. Chapter 5 is devoted to this unique organ system.

Mucous Membranes

Mucous membranes, or **mucosae** (mu-ko'se), line body cavities that open to the exterior, such as those of the hollow organs of the digestive, respiratory, and urogenital tracts (Figure 4.9b). In all cases, they are "wet," or moist, membranes bathed by secretions or, in the case of the urinary mucosa, urine. Notice that the term *mucosa* refers to the location of the epithelial membrane, *not* its cell composition, which varies. However, most mucosae contain either stratified squamous or simple columnar epithelia. The epithelial sheet is directly underlain by a layer of loose connective tissue called the **lamina propria** (lam'ĭ-nah pro'pre-ah; "one's own layer"). In some mucosae, the lamina propria rests on a third (deeper) layer of smooth muscle cells. These variations in specific mucosae are covered in later chapters.

Mucous membranes are often adapted for absorption and secretion. Although many mucosae secrete mucus, this is not a requirement. The mucosae of both the digestive and respiratory tracts secrete copious amounts of lubricating mucus; that of the urinary tract does not.

Serous Membranes

Serous membranes, or **serosae** (se-ro'se), introduced in Chapter 1 (p. 19), are the moist membranes found in closed ventral body cavities (Figure 4.9c). A serous membrane consists of simple squamous epithelium (a mesothelium) resting on a thin layer of loose connective (areolar) tissue. The mesothelial cells enrich the fluid that filters from the capillaries in the associated connective tissue with hyaluronic acid. The result is the thin, clear *serous fluid* that lubricates the facing surfaces of the parietal and visceral layers, so that they slide across each other easily.

The serosae are named according to their site and specific organ associations. For example, the serosa lining the thoracic wall and covering the lungs is the **pleura;** that enclosing the heart is the **pericardium;** and those of the abdominopelvic cavity and viscera are the **peritoneums.**

NERVOUS TISSUE

Nervous tissue is the main component of the nervous system—the brain, spinal cord, and nerves—which regulates and controls body functions. It contains two major cell types. **Neurons** are highly specialized nerve cells that generate and conduct nerve impulses (Figure 4.10). Typically, they are branching cells. Their cytoplasmic extensions, or processes, allow them to transmit electrical impulses over substantial distances within the body. The balance of nervous tissue consists of various

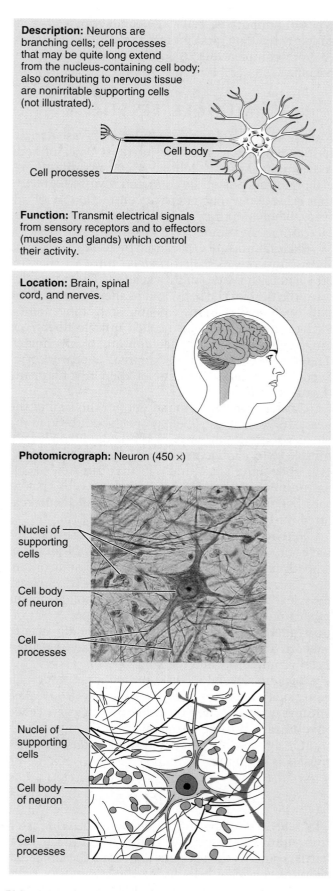

Description: Neurons are branching cells; cell processes that may be quite long extend from the nucleus-containing cell body; also contributing to nervous tissue are nonirritable supporting cells (not illustrated).

Cell processes
Cell body

Function: Transmit electrical signals from sensory receptors and to effectors (muscles and glands) which control their activity.

Location: Brain, spinal cord, and nerves.

Photomicrograph: Neuron (450 ×)

Nuclei of supporting cells
Cell body of neuron
Cell processes

Nuclei of supporting cells
Cell body of neuron
Cell processes

FIGURE 4.10 *Nervous tissue.*

types of **supporting cells,** nonconducting cells that support, insulate, and protect the delicate neurons. A more complete discussion of nervous tissue appears in Chapter 11.

MUSCLE TISSUE

Muscle tissues are highly cellular, well-vascularized tissues that are responsible for most types of body movement. Muscle cells possess **myofilaments,** elaborate versions of the *actin* and *myosin* filaments that bring about movement or contraction in all cell types. There are three kinds of muscle tissue: skeletal, cardiac, and smooth.

Skeletal muscle is packaged by connective tissue sheets into organs called *skeletal muscles* that are attached to the bones of the skeleton. These muscles form the flesh of the body, and as they contract, they pull on bones or skin, causing body movements. Skeletal muscle cells, also called **muscle fibers,** are long, cylindrical cells that contain many nuclei. Their obvious banded, or *striated,* appearance reflects the precise alignment of their myofilaments (Figure 4.11a).

Cardiac muscle is found only in the wall of the heart. Its contractions help propel blood through the blood vessels to all parts of the body. Like skeletal muscle cells, cardiac muscle cells are striated. However, they differ structurally in that cardiac cells (1) are uninucleate and (2) are branching cells that fit together tightly at unique junctions called **intercalated** (in-ter'kah-la"ted) **discs** (Figure 4.11b).

Smooth muscle is so named because its cells have no externally visible striations. Individual smooth muscle cells are spindle shaped and contain one centrally located nucleus (Figure 4.11c). Smooth muscle is found mainly in the walls of hollow organs other than the heart (digestive and urinary tract organs, uterus, and blood vessels). It generally acts to squeeze substances through these organs by alternately contracting and relaxing.

Because skeletal muscle contraction is under our conscious control, skeletal muscle is often called **voluntary muscle,** and the other two types are called **involuntary muscle.** Skeletal muscle and smooth muscle are described in detail in Chapter 9; cardiac muscle is discussed in Chapter 19.

TISSUE REPAIR

The body has many techniques for protecting itself from uninvited "guests" or injury. Intact mechanical barriers such as the skin and mucosae, the ciliary activity of epithelial cells lining the respiratory tract, and a strong acid (chemical barrier) produced by stomach glands represent three defenses exerted at the body's external boundaries. When tissue injury

(a) Skeletal muscle

Description: Long, cylindrical, multinucleate cells; obvious striations.

Function: Voluntary movement; locomotion; manipulation of the environment; facial expression; voluntary control.

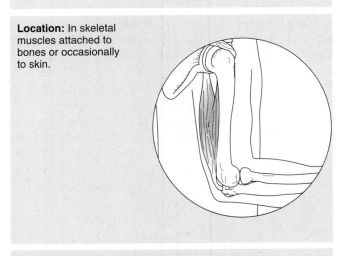

Location: In skeletal muscles attached to bones or occasionally to skin.

Photomicrograph: Skeletal muscle (approx. 130×). Notice the obvious banding pattern and the fact that these large cells are multinucleate.

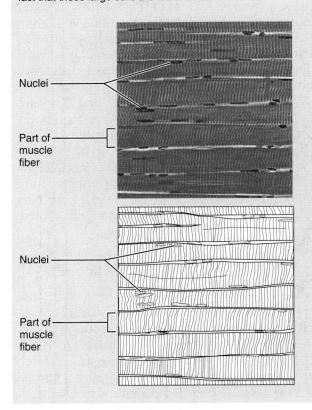

Nuclei

Part of muscle fiber

Nuclei

Part of muscle fiber

FIGURE 4.11 *Muscle tissues (a, b, and c).*

(b) Cardiac muscle

Description: Branching, striated, generally uninucleate cells that interdigitate at specialized junctions (intercalated discs).

Function: As it contracts, it propels blood into the circulation; involuntary control.

Location: The walls of the heart.

Photomicrograph: Cardiac muscle (275×); notice the striations, branching of cells, and the intercalated discs.

Intercalated discs

Nucleus

Intercalated discs

Nucleus

(c) Smooth muscle

Description: Spindle-shaped cells with central nuclei; cells arranged closely to form sheets; no striations.

Function: Propels substances or objects (foodstuffs, urine, a baby) along internal passageways; involuntary control.

Location: Mostly in the walls of hollow organs.

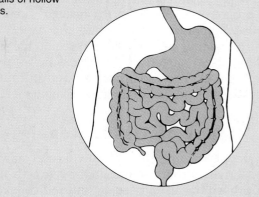

Photomicrograph: Sheet of smooth muscle (approx. 300×).

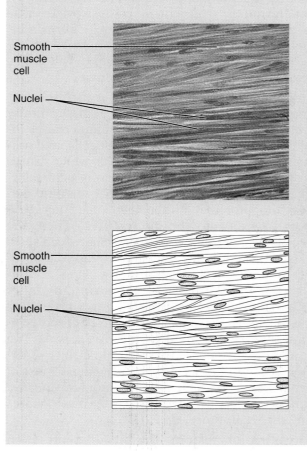

Smooth muscle cell

Nuclei

Smooth muscle cell

Nuclei

occurs these barriers are penetrated; this stimulates the body's inflammatory and immune responses, which wage their battles largely in the connective tissues of the body. The inflammatory response is a relatively nonspecific reaction that develops quickly and occurs whenever and wherever tissues are injured. Essentially, inflammation acts to get rid of the harmful agent, prevent further injury, and restore the tissue to a healthy condition. The immune response, on the other hand, is extremely specific, but takes longer to swing into action. Cells of the immune system, programmed to identify foreign substances such as microorganisms, toxins, and cancer cells, mount a vigorous attack against specific recognized invaders, either directly or by releasing antibodies into the blood. The inflammatory and immune responses are considered in detail in Chapter 22.

Steps of Tissue Repair

Tissue repair requires that cells divide and crawl (migrate), activities that are initiated by growth factors (wound hormones) released by injured cells. It occurs in two major ways: by regeneration and by fibrosis. Which of these occurs depends on (1) the type of tissue damaged and (2) the severity of the injury. **Regeneration** is replacement of destroyed tissue with the same kind of tissue, whereas **fibrosis** involves proliferation of fibrous connective tissue called **scar tissue.** In skin, the tissue we will use as our example, repair involves both activities.

1. Inflammation sets the stage. Let us briefly examine the inflammatory events that tissue injury sets into motion. First, the tissue trauma causes injured tissue cells, macrophages, mast cells, and others to release inflammatory chemicals which causes the capillaries to dilate and become very permeable. This allows white blood cells and plasma fluid rich in clotting proteins, antibodies, and other substances to seep into the injured area. Then, the leaked clotting proteins construct a clot, which stops the loss of blood, holds the edges of the wound together, and effectively walls in, or isolates, the injured area, preventing bacteria, toxins, or other harmful substances from spreading to surrounding tissues (Figure 4.12a). The part of the clot exposed to air quickly dries and hardens, forming a scab. The inflammatory events leave excess fluid, bits of destroyed cells, and other debris in the area. Most of this material is eventually removed from the area via lymphatic vessels or phagocytized by macrophages during the repair process.

2. Organization restores the blood supply. Even while the inflammatory process is going on, **organization,** the first phase of tissue repair, begins. During organization the blood clot is replaced by granulation tissue (Figure 4.12b). **Granulation tissue** is a delicate pink tissue composed of several elements. It contains capillaries that grow in from nearby areas and lay down a new capillary bed. Granulation tissue is actually named for these capillaries, which protrude nublike from its surface, giving it a granular appearance. These capillaries are fragile and bleed freely, as demonstrated when someone picks at a scab. Also present in granulation tissue are proliferating fibroblasts that produce growth factors as well as new collagen fibers to bridge the gap. Some of these fibroblasts have contractile properties that pull the margins of the wound together. As organization proceeds, macrophages digest the original blood clot and the deposit of collagen fibers continues. The granulation tissue, destined to become scar tissue (a permanent fibrous tissue patch), is highly resistant to infection because it produces bacteria-inhibiting substances.

3. Regeneration and fibrosis effect permanent repair. During organization, the surface epithelium begins to *regenerate* (see Figure 4.12b), growing under the scab, which soon detaches. As the fibrous tissue beneath matures and contracts, the regenerating epithelium thickens until it finally resembles that of the adjacent skin (Figure 4.12c). The end result is a fully regenerated epithelium, and an underlying area of scar tissue. The scar may be invisible, or visible as a thin white line, depending on the severity of the wound.

The repair process described above follows healing of a wound (cut, scrape, puncture) that breaches an epithelial barrier. In pure *infections* (a pimple or sore throat), healing is solely by regeneration. There is usually no clot formation or scarring. Only severe (destructive) infections lead to scarring.

The capacity for regeneration varies widely among the different tissues. Epithelial tissues regenerate extremely well, as do bone, areolar connective tissue, dense irregular connective tissue, and blood-forming tissue. Smooth muscle and dense regular connective tissue have a moderate capacity for regeneration, but skeletal muscle and cartilage have a weak regenerative capacity. Cardiac muscle and the nervous tissue in the brain and spinal cord have no *functional* regenerative capacity; hence they are routinely replaced by scar tissue. However, recent studies have shown that some unexpected (and highly selective) cellular division occurs in both these tissues after damage, and efforts are under way to coax them to regenerate better.

In nonregenerating tissues and in exceptionally severe wounds, fibrosis totally replaces the lost tissue. Over a period of months, the fibrous mass shrinks and becomes more and more compact. The resulting scar appears as a pale, often shiny area composed mostly of collagen fibers. Scar tissue is very strong, but it lacks the flexibility and elasticity of most normal tissues. Also, it cannot perform the normal functions of the tissue it has replaced.

? *(1) What is the makeup of granulation tissue formed at the injured site?*
(2) Where do the new capillaries that invade the injured area come from?
(3) How would this pictorial series look different if repair by regeneration was not possible?

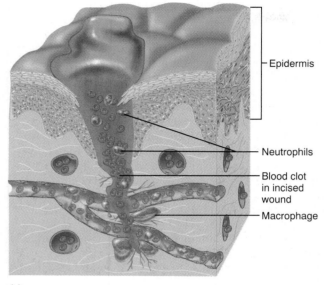

Epidermis

Neutrophils

Blood clot in incised wound

Macrophage

(a)

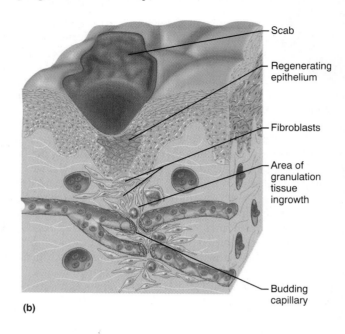

Scab

Regenerating epithelium

Fibroblasts

Area of granulation tissue ingrowth

Budding capillary

(b)

FIGURE 4.12 *Tissue repair of a nonextensive skin wound: regeneration and fibrosis.* (a) Severed blood vessels bleed. Inflammatory chemicals are released, and local blood vessels dilate and become more permeable, allowing white blood cells, fluid, clotting proteins, and other plasma proteins to invade the injured site. Clotting proteins initiate clotting; surface dries and forms a scab. **(b)** Granulation tissue is formed. Capillary buds invade the clot, restoring a vascular supply. Fibroblasts invade the region and secrete soluble collagen, which constructs collagen fibers that bridge the gap. Macrophages phagocytize dead and dying cell debris. Surface epithelial cells proliferate and migrate over the granulation tissue. **(c)** Approximately one week later, the fibrosed area (scar) has contracted and regeneration of the epithelium is in progress. The scar tissue may or may not be visible beneath the epidermis.

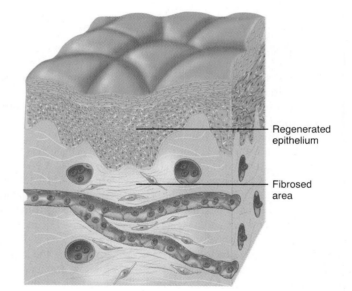

Regenerated epithelium

Fibrosed area

(c)

![Homeostatic Imbalance icon] **Homeostatic Imbalance**

Scar tissue that forms in the wall of the urinary bladder, heart, or other muscular organ may severely hamper the function of that organ. The normal shrinking of the scar reduces the internal volume of an organ and may hinder or even block movement of substances through a hollow organ. Scar tissue ham-

pers muscle's ability to contract and may interfere with its normal excitation by the nervous system. In the heart, these problems may cause progressive heart failure. In irritated visceral organs, particularly following abdominal surgery, *adhesions* may form as the newly forming scar tissue connects adjacent organs together. Such adhesions can prevent the normal shifting about (churning) of loops of the intestine, dangerously obstructing the flow of foodstuffs through it. Adhesions can also restrict heart movements and immobilize joints. ■

(1) *Budding capillaries and soft connective tissue.* (2) *They bud from uninjured blood vessels in the area.* (3) *The epidermis would not be continuous. The tissue in the injured region would be totally replaced by scar tissue.*

A CLOSER LOOK   Cancer—The Intimate Enemy

The word **cancer** elicits dread in nearly everyone. Why does cancer strike some and not others?

Although formerly perceived as an example of rapid disorganized cell growth, this disease is now known to be a logical, coordinated process in which a precise sequence of tiny alterations changes a normal cell into a killer. Let's take a closer look at what cancer really is.

When cells fail to honor normal controls of cell division and multiply excessively, an abnormal mass of proliferating cells called a *neoplasm* (ne'o-plazm; "new growth") results. Neoplasms are classified as benign ("kindly") or malignant ("bad"). A **benign neoplasm,** commonly called a *tumor,* is strictly a local affair. Its cells remain compacted, are often encapsulated, tend to grow slowly, and seldom kill their hosts if they are removed before they compress vital organs. In contrast, cancers are **malignant neoplasms,** nonencapsulated masses that grow relentlessly and may become killers. Their cells resemble immature cells, and they invade their surroundings rather than pushing them aside, as reflected in the name *cancer,* from the Latin word for "crab." Malignant cells can also break away from the parent mass, the *primary tumor,* and travel via blood or lymph to other body organs, where they form *secondary cancer masses.* This capability for traveling to other parts of the body is called **metastasis** (mĕ-tas'tah-sis). Metastasis and invasiveness distinguish cancer cells from the cells of benign neoplasms. Cancer cells consume an exceptional amount of the body's nutrients, leading to loss of weight and tissue wasting that contribute to death.

Mechanisms of Carcinogenesis

What causes a normal cell to **transform** or change into a cancerous one? It is well known that some physical factors (radiation, mechanical trauma), certain viral infections, and many chemicals (tobacco tars, saccharine, some natural food chemicals) can act as **carcinogens** (cancer-causers). What these factors have in common is that they all cause *mutations*— changes in DNA that alter the expression of certain genes. However, not all carcinogens do damage because most are eliminated by peroxisomal or lysosomal enzymes or by the immune system. Furthermore, *one* mutation usually isn't enough; it takes several genetic changes to transform a normal cell to a cancer cell.

A clue to the role of genes in cancer was provided by the discovery of **oncogenes** (Greek *onco* = tumor), or cancer-causing genes in rapidly spreading cancers. **Proto-oncogenes,** benign forms of oncogenes in normal cells, were discovered later. Proto-oncogenes code for proteins that are essential for cell division, growth, and cellular adhesion, among other things. Many have fragile sites that break when exposed to carcinogens, causing their conversion to oncogenes. Failure to code for certain proteins may lead to loss of an enzyme that controls an important metabolic process. "Switching on" of dormant genes that allow cells to become invasive and metastasize is also associated with oncogenes.

Oncogenes have been detected in only 15% to 20% of human cancers, so investigators were not too surprised by the discovery of **tumor suppressor genes,** or **antioncogenes** which work to suppress cancer by inactivating carcinogens, aiding DNA repair, or enhancing the ability of the immune system. In fact, malfunction or loss of two tumor suppressor genes—*p53* and *p16*—has been found in over half of all cancers. This is not surprising when you learn that *p53* prompts most cells to make proteins that "put the brakes on" cell division. Hence, its impairment invites uncontrolled division and cancer. Whatever genetic factor is at work, the "seeds" of cancer do appear to be in our own genes. Thus, cancer is an intimate enemy indeed.

Some of the mutations involved in colorectal cancer, one of the best-understood human cancers, are listed in the accompanying illustration. As with most cancers, the development of a metastasis is a gradual process. One of the first signs is the appearance of a polyp, a small benign growth consisting of apparently normal mucosa cells. As cell division continues, the growth enlarges, becoming an adenoma. As various tumor-suppressor genes are inactivated and the k-*ras* oncogene is mobilized, the mutations pile up and the adenoma becomes increasingly abnormal. The final consequence is colon carcinoma, a form of cancer that metastasizes quickly.

Cancer Prevalence, Types, and Management

Almost half of all Americans develop cancer in their lifetime and a fifth of us will die of it. Cancer can arise from almost any cell type, but the most common cancers originate in the skin, lung, colon, breast, and prostate gland. Although stomach and colon cancer incidence is down, skin and lymphoid cancer is up, and the overall rate of cancer deaths has actually increased over the past few decades despite a rise in expenditures for research and treatments.

Many forms of cancer are preceded by observable lumps or other structural changes in the tissue. An example is the development of *leukoplakia,* white patches in the mouth that may result from chronic irritation of badly fitting dentures or heavy smoking. Although these lesions sometimes progress to cancer, in many cases they remain stable or even revert to a normal condition if the environmental stimulus is removed.

Cancer management techniques vary, but the usual sequence is:

1. Diagnosis. Screening procedures, such as having a *mammogram* or examining one's breasts or testicles for lumps, and checking fecal samples for blood, aid in early detection of cancers. However, most cancers are diagnosed only after they have begun to cause symptoms (pain, bloody discharge, lump, etc.) and the diagnostic method most used is the *biopsy.* In a **biopsy,** a tissue sample is surgically removed

Development of Colon Cancer

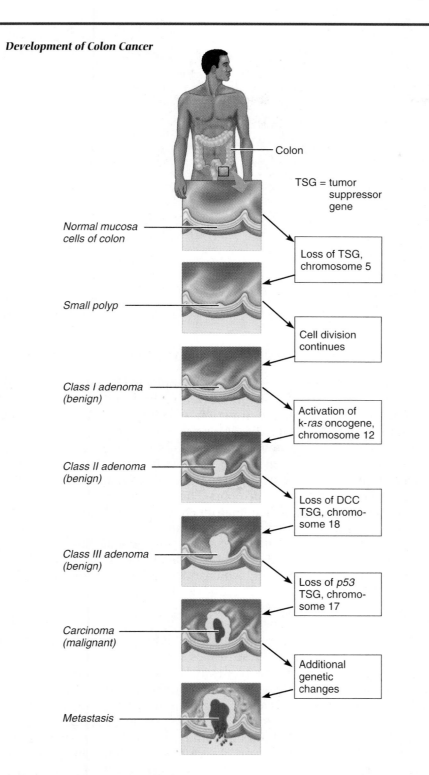

Colon

TSG = tumor suppressor gene

Normal mucosa cells of colon

Loss of TSG, chromosome 5

Small polyp

Cell division continues

Class I adenoma (benign)

Activation of k-*ras* oncogene, chromosome 12

Class II adenoma (benign)

Loss of DCC TSG, chromosome 18

Class III adenoma (benign)

Loss of *p53* TSG, chromosome 17

Carcinoma (malignant)

Additional genetic changes

Metastasis

from 1 to 4 according to the probability of cure (stage 1 has the best probability, stage 4 the worst).

3. Treatment. Most cancers are removed surgically if possible. To destroy metastasized cells, surgery is commonly followed by radiation therapy (X irradiation and/or treatment with radioisotopes) and chemotherapy (treatment with cytotoxic drugs). Chemotherapy is beset with the problem of resistance. Some cancer cells can eject the drugs and these cells proliferate, forming new tumors that are invulnerable to chemotherapy. Furthermore, anticancer drugs have unpleasant side effects because they kill *all* rapidly dividing cells, including normal ones. The side effects include nausea, vomiting, and hair loss. X rays also have side effects because, in passing through the body, they kill healthy cells that lie in the path of the cancer cells.

Treatment cures about half of all cancer cases. However, some cancers have very low survival rates; these include cancers of the lungs, liver, and ovaries.

Traditional cancer treatments—"cut, burn, and poison"—are widely recognized to be crude and painful. Promising new methods focus on delivering anticancer drugs more exclusively and precisely to the cancer (via monoclonal antibodies which respond to a single protein on a cancerous cell) and on increasing the immune system's effectiveness against cancer cells. The most recent research seeks to fix defective tumor-suppressor genes and oncogenes in cancer cells, to choke off tumors by inhibiting their ability to attract a rich blood supply (now testing *endostatin*), to use viruses to target and destroy cancer cells, and to signal cancer cells to commit suicide by apoptosis.

We should end on another positive note. Although average survival rates have not increased, the quality of life of cancer patients has improved in the last decade. The treatment of cancer-associated pain is better, and discomfort from the side-effects of chemotherapy has been reduced by antinausea drugs and other helpful medicines.

and examined microscopically for malignant cells. Increasingly, diagnosis is made by chemical or genetic analysis of the sample. Typing cancer cells by what genes are switched on or off tells clinicians what drugs to use. For example, taxol, quite successfully used against breast and ovarian cancer, only works against tumors with a specific genotype (genetic makeup).

Large tumors are recognized by medical imaging techniques (MRI, CT).

2. Staging. Several techniques (physical and histological examinations, lab tests, and imaging techniques) are used to determine the extent of the disease (size of the neoplasm, degree of metastasis, etc.). Then, the cancer is assigned a stage

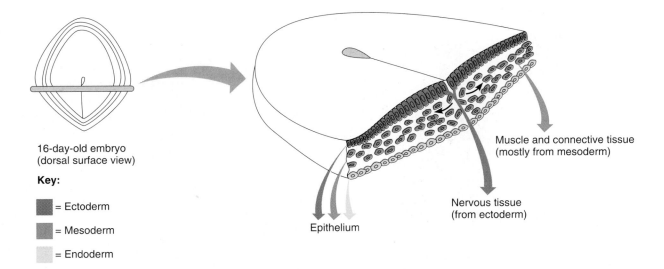

FIGURE 4.13 *Embryonic germ layers and the primary tissue types they produce.* The three embryonic layers collectively form the very early embryonic body.

DEVELOPMENTAL ASPECTS OF TISSUES

One of the first events of embryonic development is the formation of the three **primary germ layers,** which lie one atop the next like a three-layered cellular pancake. From superficial to deep, these layers are the **ectoderm, mesoderm** (mez′o-derm), and **endoderm** (Figure 4.13). These primary germ layers then specialize to form the four primary tissues from which all body organs are derived. Epithelial tissues form from all three germ layers: most mucosae from endoderm; endothelium and mesothelium from mesoderm; and the epidermis from ectoderm. Muscle and connective tissues develop from mesoderm, and nervous tissue arises from ectoderm.

By the end of the second month of development, the primary tissues have appeared, and all major organs are in place. In general, tissue cells remain mitotic and produce the rapid growth that occurs before birth. The division of nerve cells, however, stops or nearly stops during the fetal period. After birth, the cells of most other tissues continue to divide until adult body size is achieved. Cellular division then slows greatly, although many tissues retain the ability to regenerate. In adulthood, only epithelia and blood-forming tissues are highly mitotic. Some tissues that regenerate through life, such as the glandular cells of the liver, do so through division of their mature (specialized) cells. Others, like the epidermis of the skin and cells lining the intestine, have abundant *stem cells,* relatively undifferentiated cells that divide as necessary to produce new cells.

Given good nutrition, good circulation, and relatively infrequent wounds and infections, our tissues normally function efficiently through youth and middle age. But with increasing age, the epithelia thin and are more easily breached. The amount of collagen in the body declines, making tissue repair less efficient, and bone, muscle, and nervous tissues begin to atrophy. These events are due partly to decreased circulatory efficiency, which reduces delivery of nutrients to the tissues, but in some cases, diet is a contributing factor. As income declines or as chewing becomes more difficult, older people tend to eat soft foods, which may be low in protein and vitamins. As a result, tissue health suffers.

As we have seen, body cells combine to form four discrete tissue types: epithelial, connective, muscle, and nervous tissues. The cells making up each of these tissues share certain features but are by no means identical. They "belong" together because they have basic functional similarities. The connective tissues assume many guises, but perhaps the most versatile cells are those of epithelium: They protect our outer and inner surfaces, permit us to obtain oxygen, absorb vital nutrients into the blood, and allow our kidneys to excrete wastes. The important concept to carry away with you is that tissues, despite their unique abilities, cooperate to keep the body safe, healthy, and whole.

RELATED CLINICAL TERMS

Adenoma (ad"ĕ-no'mah; *aden* = gland, *oma* = tumor) Any neoplasm of glandular epithelium, benign or malignant. The malignant type is more specifically called *adenocarcinoma*.

Autopsy (aw'top-se) Examination of the body, its organs, and its tissues after death to determine the actual cause of death; also called postmortem examination and necropsy.

Carcinoma (kar"sĭ-no'mah; *karkinos* = crab, cancer) Cancer arising in an epithelium; accounts for 90% of human cancers.

Fat embolism Freely floating fat globule in the bloodstream; may result from extensive injury to fatty subcutaneous tissue or a bone fracture in which the fat in the bone marrow cavity is released. The danger is that the fatty masses may obstruct blood circulation to vital organs (heart, lungs, brain).

Healing by first intention The simplest type of healing; occurs when the edges of the wound are brought together by sutures, staples, or other means used to close surgical incisions. Only small amounts of granulation tissue need be formed.

Healing by second intention The wound edges remain separated, and the gap is bridged by relatively large amounts of granulation tissue; the manner in which unattended wounds heal. Healing is slower than in wounds in which the edges are brought together and larger scars result.

Keloid (ke'loid) Abnormal proliferation of connective tissue during healing of skin wounds; results in large, unsightly mass of scar tissue at the skin surface.

Lesion (le'zhun; "wound") Any injury, wound, or infection that affects tissue over an area of a definite size (as opposed to being widely spread throughout the body).

Marfan's syndrome Genetic disease resulting in abnormalities of connective tissues due to a defect in fibrillin, a protein that is associated with elastin in elastic fibers. Clinical signs include loose-jointedness, long limbs and spiderlike fingers and toes, visual problems, and weakened blood vessels (especially the aorta) due to poor connective tissue reinforcement.

Pathology (pah-thol'o-je) Scientific study of changes in organs and tissues produced by disease.

Pus A collection of tissue fluid, bacteria, dead and dying tissue cells, white blood cells, and macrophages in an inflamed area.

Sarcoma (sar-ko'mah; *sarkos* = flesh; *oma* = tumor) Cancer arising in the mesenchyme-derived tissues, that is, in connective tissues and muscle.

Scurvy A nutritional deficiency caused by lack of adequate vitamin C needed to synthesize collagen; signs and symptoms include blood vessel disruption, delay in wound healing, weakness of scar tissue, and loosening of teeth.

VAC (vacuum-assisted closure) Innovative healing process for open-skin wounds and skin ulcers. Often induces healing when all other methods fail. Involves covering the wound with a special sponge, and then applying suction through the sponge. In response to the subsequent skin stretching, fibroblasts in the wound form more collagen tissue and new blood vessels proliferate, bringing more blood into the injured area, which also promotes healing.

CHAPTER SUMMARY

Media study tools that could provide you additional help in reviewing specific key topics of Chapter 4 are referenced below. **SP** = *Study Partner;* **IP** = *Interactive Physiology.*

Tissues are collections of structurally similar cells with related functions. The four primary tissues are epithelial, connective, muscle, and nervous tissues.

Epithelial Tissue   (pp. 115–124)

1. Epithelial tissue is the covering, lining, and glandular tissue of the body. Its functions include protection, absorption, secretion, filtration, and sensory reception.

Special Characteristics of Epithelium   (pp. 115–116)

2. Epithelial tissues exhibit a huge degree of cellularity, specialized contacts, polarity, avascularity, support from connective tissue, and high regenerative capacity.

Classification of Epithelia   (pp. 116–121)

3. Epithelium is classified by arrangement as simple (one layer) or stratified (more than one) and by cell shape as squamous, cuboidal, or columnar. The terms denoting cell shape and arrangement are combined to describe the epithelium fully.

4. Simple squamous epithelium is a single layer of squamous cells. Highly adapted for filtration and exchange of substances. Forms walls of air sacs of the lungs and lines blood vessels. Contributes to serosae as mesothelium and forms the lining of all hollow circulatory system organs as endothelium.

5. Simple cuboidal epithelium is commonly active in secretion and absorption. Found in glands and in kidney tubules.

6. Simple columnar epithelium, specialized for secretion and absorption, consists of a single layer of tall columnar cells that exhibit microvilli and often goblet cells. Lines most of the digestive tract.

7. Pseudostratified columnar epithelium is a simple columnar epithelium that appears stratified. Its ciliated variety, rich in goblet cells, lines most of the upper respiratory passages.

8. Stratified squamous epithelium is multilayered; cells at the free edge are squamous. It is adapted to resist abrasion. It lines the esophagus and vagina; its keratinized variety forms the skin epidermis.

9. Stratified columnar epithelia are rare in the body, and are found chiefly in ducts of large glands.

10. Transitional epithelium is a modified stratified squamous epithelium, adapted for responding to stretch. It lines hollow urinary system organs.

SP Exercise: Chapter 3, Classes of Epithelial Tissues.

Glandular Epithelia   (pp. 122–124)

11. A gland is one or more cells specialized to secrete a product.

12. On the basis of site of product release, glands are classified as exocrine or endocrine. Glands are classified structurally as multicellular or unicellular.

13. Unicellular glands, also called goblet cells, are mucus-secreting single-celled glands.

14. Multicellular exocrine glands are classified according to duct structure as simple or compound, and according to the structure of their secretory parts as tubular, alveolar, or tubuloalveolar.

15. Multicellular exocrine glands of humans are classified functionally as merocrine or holocrine.

Connective Tissue   (pp. 124–134)

1. Connective tissue is the most abundant and widely distributed tissue of the body. Its functions include support, protection, binding, insulation, and transportation (blood).

Common Characteristics of Connective Tissue   (p. 124)

2. Connective tissues originate from embryonic mesenchyme and exhibit matrix. Depending on type, a connective tissue may be well vascularized (most), poorly vascularized (dense connective tissue), or avascular (cartilage).

Structural Elements of Connective Tissue   (pp. 125–127)

3. The structural elements of all connective tissues are extracellular matrix and cells.

4. Extracellular matrix consists of ground substance and fibers. It may be fluid, gel-like, or firm.

5. Each connective tissue type has a primary cell type that can exist as a mitotic, matrix-secreting cell (blast) or as a mature cell (cyte) responsible for maintaining the matrix. The chief cell type of connective tissue proper is the fibroblast; that of cartilage is the chondroblast; that of bone is the osteoblast; and that of blood-forming tissue is the hematopoietic stem cell.

Types of Connective Tissue   (pp. 127–134)

6. Embryonic connective tissue is called mesenchyme.

7. Connective tissue proper consists of loose and dense varieties. The loose connective tissues are

- Areolar: semifluid ground substance; all three fiber types loosely interwoven; a variety of cells; forms soft packing around body organs and the lamina propria; the prototype.
- Adipose: consists largely of adipocytes; scant matrix; insulates and protects body organs; provides reserve energy fuel. Brown fat, present only in infants, is more important for generating body heat.
- Reticular: finely woven reticular fibers in soft ground substance; the stroma of lymphoid organs and bone marrow.

8. Dense connective tissue proper includes

- Dense regular: dense parallel bundles of collagen fibers; few cells, little ground substance; high tensile strength; forms tendons, ligaments, aponeuroses; in cases where this tissue also contains numerous elastic fibers it is called elastic connective tissue.
- Dense irregular: like regular variety, but fibers are arranged in different planes; resists tension exerted from

many different directions; forms the dermis of the skin and organ capsules.

9. Cartilage exists as

- Hyaline: firm ground substance containing collagen fibers; resists compression well; found in fetal skeleton, at articulating surfaces of bones, and trachea; most abundant type.
- Elastic cartilage: elastic fibers predominate; provides flexible support of the external ear and epiglottis.
- Fibrocartilage: coarse parallel collagen fibers; provides support with compressibility; forms intervertebral discs and knee cartilages.

10. Bone (osseous tissue) consists of a hard, collagen-containing matrix embedded with calcium salts; forms the bony skeleton.

11. Blood consists of blood cells in a fluid matrix (plasma).

Epithelial Membranes: Coverings and Linings   (pp. 134–137)

1. Epithelial membranes are simple organs, consisting of an epithelium bound to an underlying connective tissue layer. Include mucosae, serosae, and the cutaneous membrane.

Nervous Tissue   (p. 137–138)

1. Nervous tissue forms organs of the nervous system. It is composed of neurons and supporting cells.

2. Neurons are branching cells that receive and transmit electrical impulses; involved in body regulation.

IP Nervous System I; topic: Anatomy Review, pages 1, 3.

Muscle Tissue   (pp. 138–139)

1. Muscle tissue consists of elongated cells specialized to contract and cause movement.

2. Based on structure and function, there are

- Skeletal muscle: attached to and moves the bony skeleton; cells are cylindrical and striated.
- Cardiac muscle: forms the walls of the heart; pumps blood; cells are branched and striated.
- Smooth muscle: in the walls of hollow organs; propels substances through the organs; cells are spindle shaped and lack striations.

SP Exercise: Chapter 4, Tissue Types; Blood Connective Tissue; IP Activity.

Tissue Repair   (pp. 138, 140–141)

1. Inflammation is the body's response to injury. Tissue repair begins during the inflammatory process. It may lead to regeneration, fibrosis, or both.

2. Tissue repair begins with organization, during which the blood clot is replaced by granulation tissue. If the wound is small and the damaged tissue is actively mitotic, the tissue will regenerate and cover the fibrous tissue. When a wound is extensive or the damaged tissue amitotic, it is repaired only by fibrous connective (scar) tissue.

SP Case Study: Cirrhosis of the Liver.

Developmental Aspects of Tissues   (p. 144)

1. Epithelium arises from all three primary germ layers (ectoderm, mesoderm, endoderm); muscle and connective tissue from mesoderm; and nervous tissue from ectoderm.

2. The decrease in mass and viability seen in most tissues during old age often reflects circulatory deficits or poor nutrition.

REVIEW QUESTIONS

Multiple Choice/Matching

1. Use the key to classify each of the following described tissue types into one of the four major tissue categories.

Key: **(a)** connective tissue   **(c)** muscle
       (b) epithelium   **(d)** nervous tissue

_____ Tissue type composed largely of nonliving extracellular matrix; important in protection and support

_____ The tissue immediately responsible for body movement

_____ The tissue that enables us to be aware of the external environment and to react to it

_____ The tissue that lines body cavities and covers surfaces

2. An epithelium that has several layers, with an apical layer of flattened cells, is called (choose all that apply): (a) ciliated, (b) columnar, (c) stratified, (d) simple, (e) squamous.

3. Match the epithelial types named in column B with the appropriate description(s) in column A.

Column A	Column B
_____ Lines most of the digestive tract	**(a)** pseudostratified ciliated columnar
_____ Lines the esophagus	**(b)** simple columnar
_____ Lines much of the respiratory tract	**(c)** simple cuboidal
_____ Forms the walls of the air sacs of the lungs	**(d)** simple squamous
	(e) stratified columnar
_____ Found in urinary tract organs	**(f)** stratified squamous
_____ Endothelium and mesothelium	**(g)** transitional

4. The gland type that secretes products such as milk, saliva, bile, or sweat through a duct is (a) an endocrine gland, (b) an exocrine gland.

5. The membrane lining body cavities that open to the exterior is a(n) (a) endothelium, (b) cutaneous membrane, (c) mucous membrane, (d) serous membrane.

6. Scar tissue is a variety of (a) epithelium, (b) connective tissue, (c) muscle tissue, (d) nerve tissue, (e) all of these.

Short Answer Essay Questions

7. Define tissue.

8. Name four important functions of epithelial tissue and provide at least one example of a tissue that exemplifies each function.

9. Describe the criteria used to classify covering and lining epithelia.

10. Explain the functional classification of multicellular exocrine glands and supply an example for each class.

11. Name four important functions of connective tissue and provide examples from the body that illustrate each function.

12. Name the primary cell type in connective tissue proper; in cartilage; in bone.

13. Name the two major components of matrix and, if applicable, subclasses of each component.

14. Matrix is extracellular. How does the matrix get to its characteristic position?

15. Name the specific connective tissue type found in the following body locations: (a) forming the soft packing around organs, (b) supporting the ear pinna, (c) forming "stretchy" ligaments, (d) first connective tissue in the embryo, (e) forming the intervertebral discs, (f) covering the ends of bones at joint surfaces, (g) main component of subcutaneous tissue.

16. What is the function of macrophages?

17. Differentiate clearly between the roles of neurons and the supporting cells of nervous tissue.

18. Compare and contrast skeletal, cardiac, and smooth muscle tissue relative to structure, body location, and specific function.

19. Describe the process of tissue repair, making sure you indicate factors that influence this process.

20. Name the three embryonic germ layers, and indicate which primary tissue classes derive from each layer.

Critical Thinking and Clinical Application Questions

1. John has sustained a severe injury during football practice and is told that he has a torn knee cartilage. Can he expect a quick, uneventful recovery? Explain your response.

2. The epidermis (epithelium of the cutaneous membrane or skin) is a keratinized stratified squamous epithelium. Explain why that epithelium is much better suited for protecting the body's external surface than a mucosa consisting of a simple columnar epithelium would be.

3. Your friend is trying to convince you that if the ligaments binding the bones together at your freely movable joints (such as your knee, shoulder, and hip joints) contained more elastic fibers, you would be much more flexible. Although there is *some* truth to this statement, such a condition would present some serious problems. Why?

4. In adults, over 90% of all cancers are either adenomas (adenocarcinomas) or carcinomas. In fact, cancers of the skin, lung, colon, breast, and prostate are all in these categories. Which one of the four basic tissue types gives rise to most cancers? Why do you think this is so?

5. Cindy, an overweight high school student, is overheard telling her friend that she's going to research how she can transform some of her white fat to brown fat. What is her rationale here (assuming it is possible)?

5 THE INTEGUMENTARY SYSTEM

The Skin (pp. 149–156)

1. Name the tissue types composing the epidermis and dermis. List the major layers of each and describe the functions of each layer.

2. Describe the factors that normally contribute to skin color. Briefly describe how changes in skin color may be used as clinical signs of certain disease states.

Appendages of the Skin (pp. 156–161)

3. Compare the structure and locations of sweat and oil glands. Also compare the composition and functions of their secretions.

4. Compare and contrast eccrine and apocrine glands.

5. List the parts of a hair follicle and explain the function of each part. Also describe the functional relationship of arrector pili muscles to the hair follicle.

6. Name the regions of a hair and explain the basis of hair color. Describe the distribution, growth, and replacement of hairs and the changing nature of hair during the life span.

7. Describe the structure of nails.

Functions of the Integumentary System (pp. 161–163)

8. Describe how the skin accomplishes at least five different functions.

Homeostatic Imbalances of Skin (pp. 163–165)

9. Summarize the characteristics of the three major types of skin cancers.

10. Explain why serious burns are life threatening. Describe how to determine the extent of a burn and differentiate first-, second-, and third-degree burns.

Developmental Aspects of the Integumentary System (pp. 165, 168)

11. Briefly describe the changes that occur in the skin from birth to old age. Attempt to explain the causes of such changes.

Would you be enticed by an advertisement for a coat that is waterproof, stretchable, washable, and permanent-press, that automatically repairs small cuts, rips, and burns, and that is guaranteed to last a lifetime with reasonable care? Sounds too good to be true, but you already have such a coat—your skin. The skin and its derivatives (sweat and oil glands, hairs, and nails) make up a complex set of organs that serves several functions, mostly protective. Together, these organs form the **integumentary** (in-teg″u-men′tar-e) **system.**

THE SKIN

The skin ordinarily receives very little respect from its inhabitants, but architecturally it is a marvel. It covers the entire body, has a surface area of 1.2 to 2.2 square meters, weighs 4 to 5 kilograms (4–5 kg = 9–11 pounds), and accounts for about 7% of total body weight in the average adult. The skin is also called the **integument,** which simply means "covering," but its functions go well beyond serving as a large, opaque bag for the body contents. It is pliable yet tough, allowing it to take constant punishment from external agents. Without our skin, we would quickly fall prey to bacteria and perish from water and heat loss.

The skin, which varies in thickness from 1.5 to 4.0 millimeters (mm) or more in different parts of the body, is composed of two distinct regions, the *epidermis* (ep″ĭ-der′mis) and the *dermis* (Figure 5.1). The epidermis (*epi* = upon), composed of epithelial cells, is the outermost protective shield of the body. The underlying dermis, making up the bulk of the skin, is a tough leathery layer composed of fibrous connective tissue. Only the dermis is vascularized. Nutrients reach the epidermis by diffusing through the tissue fluid from blood vessels in the dermis.

The subcutaneous tissue just deep to the skin is known as the **hypodermis** or **superficial fascia** (see Figure 5.3, p. 153). Strictly speaking, the hypodermis is not really part of the skin, but it shares some of the skin's protective functions. The hypodermis consists of mostly adipose plus some areolar connective tissue. Besides storing fat, the hypodermis anchors the skin to the underlying structures (mostly to muscles), but loosely enough that the skin can slide relatively freely over those structures. Sliding skin protects us by ensuring that many blows just glance off our bodies. Because of its fatty composition, the hypodermis also acts as a shock absorber and an insulator that prevents heat loss from the body. The hypodermis thickens markedly when one gains weight. In females, this "extra" subcutaneous fat accumulates first in the thighs and breasts, but in males it first collects in the anterior abdomen (as a "beer belly").

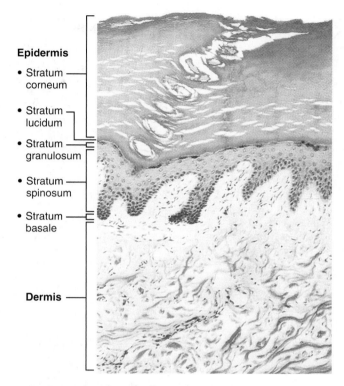

What primary tissue type gives rise to the epidermis? To the dermis?

Epidermis
• Stratum corneum
• Stratum lucidum
• Stratum granulosum
• Stratum spinosum
• Stratum basale

Dermis

FIGURE 5.1 *Photomicrograph showing layers of the skin (150 ×).* From *Gray's Anatomy,* Henry Gray. Churchill Livingstone, UK.

Epidermis

Structurally, the **epidermis** is a keratinized stratified squamous epithelium consisting of four distinct cell types and four or five distinct layers.

Cells of the Epidermis

The cells populating the epidermis include *keratinocytes, melanocytes, Merkel cells,* and *Langerhans' cells* (Figure 5.2). Most epidermal cells are keratinocytes, so we will consider them first. The chief role of **keratinocytes** (kĕ-rat′ĭ-no-sītz″; "keratin cells") is to produce **keratin,** the fibrous protein that helps give the epidermis its protective properties (*kera* = horn in Greek). Tightly connected to one another by desmosomes, the keratinocytes arise in the deepest part of the epidermis from a layer of cells (stratum basale) that undergo almost continuous mitosis. As these cells are pushed upward by the production of new cells beneath them, they make the keratin that eventually dominates their cell contents. By the time the keratinocytes reach the free

The epidermis is derived from epithelium, the dermis from connective tissue.

? *Why are the desmosomes connecting the keratino-cytes so important?*

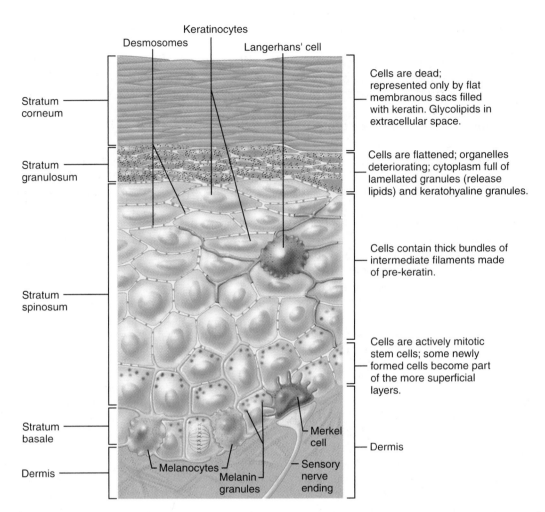

Keratinocytes

Desmosomes   Langerhans' cell

Stratum corneum — Cells are dead; represented only by flat membranous sacs filled with keratin. Glycolipids in extracellular space.

Stratum granulosum — Cells are flattened; organelles deteriorating; cytoplasm full of lamellated granules (release lipids) and keratohyaline granules.

Stratum spinosum — Cells contain thick bundles of intermediate filaments made of pre-keratin.

Cells are actively mitotic stem cells; some newly formed cells become part of the more superficial layers.

Stratum basale

Dermis

Melanocytes   Melanin granules   Sensory nerve ending   Merkel cell   Dermis

FIGURE 5.2 *Diagram showing the main features—layers and relative distribution of the different cell types—in epidermis of thin skin.* The keratinocytes (tan) form the bulk of the epidermis. Less numerous are the melanocytes (gray), which produce the pigment melanin; Langerhans' cells (blue), which function as macrophages; and Merkel cells (purple). A sensory nerve ending (yellow), extending to the Merkel cell from the dermis (pink), is shown associated with the Merkel cell forming a Merkel disc (touch receptor). Notice that the keratinocytes, but not the other cell types, are joined by numerous desmosomes (indicated by connections between membranes of adjacent cells). The stratum lucidum, present in thick skin, is not illustrated here.

surface of the skin, they are dead, scalelike structures that are little more than keratin-filled plasma membranes. Millions of these dead cells rub off every day, giving us a totally new epidermis every 25 to 45 days—the time from the "birth" of a keratinocyte to its final wearing away. In body areas regularly subjected to friction, such as the hands and feet, both cell production and keratin formation are accelerated. Persistent friction (from a poorly fitting shoe, for example) causes a thickening of the epidermis called a *callus*.

Melanocytes (mel′ah-no-sītz), the spider-shaped epithelial cells that synthesize the pigment **melanin** (mel′ah-nin; *melan* = black), are found in the deepest layer of the epidermis (Figure 5.2). As melanin is made and accumulated in membrane-bound granules called *melanosomes*, it is moved to the ends of the melanocyte's processes (the "spider arms") and then is periodically transferred to nearby keratinocytes. The melanin granules accumulate on the superficial, or "sunny," side of the keratinocyte nucleus, forming a pigment shield that protects the nucleus from the damaging effects of ultraviolet (UV) radiation in sunlight.

The star-shaped **Langerhans'** (lahng′er-hanz) **cells** arise from bone marrow and migrate to the epi-

The desmosomes connecting the keratinocytes are important to maintain continuity of the epidermis.

dermis. Also called **epidermal dendritic cells,** these cells are macrophages that help to activate our immune system as described later in this chapter (p. 162). Their slender processes extend among the surrounding keratinocytes, forming a more or less continuous network (Figure 5.2).

Occasional **Merkel cells** are present at the epidermal-dermal junction. Shaped like a spiky hemisphere (Figure 5.2), each Merkel cell is intimately associated with a disclike sensory nerve ending. The combination, called a *Merkel disc,* functions as a sensory receptor for touch.

Layers of the Epidermis

Variation in thickness of the epidermis determines if skin is *thick* or *thin.* In **thick skin,** which covers the palms, fingertips, and soles of the feet, the epidermis consists of five layers, or *strata* (stra'tah; "bed sheets") (see Figures 5.1 and 5.3). From deep to superficial, these layers are stratum basale, stratum spinosum, stratum granulosum, stratum lucidum, and stratum corneum. In **thin skin,** which covers the rest of the body, the stratum lucidum is absent and the other strata are thinner (Figure 5.2).

Stratum Basale (Basal Layer) The **stratum basale** (stra'tum ba-sǎ'le), the deepest epidermal layer, is firmly attached to the underlying dermis along a wavy borderline that reminds one of corrugated cardboard. For the most part, it consists of a single row of cells representing the youngest keratinocytes. The many mitotic nuclei seen in this layer reflect the rapid division of these cells and account for its alternate name, **stratum germinativum** (jer'mǐ-nǎ"tiv-um; "germinating layer").

Some 10 to 25% of the cells in the stratum basale are melanocytes, and their branching processes extend among the surrounding cells reaching well into the more superficial stratum spinosum layer. Occasional Merkel cells are also seen in this stratum.

Stratum Spinosum (Prickly Layer) The **stratum spinosum** (spi'no-sum; "prickly") is several cell layers thick. These cells contain a weblike system of intermediate filaments, which span their cytosol to attach to desmosomes. These intermediate filaments consist mainly of tension-resisting bundles of pre–keratin filaments. The keratinocytes in this layer flatten and appear somewhat irregular (spiny) in shape, causing them to be called *prickle cells.* However, the spines do not exist in the living cells; they arise during tissue preparation when these cells shrink but their numerous desmosomes hold tight. Scattered among the keratinocytes are melanin granules and Langerhans' cells, which are most abundant in this epidermal layer.

Stratum Granulosum (Granular Layer) The thin **stratum granulosum** (gran"u-lo'sum) consists of three to five cell layers in which a drastic change in keratinocyte appearance occurs. These cells continue to flatten, their nuclei and organelles begin to disintegrate, and they accumulate *keratohyaline granules* and *lamellated granules.* The keratohyaline (ker"ah-to-hi'ah-lin) granules help to form keratin in the upper layers, as we will see. The lamellated (lam'ǐ-la-ted; "plated") granules contain a waterproofing glycolipid that is spewed into the extracellular space and is a major factor in slowing water loss across the epidermis. The plasma membranes of these cells thicken as cytosol proteins bind to the inner membrane face and lipids released by the lamellated granules coat their external surfaces. This makes them more resistant to destruction, so you might say that the keratinocytes are "toughening up" to make the outer strata the strongest skin region.

Like all epithelia, the epidermis relies on capillaries in the underlying connective tissue (the dermis) for its nutrients. Above the stratum granulosum, the epidermal cells are too far from the dermal capillaries, so they die. This is a completely normal sequence of events.

Stratum Lucidum (Clear Layer) Through the light microscope, the **stratum lucidum** (loo'sid-um) appears as a thin translucent band just above the stratum granulosum (Figure 5.3). It consists of a few rows of clear, flat, dead keratinocytes with indistinct boundaries. Here, or in the stratum corneum above, the gummy substance of the keratohyalin granules clings to the keratin filaments in the cells, causing them to aggregate in parallel arrays. As mentioned before, the stratum lucidum is present only in thick skin.

Stratum Corneum (Horny Layer) The outermost **stratum corneum** (kor'ne-um) is a broad zone 20 to 30 cell layers thick that accounts for up to three-quarters of the epidermal thickness. Keratin and the thickened plasma membranes of cells in this stratum protect the skin against abrasion and penetration, and the glycolipid between its cells waterproofs this layer. Hence, the stratum corneum provides a durable "overcoat" for the body, protecting deeper cells from the hostile external environment (air) and from water loss, and rendering the body relatively insensitive to biological, chemical, and physical assaults. It is amazing that a layer of dead cells can still play so many roles.

The shingle-like cell remnants of the stratum corneum are referred to as *cornified,* or *horny, cells* (*cornu* = horn). They are familiar to everyone as the dandruff shed from the scalp and the flakes that slough off dry skin. (The average person sheds 18 kg

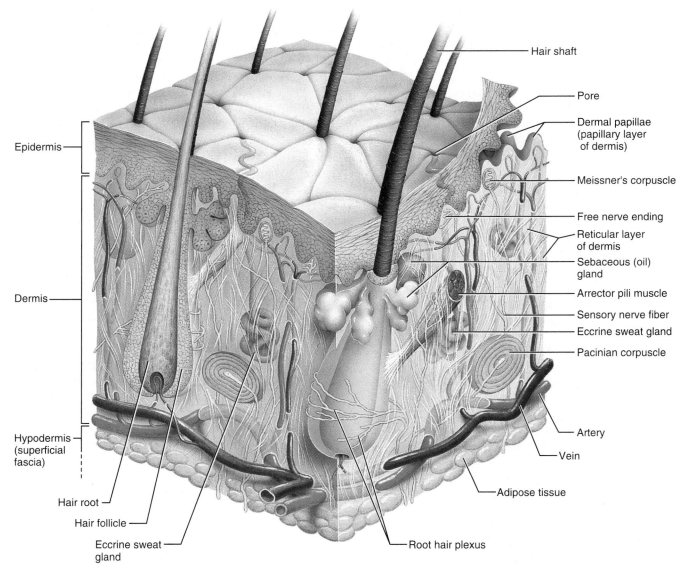

FIGURE 5.3   *Skin structure.* Three-dimensional view of the skin and underlying subcutaneous tissue. The epidermal and dermal layers have been pulled apart at the right corner to reveal the dermal papillae.

[40 pounds] of these skin flakes in a lifetime, providing a lot of fodder for the dust mites that inhabit our homes and bed linens.) The common saying "Beauty is only skin deep" is especially interesting in light of the fact that nearly everything we see when we look at someone is dead!

Dermis

The **dermis** (*derm* = skin), the second major skin region, is strong, flexible connective tissue. Its cells are typical of those found in any connective tissue proper: fibroblasts, macrophages, and occasional mast cells and white blood cells, and its semifluid matrix is heavily embedded with collagen, elastin, and reticular fibers. The dermis binds the entire body together like a body stocking. It is your "hide" and corresponds exactly to animal hides used to make leather products.

The dermis is richly supplied with nerve fibers (many equipped with sensory receptors), blood vessels, and lymphatic vessels. The major portions of hair follicles, as well as oil and sweat glands, are derived from epidermal tissue but reside in the dermis, as described later.

The dermis has two layers: the papillary and reticular. The thin superficial **papillary** (pap′il-er-e) **layer** is areolar connective tissue in which the collagen and elastin fibers form a loosely woven mat that is heavily invested with blood vessels. Its superior surface is thrown into peglike projections called **dermal papillae** (pah-pil′e; *papill* = nipple) that indent the overlying epidermis (see Figure 5.3). Many dermal papillae contain capillary loops; others house free nerve endings (pain receptors) and touch receptors called *Meissner's corpuscles* (mīs′nerz kor′pus-lz). On the palms of the hands and soles of the feet, these papillae lie atop larger mounds called *dermal ridges*, which in turn elevate the overlying epidermis to form *epidermal ridges* which increase friction and enhance the gripping ability of the fingers and feet. Epidermal ridge patterns are genetically determined and unique to each of us. Because sweat pores open along their crests, our fingertips leave identifying films of sweat called **fingerprints** on almost anything they touch.

The deeper **reticular layer,** accounting for about 80% of the thickness of the dermis, is dense irregular connective tissue. Its extracellular matrix contains thick bundles of interlacing collagen fibers that run in various planes; however, most run parallel to the skin surface. Separations, or less dense regions, between these bundles form *lines of cleavage,* or *tension lines,* in the skin. These externally invisible lines tend to run longitudinally in the skin of the head and limbs and in circular patterns around the neck and trunk. Cleavage lines are important to both surgeons and their patients. When an incision is made *parallel* to these lines, the skin gapes less and heals more readily than when the incision is made *across* cleavage lines.

The collagen fibers of the dermis give skin strength and resiliency that prevent most jabs and scrapes from penetrating the dermis. In addition, collagen binds water, helping to keep skin hydrated. Elastin fibers provide the stretch-recoil properties of skin.

In addition to the epidermal ridges and tension lines, a third type of skin marking, *flexure lines,* reflects dermal modifications. Flexure lines are dermal folds that typically occur at or near joints, where the dermis is tightly secured to deeper structures (notice the deep creases on your palms). Since the skin cannot slide easily to accommodate joint movement in such regions, the dermis folds and deep skin creases form. Flexure lines are also visible on the wrists, fingers, soles, and toes.

▨ *Homeostatic Imbalance*

Extreme stretching of the skin, such as occurs during pregnancy, can tear the dermis. Dermal tearing is indicated by silvery white scars called *striae* (stri′e; "streaks"), commonly called "stretch marks." Short-term but acute trauma (as from a burn or wielding a hoe) can cause a **blister,** the separation of the epidermal and dermal layers by a fluid-filled pocket. ∎

Skin Color

Three pigments contribute to skin color: melanin, carotene, and hemoglobin. Of these, only melanin is made in the skin. **Melanin,** a polymer made of tyrosine amino acids, ranges in color from yellow to reddish-brown to black. Its synthesis depends on an enzyme in melanocytes called tyrosinase (ti-ro′sĭ-nās) and, as noted earlier, it passes from melanocytes to the basal keratinocytes. Since all humans have the same relative number of melanocytes, individual and racial differences in skin coloring reflect the relative kind and amount of melanin made and retained. Melanocytes of black- and brown-skinned people produce many more and darker melanosomes than those of fair-skinned individuals, and their keratinocytes retain it longer. *Freckles* and *pigmented moles* are local accumulations of melanin. Melanocytes are stimulated to greater activity when we expose our skin to sunlight. Prolonged sun exposure causes a substantial melanin buildup, which helps protect the DNA of viable skin cells from UV radiation by absorbing the light and dissipating the energy as heat. Indeed, the initial signal for speeding up melanin synthesis seems to be a faster rate of repair of photodamaged DNA. In all but the darkest people, this response causes visible darkening of the skin (a tan).

▨ *Homeostatic Imbalance*

Despite melanin's protective effects, excessive sun exposure eventually damages the skin. It causes clumping of elastin fibers, leading to leathery skin; temporarily depresses the immune system; and can alter the DNA of skin cells and in this way lead to skin cancer. The fact that dark-skinned people seldom have skin cancer attests to melanin's effectiveness as a natural sunscreen.

A CLOSER LOOK — Illuminating the Skin

Human skin, only a few hundredths of an inch thick, is both highly impermeable and highly conspicuous. These qualities make life miserable for those with rosacea, large port wine stains, or tattoos that over time have become inappropriate (such as a gang insignia). Let's take a look at these conditions and how laser beams have proved to be their "miracle cure" in many cases. But first a few words about just what lasers are and what they can do.

Anyone who has ever played with a magnifying glass on a bright sunny day knows what a potent force light is when properly focused. A laser beam is about as focused as light gets—it is a single pure color with all the wavelengths in sync. Shine a laser beam on your skin and your skin cooks, because when the light beam hits pigmented cells, its energy is absorbed and converted to heat. Sadly for the first experimental subjects, that's exactly what the blue-green light of the first lasers used (the argon lasers) did when aimed at birthmarks and swollen veins. In 1989, a safer laser (the yellow light pulsed dye laser) was developed. Its yellow light is readily absorbed by the hemoglobin of the blood but not by other pigments, and if the light is pulsed, blood vessels, but not the surrounding tissues, are destroyed. Since then other lasers have come along to treat specific conditions. Although the skin may blister, scars rarely form and these newer lasers

are gentle enough to use even on a baby's skin. However, laser therapy is painful (each pulse feels like a hot needle) and expensive, costing up to $400 per square inch treated. And now to the conditions that seem to beg for laser therapy.

Rosacea

Rosacea (ro"za'shă) begins meekly enough with a sudden flushing of the skin as the dermal blood vessels of the face (particularly of the cheeks and nose) become engorged with blood. Although the flushing quickly fades, it recurs and lasts longer each time. Eventually, it becomes a persistent condition in which the facial skin becomes red and inflamed, and marked by whiteheadlike bumps and spidery blood vessels. Rosacea strikes people from all walks of life, from funnyman W. C. Fields to ordinary folks, and while it is far from fatal, it can be disfiguring and psychologically devastating. It tends to strike between the ages of 30 and 50, and though it is more common in women, its symptoms are more acute in men. Through no fault of their own, suffering patients with "fluorescent red" faces are stigmatized as alcoholics or as having poor hygiene. Alcohol does worsen the condition by causing blood to rush to the face, but so do hot spicy foods, hot baths, and sun exposure.

Those searching for a cause have blamed and then exonerated everything from emotional disturbances to skin mites. The most likely suspect is *H. pylori*, the bacterium that

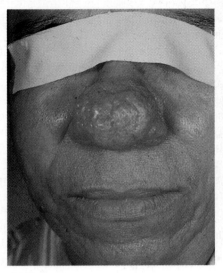

(a) Rosacea

causes peptic ulcers. Although acne medications and cortisone seem to help initially, they make subsequent flare-ups even worse. Many people with rosacea are helped by antibiotic therapy, which may be taken orally (tetracycline or erythromycin) or applied topically (*Cleocin* swabs or *Metrogel*). Untreated, rosacea takes a heavy toll. Gradually and painlessly, the skin becomes disfigured by patches of swollen veins and clusters of pustules. Most dramatic is *rhinophyma* (see photo a), the red, bulbous nose that disfigured W. C. Fields. For the more severe cases, laser beams are used to destroy the swollen blood vessels, shave away the excess tissue, and then sculpt what is left to match a "before" photograph—all with minimum blood loss and little or no scarring.

Ultraviolet radiation has other consequences as well. Many chemicals induce photosensitivity; that is, they heighten the skin's sensitivity to UV radiation, setting sun worshippers up for a skin rash the likes of which they would rather not see. Such substances include some antibiotic and antihistamine drugs, many chemicals in perfumes and detergents, and a chemical found in limes and celery. Small, itchy, blisterlike lesions erupt all over the body; then the peeling begins, in sheets! ■

Carotene (kar'o-tēn) is a yellow to orange pigment found in certain plant products such as carrots. It tends to accumulate in the stratum corneum and in fatty tissue of the hypodermis. Its color is most obvious in the palms and soles, where the stratum corneum is thickest (for example the skin of the heels), and most intense when large amounts of carotene-rich foods are eaten. However, the yellowish tinge of the skin of some Asian peoples is due to variations in melanin, not to carotene.

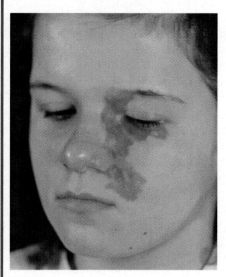

(b) Port wine stain

Port Wine Stains

Blazing red birthmarks, called *port wine stains*, are often mistaken for burns, so it's no coincidence that their technical name is *nevis flammeus* (*flamma* = flame). Folklore abounds about these colorful birthmarks. In some Central American countries, it is considered the "mark of the devil"; and the gypsies believed that a child born with a port wine stain bore the sign of its mother's guilt. The truth is that no one knows what causes these bright pink to purple blotches on the skin (see photo b), but they appear to be souvenirs of a small glitch during fetal development.

A port wine stain consists of an abnormally dense collection of dermal blood vessels, often on the face or neck. As the body grows, so does the birthmark, and in about a third of cases, the port wine stain darkens with age and develops a bumpy texture due to nodules of blood vessels that bleed profusely if cut. However, like rosacea, the most serious consequences tend to be psychological (often leading to withdrawal) because of a constant barrage of questions such as "Does it hurt? Is it catching?" and the like.

Although reverse tattoos (injections of skin-colored dye to mask the birthmark) work, they eventually fade. Radiation therapy and skin grafts, once the major means of treatment, are no longer used in the United States. Lasers, however, have made real progress against the birthmarks, particularly in adults who are better able to withstand the discomfort of treatment. Although laser therapy to remove a port wine stain the size of a small tortilla can cost thousands of dollars, the price is worth it for those who consider their birthmarks to be disfiguring.

Tattoos

Tattoos are made by using a needle to deposit pigment in the skin dermis. Dermal tattooing is believed to have originated around 8000 BC. Much more recently some males have used tattoos as a symbol of belonging to a group, such as the military or a particular street gang, while others simply view them as a symbol of individuality. In recent years, females have increasingly sought tattoos as a means of expression and for cosmetic purposes; permanent eyeliner and tattooed liplines account for more then 125,000 tattoos a year.

So should your buddy get one of the newer and trendy multicolored tattoos? Should your sister opt for permanent eyeliner (a procedure called blepharopigmentation)? What if the tattoo becomes unfashionable or the pigment migrates? Until very recently, once you had 'em, you had 'em, because attempts at removal—dermabrasion, cryosurgery (freezing the tissue), or applying caustic chemicals—left nasty scars, some as objectionable as the original tattoo. The lasers of today have no problem removing the "older" tattoos, that is, the blue and black tattoos applied a generation ago. The newer, colored tattoos are another matter entirely and require that several different lasers be used to remove them. Some of the newest pulsing lasers developed (Nd:YAG, alexandrite, and others) are effective in removing specific tattoo pigments. Complete removal requires several (seven to nine) treatments spaced about one month apart, at a pain cost about equal to that experienced in obtaining the tattoo in the first place.

But what about the noncosmetic problems? Tattooing is virtually unregulated. The U.S. Food and Drug Administration does exert some control over the composition of the pigments used (just as it does over any material injected under the skin), but the safety of those pigments is not well established. Furthermore, states vary in their tattoo regulations from entirely forbidding the practice to having no statutory regulations at all. Needles are used, bleeding occurs, and training requirements for practitioners are minimal. Even with the new lasers, is it worth the risk?

The pinkish hue of fair skin reflects the crimson color of oxygenated **hemoglobin** (he′mo-glo″bin) in the red blood cells circulating through the dermal capillaries. Because Caucasian skin contains only small amounts of melanin, the epidermis is nearly transparent and allows hemoglobin's color to show through.

⚑ *Homeostatic Imbalance*

When hemoglobin is poorly oxygenated, both the blood and the skin of Caucasians appear blue, a condition called *cyanosis* (si″ah-no′sis; *cyan* = dark blue). Skin often becomes cyanotic during heart failure and severe respiratory disorders. In dark-skinned individuals, the skin does not appear cyanotic because of the masking effects of melanin, but cyanosis is apparent in their mucous membranes and nail beds (the same sites where the red cast of normally oxygenated blood is visible).

The skin color of some people is also influenced by emotional stimuli, and many alterations in skin color signal certain disease states:

- *Redness, or erythema* (er"ĭ-the'mah): Reddened skin may indicate embarrassment (blushing), fever, hypertension, polycythemia, inflammation, or allergy.

- *Pallor, or blanching:* During fear, anger, and certain other types of emotional stress, some people become pale. Pale skin may also signify anemia or low blood pressure.

- *Jaundice* (jawn'dis), or *yellow cast:* An abnormal yellow skin tone usually signifies a liver disorder, in which yellow bile pigments accumulate in the blood and are deposited in body tissues. (Normally, the liver cells secrete the bile pigments [bilirubin] as a component of bile.)

- *Bronzing:* A bronze, almost metallic appearance of the skin is a sign of Addison's disease, hypofunction of the adrenal cortex.

- *Black-and-blue marks,* or *bruises:* Black-and-blue marks reveal where blood has escaped from the circulation and clotted beneath the skin. Such clotted blood masses are called *hematomas* (he"mah-to'mah; "blood swelling"). ▪

APPENDAGES OF THE SKIN

Along with the skin itself, the integumentary system includes several derivatives of the epidermis. These **skin appendages** include the hair follicles and hair, nails, sweat glands, and sebaceous (oil) glands. Each of these plays a unique role in maintaining body homeostasis.

Sweat (Sudoriferous) Glands

Sweat glands, also called **sudoriferous** (su"do-rif'er-us; *sudor* = sweat) **glands,** are distributed over the entire skin surface except the nipples and parts of the external genitalia. Their number is staggering—more than 2.5 million per person. There are two types of sweat glands: eccrine and apocrine.

Eccrine (ek'rin; "secreting") **sweat glands,** also called **merocrine sweat glands,** are far more numerous and are particularly abundant on the palms, soles of the feet, and forehead. Each is a simple, coiled, tubular gland. The secretory part lies coiled in the dermis; the duct extends to open in a funnel-shaped **pore** (*por* = channel) at the skin surface (see Figure 5.4b). (These sweat pores are different from the so-called pores of one's complexion, which are actually the external outlets of hair follicles.)

Eccrine gland secretion, commonly called **sweat,** is a hypotonic filtrate of the blood that passes through the secretory cells of the sweat glands and is released by exocytosis. It is 99% water, with some salts (mostly sodium chloride), vitamin C, antibodies, traces of metabolic wastes (urea, uric acid, ammonia), and lactic acid (the chemical that attracts mosquitoes). The exact composition depends on heredity and diet. Small amounts of ingested drugs may also be excreted by this route. Normally, sweat is acidic with a pH between 4 and 6.

Sweating is regulated by the sympathetic division of the autonomic nervous system, over which we have little control. Its major role is to prevent overheating of the body. Heat-induced sweating begins on the forehead and then spreads inferiorly over the remainder of the body. Emotionally induced sweating—the so-called "cold sweat" brought on by fright, embarrassment, or nervousness—begins on the palms, soles, and axillae (armpits) and then spreads to other body areas.

Apocrine (ap'o-krin) **sweat glands**[*] are largely confined to the axillary and anogenital areas. They are larger than eccrine glands, and their ducts empty into hair follicles. Apocrine secretion contains the same basic components as true sweat, plus fatty substances and proteins. Consequently, it is quite viscous and sometimes has a milky or yellowish color. The secretion is odorless, but when its organic molecules are decomposed by bacteria on the skin, it takes on a musky and generally unpleasant odor, the basis of body odor.

Apocrine glands, which begin to function at puberty under the influence of androgens, have little role to play in thermoregulation. Their precise function is not yet known, but they are activated by sympathetic nerve fibers during pain and stress. Because their activity is increased by sexual foreplay, and they enlarge and recede with the phases of a woman's menstrual cycle, they may be analogous to the sexual scent glands of other animals.

Ceruminous (sĕ-roo'mĭ-nus; *cera* = wax) **glands** are modified apocrine glands found in the lining of the external ear canal. They secrete a sticky, bitter substance called *cerumen,* or earwax, that is thought to deter insects and block entry of foreign material.

---

[*]The term *apocrine sweat glands* is a misnomer. These glands were originally thought to release their product by the apocrine mode (in which the apical surface pinches off) and were named accordingly. Subsequent studies proved them to be merocrine glands, which release their product by exocytosis like the eccrine sweat glands.

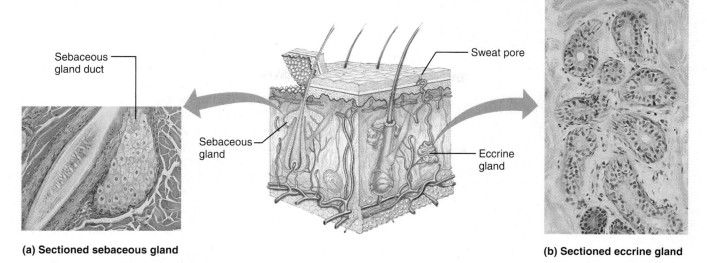

(a) Sectioned sebaceous gland

(b) Sectioned eccrine gland

FIGURE 5.4 *Cutaneous glands.* **(a)** Photomicrograph of a sebaceous gland (104×). **(b)** Photomicrograph of eccrine sweat glands (148×).

Mammary glands, another variety of specialized sweat glands, secrete milk. Although they are properly part of the integumentary system, we consider the mammary glands in Chapter 28, along with female reproductive organs.

Sebaceous (Oil) Glands

The **sebaceous** (se-ba′shus; "greasy") **glands,** or **oil glands** (see Figure 5.4a), are simple alveolar glands with little or no visible lumen that are found all over the body except on the palms and soles. They are small on the body trunk and limbs, but quite large on the face, neck, and upper chest. These glands secrete an oily secretion called **sebum** (se′bum). The central cells of the alveoli accumulate oily lipids until they become so engorged that they burst, so functionally these glands are *holocrine glands* (see p. 124). The accumulated lipids and cell fragments constitute sebum. Sebum is usually secreted into a hair follicle, or occasionally to a pore on the skin surface. Sebum softens and lubricates the hair and skin, prevents hair from becoming brittle, and slows water loss from the skin when the external humidity is low. Perhaps even more important is its *bactericidal* (bacterium-killing) action.

The secretion of sebum is stimulated by hormones, especially androgens. These glands are relatively inactive during childhood but are activated in both sexes during puberty, when androgen production begins to rise.

Homeostatic Imbalance

If a sebaceous gland duct becomes blocked by accumulated sebum, a *whitehead* appears on the skin surface. If the material oxidizes and dries, it darkens to form a *blackhead. Acne* is an active inflammation of the sebaceous glands accompanied by "pimples" (pustules or cysts) on the skin. It is usually caused by bacterial infection, particularly by staphylococcus, and can be mild or extremely severe, leading to permanent scarring. **Seborrhea** (seb″o-re′ah; "fast-flowing sebum"), known as "cradle cap" in infants, is caused by overactive sebaceous glands. It begins on the scalp as pink, raised lesions that gradually become yellow to brown and begin to slough off oily scales. Careful washing to remove the excessive oil often helps. ■

Hairs and Hair Follicles

Although hair helps to keep other mammals warm, our sparse body hair is far less luxuriant and useful. Even so, millions of hairs are distributed over our entire skin surface except our palms, soles, lips, nipples, and parts of the external genitalia (the head of the penis for example). The main function of our body hair is to sense insects on the skin before they sting us. Hair on the scalp guards the head against physical trauma, heat loss, and sunlight. Eyelashes shield the eyes, and nose hairs filter large particles like lint and insects from the air we inhale.

Structure of a Hair

Hairs, or **pili** (pi′li), are flexible strands produced by hair follicles that consist largely of dead, keratinized cells. The **hard keratin** that dominates hairs has two advantages over the **soft keratin** that is found in typical epidermal cells: (1) It is tougher and more durable, and (2) its individual cells do not flake off.

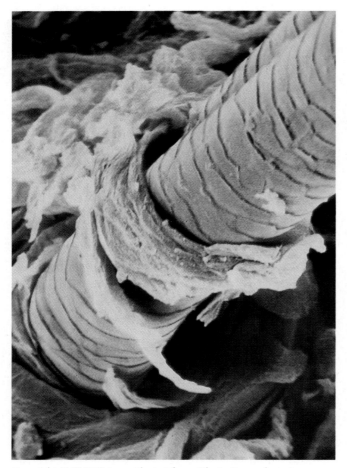

FIGURE 5.5   *Scanning electron micrograph showing a hair shaft emerging from a follicle at the epidermal surface.* Notice how the scalelike cells of the cuticle overlap one another (1500 ×).

The chief regions of a hair are the *shaft*, which projects from the skin (Figure 5.5), and the *root*, the part embedded in the skin (Figure 5.6). If the shaft is flat and ribbonlike in cross section, the hair is kinky; if it is oval, the hair is silky and wavy; if it is perfectly round, the hair is straight and tends to be coarse.

A hair has three concentric layers of keratinized cells (Figure 5.6b and c). Its central core, the *medulla* (mĕ-dul'ah; "middle"), consists of large cells and air spaces. The medulla is absent in fine hairs. The *cortex*, a bulky layer surrounding the medulla, consists of several layers of flattened cells. The outermost **cuticle** is formed from a single layer of cells that overlap one another from below like shingles on a roof (Figure 5.5). This arrangement helps to keep neighboring hairs apart so that the hair does not mat. (Hair conditioners smooth out the rough surface of the cuticle and make our hair look shiny.) The most heavily keratinized part of the hair, the cuticle, provides strength and helps keep the inner layers tightly compacted. Because it is subjected to the most abrasion, the cuticle tends to wear away at the tip of the

hair shaft, allowing the keratin fibrils in the cortex and medulla to frizz out, creating "split ends."

Hair pigment is made by melanocytes at the base of the hair follicle and transferred to the cortical cells. Various proportions of melanins of different colors (yellow, rust, brown, and black) combine to produce hair color from blond to pitch black. Additionally, red hair is colored by the iron-containing pigment called *trichosiderin*. Gray or white hair results from decreased melanin production (mediated by delayed-action genes) and from the replacement of melanin by air bubbles in the hair shaft.

Structure of a Hair Follicle

Hair follicles (*folli* = bag) extend from the epidermal surface into the dermis. In the scalp, they may even extend into the hypodermis. The deep end of the follicle is expanded, forming a **hair bulb** (Figure 5.6d and e). A knot of sensory nerve endings called a **root hair plexus** wraps around each hair bulb (see Figure 5.3), and bending the hair stimulates these endings. Consequently, our hairs act as sensitive touch receptors.

■ Feel the tickle as you run your hand over the hairs on your forearm.

A *hair papilla*, a nipplelike bit of dermal tissue, protrudes into the hair bulb. This papilla contains a knot of capillaries that supplies nutrients to the growing hair and signals it to grow. Except for its specific location, this papilla is similar to the dermal papillae underlying other epidermal regions.

The wall of a hair follicle is composed of an outer **connective tissue root sheath,** derived from the dermis, a thickened basement membrane called the *glassy membrane*, and an inner **epithelial root sheath,** derived mainly from an invagination of the epidermis (Figure 5.6d and e). The epithelial root sheath, which has external and internal parts, thins as it approaches the hair bulb, so that only a single layer of epithelial cells covers the papilla. However, the cells that comprise the **hair matrix,** or actively dividing area of the hair bulb that produces the hair, originate in a region called the *hair bulge* located a fraction of a millimeter above the hair bulb. When chemical signals diffusing from the papilla reach the hair bulge, some of its cells migrate toward the papilla, where they divide to produce the hair cells. As new hair cells are produced by the matrix, the older part of the hair is pushed upward, and its fused cells become increasingly keratinized and die.

Associated with each hair follicle is a bundle of smooth muscle cells called an **arrector pili** (ah-rek'tor pi'li; "raiser of hair") muscle. As you can see in Figure 5.3, most hair follicles approach the skin surface at a slight angle. The arrector pili muscles are attached in such a way that their contraction

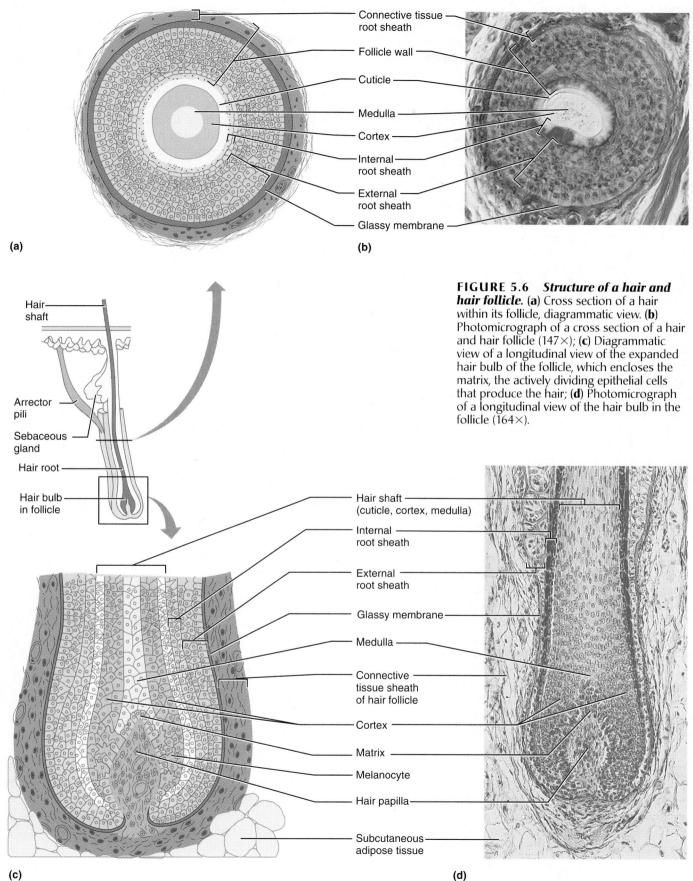

(a)

(b)

Connective tissue root sheath
Follicle wall
Cuticle
Medulla
Cortex
Internal root sheath
External root sheath
Glassy membrane

FIGURE 5.6 *Structure of a hair and hair follicle.* (**a**) Cross section of a hair within its follicle, diagrammatic view. (**b**) Photomicrograph of a cross section of a hair and hair follicle (147×); (**c**) Diagrammatic view of a longitudinal view of the expanded hair bulb of the follicle, which encloses the matrix, the actively dividing epithelial cells that produce the hair; (**d**) Photomicrograph of a longitudinal view of the hair bulb in the follicle (164×).

Hair shaft
Arrector pili
Sebaceous gland
Hair root
Hair bulb in follicle

(c)

(d)

Hair shaft (cuticle, cortex, medulla)
Internal root sheath
External root sheath
Glassy membrane
Medulla
Connective tissue sheath of hair follicle
Cortex
Matrix
Melanocyte
Hair papilla
Subcutaneous adipose tissue

pulls the hair follicle into an upright position and dimples the skin surface to produce goose bumps in response to cold external temperatures or fear. Although this "hair-raising" response is not very useful to humans, with our short sparse hairs, it is an important heat retention and protection mechanism in other animals. Furry animals can stay warmer by trapping a layer of insulating air in their fur; and a scared animal with its hair on end looks larger and more formidable to its enemy.

Types and Growth of Hair

Hairs come in various sizes and shapes, but as a rule, they can be classified as vellus or terminal. The body hair of children and adult females is of the pale, fine **vellus** (vel'us; *vell* = wool, fleece) **hair** variety. The coarser, longer hair of the eyebrows and scalp is **terminal hair;** terminal hair may also be darker. At puberty, terminal hairs appear in the axillary and pubic regions of both sexes and on the face and chest (and typically the arms and legs) of males. These terminal hairs grow in response to the stimulating effects of male sex hormones called *androgens* (of which *testosterone* is the most important).

Hair growth and density are influenced by many factors, but most importantly by nutrition and hormones. As a rule, poor nutrition means poor hair growth, whereas conditions that increase the local dermal blood flow (such as chronic physical irritation or inflammation) may enhance local hair growth. Many old-time bricklayers who carried their hod on one shoulder all the time developed one hairy shoulder. As noted, testosterone also encourages hair growth, and when male hormones are present in large amounts, terminal hair growth is luxuriant. Undesirable hair growth (such as on a woman's upper lip) may be arrested by *electrolysis* or laser treatments, which use electricity or light energy, respectively, to destroy the hair roots.

⚠ *Homeostatic Imbalance*

In women, small amounts of androgens are normally produced by both the ovaries and the adrenal glands. Excessive hairiness, or *hirsutism* (her'soot-izm; *hirsut* = hairy), as well as other signs of masculinization, may result from an adrenal gland or ovarian tumor that secretes abnormally large amounts of androgens. Since few women want a beard or hairy chest, such tumors are surgically removed as soon as possible. ■

The rate of hair growth varies from one body region to another and with sex and age, but it averages 2 mm per week. Each follicle goes through *growth cycles*. In each cycle, an active growth phase is followed by a resting phase, when the hair matrix is in-active and the follicle base and hair bulb atrophy somewhat. After the resting phase, the matrix proliferates again and forms a new hair to replace the old one that has fallen out or will be pushed out by the new hair.

The life span of hairs varies. The follicles of the scalp remain active for years (the average is four years) before becoming inactive for a few months. Because only a small percent of the hair follicles are in the shedding phase at any one time, we lose an average of 90 scalp hairs daily. The follicles of the eyebrow hairs remain active for only three to four months, which explains why your eyebrows are never as long as the hairs on your head.

Hair Thinning and Baldness

Given ideal conditions, hair grows fastest from the teen years to the 40s; then its growth slows. The fact that hairs are not replaced as fast as they are shed leads to hair thinning and some degree of baldness, or **alopecia** (al"o-pe'she-ah), in both sexes. Much less dramatic in women, the process usually begins at the anterior hairline and progresses posteriorly. Coarse terminal hairs are replaced by vellus hairs, and the hair becomes increasingly wispy.

True, or *frank, baldness* is a different story entirely. The most common type, **male pattern baldness,** is a genetically determined, sex-influenced condition. It is thought to be caused by a delayed-action gene that "switches on" in adulthood and changes the response of the hair follicles to testosterone. As a result, the follicular growth cycles become so short that many hairs never even emerge from their follicles before shedding, and those that do are fine vellus hairs that look like peach fuzz in the "bald" area. Until recently, the only cure for male pattern baldness has been drugs that inhibit testosterone production, but they also cause loss of sex drive—a trade-off few men would select. Quite by accident, it was discovered that minoxidil, a drug used to reduce high blood pressure by causing blood vessels to dilate, has an interesting side effect in some bald men; it stimulates hair regrowth. Although its results are variable, minoxidil is available over the counter in dropper bottles or spray form for application to the scalp. Finasteride, according to some the most promising cure ever developed for male pattern baldness, hit pharmacy shelves in early 1998 and has had moderate success. Available only by prescription in once-a-day pill form, it must be taken for the rest of one's life. Once the patient stops taking it, all of the new growth falls out, and so does what remains of the "old" hair.

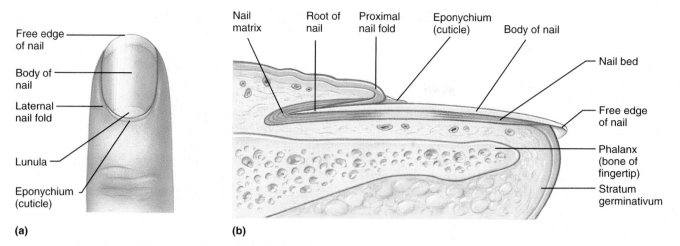

FIGURE 5.7 *Structure of a nail.* (**a**) Surface view of the distal part of a finger showing nail parts. (**b**) Sagittal section of the fingertip. The nail matrix that forms the nail lies beneath the lunula; the epidermis of the nail bed underlies the nail.

Homeostatic Imbalance

Hair thinning can be induced by a number of factors that upset the normal balance between hair loss and replacement. Outstanding examples are acutely high fever, surgery, or severe emotional trauma, and certain drugs (excessive vitamin A, some antidepressants, and most chemotherapy drugs). Protein-deficient diets and lactation lead to hair thinning because new hair growth stops when protein needed for keratin synthesis is not available or is being used for milk production. In all of these cases, hair regrows if the cause of thinning is removed or corrected. However, hair loss due to severe burns, excessive radiation, or other factors that eliminate the follicles, is permanent. ■

Nails

A **nail** is a scalelike modification of the epidermis that forms a clear protective covering on the dorsal surface of the distal part of a finger or toe (Figure 5.7). Nails, which correspond to the hooves or claws of other animals, are particularly useful as "tools" to help pick up small objects and to scratch an itch. Like hairs, nails contain hard keratin. Each nail has a *free edge,* a *body* (visible attached portion), and a proximal *root* (embedded in the skin). The deeper layers of the epidermis extend beneath the nail as the *nail bed;* the nail itself corresponds to the superficial keratinized layers. The thickened proximal portion of the nail bed, called the **nail matrix,** is responsible for nail growth. As the nail cells produced by the matrix become heavily keratinized, the nail body slides distally over the nail bed.

Nails normally appear pink because of the rich bed of capillaries in the underlying dermis. However, the region that lies over the thick nail matrix appears as a white crescent called the *lunula* (lu'nu-lah; "little moon"). The proximal and lateral borders of the nail are overlapped by skin folds, called **nail folds.** The proximal nail fold projects onto the nail body as the **cuticle** or **eponychium** (ep"o-nik'e-um; "on the nail").

FUNCTIONS OF THE INTEGUMENTARY SYSTEM

The skin and its derivatives perform a variety of functions that affect body metabolism, and prevent external factors from upsetting body homeostasis. Given its superficial location it is our most vulnerable organ system, exposed to bacteria, abrasion, temperature extremes, and harmful chemicals.

Protection

The skin constitutes at least three types of barriers: chemical, physical, and biological.

The **chemical barriers** include skin secretions and melanin. Although the skin's surface teems with bacteria, the low pH of skin secretions, or the so-called **acid mantle,** retards their multiplication. In addition, many bacteria are killed outright by bactericidal substances in sebum. Skin cells also secrete a natural antibiotic called *human defensin* that literally punches holes in bacteria, making them look like sieves. As discussed earlier, melanin provides a chemical pigment shield to prevent UV damage to the viable skin cells.

Physical, or **mechanical, barriers** are provided by the continuity of skin and the hardness of its keratinized cells. As a physical barrier, the skin is a remarkable compromise. A thicker epidermis would be more impenetrable, but we would pay the price in loss of suppleness and agility. Epidermal continuity works hand in hand with the acid mantle to ward off bacterial invasion. The waterproofing glycolipids of the epidermis effectively block the diffusion of water and water-soluble substances between cells, preventing both their loss from and entry into the body through the skin. Substances that *do* penetrate the skin in limited amounts include (1) *lipid-soluble substances*, such as oxygen, carbon dioxide, fat-soluble vitamins (A, D, E, and K), and steroids; (2) *oleoresins* (o"le-o-rez'inz) of certain plants, such as poison ivy and poison oak; (3) *organic solvents*, such as acetone, dry-cleaning fluid, and paint thinner, which dissolve the cell lipids; (4) *salts of heavy metals*, such as lead, mercury, and nickel; and (5) drug agents called *penetration enhancers* (dialkylamino acetates) that help ferry other drugs into the body.

Homeostatic Imbalance

Organic solvents and heavy metals are devastating to the body and can be lethal. Passage of organic solvents through the skin into the blood can cause the kidneys to shut down and can also cause brain damage. Absorption of lead results in anemia and neurological defects. These substances should never be handled with bare hands. ■

Biological barriers include the Langerhans' cells of the epidermis and macrophages in the dermis. Langerhans' cells are active elements of the immune system. For the immune response to be activated, the foreign substances, or *antigens*, must be presented to specialized white blood cells called lymphocytes. In the epidermis, it is the Langerhans' cells that play this role. Dermal macrophages constitute a second line of defense to dispose of viruses and bacteria that have managed to penetrate the epidermis. They, too, act as antigen "presenters."

Body Temperature Regulation

The body works best when its temperature remains within homeostatic limits. Like car engines, we need to get rid of the heat generated by our internal reactions. As long as the external temperature is lower than body temperature, the skin surface loses heat to the air and to cooler objects in its environment, just as a car radiator loses heat to the air and other nearby engine parts.

Under normal resting conditions, and as long as the environmental temperature is below 31–32°C (88°–90°F), sweat glands continuously secrete unnoticeable amounts of sweat (about 500 ml [0.5 L] of sweat per day). When body temperature rises, dermal blood vessels dilate and the sweat glands are stimulated into vigorous secretory activity. Sweat becomes noticeable and can account for the loss of up to 12 L of body water in one day. Evaporation of sweat from the skin surface dissipates body heat and efficiently cools the body, thus preventing overheating.

When the external environment is cold, dermal blood vessels constrict. This causes the warm blood to bypass the skin temporarily and allows skin temperature to drop to that of the external environment. Once this has happened, passive heat loss from the body is slowed, thus conserving body heat. Body temperature regulation is discussed in greater detail in Chapter 25.

Cutaneous Sensation

The skin is richly supplied with **cutaneous sensory receptors,** which are actually part of the nervous system. The cutaneous receptors are classified as *exteroceptors* (ek"ster-o-sep'torz) because they respond to stimuli arising outside the body. For example, Meissner's corpuscles (in the dermal papillae) and Merkel discs allow us to become aware of a caress or the feel of our clothing against our skin, whereas Pacinian receptors (in the deeper dermis or hypodermis) alert us to bumps or contacts involving deep pressure. Root hair plexuses report on wind blowing through our hair and a playful tug on a pigtail. Painful stimuli (irritating chemicals, extreme heat or cold, and others) are sensed by bare nerve endings that meander throughout the skin. Detailed discussion of these cutaneous receptors is deferred to Chapter 13, but those mentioned above are illustrated in Figure 5.3.

Metabolic Functions

When sunlight bombards the skin, modified cholesterol molecules circulating through dermal blood vessels of the skin are converted to a vitamin D precursor, and transported via the blood to other body areas to play various roles in calcium metabolism. For example, calcium cannot be absorbed from the digestive tract without vitamin D.

Besides synthesizing vitamin D, the epidermis has a host of other metabolic functions. It makes chemical conversions that supplement those of the liver—for example, keratinocyte enzymes can "disarm" many cancer-causing chemicals that penetrate the epidermis. Conversely, they can convert some harmless chemicals into carcinogens. Keratinocytes also activate some steroid hormones; for example, they can transform cortisone applied topically to irritated skin into hydrocortisone, a potent anti-inflammatory drug. Skin cells also make several biologically important proteins, including collagenase, an enzyme that aids the natural turnover of collagen (and deters wrinkles).

Blood Reservoir

The dermal vascular supply is quite extensive and can hold large volumes of blood (about 5% of the body's entire blood volume). When other body organs, such as vigorously working muscles, need a greater blood supply, the nervous system constricts the dermal blood vessels. This shunts more blood into the general circulation, making it available to the muscles and other body organs.

Excretion

Limited amounts of nitrogen-containing wastes (ammonia, urea, and uric acid) are eliminated from the body in sweat, although most such wastes are excreted in urine. Profuse sweating is an important avenue for water and salt (sodium chloride) loss.

HOMEOSTATIC IMBALANCES OF SKIN

When skin rebels, it is quite a visible revolution. Loss of homeostasis in body cells and organs reveals itself on the skin in ways that are sometimes almost too incredible to believe. The skin can develop more than 1000 different conditions and ailments. The most common skin disorders are bacterial, viral, or yeast infections. A number of these are summarized in Related Clinical Terms on pp. 168–169. Less common, but far more damaging to body well-being, are skin cancer and burns, considered next.

Skin Cancer

Most tumors that arise in the skin are benign and do not spread (metastasize) to other body areas. (A wart, a neoplasm caused by a virus, is one example.) However, some skin tumors are malignant, or cancerous, and invade other body areas. A crucial risk factor for the nonmelanoma skin cancers is overexposure to the UV radiation in sunlight, which appears to disable a tumor suppressor gene (*p53* or the patched [*ptc*] gene). In limited numbers of cases, however, frequent irritation of the skin by infections, chemicals, or physical trauma seems to be a predisposing factor.

Interestingly, sunburned skin accelerates its production of *fas*, a protein that causes genetically damaged skin to commit suicide, thus decreasing the risk of mutations that will cause sun–linked skin cancer. It is the death of these gene–damaged cells that causes the skin to peel after a sunburn.

Basal Cell Carcinoma

Basal cell carcinoma (kar"sǐ-no'mah) is the least malignant and most common skin cancer; over 30% of all white people get it in their lifetime. Cells of the stratum basale proliferate, invading the dermis and hypodermis. The cancer lesions occur most often on sun-exposed areas of the face and appear as shiny, dome-shaped nodules that later develop a central ulcer with a pearly, beaded edge (Figure 5.8a). Basal cell carcinoma is relatively slow-growing, and metastasis seldom occurs before it is noticed. Full cure by surgical excision is the rule in 99% of cases.

Squamous Cell Carcinoma

Squamous cell carcinoma arises from the keratinocytes of the stratum spinosum. The lesion appears as a scaly reddened papule (small, rounded elevation) that arises most often on the head (scalp, ears, and lower lip), and hands (Figure 5.8b). It tends to grow rapidly and metastasize if not removed. If it is caught early and removed surgically or by radiation therapy, chance of complete cure is good.

Melanoma

Melanoma (mel"ah-no'mah), cancer of melanocytes, is the most dangerous skin cancer. It accounts for only about 5% of skin cancers, but its incidence is increasing rapidly (by 3 to 8% per year in the United States). Melanoma can begin wherever there

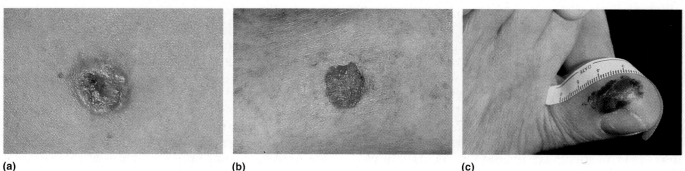

(a)   **(b)**   **(c)**

FIGURE 5.8 *Photographs of skin cancers.* (**a**) Basal cell carcinoma. (**b**) Squamous cell carcinoma. (**c**) Melanoma.

is pigment. Most such cancers appear spontaneously; about one-third develop from preexisting moles. It usually appears as a spreading brown to black patch (Figure 5.8c) that metastasizes rapidly to surrounding lymph and blood vessels. The key to surviving melanoma is early detection. The chance of survival is poor if the lesion is over 4 mm thick. The usual therapy for melanoma is wide surgical excision accompanied by immunotherapy (immunizing the body against its cancer cells).

The American Cancer Society suggests that sun worshippers regularly examine their skin for new moles or pigmented spots and apply the **ABCD rule** for recognizing melanoma. **A. Asymmetry:** The two sides of the pigmented spot or mole do not match. **B. Border irregularity:** The borders of the lesion are not smooth but exhibit indentations. **C. Color:** The pigmented spot contains several colors (blacks, browns, tans, and sometimes blues and reds). **D. Diameter:** The spot is larger than 6 mm in diameter (the size of a pencil eraser). Some experts have found that adding an **E,** for elevation above the skin surface, improves diagnosis, so they use the ABCD(E) rule.

Burns

Burns are a devastating threat to the body primarily because of their effects on the skin. A **burn** is tissue damage inflicted by intense heat, electricity, radiation, or certain chemicals, all of which denature cell proteins and cause cell death in the affected areas.

The immediate threat to life resulting from severe burns is a catastrophic loss of body fluids containing proteins and electrolytes. As fluid seeps from the burned surfaces, dehydration and electrolyte imbalance result. These, in turn, lead to renal shutdown and circulatory shock (inadequate blood circulation due to reduced blood volume). To save the patient, the lost fluids must be replaced immediately. In adults, the volume of fluid lost can be estimated by computing the percentage of body surface burned (extent of the burns) using the **rule of nines.** This method divides the body into 11 areas, each accounting for 9% of total body area, plus an additional area surrounding the genitals accounting for 1% of body surface area (Figure 5.9). This method is only approximate, so special tables are used when greater accuracy is desired (as for children, whose body proportions change rapidly).

In addition to fluid and electrolyte replacement, burn patients need thousands of extra food calories daily to replace lost proteins and allow tissue repair. No one can eat enough food to provide these calories, so burn patients are given supplementary nutrients through gastric tubes and intravenous (IV) lines. After the initial crisis has passed, infection becomes the main threat and is the leading cause of

death in burn victims. Burned skin is sterile for about 24 hours. Thereafter, bacteria, fungi, and other pathogens easily invade areas where the skin barrier is destroyed, and they multiply rapidly in the nutrient-rich environment of dead tissues and protein-containing fluid. Adding to this problem is the fact that the immune system becomes deficient within one to two days after severe burn injury.

Burns are classified according to their severity (depth) as first-, second-, or third-degree burns (Figure 5.9b–d). In **first-degree burns,** only the epidermis is damaged. Symptoms include localized redness, swelling, and pain. First-degree burns tend to heal in two to three days without special attention. Sunburn is usually a first-degree burn. **Second-degree burns** injure the epidermis and the upper region of the dermis. Symptoms mimic those of first-degree burns, but blisters also appear. Skin regeneration occurs with little or no scarring within three to four weeks if care is taken to prevent infection. First- and second-degree burns are referred to as **partial-thickness burns.**

Third-degree burns are **full-thickness burns,** involving the entire thickness of the skin. The burned area appears gray-white, cherry red, or blackened, and initially there is little or no edema. Since the nerve endings in the area have been destroyed, the burned area is not painful. Although skin regeneration might eventually occur by proliferation of epithelial cells at the edges of the burn, it is usually impossible to wait this long because of fluid loss and infection. Thus, skin grafting is usually necessary.

To prepare the burned area for grafting, the *eschar* (es'kar), or burned skin, must first be debrided (removed). To prevent infection and fluid loss, the area is then flooded with antibiotics and covered temporarily with a synthetic membrane, animal (pig) skin, cadaver skin, or "living bandage" made from the thin amniotic sac membrane that surrounds a fetus. Then healthy skin is transplanted to the burned site. Unless the graft is taken from the patient (an autograft), however, there is a good chance that it will be rejected (destroyed) by the patient's immune system. (See p. 819 in Chapter 22.) Even if the graft "takes," extensive scar tissue often forms in the burned areas.

An exciting technique is eliminating many of the traditional problems of skin grafting. Synthetic skin made of a silicone "epidermis" bound to a spongy "dermal" layer composed of collagen and ground cartilage is applied to the debrided area. In time, the patient's own dermal tissue replaces and reabsorbs the artificial one. While dermal reconstruction is going on, tiny bits of epidermal tissue are harvested from unburned parts of the patient's body. The cells are isolated and coaxed to proliferate in culture dishes. Once the new dermis is complete (usually in two to three months), the silicone sheet

? *Although the anterior head and face represent a very small proportion of the body surface, burns to this region are often much more serious than burns to the trunk. Why?*

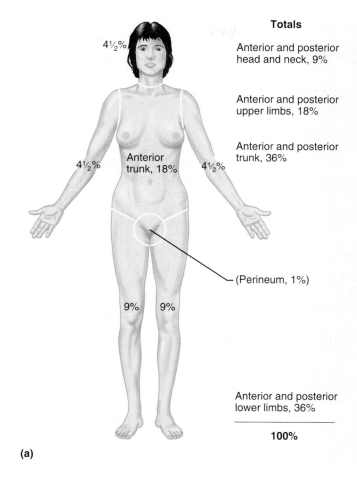

Totals

Anterior and posterior head and neck, 9%

Anterior and posterior upper limbs, 18%

Anterior and posterior trunk, 36%

4½%

Anterior trunk, 18%

4½%   4½%

(Perineum, 1%)

9%   9%

Anterior and posterior lower limbs, 36%

100%

(a)

(b)

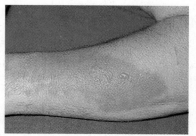

(c)

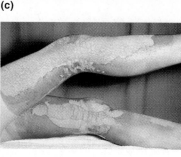

(d)

FIGURE 5.9 *Estimating the severity and extent of burns.* (a) The extent of burns is estimated by using the rule of nines. Surface area values for the anterior body surface are indicated on the human figure. Total surface area (anterior and posterior body surfaces) for each body region is indicated to the right of the figure.

Severity of burns depends on the depth of tissue damage. **(b)** In first-degree burns only the epidermis is destroyed. **(c)** In second-degree burns, the epidermis and part of the dermis are involved and blistering occurs. **(d)** In third-degree burns the epidermis and dermis (and often part of the hypodermis) are destroyed.

is stripped off, and sheets of smooth, pink, bottle-grown epidermis are grafted on the dermal surface. This artificial skin is not rejected, and it saves lives, and results in minimal scarring.

In general, burns are considered critical if any of the following conditions exists: (1) over 25% of the body has second-degree burns, (2) over 10% of the body has third-degree burns, or (3) there are third-degree burns of the face, hands, or feet. Facial burns introduce the possibility of burned respiratory passageways, which can swell and cause suffocation. Burns at joints are also troublesome because scar tissue formation can severely limit joint mobility.

Burn injury to the facial area may burn the respiratory passages, putting the patient at risk for suffocation as the burned tissues swell.

DEVELOPMENTAL ASPECTS OF THE INTEGUMENTARY SYSTEM

The epidermis develops from the embryonic ectoderm and the dermis and hypodermis develop from mesoderm. By the end of the fourth month of development, the skin is fairly well formed: The epidermis has all its strata, dermal papillae are obvious, and rudimentary epidermal derivatives are present. During the fifth and sixth months, the fetus is covered with a downy coat of delicate hairs called the **lanugo** (lah-nu′go; "wool") **coat.** This hairy cloak is shed by the seventh month, and vellus hairs make their appearance.

MAKING CONNECTIONS

SYSTEM CONNECTIONS: Homeostatic Interrelationships Between the Integumentary System and Other Body Systems

Endocrine System

- Skin protects endocrine organs; converts some hormones to their active forms
- Androgens produced by the endocrine system activate sebaceous glands and are involved in regulation of hair growth

Cardiovascular System

- Skin protects cardiovascular organs; prevents fluid loss from body; serves as blood reservoir
- Cardiovascular system transports oxygen and nutrients to skin and removes wastes from skin; provides substances needed by skin glands to make their secretions

Lymphatic System/Immunity

- Skin protects lymphatic organs; prevents pathogen invasion; Langerhans' cells and macrophages help activate the immune system
- Lymphatic system prevents edema by picking up excessive leaked fluid; immune system protects skin cells

Respiratory System

- Skin protects respiratory organs
- Respiratory system furnishes oxygen to skin cells and removes carbon dioxide via gas exchange with blood

Digestive System

- Skin protects digestive organs; provides vitamin D needed for calcium absorption; performs some of the same chemical conversions as liver cells
- Digestive system provides needed nutrients to the skin

Urinary System

- Skin protects urinary organs; excretes salts and some nitrogenous wastes in sweat
- Urinary system activates vitamin D precursor made by keratinocytes; disposes of nitrogenous wastes of skin metabolism

Reproductive System

- Skin protects reproductive organs; cutaneous receptors respond to erotic stimuli; highly modified sweat glands (mammary glands) produce milk. During pregnancy, skin stretches to accommodate growing fetus; changes in skin pigmentation may occur

Skeletal System

- Skin protects bones; skin synthesizes vitamin D that bones need for normal calcium absorption and deposit of bone (calcium) salts, which make bones hard
- Skeletal system provides support for skin

Muscular System

- Skin protects muscles
- Active muscles generate large amounts of heat, which increases blood flow to the skin and may promote activation of sweat glands of skin

Nervous System

- Skin protects nervous system organs; cutaneous sensory receptors located in skin (see Figure 5.3)
- Nervous system regulates diameter of blood vessels in skin; activates sweat glands, contributing to thermoregulation; interprets cutaneous sensation; activates arrector pili muscles

CLOSER CONNECTIONS

THE INTEGUMENTARY SYSTEM and Interrelationships with the Nervous, Cardiovascular, and Lymphatic/Immune Systems

First and foremost, our skin is a barrier. Like the skin of a grape, it keeps its contents juicy and whole. The skin, however, is a master at self (wound) repair, and interacts intimately with other body systems by making vitamin D (necessary for hard bones) and other potent molecules, all the while protecting deeper tissues from damaging external agents. Perhaps the most crucial roles of the skin in terms of overall body homeostasis are those it plays with the nervous, cardiovascular, and lymphatic systems. These interactions are detailed next.

Nervous System

The skin houses the tiny sensory receptors that provide a great deal of information about our external environment— its temperature, the pressure exerted by objects, and the presence of dangerous substances. What if we stepped on broken glass or hot pavement but did not have neural monitors in our skin? If we did not actually see, hear, taste, or smell such a damaging event, no reports would be sent to the nervous system. Consequently, the nervous system would be left "in the dark," unable to evaluate the need for a response and to order the steps needed to protect us from further damage or to get first aid. Thus, the whole body benefits from the skin's interaction with the nervous system.

Nervous and Cardiovascular System

Skin provides the site both for sensing external temperature and for responding to temperature changes. Dermal blood vessels (cardiovascular system organs) and sweat glands (controlled by the nervous system) play crucial roles in thermoregulation. So too do blood and the hot and cold receptors in the skin. When we are chilled, our blood loses heat to internal organs and cools. This alerts the nervous system to retain heat by constricting blood vessels. When body and blood temperature rises, dermal vessels dilate and sweating begins. Thermoregulation is vital: When the body overheats, life-threatening changes occur. Chemical reactions speed up, and if excessive, they destroy vital proteins and cells die. Cold has the opposite effect; cellular activity slows and ultimately stops.

Lymphatic System/Immunity

The role of the skin in immunity is very complex. Keratinocytes in the skin manufacture interferons (proteins that block viral infection) and other proteins important to the immune response. Langerhans' cells in the skin interact with antigens (foreign substances) that have penetrated the stratum corneum. The Langerhans' cells then migrate to lymphatic organs, where they present bits of the antigens to cells that will mount the immune response against them. This "messenger" function alerts the immune system early on to the presence of pathogens in the body.

Even a mild sunburn disrupts the normal immune response because UV radiation disables the skin's presenter cells. This may explain why many people infected by the cold sore virus tend to have a cold sore eruption after sun exposure.

CLINICAL CONNECTIONS

Integumentary System

A terrible collision between a trailer truck and a bus has occurred on Route 91. Several of the passengers are rushed to area hospitals for treatment. A few of these people will be followed in clinical case studies, which will continue from system to system.

Case study: Examination of Mrs. DeStephano, a 45-year-old woman, reveals several impairments of homeostasis. Relative to her integumentary system, the following comments are noted on her chart:

- Epidermal abrasions of the right arm and shoulder
- Severe lacerations of the right cheek and temple
- Cyanosis apparent

The lacerated areas are cleaned, sutured, and bandaged by the emergency room (ER) personnel, and Mrs. De-Stephano is admitted for further tests.

Relative to her signs:

1. What protective mechanisms are impaired or deficient in the abraded areas?

2. Assuming that bacteria are penetrating the dermis in these areas, what remaining skin defense mechanisms might act to prevent further bacterial invasion?

3. What benefit is conferred by suturing the lacerations? (Hint: see Chapter 4, p. 145)

4. Mrs. DeStephano's cyanotic skin may hint at what additional problem (and impairment of what body systems or functions)?

(Answers in Appendix F)

When a baby is born, its skin is covered with **vernix caseosa** (ver′niks kă-se-o′sah; "varnish of cheese"), a white, cheesy-looking substance produced by the sebaceous glands that protects the fetus's skin within the water-filled amnion. The newborn's skin is very thin and often has accumulations in the sebaceous glands on the forehead and nose that appear as small white spots called *milia* (mil′e-ah). These normally disappear by the third week after birth. During infancy and childhood, the skin thickens, and more subcutaneous fat is deposited. Although we all have approximately the same number of sweat glands, the number that begin to function increases in the first two years after birth and is determined by climate. Hence, people who grow up in hot climates have more active sweat glands than those raised in cooler areas of the world.

During adolescence, the skin and hair become oilier as sebaceous glands are activated, and acne may appear. Acne generally subsides in early adulthood, and skin reaches its optimal appearance when we reach our 20s and 30s. Thereafter, the skin starts to show the effects of cumulative environmental assaults (abrasion, wind, sun, chemicals). Scaling and various kinds of skin inflammation, or **dermatitis** (der″mah-ti′tis), become more common.

As old age approaches, the rate of epidermal cell replacement slows, the skin thins, and its susceptibility to bruises and other types of injury increases. All of the lubricating substances produced by the skin glands that make young skin so soft start to become deficient. As a result, the skin becomes dry and itchy. However, those with naturally oily skin seem to postpone this dryness until later in life. Elastic fibers begin to clump, and collagen fibers become fewer and stiffer. The subcutaneous fat layer diminishes, leading to the intolerance to cold so common in elderly people. The decreasing elasticity of the skin, along with the loss of subcutaneous tissue, inevitably leads to wrinkling. Decreasing numbers of melanocytes and Langerhans' cells enhance the risk and incidence of skin cancer in this age group. As a rule, redheads and fair-haired individuals, who have less melanin to begin with, show age-related changes more rapidly than do those with darker skin and hair.

By the age of 50 years, the number of active hair follicles is less than one-third and continues to decline, resulting in hair thinning. Hair loses its luster in old age, and the delayed-action genes responsible for graying and male pattern baldness become active.

Although there is no known way to avoid the aging of the skin, one of the best ways to slow the process is to shield your skin from the sun by means of sunscreens and protective clothing. Remember, the same sunlight that produces that fashionable tan also causes the sagging, blotchy, wrinkled skin of old age complete with pigmented "liver spots." (It is instructive to note that aged skin that has been protected from the sun has lost some elasticity and is thinned as noted, but it remains unwrinkled and unmarked.) Much of this havoc is due to UV activation of enzymes called matrix metalloproteinases, which degrade collagen and other dermal components. A drug called tretinoin related to vitamin A inhibits these enzymes and is now being used in some skin creams to slow photo-aging. Good nutrition, plenty of fluids, and cleanliness may also delay the process.

* * *

The skin is only about as thick as a paper towel—not too impressive as organ systems go. Yet, when it is severely damaged, nearly every body system reacts. Metabolism accelerates or may be impaired, immune system changes occur, bones may soften, the cardiovascular system may fail—the list goes on and on. On the other hand, when the skin is intact and performing its functions, the body as a whole benefits. Homeostatic interrelationships between the integumentary system and other organ systems are summarized in *Making Connections* on p. 166–167.

RELATED CLINICAL TERMS

Albinism (al′bĭ-nizm; *alb* = white) Inherited condition in which melanocytes do not synthesize melanin owing to a lack of tyrosinase. An albino's skin is pink, the hair pale or white, and the irises of the eyes unpigmented or poorly so.

Boils and carbuncles (kar′bung-klz; "little glowing embers") Inflammation of hair follicles and sebaceous glands in which an infection has spread to the underlying hypodermis; common on dorsal neck. Carbuncles are composite boils. A common cause is bacterial infection.

Cold sores (fever blisters) Small fluid-filled blisters that itch and smart; usually occur around the lips and in the mucosa of the mouth; caused by a herpes simplex infection. The virus localizes in a cutaneous nerve, where it remains dormant until activated by emotional upset, fever, or UV radiation.

Contact dermatitis Itching, redness, and swelling, progressing to blister formation; caused by exposure of the skin to chemicals (e.g., poison ivy) that provoke an allergic response in sensitive individuals.

Decubitus (de-ku′bĭ-tus) **ulcer** Localized breakdown and ulceration of skin due to interference with its blood supply. Usually occurs over a bony prominence, such as the hip or heel, that is subjected to continuous pressure; also called a bedsore.

Dermatology The branch of medicine that studies and treats disorders of the skin.

Epidermolysis bullosa (EB) A group of hereditary disorders characterized by inadequate or faulty synthesis of keratin, collagen, and/or basement membrane "cement" that results in lack of cohesion between layers of the skin and mucosa; a simple touch causes layers to separate and blister. For this reason, EB victims are called "touch-me-nots." In severe cases fatal blistering occurs in major vital organs. Because the blisters rupture easily, victims suffer frequent infections; treatments are aimed at relieving the symptoms and preventing infection.

Impetigo (im″pĕ-ti′go; *impet* = an attack) Pink, fluid-filled, raised lesions (common around the mouth and nose) that develop a yellow crust and eventually rupture; caused by staphylococcus infection; contagious; common in school-age children.

Mongolian spot Blue-black spot appearing in the skin of the sacral region; results from the unusual presence of a cluster of melanocytes in the dermis; the light-scattering effect of the epidermis causes these dermal melanin-containing regions to appear blue; disappears by age 3 or 4.

Porphyria (por-fer′e-ah; "purple") An inherited condition in which certain enzymes essential to form the heme of hemoglobin of blood are lacking; without these enzymes metabolic intermediates of the heme pathway called porphyrins build up, spill into the circulation, and eventually cause lesions throughout the body, especially when exposed to sunlight. The skin becomes lesioned and scarred; fingers, toes, and nose are disfigured; gums degenerate and teeth become prominent; believed to be the basis of folklore about vampires.

Psoriasis (so-ri′ah-sis) A chronic condition characterized by reddened epidermal lesions covered with dry silvery scales. When severe, it may be disfiguring and debilitating. Cause is unknown but an autoimmune attack is suspect; attacks are often triggered by trauma, infection, hormonal changes, and stress.

Vitiligo (vit″ĭ-li′go; *viti* = a vine, winding) The most prevalent skin pigmentation disorder; characterized by a loss of melanocytes and uneven dispersal of melanin, i.e., unpigmented skin regions (light spots) surrounded by normally pigmented areas.

CHAPTER SUMMARY

The Skin   (pp. 149–156)

1. The skin, or integument, is composed of two discrete tissue layers, an outer epidermis and a deeper dermis, resting on subcutaneous tissue, the hypodermis.

Epidermis   (pp. 149–152)

2. The epidermis is an avascular, keratinized sheet of stratified squamous epithelium. Most epidermal cells are keratinocytes. Scattered among the keratinocytes in the deepest epidermal layers are melanocytes, Merkel cells, and Langerhans' cells.

3. From deep to superficial, the strata, or layers of the epidermis, are the basale, spinosum, granulosum, lucidum, and corneum. The stratum lucidum is absent in thin skin. The actively mitotic stratum basale is the source of new cells for epidermal growth. The most superficial layers are increasingly keratinized and less viable.

SP **Exercise: Layers of the Epidermis**

Dermis   (pp. 152–153)

4. The dermis, composed mainly of dense, irregular connective tissue, is well supplied with blood vessels, lymphatic vessels, and nerves. Cutaneous receptors, glands, and hair follicles reside within the dermis.

5. The more superficial papillary layer exhibits dermal papillae that protrude into the epidermis above and the epidermal ridges that produce fingerprints.

6. In the deeper, thicker reticular layer, the connective tissue fibers are much more densely interwoven. Less dense regions between the collagen bundles produce cleavage, or tension, lines in the skin. Points of tight dermal attachment to the hypodermis produce dermal folds, or flexure lines.

Skin Color   (pp. 153–156)

7. Skin color reflects the amount of pigments (melanin and carotene) in the skin and the oxygenation level of hemoglobin in blood.

8. Melanin production is stimulated by exposure to ultraviolet radiation in sunlight. Melanin, produced by melanocytes and transferred to keratinocytes, protects the keratinocyte nuclei from the damaging effects of UV radiation.

9. Skin color is affected by emotional state. Alterations in normal skin color (jaundice, bronzing, erythema, and others) may indicate certain disease states.

SP **Exercise: Skin Structure**

Appendages of the Skin   (pp. 156–161)

1. Skin appendages, which derive from the epidermis, include hairs and hair follicles, nails, and glands (sweat and sebaceous).

Sweat (Sudoriferous) Glands   (pp. 156–157)

2. Eccrine (merocrine) sweat glands, with a few exceptions, are distributed over the entire body surface. Their primary function is thermoregulation. They are simple coiled tubular glands that secrete a salt solution containing small amounts of other solutes. Their ducts usually empty to the skin surface via pores.

3. Apocrine sweat glands, which may function as scent glands, are found primarily in the axillary and anogenital areas. Their secretion is similar to eccrine secretion, but it also contains proteins and fatty substances on which bacteria thrive.

Sebaceous (Oil) Glands   (p. 157)

4. Sebaceous glands occur all over the body surface except for the palms and soles. They are simple alveolar glands; their oily holocrine secretion is called sebum. Sebaceous gland ducts usually empty into hair follicles.

5. Sebum lubricates the skin and hair, prevents water loss from the skin, and acts as a bactericidal agent. Sebaceous glands are activated (at puberty) and controlled by androgens.

Hairs and Hair Follicles   (pp. 157–161)

6. A hair, produced by a hair follicle, consists of heavily keratinized cells. A typical hair has a central medulla, a cortex, and an outer cuticle and root and shaft portions. Hair color reflects the amount and kind of melanin present.

7. A hair follicle consists of an inner epidermal root sheath, enclosing the matrix (region of the hair bulb that produces the hair), and an outer connective tissue sheath derived from the dermis. A hair follicle is richly vascularized and well supplied with nerve fibers. Arrector pili muscles pull the follicles into an upright position and produce goose bumps.

8. Except for hairs of the scalp and around the eyes, hairs formed initially are fine vellus hairs; at puberty, under the influence of androgens, coarser, darker terminal hairs appear in the axillae and genital region.

9. The rate of hair growth varies in different body regions and with sex and age. Differences in life span of hairs account for differences in length on different body regions. Hair thinning reflects factors that lengthen follicular resting periods, age-related atrophy of hair follicles, and a delayed-action gene.

SP **Exercise: Hair Structure**

Nails   (p. 161)

10. A nail is a scalelike modification of the epidermis that covers the dorsum of a finger (or toe) tip. The actively growing region is the nail matrix.

Functions of the Integumentary System   (pp. 161–163)

1. Protection. The skin protects by chemical barriers (the antibacterial nature of sebum and the acid mantle, and melanin), physical barriers (the hardened keratinized and lipid-rich surface), and biological barriers (phagocytes).

2. Body temperature regulation. The skin vasculature and sweat glands, regulated by the nervous system, play an important role in maintaining body temperature homeostasis.

3. Cutaneous sensation. Cutaneous sensory receptors respond to temperature, touch, pressure, and pain stimuli.

4. Metabolic functions. Vitamin D is synthesized from cholesterol by epidermal cells. Skin cells also play a role in some chemical conversions.

5. Blood reservoir. The extensive vascular supply of the dermis allows the skin to act as a blood reservoir.

6. Excretion. Sweat contains small amounts of nitrogenous wastes and plays a minor role in excretion.

Homeostatic Imbalances of Skin   (pp. 163–165)

1. The most common skin disorders result from infections.

Skin Cancer   (pp. 163–164)

2. The most common cause of skin cancer is exposure to ultraviolet radiation.

3. Basal cell carcinoma and squamous cell carcinoma are cured if they are removed before metastasis. Melanoma, a cancer of melanocytes, is less common but dangerous.

Burns   (pp. 164–165)

4. In severe burns, the initial threat is loss of protein- and electrolyte-rich body fluids, which may lead to circulatory collapse. The second threat is overwhelming bacterial infection.

5. The extent of a burn may be evaluated by using the rule of nines (see p. 165). The severity of burns is indicated by the terms first degree, second degree, and third degree. Third-degree burns require grafting for successful recovery.

SP **Case Study: Burns**

Developmental Aspects of the Integumentary System   (pp. 165, 168)

1. The epidermis develops from embryonic ectoderm; the dermis (and hypodermis) develop from mesoderm.

2. The fetus exhibits a downy lanugo coat. Fetal sebaceous glands produce vernix caseosa, which helps protect the fetus's skin from its watery environment.

3. A newborn's skin is thin. During childhood the skin thickens and more subcutaneous fat is deposited. At puberty, sebaceous glands are activated and terminal hairs appear in greater numbers.

4. In old age, the rate of epidermal cell replacement declines and the skin and hair thin. Skin glands become less active. Loss of collagen and elastin fibers and subcutaneous fat leads to wrinkling; delayed-action genes cause graying and balding. Photodamage is a major cause of skin aging.

REVIEW QUESTIONS

Multiple Choice/Matching

1. Which epidermal cell type is most numerous? (a) keratinocyte, (b) melanocyte, (c) Langerhans' cell, (d) Merkel cell.

2. Which is probably a macrophage? (a) a keratinocyte, (b) a melanocyte, (c) a Langerhans' cell, (d) a Merkel cell.

3. The epidermis provides a physical barrier due largely to the presence of (a) melanin, (b) carotene, (c) collagen, (d) keratin.

4. Skin color is determined by (a) the amount of blood, (b) pigments, (c) oxygenation level of the blood, (d) all of these.

5. The sensations of touch and pressure are picked up by receptors located in (a) the stratum germinativum, (b) the dermis, (c) the hypodermis, (d) the stratum corneum.

6. Which is not a true statement about the papillary layer of the dermis? (a) It produces the pattern for fingerprints, (b) it is most responsible for the toughness of the skin, (c) it contains nerve endings that respond to stimuli, (d) it is highly vascular.

7. Skin surface markings that reflect points of tight dermal attachment to underlying tissues are called (a) tension lines, (b) papillary ridges, (c) flexure lines, (d) epidermal papillae.

8. Which of the following is not an epidermal derivative? (a) hair, (b) sweat gland, (c) sensory receptor, (d) sebaceous gland.

9. You can cut hair without feeling pain because (a) there are no nerves associated with hair, (b) the shaft of the hair consists of dead cells, (c) hair follicles develop from epidermal cells and the epidermis lacks a nerve supply, (d) hair follicles have no source of nourishment and therefore cannot respond.

10. An arrector pili muscle (a) is associated with each sweat gland, (b) can cause the hair to stand up straight, (c) enables each hair to be stretched when wet, (d) provides new cells for continued growth of its associated hair.

11. The product of this type of sweat gland includes protein and lipid substances that become odoriferous as a result of bacterial action: (a) apocrine gland, (b) eccrine gland, (c) sebaceous gland, (d) pancreatic gland.

12. Sebum (a) lubricates the surface of the skin and hair, (b) consists of dead cells and fatty substances, (c) in excess may cause seborrhea, (d) all of these.

13. The rule of nines is helpful clinically in (a) diagnosing skin cancer, (b) estimating the extent of a burn, (c) estimating how serious a cancer is, (d) preventing acne.

Short Answer Essay Questions

14. Which epidermal cells are also called prickle cells? Which contain keratohyalin and lamellated granules?

15. Is a bald man really hairless? Explain.

16. Both newborn infants and aged individuals have very little subcutaneous tissue. How does this affect their sensitivity to cold environmental temperature?

17. You go to the beach to swim on an extremely hot, sunshiny summer afternoon. Describe two ways in which your integumentary system acts to preserve homeostasis during your outing.

18. Distinguish clearly between first-, second-, and third-degree burns.

19. Describe the process of hair formation, and list several factors that may influence (a) growth cycles and (b) hair texture.

20. What is cyanosis and what does it reflect?

21. Why does skin wrinkle and what factors accelerate the wrinkling process?

22. Explain each of these familiar phenomena in terms of what you learned in this chapter: (a) pimples, (b) dandruff, (c) greasy hair and "shiny nose," (d) stretch marks from gaining weight, (e) freckles, (f) leaving fingerprints.

23. Count Dracula, the most famous vampire, rumored to have killed at least 200,000 people, was based on a real person who lived in eastern Europe about 600 years ago. He was indeed a "monster," even though he was not a real vampire. The historical Count Dracula may have suffered from which of the following? (a) porphyria, (b) EB, (c) halitosis, (d) vitiligo. Explain your answer.

24. Why are there no skin cancers that originate from stratum corneum cells?

25. A man got his finger caught in a machine at the factory. The damage was less serious than expected, but nonetheless the entire nail was torn off his right index finger. The parts lost were the body, root, bed, matrix, and eponychium of the nail. First, define each of these parts. Then, tell if this nail is likely to grow back.

26. On a diagram of the human body, mark off various regions according to the rule of nines. What percent of the total body surface is affected if the skin over the following body parts is burned? (a) the entire posterior trunk and buttocks, (b) an entire lower limb, (c) the entire front of the left upper limb.

Critical Thinking and Clinical Application Questions

1. Dean, a 40-year-old aging beach boy, is complaining to you that although his suntan made him popular when he was young, now his face is all wrinkled, and he has several darkly pigmented moles that are growing rapidly and are as big as large coins. He shows you the moles, and immediately you think "ABCD." What does that mean and why should he be concerned?

2. Victims of third-degree burns demonstrate the loss of vital functions performed by the skin. What are the two most important problems encountered clinically with such patients? Explain each in terms of the absence of skin.

3. A 30-year-old mental patient named Thelma has an abnormal growth of hair on the dorsum of her right index finger. The orderly comments that she gnaws on that finger continuously. What do you think is the relationship between Thelma's gnawing activity and her hairy finger?

4. A model is concerned about a new scar on her abdomen. She tells her surgeon that there is practically no scar from the appendix operation done when she was 16, but this new gallbladder scar is "gross." Her appendectomy scar is small and obliquely located on the inferior abdominal surface—it is very indistinct. By contrast, the gallbladder scar is large and lumpy and runs at right angles to the central axis of the body trunk. Can you explain why the scars are so different?

6 BONES AND SKELETAL TISSUES

Skeletal Cartilages (p. 173)

1. Describe the functional properties of each of the three types of cartilage tissue.
2. Locate the major cartilages of the adult skeleton.
3. Explain how cartilage grows.

Classification of Bones (pp. 174–176)

4. Name the major regions of the skeleton and describe their relative functions.
5. Compare and contrast the structure of the four bone classes and provide examples of each class.

Functions of Bones (p. 176)

6. List and describe five important functions of bones.

Bone Structure (pp. 176–181)

7. Describe the gross anatomy of a typical long bone and flat bone. Indicate the locations and functions of red and yellow marrow, articular cartilage, periosteum, and endosteum.

8. Indicate the functional importance of bone markings.
9. Describe the histology of compact and spongy bone.
10. Discuss the chemical composition of bone and the relative advantages conferred by its organic and its inorganic components.

Bone Development (pp. 181–184)

11. Compare and contrast the two types of bone formation: intramembranous and endochondral ossification.
12. Describe the process of long bone growth that occurs at the epiphyseal plates.

Bone Homeostasis: Remodeling and Repair (pp. 184–189)

13. Compare the locations and functions of the osteoblasts, osteocytes, and osteoclasts in bone remodeling.
14. Explain how hormones and physical stress regulate bone remodeling.
15. Describe the steps of fracture repair.

Homeostatic Imbalances of Bone (pp. 189–191, 194)

16. Contrast the disorders of bone remodeling seen in osteoporosis, osteomalacia, and Paget's disease.

Developmental Aspects of Bones: Timing of Events (p. 194)

17. Describe the timing and cause of changes in bone architecture and bone mass throughout life.

At one time or another, all of us have heard expressions like "bone tired" and "bag of bones"—pretty unflattering and inaccurate images of one of our most phenomenal tissues and our main skeletal elements. Our brains, not our bones, convey feelings of fatigue. As for "bag of bones," they are indeed more prominent in some of us, but without them to form our internal skeleton we would creep along the ground like slugs, lacking any definite shape or form. Along with its bones, the skeleton contains resilient cartilages. In this chapter, we briefly discuss the skeletal cartilages. However, the major focus here is the structure and function of bone tissue and the dynamics of its formation and remodeling throughout life. Details on individual bones of the skeleton and joints that allow skeletal mobility are the topics of Chapters 7 and 8.

SKELETAL CARTILAGES

Although our skeleton is initially made up of cartilages and fibrous membranes, most of these early supports are soon replaced by bone. The few cartilages that remain in adults are found mainly in regions where more flexible skeletal tissues are needed, as described below.

Basic Structure, Types, and Locations

A **skeletal cartilage** is made of some variety of *cartilage tissue,* which consists primarily of water. The high water content of cartilage accounts for its resilience, that is, its ability to spring back to its original shape after being compressed. The cartilage, which contains no nerves or blood vessels, is surrounded by a layer of dense irregular connective tissue, the *perichondrium* (per"ĭ-kon'drĭ-um; "around the cartilage"). The perichondrium acts like a girdle to resist outward expansion when the cartilage is compressed. Additionally, the perichondrium is the source of blood vessels from which nutrients diffuse through the matrix to reach the chondrocytes. This mode of nutrient delivery limits cartilage thickness.

As described in Chapter 4, there are three types of cartilage tissue in the body: hyaline, elastic, and fibrocartilage. All three types have the same basic components—cells called *chondrocytes,* encased in small cavities (lacunae) within an *extracellular matrix* containing a jellylike ground substance and fibers. The skeletal cartilages contain representatives from all three tissue groups (Figure 6.1).

Hyaline Cartilages

Hyaline cartilages, which look like frosted glass when freshly exposed, provide support with flexibility and resilience. They are the most abundant skeletal cartilages. When viewed under the microscope, their chondrocytes appear spherical (see Figure 4.8g), and the only fiber type in their matrix is fine collagen fibers (which, however, are not detectable). Skeletal hyaline cartilages include (1) articular cartilages, which cover the ends of most bones at movable joints; (2) costal cartilages, which connect the ribs to the sternum (breastbone); (3) respiratory cartilages, which form the skeleton of the larynx (voicebox), and reinforce other passageways of the respiratory system; and (4) the nasal cartilages which support the external nose.

Elastic Cartilages

Elastic cartilage looks very much like hyaline cartilage (see Figure 4.8h). However, it contains more stretchy elastic fibers, and so it is better able to stand up to repeated bending. The flexible **elastic cartilages** are found in only two skeletal locations (Figure 6.1)—supporting the external ear, and forming the epiglottis (the flap that bends to cover the opening of the larynx each time we swallow).

Fibrocartilages

Fibrocartilage is an unusual tissue that is highly compressible and has great tensile strength. It is the perfect intermediate between hyaline and elastic cartilages. Microscopically it consists of roughly parallel rows of chondrocytes alternating with thick collagen fibers (see Figure 4.8i). Fibrocartilages occur in sites that are subjected to both heavy pressure and stretch, such as the padlike cartilages (menisci) of the knee and the discs between the vertebrae of the spine (Figure 6.1).

Growth of Cartilage

Cartilage grows in two ways. In **appositional** (ap"o-zish'un-al) **growth** (literally, "growth from outside"), cartilage-forming cells in the surrounding perichondrium secrete new matrix against the external face of the existing cartilage tissue. In **interstitial** (in"ter-stish'al **growth** ("growth from inside"), the lacunae-bound chondrocytes inside the cartilage divide and secrete new matrix, expanding the cartilage from within. Typically, cartilage growth ends during adolescence when the skeleton itself stops growing.

Under certain conditions—during normal bone growth and aging—calcium salts may be deposited in the matrix of cartilage. Note, however, that calcified cartilage is *not* bone; cartilage and bone are always distinct tissues.

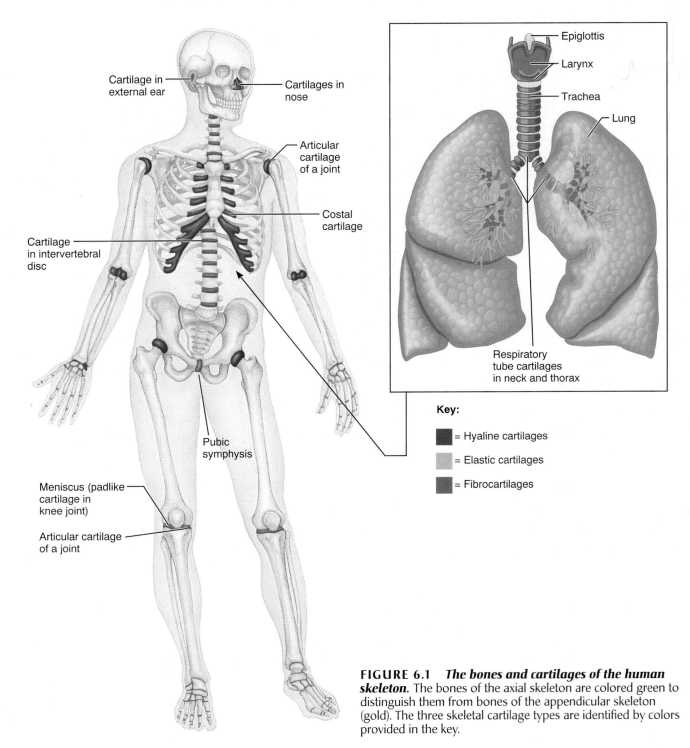

FIGURE 6.1 *The bones and cartilages of the human skeleton.* The bones of the axial skeleton are colored green to distinguish them from bones of the appendicular skeleton (gold). The three skeletal cartilage types are identified by colors provided in the key.

CLASSIFICATION OF BONES

In order to describe them more easily, the 206 named bones of the human skeleton are grouped into the axial and appendicular skeletons. The **axial skeleton** forms the long axis of the body and includes the bones of the skull, vertebral column, and rib cage. Generally speaking these bones are most involved in protecting, supporting, or carrying other body parts. The **appendicular** (ap″en-dik′u-lar) **skeleton** consists of the bones of the upper and lower limbs and the girdles (shoulder bones and hip bones) that attach the limbs to the axial skeleton. Bones of the limbs help to get us from place to place (locomotion) and to manipulate our environment. An overview of the human skeleton appears in Figure 6.1.

Bones come in many sizes and shapes. For example, the tiny pisiform bone of the wrist is the size and shape of a pea, whereas the femur (thigh bone) is nearly 2 feet long in some people and has a large ball-shaped head. The unique shape of each bone fulfills a particular need. The femur, for example, withstands great weight and pressure, and its hollow-cylinder design provides maximum strength with minimum weight.

Bones are classified by their shape as long, short, flat, and irregular (Figure 6.2).

1. Long bones. As their name suggests, long bones are considerably longer than they are wide. A long bone has a shaft plus two ends. All bones of the limbs, except the patella and those of the wrist and ankle, are long bones (see Figure 6.2a). Notice that these bones are named for their elongated shape, *not* their overall size. The three bones in each of your fingers are long bones, even though they are very small.

2. Short bones. Short bones are roughly cube shaped. The bones of the wrist and ankle are examples (see Figure 6.2b).

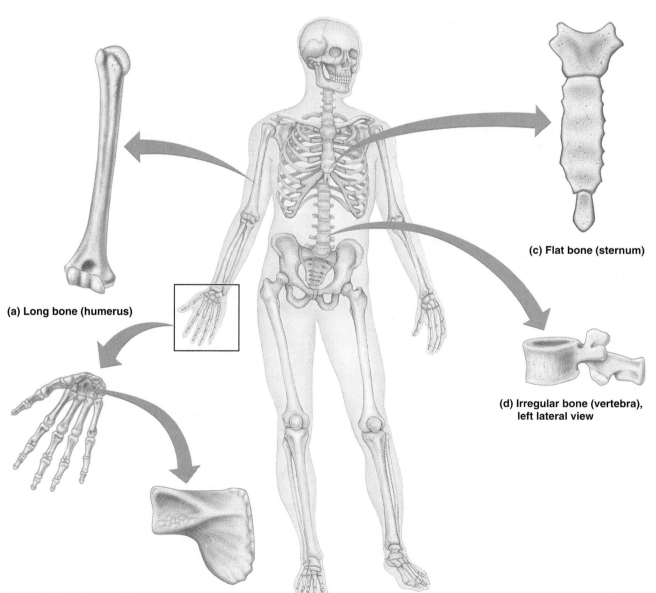

(a) Long bone (humerus)

(b) Short bone (trapezoid)

(c) Flat bone (sternum)

(d) Irregular bone (vertebra), left lateral view

FIGURE 6.2   *Classification of bones on the basis of shape.*

Sesamoid (ses'ah-moid; "shaped like a sesame seed") **bones** are a special type of short bone that form within a tendon (for example, the kneecap, or patella). They vary in size and number in different individuals. Some clearly act to alter the direction of pull of a tendon; the function of others is not known.

3. Flat bones. Flat bones are thin, flattened, and usually a bit curved. The sternum (breastbone), scapulae (shoulder blades), ribs, and most skull bones are flat bones (see Figure 6.2c).

4. Irregular bones. Irregular bones have complicated shapes that fit none of the preceding classes. Examples include the vertebrae and the hip bones (see Figure 6.2d).

FUNCTIONS OF BONES

Besides contributing to body shape and form, our bones perform several important functions:

1. Support. Bones provide a hard framework that supports the body and cradles its soft organs. For example, bones of lower limbs act as pillars to support the body trunk when we stand, and the rib cage supports the thoracic wall.

2. Protection. The fused bones of the skull provide a protective case for the brain. The vertebrae surround the spinal cord, and the rib cage helps protect the vital organs of the thorax.

3. Movement. Skeletal muscles, which attach to bones by tendons, use the bones as levers to move the body and its parts. As a result, we can walk, grasp objects, and breathe. The arrangement of bones and the design of joints determine the types of movement possible.

4. Mineral storage. Bone serves as a reservoir for minerals, the most important of which are calcium and phosphate. The stored minerals are released into the bloodstream as ions for distribution to all parts of the body as needed. Indeed, "deposits" and "withdrawals" of minerals to and from the bones go on almost continuously.

5. Blood cell formation. The bulk of blood cell formation, or *hematopoiesis* (hem"ah-to-poi-e'sis), occurs within the marrow cavities of certain bones.

BONE STRUCTURE

Bones are *organs*. (Recall that an organ contains several different tissues.) Although bone (osseous) tissue dominates bones, they also contain nervous tissue in their nerves, cartilage in their articular cartilages, fibrous connective tissue lining their cavities, and muscle and epithelial tissues in their blood vessels. We will consider bone structure at three structural levels: the gross, microscopic, and chemical levels.

Gross Anatomy

Bone Textures: Compact and Spongy Bone

Every bone of the skeleton has a dense outer layer that looks smooth and solid to the naked eye. This external layer is **compact bone** (Figures 6.3 and 6.4). Internal to this is **spongy bone** (also called *cancellous bone*), a honeycomb of small needle-like or flat pieces called *trabeculae* (trah-bek'u-le; "little beams"). In this network, the open spaces between the trabeculae are filled with red or yellow bone marrow in living bones.

Structure of a Typical Long Bone

With few exceptions, all long bones have the same general structure (Figure 6.3).

Diaphysis The tubular **diaphysis** (di-af'ĭ-sis; *dia* = through, *physis* = growth), or shaft, forms the long axis of the bone. It is constructed of a relatively thick *collar* of compact bone that surrounds a central **medullary cavity** (med'u-lar-e; "middle") or *marrow cavity*. In adults, the medullary cavity contains fat (yellow marrow) and is also called the **yellow bone marrow cavity.**

Epiphyses The **epiphyses** (e-pif'ĭ-sēz) are the bone ends (*epi* = upon). In many cases, they are more expanded than the diaphysis. Compact bone forms the exterior of epiphyses; their interior contains spongy bone. The joint surface of each epiphysis is covered with a thin layer of **articular** (hyaline) **cartilage,** which cushions the opposing bone ends during joint movement and absorbs stress. Between the diaphysis and each epiphysis of an adult long bone is an **epiphyseal line.** This line is a remnant of the epiphyseal plate, a disc of hyaline cartilage that grows during childhood to lengthen the bone (Figures 6.8 and 6.9 on pp. 183 and 184).

Membranes The external surface of the entire bone except the joint surfaces of the epiphyses is covered and protected by glistening white, double-layered membrane called the **periosteum** (per"e-os'te-um; *peri* = around, *osteo* = bone). The outer *fibrous layer* is dense irregular connective tissue. Its inner

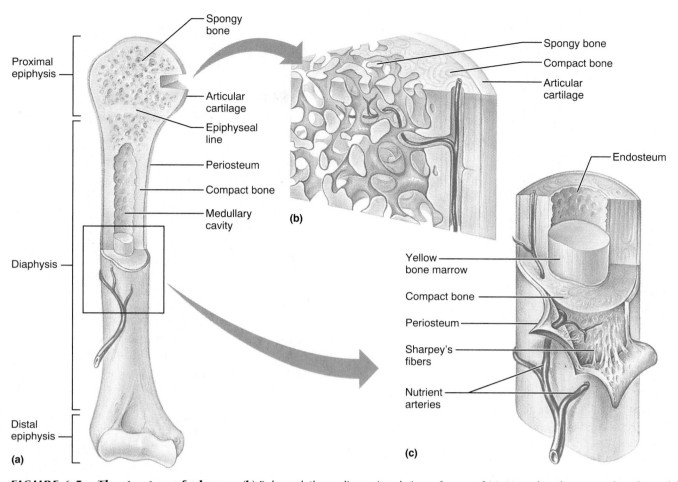

FIGURE 6.3 *The structure of a long bone (humerus of arm).* (**a**) Anterior view with bone sectioned frontally to show the interior at the proximal end.

(**b**) Enlarged, three-dimensional view of spongy bone and compact bone of the epiphysis of (a). (**c**) Enlarged cross-sectional view of the shaft (diaphysis)

of (a). Note that the external surface of the diaphysis is covered by periosteum, but the articular surface of the epiphysis is covered with hyaline cartilage.

osteogenic layer, abutting the bone surface, consists primarily of bone-forming cells, or **osteoblasts** (os'te-o-blasts; "bone germinators"), and bone-destroying cells, or **osteoclasts** ("bone breakers").

The periosteum is richly supplied with nerve fibers, lymphatic vessels, and blood vessels, which enter the bone of the shaft via a **nutrient foramen** (fo-ra'min; "opening").

The periosteum is secured to the underlying bone by **Sharpey's fibers** (Figure 6.3), tufts of collagen fibers that extend from the fibrous layer into the bone matrix. The periosteum also provides insertion or anchoring points for tendons and ligaments. At these points the Sharpey's fibers are exceptionally dense.

Internal bone surfaces are covered with a delicate connective tissue membrane called the **endosteum** (en-dos'te-um; "within the bone"). The endosteum (Figure 6.3) covers the trabeculae of spongy bone in the marrow cavities and lines the canals that pass through compact bone. Like the periosteum, the endosteum contains both osteoblasts and osteoclasts.

Structure of Short, Irregular, and Flat Bones

Like long bones, short, irregular, and flat bones share a simple design: They consist of thin plates of periosteum-covered compact bone on the outside and endosteum-covered spongy bone within. However, these bones are not cylindrical and so they have no shaft or epiphyses. They contain bone marrow (between their trabeculae), but no marrow cavity is present.

Figure 6.4 shows a typical flat bone of the skull. In flat bones, the internal layer of spongy bone is called the **diploë** (dip'lo-e; "folded") and the whole arrangement resembles a stiffened sandwich.

Location of Hematopoietic Tissue in Bones

Because the hematopoietic tissue, **red marrow,** is typically found within the cavities of spongy bone of long bones and in the diploë of flat bones, these cavities are often referred to as **red marrow cavities.** In

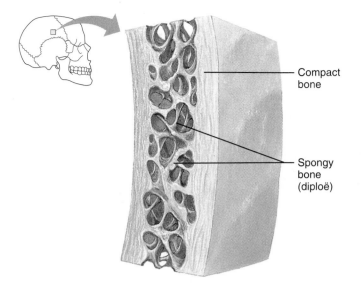

FIGURE 6.4 *Structure of a flat bone.* Flat bones, such as most bones of the skull, consist of a layer of spongy bone (the diploë) sandwiched between two thin layers of compact bone.

newborn infants, the medullary cavity and all areas of spongy bone contain red bone marrow. In most adult long bones, the fat-containing medullary cavity extends well into the epiphysis, and little red marrow is present in the spongy bone cavities. Hence, blood cell production in adult long bones routinely occurs only in the head of the femur and humerus (the long bone of the arm). The red marrow found in the diploë of flat bones (such as the sternum) and in some irregular bones (such as the hip bone) is much more active in hematopoiesis, and these are the sites routinely used for obtaining red marrow samples when problems with the blood-forming tissue are suspected. However, yellow marrow of the medullary cavity can revert to red marrow if a person becomes very anemic and needs enhanced red blood cell production.

Microscopic Structure of Bone

Compact Bone

Viewed by the unaided eye, compact bone looks dense and solid. However, a microscope reveals that it is riddled with passageways that serve as conduits for nerves, blood vessels, and lymphatic vessels (Figure 6.5). The structural unit of compact bone is called the **osteon** (os'te-on) or **Haversian** (ha-ver'shan) **system.** Each osteon is an elongated cylinder oriented parallel to the long axis of the bone. Functionally, osteons are tiny weight-bearing pillars. As shown in the "exploded" view in Figure 6.6, an osteon is a group of hollow tubes of bone matrix, one placed outside the next like the growth rings of a tree

trunk. Each of the matrix tubes is a **lamella** (lah-mel'ah; "little plate"), so compact bone is often called **lamellar bone.** Although all of the collagen fibers in a particular lamella run in a single direction, the collagen fibers in adjacent lamellae always run in opposite directions. This alternating pattern is beautifully designed to withstand torsion stresses—the adjacent lamellae reinforce one another to resist twisting. You can think of the osteon's design as a "twister resister." Collagen fibers are not the only part of bone lamellae that are beautifully ordered. The tiny bone salt crystals align with the collagen fibers and thus also alternate their direction in adjacent lamellae.

Running through the core of each osteon is the **central,** or **Haversian, canal.** This canal contains small blood vessels and nerve fibers that serve the needs of the osteon's cells. Canals of a second type called **perforating,** or **Volkmann's** (folk'mahnz), **canals,** lie at right angles to the long axis of the bone and connect the blood and nerve supply of the periosteum to those in the central canals and the medullary cavity (see Figure 6.5a). Like all internal bone cavities, these canals are lined with endosteum.

Spider-shaped **osteocytes** (mature bone cells) occupy small cavities, or **lacunae** (lah-ku'ne), at the junctions of the lamellae. Hairlike canals called **canaliculi** (kan"ah-lik'u-li) connect the lacunae to each other and to the central canal. The manner in which these canaliculi are formed is interesting. In forming bone, osteoblasts secreting the bone matrix maintain contact with one another by tentacle-like projections containing gap junctions. Then, as the matrix hardens and the maturing cells become trapped within it, an entire system of tiny canals (filled with tissue fluid and containing the osteocyte extensions) is formed. The canaliculi tie all the osteocytes in an osteon together, permitting nutrients and wastes to be easily relayed from one osteocyte to the next throughout the osteon. Although bone matrix is hard and impermeable to nutrients, its canaliculi and cell-to-cell relays (via gap junctions) allow bone cells to be very well nourished. The function of osteocytes is to maintain the bone matrix. If they die, the surrounding matrix is resorbed.

Not all lamellae of compact bone are part of osteons. Lying between intact osteons are incomplete lamellae called **interstitial** (in"ter-stish'al) **lamellae** (Figure 6.5c). These fill the gaps between forming osteons or are remnants of osteons that have been cut through by bone remodeling (discussed later). Additionally, there are **circumferential lamellae,** located just deep to the periosteum and just superficial to the endosteum, that extend around the entire circumference of the shaft. Circumferential lamellae effectively resist twisting of the long bone as a whole.

? *What membrane lines the internal canals and covers the trabeculae of spongy bone?*

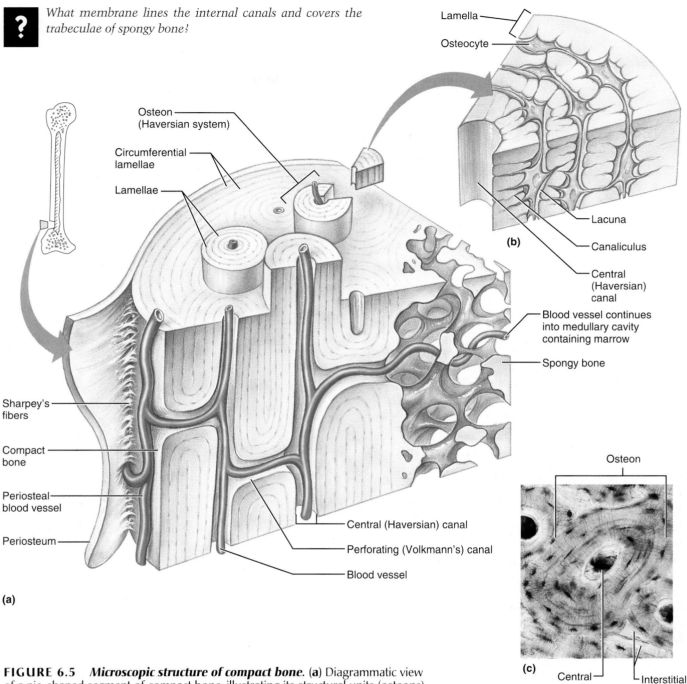

FIGURE 6.5 *Microscopic structure of compact bone.* (**a**) Diagrammatic view of a pie-shaped segment of compact bone, illustrating its structural units (osteons). (**b**) This inset shows a more highly magnified view of a portion of one osteon. Note the position of osteocytes in lacunae (cavities in the matrix). (**c**) Photomicrograph of a cross-sectional view of one osteon and portions of others (144 ×).

Spongy Bone

In contrast to compact bone, spongy bone, consisting of trabeculae, looks like a poorly organized, even haphazard, tissue (see Figure 6.4 and Figure 6.3b). However, the trabeculae align precisely along lines of stress and help the bone to resist that stress as much as possible. Thus, the tiny bone struts are as carefully positioned as the flying buttresses that support the walls of a Gothic cathedral.

Only a few cell layers thick, trabeculae contain irregularly arranged lamellae and osteocytes interconnected by canaliculi. No osteons are present. Nutrients reach the osteocytes of spongy bone by diffusing through the canaliculi from capillaries in the endosteum surrounding the trabeculae.

Endosteum.

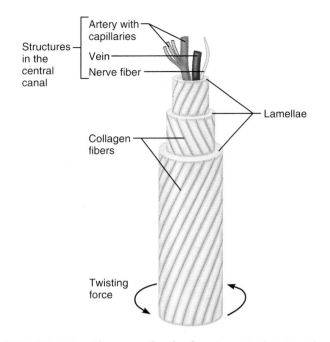

Structures in the central canal:
- Artery with capillaries
- Vein
- Nerve fiber

Lamellae

Collagen fibers

Twisting force

FIGURE 6.6 *Diagram of a single osteon.* In this view, the osteon is drawn as if pulled out like a telescope to illustrate the individual lamellae composing it. The slanted lines within each lamella indicate the direction in which its collagen fibers run within the matrix.

Chemical Composition of Bone

Bone has both organic and inorganic components. Its *organic components* include the cells (osteoblasts, osteocytes, and osteoclasts) and **osteoid** (os'te-oid), the organic part of the matrix. Osteoid, which makes up approximately one-third of the matrix, includes ground substance (composed of proteoglycans and glycoproteins) and collagen fibers, all of which are made and secreted by osteoblasts. These organic substances, particularly collagen, not only contribute to a bone's structure but also to the flexibility and great tensile strength that allow the bone to resist stretch and twisting.

The balance of bone tissue (65% by mass) consists of inorganic **hydroxyapatites** (hi-drok"se-ap'ah-tītz), or *mineral salts*, largely calcium phosphates. Calcium salts are present in the form of tiny crystals that lie around the collagen fibers in the extracellular matrix. The crystals are tightly packed and account for the most notable characteristic of bone—its exceptional hardness, which allows it to resist compression.

The proper combination of organic and inorganic matrix elements allows bones to be exceedingly durable and strong without being brittle. It is surprising to hear that healthy bone is half as strong as steel in resisting compression and fully as strong as steel in resisting tension.

TABLE 6.1	Bone Markings

Name of bone marking	Description
Projections that are sites of muscle and ligament attachment	
Tuberosity (too"bĕ-ros'ĭ-te)	Large rounded projection; may be roughened
Crest	Narrow ridge of bone; usually prominent
Trochanter (tro-kan'ter)	Very large, blunt, irregularly shaped process (The only examples are on the femur.)
Line	Narrow ridge of bone; less prominent than a crest
Tubercle (too'ber-kl)	Small rounded projection or process
Epicondyle (ep"ĭ-kon'dīl)	Raised area on or above a condyle
Spine	Sharp, slender, often pointed projection
Process	Any bony prominence
Projections that help to form joints	
Head	Bony expansion carried on a narrow neck
Facet	Smooth, nearly flat articular surface
Condyle (kon'dīl)	Rounded articular projection
Ramus (ra'mus)	Armlike bar of bone
Depressions and openings allowing blood vessels and nerves to pass	
Meatus (me-a'tus)	Canal-like passageway
Sinus	Cavity within a bone, filled with air and lined with mucous membrane
Fossa (fos'ah)	Shallow, basinlike depression in a bone, often serving as an articular surface
Groove	Furrow
Fissure	Narrow, slitlike opening
Foramen (fo-ra'men)	Round or oval opening through a bone

Because of bone salts, bones last long after death and provide an enduring "monument." In fact, skeletal remains many centuries old have revealed the shapes, sizes, races, and sexes of ancient peoples, the kinds of work they did, and many of the ailments they suffered (arthritis, for example).

Bone Markings

The external surfaces of bones are rarely smooth and featureless. Instead, they display bulges, depressions, and holes, which serve as sites of muscle, ligament, and tendon attachment, as joint surfaces, or as conduits for blood vessels and nerves. These **bone markings** are named in different ways. Projections that grow outward from the bone surface include heads, trochanters, spines, and others, and each has certain distinguishing features and functions. Depressions and openings include fossae, sinuses, foramina, and grooves. The most important types of bone markings are described in Table 6.1. You should familiarize yourself with these terms because you will meet them again as identifying marks of the individual bones described in Chapter 7 and studied in the lab.

BONE DEVELOPMENT

Osteogenesis (os″te-o-jen′ĕ-sis) and **ossification** are synonyms indicating the process of bone tissue formation (*os* = bone, *genesis* = beginning). In embryos this process leads to the *formation of the bony skeleton.* Bone growth, another form of ossification, goes on until early adulthood as we continue to increase in size. Bones are capable of growing in thickness throughout life. However, ossification in adults serves mainly for *remodeling* and repair of bones.

Formation of the Bony Skeleton

Before week 8, the skeleton of a human embryo is constructed entirely from fibrous membranes and hyaline cartilage. Bone tissue begins to develop at about this time and eventually replaces most of the existing fibrous or cartilage structures. When a bone develops from a fibrous membrane, the process is called *intramembranous ossification,* and the bone is called a **membrane bone.** Bone formation that occurs by replacing hyaline cartilage is called *endochondral ossification* (*endo* = within; *chondro* = cartilage), and the resulting bone is called a **cartilage,** or **endochondral, bone.**

Intramembranous Ossification

Intramembranous ossification results in the formation of the cranial bones of the skull and the clavicles. Notice that all bones formed by this route are flat bones. Fibrous connective tissue membranes formed by *mesenchymal cells* serve as the initial supporting structures on which ossification begins to occur at about the eighth week of development. Essentially, the process involves four major steps, depicted in Figure 6.7. The final result is a flat bone, as illustrated earlier in Figure 6.4 (p. 178).

FIGURE 6.7 *The stages in intramembranous ossification.* Diagrams 3 and 4 represent much lower magnification than the diagrams preceding them.

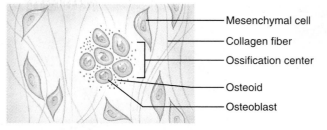

① **An ossification center appears in the fibrous connective tissue membrane.**
• Selected centrally located mesenchymal cells cluster and differentiate into osteoblasts, forming an ossification center.

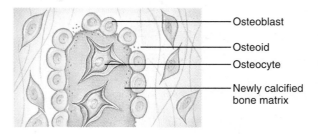

② **Bone matrix (osteoid) is secreted within the fibrous membrane.**
• Osteoblasts begin to secrete osteoid, which is mineralized within a few days.
• Trapped osteoblasts become osteocytes.

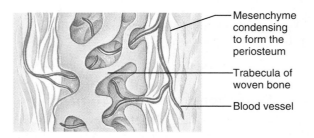

③ **Woven bone and periosteum form.**
• Accumulating osteoid is laid down between embryonic blood vessels, which form a random network. The result is a network (instead of lamellae) of trabeculae (woven bone).
• Vascularized mesenchyme condenses on the external face of the woven bone and becomes the periosteum.

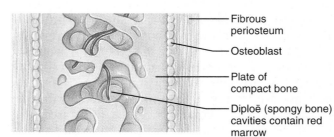

④ **Bone collar of compact bone forms and red marrow appears.**
• Trabeculae just deep to the periosteum thicken, forming a woven bone collar that is later replaced with mature lamellar bone.
• Spongy bone (diploë), consisting of distinct trabeculae, persists internally and its vascular tissue becomes red marrow.

? *What is the composition of the periosteal bud?*

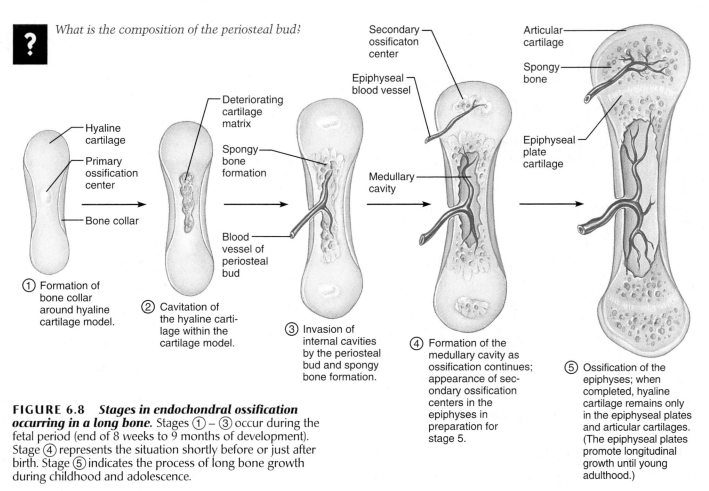

FIGURE 6.8   *Stages in endochondral ossification occurring in a long bone.* Stages ① – ③ occur during the fetal period (end of 8 weeks to 9 months of development). Stage ④ represents the situation shortly before or just after birth. Stage ⑤ indicates the process of long bone growth during childhood and adolescence.

Endochondral Ossification

Essentially all bones of the skeleton from the base of the skull down (except for the clavicles) form by the process of **endochondral** (en"do-kon'dral) **ossification.** This process, which begins in the second month of development, uses hyaline cartilage "bones" formed earlier as models, or patterns, for bone construction. It is more complex than intramembranous ossification because the hyaline cartilage must be broken down as ossification proceeds. We will use a forming long bone as our example.

The formation of a long bone typically begins in the center of the hyaline cartilage shaft at a region called the **primary ossification center.** First, the perichondrium (the fibrous membrane covering the hyaline cartilage "bone") becomes infiltrated with blood vessels, converting it to a vascularized periosteum. As a result of this change in nutrition, the underlying mesenchymal (osteoprogenitor) cells specialize into osteoblasts. The stage is now set for ossification to begin, as illustrated in Figure 6.8 and described next.

① **A bone collar forms around the diaphysis of the hyaline cartilage model.** Osteoblasts of the newly converted periosteum secrete osteoid against the hyaline cartilage diaphysis, encasing it externally in a bone collar.

② **Cartilage in the center of the diaphysis calcifies and then cavitates.** As the bone collar forms externally, the chondrocytes within the shaft hypertrophy (enlarge) and signal the surrounding cartilage matrix to calcify. Because calcified cartilage matrix is impermeable to diffusing nutrients, the chondrocytes die and the matrix they maintained begins to deteriorate. Although this opens up cavities, the deteriorating hyaline cartilage model is stabilized externally by the bone collar. Elsewhere, the cartilage remains healthy and continues to grow briskly, causing the entire cartilage model to elongate.

③ **The periosteal bud invades the internal cavities and spongy bone forms.** In the third month of development, the forming cavities are invaded by a collection of elements called the **periosteal bud.** Most importantly, the bud contains a nutrient artery and vein, lymphatics, nerve fibers, red marrow elements, osteoblasts, and osteoclasts. The entering osteoclasts partially erode the calcified cartilage ma-

Blood vessels, nerve fibers, lymph vessels, osteoblasts, osteoclasts, and red marrow elements.

trix, and the osteoblasts secrete osteoid around the remaining fragments of hyaline cartilage, forming bone-covered cartilage trabeculae. In this way, the earliest version of spongy bone in a developing long bone forms.

④ **The diaphysis elongates and a medullary cavity forms.** As the primary ossification center enlarges, spreading proximally and distally, osteoclasts break down the newly formed spongy bone and open up a medullary cavity in the center of the shaft. The formation of this cavity is the final step in ossification of the shaft. Throughout the fetal period, the rapidly growing epiphyses consist only of cartilage, and the hyaline cartilage models continue to elongate by division of viable cartilage cells at the epiphyses. Since cartilage is calcifying, being eroded, and then being replaced by bony spicules on the epiphyseal surfaces facing the medullary cavity, ossification "chases" cartilage formation along the length of the shaft. This process is explained in more detail below. (See Growth in Length of Long Bones.)

⑤ **The epiphyses ossify.** When we are born, most of our long bones have a bony diaphysis surrounding remnants of spongy bone, a widening medullary cavity, and two cartilaginous epiphyses. Shortly before or after birth, **secondary ossification centers** appear in one or both epiphyses, and the epiphyses gain bony tissue. (Typically, the large long bones form secondary centers in both epiphyses, whereas the small long bones form only one secondary ossification center.) The cartilage in the center of the epiphysis calcifies and deteriorates, opening up cavities that allow entry of a periosteal bud. Then bone trabeculae appear, just as they did earlier in the primary ossification center. (In short bones, only the primary ossification center is formed. Most irregular bones develop from several distinct ossification centers.)

Secondary ossification reproduces almost exactly the events of primary ossification, except that the spongy bone in the interior is retained and no medullary cavity forms in the epiphyses. When secondary ossification is complete, hyaline cartilage remains only at two places: (1) on the epiphyseal surfaces, as the *articular cartilages,* and (2) at the junction of the diaphysis and epiphysis, where it forms the **epiphyseal plates,** or *growth plates.*

Postnatal Bone Growth

During infancy and youth, long bones lengthen entirely by interstitial growth of the epiphyseal plates, and all bones grow in thickness by appositional growth. Most bones stop growing during adolescence or in early adulthood. However, some facial bones, such as those of the nose and lower jaw, continue to grow almost imperceptibly throughout life.

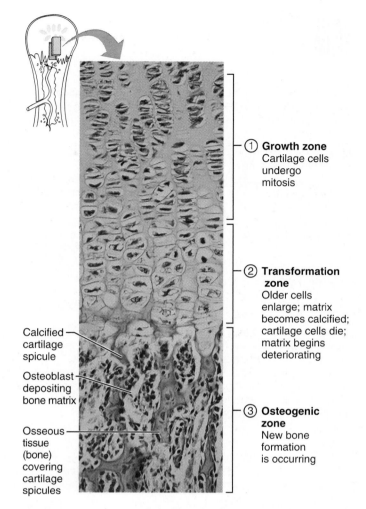

① **Growth zone** Cartilage cells undergo mitosis

② **Transformation zone** Older cells enlarge; matrix becomes calcified; cartilage cells die; matrix begins deteriorating

③ **Osteogenic zone** New bone formation is occurring

Calcified cartilage spicule

Osteoblast depositing bone matrix

Osseous tissue (bone) covering cartilage spicules

FIGURE 6.9 *Growth in length of a long bone.* The region of the epiphyseal plate closest to the epiphysis (distal face) contains resting cartilage cells. As shown in the photomicrograph (250 ×), the cells of the epiphyseal plate proximal to the resting cartilage area are arranged in three functionally different zones—growth, transformation, and osteogenic.

Growth in Length of Long Bones

The process of longitudinal bone growth mimics many of the events of endochondral ossification. On the side of the epiphyseal plate closest to the epiphysis, the cartilage is relatively quiescent and inactive. But the epiphyseal plate cartilage abutting the shaft organizes into a pattern that allows fast, efficient growth. The cartilage cells here form tall columns, like coins in a stack. The cells at the "top" (epiphyseal side) of the stack (zone 1, the *growth zone* in Figure 6.9.) divide quickly, pushing the epiphysis away from the diaphysis, causing the entire long bone to lengthen (see left side of Figure 6.10).

Meanwhile, the older chondrocytes in the stack, which are closer to the shaft (zone 2, the *transformation zone* in Figure 6.9), hypertrophy, and their lacunae erode and enlarge. Subsequently, the surrounding cartilage matrix calcifies and these

? *What do you think a long bone would look like at the end of adolescence if remodeling did not occur?*

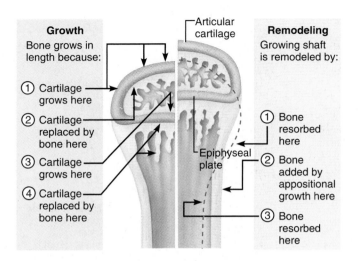

Growth
Bone grows in length because:

① Cartilage grows here

② Cartilage replaced by bone here

③ Cartilage grows here

④ Cartilage replaced by bone here

Articular cartilage

Epiphyseal plate

Remodeling
Growing shaft is remodeled by:

① Bone resorbed here

② Bone added by appositional growth here

③ Bone resorbed here

FIGURE 6.10 *Long bone growth and remodeling during youth.* The events indicated at the left depict the process of endochondral ossification that occurs at the articular cartilages and epiphyseal plates as the bone grows in length. The events indicated at the right reveal the process of bone remodeling that occurs during long bone growth to maintain proper bone proportions.

chondrocytes die and deteriorate. This leaves long spicules of calcified cartilage at the epiphysis-diaphysis junction, which look like stalactites hanging from the roof of a cave. These become part of zone 3, the *osteogenic zone,* and are invaded by marrow elements from the medullary cavity. The cartilage spicules are partly eroded by osteoclasts and then quickly covered with bone matrix by osteoblasts, forming spongy bone. The tips of the spongy bone spicules are eventually digested by osteoclasts; thus, the medullary cavity also grows longer as the long bone lengthens. During the growth period, the epiphyseal plate maintains a constant thickness because the rate of cartilage growth on its epiphyseal face is balanced by its replacement with bony tissue on its diaphyseal face.

Longitudinal growth is accompanied by almost continuous remodeling of the epiphyseal ends to maintain the proper proportions between the diaphysis and epiphyses (see Figure 6.10). Bone remodeling, involving both new bone formation and bone resorption (destruction), is described in more detail below in conjunction with the changes that occur in adult bones.

As adolescence draws to an end, the chondroblasts of the epiphyseal plates divide less often and the plates become thinner and thinner until they are entirely replaced by bone tissue. Longitudinal bone growth ends when the bone of the epiphysis and diaphysis fuses. This process, called *epiphyseal plate closure,* happens at about 18 years of age in females and 21 years of age in males. However, a bone can still increase in diameter or thickness by appositional growth if it is severely stressed by excessive muscle activity or body weight.

Growth in Width (Thickness)

Growing bones widen as they lengthen. As with cartilages, bones increase in thickness or, in the case of long bones, diameter, by the process of appositional growth. Osteoblasts beneath the periosteum secrete bone matrix on the external bone surface as osteoclasts on the endosteal surface of the diaphysis remove bone (see Figure 6.10). However, there is normally slightly less breaking down than building up. This unequal process produces a thicker, stronger bone but prevents it from becoming too heavy.

Hormonal Regulation of Bone Growth During Youth

The growth of bones that occurs throughout youth is exquisitely controlled by a symphony of hormones. During infancy and childhood, the single most important stimulus of epiphyseal plate activity is *growth hormone* released by the anterior pituitary gland. Thyroid hormones (T_3 and T_4) modulate the activity of growth hormone, ensuring that the skeleton has proper proportions as it grows. At puberty, male and female sex hormones (testosterone and estrogens, respectively) are released in increasing amounts. Initially these sex hormones promote the growth spurt typical of the adolescent, as well as the masculinization or feminization of specific parts of the skeleton. Later they induce epiphyseal plate closure, ending longitudinal bone growth.

Excesses or deficits of any of these hormones can result in obviously abnormal skeletal growth. For example, hypersecretion of growth hormone in children results in excessive height (gigantism), and deficits of growth hormone or thyroid hormone produce characteristic types of dwarfism. Hormone activity is considered in detail in Chapter 17.

BONE HOMEOSTASIS: REMODELING AND REPAIR

Bones appear to be the most lifeless of body organs, and may even summon images of a graveyard. But as you have just learned, appearances can be deceiving. Bone is a dynamic and active tissue, and small-scale changes in bone architecture occur continually. Every week we recycle 5 to 7% of our bone mass, and as much as half a gram of calcium may enter or leave the adult skeleton each day! Spongy bone is replaced every 3 to 4 years; compact bone, every 10 years or

Its shaft would still be its original length, and it would have two very elongated and expanded ends (epiphyses).

so. And when we break bones—the commonest disorder of bone homeostasis—they undergo a remarkable process of self-repair that we will describe shortly.

Bone Remodeling

In the adult skeleton, bone deposit and bone resorption (removal) occur at the periosteal and endosteal surfaces. Together, the two processes constitute **bone remodeling,** and they are coupled and coordinated by "packets" of adjacent osteoblasts and osteoclasts called *remodeling units.* In healthy young adults, total bone mass remains constant, an indication that the rates of bone deposit and resorption are essentially equal. The remodeling process is not uniform, however, and some bones (or bone areas) are very actively remodeled while others are not. For example, the distal part of the femur, or thigh bone, is fully replaced every five to six months, whereas its shaft is altered much more slowly.

Bone deposit occurs where bone is injured or added bone strength is required. For optimal bone deposit, a healthy diet rich in proteins, vitamin C (for collagen synthesis), vitamin D (essential for absorption of dietary calcium), vitamin A (needed for a balance between deposit and removal of bone), and several minerals (calcium, phosphorus, magnesium, and manganese, to name a few) is essential.

Sites of new matrix deposit by osteocytes are revealed by the presence of an **osteoid seam,** an unmineralized band of gauzy-looking bone matrix 10 to 12 μm wide. Between the osteoid seam and the older mineralized bone, there is an abrupt transition called the **calcification front.** Because the osteoid seam is always of constant width and the change from unmineralized to mineralized matrix is sudden, it seems that the osteoid must mature for about a week before it can be calcified. The precise trigger for calcification of the matrix is still controversial. However, one critical factor is the product of the local concentrations of calcium and phosphate (P_i) ions (the $Ca^{2+} \cdot P_i$ product). When this product reaches a certain level, tiny crystals of hydroxyapatite form spontaneously and then catalyze further crystallization of calcium salts in the area. Other factors involved are matrix proteins that bind and concentrate calcium, and a rich supply of the enzyme **alkaline phosphatase** (shed by the osteoblasts), which is essential for mineralization. Once proper conditions are present, calcium salt deposit occurs all at once and with great precision throughout the "matured" matrix, as the calcium salt crystals pack together beside and within the collagen fibers in a pattern that reduces the risk of cracks in a stressed bone.

Bone resorption is accomplished by osteoclasts, giant multinucleate cells that arise from the same *hematopoietic stem cells* that differentiate into macrophages. Osteoclasts move along a bone surface, digging pits or grooves called *resorption bays* as they break down the bone matrix. Their area that touches the bone is highly folded to form a ruffled membrane that clings tightly to the bone, sealing off the area of bone destruction. The ruffled border secretes (1) *lysosomal enzymes* that digest the organic matrix and (2) *acids* that convert the calcium salts into soluble forms that pass easily into solution. Osteoclasts may also phagocytize the demineralized matrix and dead osteocytes. The digested matrix end products and dissolved minerals are then endocytosed, transported across the osteoclast (by transcytosis) and released at the opposite side where they enter the interstitial fluid and then the blood.

Control of Remodeling

The extensive remodeling that goes on continuously in the skeleton is regulated by two control loops that serve different "masters." One is a negative feedback hormonal mechanism that maintains Ca^{2+} homeostasis in the blood. The other involves responses to mechanical and gravitational forces acting on the skeleton.

The hormonal mechanism described below becomes much more meaningful when you understand calcium's importance in the body. Ionic calcium (Ca^{2+}) is necessary for an amazing number of physiological processes, including transmission of nerve impulses, muscle contraction, blood coagulation, secretion by glands and nerve cells, and cell division. The human body contains some 1200–1400 g of calcium, over 99% of which is present as bone minerals. Most of the balance is within body cells. Less than 1.5 g is present in blood, and blood Ca^{2+} levels are normally maintained within the very narrow range of 9–11 mg per 100 ml of blood by the hormonal control loop. Calcium is absorbed from the intestine under the control of vitamin D metabolites. The daily requirement of calcium is 400–800 mg from birth until the age of 10, and 1200–1500 mg from ages 11 to 24.

Hormonal Mechanism The hormonal mechanism reflects the interaction of **parathyroid hormone (PTH),** produced by the parathyroid glands, and **calcitonin** (kal"sĭ-to'nin), produced by parafollicular cells (C cells) of the thyroid gland (Figure 6.11). PTH is released when blood levels of ionic calcium decline. It stimulates osteoclasts to resorb bone, releasing calcium to the blood. Osteoclasts are no respecters of matrix age. When activated, they break down both old and fairly new matrix. Only osteoid, which lacks calcium salts, escapes digestion. As blood concentrations of calcium rise, the stimulus for PTH release ends.

Calcitonin, secreted when blood calcium levels rise, inhibits bone resorption and encourages calcium

? *What is the result of the negative feedback of calcitonin?*

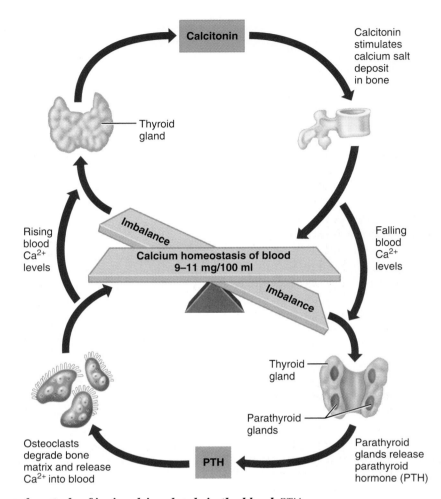

FIGURE 6.11 *Hormonal controls of ionic calcium levels in the blood.* PTH and calcitonin operate in negative feedback control systems that influence each other.

salt deposit in bone matrix, effectively reducing blood calcium levels (Figure 6.11). As blood calcium levels fall, calcitonin release wanes as well.

These hormonal controls act to maintain blood calcium homeostasis, rather than the skeleton's strength or well-being. In fact, if blood calcium levels are low for an extended time, the bones will become so demineralized that they develop large, punched out–looking holes. Thus, the bones serve as a storehouse from which ionic calcium is drawn as needed to meet the body's needs.

🔺 *Homeostatic Imbalance*

Minute changes from the homeostatic range for blood calcium can lead to severe neuromuscular problems ranging from hyperexcitability (when blood calcium levels are too low) to nonresponsive-

ness and inability to function (with high blood levels of Ca^{2+}). In addition, sustained *hypercalcemia* (hi″per-kal-se′me-ah), high blood levels of Ca^{2+}, can lead to undesirable deposits of calcium salts in the blood vessels, kidneys, and other soft organs, which may hamper the functioning of these organs. ■

Response to Mechanical Stress Mechanical stress (muscle pull) and gravity also promote skeletal remodeling. However, this set of controls serves the needs of the skeleton itself, keeping the bones strong where stressors are acting. *Wolff's* law holds that a bone grows or remodels in response to the forces or demands placed on it. The first thing to understand is that a bone's anatomy reflects the common stresses it encounters. For example, bones are loaded (stressed) when weight bears down on them or muscles pull on them. This loading is usually off center, however, and tends to *bend* the bone. Bending compresses the bone on one side and subjects it to tension (stretches it) on the other (Figure 6.12). But both forces are minimal toward the center of the bone (they cancel each other out), so a bone can "hollow out" for lightness (using spongy bone instead of compact bone) without jeopardy.

Other observations explained by this theory include these: (1) long bones are thickest midway along the shaft, exactly where bending stresses are greatest (bend a stick and it will split near the middle); (2) curved bones are thickest where they are most likely to buckle; (3) the trabeculae of spongy bone form trusses, or struts, along lines of compression; and (4) large, bony projections occur where heavy, active muscles attach. (The bones of weight lifters have enormous thickenings at the attachment sites of the most used muscles.) Wolff's law also explains the featureless bones of the fetus and the atrophied bones of bedridden people—situations in which bones are not stressed.

How do mechanical forces communicate with the cells responsible for remodeling? The mechanisms by which bone responds to mechanical stimuli are still uncertain. We do know that deforming a bone produces an electrical current. Since compressed and stretched regions are oppositely charged, it has been suggested that electrical signals direct the remodeling process. This principle and some of the devices currently used to speed bone repair and the healing of fractures are discussed in *A Closer Look*.

The skeleton is continuously subjected to both hormonal influences and mechanical forces. At the risk of constructing too large a building on too small a foundation, we can speculate that the hormonal loop determines *whether* and *when* remodeling will occur in response to changing blood calcium levels, and mechanical stress determines *where* remodeling occurs. For example, if bone must be broken down to increase blood calcium levels, PTH is released and targets the osteoclasts. However, mechanical forces determine *which* osteoclasts are most sensitive to PTH stimulation, so that bone in the *least* stressed areas (which is temporarily dispensable) is broken down.

Repair of Fractures

Despite their remarkable strength, bones are susceptible to **fractures,** or breaks. During youth, most fractures result from exceptional trauma that twists or smashes the bones (sports injuries, automobile accidents, and falls, for example). In old age, bones thin and weaken, and fractures occur more often.

Fractures may be classified in different ways: (1) by position of the bone ends after fracture (in **nondisplaced fractures** the bone ends retain their normal position; in **displaced fractures** the bone ends are out of normal alignment); (2) by complete-

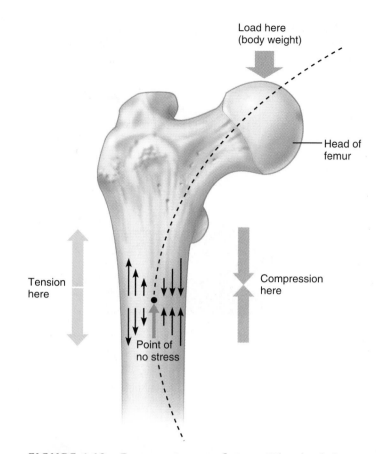

FIGURE 6.12 *Bone anatomy and stress.* When loaded, bones are subjected to bending stress. Using the femur of the thigh as an example, we show that body weight transmitted to the head of the femur threatens to bend the bone along the indicated arc. This bending compresses the bone on one side (converging arrows), and stretches it promoting tension on the other side (diverging arrows). These two forces cancel each other internally, so much less bone material is needed internally than superficially.

ness of the break (if the bone is broken through it is a **complete fracture;** if not, the fracture is **incomplete**); (3) by the orientation of the break to the long axis of the bone (if it parallels the long axis, the fracture type is **linear;** if it is perpendicular to the bone's long axis, it is a **transverse fracture**); and (4) by whether the bone ends penetrate the skin (if so, the fracture is called an **open** *(compound)* **fracture;** if not, it is a **closed** *(simple)* **fracture**). Several common types of fractures are illustrated and explained in Table 6.2.

A fracture is treated by *reduction,* the realignment of the broken bone ends. In **closed reduction,** the bone ends are coaxed back into position by the physician's hands. In **open reduction,** the bone ends are secured together surgically with pins or wires. After the broken bone is reduced, it is immobilized by a cast or traction to allow the healing process to begin. The healing time of a simple fracture is six to eight weeks, but it is much longer for large, weight-bearing

TABLE 6.2	Common Types of Fractures		
Fracture type	*Description and comments*	*Fracture type*	*Description and comments*
Comminuted	Bone fragments into 3 or more pieces Particularly common in the aged, whose bones are more brittle	Compression	Bone is crushed Common in porous bones (i.e., osteoporotic bones) subjected to extreme trauma, i.e., as in a fall
Spiral	Ragged break occurs when excessive twisting forces are applied to a bone Common sports fracture	Epiphyseal	Epiphysis separates from the diaphysis along the epiphyseal plate Tends to occur where cartilage cells are dying and calcification of the matrix is occurring
Depressed	Broken bone portion is pressed inward Typical of skull fracture	Greenstick	Bone breaks incompletely, much in the way a green twig breaks. Only one side of the shaft breaks; the other side bends Common in children, whose bones have relatively more organic matrix and are more flexible than those of adults

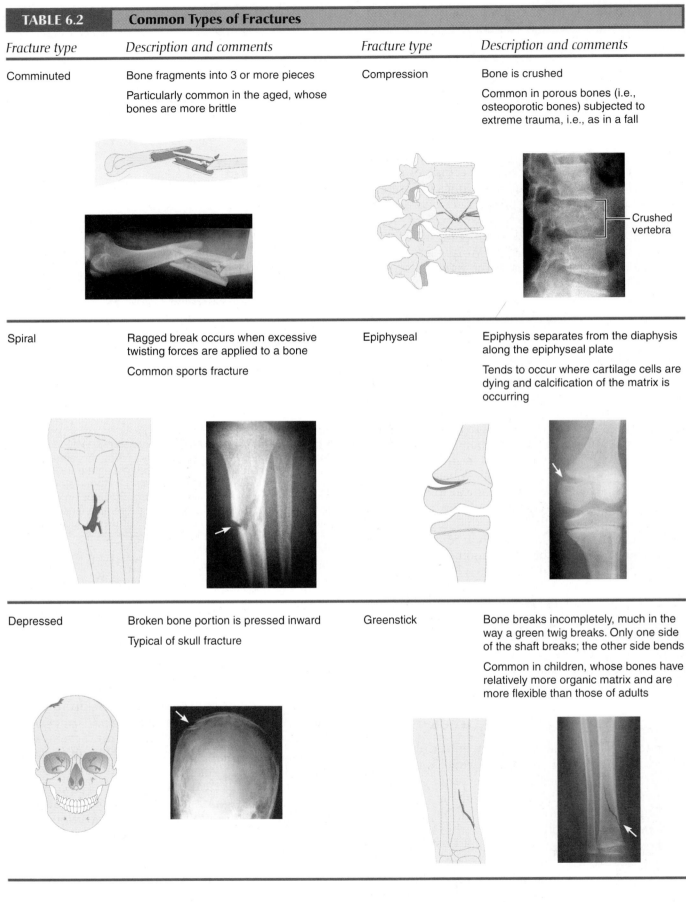

Crushed vertebra

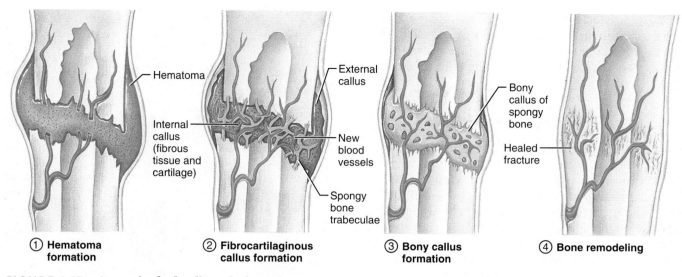

① **Hematoma formation**

② **Fibrocartilaginous callus formation**

③ **Bony callus formation**

④ **Bone remodeling**

FIGURE 6.13 *Stages in the healing of a bone fracture.*

bones and for bones of elderly people (because of their poorer circulation).

The repair process in a simple fracture involves four major phases (Figure 6.13).

① **Hematoma formation.** When a bone breaks, blood vessels in the bone and periosteum, and perhaps in surrounding tissues, are torn and hemorrhage. As a result, a **hematoma** (he"mah-to'mah), a mass of clotted blood, forms at the fracture site. Soon, bone cells deprived of nutrition begin to die, and the tissue at the site becomes swollen, painful, and obviously inflamed.

② **Fibrocartilaginous callus formation.** Within a few days, soft *granulation tissue,* also called the *soft callus* (kal'us; "hard skin") forms. Several events help to form the granulation tissue: Capillaries grow into the hematoma and phagocytic cells invade the area and begin to clean up the debris. Meanwhile, fibroblasts and osteoblasts migrate into the fracture site from the nearby periosteum and endosteum and begin reconstructing the bone. The fibroblasts produce collagen fibers that span the break and connect the broken bone ends, and some differentiate into chondroblasts that secrete cartilage matrix. Within this mass of repair tissue, osteoblasts begin forming spongy bone, but those farthest from the capillary supply secrete an externally bulging cartilaginous matrix that later calcifies. This entire mass of repair tissue, now called the **fibrocartilaginous callus,** splints the broken bone.

③ **Bony callus formation.** As a result of the labors of the osteoblasts, new bone trabeculae begin to appear in the fibrocartilaginous callus and gradually convert it to a **bony (hard) callus** of spongy (actually, woven) bone. Bony callus formation begins 3 to 4

weeks after injury and continues until a firm union is formed 2 to 3 months later.

④ **Bone remodeling.** Beginning during bony callus formation and continuing for several months after, the bony callus is remodeled. The excess material on the bone shaft exterior and within the medullary cavity is removed, and compact bone is laid down to reconstruct the shaft walls. The final structure of the remodeled area resembles that of the original unbroken bony region because it responds to the same set of mechanical stressers.

HOMEOSTATIC IMBALANCES OF BONE

Imbalances between bone formation and bone resorption underlie nearly every disease that influences the adult skeleton.

Osteomalacia and Rickets

Osteomalacia (os"te-o-mah-la'she-ah; "soft bones") includes a number of disorders in which the bones are inadequately mineralized. Osteoid is produced, but calcium salts are not deposited, so bones soften and weaken. The main symptom is pain when weight is put on the affected bones.

Rickets, the analogous disease in children, is accompanied by many of the same signs and symptoms. Because young bones are still growing rapidly, rickets is much more severe than adult osteomalacia. Bowed legs and deformities of the pelvis, skull, and rib cage are common. Because the epiphyseal plates cannot be calcified, they continue to widen, and the ends of long bones become visibly enlarged and abnormally long.

Osteomalacia and rickets are caused by insufficient calcium in the diet or vitamin D deficiency. Hence, drinking vitamin D–fortified milk and exposing the skin to sunlight usually cure these disorders.

Osteoporosis

For most of us, the phrase "bone problems of the elderly" brings to mind the stereotype of a victim of osteoporosis—a hunched-over old woman shuffling behind her walker. **Osteoporosis** (os"te-o-po-ro'sis) refers to a group of diseases in which bone resorption outpaces bone deposit. The composition of the matrix remains normal but bone mass is reduced, and the bones become more porous and lighter (Figure 6.14). Even though osteoporosis affects the entire skeleton, the spongy bone of the spine is most vulnerable, and compression fractures of the vertebrae are common. The femur (thigh bone), particularly its neck, is also very susceptible to fracture (called a *broken hip*) in people with osteoporosis.

Osteoporosis occurs most often in the aged, particularly in postmenopausal women, but cessation of sex hormone synthesis can promote the disorder in both sexes. Both estrogen and testosterone help to maintain the health and normal density of the skeleton by restraining osteoclast activity and by promoting deposit of new bone. After menopause, however, estrogen secretion wanes and estrogen deficiency is strongly implicated in osteoporosis in older women. Other factors that contribute to osteoporosis include insufficient exercise to stress the bones, a diet poor in calcium and protein, abnormal vitamin D receptors, smoking (which reduces estrogen levels), and hormone-related conditions (such as the intake of corticosteroid drugs, hyperthyroidism, and diabetes mellitus). In addition, osteoporosis can occur at any age as a result of immobility.

Osteoporosis has traditionally been treated with calcium and vitamin D supplements, increased weight-bearing exercise, and *hormone (estrogen) replacement therapy (HRT)*. Frustratingly, HRT only slows the loss of bone; it does not reverse it. For those who can't or won't take estrogen, newer drugs are available. These include *Fosamax* (alendronate), a drug that suppresses osteoclast activity and shows promise in reversing osteoporosis in the spine; and *Evista* (raloxifene), dubbed "estrogen light" because it mimics estrogen's beneficial bone sparing properties without targeting the uterus or breast. Although not a substitute for HRT, estrogenic compounds in soy protein (principally the isoflavones daidzein and genistein) offer a good addition or adjunct for some.

Osteoporosis can be prevented (or at least delayed). The first requirement is to get enough calcium while your bones are still increasing in density

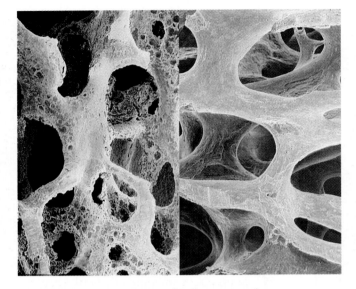

FIGURE 6.14 *Osteoporosis: the contrasting architecture of osteoporotic bone (left) and normal bone (right).*

(long bones reach their peak density between the ages of 35 and 40; bones with relatively more spongy bone are densest between the ages of 25 and 30). Second, drinking fluoridated water hardens bones (as well as teeth). Finally, getting plenty of weight-bearing exercise (walking, jogging, tennis, etc.) during youth and throughout life will increase bone mass above normal values and provide a greater buffer against age-related bone loss.

Paget's Disease

Often discovered by accident when X rays are taken for some other reason, **Paget's** (paj'ets) **disease** is characterized by excessive bone formation and breakdown. Newly formed bone, called *Pagetic bone*, is hastily made and has an abnormally high ratio of woven bone to compact bone. This, along with reduced mineralization, causes a spotty weakening of the bones. Late in the disease, osteoclast activity wanes, but osteoblasts continue to work, often forming irregular bone thickenings or filling the marrow cavity with Pagetic bone.

Paget's disease may affect any part of the skeleton, but it is usually a localized condition. The spine, pelvis, femur, and skull are most often involved and become increasingly deformed and painful. It rarely occurs before the age of 40 years, and it affects about 3% of North American elderly people. Its cause is unknown, but it may be initiated by a virus (such as the virus that causes distemper in dogs). Drug therapies for this ailment include *Didronate* (etidronate), used for three to six months at a time; calcitonin (now administered by a nasal

Text continues on page 194.

Although bones have remarkable self-regenerative powers, some conditions frustrate even their most heroic efforts. Outstanding examples include massive shattering (as in automobile accidents), poor circulation in old bones, and certain birth defects. However, recent medical advances are allowing the medical profession to deal with healing problems that bones cannot handle themselves. Let's take a look.

1. Electrical stimulation of fracture sites has dramatically increased the speed and completeness of healing in large or slowly healing fractures. For years it's been known that bone tissue is deposited in regions of negative electrical charge and absorbed in areas of positive charge, but just how electricity promotes healing is still not completely known. The theory is that the electrical fields prevent parathyroid hormone (PTH) from stimulating the bone-absorbing osteoclasts at the fracture site, thus allowing increased accumulation of bony tissue. Another theory has it that the electrical fields induce production of growth factors that stimulate the osteoblasts.

2. Ultrasound treatments. Ultrasound, one of the imaging techniques described in Chapter 1, also helps speed human bone repair. Daily exposure to pulsed low-intensity ultrasound waves reduces the healing time of broken arms and shinbones by 35 to 45%. It seems to work by stimulating the cartilage cells to make callus.

3. The free vascular fibular graft technique has proven to be a boon to the management of non-union fractures, in which the two parts of a split bone fail to join. It uses pieces of the fibula (the sticklike bone of the leg) to replace missing or severely damaged bone. In the past, extensive bone grafts (typically using pieces of bone taken from the hip) usually failed because a blood supply could not reach their interior, necessitating eventual amputation. This new technique grafts normal blood vessels along with the bone section. Subsequent remodeling

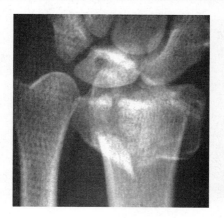

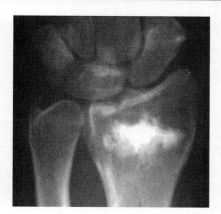

leads to a good replica of the bone normally present.

4. Bone substitutes. Desperate for bone fillers, surgeons have traditionally resorted to using crushed bone from cadavers or synthetic materials. Crushed bone induces the body to form new bone of its own. When mixed with water, it can be molded to the shape of the desired bone or packed into small, difficult-to-reach spaces. Thus, it avoids the often painful and time-consuming procedure of bone grafting. The same can be said for most synthetic bone substitutes; however, both approaches have shortcomings. Cadaveric bone carries with it a small (but real) risk of hepatitis or HIV infection. Additionally, as foreign tissue, it may be rejected by the immune system. Synthetics tend to provoke inflammation, which hinders healing and promotes infection.

Currently in clinical trials is a particularly exciting bone "stand-in" called **sea bone** (made from coral), which avoids both problems. The coral is heat-treated to kill the living coral cells and convert its mineral structure (of calcium carbonates) to hydroxyapatite, the salt in bone. The brittle coral graft is then carved to the desired shape, coated with a natural substance that enhances bone growth, called *bone morphogenic protein (BMP)*, and implanted. Osteoblasts and blood vessels migrate from the adjacent natural bone into the coral implant, gradually replacing it with living bone.

Research on bone replacement materials has also produced several

types of **artificial bone.** One made of a biodegradable ceramic substance known as TCP is soft enough to be shaped, but it is not very strong. Its biggest application has been to replace parts of non-weight-bearing bones, such as skull bones.

Unlike the bone substitutes described above, Norian SRS, a new bone cement made of calcium phosphate, provides immediate structural support to fractured or osteoporotic sites. Mixed at the time of surgery, Norian SRS starts out as a paste that can be injected into areas of damaged bone to create an "internal cast." The paste hardens in minutes and cures into a substance with greater compressive strength than spongy bone. Because its crystalline structure and chemistry is the same as that of natural bone, it is gradually remodeled and replaced by host bone cells. A limitation of Norian SRS is that it can be used only on the ends of long bones, because it cannot resist the twisting and flexing experienced by the central shaft.

Perhaps the most promising product being tested is Osteomedica's bioceramic product. Made from synthesized hydroxyapatite, it is stronger than any of the bone replacement materials discussed above. Molded or carved into long pieces, it works as an artificial bone graft. Thus far in animal tests, it has been shown to bear weight well for several months after the initial healing and remodeling.

MAKING CONNECTIONS

SYSTEM CONNECTIONS: Homeostatic Interrelationships Between the Skeletal System and Other Body Systems

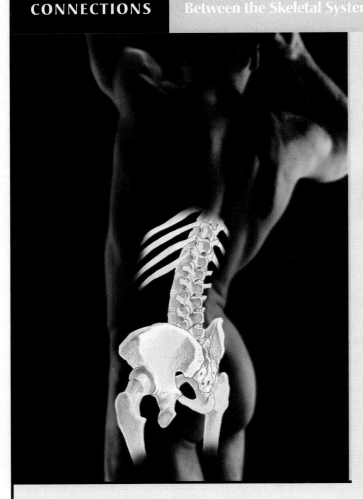

Endocrine System

- Skeletal system provides some bony protection; stores calcium needed for second messenger signaling mechanisms
- Hormones regulate uptake and release of calcium from bone; hormones promote long bone growth and maturation

Cardiovascular System

- Bone marrow cavities provide site for blood cell formation; matrix stores calcium needed for cardiac muscle activity
- Cardiovascular system delivers nutrients and oxygen to bones; carries away wastes

Lymphatic System/Immunity

- Skeletal system provides some protection to lymphatic organs; bone marrow is site of origin for lymphocytes involved in immune response
- Lymphatic system drains leaked tissue fluids; immune cells protect against pathogens

Respiratory System

- Skeletal system protects lungs by enclosure (rib cage)
- Respiratory system provides oxygen; disposes of carbon dioxide

Digestive System

- Skeletal system provides some bony protection to intestines, pelvic organs, and liver
- Digestive system provides nutrients needed for bone health and growth

Urinary System

- Skeletal system protects pelvic organs (bladder, etc.)
- Urinary system activates vitamin D; disposes of nitrogenous wastes

Reproductive System

- Skeletal system protects some reproductive organs by enclosure
- Gonads produce hormones that influence form of skeleton and epiphyseal closure

Integumentary System

- Skeletal system provides support for body organs including the skin
- Skin provides vitamin D needed for proper calcium absorption and use

Muscular System

- Skeletal system provides levers plus ionic calcium for muscle activity
- Muscle pull on bones increases bone strength and viability; helps determine bone shape

Nervous System

- Skeletal system protects brain and spinal cord; depot for calcium ions needed for neural function
- Nerves innervate bone and joint capsules, providing for pain and joint sense

CLOSER CONNECTIONS

THE SKELETAL SYSTEM and Interrelationships with the Muscular, Endocrine, and Integumentary Systems

Our skeleton supports us, protects our "innards" (the protection our brain gets from the skull is indispensable), gives us stature (for some reason, tall people get more respect), contributes to our shape (women *are* shaped differently than men), and allows us to move. Obviously, the skeletal system has important interactions with many other body systems, not the least of which are the endocrine and integumentary systems. However, its most intimate and mutually beneficial relationship is with the muscular system. Thus, we will consider that first.

Muscular System

The codependence of the skeletal and muscular systems is striking—as one system goes, so goes the other. If we participate in weight-bearing exercise (run, play tennis, do aerobics) regularly, our muscles become more efficient, and exert more force on our bones. As a result, our bones stay healthy and strong, and increase their mass to assume the added stress.

Since both spongy and compact bone reach peak density during midlife, weight-bearing exercise during youth is important, especially in females who have less bone mass than males and lose it faster.

Regular exercise also stretches the connective tissues binding bones to muscles and to other bones, and reinforcing joints. Since this increases overall flexibility, we have fewer injuries, allowing us to stay active well into old age. (Pain makes couch potatoes.)

Endocrine System

Although mechanical factors are undeniably important in shaping the skeleton and helping to keep it strong, hormones acting individually and in concert direct skeletal growth during youth, and enhance (or impair) skeletal strength in adults. Growth hormone is essential for normal skeletal growth and maintenance throughout life, whereas thyroid and sex hormones ensure that normal skeletal proportions are established during childhood and adolescence. Conversely, PTH and calcitonin serve not the skeleton but a different master—homeostasis of blood calcium levels. Regardless of the precise skeletal aspect governed by hormones, any interference with their normal functioning is soon apparent as a skeletal abnormality or malproportion.

Integumentary System

Under normal circumstances, the skeletal system is absolutely dependent on the integumentary system (the skin) for the calcium that keeps the bones hard and strong. The relationship is indirect to be sure: In the presence of sunlight, a vitamin D precursor is produced in the dermal capillary blood. It is activated elsewhere, and (among its other roles) forms part of the carrier system that absorbs calcium from ingested foods into the blood. Because calcium is required for so many body functions (see p. 185), and bones provide the "calcium bank," the bones become increasingly soft and weak in the absence of vitamin D because no daily rations of calcium are allowed to enter the blood from the digestive tract.

CLINICAL CONNECTIONS

Skeletal System

Case study: Remember Mrs. DeStephano? When we last heard about her she was being admitted for further studies. Relative to her skeletal system, the following notes have been added to her chart.

- Fracture of superior region of right tibia (shinbone of leg); skin lacerated; area cleaned and protruding bone fragments subjected to internal reduction and casted
- Nutrient artery of tibia damaged
- Medial meniscus (hyaline cartilage disc) of right knee joint crushed; knee joint inflamed and painful

Relative to these notes:

1. What type of fracture does Mrs. DeStephano have?

2. What problems can be predicted with such fractures and how are they treated?

3. What is internal reduction? Why was a cast applied?

4. Given an uncomplicated recovery, approximately how long should it take before Mrs. DeStephano has a good solid bony callus?

5. What complications might be predicted by the fact that the nutrient artery is damaged?

6. What new techniques might be used to enhance fracture repair if healing is delayed or impaired?

7. How likely is it that Mrs. DeStephano's knee cartilage will regenerate? Why? What measures will probably have to be taken to remedy this problem?

(Answers in Appendix F)

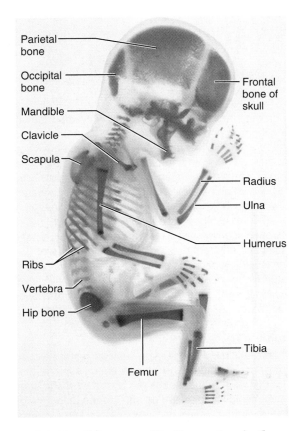

Parietal bone
Occipital bone
Mandible
Clavicle
Scapula
Frontal bone of skull
Radius
Ulna
Humerus
Ribs
Vertebra
Hip bone
Tibia
Femur

FIGURE 6.15 *Primary ossification centers in the skeleton of a 12-week-old fetus are indicated by the darker areas.*

inhaler); and the newer Fosamax, which is successful in preventing bone breakdown.

DEVELOPMENTAL ASPECTS OF BONES: TIMING OF EVENTS

The bones are on a precise schedule from the time they form until death. The mesoderm germ layer gives rise to embryonic mesenchymal cells, which in turn produce the membranes and cartilages that form the embryonic skeleton. These structures then ossify according to an amazingly predictable timetable that allows fetal age to be determined easily from X rays or sonograms. Although each bone has its own developmental schedule, most long bones begin ossifying by 8 weeks and have obvious primary ossification centers by 12 weeks (Figure 6.15).

At birth, most long bones of the skeleton are well ossified except for their epiphyses. After birth, secondary ossification centers develop in a predictable sequence. The epiphyseal plates persist and provide for long bone growth during childhood and the sex hormone–mediated growth spurt at adolescence. By the age of 25 years, nearly all bones are completely ossified and skeletal growth ceases.

In children and adolescents, bone formation exceeds bone resorption; in young adults, these processes are in balance; in old age, resorption predominates. Despite the environmental factors (listed above) that influence bone density, genetics still plays the major role in determining one's bone density (or the relative amount of bone thinning that will occur). A single gene that codes for vitamin D's cellular docking site helps determine both the tendency to accumulate bone mass during early life and one's risk of osteoporosis later in life. Beginning in the fourth decade of life, bone mass decreases with age; the only exception appears to be in bones of the skull. Among young adults, skeletal mass is generally greater in males than in females, and in blacks than in whites. As we age, the rate of bone loss is faster in whites than in blacks (who have greater bone density to begin with) and in females than in males. Qualitative changes also occur: More osteons remain incompletely formed and mineralization is less complete. The amount of nonviable bone increases, reflecting a diminished blood supply to the bones in old age.

* * *

Skeletal cartilages and bones—their architecture, composition, and dynamic nature—have been examined in some detail in this chapter. We have also discussed the role of bones in maintaining overall body homeostasis, as summarized in *Making Connections*. Now we are ready to look at the individual bones of the skeleton and how they contribute to its functions, both collectively and individually.

RELATED CLINICAL TERMS

Achondroplasia (a-kon"dro-pla′ze-ah; *a* = without; *chondro* = cartilage; *plasi* = mold, shape) A congenital condition involving defective cartilage and endochondral bone growth so that the limbs are too short but the membrane bones are of normal size; a type of dwarfism.

Bony spur Abnormal projection from a bone due to bone overgrowth; common in aging bones.

Ostealgia (os"te-al′je-ah; *algia* = pain) Pain in a bone.

Osteitis (os"te-i′tis; *itis* = inflammation) Inflammation of bony tissue.

Osteomyelitis (os"te-o-mi′ĕ-li′tis) Inflammation of bone and bone marrow caused by pus-forming bacteria that enter the body via a wound (e.g., compound bone fracture), or spread from an infection near the bone. Commonly affects the long bones, causing acute pain and fever. May result in joint stiffness, bone destruction, and shortening of a limb. Treatment involves antibiotic therapy, draining of any abscesses (local collections of pus) formed, and removal of dead bone fragments (which prevent healing).

Osteosarcoma (os′te-o-sar-ko′mah) A form of bone cancer typically arising in a long bone of a limb and most often in those 10–25 years of age. Grows aggressively, painfully eroding the bone; tends to metastasize to the lungs and cause secondary lung tumors. Usual treatment is amputation of the affected bone or limb, followed by chemotherapy and surgical removal of any metastases. Survival rate is about 50% if detected early.

Pathologic fracture Fracture in a diseased bone involving slight (coughing or a quick turn) or no physical trauma. For example, a broken hip due to osteoporosis: The weakened bone breaks, causing the person to fall.

Traction ("pulling") Placing sustained tension on a body region to keep the parts of a fractured bone in proper alignment; also prevents spasms of skeletal muscles, which would separate the fractured bone ends or crush the spinal cord in the case of vertebral column fractures.

CHAPTER SUMMARY

Skeletal Cartilages (p. 173)

Basic Structure, Types, and Locations (p. 173)

1. A skeletal cartilage exhibits chondrocytes housed in lacunae (cavities) within the extracellular matrix (ground substance and fibers). It contains large amounts of water (which accounts for its resilience), lacks nerve fibers, is avascular, and is surrounded by a fibrous perichondrium that resists expansion.

2. Hyaline cartilages appear glassy; the fibers are collagenic. They provide support with flexibility and resilience and are the most abundant skeletal cartilages, accounting for the articular, costal, respiratory, and nasal cartilages.

3. Elastic cartilages contain abundant elastic fibers, in addition to collagenic fibers, and are more flexible than hyaline cartilages. They support the outer ear and epiglottis.

4. Fibrocartilages, which contain thick collagen fibers, are the most compressible cartilages and are resistant to stretch. They form vertebral discs and knee joint cartilages.

Growth of Cartilage (p. 173)

5. Cartilages grow from within (interstitial growth) and by addition of new cartilage tissue at the periphery (appositional growth).

Classification of Bones (pp. 174–176)

1. Bones are classified as long, short, flat, or irregular on the basis of their shape and their proportion of compact or spongy bone.

Functions of Bones (p. 176)

1. Bones give the body shape; protect and support body organs; provide levers for muscles to pull on; store calcium and other minerals; and are the site of blood cell production.

SP Exercise: Classification of Bones.

Bone Structure (pp. 176–181)

Gross Anatomy (pp. 176–178)

1. A long bone is composed of a diaphysis (shaft) and epiphyses (ends). The medullary cavity of the diaphysis contains yellow marrow; the epiphyses contain spongy bone. The epiphyseal line is the remnant of the epiphyseal plate. Periosteum covers the diaphysis; endosteum lines inner bone cavities. Hyaline cartilage covers joint surfaces.

2. Flat bones consist of two thin plates of compact bone enclosing a diploë (spongy bone layer). Short and irregular bones resemble flat bones structurally.

3. In adults, hematopoietic tissue is found within the diploë of flat bones and occasionally within the epiphyses of long bones. In infants red marrow is also found in the medullary cavity.

Microscopic Structure of Bone (pp. 178–179)

4. The structural unit of compact bone, the osteon, consists of a central canal surrounded by concentric lamellae of bone matrix. Osteocytes, embedded in lacunae, are connected to each other and the central canal by canaliculi.

5. Spongy bone has slender trabeculae containing irregular lamellae, which enclose red marrow–filled cavities.

Chemical Composition of Bone (p. 180)

6. Bone is composed of living cells (osteoblasts, osteocytes, and osteoclasts) and matrix. The matrix includes organic substances that are secreted by osteoblasts and give the bone tensile strength. Its inorganic components, the hydroxyapatites (calcium salts), make bone hard.

Bone Markings (pp. 180–181)

7. Bone markings are important anatomical landmarks that reveal sites of muscle attachment, points of articulation, and sites of blood vessels and nerve passage.

SP Exercise: Bone Structure.

Bone Development (pp. 181–184)

Formation of the Bony Skeleton (pp. 181–183)

1. Intramembranous ossification forms the clavicles and most skull bones. The ground substance of the bone matrix is deposited between collagen fibers within the fibrous membrane to form spongy bone. Eventually, compact bone plates enclose the diploë.

2. Most bones are formed by endochondral ossification of a hyaline cartilage model. Osteoblasts beneath the periosteum secrete bone matrix on the cartilage model, forming the bone collar. Deterioration of the cartilage model internally opens up cavities, allowing periosteal bud entry. Bone matrix is deposited around the cartilage remnants but is later broken down.

Postnatal Bone Growth (pp. 183–184)

3. Long bones increase in length by interstitial growth of the epiphyseal plate cartilage and its replacement by bone.

4. Appositional growth increases bone diameter/thickness.

SP Case Study: Craniosynostosis.

Bone Homeostasis: Remodeling and Repair (pp. 184–189)

Bone Remodeling (pp. 185–187)

1. New bone is continually deposited and resorbed in response to hormonal and mechanical stimuli. Together these processes constitute bone remodeling.

2. An osteoid seam appears at areas of new bone formation; calcium salts are deposited a few days later.

3. Multinucleate osteoclasts release lysosomal enzymes and acids on bone surfaces to be resorbed. The dissolved products are transcytosed to the opposite face of the osteoclast for release to the extracellular fluid.

4. The hormonal mechanism of bone remodeling serves blood calcium homeostasis. When blood calcium levels decline, PTH is released and stimulates osteoclasts to digest bone matrix, releasing ionic calcium. When blood calcium levels rise, calcitonin is released, stimulating removal of calcium from the blood. Mechanical stress and gravity acting on the skeleton help maintain skeletal strength. Bones thicken, develop heavier prominences, or rearrange their trabeculae in sites where stressed.

Repair of Fractures (pp. 187–189)

5. Fractures are treated by open or closed reduction. The healing process involves formation of a hematoma, a fibrocartilaginous callus, a bony callus, and bone remodeling, in succession.

SP Exercise: Ionic Calcium Levels in the Blood.

Homeostatic Imbalances of Bone (pp. 189–191, 194)

1. Imbalances between bone formation and resorption underlie all skeletal disorders.

2. Osteomalacia and **rickets** occur when bones are inadequately mineralized. The bones become soft and deformed. The most frequent cause is inadequate vitamin D.

3. Osteoporosis is any condition in which bone breakdown outpaces bone formation, causing bones to become weak and porous. Postmenopausal women are particularly susceptible.

4. Paget's disease is characterized by excessive and abnormal bone remodeling.

SP Case Study: Skeletal Disorder.

Developmental Aspects of Bones: Timing of Events (p. 194)

1. Osteogenesis is predictable and precisely timed.

2. Longitudinal long bone growth continues until the end of adolescence. Skeletal mass increases dramatically during puberty and adolescence, when formation exceeds resorption.

3. Bone mass is fairly constant in young adulthood, but beginning in the 40s, bone resorption exceeds formation.

REVIEW QUESTIONS

Multiple Choice/Matching

1. Which is a function of the skeletal system? (a) support, (b) hematopoietic site, (c) storage, (d) providing levers for muscle activity, (e) all of these.

2. A bone with approximately the same width, length, and height is most likely (a) a long bone, (b) a short bone, (c) a flat bone, (d) an irregular bone.

3. The shaft of a long bone is properly called the (a) epiphysis, (b) periosteum, (c) diaphysis, (d) compact bone.

4. Sites of hematopoiesis include all but (a) red marrow cavities of spongy bone, (b) the diploë of flat bones, (c) medullary cavities in bones of infants, (d) medullary cavities in bones of a healthy adult.

5. An osteon has (a) a central canal carrying blood vessels, (b) concentric lamellae, (c) osteocytes in lacunae, (d) canaliculi that connect lacunae to the central canal, (e) all of these.

6. The organic portion of matrix is important in providing *all but* (a) tensile strength, (b) hardness, (c) ability to resist stretch, (d) flexibility.

7. The flat bones of the skull develop from (a) areolar tissue, (b) hyaline cartilage, (c) fibrous connective tissue, (d) compact bone.

8. The remodeling of bone is a function of which cells? (a) chondrocytes and osteocytes, (b) osteoblasts and osteoclasts, (c) chondroblasts and osteoclasts, (d) osteoblasts and osteocytes.

9. Bone growth during childhood and in adults is regulated and directed by (a) growth hormone, (b) thyroxine, (c) sex hormones, (d) mechanical stress, (e) all of these.

10. Where within the epiphyseal plate are the *dividing* cartilage cells located? (a) nearest the shaft, (b) in the marrow cavity, (c) farthest from the shaft, (d) in the primary ossification center.

11. Wolff's law is concerned with (a) calcium homeostasis of the blood, (b) the thickness and shape of a bone being determined by mechanical and gravitational stresses placed on it, (c) the electrical charge on bone surfaces.

12. Formation of the bony callus in fracture repair is followed by (a) hematoma formation, (b) fibrocartilaginous callus formation, (c) bone remodeling to convert woven bone to compact bone, (d) formation of granulation tissue.

13. A fracture in which the bone ends are incompletely separated is (a) greenstick, (b) compound, (c) simple, (d) comminuted, (e) compression.

14. The disorder in which bones are porous and thin but bone composition is normal is (a) osteomalacia, (b) osteoporosis, (c) Paget's disease.

Short Answer Essay Questions

15. Compare bone to cartilage tissue relative to its resilience, speed of regeneration, and access to nutrients.

16. Describe in proper order the events of endochondral ossification.

17. Compare compact and spongy bone in macroscopic appearance, microscopic structure, and relative location.

18. As we grow, our long bones increase in diameter, but the thickness of the bony collar of the shaft remains relatively constant. Explain this phenomenon.

19. Describe the process of new bone formation in an adult bone. Use the terms *osteoid seam* and *calcification front* in your discussion.

20. Compare and contrast controls of bone remodeling exerted by hormones and by mechanical and gravitational forces relative to the actual purpose of each control system and changes in bone architecture that might occur.

21. (a) During what period of life does skeletal mass increase dramatically? Begin to decline? (b) Why are fractures most common in elderly individuals? (c) Why are greenstick fractures most common in children?

Critical Thinking and Clinical Application Questions

1. Following a motorcycle accident, a 22-year-old man was rushed to the emergency room. X rays revealed a spiral fracture of his right tibia (main bone of the leg). Two months later, X rays revealed good bony callus formation. What is bony callus?

2. Mrs. Abbruzzo brought her 4-year-old daughter to the doctor, complaining that she didn't "look right." The child's forehead was enlarged, her rib cage was knobby, and her lower limbs were bent and deformed. X rays revealed very thick epiphyseal plates. Mrs. Abbruzzo was advised to increase dietary amounts of vitamin D and milk and to "shoo" the girl outside to play in the sun. Considering the child's signs and symptoms, what disease do you think she has? Explain the doctor's instructions.

3. You overhear some anatomy students imagining out loud what their bones would look like if they had compact bone on the inside and spongy bone on the outside, instead of the other way around. You tell them that such imaginary bones would be poorly designed mechanically and would break easily. Explain your reason for saying this.

4. Why do you think wheelchair-bound people with paralyzed lower limbs have thin, weak bones of the leg and thigh?

5. Jay Beckenstein went to weight-lifting camp in the summer between seventh and eighth grade. He noticed that the camp trainer put tremendous pressure on him and his friends to improve their strength. After an especially vigorous workout, Jay's arm felt extremely sore and weak around the elbow. He went to the camp doctor, who took X rays and then told him that the injury was serious, for the "end of his upper arm bone was starting to twist off." What had happened? Could the same thing happen to Jay's 23-year-old sister, Trixie, who was also starting a program of weight lifting? Why or why not?

6. After being picked up from the Atlantic Ocean after "splashdown," the American astronaut team was brought to the Naval Hospital for checkups. X rays revealed decreased bone mass in all of them. Isn't this surprising in view of the fact that they do exercises in the capsule while in space? Why or why not?

7 THE SKELETON

The word **skeleton** comes from the Greek word meaning "dried-up body" or "mummy," a rather unflattering description. Nonetheless, the human skeleton is a triumph of design and engineering that puts most skyscrapers to shame. It is strong, yet light, and almost perfectly adapted for the protective, locomotor, and manipulative functions it performs.

The skeleton, composed of bones, cartilages, joints, and ligaments, accounts for about 20% of body mass (about 30 pounds worth in a 160-pound person). Bones make up most of the skeleton. Cartilages occur only in isolated areas, such as the nose, parts of the ribs, and the joints. Ligaments connect bones and reinforce joints, allowing required movements while restricting motions in other directions. Joints, the junctions between bones, provide for the remarkable mobility of the skeleton. Joints and ligaments are discussed separately in Chapter 8.

■ PART 1 ■
THE AXIAL SKELETON

As described in Chapter 6, the skeleton is divided into *axial* and *appendicular* portions (see Figure 6.1, p. 174) for descriptive purposes. The **axial skeleton** is structured from 80 bones segregated into three major regions: the *skull, vertebral column,* and *bony thorax* (Figure 7.1). This part of the skeleton supports the head, neck, and trunk, and it protects the brain, spinal cord, and the organs in the thorax.

THE SKULL

The **skull** is the body's most complex bony structure. It is formed by *cranial* and *facial bones,* 22 in all. The cranial bones, or **cranium** (kra'ne-um), enclose and protect the fragile brain and furnish a site for attachment of head and neck muscles. The facial bones (1) form the framework of the face, (2) contain cavities for the special sense organs of sight, taste, and smell, (3) provide openings for passage of air and food, (4) secure the teeth, and (5) anchor the facial muscles of expression, which we use to show our feelings. As you will see, the individual skull bones are well suited to their assignments (see Table 7.1 on pp. 213–214).

Most skull bones are flat bones. Except for the mandible, which is connected to the rest of the skull by a freely movable joint, all bones of the adult skull are firmly united by interlocking joints called **sutures** (soo'cherz). The suture lines have a saw-toothed or serrated appearance. The major skull sutures, the *coronal, sagittal, squamous,* and *lambdoid sutures,* connect cranial bones (see Figures 7.2b and 7.3a). Most other skull sutures connect facial bones and are named according to the specific bones they connect.

Overview of Skull Geography

It is worth surveying basic skull "geography" before describing the individual bones. With the lower jaw removed, the skull resembles a lopsided, hollow, bony sphere. The facial bones form its anterior aspect, and the cranium forms the rest of the skull (see Figure 7.3a). The cranium can be divided into a vault and a base. The *cranial vault,* also called the *calvaria* (kal-va're-ah), forms the superior, lateral, and posterior aspects of the skull, as well as the forehead. The *cranial base,* or *floor,* forms the skull's inferior aspect. Internally, prominent bony ridges divide the base into three distinct "steps" or fossae—the *anterior, middle,* and *posterior cranial fossae* (see Figure 7.4c on p. 204). The brain sits snugly in these cranial fossae, completely enclosed by the cranial vault. Overall, the brain is said to occupy the *cranial cavity.*

In addition to the large cranial cavity, the skull has many smaller cavities. These include the middle and inner ear cavities (carved into the lateral side of its base) and, anteriorly, the nasal cavity and the orbits. The *orbits* house the eyeballs. Several bones of the skull contain air-filled sinuses. Sinuses lighten the skull and have additional roles that we will describe shortly.

The skull also has about 85 named openings (foramina, canals, fissures, etc.). The most important of these provide passageways for the spinal cord, the major blood vessels serving the brain, and the 12 pairs of cranial nerves (numbered I through XII), which transmit impulses to and from the brain.

As you read about the bones of the skull, locate each bone on the different skull views in Figures 7.2, 7.3, and 7.4. The skull bones and their important markings are also summarized in Table 7.1. The color-coded box beside a bone's name in the table corresponds to the color of that bone in the figures.

Cranium

The eight cranial bones are the paired parietal and temporal bones and the unpaired frontal, occipital, sphenoid, and ethmoid bones. Together, these construct the brain's protective bony "helmet." Because its superior aspect is curved, the cranium is self-bracing. This allows the bones to be thin, and, like an eggshell, the cranium is remarkably strong for its weight.

Frontal Bone
The shell-shaped **frontal bone** (Figures 7.2a, 7.3, and 7.4b) forms the anterior portion of the cranium.

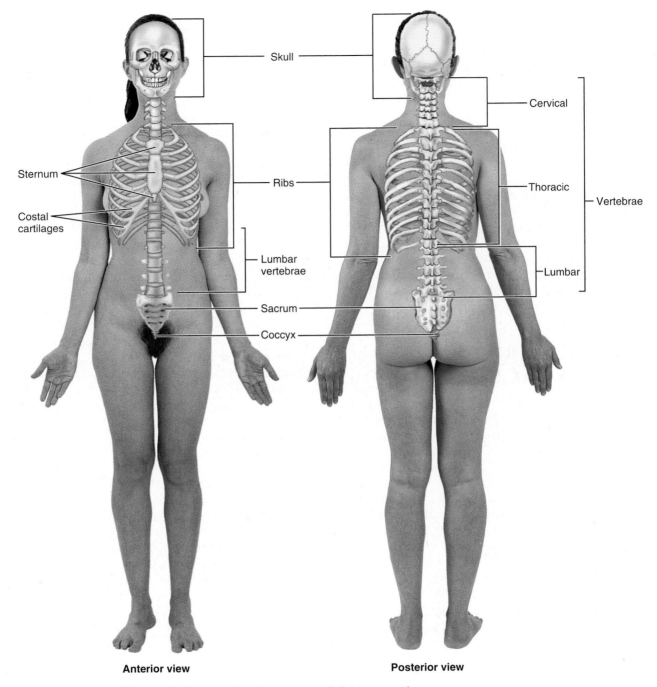

Anterior view   **Posterior view**

FIGURE 7.1 *Bones of the axial skeleton.* The three major subdivisions are the skull, bony thorax (sternum and ribs), and vertebral column (vertebrae, sacrum, and coccyx).

It articulates posteriorly with the paired parietal bones via the prominent *coronal suture.*

The most anterior part of the frontal bone is the vertical *frontal squama,* commonly called the *forehead.* The frontal squama ends inferiorly at the **supraorbital margins,** the thickened superior margins of the orbits that lie under the eyebrows. From here, the frontal bone extends posteriorly, forming the superior wall of the **orbits** and most of the **anterior cranial fossa** (see Figure 7.4b and c). This fossa

supports the frontal lobes of the brain. Each supraorbital margin is pierced by a **supraorbital foramen (notch),** which allows the supraorbital artery and nerve to pass to the forehead.

The smooth portion of the frontal bone between the orbits is the **glabella** (glah-bel'ah). Just inferior to this the frontal bone meets the nasal bones at the *frontonasal suture* (see Figure 7.2a). The areas lateral to the glabella are riddled internally with sinuses, called the **frontal sinuses** (see Figures 7.3b and 7.11).

(1) *Which of the bones illustrated in view (a) are cranial bones? **(2)** Which two terms are used interchangeably to indicate the entire group of cranial bones?*

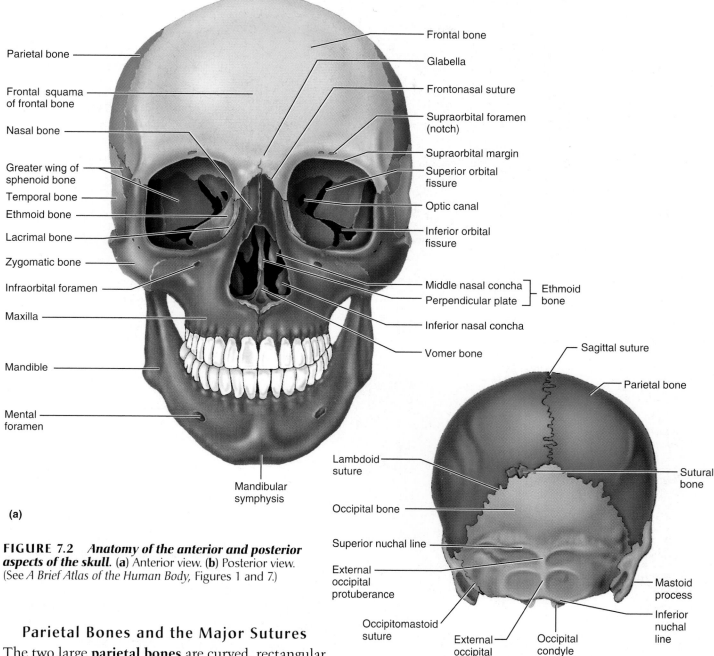

FIGURE 7.2 *Anatomy of the anterior and posterior aspects of the skull.* (a) Anterior view. (b) Posterior view. (See *A Brief Atlas of the Human Body,* Figures 1 and 7.)

Parietal Bones and the Major Sutures

The two large **parietal bones** are curved, rectangular bones that form most of the superior and lateral aspects of the skull; hence they form the bulk of the cranial vault. The four largest sutures occur where the parietal bones articulate (form a joint) with other cranial bones:

1. The **coronal** (kŏ-ro′nul) **suture,** where the parietal bones meet the frontal bone anteriorly (Figure 7.3)

2. The **sagittal suture,** where the right and left parietal bones meet superiorly at the cranial midline (see Figure 7.2b)

3. The **lambdoid** (lam′doid) **suture,** where the parietal bones meet the occipital bone posteriorly (see Figures 7.2b and 7.3)

4. The **squamous** (or **squamosal**) **suture** (one on each side), where a parietal and temporal bone meet on the lateral aspect of the skull (see Figure 7.3)

(1) The temporals, parietals, frontal, and sphenoid are all cranial bones. (2) The cranial vault or calvaria.

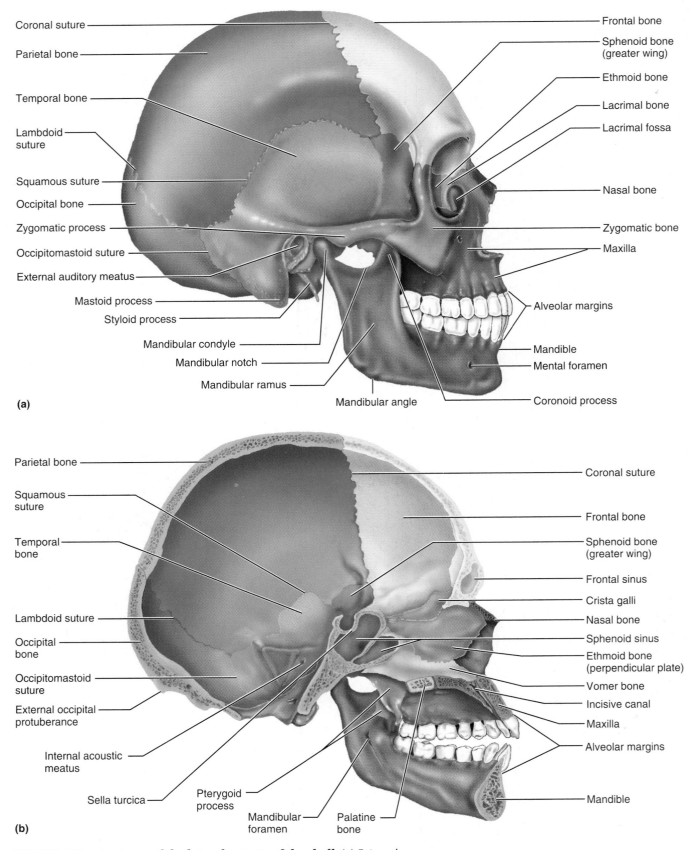

Coronal suture

Parietal bone

Temporal bone

Lambdoid suture

Squamous suture

Occipital bone

Zygomatic process

Occipitomastoid suture

External auditory meatus

Mastoid process

Styloid process

Mandibular condyle

Mandibular notch

Mandibular ramus

Mandibular angle

Frontal bone

Sphenoid bone (greater wing)

Ethmoid bone

Lacrimal bone

Lacrimal fossa

Nasal bone

Zygomatic bone

Maxilla

Alveolar margins

Mandible

Mental foramen

Coronoid process

(a)

Parietal bone

Squamous suture

Temporal bone

Lambdoid suture

Occipital bone

Occipitomastoid suture

External occipital protuberance

Internal acoustic meatus

Sella turcica

Pterygoid process

Mandibular foramen

Palatine bone

Coronal suture

Frontal bone

Sphenoid bone (greater wing)

Frontal sinus

Crista galli

Nasal bone

Sphenoid sinus

Ethmoid bone (perpendicular plate)

Vomer bone

Incisive canal

Maxilla

Alveolar margins

Mandible

(b)

FIGURE 7.3 *Anatomy of the lateral aspects of the skull.* (a) External anatomy of the right lateral aspect of the skull. (b) Midsagittal view of the skull showing the internal anatomy of the left side of the skull. (See *A Brief Atlas of the Human Body*, Figures 2 and 3.)

Occipital Bone

The **occipital** (ok-sip'ĭ-tal) **bone** forms most of the skull's posterior wall and base. It articulates anteriorly with the paired parietal and temporal bones via the *lambdoid* and *occipitomastoid sutures,* respectively (Figure 7.3). It also joins with the sphenoid bone in the cranial floor via a plate called the basioccipital, which bears a midline projection called the *pharyngeal* (fah-rin'je-ul) *tubercle* (Figure 7.4a).

Internally, the occipital bone forms the walls of the **posterior cranial fossa** (Figure 7.4b and c), which supports the cerebellum of the brain. In the base of the occipital bone is the **foramen magnum,** literally, "large hole." Through this opening, the inferior part of the brain connects with the spinal cord. The foramen magnum is flanked laterally by two **occipital condyles** (see Figure 7.4a). The rockerlike occipital condyles articulate with the first vertebra of the spinal column in a way that permits a nodding movement of the head. Hidden medial and superior to each occipital condyle is a **hypoglossal canal** (Figure 7.4b), through which a nerve of the same name passes.

Just superior to the foramen magnum is a median protrusion called the **external occipital protuberance** (see Figures 7.2b, 7.3b, and 7.4a). You can feel this knoblike projection just below the most bulging part of your posterior skull. A number of inconspicuous ridges, the *external occipital crest* and the *superior* and *inferior nuchal* (nu'kal) *lines,* mark the occipital bone near the foramen magnum. The external occipital crest secures the *ligamentum nuchae* (lig"ah-men'tum noo'ke; *nucha* = back of the neck), an elastic ligament that connects the vertebrae of the neck to the skull. The nuchal lines, and the bony regions between them, anchor many neck and back muscles. The superior nuchal line marks the upper limit of the neck.

Temporal Bones

The two **temporal bones** are best viewed on the lateral skull surface (see Figure 7.3). They lie inferior to the parietal bones and meet them at the squamosal sutures. The temporal bones form the inferolateral aspects of the skull and parts of the cranial floor. The use of the terms *temple* and *temporal,* from the Latin word *temporum,* meaning "time," came about because gray hairs, a sign of time's passing, usually appear first in the temple areas.

Each temporal bone has a complicated shape and is described in terms of its four major regions, the *squamous, tympanic, mastoid,* and *petrous regions.* The flaring **squamous region** abuts the squamous suture. It has a barlike **zygomatic process** that meets the zygomatic bone of the face anteriorly. Together, these two bony structures form the **zygomatic arch,** which you can feel as the projection of your cheek (*zygoma* = cheekbone). The small, oval **mandibular** (man-dib'u-lar) **fossa** on the inferior surface of the zygomatic process receives the condyle of the mandible (lower jawbone), forming the freely movable *temporomandibular joint.*

The **tympanic** (tim-pan'ik; "eardrum") **region** (Figure 7.5) of the temporal bone surrounds the **external auditory (acoustic) meatus,** or external ear canal, through which sound enters the ear. The external auditory meatus and the eardrum at its deep end are part of the *external ear.* In a dried skull, the eardrum has been removed; thus, part of the middle ear cavity deep to the external meatus can also be seen. Below the external auditory meatus is the needle-like **styloid** (sti'loid; "stakelike") **process,** an attachment point for several muscles of the tongue and neck and for a ligament that secures the hyoid bone of the neck to the skull (see Figure 7.12).

The **mastoid** (mas'toid) **region** of the temporal bone exhibits the conspicuous **mastoid process,** an anchoring site for some neck muscles (see Figures 7.3a, 7.4a, and 7.5). The mastoid process can be felt as a lump just posterior to the ear. The **stylomastoid foramen,** between the styloid and mastoid processes, allows cranial nerve VII (the facial nerve) to leave the skull (see Figure 7.4a).

◩ *Homeostatic Imbalance*

The mastoid process is full of air cavities, the **mastoid sinuses,** or **air cells.** Its position adjacent to the middle ear cavity (a high-risk area for infections spreading from the throat) makes it extremely susceptible to infection itself. A mastoid sinus infection, or *mastoiditis,* is notoriously difficult to treat. Since the mastoid air cells are separated from the brain by only a very thin bony plate, mastoid infections may spread to the brain as well. Surgical removal of the mastoid process was once the best way to prevent life-threatening brain inflammations in people susceptible to repeated bouts of mastoiditis. Today, antibiotic therapy is the treatment of choice. ■

The deep **petrous** (pet'rus) **region** of the temporal bone contributes to the cranial base (see Figure 7.4). It looks like a miniature mountain ridge (*petrous* = rocky) between the occipital bone posteriorly and the sphenoid bone anteriorly. The posterior slope of this ridge lies in the posterior cranial fossa; the anterior slope is in the middle cranial fossa. Together, the sphenoid bone and the petrous portions of the temporal bones construct the **middle cranial fossa** (see Figure 7.4b and c), which supports the temporal lobes of the brain. Housed within the petrous region are the *middle* and *inner ear cavities,* which contain the sensory receptors for hearing and balance.

Several foramina penetrate the bone of the petrous region (see Figure 7.4a). The large **jugular**

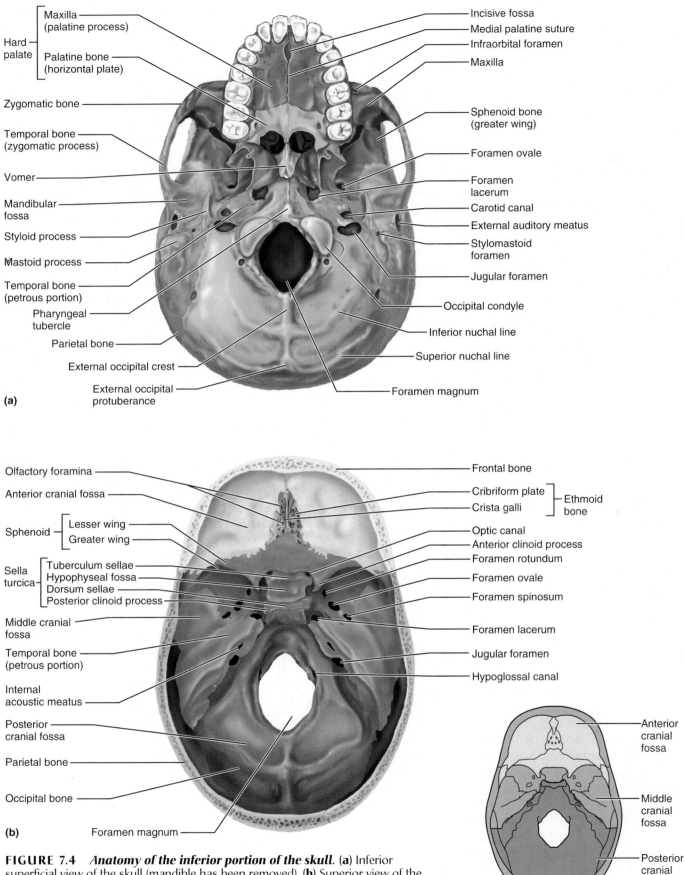

(a)

Hard palate
- Maxilla (palatine process)
- Palatine bone (horizontal plate)

Zygomatic bone

Temporal bone (zygomatic process)

Vomer

Mandibular fossa

Styloid process

Mastoid process

Temporal bone (petrous portion)

Pharyngeal tubercle

Parietal bone

External occipital crest

External occipital protuberance

Incisive fossa

Medial palatine suture

Infraorbital foramen

Maxilla

Sphenoid bone (greater wing)

Foramen ovale

Foramen lacerum

Carotid canal

External auditory meatus

Stylomastoid foramen

Jugular foramen

Occipital condyle

Inferior nuchal line

Superior nuchal line

Foramen magnum

(b)

Olfactory foramina

Anterior cranial fossa

Sphenoid
- Lesser wing
- Greater wing

Sella turcica
- Tuberculum sellae
- Hypophyseal fossa
- Dorsum sellae
- Posterior clinoid process

Middle cranial fossa

Temporal bone (petrous portion)

Internal acoustic meatus

Posterior cranial fossa

Parietal bone

Occipital bone

Foramen magnum

Frontal bone

Cribriform plate
Crista galli
Ethmoid bone

Optic canal

Anterior clinoid process

Foramen rotundum

Foramen ovale

Foramen spinosum

Foramen lacerum

Jugular foramen

Hypoglossal canal

(c)

Anterior cranial fossa

Middle cranial fossa

Posterior cranial fossa

FIGURE 7.4 *Anatomy of the inferior portion of the skull.* **(a)** Inferior superficial view of the skull (mandible has been removed). **(b)** Superior view of the floor of the cranial cavity (calvaria has been removed). **(c)** Schematic view of the cranial cavity floor showing the extent of its major fossae. (See *A Brief Atlas of the Human Body,* Figures 4 and 5.)

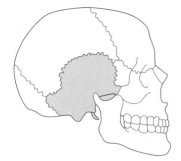

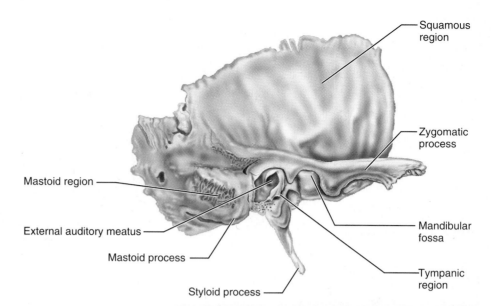

Squamous region

Zygomatic process

Mastoid region

External auditory meatus

Mastoid process

Styloid process

Mandibular fossa

Tympanic region

FIGURE 7.5 *The temporal bone.* Right lateral view. (See *A Brief Atlas of the Human Body,* Figures 3 and 8.)

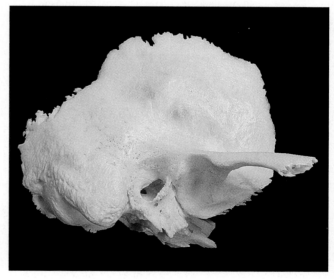

foramen at the junction of the occipital and petrous temporal bones allows passage of the internal jugular vein and three cranial nerves. The **carotid** (kar-rot'id) **canal,** just anterior to the jugular foramen, transmits the internal carotid artery into the cranial cavity. The two internal carotid arteries supply blood to over 80% of the cerebral hemispheres of the brain; their closeness to the inner ear cavities explains why, during excitement or exertion, we sometimes hear our rapid pulse as a thundering sound in the head. The **foramen lacerum** (la'ser-um) is a jagged opening (*lacerum* = torn or lacerated) between the petrous temporal bone and the sphenoid bone. It is almost completely closed by cartilage in a living person, but it is conspicuous in a dried skull, and students usually ask its name. The **internal acoustic (auditory) meatus,** positioned superolateral to the jugular foramen (see Figures 7.3b and 7.4b), transmits cranial nerves VII and VIII.

Sphenoid Bone

The butterfly-shaped **sphenoid** (sfe'noid) **bone** spans the width of the middle cranial fossa (see Figure 7.4b). The sphenoid is considered the keystone of the cranium because it forms a central wedge that articulates with all other cranial bones. It consists of a central body and three pairs of processes: the greater wings, lesser wings, and pterygoid (ter'ĭ-goid) processes (Figure 7.6). Within the **body** of the sphenoid are the paired **sphenoid sinuses** (see Figures 7.3b and 7.11).

The superior surface of the body bears a saddle-shaped prominence, the **sella turcica** (sel'ah ter'sĭ-kah), meaning "Turk's saddle." The seat of this saddle, called the **hypophyseal fossa,** forms a snug enclosure for the pituitary gland (hypophysis). This fossa is abutted anteriorly and posteriorly by the **tuberculum sellae** and the **dorsum sellae,** respectively. The dorsum sellae terminates laterally in the **posterior clinoid processes.**

The **greater wings** project laterally from the body, forming parts of (1) the middle cranial fossa (see Figure 7.4b and c), (2) the dorsal walls of the orbits (see Figure 7.2), and (3) the external wall of the skull, where they are seen as flag-shaped, bony areas medial to the zygomatic arch (see Figure 7.3). The

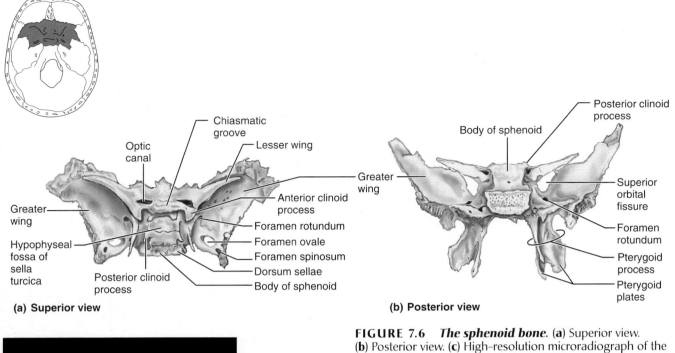

(a) **Superior view**

(b) **Posterior view**

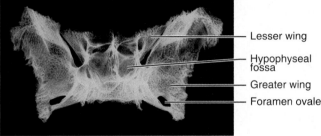

(c)

FIGURE 7.6 *The sphenoid bone.* (a) Superior view. (b) Posterior view. (c) High-resolution microradiograph of the sphenoid bone. Note in particular its body, contained sinuses, the greater and lesser wings, and the pterygoid processes. (See *A Brief Atlas of the Human Body*, Figures 3 and 9.)

hornlike **lesser wings** form part of the floor of the anterior cranial fossa (see Figure 7.4b and c) and part of the medial walls of the orbits. These terminate medially in the **anterior clinoid processes,** which provide an anchoring site for securing the brain within the skull. The trough-shaped **pterygoid processes** project inferiorly from the junction of the body and greater wings (see Figure 7.6b). They anchor the pterygoid muscles, which are important in chewing.

A number of openings in the sphenoid bone are visible in Figures 7.4b and 7.6. The **optic canals,** connected by the *chiasmatic groove,* lie anterior to the sella turcica; they allow the optic nerves to pass to the eyes. On each side of the sphenoid body is a crescent-shaped row of four openings. The anteriormost of these, the **superior orbital fissure,** is a long slit between the greater and lesser wings. It allows cranial nerves that control eye movements (III, IV, VI) to enter the orbit. This fissure is most obvious in an anterior view of the skull (see Figure 7.2). The foramen rotundum and foramen ovale provide passageways for branches of cranial nerve V (namely the maxillary and mandibular nerves) to reach the face. The **foramen rotundum** is in the medial part of the greater wing and is usually oval, despite its name meaning "round opening." The **foramen ovale** (o-vă′le), a large, oval foramen posterior to the foramen rotundum, is visible in an inferior view of the skull (see Figure 7.4a). Posterolateral to the foramen ovale is the small **foramen spinosum** (Figure 7.4b); it transmits the *middle meningeal artery,* which serves the internal faces of some of the cranial bones.

Ethmoid Bone

Like the temporal and sphenoid bones, the delicate **ethmoid bone** has a complex shape (Figure 7.7). Lying between the sphenoid and the nasal bones of the face, it is the most deeply situated bone of the skull. It forms most of the bony area between the nasal cavity and the orbits.

Of what importance is the crista galli?

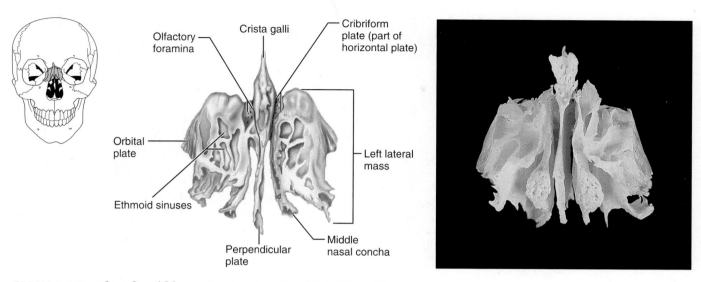

Olfactory foramina
Crista galli
Cribriform plate (part of horizontal plate)
Orbital plate
Left lateral mass
Ethmoid sinuses
Middle nasal concha
Perpendicular plate

FIGURE 7.7 *The ethmoid bone.* Anterior view. (See *A Brief Atlas of the Human Body,* Figures 3 and 10.)

The superior surface of the ethmoid, called the **cribriform** (krib′rĭ-form) **plate** (see also Figure 7.4b), helps form the roof of the nasal cavities and the floor of the anterior cranial fossa. The cribriform plate is punctured by tiny holes (*cribr* = sieve) called *olfactory foramina* that allow the olfactory nerves to pass from the smell receptors in the nasal cavities to the brain. Projecting superiorly from the midline of the cribriform plate is a triangular process called the **crista galli** (kris′tah gah′le; "rooster's comb"). The outermost covering of the brain (the dura mater) attaches to the crista galli and helps secure the brain in the cranial cavity.

The **perpendicular plate** of the ethmoid bone projects inferiorly from the cribriform plate and forms the superior part of the nasal septum, which divides the nasal cavity into right and left halves (see Figure 7.3b). Flanking the perpendicular plate on each side is a **lateral mass** riddled with the **ethmoid sinuses** (see Figures 7.7 and 7.11) for which the bone itself is named (*ethmos* = sieve). Extending medially from the lateral masses are the delicately coiled **superior** and **middle nasal conchae** (kong′ke; *concha* = shell), or **turbinates,** which protrude into the nasal cavity (Figures 7.7 and 7.10a). The lateral surfaces of the ethmoid's lateral masses are called **orbital plates** because they contribute to the medial walls of the orbits.

Sutural Bones

Sutural bones, also called **Wormian bones,** are tiny irregularly shaped bones or bone clusters that appear within sutures, most often in the lambdoid suture (see Figure 7.2b). Structurally unimportant, their number varies, and not all skulls exhibit them. Sutural bones represent additional ossification centers that appeared when the skull was expanding very rapidly during fetal development.

Facial Bones

The facial skeleton is made up of 14 bones (see Figures 7.2a and 7.3a), of which only the mandible and the vomer are unpaired. The maxillae, zygomatics, nasals, lacrimals, palatines, and inferior conchae are paired bones. As a rule, the facial skeleton of men is more elongated than that of women; therefore, women's faces tend to be rounder and less angular.

Mandible

The U-shaped **mandible** (man′dĭ-bl), or lower jawbone (Figures 7.2, 7.3, and 7.8a), is the largest, strongest bone of the face. It has a *body,* which forms the chin, and two upright *rami* (*rami* = branches). Each ramus meets the body posteriorly at a **mandibular angle.** At the superior margin of each ramus are two processes separated by the **mandibular notch.** The anterior **coronoid** (kor′o-noid;

It is the site to which the membranes surrounding the brain attach to secure the brain in the cranial cavity.

? *Which of the bones illustrated is the keystone of the facial skeleton? Which contributes to the hard palate?*

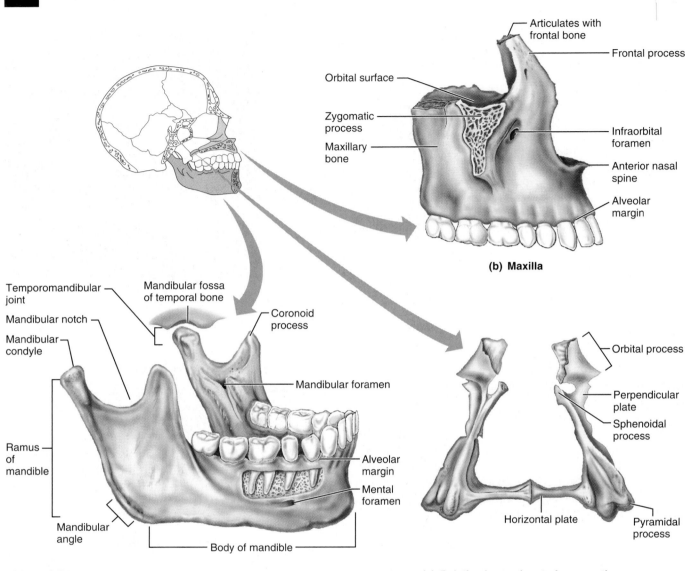

(a) **Mandible**

(b) **Maxilla**

(c) **Palatine bones (posterior aspect)**

FIGURE 7.8   *Detailed anatomy of some isolated facial bones*. (a) The mandible (and its articulation with the temporal bone). **(b)** The maxilla (and its sites of articulation with the frontal and zygomatic bones). **(c)** Posterior view of the palatine bones. Note that the bones in (c) are enlarged more than those in (a) and (b). (See *A Brief Atlas of the Human Body*, Figures 11, 12, and 13.)

"crown-shaped") **process** is an insertion point for the large temporalis muscle that elevates the lower jaw during chewing. The posterior **mandibular condyle** articulates with the mandibular fossa of the temporal bone, forming the *temporomandibular joint* on the same side.

The mandibular **body** anchors the lower teeth. Its superior border, called the **alveolar** (al-ve′o-lar)

margin, contains the sockets (*alveoli*) in which the teeth are embedded. In the midline of the mandibular body is a slight depression, the **mandibular symphysis** (sim′fih-sis), which indicates the line of fusion of the two mandibular bones during infancy (see Figure 7.2a).

Large **mandibular foramina,** one on the medial surface of each ramus, permit the nerves responsible for tooth sensation to pass to the teeth in the lower jaw. Dentists inject Novocain into these foramina to prevent pain while working on the lower teeth. The **mental foramina,** openings on the lateral aspects of

The maxilla is the keystone of the facial skeleton and forms the anterior portion of the hard palate. The palatine bones form the posterior hard palate.

the mandibular body, allow blood vessels and nerves to pass to the skin of the chin (*ment* = chin) and lower lip.

Maxillary Bones

The **maxillary bones,** or **maxillae** (mak-sil′le; "jaws") (see Figures 7.2 to 7.4 and 7.8b), are fused medially. They form the upper jaw and the central portion of the facial skeleton. All facial bones except the mandible articulate with the maxillae. Hence, the maxillae are considered the keystone bones of the facial skeleton.

The maxillae carry the upper teeth in their **alveolar margins.** Just inferior to the nose the maxillae meet medially, forming the pointed *anterior nasal spine* at their junction. The **palatine** (pă′lah-tīn) **processes** of the maxillae project posteriorly from the alveolar margins and fuse medially, forming the anterior two-thirds of the hard palate, or bony roof of the mouth (see Figures 7.3b and 7.4a). Just posterior to the teeth is a midline foramen, called the **incisive fossa,** which serves as a passageway for blood vessels and nerves.

The **frontal processes** extend superiorly to the frontal bone, forming part of the lateral aspects of the bridge of the nose (see Figures 7.2a and 7.8b). The regions that flank the nasal cavity laterally contain the **maxillary sinuses** (see Figure 7.11), the largest of the paranasal sinuses. They extend from the orbits to the upper teeth. Laterally, the maxillae articulate with the zygomatic bones via their **zygomatic processes.**

The **inferior orbital fissure** is located deep within the orbit (see Figure 7.2a) at the junction of the maxilla with the greater wing of the sphenoid. It permits the zygomatic nerve, the maxillary nerve (a branch of cranial nerve V), and blood vessels to pass to the face. Just below the eye socket on each side is an **infraorbital foramen** that allows the infraorbital nerve and artery to reach the face.

Zygomatic Bones

The irregularly shaped **zygomatic bones** (see Figures 7.2a and 7.3a) are commonly called the cheekbones (*zygoma* = cheekbone). They articulate with the zygomatic processes of the temporal bones posteriorly, the zygomatic process of the frontal bone superiorly, and with the zygomatic processes of the maxillae anteriorly. The zygomatic bones form the prominences of the cheeks and part of the inferolateral margins of the orbits.

Nasal Bones

The thin, basically rectangular **nasal** (na′zal) **bones** are fused medially, forming the bridge of the nose (see Figures 7.2a and 7.3a). They articulate with the frontal bone superiorly, the maxillary bones laterally, and the perpendicular plate of the ethmoid bone posteriorly. Inferiorly they attach to the cartilages that form most of the skeleton of the external nose.

Lacrimal Bones

The delicate fingernail-shaped **lacrimal** (lak′rĭ-mal) **bones** contribute to the medial walls of each orbit (see Figures 7.2a and 7.3a). They articulate with the frontal bone superiorly, the ethmoid bone posteriorly, and the maxillae anteriorly. Each lacrimal bone contains a deep groove that helps form a **lacrimal fossa.** The lacrimal fossa houses the *lacrimal sac,* part of the passageway that allows tears to drain from the eye surface into the nasal cavity (*lacrima* = tears).

Palatine Bones

Each L-shaped **palatine bone** is fashioned from two bony plates, the *horizontal* and *perpendicular,* and has three important articular processes, the *pyramidal, sphenoidal,* and *orbital* (Figures 7.3b, 7.4a, and 7.8c). The **horizontal plates** complete the posterior portion of the hard palate. The superiorly projecting **perpendicular (vertical) plates** form part of the posterolateral walls of the nasal cavity and a small part of the orbits.

Vomer

The slender, plow-shaped **vomer** (vo′mer) lies in the nasal cavity, where it forms part of the nasal septum (see Figures 7.2a and 7.10b). It is discussed below in connection with the nasal cavity.

Inferior Nasal Conchae

The paired **inferior nasal conchae** are thin, curved bones in the nasal cavity. They project medially from the lateral walls of the nasal cavity, just inferior to the middle nasal conchae of the ethmoid bone (see Figures 7.2a and 7.10a). They are the largest of the three pairs of conchae and, like the others, they form part of the lateral walls of the nasal cavity.

Special Characteristics of the Orbits and Nasal Cavity

Two restricted skull regions, the orbits and the nasal cavity, are formed from an amazing number of bones. Thus, even though the individual bones forming these structures have been described, a brief summary is provided here to pull the parts together.

The Orbits

The **orbits** are bony cavities within which the eyes are firmly encased and cushioned by fatty tissue.

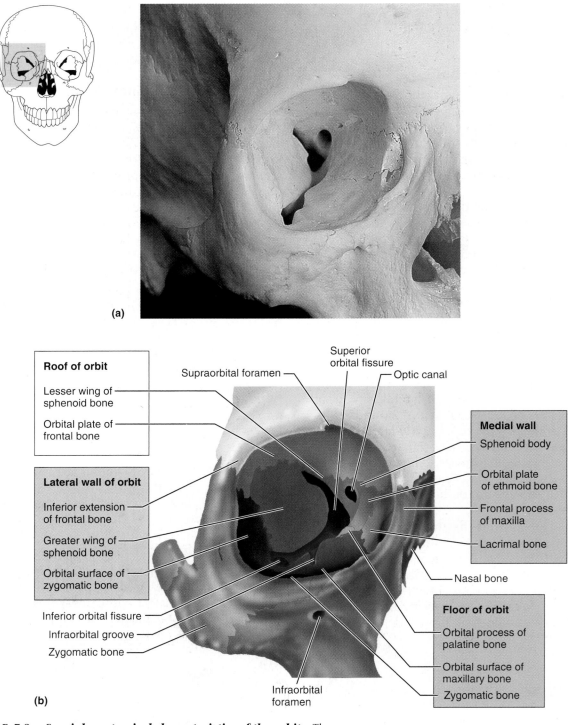

(a)

Roof of orbit

Lesser wing of
sphenoid bone

Orbital plate of
frontal bone

Supraorbital foramen

Superior
orbital fissure

Optic canal

Medial wall

Sphenoid body

Orbital plate
of ethmoid bone

Frontal process
of maxilla

Lacrimal bone

Nasal bone

Lateral wall of orbit

Inferior extension
of frontal bone

Greater wing of
sphenoid bone

Orbital surface of
zygomatic bone

Inferior orbital fissure

Infraorbital groove

Zygomatic bone

Floor of orbit

Orbital process of
palatine bone

Orbital surface of
maxillary bone

Zygomatic bone

Infraorbital
foramen

(b)

FIGURE 7.9   *Special anatomical characteristics of the orbits.* The
contribution of each of the seven bones forming the right orbit is illustrated.
(a) Photograph. **(b)** Diagrammatic view. (See *A Brief Atlas of the Human Body,* Figure
14.)

The muscles that move the eyes and the tear-
producing lacrimal glands are also housed within the
orbits. The walls of each orbit are formed by parts of
seven bones—the frontal, sphenoid, zygomatic,
maxilla, palatine, lacrimal, and ethmoid bones.

Their relationships are shown in Figure 7.9. Also
seen in the orbits are the superior and inferior orbital
fissures and the optic canals, described earlier.

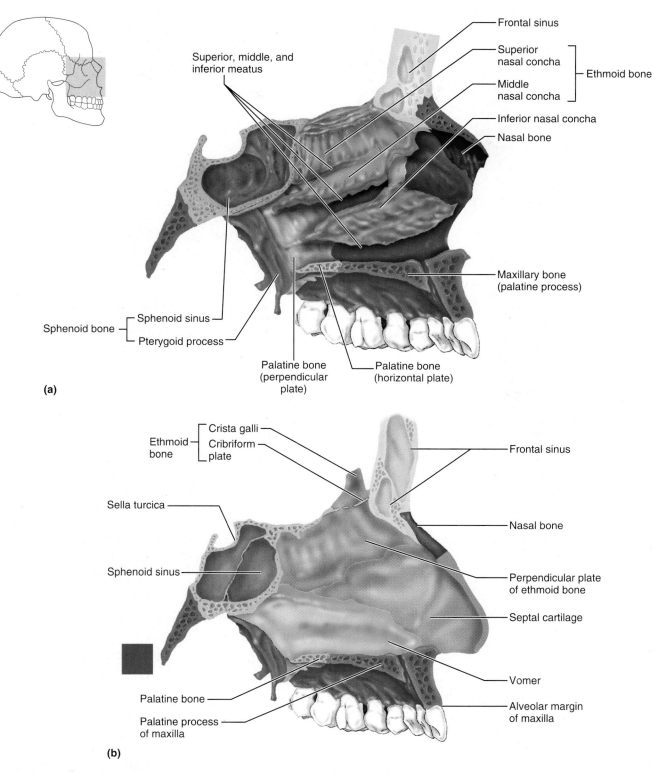

FIGURE 7.10   *Special anatomical characteristics of the nasal cavity.* **(a)** Bones forming the left lateral wall of the nasal cavity. **(b)** The contribution of the ethmoid, vomer, and cartilage to the nasal septum. (See *A Brief Atlas of the Human Body,* Figure 15.)

The Nasal Cavity

The **nasal cavity** is constructed of bone and hyaline cartilage (Figure 7.10a). The *roof* of the nasal cavity is formed by the cribriform plate of the ethmoid. The *lateral walls* are largely shaped by the superior and middle conchae of the ethmoid bone, the perpendicular plates of the palatine bones, and the inferior nasal conchae. The depressions under cover of the conchae on the lateral walls are called *meatuses* (*meatus* = passage), so there are superior, middle,

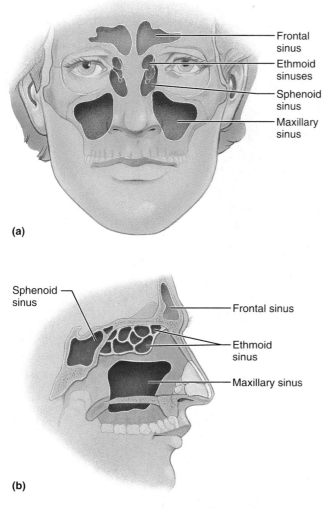

(a)

(b)

FIGURE 7.11 *Paranasal sinuses, diagrammatic view.* (**a**) Anterior aspect. (**b**) Medial aspect.

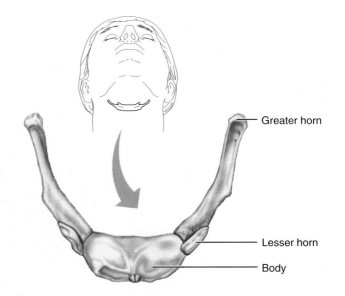

FIGURE 7.12 *Anatomical location and structure of the hyoid bone.* The hyoid bone is suspended in the mid-anterior neck by ligaments attached to the lesser horns (cornua) and the styloid processes of the temporal bones.

Paranasal Sinuses

Five skull bones—the frontal, sphenoid, ethmoid, and paired maxillary bones—contain mucosa-lined, air-filled sinuses that cause them to look rather moth-eaten in an X-ray image. These particular sinuses are called **paranasal sinuses** because they cluster around the nasal cavity (Figure 7.11). Small openings connect the sinuses to the nasal cavity and act as "two-way streets"; air enters the sinuses from the nasal cavity and mucus formed by the sinus mucosae drains into the nasal cavity. The mucosae of the sinuses also help to warm and humidify inspired air. The paranasal sinuses lighten the skull and enhance the resonance of the voice.

The Hyoid Bone

Though not really part of the skull, the **hyoid** (hi'oid; "U-shaped") **bone** (Figure 7.12) lies just inferior to the mandible in the anterior neck. The hyoid bone is unique in that it is the only bone of the body that does not articulate directly with any other bone. Instead, it is anchored by the narrow *stylohyoid ligaments* to the styloid processes of the temporal bones. Horseshoe-shaped, with a *body* and two pairs of *horns,* or *cornua,* the hyoid bone acts as a movable base for the tongue. Its body and greater horns are attachment points for neck muscles that raise and lower the larynx during swallowing and speech.

⋆ ⋆ ⋆

Table 7.1 summarizes the bones of the skull.

and inferior meatuses. The *floor* of the nasal cavity is formed by the palatine processes of the maxillae and the palatine bones. The nasal cavity is divided into right and left parts by the *nasal septum.* The bony portion of the septum is formed by the vomer inferiorly and the perpendicular plate of the ethmoid bone superiorly (Figure 7.10b). A sheet of cartilage called the *septal cartilage* completes the septum anteriorly.

The nasal septum and conchae are covered with a mucus-secreting mucosa. This mucosa moistens and warms the entering air and helps cleanse it of debris. The scroll-shaped conchae increase the turbulence of air flowing through the nasal cavity. This swirling forces more of the inhaled air into contact with the warm, damp mucosa and encourages trapping of airborne particles (dust, pollen, bacteria) in the sticky mucus.

TABLE 7.1	Bones of the Skull

Bone*	Comments	Important markings
Cranial Bones		
Frontal (1) (Figures 7.2a, 7.3, and 7.4b)	Forms forehead; superior part of orbits and anterior cranial fossa; contains sinuses	**Supraorbital foramina (notches):** allow the supraorbital arteries and nerves to pass
Parietal (2) (Figures 7.2 and 7.3)	Forms most of the superior and lateral aspects of the skull	
Occipital (1) (Figures 7.2b, 7.3, and 7.4)	Forms posterior aspect and most of the base of the skull	**Foramen magnum:** allows the spinal cord to issue from the brain stem to enter the vertebral canal
		Hypoglossal canal: allows passage of the hypoglossal nerve (cranial nerve XII)
		Occipital condyles: articulate with the atlas (first vertebra)
		External occipital protuberance and **nuchal lines:** sites of muscle attachment
		External occipital crest: attachment site of ligamentum nuchae
Temporal (2) (Figures 7.3, 7.4, and 7.5)	Forms inferolateral aspects of the skull and contributes to the middle cranial fossa; has squamous, mastoid, tympanic, and petrous regions	**Zygomatic process:** helps to form the zygomatic arch, which forms the prominence of the cheek
		Mandibular fossa: articular point of the mandibular condyle
		External auditory (acoustic) meatus: canal leading from the external ear to the eardrum
		Styloid process: attachment site for hyoid bone and several neck muscles
		Mastoid process: attachment site for several neck and tongue muscles
		Stylomastoid foramen: allows cranial nerve VII (facial nerve) to pass
		Jugular foramen: allows passage of the internal jugular vein and cranial nerves IX, X, and XI
		Internal acoustic meatus: allows passage of cranial nerves VII and VIII
		Carotid canal: allows passage of the internal carotid artery
Sphenoid (1) (Figures 7.2a, 7.3, 7.4b, and 7.6)	Keystone of the cranium; contributes to the middle cranial fossa and orbits; main parts are the body, greater wings, lesser wings, and pterygoid processes	**Sella turcica:** hypophyseal fossa portion is the seat of the pituitary gland
		Optic canals: allow passage of cranial nerve II and the ophthalmic arteries
		Superior orbital fissures: allow passage of cranial nerves III, IV, VI, part of V (ophthalmic division), and ophthalmic vein
		Foramen rotundum: allows passage of the maxillary division of cranial nerve V
		Foraman ovale: allows passage of the mandibular division of cranial nerve V
		Foramen spinosum: allows passage of the middle meningeal artery

*The color code beside each bone name corresponds to the bone's color in Figures 7.2 to 7.10. The number in parentheses ( ) following the bone name denotes the total number of such bones in the body.

Table continues on page 214.

TABLE 7.1 **Bones of the Skull** *(continued)*

Bone*	Comments	Important markings
Cranial Bones		
Ethmoid (1) (Figures 7.2a, 7.3, 7.4b, 7.7, and 7.10)	Helps to form the anterior cranial fossa; forms part of the nasal septum and the lateral walls and roof of the nasal cavity; contributes to the medial wall of the orbit	**Crista galli:** attachment point for the falx cerebri, a dural membrane **Cribriform plate:** allows passage of nerve filaments of the olfactory nerves (cranial nerve I) **Superior** and **middle nasal conchae:** form part of lateral walls of nasal cavity; increase turbulence of air flow
Ear ossicles (malleus, incus, and stapes) (2 each)	Found in middle ear cavity; involved in sound transmission; see Chapter 16	
Facial Bones		
Mandible (1) (Figures 7.2a, 7.3, and 7.8a)	The lower jaw	**Coronoid processes:** insertion points for the temporalis muscles **Mandibular condyles:** articulate with the temporal bones in freely movable joints (temporomandibular joints) of the jaw **Mandibular symphysis:** medial fusion point of the mandibular bones **Alveoli:** sockets for the teeth **Mandibular foramina:** permit the inferior alveolar nerves to pass **Mental foramina:** allow blood vessels and nerves to pass to the chin and lower lip
Maxilla (2) (Figures 7.2a, 7.3, 7.4, and 7.8b)	Keystone bones of the face; form the upper jaw and parts of the hard palate, orbits, and nasal cavity walls	**Alveoli:** sockets for the teeth **Zygomatic processes:** help form the zygomatic arches **Palatine processes:** form the anterior hard palate; meet medially in middle palatine suture **Frontal processes:** form part of lateral aspect of bridge of nose **Incisive fossa:** permits blood vessels and nerves to pass through hard palate (fused palatine processes) **Inferior orbital fissures:** permit maxillary branch of cranial nerve V, the zygomatic nerve, and blood vessels to pass **Infraorbital foramen:** allows passage of infraorbital nerve to skin of face
Zygomatic (2) (Figures 7.2a, 7.3a, and 7.4a)	Form the cheek and part of the orbit	
Nasal (2) (Figures 7.2a and 7.3)	Construct the bridge of the nose	
Lacrimal (2) (Figures 7.2a and 7.3a)	Form part of the medial orbit wall	**Lacrimal fossa:** houses the lacrimal sac, which helps to drain tears into the nasal cavity
Palatine (2) (Figures 7.3b, 7.4a, and 7.8c)	Form posterior part of the hard palate and a small part of nasal cavity and orbit walls	
Vomer (1) (Figures 7.2a and 7.10)	Part of the nasal septum	
Inferior nasal concha (2) (Figures 7.2a and 7.10a)	Form part of the lateral walls of the nasal cavity	

THE VERTEBRAL COLUMN
General Characteristics

Some people think of the **vertebral column** as a rigid supporting rod, but this is inaccurate. The vertebral column, also called the **spine**, is formed from 26 irregular bones connected in such a way that a flexible, curved structure results (Figure 7.13). Serving as the axial support of the trunk, the spine extends from the skull to the pelvis, where it transmits the weight of the trunk to the lower limbs. It also surrounds and protects the delicate spinal cord and provides attachment points for the ribs and for the muscles of the back and neck. In the fetus and infant, the vertebral column consists of 33 separate bones, or **vertebrae** (ver'tĕbre). Inferiorly, nine of these eventually fuse to form two composite bones, the sacrum and the tiny coccyx. The remaining 24 bones persist as individual vertebrae separated by intervertebral discs.

Divisions and Curvatures

The vertebral column is about 70 cm (28 inches) long in an average adult and has five major divisions (see Figure 7.13). The seven vertebrae of the neck are the **cervical** (ser'vi-kal) **vertebrae,** the next 12 are the **thoracic** (tho-ras'ik) **vertebrae,** and the five supporting the lower back are the **lumbar** (lum'bar) **vertebrae.** Remembering common meal times—7 AM, 12 noon, and 5 PM—will help you recall the number of bones in these three regions of the spine. The vertebrae become progressively larger from the cervical to the lumbar region, as they must support greater and greater weight.

Inferior to the lumbar vertebrae is the **sacrum** (sa'krum), which articulates with the hip bones of the pelvis. The terminus of the vertebral column is the tiny **coccyx** (kok'siks). All of us have the same number of cervical vertebrae. Variations in numbers of vertebrae in other regions occur in about 5% of people.

When you view the vertebral column from the side, you can see the four curvatures that give it its S, or sinusoid, shape. The **cervical** and **lumbar curvatures** are concave posteriorly; the **thoracic** and **sacral curvatures** are convex posteriorly. These curvatures increase the resilience and flexibility of the spine, allowing it to function like a spring rather than as a rigid rod.

Homeostatic Imbalance

There are several types of abnormal spinal curvatures. Some are congenital (present at birth); others result from disease, poor posture, or unequal muscle pull on the spine. *Scoliosis* (sko"le-o'sis), literally,

Why would it be anatomically "stupid" to have the sturdiest vertebrae in the cervical instead of the lumbar region of the spine?

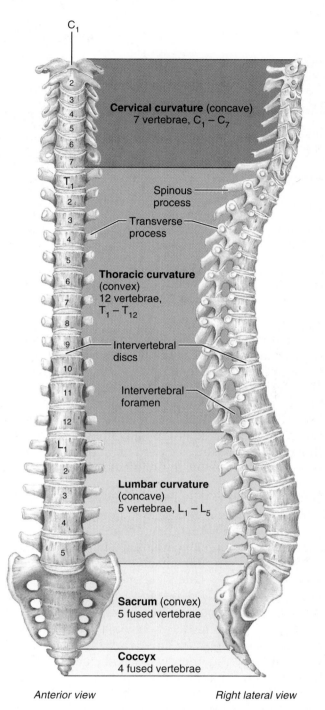

FIGURE 7.13 *The vertebral column.* Notice the curvatures in the lateral view. (The terms *convex* and *concave* refer to the curvature of the posterior aspect of the vertebral column.) (See *A Brief Atlas of the Human Body,* Figure 17.)

Because the lumbar region bears the greatest weight, its vertebrae are the sturdiest.

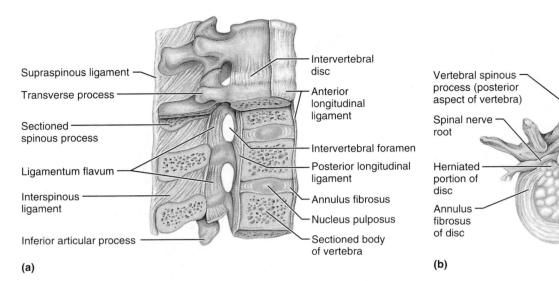

Supraspinous ligament

Transverse process

Sectioned spinous process

Ligamentum flavum

Interspinous ligament

Inferior articular process

Intervertebral disc

Anterior longitudinal ligament

Intervertebral foramen

Posterior longitudinal ligament

Annulus fibrosus

Nucleus pulposus

Sectioned body of vertebra

(a)

Vertebral spinous process (posterior aspect of vertebra)

Spinal nerve root

Herniated portion of disc

Annulus fibrosus of disc

Spinal cord

Nucleus pulposus of disc

(b)

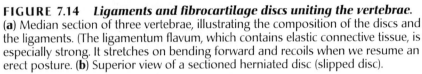

FIGURE 7.14   *Ligaments and fibrocartilage discs uniting the vertebrae.*
(a) Median section of three vertebrae, illustrating the composition of the discs and the ligaments. (The ligamentum flavum, which contains elastic connective tissue, is especially strong. It stretches on bending forward and recoils when we resume an erect posture. **(b)** Superior view of a sectioned herniated disc (slipped disc).

"twisted disease," is an abnormal *lateral* curvature that occurs most often in the thoracic region. It is quite common during late childhood, particularly in girls, for some unknown reason. Other, more severe cases result from abnormal vertebral structure, lower limbs of unequal length, or muscle paralysis. If muscles on one side of the body are nonfunctional, those of the opposite side exert an unopposed pull on the spine and force it out of alignment. Scoliosis is treated (with body braces or surgically) before growth ends to prevent permanent deformity and breathing difficulties due to a compressed lung.

Kyphosis (ki-fo'sis), or hunchback, is a *dorsally* exaggerated *thoracic* curvature. It is particularly common in aged individuals because of osteoporosis, but may also reflect tuberculosis of the spine, rickets, or osteomalacia. *Lordosis*, or swayback, is an accentuated *lumbar* curvature. It, too, can result from spinal tuberculosis or osteomalacia. Temporary lordosis is common in those carrying a "large load up front," such as men with "potbellies" and pregnant women. In an attempt to preserve their center of gravity, these individuals automatically throw back their shoulders, accentuating their lumbar curvature. ■

Ligaments

Like a tall, tremulous TV transmitting tower, the vertebral column cannot possibly stand upright by itself. It must be held in place by an elaborate system of cablelike supports. In the case of the vertebral column, straplike ligaments and the trunk muscles as-

sume this role. The trunk muscles are discussed in Chapter 10. The major supporting ligaments are the **anterior** and **posterior longitudinal ligaments** (Figure 7.14). These run as continuous bands down the front and back surfaces of the spine from the neck to the sacrum. The broad anterior ligament is strongly attached to both the bony vertebrae and the discs. Along with its supporting role, it prevents hyperextension of the spine (bending too far backward). The posterior ligament, which resists hyperflexion of the spine (bending too sharply forward), is narrow and relatively weak. It attaches only to the discs. Short ligaments connect each vertebra to those immediately above and below.

Intervertebral Discs

Each **intervertebral disc** is a cushionlike pad composed of two parts. The inner gelatinous **nucleus pulposus** (pul-po'sus; "pulp") acts like a rubber ball to give the disc its elasticity and compressibility. Surrounding the spherical nucleus pulposus is a strong collar composed of collagen fibers superficially and fibrocartilage (internally), the **annulus fibrosus** (an'u-lus fi-bro'sus; "ring of fibers") (see Figure 7.14a). The annulus fibrosus contains the nucleus pulposus, limiting its expansion when the spine is compressed. It also holds together successive vertebrae and resists tension in the spine.

The intervertebral discs act as shock absorbers during walking, jumping, and running, and they allow the spine to flex and extend, and to a lesser extent to bend laterally. At points of compression, the discs flatten and bulge out a bit between the vertebrae. The discs are thickest in the lumbar and cervical regions, which enhances the flexibility of these regions. Collectively the discs account for about 25% of the height of the vertebral column. They flatten somewhat during the course of the day, so we are always a few centimeters shorter at night than when we awake in the morning.

◪ Homeostatic Imbalance

Severe or sudden physical trauma to the spine—for example, from bending forward while lifting a heavy object—may result in herniation of one or more discs. A **herniated (prolapsed) disc** (commonly called a *slipped disc*) usually involves rupture of the annulus fibrosus followed by protrusion of the spongy nucleus pulposus through the annulus (see Figure 7.14b). If the protrusion presses on the spinal cord or on spinal nerves exiting from the cord, numbness or excruciating pain may result. Herniated discs are generally treated with moderate exercise, massage, heat therapy, and painkillers. If this fails, the protruding disc may have to be removed surgically and a bone graft done to fuse the adjoining vertebrae. For those preferring to avoid general anesthesia, the disc can be partially vaporized with a laser in an outpatient procedure called percutaneous laser disc decompression that takes only 30 to 40 minutes. The patient leaves with only an adhesive bandage to mark the spot. ∎

General Structure of Vertebrae

All vertebrae have a common structural pattern (Figure 7.15). Each vertebra consists of a **body,** or **centrum,** anteriorly and a **vertebral arch** posteriorly. The disc-shaped body is the weight-bearing region. Together, the body and vertebral arch enclose an opening called the **vertebral foramen.** Successive vertebral foramina of the articulated vertebrae form the long **vertebral canal,** through which the spinal cord passes.

The vertebral arch is a composite structure formed by two pedicles and two laminae. The **pedicles** (ped′ĭ-kel; "little foot"), short bony pillars projecting posteriorly from the vertebral body, form the sides of the arch. The **laminae** (lam′ĭ-ne) are flattened plates that fuse in the median plane, completing the arch posteriorly. Seven processes project from the vertebral arch. The **spinous process** is a median posterior projection arising at the junction of the two laminae. A **transverse process** extends laterally from each side of the vertebral arch. The spinous and transverse processes are attachment sites for muscles that move the vertebral column and for ligaments that stabilize it. The paired **superior** and **in-**

? *What structural parts of the vertebra are sites of muscle attachment?*

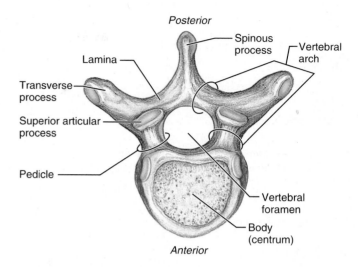

FIGURE 7.15 *Structure of a typical vertebra.* Superior view (inferior articulating surfaces not shown).

ferior articular processes protrude superiorly and inferiorly, respectively, from the pedicle-lamina junctions. The smooth joint surfaces of the articular processes, called *facets* ("little faces"), are covered with hyaline cartilage. The inferior articular processes of each vertebra form movable joints with the superior articular processes of the vertebra immediately below. Thus, successive vertebrae join both at their bodies and at their articular processes.

The pedicles have notches on their superior and inferior borders, providing lateral openings between adjacent vertebrae called **intervertebral foramina** (see Figure 7.13). The spinal nerves issuing from the spinal cord pass through these foramina.

Regional Vertebral Characteristics

In addition to the common structural features just described, vertebrae exhibit variations that allow the different regions of the spine to perform slightly different functions and movements. In general, the types of movements that can occur between vertebrae are (1) flexion and extension (anterior bending and posterior straightening of the spine), (2) lateral flexion (bending the *upper body* to the right or left), and (3) rotation (in which vertebrae rotate on one another in the longitudinal axis of the spine. The

TABLE 7.2	Regional Characteristics of Cervical, Thoracic, and Lumbar Vertebrae		
Characteristic	Cervical (3–7)	Thoracic	Lumbar
Body	Small, wide side to side	Larger than cervical; heart shaped; bears two costal demifacets	Massive; kidney shaped
Spinous process	Short; bifid; projects directly posteriorly	Long; sharp; projects inferiorly	Short; blunt; projects directly posteriorly
Vertebral foramen	Triangular	Circular	Triangular
Transverse processes	Contain foramina	Bear facets for ribs (except T_{11} and T_{12})	Thin and tapered
Superior and inferior articulating processes	Superior facets directed superoposteriorly	Superior facets directed posteriorly	Superior facets directed posteromedially (or medially)
	Inferior facets directed inferoanteriorly	Inferior facets directed anteriorly	Inferior facets directed anterolaterally (or laterally)
Movements allowed	Flexion and extension; lateral flexion; rotation; the spine region with the greatest range of movement	Rotation; lateral flexion possible but limited by ribs; flexion and extension prevented	Flexion and extension; some lateral flexion; rotation prevented

Superior View

Right Lateral View

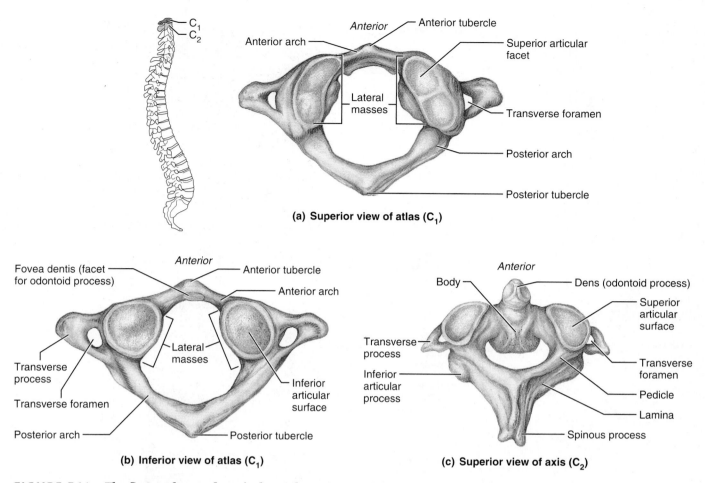

FIGURE 7.16 *The first and second cervical vertebrae.* (See *A Brief Atlas of the Human Body*, Figure 18.)

regional vertebral characteristics described in this section are illustrated and summarized in Table 7.2.

Cervical Vertebrae

The seven **cervical vertebrae**, identified as C_1–C_7, are the smallest, lightest vertebrae. The first two (C_1 and C_2) are unusual and we will skip them for the moment. The "typical" cervical vertebrae (C_3–C_7) have the following distinguishing features:

1. The body is oval—wider from side to side than in the anteroposterior dimension.

2. Except in C_7, the spinous process is short, projects directly back, and is *bifid* (bi'fid), or split at its tip.

3. The vertebral foramen is large and generally triangular.

4. Each transverse process contains a **transverse foramen** through which the vertebral arteries pass to service the brain.

The spinous process of C_7 is not bifid and is much larger than those of the other cervical vertebrae (see Figure 7.17a). Because its spinous process is visible through the skin, C_7 can be used as a landmark for counting the vertebrae and is called the **vertebra prominens** ("prominent vertebra").

The first two cervical vertebrae, the atlas and the axis, have no intervertebral disc between them, and they are highly modified, reflecting their special functions. The **atlas** (C_1) has no body and no spinous process (Figure 7.16a and b). Essentially, it is a ring of bone consisting of *anterior* and *posterior arches* and a *lateral mass* on each side. Each lateral mass has articular facets on both its superior and inferior surfaces. The superior articular facets receive the occipital condyles of the skull; thus, they "carry" the skull, just as Atlas supported the heavens in Greek mythology. These joints allow you to nod "yes." The inferior articular surfaces form joints with the axis (C_2) below.

The **axis,** which has a body, spine, and the other typical vertebral processes, is not as specialized as the atlas. In fact, its only unusual feature is the knoblike **dens** (denz; "tooth"), or **odontoid** (o-don'-toid) **process,** projecting superiorly from its body.

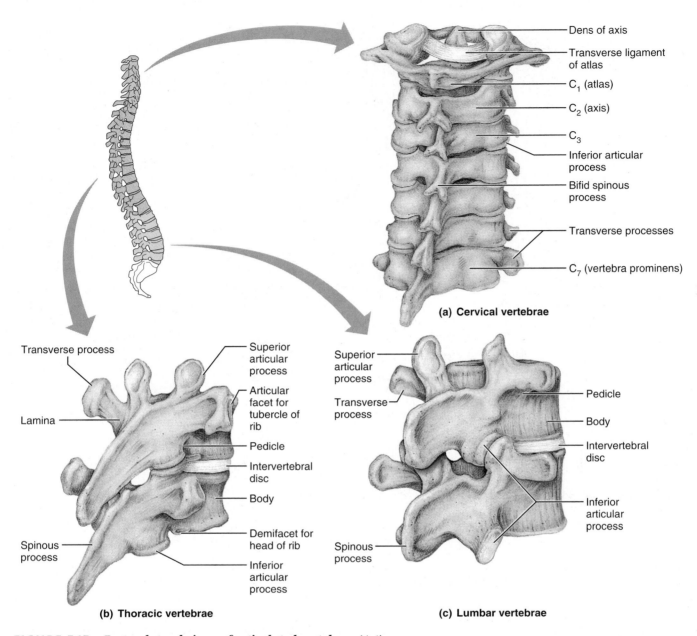

(a) Cervical vertebrae

— Dens of axis
— Transverse ligament of atlas
— C₁ (atlas)
— C₂ (axis)
— C₃
— Inferior articular process
— Bifid spinous process
— Transverse processes
— C₇ (vertebra prominens)

(b) Thoracic vertebrae

Transverse process
Lamina
Spinous process
Superior articular process
Articular facet for tubercle of rib
Pedicle
Intervertebral disc
Body
Demifacet for head of rib
Inferior articular process

(c) Lumbar vertebrae

Superior articular process
Transverse process
Spinous process
Pedicle
Body
Intervertebral disc
Inferior articular process

FIGURE 7.17 *Posterolateral views of articulated vertebrae.* Notice the bulbous tip on the spinous process of C₇, the vertebra prominens. (See *A Brief Atlas of the Human Body,* Figures 19, 20, and 21.)

The dens is actually the "missing" body of the atlas, which fuses with the axis during embryonic development. Cradled in the anterior arch of the atlas by the transverse ligaments (see Figure 7.17a), the dens acts as a pivot for the rotation of the atlas. Hence, this joint allows you to rotate your head from side to side to indicate "no."

Thoracic Vertebrae

The 12 **thoracic vertebrae** (T_1–T_{12}) all articulate with the ribs (see Table 7.2 and Figure 7.17b). However, the first looks much like C_7, and the last four show a progression toward lumbar vertebral structure. The thoracic vertebrae increase in size from the first to the last. Unique characteristics of these vertebrae include the following:

1. The body is roughly heart shaped. It typically bears two *facets*, or more typically *demifacets* (half-facets), on each side, one at the superior edge and the other at the inferior edge. The demifacets receive the heads of the ribs. (However, the bodies of T_{10}–T_{12} vary from this pattern by having only a single facet to receive their respective ribs.)

2. The vertebral foramen is circular.

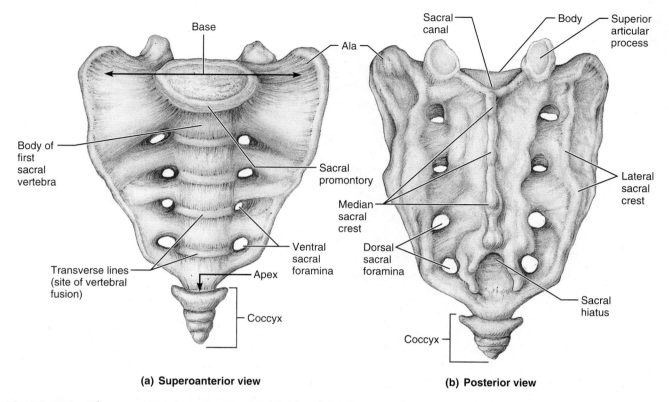

(a) Superoanterior view

(b) Posterior view

FIGURE 7.18 *The sacrum and coccyx.* (See *A Brief Atlas of the Human Body,* Figure 22.)

3. The spinous process is long and points sharply inferiorly.

4. With the exception of T_{11} and T_{12}, the transverse processes have facets that articulate with the *tubercles* of the ribs.

5. The superior and inferior articular facets lie mainly in the frontal plane, a situation that essentially prevents flexion and extension, but which allows rotation of this region of the spine.

Lumbar Vertebrae

The lumbar region of the vertebral column, commonly referred to as the small of the back, receives the most stress. The enhanced weight-bearing function of the five **lumbar vertebrae** (L_1–L_5) is reflected in their sturdier structure. Their bodies are massive and kidney shaped (see Table 7.2 and Figure 7.17c). Other characteristics typical of these vertebrae:

1. The pedicles and laminae are shorter and thicker than those of other vertebrae.

2. The spinous processes are short, flat, and hatchet shaped and are easily seen when a person bends forward. These processes are robust and project directly backward, adaptations for the attachment of the large back muscles.

3. The vertebral foramen is triangular.

4. The orientation of the facets of the articular processes of the lumbar vertebrae differs substantially from that of the other vertebra types (see Table 7.2). These modifications lock the lumbar vertebrae together and provide stability by preventing rotation of the lumbar spine. However, flexion and extension are possible.

Sacrum

The triangular **sacrum** (Figure 7.18), which shapes the posterior wall of the pelvis, is formed by five fused vertebrae (S_1–S_5) in adults. It articulates superiorly (via its **superior articular processes**) with L_5 and inferiorly with the coccyx. Laterally, the sacrum articulates, via its **auricular surfaces,** with the two hip bones to form the **sacroiliac** (sa″kro-il′e-ak) **joints** of the pelvis.

The **sacral promontory** (prom′on-tor″e), the anterosuperior margin of the first sacral vertebra, bulges anteriorly into the pelvic cavity. The body's center of gravity lies about 1 cm posterior to this landmark. Four ridges, the **transverse lines,** cross its concave anterior aspect, marking the lines of fusion of the sacral vertebrae. The **ventral sacral foramina** penetrate the sacrum at the lateral ends of these ridges and transmit blood vessels and nerves. The regions lateral to these foramina expand superiorly, forming winglike extensions, the **alae.**

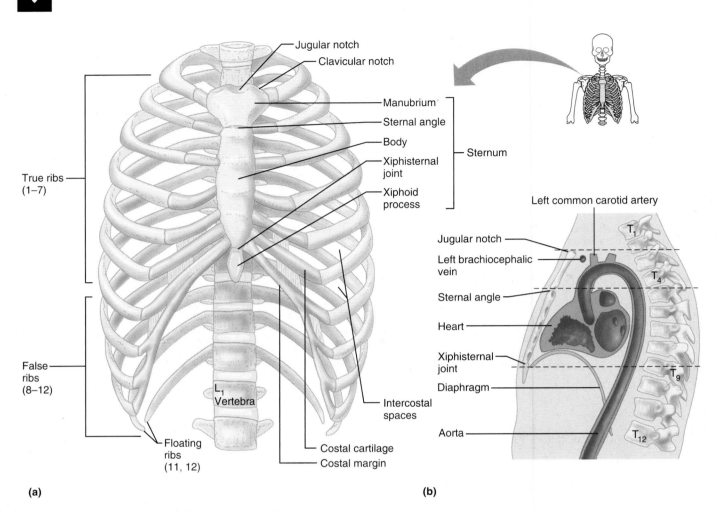

FIGURE 7.19   *The bony thorax.* (a) Skeleton of the bony thorax, anterior view (costal cartilages are shown in blue). **(b)** Left lateral view of the thorax, illustrating the relationship of the surface anatomical landmarks of the thorax to the vertebral column (thoracic portion). (See *A Brief Atlas of the Human Body,* Figure 23.)

In its dorsal midline the sacral surface is roughened by the **median sacral crest** (the fused spinous processes of the sacral vertebrae). This is flanked laterally by the **dorsal sacral foramina,** and then the **lateral sacral crests** (remnants of the transverse processes of S_1–S_5).

The vertebral canal continues inside the sacrum as the **sacral canal.** Since the laminae of the fifth (and sometimes the fourth) sacral vertebrae fail to fuse medially, an enlarged external opening called the **sacral hiatus** (hi-a′tus; "gap") is obvious at the inferior end of the sacral canal.

Coccyx

The **coccyx,** our tailbone, is a small triangular bone (Figure 7.18). It consists of four (or in some cases

three or five) vertebrae fused together. The coccyx articulates superiorly with the sacrum. (The name *coccyx* is from the Greek word meaning "cuckoo" and was so named because of its fancied resemblance to a bird's beak.) Except for the slight support the coccyx affords the pelvic organs, it is a nearly useless bone. Occasionally, a baby is born with an unusually long coccyx. In most such cases, this bony "tail" is discreetly snipped off by the physician.

⋆   ⋆   ⋆

The characteristics of the regional vertebrae are summarized in Table 7.2.

THE BONY THORAX

Anatomically, the thorax is the chest, and its bony underpinnings are called the **bony thorax** or **thoracic cage.** Elements of the bony thorax include the thoracic vertebrae dorsally, the ribs laterally, and the sternum and costal cartilages anteriorly. The costal cartilages secure the ribs to the sternum (Figure 7.19a). Roughly cone shaped with its broad dimension inferiorly, the bony thorax forms a protective cage around the vital organs of the thoracic cavity (heart, lungs, and great blood vessels), supports the shoulder girdles and upper limbs, and provides attachment points for many muscles of the neck, back, chest, and shoulders. In addition, the *intercostal spaces* between the ribs are occupied by the intercostal muscles, which lift and depress the thorax during breathing.

Sternum

The **sternum** (breastbone) lies in the anterior midline of the thorax. Vaguely resembling a dagger, it is a flat bone approximately 15 cm (6 inches) long. It results from the fusion of three bones: the manubrium, the body, and the xiphoid process. The **manubrium** (mah-nu'bre-um), the superior portion, is shaped like the knot in a necktie. It articulates via its **clavicular** (klah-vik'u-lar) **notches** with the clavicles (collarbones) laterally; it also articulates with the first two pairs of ribs. The **body,** or midportion, forms the bulk of the sternum. The sides of the body are notched where it articulates with the cartilages of the second to seventh ribs. The **xiphoid** (zif'oid; "swordlike") **process** forms the inferior end of the sternum. This small, variably shaped process is a plate of hyaline cartilage in youth, but it is usually ossified in adults. The xiphoid process articulates only with the sternal body and serves as an attachment point for some abdominal muscles.

⚠ *Homeostatic Imbalance*

In some people, the xiphoid process projects dorsally. This may present a problem because blows to the chest can push such a xiphoid into the underlying heart or liver, causing massive hemorrhage. ■

The sternum has three important anatomical landmarks: the jugular notch, the sternal angle, and the xiphisternal joint (see Figure 7.19). The easily palpated **jugular (suprasternal) notch** is the central indentation in the superior border of the manubrium. It is generally in line with the disc between the second and third thoracic vertebrae and

the point where the left common carotid artery issues from the aorta (see Figure 7.19b). The **sternal angle** can be felt as a horizontal ridge across the front of the sternum, where the manubrium joins the sternal body. This cartilaginous joint acts like a hinge, allowing the sternal body to swing forward when we inhale. The sternal angle is in line with the disc between the fourth and fifth thoracic vertebrae and at the level of the second pair of ribs. It is a handy reference point for finding the second rib and thus for counting the ribs during a physical examination and for listening to sounds made by specific heart valves. The **xiphisternal** (zif"ĭ-ster'nul) **joint** is the point where the sternal body and xiphoid process fuse. It lies opposite the ninth thoracic vertebra.

Ribs

Twelve pairs of **ribs** form the flaring sides of the thoracic cage (Figure 7.19a). All ribs attach posteriorly to the thoracic vertebrae (bodies and transverse processes) and curve inferiorly toward the anterior body surface. The superior seven rib pairs attach directly to the sternum by individual costal cartilages (bars of hyaline cartilage). These are **true** or **vertebrosternal** (ver"tĕ-bro-ster'nal) **ribs.** (Notice that the anatomical name indicates the two attachment points of a rib—the posterior attachment given first.)

The remaining five pairs of ribs are called **false ribs;** they either attach indirectly to the sternum or entirely lack a sternal attachment. Rib pairs 8–10 attach to the sternum indirectly; each joins the costal cartilage immediately above it. These ribs are also called **vertebrochondral** (ver"tĕ-bro-kon'dral) **ribs.** The inferior margin of the rib cage, or **costal margin,** is formed by the costal cartilages of ribs 7–10. Rib pairs 11 and 12 are called **vertebral ribs** or **floating ribs** because they have no anterior attachments. Instead, their costal cartilages lie embedded in the muscles of the lateral body wall. The ribs increase in length from pair 1 to pair 7, then decrease in length from pair 8 to pair 12.

The first pair of ribs is quite atypical. They are flattened superiorly to inferiorly and are quite broad, forming a horizontal table that supports the subclavian blood vessels that serve the upper limbs. There are also other exceptions to the typical rib pattern. Ribs 1 and 10–12 articulate with only one vertebral body, and ribs 11 and 12 do not articulate with a vertebral transverse process. Except for the first rib, which lies deep to the clavicle, the ribs are easily felt in people of normal weight.

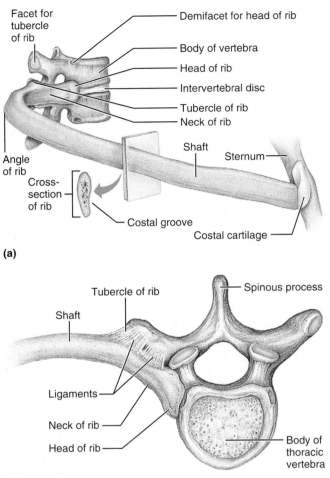

FIGURE 7.20 *Structure of a "typical" true rib and its articulations.* (a) Vertebral and sternal articulations of a typical true rib. (b) Superior view of the articulation between a rib and a thoracic vertebra. (See *A Brief Atlas of the Human Body,* Figure 23e and f.)

A typical rib is a bowed flat bone (Figure 7.20). The bulk of a rib is simply called the *shaft*. Its superior border is smooth, but its inferior border is sharp and thin and has a *costal groove* on its inner face that lodges the intercostal nerves and blood vessels. In addition to the shaft, each rib has a head, neck, and tubercle. The wedge-shaped *head*, the posterior-most end, articulates with the vertebral bodies by two facets: One joins the body of the same-numbered thoracic vertebra, the other articulates with the body of the vertebra immediately superior. The *neck* is the constricted portion of the rib just beyond the head. Lateral to this, the knoblike *tubercle* articulates with the transverse process of the same-numbered thoracic vertebra. Beyond the tubercle, the shaft angles sharply forward (at the angle of the rib) and then extends to attach to its costal cartilage anteriorly. The costal cartilages provide secure but flexible rib attachments to the sternum.

■ PART 2 ■
THE APPENDICULAR SKELETON

Bones of the limbs and their girdles are collectively called the **appendicular skeleton** because they are *appended to* the axial skeleton that forms the longitudinal axis of the body (see Figure 7.21). The yoke-like *pectoral* (pek'tor-al; "chest") *girdles* attach the upper limbs to the body trunk. The more sturdy *pelvic girdle* secures the lower limbs. Although the bones of the upper and lower limbs differ in their functions and mobility, they have the same fundamental plan: Each limb is composed of three major segments connected by movable joints.

The appendicular skeleton enables us to carry out the movements typical of our freewheeling and manipulative lifestyle. Each time we take a step, throw a ball, or pop a caramel into our mouth, we are making good use of our appendicular skeleton.

THE PECTORAL (SHOULDER) GIRDLE

The **pectoral,** or **shoulder, girdle** consists of the *clavicle* (klav'ĭ-kl) anteriorly and the *scapula* (skap'u-lah) posteriorly (Figure 7.22 and Table 7.3 on p. 227). The paired pectoral girdles and their associated muscles form your shoulders. Although the term *girdle* usually signifies a beltlike structure encircling the body, a single pectoral girdle, or even the pair, does not quite satisfy this description. Anteriorly, the medial end of each clavicle joins the sternum; the distal ends of the clavicles meet the scapulae laterally. However, the scapulae fail to complete the ring posteriorly, because their medial borders do not join each other or the axial skeleton. Instead, the scapulae are attached to the thorax and vertebral column only by the muscles that clothe their surfaces.

The pectoral girdles attach the upper limbs to the axial skeleton and provide attachment points for many of the muscles that move the upper limbs. These girdles are very light and allow the upper limbs a degree of mobility not seen anywhere else in the body. This mobility is due to the following factors:

1. Because only the clavicle attaches to the axial skeleton, the scapula can move quite freely across the thorax, allowing the arm to move with it.

2. The socket of the shoulder joint (the scapula's glenoid cavity) is shallow and poorly reinforced, so it does not restrict the movement of the humerus (arm bone). Although this arrangement is good for flexibility, it is bad for stability: Shoulder dislocations are fairly common.

? *The skeleton shown here is incomplete. What major bony regions are missing?*

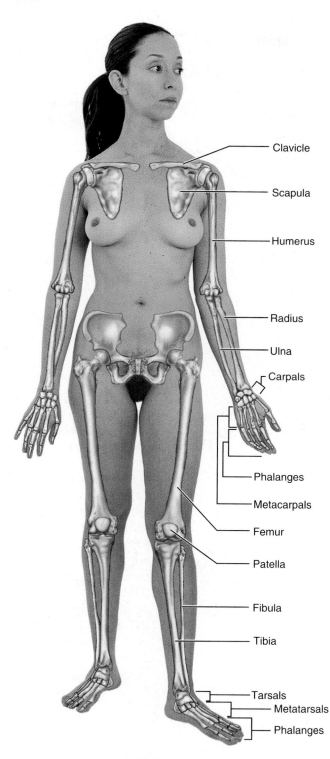

FIGURE 7.21   *The appendicular skeleton.*

Clavicles

The **clavicles** ("little keys"), or collarbones, are slender, doubly curved long bones that can be felt along their entire course as they extend horizontally across the superior thorax (Figures 7.22a–c and 7.25). Each clavicle is cone shaped at its medial **sternal end,** which attaches to the sternal manubrium, and flattened at its lateral **acromial** (ah-kro′me-al) **end,** which articulates with the scapula. The medial two-thirds of the clavicle is convex anteriorly; its lateral third is concave anteriorly. Its superior surface is smooth, but the inferior surface is ridged and grooved.

Besides providing attachment points for many muscles, the clavicles act as braces: They hold the scapulae and arms out laterally, away from the narrower superior part of the thorax. This bracing function becomes obvious when a clavicle is fractured: The entire shoulder region collapses medially. The clavicles also transmit compression forces from the upper limbs to the axial skeleton, as when someone pushes a car to a gas station.

The clavicles are not very strong and are likely to fracture, for example, when a person uses outstretched arms to break a fall. The curves in the clavicle ensure that it usually fractures anteriorly (outward). If it were to collapse posteriorly (inward), bone splinters would damage the subclavian artery, which passes just deep to the clavicle to serve the upper limb. The clavicles are exceptionally sensitive to muscle pull and become noticeably larger and stronger in those who perform manual labor or athletics involving the shoulder and arm muscles.

Scapulae

The **scapulae,** or *shoulder blades,* are thin, triangular flat bones (Figures 7.22d–f and 7.25). Interestingly, their name derives from a word meaning "spade" or "shovel," for ancient cultures made spades from the shoulder blades of animals. The scapulae lie on the dorsal surface of the rib cage, between ribs 2 and 7. Each scapula has three borders. The **superior border** is the shortest, sharpest border. The **medial,** or *vertebral,* **border** parallels the vertebral column. The thick **lateral,** or *axillary,* **border** abuts the armpit and ends superiorly in a small, shallow fossa, the **glenoid** (gle′noid) **cavity.** This cavity articulates with the humerus of the arm, forming the shoulder joint.

Like all triangles, the scapula has three corners or *angles.* The superior scapular border meets the medial border at the **superior angle** and the lateral border at the **lateral angle.** The medial and lateral borders join at the **inferior angle.** The inferior angle moves extensively as the arm is raised and lowered, and is an important landmark for studying scapular movements.

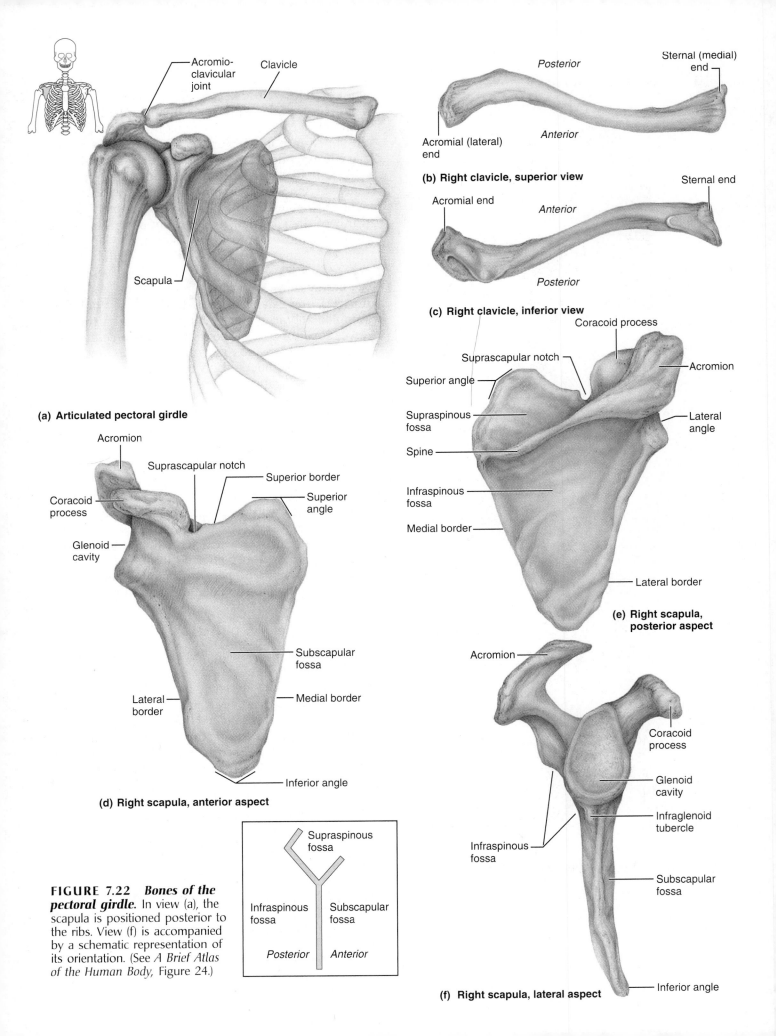

(a) Articulated pectoral girdle

(b) Right clavicle, superior view

(c) Right clavicle, inferior view

(d) Right scapula, anterior aspect

(e) Right scapula, posterior aspect

(f) Right scapula, lateral aspect

FIGURE 7.22 Bones of the pectoral girdle. In view (a), the scapula is positioned posterior to the ribs. View (f) is accompanied by a schematic representation of its orientation. (See *A Brief Atlas of the Human Body,* Figure 24.)

TABLE 7.3	Bones of the Appendicular Skeleton Part 1: Pectoral Girdle and Upper Limb			
Body region	*Bones **	*Illustration*	*Location*	*Markings*
Pectoral girdle (Figures 7.22, 7.25)	Clavicle (2)		Clavicle is in superoanterior thorax; articulates medially with sternum and laterally with scapula	Acromial end; sternal end
	Scapula (2)		Scapula is in posterior thorax; forms part of the shoulder; articulates with humerus and clavicle	Glenoid cavity; spine; acromion; coracoid process; infraspinous, supraspinous, and subscapular fossae
Free upper limb Arm (Figures 7.23, 7.25)	Humerus (2)		Humerus is the sole bone of arm; between scapula and elbow	Head; greater and lesser tubercles; intertubercular groove; deltoid tuberosity; trochlea; capitulum; coronoid and olecranon fossae; radial groove; epicondyles
Forearm (Figures 7.24, 7.25)	Ulna (2)		Ulna is the medial bone of forearm between elbow and wrist; forms elbow joint	Coronoid process; olecranon process; radial notch; trochlear notch; styloid process; head
	Radius (2)		Radius is lateral bone of forearm; carries wrist	Radial tuberosity; styloid process; head; ulnar notch
Hand (Figure 7.26)	8 Carpals (16) • scaphoid • lunate • triquetral • pisiform • trapezium • trapezoid • capitate • hamate		Carpals form a bony crescent at the wrist; arranged in two rows of four bones each	
	5 Metacarpals (10)		Metacarpals form the palm; one in line with each digit	
	14 Phalanges (28) • distal • middle • proximal		Phalanges form the fingers; three in digits 2–5; two in digit 1 (the thumb)	

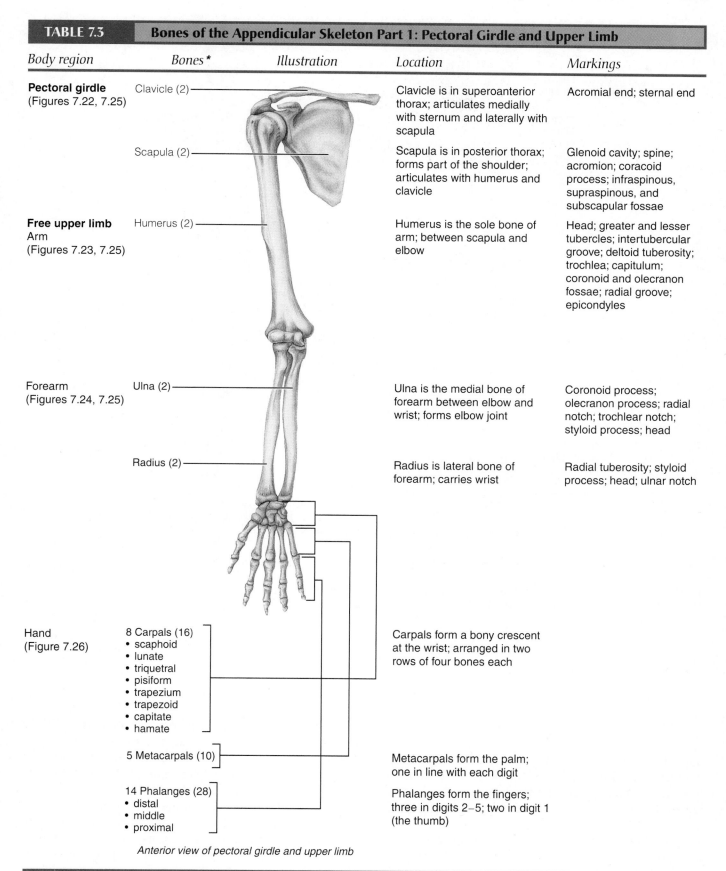

Anterior view of pectoral girdle and upper limb

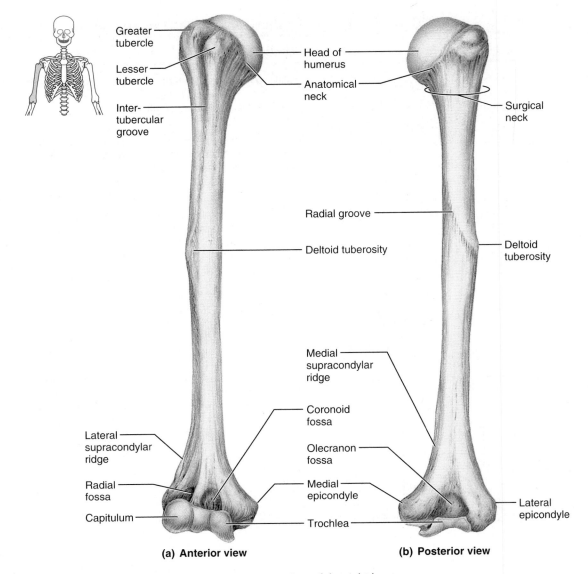

Greater tubercle

Lesser tubercle

Inter-tubercular groove

Head of humerus

Anatomical neck

Surgical neck

Radial groove

Deltoid tuberosity

Deltoid tuberosity

Medial supracondylar ridge

Lateral supracondylar ridge

Coronoid fossa

Olecranon fossa

Radial fossa

Capitulum

Medial epicondyle

Trochlea

Lateral epicondyle

(a) Anterior view

(b) Posterior view

FIGURE 7.23   *The humerus of the arm.* (a) Anterior view of the right humerus. (b) Posterior view of the right humerus. (See *A Brief Atlas of the Human Body,* Figure 25.)

The anterior, or costal, surface of the scapula is concave and relatively featureless. Its posterior surface bears a prominent **spine** that is easily felt through the skin. The spine ends laterally in an enlarged, roughened triangular projection called the **acromion** (ah-kro′me-on), literally, the "point of the shoulder." The acromion articulates with the acromial end of the clavicle, forming the **acromioclavicular joint.**

Projecting anteriorly from the superior scapular border is the **coracoid** (kor′ah-coid) **process;** *corac* means "beaklike," but this process looks more like a bent little finger. The coracoid process helps to anchor the biceps muscle of the arm. It is bounded by the **suprascapular notch** (a nerve passage) medially and by the glenoid cavity laterally.

Several large fossae appear on both sides of the scapula and are named according to location. The **infraspinous** and **supraspinous fossae** are respectively inferior and superior to the spine. The **subscapular fossa** is the shallow concavity formed by the entire anterior scapular surface.

THE UPPER LIMB

Thirty separate bones form the skeletal framework of each upper limb (see Figures 7.23 to 7.26, and Table 7.3 on p. 227). Each of these bones may be described regionally as a bone of the arm, forearm, or hand. (Keep in mind that anatomically the "arm" is only that part of the upper limb between the shoulder and elbow.)

Arm

The **humerus** (hu′mer-us), the sole bone of the arm, is a typical long bone (Figures 7.23 and 7.25). The largest, longest bone of the upper limb, it articulates with the scapula at the shoulder and with the radius and ulna (forearm bones) at the elbow.

At the proximal end of the humerus is its smooth, hemispherical **head,** which fits into the glenoid cavity of the scapula in a manner that allows the arm to hang freely at one's side. Immediately inferior to the head is a slight constriction, the **anatomical neck.** Just inferior to this are the lateral **greater tubercle** and the more medial **lesser tubercle,** separated by the **intertubercular,** or *bicipital* (bi-sip′ĭ-tal), **groove.** The tubercles are sites where muscles attach. The intertubercular groove guides a tendon of the biceps muscle of the arm to its attachment point at the rim of the glenoid cavity. Just distal to the tubercles is the **surgical neck,** so named because it is the most frequently fractured part of the humerus. About midway down the shaft on its lateral side is the **deltoid tuberosity.** This V-shaped, roughened area is the attachment site for the deltoid muscle of the shoulder. Nearby, the **radial groove** runs obliquely down the posterior aspect of the shaft. This groove marks the course of the radial nerve, an important nerve of the upper limb.

At the distal end of the humerus are two condyles, a medial **trochlea** (trok′le-ah; "pulley"), which looks like an hourglass tipped on its side, and the lateral ball-like **capitulum** (kah-pit′u-lum). These condyles articulate with the ulna and the radius, respectively. The condyle pair is flanked by the **medial** and **lateral epicondyles** (muscle attachment sites). Directly above these epicondyles are the supracondylar ridges. The ulnar nerve runs behind the medial epicondyle and is responsible for the painful, tingling sensation you experience when you hit your "funny bone."

Superior to the trochlea on the anterior surface is the **coronoid fossa;** on the posterior surface is the deeper **olecranon** (o-lek′rah-non) **fossa.** These two depressions allow the corresponding processes of the ulna to move freely when the elbow is flexed and extended. A small **radial fossa,** lateral to the coronoid fossa, receives the head of the radius when the elbow is flexed.

Forearm

Two parallel long bones, the radius and the ulna, form the skeleton of the forearm, or *antebrachium* (an″te-bra′ke-um) (Figures 7.24 and 7.25). Unless a person's forearm muscles are very bulky, these bones are easily palpated along their entire length. Their proximal ends articulate with the humerus; their distal ends form joints with bones of the wrist. The radius and ulna articulate with each other both proximally and distally at small **radioulnar** (ra″de-o-ul′nar) **joints,** and they are connected along their entire length by a flexible **interosseous** (in″ter-os′e-us) **membrane.** In the anatomical position, the radius lies laterally (on the thumb side) and the ulna medially. However, when you rotate your forearm so that the palm faces posteriorly (a movement called pronation), the distal end of the radius crosses over the ulna and the two bones form an X.

Ulna

The **ulna** (ul′nah; "elbow") is slightly longer than the radius. It has the main responsibility for forming the elbow joint with the humerus. Its proximal end looks like the adjustable end of a monkey wrench; it bears two prominent processes, the **olecranon** (elbow) and **coronoid processes,** separated by a deep concavity, the **trochlear notch** (see Figures 7.24 and 7.25). Together, these two processes grip the trochlea of the humerus, forming a hinge joint that allows the forearm to be bent upon the arm (flexed), then straightened again (extended). When the forearm is fully extended, the olecranon process "locks" into the olecranon fossa, keeping the forearm from hyperextending (moving posteriorly beyond the elbow joint). The posterior olecranon process forms the angle of the elbow when the forearm is flexed and is the bony part that rests on the table when you lean on your elbows. On the lateral side of the coronoid process is a small depression, the **radial notch,** where the ulna articulates with the head of the radius.

Distally the ulnar shaft narrows and ends in a knoblike **head.** Medial to the head is a **styloid** ("stake-shaped") **process,** from which a ligament runs to the wrist. The ulnar head is separated from the bones of the wrist by a disc of fibrocartilage and plays little or no role in hand movements.

(1) Which of these bones "carries" the hand? (2) Which plays the major role in forming the elbow joint with the humerus?

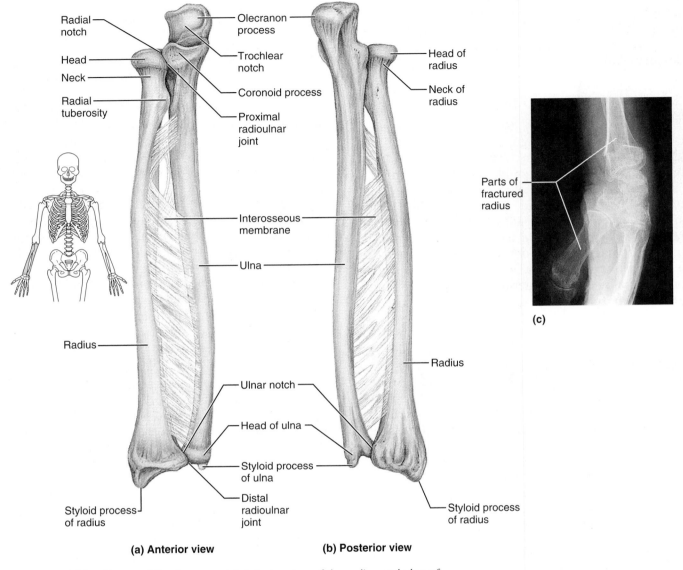

(a) Anterior view **(b) Posterior view**

FIGURE 7.24 Bones of the forearm. (a) Anterior view of the radius and ulna of the right forearm, anatomical position. Interosseous membrane also shown. **(b)** Posterior view of the radius and ulna of the right forearm. **(c)** X ray of Colle's fracture of the radius. (See *A Brief Atlas of the Human Body,* Figure 26.)

Radius

The **radius** ("rod") is thin at its proximal end and widened distally—the opposite of the ulna. The **head** of the radius is shaped somewhat like the head of a nail (see Figures 7.24 and 7.25). The superior surface of this head is concave, and it articulates with the capitulum of the humerus. Medially, the head articulates with the radial notch of the ulna. Just inferior to the head is a rough projection, the **radial tuberosity,** which anchors the biceps muscle of the arm. Distally, where the radius is expanded, it has a medial **ulnar notch,** which articulates with the ulna, and a lateral **styloid process** (an anchoring site for ligaments that run to the wrist). Between these two markings, the radius is concave where it articulates with carpal bones of the wrist. While the ulna contributes more heavily to the elbow joint,

(1) The radius carries the hand. (2) The ulna forms the elbow joint with the humerus.

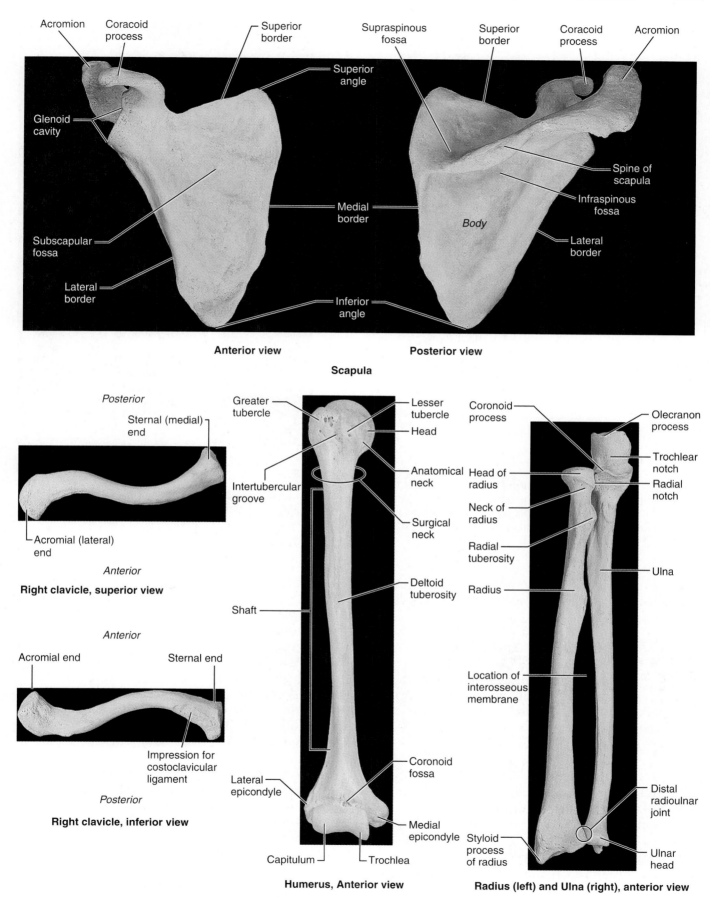

Anterior view | Posterior view

Scapula

Posterior

Right clavicle, superior view

Right clavicle, inferior view

Humerus, Anterior view

Radius (left) and Ulna (right), anterior view

FIGURE 7.25 *Photographs of selected bones of the pectoral girdle and right upper limb.* (See *A Brief Atlas of the Human Body,* Figures 24, 25, and 26.)

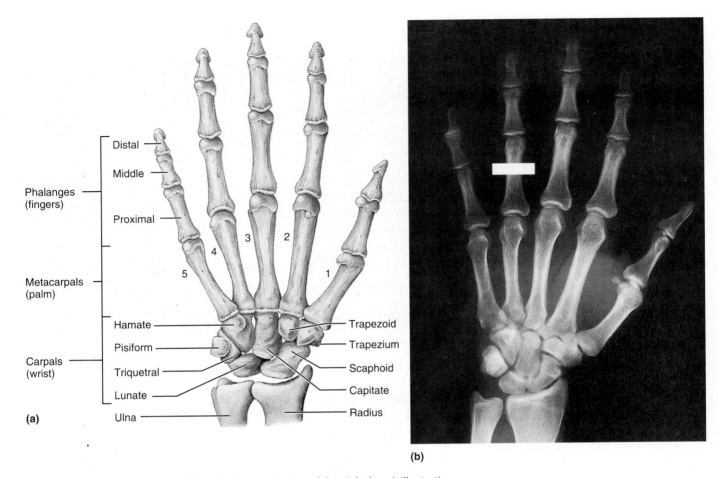

FIGURE 7.26   Bones of the hand. (a) Ventral view of the right hand, illustrating the anatomical relationships of the carpals, metacarpals, and phalanges. (b) X ray of the right hand. Notice the white bar on the proximal phalanx of finger 4, showing the position at which a ring would be worn. (See *A Brief Atlas of the Human Body,* Figure 27.)

the radius is the major forearm bone contributing to the wrist joint. When the radius moves, the hand moves with it.

🔺 Homeostatic Imbalance

Colle's fracture is a break in the distal end of the radius. It is a common fracture when a falling person attempts to break his or her fall with outstretched hands (see Figure 7.24c). ■

Hand

The skeleton of the hand (Figure 7.26) includes the bones of the *carpus* (wrist); the bones of the *metacarpus* (palm); and the *phalanges* (bones of the fingers).

Carpus (Wrist)

A "wrist" watch is actually worn on the distal forearm (over the lower ends of the radius and ulna), not on the wrist at all. The true wrist, or **carpus,** is the proximal part of the structure we generally call our "hand." The carpus consists of a group of eight marble-size short bones, or **carpals** (kar′palz), closely united by ligaments. Because gliding movements occur between these bones, the carpus as a whole is quite flexible. The carpals are arranged in two irregular rows of four bones each (Figure 7.26). In the proximal row (lateral to medial) are the **scaphoid** (skaf′oid), **lunate** (lu′nāt), **triquetral** (tri-kwe′trul), and **pisiform** (pi′sĭ-form). Only the scaphoid and lunate articulate with the radius to form the wrist joint. The carpals of the distal row (lateral to medial) are the **trapezium** (trah-pe′ze-um), **trapezoid** (tra′peh-zoid), **capitate,** and **hamate** (ham′āt). There are numerous memory-jogging phrases to help you recall the carpals in the order given above. If you don't have one, try: **S**ally **l**eft **t**he **p**arty **t**o **t**ake **C**athy **h**ome. As with all such memory jogs, the first letter of each word is the first letter of the term you need to remember.

△ *Homeostatic Imbalance*

The carpus is concave anteriorly and a ligament roofs over this concavity, forming the notorious *carpal tunnel*. Besides the median nerve (which supplies the lateral side of the hand), several long muscle tendons crowd into this tunnel. Overuse and inflammation of the tendons cause them to swell, compressing the median nerve, which causes numbness of the areas served. Those who repeatedly flex their wrists and fingers, such as those who work at computer keyboards all day, are particularly susceptible to this nerve impairment, called *carpal tunnel syndrome*. ■

Metacarpus (Palm)

Five **metacarpal bones** radiate from the wrist like spokes to form the palm of the hand (*meta* = beyond). These small long bones are not named, but instead are numbered 1 to 5 from thumb to little finger. The **bases** of the metacarpals articulate with the carpals proximally and each other medially and laterally (see Figure 7.26). Their bulbous **heads** articulate with the proximal phalanges of the fingers. When you clench your fist, the heads of the metacarpals become prominent as your *knuckles*.

Metacarpal 1, associated with the thumb, is the shortest and most mobile. It occupies a more anterior position than the other metacarpals. Consequently, the joint between metacarpal 1 and the trapezium is a unique saddle joint that allows *opposition*, the action of touching your thumb to the tips of your other fingers.

Phalanges (Fingers)

The **fingers,** or **digits** of the upper limb, are numbered 1 to 5 beginning with the thumb, or **pollex** (pol′eks). In most people, the third finger is the longest. Each hand contains 14 miniature long bones called **phalanges** (fah-lan′jēz). Except for the thumb, each finger has three phalanges: *distal, middle*, and *proximal*. The thumb has no middle phalanx. (Phalanx [fa′langks] is the singular term for phalanges.)

THE PELVIC (HIP) GIRDLE

The **pelvic girdle,** or **hip girdle,** attaches the lower limbs to the axial skeleton, transmits the weight of the upper body to the lower limbs, and supports the visceral organs of the pelvis (Figures 7.27 and 7.30 and Table 7.4, p. 236). Whereas the pectoral girdle is sparingly attached to the thoracic cage, the pelvic girdle is secured to the axial skeleton by some of the strongest ligaments in the body. And whereas the glenoid cavity of the scapula is shallow, the corresponding sockets of the pelvic girdle are deep and cuplike and firmly secure the head of the femur in place. Thus, even though both the shoulder and hip joints are ball-and-socket joints, very few of us can wheel or swing our legs and arms about with the same degree of freedom; the pelvic girdle lacks the mobility of the pectoral girdle but is far more stable.

The pelvic girdle is formed by a pair of **hip bones** (Figure 7.27), each also called an **os coxae** (ahs kok′se), or **coxal** (*coxa* = hip) **bone.** Each hip bone unites with its partner anteriorly and with the sacrum posteriorly. The deep, basinlike structure formed by the hip bones, together with the sacrum and coccyx, is called the **bony pelvis.**

Each large, irregularly shaped hip bone consists of three separate bones during childhood: the ilium, ischium, and pubis. In adults, these bones are firmly fused and their boundaries are indistinguishable. Their names are retained, however, to refer to different regions of the composite hip bone. At the point of fusion of the ilium, ischium, and pubis is a deep hemispherical socket called the **acetabulum** (as″ĕ-tab′u-lum; "vinegar cup") on the lateral surface of the pelvis (see Figure 7.27b). The acetabulum receives the head of the femur, or thigh bone, at this *hip joint.*

Ilium

The **ilium** (il′e-um; "flank") is a large flaring bone that forms the superior region of a coxal bone. It consists of a **body** and a superior winglike portion called the **ala** (a′lah). When you rest your hands on your hips, you are resting them on the thickened superior margins of the alae, the **iliac crests.** Thickest at the **tubercle of the iliac crest,** each iliac crest ends anteriorly in the blunt **anterior superior iliac spine** and posteriorly in the sharp **posterior superior iliac spine.** Located below these are the less prominent **anterior** and **posterior inferior iliac spines.** All of these spines are attachment points for the muscles of the trunk, hip, and thigh. The anterior superior iliac spine is an especially important anatomical landmark. It is easily felt through the skin and is visible in thin people. The posterior superior iliac spine is difficult to palpate, but its position is revealed by a skin dimple in the sacral region.

Just inferior to the posterior inferior iliac spine, the ilium indents deeply to form the **greater sciatic** (si-at′ik) **notch,** through which the thick cordlike sciatic nerve passes to enter the thigh. The broad posterolateral surface of the ilium, the **gluteal** (gloo′te-al) **surface,** is crossed by three ridges, the **posterior, anterior,** and **inferior gluteal lines,** to which the gluteal (buttock) muscles attach.

? *What bone marking identified in Figure 7.27 allows passage of the sciatic nerve and artery?*

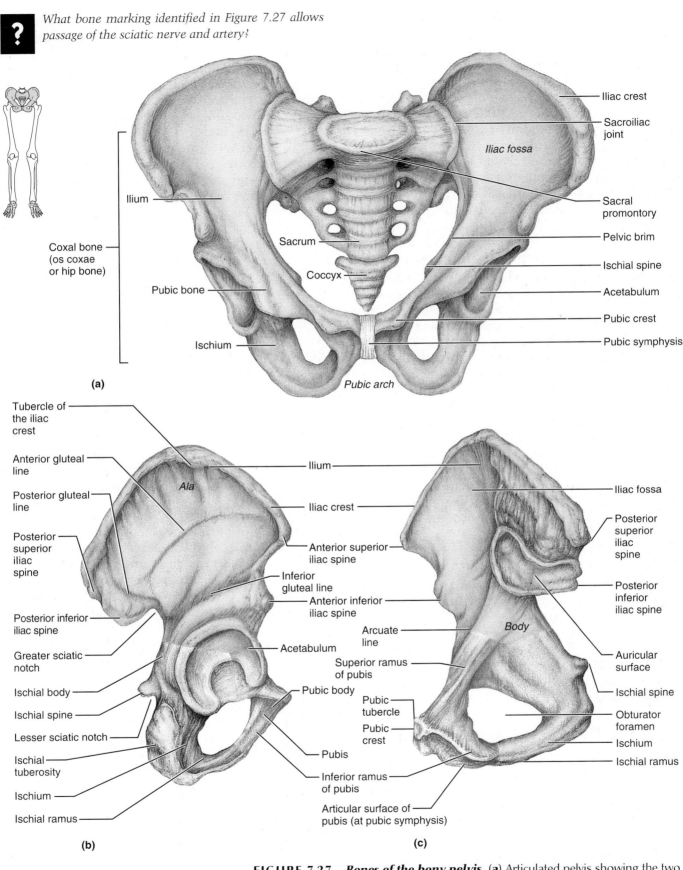

(a)

(b)

(c)

FIGURE 7.27   *Bones of the bony pelvis.* **(a)** Articulated pelvis showing the two hip (coxal) bones (which together form the pelvic girdle) and the sacrum. **(b)** Lateral view of right hip bone showing the point of fusion of the ilium (gold), ischium (blue), and pubic (pink) bones at the acetabulum. **(c)** Right hip bone, medial view. (See *A Brief Atlas of the Human Body,* Figure 28.)

The internal surface of the iliac ala exhibits a concavity called the **iliac fossa.** Posterior to this, the roughened **auricular** (aw-rik'u-lar; "ear-shaped") **surface** articulates with the same-named surface of the sacrum, forming the *sacroiliac joint.* The weight of the body is transmitted from the spine to the pelvis through the sacroiliac joints. Running inferiorly and anteriorly from the auricular surface is a robust ridge called the **arcuate** (ar'ku-at; "bowed") **line.** The arcuate line helps define the **pelvic brim,** the superior margin of the *true pelvis,* which we will discuss shortly. Anteriorly, the body of the ilium joins the ischium and pubis.

Ischium

The **ischium** (is'ke-um; "hip") forms the posteroinferior part of the hip bone (see Figures 7.27 and 7.30). Roughly L- or arc-shaped, it has a thicker, superior **body** adjoining the ilium and a thinner, inferior **ramus** (*ramus* = branch). The ramus joins the pubis anteriorly. The ischium has three important markings. Its **ischial spine** projects medially into the pelvic cavity and serves as a point of attachment of the *sacrospinous ligament* from the sacrum. Just inferior to the ischial spine is the **lesser sciatic notch.** A number of nerves and blood vessels pass through this notch to supply the perineum (anogenital area). The inferior surface of the ischial body is rough and grossly thickened as the **ischial tuberosity** (see Figure 7.27b). When we sit, our weight is borne entirely by the ischial tuberosities, which are the strongest parts of the hip bones. A massive ligament runs from the sacrum to each ischial tuberosity. This *sacrotuberous ligament* (not illustrated) helps hold the pelvis together.

Pubis

The **pubis** (pu'bis; "sexually mature"), or **pubic bone,** forms the anterior portion of the hip bone (see Figure 7.27). In the anatomical position, it lies nearly horizontally and the bladder rests upon it. Essentially, the pubis is V-shaped with **superior** and **inferior rami** issuing from a flattened medial **body.** The body of the pubis lies medially and its anterior border is thickened to form a **pubic crest.** At the lateral end of the pubic crest is the **pubic tubercle,** one of the attachments for the inguinal ligament. As the two rami of the pubic bone run laterally to join with the body and ramus of the ischium, they define a large opening in the hip bone, the **obturator** (ob"tu-ra'tor) **foramen,** through which a few blood vessels and nerves pass. Although the obturator foramen is large, it is nearly closed by a fibrous membrane in life (*obturator* = closed up).

The bodies of the two pubic bones are joined by a fibrocartilage disc, forming the midline **pubic symphysis joint.** Inferior to this joint, the inferior pubic rami angle laterally, forming an inverted V-shaped arch called the **pubic arch** (Figure 7.27). The acuteness of the pubic arch helps to differentiate the male from the female pelvis (Table 7.4).

Pelvic Structure and Childbearing

So striking are the differences between the male and female pelves that a trained anatomist can immediately determine the sex of a skeleton from a casual examination of the pelvis. The female pelvis reflects modifications for childbearing: It tends to be wider, shallower, lighter, and rounder than that of a male. The female pelvis not only accommodates a growing fetus, but it must be large enough to allow the infant's relatively large head to exit at birth. The major differences between the "typical" male and female pelves are summarized and illustrated in Table 7.4.

The pelvis is said to have a false (greater) pelvis and a true (lesser) pelvis separated by the **pelvic brim,** a continuous oval ridge that runs from the pubic crest through the arcuate line and sacral promontory (Figure 7.27a). The **false pelvis** is that portion superior to the pelvic brim; it is bounded by the alae of the ilia laterally and the lumbar vertebrae posteriorly. The false pelvis is really part of the abdomen and helps support the abdominal viscera. It does not restrict childbirth in any way. The **true pelvis** is the region inferior to the pelvic brim that is almost entirely surrounded by bone. It forms a deep "bowl" containing the pelvic organs. Its dimensions, particularly those of its *inlet* and *outlet,* are critical to the uncomplicated delivery of a baby, and they are carefully measured by an obstetrician.

The **pelvic inlet** *is* the **pelvic brim,** and its widest dimension is from right to left along the frontal plane (see Table 7.4). As labor begins, an infant's head typically enters the inlet with its forehead facing one ilium and its occiput facing the other. A sacral promontory that is particularly large can impair the infant's entry into the true pelvis. The **pelvic outlet,** illustrated in the photos at the bottom of Table 7.4, is the inferior margin of the true pelvis. It is bounded anteriorly by the pubic arch, laterally by the ischia, and posteriorly by the sacrum and coccyx. Both the coccyx and the ischial spines protrude into the outlet opening, so a sharply angled coccyx or unusually large spines can interfere with delivery. The largest dimension of the outlet is the anteroposterior diameter. Generally, after the baby's head passes through the inlet, it rotates so that the forehead faces posteriorly and the occiput anteriorly. Thus, during birth, the infant's head makes a quarter turn to follow the widest dimensions of the true pelvis.

TABLE 7.4	Comparison of the Male and Female Pelves	
Characteristic	*Female*	*Male*
General structure and functional modifications	Tilted forward; adapted for childbearing; true pelvis defines the birth canal; cavity of the true pelvis is broad, shallow, and has a greater capacity	Tilted less far forward; adapted for support of a male's heavier build and stronger muscles; cavity of the true pelvis is narrow and deep
Bone thickness	Less; bones lighter, thinner, and smoother	Greater; bones heavier and thicker, and markings are more prominent
Acetabula	Smaller; farther apart	Larger; closer
Pubic arch/angle	Broader (80–90°); more rounded	More acute (50–60°)
Anterior view	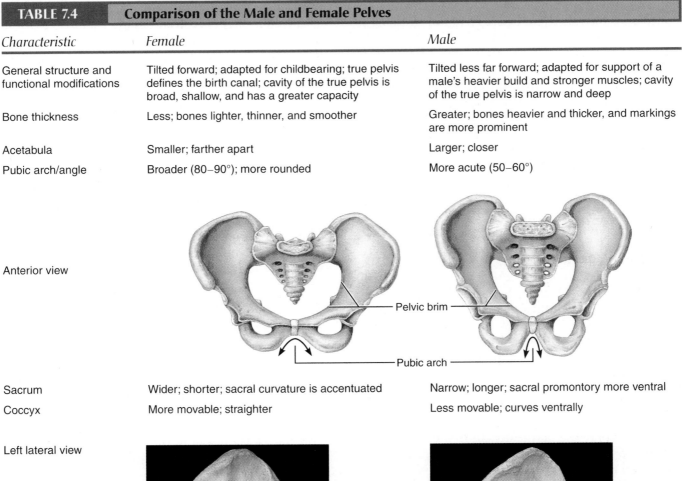	
Sacrum	Wider; shorter; sacral curvature is accentuated	Narrow; longer; sacral promontory more ventral
Coccyx	More movable; straighter	Less movable; curves ventrally
Left lateral view	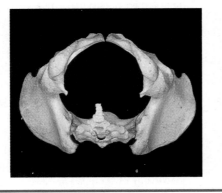	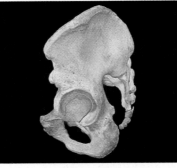
Pelvic inlet (brim)	Wider; oval from side to side	Narrow; basically heart shaped
Pelvic outlet	Wider; ischial tuberosities shorter, farther apart and everted	Narrower; ischial tuberosities longer, sharper, and point more medially
Posteroinferior view	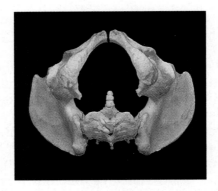	

Pelvic brim

Pubic arch

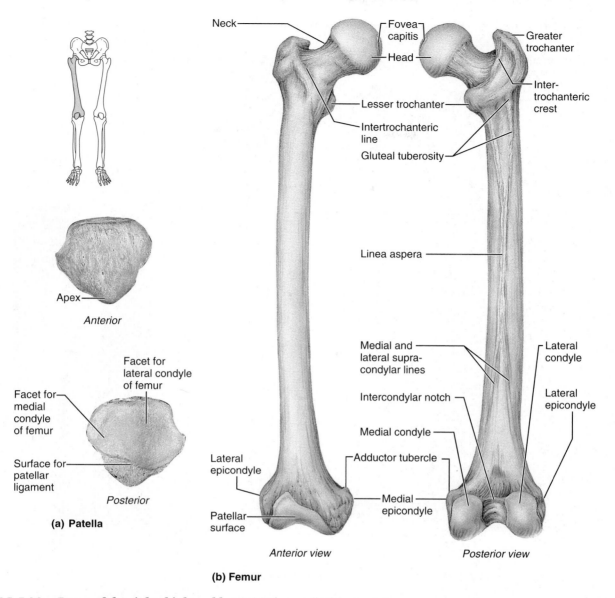

Neck
Fovea capitis
Head
Greater trochanter
Lesser trochanter
Inter-trochanteric crest
Intertrochanteric line
Gluteal tuberosity
Linea aspera
Apex
Anterior
Facet for lateral condyle of femur
Facet for medial condyle of femur
Surface for patellar ligament
Posterior
(a) Patella
Medial and lateral supra-condylar lines
Lateral condyle
Intercondylar notch
Lateral epicondyle
Medial condyle
Lateral epicondyle
Adductor tubercle
Medial epicondyle
Patellar surface
Anterior view
Posterior view
(b) Femur

FIGURE 7.28 *Bones of the right thigh and knee.* (**a**) The patella (kneecap). (**b**) The femur (thigh bone). (See *A Brief Atlas of the Human Body,* Figure 29.)

THE LOWER LIMB

The lower limbs carry the entire weight of the erect body and are subjected to exceptional forces when we jump or run. Thus, it is not surprising that the bones of the lower limbs are thicker and stronger than comparable bones of the upper limbs. The three segments of each lower limb are the thigh, the leg, and the foot (see Table 7.5 on p. 241).

Thigh

The **femur** (fe′mur; "thigh"), the single bone of the thigh (Figures 7.28 and 7.30), is the largest, longest, strongest bone in the body. Its durable structure reflects the fact that the stress on the femur during vigorous jumping can reach 280 kg/cm^2 (about 2 tons per square inch)! The femur is clothed by bulky muscles that prevent us from palpating its course down the length of the thigh. Its length is roughly one-quarter of a person's height.

Proximally, the femur articulates with the hip bone and then courses medially as it descends toward the knee. This arrangement allows the knee joints to be closer to the body's center of gravity and provides for better balance. The medial course of the two femurs is more pronounced in women because of their wider pelvis.

The ball-like **head** of the femur has a small central pit called the **fovea capitis** (fo′ve-ah kă′pĭ-tis; "pit of the head"). A short ligament, the *ligament of the head of the femur,* also called the *ligamentum teres* (tĕ′rēz), runs from this pit to the acetabulum, where it helps secure the femur. The head is carried on a *neck* that angles laterally to join the shaft. This arrangement reflects the fact that the femur articulates with the lateral aspect (rather than the inferior region) of the pelvis. The neck is the weakest part of the femur and is often fractured, an injury commonly

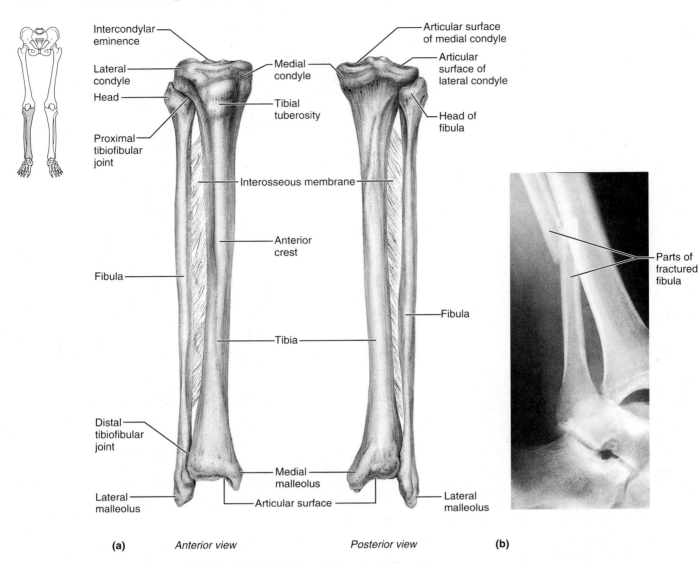

FIGURE 7.29 *The tibia and fibula of the right leg.* **(a)** Anterior (left) and posterior (right) views. **(b)** X ray of Pott's fracture of the fibula. (See *A Brief Atlas of the Human Body*, Figure 30.)

called a broken hip. At the junction of the shaft and neck are the lateral **greater trochanter** (tro-kan′ter) and posteromedial **lesser trochanter.** These projections serve as sites of attachment for thigh and buttock muscles. The two trochanters are connected by the **intertrochanteric line** anteriorly and by the prominent **intertrochanteric crest** posteriorly.

Inferior to the intertrochanteric crest on the posterior shaft is the **gluteal tuberosity,** which blends into a long vertical ridge, the **linea aspera** (lin′e-ah as′per-ah; "rough line"), inferiorly. Distally, the linea aspera appears to diverge, forming the medial and lateral supracondylar lines. All of these markings are sites of muscle attachment. Except for the linea aspera, the femur shaft is smooth and rounded.

Distally, the femur broadens and ends in the wheel-like **lateral** and **medial condyles,** which articulate with the tibia of the leg. The **medial** and **lat-**eral epicondyles (sites of muscle attachment) flank the condyles superiorly. On the superior part of the medial epicondyle is a bump, the **adductor tubercle.** The smooth **patellar surface,** between the condyles on the anterior femoral surface, articulates with the patella (pah-tel′ah), or kneecap (see Figures 7.28a, 7.30, and 7.21).

Between the condyles on the posterior aspect of the femur is the deep, U-shaped **intercondylar notch,** and superior to that on the shaft is the smooth popliteal surface.

The **patella** (pah-tel′ah; "small pan") is a triangular sesamoid bone enclosed in the (quadriceps) tendon that secures the anterior thigh muscles to the tibia. It protects the knee joint anteriorly and improves the leverage of the thigh muscles acting across the knee.

Leg

Two parallel bones, the tibia and fibula, form the skeleton of the leg, the region of the lower limb between the knee and the ankle (Figures 7.29 and 7.30). These two bones are connected by an *interosseous membrane* and articulate with each other both proximally and distally. However, unlike the joints between the radius and ulna of the forearm, the *tibiofibular* (tib"e-o-fib'u-lar) *joints* of the leg allow essentially no movement. The bones of the leg thus form a less flexible but stronger and more stable limb than those of the forearm. The large medial tibia articulates proximally with the femur to form the modified hinge joint of the knee and distally with the talus bone of the foot at the ankle. The fibula, by contrast, does not contribute to the knee joint and merely helps stabilize the ankle joint.

Tibia

The **tibia** (tib'e-ah; "shinbone") receives the weight of the body from the femur and transmits it to the foot. It is second only to the femur in size and strength. At its broad proximal end are the concave **medial** and **lateral condyles,** which look like two huge checkers lying side by side. These are separated by an irregular projection, the **intercondylar eminence.** The tibial condyles articulate with the corresponding condyles of the femur. The inferior region of the lateral tibial condyle bears a facet that indicates the site of the *proximal tibiofibular joint.* Just inferior to the condyles, the tibia's anterior surface displays the rough **tibial tuberosity,** to which the patellar ligament attaches.

The tibial shaft is triangular in cross section. Its anterior border is the sharp **anterior crest.** Neither this crest nor the tibia's medial surface is covered by muscles, so they can be felt just deep to the skin along their entire length. The anguish of a "bumped" shin is an experience familiar to nearly everyone. Distally the tibia is flat where it articulates with the talus bone of the foot. Medial to that joint surface is an inferior projection, the **medial malleolus** (mah-le'o-lus; "little hammer"), which forms the medial bulge of the ankle. The **fibular notch,** on the lateral surface of the tibia, participates in the *distal tibiofibular joint.*

Fibula

The **fibula** (fib'u-lah; "pin") is a sticklike bone with slightly expanded ends. It articulates proximally and distally with the lateral aspects of the tibia. Its proximal end is its **head;** its distal end is the **lateral malleolus.** The lateral malleolus forms the conspicuous lateral ankle bulge and articulates with

the talus. The fibular shaft is heavily ridged and appears to have been twisted a quarter turn. The fibula does not bear weight, but several muscles originate from it.

Homeostatic Imbalance

A *Pott's fracture* occurs at the distal end of the fibula, the tibia, or both. It is a common sports injury. (See Figure 7.29c.) ■

Foot

The skeleton of the foot includes the bones of the *tarsus,* the bones of the *metatarsus,* and the *phalanges,* or toe bones (Figure 7.31). The foot has two important functions: It supports our body weight, and it acts as a lever to propel the body forward when we walk and run. A single bone could serve both purposes, but it would adapt poorly to uneven ground. Segmentation makes the foot pliable, avoiding this problem.

Tarsus

The **tarsus** is made up of seven **tarsal** (tar'sal) **bones** that form the posterior half of the foot. It corresponds to the carpus of the hand. Body weight is carried primarily by the two largest, most posterior tarsals: the **talus** (ta'lus; "ankle"), which articulates with the tibia and fibula superiorly, and the strong **calcaneus** (kal-ka'ne-us; "heel bone"), which forms the heel of the foot and carries the talus on its superior surface. The thick *calcaneal,* or *Achilles, tendon* of the calf muscles attaches to the posterior surface of the calcaneus. The part of the calcaneus that touches the ground is the *tuber calcanei* (kal-ka'ne-i), the calcaneal tuberosity, and its shelflike projection that supports part of the talus is the *sustentaculum tali* (sus"ten-tak'u-lum ta'le; "supporter of the talus"). The remaining tarsals are the lateral **cuboid,** the medial **navicular** (nah-vik'u-lar), and the anterior **medial, intermediate,** and **lateral cuneiform** (ku-ne'ĭ-form; "wedge-shaped") **bones.** The cuboid and cuneiform bones articulate with the metatarsal bones anteriorly.

Metatarsus

The **metatarsus** consists of five small long bones called **metatarsal bones.** These are numbered 1 to 5 beginning on the medial (great toe) side of the foot. The first metatarsal is short and thick and its plantar surface rests on paired sesamoid bones (not shown), which play an important role in supporting body weight. The arrangement of the metatarsals is more parallel than that of the metacarpals of the hands. Distally, where the metatarsals articulate with the proximal phalanges of the toes, the enlarged head of the first metatarsal forms the "ball" of the foot.

Text continues on page 242.

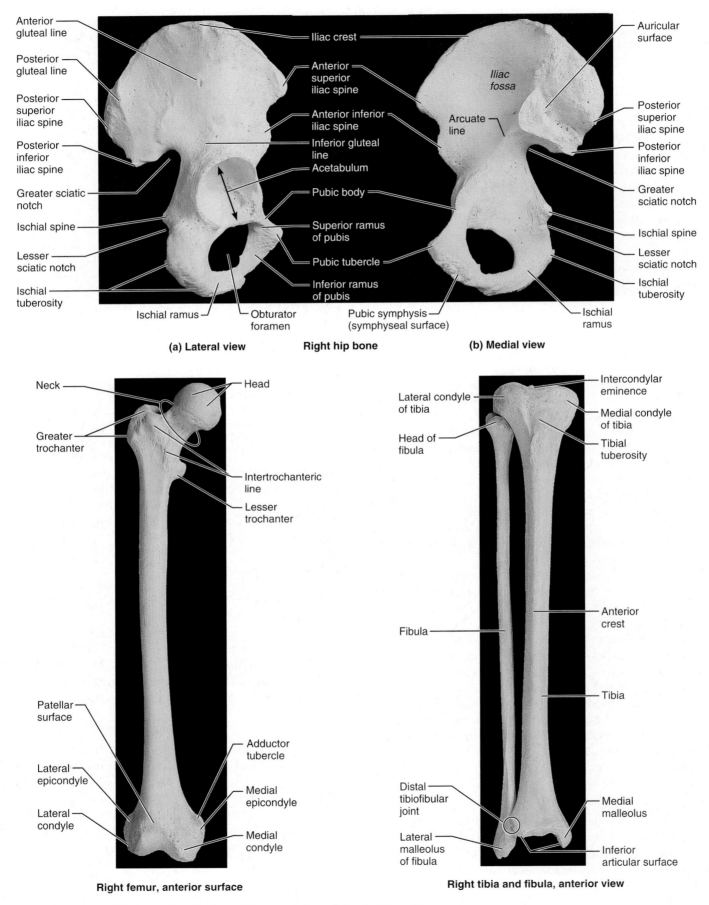

(a) Lateral view   **Right hip bone**   **(b) Medial view**

Right femur, anterior surface

Right tibia and fibula, anterior view

FIGURE 7.30 *Photographs of selected bones of the pelvic girdle and lower limb.* (See *A Brief Atlas of the Human Body,* Figures 28, 29, and 30.)

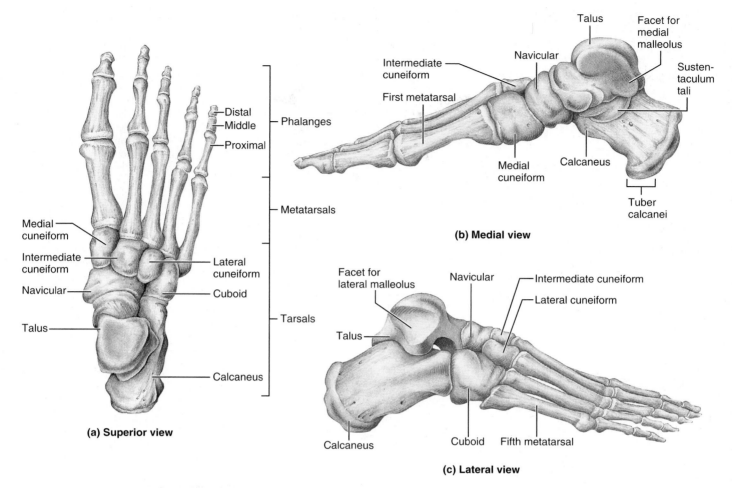

(a) Superior view

(b) Medial view

(c) Lateral view

FIGURE 7.31 **Bones of the right foot.** (See *A Brief Atlas of the Human Body,* Figure 31a, c, and d.)

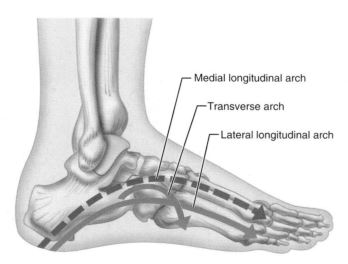

FIGURE 7.32 **Arches of the foot.**

Phalanges (Toes)

The 14 phalanges of the toes are a good deal smaller than those of the fingers and thus are less nimble. But their general structure and arrangement are the same. There are three phalanges in each digit except for the great toe, the **hallux.** The hallux has only two, proximal and distal.

Arches of the Foot

A segmented structure can hold up weight only if it is arched. The foot has three arches: two *longitudinal arches* (the *medial* and *lateral*) and one *transverse arch* (Figure 7.32), which account for its awesome strength. These arches are maintained by the interlocking shapes of the foot bones, by strong ligaments, and by the pull of some tendons during muscle activity. The ligaments and muscle tendons provide a certain amount of springiness. In general, the arches "give" or stretch slightly when weight is applied to the foot and spring back when the weight is removed, which makes walking and running more economical in terms of energy use than would otherwise be the case.

TABLE 7.5		Bones of the Appendicular Skeleton Part 2: Pelvic Girdle and Lower Limb		
Body region	*Bones**	*Illustration*	*Location*	*Markings*
Pelvic girdle (Figures 7.27, 7.30)	Coxal (2) (hip)		Each hip bone is formed by the fusion of an ilium, an ischium, and a pubic bone; the hip bones fuse anteriorly at the pubic symphysis and form sacroiliac joints with the sacrum posteriorly; girdle consisting of both hip bones is basinlike	Iliac crest; anterior and posterior iliac spines; auricular surface; greater and lesser sciatic notches; obturator foramen; ischial tuberosity and spine; acetabulum; pubic arch; pubic crest; pubic tubercle
Free lower limb Thigh (Figures 7.28b, 7.30)	Femur (2)		Femur is the only bone of thigh; between hip joint and knee; largest bone of the body	Head; greater and lesser trochanters; neck; lateral and medial condyles and epicondyles; adductor tubercle; gluteal tuberosity; linea aspera
Kneecap (Figure 7.28a)	Patella (2)		Patella is a sesamoid bone lodged in the tendon of the quadriceps (anterior thigh) muscles	
Leg (Figures 7.29, 7.30)	Tibia (2)		Tibia is the larger and more medial bone of leg; between knee and foot	Medial and lateral condyles; tibial tuberosity; anterior crest; medial malleolus
	Fibula (2)		Fibula is the lateral bone of leg; sticklike	Head; lateral malleolus
Foot (Figure 7.31)	7 Tarsals (14) • talus • calcaneus • navicular • cuboid • lateral cuneiform • intermediate cuneiform • medial cuneiform		Seven tarsal bones form the proximal part of the foot; the talus articulates with the leg bones at the ankle joint; the calcaneus, the largest tarsal, forms the heel	
	5 Metatarsals (10)		Metatarsals are five bones numbered 1–5 from the great toe	
	14 Phalanges (28) • distal • middle • proximal		Phalanges form the toes; three in digits 2–5, two in digit 1 (the great toe)	

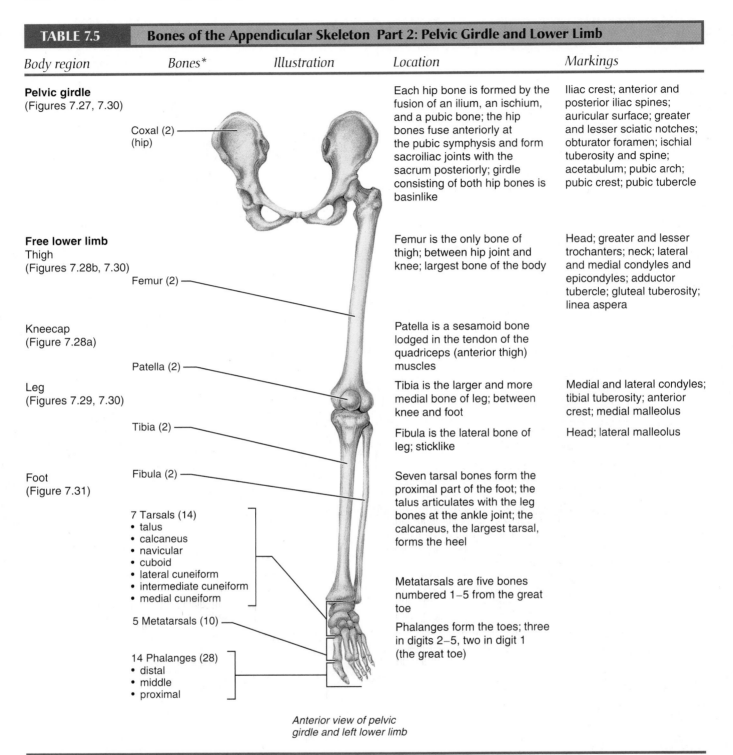

Anterior view of pelvic girdle and left lower limb

*The number in parentheses ( ) following the bone name denotes the total number of such bones in the body.

If you examine your wet footprints, you will see that the medial margin from the heel to the head of the first metatarsal leaves no print. This is because the **medial longitudinal arch** curves well above the ground. The talus is the keystone of this arch, which originates at the calcaneus, rises to the talus, and then descends to the three medial metatarsals. The **lateral longitudinal arch** is very low. It elevates the lateral part of the foot just enough to redistribute some of the weight to the calcaneus and the head of the fifth metatarsal (to the ends of the arch). The cuboid is the keystone bone of this arch. The two longitudinal arches serve as pillars for the **transverse arch,** which runs obliquely from one side of the foot to the other, following the line of the joints between the tarsals and metatarsals. Together, the arches of the foot form a half-dome that distributes about half our standing and walking weight to the heel bones and half to the heads of the metatarsals.

⚠ *Homeostatic Imbalance*

Standing immobile for extended periods places excessive strain on the tendons and ligaments of the feet (because the muscles are inactive) and can result in fallen arches, or "flat feet," particularly if one is overweight. Running on hard surfaces can also cause arches to fall unless one wears shoes that give proper arch support. ■

*   *   *

The bones of the thigh, leg, and foot are summarized in Table 7.5.

DEVELOPMENTAL ASPECTS OF THE SKELETON

The membrane bones of the skull start to ossify late in the second month of development. The rapid deposit of bone matrix at the ossification centers produces cone-shaped protrusions in the developing bones. At birth, the skull bones are still incomplete and are connected by as yet unossified remnants of fibrous membranes called **fontanels** (fon"tah-nelz') (Figure 7.33). The fontanels allow the infant's head to be compressed slightly during birth, and they accommodate brain growth in the fetus and infant. A baby's pulse can be felt surging in these "soft spots"; hence their name (*fontanel* = little fountain). The large, diamond-shaped *anterior fontanel* is palpable for 1½ to 2 years after birth. The others are replaced by bone by the end of the first year.

⚠ *Homeostatic Imbalance*

Several congenital abnormalities may distort the skull. Most common is *cleft palate,* a condition in which the right and left halves of the palate fail to

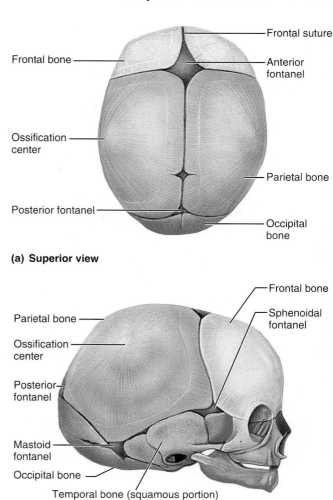

(a) Superior view

(b) Lateral view

FIGURE 7.33   *The fetal skull.* Notice that the infant's skull has more bones than that of an adult. (See *A Brief Atlas of the Human Body,* Figure 16.)

fuse medially. The persistent opening between the oral and nasal cavities interferes with sucking and can lead to aspiration (inhalation) of food into the lungs and *aspiration pneumonia.* ■

The skeleton changes throughout life, but the changes in childhood are most dramatic. At birth, the baby's cranium is huge relative to its face, and several bones are still unfused (e.g., the mandible and maxillae). The maxillae and mandible are foreshortened, and the contours of the face are flat (Figure 7.34). By 9 months after birth, the cranium is already half of its adult size (volume) because of the rapid growth of the brain. By 8 to 9 years, the cranium has almost reached adult proportions. However, between the ages of 6 and 13, the head appears to enlarge substantially as the face literally grows out from the skull. The jaws, cheekbones, and nose become more prominent. These facial changes are correlated with the expansion of the nose and paranasal sinuses, and development of the permanent teeth. Figure 7.34 tracks how differential bone growth alters body proportions throughout life.

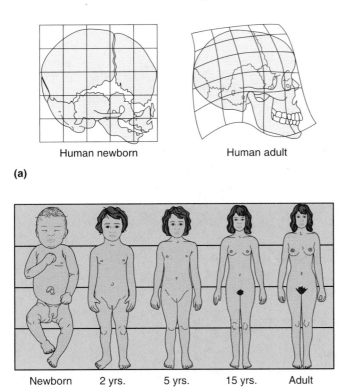

(a)

Human newborn   Human adult

Newborn   2 yrs.   5 yrs.   15 yrs.   Adult

(b)

FIGURE 7.34 *Differences in the growth rates for some parts of the body compared to others determine body proportions.* **(a)** Differential growth transforms the rounded, foreshortened skull of a newborn to the sloping skull characteristic of adults. **(b)** During growth of a human, the arms and legs grow faster than the head and trunk, as can be seen in this conceptualization of different-aged individuals all drawn at the same height.

Only the thoracic and sacral curvatures are present at birth. These so-called **primary curvatures** are convex posteriorly, and an infant's spine arches, like that of a four-legged animal. The **secondary curvatures**—cervical and lumbar—are convex anteriorly and are associated with a child's development. They result from reshaping of the intervertebral discs rather than from modifications of the vertebrae. The cervical curvature is present before birth but is not pronounced until the baby starts to lift its head (at about 3 months). The lumbar curvature develops when the baby begins to walk (at about 12 months). The lumbar curvature positions the weight of the trunk over the body's center of gravity, providing optimal balance when standing.

Vertebral problems (scoliosis or lordosis) may appear during the early school years, when rapid growth of the limb bones stretches many muscles. During the preschool years, lordosis is often present, but this is usually rectified as the abdominal muscles become stronger and the pelvis tilts forward. The thorax grows wider, but a true "military posture" (head erect, shoulders back, abdomen in, and chest out) does not develop until adolescence.

 ## Homeostatic Imbalance

The appendicular skeleton can also suffer from a number of congenital abnormalities. One that is fairly common and quite severe is *dysplasia* (dis-pla′ze-ah; "bad formation") *of the hip.* The acetabulum forms incompletely or the ligaments of the hip joint are loose, so the head of the femur tends to slip out of its socket. Early treatment is essential to prevent permanent crippling. ■

During youth, growth of the skeleton not only increases overall body height but also changes body proportions (Figure 7.34). At birth, the head and trunk are approximately 1½ times as long as the lower limbs. The lower limbs grow more rapidly than the trunk from this time on, and by the age of ten, the head and trunk are approximately the same height as the lower limbs, a condition that persists thereafter. During puberty, the female pelvis broadens in preparation for childbearing, and the entire male skeleton becomes more robust. Once adult height is reached, a healthy skeleton changes very little until late middle age.

Old age affects many parts of the skeleton, especially the spine. As the discs become thinner, less hydrated, and less elastic, the risk of disc herniation increases. By 55 years, a loss of several centimeters in stature is common. Further shortening can be produced by osteoporosis of the spine or by kyphosis (called "dowager's hump" in the elderly). What was done during youth may be undone in old age as the vertebral column gradually resumes its initial arc shape.

The thorax becomes more rigid with age, largely because the costal cartilages ossify. This loss of rib cage elasticity causes shallow breathing, which leads to less efficient gas exchange.

All bones, you will recall, lose mass with age. Cranial bones lose less mass than most, but changes in facial contours with age are common. As the bony tissue of the jaws declines, the jaws look small and childlike once again. If the elderly person loses his or her teeth, this loss of bone from the jaws is accelerated, because the alveolar region bone is resorbed. As bones become more porous, they are more likely to fracture, especially the vertebrae and the neck of the femur.

* * *

Our skeleton is a marvelous substructure, to be sure, but it is much more than that. It is a protector and supporter of other body systems, and without it (and the joints considered in Chapter 8), our muscles would be almost useless. The homeostatic relationships between the skeletal system and other body systems are illustrated in *Making Connections* in Chapter 6 (pp. 192–193).

━━━━━━━━ **RELATED CLINICAL TERMS** ━━━━━━━━

Chiropractic (ki″ro-prak′tik) A system of treating disease by manipulating the vertebral column based on the theory that most diseases are due to pressure on nerves caused by faulty bone alignment; a specialist in this field is a chiropractor.

Clubfoot A relatively common congenital defect in which the soles of the feet face medially and the toes point inferiorly; may be genetically induced or reflect an abnormal position of the foot during fetal development.

Laminectomy Surgical removal of a vertebral lamina; most often done to relieve the symptoms of a ruptured disc.

Orthopedist (or″tho-pe′dist) or **orthopedic surgeon** A physician who specializes in restoring lost skeletal system function or repairing damage to bones and joints.

Ostealgia (os″te-al′je-ah) Pain in a bone.

Pelvimetry Measurement of the dimensions of the inlet and outlet of the pelvis, usually to determine whether the pelvis is of adequate size to allow normal delivery of a baby.

Spina bifida (spi′nah bĭ′fĭ-dah; "cleft spine") Congenital defect of the vertebral column in which one or more of the vertebral arches are incomplete; ranges in severity from inconsequential to severe conditions that impair neural functioning and encourage nervous system infections.

Spinal fusion Surgical procedure involving insertion of bone chips (or crushed bone) to immobilize and stabilize a specific region of the vertebral column, particularly in cases of vertebral fracture and herniated discs.

━━━━━━━━ **CHAPTER SUMMARY** ━━━━━━━━

Media study tools that could provide you additional help in reviewing specific key topics of Chapter 7 are referenced below.

1. The axial skeleton forms the longitudinal axis of the body. Its principal subdivisions are the skull, vertebral column, and bony thorax. It provides support and protection (by enclosure).

2. The appendicular skeleton consists of the bones of the pectoral and pelvic girdles and the limbs. It allows mobility for manipulation and locomotion.

Part 1: The Axial Skeleton

The Skull   (pp. 199–214)

1. The skull is formed by 22 bones. The cranium forms the vault and base of the skull, which protect the brain. The facial skeleton provides openings for the respiratory and digestive passages and attachment points for facial muscles.

2. Except for the temporomandibular joints, all bones of the adult skull are joined by immovable sutures.

3. Cranium. The eight bones of the cranium include the paired parietal and temporal bones and the single frontal, occipital, ethmoid, and sphenoid bones (see Table 7.1, p. 213).

4. Facial bones. The 14 bones of the face include the paired maxillae, zygomatics, nasals, lacrimals, palatines, and inferior conchae and the single mandible and vomer bones (Table 7.1).

5. Orbits and nasal cavity. Both the orbits and the nasal cavities are complicated bony regions formed of several bones.

6. Paranasal sinuses. Paranasal sinuses occur in the frontal, ethmoid, sphenoid, and maxillary bones.

7. Hyoid bone. The hyoid bone, supported in the neck by ligaments, serves as an attachment point for tongue and neck muscles.

SP Exercise: Anatomy of the Skull.

SP Case Study: Craniosunostosis.

The Vertebral Column   (pp. 215–222)

1. General characteristics. The vertebral column includes 24 movable vertebrae (7 cervical, 12 thoracic, and 5 lumbar) and the sacrum and coccyx.

2. The fibrocartilage intervertebral discs act as shock absorbers and provide flexibility to the vertebral column.

3. The primary curvatures of the vertebral column are the thoracic and sacral; the secondary curvatures are the cervical and lumbar. Curvatures increase spine flexibility.

4. General structure of vertebrae. With the exception of C_1 and C_2, all vertebrae have a body, two transverse processes, two superior and two inferior articular processes, a spinous process, and a vertebral arch.

5. Regional vertebral characteristics. Special features distinguish the regional vertebrae (see Table 7.2, p. 218).

SP Exercise: The Vertebral Column.

The Bony Thorax   (pp. 223–224)

1. The bones of the thorax include the 12 rib pairs, the sternum, and the thoracic vertebrae. The bony thorax protects the organs of the thoracic cavity.

2. Sternum. The sternum consists of the fused manubrium, body, and xiphoid process.

3. Ribs. The first seven rib pairs are called true ribs; the rest are called false ribs. Ribs 11 and 12 are floating ribs.

Part 2: The Appendicular Skeleton

The Pectoral (Shoulder) Girdle*   (pp. 224–228)

1. Each pectoral girdle consists of one clavicle and one scapula. The pectoral girdles attach the upper limbs to the axial skeleton.

2. Clavicles. The clavicles hold the scapulae laterally away from the thorax. The sternoclavicular joints are the only attachment points of the pectoral girdle to the axial skeleton.

3. Scapulae. The scapulae articulate with the clavicles and with the humerus bones of the arms.

The Upper Limb*   (pp. 229–233)

1. Each upper limb consists of 30 bones and is specialized for mobility.

2. Arm/forearm/hand. The skeleton of the arm is composed solely of the humerus; the skeleton of the forearm is composed of the radius and ulna; and the skeleton of the hand consists of the carpals, metacarpals, and phalanges.

The Pelvic (Hip) Girdle*   (pp. 233–236)

1. The pelvic girdle, a heavy structure specialized for weight bearing, is composed of two hip bones that secure the lower limbs to the axial skeleton. Together with the sacrum, the hip bones form the basinlike bony pelvis.

2. Each hip bone consists of three fused bones: ilium, ischium, and pubis; the acetabulum occurs at the point of fusion.

3. **Ilium/ischium/pubis.** The ilium is the superior flaring portion of the hip bone. Each ilium forms a secure joint with the sacrum posteriorly. The ischium is a curved bar of bone; we sit on the ischial tuberosities. The V-shaped pubic bones articulate anteriorly at the pubic symphysis.

4. **Pelvic structure and childbearing.** The male pelvis is deep and narrow with larger, heavier bones than those of the female. The female pelvis, which forms the birth canal, is shallow and wide.

The Lower Limb*   (pp. 237–243)

1. Each lower limb consists of the thigh, leg, and foot and is specialized for weight bearing and locomotion.

2. **Thigh.** The femur is the only bone of the thigh. Its ball-shaped head articulates with the acetabulum.

3. **Leg.** The bones of the leg are the tibia, which participates in forming both the knee and ankle joints, and the fibula.

4. **Foot.** The bones of the foot include the tarsals, metatarsals, and phalanges. The most important tarsals are the calcaneus (heel bone) and the talus, which articulates with the tibia superiorly.

5. The foot is supported by three arches (lateral, medial, and transverse) that distribute body weight to the heel and ball of the foot.

SP Exercise: The Skeleton.

SP Case Study: Skeletal Disorders.

Developmental Aspects of the Skeleton   (pp. 243–244)

1. Fontanels, which allow brain growth and ease birth passage, are present in the skull at birth. Growth of the cranium after birth is related to brain growth; increase in size of the facial skeleton follows tooth development and enlargement of nose and sinus cavities.

2. The vertebral column is C-shaped at birth (thoracic and sacral curvatures are present); the secondary curvatures form when the baby begins to lift its head and walk.

3. Long bones continue to grow in length until late adolescence. The head and torso, initially 1½ times the length of the lower limbs, equal their length by the age of ten.

4. Changes in the female pelvis (preparatory for childbirth) occur during puberty.

5. Once at adult height, the skeleton changes little until late middle age. With old age, the intervertebral discs thin; this, along with osteoporosis, leads to a gradual loss in height and increased risk of disc herniation. Loss of bone mass increases the risk of fractures, and thoracic cage rigidity promotes breathing difficulties.

_____

*For associated bone markings, see the pages indicated in the section heads.

REVIEW QUESTIONS

Multiple Choice/Matching

1. Using the letters from column B, match the bone descriptions in column A. (Note that some require more than a single choice.)

Column A	Column B
_____ (1) connected by the coronal suture	(a) ethmoid
_____ (2) keystone bone of cranium	(b) frontal
_____ (3) keystone bone of the face	(c) mandible
_____ (4) form the hard palate	(d) maxillary
_____ (5) allows the spinal cord to pass	(e) occipital
_____ (6) forms the chin	(f) palatine
_____ (7) contain paranasal sinuses	(g) parietal
_____ (8) contains mastoid sinuses	(h) sphenoid
	(i) temporal

2. Match the key terms with the bone descriptions that follow.

Key: **(a)** clavicle   **(b)** ilium   **(c)** ischium   **(d)** pubis   **(e)** sacrum   **(f)** scapula   **(g)** sternum

_____ (1) bone of the axial skeleton to which the pectoral girdle attaches

_____ (2) markings include glenoid cavity and acromion

_____ (3) features include the ala, crest, and greater sciatic notch

_____ (4) doubly curved; acts as a shoulder strut

_____ (5) pelvic girdle bone that articulates with the axial skeleton

_____ (6) the "sit-down" bone

_____ (7) anteriormost bone of the pelvic girdle

_____ (8) part of the vertebral column

3. Use key choices to identify the bone descriptions that follow.

Key: **(a)** carpals   **(b)** femur   **(c)** fibula   **(d)** humerus   **(e)** radius   **(f)** tarsals   **(g)** tibia   **(h)** ulna

_____ (1) articulates with the acetabulum and the tibia

_____ (2) forms the lateral aspect of the ankle

_____ (3) bone that "carries" the hand

_____ (4) the wrist bones

_____ (5) end shaped like a monkey wrench

_____ (6) articulates with the capitulum of the humerus

_____ (7) largest bone of this "group" is the calcaneus

Short Answer Essay Questions

4. Name the cranial and facial bones and compare and contrast the functions of the cranial and facial skeletons.

5. How do the relative proportions of the cranium and face of a fetus compare with those of an adult skull?

6. Name and diagram the normal vertebral curvatures. Which are primary and which are secondary curvatures?

7. List at least two specific anatomical characteristics each for typical cervical, thoracic, and lumbar vertebrae that would allow *anyone* to identify each type correctly.

8. (a) What is the function of the intervertebral discs? (b) Distinguish between the annulus fibrosis and nucleus pulposus regions of a disc. (c) Which provides durability and strength? (d) Which provides resilience? (e) Which part herniates in a "slipped" disc?

9. Name the major components of the bony thorax.

10. (a) What is a true rib? A false rib? (b) Is a floating rib a true rib or a false rib? (c) Why are floating ribs easily broken?

11. The major function of the shoulder girdle is flexibility. What is the major function of the pelvic girdle? Relate these functional differences to anatomical differences seen in these girdles.

12. List three important differences between the male and female pelvis.

13. Describe the function of the arches of the foot.

14. Briefly describe the anatomical characteristics and impairment of function seen in cleft palate and hip dysplasia.

15. Compare a young adult skeleton to that of an extremely aged person relative to bone mass in general and the bony structure of the skull, thorax, and vertebral column in particular.

16. Peter Howell, a teaching assistant in the anatomy class, picked up a hip bone and pretended it was a telephone. He held the big hole in this bone right up to his ear and said, "Hello, obturator, obturator (operator, operator)." Name the structure he was helping the student to learn.

Critical Thinking and Clinical Application Questions

1. Justiniano worked in a poultry-packing plant where his job was cutting open chickens and stripping out their visceral organs. After work, he typed for long hours on his computer keyboard, writing a book about his work in the plant. Soon, his wrist and hand began to hurt whenever he flexed it, and he began to awaken at night with pain and tingling on the thumb-half of his hand. What was his probable condition?

2. Tony Bowers, an exhausted biology student, was attending a lecture. After 30 minutes or so, he lost interest and began to doze. As the lecture ended, the hubbub aroused him and he let go with a tremendous yawn. To his great distress, he couldn't close his mouth—his lower jaw was "stuck" open. What do you think had happened?

3. Ralph had polio as a boy and was partially paralyzed in one lower limb for over a year. Although no longer paralyzed, he now has a severe lateral curvature of the lumbar spine. Explain what has happened and identify his condition.

4. Mary's grandmother slipped on a scatter rug and fell heavily to the floor. Her left lower limb was laterally rotated and noticeably shorter than the right, and when she attempted to get up, she winced with pain. Mary surmised that her grandmother might have "fractured her hip," which later proved to be true. What bone was probably fractured and at what site? Why is a "fractured hip" a common type of fracture in the elderly?

8 JOINTS

1. Define joint or articulation.

Classification of Joints (p. 249)

2. Classify joints structurally and functionally.

Fibrous Joints (pp. 249–250)

3. Describe the general structure of fibrous joints. Name and give an example of each of the three common types of fibrous joints.

Cartilaginous Joints (p. 250)

4. Describe the general structure of cartilaginous joints. Name and give an example of each of the two common types of cartilaginous joints.

Synovial Joints (pp. 251–268)

5. Describe the structural characteristics shared by all synovial joints.

6. List three natural factors that stabilize synovial joints.

7. Compare the structures and functions of bursae and tendon sheaths.

8. Name and describe (or perform) the common body movements.

9. Name six types of synovial joints based on the movement(s) allowed. Provide examples of each type.

10. Describe the elbow, knee, hip, and shoulder joints. Consider in each case the articulating bones, anatomical characteristics of the joint, movements allowed, and relative joint stability.

Homeostatic Imbalances of Joints (pp. 268–272)

11. Name the most common joint injuries and discuss the symptoms and problems associated with each.

12. Compare and contrast the common types of arthritis.

Developmental Aspects of Joints (p. 272)

13. Discuss briefly the factors that promote or disturb joint homeostasis.

The graceful movements of ballet dancers and the rough-and-tumble grapplings of football players attest to the great variety of motion allowed by **joints,** or **articulations**—the sites where two or more bones meet. Our joints have two fundamental functions: They give our skeleton mobility, and they hold it together, sometimes playing a protective role in the process.

Joints are the weakest parts of the skeleton. Nonetheless, their structure resists various forces, such as crushing or tearing, that threaten to force them out of alignment.

CLASSIFICATION OF JOINTS

Joints are classified by structure and by function. The *structural classification* focuses on the material binding the bones together and whether or not a joint cavity is present. Structurally, there are *fibrous, cartilaginous,* and *synovial joints* (Table 8.1 on p. 251).

The *functional classification* is based on the amount of movement allowed at the joint. On this basis, there are **synarthroses** (sin″ar-thro′sēz), which are immovable joints (*syn* = together; *arthro* = joint); **amphiarthroses** (am″fe-ar-thro′sēz), slightly movable joints (*amphi* = on both sides); and **diarthroses** (di″ar-thro′sēz), or freely movable joints (*dia* = through, apart). Freely movable joints predominate in the limbs, whereas immovable and slightly movable joints are largely restricted to the axial skeleton.

In general, fibrous joints are immovable, and synovial joints are freely movable. However, cartilaginous joints have both rigid and slightly movable examples. Since the structural categories are more clear-cut, we will use the structural classification in this discussion, indicating functional properties where appropriate. The specific characteristics of selected body joints are summarized in Table 8.2 (pp. 261–262).

FIBROUS JOINTS

In **fibrous joints,** the bones are joined by fibrous tissue; no joint cavity is present. The amount of movement allowed depends on the length of the connective tissue fibers uniting the bones. Although a few are slightly movable, most fibrous joints are immovable or only slightly movable. The three types of fibrous joints are *sutures, syndesmoses,* and *gomphoses.*

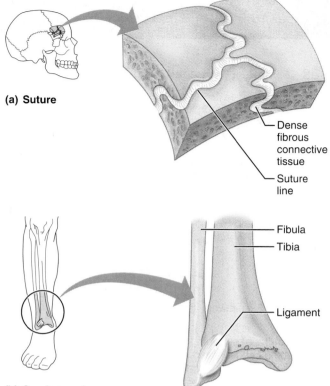

(a) Suture

Dense fibrous connective tissue
Suture line

Fibula
Tibia

Ligament

(b) Syndesmosis

FIGURE 8.1 *Fibrous joints.* **(a)** In sutures of the skull, interconnecting connective tissue fibers are very short, and the bone edges interlock so that the joint is immovable (synarthrotic). **(b)** In the syndesmosis at the distal tibiofibular joint, the fibrous tissue (ligament) connecting the bones is longer than that in sutures; "give," but no true movement, is permitted. (Gomphosis, the third type of fibrous joint, is not illustrated here.)

Sutures

Sutures, literally "seams," occur only between bones of the skull (Figure 8.1a). The wavy articulating bone edges interlock and the junction is completely filled by a minimal amount of very short connective tissue fibers that are continuous with the periosteum. The result is nearly rigid splices that bind the bones tightly together, yet allow the bones to grow at their edges during youth. During middle age, the fibrous tissue ossifies and the skull bones fuse into a single unit. At this stage, the sutures are more precisely called **synostoses** (sin″os-to′sēz), literally, "bony junctions." Because movement of the cranial bones would damage the brain, the immovable nature of sutures is a protective adaptation.

? *Which one of the three joint types shown here has greater flexibility than the other two and why?*

synovial fluid. They are common in sites where ligaments, muscles, skin, tendons, or bones rub together. Many people have never heard of a bursa, but most have heard of bunions (bun'yunz). A *bunion* is an enlarged bursa at the base of the big toe, swollen from rubbing of a tight or poorly fitting shoe.

A **tendon sheath** is essentially an elongated bursa that wraps completely around a tendon subjected to friction, like a bun around a hot dog.

Factors Influencing the Stability of Synovial Joints

Because joints are constantly stretched and compressed, they must be stabilized so that they do not dislocate (come out of alignment). The stability of a synovial joint depends chiefly on three factors: the shapes of the articular surfaces; the number and positioning of ligaments; and muscle tone.

Articular Surfaces

The shapes of articular surfaces determine what movements are possible at a joint but, surprisingly, articular surfaces play a minor role in joint stability. Many joints have shallow sockets or noncomplementary articulating surfaces ("misfits") that actually hinder joint stability. But when articular surfaces are large and fit snugly together, or when the socket is deep, stability is vastly improved. The ball and deep socket of the hip joint provide the best example of a joint made extremely stable by the shape of its articular surfaces.

Ligaments

The capsules and ligaments of synovial joints unite the bones and prevent excessive or undesirable motion. As a rule, the more ligaments a joint has, the stronger it is. However, when other stabilizing factors are inadequate, undue tension is placed on the ligaments and they stretch. Stretched ligaments stay stretched, like taffy, and a ligament can stretch only about 6% of its length before it snaps. Thus, when ligaments are the major means of bracing a joint, the joint is not very stable.

Muscle Tone

For most joints, the muscle tendons that cross the joint are the most important stabilizing factor. These tendons are kept taut at all times by the tone of their muscles. (Muscle tone is defined as low levels of contractile activity in relaxed muscles that keep the muscles healthy and ready to react to stimulation.) Muscle tone is extremely important in reinforcing the shoulder and knee joints and arches of the foot.

Movements Allowed by Synovial Joints

Every skeletal muscle of the body is attached to bone or other connective tissue structures at no fewer than two points. The muscle's **origin** is attached to the immovable (or less movable) bone. Its other end, the **insertion,** is attached to the movable bone. Body movement occurs when muscles contract across joints and their insertion moves toward their origin. The movements can be described in directional terms relative to the lines, or *axes,* around which the body part moves and the planes of space along which the movement occurs, that is, along the transverse, frontal, or sagittal plane. (These planes were described in Chapter 1.)

Range of motion allowed by synovial joints varies from **nonaxial movement** (slipping movements only, since there is no axis around which movement can occur) to **uniaxial movement** (movement in one plane) to **biaxial movement** (movement in two planes) to **multiaxial movement** (movement in or around all three planes of space and axes). Range of motion varies greatly in different people. In some, such as trained gymnasts or acrobats, range of joint movement may be extraordinary. The ranges of motion at the major joints are given in Table 8.2.

There are three general types of movements: *gliding angular movements,* and *rotation.* The most common body movements allowed by synovial joints are described next and illustrated in Figure 8.5.

Gliding Movements

Gliding movements (Figure 8.5a) are the simplest joint movements. One flat, or nearly flat, bone surface glides or slips over another similar surface. Glid-

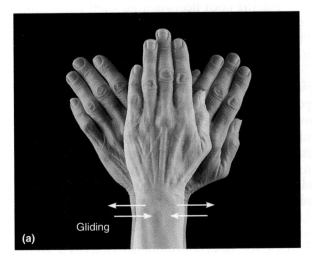

(a) Gliding

FIGURE 8.5 *Movement allowed by synovial joints.* **(a)** Gliding movements.

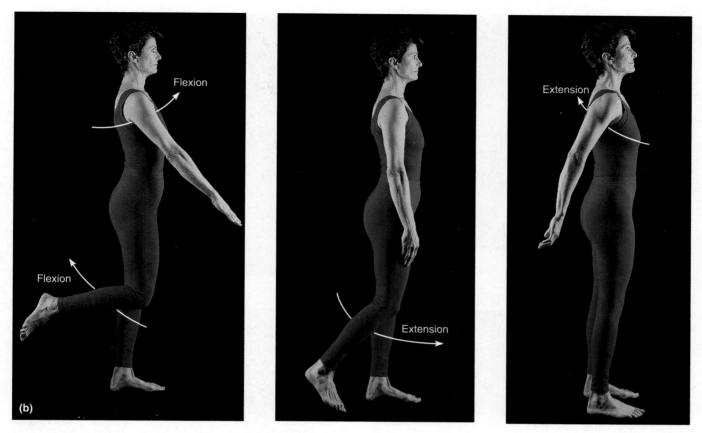

FIGURE 8.5 *(Continued)* *Movement allowed by synovial joints.*
(**b**) Angular movements.

ing movements occur at the intercarpal and inter-tarsal joints, and between the flat articular processes of the vertebrae (see Table 8.2 on pages 261–262).

Angular Movements

Angular movements (Figure 8.5b–f) increase or decrease the angle between two bones. These movements may occur in any plane of the body and include flexion, extension, hyperextension, abduction, adduction, and circumduction.

Flexion Flexion (flek′shun) is a bending movement that *decreases the angle* of the joint and brings the articulating bones closer together. The movement usually occurs along the sagittal plane. Examples include bending the head forward on the chest (Figure 8.5d) and bending the knee or the body trunk from a straight to an angled position (Figure 8.5b and c). As a less obvious example, the arm is flexed at the shoulder when the arm is lifted in an anterior direction (Figure 8.5b).

Extension Extension is the reverse of flexion and occurs at the same joints. It involves movement along the sagittal plane that *increases the angle* between the articulating bones, such as straightening a flexed neck, body trunk, elbow, or knee (Figure 8.5b–d). Bending the head backward beyond its

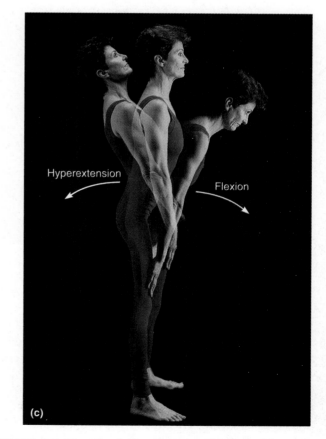

FIGURE 8.5 *(Continued)* *Movements allowed by synovial joints.* (**c**) Angular movements.

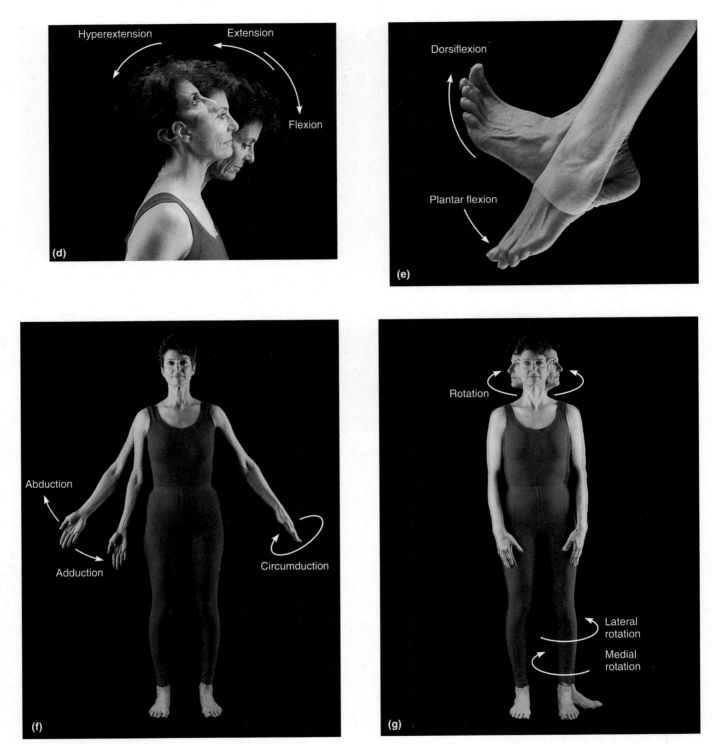

FIGURE 8.5 (Continued) *Movements allowed by synovial joints.*
(**d–f**) Angular movements, continued. (**g**) Rotation.

straight (upright) position is called **hyperextension.** At the shoulder, extension carries the arm to a point posterior to the shoulder joint.

Dorsiflexion and Plantar Flexion of the Foot The up-and-down movements of the foot at the ankle joint are given more specific names (Figure 8.5e).

Lifting the foot so that its superior surface approaches the shin is called **dorsiflexion** (corresponds to extension of the wrist), whereas depressing the foot (pointing the toes) is called **plantar flexion** (corresponds to flexion of the wrist).

Abduction Abduction ("moving away") is movement of a limb *away* from the midline or median plane of the body, along the frontal plane. Raising the arm (Figure 8.5f) or thigh laterally is an example of abduction. When the term is used to indicate the movement of the fingers or toes, it means spreading them apart. In this case "midline" is the longest digit: the third finger or second toe. Notice, however, that lateral bending of the trunk away from the body midline in the frontal plane is called *lateral flexion*, not abduction.

Adduction Adduction ("moving toward") is the opposite of abduction, so it is the movement of a limb *toward* the body midline or, in the case of the digits, toward the midline of the hand or foot (Figure 8.5f).

Circumduction Circumduction (Figure 8.5f) is moving a limb so that it describes a cone in space (*circum* = around; *duco* = to draw). The distal end of the limb moves in a circle, while the point of the cone (the shoulder or hip joint) is more or less stationary. A pitcher winding up to throw a ball is actually circumducting his or her pitching arm. Because circumduction consists of flexion, abduction, extension, and adduction performed in succession, it is the quickest way to exercise the many muscles that move the hip and shoulder ball-and-socket joints.

Rotation

Rotation is the turning of a bone around its own long axis. It is the only movement allowed between the first two cervical vertebrae and is common at the hip (Figure 8.5g) and shoulder joints. Rotation may be directed toward the midline or away from it. For example, in *medial rotation* of the thigh, the femur's anterior surface moves toward the median plane of the body; *lateral rotation* is the opposite movement.

Special Movements

Certain movements do not fit into any of the above categories and occur at only a few joints. Some of these special movements are illustrated in Figure 8.6.

Supination and Pronation The terms **supination** (soo"pĭ-na'shun; "turning backward") and **pronation** (pro-na'shun; "turning forward") refer to the movements of the radius around the ulna (Figure 8.6a). Rotating the forearm laterally so that the palm faces anteriorly or superiorly is supination. In the anatomical position, the hand is supinated and the radius and ulna are parallel. In pronation, the forearm rotates medially and the palm faces posteriorly or inferiorly. Pronation moves the distal end of the radius across the ulna so that the two bones form an X. This is the forearm's position when we are standing in a relaxed manner. Pronation is a much weaker movement than supination.

Tricks to help you keep these terms straight: If you lifted a cup of soup up to your mouth *on your palm*, you would be supinating ("soup"-inating), and a *pro* basketball player pronates his or her forearm to dribble the ball.

Inversion and Eversion Inversion and **eversion** are special movements of the foot (Figure 8.6b). In inversion, the sole of the foot turns medially. In eversion, the sole faces laterally.

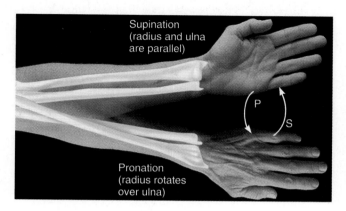

Which hand position—pronation or supination—is characteristic of the anatomical position?

Supination (radius and ulna are parallel)

Pronation (radius rotates over ulna)

(a) Supination (S) and pronation (P)

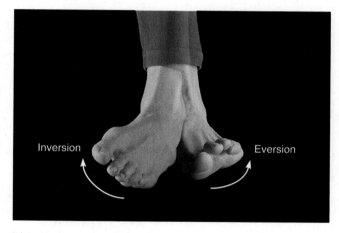

Inversion

Eversion

(b) Inversion and eversion

FIGURE 8.6 ***Special body movements.*** **(a)** Supination and pronation. **(b)** Inversion and eversion.

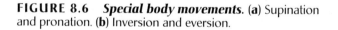

Supination.

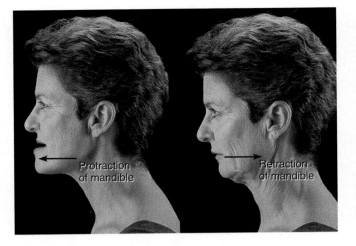

(c) Protraction and retraction

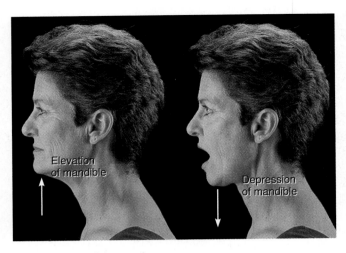

(d) Elevation and depression

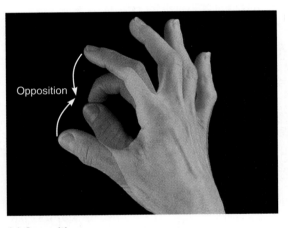

(e) Opposition

FIGURE 8.6   *Special body movements (Continued).*
(**c**) Protraction and retraction. (**d**) Elevation and depression.
(**e**) Opposition.

Protraction and Retraction Nonangular anterior and posterior movements in a transverse plane are called **protraction** and **retraction,** respectively (Figure 8.6c). The mandible is protracted when you jut out your jaw and retracted when you move it back to its original position.

Elevation and Depression Elevation means lifting a body part superiorly (Figure 8.6d). For example, the scapulae are elevated when you shrug your shoulders. Moving the elevated part inferiorly is **depression.** During chewing, the mandible is alternately elevated and depressed.

Opposition The saddle joint between metacarpal 1 and the carpals allows a movement called **opposition** of the thumb (Figure 8.6e). This movement is the action taken when you touch your thumb to the tips of the other fingers on the same hand. It is opposition that makes the human hand such a fine tool for grasping and manipulating objects.

Types of Synovial Joints

Although all synovial joints have structural features in common, they do not have a common structural plan. Based on the shape of their articular surfaces, which in turn determine the movements allowed, synovial joints can be classified further into six major categories—plane, hinge, pivot, condyloid, saddle, and ball-and-socket joints (Figure 8.7).

Plane Joints

In **plane joints** (Figure 8.7a) the articular surfaces are essentially flat, and they allow only short slipping or gliding movements. Examples are the gliding joints introduced earlier—the intercarpal and intertarsal joints, and the joints between vertebral articular processes. Gliding does not involve rotation around any axis, and gliding joints are the only examples of nonaxial joints.

Hinge Joints

In **hinge joints** (Figure 8.7b), a cylindrical projection of one bone fits into a trough-shaped surface on another. Motion is along a single plane and resembles that of a mechanical hinge. The uniaxial hinge joints permit flexion and extension only, typified by bending and straightening the elbow and interphalangeal joints.

FIGURE 8.7 *Types of synovial joints.* Dashed lines indicate the articulating bones. (**a**) Plane joint (e.g., intercarpal and intertarsal joints). (**b**) Hinge joint (e.g., elbow joints and interphalangeal joints).

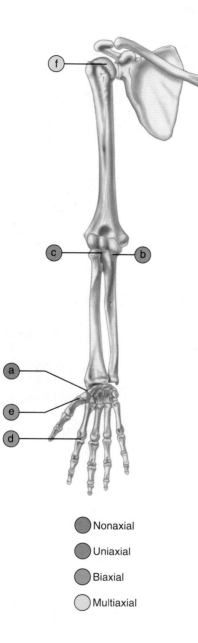

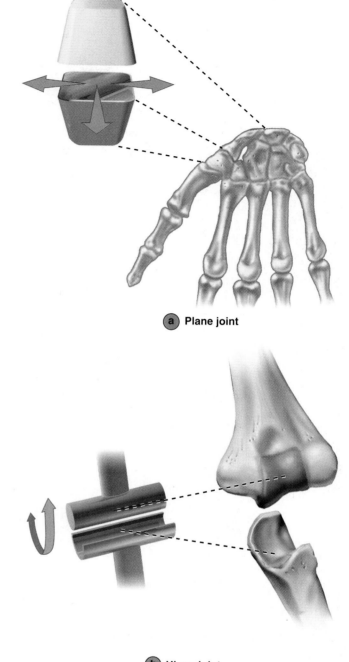

- ● Nonaxial
- ● Uniaxial
- ● Biaxial
- ○ Multiaxial

(a) **Plane joint**

(b) **Hinge joint**

Pivot Joints

In a **pivot joint** (Figure 8.7c), the rounded end of one bone protrudes into a "sleeve" or ring composed of bone (and possibly ligaments) of another. The only movement allowed is uniaxial rotation of one bone around its own long axis. An example is the joint between the atlas and dens of the axis, which allows you to move your head from side to side to indicate "no." Another is the proximal radioulnar joint, where the head of the radius rotates within a ringlike ligament secured to the ulna.

Condyloid Joints

In **condyloid** (kon′dĭ-loid; "knuckle-like"), or **ellipsoidal, joints,** the oval articular surface of one bone fits into a complementary depression in another (Figure 8.7d). The important characteristic is that both articulating surfaces are oval. The biaxial condyloid joints permit all *angular* motions, that is, flexion and extension, abduction and adduction, and circumduction. The radiocarpal (wrist) joints and

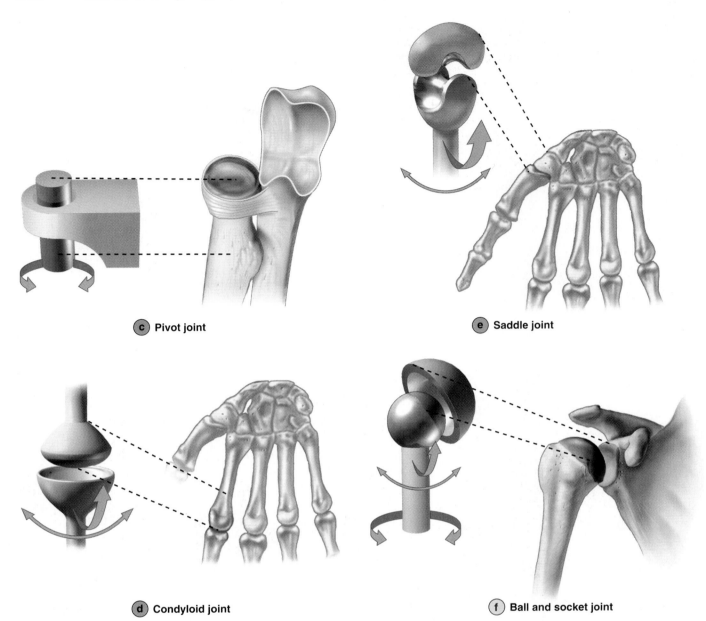

FIGURE 8.7   *(Continued) Types of synovial joints* . (**c**) Pivot joint (e.g., proximal radioulnar joint). (**d**) Condyloid joint (e.g., metacarpophalangeal joints). (**e**) Saddle joint (e.g., carpometacarpal joint of the thumb). (**f**) Ball-and-socket joint (e.g., shoulder joint).

the metacarpophalangeal (knuckle) joints are typical condyloid joints.

Saddle Joints

Saddle joints (Figure 8.7e) resemble condyloid joints, but they allow greater freedom of movement. Each articular surface has *both* concave and convex areas; that is, it is shaped like a saddle. The articular surfaces then fit together, concave to convex surfaces. The most clear-cut examples of saddle joints in the body are the carpometacarpal joints of the

thumbs, and the movements allowed by these joints are clearly demonstrated by twiddling your thumbs.

Ball-and-Socket Joints

In **ball-and-socket joints** (Figure 8.7f), the spherical or hemispherical head of one bone articulates with the cuplike socket of another. These joints are multiaxial and the most freely moving synovial joints. Universal movement is allowed (that is, in all axes and planes, including rotation). The shoulder and hip joints are examples.

TABLE 8.2   Structural and Functional Characteristics of Body Joints

Illustration	Joint	Articulating bones	Structural type*	Functional type; movements allowed
	Skull	Cranial and facial bones	Fibrous; suture	Synarthrotic; no movement
	Temporo-mandibular	Temporal bone of skull and mandible	Synovial; modified hinge† (contains articular disc)	Diarthrotic; gliding and uniaxial rotation; slight lateral movement, elevation, depression, protraction, and retraction of mandible
	Atlanto-occipital	Occipital bone of skull and atlas	Synovial; condyloid	Diarthrotic; biaxial; flexion, extension, abduction, adduction, circumduction of head on neck
	Atlantoaxial	Atlas (C_1) and axis (C_2)	Synovial; pivot	Diarthrotic; uniaxial; rotation of the head
	Intervertebral	Between adjacent vertebral bodies	Cartilaginous; symphysis	Amphiarthrotic; slight movement
	Intervertebral	Between articular processes	Synovial; plane	Diarthrotic; gliding
	Vertebrocostal	Vertebrae (transverse processes or bodies) and ribs	Synovial; plane	Diarthrotic; gliding of ribs
	Sternoclavicular	Sternum and clavicle	Synovial; shallow saddle (contains articular disc)	Diarthrotic; multiaxial (allows clavicle to move in all axes)
	Sternocostal	Sternum and rib 1	Cartilaginous; synchondrosis	Synarthrotic; no movement
	Sternocostal	Sternum and ribs 2–7	Synovial; double plane	Diarthrotic; gliding
	Acromio-clavicular	Acromion of scapula and clavicle	Synovial; plane	Diarthrotic; gliding and rotation of scapula on clavicle
	Shoulder (glenohumeral)	Scapula and humerus	Synovial; ball and socket	Diarthrotic; multiaxial; flexion, extension, abduction, adduction, circumduction, rotation of humorus
	Elbow	Ulna (and radius) with humerus	Synovial; hinge	Diarthrotic; uniaxial; flexion, extension of forearm
	Radioulnar (proximal)	Radius and ulna	Synovial; pivot	Diarthrotic; uniaxial; rotation of radius around long axis of forearm to allow pronation and supination
	Radioulnar (distal)	Radius and ulna	Synovial; pivot (contains articular disc)	Diarthrotic; uniaxial; rotation (convex head of ulna rotates in ulnar notch of radius)
	Wrist (radiocarpal)	Radius and proximal carpals	Synovial; condyloid	Diarthrotic; biaxial; flexion, extension, abduction, adduction, circumduction of hand
	Intercarpal	Adjacent carpals	Synovial; plane	Diarthrotic; gliding
	Carpometacarpal of digit 1 (thumb)	Carpal (trapezium) and metacarpal 1	Synovial; saddle	Diarthrotic; biaxial; flexion, extension, abduction, adduction, circumduction, opposition of metacarpal 1
	Carpometacarpal of digits 2–5	Carpal(s) and metacarpal(s)	Synovial; plane	Diarthrotic; gliding of metacarpals
	Knuckle (metacarpo-phalangeal)	Metacarpal and proximal phalanx	Synovial; condyloid	Diarthrotic; biaxial; flexion, extension, abduction, adduction, circumduction of fingers
	Finger (interphalangeal)	Adjacent phalanges	Synovial; hinge	Diarthrotic; uniaxial; flexion, extension of fingers

➤

TABLE 8.2	Structural and Functional Characteristics of Body Joints *continued*

Illustration	Joint	Articulating bones	Structural type*	Functional type; movements allowed
	Sacroiliac	Sacrum and coxal bone	Synovial; plane	Diarthrotic; little movement, slight gliding possible (more during pregnancy)
	Pubic symphysis	Pubic bones	Cartilaginous; symphysis	Amphiarthrotic; slight movement (enhanced during pregnancy)
	Hip (coxal)	Hip bone and femur	Synovial; ball and socket	Diarthrotic; multiaxial; flexion, extension, abduction, adduction, rotation, circumduction of thigh
	Knee (tibiofemoral)	Femur and tibia	Synovial; modified hinge† (contains articular discs)	Diarthrotic; biaxial; flexion, extension of leg, some rotation allowed
	Knee (femoropatellar)	Femur and patella	Synovial; plane	Diarthrotic; gliding of patella
	Tibiofibular	Tibia and fibula (proximally)	Synovial; plane	Diarthrotic; gliding of fibula
	Tibiofibular	Tibia and fibula (distally)	Fibrous; syndesmosis	Synarthrotic; slight "give" during dorsiflexion
	Ankle	Tibia and fibula with talus	Synovial; hinge	Diarthrotic; uniaxial; dorsiflexion, and plantar flexion of foot
	Intertarsal	Adjacent tarsals	Synovial; plane	Diarthrotic; gliding; inversion and eversion of foot
	Tarsometatarsal	Tarsal(s) and metatarsal(s)	Synovial; plane	Diarthrotic; gliding of metatarsals
	Metatarso-phalangeal	Metatarsal and proximal phalanx	Synovial; condyloid	Diarthrotic; biaxial; flexion, extension, abduction, adduction, circumduction of great toe
	Toe (interpha-langeal)	Adjacent phalanges	Synovial; hinge	Diarthrotic; uniaxial; flexion; extension of toes

*Fibrous joints indicated by orange circles; cartilaginous joints by blue circles; synovial joints by purple circles.

†These modified hinge joints are structurally bicondylar.

Selected Synovial Joints

In this section, we examine four joints (shoulder, elbow, hip, and knee) in detail. All of these joints have the five distinguishing characteristics of synovial joints, and we will not discuss these common features again. Rather, we will emphasize the unique structural features, functional abilities, and, in certain cases, functional weaknesses of each of these joints.

Shoulder (Glenohumeral) Joint

In the shoulder joint, stability has been sacrificed to provide the most freely moving joint of the body. The shoulder joint is a ball-and-socket joint. The large hemispherical head of the humerus fits in the small, shallow glenoid cavity of the scapula (Figure 8.8), like a golf ball sitting on a tee. Although the glenoid cavity is slightly deepened by a rim of fibrocartilage, called the **glenoid labrum** (*labrum* = lip), it is only about one-third the size of the humeral head and contributes little to joint stability.

The articular capsule enclosing the joint cavity (from the margin of the glenoid cavity to the anatomical neck of the humerus) is remarkably thin and loose (qualities that contribute to this joint's freedom of movement). The few ligaments reinforcing the shoulder joint are located primarily on its anterior aspect. The superiorly located **coracohumeral** (kor′ah-ko-hu′mer-ul) **ligament** provides the only strong thickening of the capsule and helps support

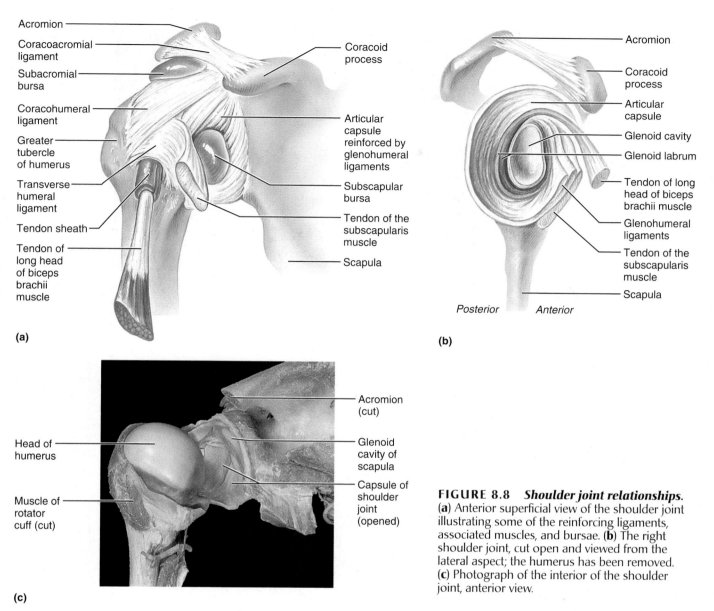

(a)

(b)

Posterior   Anterior

(c)

Head of humerus

Muscle of rotator cuff (cut)

Acromion (cut)

Glenoid cavity of scapula

Capsule of shoulder joint (opened)

FIGURE 8.8   *Shoulder joint relationships.* (a) Anterior superficial view of the shoulder joint illustrating some of the reinforcing ligaments, associated muscles, and bursae. (b) The right shoulder joint, cut open and viewed from the lateral aspect; the humerus has been removed. (c) Photograph of the interior of the shoulder joint, anterior view.

the weight of the upper limb. Three **glenohumeral** (glĕ"no-hu'mer-ul) **ligaments,** strengthen the front of the capsule somewhat but are weak and may even be absent.

Muscle tendons that cross the shoulder joint contribute most to this joint's stability. The "super-stabilizer" is the tendon of the long head of the biceps brachii muscle of the arm (Figure 8.8a). This tendon attaches to the superior margin of the glenoid labrum, travels through the joint cavity, and then runs within the intertubercular groove of the humerus. It secures the head of the humerus tightly against the glenoid cavity. Four other tendons (and the associated muscles) make up the **rotator cuff.** This cuff encircles the shoulder joint and blends with the articular capsule. The muscles include the subscapularis, supraspinatus, infraspinatus, and teres minor. (The rotator cuff muscles are illustrated

in Figure 10.14, p. 354.) The rotator cuff can be severely stretched when the arm is vigorously circumducted; this is a common injury of baseball pitchers. As noted in Chapter 7, shoulder dislocations are fairly common. Since the shoulder's reinforcements are weakest anteriorly and inferiorly, the humerus tends to dislocate in the forward and downward direction.

Hip (Coxal) Joint

The hip joint, like the shoulder joint, is a ball-and-socket joint. It has a good range of motion, but not nearly as wide as the shoulder's range. Movements occur in all possible planes but are limited by the joint's strong ligaments and its deep socket. The hip joint is formed by the articulation of the spherical head of the femur with the deeply cupped acetabulum of the hip bone (Figure 8.9). The depth of the

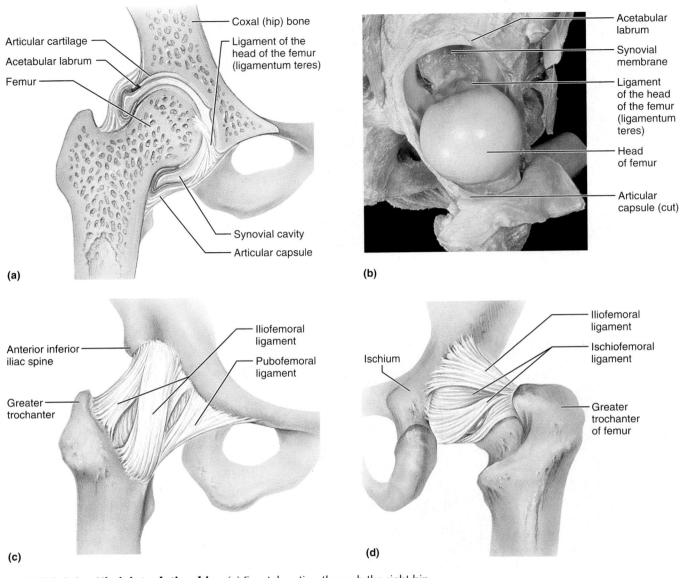

FIGURE 8.9 *Hip joint relationships.* **(a)** Frontal section through the right hip joint. **(b)** Photograph of the interior of the hip joint, lateral view. **(c)** Anterior superficial view of the right hip joint. **(d)** Posterior superficial view of the right hip joint.

acetabulum is enhanced by a circular rim of fibrocartilage called the **acetabular** (as″ĕ-tab′u-lar) **labrum** (Figure 8.9a and b). The labrum has a diameter less than that of the head of the femur, and these articular surfaces fit snugly together, so hip joint dislocations are rare.

The thick articular capsule extends from the rim of the acetabulum to the neck of the femur and completely encloses the joint. Several strong ligaments reinforce the capsule of the hip joint. These include the **iliofemoral** (il″e-o-fem′o-ral) **ligament,** a strong V-shaped ligament anteriorly; the **pubofemoral** (pu″bo-fem′o-ral) **ligament,** a triangular thickening of the inferior part of the capsule; and the **ischiofemoral** (is″ke-o-fem′o-ral) **ligament,** a spiraling posteriorly located ligament. These ligaments are arranged in such a way that they "screw"

the femur head into the acetabulum when a person stands up straight, thereby providing more stability.

The **ligament of the head of the femur,** also called the **ligamentum teres,** is a flat intracapsular band that runs from the femur head to the lower lip of the acetabulum. This ligament is slack during most hip movements, so it is not important in stabilizing the joint. In fact, its mechanical function (if any) is unclear, but it does contain an artery that helps supply the head of the femur. Damage to this artery may lead to severe arthritis of the hip joint.

Muscle tendons that cross the joint and the bulky hip and thigh muscles that surround it contribute to its stability and strength. However, the deep socket that securely encloses the femoral head and the strong ligaments contribute most to the stability of the hip joint.

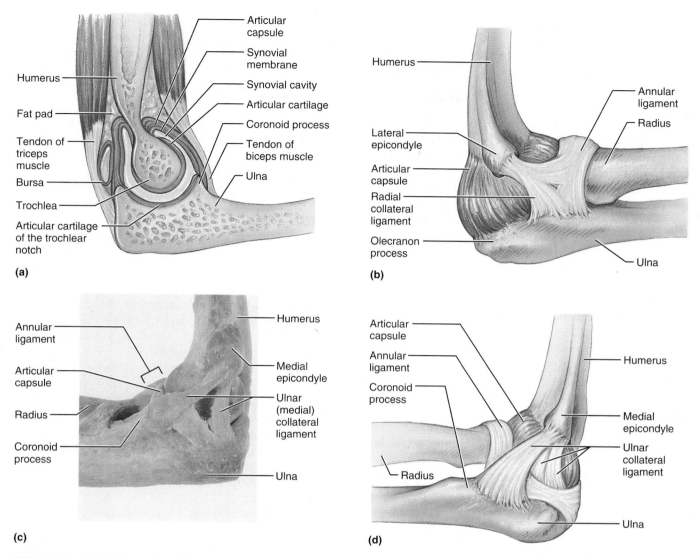

FIGURE 8.10 *Elbow joint relationships.* (**a**) Midsagittal section of right elbow joint. (**b**) Lateral view of right elbow joint. (**c**) Photograph of right elbow joint showing the important reinforcing ligaments, medial view. (**d**) Medial view of right elbow joint.

Elbow Joint

Our upper limbs are flexible extensions that permit us to reach out and manipulate things in our environment. Besides the shoulder joint, the most prominent of the upper limb joints is the elbow. The elbow joint provides a stable and smoothly operating hinge that allows flexion and extension only (Figure 8.10). Within the joint, both the radius and ulna articulate with the condyles of the humerus, but it is the close gripping of the trochlea by the ulna's trochlear notch that forms the "hinge" and stabilizes this joint. A relatively lax articular capsule extends inferiorly from the humerus to the ulna and to the **annular** (an'u-lar) **ligament** surrounding the head of the radius.

Anteriorly and posteriorly, the articular capsule is thin and allows substantial freedom for elbow flexion and extension. However, side-to-side movements are restricted by two strong capsular ligaments: the **ulnar collateral ligament** medially, and the **radial collateral ligament,** a triangular ligament on the lateral side. Additionally, tendons of several arm muscles, such as the biceps and triceps, cross the elbow joint and provide security.

The radius is a passive "onlooker" in the angular elbow movements. However, its head rotates within the annular ligament during supination and pronation of the forearm.

Knee Joint

The knee joint is the largest and most complex joint in the body (Figure 8.11). It allows extension, flexion, and some rotation. Despite its single joint cavity, the knee consists of three joints in one: an intermediate one between the patella and the lower end of the femur (the **femoropatellar joint**), and lateral and medial joints (collectively known as the **tibiofemoral joint**) between the femoral condyles above and the C-shaped **menisci,** or *semilunar cartilages,* of the tibia below. Besides deepening the shallow tibial articular surfaces, the menisci help prevent side-to-side rocking of the femur on the tibia and absorb shock transmitted to the knee joint. (However, the menisci are attached only at their outer margins and are frequently torn free.)

The tibiofemoral joint acts primarily as a hinge, permitting flexion and extension. However, it is structurally a bicondylar joint. Some rotation is possible when the knee is partly flexed, but when it is extended, side-to-side movements and rotation are strongly resisted by ligaments and the menisci. The femoropatellar joint is a plane joint, and the patella glides across the distal end of the femur during knee movements.

The knee joint is unique in that its joint cavity is only partially enclosed by a capsule. The relatively thin articular capsule is present only on the sides and posterior aspects of the knee, where it covers the bulk of the femoral and tibial condyles. Anteriorly, where the capsule is absent, three broad ligaments run from the patella to the tibia below. These are the **patellar ligament** flanked by the **medial** and **lateral patellar retinacula** (ret″ĭ-nak′u-lah; "retainers"), which merge imperceptibly into the articular capsule on each side (Figure 8.11c). The patellar ligament is actually a continuation of the tendon of the bulky quadriceps muscle of the anterior thigh. The retinacula are expansions of the same tendon. Physicians tap the patellar ligament to test the knee-jerk reflex.

The synovial cavity of the knee joint has a complicated shape, with several extensions that lead into "blind alleys." At least a dozen bursae are associated with this joint, some of which are shown in Figure 8.11a. For example, notice the *subcutaneous prepatellar bursa,* which is often injured when the knee is bumped.

Both extra- and intracapsular ligaments stabilize and strengthen the capsule of the knee joint. The **extracapsular ligaments** all act to prevent hyperextension of the knee and are stretched taut when the knee is extended. These include the following:

1. The **fibular** and **tibial collateral ligaments** are critical in preventing lateral or medial rotation when the knee is extended. The pencil-like fibular collateral ligament extends from the lateral epicondyle of the femur to the head of the fibula. The broad tibial collateral ligament runs from the medial epicondyle of the femur to the medial condyle of the tibial shaft below and is fused to the medial meniscus.

2. The **oblique popliteal** (pop″lĭ-te′al) **ligament** is actually part of the tendon of the semimembranosus muscle that crosses the posterior aspect of the knee joint (Figure 8.11e).

3. The **arcuate popliteal ligament** arcs superiorly from the head of the fibula and reinforces the joint capsule posteriorly.

The **intracapsular ligaments** are called *cruciate* (kroo′she-āt) *ligaments* because they cross each other, forming an X (*cruci* = cross) within the notch between the femoral condyles. They help to prevent anterior-posterior displacement of the articular surfaces and to secure the articulating bones when we stand (see Figure 8.11b). Although these ligaments are within the joint capsule, they are *outside* the synovial cavity and synovial membrane nearly covers their surfaces. Note that the two cruciate ligaments both run up to the femur and are named for their *tibial* attachment site. The **anterior cruciate ligament** attaches to the *anterior* intercondylar area of the tibia. From there it passes posteriorly, laterally, and upward to attach to the femur on the medial side of its lateral condyle. This ligament prevents forward sliding of the tibia on the femur and checks hyperextension of the knee. It is somewhat lax when the knee is flexed and taut when the knee is extended. The stronger **posterior cruciate ligament** is attached to the *posterior* intercondylar area of the tibia and passes anteriorly, medially, and upward to attach to the lateral side of the medial femoral condyle. This ligament prevents backward displacement of the tibia or forward sliding of the femur.

The knee capsule is heavily reinforced by muscle tendons. Most important are the strong tendons of the quadriceps muscles of the anterior thigh and the tendon of the semimembranosus muscle posteriorly. Since the muscles associated with the joint are the main knee stabilizers, the greater their strength and tone, the less the chance of knee injury.

The knees have a built-in locking device that provides steady support for the body in the standing position. As we begin to stand up, the wheel-shaped femoral condyles roll like ball bearings across the flat condyles of the tibia and the flexed leg begins to extend at the knee. Because the lateral femoral condyle stops rolling before the medial condyle stops, the femur *spins* (rotates) medially on the tibia, until all major ligaments of the knee are twisted and taut and the menisci are compressed. The tension in the ligaments effectively locks the joint into a rigid structure

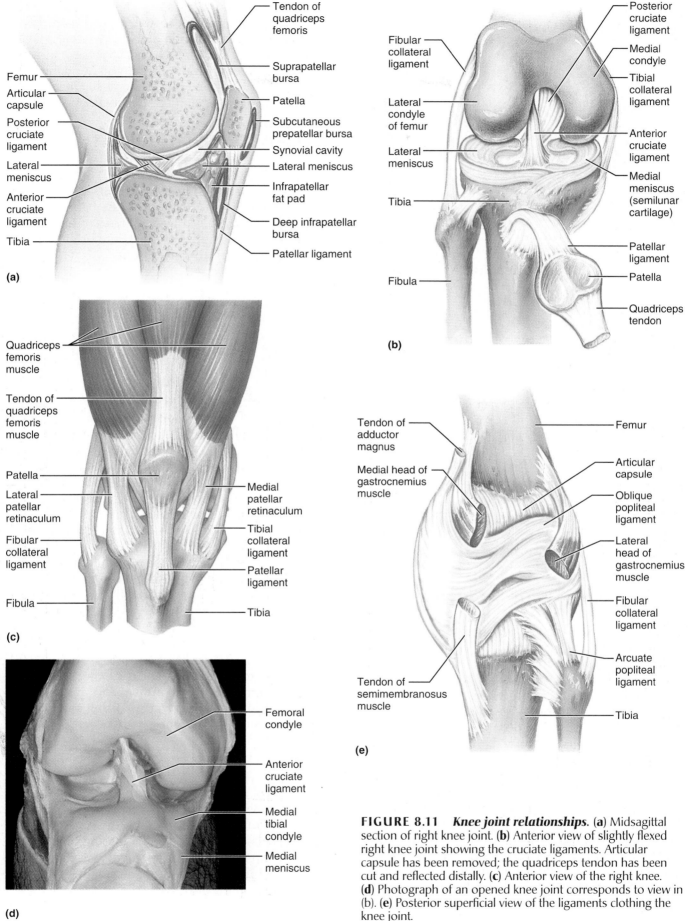

(a)

Tendon of quadriceps femoris
Suprapatellar bursa
Patella
Subcutaneous prepatellar bursa
Synovial cavity
Lateral meniscus
Infrapatellar fat pad
Deep infrapatellar bursa
Patellar ligament

Femur
Articular capsule
Posterior cruciate ligament
Lateral meniscus
Anterior cruciate ligament
Tibia

(b)

Fibular collateral ligament
Lateral condyle of femur
Lateral meniscus
Tibia
Fibula

Posterior cruciate ligament
Medial condyle
Tibial collateral ligament
Anterior cruciate ligament
Medial meniscus (semilunar cartilage)
Patellar ligament
Patella
Quadriceps tendon

(c)

Quadriceps femoris muscle
Tendon of quadriceps femoris muscle
Patella
Lateral patellar retinaculum
Fibular collateral ligament
Fibula

Medial patellar retinaculum
Tibial collateral ligament
Patellar ligament
Tibia

(d)

Femoral condyle
Anterior cruciate ligament
Medial tibial condyle
Medial meniscus

(e)

Tendon of adductor magnus
Medial head of gastrocnemius muscle
Tendon of semimembranosus muscle

Femur
Articular capsule
Oblique popliteal ligament
Lateral head of gastrocnemius muscle
Fibular collateral ligament
Arcuate popliteal ligament
Tibia

FIGURE 8.11 Knee joint relationships. (a) Midsagittal section of right knee joint. **(b)** Anterior view of slightly flexed right knee joint showing the cruciate ligaments. Articular capsule has been removed; the quadriceps tendon has been cut and reflected distally. **(c)** Anterior view of the right knee. **(d)** Photograph of an opened knee joint corresponds to view in (b). **(e)** Posterior superficial view of the ligaments clothing the knee joint.

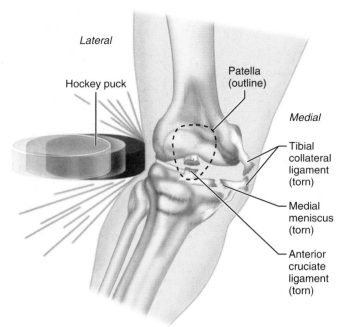

FIGURE 8.12 *A common knee injury.* Anterior view of a knee being hit by a hockey puck. In sports, most knee injuries result from blows to the lateral side. By separating the femur from the tibia medially, such blows tear both the tibial collateral ligament and the medial meniscus because the two are attached. The anterior cruciate ligament also tears.

that cannot be flexed again until it is unlocked. This unlocking is accomplished by the popliteus muscle (see Table 10.15, p. 371), which rotates the femur laterally on the tibia, causing the ligaments to become untwisted and slack.

Homeostatic Imbalance

Of all body joints, the knees are most susceptible to sports injuries because of their high reliance on nonarticular factors for stability and the fact that they carry the body's weight. The knee can absorb a vertical force equal to nearly seven times body weight. However, it is very vulnerable to *horizontal* blows, such as those that occur during blocking and tackling in football. Most dangerous are *lateral* blows to the extended knee. These forces tear the tibial collateral ligament and the medial meniscus attached to it, as well as the anterior cruciate ligament (Figure 8.12). It is estimated that 50% of all professional football players have serious knee injuries during their careers.

Although less devastating than the injury just described, injuries that affect only the anterior cruciate ligament (ACL) are becoming more common,

particularly as women's sports become more vigorous and competitive. Most ACL injuries occur when a runner changes direction quickly, twisting a hyperextended knee. A torn ACL heals poorly, so repair usually requires a ligament graft using connective tissue taken from one of the larger ligaments (e.g., patellar, Achilles, or semitendinosus). ∎

HOMEOSTATIC IMBALANCES OF JOINTS

Few of us pay attention to our joints unless something goes wrong with them. Joint pain and malfunction can be caused by a number of factors, but most joint problems result from injuries and inflammatory or degenerative conditions.

Common Joint Injuries

For most of us, sprains and dislocations are the most common trauma-induced joint injuries, but cartilage injuries are equally threatening to athletes.

Sprains

In a **sprain,** the ligaments reinforcing a joint are stretched or torn. The lumbar region of the spine, the ankle, and the knee are common sprain sites. Partially torn ligaments will repair themselves, but they heal slowly because ligaments are so poorly vascularized. Sprains tend to be painful and immobilizing. Completely ruptured ligaments require prompt surgical repair because inflammation in the joint will break down the neighboring tissues and turn the injured ligament to "mush." Surgical repair can be difficult: A ligament consists of hundreds of fibrous strands, and sewing one back together has been compared to trying to sew two hairbrushes together.

When important ligaments are too severely damaged to be repaired, they must be removed and replaced with grafts or substitute ligaments. For example, a piece of tendon from a muscle, or woven collagen bands, can be stapled to the articulating bones. Alternatively, carbon fibers are implanted in the torn ligament to form a supporting mesh, which is then invaded by fibroblasts that reconstruct the ligament.

Cartilage Injuries

Many aerobics devotees, encouraged to "feel the burn" during their workout, may feel the snap and pop of their overstressed cartilage instead. Although most cartilage injuries involve tearing of the knee menisci, overuse damage to the articular cartilages of other joints is becoming increasingly common in competitive young athletes.

Cartilage is avascular and it rarely can obtain sufficient nourishment to repair itself; thus, it usually stays torn. Because cartilage fragments (called *loose bodies*) can interfere with joint function by causing the joint to lock or bind, most sports physicians recommend that the central (nonvascular) part of a damaged cartilage be removed. Today, this can be done by **arthroscopic** (ar-thro-skop′ik; "looking into joints") **surgery,** a procedure which enables patients to be out of the hospital the same day. The arthroscope, a small instrument bearing a tiny lens and fiber-optic light source, enables the surgeon to view the joint interior (Figure 8.13), repair a ligament, or remove cartilage fragments through one or more tiny slits, minimizing tissue damage and scarring. Removal of part of a meniscus does not severely impair knee joint mobility, but the joint is definitely less stable.

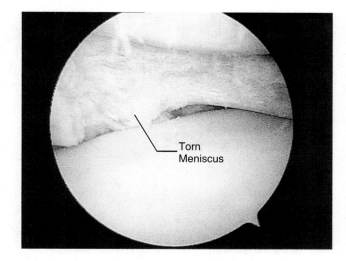

FIGURE 8.13 *Arthroscopic photograph of a torn medial meniscus (Courtesy of the author's tennis game).*

Dislocations

A **dislocation (luxation)** occurs when bones are forced out of alignment. It is usually accompanied by sprains, inflammation, and joint immobilization. Dislocations may result from serious falls and are common contact sports injuries. Joints of the jaw, shoulders, fingers, and thumbs are most commonly dislocated. Like fractures, dislocations must be *reduced;* that is, the bone ends must be returned to their proper positions by a physician. **Subluxation** is a partial dislocation of a joint.

Repeat dislocations of the same joint are common because the initial dislocation stretches the joint capsule and ligaments. The resulting loose capsule provides poor reinforcement for the joint.

Inflammatory and Degenerative Conditions

Inflammatory conditions that affect joints include bursitis, tendonitis, and various forms of arthritis.

Bursitis and Tendonitis

Bursitis is an inflammation of a bursa and is usually caused by a blow or friction. Falling on one's knee may result in a painful bursitis of the prepatellar bursa, known as *housemaid's knee* or *water on the knee.* Prolonged leaning on one's elbows may damage the bursa close to the olecranon process, producing *student's elbow,* or *olecranon bursitis.* Severe cases are treated by injecting anti-inflammatory drugs into the bursa. If excessive fluid accumulates, removing some fluid by needle aspiration may relieve the pressure.

Tendonitis is inflammation of tendon sheaths, typically caused by overuse. Its symptoms (pain and swelling) and treatment (rest, ice, and anti-inflammatory drugs) mirror those of bursitis.

Arthritis

The term **arthritis** describes over 100 different types of inflammatory or degenerative diseases that damage the joints. In all its forms, arthritis is the most widespread crippling disease in the United States. One out of seven Americans suffers its ravages. All forms of arthritis have, to a greater or lesser degree, the same initial symptoms: pain, stiffness, and swelling of the joint.

Acute forms of arthritis usually result from bacterial invasion and are treated with antibiotics. The synovial membrane thickens and fluid production decreases, causing increased friction and pain. Chronic forms of arthritis include osteoarthritis, rheumatoid arthritis, and gouty arthritis.

Osteoarthritis Osteoarthritis **(OA)** is the most common chronic arthritis. A chronic (long-term) degenerative condition, OA is often called "wear-and-tear arthritis." OA is most prevalent in the aged and is probably related to the normal aging process (although it is seen occasionally in younger people and some forms have a genetic basis). More women than men are affected, but 85% of all Americans develop this condition.

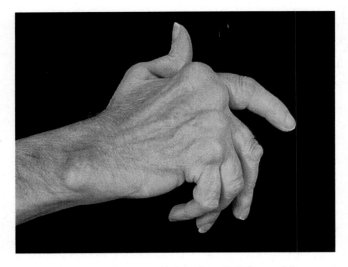

FIGURE 8.14 *Photograph of a hand deformed by rheumatoid arthritis.*

Current theory holds that normal joint use prompts the release of (metalloproteinase) enzymes that break down articular cartilage. In healthy individuals, this damaged cartilage is eventually replaced, but in people with OA, more cartilage is destroyed than replaced. Although its specific cause is unknown, OA may reflect the cumulative effects of years of compression and abrasion acting at joint surfaces, causing excessive amounts of the cartilage-destroying enzymes to be released. The result is softened, roughened, cracked, and eroded articular cartilages.

As the disease progresses, the exposed bone tissue thickens and forms bony spurs that enlarge the bone ends and may restrict joint movement. Patients complain of stiffness on arising that lessens somewhat with activity. The affected joints may make a crunching noise as they move. This sound, called *crepitus* (krep'ĭ-tus), results as the roughened articular surfaces rub together. The joints most often affected are those of the cervical and lumbar spine and the fingers, knuckles, knees, and hips.

The course of osteoarthritis is usually slow and irreversible. In many cases, its symptoms are controllable with a mild pain reliever like aspirin or acetaminophen, along with moderate activity to keep the joints mobile. Magnetic therapy (assumed to stimulate the growth and repair of articular cartilage) is reported to provide significant relief to about 70% of the patients treated. Glucosamine sulfate, a nutritional supplement used by veterinarians to treat arthritis in animals, appears to decrease pain and inflammation and may help to preserve the articular cartilage. Osteoarthritis is rarely crippling, but it can be, particularly when the hip or knee joints are involved.

Although still incompletely documented, another natural remedy sold over the counter as a dietary supplement called SAM-e (for S-adenosylmethionine) has been creating a stir since it became available for sale in the U.S. early in 1999. Although touted for depression, SAM-e's best documented claim is its use for treating arthritis. Instead of promoting gastrointestinal bleeding and destroying cartilage, which all anti-inflammatory medications (ibuprofin and others) do, it builds up cartilage matrix and appears to regenerate the tissue, or at least stems the tide. SAM-e also produces a modest reduction in stiffness and the pain of osteoarthritis.

SAM-e is produced in the body from ATP and the amino acid methionine, which is found in protein-rich foods and is believed to have properties that affect mood and mental functions. Unlike other natural remedies that have become hits without strong scientific research behind them—St. John's wort for depression and glucosamine for arthritis, for example—SAM-e's debut is supported by some top scientists.

Rheumatoid Arthritis Rheumatoid (roo'mah-toid) **arthritis (RA)** is a chronic inflammatory disorder with an insidious onset. Though it usually arises between the ages of 40 and 50, it may occur at any age. It affects three times as many women as men. While not as common as osteoarthritis, rheumatoid arthritis causes disability in millions (Figure 8.14). It occurs in more than 1% of Americans.

In the early stages of RA, joint tenderness and stiffness are common. Many joints, particularly the small joints of the fingers, wrists, ankles, and feet, are afflicted at the same time and bilaterally. For example, if the right elbow is affected, most likely the left elbow is also affected. The course of RA is variable and marked by flare-ups (exacerbations) and remissions (*rheumat* = susceptible to change). Other manifestations include anemia, osteoporosis, muscle atrophy, and cardiovascular problems.

RA is an *autoimmune disease*—a disorder in which the body's immune system attacks its own tissues. The initial trigger for this reaction is unknown, but the streptococcus bacterium and viruses have been suspect. Perhaps these microorganisms bear molecules similar to some naturally present in the joints, and the immune system, once activated, attempts to destroy both.

A CLOSER LOOK

Joints: From Knights in Shining Armor to Bionic Man

There is a stark contrast between the time it took to develop joints for medieval suits of armor and the modern development of joint prostheses (artificial joints). The greatest challenge faced by visionary men of the Middle Ages and Renaissance was to design armor joints that allowed mobility while still protecting the human joints beneath. The ball-and-socket joints that they found so difficult to protect were the first to be engineered by modern-day "visionaries" for implantation into the body.

The history of joint prostheses is less than 60 years old. It dates back to the 1940s and 1950s, when World War II and the Korean War produced large numbers of wounded who needed artificial limbs and joints. Today, over a third of a million American arthritis patients receive total joint replacements each year, mostly because of the destructive effects of osteoarthritis or rheumatoid arthritis.

To produce strong, mobile, lasting joints, a substance was needed that was strong, nontoxic to the body, and resistant to the corrosive effects of organic acids in blood. In 1963, Sir John Charnley, an English orthopedic surgeon, performed the first total hip replacement and revolutionized the therapy of arthritic hips. His device consisted of a metal ball on a stem and a cup-shaped polyethylene plastic socket anchored to the pelvis by methyl methacrylate cement. This cement proved to be exceptionally strong and relatively problem free. Hip prostheses were followed by knee prostheses, but it was not until ten years later that total knee joint replacements that operated smoothly became available. Now, the metal parts of the prostheses are strong cobalt and titanium alloys, and the number of knee replacements equals the number of hip replacements.

Replacements are now available for many other joints, including the fingers, elbows, and shoulders. Total hip and knee replacements last about 10 to 15 years in elderly patients who do not excessively stress the joint. Most such operations are done to reduce pain and restore about 80% of original joint function.

Replacement joints are not yet strong or durable enough for young, active people, but making them so is a major goal of current research. The problem is that the prostheses work loose over time. Effort is being made to minimize this problem by producing a better fit between implant and bone. In fact, the first-ever robot surgeon, named Robodoc, is being developed to drill a better-fitting hole into the top part of the femur's shaft to receive the femoral prosthesis in hip surgery. Ways are also being explored to get the bone to grow so that it binds strongly to the implant (in cementless prostheses), and to improve the strength of the cement in typical prostheses (simply eliminating air bubbles from the cement seems to increase its durability).

Dramatic changes are occurring in the way artificial joints are made. CAD/CAM (computer-aided design and computer-aided manufacturing) techniques are being used to design and create individualized joints (see photo). Fed the patient's X rays and information about the patient's problems, the computer draws from a database of hundreds of normal joints and generates possible designs and modifications for a joint prosthesis. Once the best design is selected, the computer produces a program to direct the machines that shape the joint prosthesis. Time and cost have been dramatically reduced by CAD/CAM computerization.

Joint replacement therapy is coming of age, but perhaps equally exciting is research into the ability of joint tissues to regenerate. We now know that a patient's own chondrocytes cultured in the lab and then grafted onto his or her eroded joint surfaces can produce enough new cartilage to fill small holes and breaks in the articular cartilage of the knee. This offers hope for younger patients, since it could stave off the need for a joint prosthesis for several years.

And so, through the centuries, the focus has shifted from jointed armor to artificial joints that can be put inside the body to restore lost joint function. Modern technology has in fact accomplished what the armor designers of the Middle Ages never even dreamed of.

RA begins with inflammation of the synovial membrane (*synovitis*) of the affected joints. Inflammatory cells (lymphocytes, neutrophils, and others) migrate into the joint cavity from the blood and synovial fluid accumulates, causing joint swelling. In time, the inflamed synovial membrane thickens into a **pannus** ("rag"), an abnormal tissue that clings to the articular cartilages. The pannus erodes the cartilage (and sometimes the underlying bone) and eventually scar tissue forms and connects the bone ends. Later this scar tissue ossifies, and the bone ends fuse together, immobilizing the joint. This end condition, called *ankylosis* (ang'kĭ-lo'sis; "stiff condition"), often produces bent, deformed fingers (see Figure 8.14). Not all cases of RA progress to the severely crippling ankylosis stage, but all cases do involve restriction of joint movement and extreme pain.

A wonder drug for RA sufferers is still undiscovered. Currently the pendulum is swinging from conservative RA therapy utilizing aspirin, long-term antibiotic therapy, and physical therapy to a more progressive treatment course using anti-inflammatory drugs or immunosuppressants. Particularly promising is *Embrel*, the first in a class of drugs called *biologic response modifiers*. It absorbs excess tumor necrosis factor (a chemical released by immune cells that promotes inflammation) in the bloodstream before it can make its way to the joints to do damage. Joint prostheses, if available, are the last resort for severely crippled RA patients (see *A Closer Look*).

Gouty Arthritis Uric acid, a normal waste product of nucleic acid metabolism, is ordinarily excreted in urine without any problems. However, when blood levels of uric acid rise excessively (due to its excessive production or slow excretion), it may be deposited as needle-shaped urate crystals in the soft tissues of joints. An inflammatory response follows, leading into an agonizingly painful attack of **gouty** (gow'te) **arthritis,** or gout. The initial attack typically affects one joint, often at the base of the great toe.

Gout is far more common in males than in females because males naturally have higher blood levels of uric acid. Because gout seems to run in families, genetic factors are definitely implicated.

Untreated gout can be very destructive; the articulating bone ends fuse and immobilize the joint.

Fortunately, several drugs (colchicine, nonsteroidal anti-inflammatory drugs, glucocorticoids, and others) that terminate or prevent gout attacks are available. Patients are advised to avoid alcohol excess (which promotes uric acid overproduction), and foods high in purine-containing nucleic acids, such as liver, kidneys, and sardines.

DEVELOPMENTAL ASPECTS OF JOINTS

As bones form from mesenchyme in the embryo, the joints develop in parallel. By 8 weeks, the synovial joints resemble adult joints in form and arrangement.

Injuries aside, relatively few interferences with joint function occur until late middle age. Eventually, however, advancing years do take their toll. The ligaments and tendons that connect muscles and bones shorten and weaken. The intervertebral discs become more likely to rupture (herniate), and osteoarthritis rears its ugly head. Virtually everyone has osteoarthritis to some degree by the time they are in their 70s. The middle years also see an increased incidence of rheumatoid arthritis.

Exercise that coaxes joints through their full range of motion, such as regular stretching and aerobics, is the key to postponing the immobilizing effects of aging on ligaments and tendons, to keeping cartilages well nourished, and to strengthening the muscles that stabilize the joints. The key word for exercising is "prudently," because excessive or abusive use of the joints guarantees early onset of osteoarthritis. The buoyancy of water relieves much of the stress on weight-bearing joints, and people who swim or exercise in a pool often retain good joint function as long as they live. As with so many medical problems, it is easier to prevent joint problems than to cure or correct them.

* * *

The importance of joints is obvious: The skeleton's ability to protect other organs and to move smoothly reflects their presence. Now that we are familiar with joint structure and with the movements that joints allow, we are ready to consider how the muscles attached to the skeleton cause body movements by acting across its joints.

RELATED CLINICAL TERMS

Ankylosing spondylitis (ang′kĭ-lōz″ing spon″dĭ-li′tis; *ankyl* = crooked, bent; *spondyl* = vertebra) A variant of rheumatoid arthritis that chiefly affects males; it usually begins in the sacroiliac joints and progresses superiorly along the spine. The vertebrae become interconnected by fibrous tissue, causing the spine to become rigid ("poker back").

Arthrology (ar-throl′o-je; *logos* = study) The study of joints.

Chondromalacia (kon-dro-mal-a′sĭ-ah) **patellae** ("softening of cartilage by the patella") Damage and softening of the articular cartilages on the posterior patellar surface and the anterior surface of the distal femur; most often seen in adolescent athletes. Produces a sharp pain in the knee when the leg is extended (in climbing stairs, for example). May result when the quadriceps femoris, the main group of muscles on the anterior thigh, pulls unevenly on the patella, persistently rubbing it against the femur in the knee joint; often corrected by exercises that strengthen weakened parts of the quadriceps muscles.

Rheumatism A term used by laypeople to indicate disease involving muscle or joint pain; consequently may be used to apply to arthritis, bursitis, etc.

Synovitis (sin″o-vi′tis) Inflammation of the synovial membrane of a joint. In healthy joints, only small amounts of synovial fluid are present, but synovitis causes copious amounts to be produced, leading to swelling and limitation of joint movement.

CHAPTER SUMMARY

Media study tools that could provide you additional help in reviewing specific key topics of Chapter 8 are referenced below.

1. Joints, or articulations, are sites where bones meet. Their functions are to hold bones together and to allow various degrees of skeletal movement.

Classification of Joints   (p. 249)

1. Joints are classified structurally as fibrous, cartilaginous, or synovial. They are classed functionally as synarthrotic, amphiarthrotic, or diarthrotic.

SP Exercise: Structure and Function of Body Joints.

Fibrous Joints   (pp. 249–250)

1. Fibrous joints occur where bones are connected by fibrous tissue; no joint cavity is present. Nearly all fibrous joints are synarthrotic.

2. **Sutures/syndesmoses/gomphoses.** The major types of fibrous joints are sutures, syndesmoses, and gomphoses.

Cartilaginous Joints   (p. 250)

1. In cartilaginous joints, the bones are united by cartilage; no joint cavity is present.

2. **Synchondroses/symphyses.** Cartilaginous joints include synchondroses and symphyses. Synchondroses are synarthrotic; all symphyses are amphiarthrotic.

Synovial Joints   (pp. 251–268)

1. Most body joints are synovial joints, all of which are diarthrotic.

General Structure   (pp. 251–253)

2. All synovial joints have a joint cavity enclosed by a fibrous capsule lined with synovial membrane and reinforced by ligaments, articulating bone ends covered with articular cartilage, and synovial fluid in the joint cavity. Some (e.g., the knee) contain fibrocartilage discs that absorb shock.

SP Exercise: General Structure of a Synovial Joint.

Bursae and Tendon Sheaths   (pp. 253–254)

3. Bursae are fibrous sacs lined with synovial membrane and containing synovial fluid. Tendon sheaths are similar to bursae but are cylindrical structures that surround muscle tendons. Both allow adjacent structures to move smoothly over one another.

Factors Influencing the Stability of Synovial Joints   (p. 254)

4. Articular surfaces providing the most stability have large surfaces and deep sockets and fit snugly together.

5. Ligaments prevent undesirable movements and reinforce the joint.

6. The tone of muscles whose tendons cross the joint is the most important stabilizing factor in many joints.

Movements Allowed by Synovial Joints   (pp. 254–258)

7. When a skeletal muscle contracts, the insertion (movable attachment) moves toward the origin (immovable attachment). Three common types of movements can occur when muscles contract across joints: (a) gliding movements, (b) angular movements (which include flexion, extension, abduction, adduction, and circumduction), and (c) rotation.

8. Special movements include supination and pronation, inversion and eversion, protraction and retraction, elevation and depression, and opposition.

SP Exercise: Special Body Movements.

Types of Synovial Joints   (pp. 258–262)

9. Synovial joints differ in their range of motion. Motion may be nonaxial (gliding), uniaxial (in one plane), biaxial (in two planes), or multiaxial (in all three planes).

10. The six major categories of synovial joints are plane joints (movement nonaxial), hinge joints (uniaxial), pivot joints (uniaxial, rotation permitted), condyloid joints (biaxial with angular movements in two planes), saddle joints (biaxial, like condyloid joints, but with freer movement), and ball-and-socket joints (multiaxial and rotational movement).

Selected Synovial Joints   (pp. 262–268)

11. The shoulder joint is a ball-and-socket joint formed by the glenoid cavity of the scapula and the humeral head. The most freely movable joint of the body, it allows all angular and rotational movements. Its articular surfaces are shallow. Its capsule is lax and poorly reinforced by ligaments. The tendons of the biceps brachii and rotator cuff muscles help to stabilize it.

12. The hip joint is a ball-and-socket joint formed by the acetabulum of the coxal bone and the femoral head. It is highly adapted for weight bearing. Its articular surfaces are deep and secure. Its capsule is heavy and strongly reinforced by ligaments.

13. The elbow is a hinge joint in which the ulna (and radius) articulates with the humerus, allowing flexion and extension. Its articular surfaces are highly complementary and are the most important factor contributing to joint stability.

14. The knee joint is the largest joint in the body. It is a hinge joint formed by the articulation of the tibial and femoral condyles (and anteriorly by the patella and patellar surface of the femur). Extension, flexion, and (some) rotation are allowed. Its articular surfaces are shallow and condyloid. C-shaped menisci deepen the articular surfaces. The joint cavity is enclosed by a capsule only on the sides and posterior aspects. Several extracapsular ligaments and the intracapsular anterior and posterior cruciate ligaments help prevent displacement of the joint surfaces. Muscle tone of the quadriceps and semimembranosus muscles is important in knee stability.

Homeostatic Imbalances of Joints (pp. 268–272)

Common Joint Injuries (pp. 268–269)

1. Sprains involve stretching or tearing of joint ligaments. Because ligaments are poorly vascularized, healing is slow.

2. Cartilage injuries, particularly of the knee, are common in contact sports and may result from excessive twisting or high pressure. The avascular cartilage is unable to repair itself.

3. Dislocations involve displacement of the articular surfaces of bones. They must be reduced.

Inflammatory and Degenerative Conditions (pp. 269–272)

4. Bursitis and tendonitis are inflammations of a bursa and a tendon sheath, respectively.

5. Arthritis is joint inflammation or degeneration accompanied by stiffness, pain, and swelling. Acute forms generally result from bacterial infection. Chronic forms include osteoarthritis, rheumatoid arthritis, and gouty arthritis.

6. Osteoarthritis is a degenerative condition most common in the aged. Weight-bearing joints are most affected.

7. Rheumatoid arthritis, the most crippling arthritis, is an autoimmune disease involving severe inflammation of the joints.

8. Gouty arthritis, or gout, is joint inflammation caused by the deposit of urate salts in soft joint tissues.

Developmental Aspects of Joints (p. 272)

1. Joints form from mesenchyme and in tandem with bone development in the embryo.

2. Excluding traumatic injury, joints usually function well until late middle age, at which time symptoms of connective tissue stiffening and osteoarthritis begin to appear. Prudent exercise delays these effects, whereas excessive exercise promotes the early onset of arthritis.

SP Case Study: Craniosynostosis.

REVIEW QUESTIONS

Multiple Choice/Matching

1. Match the key terms to the appropriate descriptions.

Key: **(a)** fibrous joints   **(b)** cartilaginous joints
     (c) synovial joints

_____ (1) exhibit a joint cavity

_____ (2) types are sutures and syndesmoses

_____ (3) bones connected by collagen fibers

_____ (4) types include synchondroses and symphyses

_____ (5) all are diarthrotic

_____ (6) many are amphiarthrotic

_____ (7) bones connected by a disc of hyaline cartilage or fibrocartilage

_____ (8) nearly all are synarthrotic

_____ (9) shoulder, hip, jaw, and elbow joints

2. Freely movable joints are (a) synarthroses, (b) diarthroses, (c) amphiarthroses.

3. Anatomical characteristics of a synovial joint include (a) articular cartilage, (b) a joint cavity, (c) an articular capsule, (d) all of these.

4. Factors that influence the stability of a synovial joint include (a) shape of articular surfaces, (b) presence of strong reinforcing ligaments, (c) tone of surrounding muscles, (d) all of these.

5. The following description—"Articular surfaces deep and secure; capsule heavily reinforced by ligaments and muscle tendons; extremely stable joint"—best describes (a) the elbow joint, (b) the hip joint, (c) the knee joint, (d) the shoulder joint.

6. Ankylosis means (a) twisting of the ankle, (b) tearing of ligaments, (c) displacement of a bone, (d) immobility of a joint due to fusion of its articular surfaces.

7. An autoimmune disorder in which joints are affected bilaterally and which involves pannus formation and gradual joint immobilization is (a) bursitis, (b) gout, (c) osteoarthritis, (d) rheumatoid arthritis.

Short Answer Essay Questions

8. Define joint.

9. Discuss the relative value (to body homeostasis) of immovable, slightly movable, and freely movable joints.

10. Compare the structure, function, and common body locations of bursae and tendon sheaths.

11. Joint movements may be nonaxial, uniaxial, biaxial, or multiaxial. Define what each of these terms means.

12. Compare and contrast the paired movements of flexion and extension with adduction and abduction.

13. How does rotation differ from circumduction?

14. Name two types of uniaxial, biaxial, and multiaxial joints.

15. What is the specific role of the menisci of the knee? of the anterior and posterior cruciate ligaments?

16. The knee has been called "a beauty and a beast." Provide several reasons that might explain the negative (beast) part of this description.

17. Why are sprains and cartilage injuries a particular problem?

18. List the functions of the following elements of a synovial joint: fibrous part of the capsule, synovial fluid, articular disc.

 Critical Thinking and Clinical Application Questions

1. Sophie worked as a cleaning woman for 30 years so she could send her two children to college. Several times, she had been forced to call her employers to tell them she could not come in to work because one of her kneecaps was swollen and painful. What is Sophie's condition, and what probably caused it?

2. As Harry was jogging down the road, he tripped and his left ankle twisted violently to the side. When he picked himself up, he was unable to put any weight on that ankle. The diagnosis was severe dislocation and sprains of the left ankle. The orthopedic surgeon stated that she would perform a closed reduction of the dislocation and attempt ligament repair by using arthroscopy. (a) Is the ankle joint normally a stable joint? (b) What does its stability depend on? (c) What is a closed reduction? (d) Why is ligament repair necessary? (e) What does arthroscopy entail? (f) How will the use of this procedure minimize Harry's recuperation time (and suffering)?

3. Mrs. Bell, a 45-year-old woman, appeared at her physician's office complaining of unbearable pain in the distal interphalangeal joint of her right great toe. The joint was red and swollen. When asked about previous episodes, she recalled a similar attack two years earlier that disappeared as suddenly as it had come. Her diagnosis was arthritis. (a) What type? (b) What is the precipitating cause of this particular type of arthritis?

9 MUSCLES AND MUSCLE TISSUE

Because flexing muscles look like mice scurrying beneath the skin, some scientist long ago dubbed them *muscles,* from the Latin word *mus* meaning "little mouse." Indeed, the rippling muscles of professional boxers or weight lifters are often the first thing that comes to mind when one hears the word *muscle.* But muscle is also the dominant tissue in the heart and in the walls of other hollow organs of the body. In all its forms, it makes up nearly half the body's mass. Its most distinguishing functional characteristic is its ability to transform chemical energy (in the form of ATP) into directed mechanical energy. In so doing, muscle becomes capable of exerting force.

OVERVIEW OF MUSCLE TISSUES

Muscle Types

The three types of muscle tissue are *skeletal, cardiac,* and *smooth.* These muscle tissues differ in the structure of their cells, their body location, their function, and the means by which they are activated to contract. Before exploring their differences, however, let us look at some of their similarities. First, skeletal and smooth muscle cells (but not cardiac muscle cells) are elongated and, for this reason, are called **muscle fibers.** Second, muscle contraction depends on two kinds of **myofilaments,** muscle equivalents of the actin- or myosin-containing microfilaments described in Chapter 3. As you will recall, these two proteins play roles in motility and shape changes in virtually every cell in the body, but this property reaches its highest development in the contractile muscle fibers. The third and final similarity concerns terminology: Whenever you see the prefixes **myo** or **mys** (both are word roots meaning "muscle") and **sarco** (flesh), the reference is to muscle. For example, the plasma membrane of muscle fibers is called the *sarcolemma* (sar"ko-lem'ah), literally "muscle" (sarco) "husk" (lemma), and muscle fiber cytoplasm is called *sarcoplasm.* Now we are ready to describe the three types of muscle tissue individually.

Skeletal muscle tissue is packaged into the *skeletal muscles* that attach to and cover the bony skeleton. Skeletal muscle fibers are the longest of the muscle cell types, have obvious stripes called **striations,** and can be controlled voluntarily. Although it is often activated by reflexes without our willed "command," skeletal muscle is called **voluntary muscle** because it is the *only type subject to* conscious control. When you think of skeletal muscle tissue, the key words to keep in mind are *skeletal, striated,* and *voluntary.* Skeletal muscle can contract rapidly, but it tires easily and must rest after short

periods of activity. These muscles, which are responsible for overall body mobility, can exert tremendous power, a fact revealed by reports of people lifting cars to save their loved ones. Skeletal muscle is also remarkably adaptable. For example, your hand muscles can exert a force of a fraction of an ounce to pick up a dropped paper clip and the same muscles can exert a force of about 70 pounds to pick up this book!

Cardiac muscle tissue occurs only in the heart (the body's blood pump), where it constitutes the bulk of the heart walls. Like skeletal muscle, cardiac muscle is striated, but it is not voluntary. Most of us have no conscious control over how fast our heart beats. Key words to remember for this muscle type are *cardiac, striated,* and *involuntary.* Cardiac muscle usually contracts at a fairly steady rate set by the heart's pacemaker, but neural controls allow the heart to "shift into high gear" for brief periods, as when you race across the tennis court to make that overhead smash.

Smooth muscle tissue is found in the walls of hollow visceral organs, such as the stomach, urinary bladder, and respiratory passages. Its role is to force fluids and other substances through internal body channels. It has no striations, and, like cardiac muscle, it is not subject to voluntary control. We can describe smooth muscle tissue most precisely as *visceral, nonstriated,* and *involuntary.* Contractions of smooth muscle fibers are slow and sustained. If skeletal muscle is like a speedy windup car that quickly runs down, then smooth muscle is like a steady, heavy-duty engine that lumbers along tirelessly.

Skeletal muscle and smooth muscle are discussed in this chapter. Cardiac muscle is discussed in Chapter 19 (The Heart). Table 9.4 on pp. 310–311 summarizes the most important characteristics of each type of muscle tissue.

Muscle Functions

Muscle performs four important functions for the body: It produces movement, maintains posture, stabilizes joints, and generates heat.

Producing Movement

Just about all movements of the human body and its parts are a result of muscle contraction. Skeletal muscles are responsible for all locomotion and manipulation, and they enable you to respond quickly to changes in the external environment. For example, they enable you to jump out of the way of a runaway car, direct your eyeballs, and express joy or rage with facial expressions.

The coursing of blood through your body reflects the work of the rhythmically beating cardiac muscle of your heart and the smooth muscle in the walls of your blood vessels, which helps maintain blood pressure. Smooth muscle in organs of the digestive, urinary, and reproductive tracts propels, or squeezes, substances (foodstuffs, urine, a baby) through the organs and along the tract.

Maintaining Posture

We are rarely aware of the workings of the skeletal muscles that maintain body posture. Yet these muscles function almost continuously, making one tiny adjustment after another that enables us to maintain an erect or seated posture despite the never-ending downward pull of gravity.

Stabilizing Joints

Even as muscles pull on bones to cause movements, they stabilize the joints of the skeleton. As described in Chapter 8, skeletal muscles help stabilize and strengthen joints.

Generating Heat

Finally, because no "machine" operates with perfect efficiency, muscles generate heat as they contract. This heat is vitally important in maintaining normal body temperature. Since skeletal muscle accounts for at least 40% of body mass, it is the muscle type most responsible for generating heat.

Functional Characteristics of Muscle

Muscle tissue is endowed with some special functional properties that enable it to perform its duties. These properties include excitability, contractility, extensibility, and elasticity.

Excitability, or **irritability,** is the ability to receive and respond to a stimulus. (A *stimulus* is an environmental change; it may arise either inside or outside the body.) In the case of muscle, the stimulus is usually a chemical—for example, a neurotransmitter released by a nerve cell, a hormone, or a local change in pH. The response is generation of an electrical impulse that passes along the sarcolemma and causes the muscle cells to contract.

Contractility is the ability to shorten forcibly when adequately stimulated. This property sets muscle apart from all other tissue types.

Extensibility is the ability to be stretched or extended. Muscle fibers shorten when contracting, but they can be stretched, even beyond their resting length, when relaxed.

Elasticity is the ability of a muscle fiber to recoil and resume its resting length after being stretched.

In the following section, we examine the structure and functioning of skeletal muscle in detail. Then we consider smooth muscle more briefly, largely by comparing it with skeletal muscle. Because cardiac muscle is described in Chapter 19, its treatment in this chapter is limited to the summary of its characteristics provided in Table 9.4.

SKELETAL MUSCLE

The levels of skeletal muscle organization, gross to microscopic, are summarized in Table 9.1 on p. 282.

Gross Anatomy of a Skeletal Muscle

Each **skeletal muscle** is a discrete organ, made up of several kinds of tissues. Although muscle fibers predominate, blood vessels, nerve fibers, and substantial amounts of connective tissue are also present. A skeletal muscle's shape and its attachments within the body can be examined easily without the help of a microscope.

Connective Tissue Wrappings

In an intact muscle, the individual muscle fibers are wrapped and held together by several different layers of connective tissue. We will consider these sheaths from external to internal (Figure 9.1).

1. Epimysium. An "overcoat" of dense irregular connective tissue surrounds the whole muscle. This coat is the epimysium (ep"i-mis'e-um), a name that means "outside the muscle." Sometimes, the epimysium blends with the deep fascia that lies between neighboring muscles.

2. Perimysium and **fascicles.** Within each skeletal muscle, the muscle fibers are grouped into fascicles (fas'i-klz; "bundles") that resemble bundles of sticks. Surrounding each fascicle is a layer of fibrous connective tissue called perimysium (per"i-mis'e-um; "around the muscle [fascicles]").

3. Endomysium. Within a fascicle, each muscle fiber is surrounded by a fine sheath of connective tissue consisting mostly of reticular fibers. This is the endomysium (en"do-mis'e-um; "within the muscle").

As shown in Figure 9.1, all of these connective tissue sheaths are continuous with one another as well as with the tendons that join muscles to bones. Therefore, when muscle fibers contract, they pull on these sheaths, which in turn transmit the force to the bone to be moved.

The connective tissue coverings support each cell, reinforce the muscle as a whole, and contribute

Consider the term epimysium. *What is the meaning of* epi? *Of* mys? *How do these word stems relate to the role and position of this connective tissue sheath?*

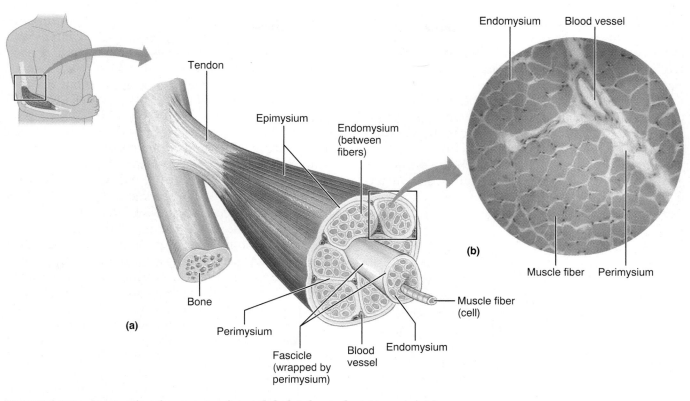

FIGURE 9.1 *Connective tissue wrappings of skeletal muscle.* **(a)** In a skeletal muscle, each muscle fiber is wrapped with a delicate connective tissue sheath, the endomysium. Bundles, or fascicles, of muscle fibers are bound by a collagenic sheath called a perimysium. The entire muscle is strengthened and wrapped by a coarse epimysium sheath. **(b)** Photomicrograph of a cross section of part of a skeletal muscle (73×).

to the natural elasticity of muscle tissue. These sheaths also provide entry and exit routes for the blood vessels and nerve fibers that serve the muscle.

Nerve and Blood Supply

The normal activity of a skeletal muscle is absolutely dependent on nerves and its rich blood supply. In general, each muscle is served by one nerve, an artery, and one or more veins, all of which enter or exit near the central part of the muscle and branch profusely through its connective tissue septa. Unlike cells of the other muscle tissues, which can contract in the total absence of nerve stimulation, each skeletal muscle fiber is supplied with a nerve ending that controls its activity. The structure of the nerve-muscle junction is discussed shortly.

Contracting muscle fibers use huge amounts of energy, which requires a more or less continuous delivery of oxygen and nutrients by way of arteries. Muscle cells also give off large amounts of metabolic wastes that must be removed through veins if contraction is to remain efficient. Muscle capillaries, the smallest of the blood vessels, are long and winding, a feature that accommodates changes in muscle length. They straighten when the muscle is stretched and contort when the muscle contracts.

Attachments

Recall from Chapter 8 that (1) most muscles span joints and are attached to bones (or other structures) in at least two places and (2) when a muscle contracts, the movable bone, the muscle's **insertion,** moves toward the immovable or less movable bone, the muscle's **origin.** In the muscles of the limbs, the origin typically lies proximal to the insertion.

Muscle attachments, whether origin or insertion, may be direct or indirect. In **direct,** or **fleshy, attachments,** the epimysium of the muscle is fused to the periosteum of a bone or perichondrium of a

epi = *outer,* mys = *muscle. The epimysium covers the external surface of the muscle.*

cartilage. In **indirect attachments,** the muscle's connective tissue wrappings extend beyond the muscle as a ropelike tendon or a sheetlike **aponeurosis** (ap"o-nu-ro'sis). The tendon or aponeurosis anchors the muscle to the connective tissue covering of a skeletal element (bone or cartilage) or to the fascia of other muscles. The temporalis muscle of the head has both direct and indirect (tendinous) attachments (see Figure 10.6, p. 335).

Of the two, indirect attachments are much more common in the body because of their durability and small size. Since tendons are mostly tough collagenic fibers, they can cross rough bony projections which would tear apart the more delicate muscle tissues. Because of their relatively small size, more tendons than fleshy muscles can pass over a joint—thus, tendons also conserve space.

Microscopic Anatomy of a Skeletal Muscle Fiber

Each skeletal muscle fiber is a long cylindrical cell with multiple oval nuclei arranged just beneath its **sarcolemma** or plasma membrane surface (see Figure 9.2a and Table 9.1 on p. 282). Skeletal muscle fibers are huge cells. Their diameter typically ranges from 10 to 100 μm—up to ten times that of an average body cell—and their length is phenomenal, some hundreds of centimeters long. Their large size and multinucleate condition are less surprising once you learn that each skeletal muscle fiber is actually a *syncytium* (sin-sish'e-um; "fused cells") produced by the fusion of hundreds of embryonic cells.

The **sarcoplasm** of a muscle fiber is similar to the cytoplasm of other cells, but it contains unusually large amounts of *glycosomes* (stored glycogen) and a unique oxygen-binding protein called *myoglobin* that is not found in other cell types. **Myoglobin,** a red pigment that stores oxygen within the muscle cells, is similar to hemoglobin, the pigment that transports oxygen in blood. The usual organelles are present, as well as some that are highly modified in muscle fibers. These are the myofibrils and sarcoplasmic reticulum. The T tubules are unique modifications of the muscle fiber's sarcolemma. Let's look at these special structures more closely.

Myofibrils

When viewed at high magnification, each muscle fiber is seen to contain a large number of rodlike **myofibrils** that run in parallel fashion and extend the entire length of the cell (see Figure 9.2b). With a diameter of 1–2 μm each, the myofibrils are so densely packed that mitochondria and other organelles appear to be squeezed between them. Hundreds to thousands of myofibrils are in a single mus-

cle fiber, depending on its size, and they account for about 80% of cellular volume. The myofibrils are the contractile elements of skeletal muscle cells.

Striations, Sarcomeres, and Myofilaments
Striations, a repeating series of dark **A bands** and light **I bands,** are evident along the length of each myofibril. In an intact muscle fiber, the A and I bands are nearly perfectly aligned with one another and this gives the cell as a whole its striped (striated) appearance.

As illustrated in Figure 9.2c, each A band has a lighter stripe in its midsection called the **H zone** (*H* stands for *helle,* which means "bright"). The H zones are visible only in relaxed muscle fibers for reasons that will soon become obvious. To make things a bit more complicated, each H zone is bisected by a dark line called the **M line.** The I bands also have a midline interruption, a darker area called the **Z disc.** A **sarcomere** (sar'ko-měr; literally, "muscle segment") is the region of a myofibril between two successive Z discs (see Figure 9.2c). Averaging 2 μm long, the sarcomere is the smallest contractile unit of a muscle fiber. Thus, the *functional units* of skeletal muscle are actually very minute portions of the myofibrils, and the myofibrils are chains of sarcomeres aligned end-to-end like boxcars in a train.

If we examine the *banding pattern* of a myofibril at the molecular level, we see that it arises from an orderly arrangement of two types of even smaller structures, called **filaments** or **myofilaments,** within the sarcomeres (Figure 9.2d). The central **thick filaments** extend the entire length of the A band. The more lateral **thin filaments** extend across the I band and partway into the A band. The Z disc, also called a *Z line,* is a coin-shaped sheet composed of proteins called *connectins.* The Z disc anchors the thin filaments and also connects each myofibril to the next throughout the width of the muscle cell. The H zone of the A band appears less dense because the thin filaments do not overlap the thick ones in this region. The M line in the center of the H zone is slightly darker because of the presence of fine protein (*desmin*) strands that hold adjacent thick filaments together in that area. The third type of filament illustrated in Figure 9.2d, the *elastic filament,* is described in the next section.

A longitudinal view of the myofilaments, such as that in Figure 9.2d, is a bit misleading because it gives the impression that each thick filament interdigitates with only four thin filaments. In areas where thick and thin filaments overlap, each thick filament is actually surrounded by a hexagonal arrangement of six thin filaments.

Ultrastructure and Molecular Composition of the Myofilaments
Thick filaments (about 16 nm in diameter) are composed primarily of the protein

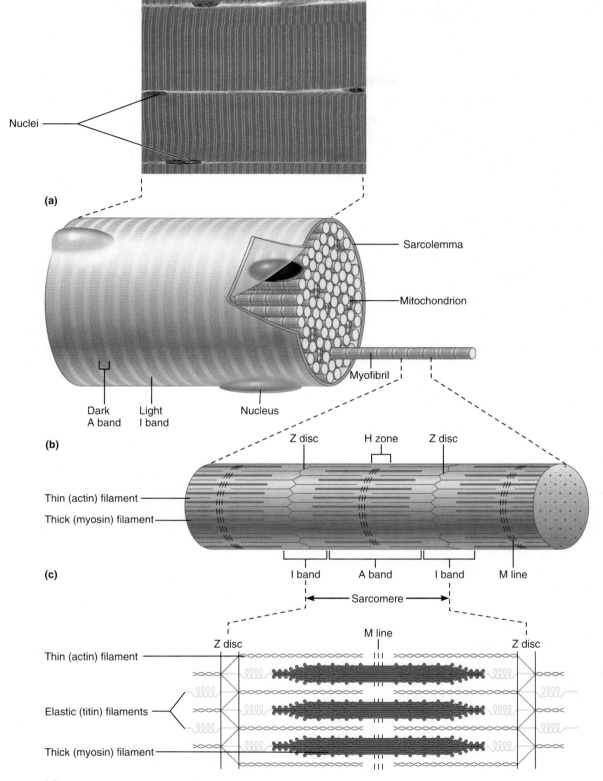

FIGURE 9.2 *Microscopic anatomy of a skeletal muscle fiber.*
(a) Photomicrograph of portions of two isolated muscle fibers (540×). Notice the obvious cross striations (alternating light and dark bands). (b) Diagram of part of a muscle fiber (cell) showing the myofibrils it contains. One myofibril is shown extending from the cut end of the muscle fiber. (c) A small portion of one myofibril is enlarged to show the myofilaments responsible for the banding pattern. Each sarcomere, or contractile unit, extends from one Z disc to the next. (d) Enlargement of one sarcomere (sectioned lengthwise). Notice the myosin heads on the thick filaments.

TABLE 9.1	Structure and Organizational Levels of Skeletal Muscle	
Structure and organizational level	*Description*	*Connective tissue wrappings*
Muscle (organ) 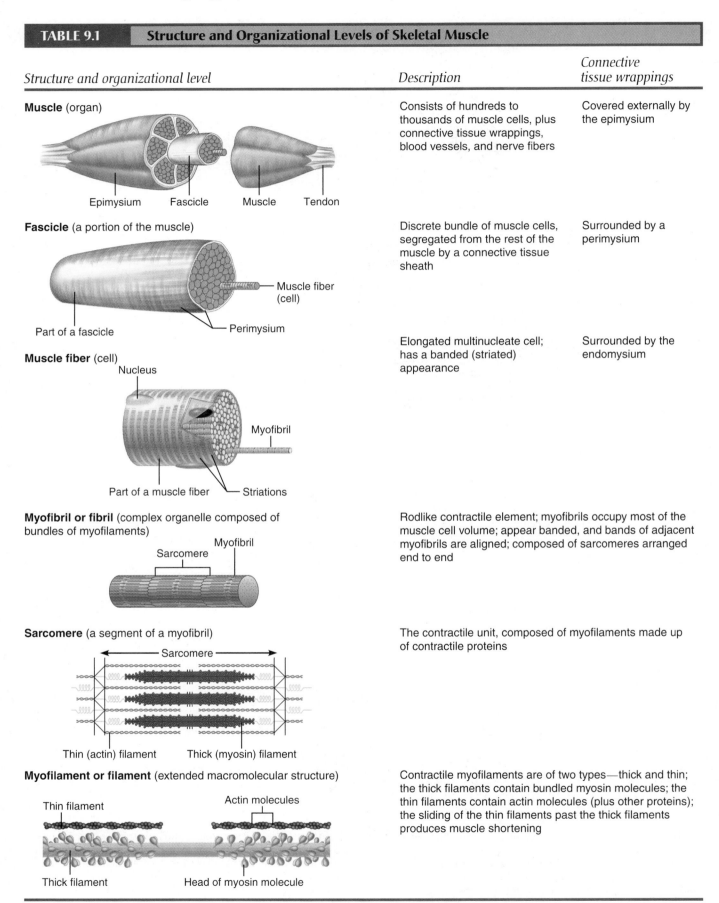 Epimysium Fascicle Muscle Tendon	Consists of hundreds to thousands of muscle cells, plus connective tissue wrappings, blood vessels, and nerve fibers	Covered externally by the epimysium
Fascicle (a portion of the muscle) Part of a fascicle Muscle fiber (cell) Perimysium	Discrete bundle of muscle cells, segregated from the rest of the muscle by a connective tissue sheath	Surrounded by a perimysium
Muscle fiber (cell) Nucleus Myofibril Part of a muscle fiber Striations	Elongated multinucleate cell; has a banded (striated) appearance	Surrounded by the endomysium
Myofibril or fibril (complex organelle composed of bundles of myofilaments) Sarcomere Myofibril	Rodlike contractile element; myofibrils occupy most of the muscle cell volume; appear banded, and bands of adjacent myofibrils are aligned; composed of sarcomeres arranged end to end	
Sarcomere (a segment of a myofibril) Sarcomere Thin (actin) filament Thick (myosin) filament	The contractile unit, composed of myofilaments made up of contractile proteins	
Myofilament or filament (extended macromolecular structure) Thin filament Actin molecules Thick filament Head of myosin molecule	Contractile myofilaments are of two types—thick and thin; the thick filaments contain bundled myosin molecules; the thin filaments contain actin molecules (plus other proteins); the sliding of the thin filaments past the thick filaments produces muscle shortening	

myosin (Figure 9.3a). Each myosin molecule has a rodlike *tail* terminating in two globular *heads*. The tail consists of two interwoven *heavy* polypeptide chains; its heads are the ends of the heavy chains, each made "fat" by the attachment of two smaller *light* polypeptide chains. Sometimes called **cross bridges,** the heads are the "business end" of myosin because they link the thick and thin myofilaments together during contraction. As explained shortly, these cross bridges generate the tension developed by a muscle cell when it contracts. Each thick filament within a sarcomere contains about 200 myosin molecules. As shown in Figure 9.3b and d, the myosin molecules are bundled together with their tails forming the central part of the filament and their heads facing outward and in opposite directions at each end. As a result, the central portion of a thick filament is smooth, but its ends are studded with a staggered array of myosin heads. Besides bearing actin binding sites, the heads of myosin molecules contain ATP binding sites and ATPase enzymes that split ATP to generate energy for muscle contraction.

The thin filaments (7–8 nm thick) are composed chiefly of **actin** (Figure 9.3c). The polypeptide subunits of actin, called *globular actin* or *G actin,* bear the active sites to which the myosin cross bridges attach during contraction. The generally kidney bean-shaped G actin monomers are polymerized into long actin filaments. The backbone of each thin filament appears to be formed by an actin filament that coils back on itself, forming a helical structure that looks like a twisted double strand of pearls. Several regulatory proteins are also present in the thin filament. Two strands of **tropomyosin** (tro"po-mi'o-sin), a rod-shaped protein, spiral about the actin core and help to stiffen it. Successive tropomyosin molecules are arranged end-to-end along the actin filaments, and in a relaxed muscle fiber, they block actin's active sites so that the myosin heads cannot bind to the thin filaments. The other major protein in the thin filament, **troponin** (tro'po-nin), is actually a three-polypeptide complex. One of these

FIGURE 9.3 *Myofilament composition in skeletal muscle.* **(a)** An individual myosin molecule has a stalklike tail, from which two "heads" protrude. **(b)** Each thick filament consists of many myosin molecules, whose heads protrude at opposite ends of the filament, as shown in (d). **(c)** A thin filament contains a strand of actin twisted into a helix. Each strand is made up of G actin subunits. Tropomyosin molecules coil around the actin filament, helping to reinforce them. A troponin complex is attached to each tropomyosin molecule. **(d)** Arrangement of the filaments in a sarcomere (longitudinal view). In the center of the sarcomere, the thick elements are devoid of myosin heads, but at points of thick and thin filament overlap, the heads extend toward the actin, with which they interact during contraction. **(e)** Transmission electron micrograph of part of a sarcomere clearly showing the myosin heads (cross bridges) that generate the contractile force.

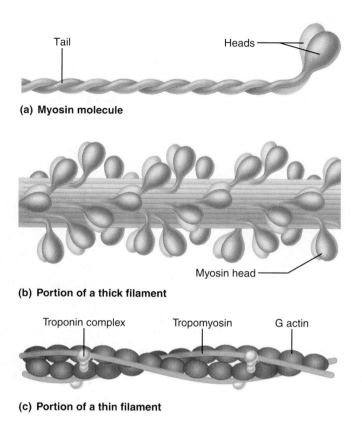

Tail   Heads

(a) Myosin molecule

Myosin head

(b) Portion of a thick filament

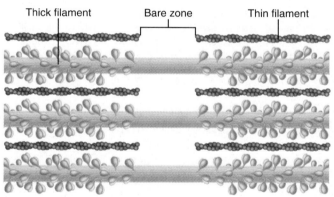

Troponin complex   Tropomyosin   G actin

(c) Portion of a thin filament

Thick filament   Bare zone   Thin filament

(d) Longitudinal section of filaments within one sarcomere of a myofibril

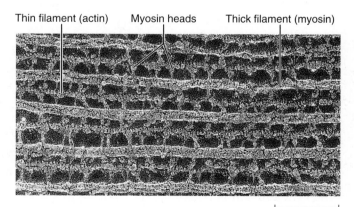

Thin filament (actin)   Myosin heads   Thick filament (myosin)

100 nm

(e) Transmission electron micrograph of part of a sarcomere

? *Which of the following structures shown in this diagram would contain the highest concentration of Ca²⁺ in a resting muscle cell? T tubule, mitochondrion, or SR?*

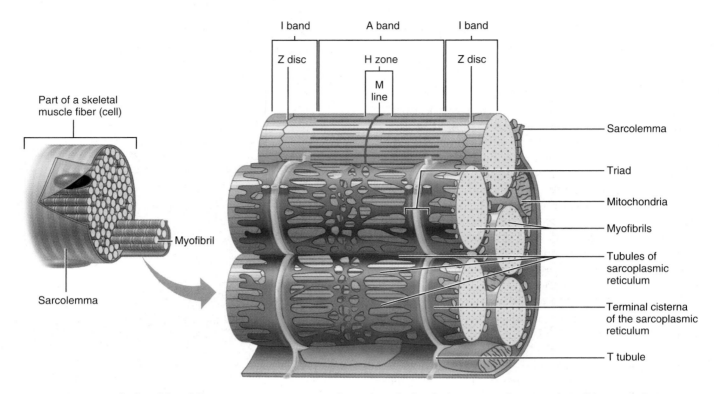

FIGURE 9.4 **Relationship of the sarcoplasmic reticulum and T tubules to the myofibrils of skeletal muscle.** The tubules of the sarcoplasmic reticulum encircle each myofibril like a "holey" sleeve. The tubules fuse to form communicating channels at the level of the H zone and abutting the A–I junctions, where the saclike elements formed are called terminal cisternae. The T tubules, inward invaginations of the sarcolemma, run deep into the cell between the terminal cisternae at the A–I junctions. Sites of close contact of these three elements (terminal cisterna, T tubule, and terminal cisterna) are called triads.

polypeptides (TnI) is an inhibitory subunit that binds to actin. Another (TnT) binds to tropomyosin and helps to position it on actin. The third (TnC) binds calcium ions. Both troponin and tropomyosin help to control the myosin-actin interactions involved in contraction.

The discoveries of additional types of muscle filaments during the past decade or so demand that the established view of striated muscle as a two-filament system be rewritten. One of these filament types, the **elastic filament** referred to earlier, is composed of the giant protein **titin**, which extends from the Z disc to the thick filament, and then actually runs within the latter to attach to the M line. It has two basic functions: (1) It holds the thick filaments in place in the sarcomere, thus maintaining the organization of the A band. (2) The part of the titin that spans the I bands is extensible, and this region unfolds when the muscle is stretched and then recoils when the tension is released, contributing to a muscle cell's ability to spring back into shape after being stretched. Titin does not resist stretching in the or-

dinary range of extension, but it stiffens the more it uncoils; therefore it strongly resists excessive stretching that might pull the sarcomeres apart.

Sarcoplasmic Reticulum and T Tubules

Skeletal muscle fibers (cells) contain two sets of intracellular tubules that participate in regulation of muscle contraction: (1) the sarcoplasmic reticulum and (2) the T tubules (Figure 9.4).

Sarcoplasmic Reticulum The **sarcoplasmic reticulum (SR)** is an elaborate smooth endoplasmic reticulum (see pp. 85 and 87). Its interconnecting tubules surround each myofibril like the sleeve of a loosely crocheted sweater surrounds your arm. Most of these tubules run longitudinally along the myofibril. Others form larger, perpendicular cross channels, called **terminal cisternae** ("end sacs"), at the A–I junctions. These terminal cisternae occur in pairs.

The major role of the SR is to regulate intracellular levels of ionic calcium: It stores calcium and releases it on demand when the muscle fiber is stimulated to contract. As you will see, calcium provides the final "go" signal for contraction.

The SR, which is a calcium storage depot.

T Tubules At each A band–I band junction, the sarcolemma of the muscle cell penetrates into the cell interior to form an elongated tube called the **transverse tubule** or **T tubule,** the lumen of which is continuous with the extracellular space. As each T tubule protrudes deep into the cell, it runs between the paired terminal cisternae of the SR so that **triads,** successive groupings of the three membranous structures (terminal cisterna, T tubule, terminal cisterna), are formed (Figure 9.4). As they pass from one myofibril to the next, the T tubules also encircle each sarcomere.

Muscle contraction is ultimately controlled by nerve-initiated impulses that travel along the sarcolemma of the muscle cell. Since T tubules are continuations of the sarcolemma, they can conduct impulses to the deepest regions of the muscle cell and to every sarcomere. These impulses signal for the release of calcium from the adjacent terminal cisternae. Thus, the T tubules can be thought of as a rapid telegraphy system that ensures that every myofibril in the muscle fiber contracts at virtually the same time.

Triad Relationships The roles of the T tubules and SR in providing signals for contraction are tightly linked. At the triads, where these organelles come into closest contact, we see something that resembles a *double zipper* of integral proteins protruding into the intermembrane space. The protruding integral proteins of the T tubule act as voltage sensors; and those of the SR (called foot proteins) are receptors that regulate the release of Ca^{2+} from the SR cisternae. We will return to consider their interaction shortly.

Contraction of a Skeletal Muscle Fiber

Although we almost always think "shortening" when we hear the word **contraction,** the true physiological meaning of the term refers only to the activation of myosin's cross bridges, which are the force-generating sites. Shortening occurs when the force or tension generated by the cross bridges on the thin filaments is more than the forces opposing shortening. Contraction ends when the cross bridges become inactive and the tension generated declines, inducing **relaxation** of the muscle fiber.

Sliding Filament Mechanism of Contraction

As we introduce the contractile process we will focus on contractions that *do* produce active shortening of the muscle cell. The **sliding filament theory of contraction,** proposed in 1954 by Hugh Huxley, states that during contraction the thin filaments slide past the thick ones so that the actin and myosin filaments overlap to a greater degree (Figure 9.5). In a relaxed

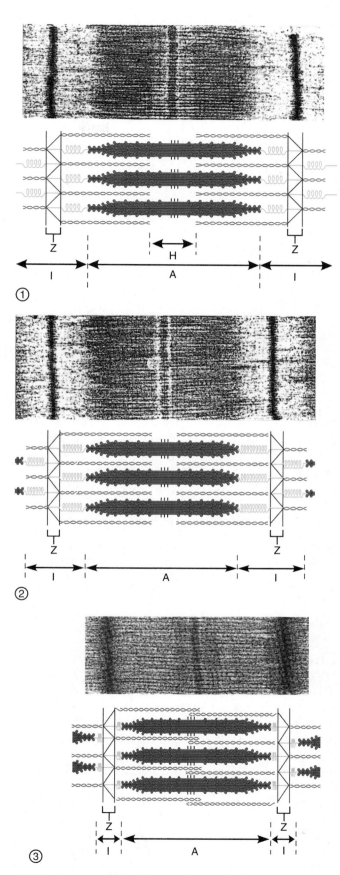

FIGURE 9.5 *Sliding filament model of contraction.* The numbers indicate the sequence of events, with ① being relaxed and ③ fully contracted. At full contraction, the Z discs abut the myosin filaments and the actin filaments overlap each other. The photomicrographs (top view in each case) show enlargements of 25,000×.

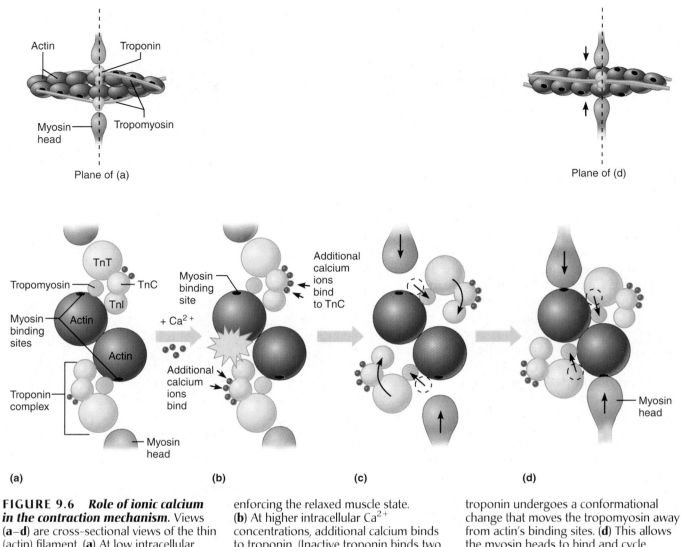

(a)   **(b)**   **(c)**   **(d)**

FIGURE 9.6 *Role of ionic calcium in the contraction mechanism.* Views **(a–d)** are cross-sectional views of the thin (actin) filament. **(a)** At low intracellular Ca^{2+} concentration, tropomyosin blocks the binding sites on actin, preventing the attachment of myosin cross bridges and enforcing the relaxed muscle state. **(b)** At higher intracellular Ca^{2+} concentrations, additional calcium binds to troponin. (Inactive troponin binds two Ca^{2+}; calcium-activated troponin binds an additional two Ca^{2+} at a separate regulatory site.) **(c)** Calcium-activated troponin undergoes a conformational change that moves the tropomyosin away from actin's binding sites. **(d)** This allows the myosin heads to bind and cycle, permitting contraction (sliding of the thin filaments by the myosin cross bridges) to begin.

muscle fiber, the thick and thin filaments overlap only slightly. But when muscle fibers are stimulated adequately by the nervous system, the cross bridges latch on to myosin binding sites on actin in the thin filaments, and the sliding begins. Each cross bridge attaches and detaches several times during a contraction, acting like a tiny ratchet to generate tension and propel the thin filaments toward the center of the sarcomere. As this event occurs simultaneously in sarcomeres throughout the cell, the muscle cell shortens. Notice that as the thin filaments slide centrally, the Z discs to which they are attached are pulled *toward* the thick filaments. Overall, the distance between successive Z discs is reduced, the I bands shorten, the H zones disappear, and the A bands move closer together but do not change in length.

Cross bridge attachment to actin requires calcium ions, and the nerve impulse leading to contraction causes an increase in calcium ions within the muscle cell. We discuss the nerve impulse and the calcium fluxes later; here let us concentrate on the contraction mechanism itself. When intracellular calcium levels are low, the muscle cell is relaxed, and the active (myosin binding) sites on actin are physically blocked by tropomyosin molecules (Figure 9.6a). As calcium ions become available, they bind to regulatory sites on troponin (Figure 9.6b), causing it to change shape. This event moves tropomyosin deeper into the groove of the actin helix and away from the myosin binding sites (Figure 9.6c and d). Thus, the tropomyosin "blockade" is removed when calcium is present.

Once binding sites on actin are exposed, the following events occur in rapid succession (Figure 9.7).

1. Cross bridge attachment. The activated myosin heads are strongly attracted to the exposed binding sites on actin and cross bridge attachment occurs.

? *What would be the result if the muscle fiber suddenly ran out of ATP when the sarcomeres had only partially contracted?*

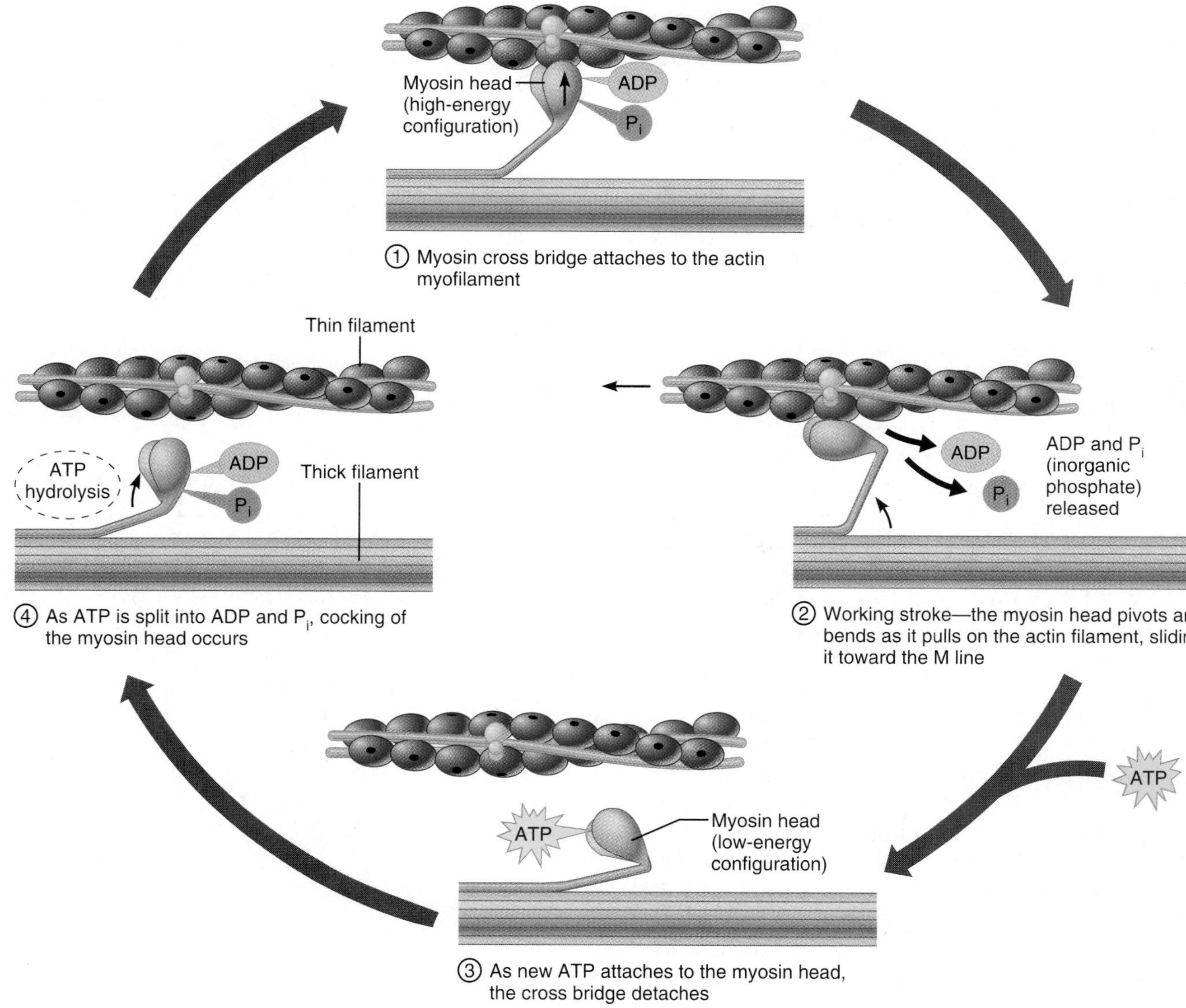

Myosin head
(high-energy
configuration)

ADP

P$_i$

① Myosin cross bridge attaches to the actin myofilament

Thin filament

ATP
hydrolysis

ADP

P$_i$

Thick filament

④ As ATP is split into ADP and P$_i$, cocking of the myosin head occurs

ADP

P$_i$

ADP and P$_i$
(inorganic phosphate)
released

② Working stroke—the myosin head pivots and bends as it pulls on the actin filament, sliding it toward the M line

ATP

ATP

Myosin head
(low-energy
configuration)

③ As new ATP attaches to the myosin head, the cross bridge detaches

FIGURE 9.7 *Sequence of events involved in the sliding of the actin filaments during contraction.* A small section of adjacent thick and thin filaments is used to illustrate the interactions that occur between the two types of myofilaments. These events occur only in the presence of ionic calcium (Ca^{2+}), which releases tropomyosin's blockade of actin's active sites.

2. The working (power) stroke. As a myosin head binds, it pivots, changing from its high-energy configuration to its bent, low-energy shape, which pulls on the thin filament, sliding it toward the center of the sarcomere. At the same time, ADP and inorganic phosphate (P$_i$) generated during the *prior* contraction cycle are released from the myosin head.

3. Cross bridge detachment. As a new ATP molecule binds to the myosin head, myosin's hold on actin loosens and the cross bridge detaches from actin.

4. "Cocking" of the myosin head. Hydrolysis of ATP to ADP and P$_i$ by ATPase provides the energy needed to return the myosin head to its high-energy, or "cocked," position. This provides the potential energy needed for its *next* sequence of attachment

The sarcomeres would remain in their partially contracted state.

and working stroke. (The ADP and P_i remain attached to the myosin head during this phase.) At this point, the cycle is back where it started. The myosin head is in its upright high-energy configuration, ready to take another "step" and attach to an actin site farther along the thin filament. This "walking" of the myosin cross bridges along the adjacent thin filaments during muscle shortening is much like a centipede's gait. Some myosin heads ("legs") are always in contact with actin (the "ground"), so that the thin filaments cannot slide backward as this cycle is repeated again and again during contraction.

A single working stroke of all the cross bridges in a muscle results in a shortening of only about 1%. Because contracting muscles routinely shorten 30 to 35% of their total resting length, each myosin cross bridge must attach and detach many times during a single contraction. It is likely that only half of the myosin heads of a thick filament are actively exerting a pulling force at the same instant; the balance are randomly seeking their next binding site. Sliding of thin filaments continues as long as the calcium signal and adequate ATP are present. As the Ca^{2+} pumps of the SR reclaim calcium ions from the sarcoplasm and troponin again changes shape, the actin active sites are covered by tropomyosin, the contraction ends, and the muscle fiber relaxes.

◿ Homeostatic Imbalance

Rigor mortis (death rigor) illustrates the fact that cross bridge detachment is ATP driven. Most muscles begin to stiffen 3 to 4 hours after death. Peak rigidity occurs at 12 hours and then gradually dissipates over the next 48 to 60 hours. Dying cells are unable to exclude calcium (which is in higher concentration in the extracellular fluid), and the calcium influx into muscle cells promotes the binding of myosin cross bridges. However, shortly after breathing stops, ATP synthesis ceases, and cross bridge detachment is impossible. Actin and myosin become irreversibly cross-linked, producing the stiffness of dead muscle. Rigor mortis disappears as muscle proteins break down several hours after death. ■

Regulation of Contraction

For a skeletal muscle fiber to contract, it must be stimulated by a nerve ending and must propagate an electrical current, or *action potential,* along its sarcolemma. This electrical event causes the short-lived rise in intracellular calcium ion levels that is the final trigger for contraction. The series of events linking the electrical signal to contraction is called *excitation-contraction coupling.*

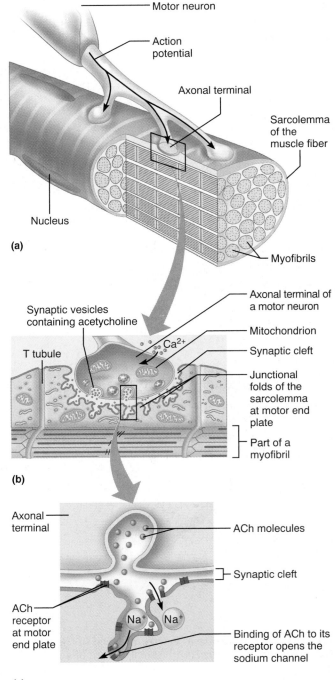

FIGURE 9.8   *The neuromuscular junction.* (a) Axonal ending of a motor neuron forming a neuromuscular junction with a muscle fiber. (b) The axonal terminal contains vesicles filled with the neurotransmitter acetylcholine (ACh), which is released when the action potential reaches the axonal terminal. The sarcolemma is highly folded adjacent to the synaptic cleft, and acetylcholine receptors are present in these junctional folds. (c) Acetylcholine diffuses across the synaptic cleft and attaches to ACh receptors on the sarcolemma, opening Na^+ channels and initiating depolarization of the sarcolemma.

The Neuromuscular Junction and the Nerve Stimulus Skeletal muscle cells are stimulated by *motor neurons* of the somatic (voluntary) division of

the nervous system. Although these motor neurons "reside" in the brain or spinal cord, their long thread-like extensions called *axons* travel, bundled within nerves, to the muscle cells they serve. The axon of each motor neuron divides profusely as it enters the muscle, and each of these axonal endings forms a branching **neuromuscular junction** with a single muscle fiber (Figure 9.8). As a rule, each muscle fiber has only one neuromuscular junction, located in its approximate middle. Although the axonal ending and the muscle fiber are exceedingly close, they remain separated by a space called the **synaptic cleft.** The synaptic cleft is filled with a gel-like extracellular substance rich in glycoproteins. Within the flattened moundlike axonal ending are **synaptic vesicles,** small membranous sacs containing a neurotransmitter called **acetylcholine** (as"ĕ-til-ko'lēn), or **ACh.** The **motor end plate,** the specific trough-like part of the muscle fiber's sarcolemma that helps form the neuromuscular junction, is highly folded. These *junctional folds* provide a large surface area for the millions of **ACh receptors** located there.

When a nerve impulse reaches the end of an axon, voltage-regulated calcium channels in its membrane open, allowing Ca^{2+} to flow in from the extracellular fluid. The presence of calcium inside the axon terminal causes some of the synaptic vesicles to fuse with the axonal membrane and release ACh into the synaptic cleft by *exocytosis.* ACh diffuses across the cleft and attaches to the flowerlike ACh receptors on the sarcolemma. The electrical events triggered in a sarcolemma when ACh binds are similar to those that take place in excited nerve cell membranes. We outline these events here and then consider them in more detail in Chapter 11.

⚠ *Homeostatic Imbalance*

Many toxins, drugs, and diseases interfere with events occurring at the neuromuscular junction. For example, *myasthenia gravis* (*asthen* = weakness; *gravi* = heavy), a disease characterized by drooping of the upper eyelids, difficulty in swallowing and talking, and generalized muscle weakness, involves a shortage of ACh receptors. The blood contains antibodies to ACh receptors, which suggests that myasthenia gravis is an autoimmune disease. Although normal numbers of receptors are initially present, they appear to be destroyed as the disease progresses.

Curare, the arrowhead poison used by natives of South America, binds to ACh receptors and blocks ACh attachment. Thus, even though the nervous system is still releasing ACh (the "go" signal), the muscles are unable to respond and respiratory arrest occurs. Curare and similar drugs are used clinically in small amounts to allow intubation and prevent muscular movements during surgery. ∎

Generation of an Action Potential Across the Sarcolemma Like the plasma membranes of all cells, a resting sarcolemma is *polarized* (Figure 9.9a). That is, a voltmeter would show there is a voltage across the membrane and the inside is negative. (The resting membrane potential is described in Chapter 3, pp. 81–82.) The binding of ACh molecules to ACh receptors on the sarcolemma opens *chemically regulated* Na^+ channels housed within the ACh receptors and leads to transient permeability changes in the sarcolemma. The result is a change in membrane potential (voltage) such that the muscle cell interior becomes slightly *less negative,* an event called **depolarization.** Initially, depolarization is a local event (occurring only at the receptor site), but if the nerve stimulus is strong enough, an action potential is generated that passes in *all directions* from the neuromuscular junction across the sarcolemma, just as ripples move away from a pebble dropped into a stream.

The **action potential** is the result of a predictable sequence of electrical changes that occurs along the sarcolemma (Figure 9.9). Essentially, three steps are involved.

1. First, the membrane is depolarized. This is a consequence of sodium (Na^+) channels opening and Na^+ entering the cell (see Figure 9.9b).

2. During step 2, the action potential is propagated as the local depolarization wave spreads to the adjacent areas of the sarcolemma and opens *voltage-regulated* sodium channels there (see Figure 9.9c). Sodium ions, normally restricted from entering, now diffuse into the cell following their electrochemical gradient.

3. Step 3 is **repolarization,** which restores the sarcolemma to its initial polarized state. The repolarization wave, which quickly follows the depolarization wave, is a consequence of Na^+ channels closing and K^+ channels opening. Since the potassium ion concentration is substantially higher inside the cell than in the extracellular fluid, K^+ diffuses rapidly out of the muscle fiber (Figure 9.9d).

During repolarization, a muscle fiber is said to be in a **refractory period,** because the cell cannot be stimulated again until repolarization is complete. Notice that repolarization only restores the *electrical conditions* of the resting (polarized) state. The ATP-dependent Na^+-K^+ pump must be "revved up" to restore the ionic conditions of the resting state, but several contractions can occur before ionic imbalances interfere with contractile activity.

Once initiated, the action potential is unstoppable and ultimately results in contraction of the muscle cell. Although the action potential itself is very brief (1–2 milliseconds [ms]), the contraction phase of a muscle fiber may persist for 100 ms

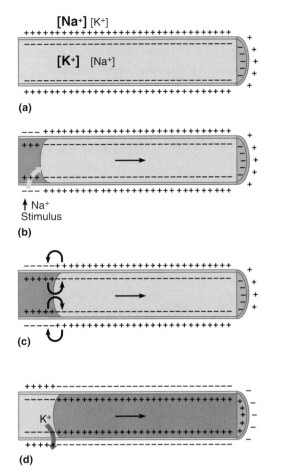

(a) Electrical conditions of a resting (polarized) sarcolemma.
The outside face is positive, while the inside face is negative. The predominant extracellular ion is sodium (Na$^+$); the predominant intracellular ion is potassium (K$^+$). The sarcolemma is relatively impermeable to both ions.

(b) Step 1: Depolarization and generation of the action potential.
Stimulation by a motor nerve fiber (via the release of acetylcholine) causes that patch of sarcolemma to become permeable to sodium (sodium channels open). As sodium ions diffuse rapidly into the cell, the resting potential is decreased (i.e., depolarization occurs). If the stimulus is strong enough, an action potential is initiated.

(c) Step 2: Propagation of the action potential.
The positive charge inside the initial patch of sarcolemma changes the permeability of an adjacent patch, opening voltage-regulated Na$^+$ channels there. Consequently the membrane potential in that region decreases and depolarization occurs there as well. Thus, the action potential travels rapidly over the entire sarcolemma.

(d) Step 3: Repolarization.
Immediately after the depolarization wave passes, the sarcolemma's permeability changes once again: Na$^+$ channels close and K$^+$ channels open, allowing K$^+$ to diffuse from the cell. This restores the electrical conditions of the resting (polarized) state. Repolarization occurs in the same direction as depolarization, and must occur before the muscle fiber can be stimulated again. The ionic concentrations of the resting state are restored later by the sodium-potassium pump.

FIGURE 9.9 *Summary of events in the generation and propagation of an action potential in a skeletal muscle fiber.*

or more and far outlasts the electrical event that triggers it because active transport of Ca^{2+} back into the SR takes substantially longer then its release.

Destruction of Acetylcholine After ACh is released from the motor neuron and binds to the ACh receptors, it is swiftly destroyed by **acetylcholinesterase** (as″ĕ-til-ko″lin-es′ter-ās), or **AChE,** an enzyme located on the sarcolemma at the neuromuscular junction and in the synaptic cleft. This prevents continued muscle fiber contraction in the absence of additional nervous system stimulation.

Excitation-Contraction Coupling Excitation-contraction coupling is the sequence of events by which transmission of an action potential along the sarcolemma leads to the sliding of myofilaments. The action potential is very brief and ends well be-

fore any signs of contraction are obvious. The period from action potential initiation to the beginning of mechanical activity (shortening) is called the *latent period* (*laten* = hidden). The events of excitation-contraction coupling occur during this period. As you will see, the electrical signal does not act directly on the myofilaments; rather, it causes the rise in intracellular calcium ion concentration that allows the filaments to slide (Figure 9.10).

Excitation-contraction coupling consists of the following steps:

① The action potential propagates along the sarcolemma and down the T tubules.

② Transmission of the action potential past the triad regions ultimately causes the terminal cisternae of the SR to release calcium ions into the sarcoplasm, where Ca^{2+} becomes available to the myofilaments. The "double zippers" (p. 285) at the T-SR junctions of the triads are involved in this action as follows: The protein particles on the T tubule side are sensitive to

(1) *Why is Ca²⁺ called the final trigger for contraction?* **(2)** *What constitutes the initial trigger?*

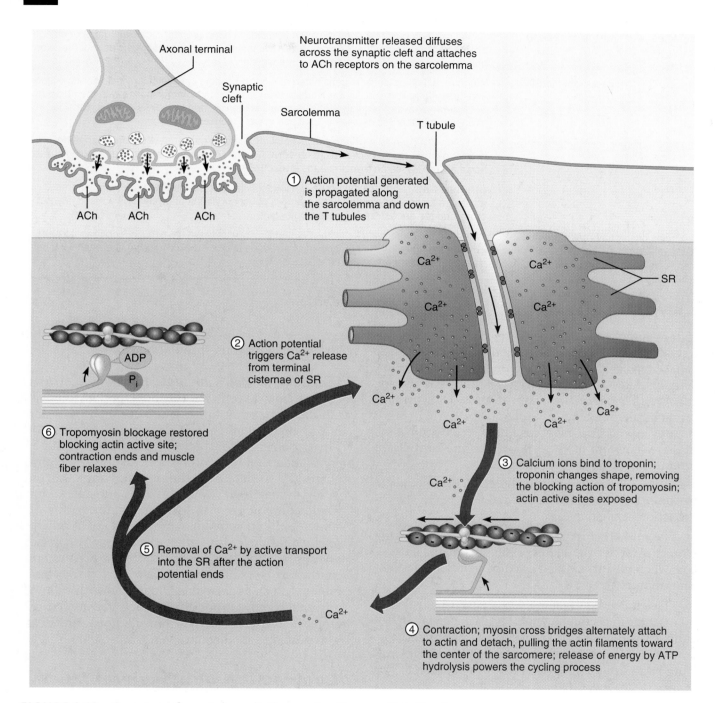

Neurotransmitter released diffuses across the synaptic cleft and attaches to ACh receptors on the sarcolemma

Axonal terminal

Synaptic cleft

Sarcolemma

T tubule

ACh   ACh   ACh

① Action potential generated is propagated along the sarcolemma and down the T tubules

Ca^{2+} Ca^{2+} SR

Ca^{2+} Ca^{2+}

② Action potential triggers Ca^{2+} release from terminal cisternae of SR

Ca^{2+} Ca^{2+} Ca^{2+}

ADP P_i

Ca^{2+} Ca^{2+}

⑥ Tropomyosin blockage restored blocking actin active site; contraction ends and muscle fiber relaxes

③ Calcium ions bind to troponin; troponin changes shape, removing the blocking action of tropomyosin; actin active sites exposed

Ca^{2+}

⑤ Removal of Ca^{2+} by active transport into the SR after the action potential ends

Ca^{2+}

④ Contraction; myosin cross bridges alternately attach to actin and detach, pulling the actin filaments toward the center of the sarcomere; release of energy by ATP hydrolysis powers the cycling process

FIGURE 9.10 *Sequence of events in excitation-contraction coupling.* Events ① through ⑤ indicate the sequence of events in the coupling process. As shown in the clockwise flow of events, contraction continues until the calcium signal ends ⑥.

(1) *Because calcium binding to troponin frees the actin active sites to bind with myosin cross bridges.* **(2)** *The initial trigger is neurotransmitter binding, which initiates generation of an action potential along the sarcolemma.*

TABLE 9.2	Roles of Ionic Calcium (Ca²⁺) in Muscle Contraction

Role	Mechanism
Promotes neurotransmitter release	As the nerve impulse reaches the axon terminal, voltage-regulated calcium gates open allowing Ca^{2+} to enter the terminal, which triggers fusion of synaptic vesicles with the axonal membrane and exocytosis of the neurotransmitter.
Triggers Ca^{2+} release from the SR	When an action potential is transmitted along the T tubules of skeletal muscle fibers, voltage sensors (integral proteins in the T tubule wall) respond by communicating with the adjoining SR foot proteins that regulate Ca^{2+} channels in the SR membrane. As a result, Ca^{2+} is released from the SR, causing a local rise in Ca^{2+} concentration.
Triggers sliding of myofilaments and ATPase activity	(1) When Ca^{2+} binds to troponin in skeletal and cardiac muscles, structural changes occur in troponin that expose active sites on actin. As a result, myosin cross bridge binding can occur and ATPases on myosin heads are activated.
	(2) When Ca^{2+} binds to calmodulin (an intracellular calcium-binding protein) in smooth muscle, a kinase is activated that catalyzes phosphorylation of myosin. As a result, myosin cross bridges are activated and sliding begins.
Promotes glycogen breakdown and ATP synthesis	When Ca^{2+} binds to and activates calmodulin, the activated calmodulin mobilizes a kinase enzyme that initiates the breakdown of glycogen to glucose. The muscle fiber then metabolizes glucose to produce ATP for muscular work.

voltage and change their conformation (3-D shape) in response to the arrival of the action potential. Their voltage-regulated change is communicated to the underlying SR foot, which in turn undergoes shape changes that open its calcium channel (and perhaps mechanically linked Ca^{2+} channels nearby). Since these events occur at every triad in the cell, within 1 ms massive amounts of Ca^{2+} flood into the sarcoplasm from the SR cisternae (see Table 9.2).

③ As already described, some of this calcium binds to troponin, which changes shape and removes the blocking action of tropomyosin.

④ The myosin cross bridges attach and pull the thin filaments toward the center of the sarcomere. This occurs when the intracellular calcium is present at a concentration of about 10^{-5} M.

⑤ The short-lived calcium signal ends, usually within 30 ms after the action potential is over. The fall in calcium levels reflects the operation of a continuously active, ATP-dependent calcium pump that moves calcium back into the tubules of the SR to be stored once again.

⑥ When intracellular calcium levels drop too low to allow contraction, the tropomyosin blockade is reestablished and myosin ATPases are inhibited. Cross bridge activity ends and relaxation occurs.

This entire sequence of events is repeated when another nerve impulse arrives at the neuromuscular junction. When the impulses are delivered very rapidly, intracellular calcium levels increase greatly due to successive "puffs" or rounds of calcium release from the SR. In such cases, the muscle cells do not completely relax between successive stimuli and

contraction is stronger and more sustained (within limits) until nervous stimulation ceases.

Summary of Roles of Ionic Calcium in Muscle Contraction Except for the brief period following muscle cell excitation, calcium ion concentrations in the sarcoplasm are kept almost undetectably low. There is a reason for this: ATP provides the cell's energy source and its hydrolysis yields inorganic phosphates (P_i). If the intracellular level of ionic calcium were always high, calcium and phosphate ions would combine to form hydroxyapatite crystals, the stony-hard salts found in bone matrix. Such calcified cells would die. Also, since calcium's physiological roles are so vital (see Table 9.2), its cytoplasmic concentration is exquisitely regulated by intracellular proteins, such as **calsequestrin** (found within the SR cisternae) and **calmodulin**, which can alternately bind calcium (removing it from solution) and release it to provide a metabolic signal.

Contraction of a Skeletal Muscle

In its relaxed state, a muscle is unimpressive. It is soft and not at all what you would expect of a prime mover of the body. However, within a few milliseconds, it can become a hard elastic structure with dynamic characteristics that intrigue not only biologists but engineers and physicists as well.

Now we are ready to consider muscle contraction on the organ level. But before we lunge into this section, let's review a few facts and lay a few ground rules for best understanding of muscle mechanics.

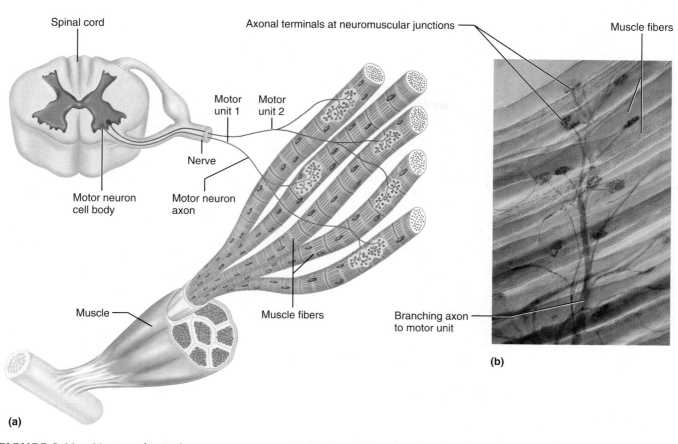

FIGURE 9.11 *Motor units.* Each motor unit consists of a motor neuron and all of the muscle fibers it innervates. **(a)** Schematic view of portions of two motor units. The cell bodies of the motor neurons reside in the spinal cord, and their axons extend to the muscle. Within the muscle, each axon divides into a number of axonal terminals, which are distributed to muscle fibers scattered throughout the muscle. **(b)** Photomicrograph of a portion of a motor unit (110×). Notice the diverging axonal terminals and the neuromuscular junctions with the muscle fibers.

1. The principles of contraction of a muscle fiber (cell) and of a skeletal muscle consisting of a huge number of cells are pretty much the same.

2. The force exerted on an object by a contracting muscle is called the **muscle tension,** and the opposing force exerted on the muscle by the weight of the object to be moved is called the **load.**

3. A contracting muscle does not always move the load (shorten). If tension develops but the load is not moved, the contraction is called *isometric* ("same measure"). If the tension developed overcomes the load and muscle shortening occurs, the contraction is *isotonic.* These major types of contraction will be described in detail shortly, but for now the important thing to remember when reading the graphs is that *increasing muscle tension* is measured in isometric contractions, whereas the *amount of shortening* (distance in millimeters) the muscle undergoes is recorded by the tracing in isotonic contractions.

4. A skeletal muscle contracts with varying force and for different periods of time. To understand how this occurs, we must look at the nerve-muscle functional unit called a *motor unit* and investigate how a muscle responds to stimuli of varying frequencies and intensities. These are our next topics.

The Motor Unit

Each muscle is served by at least one motor nerve, which contains hundreds of motor neuron axons. As an axon enters a muscle, it branches into a number of axonal terminals, each of which forms a neuromuscular junction with a single muscle fiber. A motor neuron and all the muscle fibers it supplies is called a **motor unit** (Figure 9.11). When a motor neuron fires (transmits an electrical impulse), all the muscle fibers that it innervates respond by contracting. The number of muscle fibers per motor unit may be as high as several hundred or as few as four. Muscles that exert very fine control (such as the muscles controlling movements of the fingers and eyes) have small motor units. By contrast, large, weight-bearing muscles, whose movements are less precise (such as the hip muscles), have large motor units. The muscle fibers in a single motor unit are not clustered together but are spread throughout the muscle. As a result, stimulation of a single motor unit causes a weak contraction of the *entire* muscle.

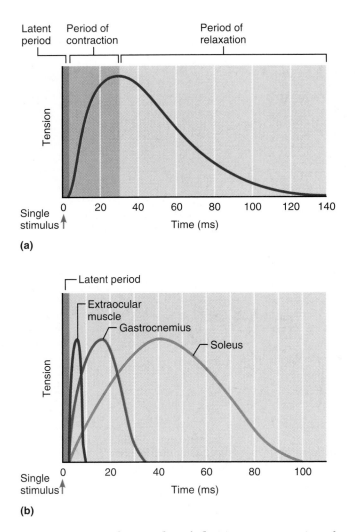

FIGURE 9.12   *The muscle twitch.* (a) Myogram tracing of an isometric twitch contraction, showing its three phases: the latent period, the period of contraction, and the period of relaxation. (b) Comparison of the twitch responses of the extraocular, gastrocnemius, and soleus muscles.

The Muscle Twitch and Development of Muscle Tension

Muscle contraction is easily investigated in a laboratory setting using an isolated muscle. The muscle is attached to an apparatus that produces a **myogram**, a graphic recording of mechanical contractile activity. (The line recording the activity is called a *tracing.*)

The response of a muscle to a single brief threshold stimulus is called a **muscle twitch.** The muscle contracts quickly and then relaxes. A twitch may be strong or weak, depending on the number of motor units activated, but in any twitch myogram, three distinct phases are obvious (Figure 9.12a).

1. Latent period. The latent period is the first few milliseconds following stimulation when excitation-contraction coupling is occurring. During this pe-

riod, muscle tension is beginning to increase but no response is seen on the myogram.

2. Period of contraction. The period of contraction is when cross bridges are active, from the onset to the peak of tension development, and the myogram tracing rises to a peak. This period lasts 10–100 ms. If the tension (pull) becomes great enough to overcome the resistance of an attached load, the muscle shortens.

3. Period of relaxation. The period of contraction is followed by the period of relaxation. This final phase, lasting 10–100 ms, is initiated by reentry of Ca^{2+} into the SR. Since contractile force is no longer being generated, muscle tension decreases to zero and the tracing returns to the baseline. If the muscle has shortened during contraction, it now returns to its initial length.

As you can see in Figure 9.12b, twitch contractions of some muscles are rapid and brief, as with the muscles controlling eye movements. In contrast, the fleshy calf muscles (gastrocnemius and soleus) contract more slowly and remain contracted for much longer periods. These differences between muscles reflect metabolic properties of the myofibrils and enzyme variations.

Graded Muscle Responses

Although *muscle twitches*—the single, jerky contractions observed in the laboratory—may occur as a result of certain neuromuscular problems, this is *not* the way our muscles normally operate in the body. Instead, our muscle contractions are relatively smooth and vary in strength as different demands are placed on them. Variations in the degree of muscle contraction (an obvious requirement for proper control of skeletal movement) are referred to as **graded responses.** In general, muscle contraction can be graded in two ways: (1) by changing the frequency (speed) of stimulation and (2) by changing the strength of the stimulus.

Muscle Response to Frequency of Stimulation: Wave Summation and Tetanus If two identical stimuli (electrical shocks or nerve impulses) are delivered to a muscle in rapid succession, the second twitch will be stronger than the first. On the myogram it will appear to ride on the shoulders of the first contraction (Figure 9.13). This phenomenon, called **wave,** or **temporal, summation,** occurs because a second contraction occurs before the muscle has completely relaxed. Since the muscle is already partially contracted and more calcium is being released to replace that being reclaimed by the SR, tension produced during the sec-

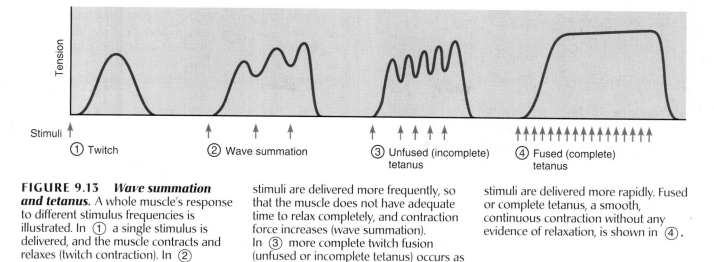

① Twitch ② Wave summation ③ Unfused (incomplete) tetanus ④ Fused (complete) tetanus

FIGURE 9.13 *Wave summation and tetanus.* A whole muscle's response to different stimulus frequencies is illustrated. In ① a single stimulus is delivered, and the muscle contracts and relaxes (twitch contraction). In ② stimuli are delivered more frequently, so that the muscle does not have adequate time to relax completely, and contraction force increases (wave summation). In ③ more complete twitch fusion (unfused or incomplete tetanus) occurs as stimuli are delivered more rapidly. Fused or complete tetanus, a smooth, continuous contraction without any evidence of relaxation, is shown in ④.

ond contraction causes more shortening than the first. In other words, the contractions are summed. (However, the refractory period is *always* honored. Thus, if a second stimulus is delivered before repolarization is complete, no summation occurs.) If the stimulus strength or voltage is held constant and the muscle is stimulated at an increasingly faster rate, the relaxation time between the twitches becomes shorter and shorter, the concentration of Ca^{2+} in the sarcoplasm higher and higher, and the degree of summation greater and greater, progressing to a sustained but quivering contraction referred to as **unfused** or **incomplete tetanus.** Finally, all evidence of muscle relaxation disappears and the contractions fuse into a smooth, sustained contraction called **fused** or **complete tetanus** (tet′ah-nus; *tetan* = rigid, tense). (Tetanus is often confused with the bacterial disease of the same name.)

Vigorous muscle activity cannot continue indefinitely. Prolonged tetanus inevitably leads to *muscle fatigue,* a situation in which the muscle is unable to contract and its tension drops to zero. A key factor in muscle fatigue is the muscle's inability to produce enough ATP to power contraction. This and other factors causing muscle fatigue are discussed later in the chapter.

Muscle Response to Stronger Stimuli: Multiple Motor Unit Summation

Although wave summation contributes to contractile force, its primary function is to produce smooth, continuous muscle contractions by rapidly stimulating a specific number of muscle cells. The force of contraction is controlled more precisely by **multiple motor unit summation.** In the laboratory this phenomenon, also called **recruitment,** is achieved by delivering shocks of increasing voltage to the muscle, calling more and more muscle fibers into play. The stimulus at which the first observable muscle contraction occurs is called the **threshold stimulus.** Beyond this point, the muscle contracts more and more vigorously as the stimulus strength is increased. The **maximal stimulus** is the strongest stimulus that produces increased contractile force. It represents the point at which all the muscle's motor units are recruited. Increasing the stimulus intensity beyond the maximal stimulus does not produce a stronger contraction. In the body, the same phenomenon is caused by neural activation of an increasingly large number of motor units serving the muscle.

In weak and precise muscle contractions, relatively few motor units are stimulated. Conversely, when the muscle contracts forcibly many motor units are activated. This explains how the same hand that pats your cheek can deliver a stinging slap. In any muscle, the smallest motor units (those with the fewest muscle fibers) are controlled by the most excitable motor neurons. These motor units tend to be activated first. The larger motor units, controlled by less excitable neurons, are activated only when a stronger contraction is necessary.

Although *all* the motor units of a muscle may be recruited simultaneously to produce an exceptionally strong contraction, more commonly motor units are activated asynchronously in the body. Some are in tetanus (most commonly unfused tetanus) while others are resting and recovering. This technique helps to prolong a strong contraction by preventing or delaying fatigue. It also explains how weak contractions promoted by infrequent stimuli can remain smooth.

Treppe: The Staircase Effect

When a muscle begins to contract, its contractions may be only half as strong as those that occur later in response to stimuli of the same strength. Thus,

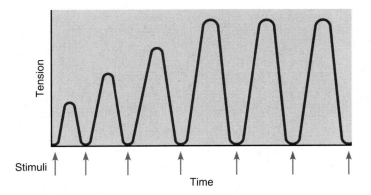

FIGURE 9.14 *Myogram of treppe (the staircase effect).* Notice that although the stimuli delivered to the muscle are of the same intensity and the muscle is not being stimulated rapidly, the first few contractile responses are increasingly stronger.

the tracing shows a staircase pattern called **treppe** (trep'ĕ) (Figure 9.14). Treppe probably reflects the increasing availability of Ca^{2+} in the sarcoplasm; the greater numbers of Ca^{2+} ions expose more active sites on the actin filaments for cross bridge attachment. Additionally, as the muscle begins to work and liberates heat, its enzyme systems become more efficient. Together, these factors produce a slightly stronger contraction with each successive stimulus during the initial phase of muscle activity. This is the basis of the warm-up period required of athletes.

Muscle Tone

Skeletal muscles are described as "voluntary muscles," but even relaxed muscles are almost always in a slightly contracted state, a phenomenon called **muscle tone.** This is due to spinal reflexes that activate first one group of motor units and then another in response to activation of stretch receptors in the muscles and tendons. (These receptors and their activity are described in Chapter 13.) Muscle tone does not produce active movements, but it keeps the muscles firm, healthy, and ready to respond to stimulation. Skeletal muscle tone also helps stabilize joints and maintain posture.

Isotonic and Isometric Contractions

As noted earlier, there are two main categories of contractions—*isotonic* and *isometric*. In **isotonic** (*iso* = same; *ton* = tension) **contractions,** the muscle changes in length (decreasing the angle at the joint) and moves the load. Once sufficient tension has developed to lift the load, the tension remains relatively constant through the rest of the contractile period (Figure 9.15a).

Isotonic contractions come in two "flavors"—*concentric* and *eccentric*. **Concentric contractions** in which the muscle *shortens* and does work—picking up a book or kicking a ball, for instance—are probably more familiar. However, **eccentric contractions,** in which the muscle contracts as it *lengthens*, are equally important for coordination and purposeful movements. Eccentric contractions occur in your calf muscle, for example, as you walk up a long, steep hill. Eccentric contractions are about 50% more forceful than concentric ones at the same load and more often cause delayed-onset muscle soreness. (Consider how your stretched calf muscles *feel* the day after hiking up that long, steep hill.) Just why this is so is unclear, but it may be that the muscle stretching that occurs during such contractions causes microtears in the muscles.

Squats, or deep knee bends, provide a simple example of how concentric and eccentric contractions work together in our everyday activities. As the knees flex, the powerful quadriceps muscles of the anterior thigh lengthen (are stretched), but they also contract (eccentrically) at the same time to counteract the force of gravity and control the descent of the torso ("muscle braking") to prevent joint injury. Raising the body back to its starting position requires that the same muscles (quadriceps) contract concentrically as they shorten to extend the knees again. As you can see, eccentric contractions put the body in position to contract concentrically. All jumping and throwing activities involve both types of contraction to some degree.

Effective movement also requires isometric (*metric* = measure) contractions. In **isometric contractions,** tension continues to increase to the muscle's peak tension-producing capacity, but the muscle *neither shortens nor lengthens* (Figure 9.15b). Isometric contractions occur when a muscle attempts to move a load that is greater than the force (tension) the muscle is able to develop—such as your arm muscles when you try to lift a piano singlehandedly. Muscles that act primarily to maintain upright posture or to hold joints in stationary positions while movements occur at other joints are contracting isometrically. In our knee bend example, the anterior thigh muscles contract isometrically at two points: (1) when the squat position is held for a few seconds, both the anterior and posterior thigh muscles contract isometrically to hold the knee in the flexed position; and (2) as one begins to rise to the upright position, the quadriceps muscles contract isometrically until their tension exceeds the load (weight of the upper body) and then muscle shortening (concentric contraction) begins. So the quadriceps contractile sequence for a deep knee bend from

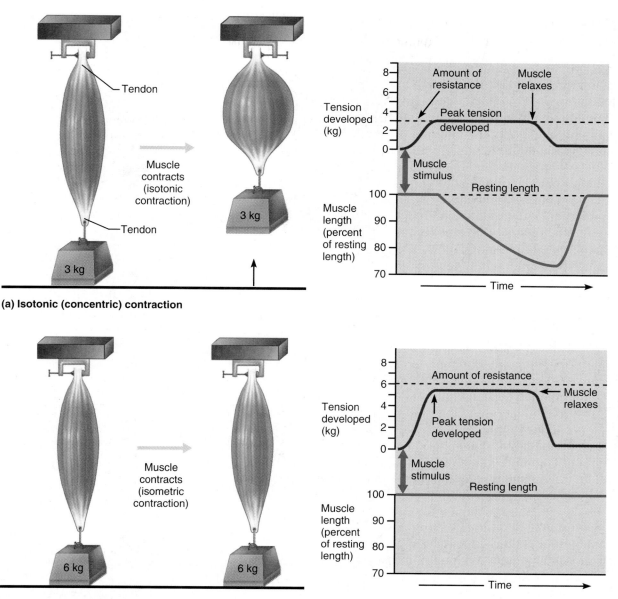

(a) Isotonic (concentric) contraction

(b) Isometric contraction

FIGURE 9.15   *Isotonic (concentric) and isometric contractions.* **(a)** On stimulation, this muscle develops enough tension (force) to lift the load (weight). Once the resistance is overcome, tension remains constant for the rest of the contraction and the muscle shortens. This is a concentric isotonic contraction. **(b)** This muscle is attached to a weight that exceeds its peak tension-developing capabilities. When stimulated, the tension increases to the muscle's peak tension-developing capability, but the muscle does not shorten (it remains at its resting length). This is an isometric contraction.

start to finish is (1) knee flex (eccentric), (2) hold squat position (isometric), (3) knee extend (isometric, then concentric). Of course, this does not even begin to consider the isometric contractions of the trunk muscles that maintain a relatively vertical trunk posture during the movement.

Electrochemical and mechanical events occurring within a muscle are identical in both isotonic and isometric contractions. However, the result is different. In isotonic contractions, the thin filaments are sliding; in isometric contractions, the cross bridges are generating force but are not moving the thin filaments. (You could say that they are "spinning their wheels" on the same actin binding site.)

Muscle Metabolism

Providing Energy for Contraction

Stored ATP   As a muscle contracts, ATP provides the energy for cross bridge movement and detachment and for operation of the calcium pump. Surprisingly, muscles store very limited reserves of ATP—4 to 6 seconds' worth at most, just enough to get you going.

Since ATP is the *only* energy source used directly for contractile activities, it must be regenerated as fast as it is broken down if contraction is to continue. Fortunately, after ATP is hydrolyzed to ADP and inorganic phosphate, it is regenerated within a

? *Which of these energy-producing pathways would predominate in the leg muscles of a long-distance cyclist?*

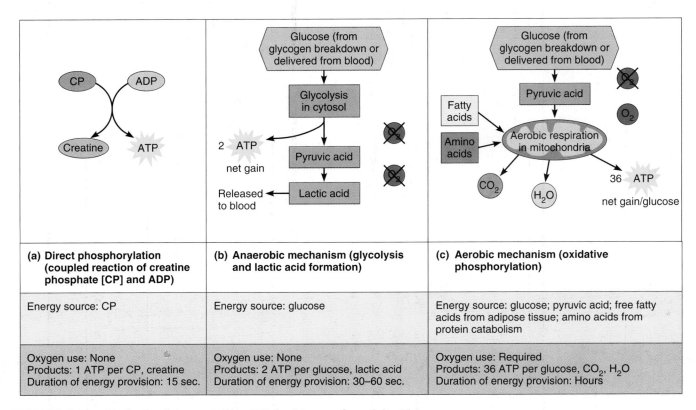

(a) **Direct phosphorylation (coupled reaction of creatine phosphate [CP] and ADP)**	(b) **Anaerobic mechanism (glycolysis and lactic acid formation)**	(c) **Aerobic mechanism (oxidative phosphorylation)**
Energy source: CP	Energy source: glucose	Energy source: glucose; pyruvic acid; free fatty acids from adipose tissue; amino acids from protein catabolism
Oxygen use: None Products: 1 ATP per CP, creatine Duration of energy provision: 15 sec.	Oxygen use: None Products: 2 ATP per glucose, lactic acid Duration of energy provision: 30–60 sec.	Oxygen use: Required Products: 36 ATP per glucose, CO_2, H_2O Duration of energy provision: Hours

FIGURE 9.16 *Methods of regenerating ATP during muscle activity.* The fastest mechanism is direct phosphorylation (**a**); the slowest is the aerobic mechanism (**c**).

fraction of a second by three pathways (Figure 9.16): (1) by interaction of ADP with creatine phosphate, (2) from stored glycogen via the anaerobic pathway called anaerobic glycolysis, and (3) by aerobic respiration. The second and third metabolic pathways are used to produce ATP in all body cells. These pathways, just touched on here, are described in detail in Chapter 25.

Direct Phosphorylation of ADP by Creatine Phosphate As we begin to exercise vigorously, ATP stored in working muscles is consumed within a few twitches. Then **creatine** (kre′ah-tin) **phosphate (CP)**, a unique high-energy molecule stored in muscles, is tapped to regenerate ATP while the metabolic pathways are adjusting to the suddenly higher demands for ATP. The reaction that occurs couples CP with ADP. The overall result is almost instant trans-

fer of energy and a phosphate group from CP to ADP to form ATP:

Creatine phosphate + ADP → creatine + ATP

Muscle cells store much more creatine phosphate than ATP, and the coupled reaction, catalyzed by the enzyme **creatine kinase,** is so efficient that the amount of ATP in muscle cells changes very little during the initial period of contraction.

Together, stored ATP and CP provide for maximum muscle power for about 15 seconds—long enough to energize a 100-meter dash. The coupled reaction is readily reversible, and CP reserves are replenished during periods of inactivity, when the muscle fibers are producing more ATP than they need by other pathways.

Anaerobic Glycolysis and Lactic Acid Formation
Even as stored ATP and CP are used, more ATP is generated by catabolism of glucose obtained from the blood or by breakdown of glycogen stores in the

(c) The aerobic pathway.

muscle. The initial phase of glucose respiration is **glycolysis** (gli-kol'ĭ-sis; "sugar splitting"). During glycolysis, glucose is broken down to two *pyruvic acid* molecules and some of the energy released is captured to form small amounts of ATP (2 per glucose). This pathway occurs in both the presence *and* absence of oxygen, but since it does not *use* oxygen, it is an *anaerobic* (an-a'er-ob-ik) pathway (see Figure 9.16b). Ordinarily, pyruvic acid then enters the oxygen-requiring aerobic pathways in the mitochondria and reacts with oxygen to produce still more ATP, as described below. In such cases, since oxygen is ultimately used, the respiration process as a whole is described as *aerobic*. But when muscles contract vigorously and contractile activity reaches about 70% of the maximum possible (e.g., running 600 meters with maximal effort), the bulging muscles compress the blood vessels within them, impairing blood and hence oxygen delivery. Under these conditions, most of the pyruvic acid produced during glycolysis is converted into **lactic acid,** and the overall process is referred to as **anaerobic glycolysis.** Thus, during oxygen deficit, lactic acid, rather than carbon dioxide and water, is the end product of cellular metabolism of glucose. Most of the lactic acid diffuses out of the muscles into the bloodstream and is completely washed out of the muscle tissue within 30 minutes after exercise stops. Subsequently, the lactic acid is picked up by liver, heart, or kidney cells, which can use it as an energy source. Additionally, liver cells can reconvert it to pyruvic acid or glucose and release it back into the bloodstream for muscle use, a shuffling process called the *Cori cycle* (Figure 9.17), or convert it to glycogen for storage.

The anaerobic pathway harvests only about 5% as much ATP from each glucose molecule as the aerobic pathway; however, it produces ATP about 2½ times faster. Thus, when large amounts of ATP are needed for moderate periods (30–40 seconds) of strenuous muscle activity, glycolysis can provide most of the ATP needed as long as the required fuels and enzymes are available. Together, stored ATP and CP and the glycolysis–lactic acid system can support strenuous muscle activity for nearly a minute.

Although anaerobic glycolysis is very effective in providing the energy to sustain short-term vigorous exercise, it has shortcomings. Huge amounts of glucose are used to produce relatively small harvests of ATP, and the accumulating lactic acid contributes to muscle fatigue and is at least partially responsible for the muscle soreness resulting from intense exercise.

Aerobic Respiration During rest and light to moderate exercise, even if prolonged, 95% of the ATP used for muscle activity comes from aerobic respira-

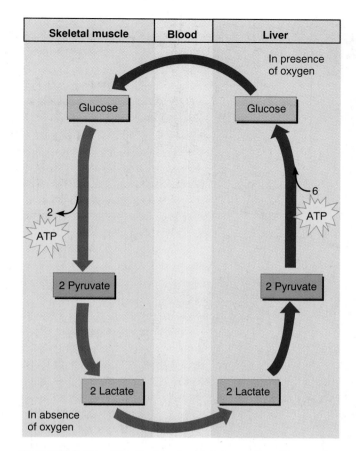

FIGURE 9.17 *The Cori cycle.* Lactate produced in muscle is tranported to the liver for resynthesis of glucose.

tion. **Aerobic respiration** occurs in the mitochondria, requires oxygen, and involves a sequence of chemical reactions in which the bonds of fuel molecules are broken and the energy released is used to make ATP. These reactions are known collectively as **oxidative phosphorylation** (fos"for-ĭ-la'shun).

During aerobic respiration, glucose is broken down entirely, yielding water, carbon dioxide, and large amounts of ATP as the final products (Figure 9.16c).

Glucose + oxygen → carbon dioxide + water + ATP

The carbon dioxide released diffuses out of the muscle tissue into the blood and is removed from the body by the lungs.

As exercise begins, muscle glycogen provides most of the fuel. After that, bloodborne glucose, pyruvic acid from glycolysis, free fatty acids, and in some cases even amino acids are the major sources of fuels for oxidation. The aerobic mechanism provides a high yield of ATP (about 36 per glucose oxidized), but it is relatively sluggish because of its many steps and it requires continuous delivery of oxygen and nutrient fuels to keep it going.

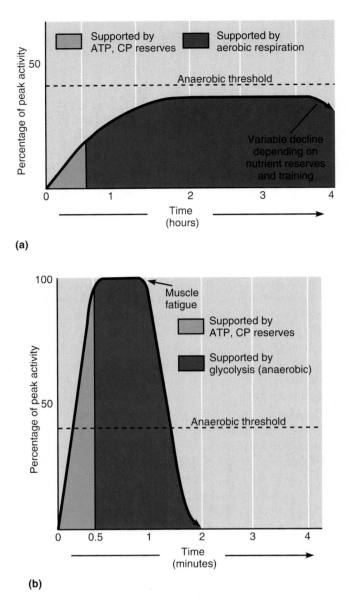

(a)

(b)

FIGURE 9.18 *Energy systems used during sports activities.* (a) Muscular activity can be continued for prolonged periods when ATP demands are maintained below the anaerobic threshold. (b) During peak activity levels, skeletal muscles rely largely on glycolysis (anaerobic respiration) for ATP synthesis. The initial activity burst is energized by ATP and CP reserves. Muscles operating at peak levels fatigue rapidly as lactic acid accumulates.

As long as it has enough oxygen, a muscle cell will form ATP by aerobic reactions. However, when exercise demands begin to exceed the ability of the muscle cells to carry out the necessary reactions quickly enough, glycolysis begins to contribute more and more of the total ATP generated. If the anaerobic mechanism is allowed to continue and accelerate, it eventually and typically leads to lactic acid accumulation and muscle fatigue.

Energy Systems Used During Sports Activities Exercise physiologists have been able to estimate the relative importance of each energy-producing system to athletic performance. Activities that require a surge of power but last only a few seconds, such as weight lifting, diving, and sprinting, rely entirely on ATP and CP stores. (Other energy-producing mechanisms do not have a chance to get going before the activity is over.) The more on-and-off or burstlike activities of tennis, soccer, and a 100-meter swim appear to be fueled almost entirely by the anaerobic mechanism producing lactic acid. Prolonged activities such as marathon runs and jogging, where endurance rather than power is the goal, depend mainly on aerobic mechanisms.

Anaerobic and aerobic pathways are closely intertwined in muscle metabolism. The anaerobic pathways often kick in temporarily early in the exercise session when the aerobic machinery is still attempting to reach full efficiency and CP reserves are exhausted. Then aerobic mechanisms take over *unless* the activity is so strenuous or prolonged that they cannot keep pace. The length of time a muscle can continue to contract using aerobic pathways is called **aerobic endurance,** and the point at which muscle metabolism converts to anaerobic glycolysis is called **anaerobic threshold.** As long as ATP demands are kept below the anaerobic threshold, light to moderate muscular activity can continue for several hours in well-conditioned individuals (Figure 9.18). By contrast, muscles working at peak levels using anaerobic glycolysis fatigue after 1–2 minutes.

Muscle Fatigue

The relatively large amounts of glycogen stored in muscle cells provide some independence from blood-delivered glucose, but with continued exertion even those reserves are exhausted. When ATP production fails to keep pace with ATP use, muscles contract less and less effectively. Ultimately muscle fatigue sets in and the muscle stops contracting, even though it may still be receiving stimuli. **Muscle fatigue** is a state of *physiological inability to contract.* This is quite different from psychological fatigue, in which we voluntarily stop exercising when we feel tired. In psychological fatigue, the flesh (muscles) is still willing, but the spirit is not! It is the will to win in the face of psychological fatigue that sets athletes apart. Notice that muscle fatigue results from a *relative deficit* of ATP, not its total absence. When no ATP is available, **contractures,** or states of continuous contraction, result because the cross bridges are unable to detach (not unlike the scenario of rigor mortis). Writer's cramp is a familiar example of temporary contractures.

Excessive accumulation of lactic acid and ionic imbalances also contribute to muscle fatigue. Lactic acid, which causes muscle pH to drop (and the muscles to ache), causes extreme fatigue and so limits the usefulness of the anaerobic mechanism for ATP production. As action potentials are transmitted, potassium is lost from the muscle cells, and excess sodium enters. So long as ATP is available to energize the Na^+-K^+ *pump*, these slight ionic imbalances are corrected. However, without ATP, the pump is inactive and muscle cells become nonresponsive. In general, intense exercise of short duration produces fatigue rapidly but recovery is also rapid. By contrast, the slow-developing fatigue of prolonged low-intensity exercise may require several hours for complete recovery.

Oxygen Debt

Whether or not fatigue occurs, vigorous exercise causes a muscle's chemistry to change dramatically. For a muscle to return to its resting state, its oxygen reserves must be replenished, the accumulated lactic acid must be reconverted to pyruvic acid, glycogen stores must be replaced, and ATP and creatine phosphate reserves must be resynthesized. Additionally, the liver must convert any persisting lactic acid released into the blood during muscle activity to glucose or glycogen (see Figure 9.17). During anaerobic muscle contraction, all of these oxygen-requiring activities occur more slowly and are (at least partially) deferred until oxygen is again available. Thus, we say an *oxygen debt* is incurred, which must be repaid. **Oxygen debt** is defined as the extra amount of oxygen that must be taken in by the body for these restorative processes, and it represents the difference between the amount of oxygen needed for totally aerobic muscle activity and the amount actually used. All nonaerobic sources of ATP used during muscle activity contribute to this debt.

A simple example will help illustrate oxygen debt. If you were to run the 100-yard dash in 12 seconds, your body would require approximately 6 L of oxygen for totally aerobic respiration. However, the VO$_2$ max (the amount of oxygen that could actually be delivered to and used by your muscles) during that 12-second interval would be about 1.2 L, far short of the required amount. Thus, you would have incurred an oxygen debt of about 4.8 L, which would be repaid by rapid deep breathing for some time after exertion ended. This heavy breathing is triggered primarily by high levels of lactic acid in the blood, which indirectly stimulates the respiratory center of the brain. The amount of oxygen used during exertion depends on several factors, including age, size, athletic training, and health. In general, the more exercise to which a person is accustomed, the higher his or her oxygen use (and anaerobic threshold) during exercise and the lower the oxygen debt incurred. For example, the VO$_2$ max of most athletes is at least 10% greater than that of a sedentary person, and that of a trained marathon runner may be as much as 50% greater.

Heat Production During Muscle Activity

Only about 40% of the energy released during muscle contraction is actually converted to useful work. The rest is given off as heat, which has to be dealt with if body homeostasis is to be maintained. When you exercise vigorously, you start to feel uncomfortably hot as your blood is warmed by the liberated heat. Ordinarily, heat buildup is prevented from reaching dangerous levels by several homeostatic mechanisms, including radiation of heat from the skin surface and sweating. Shivering represents the opposite end of homeostatic balance, in which muscle contractions are used to produce more heat. Body temperature regulation is described in Chapter 25.

Force, Velocity, and Duration of Muscle Contraction

The main factors affecting the force, velocity, and duration of muscle contraction are described next and summarized in Figure 9.19.

Force of Contraction

The force of muscle contraction is affected by (1) the number of muscle fibers contracting, (2) the relative size of the muscle, (3) series-elastic elements, and (4) the degree of muscle stretch. Let's briefly examine the role of each of these factors.

Number of Muscle Fibers Stimulated As already discussed, the more motor units that are recruited, the greater the muscle force will be.

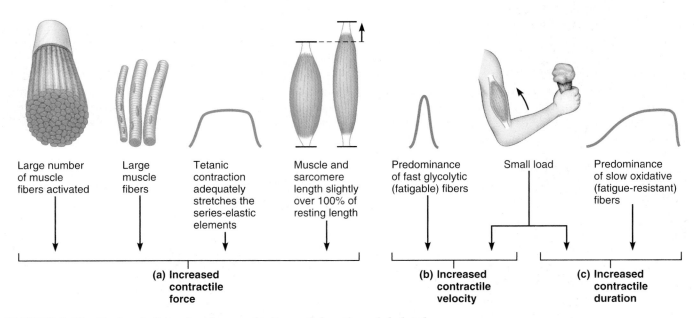

FIGURE 9.19 *Factors influencing force, velocity, and duration of skeletal muscle contraction.*

Relative Size of the Muscle The bulkier the muscle (the greater its cross-sectional area), the more tension it can develop and the greater its strength. Regular exercise increases muscle force by causing muscle cells to *hypertrophy* (hi-per′tro-fe) or increase in size.

Series-Elastic Elements Muscle shortening in and of itself does not promote movement of body parts. To do useful work, (1) a muscle must be attached to other movable structures and (2) its connective tissue coverings and tendons must be pulled taut. Then, when tension generated by cross bridge activity is transmitted to the load (the muscle insertion), movement occurs. All of the *noncontractile structures* of muscles are collectively called **series-elastic elements** because they are able to stretch and recoil. The force—the **internal tension**—generated by the contractile elements (myofibrils) stretches the series-elastic elements. These in turn transfer their tension, called the **external tension,** to the load, and when the contraction ends, their recoil helps to return the muscle to its normal resting length.

The point here is that time is required to take up slack and stretch the series-elastic elements, and while this is happening, the stretching force (internal tension) is already beginning to decline. So, in the brief twitch contractions, the external tension is always less than the internal tension (Figure 9.20a). However, when a muscle is stimulated to tetanus, more time is available to stretch the series-elastic elements, and external tension approaches that gener-

ated by the cross bridges (Figure 9.20b). So, the more rapidly a muscle is stimulated, the greater the force it exerts.

Degree of Muscle Stretch The optimal resting length for muscle fibers is the length at which they can generate maximum force (Figures 9.19 and 9.21). The ideal **length-tension relationship** within a sarcomere occurs when a muscle is slightly stretched and the actin and myosin filaments barely overlap, because this permits sliding along nearly the entire length of the actin filaments. If a muscle fiber is stretched to the extent that the myofilaments do not overlap at all (Figure 9.21c), the cross bridges have nothing to attach to and cannot generate tension. Alternatively, if the sarcomeres are so compressed and cramped that the Z discs abut the thick myofilaments and the thin filaments touch and interfere with one another (Figure 9.21a), little or no further shortening can occur.

Identical relationships exist in a whole muscle. A severely stretched muscle (say one at 180% of its optimal length) cannot develop any tension. Likewise, once a muscle contracts to 75% of its resting length, further shortening is limited. Muscles, like muscle fibers, are at an optimal operational length from about 80% to about 120% of their normal resting length. In the body, skeletal muscles are maintained near that optimum by the way they are attached to bones; that is, the joints normally prevent bone movements that would stretch attached muscles beyond their optimal range.

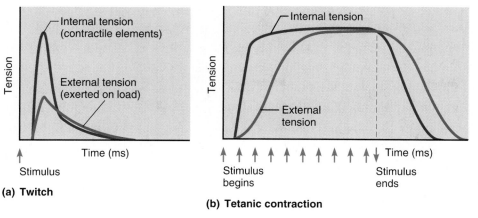

(a) **Twitch**

(b) **Tetanic contraction**

FIGURE 9.20 *Relationship of stimulus frequency to external tension exerted on the load.* **(a)** During a single-twitch contraction, the internal tension developed by the cross bridges peaks and begins to drop well before the series–elastic elements are stretched to equal tension. As a result, the external tension exerted on the load is always less than the internal tension. **(b)** When the muscle is stimulated tetanically, the internal tension lasts long enough for the series–elastic components to be stretched to similar tension, so the external tension approaches and finally equals the internal tension.

Velocity and Duration of Contraction

Muscles vary in how fast they can contract and how long they can continue to contract before they fatigue. These characteristics are influenced by muscle fiber type, load, and recruitment.

Muscle Fiber Type There are several ways of classifying muscle fibers (some students say enough ways to make your head spin), but the task of learning about these classes will be easier if you initially pay attention to just two major functional characteristics of the muscle fibers:

- **Speed of contraction.** On the basis of speed of shortening or contraction, there are **slow fibers** and **fast fibers.** The difference in speed of these fibers reflects how fast their myosin ATPases split ATP. Obviously, the fast fibers have more efficient ATPases.

- **The major pathways for forming ATP.** The cells that mostly rely on the oxygen-using aerobic pathways for ATP generation are called **oxidative fibers,** whereas those that rely more on anaerobic glycolysis are **glycolytic fibers.**

On the basis of these two criteria, we can then assign skeletal muscle cells to one of three categories: **slow oxidative fibers, fast oxidative fibers,** and **fast glycolytic fibers.** The details about each of these groups of muscle fibers are given in Table 9.3, but a word to the wise: Do not approach this information by rote memorization—you'll just get frustrated. Instead, start with what you know for any category and see how the characteristics listed support that. For example, a fiber that is a *slow oxidative fiber*

- Contracts relatively slowly because its myosin ATPases are slow (a criterion)

- Depends on oxygen delivery and aerobic mechanisms (high oxidative capacity—a criterion)

- Is fatigue resistant and has high endurance (typical of a cell that depends on aerobic metabolism)

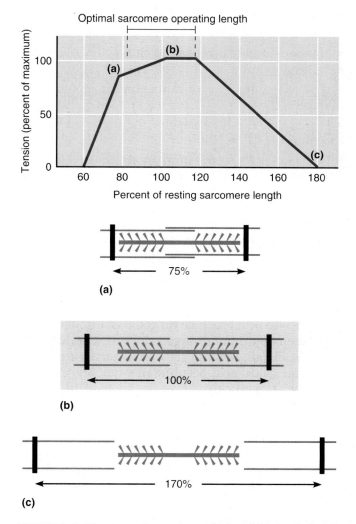

FIGURE 9.21 *Length-tension relationships in skeletal muscles.* Maximum force generation is possible when the muscle is a little over 100% of its resting length. Increases and decreases beyond the optimal range result in decreased force and finally an inability to generate tension. Depicted here is the relative sarcomere length in a muscle that is **(a)** strongly contracted, **(b)** at normal resting length, and **(c)** excessively stretched.

FIGURE 9.22 *Influence of load on velocity and duration of contraction.* (**a**) Relationship of load to degree and duration of contraction (shortening). (**b**) Relationship of load to velocity of shortening. As the load increases, the speed of contraction decreases.

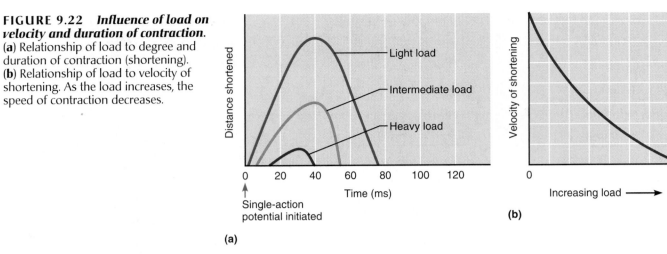

- Is thin (a large amount of cytoplasm impedes diffusion of O_2 and nutrients from the blood)

- Has relatively little power (a thin cell can contain only limited numbers of myofibrils)

- Has a large number of mitochondria (actual sites of oxygen use)

- Has a rich capillary supply (the better to deliver bloodborne O_2)

- Is red (its color stems from an abundant supply of myoglobin, an oxygen-binding pigment unique to muscle cells that stores O_2 reserves in the cell and aids diffusion of O_2 through the cell)

Conversely, a cell that is a *fast glycolytic fiber* contracts rapidly (Figure 9.19) and depends on plentiful glycogen reserves for fuel rather than a blood-delivered supply and does not use oxygen (Table 9.3). Consequently, compared to the oxidative cell, it has few mitochondria, little myoglobin (and so is white), and tends to be a much larger cell (because it doesn't depend on continuous oxygen and nutrient diffusion from the blood). Because glycogen reserves are short-lived and lactic acid accumulates quickly in these cells, they tire quickly and hence are fatigable fibers. However, their large diameter reflects their plentiful contractile filaments that allow them to contract powerfully before they "poop out." Thus, the fast glycolytic fibers are best suited for short-term, rapid, intense movements (moving furniture across the room, for example). Other characteristics, plus the details about the intermediate muscle fiber type called the fast oxidative fiber, are listed in Table 9.3.

Although some muscles have a predominance of one fiber type, most body muscles contain a mixture of fiber types, which gives them a range of contractile speeds and fatigue resistance. For example, a muscle in the calf of the leg can propel us in a sprint (using mostly its white fast glycolytic fibers), or in long-distance races (making good use of its intermediate fast oxidative fibers), or may simply help maintain our standing posture (using slow oxidative fibers). As might be expected, all muscle fibers in a particular motor unit are of the same type.

Although everyone's muscles contain mixtures of the three fiber types, some people have relatively more of one variety. These differences are genetically controlled and no doubt determine athletic capabilities to a large extent. For example, muscles of marathon runners have a high percentage of slow oxidative fibers (about 80%), while those of sprinters contain a higher percentage (about 60%) of the fast oxidative fibers. Weight lifters appear to have approximately equal amounts of fast glycolytic and slow oxidative fibers.

Load Because muscles are attached to bones, they are always pitted against some resistance, or load, when they contract. As you might expect, they contract fastest when there is no added load on them. The greater the load, the longer the latent period, the slower the contraction, and the shorter the duration of contraction (Figure 9.22). If the load exceeds the muscle's ability to move it, the speed of shortening is zero and the contraction is isometric.

Recruitment Just as many hands on a project can get the work done more quickly and also can keep working longer, the more motor units that are contracting, the faster and more prolonged a contraction can be.

Effect of Exercise on Muscles

The amount of work a muscle does is reflected in changes in the muscle itself. When used actively or strenuously, muscles may increase in size or strength or become more efficient and fatigue resistant. But muscle inactivity, whatever the cause, *always* leads to muscle weakness and wasting.

TABLE 9.3	Structural and Functional Characteristics of the Three Types of Skeletal Muscle Fibers		
	Slow oxidative fibers	*Fast oxidative fibers*	*Fast glycolytic fibers*
Metabolic characteristics:			
Speed of contraction	Slow	Fast	Fast
Myosin ATPase activity	Slow	Fast	Fast
Primary pathway for ATP synthesis	Aerobic	Aerobic	Anaerobic glycolysis
Myoglobin content	High	High	Low
Glycogen stores	Low	Intermediate	High
Rate of fatigue	Slow (fatigue-resistant)	Intermediate (moderately fatigue-resistant)	Fast (fatigable)
Activities best suited for:			
	Endurance-type activities— e.g., running a marathon; maintaining posture (antigravity muscles)	Sprinting, walking	Short-term intense or powerful movements, e.g., hitting a baseball
Structural characteristics:			
Color	Red	Red to pink	White (pale)
Fiber diameter	Small	Intermediate	Large
Mitochondria	Many	Many	Few
Capillaries	Many	Many	Few

Adaptations to Exercise

Aerobic, or **endurance, exercise** such as swimming, jogging, fast walking, and biking results in several recognizable changes in skeletal muscles. Capillaries surrounding the muscle fibers, as well as mitochondria within them, increase in number, and the fibers synthesize more myoglobin. These changes occur in all fiber types, but they are most dramatic in the slow oxidative fibers that depend primarily on aerobic pathways. The changes result in more efficient muscle metabolism and in greater endurance, strength, and resistance to fatigue.

However, aerobic exercise benefits far more than the skeletal muscles. It makes overall body metabolism and neuromuscular coordination more efficient, improves gastrointestinal mobility (and elimination), and enhances the strength of the skeleton. Aerobic exercise also promotes changes in cardiovascular and respiratory system functioning that improve the delivery of oxygen and nutrients to all body tissues. The heart hypertrophies and develops a greater stroke volume (more blood is pumped out with each beat), fatty deposits are cleared from blood vessel walls, and gas exchange in the lungs becomes more efficient.

The moderately weak but sustained type of muscle activity required for endurance exercise does not promote significant skeletal muscle hypertrophy, even though the exercise may go on for hours. Muscle hypertrophy, illustrated by the bulging biceps and chest muscles of a professional weight lifter, results mainly from high-intensity **resistance exercise** (typically under anaerobic conditions) such as weight lifting or isometric exercise, in which the muscles are pitted against high-resistance or immovable forces. Here strength, not stamina, is important; a few minutes every other day is sufficient. (Indeed, even a proverbial weakling can put on 50% more muscle within a year.) The increased muscle bulk largely reflects increases in the size of individual muscle fibers (particularly the fast glycolytic variety) rather than an increased number of muscle fibers. (However, some of the increased muscle size may result from longitudinal splitting or tearing of the enlarged fibers and subsequent growth of these "split" cells, or conversely it may arise from the proliferation and fusion of satellite cells (see p. 313). The controversy is still raging.) Vigorously stressed muscle fibers contain more mitochondria, form more myofilaments and myofibrils, and store more glycogen.

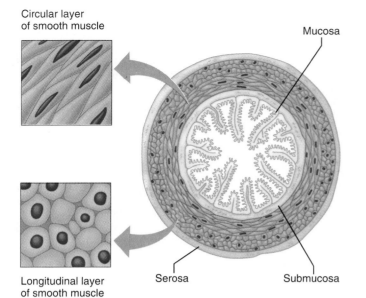

Circular layer
of smooth muscle

Mucosa

Longitudinal layer
of smooth muscle

Serosa

Submucosa

FIGURE 9.23 *Arrangement of smooth muscle in the walls of hollow organs.* As seen in this simplified cross-sectional view of the intestine, two smooth muscle layers are present (one circular and the other longitudinal) and run at right angles to each other.

The amount of connective tissue between the cells also increases. Collectively these changes promote significant increases in muscle strength and size.

Resistance training can produce magnificently bulging muscles, but if training is done unwisely, some muscles may be developed more than others. Since muscles work in antagonistic pairs (or groups), opposing muscles must be equally strong to work together smoothly. When muscle training is not balanced, individuals can become *muscle-bound;* they lack flexibility, have a generally awkward stance, and are unable to make full use of their muscles.

Endurance and resistance exercises produce different patterns of muscular response, so it is important to know what your exercise goals are. Lifting weights will not improve your endurance for a triathlon. By the same token, jogging will do little to improve muscle definition for competition in the Mr. or Ms. Muscle contest, nor will it enhance your strength for moving furniture. **Cross training,** which alternates aerobic activities with anaerobic ones, provides the best program for optimal health.

Homeostatic Imbalance
Disuse Atrophy

To remain healthy, muscles must be active. Complete immobilization due to enforced bed rest or loss of neural stimulation results in *disuse atrophy* (degeneration and loss of mass), which begins almost as soon as the muscles are immobilized. Under such conditions, muscle strength can decrease at the rate of 5% per day!

Even at rest, muscles receive weak intermittent stimuli from the nervous system. When totally deprived of neural stimulation, the result is devastating; the paralyzed muscle may ultimately atrophy to about one-quarter of its initial size. Lost muscle tissue is replaced by fibrous connective tissue, making muscle rehabilitation impossible. Atrophy of a denervated muscle may be delayed by electrically stimulating the muscle periodically while waiting to see whether the damaged nerve fibers regenerate. ■

Training Smart to Prevent Overuse Injuries

Don't expect to get in shape by playing a sport; you've *got to get in shape to play* the sport. Regardless of your choice—running, lifting weights, or tennis, for example—exercise stresses muscles. Muscle fibers tear, tendons stretch, and accumulation of lactic acid in the muscle causes pain.

Effective training walks a fine line between working hard enough to improve and preventing overuse injuries. Whatever the activity, there are certain basic guidelines for training: (1) stretching immediately after warming up the muscle is an absolute requirement, and (2) exercise gains adhere to the *overload principle.* Forcing a muscle to work hard promotes increased muscle strength and endurance, and as muscles adapt to the increased demands, they must be overloaded even more to produce further gains. To become faster, you must train at an increasingly fast pace. A heavy-workout day should be followed by one of rest or an easy workout to allow the muscles to recover and repair themselves. Doing too much too soon, or ignoring the warning signs of muscle or joint pain, increases the risk of **overuse injuries** that may prevent future sports activities, or even lead to lifetime disability. Changing or restricting activity, using ice packs to prevent or reduce inflammation, and taking nonsteroidal anti-inflammatory drugs (aspirin, ibuprofen, acetaminophen) to reduce pain are the main treatment modes for virtually all overuse injuries.

SMOOTH MUSCLE

Except for the heart, which is made up of cardiac muscle, the muscle in the walls of hollow organs is almost entirely smooth muscle. Although the chemical and mechanical events of contraction are essentially the same in all muscle tissues, smooth muscle is distinctive in several ways (see Table 9.4 on pp. 310–311).

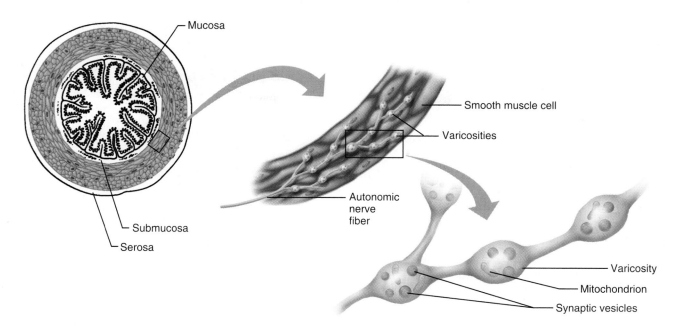

FIGURE 9.24   *Innervation of smooth muscle.* Most smooth muscle cells (the cells in single-unit muscle, discussed later) are innervated by autonomic nervous system fibers that release their neurotransmitters from varicosities into a wide synaptic cleft (a diffuse junction).

Arrangement and Microscopic Structure of Smooth Muscle Fibers

Smooth muscle fibers are spindle-shaped cells, each with one centrally located nucleus (Figure 9.23). Typically, they have a diameter of 2–10 μm and a length of several hundred μm. Skeletal muscle fibers are some 20 times wider and thousands of times longer.

Smooth muscle lacks the coarser connective tissue sheaths seen in skeletal muscle. However, a small amount of fine connective tissue (endomysium), secreted by the smooth muscles themselves and containing blood vessels and nerves, is found between the smooth muscle fibers. Most smooth muscle fibers are organized into sheets of closely apposed fibers. These sheets occur in the walls of all but the smallest blood vessels and in the walls of hollow organs of the respiratory, digestive, urinary, and reproductive tracts. In most cases, two sheets of smooth muscle are present, with their fibers oriented at right angles to each other (Figure 9.23). In one of these, the *longitudinal layer*, the muscle fibers run parallel to the long axis of the organ. Consequently, when the organ contracts, it dilates and shortens. In the other, the *circular layer*, the fibers run around the circumference of the organ. Contraction of this layer constricts the lumen (cavity) of the organ and causes the organ to elongate. The alternating contraction and relaxation of these opposing layers mixes substances in the lumen and squeezes them through the organ's internal pathway. This phenomenon is called **peristalsis** (per"i-stal'sis; "around contraction"). Contraction of smooth muscle in the rectum, urinary bladder, and uterus helps those organs to expel their contents. Smooth muscle contraction also accounts for the wheeze of asthma and for stomach cramps.

Smooth muscle lacks the highly structured neuromuscular junctions of skeletal muscle. Instead, the innervating nerve fibers, which are part of the autonomic nervous system (described in Chapter 14), have numerous bulbous swellings, called **varicosities** (Figure 9.24). The varicosities release neurotransmitter into a wide synaptic cleft in the general area of the smooth muscle cells. Such junctions are called **diffuse junctions.**

Intracellularly, the sarcoplasmic reticulum of smooth muscle fibers is less developed than that of skeletal muscle and lacks a specific pattern relative to the myofilaments. Some of the SR tubules of the smooth muscle directly touch the plasma membrane at several sites, forming what resemble half-triads that may act to couple the action potential to calcium release from the SR. T tubules are notably absent, but the plasma membrane of the smooth muscle fiber has multiple pouchlike infoldings called **caveoli** that enclose bits of extracellular fluid and allow a high concentration of Ca^{2+} to be sequestered close to the membrane. Consequently, when calcium channels open, Ca^{2+} influx can occur rapidly. Although the SR *does* release some of the calcium ions that trigger contraction, most gain entry via calcium channels directly from the extracellular space. Contraction ends when calcium is actively transported into the SR and out of the cell.

No striations are visible, as the name *smooth muscle* indicates, and no sarcomeres are present. Smooth muscle fibers *do* contain interdigitating thick and thin filaments, but the thick filaments are much longer in smooth muscle than in skeletal muscle. The proportion and organization of the myofilaments are also different:

1. The ratio of thick to thin actin filaments is much lower in smooth muscle than in skeletal muscle (1:13 compared to 1:2). However, the thick filaments of smooth muscle contain actin-gripping heads along their *entire length*, a feature that allows this class of muscles to be as powerful as skeletal muscle, size for size.

2. Tropomyosin is associated with the thin filaments, but no troponin complex is present.

3. The thick and thin filaments are arranged "on the bias" (diagonally) within the cell so that they appear to spiral down the long axis of a smooth muscle cell like the stripes on a barber pole. Because of this arrangement, the smooth muscle cells contract in a corkscrew-like way (see Figure 9.25b).

4. Smooth muscle fibers contain longitudinal bundles of noncontractile *intermediate filaments* that resist tension. These attach to dark-staining **dense bodies** at regular intervals. The dense bodies, which are also tethered to the sarcolemma, act as anchoring points for groups of thin filaments; they therefore correspond to the Z discs of skeletal muscle. The intermediate filament-dense body network forms a strong, cable-like intracellular cytoskeleton that harnesses the pull generated by the sliding of the myofilaments during contraction (Figure 9.25). During contraction, the cellular areas anchored to the sarcolemma between the dense bodies bulge outward, giving the cell a puffy appearance (see Figure 9.25b). Dense bodies at the sarcolemma surface also bind the muscle cell to the connective tissue fibers outside the cell (endomysium), an arrangement that transmits the pulling force to the surrounding connective tissue.

Contraction of Smooth Muscle

Mechanism and Characteristics of Contraction

In most cases, adjacent smooth muscle cells exhibit slow, synchronized contractions; the whole sheet responds to a stimulus in unison. This phenomenon

Intermediate filament bundles attached to dense bodies

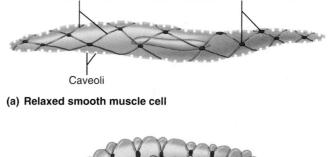
Caveoli

(a) Relaxed smooth muscle cell

(b) Contracted smooth muscle cell

FIGURE 9.25 *Intermediate filaments and dense bodies of smooth muscle fibers.* Intermediate filaments and dense bodies of smooth muscle fibers harness the pull generated by the activity of myosin cross bridges. Intermediate filaments attach to dense bodies scattered throughout the sarcoplasm and occasionally anchor to the sarcolemma. (**a**) A relaxed smooth muscle cell. (**b**) A contracted smooth muscle cell. (Caveoli are not illustrated in [b]).

reflects electrical coupling of smooth muscle cells by *gap junctions*, specialized cell connections described in Chapter 3. Skeletal muscle cells are electrically isolated from one another, each stimulated to contract by its own neuromuscular junction. By contrast, gap junctions allow smooth muscles to transmit action potentials from cell to cell. Some smooth muscle fibers in the stomach and small intestine act as *pacemaker cells* and, once excited, they act as "drummers" to set the contractile pace for the entire muscle sheet. Additionally, some of these pacemakers are self-excitatory, that is, they depolarize spontaneously in the absence of external stimuli. However, both the rate and intensity of smooth muscle contraction may be modified by neural and chemical stimuli.

The contraction mechanism in smooth muscle is like that of skeletal muscle in the following ways: (1) actin and myosin interact by the sliding filament mechanism; (2) the final trigger for contraction is a rise in the intracellular calcium ion level; and (3) the sliding process is energized by ATP.

During excitation-contraction coupling, Ca^{2+} is released by the tubules of the SR, but, as mentioned above, it also moves into the cell from the extracellular space. Ca^{2+} serves the same triggering role in all muscle types, but the activation mechanism is different in smooth muscle. Here calcium interacts with regulatory molecules, in this case **calmodulin,** a calcium-binding protein, and a kinase enzyme called **myosin light chain kinase,** that are part of the *thick* filaments in order to activate myosin. The thin filaments lack troponin and so are always ready for contraction. Apparently the sequence of events is:

1. Ionic calcium binds to calmodulin, activating it.

2. Activated calmodulin activates the kinase enzyme.

3. The activated kinase catalyzes transfer of phosphate from ATP to myosin cross bridges.

4. Phosphorylated cross bridges interact with actin of the thin filaments, producing shortening.

5. As with skeletal muscle, smooth muscle relaxes when intracellular Ca^{2+} levels drop.

Contraction of smooth muscle is slow, sustained, and resistant to fatigue. It takes 30 times longer to contract and relax than does skeletal muscle and can maintain the same contractile tension for prolonged periods at less than 1% of the energy cost. At least part of the striking energy economy of smooth muscle reflects the sluggishness of its ATPases (the myosin light chain kinases) compared to those in skeletal muscle. Moreover, smooth muscle myofilaments may latch together during prolonged contractions, saving energy in that way as well.

The ATP-efficient contraction of smooth muscle is extremely important to overall body homeostasis. The smooth muscle in small arterioles and other visceral organs routinely maintains a moderate degree of contraction, called *smooth muscle tone,* day in and day out without fatiguing. Since smooth muscle has low energy requirements, its fibers contain relatively few mitochondria, and most ATP is made via anaerobic pathways.

Regulation of Contraction

Neural Regulation In some cases, the activation of smooth muscle by a neural stimulus is identical to that described for skeletal muscle. An action potential is generated by neurotransmitter binding, and is coupled to a rise in calcium ions in the sarcoplasm. However, not all neural signals received by smooth muscle result in action potential generation. Some types of smooth muscle respond with graded potentials (local electrical signals) only. Furthermore, not all smooth muscle activation results from neural signals.

All somatic nerve endings, that is, nerve endings that serve skeletal muscle, release acetylcholine, which always excites the skeletal muscle. However, different autonomic nerves serving the smooth mus-

cle of visceral organs release different neurotransmitters, each of which may excite or inhibit a particular group of smooth muscle cells. The effect of a specific neurotransmitter on a given smooth muscle cell depends on the type of receptor molecules on the sarcolemma. For example, when acetylcholine binds to ACh receptors on smooth muscle in the bronchioles (small air passageways of the lungs), the muscle contracts strongly, narrowing the bronchioles. When norepinephrine, released by a different type of autonomic fiber, binds to norepinephrine receptors on the *same* smooth muscle cells, the effect is inhibitory; the muscle relaxes, dilating the air passageways. However, when norepinephrine binds to smooth muscle in the walls of most blood vessels, it stimulates the smooth muscle cells to contract and constrict the vessel.

Local Factors Some smooth muscle layers have no nerve supply at all; instead they depolarize spontaneously or in response to chemical stimuli. Others respond to both neural and chemical stimuli. Chemical factors that cause smooth muscle contraction or relaxation without an action potential (by enhancing or inhibiting calcium ion entry into the sarcoplasm) include certain hormones, lack of oxygen, excess carbon dioxide, and low pH. The direct response of smooth muscle to these chemical stimuli alters smooth muscle activity according to local tissue needs and probably is most responsible for smooth muscle tone. For example, the hormone gastrin stimulates contraction of stomach smooth muscle so that it can churn foodstuffs more efficiently. We will consider how the smooth muscle of specific organs is activated as we discuss each organ in subsequent chapters.

Special Features of Smooth Muscle Contraction

Smooth muscle is intimately involved with the functioning of most hollow organs and has a number of unique characteristics. Some of these—smooth muscle tone; slow, prolonged contractile activity; and low energy requirements—have already been considered. But smooth muscle also responds differently to stretch and can shorten more than other muscle types. Let's take a look.

Response to Stretch When cardiac muscle is stretched, it responds with more vigorous contractions. So does skeletal muscle up to a point (about 120% of resting length). Stretching of smooth muscle also provokes contraction, which automatically

TABLE 9.4	Comparison of Skeletal, Cardiac, and Smooth Muscle		
Characteristic	*Skeletal*	*Cardiac*	*Smooth*
Body location	Attached to bones or (some facial muscles) to skin	Walls of the heart	Single-unit muscle in walls of hollow visceral organs (other than the heart); multiunit muscle in intrinsic eye muscles
Cell shape and appearance	Single, very long, cylindrical, multinucleate cells with very obvious striations	Branching chains of cells; uni- or binucleate; striations	Single, fusiform, uninucleate; no striations
Connective tissue components	Epimysium, perimysium, and endomysium	Endomysium attached to fibrous skeleton of heart	Endomysium
Presence of myofibrils composed of sarcomeres	Yes	Yes, but myofibrils are of irregular thickness	No, but actin and myosin filaments are present throughout; dense bodies anchor actin filaments
Presence of T tubules and site of invagination	Yes; two in each sarcomere at A-I junctions	Yes; one in each sarcomere at Z disc; larger diameter than those of skeletal muscle	No; only caveoli
Elaborate sarcoplasmic reticulum	Yes	Less than skeletal muscle (1–8% of cell volume); scant terminal cisternae	Equivalent to cardiac muscle (1–8% of cell volume); some SR contacts the sarcolemma

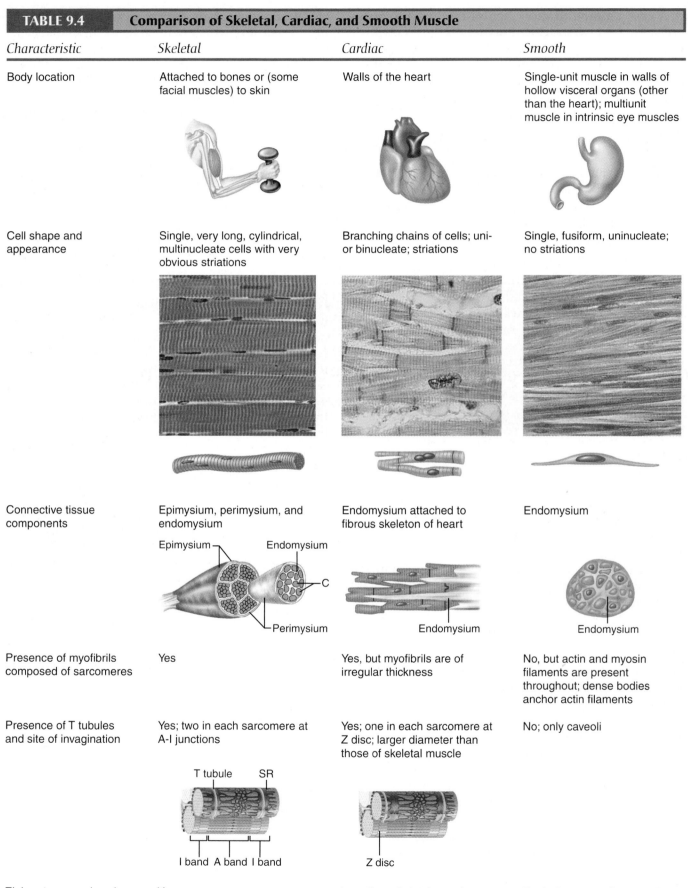

TABLE 9.4	Comparison of Skeletal, Cardiac, and Smooth Muscle *(continued)*		
Characteristic	*Skeletal*	*Cardiac*	*Smooth*
Presence of gap junctions	No	Yes; at intercalated discs	Yes; in single-unit muscle
Cells exhibit individual neuromuscular junctions	Yes	No	Not in single-unit muscle; yes in multiunit muscle
Regulation of contraction	Voluntary via axonal endings of the somatic nervous system	Involuntary; intrinsic system regulation; also autonomic nervous system controls; hormones; stretch	Involuntary; autonomic nerves, hormones, local chemicals; stretch
Source of Ca²⁺ for calcium pulse	Sarcoplasmic reticulum (SR)	SR and from extracellular fluid	SR and from extracellular fluid
Site of calcium regulation	Troponin on actin-containing thin filaments	Troponin on actin-containing thin filaments	Calmodulin on myosin-containing thick filaments
Presence of pacemaker(s)	No	Yes	Yes (in single-unit muscle only)
Effect of nervous system stimulation	Excitation	Excitation or inhibition	Excitation or inhibition
Speed of contraction	Slow to fast	Slow	Very slow
Rhythmic contraction	No	Yes	Yes in single-unit muscle
Response to stretch	Contractile strength increases with degree of stretch (to a point)	Contractile strength increases with degree of stretch	Stress-relaxation response
Respiration	Aerobic and anaerobic	Aerobic	Mainly anaerobic

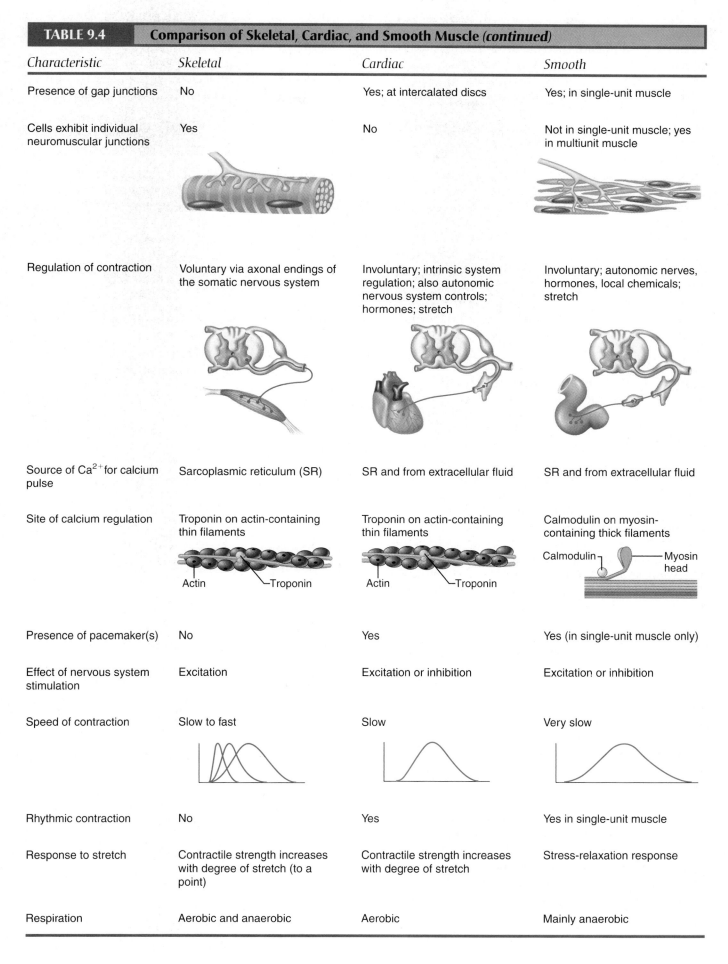

moves substances along an internal tract. However, the increased tension persists only briefly; soon the muscle adapts to its new length and relaxes, while still retaining the ability to contract on demand. This response, called the **stress-relaxation response,** allows a hollow organ to fill or expand slowly (within certain limits) to accommodate a greater volume without promoting contractions that would expel their contents. This is an important attribute, because organs such as the stomach and bladder must be able to store their contents temporarily. If this did not occur, the stretching of your stomach and intestines as you ate would lead to vigorous contractions that would rush the foodstuffs through your digestive tract, providing insufficient time for digestion and absorption of the nutrients. Likewise, your urinary bladder would be unable to store the continuously made urine until it was convenient to empty your bladder. (You would spend all your time in the bathroom while you quietly starved to death!)

Length and Tension Changes Smooth muscle stretches much more than skeletal muscle and generates more tension than skeletal muscles stretched to a comparable extent. As Figure 9.21c shows, precise, highly organized sarcomeres limit how far a skeletal muscle can be stretched before it is unable to generate force. In contrast, the lack of sarcomeres and the irregular, overlapping arrangement of smooth muscle filaments allow them to generate considerable force, even when they are substantially stretched. The total length change that skeletal muscles can undergo and still function with good to fair efficiency is about 60% (from 30% shorter to 30% longer than resting length), but smooth muscle can contract from twice its normal length to half its normal (resting) length—a total length change of 150%. This allows hollow organs to tolerate tremendous changes in volume without becoming flabby when they empty.

Hyperplasia Besides the ability to hypertrophy (increase in cell size), which is common to all muscle cells, certain smooth muscle fibers can divide to increase their numbers, that is, they undergo *hyperplasia* (hi"per-pla′ze-ah). Excellent examples are seen in the response of the uterus to estrogen. At puberty, girls' plasma estrogen levels rise. As estrogen binds to uterine smooth muscle receptors, it stimulates the synthesis of more uterine smooth muscle, causing the uterus to grow to adult size. During pregnancy, high blood levels of estrogen stimulate uterine hyperplasia to accommodate the increasing size of the fetus.

◮ *Homeostatic Imbalance*

Proliferative disorders of smooth muscle may underlie *restenosis*, the reblockage of narrowed coronary arteries that have been dilated by angioplasty. (See pp. 722–723.) ∎

Types of Smooth Muscle

The smooth muscle in different body organs varies substantially in its (1) fiber arrangement and organization, (2) responsiveness to various stimuli, and (3) innervation. However, for simplicity, smooth muscle is usually categorized into two major types: *single-unit* and *multiunit smooth muscle.*

Single-Unit Smooth Muscle

Single-unit smooth muscle, commonly called **visceral muscle,** is far more common. Its cells (1) contract as a unit and rhythmically, (2) are electrically coupled to one another by *gap junctions* (see Figure 9.24), and (3) often exhibit spontaneous action potentials. All the smooth muscle characteristics described so far pertain to single-unit smooth muscle. Thus, the cells of single-unit smooth muscle are arranged in opposing sheets, exhibit the stress-relaxation response, and so on.

Multiunit Smooth Muscle

The smooth muscles in the large airways to the lungs and in large arteries, the arrector pili muscles attached to hair follicles, and the internal eye muscles that adjust your pupil size and allow you to focus visually are all examples of **multiunit smooth muscle.**

In contrast to what we see in single-unit muscle, gap junctions are rare, and spontaneous and synchronous depolarizations are infrequent. Like skeletal muscle, multiunit smooth muscle

1. Consists of muscle fibers that are structurally independent of each other

2. Is richly supplied with nerve endings, each of which forms a motor unit with a number of muscle fibers

3. Responds to neural stimulation with graded contractions

However, while skeletal muscle is served by the somatic (voluntary) division of the nervous system, multiunit smooth muscle (like single-unit smooth muscle) is innervated by the autonomic (involuntary) division and is also responsive to hormonal controls.

DEVELOPMENTAL ASPECTS OF MUSCLES

With rare exceptions, all muscle tissues develop from embryonic mesoderm cells called **myoblasts.** The multinucleate skeletal muscle fibers form by the fusion of many myoblasts. Ordinarily, skeletal muscle fibers are contracting by week 7 when the embryo is only about 1 inch long. Initially ACh receptors "sprout" over the entire surface of the developing myoblasts. But as spinal nerves invade the muscle masses, the nerve endings target individual myoblasts and release a growth factor called *agrin.* Agrin stimulates clustering of ACh receptors at the newly forming motor end plates. Electrical activity of the neurons serving the muscle fibers also plays a critical role in muscle fiber maturation. As the muscle fibers are brought under the control of the somatic nervous system, the relative number of fast and slow contractile fiber types is determined as well.

Myoblasts producing cardiac and smooth muscle cells do not fuse. However, both develop gap junctions at a very early embryonic stage. Cardiac muscle is pumping blood just 3 weeks after fertilization.

Specialized skeletal and cardiac muscle cells become amitotic early on but retain the ability to lengthen and thicken in a growing child and to hypertrophy in adults. However, myoblastlike cells called *satellite cells* associated with skeletal muscle help to repair injured fibers and allow *very limited* regeneration of dead skeletal muscle fibers. Cardiac muscle totally lacks satellite cells. It was thought to have no regenerative capability whatsoever, but recent studies suggest that cardiac cells do divide at a modest rate. Nonetheless, injured heart muscle is repaired mostly by scar tissue. Smooth muscles have a good ability to regenerate throughout life.

At birth, a baby's movements are uncoordinated and largely reflexive. Muscular development reflects the level of neuromuscular coordination, which develops in a head-to-toe and proximal-to-distal direction. A baby can lift its head before it can walk, and gross movements precede fine ones. All through childhood, our control of our skeletal muscles becomes more and more sophisticated. By midadolescence, we have reached the peak of *natural* neural control of muscles, and can either accept that level of development or improve it by athletic or other types of training.

A frequently asked question is whether the difference in strength between women and men has a biological basis. It does. Individuals vary, but on average, women's skeletal muscles make up approximately 36% of body mass, whereas men's account for about 42%. Men's greater muscular development

is due primarily to the effects of testosterone on skeletal muscle, not to the effects of exercise. Body strength per unit muscle mass, however, is the same in both sexes. Strenuous muscle exercise causes more muscle enlargement in males than in females, again because of the influence of male sex hormone. Some athletes take large doses of synthetic male sex hormones ("steroids") to increase their muscle mass. This illegal and dangerous practice is discussed in *A Closer Look* on p. 314.

Because of its rich blood supply, skeletal muscle is amazingly resistant to infection throughout life and, given good nutrition and moderate exercise, relatively few problems afflict skeletal muscles. However, muscular dystrophy is a serious condition that deserves more than a passing mention.

Homeostatic Imbalance

The term **muscular dystrophy** refers to a group of inherited muscle-destroying diseases that generally appear during childhood. The affected muscles enlarge due to fat and connective tissue deposit, but the muscle fibers themselves degenerate and atrophy.

The most common and serious form is **Duchenne muscular dystrophy (DMD),** which is inherited as a sex-linked recessive disease: Females carry and transmit the abnormal gene, which is expressed almost exclusively in males (1 in every 3500 births). This tragic disease is usually diagnosed when the boy is between two and ten years old. Active, normal-appearing children become clumsy and fall frequently as their muscles weaken. The disease progresses relentlessly from the extremities upward, finally affecting the head and chest muscles. Victims rarely live beyond their early 20s due to respiratory failure.

Recent research has pinned down the cause of DMD: The diseased muscle fibers lack a cytoplasmic protein, called *dystrophin,* that links the cytoskeleton to the extracellular matrix and helps to stabilize the sarcolemma. Although the precise defect is still unknown, problems in regulating calcium entry are suspect.

There is still no cure for Duchenne, but one promising new technique, called *myoblast transfer therapy,* involves injecting diseased muscle with healthy myoblast cells that fuse with the unhealthy ones. Once inside the cells, the normal gene thus provided allows the fibers to produce dystrophin and so to grow normally. Clinical trials of this technique on humans are showing limited success. Another approach is to inject *plasmids,* tiny circlets of DNA containing a pared-down version of the dystrophin

A CLOSER LOOK   Athletes Looking Good and Doing Better with Anabolic Steroids?

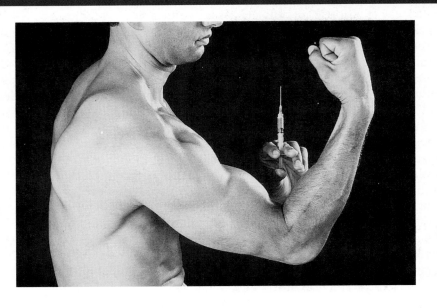

Society loves a winner and top athletes reap large social and monetary rewards. Thus, it is not surprising that some will grasp at anything that will increase their performance—including anabolic steroids. Anabolic steroids, variants of the male sex hormone testosterone engineered by pharmaceutical companies, were introduced in the 1950s to treat victims of anemia and certain muscle-wasting diseases and to prevent muscle atrophy in patients immobilized after surgery. Testosterone is responsible for the increase in muscle and bone mass and other physical changes that occur during puberty and convert boys into men. Convinced that megadoses of the steroids could produce enhanced masculinizing effects in grown men, many athletes and bodybuilders were using the steroids by the early 1960s, and the practice is still going strong today. Indeed, it has been estimated that nearly one in every ten young men has tried steroids, so use is no longer confined to athletes looking for the edge.

It has been difficult to determine the incidence of anabolic steroid use among athletes because the use of drugs has been banned by most international competitions, and users (and prescribing physicians or drug dealers) are naturally reluctant to talk about it. Nonetheless, there is little question that many professional bodybuilders and athletes competing in events that require muscle strength (e.g., shot put, discus throwing, and weight lifting) are heavy users. Sports figures such as football players have also admitted to using steroids as an adjunct to training, diet, and psychological preparation for games. Advantages of anabolic steroids cited by athletes include enhanced muscle mass and strength, increased oxygen-carrying capability owing to greater red blood cell volume, and an increase in aggressive behavior.

Typically, bodybuilders who use steroids combine high doses (up to 200 mg/day) with heavy resistance training. Intermittent use begins several months before an event, and commonly entails the use of many anabolic steroid supplements (a method called stacking). Injected or transdermal (taken via a skin patch) steroid doses are increased gradually as the competition nears.

But do the drugs do all that is claimed for them? Research studies have reported increases in isometric strength and a rise in body weight in steroid users. While these are results weight lifters dream about, there is a hot dispute over whether this also translates into athletic performance requiring the fine muscle coordination and endurance needed by runners, etc. The "jury is still out" on this question, but if you ask users, the answer will most likely be a resounding yes.

Do the proclaimed advantages conferred by steroid use outweigh the risks? Absolutely not. Physicians say they cause bloated faces (Cushingoid sign of steroid excess); shriveled testes and infertility; damage to the liver that promotes liver cancer; and changes in blood cholesterol levels (which may predispose long-term users to coronary heart disease). The psychiatric hazards of anabolic steroid use may be equally threatening: Recent studies have indicated that one-third of users have serious mental problems. Manic behavior in which the users undergo Jekyll-Hyde personality swings and become extremely violent (termed the 'roid rage) is common, as are depression and delusions.

A recent arrival on the scene, sold over the counter and touted as a "nutritional performance-enhancer," is androstenedione, which is converted to testosterone in the body. Though it is taken orally (and much of it is destroyed by the liver soon after ingestion), the few milligrams that survive temporarily boost testosterone levels. Reports of its use by baseball great Mark McGwire in the summer of '98, and of athletic wanna-bes from the fifth grade up recently sweeping the supplement off the drugstore shelves, are troubling, particularly since it is not regulated by the FDA and its long-term effects are unpredictable and untested.

The question of why some athletes use these drugs is easy to answer. Some admit to a willingness to do almost anything to win, short of killing themselves. Are they unwittingly doing this as well?

gene, into affected muscles. In initial tests on mice, about 1% of the muscle cells took up the minigenes and made functional dystrophin protein. The large size of human muscles presents a huge challenge to both the cell and gene transfer therapies. A different approach being tested is coaxing dystrophic muscles to produce more *utrophin*, a closely related protein that can compensate for dystrophin deficiency in mice.

Thus far, the only medication that has improved strength and muscle function is the steroid prednisone; however, it is not curative. In mid–1999 a surprising study reported that gentamicin, a common antibiotic, may be an effective weapon against muscular dystrophy. In animal experiments, injection of the antibiotic overrode the effect of a genetic error (a mutation called a *premature stop codon*, that is the cause of about 15% of DMD cases) enabling the mice to produce dystrophin and halting disease progress. ■

As we age, the amount of connective tissue in our skeletal muscles increases, the number of muscle fibers decreases, and the muscles become stringier, or more sinewy. Because skeletal muscles form so much of the body mass, body weight and muscle strength decline in tandem. Muscle strength has usually decreased by about 50% by the age of 80 years. This "flesh wasting" condition, called *sarcopenia* (sar"ko-pe'ne-ah) has serious health implications for the elderly, particularly because falling becomes a common event. Fortunately, regular exercise helps reverse sarcopenia, and frail elders who begin to "pump iron" (use leg and hand weights) can rebuild muscle mass and dramatically increase their strength. Performing those lifting exercises more rapidly can improve one's ability to carry out the "explosive" movements needed to rise from a chair or catch one's balance.

Muscles can also suffer indirectly. Aging of the cardiovascular system affects nearly every organ in the body, and muscles are no exception. As atherosclerosis takes its toll and begins to block distal arteries, a circulatory condition called *intermittent claudication* (klaw'dĭ-ka'shun; "limping") occurs in some individuals. This condition restricts blood delivery to the legs, leading to excruciating pains in the leg muscles during walking, forcing the person to stop and rest to get relief.

Smooth muscle is remarkably trouble-free. The few problems that impair its functioning stem from external irritants. In the gastrointestinal tract, irritation might result from ingestion of excess alcohol, spicy foods, or bacterial infection. Under such conditions, smooth muscle mobility increases in an attempt to rid the body of irritating agents, and diarrhea or vomiting occurs. Specific clinical problems of the heart are considered in Chapter 19.

⋆ ⋆ ⋆

The capacity for movement is a property of all cells but, with the exception of muscle, these movements are largely restricted to intracellular events. Skeletal muscles, the major focus of this chapter, permit us to interact with our external environment in an amazing number of ways, but they also contribute to our internal homeostasis as summarized in *Making Connections* (pp. 316–317). We have covered muscle structure from the gross to molecular levels in this chapter and have considered its physiology in some detail. Chapter 10 continues from this point to explain how muscles interact with bones and with each other, and then describes the individual skeletal muscles that contribute to the muscular system of the body.

Endocrine System

- Growth hormone and androgens influence skeletal muscle strength and mass; other hormones help regulate cardiac and smooth muscle activity

Cardiovascular System

- Skeletal muscle activity increases efficiency of cardiovascular functioning; helps prevent atherosclerosis and causes cardiac hypertrophy
- Cardiovascular system delivers needed oxygen and nutrients to muscles

Lymphatic System/Immunity

- Physical exercise may enhance or depress immunity depending on its intensity
- Lymphatic vessels drain leaked tissue fluids; immune system protects muscles from disease

Respiratory System

- Muscular exercise increases respiratory capacity
- Respiratory system provides oxygen and disposes of carbon dioxide

Digestive System

- Physical activity increases gastrointestinal mobility when at rest
- Digestive system provides nutrients needed for muscle health; liver metabolizes lactic acid

Urinary System

- Physical activity promotes normal voiding behavior; skeletal muscle forms the voluntary sphincter of the urethra
- Urinary system disposes of nitrogenous wastes

Reproductive System

- Skeletal muscle helps support pelvic organs (e.g., uterus); assists erection of penis and clitoris
- Testicular androgen promotes increased skeletal muscle size

Integumentary System

- Muscular exercise enhances circulation to skin and improves skin health; exercise also increases body heat, which the skin helps dissipate
- Skin protects the muscles by external enclosure

Skeletal System

- Skeletal muscle activity maintains bone health and strength
- Bones provide levers for muscle activity

Nervous System

- Facial muscle activity allows emotions to be expressed
- Nervous system stimulates and regulates muscle activity

THE MUSCULAR SYSTEM and Interrelationships with the Cardio-vascular, Endocrine, Lymphatic/Immunity, and Skeletal Systems

Our skeletal muscles are a marvel. In the well conditioned, they ripple with energy. In those of us who are less athletic, they still allow us to perform the rather remarkable tasks of moving and getting around. No one would argue that the nervous system is indispensable for activating muscles to contract and for keeping them healthy via tone. One func-tional-clinical example of intrinsically important system interactions is that most premature aging results from inactivity and related diseases. But we don't have to slow up during old age. The flesh is still willing and muscle is responsive to exercise throughout life. Let's look at how skeletal muscle activity affects other body systems from the vantage point of overall health and disease prevention.

Cardiovascular System

The single most important barometer of how "well" we age is the health of our cardiovascular system. More than any other factor, exercise helps to maintain that health. Any-thing that gets you huffing and puffing on a regular basis, be it racquetball or a vigorous walk, helps to keep heart muscle healthy and strong. It also keeps blood vessels clear, delaying atherosclerosis and helping to prevent the most common type of high blood pressure and hyper-tensive heart disease—ailments that can lead to the de-terioration of heart muscle and the kidneys. Uncluttered blood vessels also stave off painful or disabling intermittent claudication in which muscle pain due to ischemia hinders walking. Furthermore, regular exercise boosts blood levels of clot-busting enzymes, helping to ward off heart attacks and strokes, other scourges of old age.

Endocrine System

Muscle has a high rate of metabolism, and even at rest it uses much more energy than does fat. Consequently, exercise that builds moderate muscle mass helps to keep weight down and prevents obesity. Obesity is one of the

risk factors for development of age-related diabetes mellitus—the metabolic disorder in which body cells are unresponsive to insulin (secreted by the pancreas) and, hence, unable to utilize glucose normally. Since exercise helps to maintain normal cellular responses to insulin, it helps prevent the cardiovascular consequences (blood vessel problems) of that disease. On the other hand, several hormones, including growth hormone, thyroid hormone, and sex hormones, are essential for normal development and maturation of the skeletal muscles.

Lymphatic System/Immunity

Physical exercise has a marked effect on immunity. Moderate or mild exercise causes a temporary rise in the number of phagocytes, T cells (a particular group of white blood cells), and antibodies, all of which populate lym-phatic organs and mount the attack against infectious disease. By contrast, strenuous exercise depresses the immune system. The way in which exercise affects immunity—including these seemingly contradictory observations—is still a mystery, but the so-called stress hormones are probably involved. Like major stressors such as surgery and serious burns, strenuous exercise increases blood levels of stress hormones such as epi-nephrine and glucocorticoids. These hormones depress the immune system during severe stress. This is thought to be a protective mechanism—a way of preventing large numbers of slightly damaged cells from being rejected.

Skeletal System

Last, but not least, weight-bearing exercise of the skeletal muscles promotes skeletal strength and helps prevent osteoporosis. Since osteoporosis severely detracts from the quality of life by increasing risk of fractures, this is an extremely important interaction. But without bones, muscles would be ineffective in causing body movement.

Muscular System

Case study: Let's continue our tale of Mrs. DeStephano's medical problems, this time looking at the notes made detailing observations of her skeletal musculature.

- Severe lacerations of the muscles of the right leg and knee
- Damage to the blood vessels serving the right leg and knee
- Transection of the sciatic nerve (the large nerve serving most of the lower limb), just above the right knee

Her physician orders daily passive range-of-motion (ROM) exercise and electrical stimulation for her right leg and a diet high in protein, carbohydrates, and vitamin C.

1. Describe the step-by-step process of wound healing that will occur in her fleshy (muscle) wounds, and note the consequences of the specific restorative process that occurs.

2. What complications in healing can be anticipated owing to vascular (blood vessel) damage in the right leg?

3. What complications in muscle structure and function result from transection of the sciatic nerve? Why are passive ROM and electrical stimulation of her right leg muscles ordered?

4. Explain the reasoning behind the dietary recom-mendations.

(Answers in Appendix F)

RELATED CLINICAL TERMS

Fibromyositis (*fibro* = fiber; *itis* = inflammation) A group of conditions involving chronic inflammation of a muscle, its connective tissue coverings and tendons, and capsules of nearby joints. Symptoms are nonspecific and involve varying degrees of tenderness associated with specific trigger points.

Hernia Protrusion of an organ through its body cavity wall; may be congenital (owing to failure of muscle fusion during development), but most often is caused by heavy lifting or obesity and subsequent muscle weakening.

Myalgia (mi-al′je-ah; *algia* = pain) Muscle pain resulting from any muscle disorder.

Myofascial pain syndrome Pain caused by a tightened band of muscle fibers, which twitch when the skin over them is touched. Mostly associated with overused or strained postural muscles.

Myopathy (mi-op′ah-the; *path* = disease, suffering) Any disease of muscle.

Myotonic dystrophy A form of muscular dystrophy that is less common than DMD; in the U.S. it affects about 14 of 100,000 people. Symptoms include a gradual reduction in muscle mass and control of the skeletal muscles, abnormal heart rhythm, and diabetes mellitus. May appear any time between birth and age 60; not sex-linked. Underlying genetic defect is multiple repeats of a particular gene on chromosome 19. Because the number of repeats tends to increase from generation to generation, subsequent generations develop more severe symptoms. No effective treatment.

RICE Acronym for *rest, ice, compression,* and *elevation,* the standard treatment for a pulled muscle, or excessively stretched tendons or ligaments.

Spasm A sudden, involuntary smooth or skeletal muscle twitch ranging in severity from merely irritating to very painful; may be due to chemical imbalances. In spasms of the eyelid or facial muscles, called tics, psychological factors have been implicated. Stretching and massaging the affected area may help to end the spasm. A **cramp** is a prolonged spasm that causes a muscle to become taut and painful; common in calf, thigh, and hip muscles, and usually occurs at night or after exercise.

Strain Commonly called a "pulled muscle," a strain is excessive stretching and possible tearing of a muscle due to muscle overuse or abuse; the injured muscle becomes painfully inflamed (myositis), and adjacent joints are usually immobilized.

Tetanus (1) A state of sustained contraction of a muscle that is a normal aspect of skeletal muscle functioning. (2) An acute infectious disease caused by the anaerobic bacterium *Clostridium tetani* and resulting in persistent painful spasms of some of the skeletal muscles. Typically begins gradually with stiffness of jaw and neck muscles and progresses to fixed rigidity of the jaws (lockjaw) and spasms of trunk and limb muscles; usually fatal due to respiratory failure or exhaustion.

CHAPTER SUMMARY

Media study tools that could provide you additional help in reviewing specific key topics of Chapter 9 are referenced below. **SP** = *Study Partner;* **IP** = *Interactive Physiology.*

Overview of Muscle Tissues   (pp. 277–278)

Muscle Types   (p. 277)

1. Skeletal muscle is attached to the skeleton, is striated, and can be controlled voluntarily.

2. Cardiac muscle forms the heart, is striated, and is controlled involuntarily.

3. Smooth muscle, located chiefly in the walls of hollow organs, is controlled involuntarily. Its fibers are not striated.

SP Exercise: Chapter 4, Comparison of Muscle Cells.

Muscle Functions   (pp. 277–278)

4. Muscles move internal and external body parts, maintain posture, stabilize joints, and generate heat.

Functional Characteristics of Muscle   (p. 278)

5. Special functional characteristics of muscle include excitability, contractility, extensibility, and elasticity.

Skeletal Muscle   (pp. 278–306)

Gross Anatomy of a Skeletal Muscle   (pp. 278–280)

1. Skeletal muscle fibers (cells) are protected and strengthened by connective tissue coverings. Deep to superficial, these are endomysium, perimysium, and epimysium.

2. Skeletal muscle attachments (origins/insertions) may be direct or indirect via tendons or aponeuroses. Indirect attachments withstand friction better.

Microscopic Anatomy of a Skeletal Muscle Fiber (pp. 280–285)

3. Skeletal muscle fibers are long, striated, and multinucleate.

4. Myofibrils are contractile elements that occupy most of the cell volume. Their banded appearance results from a regular alternation of dark (A) and light (I) bands. Myofibrils are chains of sarcomeres; each sarcomere contains thick (myosin) and thin (actin) myofilaments arranged in a regular array. The heads of myosin molecules form cross bridges that interact with the thin filaments.

5. The sarcoplasmic reticulum (SR) is a system of membranous tubules surrounding each myofibril. Its function is to release and then sequester calcium ions.

 Exercise: Chapter 9, Organizational Levels of Skeletal Muscle.

6. T tubules are invaginations of the sarcolemma that run between the terminal cisternae of the SR. They allow the electrical stimulus to be delivered quickly to deep cell regions.

Contraction of a Skeletal Muscle Fiber   (pp. 285–292)

7. Muscle contraction is defined as the generation of force (tension) by the myosin cross bridges. Shortening of the muscle may or may not occur.

8. According to the sliding filament theory, the thin filaments are pulled toward the sarcomere centers by cross bridge (myosin head) activity of the thick filaments.

SP Exercise: Chapter 9, Steps of Cross-Bridge Cycling.

9. Sliding of the filaments is triggered by a rise in intracellular calcium ion levels. Troponin binding of calcium moves tropomyosin away from myosin binding sites on actin, allowing cross bridge binding. Myosin ATPases split ATP, which energizes the working strokes and is necessary for cross bridge detachment.

10. Regulation of skeletal muscle cell contraction involves (a) generation and transmission of an action potential along the sarcolemma and (b) excitation-contraction coupling.

11. The action potential is set up when acetylcholine released by a nerve ending binds to ACh receptors on the sarcolemma, causing changes in membrane permeability that allow ion flows that depolarize and then repolarize the membrane. Once initiated, the action potential is self-propagating and unstoppable.

SP Case Study: Neuromuscular Dysfunction.

12. In excitation-contraction coupling, the action potential is propagated down the T tubules, causing calcium to be released from the SR into the cell interior. Calcium initiates cross bridge activity and sliding of the filaments. Cross bridge activity ends when calcium is pumped back into the SR.

IP Muscular System CD-ROM; Topic: Sliding Filament Theory, pages 18–29.

Contraction of a Skeletal Muscle (pp. 292–297)

13. A motor unit is one motor neuron and all the muscle cells it innervates. The neuron's axon has several branches, each of which forms a neuromuscular junction with one muscle cell.

SP Exercise: Chapter 9, The Neuromuscular Junction; Case Study: Neuromuscular Dysfunction.

14. A skeletal muscle's response to a single brief threshold stimulus is a twitch. A twitch has three phases: the latent period (preparatory events occurring), the period of contraction (the muscle tenses and may shorten), and the period of relaxation (muscle tension declines and the muscle resumes its resting length).

15. Graded responses of muscles to rapid stimuli are wave summation and unfused and fused tetanus. A graded response to increasingly strong stimuli is multiple motor unit summation.

16. Isotonic contractions occur when the muscle shortens (concentric contraction) or lengthens (eccentric contraction) as the load is moved. Isometric contractions occur when muscle tension produces neither shortening nor lengthening.

IP Muscular System CD-ROM; Topic: Contraction of Motor Units, pages 1–11.

Muscle Metabolism (pp. 297–301)

17. The energy source for muscle contraction is ATP, obtained from a coupled reaction of creatine phosphate with ADP and from aerobic and anaerobic metabolism of glucose. When ATP use exceeds its production, muscle fatigue occurs.

18. When ATP is produced by nonaerobic pathways, lactic acid accumulates and an oxygen debt occurs. To return the muscles to their resting state, ATP must be produced aerobically and used to regenerate creatine phosphate and glycogen reserves and to oxidize accumulated lactic acid.

19. Only about 40% of energy released during ATP hydrolysis powers contractile activity. The rest is liberated as heat.

IP Muscular System CD-ROM; Topic: Muscle Metabolism, pages 1–7.

Force, Velocity, and Duration of Muscle Contraction (pp. 301–304)

20. The force of muscle contraction is affected by the number and size of contracting muscle cells (the more and the larger the cells, the greater the force), the series-elastic elements, and the degree of muscle stretch.

21. In twitch contractions, the external tension exerted on the load is always less than the internal tension. When a muscle is tetanized, the external tension equals the internal tension.

22. When the thick and thin filaments are slightly overlapping, the muscle can generate maximum force. With excessive increase or decrease in muscle length, force declines.

23. Factors determining the velocity and duration of muscle contraction include the load (the greater the load, the slower the contraction) and muscle fiber types.

24. There are three types of muscle fibers: (1) fast glycolytic (fatigable) fibers, (2) slow oxidative (fatigue-resistant) fibers, and (3) intermediate fast oxidative (fatigue-resistant) fibers. Most muscles contain a mixture of fiber types.

Effect of Exercise on Muscles (pp. 304–306)

25. Regular aerobic exercise results in increased efficiency, endurance, strength, and resistance to fatigue of skeletal muscles and more efficient cardiovascular, respiratory, and neuromuscular functioning.

26. Resistance exercises cause skeletal muscle hypertrophy and large gains in skeletal muscle strength.

27. Complete immobilization of muscles leads to muscle weakness and severe atrophy.

28. Improper training and excessive exercise result in overuse injuries, which may be disabling.

Smooth Muscle (pp. 306–312)

Arrangement and Microscopic Structure of Smooth Muscle Fibers (pp. 307–308)

1. Smooth muscle fibers are spindle shaped and uninucleate; they display no striations.

2. Smooth muscle cells are most often arranged in sheets. They lack elaborate connective tissue coverings.

3. The SR is poorly developed; T tubules are absent. Actin and myosin filaments are present, but sarcomeres are not. Intermediate filaments and dense bodies form an intracellular network that harnesses the pull generated during cross bridge activity and transfers it to the extra-cellular matrix.

Contraction of Smooth Muscle (pp. 308–312)

4. Smooth muscle fibers may be electrically coupled by gap junctions; the pace of contraction may be set by pacemaker cells.

5. Smooth muscle contraction is energized by ATP and is activated by a calcium pulse. However, calcium binds (to calmodulin) on the thick filaments rather than to troponin on the thin filaments.

6. Smooth muscle contracts for extended periods at low energy cost and without fatigue.

7. Neurotransmitters of the autonomic nervous system may inhibit or stimulate smooth muscle fibers. Smooth muscle contraction may also be initiated by pacemaker cells, hormones, or other local chemical factors that influence intracellular calcium levels, and by mechanical stretch.

8. Special features of smooth muscle contraction include the stress-relaxation response, the ability to generate large amounts of force when extensively stretched, and hyperplasia under certain conditions.

Types of Smooth Muscle   (p. 312)

9. Single-unit smooth muscle has electrically coupled fibers that contract synchronously and often spontaneously.

10. Multiunit smooth muscle has independent, well-innervated fibers that lack gap junctions and pacemaker cells. Stimulation is via autonomic nerves (or hormones). Multiunit muscle contractions are rarely synchronous.

Developmental Aspects of Muscles   (pp. 313, 315)

1. Muscle tissue develops from embryonic mesoderm cells called myoblasts. Skeletal muscle fibers are formed by the fusion of several myoblasts. Smooth and cardiac cells develop from single myoblasts and display gap junctions.

2. For the most part, specialized skeletal and cardiac muscle cells lose their ability to divide but retain the ability to hypertrophy. Smooth muscle regenerates well and undergoes hyperplasia.

3. Skeletal muscle development reflects maturation of the nervous system and occurs in cephalocaudal and proximal-to-distal directions. Peak development of natural neuromuscular control is achieved in midadolescence.

4. Women's muscles account for about 36% of their total body weight and men's for about 42%, a difference due chiefly to the effects of male hormones on skeletal muscle growth.

5. Skeletal muscle is richly vascularized and quite resistant to infection, but in old age, skeletal muscles become fibrous, decline in strength, and atrophy unless an appropriate exercise regimen is followed.

REVIEW QUESTIONS

Multiple Choice/Matching

1. The connective tissue covering that encloses the sarcolemma of an individual muscle fiber is called the (a) epimysium, (b) perimysium, (c) endomysium, (d) periosteum.

2. A fascicle is a (a) muscle, (b) bundle of muscle fibers enclosed by a connective tissue sheath, (c) bundle of myofibrils, (d) group of myofilaments.

3. Thick and thin myofilaments have different compositions. For each descriptive phrase, indicate whether the filament is (a) thick or (b) thin.

_____ (1) contains actin

_____ (2) contains ATPases

_____ (3) attaches to the Z disc

_____ (4) contains myosin

_____ (5) contains troponin

_____ (6) does not lie in the I band

4. The function of the T tubules in muscle contraction is to (a) make and store glycogen, (b) release Ca^{2+} into the cell interior and then pick it up again, (c) transmit the action potential deep into the muscle cells, (d) form proteins.

5. The sites where the motor nerve impulse is transmitted from the nerve endings to the skeletal muscle cell membranes are the (a) neuromuscular junctions, (b) sarcomeres, (c) myofilaments, (d) Z discs.

6. Contraction elicited by a single brief stimulus is called (a) a twitch, (b) wave summation, (c) multiple motor unit summation, (d) tetanus.

7. A smooth, sustained contraction resulting from very rapid stimulation of the muscle, in which no evidence of relaxation is seen, is called (a) a twitch, (b) wave summation, (c) multiple motor unit summation, (d) fused tetanus.

8. Characteristics of isometric contractions include all but (a) shortening, (b) increased muscle tension throughout, (c) absence of shortening, (d) use in resistance training.

9. During muscle contraction, ATP is provided by (a) a coupled reaction of creatine phosphate with ADP, (b) aerobic respiration of glucose, and (c) anaerobic glycolysis.

_____ (1) Which provides ATP fastest?

_____ (2) Which does (do) not require that oxygen be available?

_____ (3) Which (aerobic or anaerobic pathway) provides the highest yield of ATP per glucose molecule?

_____ (4) Which results in the formation of lactic acid?

_____ (5) Which has carbon dioxide and water products?

_____ (6) Which is most important in endurance sports?

10. The neurotransmitter released by somatic motor neurons is (a) acetylcholine, (b) acetylcholinesterase, (c) norepinephrine.

11. The ions that enter the muscle cell during action potential generation are (a) calcium ions, (b) chloride ions, (c) sodium ions, (d) potassium ions.

12. Myoglobin has a special function in muscle tissue. It (a) breaks down glycogen, (b) is a contractile protein, (c) holds a reserve supply of oxygen in the muscle.

13. Aerobic exercise results in all of the following except (a) increased cardiovascular system efficiency, (b) more mitochondria in the muscle cells, (c) increased size and strength of existing muscle cells, (d) increased neuromuscular system coordination.

14. The smooth muscle type found in the walls of digestive and urinary system organs and that exhibits gap junctions and pacemaker cells is (a) multiunit, (b) single-unit.

Short Answer Essay Questions

15. Name and describe the four special functional characteristics of muscle that are the basis for muscle response.

16. Distinguish between (a) direct and indirect muscle attachments and (b) a tendon and an aponeurosis.

17. (a) Describe the structure of a sarcomere and indicate the relationship of the sarcomere to the myofilament. (b) Explain the sliding filament theory of contraction using appropriately labeled diagrams of a relaxed and a contracted sarcomere.

18. What is the importance of acetylcholinesterase in muscle cell contraction?

19. Explain how a slight (but smooth) contraction differs from a vigorous contraction of the same muscle using the understandings of multiple motor unit summation.

20. Explain what is meant by the term excitation-contraction coupling.

21. Define motor unit.

22. Describe the three distinct types of skeletal muscle fibers.

23. True or false: Most muscles contain a predominance of one skeletal muscle fiber type. Explain the reasoning behind your choice.

24. Describe the cause(s) of muscle fatigue and define this term clearly.

25. Define oxygen debt.

26. Name four factors that influence contractile force and two that influence velocity and duration of contraction.

27. Smooth muscle has some unique properties, such as low energy usage, ability to maintain contraction over long periods, and the stress-relaxation response. Tie these properties to the function of smooth muscle in the body.

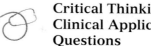

Critical Thinking and Clinical Application Questions

1. Diego was seriously out of shape the day he joined his friends for a game of touch football. While he was running pell-mell for the ball, his left calf began to hurt. He went to the clinic the next day and was told he had a strain. Diego insisted that this must be wrong, because his joints did not hurt. Clearly, Diego was confusing a *strain* with a *sprain*. Explain the difference.

2. Jim Fitch decided that his physique left much to be desired, so he joined a local health club and began to "pump iron" three times weekly. After three months of training, during which he was able to lift increasingly heavier weights, he noticed that his arm and chest muscles were substantially larger. Explain the structural and functional basis of these changes.

3. When a suicide victim was found, the coroner was unable to remove the drug vial clutched in his hand. Explain the reasons for this. If the victim had been discovered three days later, would the coroner have had the same difficulty? Explain.

4. Kristin, a dedicated sprinter, knows that the best way to treat a muscle pull is through "RICE." What does this mean?

5. When Eric returned from jogging, he was breathing heavily, sweating profusely, and complained that his legs ached and felt weak. His wife poured him a sports drink and urged him to take it easy until he could "catch his breath." On the basis of what you have learned about muscle energy metabolism, respond to the following questions.

- Why is Eric breathing heavily?

- What ATP harvesting pathway have his working muscles been using that leads to such a breathng pattern?

- What metabolic product(s) might account for his sore muscles and his feeling of muscle weakness?

10 THE MUSCULAR SYSTEM

Interactions of Skeletal Muscles in the Body (p. 323)

1. Describe the function of prime movers, antagonists, synergists, and fixators, and describe how each promotes normal muscular function.

Naming Skeletal Muscles (pp. 323–324)

2. List the criteria used in naming muscles. Provide an example to illustrate the use of each criterion.

Muscle Mechanics: Importance of Fascicle Arrangement and Leverage (pp. 324–330)

3. Define lever, and explain how a lever operating at a mechanical advantage differs from one operating at a mechanical disadvantage.

4. Name the three types of lever systems and indicate the arrangement of elements (effort, fulcrum, and load) in each. Also note the advantages of each type of lever system.

5. Name the common patterns of fascicle arrangement and relate these to power generation by muscles.

Major Skeletal Muscles of the Body (pp. 330–381)

6. Name and identify the muscles described in Tables 10.1 to 10.17. State the origin, insertion, and action of each.

As we have seen, the human body enjoys an incredibly wide range of movements. The gentle blinking of your eye, standing on tiptoe, and wielding a sledgehammer are just a small sample of the different activities promoted by the muscular system. Although muscle tissue includes all contractile tissues (skeletal, cardiac, and smooth muscle), when we study the muscular system, **skeletal muscles** take center stage. These muscular "machines" that enable us to perform so many different activities are the focus of this chapter. Before describing the individual muscles in detail, we will describe the manner in which muscles "play" with or against each other to bring about movements, consider the criteria used for naming muscles, and explain the principles of leverage.

INTERACTIONS OF SKELETAL MUSCLES IN THE BODY

The arrangement of body muscles permits them to work either together or in opposition to achieve a wide variety of movements. As you eat, for example, you alternately raise your fork to your lips and lower it to your plate, and both sets of actions are accomplished by your arm and hand muscles. But muscles can only *pull*; they never *push*. Generally as a muscle shortens, its *insertion* (attachment on the movable bone) moves toward its *origin* (its fixed or immovable point of attachment). Thus, whatever one muscle (or muscle group) can do, there is another muscle or group of muscles that "undoes" the action.

Muscles can be classified into four *functional* groups: prime movers, antagonists, synergists, and fixators. A muscle that provides the major force for producing a specific movement is called a **prime mover,** or **agonist** (ag'o-nist; "leader"), of that movement. The biceps brachii muscle, which fleshes out the anterior arm (and inserts on the radius), is a prime mover of elbow flexion.

Muscles that oppose, or reverse, a particular movement are called **antagonists** (an-tag'o-nists). When a prime mover is active, the antagonist muscles are often stretched and may be relaxed. Antagonists can also help to regulate the action of a prime mover by contracting (eccentrically) to provide some resistance, thus helping to prevent overshoot or to slow or stop the movement. As you might expect, a prime mover and its antagonist are located on opposite sides of the joint across which they act. Antagonists can also be prime movers in their own right. For example, flexion of the forearm by the biceps brachii muscle of the arm is antagonized by the triceps brachii, the prime mover for extending the forearm.

In addition to agonists and antagonists, most movements involve the action of one or more **synergists** (sin'er-jists; *syn* = together, *erg* = work). Synergists help prime movers by (1) adding a little extra force to the same movement or (2) reducing undesirable or unnecessary movements that might occur as the prime mover contracts. This latter function deserves more explanation. When a muscle crosses two or more joints, its contraction causes movement at all of the spanned joints unless other muscles act as joint stabilizers. For example, the finger flexor muscles cross both the wrist and the phalangeal joints, but you can make a fist without bending your wrist because synergistic muscles stabilize the wrist. Additionally, as some flexors act, undesirable rotatory movements may occur; synergists can prevent this movement, allowing all of the prime mover's force to be exerted in the desired direction.

When synergists immobilize a bone, or a muscle's origin, they are more specifically called **fixators** (fik'sa-terz). Recall from Chapter 7 that the scapula is held to the axial skeleton only by muscles and is quite freely movable. The fixator muscles that run from the axial skeleton to the scapula can immobilize the scapula so that only the desired movements occur at the mobile shoulder joint. Additionally, muscles that help to maintain upright posture are fixators.

In summary, although prime movers seem to get all the credit for causing certain movements, antagonistic and synergistic muscles are also important in producing smooth, coordinated, and precise movements. Furthermore, a muscle may act as a prime mover in one movement, an antagonist for another movement, a synergist for a third movement, and so on.

NAMING SKELETAL MUSCLES

Skeletal muscles are named according to a number of criteria, each of which describes the muscle in some way. Paying close attention to these cues can simplify the task of learning muscle names and actions.

1. Location of the muscle. Some muscle names indicate the bone or body region with which the muscle is associated. For example, the temporalis (tem"por-ă'lis) muscle overlies the temporal bone, and intercostal (*costal* = rib) muscles run between the ribs.

2. Shape of the muscle. Some muscles are named for their distinctive shapes. For example, the deltoid (del'-toid) muscle is roughly triangular (*deltoid* = triangle), and together the right and left trapezius (trah-pe'ze-us) muscles form a trapezoid.

3. **Relative size of the muscle.** Terms such as *maximus* (largest), *minimus* (smallest), *longus* (long), and *brevis* (short) are often used in muscle names—as in gluteus maximus and gluteus minimus (the large and small gluteus muscles, respectively).

4. **Direction of muscle fibers.** The names of some muscles reveal the direction in which their fibers (and fascicles) run in reference to some imaginary line, usually the midline of the body or the longitudinal axis of a limb bone. In muscles with the term *rectus* (straight) in their names, the fibers run parallel to that imaginary line (axis), whereas the terms *transversus* and *oblique* indicate that the muscle fibers run respectively at right angles and obliquely to that line. Specific examples include the rectus femoris (straight muscle of the thigh, or femur) and transversus abdominis (transverse muscle of the abdomen).

5. **Number of origins.** When the term *biceps, triceps,* or *quadriceps* forms part of a muscle's name, you can assume that the muscle has two, three, or four origins, respectively. For example, the biceps brachii (bra′ke-i) muscle of the arm has two origins, or *heads.*

6. **Location of the attachments.** Some muscles are named according to their points of origin and insertion. The origin is always named first. For instance, the sternocleidomastoid (ster″no-kli″do-mas′toid) muscle of the neck has a dual origin on the sternum (*sterno*) and clavicle (*cleido*), and it inserts on the *mastoid* process of the temporal bone.

7. **Action.** When muscles are named for their action, action words such as *flexor, extensor,* or *adductor* appear in the muscle's name. For example, the adductor longus, located on the medial thigh, brings about thigh adduction, and the supinator (soo′pi-na″tor) muscle supinates the forearm.

Often, several criteria are combined in the naming of a muscle. For instance, the name *extensor carpi radialis longus* tells us the muscle's action (extensor), what joint it acts on (*carpi* = wrist), and that it lies close to the radius of the forearm (radialis); it also hints at its size (longus) relative to other wrist extensor muscles. Unfortunately, not all muscle names are this descriptive.

MUSCLE MECHANICS: IMPORTANCE OF FASCICLE ARRANGEMENT AND LEVERAGE

Most factors contributing to muscle force and speed (load, fiber type, etc.) have already been covered in Chapter 9 with two important exceptions—the patterns of fascicle arrangement in muscles and lever systems (partnerships between the muscular and skeletal systems). These are attended to next.

Arrangement of Fascicles

All skeletal muscles consist of fascicles, but fascicle arrangements vary. This results in muscles with different shapes and functional capabilities. The most common patterns of fascicle arrangement are parallel, pennate, convergent, and circular (Figure 10.1).

In a **parallel** arrangement, the long axes of the fascicles run parallel to the long axis of the muscle. Such muscles are either *straplike* (Figure 10.1c), or spindle shaped with an expanded belly (midsection), like the biceps brachii muscle of the arm (Figure 10.1f). However, some authorities classify the spindle-shaped muscles into a separate class as **fusiform muscles.** This is the approach used here.

In a **pennate** (pen′āt) pattern, the fascicles are short and they attach obliquely (*penna* = feather) to a central tendon that runs the length of the muscle. If, as seen in the extensor digitorum muscle of the leg, the fascicles insert into only one side of the tendon, the muscle is *unipennate* (Figure 10.1d). If the fascicles insert into the tendon from opposite sides, so that the muscle's "grain" resembles a feather, the arrangement is *bipennate* (Figure 10.1g). The rectus femoris of the thigh is bipennate. A *multipennate* arrangement looks like many feathers situated side by side, with all their quills inserted into one large tendon. The deltoid muscle, which forms the roundness of the shoulder, is multipennate (Figure 10.1e).

A **convergent** muscle has a broad origin, and its fascicles *converge* toward a single tendon of insertion. Such a muscle is triangular or fan shaped. The pectoralis major muscle of the anterior thorax has a convergent pattern (Figure 10.1b).

The fascicular pattern is **circular** when the fascicles are arranged in concentric rings (Figure 10.1a). Muscles with this arrangement surround external body openings, which they close by contracting. A general term for such muscles is *sphincters* (literally, "squeezers"). Examples are the orbicularis muscles surrounding the eyes and the mouth.

The arrangement of a muscle's fascicles determines its range of motion and power. Since skeletal muscle fibers shorten to about 70% of their resting length when they contract, the longer and the more nearly parallel the muscle fibers are to a muscle's long axis, the more the muscle can shorten. Muscles with parallel fascicle arrangement provide the greatest degree of shortening, but they are not usually very powerful. Muscle power depends more on the total number of muscle cells in the muscle; the greater the number, the greater the power. The stocky bipennate and multipennate muscles, which "pack in" the most fibers, shorten very little but are very powerful.

? *Of the muscles illustrated, which could shorten most? Which two would probably be most powerful? Why?*

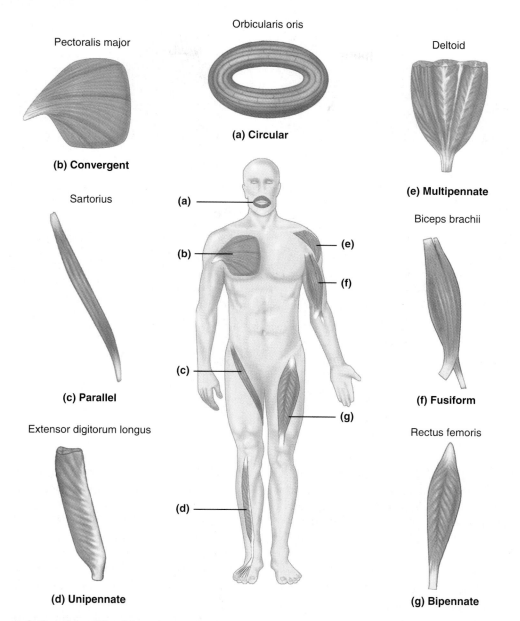

Pectoralis major

Orbicularis oris

Deltoid

(a) Circular

(b) Convergent

(e) Multipennate

Sartorius

(a)

Biceps brachii

(b)

(e)

(f)

(c) Parallel

(c)

(f) Fusiform

Extensor digitorum longus

(g)

Rectus femoris

(d)

(d) Unipennate

(g) Bipennate

FIGURE 10.1 *Relationship of fascicle arrangement to muscle structure.*

Lever Systems: Bone-Muscle Relationships

The operation of most skeletal muscles involves the use of leverage and **lever systems.** A **lever** is a rigid bar that moves on a fixed point, or **fulcrum,** when a force is applied to it. The applied force, or **effort,** is used to move a resistance, or **load.** In your body, your joints are the fulcrums, and your bones act as levers. Muscle contraction provides the effort which is applied at the muscle's insertion point on a bone. The load that is moved is the bone itself, along with overlying tissues and anything else you are trying to move with that particular lever.

The biceps brachii and sartorius could shorten most because they have the longest fibers and fibers that run generally parallel to the long axis of the bone with which it is closely associated. The pectoralis major, deltoid, and rectus femoris would be the most powerful because they are the most fleshy.

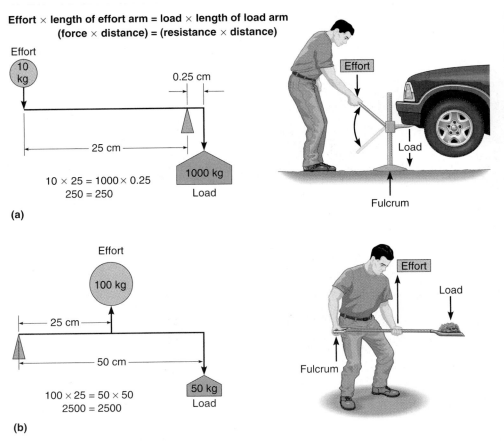

Effort × length of effort arm = load × length of load arm
(force × distance) = (resistance × distance)

$10 \times 25 = 1000 \times 0.25$
$250 = 250$

(a)

$100 \times 25 = 50 \times 50$
$2500 = 2500$

(b)

FIGURE 10.2 *Lever systems operating at a mechanical advantage and a mechanical disadvantage.* The equation at the top expresses the relationships among the forces and distances in any lever system. (**a**) 10 kg of force (the effort) is used to lift a 1000-kg car (the load). This lever system, which employs a jack, operates at a mechanical advantage: The load lifted is greater than the applied muscular effort. (**b**) Using a shovel to lift dirt is an example of a lever system that operates at a mechanical disadvantage. A muscular force (effort) of 100 kg is used to lift 50 kg of dirt (the load). Levers operating at a mechanical disadvantage are common in the body because they permit a muscle to be inserted close to the load and provide rapid contractions that have a wide range of motion.

A lever allows a given effort to move a heavier load, or to move a load farther or faster, than it otherwise could. If, as shown in Figure 10.2a, the load is close to the fulcrum and the effort is applied far from the fulcrum, a small effort exerted over a relatively large distance can be used to move a large load over a small distance. Such a lever is said to operate at a **mechanical advantage** and is commonly called a *power lever.* For example, as shown to the right in Figure 10.2a, a man can lift a car with such a lever, in this case, a jack. The car moves up only a small distance with each downward "push" of the jack handle, but relatively little muscle effort is needed. If, on the other hand, the load is far from the fulcrum and the effort is applied near the fulcrum, the force exerted by the muscle must be greater than the load moved or supported (Figure 10.2b). This lever system operates at a **mechanical disadvantage** and is a *speed lever.* These levers are useful because they allow the load to move rapidly through a large distance. Wielding a shovel is an example. As you can see, small differences in the site of a muscle's insertion (relative to the fulcrum or joint) can translate into large differences in the amount of force a muscle must generate to move a given load or resistance. Regardless of type, all levers follow the same basic principle: effort farther than load from fulcrum = mechanical advantage; effort nearer than load to fulcrum = mechanical disadvantage.

Depending on the relative position of the three elements—effort, fulcrum, and load—a lever belongs to one of three classes. In **first-class levers,** the effort is applied at one end of the lever and the load is at the other, with the fulcrum somewhere between (Figure 10.3a). Seesaws and scissors are first-class levers. First-class leverage also occurs when you lift your head off your chest. Some first-class levers in the body operate at a mechanical advantage (effort is farther from the joint than the load and is less than the load to be moved), but others, such as the action of the triceps muscle in extending the forearm against resistance, operate at a mechanical disadvantage (effort is closer to the joint and greater than the load).

In a **second-class lever,** the effort is applied at one end of the lever and the fulcrum is located at the other, with the load between them (Figure 10.3b). A wheelbarrow demonstrates this type of lever system. Second-class levers are uncommon in the body; the best example is the act of standing on your toes. Joints in the ball of the foot act together as the fulcrum, the load is the entire body weight, and the calf muscles exert the effort, pulling the heel superiorly.

? *Which of these lever systems as demonstrated at far left in the diagrams would be the fastest lever?*

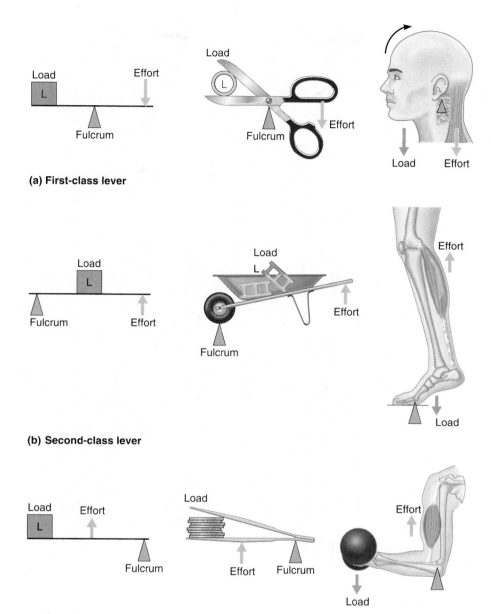

(a) First-class lever

(b) Second-class lever

(c) Third-class lever

FIGURE 10.3 *Lever systems.* **(a)** In a first-class lever, the arrangement of the elements is load-fulcrum-effort. Scissors are in this lever class. In the body, a first-class lever system is activated when you raise your head off your chest. The posterior neck muscles provide the effort, the atlanto-occipital joint is the fulcrum, and the weight to be lifted is the facial skeleton. **(b)** In second-class levers, the arrangement is fulcrum-load-effort, as exemplified by a wheelbarrow. In the body, second-class leverage is exerted when you stand on tiptoe. The joints of the ball of the foot are the fulcrum, the weight of the body is the load, and the effort is exerted by the calf muscles pulling upward on the heel (calcaneus) of the foot. **(c)** In a third-class lever, the arrangement is load-effort-fulcrum. A pair of tweezers or forceps uses this type of leverage. The operation of the biceps brachii muscle in flexing the forearm exemplifies third-class leverage. The load is the hand and distal end of the forearm, the effort is exerted on the proximal radius of the forearm, and the fulcrum is the elbow joint.

(c) The third-class lever

Which two muscles identified in (a) are used to do biceps curls (flex the elbow)? Which would be active during abdominal crunches (sit-ups)?

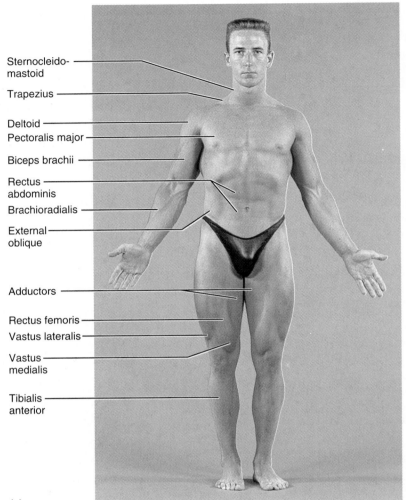

Sternocleido-
mastoid

Trapezius

Deltoid

Pectoralis major

Biceps brachii

Rectus
abdominis

Brachioradialis

External
oblique

Adductors

Rectus femoris

Vastus lateralis

Vastus
medialis

Tibialis
anterior

(a)

FIGURE 10.4 *Anterior view of superficial muscles of the body.* **(a)** Photograph of surface anatomy. **(b)** Diagrammatic view. The abdominal surface has been partially dissected on the right side to show somewhat deeper muscles. (See *A Brief Atlas of the Human Body,* Figures 61–64.)

All second-class levers in the body work at a mechanical advantage because the muscle insertion is always farther from the fulcrum than is the load to be moved. Second-class levers are levers of strength, but speed and range of motion are sacrificed for that strength.

In **third-class levers,** the effort is applied between the load and the fulcrum (Figure 10.3c). These levers operate with great speed and *always* at a mechanical disadvantage. Tweezers or forceps provide this type of leverage. Most skeletal muscles of the body act in third-class lever systems. As illustrated by the activity of the biceps muscle of the arm, the fulcrum is the elbow joint, the force is exerted on the proximal radius, and the load to be lifted is the distal forearm (and anything carried in the hand). Third-class lever systems permit a muscle to be inserted very close to the joint across which movement occurs, which allows rapid, extensive movements with relatively little shortening of the muscle. Muscles involved in third-class levers tend to be thicker and more powerful.

In conclusion, differences in the positioning of the three elements modify muscle activity with respect to (1) speed of contraction, (2) range of movement, and (3) the weight of the load that can be lifted. In lever systems that operate at a mechanical

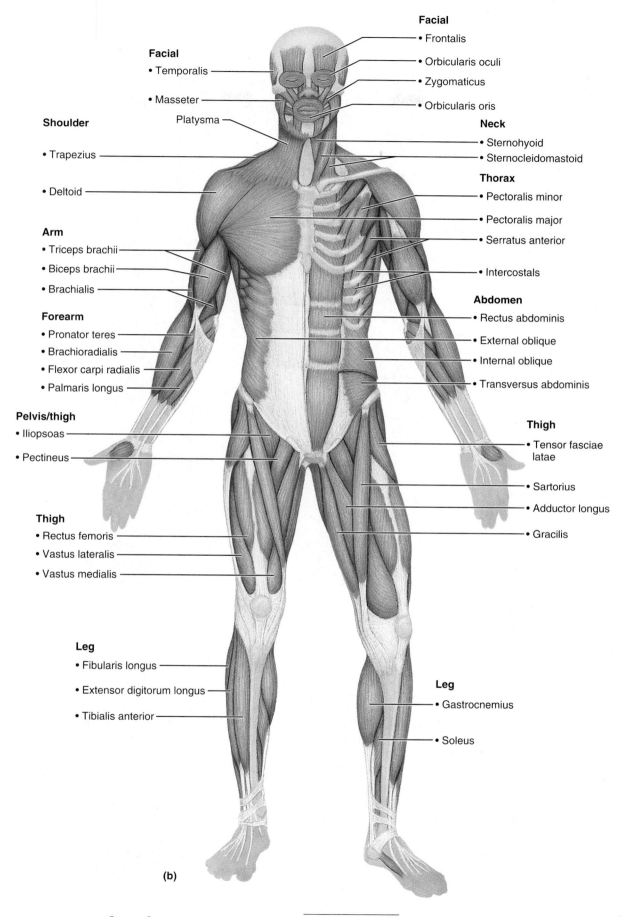

Facial
- Frontalis
- Orbicularis oculi
- Zygomaticus
- Orbicularis oris

Facial
- Temporalis
- Masseter

Platysma

Shoulder
- Trapezius
- Deltoid

Arm
- Triceps brachii
- Biceps brachii
- Brachialis

Forearm
- Pronator teres
- Brachioradialis
- Flexor carpi radialis
- Palmaris longus

Pelvis/thigh
- Iliopsoas
- Pectineus

Thigh
- Rectus femoris
- Vastus lateralis
- Vastus medialis

Leg
- Fibularis longus
- Extensor digitorum longus
- Tibialis anterior

Neck
- Sternohyoid
- Sternocleidomastoid

Thorax
- Pectoralis minor
- Pectoralis major
- Serratus anterior
- Intercostals

Abdomen
- Rectus abdominis
- External oblique
- Internal oblique
- Transversus abdominis

Thigh
- Tensor fasciae latae
- Sartorius
- Adductor longus
- Gracilis

Leg
- Gastrocnemius
- Soleus

(b)

Brachioradialis and biceps brachii would be used to do curls. The external obliques and rectus abdominis muscle would be working during crunches.

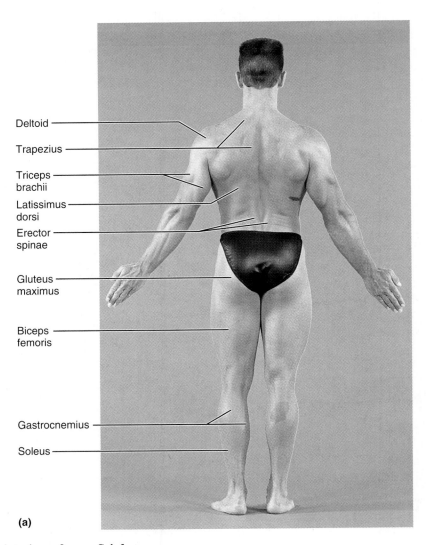

Deltoid

Trapezius

Triceps brachii

Latissimus dorsi

Erector spinae

Gluteus maximus

Biceps femoris

Gastrocnemius

Soleus

(a)

FIGURE 10.5   *Posterior view of superficial muscles of the body.* **(a)** Photograph of surface anatomy. **(b)** Diagrammatic view.

disadvantage (speed levers), force is lost but speed and range of movement are gained, and this can be a distinct benefit. Systems that operate at a mechanical advantage (power levers) are slower, more stable, and used where strength is a priority.

MAJOR SKELETAL MUSCLES OF THE BODY

The grand plan of the muscular system is all the more impressive because of the sheer number of skeletal muscles in the body—there are over 600 of them! Obviously, trying to remember all the names, locations, and actions of these muscles is a monumental task. Take heart; only the principal muscles (approximately 125 pairs of them) are considered here. Although this number is far fewer than 600, the job of learning about all these muscles will still require a concerted effort on your part. Memorization will be easier if you can apply what you have learned in a practical, or clinical, way; that is, with a *functional anatomy focus*. Once you are satisfied that you have learned the name of a muscle and can identify it on a cadaver, model, or diagram, you must

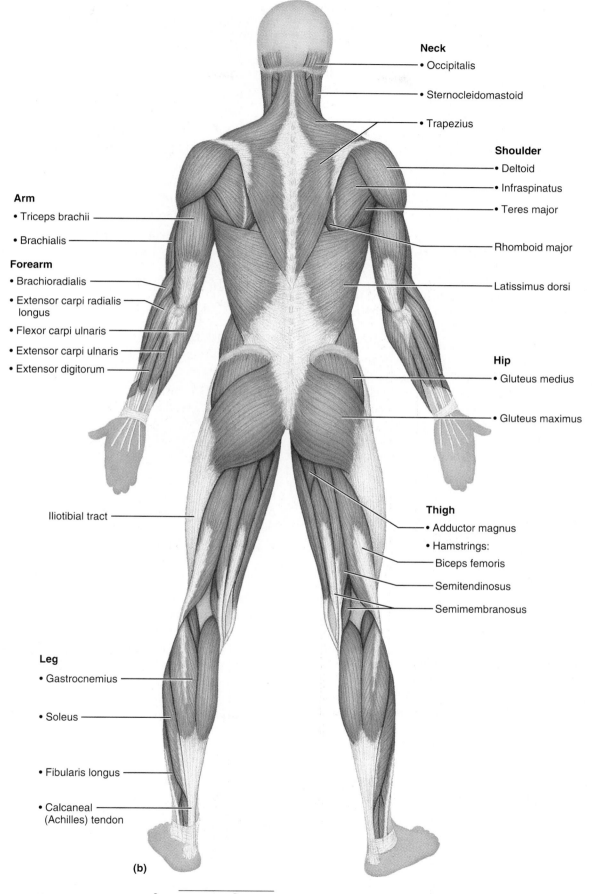

Neck
- Occipitalis
- Sternocleidomastoid
- Trapezius

Shoulder
- Deltoid
- Infraspinatus
- Teres major

Rhomboid major

Latissimus dorsi

Arm
- Triceps brachii
- Brachialis

Forearm
- Brachioradialis
- Extensor carpi radialis longus
- Flexor carpi ulnaris
- Extensor carpi ulnaris
- Extensor digitorum

Hip
- Gluteus medius
- Gluteus maximus

Iliotibial tract

Thigh
- Adductor magnus
- Hamstrings:
- Biceps femoris
- Semitendinosus
- Semimembranosus

Leg
- Gastrocnemius
- Soleus
- Fibularis longus
- Calcaneal (Achilles) tendon

(b)

The trapezius. The trapezius and rhomboids.

then flesh out your learning by asking yourself, "What does it do?" (It might be a good idea to review body movements (pp. 255–258) before you get to this point.)

In the tables that follow, the muscles of the body have been grouped by function and by location, roughly from head to foot. Each table is keyed to a particular figure (or group of figures) illustrating the muscles it describes. The legend at the beginning of each table provides an orienting overview of the types of movements effected by the muscles listed and gives pointers on the way those muscles interact with one another. The table itself describes each muscle's shape, location relative to other muscles, origin and insertion, primary actions, and innervation. (Since some instructors want students to defer learning the muscle innervations until the nervous system has been studied, you might want to check on what is expected of you in this regard.) Muscle innervations are described in detail in Chapters 13 and 15.

As you consider each individual muscle, be alert to the information its name provides. After reading its entire description, identify the muscle on the corresponding figure and, in the case of superficial muscles, also on Figure 10.4 or 10.5. This will help you to link the descriptive material in the table to a visual image of the muscle's location in the body. Also try to relate a muscle's attachments and location to the actions allowed by the joint(s) it acts across. This will focus your attention on functional details that often escape student awareness. For example, both the elbow and knee joints are hinge joints that allow flexion and extension. However, the knee flexes to the dorsum of the body (the calf moves toward the posterior thigh), whereas elbow flexion carries the forearm toward the anterior aspect of the arm. Therefore, leg flexors are located on the posterior thigh, while forearm flexors are found on the anterior aspect of the humerus.

Finally, keep in mind that the *best* way to learn muscle actions is to act out their movements yourself while feeling for the muscles contracting (bulging) beneath your skin.

The following list summarizes the organization and sequence of the tables in this chapter:

TABLE 10.1	Muscles of the Head, Part I: Facial Expression (Figure 10.6)

The superficial muscles of the head that promote facial expression include those of the scalp and face. They are highly variable in shape and strength, and adjacent muscles tend to be fused. They are unusual muscles in that they insert into skin (or other muscles), not bones. In the scalp, the main muscle is the **epicranius**, which has distinct anterior and posterior parts; the lateral scalp muscles are vestigial in humans. Muscles clothing the facial skeleton lift the eyebrows, flare the nostrils, open and close the eyes and mouth, and provide one of the best tools for influencing others—the smile. The tremendous importance of facial muscles in nonverbal communication becomes especially clear when they are paralyzed, as in some stroke victims. All muscles listed in this table are innervated by *cranial nerve VII (the facial nerve)*. The external muscles of the eye, which are contained within the orbit and act to direct the eyeball, and the levator palpebrae superioris muscles that raise the eyelids are described in Chapter 16.

Muscle	Description	Origin (O) and Insertion (I)	Action	Nerve Supply
Muscles of the Scalp				
Epicranius (occipitofrontalis) (ep″ĭ-kra′ne-us; ok-sip″ĭ-to-fron-ta′lis) (*epi* = over; *cran* = skull)	Bipartite muscle consisting of the frontalis and occipitalis muscles connected by a cranial aponeurosis, the galea aponeurotica; the alternate actions of these two muscles pull scalp forward and backward			
■ **Frontalis** (fron-ta′lis) (*front* = forehead)	Covers forehead and dome of skull; no bony attachments	O—galea aponeurotica I—skin of eyebrows and root of nose	With aponeurosis fixed, raises the eyebrows (as in surprise); wrinkles forehead skin horizontally	Facial nerve (cranial VII)
■ **Occipitalis** (ok-sip″ĭ-tal′is) (*occipito* = base of skull)	Overlies posterior occiput; by pulling on the galea, fixes origin of frontalis	O—occipital and temporal bones I—galea aponeurotica	Fixes aponeurosis and pulls scalp posteriorly	Facial nerve
Muscles of the Face				
Corrugator supercilii (kor′ah-ga-ter soo″per-sĭ′le-i) (*corrugo* = wrinkle; *supercilium* = eyebrow)	Small muscle; activity associated with that of orbicularis oculi	O—arch of frontal bone above nasal bone I—skin of eyebrow	Draws eyebrows together and inferiorly; wrinkles skin of forehead vertically (as in frowning)	Facial nerve
Orbicularis oculi (or-bik′u-lar-is ok′u-li) (*orb* =circular; *ocul* = eye)	Thin, tripartite sphincter muscle of eyelid; surrounds rim of the orbit	O—frontal and maxillary bones and ligaments around orbit I—tissue of eyelid	Protects eyes from intense light and injury; various parts can be activated individually; produces blinking, squinting, and draws eyebrows inferiorly	Facial nerve

(Table continues on page 334)

TABLE 10.1	Muscles of the Head, Part I: Facial Expression (Figure 10.6) *(continued)*			
Muscle	Description	Origin (O) and Insertion (I)	Action	Nerve Supply
Muscles of the Face				
Zygomaticus (zi-go-mat′ĭ-kus), major and minor (*zygomatic* = cheekbone)	Muscle pair extending diagonally from cheekbone to corner of mouth	O—zygomatic bone I—skin and muscle at corner of mouth	Raises lateral corners of mouth upward (smiling muscle)	Facial nerve
Risorius (ri-zor′e-us) (*risor* = laughter)	Slender muscle inferior and lateral to zygomaticus	O—lateral fascia associated with masseter muscle I—skin at angle of mouth	Draws corner of lip laterally; tenses lips; synergist of zygomaticus	Facial nerve
Levator labii superioris (lĕ-va′tor la′be-i soo-per″e-or′is) (*leva* = raise; *labi* = lip; *superior* = above, over)	Thin muscle between orbicularis oris and inferior eye margin	O—zygomatic bone and infraorbital margin of maxilla I—skin and muscle of upper lip	Opens lips; raises and furrows the upper lip	Facial nerve
Depressor labii inferioris (de-pres′or la′be-i in-fer″e-or′is) (*depressor* = depresses; *infer* = below)	Small muscle running from mandible to lower lip	O—body of mandible lateral to its midline I—skin and muscle of lower lip	Draws lower lip inferiorly (as in a pout)	Facial nerve
Depressor anguli oris (ang′gu-li or-is) (*angul* = angle, corner; *or* = mouth)	Small muscle lateral to depressor labii inferioris	O—body of mandible below incisors I—skin and muscle at angle of mouth below insertion of zygomaticus	Zygomaticus antagonist; draws corners of mouth downward and laterally (as in a "tragedy mask" grimace)	Facial nerve
Orbicularis oris	Complicated, multilayered muscle of the lips with fibers that run in many different directions; most run circularly	O—arises indirectly from maxilla and mandible; fibers blended with fibers of other facial muscles associated with the lips I—encircles mouth; inserts into muscle and skin at angles of mouth	Closes lips; purses and protrudes lips; kissing and whistling muscle	Facial nerve
Mentalis (men-ta′lis) (*ment* = chin)	One of the muscle pair forming a V-shaped muscle mass on chin	O—mandible below incisors I—skin of chin	Protrudes lower lip; wrinkles chin	Facial nerve
Buccinator (bu′sĭ-na″ter) (*bucc* = cheek or "trumpeter")	Thin, horizontal cheek muscle; principal muscle of cheek; deep to masseter (see also Figure 10.7)	O—molar region of maxilla and mandible I—orbicularis oris	Draws corner of mouth laterally; compresses cheek (as in whistling and sucking); holds food between teeth during chewing; well developed in nursing infants	Facial nerve
Platysma (plah-tiz′mah) (*platy* = broad, flat)	Unpaired, thin, sheetlike superficial neck muscle; not strictly a head muscle, but plays a role in facial expression	O—fascia of chest (over pectoral muscles and deltoid) I—lower margin of mandible, and skin and muscle at corner of mouth	Helps depress mandible; pulls lower lip back and down, i.e., produces downward sag of mouth; tenses skin of neck (e.g., during shaving)	Facial nerve

TABLE 10.1	**Muscles of the Head, Part I: Facial Expression (Figure 10.6) (continued)**

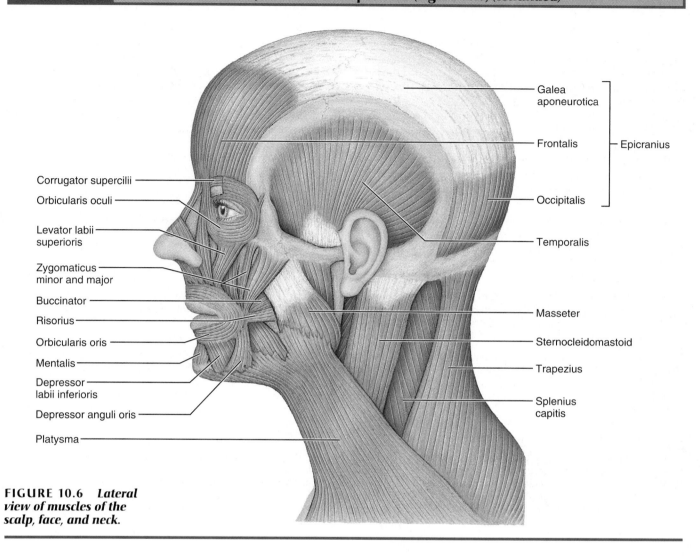

FIGURE 10.6 Lateral view of muscles of the scalp, face, and neck.

TABLE 10.2	Muscles of the Head, Part II: Mastication and Tongue Movement (Figure 10.7)

Four pairs of muscles are involved in mastication (chewing and biting) activities, and all are innervated by the mandibular division of *cranial nerve V* (the *trigeminal nerve*). The prime movers of jaw closure (and biting) are the powerful **masseter** and **temporalis** muscles, which can be palpated easily when the teeth are clenched. Grinding movements are provided for by the **pterygoid** muscles. The **buccinator** muscles (see Table 10.1) also play a role in chewing. Normally, gravity is sufficient to depress the mandible, but if there is resistance to jaw opening, neck muscles such as the digastric and mylohyoid muscles (see Table 10.3) are activated.

The tongue is composed of muscle fibers peculiar to itself that curl, squeeze, and fold the tongue during speaking and chewing. These **intrinsic tongue muscles,** arranged in several planes, change the shape of the tongue and contribute to its exceptional nimbleness, but they do not really move the tongue. They are considered in Chapter 24 with the digestive system. Only the **extrinsic tongue muscles,** which serve to anchor and move the tongue, are considered in this table. The extrinsic tongue muscles are all innervated by *cranial nerve XII* (the *hypoglossal nerve*).

Muscle	Description	Origin (O) and Insertion (I)	Action	Nerve Supply
Muscles of Mastication				
Masseter (mah-se′ter) (*maseter* = chewer)	Powerful muscle that covers lateral aspect of mandibular ramus	O—zygomatic arch and maxilla I—angle and ramus of mandible	Prime mover of jaw closure; elevates mandible	Trigeminal nerve (cranial V)
Temporalis (tem″por-ă′lis) (*tempora* = time; pertaining to the temporal bone)	Fan-shaped muscle that covers parts of the temporal, frontal, and parietal bones	O—temporal fossa I—coronoid process of mandible via a tendon that passes deep to zygomatic arch	Closes jaw; elevates and retracts mandible; synergist of pterygoids (see below) in side-to-side movements; maintains position of the mandible at rest	Trigeminal nerve
Medial pterygoid (me′de-ul ter′ĭ-goid) (*medial* = toward median plane; *pterygoid* = winglike)	Deep two-headed muscle that runs along internal surface of mandible and is largely concealed by that bone	O—medial surface of lateral pterygoid plate of sphenoid bone, maxilla, and palatine bone I—medial surface of mandible near its angle	Synergist of temporalis and masseter muscles in elevation of the mandible; acts with the lateral pterygoid muscle to protrude mandible and to promote side-to-side side (grinding) movements	Trigeminal nerve
Lateral pterygoid (*lateral* = away from median plane)	Deep two-headed muscle; lies superior to medial pterygoid muscle	O—greater wing and lateral pterygoid plate of sphenoid bone I—condyle of mandible and capsule of temporomandibular joint	Protrudes mandible (pulls it anteriorly); provides forward sliding and side-to-side grinding movements of the lower teeth	Trigeminal nerve
Buccinator	See Table 10.1	See Table 10.1	Trampoline-like action of buccinator muscles helps keep food between grinding surfaces of teeth during chewing	Facial nerve (cranial VII)
Muscles Promoting Tongue Movements (Extrinsic Muscles)				
Genioglossus (je″ne-o-glah′sus) (*geni* = chin; *glossus* = tongue)	Fan-shaped muscle; forms bulk of inferior part of tongue; its attachment to mandible prevents tongue from falling backward and obstructing respiration	O—internal surface of mandible near symphysis I—inferior aspect of the tongue and body of hyoid bone	Primarily protrudes tongue, but can depress or act in concert with other extrinsic muscles to retract tongue	Hypoglossal nerve (cranial XII)

TABLE 10.2	Muscles of the Head, Part II: Mastication and Tongue Movement (Figure 10.7) *(continued)*			
Muscle	*Description*	*Origin (O) and Insertion (I)*	*Action*	*Nerve Supply*
Muscles of Mastication				
Hyoglossus (hi'o-glos"us) (*hyo* = pertaining to hyoid bone)	Flat, quadrilateral muscle	O—body and greater horn of hyoid bone I—inferolateral tongue	Depresses tongue and draws its sides downward	Hypoglossal nerve
Styloglossus (sti-lo-glah'sus) (*stylo* = pertaining to styloid process)	Slender muscle running superiorly to and at right angles to hyoglossus	O—styloid process of temporal bone I—lateral inferior aspect of tongue	Retracts (and elevates) tongue	Hypoglossal nerve

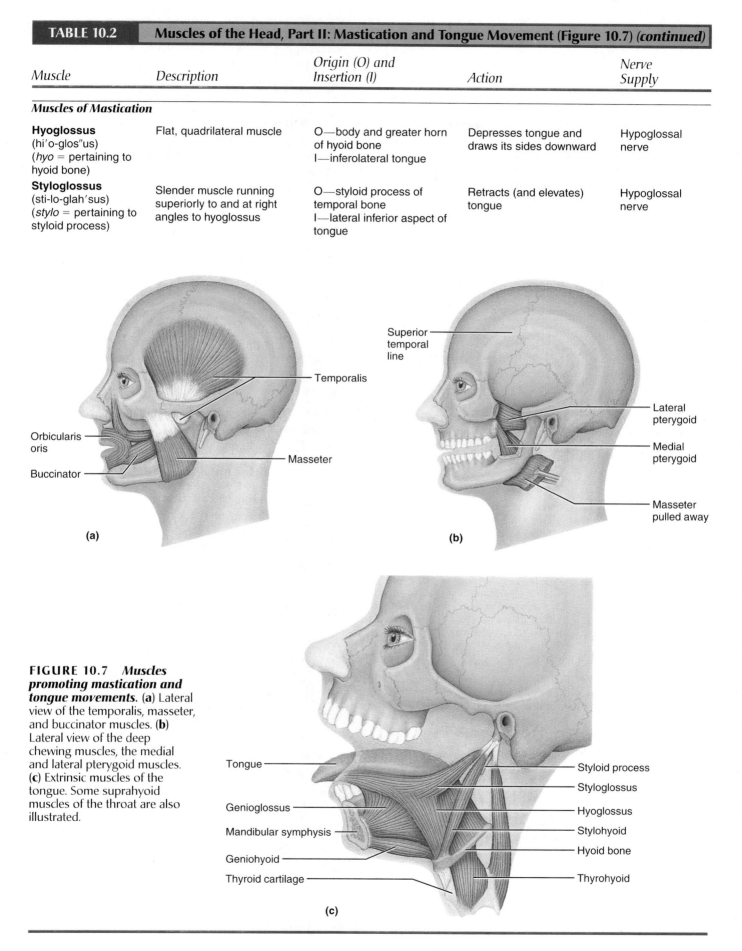

FIGURE 10.7 *Muscles promoting mastication and tongue movements.* (a) Lateral view of the temporalis, masseter, and buccinator muscles. **(b)** Lateral view of the deep chewing muscles, the medial and lateral pterygoid muscles. **(c)** Extrinsic muscles of the tongue. Some suprahyoid muscles of the throat are also illustrated.

TABLE 10.3   Muscles of the Anterior Neck and Throat: Swallowing (Figure 10.8)

The neck is divided into two triangles (anterior and posterior) by the sternocleidomastoid muscle (Figure 10.8a). This table considers the muscles of the *anterior* triangle, which are divided into **suprahyoid** and **infrahyoid muscles** (above and below the hyoid bone respectively). Most of these muscles are deep (throat) muscles involved in the coordinated muscle movements that occur during swallowing.

Swallowing begins when the tongue and buccinator muscles of the cheeks squeeze the food back along the roof of the mouth toward the pharynx. Then a rapid series of muscular movements in the posterior mouth and pharynx complete the process. Events of swallowing include: (1) Widening the pharynx to receive the food, and closing the respiratory passageway (larynx) anteriorly to prevent aspiration of food. This is accomplished by the *suprahyoid muscles*, which pull the hyoid bone upward and forward toward the mandible. Since the hyoid bone is attached by the thyrohyoid membrane to the larynx, the larynx is also pulled upward and forward, a maneuver that widens the pharynx and closes the respiratory passageway. (2) Closing off the nasal passages to prevent food from entering the superior nasal cavity due to the activation of small muscles that elevate the soft palate. (These muscles, the *tensor* and *levator veli palatini*, are not described in the table but are illustrated in Figure 10.8b.) (3) Food is propelled through the pharynx into the esophagus inferiorly by the **pharyngeal constrictor muscles.** (4) Returning the hyoid bone and larynx to their more inferior positions after swallowing, promoted by the *infrahyoid muscles*.

Muscle	Description	Origin (O) and Insertion (I)	Action	Nerve Supply
SUPRAHYOID MUSCLES (soo″prah-hi′oid)	Muscles that help form floor of oral cavity, anchor tongue, elevate hyoid, and move larynx superiorly during swallowing; lie superior to hyoid bone			
Digastric (di-gas′trik) (*di* = two; *gaster* = belly)	Consists of two bellies united by an intermediate tendon, forming a V shape under the chin	O—lower margin of mandible (anterior belly) and mastoid process of the temporal bone (posterior belly) I—by a connective tissue loop to hyoid bone	Acting in concert, the digastric muscles elevate hyoid bone and steady it during swallowing and speech; acting from behind, they open mouth and depress mandible	Mandibular branch of trigeminal nerve (cranial V) for anterior belly; facial nerve (cranial VII) for posterior belly
Stylohyoid (sti″lo-hi′oid) (also see Figure 10.7)	Slender muscle below angle of jaw; parallels posterior belly of diagastric muscle	O—styloid process of temporal bone I—hyoid bone	Elevates and retracts hyoid, thereby elongating floor of mouth during swallowing	Facial nerve
Mylohyoid (mi″lo-hi′oid) (*myle* = molar)	Flat, triangular muscle just deep to digastric muscle; this muscle pair forms a sling that forms the floor of the anterior mouth	O—medial surface of mandible I—hyoid bone and median raphe	Elevates hyoid bone and floor of mouth, enabling tongue to exert backward and upward pressure that forces food bolus into pharynx	Mandibular branch of trigeminal nerve
Geniohyoid (je′ne-o-hy″oid) (also see Figure 10.7) (*geni* = chin)	Narrow muscle in contact with its partner medially; runs from chin to hyoid bone	O—inner surface of mandibular symphysis I—hyoid bone	Pulls hyoid bone superiorly and anteriorly, shortening floor of mouth and widening pharynx for receiving food	First cervical spinal nerve via hypoglossal nerve (cranial XII)
INFRAHYOID MUSCLES	Straplike muscles that depress the hyoid bone and larynx during swallowing and speaking (see also Figure 10.9c)			
Sternohyoid (ster″no-hi′oid) (*sterno* = sternum)	Most medial muscle of the neck; thin; superficial except inferiorly, where covered by sternocleidomastoid	O—manubrium and medial end of clavicle I—lower margin of hyoid bone	Depresses larynx and hyoid bone if mandible is fixed; may also flex skull	Cervical spinal nerves 1–3 (C_1–C_3) through ansa cervicalis (slender nerve root in cervical plexus)
Sternothyroid (ster″no-thi′roid) (*thyro* = thyroid cartilage)	Lateral and deep to sternohyoid	O—posterior surface of manubrium of sternum I—thyroid cartilage	Pulls thyroid cartilage (plus larynx and hyoid bone) inferiorly	As for sternohyoid

| TABLE 10.3 | Muscles of the Anterior Neck and Throat: Swallowing (Figure 10.8) *(continued)* |

Muscle	Description	Origin (O) and Insertion (I)	Action	Nerve Supply
Omohyoid (o″mo-hi′oid) (*omo* = shoulder)	Straplike muscle with two bellies united by an intermediate tendon; lateral to sternohyoid	O—superior surface of scapula I—hyoid bone, lower border	Depresses and retracts hyoid bone	As for sternohyoid
Thyrohyoid (thi″ro-hi′oid) (also see Figure 10.7)	Appears as a superior continuation of sternothyroid muscle	O—thyroid cartilage I—hyoid bone	Depresses hyoid bone and elevates larynx if hyoid is fixed	First cervical nerve via hypoglossal
Pharyngeal constrictor muscles—superior, middle, and inferior (far-rin′je-al)	Composite of three paired muscles whose fibers run circularly in pharynx wall; arranged so that the superior muscle is innermost and inferior one is outermost; substantial overlap	O—attached anteriorly to mandible and medial pterygoid plate (superior), hyoid bone (middle), and laryngeal cartilages (inferior) I—posterior median raphe of pharynx	Working as a group and in sequence, all constrict pharynx during swallowing, which propels a food bolus to esophagus (via a massagelike action called peristalsis)	Pharyngeal plexus [branches of vagus (X)]

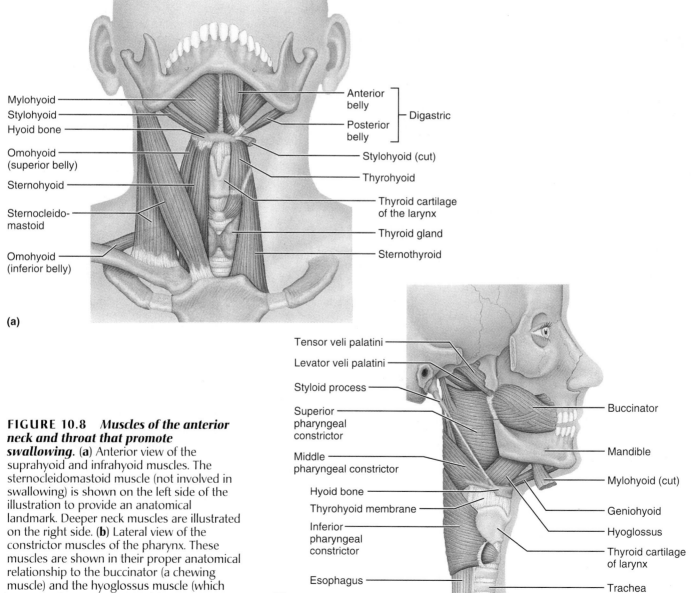

FIGURE 10.8   *Muscles of the anterior neck and throat that promote swallowing.* (a) Anterior view of the suprahyoid and infrahyoid muscles. The sternocleidomastoid muscle (not involved in swallowing) is shown on the left side of the illustration to provide an anatomical landmark. Deeper neck muscles are illustrated on the right side. (b) Lateral view of the constrictor muscles of the pharynx. These muscles are shown in their proper anatomical relationship to the buccinator (a chewing muscle) and the hyoglossus muscle (which promotes tongue movements).

TABLE 10.4	Muscles of the Neck and Vertebral Column: Head and Trunk Movements (Figure 10.9)

Head movements. The head is moved by muscles originating from the axial skeleton. The major head flexors are the **sternocleidomastoid muscles,** but the suprahyoid and infrahyoid muscles described in Table 10.3 act as synergists in this action. Lateral head movements are effected by the sternocleidomastoids and a number of deeper neck muscles, including the **scalenes,** and by several straplike muscles of the vertebral column at the back of the neck. Head extension is aided by the superficial trapezius muscles of the back, but the **splenius** muscles deep to the trapezius muscles bear most of the responsibility for head extension.

Trunk movements. Trunk extension is effected by the *deep* or *intrinsic back muscles* associated with the bony vertebral column, muscles that also play an important role in maintaining the normal curvatures of the spine. Muscles of the thorax that connect adjacent ribs (plus the diaphragm) are involved in breathing movements (see Table 10.5), while the superficial back muscles are concerned primarily with movements of the shoulder girdle and upper limbs (see Tables 10.8 and 10.9).

The deep muscles of the back form a broad, thick column extending from the sacrum to the skull. Many muscles of varying length contribute to this mass. It helps to regard each of these individual muscles as a string that when pulled causes one or several vertebrae to extend or to rotate on the vertebrae below. The largest of the deep back muscle groups is the **erector spinae** (literally, "spine erectors"). Since the origins and insertions of the different muscle groups overlap extensively, entire regions of the vertebral column can be moved simultaneously and smoothly. Acting in concert, the deep back muscles extend (or hyperextend) the spine, but contraction of the muscles on only one side causes lateral bending of the back, neck, or head. Lateral flexion is automatically accompanied by some degree of rotation of the vertebral column. During vertebral movements, the articular facets of the vertebrae glide on each other.

In addition to the long muscles of the back, there are a number of short muscles that extend from one vertebra to the next (Figure 10.9e). These small muscles (the rotatores, multifidus, interspinales, and intertransversarii muscles) act primarily as synergists in extension and rotation of the spine and as spine stabilizers. They are not described in the table but you can deduce their actions by examining their origins and insertions in the figure.

Trunk muscles also maintain the normal curvatures of the spine, acting as postural muscles. It is these deep trunk *extensors* that are considered in this table. The more superficial muscles, which have other functions, are considered in subsequent tables. For example, the muscles of the abdominal wall also help to effect movements of the vertebral column and body trunk. These muscles, which cause trunk *flexion*, are described in Table 10.6.

Muscle	Description	Origin (O) and Insertion (I)	Action	Nerve Supply
ANTEROLATERAL NECK MUSCLES (Figure 10.9a and c)				
Sternocleidomastoid (ster″no-kli″do-mas′toid) (*sterno* = breastbone; *cleido* = clavicle; *mastoid* = mastoid process)	Two-headed muscle located deep to platysma on anterolateral surface of neck; fleshy parts on either side of neck delineate limits of anterior and posterior triangles; key muscular landmark in neck; spasms of one of these muscles may cause torticollis (wryneck)	O—manubrium of sternum and medial portion of clavicle I—mastoid process of temporal bone and superior nuchal line of occipital bone	Prime mover of active head flexion; simultaneous contraction of both muscles causes neck flexion, generally against resistance as when one raises head when lying on back; acting alone, each muscle rotates head toward shoulder on opposite side and tilts or laterally flexes head to its own side	Accessory nerve (cranial nerve XI) and branches of cervical spinal nerves 2–4
Scalenes (ska′lēnz)— anterior, middle, and posterior (*scalene* = uneven)	Located more laterally than anteriorly on neck; deep to platysma and sternocleidomastoid	O—transverse processes of cervical vertebrae I—anterolaterally on first two ribs	Elevate first two ribs (aid in inspiration); flex and rotate neck	Cervical spinal nerves

TABLE 10.4	Muscles of the Neck and Vertebral Column: Head and Trunk Movements (Figure 10.9) *(continued)*				

Muscle	Description	Origin (O) and Insertion (I)	Action	Nerve Supply
INTRINSIC MUSCLES OF THE BACK (Figure 10.9b–e)				
Splenius (sple′ne-us)— (capitis and cervicis portions (kah-pit′us; ser-vis′us) (*splenion* = bandage; *caput* = head; *cervi* = neck) (Figures 10.9b and 10.6)	Broad bipartite superficial muscle (capitis and cervicis parts) extending from upper thoracic vertebrae to skull; capitis portion known as "bandage muscle" because it covers and holds down deeper neck muscles	O—ligamentum nuchae,* spinous processes of vertebrae C_7–T_6 I—mastoid process of temporal bone and occipital bone (capitis); transverse processes of C_2-C_4 vertebrae (cervicis)	Act as a group to extend or hyperextend head; when splenius muscles on one side are activated, head is rotated and bent laterally toward same side	Cervical spinal nerves (dorsal rami)

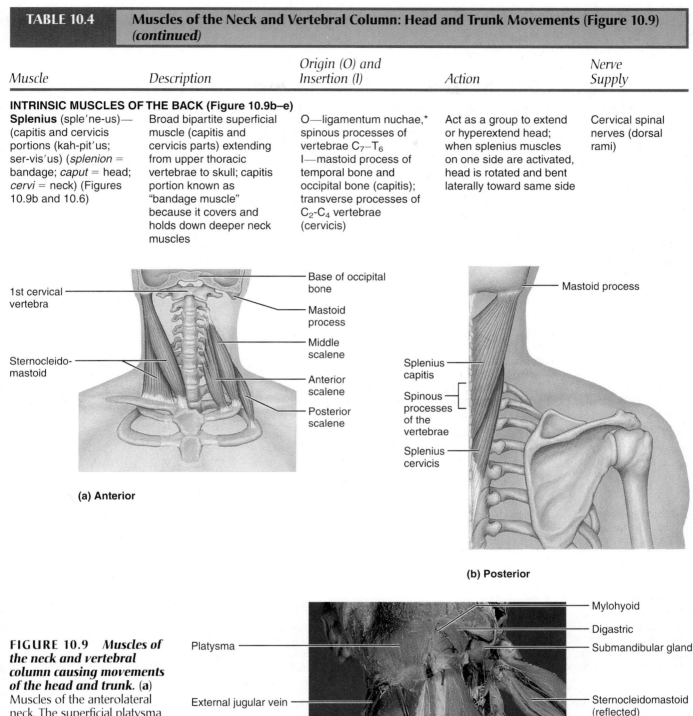

1st cervical vertebra

Sternocleido-mastoid

Base of occipital bone

Mastoid process

Middle scalene

Anterior scalene

Posterior scalene

(a) Anterior

Mastoid process

Splenius capitis

Spinous processes of the vertebrae

Splenius cervicis

(b) Posterior

FIGURE 10.9 *Muscles of the neck and vertebral column causing movements of the head and trunk.* **(a)** Muscles of the anterolateral neck. The superficial platysma muscle and the deeper neck muscles have been removed to show clearly the origins and insertions of the sternocleido-mastoid and scalene muscles. **(b)** Deep muscles of the posterior neck. **(c)** Photograph of the anterior and lateral regions of the neck. The fascia has been partially removed (left side of photograph) to expose the sternocleidomastoid muscle.

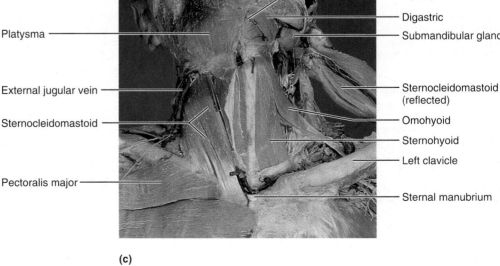

Platysma

External jugular vein

Sternocleidomastoid

Pectoralis major

Mylohyoid

Digastric

Submandibular gland

Sternocleidomastoid (reflected)

Omohyoid

Sternohyoid

Left clavicle

Sternal manubrium

(c)

*The ligamentum nuchae (lig″ah-men′tum noo′ke) is a strong, elastic ligament extending from the occipital bone of the skull along the tips of the spinous processes of the cervical vertebrae. It binds the cervical vertebrae together and inhibits excessive head and neck flexion, thus preventing damage to the spinal cord in the vertebral canal.

TABLE 10.4	Muscles of the Neck and Vertebral Column: Head and Trunk Movements (Figure 10.9) *(continued)*			
Muscle	Description	Origin (O) and Insertion (I)	Action	Nerve Supply
Erector spinae (e-rek′tor spi′ne) Also called **sacrospinalis** (Figure 10.9d, left side)	Prime mover of back extension; erector spinae muscles, each consisting of three columns—the iliocostalis, longissimus, and spinalis muscles—form intermediate layer of intrinsic back muscles; erector spinae provide resistance that helps control action of bending forward at the waist and act as powerful extensors to promote return to erect position; during full flexion (i.e., when touching fingertips to floor), erector spinae are relaxed and strain is borne entirely by ligaments of back; on reversal of the movement, these muscles are initially inactive, and extension is initiated by hamstring muscles of thighs and gluteus maximus muscles of buttocks. As a result of this peculiarity, lifting a load or moving suddenly from a bent-over position is potentially dangerous (in terms of possible injury) to muscles and ligaments of back and intervertebral discs; erector spinae muscles readily go into painful spasms following injury to back structures			
■ **Iliocostalis** (il″e-o-kos-tă′lis)— lumborum), thoracis, and cervicis portions (lum′bor-um; tho-ra′sis) (*ilio* = ilium; *cost* = rib)	Most lateral muscle group of erector spinae muscles; extend from pelvis to neck	O—iliac crests (lumborum); inferior 6 ribs (thoracis); ribs 3 to 6 (cervicis) I—angles of ribs (lumborum and thoracis); transverse processes of cervical vertebrae C_6–C_4 (cervicis)	Extend vertebral column, maintain erect posture; acting on one side, bend vertebral column to same side	Spinal nerves (dorsal rami)
■ **Longissimus** (lon-jis′ĭ-mus)— thoracis, cervicis, and capitis parts (*longissimus* = longest)	Intermediate tripartite muscle group of erector spinae; extend by many muscle slips from lumbar region to skull; mainly pass between transverse processes of the vertebrae	O—transverse processes of lumbar through cervical vertebrae I—transverse processes of thoracic or cervical vertebrae and to ribs superior to origin as indicated by name; capitis inserts into mastoid process of temporal bone	Thoracis and cervicis act together to extend vertebral column and acting on one side, bend it laterally; capitis extends head and turns the face toward same side	Spinal nerves (dorsal rami)
■ **Spinalis** (spi-nă′lis)— thoracis and cervicis parts (*spin* = vertebral column, spine)	Most medial muscle column of erector spinae; cervicis usually rudimentary and poorly defined	O—spines of upper lumbar and lower thoracic vertebrae I—spines of upper thoracic and cervical vertebrae	Extends vertebral column	Spinal nerves (dorsal rami)
Semispinalis (sem′e-spĭ-nă′lis)— thoracis, cervicis, and capitis regions (*semi* = half; *thorac* = thorax) (Figure 10.9d, right side)	Composite muscle forming part of deep layer of intrinsic back muscles; extends from thoracic region to head	O—transverse processes of C_7–T_{12} I—occipital bone (capitis) and spinous processes of cervical (cervicis) and thoracic vertebrae T_1–T_4 (thoracis)	Extends vertebral column and head and rotates them to opposite side; acts synergistically with sternocleidomastoid muscles of opposite side	Spinal nerves (dorsal rami)
Quadratus lumborum (kwod-ra′tus lum-bor′um) (*quad* = four-sided; *lumb* = lumbar region) (See also Figure 10.19a)	Fleshy muscle forming part of posterior abdominal wall	O—iliac crest and lumbar fascia I—transverse processes of upper lumbar vertebrae and lower margin of 12th rib	Flexes vertebral column laterally when acting separately; when pair acts jointly, lumbar spine is extended and 12th rib is fixed; maintains upright posture; assists in forced inspiration	T_{12} and upper lumbar spinal nerves (ventral rami)

TABLE 10.4	Muscles of the Neck and Vertebral Column: Head and Trunk Movements (Figure 10.9) (*continued*)

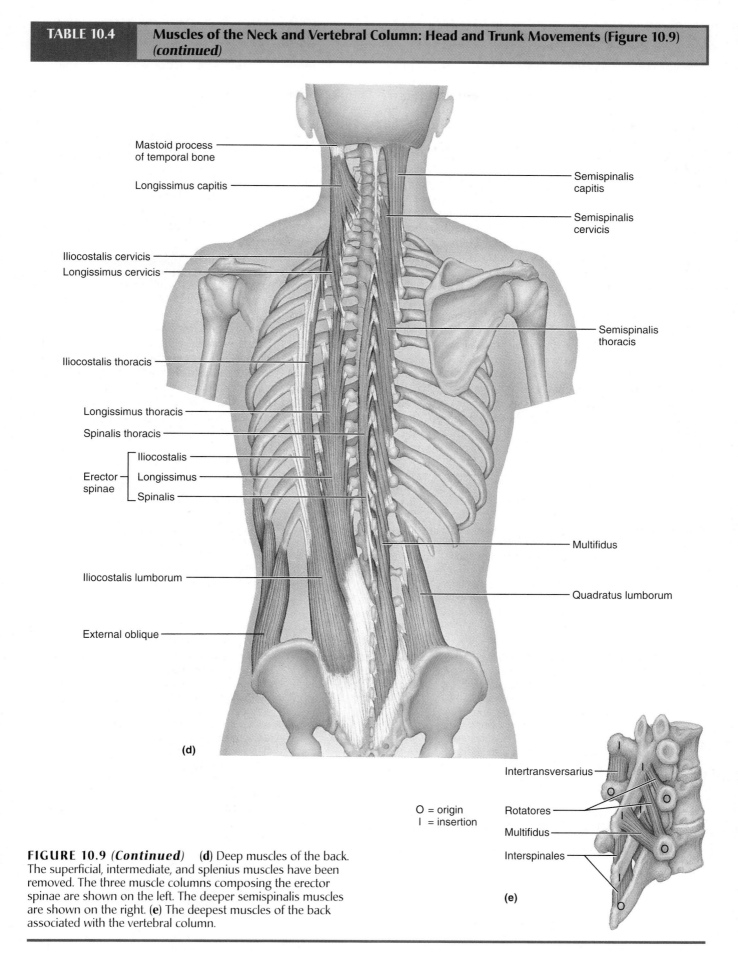

FIGURE 10.9 (*Continued*) (**d**) Deep muscles of the back. The superficial, intermediate, and splenius muscles have been removed. The three muscle columns composing the erector spinae are shown on the left. The deeper semispinalis muscles are shown on the right. (**e**) The deepest muscles of the back associated with the vertebral column.

TABLE 10.5	Muscles of the Thorax: Breathing (Figure 10.10)

The primary function of the deep muscles of the thorax is to promote movements necessary for breathing. Breathing consists of two phases—inspiration, or inhaling, and expiration, or exhaling—brought about by the alternate increase and decrease in the volume of the thoracic cavity.

Two main layers of muscles help form the anterolateral wall of the thorax. However, unlike the extensive muscles that form the abdominal wall, those of the thorax are very short, extending only from one rib to the next. On contraction, they draw the somewhat flexible ribs closer together. The **external intercostal muscles** form the more superficial layer. They lift the rib cage, an action that increases the anterior to posterior and side-to-side dimensions of the thorax, and are considered inspiratory muscles. The **internal intercostal muscles** form the deeper layer and aid active (forced) expiration by depressing the rib cage. (However, quiet expiration is largely a passive phenomenon, resulting from relaxation of the external intercostals and diaphragm and elastic recoil of the lungs.)

The **diaphragm,** the most important muscle of inspiration, forms a muscular partition between the thoracic and abdominopelvic cavities. In the relaxed state, the diaphragm is dome shaped, but when it contracts it moves inferiorly and flattens, increasing the volume of the thoracic cavity. The alternating contraction and relaxation of the diaphragm causes pressure changes in the abdominopelvic cavity below that facilitate the return of blood to the heart. In addition to its rhythmic contractions during respiration, one can also contract the diaphragm voluntarily to push down on the abdominal viscera and increase the pressure in the abdominal cavity to help evacuate pelvic organ contents (urine, feces, or a baby) or during weight lifting. When one takes a deep breath to fix the diaphragm, the abdomen becomes a firm pillar that will not buckle under the weight being lifted. Needless to say, it is important to have good control of the bladder and anal sphincters during such maneuvers.

With the exception of the diaphragm, which is served by the *phrenic nerves,* the muscles listed in this table are served by the *intercostal nerves* (anterior rami of the first 11 thoracic spinal nerves), which run between the ribs.

Forced breathing calls into play a number of other muscles that insert into the ribs; e.g., during forced inspiration the scalene and sternocleidomastoid muscles of the neck help lift the ribs. Forced expiration is aided by muscles that pull the ribs inferiorly (quadratus lumborum) and those that push the diaphragm superiorly by compressing the abdominal contents (abdominal wall muscles). Mechanics of respiration are considered in Chapter 23.

Muscle	Description	Origin (O) and Insertion (I)	Action	Nerve Supply
External intercostals (in″ter-kos′talz) (*external* = toward the outside; *inter* = between; *cost* = rib)	11 pairs lie between ribs; fibers run obliquely (down and forward) from each rib to rib below; in lower intercostal spaces, fibers are continuous with external oblique muscle forming part of abdominal wall	O—inferior border of rib above I—superior border of rib below	With first ribs fixed by scalene muscles, pull ribs toward one another to elevate rib cage; aids in inspiration; synergists of diaphragm	Intercostal nerves
Internal intercostals (*internal* = toward the inside, deep)	11 pairs lie between ribs; fibers run deep to and at right angles to those of external intercostals (i.e., run downward and posteriorly); lower internal intercostal muscles are continuous with fibers of internal oblique muscle of abdominal wall	O—superior border of rib below I—inferior border (costal groove) of rib above	With 12th ribs fixed by quadratus lumborum, muscles of posterior abdominal wall, and oblique muscles of the abdominal wall, they draw ribs together and depress rib cage; aid in forced expiration; antagonistic to external intercostals	Intercostal nerves
Diaphragm (di′ah-fram) (*dia* = across; *phragm* = partition)	Broad muscle pierced by the aorta, inferior vena cava, and esophagus, forms floor of thoracic cavity; in relaxed state is dome shaped; fibers converge from margins of thoracic cage toward a boomerang-shaped central tendon	O—inferior, internal surface of rib cage and sternum, costal cartilages of last six ribs and lumbar vertebrae I—central tendon	Prime mover of inspiration; flattens on contraction, increasing vertical dimensions of thorax; when strongly contracted, dramatically increases intra-abdominal pressure	Phrenic nerves

TABLE 10.5   **Muscles of the Thorax: Breathing (Figure 10.10) (continued)**

FIGURE 10.10 *Muscles of respiration.* **(a)** Deep muscles of the thorax. The external intercostals (inspiratory muscles) are shown on the left and the internal intercostals (expiratory muscles) are shown on the right. These two muscle layers run obliquely and at right angles to each other. **(b)** Inferior view of the diaphragm, the prime mover of inspiration. Notice that its muscle fibers converge toward a central tendon, an arrangement that causes the diaphragm to flatten and move inferiorly as it contracts. **(c)** Photograph of the diaphragm, superior view.

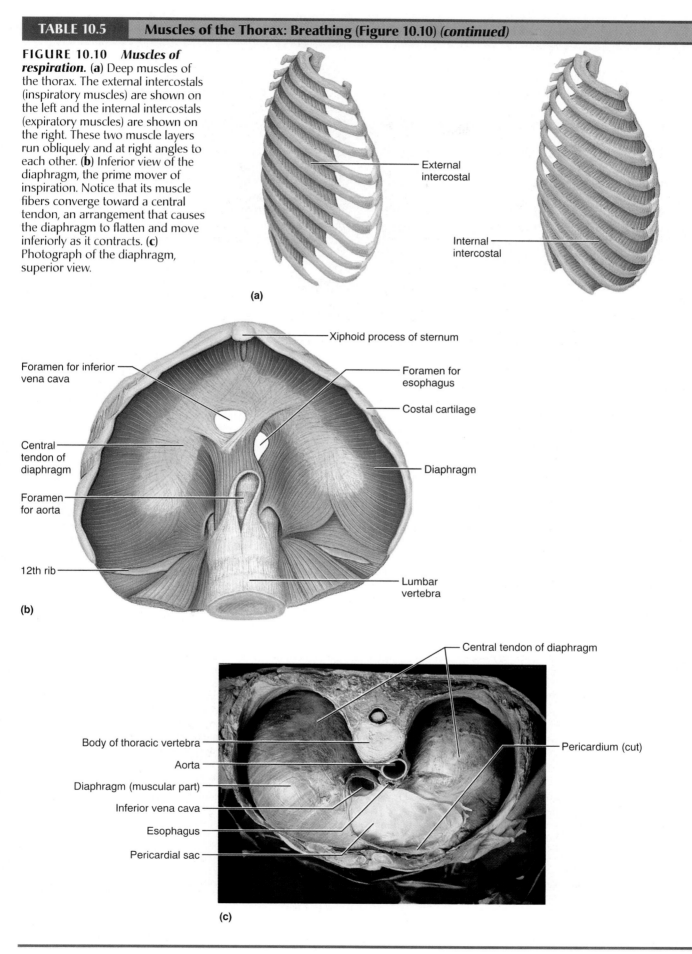

External intercostal

Internal intercostal

(a)

Xiphoid process of sternum

Foramen for inferior vena cava

Foramen for esophagus

Costal cartilage

Central tendon of diaphragm

Diaphragm

Foramen for aorta

12th rib

Lumbar vertebra

(b)

Central tendon of diaphragm

Body of thoracic vertebra

Aorta

Diaphragm (muscular part)

Inferior vena cava

Esophagus

Pericardial sac

Pericardium (cut)

(c)

TABLE 10.6	**Muscles of the Abdominal Wall: Trunk Movements and Compression of Abdominal Viscera (Figure 10.11)**

Unlike the thorax, the anterior and lateral abdominal wall has no bony reinforcements (ribs). Instead, it is a composite of four paired muscles, their investing fasciae, and their aponeuroses. Three broad flat muscle pairs, layered one atop the next, form the lateral abdominal walls: The fibers of the **external oblique muscle** run inferomedially and at right angles to those of the **internal oblique,** which immediately underlies it. The fibers of the deep **transversus abdominis muscle** run horizontally across the abdomen at an angle to both. This alternation of fascicle directions is similar to the construction of plywood (which is made of sheets with different grains) and provides great strength. These three muscles blend into broad insertion aponeuroses anteriorly. The aponeuroses, in turn, enclose a fourth muscle pair, the straplike **rectus abdominis muscles** medially, and then fuse forming the **linea alba** ("white line"), a tendinous raphe (seam) that runs from the sternum to the pubic symphysis. Enclosing the rectus abdominis muscle within the aponeuroses prevents them from "bowstringing" (protruding anteriorly). The quadratus lumborum muscles of the *posterior* abdominal wall are covered in Table 10.4.

The abdominal muscles protect and support the viscera most effectively when they are well toned. When weak or severely stretched (as during pregnancy),

they allow the abdomen to become pendulous (i.e., to form a potbelly). Additional functions include lateral flexion and rotation of the trunk and anterior flexion of the trunk against resistance (as in sit-ups). During quiet expiration, the abdominal muscles relax, allowing the abdominal viscera to be pushed inferiorly by the descending diaphragm. When all these abdominal muscles contract in unison, several different activities may be effected, depending on which other muscles are activated simultaneously. For example, when all the abdominal muscles are contracted, the ribs are pulled inferiorly and the abdominal contents compressed. This pushes the visceral organs upward on the diaphragm, aiding forced expiration. When the abdominal muscles contract with the diaphragm and the glottis is closed (an action called the Valsalva maneuver), the increased intra-abdominal pressure helps to promote urination, defecation, childbirth, vomiting, coughing, screaming, sneezing, burping, and nose blowing. (Next time you perform one of these activities, feel your abdominal muscles contract under your skin.) These muscles also contract during heavy lifting—sometimes so forcefully that hernias result. Contraction of the abdominal muscles along with contraction of the deep back muscles helps prevent hyperextension of the spine and splints the entire body trunk.

Muscle	Description	Origin (O) and Insertion (I)	Action	Nerve Supply
MUSCLES OF THE ANTERIOR AND LATERAL ABDOMINAL WALL				
	Four paired flat muscles; very important in supporting and protecting abdominal viscera and play an important role in promoting movement of vertebral column (flexion and lateral bending)			
Rectus abdominis (rek′tus ab-dom′ĭ-nis) (*rectus* = straight; *abdom* = abdomen)	Medial superficial muscle pair; extend from pubis to rib cage; ensheathed by aponeuroses of lateral muscles; segmented by 3 tendinous intersections	O—pubic crest and symphysis I—xiphoid process and costal cartilages of ribs 5–7	Flex and rotate lumbar region of vertebral column; fix and depress ribs, stabilize pelvis during walking, increase intra-abdominal pressure	Intercostal nerves (T_6 or T_7–T_{12})
External oblique (o-blēk′) (*external* = toward outside; *oblique* = running at an angle)	Largest and most superficial of the three lateral muscles; fibers run downward and medially (same direction outstretched fingers take when hands put into pants pockets); aponeurosis turns under inferiorly, forming inguinal ligament	O—by fleshy strips from outer surfaces of lower eight ribs I—most fibers insert anteriorly via a broad aponeurosis into linea alba; some into pubic crest and tubercle and iliac crest	When pair contracts simultaneously, aid rectus abdominis muscles in flexing vertebral column and in compressing abdominal wall and increasing intra-abdominal pressure; acting individually, aid muscles of back to trunk rotation and lateral flexion	Intercostal nerves (T_7–T_{12})
Internal oblique (*internal* = toward the inside; deep)	Most fibers run upward and medially; however the muscle fans so its inferior fibers run downward and medially	O—lumbar fascia, iliac crest, and inguinal ligament I—linea alba, pubic crest, last three or four ribs, and costal margin	As for external oblique	Intercostal nerves (T_7–T_{12}) and L_1
Transversus abdominis (trans-ver′sus) (*transverse* = running straight across)	Deepest (innermost) muscle of abdominal wall; fibers run horizontally	O—inguinal ligament, lumbar fascia, cartilages of last six ribs; iliac crest I—linea alba, pubic crest	Compresses abdominal contents	Intercostal nerves (T_7–T_{12}) and L_1

TABLE 10.6	Muscles of the Abdominal Wall: Trunk Movements and Compression of Abdominal Viscera (Figure 10.11) *(continued)*

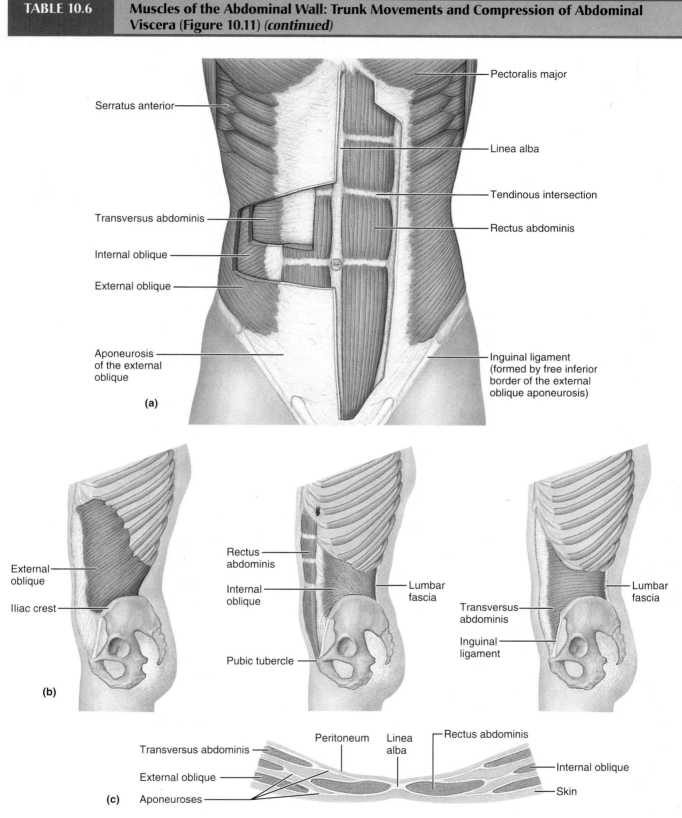

FIGURE 10.11 *Muscles of the abdominal wall.* (**a**) Anterior view of the muscles forming the anterolateral abdominal wall. The superficial muscles have been partially cut away on the left side of the diagram to reveal the deeper muscles. (**b**) Lateral view of the trunk, illustrating the fiber direction and attachments of the external and internal obliques, and transversus abdominis muscles. Although shown in the same view, the rectus abdominis is deep to the fascia of the internal oblique muscle. (**c**) Transverse section through the anterolateral abdominal wall (midregion), showing how the aponeuroses of the lateral abdominal muscles contribute to the rectus abdominis sheath.

TABLE 10.7	Muscles of the Pelvic Floor and Perineum: Support of Abdominopelvic Organs (Figure 10.12)

Two paired muscles, the **levator ani** and **coccygeus**, form the funnel-shaped pelvic floor, or **pelvic diaphragm.** These muscles (1) close the inferior outlet of the pelvis, (2) support the pelvic floor and elevate it to help release feces, and (3) resist increased intra-abdominal pressure (which would expel contents of the bladder, rectum, and uterus). The pelvic diaphragm is pierced by the rectum and urethra (urinary tube), and by the vagina in females. The body region inferior to the pelvic diaphragm is the *perineum.* The relationships of the perineum are a bit complex and worth explaining. Inferior to the muscles of the pelvic floor, and stretching between the two sides of the pubic arch in the anterior half of the perineum, is the **urogenital diaphragm.** This thin triangular sheet of muscle contains the **sphincter urethrae.** This sphincter muscle, which surrounds the urethra, allows voluntary control of urination when it is inconvenient. Superficial to the urogenital diaphragm, and covered by the skin of the perineum, is the *superficial space,* which contains muscles (**ischiocavernosus** and **bulbospongiosus**) that help maintain erection of the penis and clitoris. In the posterior half of the perineum lies the **external anal sphincter,** a sphincter that encircles the anus and allows voluntary control of defecation. Just anterior to this sphincter is the **central tendon of the perineum,** a strong tendon into which many of the perineal muscles insert.

Muscle	Description	Origin (O) and Insertion (I)	Action	Nerve Supply
MUSCLES OF THE PELVIC DIAPHRAGM (Figure 10.12a)				
Levator ani (lĕ-va′tor a′ne) (*levator* = raises; *ani* = anus)	Broad, thin, tripartite muscle (pubococcygeus, puborectalis, and iliococcygeus parts); its fibers extend inferomedially, forming a muscular "sling" around male prostate (or female vagina), urethra, and anorectal junction before meeting in the median plane	O—extensive linear origin inside pelvis from pubis to ischial spine I—inner surface of coccyx, levator ani of opposite side, and (in part) into the structures that penetrate it	Supports and maintains position of pelvic viscera; resists downward thrusts that accompany rises in intrapelvic pressure during coughing, vomiting, and expulsive efforts of abdominal muscles; forms sphincters at anorectal junction and vagina; lifts anal canal during defecation	S$_4$ and inferior rectal nerve (branch of pudendal nerve)
Coccygeus (kok-sij′e-us) (*coccy* = coccyx)	Small triangular muscle lying posterior to levator ani; forms posterior part of pelvic diaphragm	O—spine of ischium I—sacrum and coccyx	Supports pelvic viscera; supports coccyx and pulls it forward after it has been reflected posteriorly by defecation and childbirth	S$_4$ and S$_5$
MUSCLES OF THE UROGENITAL DIAPHRAGM (Figure 10.12b)				
Deep transverse perineus (per″ĭ-ne′us) (*deep* = far from surface; *transverse* = across; *perine* = near anus)	Together the pair spans distance between ischial rami; in females, lies posterior to vagina	O—ischial rami I—midline central tendon of perineum; some fibers into vaginal wall in females	Supports pelvic organs; steadies central tendon	Pudendal nerve
Sphincter urethrae (*sphin* = squeeze)	Muscle encircling urethra and vagina (female)	O—ischiopubic rami I—midline raphe	Constricts urethra; helps support pelvic organs	Pudendal nerve
MUSCLES OF THE SUPERFICIAL SPACE (Figure 10.12c)				
Ischiocavernosus (is′ke-o-kav′ern-o′sus) (*ischi* = hip; *caverna* = hollow chamber)	Runs from pelvis to base of penis or clitoris	O—ischial tuberosities I—crus of corpus cavernosa of male penis or female clitoris	Retards venous drainage and maintains erection of penis or clitoris	Pudendal nerve
Bulbospongiosus (bul″bo-spun″je-o′sus) (*bulbon* = bulb; *spongio* = sponge)	Encloses base of penis (bulb) in males and lies deep to labia in females	O—central tendon of perineum and midline raphe of male penis I—anteriorly into corpus cavernosa of penis or clitoris	Empties male urethra; assists in erection of penis in males and of clitoris in females	Pudendal nerve
Superficial transverse perineus (superficial 5 closer to surface)	Paired muscle bands posterior to urethral (and in females, vaginal) opening; variable; sometimes absent	O—ischial tuberosity I—central tendon of perineum	Stabilizes and strengthens midline tendon of perineum	Pudendal nerve

TABLE 10.7	Muscles of the Pelvic Floor and Perineum: Support of Abdominopelvic Organs (Figure 10.12) *(continued)*

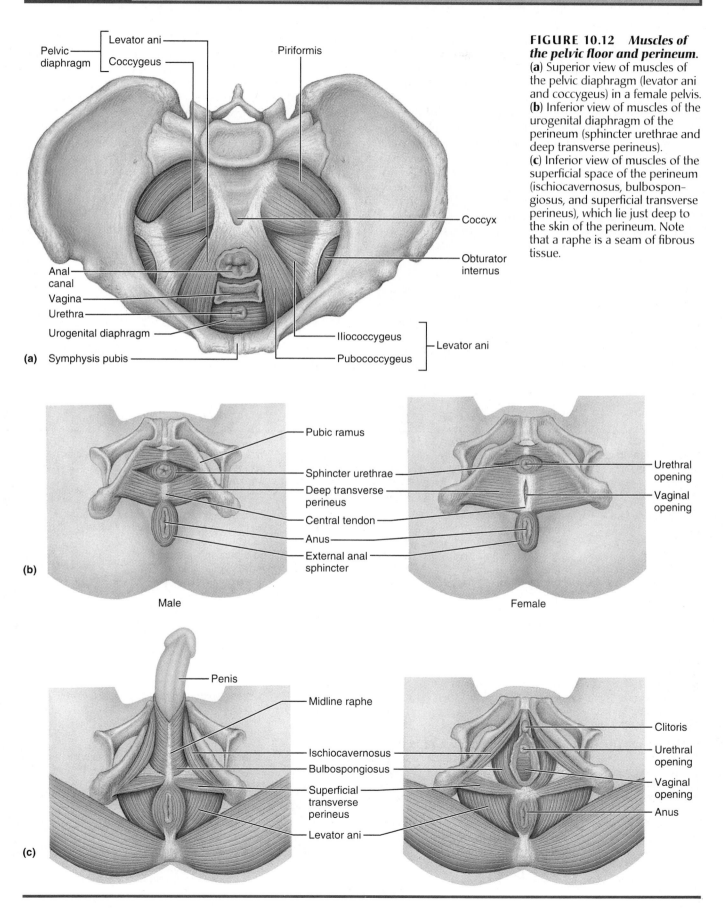

FIGURE 10.12 *Muscles of the pelvic floor and perineum.* (**a**) Superior view of muscles of the pelvic diaphragm (levator ani and coccygeus) in a female pelvis. (**b**) Inferior view of muscles of the urogenital diaphragm of the perineum (sphincter urethrae and deep transverse perineus). (**c**) Inferior view of muscles of the superficial space of the perineum (ischiocavernosus, bulbospongiosus, and superficial transverse perineus), which lie just deep to the skin of the perineum. Note that a raphe is a seam of fibrous tissue.

(a)

Pelvic diaphragm — Levator ani / Coccygeus
Piriformis
Coccyx
Obturator internus
Anal canal
Vagina
Urethra
Urogenital diaphragm
Symphysis pubis
Iliococcygeus / Pubococcygeus — Levator ani

(b)

Pubic ramus
Sphincter urethrae
Deep transverse perineus
Central tendon
Anus
External anal sphincter
Urethral opening
Vaginal opening
Male   Female

(c)

Penis
Midline raphe
Ischiocavernosus
Bulbospongiosus
Superficial transverse perineus
Levator ani
Clitoris
Urethral opening
Vaginal opening
Anus

TABLE 10.8	Superficial Muscles of the Anterior and Posterior Thorax: Movements of the Scapula (Figure 10.13)

Most superficial thorax muscles are *extrinsic shoulder muscles,* which run from the ribs and vertebral column to the shoulder girdle. They act to fix the scapula to the wall of the thorax or move the scapula to effect arm movements. The muscles of the anterior thorax include the **pectoralis major, pectoralis minor, serratus anterior,** and **subclavius.** Except for the pectoralis major, which inserts into the humerus, all muscles of the anterior group insert into the pectoral girdle. Extrinsic muscles of the posterior thorax include the **latissimus dorsi** and **trapezius muscles** superficially and the underlying **levator scapulae** and **rhomboids.** The latissimus dorsi, like the pectoralis major muscles anteriorly, insert into the humerus and are more concerned with movements of the arm than of the scapula. We will defer consideration of these two muscle pairs to Table 10.9 (arm-moving muscles).

The important movements of the pectoral girdle involve displacements of the scapula, i.e., its elevation and depression, rotation, lateral (forward) movements, and medial (backward) movements. The clavicles rotate around their own axes during scapular movements, and provide both stability and precision to scapular movements.

Except for the serratus anterior, the anterior muscles stabilize and depress the shoulder girdle. Thus, most scapular movements are promoted by the serratus anterior muscles anteriorly and by the posterior muscles. The arrangement of muscle attachments to the scapula is such that one muscle cannot bring about a simple (linear) movement on its own. To elevate or depress the scapula or to effect other scapular movements, several muscles must act in combination.

The prime movers of shoulder elevation are the trapezius and levator scapulae. When acting together to shrug the shoulder, their opposite rotational effects counterbalance each other. The scapula is depressed largely by gravity (weight of the arm), but when it is depressed against resistance, the trapezius and serratus anterior (along with the latissimus dorsi, Table 10.9) are active. Forward movements (abduction) of the scapula on the thorax wall, as in pushing or punching movements, mainly reflect serratus anterior activity. Retraction (adduction) of the scapula is effected by the trapezius and rhomboids. Although the serratus anterior and trapezius muscles are antagonists in forward/backward movements of the scapulae, they act together to coordinate *rotational* scapular movements.

Muscle	Description	Origin (O) and Insertion (I)	Action	Nerve Supply
MUSCLES OF THE ANTERIOR THORAX (Figure 10.13a)				
Pectoralis minor (pek"to-ra'lis mi'nor) (*pectus = chest; breast; minor = lesser*)	Flat, thin muscle directly beneath and obscured by pectoralis major	O—anterior surfaces of ribs 3–5 (or 2–4) I—coracoid process of scapula	With ribs fixed, draws scapula forward and downward; with scapula fixed, draws rib cage superiorly	Both pectoral nerves (C_6–C_8)
Serratus anterior (ser-a'tus) (*serratus = saw*)	Lies deep to scapula, beneath and inferior to pectoral muscles on lateral rib cage; forms medial wall of axilla; origins have serrated, or sawtooth, appearance; paralysis results in "winging" of vertebral border of scapula away from chest wall, making arm elevation impossible	O—by a series of muscle slips from ribs 1–8 (or 9) I—entire anterior surface of vertebral border of scapula	Prime mover to protract and hold scapula against chest wall; rotates scapula so that its inferior angle moves laterally and upward; raises point of shoulder; important role in abduction and raising of arm and in horizontal arm movements (pushing, punching); called "boxer's muscle"	Long thoracic nerve (C_5–C_7)
Subclavius (sub-kla've-us) (*sub = under, beneath; clav = clavicle*)	Small cylindrical muscle extending from rib 1 to clavicle	O—costal cartilage of rib 1 I—groove on inferior surface of clavicle	Helps stabilize and depress pectoral girdle	Nerve to subclavius (C_5 and C_6)
MUSCLES OF THE POSTERIOR THORAX (Figure 10.13b)				
Trapezius (trah-pe'ze-us) (*trapezion = irregular four-sided figure*)	Most superficial muscle of posterior thorax; flat, and triangular in shape; upper fibers run inferiorly to scapula; middle fibers run horizontally to scapula; lower fibers run superiorly to scapula	O—occipital bone, ligamentum nuchae, and spines of C_7 and all thoracic vertebrae I—a continuous insertion along acromion and spine of scapula and lateral third of clavicle	Stabilizes, raises, retracts, and rotates scapula; middle fibers retract (adduct) scapula; superior fibers elevate scapula or can help extend head with scapula fixed; inferior fibers depress scapula (and shoulder)	Accessory nerve (cranial nerve XI); C_3 and C_4

TABLE 10.8	Superficial Muscles of the Anterior and Posterior Thorax: Movements of the Scapula (Figure 10.13) *(continued)*			
Muscle	*Description*	*Origin (O) and Insertion (I)*	*Action*	*Nerve Supply*
Levator scapulae (skap′u-le) (*levator* = raises)	Located at back and side of neck, deep to trapezius; thick, straplike muscle border of scapula	O—transverse processes of C_1–C_4 I—medial border of the scapula, superior to the spine	Elevates/adducts scapula in concert with superior fibers of trapezius; tilts glenoid cavity downward when scapula is fixed, flexes neck to same side	Cervical spinal nerves and dorsal scapular nerve (C_3–C_5)
Rhomboids (rom′boidz)—major and minor (*rhomboid* = diamond shaped)	Two rectangular muscles lying deep to trapezius and inferior to levator scapulae; rhomboid minor is the more superior muscle	O—spinous processes of C_7 and T_1 (minor) and spinous processes of T_2–T_5 (major) I—medial border of scapula	Act together (and with middle trapezius fibers) to retract scapula, thus "squaring shoulders"; rotate glenoid cavity of scapula downward (as when arm is lowered against resistance; e.g., paddling a canoe); stabilize scapula	Dorsal scapular nerve (C_4 and C_5)

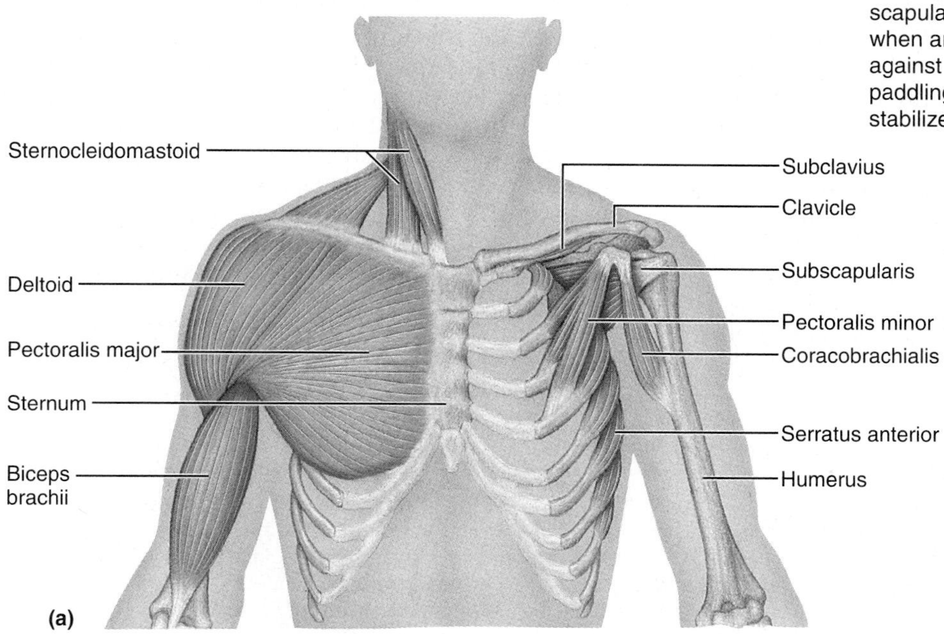

FIGURE 10.13 ***Superficial muscles of the thorax and shoulder acting on the scapula and arm.*** (**a**) Anterior view. The superficial muscles, which effect arm movements, are shown on the left side of the illustration. These muscles are removed on the right to show the muscles that stabilize or move the pectoral girdle. (**b**) Posterior view. The superficial muscles are shown for the left side of the illustration, with a corresponding photograph. The superficial muscles are removed on the right side to reveal the deeper muscles acting on the scapula and the rotator cuff muscles that help stabilize the shoulder joint.*

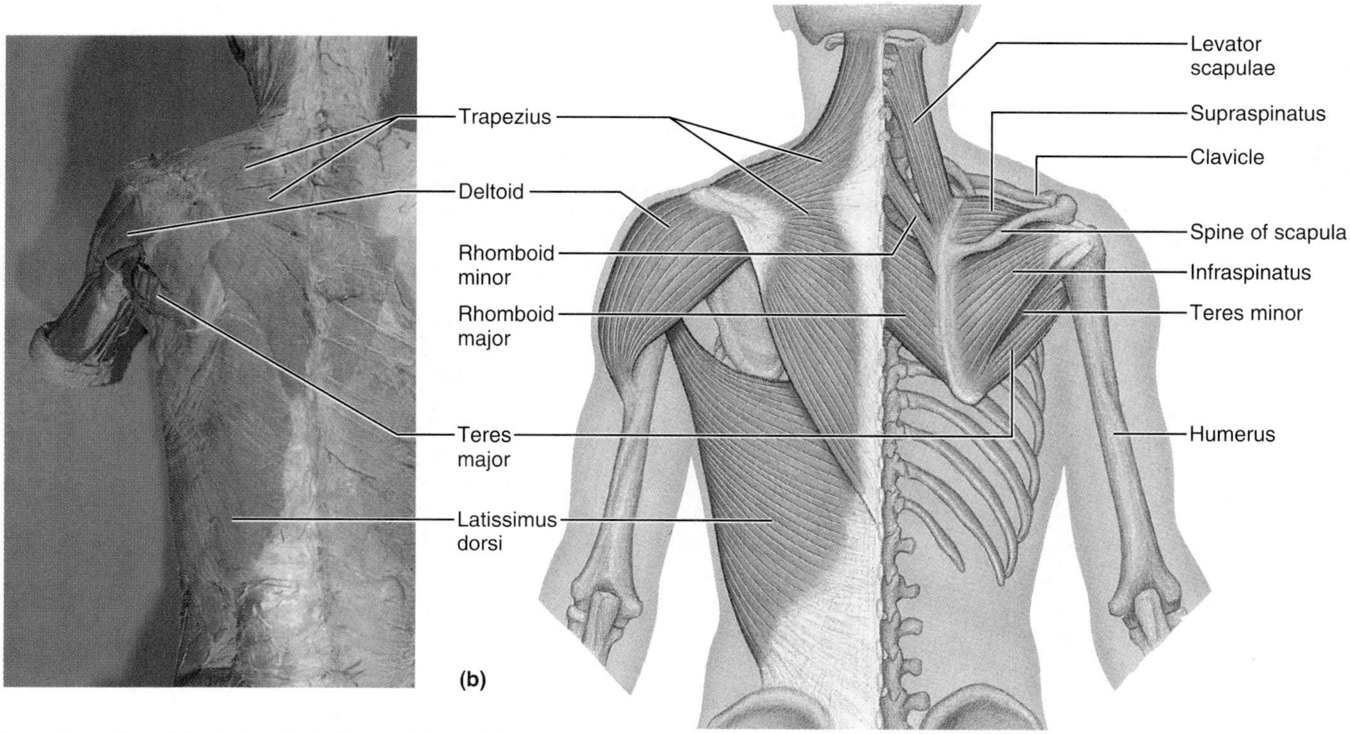

*See *A Brief Atlas of the Human Body,* Figures 6.1 and 6.2.

TABLE 10.9	Muscles Crossing the Shoulder Joint: Movements of the Arm (Figure 10.14)

Recall that the ball-and-socket shoulder joint is the most flexible joint in the body, but pays the price of instability. A total of nine muscles cross each shoulder joint to insert on the humerus. All muscles acting on the humerus originate from the pectoral girdle; however, the latissimus dorsi and pectoralis major also originate on the axial skeleton.

Of these nine muscles, only the superficial **pectoralis major, latissimus dorsi,** and **deltoid muscles** are prime movers of arm movements. The remaining six are synergists and fixators. Four of these, the **supraspinatus, infraspinatus, teres minor,** and **subscapularis,** are known as *rotator cuff muscles.* They originate on the scapula, and their tendons blend with the fibrous capsule of the shoulder joint en route to the humerus. Although the rotator cuff muscles act as synergists in the angular and rotational movements of the arm, their main function is to reinforce the capsule of the shoulder joint to prevent dislocation of the humerus. The remaining two muscles, the small **teres major** and **coracobrachialis,** cross the shoulder joint but do not contribute to its reinforcement.

Generally speaking, muscles that originate *anterior* to the shoulder joint (pectoralis major, coracobra-chialis, and anterior fibers of the deltoid) *flex* the arm, i.e., lift it anteriorly, usually in the sagittal plane. The prime mover of arm flexion is the pectoralis major. The biceps brachii of the arm (see Table 10.10) also assists in this action. Muscles originating *posterior* to the shoulder joint extend the arm. These include the latissimus dorsi and posterior fibers of the deltoid muscles (both prime movers of arm extension) and the teres major. Thus, the latissimus dorsi and pectoralis muscles are *antagonists* of one another in the flexion-extension movements of the arm.

The middle region of the fleshy deltoid muscle of the shoulder is the prime mover of arm abduction. It is antagonized in this function by the pectoralis major anteriorly and latissimus dorsi posteriorly. Depending on their precise location and insertion points, the various muscles acting on the humerus also promote lateral and medial rotation of the shoulder joint. Since the interactions of these nine muscles are complex and each muscle contributes to more than a single movement, a summary of muscles contributing to the various angular and rotational movements of the humerus is provided in Table 10.12 (Part I).

Muscle	Description	Origin (O) and Insertion (I)	Action	Nerve Supply
Pectoralis major (pek″to-ra′lis ma′jer) (*pectus* = breast, chest; *major* = larger)	Large, fan-shaped muscle covering upper portion of chest; forms anterior axillary fold; divided into clavicular and sternal parts	O—sternal end of clavicle, sternum, cartilage of ribs 1–6 (or 7), and aponeurosis of external oblique muscle I—fibers converge to insert by a short tendon into intertubercular groove of humerus	Prime mover of arm flexion; rotates arm medially; adducts arm against resistance; with scapula (and arm) fixed, pulls rib cage upward, thus can help in climbing, throwing, pushing, and in forced inspiration	Lateral and medial pectoral nerves (C_5–C_8, and T_1)
Latissimus dorsi (lah-tis′ĭ-mus dor′si) (*latissimus* = widest; *dorsi* = back)	Broad, flat, triangular muscle of lower back (lumbar region); extensive superficial origins; covered by trapezius superiorly; contributes to the posterior wall of axilla	O—indirect attachment via lumbodorsal fascia into spines of lower six thoracic vertebrae, lumbar vertebrae, lower 3 to 4 ribs, and iliac crest I—spirals around teres major to insert in floor of intertubercular groove of humerus	Prime mover of arm extension; powerful arm adductor; medially rotates arm at shoulder; depresses scapula; because of its power in these movements, it plays an important role in bringing the arm down in a power stroke, as in striking a blow, hammering, swimming, and rowing; with arms fixed overhead, it pulls the rest of the body upward and forward	Thoracodorsal nerve (C_6–C_8)

TABLE 10.9	Muscles Crossing the Shoulder Joint: Movements of the Arm (Figure 10.14) *(continued)*			
Muscle	*Description*	*Origin (O) and Insertion (I)*	*Action*	*Nerve Supply*
Deltoid (del′toid)) (*delta* = triangular)	Thick, multipennate muscle forming rounded shoulder muscle mass; responsible for roundness of shoulder; a site commonly used for intramuscular injection, particularly in males, where it tends to be quite fleshy	O—embraces insertion of the trapezius; lateral third of clavicle; acromion and spine of scapula I—deltoid tuberosity of humerus	Prime mover of arm abduction when all its fibers contract simultaneously; antagonist of pectoralis major and latissimus dorsi, which adduct the arm; if only anterior fibers are active, can act powerfully in flexion and medial rotation of humerus, therefore synergist of pectoralis major; if only posterior fibers are active, effects extension and lateral rotation of arm; active during rhythmic arm swinging movements during walking	Axillary nerve (C_5 and C_6)
Subscapularis (sub-scap″u-lar′is) (*sub* = under; *scapular* = scapula)	Forms part of posterior wall of axilla; tendon of insertion passes in front of shoulder joint; a rotator cuff muscle	O—subscapular fossa of scapula I—lesser tubercle of humerus	Chief medial rotator of humerus; assisted by pectoralis major; helps to hold head of humerus in glenoid cavity, thereby stabilizing shoulder joint	Subscapular nerves (C_5–C_7)
Supraspinatus (soo″prah-spi-nah′tus) (*supra* = above, over; *spin* = spine)	Named for its location on posterior aspect of scapula; deep to trapezius; a rotator cuff muscle	O—supraspinous fossa of scapula I—superior part of greater tubercle of humerus	Stabilizes shoulder joint; helps to prevent downward dislocation of humerus, as when carrying a heavy suitcase; assists in abduction	Suprascapular nerve
Infraspinatus (in″frah-spi-nah′tus) (*infra* = below)	Partially covered by deltoid and trapezius; named for its scapular location; a rotator cuff muscle	O—infraspinous fossa of scapula I—greater tubercle of humerus posterior to insertion of supraspinatus	Helps to hold head of humerus in glenoid cavity, stabilizing the shoulder joint; rotates humerus laterally	Suprascapular nerve
Teres minor (te′rēz) (*teres* = round; *minor* = lesser)	Small, elongated muscle; lies inferior to infraspinatus and may be inseparable from that muscle; a rotator cuff muscle	O—lateral border of dorsal scapular surface I—greater tubercle of humerus inferior to infraspinatus insertion	Same action(s) as infraspinatus muscle	Axillary nerve
Teres major	Thick, rounded muscle; located inferior to teres minor; helps to form posterior wall of axilla (along with latissimus dorsi and subscapularis)	O—posterior surface of scapula at inferior angle I—intertubercular groove of the humerus; insertion tendon fused with that of latissimus dorsi	Posteromedially extends, medially rotates, and adducts humerus; synergist of latissimus dorsi	Lower subscapular nerve (C_6 and C_7)
Coracobrachialis (kor″ah-ko-bra″ke-al′is) (*coraco* = coracoid; *brachi* = arm)	Small, cylindrical muscle	O—coracoid process of scapula I—medial surface of humerus shaft	Flexion and adduction of the humerus; synergist of pectoralis major	Musculocutaneous nerve (C_5–C_7)

TABLE 10.9   **Muscles Crossing the Shoulder Joint: Movements of the Arm (Figure 10.14) (continued)**

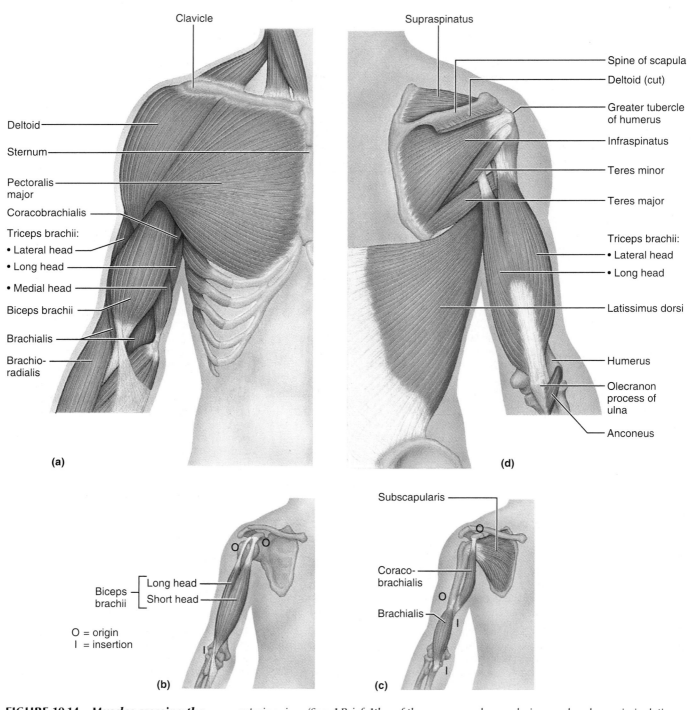

FIGURE 10.14 *Muscles crossing the shoulder and elbow joint, respectively causing movements of the arm and forearm.* (a) Superficial muscles of the anterior thorax, shoulder, and arm, anterior view. (See *A Brief Atlas of the Human Body*, Figures 63 and 64.) (b) The biceps brachii muscle of the anterior arm, shown in isolation. (c) The brachialis muscle and the coracobrachialis and subscapularis muscles shown in isolation. (d) The extent of the triceps brachii muscle of the posterior arm, shown in relation to the deep scapular muscles. The deltoid muscle of the shoulder has been removed.

TABLE 10.10	**Muscles Crossing the Elbow Joint: Flexion and Extension of the Forearm (Figure 10.14)**

Muscles fleshing out the arm cross the elbow joint to insert on the forearm bones. Since the elbow is a hinge joint, movements promoted by these muscles are limited almost entirely to flexion and extension of the forearm. Walls of fascia divide the arm into two muscle compartments—the *posterior extensors* and *anterior flexors.* The prime mover of forearm extension is the bulky **triceps brachii** muscle, which forms nearly the entire musculature of the posterior compartment. It is assisted (minimally) by the tiny **anconeus** muscle, which just about spans the elbow joint posteriorly.

All anterior arm muscles cause elbow flexion. In order of decreasing strength, these are the **brachialis, biceps brachii,** and **brachioradialis.** The brachialis and biceps are inserted (respectively) into the ulna and radius and contract simultaneously during flexion; they are the chief forearm flexors. The biceps brachii, a muscle that bulges when the forearm is flexed, is familiar to almost everyone; the brachialis, which lies deep to the biceps, is less known, but is as important in flexing the elbow. The biceps muscle also supinates the forearm and is ineffective in flexing the elbow when the forearm *must* stay pronated. (This is why doing chin-ups with palms facing anteriorly is harder than with palms facing posteriorly.) Since the brachioradialis arises from the distal humerus and inserts on the distal forearm, it resides mainly in the forearm (rather than in the arm as do the rest of this group). Because its force is exerted far from the fulcrum, the brachioradialis is a weak forearm flexor. It becomes active only when the elbow has been partially flexed and is semi-pronated.

The actions of the muscles described here are summarized in Table 10.12 (Part II).

Muscle	Description	Origin (O) and Insertion (I)	Action	Nerve Supply
POSTERIOR MUSCLES				
Triceps brachii (tri′seps bra′ke-i) (*triceps* = three heads; *brachi* = arm)	Large fleshy muscle; the only muscle of posterior compartment of arm; three-headed origin; long and lateral heads lie superficial to medial head	O—long head: infraglenoid tubercle of scapula; lateral head: posterior shaft of humerus; medial head: posterior humeral shaft distal to radial groove I—by common tendon into olecranon process of ulna	Powerful forearm extensor (prime mover, particularly medial head); antagonist of forearm flexors; long and lateral heads mainly active in extension against resistance; long head tendon may help stabilize shoulder joint and assist in arm adduction	Radial nerve (C_6–C_8)
Anconeus (an-ko′ne-us) (*ancon* = elbow) (see Figure 10.16)	Short triangular muscle; closely associated (blended) with distal end of triceps on posterior humerus	O—lateral epicondyle of humerus I—lateral aspect of olecranon process of ulna	Abducts ulna during forearm pronation; synergist of triceps brachii in elbow extension	Radial nerve
ANTERIOR MUSCLES				
Biceps brachii (bi′seps) (*biceps* = two heads)	Two-headed fusiform muscle; bellies unite as insertion point is approached; tendon of long head helps stabilize shoulder joint	O—short head: coracoid process; long head: tubercle above glenoid cavity and lip of glenoid cavity; tendon of long head runs within capsule and descends into intertubercular groove of humerus I—by common tendon into radial tuberosity	Flexes elbow joint and supinates forearm; these actions usually occur at same time (e.g., when you open a bottle of wine, it turns the corkscrew and pulls the cork); weak flexor of arm at shoulder	Musculocutaneous nerve (C_5 and C_6)
Brachialis (bra′ke-al-is)	Strong muscle that is immediately deep to biceps brachii on distal humerus	O—front of distal humerus; embraces insertion of deltoid muscle I—coronoid process of ulna	A major forearm flexor (lifts ulna as biceps lifts the radius)	Musculocutaneous nerve
Brachioradialis (bra″ke-o-ra″de-a′lis) (*radi* = radius, ray) (also see Figure 10.15)	Superficial muscle of lateral forearm; forms lateral boundary of antecubital fossa; extends from distal humerus to distal forearm	O—lateral supracondylar ridge at distal end of humerus I—base of styloid process of radius	Synergist in forearm flexion; acts to best advantage when forearm is partially flexed and semi-pronated; during rapid flexion *and* extension, stabilizes the elbow	Radial nerve (an important exception: the radial nerve typically serves extensor muscles)

The forearm muscles are functionally divided into two approximately equal groups: those causing wrist movements and those moving the fingers and thumb. In most cases, their fleshy portions contribute to the roundness of the proximal forearm and then they taper to long insertion tendons. Their insertions are securely anchored by strong ligaments called **flexor** and **extensor retinacula** ("retainers"). These "wrist bands" keep the tendons from jumping outward when tensed. Crowded together in the wrist and palm, the muscle tendons are surrounded by slippery tendon sheaths that minimize friction as they slide against one another.

Although many of the forearm muscles actually arise from the humerus (and thus cross both the elbow and wrist joints), their actions on the elbow are slight. Flexion and extension are the movements typically effected at both the wrist and finger joints. In addition, the wrist can be abducted and adducted.

The forearm muscles are subdivided by fascia into two main compartments (the *anterior flexors* and *posterior extensors*), and each compartment has superficial and deep muscle layers. Most flexors in the anterior compartment arise from a common tendon on the humerus and are innervated largely by the median nerve. Though most anterior forearm muscles are wrist or finger *flexors*, two muscles of this group are not flexors but pronators, the **pronator teres** and **pronator quadratus.** Pronation is one of the most important forearm movements.

Muscles of the posterior compartment extend the wrist and fingers. An exception is the **supinator** muscle, which assists the biceps brachii muscle of the arm in supinating the forearm. (Also residing in the posterior compartment is the brachioradialis muscle, the weak elbow flexor considered in Table 10.10.) Like those of the anterior compartment, most muscles of the posterior compartment arise from a common origin tendon on the humerus; however, the location on the humerus differs. All posterior forearm muscles are supplied by the radial nerve.

Although the hand performs many different movements, it contains relatively few of the muscles that control those movements. As described above, most muscles that move the hand are located in the forearm and "operate" the fingers via their long tendons, like operating a puppet by strings. This design makes the hand less bulky and enables it to perform finer movements. The hand movements promoted by the forearm muscles are assisted and made more precise by the small *intrinsic* muscles of the hand. These are considered separately in Table 10.13.

The locations and actions of the forearm muscles are the focus of this table. Since the forearm muscles are numerous and their interactions varied, a summary of their actions is provided in Table 10.12 (Parts II and III).

Muscle	Description	Origin (O) and Insertion (I)	Action	Nerve Supply
PART I: ANTERIOR MUSCLES (Figure 10.15)	These eight muscles of the anterior fascial compartment are listed from the lateral to the medial aspect. Most arise from a common flexor tendon attached to the medial epicondyle of the humerus and have additional origins as well. Most of the tendons of insertion of these flexors are held in place at the wrist by a thickening of deep fascia called the *flexor retinaculum*.			
SUPERFICIAL MUSCLES				
Pronator teres (pro'na'tor te'rēz) (*pronation* = turning palm posteriorly, or down; *teres* = round)	Two-headed muscle; seen in superficial view between proximal margins of brachioradialis and flexor carpi radialis; forms medial boundary of antecubital fossa	O—medial epicondyle of humerus; coronoid process of ulna I—by common tendon into lateral radius, midshaft	Pronates forearm; weak flexor of elbow	Median nerve
Flexor carpi radialis (flek'sor kar'pe ra''de-a'lis) (*flex* = decrease angle between two bones; *carpi* = wrist; *radi* = radius)	Runs diagonally across forearm; midway, its fleshy belly is replaced by a flat tendon that becomes cordlike at wrist	O—medial epicondyle of humerus I—base of second and third metacarpals; insertion tendon easily seen and provides guide to position of radial artery (used for pulse taking) at wrist	Powerful flexor of wrist; abducts hand; weak synergist of elbow flexion	Median nerve
Palmaris longus (pahl-ma'ris lon'gus) (*palma* = palm; *longus* = long)	Small fleshy muscle with a long insertion tendon; often absent; may be used as guide to find median nerve that lies lateral to it at wrist	O—medial epicondyle of humerus I—palmar aponeurosis; skin and fascia of palm	Weak wrist flexor; tenses skin and fascia of palm during hand movements; weak synergist for elbow flexion	Median nerve
Flexor carpi ulnaris (ul-na'ris) (*ulnar* = ulna)	Most medial muscle of this group; two-headed; ulnar nerve lies lateral to its tendon	O—medial epicondyle of humerus; olecranon process and posterior surface of ulna I—pisiform and hamate bones and base of fifth metacarpal	Powerful flexor of wrist; also adducts hand in concert with extensor carpi ulnaris (posterior muscle); stabilizes wrist during finger extension	Ulnar nerve (C_7 and C_8)

TABLE 10.11	Muscles of the Forearm: Movements of the Wrist, Hand, and Fingers (Figures 10.15 and 10.16) *(continued)*			
Muscle	*Description*	*Origin (O) and Insertion (I)*	*Action*	*Nerve Supply*
Flexor digitorum superficialis (dĭ″ji-tor′um soo″per-fish″e-a′lis) (*digit* =finger, toe; *superficial* = close to surface)	Two-headed muscle; more deeply placed (therefore, actually forms an intermediate layer); overlain by muscles above but visible at distal end of forearm	O—medial epicondyle of humerus, coronoid process of ulna; shaft of radius I—by four tendons into middle phalanges of fingers 2–5	Flexes wrist and middle phalanges of fingers 2–5; the important finger flexor when speed and flexion against resistance are required	Median nerve (C_7, C_8, and T_1)
DEEP MUSCLES				
Flexor pollicis longus (pah′lĭ-kis) (*pollix* = thumb)	Partly covered by flexor digitorum superficialis; parallels flexor digitorum profundus laterally	O—anterior surface of radius and interosseous membrane I—distal phalanx of thumb	Flexes distal phalanx of thumb	Branch of median nerve (C_8, T_1)
Flexor digitorum profundus (pro-fun′dus) (*profund* = deep)	Extensive origin; overlain entirely by flexor digitorum superficialis	O—coronoid process, anteromedial surface of ulna, and interosseous membrane I—by four tendons into distal phalanges of fingers 2–5	Slow-acting flexor of any or all fingers; assists in flexing wrist; the only muscle that can flex distal interphalangeal joints	Medial half by ulnar nerve; lateral half by median nerve

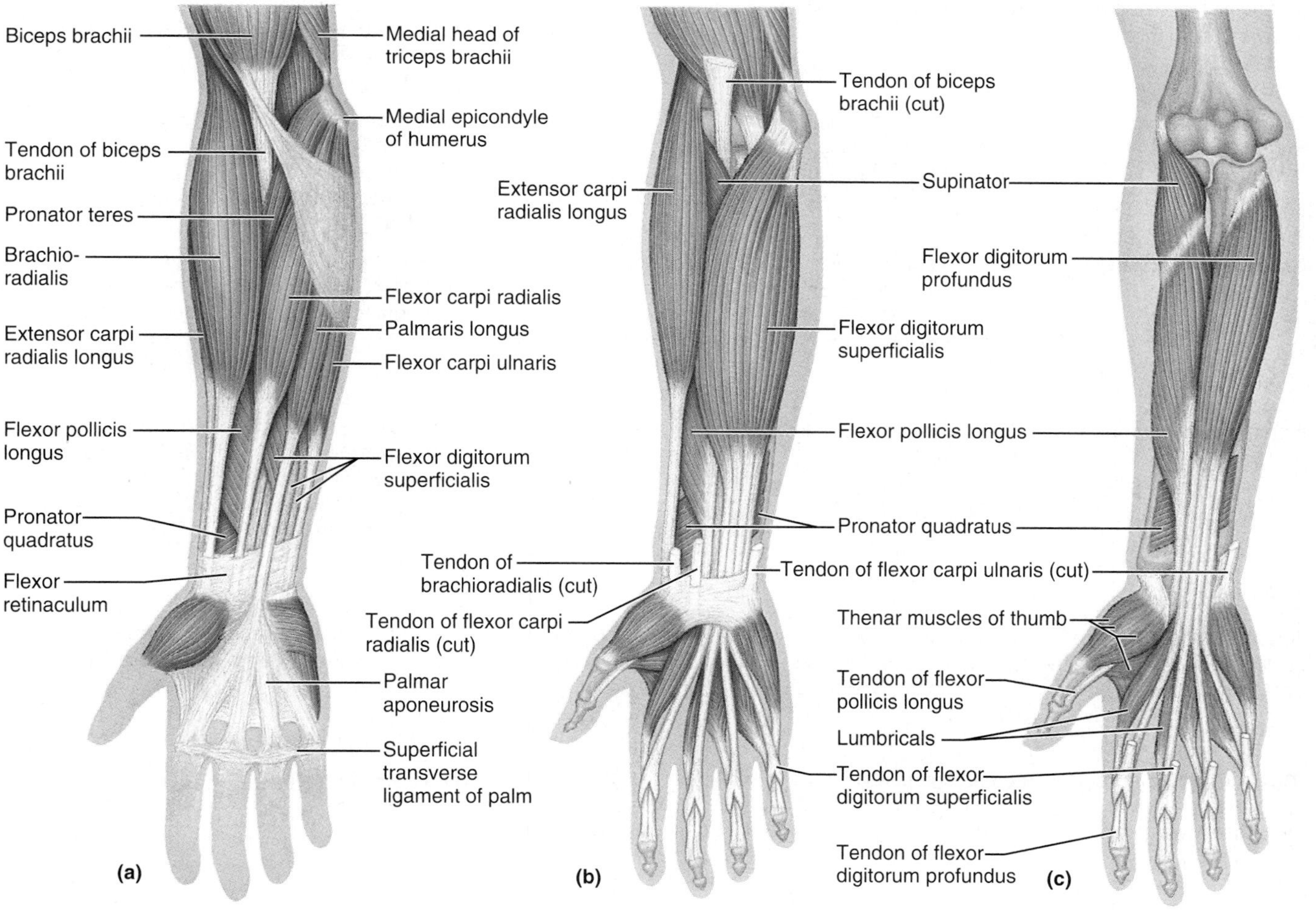

(a)

Biceps brachii
Tendon of biceps brachii
Pronator teres
Brachioradialis
Extensor carpi radialis longus
Flexor pollicis longus
Pronator quadratus
Flexor retinaculum

Medial head of triceps brachii
Medial epicondyle of humerus
Flexor carpi radialis
Palmaris longus
Flexor carpi ulnaris
Flexor digitorum superficialis
Palmar aponeurosis
Superficial transverse ligament of palm

(b)

Tendon of biceps brachii (cut)
Extensor carpi radialis longus
Flexor digitorum superficialis
Tendon of brachioradialis (cut)
Tendon of flexor carpi radialis (cut)

(c)

Supinator
Flexor digitorum profundus
Flexor pollicis longus
Pronator quadratus
Tendon of flexor carpi ulnaris (cut)
Thenar muscles of thumb
Tendon of flexor pollicis longus
Lumbricals
Tendon of flexor digitorum superficialis
Tendon of flexor digitorum profundus

FIGURE 10.15 Muscles of the anterior compartment of the forearm acting on the right wrist and fingers. (a) Superficial view. (b) The brachioradialis, flexors carpi radialis and ulnaris, and palmaris longus muscles have been removed to reveal the position of the flexor digitorum superficialis. (c) Deep muscles of the anterior compartment. The lumbricals and thenar muscles (intrinsic hand muscles) are also illustrated. (See *A Brief Atlas of the Human Body,* Figure 65a.)

TABLE 10.11	Muscles of the Forearm: Movements of the Wrist, Hand, and Fingers (Figures 10.15 and 10.16) *(continued)*

Muscle	Description	Origin (O) and Insertion (I)	Action	Nerve Supply
Pronator quadratus (kwod-ra′tus) (*quad* = square, four-sided)	Deepest muscle of distal forearm; passes downward and laterally; only muscle that arises solely from ulna and inserts solely into radius	O—distal portion of anterior ulnar shaft I—distal surface of anterior radius	Prime mover of forearm pronation; acts with pronator teres; also helps hold ulna and radius together	Median nerve (C_8 and T_1)
PART II: POSTERIOR MUSCLES (Figure 10.16)	These muscles of the posterior fascial compartment are listed from the lateral to the medial aspect. They are all innervated by the radial nerve or its branches. More than half of the posterior compartment muscles arise from a common extensor origin tendon attached to the posterior surface of the lateral epicondyle of the humerus and adjacent fascia. The extensor tendons are held in place at the posterior aspect of the wrist by the *extensor retinaculum,* which prevents "bowstringing" of these tendons when the wrist is hyperextended. The extensor muscles of the fingers end in a broad hood over the dorsal side of the digits, the extensor expansion.			

SUPERFICIAL MUSCLES

Brachioradialis (see Table 10.10)

Muscle	Description	Origin (O) and Insertion (I)	Action	Nerve Supply
Extensor carpi radialis longus (ek-sten′sor) (*extend* = increase angle between two bones)	Parallels brachioradialis on lateral forearm, and may blend with it	O—lateral supracondylar ridge of humerus I—base of second metacarpal	Extends wrist in conjunction with the extensor carpi ulnaris and abducts wrist in conjunction with the flexor carpi radialis	Radial nerve (C_6 and C_7)
Extensor carpi radialis brevis (brĕ′vis) (*brevis* = short)	Somewhat shorter than extensor carpi radialis longus and lies deep to it	O—lateral epicondyle of humerus I—base of third metacarpal	Extends and abducts wrist; acts synergistically with extensor carpi radialis longus to steady wrist during finger flexion	Deep branch of radial nerve
Extensor digitorum	Lies medial to extensor carpi radialis brevis; a detached portion of this muscle, called *extensor digiti minimi,* extends little finger	O—lateral epicondyle of humerus I—by four tendons into extensor expansions and distal phalanges of fingers 2–5	Prime mover of finger extension; extends wrist; can abduct (flare) fingers	Posterior interosseous nerve, a branch of radial nerve (C_5 and C_6)
Extensor carpi ulnaris	Most medial of superficial posterior muscles; long, slender muscle	O—lateral epicondyle of humerus and posterior border of ulna I—base of fifth metacarpal	Extends wrist in conjunction with the extensor carpi radialis and adducts wrist in conjunction with flexor carpi ulnaris	Posterior interosseous nerve

DEEP MUSCLES

Muscle	Description	Origin (O) and Insertion (I)	Action	Nerve Supply
Supinator (soo″pĭ-na′tor) (*supination* = turning palm anteriorly or upward)	Deep muscle at posterior aspect of elbow; largely concealed by superficial muscles	O—lateral epicondyle of humerus; proximal ulna I—proximal end of radius	Assists biceps brachii to forcibly supinate forearm; works alone in slow supination; antagonist of pronator muscles	Posterior interosseous nerve
Abductor pollicis longus (ab-duk′tor) (*abduct* = movement away from median plane)	Lateral and parallel to extensor pollicis longus; just distal to supinator	O—posterior surface of radius and ulna; interosseous membrane I—base of first meta-carpal and trapezium	Abducts and extends thumb	Posterior interosseous nerve
Extensor pollicis brevis and **longus**	Deep muscle pair with a common origin and action; overlain by extensor carpi ulnaris	O—dorsal shaft of radius and ulna; interosseous membrane I—base of proximal (brevis) and distal (longus) phalanx of thumb	Extends thumb	Posterior interosseous nerve

TABLE 10.11	Muscles of the Forearm: Movements of the Wrist, Hand, and Fingers (Figures 10.15 and 10.16) *(continued)*

Muscle	Description	Origin (O) and Insertion (I)	Action	Nerve Supply
(in'dĭ-kis) (*indicis* = index finger)	Tiny muscle arising close to wrist	O—posterior surface of distal ulna; interosseous membrane I—extensor expansion of index finger; joins tendon of extensor digitorum	Extends index finger and assists in wrist extension	Posterior interosseous nerve

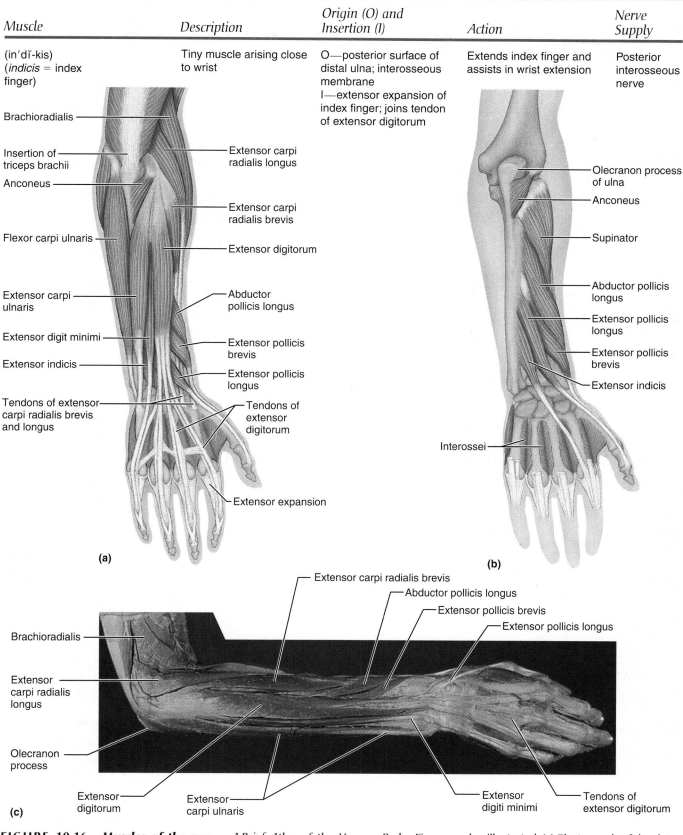

(a)

(b)

(c)

FIGURE 10.16 *Muscles of the posterior compartment of the right forearm acting on the wrist and fingers.* (a) Superficial muscles, posterior view. (See *A Brief Atlas of the Human Body,* Figure 65b.) (b) Deep posterior muscles, superficial muscles removed. The interossei, the deepest layer of intrinsic hand muscles, are also illustrated. (c) Photograph of the deep posterior muscles.

TABLE 10.12	Summary of Actions of Muscles Acting on the Arm, Forearm, and Hand (Figure 10.17)

Part I: Muscles Acting on the Arm (Humerus) (PM = prime mover)

Actions at the shoulder

	Flexion	Extension	Abduction	Adduction	Medial rotation	Lateral rotation
Pectoralis major	X (PM)			X (PM)	X	
Latissimus dorsi		X (PM)		X (PM)	X	
Deltoid	X (PM) (anterior fibers)	X (PM) (posterior fibers)	X (PM)		X (anterior fibers)	X (posterior fibers)
Subscapularis					X (PM)	
Supraspinatus			X			
Infraspinatus						X (PM)
Teres minor				X (weak)		X (PM)
Teres major		X		X	X	
Coracobrachialis	X			X		
Biceps brachii	X					
Triceps brachii				X		

Part II: Muscles Acting on the Forearm

Actions

	Elbow flexion	Elbow extension	Pronation	Supination
Biceps brachii	X (PM)			X
Triceps brachii		X (PM)		
Anconeus		X		
Brachialis	X (PM)			
Brachioradialis	X			
Pronator teres	X (weak)		X	
Pronator quadratus			X (PM)	
Supinator				X

Part III: Muscles Acting on the Wrist and Fingers

	Actions on the wrist				Actions on the fingers	
	Flexion	Extension	Abduction	Adduction	Flexion	Extension
Anterior compartment: Flexor carpi radialis	X (PM)		X			
Palmaris longus	X (weak)					
Flexor carpi ulnaris	X (PM)			X		
Flexor digitorum superficialis	X (PM)				X	
Flexor pollicis longus					X (thumb)	
Flexor digitorum profundus	X				X	
Posterior compartment: Extensor carpi radialis longus and brevis		X	X			
Extensor digitorum		X (PM)				X (and abducts)
Extensor carpi ulnaris		X		X		
Abductor pollicis longus			X			(abducts thumb)
Extensor pollicis longus and brevis						X (thumb)
Extensor indicis						X (index finger)

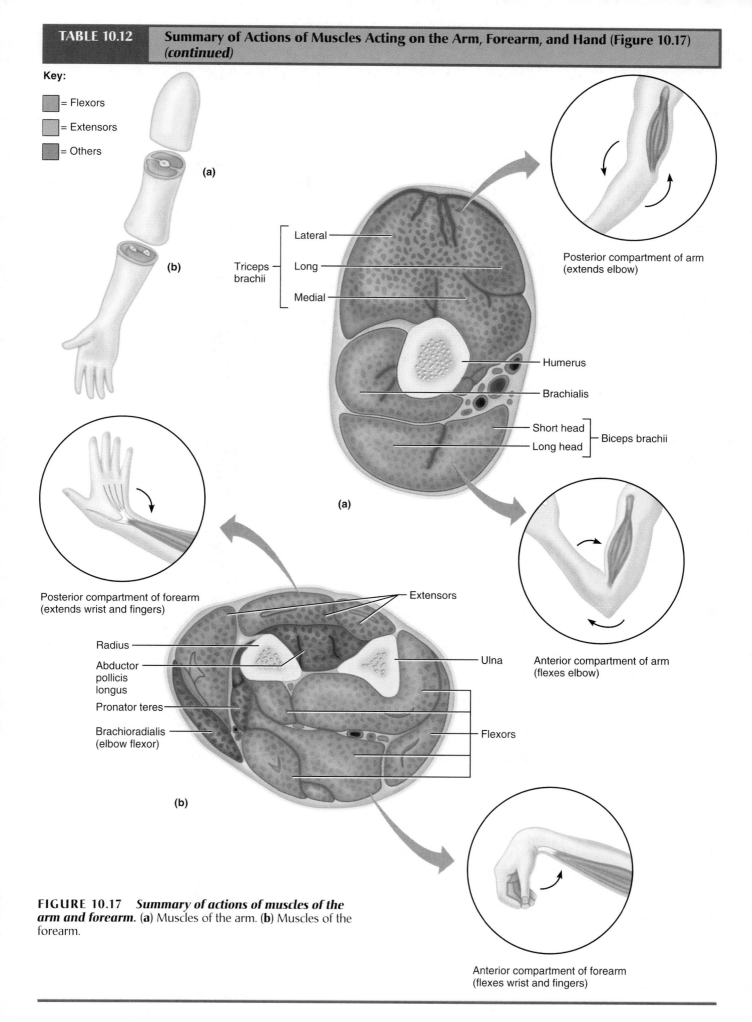

Key:
- ▢ = Flexors
- ▢ = Extensors
- ▢ = Others

(a)

(b)

Posterior compartment of arm (extends elbow)

Triceps brachii
- Lateral
- Long
- Medial

Humerus

Brachialis

Short head
Long head
} Biceps brachii

(a)

Anterior compartment of arm (flexes elbow)

Posterior compartment of forearm (extends wrist and fingers)

Extensors

Radius

Abductor pollicis longus

Pronator teres

Brachioradialis (elbow flexor)

Ulna

Flexors

(b)

Anterior compartment of forearm (flexes wrist and fingers)

FIGURE 10.17 *Summary of actions of muscles of the arm and forearm.* (**a**) Muscles of the arm. (**b**) Muscles of the forearm.

TABLE 10.13	**Intrinsic Muscles of the Hand: Fine Movements of the Fingers (Figure 10.18)**

This table considers the small muscles that lie entirely in the hand. All are in the palm, none on the hand's dorsal side. All move the metacarpals and fingers. Small, weak muscles, they mostly control precise movements (such as threading a needle), leaving the powerful movements of the fingers ("power grip") to the forearm muscles.

The intrinsic muscles include the main abductors and adductors of the fingers, as well as muscles that produce the movement of opposition—moving the thumb toward the little finger—that enables one to grip objects in the palm (the handle of a hammer, for example). Many palm muscles are specialized to move the thumb, and many move the little finger. Thumb movements are defined differently from movements of other fingers because the thumb lies at a right angle to the rest of the hand. The thumb flexes by bending medially along the palm, not anteriorly, as do the other fingers. (Start with your hand in the anatomical position or this will not be clear!) The thumb extends by pointing laterally (as in hitchhiking), not posteriorly, as do the other fingers. To abduct the fingers is to splay them laterally, but to abduct the thumb is to point it anteriorly. Adduction of the thumb brings it back posteriorly.

The intrinsic muscles of the palm are divided into three groups, those in (1) the *thenar eminence* (ball of the thumb); (2) the *hypothenar eminence* (ball of the little finger); and (3) the *midpalm*. Thenar and hypothenar muscles are almost mirror images of each other, each containing a small flexor, an abductor, and an opponens muscle. The midpalmar muscles, called **lumbricals** and **interossei**, extend our fingers at the interphalangeal joints. The interossei are also the main finger abductors and adductors.

Muscle	Description	Origin (O) and Insertion (I)	Action	Nerve Supply
THENAR MUSCLES IN BALL OF THUMB (the′nar) (*thenar* = palm)				
Abductor pollicis brevis (*pollex* = thumb)	Lateral muscle of thenar group; superficial	O—flexor retinaculum and nearby carpals I—lateral base of thumb's proximal phalanx	Abducts thumb (at carpometacarpal joint)	Median nerve ($C_8 T_1$)
Flexor pollicis brevis	Medial and deep muscle of thenar group	O—flexor retinaculum and trapezium I—lateral side of base of proximal phalanx of thumb	Flexes thumb (at carpometacarpal and metacarpophalangeal joints)	Median (or occasionally ulnar) nerve (C_8, T_1)
Opponens pollicis (o-pōn′enz) (*opponens* = opposition)	Deep to abductor pollicis brevis, on metacarpal 1	O—flexor retinaculum and trapezium I—whole anterior side of metacarpal 1	Opposition: moves thumb to touch tip of little finger	Median (or occasionally ulnar) nerve
Adductor pollicis	Fan-shaped with horizontal fibers; distal to other thenar muscles; oblique and transverse heads	O—capitate bone and bases of metacarpals 2–4; front of metacarpal 3 I—medial side of base of proximal phalanx of thumb	Adducts and helps to oppose thumb	Ulnar nerve (C_8, T_1)
HYPOTHENAL MUSCLES IN BALL OF LITTLE FINGER				
Abductor digiti minimi (dĭ′jĭ-ti min′ĭ-mi) (*digiti minimi* = little finger)	Medial muscle of hypothenar group; superficial	O—pisiform bone I—medial side of proximal phalanx of little finger	Abducts little finger at metacarpophalangeal joint	Ulnar nerve
Flexor digiti minimi brevis	Lateral deep muscle of hypothenar group	O—hamate bone and flexor retinaculum I—same as abductor digiti minimi	Flexes little finger at metacarpophalangeal joint	Ulnar nerve
Opponens digiti minimi	Deep to abductor digiti minimi	O—same as flexor digiti minimi brevis I—most of length of medial side of metacarpal 5	Helps in opposition: brings metacarpal 5 toward thumb to cup the hand	Ulnar nerve

| TABLE 10.13 | Intrinsic Muscles of the Hand: Fine Movements of the Fingers (Figure 10.18) *(continued)* |

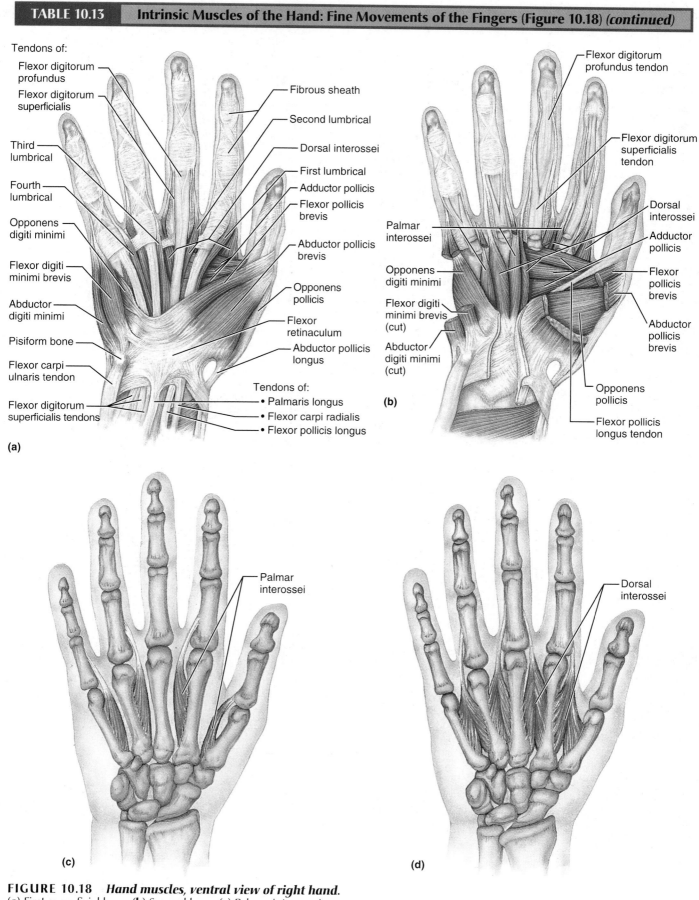

Tendons of:

Flexor digitorum profundus

Flexor digitorum superficialis

Third lumbrical

Fourth lumbrical

Opponens digiti minimi

Flexor digiti minimi brevis

Abductor digiti minimi

Pisiform bone

Flexor carpi ulnaris tendon

Flexor digitorum superficialis tendons

Fibrous sheath

Second lumbrical

Dorsal interossei

First lumbrical

Adductor pollicis

Flexor pollicis brevis

Abductor pollicis brevis

Opponens pollicis

Flexor retinaculum

Abductor pollicis longus

Tendons of:
• Palmaris longus
• Flexor carpi radialis
• Flexor pollicis longus

(a)

Flexor digitorum profundus tendon

Flexor digitorum superficialis tendon

Palmar interossei

Dorsal interossei

Opponens digiti minimi

Adductor pollicis

Flexor digiti minimi brevis (cut)

Flexor pollicis brevis

Abductor digiti minimi (cut)

Abductor pollicis brevis

Opponens pollicis

Flexor pollicis longus tendon

(b)

Palmar interossei

(c)

Dorsal interossei

(d)

FIGURE 10.18 *Hand muscles, ventral view of right hand.*
(**a**) First superficial layer. (**b**) Second layer. (**c**) Palmar interossei (isolated). (**d**) Dorsal interossei (isolated).

TABLE 10.13	Intrinsic Muscles of the Hand: Fine Movements of the Fingers (Figure 10.18) *(continued)*			
Muscle	*Description*	*Origin (O) and Insertion (I)*	*Action*	*Nerve Supply*
MIDPALMAR MUSCLES				
Lumbricals (lum′brĭ-klz) (*lumbric* = earthworm)	Four worm-shaped muscles in palm, one to each finger (except thumb); odd because they originate from the tendons of another muscle	O—lateral side of each tendon of flexor digitorum profundus in palm I—lateral edge of extensor expansion on first phalanx of fingers 2–5	Flex fingers at metacarpophalangeal joints but extend fingers at interphalangeal joints	Median nerve (lateral two) and ulnar nerve (medial two)
Palmar interossei (in″ter-os′e-i) (*interossei* = between bones)	Four long, cone-shaped muscles in the spaces between the metacarpals; lie ventral to the dorsal interossei	O—the side of each metacarpal that faces the midaxis of the hand (metacarpal 3, where it's absent) I—extensor expansion on first phalanx of each finger (except finger 3), on side facing midaxis of hand	Adductors of fingers: pull fingers in toward third digit; act with lumbricals to extend fingers at interphalangeal joints and flex them at metacarpophalangeal joints	Ulnar nerve
Dorsal interossei	Four bipennate muscles filling spaces between the metacarpals; deepest palm muscles, also visible on dorsal side of hand (Figure 10.15b)	O—sides of metacarpals I—extensor expansion over first phalanx of fingers 2–4 on side opposite midaxis of hand (finger 3), but on *both* sides of finger 3	Abduct (diverge) fingers; extend fingers at interphalangeal joints and flex them at metacarpophalangeal joints	Ulnar nerve

The muscles fleshing out the thigh are difficult to segregate into groups on the basis of action. Some thigh muscles act only at the hip joint, others only at the knee, while still others act at both joints. Attempts to classify these muscles solely on the basis of location is equally frustrating because different muscles in a particular location often have very different actions. However, the *most anterior* muscles of the hip and thigh tend to flex the femur at the hip and to extend the leg at the knee—producing the foreswing phase of walking. The *posterior* muscles of the hip and thigh, by contrast, mostly extend the thigh and flex the leg—the backswing phase of walking. The third group of muscles in this region, the *medial,* or *adductor,* muscles, all adduct the thigh; they have no effect on the leg. In the thigh, the anterior, posterior, and adductor muscles are separated by walls of fascia into *anterior, posterior,* and *medial compartments* (see Figure 10.24a). The deep fascia of the thigh, the *fascia lata,* surrounds and encloses all three groups of muscles like a support stocking.

Movements of the thigh (occurring at the hip joint) are accomplished largely by muscles anchored to the pelvic girdle. Like the shoulder joint, the hip joint is a ball-and-socket joint permitting flexion, extension, abduction, adduction, circumduction, and rotation. Muscles effecting these movements of the thigh are among the most powerful muscles of the body. For the most part, the thigh *flexors* pass in front of the hip joint. The most important of these are the **iliopsoas, tensor fasciae latae,** and **rectus femoris.** They are assisted in this action by the **adductor muscles** of the medial thigh and the straplike **sartorius.** The prime mover of thigh flexion is the iliopsoas.

Thigh *extension* is effected primarily by the massive **hamstring muscles** of the posterior thigh. However, during forceful extension, the **gluteus maximus** of the buttock is activated. Buttock muscles that lie lateral to the hip joint (**gluteus medius** and **minimus**) *abduct* the thigh and rotate it medially. Medial rotation is opposed by six small deep muscles of the gluteal region called the **lateral rotators.** Thigh adduction is the role of the adductor muscles of the medial thigh. Abduction and adduction of the thighs are extremely important during walking to keep the body's weight balanced over the limb that is on the ground.

At the knee joint, flexion and extension are the main movements. The sole knee *extensor* is the **quadriceps femoris** muscle of the anterior thigh, the most powerful muscle in the body. The quadriceps is antagonized by the hamstrings of the posterior compartment, which are prime movers of knee flexion and also act to rotate the leg when the knee is semiflexed.

The actions of the muscles presented here are further summarized in Table 10.16 (Part I).

Muscle	Description	Origin (O) and Insertion (I)	Action	Nerve Supply
PART I: ANTERIOR AND MEDIAL MUSCLES (Figure 10.19) ORIGIN ON THE PELVIS				
Iliopsoas (il″e-o-so′us)	Iliopsoas is a composite of two closely related muscles (iliacus and psoas major) whose fibers pass under the inguinal ligament (see Table 10.6 [external oblique muscle] and Figure 10.11) to insert via a common tendon on the femur.			
▪ **Iliacus** (il-e-ak′us) (*iliac* = ilium)	Large, fan-shaped, more lateral muscle	O—iliac fossa and crest, lateral sacrum I—femur on and immediately below lesser trochanter of femur via iliopsoas tendon	Iliopsoas is the prime mover for flexing thigh or for flexing trunk on thigh during a bow	Femoral nerve (L_2 and L_3)
▪ **Psoas** (so′us) **major** (*psoa* = loin muscle; *major* = larger)	Longer, thicker, more medial muscle of the pair (Butchers refer to this muscle as the tenderloin)	O—by fleshy slips from transverse processes, bodies, and discs of lumbar vertebrae and T_{12} I—lesser trochanter of femur via iliopsoas tendon	As above; also effects lateral flexion of vertebral column; important postural muscle	Ventral rami L_1–L_3
Sartorius (sar-tor′e-us) (*sartor* = tailor)	Straplike superficial muscle running obliquely across anterior surface of thigh to knee; longest muscle in body; crosses both hip and knee joints	O—anterior superior iliac spine I—winds around medial aspect of knee and inserts into medial aspect of proximal tibia	Flexes, abducts, and laterally rotates thigh; flexes knee (weak); known as "tailor's muscle" because it helps effect cross-legged position in which tailors are often depicted	Femoral nerve

TABLE 10.14   **Muscles Crossing the Hip and Knee Joints: Movements of the Thigh and Leg (Figures 10.19 and 10.20) (continued)**

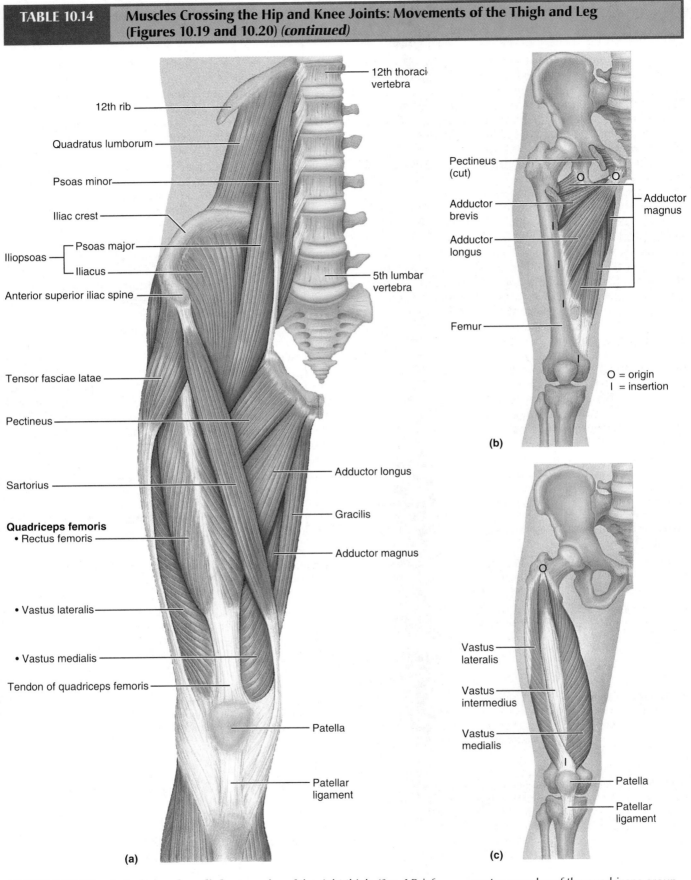

(a)

(b)

O = origin
I = insertion

(c)

FIGURE 10.19 *Anterior and medial muscles promoting movements of the thigh and leg.* (a) Anterior view of the deep muscles of the pelvis and superficial muscles of the right thigh. (See *A Brief Atlas of the Human Body,* Figure 68.) (b) Adductor muscles of the medial compartment of the thigh isolated. (c) The vastus muscles of the quadriceps group. The rectus femoris muscle of the quadriceps group and surrounding muscles have been removed.

TABLE 10.14	Muscles Crossing the Hip and Knee Joints: Movements of the Thigh and Leg (Figures 10.19 and 10.20) *(continued)*			
Muscle	*Description*	*Origin (O) and Insertion (I)*	*Action*	*Nerve Supply*

MUSCLES OF THE MEDIAL COMPARTMENT OF THE THIGH

Adductors (ah-duk′torz)	Large muscle mass consisting of three muscles (magnus, longus, and brevis) forming medial aspect of thigh; arise from inferior part of pelvis and insert at various levels on femur; all used in movements that press thighs together, as when astride a horse; important in pelvic tilting movements that occur during walking and in fixing the hip when the knee is flexed and the foot is off the ground; entire group innervated by obturator nerve. Strain or stretching of this muscle group is called a "pulled groin"			
■ **Adductor magnus** (mag′nus) (*adduct* = move toward midline; *magnus* = large)	A triangular muscle with a broad insertion; is a composite muscle that is part adductor and part hamstring in action	O—ischial and pubic rami and ischial tuberosity I—linea aspera and adductor tubercle of femur	Anterior part adducts and medially rotates and flexes thigh; posterior part is a synergist of hamstrings in thigh extension	Obturator nerve and sciatic nerve (L_2–L_4)
■ **Adductor longus** (*longus* = long)	Overlies middle aspect of adductor magnus; most anterior of adductor muscles	O—Pubis near pubic symphysis I—linea aspera	Adducts, flexes, and medially rotates thigh	Anterior division of obturator nerve (L_{2-4})
■ **Adductor brevis** (*brevis* = short)	In contact with obturator externus muscle; largely concealed by adductor longus and pectineus	O—body and inferior ramus of pubis I—linea aspera above adductor longus	Adducts and medially rotates thigh	Obturator nerve
Pectineus (pek-tin′e-us) (*pecten* = comb)	Short, flat muscle; overlies adductor brevis on proximal thigh; abuts adductor longus medially	O—pectineal line of pubis (and superior ramus) I—inferior from lesser trochanter to the linea aspera of posterior aspect of femur	Adducts, flexes, and medially rotates thigh	Femoral and sometimes obturator nerve
Gracilis (grah-sĭ′lis) (*gracilis* = slender)	Long, thin, superficial muscle of medial thigh	O—inferior ramus and body of pubis and adjacent ischial ramus I—medial surface of tibia just inferior to its medial condyle	Adducts thigh, flexes, and medially rotates leg, especially during walking	Obturator nerve

MUSCLES OF THE ANTERIOR COMPARTMENT OF THE THIGH

Quadriceps femoris (kwod′rĭ-seps fem′o-ris)	Quadriceps femoris arises from four separate heads (*quadriceps* = four heads) that form the flesh of front and sides of thigh; these heads (rectus femoris, and lateral, medial, and intermediate vastus muscles) have a common insertion tendon, the quadriceps tendon, which inserts into the patella and then via the patellar ligament into tibial tuberosity. The quadriceps is a powerful knee extensor used in climbing, jumping, running, and rising from seated position; group is innervated by femoral nerve; the tone of quadriceps plays important role in strengthening knee joint			
■ **Rectus femoris** (rek′tus) (*rectus* = straight; *femoris* = femur)	Superficial muscle of anterior thigh; runs straight down thigh; longest head and only muscle of group to cross hip joint	O—anterior inferior iliac spine and superior margin of acetabulum I—patella and tibial tuberosity via patellar ligament	Extends knee and flexes thigh as hip	Femoral nerve (L_2–L_4)
■ **Vastus lateralis** (vas′tus lat″er-a′lis) (*vastus* = large; *lateralis* = lateral)	Largest head of the group, forms lateral aspect of thigh; a common intramuscular injection site, particularly in infants (who have poorly developed buttock and arm muscles)	O—greater trochanter, intertrochanteric line, linea aspera I—as for rectus femoris	Extends and stabilizes knee	Femoral nerve
■ **Vastus medialis** (me″de-a′lis) (*medialis* = medial)	Forms inferomedial aspect of thigh	O—linea aspera, intertrochanteric line I—as for rectus femoris	Extends knee; inferior fibers stabilize patella	Femoral nerve

➤

TABLE 10.14	Muscles Crossing the Hip and Knee Joints: Movements of the Thigh and Leg (Figures 10.19 and 10.20) (continued)			
Muscle	Description	Origin (O) and Insertion (I)	Action	Nerve Supply
■ **Vastus intermedius** (in"ter-me′de-us) (*intermedius* = intermediate)	Obscured by rectus femoris; lies between vastus lateralis and vastus medialis on anterior thigh	O—anterior and lateral surfaces of proximal femur shaft I—as for rectus femoris	Extends knee	Femoral nerve
Tensor fasciae latae (ten′sor fä′she-e la′te) (*tensor* = to make tense; *fascia* = band; *lata* = wide)	Enclosed between fascia layers of anterolateral aspect of thigh; functionally associated with medial rotators and flexors of thigh	O—anterior aspect of iliac crest and anterior superior iliac spine I—iliotibial tract*	Flexes and abducts thigh (thus a synergist of the iliopsoas and gluteus medius and minimus muscles); rotates thigh medially; steadies trunk on thigh by making iliotibial tract taut	Superior gluteal nerve (L_4 and L_5)

PART II: POSTERIOR MUSCLES (Figure 10.20)
GLUTEAL MUSCLES—ORIGIN ON PELVIS

Gluteus maximus (gloo′te-us mak′sĭ-mus) (*glutos* = buttock; *maximus* = largest)	Largest and most superficial of gluteus muscles; forms bulk of buttock mass; fibers are thick and coarse; important site of intramuscular injection (dorsal gluteal site); overlies large sciatic nerve; covers ischial tuberosity only when standing; when sitting, moves superiorly, leaving ischial tuberosity exposed in the subcutaneous position	O—dorsal ilium, sacrum, and coccyx I—gluteal tuberosity of femur; iliotibial tract	Major extensor of thigh; complex, powerful, and most effective when thigh is flexed and force is necessary, as in rising from a forward flexed position and in thrusting the thigh posteriorly in climbing stairs and running; generally inactive during standing and walking; laterally rotates and abducts thigh	Inferior gluteal nerve (L_5, S_1, and S_2)
Gluteus medius (me′de-us) (*medius* = middle)	Thick muscle largely covered by gluteus maximus; important site for intramuscular injec-ions (ventral gluteal site); considered safer than dorsal gluteal site because there is less chance of injuring sciatic nerve	O—between anterior and posterior gluteal lines on lateral surface of ilium I—by short tendon into lateral aspect of greater trochanter of femur	Abducts and medially rotates thigh; steadies pelvis; its action is extremely important in walking; e.g., muscle of limb planted on ground tilts or holds pelvis in abduction so that pelvis on side of swinging limb does not sag; the foot of swinging limb can thus clear the ground	Superior gluteal nerve (L_5, S_1)
Gluteus minimus (mĭ′nĭ-mus) (*minimus* = smallest)	Smallest and deepest of gluteal muscles	O—between anterior and inferior gluteal lines on external surface of ilium I—anterior border of greater trochanter of femur	As for gluteus medius	Superior gluteal nerve (L_5, S_1)
LATERAL ROTATORS **Piriformis** (pir′ĭ-form-is) (*piri* = pear; *forma* = shape)	Pyramidal muscle located on posterior aspect of hip joint; inferior to gluteus minimus; issues from pelvis via greater sciatic notch	O—anterolateral surface of sacrum (opposite greater sciatic notch) I—superior border of greater trochanter of femur	Rotates extended thigh laterally; since inserted above head of femur, can also assist in abduction of thigh when hip is flexed; stabilizes hip joint	S_1 and S_2, L_5

*The iliotibial tract is a thickened lateral portion of the *fascia lata* (the fascia that ensheathes all the muscles of the thigh). It extends as a tendinous band from the iliac crest to the knee (see Figure 10.20a).

TABLE 10.14	**Muscles Crossing the Hip and Knee Joints: Movements of the Thigh and Leg (Figures 10.19 and 10.20) (continued)**

| **Obturator externus** (ob"tu-ra'tor ek-ster'nus) (*obturator* = obturator foramen; *externus* = outside) | Flat, triangular muscle deep in upper medial aspect of thigh | O—outer surfaces of obturator membrane, pubis, and ischium, margins of obturator foramen I—by a tendon into trochanteric fossa of posterior femur | As for piriformis | Obturator nerve |

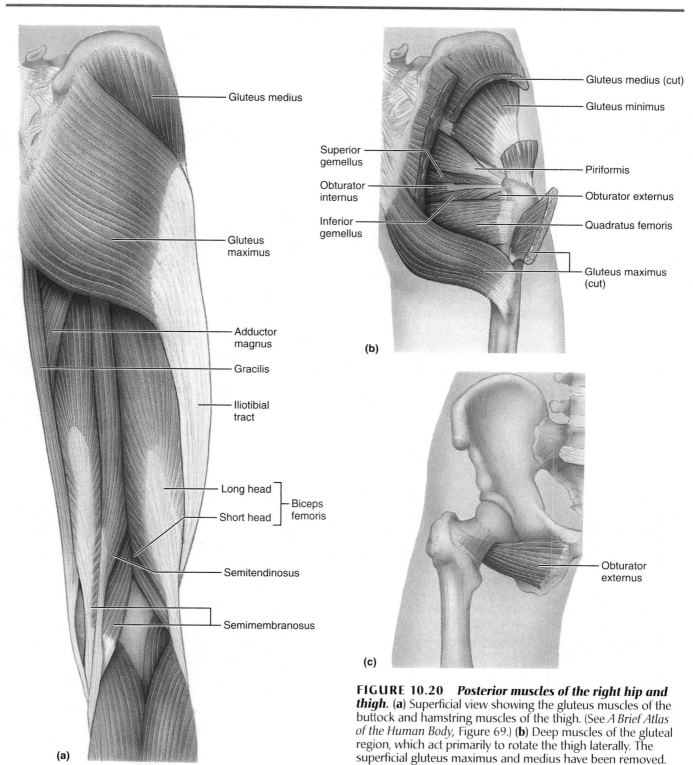

(a)

(b)

(c)

FIGURE 10.20 *Posterior muscles of the right hip and thigh.* **(a)** Superficial view showing the gluteus muscles of the buttock and hamstring muscles of the thigh. (See *A Brief Atlas of the Human Body*, Figure 69.) **(b)** Deep muscles of the gluteal region, which act primarily to rotate the thigh laterally. The superficial gluteus maximus and medius have been removed. **(c)** Anterior view of the isolated obturator externus muscle.

TABLE 10.14	Muscles Crossing the Hip and Knee Joints: Movements of the Thigh and Leg (Figures 10.19 and 10.20) *(continued)*			

Muscle	Description	Origin (O) and Insertion (I)	Action	Nerve Supply
Obturator internus (in-ter′nus) (*internus* = inside)	Surrounds obturator foramen within pelvis; leaves pelvis via lesser sciatic notch and turns acutely forward to insert in femur	O—inner surface of obturator membrane, greater sciatic notch, and margins of obturator foramen I—greater trochanter in front of piriformis	As for piriformis	L_5 and S_1
Gemellus (jĕ-mě′lis)— superior and inferior (*gemin* = twin, double; *superior* = above; *inferior* = below)	Two small muscles with common insertions and actions; considered extrapelvic portions of obturator internus	O—ischial spine (superior); ischial tuberosity (inferior) I—greater trochanter of femur	As for piriformis	L_5 and S_1
Quadratus femoris (*quad* = four-sided square)	Short, thick muscle; most inferior of lateral rotator muscles; extends laterally from pelvis	O—ischial tuberosity I—trochanteric crest of femur	Rotates thigh laterally and stabilizes hip joint	L_5 and S_1

MUSCLES OF THE POSTERIOR COMPARTMENT OF THE THIGH

Hamstrings	The hamstrings are fleshy muscles of the posterior thigh (biceps femoris, semitendinosus, and semimembranosus); they cross both the hip and knee joints and are prime movers of thigh extension and knee flexion; group has a common origin site and is innervated by sciatic nerve; ability of hamstrings to act on one of the two joints spanned depends on which joint is fixed; i.e., if knee is fixed (extended), they promote hip extension; if hip is extended, they promote knee flexion; however, when hamstrings are stretched, they tend to restrict full accomplishment of antagonistic movements; e.g., if knees are fully extended, it is difficult to flex the hip fully (and touch your toes), and when the thigh is fully flexed as in kicking a football, it is almost impossible to extend the knee fully at the same time (without considerable practice); name of this muscle group comes from old butchers' practice of using their tendons to hang hams for smoking; "pulled hamstrings" are common sports injuries in those who run very hard, e.g., football halfbacks			
■ **Biceps femoris** (*biceps* = two heads)	Most lateral muscle of the group; arises from two heads	O—ischial tuberosity (long head); linea aspera and distal femur (short head) I—common tendon passes downward and laterally (forming lateral border of popliteal fossa) to insert into head of fibula and lateral condyle of tibia	Extends thigh and flexes knee; laterally rotates leg, especially when knee is flexed	Sciatic nerve (L_5–S_2)
■ **Semitendinosus** (sem″e-ten″dĭ-no′sus) (*semi* = half; *tendinosus* = tendon)	Lies medial to biceps femoris; although its name suggests that this muscle is largely tendinous, it is quite fleshy; its long slender tendon begins about two-thirds of the way down thigh	O—ischial tuberosity in common with long head of biceps femoris I—medial aspect of upper tibial shaft	Extends thigh at hip; flexes knee; with semimembranosus, medially rotates leg	Sciatic nerve
■ **Semimembranosus** (sem″e-mem″brah-no′sus) (*membranosus* = membrane)	Deep to semitendinosus	O—ischial tuberosity I—medial condyle of tibia; via oblique popliteal ligament to lateral condyle of femur	Extends thigh and flexes knee; medially rotates leg	Sciatic nerve

TABLE 10.15	Muscles of the Leg: Movements of the Ankle and Toes (Figures 10.21 to 10.23)

The deep fascia of the leg is continuous with the fascia lata that ensheathes the thigh. Like a tight knee sock beneath the skin, the leg fascia binds the leg muscles tightly, helping to prevent excessive swelling of muscles during exercise and aiding venous return. Its inward extensions segregate the leg muscles into *anterior, lateral,* and *posterior compartments* (see Figure 10.24b), each with its own nerve and blood supply. Distally the leg fascia thickens to form the **flexor, extensor,** and **fibular** (or **peroneal**) **retinaculae,** which secure the muscle tendons in place as they cross the ankle to run to the foot.

Depending on their location and placement, the various muscles of the leg promote movements at the ankle joint (dorsiflexion and plantar flexion), the intertarsal joints (inversion and eversion of the foot), and/or the toes (flexion and extension). Muscles in the *anterior extensor compartment* of the leg—**tibialis anterior, extensor digitorum longus, extensor hallucis longus,** and **fibularis tertius**—are primarily toe extensors and ankle dorsiflexors. Although dorsiflexion is

not a powerful movement, it is extremely important in preventing the toes from dragging during walking. Lateral compartment muscles are the **fibular** or **peroneal** (*peron* = fibula) **muscles** that plantar flex and evert the foot. Muscles of the *posterior flexor compartment*—**gastrocnemius, soleus, tibialis posterior, flexor digitorum longus,** and **flexor hallucis longus**—primarily plantar flex the foot and flex the toes. Plantar flexion is the most powerful movement of the ankle (and foot) because it lifts the entire weight of our body. It is essential for standing on tiptoe and provides the necessary forward thrust when walking and running. The **popliteus muscle,** which crosses the knee, is important in "unlocking" the extended knee in preparation for flexion.

The tiny intrinsic muscles of the sole of the foot (lumbricals, interossei, and others) are considered separately in Table 10.17.

The actions of the muscles in this table are summarized in Table 10.16 (Part II).

Muscle	Description	Origin (O) and Insertion (I)	Action	Nerve Supply
PART I: MUSCLES OF THE ANTERIOR COMPARTMENT (Figures 10.21 and 10.22)				
	All muscles of the anterior compartment are dorsiflexors of the ankle and have a common innervation, the deep fibular nerve. Paralysis of the anterior muscle group causes *foot drop*, which requires that the leg be lifted unusually high during walking to prevent tripping over one's toes. "Shin splints" is a painful inflammatory condition of the muscles of the anterior compartment.			
Tibialis anterior (tib"e-a′lis) (*tibial* = tibia; *anterior* = toward the front)	Superficial muscle of anterior leg; laterally parallels sharp anterior margin of tibia	O—lateral condyle and upper ⅔ of tibial shaft; interosseous membrane I—by tendon into inferior surface of medial cuneiform and first metatarsal bone	Prime mover of dorsiflexion; inverts foot; assists in supporting medial longitudinal arch of foot	Deep fibular nerve (L_4 and L_5)
Extensor digitorum longus (*extensor* = increases angle at a joint; *digit* = finger or toe; *longus* = long)	Unipennate muscle on anterolateral surface of leg; lateral to tibialis anterior muscle	O—lateral condyle of tibia; proximal ¾ of fibula; interosseous membrane I—middle and distal phalanges of toes 2–5 via extensor expansion	Prime mover of toe extension (acts mainly at metatarsophalangeal joints); dorsiflexes foot (in conjunction with the tibialis anterior and extensor hallucis longus)	Deep fibular nerve (L_5 and S_1)
Fibularis (peroneus) tertius (fib-u-lar′ris ter′shus) (*perone* = fibula; *tertius* = third)	Small muscle; usually continuous and fused with distal part of extensor digitorum longus; not always present	O—distal anterior surface of fibula and interosseous membrane I—tendon inserts on dorsum of fifth metatarsal	Dorsiflexes and everts foot	Deep fibular nerve (L_5 and S_1)
Extensor hallucis (hal′u-sis) **longus** (*hallux* = great toe)	Deep to extensor digitorum longus and tibialis anterior; narrow origin	O—anteromedial fibula shaft and interosseous membrane I—tendon inserts on distal phalanx of great toe	Extends great toe; dorsiflexes foot	Deep fibular nerve (L_5 and S_1)

▶

TABLE 10.15   **Muscles of the Leg: Movements of the Ankle and Toes (Figures 10.21 to 10.23)** *(continued)*

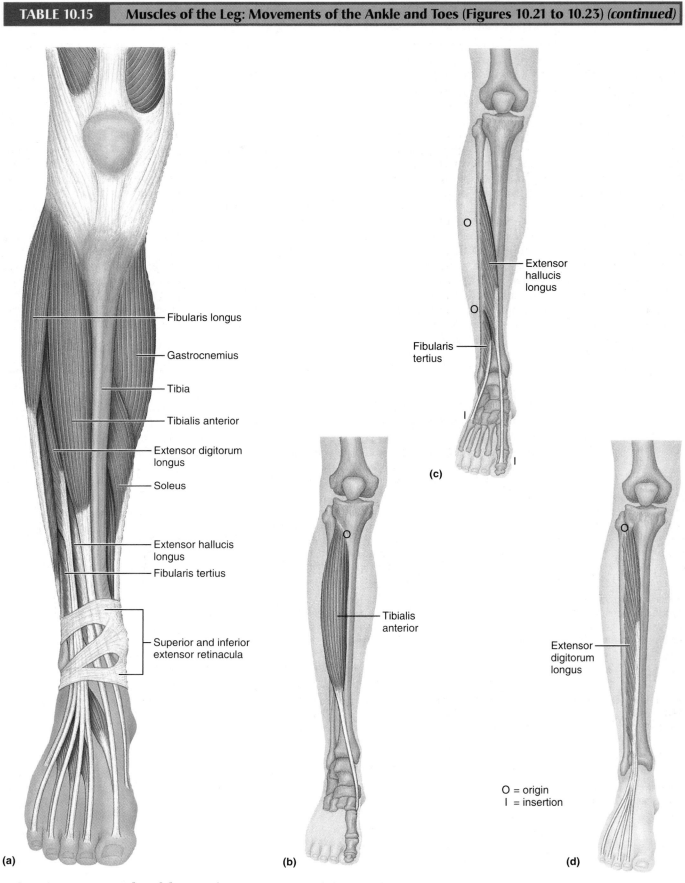

FIGURE 10.21   *Muscles of the anterior compartment of the right leg.*
(**a**) Superficial view of anterior leg muscles. (**b–d**) Some of the same muscles shown
in isolation to allow visualization of their origins and insertions.

TABLE 10.15	Muscles of the Leg: Movements of the Ankle and Toes (Figures 10.21 to 10.23) *(continued)*

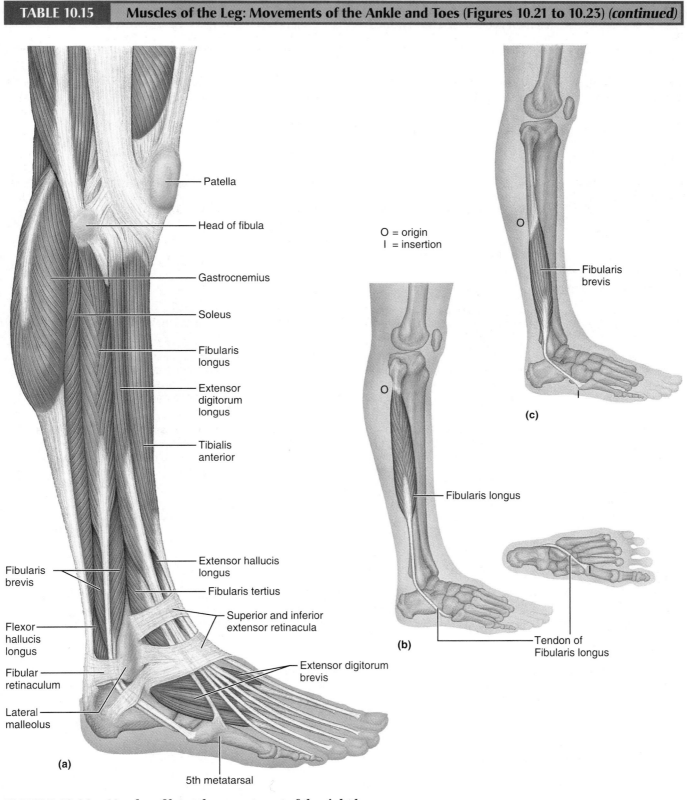

O = origin
I = insertion

Patella

Head of fibula

Gastrocnemius

Soleus

Fibularis longus

Extensor digitorum longus

Tibialis anterior

Extensor hallucis longus

Fibularis tertius

Superior and inferior extensor retinacula

Fibularis brevis

Flexor hallucis longus

Fibular retinaculum

Lateral malleolus

Extensor digitorum brevis

5th metatarsal

(a)

Fibularis brevis

Fibularis longus

(b)

(c)

Tendon of Fibularis longus

FIGURE 10.22 *Muscles of lateral compartment of the right leg.*
(**a**) Superficial view of lateral aspect of the leg, illustrating positions of lateral compartment muscles (fibularis longus and brevis) relative to anterior and posterior leg muscles. (**b**) Isolated view of fibularis longus; inset illustrates the insertion of the fibularis longus on the plantar surface of the foot. (**c**) Isolated view of the fibularis brevis muscle.

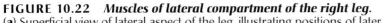

Muscle	Description	Origin (O) and Insertion (I)	Action	Nerve Supply
PART II: MUSCLES OF THE LATERAL COMPARTMENT (Figures 10.22 and 10.23) These muscles have a common innervation, the superficial fibular nerve. Besides plantar flexion and foot eversion, these muscles stabilize the lateral ankle and lateral longitudinal arch of the foot.				
Fibularis (peroneus) longus (See also Figure 10.21)	Superficial lateral muscle; overlies fibula	O—head and upper portion of lateral aspect of fibula I—by long tendon that curves under foot to first metatarsal and medial cuneiform	Plantar flexes and everts foot; may help keep foot flat on ground	Superficial fibular nerve ($L_5 - S_2$)
Fibularis (peroneus) brevis (*brevis* = short)	Smaller muscle; deep to fibularis longus; enclosed in a common sheath	O—distal fibula shaft I—by tendon running behind lateral malleolus to insert on proximal end of fifth metatarsal	Plantar flexes and everts foot	Superficial fibular nerve
PART III: MUSCLES OF THE POSTERIOR COMPARTMENT (Figure 10.23) The muscles of the posterior compartment have a common innervation, the tibial nerve. They act in concert to plantar flex the ankle.				
SUPERFICIAL MUSCLES				
Triceps surae (tri"seps sur'e) (See also Figure 10.22)	Refers to muscle pair (gastrocnemius and soleus) that shapes the posterior calf and inserts via a common tendon into the calcaneus of the heel; this calcaneal or Achilles tendon is the largest tendon in the body; prime movers of ankle plantar flexion			
▪ **Gastrocnemius** (gas"truk-ne'me-us) (*gaster* = belly; *kneme* = leg)	Superficial muscle of pair; two prominent bellies that form proximal curve of calf	O—by two heads from medial and lateral condyles of femur I—posterior calcaneus via calcaneal tendon	Plantar flexes foot when knee is extended; since it also crosses knee joint, it can flex knee when foot is dorsiflexed	Tibial nerve (S_1, S_2)
▪ **Soleus** (so'le-us) (*soleus* = fish)	Broad, flat muscle, deep to gastrocnemius on posterior surface of calf	O—extensive cone-shaped origin from superior tibia, fibula, and interosseous membrane I—as for gastrocnemius	Plantar flexes foot; important locomotor and postural muscle during walking, running, and dancing	Tibial nerve
Plantaris (plan-tar'is) (*planta* = sole of foot)	Generally a small feeble muscle, but varies in size and extent; may be absent	O—posterior femur above lateral condyle I—via a long, thin tendon into calcaneus or calcaneal tendon	Assists in knee flexion and plantar flexion of foot	Tibial nerve
DEEP MUSCLES				
Popliteus (pop-lit'e-us) (*poplit* = back of knee)	Thin, triangular muscle at posterior knee; passes downward and medially to tibial surface	O—lateral condyle of femur and lateral meniscus I—proximal tibia	Flexes and rotates leg medially to unlock knee from full extension; with tibia fixed, rotates thigh laterally	Tibial nerve (L_4–S_1)
Flexor digitorum longus (*flexor* = decreases angle at a joint)	Long, narrow muscle; runs medial to and partially overlies tibialis posterior	O—extensive origin on the posterior tibia I—tendon runs behind medial malleolus and splits to insert into distal phalanges of toes 2–5	Plantar flexes and inverts foot; flexes toes; helps foot "grip" ground	Tibial nerve (S_2 and S_3)
Flexor hallucis longus (See also Figure 10.22)	Bipennate muscle; lies lateral to inferior aspect of tibialis posterior	O—mid shaft of fibula; interosseous membrane I—tendon runs under foot to distal phalanx of great toe	Plantar flexes and inverts foot; flexes great toe at all joints; "push off" muscle during walking	Tibial nerve (S_2 and S_3)
Tibialis posterior (*posterior* = toward the back)	Thick, flat muscle deep to soleus; placed between posterior flexors	O—extensive origin from superior tibia and fibula and interosseous membrane I—tendon passes behind medial malleolus and under arch of foot; inserts into several tarsals and metatarsals 2–4	Prime mover of foot inversion; plantar flexes foot; stabilizes medial longitudinal arch of foot (as during ice skating)	Tibial nerve (L_4 and L_5)

| TABLE 10.15 | Muscles of the Leg: Movements of the Ankle and Toes (Figures 10.21 to 10.23) *(continued)* |

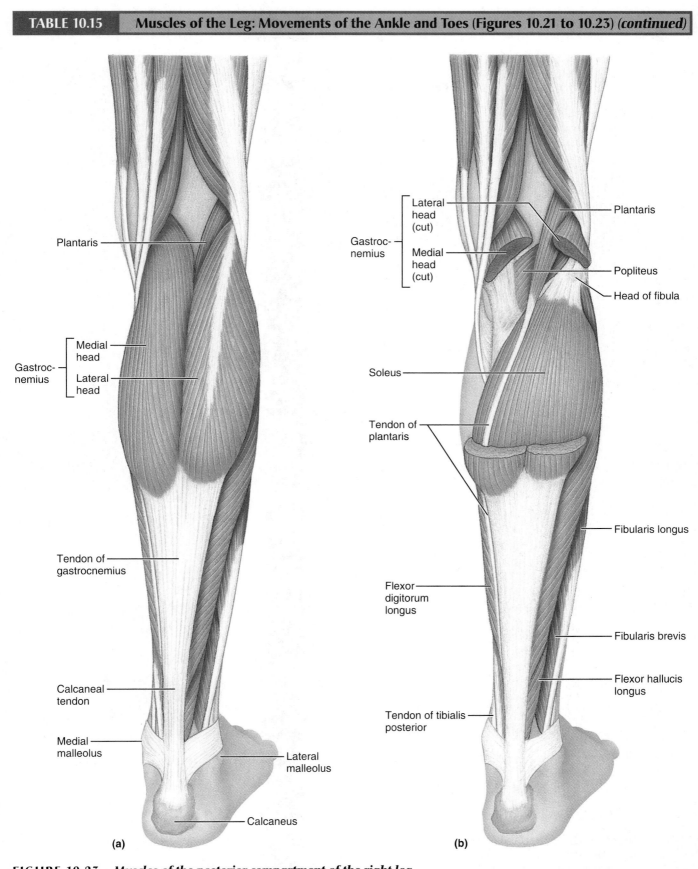

(a)

(b)

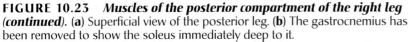

FIGURE 10.23 *Muscles of the posterior compartment of the right leg (continued).* **(a)** Superficial view of the posterior leg. **(b)** The gastrocnemius has been removed to show the soleus immediately deep to it.

TABLE 10.15 | **Muscles of the Leg: Movements of the Ankle and Toes (Figures 10.21 to 10.23)** *(continued)*

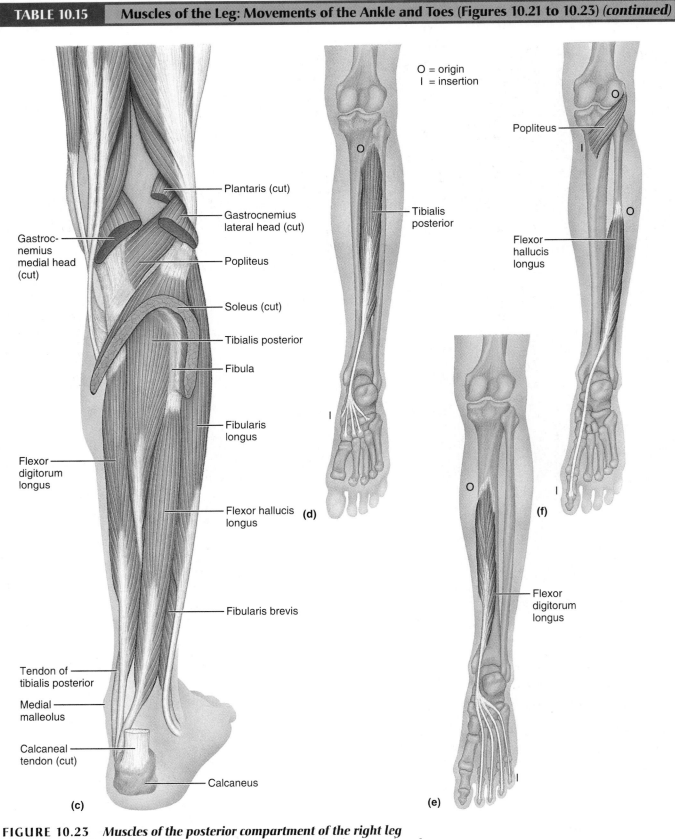

FIGURE 10.23 *Muscles of the posterior compartment of the right leg (continued).* (**c**) The triceps surae has been removed to show the deep muscles of the posterior compartment. (**d–f**) Individual deep muscles are shown in isolation.

TABLE 10.16   **Summary of Actions of Muscles Acting on the Thigh, Leg, and Foot (Figure 10.24)**

Part I: Muscles Acting on the Thigh and Leg (PM = prime mover)	Actions at the hip joint						Actions at the knee	
	Flexion	Extension	Abduction	Adduction	Medial rotation	Lateral rotation	Flexion	Extension
Anterior and medial muscles:								
Iliopsoas	X (PM)							
Sartorius	X		X			X	X	
Adductor magnus		X		X	X			
Adductor longus	X			X	X			
Adductor brevis	X			X	X			
Pectineus	X			X	X			
Gracilis				X	X		X	
Rectus femoris	X							X (PM)
Vastus muscles								X (PM)
Tensor fasciae latae	X		X		X			X
Posterior muscles:								
Gluteus maximus		X (PM)	X			X		
Gluteus medius			X (PM)		X			
Gluteus minimus			X		X			
Piriformis			X			X		
Obturator internus						X		
Obturator externus						X		
Gemelli						X		
Quadratus femoris						X		
Biceps femoris		X (PM)				X	X (PM)	
Semitendinosus		X			X		X (PM)	
Semimembranosus		X			X		X (PM)	
Gastrocnemius							X	
Plantaris							X	
Popliteus						X	X (and rotates leg medially)	

Part II: Muscles Acting on the Ankle and Toes	Actions at the ankle joint				Actions at the toes	
	Plantar flexion	Dorsiflexion	Inversion	Eversion	Flexion	Extension
Anterior compartment:						
Tibialis anterior		X (PM)	X			
Extensor digitorum longus		X				X (PM)
Fibularis tertius		X		X		
Extensor hallucis longus		X	X (weak)			X (great toe)
Lateral compartment:						
Fibularis longus and brevis	X			X		
Posterior compartment:						
Gastrocnemius	X (PM)					
Soleus	X (PM)					
Plantaris	X					
Flexor digitorum longus	X		X		X (PM)	
Flexor hallucis longus	X		X		X (great toe)	
Tibialis posterior	X		X (PM)			

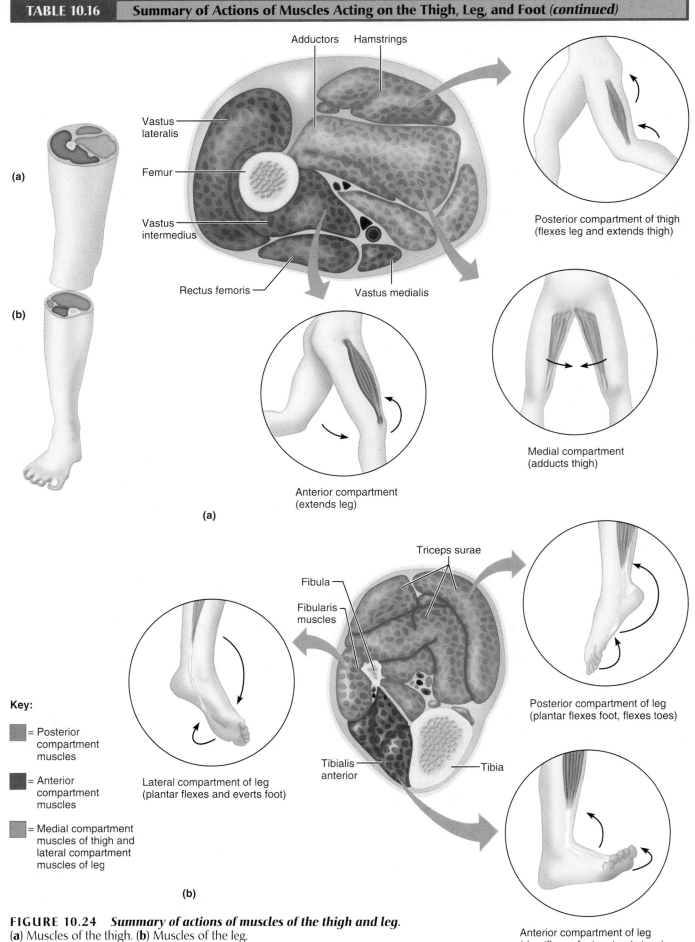

Posterior compartment of thigh
(flexes leg and extends thigh)

Adductors   Hamstrings

Vastus lateralis

Femur

Vastus intermedius

Rectus femoris   Vastus medialis

Anterior compartment
(extends leg)

Medial compartment
(adducts thigh)

(a)

Triceps surae

Fibula

Fibularis muscles

Tibialis anterior   Tibia

Posterior compartment of leg
(plantar flexes foot, flexes toes)

Lateral compartment of leg
(plantar flexes and everts foot)

Key:

= Posterior compartment muscles

= Anterior compartment muscles

= Medial compartment muscles of thigh and lateral compartment muscles of leg

(b)

Anterior compartment of leg
(dorsiflexes foot, extends toes)

FIGURE 10.24   *Summary of actions of muscles of the thigh and leg.*
(**a**) Muscles of the thigh. (**b**) Muscles of the leg.

TABLE 10.17	Intrinsic Muscles of the Foot: Toe Movement and Arch Support (Figure 10.25)

The intrinsic muscles of the foot help to flex, extend, abduct, and adduct the toes. Collectively, along with the tendons of some leg muscles that enter the sole, the foot muscles help support the arches of the foot. There is a single muscle on the foot's dorsum (superior aspect), and several muscles on the plantar aspect (the sole). The plantar muscles occur in four layers, from superficial to deep. Overall, the foot muscles are remarkably similar to those in the palm of the hand.

Muscle	Description	Origin (O) and Insertion (I)	Action	Nerve Supply
MUSCLE ON DORSUM OF FOOT				
Extensor digitorum brevis (Figure 10.22a)	Small, four-part muscle on dorsum of foot; deep to the tendons of extensor digitorum longus; corresponds to the extensor indicis and extensor pollicis muscles of forearm	O—anterior part of calcaneus bone; extensor retinaculum I—base of proximal phalanx of great toe; extensor expansions on toes 2–5	Helps extend toes at metatarsophalangeal joints	Deep fibular nerve (S_1 and S_2)
MUSCLES ON SOLE OF FOOT (Figure 10.25)				
First Layer (most superficial)				
■ **Flexor digitorum brevis**	Bandlike muscle in middle of sole; corresponds to flexor digitorum superficialis of forearm and inserts into digits in the same way	O—tuber calcanei I—middle phalanx of toes 2–4	Helps flex toes	Medial plantar nerve (S_2 and S_3)
■ **Abductor hallucis** (hal'yu-kis) (*hallux* = the big toe)	Lies medial to flexor digitorum brevis (recall the similar thumb muscle, abductor pollicis brevis)	O—tuber calcanei and flexor retinaculum I—proximal phalanx of great toe, medial side, in the tendon of flexor hallucis brevis (see below)	Abducts great toe	Medial plantar nerve
■ **Abductor digiti minimi**	Most lateral of the three superficial sole muscles; (recall similar abductor muscle in palm)	O—tuber calcanei I—lateral side of base of little toe's proximal phalanx	Abducts and flexes little toe	Lateral plantar nerve (S_2 and S_3)
Second Layer				
■ **Flexor accessorius (quadratus plantae)**	Rectangular muscle just deep to flexor digitorum brevis in posterior half of sole; two heads (see also Figure 10.25c)	O—medial and lateral sides of calcaneus I—tendon of flexor digitorum longus in midsole	Straightens out the oblique pull of flexor digitorum longus	Lateral plantar nerve
■ **Lumbricals**	Four little "worms" (like lumbricals in hand)	O—from each tendon of flexor digitorum longus I—extensor expansion on proximal phalanx of toes 2–5, medial side	By pulling on extensor expansion, flex toes at metatarsophalangeal joints and extend toes at interphalangeal joints	Medial plantar nerve (first lumbrical) and lateral plantar nerve (second to fourth lumbrical)

TABLE 10.17   Intrinsic Muscles of the Foot: Toe Movement and Arch Support (Figure 10.25) (continued)

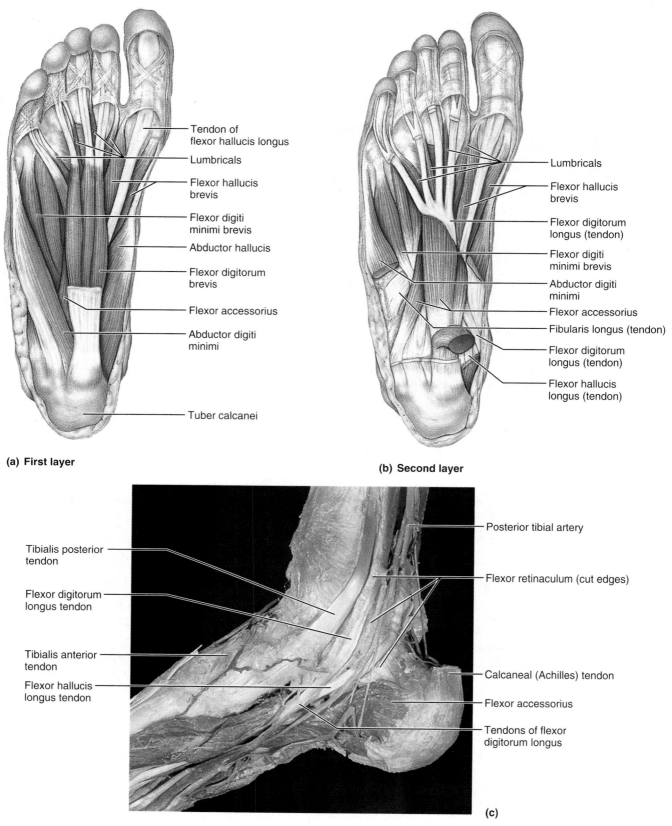

(a) First layer

- Tendon of flexor hallucis longus
- Lumbricals
- Flexor hallucis brevis
- Flexor digiti minimi brevis
- Abductor hallucis
- Flexor digitorum brevis
- Flexor accessorius
- Abductor digiti minimi
- Tuber calcanei

(b) Second layer

- Lumbricals
- Flexor hallucis brevis
- Flexor digitorum longus (tendon)
- Flexor digiti minimi brevis
- Abductor digiti minimi
- Flexor accessorius
- Fibularis longus (tendon)
- Flexor digitorum longus (tendon)
- Flexor hallucis longus (tendon)

- Posterior tibial artery
- Tibialis posterior tendon
- Flexor digitorum longus tendon
- Flexor retinaculum (cut edges)
- Tibialis anterior tendon
- Flexor hallucis longus tendon
- Calcaneal (Achilles) tendon
- Flexor accessorius
- Tendons of flexor digitorum longus

(c)

FIGURE 10.25   Muscles of the right foot, plantar and medial aspects.

TABLE 10.17	Intrinsic Muscles of the Foot: Toe Movement and Arch Support (Figure 10.25) *(continued)*

Third Layer

■ **Flexor hallucis brevis**	Covers metatarsal 1; splits into two bellies—recall flexor pollicis brevis of thumb (see Figure 10.25d)	O—lateral cuneiform and cuboid bones I—via two tendons onto base of the proximal phalanx of great toe	Flexes great toe's metatarsophalangeal joint	Medial plantar nerve
■ **Adductor hallucis**	Oblique and transverse heads; deep to lumbricals (recall adductor pollicis in thumb)	O—from bases of metatarsals 2–4 and from fibularis longus tendon sheath (oblique head); from a ligament across metatarsophalangeal joints (transverse head) I—base of proximal phalanx of great toe, lateral side	Helps maintain the transverse arch of foot; weak adductor of great toe	Lateral plantar nerve (S_2 and S_3)
■ **Flexor digiti minimi brevis**	Covers metatarsal 5 (recall same muscle in hand)	O—base of metatarsal 5 and tendon sheath of fibularis longus I—base of proximal phalanx of toe 5	Flexes little toe at metatarsophalangeal joint	Lateral plantar nerve
Fourth Layer (deepest) ■ **Plantar (3) and dorsal interossei (4)**	Similar to the palmar and dorsal interossei of hand in locations, attachments, and actions; however, the long axis of foot around which these muscles orient is the second digit, not the third	See palmar and dorsal interossei (Table 10.13)	See palmar and dorsal interossei (Table 10.13)	Lateral plantar nerve

(d) Third layer

- Adductor hallucis (transverse head)
- Adductor hallucis (oblique head)
- Interosseous muscles
- Flexor hallucis brevis
- Flexor digiti minimi brevis
- Fibularis longus (tendon)
- Flexor accessorius
- Flexor digitorum longus (tendon)
- Flexor hallucis longus (tendon)

Plantar interossei

(e) Fourth layer: plantar interossei

Dorsal interossei

(f) Fourth layer: dorsal interossei

FIGURE 10.25 (Continued) *Muscles of the right foot, plantar aspect.*

RELATED CLINICAL TERMS

Charley horse A muscle contusion, i.e., tearing of muscle followed by bleeding into the tissues (hematoma) and severe, prolonged pain; a common contact sports injury; a charley horse of the quadriceps muscle of the thigh occurs frequently in football players.

Electromyography Recording and interpretation of graphic records of the electrical activity of contracting muscles. Electrodes inserted into the muscles record the impulses that pass over muscle-cell membranes to stimulate contraction. The best and most important technique for determining the functions of muscles and muscle groups.

Hernia An abnormal protrusion of abdominal contents (typically coils of the small intestine) through a weak point in the muscles of the abdominal wall. Most often caused by increased intra-abdominal pressure during lifting or straining. The hernia penetrates the muscle wall but not the skin and so appears as a visible bulge in the body surface. Common abdominal hernias include the inguinal and umbilical hernias.

Quadriceps and hamstring strains Also called quad and hamstring pulls, these conditions involve tearing of these muscles or their tendons; happen mainly in athletes who do not warm up properly and then fully extend their hip (quad pull) or knee (hamstring pull) quickly or forcefully (e.g., sprinters, tennis players); not painful at first, but pain intensifies within 3 to 6 hours (30 minutes if the tearing is severe). After a week of rest, stretching is the best therapy.

Rupture of the calcaneal tendon Although the calcaneal (Achilles) tendon is the largest, strongest tendon in the body, its rupture is surprisingly common, particularly in older men as a result of stumbling and in young sprinters when the tendon is traumatized during the takeoff; the rupture is followed by abrupt pain; a gap is seen just above the heel, and the calf bulges as the triceps surae are released from their insertion; plantar flexion is not possible, but dorsiflexion is exaggerated.

Shin splints Common term for pain in the anterior compartment of the leg caused by irritation of the anterior tibialis muscle as might follow extreme or unusual exercise without adequate prior conditioning; as the inflamed muscle swells, its circulation is impaired by the tight fascial wrappings, leading to pain and tenderness to touch. However, the term is used loosely to indicate a variety of ailments ranging from stress fracture of the tibia, to inflammation of shin fascia, to muscle tears.

Tennis elbow Tenderness due to trauma or overuse of the tendon of origin of the forearm extensor muscles at the lateral epicondyle of the humerus. Caused and aggravated when these muscles contract forcefully to extend the hand at the wrist—as in executing a tennis backhand or lifting a loaded snow shovel. Despite its name, tennis elbow does not involve the elbow joint; most cases caused by work activities.

Torticollis (tor"ti-kol'is; *tort* = twisted) A twisting of the neck in which there is a chronic rotation and tilting of the head to one side, due to injury of the sternocleidomastoid muscle on one side; also called wryneck; sometimes present at birth when the muscle fibers are torn during difficult delivery; exercise that stretches the affected muscle is the usual treatment.

CHAPTER SUMMARY

Media study tools that could provide you additional help in reviewing specific key topics of Chapter 10 are referenced below. **SP** *= Study Partner;* **IP** *= Interactive Physiology.*

Interactions of Skeletal Muscles in the Body (p. 323)

1. Skeletal muscles are arranged in opposing groups across body joints so that one group can reverse or modify the action of the other.

SP Exercise: Chapter 10, Lever Systems.

2. Muscles are classified as prime movers or agonists (bear the chief responsibility for producing movement), antagonists (reverse, or oppose, the action of another muscle), synergists (aid a prime mover by effecting the same action, stabilizing joints, or preventing undesirable movements), and fixators (function to immobilize a bone or a muscle's origin).

Naming Skeletal Muscles (pp. 323–324)

1. Criteria used to name muscles include a muscle's location, shape, relative size, fiber (fascicle) direction, number of origins, attachment sites (origin/insertion), and action. Several criteria are combined to name some muscles.

Muscle Mechanics: Importance of Leverage and Fascicle Arrangement (pp. 324–330)

1. A lever is a bar that moves on a fulcrum. When an effort is applied to the lever, a load is moved. In the body, bones are the levers, joints are the fulcrums, and the effort is exerted by skeletal muscles at their insertions.

SP Exercise: Chapter 10, Lever Systems.

2. When the effort is farther from the fulcrum than is the load, the lever operates at a mechanical advantage (is slow and strong). When the effort is exerted closer to the fulcrum than is the load, the lever operates at a mechanical disadvantage (is fast and promotes a large degree of movement).

3. First-class levers (effort-fulcrum-load) may operate at a mechanical advantage or disadvantage. Second-class levers (fulcrum-load-effort) all operate at a mechanical advantage. Third-class levers (fulcrum-effort-load) always operate at a mechanical disadvantage.

4. Common patterns of fascicle arrangement are parallel, fusiform, pennate, convergent, and circular. Muscles with fibers that run parallel to the long axis of the muscle shorten most; stocky pennate muscles shorten little but are the most powerful muscles.

Major Skeletal Muscles of the Body (pp. 330–381)

1. Muscles of the head that produce facial expression tend to be small and to insert into soft tissue (skin and other muscles) rather than into bone. These muscles open and close the eyes and mouth, compress the cheeks, allow smiling and other types of facial language (see Table 10.1*).

2. Muscles of the head involved in mastication include the masseter and temporalis that elevate the mandible and two deep muscle pairs that promote grinding and sliding jaw movements (see Table 10.2*). Extrinsic muscles of the tongue anchor the tongue and control its movements.

3. Deep muscles of the anterior neck promote swallowing movements, including elevation/depression of the hyoid bone, closure of the respiratory passages, and peristalsis of the pharynx (see Table 10.3*).

4. Neck muscles and deep muscles of the vertebral column promote head and trunk movements (see Table 10.4*). The deep muscles of the posterior trunk can extend large regions of the vertebral column (and head) simultaneously. Head flexion and rotation are effected by the anteriorly located sternocleidomastoid and scalene muscles.

5. Movements of quiet breathing are promoted by the diaphragm and the external intercostal muscles of the thorax (see Table 10.5*). Downward movement of the diaphragm increases intra-abdominal pressure.

6. The four muscle pairs forming the abdominal wall are layered like plywood to form a natural muscular girdle that protects, supports, and compresses abdominal contents. These muscles also flex and laterally rotate the trunk (see Table 10.6*).

7. Muscles of the pelvic floor and perineum (see Table 10.7*) support the pelvic viscera, resist increases in intra-abdominal pressure, inhibit urination and defecation, and aid erection.

8. Except for the pectoralis major and the latissimus dorsi, the superficial muscles of the thorax act to fix or promote movements of the scapula (see Table 10.8*).

Scapular movements are effected primarily by posterior thoracic muscles.

9. Nine muscles cross the shoulder joint to effect movements of the humerus (see Table 10.9*). Of these, seven originate on the scapula and two arise from the axial skeleton. Four muscles contribute to the "rotator cuff" helping to stabilize the multiaxial shoulder joint. Generally speaking, muscles located anteriorly flex, rotate, and adduct the arm. Those located posteriorly extend, rotate, and adduct the arm. The deltoid muscle of the shoulder is the prime mover of shoulder abduction.

10. Muscles causing movements of the forearm form the flesh of the arm (see Table 10.10*). Anterior arm muscles are forearm flexors; posterior muscles are forearm extensors.

11. Movements of the wrist, hand, and fingers are effected mainly by muscles originating on the forearm (see Table 10.11*). Except for the two pronator muscles, the anterior forearm muscles are wrist and/or finger flexors; those of the posterior compartment are wrist and finger extensors.

12. The intrinsic muscles of the hand aid in precise movements of the fingers (Table 10.13*) and in opposition, which helps us grip things in our palms. These small muscles are divided into thenar, hypothenar, and mid-palmar groups.

13. Muscles crossing the hip and knee joints effect thigh and leg movements (see Table 10.14*). Anteromedial muscles include thigh flexors and/or adductors and knee extensors. Muscles of the posterior gluteal region extend and rotate the thigh. Posterior thigh muscles extend the hip and flex the knee.

14. Muscles in the leg act on the ankle and toes (see Table 10.15*). Anterior compartment muscles are largely ankle dorsiflexors. Lateral compartment muscles are plantar flexors and foot everters. Those of the posterior leg are plantar flexors. Intrinsic foot muscles support foot arches and help effect toe movements.

15. The intrinsic muscles of the foot (Table 10.17*) support the foot arches and help move the toes. Most occur in the sole, arranged in four layers. They resemble the small muscles in the palm of the hand.

---

*See specific table cited for detailed description of each muscle in the group.

REVIEW QUESTIONS

Multiple Choice/Matching

1. A muscle that assists an agonist by causing a like movement or by stabilizing a joint over which an agonist acts is (a) an antagonist, (b) a prime mover, (c) a synergist, (d) an agonist.

2. A muscle in which the fibers are arranged at an angle to a central longitudinal tendon has a _____ arrangement. (a) circular, (b) longitudinal, (c) pennate, (d) parallel.

3. Match the muscle names in column B to the facial muscles described in column A.

Column A	Column B
_____ (1) squints the eyes	(a) corrugator supercilii
_____ (2) raises the eyebrows	(b) depressor anguli oris
_____ (3) smiling muscle	(c) frontalis
_____ (4) puckers the lips	(d) occipitalis
_____ (5) pulls the scalp posteriorly	(e) orbicularis oculi
	(f) orbicularis oris
	(g) zygomaticus

4. The prime mover of inspiration is the (a) diaphragm, (b) internal intercostals, (c) external intercostals, (d) abdominal wall muscles.

5. The arm muscle that both flexes the elbow and supinates the forearm is the (a) brachialis, (b) brachioradialis, (c) biceps brachii, (d) triceps brachii.

6. The chewing muscles that protrude the mandible and produce side-to-side grinding movements are the (a) buccinators, (b) masseters, (c) temporalis, (d) pterygoids.

7. Muscles that depress the hyoid bone and larynx include all but the (a) sternohyoid, (b) omohyoid, (c) geniohyoid, (d) sternothyroid.

8. Intrinsic muscles of the back that promote extension of the spine (or head) include all but (a) splenius muscles, (b) semispinalis muscles, (c) scalene muscles, (d) erector spinae.

9. Several muscles act to move and/or stabilize the scapula. Which of the following are small rectangular muscles that square the shoulders as they act together to retract the scapula? (a) levator scapulae, (b) rhomboids, (c) serratus anterior, (d) trapezius.

10. The quadriceps include all but (a) vastus lateralis, (b) vastus intermedius, (c) vastus medialis, (d) biceps femoris, (e) rectus femoris.

11. A prime mover of hip flexion is the (a) rectus femoris, (b) iliopsoas, (c) vasti muscles, (d) gluteus maximus.

12. The prime mover of hip extension *against* resistance is the (a) gluteus maximus, (b) gluteus medius, (c) biceps femoris, (d) semimembranosus.

13. Muscles that cause plantar flexion include all but the (a) gastrocnemius, (b) soleus, (c) tibialis anterior, (d) tibialis posterior, (e) peroneus muscles.

14. In walking, which two lower limb muscles keep the forward-swinging foot from dragging on the ground? (a) pronator teres and popliteus, (b) flexor digitorum longus and popliteus, (c) adductor longus and abductor digiti minimi in foot, (d) gluteus medius and tibialis anterior.

15. List four criteria used in naming muscles, and name a specific muscle that illustrates each criterion. (Do not use the examples given in the text.)

16. Which of the following is a large, deep muscle that protracts the scapula during punching? (a) serratus anterior, (b) rhomboids, (c) levator scapulae, (d) subscapularis.

Short Answer Essay Questions

17. Name four criteria used in naming muscles, and provide an example (other than those used in the text) that illustrates each criterion.

18. Differentiate between the arrangement of elements (load, fulcrum, and effort) in first-, second-, and third-class levers.

19. What does it mean when we say that a lever operates at a mechanical disadvantage, and what benefits does such a lever system provide?

20. What muscles act to propel a food bolus down the length of the pharynx to the esophagus?

21. Name and describe the action of muscles used to shake your head no. To nod yes.

22. (a) Name the four muscle pairs that act in unison to compress the abdominal contents. (b) How does their arrangement (fiber direction) contribute to the strength of the abdominal wall? (c) Which of these muscles can effect lateral rotation of the spine? (d) Which can act alone to flex the spine?

23. List all (six) possible movements that can occur at the shoulder joint and name the prime mover(s) of each movement. Then name their antagonists.

24. (a) Name two forearm muscles that are powerful extensors and abductors of the wrist. (b) Name the sole forearm muscle that can flex the distal interphalangeal joints.

25. Name the muscles usually grouped together as the lateral rotators of the hip.

26. Name three thigh muscles that help you keep your seat astride a horse.

27. (a) Name three muscles or muscle groups used as sites for intramuscular injections. (b) Which of these is used most often in infants, and why?

28. Name two muscles in each of the following compartments or regions: (a) thenar eminence (ball of thumb), (b) posterior compartment of forearm, (c) anterior compartment of forearm—deep muscle group, (d) anterior muscle group in the arm, (e) muscles of mastication, (f) third muscle layer of the foot, (g) posterior compartment of leg, (h) medial compartment of thigh, (i) posterior compartment of thigh.

Critical Thinking and Clinical Application Questions

1. Assume you have a 10-lb weight in your right hand. Explain why it is easier to flex the right elbow when your forearm is supinated than when it is pronated.

2. When Mrs. O'Brien returned to her doctor for a follow-up visit after childbirth, she complained that she was having problems controlling her urine flow (was incontinent) when she sneezed. The physician asked his nurse to give Mrs. O'Brien instructions on how to perform exercises to strengthen the muscles of the pelvic floor. To what muscles was he referring?

3. Mr. Ahmadi, an out-of-shape 45-year-old man, was advised by his physician to lose weight and to exercise on a regular basis. He followed his diet faithfully and began to jog daily. One day, while on his morning jog, he heard a snapping sound that was immediately followed by a severe pain in his right lower calf. When examined, a gap was seen between his swollen upper calf region and his heel, and he was unable to plantar flex that ankle. What do you think happened? Why was the upper part of his calf swollen?

4. What is the functional reason why the muscle group on the dorsal leg (calf) is so much larger than the muscle group in the ventral region of the leg?

11 FUNDAMENTALS OF THE NERVOUS SYSTEM AND NERVOUS TISSUE

1. List the basic functions of the nervous system.

Organization of the Nervous System (pp. 387–388)

2. Explain the structural and functional divisions of the nervous system.

Histology of Nervous Tissue (pp. 388–396)

3. List the types of supporting cells and cite their functions.

4. Define neuron, describe its important structural components, and relate each to a functional role.

5. Differentiate between a nerve and a tract, and between a nucleus and a ganglion.

6. Explain the importance of the myelin sheath and describe how it is formed in the central and peripheral nervous systems.

7. Classify neurons structurally and functionally.

Neurophysiology (pp. 396–419)

8. Define resting membrane potential and describe its electrochemical basis.

9. Compare and contrast graded and action potentials.

10. Explain how action potentials are generated and propagated along neurons.

11. Define absolute and relative refractory periods.

12. Define saltatory conduction and contrast it to conduction along unmyelinated fibers.

13. Define synapse. Distinguish between electrical and chemical synapses structurally and in their mechanisms of information transmission.

14. Distinguish between excitatory and inhibitory postsynaptic potentials.

15. Describe how synaptic events are integrated and modified.

16. Define neurotransmitter and name several classes of neurotransmitters.

Basic Concepts of Neural Integration (pp. 419–421)

17. Describe common patterns of neuronal organization and processing.

18. Distinguish in a general sense between serial and parallel processing.

Developmental Aspects of Neurons (pp. 422–424)

19. Describe the role of astrocytes and nerve cell adhesion molecules (N-CAMs) in neuronal differentiation.

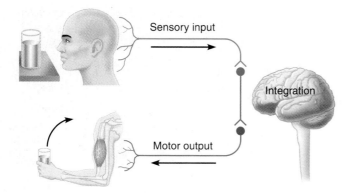

FIGURE 11.1   *The nervous system's functions.*

You are driving down the freeway, and a horn blares to your right. You swerve to your left. Charlie leaves a note on the kitchen table: "See you later—have the stuff ready at 6." You know that the "stuff" is chili with taco chips. You are dozing, and your infant son makes a soft cry. Instantly, you awaken. What do all these events have in common? They are all everyday examples of the functioning of your nervous system, which has your body cells humming with activity nearly all the time.

The **nervous system** is the master controlling and communicating system of the body. It is responsible for all behavior—every thought, action, and emotion reflects its activity. Its cells communicate by electrical signals, which are rapid, specific, and usually cause almost immediate responses.

The nervous system has three overlapping functions (Figure 11.1): (1) It uses its millions of sensory receptors to monitor changes occurring both inside and outside the body. These changes are called *stimuli* and the gathered information is called **sensory input.** (2) It processes and interprets the sensory input and decides what should be done at each moment—a process called **integration.** (3) It causes a *response* by activating *effector organs*, our muscles or glands; the response is called **motor output.** An example will illustrate how these functions work together. When you are driving and see a red light ahead (sensory input), your nervous system integrates this information (red light means "stop"), and your foot goes for the brake (motor output).

This chapter begins with a brief overview of the organization of the nervous system. It then focuses on the functional anatomy and arrangement of nervous tissue, especially that of nerve cells, or *neurons,* which are the key to neural communication.

ORGANIZATION OF THE NERVOUS SYSTEM

We have only one highly integrated nervous system. However, for convenience, it can be divided into two principal parts (Figure 11.2). The **central nervous system (CNS)** consists of the *brain* and *spinal cord,* which occupy the dorsal body cavity. The CNS is the integrating and command center of the nervous system. It interprets incoming sensory information and dictates motor responses based on past experience, reflexes, and current conditions. The **peripheral nervous system (PNS),** the part of the nervous system *outside* the CNS, consists mainly of the nerves that extend from the brain and spinal cord. *Spinal nerves* carry impulses to and from the spinal cord; *cranial nerves* carry impulses to and from the brain. These peripheral nerves serve as the communication lines that link all parts of the body to the central nervous system.

The peripheral nervous system has two functional subdivisions (see Figure 11.2). The **sensory,** or **afferent** (af'er-ent; "carrying toward"), **division** consists of nerve fibers that convey impulses *to* the central nervous system from sensory receptors located throughout the body. Sensory fibers conveying impulses from the skin, skeletal muscles, and joints are called *somatic afferent fibers* (soma = body), whereas those transmitting impulses from the visceral organs (organs within the ventral body cavity) are called *visceral afferent fibers.* The sensory division keeps the CNS constantly informed of events going on both inside and outside the body.

The **motor,** or **efferent** (ef'er-ent; "carrying away"), **division** transmits impulses *from* the CNS to effector organs, the muscles and glands. These impulses activate muscles to contract and glands to secrete; that is, they *effect* (bring about) a motor response.

The motor division also has two main parts (see Figure 11.2):

1. The **somatic nervous system** is composed of somatic motor nerve fibers that conduct impulses from the CNS to skeletal muscles. It is often referred to as the **voluntary nervous system** because it allows us to consciously control our skeletal muscles.

2. The **autonomic nervous system (ANS)** consists of visceral motor nerve fibers that regulate the activity of smooth muscles, cardiac muscles, and glands. *Autonomic* literally means "a law unto itself," and because we generally cannot control activities such as the pumping of our hearts or the movement of

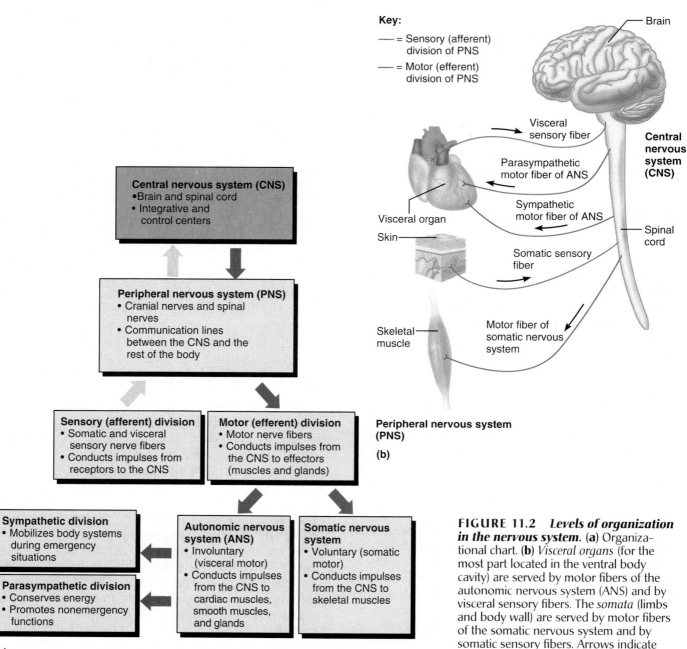

Key:

—— = Sensory (afferent) division of PNS

—— = Motor (efferent) division of PNS

Central nervous system (CNS)
- Brain and spinal cord
- Integrative and control centers

Peripheral nervous system (PNS)
- Cranial nerves and spinal nerves
- Communication lines between the CNS and the rest of the body

Sensory (afferent) division
- Somatic and visceral sensory nerve fibers
- Conducts impulses from receptors to the CNS

Motor (efferent) division
- Motor nerve fibers
- Conducts impulses from the CNS to effectors (muscles and glands)

Sympathetic division
- Mobilizes body systems during emergency situations

Parasympathetic division
- Conserves energy
- Promotes nonemergency functions

Autonomic nervous system (ANS)
- Involuntary (visceral motor)
- Conducts impulses from the CNS to cardiac muscles, smooth muscles, and glands

Somatic nervous system
- Voluntary (somatic motor)
- Conducts impulses from the CNS to skeletal muscles

(a)

Brain

Visceral sensory fiber

Parasympathetic motor fiber of ANS

Central nervous system (CNS)

Sympathetic motor fiber of ANS

Visceral organ

Skin

Somatic sensory fiber

Spinal cord

Skeletal muscle

Motor fiber of somatic nervous system

Peripheral nervous system (PNS)

(b)

FIGURE 11.2 *Levels of organization in the nervous system.* (a) Organizational chart. (b) *Visceral organs* (for the most part located in the ventral body cavity) are served by motor fibers of the autonomic nervous system (ANS) and by visceral sensory fibers. The *somata* (limbs and body wall) are served by motor fibers of the somatic nervous system and by somatic sensory fibers. Arrows indicate the direction of nerve impulses.

food through our digestive tracts, the ANS is also referred to as the **involuntary nervous system.** As indicated in Figure 11.2 and described in Chapter 14, the ANS has two functional subdivisions, the **sympathetic** and the **parasympathetic,** which typically bring about opposite effects on the activity of the same visceral organs. What one subdivision stimulates, the other inhibits.

HISTOLOGY OF NERVOUS TISSUE

The nervous system consists mostly of nervous tissue, which is highly cellular. For example, less than

20% of the CNS is extracellular space, which means that the cells there are densely packed and tightly intertwined. Although it is very complex, nervous tissue is made up of just two principal types of cells: (1) *neurons,* the excitable nerve cells that transmit electrical signals, and (2) *supporting cells,* smaller cells that surround and wrap the more delicate neurons. Together the two types of cells form the structures of both the central and peripheral nervous systems.

Supporting Cells (Neuroglia)

All neurons associate closely with nonnervous **supporting cells,** also called **neuroglia** (nu-rog'le-ah; "nerve glue") or simply **glial** (gle'al) **cells.** There are

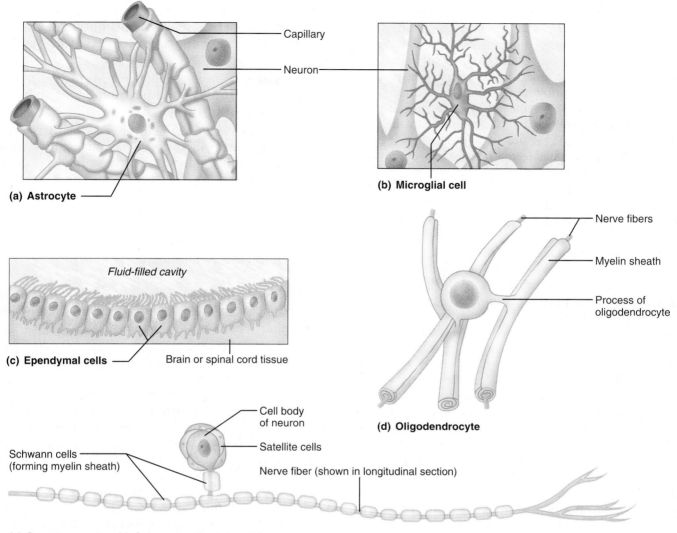

FIGURE 11.3 *Supporting cells of nervous tissue.* (a–d) Types of supporting cells found in the central nervous system. Notice in (d) that it is the processes of the oligodendrocytes that form the myelin sheaths around nerve fibers in the CNS. (e) The relationships of Schwann cells (myelinating cells) and satellite cells to a sensory neuron (nerve cell) in the peripheral nervous system.

six types of neuroglia—four in the CNS and two in the PNS (Figure 11.3). Each type has a unique function, but in general, these cells provide a supportive scaffolding for the neurons. Selected supporting cells segregate and insulate neurons so that electrical activities of adjacent neurons don't interfere with each other. Others produce chemicals that guide young neurons to the proper connections, and promote neuron health and growth.

Supporting Cells in the CNS

The supporting cells in the CNS include *astrocytes, microglia, ependymal cells,* and *oligodendrocytes.* Like neurons, most glial cells have branching processes and a central cell body (Figure 11.3a–d). Neuroglia can be distinguished, however, by their much smaller size and by their darker-staining nu-

clei. They outnumber neurons in the CNS by about 10 to 1, and make up about half the mass of the brain.

The star-shaped **astrocytes** (as'tro-sītz; "star cells") are the most abundant and the most versatile glial cells. Their numerous radiating processes cling to neurons and cover nearby capillaries, supporting and bracing the neurons and anchoring them to their nutrient supply lines, the blood capillaries (Figure 11.3a). Astrocytes have roles to play in making exchanges between capillaries and neurons, guiding the migration of young neurons, and helping to determine capillary permeability. Astrocytes also control the chemical environment around neurons, most importantly "mopping up" leaked potassium ions (K^+) and recapturing (and recycling) released neurotransmitters. Glucose provides the fuel source for producing the ATP needed to carry out their activities. As described later, the electrical charge and ion types outside nerve fibers must be just right for nerve impulses to be conducted properly. Furthermore, astrocytes (connected together by gap junctions) have been shown to signal each other (and perhaps neurons) via intracellular calcium pulses.

Microglia (mi-krog'le-ah) are small ovoid cells with relatively long "thorny" processes (Figure 11.3b). Their branches touch nearby neurons, monitoring their health, and when they sense that certain neurons are injured or in other trouble, the microglia round up and migrate toward them. Where invading microorganisms or dead neurons are present, the microglia transform into a special type of macrophage that phagocytizes the microorganisms or neuronal debris. This protective role of the microglia is important because cells of the immune system are denied access to the CNS.

Ependymal (ĕ-pen'dĭ-mul; "wrapping garment") **cells** range in shape from squamous to columnar, and many are ciliated. They line the central cavities of the brain and the spinal cord, where they form a fairly permeable barrier between the cerebrospinal fluid that fills those cavities and the tissue fluid bathing the cells of the CNS. The beating of their cilia helps to circulate the cerebrospinal fluid that cushions the brain and spinal cord (Figure 11.3c).

Though they also branch, the **oligodendrocytes** (ol"ĭ-go-den'dro-sīts) have fewer branches (*oligo* = few; *dendr* = branch) than the astrocytes. Oligodendrocytes line up along the thicker neuron fibers in the CNS and wrap their cytoplasmic extensions tightly around the fibers, producing insulating coverings called *myelin sheaths* (Figure 11.3d).

Supporting Cells in the PNS

The two kinds of supporting cells in the PNS—*satellite cells* and *Schwann cells*—differ mainly in location. **Satellite cells** surround neuron cell bodies within ganglia (Figure 11.3e), but their function is still largely unknown. Their name comes from a fancied resemblance to the moons (satellites) around a planet.

Schwann cells (also called *neurolemmocytes*) surround and form myelin sheaths around the larger nerve fibers in the peripheral nervous system (Figures 11.3e and 11.4b). Hence, they are functionally similar to oligodendrocytes. (The formation of myelin sheaths is described later in this chapter.) Schwann cells are vital to the process of peripheral nerve fiber regeneration.

Neurons

The billions of **neurons,** also called **nerve cells,** are the structural units of the nervous system. Neurons are highly specialized cells that conduct messages in the form of nerve impulses from one part of the body to another. Besides their ability to conduct nerve impulses, neurons have some other special characteristics:

1. They have *extreme longevity.* Given good nutrition, neurons can live and function optimally for a lifetime (over 100 years).

2. They are *amitotic.* As neurons assume their roles as communicating links of the nervous system, they lose their ability to divide. We pay a high price for this neuron feature because they cannot replace themselves if destroyed. There *are* exceptions to this rule. For example, olfactory neurons and some hippocampal neurons retain the ability to reproduce themselves throughout life. (The hippocampus is a brain region involved in memory.)

3. They have an exceptionally *high metabolic rate* and require continuous and abundant supplies of oxygen and glucose. Neurons cannot survive for more than a few minutes without oxygen.

Neurons are typically large, complex cells. Although they vary in structure, they all have a *cell body* from which one or more slender *processes* project (Figure 11.4). The plasma membrane of neurons is the site of electrical signaling, and it plays a crucial role in cell-to-cell interactions that occur during development. Most neurons have three functional components in common (Table 11.1, p. 395): (1) a *receptive,* or *input, region;* (2) a *conducting component;* and (3) a *secretory,* or *output, component.* As described shortly, each of these functional components is associated with a particular region of a neuron's anatomy.

Cell Body

The **cell body** consists of a transparent, spherical nucleus with a conspicuous nucleolus surrounded by cytoplasm. Also called the **perikaryon** (*peri* = around, *kary* = nucleus) or **soma,** it ranges in diameter from 5 to 140 μm. The cell body is the major *biosynthetic center* of a neuron. Except for centrioles, it contains the usual organelles. (Centrioles play an important role in forming the mitotic spindle. Their apparent absence reflects the amitotic nature of most neurons.)

A neuron's protein- and membrane-making machinery, consisting of clustered free ribosomes and rough endoplasmic reticulum (ER), is probably the most active and best developed of any cell in the body. This rough ER, referred to as **Nissl** (nis'l) **bodies** or **chromatophilic** (literally, "color loving") **substance,** stains darkly with basic dyes. The Golgi apparatus is also well developed and forms an arc or complete circle around the nucleus. Mitochondria are scattered everywhere among the other organelles. Microtubules and **neurofibrils,** bundles of intermediate filaments (*neurofilaments*), which are

Neurons with the longest axons tend to have large cell bodies, while those with short axons usually have small cell bodies. Can you think of an explanation for this relationship?

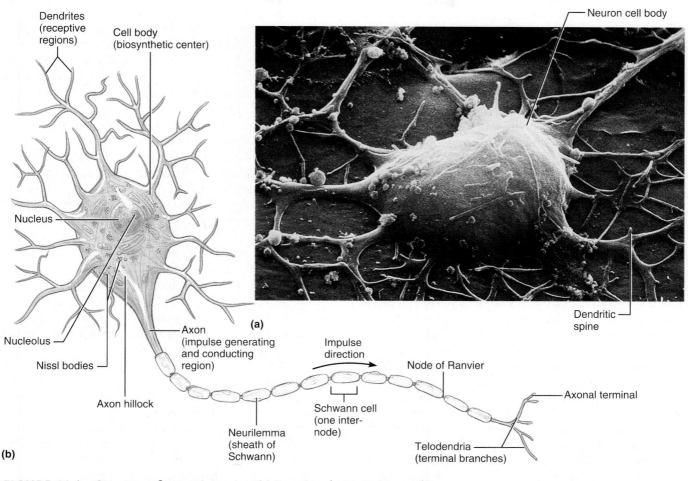

FIGURE 11.4 *Structure of a motor neuron.* (**a**) Scanning electron micrograph showing the neuron cell body and dendrites with obvious dendritic spines (5500 ×). (**b**) Diagrammatic view.

important in maintaining cell shape and integrity, are seen throughout the cell body. The cell body of some neurons also contains pigment inclusions. For example, some contain a black melanin, a red iron-containing pigment, or a golden-brown pigment called *lipofuscin* (lip″o-fu′sin). Lipofuscin, a harmless by-product of lysosomal activity, is sometimes called the "aging pigment" because it accumulates in neurons of elderly individuals.

The cell body is the focal point for the outgrowth of neuron processes during embryonic development. In most neurons, the plasma membrane of the cell body also acts as *part of the receptive surface* that receives information from other neurons.

Most neuron cell bodies are located within the CNS, where they are protected by the bones of the skull and vertebral column. Clusters of cell bodies in the CNS are called **nuclei,** whereas the far fewer collections of cell bodies that lie along the nerves in the PNS are called **ganglia** (gang′gle-ah; *ganglion* = knot on a string, swelling).

Processes

Armlike **processes** extend from the cell body of all neurons. The brain and spinal cord (CNS) contain both neuron cell bodies and their processes. The PNS, for the most part, consists chiefly of neuron processes. Bundles of neuron processes are called **tracts** in the CNS and **nerves** in the PNS.

The longer the neuron, the larger the cell body needed to service it.

The two types of neuron processes, *dendrites* and *axons* (ak'sonz), differ from each other in the structure and function of their plasma membranes. The convention is to describe these processes using a motor neuron as an example of a typical neuron. We shall follow this practice, but keep in mind that many sensory neurons and some tiny CNS neurons differ from the "typical" pattern we present here.

Dendrites **Dendrites** of motor neurons are short, tapering, diffusely branching extensions. Typically, motor neurons have hundreds of lacy dendrites clustering close to the cell body. Virtually all organelles present in the cell body also occur in dendrites. Dendrites correspond to the main **receptive** or **input regions** (see Table 11.1, p. 395). They provide an enormous surface area for receiving signals from other neurons. In many brain areas, the finer dendrites are highly specialized for information collection. They bristle with thorny appendages with bulbous ends called *dendritic spines* (see Figure 11.4a), which represent points of close contact (synapses) with other neurons. Dendrites convey incoming messages *toward* the cell body. These electrical signals are *not* nerve impulses (action potentials) but are short-distance signals called *graded potentials,* as described shortly.

The Axon Each neuron has a single **axon** (*axo* = axis, axle). The initial region of the axon arises from a cone-shaped area of the cell body called the **axon hillock** ("little hill") and then narrows to form a slender process that is uniform in diameter for the rest of its length (Figure 11.4). In some neurons, the axon is very short or absent; in others it is long and accounts for nearly the entire length of the neuron. For example, axons of motor neurons controlling the skeletal muscles of your great toe extend from the lumbar region of your spine to your foot, a distance of a meter or more (3–4 feet), making them the longest cells in the body. Any long axon is called a **nerve fiber.**

Each neuron has only one axon, but axons may give off occasional branches along their length. These rare branches, called **axon collaterals,** extend from the axon at more or less right angles. Whether an axon is undivided or has collaterals, it usually branches profusely at its end (terminus): 10,000 or more of these **terminal branches** (see Figure 11.4) per neuron is not unusual. The knoblike distal endings of the terminal branches are variously called **axonal terminals, synaptic knobs,** or **boutons** (bootonz; "buttons"). Take your pick!

Two of the three functional components of neurons introduced earlier are part and parcel of axons. Functionally, axons are the **conducting component** of the neuron (see Table 11.1). They *generate nerve impulses* and *transmit them,* typically away from the cell body. In motor neurons, the nerve impulse is generated at the junction of the axon hillock and

axon (which is therefore the *trigger zone*) and conducted along the axon to the axonal terminals. These terminals represent the **secretory component** of the neuron. When the impulse reaches the axonal terminals, it causes chemicals (*neurotransmitters*) stored in vesicles there to be released into the extracellular space. Neurotransmitters excite or inhibit neurons (or effector cells) with which the axon is in close contact. Because each neuron both receives signals from and sends signals to scores of other neurons, it carries on "conversations" with many different neurons at the same time.

An axon contains the same organelles found in the dendrites and cell body with two very important exceptions—it lacks the structures involved with protein synthesis and packaging—Nissl bodies and a Golgi apparatus. Consequently, an axon depends on its cell body to renew the necessary proteins and membrane components, and on efficient transport mechanisms to distribute them. Axons quickly decay if cut or severely damaged. Because axons are often very long, moving molecules along their length might appear to be a problem. However, through the cooperative effort of several types of cytoskeletal elements (microtubules, actin filaments, etc.), substances travel continuously both away from and toward the cell body along the axon.

Substances moved toward the axonal terminals (in the anterograde direction) include mitochondria, cytoskeletal elements, membrane components to be used for renewing the plasma membrane of the axon, or **axolemma** (ak"so-lem'ah), and enzymes needed for synthesis of certain neurotransmitter chemicals. (Some neurotransmitters are synthesized in the soma and then transported to the axonal terminals.)

Substances transported in the opposite (retrograde) direction are mostly organelles being returned to the cell body for degradation or recycling. The reverse transport process is also an important means of intracellular communication for "advising" the cell body of conditions at the axonal terminals, and for delivering vesicles containing signal molecules (like nerve growth factor, which activates certain nuclear genes promoting growth) to the soma.

There are two or three different transport mechanisms, the fastest of which is ATP dependent, bidirectional, and uses a "motor" protein (an ATPase called *kinesin*). This protein propels membranous particles along the microtubules like trains along tracks.

Homeostatic Imbalance

Certain viruses and bacterial toxins that damage neural tissues use retrograde axonal transport to reach the cell body. This transport mechanism has been demonstrated for polio, rabies, and herpes simplex viruses and for tetanus toxin. Its use as a tool to

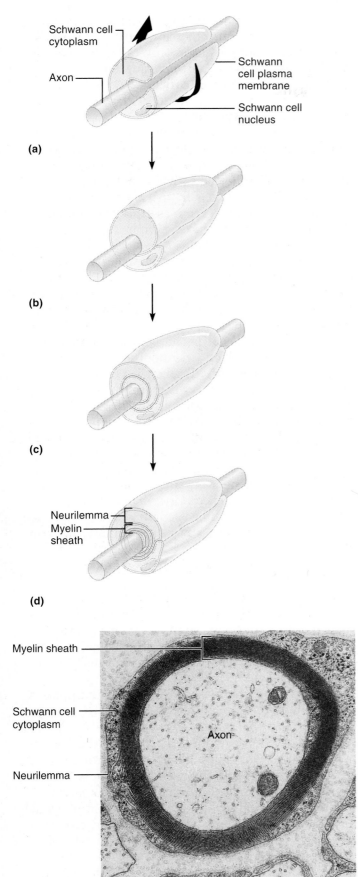

Schwann cell cytoplasm

Axon

Schwann cell plasma membrane

Schwann cell nucleus

(a)

(b)

(c)

Neurilemma

Myelin sheath

(d)

Myelin sheath

Schwann cell cytoplasm

Axon

Neurilemma

(e)

FIGURE 11.5 *Relationship of Schwann cells to axons in the peripheral nervous system.* **(a–d)** Myelination of a nerve fiber (axon). As illustrated, a Schwann cell envelops an axon in a trough. It then begins to rotate around the axon, enveloping it loosely in successive layers of its plasma membrane. Eventually, the Schwann cell cytoplasm is forced from between the membranes and comes to lie peripherally just beneath the exposed portion of the Schwann cell plasma membrane. The tight membrane wrappings surrounding the axon form the myelin sheath; the area of Schwann cell cytoplasm and its exposed membrane is referred to as the neurilemma. **(e)** Electron micrograph of a myelinated axon, cross-sectional view (20,000 ×).

treat genetic diseases by introducing viruses containing "corrected" genes is under investigation. ■

Myelin Sheath and Neurilemma Many nerve fibers, particularly those that are long or large in diameter, are covered with a whitish, fatty (protein-lipoid), segmented sheath called the **myelin** (mi′ĕ-lin) **sheath.** Myelin protects and electrically insulates fibers from one another, and it increases the speed of transmission of nerve impulses. **Myelinated fibers** (axons bearing a myelin sheath) conduct nerve impulses rapidly, whereas **unmyelinated fibers** conduct impulses quite slowly. (The order of difference may be more than 150 times—from 150 meters per second [m/s] to less than 1 m/s.) Note that myelin sheaths are associated only with axons. Dendrites are *always* unmyelinated.

Myelin sheaths in the peripheral nervous system are formed by Schwann cells. The Schwann cells indent to receive the axon and then wrap themselves around it in a jelly roll fashion (Figure 11.5). Initially the wrapping is loose, but the Schwann cell cytoplasm is gradually squeezed from between the membrane layers. When the wrapping process is complete, many concentric layers of Schwann cell plasma membrane enclose the axon, much like gauze wrapped around an injured finger. This tight coil of wrapped membranes *is* the myelin sheath, and its thickness depends on the number of spirals. Plasma membranes of myelinating cells contain much less protein than the plasma membranes of most body cells. Channel and carrier proteins are notably absent, a characteristic that makes myelin sheaths exceptionally good electrical insulators. Another unique characteristic of these membranes is the presence of specific protein molecules that interlock to form a sort of molecular Velcro (myelin glue) between the adjacent myelin membranes.

The nucleus and most of the cytoplasm of the Schwann cell end up as a bulge just external to the myelin sheath. This portion of the Schwann cell, which includes the exposed part of its plasma membrane, is called the **neurilemma** ("neuron husk"). Adjacent Schwann cells along an axon do not touch one another, so there are gaps in the sheath. These gaps, called **nodes of Ranvier** (ran'vēr) or **neurofibral nodes,** occur at regular intervals (about 1 mm apart) along the myelinated axon. It is at these nodes that axon collaterals can emerge from the axon.

Sometimes Schwann cells surround peripheral nerve fibers but the coiling process does not occur. In such instances, a single Schwann cell can partially enclose 15 or more axons, each of which occupies a separate recess in the Schwann cell surface. Nerve fibers associated with Schwann cells in this manner are said to be *unmyelinated* and are typically thin fibers.

Both myelinated and unmyelinated axons are also found in the central nervous system. However, it is oligodendrocytes that form CNS myelin sheaths (see Figure 11.3d). In contrast to Schwann cells, each of which forms only one segment (internode) of a myelin sheath, oligodendrocytes have multiple flat processes that can coil around as many as 60 different axons at the same time. Nodes of Ranvier are present, but are more widely spaced than in the PNS. CNS myelin sheaths lack a neurilemma because cell extensions are doing the coiling and the squeezed-out cytoplasm is forced not peripherally but back toward the centrally located nucleus. As in the PNS, the thinnest axons are unmyelinated. These unmyelinated axons are covered by the long extensions of adjacent glial cells.

Regions of the brain and spinal cord containing dense collections of myelinated fibers are referred to as **white matter** and are primarily fiber tracts. **Gray matter** contains mostly nerve cell bodies and unmyelinated fibers.

Classification of Neurons

Neurons are classified structurally and functionally. We will describe both classifications here, but we will use the functional classification in most discussions.

Structural Classification Neurons are grouped structurally according to the number of processes extending from their cell body. Three major neuron groups make up this classification: multipolar (*polar* = end, pole), bipolar, and unipolar neurons (Table 11.1).

Multipolar neurons have three or more processes. They are the most common neuron type in humans (over 99% of neurons belong to this class) and the major neuron type in the central nervous system. Although the "typical" multipolar neuron exhibits numerous branching dendrites and one axon, some neurons of this class lack an axon entirely and have only dendrites.

Bipolar neurons have two processes—an axon and a dendrite—that extend from opposite sides of the cell body. These rare neurons are found only in some of the special sense organs, where they act as receptor cells. Examples include some neurons in the retina of the eye and in the olfactory mucosa.

Unipolar neurons have a single short process that emerges from the cell body and divides T-like into proximal and distal branches. The more distal process of the neuron, which is often associated with a sensory receptor, is commonly called the **peripheral process,** whereas that entering the CNS is referred to as the **central process** (see Table 11.1). Unipolar neurons are more accurately called **pseudounipolar neurons** (*pseudo* = false) because they originate as bipolar neurons. Then, during early embryonic development, the two processes converge and partially fuse to form the short single process that issues from the cell body. Unipolar neurons are found chiefly in ganglia in the peripheral nervous system, where they function as sensory neurons.

The fact that the fused peripheral and central processes of unipolar neurons are continuous and function as a single fiber might make you wonder whether they are axons or dendrites. The central process is definitely an axon because it conducts impulses away from the cell body (one definition of an axon). However, the peripheral process is perplexing. Facts that favor classifying it as an axon are: (1) It generates and conducts an impulse (the functional definition of an axon); (2) when large, it is heavily myelinated; and (3) it has a uniform diameter and is indistinguishable microscopically from an axon. However, the older definition of a dendrite as a process that transmits the impulse *toward* the cell body keeps interfering with that conclusion. What is it? In this book, we have chosen to emphasize the newer impulse generating and transmitting definition of the axon. Hence, *for unipolar neurons,* we will refer to the combined length of the peripheral and central process as an axon and will use the term "dendrites" to refer only to the small receptive branches at the end of the peripheral process.

Functional Classification The functional classification scheme groups neurons according to the direction in which the nerve impulse travels relative to the central nervous system. Based on this criterion, there are sensory neurons, motor neurons, and interneurons (Table 11.1).

Sensory, or **afferent, neurons** transmit impulses from sensory receptors in the skin or internal organs *toward* or *into* the central nervous system. Except for the bipolar neurons found in some of the special sense organs, virtually all sensory neurons of the body are unipolar, and their cell bodies are located in sensory ganglia *outside* the CNS. Functionally, only the most distal parts of these unipolar neurons act as impulse receptor sites and the peripheral processes are often very long. For example, fibers carrying sensory impulses from the skin of your great toe travel for more than a meter before they reach their cell bodies in a ganglion close to the spinal cord.

Although the dendritic endings of some primary sensory neurons are naked, in which case the dendrites themselves function as sensory receptors,

TABLE 11.1	Comparison of Structural Classes of Neurons	
	Neuron Type	
Multipolar	*Bipolar*	*Unipolar (pseudounipolar)*

Structural Class: Neuron Type According to the Number of Processes Extending from the Cell Body

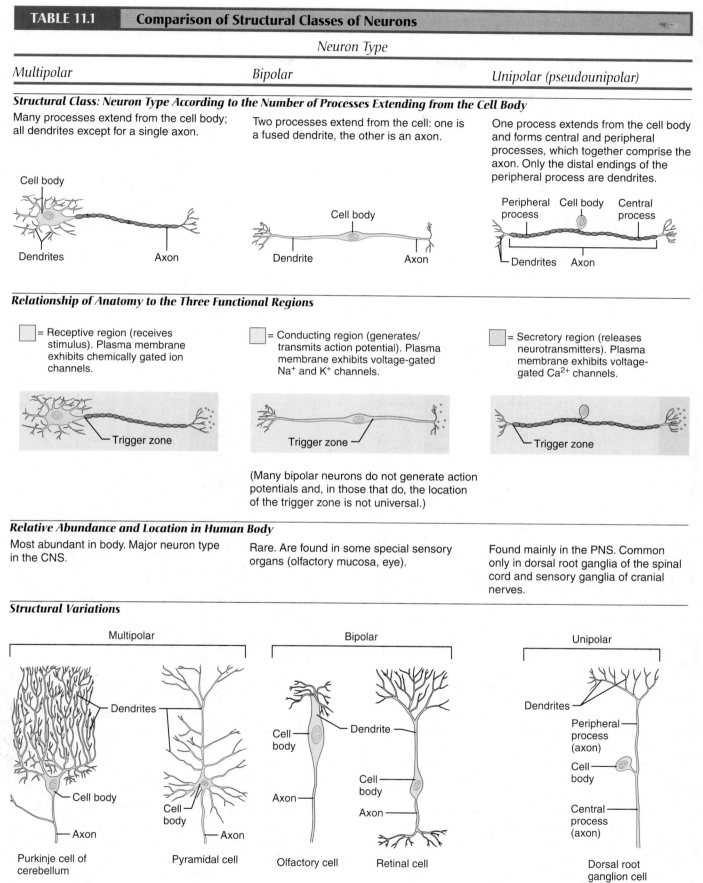

Many processes extend from the cell body; all dendrites except for a single axon.

Two processes extend from the cell: one is a fused dendrite, the other is an axon.

One process extends from the cell body and forms central and peripheral processes, which together comprise the axon. Only the distal endings of the peripheral process are dendrites.

Cell body

Dendrites — Axon

Cell body

Dendrite — Axon

Peripheral process   Cell body   Central process

Dendrites Axon

Relationship of Anatomy to the Three Functional Regions

☐ = Receptive region (receives stimulus). Plasma membrane exhibits chemically gated ion channels.

☐ = Conducting region (generates/transmits action potential). Plasma membrane exhibits voltage-gated Na⁺ and K⁺ channels.

☐ = Secretory region (releases neurotransmitters). Plasma membrane exhibits voltage-gated Ca²⁺ channels.

Trigger zone

Trigger zone

Trigger zone

(Many bipolar neurons do not generate action potentials and, in those that do, the location of the trigger zone is not universal.)

Relative Abundance and Location in Human Body

Most abundant in body. Major neuron type in the CNS.

Rare. Are found in some special sensory organs (olfactory mucosa, eye).

Found mainly in the PNS. Common only in dorsal root ganglia of the spinal cord and sensory ganglia of cranial nerves.

Structural Variations

Multipolar

Bipolar

Unipolar

Dendrites

Dendrites

Cell body

Axon

Cell body

Cell body

Axon

Dendrite

Cell body

Axon

Dendrites

Peripheral process (axon)

Cell body

Central process (axon)

Purkinje cell of cerebellum

Pyramidal cell

Olfactory cell

Retinal cell

Dorsal root ganglion cell

TABLE 11.1	**Comparison of Structural Classes of Neurons (continued)**

Neuron Type

Multipolar	*Bipolar*	*Unipolar (pseudounipolar)*

Functional Class: Neuron Type According to Direction of Impulse Conduction

1. Some multipolar neurons are **motor neurons** that conduct impulses along the efferent pathways from the CNS to an effector (muscle/gland).

2. Some multipolar neurons are higher level sensory neurons that convey sensory input from the first-order sensory neurons to higher CNS levels.

3. Most multipolar neurons are **interneurons (association neurons)** that conduct impulses within the CNS; may be one of a chain of CNS neurons, or single neuron connecting sensory and motor neurons.

Essentially all bipolar neurons are **sensory neurons** that are located in some special sense organs. For example, bipolar cells of

the retina are involved with the transmission of visual inputs from eye to the brain (via an intermediate chain of neurons).

Most unipolar neurons are **sensory neurons** that conduct impulses along afferent pathways to the CNS for interpretation. (These sensory neurons are primary or first-order sensory neurons.)

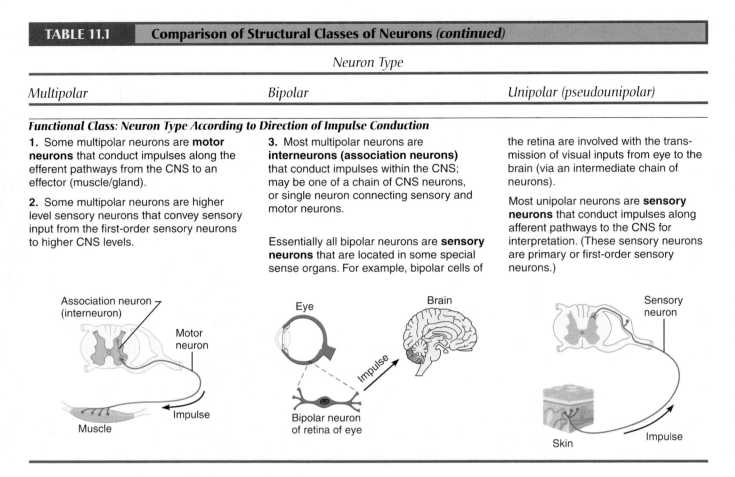

many bear receptors that include other cell types. The various types of general sensory receptor end organs, such as those of the skin, are described in Chapter 13. The special sensory receptors (of the ear, eye, etc.) are the topic of Chapter 16.

Motor, or **efferent, neurons** carry impulses *away from* the CNS to the effector organs (muscles and glands) of the body periphery. Motor neurons are multipolar, and except for some neurons of the autonomic nervous system, their cell bodies are located in the CNS.

Interneurons, or **association neurons,** lie between motor and sensory neurons in neural pathways and shuttle signals through CNS pathways where integration occurs. Most interneurons are confined entirely within the CNS. They make up over 99% of the neurons of the body, including most of those in the central nervous system. Almost all interneurons are multipolar, but there is considerable diversity both in their size and fiber-branching patterns. The Purkinje and pyramidal cells illustrated as structural variations in Table 11.1 are just two examples of their variety.

NEUROPHYSIOLOGY

Neurons are highly *irritable* (responsive to stimuli). When a neuron is adequately stimulated, an electri-

cal impulse is conducted along the length of its axon. This response, called the *action potential (nerve impulse),* is always the same, regardless of the source or type of stimulus, and it underlies virtually all functional activities of the nervous system.

In this section, we will consider how neurons become excited or inhibited and how they communicate with other cells in the body. However, we first need to explore some basic principles of electricity and to revisit the resting membrane potential.

Basic Principles of Electricity

The human body as a whole is electrically neutral; it has the same number of positive and negative charges. However, there are areas where one type of charge predominates, making such regions positively or negatively charged. Because opposite charges attract each other, energy must be used (work must be done) to separate them. On the other hand, when opposite charges come together, energy is liberated that can be used to do work. The stored energy of a battery, for example, is released when the two points are connected (the circuit is completed), allowing electrons to flow from the negatively charged area to the positively charged area. Thus, separated electrical charges of opposite sign have potential energy.

Some Definitions: Voltage, Resistance, Current

The measure of potential energy generated by separated charge is called **voltage** and is measured in either *volts* or *millivolts* (1 mV = 0.001 V). Voltage is always measured between two points and is called the **potential difference** or simply the **potential** between these two points. The greater the difference in charge between two points, the higher the voltage.

The flow of electrical charge from one point to another is called a **current,** and it can be used to do work—for example, to power a flashlight. The amount of charge that moves between the two points depends on two factors: the voltage and the resistance. **Resistance** is hindrance to charge flow provided by intervening substances through which the current must pass. Substances with high electrical resistance are called *insulators;* those with low resistance are called *conductors.*

The relationship between voltage, current, and resistance is given by **Ohm's law:**

$$\text{Current } (I) = \frac{\text{voltage } (V)}{\text{resistance } (R)}$$

As you can see, the current (I) is directly proportional to the voltage: The greater the voltage (potential difference), the greater the current. And no net current flows between points that have the same potential. Ohm's law also tells us that current is inversely related to the resistance: The greater the resistance, the smaller the current.

In the body, electrical currents reflect the flow of ions (rather than free electrons) across cellular membranes. (There are no free electrons "running around" in a living system.) As described in Chapter 3, there is a slight difference in the numbers of positive and negative ions on the two sides of cellular plasma membranes (there is a charge separation), so there is a potential across those membranes. The resistance to current flow is provided by the plasma membranes themselves.

Role of Membrane Ion Channels

Plasma membranes are peppered with a variety of *ion channels* made up of membrane proteins. Some of these membrane channels are *passive,* or *leakage, channels* that are always open. Others are *active,* or *gated, channels* (Figure 11.6). Gated channels have a molecular "gate," usually one or more protein molecules, that can change shape to open or close the channel in response to various signals. **Chemically gated,** or **transmitter-gated, channels** open when the appropriate neurotransmitter binds. **Voltage-gated channels** open and close in response to changes in the membrane potential. We will speak more of this later. Each type of channel is selective as

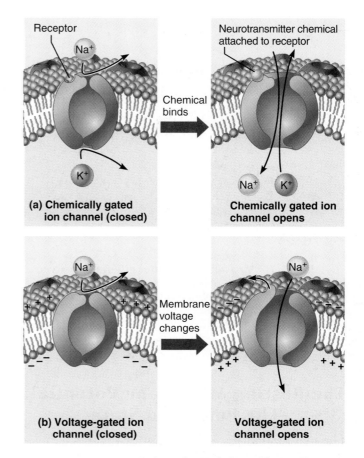

FIGURE 11.6   *Operation of gated channels.* (a) Chemically (ligand) gated channels (a Na$^+$-K$^+$ channel in this example) open when the appropriate neurotransmitter binds to the receptor, allowing simultaneous movement of Na$^+$ and K$^+$ through the channel. **(b)** Voltage-gated channels open or close in response to changes in membrane voltage. In this example, a Na$^+$ channel opens as the membrane interior changes from negative to positive.

to the type of ion (or ions) it allows to pass. For example, a potassium ion channel generally allows only potassium ions to pass.

When gated ion channels are open, ions diffuse quickly across the membrane following their electrochemical gradients, creating electrical currents and voltage changes across the membrane according to the rearranged equation

$$\text{Voltage } (V) = \text{current } (I) \times \text{resistance } (R)$$

Ions move along *chemical gradients* when they diffuse passively from an area of their higher concentration to an area of lower concentration, and along *electrical gradients* when they move toward an area of opposite electrical charge. Together, electrical and chemical gradients constitute the **electrochemical gradient.** It is these ion flows along electrochemical gradients that underlie all electrical phenomena in neurons.

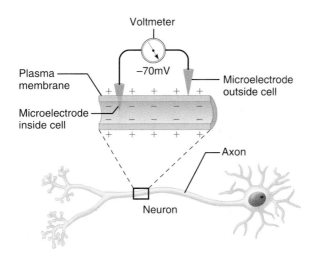

FIGURE 11.7 *Measuring potential difference between two points in neurons.* When one electrode of a voltmeter is placed on the external membrane and the other electrode is inserted just inside the membrane, a voltage (membrane potential) of approximately −70 mV (inside negative) is recorded.

The Resting Membrane Potential: The Polarized State

The potential difference between two points is measured by using two microelectrodes connected to a voltmeter (Figure 11.7). When one microelectrode is inserted into the neuron and the other rests on its outside surface, a voltage across the membrane of approximately −70 mV is recorded. The minus sign indicates that the cytoplasmic side (inside) of a neu-

ron's membrane is negatively charged with respect to the outside. This potential difference or voltage in a resting neuron (V_r) is called the **resting membrane potential,** and the membrane is said to be **polarized.** The value of the resting membrane potential varies (from −40 mV to −90 mV) in different types of neurons. (The range is even greater if all cell types are considered.)

The V_r exists only across the membrane; that is, the bulk solutions inside and outside the cell are electrically neutral. As shown in Figure 11.8, the resting membrane potential is generated by differences in the ionic makeup of the intracellular and extracellular fluids. The cell cytosol contains a lower concentration of sodium (Na^+) and a higher concentration of potassium (K^+) than the extracellular fluid surrounding it. Although there are many other solutes (glucose, urea, and other ions) in both fluids, sodium and potassium play the most important roles in generating the membrane potential. In the extracellular fluid, the positive charges of sodium and other cations are balanced chiefly by chloride ions (Cl^-). Negatively charged (anionic) proteins (A^-) help to balance the positive charges of intracellular cations (primarily K^+).

These ionic differences are the consequence of (1) the differential permeability of the plasma membrane to sodium and potassium ions and (2) the operation of the sodium-potassium pump, which actively transports Na^+ out of the cell and K^+ into the cell (see Figure 11.8). At rest the membrane is impermeable to the large anionic cytoplasmic proteins,

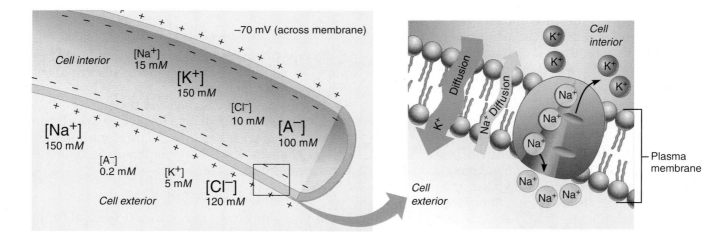

FIGURE 11.8 *The basis of the resting membrane potential.* Approximate ion concentrations of sodium [Na^+], potassium [K^+], chloride [Cl^-], and protein anions [A^-] in the intracellular and extracellular fluids of mammalian cells are indicated in millimoles per liter (m**M**). Although many other ions are present (e.g., Ca^{2+}, PO_4^{3-}, and HCO_3^-), those shown here are most

significant in terms of neural function. Steady diffusion of K^+ to the cell exterior is strongly promoted by its concentration (chemical) gradient. Na^+ is strongly attracted to the cell interior by its concentration gradient, but it is less able to cross the cell membrane. The relative membrane permeability to Na^+ and K^+ is indicated by diffusion arrows of different thickness. The resulting net outward

diffusion of positive charge leads to a state of relative negativity on the inner membrane face. This membrane potential (−70 mV, inside negative) is maintained by the sodium–potassium pump, which transports $3Na^+$ out of the cell for each $2K^+$ transported back into the cell. Notice that the electrical gradient thus maintained resists K^+ efflux but enhances the drive for Na^+ entry.

? Hyper *means greater than or more than. When we say the membrane becomes hyperpolarized, what value has increased or become greater?*

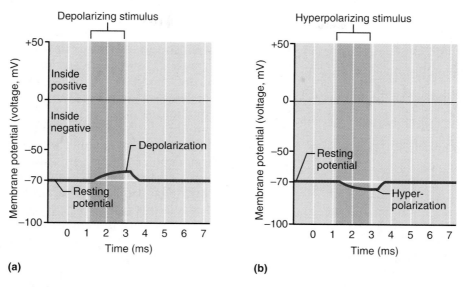

FIGURE 11.9 *Depolarization and hyperpolarization of the membrane.* The resting membrane potential is approximately −70 mV (inside negative) in neurons. Changes in this potential result in depolarization or hyperpolarization of the membrane. **(a)** In depolarization, the membrane potential moves toward 0 mV, the inside becoming less negative (more positive). **(b)** In hyperpolarization, the membrane potential increases, the inside becoming more negative.

slightly permeable to sodium, approximately 75 times more permeable to potassium than to sodium, and quite freely permeable to chloride ions. These resting permeabilities reflect the properties of the passive (leakage) ion channels present in the membrane. Potassium ions diffuse out of the cell along their *concentration gradient* much more easily and quickly than sodium ions can enter the cell along theirs. Thus a slightly greater number of positive ions diffuse out than in, leaving the cell interior with a slight excess of negative charge, which results in an *electrical gradient* that produces the resting membrane potential. Because some K^+ is always leaking out of the cell and Na^+ is always leaking in, it would appear that the concentration gradients would eventually "run down," resulting in equal concentrations of Na^+ and K^+ inside and outside the cell. This does not happen, because the ATP-driven sodium-potassium pump first ejects $3Na^+$ from the cell and then transports $2K^+$ back into the cell. Thus, the sodium-potassium pump stabilizes the resting membrane potential by serially pumping Na^+ and K^+ to maintain the diffusion gradients for sodium and potassium.

Membrane Potentials That Act as Signals

Neurons use changes in their membrane potential as communication signals for receiving, integrating,

and sending information. A change in membrane potential can be produced by (1) anything that changes membrane permeability to any type of ion or (2) anything that alters ion concentrations on the two sides of the membrane. Two types of signals are produced by a change in the membrane potential: *graded potentials*, which signal over short distances, and *action potentials*, which are long-distance signals.

The terms *depolarization* and *hyperpolarization* are used frequently in the following sections to describe membrane potential changes *relative to the resting membrane potential*, so it is important to understand these terms clearly. **Depolarization** is a reduction in membrane potential: The inside of the membrane becomes *less negative* (closer to zero) than the resting potential (Figure 11.9a). For instance, a change from a resting potential of −70 mV to a potential of −65 mV is a depolarization. By convention, depolarization also includes events in which the membrane potential reverses and moves above zero to become positive.

Hyperpolarization occurs when the membrane potential or voltage increases, becoming *more negative* than the resting potential. For example, a change from −70 mV to −75 mV (increased negativity inside) is hyperpolarization (Figure 11.9b). As

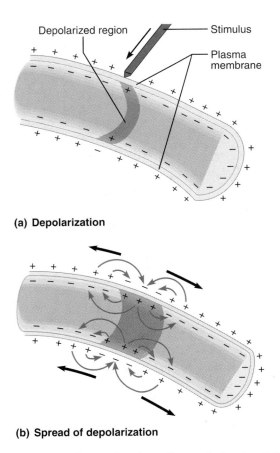

(a) Depolarization

(b) Spread of depolarization

FIGURE 11.10 *The mechanism of a graded potential.* (**a**) A small patch of the membrane has become depolarized, resulting in a change in polarity at that point. (**b**) As positive ions flow toward negative areas (and negative ions flow toward adjacent more positive areas), local currents are created that depolarize adjacent membrane areas and allow the wave of depolarization to spread.

described shortly, depolarization increases the probability of producing nerve impulses, whereas hyperpolarization reduces this probability.

Graded Potentials

Graded potentials are short-lived, local changes in membrane potential that can be either depolarizations or hyperpolarizations. These changes cause current flows that decrease with distance. Graded potentials are called "graded" because their magnitude varies directly with the strength of the stimulus. The stronger the stimulus, the more the voltage changes and the farther the current flows.

Graded potentials are triggered by some change (a stimulus) in the neuron's environment that causes gated ion channels to open. Graded potentials are given different names, depending on where they occur and the functions they perform. For example, when the receptor of a sensory neuron is excited by some form of energy (heat, light, or others),

the resulting graded potential is called a *generator potential*. Generator potentials are considered in Chapter 15. When the stimulus is a neurotransmitter released by another neuron, the graded potential is called a *postsynaptic potential*, because the neurotransmitter is released into a fluid-filled gap called a synapse and influences the neuron beyond (post) the synapse. Postsynaptic potentials are examined later in this chapter.

Fluids inside and outside cells are fairly good conductors, and current, carried by ions, flows through them whenever voltage changes occur. Let us assume that a small area of the membrane has been depolarized by a stimulus (Figure 11.10a). Current will flow on both sides of the membrane between the depolarized (active) membrane area and the adjacent polarized (resting) areas. Positive ions migrate toward more negative areas (the direction of cation movement indicates the direction of current flow), and negative ions simultaneously move in the opposite direction, toward more positive areas (Figure 11.10b). Thus, inside the cell, positive ions (mostly K^+) move away from the active area and accumulate on the neighboring membrane areas, where they displace negative ions. Meanwhile, positive ions on the outer membrane face are moving toward the active region of reversed membrane polarity (the depolarized region), which is momentarily less positive. As they move, their "places" on the membrane become occupied by negative ions (such as Cl^- and HCO^{3-}), sort of like ionic musical chairs. Thus, at the neighboring region, the inside becomes less negative and the outside becomes less positive; that is, the neighboring membrane is depolarized.

To simplify the explanation, virtually all diagrams that illustrate local currents give the impression that the circuit is completed by ions passing into and out of the cell through the membrane. This is *not* the case. Only the inward current across the membrane is caused by the flow of ions through gated channels. The outward current, the so-called *capacitance current*, reflects changes in the charge distribution as the ions migrate *along* the two membrane faces as described above. The capacitance current reflects the fact that the fatty membrane interior is a poor conductor of current. That is, it is a *capacitor* that *temporarily* stores the charge, forcing the ions of opposite charge to accumulate opposite each other on either side of the membrane.

As just explained, the flow of current to adjacent membrane areas changes the membrane potential there as well. However, the plasma membrane is permeable like a leaky water hose, and most of the charge is quickly lost through the membrane. Consequently, the current flow is *decremental*; that is, it

? *Since it is obvious from this figure that current flow quickly falls off with distance, which synaptic contacts—those on distal parts of dendrites, those on proximal parts of dendrites, or those on the cell body—would have the greatest potential for producing an action potential?*

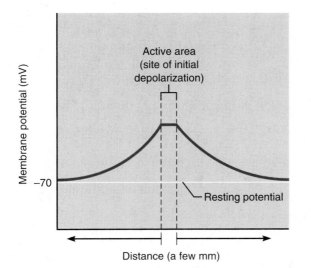

FIGURE 11.11 *Changes in membrane potential produced by a depolarizing graded potential.* Such voltage changes are decremental because the current is quickly dissipated by loss of ions through the "leaky" plasma membrane. Consequently, graded potentials are short-distance signals.

dies out within a few millimeters of its origin (Figure 11.11). Because the current dissipates quickly and dies out with increasing distance, graded potentials can act as signals only over very short distances. Nonetheless, they are essential in initiating action potentials, the long-distance signals.

Action Potentials

The principal way neurons communicate is by generating and propagating **action potentials (APs)**, and for the most part, only cells with *excitable membranes*—neurons and muscle cells—can generate action potentials. As illustrated in Figure 11.12a, an action potential is a brief reversal of membrane potential with a total amplitude (change in voltage) of about 100 mV (from −70 mV to +30 mV). The whole event is over in a few milliseconds. Unlike graded potentials, action potentials do not decrease in strength with distance.

The events of action potential generation and transmission are identical in skeletal muscle cells

and neurons. However, in a neuron, a transmitted action potential is also called a **nerve impulse.** A neuron transmits a nerve impulse only when it is adequately stimulated. The stimulus changes the permeability of the neuron's membrane by opening specific voltage-gated channels that are located on axons. *Only axons are capable of generating action potentials.* These channels open and close in response to changes in the membrane potential and are activated by local currents (graded potentials) that spread toward the axon along the dendritic and cell body membranes. In many neurons, the transition from the local graded potential to the action potential takes place at the axon hillock. In sensory neurons, the action potential is generated by the peripheral (axonal) process just proximal to the receptor region. However, for simplicity, we will just use the term *axon* in our discussion.

Generation of an Action Potential Generating an action potential involves three consecutive but overlapping changes in membrane permeability due to opening and closing of active ion gates, all induced by depolarization of the axonal membrane (Figure 11.12). Sequentially, these permeability changes are first a transient increase in Na$^+$ permeability, followed by restoration of Na$^+$ impermeability, and then a short-lived increase in K$^+$ permeability. The first two permeability changes occur during the *depolarization phase* of action potential generation, which is indicated by the upward-rising part of the AP curve or spike. The third permeability change is responsible for both the *repolarization* (the downward part of the AP spike) and *undershoot phases* shown in Figure 11.12a. Let's examine each of these phases more carefully—we will start with a neuron in the resting (polarized) state.

① **Resting state: Voltage-gated channels closed.** In the resting state, virtually all the voltage-gated Na$^+$ and K$^+$ channels are closed. However, small amounts of K$^+$ are passing outward via leakage channels and even smaller amounts of Na$^+$ are diffusing in.

Each Na$^+$ channel actually has two voltage-sensitive gates: an *activation gate* that is closed when resting and responds to depolarization by opening rapidly, and an *inactivation gate* that is open at rest and responds to depolarization by closing slowly. Thus, *depolarization opens and then closes the sodium channels.* Both gates must be open for the channel to allow Na$^+$ entry, but the closing of *either* gate effectively closes the channel. By contrast, each active potassium channel has a single voltage-sensitive gate; it is closed when resting and opens slowly in response to depolarization.

② **Depolarizing phase: Increase in sodium permeability and reversal of the membrane potential.** As the axonal membrane is depolarized by local currents, the sodium channel activation gates open quickly and

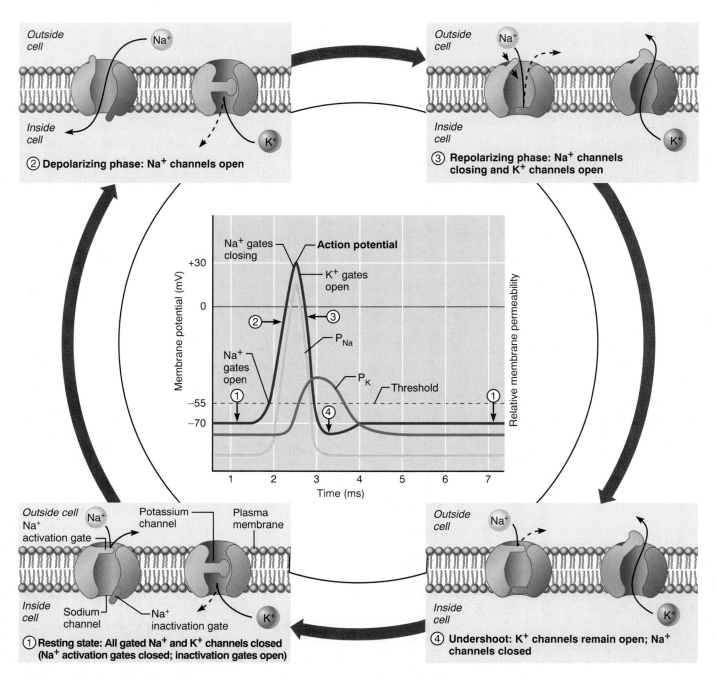

FIGURE 11.12 *The phases of the action potential and the role of gated ion channels in those phases.* The action potential scan, which is shown in the center of the diagram, can be divided into four phases, during which the voltage-sensitive gates controlling the sodium and potassium channels are in different states and membrane permeability to sodium (P_{Na}) and potassium (P_K) is changing. These phases are illustrated clockwise from ① to ④. ① In the resting state, neither channel is open. (Sodium inactivation gates are open, but activation gates are closed.) ② During the depolarizing phase of the action potential, sodium gates are open, but potassium channels remain closed. ③ During the repolarizing phase of the action potential, sodium inactivation gates close the sodium channels, and potassium channels open. ④ During the undershoot (hyperpolarization), both gates of the sodium channels have closed, but potassium channels remain open temporarily because the relatively slow gates of those channels have not had time to respond to the repolarization of the membrane. Within another millisecond or two, the resting state ① is restored, and the system is ready to respond to the next stimulus.

sodium rushes into the cell at the point of depolarization. This influx of positive charge depolarizes that local "patch" of membrane still further, opening more activation gates so that the cell interior becomes progressively less negative. When depolarization of the membrane at the stimulation site reaches a certain critical level called **threshold** (often between -55 and -50 mV), depolarization becomes self-generating, urged on by the positive feedback. That is, after being initiated by the stimulus, depolarization is driven by the ionic currents created by sodium ion influx. As more sodium enters, the membrane depolarizes further and opens still more activation gates until all sodium channels are open. At this point, Na^+ permeability is about 1000 times greater than it is in a resting neuron. As a result, the membrane potential becomes less and less negative and then overshoots to about $+30$ mV as sodium ions rush inward along their electrochemical gradient. This rapid depolarization and polarity reversal produce the sharply upward, or rising, *spike* of the action potential (see Figure 11.12a).

Earlier, we stated that membrane potential depends on membrane permeability, but here we are saying that membrane permeability depends on membrane potential. Can both statements be true? As a matter of fact they are, because these two distinct relationships interlock to establish a *positive feedback* cycle (increased Na^+ permeability due to increased channel openings leads to greater depolarization, which in turn leads to increased Na^+ permeability, and so on). It is this explosive positive feedback cycle that is responsible for the rising (depolarizing) phase of action potentials and is the property that puts the "action" in the action potential.

③ **Repolarizing phase: Decrease in sodium permeability.** The explosively rising phase of the action potential (period of sodium permeability) persists for only about 1 ms and is self-limiting. As the membrane potential passes 0 mV and becomes increasingly positive, the positive intracellular charge resists further sodium entry. In addition, the slow inactivation gates of the sodium channels have now had time to respond and begin to close after a few milliseconds of depolarization. As a result, the membrane permeability to sodium declines to resting levels, and the net influx of sodium finally stops completely (see Figure 11.12). As a result, the AP spike stops rising and reverses direction.

Repolarizing phase: Increase in potassium permeability. As sodium entry declines, the slow voltage-sensitive potassium gates open and potassium rushes out of the cell, following its electrochemical gradient (see Figure 11.12b). Consequently, internal negativity of the resting neuron is restored, an event called **repolarization** (see Figure 11.12a). Both the abrupt decline in sodium permeability and the increased permeability to potassium contribute to the repolarization process.

④ **Undershoot: Potassium permeability continues.** Because the potassium gates are sluggish gates that are slow to respond to the depolarization signal, the period of increased K^+ permeability typically lasts longer than needed to restore the resting polarized state. As a result of the excessive K^+ efflux, an *after-hyperpolarization* called the **undershoot** is seen on the recording as a slight dip following the action potential spike. Notice that both the activation and inactivation gates of the sodium channels are still closed during the undershoot; hence the neuron is insensitive to a stimulus and depolarization at this time.

Although repolarization restores resting electrical conditions, it does *not* restore the original ionic distributions of the resting state. The redistribution is accomplished by "revving up" the **sodium-potassium pump** following repolarization. While it might appear that tremendous numbers of sodium and potassium ions change places during action potential generation, this is not the case. Only small amounts of sodium and potassium cross the membrane. (The Na^+ influx required to reach threshold produces only a 0.012% change in cellular Na^+ concentration.) Because an axonal membrane has thousands of sodium-potassium pumps, these small ionic changes are quickly corrected.

Propagation of an Action Potential If an action potential is to serve as the neuron's signaling device, it must be **propagated** (sent or transmitted) along the axon's entire length (Figure 11.13). As we have seen, the action potential is generated by the influx of sodium ions, and the local patch of axonal membrane undergoes polarity reversal: The inside becomes positive, and the outside becomes negative. The positive ions in the axoplasm move laterally from the area of polarity reversal toward the membrane area that is still negative (polarized), and those in the extracellular fluid migrate back toward the area of greatest negative charge (the area of polarity reversal) to complete the circuit.

This establishes local current flows that depolarize adjacent membrane areas in the forward direction (away from the origin of the nerve impulse), which opens voltage-gated channels and triggers an action potential there. Because the area in the opposite direction has just generated an action potential, the sodium gates are closed—no new action potential is generated there. Thus, the impulse propagates away from its point of origin. (If an *isolated* axon is stimulated by an electrode, or an axon is stimulated at a node of Ranvier, the nerve impulse will move away from the point of stimulus in *all* directions along the membrane.) In the body, action potentials are initiated at one end of the axon and conducted away from that point toward the axon's terminals. Once initiated, an action potential is a *self-propagating* process that continues along the axon at a constant velocity—something like a domino effect.

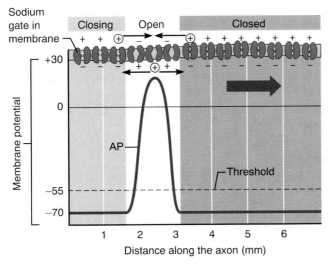

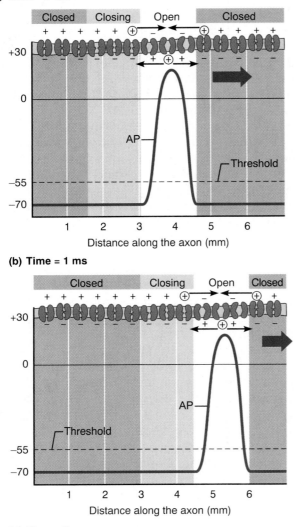

(a) Time = 0 ms

(b) Time = 1 ms

(c) Time = 2 ms

FIGURE 11.13 *Propagation of an action potential* (**AP**). The propagating action potential along the axon is shown at time 0 ms, 1 ms, and 2 ms. The sodium gates are labeled closed, open, or closing. Small circular arrows indicate local currents generated by the movement of positive ions. The large arrows indicate the direction of action potential propagation. Not depicted, the current flows created by the opening of potassium channels (and subsequent repolarization) occur where sodium gates are shown as closing.

Following depolarization, each segment of axonal membrane repolarizes, which restores the resting membrane potential in that region. Since these electrical changes also set up local current flows, the repolarization wave chases the depolarization wave down the length of the axon. The propagation process just described occurs on unmyelinated axons. The unique propagation process that occurs in myelinated axons, called *saltatory conduction*, is described shortly.

Although the phrase *conduction of a nerve impulse* is commonly used, nerve impulses are not really conducted in the same way that an insulated wire conducts current. In fact, neurons are fairly poor conductors of electrical current, and local current flows decline with distance because the charges leak through the membrane. The expression *propagation of a nerve impulse* is more accurate, because the action potential is *regenerated anew* at each membrane patch, and every subsequent action potential is identical to that generated initially.

Threshold and the All-or-None Phenomenon Not all local depolarization events produce action potentials. The depolarization must reach threshold values if an axon is to "fire." What determines the *threshold point*? One explanation is that threshold is the membrane potential at which the outward current carried by K^+ is exactly equal to the inward current created by Na^+. Threshold is typically reached when the membrane has been depolarized by 15 to 20 mV from the resting value. It seems to represent an unstable equilibrium state at which one of two things can happen. If an extra sodium ion enters, further depolarization occurs, opening more sodium channels and allowing more sodium ions entry. If, on the other hand, another potassium ion leaves, the membrane potential is driven away from threshold, sodium channels close, and potassium ions continue to diffuse outward until the potential returns to its resting value.

Recall that the local depolarizations are graded potentials and that their magnitude increases with increasing stimulus intensity. Brief weak stimuli (*subthreshold stimuli*) produce subthreshold depolarizations that are not translated into nerve impulses. On the other hand, stronger *threshold stimuli* produce depolarizing currents that push the membrane potential toward and beyond the threshold voltage. As a result, sodium permeability is increased to such an extent that sodium entry "swamps" (exceeds) the outward movement of potassium ions, allowing the positive feedback cycle to become established and generating an action potential. The critical factor here is the total amount of current that flows through the membrane during a stimulus (electrical charge × time). Strong stimuli

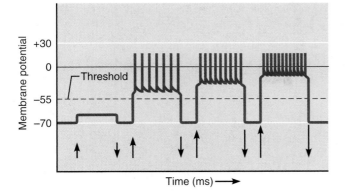

FIGURE 11.14 *Relationship between stimulus strength, the local potential produced, and action potential frequency.* In the tracing, action potentials are shown as vertical lines rather than the typical spike tracings. Upward arrows (↑) indicate points of stimulus application; downward arrows (↓) indicate stimulus cessation; length of arrows indicate strength of stimulus. Notice that a sub-threshold stimulus does not generate an action potential at all, but once threshold voltage is reached, the stronger the stimulus, the more frequently action potentials are generated.

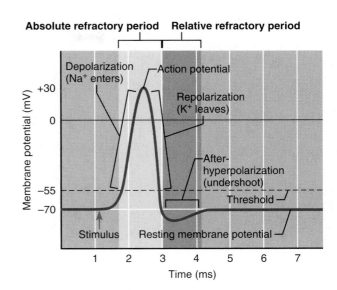

What causes the undershoot?

FIGURE 11.15 *Recording of an action potential indicating the timing of the absolute and relative refractory periods.*

depolarize the membrane to threshold quickly. Weaker stimuli must be applied for longer periods to provide the crucial amount of current flow. Very weak stimuli fail to trigger an action potential because the local current flows they produce are so slight that they dissipate long before threshold is reached.

The action potential is an **all-or-none phenomenon;** it either happens completely or it doesn't happen at all. It might help to compare the generation of the action potential to lighting a match under a small dry twig. The changes occurring where the twig is being heated can be compared to the change in membrane permeability that initially allows more sodium ions to enter the cell. When that part of the twig becomes hot enough (when enough sodium ions have entered the cell), the critical flash point (threshold) is reached and the flame will consume the entire twig, even if you then blow out the match (the action potential will be generated and propagated whether or not the stimulus continues). But if the match is extinguished just before the twig has reached the critical temperature, ignition will not take place. Likewise, if too few sodium ions enter the cell to achieve threshold, no action potential will occur.

Coding for Stimulus Intensity

Once generated, all action potentials are independent of stimulus strength, and all action potentials are alike. So how can the CNS determine whether a particular stimulus is intense or weak—information it needs to initiate an appropriate response? The answer is really quite simple: Strong stimuli cause nerve impulses to be generated more *often* in a given time interval than

do weak stimuli (Figure 11.14). Thus, stimulus intensity is coded for by the number of impulses generated per second—that is, by the *frequency of impulse transmission*—rather than by increases in the strength (amplitude) of the individual action potentials.

Absolute and Relative Refractory Periods

When a patch of neuron membrane is generating an action potential and its sodium channels are open, the neuron is incapable of responding to another stimulus, no matter how strong. This period from the opening of the activation gates of the Na^+ channels to the closing of the inactivation gates, called the **absolute refractory period** (see Figure 11.15), ensures that each action potential is a separate, *all-or-none event* and enforces one-way transmission of the action potential.

The **relative refractory period** is the interval following the absolute refractory period when the sodium gates are closed and most have returned to their resting state, the potassium gates are open, and repolarization is occurring (Figure 11.15). During this time, the axon's threshold for impulse generation is substantially elevated. A threshold stimulus

---

Undershoot occurs because potassium channel gates are slow to respond to the membrane depolarization. Hence more K^+ leaves the cell than is absolutely necessary to restore the neuron to its V_r.

How does the location of voltage-gated Na$^+$ channels differ in a myelinated axon (as depicted here) and in an unmyelinated axon?

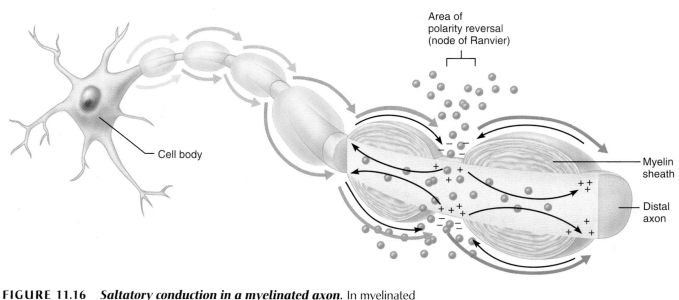

Area of
polarity reversal
(node of Ranvier)

Cell body

Myelin
sheath

Distal
axon

FIGURE 11.16 *Saltatory conduction in a myelinated axon.* In myelinated fibers, the local current flows (small black arrows in longitudinally sectioned axon and immediately outside the myelin sheath) give rise to a propagated action potential (large pink to red arrows) that appears to jump from node to node. Notice that while action potentials are generated only at the nodes, the current flows along the axon from node to node.

won't trigger an action potential during the relative refractory period, but an exceptionally strong stimulus can reopen the sodium gates and allow another impulse to be generated. Thus, by intruding into the relative refractory period, strong stimuli cause more frequent generation of action potentials.

Conduction Velocities of Axons Conduction velocities of neurons vary widely. Fibers that transmit impulses most rapidly (100 m/s or more) are found in neural pathways where speed is essential, such as those that mediate some postural reflexes. More slowly conducting axons typically serve internal organs (the gut, glands, blood vessels), where slower responses are not usually a handicap. The rate of impulse propagation depends largely on two factors: axon diameter and degree of myelination.

1. Influence of axon diameter. The diameter of different axons varies considerably and, as a general rule, the larger the axon's diameter, the faster it conducts impulses. This is because larger axons offer less resistance to the flow of local current, so adja-

cent areas of the membrane can more quickly be brought to threshold.

2. Influence of a myelin sheath. On unmyelinated axons, action potentials are generated at sites immediately adjacent to each other and conduction is relatively slow, a type of AP propagation called **continuous conduction.** The presence of a myelin sheath dramatically increases the rate of impulse propagation because myelin acts as an insulator to prevent almost all leakage of charge from the axon. Current can pass through the membrane of a myelinated axon *only* at the nodes of Ranvier, where the myelin sheath is interrupted and the axon is bare, and essentially all the voltage-regulated sodium channels are concentrated at the nodes. Thus, when an action potential is generated in a myelinated fiber, the local depolarizing current does not dissipate through the adjacent (nonexcitable) membrane regions but instead is maintained and moves to the next node, a distance of approximately 1 mm, where it triggers another action potential. Consequently, action potentials are triggered only at the nodes, a type of conduction called **saltatory conduction** (*saltare* = to leap) because the electrical signal jumps from node to node along the axon (Figure 11.16). Saltatory conduction is much faster than the conduction produced by the continuous spread of local depolarizing currents along unmyelinated membranes.

In a myelinated axon, voltage-gated sodium channels are located only at the nodes of Ranvier, as opposed to being located along the entire length of an unmyelinated axon.

◭ *Homeostatic Imbalance*

The importance of myelin to nerve transmission is painfully clear to those with demyelinating diseases such as **multiple sclerosis (MS).** This autoimmune disease affects mostly young adults. Common symptoms are visual disturbances (including blindness), problems controlling muscles (weakness, clumsiness, and ultimately paralysis), and urinary incontinence. In this disease, nerve fibers are severed, and myelin sheaths in the CNS are gradually destroyed, reduced to nonfunctional hardened lesions called *scleroses.* The loss of myelin (due to the immune system's attack on myelin proteins) results in such substantial shunting and short-circuiting of the current that successive nodes are excited more and more slowly, and eventually impulse conduction ceases. However, the axons themselves are not damaged and growing numbers of sodium channels appear spontaneously in the demyelinated fibers. This may account for the remarkably variable cycles of relapse (disability) and remission (temporary recovery) that are typical of this disease.

The anti-inflammatory steroid methylprednisolone, and more recently, injections of beta interferon (a hormonelike substance released by cells of the immune system), reduce the frequency of attacks but do not eliminate symptoms during attacks. An experimental procedure in which MS patients had their blood removed, mixed with new plasma, and then returned to their bodies exhibited moderate to marked improvement including some regain in limb function. ■

Nerve fibers may be classified as group A, B, or C fibers based on their diameter, degree of myelination, and speed of conduction. **Group A fibers** are mostly somatic sensory and motor fibers serving the skin, skeletal muscles, and joints. They have the largest diameter and thick myelin sheaths, and conduct impulses at speeds ranging up to 150 m/s (over 300 mph). Autonomic nervous system motor fibers serving the visceral organs; visceral sensory fibers; and the smaller somatic sensory fibers transmitting afferent impulses from the skin (such as pain and small touch fibers) belong in the B and C fiber groups. **Group B fibers,** lightly myelinated fibers of intermediate diameter, transmit impulses at an average rate of 15 m/s (about 40 mph). **Group C fibers** have the smallest diameter and are unmyelinated. Hence, these fibers are incapable of saltatory conduction and conduct impulses at a leisurely pace—1 m/s (2 mph) or less.

◭ *Homeostatic Imbalance*

A number of chemical and physical factors impair the conduction of impulses. Although their mechanisms of action differ, alcohol, sedatives, and injected anesthetics all block nerve impulses by reducing membrane permeability to ions, mainly sodium. As we have seen, no sodium entry—no action potential.

Cold and continuous pressure interrupt blood circulation (and hence the delivery of oxygen and nutrients) to the neuronal processes, impairing their ability to conduct impulses. For example, your fingers get numb when you hold an ice cube for more than a few seconds. Likewise, when you sit on your foot, it "goes to sleep." When you remove the pressure, the impulses begin to be transmitted once again, leading to an unpleasant prickly feeling. ■

The Synapse

The operation of the nervous system depends on the flow of information through chains of neurons functionally connected by synapses. A **synapse** (sin′aps), from the Greek *syn,* meaning "to clasp or join," is a unique junction that mediates information transfer from one neuron to the next or from a neuron to an effector cell—its where the action is.

Most synapses occur between the axonal endings of one neuron and the dendrites or the cell bodies (soma) of other neurons and are called **axodendritic** or **axosomatic synapses,** respectively. Less common (and far less understood) are the synapses between axons (*axoaxonic*), between dendrites (*dendrodendritic*), or between dendrites and cell bodies (*dendrosomatic*). Some of these synaptic patterns are illustrated in Figure 11.17.

The neuron conducting impulses toward the synapse is called the **presynaptic neuron.** The neuron that transmits the electrical signal away from the synapse is called the **postsynaptic neuron.** At a *given* synapse, the presynaptic neuron is the information sender, and the postsynaptic neuron is the information receiver. As you might anticipate, most neurons function both as presynaptic and postsynaptic neurons. Neurons have anywhere from 1000 to 10,000 axonal terminals making synapses and are stimulated by an equal number of other neurons. In the body periphery, the postsynaptic cell may be either another neuron or an effector cell (a muscle cell or gland cell).

There are two varieties of synapses: *electrical* and *chemical.* These are described next.

Electrical Synapses

Electrical synapses, the less common variety, correspond to the gap junctions found between certain other body cells (see Figure 3.4, p. 71). They contain protein channels that interconnect the cytoplasm of adjacent neurons, and allow the current-carrying ions to flow directly from one neuron to the next.

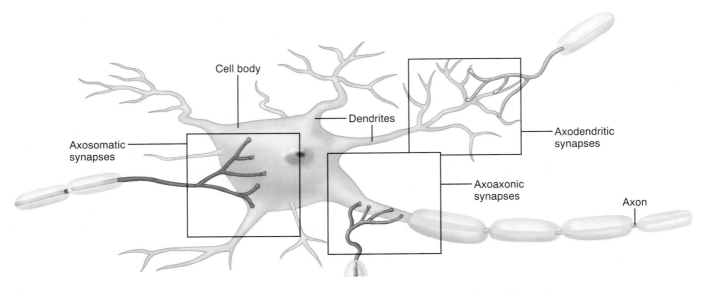

(a)

FIGURE 11.17 *Examples of types of chemical synapses.* **(a)** Axodendritic, axosomatic, and axoaxonic synapses. Axonal terminals of the presynaptic neurons are indicated in different colors at the different synapse types. **(b)** Scanning electron micrograph of incoming fibers to axosomatic synapses (16,500×)

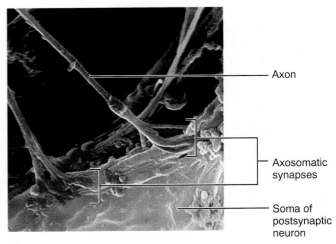

(b)

Neurons joined in this way are said to be *electrically coupled,* and transmission across these synapses is very rapid. Depending on the nature of the synapse, communication may be unidirectional or bidirectional.

A key feature of electrical synapses is that they provide a simple means of synchronizing the activity of all interconnected neurons. In adults, electrical synapses are found in regions of the brain responsible for certain stereotyped movements, such as the normal jerky movements of the eyes. They are far more abundant in embryonic nervous tissue, where they permit exchange of guiding clues during early neuronal development so that neurons can connect properly with one another. As nervous system development continues, most electrical synapses are replaced by chemical synapses.

Chemical Synapses

In contrast to electrical synapses, which are specialized to allow the flow of ions between neurons, **chemical synapses** are specialized for release and reception of chemical **neurotransmitters.** Neurotransmitters (described in detail shortly) act to open or close ion channels that influence membrane permeability and, consequently, membrane potential.

A typical chemical synapse is made up of two parts: (1) a knoblike *axonal terminal* of the presynaptic (transmitting) neuron, which contains many tiny, membrane-bounded sacs called **synaptic vesicles,** each containing thousands of neurotransmitter molecules; and (2) a *receptor region* on the membrane of a dendrite or the cell body of the postsynaptic neuron, which bears neurotransmitter receptors, often of several types (Figure 11.18).

Although close, the presynaptic and postsynaptic membranes are always separated by the **synaptic cleft,** a fluid-filled space approximately 30 to 50 nm (about one-millionth of an inch) wide. Because the current from the presynaptic membrane dissipates in the fluid-filled cleft, chemical synapses effectively prevent a nerve impulse from being *directly* transmitted from one neuron to another. Instead, transmission of signals across these synapses is a *chemical event* that depends on the release, diffusion, and receptor binding of neurotransmitter molecules and results in *unidirectional communication* between neurons. Thus, while transmission of nerve impulses along an axon and across electrical synapses is a purely electrical event, chemical synapses convert the electrical signals into chemical signals (neurotransmitters) that travel across the synapse to the postsynaptic cells, where they are converted back into electrical signals.

> **?** *Why may such axonal terminals be referred to as "biological transducers"?*

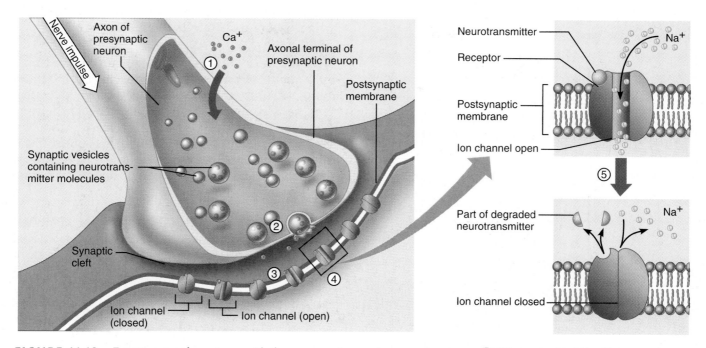

FIGURE 11.18 *Events occurring at a chemical synapse in response to depolarization of the axonal terminal.* ① Arrival of the depolarization wave (nerve impulse) results in the opening of calcium gates and calcium (Ca²⁺) influx into the axonal terminal. ② Calcium ions promote the fusion of synaptic vesicles with the presynaptic membrane and exocytosis of neurotransmitter. ③ The neurotransmitter diffuses across the synaptic cleft and attaches to receptors on the postsynaptic membrane. ④ Binding of neurotransmitter opens the ion channels, which results in voltage changes in the postsynaptic membrane.

⑤ Effects are short-lived because the neurotransmitter is quickly destroyed by enzymes or taken back up into the presynaptic terminal, which closes the ion channels and terminates the synaptic response.

Information Transfer Across Chemical Synapses

When a nerve impulse reaches the axon terminal, it sets into motion a chain of events that triggers neurotransmitter release. The neurotransmitter crosses the synaptic cleft and, on binding to receptors on the postsynaptic membrane, causes changes in the postsynaptic membrane permeability. The successive events are as follows (Figure 11.18):

① **Calcium gates open in the presynaptic axonal terminal.** When the nerve impulse reaches the axon terminal, depolarization of the membrane opens not only Na⁺ channels but voltage-regulated Ca²⁺ channels as well. During the brief time the calcium gates are open, Ca²⁺ floods into the terminal from the extracellular fluid.

② **Neurotransmitter is released by exocytosis.** The surge of free calcium into the axonal terminal acts as an intracellular messenger, directing synaptic vesicles to fuse with the axonal membrane and empty their contents by exocytosis into the synaptic cleft. The Ca²⁺ is then quickly removed, either taken up into the mitochondria or ejected to the outside by an active calcium membrane pump.

③ **Neurotransmitter binds to postsynaptic receptors.** The neurotransmitter diffuses across the synaptic cleft and binds reversibly to specific protein receptors that are densely clustered on the postsynaptic membrane.

④ **Ion channels open in the postsynaptic membrane.** As the receptor proteins bind neurotransmitter molecules, their three-dimensional shape changes. This causes ion channels to open, and the resulting current flows produce local changes in the membrane potential. Depending on the receptor protein to which the neurotransmitter binds and the type of channel the receptor controls, the postsynaptic neuron may be excited or inhibited.

For each nerve impulse reaching the presynaptic terminal, many vesicles (perhaps 300) are emptied into the synaptic cleft. The higher the frequency of impulses reaching the terminals (that is, the more

Because they change the signal from an electrical current to a chemical signal (neurotransmitter), which in turn initiates an electrical current in the postsynaptic neuron.

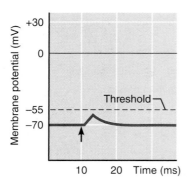

(a) Excitatory postsynaptic potential (EPSP)

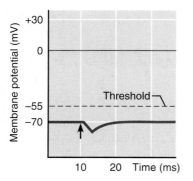

(b) Inhibitory postsynaptic potential (IPSP)

FIGURE 11.19 *Postsynaptic potentials.* **(a)** An excitatory postsynaptic potential (EPSP) is a local depolarization of the postsynaptic membrane that brings the neuron closer to threshold for AP generation. It is mediated by neurotransmitter binding that opens channels allowing the simultaneous passage of sodium and potassium through the postsynaptic membrane. **(b)** An inhibitory postsynaptic potential results in hyperpolarization of the postsyaptic neuron and drives the neuron away from the threshold for firing. It is mediated by neurotransmitter binding that opens potassium or chloride gates or both. The black vertical arrows represent stimulation.

intense the stimulus), the greater the total number of synaptic vesicles that fuse and spill their contents, and the greater the effect on the postsynaptic cell.

Termination of Neurotransmitter Effects
As long as a neurotransmitter is bound to a postsynaptic receptor, it continues to produce its effects on membrane permeability and blocks reception of additional "messages" from presynaptic neurons. Thus, some means of "wiping the postsynaptic slate clean" is necessary. The effects of neurotransmitters last a few milliseconds before being terminated by one of three mechanisms. Depending on the particular neurotransmitter, the mechanism may be

1. Degradation by enzymes associated with the postsynaptic membrane or present in the synapse (see Figure 11.18, step 5). This is the case for acetylcholine.

2. Removal from the synapse by reuptake by astrocytes or the presynaptic terminal, where it is stored or destroyed by enzymes, as with norepinephrine.

3. Diffusion of the neurotransmitter away from the synapse, ending its effects there.

Synaptic Delay Although some neurons can transmit impulses at rates that approach 100 m/s (250 mph), neural transmission across a chemical synapse is comparatively slow and reflects the time required for neurotransmitter release, diffusion across the synapse, and binding to receptors. Typically, this **synaptic delay**, which lasts 0.3–5.0 ms, is the *rate-limiting* (slowest) step of neural transmission. Synaptic delay helps explain why transmission along short neural pathways involving only two or three neurons occurs rapidly, while that along the multisynaptic pathways typical of higher mental functioning occurs much more slowly. However, in practical terms these differences are not noticeable.

Postsynaptic Potentials and Synaptic Integration

Many receptors present on postsynaptic membranes at chemical synapses are specialized to open ion channels, thereby converting chemical signals into electrical signals. However, unlike the voltage-gated ion channels responsible for action potentials, these chemically gated channels are relatively insensitive to changes in membrane potential. Consequently, channel opening at postsynaptic membranes cannot possibly become a self-amplifying or self-generating event. Instead, neurotransmitter receptors mediate local changes in membrane potential (voltage) that are *graded* according to the amount of transmitter released and the time it remains in the area.

Chemical synapses are divided into two types, excitatory and inhibitory, according to how they affect the membrane potential of the postsynaptic neuron (Figure 11.19). *Excitatory postsynaptic potentials (EPSPs)* occur at excitatory synapses and *inhibitory postsynaptic potentials (IPSPs)* are generated at inhibitory synapses. Action potentials are compared with postsynaptic potentials in Table 11.2.

Excitatory Synapses and EPSPs
At excitatory synapses, neurotransmitter binding causes depolarization of the postsynaptic membrane. However, in contrast to what happens on axonal membranes, only a single type of channel opens on postsynaptic membranes (membranes of dendrites and neuronal cell bodies). This channel allows

TABLE 11.2	**Comparison of Action Potentials with Postsynaptic Potentials**		
Characteristic	*Action potential*	*EPSP*	*IPSP*
Function	Long-distance signaling; constitutes the nerve impulse	Short-distance signaling; depolarization that spreads to axon hillock; moves membrane potential *toward* threshold for generation of action potential	Short-distance signaling; hyperpolarization that spreads to axon hillock; moves potential *away from* threshold for generation of action potential
Stimulus for opening of ionic gates	Voltage (depolarization)	Chemical (neurotransmitter)	Chemical (neurotransmitter)
Initial effect of stimulus	First opens sodium gates, then potassium gates	Opens channels that allow simultaneous sodium and potassium fluxes	Opens potassium or chloride channels or both
Repolarization	Voltage regulated; closing of sodium gates followed by opening of potassium gates	Dissipation of membrane charges with time and distance	
Conduction distance	Not conducted by local current flows; continually regenerated (propagated) along entire axon; intensity does not decline with distance	1–2 mm; local electrical events; intensity declines with distance	
Positive feedback cycle	Present	Absent	Absent
Peak membrane potential	+40 to +50 mV	0 mV	Becomes hyperpolarized; moves toward −90 mV
Summation	None; an all-or-none phenomenon	Present; produces graded depolarization	Present; produces graded hyperpolarization
Refractory period	Present	Absent	Absent

sodium and potassium ions to diffuse *simultaneously* through the membrane in opposite directions (rather than first opening sodium gates and later potassium gates). Although this may appear to be self-defeating when depolarization is the goal, remember that the electrochemical gradient for sodium is much steeper than that for potassium. Hence, the influx of Na^+ is greater than the efflux of K^+, and *net* depolarization occurs.

If enough neurotransmitter binds, depolarization of the postsynaptic membrane can successfully reach 0 mV, which is well above an axon's threshold (about −50 mV) for "firing off" an action potential. However, *postsynaptic membranes do not generate action potentials; only axons* (fibers with voltage-gated channels) *have this capability*. The dramatic polarity reversal seen in axons never occurs in membranes containing *only* chemically gated channels because the opposite movements of K^+ and Na^+ prevent accumulation of excessive positive charge inside the cell. Hence, instead of action potentials, local graded depolarization events called **excitatory postsynaptic potentials (EPSPs)** occur at excitatory postsynaptic membranes (see Figure 11.19a). Each EPSP lasts a few milliseconds and then the membrane returns to its resting potential. The only function of EPSPs is to help trigger an action potential

distally at the axon hillock of the postsynaptic neuron. Although current flows created by individual EPSPs decline with distance, they can and often *do* spread all the way to the axon hillock. If currents reaching the axon hillock are strong enough to depolarize the axon to threshold, axonal voltage-gated channels open and an action potential is generated.

Inhibitory Synapses and IPSPs

Binding of neurotransmitters at inhibitory synapses *reduces* a postsynaptic neuron's ability to generate an action potential. Most inhibitory neurotransmitters induce hyperpolarization of the postsynaptic membrane by making the membrane more permeable to potassium ions, chloride ions, or both. Sodium ion permeability is not affected. If potassium gates are opened, potassium ions move out of the cell; if chloride gates are opened, chloride ions move in. In either case, the charge on the inner face of the membrane becomes relatively more negative. As the membrane potential increases and is driven farther from the axon's threshold, the postsynaptic neuron becomes less and less likely to "fire" and larger depolarizing currents are required to induce an action potential. Such changes in potential are called **inhibitory postsynaptic potentials (IPSPs)** (see Figure 11.19b).

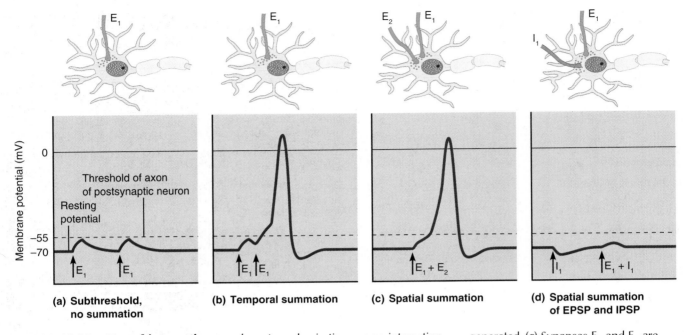

(a) Subthreshold, no summation

(b) Temporal summation

(c) Spatial summation

(d) Spatial summation of EPSP and IPSP

FIGURE 11.20 *Neural integration of EPSPs and IPSPs at the axonal membrane of the postsynaptic cell.* Synapses E_1 and E_2 are excitatory; synapse I_1 is inhibitory. Each individual impulse is itself subthreshold. **(a)** Synapse E_1 is stimulated and then is stimulated again shortly thereafter. The two EPSPs do not overlap in time, so no interaction (summation) occurs; threshold is not reached in the axon of the postsynaptic neuron. **(b)** Synapse E_1 is stimulated a second time before the initial EPSP has died away; temporal summation occurs, and the axon's threshold is reached, causing an action potential to be generated. **(c)** Synapses E_1 and E_2 are stimulated simultaneously (spatial summation), resulting in a threshold depolarization. **(d)** Synapse I_1 is stimulated, resulting in a short-lived IPSP (hyperpolarization). When E_1 and I_1 are simultaneously stimulated, the changes in potential cancel each other out.

Integration and Modification of Synaptic Events

Summation by the Postsynaptic Neuron

A single EPSP cannot induce an action potential in the postsynaptic neuron. But if thousands of excitatory axonal terminals are firing on the same postsynaptic membrane, or if a smaller number of terminals are delivering impulses very rapidly, the probability of reaching threshold depolarization increases greatly. Thus, EPSPs can add together, or **summate,** to influence the activity of a postsynaptic neuron. Nerve impulses would never be initiated if this were not so.

Two types of summation occur: temporal and spatial, and typically they occur together. However, for ease of discussion we will consider the two types of summation separately. First we will explain these events in terms of EPSPs. **Temporal summation** occurs when one or more presynaptic neurons transmit impulses in rapid-fire order (Figure 11.20b) and waves of neurotransmitter release occur in quick succession. The first impulse produces a slight EPSP, and before it dissipates, successive impulses trigger more EPSPs. These summate, producing a much greater depolarization of the postsynaptic membrane than would result from a single EPSP.

Spatial summation occurs when the postsynaptic neuron is stimulated by a large number of terminals from the same or, more commonly, different neurons at the same time. As a result, huge numbers of its receptors can bind neurotransmitter and simultaneously initiate EPSPs, which sum up and dramatically enhance the depolarization (Figure 11.20c).

Although we have focused on summation of EPSPs here, IPSPs also summate, both temporally and spatially. In this case, the postsynaptic neuron is inhibited to a greater degree.

Most neurons receive both stimulatory and inhibitory inputs from thousands of other neurons. Additionally, the same fiber may form different types of synapses (in terms of biochemical and electrical characteristics) with different types of target neurons. How is all of this conflicting information sorted out? Each neuron's axon hillock appears to act like a "bean counter" by keeping a running account of all the signals it receives (Figure 11.20d). Not only do EPSPs summate and IPSPs summate, but also EPSPs summate with IPSPs. If the stimulatory effects of EPSPs dominate the membrane potential enough to reach threshold, the neuron will fire. If, on the other hand, the summing process yields only subthreshold depolarization or hyperpolarization, the neuron fails to generate an action potential. However, partially depolarized neurons are **facilitated**—that is, more easily excited by successive depolarization events—because they are already nearer to threshold. Thus, axon hillock membranes function as *neural integrators,* and their potential at any time reflects the sum of all incoming neural information.

Since EPSPs and IPSPs are graded potentials that diminish in strength the farther they spread, the most effective synapses are those closest to the axon hillock. Synapses on distal dendrites have far less influence on firing the axon than those on the cell body.

Synaptic Potentiation Repeated or continuous use of a synapse (even for short periods) considerably enhances the presynaptic neuron's ability to excite the postsynaptic neuron, producing postsynaptic potentials much larger than might be predicted by the stimulus. This phenomenon is called **synaptic potentiation.** The presynaptic terminals at such synapses contain higher concentrations of calcium ions, which (presumably) trigger the release of more neurotransmitter, which in turn produces larger EPSPs. Furthermore, synaptic potentiation increases calcium influx via dendritic spines in the postsynaptic neuron as well. A brief high-frequency pattern of stimulation specifically activates voltage-regulated receptors called *NMDA (N-methyl-D-aspartate) receptors,* located on the postsynaptic membrane, which couple the depolarization to increased calcium entry. Theoretically, as calcium floods into the cell, it activates certain kinase enzymes that promote changes that result in more efficient responses to subsequent stimuli.

Functionally, synaptic potentiation, also called *post-tetanic potentiation,* can be viewed as a learning process that increases the efficiency of neurotransmission along a particular pathway. Indeed, the hippocampus of the brain, which plays a special role in memory and learning (see pp. 461, 552–554) exhibits especially long post-tetanic potentiations.

Another facet of synaptic potentiation that appears to occur under certain conditions and at certain brain synapses involves the dendrites as major players. Dendrites of pyramidal neurons of the cerebral cortex and hippocampal neurons have fast sodium gates and high-threshold gates for calcium currents that act in synaptic integration and influence the strength and plasticity of their synapses. Apparently as a neuron shoots off an AP, a back-propagating AP travels from the cell body to the dendrites. This current flow alters the efficiency of the synapses, promoting synaptic potentiation.

Presynaptic Inhibition and Neuromodulation Postsynaptic activity can also be influenced by events occurring at the presynaptic membrane, such as presynaptic inhibition and neuromodulation.

Presynaptic inhibition occurs when the release of excitatory neurotransmitter by one neuron is inhibited by the activity of another neuron via an axoaxonic synapse. More than one mechanism is involved, but the end result is that less neurotransmitter is released and bound, and smaller EPSPs are formed. (Notice that this is opposite to what we see with synaptic potentiation.) In contrast to postsynaptic inhibition by IPSPs, which decreases the excitability of the postsynaptic neuron, presynaptic inhibition is more like a functional synaptic "pruning." It reduces excitatory stimulation of the postsynaptic neuron by the presynaptic neuron.

Neuromodulation occurs when a neurotransmitter acts via slow changes in target cell metabolism (see p. 418), or chemicals other than neurotransmitters modify neuronal activity. Some *neuromodulators* influence the synthesis, release, degradation, or reuptake of neurotransmitter by a presynaptic neuron. Others alter the sensitivity of the postsynaptic membrane to the neurotransmitter. Many neuromodulators are hormones that act at sites relatively far from their release site.

Neurotransmitters and Their Receptors

Neurotransmitters, along with electrical signals, are the "languages" of the nervous system—the means by which each neuron communicates with others to process and send messages to the rest of the body. These versatile molecules that chemically connect neurons affect body and brain. Sleep, thought, rage, hunger, memory, movement, and even your smile reflect their "doings." Most factors that specifically affect synaptic transmission do so by enhancing or inhibiting neurotransmitter release or destruction, or by blocking their binding to receptors. Just as speech defects may hinder interpersonal communication, interferences with neurotransmitter activity may short-circuit the brain's "conversations" or internal talk (see *A Closer Look* on pp. 422–423).

At present, over 50 different neurotransmitters or neurotransmitter candidates have been identified. Although some neurons produce and release only one kind of neurotransmitter, most make two or more and may release any one or all of them. It appears that in most cases, different neurotransmitters are released at different stimulation frequencies, which avoids producing a jumble of nonsense messages. However co-release of two neurotransmitters from the same vesicles has been documented. The coexistence of more than one neurotransmitter in a single neuron makes it possible for that cell to exert several influences rather than a single discrete effect.

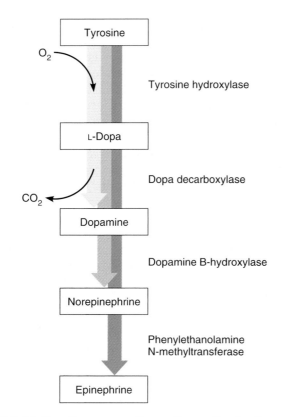

FIGURE 11.21 *Common pathway for synthesis of dopamine, norepinephrine, and epinephrine.* How far synthesis proceeds along the pathway depends on the enzymes present in the cell. The production of dopamine and norepinephrine takes place in the axonal terminals of neurons releasing these neurotransmitters. Epinephrine, a hormone, is released (along with norepinephrine) by the cells of the adrenal medulla.

Neurotransmitters are classified in two ways: chemically and functionally. Table 11.3 (p. 416) provides a fairly detailed overview of neurotransmitters, some of which are described here. No one expects you to memorize this table at this point, but it will be a handy reference for you to look back to when the neurotransmitters are mentioned in subsequent chapters.

Classification of Neurotransmitters According to Chemical Structure

Neurotransmitters fall into several chemical classes based on molecular structure.

Acetylcholine (ACh) Acetylcholine (as″ĕ-til-ko′lēn) was the first neurotransmitter to be identified. It is still the best understood because it is released at neuromuscular junctions, which are much easier to study than synapses buried within the CNS. ACh is synthesized and enclosed in synaptic vesicles within the axonal terminals in a reaction catalyzed by the enzyme *choline acetyltransferase.* Acetic acid is bound to coenzyme A (CoA) to form acetyl-CoA,

which in turn is combined with choline. Then the coenzyme is released.

$$\text{Acetyl-CoA} + \text{choline} \xrightarrow{\substack{\text{choline}\\\text{acetyltransferase}}} \text{ACh} + \text{CoA}$$

After ACh is released by the presynaptic terminal, it binds to the postsynaptic receptors briefly. Then it is released and degraded to acetic acid and choline by the enzyme **acetylcholinesterase (AChE)**, located in the synaptic cleft and on postsynaptic membranes. The released choline is recaptured by the presynaptic terminals and reused to synthesize more ACh.

Acetylcholine is released by all neurons that stimulate skeletal muscles and by some neurons of the autonomic nervous system. ACh-releasing neurons are also prevalent in the central nervous system.

Biogenic Amines The **biogenic** (bi″o-jen′ik) **amines** include the **catecholamines** (kat″ĕ-kol′ah-mēnz), such as dopamine, norepinephrine, and epinephrine, and the **indolamines**, which include serotonin and histamine. As illustrated in Figure 11.21, **dopamine** and **norepinephrine** (nor″ep-ĭ-nef′rin) **(NE)** are synthesized from the amino acid tyrosine in a common pathway consisting of several steps. Apparently, neurons contain only the enzymes needed to produce their own neurotransmitter. Thus, the sequence stops at dopamine in dopamine-releasing neurons but continues on to NE in NE-releasing neurons. The same pathway is used by the epinephrine-releasing cells of the brain and the adrenal medulla. Since epinephrine produced by the adrenal gland is released to the blood, it is a hormone as well as a neurotransmitter. **Serotonin** (se″ro-to′nin) is synthesized from the amino acid tryptophan. **Histamine** is synthesized from the amino acid histidine.

Biogenic amine neurotransmitters are broadly distributed in the brain, where they play a role in emotional behavior and help regulate the biological clock. Additionally, catecholamines (particularly NE) are released by some motor neurons of the autonomic nervous system. Imbalances of these neurotransmitters are associated with mental illness; for example, overproduction of dopamine occurs in schizophrenia. Additionally, certain psychoactive drugs (LSD and mescaline) can bind to biogenic amine receptors and induce hallucinations.

Amino Acids It has been more difficult to prove a neurotransmitter role when the suspect is an **amino acid.** ACh and biogenic amines are found only in neurons, but amino acids occur in all cells of the body and are important in many biochemical reactions. The amino acids for which a neurotransmitter role is certain include **gamma (γ)-aminobutyric acid (GABA), glycine, aspartate,** and **glutamate,** but there are others. Amino acid neurotransmitters have so far been found only in the CNS.

Peptides The **neuropeptides,** essentially strings of amino acids, include a broad spectrum of molecules with diverse effects. For example, a neuropeptide called **substance P** is an important mediator of pain signals. By contrast, **endorphins,** which include **beta endorphin, dynorphin,** and **enkephalins** (en-kef′ah-linz), act as natural opiates, reducing our perception of pain under certain stressful conditions. Enkephalin activity increases dramatically in pregnant women in labor. Endorphin release is enhanced when an athlete gets a so-called second wind and is probably responsible for the "runner's high." Additionally, some claim that the placebo effect is due to endorphin release. These pain-killing neurotransmitters remained undiscovered until investigators began to study why morphine and other opiates reduce anxiety and pain. As it turned out, these drugs attach to the same receptors that bind natural opiates, producing similar but stronger effects.

Some neuropeptides, such as somatostatin, vasoactive intestinal peptide (VIP), and cholecystokinin, are also produced by nonneural body tissues and are widespread in the gastrointestinal tract. Such peptides are commonly referred to as **gut-brain peptides.**

Novel Messengers Just a few years ago, it would have been scientific suicide to suggest that ATP, nitric oxide, and carbon monoxide—all ubiquitous molecules—might be neurotransmitters. Nonetheless, the discovery of these unlikely messengers has opened up a whole new chapter in the story of neurotransmission.

Although it was well known that **adenosine triphosphate (ATP),** the universal form of cellular energy, was stored in synaptic vesicles, its presence there was thought to promote synthesis or reuptake of other neurotransmitters. ATP is now recognized as a major neurotransmitter (perhaps the most primitive one) in both the CNS and PNS. Like glutamate and acetylcholine it produces a fast excitatory response at certain receptors (channel-linked receptors, discussed later).

Even more preposterous was the idea that **nitric oxide (NO),** a short-lived toxic gas, could be a neurotransmitter. It defies all the official descriptions of neurotransmitters—it is synthesized on demand and diffuses out of the cells making it, rather than being stored in vesicles and released by exocytosis. Instead of attaching to surface receptors, it zooms through the plasma membrane of nearby cells to bind with a peculiar intracellular receptor—iron in *guanylyl cyclase,* the enzyme that makes the second messenger *cyclic GMP.* The precise role of NO in the brain has been elusive, but when the neurotransmitter glutamate binds to NMDA receptors (a specific subset of its receptors), calcium channels open to al-

low an influx of Ca^{2+} into the cell. The calcium triggers reactions that activate *nitric oxide synthase (NOS),* the enzyme needed for NO synthesis. Many believe that NO is the retrograde messenger involved in long-term potentiation (learning and memory)—that is, it may travel back from the postsynaptic neuron to the presynaptic neuron where it activates guanylyl cyclase. Excessive release of NO is responsible for much of the brain damage seen in stroke patients (see pp. 460–461). In the myenteric plexus of the intestine, NO causes relaxation of intestinal smooth muscle.

It may be that NO is just the first of a soon-to-be-discovered class of signaling gases that pass swiftly into cells where they bind briefly to metal-containing enzymes and then vanish. **Carbon monoxide (CO),** another airy messenger, also stimulates synthesis of cyclic GMP, and some researchers speculate that CO is the main regulator of cyclic GMP levels in the brain. Though NO and CO are found in different brain regions and appear to act in different pathways, their mode of action is very similar. The proposed role of CO in neurotransmission leads to an interesting speculation: Heavy smokers who quit often bemoan the fact that they can't concentrate for weeks after. Since CO displaces oxygen in the blood, it would seem that they should think better after quitting. Perhaps the higher blood levels of CO they were accustomed to enhanced neurotransmission in certain circuits involved in logic. Time will tell.

Classification of Neurotransmitters According to Function

This text cannot begin to describe the incredible diversity of functions that neurotransmitters mediate. Therefore, we limit our discussion to two broad ways of classifying neurotransmitters according to function; more details will be added as appropriate in subsequent chapters.

Effects: Excitatory versus Inhibitory We can summarize this classification scheme by saying that some neurotransmitters are excitatory (cause depolarization), some are inhibitory (cause hyperpolarization), and others exert both effects, depending on the specific receptor types with which they interact. For example, the amino acids GABA and glycine are usually inhibitory, while glutamate is typically excitatory. On the other hand, ACh and NE each bind to at least two receptor types that cause opposite effects. For example, acetylcholine is excitatory at neuromuscular junctions with skeletal muscle and inhibitory when released on cardiac muscle. These effects of ACh and NE are described more fully in Chapter 14.

TABLE 11.3	Neurotransmitters		
Neurotransmitter	*Functional classes*	*Sites where secreted*	*Comments*

Acetylcholine

| | Excitatory to skeletal muscles; excitatory or inhibitory to visceral effectors, depending on receptors bound

Direct action at nicotinic receptors; indirect action via second messengers at muscarinic receptors | CNS: basal nuclei and some neurons of motor cortex of brain

PNS: all neuromuscular junctions with skeletal muscle; some autonomic motor endings (all preganglionic and parasympathetic postganglionic fibers) | Effects prolonged (leading to tetanic muscle spasms and neural "frying") by nerve gas and organophosphate insecticides (malathion); release inhibited by botulinus toxin and barbiturates; binding to receptors inhibited by curare (a muscle paralytic agent) and some snake venoms; decreased ACh levels in certain brain areas in Alzheimer's disease; ACh receptors destroyed in myasthenia gravis; binding of nicotine to nicotinic ACh receptors in the brain enhances excitatory neurotransmitter (glutamate, ACh) release by enhancing presynaptic Ca^{2+} levels; may account for behavioral effects of nicotine in smokers; atropine competes with ACh for binding sites. |

$$H_3C-\overset{\overset{\textstyle O}{\|}}{C}-O-CH_2-CH_2-\overset{+}{N}-(CH_3)_3$$

Biogenic Amines

| Norepinephrine | Excitatory or inhibitory, depending on receptor type bound

Indirect action via second messengers | CNS: brain stem, particularly in the locus coeruleus of the midbrain; limbic system; some areas of cerebral cortex

PNS: main neurotransmitter of ganglion cells in the sympathetic nervous system | A "feeling good" neurotransmitter; release enhanced by amphetamines; removal from synapse blocked by tricyclic antidepressants (Elavil and others) and cocaine; brain levels reduced by reserpine (an antihypertensive drug) leading to depression |

$$HO-\text{⬡}-\underset{\underset{\textstyle OH}{|}}{CH}-CH_2-NH_2$$

| Dopamine | Excitatory or inhibitory depending on the receptor type bound

Indirect action via second messengers | CNS: substantia nigra of midbrain; hypothalamus; is the principal neurotransmitter of extrapyramidal system

PNS: some sympathetic ganglia | A "feeling good" neurotransmitter; release enhanced by L-dopa and amphetamines; reuptake blocked by cocaine; deficient in Parkinson's disease; may be involved in pathogenesis of schizophrenia |

$$HO-\text{⬡}-CH_2-CH_2-NH_2$$

| Serotonin (5-HT) | Mainly inhibitory

Indirect action via second messengers; direct action at 5-HT$_3$ receptors | CNS: brain stem, especially midbrain; hypothalamus; limbic system; cerebellum; pineal gland; spinal cord | Activity blocked by LSD; may play a role in sleep, appetite, nausea, migraine headaches, and regulation of mood; drugs that block its uptake (Prozac) relieve anxiety and depression |

$$HO-\text{⬡}-\underset{\underset{\textstyle N}{\underset{\textstyle |}{}}}{\overset{\overset{\textstyle C}{\|}}{C}}-CH_2-CH_2-NH_2$$

| Histamine | Indirect action via second messengers | CNS: hypothalamus | Also released by mast cells during inflammation and acts as powerful vasodilator |

$$HC=C-CH_2-CH_2-NH_2$$

TABLE 11.3	Neurotransmitters *continued*

Amino Acids

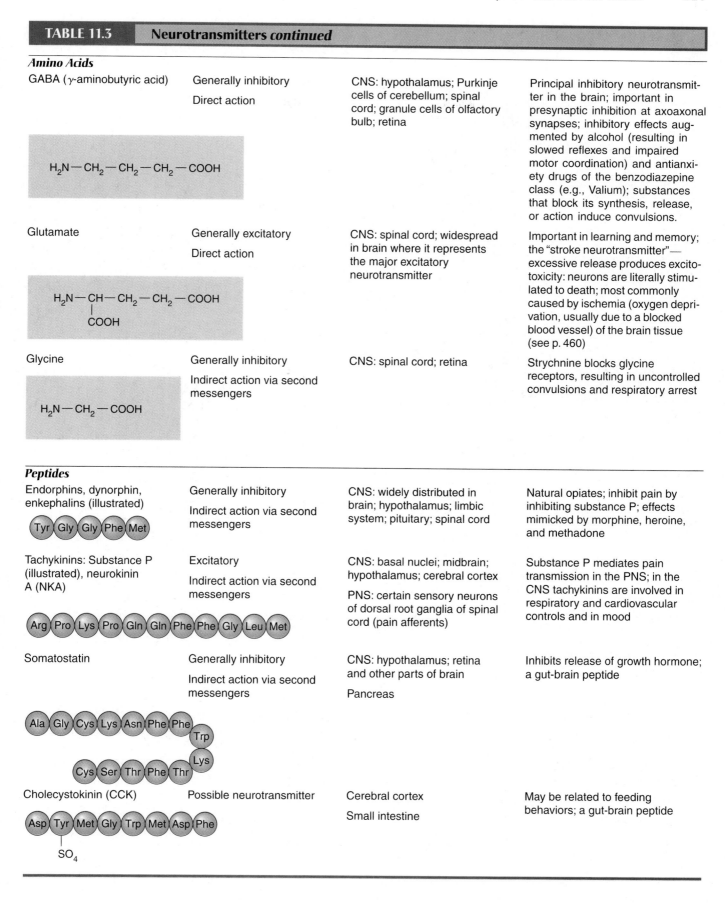

GABA (γ-aminobutyric acid)	Generally inhibitory Direct action	CNS: hypothalamus; Purkinje cells of cerebellum; spinal cord; granule cells of olfactory bulb; retina	Principal inhibitory neurotransmitter in the brain; important in presynaptic inhibition at axoaxonal synapses; inhibitory effects augmented by alcohol (resulting in slowed reflexes and impaired motor coordination) and antianxiety drugs of the benzodiazepine class (e.g., Valium); substances that block its synthesis, release, or action induce convulsions.
$H_2N—CH_2—CH_2—CH_2—COOH$			
Glutamate	Generally excitatory Direct action	CNS: spinal cord; widespread in brain where it represents the major excitatory neurotransmitter	Important in learning and memory; the "stroke neurotransmitter"—excessive release produces excitotoxicity: neurons are literally stimulated to death; most commonly caused by ischemia (oxygen deprivation, usually due to a blocked blood vessel) of the brain tissue (see p. 460)
$H_2N—CH—CH_2—CH_2—COOH$ | $COOH$			
Glycine	Generally inhibitory Indirect action via second messengers	CNS: spinal cord; retina	Strychnine blocks glycine receptors, resulting in uncontrolled convulsions and respiratory arrest
$H_2N—CH_2—COOH$			

Peptides

Endorphins, dynorphin, enkephalins (illustrated) Tyr–Gly–Gly–Phe–Met	Generally inhibitory Indirect action via second messengers	CNS: widely distributed in brain; hypothalamus; limbic system; pituitary; spinal cord	Natural opiates; inhibit pain by inhibiting substance P; effects mimicked by morphine, heroine, and methadone
Tachykinins: Substance P (illustrated), neurokinin A (NKA) Arg–Pro–Lys–Pro–Gln–Gln–Phe–Phe–Gly–Leu–Met	Excitatory Indirect action via second messengers	CNS: basal nuclei; midbrain; hypothalamus; cerebral cortex PNS: certain sensory neurons of dorsal root ganglia of spinal cord (pain afferents)	Substance P mediates pain transmission in the PNS; in the CNS tachykinins are involved in respiratory and cardiovascular controls and in mood
Somatostatin Ala–Gly–Cys–Lys–Asn–Phe–Phe–Trp–Lys–Thr–Phe–Thr–Ser–Cys	Generally inhibitory Indirect action via second messengers	CNS: hypothalamus; retina and other parts of brain Pancreas	Inhibits release of growth hormone; a gut-brain peptide
Cholecystokinin (CCK) Asp–Tyr–Met–Gly–Trp–Met–Asp–Phe | SO_4	Possible neurotransmitter	Cerebral cortex Small intestine	May be related to feeding behaviors; a gut-brain peptide

Why is cyclic AMP called a second messenger?

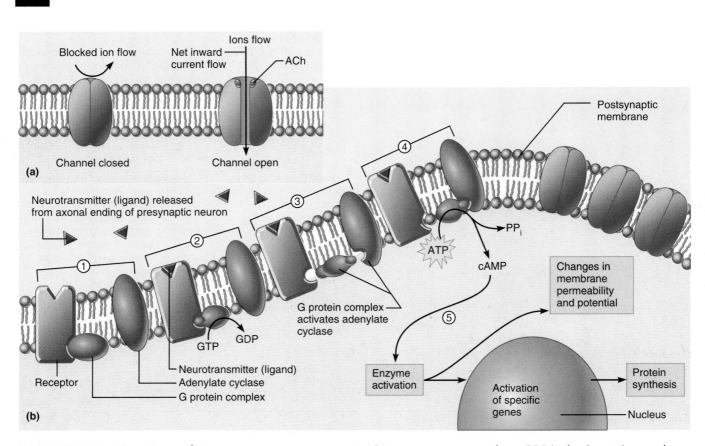

FIGURE 11.22 *Neurotransmitter receptor mechanisms.* **(a)** Channel-linked receptors (ligand-activated ion channels) open in response to ligand (ACh in this case) binding. When no ligand is bound, the channel is closed and no current passes through it. Immediately upon ligand binding, the channel opens and ions begin to flow through the channel. The precise ion current is determined by the structure and charge of the channel proteins. (In ACh channels, cations [Na^+, K^+, and Ca^{2+}] pass, resulting in a depolarizing current.) **(b)** Mechanism of a G protein–linked receptor. In the example shown, effects are mediated by cyclic AMP, which acts as a second messenger to relay the signal from the receptor to other cell components.
① Unbound receptor. ② Binding of the neurotransmitter (ligand) to the receptor results in receptor–G protein interaction and activation of the G protein. Once GTP replaces GDP in the G protein complex, ③ the G protein can interact with and activate adenylate cyclase. ④ Activated adenylate cyclase catalyzes the formation of cyclic AMP from ATP. ⑤ Cyclic AMP, acting as an intracellular second messenger, then mediates events which activate various enzymes that bring about the postsynaptic neuron's response (changes in membrane potential, protein synthesis, etc.).

Mechanism of Action: Direct versus Indirect

Neurotransmitters that open ion channels are said to act *directly*. These neurotransmitters provoke rapid responses in the postsynaptic cells by promoting changes in membrane potential. ACh and the amino acid neurotransmitters are direct-acting neurotransmitters.

The *indirectly acting* neurotransmitters promote broader, longer-lasting effects by acting through intracellular *second-messenger* molecules (typically via G protein mechanisms). In this way their mechanism of action is similar to that of many hormones. The biogenic amines and the peptides are indirect neurotransmitters.

Neurotransmitter Receptors

In Chapter 3, we introduced the various types of receptors involved in cell signaling. Now we are ready to pick up that thread again as we examine the action of receptors that bind neurotransmitters. For the most part, neurotransmitter receptors are either channel-linked receptors that mediate fast synaptic transmission or G protein–linked receptors that oversee slow synaptic responses.

The term first messenger designates the original chemical messenger (the stimulus), which in this case is the neurotransmitter. The binding of the first messenger promotes events that generate the second messenger, which acts within the cell to bring about the desired response.

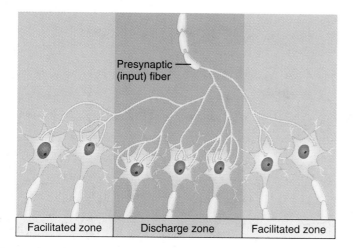

FIGURE 11.23 Simple neuronal pool. This simplified representation of a neuronal pool contains seven neurons and shows the relative position of postsynaptic neurons in the discharge and facilitated zones. Notice that the presynaptic fiber makes more synapses per neuron with neurons in the discharge zone.

Mechanism of Action of Channel-Linked Receptors Equivalent to ligand-gated ion channels, **channel-linked receptors** mediate direct transmitter action. Also called *ionotropic receptors,* these receptors are composed of several protein subunits arranged to form a "rosette" around a central pore (Figure 11.22a). As the ligand binds to one (or more) of the receptor subunits, the proteins promptly change shape. This event opens the central channel and allows ions to pass, altering the membrane potential of the target cell. Channel-linked receptors are always located precisely opposite sites of neurotransmitter release, and their ion channels open instantly upon ligand binding and remain open 1 ms or less while the ligand is bound. At excitatory receptor sites (nicotinic ACh channels and receptors for glutamate, aspartate, and ATP), the channel-linked receptors are *cation* channels that allow small cations (Na^+, K^+, and Ca^{2+}) to pass, but the greatest drive is for Na^+ entry, which contributes to membrane depolarization. Channel-linked receptors that respond to GABA and glycine, and allow K^+ or Cl^- to pass, mediate fast inhibition (hyperpolarization).

Mechanism of Action of G Protein–Linked Receptors Unlike responses to neurotransmitter binding at channel-linked receptors, which are immediate, simple, brief, and highly localized to a single postsynaptic cell, the activity mediated by **G protein–linked receptors** is indirect, slow (hundreds of ms or more), complex, prolonged, and often diffuse–ideal as a basis for some types of learning. Receptors that fall into this class are transmembrane protein complexes and include muscarinic ACh re-

ceptors and those that bind the biogenic amines and neuropeptides. When a neurotransmitter binds to a G protein–linked receptor, the G protein is activated (Figure 11.22b). Activated G proteins typically bring about their effects by controlling the production of second messengers such as **cyclic AMP, cyclic GMP, diacylglycerol,** or Ca^{2+}. These, in turn, act as go-betweens to regulate (open or close) ion channels or to activate kinase enzymes that initiate a cascade of enzymatic reactions in the target cells. Some of these modify (activate or inactivate) other proteins, including channel proteins, by attaching phosphate groups to them. Others interact with nuclear proteins that activate genes and induce synthesis of new proteins in the target cell. Since the effects produced tend to bring about widespread metabolic changes, receptors like the G protein–linked receptors are commonly called *metabotropic receptors.*

BASIC CONCEPTS OF NEURAL INTEGRATION

Until now, we have been concentrating on the activities of individual neurons or, at most, a pair of nerve cells that make synaptic contacts with each other. However, neurons function in groups, and each group contributes to still broader neural functions. Thus, the organization of the nervous system is hierarchical, or ladderlike.

Any time you have a large number of *anything*—people included—there must be *integration;* that is, the parts must be fused into a smoothly operating whole. In this section, we will begin to ascend the ladder of **neural integration,** dealing only with the first rung, *neuronal pools* and their basic patterns of communicating with other parts of the nervous system. Chapter 15 picks up the thread of neural integration once again to examine how sensory inputs interface with motor activity, and looks at the highest levels of neural integration—how we think and remember.

Organization of Neurons: Neuronal Pools

The millions of neurons in the central nervous system are organized into **neuronal pools,** functional groups of neurons that integrate incoming information received from other sources (receptors or different neuronal pools) and then forward the processed information to other destinations.

The composition of a simple type of neuronal pool is shown in Figure 11.23. In this example, one incoming presynaptic fiber branches profusely as it enters the pool and then synapses with several different neurons in the pool. When the incoming fiber

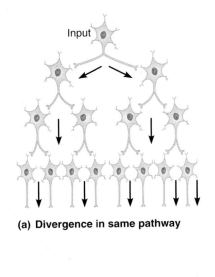

(a) Divergence in same pathway

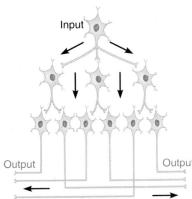

(b) Divergence to multiple pathways

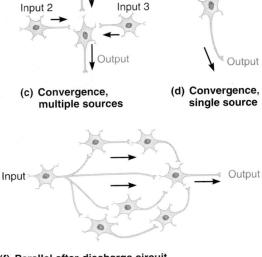

(c) Convergence, multiple sources

(d) Convergence, single source

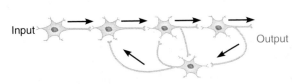

(e) Reverberating circuit

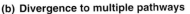

(f) Parallel after-discharge circuit

FIGURE 11.24   *Types of circuits in neuronal pools.*

is excited, it will excite some postsynaptic neurons and facilitate others. Neurons most likely to generate impulses are those closely associated with the incoming fiber, because they receive the bulk of the synaptic contacts. Those neurons are said to be in the *discharge zone* of the neuronal pool. Neurons farther away from the center are not usually excited to threshold by EPSPs induced by the incoming fiber, but they are facilitated and can easily be brought to threshold by stimuli from another source. Thus, the periphery of the pool is the *facilitated zone.* Keep in mind, however, that our figure is a gross oversimplification. Most neuronal pools consist of thousands of neurons and include inhibitory as well as excitatory neurons.

Types of Circuits

Individual neurons in a neuronal pool both send and receive information, and synaptic contacts may cause either excitation or inhibition. The patterns of synaptic connections in neuronal pools are called **circuits,** and they determine the functional capabilities of each type of pool. Four basic types of circuit patterns—diverging, converging, reverberating, and parallel after-discharge—are shown in simplified form in Figure 11.24.

In **diverging circuits,** one incoming fiber triggers responses in ever-increasing numbers of neurons farther and farther along in the circuit. Thus, diverging circuits are often *amplifying* circuits. Divergence can occur along a single pathway or along several (see Figure 11.24a and b, respectively). These circuits are common in both sensory and motor systems. For example, impulses traveling from a single neuron of the brain can activate a hundred or more motor neurons in the spinal cord and, consequently, thousands of skeletal muscle fibers.

The pattern of **converging circuits** is opposite that of diverging circuits, but they too are common in both sensory and motor pathways. In these circuits, the pool receives inputs from several presynaptic neurons, and the circuit as a whole has a funneling, or *concentrating*, effect. Incoming stimuli may converge from one or many different areas (see Figure 11.24c and d), which results in strong stimulation or inhibition. The former condition helps explain how different types of sensory stimuli can cause the same ultimate effect or reaction. For instance, seeing the smiling face of their infant, smelling the baby's freshly powdered skin, or hearing the baby gurgle can all trigger a flood of loving feelings in parents.

In **reverberating,** or **oscillating, circuits** (see Figure 11.24e), the incoming signal travels through a chain of neurons, each of which makes collateral synapses with neurons in the previous part of the pathway. As a result of the positive feedback, the impulses reverberate (are sent through the circuit again and again), giving a continuous output signal until one neuron in the circuit fails to fire. Reverberating circuits are involved in control of rhythmic activities, such as the sleep-wake cycle, breathing, and certain motor activities (such as arm swinging when walking). Some researchers believe that such circuits underlie short-term memory. Depending on the specific circuit, reverberating circuits may continue to oscillate for seconds, hours, or (in the case of the circuit controlling the rhythm of breathing) a lifetime.

In **parallel after-discharge circuits,** the incoming fiber stimulates several neurons arranged in parallel arrays that eventually stimulate a common output cell (see Figure 11.24f). Impulses reach the output cell at different times, creating a burst of impulses called an *after discharge* that lasts 15 ms or more after the initial input has ended. This type of circuit has no positive feedback, and once all the neurons have fired, circuit activity ends. Parallel after-discharge circuits may be involved in complex, exacting types of mental processing, such as working on mathematical equations or other types of problem solving.

Patterns of Neural Processing

Processing of inputs in the various circuits is both *serial* and *parallel.* In serial processing, the input travels along one pathway to a specific destination. In parallel processing, the input travels along several different pathways to be integrated in different CNS regions. Each mode of information handling has unique advantages in the overall scheme of neural functioning, but as an information processor, the brain derives its power from its ability to process in parallel.

Serial Processing

In **serial processing,** the whole system works in a predictable all-or-nothing manner. One neuron stimulates the next in sequence, which stimulates the next, and so on, eventually causing a specific, anticipated response. The most clear-cut examples of serial processing are provided by spinal reflexes, but straight-through sensory pathways from the receptors to the brain provide additional examples. Since reflexes are the functional units of the nervous system, it is important that you gain at least some understanding of them early on.

Reflexes are rapid, automatic responses to stimuli, in which a particular stimulus always causes the *same* motor response. You could say that reflex activity, which produces the simplest of behaviors, is stereotyped and dependable. For example, jerking away your hand after touching a hot object is the norm, and an object approaching the eye triggers a blink. Reflexes occur over neural pathways called **reflex arcs** that have five essential components—re-

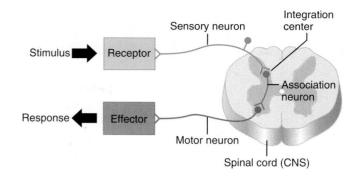

FIGURE 11.25 *A simple reflex arc.* The essential elements in a vertebrate reflex arc—receptor, sensory neuron, integration center, motor neuron, and effector—are illustrated.

ceptor, sensory neuron, CNS integration center, motor neuron, and effector (Figure 11.25). Reflexes are described in detail in Chapter 13.

Parallel Processing

In **parallel processing,** inputs are segregated into many different pathways, and information delivered by each pathway is dealt with simultaneously by different parts of the neural circuitry. For example, smelling a pickle (the input) may cause you to remember picking cucumbers on a farm; or it may remind you that you don't like pickles or that you must buy some at the market; or perhaps it will call to mind *all* these thoughts. For each person, parallel processing will trigger some pathways that are unique. The same stimulus—pickle smell, in our example—promotes many responses beyond simple awareness of the smell. Parallel processing is not repetitious because the circuits do different things with the information, and each "channel" is decoded in relation to all the others to produce a total picture.

Think, for example, about what happens when you step on something sharp when walking barefoot through the grass. The serially processed withdrawal reflex would cause instantaneous removal of your injured foot from the sharp object (painful stimulus). At the same time, pain and pressure impulses would be speeding up to the brain along parallel pathways that would allow you to decide whether to simply rub the hurt spot to soothe it or to seek first aid.

Parallel processing is also extremely important for higher level mental functioning—for putting the parts together to understand the whole. Because of parallel processing, you can recognize a dollar bill in a split second, a task that takes a serial-based computer a fairly long time. This is because parallel processing, which allows a single neuron to send information along several different pathways instead of just one, allows a large amount of information to be packed into a small volume. Hence, logic systems can work much faster.

Deep inside the hypothalamus is a small bundle of neural tissue, called the *pleasure center,* that motivates much of human behavior. The pleasure center reinforces our drives to eat, drink, and reproduce, but in doing so it also makes us perilously vulnerable. Few would deny that much of what we do and value is driven by the "pleasure principle," and herein lies the genesis of addiction.

Our ability to "feel good" involves brain neurotransmitters. For example, the ecstasy of romantic love has been described as a brain bath of norepinephrine, dopamine, and especially phenylethylamine (PEA), which target the pleasure center. Cole Porter knew what he was talking about when he wrote "I get a kick out of you," because these neurotransmitters are chemical cousins of the amphetamines. People who use "speed" (methamphetamine) artificially stimulate their brains to provide their pleasure flush. However, their pleasure is short-lived, because when the brain is flooded with neurotransmitter-like chemicals from the outside, it makes less of its own (why bother?). The wisdom of the body mediates against unnecessary effort.

Cocaine, another pleasure center titillater, has been around for a long time in its prohibitively expensive granular form, which is inhaled, or "snorted." Enter "crack"—the cheaper, much more potent, smokable form of cocaine. For ten dollars or less, a novice user can get "wasted for the night" and flooded with a rush of intense pleasure. But crack is treacherous and intensely addictive. It produces not only a higher high than the inhaled form of cocaine, but also a deeper crash that leaves the user desperate for more.

The prevailing theory of how cocaine produces its effects, at least until the early 1990s, was that the drug first helps to stimulate the pleasure center and then "squeezes it dry." Cocaine produces the "rush" by hooking up to the dopamine reuptake transporter protein, thus blocking the reabsorption of dopamine. As a result, the neurotransmitter remains in the synapse and stimulates the postsynaptic receptor cells again and again, allowing the body to feel its effects over a prolonged period. This sensation is accompanied by increases in heart rate, blood pressure, and sexual appetite. But over time—and sometimes after just one use—the effect changes. As dopamine reabsorption continues to be blocked, that which had accumulated

in the synapses is washed away and degraded, and the brain's dopamine supply becomes inadequate to maintain normal mood. The sending cells cannot produce neural dopamine fast enough to make up for its loss, and the pleasure circuits go dry. Meanwhile, the postsynaptic cells become hypersensitive, and sprout new receptors in a desperate effort to pick up dopamine signals. A vicious cycle of addiction is established. Cocaine is needed to experience pleasure, but using it further depletes the neurotransmitter supply.

Enter a new theory and a new player. It is beginning to look like another neurotransmitter, glutamate, which plays an important role in normal learning, is acting as the devil's advocate or tutor in the addiction experience. Researchers are now claiming that instead of actually producing the pleasurable feelings, dopamine works to heighten attention to external events, especially those that are rewarding, novel, or startling, thus facilitating learning. While the bursts of dopamine release caused by the drug get the brain's attention, glutamate release in the pleasure centers increases (running 50 to 100% above normal levels). Glutamate signaling seems to produce more permanent changes in the brain (synaptic potentiation) that lead

DEVELOPMENTAL ASPECTS OF NEURONS

The nervous system is covered in several chapters, so we limit our attention here to the development of neurons, beginning with the questions, How do nerve cells originate? and How do they mature?

As described in greater detail in Chapter 12, the nervous system originates from a dorsal *neural tube* and the *neural crest* (see Figure 12.2, p. 430), formed from surface ectoderm. The neural tube, whose walls begin as a layer of *neuroepithelial* cells, becomes the CNS. The neuroepithelial cells then begin a three-phase process of differentiation, which occurs largely in the second month of development. (1) They *proliferate* to produce the appropriate number

of cells needed for nervous system development. (2) The potential neurons become amitotic and *migrate* externally. (3) Once they reach characteristic positions in the forming CNS, they *differentiate* into **neuroblasts.** For phase 3 to be achieved, appropriate neurotransmitters must be synthesized, and the correct synapses must be made.

Little is known about the biochemical specialization of neurons, and synapse formation in particular poses several difficult questions. How does a neuroblast's growing axon "know" where to go—and once it gets there, where to stop and make the proper connection? The tortuous growth of an axon toward an appropriate target appears to be guided by multiple signals: a scaffold laid down by similar, older "pathfinder" neurons and orienting glial fibers;

to compulsive drug-seeking behaviors elicited by external cues. Hence, repeated exposure to cocaine seems to activate glutamate transmission and to prompt the synthesis of more glutamate receptors as cocaine addiction takes hold.

Whatever the precise mechanism, the cocaine user becomes anxious and, in a very real sense, unable to experience pleasure without cocaine. Cocaine addicts tend to lose weight, have trouble sleeping, develop cardiac and pulmonary abnormalities ("crack heart and lung"), and have frequent infections.

Cocaine addiction carries with it an out-of-control, desperate craving that is notoriously difficult to manage or treat. Drug abusers call it *jonesing.* Various drugs have been used with mixed success to treat cocaine addicts. Traditional antiaddiction drugs take so long to reduce the overpowering cravings that users commonly drop out of treatment programs. However, three new drugs are generating excitement as likely candidates to help addicts abandon their addiction. Manzindol, an appetite suppressant, binds dopamine receptors more tightly than cocaine, thus blocking cocaine's effects. Reports on buprenorphine, a pain reliever, look even brighter: Researchers found that laboratory monkeys addicted to cocaine rapidly lost interest in the drug after receiving low doses of buprenorphine.

Acamprosate, a drug approved for treating alcoholism in Europe, modifies glutamate transmission. It shows promise for treating cocaine addiction, as do catalytic anticocaine antibodies, which degrade cocaine and hence prevent it from binding and causing its effects.

Crack use in the United States is still a national problem. Cocaine-related cardiovascular deaths (due to hypertensive crisis) and strokes are increasing daily, and "crack babies" are a sad heritage.

A mixture of speed and crack called "croak" (funny or prophetic?) and a new smokable form of speed called "ice" have gained a foothold in the United States. Unlike crack, which creates a 20- to 30-minute high, ice produces a high that lasts 12 to 24 hours, followed by a crash that seemingly creates psychosis (paranoid schizophrenia). Moreover, a formaldehyde (embalming fluid)-laced marijuana known on the street as "ill face" has become popular.

As if these were not enough to contend with, a sinister kind of heroin known as *China white* that comes from the Golden Triangle of Southeast Asia is flooding the United States. Unlike the heroin forms displaced by cocaine in the 1980s, which had a purity of 5 to 10%, this heroin is up to 90% pure and it can be snorted or smoked just like cocaine. Because heroin is a depressant that does not incite the violence

that crack (a stimulant) does, the recent rise in heroin addiction went largely unnoticed. Removed from the stigma and dangers of needle injection, the pattern of China white abuse mirrors that of cocaine, first appearing in the entertainment industry and filtering down to Small Town, USA. More expensive than cocaine, it demands dramatically higher doses to achieve the same effect. Deprived of the heroin it increasingly depends on, the body goes into incapacitating nausea and convulsions; but with an overdose, the lungs fill with fluid and the user drowns.

The brain, with its complex biochemistry, always circumvents attempts to keep it in a euphoric haze. Perhaps this means that pleasure must be transient and can be experienced only against a background of its absence.

Smoking with a crack cocaine pipe.

nerve growth factor (NGF) released by astrocytes and various chemicals called *neurotropins* released by other neurons; inhibitory chemicals that cause the axon to avoid an area, such as RGM and SLIT (repulsion guiding molecules); and attractants released by the target cells. Further, a "sticky" molecule called *nerve cell adhesion molecule (N-CAM)* has been identified on skeletal muscle cells and on the surfaces of glial cells. N-CAM seems to provide some sort of nerve glue during the formative period. This molecule is so important in establishing neural patterns that when its action is blocked, developing neural tissue falls into a tangled, spaghetti-like mass, and neural function is hopelessly impaired.

The ability of an axon to interact with its environment and to grow reside in a prickly, fanlike structure at its growing tip called the **growth cone.** The membrane of the growth cone bears N-CAM molecules and has *filopodia,* oozing extensions composed mainly of actin. The filopodia elongate and then contract like inchworms to pull the growth cone along. The growth cone also engulfs molecules from its environment that are then transported to the nerve cell body and inform it of the pathway being taken for growth and synapse formation. Once the axon has reached its general target area, it must select the right site on the right target cell to form a synapse, an event that requires signaling by both presynaptic and postsynaptic cells. This is far from an exact process—trial and error occurs and gap junctions appear and disappear before permanent chemical synapses are formed.

Neurons that fail to make appropriate or functional synaptic contacts act as if they have been deprived of some essential nutrient and die. Besides cell death resulting from unsuccessful synapse formation, *programmed cell death* also appears to be a normal part of the developmental process. Of the neurons formed during the embryonic period, perhaps two-thirds die before we are born. Those that remain constitute our total neural endowment for life. The generally amitotic nature of neurons is important because their activity depends on the synapses they've formed; if neurons were to divide, their connections might be hopelessly disrupted. This aside, there *do* appear to be some specific neuronal populations that remain mitotic—notably olfactory neurons and some cells of the hippocampus, a brain region very much involved with learning and memory.

★   ★   ★

In this chapter, we have examined how the amazingly complex neurons, via their electrical and chemical signals, serve the body in a variety of ways: Some serve as "lookouts," others process information for immediate use or for future reference, and still others stimulate the body's muscles and glands into activity. With this background, we are ready to study the most sophisticated mass of neural tissue in the entire body—the brain (and its continuation, the spinal cord), the focus of Chapter 12.

RELATED CLINICAL TERMS

Neuroblastoma (nu″ro-blas-to′mah; *oma* = tumor) A malignant tumor in children; arises from cells that retain a neuroblast-like structure. These tumors sometimes arise in the brain, but most occur in the peripheral nervous system.

Neurologist (nu-rol′o-jist) A medical specialist in the study of the nervous system, its functions, and its disorders.

Neuropathy (nu-rop′ah-the) Any disease of nervous tissue, but particularly degenerative disease of nerves.

Neuropharmacology (nu″ro-far″mah-kol′o-je) Scientific study of the effects of drugs on the nervous system.

Neurotoxin Substance that is poisonous or destructive to nervous tissue, e.g., botulism and tetanus toxins.

Rabies (*rabies* = madness) A viral infection of the nervous system transferred to humans by the bites of infected mammals (dogs, bats, skunks, and so on). After entry, the virus is transported via axonal transport in peripheral nerve axons to the CNS, where it causes brain inflammation, delirium, and death. A vaccine- and antibody-based treatment is effective if given before symptoms appear; rabies is very rare in the United States.

Shingles (herpes zoster) A viral infection of sensory neurons serving the skin. Characterized by scaly, painful blisters usually confined to a narrow strip of skin, often on one side of the body trunk. Caused by the varicella-zoster virus, which causes chicken pox (generally during childhood); during this initial infection the virus is transported from the skin lesions to the sensory cell bodies in the sensory ganglia. Typically, the virus is held in check by the immune system and remains dormant until the immune system is weakened, often by stress. Then viral particles multiply and travel back to the skin, producing the characteristic rash. Attacks last several weeks, alternating between periods of healing and relapse. Seen mostly in those over 50 years old.

CHAPTER SUMMARY

Media study tools that could provide you additional help in reviewing specific key topics of Chapter 11 are referenced below. **SP** = *Study Partner;* **IP** = *Interactive Physiology.*

1. The nervous system bears a major responsibility for maintaining body homeostasis. Its chief functions are to monitor, integrate, and respond to information in the environment.

Organization of the Nervous System   (pp. 387–388)

1. The nervous system is divided anatomically into the central nervous system (brain and spinal cord) and the peripheral nervous system (cranial and spinal nerves and ganglia).

2. The major functional divisions of the nervous system are the sensory (afferent) division, which conveys impulses to the CNS, and the motor (efferent) division, which conveys impulses from the CNS.

3. The efferent division includes the somatic (voluntary) system, which serves skeletal muscles, and the autonomic (involuntary) system, which innervates smooth and cardiac muscle and glands.

Histology of Nervous Tissue   (pp. 388–396)
Supporting Cells (Neuroglia)   (pp. 388–390)

1. Supporting cells (neuroglia) segregate and insulate neurons and assist neurons in various other ways.

2. CNS supporting cells include astrocytes, microglia, ependymal cells, and oligodendrocytes. Schwann cells and satellite cells are supporting cells found in the PNS.

Neurons   (pp. 390–396)

3. Neurons have a cell body and cytoplasmic processes called axons and dendrites.

4. A bundle of nerve fibers is called a tract in the CNS and a nerve in the PNS. A collection of cell bodies is called a nucleus in the CNS and a ganglion in the PNS.

5. The cell body is the biosynthetic (and receptive) center of the neuron. Except for those found in ganglia, cell bodies are found in the CNS.

6. Some neurons have many dendrites, receptive sites that conduct signals from other neurons toward the nerve cell

body. With few exceptions, all neurons have one axon, which generates and conducts nerve impulses away from the nerve cell body. Terminal endings of axons release neurotransmitter.

SP Exercise: Chapter 11, Structure of a Motor Neuron.

7. Transport along axons occurs via different mechanisms. Best understood is the bidirectional, ATP-dependent process that moves particulate material, neurotransmitters, and enzymes toward the axonal terminals and conducts substances destined for degradation back to the cell body. It involves microtubules, microfilaments, and motor proteins.

8. Large nerve fibers (axons) are myelinated. The myelin sheath is formed in the PNS by Schwann cells and in the CNS by oligodendrocytes. The sheath has gaps called nodes of Ranvier. Unmyelinated fibers are surrounded by supporting cells, but the membrane-wrapping process does not occur.

9. Anatomically, neurons are classified according to the number of processes issuing from the cell body as multipolar, bipolar, or unipolar.

10. Functionally, neurons are classified according to the direction of nerve impulse conduction. Sensory neurons conduct impulses toward the CNS, motor neurons conduct away from the CNS, and interneurons (association neurons) lie between sensory and motor neurons in the neural pathways.

IP Nervous System I CD-ROM; Topic: Anatomy Review Pages 1–12.

Neurophysiology (pp. 396–419)

Basic Principles of Electricity (pp. 396–397)

1. The measure of the potential energy of separated electrical charges is called voltage (*V*) or potential. Current (*I*) is the flow of electrical charge from one point to another. Resistance (*R*) is hindrance to current flow. The relationship among these is given by Ohm's law: $I = V/R$.

2. In the body, electrical charges are provided by ions; cellular plasma membranes provide resistance to ion flow. The membranes contain passive (open) and active (gated) channels.

IP Nervous System I CD-ROM; Topic: Ion Channels, Pages 1–10.

The Resting Membrane Potential: The Polarized State (pp. 398–399)

3. A resting neuron exhibits a resting membrane potential, which is −70 mV (inside negative), owing to differences in sodium and potassium ion concentrations inside and outside the cell.

4. The ionic differences result from greater permeability of the membrane to potassium than to sodium and from the operation of the sodium-potassium pump, which ejects $3Na^+$ from the cell for each $2K^+$ transported in.

SP Exercise: Chapter 11, Changes in Neuronal Permeability; Case Study: Cerebrovascular Accident.

IP Nervous System I CD-ROM; Topic: The Membrane Potential, Pages 1–16.

Membrane Potentials That Act as Signals (pp. 399–407)

5. Depolarization is a reduction in membrane potential (inside becomes less negative); hyperpolarization is an increase in membrane potential (inside becomes more negative).

6. Graded potentials are small, brief, local changes in membrane potential that act as short-distance signals. The current produced dissipates with distance.

7. An action potential, or nerve impulse, is a large, but brief, depolarization signal (and polarity reversal) that underlies long-distance neural communication. It is an all-or-none phenomenon.

8. Generation of an action potential involves three phases: (1) Increase in sodium permeability and reversal of the membrane potential to approximately +30 mV (inside positive). Local depolarization opens voltage-sensitive sodium gates; at threshold, depolarization becomes self-generating (driven by sodium ion influx). (2) Decrease in sodium permeability. (3) Increase in potassium permeability. Repolarization is ongoing during phases 2 and 3.

9. In nerve impulse propagation, each action potential provides the depolarizing stimulus for triggering an action potential in the next membrane patch. Regions that have just generated action potentials are refractory; hence, the nerve impulse is propagated in one direction only.

10. If threshold is reached, an action potential is generated; if not, depolarization remains local.

11. Action potentials are independent of stimulus strength: strong stimuli cause action potentials to be generated more frequently but not with greater amplitude.

12. During the absolute refractory period, a neuron cannot respond to another stimulus because it is already generating an action potential. During the relative refractory period the neuron's threshold is elevated because repolarization is ongoing.

13. In unmyelinated fibers, action potentials are produced in a wave all along the axon, that is, by continuous conduction. In myelinated fibers, action potentials are generated only at nodes of Ranvier and are propagated more rapidly by saltatory conduction.

IP Nervous System I CD-ROM; Topic: The Action Potential, Pages 1–18.

The Synapse (pp. 407–410)

14. A synapse is a functional junction between neurons. The information-transmitting neuron is the presynaptic neuron; the neuron beyond the synapse is the postsynaptic neuron.

15. Electrical synapses allow ions to flow directly from one neuron to another; the cells are electrically coupled.

16. Chemical synapses are sites of neurotransmitter release and binding. When the impulse reaches the presynaptic axonal terminals, voltage-regulated Ca^{2+} channels open, and Ca^{2+} enters the cell and mediates neurotransmitter release. Neurotransmitters diffuse across the synaptic cleft and attach to postsynaptic membrane receptors, opening ion channels. After binding, the neurotransmitters are removed from the synapse by enzymatic breakdown or by reuptake into the presynaptic terminal or astrocytes.

SP Exercise: Chapter 11, Synaptic Transmission.

IP Nervous System II CD-ROM; Topics: Anatomy Review, Pages 1–9, Ion Channels, Pages 1–8, Synaptic Transmission, Pages 1–7.

Postsynaptic Potentials and Synaptic Integration (pp. 410–413)

17. Binding of neurotransmitter at excitatory chemical synapses results in local graded depolarizations called EPSPs, caused by the opening of channels that allow simultaneous passage of Na^+ and K^+.

18. Neurotransmitter binding at inhibitory chemical synapses results in hyperpolarizations called IPSPs, caused by the opening of K^+ or Cl^- gates or both. IPSPs drive the membrane potential farther from threshold.

19. EPSPs and IPSPs summate temporally and spatially. The membrane of the axon hillock acts as a neuronal integrator.

20. Synaptic potentiation, in which the postsynaptic neuron's response is enhanced, is produced by intense repeated stimulation. Ionic calcium appears to mediate such effects, which may be the basis of learning.

21. Presynaptic inhibition is mediated by axoaxonal synapses that reduce the amount of neurotransmitter released by the inhibited neuron. Neuromodulation occurs when chemicals (often other than neurotransmitters) alter neuronal or neurotransmitter activity.

SP Exercise: Chapter 11, Synaptic Transmission.

IP Nervous System II CD-ROM; Topic: Synaptic Potentials and Cellular Intergration, Pages 1–10.

Neurotransmitters and Their Receptors (pp. 413–419)

22. The major classes of neurotransmitters based on chemical structure are acetylcholine, biogenic amines, amino acids, and peptides.

23. Functionally, neurotransmitters are classified as (1) inhibitory or excitatory (or both) and (2) direct or indirect. Direct-acting neurotransmitters cause channel opening. Indirect-acting neurotransmitters act through second messengers and cause complex changes in target cell metabolism.

24. Neurotransmitter receptors are either channel-linked receptors that open ion channels, leading to fast changes in membrane potential, or G protein–linked receptors that oversee slow synaptic responses mediated by G proteins and intracellular second messengers. Second messengers most often activate kinases, which in turn act on ion channels or activate other proteins.

IP Nervous System II CD-ROM; Topic: Synaptic Transmission Pages 6–15.

SP Case Study: Parkinson's Disease.

Basic Concepts of Neural Integration (pp. 419–421)

Organization of Neurons: Neuronal Pools (pp. 419–420)

1. CNS neurons are organized into several types of neuronal pools, each with distinguishing patterns of synaptic connections called circuits.

Types of Circuits (pp. 420–421)

2. The four basic circuit types are diverging, converging, reverberating, and parallel after-discharge.

Patterns of Neural Processing (p. 421)

3. In serial processing, one neuron stimulates the next in sequence, producing specific, predictable responses, as in spinal reflexes. A reflex is a rapid, involuntary motor response to a stimulus.

4. Reflexes are mediated over neural pathways called reflex arcs. The minimum number of elements in a reflex arc is five: receptor, sensory neuron, integration center, motor neuron, and effector.

5. In parallel processing, which underlies complex mental functions, impulses are sent along several pathways to different integration centers.

Developmental Aspects of Neurons (pp. 422–424)

1. Neuron development involves proliferation, migration, and cellular differentiation. Cellular differentiation entails neuron specialization, specific neurotransmitter synthesis, and synapse formation.

2. Axonal outgrowth and synapse formation are guided by other neurons, glial fibers, and chemicals (such as N-CAM and nerve growth factor). Neurons that do not make appropriate synapses die, and approximately two-thirds of neurons formed in the embryo undergo programmed cell death before birth.

REVIEW QUESTIONS

Multiple Choice/Matching

1. Which of the following structures is not part of the central nervous system? (a) the brain, (b) a nerve, (c) the spinal cord, (d) a tract.

2. Match the names of the supporting cells found in column B with the appropriate descriptions in column A.

Column A	Column B
_____ **(1)** myelinates nerve fibers in the CNS	**(a)** astrocyte
_____ **(2)** lines brain cavities	**(b)** ependymal cell
_____ **(3)** myelinates nerve fibers in the PNS	**(c)** microglia
_____ **(4)** CNS phagocytes	**(d)** oligodendrocyte
_____ **(5)** may help regulate the ionic composition of the extracellular fluid	**(e)** satellite cell
	(f) Schwann cell

3. What type of current flows through the axolemma during the steep phase of repolarization? (a) chiefly a sodium current, (b) chiefly a potassium current, (c) sodium and potassium currents of approximately the same magnitude.

4. Assume that an EPSP is being generated on the dendritic membrane. Which will occur? (a) specific Na^+ gates will open, (b) specific K^+ gates will open, (c) a single type of channel will open, permitting simultaneous flow of Na^+ and K^+, (d) Na^+ gates will open first and then close as K^+ gates open.

5. The velocity of nerve impulse conduction is greatest in (a) heavily myelinated, large-diameter fibers, (b) myelinated, small-diameter fibers, (c) unmyelinated, small-diameter fibers, (d) unmyelinated, large-diameter fibers.

6. Chemical synapses are characterized by all of the following except (a) the release of neurotransmitter by the presynaptic membranes, (b) postsynaptic membranes bearing receptors that bind neurotransmitter, (c) ions flowing through protein channels from the presynaptic to the postsynaptic neuron, (d) a fluid-filled gap separating the neurons.

7. Biogenic amine neurotransmitters include all but (a) norepinephrine, (b) acetylcholine, (c) dopamine, (d) serotonin.

8. The neuropeptides that act as natural opiates are (a) substance P, (b) somatostatin, (c) cholecystokinin, (d) enkephalins.

9. Inhibition of acetylcholinesterase by poisoning blocks neurotransmission at the neuromuscular junction because (a) ACh is no longer released by the presynaptic terminal,

(b) ACh synthesis in the presynaptic terminal is blocked, (c) ACh is not degraded, hence prolonged depolarization is enforced on the postsynaptic neuron, (d) ACh is blocked from attaching to the postsynaptic ACh receptors.

10. The anatomical region of a multipolar neuron that has the lowest threshold for generating an action potential is the (a) soma, (b) dendrites, (c) axon hillock, (d) distal axon.

11. An IPSP is inhibitory because (a) it hyperpolarizes the postsynaptic membrane, (b) it reduces the amount of neurotransmitter released by the presynaptic terminal, (c) it prevents calcium ion entry into the presynaptic terminal, (d) it changes the threshold of the neuron.

12. Identify the neuronal circuits described by choosing the correct response from the key.

Key: (a) converging, **(b)** diverging, **(c)** parallel after-discharge, **(d)** reverberating

_____ **(1)** Impulses continue around and around the circuit until one neuron stops firing.

_____ **(2)** One or a few inputs ultimately influence large numbers of neurons.

_____ **(3)** Many neurons influence a few neurons.

_____ **(4)** May be involved in exacting types of mental activity.

Short Answer Essay Questions

13. Explain both the anatomical and functional divisions of the nervous system. Include the subdivisions of each.

14. (a) Describe the composition and function of the cell body. (b) How are axons and dendrites alike? In what ways (structurally and functionally) do they differ?

15. (a) What is myelin? (b) How does the myelination process differ in the CNS and PNS?

16. (a) Contrast unipolar, bipolar, and multipolar neurons structurally. (b) Indicate where each is most likely to be found.

17. What is the polarized membrane state? How is it maintained? (Note the relative roles of both passive and active mechanisms.)

18. Describe the events that must occur to generate an action potential. Indicate how the ionic gates are controlled, and explain why the action potential is an all-or-none phenomenon.

19. Since all action potentials generated by a given nerve fiber have the same magnitude, how does the CNS "know" whether a stimulus is strong or weak?

20. (a) Explain the difference between an EPSP and an IPSP. (b) What specifically determines whether an EPSP or IPSP will be generated at the postsynaptic membrane?

21. Since at any moment a neuron is likely to have thousands of neurons releasing neurotransmitters at its surface, how is neuronal activity (to fire or not to fire) determined?

22. The effects of neurotransmitter binding are very brief. Explain.

23. During a neurobiology lecture, a professor repeatedly refers to type A and type B fibers, absolute refractory period, and nodes of Ranvier. Define these terms.

24. Distinguish between serial and parallel processing.

25. Briefly describe the three stages of neuron development.

26. What factors appear to guide the outgrowth of an axon and its ability to make the "correct" synaptic contacts?

Critical Thinking and Clinical Application Questions

1. Mr. Miller is hospitalized for cardiac problems. Somehow, medical orders are mixed up and Mr. Miller is infused with a K^+-enhanced intravenous solution meant for another patient who is taking potassium-wasting diuretics (i.e., drugs that cause excessive loss of potassium from the body in urine). Mr. Miller's potassium levels are normal before the IV is administered. What do you think will happen to Mr. Miller's neuronal resting potentials? To his neurons' ability to generate action potentials?

2. General and local anesthetics block action potential generation, thereby rendering the nervous system quiescent while surgery is performed. What specific process do anesthetics impair, and how does this interfere with nerve transmission?

3. When admitted to the emergency room, John was holding his right hand, which had a deep puncture hole in its palm. He explained that he had fallen on a nail while exploring a barn. John was given an antitetanus shot to prevent neural complications. Tetanus bacteria fester in deep, dark wounds, but how do they travel in neural tissue?

4. Rochelle developed multiple sclerosis when she was 27. After eight years she had lost a good portion of her ability to control her skeletal muscles. Why did this happen?

12 THE CENTRAL NERVOUS SYSTEM

The Brain (pp. 429–461)

1. Describe the process of brain development.
2. Name the major regions of the adult brain.
3. Name and locate the ventricles of the brain.
4. List the major lobes, fissures, and functional areas of the cerebral cortex.
5. Explain lateralization of hemisphere function.
6. Differentiate between commissures, association fibers, and projection fibers.
7. Describe the general function of the basal nuclei (basal ganglia).
8. Describe the location of the diencephalon, and name its subdivisions.
9. Identify the three major regions of the brain stem, and note the functions of each area.
10. Describe the structure and function of the cerebellum.

11. Locate the limbic system and the reticular formation, and explain the role of each functional system.
12. Describe how meninges, cerebrospinal fluid, and the blood-brain barrier protect the CNS.
13. Describe the formation of cerebrospinal fluid, and follow its circulatory pathway.
14. Describe the cause (if known) and major signs and symptoms of cerebrovascular accidents, Alzheimer's disease, Huntington's disease, and Parkinson's disease.

The Spinal Cord (pp. 461–467)

15. Describe the embryonic development of the spinal cord.
16. Describe the gross and microscopic structure of the spinal cord.
17. List the major spinal cord tracts, and classify each as a motor or sensory tract.

18. Distinguish between flaccid and spastic paralysis and between paralysis and paresthesia.

Diagnostic Procedures for Assessing CNS Dysfunction (p. 468)

19. List and explain several techniques used to diagnose brain disorders.

Developmental Aspects of the Central Nervous System (pp. 468–469)

20. Indicate several maternal factors that can impair development of the nervous system in an embryo.
21. Explain the effects of aging on the brain.

Historically, the **central nervous system (CNS)**—brain and spinal cord—has been compared to the central switchboard of a telephone system that interconnects and directs a dizzying number of incoming and outgoing calls. Nowadays, many compare it to a kind of supercomputer. These analogies may explain some workings of the spinal cord, but neither does justice to the fantastic complexity of the human brain. Whether we view it as an evolved biological organ, an impressive computer, or simply a miracle, the human brain is certainly one of the most amazing things known.

During the course of animal evolution, **cephalization** (se"fah-lĭ-za'shun) has occurred. That is, there has been an elaboration of the rostral ("toward the snout"), or anterior, portion of the central nervous system, along with an increase in the number of neurons in the head. This phenomenon reaches its highest level in the human brain.

In this chapter, we examine the structure of the central nervous system and functions associated with its specific anatomical regions. Complex integrative functions, such as sensory and motor integration, sleep-wake cycles, and higher mental functions (consciousness and memory) are covered in Chapter 15.

THE BRAIN

The unimpressive appearance of the human **brain** (Figure 12.1) gives few hints of its remarkable abilities. It is about two good fistfuls of quivering pinkish gray tissue, wrinkled like a walnut, and somewhat the consistency of cold oatmeal. The average adult man's brain weighs about 1600 g (3.5 pounds); that of a woman averages 1450 g. In terms of brain weight per body weight, however, males and females have equivalent brain sizes.

Embryonic Development of the Brain

Breaking from our usual pattern, we will begin with brain embryology first, as the terminology used for the structural divisions of the adult brain is easier to follow when you understand brain development.

The earliest phase of brain development is shown in Figure 12.2. Starting in the three-week embryo, the ectoderm thickens along the dorsal midline axis of the embryo to form the **neural plate.** The neural plate then invaginates, forming a groove flanked by **neural folds.** As this **neural groove** deepens, the superior edges of the neural folds fuse, forming the

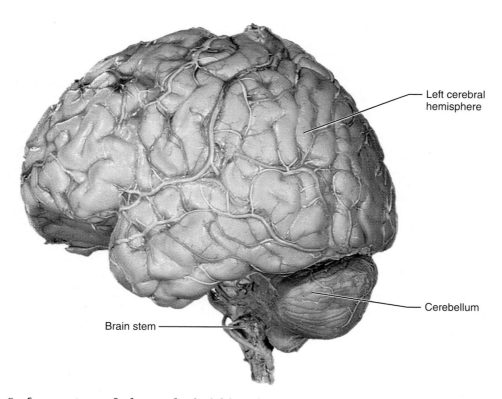

FIGURE 12.1 *Surface anatomy of a human brain,* left lateral aspect. (Veins and arteries are injected with colored latex.)

Left cerebral hemisphere

Cerebellum

Brain stem

Anterior (rostral) end

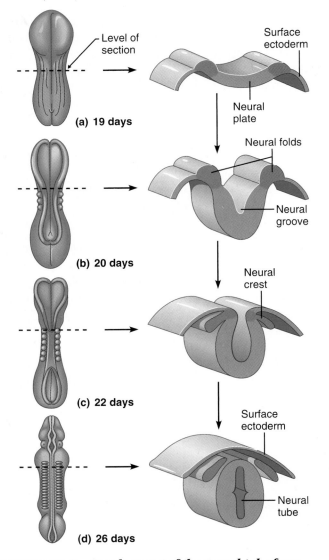

FIGURE 12.2 *Development of the neural tube from embryonic ectoderm.* Dorsal surface views of the embryo are shown at the left; transverse sections are shown at the right. **(a)** Formation of the neural plate from surface ectoderm. **(b–d)** Progressive development of the neural plate into the neural groove (flanked by neural folds) and then the neural tube. The neural tube forms CNS structures; the neural crest cells form some PNS structures.

neural tube, which soon detaches from the surface ectoderm and sinks to a deeper position. The neural tube is formed by the fourth week of pregnancy and differentiates rapidly into the CNS. The brain forms anteriorly (rostrally) and the spinal cord develops from the caudal ("toward the tail") or posterior portion of the neural tube. Small groups of neural fold cells migrate laterally from between the surface ectoderm and the neural tube, forming the **neural crest** (Figure 12.2c), which gives rise to some neurons destined to reside in ganglia.

As soon as the neural tube is formed, its anterior end immediately begins to expand and constrictions appear that mark off the three **primary brain vesicles** (Figure 12.3): the **prosencephalon** (pros″en-sef′ah-lon), or **forebrain;** the **mesencephalon** (mes″en-sef′ah-lon), or **midbrain;** and the **rhombencephalon** (romb″en-sef′ah-lon), or **hindbrain.** (Note that *encephalo* means "brain.") The remainder of the neural tube becomes the spinal cord. We will discuss its development later in the chapter.

In week 5, the primary vesicles give rise to the **secondary brain vesicles.** The forebrain divides into the **telencephalon** ("endbrain") and **diencephalon** ("interbrain"), and the hindbrain constricts, forming the **metencephalon** ("afterbrain") and **myelencephalon** ("spinal brain"). The midbrain remains undivided. Each of the five secondary brain vesicles then develops rapidly to produce the major structures of the adult brain (Figure 12.3d). The greatest change occurs in the telencephalon, which sprouts two lateral swellings that look like Mickey Mouse's ears. These become the *cerebral hemispheres*, referred to collectively as the **cerebrum** (ser′ĕ-brum). The diencephalon, also part of the forebrain, specializes to form the *hypothalamus* (hi″po-thal′ah-mus), *thalamus*, and *epithalamus*. Less dramatic changes occur in the mesencephalon, metencephalon, and myelencephalon as these regions are transformed into the *midbrain, pons, and cerebellum*, and *medulla oblongata*, respectively. All of these midbrain and hindbrain structures, except for the cerebellum, form the **brain stem.** The central cavity of the neural tube remains continuous and enlarges in four areas to form the fluid-filled *ventricles* (*ventr* = little belly) of the brain. The ventricles are described shortly.

During later development the brain continues to grow rapidly and changes occur in the relative positions of its parts. Because the brain grows more rapidly than the membranous skull that contains it, two major flexures develop—the *midbrain* and *cervical flexures*—which bend the forebrain toward the brain stem (Figure 12.4a). A second consequence of restricted space is that the cerebral hemispheres are forced to take a horseshoe-shaped course and grow posteriorly and laterally (indicated by the arrows in Figure 12.4b and c). As a result, they grow back over and almost completely envelop the diencephalon and midbrain. By week 26, the continued growth of the cerebral hemispheres causes their surfaces to crease and fold (Figure 12.4c and d), producing their typical *convolutions* and increasing their surface area, which allows more neurons to occupy the limited space.

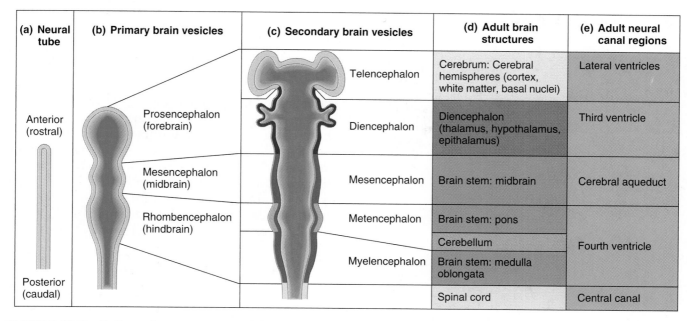

(a) Neural tube	(b) Primary brain vesicles	(c) Secondary brain vesicles	(d) Adult brain structures	(e) Adult neural canal regions
	Prosencephalon (forebrain)	Telencephalon	Cerebrum: Cerebral hemispheres (cortex, white matter, basal nuclei)	Lateral ventricles
		Diencephalon	Diencephalon (thalamus, hypothalamus, epithalamus)	Third ventricle
	Mesencephalon (midbrain)	Mesencephalon	Brain stem: midbrain	Cerebral aqueduct
	Rhombencephalon (hindbrain)	Metencephalon	Brain stem: pons	Fourth ventricle
			Cerebellum	
		Myelencephalon	Brain stem: medulla oblongata	
			Spinal cord	Central canal

Anterior (rostral) — Posterior (caudal)

FIGURE 12.3 *Embryonic development of the human brain.* **(a)** Formed by week 4, the neural tube quickly subdivides into **(b)** the primary brain vesicles, which subsequently form **(c)** the secondary brain vesicles by week 5. These differentiate into **(d)** the adult brain structures. **(e)** The adult structures derived from the neural canal.

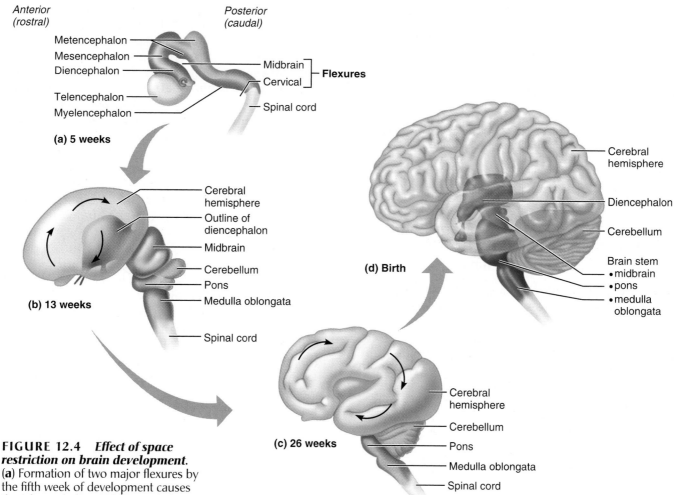

FIGURE 12.4 *Effect of space restriction on brain development.* **(a)** Formation of two major flexures by the fifth week of development causes the telencephalon and diencephalon to be angled toward the brain stem. Development of the cerebral hemispheres at **(b)** 13 weeks, **(c)** 26 weeks, and **(d)** birth. Initially, the cerebral surface is smooth; the folding begins in the sixth month and convolutions become more obvious as development continues. The posterolateral growth of the cerebral hemispheres ultimately completely encloses the diencephalon and superior aspect of the brain stem (seen through the cerebral hemispheres in this see-through view).

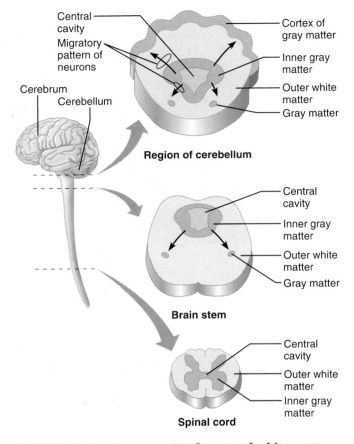

Cerebrum
Cerebellum

Region of cerebellum
- Central cavity
- Migratory pattern of neurons
- Cortex of gray matter
- Inner gray matter
- Outer white matter
- Gray matter

Brain stem
- Central cavity
- Inner gray matter
- Outer white matter
- Gray matter

Spinal cord
- Central cavity
- Outer white matter
- Inner gray matter

FIGURE 12.5 *Arrangement of gray and white matter in the CNS (highly simplified).* The diagrams represent cross sections at three different levels of the CNS. In each section, the dorsal aspect is at the top. In general, white matter lies external to gray matter; however, collections of gray matter migrate externally into the white matter in the developing brain (see arrows). The cerebrum resembles the cerebellum in its external cortex of gray matter.

Regions and Organization of the Brain

There are various approaches to describing brain structure. Some neuroanatomists (and textbooks) discuss brain anatomy in terms of the *embryonic scheme* (see Figure 12.3c), whereas in medical facilities, the names of the adult brain regions are more often heard (see Figure 12.3d). In this text, we will consider the brain in terms of the medical scheme and the regions shown in Figure 12.4d: (1) cerebral hemispheres, (2) diencephalon, (3) brain stem (midbrain, pons, and medulla), and (4) cerebellum.

The basic pattern of the CNS is revealed in spinal cord sections. It consists of a central cavity surrounded by a gray matter core, external to which is white matter (myelinated fiber tracts). The brain exhibits this basic design but it also has additional regions of gray matter that are not present in the spinal cord (Figure 12.5). Both the cerebral hemispheres and the cerebellum have an outer sheet or

"bark" of gray matter consisting of neuron cell bodies; this layer is called a *cortex*. This pattern changes with descent through the brain stem—the cortex disappears, but scattered gray matter nuclei are seen within the white matter. At the caudal end of the brain stem, the basic pattern is evident.

To help you picture the spatial relationships among the brain regions, we first consider the fluid-filled ventricles that lie deep within the brain. Then we will examine each of the brain regions from rostral to caudal. Finally, a brief summary of the brain regions is provided in Table 12.1 on p. 449.

Ventricles of the Brain

As noted earlier, the **ventricles** of the brain arise from expansions of the lumen of the embryonic neural tube. They are continuous with each other and with the central canal of the spinal cord (Figure 12.6). The hollow ventricular chambers are filled with cerebrospinal fluid and lined by *ependymal cells*, a type of neuroglia (see Figure 11.3c).

The paired **lateral ventricles**, one deep within each cerebral hemisphere, are large C-shaped chambers that reflect the pattern of cerebral growth. Anteriorly, the lateral ventricles lie close together, separated only by a thin median membrane called the **septum pellucidum** (pĕ-lu'sid-um; "transparent wall"). (See Figure 12.12, p. 443.) Each lateral ventricle communicates with the narrow **third ventricle** in the diencephalon via a channel called an **interventricular foramen** (*foramen of Monro*). The third ventricle is continuous with the **fourth ventricle** via the canal-like **cerebral aqueduct** that runs through the midbrain to connect the third and fourth ventricles. The fourth ventricle lies in the hindbrain dorsal to the pons and superior medulla. It is continuous with the central canal of the spinal cord inferiorly. Three openings mark the walls of the fourth ventricle: the paired **lateral apertures** in its side walls and the **median aperture** in its roof. These apertures connect the ventricles to the *subarachnoid* (sub"ah-rak'noid) *space*, a fluid-filled space surrounding the brain. The importance of the cerebrospinal (ser"ĕ-bro-spi'nal) fluid, which fills both the internal brain cavities and the subarachnoid space, is discussed later in the chapter.

The Cerebral Hemispheres

The **cerebral hemispheres** form the superior part of the brain (Figure 12.7). Together they account for about 83% of total brain mass and are the most conspicuous parts of an intact brain. Picture how a mushroom cap covers the top of its stalk, and you have a fairly good idea of how the paired cerebral hemispheres cover and obscure the diencephalon and the top of the brain stem (see Figure 12.4d).

? *Why are the lateral ventricles horn-shaped instead of being vertically erect structures like the third and fourth ventricles?*

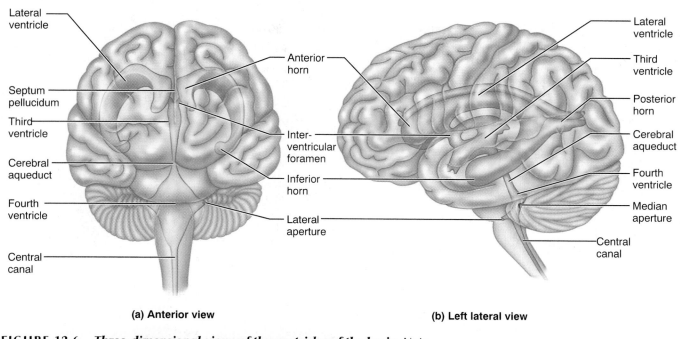

(a) Anterior view

(b) Left lateral view

FIGURE 12.6 *Three-dimensional views of the ventricles of the brain.* Note that different regions of the large lateral ventricles are indicated by the terms anterior horn, posterior horn, and inferior horn.

Nearly the entire surface of the cerebral hemispheres is marked by elevated ridges of tissue called **gyri** (ji′ri), literally "twisters," separated by shallow grooves called **sulci** (sul′ki), from the Greek meaning "furrows." (The singular forms of these terms are *gyrus* and *sulcus.*) The deeper grooves, called **fissures,** separate large regions of the brain. The more prominent gyri and sulci are similar in all people and are important anatomical landmarks. The median **longitudinal fissure** separates the cerebral hemispheres (see Figure 12.7b). Another large fissure, the **transverse fissure,** separates the cerebral hemispheres from the cerebellum below (see Figure 12.7a).

Some deep sulci divide each hemisphere into five lobes—frontal, parietal, temporal, occipital, and insula—all but the last named for the cranial bones that overlie them (see Figure 12.7a). The **central sulcus,** which lies in the frontal plane, separates the **frontal lobe** from the **parietal lobe.** Bordering the central sulcus are two important gyri, the **precentral gyrus** anteriorly and the **postcentral gyrus** posteri-

orly. More posteriorly, the **occipital lobe** is separated from the parietal lobe by the **parieto-occipital** (pah-ri″ĕ-to-ok-sip′ĭ-tal) **sulcus.** This sulcus is located on the medial surface of the hemisphere.

The deep **lateral sulcus** outlines the flaplike **temporal lobe** and separates it from the overlying parietal and frontal lobes. A fifth lobe of the cerebral hemisphere, the **insula** (in′su-lah; "island"), is buried deep within the lateral sulcus and forms part of its floor. The insula is covered by portions of the temporal, parietal, and frontal lobes (Figure 12.7a).

The cerebral hemispheres fit snugly into the skull. Rostrally, the frontal lobes lie in the anterior cranial fossa. The anterior parts of the temporal lobes fill the middle cranial fossa. The posterior cranial fossa, however, houses the brain stem and cerebellum; the occipital lobes are located well superior to that cranial fossa.

Each cerebral hemisphere has three basic regions: a superficial *cortex* of gray matter, which looks gray in fresh brain tissue; an internal *white matter;* and the *basal nuclei,* islands of gray matter situated deep within the white matter (see Figure 12.10). We consider these regions next.

Cerebral Cortex

The **cerebral cortex** is the "executive suite" of the nervous system, where our *conscious mind* is found.

The growth of the cerebral hemispheres is restricted by the presence of the forming embryonic skull such that the cerebral hemispheres are forced to grow posteriorly and inferiorly. Hence their cavities (ventricles) become horn-shaped.

FIGURE 12.7 *Lobes and fissures of the cerebral hemispheres.* (**a**) Left lateral view of the brain. Parts of the frontal and temporal lobes have been removed to reveal the lateral sulcus and insula. (**b**) Superior surface of the cerebral hemispheres.

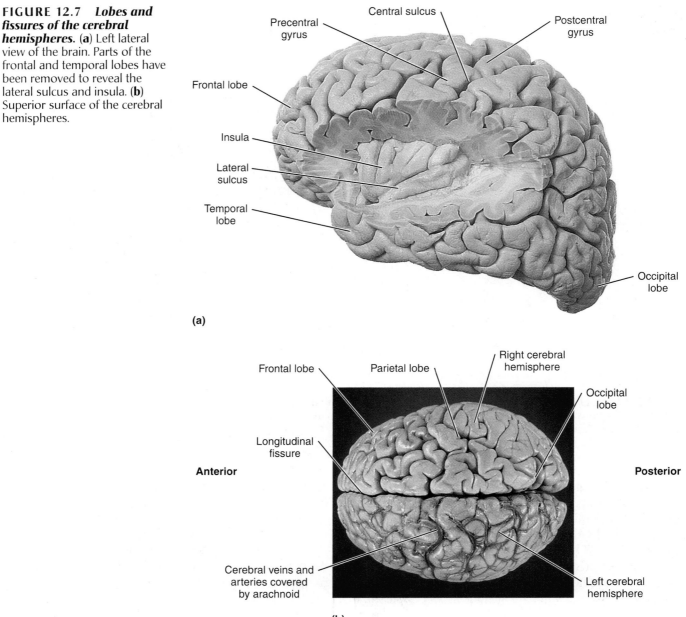

(a)

(b)

It enables us to be aware of ourselves and our sensations, to communicate, remember, and understand, and to initiate voluntary movements. Because it is composed of gray matter, the cerebral cortex consists of neuron cell bodies, dendrites, and unmyelinated axons (plus associated glia and blood vessels), but no fiber tracts. The cerebral cortex contains billions of neurons arranged in six layers, and it accounts for roughly 40% of total brain mass. Although it is only 2–4 mm (about ⅛ inch) thick, its many convolutions effectively triple its surface area.

In the late 1800s, anatomists were able to map subtle variations in its thickness and in the structure of the cerebral cortex. Most successful in these efforts was K. Brodmann, who produced an elaborate numbered mosaic of 52 different cortical areas, called **Brodmann areas,** in 1906. With a structural map emerging, early neurologists were anxious to localize *functional* regions of the cortex as well. Modern imaging techniques (PET scans to show maximal metabolic activity in the brain, or functional MRI scans to reveal blood flow) have

What anatomical landmark separates motor from sensory areas of the cerebral cortex?

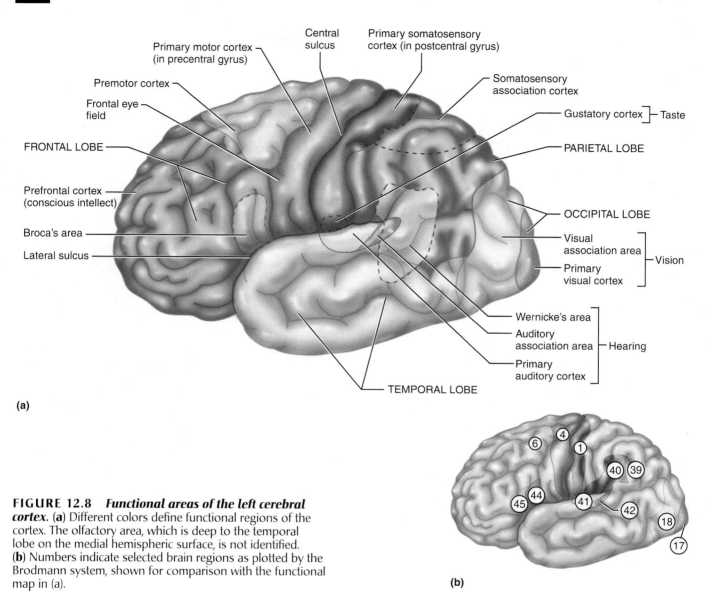

(a)

FIGURE 12.8 *Functional areas of the left cerebral cortex.* (a) Different colors define functional regions of the cortex. The olfactory area, which is deep to the temporal lobe on the medial hemispheric surface, is not identified. (b) Numbers indicate selected brain regions as plotted by the Brodmann system, shown for comparison with the functional map in (a).

(b)

shown that specific motor and sensory functions are localized in discrete cortical areas called *domains*. However, many higher mental functions, such as memory and language, appear to have overlapping domains and are spread over very large areas of the cortex. Some of the most important Brodmann areas are numbered in Figure 12.8b; however, our focus here is to examine the functional regions of the cerebral cortex only in the most general sense. But first, some generalizations about the cerebral cortex:

1. The cerebral cortex contains three kinds of functional areas: **motor areas** that control voluntary motor functions; **sensory areas** that provide for conscious awareness of sensation; and **association areas** that act mainly to integrate diverse information for purposeful action. As you read about these areas, do not confuse the sensory and motor areas of the cortex with sensory and motor neurons. All neurons in the cortex are interneurons.

2. Each hemisphere is chiefly concerned with the sensory and motor functions of the opposite (contralateral) side of the body.

3. Although largely symmetrical in structure, the two hemispheres are not entirely equal in function. Instead, there is a lateralization (specialization) of cortical functions, as discussed later.

? *Why are the motor and sensory homunculi "disformed" anatomically?*

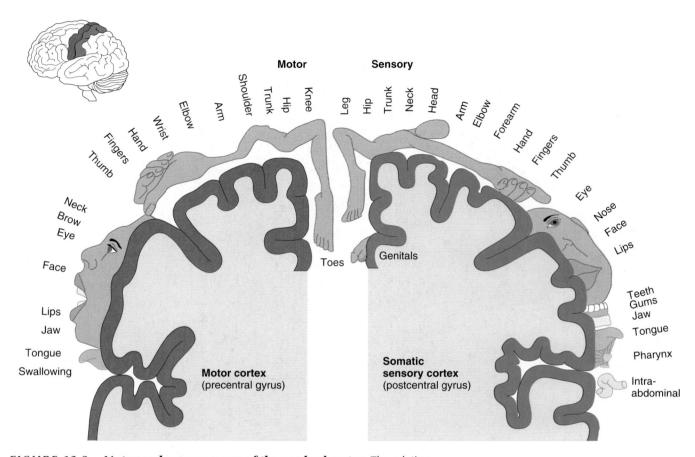

FIGURE 12.9 *Motor and sensory areas of the cerebral cortex.* The relative amount of cortical tissue devoted to each function is indicated by the amount of the gyrus occupied by the body area diagrams. The primary motor cortex in the precentral gyrus is represented on the left, and the somatic sensory cortex in the postcentral gyrus is represented on the right.

4. The final, and perhaps the most important, generalization to keep in mind is that our approach is a gross oversimplification; *no* functional area of the cortex acts alone, and conscious behavior involves the entire cortex in one way or another.

Motor Areas Cortical areas controlling motor functions lie in the posterior part of the frontal lobes. These include the primary motor cortex, the premotor cortex, Broca's area, and the frontal eye field (see Figure 12.8a).

1. Primary (somatic) motor cortex. The primary motor cortex is located in the precentral gyrus of the frontal lobe of each hemisphere (Brodmann area 4).

_____

Because the anatomical representations indicate the relative amount of the cerebral cortex dedicated to that particular body area.

Large neurons, called **pyramidal cells,** in these gyri allow us to consciously control the precise or skilled voluntary movements of our skeletal muscles. Their long axons, which project to the spinal cord, form the massive voluntary motor tracts called the **pyramidal,** or **corticospinal** (kor″tĭ-ko-spi′nal), **tracts.** All other descending motor tracts issue from brain stem nuclei and consist of chains of two, three, or more neurons.

The entire body is represented spatially in the primary motor cortex of each hemisphere. In other words, the pyramidal cells that control foot movements are in one place and those that control hand movements are in another. Such a mapping of the body in CNS structures is called **somatotopy** (so″mah-to-to′pe). As illustrated in Figure 12.9, the body is typically represented upside down—the head at the inferolateral part of the precentral gyrus, and the toes at the superomedial end. Most of the neurons in these gyri control muscles in body areas having the most precise motor control—that is, the face, tongue, and hands. Consequently, these regions of the caricature-like **motor homunculus** (ho-mung'ku-lus; "little man") drawn on the gyrus in Figure 12.9 are shown disproportionally large. The motor innervation of the body is **contralateral** (*contra* = opposite); that is, the left primary motor gyrus controls muscles on the right side of the body, and vice versa.

The motor homunculus view of the primary motor cortex implies a one-to-one correspondence between certain cortical neurons and the muscles they control, but this is somewhat misleading. Current research indicates that a given muscle is controlled by multiple spots on the cortex and that individual cortical motor neurons actually send impulses to more than one muscle. In other words, individual motor neurons control muscles that work together in a synergistic way. For example, reaching forward with one arm involves some muscles acting at the shoulder and some acting at the elbow. Thus, instead of the discrete map offered by the motor homunculus, the primary motor cortex map is an orderly but fuzzy map with neurons arranged in useful ways to control and coordinate sets of muscles. As expected, neurons controlling the arm are intermingled and overlap with those controlling interacting body areas such as the hand or shoulder. However, neurons controlling unrelated movements, such as those controlling the arm and those controlling body trunk muscles, do not cooperate in motor activity. Thus, the motor homunculus is useful to show that broad areas of the primary cortex are devoted to governing the leg, arm, torso, and head, but neuron organization within those broad areas is much more diffuse than initially imagined.

2. Premotor cortex. Just anterior to the precentral gyrus in the frontal lobe (Brodmann area 6) is the premotor cortex (see Figure 12.8). This region controls learned motor skills of a repetitious or patterned nature, such as playing a musical instrument and typing. The premotor cortex coordinates the movement of several muscle groups simultaneously or sequentially, mainly by sending activating impulses to the primary motor cortex. However, the premotor cortex also influences motor activity more directly by supplying about 15% of pyramidal tract fibers. Think of this region as the memory bank for skilled motor activities.

This area also appears to be involved in planning movements. Using highly processed sensory information received from other cortical areas, it can control voluntary actions that depend on sensory feedback, such as moving an arm through a maze to grasp a hidden object.

3. Broca's (bro'kahz) **area.** Broca's area lies anterior to the inferior region of the premotor area and overlaps Brodmann areas 44 and 45. Broca's area has long been considered to be (1) present in one hemisphere only (usually the left) and (2) a special *motor speech area* that directs the muscles of the tongue, throat, and lips involved in speech production. However, recent studies using PET scans to "light up" active areas of the brain show that this may not be the only function of Broca's area. These indicate that Broca's area becomes active as we prepare to speak and even as we think about (plan) many voluntary motor activities other than speech.

4. Frontal eye field. The frontal eye field is located partially in and anterior to the premotor cortex and superior to Broca's area. This cortical region controls voluntary movement of the eyes.

Homeostatic Imbalance

Damage to localized areas of the *primary motor cortex* (as from a stroke) paralyzes the body muscles controlled by those areas. If the lesion is in the right hemisphere, the left side of the body will be paralyzed. Only *voluntary* control is lost, however, as the muscles can still contract reflexively.

Destruction of the *premotor cortex*, or part of it, results in a loss of the motor skill(s) programmed in that particular region, but muscle strength and the ability to perform the discrete individual movements are not hindered. For example, if the premotor area controlling the flight of your fingers over the computer keyboard were damaged, you would not be able to type with your usual speed, but you could still make the same movements with your fingers. Reprogramming the skill into another set of premotor neurons would require practice, just as the initial learning process did. ■

Sensory Areas Unlike the motor areas, which are confined to the frontal lobe cortex, areas concerned with conscious awareness of sensation occur in the parietal, temporal, and occipital lobes (see Figure 12.8).

1. Primary somatosensory cortex. The primary somatosensory cortex resides in the postcentral gyrus of the parietal lobe, just posterior to the premotor cortex (Brodmann areas 1–3). Neurons in this gyrus receive information (relayed via a three-neuron synaptic chain) from the general (somatic) sensory receptors located in the skin and from proprioceptors in skeletal muscles. They then identify the body region being stimulated, an ability called **spatial discrimination**. As with the primary motor cortex, the body is represented spatially and in an upside-down fashion according to the site of stimulus input (see

Figure 12.9), and the right hemisphere receives input from the left side of the body. The amount of sensory cortex devoted to a particular body region is related to that region's sensitivity (that is, to how many receptors it has), not to the size of the body region. In humans, the face (especially the lips) and fingertips are the most sensitive body areas, hence these are the largest parts of the **somatosensory homunculus.**

2. Somatosensory association area. The somatosensory association area lies just posterior to the primary somatosensory cortex (see Figure 12.8) and has many connections with it. The major function of this area is to integrate different sensory inputs (temperature, pressure, etc.) relayed to it via the primary somatosensory cortex to produce a comprehensive understanding of an object being felt: its size, texture, and the relationship of its parts. For example, when you reach into your pocket, this area draws upon stored memories of past sensory experiences and perceives the objects you feel as coins or keys. Someone with damage to this area could not recognize these objects without looking at them.

3. Visual areas. The **primary visual (striate) cortex** is seen on the extreme posterior tip of the occipital lobe (Figure 12.8). Most of it, however, is buried in the medial aspect of the occipital lobe. The largest of all cortical sensory areas, the primary visual cortex receives visual information that originates on the retinas of the eyes. There is a map of visual space on the primary visual cortex, analogous to the body map on the somatosensory cortex. The right half of visual space is represented on the left visual cortex, the left half on the right cortex.

The **visual association area** surrounds the primary visual area and covers much of the occipital lobe (Figure 12.8). Communicating with the primary visual area, the visual association area interprets visual stimuli (color, form, and movement) using past visual experiences, enabling us to recognize a flower or a person's face and to appreciate what we are seeing. We do our "seeing" with these cortical neurons. However, recent experiments on monkeys indicate that complex visual processing involves the entire posterior half of the cerebral hemispheres. Particularly important are two visual "streams"—one running along the top of the brain that handles object identity, the other taking the lower road and focusing on object locations.

4. Auditory areas. Each **primary auditory cortex** is located in the superior margin of the temporal lobe abutting the lateral sulcus. Sound energy exciting the hearing (cochlear) receptors of the inner ear causes impulses to be transmitted to the primary auditory cortex, where they are related to pitch, rhythm, and loudness.

The more posterior **auditory association area** then permits the perception of the sound stimulus, which we "hear" as speech, a scream, music, thunder, noise, and so on. Memories of sounds heard in the past appear to be stored here for reference.

5. Olfactory (smell) cortex. The olfactory cortices are found in small areas of the frontal lobes just above the orbits, and in the medial aspects of the temporal lobes in a small region called the *piriform lobe* which is dominated by the hooklike *uncus.* Afferent fibers from the smell receptors in the superior nasal cavities send impulses along the olfactory tracts that are ultimately relayed to the olfactory cortices. The outcome is conscious awareness of different odors.

The olfactory cortex is part of the primitive **rhinencephalon** (ri″nen-sef′ah-lon; "nose brain"), which includes all parts of the cerebrum that receive olfactory signals. It includes the orbitofrontal cortex, the uncus and associated regions located on or in the medial aspects of the temporal lobes, and the protruding olfactory tracts and bulbs that extend to the nose region. During the course of evolution, most of the "old" rhinencephalon has taken on new functions concerned chiefly with emotions and memory. This "newer" emotional brain, called the *limbic system,* is considered on p. 452. The only portions of the rhinencephalon still devoted to smell in humans are the olfactory bulbs and tracts (described in Chapter 16) and the greatly reduced olfactory cortices.

6. Gustatory (taste) cortex. The gustatory (gus′tah-tor-e) cortex (see Figure 12.8a), a region involved in the perception of taste stimuli, is located in the parietal lobe just deep to the temporal lobe. Logically, it occurs at the tip of the tongue of the somatosensory homunculus.

7. Vestibular (equilibrium) cortex. It has been difficult to pin down the part of the cortex responsible for conscious awareness of balance, that is, of the position of the head in space. However, medical imaging studies now locate this region in the posterior part of the insula, deep to the temporal lobe.

Homeostatic Imbalance

Damage to the *primary visual cortex* results in functional blindness. By contrast, those with damage to the visual association area can see, but they do not comprehend what they are looking at. ∎

Association Areas Association areas include any cortical area that doesn't have the word *primary* in its name. As already described, the primary somatosensory cortex and each of the special sensory areas have nearby association areas with which they communicate. The association areas, in turn, communicate ("associate") with the motor cortex and with other sensory association areas to analyze, recognize, and act on sensory inputs in the light of past experience. Each of these areas has multiple inputs and outputs quite independent of the primary sensory and motor areas, indicating that their function is complex indeed. The remaining association areas are described next.

1. Prefrontal cortex. The prefrontal cortex, which occupies the anterior portions of the frontal lobes (see Figure 12.8), is the most complicated cortical region of all. It is involved with intellect, complex learning abilities (called cognition), recall, and personality. It is necessary for the production of abstract ideas, judgment, reasoning, persistence, long-term planning, concern for others, and conscience. These qualities all develop slowly in children, so it appears that the prefrontal cortex matures slowly and is heavily dependent on positive and negative feedback from one's social environment. The prefrontal cortex is closely linked to the emotional part of the brain (limbic system) and plays a role in intuitive judgments and mood. It is the tremendous elaboration of this region that sets human beings apart from other animals.

Tumors or other lesions of the *prefrontal cortex* may cause mental and personality disorders. Wide mood swings and loss of judgment, attentiveness, and inhibitions may occur. The affected individual may be oblivious to social restraints, perhaps becoming careless about personal appearance, or rashly attacking a 7-foot opponent rather than running. Psychoactive drugs are the therapy of choice for most such patients.

2. Language areas. A large continuous area for language comprehension and articulation surrounds the lateral sulcus in the left or language-dominant hemisphere. The best known parts of this area are: (1) **Wernicke's area** (Figure 12.8a), formerly believed to be the area responsible for understanding written and spoken language, now thought to be primarily involved in sounding out unfamiliar words; (2) Broca's area (mentioned previously) for speech production; (3) the *lateral prefrontal cortex,* involved in the more complex process of language comprehension and word analysis; (4) most of the lateral and ventral parts of the temporal lobe, which coordinate auditory and visual aspects of language as when naming objects or reading. (This is only part of the story, but you've probably heard enough.)

The corresponding areas in the right or non-language-dominant hemisphere are involved in "body language"—the nonverbal emotional (affective) components of language rather than speech mechanics. These areas allow the lilt or tone of our voice and our gestures to express our emotions when we speak, and permit us to comprehend the emotional content of what we hear. For example, a soft melodious response to your question conveys quite a different meaning than a sharp reply.

3. General (common) interpretation area. The general interpretation area is an ill-defined region encompassing parts of the temporal, parietal, and occipital lobes. It is found in one hemisphere only, usually the left. This region receives input from all the sensory association areas and integrates all incoming signals into a single thought or understanding of the situation. Suppose, for example, you drop a bottle of hydrochloric acid and it splashes on you. You see the bottle shatter; you hear the crash; you feel the burning on your skin; and you smell the acid fumes. However, these individual perceptions do not dominate your consciousness. What *does* is the overall message "danger"—by which time your leg muscles are already activated and are propelling you to the safety shower.

So the traditional story goes—but some more modern sources are abandoning this idea of a multisensory interpretation area or at least confining it to a much smaller region near the top and back of Wernicke's area. This change in theory reflects the newer studies which indicate that most of the relevant area is involved in the processing of spatial relationships.

4. Visceral association area. The cortex of the insula may be involved in conscious perception of visceral sensations (upset stomach, full bladder, and the like). However, a small part of the insular cortex functions in language.

Lateralization of Cortical Functioning We use both cerebral hemispheres for almost every activity, and the hemispheres appear nearly identical. Nonetheless, there *is* a division of labor, and each hemisphere has unique abilities not shared by its partner. This phenomenon is called **lateralization.** Although one cerebral hemisphere or the other "dominates" each task, the term **cerebral dominance** designates the hemisphere that is *dominant for language.* In most people (about 90%), the left hemisphere has greater control over language abilities, math, and logic. This so-called dominant hemisphere is working when we compose a sentence, balance a checkbook, and memorize a list of names. The other hemisphere (usually the right) is more free-spirited, involved in visual-spatial skills, intuition, emotion, and artistic and musical skills. It is the poetic, creative, and the "Ah-ha!" (insightful) side of our nature, and it is far better at recognizing faces. Most individuals with left cerebral dominance are right-handed.

In the remaining 10% of people, the roles of the hemispheres are reversed or the hemispheres share their functions equally. Typically, many right-cerebral-dominant people are left-handed and more often male. In "lefties" in whom the cerebral cortex functions bilaterally, the mutuality (or duality) of brain control sometimes results ambidexterity. In other cases, however, this phenomenon results in cerebral confusion ("Is it your turn, or mine?") and learning disabilities, such as the reading disorder *dyslexia*, in which otherwise intelligent people may reverse the order of letters in words (and of words in sentences).

The two cerebral hemispheres have perfect and almost instant communication with one another via connecting fiber tracts, as well as complete integration of their functions. Furthermore, although lateralization means that each hemisphere is better than the other at certain functions, neither side is better at everything.

Cerebral White Matter

From what has already been described, it is clear that communication within the brain is extensive. The **cerebral white matter** (see Figure 12.10) deep to the gray matter of the cortex is responsible for communication between cerebral areas and between the cerebral cortex and lower CNS centers. White matter consists largely of myelinated fibers bundled into large tracts. These fibers and the tracts they form are classified according to the direction in which they run as *commissural, association,* or *projection fibers.*

Commissures (kom′ĭ-shŭrz), composed of **commissural fibers,** connect corresponding gray areas of the two hemispheres, enabling them to function as a coordinated whole. The largest commissure is the **corpus callosum** (kah-lo′sum; "thickened body"), which lies superior to the lateral ventricles, deep within the longitudinal fissure (Figure 12.10b). Less important examples are the **anterior** and **posterior commissures.**

Association fibers connect different parts of the same hemisphere. Short association fibers connect adjacent gyri. Long association fibers are bundled into tracts and connect different cortical lobes.

Projection fibers include fibers entering the cerebral hemispheres from lower brain or cord centers, and fibers leaving the cortex to travel to lower areas. They tie the cortex to the rest of the nervous system and to the receptors and effectors of the body. In contrast to commissural and association fibers, which run horizontally, projection fibers run vertically.

At the upper limit of the brain stem, the projection fibers on each side form a compact band called the **internal capsule** (see Figures 12.10b and 12.11a) that passes between the thalamus and some of the basal nuclei. Beyond that point, they radiate fanlike through the cerebral white matter to the cortex. This distinctive arrangement of projection tract fibers is known as the **corona radiata** ("radiating crown").

Basal Nuclei

Deep within the cerebral white matter of each hemisphere is a group of subcortical nuclei called the **basal nuclei** or **basal ganglia.** * Although the definition of the precise structures forming the basal nuclei is controversial, most agree that the **caudate** (kaw′dāt) **nucleus, putamen** (pu-ta′men), and **globus pallidus** (glo′bis pal′ĭ-dus) constitute their main mass (see Figure 12.11). Together, the putamen ("pod") and globus pallidus ("pale globe") form a lens-shaped mass, called the **lentiform nucleus,** that flanks the internal capsule laterally. The comma-shaped caudate nucleus arches superiorly over the diencephalon. Collectively, the lentiform and caudate nuclei are called the **corpus striatum** (stri-a′tum) because the fibers of the internal capsule that course past and through them give them a striped appearance (see Figure 12.11a). The basal nuclei are functionally associated with the *subthalamic nuclei* (located in the lateral "floor" of the diencephalon) and the *substantia nigra* of the midbrain (Figure 12.16a).

The almond-shaped **amygdala** (ah-mig′dah-lah; "almond") sits on the tail of the caudate nucleus. Traditionally, it has been grouped with the basal nuclei, but functionally it belongs to the limbic system (see p. 452).

The striatum of the basal nuclei receives inputs from the entire cerebral cortex, as well as from other subcortical nuclei and each other. Via relays through the thalamus, the output nuclei of the basal nuclei (globus pallidus and the substantia nigra) project to the premotor and prefrontal cortices and so influence muscle movements directed by the primary motor cortex. The basal nuclei have no direct access to motor pathways.

The precise role of the basal nuclei has been elusive because of their inaccessible location and because their functions overlap to some extent with those of the cerebellum. The role of the basal nuclei in motor control is very complex and there is evidence that they also play some role in regulating attention and in cognition. The basal nuclei are particularly important in starting, stopping, and monitoring movements executed by the cortex, especially those that are relatively slow or stereotyped, such as arm-swinging during walking. They also regulate the *intensity* of these

---

*Since a nucleus is a collection of nerve cell bodies within the CNS, the term *basal nuclei* is technically correct. The more frequently used but misleading historical term *basal ganglia* is a misnomer and should be abandoned, because ganglia are PNS structures.

Commissures allow the cerebral hemispheres to "talk to each other."

How can you explain the observation that the two cerebral hemispheres have instant communication when it is quite obvious in diagram (b) that the projection fibers are part of crossed pathways and information flows to only one (the opposite) hemisphere from each side?

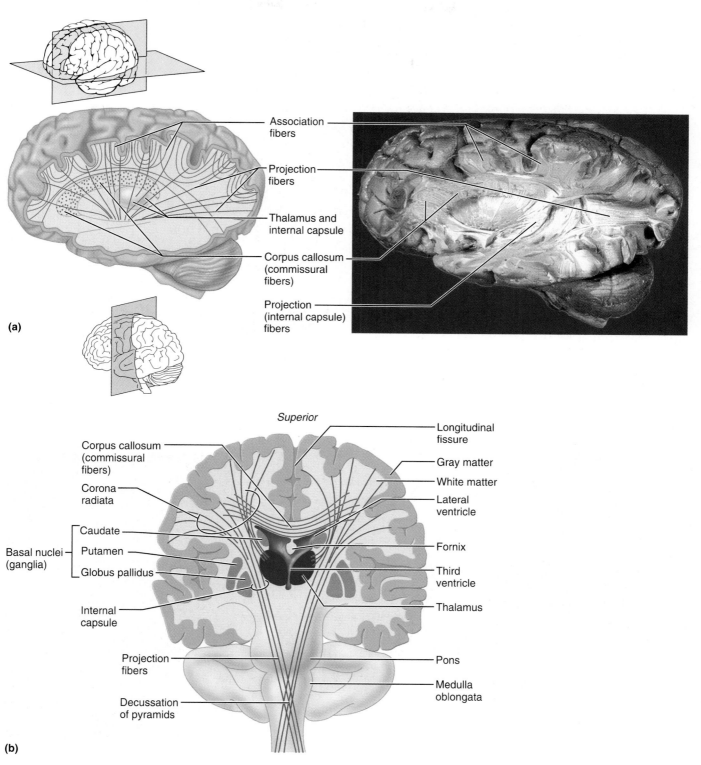

(a)

Association fibers

Projection fibers

Thalamus and internal capsule

Corpus callosum (commissural fibers)

Projection (internal capsule) fibers

Superior

Corpus callosum (commissural fibers)

Corona radiata

Basal nuclei (ganglia)
- Caudate
- Putamen
- Globus pallidus

Internal capsule

Projection fibers

Decussation of pyramids

Longitudinal fissure

Gray matter

White matter

Lateral ventricle

Fornix

Third ventricle

Thalamus

Pons

Medulla oblongata

(b)

FIGURE 12.10 *Types of fiber tracts in the white matter of the cerebral medulla.* (a) Midsagittal views of the right cerebral hemisphere showing the association tracts (tracts connecting different parts of the same hemisphere), and the corpus callosum, a commissure that connects the hemispheres. **(b)** Frontal section showing commissural fibers of the cerebrum and projection fibers that run between the cerebrum and lower CNS centers. Notice the tight band of projection fibers called the internal capsule that passes between the thalamus and the basal nuclei, and then fans out as the corona radiata.

? *Why are the lentiform nucleus and caudate nucleus collectively referred to as the corpus striatum?*

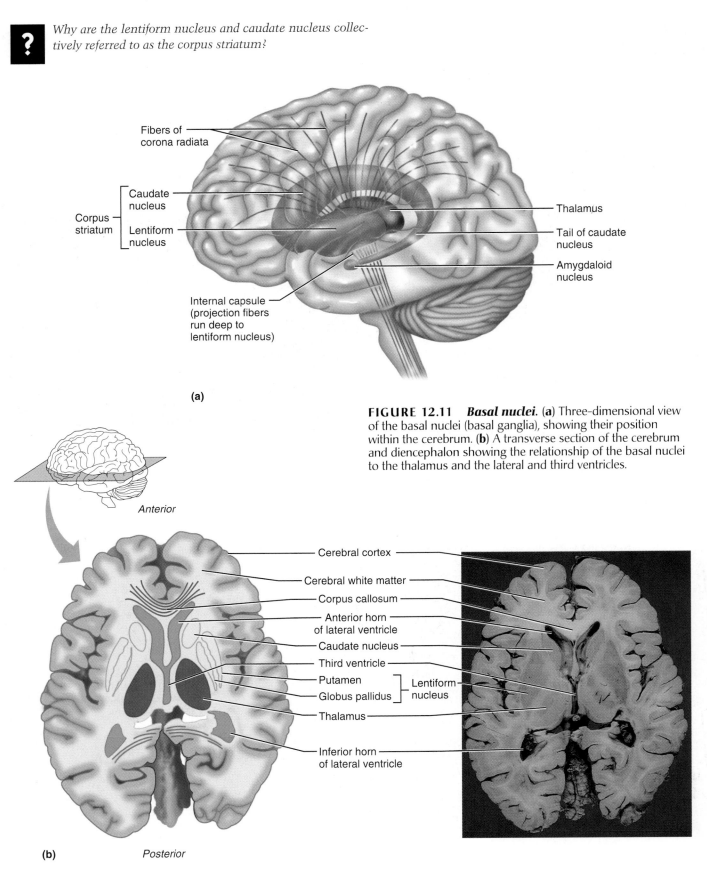

(a)

FIGURE 12.11   *Basal nuclei.* (a) Three-dimensional view of the basal nuclei (basal ganglia), showing their position within the cerebrum. **(b)** A transverse section of the cerebrum and diencephalon showing the relationship of the basal nuclei to the thalamus and the lateral and third ventricles.

(b)

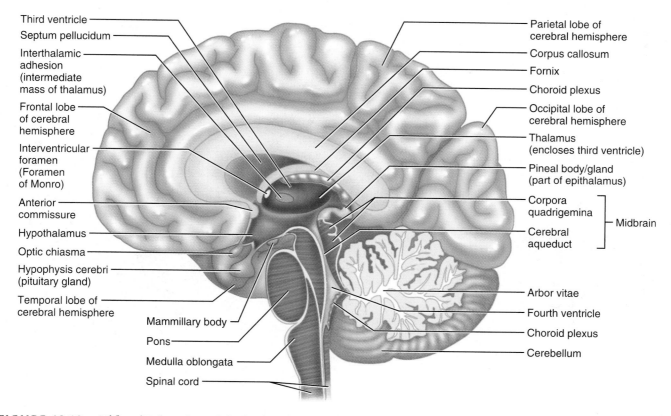

Third ventricle
Septum pellucidum
Interthalamic adhesion (intermediate mass of thalamus)
Frontal lobe of cerebral hemisphere
Interventricular foramen (Foramen of Monro)
Anterior commissure
Hypothalamus
Optic chiasma
Hypophysis cerebri (pituitary gland)
Temporal lobe of cerebral hemisphere
Mammillary body
Pons
Medulla oblongata
Spinal cord

Parietal lobe of cerebral hemisphere
Corpus callosum
Fornix
Choroid plexus
Occipital lobe of cerebral hemisphere
Thalamus (encloses third ventricle)
Pineal body/gland (part of epithalamus)
Corpora quadrigemina
Cerebral aqueduct
Midbrain
Arbor vitae
Fourth ventricle
Choroid plexus
Cerebellum

FIGURE 12.12 *Midsagittal section of the brain illustrating the diencephalon and brain stem structures.*

movements, resembling the motor drive that regulates how fast a mechanical motor works. Additionally, they inhibit antagonistic or unnecessary movements; thus their input seems necessary to our ability to perform several activities at once. Disorders of the basal nuclei result in either too much or too little movement as exemplified by Huntington's chorea and Parkinson's disease, respectively (see p. 459). The regulatory activities and role in skills memory of the basal nuclei are described in more detail in Chapter 15.

The Diencephalon

The **diencephalon** forms the central core of the forebrain and is surrounded by the cerebral hemispheres. It consists largely of three paired structures—the thalamus, hypothalamus, and epithalamus. These gray matter areas collectively enclose the third ventricle (see Figure 12.12).

The Thalamus

The egg-shaped **thalamus** makes up 80% of the diencephalon and forms the superolateral walls of the third ventricle (Figures 12.10 and 12.12). In some people, its bilateral masses of gray matter are held together by a midline connection called the **interthalamic adhesion (intermediate mass)**. *Thala-*

mus is a Greek word meaning "inner room," which well describes this deep, well-hidden brain region.

The thalamus contains about a dozen nuclei, most named according to their relative location (Figure 12.13a). Each of these nuclei has a functional specialty, and each projects fibers to and receives fibers from a specific region of the cerebral cortex. Afferent impulses from all senses and all parts of the body converge on the thalamus and synapse with at least one of its nuclei. For example, the *ventral posterior lateral nucleus* receives impulses from the general somatic sensory receptors (touch, pressure, pain, etc.), and the *lateral* and *medial geniculate* (jĕ-nik'u-lāt; "knee shaped") *bodies* are important visual and auditory relay centers, respectively. Within the thalamus, a sorting-out and information "editing" process occurs. Impulses having to do with similar functions are relayed as a group via the internal capsule to the appropriate area of the sensory cortex as well as to specific cortical association areas. As the afferent impulses reach the thalamus, we have a crude recognition of the sensation as pleasant or unpleasant. However, specific stimulus localization and discrimination occur in the cerebral cortex.

In addition to sensory inputs, virtually *all* inputs ascending to the cerebral cortex funnel through thalamic nuclei. These include impulses participating in the regulation of emotion and visceral function

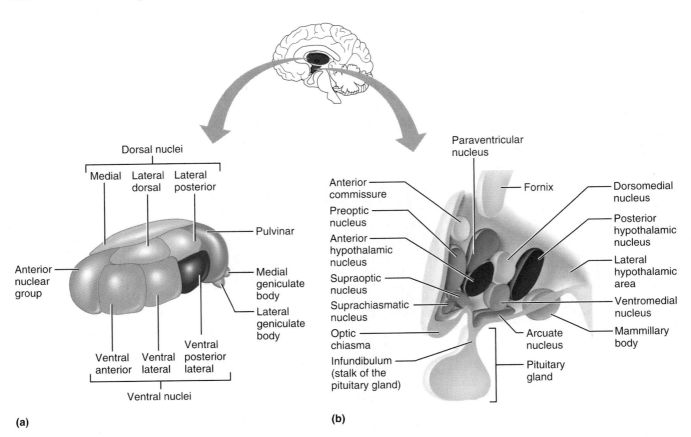

FIGURE 12.13   *Selected structures of the diencephalon.* **(a)** Thalamus showing the main thalamic nuclei. (The reticular nuclei that "cap" the thalamic nuclei laterally are depicted as curving translucent structures.) **(b)** The main hypothalamic nuclei.

from the hypothalamus (via the anterior nuclei), and some that help to direct the activity of the motor cortices from the cerebellum and basal nuclei (via the ventral lateral and ventral anterior nuclei, respectively). Several thalamic nuclei (pulvinar, lateral dorsal, and lateral posterior nuclei) are involved in integration of sensory information and project to specific association cortices. Thus the thalamus plays a key role in mediating sensation, motor activities, cortical arousal, learning, and memory. It is truly the gateway to the cerebral cortex.

The Hypothalamus

The **hypothalamus,** named for its position below *(hypo)* the thalamus, caps the top of the brain stem and forms the inferolateral walls of the third ventricle (see Figure 12.12). Merging into the midbrain inferiorly, it extends from the optic chiasma (point of crossover of the optic nerves) to the posterior margin of the mammillary bodies. The **mammillary** (mam′mil-er-e; "little breast") **bodies,** paired pealike nuclei that bulge anteriorly from the hypothalamus, are relay stations in the olfactory pathways. Between the optic chiasma and mammillary bodies is the **infundibulum** (in″fun-dib′u-lum), a stalk of hypothalamic tissue that connects the **pituitary gland** to the

base of the hypothalamus. Like the thalamus, the hypothalamus contains many functionally important nuclei (Figure 12.13b).

Despite its small size, the hypothalamus is the main visceral control center of the body and is vitally important to overall body homeostasis. Few tissues in the body escape its influence. Its chief homeostatic roles are summarized next.

1. Autonomic control center. As you will remember, the autonomic nervous system (ANS) is a system of peripheral nerves that regulates cardiac and smooth muscle and secretion by the glands. The hypothalamus regulates ANS activity by controlling the activity of centers in the brain stem and spinal cord. In this role, the hypothalamus influences blood pressure, rate and force of heartbeat, digestive tract motility, respiratory rate and depth, eye pupil size, and many other visceral activities as described in Chapter 14.

2. Center for emotional response. The hypothalamus lies at the "heart" of the limbic system (the emotional part of the brain). Nuclei involved in the perception of pleasure, fear, and rage, as well as those involved in biological rhythms and drives (such as the sex drive), are found in the hypothalamus.

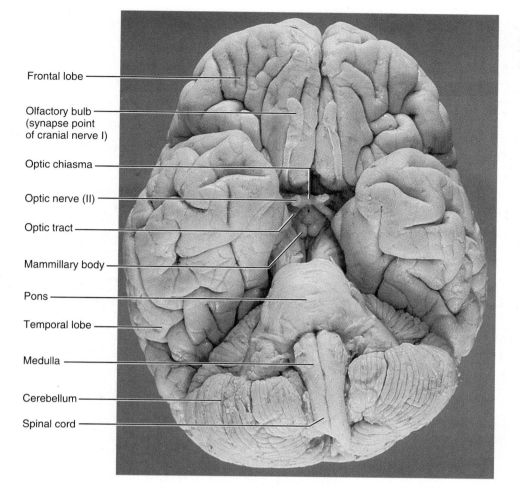

Frontal lobe

Olfactory bulb
(synapse point
of cranial nerve I)

Optic chiasma

Optic nerve (II)

Optic tract

Mammillary body

Pons

Temporal lobe

Medulla

Cerebellum

Spinal cord

FIGURE 12.14 *Ventral aspect of the human brain, showing the three regions of the brain stem.* Only a small portion of the midbrain can be seen; the rest is surrounded by other brain regions.

The hypothalamus acts through ANS pathways to initiate most physical expressions of emotion. For example, a fearful person has a pounding heart, high blood pressure, pallor, sweating, and a dry mouth.

3. Body temperature regulation. The body's thermostat is in the hypothalamus. Thermoreceptors located in other parts of the brain as well as in the body periphery, and certain hypothalamic neurons (particularly those in the *preoptic region*) monitor blood temperature. Accordingly, the hypothalamus initiates cooling (sweating) or heat-retention mechanisms (shivering) as needed to maintain relatively constant body temperature. (See Chapter 25.)

4. Regulation of food intake. In response to changing blood levels of certain nutrients (glucose and perhaps amino acids) or hormones (insulin and others), the hypothalamus regulates feelings of hunger and satiety (see Chapter 25). Its *ventromedial nuclei* appear to play a role in satiety.

5. Regulation of water balance and thirst. When body fluids become too concentrated, hypothalamic neurons called *osmoreceptors* are activated. They, in turn, excite hypothalamic nuclei that trigger the release of antidiuretic hormone (ADH) from the posterior pituitary. ADH causes the kidneys to retain water. The same conditions also stimulate hypo-

thalamic neurons in the *thirst center*, causing us to drink fluids.

6. Regulation of sleep-wake cycles. Acting with other brain regions, the hypothalamus helps regulate sleep. The hypothalamus, through the operation of its *suprachiasmatic nucleus* (our biological clock), sets the timing of the sleep cycle in response to daylight-darkness cues received from the visual pathways.

7. Control of endocrine system functioning. The hypothalamus acts as the helmsman of the endocrine system in two important ways. First, its *releasing hormones* control the secretion of hormones by the anterior pituitary gland. Second, two of its nuclei, the *supraoptic* and *paraventricular nuclei*, produce hormones called ADH and oxytocin. The endocrine roles of the hypothalamus are described in more detail in Chapter 17.

◢ *Homeostatic Imbalance*

Hypothalamic disturbances cause a number of disorders in body homeostasis, such as severe body wasting or obesity, sleep disturbances, dehydration, and a broad range of emotional imbalances. For example, infants deprived of a warm, nurturing relationship may develop sleep disorders and fail to thrive. ■

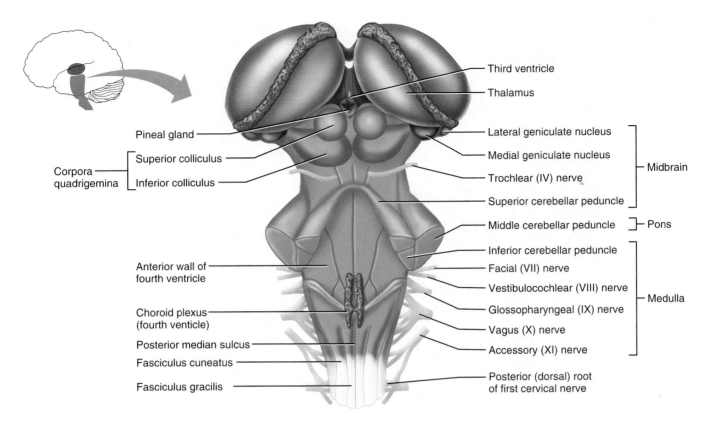

FIGURE 12.15 *Relationship of the brain stem to the diencephalon, posterior superficial view.* Cerebral hemispheres have been removed.

The Epithalamus

The **epithalamus** is the most dorsal portion of the diencephalon and forms the roof of the third ventricle. Extending from its posterior border and visible externally is the **pineal** (pin'e-al; "pine cone shaped") **body** or **gland** (see Figures 12.12 and 12.15). The pineal gland secretes the hormone *melatonin* (the sleep-inducing signal) and, along with hypothalamic nuclei, helps regulate the sleep-wake cycle and some aspects of mood (as discussed further in Chapter 17). A cerebrospinal fluid-forming structure called a **choroid plexus** (ko'roid plek'sus) is also part of the epithalamus (see Figure 12.12).

The Brain Stem

From superior to inferior, the brain stem regions are the midbrain, pons, and medulla oblongata (see Figures 12.12, 12.14, and 12.15). Each of these regions is roughly an inch long. Histologically, the organization of the brain stem is similar (but not identical) to that of the spinal cord—deep gray matter surrounded by white matter fiber tracts (see Figure 12.5). However, there are nuclei of gray matter embedded in the white matter.

Brain stem centers produce the rigidly programmed, automatic behaviors necessary for our survival. Positioned between the cerebrum and the spinal cord, the brain stem also provides a pathway for fiber tracts running between higher and lower neural centers. Additionally, brain stem nuclei are associated with 10 of the 12 pairs of cranial nerves (described in Chapter 13), so it is heavily involved with innervation of the head.

The Midbrain

The **midbrain** is located between the diencephalon and the pons more inferiorly (see Figure 12.14). On its ventral aspect the two bulging **cerebral peduncles** (pĕ-dung'klz) form vertical pillars that seem to hold up the cerebrum, hence their name meaning "little feet of the cerebrum." These peduncles contain the large pyramidal (corticospinal) motor tracts descending toward the spinal cord. The *superior cerebellar peduncles,* also fiber tracts, connect the midbrain to the cerebellum dorsally (Figure 12.15a).

Running through the midbrain is the hollow **cerebral aqueduct,** which connects the third and fourth ventricles, and delineates the cerebral peduncles ventrally from the midbrain's roof, called the *tectum.* Surrounding the aqueduct is the periaqueductal gray matter (Figure 12.16a) that is involved in pain suppression and serves as the link between the fear-perceiving amygdala and ANS pathways that control the "fight or flight" response. It also includes nuclei that control two cranial nerves, the *oculomotor* (III) and the *trochlear* (trok'le-ar) (IV) *nuclei* (see Figures 12.15 and 12.16a).

Just what are the pyramids of the medulla?

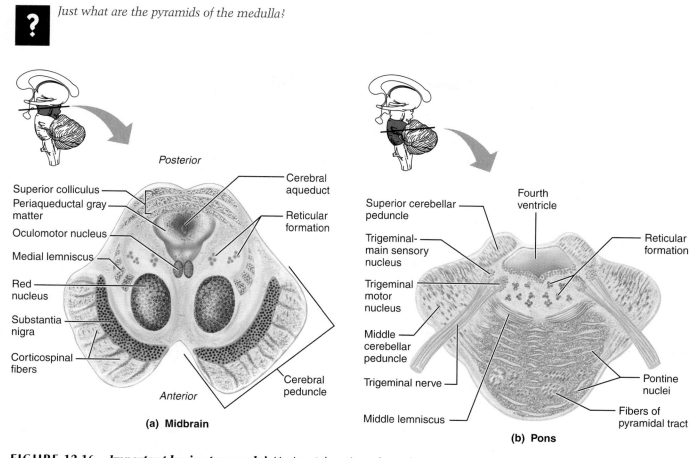

FIGURE 12.16 *Important brain stem nuclei.* Horizontal sections through **(a)** the midbrain at the level of the superior colliculi, **(b)** the pons at the level of its cranial nerve nuclei.

Nuclei are also scattered in the surrounding white matter of the midbrain. The largest of these nuclei are the **corpora quadrigemina** (kor′por-ah kwod″ri-jem′i-nah; "quadruplets"), which raise four domelike protrusions on the dorsal midbrain surface (see Figures 12.12 and 12.15). The superior pair of nuclei, the **superior colliculi** (kŏ-lik′u-li), are visual reflex centers that coordinate head and eye movements when we visually follow or track a moving object, even if we are not consciously looking at the object. The **inferior colliculi** are part of the auditory relay from the hearing receptors of the ear to the sensory cortex. They also act in reflexive responses to sound, such as in the *startle reflex,* which causes you to turn your head toward an unexpected sound.

Also embedded in the white matter of the midbrain are two pigmented nuclei, the substantia nigra and red nucleus (Figure 12.16a). The bandlike **substantia nigra** (sub-stan′she-ah ni′grah) is located deep to the cerebral peduncle. Its dark (*nigr* = black) color reflects the high content of melanin pigment, a precursor of the neurotransmitter (dopamine) released by these neurons. The substantia nigra is functionally linked to the basal nuclei (its axons project to the globus pallidus), and is considered part of

the basal nuclear complex by many authorities. Degeneration of the dopamine-releasing neurons of the substantia nigra is the ultimate cause of Parkinson's disease.

The oval **red nucleus** lies deep to the substantia nigra. Its reddish hue is due to its rich blood supply and to the presence of iron pigment in the cell bodies of its neurons. The red nuclei are relay nuclei in some descending motor pathways that effect limb flexion. The red nucleus is the largest nucleus of the *reticular formation,* a system of small nuclei scattered through the core of the brain stem (see p. 453).

The Pons

The **pons** is the bulging brain stem region wedged between the midbrain and the medulla oblongata (see Figures 12.12, 12.14, and 12.15). Dorsally, it forms part of the anterior wall of the fourth ventricle.

The corticospinal (pyramidal) tracts. The large voluntary motor tracts descending from the motor cortex.

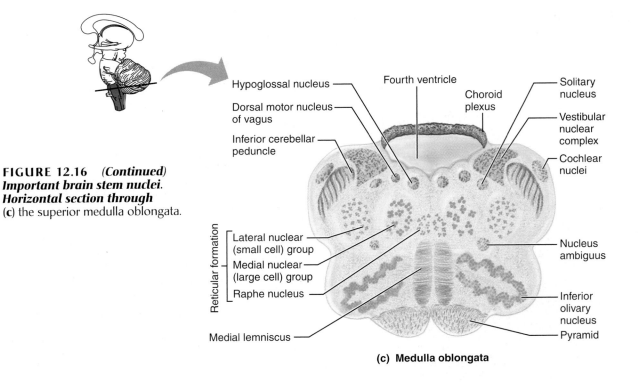

FIGURE 12.16 *(Continued)*
Important brain stem nuclei.
Horizontal section through
(**c**) the superior medulla oblongata.

(c) Medulla oblongata

As its name suggests (*pons* = bridge), the pons is chiefly composed of conduction tracts. Pons tracts course in two directions. The deep projection fibers run longitudinally and complete the pathway between higher brain centers and the spinal cord. The more superficial ventral fibers issue from numerous *pontine nuclei*, which act as relays for "conversations" between the motor cortex and the cerebellum. These fibers are oriented transversely and dorsally as the *middle cerebellar peduncles* (Figure 12.15a) and connect the pons bilaterally with the two sides of the cerebellum posteriorly.

Several cranial nerve pairs issue from pons nuclei (see Figure 12.16b). These include the *trigeminal* (tri-jem′ĭ-nal) *nerves* (V), the *abducens* (ab-du′senz) *nerves* (VI), and the *facial nerves* (VII). These cranial nerves and their functions are discussed in Chapter 13. Other important pons nuclei are part of the reticular formation. The *pneumotaxic center*, for example, is a respiratory center. Together with medullary respiratory centers, it helps to maintain the normal rhythm of breathing.

The Medulla Oblongata

The conical **medulla oblongata** (me-dul′ah ob″long-gah′tah), or simply **medulla,** is the most inferior part of the brain stem. It blends imperceptibly into the spinal cord at the level of the foramen magnum of the skull (see Figures 12.12 and 12.14). The central canal of the spinal cord continues upward into the medulla, where it broadens out to form the cavity of the fourth ventricle. Thus, both the medulla and the pons help to form the ventral wall of the fourth ventricle. (The

dorsal wall of the ventricle is formed by a thin capillary-rich membrane called a choroid plexus which abuts the cerebellum dorsally [see Figure 12.12].)

The medulla has several externally visible landmarks (see Figure 12.14). Flanking the midline on its ventral aspect are two obvious longitudinal ridges called the **pyramids,** formed by the large pyramidal (corticospinal) tracts descending from the motor cortex. Just above the medulla–spinal cord junction, most of these fibers cross over to the opposite side before continuing their descent into the spinal cord. This crossover point is called the **decussation** (de″kus-sa′shun; "a crossing") **of the pyramids.** As mentioned earlier, the consequence of this crossover is that each cerebral hemisphere chiefly controls the voluntary movements of muscles on the opposite (*contralateral*) side of the body.

Also obvious externally are the inferior cerebellar peduncles, the olives, and the inferiormost five pairs of cranial nerves. The *inferior cerebellar peduncles* are fiber tracts that connect the medulla to the cerebellum dorsally. Situated lateral to the pyramids, the **olives** are oval swellings (which *do* resemble olives) produced mainly by the underlying **inferior olivary nuclei,** wavy folds of gray matter (Figure 12.16c). The olivary nuclei relay sensory information on the state of stretch of our muscles and joints to the cerebellum. The rootlets of the *hypoglossal nerves* (XII) emerge from the groove between the pyramid and olive on each side of the brain stem. Other cranial nerves associated with the medulla are the *glossopharyngeal nerves* (IX), the *vagus nerves* (X), and portions of the *accessory nerves* (XI) (see Figure

TABLE 12.1	Functions of Major Brain Regions

Region	Function

Cerebral hemispheres (pp. 432–443)

Cortical gray matter localizes and interprets sensory inputs, controls voluntary and skilled skeletal muscle activity, and functions in intellectual and emotional processing; basal nuclei (ganglia) are subcortical motor centers important in initiation of skeletal muscle movements

Diencephalon (pp. 443–446)

Thalamic nuclei are relay stations in conduction of (1) sensory impulses to cerebral cortex for interpretation, and (2) impulses to and from cerebral motor cortex and lower (subcortical) motor centers, including cerebellum; thalamus is also involved in memory processing

Hypothalamus is chief integration center of autonomic (involuntary) nervous system; it functions in regulation of body temperature, food intake, water balance, thirst, and biological rhythms and drives; regulates hormonal output of anterior pituitary gland and is an endocrine organ in its own right (produces ADH and oxytocin); part of limbic system

Limbic system (pp. 452–453)

A functional system involving cerebral and diencephalon structures that mediates emotional response; also involved in memory processing

Brain stem (pp. 446–450)

Midbrain: Conduction pathway between higher and lower brain centers (e.g., cerebral peduncles contain the fibers of the pyramidal tracts); its superior and inferior colliculi are visual and auditory reflex centers; substantia nigra and red nuclei are subcortical motor centers; contains nuclei for cranial nerves III and IV

Pons: Conduction pathway between higher and lower brain centers; pontine nuclei relay information from the cerebrum to the cerebellum; its respiratory nuclei cooperate with the medullary respiratory centers to control respiratory rate and depth; houses nuclei of cranial nerves V–VII

Medulla oblongata: Conduction pathway between higher brain centers and spinal cord; site of decussation of the pyramidal tracts; houses nuclei of cranial nerves VIII–XII; contains nuclei cuneatus and gracilis (synapse points of ascending sensory pathways transmitting sensory impulses from skin and proprioceptors), and visceral nuclei controlling heart rate, blood vessel diameter, respiratory rate, vomiting, coughing, etc.; its inferior olivary nuclei provide the sensory relay to the cerebellum

Reticular formation (p. 453)

A functional brain stem system that maintains cerebral cortical alertness (reticular activating system) and filters out repetitive stimuli; its motor nuclei help regulate skeletal and visceral muscle activity

Cerebellum (pp. 450–451)

Processes information from cerebral motor cortex and from proprioceptors and visual and equilibrium pathways, and provides "instructions" to cerebral motor cortex and subcortical motor centers that result in proper balance and posture and smooth, coordinated skeletal muscle movements

12.15). Additionally, the fibers of the *vestibulocochlear* (ves-tib"u-lo-kok'le-ar) *nerves* (VIII) synapse with the **cochlear nuclei** (auditory relays), and with numerous vestibular nuclei in both the pons and medulla (see Figure 12.16c). Collectively, the vestibular nuclei, called the **vestibular nuclear complex,** mediate responses that maintain equilibrium.

Also housed within the medulla are several nuclei associated with ascending sensory tracts. The most prominent of these are the dorsally located **nucleus gracilis** (grah-si'lis) and **nucleus cuneatus** (ku'ne-āt-us), associated with the **medial lemniscal tract** (see Figure 12.16c and Figure 15.2 on p. 538). These serve as relay nuclei in a pathway by which general somatic sensory information ascends from the spinal cord to the somatosensory cortex.

The small size of the medulla belies its crucial role as an autonomic reflex center involved in maintaining body homeostasis. Important visceral motor nuclei found in the medulla include the following:

1. The cardiovascular center. The cardiovascular center includes cardiac and vasomotor centers. The *cardiac center* adjusts the force and rate of heart contraction to meet the body's needs. The *vasomotor center* regulates blood pressure by acting on smooth muscle in the walls of blood vessels to effect changes in blood vessel diameter.

2. The respiratory centers. The medullary respiratory centers control the rate and depth of breathing and (in a negative feedback interaction with centers of the pons) maintain respiratory rhythm.

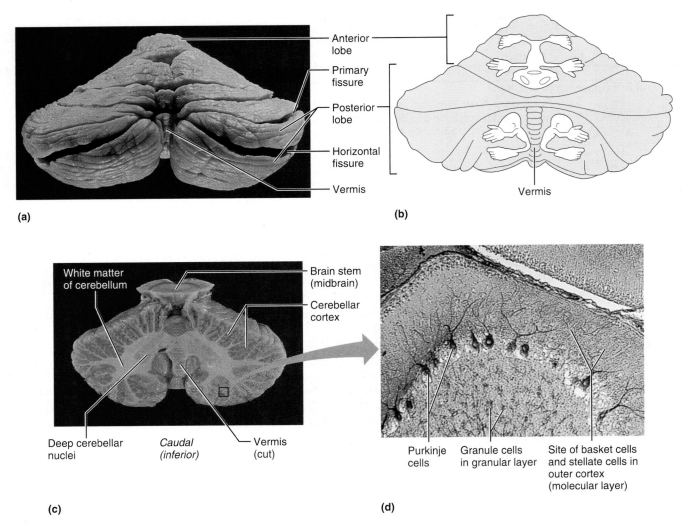

(a)

(b)

(c)

(d)

FIGURE 12.17 *Cerebellum.*
(**a**) Photograph of the posterior aspect of the cerebellum. The flocculonodular lobe, which lies deep to the vermis, is not visible in this view. (**b**) Positioning of overlapping motor and sensory maps of the body (in the form of three homunculi) in the cerebellum. (**c**) Posterior view of the cerebellum sectioned frontally to reveal its three layers. (**d**) Photomicrograph of a small portion of the cerebellar cortex showing the relative positions of the granule cells, Purkinje cells, and stellate and basket cells (115×).

3. Various other centers. Additional centers regulate activities such as vomiting, hiccuping, swallowing, coughing, and sneezing.

Notice that many functions listed above were also attributed to the hypothalamus (pp. 444–445). The overlap is easily explained. The hypothalamus exerts its control over many visceral functions by relaying its instructions through medullary reticular centers, which carry them out.

The Cerebellum

The cauliflower-like **cerebellum** (ser″ĕ-bel′um), literally, our "small brain," is exceeded in size only by the cerebrum. It accounts for about 11% of total brain mass. The cerebellum is located dorsal to the pons and medulla (and to the intervening fourth ventricle). It protrudes under the occipital lobes of the cerebral hemispheres, from which it is separated by the transverse fissure (see Figure 12.7).

The cerebellum processes inputs received from the cerebral motor cortex, various brain stem nuclei, and sensory receptors to provide the precise timing and appropriate patterns of skeletal muscle contraction for the smooth, coordinated movements and agility needed for our daily living—driving, typing, and for some of us, playing the tuba. Cerebellar activity occurs subconsciously; that is, we have no awareness of its functioning.

Anatomy of the Cerebellum

The cerebellum is bilaterally symmetrical; its two apple-sized **cerebellar hemispheres** are connected medially by the wormlike **vermis** (Figure 12.17). Its surface is heavily convoluted, with fine pleatlike gyri known as **folia** ("leaves") that are transversely oriented. Deep fissures subdivide each hemisphere into three lobes: the **anterior, posterior,** and **flocculo-**

nodular (flok"u-lo-nod'u-lar) **lobes.** The small propeller-shaped flocculonodular lobes, which are situated deep to the vermis and posterior lobe, cannot be seen in a surface view.

Like the cerebrum, the cerebellum has a thin outer cortex of gray matter, internal white matter, and small, deeply situated, paired masses of gray matter, the most familiar of which are the *dentate nuclei.* Several types of neurons populate the cerebellar cortex, including stellate, basket, granule, and Purkinje cells (Figure 12.17d). Of these, the large **Purkinje cells,** with their extensively branched dendrites, are the only cortical neurons that send their axons through the white matter to synapse with the central nuclei of the cerebellum. The distinctive pattern of white matter in the cerebellum resembles a branching tree, a pattern fancifully called the **arbor vitae** (ar'bor vi'te), literally, "tree of life" (see Figure 12.12 and 12.17c).

The anterior and posterior lobes of the cerebellum, which act to coordinate body movements, have completely overlapping sensory and motor maps of the entire body (see the three homunculi in Figure 12.17b). The medial portions influence the motor activities of the trunk and girdle muscles. The intermediate parts of each hemisphere are more concerned with the distal parts of the limbs and skilled movements. The lateralmost parts of each hemisphere integrate information from the association areas of the cerebral cortex and appear to play a role in planning rather than executing movements. The small flocculonodular lobes receive inputs from the equilibrium apparatus of the inner ears, and adjust posture to maintain balance.

Cerebellar Peduncles

As noted earlier, three paired fiber tracts—the cerebellar peduncles—connect the cerebellum to the brain stem (see Figure 12.15). Unlike the contralateral fiber distribution to and from the cerebral cortex, virtually all fibers entering and leaving the cerebellum are **ipsilateral** (*ipsi* = same)—from and to the *same* side of the body. The **superior cerebellar peduncles** connecting the cerebellum and midbrain carry instructions from neurons in the deep cerebellar nuclei to the cerebral motor cortex via thalamic relays. Like the basal nuclei, the cerebellum has no *direct* connections with the cerebral cortex.

The **middle cerebellar peduncles** connect the pons and cerebellum. They carry one-way communication from the pons to the cerebellum, advising the cerebellum of voluntary motor activities initiated by the motor cortex (via relays in the pontine nuclei). The **inferior cerebellar peduncles** connect the medulla and the cerebellum. These peduncles convey sensory information to the cerebellum from (1) muscle proprioceptors throughout the body and (2) the vestibular nuclei of the brain stem, which are concerned with equilibrium and balance.

Cerebellar Processing

The functional scheme of cerebellar processing for motor activity seems to be as follows:

1. The frontal motor association area of the cerebral cortex, via collateral fibers of the pyramidal tracts, notifies the cerebellum of its intent to initiate voluntary muscle contractions.

2. At the same time, the cerebellum receives information from proprioceptors throughout the body (regarding the tension in the muscles and tendons, and joint position) and from visual and equilibrium pathways. This information enables it to evaluate body position and momentum, that is, where the body is and where it is going.

3. The cerebellar cortex calculates the best way to coordinate the force, direction, and extent of muscle contraction to prevent overshoot, maintain posture, and ensure smooth, coordinated movements.

4. Then, via the superior peduncles, the cerebellum dispatches its "blueprint" for coordinating movement to the cerebral motor cortex. Cerebellar fibers also send information to brain stem nuclei, such as the *red nuclei* of the midbrain, which in turn influence motor neurons of the spinal cord.

The cerebellum can be compared to the control system of an automatic pilot. Just as an automatic pilot compares a plane's instrument settings with its actual course, the cerebellum continually compares the higher brain's intention with the body's performance and sends out messages to initiate the appropriate corrective measures. In this way, it helps promote smooth voluntary movements that are precise and economical in terms of muscular effort. Cerebellar injury results in loss of muscle tone and clumsy, unsure movements, and sometimes even impaired thoughts about movements. These disturbances are described in more detail later in this chapter.

Cognitive Function of the Cerebellum

Recent findings indicate that the cerebellum participates in "thinking," that is, it plays a role in cognition. Functional imaging studies indicate that the cerebellum plays a role in language and problem solving, and researchers surmise that the cerebellum's overall cognitive function may be to recognize and predict sequences of events so that it may adjust for the multiple forces on a limb during complex movements involving several joints.

Functional Brain Systems

Functional brain systems are networks of neurons that work together but span relatively large distances

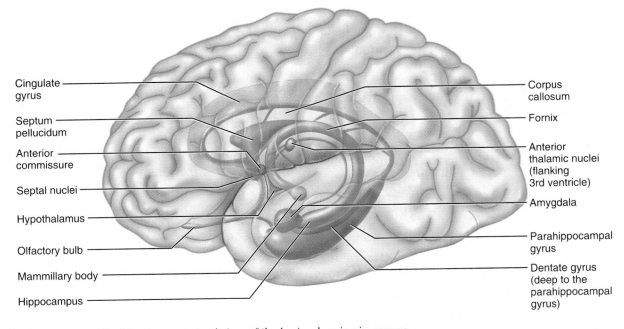

FIGURE 12.18 *The limbic system.* Lateral view of the brain, showing in orange some of the structures that constitute the limbic system, the emotional–visceral brain. The brain stem is not illustrated.

within the brain, so they cannot be localized to specific brain regions. The *limbic system,* spread widely throughout the forebrain, and the *reticular formation,* which traverses the brain stem, are excellent examples of functional systems (see Table 12.1, page 449).

The Limbic System

The **limbic system** is a group of structures located on the medial aspect of each cerebral hemisphere and diencephalon. Its cerebral structures encircle (*limbus* = ring) the upper part of the brain stem (Figure 12.18), and include parts of the rhinencephalon (the *septal nuclei, cingulate gyrus, parahippocampal gyrus, dentate gyrus,* and C-shaped *hippocampus*), and part of the *amygdala.* In the diencephalon, the main limbic structures are the *hypothalamus* and the *anterior nucleus of the thalamus.* The **fornix** ("arch") and other fiber tracts link these limbic system regions together.

The limbic system is our *emotional,* or *affective* (feelings), *brain.* Two parts seem especially important in emotions—the amygdala and the anterior part of the **cingulate gyrus.** The amygdala recognizes angry or fearful facial expressions, assesses danger, and elicits the fear response. The cingulate gyrus plays a role in expressing our emotions through gestures and resolving mental conflicts when we are frustrated.

Odors often trigger emotional reactions and memories, which reflect the origin of much of this system in the primitive "smell brain" (rhinencephalon). Our reactions to odors are rarely neutral (a skunk smells *bad* and repulses us), and odors often recall memories of emotionally laden events.

Extensive connections between the limbic system and lower and higher brain regions allow the system to integrate and respond to a wide variety of environmental stimuli. Most limbic system output is relayed through the hypothalamus. Because the hypothalamus is the neural clearinghouse for both autonomic (visceral) function and emotional response, it is not surprising that some people under acute or unrelenting emotional stress fall prey to visceral illnesses, such as high blood pressure and heartburn. Such emotion-induced illnesses are called **psychosomatic illnesses.** The most extreme consequence of severe emotional upheaval (paralyzing fear, extreme joy, or overwhelming grief, for example) is cardiac arrest.

The limbic system also interacts with the prefrontal lobes, so there is an intimate relationship between our feelings (mediated by the emotional brain) and our thoughts (mediated by the cognitive brain). As a result, we (a) react emotionally to things we consciously understand to be happening, and (b) are consciously aware of the emotional richness of our lives. Communication between the cerebral cortex and the limbic system explains why emotions sometimes override logic and, conversely, why reason can stop us from expressing our emotions in inappropriate situations. As described in Chapter 15, particular limbic system structures—the **hippocampal struc-**

tures and amygdala—also play an important role in converting new information into long-term memories.

The Reticular Formation

The **reticular formation** extends through the central core of the medulla oblongata, pons, and midbrain (Figure 12.19). It is composed of loosely clustered neurons in what is otherwise white matter. These reticular neurons form three broad columns along the length of the brain stem (see Figure 12.16c): (1) the midline **raphe** (ra′fe) **nuclei** (*raphe* = seam or crease), which are flanked laterally by (2) the **medial (large cell) group** and then (3) the **lateral (small cell) group of nuclei.** You will be hearing more about some of these nuclei when we discuss sleep in Chapter 15.

The outstanding feature of the reticular neurons is their far-flung axonal connections. Individual reticular neurons project to the hypothalamus, thalamus, cerebellum, and spinal cord. Such widespread connections make reticular neurons ideal for governing the arousal of the brain as a whole. For example, certain reticular neurons, unless inhibited by other brain areas, send a continuous stream of impulses (via thalamic relays) to the cerebral cortex, which keeps the cortex alert and conscious and enhances its excitability. This particular arm of the reticular formation is known as the **reticular activating system (RAS).** Impulses from all the great ascending sensory tracts synapse with RAS neurons, keeping them active and enhancing their arousing effect on the cerebrum. (This may help explain why many students, stimulated by a bustling environment, like to study in a crowded cafeteria.) The RAS also acts like a filter for this flood of sensory inputs. Repetitive, familiar, or weak signals are damped (filtered out), but unusual, significant, or strong impulses do reach our consciousness. For example, you are probably unaware of your watch encircling your wrist, but you would immediately notice it if the clasp broke. Between them, the RAS and the cerebral cortex disregard perhaps 99% of all sensory stimuli as unimportant. If this did not occur, the sensory overload would drive us crazy. The drug LSD interferes with these sensory dampers, promoting an often overwhelming sensory overload.

■ Take a moment to become aware of all the stimuli in your environment. Notice all the colors, shapes, odors, sounds, and so on. How many of these sensory stimuli are you *usually* aware of?

The RAS is inhibited by sleep centers located in the hypothalamus and other neural regions, and is depressed by alcohol, sleep-inducing drugs, and tranquilizers. Severe injury to this system, as might follow a knockout punch that twists the brain stem, re-

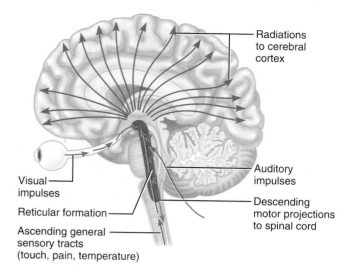

FIGURE 12.19 *The reticular formation.* The reticular formation extends the length of the brain stem. A portion of this formation, the reticular activating system (RAS), maintains alert wakefulness of the cerebral cortex. Ascending arrows on the diagram indicate input of sensory systems to the RAS and then reticular output via thalamic relays to the cerebral cortex. Other reticular nuclei are involved in the coordination of muscle activity. Their output is indicated by the arrow descending the brain stem.

sults in permanent unconsciousness (irreversible *coma*).

The reticular formation also has a *motor* arm. Some of its motor nuclei project to motor neurons in the spinal cord via the *reticulospinal tracts,* and help control skeletal muscles during coarse movements of the limbs. Other reticular motor nuclei, such as the vasomotor, cardiac, and respiratory centers of the medulla, are autonomic centers that regulate visceral motor functions.

Protection of the Brain

Nervous tissue is soft and delicate, and the largely irreplaceable neurons are injured by even slight pressure. However, the brain is protected by bone (the skull), membranes (the meninges), and a watery cushion (cerebrospinal fluid). Furthermore, the brain is protected from harmful substances in the blood by the so-called blood-brain barrier. We described the cranium, the brain's bony encasement, in Chapter 7. Here we will attend to the other protective devices.

Meninges

The **meninges** (mĕ-nin′jēz; *mening* = membrane) are three connective tissue membranes that lie just external to the central nervous system organs. They (1) cover and protect the CNS, (2) protect blood vessels and enclose venous sinuses, (3) contain

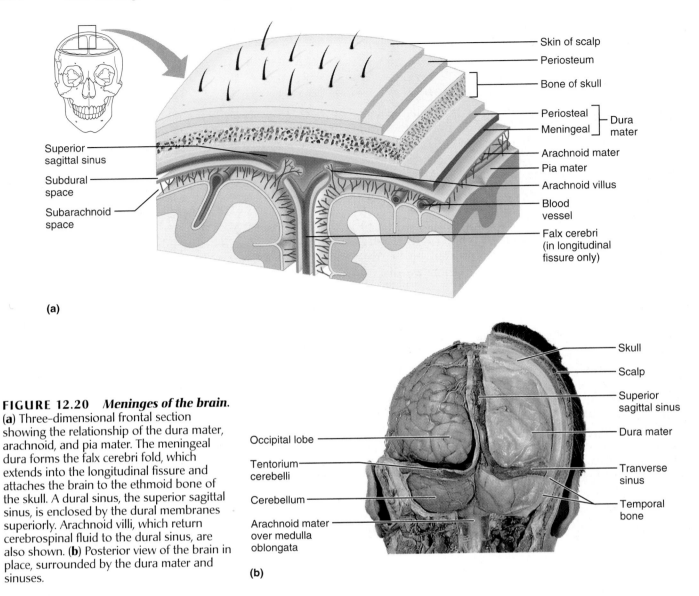

Skin of scalp
Periosteum
Bone of skull
Periosteal ⎤ Dura
Meningeal ⎦ mater
Arachnoid mater
Pia mater
Arachnoid villus
Blood vessel
Falx cerebri (in longitudinal fissure only)

Superior sagittal sinus
Subdural space
Subarachnoid space

(a)

Skull
Scalp
Superior sagittal sinus
Dura mater
Tranverse sinus
Temporal bone

Occipital lobe
Tentorium cerebelli
Cerebellum
Arachnoid mater over medulla oblongata

(b)

FIGURE 12.20 *Meninges of the brain.* **(a)** Three-dimensional frontal section showing the relationship of the dura mater, arachnoid, and pia mater. The meningeal dura forms the falx cerebri fold, which extends into the longitudinal fissure and attaches the brain to the ethmoid bone of the skull. A dural sinus, the superior sagittal sinus, is enclosed by the dural membranes superiorly. Arachnoid villi, which return cerebrospinal fluid to the dural sinus, are also shown. **(b)** Posterior view of the brain in place, surrounded by the dura mater and sinuses.

cerebrospinal fluid, and (4) form partitions within the skull. From external to internal, the meninges (singular, **meninx**) are the dura mater, arachnoid, and pia mater (Figure 12.20).

The Dura Mater The leathery **dura mater** (du'rah ma'ter), meaning "tough mother," is the strongest of the meninges. Where it surrounds the brain, it is a two-layered sheet of fibrous connective tissue. Its more superficial layer, the *periosteal layer*, is attached to the inner surface of the skull (the periosteum). Note that there is no dural periosteal layer surrounding the spinal cord. The deeper *meningeal layer* forms the true external covering of the brain and continues caudally in the vertebral canal as the dural sheath of the spinal cord. The brain's two dural layers are fused together except in certain areas, where they separate to enclose **dural sinuses** that collect venous blood from the brain and direct it into the internal jugular veins of the neck.

In several places, the meningeal dura mater extends inward to form flat partitions that subdivide the cranial cavity. These **dural septa,** which limit excessive movement of the brain within the cranium, include the following (Figure 12.21):

- **Falx cerebri** (falks ser'ĕ-bri). A large sickle-shaped (*falks* = sickle) fold that dips into the longitudinal fissure between the cerebral hemispheres. Anteriorly, it attaches to the crista galli of the ethmoid bone.

- **Falx cerebelli** (ser"ĕ-bel'i). Continuing inferiorly from the posterior falx cerebri, this small midline partition runs along the vermis of the cerebellum.

- **Tentorium** (ten-to're-um) **cerebelli.** Resembling a tent over the cerebellum, this nearly horizontal dural fold extends into the transverse fissure between the cerebral hemispheres (which it helps to support) and the cerebellum. (*Tentorium* means "tent.")

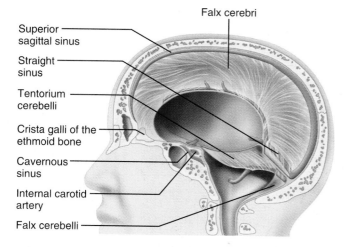

Superior
sagittal sinus

Straight
sinus

Tentorium
cerebelli

Crista galli of the
ethmoid bone

Cavernous
sinus

Internal carotid
artery

Falx cerebelli

Falx cerebri

FIGURE 12.21 *Partitioning folds of dura mater in the cranial cavity.* Note the falx cerebri, falx cerebelli, and tentorium cerebelli. Some dural sinuses are also shown.

The Arachnoid Mater The middle meninx, the **arachnoid mater,** or simply the **arachnoid** (ah-rak′noid), forms a loose brain covering, never dipping into the sulci at the cerebral surface (Figure 12.20a). It is separated from the dura mater by a narrow serous cavity, the **subdural space,** which contains a film of fluid. Beneath the arachnoid membrane is the wide **subarachnoid space.** Weblike extensions span this space and secure the arachnoid mater to the underlying pia mater. (*Arachnida* means "spider," and this membrane was named for its weblike extensions.) The subarachnoid space is filled with cerebrospinal fluid and also contains the largest blood vessels serving the brain. Since the arachnoid is fine and elastic, these blood vessels are rather poorly protected.

Knoblike projections of the arachnoid called **arachnoid villi** (vil′i) protrude superiorly through the overlying dura mater and into the superior sagittal sinus (see Figure 12.20). Cerebrospinal fluid is absorbed into the venous blood of the sinus by these valvelike villi.

The Pia Mater The **pia** (pi′ah) **mater,** meaning "gentle mother," is composed of delicate connective tissue and is richly invested with tiny blood vessels. It is the only meninx that clings tightly to the brain, following its every convolution. Small arteries entering the brain tissue carry ragged sheaths of pia mater inward with them for short distances.

⚠ Homeostatic Imbalance

Meningitis, an inflammation of the meninges, is a serious threat to the brain because a bacterial or viral meningitis may spread into the nervous tissue of the CNS. This condition of brain inflammation is called *encephalitis* (en′sef-ah-li′tis). Meningitis is usually diagnosed by taking a sample of cerebrospinal fluid from the subarachnoid space and examining it for microbes. ∎

Cerebrospinal Fluid

Cerebrospinal fluid (CSF), found in and around the brain and spinal cord, forms a liquid cushion that gives buoyancy to the CNS organs. By floating the jellylike brain, the CSF effectively reduces brain weight by 97% and prevents the brain from crushing under its own weight. CSF also protects the brain and spinal cord from blows and other trauma. Additionally, although the brain has a rich blood supply, cerebrospinal fluid helps to nourish the brain, and there is some evidence that it carries chemical signals (such as hormones and sleep- and appetite-inducing molecules) from one part of the brain to another.

CSF is a watery "broth" similar in composition to blood plasma, from which it arises. However, it contains less protein and its ion concentration is different. For example, CSF contains more sodium, chloride, and hydrogen ions than blood plasma, and fewer calcium and potassium ions. CSF composition, particularly its pH, is important in the control of cerebral blood flow and breathing, as described in later chapters.

The **choroid plexuses** that hang from the roof of each ventricle (Figure 12.22a) form CSF. These plexuses are frond-shaped clusters of broad, thin-walled capillaries (*plex* = interwoven) enclosed by pia mater and then a layer of ependymal cells lining the ventricles. The capillaries of the choroid plexuses are fairly permeable, and tissue fluid filters continuously from the bloodstream. However, the choroid plexus ependymal cells are joined by tight junctions and have ion pumps that allow them to modify this filtrate by actively transporting only certain ions across their membranes into the CSF pool. This sets up ionic gradients that cause water to diffuse into the ventricles as well. In adults, the total CSF volume of about 150 ml (about half a cup) is replaced every 8 hours or so; hence about 500 ml of CSF are formed daily. The choroid plexuses also help to cleanse the CSF by removing waste products and other unnecessary solutes.

Once produced, CSF moves freely through the ventricles. Some CSF circulates into the central canal of the spinal cord, but most enters the subarachnoid space via the lateral and medial apertures in the walls of the fourth ventricle (Figure 12.22b). The constant motion of the CSF is aided by movement of the long microvilli of the ependymal cells lining the ventricles. In the subarachnoid space, CSF bathes the outer surfaces of the brain and cord and then returns to the blood in the dural sinuses via the arachnoid villi.

? *What structures return CSF to the bloodstream?*

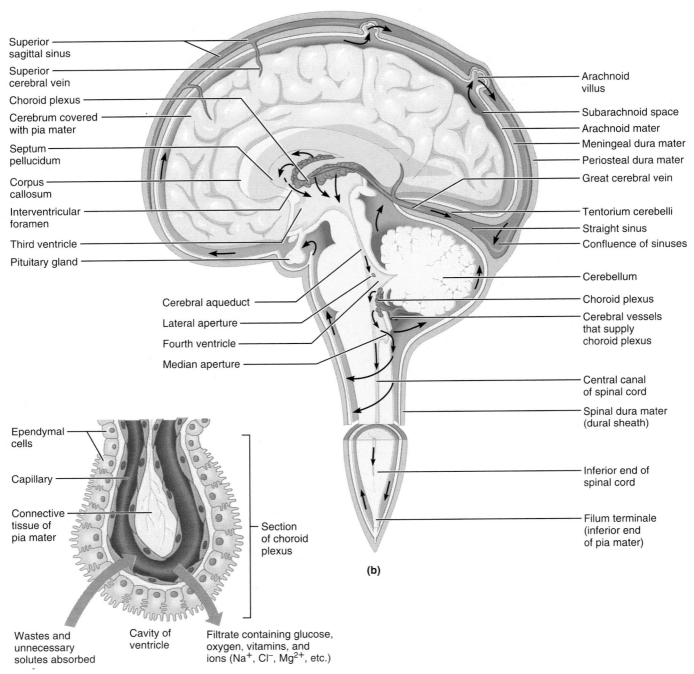

Superior sagittal sinus

Superior cerebral vein

Choroid plexus

Cerebrum covered with pia mater

Septum pellucidum

Corpus callosum

Interventricular foramen

Third ventricle

Pituitary gland

Arachnoid villus

Subarachnoid space

Arachnoid mater

Meningeal dura mater

Periosteal dura mater

Great cerebral vein

Tentorium cerebelli

Straight sinus

Confluence of sinuses

Cerebellum

Choroid plexus

Cerebral vessels that supply choroid plexus

Central canal of spinal cord

Spinal dura mater (dural sheath)

Inferior end of spinal cord

Filum terminale (inferior end of pia mater)

Cerebral aqueduct

Lateral aperture

Fourth ventricle

Median aperture

(b)

Ependymal cells

Capillary

Connective tissue of pia mater

Section of choroid plexus

Wastes and unnecessary solutes absorbed

Cavity of ventricle

Filtrate containing glucose, oxygen, vitamins, and ions (Na⁺, Cl⁻, Mg²⁺, etc.)

(a)

FIGURE 12.22 *Formation, location, and circulation of CSF.* **(a)** Choroid plexus forms CSF. Each choroid plexus consists of a knot of porous capillaries surrounded by a single layer of ependymal cells joined by tight junctions and bearing long microvilli with blunted ends. Although the filtrate moves easily from the capillaries, it must move through and be processed by the ependymal cells before it is allowed to enter the ventricles as cerebrospinal fluid. **(b)** Location and circulatory pattern of CSF. Direction of circulation is indicated by the arrows. (The relative position of the right lateral ventricle is indicated by the pale blue area deep to the corpus callosum and septum pellucidum.)

⚠ *Homeostatic Imbalance*

Ordinarily, CSF is produced and drained at a constant rate. However, if something (such as a tumor) obstructs its circulation or drainage, it accumulates and exerts pressure on the brain. This condition is called *hydrocephalus* ("water on the brain"). Hydrocephalus in a newborn baby causes its head to enlarge (Figure 12.23); this is possible because the skull bones have not yet fused. In adults, however, hydrocephalus is more likely to result in brain damage because the skull is rigid and hard, and accumulating fluid compresses the blood vessels serving the brain and crushes the soft nervous tissue. Hydrocephalus is treated by inserting a shunt in the ventricles that drains off the excess fluid into a vein in the neck. ∎

Blood–Brain Barrier

The **blood-brain barrier** is a protective mechanism that helps maintain a stable environment for the brain. No other body tissue is so absolutely dependent on a constant internal milieu as is the brain. In other body regions, the extracellular concentrations of hormones, amino acids, and ions are in constant flux, particularly after eating or exercise. If the brain were exposed to such chemical variations, the neurons would fire uncontrollably, because some hormones and amino acids serve as neurotransmitters and certain ions (particularly potassium) modify the threshold for neuronal firing.

Bloodborne substances within the brain's capillaries are separated from the extracellular space and neurons by

1. The continuous endothelium of the capillary wall

2. A relatively thick basal lamina surrounding the external face of the capillary

3. The bulbous "feet" of the astrocytes that cling to the capillaries.

The capillary endothelial cells are almost seamlessly joined all around by *tight junctions* (see Chapter 20), making them the least permeable capillaries in the entire body, a characteristic that constitutes most (if not all) of the blood-brain barrier. Although it was formerly assumed that the astrocytes contribute to the blood-brain barrier, their end feet are too far apart to form any true seal. It now appears that the major role of the astrocyte end feet is to provide the signals that stimulate the endothelial cells of the brain capillaries to *form* the tight junctions characteristic of the blood-brain barrier.

The blood-brain barrier is a selective, rather than an absolute, barrier. Nutrients, such as glucose, essential amino acids, and some electrolytes, move passively by facilitated diffusion through the endothelial cell membranes. Bloodborne metabolic

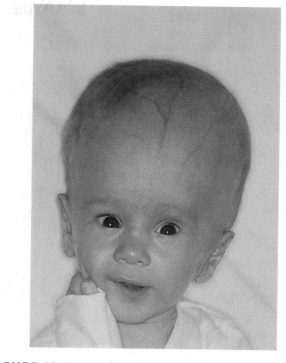

FIGURE 12.23 *Hydrocephalus in the newborn.*

wastes, such as urea and creatinine, as well as proteins, certain toxins, and most drugs, are prevented from entering brain tissue. Small nonessential amino acids and potassium ions not only are prevented from entering the brain, but also are actively pumped from the brain across the capillary endothelium.

The barrier is ineffective against fats, fatty acids, oxygen and carbon dioxide, and other fat-soluble molecules that diffuse easily through all plasma membranes. This explains why bloodborne alcohol, nicotine, and anesthetics can affect the brain.

The structure of the blood-brain barrier is not completely uniform. As noted above, the capillaries of the choroid plexuses are very porous, but the ependymal cells surrounding them have tight junctions. In some brain areas, particularly some surrounding the third and fourth ventricles, the blood-brain barrier is entirely absent and the capillary endothelium is quite permeable, allowing bloodborne molecules easy access to the neural tissue. One such region is the vomiting center of the brain stem, which monitors the blood for poisonous substances. Another is in the hypothalamus, which regulates water balance, body temperature, and many metabolic activities; lack of a blood-brain barrier here is essential to allow the hypothalamus to sample the chemical composition of the blood. The blood-brain barrier is incomplete in newborn and premature infants, and potentially toxic substances

It's naive to think that scientists aren't as competitive as corporate raiders or professional athletes. Nothing gets their juices flowing as quickly as the possibility of discovering the gene(s) responsible for some dreaded human disease. Although there are certainly many other examples, the degenerative diseases of the CNS—Alzheimer's disease, Parkinson's disease, and Huntington's disease—provide a fertile hunting ground and battleground. It is these three scourges that we look at here.

Alzheimer's Disease

Alzheimer's (altz'hi-merz) **disease (AD)** is a progressive degenerative disease of the brain that ultimately results in dementia (mental deterioration). Stroke patients represent nearly half of people in nursing homes. Alzheimer's patients account for most of the remaining half. Although usually seen in elderly people, Alzheimer's disease may begin in middle age. Between 5 and 15% of those over 65 develop this condition, and up to half of those over 85 die from it.

Its victims exhibit memory loss (particularly for recent events), shortened attention span and disorientation, and eventual language loss. Formerly good-natured individuals become irritable (and often combative), moody, and confused. Ultimately, hallucinations occur. These changes occur over a period of several years, during which time the family members watch the person they love "disappear."

Alzheimer's disease is associated with a shortage of ACh and with structural changes in the brain, particularly in the associative areas of the cerebral cortex and the hippocampus, areas involved with cognitive

functions and memory. The gyri shrink and many neurons are lost. Some cases of AD run in families, and several genetic mutations appear to be involved. (1) Individuals with two genes for APO-E4 get the disease earlier than those with one, who get it earlier than those with none. (APO-E4 directs synthesis of a protein called apolipoprotein E, which helps process cholesterol, plays a role in atherosclerosis, and is linked to at least one type of heart disease.) (2) Mutant A_2M genes (inhibitors of many body proteases) pose a significant risk factor for AD. (3) Mutations in the genes coding for presenilin proteins 1 and 2 (components of the ER membrane) can be damaging because these proteins upset Ca^{2+} homeostasis, causing neuronal apoptosis.

Microscopic examinations of brain tissue reveal the presence of **senile plaques** (aggregations of cells and degenerated fibers around a *beta amyloid peptide* core) littering the brain like shrapnel between the neurons. They also show **neurofibrillar tangles** (twisted fibrils within neuron cell bodies).

It has been frustratingly difficult for researchers to uncover how beta amyloid peptide acts as a neurotoxin, particularly since it is also present in healthy brain cells (but in lower amounts). Essentially, beta amyloid is a lysosomal breakdown product of a normal cell surface protein called amyloid precursor protein (APP). All cells break down APP via two pathways: one that yields harmless protein fragments, and the other (lysosomal pathway) that cleaves beta amyloid from APP. Just what tips the biochemical pathways off balance to favor production of beta amyloid is not understood, but it is known that this tiny peptide does its damage by enhancing calcium influx mediated by glutamate into neurons. Thus, as in stroke patients, the unholy trio of glutamate-NMDA receptors-Ca^{2+} appears to be responsible for at least part of the neural havoc that occurs.

Another line of research has implicated a protein called *tau*, which appears to function like railroad ties to bind microtubule "tracks" together. (Recall that microtubules are essential for intracellular transport, a crucial function in neurons.) When tau mutates and is inappropriately phosphorylated as in the brains of Alzheimer's victims, it abandons its microtubule stabilizing role and grabs onto other tau molecules, forming the neurofibrillary tangles. Once calcium enters the cell it triggers the cascade that leads to phosphorylation of tau, breakdown of microtubules, and formation of the spaghetti-like neurofibrillary tangles. Presumably, the beta amyloid and tau lines of investigation will eventually merge and point to the treatment for this terrible disorder, but other than drugs (tacrine, donepezil) that ease symptoms by inhibiting the breakdown of ACh, no one has the answer yet. However, the answer may be in the wings—because recent studies indicate that immunization with β-amyloid (the protein fragment in plaques) reverses or prevents plaque formation and neuronal damage.

Parkinson's Disease

Typically striking people in their 50s and 60s, **Parkinson's disease** results from a degeneration of the dopamine-releasing neurons of the substantia nigra, a brain stem nucleus that projects to the corpus striatum. As the substantia nigra neurons deteriorate, the dopamine-deprived basal nuclei they target become overactive, causing the well-known symptoms of the disease. Afflicted individuals have a persistent tremor at rest (exhibited by head nodding and a "pill-rolling" movement of their fingers), a forward-bent walking posture and shuffling gait, and a stiff facial expression, and they are slow in initiating and executing movement. It appears that the globus pallidus is overactive and, consequently, inhibits the motor cortex. The cause of Parkinson's disease is unknown, but environmental chemicals (such as pesticides and herbicides) may increase the risk of developing the disorder.

Used since the 1960s, the drug *L-dopa,* a precursor of dopamine, often helps to alleviate some of the symptoms. However, L-dopa is not curative, and some believe that its chronic use may ultimately contribute to the degenerative process. L-dopa also has undesirable side effects: severe nausea, dizziness, and, in some, liver damage. Presently, a combination drug therapy is the most-used treatment of Parkinson's disease. By using (1) L-dopa (to replace dopamine) in combination with carbidopa, which blocks the conversion of L-dopa to dopamine outside the brain (an event that leads to the side effects) and (2) artificial dopamine agonists (bromo-criptine, ropinorole, and others),

which bind to selected dopamine receptors in the brain and so mimic the effects of dopamine, lower doses of both drug classes can be used and side effects minimized. Deep brain (thalamic) stimulation via implanted electrodes has proved helpful in alleviating tremors but little else.

More promising for long-term results are the intrabrain transplants of fetal substantia nigra tissue, genetically engineered adult nigral cells, and fetal pig dopamine-producing cells, all of which have produced regression of disease symptoms. However, the use of fetal tissue is controversial and riddled with ethical and legal roadblocks, making use of the modified adult nigral cells and fetal pig cells more attractive alternatives. Better yet (it seems) is the use of self cells taken from the carotid bodies (tiny oxygen sensors located in the neck). Although the cells do not multiply after transplantation to the brain, they do produce dopamine and growth inducing chemicals that prod the brain cells to produce more dopamine.

Newer to the drug therapy scene is deprenyl. When given early on in the disease, deprenyl slows the neurological deterioration to some extent and delays the need to administer L-dopa for as long as 18 months. Additionally, tropic factors are in clinical tests. GDNF (glial-cell-derived neurotropic factor) appears to protect the dopamine-releasing neurons and to actually reverse progress of the disease; treated animals show marked improvement in mobility and balance.

Huntington's Disease

Huntington's disease is a hereditary disorder that strikes during middle age and leads to massive degeneration of the basal nuclei and later of the cerebral cortex. Its initial symptoms in many of the afflicted are wild, jerky, and almost continuous "flapping"

movements called *chorea* (Greek for "dance"). Although the movements appear to be voluntary, they are not. These hyperkinetic manifestations are essentially the opposite of those of Parkinson's disease (impairments involve overstimulation instead of inhibition of the motor drive), and Huntington's disease is usually treated with drugs that block, rather than enhance, dopamine's effects. As with Parkinson's disease, fetal tissue implants may provide promise for its treatment in the future.

Huntington's disease is progressive and usually fatal within 15 years of onset of symptoms. The disease—now known to result from a "genetic stutter," an explosive repeat of three bases of DNA (CAG) near the end of chromosome 4—appears to impair the activity of a key enzyme (GAPDH) in glycolysis (see Chapter 25), thereby stunting energy (ATP) production in the affected neurons. Late in the disease, marked mental deterioration (dementia) occurs.

Commonalities

Both the research findings and the treatments for these three dreaded brain disorders are converging. Although abnormal protein aggregates (neurofibrillary tangles of tau) are most obvious in Alzheimer's disease, similar aggregates have also been found in the affected CNS neurons of Parkinson's and Huntington's disease patients (Lewy bodies and clumps of a glutamine-rich enzyme called HPRT, respectively). Fetal cell implants, and the even more recent human stem cell implants, may ultimately provide the best treatment avenue for all three disorders.

Also, if investigators can figure out why parts of the brain do or don't grow (the hippocampus *does* grow new neurons), we may finally have the means of inducing cell proliferation to replace the neurons laid low by these dread diseases.

can readily enter the CNS and cause problems not seen in adults.

Stress markedly increases the ability of chemicals to penetrate the blood-brain barrier. Moreover, injury to the brain, whatever the cause, may result in a localized breakdown of the blood-brain barrier. Most likely, this reflects some change in the capillary endothelial cells or their tight junctions. This supposition is borne out by a procedure that infuses a concentrated solution of mannitol (sugar) prior to infusing chemotherapeutic drugs. The mannitol causes the capillary endothelial cells to shrink, opening up gaps between their tight junctions that allow the drugs to breach the blood-brain barrier and gain access to brain tumors.

Homeostatic Imbalances of the Brain

Brain dysfunctions are unbelievably varied and extensive. We have mentioned some of them already; degenerative brain disorders are covered in *A Closer Look*, and we will discuss developmental problems in the final section of the chapter. Here, we will focus on traumatic brain injuries and cerebrovascular accidents.

Traumatic Brain Injuries

Head injuries are the leading cause of accidental death in the United States. Consider, for example, what happens if you forget to fasten your seat belt and then rear-end another car. Your head is moving and then is suddenly stopped as it hits the windshield. Brain damage is caused not only by localized injury at the site of the blow (the *coup* injury), but also by the ricocheting effect as the brain hits the opposite end of the skull (the *contrecoup* injury).

A **concussion** occurs when brain injury is slight and the symptoms are mild and short-lived. The victim may be dizzy, "see stars," or lose consciousness briefly, but there is no permanent neurological damage.

The result of marked brain tissue destruction is a **contusion.** In cortical contusions, the individual may remain conscious, but severe brain stem contusions always cause unconsciousness (coma), lasting from hours to a lifetime because of injury to the reticular activating system.

Following head blows, death may result from **subdural** or **subarachnoid hemorrhage** (bleeding from ruptured vessels into those spaces). Individuals who are initially lucid following head trauma and then begin to deteriorate neurologically are, in all probability, hemorrhaging intracranially. Accumulating blood within the skull increases intracranial pressure and compresses brain tissue. If the pressure is sufficient to force the brain stem inferiorly through the foramen magnum, control of blood pressure, heart rate, and respiration is lost. Intracranial hemorrhages are treated by surgical removal of the hematoma (localized blood mass) and repair of the ruptured blood vessels.

Another consequence of traumatic head injury is **cerebral edema,** or swelling of the brain. This results both from exudate formation during the inflammatory response and from water uptake by brain tissue. Anti-inflammatory drugs such as prednisone and other steroids are routinely administered to head injury patients in an attempt to prevent cerebral edema from aggravating brain injury.

Cerebrovascular Accidents

The single most common nervous system disorder and the third leading cause of death in the United States are **cerebrovascular** (ser″ĕ-bro-vas′ku-lar) **accidents (CVAs),** also called *strokes* or *brain attacks.* Strokes occur when blood circulation to a brain area is blocked and brain tissue dies. (Deprivation of blood supply to *any* tissue is called **ischemia** [is-ke′me-ah; "to hold back blood"] and results in deficient oxygen and nutrient delivery to cells.) The most common cause of CVA is blockage of a cerebral artery by a blood clot. Other causes include compression of brain tissue by hemorrhage or edema and progressive narrowing of brain vessels by atherosclerosis.

Eighty percent of those who suffer massive *cerebral hemorrhage* die during the initial acute attack; the mortality is far lower in those with nonhemorrhagic strokes. Those who survive a CVA are typically paralyzed on one side of the body and many exhibit sensory deficits. If the language areas are damaged, difficulties in understanding or vocalizing speech develop. Whatever the cause, the nature of the neurological deficits that follow depends on the area and extent of brain damage that occurred. Additionally, those suffering CVAs from blood clots are likely to have recurrent clotting problems—and more CVAs. Fewer than 35% of those that survive a CVA are alive three years later. Even so, the picture is not hopeless. Some patients recover at least part of their lost faculties, because undamaged neurons sprout new branches that spread into the area of injury and take over some lost functions. However, neuronal spread is limited to only about 5 mm (0.2 inch); thus, large lesions always result in less than complete recovery. Physical therapy is usually started as soon as possible to prevent muscle contractures (abnormal shortening of the paralyzed muscles which occurs due to lack of exercise).

Not all strokes are "completed." Temporary episodes of reversible cerebral ischemia, called **transient ischemic attacks (TIAs),** are common. TIAs last from 5 to 50 minutes and are characterized by temporary numbness, paralysis, and impaired speech. While these deficits are not permanent, TIAs do constitute "red flags" that warn of impending, more serious CVAs. If left untreated, about one-third of patients with TIAs eventually suffer a completed stroke. Cumulative trials have strongly indicated that 300 mg of aspirin daily effectively reduces the risk of both TIAs and CVAs.

Historically, medicine has been able to do little for a stroke patient, but recent findings on the mechanism of cell death following a stroke are particularly exciting. Strokes are a lot like undersea earthquakes. It's not the initial temblor that does most of the damage, it's the tsunami wave that floods the coast later. Similarly, the initial vascular blockage during a stroke is not usually disastrous because there are many blood vessels in the brain that can pick up the slack; it's the events that lead to killing of neurons not in the initial ischemic zone that wreak the most havoc. Let's see how this happens.

Apparently the main culprit is *glutamate,* an excitatory neurotransmitter that helps shape the processes of learning and memory. Normally, glutamate binding to NMDA (*N*-methyl D-aspartate) receptors (named for the chemical used to identify them) opens specific gates that allow calcium ions (which set off signaling cascades that induce changes essential for learning or potentiation) to enter the stimulated neuron. After brain injury, neurons totally deprived of oxygen begin to disintegrate, unleashing the cellular equivalent of "buckets" of glutamate that acts as an *excitotoxin* to mediate both immediate and delayed changes in ion transport by activating more NMDA receptors. The acute change, which occurs within minutes, consists of neuronal uptake of Na^+, Cl^-, and water. The delayed changes, which take place over the next few hours, involve unregulated calcium influx and generation of free radicals including NO that kill or injure thousands of surrounding, still-healthy neurons. As more and more newly oxygen-deprived neurons become overexcited and themselves release glutamate, the chain reaction leading to even more cell death is reinforced. This in turn activates the microglia, which phagocytize dead and injured neurons, and initiates inflammation, which compounds the damage. These new insights have launched an intense search for neuroprotective drugs, such as NMDA and NO antagonists and calcium channel blockers, that can prevent this neurotoxic domino effect and improve the outlook for future stroke victims. Additional therapies include those that prevent or dissolve blood clots (Coumadin, tissue plasminogen activator [tPA], and others). Indeed, clinical trials have shown that if tPA, a clot-busting drug, is given within three hours of stroke onset, the number of

stroke patients that walk away without permanent disability is boosted by 50%.

An astonishing and still experimental procedure initiated in 1997 called *retrograde venous neuroperfusion* has been producing fairly good results in trials. Blood from a large artery in the groin is drawn through a mechanical pump, guided into a pair of veins at the back of the brain, and then pumped "backward" in the brain toward the affected artery. This maintains brain circulation while emergency measures are started (surgery or tPA administration). This "backdoor approach" to stroke treatment has allowed some lucky patients to walk out of the hospital within five days of their CVA with no permanent disability. Another radical approach is injecting LBS-Neurons (immature neurons grown from human cancer cells) into stroke-damaged brain regions in hope that they will take on properties of nearby mature neurons. Although the injected neurons have been rendered amitotic by chemical treatment, it is not known whether they will remain so, and the treatment is highly controversial.

THE SPINAL CORD

Gross Anatomy and Protection of the Spinal Cord

The spinal cord, enclosed within the vertebral column, extends from the foramen magnum of the skull to the level of the first or second lumbar vertebra, just inferior to the ribs (Figure 12.24). About 42 cm (17 inches) long and 1.8 cm (3/4 of an inch) thick, the glistening-white **spinal cord** provides a two-way conduction pathway to and from the brain. It is a major reflex center: Spinal reflexes are initiated and completed at the spinal cord level. We discuss reflex functions and motor activity of the cord in subsequent chapters. In this section we focus on the anatomy of the cord and on the location and naming of its ascending and descending tracts.

Like the brain, the spinal cord is protected by bone, meninges, and cerebrospinal fluid. The single-layered dura mater of the spinal cord, called the **spinal dural sheath** (see Figures 12.22b and 12.27), is not attached to the bony walls of the vertebral column. Between the bony vertebrae and the dural sheath is a rather large **epidural space** filled with a soft padding of fat and a network of veins. Cerebrospinal fluid fills the subarachnoid space between the *arachnoid* and *pia mater* meninges. Inferiorly, the dural and arachnoid membranes extend to the level of S_2, well beyond the end of the spinal cord. The spinal cord typically ends at L_1 (Figure 12.24a). Thus, the subarachnoid space within the meningeal sac inferior to that point provides a nearly ideal spot for removing cerebrospinal fluid for testing, a

? *What is the explanation for the cervical and lumbar enlargements of the cord?*

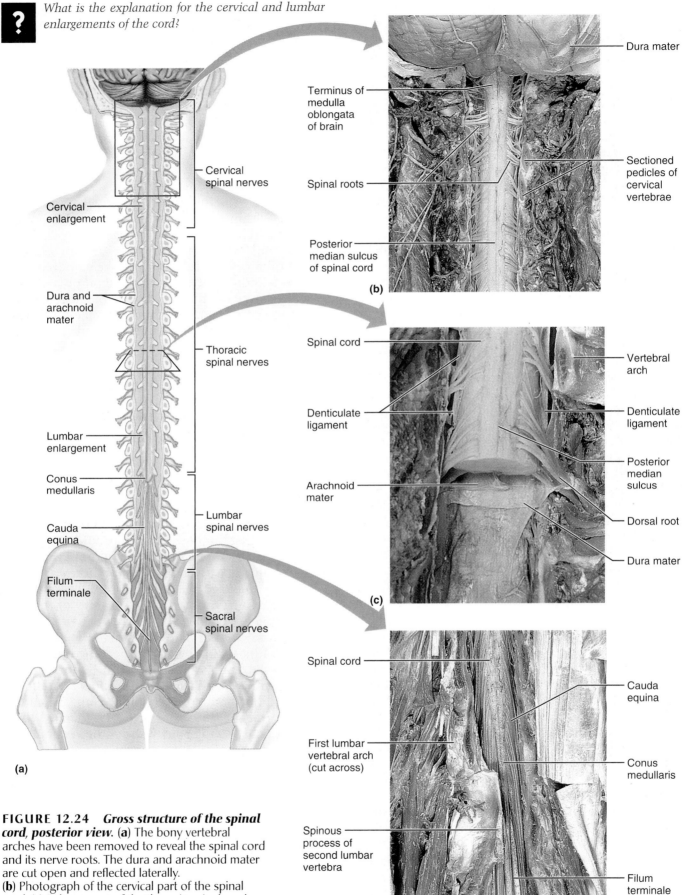

Cervical spinal nerves

Cervical enlargement

Dura and arachnoid mater

Thoracic spinal nerves

Lumbar enlargement

Conus medullaris

Cauda equina

Filum terminale

Lumbar spinal nerves

Sacral spinal nerves

(a)

Dura mater

Terminus of medulla oblongata of brain

Spinal roots

Sectioned pedicles of cervical vertebrae

Posterior median sulcus of spinal cord

(b)

Spinal cord

Vertebral arch

Denticulate ligament

Denticulate ligament

Arachnoid mater

Posterior median sulcus

Dorsal root

Dura mater

(c)

Spinal cord

Cauda equina

First lumbar vertebral arch (cut across)

Conus medullaris

Spinous process of second lumbar vertebra

Filum terminale

(d)

FIGURE 12.24   Gross structure of the spinal cord, posterior view. (a) The bony vertebral arches have been removed to reveal the spinal cord and its nerve roots. The dura and arachnoid mater are cut open and reflected laterally.
(b) Photograph of the cervical part of the spinal cord. **(c)** Enlargement of the thoracic spinal cord, showing the denticulate ligaments. **(d)** Inferior end of the spinal cord, showing the conus medullaris, cauda equina, and filum terminale.

The nerves serving the limbs arise in these areas.

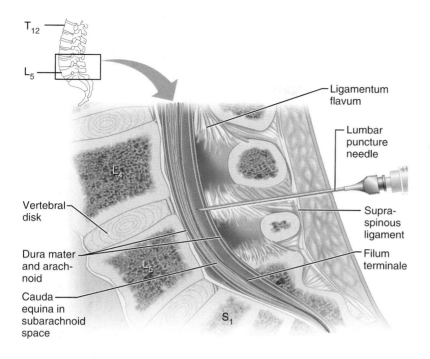

FIGURE 12.25   *Diagrammatic view of a lumbar tap.*

procedure called a **lumbar puncture** or **tap** (Figure 12.25). Because the spinal cord is absent there and the delicate nerve roots drift away from the point of needle insertion, there is little or no danger of damaging the cord (or spinal roots) beyond L₃.

Inferiorly, the spinal cord terminates in a tapering cone-shaped structure called the **conus medullaris** (ko'nus me"dul-ar'is). A fibrous extension of the pia mater, the **filum terminale** (fi'lum ter"mĭ-nah'le; "terminal filament") extends inferiorly from the conus medullaris to the posterior surface of the coccyx, where it anchors the spinal cord in place so it is not jostled by body movements (Figure 12.24d). Furthermore, the spinal cord is secured to the bony walls of the vertebral canal throughout its length by saw-toothed shelves of pia mater (Figure 12.24c) called **denticulate ligaments** (den-tik'u-lāt; "toothed").

In humans, 31 pairs of *spinal nerves* attach to the cord by paired roots and exit from the vertebral column via the intervertebral foramina to travel to the body regions they serve. Just as the vertebral column is segmented, so too is the spinal cord (however, neural spinal segments are not visible externally). Each segment of the cord is defined by a pair of spinal nerves that lie superior to their corresponding vertebra. The spinal cord is about the width of a thumb for most of its length, but it has obvious enlargements in the cervical and lumbosacral regions, where the nerves serving the upper and lower limbs arise. These enlargements are the **cervical** and **lumbar enlargements,** respectively (see Figure 12.24a). Because the cord does not reach the end of the vertebral column, the lumbar and sacral spinal nerve roots angle sharply downward and travel inferiorly

through the vertebral canal for some distance before reaching their intervertebral foramina. The collection of nerve roots at the inferior end of the vertebral canal is named the **cauda equina** (kaw'da e-kwi'nuh) because of its resemblance to a horse's tail. This rather strange arrangement reflects the fact that during fetal development, the vertebral column grows more rapidly than does the spinal cord, forcing the lower spinal nerve roots to "chase" their exit points inferiorly through the vertebral canal.

Embryonic Development of the Spinal Cord

The spinal cord develops from the caudal portion of the embryonic neural tube (see Figures 12.2 and 12.3 on pp. 430 and 431). By the sixth week, each side of the developing cord has two recognizable clusters of neuroblasts that have migrated outward from the original neural tube: a dorsal **alar** (a'lar) **plate** and a ventral **basal plate** (Figure 12.26).

Alar plate neuroblasts become interneurons. The basal plate neuroblasts develop into motor neurons and sprout axons that grow out to the effector organs. Axons that emerge from alar plate cells (and some basal plate cells) form the white matter of the cord by growing outward along the length of the CNS. As development progresses, these plates expand dorsally and ventrally to produce the H-shaped central mass of gray matter of the adult spinal cord. Neural crest cells that come to lie alongside the cord form the *dorsal root ganglia* containing sensory neuron cell bodies; these send their axons into the dorsal aspect of the cord.

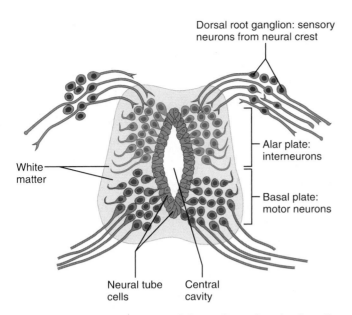

FIGURE 12.26 *Structure of the embryonic spinal cord.* At six weeks of development, aggregations of gray matter called the alar plates (future interneurons) and basal plates (future motor neurons) have formed, and dorsal root ganglia (future sensory neurons) have arisen from neural crest cells.

Cross-Sectional Anatomy of the Spinal Cord

The spinal cord is somewhat flattened from front to back and two grooves mark its surface: the **anterior median fissure** and the more shallow **posterior median sulcus** (see Figure 12.27b). These grooves run the length of the cord and partially divide it into right and left halves. As mentioned earlier, the gray matter of the cord is located in its core, the white matter outside.

Gray Matter and Spinal Roots

As in other regions of the central nervous system, the gray matter of the cord consists of a mixture of neuron cell bodies, their unmyelinated processes, and neuroglia. All neurons whose cell bodies are in the gray matter of the cord are multipolar neurons.

The gray matter of the cord looks like the letter H or like a butterfly in cross section (Figure 12.27b). It consists of mirror-image lateral gray masses connected by a cross-bar of gray matter, called the **gray commissure,** that encloses the central canal. The two posterior projections of the gray matter are the **posterior (dorsal) horns;** the anterior pair are the **anterior (ventral) horns.** An additional pair of gray matter columns, the small **lateral horns,** is present in the thoracic and superior lumbar segments of the cord.

The posterior horns consist entirely of interneurons. The anterior horns also have some interneurons but mainly house nerve cell bodies of somatic motor neurons. These send their axons out via the **ventral roots** of the spinal cord (Figure 12.27b) to the skeletal muscles (their effector organs). The amount of ventral gray matter present at a given level of the spinal cord reflects the amount of skeletal muscle innervated at that particular level. Thus, the anterior horns are largest in the limb-innervating cervical and lumbar regions of the cord and are responsible for the cord enlargements seen in those regions.

The lateral horn neurons are autonomic (sympathetic division) motor neurons that serve visceral organs. Their axons leave the cord via the ventral root along with those of the somatic motor neurons. Since the ventral roots contain both somatic and autonomic efferents, they serve both motor divisions of the peripheral nervous system (Figure 12.28).

Afferent fibers carrying impulses from peripheral sensory receptors form the **dorsal roots** of the spinal cord (see Figure 12.27). The nerve cell bodies of the as-

FIGURE 12.27
Anatomy of the spinal cord. **(a)** Cross section through the spinal cord illustrating its relationship to the surrounding vertebral column.

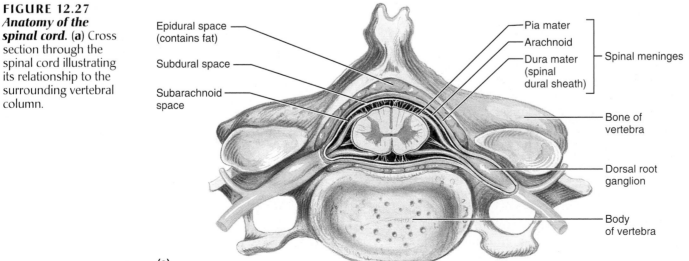

(a)

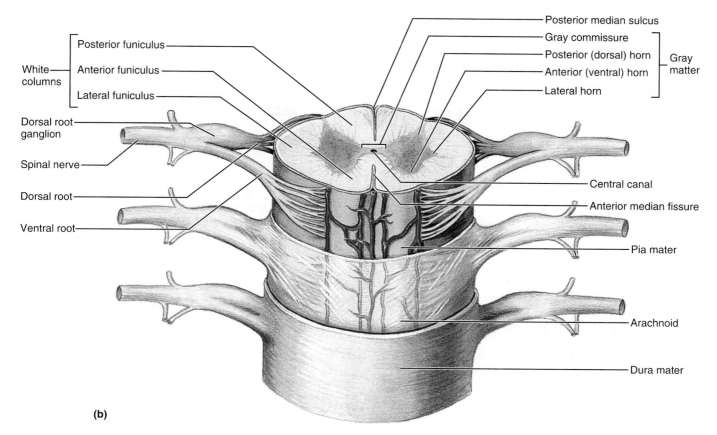

(b)

FIGURE 12.27 *(Continued) Anatomy of the spinal cord.* **(b)** Three-dimensional view of the adult spinal cord and its meningeal coverings.

sociated sensory neurons are found in an enlarged region of the dorsal root called the **dorsal root ganglion** or **spinal ganglion.** After entering the cord, their axons may take a number of routes. For example, some enter the posterior white matter of the cord directly and travel to synapse at higher cord or brain levels. Others synapse with interneurons in the posterior horns of the spinal cord gray matter at their entry level.

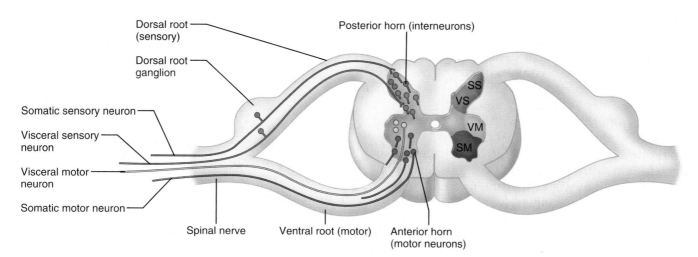

FIGURE 12.28 *Organization of the gray matter of the spinal cord.* In terms of processing sensory and motor activity, the gray matter of the spinal cord can be divided into a sensory half dorsally and a motor half ventrally. SS: interneurons receiving input from somatic sensory neurons. VS: interneurons receiving input from visceral sensory neurons. VM: visceral motor (autonomic) neurons. SM: somatic motor neurons. Note that the dorsal and ventral roots are part of the PNS, not of the spinal cord.

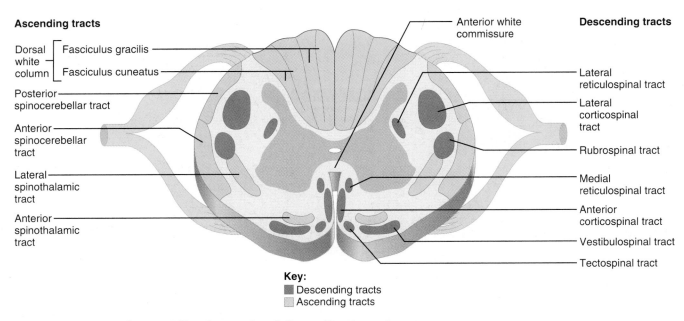

Ascending tracts

Dorsal white column — Fasciculus gracilis
Fasciculus cuneatus

Posterior spinocerebellar tract

Anterior spinocerebellar tract

Lateral spinothalamic tract

Anterior spinothalamic tract

Anterior white commissure

Descending tracts

Lateral reticulospinal tract

Lateral corticospinal tract

Rubrospinal tract

Medial reticulospinal tract

Anterior corticospinal tract

Vestibulospinal tract

Tectospinal tract

Key:
Descending tracts
Ascending tracts

FIGURE 12.29 *Major ascending (sensory) and descending (motor) tracts of the spinal cord, cross-sectional view.* Ascending tracts are labeled only on the left; descending tracts are labeled on the right side of the figure.

The dorsal and ventral roots are very short and fuse laterally to form the **spinal nerves.** The spinal nerves, which are part of the peripheral nervous system, are considered further in Chapter 13.

The spinal gray matter can be divided further according to its neurons' relative involvement in the innervation of the somatic and visceral regions of the body. The following four zones are evident within this gray matter (Figure 12.28): **somatic sensory (SS), visceral** (autonomic) **sensory (VS), visceral motor (VM), somatic motor (SM).**

White Matter

The white matter of the spinal cord is composed of myelinated and unmyelinated nerve fibers that allow communication between different parts of the spinal cord and between the cord and brain. These fibers run in three directions: (1) *ascending*—up to higher centers (sensory inputs), (2) *descending*—down to the cord from the brain or within the cord to lower levels (motor outputs), and (3) *transversely*—across from one side of the cord to the other (commissural fibers). The vertical (ascending and descending) tracts make up most of the white matter.

The white matter on each side of the cord is divided into three **white columns,** or **funiculi** (fu-nik'u-li; "long ropes"). These are named according to their position as the **posterior, lateral,** and **anterior funiculi** (see Figure 12.27b). Each funiculus contains several fiber tracts, and each tract is made up of axons with similar destinations and functions. With a few exceptions, the names of the spinal tracts reveal both their origin and destination. The principal ascending and descending tracts of the spinal

cord are illustrated schematically in Figure 12.29.

All major spinal tracts are actually *part of multineuron pathways* that connect the brain to the body periphery. These great ascending and descending pathways contain not only spinal cord neurons but also parts of peripheral neurons and neurons in the brain. In Chapter 15, we get specific about the individual spinal tracts; here we will just make some generalizations about them and the pathways to which they contribute:

1. Most pathways cross over from one side of the CNS to the other (decussate) at some point along their journey.

2. Most consist of a chain of two or three neurons that contribute to successive tracts of the pathway.

3. Most exhibit *somatotopy*, a precise spatial relationship among the tract fibers that reflects the orderly mapping of the body. For example, in an ascending sensory tract, somatotopy refers to the fact that fibers transmitting inputs from sensory receptors in the superior parts of the body lie lateral to those conveying sensory information from inferior body regions.

4. All pathways and tracts are paired (right and left) with a member of the pair present on each side of the spinal cord or brain.

Spinal Cord Trauma and Disorders

Spinal Cord Trauma

The spinal cord is elastic, stretching with every turn of the head or bend of the trunk, but it is exquisitely sensitive to direct pressure. Any localized damage to the spinal cord or spinal roots leads to some form of functional loss, either **paralysis** (loss of motor function) or **paresthesias** (par″es-the′ze-ahz) (sensory losses). Severe damage to ventral root or anterior horn cells results in a **flaccid** (flak′sid) **paralysis** of the skeletal muscles served. Because the lower motor neurons are damaged, nerve impulses do not reach these muscles, which consequently cannot move either voluntarily or involuntarily. Without stimulation, the muscles begin to atrophy. However, when only the upper motor neurons of the primary motor cortex are damaged, **spastic paralysis** occurs. In this case, the spinal motor neurons remain intact and the muscles continue to be stimulated irregularly by spinal reflex activity. Thus, the muscles remain healthy longer, but their movements are no longer subject to voluntary control. In many such cases, the muscles become permanently shortened and fibrotic in time. Patients with spinal cord injuries do better if they receive massive doses of the steroid prednisolone within eight hours of injury. By reducing inflammation and blocking the cascade of harmful biochemical reactions that produce the so-called second tier of spinal cord injury, it is possible to significantly reduce neurological damage and paralysis in patients who previously were deemed beyond help.

Transection (cross sectioning) of the spinal cord at any level results in total motor and sensory loss in body regions inferior to the site of damage. If the transection occurs between T_1 and L_1, both lower limbs are affected, resulting in **paraplegia** (par″ah-ple′je-ah). If the injury occurs in the cervical region, all four limbs are affected and **quadriplegia** occurs. *Hemiplegia*, paralysis of one side of the body, usually reflects brain, rather than spinal cord, injury.

Anyone with traumatic spinal cord injury must be watched for symptoms of **spinal shock,** a transient period of functional loss that follows injury to the cord. Spinal shock results in immediate depression of all reflex activity caudal to the lesion site. Bowel and bladder reflexes stop, blood pressure falls, and all muscles (somatic and visceral alike) below the injury are paralyzed and insensitive. Since the individual does not perspire in the paralyzed body region, a fever may develop. Neural function usually returns within a few hours following injury. If function does not resume within 48 hours, paralysis is permanent in most cases. Over 10,000 Americans are paralyzed yearly in automotive or sports accidents.

Poliomyelitis

Poliomyelitis (po″le-o-mi″ĕ-li′tis; *polio* = gray matter; *myelitis* = inflammation of the spinal cord) results from destruction of anterior horn motor neurons by the poliovirus. Early symptoms include fever, headache, muscle pain and weakness, and loss of certain somatic reflexes. Later, paralysis develops and the muscles served atrophy. The victim may die from paralysis of the respiratory muscles or from cardiac arrest if neurons in the medulla oblongata are destroyed. In most cases, the poliovirus enters the body in feces-contaminated water (such as might occur in a public swimming pool), and the incidence of the disease has traditionally been highest in children and during the summer months. Fortunately, Salk and Sabin polio vaccines have nearly eliminated this disease.

However, many of the "recovered" survivors of the great polio epidemic of the late 1940s and 1950s have begun to experience extreme lethargy, sharp burning pains in their muscles, and progressive muscle weakness and atrophy. Initially dismissed as the flu or psychosomatic illness, these disturbing symptoms are now referred to as **postpolio syndrome.** The cause of postpolio syndrome is not known, but a likely explanation is that its victims, like all of us, continue to lose neurons throughout life. While a healthy nervous system can recruit nearby neurons to compensate for the losses, polio survivors have already drawn on that "pool" and have few neurons left to take over. Ironically, those who worked hardest to overcome their disease are its newest victims.

Amyotrophic Lateral Sclerosis

Amyotrophic (ah-mi″o-trof′ik) *lateral sclerosis (ALS)*, also called Lou Gehrig's disease, is a devastating neuromuscular condition that involves progressive destruction of anterior horn motor neurons and fibers of the pyramidal tract. As the disease progresses, the sufferer loses the ability to speak, swallow, and breathe. Death typically occurs within five years. Lou Gehrig's disease seems to be linked to abnormal genes. For example, defects of the EAAT2 gene, which codes for the glutamate transporter that allows glial cells to "sop up" or reclaim released glutamate neurotransmitter, can account for neuronal death typical of this disorder because exposure of neurons to excessive amounts of glutamate is neurotoxic (see p. 461). Additionally, some cases have been found to be due to a defect in the gene that directs production of superoxide dismutase (SOD), an enzyme that protects cells from the ravages of free radicals (see Chapter 3, p. 109). It is not known whether defects in this gene also affect the glutamate transporter, but the mutant gene does somehow trigger events that lead to nerve cell death. Riluzole, the only advance in treatment of ALS in 50 years, appears to prolong life for ALS victims by indirectly inhibiting free radical formation by inhibiting the release of glutamate.

DIAGNOSTIC PROCEDURES FOR ASSESSING CNS DYSFUNCTION

Anyone who has had a routine physical examination is familiar with the reflex tests done to assess neural function. A tap with a reflex hammer stretches your quadriceps tendon and your anterior thigh muscles contract, which results in the knee-jerk response. This response shows that the spinal cord and upper brain centers are functioning normally. Abnormal reflex test responses may indicate serious disorders such as intracranial hemorrhage, multiple sclerosis, or hydrocephalus and suggest that other more sophisticated neurological tests are needed to identify the problem.

Pneumoencephalography (nu"mo-en-sef"ah-log'-rah-fe) provides a fairly clear X-ray picture of the brain ventricles and has been the procedure of choice for diagnosing hydrocephalus. Cerebrospinal fluid is withdrawn, and air injected into the subarachnoid space floats upward into the ventricles, allowing them to be visualized.

A *cerebral angiogram* (an'je-o-gram") is commonly ordered to assess the condition of the cerebral arteries serving the brain (or the carotid arteries of the neck, which feed most of those vessels) in individuals who have suffered a stroke or TIA. A dye is injected into an artery and allowed to disperse in the brain. Then an X ray is taken of the arteries of interest. The dye allows visualization of arteries narrowed by arteriosclerosis or blocked by blood clots.

The new imaging techniques described in Chapter 1 (pp. 20–21) have revolutionized the diagnosis of brain lesions. Together the various *CT* and *MRI scanning techniques* allow most tumors, intracranial lesions, multiple sclerosis plaques, and areas of dead brain tissue (infarcts) to be identified quickly. PET scans can localize brain lesions that generate seizures (epileptic tissue) and diagnose Alzheimer's disease.

DEVELOPMENTAL ASPECTS OF THE CENTRAL NERVOUS SYSTEM

Established during the first month of development, the brain and spinal cord continue to grow and mature throughout the prenatal period. During development, gender-specific areas appear in both the brain and the spinal cord. For example, certain hypothalamic nuclei concerned with regulating typical male sexual behavior and clusters of neurons in the spinal cord that serve the external genitals are much larger in males. Females have a larger corpus callosum, and gender differences are seen in the language and auditory area of the cerebral cortex. The key to this CNS gender-specific development is whether or not testosterone is being secreted by the fetus. If it is, the male pattern develops.

Maternal exposure to radiation, various drugs (alcohol, opiates, and others), and infections can harm a developing infant's nervous system, particularly during the initial formative stages. For example, rubella (German measles) often leads to deafness and other types of CNS damage in the newborn. Smoking decreases the amount of oxygen in the blood, and lack of oxygen for even a few minutes causes death of neurons. Hence, a smoking mother may be sentencing her infant to some degree of brain damage.

Homeostatic Imbalance

In difficult deliveries, a temporary lack of oxygen may lead to **cerebral palsy,** but any of the factors listed above may also be a cause. Cerebral palsy is a neuromuscular disability in which the voluntary muscles are poorly controlled or paralyzed as a result of brain damage. In addition to spasticity, speech difficulties, and other motor impairments, about half of cerebral palsy victims have seizures, half are mentally retarded, and about a third have some degree of deafness. Visual impairments are also common. Cerebral palsy does not get worse over time, but its deficits are irreversible. It is the largest single cause of crippling in children, affecting six out of every thousand births.

The CNS is plagued by a number of other congenital malformations triggered by genetic or environmental factors during early development. The most serious are congenital hydrocephalus (see p. 457), anencephaly, and spina bifida.

In **anencephaly** (an"en-sef'ah-le; "without brain") the cerebrum and part of the brain stem never develop, presumably because the neural folds fail to fuse rostrally. The child is totally vegetative, unable to see, hear, or process sensory inputs. Muscles are flaccid, and no voluntary movement is possible. Mental life as we know it does not exist. Mercifully, death occurs soon after birth.

Spina bifida (spi'nah bif'ĭ-dah; "forked spine") results from incomplete formation of the vertebral arches and typically involves the lumbosacral region. The technical definition is that laminae and spinous processes are missing on at least one vertebra. If the condition is severe, neural deficits occur as well. *Spina bifida occulta*, the least serious type, involves one or only a few vertebrae and causes no neural problems. Other than a small dimple or tuft of hair over the site of nonfusion, it has no external manifestations. In the more common and severe form, *spina bifida cystica*, a saclike cyst protrudes dorsally from the child's spine. The cyst may contain meninges and cerebrospinal fluid (a *meningocele* [mĕ-ning'go-sēl]), or even portions of the spinal cord

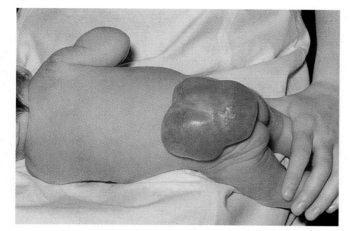

FIGURE 12.30 *Newborn with a lumbar myelomeningocele.*

and spinal nerve roots (a *myelomeningocele*, "spinal cord in a meningeal sac," see Figure 12.30). The larger the cyst and the more neural structures it contains, the greater the neurological impairment. In the worst case, where the inferior spinal cord is rendered functionless, bowel incontinence, bladder muscle paralysis (which predisposes the infant to urinary tract infection and renal failure), and lower limb paralysis occur. Problems with infection are continual because the cyst wall is thin and porous and tends to rupture or leak. Spina bifida cystica is accompanied by hydrocephalus in 90% of all cases.

Up to 70% of cases of spina bifida are caused by inadequate amounts of the B vitamin folate in the maternal diet, a finding that has led to folate supplementation of all bread, flour, and pasta products sold in the United States. Additionally, another B vitamin called inositol further reduces the incidence of spina bifida in experimental mice, and this is expected to be true for humans as well. ■

One of the last CNS areas to mature is the hypothalamus. Since the hypothalamus contains body temperature regulatory centers, premature babies have problems controlling their loss of body heat and must be kept in temperature-controlled environments. PET scans have revealed that the thalamus and somatosensory cortex are active in a five-day-old baby, but that the visual cortex is not. This explains why infants of this age respond to touch but have poor vision. By 11 weeks, more of the cortex is active, and the baby can reach for a rattle. By eight months, the cortex is very active and the child can think about what he or she sees. Growth and maturation of the nervous system continue throughout childhood and largely reflect progressive myelination. As described in Chapter 9, neuromuscular coordination progresses in a superior-to-inferior direction and in a proximal-to-distal direction, and we know that myelination also occurs in this sequence.

The brain reaches its maximum weight in the young adult. Over the next 60 years or so, neurons are damaged and die, and brain weight and volume steadily decline. However, the number of neurons lost over the decades is normally only a small percentage of the total, and the remaining neurons can change their synaptic connections, providing for continued learning throughout life.

Although age brings some cognitive declines in spatial ability, speed of perception, decision making, reaction time, and memory, these losses are not significant in the *healthy individual* until after the seventh decade. Then the brain becomes increasingly fragile, presumably due to an increase in the density of Ca^{2+} channels in aging neurons (as we have seen, elevated Ca^{2+} levels are neurotoxic), and a fairly rapid decline occurs in some, but not all, of these abilities. Ability to build on experience, mathematical skills, and verbal fluency do not decline with age, and many people continue to enjoy intellectual lives and to work at mentally demanding tasks their entire life. Fewer than 5% of people over 65 years old demonstrate true senility. Sadly, there are many cases of "reversible senility" caused by low blood pressure, constipation, poor nutrition, prescription drug effects, dehydration, and hormone imbalances that go undiagnosed. The best way to maintain one's mental abilities in old age may be to seek regular medical checkups throughout life.

Although eventual shrinking of the brain is normal, alcoholics and professional boxers accelerate the process. Whether a boxer wins the match or not, the likelihood of brain damage and atrophy increases with every blow. Everyone recognizes that alcohol profoundly affects both the mind and the body. However, these effects may not be temporary. CT scans of alcoholics reveal a reduction in brain size and density appearing at a fairly early age. Both boxers and alcoholics exhibit signs of senility unrelated to the aging process.

＊   ＊   ＊

The human cerebral hemispheres—our "thinking caps"—are awesome in their complexity. But no less amazing are the brain regions that oversee our subconscious, autonomic body functions—the diencephalon and brain stem—particularly when you consider their relatively insignificant size. The spinal cord, which acts as a reflex center and a communication link between the brain and body periphery, is equally important to body homeostasis.

A good deal of new terminology has been introduced in this chapter, and much of it will come up again in one or more of the remaining nervous system chapters. Chapter 13, your next challenge, considers the structures of the peripheral nervous system that work hand in hand with the CNS to keep it informed and to deliver its orders to the effectors of the body.

RELATED CLINICAL TERMS

Cordotomy (kor-dot′o-me) A procedure in which a tract in the spinal cord is severed surgically; usually done to relieve unremitting pain.

Encephalopathy (en-sef″ah-lop′ah-the; *enceph* = brain; *path* = disease) Any disease or disorder of the brain.

Functional brain disorders Psychological disorders for which no structural cause can be found; include neuroses and psychoses (see entries at right).

Microencephaly (mi″kro-en-sef′ah-le; *micro* = small) Congenital condition involving the formation of a small brain, as evidenced by reduced skull size; most microencephalic children are retarded.

Myelitis (mi″ĕ-li′tis; *myel* = spinal cord; *itis* = inflammation) Inflammation of the spinal cord.

Myelogram (*gram* = recording) X ray of the spinal cord after injection of a contrast medium.

Neuroses (nu-ro′sēs) The less debilitating class of functional brain disorders; examples include severe anxiety (panic attacks), phobias (irrational fears), and obsessivecompulsive behaviors (e.g., washing one's hands every few minutes); the affected individual, however, retains contact with reality.

Pallidectomy ("cutting the pallidus") Surgically lesioning part of the globus pallidus of the basal nuclei to relieve some symptoms of Parkinson's disease.

Psychoses (si-ko′sēs) Class of functional brain disorders in which the affected individuals detach themselves from reality and exhibit bizarre behaviors; the *legal* word for psychotic behavior is *insanity*; the psychoses include *schizophrenia* (skit-so-fre′ne-ah), *manic-depressive disease,* and *depression.*

CHAPTER SUMMARY

Media study tools that could provide you additional help in reviewing specific key topics of Chapter 12 are referenced below. **SP** = *Study Partner;* **IP** = *Interactive Physiology.*

The Brain (pp. 429–461)

1. The brain provides for voluntary movements, interpretation and integration of sensation, consciousness, and cognitive function.

Embryonic Development of the Brain (pp. 429–431)

2. The brain develops from the rostral portion of the embryonic neural tube.

3. Early brain development yields the three primary brain vesicles: the prosencephalon (cerebral hemispheres and diencephalon), mesencephalon (midbrain), and rhombencephalon (pons, medulla, and cerebellum).

4. Cephalization results in the envelopment of the diencephalon and superior brain stem by the cerebral hemispheres.

Regions and Organization of the Brain (p. 432)

5. In a widely used system, the adult brain is divided into the cerebral hemispheres, diencephalon, brain stem, and cerebellum.

6. The cerebral hemispheres and cerebellum have gray matter nuclei surrounded by white matter and an outer cortex of gray matter. The diencephalon and brain stem lack a cortex.

Ventricles of the Brain (p. 432)

7. The brain contains four ventricles filled with cerebrospinal fluid. The lateral ventricles are in the cerebral hemispheres; the third ventricle is in the diencephalon; the fourth ventricle is in the brain stem and connects with the central canal of the spinal cord.

SP Exercise: Chapter 12, Ventricles of the Brain.

The Cerebral Hemispheres (pp. 432–443)

8. The two cerebral hemispheres exhibit gyri, sulci, and fissures. The longitudinal fissure partially separates the hemispheres; other fissures or sulci subdivide each hemisphere into lobes.

SP Exercise: Chapter 12, Structures of the Brain.

9. Each cerebral hemisphere consists of the cerebral cortex, the cerebral white matter, and the basal nuclei (ganglia).

10. Each cerebral hemisphere receives sensory impulses from, and dispatches motor impulses to, the opposite side of the body. The body is represented in an upside-down fashion in the sensory and motor cortices.

11. Functional areas of the cerebral cortex include (1) motor areas: primary motor and premotor areas of the frontal lobe, the frontal eye field, and Broca's area in the frontal lobe of one hemisphere (usually the left); (2) sensory areas: primary somatosensory cortex, somatosensory association cortex, and gustatory area, in the parietal lobe; visual areas in the occipital lobe; olfactory and auditory areas in the temporal lobe; and a vestibular area in the insula; (3) association areas: prefrontal cortex in the frontal lobe; general interpretation area at junction of temporal, parietal, and occipital lobe on one side (usually the left); language areas, including the lateral prefrontal cortex, a large portion of the temporal lobe, Broca's area, Wernicke's area in one hemisphere (temporal lobe) only, usually the left, and affective language areas in one hemisphere (usually the right).

12. The cerebral hemispheres show lateralization of cortical function. In most people, the left hemisphere is dominant (i.e., specialized for language and mathematical skills); the right hemisphere is more concerned with visual-spatial skills and creative endeavors.

13. Fiber tracts of the cerebral white matter include commissures, association fibers, and projection fibers.

14. The paired basal nuclei (also called basal ganglia) include the lentiform nucleus (globus pallidus and putamen) and caudate nucleus. The basal nuclei are subcortical nuclei that help control muscular movements. Functionally they are closely associated with the substantia nigra of the midbrain.

The Diencephalon (pp. 443–446)

15. The diencephalon consists of the thalamus, hypothalamus, and epithalamus and encloses the third ventricle.

16. The thalamus is the major relay station for (1) sensory impulses ascending to the sensory cortex, (2) inputs from subcortical motor nuclei and the cerebellum traveling to the cerebral motor cortex, and (3) impulses traveling to association cortices from lower centers.

17. The hypothalamus is an important autonomic nervous system control center and a pivotal part of the limbic system. It maintains water balance and regulates thirst, eating behavior, gastrointestinal activity, body temperature, and the activity of the anterior pituitary gland.

18. The epithalamus consists of the pineal gland and the choroid plexus of the third ventricle.

The Brain Stem   (pp. 446–450)

19. The brain stem includes the midbrain, pons, and medulla oblongata.

20. The midbrain contains the corpora quadrigemina (visual and auditory reflex centers), the red nucleus (subcortical motor centers), and the substantia nigra. The periaqueductal gray matter elicits the fear response and contains the motor nuclei of cranial nerves III and IV. The cerebral peduncles on its ventral face house the pyramidal fiber tracts. The midbrain surrounds the cerebral aqueduct.

21. The pons is mainly a conduction area. Its nuclei contribute to regulation of respiration and cranial nerves V–VII.

22. The pyramids (descending corticospinal tracts) form the ventral face of the medulla oblongata; these fibers cross over (decussation of the pyramids) before entering the spinal cord. Important nuclei in the medulla regulate respiratory rhythm, heart rate, and blood pressure and serve cranial nerves VIII–XII. The olivary nucleus and cough, sneezing, swallowing, and vomiting centers are in the medulla.

The Cerebellum   (pp. 450–451)

23. The cerebellum consists of two hemispheres, marked by convolutions and separated by the vermis. It is connected to the brain stem by superior, middle, and inferior peduncles.

24. The cerebellum processes and interprets impulses from the motor cortex and sensory pathways and coordinates motor activity so that smooth, well-timed movements occur. It also plays a poorly understood role in cognition.

Functional Brain Systems   (pp. 451–453)

25. The limbic system consists of numerous structures that encircle the diencephalon. It is the "emotional-visceral brain." It also plays a role in memory.

26. The reticular formation includes nuclei spanning the length of the brain stem. It maintains the alert state of the cerebral cortex (RAS) and its motor nuclei serve both somatic and visceral motor activities.

Protection of the Brain   (pp. 453–457, 460)

27. The brain is protected by bone, meninges, cerebrospinal fluid, and the blood-brain barrier.

28. The meninges from superficial to deep are the dura mater, the arachnoid mater, and the pia mater. They enclose the brain and spinal cord and their blood vessels. Inward folds of the inner layer of the dura mater secure the brain to the skull.

SP Exercise: Chapter 12, Meninges of the Brain.

29. Cerebrospinal fluid (CSF), formed by the choroid plexuses from blood plasma, circulates through the ventricles and into the subarachnoid space. It returns to the dural venous sinuses via the arachnoid villi. CSF supports and cushions the brain and cord and helps to nourish them.

30. The blood-brain barrier reflects the relative impermeability of the epithelium of capillaries of the brain. It allows water, respiratory gases, essential nutrients, and fat-soluble molecules to enter the neural tissue, but prevents entry of other, water-soluble, potentially harmful substances.

Homeostatic Imbalances of the Brain   (pp. 460–461)

31. Head trauma may cause brain injuries called concussions (reversible damage) or contusions (irreversible damage). When the brain stem is affected, unconsciousness (temporary or permanent) occurs. Trauma-induced brain injuries may be aggravated by intracranial hemorrhage or cerebral edema, both of which compress brain tissue.

32. Cerebrovascular accidents (strokes) result when blood circulation to brain neurons is impaired and brain tissue dies. The result may be hemiplegia, sensory deficits, or speech impairment.

33. Alzheimer's disease is a degenerative brain disease in which beta amyloid peptide deposits and neurofibrillar tangles appear. Marked by a deficit of ACh, it results in slow, progressive loss of memory and motor control and increasing dementia.

34. Parkinson's disease and Huntington's disease are neurodegenerative disorders of the basal nuclei. Both involve abnormalities of the neurotransmitter dopamine (too little or too much secreted) and are characterized by abnormal movements.

SP Case Study: Parkinson's Disease.

The Spinal Cord   (pp. 461–467)

Gross Anatomy and Protection of the Spinal Cord (pp. 461–463)

1. The spinal cord, a two-way impulse conduction pathway and a reflex center, resides within the vertebral column and is protected by meninges and cerebrospinal fluid. It extends from the foramen magnum to the end of the first lumbar vertebra.

2. Thirty-one pairs of spinal nerve roots issue from the cord. The cord is enlarged in the cervical and lumbar regions, where spinal nerves serving the limbs arise.

Embryonic Development of the Spinal Cord (pp. 463–464)

3. The spinal cord develops from the neural tube. Its gray matter forms from the alar and basal plates. Fiber tracts form the outer white matter. The neural crest forms the sensory (dorsal root) ganglia.

SP Case Study: Congenital Defect.

Cross-Sectional Anatomy of the Spinal Cord (pp. 464–466)

4. The central gray matter of the cord is H shaped. Anterior horns mainly contain somatic motor neurons. Lateral horns contain visceral (autonomic) motor neurons. Posterior horns contain interneurons.

5. Axons of neurons of the lateral and anterior horns emerge in common from the cord via the ventral roots. Axons of sensory neurons (with cell bodies located in the

dorsal root ganglion) enter the posterior aspect of the cord and form the dorsal roots. The ventral and dorsal roots combine to form the spinal nerves.

6. Each side of the white matter of the cord has posterior, lateral, and anterior columns (funiculi), and each funiculus contains a number of ascending and descending tracts. All tracts are paired and most decussate.

7. Ascending (sensory) tracts include the fasciculi gracilis and cuneatus, spinothalamic tracts, and spinocerebellar tracts.

8. Descending tracts include the pyramidal tracts (anterior and lateral corticospinal tracts) and a number of motor tracts originating from subcortical motor nuclei.

Spinal Cord Trauma and Disorders   (pp. 466–467)

9. Injury to the anterior horn neurons or the ventral roots results in flaccid paralysis. (Injury to the upper motor neurons in the brain results in spastic paralysis.) If the dorsal roots or sensory tracts are damaged, paresthesias occur.

10. Poliomyelitis results from inflammation and destruction of the anterior horn neurons by the poliovirus. Paralysis and muscle atrophy ensue.

11. Amyotrophic lateral sclerosis results from destruction of the anterior horn neurons and the pyramidal tract. The victim loses the ability to swallow, speak, and breathe. Death occurs within five years.

Diagnostic Procedures for Assessing CNS Dysfunction   (p. 468)

1. Diagnostic procedures used to assess neurological condition and function range from routine reflex testing to sophisticated techniques such as pneumoencephalography, cerebral angiography, CT scans, MRI scans, and PET scans.

Developmental Aspects of the Central Nervous System   (pp. 468–469)

1. Maternal and environmental factors may impair embryonic brain development, and oxygen deprivation destroys brain cells. Severe congenital brain diseases include cerebral palsy, anencephaly, hydrocephalus, and spina bifida.

2. Premature babies have trouble regulating body temperature because the hypothalamus is one of the last brain areas to mature prenatally.

3. Development of motor control indicates progressive myelination and maturation of a child's nervous system.

4. Brain growth ends in young adulthood. Neurons die throughout life and are not replaced; thus, brain weight and volume decline with age.

5. Healthy elders maintain nearly optimal intellectual function. Disease—particularly cardiovascular disease—is the major cause of declining mental function with age.

REVIEW QUESTIONS

Multiple Choice/Matching

1. The primary motor cortex, Broca's area, and the premotor area are located in which lobe? (a) frontal, (b) parietal, (c) temporal, (d) occipital.

2. The innermost layer of the meninges, delicate and closely apposed to the brain tissue, is the (a) dura mater, (b) corpus callosum, (c) arachnoid, (d) pia mater.

3. Cerebrospinal fluid is formed by (a) arachnoid villi, (b) the dura mater, (c) choroid plexuses, (d) all of these.

4. A patient has suffered a cerebral hemorrhage that has caused dysfunction of the precentral gyrus of his right cerebral cortex. As a result, (a) he cannot voluntarily move his left arm or leg, (b) he feels no sensation on the left side of his body, (c) he feels no sensation on his right side.

5. Choose the proper term from the key to respond to the statements describing various brain areas.

Key:

(a) cerebellum   (d) corpus striatum   (g) midbrain
(b) corpora quadrigemina (e) hypothalamus   (h) pons
(c) corpus callosum   (f) medulla   (i) thalamus

_____ **(1)** basal nuclei involved in fine control of motor activities

_____ **(2)** region where there is a gross crossover of fibers of descending pyramidal tracts

_____ **(3)** control of temperature, autonomic nervous system reflexes, hunger, and water balance

_____ **(4)** houses the substantia nigra and cerebral aqueduct

_____ **(5)** relay stations for visual and auditory stimuli input; found in midbrain

_____ **(6)** houses vital centers for control of the heart, respiration, and blood pressure

_____ **(7)** brain area through which all the sensory input must travel to get to the cerebral cortex

_____ **(8)** brain area most concerned with equilibrium, body posture, and coordination of motor activity

6. Ascending pathways in the spinal cord convey (a) motor impulses, (b) sensory impulses, (c) commissural impulses, (d) all of these.

7. Destruction of the anterior horn cells of the spinal cord results in loss of (a) integrating impulses, (b) sensory impulses, (c) voluntary motor impulses, (d) all of these.

8. Fiber tracts that allow neurons within the same cerebral hemisphere to communicate are (a) association tracts, (b) commissures, (c) projection tracts.

9. A number of brain structures are listed below. If an area is primarily gray matter, write **a** in the answer blank; if mostly white matter, respond with **b**.

_____ **(1)** cerebral cortex
_____ **(2)** corpus callosum and corona radiata
_____ **(3)** red nucleus
_____ **(4)** medial and lateral nuclear groups
_____ **(5)** medial lemniscal tract
_____ **(6)** cranial nerve nuclei
_____ **(7)** spinothalamic tract
_____ **(8)** fornix
_____ **(9)** cingulate and precentral gyri

10. A professor unexpectedly blew a loud horn in his anatomy and physiology class. The students looked up, startled. The reflexive movements of their eyes were mediated by the (a) cerebral cortex, (b) inferior olives, (c) raphe nuclei, (d) superior colliculi, (e) nucleus gracilis.

Short Answer Essay Questions

11. Make a diagram showing the three primary (embryonic) brain vesicles. Name each and then use clinical terminology to name the resulting adult brain regions.

12. (a) What is the advantage of having a cerebrum that is highly convoluted? (b) What term is used to indicate its grooves? Its outward folds? (c) What groove divides the cerebrum into two hemispheres? (d) What divides the parietal from the frontal lobe? The parietal from the temporal lobe?

13. (a) Make a rough drawing of the lateral aspect of the left cerebral hemisphere. (b) You may be thinking, "But I just can't draw!" So, name the hemisphere involved with most people's ability to draw. (c) On your drawing, locate the following areas and provide the major function of each: primary motor cortex, premotor cortex, somatosensory association area, primary sensory area, visual and auditory areas, prefrontal cortex, Wernicke's and Broca's areas.

14. (a) What does lateralization of cortical functioning mean? (b) Why is the term *cerebral dominance* a misnomer?

15. (a) What is the function of the basal nuclei? (b) Which basal nuclei form the lentiform nucleus? (c) Which arches over the diencephalon?

16. (a) Explain how the cerebellum is physically connected to the brain stem. (b) List ways in which the cerebellum is very similar to the cerebrum.

17. Describe the role of the cerebellum in maintaining smooth, coordinated skeletal muscle activity.

18. (a) Where is the limbic system located? (b) What structures make up this system? (c) How is the limbic system important in behavior?

19. (a) Localize the reticular formation in the brain. (b) What does RAS mean, and what is its function?

20. List four ways in which the CNS is protected.

21. (a) How is cerebrospinal fluid formed and drained? Follow its pathway within and around the brain. (b) What happens if CSF is not drained properly? Why is this consequence more harmful in adults?

22. What constitutes the blood-brain barrier?

23. A brain surgeon is about to make an incision. Name all the tissue layers that he cuts through from the skin to the brain.

24. (a) Differentiate between a concussion and a contusion. (b) Why do severe brain stem contusions result in unconsciousness?

25. Describe the spinal cord, depicting its extent, its composition of gray and white matter, and its spinal roots.

26. Differentiate between spastic and flaccid paralysis.

27. How do the conditions paraplegia, hemiplegia, and quadriplegia differ?

28. (a) Define cerebrovascular accident or CVA. (b) Describe its possible causes and consequences.

29. (a) What factors account for brain growth after birth? (b) List some structural brain changes observed with aging.

Critical Thinking and Clinical Application Questions

1. A 10-month-old infant has an enlarging head circumference and delayed overall development. She has a bulging anterior fontanel and her CSF pressure is elevated. Based on these findings, answer the following questions: (a) What are the possible cause(s) of an enlarged head? (b) What tests might be helpful in obtaining information about this infant's problem? (c) Assuming the tests conducted showed the cerebral aqueduct to be constricted, which ventricles or CSF-containing areas would you expect to be enlarged? Which would likely not be visible? Respond to the same questions based on a finding of obstructed arachnoid villi.

2. Mrs. Jones has had a progressive decline in her mental capabilities in the last five or six years. At first her family attributed her occasional memory lapses, confusion, and agitation to grief over her husband's death six years earlier. When examined, Mrs. Jones was aware of her cognitive problems and was shown to have an IQ score approximately 30 points less than would be predicted by her work history. A CT scan showed diffuse cerebral atrophy. The physician prescribed a mild tranquilizer for Mrs. Jones and told her family that there was little else he could recommend. What is Mrs. Jones's problem?

3. Robert, a brilliant computer analyst, suffered a blow to his anterior skull from a falling rock while mountain climbing. Shortly thereafter, it was obvious to his coworkers that his behavior had undergone a dramatic change. Although previously a smart dresser, he was now unkempt. One morning, he was observed defecating into the wastebasket. His supervisor ordered Robert to report to the company's doctor immediately. What region of Robert's brain was affected by the cranial blow?

4. Mrs. Adams is ready to deliver her first baby. Unhappily, the baby appears to have a myelomeningocele. Would a vaginal or surgical (C-section) delivery be more appropriate and why?

5. The medical chart of a 68-year-old man includes the following notes: "Slight tremor of right hand at rest; stony facial expression; difficulty in initiating movements." (a) Based on your present knowledge, what is the diagnosis? (b) What brain areas are most likely involved in this man's disorder, and what is the deficiency? (c) How is this condition currently treated?

13 THE PERIPHERAL NERVOUS SYSTEM AND REFLEX ACTIVITY

Overview of the Peripheral Nervous System (pp. 475–483)

1. Define peripheral nervous system and list its components.
2. Classify sensory receptors according to structure, stimulus detected, and body location.
3. Define nerve and describe the general structure of a nerve.
4. Distinguish between sensory, motor, and mixed nerves.
5. Define ganglion and indicate the general location of ganglia in the body periphery.
6. Describe the process of nerve fiber regeneration.
7. Compare and contrast the motor endings of somatic and autonomic nerve fibers.

Cranial Nerves (pp. 483–491)

8. Name the 12 pairs of cranial nerves and describe the body region and structures innervated by each.

Spinal Nerves (pp. 491–502)

9. Describe the formation of a spinal nerve, and distinguish between spinal roots and rami. Describe the general distribution of the ventral and dorsal rami.
10. Define plexus. Name the major plexuses, their origin sites, and the major nerves arising from each, and describe the distribution and function of the peripheral nerves.
11. Explain Hilton's law of joint innervation.

Reflex Activity (pp. 502–507)

12. Distinguish between autonomic and somatic reflexes.
13. Compare and contrast stretch, flexor, and crossed extensor reflexes.

Developmental Aspects of the Peripheral Nervous System (p. 508)

14. Describe the developmental relationship between the segmented arrangement of peripheral nerves, skeletal muscles, and skin dermatomes.

The human brain, for all its sophistication, would be useless without its links to the outside world. When blindfolded volunteers were suspended in warm water in a sensory deprivation tank (a situation that limited sensory inputs), they began to hallucinate: One saw charging herds of pink and purple elephants, another heard a singing chorus, still others had taste hallucinations. Our very sanity depends on a continual flow of information from the outside.

No less important to our well-being are the barrages of orders sent from the CNS to voluntary muscles and other effectors of the body. This almost continual flow of nerve impulses allows us to move and to take care of our own needs. The frustration of paralysis victims (who can no longer dispatch the motor output that controls their muscles) is very real and unrelenting. The **peripheral nervous system (PNS)** provides these links from and to the real world. Its ghostly white nerves thread through virtually every part of the body, enabling the CNS to receive information and to carry out its decisions.

The peripheral nervous system includes all neural structures outside the brain and spinal cord, that is, the *sensory receptors,* peripheral *nerves* and their associated *ganglia,* and efferent *motor endings.* Its basic components are diagrammed in Figure 13.1. We will begin this chapter with an overview of the functional anatomy of each PNS element. Then we will consider the distribution and function of cranial and spinal nerves. Finally, we will examine the components of reflex arcs before looking at the way some important somatic reflexes, played out almost entirely in PNS structures, help to maintain homeostasis.

OVERVIEW OF THE PERIPHERAL NERVOUS SYSTEM

Sensory Receptors

Sensory receptors are structures that are specialized to respond to changes in their environment; such environmental changes are called **stimuli.** Typically, activation of a sensory receptor by an adequate stimulus results in local depolarizations or graded potentials that, in turn, trigger nerve impulses along the afferent fibers coursing to the CNS. (This process is discussed on pp. 536–541.) As described in more detail in the same chapter (Chapter 15), *sensation* (awareness of the stimulus) and *perception* (interpretation of the meaning of the stimulus) occur in the brain. But we are getting ahead of ourselves here. For now, let us just examine how sensory receptors

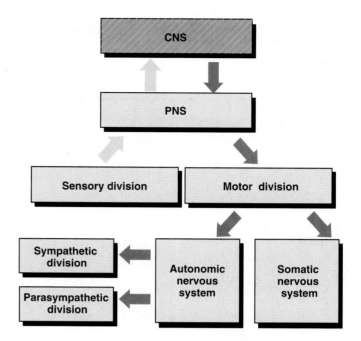

FIGURE 13.1   Place of the PNS in the structural organization of the nervous system.

are classified. Basically, there are three ways to classify sensory receptors: (1) by the type of stimulus they detect; (2) by their body location; and (3) by the relative complexity of their structure.

Classification by Stimulus Type Detected

The receptor classes named according to the stimuli that activate them are easy to remember because the class name usually indicates the stimulus.

1. Mechanoreceptors generate nerve impulses when they, or adjacent tissues, are deformed by mechanical forces such as touch, pressure (including blood pressure), vibration, stretch, and itch.

2. Thermoreceptors are sensitive to temperature changes.

3. Photoreceptors, such as those of the retina of the eye, respond to light energy.

4. Chemoreceptors respond to chemicals in solution (molecules smelled or tasted, or changes in blood chemistry).

5. Nociceptors (no"se-sep'torz; *noci* = harm) respond to potentially damaging stimuli that result in pain.

Overstimulation of any receptor is painful, and essentially all receptor types function as nociceptors at one time or another. For example, searing heat, extreme cold, excessive pressure, and inflammatory chemicals are all interpreted as painful.

TABLE 13.1		General Sensory Receptors Classified by Structure and Function	
Anatomical class (structure)	Illustration	Functional classes according to location (L) and stimulus type (S)	Body location
Unencapsulated			
Free dendritic nerve endings of sensory neurons		L: Exteroceptors, interoceptors, and proprioceptors S: Nociceptors (pain), thermoreceptors (heat and cold), possible mechanoreceptors (pressure)	Most body tissues; most dense in connective tissues (ligaments, tendons, dermis, joint capsules, periostea) and epithelia (epidermis, cornea, mucosae, and glands)
Modified free dendritic endings: Merkel discs		L: Exteroceptors S: Mechanoreceptors (light pressure); slowly adapting	Basal layer of epidermis of skin
Root hair plexuses		L: Exteroceptors S: Mechanoreceptors (hair deflection); rapidly adapting	In and surrounding hair follicles
Encapsulated			
Meissner's corpuscles		L: Exteroceptors S: Mechanoreceptors (light pressure, discriminative touch, vibration of low frequency)	Dermal papillae of hairless skin, particularly nipples, external genitalia, fingertips, soles of feet, eyelids
Krause's end bulbs		L: Exteroceptors S: Mechanoreceptors (probably modified Meissner's corpuscles)	Connective tissue of mucosae (mouth, conjunctiva of eye) and of hairless skin near body openings (lips)

Classification by Location

Three receptor classes are recognized according to their location or the location of the stimulus to which they respond.

1. Exteroceptors (ek"ster-o-sep'torz) are sensitive to stimuli arising outside the body (*extero* = outside). As might be expected, most exteroceptors are near or at the body surface. They include touch, pressure, pain, and temperature receptors in the skin and most receptors of the special sense organs.

2. Interoceptors (in"ter-o-sep'torz), also called **visceroceptors,** respond to stimuli arising from within the body (*intero* = inside), such as from the internal viscera and blood vessels. The different interoceptors monitor a variety of stimuli, including chemical changes, stretching of tissues, and temperature. Sometimes their activity causes us to feel pain, dis-

comfort, hunger, or thirst. However, we are usually unaware of the workings of our visceral receptors.

3. Proprioceptors (pro"pre-o-sep'torz), like interoceptors, respond to internal stimuli; however, their location (in the musculoskeletal organs) is much more restricted. Proprioceptors occur in skeletal muscles, tendons, joints, and ligaments and connective tissue coverings of bones and muscles. (Some authorities also include the equilibrium receptors of the inner ear in this class.) Proprioceptors constantly advise the brain of our own movements (*propria* = one's own) by monitoring the degree of stretch of the organs they occupy.

Classification by Structural Complexity

On the basis of overall receptor structure, there are **simple** and **complex receptors,** but the overwhelm-

TABLE 13.1		General Sensory Receptors Classified by Structure and Function (*continued*)	
Anatomical class (structure)	*Illustration*	*Functional classes according to location (L) and stimulus type (S)*	*Body location*
Pacinian corpuscles		L: Exteroceptors, interoceptors, and some proprioceptors S: Mechanoreceptors (deep pressure, stretch, vibration of high frequency); rapidly adapting	Subcutaneous tissue of the skin; periostea, mesentery, tendons, ligaments, joint capsules, most abundant on fingers, soles of feet, external genitalia, nipples
Ruffini's corpuscles		L: Exteroceptors and proprioceptors S: Mechanoreceptors (deep pressure and stretch); slowly adapting	Deep in dermis, hypodermis, and joint capsules
Muscle spindles	Intrafusal fibers	L: Proprioceptors S: Mechanoreceptors (muscle stretch)	Skeletal muscles, particularly those of the extremities
Golgi tendon organs		L: Proprioceptors S: Mechanoreceptors (tendon stretch)	Tendons
Joint kinesthetic receptors (Pacinian and Ruffini corpuscles, free dendritic endings, and receptors resembling Golgi tendon organs)		L: Proprioceptors S: Mechanoreceptors and nociceptors	Joint capsules of synovial joints

ing majority are simple receptors. The simple receptors are modified dendritic endings of sensory neurons. They are found in the skin, mucous membranes, muscles, and connective tissues, and monitor most types of general sensory information. By contrast, the complex receptors are actually **sense organs,** localized collections of cells (usually of many types) working together to accomplish a specific receptive process. Complex receptors are associated with the **special senses** (vision, hearing, smell, and taste). For example, the eye is composed not only of sensory neurons but also of nonneural cells that form its supporting wall, lens, and other associated structures. Though the complex sense organs (described in detail in Chapter 16) are most familiar to us, the simple sensory receptors associated with the **general senses** are no less important. Here we will concentrate on the structure and function of these tiny sentinels that keep the CNS apprised of what is happening both deep within the body and at its surfaces.

The widely distributed **general sensory receptors** are involved in tactile sensation (a mix of touch, pressure, stretch, and vibration), temperature monitoring (heat and cold), and pain, as well as the "muscle sense" provided by proprioceptors. Anatomically, these receptors are either (1) *free dendritic endings* or (2) *encapsulated dendritic endings.* The general sensory receptors are illustrated and classified in Table 13.1 according to both the structural and functional schemes. However, it should be pointed out that the functional classification is still controversial because many of these receptors respond to similar stimuli, and individual receptors may monitor several different kinds of stimuli.

Free Dendritic Endings Free, or **naked, dendritic endings** of sensory neurons are present nearly everywhere in the body, but they are particularly abundant in epithelia and connective tissues. Most of these sensory fibers are unmyelinated, all have a small diameter, and their distal endings usually have small

Text continues on page 480.

Pain is a four-letter word that means suffering. It is also a primitive experience that we humans share with nearly all animals. Almost everyone has suffered the smarting of a bee sting and the cruel persistence of a headache. Although pain warns of actual or potential tissue damage, its protective value is difficult to appreciate when we are its victims.

Clinically, pain is measured by roundabout techniques, such as tapping the painful spot and observing the patient's degree of wincing. This seems a bit strange, given that the other most common sign of disease or injury—fever—is readily measured with a thermometer.

Pain Reception

The principal pain receptors are the several million bare sensory nerve endings that weave through all the tissues and organs of the body (except the brain) and respond to noxious stimuli—a catchall term for anything damaging to tissues. Wherever body tissue is injured, the damaged cells release an "inflammatory soup" of chemicals including bradykinins, the most potent pain-producing chemicals known. Bradykinins, in turn, unleash the production of an avalanche of inflammatory chemicals, such as histamine and prostaglandins, that initiate healing.

Another surprising entry into the arena as a pain-causing chemical is ATP, that ubiquitous energy source for all cells. It now appears that ATP released from injured cells into the extracellular space may stimulate certain receptors on small-diameter C fibers, thus initiating pain signals.

Pain Transmission and Perception

Clinically, pain is classified as *somatic* or *visceral*. Somatic pain, arising from the skin, muscles, or joints, can be either superficial or deep. Superficial somatic pain is a sharp, pricking pain that often causes us to cry out. Localizable to the skin epidermis or mucosae, it tends to be brief. This type of pain is transmitted along finely myelinated A delta (δ) fibers at the rate of 12 to 80 m/s (up to 250 ft/s). Deep somatic pain is more likely to be burning, itching, or aching pain; it results from stimulation of pain receptors in the deep skin layers, muscles, or joints. Deep somatic pain is both more diffuse and longer lasting than superficial somatic pain, and it *always* indicates tissue destruction. Impulses from the deep pain receptors are transmitted slowly (0.4 to over 1 m/s, or up to 3.5 ft/s) along the small, unmyelinated C fibers.

Visceral pain results from noxious stimulation of receptors in the organs of the thorax and abdominal cavity. Like deep somatic pain, it is usually a dull ache, a burning feeling, or a gnawing pain. Important stimuli for visceral pain are extreme stretching of tissue, ischemia, irritating chemicals, and muscle spasms. Because visceral pain inputs follow the same pathways as somatic pain (convergence), projection by the brain may cause visceral pain to be perceived as somatic in origin. This phenomenon, *referred pain,* is described in Chapter 14.

Both the superficial somatic pain fibers and deep somatic/visceral pain fibers synapse with interneurons in the dorsal horns of the spinal cord. Transmission of pain impulses to the CNS along the sensory neurons causes release of glutamate and *substance P,* the pain neurotransmitter, into the synaptic cleft. On binding to receptors that allow Na^+ entry to and K^+ loss from the postsynaptic neuron, glutamate causes its depolarization. Depolarization in turn makes NDMA receptors on the interneuron more receptive to glutamate binding, which allows Ca^{2+} entry and heightens pain sensitivity via second-messenger mechanisms. Substance P's effects are more diffuse, but basically it affects large numbers of dorsal horn interneurons.

What happens next is unclear, but it appears that the axons of most of these interneurons cross the cord and enter the lateral spinothalamic tracts, which ascend to the thalamus, most importantly its ventral posterior lateral (VPL) nucleus. From the thalamus, impulses are relayed to the somatosensory cortex for interpretation.

Some fibers "take the express"— that is, they ascend directly to the thalamus, allowing the sensory cortex to rapidly analyze what hurts and how much. Other fibers of the spinothalamic tracts "take the local," making abundant synapses in the brain stem and in the hypothalamus and other limbic system structures (particularly the anterior cingulate gyrus) before reaching the thalamus. This second set of pathways mediates arousal and emotional reactions to pain, and has longer-lasting effects on the CNS.

We all have the same **pain threshold;** that is, we perceive it at the same stimulus intensity. For example, heat is perceived as painful at 44°–46°C, the range at which it begins to damage tissue. However, reactions to pain, or **pain tolerance,** vary widely and are heavily influenced by persistent pain stimuli (they decrease it) as well as cultural and psychological factors. When we say that someone is "very sensitive to pain," we are referring to that person's pain tolerance rather than to their pain threshold. Pain tolerance seems to increase with age, and is modified by emotions and mental state. In a disaster, many who feel no pain while struggling to help others may actually be severely injured.

Normally a steady state is maintained that correlates injury and pain. However, long-lasting or very intense pain inputs can disrupt this system, leading to chronic pain. **NMDA receptors,** the same receptors that help strengthen neural connections during certain kinds of learning, appear to play a similar role in **hyperalgesia,** or pain amplification, in the spinal cord. (Hyperalgesia is familiar to anyone with a bad sunburn who steps into a warm shower.) Normal pain input doesn't turn on the spinal cord NMDA receptors, but repeated pain transmission does. Once turned on, the NMDA receptors sensitize the spinal cord neurons to future signals. NMDA receptor blockers prevent hyperalgesia while leaving the normal pain pathways intact, a phenomenon which suggests a drug therapy for managing chronic and phantom pain.

Pain Modulation and Analgesia

The extraordinary plasticity of human pain suggests that natural neural mechanisms must exist to modulate pain transmission and perception. One of the most significant advances in the understanding and management of pain was the publication of Melzack and Wall's *gate control theory of pain* in the 1960s. It suggested that:

1. A pain "gate" exists in the dorsal horn (substantia gelatinosa) where impulses from small unmyelinated pain fibers and large touch (A beta) fibers enter the cord.

2. If impulses along the pain fibers outnumber those transmitted along the touch fibers, the gate opens and pain impulses are transmitted. If the reverse is true, the gate is closed by enkephalin-releasing interneurons in the spinal cord that inhibit transmission of both touch and pain impulses, thus reducing pain perception.

Although several studies have failed to support the gate theory of pain control as originally proposed, it fostered a generation of pain research, and some of its clinical predictions are useful. For example, it's been found that (1) threshold stimulation of the large touch fibers (as by massage) results in a burst of

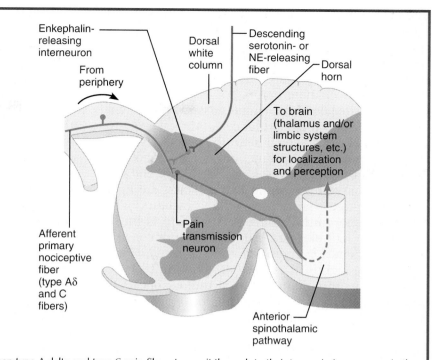

When type A delta and type C pain fibers transmit through to their transmission neurons in the spinothalamic pathway, pain impulses are transmitted to the cerebral cortex. Descending control of pain transmission (analgesia) is mediated by descending central fibers that synapse with small enkephalin-releasing interneurons in the dorsal horn that make inhibitory synapses with the afferent pain fibers. Activation of these interneurons inhibits pain transmission by preventing their release of substance P.

firing in the substantia gelatinosa cells, followed by a brief period of inhibited pain transmission (it *does* close the pain "gate"), and (2) it has been amply proven that direct stimulation, or even transcutaneous electrical nerve stimulation (TENS), of dorsal column (large-diameter touch) fibers *does* provide extended pain relief (however, the precise mechanisms of these effects are still unknown).

For some time, it has been known that our natural opiates (beta endorphins and enkephalins) are released in the brain when we are in pain and act to reduce its perception. Hypnosis, natural childbirth techniques, morphine, and stimulus-induced analgesia all tap into these natural-opiate pathways, which originate in certain brain regions. These regions, which include the *periventricular gray matter* of the hypothalamus and the *periaqueductal gray matter* of the midbrain, oversee

descending pain suppressor fibers that synapse in the dorsal horns. When transmitting, these fibers (most importantly some from the medullary raphe magnus) produce analgesia, presumably by synapsing with opiate (enkephalin) releasing interneurons that in turn actively inhibit forward transmission of pain inputs (see the diagram). The mechanism of this inhibition appears to be that enkephalin blocks Ca^{2+} influx into the sensory terminals, thereby blocking their release of substance P. However, this is only one mechanism of pain modulation. A variety of other neurotransmitter receptor systems in the dorsal horn also regulate pain perception. For example, baclofen (a GABA receptor agonist) and somatostatin promote analgesia in morphine-tolerant patients. You could say that such pain-reducing techniques—drugs and electrical stimulation and massage—work because the body allows it!

knoblike swellings. They respond chiefly to pain and temperature, but some respond to tissue movements caused by pressure as well. Thus, although the bare nerve endings are considered "the" pain receptors (nociceptors) that warn of tissue injury, they also function as mechanoreceptors.

Certain free dendritic endings associate with enlarged, disc-shaped epidermal cells (*Merkel cells*) to form **Merkel discs,** which lie in the deeper layers of the skin epidermis and function as light touch receptors. **Root hair plexuses,** free dendritic endings that wrap basketlike around hair follicles, are light touch receptors that detect bending of hairs. The tickle of a mosquito landing on your skin is mediated by root hair plexuses. Another receptor consisting of free dendritic endings is the *itch receptor.* Located in the dermis, it escaped detection until 1997 because of its thin diameter. Its stimulus appears to be a chemical—such as *histamine* or *bradykinin*—present at inflamed sites.

Encapsulated Dendritic Endings

All **encapsulated dendritic endings** consist of one or more fiber terminals of sensory neurons enclosed in a connective tissue capsule. Virtually all the encapsulated receptors are mechanoreceptors; however, they vary greatly in shape, size, and distribution in the body.

Meissner's corpuscles (mīs′nerz kor′pus″lz) are small receptors in which a few spiraling dendrites are surrounded by Schwann cells and then by a thin egg-shaped capsule of connective tissue. Meissner's corpuscles are found just beneath the skin epidermis in the dermal papillae and are especially numerous in sensitive and hairless skin areas such as the nipples, fingertips, and soles of the feet. Also called **tactile corpuscles,** they are receptors for discriminative touch, and apparently play the same role in light touch reception in hairless skin as do the root hair plexuses in hairy skin.

Krause's end bulbs are considered to be a variation of Meissner's corpuscles. However, while Meissner's corpuscles are found in the skin, Krause's end bulbs are most abundant in mucous membranes, in the lining of the mouth for example. For this reason, they are also called **mucocutaneous corpuscles.**

Pacinian corpuscles, also called **large lamellated corpuscles,** are scattered deep in the dermis, and in subcutaneous tissue underlying the skin. Although they are mechanoreceptors stimulated by deep pressure, they respond only when the pressure is first applied. Thus, Pacinian receptors are best suited to monitor vibration (an "on/off" pressure stimulus). They are the largest of the corpuscular receptors. Some are over 3 mm long and half as wide and are visible to the naked eye as white, egg-shaped bodies. In section, a Pacinian corpuscle resembles a cut onion. Its single dendrite is surrounded by up to 60 layers of flattened Schwann cells, which in turn are enclosed by a connective tissue capsule.

Ruffini's corpuscles, which lie in the dermis, subcutaneous tissue, and joint capsules, contain a spray of dendritic endings enclosed by a flattened capsule. They bear a striking resemblance to Golgi tendon organs (which monitor tendon stretch) and they probably play a similar role in other dense connective tissues where they respond to deep and *continuous* pressure.

Muscle spindles (neuromuscular spindles) are fusiform (spindle-shaped) proprioceptors found throughout the perimysia of skeletal muscles. Each muscle spindle consists of a bundle of modified skeletal muscle fibers, called **intrafusal** (in″trah-fu′zal; *intra* = within, *fusal* = the spindle) **fibers,** enclosed in a connective tissue capsule (see Table 13.1). The intrafusal fibers detect when a muscle is stretched and initiate a reflex that resists the stretch. Details of muscle spindle innervation are considered later in this chapter when the stretch reflex is described (see pp. 503–504).

Golgi tendon organs are proprioceptors located in tendons, close to the point of skeletal muscle insertion. They consist of small bundles of tendon (collagen) fibers enclosed in a layered capsule, with dendrites coiling between and around the fibers. Golgi tendon organs are stimulated when the associated muscle contracts and stretches the tendon; hence, their activity is functionally associated with that of the muscle spindles. When Golgi tendon organs are activated, the contracting muscle is inhibited, which causes it to relax, ending the stimulation of the Golgi tendon organs.

Joint kinesthetic (kin″es-thet′ik) **receptors** are proprioceptors that monitor stretch in the articular capsules that enclose synovial joints. At least four different receptor types (see Table 13.1) contribute to this receptor category. Together these receptors provide information on joint position and motion (*kines* = movement), a sensation of which we are highly conscious.

■ Close your eyes and flex and extend your fingers—you can *feel* exactly which joints are moving.

Nerves and Associated Ganglia

Structure and Classification

A **nerve** is a cordlike organ that is part of the peripheral nervous system. Nerves vary in size, but every nerve consists of parallel bundles of peripheral axons (some myelinated and some not) enclosed by successive wrappings of connective tissue (Figure 13.2).

Within a nerve, each axon is surrounded by **endoneurium** (en″do-nu′re-um), a delicate layer of loose connective tissue that also encloses the fiber's associated myelin or neurilemma sheath. Groups of fibers are bound into bundles or **fascicles** by a coarser connective tissue wrapping, the **perineurium.** Finally, all the fascicles are enclosed by a tough fibrous sheath, the **epineurium,** to form the nerve. Neuron processes actually constitute only a small fraction of a nerve's bulk. The balance consists chiefly of myelin, the protective connective tissue wrappings, blood vessels, and lymphatic vessels.

Recall that the PNS is divided into *sensory* (afferent) and *motor* (efferent) divisions. Hence, nerves are classified according to the direction in which they transmit impulses. Nerves containing both sensory and motor fibers and transmitting impulses both to and from the central nervous system are called **mixed nerves.** Nerves that carry impulses toward the CNS only are **sensory (afferent) nerves;** and nerves carrying impulses only away from the CNS are **motor (efferent) nerves.** Most nerves are mixed. Purely sensory or motor nerves are extremely rare.

Since mixed nerves often carry both somatic and autonomic (visceral) nervous system fibers, the fibers within them may be classified further according to the region they innervate as *somatic afferent, somatic efferent, visceral afferent,* and *visceral efferent.*

For convenience, the peripheral nerves are classified as *cranial nerves* or *spinal nerves,* depending on whether they arise from the brain or spinal cord, respectively. Although autonomic efferents of cranial nerves are mentioned, this chapter focuses on somatic functions. The discussion of the autonomic nervous system and the visceral functions it serves is deferred to Chapter 14.

Ganglia are collections of neuron cell bodies associated with nerves in the PNS. Ganglia associated with *afferent* nerve fibers contain cell bodies of sensory neurons. (These are the *dorsal root ganglia* studied in Chapter 12.) Ganglia associated with *efferent* nerve fibers mostly contain cell bodies of autonomic motor neurons. These more complicated autonomic ganglia are described in Chapter 14.

Regeneration of Nerve Fibers

Damage to nervous tissue is serious because, as a rule, mature neurons do not divide. If the damage is severe or close to the cell body, the entire neuron may die, and other neurons that are normally stimulated by its axon may die as well. However, if the cell body remains intact, cut or compressed axons on peripheral nerves can regenerate successfully.

Almost immediately after a peripheral axon has been severed or crushed, the separated ends seal themselves off and then swell as substances being transported along the axon begin to accumulate in

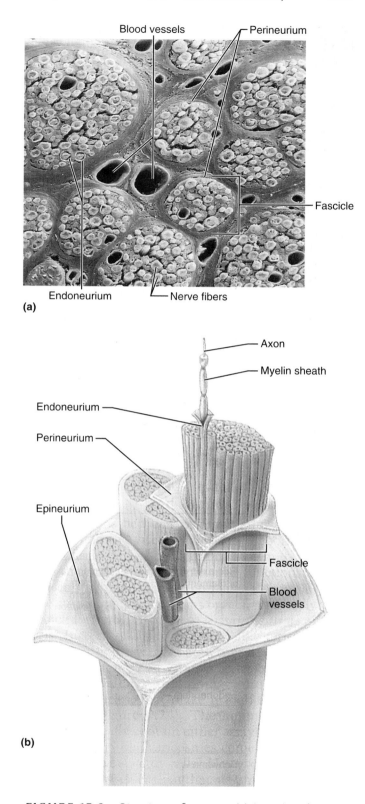

FIGURE 13.2 *Structure of a nerve.* **(a)** Scanning electron micrograph of a cross section of a portion of a nerve (400 ×). (© R. G. Kessel and R. H. Kardon, *Tissues and Organs: A Text-Atlas of Scanning Electron Microscopy,* W. H. Freeman and Company, 1979, all rights reserved.) **(b)** Three-dimensional view of a portion of a nerve, showing connective tissue wrappings.

TABLE 13.2	Cranial Nerves *(continued)*

II The Optic Nerves

Origin and course: Fibers arise from retina of eye to form optic nerve, which passes through optic foramen of orbit; the optic nerves converge to form the optic chiasma (ki-az'mah) where fibers partially cross over, continue on as optic tracts, enter thalamus, and synapse there; thalamic fibers run (as the optic radiation) to occipital (visual) cortex, where visual interpretation occurs; see also Figure 16.22

Function: Purely sensory; carry afferent impulses for vision

Clinical testing: Vision and visual field are determined with eye chart and by testing the point at which the person first sees an object (finger) moving into the visual field. Fundus of eye viewed with ophthalmoscope to detect papilledema (swelling of optic disc, or site where the optic nerve leaves the eyeball), as well as for routine examination of the optic disc and retinal blood vessels.

Homeostatic imbalance: Damage to optic nerve results in blindness in eye served by nerve; damage to visual pathway beyond the optic chiasma results in partial visual losses; visual defects are called *anopsias* (ah-nop'se-ahz). ■

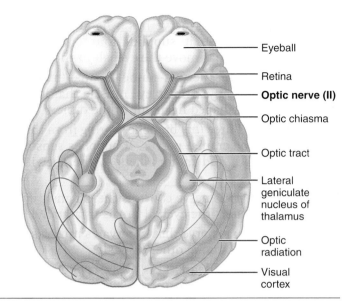

Eyeball
Retina
Optic nerve (II)
Optic chiasma
Optic tract
Lateral geniculate nucleus of thalamus
Optic radiation
Visual cortex

III The Oculomotor (ok"u-lo-mo'tor) Nerves

Origin and course: Fibers extend from ventral midbrain (near its junction with pons) and pass through bony orbit, via superior orbital fissure, to eye

Function: Chiefly motor nerves (*oculomotor* = motor to the eye); contain a few proprioceptive afferents; each nerve includes:

- Somatic motor fibers to four of the six extrinsic eye muscles (inferior oblique and superior, inferior, and medial rectus muscles) that help direct eyeball, and to levator palpebrae superioris muscle, which raises upper eyelid
- Parasympathetic (autonomic) motor fibers to constrictor muscles of iris, which cause pupil to constrict, and to ciliary muscle, controlling lens shape for visual focusing
- Sensory (proprioceptor) afferents, which run from same four extrinsic eye muscles to midbrain

Clinical testing: Pupils are examined for size, shape, and equality. Pupillary reflex is tested with penlight (pupils should constrict when illuminated). Convergence for near vision is tested, as is subject's ability to follow objects with the eyes.

Homeostatic imbalance: In oculomotor nerve paralysis, eye cannot be moved up, down, or inward, and at rest, eye rotates laterally (*external strabismus* [strah-biz'mus]) because the actions of the two extrinsic eye muscles not served by cranial nerve III are unopposed; upper eyelid droops *(ptosis)*, and the person has double vision and trouble focusing on close objects. ■

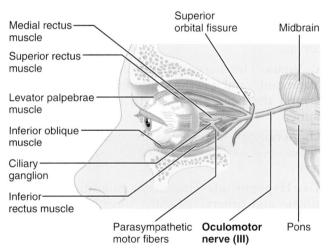

Medial rectus muscle
Superior rectus muscle
Levator palpebrae muscle
Inferior oblique muscle
Ciliary ganglion
Inferior rectus muscle
Superior orbital fissure
Midbrain
Parasympathetic motor fibers
Oculomotor nerve (III)
Pons

IV The Trochlear (trok'le-ar) Nerves

Origin and course: Fibers emerge from dorsal midbrain and course ventrally around midbrain to enter orbits through *superior orbital fissures* along with oculomotor nerves

Function: Primarily motor nerves; supply somatic motor fibers to (and carry proprioceptor fibers from) one of the extrinsic eye muscles, the superior oblique muscle

Clinical testing: Tested in common with cranial nerve III.

Homeostatic imbalance: Trauma to, or paralysis of, a trochlear nerve results in double vision and reduced ability to rotate eye inferolaterally. ■

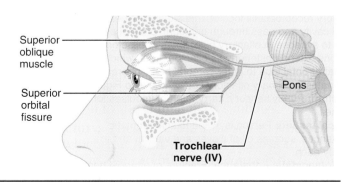

Superior oblique muscle
Superior orbital fissure
Pons
Trochlear nerve (IV)

TABLE 13.2	Cranial Nerves *(continued)*

V *The Trigeminal Nerves*

Largest of cranial nerves; fibers extend from pons to face, and form three divisions (*trigemina* = threefold): ophthalmic, maxillary, and mandibular divisions; major general sensory nerves of face; transmit afferent impulses from touch, temperature, and pain receptors; cell bodies of sensory neurons of all three divisions are located in large *trigeminal* (also called *semilunar* or *gasserian*) *ganglion;* the mandibular division also contains some motor fibers that innervate chewing muscles.

Dentists desensitize upper and lower jaws by injecting local anesthetic (such as Novocain) into alveolar branches of maxillary and mandibular divisions, respectively; since this blocks pain-transmitting fibers of teeth, the surrounding tissues become numb.

	Ophthalmic division (V₁)	Maxillary division (V₂)	Mandibular division (V₃)
Origin and course	Fibers run from face to pons via superior orbital fissure	Fibers run from face to pons via foramen rotundum	Fibers pass through skull via foramen ovale
Function	Conveys sensory impulses from skin of anterior scalp, upper eyelid, and nose, and from nasal cavity mucosa, cornea, and lacrimal gland	Conveys sensory impulses from nasal cavity mucosa, palate, upper teeth, skin of cheek, upper lip, lower eyelid	Conveys sensory impulses from anterior tongue (except taste buds), lower teeth, skin of chin, temporal region of scalp; supplies motor fibers to, and carries proprioceptor fibers from, muscles of mastication
Clinical testing	Corneal reflex tested: touching cornea with wisp of cotton should elicit blinking	Sensations of pain, touch, and temperature are tested with safety pin and hot and cold objects	Motor branch assessed by asking person to clench his teeth, open mouth against resistance, and move jaw side to side

⚠ **Homeostatic imbalance:** *Tic douloureux* (tik doo″loo-roo′), or *trigeminal neuralgia* (nu-ral′je-ah), caused by inflammation of trigeminal nerve, is widely considered to produce most excruciating pain known; the stabbing pain lasts for a few seconds to a minute, but it can be relentless, occurring a hundred times a day (the term *tic* refers to victim's wincing during pain); usually provoked by some sensory stimulus, such as brushing teeth or even a passing breeze hitting the face, but seems to be caused by pressure on the trigeminal nerve root; analgesics and Tegretol (an anticonvulsant) only partially effective; in severe cases, nerve is cut proximal to trigeminal ganglion; this relieves the agony, but also results in loss of sensation on that side of face. ■

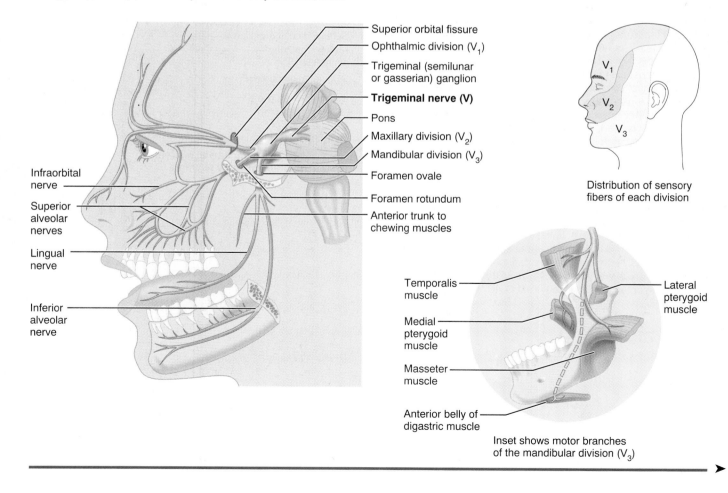

Superior orbital fissure
Ophthalmic division (V₁)
Trigeminal (semilunar or gasserian) ganglion
Trigeminal nerve (V)
Pons
Maxillary division (V₂)
Mandibular division (V₃)
Foramen ovale
Foramen rotundum
Anterior trunk to chewing muscles

Infraorbital nerve
Superior alveolar nerves
Lingual nerve
Inferior alveolar nerve

V₁
V₂
V₃
Distribution of sensory fibers of each division

Temporalis muscle
Medial pterygoid muscle
Masseter muscle
Anterior belly of digastric muscle
Lateral pterygoid muscle

Inset shows motor branches of the mandibular division (V₃)

TABLE 13.2	Cranial Nerves *(continued)*

VI The Abducens (ab-du'senz) Nerves

Origin and course: Fibers leave inferior pons and enter orbit via superior orbital fissure to run to eye

Function: Mixed nerve, primarily motor; supplies somatic motor fibers to lateral rectus muscle, an extrinsic muscle of the eye; conveys proprioceptor impulses from same muscle to brain.

Clinical testing: Tested in common with cranial nerve III.

⚡ **Homeostatic imbalance:** In abducens nerve paralysis, eye cannot be moved laterally; at rest, affected eyeball rotates medially *(internal strabismus).* ■

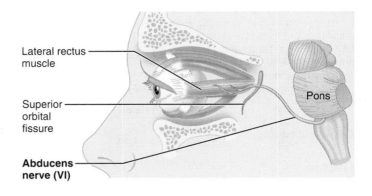

VII The Facial Nerves

Origin and course: Fibers issue from pons, just lateral to abducens nerves (see Figure 13.4), enter temporal bone via *internal acoustic meatus,* and run within bone (and through inner ear cavity) before emerging through *stylomastoid foramen;* nerve then courses to lateral aspect of face

Function: Mixed nerves that are the chief motor nerves of face; have five major branches: temporal, zygomatic, buccal, mandibular, and cervical (see **c**)

■ Convey motor impulses to skeletal muscles of face (muscles of facial expression), except for chewing muscles served by trigeminal nerves, and transmit proprioceptor impulses from same muscles to pons (see **b**)

■ Transmit parasympathetic (autonomic) motor impulses to lacrimal (tear) glands, nasal and palatine glands, and submandibular and sublingual salivary glands. Some of the cell bodies of these parasympathetic motor neurons are in *pterygopalatine* (ter"eh-go-pal'ah-tīn) and *submandibular ganglia* on the trigeminal nerve (see **a**)

■ Convey sensory impulses from taste buds of anterior two-thirds of tongue; cell bodies of theses sensory neurons are in *geniculate ganglion* (see **a**)

Clinical testing: Anterior two-thirds of tongue is tested for ability to taste sweet (sugar), salty, sour (vinegar), and bitter (quinine) substances. Symmetry of face is checked. Subject is asked to close eyes, smile, whistle, and so on. Tearing is assessed with ammonia fumes.

⚡ **Homeostatic imbalance:** *Bell's palsy,* characterized by paralysis of facial muscles on affected side and partial loss of taste sensation, may develop rapidly (often overnight); caused by herpes simplex I viral infection, which causes swelling and inflammation of facial nerve; lower eyelid droops, corner of mouth sags (making it difficult to eat or speak normally), tears drip continuously from eye and eye cannot be completely closed (conversely, dry-eye syndrome may occur); condition may disappear spontaneously without treatment. ■

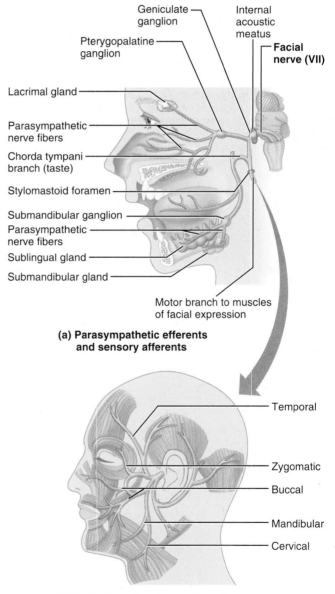

(a) Parasympathetic efferents and sensory afferents

(b) Motor branches to muscles of facial expression and scalp muscles

TABLE 13.2	Cranial Nerves (continued)

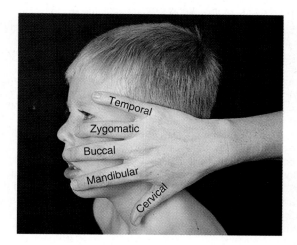

(c) A simple method of remembering the courses of the five major motor nerves of the face

VIII The Vestibulocochlear (ves-tib″u-lo-kok′le-ar) Nerves

Origin and course: Fibers arise from hearing and equilibrium apparatus located within inner ear of temporal bone and pass through internal acoustic meatus to enter brain stem at pons-medulla border; afferent fibers from hearing receptor in cochlea form the *cochlear division;* those from equilibrium receptors in semicircular canals and vestibule form the *vestibular division* (vestibular nerve); the two divisions merge to form vestibulocochlear nerve; see also Figure 16.25

Function: Purely sensory; vestibular branch transmits afferent impulses for sense of equilibrium, and sensory nerve cell bodies are located in *vestibular ganglia;* cochlear branch transmits afferent impulses for sense of hearing, and sensory nerve cell bodies are located in *spiral ganglion* within cochlea

Clinical testing: Hearing is checked by air and bone conduction using tuning fork.

Homeostatic imbalance: Lesions of cochlear nerve or cochlear receptors result in *central* or *nerve deafness,* whereas damage to vestibular division produces dizziness, rapid involuntary eye movements, loss of balance, nausea and vomiting. ■

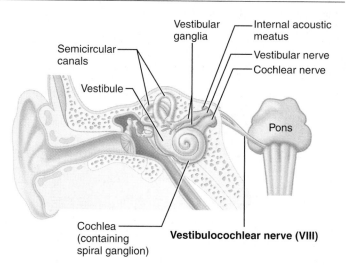

Vestibulocochlear nerve (VIII)

IX The Glossopharyngeal (glos″o-fah-rin′je-al) Nerves

Origin and course: Fibers emerge from medulla and leave skull via *jugular foramen* to run to throat

Function: Mixed nerves that innervate part of tongue and pharynx; provide motor fibers to, and carry proprioceptor fibers from, a superior pharyngeal muscle called the *stylopharyngeus,* which elevates the pharynx in swallowing; provide parasympathetic motor fibers to parotid salivary gland (some of the nerve cell bodies of these parasympathetic motor neurons are located in *otic ganglion*)

Sensory fibers conduct taste and general sensory (touch, pressure, pain) impulses from pharynx and posterior tongue, from chemoreceptors in the carotid body (which monitor O_2 and CO_2 levels in the blood and help regulate respiratory rate and depth), and from pressure receptors of carotid sinus (which help to regulate blood pressure by providing feedback information); sensory neuron cell bodies are located in *superior* and *inferior* ganglia

Clinical testing: Position of the uvula is checked. Gag and swallowing reflexes are checked. Subject is asked to speak and cough. Posterior third of tongue may be tested for taste.

⚠ **Homeostatic imbalance:** Injury or inflammation of glossopharyngeal nerves impairs swallowing and taste, particularly for sour and bitter substances. ■

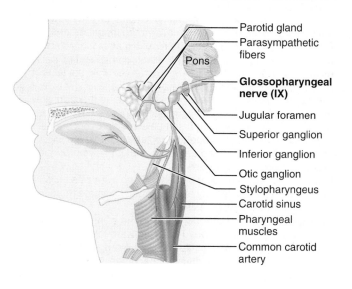

Glossopharyngeal nerve (IX)

TABLE 13.2	Cranial Nerves *(continued)*

X *The Vagus* (va′gus) *Nerves*

Origin and course: The only cranial nerves to extend beyond head and neck region; fibers emerge from medulla, pass through skull via jugular foramen, and descend through neck region into thorax and abdomen; see also Figure 14.4

Function: Mixed nerves; nearly all motor fibers are parasympathetic efferents, except those serving skeletal muscles of pharynx and larynx (involved in swallowing); parasympathetic motor fibers supply heart, lungs, and abdominal viscera and are involved in regulation of heart rate, breathing, and digestive system activity; transmit sensory impulses from thoracic and abdominal viscera, from the carotid sinus (pressoreceptor for blood pressure) and the carotid and aortic bodies (chemoreceptors for respiration), and taste buds of posterior tongue and pharynx; carry proprioceptor fibers from muscles of larynx and pharynx

Clinical testing: As for cranial nerve IX (IX and X are tested in common, since they both innervate muscles of throat and mouth).

⚠ **Homeostatic imbalance:** Since nearly all muscles of the larynx ("voice box") are innervated by laryngeal branches of the vagus, vagal nerve paralysis can lead to hoarseness or loss of voice; other symptoms are difficult swallowing and impaired digestive system mobility. Total destruction of both vagus nerves is incompatible with life, because these parasympathetic nerves are crucial in maintaining normal state of visceral organ activity; without their influence, the activity of the sympathetic nerves, which mobilize and accelerate vital body processes (and shut down digestion), would be unopposed. ■

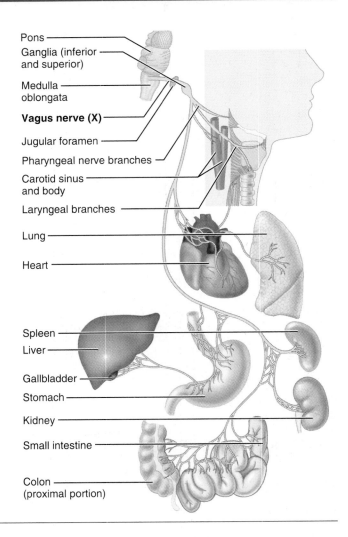

XI *The Accessory Nerves*

Origin and course: Unique in that they are formed by union of a *cranial root* and a *spinal root;* cranial root emerges from lateral aspect of medulla of brain stem; spinal root arises from superior region (C₁–C₅) of spinal cord. Spinal portion passes upward along spinal cord, enters skull via foramen magnum, and temporarily joins cranial root; the resulting accessory nerve exits from skull through *jugular foramen;* then cranial and spinal fibers diverge; cranial root fibers join vagus nerve, while spinal root runs to the largest neck muscles

Function: Mixed nerves, but primarily motor in function; cranial division joins with fibers of vagus nerve (X) to supply motor fibers to larynx, pharynx, and soft palate; spinal root supplies motor fibers to trapezius and sternocleidomastoid muscles, which together move head and neck, and conveys proprioceptor impulses from same muscles

Clinical testing: Sternocleidomastoid and trapezius muscles are checked for strength by asking person to rotate head and shrug shoulders against resistance.

⚠ **Homeostatic imbalance:** Injury to the spinal root of one accessory nerve causes head to turn toward injury side as result of sternocleidomastoid muscle paralysis; shrugging of that shoulder (role of trapezius muscle) becomes difficult. ■

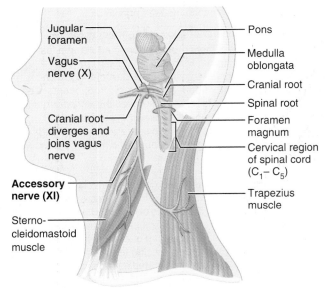

TABLE 13.2	Cranial Nerves (continued)

XII The Hypoglossal (hi"po-glos'al) Nerves

Origin and course: As their name implies (*hypo* = beneath; *glossal* = tongue), hypoglossal nerves mainly serve the tongue; fibers arise by a series of roots from medulla and exit from skull via *hypoglossal canal* to travel to tongue; see also Figure 13.4

Function: Mixed nerves, but primarily motor in function; carry somatic motor fibers to intrinsic and extrinsic muscles of tongue, and proprioceptor fibers from same muscles to brain stem; hypoglossal nerve control allows not only food mixing and manipulation by tongue during chewing, but also tongue movements that contribute to swallowing and speech.

Clinical testing: Person is asked to protrude and retract tongue. Any deviations in position are noted.

⚠ Homeostatic imbalance: Damage to hypoglossal nerves causes difficulties in speech and swallowing; if both nerves are impaired, the person cannot protrude tongue; if only one side is affected, tongue deviates (leans) toward affected side; eventually paralyzed side begins to atrophy. ■

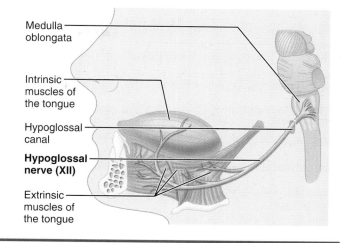

Medulla oblongata

Intrinsic muscles of the tongue

Hypoglossal canal

Hypoglossal nerve (XII)

Extrinsic muscles of the tongue

SPINAL NERVES

General Features of Spinal Nerves

Thirty-one pairs of **spinal nerves,** each containing thousands of nerve fibers, arise from the spinal cord and supply all parts of the body except the head and some areas of the neck. All are mixed nerves. As illustrated in Figure 13.5, the spinal nerves are named according to their point of issue from the spinal cord. There are 8 pairs of cervical spinal nerves (C_1-C_8), 12 pairs of thoracic nerves (T_1-T_{12}), 5 pairs of lumbar nerves (L_1-L_5), 5 pairs of sacral nerves (S_1-S_5), and 1 pair of tiny coccygeal nerves (designated C_0).

Notice that there are eight pairs of cervical nerves but only seven cervical vertebrae. This "discrepancy" is easily explained. The first seven pairs of cervical nerves exit the vertebral canal *superior to* the vertebrae for which they are named. C_8, on the other hand, emerges *inferior to* the seventh cervical vertebra (between C_7 and T_1). Below the cervical level, each spinal nerve leaves the vertebral column *inferior to* the same-numbered vertebra.

Why are the lumbar and cervical regions of the spinal cord enlarged?

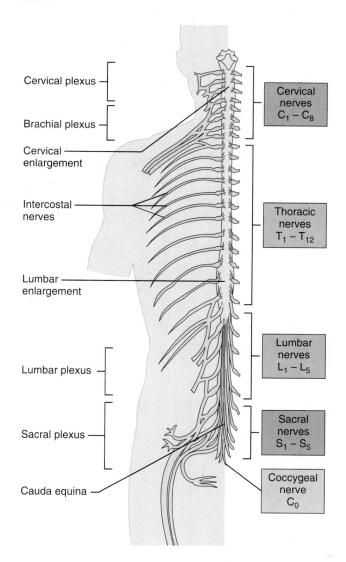

Cervical plexus

Brachial plexus

Cervical enlargement

Intercostal nerves

Lumbar enlargement

Lumbar plexus

Sacral plexus

Cauda equina

Cervical nerves $C_1 - C_8$

Thoracic nerves $T_1 - T_{12}$

Lumbar nerves $L_1 - L_5$

Sacral nerves $S_1 - S_5$

Coccygeal nerve C_0

The peripheral nerves serving the limbs arise in the lumbar and cervical regions.

FIGURE 13.5 *Distribution of spinal nerves, posterior view.* Note that the spinal nerves are named according to their points of issue.

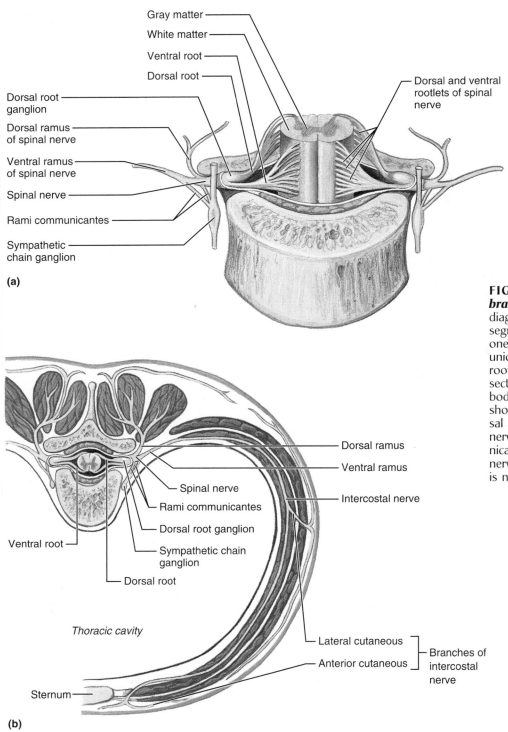

Gray matter

White matter

Ventral root

Dorsal root

Dorsal and ventral rootlets of spinal nerve

Dorsal root ganglion

Dorsal ramus of spinal nerve

Ventral ramus of spinal nerve

Spinal nerve

Rami communicantes

Sympathetic chain ganglion

(a)

Dorsal ramus

Ventral ramus

Spinal nerve

Rami communicantes

Dorsal root ganglion

Sympathetic chain ganglion

Ventral root

Dorsal root

Intercostal nerve

Thoracic cavity

Lateral cutaneous

Anterior cutaneous

Branches of intercostal nerve

Sternum

(b)

FIGURE 13.6 *Formation and branches of spinal nerves.* (a) A diagrammatic view of a spinal cord segment, illustrating the formation of one pair of spinal nerves by the union of the ventral and dorsal roots of the spinal cord. **(b)** Cross-sectional view of the left side of the body at the level of the thorax, showing the distribution of the dorsal and ventral rami of the spinal nerve. Notice also the rami communicantes branches of the spinal nerve. (The small meningeal branch is not illustrated.)

As we mentioned in Chapter 12, each spinal nerve connects to the spinal cord by two roots, the **dorsal** and **ventral roots.** Each root forms from a series of **rootlets** that attach along the whole length of the corresponding spinal cord segment (Figure 13.6). The *ventral roots* contain *motor* (efferent) fibers, arising from anterior horn motor neurons, that extend to and innervate the skeletal muscles. (The autonomic nervous system efferents also contained in the ventral roots are described in Chapter 14.) *Dorsal roots* contain *sensory* (afferent) fibers, arising from sensory neurons in the dorsal root ganglia, that conduct impulses from peripherally located receptors to the spinal cord.

The spinal roots pass laterally from the cord and unite just distal to the dorsal root ganglion to form a spinal nerve before emerging from the vertebral column via their respective intervertebral foramina. The motor and sensory fibers mingle together in the spinal nerve, so the spinal nerve proper contains both efferent and afferent fibers. The length of the spinal roots increases progressively from the superior to the inferior aspect of the cord. In the cervical region, the roots are short and run horizontally, but the roots of the lumbar and sacral nerves extend inferiorly for some distance through the lower vertebral canal as the *cauda equina* before exiting the vertebral column (see Figure 13.5).

A spinal nerve is quite short (only 1–2 cm) because almost immediately after emerging from its foramen, it divides into a small **dorsal ramus,** a larger **ventral ramus** (ra'mus; "branch"), and a tiny **meningeal** (mĕ-nin'je-al) **branch** that reenters the vertebral canal to innervate the meninges and blood vessels within. Each ramus, like the spinal nerve itself, is mixed. Finally, joined to the base of ventral rami of spinal nerves in the thoracic region, there are the **rami communicantes.** These rami, which contain autonomic (visceral) nerve fibers, are considered in detail in Chapter 14.

The rest of this section focuses on the spinal nerve rami and their main branches, which supply the entire somatic region of the body (voluntary muscles and skin) from the neck down. The dorsal rami supply the posterior body trunk. The thicker ventral rami supply the rest of the body trunk and the limbs. However, before getting into rami distribution, let's review the difference between roots and rami: Roots lie medial to and form the spinal nerves, and each root is strictly sensory *or* motor in function. Rami lie distal to and are lateral branches of the spinal nerves and, like spinal nerves, carry both sensory and motor fibers.

Innervation of Specific Body Regions

This section explores how the rami and their branches innervate the following regions of the body: the back, the thorax and abdominal wall, the neck, the limbs, the joints, and the skin. Before getting into the specifics, however, some important points about the ventral rami of the spinal nerves need to be understood.

Except for T_2–T_{12}, all ventral rami branch and join one another lateral to the vertebral column, forming complicated **nerve plexuses** (see Figure 13.5). Such interlacing nerve networks occur in the cervical, brachial, lumbar, and sacral regions and primarily serve the limbs. (Note that only ventral [not dorsal] rami form plexuses.) Within the plexuses, fibers from the different ventral rami crisscross each other and become redistributed so that (1) each resulting branch of the plexus contains fibers from several different spinal nerves and (2) fibers from each ventral ramus travel to the body periphery via several different routes or branches. Thus, each muscle in a limb receives its nerve supply from more than one spinal nerve. An advantage of this fiber regrouping is that damage to one spinal segment or root cannot completely paralyze any limb muscle.

Throughout this section, we will be mentioning major groups of skeletal muscles served. For more specific information on muscle innervations, see Tables 10.1–10.17 on pp. 334–381.

The Back

The innervation of the posterior body trunk by the dorsal rami follows a neat, segmented plan. Via its several branches (see Figure 13.6), each dorsal ramus innervates the narrow strip of muscle (and skin) in line with its emergence point from the spinal column.

The Anterolateral Thorax and Abdominal Wall

Only in the thorax are the ventral rami arranged in a simple segmental pattern corresponding to that of the dorsal rami. The ventral rami of T_1–T_{12} mostly course anteriorly, deep to each rib, as the **intercostal nerves.** Along their course, the intercostal nerves give off *cutaneous branches* to the skin (see Figure 13.6b). Two nerves in this thoracic series are unusual: the tiny T_1 (most fibers enter the brachial plexus) and T_{12}, which lies inferior to the twelfth rib, making it a **subcostal nerve.** The intercostal nerves and their branches supply the intercostal muscles lying between the ribs, the muscle and skin of the anterolateral thorax, and most of the abdominal wall.

TABLE 13.3	Branches of the Cervical Plexus (See Figure 13.7)	

Nerves	Spinal roots (ventral rami)	Structures served
Cutaneous Branches (Superficial)		
Lesser occipital	C_2 (C_3)	Skin on posterolateral aspect of neck
Greater auricular	C_2, C_3	Skin of ear, skin over parotid gland
Transverse cutaneous	C_2, C_3	Skin on anterior and lateral aspect of neck
Supraclavicular (anterior, middle, and posterior)	C_3, C_4	Skin of shoulder and anterior aspect of chest
Motor Branches (Deep)		
Ansa cervicalis (superior and inferior roots)	C_1–C_3	Infrahyoid muscles of neck (omohyoid, sternohyoid, and sternothyroid)
Segmental and other muscular branches	C_1–C_5	Deep muscles of neck (geniohyoid and thyrohyoid) and portions of scalenes, levator scapulae, trapezius, and sternocleidomastoid muscles
Phrenic	C_3–C_5	Diaphragm (sole motor nerve supply)

Cervical Plexus and the Neck

The **cervical plexus** is buried deep in the neck under the sternocleidomastoid muscle. This plexus is formed by the ventral rami of the first four cervical nerves (Figure 13.7). The branches of the looping cervical plexus are summarized in Table 13.3. Most branches are **cutaneous nerves** (nerves that supply only the skin) that transmit sensory impulses from the skin of the neck, the ear area, the back of the head, and the shoulder. Other branches innervate muscles of the anterior neck.

The single most important nerve from this plexus is the **phrenic** (fren'ik) **nerve** (which receives its major input from C_3 and C_4). The phrenic nerve runs inferiorly through the thorax and supplies both motor and sensory fibers to the diaphragm (*phren* = diaphragm). The diaphragm is the chief muscle causing breathing movements (see Chapter 23).

⚠ *Homeostatic Imbalance*

Irritation of the phrenic nerve causes spasms of the diaphragm, or hiccups. If both phrenic nerves are severed, or if the C_3–C_5 region of the spinal cord is crushed or destroyed, the diaphragm is paralyzed and respiratory arrest occurs. Victims are kept alive by mechanical respirators that force air into their lungs and do their breathing for them. ■

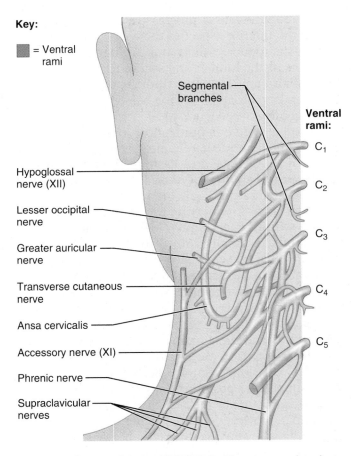

FIGURE 13.7   *The cervical plexus.* The nerves colored gray connect to the plexus but do not belong to it. (See Table 13.3.)

TABLE 13.4	Branches of the Brachial Plexus (See Figure 13.8)	

Nerves	Cord and spinal roots (ventral rami)	Structures served
Axillary	Posterior cord (C_5, C_6)	Muscular branches: deltoid and teres minor muscles Cutaneous branches: some skin of shoulder region
Musculocutaneous	Lateral cord (C_5–C_7)	Muscular branches: flexor muscles in anterior arm (biceps brachii, brachialis, coracobrachialis) Cutaneous branches: skin on anterolateral forearm (extremely variable)
Median	By two branches, one from medial cord (C_8, T_1) and one from the lateral cord (C_5–C_7)	Muscular branches to flexor group of anterior forearm (palmaris longus, flexor carpi radialis, flexor digitorum superficialis, flexor pollicis longus, lateral half of flexor digitorum profundus, and pronator muscles); intrinsic muscles of lateral palm and digital branches to the fingers Cutaneous branches: skin of lateral two-thirds of hand, palm side and dorsum of fingers 2 and 3
Ulnar	Medial cord (C_8, T_1)	Muscular branches: flexor muscles in anterior forearm (flexor carpi ulnaris and medial half of flexor digitorum profundus); most intrinsic muscles of hand Cutaneous branches: skin of medial third of hand, both anterior and posterior aspects
Radial	Posterior cord (C_5–C_8, T_1)	Muscular branches: posterior muscles of arm, forearm, and hand (triceps brachii, anconeus, supinator, brachioradialis, extensors carpi radialis longus and brevis, extensor carpi ulnaris, and several muscles that extend the fingers) Cutaneous branches: skin of posterolateral surface of entire limb (except dorsum of fingers 2 and 3)
Dorsal scapular	Branches of C_5 rami	Rhomboid muscles and levator scapulae
Long thoracic	Branches of C_5–C_7 rami	Serratus anterior muscle
Subscapular	Posterior cord; branches of C_5 and C_6 rami	Teres major and subscapular muscles
Suprascapular	Upper trunk (C_5, C_6)	Shoulder joint; supraspinatus and infraspinatus muscles
Pectoral (lateral and medial)	Branches of lateral and medial cords (C_5–T_1)	Pectoralis major and minor muscles

Brachial Plexus and the Upper Limb

The large, important **brachial plexus** is situated partly in the neck and partly in the axilla. It gives rise to virtually all the nerves that innervate the upper limb (Table 13.4). It can be palpated (felt) in a living person just superior to the clavicle at the lateral border of the sternocleidomastoid muscle.

This plexus is formed by intermixing of the ventral rami of the inferior four cervical nerves (C_5–C_8) and most of the T_1 ramus. Additionally, it often receives fibers from C_4 or T_2 or both.

The brachial plexus is very complex (some consider it to be the anatomy student's nightmare). Perhaps the simplest approach is to master the terms used for its four major groups of branches: From medial to lateral, these are the (1) ventral rami, misleadingly called *roots*, which form (2) *trunks*, which form (3) *divisions*, which in turn form (4) *cords*. These elements are shown in Figure 13.8a and b. You might want to use the saying "**R**eally **t**ired? **D**rink **c**offee" to help you remember this sequence of brachial plexus branching.

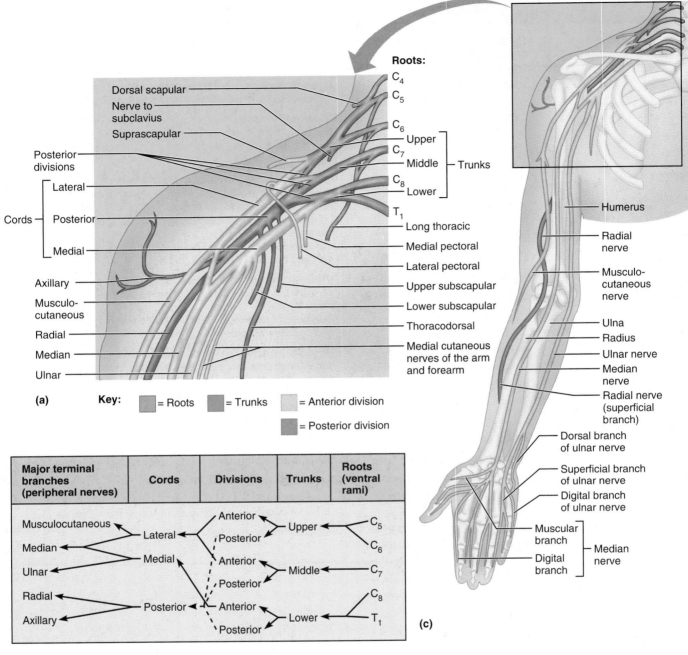

Key:
| ■ = Roots | ■ = Trunks | ▢ = Anterior division |
| | ■ = Posterior division | |

(a)

Major terminal branches (peripheral nerves)	Cords	Divisions	Trunks	Roots (ventral rami)

(b)

(c)

FIGURE 13.8 *The brachial plexus.* (a) Roots, trunks, divisions, and cords of the brachial plexus. (b) Flowchart of consecutive branches formed in the brachial plexus from the spinal roots (ventral rami) to the major nerves formed from the cords. (c) Distribution of the major peripheral nerves of the upper limb. (See Table 13.4.)

The five **roots** (ventral rami C_5–T_1) of the brachial plexus lie deep to the sternocleidomastoid muscle. At the lateral border of that muscle, these roots unite to form **upper, middle,** and **lower trunks,** each of which almost immediately divides into an **anterior** and a **posterior division.** These divisions generally indicate which fibers ultimately serve the front or back of the limb. The divisions pass deep to the clavicle and enter the axilla, where they give rise to three large fiber bundles called the **lateral, medial,** and **posterior cords.** (The cords are named for their relationship to the axillary artery, which runs through the axilla.) All along the plexus, small nerves branch off. These supply the muscles and skin of the shoulder and superior thorax.

Homeostatic Imbalance

Injuries to the brachial plexus are common; when severe, they cause weakness or paralysis of the entire upper limb. Such injuries may occur when the upper limb is pulled hard and the plexus is stretched (as when a football tackler yanks the arm of the halfback), and by blows to the top of the shoulder that force the humerus inferiorly (as when a cyclist is pitched headfirst off his motorcycle and grinds his shoulder into the pavement). ■

The brachial plexus ends within the axilla, where its three cords wind along the axillary artery and then give rise to the main nerves of the upper limb. Five of these nerves are especially important: the axillary, musculocutaneous, median, ulnar, and radial nerves. These nerves are illustrated in Figure 13.8c. Their distribution and targets are described briefly below, and in more detail in Table 13.4.

The **axillary nerve** branches off the posterior cord and runs posterior to the surgical neck of the humerus. It innervates the deltoid and teres minor muscles and the skin and joint capsule of the shoulder.

The **musculocutaneous nerve,** the major end branch of the lateral cord, courses inferiorly within the anterior arm, supplying motor fibers to the biceps brachii and brachialis muscles. Distal to the elbow, it provides for cutaneous sensation of the lateral forearm.

The **median nerve** descends through the arm to the anterior forearm, where it gives off branches to the skin and to most of the flexor muscles. On reaching the hand, it innervates five intrinsic muscles of the lateral palm. The median nerve activates muscles that pronate the forearm, flex the wrist and fingers, and oppose the thumb.

Homeostatic Imbalance

Median nerve injury makes it difficult to use the pincer grasp (oppose the thumb and index finger) and thus to pick up small objects. This nerve runs down the midline of the forearm and wrist, so it is a frequent casualty of wrist-slashing suicide attempts. ■

The **ulnar nerve** branches off the medial cord of the plexus. It descends along the medial aspect of the arm toward the elbow, swings behind the medial epicondyle, and then follows the ulna along the medial forearm. There it supplies the flexor carpi ulnaris and the medial part of the flexor digitorum profundus (the flexors not supplied by the median nerve). It continues into the hand, where it innervates most intrinsic hand muscles and the skin of the medial

aspect of the hand. The ulnar nerve produces wrist and finger flexion and adduction and abduction of the medial fingers (with the median nerve).

Homeostatic Imbalance

Where it takes a superficial course, the ulnar nerve is very vulnerable to injury. Striking the "funny bone"—the spot where this nerve rests against the medial epicondyle—causes tingling of the little finger. Severe or chronic damage can lead to sensory loss, paralysis, and atrophy of the muscles served by the ulnar nerve. Affected individuals have trouble making a fist and gripping objects. The little and ring fingers become hyperextended at the knuckles and flexed at the distal interphalangeal joints, so the hand contorts into a *clawhand.* ■

The **radial nerve,** the largest branch of the brachial plexus, is a continuation of the posterior cord. This nerve wraps around the humerus (in the radial groove), runs anteriorly around the lateral epicondyle at the elbow, where it divides into a superficial branch that follows the lateral edge of the radius to the hand and a deep branch (not illustrated) that runs posteriorly. It supplies the posterior skin of the limb along its entire course. Its motor branches innervate essentially all the extensor muscles of the upper limb. The radial nerve produces elbow extension, supination of the forearm, extension of the wrist and fingers, and abduction of the thumb.

Homeostatic Imbalance

Trauma to the radial nerve results in *wrist drop,* an inability to extend the hand at the wrist. Improper use of a crutch or "Saturday night paralysis," in which an intoxicated person falls asleep with an arm draped over the back of a chair or sofa edge, cause radial nerve compression and ischemia. ■

Lumbosacral Plexus and the Lower Limb

The sacral and lumbar plexuses overlap substantially, and since many fibers of the lumbar plexus contribute to the sacral plexus via the **lumbosacral trunk,** the two plexuses are often referred to as the **lumbosacral plexus.** Although the lumbosacral plexus mainly serves the lower limb, it also sends some branches to the abdomen, pelvis, and buttock.

Lumbar Plexus The **lumbar plexus** arises from the first four lumbar spinal nerves (Figure 13.9) and lies within the psoas major muscle. Its proximal branches innervate parts of the abdominal wall muscles and the psoas muscle, but the major branches of

TABLE 13.5	Branches of the Lumbar Plexus (See Figure 13.9)	
Nerves	**Spinal roots (ventral rami)**	**Structures served**
Femoral	L₂–L₄	Skin of anterior and medial thigh via *anterior femoral cutaneous* branch; skin of medial leg and foot, hip and knee joints via *saphenous* branch; motor to anterior muscles (quadriceps and sartorius) of thigh; pectineus, iliacus
Obturator	L₂–L₄	Motor to adductor magnus (part), longus, and brevis muscles, gracilis muscle of medial thigh, obturator externus; sensory for skin of medial thigh and for hip and knee joints
Lateral femoral cutaneous	L₂, L₃	Skin of lateral thigh; some sensory branches to peritoneum
Iliohypogastric	L₁	Skin of lower abdomen, lower back, and hip; muscles of anterolateral abdominal wall (obliques and transversus) and pubic region
Ilioinguinal	L₁	Skin of external genitalia and proximal medial aspect of the thigh; inferior abdominal muscles
Genitofemoral	L₁, L₂	Skin of scrotum in males, of labia majora in females, and of anterior thigh inferior to middle portion of inguinal region; cremaster muscle in males

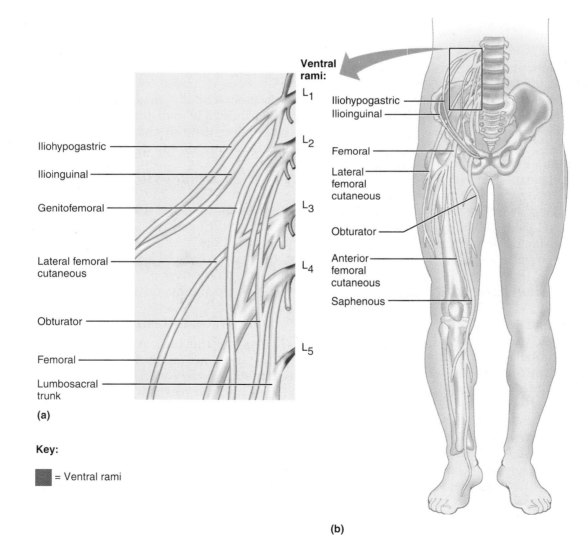

FIGURE 13.9 *The lumbar plexus.* **(a)** Spinal roots (anterior rami) and major branches of the lumbar plexus. **(b)** Distribution of the major peripheral nerves of the lumbar plexus in the lower limb (anterior view). (See Table 13.5.)

this plexus descend to innervate the anterior and medial thigh. The **femoral nerve,** the largest terminal nerve of this plexus, runs deep to the inguinal ligament to enter the thigh and then divides into a number of large branches. The motor branches innervate the anterior thigh muscles (quadriceps), which are the principal thigh flexors and knee extensors. The cutaneous branches serve the skin of the anterior thigh and the medial surface of the leg from the knee to the foot. The **obturator** (ob"tu-ra'tor) **nerve** enters the medial thigh via the obturator foramen and innervates the adductor muscles. These and other smaller branches of the lumbar plexus are summarized in Table 13.5.

◪ Homeostatic Imbalance

Compression of the spinal roots of the lumbar plexus, as by a herniated disc, results in gait problems because the femoral nerve serves the prime movers of both hip flexion and knee extension. Anesthesia of the anterior thigh and pain in the medial thigh occur if the obturator nerve is impaired. ■

Sacral Plexus The **sacral plexus** arises from spinal nerves L_4–S_4 and lies immediately caudal to the lumbar plexus (Figure 13.10). Some fibers of the lumbar plexus contribute to this plexus via the *lum-*

bosacral trunk, as mentioned earlier. The sacral plexus has about a dozen named branches. Around half of these serve the buttock and lower limb; the others innervate pelvic structures and the perineum. The most important branches are described here. Table 13.6 summarizes all but the smallest branches.

The largest branch of the sacral plexus is the **sciatic** (si-at'ik) **nerve,** the thickest and longest nerve in the body. It supplies nearly the entire lower limb, except the anteromedial thigh.

Actually two nerves (the tibial and common fibular) wrapped in a common sheath, the sciatic nerve leaves the pelvis via the greater sciatic notch. It courses deep to the gluteus maximus muscle and enters the posterior thigh just medial to the hip joint (*sciatic* = of the hip). There it gives off motor branches to the hamstring muscles (all thigh extensors and knee flexors) and to the adductor magnus. Immediately above the knee, its two divisions diverge.

The **tibial nerve** courses through the popliteal fossa (the region just posterior to the knee joint) and supplies the posterior compartment muscles of the leg and the skin of the posterior calf and sole of the foot. Important branches of the tibial nerve are the **sural,** which serves the skin of the posterolateral leg, and the **plantar nerves,** which serve most of the foot.

TABLE 13.6	Branches of the Sacral Plexus (See Figure 13.10)	
Nerves	*Spinal roots (ventral rami)*	*Structures served*
Sciatic nerve	L_4, L_5, S_1–S_3	Composed of two nerves (tibial and common peroneal) in a common sheath that diverge just proximal to the knee
▪ Tibial (including sural branch and medial and lateral plantar branches)	L_4–S_3	Cutaneous branches: to skin of posterior surface of leg and sole of foot Motor branches: to muscles of back of thigh, leg, and foot (hamstrings [except short head of biceps femoris], posterior part of adductor magnus, triceps surae, tibialis posterior, popliteus, flexor digitorum longus, flexor hallucis longus, and intrinsic muscles of foot)
▪ Common fibular (superficial and deep branches)	L_4–S_2	Cutaneous branches: to skin of anterior surface of leg and dorsum of foot Motor branches: to short head of biceps femoris of thigh, peroneal muscles of lateral compartment of leg, tibialis anterior, and extensor muscles of toes (extensor hallucis longus, extensors digitorum longus and brevis)
Superior gluteal	L_4, L_5, S_1	Motor branches: to gluteus medius and minimus and tensor fasciae latae
Inferior gluteal	L_5–S_2	Motor branches: to gluteus maximus
Posterior femoral cutaneous	S_1–S_3	Skin of buttock, posterior thigh, and popliteal region; length variable; may also innervate part of skin of calf and heel
Pudendal	S_2–S_4	Supplies most of skin and muscles of perineum (region encompassing external genitalia and anus and including clitoris, labia, and vaginal mucosa in females, and scrotum and penis in males); external anal sphincter

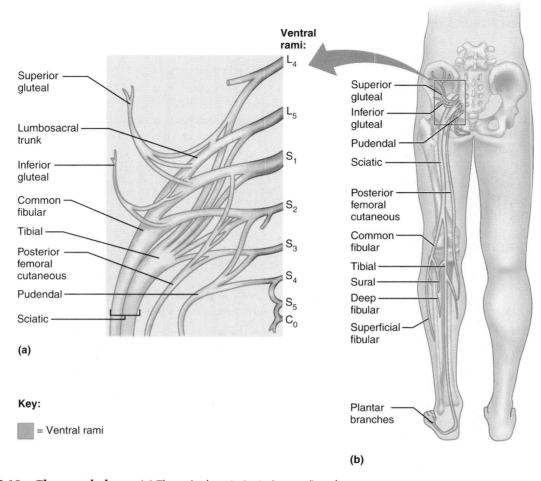

FIGURE 13.10 *The sacral plexus.* **(a)** The spinal roots (anterior rami) and major branches of the sacral plexus. **(b)** Distribution of the major peripheral nerves of the sacral plexus in the lower limb (posterior view). (See Table 13.6.)

The **common fibular (peroneal) nerve** descends from its point of origin, wraps around the head of the fibula (fibula = *perone*), and then divides into superficial and deep branches. These branches innervate the knee joint, the skin of the lateral calf and dorsum of the foot, and the muscles of the anterolateral leg (the extensors that dorsiflex the foot).

The next largest branches of the sacral plexus are the **superior** and **inferior gluteal nerves.** Together, they innervate the buttock (gluteal) and tensor fasciae latae muscles. The **pudendal** (pu-den'dal; "shameful") **nerve** innervates the muscles and skin of the perineum, mediates the act of erection, and is involved in voluntary control of urination (see Table 10.7, p. 348). Other branches of the sacral plexus supply the thigh rotators and muscles of the pelvic floor.

Homeostatic Imbalance

Injury to the proximal part of the sciatic nerve, as might follow a fall, disc herniation, or improper ad-

ministration of an injection into the buttock, results in a number of lower limb impairments, depending on the precise nerve roots injured. *Sciatica* (si-at'ĭ-kah), characterized by stabbing pain radiating over the course of the sciatic nerve, is common. When the sciatic nerve is completely transected, the leg is nearly useless. The leg cannot be flexed (because the hamstrings are paralyzed), and all foot and ankle movements are lost. The foot drops into plantar flexion (dangles), a condition called *footdrop.* Recovery from sciatic nerve injury is usually slow and incomplete.

If the lesion occurs below the knee, thigh muscles are spared. When the tibial nerve is injured, the paralyzed calf muscles cannot plantar flex the foot and a shuffling gait develops. The common fibular nerve is susceptible to injury largely because of its superficial location at the head and neck of the fibula. Even a tight leg cast, or remaining too long in a side-lying position on a firm mattress, can compress this nerve and cause footdrop. ■

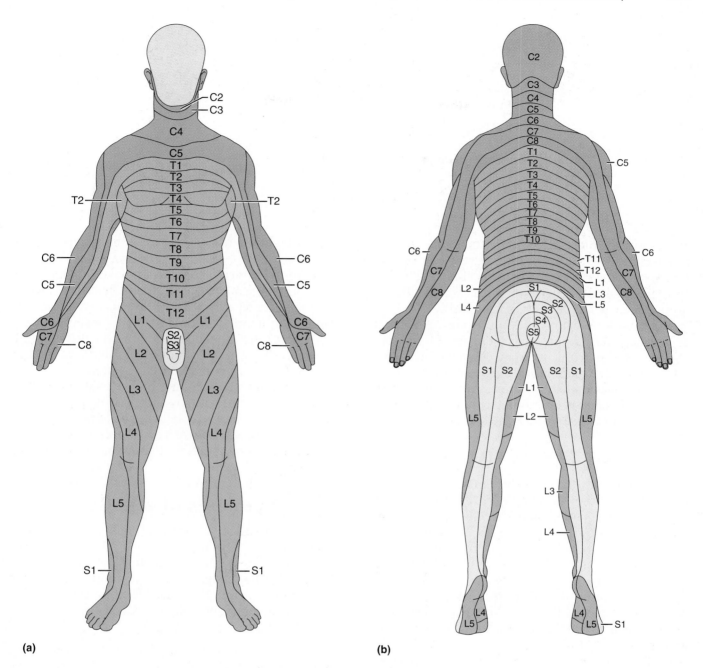

(a)   **(b)**

FIGURE 13.11 *Dermatomes (skin segments) of the body are related to the sensory innervation regions of the spinal nerves.* All spinal nerves but C_1 participate in the innervation of the dermatomes. (**a**) Anterior view. (**b**) Posterior view.

Innervation of Joints

The easiest way to remember which nerves serve each synovial joint is to use **Hilton's law,** which says that *any nerve serving a muscle producing movement at a joint also innervates the joint itself and the skin over the joint.* Hence, once you learn which nerves serve the various major muscles and muscle groups, no *new* learning is necessary. For example, the knee is crossed by the quadriceps, gracilis, and hamstring muscles. The nerves to these muscles are the femoral nerve anteriorly and branches of the sciatic and obturator nerves posteriorly. Consequently, these nerves innervate the knee joint as well.

Innervation of Skin: Dermatomes

The area of skin innervated by the cutaneous branches of a single spinal nerve is called a **dermatome** (der'mah-tōm; "skin segment"). All spinal nerves except C_1 participate in dermatomes. Adjacent dermatomes on the body trunk are fairly uniform in width, almost horizontal, and in direct line with their spinal nerves (Figure 13.11). The arrangement of the dermatomes in the limbs is less obvious. (It is also less well worked out and different clinicians have mapped a variety of areas for the same dermatomes.) The skin of the upper limbs is supplied by ventral rami of C_5–T_1 (or T_2). The lumbar

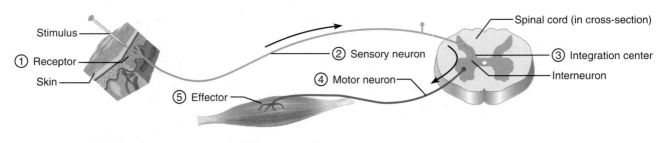

FIGURE 13.12 *The basic components of all human reflex arcs:* Receptor, sensory neuron, integration center (one or more synapses in the CNS), motor neuron, and effector. (The reflex arc illustrated is polysynaptic.)

nerves supply most of the anterior surfaces of the thighs and legs, and the sacral nerves serve most of the posterior surfaces of the lower limbs. (This distribution basically reflects the areas supplied by the lumbar and sacral plexuses, respectively.)

Dermatome regions are not as cleanly separate as a typical dermatome map indicates. Dermatome regions of the trunk overlap considerably (about 50%); thus, destruction of a single spinal nerve will not result in complete numbness anywhere. In the limbs, the overlap is less complete and some skin regions are innervated by just one spinal nerve.

REFLEX ACTIVITY

Many of the body's control systems belong to a general category known as reflexes. In the most restricted definition of the term, a **reflex** is a rapid, predictable motor response to a stimulus. It is unlearned, unpremeditated, and involuntary. Such basic reflexes may be considered to be built into our neural anatomy.

In many cases, we are aware of the final response of the reflex activity. If you splash a pot of boiling water on your arm, you are likely to drop the pot instantly and involuntarily before you have any sensation of pain. This response is triggered by a spinal reflex without any help from the brain. But the pain signals picked up by the interneurons of the spinal cord are quickly transmitted to the brain, so that within the next few seconds you *do* become aware of pain, and you also know what happened to cause it. Recall the discussion in Chapter 11 about serial and parallel processing. Here we have a good example of how they work together. The serial processing phase is the withdrawal reflex mediated by the spinal cord. Pain awareness reflects parallel processing of the sensory input.

In other cases, reflex activities go on without any awareness on our part. This is typical of many visceral activities, which are regulated by the subconscious lower regions of the CNS, specifically the brain stem and spinal cord.

In addition to these basic, inborn types of reflexes, there are *learned,* or *acquired, reflexes* that result from practice or repetition. Take, for instance, the complex sequence of reactions that occurs when an experienced driver drives a car. The process is largely automatic, but only because substantial time and effort were expended to acquire the driving skill. However, most reflex actions are subject to modification by learning and conscious effort. If a three-year-old child was standing by your side when you scalded your arm, you most likely would set the pot down (rather than just letting go) because of your conscious recognition of the danger to the child. Thus, there is no clear-cut distinction between basic and learned reflexes.

Components of a Reflex Arc

As you learned in Chapter 11, reflexes occur over highly specific neural paths called **reflex arcs.** However, in order to understand the body reflexes described in this chapter, we have to examine reflex arcs in more detail. Basically, all reflex arcs have five essential components (Figure 13.12):

① The **receptor** is the site of the stimulus action.

② The **sensory neuron** transmits the afferent impulses to the CNS.

③ The **integration center** in the simplest reflex arcs may be a single synapse between the sensory neuron and a motor neuron (**monosynaptic reflex**). In more complex reflexes, it involves multiple synapses with chains of interneurons (**polysynaptic reflex**). The integration center is always within the CNS, exemplified by the spinal cord in Figure 13.12.

④ The **motor neuron** conducts efferent impulses from the integration center to an effector organ.

⑤ The **effector** is the muscle fiber or gland cell that responds to the efferent impulses in a characteristic way (by contracting or secreting).

Reflexes are classified functionally as **somatic reflexes** if they activate skeletal muscle, and as **autonomic (visceral) reflexes** when visceral effectors (smooth or cardiac muscle or glands) are activated. The next section describes some common types of somatic reflexes mediated by the spinal cord. Autonomic reflexes are considered in later chapters with the visceral processes they help to regulate.

Spinal Reflexes

Somatic reflexes mediated by the spinal cord are called **spinal reflexes.** Many spinal reflexes occur without the involvement of higher brain centers. These reflexes work equally well in decerebrate animals (those in which the brain has been destroyed) as long as the spinal cord is functional. Other spinal reflexes require the activity of the brain for their successful completion. Additionally, the brain is "advised" of most types of spinal cord reflex activity and can then facilitate or inhibit them. Moreover, continuous facilitating signals from the brain are required for normal cord activity because, as noted in Chapter 12, if the spinal cord is suddenly transected, *spinal shock* occurs (all functions controlled by the cord are immediately depressed).

Testing of somatic reflexes is important clinically to assess the condition of the nervous system. Exaggerated, distorted, or absent reflexes indicate degeneration or pathology of specific regions of the nervous system, often before other signs are apparent.

Stretch and Deep Tendon Reflexes

For skeletal muscles to perform normally, (1) the brain must be continually informed of the current state of the muscles, and (2) the muscles must have healthy *tone* (resistance to active or passive stretch at rest). The first requirement depends on transmission of information from *muscle spindles* and *Golgi tendon organs* (proprioceptors located in the skeletal muscles and their associated tendons) to the cerebellum and cerebral cortex. Normal muscle tone, the second requirement, depends on **stretch reflexes** initiated by the muscle spindles, which monitor changes in muscle length. These processes are important to normal skeletal muscle function, posture, and locomotion.

Functional Anatomy of Muscle Spindles Before we examine the functional role of muscle spindles, we need to comment briefly on their unique nature. Each muscle spindle consists of 3 to 10 **intrafusal** (*intra* = within; *fusal* = the spindle) **muscle fibers** enclosed in a connective tissue capsule (see Figure 13.13). These modified skeletal muscle fibers are less than one-quarter of the size of the **extrafusal muscle fibers** (the effector fibers of the muscle). The central regions of the intrafusal fibers, which lack

There is a major *difference in the functions of the intrafusal and extrafusal muscle fibers. What is it?*

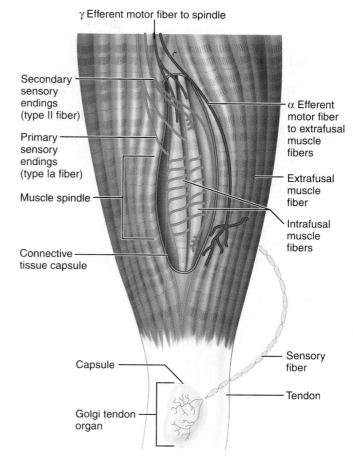

FIGURE 13.13 ***Anatomy of the muscle spindle and Golgi tendon organ.*** Notice the afferent fibers from and efferent fibers to the muscle spindle.

myofilaments and are noncontractile, serve as the receptive surfaces of the spindle. These regions are wrapped by two types of afferent endings that send sensory inputs to the CNS. The **primary sensory endings** of large **type Ia fibers,** which innervate the center of the spindle, are stimulated by both the rate and amount of stretch. The **secondary sensory endings** of small **type II fibers** supply the ends of the spindle and are stimulated only by degree of stretch. The contractile regions of the intrafusal muscle fibers are limited to their ends because these are the only cellular areas that contain actin and myosin myofilaments. These regions are innervated by

The extrafusal fibers are effectors; they contract, shortening the muscle, when adequately stimulated. The intrafusal fibers are part of a receptor apparatus that provides input that helps to regulate the activity of the muscle.

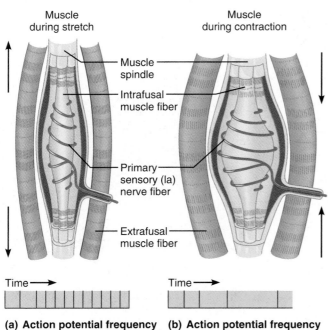

Muscle
during stretch

Muscle
during contraction

Muscle
spindle

Intrafusal
muscle fiber

Primary
sensory (Ia)
nerve fiber

Extrafusal
muscle fiber

Time →

Time →

**(a) Action potential frequency
increases during stretch**

**(b) Action potential frequency
declines during contraction**

FIGURE 13.14 *Operation of the muscle spindle.*
(a) Stretching the muscle activates the muscle spindle and
causes an increased rate of action potential generation in the
associated sensory (Ia) fiber. **(b)** Contraction of the muscle
reduces tension on the muscle spindle and lowers the rate of
action potential generation. Arrows indicate the direction of
force on the muscle spindles.

gamma (γ) efferent fibers that arise from small motor neurons in the ventral horn of the spinal cord. These motor fibers, which maintain the sensitivity of the spindle (as described shortly), are distinct from the **alpha (α) efferent fibers** of the large **alpha (α) motor neurons** that stimulate the extrafusal muscle fibers to contract.

The Stretch Reflex Stretching (and exciting) the muscle spindle occurs in one of two ways: (1) by applying an external force that lengthens the entire muscle, such as occurs when we carry a heavy weight or when antagonistic muscles contract *(external stretch)*, or (2) by activating the γ motor neurons that stimulate the distal ends of the intrafusal fibers to contract, thus stretching the middle of the spindle *(internal stretch)*. Whatever the stimulus, when the spindles are activated the associated sensory neurons transmit impulses at a higher frequency to the spinal cord (Figure 13.14a). There the sensory neurons synapse directly with α motor neurons, which in turn rapidly excite the extrafusal muscle fibers of the stretched muscle (Figure 13.15a). The reflexive muscle contraction that follows (an example of serial processing) resists further stretching of the muscle. Branches of the afferent fibers also synapse with interneurons

that inhibit motor neurons controlling antagonistic muscles (parallel processing). The resulting inhibition is called **reciprocal inhibition.** Consequently, the stretch stimulus causes the antagonists to relax so that they cannot resist the shortening of the "stretched" muscle caused by the main reflex arc.

While this spinal reflex is occurring, information on muscle length and the velocity of muscle shortening is also being relayed (mainly via the dorsal white columns) to higher brain centers (more parallel processing). This sequence of events is the way muscle tone is maintained and adjusted reflexively to the requirements of posture and movement. If either the afferent or efferent fibers are cut, the muscle immediately loses its tone and becomes flaccid. The stretch reflex is most important in the large extensor muscles that sustain upright posture, such as the quadriceps, and in postural muscles of the trunk. For example, contractions of the postural muscles of the spine are almost continuously regulated by stretch reflexes initiated first on one side of the spine and then on the other.

Although the stretch reflex is essential for normal muscle tone and activity, it essentially never acts alone. It is always accompanied by the **gamma motor neuron reflex arc.** There is a good reason for this: As the muscle shortens, the spindle's rate of firing declines; this in turn reduces impulse generation by the alpha motor neurons (see Figure 13.14b). Thus, the stretch reflex alone would result in crude, jerky muscle contractions. However, this is not what we see because contractions promoted by stretch reflexes are refined and smoothed out by gamma motor neuron reflex arcs, which regulate the response of the intrafusal muscle fibers of the muscle spindle. Descending fibers of motor pathways synapse with both α and γ motor neurons, and motor impulses are sent to the large extrafusal fibers and to intrafusal fibers of the muscle spindles at the same time. Stimulating the intrafusal fibers maintains the muscle spindle's tension (and sensitivity) during muscle contraction, so that the brain continues to be notified of conditions in the muscle. Without such a system, information on muscle length and its rate of change would cease to flow from contracting muscles.

The motor supply to the muscle spindle also allows us to voluntarily control the stretch reflex response and the firing rate of α motor neurons by stimulating or inhibiting the γ neurons. When the γ neurons are vigorously stimulated by impulses descending from the brain, the spindle is stretched and highly sensitive, and force of muscle contraction is maintained or increased. When the γ motor neurons are inhibited, the muscle spindle resembles a loose rubber band and is nonresponsive, and the extrafusal muscles relax.

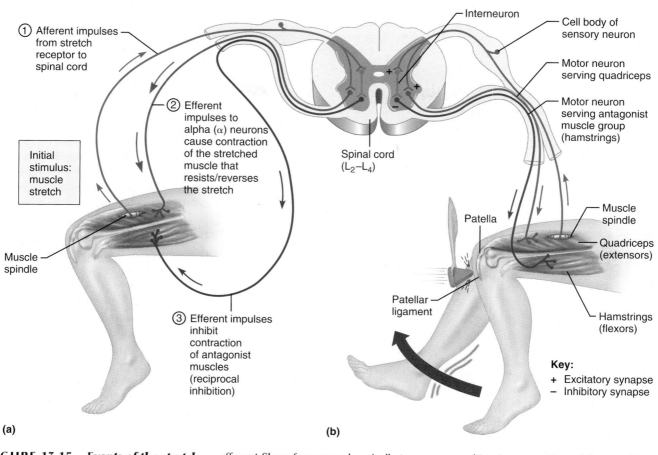

① Afferent impulses from stretch receptor to spinal cord

Interneuron

Cell body of sensory neuron

② Efferent impulses to alpha (α) neurons cause contraction of the stretched muscle that resists/reverses the stretch

Motor neuron serving quadriceps

Motor neuron serving antagonist muscle group (hamstrings)

Initial stimulus: muscle stretch

Spinal cord (L₂–L₄)

Patella

Muscle spindle

Quadriceps (extensors)

Muscle spindle

③ Efferent impulses inhibit contraction of antagonist muscles (reciprocal inhibition)

Patellar ligament

Hamstrings (flexors)

Key:
+ Excitatory synapse
– Inhibitory synapse

(a)   **(b)**

FIGURE 13.15   *Events of the stretch reflex.* (a) Events of the stretch reflex by which muscle stretch is damped. The events are shown in circular fashion. ① Stretching the muscle activates a muscle spindle. ② Impulses transmitted by afferent fibers from muscle spindle to alpha (α) motor neurons in the spinal cord cause the stretched muscle to contract. ③ Impulses transmitted by afferent fibers from muscle spindle to interneurons in the spinal cord result in reciprocal inhibition of the antagonist muscle. **(b)** The patellar reflex. Tapping the patellar tendon excites muscle spindles in the quadriceps muscle. Afferent impulses travel to the spinal cord, where synapses occur with motor neurons and interneurons. The motor neurons send activating impulses to the same muscle, resulting in contraction of the quadriceps muscle, which causes extension of the knee and a forward movement of the foot, counteracting the initial stretch. The interneurons make inhibitory synapses with anterior horn neurons that serve the antagonistic muscle group (hamstrings), preventing it from resisting the contraction.

The ability to modify the stretch reflex is important in many situations. For example, if you want to wind up to pitch a baseball, it is essential to suppress the stretch reflex so that your muscles can produce a large degree of motion. On the other hand, when maximum force is desired, the muscle should be stretched as much and as quickly as possible just before the movement so that a volley of involuntary *(γ)* efferent impulses reaches the muscle just as the voluntary command is given. This advantage is demonstrated by the crouch that athletes assume just before jumping or running.

The most familiar clinical example of the stretch reflex is the **patellar** (pah-tel'ar), or **knee-jerk, reflex,** elicited by striking the patellar tendon with a reflex hammer (Figure 13.15b). The sudden jolt to the tendon (1) stretches the quadriceps muscle, (2) stim-ulates the muscle spindles, and (3) results in contraction of the quadriceps muscle and inhibition of its antagonist, the hamstrings. Stretch reflexes can be elicited in any skeletal muscle of the body by suddenly striking its tendon or the muscle itself. All stretch reflexes are **monosynaptic** and **ipsilateral** (*ipsi* = same; *lateral* = side); that is, they involve a single synapse and motor activity on the same side of the body. Although stretch reflexes *themselves* are monosynaptic, the reflex arcs that inhibit the antagonistic muscles involve *more* than two neurons and one synapse; that is, they are polysynaptic.

A positive knee jerk (or a positive result for any stretch reflex test) provides two important pieces of information. First, it proves that the sensory and motor connections between that muscle and the spinal cord are intact. Second, the vigor of the motor

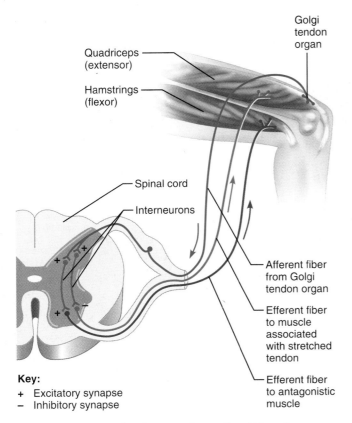

Golgi tendon organ

Quadriceps (extensor)

Hamstrings (flexor)

Spinal cord

Interneurons

Afferent fiber from Golgi tendon organ

Efferent fiber to muscle associated with stretched tendon

Efferent fiber to antagonistic muscle

Key:
+ Excitatory synapse
– Inhibitory synapse

FIGURE 13.16   *The deep tendon reflex.* When the quadriceps muscle of the thigh contracts, the tension in the patellar region increases, activating Golgi tendon organs in the tendon. Afferent neurons associated with the Golgi tendon organs synapse with interneurons in the spinal cord that inhibit motor neurons serving the contracting muscle and stimulate motor neurons activating the antagonistic muscle groups (the hamstrings). As a result, the contracting muscle relaxes and lengthens as the antagonist contracts and shortens.

response indicates the degree of excitability of the spinal cord. When the spinal motor neurons are highly facilitated by impulses descending from higher centers, just touching the muscle tendon produces a vigorous reflex response. On the other hand, if the lower motor neurons are bombarded by inhibitory signals, even pounding on the tendon may fail to cause the reflex response.

Homeostatic Imbalance

Stretch reflexes tend to be hypoactive or absent in cases of peripheral nerve damage or ventral horn injury involving the tested area. They are absent in those with chronic diabetes mellitus and neurosyphilis and during coma. However, these reflexes are hyperactive when lesions of the corticospinal tract reduce the inhibitory effect of the brain on the spinal cord (as in polio and stroke patients). ■

The Deep Tendon Reflex While stretch reflexes cause muscle contraction in response to increased muscle length (stretch), the polysynaptic **deep tendon reflexes** produce exactly the opposite effect: muscle relaxation and lengthening in response to the muscle's contraction (Figure 13.16). When muscle tension increases moderately during muscle contraction or passive stretching, Golgi tendon receptors in the muscle's tendon are activated, and afferent impulses are transmitted to the spinal cord. From there this information is sent to the cerebellum, where it is used to help adjust muscle tension appropriately. Simultaneously, motor neurons in spinal cord circuits supplying the contracting muscle are inhibited and antagonist muscles are activated, a phenomenon called **reciprocal activation.** As a result, the contracting muscle relaxes as its antagonist is activated.

Golgi tendon organs help ensure smooth onset and termination of muscle contraction and are particularly important in activities involving rapid switching between flexion and extension, such as running. Notice that the Golgi tendon organs are also stimulated during a clinically induced stretch reflex, such as the flexor reflex, but the brevity of the stimulus and the fact that the stretched muscle is already relaxed prevent it from being inhibited by the deep tendon reflex.

The Flexor Reflex

The **flexor,** or **withdrawal, reflex** is initiated by a painful stimulus (actual or perceived) and causes automatic withdrawal of the threatened body part from the stimulus (see Figure 13.17). The response that occurs when you prick your finger is a good example. So, too, is the trunk flexion that occurs when someone pretends to "throw a punch" at your abdomen. Flexor reflexes are ipsilateral, polysynaptic reflexes that involve more than one spinal cord segment—a necessity when several muscles must be recruited to withdraw the injured body part. Since flexor reflexes are protective reflexes important to our survival, they override the spinal pathways and prevent any other reflexes from using them at the same time.

The Crossed Extensor Reflex

The **crossed extensor reflex** is a complex spinal reflex consisting of an ipsilateral withdrawal reflex and a contralateral (opposite side) extensor reflex. Incoming afferent fibers synapse with interneurons that control the flexor withdrawal response on the same side of the body and with other interneurons that control the extensor muscles on the opposite side. This reflex is quite obvious when someone unexpectedly grabs for your arm (Figure 13.17). A more common example occurs when you step barefoot on broken glass. The ipsilateral response causes rapid lifting of the cut foot, while the contralateral re-

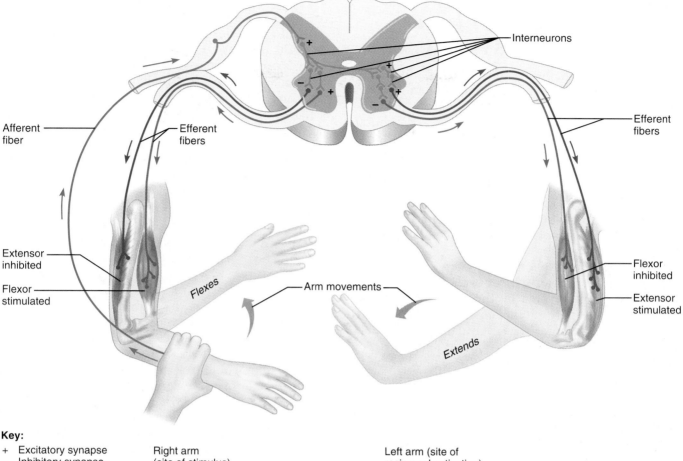

Key:
+ Excitatory synapse   Right arm
− Inhibitory synapse   (site of stimulus)

Left arm (site of
reciprocal activation)

FIGURE 13.17 *The crossed extensor reflex.* The crossed extensor reflex causes reflex withdrawal of the body part on the stimulated side (flexor reflex) as shown on the left side of the diagram, and extension of the muscles on the opposite side of the body (extensor reflex) as shown on the right side of the diagram. The example illustrated might occur if a stranger suddenly grabbed your arm. The flexor reflex would cause immediate withdrawal of your clutched arm, while the extensor reflex would simultaneously push the intruder away with the other arm by activating extensor muscles on the opposite side of the body.

sponse activates the extensor muscles of the opposite leg to support the weight suddenly shifted to it. Crossed extensor reflexes are particularly important in maintaining balance.

Superficial Reflexes

Superficial reflexes are elicited by gentle cutaneous stimulation, such as that produced by stroking the skin with a tongue depressor or a blunt key. The superficial reflexes are clinically important reflexes that depend both on functional upper motor pathways and on cord-level reflex arcs. The best known of these are the plantar and abdominal reflexes.

The **plantar reflex** tests the integrity of the spinal cord from L_4 to S_2 and indirectly determines if the corticospinal tracts are functioning properly. It is elicited by drawing a blunt object downward along the lateral aspect of the plantar surface (sole) of the foot. The normal response is a downward flexion

(curling) of the toes. However, if the primary motor cortex or corticospinal tract is damaged, the plantar reflex is replaced by an abnormal reflex called **Babinski's sign,** in which the great toe dorsiflexes and the smaller toes fan laterally. Infants exhibit Babinski's sign until they are about a year old because their nervous systems are incompletely myelinated before that age. Despite its clinical significance, the physiological mechanism of Babinski's sign is not understood.

Stroking the skin of the lateral abdomen above, to the side, or below the umbilicus may induce a reflex contraction of the abdominal muscles in which the umbilicus moves toward the stimulated site. These reflexes, called **abdominal reflexes,** check the integrity of the spinal cord and ventral rami from T_8 to T_{12}. The abdominal reflexes vary in intensity in different people. They are absent when corticospinal tract lesions are present.

DEVELOPMENTAL ASPECTS OF THE PERIPHERAL NERVOUS SYSTEM

As described in Chapter 9, most skeletal muscles of the body derive from paired blocks of mesoderm (somites) distributed segmentally down the postero-medial aspect of the embryonic body. The spinal nerves branch from the developing spinal cord and adjacent neural crest and exit between the forming vertebrae, and each becomes associated with the adjacent muscle mass. The spinal nerves supply both sensory and motor fibers to the developing muscles and help to direct their maturation. Cranial nerves innervate muscles of the head in a comparable manner.

The distribution of cutaneous nerves to the skin follows a similar pattern. Most of the scalp and skin of the face is innervated by the trigeminal nerves. Spinal nerves supply cutaneous branches to specific (adjacent) dermatomes that eventually become dermal segments. Thus, the distribution and growth of the spinal nerves correlate with the segmented body plan, which is established by the fourth week of embryonic development.

Growth of the limbs and unequal growth of other body areas result in an adult pattern of dermatomes with unequal sizes and shapes and varying degrees of overlap (see Figure 13.11, p. 501). Because embryonic muscle cells migrate extensively, much of the early segmental pattern is lost. Understanding the general pattern of sensory nerve distribution is critical to physicians. For example, in areas of substantial dermatome overlap, two or three spinal nerves must be blocked to perform local surgery.

Sensory receptors atrophy to some degree with age, and there is some lessening of muscle tone in the face and neck. Reflexes occur a bit more slowly during old age. However, this seems to reflect a general loss of neurons, fewer synapses per neuron, and a slowdown in central processing rather than any major changes in the peripheral nerve fibers. In fact, the peripheral nerves remain viable and normally functional throughout life unless subjected to traumatic injury or ischemia. The most common symptom of ischemia (deprivation of blood supply) is a tingling sensation or numbness in the affected region.

* * *

The peripheral nervous system, composed primarily of nerves, is an essential part of any functional nervous system. Without it, the central nervous system would lack its rich bank of information about events of the external and internal environments. Now that we have connected the CNS to both of these environments, we are ready to consider the autonomic nervous system, a special subdivision of the peripheral nervous system and the topic of Chapter 14.

RELATED CLINICAL TERMS

Analgesia (an″al-je′ze-ah; *an* = without; *algos* = pain) Reduced ability to feel pain, not accompanied by loss of consciousness. An analgesic is a pain-relieving drug.

Nerve conduction studies Diagnostic tests that assess the integrity of nerves as indicated by their conduction velocities; the nerve is stimulated separately at two points, and the time required for the stimulus to reach the muscle it serves is recorded; used to assess suspected peripheral neuropathies.

Neuralgia (nu-ral′je-ah; *neuro* = nerve) Sharp spasmlike pain along the course of one or more nerves; usually caused by inflammation or injury to the nerve(s).

Neuritis (nu-ri′tis) Inflammation of a nerve; there are many forms with different effects (e.g., increased or decreased nerve sensitivity, paralysis of structure served, and pain).

Paresthesia (par″es-the′ze-ah) An abnormal sensation (burning, numbness, tingling) resulting from a disorder of a sensory nerve.

Shingles A viral inflammation caused by *herpes zoster* (the chicken pox virus), which invades the dorsal roots of the spinal nerves. Indicated by blisterlike skin lesions along the course of one or more dermatomes. Commonly due to reactivation of the (latent) virus acquired during childhood disease when the immune system is weakened. The reactivated virus is transported along the sensory fibers associated with the infected dorsal root ganglion, causing pain and a characteristic rash.

CHAPTER SUMMARY

Media study tools that could provide you additional help in reviewing specific key topics of Chapter 13 are referenced below. **SP** = *Study Partner;* **IP** = *Interactive Physiology.*

Overview of the Peripheral Nervous System (pp. 475–483)

1. The peripheral nervous system consists of sensory receptors, nerves conducting impulses to and from the CNS, their associated ganglia, and motor endings.

Sensory Receptors (pp. 475–480)

2. Sensory receptors are specialized to respond to environmental changes (stimuli).

3. Sensory receptors include the simple (general) receptors for pain, touch, pressure, and temperature found in the skin, as well as those found in skeletal muscles and tendons and in the visceral organs. Complex receptors (sense organs), consisting of sensory receptors and other cells, serve the special senses (vision, hearing, equilibrium, smell, and taste).

4. Receptors are classified according to stimulus detected as mechanoreceptors, thermoreceptors, photoreceptors, chemoreceptors, and nociceptors, and according to location as exteroceptors, interoceptors, and proprioceptors.

5. The general sensory receptors are classified structurally as free or encapsulated dendritic endings of sensory neurons. The free endings are mainly receptors for temperature and pain, although two are for light touch (Merkel discs and root hair plexuses). The encapsulated dendritic endings, which are mechanoreceptors, include Meissner's corpuscles, Pacinian corpuscles, Krause's end bulbs, Ruffini's corpuscles, muscle spindles, Golgi tendon organs, and joint kinesthetic receptors.

Nerves and Associated Ganglia (pp. 480–483)

6. A nerve is a bundle of neuron fibers in the PNS. Each fiber is enclosed by an endoneurium, fascicles of fibers are wrapped by a perineurium, and the whole nerve is bundled by the epineurium.

7. Nerves are classified according to the direction of impulse conduction of their fibers as sensory, motor, or mixed; most nerves are mixed. The efferent fibers may be somatic or autonomic.

8. Ganglia are collections of nerve cell bodies associated with nerves. Examples are the dorsal root (sensory) ganglia and autonomic (motor) ganglia.

9. Injured PNS fibers may regenerate if macrophages enter the area, phagocytize the debris, and release chemicals that cause axonal regrowth and promote Schwann cell proliferation. Schwann cells then form a channel to guide axon sprouts to their original contacts. Fibers in the CNS do not normally regenerate because the oligodendrocytes fail to aid the process, macrophages are largely excluded, and growth-inhibiting proteins are present in the surrounding tissue.

Motor Endings (p. 483)

10. Motor endings of somatic nerve fibers (axonal terminals) help to form elaborate neuromuscular junctions with skeletal muscle cells. Axonal terminals contain synaptic vesicles filled with acetylcholine, which (when released) signals the muscle cell to contract. An elaborate basal lamina fills the synaptic cleft.

11. Autonomic motor endings, called varicosities, are functionally similar, but structurally simpler, beaded terminals that innervate smooth muscle and glands. They do not form specialized neuromuscular junctions and generally a wider synaptic cleft separates them from their effector cells.

SP Case Study: Neuromuscular Dysfunction.

Cranial Nerves (pp. 483–491)

1. Twelve pairs of cranial nerves originate from the brain and issue through the skull to innervate the head and neck. Only the vagus nerves extend into the thoracic and abdominal cavities.

2. Cranial nerves are numbered from anterior to posterior in order of emergence from the brain. Their names reflect structures served or function or both. The cranial nerves include:

- The olfactory nerves (I): purely sensory; concerned with the sense of smell.

- The optic nerves (II): purely sensory; transmit visual impulses from the retina to the thalamus.

- The oculomotor nerves (III): primarily motor; emerge from the midbrain and serve four extrinsic eye muscles, the levator palpebrae superioris of the eyelid, and the intrinsic ciliary muscle of the eye and constrictor fibers of the iris; also carry proprioceptive impulses from the skeletal muscles served.

- The trochlear nerves (IV): primarily motor; emerge from the dorsal midbrain and carry motor and proprioceptor impulses to and from superior oblique muscles of the eyeballs.

- The trigeminal nerves (V): mixed nerves; emerge from the lateral pons; the major general sensory nerves of the face; each has three sensory divisions—ophthalmic, maxillary, and mandibular; the mandibular branch also contains motor fibers that innervate the chewing muscles.

- The abducens nerves (VI): primarily motor; emerge from the pons and serve the motor and proprioceptive functions of the lateral rectus muscles of the eyeballs.

- The facial nerves (VII): mixed nerves; emerge from the pons; major motor nerves of the face; also carry sensory impulses from the taste buds of anterior two-thirds of the tongue.

- The vestibulocochlear nerves (VIII): purely sensory; transmit impulses from the hearing and equilibrium receptors of the inner ears.

- The glossopharyngeal nerves (IX): mixed nerves; issue from the medulla; transmit sensory impulses from the taste buds of the posterior tongue, from the pharynx, and from chemo- and pressoreceptors of the carotid bodies and sinuses; innervate some pharyngeal muscles and parotid glands.

- The vagus nerves (X): mixed nerves; arise from the medulla; almost all motor fibers are autonomic parasympathetic fibers; motor efferents to, and sensory fibers from, the pharynx, larynx, and visceral organs of the thoracic and abdominal cavities.

- The accessory nerves (XI): motor only; consist of a cranial root arising from the medulla and a spinal root arising from the cervical spinal cord; cranial root supplies motor fibers to the pharynx and larynx; spinal root supplies somatic efferents to the trapezius and sternocleidomastoid muscles of the neck and carries proprioceptor afferents from the same muscles.

- The hypoglossal nerves (XII): motor only; issue from the medulla; carry somatic motor efferents to, and proprioceptive fibers from, the tongue muscles.

SP Exercise: Location of the Cranial Nerves.

Spinal Nerves (pp. 491–502)

General Features of Spinal Nerves (pp. 491–493)

1. The 31 pairs of spinal nerves (all mixed nerves) are numbered successively according to the region of the spinal cord from which they issue.

2. Spinal nerves are formed by the union of dorsal and ventral roots of the spinal cord and are short, confined to the intervertebral foramina.

3. Branches of each spinal nerve include dorsal and ventral rami, a meningeal branch, and rami communicantes (ANS branches).

SP Exercise: Formation and Branches of Spinal Nerves.

Innervation of Specific Body Regions (pp. 493–502)

4. Ventral rami, except T_2–T_{12}, form plexuses that serve the limbs.

5. Dorsal rami serve the muscles and skin of the posterior body trunk. T_2–T_{12} ventral rami give rise to intercostal nerves that serve the thorax wall and abdominal surface.

6. The cervical plexus (C_1–C_4) innervates the muscles and skin of the neck and shoulder. Its phrenic nerve serves the diaphragm.

7. The brachial plexus serves the shoulder, some thorax muscles, and the upper limb. It arises primarily from C_5–T_1. Proximal to distal, the brachial plexus has roots, trunks, divisions, and cords. The main nerves arising from the cords are the axillary, musculocutaneous, median, radial, and ulnar nerves.

8. The lumbar plexus (L_1–L_4) provides the motor supply to the anterior and medial thigh muscles and the cutaneous supply to the anterior thigh and part of the leg. Its chief nerves are the femoral and obturator.

9. The sacral plexus (L_4–S_4) supplies the posterior muscles and skin of the lower limb. Its principal nerve is the large sciatic nerve composed of the tibial and common fibular (peroneal) nerves.

10. Joints are innervated by the same nerves that serve the muscles acting at the joint. All spinal nerves except C_1 innervate specific segments of the skin called dermatomes.

SP Exercise: Distribution of Spinal Nerves.

Reflex Activity (pp. 502–507)

Components of a Reflex Arc (pp. 502–503)

1. A reflex is a rapid, involuntary motor response to a stimulus. The reflex arc has five elements: receptor, sensory neuron, integration center, motor neuron, and effector.

Spinal Reflexes (pp. 503–507)

2. Testing of somatic spinal reflexes provides information on the integrity of the reflex pathway and the degree of excitability of the spinal cord.

3. Somatic spinal reflexes include stretch, deep tendon, flexor, crossed extensor, and superficial reflexes.

4. A stretch reflex, initiated by stretching of muscle spindles, causes contraction of the stimulated muscle and inhibits its antagonist. It is monosynaptic and ipsilateral. Stretch reflexes maintain muscle tone and body posture.

5. Deep tendon reflexes, initiated by stimulation of Golgi tendon organs by increased muscle tension, are polysynaptic reflexes. They cause relaxation of the stimulated muscle and contraction of its antagonist.

6. Flexor reflexes are initiated by painful stimuli. They are polysynaptic, ipsilateral reflexes that are protective in nature.

7. Crossed extensor reflexes consist of an ipsilateral flexor reflex and a contralateral extensor reflex.

8. Superficial reflexes (e.g., the plantar and abdominal reflexes) are elicited by cutaneous stimulation. They require functional cord reflex arcs and corticospinal pathways.

Developmental Aspects of the Peripheral Nervous System (p. 508)

1. Each spinal nerve provides the sensory and motor supply of an adjacent muscle mass (destined to become skeletal muscles) and the cutaneous supply of a dermatome (skin segment).

2. Reflexes decline in speed with age; this probably reflects neuronal loss or sluggish CNS integration circuits.

REVIEW QUESTIONS

Multiple Choice/Matching

1. The large onion-shaped receptors that are found deep in the dermis and in subcutaneous tissue and that respond to deep pressure are (a) Merkel discs, (b) Pacinian receptors, (c) free nerve endings, (d) Krause's end bulbs.

2. Proprioceptors include all of the following except (a) muscle spindles, (b) Golgi tendon organs, (c) Merkel discs, (d) joint kinesthetic receptors.

3. The connective tissue sheath that surrounds a fascicle of nerve fibers is the (a) epineurium, (b) endoneurium, (c) perineurium, (d) neurilemma.

4. Match the receptor type from column B to the correct description in column A.

Column A	Column B
_____ (1) pain, itch, and temperature receptors	(a) Ruffini's corpuscle
_____ (2) looks and functions like a Meissner's corpuscle but found in mucosae	(b) Golgi tendon organ
_____ (3) contains intrafusal fibers and type Ia and II sensory endings	(c) muscle spindle
_____ (4) discriminative touch receptor in hairless skin (fingertips)	(d) Krause's end bulb
_____ (5) contains dendrites wrapped around thick collagen bundles	(e) free dendritic endings
	(f) Pacinian corpuscle
	(g) Meissner's corpuscle

_____ (6) rapidly adapting deep-pressure receptor
_____ (7) slowly adapting deep-pressure receptor

5. Match the names of the cranial nerves in column B to the appropriate description in column A.

Column A	Column B
_____ (1) causes pupillary constriction	(a) abducens
_____ (2) is the major sensory nerve of the face	(b) accessory
_____ (3) serves the sternocleidomastoid and trapezius muscles	(c) facial
_____ (4) are purely sensory (three nerves)	(d) glossopharyngeal
_____ (5) serves the tongue muscles	(e) hypoglossal
_____ (6) allows you to chew your food	(f) oculomotor
_____ (7) is impaired in tic douloureux	(g) olfactory
_____ (8) helps to regulate heart activity	(h) optic
_____ (9) helps you to hear and to maintain your balance	(i) trigeminal
_____ (10) contain parasympathetic motor fibers (four nerves)	(j) trochlear
	(k) vagus
	(l) vestibulocochlear

6. For each of the following muscles or body regions, identify the plexus and the peripheral nerve(s) (or branch of one) involved. Use choices from keys A and B.

—; — **(1)** the diaphragm

—; — **(2)** muscles of the posterior thigh and leg

—; — **(3)** anterior thigh muscles

—; — **(4)** medial thigh muscles

—; — **(5)** anterior arm muscles that flex the forearm

—; — **(6)** muscles that flex the wrist and digits (two nerves)

—; — **(7)** muscles that extend the wrist and digits

—; — **(8)** skin and extensor muscles of the posterior arm

—; — **(9)** fibularis muscles, tibialis anterior, and toe extensors

—; —, —, —, — **(10)** elbow joint

Key A: Plexuses
(a) brachial
(b) cervical
(c) lumbar
(d) sacral

Key B: Nerves
(1) common fibular
(2) femoral
(3) median
(4) musculocutaneous
(5) obturator
(6) phrenic
(7) radial
(8) tibial
(9) ulnar

7. Characterize each receptor activity described below by choosing the appropriate letter and number from keys A and B.

Key A: (a) exteroceptor
(b) interoceptor
(c) proprioceptor

Key B: (1) chemoreceptor
(2) mechanoreceptor
(3) nociceptor
(4) photoreceptor
(5) thermoreceptor

—, — You are enjoying an ice cream cone.

—, — You have just scalded yourself with hot coffee.

—, — The retinas of your eyes are stimulated.

—, — You bump (lightly) into someone.

—, — You have been lifting weights and experience a sensation in your upper limbs.

8. An ipsilateral reflex that causes rapid withdrawal of a body part from a painful stimulus is the (a) crossed extensor, (b) flexor, (c) Golgi tendon, (d) muscle stretch.

Short Answer Essay Questions

9. What is the functional relationship of the peripheral nervous system to the central nervous system?

10. List the structural components of the peripheral nervous system, and describe the function of each component.

11. Explain why damage to peripheral nerve fibers is often reversible, whereas damage to CNS fibers rarely is.

12. (a) Describe the formation and composition of a spinal nerve. (b) Name the branches of a spinal nerve (other than the rami communicantes), and indicate their distribution.

13. (a) Define plexus. (b) Indicate the spinal roots of origin of the four major nerve plexuses, and name the general body regions served by each.

14. Differentiate between ipsilateral and contralateral reflexes.

15. What is the homeostatic value of flexor reflexes?

16. Compare and contrast flexor and crossed extensor reflexes.

17. Explain how a crossed extensor reflex exemplifies both serial and parallel processing.

18. What clinical information can be gained by conducting somatic reflex tests?

19. What is the structural and functional relationship between spinal nerves, skeletal muscles, and dermatomes?

Critical Thinking and Clinical Application Questions

1. In 1962 a boy playing in a train yard fell under a train. His right arm was cut off cleanly by the train wheel. Surgeons reattached the arm, sewing nerves and vessels back together. The boy was told he should eventually regain the use of his arm but that it would never be strong enough to pitch a baseball. Explain why full recovery of strength was unlikely.

2. Jefferson, a football quarterback, suffered torn menisci in his right knee joint when tackled from the side. The same injury crushed his common fibular nerve against the head of the fibula. What locomotor problems did Jefferson have after this?

3. As Harry fell off a ladder, he grabbed a tree branch with his right hand, but unfortunately lost his grip and fell heavily to the ground. Days later, Harry complained that his upper limb was numb. What was damaged in his fall?

4. Mr. Frank, a former stroke victim who had made a remarkable recovery, suddenly began to have problems reading. He complained of seeing double and also had problems navigating steps. He was unable to move his left eye downward and laterally. What cranial nerve was the site of lesion? (Right or left?)

5. One of a group of rabbit hunters was accidentally sprayed with buckshot in both of his gluteal prominences. When his companions saw that he would survive, they laughed and joked about where he had been shot. They were horrified and ashamed a week later when it was announced that their friend would be permanently paralyzed and without sensation in both legs from the knee down, as well as on the back of his thighs. What had happened?

14 THE AUTONOMIC NERVOUS SYSTEM

In its never-ending attempt to maintain balance, the body is far more sensitive to changing events than any human invention or cultural institution. Although all body systems contribute, the stability of our internal environment depends to a large extent on the orchestrations of the **autonomic nervous system (ANS),** the system of motor neurons that innervates smooth and cardiac muscle and glands (see Figure 14.1). At every moment, signals flood from visceral organs into the central nervous system, and autonomic nerves make adjustments as necessary to ensure optimal support for body activities. In response to changing conditions, the ANS shunts blood to more "needy" areas, speeds or slows heart rate, adjusts blood pressure and body temperature, and increases or decreases stomach secretions.

Most of this fine-tuning occurs without our awareness or attention. Can you tell when your arteries are constricting or when your pupils are dilating? Maybe not, but if you've ever been stuck in a grocery store check-out line, and your full bladder was contracting as if it had a mind of its own, you've been very aware of its activity. These functions, those that we're aware of and those that occur without our awareness or attention, are controlled by the autonomic nervous system. Indeed, as the term *autonomic* (*auto* = self; *nom* = govern) implies, this motor subdivision of the peripheral nervous system has a certain amount of functional independence. The autonomic nervous system is also called the **involuntary nervous system,** which reflects its subconscious control, and the **general visceral motor system,** which indicates the location of most of its effectors.

INTRODUCTION TO THE AUTONOMIC NERVOUS SYSTEM
Comparison of the Somatic and Autonomic Nervous Systems

Our previous discussions of motor nerves have focused largely on the activity of the somatic nervous system. So, before describing autonomic nervous system anatomy, we will point out the major differences between the somatic and autonomic divisions, as well as some areas of functional overlap. Although both systems have motor fibers, the somatic and autonomic nervous systems differ (1) in their effectors, (2) in their efferent pathways, and (3) to some degree in target organ responses to their neurotransmitters. The differences are summarized in Figure 14.2.

Effectors

The somatic nervous system stimulates skeletal muscles, whereas the ANS innervates cardiac and smooth muscle and glands. The differing physiology of the effector organs themselves accounts for most

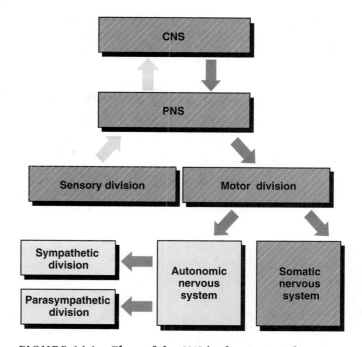

FIGURE 14.1 *Place of the ANS in the structural organization of the nervous system.*

of the remaining differences between somatic and autonomic effects on their target organs.

Efferent Pathways and Ganglia

In the somatic nervous system, cell bodies of the motor neurons are in the central nervous system, and their axons extend in spinal nerves all the way to the skeletal muscles they serve. Somatic motor fibers are typically thick, heavily myelinated type A fibers that conduct nerve impulses very rapidly.

In contrast, the motor unit of the ANS is a *two-neuron chain.* The cell body of the first neuron, the **preganglionic neuron,** resides in the brain or spinal cord. Its axon, called the **preganglionic axon,** synapses with the second motor neuron, the **ganglionic neuron,** in an **autonomic ganglion** outside the central nervous system. The axon of the ganglionic neuron, called the **postganglionic axon,** then extends to the effector organ. If you take the time to think about the meanings of all these terms while referring to Figure 14.2, understanding the rest of the chapter will be much easier. The preganglionic axons are lightly myelinated, thin fibers; the postganglionic axons are even thinner and are unmyelinated. Consequently, conduction through the autonomic efferent chain is slower than conduction in the somatic motor system. Many pre- and postganglionic fibers are incorporated into spinal or cranial nerves for most of their course.

It is important to keep in mind that autonomic ganglia are *motor* ganglia, containing the cell bodies of motor neurons. Technically, they are sites of

> **?** *Why are the axons of the first and second neurons of the two-neuron motor autonomic chain named the preganglionic and postganglionic axons respectively?*

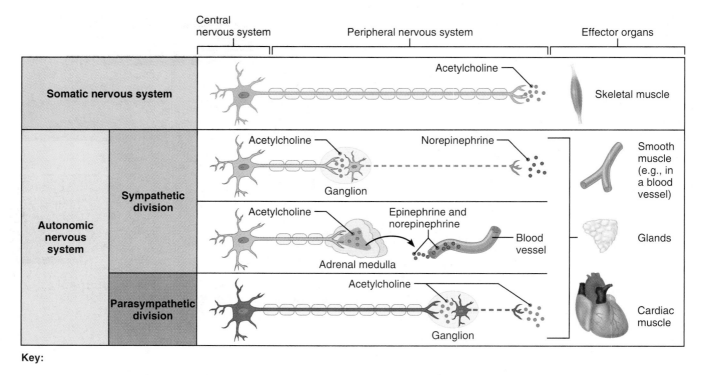

Key:

▬▬ = Preganglionic axons (sympathetic)   ─ ─ ─ = Postganglionic axons (sympathetic)   ⊜ = Myelination   ▬▬ = Preganglionic axons (parasympathetic)   ─ ─ ─ = Postganglionic axons (parasympathetic)

FIGURE 14.2 *Comparison of the somatic and autonomic nervous systems.* *Somatic Division:* Axons of somatic motor neurons extend from the CNS to their effectors (skeletal muscle cells). These axons are typically heavily myelinated. Somatic motor neurons release acetylcholine, and the effect is always stimulatory.
Autonomic Division: Axons of most preganglionic neurons run from the CNS to synapse in a peripheral autonomic ganglion with a ganglionic neuron. A few sympathetic preganglionic axons synapse with cells of the adrenal medulla. Postganglionic axons run from the ganglion to the effectors (cardiac and smooth muscle fibers and glands). Preganglionic axons are lightly myelinated; postganglionic axons are unmyelinated. All preganglionic fibers release acetylcholine; all parasympathetic postganglionic fibers release acetylcholine; most sympathetic postganglionic fibers release norepinephrine. Stimulated adrenal medullary cells release norepinephrine and epinephrine into the blood. Autonomic effects are stimulatory or inhibitory, depending on the postganglionic neurotransmitter released and the receptor types on the effectors.

synapse and information transmission from preganglionic to ganglionic neurons. Also, notice that the somatic motor division *lacks* ganglia entirely. The dorsal root ganglia are part of the sensory, not the motor, division of the peripheral nervous system.

Neurotransmitter Effects

All somatic motor neurons release acetylcholine at their synapses with their effectors, the skeletal muscle fibers. The effect is always *excitatory*, and if stimulation reaches threshold, the skeletal muscle fiber contracts.

Neurotransmitters released onto visceral effector organs by postganglionic autonomic fibers include **norepinephrine** secreted by most sympathetic fibers, and **acetylcholine** released by parasympathetic fibers. Depending on the receptors present on the target organ (see Figure 14.2 and Table 14.3 on p. 524), its response to these neurotransmitters may be either excitation or inhibition.

Overlap of Somatic and Autonomic Function

Higher brain centers regulate and coordinate both somatic and visceral motor activities, and nearly all spinal nerves (and many cranial nerves) contain both somatic and autonomic fibers. Moreover, most of the body's adaptations to changing internal and external conditions involve both skeletal muscle activity and enhanced responses of certain visceral organs. For example, when skeletal muscles are working hard, they need more oxygen and glucose, and autonomic control mechanisms speed up the heart and breathing to meet these needs and maintain homeostasis. The autonomic nervous system is only one part of our highly integrated nervous system, but according to convention we will consider it an individual entity and describe its role in isolation in the sections that follow.

The Divisions of the Autonomic Nervous System

The two arms of the autonomic nervous system, the *parasympathetic* and *sympathetic divisions,* generally serve the same visceral organs but cause essentially opposite effects. If one division stimulates certain smooth muscles to contract or a gland to secrete, the other division inhibits that action. Through this process of **dual innervation,** the two divisions counterbalance each other's activities to keep body systems running smoothly. The sympathetic part mobilizes the body during extreme situations (such as fear, exercise, or rage), whereas the parasympathetic arm allows us to unwind as it performs maintenance activities and conserves body energy. Let's elaborate on these functional differences by focusing briefly on situations in which each division is exerting primary control.

Role of the Parasympathetic Division

The **parasympathetic division** is most active in nonstressful situations. This division, sometimes called the "resting and digesting" system, is chiefly concerned with keeping body energy use as low as possible, even as it directs vital "housekeeping" activities like digestion and elimination of feces and urine. (This explains why it is a good idea to relax after a heavy meal: so that digestion is not interfered with by sympathetic activity.) Its activity is best illustrated in a person who relaxes after a meal and reads the newspaper. Blood pressure, heart rate, and respiratory rate are regulated at low normal levels, the gastrointestinal tract is actively digesting food, and the skin is warm (indicating that there is no need to divert blood to skeletal muscles or vital organs). The eye pupils are constricted to protect the retinas from excessive damaging light, and the lenses of the eyes are accommodated for close vision.

Role of the Sympathetic Division

The **sympathetic division** is often referred to as the "fight-or-flight" system. Its activity is evident when we are excited or find ourselves in emergency or threatening situations, such as being frightened by street toughs late at night. A pounding heart; rapid, deep breathing; cold, sweaty skin; and dilated eye pupils are sure signs of mobilization of the sympathetic nervous system. Not as obvious, but equally characteristic, are changes in brain wave patterns and in the electrical resistance of the skin (galvanic skin resistance)—events that are most frequently recorded during lie detector examinations.

During exercise or any type of vigorous physical activity, the sympathetic division also promotes a number of other adjustments. Visceral (and perhaps cutaneous) blood vessels are constricted, and blood is shunted to active skeletal muscles and to the vigorously working heart. The bronchioles in the lungs dilate, increasing ventilation (and ultimately increasing oxygen delivery to the body cells), and the liver releases more sugar (glucose) into the blood to provide for the increased energy needs of body cells. At the same time, temporarily nonessential activities, such as gastrointestinal and urinary tract motility, are damped. If you are running from a mugger, digesting lunch can wait! It is far more important that your muscles be provided with everything they need to get you out of danger's way.

The sympathetic division generates a head of steam that enables the body to cope rapidly with situations that threaten homeostasis. Its function is to provide the optimal conditions for an appropriate response to some threat, whether that response is to run, to see better, or to think more clearly.

An easy way to remember the most important roles of the two ANS divisions is to think of the parasympathetic division as the **D** (digesting, defecation, and diuresis [urination]) division and the sympathetic division as the **E** (exercise, excitement, emergency, and embarrassment) division. Remember, however, while it is easiest to think of the sympathetic and parasympathetic divisions as working in an all-or-none fashion, this is rarely the case. A dynamic antagonism exists between the two divisions, and fine adjustments are made continuously by both. We will go into more detail on these matters later.

TABLE 14.1	Anatomical and Physiological Differences Between the Parasympathetic and Sympathetic Divisions	
Characteristic	*Parasympathetic*	*Sympathetic*
Origin	Craniosacral outflow: brain stem nuclei of cranial nerves III, VII, IX, and X; spinal cord segments S_2–S_4	Thoracolumbar outflow: lateral horn of gray matter of spinal cord segments T_1–L_2
Location of ganglia	Ganglia in (intramural) or close to visceral organ served	Ganglia within a few centimeters of CNS: alongside vertebral column (chain or paravertebral ganglia) and anterior to vertebral column (collateral or prevertebral ganglia)
Relative length of pre- and postganglionic fibers	Long preganglionic; short postganglionic	Short preganglionic; long postganglionic
Rami communicantes	None	Gray and white rami communicantes. White rami contain myelinated preganglionic fibers; gray contain unmyelinated postganglionic fibers
Degree of branching of preganglionic fibers	Minimal	Extensive
Functional role	Maintenance functions; conserves and stores energy	Prepares body to cope with emergencies and intense muscular activity
Neurotransmitters	All fibers release ACh (cholinergic fibers)	All preganglionic fibers release ACh; most postganglionic fibers release norepinephrine (adrenergic fibers); postganglionic fibers serving sweat glands and some blood vessels of skeletal muscles release ACh; neurotransmitter activity augmented by release of adrenal medullary hormones (norepinephrine and epinephrine)

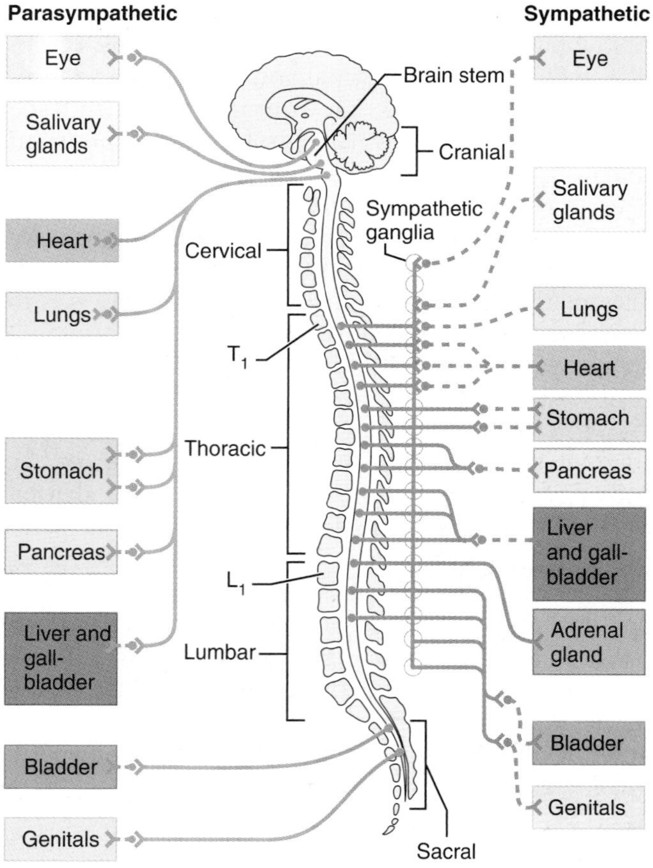

FIGURE 14.3 *Overview of the subdivisions of the autonomic nervous system.* Sites of origin of their nerves and major organs served are indicated. Synapse sites indicate the relative locations of the sympathetic and parasympathetic ganglia.

ANATOMY OF THE AUTONOMIC NERVOUS SYSTEM

The sympathetic and parasympathetic divisions are distinguished by

1. Their unique origin sites: Parasympathetic fibers emerge from the brain and sacral spinal cord, whereas sympathetic fibers originate from the thoracolumbar region of the spinal cord.

2. The different lengths of their fibers: The parasympathetic division has long preganglionic and short postganglionic fibers, whereas the sympathetic has the opposite condition.

3. The location of their ganglia: Most parasympathetic ganglia are located in the visceral effector organs, whereas sympathetic ganglia lie close to the spinal cord.

These and other anatomical considerations are summarized in Table 14.1 and illustrated in overview fashion in Figure 14.3.

Parasympathetic (Craniosacral) Division

We begin our exploration of the autonomic nervous system with the anatomically simpler parasympathetic division. Because its preganglionic fibers spring from opposite ends of the central nervous system—the brain stem and the sacral region of the spinal cord—it is also called the **craniosacral division** (see Figure 14.4). The preganglionic axons extend from the CNS nearly all the way to the structures to be innervated. There they synapse with ganglionic neurons located in **terminal** or **intramural** (in"trah-mu'ral) **ganglia** very close to or within the target organs. Very short postganglionic axons issue from the terminal ganglia and synapse with effector cells in their immediate area.

Cranial Outflow

Several cranial nerves contain the parasympathetic outflow (Figure 14.4). More specifically, preganglionic fibers run in the oculomotor, facial, glossopharyngeal, and vagus nerves. Their cell bodies lie in associated motor cranial-nerve nuclei in the brain stem (see Figure 12.16). The precise locations of the preganglionic and ganglionic neurons of the cranial parasympathetics are described next.

1. Oculomotor nerves (III). The parasympathetic fibers of the oculomotor nerves innervate smooth muscles in the eyes that cause the pupils to constrict and the lenses of the eyes to bulge—actions that allow focusing on close objects in the field of vision. The preganglionic axons found in the oculomotor nerves issue from the **accessory oculomotor nuclei** in the midbrain. The cell bodies of the ganglionic neurons are in the **ciliary ganglia** within the eye orbits (see Table 13.2, p. 486).

2. Facial nerves (VII). The parasympathetic fibers of the facial nerves stimulate the secretory activity of many large glands in the head. The pathway that activates the nasal glands and the lacrimal glands of the eyes originates in the *lacrimal nuclei* of the pons. The preganglionic fibers then run to synapse with ganglionic neurons in the **pterygopalatine** (ter"eh-go-pal'ah-tīn) **ganglia** just posterior to the maxillae. The preganglionic neurons that stimulate the submandibular and sublingual salivary glands originate in the *superior salivatory nuclei* of the pons and synapse with ganglionic neurons in the **submandibular ganglia**, deep to the mandibular angles (see Table 13.2, p. 488).

3. Glossopharyngeal nerves (IX). The parasympathetics in the glossopharyngeal nerves originate in the *inferior salivatory nuclei* of the medulla and

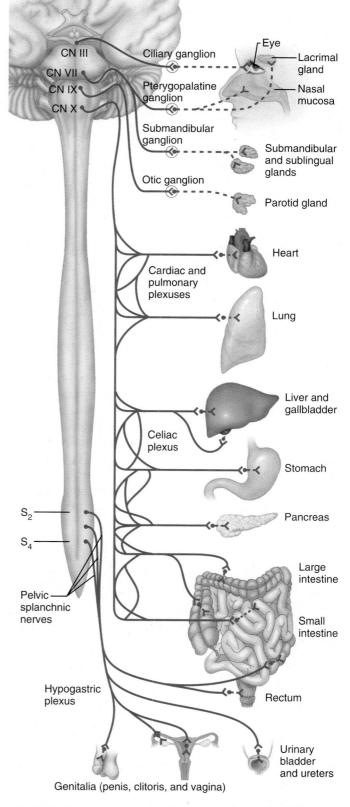

FIGURE 14.4 *Parasympathetic (craniosacral) division of the autonomic nervous system.* Solid lines indicate preganglionic nerve fibers. Dashed lines indicate postganglionic fibers. Terminal ganglia of the vagus nerve and pelvic splanchnic nerves fibers are not shown; most of these ganglia are located in or on the target organ. (Note: CN = cranial nerve.)

synapse in the **otic ganglia,** located just inferior to the foramen ovale of the skull. The postganglionic fibers then course to and activate the parotid salivary glands, which lie anterior to the ears (see Table 13.2, p. 489).

These three pairs of cranial nerves (III, VII, and IX) supply the entire parasympathetic innervation of the head. However, only the *preganglionic fibers* lie within these three pairs of cranial nerves—postganglionic fibers do not. The distal ends of the preganglionic fibers "jump over" to branches of the *trigeminal nerves (V)* to synapse; then the postganglionic fibers travel in the trigeminal nerves to reach the face. This "hitchhiking" takes advantage of the fact that the trigeminal nerves have the broadest facial distribution of all the cranial nerves and thus are well suited for this "delivery" role to the widely separated glands and smooth muscles in the head.

4. Vagus nerves (X). The remainder and major portion of the parasympathetic cranial outflow is via the vagus (X) nerves. Between them, the two vagus nerves account for about 90% of all preganglionic parasympathetic fibers in the body. They provide fibers to the neck and contribute to nerve plexuses (interweaving networks of nerves) that serve virtually every organ in the thoracic and abdominal cavities. The vagal nerve fibers (preganglionic axons) arise mostly from the *dorsal motor nuclei* of the medulla and terminate by synapsing in terminal ganglia that are usually located within the walls of the target organ. Most of these terminal ganglia are not individually named; instead they are collectively called *intramural ganglia,* literally ganglia "within the walls." As the vagus nerves pass into the thorax, they send branches to the following autonomic plexuses located within or near the organs served:

■ **Cardiac plexuses,** supplying fibers to the heart that slow heart rate

■ **Pulmonary plexuses,** serving the lungs and bronchi

■ **Esophageal** (ĕ-sof"ah-je'al) **plexuses,** supplying the esophagus

When the main trunks of the vagus nerves reach the esophagus, their fibers intermingle and form the **anterior** and **posterior vagal trunks,** each containing fibers from both vagus nerves. These vagal trunks then "ride" the esophagus down to enter the abdominal cavity. There they send fibers *through* the large **aortic plexus** (formed by a number of smaller plexuses [e.g., *celiac, superior mesenteric,* and *hypogastric*] that run along the aorta) before giving off branches to the abdominal viscera. Abdominal organs receiving vagal innervation include the liver, gallbladder, stomach, small intestine, kidneys, pancreas, and the proximal half of the large intestine.

The rest of the large intestine and the pelvic organs are served by the sacral outflow.

Sacral Outflow

The sacral outflow arises from neurons located in the lateral gray matter of spinal cord segments S_2–S_4. The axons of these neurons run in the ventral roots of the spinal nerves to the ventral rami and then branch off to form the **pelvic (splanchnic) nerves** (see Figure 14.4), which pass through the **inferior hypogastric (pelvic) plexus** in the pelvic floor. Some preganglionic fibers synapse with ganglia in this plexus, but most synapse in intramural ganglia located within the walls of the following organs: distal half of the large intestine, the urinary bladder, ureters, and reproductive organs (e.g., uterus and external genitalia).

Sympathetic (Thoracolumbar) Division

The sympathetic division is decidedly more complex than the parasympathetic division, partly because it innervates more organs. It not only supplies the visceral organs in the internal body cavities but also all visceral structures in the superficial (somatic) part of the body. This sounds impossible, but there is an explanation—some glands and smooth muscle structures in the soma (sweat glands and the hair-raising arrector pili muscles of the skin) require autonomic innervation and are served only by sympathetic fibers. Moreover, all arteries and veins (be they deep or superficial) have smooth muscle in their walls that is innervated by sympathetic fibers. But these matters will be explained in more detail later—let us get on with the anatomy of the sympathetic division.

All preganglionic fibers of the sympathetic division arise from cell bodies of preganglionic neurons located in spinal cord segments T_1 through L_2 (see Figure 14.5). For this reason, the sympathetic division is also referred to as the **thoracolumbar** (tho"rah-ko-lum'bar) **division.** The presence of numerous preganglionic sympathetic neurons in the gray matter of the spinal cord produces the **lateral horns**—the so-called **visceral motor zones** (see Figures 12.27b, p. 465, and 12.28, p. 465). The lateral horns protrude just posterolateral to the ventral horns that house motor neurons of the somatic motor division. (Parasympathetic preganglionic neurons in the sacral cord are far less abundant than the comparable sympathetic neurons in the thoracolumbar regions, and *lateral horns are absent* in the sacral region of the spinal cord. This is a major anatomical difference between the two divisions.)

After leaving the cord via the ventral root, the preganglionic sympathetic fibers pass through a

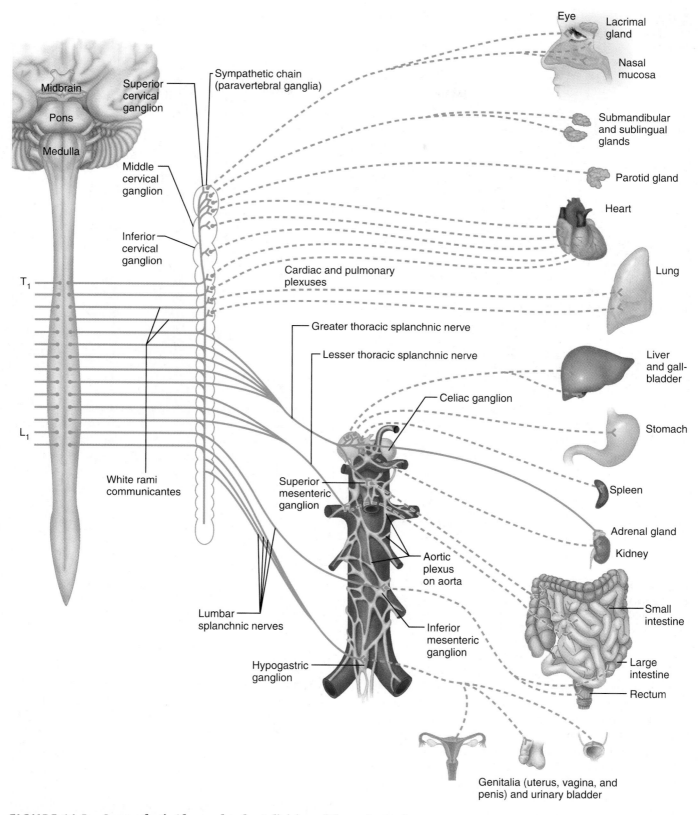

FIGURE 14.5 *Sympathetic (thoracolumbar) division of the autonomic nervous system.* Solid lines indicate preganglionic fibers; dashed lines indicate postganglionic fibers. The least thoracic and sacral splanchnic nerves are not shown.

? *On what basis are the rami communicantes distinguished as gray or white?*

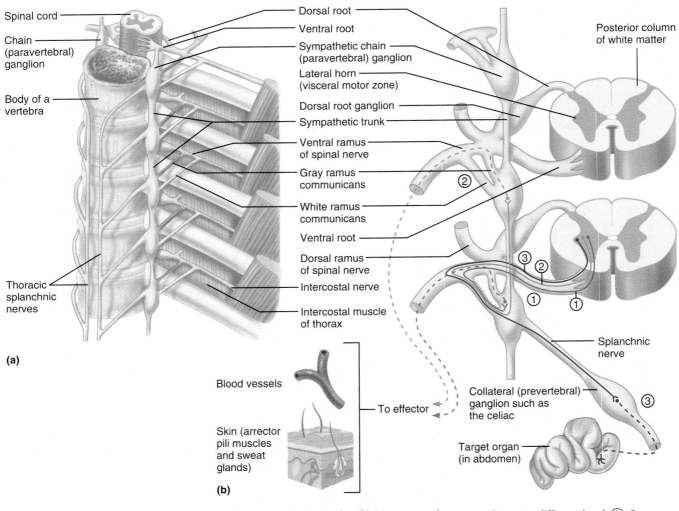

(a)

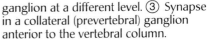

(b)

FIGURE 14.6 *Sympathetic trunks and pathways.* **(a)** The organs of the anterior thorax have been removed to allow visualization of the chain ganglia of the sympathetic trunks. **(b)** Diagrammatic view of sympathetic pathway. ① Synapse in a chain (paravertebral) ganglion at the same level. ② Synapse in a chain ganglion at a different level. ③ Synapse in a collateral (prevertebral) ganglion anterior to the vertebral column.

white ramus communicans (plural, **rami communicantes** [kom-mu′nĭ-kan″tēz]) to enter an adjoining **chain (paravertebral) ganglion** forming part of the **sympathetic trunk** or **chain** (Figure 14.6). Looking much like strands of glistening white beads, the sympathetic trunks flank each side of the vertebral column. Thus, these ganglia are named for their location.

Although the sympathetic trunks extend from the neck to the pelvis, sympathetic fibers arise only from the thoracic and lumbar spinal cord segments, as shown in Figure 14.5. The ganglia vary in size, position, and number, but typically there are 23 ganglia in each sympathetic chain—3 cervical, 11 thoracic, 4 lumbar, 4 sacral, and 1 coccygeal.

Once a preganglionic axon reaches a paravertebral ganglion, one of three things can happen to it:

1. It can synapse with a ganglionic neuron within the same chain ganglion (see pathway ① in Figure 14.6).

2. It can ascend or descend the sympathetic chain to synapse in another chain ganglion (see pathway ② in Figure 14.6). (It is these fibers running from one ganglion to another that connect the ganglia into the sympathetic trunk.)

3. It can pass through the chain ganglion and emerge from the sympathetic chain without synapsing (see pathway ③ in Figure 14.6).

The gray rami are unmyelinated. The white rami are myelinated, which gives them a distinguishing white color.

Preganglionic fibers exhibiting the last pattern help to form the **splanchnic** (splank'nik) **nerves,** which synapse with **prevertebral,** or **collateral, ganglia** located anterior to the vertebral column. Unlike chain (paravertebral) ganglia, the prevertebral ganglia

1. Are neither paired nor segmentally arranged
2. Occur only in the abdomen and pelvis

Notice that regardless of where the synapse occurs, all sympathetic ganglia are close to the spinal cord, and the postganglionic fibers, which run from the ganglion to the organs they supply, are typically much longer than the preganglionic fibers. Recall that the opposite condition occurs in the parasympathetic division. As you will soon see, these anatomical distinctions are functionally important as well.

Pathways with Synapses in a Chain Ganglion

When synapses are made in the chain ganglia, the postganglionic axons enter the ventral (or dorsal) ramus of the adjoining spinal nerves by way of communicating branches called **gray rami communicantes** (see Figure 14.6). From there they travel via branches of the rami to the sweat glands and arrector pili muscles of the skin. Anywhere along their path, the postganglionic axons may transfer over to nearby blood vessels and then innervate the vascular smooth muscle all the way to their final branches.

Notice that the naming of the rami communicantes as *white* or *gray*, which reflects their appearance, reveals whether or not their fibers are myelinated (and has *no* relationship to the white and gray matter of the CNS). Preganglionic fibers composing the white rami are myelinated. Postganglionic axons forming the gray rami are not.

The white rami, which carry preganglionic axons to the sympathetic trunks, are found only in the T_1–L_2 cord segments, the regions of sympathetic outflow. However, gray rami carrying postganglionic fibers headed for the periphery issue from *every* chain ganglion from the cervical to the sacral region. This allows sympathetic output to reach all parts of the body. Because parasympathetic fibers never run in spinal nerves, *rami communicantes are associated only with the sympathetic division.*

Pathways to the Head Sympathetic preganglionic fibers serving the head emerge from the first four spinal cord segments (T_1–T_4) and ascend the sympathetic chain to synapse with ganglionic neurons in the large **superior cervical ganglion.** This ganglion contributes sympathetic fibers that run with several cranial nerves and with the upper three or four cervical spinal nerves. Besides serving the skin and blood vessels of the head, its fibers stimulate the

Spinal cord segment	Organs served
T_1—T_5	Head and neck, heart
T_2—T_4	Bronchi and lungs
T_2—T_5	Upper limb
T_5—T_6	Esophagus
T_6—T_{10}	Stomach, spleen, pancreas
T_7—T_9	Liver
T_9—T_{10}	Small intestine
T_{10}—L_1	Kidney, reproductive organs (uterus, testis, ovary, etc.)
T_{10}—L_2	Lower limb
T_{11}—L_2	Large intestine, ureter, urinary bladder

TABLE 14.2 Segmental Sympathetic Supplies

dilator muscles of the irises of the eyes, inhibit the nasal and salivary glands (the reason our mouth goes dry when we are scared), and innervate the smooth (tarsal) muscle which lifts the upper eyelid. The superior cervical ganglion also gives off direct branches to the heart.

Pathways to the Thorax Sympathetic preganglionic fibers innervating the thoracic organs originate at T_1–T_6. From there the preganglionic fibers run to synapse in the cervical chain ganglia. Postganglionic fibers emerging from the **middle** and **inferior cervical ganglia** enter cervical nerves C_4–C_8. Some of these fibers innervate the heart via the cardiac plexus, and some innervate the thyroid gland, but most serve the skin. Additionally, some of the T_1–T_6 preganglionic fibers synapse in the nearest chain ganglion, and the postganglionic fibers pass directly to the organ served. Fibers to the heart, aorta, lungs, and esophagus take this direct route. Along the way, they run into the plexuses associated with those organs. Sympathetic innervation by spinal cord segment is summarized in Table 14.2.

Pathways with Synapses in a Collateral (Prevertebral) Ganglion

The preganglionic fibers from T_5 down synapse in collateral ganglia; thus these fibers enter and leave the sympathetic chains without synapsing. They form several nerves called splanchnic nerves, including the **thoracic splanchnic nerves** (greater, lesser, and least) and the **lumbar** and **sacral splanchnics.** The splanchnic nerves contribute to a number of interweaving nerve plexuses known collectively as the **abdominal aortic** (a-or'tik) **plexus,** which clings to the surface of the abdominal aorta (see Figure 14.5). This complex plexus contains several ganglia, some large and some small, that together serve the entire abdominopelvic viscera (*splanchni* = viscera). From the superior to the inferior aspect, the most important of these ganglia (and related subplexuses) are the **celiac, superior mesenteric, inferior mesenteric,** and **hypogastric,** named for the arteries with which they most closely associate. Postganglionic fibers issuing from these ganglia generally travel to their target organs in the company of arteries serving these organs.

Pathways to the Abdomen Sympathetic innervation of the abdomen is via preganglionic fibers from T$_5$ to L$_2$, which travel in the *thoracic splanchnic nerves* to synapse mainly at the celiac and superior mesenteric ganglia. Postganglionic fibers issuing from these ganglia serve most of the abdominal viscera: stomach, intestines (except the distal half of the large intestine), liver, spleen, and kidneys.

Pathways to the Pelvis Preganglionic fibers innervating the pelvis originate from T$_{10}$ to L$_2$ and then descend in the sympathetic trunk to the lumbar and sacral chain ganglia. Some fibers synapse there, but most go directly out from the cord via the *lumbar* and *sacral splanchnic nerves*, which send the bulk of their fibers to the inferior mesenteric and hypogastric ganglia. From these ganglia, postganglionic fibers serving the distal half of the large intestine, the urinary bladder, and the pelvic reproductive organs issue. For the most part, sympathetic fibers *inhibit* the activity of the muscles and glands in these visceral organs.

Pathways with Synapses in the Adrenal Medulla

Some fibers traveling in the thoracic splanchnic nerves pass through the celiac ganglion without synapsing and terminate by synapsing with the hormone-producing medullary cells of the adrenal gland (see Figure 14.5). When stimulated by the preganglionic fibers, the medullary cells secrete **norepinephrine** and **epinephrine** (also called *noradrenaline* and *adrenaline*, respectively) into the blood, producing the widespread excitatory effects we have all felt as a "surge of adrenaline." Embryologically, sympathetic ganglia and the adrenal medulla arise from the same tissue. For this reason, the adrenal medulla is sometimes viewed as a "misplaced" sympathetic ganglion, and its hormone-releasing cells, although lacking nerve processes, are considered equivalent to ganglionic sympathetic neurons.

 According to the figure legend, the afferent neuron shown in this diagram reaches the CNS via a spinal nerve. How would you know that if it was not mentioned?

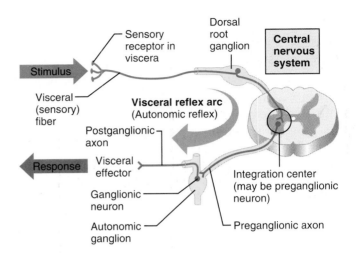

FIGURE 14.7 *Visceral reflexes.* Visceral reflexes have the same elements as somatic reflexes but they always occur over polysynaptic pathways because of the two-motor-neuron efferent pathway. The integration center may involve a dorsal horn interneuron as shown, or may just be a synapse with a preganglionic neuron. The visceral afferent fibers are found both in spinal nerves (as depicted here) and in autonomic nerves.

Visceral Reflexes

Because most anatomists consider the autonomic nervous system to be a visceral motor system, the presence of sensory fibers (mostly visceral pain afferents) in autonomic nerves is often overlooked. However, the **visceral sensory neurons,** which send information concerning chemical changes, stretch, and irritation of the viscera, are the first link in autonomic reflexes. **Visceral reflex arcs** have essentially the same components as somatic reflex arcs— receptor, sensory neuron, integration center, motor neuron, and effector—except that a visceral reflex arc has a *two-neuron* motor chain (Figure 14.7).

Nearly all the sympathetic and parasympathetic fibers described thus far are accompanied by afferent fibers conducting sensory impulses from glands or muscular structures. Thus, peripheral processes of visceral sensory neurons are found in cranial nerves VII, IX, and X, the splanchnic nerves, and the sympathetic trunk, as well as in spinal nerves. Like sensory neurons serving somatic structures (skeletal muscles and skin), the cell bodies of visceral sensory

The visceral afferent fiber enters the spinal cord (CNS) via a dorsal root ganglion. Dorsal root ganglia are found only on the dorsal root of spinal nerves.

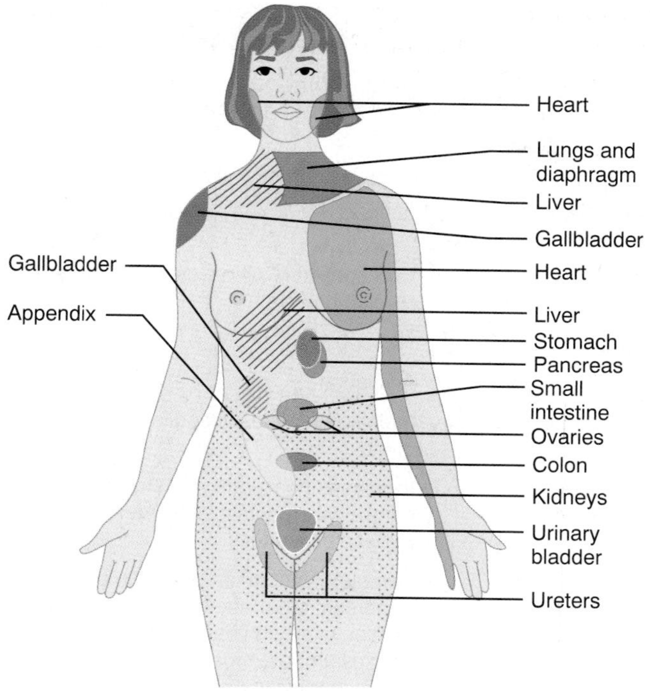
Heart
Lungs and diaphragm
Liver
Gallbladder
Heart
Liver
Stomach
Pancreas
Small intestine
Ovaries
Colon
Kidneys
Urinary bladder
Ureters
Gallbladder
Appendix

FIGURE 14.8 ***Referred pain.*** Anterior cutaneous areas to which pain from certain visceral organs is referred.

neurons are located in the sensory ganglia of associated cranial nerves or in the dorsal root ganglia of the spinal cord. However, visceral sensory neurons are also found within sympathetic ganglia where synapses with preganglionic neurons occur. Furthermore, complete three-neuron reflex arcs (with sensory, motor, and intrinsic neurons) exist entirely within the walls of the gastrointestinal tract. Neurons composing these reflex arcs make up the *enteric nervous system,* which plays an important role in controlling gastrointestinal tract activity.

The fact that visceral pain afferents travel along the same pathways as somatic pain fibers helps to explain the phenomenon of **referred pain,** in which pain stimuli arising in the viscera are perceived as somatic in origin. For example, a heart attack may produce a sensation of pain that radiates to the superior thoracic wall and along the medial aspect of the left arm. Since the same spinal segments (T_1–T_5) innervate both the heart and the regions to which pain signals from heart tissue are referred, the brain interprets most such inputs as coming from the more common somatic pathway. Cutaneous areas to which visceral pain is commonly referred are shown in Figure 14.8. (Somatic and visceral pain are described in more detail in *A Closer Look* in Chapter 13.)

PHYSIOLOGY OF THE AUTONOMIC NERVOUS SYSTEM

Neurotransmitters and Receptors

Acetylcholine (ACh) and *norepinephrine (NE)* are the major neurotransmitters released by ANS neurons. Acetylcholine, the same neurotransmitter secreted by somatic motor neurons, is released by (1) all preganglionic axons of the ANS and (2) all parasympathetic postganglionic axons at synapses with their effectors (see Figure 14.2 and Table 14.1). ACh-releasing fibers are called **cholinergic** (ko″lin-er′jik) **fibers.**

In contrast, most sympathetic postganglionic axons release NE and are classified as **adrenergic** (ad″ren-er′jik) **fibers.** The only exceptions are sympathetic postganglionic fibers innervating the sweat glands of the skin, and some blood vessels within skeletal muscles and the external genitalia, all of which secrete ACh. As mentioned earlier, NE was originally named *noradrenaline,* and the newer term *norepinephrine* has never completely replaced the old. NE-releasing fibers continue to be called adrenergic fibers (perhaps because the term is easier to pronounce than "epinephrinergic fibers").

Unfortunately for memorization purposes, the effects of ACh and NE on their effectors are not consistently either excitation or inhibition. The response of visceral effectors to these neurotransmitters depends not only on the neurotransmitters themselves but also on the receptors to which they attach. The two or more kinds of receptors that exist for each of the autonomic neurotransmitters allow them to exert different effects (activation or inhibition) at different body targets (Table 14.3).

Cholinergic Receptors

The two types of receptors that bind ACh are named for certain drugs that bind to them and mimic acetylcholine's effects. The first of these receptors identified were the **nicotinic** (nik″o-tin′ik) **receptors.** A mushroom poison, *muscarine* (mus′kah-rin), activates a different set of ACh receptors, named **muscarinic receptors.** All acetylcholine receptors are either nicotinic or muscarinic.

Nicotinic receptors are found on (1) the motor end plates of skeletal muscle cells (which, however, are somatic and not autonomic targets), (2) *all* ganglionic neurons, both sympathetic and parasympathetic, and (3) the hormone-producing cells of the adrenal medulla. The effect of ACh binding to nicotinic receptors is *always* stimulatory and results in excitation of the neuron or effector cell.

TABLE 14.3	**Cholinergic and Adrenergic Receptors**		
Neurotransmitter	*Receptor type*	*Major locations*	*Effect of binding*
Acetylcholine	**Cholinergic**		
	Nicotinic	All ganglionic neurons; adrenal medullary cells (also neuromuscular junctions of skeletal muscle)	Excitation
	Muscarinic	All parasympathetic target organs	Excitation in most cases; inhibition of cardiac muscle
		Selected sympathetic targets:	
		■ Eccrine sweat glands	Activation
		■ Blood vessels in skeletal muscles	Inhibition (causes vasodilation)
Norepinephrine (and epinephrine released by adrenal medulla)	**Adrenergic**		
	β_1	Heart and coronary blood vessels predominantly, but also kidneys, liver, and adipose tissue	Increases heart rate and strength; dilates coronary arterioles; stimulates renin release by kidneys
	β_2	Lungs and most other sympathetic target organs; abundant on blood vessels serving the heart	Stimulates secretion of insulin by pancreas; other effects mostly inhibitory: dilation of blood vessels and bronchioles; relaxes smooth muscle walls of digestive and urinary visceral organs; relaxes pregnant uterus
	β_3	Adipose tissue	Stimulates lipolysis by fat cells
	α_1	Most importantly blood vessels serving the skin, mucosae, abdominal viscera, kidneys, and salivary glands; but virtually all sympathetic target organs except heart	Constricts blood vessels and visceral organ sphincters; dilates pupils of the eyes
	α_2	Membrane of adrenergic axon terminals; blood platelets	Mediates inhibition of NE release from adrenergic terminals; promotes blood clotting

Muscarinic receptors occur on all effector cells stimulated by postganglionic cholinergic fibers—that is, all parasympathetic target organs and a few sympathetic targets, such as eccrine sweat glands and some blood vessels of skeletal muscles. The effect of ACh binding to muscarinic receptors is inhibitory or stimulatory, depending on the target organ. For example, binding of ACh to cardiac muscle receptors slows heart activity, whereas ACh binding to receptors on smooth muscle of the gastrointestinal tract increases its motility.

Adrenergic Receptors

There are also two major classes of adrenergic (NE-binding) receptors: **alpha** (α) and **beta** (β). Organs that respond to NE (or epinephrine) display one or both types of receptor. In general, NE or epinephrine binding to α receptors is stimulatory, while their binding to β receptors is inhibitory. But there are notable exceptions. For example, binding of NE to the β receptors of cardiac muscle prods the heart into more vigorous activity. These differences reflect the fact that both α and β receptors have two or three re-

ceptor subclasses (α_1 and α_2; β_1, β_2, and β_3), and each receptor type tends to predominate in certain target organs. An overview of the locations and effects of these receptor subclasses is provided in Table 14.3.

The Effects of Drugs

Knowing the locations of the cholinergic and adrenergic receptor subtypes is important clinically because it allows specific drugs to be prescribed to obtain the desired blocking or stimulatory effects on selected target organs. For example, atropine, an anticholinergic drug that blocks parasympathetic effects, is routinely administered before surgery to prevent salivation and dry up respiratory system secretions during surgery. It is also used by ophthalmologists to dilate the pupils for eye examination. The anticholinesterase drug *neostigmine* inhibits the enzyme acetylcholinesterase, thus preventing enzymatic breakdown of ACh and allowing it to accumulate in the synapses. This drug is used to treat myasthenia gravis, a condition in which skeletal muscle activity is impaired for lack of ACh stimulation.

Drugs called *tricyclic antidepressants* (Elavil and Sinequan) help to relieve depression because they prolong the activity of NE on the postsynaptic membrane. As described in Chapter 11 (Table 11.3, p. 416), NE is one of our "feeling good" neurotransmitters. Hundreds of over-the-counter drugs used to treat colds, coughs, allergies, and nasal congestion contain sympathomimetics (ephedrine, phenylephrine, and others), drugs that stimulate α-adrenergic receptors. Much of the research going on in pharmaceutical companies is directed toward finding drugs that affect only one subclass of receptor without upsetting the whole adrenergic or cholinergic system. An important breakthrough was finding adrenergic blockers that attach mainly to the β_1 receptors of cardiac muscle. Also called *beta-blockers*, acebutolol (Sectral), metoprolol (Lopressor), and other drugs of their class are used when the goal is to reduce heart rate and prevent arrhythmias (irregular heartbeats) in cardiac patients without interfering with other sympathetic effects.

Interactions of the Autonomic Divisions

As mentioned earlier, most visceral organs are innervated by both sympathetic and parasympathetic fibers—that is, they receive *dual innervation*. Normally, both ANS divisions are partially active, producing a dynamic antagonism that allows visceral activity to be very precisely controlled. However, one division or the other usually exerts the predominant effects in given circumstances, and in a few cases, the autonomic divisions actually cooperate with one another. The major effects of the two divisions are summarized in Table 14.4.

Antagonistic Interactions

Antagonistic effects, which were described earlier, are most clearly seen on the activity of the heart, respiratory system, and gastrointestinal organs. In a fight-or-flight situation, the sympathetic division increases your respiratory and heart rates while inhibiting digestion and elimination. When the emergency is over, the parasympathetic division restores heart and breathing rates to resting levels and then attends to processes that refuel body cells and discard wastes.

Sympathetic and Parasympathetic Tone

Although we have described the parasympathetic division as the "resting and digesting" division, the sympathetic division is the major actor in controlling blood pressure, even at rest. With few exceptions, the vascular system is entirely innervated by sympathetic fibers that keep the blood vessels in a continual state of partial constriction called **sympathetic** or **vasomotor tone.** When faster blood delivery is needed, these fibers deliver impulses more rapidly, causing blood vessels to constrict and blood pressure to rise. Conversely, when blood pressure is to be decreased, the vessels are prompted to dilate. *Alpha-blocker drugs* (such as phentolamine) that interfere with the activity of these **vasomotor fibers** are often used to treat hypertension. During circulatory shock (inadequate blood delivery to body tissues), or when more blood is needed to meet the increased needs of working skeletal muscles, blood vessels serving the skin and abdominal viscera are strongly constricted. This blood "shunting" helps to maintain circulation to vital organs or enhance blood delivery to skeletal muscles.

On the other hand, parasympathetic effects normally dominate the heart and smooth muscle of digestive and urinary tract organs. Thus, these organs exhibit **parasympathetic tone.** The parasympathetic division slows the heart and dictates the normal activity levels of the digestive and urinary tracts. However, the sympathetic division can override these parasympathetic effects during times of stress. Drugs that block parasympathetic responses increase heart rate and cause fecal and urinary retention. Except for the adrenal glands and sweat glands of the skin, most glands are activated by parasympathetic fibers.

Cooperative Effects

The best example of cooperative ANS effects is seen in controls of the external genitalia. Parasympathetic stimulation causes vasodilation of blood vessels in the external genitalia and is responsible for erection of the male penis or female clitoris during sexual excitement. (This may help explain why sexual performance is sometimes impaired when people are anxious or upset and the sympathetic division is in charge.) Sympathetic stimulation then causes ejaculation of semen by the male or reflex peristalsis of a female's vagina.

Unique Roles of the Sympathetic Division

The sympathetic division regulates many functions that are not subject to parasympathetic influence. For example, the adrenal medulla, the sweat glands and arrector pili muscles of the skin, the kidneys, and most blood vessels receive only sympathetic fibers. It is easy to remember that the sympathetic system innervates these structures because most of us sweat under stress, our scalp "prickles" during fear, and our blood pressure skyrockets (from widespread vasoconstriction) when we get excited. We have already described how sympathetic control of blood vessels regulates blood pressure and shunting

TABLE 14.4	Effects of the Parasympathetic and Sympathetic Divisions on Various Organs	
Target organ or system	Parasympathetic effects	Sympathetic effects
Eye (iris)	Stimulates constrictor muscles; constricts eye pupils	Stimulates dilator muscles; dilates eye pupils
Eye (ciliary muscle)	Stimulates muscles, which results in bulging of the lens for accommodation and close vision	No effect (no innervation)
Glands (nasal, lacrimal, salivary, gastric, pancreas)	Stimulates secretory activity	Inhibits secretory activity; causes vasoconstriction of blood vessels supplying the glands
Sweat glands	No effect (no innervation)	Stimulates copious sweating (cholinergic fibers)
Adrenal medulla	No effect (no innervation)	Stimulates medulla cells to secrete epinephrine and norepinephrine
Arrector pili muscles attached to hair follicles	No effect (no innervation)	Stimulates to contract (erects hairs and produces "goosebumps")
Heart muscle	Decreases rate; slows and steadies heart	Increases rate and force of heartbeat
Heart: coronary blood vessels	Constricts coronary vessels	Causes vasodilation
Bladder/urethra	Causes contraction of smooth muscle of bladder wall; relaxes urethral sphincter; promotes voiding	Causes relaxation of smooth muscle of bladder wall; constricts urethral sphincter; inhibits voiding
Lungs	Constricts bronchioles	Dilates bronchioles and mildly constricts blood vessels
Digestive tract organs	Increases motility (peristalsis) and amount of secretion by digestive organs; relaxes sphincters to allow movement of foodstuffs along tract	Decreases activity of glands and muscles of digestive system and constricts sphincters (e.g., anal sphincter)
Liver	No effect (no innervation)	Epinephrine stimulates liver to release glucose to blood
Gallbladder	Excites (gallbladder contracts to expel bile)	Inhibits (gallbladder is relaxed)
Kidney	No effect (no innervation)	Causes vasoconstriction; decreases urine output; promotes renin formation
Penis	Causes erection (vasodilation)	Causes ejaculation
Vagina/clitoris	Causes erection (vasodilation) of clitoris	Causes reverse peristalsis (contraction) of vagina
Blood vessels	Little or no effect	Constricts most vessels and increases blood pressure; constricts vessels of abdominal viscera and skin to divert blood to muscles, brain, and heart when necessary; dilates vessels of the skeletal muscles (via cholinergic fibers and epinephrine) during exercise
Blood coagulation	No effect (no innervation)	Increases coagulation
Cellular metabolism	No effect (no innervation)	Increases metabolic rate
Adipose tissue	No effect (no innervation)	Stimulates lipolysis (fat breakdown)
Mental activity	No effect (no innervation)	Increases alertness

of blood in the vascular system. We will now mention several other uniquely sympathetic functions.

Thermoregulatory Responses to Heat The sympathetic division mediates reflexes that help to regulate body temperature. For example, applying heat to the skin causes reflex dilation of blood vessels in that area. When systemic body temperature is elevated, sympathetic nerves effect widespread dilation of the skin vasculature, allowing the skin to become flushed with warm blood, and activate the sweat glands to help cool the body. Conversely, when body temperature falls, skin blood vessels are constricted, and blood is restricted to deeper, more vital organs.

Release of Renin from the Kidneys Sympathetic impulses stimulate the kidneys to release renin, an enzyme that promotes an increase in blood pressure. This renin-angiotensin mechanism, which helps maintain fluid balance in the body, is described in Chapter 26.

Metabolic Effects Through both direct neural stimulation and release of adrenal medullary hormones, the sympathetic division promotes a number of metabolic effects not reversed by parasympathetic activity. It (1) increases the metabolic rate of body cells; (2) raises blood glucose levels; (3) mobilizes fats for use as fuels; and (4) increases mental alertness by stimulating the reticular activating system (RAS) of the brain stem. These hormones also cause skeletal muscle to contract more strongly and quickly. As a side effect, the muscle spindles are stimulated more often and, consequently, nerve impulses traveling to the muscles occur more synchronously. These neural bursts, which put muscle contractions on a "hair trigger," are great if you have to make a quick jump or run, but they can be embarrassing or even disabling to the nervous musician or surgeon.

Localized Versus Diffuse Effects

In the parasympathetic division, one preganglionic neuron synapses with one (or at most a few) ganglionic neurons. Additionally, all parasympathetic fibers release ACh, which is quickly destroyed (hydrolyzed) by acetylcholinesterase. Consequently, the parasympathetic division exerts short-lived, highly localized control over its effectors. In contrast, preganglionic sympathetic axons branch profusely as they enter the sympathetic trunk and synapse with many ganglionic neurons at several levels. Thus, when the sympathetic division is activated, it responds in a diffuse and highly interconnected way. Indeed, the literal translation of sympathetic (*sym* = together; *pathos* = feeling) relates to the bodywide mobilization it provokes.

Effects produced by sympathetic activation are also much longer-lasting than those of parasympathetic activation. This reflects three phenomena: (1) NE is inactivated more slowly than ACh because it must be taken back up into the presynaptic ending before it is hydrolyzed (or stored); (2) because NE is an indirectly acting neurotransmitter that acts through a second-messenger system (see p. 418), it exerts its effects more slowly than ACh, which acts directly; and (3) when the sympathetic division is mobilized, norepinephrine and epinephrine are secreted into the blood by adrenal medullary cells. Although epinephrine is more potent at increasing heart rate and raising blood sugar levels and metabolic rate, these hormones have essentially the same effects as NE released by sympathetic neurons. In fact, circulating adrenal medullary hormones produce 25 to 50% of all the sympathetic effects acting on the body at a given time. These effects continue for several minutes until the hormones are destroyed by the liver. Thus, although sympathetic nerve impulses act only briefly, the hormonal effects they provide linger. The widespread and prolonged effect of sympathetic activation helps explain why we need time to "come down" after an extremely stressful experience.

Control of Autonomic Functioning

Although the ANS is not usually considered to be under voluntary control, its activity *is* regulated. Several levels of CNS controls exist—in the spinal cord, brain stem, hypothalamus, and cerebral cortex. In general, the hypothalamus is like a "head ganglion" at the top of the ANS control hierarchy. From there, orders flow to lower and lower CNS centers for execution. Although the cerebral cortex may modify the workings of the ANS, it does so at the subconscious level and by acting through limbic system structures on hypothalamic centers (Figure 14.9).

Brain Stem and Spinal Cord Controls

The hypothalamus is the "boss," but the reticular formation of the brain stem appears to exert the most *direct* influence over autonomic functions. For example, certain motor centers in the ventrolateral medulla reflexively regulate heart rate and blood vessel diameter (*cardiac* and *vasomotor centers*) and respiration. Other medullary regions oversee certain gastrointestinal activities. Most sensory impulses involved in eliciting these autonomic reflexes reach the brain stem via vagus nerve afferents. The pons contains *respiratory centers* that interact with those of the medulla, and there are midbrain centers (*oculomotor nuclei*) concerned with controlling the size of the eye pupils. The defecation and micturition reflexes that promote emptying of the rectum and bladder are integrated at the spinal cord level but are subject to conscious inhibition. All of these autonomic reflexes are described in later chapters in relation to the organ systems they serve.

Hypothalamic Controls

As noted, the main integration center of the autonomic nervous system is the hypothalamus. Medial and anterior hypothalamic regions appear to direct parasympathetic functions, whereas lateral and posterior areas direct sympathetic functions. These centers exert their effects both directly and via relays through the *reticular formation*, which in turn influences the preganglionic motor neurons in

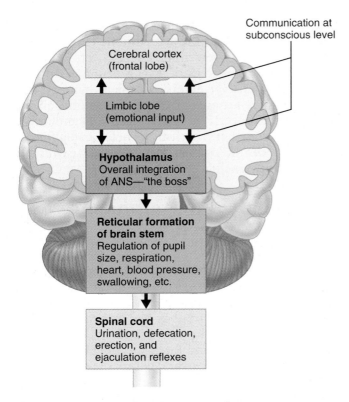

Communication at subconscious level

Cerebral cortex (frontal lobe)

Limbic lobe (emotional input)

Hypothalamus
Overall integration of ANS—"the boss"

Reticular formation of brain stem
Regulation of pupil size, respiration, heart, blood pressure, swallowing, etc.

Spinal cord
Urination, defecation, erection, and ejaculation reflexes

FIGURE 14.9 *Levels of ANS control.* The hypothalamus stands at the top of the control hierarchy as the integrator of ANS activity. However, subconscious cerebral inputs via limbic lobe connections influence hypothalamic functioning.

the brain stem and spinal cord (Figure 14.9). The hypothalamus contains centers that coordinate heart activity, blood pressure, body temperature, water balance, and endocrine activity. It also contains centers that help mediate various emotional states (rage, pleasure) and biological drives (thirst, hunger, sex). (The relationship of certain drug abuse behaviors to hypothalamic pleasure centers is the topic of *A Closer Look* in Chapter 11.)

Via its associations with the periaqueductal gray matter and the amygdala, the hypothalamus also mediates our reactions to fear. Emotional responses of the limbic lobe of the cerebrum to danger and stress signal the hypothalamus to activate the sympathetic system to fight-or-flight status. Thus, the hypothalamus serves as the keystone of the emotional and visceral brain, and through its centers emotions influence autonomic nervous system functioning and behavior.

Cortical Controls

It was originally believed that the autonomic nervous system is not subject to voluntary controls. However, we have all had occasions when remembering a frightening event made our heart race (sympathetic response) or just the thought of a favorite

food (pecan pie for example) made our mouth water (a parasympathetic response). These inputs converge on the hypothalamus through its connections to the limbic lobe.

Additionally, studies of meditating individuals and biofeedback have shown that voluntary cortical control of visceral activities is also possible—a capability that is untapped by most people.

Influence of Biofeedback on Autonomic Function
During **biofeedback training,** subjects are connected to monitoring devices that provide an awareness of what is happening within their body, which is called **biofeedback.** The devices detect and amplify changes in physiological processes such as heart rate, blood pressure, and skeletal muscle tone, and these data are then "fed back" to the subject in the form of flashing lights or audible tones. Subjects are asked to try to alter or control some "involuntary" function by concentrating on calming, pleasant thoughts. The monitor allows them to identify changes in the desired direction, so they can gradually recognize the feelings associated with these changes and learn to produce the changes at will.

Biofeedback techniques have been quite successful in helping individuals plagued by migraine headaches. They are also being used by cardiac patients to manage stress, and so reduce their risk of heart attack. However, biofeedback training is a time-consuming and often frustrating process, and the training equipment is expensive and difficult to use. Furthermore, not everyone can master this technique.

HOMEOSTATIC IMBALANCES OF THE AUTONOMIC NERVOUS SYSTEM

Because the autonomic nervous system is involved in nearly every important process that goes on within the body, it is not surprising that abnormalities of autonomic functioning can have far-reaching effects. Such abnormalities can threaten life itself.

Most autonomic disorders reflect exaggerated or deficient controls of smooth muscle activity. Of these, the most devastating involve blood vessels and include conditions such as hypertension, Raynaud's disease, and the mass reflex reaction.

Hypertension, or high blood pressure, may result from an overactive sympathetic vasoconstrictor response promoted by continual high levels of stress. Hypertension, discussed in Chapter 20, is always serious because (1) it increases the work load on the heart, which may precipitate heart disease, and (2) it increases the wear and tear on the artery walls. Stress-induced hypertension is treated with adrenergic receptor-blocking drugs.

Raynaud's disease is characterized by intermittent attacks during which the skin of the fingers and toes becomes pale, then cyanotic and painful. Commonly provoked by exposure to cold or emotional stress, it is an exaggerated vasoconstriction response in the affected body regions. The severity of this disease ranges from merely uncomfortable to such severe constriction of blood vessels that ischemia and gangrene (tissue death) results. To treat severe cases, preganglionic sympathetic fibers serving the affected regions are severed (a procedure called *sympathectomy*). The involved vessels then dilate, reestablishing adequate blood delivery to the region.

The *mass reflex reaction* is a life-threatening condition involving uncontrolled activation of both autonomic and somatic motor neurons. It occurs in a majority of individuals with quadriplegia and in others with spinal cord injuries above the level of T_6. The initial cord injury is followed by *spinal shock* (see p. 467). When reflex activity returns, it is usually exaggerated because of lack of inhibitory input from higher centers. Then episodes of mass reflex, caused by surges of nervous output from large regions of the spinal cord, begin to occur. The usual trigger for the mass reflex is a painful stimulus to the skin or overfilling of a visceral organ, such as the bladder. The body goes into flexor spasms, the colon and bladder empty, and profuse sweating begins. Arterial blood pressure rises to life-threatening levels (200 mm Hg or more), which may rupture a blood vessel in the brain, precipitating a stroke. The precise mechanism of the mass reflex is unknown, but it has been envisioned as a type of epilepsy of the spinal cord.

DEVELOPMENTAL ASPECTS OF THE AUTONOMIC NERVOUS SYSTEM

Preganglionic neurons of the autonomic nervous system derive from the embryonic *neural tube*, as do somatic motor neurons. ANS structures in the PNS—ganglionic neurons, the adrenal medulla, and all autonomic ganglia—derive from the *neural crest* (along with all sensory neurons) (see Figure 12.2). The neural crest cells appear to reach their ultimate destinations by migrating along growing axons. Forming ganglia receive axons from preganglionic neurons within the cord or brain and in turn send their axons to synapse with their effector cells in the body periphery. This process appears to be guided by

nerve growth factor (NGF), a protein secreted by the target cells of the postganglionic axons.

Homeostatic Imbalance

Congenital abnormalities of the autonomic nervous system are rare, but they do occur. Perhaps the most notable example is *Hirschsprung's disease (congenital megacolon)*, in which parasympathetic innervation of the distal part of the large intestine (colon) fails to develop normally. As a result, the distal colon is immobile, and as feces accumulate proximal to the inactive bowel segment, the area becomes enormously distended. The condition is corrected surgically by removing the inactive bowel segment. ∎

During youth, impairments of autonomic nervous function are usually due to injuries to the spinal cord or autonomic nerves. In old age the efficiency of the autonomic nervous system begins to decline. At least part of the problem appears to be due to structural changes (bloating) of some (neuropeptide Y-releasing) preganglionic axonal terminals, which become congested with neurofilaments with age.

Many elderly people complain of constipation (due to reduced gastrointestinal tract motility), and diminished tear formation leads to dry eyes and frequent eye infections. Additionally, when they stand up they may have fainting episodes due to **orthostatic hypotension** (*ortho* = straight; *stat* = standing), a form of low blood pressure that occurs because the aging sympathetic vasoconstrictor centers respond more slowly to position changes. These problems are distressing, but they are not usually life threatening, and most can be managed by lifestyle changes or artificial aids. For example, changing position slowly gives the sympathetic nervous system more time to adjust the blood pressure; drinking ample fluids helps relieve constipation; and eye drops (artificial tears) are available for the dry-eye problem.

* * *

In this chapter, we have described the structure and function of the autonomic nervous system, one arm of the motor division of the peripheral nervous system. Because virtually every organ system still to be considered depends on autonomic controls to function normally, you will be hearing more about the autonomic nervous system in the chapters that follow. Now that we have explored most of the nervous system, this is a good time to examine how it interacts with the rest of the body as summarized in the *Making Connections* feature, pp. 530–531.

MAKING CONNECTIONS

SYSTEM CONNECTIONS: Homeostatic Interrelationships Between the Nervous System and Other Body Systems

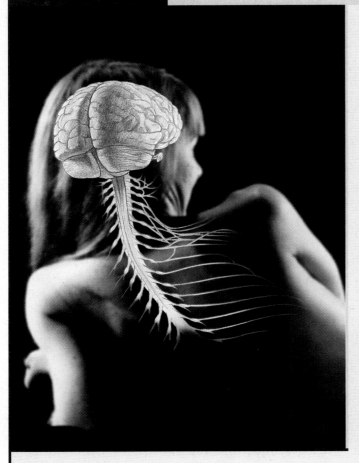

Endocrine System

- Sympathetic division of the ANS activates the adrenal medulla; hypothalamus helps regulate the activity of the anterior pituitary gland and produces two hormones
- Hormones influence neuronal metabolism

Cardiovascular System

- ANS helps regulate heart rate and blood pressure
- Cardiovascular system provides blood containing oxygen and nutrients to the nervous system; carries away wastes

Lymphatic System/Immunity

- Nerves innervate lymphoid organs; the brain plays a role in regulating immune function
- Lymphatic vessels carry away leaked tissue fluids from tissues surrounding nervous system structures; immune elements protect all body organs from pathogens (CNS has additional mechanisms as well)

Respiratory System

- Nervous system regulates respiratory rhythm and depth
- Respiratory system provides life-sustaining oxygen; disposes of carbon dioxide

Digestive System

- ANS (particularly the parasympathetic division) regulates digestive mobility and glandular activity
- Digestive system provides nutrients needed for neuronal health

Urinary System

- ANS regulates bladder emptying and renal blood pressure
- Kidneys help to dispose of metabolic wastes and maintain proper electrolyte composition and pH of blood for neural functioning

Reproductive System

- ANS regulates sexual erection and ejaculation in males; erection of the clitoris in females
- Testosterone causes masculinization of the brain and underlies sex drive and aggressive behavior

Integumentary System

- Sympathetic division of the ANS regulates sweat glands and blood vessels of skin (therefore heat loss/retention)
- Skin serves as heat loss surface

Skeletal System

- Nerves innervate bones
- Bones serve as depot for calcium needed for neural function; skeletal system protects CNS structures

Muscular System

- Somatic division of nervous system activates skeletal muscles; maintains muscle health
- Skeletal muscles are the effectors of the somatic division

CLOSER CONNECTIONS

THE NERVOUS SYSTEM and Interrelationships with the Muscular, Respiratory, and Digestive Systems

The nervous system pokes its figurative nose into the activities of virtually every organ system of the body. Hence, trying to choose the most significant interactions comes pretty close to a lesson in futility. Which is more important: digestion, elimination of wastes, or mobility? Your answer probably depends on whether you are hungry or need to make a "pit stop" at the bathroom when you read this. Since neural interactions with several other systems are thoroughly plumbed in chapters to come, here we will look at the all-important interactions between the nervous system and the muscular, respiratory, and digestive systems.

Muscular System

Most simply said, the muscular system would cease to function without the nervous system. Unlike the visceral muscles or cardiac muscle, both of which have other controlling systems, somatic motor fibers are *it* for skeletal muscle activation and regulation. Somatic nerve fibers "tell" skeletal muscles not only when to contract but also how strongly. Additionally, as the nervous system makes its initial synapses with skeletal muscle fibers, it determines their fate as fast or slow fibers, which forever after affects our potential for muscle speed and endurance. The interactions of the various brain regions (basal nuclei, cerebellum, premotor cortex, etc.) and inputs of stretch receptors also determine our grace—how smooth and coordinated we are. Nonetheless, keep in mind that as long as the skeletal muscle effector cells are healthy, they help determine the viability of the neurons synapsing with them. The relationship is truly synergistic.

Respiratory System

Another system that depends entirely on the nervous system for its function is the respiratory system, which continuously refreshes the blood with oxygen and unloads the carbon dioxide waste to the sea of air that surrounds us. Neural centers in the medulla and pons both initiate and maintain the tidelike rhythm of air flushing into and out of our lungs by activating skeletal muscles that change the volume (thus the gas pressure) within the lungs. If some of those CNS centers are damaged, reflexes set up by stretch receptors in the lungs can still maintain the vital function of breathing.

Digestive System

Although the digestive system responds to many different types of controls—for example, hormones, local pH, and irritating chemicals or bacteria—the parasympathetic nervous system is crucial to its normal functioning. Without the parasympathetic inputs, sympathetic neural activity, which inhibits normal digestion (and hence nourishment of the body), would be unopposed. So important are para-sympathetic controls that some of the parasympathetic neurons are actually located in the walls of the digestive organs, in the so-called intrinsic plexuses. Thus, even if all extrinsic controls are severed, the intrinsic mechanisms can still maintain this crucial body function. The role the digestive system plays for the nervous system is the same one it offers to all body systems—it sees that ingested foodstuff gets digested and loaded into the blood for cell use.

CLINICAL CONNECTIONS

Nervous System

Case Study: On arrival at Holyoke Hospital, Jimmy Chin, a 10-year-old boy, is immobilized on a rigid stretcher so that he is unable to move his head or trunk. The para-medics report that when they found him some 50 feet from the bus, he was awake and alert, but crying and complaining that he couldn't "get up to find his mom" and that he had a "wicked headache." He has severe bruises on his lumbar region and head, and lacerations of his back and scalp. His blood pressure is low, body tempera-ture elevated (103°F), and his lower limbs paralyzed and insensitive to painful stimuli. Although still alert on arrival, Jimmy soon becomes drowsy and incoherent and begins to drift in and out of unconsciousness.

Jimmy is immediately scheduled for a CT scan, an operating room is reserved, and an intravenous infusion of dexamethasone, an anti-inflammatory steroid, is started. Relative to Jimmy's condition:

1. Why were his head and torso immobilized for transport to the hospital?

2. What do his worsening neurological signs (drowsiness, incoherence, etc.) probably indicate? (Relate this to the type of surgery that will be performed.)

3. Why was dexamethasone ordered?

4. The CT scan shows no permanent damage to Jimmy's spinal cord, so how can his lower limb paralysis be explained?

Some 24 hours after his brain surgery, Jimmy's lower limb paralysis ends, but then as reflex activity returns, his body begins to go into uncontrolled flexor spasms, and he is incontinent. When examined, he is sweating intensely (diaphoretic) and his blood pressure is way above normal.

5. What is this condition called and what precipitates it?

6. How does Jimmy's excessively high blood pressure put him at risk?

(Answers in Appendix F)

RELATED CLINICAL TERMS

Achalasia (ak"ah-la'ze-ah; *a* = not, without; *chalas* = relaxed) A condition in which the esophagus loses its ability to propel food inferiorly and the sphincter at the esophagus-stomach junction does not relax with swallowing, blocking food passage into the stomach; the distal esophagus becomes enormously dilated and vomiting is common. Cause is unknown; may reflect a congenital deficiency of parasympathetic ganglionic neurons in the esophageal wall, or may be a secondary result of fibrosis of the esophagus (as in scleroderma).

Atonic bladder (ah-ton'ik; *a* = without; *ton* = tone, tension) A condition in which the urinary bladder becomes flaccid and overfills, allowing urine to dribble through the sphincters; results from temporary loss of the micturition reflex following spinal cord injury.

Horner's syndrome A condition due to damage to the superior sympathetic trunk on one side of the body; the affected person exhibits drooping of the upper eyelid (ptosis) and constricted pupil, and does not sweat on the affected side of the head.

Vagotomy (va-got'o-me) Cutting or severing of the vagus nerve, often done to decrease secretion of gastric juice in those with peptic ulcers.

CHAPTER SUMMARY

Media study tools that could provide you additional help in reviewing specific key topics of Chapter 14 are referenced below. **SP** = *Study Partner;* **IP** = *Interactive Physiology.*

1. The autonomic nervous system is the motor division of the PNS that controls visceral activities, with the goal of maintaining internal homeostasis.

Introduction to the Autonomic Nervous System (pp. 513–515)

Comparison of the Somatic and Autonomic Nervous Systems (pp. 513–515)

1. The somatic (voluntary) nervous system provides motor fibers to skeletal muscles. The autonomic (involuntary or visceral motor) nervous system provides motor fibers to smooth and cardiac muscles and glands.

SP Exercise: Chapter 14, The Somatic and Autonomic Nervous Systems.

2. In the somatic division, a single motor neuron forms the efferent pathway from the CNS to the effectors. The efferent pathway of the autonomic division consists of a two-neuron chain: the preganglionic neuron in the CNS and the ganglionic neuron in a ganglion.

3. Acetylcholine, the neurotransmitter of somatic motor neurons, is stimulatory to skeletal muscle fibers. Neurotransmitters released by autonomic motor neurons (acetylcholine and norepinephrine) may cause excitation or inhibition.

IP Nervous System II CD-ROM; Topic: Synaptic Transmission, pages 8–11.

The Divisions of the Autonomic Nervous System (p. 515)

4. The autonomic nervous system consists of two divisions, the parasympathetic and sympathetic, which normally exert antagonistic effects on many of the same target organs.

5. The parasympathetic division (the resting-digesting system) conserves body energy and maintains body activities at basal levels.

6. Parasympathetic effects include pupillary constriction, glandular secretion, increased digestive tract mobility, and smooth muscle activity leading to elimination of feces and urine.

7. The sympathetic division activates the body under conditions of emergency and is called the fight-or-flight system.

8. Sympathetic responses include dilated pupils, increased heart and respiratory rates, increased blood pressure, dilation of the bronchioles of the lungs, increased blood glucose levels, and sweating. During exercise, sympathetic vasoconstriction shunts blood from the skin and digestive viscera to the heart, brain, and skeletal muscles.

SP Exercise: Chapter 13, Parasympathetic and Sympathetic Division of the ANS.

Anatomy of the Autonomic Nervous System (pp. 516–523)

Parasympathetic (Craniosacral) Division (pp. 517–518)

1. Parasympathetic preganglionic neurons arise from the brain stem and from the sacral (S_2–S_4) region of the cord.

2. Preganglionic fibers synapse with ganglionic neurons in intramural or terminal ganglia located in or close to their effector organs. Preganglionic fibers are long; post-ganglionic fibers are short.

3. Cranial fibers arise in the brain stem nuclei of cranial nerves III, VII, IX, and X and synapse in ganglia of the head, thorax, and abdomen. The vagus nerve serves virtually all organs of the thoracic and abdominal cavities.

4. Sacral fibers (S_2–S_4) issue from the lateral region of the cord and form pelvic splanchnic nerves that innervate the pelvic viscera. The preganglionic axons do not travel within rami communicantes or spinal nerves.

Sympathetic (Thoracolumbar) Division (pp. 518–522)

5. Preganglionic sympathetic neurons arise from the lateral horn of the spinal cord from the level of T_1 to L_2.

6. Preganglionic axons leave the cord via white rami communicantes and enter the chain (paravertebral) ganglia in the sympathetic trunk. An axon may synapse in a chain ganglion at the same or at a different level, or it may issue from the sympathetic chain without synapsing. Preganglionic fibers are short; postganglionic fibers are long.

7. When the synapse occurs in a chain ganglion, the postganglionic fiber may enter the spinal nerve ramus via the gray ramus communicans to travel to the body periphery. Postganglionic fibers issuing from the cervical ganglia also serve visceral organs and blood vessels of the head, neck, and thorax.

8. When synapses do not occur in the chain ganglia, the preganglionic fibers form splanchnic nerves (thoracic, lum-

bar, and sacral). Most splanchnic nerve fibers synapse in collateral ganglia, and the postganglionic fibers serve the abdominal viscera. Some splanchnic nerve fibers synapse with cells of the adrenal medulla.

Visceral Reflexes (pp. 522–523)

9. Visceral reflex arcs have the same components as somatic reflexes.

10. Cell bodies of visceral sensory neurons are located in dorsal root ganglia, sensory ganglia of cranial nerves, or autonomic ganglia. Visceral afferents are found in spinal nerves and in virtually all autonomic nerves.

Physiology of the Autonomic Nervous System (pp. 523–528)

Neurotransmitters and Receptors (pp. 523–524)

1. Two major neurotransmitters, acetylcholine (ACh) and norepinephrine (NE), are released by autonomic motor neurons. On the basis of the neurotransmitter released, the fibers are classified as cholinergic or adrenergic.

SP Exercise: Chapter 15, Norepinephrine and Its Receptors; Case Study: Cardiac Arrythmia.

2. ACh is released by all preganglionic fibers and all parasympathetic postganglionic fibers. NE is released by all sympathetic postganglionic fibers except those serving the sweat glands of the skin, some blood vessels within skeletal muscles, and the external genitalia (those fibers secrete ACh).

3. Neurotransmitter effects depend on the receptors to which the transmitter binds. Cholinergic (ACh) receptors are classified as nicotinic and muscarinic. Adrenergic (NE) receptors are classified as α_1 and α_2 and β_1, β_2, and β_3.

IP Nervous System II CD-ROM; Topic: Synaptic Transmission, pages 8-11, 14.

SP Exercise: Chapter 14, Synaptic Transmission.

The Effects of Drugs (pp. 524–525)

4. Drugs that mimic, enhance, or inhibit the action of ANS neurotransmitters are used to treat conditions caused by excessive, inadequate, or inappropriate ANS functioning. Some drugs bind with only one receptor subtype, allowing specific ANS-mediated activities to be enhanced or blocked.

SP Case Study: Cardiac Arrythmia.

Interactions of the Autonomic Divisions (pp. 525–527)

5. Most visceral organs are innervated by both divisions; they interact in various ways but usually exert a dynamic antagonism. Antagonistic interactions mainly involve the heart, respiratory system, and gastrointestinal organs.

Sympathetic activity increases heart and respiratory system activity and depresses gastrointestinal activity. Parasympathetic activity reverses these effects.

6. Most blood vessels are innervated only by sympathetic fibers and exhibit vasomotor tone. Parasympathetic activity dominates the heart and muscles of the gastrointestinal tract (which normally exhibit parasympathetic tone) and glands.

7. The two ANS divisions exert cooperative effects on the external genitalia.

8. Roles unique to the sympathetic division are blood pressure regulation, shunting of blood in the vascular system, thermoregulatory responses, stimulation of renin release by the kidneys, and metabolic effects.

9. Activation of the sympathetic division causes widespread, long-lasting mobilization of the fight-or-flight response. Parasympathetic effects are highly localized and short lived.

Control of Autonomic Functioning (pp. 527–528)

10. Autonomic function is controlled at several levels: (1) Reflex activity is mediated by the spinal cord and brain stem (particularly medullary) centers; (2) hypothalamic integration centers interact with both higher and lower centers to orchestrate autonomic, somatic, and endocrine responses; and (3) cortical centers influence autonomic functioning via connections with the limbic system; conscious controls of autonomic function are rare but possible, as illustrated by biofeedback training.

SP Case Study: Cardiac Arrythmia.

Homeostatic Imbalances of the Autonomic Nervous System (pp. 528–529)

1. Most autonomic disorders reflect problems with smooth muscle control. Abnormalities in vascular control, such as occur in hypertension, Raynaud's disease, and the mass reflex reaction, are most devastating.

Developmental Aspects of the Autonomic Nervous System (p. 529)

1. Preganglionic neurons develop from the neural tube; ganglionic neurons develop from the embryonic neural crest.

2. Hirschsprung's disease involves functional blockage of the large intestine due to failure of parasympathetic innervation.

3. The efficiency of the autonomic nervous system declines in old age, as reflected by decreased glandular secretory activity, decreased gastrointestinal motility, and slowed sympathetic vasomotor responses to changes in position.

REVIEW QUESTIONS

Multiple Choice/Matching

1. All of the following characterize the ANS except (a) two-neuron efferent chain, (b) presence of nerve cell bodies in the CNS, (c) presence of nerve cell bodies in the ganglia, (d) innervation of skeletal muscles.

2. Relate each of the following terms or phrases to either the sympathetic (S) or parasympathetic (P) division of the autonomic nervous system:

_____ **(1)** short preganglionic, long postganglionic fibers

_____ **(2)** intramural ganglia

_____ **(3)** craniosacral outflow

_____ **(4)** adrenergic fibers

_____ **(5)** cervical ganglia

_____ **(6)** otic and ciliary ganglia

_____ **(7)** more specific control

_____ **(8)** increases heart rate, respiratory rate, and blood pressure

_____ **(9)** increases gastric motility and secretion of lacrimal, salivary, and digestive juices

_____ (10) innervates blood vessels
_____ (11) most active when you are swinging in a hammock
_____ (12) active when you are running in the Boston Marathon

3. Preganglionic neurons develop from (a) neural crest cells, (b) neural tube cells, (c) alar plate cells, (d) endoderm.

4. The white rami communicantes contain what kind of fibers? (a) preganglionic parasympathetic, (b) postganglionic parasympathetic, (c) preganglionic sympathetic, (d) postganglionic sympathetic.

5. Prevertebral sympathetic ganglia are involved with the innervation of the (a) abdominal organs, (b) thoracic organs, (c) head, (d) arrector pill, (e) all of these.

Short Answer Essay Questions

6. Briefly explain why the following terms are sometimes used to refer to the autonomic nervous system: involuntary nervous system and emotional-visceral system.

7. Describe the anatomical relationship of the white and gray rami to the spinal nerve, and indicate the kind of fibers found in each ramus type.

8. Indicate the results of sympathetic activation of the following structures: sweat glands, eye pupils, adrenal medulla, heart, lungs, liver, blood vessels of vigorously working skeletal muscles, blood vessels of digestive viscera, salivary glands.

9. Which of the effects listed in response to question 6 would be reversed by parasympathetic activity?

10. Which ANS fibers release acetylcholine? Which release norepinephrine?

11. Describe the meaning and importance of sympathetic tone and parasympathetic tone.

12. List the receptor subtypes for ACh and NE, and indicate the major sites where each type is found.

13. What area of the brain is most directly involved in mediating autonomic reflexes?

14. Describe the importance of the hypothalamus in controlling the autonomic nervous system.

15. Describe the basis and uses of biofeedback training.

16. What manifestations of decreased autonomic nervous system efficiency are seen in elderly individuals?

17. Ganglionic neurons were initially called postganglionic neurons. Why is this a misnomer?

Critical Thinking and Clinical Application Questions

1. Mr. Johnson has been suffering from functional urinary retention and a hypoactive urinary bladder. Bethanechol, a drug that mimics acetylcholine's autonomic effects, is prescribed to manage his problem. First explain the rationale for prescribing bethanechol, and then predict which of the following adverse effects Mr. Johnson might experience while taking this drug (select all that apply): dizziness, low blood pressure, deficient tear formation, wheezing, increased mucus production in bronchi, deficient salivation, diarrhea, cramping, excessive sweating, undesirable erection of penis.

2. Mr. Jake was admitted to the hospital with excruciating pain in his left shoulder and arm. He was found to have suffered a heart attack. Explain the phenomenon of referred pain as exhibited by Mr. Jake.

3. A 32-year-old woman complains that she has been experiencing aching pains in the medial two fingers of both hands and that during such episodes, the fingers become blanched and then blue. Her history is taken, and it is noted that she is a heavy smoker. The physician advises her that she must stop smoking and states that he will not prescribe any medication until she has discontinued smoking for a month. What is this woman's problem, and why was she told to stop smoking?

4. Two-year-old Jimmy has a swollen abdomen and is continuously constipated. On examination, a mass was felt over the descending colon, and the X ray showed the colon to be greatly distended in that region. What do you think is wrong with Jimmy, and how does this relate to the innervation of the colon?

5. As the aroma of freshly brewed coffee drifted by dozing Henry's nose, his mouth started to water and his stomach began to rumble. Explain his reactions in terms of ANS activity.

NEURAL INTEGRATION 15

Sensory Integration: From Reception to Perception (pp. 536–541)

1. List the three levels of sensory integration.
2. Describe receptor and generator potentials and define adaptation.
3. Describe the role of receptors in sensory processing.
4. List the major ascending spinal cord tracts, and describe each in terms of its termination and function.
5. Compare and contrast specific and nonspecific ascending somatosensory pathways.
6. Describe the main features of perceptual processing of sensory inputs.

Motor Integration: From Intention to Effect (pp. 541–546)

7. Describe the levels of the motor control hierarchy.
8. Define central pattern generator and command neuron.
9. List the major descending spinal cord tracts and describe each in terms of its origin and function.
10. Compare the roles of the direct and indirect systems in controlling motor activity.
11. Explain the function of the cerebellum and basal nuclei in somatic sensory and motor integration.
12. Describe symptoms of cerebellar and basal nuclear disease.

Higher Mental Functions (pp. 546–554)

13. Define EEG and distinguish between alpha, beta, theta, and delta brain waves.
14. Compare and contrast the events and importance of slow-wave and REM sleep, and indicate how their patterns change through life.
15. Describe consciousness clinically.
16. Compare and contrast the stages and categories of memory.
17. Indicate the major brain structures believed to be involved in fact and skill memories and describe their relative roles.

A lthough we seldom look at nervous system functioning in its entirety, all of its activities —sensory, integrative, and motor alike—go on simultaneously. Consider, for example, a leisurely drive with a friend. Sensory inputs, such as what you see in your visual field and the pressure of the pedals on your feet, help keep your CNS constantly informed about what is going on both inside and outside your body. Your muscles respond to CNS commands to brake or accelerate, and all the while, you carry on an animated conversation with your passenger. This chapter analyzes neural events that underlie such daily activities.

The immediacy of sensory experiences makes them somehow more understandable than other types of neural functions, so they are good starting points for our studies of neural integration. After considering sensory integration, we will examine how some of the motor activities we take so much for granted (like walking and maintaining our upright posture) are accomplished. The final section of this chapter looks at how we think and remember— nebulous topics, to be sure, but fertile ground for speculation.

SENSORY INTEGRATION: FROM RECEPTION TO PERCEPTION

Our survival depends not only on **sensation** (awareness of changes in the internal and external environments) but also on **perception** (conscious interpretation of those stimuli). For example, a pebble kicked up into my shoe causes the *sensation* of localized deep pressure, but my *perception* of it is an awareness of discomfort. Perception in turn determines how we will respond to stimuli. In the case of the pebble-in-my-shoe example, I'm taking off my shoe to get rid of the pesky pebble in a hurry.

In this section, we will follow sensory inputs from the sensory receptors to the somatosensory cortex and examine the role of the neural structures at each level of the pathway.

General Organization of the Somatosensory System

The **somatosensory system,** or that part of the sensory system serving the body wall and limbs, receives inputs from exteroceptors, proprioceptors, and interoceptors; consequently, it transmits information about several different sensory modalities.

There are three main levels of neural integration in the somatosensory (or any sensory) system:

① **Receptor level,** corresponding to the sensor receptors.

② **Circuit level,** corresponding to the ascending pathways.

③ **Perceptual level,** corresponding to the neuronal circuits in the cerebral cortex (see Figure 15.1). Although sensory input is generally relayed toward the head, it is also processed along the way.

Processing at the Receptor Level

Information about our external and internal worlds presents itself as different forms of energy: sound, mechanical, chemical, and so on. The sensory receptors are specialized to respond to these energetic stimuli. However, before they can communicate with other neurons, they must translate the infor-

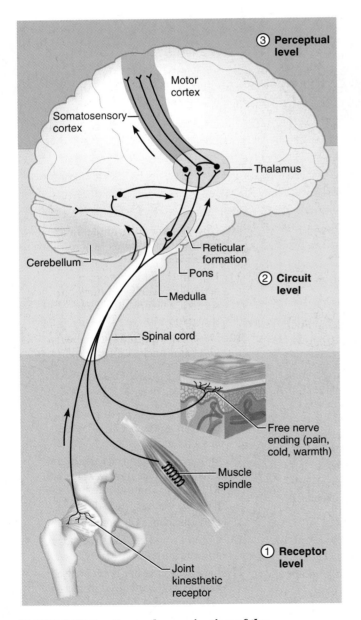

FIGURE 15.1 *General organization of the somatosensory system.* The three basic levels of neural integration are ① the receptor level, ② the circuit level, and ③ the perceptual level. The circuit level involves all CNS centers except for the somatosensory cortex in the cerebrum.

mation of their stimulus into nerve impulses, the universal neural language. The process by which a stimulus is transformed into an electrical response at a receptor is called **transduction.** Let's look at how this occurs.

Transduction by Sensory Receptors As the receptor absorbs stimulus energy, changes in the permeability of the receptor membrane occur that open or close ion channels. The result is a local graded potential, called a **receptor potential,** which is essentially the same as an EPSP (p. 410) generated at a postsynaptic membrane in response to binding of neurotransmitters. In both cases, one type of ion channel is opened that allows ion fluxes (usually of sodium and potassium) across the membrane, and summation of local potentials occurs. Depolarization of the afferent fiber is called a **generator potential** if the receptor is a separate cell (like a hair cell of the ear's hearing receptor) that depolarizes and releases a neurotransmitter. The neurotransmitter, *in turn,* excites the associated afferent neuron.

If the receptor potential is at or above threshold strength when it reaches the *voltage-gated* sodium ion channels on the axon (these are usually just proximal to the receptor membrane, often at the first node of Ranvier), those sodium channels will be opened and an action potential will be generated and propagated to the CNS.

Nerve impulse transmission to the CNS will continue as long as a threshold stimulus is applied. Since the strength of the stimulus is encoded in the frequency of impulses sent, stronger stimuli cause more impulses per second to reach the CNS, which enables the integrative centers to determine the intensity of stimuli acting on the body.

Adaptation of Sensory Receptors Although receptor potentials are graded potentials that vary with stimulus intensity and can summate, an unusual phenomenon called **adaptation** occurs in certain sensory receptors when they are subjected to an unchanging stimulus. The receptor membranes become less responsive with time and, as a result, the receptor potentials decline in frequency, or stop completely. Some receptors, such as most of those responding to pressure, touch, and smell, adapt rapidly. This explains why we are not usually aware (after a short period) of the feel of our clothing against our skin. However, some receptors adapt slowly, and others do not adapt at all. Pain receptors and proprioceptors, for example, respond more or less continuously to threshold stimuli. Since pain typically warns of tissue damage and proprioceptive sense is essential to balance and coordinated skeletal muscle activity, the nonadaptability of these receptors is important. Slowly adapting receptors include

Merkel's discs, Ruffini's corpuscles, and some interoceptors that respond to changing levels of chemicals in the blood.

Processing at the Circuit Level

Neuronal Composition of the Sensory Pathway to the Brain The ascending pathways conduct sensory impulses upward, typically through chains of three successive neurons (first-, second-, and third-order neurons) to various areas of the brain (Figure 15.2). (Note that both second- and third-order neurons are interneurons.)

- **First-order neurons,** whose cell bodies reside in a ganglion (dorsal root or cranial), conduct impulses from the cutaneous receptors of the skin and from proprioceptors to the spinal cord or brain stem, where they synapse with second-order neurons. Impulses from the facial area are transmitted by cranial nerves; spinal nerves conduct somatic sensory impulses from the rest of the body to the CNS.

- **Second-order neurons,** whose cell bodies reside in the dorsal horn of the spinal cord or in medullary nuclei, transmit impulses to the thalamus or to the cerebellum where they synapse.

- **Third-order neurons** are located in the thalamus and conduct impulses to the somatosensory cortex of the cerebrum. (There are no third-order neurons in the cerebellum.)

The Main Ascending Pathways The central processes of first-order neurons branch diffusely as they enter the spinal cord or medulla. Some branches take part in local spinal cord reflexes, directly initiating motor activities. Others synapse with second-order neurons in the cord or continue upward to synapse in medullary nuclei. Small-diameter pain fibers synapse with *substantia gelatinosa neurons* in the dorsal horn, whereas the large myelinated fibers from pressure and touch receptors make collateral synapses with interneurons in the dorsal horns that allow pain inputs to be squelched at the circuit level (see *A Closer Look*, pp. 478–479).

In general, somatosensory information is conveyed along three main pathways on each side of the spinal cord. Two of these pathways (the *nonspecific* and *specific ascending pathways*) transmit impulses to the sensory cortex. Collectively the inputs of these sister tracts provide us with what is referred to as *discriminative touch* and *conscious proprioception.* Both of these pathways cross over (decussate)—the first in the spinal cord and the second in the medulla.

The third pathway (consisting of the *spinocerebellar tracts*) tracks to the cerebellum, and hence does not contribute to sensory perception. Let's examine these pathways more closely.

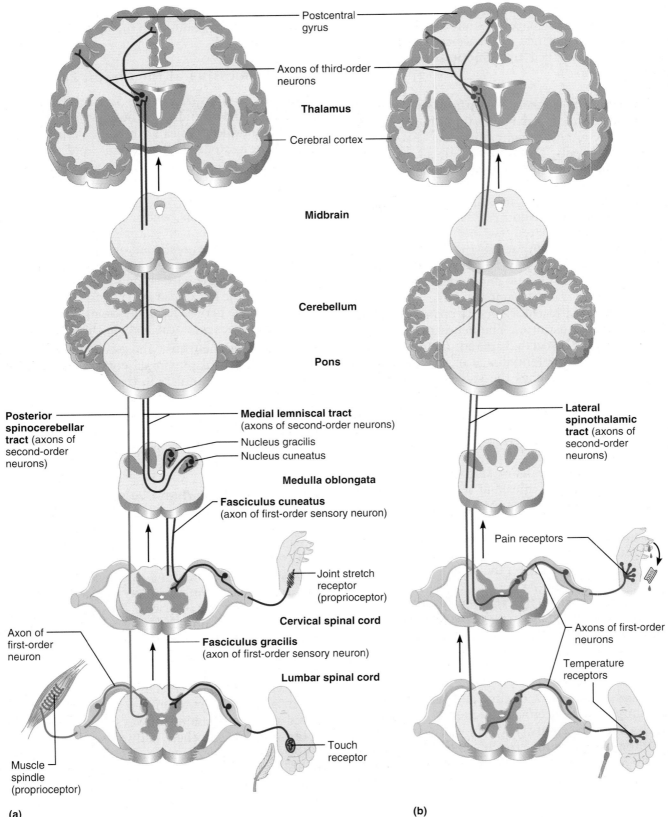

FIGURE 15.2 *Pathways of selected ascending spinal cord tracts.* (**a**) Right: Specific ascending pathways for discriminative touch and conscious proprioception within the fasciculus gracilis and fasciculus cuneatus tracts and their continuation, the medial lemniscal tracts. Left: The posterior spinocerebellar tract (tract extends only to cerebellum). (**b**) Nonspecific ascending pathways for pain, temperature, and crude touch within the lateral spinothalamic tract of the anterolateral pathways. The entire ascending pathway is shown in each case.

1. Nonspecific ascending pathways. The evolutionarily older **nonspecific ascending pathways** are so named because they receive inputs from many different types of sensory receptors and make multiple synapses in the brain stem. These pathways, also called the **anterolateral pathways** (because they are located in the anterior and lateral white columns of the spinal cord), are largely formed by the **lateral** and **anterior spinothalamic tracts** (see Table 15.1 and Figure 15.2b). Crossover of fibers occurs in the spinal cord.

Most of these fibers transmit pain, temperature, and coarse touch impulses, sensations that we are aware of but have difficulty localizing precisely on the body surface. After crossing over in the spinal cord, spinothalamic tract fibers synapse in several thalamic nuclei, but en route they give off numerous branches to the reticular formation. The reticular neurons, in turn, form synapses with most parts of the brain. Because a given neuron in the pathway may be receiving inputs initiated by several different types of stimuli, the information sent to "headquarters" is fairly general and nondiscriminatory, such as "something is happening but can't really tell you what."

The nonspecific somatosensory system is heavily involved in emotional aspects of perception (pleasure, aversion, arousal, and pain perception). It also participates in certain higher-level motor reflexes and *orienting responses* (such as turning the head toward a novel stimulus).

2. Specific ascending pathways. Also called the **lemniscal** (lem-nis′kul; *lemnisc* = a ribbon) **system,** the **specific ascending pathways** mediate precise, straight-through transmission of inputs from a single type (or a few related types) of sensory receptor that *can* be localized precisely on the body surface. These pathways are formed by the paired tracts of the **dorsal white column—fasciculus cuneatus** and **fasciculus gracilis—**of the spinal cord and the **medial lemniscal tracts.** The latter arise in the medulla and terminate in the *ventral posterior nuclei* of the thalamus (see Table 15.1 and Figure 15.2a). From the thalamus, impulses are forwarded to specific areas of the somatosensory cortex. The trigeminal nerve (cranial nerve V) also feeds into these pathways. The specific pathways send information about discriminative touch (tactile discrimination), pressure, vibration, and conscious proprioception (limb and joint position) to the cortex. These pathways send some fibers to the reticular nuclei, so they also contribute to the arousal mechanism.

The specific and nonspecific divisions of the somatosensory system are *parallel pathways* that are activated simultaneously, and their interactions with each other and the cerebral cortex are numerous and overlapping. However, parallel pathways offer advantages in that they (1) add richness to our perceptions by allowing the same information to be handled in different ways, and (2) provide insurance—if one pathway is damaged, the other still provides some of the same inputs.

3. Spinocerebellar tracts. The last pair of ascending pathways, the **anterior** and **posterior spinocerebellar tracts,** convey information from proprioceptors (muscle or tendon stretch) to the cerebellum, which uses this information to coordinate skeletal muscle activity. Since the somatosensory cortex is not their terminus, these pathways do *not* contribute to conscious sensation. The fibers of the spinocerebellar pathways either do not decussate or else cross over twice (thus "undoing" the decussation).

These ascending pathways and the spinal cord fiber tracts contributing to them are described in more detail in Table 15.1. Four of these tracts are illustrated in Figure 15.2. Notice that only the fasciculi cuneatus and gracilis (specific ascending pathway) are formed by the axons of primary sensory (first-order) neurons. The other ascending tracts described in Table 15.1 are formed by axons of second-order neurons in the spinal cord with which the transmitting sensory neurons synapse.

Recall that our ability to identify and appreciate the kind of sensation being transmitted depends on the specific location of the target neurons in the cerebral sensory cortex, with which the ascending pathway fibers synapse, and not on the nature of the message, which is *always* action potentials. Each sensory nerve fiber is analogous to a "labeled line" in a telephone system that tells the brain "who" is calling, whether it is a taste bud or a pressure receptor. Furthermore, when a sensory neuron is excited, the brain interprets its activity as a specific sensory modality, regardless of how the receptor is activated. For example, if we press a pressure receptor in the index fingertip, jolt it with an electrical shock, or electrically stimulate the area of the somatosensory cortex that recognizes it, the result will be the same: We will perceive deep touch or pressure and will interpret it as coming from the index fingertip. This phenomenon, by which the brain refers sensations to their usual point of stimulation, is called **projection.**

Processing at the Perceptual Level

Perception, the final stage of sensory processing, allows us to be aware of stimuli and to discriminate their characteristics. As sensory information reaches the thalamus, origins of the sensory input are roughly localized, and their modalities are perceived in a crude sense. But full appreciation of their qualities and sharp localization awaits the arrival of the input at the somatosensory cortex.

TABLE 15.1	Major Ascending (Sensory) Pathways and Spinal Cord Tracts			
Spinal cord tract	*Location (funiculus)*	*Origin*	*Termination*	*Function*
Specific Ascending (Lemniscal) Pathways				
Fasciculus cuneatus and fasciculus gracilis (dorsal white column)	Posterior	Central axons of sensory (first-order) neurons enter dorsal root of the spinal cord and branch; branches enter posterior white column on same side without synapsing	By synapse with second-order neurons in nucleus cuneatus and nucleus gracilis in medulla; fibers of medullary neurons cross over and ascend in medial lemniscal tracts to thalamus, where they synapse with third-order neurons; thalamic neurons then transmit impulses to somatosensory cortex	Both tracts transmit sensory impulses from general sensory receptors of skin and proprioceptors, which are interpreted as discriminative touch, pressure, and "body sense" (limb and joint position) in opposite somatosensory cortex; cuneatus transmits afferent impulses from upper limbs, upper trunk, and neck; it is not present in spinal cord below level of T_6; gracilis carries impulses from lower limbs and inferior body trunk
Nonspecific Ascending (Anterolateral) Pathways				
Lateral spinothalamic	Lateral	Association (second-order) neurons of posterior horn; fibers cross to opposite side before ascending	By synapse with third-order neurons in thalamus; impulses then conveyed to somatosensory cortex by thalamic neurons	Transmits impulses concerned with pain and temperature to opposite side of brain; eventually interpreted in somatosensory cortex
Anterior spinothalamic	Anterior	Association neurons in posterior horns; fibers cross to opposite side before ascending	By synapse with third-order neurons in thalamus; impulses eventually conveyed to somato-sensory cortex by thalamic neurons	Transmits impulses concerned with crude touch and pressure to opposite side of brain for interpretation by somatosensory cortex
Spinocerebellar Pathways				
Posterior spinocerebellar*	Lateral (posterior part)	Association (second-order) neurons in posterior horn on same side of cord; fibers ascend without crossing	By synapse in cerebellum	Transmits impulses from trunk and lower limb proprioceptors on one side of body to same side of cerebellum; subconscious proprioception
Anterior spinocerebellar*	Lateral (anterior part)	Association (second-order) neurons of posterior horn; contains crossed fibers that cross back to the opposite side in the pons	By synapse in cerebellum	Transmits impulses from the trunk and lower limb on the same side of body to cerebellum; subconscious proprioception

*The posterior and anterior spinocerebellar tracts carry information from the lower limbs and trunk only. The corresponding tracts for the upper limb and neck (rostral spinocerebellar and others) are beyond the scope of this book.

The thalamus projects fibers (sorted out by sensation type) to the primary somatosensory cortex and to sensory association areas, first those restricted to the same modality and then to those considering more than one. This input flow allows parallel processing of the various somatosensory inputs, and the result is an internal, conscious image of the stimulus.

Sensory input generally evokes a behavioral response. In human beings, however, a response is not obligatory. When cortical processing produces a conscious image of the stimulus, we can either act or not act on the basis of the image's information. The choice we make depends, of course, on our past experience with similar sensory inputs.

Main Aspects of Sensory Perception

The main aspects of sensory perception are detection, magnitude estimation, spatial discrimination, feature abstraction, quality discrimination, and pattern recognition. We examine each of these next.

Perceptual Detection Detecting that a stimulus has occurred is the simplest level of perception. As a general rule, several receptor impulses must be summated for **perceptual detection** to occur.

Magnitude Estimation The ability to detect how *much* of the stimulus is acting on the body is **magnitude estimation.** Because of frequency coding, perception increases as stimulus intensity increases.

Spatial Discrimination Spatial discrimination allows us to identify the site or pattern of stimulation. A common tool for studying this quality in the laboratory is the **two-point discrimination test.** The test determines how close together two points on the skin can be and still be perceived as two points by the somatosensory cortex when stimulated, rather than as one. This test provides a crude map of the density of tactile receptors in the various regions of the skin. The distance between perceived points varies from less than 5 mm on highly sensitive body areas (tip of the tongue and of the index finger) to more than 50 mm on less sensitive areas (back and posterior neck).

Feature Abstraction Sensation usually involves an interplay of several stimulus properties. For example, one touch tells us that velvet is warm, compressible, and smooth but not completely continuous. This implies that perception is tuned to coordinated sets of several stimulus properties, called *features*. The mechanism by which a neuron or circuit is tuned to one feature in preference to others is called **feature abstraction.** When we run our fingers over a slab of marble, we may notice first its coolness, then its hardness, then its smoothness—each a feature that contributes to our perception of the marble surface. Feature abstraction enables us to identify a substance or object that has specific texture or shape.

Quality Discrimination Each sensory modality has several **qualities,** or submodalities. For example, the submodalities of taste are the qualities sweet, salt, bitter, and sour. **Quality discrimination** is the ability to differentiate the submodalities of a particular sensation.

Quality discrimination may be analytic or synthetic. In **analytic discrimination,** each quality retains its individual nature. If we mix sugar and salt, analytic discrimination allows us to taste each quality; they do not merge into a new sensation. On the other hand, chocolate is a mixture of qualities (sweet, bitter, and perhaps some salt), but our perception of it is *synthesized* from the primary qualities and is distinct from them. **Synthetic discrimination** is also very important in vision. Our color photoreceptors respond chiefly to red, blue, and green wavelengths of light, yet we are able to see yellow, purple, and orange when different numbers of each type of color receptor are stimulated. Vision and olfaction use only synthetic discrimination.

Pattern Recognition Our ability to take in the scene around us and recognize a familiar pattern, an unfamiliar one, or one that has special significance for us is **pattern recognition.** For example, a figure made of dots may be recognized as a familiar face, and when we listen to music, we hear the melody, not just a string of notes.

MOTOR INTEGRATION: FROM INTENTION TO EFFECT

In the motor system, we have effectors (muscle fibers) instead of sensory receptors, descending efferent circuits instead of ascending afferent circuits, and motor behavior instead of perception. However, like sensory systems, the basic mechanisms of motor systems operate at three levels. These three levels form a hierarchy of controls of motor activity.

Levels of Motor Control

The idea of multiple and successive levels of motor control, or a **motor hierarchy,** was proposed in 1873 by the English neurologist John Jackson. According to Jackson, the reflex activity of the spinal cord and brain stem is the most elementary level of motor control. The next is in the cerebellum, and the highest level is in the motor cortex of the cerebrum.

Modern research has revealed that information flow in the motor control hierarchy is somewhat different. The cerebral cortex is the instrument of volition, and is at the highest level of the conscious motor pathways, but it is *not* the ultimate planner and coordinator of complex motor activities. The cerebellum and basal nuclei (ganglia) play this role and are therefore put at the top of the functional motor control hierarchy. Motor control exerted by lower levels is mediated by *reflex arcs* in some cases, but complex motor behavior, such as walking and swimming, appears to depend on fixed-action patterns. **Fixed-action patterns** are stereotyped sequential motor actions triggered internally or by appropriate environmental stimuli. Once triggered, the entire

Which control level is the level of volition?

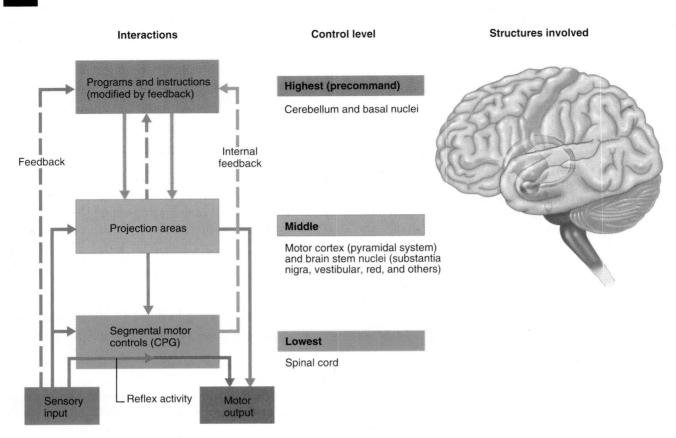

FIGURE 15.3 *Hierarchy of motor control.*

sequence is released in an all-or-none fashion. Currently, we define three levels of motor control: the *segmental level,* the *projection level,* and the *programs/instructions level* (Figure 15.3) as described in detail next.

The Segmental Level

The lowest level of the motor hierarchy, the **segmental level,** consists of the **segmental circuits of the spinal cord.** A segmental circuit activates a network of anterior horn neurons of a single cord segment, causing them to stimulate a specific group of muscle fibers. Those circuits that control locomotion and other specific and oft-repeated motor activity are called **central pattern generators (CPGs).**

Most of our information about motor control comes from locomotion studies done over the last century. Early investigators found that segments of cat nerve cords, freed of all nerve connections and placed in a physiological solution, generated rhythmic bursts of motor impulses that would excite extensors and flexors in the proper sequence and rhythm to produce normal stepping movements in an intact ani-

mal. This was surprising—a nerve cord, sitting in a dish with *no* sensory input, was continuing to "regulate locomotion." That revelation led to the concept of the CPG. The mechanism of the CPG is not yet clear, but it probably involves spinal cord neurons arranged in reverberating circuits (see Figure 11.24e, p. 420). It is assumed that human locomotion is controlled in much the same way and that the pattern for walking is imprinted in our spinal cords. It is inherited, not learned, and is present even in infants.

What controls a CPG (or other segmental circuit) so that it produces a variety of purposeful movements? Clearly, we are not locomoting all the time, and when we are it may be an amble, a jog, or a flat-out sprint. The most accepted theory is that the segmental apparatus is turned on or off and modulated by a "switch" located in the higher neural centers. This switch is apparently made up of brain stem interneurons called *command neurons* that form part of the projection level.

The Projection Level

The spinal cord is under the direct control of the **projection level** of motor control. The projection level consists of cortical motor areas that produce the direct

TABLE 15.2	Major Descending (Motor) Pathways and Spinal Cord Tracts			
Spinal cord tract	Location (funiculus)	Origin	Termination	Function
Direct (Pyramidal)				
Lateral corticospinal	Lateral	Pyramidal neurons of motor cortex of the cerebrum; decussate in pyramids of medulla	Anterior horn interneurons that influence motor neurons and occasionally directly with anterior horn motor neurons	Transmits motor impulses from cerebrum to spinal cord motor neurons (which activate skeletal muscles on opposite side of body); voluntary motor tract
Anterior corticospinal	Anterior	Pyramidal neurons of motor cortex; fibers cross over at the spinal cord level	Anterior horn (as above)	Same as lateral corticospinal tract
Indirect (Extrapyramidal) Pathways				
Tectospinal	Anterior	Superior colliculus of midbrain of brain stem (fibers cross to opposite side of cord)	Anterior horn (as above)	Transmits motor impulses from midbrain that are important for coordinated movement of head and eyes toward visual targets
Vestibulospinal	Anterior	Vestibular nuclei in medulla of brain stem (fibers descend without crossing)	Anterior horn (as above)	Transmits motor impulses that maintain muscle tone and activate ipsilateral limb and trunk extensor muscles and muscles that move head; in this way helps maintain balance during standing and moving
Rubrospinal	Lateral	Red nucleus of midbrain of brain stem (fibers cross to opposite side just inferior to the red nucleus)	Anterior horn (as above)	In experimental animals, transmits motor impulses concerned with muscle tone of distal limb muscles (mostly flexors) on opposite side of body; function in humans is controversial: may be vestigial with function largely assumed by corticospinal tracts
Reticulospinal (anterior, medial, and lateral)	Anterior and lateral	Reticular formation of brain stem (medial nuclear group of pons and medulla); both crossed and uncrossed fibers	Anterior horn (as above)	Transmits impulses concerned with muscle tone and many visceral motor functions; may control most unskilled movements

(pyramidal) system, and of brain stem motor areas that oversee the indirect (multineuronal) system. Axons of these neurons *project* to the spinal cord, where they help control reflex and fixed-action pattern activities and produce discrete voluntary movements. As previously noted, the projection level houses the command neurons that modify and control the segmental apparatus. In order to understand the role of this level, we must discuss the descending motor pathways that connect the projection and segmental levels.

Descending (Motor) Pathways and Tracts The descending tracts that deliver efferent impulses from the brain to the spinal cord are divided into two groups: (1) the *direct pathways* equivalent to the pyramidal tracts and (2) the *indirect pathways*, essentially all others. Motor pathways involve two neurons, referred to as the upper and lower motor neurons. The pyramidal cells of the motor cortex, as well as the neurons in subcortical motor nuclei that give rise to other descending motor pathways, are called **upper motor neurons.** The anterior horn motor neurons, which actually innervate the skeletal muscles (their effectors), are called **lower motor neurons.** These tracts are briefly overviewed here; most of the detailed information is provided in Table 15.2.

The Direct (Pyramidal) System The direct pathways originate with the pyramidal neurons located in the precentral gyri (plus some premotor area neurons of the frontal lobes). These neurons send impulses through the brain stem via the large **pyramidal (corticospinal) tracts** (see Table 15.2 and Figure 15.4a). The direct pathways are so called because their axons descend without synapsing from the pyramidal neurons to the spinal cord. There they synapse with interneurons or directly with anterior horn motor neurons. Stimulation of the anterior horn neurons activates the skeletal muscles with which they are associated. As the pyramidal tracts descend through the brain stem, they send collaterals to brain stem motor nuclei, basal nuclei, and the cerebellum. Part of the direct pathway called the **corticobulbar tracts** innervates the cranial nerve nuclei of the brain stem.

Although the axons of the cerebrospinal tracts act on *all* anterior horn cells (α and γ alike), they influence mainly the most lateral alpha motor neurons that control skeletal muscles of the distal extremities. Hence the direct pathway primarily regulates fast and fine (or skilled) movements such as those performed by the fingers, like threading a needle and writing.

Homeostatic Imbalance

Lesions of the pyramidal tracts in the cerebral cortex or the medullary pyramids result in *hypotonia,* some degree of weakness or spastic paralysis of the distal limb muscles such that it is difficult to release one's grip and all finger movements are slower and weaker. Other consequences include loss of the abdominal reflex on one side and appearance of Babinski's sign (see p. 507). ■

The Indirect (Extrapyramidal) System The indirect system includes brain stem motor nuclei and *all motor pathways except* the pyramidal pathways. It is the home of the command neurons mentioned earlier. These remaining descending tracts include the *rubrospinal* (roo″bro-spi′nal), *vestibulospinal, reticulospinal,* and *tectospinal tracts,* which originate in different subcortical motor nuclei of the brain stem. These tracts were formerly lumped together as the **extrapyramidal tracts,** or **system,** because their nuclei of origin were presumed to be "independent of" (extra to) the pyramidal tracts. This term is still widely used clinically; however, pyramidal tract neurons are now known to project to and influence the activity of most "extrapyramidal" nuclei, so modern anatomists prefer to use the term **indirect,** or **multineuronal pathways,** or even just the names of the individual motor pathways.

These motor pathways are very complex and multisynaptic. They are most involved in regulating (1) the axial muscles that maintain balance and posture, (2) the muscles controlling coarse movements of the proximal portions of the limbs, and (3) head, neck, and eye movements that act to follow objects in the visual field. Many of the activities controlled by subcortical motor nuclei are heavily dependent on reflex activity. One of these tracts, the rubrospinal tract, is illustrated in Figure 15.4b.

The **reticular nuclei,** which transmit via the descending **reticulospinal tracts** (see Table 15.2), have complex and contrasting functions. Overall, they maintain balance by varying the tone of postural muscles. The **vestibular nuclei** receive inputs from the equilibrium apparatus of the ears and from the cerebellum. Impulses conducted downward along the **vestibulospinal tracts** are important in controlling the segmental (spinal cord) apparatus during standing—that is, in supporting the body against gravity. The **red nuclei** (at least in other animals) send facilitating impulses via the **rubrospinal tracts** to motor neurons controlling flexor muscles, whereas the **superior colliculi** and the **tectospinal tracts** mediate head movements in response to visual stimuli. Interactions between these brain stem nuclei presumably control the CPGs of the cord during locomotion and other rhythmic activities, such as arm-swinging and scratching.

The **command neurons,** found in the brain stem nuclei, are higher-level interneurons that mediate many of these activities. That is, they function to start, stop, or modify the basic rhythm of CPGs of the spinal cord or of other segmental circuits.

Notice that the cerebral motor cortex can bypass the segmental apparatus of the spinal cord to directly activate the anterior horn motor neurons. Additionally, information continually feeds back from every motor system level. Projection motor pathways not only convey information to lower motor neurons, but also send a copy of that information (*internal feedback*) back to higher command levels, continually informing them of what is happening. The direct and indirect systems provide separate and parallel pathways for controlling the spinal cord, but these systems are interrelated at all levels.

The Programs and Instructions Level

Two other vast systems of brain neurons, located in the cerebellum and basal nuclei, regulate motor activity. They precisely start or stop movements, coordinate movements with posture, block unwanted movements, and monitor muscle tone. These systems, collectively called **precommand areas,** *control the outputs* of the cortex and brain stem motor centers and stand at the highest level of the motor

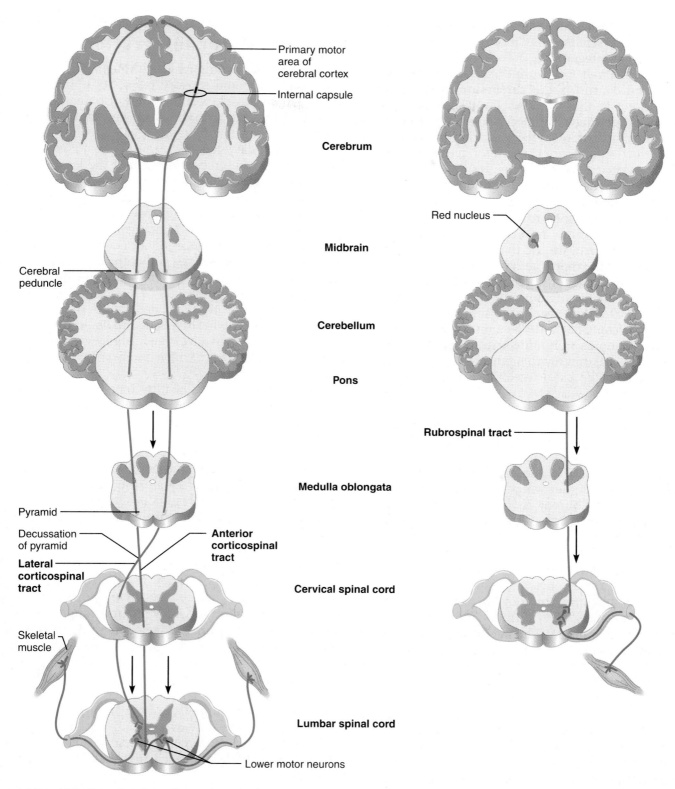

(a) Pyramidal (lateral and anterior corticospinal) tracts

(b) Rubrospinal tract (an extrapyramidal tract)

FIGURE 15.4 *Pathways of selected descending spinal cord tracts.* (**a**) Direct pathways, that is, the pyramidal tracts (lateral and anterior corticospinal tracts), carrying motor impulses to skeletal muscles. (**b**) The rubrospinal tract, one of the indirect or extrapyramidal tracts, helps regulate tone of muscles on opposite side of body.

hierarchy, the **programs and instructions level.** At this level, the sensory and motor systems finally merge and are integrated.

The key center for "on-line" sensorimotor integration and control is the **cerebellum.** Remember from Chapter 12 that the cerebellum is the ultimate target of ascending proprioceptor, tactile, equilibrium, and visual inputs—feedback that it needs for rapid correction of "errors" in motor activity. It also receives information from the motor cortex via branches from descending pyramidal tracts, and from various brain stem nuclei. Since the cerebellum lacks direct connections to the spinal cord, it acts through the projection areas of the brain stem on motor pathways and via the thalamus on the motor cortex to fine-tune motor activity.

Like the cerebellum, the **basal nuclei** are at the crossroads of many afferent and efferent pathways. However, they do not receive sensory fibers, nor do they send efferent fibers to the spinal cord. Instead, they receive inputs from *all* cortical areas and send their output back mainly to premotor and prefrontal cortical areas. The basal nuclei also have complex interconnections with each other as well as with brain stem neurons that can initiate different "motor programs." Hence, as compared to the cerebellum, the basal nuclei appear to be involved in more complex aspects of motor control. Under resting conditions, the basal nuclei inhibit various motor centers of the brain, but when the motor centers are released from inhibition, coordinated motions can begin. Cells in both the basal nuclei and the cerebellum discharge in advance of willed movements.

Once the motor association areas of your frontal lobes (Broca's area and others) have indicated their intent to perform a movement such as drumming your fingers, planning the movement occurs in the precommand areas. The primary motor cortex is quiet during this phase. The lateral portions of the cerebellar hemispheres (and the dentate nuclei) and the caudate and putamen nuclei of the basal nuclei are most involved in this unconscious planning. When you actually move your fingers, both the precommand areas and the primary motor cortex are active. The precommand areas provide the "orders," and then the motor cortex executes those orders by sending activating commands to the appropriate muscle groups. At the risk of oversimplifying, it appears that the cortex says, "I want to do this," and then lets the precommand areas take over to provide the proper timing and patterns to execute the movements desired. The programs of the precommand areas control the motor cortex and provide its readiness to initiate a voluntary act. The conscious cortex then chooses to act or not to act, but the groundwork has already been laid.

Homeostatic Imbalances of Motor Integration

The cerebellum is amazing. Although it contains complete motor and sensory maps, its injury produces neither muscle weakness nor disorders of perception. Cerebellar damage affects the same side of the body (produces *ipsilateral* signs and symptoms) and cerebellar disorders fall into three major groups, categorized by the precise location of the lesion: (1) disorders of synergy and muscle tone, (2) disturbances of equilibrium, and (3) speech disorders.

Synergy (sin'er-je; *synerg* = work together) is the coordination of agonist and antagonist muscles by the cerebellum to produce smooth, well-controlled movements. When synergy is disrupted, ataxia (ah-tak'se-ah) results. **Ataxia** sufferers exhibit slow, tentative, fumbling movements. They are unable to touch their finger to their nose with their eyes closed—a feat that healthy individuals accomplish easily. Typically, they have a wide stance and an unsteady "drunken" gait, which predisposes them to falling.

When muscles have healthy tone, manipulated limbs offer some resistance to movement. When the cerebellum is impaired, muscle tone declines, leading to **lack of check,** or overshoot (a rapidly moving limb is unable to stop quickly and sharply). Cerebellar damage may also produce equilibrium problems and speech difficulties. The so-called **scanning speech** that sometimes results is slurred and somewhat slow and singsong in nature. Alternatively, the voice may waver between a shout and a whisper.

Characteristic symptoms of basal nuclei disorders are collectively called **dyskinesia** (dis-ki-ne'ze-ah; "bad movements"). They include disorders of muscle tone and posture, and involuntary movements. The abnormal movements range from *tremor,* to slow, writhing movements of the fingers and hands (*athetosis*), to violent, flailing movements of the limbs (*ballism*). The most familiar basal nuclear diseases are Parkinson's disease and Huntington's disease, both neurodegenerative diseases involving disturbances of the "motor drive"— the neural circuits that oversee smooth initiation and execution of movement. These disorders and their treatment are discussed in *A Closer Look* in Chapter 12.

HIGHER MENTAL FUNCTIONS

During the last four decades, an exciting exploration of our "inner space" has been going on that has nearly escaped public notice. Psychology and biology have begun to converge in investigations of higher mental functions, or what we commonly call *the mind,* including consciousness, memory, and rea-

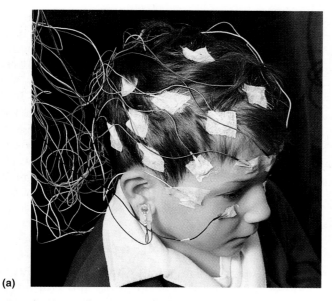

(a)

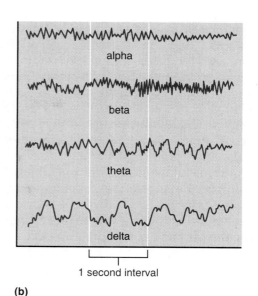

(b)

FIGURE 15.5 *Electroencephalography and brain waves.* (a) To obtain a recording of brain wave activity (an EEG), electrodes are positioned on the patient's scalp and attached to a recording device called an electroencephalograph. **(b)** Typical EEGs. Alpha waves are typical of the awake but relaxed state; beta waves occur in the awake, alert state; theta waves are common in children but not in normal awake adults; delta waves occur during deep sleep.

soning. But researchers in the field of cognition are still struggling to understand how the mind's presently incomprehensible qualities might spring from living tissue and electrical impulses. Souls and synapses are hard to reconcile!

In Chapter 11, we described the simple neuronal pools that provide the basis for neural activity, and in this chapter, we have so far considered the more complex mechanisms of sensorimotor integration. Now we will take a giant step to the highest levels of mental functioning. Since brain waves reflect the electrical activity on which such functioning is based, we will consider them first, along with the related topics of consciousness and sleep. We will then (hesitantly) examine the topic of memory, an area in which we are forced to hypothesize or speculate about what must (or should) happen.

Brain Wave Patterns and the EEG

Normal brain function involves continuous electrical activity of neurons. An **electroencephalogram** (e-lek″tro-en-sef′ah-lo-gram), or **EEG,** records some aspects of this activity. An EEG is made by placing electrodes at various locations on the scalp and then connecting the electrodes to an apparatus that measures electrical potential differences between various cortical areas (Figure 15.5a). The patterns of neuronal electrical activity recorded are called **brain waves.**

Because people differ genetically, and because everything we have ever experienced has left its imprint in our brain, each of us has a brain wave pattern that is as unique as our fingerprints. However, for simplicity we can group brain waves into four frequency classes (Figure 15.5b).

- **Alpha waves** are low-amplitude, slow, synchronous waves with an average frequency of 8–13 Hz (hertz, or cycles per second). In most cases, they indicate a brain that is "idling"—a calm, relaxed state of wakefulness.

- **Beta waves** are also rhythmic, but they are more irregular than alpha waves and have a higher frequency (14–25 Hz). Beta waves occur when we are awake and mentally alert, as when we concentrate on some problem or visual stimulus.

- **Theta waves** are still more irregular, and they have a frequency of 4–7 Hz. Though common in children, theta waves are considered abnormal in adults who are awake.

- **Delta waves** are high-amplitude waves with a frequency of 4 Hz or less. They are seen during deep sleep and when the reticular activating system is damped, such as during anesthesia. In awake adults, they indicate brain damage.

Brain waves have a normal frequency range of 1–30 Hz, a dominant rhythm of 10 Hz, and an average amplitude of 20–100 μV. The amplitude reflects the number of neurons firing together in synchrony, not the degree of electrical activity of individual neurons. When the brain is active, brain waves are complex and low amplitude. But when the brain is inactive, as during sleep, neurons tend to fire synchronously, producing similar, high-amplitude brain waves.

Brain waves change with age, sensory stimuli, brain disease, and the chemical state of the body. EEGs are used for diagnosis and localization of many types of brain lesions, such as tumors, infarcts, infections, abscesses, and epileptic lesions. Interference with cerebral cortical functions is suggested by brain waves that are too fast or too slow, and unconsciousness occurs at both extremes. Brain-depressing drugs and coma produce brain wave patterns that are abnormally slow, whereas fright, various types of drug intoxication, and epilepsy are associated with excessively fast brain waves. Because spontaneous brain waves are always present, even during unconsciousness and coma, their absence—called a "flat EEG"—is clinical evidence of brain death.

Abnormal Electrical Activity of the Brain: Epilepsy

Almost without warning, a victim of epilepsy may lose consciousness and fall stiffly to the ground, body wracked by uncontrollable jerking. These seizures, called **epileptic seizures,** reflect abnormal electrical discharges of groups of brain neurons, and while their uncontrolled activity is occurring, no other messages can get through. Epilepsy is not associated with, nor does it cause, intellectual impairments. Some cases of epilepsy are induced by genetic factors, but it can also result from brain injuries caused by blows to the head, stroke, infections, abnormal immune system activity, high fever, or tumors. Epilepsy affects about 1% of the population.

Epileptic seizures can vary tremendously in their expressions and severity. *Absence seizures,* formerly known as *petit mal,* are mild forms in which the facial muscles twitch and the expression goes blank for a few seconds. These are typically seen in young children and usually disappear by the age of ten. Another form, sometimes misdiagnosed as emotional or mental illness, is specifically called *temporal lobe epilepsy* (for the source of the hyperactivity); the victim loses contact with reality briefly and may experience hallucinations, flashbacks, and emotional outbursts or rampages. In the most severe, convulsive form of epileptic seizures, *grand mal seizures,* the person loses consciousness. Bones are often broken during the intense convulsions that take place, showing the incredible strength of the muscle contractions that occur. Loss of bowel and bladder control and severe biting of the tongue are common. The seizure lasts for a few minutes, then the muscles relax and the person awakens but remains disoriented for several minutes after. Many seizure sufferers experience a sensory hallucination, such as a taste, smell, or flashes of light, just before the seizure begins. This phenomenon, called an **aura,** is helpful because it gives the person time to lie down and avoid falling to the floor.

Epilepsy can usually be controlled by anticonvulsive drugs. Currently, nonsedating drugs (such as valproic acid, which enhances the amount of the inhibitory neurotransmitter GABA) are preferred over the sedating ones (e.g., phenobarbital). New on the scene, providing hope for those with seizures that cannot be controlled by medication or surgery, is the *vagus nerve stimulator.* This pocket watch-size "pacemaker for the brain" is implanted under the skin of the chest. Its leads (entwined around the vagus nerve) deliver pulses to the brain at predetermined intervals to keep the electrical activity of the brain from becoming chaotic.

Consciousness

Consciousness encompasses conscious perception of sensations, voluntary initiation and control of movement, and capabilities associated with higher mental processing (memory, logic, judgment, perseverance, and so on). Clinically, consciousness is defined on a continuum that grades levels of behavior in response to stimuli as (1) *alertness,* (2) *drowsiness* or *lethargy* (which proceeds to sleep), (3) *stupor,* and (4) *coma.* Alertness is the highest state of consciousness and cortical activity, whereas coma is the most depressed state. But when we get away from the clinical definition, consciousness is very difficult to define. A sleeping person obviously lacks something that he or she has when awake, and we call this "something" consciousness. Human consciousness is still a mystery, and most of what you read about it is likely to be speculation. The current suppositions about consciousness are as follows:

1. Consciousness involves simultaneous activity of large areas of the cerebral cortex. Localized damage to any specific region of the cerebral cortex does *not* destroy consciousness.

2. Consciousness is superimposed on other types of neural activity. At any time, specific neurons and neuronal pools are involved both in localized activities (such as motor control or sensory perception) and in mediating cognitive behaviors.

3. Consciousness is holistic and totally interconnected. Information for "thought" can be claimed from many locations in the cerebrum simultaneously. For example, retrieval of a specific memory can be triggered by one of several routes—a smell, a place, a particular person, and so on. The cortical cross-file is enormous.

Homeostatic Imbalance

Except for sleep, unconsciousness is always a signal that brain function is impaired. A brief loss of consciousness is called **fainting** or **syncope** (sing'ko-pe;

"cut short"). Most often, it indicates inadequate cerebral blood flow due to low blood pressure, as might follow hemorrhage or sudden emotional stress. Fainting is usually preceded by a feeling of light-headedness as blood pressure falls to critically low levels.

When a person is totally unresponsive to sensory stimuli for an extended period, the condition is called **coma.** Coma is *not* deep sleep. During sleep, the brain is active and oxygen consumption resembles that of the waking state. In contrast, oxygen use is always below resting levels in coma patients.

Blows to the head may induce coma by causing widespread cerebral or brain stem trauma, resulting from the damage done by the blow itself or from hemorrhage or edema. Coma may also be produced by tumors or infections that invade the brain stem. Metabolic disturbances such as hypoglycemia (abnormally low blood sugar levels), drug overdose, or liver or kidney failure interfere with overall brain function and can result in coma. Cerebral infarctions, unless massive and accompanied by extreme swelling of the brain, rarely cause coma.

When the brain has suffered irreparable damage, irreversible coma occurs, even though life-support measures may have restored vitality to other body organs. The result is **brain death,** a dead brain in an otherwise living body. Physicians must determine whether a patient in an irreversible coma is legally alive or dead, because life support can be removed only after death. ■

Sleep and Sleep-Awake Cycles

Sleep is defined as a state of changed consciousness, or partial unconsciousness, from which a person can be aroused by stimulation. This distinguishes sleep from *coma,* a state of unconsciousness from which a person *cannot* be aroused by even the most vigorous stimuli. Cortical activity is depressed during sleep, but brain stem functions, such as control of respiration, heart rate, and blood pressure, continue. Even environmental monitoring continues to some extent, as illustrated by the fact that strong stimuli ("things that go bump in the night") immediately arouse us. In fact, people who sleepwalk can avoid objects and navigate stairs while truly asleep.

Types of Sleep

The two major types of sleep, which alternate through most of the sleep cycle, are **non-rapid eye movement (NREM) sleep** and **rapid eye movement (REM) sleep.** The types of sleep are defined in terms of their EEG patterns. During the first 30 to 45 minutes of the sleep cycle, we pass through the four stages of NREM sleep, culminating in **slow-wave sleep.** As we pass through these stages and slip into

deeper and deeper sleep, the frequency of the EEG pattern declines, but the wave amplitude increases. REM sleep, recognized by an EEG pattern more typical of the awake state, does not appear until after slow-wave sleep has been achieved.

NREM Sleep All stages of sleep other than REM are collectively referred to as non-rapid eye movement (NREM) sleep. The four stages of NREM sleep, in which both the frequency of brain waves and vital signs decline, are summarized in Table 15.3.

REM Sleep About 90 minutes after sleep begins, the EEG pattern changes abruptly. It becomes very irregular and appears to backtrack quickly through the stages until alpha waves, small fast waves characteristic of stage 1 and indicating the onset of REM, appear (Table 15.3). This brain wave change is coupled with increases in body temperature, heart rate, respiratory rate, and blood pressure and a decrease in gastrointestinal motility. REM sleep is also called **paradoxical sleep** because its EEG pattern is more typical of the awake state. Oxygen use by the brain is tremendous during REM—greater than during the awake state.

Although the eyes move rapidly under the lids during REM, most of the body's skeletal muscles are actively inhibited and go limp. This temporary paralysis prevents us from acting out our dreams. Most dreaming occurs during REM sleep, and some suggest that the flitting eye movements are following the visual imagery of our dreams. (Note, however, that most nightmares and night terrors occur during stages 3 and 4 of NREM sleep.) In adolescent and adult males, REM episodes are frequently associated with erections of the penis. The threshold for arousal to consciousness is highest during REM sleep. On the other hand, the sleeper is more likely to awaken spontaneously and to remember his or her dreams during this time.

Sleep Patterns

Although most people spend about a third of their lives asleep, we know little about sleep's biological basis. But we do know that the alternating cycles of sleep and wakefulness reflect a natural *circadian,* or 24-hour, *rhythm.* In the awake state, alertness of the cerebral cortex (the "conscious brain") is mediated by the *reticular activating system,* or *RAS* (see Figure 12.19, p. 453). When RAS activity declines, cerebral cortical activity declines as well; thus, lesions of the RAS nuclei result in unconsciousness. However, sleep is much more than simply turning off the arousal (RAS) mechanism. The brain is actively guided into sleep. RAS centers not only help maintain the awake state but also mediate some sleep stages, especially dreaming sleep. The hypothalamus

TABLE 15.3	Types and Stages of Sleep	

Non-rapid Eye Movement (NREM) Sleep

Stage 1		Eyes are closed and relaxation begins. Thoughts flit in and out and a drifting sensation occurs; vital signs (body temperature, respiration, pulse, and blood pressure) are normal; the EEG pattern shows alpha waves; if stimulated, arousal is immediate.
Stage 2		The EEG pattern becomes more irregular; *sleep spindles*—sudden, short high-voltage wave bursts occurring at 12–14 Hz—appear, and arousal is more difficult.
Stage 3		Sleep deepens, and theta and delta waves appear; vital signs begin to decline, and the skeletal muscles are very relaxed; dreaming is common; usually reached about 20 minutes after the onset of stage 1.
Stage 4		Called slow-wave sleep because the EEG pattern is dominated by delta waves (1–4 Hz); vital signs reach their lowest normal levels, and digestive system motility increases; skeletal muscles are relaxed, but normal sleepers turn approximately every 20 minutes; arousal is difficult; bed-wetting and sleepwalking occur during this phase.

Rapid Eye Movement (REM) Sleep

		EEG pattern reverts through the NREM stages to the stage 1 pattern. Vital signs increase and digestive system activity declines; skeletal muscles (except for ocular muscles) are inhibited; the stage of most dreaming.

is responsible for the timing of the sleep cycle: Its teardrop-shaped *suprachiasmatic nucleus* (our biological clock) regulates its *preoptic nucleus* (a sleep-inducing center).

In young and middle-aged adults, a typical night's sleep alternates between REM and NREM sleep (Figure 15.6). Following each REM episode, the sleeper descends to stage 4 sleep again. REM recurs about every 90 minutes, with each REM period getting longer. The first REM lasts 5–10 minutes and the final one from 20 to as much as 50 minutes. Hence, our longest dreams are in the "wee hours" of the morning. When the neurons of the dorsal raphe nuclei of the midbrain reticular formation fire at maximal rates, we awaken for the day. According to some researchers, this event is rigidly tied to a rise in core body temperature.

Key generators of REM sleep are reticular cells in the lateral pontine and medial medullary areas. The activity of ACh-secreting cells of the giant cell nucleus, part of the pons reticular formation, is believed to cause the arousal pattern of REM sleep, and norepinephrine released by the locus coeruleus of the pons is thought to induce the transient paralysis

of REM sleep. NREM sleep areas include the nucleus of the solitary tract, the reticular nucleus of the thalamus, the hypothalamus, and the basal forebrain.

Importance of Sleep

Slow-wave and REM sleep seem to be important in different ways. Shakespeare called sleep "nature's soft nurse," referring to its restorative effects on the body. Slow-wave sleep is presumed to be the restorative stage—the time when most neural mechanisms can wind down to basal levels. When we are deprived of sleep, we spend more time than usual in slow-wave sleep during the next sleep episode.

Those persistently deprived of REM sleep become moody and depressed, and exhibit various personality disorders. REM sleep may give the brain an opportunity to analyze the day's events and to work through emotional problems in dream imagery. Sigmund Freud saw dreams as symbols of frustrated desires and attempts to fulfill wishes. Another idea is that REM sleep is reverse learning. According to this hypothesis, accidental, repetitive, and meaningless communications continually occur, and they must be eliminated from the neural networks by dreaming if the cortex is to remain a well-behaved and efficient thinking system. In other words, we dream to forget.

Alcohol and most sleep medications (barbiturates and others) suppress REM sleep but not slow-wave sleep. On the other hand, certain tranquilizers, such as diazepam (Valium) and chlordiazepoxide (Librium), reduce slow-wave sleep much more than REM sleep.

Whatever its importance, the daily sleep requirement declines steadily from 16 hours or so in infants to approximately 7 hours in early adulthood. It then levels off before declining once again in old age. Sleep patterns also change throughout life. REM sleep occupies about half of the total sleeping time in infants and then declines until the age of ten years, when it stabilizes at about 25% (approximately 1.5 to 2 hours per night). In contrast, stage 4 sleep declines steadily from birth and often disappears completely in those over 60. The elderly awaken more frequently during the night because of this perpetually light sleep.

Homeostatic Imbalances of Sleep

Three clinically important disorders of sleep are narcolepsy, insomnia, and sleep apnea. People with **narcolepsy** lapse abruptly into sleep from the awake state. These sleep episodes last about 15 minutes, can occur without warning at any time, and are often triggered by a pleasurable event—a good joke, a game of poker, a sports event. This condition can be extremely hazardous when the person must drive, operate machinery, or take a bath. In all daytime episodes, narcoleptics show the typical EEG of REM sleep. During the evening sleep cycle, however, they spend much less time in REM than normal.

Insomnia is a chronic inability to obtain the *amount* or *quality* of sleep needed to function adequately during the daytime. Sleep requirements vary from four to nine hours a day among healthy people, so there is no way to determine the "right" amount. Self-professed insomniacs tend to exaggerate the extent of their sleeplessness, and many have a notorious tendency toward self-medication and barbiturate abuse.

True insomnia often reflects normal age-related changes. In travelers, it may be caused by *jet lag*; but perhaps the most common cause of insomnia is psychological disturbance. We have difficulty falling asleep when we are anxious or upset, and depression is often accompanied by early awakening.

Sleep apnea, a temporary cessation of breathing during sleep, is a scary situation. The victim awakes abruptly due to hypoxia—a condition that may occur repeatedly throughout the night. Its cause varies and is uncertain. However, it tends to be more common in the elderly (the stimulating effects of carbon dioxide buildup on breathing decline with age), and in those with some sort of blockage of the upper air-

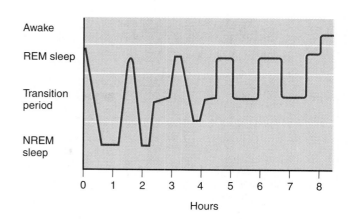

FIGURE 15.6 *Sleep pattern of a typical young adult.*

way (tonsillitis, for example). Sleep apnea is enhanced by certain drugs (alcohol), high blood pressure, obesity, and conditions that affect CNS respiratory centers. Treatment is focused on treating the cause if known (i.e., losing weight if obese, removal of tonsils, drugs to reduce hypertension). Sold without a prescription, adhesive nasal patches that hold the nostrils wide open during sleep help some.

Memory

Memory is the storage and retrieval of information or, more simply, the wonderful ability to carry around the past inside our heads. Memories are essential for learning and incorporating our experiences into behavior and are part and parcel of our consciousness. Stored somewhere in your 3 pounds of wrinkled brain are zip codes, the face of your grandfather, and the taste of yesterday's pizza. In essence, your memories reflect your lifetime.

Most of what is known about memory and learning can be summarized in three principles: (1) Memory storage occurs in stages and is continually changing; (2) the hippocampus and surrounding structures play unique roles in memory processing; and (3) memory traces—the chemical or structural changes that encode memory—are widely distributed in the brain.

Stages of Memory

Memory storage involves two distinct stages: short-term memory and long-term memory (Figure 15.7). **Short-term memory (STM),** also called *working memory,* is a fleeting memory of the events that continually parade before you. Lasting for seconds to a few hours, it is the preliminary step to long-term memory, as well as the power that lets you look up a telephone number, dial it, and then never think of it again. The capacity of STM is limited to seven or eight chunks of information, such as the digits of a telephone number or the sequence of words in an elaborate sentence.

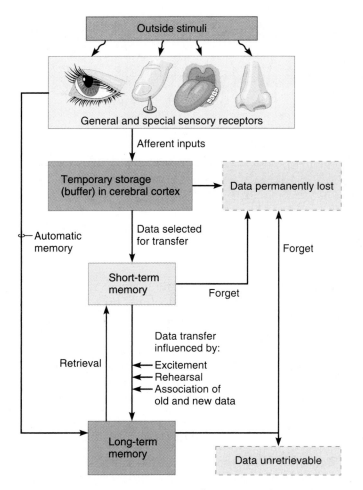

FIGURE 15.7 *Memory processing.* Sensory inputs are processed by the cerebral cortex (indicated as the temporary storage area), which selects what is to be sent to short-term memory. Rehearsal and other factors help the transfer of information from short-term memory to long-term memory. For long-term memory to become permanent, consolidation must occur. Some information not attended to consciously goes directly into long-term memory (automatic memory).

In contrast to STM, **long-term memory (LTM)** seems to have a limitless capacity. Although our short-term memory devices cannot recall numbers much longer than a telephone number, we can remember scores of telephone numbers by committing them to LTM. However, our ability to store and to retrieve information declines with aging. Thus, long-term memories too can be forgotten, and our memory bank continually changes with time.

We do not remember or even consciously notice much of what is going on around us. As sensory inputs flood into our cerebral cortex, they are processed, and some of this information is selected for transfer to STM (Figure 15.7). STM serves as a sort of temporary holding bin for data that we may or may not want to retain. The transfer of information from STM to LTM is affected by many factors, including:

1. Emotional state. We learn best when we are alert, motivated, and aroused. For example, when we witness shocking events, transferal is almost immediate. Norepinephrine is involved in memory processing of emotionally charged events and more NE is released when we are excited or "stressed out," which helps to explain this phenomenon.

2. Rehearsal. An opportunity to repeat or rehearse the material enhances memory.

3. Association of "new" information with "old" information already stored in LTM. A staunch football fan can tell you who did what in all of the important plays of a particular football game, whereas someone unfamiliar with the sport would find them difficult to understand and therefore to remember.

4. Automatic memory. Not all impressions that become part of LTM are consciously formed. For example, a student concentrating on a lecturer's speech may also record an automatic memory of the pattern of the lecturer's tie.

Memories transferred to LTM take time to become permanent. The process of **memory consolidation** apparently involves fitting new facts into the various categories of knowledge already stored in the cerebral cortex.

Categories of Memory

The brain distinguishes between factual knowledge and skills, and we process and store these different kinds of information in different ways. **Fact (declarative) memory** entails learning explicit information, such as names, faces, words, and dates. It is related to our conscious thoughts and our ability to manipulate symbols and language. When fact memories are committed to LTM, they are usually filed along with the context in which they were learned. For instance, when you think of your new friend Joe, you probably picture him at the basketball game where you met him.

Skill (procedural) memory is less conscious learning and usually involves motor activities. It is acquired only through practice, as when we learn to ride a bike or play a bass fiddle. Skill memories do not preserve the circumstances of learning; in fact, they are best remembered in the doing. A batter does not consciously recall the moves needed to properly time and swing the bat, nor do you have to think through how to tie your shoes. Once learned, skill memories are hard to unlearn.

Brain Structures Involved in Memory

Much of what we know about learning and memory comes from two sources: experiments with macaque monkeys and studies of amnesia in humans. Such

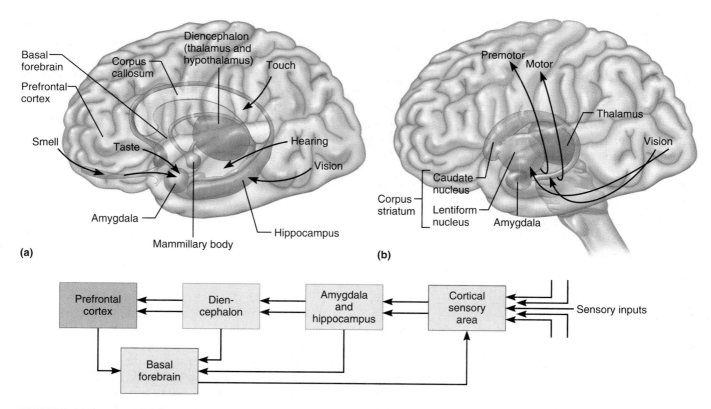

FIGURE 15.8 *Proposed memory circuits.* **(a)** The essential structures of the fact memory circuits. The flowchart below indicates how these structures may interact in memory formation. Sensory inputs from the cortex flow through parallel circuits, one involving the hippo‑campus and the other the amygdala. Both circuits encompass parts of the dience‑phalon, basal forebrain, and prefrontal cortex. The basal forebrain feeds back to the sensory cortex, closing the memory loop. **(b)** Major structures involved in skills memory. The corpus striatum mediates the automatic connections between a stimulus (visual here) and a motor response.

studies have revealed that different brain structures are involved in the two stages of memory (Figure 15.8).

It appears that specific pieces of each memory are stored near regions of the brain that need them so that new inputs can be quickly associated with the old. Accordingly, visual memories would be stored in the occipital cortex, memories of music in the temporal cortex, and so on. Thus, your memory of a dear aunt—the scent of her cologne, the softness of her hands—is found in bits and pieces scattered all over your cortex.

But how are the memory connections made? Research indicates that the crucial structures for fact memory are the *hippocampus* and the *amygdala* (both part of the limbic system), the *diencephalon* (specific areas of the thalamus and hypothalamus), the *ventromedial prefrontal cortex* (a cortical region tucked beneath the front of the brain), and the *basal forebrain* (a cluster of ACh‑secreting neurons ante‑rior to the hypothalamus) (see Figure 15.8a).

The proposed scheme of information flow is as follows. When a sensory perception is formed in the cerebral sensory cortex, the cortical neurons dis‑patch impulses along two parallel circuits to the hip‑pocampus and the amygdala, both of which have connections with the diencephalon, basal forebrain, and prefrontal cortex. The basal forebrain then closes the memory loop by sending impulses back to the sensory cortical areas initially forming the per‑ception. This feedback presumably causes changes that transform the new perception into a more durable memory. This suggests that the subcortical structures make the initial connections between stored memories and the new perception by access‑ing cortical regions where the LTM is stored in a kind of "conference call" until the new information can be consolidated. Later recall of the new memory occurs when the same cortical neurons are stimu‑lated.

The prefrontal cortex seems to be necessary for retrieving fact learning from long‑term storage de‑pots elsewhere in the brain for use in present tasks. The hippocampus oversees the circuitry for learning and remembering spatial relationships. The amyg‑dala, with its widespread connections to all cortical sensory areas and the emotional response centers of the hypothalamus, seems to be responsible for asso‑ciating memories formed through different senses and linking them to emotional states generated in the hypothalamus.

⬜ *Homeostatic Imbalance*

Damage to either the hippocampus or amygdala results in only slight memory loss, but bilateral destruction of both structures causes global (widespread) amnesia. Consolidated memories are not lost, but new sensory inputs cannot be associated with old, and the person lives in the here and now from that point on. This phenomenon is called *anterograde* (an'ter-o-grād") *amnesia,* in contrast to *retrograde amnesia,* which is the loss of memories formed in the distant past. You could carry on an animated conversation with a person with anterograde amnesia, excuse yourself, return five minutes later, and that person would not remember you. ■

Individuals suffering from anterograde amnesia can still learn skills. Consequently, a second learning circuit, independent of the pathways used in fact memory, has been suggested (see Figure 15.8b). It appears that the cerebral cortex, activated by sensory inputs, signals the *corpus striatum* (lentiform and caudate nuclei of the basal nuclei) of its intent to mobilize a skill memory. The corpus striatum then communicates with one or more of the brain stem nuclei as well as with the cerebral cortex to promote the desired movement. Thus, the corpus striatum forms the link between a perceived stimulus and a motor response. Because conditioning of voluntary muscle contractions is involved in many skills, the cerebellum also plays a role in skill memory. Some authorities prefer to call these pathways the **habit system,** rather than memory in the true sense.

Mechanisms of Memory

In the early 1900s, Karl Lashley, a pioneer neuropsychologist, set out to find the **engram,** the hypothetical unit of memory or permanent memory trace. Now, a century later, the mysterious engram still eludes us. Because STM is so fleeting, it is not believed to involve permanent changes in neural circuits. If an engram exists, it will probably be found in long-term memory depots. Experimental studies reveal that during learning, (1) neuronal RNA content is altered and newly synthesized mRNAs are delivered to axons and dendrites, (2) dendritic spines change in shape, (3) unique extracellular proteins are deposited at synapses involved in LTM, (4) presynaptic terminals may increase in number and size, and (5) more neurotransmitter is released by the presynaptic neurons.

Human memory is notoriously difficult to study.

However, the growing understanding of **NMDA receptors,** named after the chemical *N*-methyl-D-aspartate used to detect them, is changing this picture. As mentioned in Chapter 11, these unique receptors act as calcium channels and mediate synaptic long-term potentiation (LTP), the persistent changes in synaptic strength first identified in hippocampal memory pathways that use the amino acid glutamate as their transmitter.

Postsynaptic cells in which LTP can be induced bear two types of glutamate receptors—non-NMDA and NMDA channels. Resting NMDA channels bind magnesium ions (Mg^{2+}) that block calcium passage through the channel. Binding of glutamate to non-NMDA receptors (essentially chemically gated sodium channels) during high-intensity stimulation causes membrane depolarization that expels Mg^{2+} from the NMDA receptors and allows glutamate to bind to them, causing calcium channels to open.

Thus, the NMDA receptors are activated by both voltage (a depolarizing current) and a chemical (subsequent binding of glutamate). These signals, which apparently must occur sequentially ("click-click"), lead to Ca^{2+} influx, which provides the signal necessary for the induction of LTP. Details of events from this point on are murky, but Ca^{2+} appears to promote activation of several enzymes in the postsynaptic cells, including proteases and protein kinases. It's a good bet that at least some of these "reshaped proteins" bring about changes in the NMDA receptors that increase their sensitivity to glutamate.

Despite substantial research into the mystery of memory, our understanding of it is still wispy, and each new discovery inevitably creates new questions. The "mind" within the brain is always just a bit beyond our grasp.

* * *

In this chapter, we have brought together most of the neural machinery studied in earlier chapters and have examined how sensory inputs are integrated with motor activity in the central nervous system. We have also taken a look at the most complex aspects of neural integration, which allow us to think and remember. Now that the nervous system has been viewed as an integrated whole, we are ready to consider the special senses in Chapter 16.

RELATED CLINICAL TERMS

Apraxia (ah-prak'se-ah) Loss of the voluntary ability to perform skilled motor activities, which, however, can still occur automatically. Motor apraxia of speech is sometimes called **motor aphasia**; the individual is unable to speak in grammatical sentences due to lesions of Broca's area or the surrounding portions of the more anterior frontal cortex; common in stroke victims who have right-sided paralysis.

Dysarthria (dis-ar'three-ah) Disorders in the motor pathways resulting in weakness, uncoordinated motion, and characteristic speech disturbances by altering respiration or speech articulation or rhythm. For example, lesions of cranial nerves IX, X, and XII result in nasal, breathy speech, and lesions in upper motor pathways produce a hoarse, strained voice.

Dystonia (dis-to'ne-ah) Impairment of muscle tone.

Hypersomnia (*hyper* = excess; *somnus* = sleep) A condition in which affected individuals sleep as much as 15 hours daily.

Myoclonus (mi"o-klo'nus; *myo* = muscle; *clon* = violent motion, tumult) Sudden contraction of a muscle or muscle part, usually involving muscles of the limbs; myoclonal jerks occurring in normal individuals as they are falling asleep are thought to reflect a fleeting reactivation of the RAS; may also be due to diseases of the reticular formation or cerebellum.

Phantom pain A phenomenon in which amputees have chronic pain in their nonexistent (amputated) limb; although it is not well understood, apparently the cut sensory nerves that previously transmitted inputs from the limb are hyperalgesic and continue to respond to the trauma of amputation, and the brain interprets (projects) these painful stimuli as coming from the missing limb.

Tabes dorsalis (ta'bēz dor-sa'lis) A slowly progressive condition caused by deterioration of the posterior white matter tracts (gracilis and cuneatus) and associated dorsal roots; a late sign of the neurological damage caused by the syphilis bacterium. Because joint proprioceptor tracts are destroyed, affected individuals have poor muscle coordination and an unstable gait. Bacterial invasion of the sensory (dorsal) roots results in pain, which ends when dorsal root destruction is complete.

CHAPTER SUMMARY

Media study tools that could provide you additional help in reviewing specific key topics of Chapter 15 are referenced below. **SP** = *Study Partner;* **IP** = *Interactive Physiology.*

Sensory Integration: From Reception to Perception (pp. 536–541)

1. Sensation is awareness of internal and external stimuli; perception is conscious interpretation of those stimuli.

General Organization of the Somatosensory System (pp. 536–541)

2. The three levels of sensory integration are the receptor, circuit, and perceptual levels. These levels are functions of the sensory receptors, the ascending pathways, and the cerebral cortex, respectively.

3. Sensory receptors transduce (convert) stimulus energy via receptor or generator potentials into action potentials. Stimulus strength is frequency coded. Adaptation (decreased response to a continuous or unchanging stimulus) is seen in all general receptors except pain and proprioceptors.

SP Exercise: Chapter 15, Synaptic Potentials and Cellular Integration.

4. Some sensory fibers entering the spinal cord act in local reflex arcs. Some synapse with the dorsal horn neurons (nonspecific ascending pathways), and others continue upward to synapse in medullary nuclei (specific ascending pathways). Second-order neurons of both specific and nonspecific ascending pathways terminate in the thalamus.

5. The specific sensory pathway consists of the dorsal white column (fasciculus cuneatus, fasciculus gracilis) and the medial lemniscal tracts, which are concerned with straight-through, precise transmission of one or a few related sensory modalities. The nonspecific pathway (the spinothalamic tracts) is a multimodal pathway that permits brain stem processing of ascending impulses.

6. The spinocerebellar tracts, which terminate in the cerebellum, serve muscle sense, not conscious sensory perception.

7. The origins and modalities of sensory impulses are crudely identified in the thalamus and then projected to the somatosensory cortex and other sensory association areas.

8. Perception—the internal, conscious image of the stimulus that serves as the basis for response—is the result of cortical processing. Regions of the somatosensory cortex devoted to a particular body area reflect the number of receptors in that area.

9. The major aspects of sensory perception are perceptual detection, magnitude estimation, spatial discrimination, feature abstraction, quality discrimination, and pattern recognition.

SP Case Study: Cerebrovascular Accident.

SP Exercise: Chapter 15, Two-Point Threshold in the Adult; Case Study: Cerebrovascular Accident.

Motor Integration: From Intention to Effect (pp. 541–546)

1. Motor mechanisms operate at the level of the effectors (muscle fibers), descending circuits, and control levels of motor behavior.

Levels of Motor Control (pp. 541–546)

2. The motor control hierarchy consists of the segmental level, the projection level, and the programs and instructions level.

SP Exercise: Chapter 15, Hierarchy of Motor Control.

3. The segmental level is the spinal cord circuitry that activates anterior horn motor neurons to stimulate the muscles. It directly controls reflexes and fixed-action patterns. Segmental circuits controlling locomotion are central pattern generators (CPGs).

4. The projection level consists of descending fibers that project to and control the segmental level. These fibers issue from the brain stem motor areas (indirect [extrapyramidal] system) and cortical motor areas (direct [pyramidal] system). Command neurons in the brain stem appear to turn CPGs on and off, or to modulate them.

5. The programs and instructions level consists of the cerebellum and basal nuclei. These constitute the precommand areas that subconsciously integrate mechanisms mediated by the projection level.

SP Exercise: Hierarchy of Motor Control.

Homeostatic Imbalances of Motor Integration (p. 546)

6. Cerebellar deficits cause ipsilateral symptoms, which include disorders in synergy and muscle tone, equilibrium problems, and speech problems. Ataxia is the most common symptom.

SP Case Study: Cerebrovascular Accident.

7. Basal nuclear deficits result in abnormal muscle tone and involuntary movements (tremors, athetosis, and ballism). Parkinson's disease and Huntington's disease are examples.

SP Case Study 15: Parkinson's Disease.

Higher Mental Functions (pp. 546–554)

Brain Wave Patterns and the EEG (pp. 547–548)

1. Patterns of electrical activity of the brain are called brain waves; a record of this activity is an electroencephalogram (EEG). Brain wave patterns, identified by their frequencies, include alpha, beta, theta, and delta waves.

2. Epilepsy results from abnormal electrical activity of brain neurons. Involuntary muscle contractions and sensory auras are typical during such seizures.

Consciousness (pp. 548–549)

3. Consciousness includes sensory perception, initiation and control of voluntary movement, and higher mental processing capabilities. It is described clinically on a continuum from alertness to drowsiness to stupor and finally to coma.

4. Human consciousness is thought to involve holistic information processing, which is (1) not localizable, (2) superimposed on other types of neural activity, and (3) totally interconnected.

5. Fainting (syncope) is a temporary loss of consciousness that usually reflects inadequate blood delivery to the brain. Coma is loss of consciousness in which the victim is unresponsive to stimuli.

Sleep and Sleep-Awake Cycles (pp. 549–551)

6. Sleep is a state of altered consciousness from which one can be aroused by stimulation. The two major types of sleep are non-rapid eye movement (NREM) sleep and rapid eye movement (REM) sleep.

7. During stages 1–4 of NREM sleep, brain waves become more irregular and increase in amplitude until delta wave sleep (stage 4) is achieved. REM sleep is indicated by a return to a stage 1 EEG. During REM, the eyes move rapidly under the lids. NREM and REM sleep alternate throughout the night.

8. Slow-wave sleep (stage 4 of NREM) appears to be restorative. REM sleep is important for emotional stability.

9. REM occupies half of an infant's sleep time and then declines to about 25% of sleep time by the age of ten years. Time spent in slow-wave sleep declines steadily throughout life.

10. Narcolepsy is involuntary lapses into sleep that occur without warning during waking periods. Insomnia is a chronic inability to obtain the amount or quality of sleep needed to function adequately. Sleep apnea is a temporary cessation of breathing during sleep due to hypoxia.

Memory (pp. 551–554)

11. Memory is the ability to recall one's thoughts. It is essential for learning and is part of consciousness.

12. Memory storage has two stages: short-term memory (STM) and long-term memory (LTM). Transfer of information from STM to LTM takes minutes to hours, but more time is required for LTM consolidation.

13. Fact memory is the ability to learn and consciously remember information. Skill memory is the learning of motor skills, which are then performed without conscious thought.

14. Fact (declarative) memory appears to involve the hippocampus, amygdala, diencephalon, basal forebrain, and prefrontal cortex. Skill (procedural) memory pathways are mediated by the corpus striatum.

15. The nature of memory traces in the human brain is not fully known, but NMDA receptors (essentially calcium channels), activated sequentially by depolarization and glutamate binding, play a major role in long-term potentiation (LTP). The calcium influx that follows NMDA receptor activation mobilizes enzymes that mediate events necessary for memory consolidation.

SP Exercise: Synaptic Potentials and Cellular Integration.

REVIEW QUESTIONS

Multiple Choice/Matching

1. All of the following descriptions refer to specific ascending pathways except one: (a) they include the fasciculus gracilis and fasciculus cuneatus, which terminate in the thalamus; (b) they include a chain of three neurons; (c) their connections are diffuse and polymodal; (d) they are concerned with precise transmission of one or a few related sensory modalities.

2. The aspect of sensory perception by which the cerebral cortex identifies the site or pattern of stimulation is (a) perceptual detection, (b) feature abstraction, (c) pattern recognition, (d) spatial discrimination.

3. The neural machinery of the spinal cord is at the (a) programs level, (b) projection level, (c) segmental level.

4. Which of the following is part of the indirect (multineuronal) system? (a) vestibular nuclei, (b) red nucleus, (c) superior colliculi, (d) reticular nuclei, (e) all of these.

5. Brain waves typical of the alert, wide-awake state are (a) alpha, (b) beta, (c) delta, (d) theta.

6. Identify the stage of sleep described by using choices from the key. (Note that responses a–d refer to NREM sleep.)

Key: (a) stage 1, **(b)** stage 2, **(c)** stage 3, **(d)** stage 4, **(e)** REM

_____ (1) the stage when vital signs (blood pressure, heart rate, and body temperature) reach their lowest levels

_____ (2) indicated by movement of the eyes under the lids; dreaming occurs

_____ (3) when sleepwalking is likely to occur

_____ (4) when the sleeper is very easily awakened; EEG shows alpha waves

Short Answer Essay Questions

7. Differentiate clearly between sensation and perception.

8. How do synthetic and analytic quality discrimination differ?

9. Central pattern generators (CPGs) are found at the segmental level of motor control. (a) What is the job of the CPGs? (b) What controls them, and where is this control localized?

10. Make a diagram of the hierarchy of motor control. Position the CPGs, the command neurons, and the cerebellum and basal nuclei in this scheme.

11. How do the types of motor activity controlled by the direct (pyramidal) and indirect systems differ?

12. Describe the functional problems that would be experienced by a person in which these fiber tracts have been cut: (a) lateral spinothalamic, (b) anterior and posterior spinocerebellar, (c) tectospinal.

13. Why are the cerebellum and basal nuclei called *pre* command areas?

14. Define EEG.

15. How do sleep patterns, extent of sleeping time, and amount of time spent in REM and NREM sleep change through life?

16. (a) Define narcolepsy and insomnia. (b) Why is narcolepsy thought to reflect a problem with the sleep control centers? (c) Why is narcolepsy a dangerous condition?

17. (a) Define epilepsy. (b) Which form of this disorder is least distressing? Which is most distressing? Explain.

18. What is an aura?

19. What is holistic information processing, and what are its essential characteristics?

20. Compare and contrast short-term memory (STM) and long-term memory (LTM) relative to storage capacity and duration of the memory.

21. (a) Name several factors that can enhance transfer of information from STM to LTM. (b) Define memory consolidation.

22. Compare and contrast fact and skill memory relative to the types of things remembered and the importance of conscious retrieval.

23. Explain why consciousness, memory, and language are inseparable.

Critical Thinking and Clinical Application Questions

1. Soon after Mr. Camors noted a weakness in his right arm, he was unable to speak or move the arm. During the examination, he seemed to understand what was said to him but could not respond appropriately—his words were garbled, and his right cheek and the right side of his mouth drooped. Although lower limb strength and reflexes were normal, reflexes in his right arm were hypoactive. No sensory losses or gait problems were noted. Given this situation, answer the following questions: (a) Specifically, what motor system (direct or indirect) and what brain site(s) are involved? (b) How do the abdominal and plantar reflexes change with lesions in this system? (c) What is Mr. Camors's speech disorder called?

2. Cynthia, a 16-year-old girl, was rushed to the hospital after taking a bad spill off the parallel bars. After a complete neurological workup, her family was told that she would be permanently paralyzed from the waist down. The neurologist then outlined for Cynthia's parents the importance of preventing complications in such cases. Common complications include urinary infection, bed sores, and muscular spasms. Using your knowledge of neuroanatomy, explain the underlying reasons for these complications.

3. You are at a party at Mary's house. After being blindfolded, an object (a key *or* a rabbit's foot) is placed in your hand. What spinal tracts carry the signals to the cortex that will allow you to differentiate between these objects, and what aspects of sensory perception are operating?

16 THE SPECIAL SENSES

The Chemical Senses: Taste and Smell (pp. 559–564)

1. Describe the location, structure, and afferent pathways of taste and smell receptors, and explain how these receptors are activated.

The Eye and Vision (pp. 564–586)

2. Describe the structure and function of accessory eye structures, eye tunics, lens, and humors of the eye.

3. Trace the pathway of light through the eye to the retina, and explain how light is focused for distant and close vision.

4. Describe the events involved in the stimulation of photoreceptors by light, and compare and contrast the roles of rods and cones in vision.

5. Note the cause and consequences of astigmatism, cataract, glaucoma, hyperopia, myopia, and color blindness.

6. Compare and contrast light and dark adaptation.

7. Trace the visual pathway to the optic cortex, and briefly describe the process of visual processing.

The Ear: Hearing and Balance (pp. 586–601)

8. Describe the structure and general function of the outer, middle, and inner ears.

9. Describe the sound conduction pathway to the fluids of the inner ear, and follow the auditory pathway from the organ of Corti to the temporal cortex.

10. Explain how one is able to differentiate pitch and loudness of sounds and to localize the source of sounds.

11. Explain how the balance organs of the semicircular canals and the vestibule help to maintain dynamic and static equilibrium.

12. List possible causes and symptoms of otitis media, deafness, Ménière's syndrome, and motion sickness.

Developmental Aspects of the Special Senses (pp. 601–602)

13. List changes that occur in the special sense organs with aging.

People are responsive creatures. The aroma of freshly baked bread makes our mouths water. A sudden clap of thunder makes us jump. These irritants and many others are the stimuli that continually greet us and are interpreted by our nervous systems.

We are usually told that we have five senses: touch, taste, smell, sight, and hearing. Actually, touch reflects the activity of the general senses that we considered in Chapter 13. The other four traditional senses—*smell, taste, sight,* and *hearing*—are called **special senses.** Receptors for a fifth special sense, *equilibrium,* are housed in the ear, along with the organ of hearing. In contrast to the widely distributed general receptors (most of which are modified dendrites of sensory neurons), the **special sensory receptors** are distinct *receptor cells.* These receptor cells are confined to the head region and are highly localized, either housed within complex sensory organs (eyes and ears) or within distinct epithelial structures (taste buds and olfactory epithelium).

This chapter considers the functional anatomy of each of the five special senses. But keep in mind that our perceptions of sensory inputs are overlapping. What we finally experience—our "feel" of the world—is a blending of stimulus effects.

THE CHEMICAL SENSES: TASTE AND SMELL

Taste and smell are gritty, primitive senses that alert us to whether that "stuff" nearby (or in our mouth) is to be savored or avoided. The receptors for taste (gustation) and smell (olfaction) are classified as **chemoreceptors** because they respond to chemicals in an aqueous solution. Taste receptors are excited by food chemicals dissolved in saliva; smell receptors, by airborne chemicals that dissolve in fluids coating nasal membranes. The receptors for taste and smell complement each other and respond to many of the same stimuli.

Taste Buds and the Sense of Taste

The word *taste* comes from the Latin *taxare,* meaning "to touch, estimate, or judge." When we taste things, we are in fact intimately testing or judging our environment, and the sense of taste is considered by many to be the most pleasurable of the special senses.

Localization and Structure of Taste Buds

The **taste buds,** the sensory receptor organs for taste, are located primarily in the oral cavity. Of our 10,000 or so taste buds, most are on the tongue. A few are scattered on the soft palate, inner surface of the cheeks, pharynx, and epiglottis of the larynx.

Most taste buds are found in **papillae** (pah-pil'e), peglike projections of the tongue mucosa that give the tongue surface a slightly abrasive feel. These papillae are of three major types: filiform, fungiform, and circumvallate, and the latter two types house most of the taste buds. The mushroom-shaped **fungiform** (fun'jĭ-form) **papillae** are scattered over the entire tongue surface, but are most abundant at its tip and along its sides (Figure 16.1a). The round **circumvallate** (ser"kum-val'āt), or simply **vallate, papillae** are the largest and least numerous papillae; 7 to 12 of these form an inverted V at the back of the tongue (see Figure 16.1a). Taste buds are located in the epithelium of the side walls of the circumvallate papillae (Figure 16.1b) and on the tops of the fungiform papillae.

Each gourd-shaped taste bud consists of 40 to 100 *epithelial cells* of three major types: supporting cells, receptor cells, and basal cells (Figure 16.1c). **Supporting cells** form the bulk of the taste bud. They insulate the receptor cells called **gustatory,** or **taste, cells** from each other and from the surrounding tongue epithelium. Long microvilli called **gustatory hairs** project from the tips of both cell types and extend through a **taste pore** to the surface of the epithelium, where they are bathed by saliva. Apparently, the gustatory hairs are the sensitive portions *(receptor membranes)* of the gustatory cells. Coiling intimately around the gustatory cells are sensory dendrites that represent the initial part of the gustatory pathway to the brain. Each afferent fiber receives signals from several receptor cells within the taste bud. Because of their location, taste bud cells are subjected to huge amounts of friction and are routinely burned by hot foods. Luckily, they are among the most dynamic cells in the body, and they are replaced every seven to ten days. The **basal cells** act as stem cells, dividing and differentiating into supporting cells, which in turn give rise to new gustatory cells.

Basic Taste Sensations

Normally, our taste sensations are complicated mixtures of qualities. However, when taste is tested with pure chemical compounds, taste sensations can all be grouped into one of four basic qualities:* *sweet,*

---

*A fifth distinct taste sensation elicited by glutamate has recently been identified and named *umami* (Japanese: "deliciousness"). Glutamate appears to be responsible for the "beef taste" of steak and enhances the other taste qualities. It is the taste of the food additive monosodium glutamate. Taste receptors responding to umami are located on the posterior pharynx.

Think back to your studies of the cranial nerves as you view Figure 16.1a. Which taste sensitivity would be most impaired by injury to the glossopharyngeal nerve?

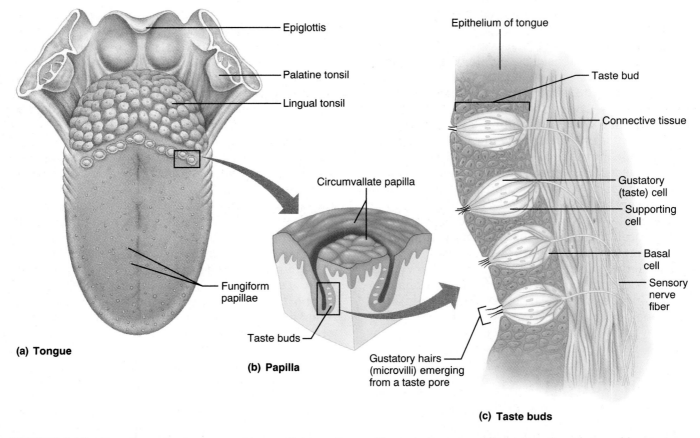

(a) Tongue

Epiglottis

Palatine tonsil

Lingual tonsil

Circumvallate papilla

Fungiform papillae

Taste buds

(b) Papilla

Gustatory hairs (microvilli) emerging from a taste pore

Epithelium of tongue

Taste bud

Connective tissue

Gustatory (taste) cell

Supporting cell

Basal cell

Sensory nerve fiber

(c) Taste buds

FIGURE 16.1 *Location, structure, and taste sensitivity of taste buds on the tongue.* **(a)** Taste buds on the tongue are associated with fungiform and vallate (circumvallate) papillae, peglike projections of the tongue mucosa. **(b)** A sectioned circumvallate papilla shows the positioning of the taste buds in its lateral walls. **(c)** An enlarged view of four taste buds, illustrating gustatory (taste), supporting, and basal cells.

sour, salty, and *bitter.* The sweet taste is elicited by many organic substances including sugars, saccharin, alcohols, some amino acids, and some lead salts (such as those found in lead paint). Sour taste is produced by acids, specifically their hydrogen ions (H^+) in solution. Salty taste is produced by metal ions (inorganic salts); table salt (sodium chloride) tastes the "saltiest." The bitter taste is elicited by alkaloids (such as quinine, nicotine, caffeine, morphine, and strychnine) as well as a number of nonalkaloid substances, such as aspirin. However, these differences are not absolute: Most taste buds respond to two, three, or all four taste qualities, and many substances produce a mixture of the basic taste sensations. Furthermore, some substances change in flavor as they move through the mouth. Saccharin is

initially tasted as sweet, but acquires a bitter aftertaste at the back of the tongue.

Although taste maps that spatially assign sweet receptors to the tip of the tongue, salty and sour receptors to the sides, and bitter receptors to the back are common in the literature, researchers have known for years that these mapped areas are misrepresentations, or are dubious at the very least. In reality, there are only slight differences in the localization of specific taste receptors in different regions of the tongue, and all modalities of taste can be elicited from all areas that contain taste buds.

Taste likes and dislikes have homeostatic value. A liking for sugar and salt helps satisfy the body's need for carbohydrates and minerals (as well as some amino acids). Many sour, naturally acidic foods (such as oranges, lemons, and tomatoes) are rich sources of vitamin C, an essential vitamin. Because many natural poisons and spoiled foods are bitter, our dislike for bitterness is protective. The position of the bitter receptors on the most posterior part of the tongue seems strange, however, because usually by the time we actually taste a bitter substance, we have swallowed some of it. However, many fruits and vegetables (e.g., grapefruit, kale, broccoli, and tofu) contain bitter chemicals that are powerful anticancer agents. Although many people learn to enjoy these and other bitter foods such as olives and quinine water, most animals scrupulously avoid bitter substances.

Physiology of Taste

Activation of Taste Receptors For a chemical to be tasted it must dissolve in saliva, diffuse into the taste pore, and contact the gustatory hairs. The gustatory cells contain synaptic vesicles, and binding of the food chemical to the gustatory cell membrane induces a depolarizing potential that causes the release of the synaptic vesicle contents, neurotransmitter. Binding of the neurotransmitter to the associated sensory dendrites triggers generator potentials that elicit action potentials in these fibers.

The different gustatory cells have different thresholds for activation. The bitter receptors (in line with their protective nature) detect substances present in minute amounts. Sour receptors are less sensitive; the sweet and salty receptors are least sensitive. Taste receptors adapt rapidly, with partial adaptation in 3–5 seconds and complete adaptation in 1–5 minutes.

Mechanism of Taste Transduction Transduction, as you may recall, is the process by which stimulus energy is converted into a nerve impulse. The mechanism of taste transduction is only partly known, but each taste quality appears to have its own special mechanism. Ionic stimuli act directly at apically located ion channels to depolarize the taste receptor cells. Salty taste seems to be due to Na^+ influx through sodium channels followed by Ca^{2+} influx; sour is mediated by H^+ and blockade of K^+ (or Na^+) channels. Bitter and sweet responses are mediated by G protein–dependent mechanisms that act via second messengers to promote depolarization by increasing intracellular levels of Ca^{2+} (bitter) or closing K^+ channels (sweet). The G protein that mediates transduction of bitter and sweet tastes has recently been identified and named *gustducin*.

The Gustatory Pathway

As described in Chapter 13, afferent fibers carrying taste information from the tongue are found primarily in two cranial nerve pairs. A branch of the **facial nerve** (VII), the *chorda tympani,* transmits impulses from taste receptors in the anterior two-thirds of the tongue, whereas the lingual branch of the **glos-**sopharyngeal nerve** (IX) services the posterior third and the pharynx just behind. Taste impulses from the few taste buds in the epiglottis and the lower pharynx are conducted primarily by the **vagus nerve** (X). These afferent fibers synapse in the **solitary nucleus** of the medulla, and from there impulses stream to the thalamus and ultimately to the gustatory cortex in the parietal lobes. Fibers also project to the hypothalamus and limbic system structures, regions that determine our appreciation of what we are tasting.

An important role of taste is to trigger reflexes involved in digestion. As taste impulses pass through the solitary nucleus, they initiate reflexes (via synapses with parasympathetic nuclei) that increase secretion of saliva into the mouth and gastric juice into the stomach. Saliva contains mucus that moistens food and digestive enzymes that begin the digestion of starch. Acidic foods are particularly strong stimulants of the salivary reflex. Additionally, when we eat revolting or foul-tasting substances, gagging or even reflexive vomiting may be initiated.

Influence of Other Sensations on Taste

Taste is 80% smell. When the olfactory receptors in the nasal cavity are blocked by nasal congestion (or pinching your nostrils), food is bland. Without smell, our morning coffee would lack its richness and simply taste bitter.

The mouth also contains thermoreceptors, mechanoreceptors, and nociceptors, and the temperature and texture of foods can enhance or detract from their taste. "Hot" foods such as chili peppers actually bring about their pleasurable effects by exciting pain receptors in the mouth.

The Olfactory Epithelium and the Sense of Smell

Although our olfactory sense (*olfact* = to smell) is far less acute than that of many other animals, the human nose is still no slouch in picking up small differences in odors. Some people capitalize on this ability by becoming wine tasters.

Localization and Structure of Olfactory Receptors

Like taste, olfaction detects chemicals in solution. The organ of smell is a yellow-tinged patch (about 5 cm^2) of pseudostratified epithelium called the **olfactory epithelium** located in the roof of the nasal cavity (Figure 16.2). Air entering the nasal cavity must make a hairpin turn to stimulate olfactory receptors before entering the respiratory passageway below, so the human olfactory epithelium is in a poor position for doing its job. (This is why sniffing, which draws

What is the importance of the layer of mucus that coats the olfactory epithelium surface?

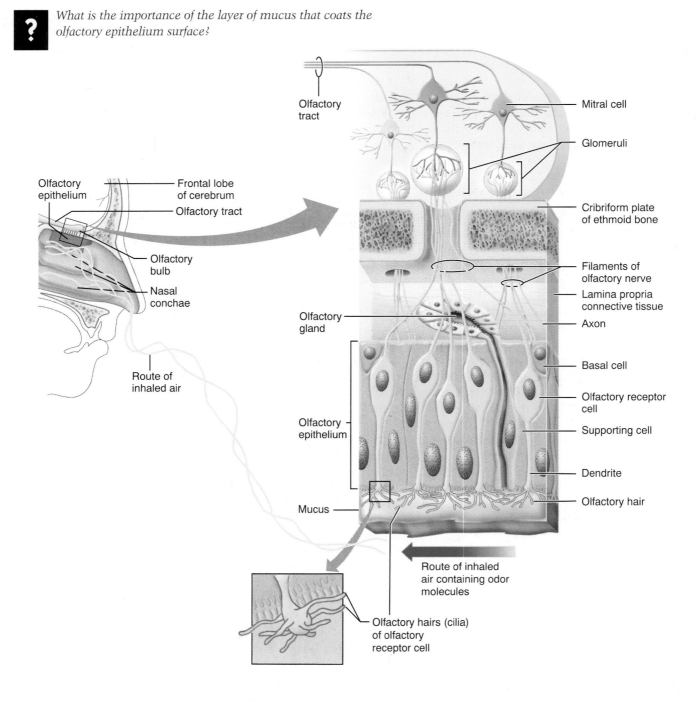

FIGURE 16.2   Olfactory receptors.
Site of olfactory epithelium in the superior aspect of the nasal cavity is shown on the left. On the right is an enlarged view illustrating the cellular composition of the olfactory epithelium and the course of the fibers of the olfactory nerve (I) through the ethmoid bone to synapse in the overlying olfactory bulb. The glomeruli and mitral cells (output cells) within the olfactory bulb are also shown.

more air superiorly across the olfactory epithelium, intensifies the sense of smell.) The olfactory epithelium covers the superior nasal concha on each side of the nasal septum, and contains millions of bowling pin–shaped **olfactory receptor cells.** These are surrounded and cushioned by columnar **supporting cells,** which make up the bulk of the penny-thin epithelial membrane (Figure 16.2). The supporting cells contain a yellow-brown pigment similar to lipofuscin, which gives the olfactory epithelium its yellow hue. At the base of the epithelium lie the short **basal cells.**

The olfactory receptor cells are unusual bipolar neurons. Each has a thin apical dendrite that terminates in a knob from which several long cilia radiate. These **olfactory cilia,** which substantially increase the receptive surface area, typically lie flat on the nasal ep-

It serves as a dissolving medium for odor molecules.

ithelium and are covered by a coat of thin mucus produced by the supporting cells and by olfactory glands in the underlying connective tissue. This mucus is a solvent that "captures" and dissolves airborne odor molecules. It is renewed continuously, flushing away "old" odor molecules so that new odor chemicals always have access to the receptor cilia. Unlike other cilia in the body, which beat rapidly and in a coordinated manner, olfactory cilia are largely nonmotile. The slender, unmyelinated axons of the olfactory receptor cells are gathered into small fascicles that collectively form the **filaments of the olfactory nerve** (cranial nerve I). They project superiorly through the openings in the cribriform plate of the ethmoid bone, where they synapse in the overlying olfactory bulbs.

Olfactory receptor cells are unique in that they are one of the few types of *neurons* that undergo noticeable turnover throughout adult life. (Remember, the taste receptors are epithelial cells.) Their typical life span is 60 days, after which they are replaced by differentiation of the basal cells in the olfactory epithelium.

Specificity of the Olfactory Receptors

Smell is difficult to research because any given odor (say, tobacco smoke) may be made up of hundreds of different chemicals. Taste has been neatly packaged into four or five taste qualities, but science has yet to discover any similar means for classifying smell. Humans can distinguish 10,000 or so chemicals, but research suggests that our olfactory receptors are stimulated by different combinations of a more limited number of olfactory qualities. Most promising in throwing light on this matter are the studies done in the early 1990s suggesting that there are at least 1000 "smell genes" that are active only in the nose. Each such gene encodes an *odorant binding protein,* a unique receptor protein, and it appears that each of the receptor proteins responds to several different odors and each odor binds to several different receptor types. However, each receptor cell has only one type of receptor protein.

Olfactory neurons are exquisitely sensitive—in some cases, just a few molecules activate them. As with taste, some of what we call smell is really pain. The nasal cavities contain pain receptors that respond to irritants such as the sharpness of ammonia, the hotness of chili peppers, and the "chill" of menthol. Impulses from these pain receptors reach the central nervous system via afferent fibers of the trigeminal nerves.

Physiology of Smell

Activation of the Olfactory Receptors For us to smell a particular chemical, it must be *volatile*—that is, it must be in the gaseous state as it enters the nasal cavity. Additionally, it must be sufficiently wa-

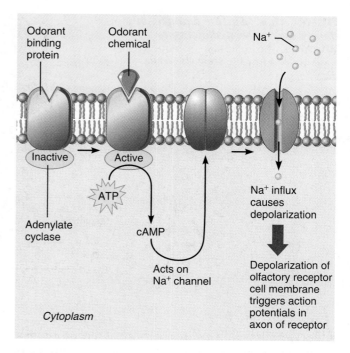

FIGURE 16.3 *Brief schematic of steps in the olfactory transduction process.* Odorant chemical binding to the G protein–associated receptor sets the cAMP second messenger system into motion, which in turn causes Na$^+$ channels to open. Influx of Na$^+$ causes depolarization.

ter soluble to dissolve in the fluid coating the olfactory epithelium. The dissolved chemicals stimulate the olfactory receptors by binding to protein receptors in the olfactory cilium membranes and opening specific ion (Na$^+$) channels. This leads to a receptor potential and ultimately (assuming threshold stimulation) to an action potential that is conducted to the first relay station in the olfactory bulb.

Mechanism of Smell Transduction The details of transduction in olfactory receptor cells are still to be fully worked out. However, there is little doubt that G proteins are involved (Figure 16.3) and that at least in some receptor cells (1) cyclic AMP is a second messenger and (2) opening of Na$^+$ channels leads to depolarization and impulse transmission.

The Olfactory Pathway

As noted above, axons of the olfactory receptor cells constitute the olfactory nerves that synapse in the overlying **olfactory bulbs,** the distal ends of the olfactory tracts (see Table 13.2). There, the filaments of the olfactory nerves synapse with **mitral** (mi'tral) **cells** (second-order neurons) in complex structures called **glomeruli** (glo-mer'u-li; "little balls") (see Figure 16.2). Axons from neurons bearing the same kind of receptor converge on a given type of glomerulus. Hence, each glomerulus represents a "file" that receives only one type of odor signal, and different

FIGURE 16.4 *Surface anatomy of the eye and associated accessory structures.* The eyelids are open more widely than normal to show the entire iris.

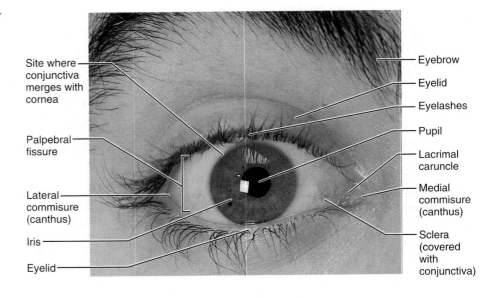

Site where conjunctiva merges with cornea

Palpebral fissure

Lateral commisure (canthus)

Iris

Eyelid

Eyebrow

Eyelid

Eyelashes

Pupil

Lacrimal caruncle

Medial commisure (canthus)

Sclera (covered with conjunctiva)

odors (which would excite a variety of receptor types) would be expected to activate different subsets of glomeruli. The mitral cells appear to play an important local integration role in olfaction; they refine the signal, amplify it, and then relay it. The olfactory bulbs also house *granule cells,* GABA-releasing cells that inhibit mitral cells (via dendrodendritic synapses), so that only highly excitatory olfactory impulses are transmitted. This inhibition is believed to contribute to the mechanism of *olfactory adaptation* and helps explain how a person working in a paper mill or sewage treatment plant can still enjoy lunch!

When the mitral cells are activated, impulses flow from the olfactory bulbs via the **olfactory tracts** (composed mainly of mitral cell axons) to two main destinations: (1) via the thalamus to the piriform lobe of the olfactory cortex and the part of the frontal lobe just above the orbit, where smells are consciously interpreted and identified, and (2) via the subcortical route to the hypothalamus, amygdala, and other regions of the limbic system, which elicits emotional responses to odors. Smells associated with danger—smoke, cooking gas, or skunk scent—trigger the sympathetic fight-or-flight response. As with tastes, appetizing odors cause increased salivation and stimulation of the digestive tract, and unpleasant odors can trigger protective reflexes such as sneezing and choking. Substances with a sharp smell, like ammonia, cause reflex cessation of breathing.

Homeostatic Imbalances of the Chemical Senses

While dysfunctions of both taste and smell do occur, most people seeking medical help for problems of the chemical senses have olfactory disorders, or *anosmias* (an-oz′me-ahz; "without smells"). Most anosmias result from head injuries that tear the olfactory nerves, the aftereffects of nasal cavity inflammation (due to a cold, an allergy, or smoking), physical obstruction of the nasal cavity (as by polyps), and aging. The culprit in one-third of all cases of chemical sense loss is zinc deficiency, and the cure is rapid once the right form of zinc supplement is prescribed. Zinc is a known growth factor for the receptors of the chemical senses.

Brain disorders can distort the sense of smell. Some people have *uncinate* (uns′ih-nāt) *fits,* olfactory hallucinations during which they experience a particular (usually unpleasant) odor, such as rotting meat. Some such cases are undeniably psychological, but many result from irritation of the olfactory pathway by brain surgery or head trauma. *Olfactory auras* experienced by some epileptics just before they have a seizure are transient uncinate fits.

THE EYE AND VISION

Vision is our dominant sense: Some 70% of all the sensory receptors in the body are in the eyes, and nearly half of the cerebral cortex is involved in some aspect of processing of visual information. The visual receptor cells (photoreceptors) sense and encode the patterns of light that enter the eye, and the brain uses these signals to fashion images of the world around us.

The adult **eye** is a sphere with a diameter of about 2.5 cm (1 inch). Only the anterior one-sixth of the eye's surface is visible (Figure 16.4); the rest is enclosed and protected by a cushion of fat and the walls of the bony orbit. The fat pad occupies nearly all of the orbit not occupied by the eye itself. The eye is a complex structure and only a small portion of its tissues are actually involved in photoreception. Before turning our attention to the eye itself, let us consider the accessory structures that protect it or aid its functioning.

Accessory Structures of the Eye

The **accessory structures** of the eye include the eyebrows, eyelids, conjunctiva, lacrimal apparatus, and extrinsic eye muscles.

Eyebrows

The **eyebrows** are short, coarse hairs that overlie the supraorbital margins of the skull (Figures 16.4 and 16.5a). They help shade the eyes from sunlight and prevent perspiration trickling down the forehead from reaching the eyes. Deep to the skin of the eyebrows are parts of the orbicularis oculi and corrugator muscles. Contraction of the orbicularis muscle depresses the eyebrow, whereas the corrugator moves the eyebrow medially.

Eyelids

Anteriorly, the eyes are protected by the mobile **eyelids** or **palpebrae** (pal'pĕ-bre). The eyelids are separated by the **palpebral fissure** ("eyelid slit") and meet at the medial and lateral angles of the eye—the **medial** and **lateral commissures (canthi)**, respectively (see Figure 16.4). The medial canthus sports a fleshy elevation called the **lacrimal caruncle** (kar'ung-kl; "a bit of flesh"). The caruncle contains sebaceous and sweat glands and produces the whitish, oily secretion (fancifully called the Sandman's eye-sand) that sometimes collects at the medial canthus, especially during sleep. In most Asian peoples, a vertical fold of skin called the *epicanthic fold* commonly appears on both sides of the nose and sometimes covers medial commissure.

The eyelids are thin, skin-covered folds supported internally by connective tissue sheets called **tarsal plates** (see Figure 16.5a). The tarsal plates also anchor the orbicularis oculi and levator palpebrae superioris muscles that run within the eyelid. The orbicularis muscle encircles the eye; when it contracts, the eye closes. Of the two eyelids, the larger upper one is much more mobile, mainly because of the **levator palpebrae superioris** muscle, which raises that eyelid to open the eye. The eyelid muscles are activated reflexively to cause blinking every 3–7 seconds and to protect the eye when it is threatened by foreign objects. Reflex blinking helps prevent desiccation (drying) of the eyes because each time we blink, accessory structure secretions (oil, mucus, and saline solution) are spread across the eyeball surface.

Projecting from the free margin of each eyelid are the **eyelashes.** The follicles of the eyelash hairs are richly innervated by nerve endings (root hair

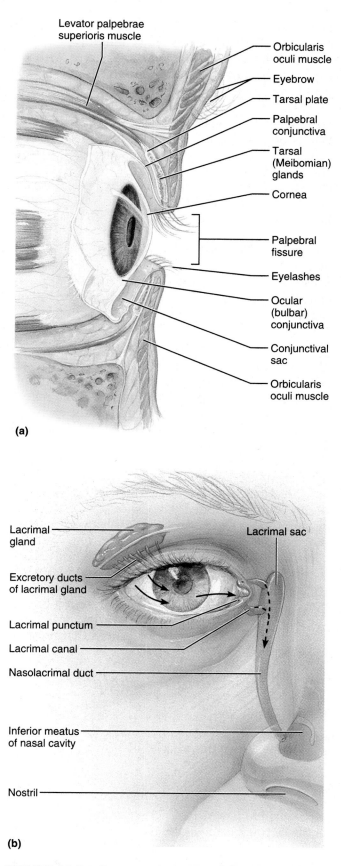

(a)

(b)

FIGURE 16.5 *Accessory structures of the eye.* **(a)** Sagittal section of the accessory structures associated with the anterior portion of the eye. **(b)** The lacrimal apparatus. Arrows indicate the direction of flow of lacrimal fluid after its release by the lacrimal gland.

plexuses), and anything that touches the eyelashes (even a puff of air) triggers reflex blinking.

Several types of glands are associated with the eyelids. The **tarsal,** or **Meibomian** (mi-bo′me-an), **glands** are embedded in the tarsal plates (see Figure 16.5a), and their ducts open at the eyelid edge just posterior to the eyelashes. These modified sebaceous glands produce an oily secretion that lubricates the eyelid and the eye and prevents the eyelids from sticking together. Associated with the eyelash follicles are a number of smaller, more typical sebaceous glands. Modified sweat glands called *ciliary glands* lie between the hair follicles (*cilium* = eyelash).

Homeostatic Imbalance

Infection of a tarsal gland results in an unsightly cyst called a *chalazion* (kah-la′ze-on; "swelling"). Inflammation of any of the smaller glands is called a *sty.* ■

Conjunctiva

The **conjunctiva** (kon″junk-ti′vah; "joined together") is a transparent mucous membrane. It lines the eyelids as the **palpebral conjunctiva** and reflects (folds back) over the anterior surface of the eyeball as the **ocular,** or **bulbar, conjunctiva** (see Figure 16.5a). The latter covers only the white of the eye, not the cornea (the clear "window" over the iris and pupil). The ocular conjunctiva is very thin, and blood vessels are clearly visible beneath it. (They are even more visible in irritated "bloodshot" eyes.) When the eye is closed, a slitlike space occurs between the conjunctiva-covered eyeball and eyelids. This so-called **conjunctival sac** is where a contact lens lies, and eye medications are often administered into its inferior recess. Although the conjunctiva protects the eye by preventing foreign objects from penetrating beyond the confines of the conjunctival sac, its major function is to produce a lubricating mucus that prevents the eyes from drying out.

Homeostatic Imbalance

Inflammation of the conjunctiva, called *conjunctivitis,* results in reddened, irritated eyes. *Pinkeye,* a conjunctival infection caused by bacteria or viruses, is highly contagious. ■

Lacrimal Apparatus

The **lacrimal** (lak′rĭ-mal; "tear") **apparatus** consists of the lacrimal gland and the ducts that drain the excess lacrimal secretions into the nasal cavity (Figure 16.5b). The **lacrimal gland** lies within the orbit above the lateral end of the eye and is visible through the conjunctiva when the lid is everted. It continually releases a dilute saline solution called **lacrimal secretion**—or, more commonly, **tears**—into the su-

perior part of the conjunctival sac through several small excretory ducts. Blinking spreads the tears downward and across the eyeball to the medial commissure, where they enter the paired **lacrimal canaliculi (canals)** via two tiny openings called **lacrimal puncta** (literally, "prick points"), visible as tiny red dots on the medial margin of each eyelid. From the lacrimal canals, the tears drain into the **lacrimal sac** and then into the **nasolacrimal duct,** which empties into the nasal cavity at the inferior nasal meatus.

Lacrimal fluid contains mucus, antibodies, and **lysozyme,** an enzyme that destroys bacteria. Thus, it cleanses and protects the eye surface as it moistens and lubricates it. When lacrimal secretion increases substantially, tears spill over the eyelids and fill the nasal cavities, causing congestion and the "sniffles." This happens when the eyes are irritated and when we are emotionally upset. In the case of eye irritation, enhanced tearing serves to wash away or dilute the irritating substance. The importance of emotionally induced tears is poorly understood. The fact that only humans shed emotional tears suggests that crying is important in reducing stress. Anyone who has had a good cry would probably agree, but this has been difficult to demonstrate scientifically.

Homeostatic Imbalance

Because the nasal cavity mucosa is continuous with that of the lacrimal duct system, a cold or nasal inflammation often causes the lacrimal mucosa to become inflamed and swell. This constricts the ducts and prevents tears from draining from the eye surface, causing "watery" eyes. ■

Extrinsic Eye Muscles

The movement of each eyeball is controlled by six straplike **extrinsic eye muscles.** These originate from the bony orbit (*extrins* = from the outside) and insert into the outer surface of the eyeball (Figure 16.6). These muscles allow the eyes to follow a moving object. They also provide external "guy-wires" that help to maintain the shape of the eyeball and hold it in the orbit.

Four of the extrinsic muscles, the *rectus muscles,* originate from a common tendinous ring, the **annular ring,** at the back of the orbit and run straight to their insertion on the eyeball. Their locations and the movements that they promote are clearly indicated by their names: **superior, inferior, lateral,** and **medial rectus muscles.** The actions of the two *oblique muscles* are less easy to deduce because they take rather strange paths through the orbit. They move the eye in the vertical plane when the eyeball is already turned medially by the rectus muscle. The **superior oblique muscle** originates in

Which three extrinsic eye muscles turn the eyeball laterally?

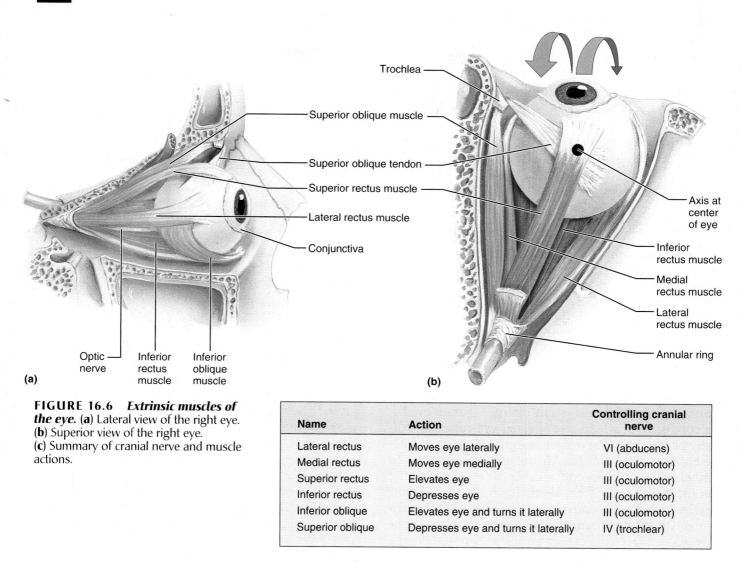

FIGURE 16.6 *Extrinsic muscles of the eye.* (**a**) Lateral view of the right eye. (**b**) Superior view of the right eye. (**c**) Summary of cranial nerve and muscle actions.

Name	Action	Controlling cranial nerve
Lateral rectus	Moves eye laterally	VI (abducens)
Medial rectus	Moves eye medially	III (oculomotor)
Superior rectus	Elevates eye	III (oculomotor)
Inferior rectus	Depresses eye	III (oculomotor)
Inferior oblique	Elevates eye and turns it laterally	III (oculomotor)
Superior oblique	Depresses eye and turns it laterally	IV (trochlear)

(c)

common with the rectus muscles, runs along the medial wall of the orbit, and then makes a right-angle turn and passes through a fibrocartilaginous loop suspended from the frontal bone called the **trochlea** (trok'le-ah; "pulley") before inserting on the superolateral aspect of the eyeball. Its contraction rotates the eye downward and somewhat laterally. The **inferior oblique muscle** originates from the medial orbit surface and runs laterally and obliquely to insert on the inferolateral eye surface. Hence, it rotates the eye up and laterally.

The four rectus muscles seem to provide all the eye movements we require—medial, lateral, superior, and inferior—so why the two oblique muscles?

The simplest way to answer this question is to point out that the superior and inferior recti cannot elevate or depress the eye *without also turning it medially* because they approach the eye from a posteromedial direction. For an eye to be *directly* elevated or depressed, the lateral pull of the oblique muscles is necessary to cancel the medial pull of the superior and inferior recti.

Except for the lateral rectus and superior oblique muscles, which are innervated respectively by the *abducens* and *trochlear nerves*, all extrinsic eye muscles are served by the *oculomotor nerves*. The actions and nerve supply of these muscles are summarized in Figure 16.6c. The courses of the associated cranial nerves are illustrated in Table 13.2.

The extrinsic eye muscles are among the most precisely and rapidly controlled skeletal muscles in

The lateral rectus and the superior and inferior obliques.

the entire body. This reflects their high axon-to-muscle-fiber ratio: The motor units of these muscles contain only 8 to 12 muscle cells and in some cases even fewer. The extrinsic muscles effect two basic types of eye movements:

1. **Saccades** (sah-kādz'; "jerks on the fibers") are small, jerky movements that quickly move the eye from one point to another, allowing the entire visual field to be seen in a short time. (The **visual field** is the field of view seen when the head is still.)

2. Slow **scanning movements** are *tracking* movements that allow us to follow a particular object moving in our visual fields or to keep our gaze focused on an object as we move our heads.

⚠ *Homeostatic Imbalance*

When movements of the external muscles of the two eyes are not perfectly coordinated, a person cannot properly focus the images of the same area of the visual field from each eye and so sees two images instead of one. This condition is called **diplopia** (dĭ-plo'pe-ah), or *double vision*. It can result from paralysis or weakness of certain extrinsic muscles, or it may be a temporary consequence of acute alcohol intoxication.

Congenital weakness of the external eye muscles may cause **strabismus** (strah-biz'mus; "cross-eyed"), a condition in which the affected eye rotates medially or laterally. To compensate, the normal and deviant eyes may alternate in focusing on objects. In other cases, only the controllable eye is used, and the brain begins to disregard inputs from the deviant eye, which then becomes functionally blind. Strabismus cannot be outgrown. Less severe cases may be treated either with eye exercises to strengthen the weak muscles or by temporarily placing a patch on the stronger eye, which forces the child to use the weaker eye. Surgery on the eye muscles is needed for unyielding conditions. ▪

Structure of the Eyeball

The eye itself, commonly called the **eyeball,** is a slightly irregular hollow sphere (Figure 16.7). Because the eyeball is shaped roughly like the globe of the earth, it is said to have poles. Its most anterior point is the **anterior pole;** its most posterior point, the **posterior pole.** Its wall is composed of three coats, or tunics, the fibrous, vascular, and sensory tunics. Its internal cavity is filled with fluids called *humors* that help to maintain its shape. The lens, the adjustable focusing apparatus of the eye, is supported vertically within the internal cavity, dividing it into *anterior* and *posterior segments*.

Tunics Forming the Wall of the Eyeball

The Fibrous Tunic The outermost coat of the eye, the **fibrous tunic,** is composed of dense avascular connective tissue. It has two obviously different regions: the sclera and the cornea. The **sclera** (skle'rah), forming the posterior portion and the bulk of the fibrous tunic, is glistening white and opaque. Seen anteriorly as the "white of the eye," the tough, tendonlike sclera (*sclera* = hard) protects and shapes the eyeball and provides a sturdy anchoring site for the extrinsic eye muscles. Posteriorly, where the sclera is pierced by the optic nerve, it is continuous with the dura mater of the brain. The anterior sixth of the fibrous tunic is modified to form the transparent **cornea,** which bulges anteriorly from its junction with the sclera. The regular arrangement of its collagen fibers makes the cornea crystal clear, allowing it to form a window that lets light enter the eye. As explained shortly, the cornea is also part of the light-bending apparatus of the eye.

The cornea is covered by epithelial sheets on both faces. The external sheet, a stratified squamous epithelium, helps to protect the cornea from abrasion. It merges with the ocular conjunctiva at the sclera-cornea junction, a region called the *limbus.* Epithelial cells that continually renew the cornea are located here. The deep *corneal endothelium,* composed of simple squamous epithelium, lines the inner face of the cornea. Its cells have active sodium pumps that maintain the clarity of the cornea: They continually eject sodium ions from the cornea into the fluids of the eyeball and water follows (along its osmotic gradients). This action prevents tissue fluid from accumulating between the tightly packed collagen fibers, which would force them apart and reduce the transparency of the cornea.

The cornea is well supplied with nerve endings, most of which are pain receptors. (For this reason, some people can never adjust to wearing contact lenses.) When the cornea is touched, blinking and increased tearing occur reflexively. Even so, the cornea is the most exposed part of the eye and is very vulnerable to damage from dust, slivers, and the like. Luckily, its capacity for regeneration and repair is extraordinary. Furthermore, it can be replaced surgically. The cornea is the only tissue in the body that can be transplanted from one person to another with little or no possibility of rejection. Because it has no blood vessels, it is beyond the reach of the immune system.

The Vascular Tunic (Uvea) The **vascular tunic,** or middle coat of the eyeball, is also called the **uvea** (u've-ah; "grape"). The pigmented uvea has three regions: choroid, ciliary body, and iris (see Figure 16.7).

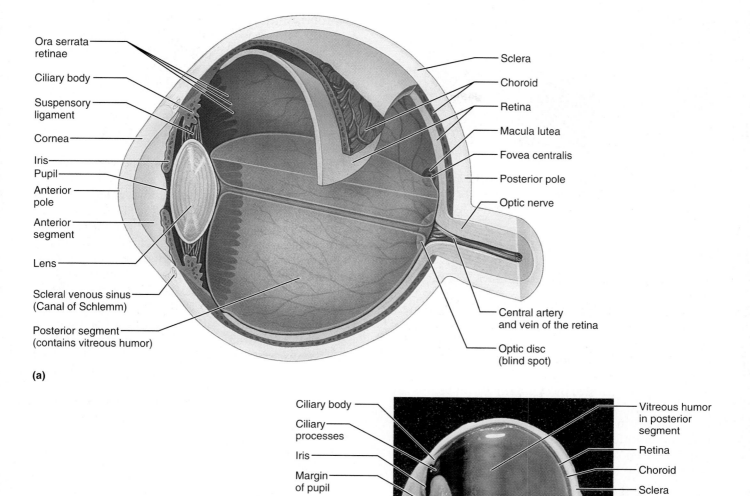

FIGURE 16.7 *Internal structure of the eye (sagittal section).* (**a**) Diagrammatic view. The vitreous humor is illustrated only in the bottom half of the eyeball. (**b**) Photograph of the human eye.

The **choroid** is a highly vascular, dark brown membrane (*choroid* = membranelike) that forms the posterior five-sixths of the uvea. Its blood vessels provide nutrition to all eye tunics. Its brown pigment, produced by melanocytes, helps absorb light, preventing it from scattering and reflecting within the eye (which would cause visual confusion). The choroid is incomplete posteriorly where the optic nerve leaves the eye. Anteriorly, it becomes the **ciliary body,** a thickened ring of tissue that encircles the lens. The ciliary body consists chiefly of interlacing smooth muscle bundles called **ciliary muscles,** which are important in controlling lens shape. Near the lens, its posterior surface is thrown into radiating folds called **ciliary processes,** which contain the

capillaries that secrete the fluid that fills the cavity of the anterior segment of the eyeball. The **suspensory ligament,** or **zonule,** extends from the ciliary processes to the lens; this halo of fine fibers encircles and helps hold the lens in its upright position in the eye.

The **iris,** the visible colored part of the eye, is the most anterior portion of the uvea. Shaped like a flattened doughnut, it lies between the cornea and the lens and is continuous with the ciliary body posteriorly. Its round central opening, the **pupil,** allows light to enter the eye. The iris is made up of two smooth muscle layers with bunches of sticky elastic fibers that congeal into a random pattern before birth. Its muscle fibers allow it to act as a reflexively

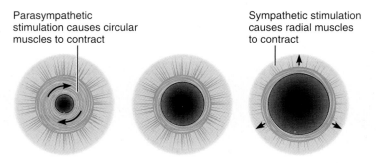

Parasympathetic stimulation causes circular muscles to contract

Sympathetic stimulation causes radial muscles to contract

FIGURE 16.8 Pupil dilation and constriction, anterior view. The circular and radial muscles of the iris respond to stimulation from different parts of the autonomic nervous system.

activated diaphragm to vary pupil size (Figure 16.8). In close vision and bright light, the circular layer muscles contract and the pupil constricts. In distant vision and dim light, the radially arranged muscles contract and the pupil dilates, allowing more light to enter the eye. As described in Chapter 14 and reviewed in Figure 16.8, pupillary dilation and constriction are controlled by sympathetic and parasympathetic fibers, respectively.

Changes in pupil size may also reflect our interests and emotional reactions to what we are seeing. Pupil dilation often occurs when the subject matter is appealing and during problem solving. (Computing your taxes should make your pupils get bigger and bigger.) On the other hand, boredom or subject matter that is personally repulsive causes pupil constriction.

Although irises come in different colors (iris = rainbow), they contain only brown pigment. When they have a lot of pigment, the eyes appear brown or black. If the amount of pigment is small and restricted to the posterior surface of the iris, the shorter wavelengths of light are scattered from the unpigmented parts, and the eyes appear blue, green, or gray. This is similar to the scattering of light that makes the sky appear blue. Most newborn babies' eyes are slate gray or blue because their iris pigment is not yet developed.

The Sensory Tunic (Retina) The innermost tunic is the delicate, two-layered **retina** (ret'ĭ-nah). The outer **pigmented layer,** a single-cell-thick lining, abuts the choroid, and extends anteriorly to cover the ciliary body and the posterior face of the iris. Its pigmented epithelial cells, like those of the choroid, absorb light and prevent it from scattering in the eye. They also act as phagocytes and store vitamin A needed by the photoreceptor cells. The transparent inner **neural (nervous) layer** extends anteriorly to the posterior margin of the ciliary body. This junction is called the **ora serrata retinae,** literally the

saw-toothed margin of the retina (see Figure 16.7). Actually an outpocketing of the brain, the retina contains millions of photoreceptors that transduce light energy as well as other neurons involved in the processing of light stimuli and glia. Although these two layers are very close together, they are not fused. Notice that while the retina is commonly called the **sensory tunic,** only its neural layer plays a direct role in vision.

From its posterior to anterior aspects, the neural layer is composed of three main types of neurons: **photoreceptors, bipolar cells,** and **ganglion cells** (Figure 16.9). Local currents produced in response to light spread from the photoreceptors (abutting the pigmented layer) to the bipolar neurons and then to the innermost ganglion cells, where action potentials are generated. The ganglion cell axons make a right-angle turn at the inner face of the retina, then leave the posterior aspect of the eye as the thick optic nerve. The retina also contains other types of neurons—horizontal cells and amacrine cells—whose processes run laterally. The role of these cells in visual processing is described shortly. The **optic disc,** where the optic nerve exits the eye, is a weak spot in the **fundus** (posterior wall) of the eye because it is not reinforced by the sclera. The optic disc is also called the **blind spot** because it lacks photoreceptors and light focused on it cannot be seen. Nonetheless, we do not usually notice these gaps in our vision because the brain uses a sophisticated process called *filling in* to deal with absence of input.

The quarter-billion photoreceptors found in the neural retinas are of two types: rods and cones. The more numerous **rods** are our dim-light and peripheral vision receptors. They are far more sensitive to light than cones are, but they do not provide either sharp images or color vision. This is why colors are indistinct and edges of objects appear fuzzy in dim light and at the edges of our visual field. **Cones,** by contrast, operate in bright light and provide high-acuity color vision. Lateral to the blind spot of each eye, and located precisely at the eye's posterior pole, is an oval region called the **macula lutea** (mak'u-lah lu'te-ah; "yellow spot") with a minute (0.4 mm) pit in its center called the **fovea centralis** (see Figure 16.7). In this region, the retinal structures abutting the vitreous humor are displaced to the sides. This allows light to pass almost directly to the photoreceptors (mostly cones) rather than through several retinal layers, and it greatly enhances visual acuity. The fovea contains only cones; the macula contains mostly cones; and from the edge of the macula toward the retina periphery, cone density declines gradually. The retina periphery contains only rods, which continuously decrease in density from there to the macula. Only the foveae (plural of fovea) have a sufficient cone density to provide detailed color

Which retinal cells are the output *cells?*

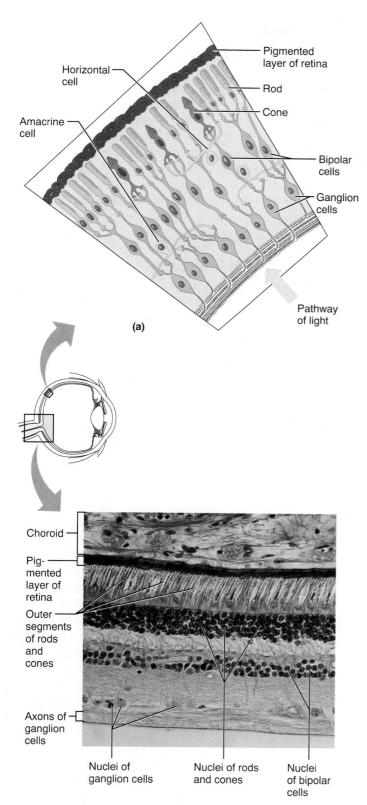

Pigmented layer of retina

Horizontal cell

Rod

Amacrine cell

Cone

Bipolar cells

Ganglion cells

Pathway of light

(a)

Choroid

Pig-mented layer of retina

Outer segments of rods and cones

Axons of ganglion cells

Nuclei of ganglion cells

Nuclei of rods and cones

Nuclei of bipolar cells

(c)

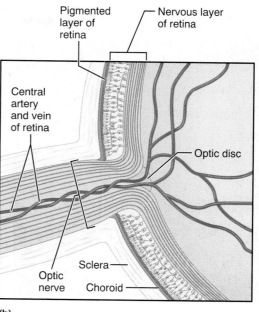

Pigmented layer of retina

Nervous layer of retina

Central artery and vein of retina

Optic disc

Sclera

Optic nerve

Choroid

(b)

FIGURE 16.9 *Microscopic anatomy of the retina.*
(**a**) Schematic view of the three major neuronal populations (photoreceptors, bipolar cells, and ganglion cells) composing the neural layer of the retina. Notice that light passes through the retina to excite the photoreceptor cells (as indicated by the yellow arrow). The flow of electrical signals occurs in the opposite direction. (**b**) Schematic view of the posterior aspect of the eyeball illustrating how the axons of the ganglion cells form the optic nerve, which leaves the back of the eyeball at the optic disc. (**c**) Photomicrograph of the retina (280×).

? *Which of the areas shown here would have the highest cone density?*

Central artery and vein
emerging from the optic disc

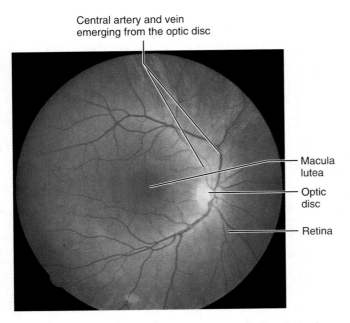

Macula lutea

Optic disc

Retina

FIGURE 16.10 *Part of the posterior wall (fundus) of the retina as seen with an ophthalmoscope.* Note the optic disc from which the retinal blood vessels radiate.

vision, so anything we wish to view critically is focused on the foveae. Because each fovea is only about the size of the head of a pin, not more than a thousandth of the entire visual field is in *hard focus* (foveal focus) at a given moment. Consequently, for us to visually comprehend a scene that is rapidly changing (as when we drive in traffic), our eyes must flick rapidly back and forth (in fast saccades movements) to provide the foveae with images of different parts of the visual field.

The neural retina receives its blood supply from two sources. The outer third (containing photoreceptors) is supplied by vessels in the choroid. The inner two-thirds is served by the **central artery** and **central vein of the retina,** which enter and leave the eye through the center of the optic nerve. Radiating outward from the optic disc, these vessels give rise to a rich vascular network that weaves through the retina's inner face (*retin* = net) and is clearly seen when the eyeball interior is examined with an ophthalmoscope (Figure 16.10). The fundus of the eye is the only place in the body where small blood vessels can be observed directly in a living person.

⚠ *Homeostatic Imbalance*

The pattern of vascularization of the retina makes it susceptible to *retinal detachment*. This condition, in which the pigmented and nervous layers separate (detach) and allow the jellylike vitreous humor to seep between them, can cause permanent blindness because it deprives the neural retina of its nutrient source. It usually happens when the retina is torn during a traumatic blow to the head or when the head stops moving suddenly and then is jerked in the opposite direction (as in bungee jumping). The symptom that victims most often describe is "a curtain being drawn across the eye," but some people see sootlike spots or light flashes. If diagnosed early, it is often possible to reattach the retina with a laser or by cryosurgery (local application of extreme cold) before photoreceptor damage becomes permanent. ■

Internal Chambers and Fluids

As noted earlier, the lens and its halolike suspensory ligament divide the eye into two segments, the anterior segment in front of the lens and the larger posterior segment behind it (Figure 16.7a). The **posterior segment** (cavity) is filled with a clear gel called **vitreous humor** (*vitre* = glassy), consisting of a network of fine collagenic fibrils embedded in a viscous ground substance that binds tremendous amounts of water. Vitreous humor (1) transmits light, (2) supports the posterior surface of the lens and holds the neural retina firmly against the pigmented layer, and (3) contributes to intraocular pressure, helping to counteract the pulling force of the extrinsic eye muscles.

The **anterior segment** (cavity) (Figure 16.11) is partially subdivided by the iris into the **anterior chamber** (between the cornea and the iris) and the **posterior chamber** (between the iris and the lens). The *entire* anterior segment is filled with **aqueous humor,** a clear fluid similar in composition to blood plasma. Unlike the vitreous humor, which forms in the embryo and lasts for a lifetime, aqueous humor forms and drains continually and is in constant motion. It filters from the capillaries of the ciliary processes into the posterior chamber and freely diffuses through the vitreous humor in the posterior segment before returning to the anterior segment. After flowing through the pupil into the anterior chamber, it drains into the venous blood via the **scleral venous sinus (canal of Schlemm),** an unusual venous channel that encircles the eye in the angle at the sclera-cornea junction. Normally, aqueous humor is produced and drained at the same rate. As a result, a constant intraocular pressure of about 16 mm Hg is maintained, which helps to support the eyeball internally. Aqueous humor supplies nutrients and oxygen to the lens and cornea, and to some

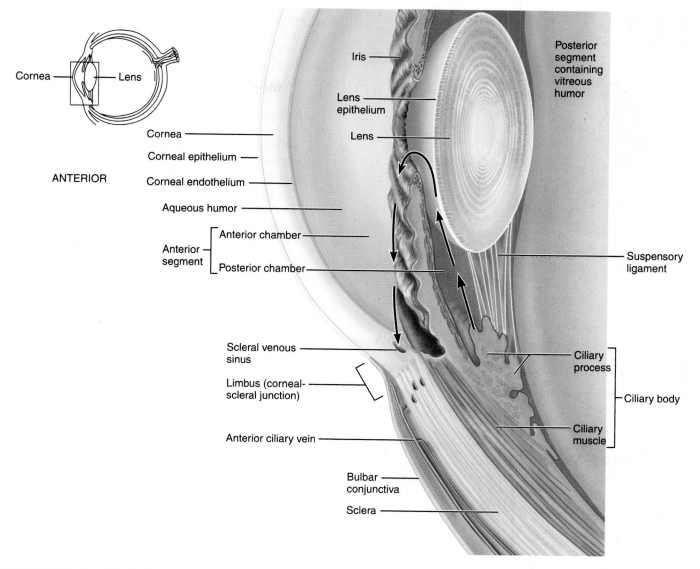

FIGURE 16.11 *Circulation of aqueous humor.* The anterior segment in front of the lens is incompletely divided into the anterior chamber (anterior to the iris) and the posterior chamber (behind the iris), which are continuous through the pupil. Aqueous humor, which fills the chambers and diffuses freely through the vitreous humor in the posterior segment, is formed by filtration from capillaries in the ciliary processes and is reabsorbed into the venous blood by the scleral venous sinus. The arrows indicate the circulation pathway.

cells of the retina, and it carries away their metabolic wastes.

Homeostatic Imbalance

If the drainage of aqueous humor is blocked, pressure within the eye may increase to dangerous levels and cause compression of the retina and optic nerve—a condition called **glaucoma** (glaw-ko'mah). The eventual result is blindness (*glaucoma* = vision growing gray) unless the condition is detected early. Unfortunately, many forms of glaucoma steal sight so slowly and painlessly that people do not realize they have a problem until the damage is done. Late signs include seeing halos around lights and blurred vision. The glaucoma ex-amination is simple. The intraocular pressure is determined by directing a puff of air at the sclera and measuring the amount of scleral deformation it causes. This exam should be done yearly after the age of 40. Treatment begins with eye drops (miotics) that increase the rate of aqueous humor drainage and progresses to surgery. ■

Lens

The **lens** is a biconvex, transparent, flexible structure that can change shape to allow precise focusing of light on the retina. It is enclosed in a thin, elastic capsule and held in place just posterior to the iris by the suspensory ligament (see Figures 16.7 and 16.11). Like the cornea, the lens is avascular; blood vessels interfere with transparency.

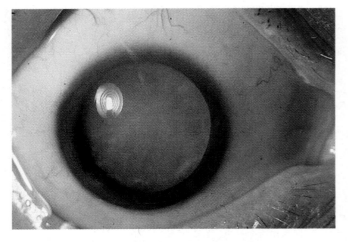

FIGURE 16.12 *Photograph of a cataract (due to a clouded lens).*

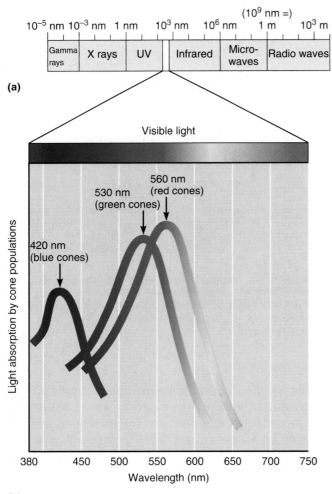

(a)

(b)

FIGURE 16.13 *The electromagnetic spectrum and cone cell sensitivities.* **(a)** The electromagnetic spectrum extends from the very short gamma waves to the long radio waves. The wavelengths of the electromagnetic spectrum are indicated in nanometers (1 nm = 10^{-9} m). The spectrum of visible light constitutes only a small portion of this energy range. **(b)** Sensitivities of the three cone types to the different wavelengths of the visible spectrum.

The lens has two regions: the lens epithelium and the lens fibers (see Figure 16.11). The **lens epithelium,** confined to the anterior lens surface, consists of cuboidal cells that eventually differentiate into the **lens fibers** that form the bulk of the lens. The lens fibers, which are packed tightly together like the layers in an onion, contain no nuclei and few organelles. They do, however, contain precisely folded proteins called **crystallins.** The structure of the crystallins makes them transparent, and their enzymatic properties allow them to convert sugar into energy for use by the lens. Since new lens fibers are continually added, the lens enlarges throughout life, becoming denser, more convex, and less elastic, all of which gradually impair its ability to focus light properly.

⬛ Homeostatic Imbalance

A **cataract** ("waterfall") is a clouding of the lens that causes the world to appear distorted, as if seen through frosted glass (Figure 16.12). Some cataracts are congenital, but most result from age-related hardening and thickening of the lens or are a secondary consequence of diabetes mellitus. Heavy smoking and frequent exposure to intense sunlight increase the risk for cataracts, whereas long-term dietary supplementation with vitamin C appears to significantly decrease the risk. Whatever the promoting factors, the *direct* cause of cataracts seems to be inadequate delivery of nutrients to the deeper lens fibers. The metabolic changes that result promote clumping of the crystallin proteins. Fortunately, the offending lens can be surgically removed and an artificial lens implanted to save the patient's sight. ■

Physiology of Vision

Overview: Light and Optics

To really comprehend the function of the eye as a photoreceptor organ, we need to have some understanding of the properties of light.

Wavelength and Color Electromagnetic radiation includes all energy waves, from the long radio waves (with wavelengths measured in meters) to the very short gamma (γ) waves and X rays with wavelengths of 1 nm and less. Our eyes respond to the part of the spectrum called **visible light,** which has a wavelength range of approximately 400–700 nm (Figure 16.13a). (1 nm = 10^{-9} m, or one-billionth of a meter.) Visible light travels in the form of waves, and its wavelengths can be measured very accurately. However, light is actually composed of small particles or packets of energy called **photons** or **quanta.** Attempts to reconcile these two findings have led to the present concept of light as packets of energy (photons) traveling in a wavelike fashion at very high speeds (186,000 miles per second or about 300,000 km/s). Thus, light can be thought of as a vibration of pure energy ("a bright wiggle") rather than a material substance.

When visible light passes through a prism, each of its component waves bends to a different degree, so that the beam of light is dispersed and a **visible spectrum,** or band of colors, is seen (see Figure 16.13). (In the same manner, the rainbow seen during a summer shower represents the collective prismatic effects of all the tiny water droplets suspended in air.) The visible spectrum progresses from red to violet. Red wavelengths are the longest and have the lowest energy, whereas the violet wavelengths are the shortest and most energetic. Objects have color because they absorb some wavelengths and reflect others. Things that look white reflect all wavelengths of light, whereas black objects absorb them all. A red apple reflects mostly red light; grass reflects more of the green.

Refraction and Lenses

Light travels in straight lines and is easily blocked by any nontransparent object. Like sound, light can reflect, or bounce, off a surface. **Reflection** of light by objects in our environment accounts for most of the light reaching our eyes.

When light is traveling in a given medium, its speed is constant. But when it passes from one transparent medium into another with a different density, its speed changes. Light speeds up as it passes into a less dense medium and slows as it passes into a denser medium. Because of these changes in speed, bending or **refraction** of a light ray occurs when it meets the surface of a different medium at an oblique angle rather than at a right angle (perpendicular). The greater the incident angle, the greater the amount of bending. Figure 16.14 demonstrates the consequence of light refraction when a spoon is placed in a half-full glass of water: The spoon appears to break at the air-water interface.

A lens is a piece of transparent material curved on one or both surfaces. Since light hits the curve at an angle, it is refracted. If the lens surface is convex, that is, thickest in the center like a camera lens, the light rays are bent so that they converge (come together) or intersect at a single point called the **focal point** (Figure 16.15). In general, the thicker (more convex) the lens, the more the light is bent and the shorter the focal distance (distance between the lens and focal point). The image formed by a convex lens, called a **real image,** is upside down and reversed from left to right. On the other hand, concave lenses, which are thicker at the edges than at the center, like those of magnifying glasses, diverge the light (bend it outward) so that the light rays move away from each other. Consequently, concave lenses prevent light from focusing and extend the focal distance.

FIGURE 16.14 *A spoon standing in a glass of water appears to be broken at the water-air interface.* This phenomenon occurs because light is bent toward the perpendicular when it travels from a less dense to a more dense medium (air to water in this example).

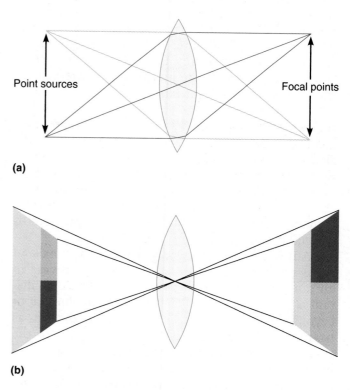

(a)

(b)

FIGURE 16.15 *Bending of light points by a convex lens.* **(a)** How two light points are focused on the opposite side of the lens. **(b)** Formation of an image by a convex lens. Notice that the image is upside down and reversed from right to left.

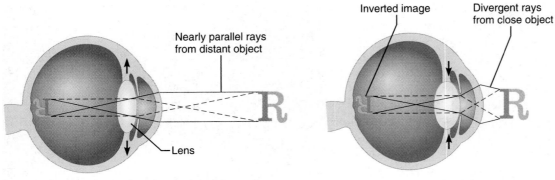

(a) Lens is flattened for distant vision

Nearly parallel rays from distant object

Lens

(b) Lens bulges for close vision

Inverted image

Divergent rays from close object

FIGURE 16.16 *Focusing for distant and close vision.* **(a)** Light from distant objects (over 6 m away) approaches as parallel rays and, in the normal eye, need not be adjusted for proper focusing on the retina. **(b)** Light from close objects (less than 6 m away) tends to diverge and lens convexity must be increased (accommodation) for proper focusing. Notice in both cases that the image formed on the retina is inverted and reversed from left to right (i.e., it is a real image).

Focusing of Light on the Retina

As light passes from air into the eye, it moves sequentially through the cornea, aqueous humor, lens, and vitreous humor, and then passes *through the entire thickness of the neural layer of the retina* to excite the photoreceptors that abut the pigmented layer (see Figures 16.7 and 16.9). During its passage, light is bent three times: as it enters the cornea and on entering and leaving the lens. The cornea is responsible for most light refraction in the eye; but its thickness is constant, so its light-bending power is unchanging. On the other hand, the lens is highly elastic, and its curvature can be actively changed to allow fine focusing of the image. The humors are of little importance in light refraction.

Focusing for Distant Vision Our eyes are best adapted ("preset to focus") for distant vision. To look at distant objects, we need only aim our eyeballs so that they are both fixated on the same spot. The **far point of vision** is that distance beyond which no change in lens shape (accommodation) is needed for focusing. For the normal or **emmetropic** (em″ĕ-tro′pik) eye, the far point is 6 m (20 feet).

Any object being viewed can be said to consist of many small points, with light radiating outward in all directions from each point. However, since distant objects appear smaller, light from an object at or beyond the far point of vision approaches the eyes as nearly parallel rays and is focused precisely on the retina by the fixed refractory apparatus (cornea and humors) and the at-rest lens (Figure 16.16a). During distant vision, the ciliary muscles are completely relaxed, and the lens (stretched flat by tension in the suspensory ligaments) is as thin as it gets. Consequently, it is at its lowest refractory power.

Focusing for Close Vision Light from objects less than 6 m away diverges as it approaches the eyes and it comes to a focal point farther from the lens. Thus, while distant vision is the natural state, close vision demands that the eye make active adjustments. To restore focus, three processes—accommodation of the lenses, constriction of the pupils, and convergence of the eyeballs—must occur simultaneously. The signal that induces this trio of reflex responses appears to be a blurring of the retinal image.

1. Accommodation of the lenses. The process that increases the refractory power of the lens so that diverging light rays are bent more sharply is called **accommodation.** This occurs as the ciliary muscles contract. This action pulls the ciliary body anteriorly and inward toward the pupil and releases tension in the suspensory ligament. No longer stretched, the elastic lens recoils and bulges, providing the shorter focal length needed to focus the image of a close object on the retina (Figure 16.16b). Contraction of the ciliary muscles is controlled mainly by the parasympathetic fibers of the oculomotor nerves.

The closest point on which we can focus clearly is called the **near point of vision,** and it represents the maximum bulge the lens can achieve. Although we can see objects closer to us than the near point, they are out of focus and fuzzy. In young adults with emmetropic vision, the near point is 10 cm (4 inches) from the eye. However, it is closer in children and gradually recedes with age, explaining why children can hold their books very close to their faces while many elderly people must hold the newspaper at arm's length. The gradual loss of accommodation with age reflects the lens's decreasing elasticity. In many people over the age of 50, the lens is nonaccommodating, a condition known as *presbyopia* (pres″be-o′pe-ah), literally "old person's vision."

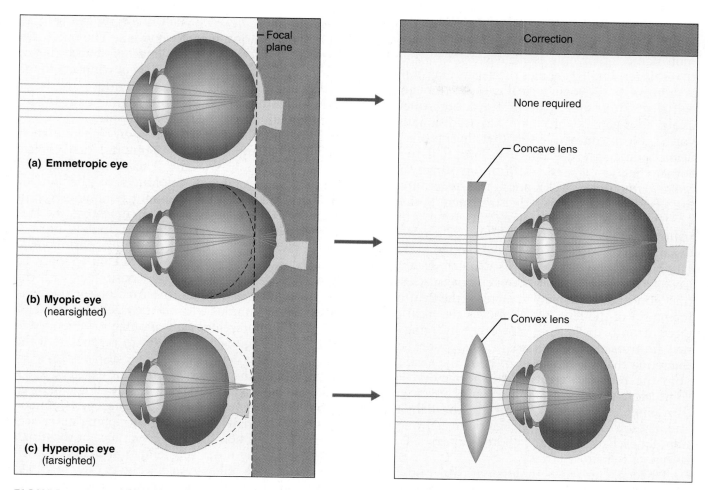

FIGURE 16.17 *Problems of refraction.* (**a**) In the emmetropic (normal) eye, light from both near and distant objects is focused properly on the retina. (**b**) In a myopic eye, light from a distant object is brought to a focal point before reaching the retina and then diverges again. (**c**) In the hyperopic eye, light from a near object is brought to a focal point behind (past) the retina. (Refractory effect of the cornea is ignored.)

2. Constriction of the pupils. The circular (constrictor) muscles of the iris enhance the effect of accommodation by reducing the size of the pupil (see Figure 16.8) toward 2 mm. This **accommodation pupillary reflex,** mediated by parasympathetic fibers of the oculomotor nerves, prevents the most divergent light rays from entering the eye. Such rays would pass through the extreme edge of the lens and would not be focused properly, and so would cause blurred vision. Pupillary constriction for close vision makes the eye act like a pinhole camera and increases the clarity and depth of focus.

3. Convergence of the eyeballs. Synchronized movements of the eyeballs (promoted by the extrinsic eye muscles) allow us to fix our gaze on something. The goal is always to keep the object being viewed focused on the retinal foveae. When we look at distant objects, both eyes are directed either straight ahead or to one side to the same degree, but when we fixate on a close object, our eyes converge. **Convergence,** controlled by somatic motor fibers of the oculomotor nerves, is the medial rotation of the eyeballs by the medial rectus muscles so that each is directed toward the object being viewed. The closer that object, the greater the degree of convergence required; when you focus on the tip of your nose, you "go cross-eyed."

Reading or other close work requires almost continuous accommodation, pupillary constriction, and convergence. This is why prolonged periods of reading tire the eye muscles and can result in *eyestrain.* When you read for an extended time, it is helpful to look up and stare into the distance occasionally. This temporarily relaxes the intrinsic eye muscles.

Homeostatic Imbalances of Refraction Visual problems related to refraction can result from a hyperrefractive (overconverging) or hyporefractive (underconverging) lens or from structural abnormalities of the eyeball.

Myopia (mi-o'pe-ah; "short vision") occurs when distant objects are focused not on, but in front of, the retina (Figure 16.17b). Myopic people can see close objects without problems because of the active ability of their lens to accommodate to the extent necessary, but distant objects are blurred. The common name for myopia is *nearsightedness.* (Notice that this terminology names the aspect of vision that is *unimpaired.*) Myopia typically results from an eyeball that is too long. One of every four American adults is myopic. Correction has traditionally involved use of concave lenses that diverge the light before it enters the eye, but procedures to flatten the cornea slightly—a painless 10-minute surgery called *radial keratotomy,* or PRK and LASIK procedures using a laser—have offered other treatment options.

Hyperopia (hy′per-o″pe-ah; "far vision"), or *farsightedness,* occurs when the parallel light rays from distant objects are focused *behind* the retina (Figure 16.17c). Hyperopic individuals can see distant objects perfectly well because their ciliary muscles contract almost continuously to increase the light-bending power of the lens, which moves the focal point forward onto the retina. However, diverging light rays from *nearby* objects are focused so far behind the retina that the lens cannot bring the focal point onto the retina even at its full refractory power. Thus, close objects appear blurry, and convex corrective lenses are needed to converge the light more strongly for close vision. Hyperopia usually results from an eyeball that is too short or a lazy lens (one with poor refractory power).

Unequal curvatures in different parts of the lens (or cornea) lead to blurry images because points of light are focused on the retina as lines (not points). This refractory problem is **astigmatism** (*astigma* = not a point). Special cylindrically ground lenses and laser procedures are used to correct this problem.

Photoreception

Once light is focused on the retina, the photoreceptors come into play. We will approach **photoreception,** the process by which the eye detects light energy, by examining several interrelated topics. First, we will describe the functional anatomy of photoreceptor cells, then the chemistry of their visual pigments and their response to light. Finally, activation of the photoreceptors will be covered.

Functional Anatomy of the Photoreceptors Although photoreceptors are modified neurons, structurally they resemble tall epithelial cells turned upside down with their "tips" immersed in the pigmented layer of the retina (Figure 16.18a). Moving from the pigmented layer into the neural layer, both the rods and cones have an **outer segment** (the receptor region) joined to an **inner segment** by a stalk containing a cilium. The inner segment connects to the *cell body,* which is continuous with an *inner fiber* bearing synaptic endings. The outer segment of the rods is slender and rod shaped (hence their name) and their inner segment connects to the cell body by the *outer fiber.* By contrast, the fatter cones have a short conical outer segment and the inner segment directly joins the cell body.

The outer segments contain an elaborate array of **visual pigments (photopigments)** that change shape as they absorb light. These are packaged in discs that develop as infoldings of the surface (plasma) membrane. This coupling of photoreceptor pigments to cellular membranes magnifies the surface area available for trapping light. In rods, most of these discs pinch off and become discontinuous—stacked like a row of pennies in a coin wrapper. In cones the disc membranes are continuous with the plasma membrane; thus, the interiors of the cone discs are continuous with the extracellular space.

The photoreceptor cells are highly vulnerable to damage and immediately begin to degenerate if the retina becomes detached. They are also destroyed by intense light, the very energy they are designed to detect. How is it, then, that we do not all gradually go blind? The answer lies in the photoreceptors' unique system for renewing their light-trapping outer segment. In rods, new discs are assembled at the proximal end of the outer segment from materials synthesized in the cell body at the end of each night. As new discs are made, they push the others peripherally. The discs at the tip of the outer segment continually fragment off and are phagocytized by cells of the pigmented layer. The tips of the outer segments of cones are also removed and renewed (but at the end of each day); however, the precise way the membrane elements migrate into the outer segment is still unclear. Like photopigment synthesis, phagocytic activity occurs in a circadian (approximately 24-hour) rhythm. This retinal "clock" is apparently "set" by varying levels of melatonin in the eye. Rod tips are phagocytized in the morning after first exposure to light. Cone phagocytosis, triggered by darkness, occurs at night.

Because the rods and each of the three cone types contain unique visual pigments, they absorb different wavelengths of light and have different thresholds for activation. Rods, for example, (1) are very sensitive (respond to very dim light), making them best suited for night vision and peripheral vision, and (2) absorb all wavelengths of visible light, but their inputs are perceived only in gray tones. Cones, on the other hand, (1) need bright light for activation (have low sensitivity), but (2) have pigments that furnish a vividly colored view of the world.

What is the advantage of having the photoreceptor pigment form part of the disc membranes?

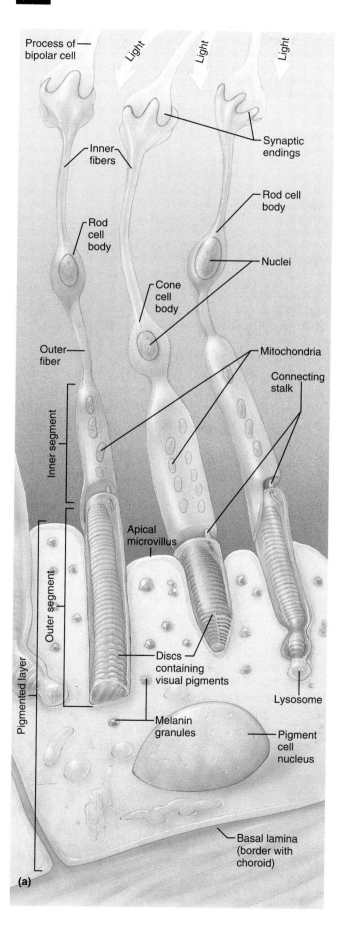

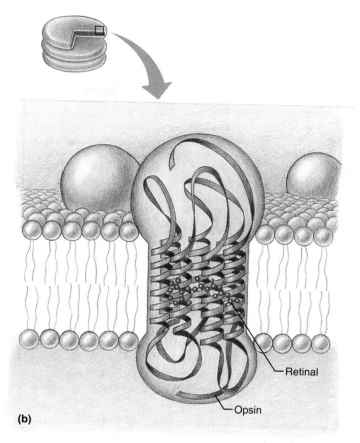

(b)

FIGURE 16.18 *Photoreceptors of the retina.* (a) Diagrammatic view of the photoreceptors (rods and cones). The relationship of the outer segments of the photoreceptors to the pigmented layer of the retina is also illustrated. Notice that the tip of the outer segment of the rod cell on the right is sloughing off. (b) Diagrammatic view of a small section of the membrane of a visual pigment–containing disc of the outer segment of a rod cell. The visual pigments consist of a light-absorbing molecule called retinal (derived from vitamin A) bound to an opsin protein. Each type of photoreceptor has a characteristic kind of opsin protein, which affects the absorption spectrum of the retinal. In rods, the whole pigment-opsin complex is called rhodopsin. Notice that the light-absorbing retinal occupies the core of the rhodopsin molecule.

In addition, rods and cones are "wired" differently to other retinal neurons. Rods participate in converging pathways, and as many as 100 rods may ultimately feed into each ganglion cell. As a result, rod effects are summated and considered collectively, resulting in vision that is fuzzy and indistinct. (The visual cortex has no way of knowing exactly *which* rods of the large number influencing a particular ganglion cell are actually activated.) In contrast, each cone in the fovea (or at most a few) has a straight-through pathway via its "own personal bipolar cell" to a ganglion cell (see Figure 16.9). Hence, each cone essentially has its own "labeled line" to

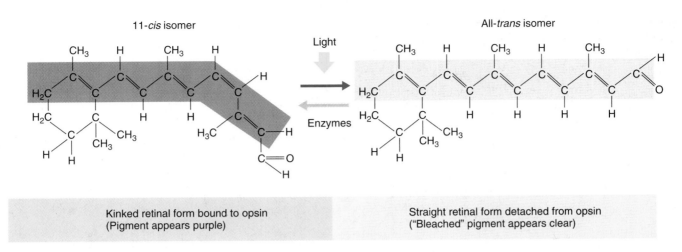

FIGURE 16.19 *Structure of retinal isomers involved in photoreception.*
When the visual pigment absorbs light, 11-*cis* retinal is transformed to the all-*trans*
isomer and the pigment is bleached.

the higher visual centers. This accounts for the detailed, high-resolution views of very small areas of
the visual field provided by cones.

Because rods are absent from the foveae and
cones do not respond to low-intensity light, we see
dimly lit objects best when we do not look at them
directly. However, without foveal focusing, or *hard
focus*, our vision is nondiscriminatory. We see only
vague outlines of objects and recognize them best
when they move. If you doubt this, go out into your
backyard on a moonlit evening and see how much
you can actually discriminate.

The Chemistry of Visual Pigments How does light
affect the photoreceptors so that it is ultimately
translated into electrical signals? The key is a light-
absorbing molecule called **retinal** that combines
with proteins called **opsins** to form four types of visual pigments. Depending on the type of opsin to
which it is bound, retinal preferentially absorbs different wavelengths of the visible spectrum. Retinal is
chemically related to vitamin A and is made from it.
The liver stores vitamin A and releases it as needed
by the photoreceptors to make their visual pigments.
The cells of the pigmented layer of the retina absorb
vitamin A from the blood and serve as the "local" vitamin A depot for the rods and cones.

Retinal can assume a variety of distinct three-
dimensional forms, each form called an isomer.
When bound to opsin, retinal has a bent, or kinked,
shape called the **11-*cis* isomer** (Figure 16.19). However, when the pigment is struck by light and absorbs
photons, retinal twists and snaps into a new configuration, its **all-*trans* isomer.** This shape-changing
process, called *isomerization,* causes retinal to detach from opsin. This is the *only* light-dependent
stage, and this simple photochemical event initiates

a whole chain of chemical and electrical reactions in
rods and cones that ultimately causes electrical impulses to be transmitted along the optic nerve.

Stimulation of the Photoreceptors

1. Excitation of rods. The visual pigment of rods is
a deep purple pigment called **rhodopsin** (ro-dop'sin;
rhodo = rose, *opsis* = vision). (Do you suppose the
person who coined the expression "looking at the
world through rose-colored glasses" knew the meaning of "rhodopsin"?) Rhodopsin molecules are
arranged in a single layer in the membranes of each
of the thousands of discs in the rods' outer segments
(see Figure 16.18b). Although rhodopsin absorbs
light throughout the entire visible spectrum, it maximally absorbs green light.

Rhodopsin forms and accumulates in the dark in
the sequence of reactions shown on the left side of
Figure 16.20. As illustrated, vitamin A is oxidized to
the 11-*cis* retinal form and then combined with
opsin to form rhodopsin. When rhodopsin absorbs
light, retinal changes shape to its all-*trans* isomer,
and the retinal-opsin combination breaks down, allowing retinal and opsin to separate. This breakdown is known as the **bleaching of the pigment.** Actually, the story is a good deal more complex than
this, and the breakdown of rhodopsin that triggers
the transduction process involves a rapid cascade of
intermediate steps (right side of Figure 16.20) that
takes just a few milliseconds. Once the light-struck
all-*trans* retinal actually detaches from opsin, it is
shuttled through the gelatinous fluid in the paper-
thin subretinal space to the pigmented epithelium
by a protein carrier. There, within the pigmented epithelial cells, retinal is reconverted by enzymes to its
11-*cis* isomer in an ATP-requiring process that takes
several minutes. Then, retinal heads "homeward"

again to the photoreceptor cells' outer segments. Rhodopsin is regenerated when 11-*cis* retinal is rejoined to opsin.

2. Excitation of cones. The chemistry of cone function is not completely worked out, but breakdown and regeneration of their visual pigments is essentially the same as for rhodopsin. However, the threshold for cone activation is much higher than that for rods because cones respond only to high-intensity (bright) light.

Visual pigments of cones, like those of rods, are a combination of retinal and opsins. However, there are three distinct types of cones, and each type is most sensitive to a different wavelength range, which reflects the specific properties of the opsin to which the retinal is bound. The three types of cone opsins differ both from the opsin of the rods and from one another. The naming of cones reflects the colors (that is, wavelengths) of light that each cone variety absorbs best and that is most efficient at causing the retinal to change shape and detach from its opsin. The blue cones respond maximally to wavelengths around 420 nm, the green cones to wavelengths of 530 nm, and the red cones to wavelengths at or close to 560 nm (see Figure 16.13b). However, as shown in the figure, their absorption spectra overlap, and perception of intermediate hues, such as orange, yellow, and purple, results from differential activation of more than one type of cone at the same time. For example, yellow light stimulates both red and green cone receptors, but if the red cones are stimulated more strongly than the green cones, we see orange instead of yellow. When all cones are stimulated equally, we see white. Color brightness and saturation (for example, red as opposed to pink) also reflects the degree to which the cone populations are stimulated.

Homeostatic Imbalance

Color blindness is due to a congenital lack of one or more of the cone types. Inherited as a sex-linked condition, it is far more common in males than in females. As many as 8–10% of males have some form of color blindness. The most common type is red-green color blindness, resulting from a deficit or absolute absence of either red or green cones. Red and green are seen as the same color—either red or green, depending on the cone type present. Many color-blind people are unaware of their condition because they have learned to rely on other cues—such as differences in intensities of the same color—to distinguish something green from something red, such as traffic signals. ■

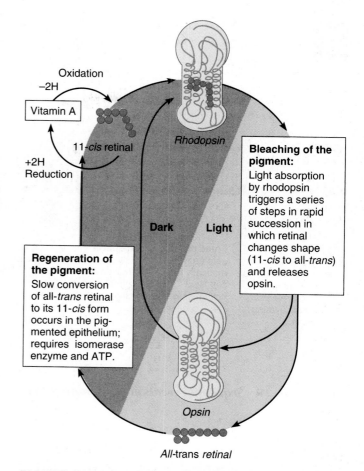

FIGURE 16.20 *Sequence of chemical events in the formation and breakdown of rhodopsin.* When struck by light, rhodopsin breaks down and is converted to its precursors via a number of intermediate steps. Absorption of light causes rapid conversion of 11-*cis* retinal to all-*trans* retinal, which separates from opsin, an event that is responsible for transduction of the light stimulus, and subsequently the separation of all-*trans* retinal and opsin. The 11-*cis* retinal is regenerated in energy (ATP)-requiring reactions by enzymatic action from all-*trans* retinal or from vitamin A and recombined with opsin to form rhodopsin.

Light Transduction in Photoreceptors How is the light stimulus transduced into an electrical event, that is, a change in membrane potential? In the dark, **cyclic GMP** (cGMP) binds to sodium channels in the outer segments, holding them open. Consequently, Na^+ continually enters the outer segment, producing a *dark current* that maintains a transmembrane potential (a *dark potential*) of about -40 mV. This depolarizing current keeps the Ca^{2+} channels *at their synaptic endings* open and permits more or less continuous neurotransmitter (glutamate) release by the photoreceptors at their synapses with the bipolar cells.

When light triggers pigment breakdown, the net effect is to "turn off" sodium entry. That is, an enzymatic cascade occurs that ultimately destroys the

> **?** *Assume that everything is proceeding normally in the scheme illustrated, but then a metabolic poison that interferes with the Na⁺-K⁺ pump is introduced. What would happen and why?*

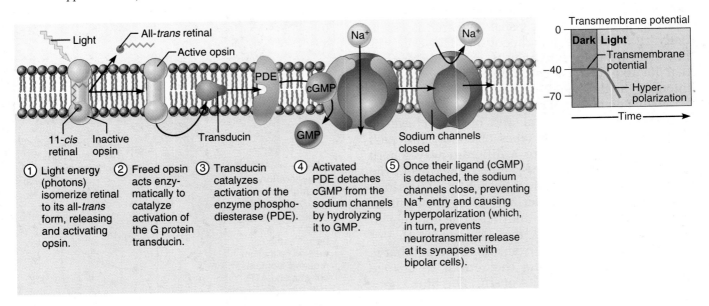

FIGURE 16.21 *Sequential events of phototransduction.*

① Light energy (photons) isomerize retinal to its all-*trans* form, releasing and activating opsin.

② Freed opsin acts enzymatically to catalyze activation of the G protein transducin.

③ Transducin catalyzes activation of the enzyme phosphodiesterase (PDE).

④ Activated PDE detaches cGMP from the sodium channels by hydrolyzing it to GMP.

⑤ Once their ligand (cGMP) is detached, the sodium channels close, preventing Na⁺ entry and causing hyperpolarization (which, in turn, prevents neurotransmitter release at its synapses with bipolar cells).

cGMP that keeps the sodium gates open in the dark. In this process, described in more detail in Figure 16.21, freed opsin (or its earlier intermediate metarhodopsin II) interacts with a G protein subunit called **transducin,** activating it. Transducin, in turn, activates *PDE (phosphodiesterase),* the enzyme that breaks down cGMP, and the sodium gates close. Sodium permeability now decreases dramatically but potassium permeability remains unchanged, so the photoreceptors develop a *hyperpolarizing receptor potential* (−70 mV) that inhibits their release of neurotransmitter. This is bewildering, to say the least. Here we have receptors built to detect light that depolarize in the dark and hyperpolarize when exposed to light! Nevertheless, as we will explain in the section called Visual Processing, turning off neurotransmitter release by hyperpolarization conveys information as effectively as depolarization.

The photoreceptors do not generate action potentials. In fact, none of the retinal neurons generate action potentials except the ganglion cells, which is consistent with their function as the output neurons of the retina. In all other cases, only local currents (graded potentials) are produced. This is not surprising if you remember that the primary function of action potentials is to carry information rapidly over

long distances. Since the retinal cells are small cells that are very close together, graded potentials can serve quite adequately as signals to regulate neurotransmitter release.

Light and Dark Adaptation Rhodopsin is amazingly sensitive; even starlight causes some of the molecules to become bleached. As long as the light is low intensity, relatively little rhodopsin is bleached and the retina continues to respond to light stimuli. However, in high-intensity light there is wholesale bleaching of the pigment, plus rhodopsin breaks down nearly as rapidly as it is made. At this point, the rods become nonfunctional and cones begin to take over. Hence, retinal sensitivity automatically adjusts to the amount of light present.

Light adaptation occurs when we move from darkness into bright light, as when leaving a movie matinee. We are momentarily dazzled—all we see is white light—because the sensitivity of the retina is still "set" for dim light. Both rods and cones are strongly stimulated, and large amounts of the photopigments are broken down almost instantaneously, producing a flood of signals that accounts for the glare. Under such conditions, compensations occur. (1) The sensitivity of the retina decreases dramatically, and (2) retinal neurons adapt rapidly, switching from the rod to the cone system. Within about 60 seconds, the cones are sufficiently excited by the bright light to take over. Visual acuity and color vision continue to improve over the next 5–10

The whole process will come to a halt because excitation depends on having the proper ionic concentrations on the two sides of the membrane.

minutes. Thus, during light adaptation, retinal sensitivity (rod function) is lost, but visual acuity is gained.

Dark adaptation is essentially the reverse of light adaptation. It occurs when we go from a well-lit area into a dark one. Initially, we see nothing but velvety blackness because (1) our cones stop functioning in low-intensity light, and (2) our rod pigments have been bleached out by the bright light, and the rods are still initially inhibited. But once we are in the dark, rhodopsin accumulates and sensitivity of the retina increases. Dark adaptation is much slower than light adaptation and can go on for hours. However, usually enough rhodopsin is accumulated within 20–30 minutes to allow adequate dim-light vision.

During these adaptations, reflexive changes also occur in pupil size. Bright light shining in one or both eyes causes both pupils to constrict (elicits the *consensual* and *pupillary light reflexes*). These pupillary reflexes are mediated by the pretectal nucleus of the midbrain and by parasympathetic fibers. In dim light, the pupils dilate, allowing more light to enter the eye interior.

Homeostatic Imbalance

Night blindness, or *nyctalopia* (nic"tă-lo'pe-uh), is a condition in which rod function is seriously hampered, impairing one's ability to drive safely at night. The most common cause of night blindness is prolonged vitamin A deficiency, which leads to rod degeneration. Vitamin A supplements restore function if they are administered before degenerative changes occur. Although visual pigment content is reduced in both rods and cones in nyctalopia, the cones usually have sufficient photopigment to respond to their intense stimulus (bright light), except in the most severe cases. On the other hand, even a modest deficiency of rhodopsin in the rods renders them nonfunctional in dim light. ■

The Visual Pathway to the Brain

As described earlier, the axons of the retinal ganglion cells issue from the back of the eyeballs in the **optic nerves** (Figure 16.22). At the X-shaped **optic chiasma** (*chiasm* = cross), fibers from the medial aspect of each eye cross over to the opposite side and then continue on via the **optic tracts.** Thus, each optic tract (1) contains fibers from the lateral (temporal) aspect of the eye on the same side and fibers from the medial (nasal) aspect of the opposite eye, and (2) carries all the information from the same half of the visual field. Also notice that, because the lens system of each eye reverses all images, the medial half of each retina receives light rays from the *temporal* (lateralmost) part of the visual field (that is, from the

far left or far right rather than from straight ahead), and the lateral half of each retina receives an image of the *nasal* (central) part of the visual field. Consequently, the left optic tract carries (and sends on) a complete representation of the right half of the visual field, and the opposite is true for the right optic tract.

The paired optic tracts sweep posteriorly around the hypothalamus and send most of their axons to synapse with neurons in the **lateral geniculate body** of the thalamus, which maintains the fiber separation established at the chiasma, but balances and combines the retinal input for delivery to the visual cortex. Axons of the thalamic neurons then project through the internal capsule to form the **optic radiation** of fibers in the cerebral white matter. These fibers project to the **primary visual cortex** in the occipital lobes, where conscious perception of visual images (seeing) occurs.

Some nerve fibers in the optic tracts send branches to the midbrain. These fibers end in the **superior colliculi,** visual reflex centers controlling the extrinsic muscles of the eyes, or in the **pretectal nuclei,** which, as already noted, mediate the pupillary light reflexes. Still other fibers of the visual pathway project to the **suprachiasmatic nucleus** of the hypothalamus, which functions as the "timer" to set our daily biorhythms. Visual inputs keep it synchronized with the natural cycles of daylight and darkness. As you can see, visual processing is a big user of both serial and parallel processing of inputs.

Stereoscopic Vision and Depth Perception

Stereoscopic vision (literally, "seeing solid") is found in humans and most other primates, as well as in predatory birds and cats. Although both eyes are set anteriorly and look in approximately the same direction, their visual fields, each about 170 degrees, overlap to a considerable extent, and each eye sees a slightly different view (see Figure 16.22). Crossing over of about half of the optic nerve fibers at the optic chiasma provides each visual cortex with the same image (or part of the visual field) as viewed by both eyes—but the two eyes behold that image from slightly different angles. In contrast, many animals (pigeons, rabbits, and others) have *panoramic vision*. Their eyes are placed more laterally on the head so that the two visual fields overlap very little, and crossover of the optic nerve fibers is almost complete. Consequently, each visual cortex receives input principally from a single eye and a totally different visual field. Stereoscopic vision provides a smaller total visual field than panoramic vision, but it also provides **depth perception,** an accurate means of locating objects in space. This faculty, also called **three-dimensional vision,** results from cortical "fusion" of the slightly different images delivered by the two eyes.

? Where on the retina of the right eye would light from an object in the nasal portion of its visual field be projected?

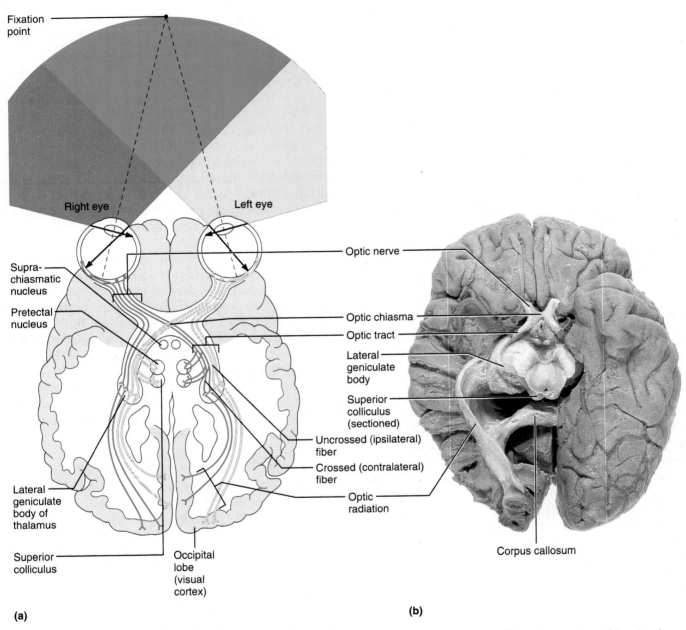

(a)

(b)

FIGURE 16.22 *Visual fields of the eyes and visual pathway to the brain, inferior view.* (a) Diagram. (b) Photograph. In (a) notice that the visual fields of the two eyes overlap considerably. Also notice the retinal sites at which a real image would be focused when both eyes are fixated on a close, pointlike object. The full lateral extension of the visual fields is not illustrated here.

Depth perception depends on the two eyes working together and accurately focusing on the object. If only one eye is used, depth perception is lost, and the person must learn to judge an object's position based on learned cues (for example, nearer objects appear larger, and parallel lines converge with distance).

To the temporal portion of the retina.

⚖ *Homeostatic Imbalance*

The relationships described above explain patterns of blindness that follow damage to different visual structures. Loss of an eye or destruction of one optic nerve eliminates true depth perception, and peripheral vision is lost on the side of damage. For example, if the "left eye" in Figure 16.22 were lost in a

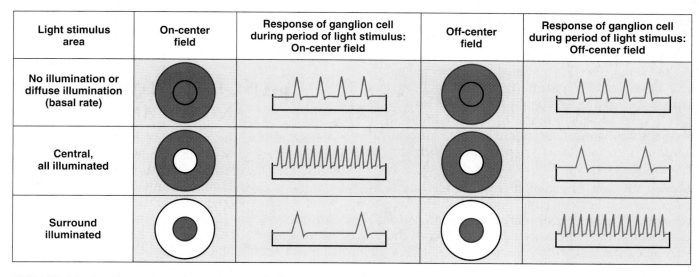

Light stimulus area	On-center field	Response of ganglion cell during period of light stimulus: On-center field	Off-center field	Response of ganglion cell during period of light stimulus: Off-center field
No illumination or diffuse illumination (basal rate)				
Central, all illuminated				
Surround illuminated				

FIGURE 16.23 *Responses of on-center and off-center retinal ganglion cells to different types of illumination.* Intermediate amounts of illumination of the central and surround regions lead to intermediate rates of impulse transmission.

hunting accident, nothing would be seen in the visual field area colored gold in that figure. If neural destruction occurs beyond the optic chiasma—in an optic tract, the thalamus, or visual cortex—then part or all of the opposite half of the visual field is lost. For example, a stroke affecting the left visual cortex leads to blindness in the right half of the visual field, but since the right (undamaged) visual cortex still receives inputs from both eyes, depth perception in the remaining half of the visual field is retained. ■

Visual Processing

How does information received by the rods and cones become vision? In the past few years, a tremendous amount of research has been devoted to this question. Some of the more basic information that has come out of these studies is presented here.

Retinal Processing The retinal ganglion cells generate action potentials at a fairly steady rate (20–30 Hz), even in the dark. Surprisingly, illuminating the entire retina evenly has no effect on that basal rate. However, the activity of individual ganglion cells changes dramatically when a tiny spot of light falls just on certain portions of their receptive field. The receptive field of a ganglion cell consists of the area of the retina that, when stimulated, influences the activity of that ganglion cell (the rods or cones that funnel their impulses to it).

When researchers mapped the receptive fields of ganglion cells receiving inputs only from rods, they found two types of doughnut-shaped (circle-within-a-circle) receptive fields (Figure 16.23). These are called **on-center** and **off-center fields,** based on

what happens to the ganglion cell when the center of its receptive field is illuminated with a spot of light. Ganglion cells with on-center fields are stimulated (depolarized) by light hitting the field center (the "doughnut hole") and are inhibited by light hitting rods in the periphery (the doughnut itself). The opposite situation occurs in ganglion cells with off-center fields. Equal illumination of "on" and "off" regions causes little change in the basal rate of ganglion cell firing.

The mechanisms of retinal processing as currently understood can be summarized briefly as follows:

1. The action of light on photoreceptors hyperpolarizes them.

2. Bipolar neurons in the "on" regions are excited (depolarize) and excite the associated ganglion cell when the rods feeding into them are illuminated (and hyperpolarized). Bipolar neurons in the "off" regions are inhibited (hyperpolarize) and inhibit the ganglion cell when the rods feeding into them are stimulated. The opposite responses of the bipolar neurons in the "on" versus the "off" regions reflect the fact that they have different receptor types for glutamate.

3. Bipolar neurons receiving signals from cones feed directly into excitatory synapses on ganglion cells. Hence, cone inputs are perceived as sharp and clear (and in color).

4. Bipolar neurons receiving inputs from rods excite amacrine cells via gap junctions. These local integrator neurons modify rod inputs and ultimately direct excitatory inputs to appropriate ganglion cells. Thus, rod inputs are not only summated but are subject to "detours" before reaching the output (ganglion) cells. Together these factors provide a more smeary picture than cone inputs.

5. Rod inputs are also modified and subjected to lateral inhibition by synaptic (gap junction) contacts with horizontal cells. Inputs from these local integrator cells allow the retina to convert inputs that are points of light into perceptually more meaningful contour information by accentuating bright/dark contrasts, or edges.

6. Two varieties of ganglion cells are identified. M cells respond best to large, moving objects at the edge of the image. P cells relay information concerning small nonmoving objects in the center of the visual field, that is, their color and details.

Thalamic Processing The lateral geniculate nuclei (LGN) of the thalamus relay information on movement, segregate the retinal axons in preparation for depth perception, emphasize visual inputs from the region of high cone density, and sharpen the contrast information received from the retina. The separation of signals from the two eyes is relayed accurately to the visual cortex.

Cortical Processing Two types of areas for processing retinal inputs are found in the visual cortex. The *primary visual cortex,* also called the **striate cortex,** is thick with fibers coming in from the LGN. This area contains an accurate topographical map of the retina; the left visual cortex receives input from the right visual field and vice versa. Visual processing here occurs at a relatively basic level, with the processing neurons responding to dark and bright edges (contrast information), and object orientation.

The striate cortex also provides form, color, and motion inputs, with temporary storage areas called "blobs," which forward information on to *visual association areas* collectively called the prestriate cortices. The **prestriate cortices** are centers that continue the processing of visual information concerned with form, color, and movement. Although the visual association areas have been localized in the occipital lobes, functional neuroimaging of humans has revealed that complex visual processing extends well forward into the temporal, parietal, and frontal lobes. Visual information proceeds anteriorly via parallel pathways through these areas in two main streams: (1) The "what" processing stream extends through the ventral part of the temporal lobe and specializes in the identification of objects in the visual field. (2) The "where" processing stream takes a dorsal path through the parietal cortex all the way to the postcentral gyrus and uses information from the primary visual cortex to assess the spatial location of objects. Output from both these regions then appears to pass to the frontal cortex, which uses that information to direct activities that, among other things, can guide movements such as reaching for a juicy peach.

THE EAR: HEARING AND BALANCE

At first glance, the machinery for hearing and balance appears very crude. Fluids must be stirred to stimulate the mechanoreceptors of the ear: Sound vibrations move fluid to stimulate hearing receptors, whereas gross movements of the head disturb fluids surrounding the balance organs. Nevertheless, our hearing apparatus allows us to hear an extraordinary range of sound, and our equilibrium receptors keep the nervous system continually informed of head movements and position. Although the organs serving these two senses are structurally interconnected within the ear, their receptors respond to different stimuli and are activated independently of one another.

Structure of the Ear

The ear is divided into three major areas: outer ear, middle ear, and inner ear (Figure 16.24). The outer and middle ear structures are involved with hearing only and are rather simply engineered. The inner ear functions in both equilibrium and hearing and is extremely complex.

The Outer (External) Ear

The **outer (external) ear** consists of the auricle and the external auditory canal. The **auricle,** or **pinna,** is what most people call the ear—the shell-shaped projection surrounding the opening of the external auditory canal. The auricle is composed of elastic cartilage covered with thin skin and an occasional hair. Its rim, the **helix,** is somewhat thicker, and its fleshy, dangling **lobule** ("earlobe") lacks supporting cartilage. The function of the auricle is to direct sound waves into the external auditory canal. Some animals (coyotes, for example) can move their auricles toward a sound source, but in humans the comparable muscles are vestigial and inefficient.

The **external auditory canal (meatus)** is a short, curved tube (about 2.5 cm long by 0.6 cm wide) that extends from the auricle to the eardrum. Near the auricle, its framework is elastic cartilage; the remainder of the canal is carved into the temporal bone. The entire canal is lined with skin bearing hairs, sebaceous glands, and modified apocrine sweat glands called **ceruminous** (sĕ-roo′mĭ-nus) **glands.** These glands secrete yellow-brown waxy **cerumen,** or earwax (*cere* = wax), which provides a sticky trap for foreign bodies and repels insects. In many people, the ear is naturally cleansed as the cerumen dries and then falls out of the external auditory canal. In others, cerumen builds up excessively and becomes compacted, a condition that can impair hearing.

Sound waves entering the external auditory canal eventually hit the **tympanic membrane,** or **eardrum** (*tympanum* = drum), the boundary between the outer and middle ears. The eardrum is a thin, translucent, connective tissue membrane, covered by skin on its external face and by a mucosa internally. It is shaped like a flattened cone, with its apex protruding medially into the middle ear. Sound waves make the eardrum vibrate. The eardrum, in turn, transfers the sound energy to the tiny bones of the middle ear and sets them into vibration.

The Middle Ear

The **middle ear,** or **tympanic cavity,** is a small, air-filled, mucosa-lined cavity within the petrous portion of the temporal bone. It is flanked laterally by the eardrum and medially by a bony wall with two openings, the superior **oval (vestibular) window** and the inferior **round (cochlear) window.** The round window is closed by the *secondary tympanic membrane.* Superiorly the tympanic cavity arches upward as the **epitympanic recess,** the "roof" of the middle ear cavity. The **mastoid antrum,** a canal in the posterior wall of the tympanic cavity, allows it to communicate with the *mastoid air cells* housed in the mastoid process. The anterior wall of the middle ear abuts the internal carotid artery (the main artery supplying the brain) and contains the opening of the **pharyngotympanic (auditory) tube.** This tube, formerly called the Eustachian tube, runs obliquely downward to link the middle ear cavity with the nasopharynx (the superiormost part of the throat), and the mucosa of the middle ear is continuous with that lining the pharynx (throat).

Normally, the pharyngotympanic tube is flattened and closed, but swallowing or yawning opens it briefly to equalize pressure in the middle ear cavity with external air pressure. This is important because the eardrum does not vibrate freely unless the pressure on both of its surfaces is the same. When the pressures are unequal, the eardrum bulges inward or outward, causing hearing difficulty (voices may sound far away) and sometimes earaches. The ear-popping sensation of the pressures equalizing is familiar to anyone who has flown in an airplane.

⚠ *Homeostatic Imbalance*

Otitis media (me'de-ah), or middle ear inflammation, is a fairly common result of a sore throat, especially in children, whose auditory tubes are shorter and run more horizontally. Otitis media is the most

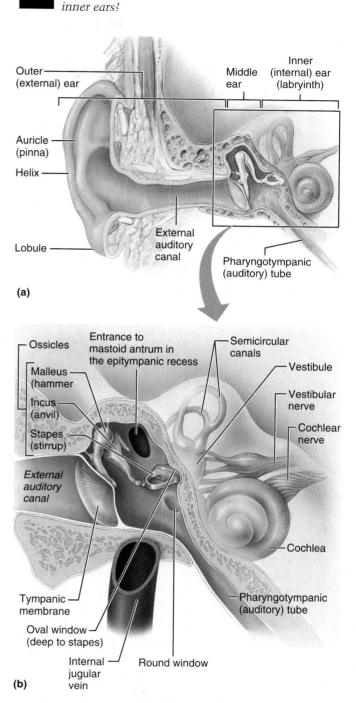

? *Besides the bony boundaries, what structure(s) separates the outer from the middle ear? The middle and inner ears?*

(a)

(b)

FIGURE 16.24 *Structure of the ear.* **(a)** The three regions of the ear. **(b)** Enlarged view of the middle and inner ear. Notice that the semicircular canals, vestibule, and cochlea of the inner ear form the bony labyrinth.

The tympanic membrane. Secondary tympanic membrane.

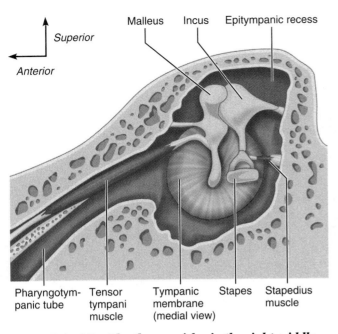

Superior

Anterior

Malleus   Incus   Epitympanic recess

Pharyngotym-   Tensor   Tympanic   Stapes   Stapedius
panic tube   tympani   membrane   muscle
              muscle   (medial view)

FIGURE 16.25 *The three ossicles in the right middle ear, medial view.*

frequent cause of hearing loss in children. In acute forms, in which infectious bacteria colonize the middle ear, the eardrum bulges and becomes inflamed and red. Most cases of otitis media are treated with antibiotics. When large amounts of fluid or pus accumulate in the cavity, an emergency *myringotomy* (lancing of the eardrum) may be required to relieve the pressure. During myringotomy a tiny tube is implanted in the eardrum, permitting pus to drain into the external ear. The tube falls out by itself within the year. Noninfectious cases of otitis media, in which clear fluid accumulates in the tympanic cavity, are often due to food allergies (typically milk or wheat). In such cases, the cure is restricting the offending food rather than taking antibiotics. ■

The tympanic cavity is spanned by the three smallest bones in the body: the **ossicles** (see Figures 16.24 and 16.25). These bones, named for their shape, are the **malleus** (mal'e-us), or **hammer;** the **incus** (ing'kus), or **anvil;** and the **stapes** (sta'pēz), or **stirrup.** The "handle" of the malleus is secured to the eardrum, and the base of the stapes fits into the oval window.

Tiny ligaments suspend the ossicles, and minisynovial joints link them together into a chain that spans the middle ear cavity. The incus articulates with the malleus laterally and the stapes medially. The ossicles transmit the vibratory motion of the eardrum to the oval window, which in turn sets the fluids of the inner ear into motion, eventually exciting the hearing receptors.

Two tiny skeletal muscles are associated with the ossicles (Figure 16.25). The **tensor tympani** (ten'sor tim'pah-ni) arises from the wall of the auditory tube and inserts on the malleus. The **stapedius** (stahpe'de-us) runs from the posterior wall of the middle ear cavity and inserts into the stapes. When the ears are assaulted by very loud sounds, these muscles contract reflexively to prevent damage to the hearing receptors. Specifically, the tensor tympani tenses the eardrum by pulling it medially; the stapedius checks vibration of the whole ossicle chain and limits the movement of the stapes in the oval window.

The Inner (Internal) Ear

The **inner (internal) ear** is also called the **labyrinth** ("maze") because of its complicated shape (see Figure 16.24). It lies deep within the temporal bone behind the eye socket and provides a secure site for all of the delicate receptor machinery housed there. The inner ear consists of two major divisions: the bony labyrinth and the membranous labyrinth. The **bony,** or **osseous, labyrinth** is a system of tortuous channels worming through the bone. Its three structurally and functionally unique regions are the *vestibule*, the *cochlea* (kok'le-ah), and the *semicircular canals*. The views of the bony labyrinth typically seen in most textbooks, including this one, are somewhat misleading because we are talking about a *cavity* here. The representation seen in Figure 16.24 can be compared to a plaster of paris cast of the cavity or hollow space inside the bony labyrinth. The **membranous labyrinth** is a continuous series of membranous sacs and ducts contained within the bony labyrinth and (more or less) following its contours (Figure 16.26).

The bony labyrinth is filled with **perilymph,** a fluid similar to cerebrospinal fluid and continuous with it. The membranous labyrinth is surrounded by and floats in the perilymph; its interior contains **endolymph,** which is chemically similar to K^+-rich intracellular fluid. These two fluids conduct the sound vibrations involved in hearing and respond to the mechanical forces occurring during changes in body position and acceleration. They have no relationship to the lymph circulating in the lymphatic vessels of the body.

The Vestibule The **vestibule** is the central eggshaped cavity of the bony labyrinth. It lies posterior

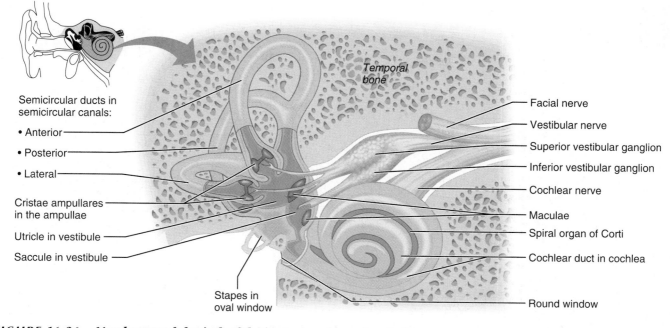

Semicircular ducts in
semicircular canals:

• Anterior

• Posterior

• Lateral

Cristae ampullares
in the ampullae

Utricle in vestibule

Saccule in vestibule

Stapes in
oval window

Temporal bone

Facial nerve

Vestibular nerve

Superior vestibular ganglion

Inferior vestibular ganglion

Cochlear nerve

Maculae

Spiral organ of Corti

Cochlear duct in cochlea

Round window

FIGURE 16.26 *Membranous labyrinth of the inner ear, shown in relation to the chambers of the bony labyrinth.* The locations of the specialized receptors for hearing (organ of Corti) and equilibrium (maculae and cristae ampullares) are also indicated.

to the cochlea, anterior to the semicircular canals, and flanks the middle ear medially. In its lateral wall is the oval window. Suspended within its perilymph and united by a small duct are two membranous labyrinth sacs, the **saccule** and **utricle** (u'tri-kl) (Figure 16.26). The smaller saccule is continuous with the membranous labyrinth extending anteriorly into the cochlea (the *cochlear duct*), whereas the utricle is continuous with the semicircular ducts extending into the semicircular canals posteriorly. The saccule and utricle house equilibrium receptor regions called *maculae* that respond to the pull of gravity and report on changes of head position.

The Semicircular Canals The **semicircular canals** lie posterior and lateral to the vestibule and each of these canals defines about 2/3 of a circle. The cavities of the bony semicircular canals project from the posterior aspect of the vestibule; each is oriented in one of the three planes of space. Accordingly, there is an *anterior, posterior,* and *lateral* semicircular canal in each inner ear. The anterior and posterior canals are oriented at right angles to each other in the vertical plane, whereas the lateral canal lies horizontally (see Figure 16.26). Snaking through each semicircular canal is a corresponding membranous **semicircular duct,** which communicates with the utricle anteriorly. Each of these ducts has an en-

larged swelling at one end called an **ampulla,** which houses an equilibrium receptor region called a *crista ampullaris* (literally, crest of the ampulla). These receptors respond to angular (rotational) movements of the head.

The Cochlea The **cochlea,** from the Latin meaning "snail," is a spiral, conical, bony chamber about half the size of a split pea. It extends from the anterior part of the vestibule and then coils for about 2½ turns around a bony pillar called the **modiolus** (mo-di'o-lus) (Figure 16.27a). Running through its center like a wedge-shaped worm is the membranous **cochlear duct,** which ends blindly at the cochlear apex. The cochlear duct houses the **spiral organ of Corti,** the receptor organ for hearing (Figure 16.27b). The cochlear duct and the **spiral lamina,** a thin shelflike extension of bone that spirals up the modiolus like the thread on a screw, together divide the cavity of the bony cochlea into three separate chambers or **scalas** (*scala* = ladder). The **scala vestibuli** (ska'lah ves-tı̆'bu-li), which lies superior to the cochlear duct, is continuous with the vestibule and abuts the oval window. The middle **scala media** is the cochlear duct itself. The **scala tympani,** which terminates at the round window, is inferior to the cochlear duct. Since the scala media is part of the membranous labyrinth, it is filled with endolymph.

? *Which scala contains endolymph?*

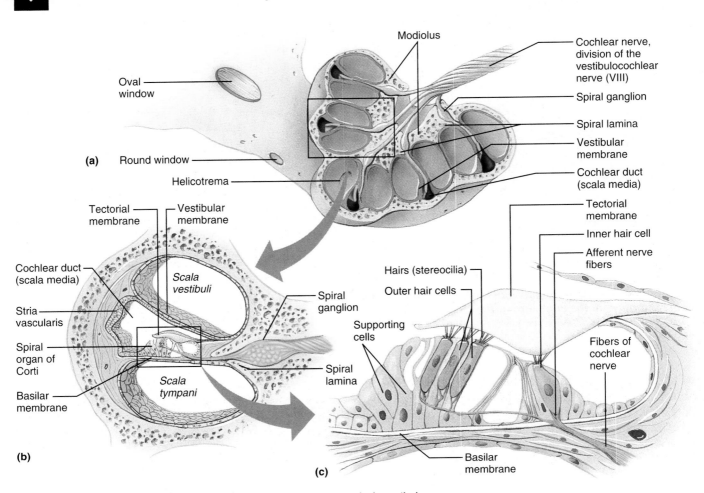

FIGURE 16.27   *Anatomy of the cochlea.* (a) A section through the coiled cochlea. (b) Magnified cross-sectional view of one turn of the cochlea, showing the relationship of the three scalae. The scalae vestibuli and tympani contain perilymph; the cochlear duct (scala media) contains endolymph. (c) Detailed structure of the spiral organ of Corti.

The scala vestibuli and the scala tympani, both part of the bony labyrinth, contain perilymph. The perilymph-containing chambers are continuous with each other at the cochlear apex, a region called the **helicotrema** (hel″ĭ-ko-tre′mah; "the hole in the spiral").

The "roof" of the cochlear duct, separating the scala media from the scala vestibuli, is the **vestibular membrane** (see Figure 16.27b). Its external wall, the **stria vascularis,** is composed of an unusual richly vascularized mucosa that secretes the

endolymph. The "floor" of the cochlear duct is composed of the bony spiral lamina and the flexible, fibrous **basilar membrane,** which supports the organ of Corti. (We will describe the organ of Corti when we discuss the mechanism of hearing.) The basilar membrane is narrow and thick near the oval window and gradually becomes wider and thinner as it approaches the cochlear apex. As you will see, its structure plays a critical role in sound reception. The *cochlear nerve,* a division of the *vestibulocochlear (eighth cranial) nerve,* runs from the organ of Corti through the modiolus on its way to the brain.

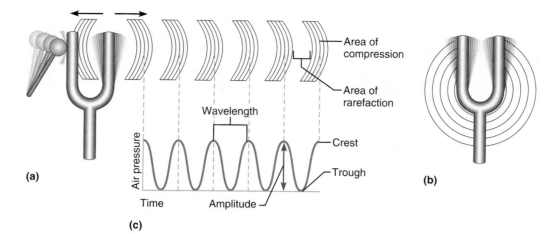

(a)

Wavelength

Air pressure

Time   Amplitude

Crest

Trough

Area of compression

Area of rarefaction

(b)

(c)

FIGURE 16.28   Sound: source and propagation. The source of sound is a vibrating object. **(a)** A tuning fork is struck on the left with a mallet. Its tines move to the right, compressing the air molecules in that region. (The tine movement is shown on the right side only and is exaggerated.) Then the tines move in the opposite direction, compressing the air there and leaving a rarefied area to the right side of the tuning fork. **(b)** In this way, sound waves, consisting of alternate areas of compression and rarefaction, are set up and radiate outward in all directions from the sound source. **(c)** The sound wave may be depicted as a sine wave. The peaks of the wave represent areas of high pressure (compressed areas), and the troughs represent areas of low pressure (rarefied areas). For any tone, the distance between two corresponding points on the wave (two crests or two troughs) is called the wavelength. The height (amplitude) of the crests is related to the energy, or intensity, of the sound wave.

Sound and Mechanisms of Hearing

The mechanics of human hearing can be summed up in a single sprawling sentence: Sounds set up vibrations in air that beat against the eardrum that pushes a chain of tiny bones that press fluid in the inner ear against membranes that set up shearing forces that pull on the tiny hair cells that stimulate nearby neurons that give rise to impulses that travel to the brain, which interprets them—and you hear. Before we unravel this intriguing sequence, let us pause to describe sound, the stimulus for hearing.

Properties of Sound

Light can be transmitted through a vacuum (for instance, outer space), but sound depends on an *elastic* medium for its transmission. Sound also travels much more slowly than light. Its speed in dry air is only about 0.2 mile per second (331 m/s), as opposed to the 186,000 miles per second (about 300,000 km/s) of light. A lightning flash is almost instantly visible, but the sound it creates (thunder) reaches our ears much more slowly. (For each second between the lightning bolt and the roll of thunder, the storm is 1/5 mile farther away.) The speed of sound is constant in a given medium. It is greatest in solids and lowest in gases, including air.

Sound is a pressure disturbance originating from a vibrating object and propagated by the molecules of the medium. Consider a sound arising from a vibrat-ing tuning fork (Figure 16.28a). If the tuning fork is struck on the left, its prongs will move first to the right, creating an area of high pressure by compressing the air molecules there. Then, as the prongs rebound to the left, the air on the left will be compressed, and the region on the right will be a rarefied, or low-pressure, area (since most of its air molecules have been pushed farther to the right). As the fork vibrates alternately from right to left, it produces a series of compressions and rarefactions, collectively called a *sound wave*, which moves outward in all directions (Figure 16.28b). However, the individual air molecules just vibrate back and forth for short distances as they bump other molecules and rebound. Because the outward-moving molecules give up kinetic energy to the molecules they bump, energy is always transferred in the direction the sound wave is traveling. Thus, with time and distance, the energy of the wave declines, and the sound dies a natural death.

We can illustrate a sound wave as an S-shaped curve, or *sine wave*, in which the compressed areas are crests and the rarefied areas are troughs (Figure 16.28c). Sound can be described in terms of two physical properties inferred from this sine wave graph: frequency and amplitude.

Frequency   The sine wave of a pure tone is *periodic*; that is, its crests and troughs repeat at definite distances. The distance between two consecutive crests (or troughs) is called the **wavelength** of the sound

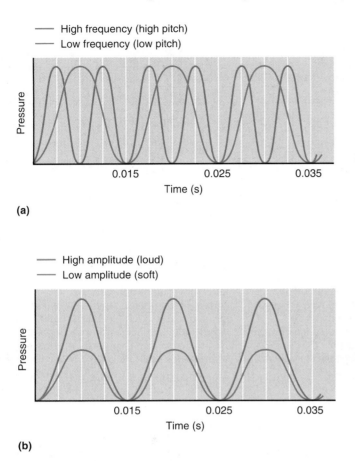

(a)

(b)

FIGURE 16.29 *Frequency and amplitude of sound waves.* **(a)** The wave shown in red has a shorter wavelength, and thus a greater frequency, than that shown in blue. The frequency of sound is perceived as pitch. **(b)** The wave shown in red has a greater amplitude (intensity) than that shown in blue and is perceived as being louder.

and is constant for a particular tone. **Frequency,** expressed in *hertz*, is defined as the number of waves that pass a given point in a given time. The shorter the wavelength, the higher the frequency of the sound (Figure 16.29a).

The frequency range of human hearing is from 20 to 20,000 Hz. Our ears are most sensitive to frequencies between 1500 and 4000 Hz and, in that range, we can distinguish frequencies differing by only 2–3 Hz. We perceive different sound frequencies as differences in **pitch:** The higher the frequency, the higher the pitch. A tuning fork produces a *tone,* a pure (but bland) sound with a single frequency, but most sounds are mixtures of several frequencies. It is this characteristic of sound, called **quality,** that enables us to distinguish between the same musical note—say, high C—sung by a soprano and played on a piano or clarinet. It is the quality of sound that provides the richness and complexity of sounds (and music) that we hear, as well as the sound input we call "noise."

Amplitude The **intensity** of a sound is related to its energy, or the pressure differences between its compressed and rarefied areas. When we represent a sound wave graphically, as in Figure 16.29b, its intensity corresponds to the **amplitude,** or height, of the sine wave crests.

Intensity is an objective, precisely measurable, physical property of sound, but **loudness** refers to our subjective interpretation of sound intensity. Because we can hear such an enormous range of intensities, from the proverbial pin drop to a steam whistle 100 trillion times as loud, sound intensity (and loudness) is measured in logarithmic units called **decibels** (des'ĭ-belz) **(dB).** On a clinical audiometer, the decibel scale is arbitrarily set to begin at 0 dB, which is the threshold of hearing (barely audible sound) for normal ears. Each 10 dB increase represents a tenfold increase in sound intensity. A sound of 10 dB has ten times more energy than one of 0 dB, and a 20 dB sound has 100 times (10 × 10) more energy than one of 0 dB. However, the same 10 dB increase represents only a twofold increase in loudness. In other words, most people would report that a 20 dB sound seems about twice as loud as a 10 dB sound. The healthy adult ear can pick up differences in sound intensity as small as 0.1 dB, and the normal range of hearing (from barely audible to the loudest sound we can process without excruciating pain) extends over a range of 120 dB. (The threshold of pain is 130 dB.)

Severe hearing loss occurs with frequent or prolonged exposure to sounds with intensities greater than 90 dB, and in the U.S., employees exposed to occupational noise over that range must wear ear (hearing) protection. This figure becomes more meaningful when you realize that a normal conversation is in the 50 dB range, a noisy restaurant has 70 dB levels, and amplified rock music is, at 120 dB or more, far above the 90 dB danger zone.

Transmission of Sound to the Inner Ear

Hearing occurs when the auditory area of the temporal lobe cortex is stimulated. However, before this can happen, sound waves must be propagated through air, membranes, bones, and fluids to reach and stimulate receptor cells in the organ of Corti (Figure 16.30).

Airborne sound entering the external auditory canal strikes the tympanic membrane and sets it vibrating at the same frequency. The distance the membrane moves in its vibratory motion varies with sound intensity. The greater the intensity, the farther the membrane is displaced. The motion of the tympanic membrane is amplified and transferred to the oval window by the ossicles. If sound hit the oval window directly, most of its energy would be reflected and lost because of the high impedance (resistance to motion) of the cochlear fluid in the inner

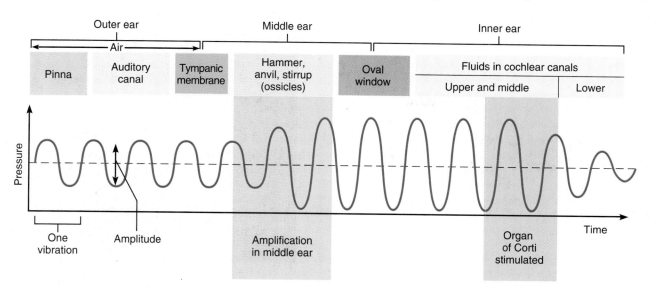

FIGURE 16.30 *Route of sound waves through the ear.* To excite the hair cells in the organ of Corti in the inner ear, sound wave vibrations must pass through air, membranes, bones, and fluids.

ear. However, the ossicle lever system acts much like a hydraulic press or piston to transfer the same total force hitting the eardrum to the oval window. Because the tympanic membrane is 17 to 20 times larger than the oval window, the pressure (force per unit area) actually exerted on the oval window is about 20 times that on the tympanic membrane. This increased pressure overcomes the impedance of cochlear fluid and sets it into wave motion. This situation can be roughly compared to the difference in pressure relayed to the floor by the broad rubber heels of a man's shoes and a woman's tiny spike heels. The man's weight—say, 70 kg (about 150 pounds)—is spread over several square inches, and his heels will not make dents in a pliable vinyl floor. But spike heels will concentrate the same 70 kg force in an area of about 2.5 cm^2 (1 square inch) and *will* dent the floor.

Resonance of the Basilar Membrane

As the stapes rocks back and forth against the oval window, it sets the perilymph in the scala vestibuli into a similar back-and-forth motion. During these movements, the basilar membrane swings up and down, causing the nearby (basal) part of the cochlear duct to oscillate in time, and a pressure wave travels through the perilymph from the basal end toward the helicotrema, much as a piece of rope held horizontally can be set into wave motion by movements initiated at one end. Sounds of very low frequency (below 20 Hz) create pressure waves that take the complete route through the cochlea—up the scala vestibuli, around the helicotrema, and back toward the round window through the scala tympani (Figure

16.31a). Such sounds do not activate the organ of Corti (are below the threshold of hearing). But sounds of higher frequency (and shorter wavelengths) create pressure waves that fail to reach the helicotrema and instead take a "shortcut" and are transmitted through the cochlear duct into the perilymph of the scala tympani. Fluids are incompressible. Think of what happens when you sit on a water bed—you sit "here" and it bulges over "there." Thus, each time the fluid adjacent to the oval window is forced medially by the stapes, the membrane of the round window bulges laterally into the middle ear cavity and acts as a pressure valve. Hence, for each *in and out* movement of the stapes at the oval window, the basilar membrane moves *down and up*, and the membrane of the round window moves *out and in*. It may help to visualize this chain of events.

As a pressure wave descends through the flexible vestibular membrane, the endolymph of the cochlear duct, and then through the basilar membrane, it sets the entire basilar membrane into vibrations. Maximal displacement of the membrane occurs where the fibers of the basilar membranes are "tuned" to a particular sound frequency (Figure 16.31b). (This characteristic of many natural substances is called *resonance.*) The fibers of the basilar membrane span its width like the strings of a harp. The fibers near the oval window (cochlear base) are short and stiff, and they resonate in response to high-frequency pressure waves (Figure 16.31b). The longer, more floppy basilar membrane fibers near the cochlear apex resonate in time with lower-frequency pressure waves. Thus, sound signals are mechanically processed by the resonance of the basilar membrane, before ever reaching the receptors.

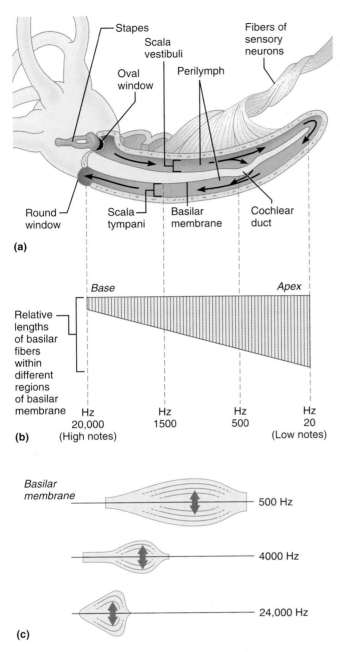

(a)

(b)

(c)

FIGURE 16.31 *Resonance of the basilar membrane and activation of the cochlear hair cells.* **(a)** The cochlea is depicted as if it is uncoiled to make the events of sound transmission occurring there easier to follow. Sound waves of low frequency that are below the level of hearing travel around the helicotrema without exciting hair cells. But sounds of higher frequency result in pressure waves that penetrate through the cochlear duct and basilar membrane to reach the scala tympani. This sets up vibrations in the basilar membrane, causing it to vibrate maximally in certain areas in response to certain frequencies of sound. **(b)** Fibers span the width of the basilar membrane. Like harp strings, these fibers vary in length, being shorter near the base of the membrane and longer near its apex. The length of the fibers "tunes" specific regions of the basilar membrane to vibrate at specific frequencies. **(c)** Different frequencies of pressure waves in the cochlea cause certain places along the basilar membrane to vibrate, stimulating particular hair cells and sensory neurons. The differential stimulation of hair cells is perceived in the brain as sound of a certain pitch.

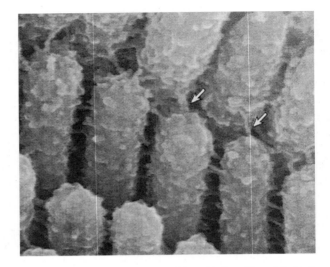

FIGURE 16.32 *Photo of cochlear hair cells with their precise arrays of stereocilia.* White arrows point to tip-links.

Excitation of Hair Cells in the Organ of Corti

The organ of Corti, which rests atop the basilar membrane, is composed of supporting cells and some 16,000 hearing receptor cells called *cochlear hair cells*. The hair cells are arranged functionally—specifically, one row of **inner hair cells** and three rows of **outer hair cells**—sandwiched between the tectorial and basilar membranes. Afferent fibers of the **cochlear nerve** (a division of the vestibulo-cochlear nerve [VIII]) are coiled about the bases of the hair cells. The "hairs" (stereocilia) of the hair cells are stiffened by actin filaments and linked together by fine fibers called *tip-links* (Figure 16.32). They protrude into the K$^+$-rich endolymph, and the longest of them are enmeshed in the overlying, gel-like **tectorial membrane** (see Figure 16.27c).

Transduction of sound stimuli occurs after the trapped stereocilia of the hair cells are "tweaked" or deflected by localized movements of the basilar membrane. Bending the cilia toward the tallest cilium puts tension on the tip-links which in turn opens cation channels in the adjacent shorter stereocilia. This results in an inward K$^+$ (and Ca^{2+}) current and a graded depolarization. Bending the cilia away from the tallest cilium allows the tip-links to relax, closes the mechanically gated ion channels, and produces a graded hyperpolarization. Depolarization and the rise in intracellular Ca^{2+} increases the hair cells' release of neurotransmitter (probably glutamate), which causes the afferent cochlear fibers to transmit a faster stream of impulses to the brain for auditory interpretation. Hyperpolarization produces the exact opposite effect. As you might guess, activation of the hair cells occurs at points of vigorous basilar membrane vibration. Hair cells nearest the oval window are activated by the highest-pitch sounds, and those located at the cochlear apex are stimulated by the lowest-frequency tones (Figure 16.31c).

Although the outer hair cells are much more numerous, they instigate relatively few action potentials. In fact, 90–95% of the sensory fibers of the spiral ganglion service the inner hair cells, which shoulder nearly the entire responsibility for sending auditory messages to the brain. By contrast, most fibers coiling around the outer hair cells are *efferent* fibers that convey messages in the opposite direction, from brain to ear. (What is the brain "telling" the outer hair cells?) Not only are the outer hair cells wired oddly, they behave oddly as well; that is, they stretch and contract (on hyperpolarization and depolarization respectively)—in a type of cellular boogie called *fast motility*. Sound activates a feedback loop from the outer hair cells to the brain stem and back that promotes the length changes in these hair cells. Their strange behavior changes the stiffness of the basilar membrane, which alters its motion and amplifies the responsiveness of the inner hair cells— kind of a cochlear tuning.

Outer hair cell mobility is also responsible for producing ear sounds that are so audible in some people that they can actually be heard by others. More precisely called *otoacoustic emissions*, these ear sounds seem to represent a kind of background noise or "line busy" signal that can interfere with auditory transmission. In clinical pediatrics, detection of spontaneous otoacoustic emissions has proven a quick, inexpensive way to screen newborns for hearing defects.

The Auditory Pathway to the Brain

The ascending auditory pathways to the brain involve several brain stem nuclei, but here we will stress only the major way stations. Impulses generated in the cochlea pass through the **spiral ganglion,** where the bipolar sensory neurons for the sense of hearing reside, and along the afferent fibers of the cochlear nerve to the **cochlear nuclei** of the medulla (Figure 16.33). From there, impulses are sent on to the **superior olivary nucleus** and then via the **lateral lemniscal tract** to the **inferior colliculus** (auditory reflex center in the midbrain). From there impulses are sent on to the **auditory cortex** (which provides for conscious awareness of sound) in the temporal lobe via relays in the **medial geniculate body** of the thalamus. The inferior colliculus also sends input to the superior colliculus of the midbrain. The colliculi act together to initiate auditory reflexes to sound, such as the startle reflex and head turning. The auditory pathway is unusual in that not all of the fibers from each ear decussate (cross over); therefore each auditory cortex receives impulses from both ears.

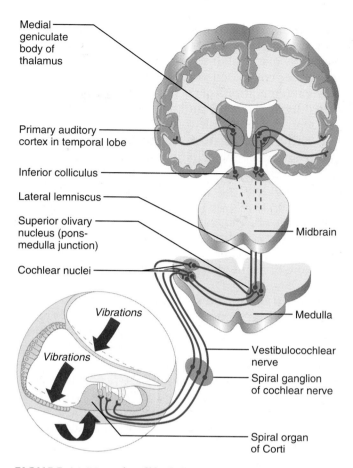

FIGURE 16.33 *Simplified diagram of the auditory pathway from the organ of Corti to the auditory cortex in the temporal lobe of the brain.* For clarity, only the pathway from the right ear is illustrated.

Auditory Processing

If you are at a Broadway musical, the sound of the instruments, the actors' voices, rustling of clothing, and closing of doors are all intermingled in your sound awareness. Yet, the auditory cortex can distinguish the separate parts of this auditory jumble. Auditory processing is analytic and parallel. Whenever the difference between sound wavelengths is sufficient for discrimination, you hear two separate and distinct tones. In fact, the analytic powers of the auditory cortex are so great that we are able to pick single instruments out of a whole orchestra.

Cortical processing of sound stimuli seems to be quite complex. For example, certain cortical cells depolarize at the beginning of a particular tone, others depolarize when the tone ends. Some cortical cells depolarize continuously, others appear to have high thresholds (low sensitivity), and so on. Here we will concentrate on the more straightforward aspects of cortical perception of pitch, loudness, and sound location.

Perception of Pitch As just explained, hair cells in different parts of the organ of Corti are activated by sound waves of different frequencies. When the sound is composed of tones of many frequencies, several populations of cochlear hair cells and cortical cells are activated simultaneously, resulting in the perception of multiple tones.

Early investigators concluded that impulses from specific hair cells are interpreted as specific pitches. However, the coding mechanism for pitch perception is much more complex and involves local processing in the cochlear nuclei and probably other areas as well. Indeed, both the cochlear nuclei and the primary auditory cortex have *tonotopic maps* that correspond spatially to precise regions of the organ of Corti and to specific frequencies.

Detection of Loudness Our perception of loudness suggests that certain cochlear cells have higher thresholds than others for responding to a tone of the same frequency. For example, some of the receptors for a tone with a frequency of 540 Hz might be stimulated by a sound wave of very low intensity. As the intensity of the sound increases, the basilar membrane vibrates more vigorously. As a result, more hair cells would begin to respond and more impulses would reach the auditory cortex and be recognized as a louder sound of the same pitch.

Localization of Sound Several brain stem nuclei (most importantly the superior olivary nuclei) process signals that help us localize a sound's source in space by means of two cues: the *relative intensity* and the *relative timing* of sound waves reaching the two ears. If the sound source is *directly* in front, in back, or over the midline of the head, the intensity and timing cues are the same for both ears. However, when sound comes from one side, the receptors of the nearer ear are activated slightly earlier and more vigorously (because of the greater intensity of the sound waves entering that ear).

Homeostatic Imbalances of Hearing

Deafness

Any hearing loss, no matter how slight, is **deafness** of some sort. Hearing loss may range from the inability to hear sound of a certain pitch or intensity to a complete inability to detect sound. Deafness is classified as conduction or sensorineural deafness, according to its cause.

Conduction deafness occurs when something hampers sound conduction to the fluids of the inner ear. The possibilities are almost limitless. For example, impacted earwax blocking the auditory canal hinders vibration of the eardrum. A *perforated* or *ruptured* eardrum prevents sound conduction from the eardrum to the ossicles. But the most common causes of conduction deafness are middle ear inflammations (otitis media) and **otosclerosis** (o"to-sklĕ-ro'sis) of the ossicles. Otosclerosis ("hardening of the ear"), a common age-related problem, occurs when overgrowth of bony tissue fuses the stapes foot plate to the oval window or welds the ossicles to one another. In such cases, sound is conducted to the receptors of that ear through vibrations of the skull bones, which is far less satisfactory. Otosclerosis is treated surgically to remove the excess tissue or to replace one or more of the ossicles or the oval window.

Sensorineural deafness results from damage to neural structures at any point from the cochlear hair cells to and including the auditory cortical cells. This type of deafness may be partial or complete and typically results from the gradual loss of the hearing receptor cells throughout life. These cells can also be destroyed at an earlier age by a single explosively loud noise or prolonged exposure to high-intensity sounds, such as rock bands or airport noise, which causes them to stiffen or tears their cilia. Degeneration of the cochlear nerve, cerebral infarcts, and tumors in the auditory cortex are other causes. For age- or noise-related cochlear damage, cochlear implants can be inserted into a drilled recess in the temporal bone. These miniature transducer devices convert sound energy into electrical signals that are delivered directly to the cochlear nerve fibers. The original cochlear implants provided less than normal hearing; for example, they made human voices sound tinny and robotlike. The modern versions have 20 electrodes, each responding to a different frequency, and are so effective that even children born deaf can hear well enough to learn to speak well.

Tinnitus

Tinnitus (tĭ-ni'tus) is a ringing or clicking sound in the ears in the absence of auditory stimuli. It is more a symptom of pathology than a disease. For example, tinnitus is one of the first symptoms of cochlear nerve degeneration. It may also result from inflammation of the middle or inner ears and is a side effect of some medications, such as aspirin.

Ménière's Syndrome

Classic **Ménière's** (men"ĕ-ār'z) **syndrome** is a labyrinth disorder that affects both the semicircular canals and the cochlea. The afflicted person has repeated attacks of vertigo, nausea, and vomiting. Balance is so disturbed that standing erect is nearly impossible. A "howling" tinnitus is common, so hearing is impaired (and ultimately lost) as well. The cause is uncertain, but it may result from distortion of the membranous labyrinth by excessive endolymph accumulation, or rupture of membranes which allows the perilymph and endolymph to mix, an event that generates ototoxins (chemicals poisonous to the ear). Mild cases can usually be managed by antimotion drugs. For more debilitating attacks, a low-salt diet and diuretics are used to decrease overall extracellular fluid volumes and, consequently, endolymph fluid volume. In severe cases, surgery may be performed to drain the excess endolymph from the inner ear. A last resort is removal of the entire malfunctioning labyrinth, which is usually deferred until hearing loss is complete.

Mechanisms of Equilibrium and Orientation

The equilibrium sense is not easy to describe because it does not "see," "hear," or "feel," but responds (frequently without our awareness) to various head movements. Furthermore, this sense depends not only on inputs from the inner ear but also on vision and on information from stretch receptors of muscles and tendons. Under normal conditions the equilibrium receptors in the semicircular canals and vestibule, collectively called the **vestibular apparatus,** send signals to the brain that initiate reflexes needed to make the simplest changes in position as well as to serve a tennis ball precisely to the right spot (your opponent's backhand). However, if the vestibular apparatus is damaged, the system can adapt and learn to function without it. Thus, it is difficult to credit any one set of receptors with our ability to maintain our orientation and balance in space. Nevertheless, we do know that the equilibrium receptors of the inner ear can be divided into two functional arms, with the receptors in the vestibule and semicircular canals monitoring **static** and **dynamic equilibrium,** respectively.

The Maculae and Static Equilibrium

The sensory receptors for static equilibrium are the **maculae** (mak′u-le; "spots"), one in each saccule wall and one in each utricle wall. These receptors monitor the position of the head in space, and in so doing, they play a key role in control of posture. They respond to *linear* acceleration forces, that is, straight-line changes in speed and direction, but not to rotation.

Anatomy of the Maculae Each macula is a flat epithelial patch containing **supporting cells** and scattered receptor cells called **hair cells** (Figure 16.34). Displacement of the head in any linear direction activates the receptors in either the utricles or the saccules. The hair cells have numerous *stereocilia* (long microvilli) and a single *kinocilium* (a true cilium)

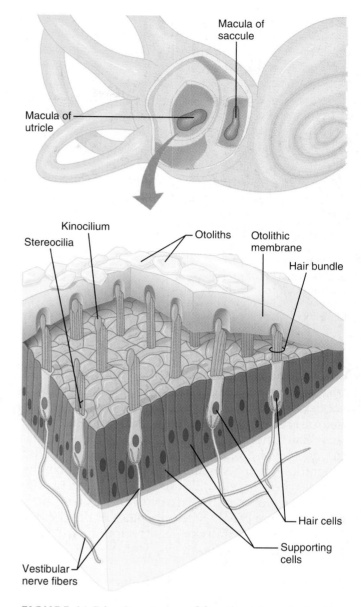

FIGURE 16.34   *Structure and function of a macula.* The "hairs" of the receptor cells of a macula project into the gelatinous otolithic membrane. Vestibular nerve fibers surround the base of the hair cells.

protruding from their apices. These "hairs" are embedded in the overlying **otolithic** (o′to-lith-ik) **membrane,** a flattened jellylike mass studded with tiny stones (calcium carbonate crystals) called **otoliths** ("ear stones"). The otoliths, though small, are dense and they increase the membrane's weight and its inertia (resistance to change in motion). In the utricle, the macula is horizontal (Figure 16.34), and the hairs are vertically oriented when the head is upright. Thus, the utricular maculae respond best to acceleration in the horizontal plane and tilting the head to the side, because vertical (up-down) movements do not displace their horizontal otolithic membrane. In the saccule, the macula is nearly vertical, and the hairs protrude horizontally into the

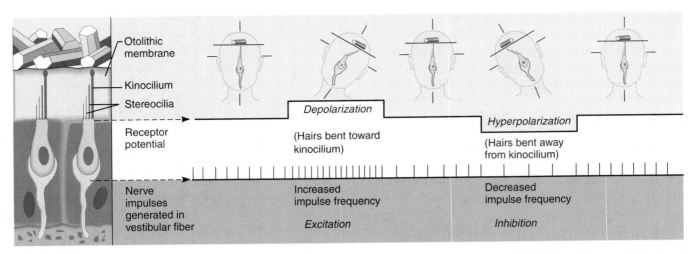

FIGURE 16.35   *The effect of gravitational pull on a macula receptor cell in the utricle.* When movement of the otolithic membrane (direction indicated by the arrow) bends the hair cells in the direction of the kinocilium, the vestibular nerve fibers depolarize and generate action potentials more rapidly. When the hairs are bent in the direction away from the kinocilium, the hair cells become hyperpolarized, and the nerve fibers send impulses at a reduced rate (i.e., below the resting rate of discharge).

otolithic membrane. The saccular maculae respond best to vertical movements, such as the sudden acceleration of an elevator. The receptor cells release neurotransmitter continually, but movement of their hairs modifies the amount they release, causing an increase or decrease in the rate of impulse generation by the **vestibular nerve** endings coiling around their bases. (Like the cochlear nerve, the vestibular nerve is a subdivision of the vestibulocochlear nerve [VIII].) The cell bodies of the sensory neurons are located in the nearby **superior** and **inferior vestibular ganglia.**

Transducing Gravity and Linear Acceleration Stimuli Let's look more closely at the happenings in the maculae that lead to sensory transduction. When your head starts or stops moving in a linear direction, inertia causes the otolithic membrane to slide backward or forward like a greased plate over the hair cells, bending the hairs. For example, when you run, the otolithic membranes of the utricle maculae lag behind, bending the hairs backward. When you suddenly stop, the otolithic membrane slides abruptly forward (just as you slide forward in your car when you brake), bending the hair cells forward. Likewise, when you nod your head or fall, the otoliths roll inferiorly, bending the hairs of the maculae in the saccules. When the hairs are bent *toward the kinocilium,* the hair cells depolarize and step up their pace of neurotransmitter release. As a result, a faster stream of impulses is sent to the brain (Figure 16.35). When the hairs are bent in the opposite direction, the receptors hyperpolarize, and neurotransmitter release and impulse generation decline. In either case, the brain is informed of the changing position of the head in space.

It is important to understand that the maculae respond *only to changes* in acceleration or velocity of head movement. Because the hair cells adapt quickly (resuming their basic level of neurotransmitter release), they do not report on unchanging head positions. Thus, the maculae help us to maintain normal head position with respect to gravity. They also contribute to dynamic equilibrium by responding to changes in linear acceleration.

The Crista Ampullaris and Dynamic Equilibrium

The receptor for dynamic equilibrium, called the **crista ampullaris,** or simply *crista,* is a minute elevation in the ampulla of each semicircular canal (see Figure 16.26). Like the maculae, the cristae are excited by head movement (acceleration and deceleration), but in this case the major stimuli are rotatory (angular) movements. When you twirl on the dance floor or suffer through a rough boat ride, these gyroscopelike receptors are working overtime. Since the semicircular canals are located in all three planes of space, all rotatory movements of the head disturb one or another *pair* of cristae (one in each ear).

Anatomy of the Crista Ampullaris Each crista is composed of supporting cells and hair cells. These hair cells, like those of the maculae, have stereocilia plus one kinocilium that project into a gel-like mass. In this case, the gelled mass is a **cupula** (ku'pu-lah), which resembles a pointed cap (Figure 16.36b and c). The cupula is a delicate, loosely organized network of gelatinous strands that radiate outward to contact the "hairs" of each hair cell. Dendrites of vestibular nerve fibers encircle the base of the hair cells.

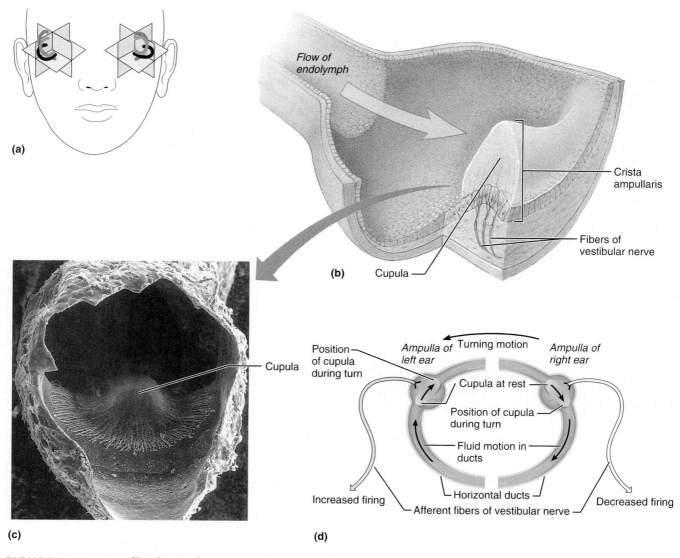

FIGURE 16.36 *Localization and structure of a crista ampullaris.* **(a)** Location of the semicircular canals within the temporal bones of the skull. **(b)** The ampulla of a semicircular duct has been sectioned to show the positioning of a crista ampullaris receptor. **(c)** Scanning EM of a crista ampullaris. (215×) **(d)** View of the horizontal ducts from above shows how paired semicircular canals work together to provide bilateral information on rotatory head movement.

Transducing Rotational Stimuli The cristae respond to *changes* in the velocity of rotatory movements of the head. Because of its inertia, the endolymph in the semicircular ducts moves briefly in the direction *opposite* the body's rotation, deforming the crista in the duct. As the hairs are bent, the hair cells depolarize and impulses reach the brain at a faster rate. Bending the cilia in the opposite direction causes hyperpolarization and reduces impulse generation. Because the axes of the hair cells in the complementary semicircular ducts are opposite, rotation in a given direction causes depolarization of the receptors in one ampulla of the pair, and hyperpolarization of the receptors in the other (Figure 16.36d).

If the body continues to rotate at a constant rate, the endolymph eventually comes to rest—it moves along at the same speed as the body and stimulation of the hair cells ends. Consequently, if we are blindfolded, we cannot tell whether we are moving at a constant speed or not moving at all after the first few seconds of rotation. However, when we suddenly stop moving, the endolymph keeps on going, in effect reversing its direction within the canal. This sudden reversal in the direction of hair bending results in membrane voltage changes in the receptor cells and modifies the rate of impulse transmission, which tells the brain that we have slowed or stopped.

The key point to remember when considering both the dynamic and the static equilibrium receptors is that the rigid bony labyrinth moves with the body, while the fluids (and gels) within the membranous labyrinth are free to move at various rates,

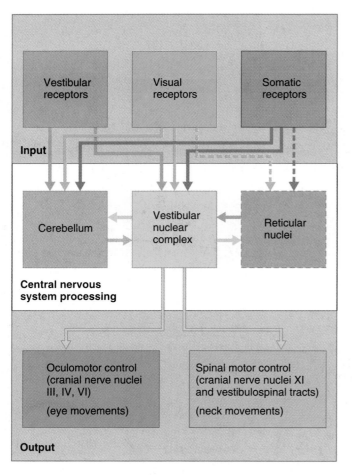

FIGURE 16.37 *Pathways of the balance and orientation system.* The three modes of input—vestibular, visual, and somatic (skin, muscle, and joint) receptors—transmit impulses to two major processing areas: the vestibular nuclei of the brain stem and the cerebellum. (The pathways to and the precise function of the reticular nuclei are unclear.) After vestibular nuclear processing, impulses are sent to one of two major output regions—areas controlling eye movements (oculomotor control) or those controlling the skeletal muscles of the neck (spinal motor control).

depending on the forces (gravity, acceleration, and so on) acting on them.

Impulses transmitted from the semicircular canals are particularly important to reflex movements of the eyes. **Vestibular nystagmus** is a complex of rather strange eye movements that occurs during and immediately after rotation. As you rotate, your eyes slowly drift in the opposite direction, as though fixed on some object in the environment. This reaction relates to the backflow of endolymph in the semicircular canals. Then, because of compensating mechanisms of the central nervous system, the eyes jump rapidly toward the direction of rotation to establish a new fixation point. These alternating eye movements continue until the endolymph comes to rest. When you stop rotating, at first your eyes continue to move in the direction of

the previous spin, and then they jerk rapidly in the opposite direction. This sudden change is caused by the change in the direction in which the cristae are bent after you stop. Nystagmus is often accompanied by vertigo.

The Equilibrium Pathway to the Brain

Our responses to body imbalance, such as during stumbling, must be fast and reflexive. By the time we "thought about" correcting our fall, we would already be on the ground. Hence, information from the balance receptors goes directly to reflex centers in the brain stem, rather than to the cerebral cortex as with the other special senses. However, the nerve pathways that connect the vestibular apparatus with the brain are complex and poorly traced. The transmission sequence begins when the hair cells in the vestibular apparatus are activated. As depicted in Figure 16.37, impulses travel initially to one of two destinations: the **vestibular nuclear complex** in the brain stem or the **cerebellum.** The vestibular nuclei, the major integrative center for balance, also receive inputs from the visual and somatic receptors, particularly from proprioceptors in neck muscles that report on the position of the head. They integrate this information and then send commands to brain stem motor centers that control the extrinsic eye muscles (cranial nerve nuclei III, IV, and VI) and reflex movements of the neck, limb, and trunk muscles (via the vestibulospinal tracts). The ensuing reflex movements of the eyes and body allow us to remain focused on the visual field and to quickly adjust our body position to maintain or regain balance.

The cerebellum also integrates inputs from the eyes and somatic receptors (as well as from the cerebrum). It coordinates skeletal muscle activity and regulates muscle tone so that head position, posture, and balance are maintained, often in the face of rapidly changing inputs. Its "specialty" is fine control of delicate postural movements and timing. Inputs from the eyes and somatic receptors also reach the reticular nuclei, but their precise function in equilibrium is unclear.

Notice that the vestibular apparatus does *not automatically compensate* for forces acting on the body. Its job is to send warning signals to the central nervous system, which initiates the appropriate compensations (righting) to keep your body balanced, your weight evenly distributed, and your eyes focused on what you were looking at when the disturbance occurred.

Homeostatic Imbalance

Responses to equilibrium signals are totally reflexive, and usually we are aware of vestibular apparatus activity only when its functioning is impaired. Equilibrium problems are usually obvious and unpleasant. Nausea, dizziness, and loss of balance are common symptoms. Also, there may be nystagmus in the absence of rotational stimuli.

Motion sickness, a common equilibrium disorder, has been difficult to explain, but it appears to be due to sensory input mismatch. For example, if you are inside a ship during a storm, visual inputs indicate that your body is fixed with reference to a stationary environment (your cabin). But as the ship is tossed about by the rough seas, your vestibular apparatus detects movement and sends impulses that "disagree" with the visual information. The brain thus receives conflicting information, and its "confusion" somehow leads to motion sickness. Warning signals, which precede nausea and vomiting, include excessive salivation, pallor, rapid deep breathing, and profuse sweating. Removal of the stimulus usually ends the symptoms. Over-the-counter antimotion drugs, such as meclizine HCl (Bonine), depress vestibular inputs and help many sufferers. They are most effective when taken "before the fact." Scopolamine-containing timed-release skin patches are also used to head off motion sickness. ■

DEVELOPMENTAL ASPECTS OF THE SPECIAL SENSES

Taste and Smell

All the special senses are functional, to a greater or lesser degree, at birth. Smell and taste are sharp, and infants relish food that adults consider bland or tasteless. Some researchers claim that smell is just as important as touch in guiding newborn infants to their mother's breast. However, very young children seem indifferent to odors and can play happily with their own feces. As they get older, their emotional responses to specific odors and foods increase.

There are few problems with the chemical senses during childhood and young adulthood. Women generally have a more acute sense of smell than men, and nonsmokers have a sharper sense of smell than smokers. Beginning in the fourth decade of life, our ability to taste and smell declines. This reflects the gradual loss of receptors, which are replaced more slowly than in younger people. Over half of people over the age of 65 years have serious problems detecting odors, which may explain their tendency to douse themselves with large amounts of cologne, or their inattention to formerly disagreeable odors. Additionally, their sense of taste is poor. This along with the decline in the sense of smell makes food taste bland and results in loss of appetite.

Vision

By the fourth week of development, the beginnings of the eyes are seen in the **optic vesicles** that protrude from the diencephalon. Soon these hollow vesicles indent to form double-layered **optic cups,** and their stalks form the optic nerves and provide a pathway for blood vessels to reach the eye interior. Once an optic vesicle reaches the overlying surface ectoderm, it induces the ectoderm to thicken and then form a **lens vesicle** that pinches off into the cavity of the optic cup, where it becomes the lens. The lining (internal layer) of the optic cup becomes the neural retina, and the outer layer forms the pigmented layer of the retina. The balance of the eye tissues and the vitreous humor are formed by mesenchymal cells derived from the mesoderm that surrounds the optic cup.

In the darkness of the uterus, the fetus cannot see. Nonetheless, even before the light-sensitive portions of the photoreceptors develop, the central nervous system connections have been made and are functional. During infancy, synaptic connections are fine-tuned, and the typical cortical fields that allow binocular vision are established.

🔁 *Homeostatic Imbalance*

Congenital problems of the eyes are relatively uncommon, but their incidence is dramatically increased by certain maternal infections, particularly rubella (German measles), occurring during the critical first three months of pregnancy. Common rubella sequels include blindness and cataracts. ■

As a rule, vision is the only special sense not fully functional at birth. Because the eyeballs are foreshortened, most babies are hyperopic. The newborn infant sees only in gray tones, eye movements are uncoordinated, and often only one eye at a time is used. The lacrimal glands are not completely developed until about two weeks after birth, so babies are tearless for this period, even though they may cry lustily. By five months, infants can follow moving objects with their eyes, but visual acuity is still poor (20/200). By the age of five years, depth perception is present and color vision is well developed. Because the eyeball has grown, visual acuity has improved to about 20/30, providing a readiness to begin reading. By first grade, the initial hyperopia has usually been replaced by emmetropia, and the eye reaches its adult size at eight to nine years of age. Emmetropia usually continues until presbyopia begins to set in around age 40 owing to decreasing lens elasticity.

With age, the lens loses its crystal clarity and discolors. As a result, it begins to scatter light, causing a glare that is distressing when driving at night.

The dilator muscles of the iris become less efficient so the pupils stay partly constricted. These two changes together decrease the amount of light reaching the retina, and visual acuity is dramatically lower in people over 70. In addition, the lacrimal glands are less active and the eyes tend to be dry and more susceptible to infection. Elderly persons are also at risk for certain conditions that cause blindness, such as macular degeneration, glaucoma, cataracts, arteriosclerosis, and diabetes mellitus.

Hearing and Balance

Development of the ear begins in the three-week embryo. The inner ears develop first, from thickenings of the surface ectoderm, called the **otic** (o'tik) **placodes,** which lie lateral to the hindbrain on each side. The otic placode invaginates, forming the **otic pit** and then the **otic vesicle,** which detaches from the surface epithelium. The otic vesicle develops into the membranous labyrinth. The surrounding mesenchyme forms the bony labyrinth.

Middle ear structures (the middle ear cavity, pharyngotympanic tube, and ossicles) develop from **pharyngeal pouches,** lateral outpocketings of the endoderm lining the pharynx.

The external auditory canal and external face of the tympanic membrane of the outer ear differentiate from the **branchial** (brang'ke-al) **groove,** an indentation of the surface ectoderm, and the auricle develops from swellings of the surrounding tissue.

Newborn infants can hear, but early responses to sound are mostly reflexive—for example, crying and clenching the eyelids in response to a startling noise. By the fourth month, infants can localize sounds and will turn to the voices of family members. Critical listening begins to occur in toddlers as they increase their vocabulary, and good language skills are very closely tied to the ability to hear well.

⧮ *Homeostatic Imbalance*

Congenital abnormalities of the ears are fairly common. Examples include partly or completely missing pinnae and closed or absent external ear canals. Maternal rubella during the first trimester commonly results in sensorineural deafness. ■

Except for ear inflammations resulting from bacterial infections or allergies, few problems affect the ears during childhood and adult life. By the 60s, however, deterioration of the organ of Corti becomes noticeable. We are born with approximately 40,000 hair cells, but their number decreases when they are damaged or destroyed by loud noises, disease, or drugs. The hair cells *are* replaced, but at such a slow rate that there really is no functional regeneration. It is estimated that if we were to live 140 years, we would have lost all of our hearing receptors.

The ability to hear high-pitched sounds leaves us first. This condition, called **presbycusis** (pres"bǐ-ku'sis), is a type of sensorineural deafness. Although presbycusis is considered a disability of old age, it is becoming much more common in younger people as our world grows noisier. Since noise is a stressor, one of its physiological consequences is vasoconstriction, and when delivery of blood to the ear is reduced, the ear becomes even more sensitive to the damaging effects of noise.

★ ★ ★

Our abilities to see, hear, taste, and smell—and some of our responses to the effects of gravity—are largely the work of our brain. However, as we have discovered in this, the last of the nervous system chapters, the large and often elaborate sensory receptor organs that serve the special senses are works of art in and of themselves.

The final chapter of this unit describes how the body's functions are controlled by chemicals called hormones in a manner quite different from what we have described for neural control.

RELATED CLINICAL TERMS

Ageusia (ah-gu'ze-ah) Loss or impairment of the taste sense.

Age-related macular degeneration (ARMD) Progressive deterioration of the retina that affects the macula lutea and leads to loss of central vision; the main cause of vision loss in those over age 65. Early stages and mild forms involve the buildup of pigments in the macula, probably because the retina's phagocytic pigment cells cannot clear away all the debris from the normal recycling of the cone membranes and lipofuscin deposits. Continued accumulation of this pigment leads to the "dry" form of ARMD, in which many of the macular photoreceptors die. More common is the "wet" form in which an overgrowth of new blood vessels invades the retina from the choroid layer. Leakage of blood

and fluids from these vessels causes scarring and detachment of the retina. The cause is unknown and ARMD is largely untreatable, although treatment (laser photocoagulation) can destroy some of the growing vessels in the "wet" form.

Blepharitis (blef"ah-ri'tis; *blephar* = eyelash; *itis* = inflammation) Inflammation of the margins of the eyelids.

Enucleation (e-nu"kle-a'shun) Surgical removal of an eyeball.

Exophthalmos (ek"sof-thal'mos; *exo* = out; *phthalmo* = the eye) Anteriorly bulging eyeballs; this condition is seen in some cases of hyperthyroidism.

Labyrinthitis Inflammation of the labyrinth.

Ophthalmology (of″thal-mol′o-je) The science that studies the eye and eye diseases. An **ophthalmologist** is a medical doctor who specializes in treating eye disorders.

Optometrist A licensed nonphysician who measures vision and prescribes corrective lenses.

Otalgia (o-tal′je-ah; *algia* = pain) Earache.

Otitis externa Inflammation and infection of the external auditory canal, caused by bacteria or fungi that enter the canal from outside, especially when the canal is moist (e.g., after swimming).

Papilledema (pap″il-ĕ-de′mah; *papill* = nipple; *edema* = swelling) Protrusion of the optic disc into the eyeball, which can be observed by ophthalmoscopic examination; caused by conditions that increase intracranial pressure.

Scotoma (sko-to′mah; *scoto* = darkness) A blind spot other than the normal (optic disc) blind spot; often reflects the presence of a brain tumor pressing on fibers of the visual pathway.

Trachoma (trah-ko′mah; *trach* = rough) A highly contagious bacterial (chlamydial) infection of the conjunctiva and cornea. Common worldwide, it blinds millions of people in poor countries of Africa and Asia; treated with eye ointments containing antibiotic drugs.

Weber's test Hearing test during which a sounding tuning fork is held to the forehead. In those with normal hearing, the tone is heard equally in both ears. The tone will be heard best in the "good" ear if sensorineural deafness is present, and in the "bad" ear if conduction deafness is present.

CHAPTER SUMMARY

Media study tools that could provide you additional help in reviewing specific key topics of Chapter 16 are referenced below. **SP** = *Study Partner;* **IP** = *Interactive Physiology.*

The Chemical Senses: Taste and Smell (pp. 559–564)

Taste Buds and the Sense of Taste (pp. 559–561)

1. The taste buds are scattered in the oral cavity and pharynx but are most abundant on the tongue papillae.

2. Gustatory cells, the receptor cells of the taste buds, have gustatory hairs (microvilli) that serve as the receptor regions. The gustatory cells are excited by the binding of chemicals to receptors on their microvilli.

3. The four basic taste qualities—sweet, sour, salty, and bitter—are sensed best at different regions on the tongue.

4. The taste sense is served by cranial nerves VII, IX, and X, which send impulses to the solitary nucleus of the medulla. From there, impulses are sent to the thalamus and the taste cortex.

The Olfactory Epithelium and the Sense of Smell (pp. 561–564)

5. The olfactory epithelium is located in the roof of the nasal cavity. The receptor cells are ciliated bipolar neurons. Their axons are the filaments of the olfactory nerve (cranial nerve I).

6. Individual olfactory neurons show a range of responsiveness to different chemicals. Olfactory cells bearing the same odorant receptors synapse in the same glomerulus type.

7. Olfactory neurons are excited by volatile chemicals that bind to receptors in the olfactory cilia.

8. Action potentials of the olfactory nerve filaments are transmitted to the olfactory bulb where the filaments synapse with mitral cells. The mitral cells send impulses via the olfactory tract to the olfactory cortex. Fibers carrying impulses from the olfactory receptors also project to the limbic system.

Homeostatic Imbalances of the Chemical Senses (p. 564)

9. Most chemical sense dysfunctions are olfactory disorders (anosmias). Common causes are nasal damage or obstruction and zinc deficiency.

The Eye and Vision (pp. 564–586)

1. The eye is enclosed in the bony orbit and cushioned by fat.

Accessory Structures of the Eye (pp. 565–568)

2. Eyebrows help to shade and protect the eyes.

3. Eyelids protect and lubricate the eyes by reflex blinking. Within the eyelids are the orbicularis oculi and levator palpebrae muscles, modified sebaceous glands, and sweat glands.

4. The conjunctiva is a mucosa that lines the eyelids and covers the anterior eyeball surface. Its mucus lubricates the eyeball surface.

5. The lacrimal apparatus consists of the lacrimal gland (which produces a saline solution containing mucus, lysozyme, and antibodies), the lacrimal canals, the lacrimal sac, and the nasolacrimal duct.

6. The extrinsic eye muscles (superior, inferior, lateral, and medial rectus and superior and inferior oblique) move the eyeballs.

Structure of the Eyeball (pp. 568–574)

7. The wall of the eyeball is made up of three layers, or tunics. The outermost fibrous tunic consists of the sclera and the cornea. The sclera protects the eye and gives it shape; the cornea allows light to enter the eye.

8. The middle pigmented vascular tunic (uvea) consists of the choroid, the ciliary body, and the iris. The choroid provides nutrients to the eye and prevents light scattering within the eye. The ciliary muscles of the ciliary body control lens shape; the iris controls the size of the pupil.

9. The sensory tunic, or retina, consists of an outer pigmented layer and an inner nervous layer. The neural layer contains photoreceptors (rods and cones), bipolar cells, and ganglion cells. Ganglion cell axons form the optic nerve, which exits via the optic disc ("blind spot").

10. The outer segments of the photoreceptors contain the light-absorbing pigment in membrane-bounded discs.

11. The posterior segment, behind the lens, contains vitreous humor, which helps support the eyeball and keep the retina in place. The anterior segment, anterior to the lens, is filled with aqueous humor, formed by capillaries in the ciliary processes and drained into the scleral venous sinus. Aqueous humor is a major factor in maintaining intraocular pressure.

12. The biconvex lens is suspended within the eye by the suspensory ligaments attached to the ciliary body. It is the only adjustable refractory structure of the eye.

SP Exercise: Chapter 16, Internal Structures of the Eye.

Physiology of Vision (pp. 574–586)

13. Light is made up of those wavelengths of the electro-magnetic spectrum that excite the photoreceptors.

14. Light is refracted (bent) when passing from one transparent medium to another of different density, or when it strikes a curved surface. Concave lenses disperse light; convex lenses converge light and bring its rays to a focal point.

15. As light passes through the eye, it is bent by the cornea and the lens and focused on the retina. The cornea accounts for most of the refraction, but the lens allows active focusing for different distances.

16. Focusing for distance vision requires no special movements of the eye structures. Focusing for close-up vision requires accommodation (bulging of the lens), pupillary constriction, and convergence of the eyeballs. All three reflexes are controlled by cranial nerve III.

17. Refractory problems include myopia, hyperopia, and astigmatism. (See pp. 577–578.)

SP Exercise: Chapter 16, Optics of the Eye.

18. Rods respond to low-intensity light and provide night and peripheral vision. Cones are bright-light, high-discrimination receptors that provide for color vision. Anything that must be viewed precisely is focused on the cone-rich fovea centralis.

19. The light-absorbing molecule retinal is combined with various opsins in the visual pigments. When struck by light, retinal changes shape (11-*cis* to all-*trans*) and releases opsin. Freed opsin activates transducin (a G protein subunit) which in turn activates PDE, an enzyme that breaks down cGMP, allowing the Na^+ gates to close. This results in hyperpolarization of the receptor cells and inhibits their release of neurotransmitter.

20. Rod visual pigment, rhodopsin, is a combination of retinal and opsin. The light-triggered changes in retinal cause hyperpolarization of the rods. Photoreceptors and bipolar cells generate local potentials only; action potentials are generated by the ganglion cells.

21. The three types of cones all contain retinal, but each has a different type of opsin. Each cone type responds maximally to one color of light: red, blue, or green. The chemistry of cone function is similar to that of rods.

22. During light adaptation, photopigments are bleached and rods are inactivated; then, as cones begin to respond to high-intensity light, high-acuity vision ensues. In dark adaptation, cones cease functioning, and visual acuity decreases; rod function begins when sufficient rhodopsin has accumulated.

23. The visual pathway to the brain begins with the optic nerve fibers (ganglion cell axons) from the retina. At the optic chiasma, fibers from the medial half of each retinal field cross over and continue on in the optic tracts to the thalamus. Thalamic neurons project to the optic cortex via the optic radiation. Fibers also project from the retina to the midbrain pretectal nuclei and the superior colliculi, and to the suprachiasmatic nucleus of the hypothalamus.

24. In stereoscopic vision, each eye receives a slightly different view of the visual field. These views are fused by the optic cortices to provide for depth perception.

25. Retinal processing involves the selective destruction of rod inputs so as to emphasize bright/dark contrasts and edges. (The horizontal cells and amacrine cells are local integrator cells of the retina that modify and process rod inputs to the ganglion cells.) Thalamic processing subserves high-acuity color vision and depth perception. Cortical processing involves neurons of the striate (primary) cortex, which receive inputs from the retinal ganglion cells, and neurons of the prestriate (association) cortices, which receive inputs from striate cortical cells and mostly integrate inputs concerned with color, form, and movement. Visual processing also proceeds anteriorly in the "what" and "where" processing streams via the temporal and parietal lobes, respectively.

SP Case Study: Cerebrovascular Accident.

The Ear: Hearing and Balance (pp. 586–601)

Structure of the Ear (pp. 586–590)

1. The auricle and external auditory canal compose the outer ear. The tympanic membrane, the boundary between the outer and middle ears, transmits sound waves to the middle ear.

2. The middle ear is a small chamber within the temporal bone, connected by the pharyngotympanic tube to the nasopharynx. The ossicles, which help to amplify sound, span the middle ear cavity and transmit sound vibrations from the eardrum to the oval window.

3. The inner ear consists of the bony labyrinth, within which the membranous labyrinth is suspended. The bony labyrinth chambers contain perilymph; the membranous labyrinth ducts and sacs contain endolymph.

4. The vestibule contains the saccule and utricle. The semicircular canals extend posteriorly from the vestibule in three planes. They contain the semicircular ducts.

5. The cochlea houses the cochlear duct (scala media), which contains the organ of Corti (hearing receptor). Within the cochlear duct, the hair (receptor) cells rest on the basilar membrane, and their hairs project into the gelatinous tectorial membrane.

SP Exercise: Chapter 16, Structure of the Ear.

Sound and Mechanisms of Hearing (pp. 591–596)

6. Sound originates from a vibrating object and travels in waves consisting of alternating areas of compression and rarefaction of the medium.

7. The distance from crest to crest on a sine wave is the sound's wavelength; the shorter the wavelength, the higher the frequency (measured in hertz). Frequency is perceived as pitch.

8. The amplitude of sound is the height of the peaks of the sine wave, which reflect the sound's intensity. Sound intensity is measured in decibels. Intensity is perceived as loudness.

9. Sound passing through the external auditory canal sets the eardrum into vibration at the same frequency. The ossicles amplify and deliver the vibrations to the oval window.

10. Pressure waves in cochlear fluids set specific basilar membrane fibers into resonance. At points of maximal membrane vibration, the hair cells of the organ of Corti are alternately depolarized and hyperpolarized by the vibratory motion. High-frequency sounds stimulate hair cells near the oval window; low-frequency sounds stimulate hair cells near the apex. Most auditory inputs are sent to the brain by inner hair cells. Outer hair cells amplify responsiveness of the inner hair cells.

11. Impulses generated along the cochlear nerve travel to the cochlear nuclei of the medulla and from there through several brain stem nuclei to the medial geniculate nucleus of the thalamus and then the auditory cortex. Each auditory cortex receives impulses from both ears.

12. Auditory processing is analytic; each tone is perceived separately. Perception of pitch is related to the position of the excited hair cells along the basilar membrane. Intensity perception reflects the fact that as sound intensity increases, basilar membrane mobility is increased and the frequency of impulse transmission to the cortex is enhanced. Cues for sound localization include the intensity and timing of sound arriving at each ear.

Homeostatic Imbalances of Hearing   (pp. 596–597)

13. Conduction deafness results from interference with conduction of sound vibrations to the fluids of the inner ear. Sensorineural deafness reflects damage to neural structures.

14. Tinnitus is an early sign of sensorineural deafness; it may also result from the use of certain drugs.

15. Ménière's syndrome is a disorder of the membranous labyrinth. Symptoms include tinnitus, deafness, and vertigo. Excessive endolymph accumulation is the suspected cause.

Mechanisms of Equilibrium and Orientation (pp. 597–601)

16. The equilibrium receptor regions of the inner ear are called the vestibular apparatus.

17. The receptors for static equilibrium are the maculae of the saccule and utricle. A macula consists of hair cells with stereocilia and a kinocilium embedded in an overlying otolithic membrane. Linear movements cause the otolithic membrane to move, pulling on the hair cells. Movements toward the kinocilium depolarize the hair cells and increase the rate of impulse generation in the vestibular nerve fibers. Movements away from the kinocilium have the opposite effect.

18. The dynamic equilibrium receptor, the crista ampullaris within each semicircular duct, responds to angular or rotatory movements in one plane. It consists of a tuft of hair cells whose microvilli are embedded in the gelatinous cupula. Rotatory movements cause the endolymph to flow in the opposite direction, bending the cupula and either exciting or inhibiting the hair cells.

19. Impulses from the vestibular apparatus are sent via vestibular nerve fibers mainly to the vestibular complex of the brain stem and the cerebellum. These centers initiate responses that fix the eyes on objects and activate muscles to maintain balance.

Developmental Aspects of the Special Senses (pp. 601–602)

Taste and Smell   (p. 601)

1. The chemical senses are sharpest at birth and gradually decline with age as replacement of the receptor cells becomes more sluggish.

Vision   (pp. 601–602)

2. Congenital eye problems are uncommon, but maternal rubella can cause blindness.

3. The eye starts as an optic vesicle, an outpocketing of the diencephalon that invaginates to form the optic cup, which becomes the retina. Overlying ectoderm folds to form the lens vesicle, which gives rise to the lens. The remaining eye tissues and the accessory structures are formed by mesenchyme.

4. The eye is foreshortened at birth and reaches adult size at the age of eight or nine years. Depth perception and color vision develop during early childhood.

5. With age, the lens loses its elasticity and clarity, there is a decline in the ability of the iris to dilate, and visual acuity decreases. The elderly are at risk for eye problems resulting from dry eyes and disease.

Hearing and Balance   (p. 602)

6. The membranous labyrinth develops from the otic placode, an ectodermal thickening lateral to the hindbrain. Mesenchyme forms the surrounding bony structures. Pharyngeal pouch endoderm, in conjunction with mesenchyme, forms middle ear structures; the outer ear is formed largely by ectoderm.

7. Congenital ear problems are fairly common. Maternal rubella can cause deafness.

8. Response to sound in infants is reflexive. By five months, an infant can locate sound. Critical listening develops in toddlers.

9. Deterioration of the organ of Corti occurs throughout life as noise, disease, and drugs destroy cochlear hair cells. Age-related loss of hearing (presbycusis) occurs in the 60s and 70s.

REVIEW QUESTIONS

Multiple Choice/Matching

1. Olfactory tract damage would probably affect your ability to (a) see, (b) hear, (c) feel pain, (d) smell.

2. Sensory impulses transmitted over the facial, glossopharyngeal, and vagus nerves are involved in the sensation of (a) taste, (b) touch, (c) equilibrium, (d) smell.

3. Taste buds are found on the (a) anterior part of the tongue, (b) posterior part of the tongue, (c) palate, (d) all of these.

4. Gustatory cells are stimulated by (a) movement of otoliths, (b) stretch, (c) substances in solution, (d) photons of light.

5. Cells in the olfactory bulb that act as local "integrators" of olfactory inputs are the (a) hair cells, (b) granule cells, (c) basal cells, (d) mitral cells, (e) supporting cells.

6. Olfactory nerve filaments are found (a) in the optic bulbs, (b) passing through the cribriform plate of the ethmoid bone, (c) in the optic tracts, (d) in the olfactory cortex.

7. The accessory glands that produce an oily secretion are the (a) conjunctiva, (b) lacrimal glands, (c) tarsal glands.

8. The portion of the fibrous tunic that is white and opaque is the (a) choroid, (b) cornea, (c) retina, (d) sclera.

9. Which sequence best describes a normal route for the flow of tears from the eyes into the nasal cavity?
(a) lacrimal canals, nasolacrimal ducts, nasal cavity,
(b) lacrimal ducts, lacrimal canals, nasolacrimal ducts,
(c) nasolacrimal ducts, lacrimal canals, lacrimal sacs.

10. Four refractory media of the eye, listed in the sequence in which they refract light, are (a) vitreous humor, lens, aqueous humor, cornea, (b) cornea, aqueous humor, lens, vitreous humor, (c) cornea, vitreous humor, lens, aqueous humor, (d) lens, aqueous humor, cornea, vitreous humor.

11. Damage to the medial recti muscles would probably affect (a) accommodation, (b) refraction, (c) convergence, (d) pupil constriction.

12. The phenomenon of light adaptation is best explained by the fact that (a) rhodopsin does not function in dim light, (b) rhodopsin breakdown occurs slowly, (c) rods exposed to intense light need time to generate rhodopsin, (d) cones are stimulated to function by bright light.

13. Blockage of the scleral venous sinus might result in (a) a sty, (b) glaucoma, (c) conjunctivitis, (d) a cataract.

14. Nearsightedness is more properly called (a) myopia, (b) hyperopia, (c) presbyopia, (d) emmetropia.

15. Of the neurons in the retina, the axons of which of these form the optic nerve? (a) bipolar neurons, (b) ganglion cells, (c) cone cells, (d) horizontal cells.

16. Which sequence of reactions occurs when a person looks at a distant object? (a) pupils constrict, suspensory ligament relaxes, lenses become less convex, (b) pupils dilate, suspensory ligament becomes taut, lenses become less convex, (c) pupils dilate, suspensory ligament becomes taut, lenses become more convex, (d) pupils constrict, suspensory ligament relaxes, lenses become more convex.

17. During embryonic development, the lens of the eye forms (a) as part of the choroid coat, (b) from the surface ectoderm overlying the optic cup, (c) as part of the sclera, (d) from mesodermal tissue.

18. The blind spot of the eye is (a) where more rods than cones are found, (b) where the macula lutea is located, (c) where only cones occur, (d) where the optic nerve leaves the eye.

19. Conduction of sound from the middle ear to the inner ear occurs via vibration of the (a) malleus against the tympanic membrane, (b) stapes in the oval window, (c) incus in the round window, (d) stapes against the tympanic membrane.

20. The transmission of sound vibrations through the inner ear occurs chiefly through (a) nerve fibers, (b) air, (c) fluid, (d) bone.

21. Which of the following statements does not correctly describe the organ of Corti? (a) sounds of high frequency stimulate hair cells at the basal end, (b) the "hairs" of the receptor cells are embedded in the tectorial membrane, (c) the basilar membrane acts as a resonator, (d) sounds of high frequency stimulate cells at the apex of the basilar membrane.

22. Pitch is to frequency of sound as loudness is to (a) quality, (b) intensity, (c) overtones, (d) all of these.

23. The structure that allows pressure in the middle ear to be equalized with atmospheric pressure is the (a) pinna, (b) pharyngotympanic tube, (c) tympanic membrane, (d) secondary tympanic membrane.

24. Which of the following is important in maintaining the balance of the body? (a) visual cues, (b) semicircular canals, (c) the saccule, (d) proprioceptors, (e) all of these.

25. Static equilibrium receptors that report the position of the head in space relative to the pull of gravity are (a) organ of Corti, (b) maculae, (c) cristae ampullares.

26. Which of the following is *not* a possible cause of conduction deafness? (a) impacted cerumen, (b) middle ear infection, (c) cochlear nerve degeneration, (d) otosclerosis.

27. Which of the following are intrinsic eye muscles? (a) superior rectus, (b) orbicularis oculi, (c) smooth muscles of the iris and ciliary body, (d) levator palpebrae superioris.

28. Which lies closest to the exact posterior pole of the eye? (a) optic nerve, (b) optic disc, (c) macula lutea, (d) point of entry of central artery into the eye.

29. Otoliths (ear stones) are (a) a cause of deafness, (b) a type of hearing aid, (c) important in equilibrium, (d) the rock-hard petrous temporal bones.

Short Answer Essay Questions

30. Name the four primary taste qualities, and indicate the regions of the tongue that are most sensitive to each quality.

31. Where are the olfactory receptors, and why is that site poorly suited for their job?

32. Why do you often have to blow your nose after crying?

33. How do rods and cones differ functionally?

34. Where is the fovea centralis, and why is it important?

35. Describe the response of rhodopsin to light stimuli. What is the outcome of this cascade of events?

36. Since there are only three types of cones, how can you explain the fact that we see many more colors?

37. Describe the effect of aging on the special sense organs.

38. Each olfactory cell responds to a single odorant molecule. True or false? Explain your choice.

Critical Thinking and Clinical Application Questions

1. During an ophthalmoscopic examination, Mrs. James was found to have bilateral papilledema. Further investigation indicated that this condition resulted from a rapidly growing intracranial tumor. First, define papilledema. Then, explain its presence in terms of Mrs. James's diagnosis.

2. Sally, a nine-year-old girl, told the clinic physician that her "ear lump hurt" and she kept "getting dizzy and falling down." As she told her story, she pointed to her mastoid process. An otoscopic examination of the external auditory canal revealed a red, swollen eardrum, and her throat was inflamed. Her condition was described as mastoiditis with secondary labyrinthitis (inflammation of the labyrinth). Describe the most likely route of infection and the infected structures in Sally's case. Also explain the cause of her dizziness and falling.

3. Mr. Gaspe appeared at the eye clinic complaining of a chip of wood in his eye. No foreign body was found, but the conjunctiva was obviously inflamed. What name is given to this inflammatory condition, and where would you look for a foreign body that has been floating around on the eye surface for a while?

4. Mrs. Orlando has been noticing flashes of light and tiny specks in her right visual field. When she begins to see a "veil" floating before her right eye, she makes an appointment to see the eye doctor. What is your diagnosis? Is the condition serious? Explain.

5. David Norris, an engineering student, has been working in a disco to earn money to pay for his education. After about eight months, he notices that he is having problems hearing high-pitched tones. What is the cause-and-effect relationship here?

6. Assume that a tumor in the pituitary gland or hypothalamus is protruding inferiorly and compressing the optic chiasma. What could be the visual outcome?

7. Four-year-old Gary Zammer is brought to the ophthalmologist for a routine checkup on his vision. Gary is an albino. How do you think albinism affects vision?

17 THE ENDOCRINE SYSTEM

Y ou don't have to watch emergency room medical shows on TV to experience action-packed drama. Inside your body, molecules and cells have dynamic adventures on microscopic levels. When insulin molecules, carried passively along in the blood, attach tenaciously to protein receptors of nearby cells, the response is dramatic: Transported glucose molecules begin to disappear into the cells, and cellular activity accelerates. Such is the power of the second great controlling system of the body, the **endocrine system.** However, the endocrine system does not act alone—it interacts with the nervous system to coordinate and integrate the activity of body cells.

The means and speed of control used by these two great regulatory systems are very different. The nervous system regulates the activity of muscles and glands via electrochemical impulses delivered by neurons; those organs respond within milliseconds. The endocrine system, on the other hand, influences the metabolic activities of cells by means of **hormones** (*hormon* = to excite), chemical messengers released to the blood to be transported throughout the body. Tissue or organ responses to hormones typically occur after a lag period of seconds or even days. But once initiated, those responses tend to be much more prolonged than those induced by the nervous system.

Hormonal targets ultimately include most cells of the body, and hormones have widespread and diverse effects. The major processes controlled and integrated by these "mighty molecules" are reproduction; growth and development; mobilizing body defenses against stressors; maintaining electrolyte, water, and nutrient balance of the blood; and regulating cellular metabolism and energy balance. As you can see, the endocrine system orchestrates processes that go on for relatively long periods, in some instances continuously. The scientific study of hormones and the endocrine organs is called **endocrinology.**

THE ENDOCRINE SYSTEM: AN OVERVIEW

Compared to other organs of the body, the organs of the endocrine system are small and unimpressive. Indeed, to collect 1 kg (2.2 lbs) of hormone-producing tissue, you would need to collect *all* of the endocrine tissue from eight or nine adults! In addition, the anatomical continuity typical of most organ systems does not exist between the endocrine organs. Instead, endocrine organs are widely scattered about the body (see Figure 17.4, p. 616).

As explained in Chapter 4, there are two kinds of glands: endocrine and exocrine. Exocrine glands have ducts through which their nonhormonal products are routed to a membrane surface. In contrast, endocrine glands, also called *ductless glands*, lack ducts. They release hormones into the surrounding tissue fluid (*endo* = within; *crine* = to secrete), and they typically have a rich vascular and lymphatic drainage to receive their hormones. Most hormone-producing cells are arranged in cords and branching networks—a situation that maximizes contact between them and the surrounding capillaries.

The **endocrine glands** of the body include the pituitary, thyroid, parathyroid, adrenal, pineal, and thymus glands. In addition, several organs of the body contain discrete areas of endocrine tissue and produce hormones as well as exocrine products. Such organs, which include the pancreas and gonads (the ovaries and testes), are also major endocrine glands. The hypothalamus also falls into this latter category. Along with its neural functions, it produces and releases hormones, so we can consider the hypothalamus a **neuroendocrine organ.**

Besides the major endocrine organs, various other tissues and organs produce hormones. For example, adipose cells release leptin, and pockets of hormone-producing cells are found in the walls of the small intestine, stomach, kidneys, and heart—organs whose chief functions have little to do with hormone production. These other hormone-producing structures are described on pp. 640–642.

⬥ *Homeostatic Imbalance*

Certain tumor cells, such as those of some cancers of the lung or pancreas, synthesize hormones identical to those made in normal endocrine glands. However, they do so in an excessive and uncontrolled fashion. ■

HORMONES

The Chemistry of Hormones

Hormones are chemical substances, secreted by cells into the extracellular fluids, that regulate the metabolic function of other cells in the body. Although a large variety of hormones are produced, nearly all of them can be classified chemically as either **amino acid–based hormones** or **steroids.**

Most hormones belong to the first group. Molecular size varies widely in this group—from the simple amino acid derivatives, which include the amines, thyroxine, and peptides (short chains of amino acids), to proteins (long polymers of amino acids). Hormones of the second group, the steroids, are synthesized from cholesterol. Of the hormones produced by the major endocrine organs, only the gonadal hormones and adrenocortical hormones are steroids.

If we also consider the **eicosanoids** (i-ko′să-noyds), which include *leukotrienes* and *prosta-glandins,* we must add a third chemical class. These *paracrines,* or *local hormones,* are biologically active lipids (made from arachidonic acid) released by nearly all cell membranes. Leukotrienes are signaling chemicals that mediate inflammation and some allergic reactions. Prostaglandins have multiple targets and effects, ranging from raising blood pressure and increasing the expulsive uterine contractions of birth (via stimulation of smooth muscle) to enhancing blood clotting and inflammation. Since the effects of eicosanoids are typically highly localized, they do not quite fit the definition of true *circulating hormones,* which influence distant targets. Hence, this class of hormonelike chemicals will not be considered further here, but its members are considered in later chapters as appropriate.

Mechanisms of Hormone Action

Hormones bring about their characteristic effects on target cells by *altering* cell activity; that is, they increase or decrease the rates of normal cellular processes. The precise response depends on the target cell type. For example, epinephrine binding to smooth muscle cells in blood vessel walls stimulates them to contract. Epinephrine binding to cells other than muscle cells may have a different effect, but it does *not* cause those cells to shorten.

A hormonal stimulus typically produces one or more of the following changes:

1. Alters plasma membrane permeability or electrical state (membrane potential) or both by opening or closing ion channels

2. Stimulates the synthesis of proteins or certain regulatory molecules (such as enzymes) within the cell

3. Activates or deactivates enzymes

4. Induces secretory activity

5. Stimulates mitosis

Two main mechanisms harness hormone receptor binding to the intracellular machinery needed for hormone action. One, used by amino acid–based hormones, involves regulatory molecules called G proteins and one or more intracellular second messengers, which mediate the target cell's response to the hormone. The other, the steroid hormone mechanism, involves direct gene (DNA) activation by the hormone itself.

Amino Acid–Based Hormones and Second-Messenger Systems

Because proteins and peptides cannot penetrate the plasma membranes of tissue cells, virtually all of the amino acid–based hormones exert their signaling effects through intracellular **second messengers** generated by hormone binding to plasma membrane receptors. Of the second messengers, **cyclic AMP,** which also mediates the effects of certain neurotransmitters, is by far the best understood, so it will receive most of our attention.

The Cyclic AMP Signaling Mechanism In this mechanism, three plasma membrane components interact to determine intracellular levels of cyclic AMP (Figure 17.1a): the hormone receptor, the signal transducer (a G protein), and the effector enzyme (adenylate cyclase). The G protein behaves like a light switch; it is "off" when GDP is bound to it, and "on" when GTP is bound.

(1) The hormone, acting as the **first messenger,** binds to its receptor. This causes the receptor to change shape, allowing it to bind the **G protein.**

(2) The G protein is activated as it binds GTP, which displaces GDP.

(3) The activated G protein (moving freely along the membrane) binds to and activates the effector enzyme **adenylate cyclase** as it hydrolyzes its bound GTP and becomes inactive once again. (The G protein possesses GTPase activity and cleaves the terminal phosphate group off GTP in much the same way that ATPase enzymes hydrolyze ATP.)

(4) Adenylate cyclase then generates the second-messenger cyclic AMP from ATP.

(5) Cyclic AMP, which is free to diffuse throughout the cell, triggers a cascade of chemical reactions in which one or more different enzymes, called **protein kinases,** are activated. The protein kinases *phosphorylate* (add a phosphate group to) different proteins, many of which are other enzymes. Phosphorylation activates some of these proteins and inhibits others, so a variety of reactions may occur in the same target cell at the same time. For example, a fat cell responds to epinephrine binding by breaking down glycogen and stored fat, reactions mediated by different enzymes.

This type of intracellular enzymatic cascade has a huge amplification effect. Each activated adenylate cyclase generates large numbers of cyclic AMP molecules, and a single kinase enzyme can catalyze literally hundreds of reactions. Hence, as the reaction cascades through one enzyme intermediate after another, the number of product molecules increases dramatically at each step. Theoretically, receptor binding of a single hormone molecule could generate millions of the final product molecules!

The sequence of biochemical reactions set into motion by cyclic AMP depends on the target cell type, the specific protein kinases it contains, and the hormone acting as first messenger. For example, in

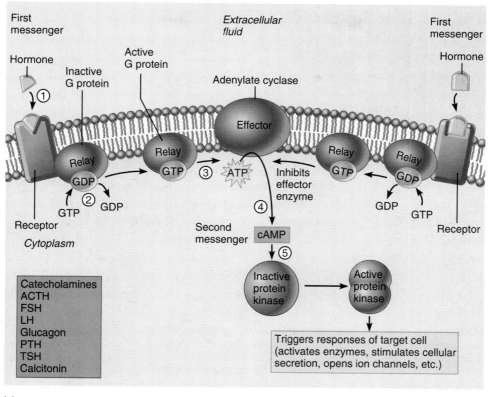

(a)

FIGURE 17.1 *Second-messenger mechanisms of amino acid–based hormones.* **(a)** Mechanisms that generate cyclic AMP by activation of adenylate cyclase are mediated by G proteins, which are activated when the hormone (first messenger) binds to plasma membrane receptors. Cyclic AMP (the second messenger) acts intracellularly to activate protein kinase enzymes that mediate the cell's responses to the hormone. A list of hormones that act through these mechanisms is provided.

thyroid cells, binding of thyroid-stimulating hormone ultimately promotes the synthesis of the thyroid hormone thyroxine; in bone and muscle cells, growth hormone binding activates anabolic reactions in which amino acids are built into tissue proteins. Notice that not all G proteins activate adenylate cyclase; some inhibit it (see Figure 17.1a), thus reducing the cytoplasmic concentration of cyclic AMP. Such opposing effects permit even slight changes in levels of antagonistic hormones to influence a target cell's activity.

Since cyclic AMP is rapidly degraded by the intracellular enzyme **phosphodiesterase**, its action persists only briefly. While at first glance this may appear to be a problem, it is quite the opposite. Because of the amplification effect just explained, most hormones need to be present only briefly to cause the desired results. Continued production of hormones then prompts continued cellular activity; no extracellular controls are necessary to stop the activity.

The PIP-Calcium Signal Mechanism Although cyclic AMP is the activating second messenger in some tissues for at least ten of the amino acid–based hormones, some of the same hormones (e.g., epinephrine) act through a different second-messenger system in other tissues. In one such mechanism, intracellular calcium ions act as the final mediator. Let's follow this process (Figure 17.1b).

① Hormone docking to the receptor causes it to bind the inactive G protein.

The G protein(s) acts as the signal transducer. Because each step in the pathway has a huge amplification effect and the number of product molecules increases dramatically at each step.

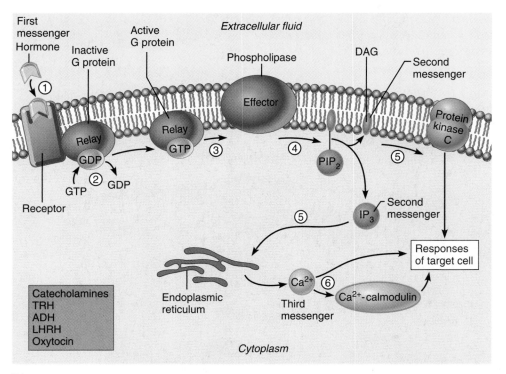

(b)

FIGURE 17.1 (continued)
Second-messenger mechanisms of amino acid—based hormones.
(b) Some mechanisms utilize Ca^{2+} as an intracellular messenger. The activated G protein stimulates phospholipase (an enzyme in the plasma membrane) which cleaves PIP into two fragments—inositol triphosphate (IP_3) and diacylglycerol (DAG), which act as intracellular second messengers to activate protein kinases and increase the cytoplasmic concentration of Ca^{2+}. Ionic calcium also acts as a second messenger to modify the activity of cellular proteins. A list of hormones that act through these mechanisms is provided.

② The G protein is activated as GTP binds, displacing GDP.

③ The activated G protein binds to and activates a membrane-bound **phospholipase** enzyme, and then becomes inactive.

④ Phospholipase splits a plasma-membrane phospholipid called **PIP₂** (phosphatidyl inositol biphosphate) into **diacylglycerol** and **IP₃** (inositol triphosphate), and both of these molecules act as second messengers.

⑤ Diacylglycerol activates specific protein kinases, whereas IP_3 triggers the release of Ca^{2+} from the endoplasmic reticulum (ER) and other intracellular storage sites (Figure 17.1b).

⑥ The liberated calcium ions, in turn, take on a second-messenger role, either by directly altering the activity of specific enzymes and plasma membrane ion (Ca^{2+}) channels or by binding to the intracellular regulatory protein **calmodulin.** Once calcium binds to calmodulin, enzymes are activated that amplify the cellular response.

Hormones known to act on their target cells via G protein receptors which act via cyclic AMP or the PIP mechanism are listed in Figure 17.1. Other amino acid–based hormones act on their target cells through different (and in some cases, unknown) mechanisms or messengers. For example, *cyclic GMP* (cyclic guanosine monophosphate) is a second messenger for selected hormones. Insulin (and other growth factors) appear to work without second messengers. The insulin receptor is a *tyrosine kinase* enzyme that is activated by autophosphorylation (addition of phosphate to several of its own tyrosines) when insulin binds. The activated receptor is recognized by intracellular *relay proteins* that are activated by the interaction and then trigger specific cell responses. In certain instances, any of the second messengers mentioned—and the hormone receptor itself—can cause changes in intracellular calcium ion levels.

Steroid Hormones and Direct Gene Activation

Being lipid soluble, steroid hormones (and, strangely, thyroid hormone, a small iodinated amine) can diffuse easily into their target cells. Once inside, they bind to an intracellular receptor that is activated by the coupling. The activated hormone-

? *What is the crucial difference between the signaling mechanism depicted here and that shown in Figure 17.1?*

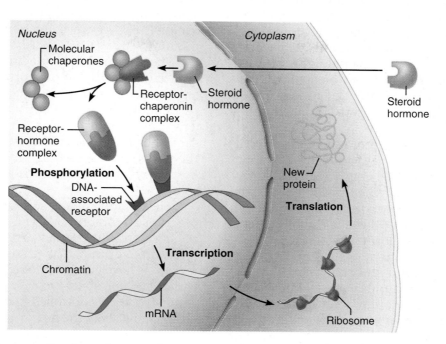

FIGURE 17.2 **Direct gene activation mechanism of steroid hormones.** Lipid-soluble steroids diffuse through the plasma membrane of the target cell and bind to intracellular receptor–chaperonin complexes located in the nucleus. Once hormone binding occurs, the chaperonin complex dissociates from the receptor and the hormone–receptor complex is activated by phosphorylation and binds to specific receptor proteins on the chromatin, initiating transcription of certain genes. The messenger RNA formed migrates into the cytoplasm, where it directs synthesis of specific proteins.

receptor complex then makes its way to the nuclear chromatin, where the hormone binds to a DNA-associated *receptor protein* specific for it. This interaction "turns on" a gene, that is, it prompts transcription of DNA to produce a messenger RNA (mRNA). The mRNA is then translated on the cytoplasmic ribosomes, producing specific protein molecules. These proteins include enzymes that promote the metabolic activities induced by that particular hormone and, in some cases, synthesis of structural proteins or proteins that are exported by the target cell. The steroidal gene activation mechanism is depicted in Figure 17.2.

In the absence of hormone, the nuclear receptors are bound into receptor-chaperonin complexes, associations that seem to silence the unbound receptors (that is, prevent them from binding to DNA) and

perhaps protect the receptor from proteolysis. (Check back to Chapter 2, p. 54, if your recall of molecular chaperones needs a jog.) When the hormone *is* present, the complex dissociates and the receptor is phosphorylated, which allows it to bind to the DNA and influence transcription.

Hormone–Target Cell Specificity

Even though all major hormones circulate to virtually all tissues, a given hormone influences the activity of only certain tissue cells, referred to as its **target cells.** For a target cell to respond to a hormone, it must have specific protein **receptors** on its plasma membrane or in its interior to which that hormone can bind. For example, receptors for adrenocorticotropic hormone (ACTH) are normally found only on certain cells of the adrenal cortex. By contrast, thyroxine is the principal hormone stimulating cellular metabolism, and nearly all body cells have thyroxine receptors.

An analogy can be drawn between the endocrine gland–target cell response and a radio's transmitter-receiver system. Let's assume that the endocrine gland is the transmitter, sending signals

In the one illustrated here, the hormone (first messenger) enters the cell nucleus and directly stimulates the desired effect by acting on DNA to cause transcription. In the Figure 17.1 mechanism, the hormone cannot enter the cell. Hence, it binds to a plasma membrane receptor that, in turn, interacts with a signal-transducing G peptide.

to the rest of the body, and the receptors on target cells are the receivers. Like a radio tuned to a single station, the receptors respond only to their particular signal, even though many other signals are present at the same time. A radio receiver responds to the radio signal by producing sound. A hormone receptor responds to hormone binding by prompting the cell to perform, or "turn on," some gene-determined "preprogrammed" function. Hence, hormones are molecular "triggers" rather than informational molecules.

Although hormone-receptor binding is the crucial first step, target cell activation by hormone-receptor interaction depends equally on three factors: (1) blood levels of the hormone, (2) the relative numbers of receptors for that hormone on or in the target cells, and (3) the *affinity* (strength) of the bond between the hormone and the receptor. Changes in all three factors occur rapidly in response to various stimuli and changes within the body. As a rule, a large number of high-affinity receptors produces a pronounced hormonal effect, whereas a smaller number of low-affinity receptors results in reduced target cell response or outright endocrine dysfunction at the same blood hormone levels.

Furthermore, receptors are dynamic structures. In some instances, the target cells form more receptors in response to rising blood levels of the specific hormones to which they respond, a phenomenon called **up-regulation.** But, in other cases, prolonged exposure to high hormone concentrations desensitizes the target cells, so that they respond less vigorously to hormonal stimulation. This **down-regulation** involves loss of receptors and prevents the target cells from overreacting to persistently high hormone levels. Hormones influence the number and affinity not only of their own receptors, but also of receptors that respond to other hormones. For example, progesterone induces a loss of estrogen receptors in the uterus, thus antagonizing estrogen's actions. On the other hand, estrogen causes the same cells to produce more progesterone receptors, enhancing their ability to respond to progesterone.

Half-Life, Onset, and Duration of Hormone Activity

Hormones are potent chemicals, and they exert profound effects on their target organs at very low concentrations. The concentration of a circulating hormone in blood at any time reflects (1) its rate of release, and (2) the speed of its inactivation and removal from the body. Some hormones are rapidly degraded by enzymes within their target cells, but most are removed from the blood by kidney and liver enzyme systems, and their breakdown products are excreted from the body in urine or, to a lesser extent, in feces. As a result, the persistence of a hormone in the blood, referred to as its **half-life,** is usually brief—from a fraction of a minute to 30 minutes.

The time required for hormone effects to appear varies greatly. Some hormones provoke target organ responses almost immediately, while others, particularly the steroid hormones, require hours to days before their effects are seen. Additionally, some hormones, such as testosterone secreted by the testes, are secreted in a relatively inactive *(prohormone)* form and must be activated in the target cells.

The duration of hormone action is limited, ranging from 20 minutes to several hours, depending on the hormone. Effects may disappear rapidly as blood levels drop, or they may persist for hours after very low levels have been reached. Because of these many variations, hormonal blood levels must be precisely and individually controlled to meet the continuously changing needs of the body.

Control of Hormone Release

The synthesis and release of most hormones are regulated by some type of **negative feedback system** (see Chapter 1, pp. 11–12). In such a system, hormone secretion is triggered by some internal or external stimulus. As hormone levels rise, they cause target organ effects and inhibit further hormone release. As a result, blood levels of many hormones vary only within a narrow "desirable" range.

Endocrine Gland Stimuli

The endocrine glands of the body are stimulated to manufacture and release their hormones by three major types of stimuli: *humoral* (hu'mer-ul), *neural,* and *hormonal.*

Humoral Stimuli Some endocrine glands secrete their hormones in direct response to changing blood levels of certain ions and nutrients. These stimuli are called *humoral stimuli* to distinguish them from hormonal stimuli, which are also bloodborne chemicals. The term *humoral* harks back to the ancient use of the term *humor* to refer to various body fluids (blood, bile, and others). This is the simplest of the endocrine control systems. For example, cells of the parathyroid glands directly monitor the concentration of calcium ions in blood, and when they detect a decline from normal Ca^{2+} values, they secrete parathyroid hormone (PTH). Since PTH acts by several routes to reverse that decline, blood Ca^{2+} levels soon rise, ending the initiative for PTH release

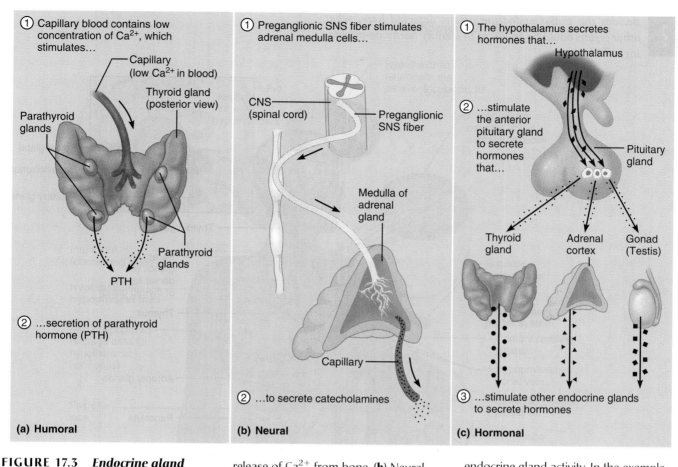

① Capillary blood contains low concentration of Ca^{2+}, which stimulates...

Capillary (low Ca^{2+} in blood)

Thyroid gland (posterior view)

Parathyroid glands

Parathyroid glands

PTH

② ...secretion of parathyroid hormone (PTH)

(a) Humoral

① Preganglionic SNS fiber stimulates adrenal medulla cells...

CNS (spinal cord)

Preganglionic SNS fiber

Medulla of adrenal gland

Capillary

② ...to secrete catecholamines

(b) Neural

① The hypothalamus secretes hormones that...

Hypothalamus

② ...stimulate the anterior pituitary gland to secrete hormones that...

Pituitary gland

Thyroid gland

Adrenal cortex

Gonad (Testis)

③ ...stimulate other endocrine glands to secrete hormones

(c) Hormonal

FIGURE 17.3 *Endocrine gland stimuli: Three different mechanisms.* (a) Humoral stimulus of endocrine gland activity. Low blood calcium levels trigger parathyroid hormone (PTH) release from the parathyroid glands. PTH causes blood calcium levels to rise by stimulating release of Ca^{2+} from bone. (b) Neural stimulus of endocrine gland activity. The stimulation of adrenal medullary cells by sympathetic nervous system (SNS) fibers triggers the release of catecholamines (epinephrine and norepinephrine) to the blood. (c) Hormonal stimulus of endocrine gland activity. In the example illustrated, hormones released by the hypothalamus stimulate the anterior pituitary to release hormones that stimulate other endocrine organs to secrete hormones.

(Figure 17.3a). Other hormones released in response to humoral stimuli include insulin, produced by the pancreas, and aldosterone, one of the adrenal cortex hormones.

Neural Stimuli In a few cases, nerve fibers stimulate hormone release. The classic example is sympathetic nervous system stimulation of the adrenal medulla to release catecholamines (norepinephrine and epinephrine) during periods of stress (Figure 17.3b). Also, oxytocin and antidiuretic hormone are released from the posterior pituitary in response to nerve impulses from hypothalamic neurons.

Hormonal Stimuli Finally, many endocrine glands release their hormones in response to hormones produced by other endocrine organs. For example, release of most anterior pituitary hormones is regulated by releasing and inhibiting hormones produced by the hypothalamus, and many anterior pi-

tuitary hormones in turn stimulate other endocrine organs to release their hormones into the blood (Figure 17.3c). As the hormones produced by the final target glands increase in the blood, they inhibit the release of anterior pituitary hormones and thus their own release. This hypothalamic–pituitary–target endocrine organ feedback loop lies at the very core of the science of endocrinology, and it will come up many times in this chapter. The hormonal mechanism promotes rhythmic hormone release, with hormone blood levels rising and falling in a specific pattern.

While these three mechanisms typify most systems that control hormone release, they are by no means all-inclusive or mutually exclusive, and some endocrine organs respond to multiple stimuli.

Nervous System Modulation

Both "turn on" factors (hormonal, humoral, and neural stimuli) and "turn off" factors (feedback

lobe plus the infundibulum make up the region called the **neurohypophysis** (nu″ro-hi-pof′ĭ-sis), a term that is commonly used (incorrectly) to indicate the posterior pituitary lobe. The **anterior lobe,** or **adenohypophysis** (ad″ĕ-no-hi-pof′ĭ-sis), is composed of glandular tissue (*adeno* = gland), and it manufactures and releases a number of hormones (Table 17.1 on pp. 622–623).

Arterial blood is delivered to the pituitary via hypophyseal branches of the internal carotid arteries. The veins leaving the pituitary drain into the dural sinuses.

Pituitary-Hypothalamic Relationships

The contrasting histology of the two pituitary lobes reflects the dual origin of this tiny gland. The posterior lobe is actually part of the brain. It derives from a downgrowth of hypothalamic (nervous) tissue and maintains its neural connection with the hypothalamus via a nerve bundle called the **hypothalamic-hypophyseal tract,** which runs through the infundibulum (Figure 17.5). This tract arises from neurons in the **supraoptic** and **paraventricular nuclei** of the hypothalamus (see Figure 12.13b, p. 444). These neurosecretory cells synthesize two **neurohormones** and transport them along their axons to the posterior pituitary. Oxytocin is made by the paraventricular neurons, and antidiuretic hormone (ADH) is synthesized by the supraoptic neurons. When these hypothalamic neurons fire, they release the stored hormones into a capillary bed in the posterior pituitary for distribution throughout the body.

In contrast, the anterior lobe originates from a superior outpocketing of the oral mucosa *(Rathke's pouch)* and is formed from the epithelial tissue. After touching the posterior lobe, the anterior lobe loses its connection with the oral mucosa and adheres to the neurohypophysis to form the composite organ. There is no direct neural connection between the anterior lobe and hypothalamus, but there is a vascular connection. Specifically, the **primary capillary plexus** in the infundibulum communicates inferiorly via the small **hypophyseal portal veins** with a **secondary capillary plexus** in the anterior lobe. The primary and secondary capillary plexuses and the intervening hypophyseal portal veins make up the **hypophyseal portal system*** (Figure 17.5). Via this portal system, **releasing** and **inhibiting hormones** secreted by neurons in the ventral hypothalamus circulate directly to the adenohypophysis, where they regulate the secretion of its hormones. All of these hypothalamic regulatory hormones are amino acid–based, but they vary widely in size from amines to polypeptides.

---

*A *portal system* is an unusual arrangement of blood vessels in which a capillary bed feeds into veins, which in turn feed into another capillary bed.

Adenohypophyseal Hormones

The adenohypophysis has traditionally been called the "master endocrine gland" because of its numerous hormonal products, many of which regulate the activity of other endocrine glands. Although the adenohypophysis has been dethroned by the hypothalamus, which is now known to control anterior pituitary activity, six distinct adenohypophyseal hormones with specific physiological actions in humans are known. All of these are proteins. In addition, a large molecule with the tongue-twisting name **pro-opiomelanocortin** (pro″o″pe-o-mah-lan′o-kor″tin) **(POMC)** has been isolated from the anterior pituitary. POMC is a *prohormone*, that is, a large precursor molecule from which other active molecules are split by enzymes. POMC is the source of adrenocorticotropic hormone, two natural opiates (an enkephalin and beta endorphin, described in Chapter 11), and *melanocyte-stimulating hormone (MSH)*. MSH stimulates melanocytes of amphibians, reptiles, and other animals to increase their synthesis of melanin pigment, but in humans its plasma levels are insignificant and it is probably more important as a CNS neurotransmitter (neuropeptide) than a hormone. Additionally, MSH release is tonically suppressed by dopamine-releasing hypothalamic neurons.

When the adenohypophysis receives an appropriate chemical stimulus from the hypothalamus, one or more of its hormones is released by certain of its cells. Although many different releasing and inhibiting hormones pass from the hypothalamus to the anterior lobe, the individual target cells of the anterior pituitary distinguish the messages directed to them and respond in kind—synthesizing and secreting the proper hormones in response to specific releasing hormones, and shutting off hormone release in response to inhibiting hormones. The releasing hormones are far more important as regulatory factors because very little hormone is stored by the secretory cells of the anterior lobe.

Four of the six anterior pituitary lobe hormones—thyroid-stimulating hormone (TSH), adrenocorticotropic hormone (ACTH), follicle-stimulating hormone (FSH), and luteinizing hormone (LH)—are **tropins** or **tropic hormones.** Tropic (*tropi* = turn on, change) hormones regulate the secretory action of other endocrine glands. All adenohypophyseal hormones affect their target cells via a cyclic AMP second-messenger system. The adenohypophyseal hormones, their effects, and their relationships to hypothalamic regulatory factors are summarized briefly in Table 17.1.

Growth Hormone (GH) Growth hormone (GH) is produced by the **somatotropic cells** or **somatotropes** of the anterior lobe. Although GH stimulates most body cells to increase in size and divide, its

FIGURE 17.6 *Proposed scheme for classifying the metabolic actions of growth hormone (GH).* The direct actions of GH on target cells stimulate the liver, skeletal muscle, bone, and cartilage cells to produce and release insulin-like growth factor 1, which mediates the indirect, largely anabolic effects of GH. GH acts directly on other tissue cells to promote fat breakdown (lipolysis) and release from adipose tissues and hinder glucose uptake from the blood by tissue cells, maintaining blood sugar at fairly high levels. (Because these actions antagonize those of the pancreatic hormone insulin, they are referred to as anti-insulin actions.) Elevated levels of GH and IGF-1 feed back to promote GHIH release (and depress GHRH release) by the hypothalamus and to inhibit GH release by the anterior pituitary gland.

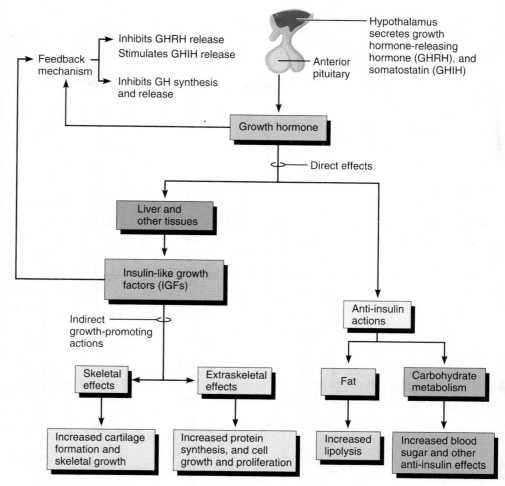

major targets are the bones and skeletal muscles. Stimulation of the epiphyseal plate leads to long bone growth; its effects on skeletal muscles promote increased muscle mass.

Essentially an anabolic hormone, GH promotes protein synthesis, and it encourages the use of fats for fuel, thus conserving glucose (Figure 17.6). Most growth-promoting effects of GH are mediated indirectly by **insulin-like growth factors (IGFs),** also known as **somatomedins** (so"mah-to-me′dinz), a family of growth-promoting proteins produced by the liver, skeletal muscle, bone, and other tissues. Specifically, IGFs (1) stimulate uptake of amino acids from the blood and their incorporation into cellular proteins throughout the body; and (2) stimulate uptake of sulfur (needed for the synthesis of chondroitin sulfate) into cartilage matrix. Acting directly, GH mobilizes fats from fat depots for transport to cells, increasing blood levels of fatty acids. It also decreases the rate of glucose uptake and metabolism. In the liver, it encourages glycogen breakdown and release of glucose to the blood. The elevation in blood sugar levels that subsequently occurs as a result of this *glucose sparing* is called the **diabetogenic**

effect of GH, because it mimics the high blood sugar levels typical of diabetes mellitus (see p. 637).

Secretion of GH is regulated chiefly by two hypothalamic hormones with antagonistic effects. **Growth hormone–releasing hormone (GHRH)** stimulates GH release, while **growth hormone-inhibiting hormone (GHIH),** also called **somatostatin** (so"mah-to-stat′in), inhibits it. Somatostatin release is (presumably) triggered by the feedback of GH and IGFs. Rising levels of GH also feed back to inhibit its own release. As indicated in Table 17.1, a number of secondary triggers such as stress, nutritional factors, and sleep patterns also influence GH release. Typically, GH secretion has a daily cycle, with the highest levels occurring during evening sleep, but the total amount secreted daily peaks during adolescence and then declines with age.

The far-reaching effects of somatostatin deserve special mention. Besides inhibiting growth hormone secretion, it blocks the release of several other adenohypophyseal hormones (see Table 17.1). Additionally, it inhibits the release of virtually all gastrointestinal and pancreatic secretions—both endocrine and exocrine.

⬥ *Homeostatic Imbalance*

Both hypersecretion and hyposecretion of GH may result in structural abnormalities. Hypersecretion in children results in **gigantism.** The person becomes abnormally tall, often reaching a height of 2.4 m (8 feet), but has relatively normal body proportions. If excessive amounts of GH are secreted after the epiphyseal plates have closed, **acromegaly** (ak"ro-meg'ah-le) results. Acromegaly, literally translated as "enlarged extremities," is characterized by overgrowth of bony areas still responsive to GH, namely bones of the hands, feet, and face (Figure 17.7). Thickening of soft tissues leads to coarse or malformed facial features and an enlarged tongue. Hypersecretion of GH usually results from an adenohypophyseal tumor that churns out excessive amounts of GH. The usual treatment is surgical removal of the tumor; however, anatomical changes that have already occurred are not reversible.

Hyposecretion of GH in adults usually causes no problems, but in rare instances the deficit is so severe that **progeria** ensues—body tissues begin to atrophy and clinical signs of premature aging appear. Growth hormone deficiency in children results in slowed long bone growth, a condition called **pituitary dwarfism.** Such individuals attain a maximum height of only 1.2 m (4 feet), but usually have fairly normal body proportions. Lack of GH is typically accompanied by deficiencies of other adenohypophyseal hormones, and if thyroid-stimulating hormone and gonadotropins are lacking, the individual will be malproportioned and will fail to mature sexually as well. Fortunately, human growth hormone is produced commercially by genetic engineering techniques and when pituitary dwarfism is diagnosed before puberty, growth hormone replacement therapy can promote nearly normal somatic growth. ∎

Thyroid-Stimulating Hormone (TSH) Thyroid-stimulating hormone (TSH), or **thyrotropin,** is a tropic hormone that stimulates normal development and secretory activity of the thyroid gland. TSH release from the **thyrotrope cells** of the anterior pituitary is triggered by the hypothalamic peptide **thyrotropin-releasing hormone (TRH).** Rising blood levels of thyroid hormones to act on both the pituitary and the hypothalamus to inhibit TSH secretion. The hypothalamus, in response, releases somatostatin, which reinforces the blockade of TSH release.

Adrenocorticotropic Hormone (ACTH) Adrenocorticotropic (ah-dre"no-kor"tǐ-ko-trōp'ik) **hormone (ACTH),** or **corticotropin,** is secreted by the **corticotrope cells** of the adenohypophysis. ACTH stimu-lates the adrenal cortex to release corticosteroid hormones, most importantly glucocorticoids that help the body to resist stressors. ACTH release, elicited by hypothalamic **corticotropin-releasing hormone (CRH),** has a daily rhythm, with levels peaking in the morning, shortly after one arises. Rising levels of glucocorticoids feed back and block secretion of CRH and consequently ACTH release. Internal and external factors that alter the normal ACTH rhythm by triggering CRH release include fever, hypoglycemia, and stressors of all types.

Gonadotropins Follicle-stimulating hormone (FSH) and **luteinizing** (lu'te-in-īz"ing) **hormone (LH),** referred to collectively as **gonadotropins,** regulate the function of the gonads (ovaries and testes). In both sexes, FSH stimulates gamete (sperm or egg) production, while LH promotes production of gonadal hormones. In females, LH works with FSH to cause maturation of an egg-containing ovarian follicle. It then independently triggers ovulation (expulsion of the egg from the follicle) and promotes synthesis and release of ovarian hormones (estrogens and progesterone). In males, LH stimulates the interstitial cells of the testes to produce the male hormone testosterone. For this reason, LH is sometimes called **interstitial** (in"ter-stish'al) **cell–stimulating hormone (ICSH)** in males.

Gonadotropins are virtually absent from the blood of prepubertal boys and girls. However, during puberty, the **gonadotrope cells** of the adenohypophysis are activated and gonadotropin levels begin to rise, causing the gonads to mature to the adult state. In both sexes, gonadotropin release by the adenohypophysis is prompted by **gonadotropin-releasing hormone (GnRH)** produced by the hypothalamus. Gonadal hormones, produced in response to the gonadotropins, feed back to suppress FSH and LH release.

Prolactin (PRL) Prolactin (PRL) is a protein hormone structurally similar to growth hormone. Produced by the **lactotropes,** it stimulates the gonads of some animals and is considered a gonadotropin by some researchers. Its only well-documented effect in humans is to stimulate milk production by the breasts (*pro* = for; *lact* = milk). However, there is some evidence that PRL enhances testosterone production in males.

Like GH, prolactin release is controlled by hypothalamic releasing and inhibiting hormones. The exact nature of **prolactin-releasing hormone (PRH)** is unknown, but it is thought to be serotonin. PRH causes prolactin synthesis and release, whereas **prolactin-inhibiting hormone (PIH),** now known to be the neurotransmitter dopamine, prevents prolactin secretion. In males, the influence of PIH predominates, but in women, prolactin levels rise and fall in

FIGURE 17.7 *An individual with acromegaly.* This condition results from hypersecretion of growth hormone in the adult. Notice the enlarged jaw, nose, and hands. Left to right, the same person is shown at age 16, age 33, and age 52.

rhythm with estrogen blood levels. Low estrogen stimulates PIH release, while high estrogen promotes release of PRH and, thus, prolactin. A brief rise in prolactin levels just before the menstrual period partially accounts for the breast swelling and tenderness some women experience at that time, but since this PRL stimulation is so brief, the breasts do not produce milk. In pregnant women, PRL blood levels rise dramatically toward the end of pregnancy, and milk production by the breasts becomes possible. After birth, the infant's suckling stimulates PRH release in the mother, encouraging continued milk production and availability.

Homeostatic Imbalance

Hypersecretion of prolactin is more common than its hyposecretion (which is not a problem in anyone except women who choose to nurse). In fact, hyperprolactinemia is the most frequent abnormality of adenohypophyseal tumors. Clinical signs include inappropriate lactation *(galactorrhea)*, lack of menses, infertility in females, and gynecomastia (breast enlargement) and impotence in males. ∎

The Posterior Pituitary and Hypothalamic Hormones

The posterior pituitary, made largely of the axons of hypothalamic neurons stores antidiuretic hormone (ADH) and oxytocin, which have been synthesized and forwarded by hypothalamic neurons of the supraoptic and paraventricular nuclei, respectively. These hormones are later released "on demand" in response to nerve impulses from the same hypothalamic neurons.

ADH and oxytocin, each composed of nine amino acids, are almost identical. They differ in only two amino acids, yet they have very different physiological effects on their target organs. ADH influences body water balance, whereas oxytocin stimulates the contraction of smooth muscle, particularly that of the uterus and breasts (see Table 17.1). Both hormones use the PIP second-messenger mechanism to exert their effects on their target cells.

Oxytocin Oxytocin (ok″sĭ-to′sin), a strong stimulant of uterine contraction, is released in significantly higher amounts during childbirth *(ocytocia = childbirth)* and in nursing women. The number of oxytocin receptors in the uterus peaks near the end of pregnancy, and uterine smooth muscle becomes more and more sensitive to the stimulatory effects of oxytocin. Stretching of the uterus and cervix as birth nears dispatches afferent impulses to the hypothalamus. The hypothalamus responds by synthesizing oxytocin and triggering its release from the neurohypophysis. As blood levels of oxytocin rise, the expulsive contractions of labor gain momentum and finally end in birth.

Oxytocin also acts as the hormonal trigger for milk ejection (the "letdown" reflex) in women whose breasts are actively producing milk in response to prolactin. Suckling causes a reflex-initiated release of oxytocin, which targets specialized myoepithelial cells surrounding the milk-producing glands. As these cells contract, milk is forced from the breast into the mouth of the suckling infant. Both of these mechanisms are *positive feedback mechanisms* and are described in more detail in Chapter 29.

TABLE 17.1	Pituitary Hormones: Regulation and Effects		
Hormone (chemical structure and cell type)	*Regulation of release*	*Target organ and effects*	*Effects of hyposecretion ↓ and hypersecretion ↑*

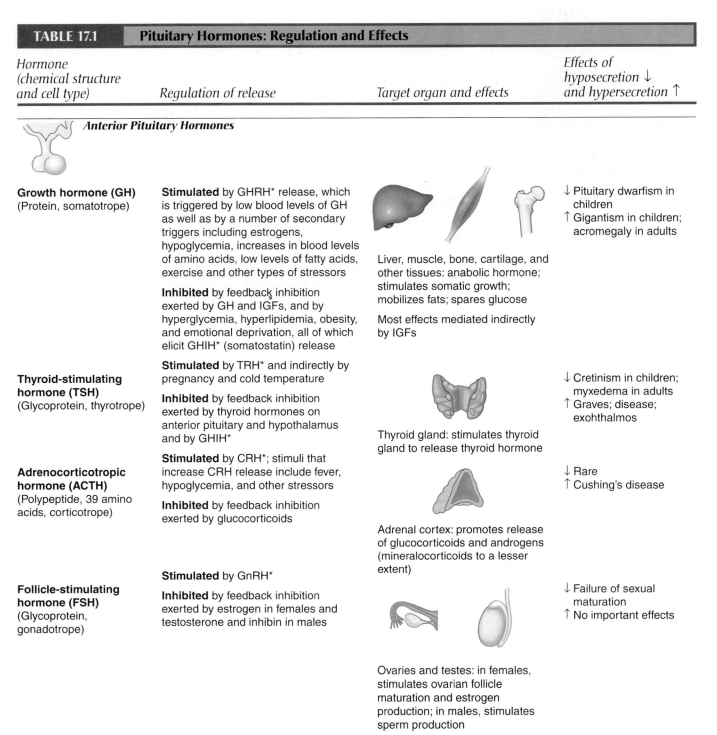

Anterior Pituitary Hormones

Growth hormone (GH) (Protein, somatotrope)	**Stimulated** by GHRH* release, which is triggered by low blood levels of GH as well as by a number of secondary triggers including estrogens, hypoglycemia, increases in blood levels of amino acids, low levels of fatty acids, exercise and other types of stressors **Inhibited** by feedback inhibition exerted by GH and IGFs, and by hyperglycemia, hyperlipidemia, obesity, and emotional deprivation, all of which elicit GHIH* (somatostatin) release	Liver, muscle, bone, cartilage, and other tissues: anabolic hormone; stimulates somatic growth; mobilizes fats; spares glucose Most effects mediated indirectly by IGFs	↓ Pituitary dwarfism in children ↑ Gigantism in children; acromegaly in adults
Thyroid-stimulating hormone (TSH) (Glycoprotein, thyrotrope)	**Stimulated** by TRH* and indirectly by pregnancy and cold temperature **Inhibited** by feedback inhibition exerted by thyroid hormones on anterior pituitary and hypothalamus and by GHIH*	Thyroid gland: stimulates thyroid gland to release thyroid hormone	↓ Cretinism in children; myxedema in adults ↑ Graves; disease; exohthalmos
Adrenocorticotropic hormone (ACTH) (Polypeptide, 39 amino acids, corticotrope)	**Stimulated** by CRH*; stimuli that increase CRH release include fever, hypoglycemia, and other stressors **Inhibited** by feedback inhibition exerted by glucocorticoids	Adrenal cortex: promotes release of glucocorticoids and androgens (mineralocorticoids to a lesser extent)	↓ Rare ↑ Cushing's disease
Follicle-stimulating hormone (FSH) (Glycoprotein, gonadotrope)	**Stimulated** by GnRH* **Inhibited** by feedback inhibition exerted by estrogen in females and testosterone and inhibin in males	Ovaries and testes: in females, stimulates ovarian follicle maturation and estrogen production; in males, stimulates sperm production	↓ Failure of sexual maturation ↑ No important effects

Both natural and synthetic oxytocic drugs (Pitocin and others) are used to induce labor or to hasten normal labor that is progressing slowly. Less frequently, oxytocics are used to stop postpartum bleeding (by compressing ruptured blood vessels at the placental site) and to stimulate the milk ejection reflex.

Until recently, oxytocin's role in males and non-pregnant, nonlactating females was unknown, but new studies reveal that this potent peptide plays a role in sexual arousal and orgasm when the body is already primed for reproduction by sex hormones. Then, it is responsible for the feeling of sexual satiety or satisfaction that results from that interaction. In nonsexual relationships, it is thought to promote nurturing and affectionate behavior, that is, it acts as a "cuddle hormone."

TABLE 17.1	*continued*		
Hormone (chemical structure and cell type)	Regulation of release	Target organ and effects	Effects of hyposecretion ↓ and hypersecretion ↑
Luteinizing hormone (LH) (Glycoprotein, gonadotrope)	**Stimulated** by GnRH* **Inhibited** by feedback inhibition exerted by estrogen and progesterone in females and testosterone in males	Ovaries and testes: in females, triggers ovulation and stimulates ovarian production of estrogen and progesterone; in males, promotes testosterone production	As for FSH
Prolactin (PRL) (Protein, lactotrope)	**Stimulated** by PRH*; PRH release enhanced by estrogens, birth control pills, opiates, and breast-feeding **Inhibited** by PIH* (dopamine)	Breast secretory tissue: promotes lactation	↓ Poor milk production in nursing women ↑ Galactorrhea; cessation of menses in females; impotence and gynecomastia in males

Posterior Pituitary Hormones (made by hypothalamic neurons and stored in posterior pituitary)

Oxytocin (Peptide, neurons in paraventricular nucleus of hypothalamus)	**Stimulated** by impulses from hypothalamic neurons in response to cervical/uterine stretching and suckling of infant at breast **Inhibited** by lack of appropriate neural stimuli	Uterus: stimulates uterine contractions; initiates labor; breast: initiates milk ejection	Unknown
Antidiuretic hormone (ADH) or **vasopressin** (Peptide, neurons in supraoptic nucleus of hypothalamus)	**Stimulated** by impulses from hypothalamic neurons in response to increased osmolarity of blood or decreased blood volume; also stimulated by pain, some drugs, low blood pressure **Inhibited** by adequate hydration of the body and by alcohol	Kidneys: stimulates kidney tubule cells to reabsorb water	↓ Diabetes insipidus ↑ Syndrome of inappropriate ADH secretion (SIADH)

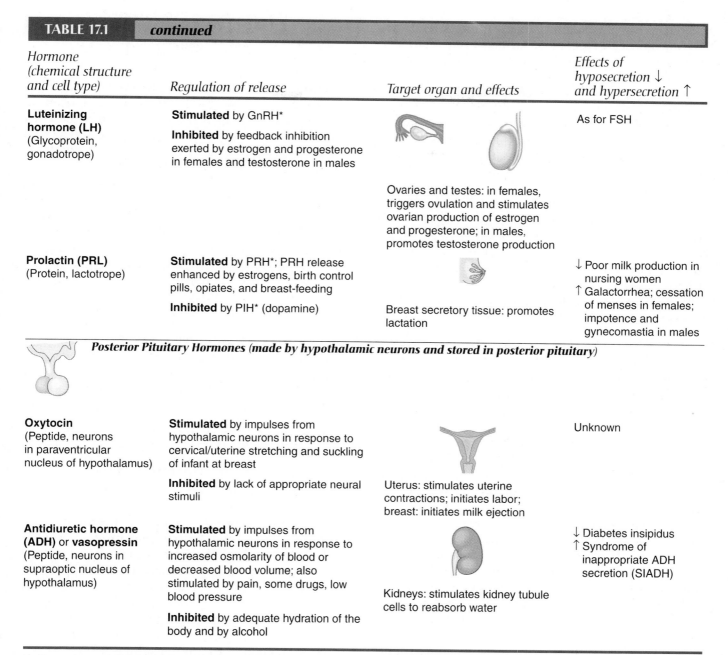

*Indicates hypothalamic releasing and inhibiting hormones: GHRH = growth hormone–releasing hormone; GHIH = growth hormone–inhibiting hormone; TRH = thyrotropin-releasing hormone; CRH = corticotropin-releasing hormone; GnRH = gonadotropin-releasing hormone; PRH = prolactin-releasing hormone; PIH = prolactin-inhibiting hormone.

Antidiuretic Hormone (ADH) Diuresis (di″u-re′sis) is urine production. Thus, an *antidiuretic* (an″tĭ-di″u-ret′ik) is a substance that inhibits or prevents urine formation. **Antidiuretic hormone (ADH)** prevents wide swings in water balance, helping to avoid dehydration or water overload. Highly specialized hypothalamic neurons, called *osmoreceptors*, continually monitor the solute concentration (and thus the water concentration) of the blood. When solutes threaten to become too concentrated (as might follow excessive perspiration or inadequate fluid intake), the osmoreceptors transmit excitatory impulses to the hypothalamic neurons in the supraoptic nucleus, which synthesize and release ADH. ADH is liberated into the blood by the neurohypophysis, and targets the kidney tubules. The tubule cells respond by reabsorbing more water from the forming urine and returning it to the bloodstream. As a result, less urine is produced and blood volume increases. As solute concentration declines, the osmoreceptors stop depolarizing, effectively ending ADH release. Other stimuli triggering ADH release include pain, low blood pressure, and certain drugs (such as nicotine, morphine, and barbiturates).

Drinking alcoholic beverages inhibits ADH secretion and causes copious urine output. The dry mouth and intense thirst of the "morning after" reflect this dehydrating effect of alcohol. As might be expected, ADH release is also inhibited by drinking excessive amounts of water. *Diuretic drugs* antagonize the effects of ADH and cause water to be flushed from the body. These drugs are used to manage some cases of hypertension and the edema (water retention in tissues) typical of congestive heart failure.

At high blood concentrations, ADH has a decided pressor effect; that is, it causes vasoconstriction, primarily of the visceral blood vessels. Under certain conditions, such as severe blood loss, exceptionally large amounts of ADH are released. This, in turn, causes a rise in systemic blood pressure. The alternative name for this hormone (**vasopressin**) reflects this particular effect.

🔺 *Homeostatic Imbalance*

The only important deficiency of ADH secretion is *diabetes insipidus*, a syndrome marked by the output of huge amounts of urine (polyuria) and intense thirst. The name of this condition (*diabetes* = overflow; *insipidus* = tasteless) distinguishes it from diabetes mellitus (*mel* = honey), in which insulin deficiency causes large amounts of blood sugar to be lost in the urine. At one time, urine was tasted to determine the cause of the polyuria.

Diabetes insipidus can be caused by a blow to the head that damages the hypothalamus or the posterior pituitary releasing site. In either case, ADH release is deficient. Though inconvenient, the condition is not serious when the thirst center is operating properly and the person drinks enough water to prevent dehydration. However, it can be life threatening in unconscious or comatose patients, so accident victims with head trauma must be carefully monitored.

Hypersecretion of ADH is seen in children with meningitis, may follow neurosurgery or head (hypothalamic) injury, or be the consequence of ectopic ADH secretion by cancer cells (particularly pulmonary cancers). It also may occur after general anesthesia or administration of certain drugs (meperidine and others). The syndrome resulting is called *syndrome of inappropriate ADH secretion (SIADH)* and is marked by overretention of fluid, headache and disorientation due to brain edema, weight gain, and hypo-osmolarity of the blood. SIADH management requires fluid restriction and careful monitoring of blood sodium levels. ■

The Thyroid Gland

Location and Structure

The butterfly-shaped thyroid gland is located in the anterior neck, on the trachea just inferior to the larynx (Figures 17.4 and 17.8a). Its two lateral **lobes** are connected by a median tissue mass called the **isthmus.** The thyroid gland is the largest pure endocrine gland in the body. Its prodigious blood supply (from the *superior* and *inferior thyroid arteries*) makes thyroid surgery a painstaking (and bloody) endeavor.

Internally, the thyroid gland is composed of hollow, spherical **follicles** (Figure 17.8b). The walls of each follicle are formed largely by cuboidal or squamous epithelial cells called *follicle cells,* which produce the glycoprotein **thyroglobulin** (thi″ro-glob′u-lin). The central cavity, or lumen, of the follicle stores **colloid,** an amber-colored, sticky material consisting of thyroglobulin molecules with attached iodine atoms. *Thyroid hormone* is derived from this iodinated thyroglobulin. The *parafollicular cells,* another population of endocrine cells found in the thyroid gland, produce *calcitonin,* an entirely different hormone (see p. 628). The parafollicular cells lie within the follicular epithelium but protrude into the soft connective tissue that separates and surrounds the thyroid follicles.

Thyroid Hormone (TH)

Thyroid hormone (TH), often referred to as the body's major metabolic hormone, is actually two active iodine-containing hormones, **thyroxine** (thirok′sin), or T_4, and **triiodothyronine** (tri″i-o″do-thi′ro-nēn), or T_3. Thyroxine is the major hormone secreted by the thyroid follicles; most triiodothyronine is formed at the target tissues by conversion of T_4 to T_3. Very much like one another, each hormone is constructed from two tyrosine amino acids linked together, but thyroxine has four bound iodine atoms, whereas triiodothyronine has three (thus, T_4 and T_3).

Except for the adult brain, spleen, testes, uterus, and the thyroid gland itself, thyroid hormone affects virtually every cell in the body. Generally speaking, it stimulates enzymes concerned with glucose oxidation. In this way, it increases basal metabolic rate (and oxygen consumption) and body heat production; this is known as the hormone's **calorigenic effect** (*calorigenic* = heat producing). Because thyroid hormone provokes an increase in the number of adrenergic receptors in blood vessels, it plays an important role in maintaining blood pressure. Additionally, it is an important regulator of tissue growth and development. It is especially critical for normal

Which of the cell populations shown in view (b) produces calcitonin?

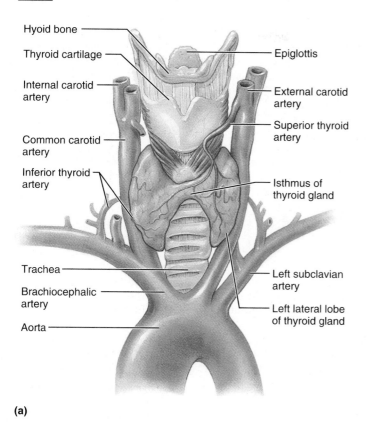

Hyoid bone

Thyroid cartilage

Internal carotid artery

Common carotid artery

Inferior thyroid artery

Trachea

Brachiocephalic artery

Aorta

Epiglottis

External carotid artery

Superior thyroid artery

Isthmus of thyroid gland

Left subclavian artery

Left lateral lobe of thyroid gland

(a)

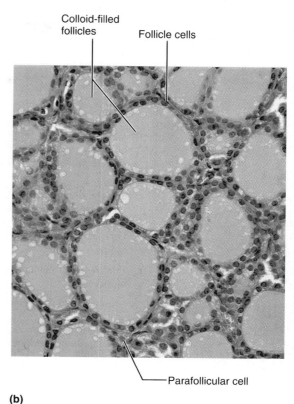

Colloid-filled follicles

Follicle cells

Parafollicular cell

(b)

FIGURE 17.8 *Gross and microscopic anatomy of the thyroid gland.*
(**a**) Location and arterial blood supply of the thyroid gland, anterior view.
(**b**) Photomicrograph of the thyroid gland (165×).

skeletal and nervous system development and maturation and reproductive capabilities. A summary of the broad effects of thyroid hormone appears in Table 17.2.

Transport and Regulation Most released T$_4$ and T$_3$ immediately bind to transport proteins, most importantly *thyroxine-binding globulins (TBGs)* produced by the liver. Both T$_4$ and T$_3$ bind to target tissue receptors, but T$_3$ binds much more avidly and is about ten times more active. Most peripheral tissues have the enzymes needed to convert T$_4$ to T$_3$, a process that entails enzymatic removal of one iodine group.

Several mechanisms of thyroid hormone activity probably exist; however, like steroid hormones, the T$_3$ form is known to enter a target cell and bind to intracellular receptors within the cell's nucleus, initiating transcription of mRNA.

Falling thyroxine blood levels trigger release of *thyroid-stimulating hormone (TSH)*, and ultimately thyroxine. Rising levels of thyroxine feed back to inhibit the hypothalamic-adenohypophyseal axis, temporarily shutting off the stimulus for TSH release. Conditions that increase body energy requirements, such as pregnancy and prolonged cold, stimulate the hypothalamus to secrete *thyrotropin-releasing hormone (TRH)*, which triggers TSH release, and in such instances, TRH overcomes the negative feedback controls. As the thyroid gland releases larger amounts of thyroid hormones, body metabolism and heat production are enhanced. Factors that inhibit TSH release include somatostatin, rising levels of glucocorticoids and sex hormones (estrogens or testosterone), and excessively high blood iodide concentration.

 *The parafollicular cells.*

⊠ *Homeostatic Imbalance*

Both overactivity and underactivity of the thyroid gland can cause severe metabolic disturbances (see

TABLE 17.2	Major Effects of Thyroid Hormone (T_4 and T_3) in the Body		
Process or system affected	Normal physiological effects	Effects of hyposecretion	Effects of hypersecretion
Basal metabolic rate (BMR)/ temperature regulation	Promotes normal oxygen use and BMR; calorigenesis; enhances effects of sympathetic nervous system	BMR below normal; decreased body temperature and cold intolerance; decreased appetite; weight gain; reduced sensitivity to catecholamines	BMR above normal; increased body temperature and heat intolerance; increased appetite; weight loss; increased sensitivity to catecholamines may lead to high blood pressure
Carbohydrate/lipid/protein metabolism	Promotes glucose catabolism; mobilizes fats; essential for protein synthesis; enhances liver's synthesis of cholesterol	Decreased glucose metabolism; elevated cholesterol/triglyceride levels in blood; decreased protein synthesis; edema	Enhanced catabolism of glucose, proteins, and fats; weight loss; loss of muscle mass
Nervous system	Promotes normal development of nervous system in fetus and infant; promotes normal adult nervous system function	In infant, slowed/deficient brain development, retardation; in adult, mental dulling, depression, paresthesias, memory impairment, hypoactive reflexes	Irritability, restlessness, insomnia, exophthalmos, personality changes
Cardiovascular system	Promotes normal functioning of the heart	Decreased efficiency of pumping action of the heart; low heart rate and blood pressure	Rapid heart rate and possible palpitations; high blood pressure; if prolonged, heart failure
Muscular system	Promotes normal muscular development, and function	Sluggish muscle action; muscle cramps; myalgia	Muscle atrophy and weakness
Skeletal system	Promotes normal growth and maturation of the skeleton	In child, growth retardation, skeletal stunting and retention of child's body proportions; in adult, joint pain	In child, excessive skeletal growth initially, followed by early epiphyseal closure and short stature; in adult, demineralization of skeleton
Gastrointestinal system	Promotes normal GI motility and tone; increases secretion of digestive juices	Depressed GI motility, tone, and secretory activity; constipation	Excessive GI motility; diarrhea; loss of appetite
Reproductive system	Promotes normal female reproductive ability and lactation	Depressed ovarian function; sterility; depressed lactation	In females, depressed ovarian function; in males, impotence
Integumentary system	Promotes normal hydration and secretory activity of skin	Skin pale, thick, and dry; facial edema; hair coarse and thick	Skin flushed, thin, and moist; hair fine and soft; nails soft and thin

Table 17.2). Hypothyroid disorders may result from some thyroid gland defect or secondarily from inadequate TSH or TRH release. They also occur when the thyroid gland is removed surgically (thyroidectomy) and when dietary iodine is inadequate.

In adults, the full-blown hypothyroid syndrome is called **myxedema** (mik″sĕ-de′mah; "mucous swelling"). Symptoms include a low metabolic rate; feeling chilled; constipation; thick, dry skin and puffy eyes; edema; lethargy; and mental sluggishness (but not mental retardation). If myxedema results from lack of iodine, the thyroid gland enlarges, a condition called **endemic** (en-dem′ik), or colloidal, **goiter.** The follicle cells produce colloid but cannot iodinate it or make functional hormones. The pituitary gland secretes increasing amounts of TSH in a

futile attempt to stimulate the thyroid to produce TH, but the only result is that the follicles accumulate more and more unusable colloid. Untreated, the thyroid cells eventually "burn out" from frantic activity, and the gland atrophies. Before the marketing of iodized salt, parts of the midwestern United States were called the "goiter belt." Because these areas had iodine-poor soil and no access to iodine-rich shellfish, goiters were very common there. Depending on the cause, myxedema can be reversed by iodine supplements or hormone replacement therapy.

Severe hypothyroidism in infants is called **cretinism** (kre′tĭ-nizm). The child has a short, disproportionate body, a thick tongue and neck, and is mentally retarded. Cretinism may reflect a genetic

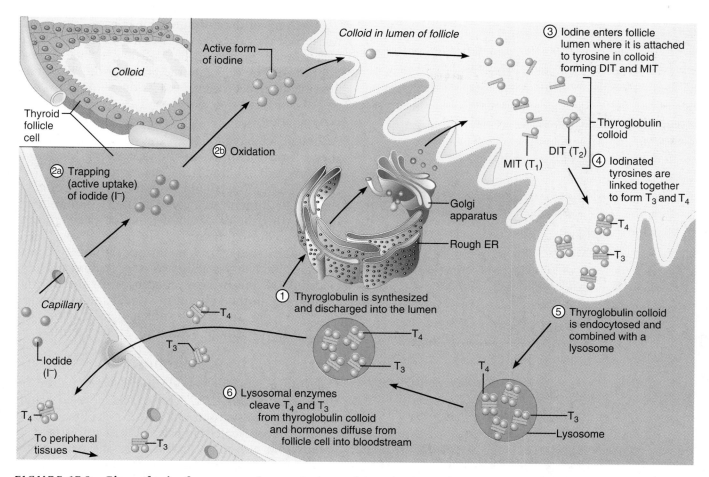

FIGURE 17.9 *Biosynthesis of thyroid hormones.* Binding of TSH to the follicle cell receptors (not illustrated) stimulates synthesis of the thyroglobulin colloid, the steps illustrated here, and the subsequent release of thyroxine (T_4) and triiodothyronine (T_3) from the cell. A complete description of these events appears in the text. (Note: Only iodinated tyrosine amino acids of the thyroglobulin colloid are illustrated. The rest of the colloid is indicated as the unstructured yellow substance.)

deficiency of the fetal thyroid gland or maternal factors, such as lack of dietary iodine. Cretinism is preventable by thyroid hormone replacement therapy, but once developmental abnormalities and mental retardation appear, they are not reversible.

The most common (and puzzling) hyperthyroid pathology is **Graves' disease.** Since patients' serum often contains abnormal antibodies that mimic TSH and continuously stimulate release of thyroid hormone, Graves' disease is believed to be an autoimmune disease. Typical symptoms include an elevated metabolic rate; sweating; rapid, irregular heartbeat; nervousness; and weight loss despite adequate food intake. *Exophthalmos*, protrusion of the eyeballs, may occur if the tissue behind the eyes becomes edematous and then fibrous. Treatments include surgical removal of the thyroid gland or ingestion of radioactive iodine (^{131}I) which selectively destroys the most active thyroid cells. ■

Synthesis Synthesis of thyroid hormone involves six interrelated processes that begin when thyroid-stimulating hormone (TSH) secreted by the anterior pituitary binds to follicular cell receptors. The following numbers correspond to steps 1–6 in Figure 17.9:

① **Formation and storage of thyroglobulin.** Thyroglobulin is synthesized on the ribosomes and then transported to the Golgi apparatus, where sugar residues are attached and the molecules are packed into vesicles. These transport vesicles move to the apex of the follicle cell, where their contents are discharged into the lumen and become part of the stored colloid.

② **Iodide trapping and oxidation to iodine.** To produce the functional iodinated hormones, the follicle cells must accumulate iodides (anions of iodine) from the blood. Since their intracellular concentration of iodides is over 30 times higher than that in blood, iodide trapping depends on active transport. Upon entry, iodides (I^-) are oxidized (by removal of electrons) and converted to iodine (I_2).

③ **Iodination.** Once formed, iodine is attached to the tyrosine amino acids forming part of the thyroglobulin colloid. This iodination reaction occurs at the apical follicle cell–colloid junction and is mediated by peroxidase enzymes.

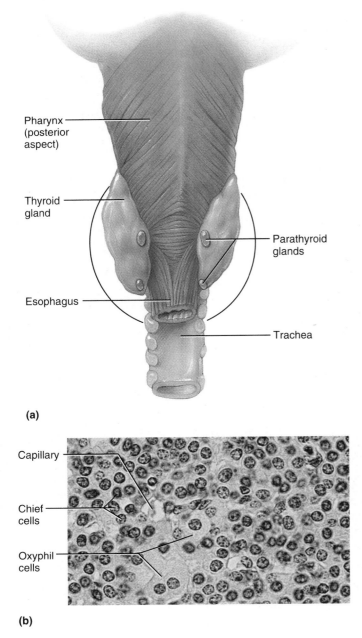

(a)

(b)

FIGURE 17.10 *The parathyroid glands.* **(a)** Location of the parathyroid glands on the posterior aspect of the thyroid gland. These tiny glands may be more inconspicuous than depicted. **(b)** Micrograph of a section of the parathyroid gland (815×).

④ **Coupling of T_2 and T_1.** Attachment of one iodine to a tyrosine produces **monoiodotyrosine (MIT or T_1)**; attachment of two iodines produces **diiodotyrosine (DIT or T_2).** Then, enzymes within the colloid link T_1 and T_2 together. Two linked DITs result in T_4, and a coupling of MIT and DIT produces T_3. At this point, however, the hormones are still part of the thyroglobulin colloid.

⑤ **Colloid endocytosis.** Hormone secretion requires that the follicle cells reclaim iodinated thyroglobulin by endocytosis and combine the vesicles with lysosomes.

⑥ **Cleavage of the hormones for release.** Within the lysosomes, the hormones are cleaved out of the colloid by lysosomal enzymes. The hormones then diffuse from the follicle cells into the bloodstream. The main hormonal product secreted is T_4. Some T_4 is converted to T_3 before secretion, but most T_3 is generated in the peripheral tissues.

Although we have followed the process of TH synthesis from beginning to end (secretion), the *initial response* to TSH binding is secretion of thyroid hormone. Then more colloid is synthesized to "restock" the follicle lumen. As a general rule, TSH levels are lower during the day, peak just before sleep, and remain high during the night.

The thyroid gland is unique among the endocrine glands in its ability to store its hormone extracellularly and in large quantities. In the normal thyroid gland, the amount of stored colloid remains relatively constant and is sufficient to provide normal levels of hormone release for several months.

Calcitonin

Calcitonin is a polypeptide hormone produced by the **parafollicular,** or **C, cells** of the thyroid gland. Because its most important effect is to lower blood calcium levels, calcitonin is a direct antagonist of parathyroid hormone, produced by the parathyroid glands. Calcitonin targets the skeleton, where it (1) inhibits osteoclast activity and hence bone resorption and release of ionic calcium from the bony matrix and (2) stimulates calcium uptake and incorporation into bone matrix. Thus, calcitonin has a bone-sparing effect.

Excessive blood levels of ionic calcium (approximately 20% above normal) act as a humoral stimulus for calcitonin release, whereas declining blood calcium levels inhibit C cell secretory activity. Calcitonin regulation of blood calcium levels is short-lived but extremely rapid.

Calcitonin appears to be important only in childhood, when the skeleton grows quickly and the bones are changing dramatically in mass, size, and shape. In adults, calcitonin is at best a weak hypocalcemic agent.

The Parathyroid Glands

The tiny, yellow-brown **parathyroid glands** are nearly hidden from view in the posterior aspect of the thyroid gland (Figure 17.10a). There are usually four of these glands, but the precise number varies. As many as eight have been reported, and some may be located in other regions of the neck or even in the thorax. The parathyroid's glandular cells are arranged in thick, branching cords containing scattered *oxyphil cells* and large numbers of the smaller chief cells (Figure 17.10b). The chief cells, also called *principal cells*, secrete PTH. The function of the oxyphil cells is unclear.

Discovery of the parathyroid glands was accidental. Years ago, surgeons were baffled by the observation that while most patients recovered uneventfully after partial (or even total) thyroid gland removal, others suffered uncontrolled muscle spasms and severe pain, and subsequently died. It was only after several such tragic deaths that the parathyroid glands were discovered and their hormonal function, quite different from that of the thyroid gland hormones, became known.

Parathyroid hormone (PTH), or **parathormone**, the protein hormone of these glands, is the single most important hormone controlling the calcium balance of the blood. Its release is triggered by falling blood calcium levels and inhibited by hypercalcemia. PTH increases ionic calcium levels in blood by stimulating three target organs: the skeleton (which contains considerable amounts of calcium salt in its matrix), the kidneys, and the intestine (Figure 17.11).

PTH release (1) stimulates osteoclasts (bone-resorbing cells) to digest some of the bony matrix and release ionic calcium and phosphates to the blood; (2) enhances reabsorption of calcium (and excretion of phosphates) by the kidneys; and (3) increases absorption of calcium by the intestinal mucosal cells. Calcium absorption by the intestine is enhanced indirectly by PTH's effect on vitamin D activation. Vitamin D is required for absorption of calcium from ingested food, but the form in which the vitamin is ingested or produced by the skin is relatively inactive. For vitamin D to exert its physiological effects, it must first be converted by the kidneys to its active vitamin D_3 form, **calcitriol** (1,25-dihydroxycholecalciferol), a transformation stimulated by PTH.

Since plasma calcium ion homeostasis is essential for so many functions, including transmission of nerve impulses, muscle contraction, and blood clotting, precise control of ionic calcium levels is critical.

Homeostatic Imbalance

Hyperparathyroidism is rare and usually the result of a parathyroid gland tumor. When it occurs, calcium is leached from the bones, and the bones soften and deform as their mineral salts are replaced by fibrous connective tissue. In *osteitis cystica fibrosa*, a severe example of this disorder, the bones have a moth-eaten appearance on X rays and tend to fracture spontaneously. The resulting abnormally elevated blood calcium level (hypercalcemia) has many outcomes, but the two most notable are (1) depression of the nervous system, which leads to abnormal reflexes and weakness of the skeletal muscles, and (2) formation of kidney stones as excess calcium salts

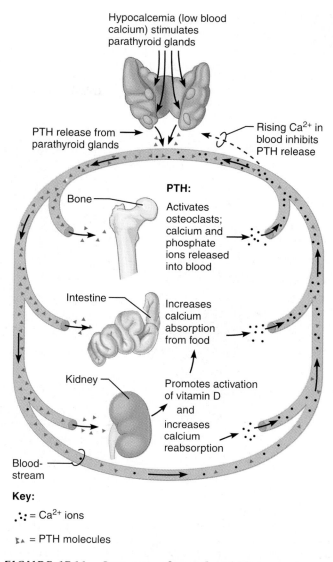

FIGURE 17.11 *Summary of parathyroid hormone's effects on bone, the intestine, and the kidneys.* Note that PTH's effect on the intestine is indirect via activated vitamin D.

precipitate in the kidney tubules. Calcium deposits may also form in soft tissues throughout the body and severely impair vital organ functioning, a condition called *metastatic calcification*.

Hypoparathyroidism, or PTH deficiency, most often follows parathyroid gland trauma or removal during thyroid surgery. However, an extended deficiency of dietary magnesium (required for PTH secretion) can cause functional hypoparathyroidism. The resulting hypocalcemia increases the excitability of neurons and accounts for the classical symptoms of *tetany* such as loss of sensation, muscle twitches, and convulsions. Untreated, the symptoms progress to spasms of the larynx, respiratory paralysis, and death. ∎

? *Suppose your professor springs a "pop" lab quiz and has adrenal cortical tissue under the microscope for identification. What information about the cortical zone cell arrangements would help you in this quiz?*

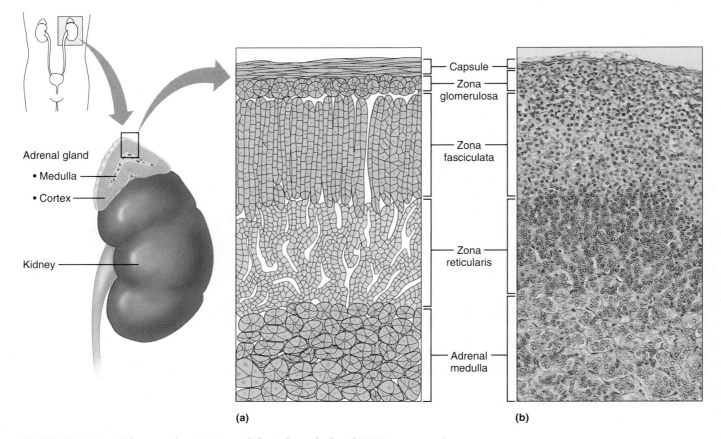

FIGURE 17.12 *Microscopic structure of the adrenal gland.* **(a)** Diagram and **(b)** photomicrograph of the zona glomerulosa, zona fasciculata, and zona reticularis of the adrenal cortex (50×). (Part of the adrenal medulla is also shown.)

Zona glomerulosa cells are in clusters, zona fasciculata cells are in parallel cords, and zona reticularis cells have a netlike arrangement.

The Adrenal (Suprarenal) Glands

The paired **adrenal** (*ad* = near; *renal* = kidney) **glands** are pyramid-shaped organs perched atop the kidneys, where they are enclosed in a fibrous capsule and a cushion of fat (see Figure 17.1). They are often referred to as the **suprarenal glands** (*supra* = above) as well.

Each adrenal gland is structurally and functionally two endocrine glands in one. The inner **adrenal medulla,** more like a "knot" of nervous tissue than a gland, acts as part of the sympathetic nervous system. The outer **adrenal cortex,** encapsulating the medullary region and forming the bulk of the gland, is glandular tissue derived from embryonic mesoderm. Each of these regions produces its own set of hormones (Table 17.3), but all adrenal hormones help us cope with "extreme" (stressful) situations.

The Adrenal Cortex

Well over two dozen *steroid* hormones, collectively called **corticosteroids,** are synthesized from cholesterol by the adrenal cortex. The pathway is multistep and involves varying intermediates depending on the hormone being formed. The structures of several corticosteroid hormones are illustrated in Table 17.3.

The large, lipid-laden cortical cells are arranged in three distinct layers or zones (Figure 17.12). The cell clusters forming the superficial **zona glomerulosa** (zo'nah glo-mer"u-lo'sah) mainly produce mineralocorticoids, hormones that help control the bal-

TABLE 17.3	Adrenal Gland Hormones: Regulation and Effects

Hormone	Structure	Regulation of release	Target organ and effects	Effects of hyposecretion ↓ and hypersecretion ↑
Adrenocortical Hormones				
Mineralocorticoids (chiefly aldosterone)	Aldosterone	Stimulated by renin-angiotensin mechanism (activated by decreasing blood volume or blood pressure), elevated K^+ or low Na^+ blood levels, and ACTH (minor influence); inhibited by increased blood volume and pressure, increased Na^+ and decreased K^+ blood levels	Kidneys: increased blood levels of Na^+ and decreased blood levels of K^+; since water reabsorption accompanies sodium retention, blood volume and blood pressure rise	↑ Aldosteronism ↓ Addison's disease
Glucocorticoids (chiefly cortisol)	Cortisol	Stimulated by ACTH; inhibited by feedback inhibition exerted by cortisol	Body cells: promote gluconeogenesis and hyperglycemia; mobilize fats for energy metabolism; stimulate protein catabolism; assist body to resist stressors; depress inflammatory and immune responses	↑ Cushing's disease ↓ Addison's disease
Gonadocorticoids (chiefly androgens such as testosterone)	Testosterone	Stimulated by ACTH; mechanism of inhibition incompletely understood, but feedback inhibition not seen	Insignificant effects in adults; may be responsible for female libido and source of estrogen after menopause	↑ Virilization of females (androgenital syndrome) ↓ No effects known
Adrenal Medullary Hormones				
Catecholamines (epinephrine and norepinephrine)	Epinephrine	Stimulated by preganglionic fibers of the sympathetic nervous system	Sympathetic nervous system target organs: effects mimic sympathetic nervous system activation; increase heart rate and metabolic rate; increase blood pressure by promoting vasoconstriction	↑ Prolonged fight-or-flight response; hypertension ↓ Unimportant

ance of minerals and water in the blood. The middle **zona fasciculata** (fah-sik"u-la'tah) cells, arranged in more or less linear cords, secrete the metabolic hormones called glucocorticoids. The cells of the innermost **zona reticularis** (rĕ-tik'u-lar-is), abutting the adrenal medulla, have a netlike arrangement. These cells produce small amounts of adrenal sex hormones, or gonadocorticoids. Although, as just described, there is a division of labor in corticosteroid production, the entire spectrum of corticosteroids is produced by all three cortical layers.

Mineralocorticoids The essential function of mineralocorticoids is regulation of the electrolyte (mineral salt) concentrations in extracellular fluids, particularly of sodium and potassium ions. The single most abundant positive ion (cation) in extracellular fluid is sodium, and while sodium is vital to homeostasis, excessive sodium intake and retention may promote high blood pressure (hypertension) in susceptible individuals. Although there are several mineralocorticoids, **aldosterone** (al-dos'ter-ōn) is the most potent and accounts for over 95% of the

Generally speaking, how does the response time for this hormonal mechanism compare to that triggered when sympathetic nerves act to control blood pressure?

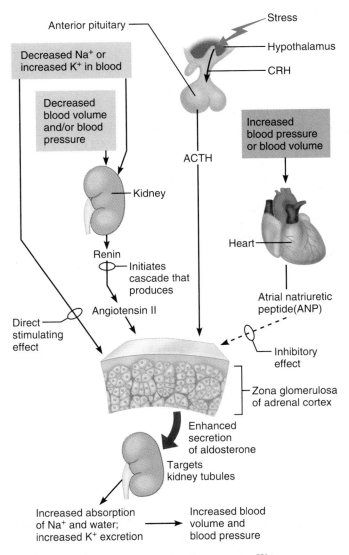

FIGURE 17.13 *Major mechanisms controlling aldosterone release from the adrenal cortex.*

mineralocorticoids produced (see Table 17.3). As you might guess, maintaining sodium ion balance is aldosterone's primary goal.

Aldosterone reduces excretion of sodium from the body. Its primary target is the distal parts of the kidney tubules, where it stimulates reabsorption of sodium ions from the forming urine and their return to the bloodstream. Aldosterone also enhances sodium ion reabsorption from perspiration, saliva,

_____

This response is much slower because the "messengers" are bloodborne hormones instead of nerve impulses and neurotransmitters.

and gastric juice. The mechanism of aldosterone activity appears to involve the synthesis of an enzyme required for sodium transport.

Regulation of a number of other ions—including potassium, hydrogen, bicarbonate, and chloride—is coupled to that of sodium; and where sodium goes, water follows—an event that leads to changes in blood volume and blood pressure. Hence, sodium ion regulation is crucial to overall body homeostasis. Briefly, aldosterone's effects on the renal tubules cause sodium and water retention accompanied by potassium ion elimination and, in some instances, alterations in the acid-base balance of the blood (by H^+ excretion). Since aldosterone's regulatory effects are brief (lasting approximately 20 minutes), plasma electrolyte balance can be very precisely controlled and modified continuously. These matters are described in greater detail in Chapter 27.

Aldosterone secretion is stimulated by a number of factors: rising blood levels of potassium ions, low blood levels of sodium, and decreasing blood volume and blood pressure. The reverse conditions inhibit aldosterone secretion. Four mechanisms regulate aldosterone secretion: the renin-angiotensin system, plasma concentrations of sodium and potassium ions, controls exerted by ACTH, and plasma concentrations of atrial natriuretic peptide (Figure 17.13). These mechanisms are described next.

1. The renin-angiotensin (re'nin an"je-o-ten'sin) **mechanism.** The renin-angiotensin mechanism, the major regulator of aldosterone release, influences both the electrolyte-water balance of the blood and blood pressure. Specialized cells of the *juxtaglomerular apparatus* in the kidneys become excited when blood pressure (or blood volume) declines or plasma osmolarity (solute concentration) drops. They respond by releasing **renin** into the blood. Renin cleaves off part of the plasma protein **angiotensinogen** (an"je-o-ten'sin-o-gen), triggering an enzymatic cascade leading to the formation of **angiotensin II**, a potent stimulator of aldosterone release by the glomerulosa cells of the adrenal cortex. However, the renin-angiotensin mechanism does much more than trigger aldosterone release, and all of its widespread effects are ultimately involved in raising the systemic blood pressure. These additional effects are described in detail in Chapters 26 and 27.

2. Plasma concentration of sodium and potassium ions. Fluctuating blood levels of sodium and potassium ions directly influence the zona glomerulosa cells. Increased potassium and decreased sodium are stimulatory; the opposite conditions are inhibitory.

3. ACTH. Under normal circumstances, ACTH released by the anterior pituitary has little or no effect on aldosterone release. However, when a person is severely stressed, the hypothalamus secretes more corticotropin-releasing hormone (CRH), and the rise in ACTH blood levels that follows steps up the rate of aldosterone secretion to a small extent. The rise in blood volume and blood pressure that results helps ensure adequate delivery of nutrients and respiratory gases during the stressful period.

4. Atrial natriuretic peptide (ANP). Atrial natriuretic peptide, a hormone secreted by the heart when blood pressure rises, fine-tunes blood pressure and sodium-water balance of the body. Its major effect is to inhibit the renin-angiotensin mechanism: It blocks renin and aldosterone secretion and inhibits other angiotensin-induced mechanisms that enhance water and sodium reabsorption. Consequently, ANP's overall influence is to decrease blood pressure by allowing Na^+ (and water) to flow out of the body in urine (*natriuretic* = producing salty urine).

Homeostatic Imbalance

Hypersecretion of aldosterone, a condition called *aldosteronism*, typically results from adrenal neoplasms. Two major sets of problems result: (1) hypertension and edema due to excessive sodium and water retention and (2) accelerated excretion of potassium ions. If potassium loss is extreme, neurons become nonresponsive and muscle weakness (and eventually paralysis) occurs. *Addison's disease,* a hyposecretory disease of the adrenal cortex, generally involves a deficient output of both mineralocorticoids and glucocorticoids, as described shortly. ■

Glucocorticoids The **glucocorticoids** influence the metabolism of most body cells and help us to resist stressors. They are absolutely essential to life. Under normal circumstances, glucocorticoids help the body adapt to intermittent food intake by keeping blood sugar levels fairly constant, and maintain blood volume by preventing the shift of water into tissue cells. However, severe stress due to hemorrhage, infections, or physical or emotional trauma evokes a dramatically higher output of glucocorticoids, which helps the body to negotiate the crisis. Glucocorticoid hormones include **cortisol (hydrocortisone)** (see Table 17.3), **cortisone,** and **corticosterone,** but only cortisol is secreted in significant amounts in humans. As for all steroid hormones, the basic mechanism of glucocorticoid activity on target cells is to modify gene activity.

Glucocorticoid secretion is regulated by a typical negative feedback system. Cortisol release is promoted by ACTH, triggered in turn by the hypothalamic releasing hormone CRH. Rising cortisol levels feed back to act on both the hypothalamus and the anterior pituitary, preventing CRH release and shut-

ting off ACTH and cortisol secretion. Cortisol secretory bursts, driven by patterns of eating and activity, occur in a definite pattern throughout the day and night. Cortisol blood levels peak shortly after we arise in the morning. The lowest levels occur in the evening just before and shortly after sleep ensues. The normal cortisol rhythm is interrupted by acute stress of any variety as the sympathetic nervous system overrides the (usually) inhibitory effects of elevated cortisol levels and triggers CRH release. The resulting increase in ACTH blood levels causes an outpouring of cortisol from the adrenal cortex.

Stress results in a dramatic rise in blood levels of glucose, fatty acids, and amino acids, all provoked by cortisol. Cortisol's prime metabolic effect is to provoke *gluconeogenesis*, that is, the formation of glucose from noncarbohydrate molecules (fats and proteins). To "save" glucose for the brain, cortisol mobilizes fatty acids from adipose tissue and encourages their increased use for energy. Under cortisol's influence, stored proteins are broken down to provide building blocks for repair or for making enzymes used in metabolic processes. Cortisol enhances epinephrine's vasoconstrictive effects, and the rise in blood pressure and circulatory efficiency that results helps ensure that these nutrients are quickly distributed to the cells.

Although *ideal amounts of glucocorticoids promote normal function*, significant anti-inflammatory and anti-immune effects are associated with cortisol excess. Excessive levels of glucocorticoids (1) depress cartilage and bone formation, (2) inhibit inflammation by stabilizing lysosomal membranes and preventing vasodilation, (3) depress the immune system, and (4) promote changes in cardiovascular, neural, and gastrointestinal function. Recognition of the effects of glucocorticoid hypersecretion has led to widespread use of glucocorticoid drugs to control symptoms of many chronic inflammatory disorders, such as rheumatoid arthritis or allergic responses. However, the use of these potent drugs at doses that are higher than normal (*pharmacological doses*) is a double-edged sword. While it relieves some of the symptoms, it also causes the undesirable effects of excessive levels of these hormones.

Homeostatic Imbalance

The pathology of cortisone excess, **Cushing's disease,** or **syndrome,** may be caused by an ACTH-releasing tumor of the pituitary; by an ACTH-releasing malignancy of the lungs, pancreas, or kidneys; or by a tumor of the adrenal cortex. However, it most often results from the clinical administration of pharmacological doses of glucocorticoid drugs. The syndrome is characterized by persistent hyperglycemia

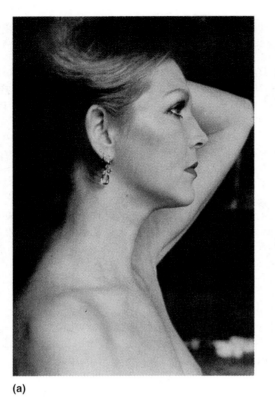
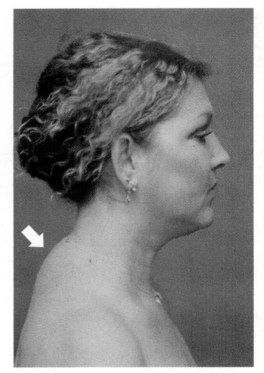

(a)

(b)

FIGURE 17.14 *Appearance of a woman (a) before and (b) during Cushing's disease.* The characteristic "buffalo hump" of fat on the woman's upper back is indicated by a white arrow.

(steroid diabetes), dramatic losses in muscle and bone protein, and water and salt retention, leading to hypertension and edema. The so-called *cushingoid signs* (Figure 17.14) include a swollen "moon" face, redistribution of fat to the abdomen and the posterior neck (causing a "buffalo hump"), a tendency to bruise, and poor wound healing. Because of enhanced anti-inflammatory effects, infections may become overwhelmingly severe before producing recognizable symptoms. Eventually, muscles weaken and spontaneous fractures force the person to become bedridden. The only treatment is removal of the cause—be it surgical removal of the offending tumor or discontinuation of the drug.

Addison's disease, the major hyposecretory disorder of the adrenal cortex, usually involves deficits in both glucocorticoids and mineralocorticoids. Its victims tend to lose weight; their plasma glucose and sodium levels drop and potassium levels rise. Severe dehydration and hypotension are common. Corticosteroid replacement therapy at physiological doses is the usual treatment. ■

Gonadocorticoids (Sex Hormones) The bulk of the **gonadocorticoids** secreted are **androgens,** or male sex hormones, the most important of which is testosterone (see Table 17.3). The adrenal cortex also makes small amounts of female hormones (estrogens). The amount of sex hormones produced by the adrenal cortex is insignificant compared to the amounts made by the gonads during late puberty and adulthood. The exact role of the adrenal sex hormones is still in question, but because adrenal androgen levels rise continuously between the ages of 7 and 13 in both boys and girls, it is assumed that they contribute to the onset of puberty and the appearance of axillary and pubic hair during that time. In adult women adrenal androgens are thought to be responsible for the sex drive, and they may be converted to estrogens after menopause when ovarian estrogens are no longer produced. Control of gonadocorticoid secretion is not completely understood. Release seems to be stimulated by ACTH, but the gonadocorticoids do not appear to exert feedback inhibition on ACTH release.

 Homeostatic Imbalance

Since androgens predominate, hypersecretion of gonadocorticoids usually causes *masculinization,* or *virilization.* In adult males, these effects may be obscured, since testicular testosterone has already produced virilization; but in prepubertal males and in females, the results can be dramatic. In the young

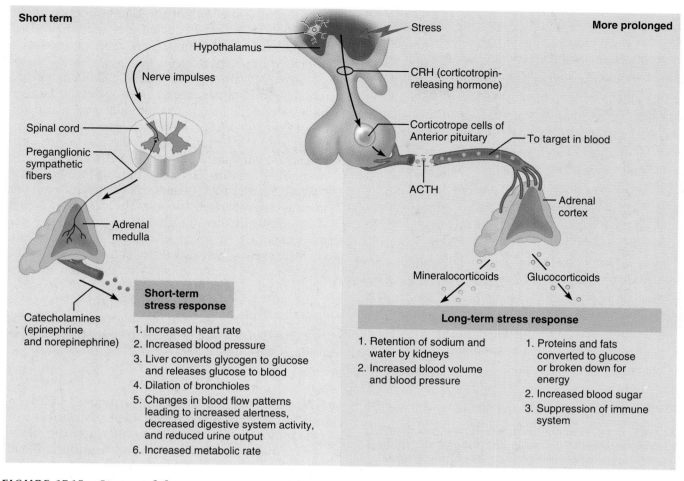

Short term ... **More prolonged**

Stress

Hypothalamus

Nerve impulses

CRH (corticotropin-releasing hormone)

Spinal cord

Corticotrope cells of Anterior pituitary

Preganglionic sympathetic fibers

ACTH

To target in blood

Adrenal medulla

Adrenal cortex

Catecholamines (epinephrine and norepinephrine)

Mineralocorticoids

Glucocorticoids

Short-term stress response

1. Increased heart rate
2. Increased blood pressure
3. Liver converts glycogen to glucose and releases glucose to blood
4. Dilation of bronchioles
5. Changes in blood flow patterns leading to increased alertness, decreased digestive system activity, and reduced urine output
6. Increased metabolic rate

Long-term stress response

1. Retention of sodium and water by kidneys
2. Increased blood volume and blood pressure

1. Proteins and fats converted to glucose or broken down for energy
2. Increased blood sugar
3. Suppression of immune system

FIGURE 17.15 *Stress and the adrenal gland.* Stressful stimuli cause the hypothalamus to activate the adrenal medulla via sympathetic nerve impulses and the adrenal cortex via hormonal signals. The medulla mediates short-term responses to stress by secreting catecholamines (epinephrine and norepinephrine). The cortex controls more prolonged responses by secreting its steroid hormones.

man, maturation of the reproductive organs and appearance of the secondary sex characteristics occur rapidly, and the sex drive emerges with a vengeance. Females develop a beard and a masculine pattern of body hair distribution, and the clitoris grows to resemble a small penis. ■

The Adrenal Medulla

The centrally located **adrenal medulla** is discussed in Chapter 14 as part of the autonomic nervous system, so it is covered only briefly here. Its spherical **chromaffin** (kro-maf'in) **cells,** which crowd around blood-filled capillaries and sinusoids, are modified ganglionic sympathetic neurons that secrete the *catecholamines* **epinephrine** and **norepinephrine** into the blood.

When the body is activated to its fight-or-flight status by some short-term stressor or emergency, the sympathetic nervous system is mobilized. As a result, blood sugar levels rise, blood vessels constrict and the heart beats faster (together raising the blood pressure), and blood is diverted from temporarily nonessential organs to the brain, heart, and skeletal muscles. At the same time, preganglionic sympathetic nerve endings that weave through the adrenal medulla signal for release of catecholamines, which reinforce and prolong the fight-or-flight response.

Unequal amounts of the two hormones are released; approximately 80% is epinephrine. With a few key exceptions, the two hormones exert the same effects (summarized in Table 14.3, p. 524). Epinephrine is the more potent stimulator of the heart and metabolic activities, while norepinephrine has the greater influence on peripheral vasoconstriction and blood pressure. Epinephrine is used clinically as a heart stimulant and to dilate the bronchioles during acute asthmatic attacks.

Unlike the adrenocortical hormones, which promote long-lasting body responses to stressors, catecholamines cause fairly brief responses. The interrelationships of the adrenal hormones and the hypothalamus, the "director" of the stress response, are depicted in Figure 17.15.

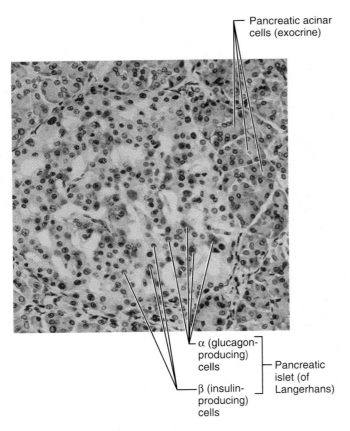

Pancreatic acinar cells (exocrine)

α (glucagon-producing) cells

β (insulin-producing) cells

Pancreatic islet (of Langerhans)

FIGURE 17.16 *Photomicrograph of* pancreatic tissue differentially stained. A pancreatic islet is surrounded by the blue–gray acinar cells, which produce the exocrine product (enzyme–rich pancreatic juice). In this preparation, the β cells of the islets that produce insulin are stained pale pink, and the α cells that produce glucagon are stained bright pink. (270×)

 Homeostatic Imbalance

Because hormones of the adrenal medulla merely intensify activities set into motion by the sympathetic nervous system neurons, their deficiency is not a problem. Unlike glucocorticoids, catecholamines are not essential for life. However, hypersecretion of catecholamines, sometimes arising from a rare chromaffin cell tumor called a *pheochromocytoma* (fe-o-kro"mo-si-to'mah), produces symptoms of massive and uncontrolled sympathetic nervous system activity—hyperglycemia, increased metabolic rate, rapid heartbeat and palpitations, hypertension, intense nervousness, and sweating. ■

The Pancreas

Located partially behind the stomach in the abdomen, the soft, triangular **pancreas** is a mixed gland composed of both endocrine and exocrine gland cells (Figure 17.4). Like the thyroid and parathyroids, it develops as an outpocketing of the epithelial lining of the gastrointestinal tract. *Acinar*

cells, forming the bulk of the gland, produce an enzyme-rich juice that is ducted into the small intestine during food digestion. This exocrine product is discussed in Chapter 24.

Scattered among the acinar cells are approximately a million **pancreatic islets (islets of Langerhans),** tiny cell clusters that produce pancreatic hormones (Figure 17.16). The islets contain two major populations of hormone-producing cells, the glucagon-synthesizing **alpha (α) cells** and the more numerous insulin-producing **beta (β) cells.** These cells act as tiny fuel sensors, secreting glucagon and insulin appropriately during the fasting and fed states. Thus insulin and glucagon are intimately but independently involved in the regulation of blood glucose levels. Their effects are opposite: Insulin is a *hypoglycemic* hormone, whereas glucagon is a *hyperglycemic* hormone (Figure 17.17). Some islet cells also synthesize other peptides in small amounts. These include *somatostatin, pancreatic polypeptide (PP),* and others. However, these will not be dealt with here.

Glucagon

Glucagon (gloo'kah-gon), a 29-amino-acid polypeptide, is an extremely potent hyperglycemic agent. One molecule of this hormone can cause the release of 100 million molecules of glucose into the blood! The major target of glucagon is the liver, where it promotes the following actions (Figure 17.17):

1. Breakdown of glycogen to glucose *(glycogenolysis)*

2. Synthesis of glucose from lactic acid and from noncarbohydrate molecules such as the glycerol portion of fats and amino acids *(gluconeogenesis)*

3. Release of glucose to the blood by the liver cells, which causes blood sugar levels to rise

A secondary effect is a fall in the amino acid concentration in the blood as the liver cells sequester them to make new glucose molecules.

Secretion of glucagon by the α cells is prompted by humoral stimuli, most importantly falling blood sugar levels. However, rising amino acid levels (as might follow a protein-rich meal) are also stimulatory. Glucagon release is suppressed by rising blood sugar levels and somatostatin. Because glucagon is such an important hyperglycemic agent, it has been speculated that people with persistently low blood sugar levels (hypoglycemics) are deficient in glucagon.

Insulin

Insulin is a small (51-amino-acid) protein consisting of two amino acid chains linked by disulfide bonds. As shown in Figure 17.18, it is initially synthesized as part of a larger polypeptide chain called **proinsulin.** The middle portion of this chain is then

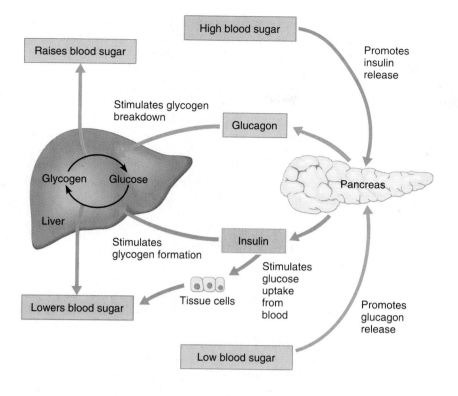

FIGURE 17.17 *Regulation of blood sugar levels by insulin and glucagon.* When blood sugar levels are high, the pancreas releases insulin. Insulin stimulates sugar uptake by cells and glycogen formation in the liver, which lowers blood sugar levels; that is, insulin exerts hypoglycemic effects. Glucagon, released when blood sugar levels are low, stimulates glycogen breakdown and thereby raises blood sugar; that is, glucagon is a hyperglycemic hormone.

excised by enzymes, releasing functional insulin. This "clipping" process occurs in the secretory vesicles just before insulin is released from the beta cell.

Insulin's effects are most obvious when we have just eaten. Its main effect is to lower blood sugar levels (Figure 17.17), but it also influences protein and fat metabolism. Circulating insulin lowers blood sugar levels by enhancing membrane transport of glucose (and other simple sugars) into body cells, especially muscle and fat cells. (It does *not* accelerate glucose entry into liver, kidney, and brain* tissue, all of which have easy access to blood glucose regardless of insulin levels.) Insulin inhibits the breakdown of glycogen to glucose and the conversion of amino acids or fats to glucose; thus, it counters any metabolic activity that would increase plasma levels of glucose. Insulin's mechanism of action is still being investigated, but as noted earlier, its receptor is a tyrosine kinase enzyme.

After glucose enters the target cells, insulin binding triggers enzymatic activities that

1. Catalyze the oxidation of glucose for ATP production

2. Join glucose together to form glycogen

3. Convert glucose to fat (particularly in adipose tissue)

As a rule, energy needs are met first, then glycogen deposit occurs. Finally, if excess glucose is still avail-

able, fat deposit occurs. Insulin also stimulates amino acid uptake and protein synthesis in muscle tissue. In summary, insulin sweeps glucose out of the blood, causing it to be used for energy or converted to other forms (glycogen or fats), and it promotes protein synthesis and fat storage.

Beta cells are stimulated to secrete insulin chiefly by elevated blood sugar levels, but rising plasma levels of amino acids and fatty acids also trigger insulin release. As body cells avidly take up sugar and other nutrients, and plasma levels of these substances drop, insulin secretion is suppressed. Other hormones also influence insulin release directly or indirectly. For example, any hyperglycemic hormone (such as glucagon, epinephrine, growth hormone, thyroxine, or glucocorticoids) called into action as blood sugar levels drop indirectly stimulates insulin release by promoting glucose entry into the bloodstream. Somatostatin depresses insulin release. Thus, blood sugar levels represent a balance of both humoral and hormonal influences. Insulin and (indirectly) somatostatin are the hypoglycemic factors that counterbalance the many hyperglycemic hormones.

Homeostatic Imbalance

Diabetes mellitus (DM) results from hyposecretion or hypoactivity of insulin. When insulin is absent or deficient, blood sugar levels remain high after a meal because glucose is unable to enter most tissue cells. Ordinarily, when blood sugar levels rise, hyperglycemic hormones are not released, but when hyperglycemia becomes excessive, the person begins to

---

*However, the cerebral cortex and hippocampus are well supplied with insulin receptors, hinting that insulin may play a different role in the brain, perhaps in cognition, memory, or neuronal growth.

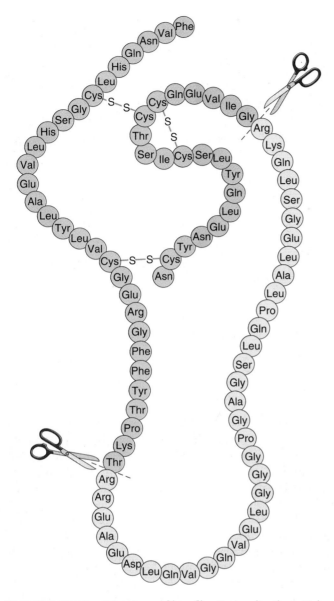

FIGURE 17.18 *Structure of insulin.* Proinsulin, the initial polypeptide product synthesized by the beta cells of the pancreas, is a single polypeptide chain with three disulfide (–S–S–) bonds. This inactive product is converted to insulin (green portion consisting of two peptide chains) by enzymatic cleavage, in which the middle portion of the polypeptide (the so-called C peptide) is cut out. Each "bead" in this model is an amino acid.

feel nauseated, which precipitates the fight-or-flight response. This results, inappropriately, in all the reactions that normally occur in the hypoglycemic (fasting) state to make glucose available—that is, glycogenolysis, lipolysis, and gluconeogenesis. Thus, the already high blood sugar levels soar even higher, and excesses of glucose begin to be lost from the body in the urine (*glycosuria*).

When sugars cannot be used as cellular fuel, more fats are mobilized, resulting in high fatty acid levels in the blood, a condition called lipidemia or lipemia. In severe cases of diabetes mellitus, blood levels of fatty acids and their metabolites (acetoacetic acid, acetone, and others) rise dramatically. The fatty acid metabolites, collectively called **ketones** (ke′tōnz) or **ketone bodies,** are strong organic acids. When they accumulate faster than they are used or excreted, the blood pH drops, resulting in **ketoacidosis,** and ketones begin to spill into the urine (*ketonuria*). Severe ketoacidosis is life threatening. The nervous system responds by initiating rapid deep breathing to blow off carbon dioxide from the blood and increase blood pH. (The physiological basis of this mechanism is explained in Chapter 23.) If untreated, ketoacidosis disrupts heart activity and oxygen transport, and severe depression of the nervous system leads to coma and finally death.

The three cardinal signs of diabetes mellitus are polyuria, polydipsia, and polyphagia. The excessive glucose in the kidney filtrate acts as an osmotic diuretic, that is, it inhibits water reabsorption by the kidney tubules, resulting in **polyuria,** a huge urine output that leads to decreased blood volume and dehydration. Serious electrolyte losses also occur as the body rids itself of excess ketones. Ketones are negatively charged and carry positive ions out with them; as a result, sodium and potassium ions are also lost from the body. Because of the electrolyte imbalance, the person gets abdominal pains and may vomit, and the stress reaction spirals even higher. Dehydration stimulates hypothalamic thirst centers, causing **polydipsia,** or excessive thirst. The final cardinal sign, **polyphagia,** refers to excessive hunger and food consumption, a sign that the person is "starving in the land of plenty." That is, although plenty of glucose is available, it cannot be used, and the body starts to utilize its fat and protein stores for energy metabolism. Figure 17.19 summarizes the consequences of insulin deficiency. DM is discussed in greater detail in *A Closer Look* on pp. 640–641.

Hyperinsulinism, or excessive insulin secretion, results in low blood sugar levels, or **hypoglycemia.** Hypoglycemia triggers the release of hyperglycemic hormones, which cause anxiety, nervousness, tremors, and a feeling of weakness. Insufficient glucose delivery to the brain causes disorientation, progressing to convulsions, unconsciousness, and even death. In rare cases, hyperinsulinism results from an islet cell tumor. More commonly, it is caused by an overdose of insulin and is easily treated by ingesting some sugar. ■

The Gonads

The male and female **gonads** (see Figure 17.4) produce steroidal sex hormones, identical in every way to those produced by adrenal cortical cells. The major distinction is the source and relative amounts produced. The paired *ovaries* are small, oval organs

Organs/tissue involved	Organ/tissue responses to insulin deficiency	Resulting condition of:		Signs and symptoms
		Blood	Urine	
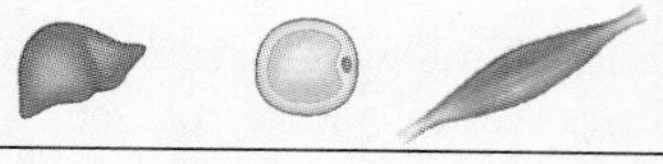 = Liver, Adipose tissue, Muscle	Decreased glucose uptake and utilization	Hyperglycemia	Glycosuria	**Polyuria** - dehydration - soft eyeballs
 = Liver	Glycogenolysis		Osmotic diuresis	**Polydipsia** Fatigue
 = Liver, Muscle	Protein catabolism and gluconeogenesis			Weight loss **Polyphagia**
= Liver, Adipose tissue	Lipolysis and ketogenesis	Lipidemia and ketoacidosis	Ketonuria	Acetone breath Hyperpnea
			Loss of Na$^+$, K$^+$; electrolyte and acid-base imbalances	Nausea/vomiting/abdominal pain Cardiac irregularities Central nervous system depression; coma

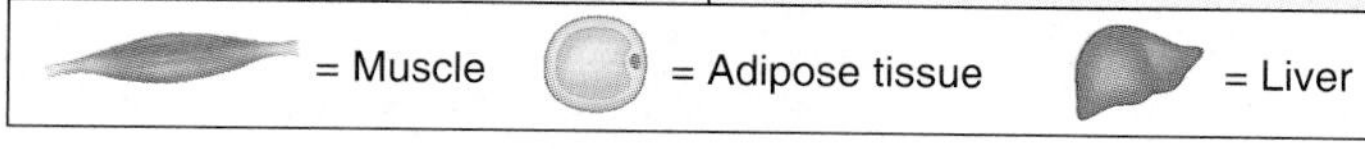

= Muscle   = Adipose tissue   = Liver

FIGURE 17.19   *Symptomatic results of insulin deficit (diabetes mellitus).*

located in the female's abdominopelvic cavity. Besides producing ova, or eggs, the ovaries produce **estrogens** and **progesterone** (pro-jes'tĕ-rōn). The estrogens alone are responsible for maturation of the reproductive organs and the appearance of the secondary sex characteristics of females at puberty. Acting with progesterone, estrogens promote breast development and cyclic changes in the uterine mucosa (the menstrual cycle).

The *testes,* located in an extra-abdominal skin pouch called the scrotum, produce sperm and male sex hormones, primarily **testosterone** (tes-tos'tĕ-rōn). During puberty, testosterone initiates the maturation of the male reproductive organs and the appearance of secondary sex characteristics and sex drive. In addition, testosterone is necessary for normal sperm production and maintains the reproductive organs in their mature functional state in adult males.

The release of gonadal hormones is regulated by gonadotropins, as described earlier. We will discuss the roles of the gonadal, placental, and gonadotropic hormones in detail in Chapters 28 and 29, where we consider the reproductive system and pregnancy.

The Pineal Gland

The tiny, pine cone–shaped **pineal gland** hangs from the roof of the third ventricle within the diencephalon (see Figure 17.4). Its secretory cells, called **pinealocytes,** are arranged in compact cords and clusters. Lying between clustered pinealocytes in adults are dense particles containing calcium salts ("brain-sand" or "pineal sand"). These salts are ra-

diopaque; hence, the pineal gland is a handy landmark for determining brain orientation in X rays.

The endocrine function of the pineal gland is still somewhat of a mystery. Although many peptides and amines (including serotonin, neuropeptide Y, norepinephrine, dopamine, and histamine) have been isolated from this minute gland, its only major secretory product is **melatonin** (mel"ah-to'nin). Melatonin concentrations in the blood rise and fall in a diurnal cycle. Peak levels occur during the night and make us drowsy; lowest levels occur during daylight hours around noon.

The pineal gland indirectly receives input from the visual pathways (retina → suprachiasmatic nucleus of hypothalamus → superior cervical ganglion → pineal gland) concerning the intensity and duration of daylight. In some animals, mating behavior and gonadal size vary with changes in the relative lengths of light and dark periods, and melatonin mediates these effects. In children, melatonin may have an antigonadotropic effect, that is, it may inhibit precocious (too early) sexual maturation and thus affect the timing of puberty.

The *suprachiasmatic (SC) nucleus* of the hypothalamus, an area referred to as our "biological clock," is richly supplied with melatonin receptors, and exposure to bright light (known to suppress melatonin secretion) can reset the clock timing. Hence, changing melatonin levels may also be a means by which the day/night cycles influence physiological processes that show rhythmic variations, such as body temperature, sleep, appetite, and hypothalamic activity in general.

Sweet Revenge: Can Biotechnology Tame the DM Monster?

Few medical breakthroughs have been as electrifying as the discovery of insulin in 1921, an event that changed diabetes mellitus from a death sentence to a survivable disease. Nonetheless, DM is still a huge health problem: Determining blood sugar levels accurately and then maintaining desirable levels sorely challenge our present biotechnology. Before we get into these specifics, let's take a closer look at the similarities and differences between type I and type II DM, the two major forms of diabetes mellitus that we have to deal with.

Type I diabetes mellitus, or *insulin-dependent diabetes mellitus (IDDM),* was formerly known as *juvenile-onset diabetes.* Some 750,000 Americans have this type of DM, in which the symptoms appear suddenly, usually before the age of 15 years. However, a long asymptomatic period during which the beta cells are systemically destroyed by an autoimmune response precedes these symptoms.

IDDM susceptibility genes have been localized on several chromosomes, indicating that type I diabetes is an example of a multigene autoimmune response. However, some investigators believe that *molecular mimicry* is at least part of the problem: Some foreign substance (for example, a virus) that has entered

the body and has been recognized as an alien (and provoked an immune response) is so similar to certain beta cell proteins that those self (beta cell) proteins are also attacked by the immune system.

Indeed, elegant studies on mice have demonstrated that T cells target a specific enzyme called *glutamate decarboxylase (GAD)* present in the beta cells. GAD converts the amino acid glutamate into GABA, a key messenger between neurons and, to a lesser extent, between pancreatic cells. Because most GAD is in the brain, essentially hidden from the immune system, these researchers suggest that the immune system may not recognize it as a self-protein. A piece of GAD closely resembles the p69 protein that beta cells display when they are infected by viruses. GAD is probably not the whole answer either, but it appears to be an important part of it, because when mice prone to develop diabetes were injected with GAD before the autoimmune attack on the pancreas began, all mice treated escaped developing diabetes.

Type I diabetics totally lack insulin activity, and their disease is difficult to control. Because of the early onset of their disease, type I diabetics typically exhibit long-term vascular and neural problems. The lipidemia and high blood cholesterol levels typical of the disease lead to severe vascular complications including atherosclerosis, strokes, heart

attacks, renal shutdown, gangrene, and blindness. Consequences of neuropathies include loss of sensation, impaired bladder function, and impotence. Female type I diabetics also tend to have lumpy breasts and to undergo premature menopause, which increases their risk for cardiac problems.

Previously, the protocol was insulin injections once or twice daily to manage ketoacidosis and, to a lesser extent, hyperglycemia. Currently, more frequent insulin injections (up to four times daily) are recommended to reduce vascular and renal complications. But insulin injections are not pleasant and many patients have resisted the more frequently scheduled shots. Insulin delivered continuously by an insulin pump worn externally has proved to be a much more desirable way to get the needed insulin, but not all patients can afford the device. Inhaled insulin-containing mist and powder are in clinical trials and will be welcomed if (and when) they become available. Although pancreatic islet cell transplants have been somewhat successful in helping type I diabetics, the need for immunosuppression is a problem, particularly in young patients. Some diabetics are managing to delay renal damage by taking drugs used to manage high blood pressure. Other than new routes to administer insulin, there is little in the pipeline for type I DM

The Thymus

Located deep to the sternum in the thorax is the lobulated **thymus gland.** Large and conspicuous in infants and children, the thymus diminishes in size throughout adulthood. By old age, it is composed largely of adipose and fibrous connective tissues.

The major hormonal product of the thymic epithelial cells is a family of peptide hormones, including **thymopoietins** and **thymosins** (thi'mo-sinz), which appear to be essential for the normal development of *T lymphocytes (T cells)* and the immune response. The role of thymic hormones is described in Chapter 22 with the discussion of the immune system.

OTHER HORMONE-PRODUCING STRUCTURES

Other hormone-producing cells occur within various organs of the body, including the following (see Table 17.4):

1. Heart. The atria of the heart contain some specialized cardiac muscle cells that secrete **atrial natriuretic peptide (ANP)** (na"tre-u-ret'ik; "producing salty urine"). This peptide hormone reduces blood volume, blood pressure, and blood sodium concentration when these parameters become elevated. ANP accomplishes this task by signaling the kidney

patients, who must receive insulin to survive.

Type II diabetes, or *non-insulin-dependent diabetes mellitus (NIDDM)* was formerly called *mature-onset diabetes* because it occurs mostly after the age of 40 years and is increasingly common with age. About 7.5 million people in the U.S. have been diagnosed with type II diabetes, and roughly the same number are believed to be undiagnosed victims. Heredity or a familial predisposition is particularly striking in this diabetic group. An estimated 25–30% of Americans carry a gene that predisposes them to NIDDM, with nonwhites affected to a much greater extent. If an identical twin has type II diabetes mellitus, the probability that the other twin will have the disease is 100%. Most type II diabetics produce insulin, but for some reason the insulin receptors are unable to respond to it, a phenomenon called **insulin resistance.** PC-1, a membrane protein whose levels are higher in those with NIDDM, may be the culprit. PC-1 inhibits the insulin receptor tyrosine kinase, but its mechanism of action is still a mystery.

Type II diabetics are almost always overweight and account for over 90% of the known cases of diabetes mellitus. The link between obesity and DM has been obscure until recently. It now appears that adipose cells of obese people overproduce a hormonelike chemical called *tumor necrosis factor-alpha*, which depresses synthesis of a protein (glut4). Since glut4 (glucose translo-

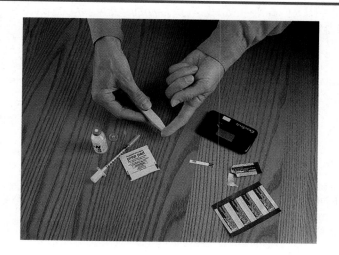

cation protein) enables glucose to pass through insulin-primed plasma membranes, cells cannot take up glucose in its absence. Ketosis is not a major problem for this group, and in many cases (and at least initially) the symptoms can be managed solely by exercise and diet (weight loss), and oral medications like Orinase, which has been around for decades. Eventually, however, most type II diabetics must inject insulin (today manufactured by recombinant DNA techniques or produced by animals, usually pigs). *Insulin resistance reducers*, drugs engineered to increase the sensitivity of insulin receptors, have begun to appear to help this group. Early on the scene was troglitazone (Rezulin); however, it was found to cause liver problems in some 2% of patients. But that early arrival was just the tip of the iceberg. Nearly all pharmaceutical companies are vying to find *the* drug to treat type II diabetes—not only to conquer insulin resistance but also to decrease the neurological and vascular consequences of the disease.

The race is also on to find an easier method of measuring blood sugar. Bloodless testing is the goal and several products are in the wings, from Diasensor (a fiber-optic probe that passes infrared light through the skin, and then analyzes its reflection), to GlucoWatch (which is worn like a wristwatch and uses a mild electric current to extract glucose painlessly onto a transdermal pad), to an implanted sensor, about three times the thickness of a human hair, that continuously monitors blood sugar levels and sets off an alarm when insulin is needed. Biotechnology is on the threshold of taking on the monster and looks like a good bet to win that battle pretty soon.

to increase its production of salty urine and by inhibiting aldosterone release by the adrenal cortex (see Figure 17.13).

2. Gastrointestinal tract. *Enteroendocrine cells* are hormone-secreting cells sprinkled within the mucosa of organs of the gastrointestinal (GI) tract (e.g., stomach, small intestine). These scattered cells release several amine and peptide hormones that help regulate a wide variety of digestive functions, some of which are summarized in Table 17.4. Enteroendocrine cells are sometimes referred to as *paraneurons* because they are similar in certain ways to neurons. Many of their hormones are chemically identical to neurotransmitters (e.g., serotonin and cholecystokinin), and in some cases, the hormones simply diffuse to and influence nearby target cells without first entering the bloodstream, that is, they act as local hormones or *paracrines*. These hormones are described in more detail in Chapter 24.

3. Placenta. Besides sustaining the fetus during pregnancy, the placenta secretes several steroid and protein hormones that influence the course of pregnancy (Chapter 29). Placental hormones include estrogens and progesterone (hormones more often associated with the ovary), as well as human chorionic gonadotropin (hCG), and others.

TABLE 17.4	Selected Examples of Hormones Produced by Organs Other Than the Major Endocrine Organs			
Source	Hormone	Chemical composition	Trigger	Target organ and effects
GI tract mucosa				
■ Stomach	Gastrin	Peptide	Secreted in response to food	Stomach: stimulates glands to release hydrochloric acid (HCl)
■ Stomach	Serotonin	Amine	Secreted in response to food	Stomach: causes contraction of stomach muscle
■ Duodenum of small intestine	Intestinal gastrin	Peptide	Secreted in response to food, especially fats	Stomach: inhibits HCl secretion and gastrointestinal tract mobility
■ Duodenum	Secretin	Peptide	Secreted in response to food	Pancreas and liver: stimulates release of bicarbonate-rich juice; stomach: inhibits secretory activity
■ Duodenum	Cholecystokinin (CCK)	Peptide	Secreted in response to food	Pancreas: stimulates release of enzyme-rich juice; gallbladder: stimulates expulsion of stored bile; sphincter of Oddi: causes sphincter to relax, allowing bile and pancreatic juice to enter duodenum
Kidney	Erythropoietin (EPO)	Glycoprotein	Erythropoietin secreted in response to hypoxia	Bone marrow: stimulates production of red blood cells
Skin (epidermal cells)	Cholecalciferol (provitamin D_3)	Steroid	Cholecalciferol activated by the kidneys to active vitamin D_3 (1,25(OH) D_3) and released in response to parathyroid hormone	Intestine: stimulates active transport of dietary calcium across intestinal cell membranes
Heart (atria)	Atrial natriuretic peptide	Peptide	Secreted in response to stretching of atria (by rising blood pressure)	Kidney: inhibits sodium ion reabsorption and renin release; adrenal cortex: inhibits secretion of aldosterone; decreases blood pressure
Adipose tissue	Leptin	Peptide	Secreted in response to fatty foods	Suppresses appetite; may play a role in triggering onset of puberty

4. Kidney. As yet unidentified cells in the kidney secrete **erythropoietin** (ĕ-rith″ro-poi′ĕ-tin; "red-maker"). This protein hormone signals the bone marrow to increase its production of red blood cells.

5. Skin. The skin produces **cholecalciferol,** an inactive form of vitamin D_3, when modified cholesterol molecules in epidermal cells are exposed to ultraviolet radiation. This compound then enters the blood via the dermal capillaries, is modified in the liver, and becomes fully activated in the kidney. The active form of vitamin D_3 (1,25-dihydroxycholecalciferol

or 1,25 (OH) D_3) is an essential part of the carrier system that intestinal cells use to absorb Ca^{2+} from the diet. Without this vitamin, the bones become weak and soft.

Adipose Tissue Adipose cells release **leptin** following their uptake of glucose and lipids, which they store as fat. Leptin binds to CNS neurons concerned with appetite control, producing a sensation of satiety.

DEVELOPMENTAL ASPECTS OF THE ENDOCRINE SYSTEM

Hormone-producing glands arise from all three embryonic germ layers. Endocrine glands derived from mesoderm produce steroid hormones. All others produce amines or amino acid and protein hormones.

Barring hypersecretory and hyposecretory disorders, most endocrine organs operate smoothly throughout life until old age. Aging may bring about changes in the rates of hormone secretion, breakdown, and excretion, or in the sensitivity of target cell receptors to their hormones. Research on typical endocrine functioning in the elderly is difficult, however, because it is frequently altered by the chronic illnesses common in that age group.

Structural changes in the anterior pituitary occur with age; the amount of connective tissue increases, vascularization decreases, and the number of hormone-secreting cells declines. This may or may not affect hormone production. For example, blood levels and the release rhythm of ACTH remain constant, whereas levels of TSH and gonadotropins increase in women with age. GH levels decline, which partially explains muscle atrophy in old age. Endocrinologists have discovered an exciting effect of GH supplementation. When aging clients received injections of genetically engineered GH in clinical tests over a period of 6 months, the injections dramatically spurred muscle growth, reduced body fat, and wiped 20 years off the elders' sagging physiques!

The adrenal gland also shows structural changes with age, but normal controls of cortisol appear to persist as long as a person is healthy and not stressed. Chronic stress, on the other hand, drives up blood levels of cortisol and appears to contribute to hippocampal (and memory) deterioration. Plasma levels of aldosterone are reduced by half in old age; however, this may reflect a decline in renin release by the kidneys, which become less responsive to renin-evoking stimuli. No age-related differences have been found in the release of catecholamines by the adrenal medulla.

The gonads, particularly the ovaries, undergo significant weight changes with age. In late middle age, the ovaries decrease in size and weight, and they become unresponsive to gonadotropins. As female hormone production declines dramatically, the ability to bear children ends, and problems associated with estrogen deficiency, such as arteriosclerosis and osteoporosis, begin to occur. Testosterone production by the testes also wanes with age, but this effect usually is not seen until very old age.

Glucose tolerance (the ability to dispose of a glucose load effectively) begins to deteriorate as early as the fourth decade of life. Blood glucose levels rise higher and return to resting levels slower in the elderly than in young adults. *Glucose tolerance tests reveal that a high proportion of elderly have borderline diabetes.* Near-normal amounts of insulin continue to be secreted by the islet cells, hence the decreasing glucose tolerance may reflect declining receptor sensitivity to insulin (pre–type II diabetes).

Thyroid hormone synthesis and release diminishes somewhat with age. Typically, the follicles are loaded with colloid in the elderly, and fibrosis of the gland occurs. Basal metabolic rate declines with age. Mild hypothyroidism is only one cause of this decline. The increase in body fat relative to muscle is equally important, since muscle tissue is more active metabolically (uses much more oxygen) than fat.

The parathyroid glands change little with age, and PTH levels remain at fairly normal values. Estrogen protects women against the demineralizing effects of PTH, but after menopause estrogen production wanes, leaving older women more vulnerable to the bone-demineralizing effects of PTH and osteoporosis.

* * *

In this chapter, we have covered the general mechanisms of hormone action and have provided an overview of the major endocrine organs, their chief targets, and their most important physiological effects, as summarized in *Making Connections* on p. 644. However, every one of the hormones discussed here comes up in at least one other chapter in this text, where its actions are described as part of the functional framework of a particular organ system. For example, the effects of PTH and calcitonin on bone mineralization are described in Chapter 6 along with the discussion of bone remodeling, and gonadal hormones take center stage as the regulators of reproductive system maturation and function in Chapters 28 and 29.

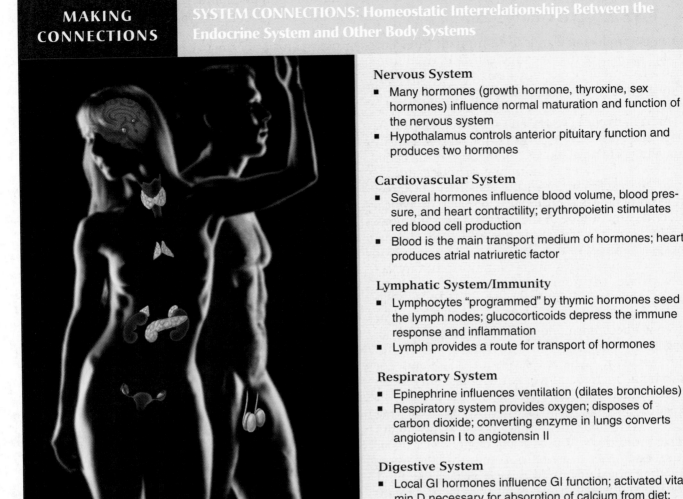

Integumentary System
- Androgens cause activation of sebaceous glands; estrogen increases skin hydration
- Skin produces cholecalciferol (provitamin D)

Skeletal System
- PTH and calcitonin regulate calcium blood levels; growth hormone, T_3, T_4, and sex hormones are necessary for normal skeletal development
- The skeleton provides some protection to endocrine organs, especially to those in the brain, chest, and pelvis

Muscular System
- Growth hormone is essential for normal muscular development; other hormones (thyroxine and catecholamines) influence muscle metabolism
- Muscular system mechanically protects some endocrine glands; muscular activity elicits catecholamine release

Nervous System
- Many hormones (growth hormone, thyroxine, sex hormones) influence normal maturation and function of the nervous system
- Hypothalamus controls anterior pituitary function and produces two hormones

Cardiovascular System
- Several hormones influence blood volume, blood pressure, and heart contractility; erythropoietin stimulates red blood cell production
- Blood is the main transport medium of hormones; heart produces atrial natriuretic factor

Lymphatic System/Immunity
- Lymphocytes "programmed" by thymic hormones seed the lymph nodes; glucocorticoids depress the immune response and inflammation
- Lymph provides a route for transport of hormones

Respiratory System
- Epinephrine influences ventilation (dilates bronchioles)
- Respiratory system provides oxygen; disposes of carbon dioxide; converting enzyme in lungs converts angiotensin I to angiotensin II

Digestive System
- Local GI hormones influence GI function; activated vitamin D necessary for absorption of calcium from diet; catecholamines influence digestive mobility and secretory activity
- Digestive system provides nutrients to endocrine organs

Urinary System
- Aldosterone and ADH influence renal function; erythropoietin released by kidneys influences red blood cell formation
- Kidneys activate vitamin D (considered a hormone)

Reproductive System
- Hypothalamic, anterior pituitary, and gonadal hormones direct reproductive system development and function; oxytocin and prolactin involved in birth and breast-feeding
- Gonadal hormones feed back to influence endocrine system function

Like most body systems, the endocrine system performs many functions that benefit the body as a whole. For example, without insulin, thyroxine, and various other metabolic hormones, body cells would be unable to get or use glucose, and would die. Likewise, total body growth is beholden to the endocrine system, which coordinates the growth spurts with increases in skeletal and muscular mass so that we don't look out of proportion most of the time. But the interactions that are most noticeable and crucial are those that the endocrine system has with the nervous and the reproductive systems, and they begin before birth.

Nervous System

The influence of hormones on behavior is striking. While we are still in the wet darkness of our mother's uterus, testosterone—or lack of it—is determining the "sex" of our brain. If testosterone is produced by the exceedingly tiny male testes, then certain areas of the brain enlarge, develop large numbers of androgen receptors, and thereafter determine the so-called masculine aspects of behavior (aggressiveness, etc.). Conversely, in the absence of testosterone, the brain is feminized. At puberty, Mom and Dad's "little angels," driven by raging hormones, turn into strangers. The surge of androgens—produced first by the adrenal cortices, and then by the maturing gonads—produces an often thoughtless aggressiveness and galloping sex drive, typically long before the cognitive abilities of the brain can rein them in.

Neural involvement in hormonal affairs is no less striking. Not only is the hypothalamus an endocrine organ in its own right, it also effectively regulates the bulk of hormonal activity via its hormonal or neural controls of the pituitary and adrenal medulla. And that's just on a normal day. The effects of trauma on the hypothalamic-pituitary axis can be far-reaching. Lack of loving care to a newborn baby results in failure to thrive; exceptionally vigorous athletic training in the pubertal female can result in bone wasting and infertility; and overwhelming, prolonged emotional stress on almost anyone can lead to Addison's disease (corticosteroid burnout). The shadow the nervous system casts over the endocrine system is long indeed.

Reproductive System

The reproductive system is totally dependent on hormones to "order up the right organs" to match our genetic sex. Testosterone secretion by the testes of male embryos directs formation of the male reproductive tract and external genitalia. Without testosterone, female structures develop—regardless of gonadal sex. The next crucial period is puberty, when gonadal sex hormone production rises and steers maturation of the reproductive organs, bringing them to their adult structure and function. Without these hormonal signals, the reproductive organs remain childlike and the person cannot produce offspring. Pregnancy invites more endocrine system interactions with the reproductive system. The placenta, a temporary endocrine organ, churns out estrogen and progesterone, which help maintain the pregnancy and prepare the mother's breasts for lactation, as well as a number of other hormones that influence maternal metabolism. During and after birth, oxytocin and prolactin take center stage to promote labor and delivery, and then milk production and ejection. Other than feedback inhibition exerted by its sex hormones on the hypothalamic-pituitary axis, the influence of reproductive organs on the endocrine system is negligible.

CLINICAL CONNECTIONS

Endocrine System

Case study: We have a new patient to consider today. Mr. Gutteman, a 70-year-old male, was brought into the ER in a comatose state and has yet to come out of it. It is obvious that he suffered severe head trauma—his scalp was badly lacerated, and he has an impacted skull fracture. His initial lab tests (blood and urine) were within normal limits. His fracture was repaired and the following orders (and others) were given:

- Check qh (every hour) and record: spontaneous behavior, level of responsiveness to stimulation, movements, pupil size and reaction to light, speech, and vital signs.
- Turn patient q4h and maintain meticulous skin care and dryness.

1. Explain the rationale behind these orders.

On the second day of his hospitalization, the aide reports that Mr. Gutteman is breathing irregularly, his skin is dry and flaccid, and that she has emptied his urine reservoir several times during the day. Upon receiving this information, the physician ordered:

- Blood and urine tests for presence of sugar and ketones
- Strict I&O (fluid intake and output recording)

Mr. Gutteman is found to be losing huge amounts of water in urine and the volume lost is being routinely replaced (via IV line). Mr. Gutteman's blood and urine tests are negative for sugar and ketones.

Relative to these findings:

2. What would you say Mr. Gutteman's hormonal problem is and what do you think caused it?

3. Is it life threatening? (Explain your answer.)

(Answers in Appendix F)

RELATED CLINICAL TERMS

Hirsutism (her′soot-izm; *hirsut* = hairy, rough) Excessive hair growth; usually refers to this phenomenon in women and reflects excessive androgen production.

Hypophysectomy (hi-pof″i-sek′to-me) Surgical removal of the pituitary gland.

Prolactinoma (pro-lak″tĭ-no′mah; *oma* = tumor) The most common type (30–40% or more) of pituitary gland tumor; evidenced by hypersecretion of prolactin and menstrual disturbances in women.

Psychosocial dwarfism Dwarfism (and failure to thrive) resulting from stress and emotional disorders that suppress hypothalamic release of growth hormone–releasing hormone and thus anterior pituitary secretion of growth hormone.

Thyroid storm (thyroid crisis) A sudden and dangerous increase in all of the symptoms of hyperthyroidism due to excessive amounts of circulating TH; symptoms of this hypermetabolic state include fever, rapid heart rate, high blood pressure, dehydration, nervousness, and tremors; precipitating factors include stressful situations, excessive intake of TH supplements, and trauma to the thyroid gland.

CHAPTER SUMMARY

Media study tools that could provide you additional help in reviewing specific key topics of Chapter 17 are referenced below. **SP** = *Study Partner;* **IP** = *Interactive Physiology.*

1. The nervous and endocrine systems are the major controlling systems of the body. The nervous system exerts rapid controls via nerve impulses; the endocrine system's effects are mediated by hormones and are more prolonged.

The Endocrine System: An Overview   (p. 609)

1. Endocrine organs are ductless, well-vascularized glands that release hormones directly into the blood or lymph. They are small and widely separated in the body.

2. The major endocrine organs are the pituitary, thyroid, parathyroid, adrenal, pineal, and thymus glands, as well as the pancreas and gonads. The hypothalamus is a neuro-endocrine organ.

3. Hormonally regulated processes include reproduction; growth and development; mobilization of body defenses to stressors; maintaining electrolyte, water, and nutrient balance; and regulating cellular metabolism.

Hormones   (pp. 609–615, 617)

The Chemistry of Hormones   (pp. 609–610)

1. Most hormones are steroids or amino acid derivatives.

Mechanisms of Hormone Action   (pp. 610–613)

2. Hormones alter cell activity by stimulating or inhibiting characteristic cellular processes.

3. Cell responses to hormone stimulation may involve changes in membrane permeability; enzyme synthesis, activation, or inhibition; secretory activity; and gene activation.

4. Second-messenger mechanisms employing intracellular messengers and transduced by G proteins are a common means by which amino acid–based hormones interact with their target cells. In the cyclic AMP system, the hormone binds to a plasma membrane receptor that couples to a G protein. When the G protein is activated it, in turn, couples to adenylate cyclase, which catalyzes the synthesis of cyclic AMP from ATP. Cyclic AMP initiates reactions that activate protein kinases and other enzymes, leading to cellular response. The phosphatidyl inositol mechanism is another important second-messenger system. Other presumed "messengers" are cyclic GMP and calcium.

5. Steroid hormones (and thyroid hormone) enter their target cells and effect responses by activating DNA, initi-ating messenger RNA formation leading to protein synthesis.

Hormone–Target Cell Specificity   (pp. 613–614)

6. The ability of a target cell to respond to a hormone depends on the presence of receptors, within the cell or on its plasma membrane, to which the hormone can bind.

7. Hormone receptors are dynamic structures. Changes in number and sensitivity of hormone receptors may occur in response to high or low levels of stimulating hormones.

Half-Life, Onset, and Duration of Hormone Activity   (p. 614)

8. Blood levels of hormones reflect a balance between secretion and degradation/excretion. The liver and kidneys are the major organs that degrade hormones; breakdown products are excreted in urine and feces.

9. Hormone half-life and duration of activity are limited and vary from hormone to hormone.

Control of Hormone Release   (pp. 614–615, 617)

10. Endocrine organs are activated to release their hormones by humoral, neural, or hormonal stimuli. Negative feedback is important in regulating hormone levels in the blood.

11. The nervous system, acting through hypothalamic controls, can in certain cases override or modulate hormonal effects.

Major Endocrine Organs   (pp. 616–640)

The Pituitary Gland (Hypophysis)   (pp. 617–624)

1. The pituitary gland hangs from the base of the brain by a stalk and is enclosed by bone. It consists of a hormone-producing glandular portion (anterior pituitary) and a neural portion (posterior pituitary), which is an extension of the hypothalamus.

2. The hypothalamus (a) regulates the hormonal output of the anterior pituitary via releasing and inhibiting hormones and (b) synthesizes two hormones that it exports to the posterior pituitary for storage and later release.

3. Four of the six adenohypophyseal hormones are tropic hormones that regulate the function of other endocrine organs. Most anterior pituitary hormones exhibit a diurnal rhythm of release, which is subject to modification by stimuli influencing the hypothalamus.

4. Growth hormone (GH) is an anabolic hormone that stimulates growth of all body tissues but especially skele-

tal muscle and bone. It may act directly or indirectly via insulin-like growth factors (IGFs). GH mobilizes fats, stimulates protein synthesis, and inhibits glucose uptake and metabolism. Secretion is regulated by growth hormone–releasing hormone (GHRH) and growth hormone–inhibiting hormone (GHIH), or somatostatin. Hypersecretion causes gigantism in children and acromegaly in adults; hyposecretion in children causes pituitary dwarfism.

5. Thyroid-stimulating hormone (TSH) promotes normal development and activity of the thyroid gland. Thyrotropin-releasing hormone (TRH) stimulates its release; negative feedback of thyroid hormone inhibits it.

6. Adrenocorticotropic hormone (ACTH) stimulates the adrenal cortex to release corticosteroids. ACTH release is triggered by corticotropin-releasing hormone (CRH) and inhibited by rising glucocorticoid levels.

7. The gonadotropins—follicle-stimulating hormone (FSH) and luteinizing hormone (LH)—regulate the functions of the gonads in both sexes. FSH stimulates sex cell production; LH stimulates gonadal hormone production. Gonadotropin levels rise in response to gonadotropin-releasing hormone (GnRH). Negative feedback of gonadal hormones inhibits gonadotropin release.

8. Prolactin (PRL) promotes milk production in humans. Its secretion is prompted by prolactin-releasing hormone (PRH) and inhibited by prolactin-inhibiting hormone (PIH).

9. The neurohypophysis stores and releases two hypothalamic hormones, oxytocin and antidiuretic hormone (ADH).

10. Oxytocin stimulates powerful uterine contractions, which trigger labor and delivery of an infant, and milk ejection in nursing women. It also appears to promote sexual arousal and nurturing. Its release is mediated reflexively by the hypothalamus and represents a positive feedback mechanism.

11. Antidiuretic hormone stimulates the kidney tubules to reabsorb and conserve water; as urine output declines, blood volume and blood pressure rise. ADH is released in response to high solute concentrations in the blood and inhibited by low solute concentrations in the blood. Hyposecretion results in diabetes insipidus.

The Thyroid Gland *(pp. 624–628)*

12. The thyroid gland is located in the anterior throat. Thyroid follicles store thyroglobulin, a colloid from which thyroid hormone is derived.

SP Exercise: Chapter 17, Hormones and their Target Cells.

13. Thyroid hormone (TH) includes thyroxine (T_4) and triiodothyronine (T_3), which increase the rate of cellular metabolism. Consequently, oxygen use and heat production rise.

14. Secretion of thyroid hormone, prompted by TSH, requires reuptake of the stored colloid by the follicle cells and splitting of the hormones from the colloid for release. Rising levels of thyroid hormone feed back to inhibit the pituitary and hypothalamus.

15. Most T_4 is converted to T_3 (the more active form) in the target tissues. These hormones appear to act via a steroidlike mechanism.

16. Hypersecretion of thyroid hormone results most importantly in Graves' disease; hyposecretion causes cretinism in infants and myxedema in adults.

17. Calcitonin, produced by the parafollicular (C) cells of the thyroid gland in response to rising blood calcium levels, depresses blood calcium levels by inhibiting bone matrix resorption and enhancing calcium deposit in bone.

The Parathyroid Glands *(pp. 628–629)*

18. The parathyroid glands, located on the dorsal aspect of the thyroid gland, secrete parathyroid hormone (PTH), which causes an increase in blood calcium levels by targeting bone, the intestine, and the kidneys. PTH is the antagonist of calcitonin.

19. PTH release is triggered by falling blood calcium levels and is inhibited by rising blood calcium levels.

20. Hyperparathyroidism results in hypercalcemia and all its effects and in extreme bone wasting. Hypoparathyroidism leads to hypocalcemia, evidenced by tetany and respiratory paralysis.

The Adrenal (Suprarenal) Glands *(pp. 630–636)*

21. The paired adrenal (suprarenal) glands sit atop the kidneys. Each adrenal gland has two functional portions, the cortex and the medulla.

22. Three groups of steroid hormones are produced by the cortex from cholesterol.

23. Mineralocorticoids (primarily aldosterone) regulate sodium ion reabsorption by the kidneys and thus indirectly regulate levels of other electrolytes that are coupled to sodium transport. Release of aldosterone is stimulated by the renin-angiotensin mechanism, rising potassium ion or falling sodium levels in the blood, and ACTH. Atrial natriuretic peptide inhibits aldosterone release.

24. Glucocorticoids (primarily cortisol) are important metabolic hormones that help the body resist stressors by increasing blood glucose, fatty acid and amino acid levels, and blood pressure. High levels of glucocorticoids depress the immune system and the inflammatory response. ACTH is the major stimulus for glucocorticoid release.

25. Gonadocorticoids (mainly androgens) are produced in small amounts throughout life.

26. Hypoactivity of the adrenal cortex results in Addison's disease. Hypersecretion can result in aldosteronism, Cushing's disease, and masculinization.

27. The adrenal medulla produces catecholamines (epinephrine and norepinephrine) in response to sympathetic nervous system stimulation. Its catecholamines enhance and prolong the fight-or-flight response to short-term stressors. Hypersecretion leads to symptoms typical of sympathetic nervous system overactivity.

SP Exercise: Chapter 17, Stress and the Adrenal Gland.

The Pancreas *(pp. 636–638)*

28. The pancreas, located in the abdomen close to the stomach, is both an exocrine and an endocrine gland. The endocrine portion (pancreatic islets) releases insulin and glucagon (plus pancreatic polypeptide and somatostatin) to the blood.

29. Glucagon, released by alpha (α) cells when blood levels of glucose are low, stimulates the liver to release glucose to the blood.

30. Insulin is released by beta (β) cells when blood levels of glucose (and amino acids) are rising. It increases the rate of glucose uptake and metabolism by most body cells. Hyposecretion of insulin results in diabetes mellitus; cardinal signs are polyuria, polydipsia, and polyphagia.

SP Exercise: Chapter 17, Consequences of Insulin Deficit; Case Study: Diabetes Mellitus.

The Gonads   (pp. 638–639)

31. The ovaries of the female, located in the pelvic cavity, release two main hormones. Secretion of estrogens by the ovarian follicles begins at puberty under the influence of FSH. Estrogens stimulate maturation of the female reproductive system and development of the secondary sex characteristics. Progesterone is released in response to high blood levels of LH. It works with estrogens in establishing the menstrual cycle.

32. The testes of the male begin to produce testosterone at puberty in response to LH (ICSH). Testosterone promotes maturation of the male reproductive organs, development of secondary sex characteristics, and production of sperm by the testes.

The Pineal Gland   (p. 639)

33. The pineal gland is located in the diencephalon. Its primary hormone is melatonin, which influences daily rhythms and may have an antigonadotropic effect in humans.

The Thymus   (p. 640)

34. The thymus gland, located in the upper thorax, declines in size and function with age. Its hormones, thymosins and thymopoietins, are important to the normal development of the immune response.

Other Hormone-Producing Structures (pp. 640–642)

1. Many body organs not normally considered endocrine organs contain isolated cell clusters that secrete hormones. Examples include the heart (atrial natriuretic peptide); gastrointestinal tract organs (gastrin, secretin, and others); the placenta (hormones of pregnancy—estrogen, progesterone, and others); the kidneys (erythropoietin); skin (cholecalciferol); and adipose tissue (leptin).

Developmental Aspects of the Endocrine System (p. 643)

1. Endocrine glands derive from all three germ layers. Those derived from mesoderm produce steroidal hormones; the others produce the amino acid–based hormones.

2. The natural decrease in function of the female's ovaries during late middle age results in menopause.

3. The efficiency of all endocrine glands seems to decrease gradually as aging occurs. This leads to a generalized increase in the incidence of diabetes mellitus and a lower metabolic rate.

REVIEW QUESTIONS

Multiple Choice/Matching

1. The major stimulus for release of parathyroid hormone is (a) hormonal, (b) humoral, (c) neural.

2. The anterior pituitary secretes all but (a) antidiuretic hormone, (b) growth hormone, (c) gonadotropins, (d) TSH.

3. A hormone *not* involved in sugar metabolism is (a) glucagon, (b) cortisone, (c) aldosterone, (d) insulin.

4. Parathyroid hormone (a) increases bone formation and lowers blood calcium levels, (b) increases calcium excretion from the body, (c) decreases calcium absorption from the gut, (d) demineralizes bone and raises blood calcium levels.

5. Choose from the following key to identify the hormones described.

Key: (a) aldosterone   (e) oxytocin
       (b) antidiuretic hormone (f) prolactin
       (c) growth hormone   (g) T_4 and T_3
       (d) luteinizing hormone   (h) TSH

_____ **(1)** important anabolic hormone; many of its effects mediated by IGFs

_____ **(2)** involved in water balance; causes the kidneys to conserve water

_____ **(3)** stimulates milk production

_____ **(4)** tropic hormone that stimulates the gonads to secrete sex hormones

_____ **(5)** increases uterine contractions during birth

_____ **(6)** major metabolic hormone(s) of the body

_____ **(7)** causes reabsorption of sodium ions by the kidneys

_____ **(8)** tropic hormone that stimulates the thyroid gland to secrete thyroid hormone

_____ **(9)** a hormone secreted by the neurohypophysis (two possible choices)

_____ **(10)** the only steroid hormone in the list

6. A hypodermic injection of epinephrine would (a) increase heart rate, increase blood pressure, dilate the bronchi of the lungs, and increase peristalsis, (b) decrease heart rate, decrease blood pressure, constrict the bronchi, and increase peristalsis, (c) decrease heart rate, increase blood pressure, constrict the bronchi, and decrease peristalsis, (d) increase heart rate, increase blood pressure, dilate the bronchi, and decrease peristalsis.

7. Testosterone is to the male as what hormone is to the female? (a) luteinizing hormone, (b) progesterone, (c) estrogen, (d) prolactin.

8. If anterior pituitary secretion is deficient in a growing child, the child will (a) develop acromegaly, (b) become a dwarf but have fairly normal body proportions, (c) mature sexually at an earlier than normal age, (d) be in constant danger of becoming dehydrated.

9. If there is adequate carbohydrate intake, secretion of insulin results in (a) lower blood sugar levels, (b) increased cell utilization of glucose, (c) storage of glycogen, (d) all of these.

10. Hormones (a) are produced by exocrine glands, (b) are carried to all parts of the body in blood, (c) remain at constant concentration in the blood, (d) affect only non-hormone-producing organs.

11. Some hormones act by (a) increasing the synthesis of enzymes, (b) converting an inactive enzyme into an active enzyme, (c) affecting only specific target organs, (d) all of these.

12. Absence of thyroxine would result in (a) increased heart rate and increased force of heart contraction, (b) depression of the CNS and lethargy, (c) exophthalmos, (d) high metabolic rate.

13. Chromaffin cells are found in the (a) parathyroid gland, (b) anterior pituitary gland, (c) adrenal gland, (d) pineal gland.

14. Atrial natriuretic hormone secreted by the heart has exactly the opposite function of this hormone secreted by the zona glomerulosa: (a) antidiuretic hormone, (b) epinephrine, (c) calcitonin, (d) aldosterone, (e) androgens.

Short Answer Essay Questions

15. Define hormone.

16. (a) Describe the body location of each of the following endocrine organs: anterior pituitary, pineal gland, pancreas, ovaries, testes, and adrenal glands. (b) List the hormones produced by each organ.

17. Name two endocrine glands (or regions) that are important in the stress response, and explain why they are important.

18. The anterior pituitary is often referred to as the master endocrine organ, but it, too, has a "master." What controls the release of anterior pituitary hormones?

19. The posterior pituitary is not really an endocrine gland. Why not? What is it?

20. A colloidal, or endemic, goiter is not really the result of malfunction of the thyroid gland. What does cause it?

21. List some problems that elderly people might have as a result of decreasing hormone production.

22. Name a hormone secreted by a muscle cell and two hormones secreted by neurons.

23. How are the hyperglycemia and lipidemia of insulin deficiency linked?

Critical Thinking and Clinical Application Questions

1. Richard Neis had symptoms of excessive secretion of PTH (high blood calcium levels), and his physicians were certain he had a parathyroid gland tumor. Yet when surgery was performed on his neck, the surgeon could not find the parathyroid glands at all. Where should the surgeon look next to find the tumorous parathyroid gland?

2. Mary Morgan has just been brought into the emergency room of City General Hospital. She is perspiring profusely and is breathing rapidly and irregularly. Her breath smells like acetone (sweet and fruity), and her blood sugar tests out at 650 mg/100 ml blood. She is in acidosis. What hormone drug should be administered, and why?

3. Johnny, a five-year-old boy, has been growing by leaps and bounds; his height is 100% above normal for his age. He has been complaining of headaches and vision problems. A CT scan reveals a large pituitary tumor. (a) What hormone is being secreted in excess? (b) What condition will Johnny exhibit if corrective measures are not taken? (c) What is the probable cause of his headaches and visual problems?

4. As Martina sat lazily scanning the newspaper, a headline caught her eye, "Anabolic steroids declared a controlled substance." Hmm, she thought, that's interesting . . . it's about time those drugs got put in the same class with heroin. That night, she awoke from a dream in a cold sweat. In her dream all her male friends were being rounded up by government drug agents and charged with illegal possession of a controlled substance. What is the connection, if any, between the headline and Martina's bizarre dream?

5. Roger Proulx has severe arthritis and has been taking prednisone (a glucocorticoid) at pharmacologic levels for two months. He isn't feeling well, complains of repeated "colds," and is extremely "puffy" (edematous). Explain the reason for these symptoms.

18 BLOOD

Which of the percentages given in this figure represents the measurement called the hematocrit?

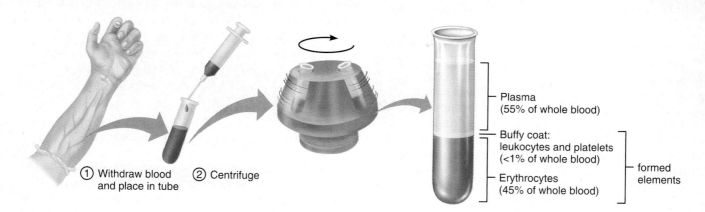

① Withdraw blood and place in tube   ② Centrifuge

Plasma (55% of whole blood)

Buffy coat: leukocytes and platelets (<1% of whole blood)

Erythrocytes (45% of whole blood)

formed elements

FIGURE 18.1 *The major components of whole blood.* When a blood sample is spun in a centrifuge, it separates into layers, the heaviest particles (erythrocytes) moving to the bottom of the test tube. The least dense component is plasma which comprises the top layer.

Blood is the river of life that surges within us, transporting nearly everything that must be carried from one place to another within the body. Long before modern medicine, blood was viewed as magical, an elixir that held the mystical force of life—because when it drained from the body, life departed as well. Today, centuries later, blood retains enormous importance in the practice of medicine. Clinicians examine it more often than any other tissue when trying to determine the cause of disease in their patients.

In this chapter, we describe the composition and functions of this life-sustaining fluid that serves as a transport "vehicle" for the organs of the cardiovascular system. To get started, we need a brief overview of blood circulation, which is initiated by the pumping action of the heart. Blood exits the *heart* via *arteries,* which branch repeatedly until they become tiny *capillaries.* By diffusing across the capillary walls, oxygen and nutrients leave the blood and enter the body tissues, and carbon dioxide and wastes move from the tissues to the bloodstream. As oxygen-deficient blood leaves the capillary beds, it flows into *veins,* which return it to the heart. The returning blood then flows from the heart to the lungs, where it picks up oxygen and then returns to the heart to be pumped throughout the body once again. Now let us look more closely at the nature of blood.

OVERVIEW: COMPOSITION AND FUNCTIONS OF BLOOD

Components

Blood is unique in that it is the only fluid tissue in the body. Although blood appears to be a thick, homogeneous liquid, the microscope reveals that it has both cellular and liquid components. Blood is a specialized type of connective tissue in which living blood cells, the **formed elements,** are suspended in a nonliving fluid matrix called **plasma** (plaz'mah). The collagen and elastic fibers typical of other connective tissues are absent from blood, but dissolved fibrous proteins become visible as fibrin strands when blood clotting occurs.

If a sample of blood is spun in a centrifuge, the heavier formed elements are packed down by centrifugal force and the less dense plasma remains at the top (Figure 18.1). Most of the reddish mass at the bottom of the tube is *erythrocytes* (ĕ-rith'ro-sīts; *erythro* = red), the red blood cells that transport oxygen. A thin, whitish layer called the **buffy coat** is present at the erythrocyte-plasma junction. This layer contains *leukocytes* (*leuko* = white), the white blood cells that act in various ways to protect the body, and *platelets,* cell fragments that help stop bleeding. Erythrocytes normally constitute about 45% of the total volume of a blood sample, a percentage known as the **hematocrit** (he-mat'o-krit; "blood fraction"). Normal hematocrit values vary. In healthy males the norm is 47% ± 5%; in females it is 42% ± 5%. Leukocytes and platelets contribute less than 1% of blood volume. Plasma makes up most of the remaining 55% of whole blood.

45%, the percentage of the erythrocytes.

Physical Characteristics and Volume

Blood is a sticky, opaque fluid with a characteristic metallic taste. As children, we discover its saltiness the first time we stick a cut finger into our mouth. Depending on the amount of oxygen it is carrying, the color of blood varies from scarlet (oxygen-rich) to a dark red (oxygen-poor). Blood is more dense (heavier) than water and about five times more viscous, largely because of its formed elements. Blood is slightly alkaline, with a pH between 7.35 and 7.45. Its temperature (38°C or 100.4°F) is always slightly higher than body temperature.

Blood accounts for approximately 8% of body weight. Its average volume in healthy adult males is 5–6 L (about 1.5 gallons), somewhat greater than in healthy adult females (4–5 L).

Functions

Blood performs a number of functions, all concerned in one way or another with substance distribution, regulating blood levels of particular substances, or body protection. These functions overlap and interact to maintain the constancy of our internal environment.

Distribution

The *distribution functions* of blood include:

■ Delivering oxygen from the lungs and nutrients from the digestive tract to all body cells.

■ Transporting metabolic waste products from cells to elimination sites (to the lungs for elimination of carbon dioxide, and to the kidneys for disposal of nitrogenous wastes in urine).

■ Transporting hormones from the endocrine organs to their target organs.

Regulation

The *regulatory functions* of blood include:

■ Maintaining appropriate body temperature by absorbing and distributing heat throughout the body and to the skin surface to encourage heat loss.

■ Maintaining normal pH in body tissues. Many blood proteins and other bloodborne solutes act as buffers to prevent excessive or abrupt changes in blood pH, which could jeopardize normal cellular activities. Additionally, blood acts as the reservoir for the body's "alkaline reserve" of bicarbonate atoms.

■ Maintaining adequate fluid volume in the circulatory system. Salts (sodium chloride and others) and blood proteins act to prevent excessive fluid loss from the bloodstream into the tissue spaces. As a result, the fluid volume in the blood vessels remains ample to support efficient blood circulation to all parts of the body.

Protection

The *protective functions* of blood include:

■ Preventing blood loss. When a blood vessel is damaged, platelets and plasma proteins initiate clot formation, halting blood loss.

■ Preventing infection. Drifting along in blood are antibodies, complement proteins, and white blood cells, all of which help defend the body against foreign invaders such as bacteria and viruses.

BLOOD PLASMA

Blood plasma is a straw-colored, sticky fluid (see Figure 18.1). Although it is mostly water (about 90%), plasma contains over 100 different dissolved solutes. These include nutrients, gases, hormones, various wastes and products of cell activity, ions, and proteins. Table 18.1 summarizes the major plasma components.

Plasma proteins, accounting for about 8% by weight of plasma volume, are the most abundant plasma solutes. Except for bloodborne hormones and gamma globulins, most plasma proteins are produced by the liver. Although plasma proteins serve a variety of functions, note that they are *not* taken up by cells to be used as fuels or metabolic nutrients as are most other plasma solutes, such as glucose, fatty acids, and oxygen. **Albumin** (al-bu'min) accounts for some 60% of plasma protein. It acts as a carrier to shuttle certain molecules through the circulation, is an important blood buffer, and is the major blood protein contributing to the plasma osmotic pressure (the pressure that helps to keep water in the bloodstream). (Sodium ions are the other major solute contributing to blood osmotic pressure.)

The makeup of plasma varies continuously as cells remove or add substances to the blood. However, assuming a healthy diet, plasma composition is kept relatively constant by various homeostatic mechanisms. For example, when blood protein levels drop undesirably, the liver makes more proteins; and when the blood starts to become too acidic (acidosis), both the respiratory system and the kidneys are called into action to restore plasma's normal, slightly alkaline pH. Body organs make dozens of adjustments, day in and day out, to maintain the many plasma solutes at life-sustaining levels. In addition to transporting various solutes around the body, plasma distributes heat (a by-product of cellular metabolism) throughout the body.

FORMED ELEMENTS

The **formed elements of blood** are *erythrocytes, leukocytes,* and *platelets.* These elements have some unusual features. (1) Two of the three are not even true cells: Erythrocytes have no nuclei or organelles,

TABLE 18.1	Composition of Plasma
Constituent	*Description and importance*
Water	90% of plasma volume; dissolving and suspending medium for solutes of blood; absorbs heat
Solutes Proteins	8% (by weight) of plasma volume
■ Albumin	60% of plasma proteins; produced by liver; exerts osmotic pressure to maintain water balance between blood and tissues
■ Globulins	36% of plasma proteins
alpha, beta	Produced by liver; are transport proteins that bind to lipids, metal ions, and fat-soluble vitamins
gamma	Antibodies released primarily by plasma cells during immune response
■ Clotting proteins	4% of plasma proteins; include fibrinogen and prothrombin produced by liver; act in blood clotting
■ Others	Metabolic enzymes, antibacterial proteins (such as complement), hormones
Nonprotein nitrogenous substances	By-products of cellular metabolism, such as lactic acid, urea, uric acid, creatinine, and ammonium salts
Nutrients (organic)	Materials absorbed from digestive tract and transported for use throughout body; include glucose and other simple carbohydrates, amino acids (digestion products of proteins), fatty acids, glycerol and triglycerides (fat products), cholesterol, and vitamins
Electrolytes	Cations include sodium, potassium, calcium, magnesium; anions include chloride, phosphate, sulfate, and bicarbonate; help to maintain plasma osmotic pressure and normal blood pH
Respiratory gases	Oxygen and carbon dioxide; some dissolved oxygen (most bound to hemoglobin inside RBCs); carbon dioxide transported bound to hemoglobin in RBCs and as bicarbonate ion dissolved in plasma

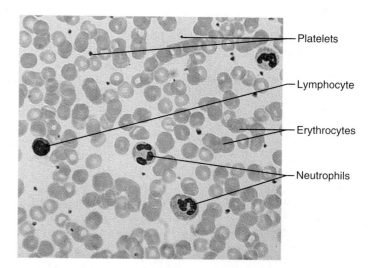

FIGURE 18.2 *Photomicrograph of a human blood smear stained with Wright's stain.*

structural and functional characteristics of the various types of formed elements.

Erythrocytes

Structural Characteristics

Erythrocytes or **red blood cells (RBCs)** are small cells, about 7.5 μm in diameter. Shaped like biconcave discs—flattened discs with depressed centers (Figure 18.3)—their thin centers appear lighter in color than their edges. Consequently, erythrocytes look like miniature doughnuts when viewed with a microscope. Mature erythrocytes are bound by a plasma membrane, but lack a nucleus (are *anucleate*) and have essentially no organelles. In fact, they are little more than "bags" of *hemoglobin (Hb)*, the RBC protein that functions in gas transport. Other proteins are present, but they function mainly to maintain the plasma membrane or promote changes in RBC shape. For example, the biconcave shape of an erythrocyte is maintained by a network of proteins, especially one called *spectrin*, attached to the cytoplasmic face of its plasma membrane. Because the spectrin net is deformable, it gives erythrocytes enough flexibility to change shape as necessary—to twist, turn, and become cup-shaped as they are carried passively along in the bloodstream and through capillaries with diameters smaller than themselves—and then to resume their biconcave shape.

The erythrocyte is a superb example of complementarity of structure and function. It picks up oxygen in the capillary beds of the lungs and releases it to tissue cells across other capillaries throughout the body. It also transports some 20% of the carbon dioxide released by tissue cells in the opposite direction—back to the lungs. Each of the erythrocyte's

and platelets are just cell fragments. Only leukocytes are complete cells. (2) Most of the formed elements survive in the bloodstream for only a few days. (3) Most blood cells do not divide. Instead, they are continuously renewed by division of cells in bone marrow, where they originate.

If you examine a stained smear of human blood under the light microscope, you will see disc-shaped red blood cells, a variety of gaudily stained spherical white blood cells, and some scattered platelets that look like debris (Figure 18.2). Erythrocytes vastly outnumber the other types of formed elements. Table 18.2 on p. 663 summarizes the important

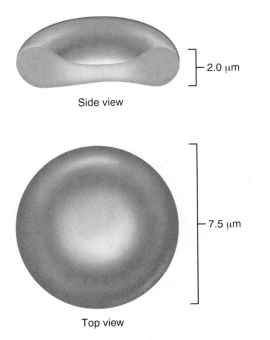

Side view

2.0 μm

Top view

7.5 μm

FIGURE 18.3 *Structure of erythrocytes.* An erythrocyte drawn in a cut side view and in a superficial surface view. Notice the distinctive biconcave shape.

structural characteristics contributes to its gas transport functions: (1) Its small size and biconcave shape provide a huge surface area relative to volume (about 30% more surface area than comparable spherical cells). Since no point within its cytoplasm is far from the surface, the biconcave disc shape is ideally suited for gas exchange. (2) Discounting its water content, an erythrocyte is over 97% hemoglobin, the molecule that binds to and transports respiratory gases. (3) Because erythrocytes lack mitochondria and generate ATP by anaerobic mechanisms, they do not consume any of the oxygen they are transporting, making them very efficient oxygen transporters indeed.

Erythrocytes are the major factor contributing to blood viscosity. Women typically have a lower red blood cell count than men (4.3–5.2 million cells per cubic millimeter [mm^3] of blood versus 5.1–5.8 million cells per cubic millimeter respectively). When the number of red blood cells increases beyond the normal range, blood viscosity rises and blood flows more slowly. Similarly, as the number of red blood cells drops below the lower end of the range, the blood thins and flows more rapidly.

Function

Erythrocytes are completely dedicated to their job of respiratory gas (oxygen and carbon dioxide) transport. Their hemoglobin binds easily and reversibly with oxygen, and most oxygen carried in blood is bound to hemoglobin. Normal values for hemoglobin are 14 to 20 grams per 100 milliliters of blood (g/100 ml) in infants, 13 to 18 g/100 ml in adult males, and 12 to 16 g/100 ml in adult females.

Hemoglobin is made up of the protein **globin** bound to the red **heme** pigment. Globin consists of four polypeptide chains—two alpha (α) and two beta (β)—each bound to a ringlike heme group. Each heme group bears an atom of iron set like a jewel in its center (Figure 18.4). Since each iron atom can combine reversibly with one molecule of oxygen, a hemoglobin molecule can transport four molecules of oxygen. A single red blood cell contains about 250 million hemoglobin molecules, so each of these tiny cells can carry about 1 billion molecules of oxygen!

The fact that hemoglobin is contained in erythrocytes, rather than existing free in plasma, prevents it (1) from breaking into fragments that would leak out of the bloodstream (through the rather porous capillary membranes) and (2) from contributing to blood viscosity and osmotic pressure.

Oxygen loading occurs in the lungs and the direction of transport is from lungs to tissue cells. As oxygen-deficient blood moves through the lungs, oxygen diffuses from the air sacs of the lungs into the blood and then into the erythrocytes, where it binds to hemoglobin. When oxygen binds to iron, the hemoglobin, now called **oxyhemoglobin,** assumes a new three-dimensional shape and becomes bright red. In the tissues, the process is reversed. Oxygen detaches from iron, hemoglobin resumes its former shape, and the resulting **deoxyhemoglobin,** or *reduced hemoglobin,* becomes dark red. The released oxygen diffuses from the blood into the tissue fluid and then into the tissue cells.

About 20% of the carbon dioxide transported in the blood combines with hemoglobin, but it binds to amino acids of globin rather than with the heme group. This formation of **carbaminohemoglobin** (kar-bam″ĭ-no-he″mo-glo′bin) occurs more readily when hemoglobin is in the reduced state (dissociated from oxygen). Carbon dioxide loading occurs in the tissues and the direction of transport is from tissues to lungs, where carbon dioxide is eliminated from the body. The mechanisms of loading and unloading these respiratory gases are described in Chapter 23.

Production of Erythrocytes

Blood cell formation is referred to as **hematopoiesis** (hem″ah-to-poi-e′sis), or **hemopoiesis** (hemo, hemato = blood; poiesis = to make). This process occurs in the **red bone marrow,** which is composed largely of a soft network of reticular connective tissue

? *How many molecules of oxygen can a hemoglobin molecule transport?*

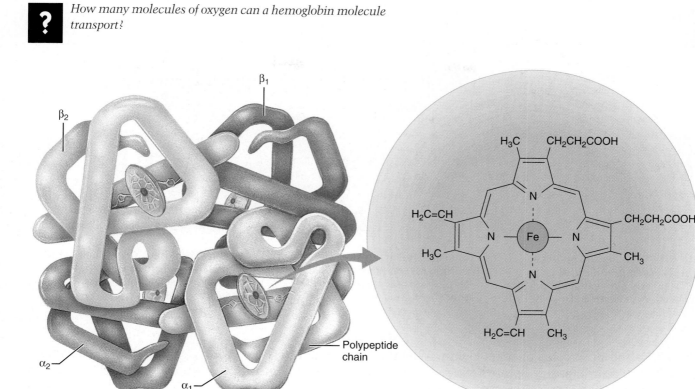

(a) Hemoglobin

(b) Iron-containing heme group

FIGURE 18.4   *Structure of hemoglobin.* **(a)** The intact hemoglobin molecule is composed of the protein globin bound to the iron-containing heme pigments. Each globin molecule has four polypeptide chains: two alpha (α) chains and two beta (β) chains. Each chain is complexed with a heme group, shown as a circular green structure with iron at its center. **(b)** Structure of a single heme group.

that borders on wide blood capillaries called *blood sinusoids.* Within this network are immature blood cells, macrophages, fat cells, and *reticular cells* (the fibroblasts that secrete the fibers). In adults, red marrow is found chiefly in the bones of the axial skeleton and girdles, and in the proximal epiphyses of the humerus and femur. Each type of blood cell is produced in different numbers in response to changing body needs and different regulatory factors. As they mature, they migrate through the thin walls of the sinusoids to enter the bloodstream. On average, the marrow turns out an ounce of new blood containing some 100 billion new cells each and every day.

Although the various formed elements have different functions, there are important similarities in their life histories. All of the formed elements arise from the same type of *stem cell,* the pleuripotential

hematopoietic stem cell, or **hemocytoblast** (*cyte* = cell, *blast* = bud), which resides in the red bone marrow. However, their maturation pathways differ; and once a cell is *committed* to a specific blood cell pathway, it cannot change. This commitment is signaled by the appearance of membrane surface receptors that respond to specific hormones or growth factors, which in turn "push" the cell toward further specialization.

Erythrocyte production, or **erythropoiesis** (ĕ-rith″ro-poi-e′sis) begins when a hemocytoblast descendant called a **myeloid stem cell** is transformed into a **proerythroblast** (Figure 18.5). Proerythroblasts, in turn, give rise to the **early (basophilic) erythroblasts** that produce huge numbers of ribosomes. During these first two phases, the cells divide many times. Hemoglobin synthesis and iron accumulation occur as the early erythroblast is transformed into a **late erythroblast** and then a **normoblast.** The "color" of the cell cytoplasm changes as the blue-staining ribosomes become

One to four O₂ per hemoglobin molecule.

What does the term "committed cell" refer to?

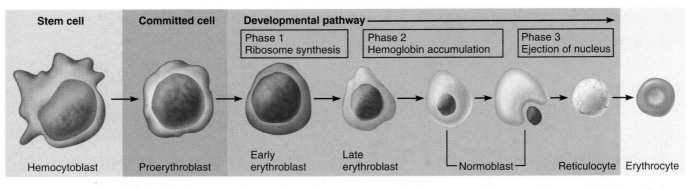

Stem cell — Hemocytoblast
Committed cell — Proerythroblast
Developmental pathway
Phase 1 Ribosome synthesis — Early erythroblast
Phase 2 Hemoglobin accumulation — Late erythroblast
Phase 3 Ejection of nucleus — Normoblast — Reticulocyte — Erythrocyte

FIGURE 18.5   *Erythropoiesis: genesis of red blood cells.*
Erythropoiesis is a sequence involving proliferation and differentiation of committed red marrow cells through the erythroblast and normoblast stages to the reticulocytes that are released into the bloodstream, and finally become erythrocytes. (The myeloid stem cell, the phase intermediate between the hemocytoblast and the proerythroblast, is not illustrated.)

masked by the pink color of hemoglobin. When a normoblast accumulates a hemoglobin concentration of about 34%, most of its organelles are ejected. Additionally, its nuclear functions end and its nucleus degenerates and is pinched off, causing the cell to collapse inward and assume the biconcave shape. The result is the *reticulocyte*. **Reticulocytes** (essentially young erythrocytes) are so named because they still contain a scant network of clumped ribosomes and rough endoplasmic reticulum. The entire process from hemocytoblast to reticulocyte takes three to five days. The reticulocytes, filled almost to bursting with hemoglobin, enter the bloodstream to begin their task of oxygen transport. Usually they become fully mature erythrocytes within two days of release as their ribosomes are degraded by intracellular enzymes. Reticulocytes account for 1–2% of all erythrocytes in the blood of healthy people. **Reticulocyte counts** are used clinically as a rough index of the *rate* of RBC formation—reticulocyte counts below or above this percentage range indicate abnormal rates of erythrocyte formation.

Regulation and Requirements for Erythropoiesis

The number of circulating erythrocytes in a given individual is remarkably constant and reflects a balance between red blood cell production and destruction. This balance is extremely important because having too few erythrocytes leads to tissue hypoxia (oxygen deprivation), whereas having too many makes the blood undesirably viscous. To ensure that the number of erythrocytes in the blood remains within the homeostatic range, new cells are produced at the incredibly rapid rate of more than 2 million per second in healthy people. This process is controlled hormonally and depends on adequate supplies of iron, amino acids, and certain B vitamins.

Hormonal Controls The direct stimulus for erythrocyte formation is provided by **erythropoietin (EPO)**, a glycoprotein hormone. Normally, a small amount of erythropoietin circulates in the blood at all times and sustains red blood cell production at a basal rate (Figure 18.6). Although the liver produces some, the kidneys play the major role in erythropoietin production. When certain kidney cells become hypoxic (i.e., have inadequate oxygen), they accelerate their release of erythropoietin. The drop in normal blood oxygen levels that triggers erythropoietin formation can result from:

1. Reduced numbers of red blood cells due to hemorrhage or excess RBC destruction

2. Reduced availability of oxygen, as might occur at high altitudes or during pneumonia

3. Increased tissue demands for oxygen (common in those who engage in aerobic exercise)

Conversely, too many erythrocytes or excessive oxygen in the bloodstream depresses erythropoietin production. The easily missed point to keep in mind is that the number of erythrocytes in the blood does *not* control the rate of erythropoiesis. Control is

---

A committed cell is a blood cell precursor whose path of specialization is determined. For example, the proerythroblast can only become an erythrocyte, not one of the leukocytes.

How does blood doping, practiced by some athletes (see p. 660), affect the negative feedback cycle outlined here?

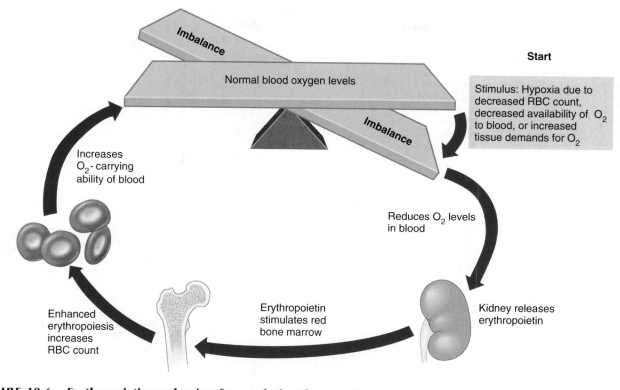

FIGURE 18.6   *Erythropoietin mechanism for regulating the rate of erythropoiesis*. Notice that increased erythropoietin release, which stimulates erythropoiesis in bone marrow, occurs when oxygen levels in the blood become inadequate to support normal cellular activity, whatever the cause.

based on their ability to transport enough oxygen to meet tissue demands.

Bloodborne erythropoietin stimulates red marrow cells that *are already committed* to becoming erythrocytes, causing them to mature more rapidly. One to two days after erythropoietin levels rise in the blood, a marked increase in the rate of reticulocyte release (and, hence, reticulocyte count) occurs. Notice that hypoxia (oxygen deficit) does not activate the bone marrow directly. Instead it stimulates the kidneys, which in turn provide the hormonal stimulus that activates the bone marrow.

Homeostatic Imbalance

Renal dialysis patients whose kidneys have failed do not produce enough erythropoietin to support nor-

mal erythropoiesis. Consequently, they routinely have red blood cell counts less than half that of healthy individuals. Recombinant (genetically engineered) erythropoietin has helped such patients immeasurably. Although created to help those in renal failure, recombinant EPO has also become a substance of abuse in athletes—particularly in professional bike racers and marathon runners seeking increased stamina and performance. However, the consequences in some cases can be deadly. By injecting EPO, healthy athletes increase their normal RBC volume from 45% to as much as 65%. Then, with the dehydration that occurs in a long race, the blood concentrates even further, becoming a thick, sticky "sludge" that can cause clotting, stroke, and even heart failure. ■

The male sex hormone *testosterone* also enhances erythropoietin production by the kidneys. Because female sex hormones do not have similar stimulatory effects, testosterone may be at least partially responsible for the higher RBC counts and hemoglobin levels seen in males. Also, a wide variety of chemicals released by leukocytes, platelets, and even reticular cells stimulates bursts of RBC production.

Blood doping, which increases the number of RBCs in the circulation and thus increases the amount of oxygen being transported, would inhibit erythropoietin release by ending the stimulus for its formation.

? *How would you expect blood levels of bilirubin to change in a person that has severe liver disease?*

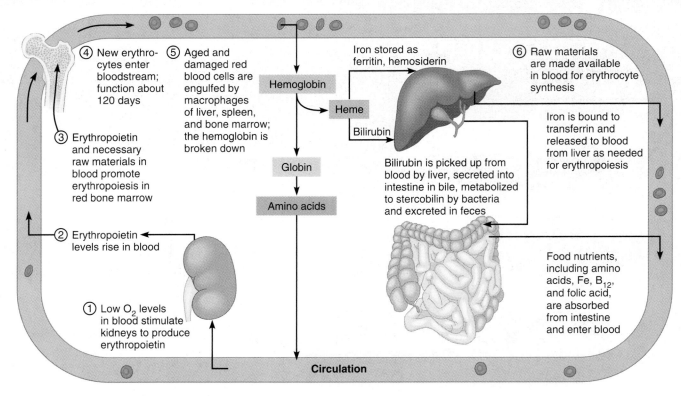

FIGURE 18.7   *Life cycle of red blood cells.*

Dietary Requirements: Iron and B-Complex Vitamins The raw materials required for erythropoiesis include the usual nutrients and structural materials—proteins, lipids, and carbohydrates. Iron is essential for hemoglobin synthesis (Figure 18.7). Iron is available from the diet, and its absorption into the bloodstream is precisely controlled by intestinal cells in response to changing body stores of iron.

Approximately 65% of the body's iron supply (about 4000 mg) is in hemoglobin. Most of the remainder is stored in the liver, spleen, and (to a much lesser extent) bone marrow. Because free iron is toxic, iron is stored inside cells as protein-iron complexes such as **ferritin** (fer'ĭ-tin) and **hemosiderin** (he"mo-sid'er-in). In blood, iron is transported loosely bound to a transport protein called **transferrin**. Developing erythrocytes take up iron as needed to form functional hemoglobin molecules. Small amounts of iron are lost each day in feces, urine, and perspiration. The average daily loss of iron is 1.7 mg

in women and 0.9 mg in men. In women, the menstrual flow accounts for the additional losses.

Two B-complex vitamins—vitamin B_{12} and folic acid—are necessary for DNA synthesis. Thus, even slight deficits jeopardize rapidly dividing cell populations, such as developing erythrocytes.

Fate and Destruction of Erythrocytes

The anucleate condition of erythrocytes carries with it some important limitations. Red blood cells are unable to synthesize proteins, to grow, or to divide. Erythrocytes become "old" as they lose their flexibility and become increasingly rigid and fragile, and their contained hemoglobin begins to degenerate. Red blood cells have a useful life span of 100 to 120 days, after which they become trapped and fragment in smaller circulatory channels, particularly in those of the spleen. For this reason, the spleen is sometimes called the "red blood cell graveyard."

Dying erythrocytes are engulfed and destroyed by macrophages. The heme of their hemoglobin is split off from globin. Its core of iron is salvaged, bound to protein (as ferritin or hemosiderin), and stored for reuse. The balance of the heme group is

It would increase because the liver's processing functions are impaired.

degraded to **bilirubin** (bil'i-roo'bin), a yellow pigment that is released to the blood and binds to albumin for transport (Figure 18.7). Bilirubin is picked up by liver cells, which in turn secrete it (in bile) into the intestine, where it is metabolized to *urobilinogen*. Most of this degraded pigment leaves the body in feces, as a brown pigment called *stercobilin*. The protein (globin) part of hemoglobin is metabolized or broken down to amino acids, which are released to the circulation (see Figure 18.7).

Erythrocyte Disorders

Most erythrocyte disorders can be classified as anemias or polycythemias. The many varieties and causes of these conditions are described next.

Anemias Anemia (ah-ne'me-ah; "lacking blood") is a condition in which the blood has an abnormally low oxygen-carrying capacity. It is a *symptom* of some disorder rather than a disease in and of itself. Its hallmark is blood oxygen levels that are inadequate to support normal metabolism. Anemic individuals are fatigued, often pale, short of breath, and chilly. Common causes of anemia include:

1. An insufficient number of red blood cells. Conditions that reduce the red blood cell count include blood loss, excessive destruction of RBCs, and bone marrow failure.

Hemorrhagic (hem"o-raj'ik) *anemias* result from blood loss. In acute hemorrhagic anemia, blood loss is rapid (as might follow a severe stab wound); it is treated by blood replacement. Slight but persistent blood loss (due to hemorrhoids or an undiagnosed bleeding ulcer, for example) causes chronic hemorrhagic anemia. Once the primary problem is resolved, normal erythropoietic mechanisms replace the deficient blood cells.

In *hemolytic* (he"mo-lit'ik) *anemias*, erythrocytes rupture, or lyse, prematurely. Hemoglobin abnormalities, transfusion of mismatched blood, and certain bacterial and parasitic infections are all possible causes.

Aplastic anemia results from destruction or inhibition of the red marrow by certain bacterial toxins, drugs, and ionizing radiation. Because marrow destruction impairs formation of *all* formed elements, anemia is just one of its signs. Defects in blood clotting and immunity are also present. Blood transfusions provide a stopgap treatment until bone marrow transplants or transfusion of umbilical blood, which contains stem cells, can be done.

2. Decreased hemoglobin content. When hemoglobin molecules are normal, but erythrocytes contain fewer than the usual number, a nutritional anemia is always suspected.

Iron-deficiency anemia is generally a secondary result of hemorrhagic anemias, but it also results from inadequate intake of iron-containing foods and impaired iron absorption. The erythrocytes produced, called **microcytes,** are small and pale. The obvious treatment is iron supplements, but if chronic hemorrhage is the cause, blood transfusions may also be needed.

When athletes exercise vigorously, their blood volume expands and can increase by as much as 15%. Since this in effect dilutes the blood components, a test for the iron content of the blood at such times would indicate iron-deficiency anemia. However, this iron deficiency illusion, called **athlete's anemia,** is quickly reversed as blood components return to physiological levels within a day or so after resuming a normal level of activity.

Pernicious anemia is due to a deficiency of vitamin B_{12}. Because meats, poultry, and fish provide ample amounts of the vitamin, diet is rarely the problem except for strict vegetarians. A substance called **intrinsic factor,** produced by the stomach mucosa, must be present for vitamin B_{12} to be absorbed by intestinal cells. In most cases of pernicious anemia, intrinsic factor is deficient. Consequently, the developing erythrocytes grow but do not divide, and large, pale cells called **macrocytes** result. Because the stomach mucosa atrophies with age, the elderly are particularly at risk for this type of anemia. Treatment involves regular intramuscular injections of vitamin B_{12}.

3. Abnormal hemoglobin. Production of abnormal hemoglobin usually has a genetic basis. Two such examples, thalassemia and sickle-cell anemia, can be serious, incurable, and sometimes fatal diseases. In both diseases the globin part of hemoglobin is abnormal and the erythrocytes produced are fragile and rupture prematurely.

Thalassemias (thal"ah-se'me-ahs; "sea blood") are typically seen in people of Mediterranean ancestry, such as Greeks and Italians. One of the globin chains is absent or faulty, and the erythrocytes are thin, delicate, and deficient in hemoglobin. The red blood cell count is generally less than 2 million cells per cubic millimeter. In most cases the reduced RBC count is not a major problem and no treatment is required. Severe cases require monthly blood transfusions.

In **sickle-cell anemia,** the havoc caused by the abnormal hemoglobin formed, *hemoglobin S (HbS)*, results from a change in just *one* of the 287 amino acids in a beta chain of the globin molecule! This alteration causes the beta chains to link together to form stiff rods under low-oxygen conditions, and as a result, hemoglobin S becomes spiky and sharp. This, in turn, causes the red blood cells to become

FIGURE 18.8 *Comparison of*
(a) normal erythrocytes to
(b) sickled erythrocytes (31,500×).

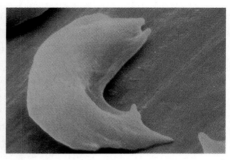

(a)   **(b)**

crescent shaped (Figure 18.8) when they unload oxygen molecules or when the oxygen content of the blood is lower than normal, as during vigorous exercise and other activities that increase metabolic rate. The stiffened and deformed erythrocytes rupture easily and tend to dam up in small blood vessels. These events interfere with oxygen delivery, leaving the victims gasping for air and in extreme pain. Bone and chest pain are particularly severe; infection and stroke are common sequels. The standard treatment for an acute sickle-cell crisis is blood transfusion.

Sickle-cell anemia occurs chiefly in black people who live in the malaria belt of Africa and among their descendants. It strikes nearly 1 of every 400 black newborns in the United States. Apparently, the gene that causes sickling of red blood cells also causes erythrocytes infected by the malaria-causing parasite to stick to the capillary walls. Then, during oxygen deficit, these abnormal erythrocytes lose potassium, an essential ingredient for survival of the parasite. Thus, individuals with the sickle-cell gene have a better chance of surviving in regions where malaria is prevalent. Only those carrying two copies of the defective gene have sickle-cell anemia. Those carrying only one sickling gene have sickle-cell trait; they generally do not display the symptoms, but can transmit the gene to their offspring.

Since fetal hemoglobin (HbF) does not "sickle," even in those destined to make HbS and have sickle-cell anemia, investigators have been looking for ways to switch the fetal hemoglobin gene back on. *Hydroxyurea*, a drug used to treat chronic leukemia, appears to do just that. This drug reduces the excruciating pain and overall severity and complications of sickle-cell anemia by 50%—so dramatically that the National Institutes of Health ended its clinical drug trials (begun in 1992) 4 months early. Additionally, several vaccines are now being clinically tested for their effectiveness against this scourge.

Polycythemia Polycythemia (pol"e-si-the'me-ah; "many blood cells") is an abnormal excess of erythrocytes that increases blood viscosity, causing it to sludge, or flow sluggishly. *Polycythemia vera*, most often a result of bone marrow cancer, is a serious condition characterized by dizziness and an exceptionally high RBC count (8–11 million cells per cubic millimeter). The hematocrit may be as high as 80% and blood volume may double, causing the vascular system to become engorged with blood and severely impairing circulation.

Secondary polycythemias result when less oxygen is available or erythropoietin production increases. The secondary polycythemia that appears in individuals living at high altitudes is a normal physiological response that compensates for the reduced atmospheric pressure and lower oxygen content of the air in such areas. RBC counts of 6–8 million per cubic millimeter are common in such people. Severe polycythemia is treated by blood dilution, that is, removing some blood and replacing it with saline.

Blood doping, practiced by some athletes competing in aerobic events, is artificially induced polycythemia. Some of the athlete's red blood cells are drawn off and then reinjected a few days before the event. Because the erythropoietin mechanism is triggered shortly after the blood is removed, the erythrocytes are quickly replaced. Then, when the stored blood is reinfused, a temporary polycythemia results. Since red blood cells carry oxygen, the additional infusion should translate into increased oxygen-carrying capacity due to a higher hematocrit, and hence greater endurance and speed. (Similar results have been produced by injecting recombinant EPO, as mentioned previously.) Other than problems that might derive from increased blood viscosity, such as temporary high blood pressure or reduced perfusion (blood delivery) to body tissues, blood doping seems to work. However, the practice is considered unethical and has been banned from the Olympic Games.

Leukocytes

General Structural and Functional Characteristics

Leukocytes (*leuko* = white), or **white blood cells (WBCs)**, are the only formed elements that are complete cells, with nuclei and the usual organelles. Leukocytes, accounting for less than 1% of total blood volume, are far less numerous than red blood cells. On average, there are 4000–11,000 WBCs per cubic millimeter of blood.

Leukocytes are crucial to our defense against disease. They form a mobile army that helps protect the body from damage by bacteria, viruses, parasites, toxins, and tumor cells. As such, they have some very special functional characteristics. Red blood cells are confined to the bloodstream, and they carry out their functions in the blood. But white blood cells are able to slip out of the capillary blood vessels—a process called **diapedesis** (di"ah-pĕ-de'sis; "leaping across")—and the circulatory system is simply their means of transport to areas of the body (largely loose connective tissues or lymphoid tissues) where they are needed to mount inflammatory or immune responses. As explained in more detail in Chapter 22, the signals that prompt WBCs to leave the bloodstream at specific locations are adhesion molecules (*selectins*) displayed by endothelial cells forming the capillary walls at sites of inflammation. Once out of the bloodstream, leukocytes move through the tissue spaces by **amoeboid motion** (they form flowing cytoplasmic extensions that move them along). By following the chemical trail of molecules released by damaged cells or other leukocytes, a phenomenon called **positive chemotaxis**, they can pinpoint and gather in large numbers at areas of tissue damage and infection to destroy foreign substances or dead cells.

Whenever white blood cells are mobilized for action, the body speeds up their production and twice the normal number may appear in the blood within a few hours. A *white blood cell count* of over 11,000 cells per cubic millimeter is called **leukocytosis.** This condition is a normal homeostatic response to bacterial or viral invasion of the body.

Leukocytes are grouped into two major categories on the basis of structural and chemical characteristics: *Granulocytes* contain specialized membrane-bound cytoplasmic granules; *agranulocytes* lack obvious granules. General information about the various leukocytes is provided next. Details of their size, concentration in blood, and percentage in the white blood cell population appear in Figure 18.9 and Table 18.2 (p. 663).

Students are often asked to list the leukocytes in order from most abundant to least abundant. The following phrase may help you with this task: Never

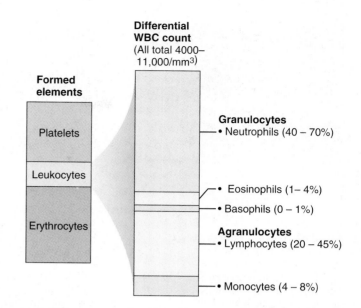

FIGURE 18.9   *Types of leukocytes and their relative percentages of the WBC population in normal blood.* (Note that the relationship of the formed elements in the left column is not shown to scale because erythrocytes comprise nearly 98% of the formed elements whereas leukocytes and platelets together account for the remaining 2+%.)

let **m**onkeys **e**at **b**ananas (neutrophils, lymphocytes, monocytes, eosinophils, basophils).

Granulocytes

The **granulocytes** (gran'u-lo-sīts), which include neutrophils, basophils, and eosinophils, are all roughly spherical in shape. They are larger and much shorter lived (in most cases) than erythrocytes. They characteristically have lobed nuclei (rounded nuclear masses connected by thinner strands of nuclear material), and their membrane-bound cytoplasmic granules stain quite specifically with Wright's stain. Functionally, all the granulocytes are phagocytes.

Neutrophils Neutrophils (nu'tro-filz), the most numerous of the white blood cells, typically account for half or more of the WBC population. Neutrophils are about twice as large as erythrocytes.

The neutrophil cytoplasm stains pale lilac and contains very fine granules (of two varieties) that are difficult to see (see Table 18.2 and Figure 18.10a). They are called neutrophils (literally, "neutral-loving") because their granules take up both *basic* (blue) and *acidic* (red) *dyes.* Together, the two types of granules give the cytoplasm a lilac color. Some of these granules contain peroxidases and other hydrolytic enzymes, and are regarded as lysosomes. Others, especially the smaller granules, contain a potent "brew" of antibiotic-like proteins, called **defensins.** Neutrophil nuclei consist of three to six

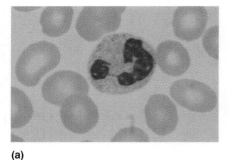

(a)

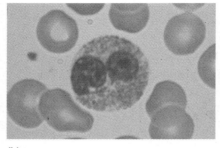

(b)

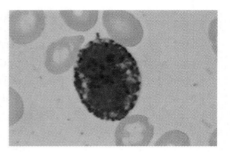

(c)

FIGURE 18.10 *Leukocytes.*
(a) Neutrophil. (b) Eosinophil.
(c) Basophil. (d) Small lymphocyte.
(e) Monocyte. In each case the
leukocytes are surrounded by
erythrocytes. (All 3350×.)

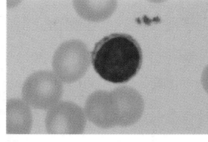

(d)

(e)

lobes. Because of this nuclear variability, they are often called **polymorphonuclear** ("many shapes of the nucleus") **leukocytes (PMNs)** or simply **polys.** (However, some authorities use this term to refer to all granulocytes.)

Neutrophils are chemically attracted to sites of inflammation and are active phagocytes. They are especially partial to bacteria and some fungi, which they can ingest and destroy. Bacterial killing is promoted by a process called a **respiratory burst.** In the respiratory burst, oxygen is actively metabolized to produce potent germ-killer oxidizing substances such as bleach and hydrogen peroxide, and defensin-mediated lysis occurs. It appears that when the granules containing defensins are merged with a microbe-containing phagosome, the defensins form peptide "spears" that pierce holes in the membrane of the ingested "foe." Neutrophils are our body's bacteria slayers, and their numbers increase explosively during acute bacterial infections such as meningitis and appendicitis.

Eosinophils Eosinophils (e"o-sin'o-filz) account for 1–4% of all leukocytes and are approximately the size of neutrophils. Their deep red nucleus usually resembles an old-fashioned telephone receiver; that is, it has two lobes connected by a broad band of nuclear material (see Table 18.2 and Figure 18.10b).

Large, coarse granules that stain from brick red to crimson with acid (eosin) dyes pack the cytoplasm. These granules are lysosome-like and filled with a unique variety of digestive enzymes. However, unlike typical lysosomes, they lack enzymes that specifically digest bacteria.

Eosinophils perform several functions, but perhaps their most important role is to lead the counterattack against parasitic worms, such as flatworms (tapeworms and flukes) and roundworms (pinworms and hookworms) that are too large to be phagocytized. These worms are ingested in food (especially raw fish [sushi]) or invade the body via the skin and then typically burrow into the intestinal or respiratory mucosae. Eosinophils tend to reside in the loose connective tissues at the same body sites, and when a parasitic worm "prey" is encountered, they gather around and release the enzymes from their cytoplasmic granules onto the parasite's surface, thus digesting it away. Eosinophils also lessen the severity of allergies by phagocytizing immune (antigen-antibody) complexes involved in allergy attacks and inactivating certain inflammatory chemicals released during allergic reactions.

Basophils Basophils are the rarest white blood cells, averaging only 0.5% of the leukocyte population (1 in 200). Basophils are the same size as or

TABLE 18.2	**Summary of Formed Elements of the Blood**				
Cell type	Illustration	Description*	Number of cells/mm3 (μl) of blood	Duration of development (D) and life span (LS)	Function
Erythrocytes (red blood cells, RBCs)		Biconcave, anucleate disc; salmon-colored; diameter 7–8 μm	4–6 million	D: 5–7 days LS: 100–120 days	Transport oxygen and carbon dioxide
Leukocytes (white blood cells, WBCs)		Spherical, nucleated cells	4,800–10,800		
Granulocytes ■ Neutrophil		Nucleus multilobed; inconspicuous cytoplasmic granules; diameter 10–14 μm	3000–7000	D: 6–9 days LS: 6 hours to a few days	Phagocytize bacteria
■ Eosinophil		Nucleus bilobed; red cytoplasmic granules; diameter 10–14 μm	100–400	D: 6–9 days LS: 8–12 days	Kill parasitic worms; destroy antigen-antibody complexes; inactivate some inflammatory chemicals of allergy
■ Basophil		Nucleus lobed; large blue-purple cytoplasmic granules; diameter 10–12 μm	20–50	D: 3–7 days LS: ? (a few hours to a few days)	Release histamine and other mediators of inflammation; contain heparin, an anticoagulant
Agranulocytes ■ Lymphocyte		Nucleus spherical or indented; pale blue cytoplasm; diameter 5–17 μm	1500–3000	D: days to weeks LS: hours to years	Mount immune response by direct cell attack or via antibodies
■ Monocyte		Nucleus U or kidney shaped; gray-blue cytoplasm; diameter 14–24 μm	100–700	D: 2–3 days LS: months	Phagocytosis; develop into macrophages in tissues
Platelets		Discoid cytoplasmic fragments containing granules; stain deep purple; diameter 2–4 μm	250,000–500,000	D: 4–5 days LS: 5–10 days	Seal small tears in blood vessels; instrumental in blood clotting

*Appearance when stained with Wright's stain.

slightly smaller than neutrophils (see Table 18.2 and Figure 18.10c). Their cytoplasm contains large, coarse histamine-containing granules that have an affinity for the basic dyes (*basophil* = "base loving") and stain purplish-black. *Histamine* is an inflammatory chemical that acts as a vasodilator (makes blood vessels dilate) and attracts other white blood cells to the inflamed site. The deep purple nucleus is generally U or S shaped with two or three conspicuous constrictions.

Granulated cells similar to basophils, called *mast cells,* are found in connective tissues. Although mast cell nuclei tend to be more oval than lobed, the cells are similar microscopically, and both cell types bind to a particular antibody (immunoglobulin E) that causes the cells to release histamine. However, they arise from different cell lines.

Agranulocytes

The **agranulocytes** include lymphocytes and monocytes, WBCs that lack *visible* cytoplasmic granules. Although they are similar structurally, they are functionally distinct and unrelated cell types. Their nuclei are typically spherical or kidney shaped.

Lymphocytes **Lymphocytes** are the second most numerous leukocytes in the blood. When stained, a typical lymphocyte has a large, dark-purple nucleus that occupies most of the cell volume. The nucleus is usually spherical but may be slightly indented, and it is surrounded by a thin rim of pale-blue cytoplasm (see Table 18.2 and Figure 18.10d). Lymphocyte diameter ranges from 5 to 17 μm, but they are often classified according to size as small (5 to 8 μm), medium (10 to 12 μm), and large (14 to 17 μm).

Although large numbers of lymphocytes exist in the body, only a small proportion of them (mostly the small lymphocytes) is found in the bloodstream. In fact, lymphocytes are so called because most are firmly enmeshed in lymphoid tissues (lymph nodes, spleen, etc.), where they play a crucial role in immunity. **T lymphocytes** (T cells) function in the immune response by acting directly against virus-infected cells and tumor cells. **B lymphocytes** (B cells) give rise to *plasma cells*, which produce **antibodies** (immunoglobulins) that are released to the blood. (The specifics of B and T lymphocyte function are described in Chapter 22.)

Monocytes **Monocytes,** which account for 4–8% of WBCs, have an average diameter of 18 μm and are the largest leukocytes. They have abundant pale-blue cytoplasm and a darkly staining purple nucleus, which is distinctively U or kidney shaped (see Table 18.2 and Figure 18.10e). Once in the tissues, monocytes differentiate into highly mobile **macrophages** with prodigious appetites. In *chronic* infections such as tuberculosis, the macrophages increase in number and are actively phagocytic, and they are crucial in the body's defense against viruses and certain intracellular bacterial parasites. As explained in Chapter 22, macrophages are also important in activating lymphocytes to mount the immune response.

Production and Life Span of Leukocytes

Like erythropoiesis, **leukopoiesis**, or the production of white blood cells, is hormonally stimulated. These hormones are glycoproteins, generally referred to as *cytokines*, that fall into two families of hematopoietic factors, **interleukins** and **colony-stimulating factors,** or **CSFs.** While the interleukins are numbered (e.g., IL-3, IL-5), most CSFs are named for the leukocyte population they stimulate—thus *granulocyte-CSF (G-CSF)* stimulates production of granulocytes. The hematopoietic factors not only prompt the white blood cell precursors to divide and mature, but also enhance the protective potency of mature leukocytes. Macrophages and T lymphocytes are the most important sources of these cytokines, but they are also produced by selected other cell types.

Apparently the stimulatory hormones are released in response to specific chemical signals. The network of chemicals that marshals an army of leukocytes to fight off some attack on the body is complex and very much tied into the immune response. Many of the hematopoietic hormones (EPO and several of the CSFs) are being used clinically to stimulate the bone marrow of cancer patients who are receiving chemotherapy (which suppresses the marrow) and of those who have received marrow transplants, and to beef up the protective responses of AIDS patients.

Figure 18.11 shows the pathways of leukocyte differentiation. An early branching of the pathway divides the **lymphoid stem cells,** which produce lymphocytes, from the **myeloid stem cells,** which give rise to all other formed elements. For each granulocyte line, the committed cells, called **myeloblasts** (mi′ĕ-lo-blasts″), accumulate lysosomes and go on to become **promyelocytes.** The distinctive granules of each granulocyte type appear next in the **myelocyte** stage; then cell division stops. In the subsequent stage, the nuclei distort and become arc-like, producing the **band cell** stage. Just before granulocytes leave the marrow and enter the circulation, their nuclei constrict, beginning the process of nuclear segmentation. The bone marrow stores mature granulocytes and usually contains 10 to 20 times more granulocytes than are found in the blood. The normal ratio of granulocytes to erythrocytes produced is about 3:1, which reflects the much shorter life span (0.5 to 9.0 days) of the granulocytes. Most granulocytes are believed to die in combat with invading microorganisms.

Despite their similar appearances, the two types of agranulocytes have very different lineages. Like granulocytes, monocytes diverge from common pleuripotent myeloid stem cells. They then progress through **monoblast** and **promonocyte** stages (see Figure 18.11). By contrast, lymphocytes derive from the lymphoid stem cell and progress through the **lymphoblast** and **prolymphocyte** stages. The promonocytes and prolymphocytes leave the bone marrow and travel to the lymphoid tissues, where their further differentiation occurs (as described in

FIGURE 18.11 ***Formation of leukocytes.*** Leukocytes arise from ancestral stem cells called hemocytoblasts. (**a–c**) Granular leukocytes develop via a sequence involving myeloblasts. The developmental pathway is common until the granules typical of each granulocyte begin to form. (**d–e**) Agranular leukocytes develop from monoblasts and lymphoblasts. Monocytes, like granular leukocytes, are progeny of the myeloid stem cell. Only lymphocytes arise via the lymphoid stem cell line. Lymphocytes released from the bone marrow are immature; their further differentiation occurs in the lymphoid organs.

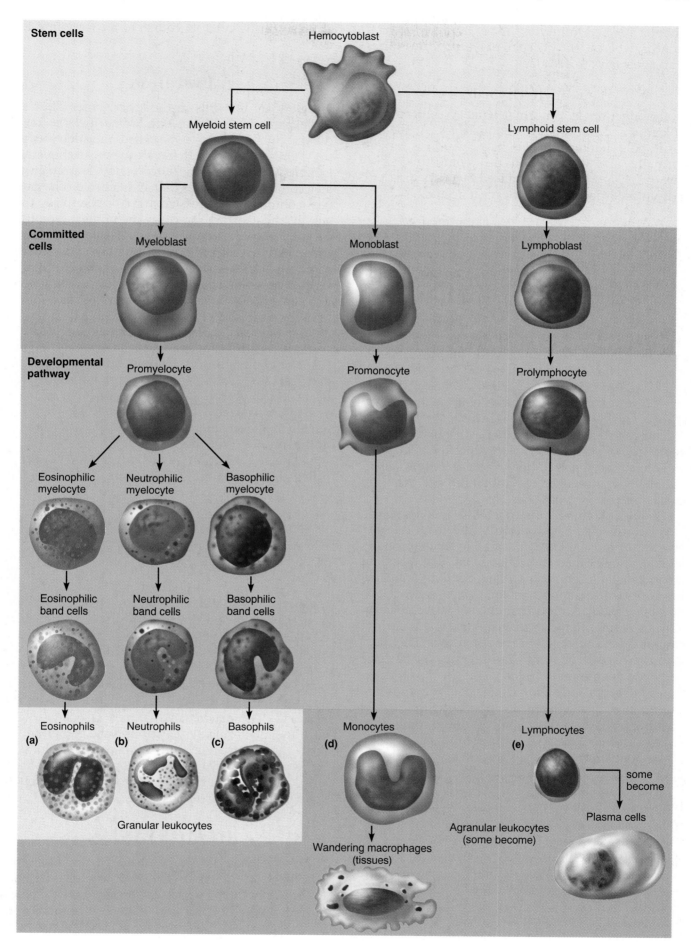

Stem cells

Hemocytoblast

Myeloid stem cell

Lymphoid stem cell

Committed cells

Myeloblast

Monoblast

Lymphoblast

Developmental pathway

Promyelocyte

Promonocyte

Prolymphocyte

Eosinophilic myelocyte

Neutrophilic myelocyte

Basophilic myelocyte

Eosinophilic band cells

Neutrophilic band cells

Basophilic band cells

Eosinophils

Neutrophils

Basophils

Monocytes

Lymphocytes

(a)

(b)

(c)

(d)

(e)

some become

Granular leukocytes

Wandering macrophages (tissues)

Agranular leukocytes (some become)

Plasma cells

Chapter 22). Monocytes may live for several months, whereas the life span of lymphocytes varies from a few days to decades.

Leukocyte Disorders

Overproduction of abnormal leukocytes occurs in leukemia and infectious mononucleosis. At the opposite pole, **leukopenia** (loo"ko-pe'ne-ah) is an abnormally low white blood cell count (*penia* = poverty). It is commonly induced by drugs, particularly glucocorticoids and anticancer agents.

Leukemias The term *leukemia*, literally "white blood," refers to a group of cancerous conditions involving white blood cells. As a rule, the renegade leukocytes are members of a single *clone* (descendants of a single cell) that remain unspecialized and mitotic, and impair normal bone marrow function. The leukemias are named according to the abnormal cell type primarily involved. For example, *myelocytic leukemia* involves descendants of the myeloblasts, whereas *lymphocytic leukemia* involves the lymphocytes. Leukemia is *acute* (quickly advancing) if it derives from blast-type cells like lymphoblasts, and *chronic* (slowly advancing) if it involves proliferation of later cell stages like myelocytes. The more serious acute forms primarily affect children. Chronic leukemia is seen more often in elderly people. Without therapy, all leukemias are fatal; only the time course differs.

In all leukemias immature WBCs flood into the bloodstream. More significantly, the bone marrow becomes almost totally occupied by the cancerous leukocytes. Because the other blood cell lines are crowded out, severe anemia and bleeding problems result. Other symptoms include fever, weight loss, and bone pain. Although tremendous numbers of leukocytes are produced, they are nonfunctional and cannot defend the body in the usual way. The most common causes of death are internal hemorrhage and overwhelming infections.

Treatments, involving irradiation and administration of antileukemic drugs to destroy the rapidly dividing cells, have successfully induced remissions (symptom-free periods) lasting from months to years. Bone marrow transplants are used in selected patients when compatible donors are available.

Infectious Mononucleosis Once called the kissing disease, *infectious mononucleosis* (glandular fever) is a highly contagious viral disease that is most often seen in children and young adults. Caused by the Epstein-Barr virus, its hallmark is excessive numbers of agranulocytes, many of which are atypical. The affected individual complains of being tired and achy, and has a chronic sore throat and a low-grade fever. There is no cure, but with rest the condition typically runs its course to recovery in a few weeks.

Platelets

Platelets are not cells in the strict sense. They are cytoplasmic fragments of extraordinarily large (up to 60 μm in diameter) cells called **megakaryocytes** (meg"ah-kar'e-o-sīts). In blood smears, each platelet exhibits a blue-staining outer region and an inner area containing granules that stain purple. The granules contain an impressive array of chemicals that act in the clotting process, including serotonin, Ca^{2+}, a variety of enzymes, ADP, and platelet-derived growth factor (PDGF). Although platelets are sometimes called **thrombocytes** (*thromb* = clot), most anatomists reserve that term for the functionally comparable *nucleated cells* of vertebrates other than mammals.

Platelets are essential for the clotting process that occurs in plasma when blood vessels are ruptured or their lining is injured. By sticking to the damaged site, platelets form a temporary plug that helps seal the break. (This mechanism is explained shortly.) Because platelets are anucleate, they age quickly and degenerate in about ten days if they are not involved in clotting.

Platelet formation is regulated by a hormone called **thrombopoietin.** Their immediate ancestral cells, the megakaryocytes, are progeny of the hemocytoblast and the myeloid stem cell, but their formation is quite unusual (Figure 18.12). In this line, repeated mitoses of the **megakaryoblast** occur, but cytokinesis does not. The final result is the megakaryocyte (literally "big nucleus cell"), a bizarre cell with a huge, multilobed nucleus and a large cytoplasmic mass. When formed, the megakaryocyte presses up against a sinusoid (the specialized type of capillary in the marrow) and sends cytoplasmic extensions through the sinusoid wall into the bloodstream. These extensions then rupture, releasing the platelet fragments like stamps being torn from a sheet of postage stamps, to seed the blood with platelets. The plasma membranes associated with each fragment quickly seal around the cytoplasm to form the grainy, roughly disc-shaped platelets (see Table 18.2), each with a diameter of 2 to 4 μm. Each cubic millimeter of blood contains between 250,000 and 500,000 of the tiny platelets.

HEMOSTASIS

Normally, blood flows smoothly past the intact unperturbed lining (endothelium) of the blood vessel walls. But if a blood vessel wall breaks, a whole series of reactions is set in motion to accomplish **hemostasis** (he"mo-sta'sis), or stoppage of bleeding (*stasis* =

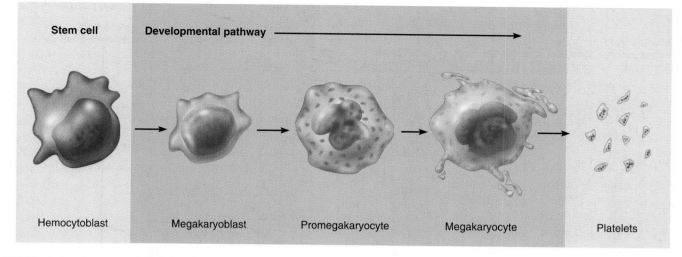

FIGURE 18.12 *Genesis of platelets.* The stem cell (hemocytoblast) gives rise to cells that undergo several mitotic divisions unaccompanied by cytoplasmic division to produce the megakaryocytes. The cytoplasm of the megakaryocyte becomes compartmentalized by membranes, and the plasma membrane then fragments, liberating the platelets. (Intermediate stages between the hemocytoblast and megakaryoblast are not illustrated.)

Labels (left to right): Stem cell — Developmental pathway → ; Hemocytoblast — Megakaryoblast — Promegakaryocyte — Megakaryocyte — Platelets

halting). Without this plug-the-hole defensive reaction, we would quickly bleed out our entire blood volume from even the smallest of cuts.

The hemostasis response, which is fast, localized, and carefully controlled, involves many blood coagulation factors normally present in plasma as well as some substances that are released by platelets and injured tissue cells. During hemostasis, three phases occur in rapid sequence: (1) vascular spasms, (2) platelet plug formation, and (3) coagulation, or blood clotting. Blood loss at the site is permanently prevented when fibrous tissue grows into the clot and seals the hole in the blood vessel.

Vascular Spasms

The immediate response to blood vessel injury is constriction of the damaged blood vessel (vasoconstriction). Factors that trigger this **vascular spasm** include direct injury to vascular smooth muscle, chemicals released by endothelial cells and platelets, and reflexes initiated by local pain receptors. Generally speaking, the spasm mechanism becomes more and more efficient as the amount of tissue damage increases, and is most effective in the smaller blood vessels. The value of the spasm response is obvious: A strongly constricted artery can significantly reduce blood loss for 20–30 minutes, allowing time for platelet plug formation and blood clotting to occur.

Platelet Plug Formation

Platelets play a key role in hemostasis by forming a plug that temporarily seals the break in the vessel wall. They also help to orchestrate subsequent events that lead to blood clot formation. These events are shown in simplified form in Figure 18.13a.

As a rule, platelets do not stick to each other or to the smooth endothelial linings of blood vessels. However, when the endothelium is damaged and underlying collagen fibers are exposed, platelets undergo some remarkable changes. They swell and form spiked processes. They also become sticky and adhere tenaciously to the exposed collagen. Once attached, the platelet granules begin to break down and release several chemicals. Some, like **serotonin**, enhance the vascular spasm. Others, like **adenosine diphosphate (ADP)**, are potent aggregating agents that attract more platelets to the area and cause them to release their contents. **Thromboxane (throm-boks'ān) A_2**, a short-lived prostaglandin derivative that is generated and released, stimulates both events. Other phospholipids are also released during the degranulation process. Thus, a positive feedback cycle that activates and attracts greater and greater numbers of platelets to the area begins and, within one minute, a platelet plug is built up, which further reduces blood loss. The platelet plug is limited to the immediate area where it is needed by PGI_2, a prostaglandin produced by the endothelial cells. Also called **prostacyclin**, PGI_2 is a strong inhibitor of platelet aggregation. Platelet plugs are loosely knit, but when reinforced by fibrin threads which act as a "molecular glue" for the aggregated platelets, they are quite effective in sealing the small tears in a blood vessel that occur with normal activity. Once the platelet plug is formed, the next stage, coagulation, comes into play.

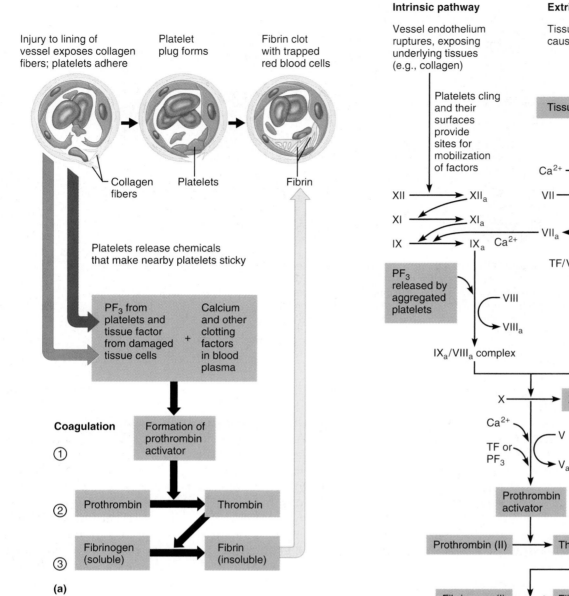

FIGURE 18.13 *Events of platelet plug formation and blood clotting.* **(a)** Simplified schematic of events. Steps numbered 1–3 represent the major events of coagulation. The color of the arrows indicates their source or destination: red from tissue, purple from platelets, and yellow to fibrin. **(b)** Detailed flowchart indicating the intermediates and events involved in platelet plug formation and the intrinsic and extrinsic mechanisms of blood clotting (coagulation). Once the prothrombin activator is present, the remaining events are common to both the extrinsic and intrinsic pathways. (The subscript "a" indicates the activated procoagulant.)

Coagulation

Coagulation or **blood clotting** (Figure 18.13a), during which blood is transformed from a liquid to a gel, is a multistep process that leads to its critically important last three phases. The final reactions are:

1. A complex substance called *prothrombin activator* is formed.

2. Prothrombin activator converts a plasma protein called *prothrombin* into *thrombin,* an enzyme.

3. Thrombin catalyzes the joining of *fibrinogen* molecules present in plasma to a *fibrin mesh,* which traps blood cells and effectively seals the hole until the blood vessel can be permanently repaired.

The complete coagulation process is much more complicated, however. Over 30 different substances are involved. Factors that enhance clot formation are called **clotting factors** or **procoagulants.** Although vitamin K is not directly involved in coagulation, this fat-soluble vitamin is required for the synthesis of four of the procoagulants made by the liver (see Table 18.3). Factors that inhibit clotting are called **anticoagulants.** Whether or not blood clots depends on a delicate balance between these two groups of factors. Normally, anticoagulants dominate and clotting is prevented; but when a vessel is ruptured, procoagulant activity in that area increases dramatically and clot formation begins. The procoagulants are numbered I to XIII (Table 18.3) according to the order of their discovery; hence the numerical order does *not* reflect the reaction sequence. Tissue factor (III) and Ca^{2+} (IV) are usually indicated by their names, rather than by numerals. Most of these factors are plasma proteins made by the liver that circulate in an inactive form in blood until mobilized.

Phase 1: Two Pathways to Prothrombin Activator

Clotting may be initiated by either the **intrinsic** or the **extrinsic pathway** (Figure 18.13b), and in the body both pathways are usually triggered by the same tissue-damaging events. Clotting of blood outside the body (such as in a test tube) is initiated *only* by the intrinsic mechanism, whereas clotting of blood that has escaped into the tissues is promoted by the extrinsic pathway. Let's examine the "why" of these differences.

A pivotal molecule in both mechanisms is PF_3, a phospholipid associated with the external surfaces of aggregated platelets. Apparently, many intermediates of both pathways can be activated only in the presence of PF_3. In the slower intrinsic pathway, all factors needed for clotting are present in (intrinsic to) the blood. By contrast, when blood is exposed to an additional factor released by injured cells called **tissue factor (TF)** or **tissue thromboplastin,** the "shortcut" extrinsic mechanism, which completely bypasses several steps of the intrinsic pathway, is triggered.

Each pathway requires ionic calcium and involves the activation of a *series* of procoagulants, each functioning as an enzyme to activate the next procoagulant in the sequence. The intermediate steps of each pathway *cascade* toward a common intermediate, factor X (Figure 18.13b). Once factor X has been activated, it complexes with calcium ions, tissue factor or PF_3, and factor V to form **prothrombin activator.** This step is usually the slowest step of the blood clotting process, but once prothrombin activator is present, the clot forms in 10 to 15 seconds.

Phase 2: Common Pathway to Thrombin

Prothrombin activator catalyzes the transformation of the plasma protein **prothrombin** to the active enzyme **thrombin.**

Phase 3: Common Pathway to the Fibrin Mesh

Thrombin catalyzes the polymerization of **fibrinogen** (another plasma protein made by the liver). As the fibrinogen molecules become aligned into long, hairlike, insoluble **fibrin** strands, they glue the platelets together and intertwine to make a web that forms the structural basis of the clot. In the presence of fibrin, plasma becomes gel-like and traps formed elements that try to pass through it (Figure 18.14).

In the presence of calcium ions, thrombin also activates **factor XIII (fibrin stabilizing factor),** a cross-linking enzyme that binds the fibrin strands tightly together and strengthens and stabilizes the clot. Clot formation is normally complete within 3 to 6 minutes after blood vessel damage. Because the extrinsic pathway involves fewer steps it is more rapid than the intrinsic pathway, and in cases of severe tissue trauma the extrinsic mechanism can promote clot formation within 15 seconds.

Clot Retraction and Repair

Within 30 to 60 minutes, the clot is stabilized further by a platelet-induced process called **clot retraction.** Platelets contain contractile proteins (actin and myosin), and they contract in much the same manner as muscle cells. As the platelets contract, they pull on the surrounding fibrin strands, squeezing **serum** (plasma minus the clotting proteins) from the mass. As a result, the clot is compacted and the ruptured edges of the blood vessel are drawn more closely together. Even as clot retraction is occurring, vessel healing is also taking place. **Platelet-derived growth factor (PDGF)** released by platelet degranulation stimulates smooth muscle and fibroblasts to divide and rebuild the wall. As fibroblasts form a connective tissue patch in the injured area, endothelial cells multiply to restore the endothelial lining.

Fibrinolysis

A clot is not a permanent solution to blood vessel injury, and a process called **fibrinolysis** removes unneeded clots when healing has occurred. Because small clots are formed continually in vessels throughout the body, this cleanup detail is crucial. Without fibrinolysis, blood vessels would gradually become completely blocked.

The critical natural "clot buster" is a fibrin-digesting enzyme called **plasmin,** which is produced when the blood protein **plasminogen** is activated. Large amounts of plasminogen are incorporated into a forming clot, where it remains inactive until

TABLE 18.3	Blood Clotting Factors (Procoagulants)		
Factor number	Factor name	Nature/origin	Function or pathway
I	Fibrinogen	Plasma protein; synthesized by liver	Common pathway; converted to fibrin, weblike substance of clot
II	Prothrombin	Plasma protein; synthesized by liver; formation requires vitamin K	Common pathway; converted to thrombin, which enzymatically converts fibrinogen to fibrin
III	Tissue factor (TF) or tissue thromboplastin	Lipoprotein complex; released from damaged tissues	Activates extrinsic pathway
IV	Calcium ions (Ca^{2+})	Inorganic ion present in plasma; acquired from diet or released from bone	Needed for essentially all stages of coagulation process
V	Proaccelerin, labile factor, or platelet accelerator	Plasma protein; synthesized in liver; also released by platelets	Both extrinsic and intrinsic mechanisms
VI	Number no longer used; substance now believed to be same as factor V		
VII	Serum prothrombin conversion accelerator (SPCA) or stable factor	Plasma protein; synthesized in liver in process that requires vitamin K	Both extrinsic and intrinsic mechanisms
VIII	Antihemophilic factor (AHF)	Globulin synthesized in liver; deficiency causes hemophilia A	Intrinsic mechanism
IX	Plasma thromboplastin component(PTC) or Christmas factor	Plasma protein; synthesized in liver; deficiency results in hemophilia B; synthesis requires vitamin K	Intrinsic mechanism
X	Stuart factor, Stuart-Prower factor, or thrombokinase	Plasma protein; synthesized in liver; synthesis requires vitamin K	Both extrinsic and intrinsic pathways
XI	Plasma thromboplastin antecedent(PTA)	Plasma protein; synthesized in liver; deficiency results in hemophilia C	Intrinsic mechanism
XII	Hageman factor, glass factor	Plasma protein; proteolytic enzyme; synthesized in the liver	Intrinsic mechanism; activates plasmin; known to be activated by contact with glass and may initiate clotting in vitro
XIII	Fibrin stabilizing factor (FSF)	Plasma protein; synthesized in liver and present in platelets	Cross-links fibrin and renders it insoluble

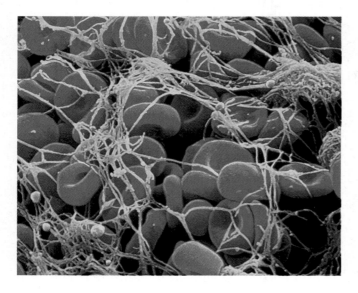

appropriate signals reach it. The presence of a clot in and around the blood vessel causes the endothelial cells to secrete **tissue plasminogen activator (tPA).** Activated factor XII and thrombin released during clotting also serve as plasminogen activators. As a result, most plasmin activity is confined to the clot, and any plasmin that strays into the plasma is quickly destroyed by circulating enzymes.

FIGURE 18.14 *Scanning electron micrograph of erythrocytes trapped in a fibrin mesh.* (3025×)

Fibrinolysis begins within two days and continues slowly over several days until the clot is finally dissolved.

Factors Limiting Clot Growth or Formation

Factors Limiting Normal Clot Growth

Once the clotting cascade has begun, it continues until a clot is formed. Normally, two important homeostatic mechanisms prevent clots from becoming unnecessarily large: (1) swift removal of clotting factors, and (2) inhibition of activated clotting factors. For clotting to occur in the first place, the concentration of activated procoagulants must reach certain critical levels. Clot formation in rapidly moving blood is usually curbed because the activated clotting factors are diluted, washed away, and prevented from accumulating. For the same reasons, further growth of a forming clot is hindered when it contacts normally flowing blood.

Other mechanisms block the final step in which fibrinogen is polymerized into fibrin by restricting thrombin to the clot or by inactivating it if it manages to escape into the general circulation. As a clot forms, almost all of the thrombin produced is adsorbed (bound) onto the fibrin threads. This is an important safeguard because thrombin also exerts positive feedback effects on the coagulation process prior to the common pathway. Not only does it speed up the production of prothrombin activator by acting indirectly through factor V, but it also accelerates the earliest steps of the intrinsic pathway by activating platelets. Thus, fibrin effectively acts as an anticoagulant to prevent enlargement of the clot and prevents thrombin from acting elsewhere. Thrombin not adsorbed to fibrin is quickly inactivated by **antithrombin III,** a protein present in plasma. Antithrombin III and **protein C,** another protein produced in the liver, also inhibit the activity of other intrinsic pathway procoagulants.

Heparin, the natural anticoagulant contained in the granules of basophils and mast cells and also produced by endothelial cells, is ordinarily secreted in small amounts into the plasma. It inhibits thrombin by enhancing the activity of antithrombin III. Heparin, like most other clotting inhibitors, also inhibits the intrinsic pathway.

Factors Preventing Undesirable Clotting

Factors that normally ward off unnecessary clotting include both structural and molecular characteristics of the endothelial lining of blood vessels. As long as the endothelium is smooth and intact, platelets are prevented from clinging and piling up. Also, the antithrombic substances—heparin and PGI_2—secreted by the endothelial cells normally prevent platelet adhesion. PGI_2 secreted by the endothelial surface actively repels the platelets as they pass by in the bloodstream. Additionally, it has been found that vitamin E quinone, a molecule formed in the body when vitamin E reacts with oxygen, is a potent anticoagulant.

Disorders of Hemostasis

Though blood clotting is one of nature's most elegant creations, it sometimes goes awry, disrupting homeostasis. The two major disorders of hemostasis are at opposite poles. **Thromboembolytic disorders** result from conditions that cause undesirable clot formation. **Bleeding disorders** arise from abnormalities that prevent normal clot formation.

Thromboembolytic Conditions

Despite the body's many safeguards, undesirable intravascular clotting, called "hemostasis in the wrong place" by some, sometimes occurs. A clot that develops and persists in an *unbroken* blood vessel is called a **thrombus.** If the thrombus is large enough, it may block circulation to the cells beyond the occlusion and lead to the death of those tissues. For example, if the blockage occurs in the coronary circulation of the heart (coronary thrombosis), the consequences may be death of heart muscle and a fatal heart attack. If the thrombus breaks away from the vessel wall and floats freely in the bloodstream, it becomes an **embolus** (plural, *emboli*). An embolus ("wedge") is usually no problem until it encounters a blood vessel that is too narrow for it to pass through. For example, emboli that become trapped in the lungs (pulmonary emboli) can dangerously impair the ability of the body to obtain oxygen. A cerebral embolus may cause a stroke. Drugs that dissolve blood clots (such as tPA) and innovative medical techniques for removing clots are described in *A Closer Look* on pp. 722–723 in Chapter 20.

Conditions that roughen the vessel endothelium, such as arteriosclerosis, severe burns, or inflammation, cause thromboembolytic disease by allowing platelets to gain a foothold. Slowly flowing blood or blood stasis is another risk factor, particularly in bedridden patients and those taking a long flight in economy-class seats. In this case, clotting factors are *not* washed away as usual and accumulate so that clot formation finally becomes possible.

A number of drugs, most importantly aspirin, heparin, and dicumarol, are used clinically to prevent undesirable clotting in patients at risk for heart attack or stroke. **Aspirin** is an antiprostaglandin drug that inhibits thromboxane A_2 formation (hence, it blocks platelet aggregation and platelet plug formation). Clinical studies of men taking low-dose aspirin (one aspirin every two days) over a period of several years demonstrated a 50% reduction in (anticipated) incidence of heart attack. Heparin (see above) is also prescribed as an anticoagulant drug, as is warfarin, an ingredient in rat poison. Administered in injectable form, heparin is the anticoagulant most used clinically (i.e., for preoperative and postoperative cardiac patients and for those receiving blood transfusions). Taken orally, **warfarin** (Coumadin) is a mainstay in the treatment of those prone to atrial fibrillation, a condition in which blood pools in the heart, to reduce the risk of stroke. It works via a different mechanism than heparin; it interferes with the action of vitamin K in the production of some procoagulants (see Impaired Liver Function below). Flavonoids, found in tea, red wine, and grape juice, have natural anticoagulant activity.

Bleeding Disorders

Anything that interferes with the clotting mechanism can result in abnormal bleeding. The most common causes are platelet deficiency (thrombocytopenia) and deficits of some of the procoagulants such as might result from impaired liver function or certain genetic conditions (hemophilias).

Thrombocytopenia A condition in which the number of circulating platelets is deficient, **thrombocytopenia** (throm"bo-si"to-pe'ne-ah) causes spontaneous bleeding from small blood vessels all over the body. Even normal movement leads to widespread hemorrhage, evidenced by many small purplish blotches, called *petechiae* (pe-te'ke-e), on the skin. Thrombocytopenia can arise from any condition that suppresses or destroys the bone marrow, such as bone marrow malignancy, exposure to ionizing radiation, or certain drugs. A platelet count of under 50,000 platelets per cubic millimeter of blood is usually diagnostic for this condition. Whole blood transfusions provide temporary relief from bleeding.

Impaired Liver Function When the liver is unable to synthesize its usual supply of procoagulants, abnormal, and often severe, bleeding occurs. The causes can range from an easily resolved vitamin K deficiency (common in newborns and after taking systemic antibiotics) to nearly total impairment of liver function (as in hepatitis or cirrhosis). Vitamin K is required by the liver cells for production of the clotting factors and, because vitamin K is produced by bacteria that reside in the intestines, dietary deficiencies are rarely a problem. However, vitamin K deficiency can occur if fat absorption is impaired, because vitamin K is a fat-soluble vitamin that is absorbed into the blood along with fats. In liver disease, the nonfunctional liver cells fail to produce not only the procoagulants but also bile, which is required for fat and vitamin K absorption.

Hemophilias The term **hemophilia** refers to several different hereditary bleeding disorders that have similar signs and symptoms. *Hemophilia A*, or *classical hemophilia*, results from a deficiency of **factor VIII (antihemophilic factor).** This is the most common type of hemophilia and accounts for 83% of cases. *Hemophilia B* results from a deficiency of factor IX. Both types are sex-linked conditions occurring primarily in males (that is, the conditions are due to recessive genes on X chromosomes; this topic is discussed in Chapter 30). *Hemophilia C*, a less severe form of hemophilia seen in both sexes, is due to a lack of factor XI. The relative mildness of this form, as compared to the A and B forms, reflects the fact that the procoagulant (factor IX) that factor XI activates may also be activated by factor VII (see Figure 18.13b). Additionally, some cases of hemophilia originally thought to be caused by a deficiency of factor XI are now being attributed to alterations of factor VII.

Symptoms of hemophilia begin early in life; even minor tissue trauma causes prolonged bleeding into tissues that can be life threatening. Commonly, the person's joints become seriously disabled and painful because of repeated bleeding into the joint cavities after exercise or trauma. Hemophilias are managed clinically by transfusions of fresh plasma or injections of the appropriate purified clotting factor. These therapies provide relief for several days but are expensive and inconvenient. Because hemophiliacs are absolutely dependent on blood transfusions or factor injections, many have become infected by the hepatitis virus and, since the early 1980s, by HIV, a blood-transmitted virus that depresses the immune system and causes AIDS. (See Chapter 22.) This problem has been resolved because of new testing methods for HIV and the availability of genetically engineered factor VIII.

TRANSFUSION AND BLOOD REPLACEMENT

Transfusion of Whole Blood

The human cardiovascular system is designed to minimize the effects of blood loss by (1) reducing the volume of the affected blood vessels, which helps to maintain normal circulation, and (2) stepping up the production of red blood cells. However, the body can compensate for only so much blood loss. Losses of 15–30% cause pallor and weakness. A loss of more than 30% of blood volume results in severe shock, which can be fatal.

Whole blood transfusions are routine when blood loss is substantial and when treating thrombocytopenia. Infusions of **packed red cells** (whole blood from which most of the plasma has been removed) are preferred to treat anemia. The usual

	Frequency (% U.S. population)				RBC antigens (ag-glutinogens)	Illustration	Plasma antibodies (agglutinins)	Blood that can be received
Blood group	White	Black	Asian	Native American				
AB	4	4	5	<1	A B		None	A, B, AB, O Universal recipient
B	11	20	27	4	B		Anti-A (a)	B, O
A	40	27	28	16	A		Anti-B (b)	A, O
O	45	49	40	79	None		Anti-A (a) Anti-B (b)	O Universal donor

TABLE 18.4 ABO Blood Groups

blood bank procedure involves collecting blood from a donor and then mixing it with an anticoagulant, such as certain citrate or oxalate salts, which prevents clotting by binding with calcium ions. The shelf life of the collected blood at 4°C is about 35 days. When freshly collected blood is to be transfused, heparin is the anticoagulant used.

Human Blood Groups

People have different blood types and transfusion of incompatible blood can be fatal. RBC plasma membranes, like those of all body cells, bear highly specific glycoproteins (antigens) at their external surfaces, which identify each of us as unique from all others. One person's RBC proteins may be recognized as foreign if transfused into someone with a different red blood cell type, and the transfused cells may be agglutinated (clumped) and destroyed. Since these RBC antigens promote agglutination, they are more specifically called **agglutinogens** (ag"loo-tin'o-jenz).

At least 30 varieties of naturally occurring RBC antigens are common in humans. Besides these, perhaps 100 others occur in individual families ("private antigens") rather than in the general population. The presence or absence of each antigen allows each person's blood cells to be classified into several different blood groups. The antigens determining the ABO and Rh blood groups cause vigorous transfusion reactions (in which the foreign erythrocytes are destroyed) when they are improperly transfused.

Thus, blood typing for these antigens is *always* done before blood is transfused. Other antigens (such as the M, N, Duffy, Kell, and Lewis factors) are mainly of legal or academic importance. Because these factors cause weak or no transfusion reactions, blood is not specifically typed for them unless the person is expected to need several transfusions, in which case the many weak transfusion reactions could have cumulative effects. Only the ABO and Rh blood groups are described here.

ABO Blood Groups As shown in Table 18.4, the **ABO blood groups** are based on the presence or absence of two agglutinogens, type A and type B. Depending on which of these a person inherits, his or her ABO blood group will be one of the following: A, B, AB, or O. The O blood group, which has neither agglutinogen, is the most common ABO blood group in white, black, Asian, and native Americans; AB, with both antigens, is the least prevalent. The presence of either the A or the B agglutinogen results in group A or B, respectively. Inheritance of the ABO blood types is an example of multiple gene inheritance; its genetic basis is explained in Chapter 30.

Unique to the ABO blood groups is the presence in the plasma of *preformed antibodies* called **agglutinins**. The agglutinins act against RBCs carrying ABO antigens that are *not* present on a person's own red blood cells. A newborn lacks these antibodies, but they begin to appear in the plasma within two months. They reach peak levels between eight and ten years of age and then slowly decline throughout the rest of life. As indicated in Table 18.4, a person with neither the A nor the B antigen (group O) possesses both anti-A and anti-B antibodies, also called *a* and *b agglutinins* respectively. Those with group A blood have anti-B antibodies, while those with group B have anti-A antibodies. Neither antibody is produced by AB individuals.

Rh Blood Groups There are at least eight different types of Rh agglutinogens, each of which is called an **Rh factor.** Only three of these, the C, D, and E antigens, are fairly common. The Rh blood typing system is so named because one Rh antigen (agglutinogen D) was originally identified in *rhesus* monkeys. Later, the same antigen was discovered in human beings. Most Americans (about 85%) are Rh^+ (Rh positive), meaning that their RBCs carry the Rh antigen.

Unlike the ABO system, anti-Rh antibodies are not spontaneously formed in the blood of Rh^- (Rh negative) individuals. However, if an Rh^- person receives Rh^+ blood, the immune system becomes sensitized and begins producing anti-Rh antibodies against the foreign antigen soon after the transfusion. Hemolysis does not occur after the first such transfusion because it takes time for the body to react and start making antibodies. But the second time, and every time thereafter, a typical transfusion reaction occurs in which the recipient's antibodies attack and rupture the donor RBCs.

⬙ *Homeostatic Imbalance*

An important problem related to the Rh factor occurs in pregnant Rh^- women who are carrying Rh^+ babies. The first such pregnancy usually results in the delivery of a healthy baby. But the mother is sensitized by her baby's Rh^+ antigens that have passed through the placenta into her bloodstream, and she will form anti-Rh antibodies unless treated with RhoGAM before or shortly after she has given birth. (The same precautions are taken in women who have miscarried or aborted the fetus.) RhoGAM is a serum containing anti-Rh agglutinins. Because it agglutinates the Rh factor, it blocks the mother's immune response and prevents her sensitization. If the mother is not treated and becomes pregnant again with an Rh^+ baby, her antibodies will cross through the placenta and destroy the baby's RBCs, producing a condition known as **hemolytic disease of the newborn,** or **erythroblastosis fetalis.** The baby becomes anemic and hypoxic. Brain damage and even death may result unless transfusions are done *before* birth

to provide the fetus with more erythrocytes for oxygen transport. Additionally, one or two *exchange transfusions* (see Related Clinical Terms, p. 678) are done after birth. The baby's Rh^+ blood is removed, and Rh^- blood is infused. Within six weeks, the transfused Rh^- erythrocytes have been broken down and replaced with the baby's own Rh^+ cells. ■

Transfusion Reactions: Agglutination and Hemolysis

When mismatched blood is infused, a **transfusion reaction** occurs in which the donor's red blood cells are avidly attacked by the recipient's plasma agglutinins. (Note that the donor's plasma antibodies may also be agglutinating the host's RBCs, but they are so diluted in the recipient's circulation that this does not usually present a serious problem.)

The initial event, agglutination of the foreign red blood cells, leads to clogging of small blood vessels throughout the body. During the next few hours, the clumped red blood cells begin to rupture or are destroyed by phagocytes, and their hemoglobin is released into the bloodstream. (When the transfusion reaction is exceptionally severe, the RBCs are lysed almost immediately.) These events lead to two easily recognized problems: (1) The oxygen-carrying capability of the transfused blood cells is disrupted, and (2) the clumping of red blood cells in small vessels hinders blood flow to tissues beyond those points. Less apparent, but more devastating, is the consequence of hemoglobin escaping into the bloodstream. Circulating hemoglobin passes freely into the kidney tubules, and in high concentration it precipitates, blocking the kidney tubules and causing renal shutdown. If shutdown is complete (acute renal failure), the person may die.

Transfusion reactions can also cause fever, chills, nausea, vomiting, and general toxicity; but in the absence of renal shutdown, these reactions are rarely lethal. Treatment of transfusion reactions is directed toward preventing kidney damage by infusing alkaline fluids to dilute and dissolve the hemoglobin and wash it out of the body. Diuretics, which increase urine output, are also given.

As indicated in Table 18.4, group O red blood cells bear neither the A nor the B antigen; thus, theoretically this blood group is the **universal donor.** Indeed, some medical centers are busily involved in enzymatically converting type B blood to type O by clipping off the extra (B-specific) sugar residue. Since group AB plasma is devoid of antibodies to both A and B antigens, group AB people are theoretically **universal recipients** and can receive blood transfusions from any of the ABO groups. However, these classifications are misleading, because they do not take into account the other agglutinogens in blood that can trigger transfusion reactions.

Because pooled blood transfusions carry the risk of transfusion reactions and transmission of life-threatening infections (particularly with HIV), public interest in **autologous transfusions** has risen. Autologous (*auto* = self) transfusions are particularly sought by those considering elective surgery when they are in no immediate danger. The patient *predonates* his or her own blood, and it is stored and immediately available if needed during or after the operation. Iron supplements are given, and as long as the patient's preoperative hematocrit is at least 30%, one unit (400–500 ml) of blood can be collected every 4 days, with the last unit taken 72 hours prior to surgery.

Blood Typing

The importance of determining the blood group of both the donor and the recipient *before* blood is transfused is glaringly obvious. The general procedure for determining ABO blood type is briefly outlined in Figure 18.15. Because it is critical that blood groups be compatible, cross matching is also done. *Cross matching* tests for agglutination of donor RBCs by the recipient's serum, and of the recipient's RBCs by the donor serum. Typing for the Rh factors is done in the same manner as ABO blood typing.

Plasma and Blood Volume Expanders

When a patient's blood volume is so low that death from shock is imminent, there may not be time to type blood, or appropriate whole blood may be unavailable. Such emergencies demand that blood *volume* be replaced immediately to restore adequate circulation. Although there is not yet a satisfactory whole blood substitute, research is ongoing as described in *A Closer Look* on p. 676.

Plasma can be administered to anyone without concern about a transfusion reaction because the antibodies it contains become harmlessly diluted in the recipient's blood. Except for red blood cells, plasma provides a complete and natural blood replacement. When plasma is not available, various colloidal solutions, or **plasma expanders,** such as *purified human serum albumin, plasminate,* and *dextran,* can be infused. All of these have osmotic properties that directly increase the fluid volume of blood. Still another option is to infuse isotonic salt solutions. *Normal saline* or a *multiple electrolyte solution* that mimics the electrolyte composition of plasma (for example, *Ringer's solution*) is a common choice.

? *Is an agglutinin a plasma membrane glycoprotein or an antibody in the plasma?*

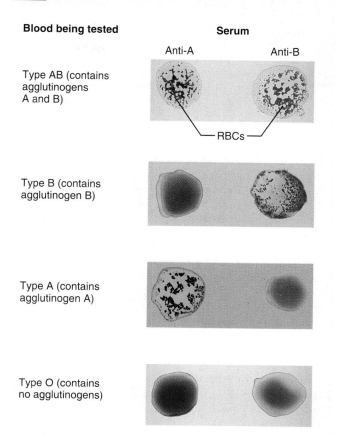

FIGURE 18.15 *Blood typing of ABO blood types.* When serum containing anti-A or anti-B agglutinins is added to a blood sample diluted with saline, agglutination will occur between the agglutinin and the corresponding agglutinogen (A or B). As illustrated, agglutination occurs with both sera in blood group AB, with anti-B serum in blood group B, with anti-A serum in blood group A, and with neither serum in blood group O.

DIAGNOSTIC BLOOD TESTS

A laboratory examination of blood yields information that can be used to evaluate a person's current state of health. For example, in some anemias, the blood is pale and yields a low hematocrit. A high fat content (*lipidemia*) gives blood plasma a yellowish hue and forecasts possible problems in those with heart disease. How well a diabetic is controlling diet and blood sugar levels is routinely determined by blood glucose tests. Infections are signaled by leukocytosis and, if severe, larger than normal buffy coats in the hematocrit.

A CLOSER LOOK — Concocting Blood: Artificial Blood Substitutes

The term "blood substitute" is somewhat misleading. Blood has many components that play such a wide variety of roles—from fighting infection to transporting oxygen—that no single artificial substitute yet engineered can fulfill all those functions. However, substitute liquids are available that can transport oxygen from the lungs through the body and can "stretch" a limited blood supply, while sidestepping transfusion reactions. An important benefit of these blood substitutes is that the recipient is not at risk for transmission of bloodborne disease factors. Four of these oxygen-transporting products—PFC-based compounds, chemically altered hemoglobin, artificial red blood cells, and Hemopure—are described briefly.

PFC-based Compounds

The main ingredient of *Fluosol*, a milky artificial blood substitute, is perfluorocarbons (PFCs), compounds similar to Teflon, the nonstick coating used in cookware. Developed in Japan,

the product was first tested in the United States in 1982. Many of the early recipients were people who needed surgery but refused blood transfusions on religious grounds.

Fluosol serves as a dissolving medium for oxygen. In order to "load" sufficient amounts of oxygen into it, patients must breathe pure oxygen by mask or must be in a hyperbaric (high-pressure) chamber. The developers claim that oxygen transported by Fluosol is used more easily by the body tissues because the slippery particles are much smaller than erythrocytes and glide through the capillaries at a faster rate. PFCs are cleared from the circulation and stored in the spleen and liver until they are exhaled as vapor by the lungs 4 to 12 hours after injection. Although initially promising for therapy of heart attack, carbon monoxide poisoning, and sickle-cell anemia, its use was clouded by research indicating that Fluosol depressed the immune system.

Currently under development is Oxyfluor, a new generation of PFC-based compounds designed to resolve the problems of Fluosol. It has a long

shelf life (two years if refrigerated) and delivers up to four times as much oxygen as the earlier versions did. However, increasing the oxygen-carrying capacity of blood can have diminishing returns because, when oxygen accumulates in body tissues, it causes tissue damage.

Chemically Altered Hemoglobin

In their search to find substitutes to boost oxygen delivery, scientists have found several ways to alter the chemistry of hemoglobin: (1) create a chemical bridge between two of its four peptide chains (cross-link it, as in HemAssist); (2) link several hemoglobin molecules together (polymerize hemoglobin, as with PolyHeme); or (3) attach polyethylene glycol (PEG) to hemoglobin to stabilize it (as with PHP). The altered hemoglobin gives up more oxygen to the tissues than normal hemoglobin, even at low temperatures (10°C). Since body temperature is routinely lowered in patients undergoing heart surgery, using the modified hemoglobin during such procedures might boost oxygen delivery to the patient's tissues. An

---

Microscopic studies of blood can reveal variations in the size and shape of erythrocytes that predict iron deficiency or pernicious anemia. Furthermore, a determination of the relative proportion of individual leukocyte types, a **differential white blood cell count,** is a valuable diagnostic tool. For example, a high eosinophil count may indicate a parasitic infection or an allergic response somewhere in the body.

A number of tests provide information on the status of the hemostasis system. The amount of prothrombin present in blood is assessed by determining the **prothrombin time,** and a **platelet count** is done when thrombocytopenia is suspected.

Two batteries of tests—a SMAC, SMA12–60, or similar series, and a **complete blood count (CBC)**—are routinely ordered during physical examinations and before hospital admissions. SMAC is a blood *chemistry* profile. The CBC includes counts of the different types of formed elements, a hematocrit,

and tests for clotting factors. Together these tests provide a comprehensive picture of one's general health status in relation to normal blood values.

A listing of the normal values for selected blood tests is provided in Appendix E.

DEVELOPMENTAL ASPECTS OF BLOOD

Before birth, there are many sites of blood cell formation—the fetal yolk sac, liver, and spleen, among others—but by the seventh month of development, the red marrow has become the primary hematopoietic area and remains so (barring serious illness) throughout life. If there is a severe need for blood cell production, however, the liver and the spleen may resume their fetal blood-forming roles. Additionally, inactive yellow bone marrow regions (essentially fatty tissue) may reconvert to active red marrow.

additional plus is that the cross-linked form of hemoglobin does not fragment in the blood (as natural hemoglobin does), so the necessity for a RBC plasma membrane "container" is eliminated.

Although test results are promising, there are still important problems to be solved. For example, potent poisons (endotoxins) produced by some bacteria tend to cling to the modified hemoglobin, and there is some evidence that the free hemoglobin provokes a generalized constriction of blood vessels, making oxygen delivery more difficult.

Neohemocytes

Researchers at the University of California at San Francisco have created artificial RBCs, which they call *neohemocytes*, by packaging natural hemoglobin molecules in fat bubbles made from phospholipids and cholesterol. The resulting "red cells" are about one-twelfth the size of human erythrocytes. Although the neohemocytes are destroyed and cleared more rapidly from the bloodstream than are real RBCs, they have a shelf life of six months (versus about 35

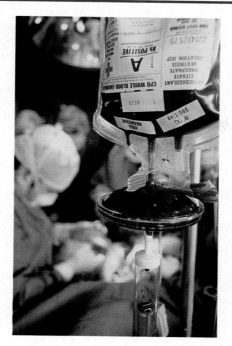

days for whole blood). This would make neohemocyte infusion a viable choice for trauma patients in immediate need of blood, but clinical trials on humans are still in the distant future.

Hemopure

Hemopure, a natural but nonhuman blood substitute recently approved by the FDA, contains purified and polymerized hemoglobin extracted from cattle blood. Because there are no plasma membranes to deal with, cross reactions should not be a problem. However, this bovine-based substitute must address the danger of transmission of "mad-cow" disease and perhaps other not yet identified diseases.

Although the FDA has encouraged the development of artificial blood for over 20 years, no marketable product, including those described above, has been approved for anything but experimental use in humans. This may change. Early in 1997, researchers at Albany Medical Center reported that they had developed a process to coat RBCs with the harmless polymer PEG to make all blood compatible with all patients. However, blood still remains a priceless commodity, and its beautiful complexity has yet to be replaced by modern medical technology.

Blood cells develop from collections of mesenchymal cells, called *blood islands*, derived from the mesoderm germ layer. The fetus forms a unique hemoglobin, **hemoglobin F**, that has a higher affinity for oxygen than does adult hemoglobin (hemoglobin A). It contains two alpha and two gamma (γ) polypeptide chains per globin molecule, instead of the paired alpha and beta chains typical of hemoglobin A. After birth, fetal erythrocytes carrying hemoglobin F are rapidly destroyed by the liver, and the baby's erythrocytes begin to produce hemoglobin A.

The most common diseases of the blood that appear during aging are chronic leukemias, anemias, and clotting disorders. However, these and most other age-related blood disorders are usually precipitated by disorders of the heart, blood vessels, or immune system. For example, the increased incidence of leukemias in old age is believed to result from the waning efficiency of the immune system, and abnormal thrombus and embolus formation reflects the progress of atherosclerosis, which roughens the blood vessel walls.

★   ★   ★

Because blood serves as the vehicle that the cardiovascular system uses to transport substances throughout the body, it could be considered the servant of the cardiovascular system. On the other hand, without blood, the normal functions of the heart and blood vessels are impossible. So perhaps the organs of the cardiovascular system, described in Chapters 19 and 20, are subservient to blood. The point of this circular thinking is that blood and the cardiovascular system organs are vitally intertwined in their common functions—to see that nutrients, oxygen, and other vital substances reach all tissue cells of the body and to relieve the cells of their wastes.

RELATED CLINICAL TERMS

Blood chemistry tests Chemical analysis of various chemical substances in the blood, e.g., glucose, iron, or calcium determinations, protein tests, bilirubin content, and pH.

Blood fraction Any one of the components of whole blood that has been separated out from the other blood components, such as platelets or clotting factors.

Bone marrow biopsy A sample of red bone marrow is obtained by needle aspiration (typically from the sternum or the iliac crest), and is examined to diagnose disorders of blood-cell formation, leukemia, various marrow infections, and anemias resulting from damage to or failure of the marrow.

Exchange transfusion Direct transfer of blood from a donor to a recipient, as the recipient's blood is being removed; a technique to treat fetal blood incompatibilities and poisoning victims.

Hematology (hem"ah-tol'o-je) Study of blood.

Hematoma (hem"ah-to'mah) Accumulated, clotted blood in the tissues usually resulting from injury; visible as "black and blue" marks or bruises; eventually absorbed naturally unless infections develop.

Hemochromatosis (he"mo-kro"mah-to'sis) An inherited disorder of iron overload in which the intestine absorbs too much iron from the diet. The iron builds up in body tissues, where it oxidizes to form compounds that poison those organs (especially joints, liver, and pancreas).

Plasmapheresis (plaz"mah-fe-re'sis) Chemical filtering technique used to cleanse plasma of specific blood components such as proteins or toxins. Most important application appears to be for removal of antibodies or immune complexes from the blood of individuals with autoimmune disorders (multiple sclerosis, myasthenia gravis, and others).

Septicemia (sep"ti-se'me-ah; *septos* = rotten) Situation of excessive and harmful levels of bacteria or their toxins in the blood; also called blood poisoning.

CHAPTER SUMMARY

Media study tools that could provide you additional help in reviewing specific key topics of Chapter 18 are referenced below. **SP** *= Study Partner;* **IP** *= Interactive Physiology.*

Overview: Composition and Functions of Blood (pp. 651–652)

Components *(p. 651)*

1. Blood is composed of formed elements and plasma. The hematocrit is a measure of one formed element, erythrocytes, as a percentage of total blood volume.

Physical Characteristics and Volume *(p. 652)*

2. Blood is a viscous, slightly alkaline fluid representing about 8% of total body weight. Blood volume of a normal adult is about 5 L.

Functions *(p. 652)*

3. Distribution functions include delivery of oxygen and nutrients to body tissues, removal of metabolic wastes, and transport of hormones.

4. Regulation functions are maintenance of body temperature, of constant blood pH, and of adequate fluid volume.

5. Protective functions include hemostasis and prevention of infection.

Blood Plasma *(p. 652)*

1. Plasma is a straw-colored, viscous fluid composed of 90% water. The remaining 10% is solutes, such as nutrients, respiratory gases, salts, hormones, and proteins. Plasma makes up 55% of whole blood.

2. Plasma proteins, most made by the liver, include albumin, globulins, and clotting proteins. Albumin is an important blood buffer and contributes to the osmotic pressure of blood.

Formed Elements *(pp. 652–666)*

1. Formed elements, accounting for 45% of whole blood, are erythrocytes, leukocytes, and platelets. All formed elements arise from hemocytoblasts in red bone marrow.

Erythrocytes *(pp. 653–660)*

2. Erythrocytes (red blood cells) are small biconcave cells containing large amounts of hemoglobin. They have no nucleus and few organelles. Spectrin allows the cells to change shape as they pass through tiny capillaries.

3. Oxygen transport is the major function of erythrocytes. In the lungs, oxygen binds to iron atoms in hemoglobin molecules, producing oxyhemoglobin. In body tissues, oxygen dissociates from iron, producing deoxyhemoglobin.

4. Red blood cells begin as hemocytoblasts and, through erythropoiesis, proceed from the proerythroblast (committed cell) stage to the erythroblast (early and late), normoblast, and reticulocyte stages. During this process, hemoglobin accumulates and the organelles and nucleus are extruded. Differentiation of reticulocytes is completed in the bloodstream.

5. Erythropoietin and testosterone enhance erythropoiesis.

6. Iron, vitamin B_{12}, and folic acid are essential for production of hemoglobin.

7. Red blood cells have a life span of approximately 120 days. Old and damaged erythrocytes are removed from the circulation by macrophages of the liver and spleen. Released iron from hemoglobin is stored as ferritin or hemosiderin to be reused. The balance of the heme group is degraded to bilirubin and secreted in bile. Amino acids of globin are metabolized or recycled.

IP Respiratory System CD-ROM; Topic: Gas Transport, pages 3–5, 11–17.

8. Erythrocyte disorders include anemias and polycythemia.

SP Exercise: Chapter 18, Regulating the Rate of Hematopoiesis; Case Study: Sickle Cell Anemia.

Leukocytes *(pp. 661–666)*

9. Leukocytes are white blood cells. All are nucleated, and all have crucial roles in defending against disease. Two main categories exist: granulocytes and agranulocytes.

10. Granulocytes include neutrophils, basophils, and eosinophils. Basophils contain histamine, which enhances

migration of leukocytes to inflammatory sites and promotes vasodilation. Neutrophils are active phagocytes. Eosinophils attack parasitic worms and their numbers increase during allergic reactions.

11. Agranulocytes have crucial roles in immunity. They include lymphocytes, the "immune cells," and monocytes, which differentiate into macrophages.

12. Leukopoiesis is directed by colony-stimulating factors and interleukins released mainly by lymphocytes and macrophages.

13. Leukocyte disorders include leukemias and infectious mononucleosis.

Platelets (p. 666)

14. Platelets are fragments of large, multinucleate megakaryocytes formed in red marrow. When a blood vessel is damaged, platelets form a plug to help prevent blood loss and play a central role in the clotting cascade.

SP Exercise: Chapter 18, Blood Components.

Hemostasis (pp. 666-672)

1. Hemostasis is prevention of blood loss. The three major phases of hemostasis are vascular spasms, platelet plug formation, and blood coagulation.

Coagulation (pp. 668-670)

2. Coagulation of blood may be initiated by either the intrinsic or the extrinsic pathway. Platelet phospholipid (PF₃) is crucial to both pathways. Tissue factor (thromboplastin) generated by tissue cell injury allows the extrinsic pathway to bypass many steps of the intrinsic pathway. A series of activated procoagulants oversees the intermediate steps of each cascade. The pathways converge as prothrombin is converted to thrombin.

Clot Retraction and Repair (p. 669)

3. After a clot is formed, clot retraction occurs. Serum is squeezed out and the ruptured vessel edges are drawn together. The vessel is repaired by smooth muscle, connective tissue, and endothelial cell proliferation and migration.

Fibrinolysis (pp. 669-671)

4. When healing is complete, clot digestion (fibrinolysis) occurs.

Factors Limiting Clot Growth or Formation (p. 671)

5. Abnormal expansion of clots is prevented by removal of coagulation factors in contact with rapidly flowing blood and by inhibition of activated blood factors. PGI₂ (prostacyclin) secreted by the endothelial cells helps prevent undesirable (unnecessary) clotting.

Disorders of Hemostasis (pp. 671-672)

6. Thromboembolytic disorders involve undesirable clot formation, which can occlude vessels.

7. Thrombocytopenia, a deficit of platelets, causes spontaneous bleeding from small blood vessels. Hemophilia is caused by a genetic deficiency of certain coagulation factors. Liver disease can also cause bleeding disorders because many coagulation proteins are formed by the liver.

Transfusion and Blood Replacement (pp. 672-675)

Transfusion of Whole Blood (pp. 672-675)

1. Whole blood transfusions are given to replace severe blood loss and to treat anemia or thrombocytopenia.

2. Blood group is based on agglutinogens (antigens) present on red blood cell membranes.

3. When mismatched blood is transfused, the recipient's agglutinins (plasma antibodies) clump the foreign RBCs; the clumped cells are then lysed. Blood vessels may be blocked by clumped RBCs; released hemoglobin may precipitate in the kidney tubules, causing renal shutdown.

4. Before whole blood can be transfused, it must be typed and cross matched so that transfusion reactions are avoided. The most important blood groups for which blood must be typed are the ABO and Rh groups.

SP Exercise: Chapter 18, Blood Types.

Plasma and Blood Volume Expanders (p. 675)

5. Transfusions of plasma alone or plasma expanders are given when rapid replacement of blood volume is necessary.

Diagnostic Blood Tests (pp. 675-676)

1. Diagnostic blood tests can provide large amounts of information about the current status of the blood and of the body as a whole.

Developmental Aspects of Blood (pp. 676-677)

1. Fetal hematopoietic sites include the yolk sac, liver, and spleen. By the seventh month of development, the red bone marrow is the primary blood-forming site.

2. Blood cells develop from blood islands derived from mesoderm. Fetal blood contains hemoglobin F. After birth, hemoglobin A is formed.

3. The major blood-related problems associated with aging are leukemia, anemia, and thromboembolytic disease.

REVIEW QUESTIONS

Multiple Choice/Matching

1. The blood volume in an adult averages approximately (a) 1 L, (b) 3 L, (c) 5 L, (d) 7 L.

2. The hormonal stimulus that prompts red blood cell formation is (a) serotonin, (b) heparin, (c) erythropoietin, (d) thrombopoietin.

3. All of the following are true of RBCs except (a) biconcave disc shape, (b) life span of approximately 120 days, (c) contain hemoglobin, (d) contain nuclei.

4. The most numerous WBC is the (a) eosinophil, (b) neutrophil, (c) monocyte, (d) lymphocyte.

5. Blood proteins play an important part in (a) blood clotting, (b) immunity, (c) maintenance of blood volume, (d) all of the above.

6. The white blood cell that releases histamine and other inflammatory chemicals is the (a) basophil, (b) neutrophil, (c) monocyte, (d) eosinophil.

7. The blood cell that is said to be immunologically competent is the (a) lymphocyte, (b) megakaryocyte, (c) neutrophil, (d) basophil.

8. The normal erythrocyte count (per cubic millimeter) for adults is (a) 3 to 4 million, (b) 4.5 to 5 million, (c) 8 million, (d) 500,000.

9. The normal pH of the blood is about (a) 8.4, (b) 7.8, (c) 7.4, (d) 4.7.

10. Suppose your blood was found to be AB positive. This means that (a) agglutinogens A and B are present on your red blood cells, (b) there are no anti-A or anti-B agglutinins in your plasma, (c) your blood is Rh^+, (d) all of the above.

Short Answer Essay Questions

11. (a) Define formed elements and list their three major categories. (b) Which is least numerous? (c) Which comprise(s) the buffy coat in a hematocrit tube?

12. Discuss hemoglobin relative to its chemical structure, its function, and the color changes it undergoes during loading and unloading of oxygen.

13. If you had a high hematocrit, would you expect your hemoglobin determination to be low or high? Why?

14. What nutrients are needed for erythropoiesis?

15. (a) Describe the process of erythropoiesis. (b) What name is given to the immature cell type released to the circulation? (c) How does it differ from a mature erythrocyte?

16. Besides the ability to move by amoeboid action, what other physiological attributes contribute to the function of white blood cells in the body?

17. (a) If you had a severe infection, would you expect your WBC count to be closest to 5000, 10,000, or 15,000 per cubic millimeter? (b) What is this condition called?

18. (a) Describe the appearance of platelets and state their major function. (b) Why should platelets not be called "cells"?

19. (a) Define hemostasis. (b) List the three major steps of coagulation. Explain what initiates each step and what the step accomplishes. (c) In what general way do the intrinsic and extrinsic mechanisms of clotting differ? (d) What ion is essential to virtually all stages of coagulation?

20. (a) Define fibrinolysis. (b) What is the importance of this process?

21. (a) How is clot overgrowth usually prevented? (b) List two conditions that may lead to unnecessary (and undesirable) clot formation.

22. How can liver dysfunction cause bleeding disorders?

23. (a) What is a transfusion reaction and why does it happen? (b) What are its possible consequences?

24. How can poor nutrition lead to anemia?

25. What blood-related problems are most common in the aged?

Critical Thinking and Clinical Application Questions

1. Cancer patients being treated with chemotherapeutic drugs designed to destroy rapidly dividing cells are monitored closely for changes in their red and white blood counts. Why so?

2. Mary Healy, a young woman with severe vaginal bleeding, is admitted to the emergency room. She is three months pregnant, and the physician is concerned about the volume of blood she is losing. (a) What type of transfusion will probably be given to this patient? (b) What blood tests will be performed before starting the transfusion?

3. Alan Forsythe, a middle-aged college professor from Boston, is in the Swiss Alps studying astronomy during his sabbatical leave. He has been there for two days and plans to stay the entire year. However, he notices that he is short of breath when he walks up steps and tires easily with any physical activity. His symptoms gradually disappear, and after two months he feels fine. Upon returning to the United States, he has a complete physical exam and is told that his erythrocyte count is higher than normal. (a) Attempt to explain this finding. (b) Will his RBC count remain at this higher-than-normal level? Why or why not?

4. A young child is diagnosed as having acute lymphocytic leukemia. Her parents cannot understand why infection is a major problem for Janie when her WBC count is so high. Can you provide an explanation for Janie's parents?

5. Mrs. Ryan, a middle-aged woman, appears at the clinic complaining of multiple small hemorrhaged spots in her skin and severe nosebleeds. While taking her history, the nurse notes that Mrs. Ryan is taking apronal (a sleeping medication) because she has problems getting to sleep at night. This drug is known to be toxic to red marrow. Using your knowledge of physiology, explain the connection between the bleeding problems and apronal.

6. A reticulocyte count indicated that 5% of Tyler's red blood cells were reticulocytes. His blood test also indicated he had polycythemia and a hematocrit of 65%. Explain the connection between these three facts.

7. In 1998, the U.S. Food And Drug Administration approved the nation's first commercial surgical glue to control bleeding during certain surgeries. Called Tisseel, it forms a flexible mesh over an oozing blood vessel to help stem bleeding within five minutes. This sealant is made from two blood proteins that naturally cause blood to clot when they latch together. Name these proteins.

THE CARDIOVASCULAR SYSTEM: THE HEART

19

Heart Anatomy   (pp. 683–693)

1. Describe the size and shape of the heart, and indicate its location and orientation in the thorax.
2. Name the coverings of the heart.
3. Describe the structure and function of each of the three layers of the heart wall.
4. Describe the structure and functions of the four heart chambers. Name each chamber and provide the name and general route of its associated great vessel(s).
5. Trace the pathway of blood through the heart.
6. Name the major branches of the coronary arteries and describe their distribution.
7. Name the heart valves and describe their location, function, and mechanism of operation.

Properties of Cardiac Muscle Fibers   (pp. 693–696)

8. Describe the structural and functional properties of cardiac muscle, and explain how it differs from skeletal muscle.
9. Briefly describe the events of cardiac muscle cell contraction.

Heart Physiology   (pp. 696–709)

10. Name the components of the conduction system of the heart, and trace the conduction pathway.
11. Draw a diagram of a normal electrocardiogram tracing; name the individual waves and intervals, and indicate what each represents. Name some abnormalities that can be detected on an ECG tracing.
12. Describe the timing and events of the cardiac cycle.

13. Describe normal heart sounds, and explain how heart murmurs differ from normal sounds.
14. Name and explain the effects of the various factors involved in regulating stroke volume and heart rate.
15. Explain the role of the autonomic nervous system in regulating cardiac output.

Developmental Aspects of the Heart   (pp. 709–711)

16. Describe fetal heart formation, and indicate how the fetal heart differs from the adult heart.
17. Provide examples of age-related changes in heart function.

? *The title for this figure points out that the heart is in the mediastinum. Just what is the mediastinum?*

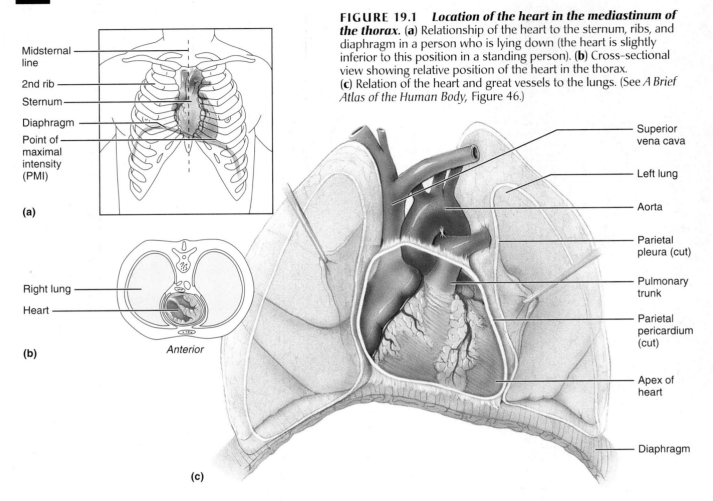

Midsternal line
2nd rib
Sternum
Diaphragm
Point of maximal intensity (PMI)

(a)

Right lung
Heart

(b)   Anterior

FIGURE 19.1 *Location of the heart in the mediastinum of the thorax.* (**a**) Relationship of the heart to the sternum, ribs, and diaphragm in a person who is lying down (the heart is slightly inferior to this position in a standing person). (**b**) Cross-sectional view showing relative position of the heart in the thorax. (**c**) Relation of the heart and great vessels to the lungs. (See *A Brief Atlas of the Human Body*, Figure 46.)

Superior vena cava
Left lung
Aorta
Parietal pleura (cut)
Pulmonary trunk
Parietal pericardium (cut)
Apex of heart
Diaphragm

(c)

The ceaselessly beating heart within the chest has intrigued people for centuries. The ancient Greeks believed the heart was the seat of intelligence. Others thought it was the source of emotions. While these theories have proven false, we do know that emotions affect heart rate. When your heart pounds or occasionally skips a beat, you become acutely aware of how much you depend on this dynamic organ for your very life.

Despite its vital importance, the heart does not work alone. Indeed, it is part of the cardiovascular system, which also includes the miles of blood vessels that run throughout the body. Day and night, tissue cells take in nutrients and oxygen and excrete wastes. Because cells can make such exchanges only with their immediate environment, some means of changing and renewing that environment is necessary to ensure a continual supply of nutrients and to prevent buildup of wastes. The cardiovascular system provides the transport system "hardware" that keeps blood continuously circulating to fulfill this critical homeostatic need.

When stripped of its romantic cloak, the **heart** is no more than the transport system pump; the hollow blood vessels are the delivery routes. Using blood as the transport medium, the heart continually propels oxygen, nutrients, wastes, and many other substances into the interconnecting blood vessels that move to and past the body cells. Blood and blood vessels are considered separately in Chapters 18 and 20, respectively. This chapter focuses on the structure and function of the heart.

The medial cavity of the thorax within which the heart, great vessels, and trachea are found.

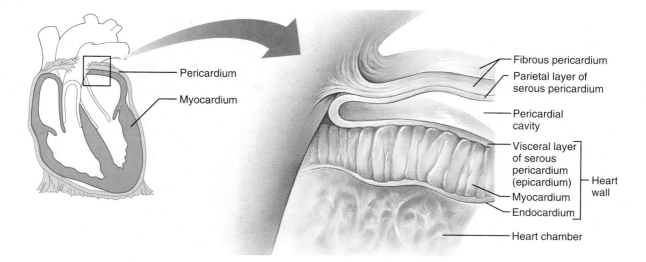

FIGURE 19.2 *The pericardial layers and layers of the heart wall.*

HEART ANATOMY

Size, Location, and Orientation

The modest size and weight of the heart belie its incredible strength and endurance. About the size of a fist, the hollow, cone-shaped heart (Figure 19.1) weighs between 250 and 350 grams—less than a pound.

Snugly enclosed within the **mediastinum** (me″de-ah-sti′num), the medial cavity of the thorax, the heart extends obliquely for 12 to 14 cm (about 5 inches) from the second rib to the fifth intercostal space (Figure 19.1a). As it rests on the superior surface of the diaphragm, the heart lies anterior to the vertebral column and posterior to the sternum. The lungs flank the heart laterally and partially obscure it. Approximately two-thirds of its mass lies to the left of the midsternal line; the balance projects to the right. Its broad, flat **base,** or posterior surface, is about 9 cm (3½ in) wide and is directed toward the right shoulder. Its **apex** points inferiorly toward the left hip. If you press your fingers between the fifth and sixth ribs just below the left nipple, you can easily feel the beating of your heart where the apex contacts the chest wall. Hence, this site is referred to as the **point of maximal intensity (PMI).**

Coverings of the Heart

The heart is enclosed in a double-walled sac called the **pericardium** (per″ĭ-kar′de-um; *peri* = around, *cardi* = heart) (Figure 19.2). The loosely fitting superficial part of this sac is the **fibrous pericardium.** This tough, dense connective tissue layer (1) protects the heart, (2) anchors it to surrounding struc-

tures, such as the diaphragm and the great vessels that issue from the heart, and (3) prevents overfilling of the heart with blood.

Deep to the fibrous pericardium is the **serous pericardium,** a thin slippery serous membrane composed of two layers. Its **parietal layer** lines the internal surface of the fibrous pericardium. At the superior margin of the heart, the parietal layer attaches to the large arteries exiting from the heart, and then turns inferiorly and continues over the external heart surface as the **visceral layer,** also called the **epicardium** ("upon the heart"). The epicardium is an integral part of the heart wall.

Between the two layers of serous pericardium is the slitlike **pericardial cavity,** which contains a film of serous fluid. The serous membranes, lubricated by the fluid, glide smoothly against one another during heart activity, allowing the mobile heart to work in a relatively friction-free environment.

◩ *Homeostatic Imbalance*

Inflammation of the pericardium, or *pericarditis* (as might follow bacterial pneumonia), hinders the production of serous fluid and roughens the serous membrane surfaces. Consequently, as the beating heart rubs against its pericardial sac, it creates a creaking sound *(pericardial friction rub)* that can be heard with a stethoscope. Pericarditis is characterized by pain deep to the sternum. Over time, it may lead to adhesions in which the pericardia stick together and impede heart activity. In severe cases of pericarditis, the situation changes and *large* amounts of inflammatory fluid seep into the pericardial cavity. This excess fluid compresses the heart, limiting its

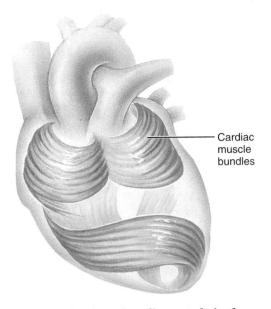

Cardiac muscle bundles

FIGURE 19.3   *Packaging of cardiac muscle in the heart.* Longitudinal view showing the spiral and circular arrangement of cardiac muscle bundles.

ability to pump blood. This condition of heart compression by fluid is called *cardiac tamponade* (tam"po-nād′), literally, "heart plug." Physicians treat cardiac tamponade by inserting a syringe into the pericardial cavity and draining off the excess fluid. ■

Layers of the Heart Wall

The heart wall is composed of three layers: a superficial epicardium, a middle myocardium, and a deep endocardium (Figure 19.2). All three layers are richly supplied with blood vessels.

The **epicardium,** or visceral layer of the serous pericardium, is often infiltrated with fat, especially in older people.

The **myocardium** ("muscle heart"), composed mainly of cardiac muscle, forms the bulk of the heart. It is the layer that actually contracts. Within this layer, the branching cardiac muscle cells are tethered to each other by crisscrossing connective tissue fibers and arranged in spiral or circular *bundles* (Figure 19.3). These interlacing bundles effectively link all parts of the heart together. The connective tissue fibers form a dense network, called the **fibrous skeleton of the heart,** that reinforces the myocardium internally and anchors the cardiac muscle fibers. This network of collagen and elastin fibers is thicker in some areas than others. For example, it constructs ropelike rings that provide additional support where the great vessels issue from the heart and around the heart valves (see also Figure 19.8, p. 691), which otherwise might eventually become stretched because of the continual stress of blood pulsing through them. Additionally, since con-

nective tissue is not electrically excitable, it limits the direct spread of action potentials across the heart to specific pathways. (The importance of this fact is discussed later in this chapter.)

The **endocardium** ("inside the heart") is a glistening white sheet of endothelium (squamous epithelium) resting on a thin connective tissue layer. Located on the inner myocardial surface, it lines the heart chambers and covers the connective tissue skeleton of the valves. The endocardium is continuous with the endothelial linings of the blood vessels leaving and entering the heart.

Chambers and Associated Great Vessels

The heart has four chambers (Figure 19.4, pp. 685–687)—two superior **atria** (a′tre-ah) and two inferior **ventricles** (ven′trĭ-klz). The internal partition that divides the heart longitudinally is called the **interatrial septum** when it separates the atria, and the **interventricular septum** (Figure 19.4e) when it separates the ventricles. The right ventricle forms most of the anterior surface of the heart. The left ventricle dominates the inferoposterior aspect of the heart and forms the heart apex. Two grooves visible on the heart surface indicate the boundaries of its four chambers and carry the blood vessels that supply the myocardium. The **atrioventricular groove** (Figure 19.4b, d), or **coronary sulcus,** encircles the junction of the atria and ventricles like a crown (*corona* = crown). The **anterior interventricular sulcus,** cradling the anterior interventricular artery (Figure 19.4a, b), marks the anterior position of the septum separating the right and left ventricles. It continues as the **posterior interventricular sulcus,** which provides a similar landmark on the heart's posteroinferior surface.

Atria: The Receiving Chambers

Except for small, wrinkled, protruding appendages called **auricles** (or′ĭ-klz; *auricle* = little ear), which increase the atrial volume somewhat, the right and left atria are remarkably free of distinguishing surface features. Internally, the atrium has two basic parts (Figure 19.4c): a smooth-walled posterior part and an anterior portion in which the walls are ridged by bundles of muscle tissue. Since the walls look as if they were raked by the tines of a comb, these muscle bundles are called **pectinate** (*pectin* = comb) **muscles.** These two atrial regions are separated by a C-shaped ridge called the *crista terminalis* ("terminal crest"). The interatrial septum bears a shallow depression, the **fossa ovalis** (o-vă′lis) (Figure 19.4c),

Left ventricle. It's the systemic pump that has to pump blood through the entire systemic circulation against high resistance.

? *Which heart chamber has the thickest walls? What is the significance of this structural difference?*

FIGURE 19.4 *Gross anatomy of the heart.*
(**a**) Photograph of anterior aspect of the heart (pericardium removed). (**b**) Anterior view. (See *A Brief Atlas of the Human Body*, Figure 47.)

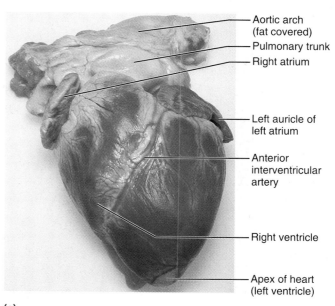

Aortic arch (fat covered)
Pulmonary trunk
Right atrium
Left auricle of left atrium
Anterior interventricular artery
Right ventricle
Apex of heart (left ventricle)

(a)

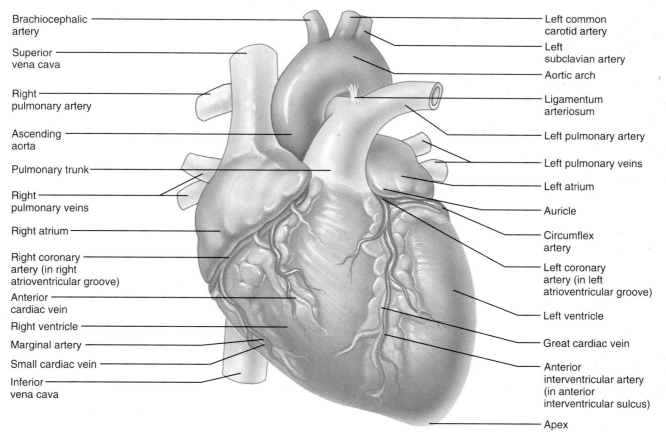

Brachiocephalic artery
Superior vena cava
Right pulmonary artery
Ascending aorta
Pulmonary trunk
Right pulmonary veins
Right atrium
Right coronary artery (in right atrioventricular groove)
Anterior cardiac vein
Right ventricle
Marginal artery
Small cardiac vein
Inferior vena cava

Left common carotid artery
Left subclavian artery
Aortic arch
Ligamentum arteriosum
Left pulmonary artery
Left pulmonary veins
Left atrium
Auricle
Circumflex artery
Left coronary artery (in left atrioventricular groove)
Left ventricle
Great cardiac vein
Anterior interventricular artery (in anterior interventricular sulcus)
Apex

(b)

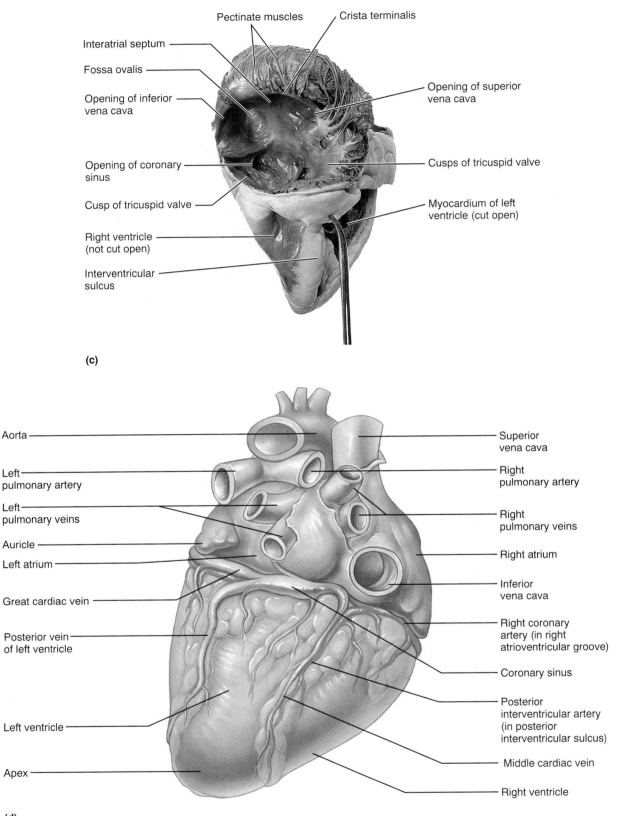

Pectinate muscles

Crista terminalis

Interatrial septum

Fossa ovalis

Opening of inferior
vena cava

Opening of superior
vena cava

Opening of coronary
sinus

Cusps of tricuspid valve

Cusp of tricuspid valve

Myocardium of left
ventricle (cut open)

Right ventricle
(not cut open)

Interventricular
sulcus

(c)

Aorta

Superior
vena cava

Left
pulmonary artery

Right
pulmonary artery

Left
pulmonary veins

Right
pulmonary veins

Auricle

Right atrium

Left atrium

Inferior
vena cava

Great cardiac vein

Right coronary
artery (in right
atrioventricular groove)

Posterior vein
of left ventricle

Coronary sinus

Posterior
interventricular artery
(in posterior
interventricular sulcus)

Left ventricle

Apex

Middle cardiac vein

Right ventricle

(d)

FIGURE 19.4 (continued) *Gross anatomy of the heart.* **(c)** Right anterior view of the internal aspect of the right atrium. The anterior wall is incised near its left margin and folded back to the right. **(d)** Posterior surface view.

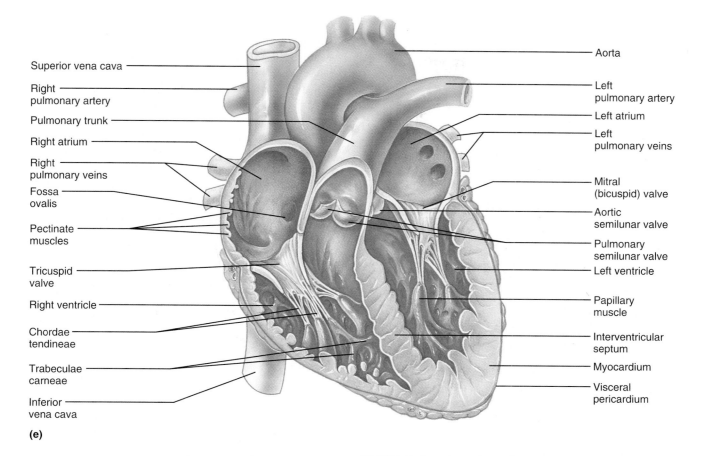

Superior vena cava

Right pulmonary artery

Pulmonary trunk

Right atrium

Right pulmonary veins

Fossa ovalis

Pectinate muscles

Tricuspid valve

Right ventricle

Chordae tendineae

Trabeculae carneae

Inferior vena cava

Aorta

Left pulmonary artery

Left atrium

Left pulmonary veins

Mitral (bicuspid) valve

Aortic semilunar valve

Pulmonary semilunar valve

Left ventricle

Papillary muscle

Interventricular septum

Myocardium

Visceral pericardium

(e)

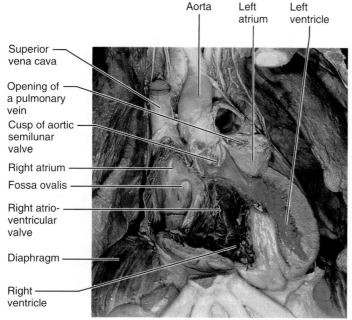

Aorta   Left atrium   Left ventricle

Superior vena cava

Opening of a pulmonary vein

Cusp of aortic semilunar valve

Right atrium

Fossa ovalis

Right atrio-ventricular valve

Diaphragm

Right ventricle

(f)

FIGURE 19.4 (continued) *Gross anatomy of the heart.* **(e)** Frontal section showing interior chambers and valves. (See *A Brief Atlas of the Human Body,* Figure 48.) **(f)** Photograph of a cadaver heart comparable to diagrammatic view shown in Figure 19.4e.

ventricles, the atria are relatively small, thin-walled chambers. As a rule, they contribute little to the propulsive pumping activity of the heart.

Blood enters the *right atrium* via three veins: (1) The **superior vena cava** returns blood from body regions superior to the diaphragm; (2) the **inferior vena cava** returns blood from body areas below the diaphragm; and (3) the **coronary sinus** collects blood draining from the myocardium itself. Four **pulmonary veins** enter the *left atrium,* which makes up most of the heart's base. The pulmonary veins transport blood from the lungs back to the heart. These vessels are best seen in a posterior view of the heart (Figure 19.4d).

Ventricles: The Discharging Chambers

Together the ventricles (*ventr* = underside) make up most of the mass of the heart. As already mentioned, the right ventricle forms most of the heart's anterior surface while the left ventricle dominates its posteroinferior surface. Marking the internal walls of the ventricular chambers are irregular ridges of muscle called **trabeculae carneae** (trah-bek'u-le kar'ne-e) (Figure 19.4e), literally, "crossbars of

that marks the spot where an opening, the *foramen ovale,* existed in the fetal heart (see Figures 19.4c, 19.4e, and 19.24, p. 709).

Functionally, the atria are receiving chambers for blood returning to the heart from the circulation (*atrium* = entry way). Because they need contract only minimally to push blood "next door" into the

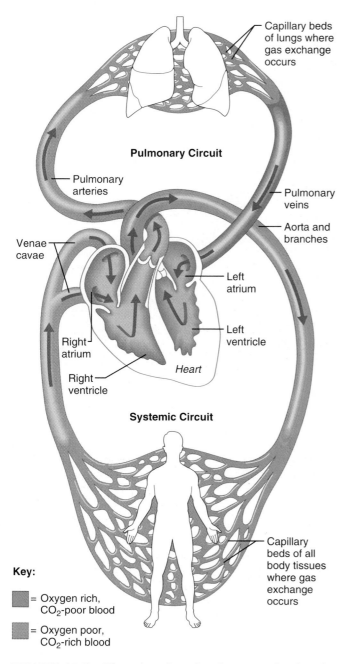

Pulmonary Circuit

Capillary beds
of lungs where
gas exchange
occurs

Pulmonary
arteries

Pulmonary
veins

Venae
cavae

Aorta and
branches

Left
atrium

Right
atrium

Left
ventricle

Right
ventricle

Heart

Systemic Circuit

Capillary
beds of all
body tissues
where gas
exchange
occurs

Key:

= Oxygen rich,
CO_2-poor blood

= Oxygen poor,
CO_2-rich blood

FIGURE 19.5 *The systemic and pulmonary circuits.* The heart is a double pump that serves two circulations. The right side of the heart pumps blood through the pulmonary circuit* (to the lungs and back to the left heart). The left heart pumps blood via the systemic circuit to all body tissues and back to the right heart. Notice that blood flowing through the pulmonary circuit gains oxygen and loses carbon dioxide as depicted by the color change from blue to red. Blood flowing through the systemic circuit loses oxygen and picks up carbon dioxide (red to blue color change).

*For simplicity, the actual number of two pulmonary arteries and four pulmonary veins has been reduced to one each.

flesh." Still other muscle bundles, the conelike **papillary muscles,** project into the ventricular cavity. As described shortly, the papillary muscles play a role in valve function.

The ventricles are the discharging chambers or actual pumps of the heart (the difference in function between the atria and the ventricles is reflected in the much more massive ventricular walls; Figure 19.4e). When the ventricles contract, blood is propelled out of the heart into the circulation. The right ventricle pumps blood into the **pulmonary trunk,** which routes the blood to the lungs where gas exchange occurs. The left ventricle ejects blood into the **aorta** (a-or'tah), the largest artery in the systemic circulation—indeed, in the body.

Pathway of Blood Through the Heart

Until the sixteenth century, it was believed that blood moved through the heart in a side-to-side fashion by seeping through pores in its midline wall or septum. We now know that the heart passages open not from side to side but vertically and that the heart is actually two side-by-side pumps, each serving a separate blood circuit (Figure 19.5). The blood vessels that carry blood to and from the lungs form the **pulmonary circuit** (*pulmonos* = lung). This circuit serves gas exchange. The blood vessels that carry the functional blood supply to and from all body tissues constitute the **systemic circuit.**

The right side of the heart is the *pulmonary circuit pump.* Blood returning from the body, which is oxygen-poor and carbon dioxide–rich (relatively speaking), enters the right atrium and passes into the right ventricle, which pumps it to the lungs via the pulmonary trunk (Figure 19.4a, b). In the lungs, the blood unloads carbon dioxide and picks up oxygen. The freshly oxygenated blood is carried by the pulmonary veins back to the left side of the heart. Notice how unique this circulation is. Typically, we think of veins as vessels that carry blood that is relatively oxygen-poor to the heart and arteries as transporters of oxygen-rich blood from the heart to the rest of the body. Exactly the opposite condition exists in the pulmonary circuit.

The left side of the heart is the *systemic circuit pump.* Freshly oxygenated blood leaving the lungs is returned to the left atrium and passes into the left ventricle, which pumps it into the aorta. From there the blood is transported via smaller systemic arteries to the body tissues, where gases and nutrients are exchanged across the capillary walls. Then the blood, once again loaded with carbon dioxide and oxygen depleted, returns through the systemic veins to the right side of the heart, where it enters the right atrium through the superior and inferior venae cavae. This whole cycle repeats itself continuously.

Although equal volumes of blood are being pumped to the pulmonary and systemic circuits at any moment, the two ventricles have very unequal workloads. The pulmonary circuit, served by the right ventricle, is a short, low-pressure circulation, whereas the systemic circuit, associated with the left ventricle, takes a long pathway through the entire body and encounters about five times as much friction, or resistance to blood flow. This functional difference is revealed in the comparative anatomy of the two ventricles (Figures 19.4e and 19.6). The walls of the left ventricle are three times as thick as those of the right ventricle, and its cavity is nearly circular. The right ventricular cavity is flattened into a crescent shape that partially encloses the left ventricle, much the way a hand might loosely grasp a clenched fist. Consequently, the left ventricle can generate much more pressure than the right and is a far more powerful pump.

Coronary Circulation

Although the heart is more or less continuously filled with blood, this contained blood provides very little nourishment to the heart tissues. (The myocardium is too thick to make diffusion a practical means of nutrient delivery.) The **coronary circulation,** the functional blood supply of the heart, is the shortest circulation in the body. The arterial supply of the coronary circulation is provided by the *right* and *left coronary arteries*, both of which arise from the base of the aorta and encircle the heart in the atrioventricular groove (Figure 19.7a). The **left coronary artery** runs toward the left side of the heart and then divides into its major branches: the **anterior interventricular artery,** which follows the anterior interventricular sulcus and supplies blood to the interventricular septum and anterior walls of both ventricles; and the **circumflex artery,** which supplies the left atrium and the posterior walls of the left ventricle.

The **right coronary artery** courses to the right side of the heart, where it also divides into two branches: The **marginal artery** serves the myocardium of the lateral part of the right side of the heart. The more important **posterior interventricular artery** runs to the heart apex and supplies the posterior ventricular walls. Near the apex of the

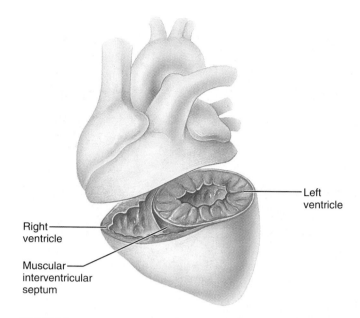

FIGURE 19.6 *Anatomical differences in the right and left ventricles.* The left ventricle has a thicker wall and its cavity is basically circular; by contrast, the right ventricle cavity is crescent shaped and wraps around the left ventricle.

heart, it merges (anastomoses) with the anterior interventricular artery. Overall the branches of the right coronary artery supply the right atrium and nearly all of the right ventricle.

The arterial supply of the heart varies considerably. For example, in 15% of people, the left coronary artery gives rise to *both* interventricular arteries; in about 4%, a single coronary artery supplies the whole heart. Additionally, there are many anastomoses among the coronary arterial branches. These fusing networks provide additional *(collateral)* routes for blood delivery to the heart muscle, which explains how the heart can receive adequate nutrition even when one of its coronary arteries is almost entirely occluded. Even so, complete blockage of a coronary artery leads to tissue death and heart attack (see below).

The coronary arteries provide an intermittent, pulsating blood flow to the myocardium. These vessels and their main branches lie in the epicardium and send branches inward (deep) to nourish the myocardium. They actively deliver blood when the heart is relaxed, but they are fairly ineffective when the ventricles are contracting because (1) they are compressed by the contracting myocardium, and (2) the entrances to the coronary arteries are partly blocked by the flaps of the open aortic semilunar valve. Although the heart represents only about 1/200 of body weight, it requires about 1/20 of the body's blood supply. As might be expected, the left ventricle receives the most plentiful blood supply.

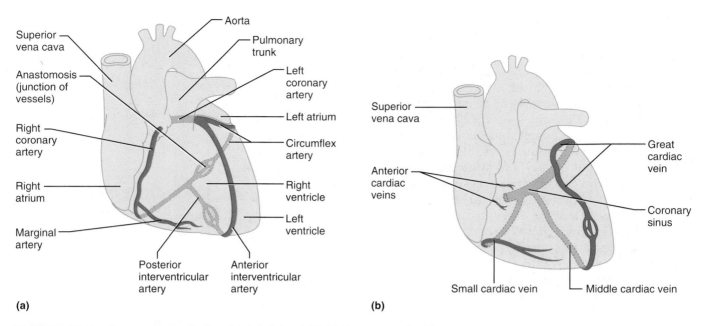

FIGURE 19.7 *Coronary circulation.* (**a**) Arterial supply. (**b**) Venous supply. (The lighter-tinted vessels in each case represent those more posterior in the heart.)

After passing through the capillary beds of the myocardium, the venous blood is collected by the **cardiac veins,** whose paths roughly follow those of the coronary arteries. The cardiac veins, in turn, join together to form an enlarged vessel called the **coronary sinus,** which empties the blood into the right atrium. The coronary sinus is obvious on the posterior aspect of the heart (Figure 19.7b). The coronary sinus has three large tributaries: the **great cardiac vein,** found in the anterior interventricular sulcus, the **middle cardiac vein** lying in the posterior interventricular sulcus, and the **small cardiac vein,** which runs along the heart's right inferior margin. Additionally, several **anterior cardiac veins** empty directly into the right atrium anteriorly.

⚠ *Homeostatic Imbalance*

Any blockage of the coronary arterial circulation can be serious; in some cases, it is fatal. **Angina pectoris** (an-ji'nah pek'tor-is), literally, "choked chest," is thoracic pain caused by a fleeting deficiency in blood delivery to the myocardium. It may result from stress-induced spasms of the coronary arteries or increased physical demands on the heart. The myocardial cells are weakened by the temporary lack of oxygen, but they do not die. Far more serious is prolonged coronary blockage, which can lead to a **myocardial infarction (MI),** commonly called a **heart attack** or a **coronary** (see *A Closer Look* on pp. 712–713). Because adult cardiac muscle is amitotic, any areas of cell death are repaired with noncontrac-

tile scar tissue. Whether or not a person survives a myocardial infarction depends on the extent and location of the damage. Damage to the left ventricle, which is the systemic pump, is most serious. ■

Heart Valves

Blood flows through the heart in one direction: from the atria to the ventricles and out the great arteries leaving the superior aspect of the heart. This one-way traffic is enforced by four heart valves: the paired atrioventricular and semilunar valves (Figures 19.4e and 19.8). The valves open and close in response to differences in blood pressure on their two sides.

Atrioventricular (AV) Valves

The two **atrioventricular (AV) valves,** one located at each atrial-ventricular junction, prevent backflow into the atria when the ventricles are contracting. The right AV valve, the **tricuspid** (tri-kus'pid) **valve,** has three flexible cusps (flaps of endocardium reinforced by connective tissue cores). The left AV valve, with two flaps, is the **bicuspid valve,** but is more commonly called the **mitral** (mi'tral) **valve** because of its resemblance to the two-sided bishop's miter or hat. Attached to each of the AV valve flaps are tiny white collagen cords called **chordae tendineae** (kor'de ten"dĭ-ne-e; "tendonous cords"), "heart strings," that anchor the cusps to the papillary muscles protruding from the ventricular walls.

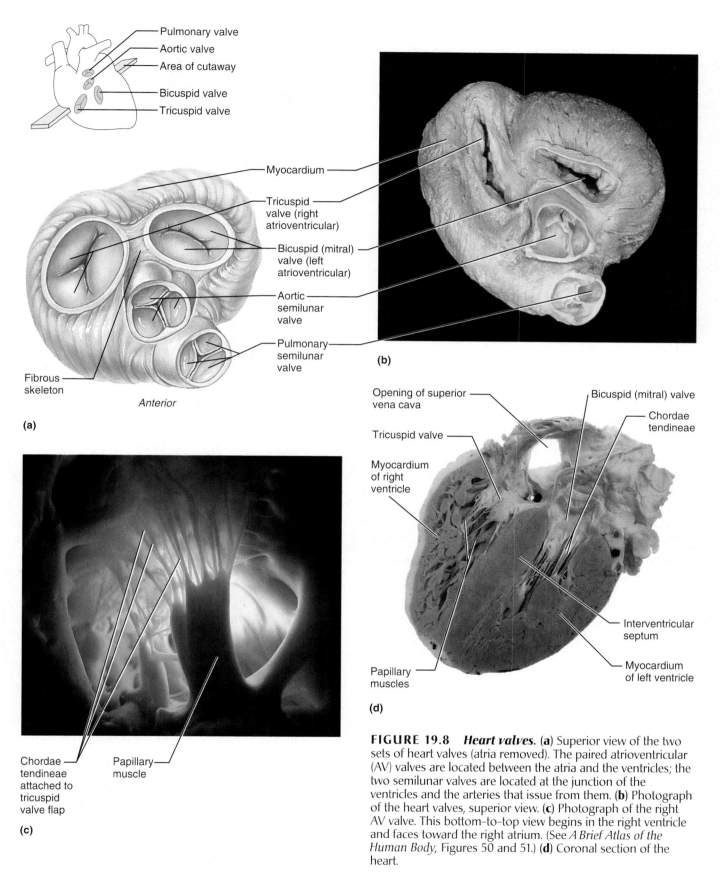

(a) Pulmonary valve
Aortic valve
Area of cutaway
Bicuspid valve
Tricuspid valve

Myocardium

Tricuspid valve (right atrioventricular)

Bicuspid (mitral) valve (left atrioventricular)

Aortic semilunar valve

Pulmonary semilunar valve

Fibrous skeleton

Anterior

(a)

(b)

Chordae tendineae attached to tricuspid valve flap

Papillary muscle

(c)

Opening of superior vena cava

Tricuspid valve

Myocardium of right ventricle

Papillary muscles

Bicuspid (mitral) valve

Chordae tendineae

Interventricular septum

Myocardium of left ventricle

(d)

FIGURE 19.8   *Heart valves.* (a) Superior view of the two sets of heart valves (atria removed). The paired atrioventricular (AV) valves are located between the atria and the ventricles; the two semilunar valves are located at the junction of the ventricles and the arteries that issue from them. **(b)** Photograph of the heart valves, superior view. **(c)** Photograph of the right AV valve. This bottom-to-top view begins in the right ventricle and faces toward the right atrium. (See *A Brief Atlas of the Human Body,* Figures 50 and 51.) **(d)** Coronal section of the heart.

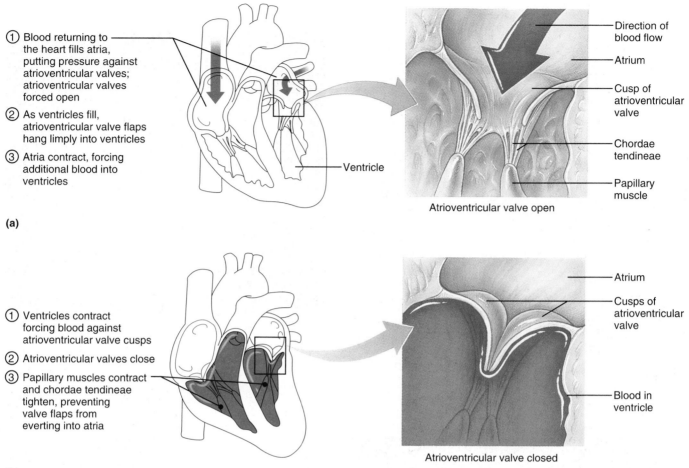

① Blood returning to the heart fills atria, putting pressure against atrioventricular valves; atrioventricular valves forced open

② As ventricles fill, atrioventricular valve flaps hang limply into ventricles

③ Atria contract, forcing additional blood into ventricles

Ventricle

Direction of blood flow

Atrium

Cusp of atrioventricular valve

Chordae tendineae

Papillary muscle

Atrioventricular valve open

(a)

① Ventricles contract forcing blood against atrioventricular valve cusps

② Atrioventricular valves close

③ Papillary muscles contract and chordae tendineae tighten, preventing valve flaps from everting into atria

Atrium

Cusps of atrioventricular valve

Blood in ventricle

Atrioventricular valve closed

(b)

FIGURE 19.9 *Operation of the atrioventricular valves of the heart.* (a) The valves open when the blood pressure exerted on their atrial side is greater than that exerted on their ventricular side. (b) The valves are forced closed when the ventricles contract and intraventricular pressure rises, moving their contained blood superiorly. The action of the papillary muscles and chordae tendineae maintains the valve flaps in the closed position.

When the heart is completely relaxed, the AV valve flaps hang limply into the ventricular chambers below; blood flows into the atria and then through the open AV valves into the ventricles. When the ventricles begin to contract, compressing the blood in their chambers, the intraventricular pressure rises, forcing the blood superiorly against the valve flaps. As a result, the valve flap edges meet, closing the valve (Figure 19.9). The chordae tendineae and the papillary muscles serve as guy-wires to anchor the valve flaps in their *closed* position. If the cusps were not anchored in this manner, they would be blown upward into the atria, in the same way that an umbrella is blown inside out by a gusty wind. The papillary muscles actually contract *before* the other ventricular musculature so that they take up the slack on the chordae tendineae before the full force of ventricular contraction hurls the blood against the AV valve flaps.

Semilunar (SL) Valves

The **aortic** and **pulmonary semilunar valves** guard the bases of the large arteries issuing from the ventricles (aorta and pulmonary trunk, respectively) and prevent backflow into the associated ventricles. Each semilunar valve is fashioned from three pocketlike cusps, each shaped roughly like a crescent moon (*semilunar* = half-moon). Their mechanism of action differs from that of the AV valves. In this case, when the ventricles are contracting and intraventricular pressure *rises above* the blood pressure in the aorta and pulmonary trunk, the semilunar valves are forced open and their cusps flatten against the arterial walls as the blood rushes past them (Figure 19.10). When the ventricles relax, and the blood (no longer propelled forward by the pressure of ventricular contraction) flows backward toward the heart, it fills the cusps and closes the valves.

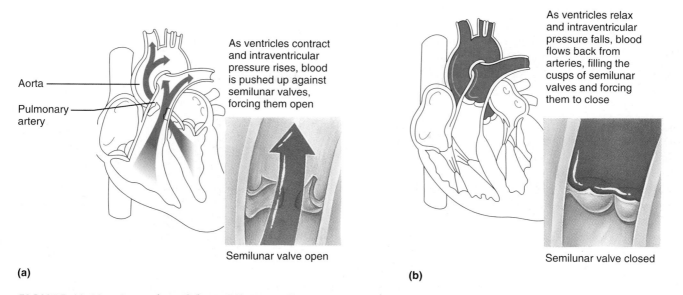

As ventricles contract and intraventricular pressure rises, blood is pushed up against semilunar valves, forcing them open

Aorta

Pulmonary artery

Semilunar valve open

(a)

As ventricles relax and intraventricular pressure falls, blood flows back from arteries, filling the cusps of semilunar valves and forcing them to close

Semilunar valve closed

(b)

FIGURE 19.10   *Operation of the semilunar valves.* (a) During ventricular contraction, the valves are open and their flaps are flattened against the artery walls. **(b)** When the ventricles relax, the backflowing blood fills the valve cusps and closes the valves.

Let us complete the valve story by mentioning what seems to be an important omission—there are no valves guarding the entrances of the venae cavae and pulmonary veins into the right and left atria, respectively. Small amounts of blood *do* spurt back into these vessels during atrial contraction, but blood backflow is largely prevented because the atrial myocardium compresses (and collapses) these venous entry points as it contracts.

Homeostatic Imbalance

Heart valves are simple devices, and the heart—like any mechanical pump—can function with "leaky" valves as long as the impairment is not too great. However, severe valve deformities can seriously hamper cardiac function. For example, an *incompetent valve* forces the heart to pump and repump the same blood because the valve does not close properly and blood backflows. In valvular *stenosis* ("narrowing"), the valve flaps become stiff (typically because of scar tissue formation following endocarditis or calcium salt deposit) and constrict the opening. This stiffness compels the heart to contract more forcibly than normal. In both instances, the work load of the heart increases and, ultimately, the heart may become severely weakened. Under such conditions, the faulty valve (most often the mitral valve) is replaced with a synthetic valve or a pig heart valve chemically treated to prevent rejection. ■

PROPERTIES OF CARDIAC MUSCLE FIBERS

Although cardiac muscle is similar to skeletal muscle, it displays some special anatomical features that reflect its unique blood-pumping role. Because skeletal muscle has already been covered in some detail in Chapter 9, it seems appropriate to compare and contrast these two muscle types.

Microscopic Anatomy

Cardiac muscle, like skeletal muscle, is striated, and its contraction is accomplished by the sliding filament mechanism. However, in contrast to the long, cylindrical, multinucleate skeletal muscle fibers, cardiac cells are short, fat, branched, and interconnected. Each fiber contains one or at most two large, pale, *centrally* located nuclei (Figure 19.11a). The intercellular spaces are filled with a loose connective tissue matrix (the *endomysium*), which contains numerous capillaries. This delicate matrix is connected, in turn, to the previously mentioned fibrous skeleton of the heart that links the cardiac cells together in a coiling array and reinforces the basketlike walls of the heart (see Figures 19.3 and 19.8a). This "skeleton" acts both as a tendon and as an insertion, giving the cardiac cells something to pull or exert their force against.

Skeletal muscle fibers are independent of one another both structurally and functionally. By contrast,

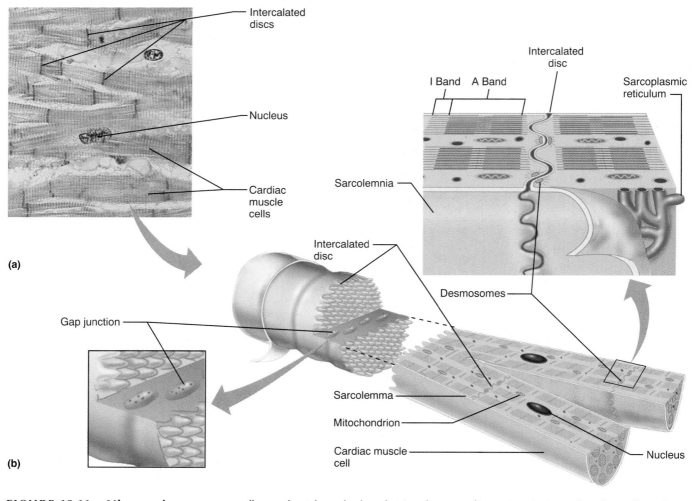

(a)

(b)

FIGURE 19.11 *Microscopic anatomy of cardiac muscle.* **(a)** Photomicrograph of cardiac muscle (465×). Notice that the cardiac muscle cells are short, branched, and striated. Also note the dark-staining intercalated discs or junctions between the adjacent cardiac cells. **(b)** Three-dimensional diagrammatic view of cardiac cells and their relationships at the intercalated discs.

the plasma membranes of adjacent cardiac cells interlock like the ribs of two sheets of corrugated cardboard at dark-staining junctions called **intercalated** (in-ter'kah-la"ted; *intercala* = insert) **discs** (Figure 19.11b). These discs contain anchoring *desmosomes* and *gap junctions* (specialized cell junctions discussed in Chapter 3). The desmosomes prevent adjacent cardiac cells from separating during contraction. Gap junctions allow ions to pass freely from cell to cell, directly transmitting the depolarizing current across the entire heart. Because cardiac cells are electrically coupled by the gap junctions, the entire myocardium *behaves* as a single coordinated unit or **functional syncytium.**

Large mitochondria account for about 25% of the volume of cardiac cells (compared to only 2% in skeletal muscle) and give the cardiac cells a high resistance to fatigue. Most of the remaining volume is occupied by myofibrils composed of fairly typical sarcomeres. The sarcomeres exhibit Z discs, A bands, and I bands, which reflect the arrangement of the thick (myosin) and thin (actin) myofilaments composing them. However, in contrast to what is seen in skeletal muscle, the myofibrils of cardiac muscle cells vary greatly in diameter and branch extensively, accommodating the abundant mitochondria that lie between them. This results in a banding pattern less dramatic than that seen in skeletal muscle.

The system for delivering Ca^{2+} is less elaborate in cardiac muscle cells. The T tubules are wider and fewer than those of skeletal muscle and they enter the cells once per sarcomere at the Z discs. (Remember that T tubules are invaginations of the sarcolemma. In skeletal muscle, the T tubules invaginate twice per sarcomere and at the A-I junctions.) The cardiac sarcoplasmic reticulum is simpler and lacks the large terminal cisternae seen in skeletal muscle. Hence, no *triads* are seen in cardiac muscle fibers.

Mechanism and Events of Contraction

Although both heart and skeletal muscle are contractile tissues, they have some fundamental differences:

1. Means of stimulation. Each skeletal muscle fiber must be independently stimulated to contract by a nerve ending. By contrast, some cardiac muscle cells are self-excitable and can initiate their own depolarization, and that of the rest of the heart, in a spontaneous and rhythmic way. This property, called **automaticity,** or **autorhythmicity,** is described in the next section.

2. Organ versus motor unit contraction. In skeletal muscle, all cells of a given motor unit (but not necessarily all motor units of the muscle) are stimulated and contract at the same time. Impulses do not spread from cell to cell. In cardiac muscle, the heart contracts as a unit or it doesn't contract at all. This is ensured by the transmission of the depolarization wave across the heart from cell to cell via ion passage through the gap junctions, which tie all cardiac muscle cells together into a single contractile unit.

3. Length of the absolute refractory period. The absolute refractory period (the inexcitable period when sodium channels are still open or are closed or inactivated) lasts approximately 250 ms in cardiac muscle cells, nearly as long as the contraction (see Figure 19.12a). Contrast this to the 1- to 2-ms refractory periods of skeletal muscle fibers, in which contractions last 20 to 100 ms. This long refractory period normally prevents tetanic contractions, which would stop the heart's pumping action.

Now that we have probed their major differences, let's look at how the contractile mechanism of these two kinds of muscle is much the same. As with skeletal muscle, heart contraction is triggered by action potentials that sweep across the muscle cell membranes. Although about 1% of the cardiac fibers are *autorhythmic* ("self-rhythm") with the special ability to depolarize spontaneously and, thus, to pace the heart, the bulk of the heart muscle is composed of *contractile muscle fibers* which are responsible for the pumping activity of the heart. The *sequence* of electrical events in these *contractile* cells is similar to that of skeletal muscle fibers:

1. Influx of sodium ions from the extracellular fluid into the cardiac cells initiates a positive feedback cycle that causes the rising phase of the action potential (and reversal of the membrane potential from -90 mV to nearly $+30$ mV) by opening **voltage-regulated fast sodium channels** (Figure 19.12). The period of increased sodium permeability is very brief, because the sodium gates are quickly inactivated and the sodium channels close.

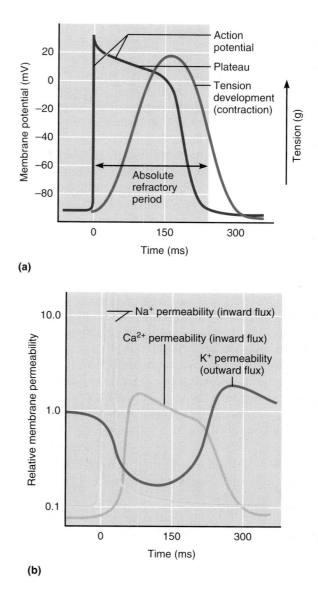

(a)

(b)

FIGURE 19.12   *Changes in membrane potential and membrane permeability during action potentials of contractile cardiac muscle cells.* (a) Relationship between the action potential (membrane potential changes), period of contraction (tension development), and absolute refractory period in a single ventricular cell. (The tracing for an atrial contractile cell is similar but the plateau phase is shorter.) (b) Membrane permeability changes during the cardiac contractile cell action potential. (The Na$^+$ permeability, which allows a rapid flux of Na$^+$ into the cell, rises to a point off the scale during the spike of the action potential.)

2. Transmission of the depolarization wave down the T tubules (ultimately) causes the sarcoplasmic reticulum (SR) to release calcium into the sarcoplasm.

3. Excitation-contraction coupling occurs as Ca^{2+} provides the signal (via troponin binding) for cross bridge activation and couples the depolarization wave to the sliding of the myofilaments.

Although the steps just listed are common to both skeletal and cardiac muscle cells, the manner in which the SR is stimulated to release Ca^{2+} differs in the two muscle types. Let's take a look.

Some 10–20% of the calcium needed for the calcium pulse that triggers contraction enters the cardiac cells from the extracellular space. Once inside, it stimulates the release of much larger amounts of Ca^{2+} (the remaining 80%) from the SR. Ionic calcium is barred from entering nonstimulated cardiac fibers, but when sodium-dependent membrane depolarization occurs, the voltage change *also* opens calcium channels that allow Ca^{2+} entry from the extracellular space. These channels are called **slow Ca^{2+} channels** because channel opening is delayed a bit (Figure 19.12b). The local influxes of Ca^{2+} through these channels trigger opening of nearby Ca^{2+}-sensitive (ryanodine) channels in the SR tubules, which liberate bursts of calcium ions ("calcium sparks") that dramatically increase the intracellular calcium concentration.

Although sodium permeability has plummeted to its resting levels and repolarization has already begun by this point, the calcium surge across the membrane prolongs the depolarization potential briefly, producing a **plateau** in the action potential tracing (Figure 19.12a). At the same time, potassium permeability decreases, which also prolongs the plateau and prevents rapid repolarization. As long as Ca^{2+} is entering the cardiac cells, the cells will continue to contract. Notice in Figure 19.12a that muscle tension develops during the plateau, and peaks just after the plateau ends. Notice also that the duration of the action potential and the contractile phase is much greater in cardiac muscle than in skeletal muscle. In skeletal muscle, the action potential typically lasts between 1 and 5 ms and the period of muscle contraction (for a single stimulus) ranges from 15 to 100 ms. In cardiac muscle, the action potential lasts 200 ms or more (because of the plateau), and tension development persists for 200 ms or more. This persistence provides the sustained contraction needed to eject blood from the heart.

Eventually (after about 200 ms), the slope of the action potential tracing falls rapidly. This results from closure of the Ca^{2+} channels and opening of voltage-regulated K^+ channels, which allows a rapid loss of potassium from the cell that restores the resting membrane potential. During repolarization, calcium ions are pumped back into the SR and the extracellular space.

Energy Requirements

Cardiac muscle has more mitochondria, which reflects its greater dependence on oxygen for its energy metabolism as compared with skeletal muscle, which can contract for prolonged periods, even during oxygen deficits, by carrying out anaerobic respiration. In contrast, the heart relies almost exclusively on aerobic respiration; hence, unlike skeletal muscle, it cannot incur much of an oxygen debt and still operate effectively.

Both types of muscle tissue use multiple fuel molecules, including glucose, fatty acids, and others. But cardiac muscle is much more adaptable and readily switches metabolic pathways to use whatever nutrient supply is available, including lactic acid generated by skeletal muscle activity. Thus, the real danger of an inadequate blood supply to the myocardium is lack of oxygen, not of nutrient fuels.

⚡ *Homeostatic Imbalance*

When a region of heart muscle is deprived of its blood supply (is ischemic), the oxygen-starved cells begin to metabolize anaerobically, producing lactic acid. The falling pH (rising H^+ level) hinders the ability of the cardiac cells to produce the ATP they need to pump Ca^{2+} into the extracellular fluid. The resulting increase in intracellular H^+ and Ca^{2+} levels causes the gap junctions (which are usually open) to close, electrically isolating the damaged cells and forcing generated action potentials to find alternate routes to reach the cardiac cells beyond them. If the ischemic area is very large, the pumping activity of the heart as a whole may be severely impaired, leading to a heart attack. ∎

HEART PHYSIOLOGY

Electrical Events

The ability of cardiac muscle to depolarize and contract is intrinsic; that is, it is a property of the heart muscle itself and does not depend on the nervous system. In fact, even if all nerve connections to the heart are severed, the heart continues to beat rhythmically, as demonstrated by transplanted hearts. Nevertheless, the healthy heart is amply supplied with nerve fibers that can alter the basic rhythm of heart activity set by intrinsic factors. Both sets of controls are examined next.

Setting the Basic Rhythm: The Intrinsic Conduction System

The independent, but coordinated, activity of the heart is a function of (1) the presence of gap junctions (mentioned above), and (2) the activity of the heart's "in-house" conduction system. The **intrinsic cardiac conduction system** consists of noncontractile cardiac cells specialized to initiate and distribute impulses throughout the heart, so that it depolarizes and contracts in an orderly, sequential manner from

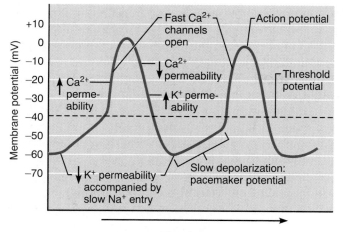

FIGURE 19.13 *Pacemaker and action potentials of autorhythmic cells of the heart.*

atria to ventricles. Thus, the heart beats as a coordinated unit. Let's look at how this system works.

Action Potential Initiation by Autorhythmic Cells

Unlike unstimulated contractile cells of the heart (and neurons and skeletal muscle fibers), the **autorhythmic cells** making up the intrinsic conduction system of the heart do *not* maintain a stable resting membrane potential. Instead, they have an **unstable resting potential** that continuously depolarizes, drifting slowly toward threshold for firing. These spontaneously changing membrane potentials are called **pacemaker potentials,** or **prepotentials,** and they initiate the action potentials (Figure 19.13) that spread throughout the heart to trigger its rhythmic contractions. The precise mechanism of the pacemaker potential is still in doubt, but it is believed to result from gradually reduced membrane permeability to K^+. Then, because Na^+ permeability is unchanged and it continues to diffuse into the cell at a slow rate, the balance between K^+ loss and Na^+ entry is upset and the membrane interior becomes less and less negative (more positive). Ultimately, when threshold (at approximately -40 mV) is reached, **fast Ca^{2+} channels** open, allowing explosive entry of Ca^{2+} (as well as some Na^+) from the extracellular space. Thus, in these autorhythmic cells, it is the influx of calcium (rather than sodium) that produces the rising phase of the action potential and reverses the membrane potential.

As in other excitable cells, the falling phase of the action potential and repolarization reflect increased K^+ permeability and efflux from the cell. Once repolarization is complete, the K^+ channels are inactivated, K^+ permeability declines, and the process of slow depolarization to threshold begins again.

Sequence of Excitation

Autorhythmic cardiac cells are localized in the following areas (Figure 19.14): ① sinoatrial (SA) node, ② atrioventricular (AV) node, ③ atrioventricular (AV) bundle (bundle of His), ④ right and left bundle branches, and ⑤ ventricular walls as Purkinje fibers. Impulses pass across the heart in the same order.

① **Sinoatrial (SA) node.** The crescent-shaped sinoatrial (si"no-a'tre-al) node is located in the right atrial wall, just inferior to the entrance of the superior vena cava. A minute cell mass with a mammoth job, the SA node typically generates impulses about 75 times every minute. (However, its inherent rate in the absence of extrinsic neural and hormonal factors is closer to 100 times per minute.) Because no other region of the conduction system or the myocardium has a faster depolarization rate, the SA sets the pace for the heart as a whole. Hence, it is the heart's **pacemaker,** and its characteristic rhythm, called **sinus rhythm,** determines heart rate.

② **Atrioventricular (AV) node.** From the SA node, the depolarization wave spreads via gap junctions throughout the atria and via the *internodal pathway* to the atrioventricular node, located in the inferior portion of the interatrial septum immediately above the tricuspid valve. At the AV node, the impulse is delayed momentarily (for about 0.1 s), allowing the atria to respond and complete their contraction before the ventricles contract. This delay largely reflects the smaller diameter of the fibers here and the fact that they have fewer gap junctions for current flow. Thus, the AV node conducts impulses more slowly than other parts of the system, just as traffic slows when cars are forced to merge into two lanes from four. Once through the AV node, the signaling impulse passes rapidly through the rest of the system (see ③–⑤).

③ **Atrioventricular (AV) bundle.** From the AV node, the impulse sweeps to the atrioventricular bundle (also called the **bundle of His**) in the superior part of the interventricular septum. Although the atria and ventricles abut each other, they are *not* connected by gap junctions. The AV bundle is the *only* electrical connection between them. The balance of the AV junction is insulated by the nonconducting fibrous skeleton of the heart.

④ **Right and left bundle branches.** The AV bundle persists only briefly before spitting into two pathways—the right and left bundle branches. The bundle branches course along the interventricular septum toward the apex of the heart.

⑤ **Purkinje fibers.** The Purkinje (pur-kin'je) fibers, essentially long strands of barrel-shaped cells with few myofibrils, complete the pathway through the interventricular septum, penetrate into the heart apex, and then turn superiorly into the ventricular walls. The

? *What ionic event produces the plateau in the AP of the contractile muscle cells?*

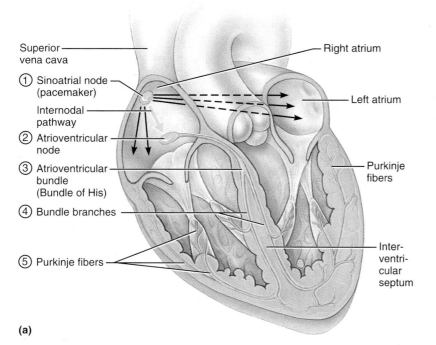

(a)

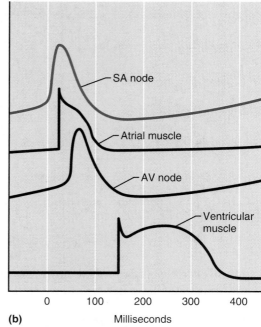

(b)

FIGURE 19.14 *The intrinsic conduction system of the heart and succession of the action potential through selected areas of the heart during one heartbeat.* (a) The depolarization wave is initiated by the SA node and then passes successively through the atrial myocardium to the AV node, the AV bundle, the right and left bundle branches, and the Purkinje fibers in the ventricular myocardium. (b) The sequence of potentials generated across the heart is shown from top to bottom beginning with the pacemaker potential generated by the SA node and ending with an action potential (with a broad plateau) typical of the contractile cardiac cells of the ventricles.

bundle branches excite the septal cells, but the bulk of ventricular depolarization depends on the large Purkinje fibers and, ultimately, on cell-to-cell transmission of the impulse via gap junctions between the ventricular muscle cells themselves. Because the left ventricle is much larger than the right, the Purkinje network is more elaborate in that side of the heart.

Interestingly, the Purkinje fibers directly supply the papillary muscles, which are excited to contract before the rest of the ventricular muscles. As mentioned before, the papillary muscles are already tightening the chordae tendineae before the full force of the ventricular contraction pushes blood against the AV valve flaps.

The total time between initiation of the impulse by the SA node and depolarization of the last of the ventricular muscle cells is approximately 0.22 s (220 ms) in a healthy human heart.

Ventricular contraction almost immediately follows the ventricular depolarization wave. A wringing contraction of the ventricles begins at the heart apex and moves toward the atria, following the direction of the excitation wave through the ventricles. This effectively ejects some of the contained blood *superiorly* into the large arteries leaving the ventricles.

Although autorhythmic cardiac cells are found in nearly all parts of the heart, their rates of spontaneous depolarization differ. For example, while the SA node normally drives the heart at a rate of 75 beats per minute, the AV node depolarizes at the average rate of about 50/min, the AV bundle and the Purkinje fibers (though they conduct very rapidly) at the rate of only about 30/min. However, these other "pacemakers" cannot dominate the heart unless everything faster than them becomes nonfunctional.

The cardiac conduction system not only coordinates and synchronizes heart activity, it also forces the heart as a whole to beat faster. Without it, the impulse would travel much more slowly through the myocardium—at the rate of 0.3 to 0.5 meters per second as opposed to several m/s in most parts of the conduction system. This would allow some muscle fibers to contract well before others, resulting in reduced pumping effectiveness.

◢ Homeostatic Imbalance

Defects in the intrinsic conduction system can cause irregular heart rhythms, or **arrhythmias** (ah-

Calcium entry.

rith'me-ahz), uncoordinated atrial and ventricular contractions, and even fibrillation. **Fibrillation** is a condition of rapid and irregular or out-of-phase contractions (see Figure 19.18d, p. 701); the heart in fibrillation has been compared with a squirming bag of worms. Fibrillating ventricles are useless as pumps; and unless the heart is defibrillated quickly, circulation stops and brain death occurs. Defibrillation is accomplished by electrically shocking the heart, which interrupts its chaotic twitching by depolarizing the entire myocardium. The hope is that "with the slate wiped clean" the SA node will begin to function normally and sinus rhythm will be reestablished.

A defective SA node may have several consequences. An **ectopic** (ek-top′ik) **focus** or abnormal pacemaker may appear and assume control of heart rate, or the AV node may become the pacemaker. The pace set by the AV node (**junctional rhythm**) is 40 to 60 beats per minute, slower than sinus rhythm but still adequate to maintain circulation. Occasionally, ectopic pacemakers appear even when the conduction system is operating normally. A small region of the heart becomes hyperexcitable, sometimes as a result of too much caffeine (several cups of coffee) or nicotine (excessive smoking), and generates impulses even more quickly than the SA node. This leads to a *premature contraction* or **extrasystole** (ek″strah-sis′to-le) before the SA node initiates the next contraction. As you might guess, **PVCs (premature ventricular contractions)** are most problematic.

Because the only route for impulse transmission from the atria to the ventricles is through the AV node, any damage to the AV node, referred to as a **heart block,** interferes with the ability of the ventricles to receive pacing impulses. In total heart block no impulses are conducted from atria to ventricles. The ventricles beat at their intrinsic rate, which is too slow to maintain adequate circulation. In such cases, a *fixed-rate* artificial pacemaker, set to discharge (deliver an impulse) at a constant rate, is usually implanted. Those suffering from partial heart block, in which some of the atrial impulses reach the ventricles, commonly receive *demand-type* pacemakers, which deliver impulses only when the heart is not transmitting on its own. ■

Modifying the Basic Rhythm: Extrinsic Innervation of the Heart

Although the basic rate of heartbeat is set by the intrinsic conduction system, fibers of the autonomic nervous system modify the marchlike beat and introduce a subtle variability from one beat to the next. The sympathetic nervous system (the "accelerator") increases both the rate and force of heartbeat; parasympathetic activation (the "brakes") slows the

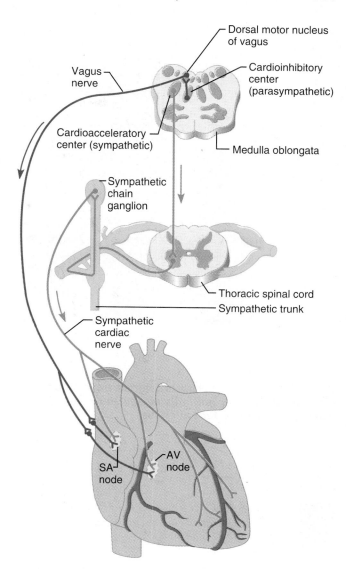

FIGURE 19.15 *Autonomic innervation of the heart.*

heart. These neural controls are explained in more detail later on pp. 706–707. Here we discuss the anatomy of the nerve supply to the heart.

The cardiac centers are located in the medulla oblongata. The sympathetic **cardioacceleratory center** projects to motor neurons in the T_1–T_5 level of the spinal cord. These, in turn, synapse with ganglionic neurons located in the cervical and upper thoracic sympathetic chain ganglia (Figure 19.15). From there, postganglionic fibers run through the cardiac plexus to the heart. The parasympathetic center, or **cardioinhibitory center,** sends impulses to the dorsal vagus nucleus in the medulla, which in turn sends inhibitory impulses to the heart via branches of the vagus nerves (cranial nerves X). Most ganglionic motor neurons lie in ganglia within the heart wall. Although these autonomic nerve fibers project to cardiac musculature throughout the heart, they project most heavily to the SA and AV nodes and the coronary arteries.

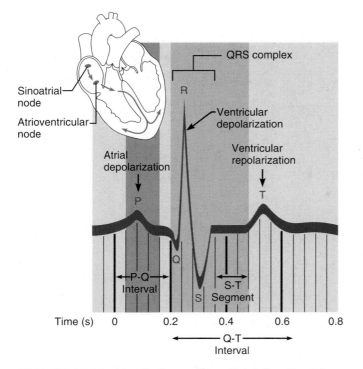

FIGURE 19.16 *An electrocardiogram tracing (lead I) illustrating the three normally recognizable deflection waves and the important intervals.*

Electrocardiography

The electrical currents generated and transmitted through the heart also spread throughout the body and can be monitored and amplified with an instrument called an **electrocardiograph.** The graphic recording of heart activity that is obtained is called an **electrocardiogram** (**ECG** or **EKG;** *kardio* is German for "heart"). An ECG is a composite of all of the APs being generated by both nodal and contractile cells at a given time and *not*, as sometimes assumed, a tracing of a single AP. Recording electrodes (leads) are positioned at various sites on the body surface; typically in a clinical setting, 12 standard leads are used to record an ECG. Three of these are bipolar leads that measure the voltage difference between the arms, or an arm and a leg, and nine are unipolar leads. Together the 12 leads provide a fairly comprehensive picture of the electrical activity of the heart.

A typical ECG consists of a series of three distinguishable waves called **deflection waves** (Figure 19.16). The first wave, the small **P wave,** lasting about 0.08 s, results from movement of the depolarization wave from the SA node through the atria. Approximately 0.1 s after the P wave begins, the atria contract.

 Since ventricular contraction follows the pattern of ventricular excitation, how can you expect the wave of contraction to move across the ventricles?

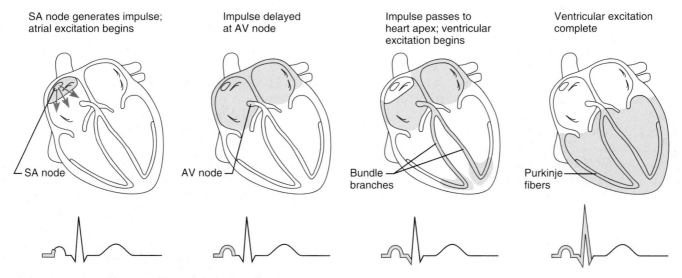

FIGURE 19.17 *The sequence of excitation of the heart related to the deflection waves of an ECG tracing.*

The large **QRS complex** results from ventricular depolarization and precedes ventricular contraction. It has a complicated shape because the paths taken by the depolarization waves through the walls of the two ventricles change continuously, producing corresponding changes in current direction. Additionally, the time required for each ventricle to depolarize depends on its relative size. Average duration of the QRS complex is 0.08 s.

The **T wave** is caused by ventricular repolarization and typically lasts about 0.16 s. Repolarization is slower than depolarization, so the T wave is more spread out and has a lower amplitude (height) than the QRS wave. Because atrial repolarization takes place during the period of ventricular excitation, its occurrence is normally obscured by the large QRS complex being recorded at the same time.

The **P-Q interval** is the time (about 0.16 s) from the beginning of atrial excitation to the beginning of ventricular excitation. It includes atrial depolarization (and contraction) as well as the passage of the depolarization wave through the rest of the conduction system. During the **S-T** segment when the action potential is in its plateau phase, the entire ventricular myocardium is depolarized. The **Q-T interval,** lasting about 0.38 s, is the period from the beginning of ventricular depolarization through their repolarization. It is sometimes referred to as the period of "electrical systole of the ventricles" because it includes the time of ventricular contraction. The relationship of the parts of an ECG tracing to movement of the action potential over the heart is illustrated in Figure 19.17.

In a healthy heart, the size, duration, and timing of the deflection waves tend to be consistent. Thus, changes in the pattern or timing of the ECG may reveal a diseased or damaged heart or problems with the heart's conduction system. For example, an enlarged R wave hints of enlarged ventricles, a flattened T wave indicates cardiac ischemia, and a prolonged Q-T interval reveals a repolarization abnormality that increases the risk of ventricular arrhythmias. A few examples of abnormalities detectable with an ECG appear in Figure 19.18.

Mechanical Events: The Cardiac Cycle

The heart undergoes some fairly dramatic writhing movements as it alternately contracts, forcing blood out of its chambers, and then relaxes, allowing its chambers to refill with blood. The terms **systole** (sis′to-le) and **diastole** (di-as′to-le) refer respectively, and literally, to these *contraction* and *relaxation* periods of heart activity. The **cardiac cycle** includes *all* events associated with the flow of blood through the

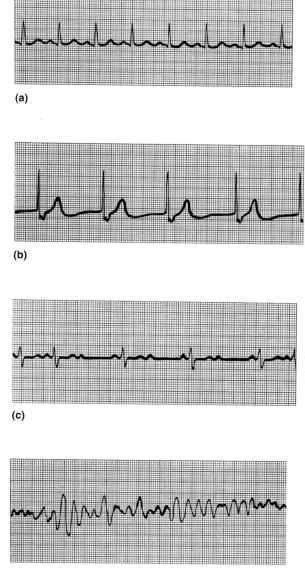

FIGURE 19.18   *Normal and abnormal ECG tracings.* (a) Normal sinus rhythm. **(b)** Junctional rhythm. SA node nonfunctional, P waves are absent, and heart rate is paced by the AV node at 40–60 beats/min. **(c)** Second-degree heart block. Some of the P waves are not conducted through the AV node; hence more P than QRS waves are seen. In cases where P waves are conducted normally, the P:QRS ratio is 1:1. In total heart block, there is no whole number ratio between the P and QRS waves, and the ventricles are no longer paced by the SA node. **(d)** Ventricular fibrillation. Chaotic, grossly irregular, bizarre ECG deflections. Seen in acute heart attack and electrical shock.

heart during one complete heartbeat, that is, atrial systole and diastole followed by ventricular systole and diastole.

The cardiac cycle is marked by a succession of pressure and blood volume changes within the heart. Although pressure changes occurring in the left side of the heart (illustrated in Figure 19.19a) are about five times greater than those occurring in the right, each ventricle pumps the same volume of blood per beat and the overall relationships are the same for both chambers.

Blood circulates endlessly, so we will arbitrarily choose a starting point to follow its route through the heart during a single cardiac cycle (Figure 19.19). We begin our explanation with the heart in total relaxation: Both the atria and ventricles are quiet, and it is mid-to-late diastole.

① **Period of ventricular filling: mid-to-late diastole.** Pressure in the heart is low, and blood returning from the circulation is flowing passively through the atria and the open AV valves into the ventricles below. The semilunar valves (aortic and pulmonary) are closed. This phase is indicated by interval 1 in Figure 19.19. Approximately 70% of ventricular filling occurs during this period, and the AV valve flaps begin to drift upward toward their closed position. (The remaining 30% is delivered to the ventricles when the atria contract toward the end of this phase.) Now the stage is set for atrial systole. Following depolarization (the P wave of the ECG), the atria contract, compressing the blood in their chambers. This causes a sudden slight rise in atrial pressure, which propels residual blood out of the atrial chambers into the ventricles. At this point the ventricles are in the last part of their resting phase (diastole) and have the maximum volume of blood they will contain in the entire cardiac cycle, a volume called the *end diastolic volume (EDV)*. Then the atria relax and the ventricles depolarize (QRS complex on the ECG). Atrial diastole persists through the rest of the cycle.

② **Ventricular systole.** As the atria relax, the ventricles begin their contraction phase. Their walls close in on the blood in their chambers, and ventricular pressure rises rapidly and sharply, closing the AV valves. For a split second, the ventricles are completely closed chambers and blood volume in the chambers remains constant. This **isovolumetric** (i″so-vol″u-met′rik) **contraction phase** is phase 2a in Figure 19.19. Ventricular pressures continue to rise and when they finally exceed those in the large arteries issuing from them, the isovolumetric stage ends as the semilunar valves are forced open and blood is expelled from the ventricles into the aorta and pulmonary trunk. During this **ventricular ejection phase** (phase 2b in Figure 19.19), the pressure in the aorta normally reaches about 120 mm Hg.

③ **Isovolumetric relaxation: early diastole.** During this brief phase following the T wave, the ventricles relax. Because the blood remaining in their chambers, referred to as the *end systolic volume (ESV)*, is no longer

compressed, ventricular pressure drops rapidly and blood in the aorta and pulmonary trunk starts to backflow toward the heart, closing the semilunar valves. Closure of the aortic semilunar valve causes a brief rise in aortic pressure as backflowing blood rebounds off the closed valve cusps. This is indicated in Figure 19.19 as the **dicrotic notch.** Once again the ventricles are totally closed chambers. This **isovolumetric relaxation** phase is indicated by interval 3 in Figure 19.19.

All during ventricular systole, the atria have been in diastole. During that time, they have been filling with blood and the intra-atrial pressure has been rising. When the pressure exerted by blood on the atrial side of the AV valves exceeds that in the ventricles below, the AV valves are forced open and ventricular filling, phase 1, begins again. Atrial pressure drops to its lowest point and ventricular pressure begins to rise, completing the cycle.

Assuming that the average heart beats approximately 75 times each minute, the length of the cardiac cycle is about 0.8 s. Of this time, atrial systole accounts for 0.1 s and ventricular systole takes 0.3 s. The remaining 0.4 s is the period of total heart relaxation, or the **quiescent period.**

Notice two important points: (1) blood flow through the heart is controlled entirely by pressure changes, and (2) blood flows along a pressure gradient, always from higher to lower pressure through any available opening. The pressure changes, in turn, reflect the alternating contraction and relaxation of the myocardium and cause the opening of the heart valves, which keeps blood flowing in the forward direction.

As already explained, the situation in the right side of the heart is essentially the same as seen in the left side *except* for the pressures seen there. The pulmonary circulation is a low-pressure circulation as evidenced by the much thinner myocardium of its right ventricle. Hence, typical systolic and diastolic pressures for the pulmonary artery are 24 and 8 mm Hg as compared to systemic aortic pressures of 120 and 80 mm Hg, respectively. However, both sides of the heart eject the same blood volume with each heartbeat.

Heart Sounds

During each cardiac cycle, two distinguishable sounds can be heard when the thorax is auscultated (listened to) with a stethoscope. These **heart sounds,** often described as lub-dup, are associated with closing of heart valves. The timing of heart sounds in the cardiac cycle is shown in Figure 19.19a.

The basic rhythm of the heart sounds is lub-dup, pause, lub-dup, pause, and so on. The pause indicates the quiescent period. The first heart

? *Are the ventricular cardiac cells in isotonic or isometric contraction during phase 2a of part (b)?*

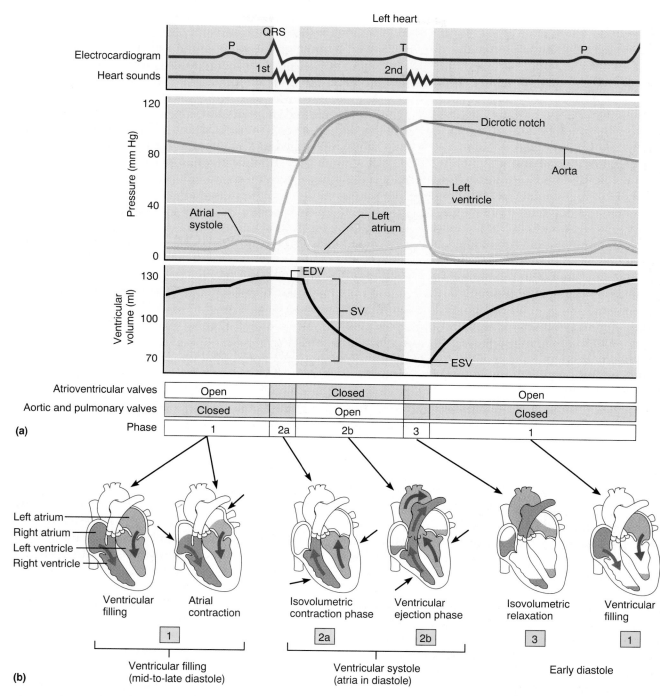

FIGURE 19.19   Summary of events occurring in the heart during the cardiac cycle. (a) Events in the left side of the heart. An ECG tracing is placed above the graphs of pressure and volume changes so that they can be related to electrical events occurring at any point. Time occurrence of heart sounds is also indicated. (b) Events of phases 1 through 3 of the cardiac cycle are depicted in diagrammatic views of the heart.

They are contracting isometrically until ventricular pressure rises above the pressure in the aorta.

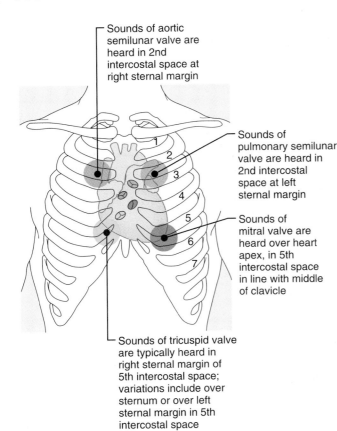

Sounds of aortic semilunar valve are heard in 2nd intercostal space at right sternal margin

Sounds of pulmonary semilunar valve are heard in 2nd intercostal space at left sternal margin

Sounds of mitral valve are heard over heart apex, in 5th intercostal space in line with middle of clavicle

Sounds of tricuspid valve are typically heard in right sternal margin of 5th intercostal space; variations include over sternum or over left sternal margin in 5th intercostal space

FIGURE 19.20 *Areas of the thoracic surface where the heart sounds can be best detected.*

sound, which occurs as the AV valves close, signifies the beginning of systole when ventricular pressure rises above atrial pressure. It tends to be louder, longer, and more resonant than the second: the short, sharp sound that is heard as the semilunar valves snap shut at the beginning of ventricular diastole. Because the mitral valve closes slightly before the tricuspid valve, and the aortic semilunar valve generally snaps shut just before the pulmonary valve, it is possible to distinguish the individual valve sounds by auscultating four specific regions of the thorax (Figure 19.20). Notice that these four points, while not directly superficial to the valves (because the sounds take oblique paths to reach the chest wall), do handily define the four corners of the normal heart. Knowing normal heart size and location is essential for recognizing an enlarged (and often diseased) heart.

⚠ *Homeostatic Imbalance*

Abnormal or unusual heart sounds are called **murmurs.** Blood flows silently as long as the flow is smooth and uninterrupted. If it strikes obstructions, its flow becomes turbulent and generates sounds, such as heart murmurs, that can be heard

with a stethoscope. Heart murmurs are fairly common in young children (and some elderly people) with perfectly healthy hearts, probably because their heart walls are relatively thin and vibrate with rushing blood. However, most often murmurs indicate valve problems. For example, if a valve is *incompetent*, a swishing sound will be heard as the blood backflows or regurgitates through the partially open valve, *after* that valve has (supposedly) closed. A stenotic valve, in which the valvular opening is narrowed, restricts blood flow *through* the valve. In the case of a stenotic aortic valve, for instance, a high-pitched sound or click can be detected when the valve should be wide open, that is, during systole just before valve closure. ■

Cardiac Output

Cardiac output (CO) is the amount of blood pumped out by *each* ventricle in 1 minute. It is the product of heart rate (HR) and stroke volume (SV). **Stroke volume** is the volume of blood pumped out by a ventricle with each beat. In general, stroke volume is correlated with force of ventricular contraction.

Using the normal resting values for heart rate (75 beats/min) and stroke volume (70 ml/beat), the average adult cardiac output can easily be computed:

CO (ml/min) = HR (75 beats/min) × SV (70 ml/beat)

Thus   CO = 5250 ml/min (5.25 L/min)

The normal adult blood volume is about 5 L (a little more than 1 gallon). Thus, the entire blood supply passes through each side of the heart once each minute. Notice that cardiac output varies directly with SV and HR. Thus CO increases when the stroke volume increases or the heart beats faster or both, and decreases when either or both of these factors decrease.

Cardiac output is highly variable and increases markedly in response to special demands, such as making a mad dash to catch the bus. The difference between resting and maximal CO is referred to as **cardiac reserve.** In nonathletic people, cardiac reserve is typically four to five times their normal CO (20–25 L/min). But in trained athletes, cardiac output during competition may reach 35 L/min (a seven-times increase). To understand how the heart accomplishes such tremendous increases in output, let's look at how stroke volume and heart rate are regulated.

Regulation of Stroke Volume

Essentially, SV represents the difference between **end diastolic volume (EDV),** the amount of blood that collects in a ventricle during diastole, and **end** systolic volume (**ESV**), the volume of blood remaining in a ventricle *after* it has contracted. The EDV, determined by the length of ventricular diastole and venous pressure (increases in either *raise* the EDV), is normally about 120 ml. The ESV, determined by arterial blood pressure (the higher the pressure the higher the ESV) and the force of ventricular contraction, is approximately 50 ml. To figure normal stroke volume, these values are simply plugged into this equation:

$$SV \text{ (ml/beat)} = EDV \text{ (120 ml)} - ESV \text{ (50 ml)}$$

$$SV = 70 \text{ ml/beat}$$

Hence, each ventricle pumps out about 70 ml of blood, or about 60% of the blood in its chambers, with each heartbeat.

So what is important here—how do we make sense out of this alphabet soup (SV, ESV, EDV)? Although several factors affect SV by causing changes in EDV or ESV, the three most important are *preload* (the amount the ventricles are stretched by contained blood), *contractility* (cardiac cell contractile force that is due to factors other than EDV), and *afterload* (back pressure exerted by blood in the large arteries leaving the heart). As described in more detail next, preload affects EDV, whereas contractility and afterload affect the ESV.

Preload: Degree of Stretch of Heart Muscle According to the **Frank-Starling law of the heart,** the critical factor controlling stroke volume is the **preload** or *degree of stretch of the cardiac muscle cells just before they contract* (Figure 19.21a). Recall that (1) stretching muscle fibers (and sarcomeres) increases the number of active cross bridge attachments that can be made between actin and myosin, and (2) the more that muscle fibers are stretched within physiological limits, the greater the force of contraction will be. Cardiac muscle, like skeletal muscle, exhibits a *length-tension relationship.* However, resting skeletal muscle fibers are kept at or near their optimal length for developing maximal tension, while resting cardiac cells are normally *shorter* than the optimal length. Therefore, increasing their length (stretching them) can produce dramatic increases in contractile force. The important factor stretching cardiac muscle is the amount of blood returning to the heart (venous return) and distending its ventricles—the EDV.

Anything that increases the volume or speed of venous return, such as a slow heart rate or exercise, increases EDV and, consequently, stroke volume and force of contraction. A slow heartbeat allows more time for ventricular filling. Exercise speeds venous return because of an increased heart rate and the squeezing action of the skeletal muscles on the veins

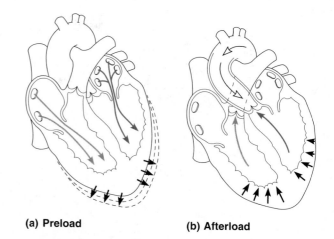

What problem might ensue relative to cardiac output if the heart beats far too rapidly for an extended period, that is, if tachycardia occurs? Why?

(a) Preload   **(b) Afterload**

FIGURE 19.21 *Preload and afterload influence stroke volume.* **(a)** Preload is related to the amount of blood stretching the ventricular fibers just before systole. **(b)** Afterload is the pressure that the ventricles must overcome to force open the aortic and pulmonary valves.

returning blood to the heart. During vigorous exercise, SV may double as a result of these factors. Conversely, low venous return, such as might result from severe blood loss or an extremely rapid heart rate, decreases EDV, causing the heart to beat less forcefully and the SV to drop (see Figure 19.23 on p. 707).

Because the systemic and pulmonary circulations are in series, this intrinsic mechanism ensures equal outputs of the ventricles and proper distribution of blood volume between the two circuits. If one side of the heart suddenly begins to pump more blood than the other, the increased venous return to the opposite ventricle forces it to pump out an equal volume, thus preventing backup or accumulation of blood in the circulation.

Contractility Although EDV is the major *intrinsic factor* influencing stroke volume, SV can also be enhanced by *extrinsic factors* that increase the *contractility* of heart muscle. **Contractility** is an increase in contractile strength that is independent of muscle stretch and the EDV. Instead, the more vigorous contractions are a direct consequence of a greater calcium influx into the cytoplasm from the extracellular fluid and the SR. Enhanced contractility results in more complete ejection of blood from the heart, hence a lower ESV and greater SV. This is the result

Pathological increase in heart rate results in too little time for ventricular filling. Therefore, CO declines and the heart beats feebly.

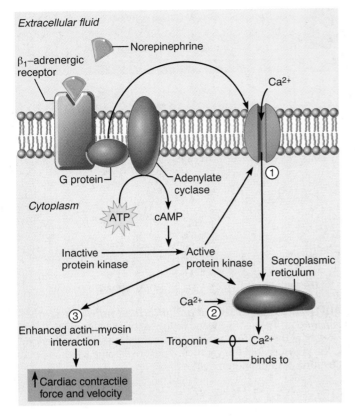

Extracellular fluid

Norepinephrine

β_1-adrenergic receptor

Ca^{2+}

G protein

Adenylate cyclase

①

Cytoplasm

ATP

cAMP

Inactive protein kinase

Active protein kinase

Sarcoplasmic reticulum

Ca^{2+} ②

③

Enhanced actin–myosin interaction

Troponin ← Ca^{2+}

binds to

↑Cardiac contractile force and velocity

FIGURE 19.22 *Mechanism by which norepinephrine influences heart contractility.* Cyclic AMP activation of protein kinases phosphorylates ① slow calcium channels promoting entry of more Ca^{2+} from the extracellular space; ② an SR protein that controls release of more Ca^{2+} by the SR; and ③ myosin, which increases the rate of myosin crossbridge cycling. In addition, phosphorylation of the SR calcium uptake pump removes calcium more rapidly from the sarcoplasm, thus speeding relaxation.

of increased sympathetic stimulation of the heart (see Figure 19.23). As noted on page 699, sympathetic fibers serve not only the intrinsic conduction system but the entire heart, and one effect of norepinephrine or epinephrine binding is to initiate a cyclic AMP second-messenger system that increases Ca^{2+} entry, which in turn promotes more cross bridge binding and enhances ventricular contractility (Figure 19.22).

Additionally, a whole battery of other chemicals influence contractility. For example, contractility is enhanced by certain hormones (glucagon, thyroxine, and epinephrine), some ions (Ca^{2+}), and certain drugs (digitalis). Factors that increase contractility are called *positive inotropic* (*ino* = muscle, fiber) *agents.* Agents or factors that impair or decrease contractility *(negative inotropic agents)* include acidosis (excess H^+), rising extracellular levels of K^+, and calcium channel blockers (e.g., verapamil).

Afterload: Back Pressure Exerted by Arterial Blood Afterload, the pressure that must be overcome for the ventricles to eject blood from the heart, is essentially the back pressure exerted on the aortic

and pulmonary valves by arterial blood (Figure 19.21b). In healthy individuals, afterload (about 80 mm Hg in the aorta and 8 mm Hg in the pulmonary trunk) is not a major determinant of stroke volume because it is relatively constant. However, in those with hypertension (high blood pressure), it is important indeed because it reduces the ability of the ventricles to eject blood. Consequently, more blood remains in the heart after systole, resulting in increased ESV and reduced stroke volume.

Regulation of Heart Rate

Given a healthy cardiovascular system, stroke volume tends to be relatively constant. However, when blood volume drops sharply or when the heart is seriously weakened, stroke volume declines and cardiac output is maintained by increasing the rate of heartbeat and contractility. Temporary stressors can also influence heart rate—and consequently cardiac output—by acting through homeostatic mechanisms induced neurally, chemically, and physically. These mechanisms are described next.

Autonomic Nervous System Regulation The most important extrinsic controls on heart rate are exerted by the autonomic nervous system. When the sympathetic nervous system is activated by emotional or physical stressors, such as fright, anxiety, excitement, or exercise (Figure 19.23), sympathetic nerve fibers release norepinephrine at their cardiac synapses. Norepinephrine (and epinephrine as well) bind to β_1 adrenergic receptors in the heart, causing threshold to be reached more quickly; thus the pacemaker fires more rapidly and the heart responds by beating faster. Sympathetic stimulation also enhances contractility by enhancing Ca^{2+} entry into the contractile cells as described above and in Figure 19.22. Since ESV falls, stroke volume does not decline, as it would if only heart *rate* were increased.

The parasympathetic division opposes sympathetic effects and effectively reduces heart rate when a stressful situation has passed. However, the parasympathetic division may be persistently activated in certain emotional conditions, such as grief and severe depression. Parasympathetic-initiated cardiac responses are mediated by acetylcholine, which hyperpolarizes the membranes of its effector cells by *opening* potassium channels. Since vagal innervation of the ventricles is sparse, parasympathetic nervous system activity has little or no effect on cardiac contractility.

Under resting conditions, both autonomic divisions continuously send impulses to the SA node of the heart, but the *dominant* influence is inhibitory. Thus, the heart is said to exhibit **vagal tone,** and heart rate is generally slower than it would be if the

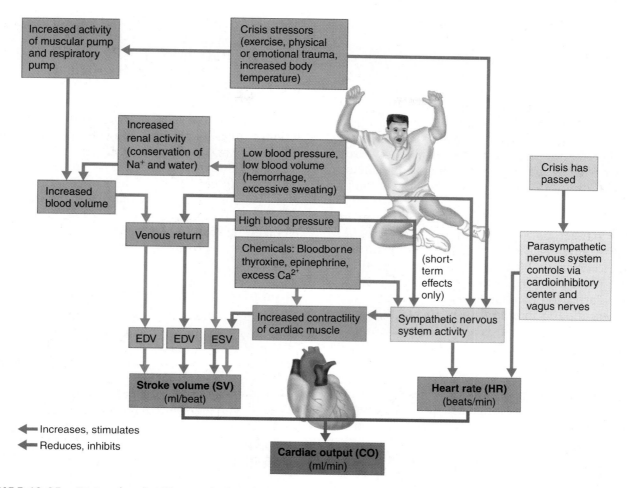

FIGURE 19.23 *Factors involved in regulation of cardiac output.*

vagal nerves were not innervating it. Cutting the vagal nerves results in an almost immediate increase in heart rate by about 25 beats/min, reflecting the inherent rate (100 beats/min) of the pacemaking SA node.

When either division of the autonomic nervous system is stimulated more strongly by sensory inputs relayed from various parts of the cardiovascular system, the alternate division is temporarily inhibited. Most such sensory input is generated by *baroreceptors (pressoreceptors)* that respond to changes in systemic blood pressure. For example, the **atrial (Bainbridge) reflex** is a sympathetic reflex initiated by an increase in venous return and blood congestion in the atria. Stretching of the atrial walls produces an increase in heart rate and force by (1) directly stimulating the SA node, and (2) stimulating baroreceptors in the atria that trigger reflexive adjustments that result in increased sympathetic stimulation of the heart.

Because increased cardiac output results in increased systemic blood pressure and vice versa, blood pressure regulation often involves reflex controls of heart rate. The neural mechanisms that regulate blood pressure are described in more detail in Chapter 20.

Chemical Regulation Chemicals normally present in the blood and other body fluids may influence heart rate (Figure 19.23), particularly if they become excessive or deficient. Some of these chemical factors are described briefly next.

1. Hormones. *Epinephrine,* a hormone liberated by the adrenal medulla during periods of sympathetic nervous system activation, produces the same cardiac effects as does norepinephrine released by the sympathetic nerves; that is, it enhances heart rate and contractility. *Thyroxine* is a thyroid gland hormone that increases metabolic rate and body heat production. When it is released in large quantities, it causes a slower but more sustained increase in heart rate than that caused by epinephrine. Since it also *enhances* the effect of epinephrine and norepinephrine on the heart, chronically hyperthyroid individuals may develop a weakened heart.

2. Ions. Physiological relationships between intracellular and extracellular ions must be maintained for normal heart function. Plasma electrolyte imbalances pose real dangers to the heart.

Homeostatic Imbalance

Reduced blood levels of ionic calcium *(hypocalcemia)* depress the heart. Conversely, *hypercalcemia* tightly couples the excitation-contraction mechanism and prolongs the plateau phase of the action potential. These events dramatically increase heart irritability, which can lead to spastic heart contractions that permit the heart little rest. Many drugs used to treat heart conditions focus on calcium transport into cardiac cells (see *A Closer Look* on pp. 712–713).

Excesses of ionic sodium and potassium are equally dangerous. Too much sodium *(hypernatremia)* inhibits transport of ionic calcium into the cardiac cells, thus blocking heart contraction. Excessive potassium *(hyperkalemia)* interferes with the depolarization mechanism by lowering the resting potential, and may lead to heart block and cardiac arrest. *Hypokalemia* is also life threatening, in that the heart beats feebly and abnormal rhythms appear. ∎

Other Factors Although they are less important than neural factors, a number of other factors, including age, gender, exercise, and body temperature, also influence heart rate. Resting heart rate is fastest in the fetus (between 140 and 160 beats/min) and gradually declines throughout life. Average heart rate is faster in females (72 to 80 beats/min) than in males (64 to 72 beats/min).

Exercise raises heart rate by acting through the sympathetic nervous system, which also increases systemic blood pressure and routes more blood to the working muscles (see Figure 19.23). However, resting heart rate in the physically fit tends to be substantially lower than in those who are out of condition. In trained athletes, it may be as slow as 40 to 60 beats/min. This apparent paradox is explained below.

Heat increases heart rate by enhancing the metabolic rate of cardiac cells. This explains the rapid, pounding heartbeat you feel when you have a high fever and also accounts, in part, for the effect of exercise on heart rate (remember, working muscles generate heat). Cold has the opposite effect: It directly decreases heart rate.

Homeostatic Imbalance

Although heart rate varies with changes in activity, marked and persistent rate changes usually signal cardiovascular disease. **Tachycardia** (tak"e-kar'de-ah; "heart hurry") is an abnormally fast heart rate, over 100 beats per minute, which may result from elevated body temperature, stress, certain drugs, or heart disease. Because tachycardia occasionally promotes fibrillation, persistent tachycardia is considered pathological.

Bradycardia (brad"e-kar'de-ah; *brady* = slow) is a heart rate slower than 60 beats/min. It may result from low body temperature, certain drugs, or parasympathetic nervous activation. It is a known, and desirable, consequence of endurance-type athletic training. As physical and cardiovascular conditioning increases, the heart hypertrophies and its stroke volume increases; thus, resting heart rate can be lower and still provide the same cardiac output. However, persistent bradycardia in poorly conditioned people may result in grossly inadequate blood circulation to body tissues, and bradycardia is a frequent warning of brain edema after head trauma. ∎

Homeostatic Imbalance of Cardiac Output

The pumping action of the heart ordinarily maintains a balance between cardiac output and venous return. Were this not so, a dangerous damming up of blood (blood congestion) would occur in the veins returning blood to the heart.

When the pumping efficiency (CO) of the heart is so low that blood circulation is inadequate to meet tissue needs, the heart is said to be in **congestive heart failure (CHF)**. Congestive heart failure is usually a progressively worsening condition that reflects weakening of the myocardium by coronary atherosclerosis, persistent high blood pressure, multiple myocardial infarcts, or cardiomyopathy. The manner in which these various conditions damage the myocardium differs, but they are all equally devastating. Let's take a look.

1. Coronary atherosclerosis, essentially a clogging of the coronary vessels with fatty buildup, impairs blood and oxygen delivery to the cardiac cells. The heart becomes increasingly hypoxic and begins to contract ineffectively.

2. Normally, pressure in the aorta during diastole is 80 mm Hg, and the left ventricle exerts only slightly over that amount of force to eject blood from its chamber. When aortic blood pressure during diastole rises to 90 mm Hg or more, the myocardium must exert more power to force the aortic valve open to pump out the same amount of blood. If this situation of enhanced afterload becomes chronic, the ESV rises and the myocardium hypertrophies. Eventually, the stress takes its toll and the myocardium becomes progressively weaker.

3. Successive myocardial infarcts depress pumping efficiency because the dead heart cells are replaced by noncontractile fibrous (scar) tissue.

4. The cause of *dilated cardiomyopathy* (kar"de-o-my-ah'path-e), or DCM, a condition in which the ventricles stretch and become flabby and the myo-

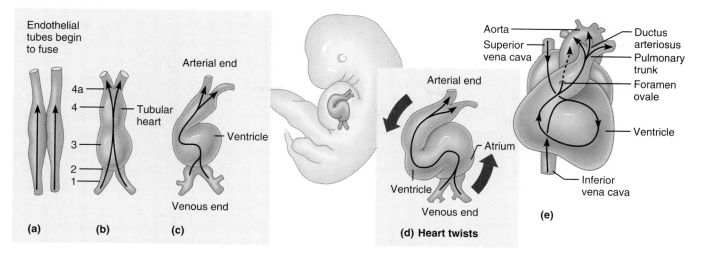

FIGURE 19.24 *Development of the human heart during week 4.* Ventral view, with the cranial direction toward the top of the figures. Arrows show the direction of blood flow. (**a**) Around day 21. (**b**) Day 22: 1 is the sinus venosus, 2 is the atrium, 3 is the ventricle, 4 is the bulbus cordis, and 4a is the truncus arteriosus. (**c**) Day 23.(**d**) Day 24. (**e**) Day 28.

cardium deteriorates, is often unknown. Drug toxicity (alcohol, cocaine, excess catecholamines, chemotherapeutic agents), hyperthyroidism, and inflammation of the heart are implicated in some cases. So too is congestive heart failure itself. The body's attempts to work harder result in increasing levels of Ca^{2+} in the cardiac cells. The calcium activates calcineurin, a calcium–sensitive enzyme which initiates a cascade that switches on genes that cause heart enlargement. Because ventricular contractility is impaired, CO is poor and the condition progressively worsens.

Because the heart is a double pump, each side can initially fail independently of the other. If the left side fails, **pulmonary congestion** occurs. The right side of the heart continues to propel blood to the lungs, but the left side does not adequately eject the returning blood into the systemic circulation. Thus, blood vessels in the lungs become engorged with blood, pressure within them increases, and fluid leaks from the circulation into the lung tissue, causing pulmonary edema. If the congestion is untreated, the person suffocates.

If the right side of the heart fails, **peripheral congestion** occurs. Blood stagnates within body organs, and pooled fluids in the tissue spaces impair the ability of body cells to obtain adequate nutrients and oxygen and to rid themselves of wastes. Edema is most noticeable in the extremities (feet, ankles, and fingers).

Failure of one side of the heart puts a greater strain on the opposite side, and ultimately the whole heart fails. A seriously weakened, or *decompensated,* heart is irreparable. Treatment is directed primarily toward (1) conserving heart energy with digitalis derivatives such as digoxin (which makes the heart

beat more slowly), (2) removing the excess leaked fluid with *diuretics* (drugs that increase excretion of Na^+ and water by the kidneys), and (3) reducing cardiac afterload by using drugs that drive down blood pressure. However, heart transplants and other surgical or mechanical remedies to replace damaged heart muscle are providing additional hope for selected cardiac patients (see *A Closer Look* on pp. 712–713).

DEVELOPMENTAL ASPECTS OF THE HEART

The human heart, derived from mesoderm, begins its existence as two simple endothelial tubes that quickly fuse to form a single chamber or heart tube that is busily pumping blood by the 23rd day of gestation (Figure 19.24). One or two days later, the heart tube exhibits four slightly bulged areas that represent the earliest heart chambers. From tail to head, following the direction of blood flow, the four chambers are:

1. Sinus venosus (ven-o'sus). This chamber initially receives all the venous blood of the embryo. It will become the smooth-walled part of the right atrium and the coronary sinus. It also gives rise to the sino-atrial node, which takes the "baton" and sets heart rate early in embryonic development.

2. Atrium. This embryonic chamber eventually becomes the pectinate muscle-ridged parts of the atria.

3. Ventricle. The strongest pumping chamber of the early heart, the ventricle gives rise to the *left* ventricle.

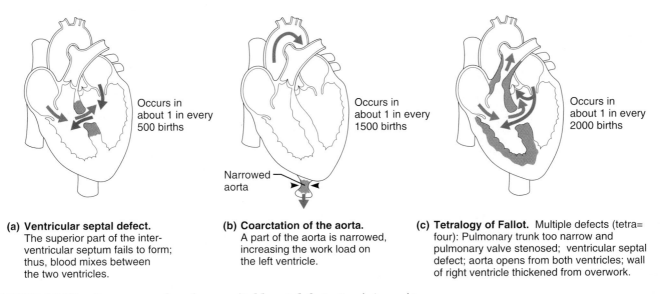

(a) Ventricular septal defect.
The superior part of the interventricular septum fails to form;
thus, blood mixes between
the two ventricles.

Occurs in
about 1 in every
500 births

Narrowed
aorta

(b) Coarctation of the aorta.
A part of the aorta is narrowed,
increasing the work load on
the left ventricle.

Occurs in
about 1 in every
1500 births

(c) Tetralogy of Fallot. Multiple defects (tetra=
four): Pulmonary trunk too narrow and
pulmonary valve stenosed; ventricular septal
defect; aorta opens from both ventricles; wall
of right ventricle thickened from overwork.

Occurs in
about 1 in every
2000 births

FIGURE 19.25   *Some examples of congenital heart defects.* Purple is used to
show the heart areas where the defects are present.

4. Bulbus cordis. This chamber, plus its cranial extension, the *truncus arteriosus,* give rise to the pulmonary trunk, first part of the aorta, and most of the
right ventricle.

During the next three weeks, the heart "tube" exhibits dramatic contortions as it undergoes right-
ward looping, and major structural changes occur
that convert it into a four-chambered organ capable
of acting as a double pump—all without missing a
beat! The ventricle moves caudally and the atrium
cranially, assuming their adult positions. The heart
divides into its four definitive chambers (via a number of stages), the midline septum forms, and the
bulbus cordis splits into the pulmonary trunk and
ascending aorta. After the second month, few
changes other than growth occur until birth.

The interatrial septum of the fetal heart is incomplete. The **foramen ovale** (literally, "oval door")
connects the two atria and allows blood entering the
right heart to bypass the pulmonary circuit and the
collapsed, nonfunctional fetal lungs. Another lung
bypass, the **ductus arteriosus,** exists between the
pulmonary trunk and the aorta. (These and other fetal circulatory bypasses are described in more detail
in Chapter 29.) At or shortly after birth, these shunts
close, completing the separation between the right
and left sides of the heart. In the adult heart, the position of the foramen ovale is revealed by the fossa
ovalis, and the **ligamentum arteriosum** is the fibrous remnant of the ductus arteriosus (see Figure
19.4b).

 Homeostatic Imbalance

Building a perfect heart is difficult. Each year about
30,000 infants are born with one or more of the 30
different **congenital heart defects,** making them
the most common of all birth defects. The most
prevalent abnormalities produce two basic kinds of
disorders in the newborn. They either (1) lead to
mixing of oxygen-poor systemic blood with oxygenated pulmonary blood (so that inadequately oxygenated blood reaches the body tissues) or (2) involve narrowed valves or vessels that greatly
increase the work load on the heart. Examples of the
first type of defect are *septal defects* (Figure 19.25a)
and *patent ductus arteriosus,* in which the connection between the aorta and pulmonary trunk remains open. *Coarctation of the aorta* (Figure
19.25b) is an example of the second type of problem.
Tetralogy of Fallot (te-tral'o-je ov fal-o'), a serious
condition in which the baby becomes cyanotic
within minutes of birth, encompasses both types of
disorders (Figure 19.25c). Modern surgical techniques can correct most of these heart defects. Most
congenital heart problems are traceable to environmental influences, such as maternal infection or
drug intake during month 2 when the major events
of heart formation occurred. ■

In the absence of congenital heart problems, the
heart functions admirably throughout a long lifetime
for most people. Homeostatic mechanisms are normally so efficient that people rarely notice when the

heart is working harder. In people who exercise regularly and vigorously, the heart gradually adapts to the increased demand by increasing in size and becoming a more efficient and powerful pump. Consequently, stroke volume increases and resting heart rate declines. Aerobic exercise also helps to clear fatty deposits from blood vessel walls throughout the body, thus retarding the process of atherosclerosis and coronary heart disease. Barring some chronic illnesses, this beneficial cardiac response to exercise persists into ripe old age.

The key word on the benefits of exercise is *regularity*. Regular exercise gradually enhances the endurance and strength of the myocardium. For example, 30 minutes a day of moderately vigorous exercise (brisk walking, biking, or yard work) offers significant health benefits to most adults. However, intermittent vigorous exercise, enjoyed by weekend athletes, may push an unconditioned heart beyond its ability to respond to the unexpected demands and bring on a myocardial infarct.

Because of the incredible amount of work the heart does over the course of a lifetime, certain structural changes are inevitable. These age-related changes affecting the heart include:

1. Sclerosis and thickening of the valve flaps. This occurs particularly where the stress of blood flow is greatest (mitral valve). Thus, heart murmurs are more common in elderly people.

2. Decline in cardiac reserve. Although the passing years seem to cause few changes in resting heart rate, the aged heart is less able to respond to both sudden and prolonged stressors that demand increased output. Sympathetic control of the heart becomes less efficient (presumably because calcium channels in sympathetic nerve endings decrease in number with age), and heart rate gradually becomes more variable. There is a decline in maximum heart rate, but this is much less of a problem in physically active seniors.

3. Fibrosis of cardiac muscle. As one ages, more and more cardiac cells die and are replaced with fibrous tissue. This reduces the stroke volume of the heart. Additionally, the nodes of the heart's conduction system sometimes become fibrosed (scarred) with age, which increases the incidence of arrhythmias and other conduction problems.

4. Atherosclerosis. Although the insidious progress of atherosclerosis begins in childhood, it is accelerated by inactivity, smoking, and stress. The most serious consequences to the heart are hypertensive heart disease and coronary artery occlusion. These conditions, in turn, increase the risk of heart attack and stroke. Although the aging process itself leads to changes in blood vessel walls that result in atherosclerosis, many investigators feel that diet, not aging, is the single most important contributor to cardiovascular disease. There is some agreement that risk is lowered if people consume less animal fat, cholesterol, and salt.

*   *   *

The heart is an exquisitely engineered double pump that operates with precision to propel blood into the large arteries leaving its chambers. However, continuous circulation of blood also depends critically on the pressure dynamics existing within the blood vessels. Chapter 20 considers the structure and function of the body's blood vessels and relates this information to the work of the heart to provide a complete picture of cardiovascular functioning.

Getting up can be perilous to your health! Even if you are a morning person, your risk of dropping dead from a heart attack is twice as high the first two hours after you climb out of bed as at any other time of the day or night. For some reason, platelets are stickier and plasma levels of stress hormones (norepinephrine and others) increase right after you arise.

When a heart attack hits you, it seems as though a lightning bolt has struck. Few events in life are more frightening. A crushing, breath-robbing pain crashes through your chest. You break into a cold sweat while nausea and a sense of doom overwhelm you. Several million Americans will suffer a heart attack this year. Perhaps one-third of them will die almost immediately as a result of it; half will be dead within the year.

But exactly what is a heart attack? Basically, it is a condition of total interruption of oxygen delivery to part of the myocardium. If the oxygen deprivation is sustained long enough, it leads to a myocardial infarction (MI), a situation in which heart muscle dies and is (eventually) replaced by scar tissue. Although it was once presumed that it was the lack of oxygen itself that injured the myocardial cells, this is apparently not so. The trouble starts when the ischemic area is reperfused with blood, and lymphocytes and other inflammatory cells congregate in the area. These "immigrants" release cytokines (signaling chemicals) that prod macrophages and neutrophils (granulocytes introduced in Chapter 18) to release a deluge of free radicals, including nitric oxide. These toxic substances depress cardiac contractility ("stunned myocardium"), and also appear to cause most of the local tissue damage. If the heart is not "tipped" into fibrillation by the insult, the patient generally survives unless damage to the myocardium is massive. The most common cause of MI is blockage of a coronary artery (particularly those narrowed by

atherosclerosis) by a thrombus or embolus, but irregular and unexplainable spasms of the coronary vessels bedevil some MI victims.

The first signs of heart disease are likely to be different in women than in men. In almost two-thirds of men, the first symptom is a heart attack or sudden death. In 56% of women, the first sign is angina, sharp squeezing chest pains that are fleeting. Heart attack victims lucky enough to reach the hospital quickly are now routinely treated with tPA or other clot-dissolving drugs, which stop many heart attacks in progress and help prevent loss of viable heart muscle.

Those who suffer angina or survive a heart attack have been given a painful warning that, if heeded, allows time to make dietary changes, take heart-saving drugs, or have coronary bypass surgery. In coronary bypass surgery, blood vessels removed from another body site are used to provide alternate routes for blood circulation in the heart. Not all people with repeated myocardial ischemia are so fortunate. For some, a fatal heart attack is the first and last symptom of heart disease. These victims of *silent ischemia* (periods when the myocardium is starved for oxygen but the person feels no pain) have no obvious symptoms of heart disease. They are at much greater risk of suffering a heart attack, which, even if not fatal, permanently damages the myocardium. What causes silent heart disease is still a mystery; some suggest that these people have abnormal pain mechanisms. Whatever the cause, a severely damaged heart typically begins its downward spiral toward congestive heart failure and death.

The key to preventing heart attacks and cardiac death is to identify those at risk, early if possible, by using sophisticated medical imaging techniques such as CT scans and MRI techniques (see pp. 20–21), as well as other diagnostic techniques. Particularly telling is presence of a specific type of low-density lipoprotein (LDL), called *lipoprotein (a)*. (LDLs are the form in which cholesterol is transported in blood to tissue cells.) About

one-quarter of all heart attacks in men younger than 60 occur in those with high plasma levels of lipoprotein (a). Unlike levels of other lipoproteins (both HDLs and LDLs), which vary widely in response to changes in diet, exercise, and drugs, lipoprotein (a) levels remain stable and positively correlated with increased incidence of clotting, vascular spasms, and the progress of atherosclerosis (see *A Closer Look* in Chapter 20), a major risk factor for heart disease.

Once the possibility of heart attack has been detected, the next step is to reduce its risk. All risk factors that can be eliminated voluntarily—excess weight, high cholesterol and saturated fat in diet, smoking, excess stress—should be attended to first. Vitamins with antioxidant abilities (vitamins E and C) are being suggested to counter free radical damage in the heart, and daily intake of 400 μg of folate and 3 mg of vitamin B_6 are suggested to minimize cardiovascular mortality.

Drug therapy is the usual next step. Traditional cardiac drugs, such as nitroglycerine (which dilates the coronary vessels) and digitalis (which slows the heart, enhancing venous return and sparing the heart's energy), are still popular. "Take one aspirin every other day" is suggested for those prone to excessive clotting, particularly those who have already had an MI or stroke. (Aspirin makes platelets less sticky.) A number of newer drugs, such as beta-blockers and calcium channel blockers, have revolutionized cardiac medicine. When norepinephrine (or epinephrine) attaches to beta receptors, heart rate and force increase. Because beta-blockers inhibit this activation by the sympathetic nervous system and successfully prevent increases in blood pressure, they reduce the strain on the heart. Calcium channel blockers such as verapamil help prevent the coronary artery spasms that often trigger heart attacks.

What are the options for someone whose heart is so devastated by disease that these measures are too little, too late? Until recently, heart transplant

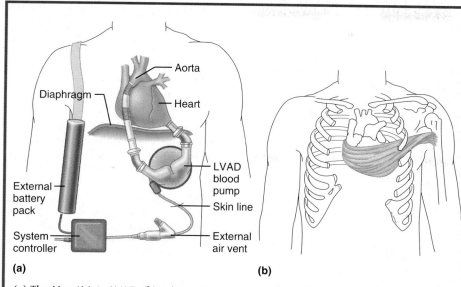

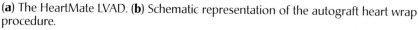

(a) The HeartMate LVAD. **(b)** Schematic representation of the autograft heart wrap procedure.

surgery has been the only hope for a dying heart. But this procedure is riddled with problems. First, an acceptable tissue match must be found. Then, after complicated and traumatic surgery, the heart recipient is dosed with drugs that suppress the immune system to thwart rejection of the donor heart. After about a year of survival, a new type of rejection may develop, in which antibodies attack the inner layer of the coronary vessels; this can lead to a fatal heart attack and it does not respond to any known immunosuppressive treatment. Despite the complications, heart transplants are becoming more and more successful. Over three-quarters of transplant patients now survive for 2 years, and half of them survive for more than 12 years.

The severe shortage of donor hearts contrasts with the great demand for them (fully one-third of patients awaiting a heart transplant in the U.S. die without receiving one). This unmet demand has created tremendous pressure to develop alternative "hearts" or treatment procedures. In December 1982, the first permanent artificial heart, called the Jarvik-7 after its developer, was implanted in Dr. Barney Clark. Barney Clark survived for only ten days with his artificial heart, and subsequent patients that received the Jarvik-7 succumbed to strokes, overwhelming infections, and kidney failure. Because

of these disappointing results, federal regulations presently allow use of an artificial heart *only* as a lifesaving stopgap measure while the patient awaits a permanent human heart transplant.

Newer to the scene and more broadly useful for temporary purposes are portable devices that take over part of the workload of an ailing heart. *LVADs* (left ventricular assist devices), such as the *HeartMate,* are mechanical pumps that are implanted in the abdomen, connected to the left ventricle, and powered by an external battery carried in a backpack or shoulder holster (see illustration a). HeartMates have functioned without problems for well over a year in patients' bodies. Indeed, several mortally ill patients who had received the device made such stunning recoveries that it could be removed. Because of these results, it has been suggested that some 20% of patients with failing hearts (particularly those too old to be considered for transplants) may be good candidates for permanently implanted LVADs.

An exciting technique is *cardio-myoplasty,* an autotransplant technique in which the patient's own skeletal muscle is used to form a living patch in the heart wall or to augment its pumping ability. The damaged portion of the heart wall is surgically removed and the wall is reinforced with latissimus dorsi muscle taken from the back (see illustration b). The

rest of the muscle patch is left attached to the posterior body wall and to its blood and nerve supply. A pacemaker is attached to the skeletal-muscle patch to stimulate it intermittently, ultimately in synchrony with the heartbeat. Because the muscle tissue is "home grown," there is obviously no problem of rejection. The major hurdle in autotransplants has been in coaxing the skeletal muscle, which is built to work in short spurts, to perform in the same manner as cardiac muscle—around the clock. But researchers are working on conditioning skeletal muscle with electrical shocks, which increases the percentage of slow oxidative, fatigue-resistant red fibers.

An unorthodox new procedure called *partial left ventriculectomy,* was developed in 1994 in a primitive hospital in the jungles of Brazil by Dr. Randas Batista. The deceptively simple surgical technique strengthens the left ventricle of an enlarged heart by cutting out a chunk of it. The heart, restored to its proper size, can beat more efficiently and with reduced tension on the ventricular wall. Although it is still experimental, many surgeons believe that this operation could be a breakthrough in treating those clients with enlarged and weakened hearts.

The final procedure in our heart-saving lineup is *transmyocardial laser revascularization (TMLR),* a newly developed therapy for those with severe coronary disease and angina who are terminal or are not good candidates for other, more demanding surgical techniques. Using a laser that operates only during diastole, the surgeon zaps 15 to 20 small channels into the wall of the left ventricular myocardium from inside the heart. The procedure seems to work by immediately increasing blood flow into the ischemic region of the myocardium, and subsequently by stimulating the coronary vessels to send more branches into the myocardium. Clinical studies indicate that TMLR reduced chest pain and improved blood flow in over 40% of those treated.

These newer procedures are revolutionizing the field of cardiac therapy, but we have yet to find that perfect cure.

RELATED CLINICAL TERMS

Asystole (a-sis'to-le) Situation in which the heart fails to contract.

Cardiac catheterization Diagnostic procedure involving passage of a fine catheter (tubing) through a blood vessel into the heart; blood samples are withdrawn, allowing oxygen content, blood flow, and pressures within the heart to be measured. Findings help to detect valve problems, heart deformities, and other heart malfunctions.

Commotio cordis ("concussion of the heart") Situation in which a relatively mild blow to the chest causes heart failure and sudden death because it occurs during a vulnerable interval (1/100 s) when the heart is repolarizing. Explains those rare instances when youngsters drop dead on the playing field after being hit in the chest by a ball.

Cor pulmonale (kor pul-mun-nā'le; *cor* = heart, *pulmo* = lung) A life-threatening condition of right-sided heart failure resulting from elevated blood pressure in the pulmonary circuit (pulmonary hypertension); acute cases may develop suddenly due to a pulmonary embolism; chronic cases are usually associated with chronic lung disorders such as emphysema.

Endocarditis (en"do-kar-di'tis) Inflammation of the endocardium, usually confined to the endocardium of the heart valves. Endocarditis often results from infection by bacteria that have entered the bloodstream but may result from fungal infection or an autoimmune response. Drug addicts may develop endocarditis by injecting themselves with contaminated needles.

Heart palpitation A heartbeat that is unusually strong, fast, or irregular so that the person becomes aware of it; may be caused by certain drugs, emotional pressures ("nervous heart"), or heart disorders.

Hypertrophic cardiomyopathy (HCM) The leading cause of death in young athletes, this condition causes the cardiac muscle cells to enlarge, thickening the heart wall. The heart pumps strongly but doesn't relax well during diastole when the heart is filling.

Mitral valve prolapse Valve disorder affecting up to 10% of the population; most often seen in young women and appears to have a genetic basis resulting in abnormal chordae tendineae or a malfunction of the papillary muscles; one or more of the flaps of the mitral valve become incompetent and billow into the left atrium during ventricular systole, allowing blood regurgitation; mitral valve prolapse is a common indication for valve replacement surgery.

Myocarditis (mi"o-kar-di'tis; *myo* = muscle, *card* = heart, *itis* = inflammation) Inflammation of the cardiac muscle layer (myocardium) of the heart; sometimes follows an untreated streptococcal infection in children; may weaken the heart and impair its ability to act as an effective pump.

CHAPTER SUMMARY

Media study tools that could provide you additional help in reviewing specific key topics of Chapter 19 are referenced below. **SP** *= Study Partner;* **IP** *= Interactive Physiology.*

Heart Anatomy (pp. 683–693)

Size, Location, and Orientation (p. 683)

1. The human heart, about the size of a clenched fist, is located obliquely within the mediastinum of the thorax.

Coverings of the Heart (pp. 683–684)

2. The heart is enclosed within a double sac made up of the outer fibrous pericardium and the inner serous pericardium (parietal and visceral layers). The pericardial cavity between the serous layers contains lubricating serous fluid.

Layers of the Heart Wall (p. 684)

3. Layers of the heart wall, from the interior out, are the endocardium, the myocardium (reinforced by a fibrous skeleton), and the epicardium (visceral layer of the serous pericardium).

Chambers and Associated Great Vessels (pp. 684–688)

4. The heart has two superior atria and two inferior ventricles. Functionally, the heart is a double pump.

5. Entering the right atrium are the superior vena cava, the inferior vena cava, and the coronary sinus. Four pulmonary veins enter the left atrium.

6. The right ventricle discharges blood into the pulmonary trunk; the left ventricle pumps blood into the aorta.

Pathway of Blood Through the Heart (pp. 688–689)

7. The right heart is the pulmonary circuit pump. Oxygen-poor systemic blood enters the right atrium, passes into the right ventricle, through the pulmonary trunk to the lungs, and back to the left atrium via the pulmonary veins.

8. The left heart is the systemic circuit pump. Oxygen-laden blood entering the left atrium from the lungs flows into the left ventricle and then into the aorta, which provides the functional supply of all body organs. Systemic veins return the oxygen-depleted blood to the right atrium.

Coronary Circulation (pp. 689–690)

9. The right and left coronary arteries branch from the aorta and via their main branches (anterior and posterior interventricular, marginal, and circumflex arteries) supply the heart itself. Venous blood, collected by the cardiac veins (great, middle, and small), is emptied into the coronary sinus.

10. Blood delivery to the myocardium occurs during heart relaxation.

Heart Valves (pp. 690–693)

11. The atrioventricular valves (tricuspid and bicuspid) prevent backflow into the atria when the ventricles are contracting; the pulmonary and aortic semilunar valves prevent backflow into the ventricles when the ventricles are relaxing.

IP Cardiovascular System CD-ROM; Topic: Anatomy Review: The Heart, pages 1–10.

Properties of Cardiac Muscle Fibers   (pp. 693–696)

Microscopic Anatomy   (pp. 693–694)

1. Cardiac muscle cells are branching, striated, generally uninucleate cells. They contain myofibrils consisting of typical sarcomeres.

2. Adjacent cardiac cells are connected by intercalated discs containing desmosomes and gap junctions. The myocardium behaves as a functional syncytium because of electrical coupling provided by gap junctions.

Mechanism and Events of Contraction   (pp. 695–696)

3. Action-potential generation in the contractile cardiac muscle mimics that of skeletal muscle. Membrane depolarization causes opening of sodium channels and sodium entry, which is responsible for the rising phase of the action potential curve. Depolarization also opens slow calcium channels; Ca^{2+} entry prolongs the period of depolarization (creates the plateau). The action potential is coupled to sliding of the myofilaments by ionic calcium released by the SR and entering from the extracellular space. Compared to skeletal muscle, cardiac muscle has a prolonged refractory period that prevents tetanization.

IP Cardiovascular System CD-ROM; Topic: Cardiac Action Potential, pages 11–18.

Energy Requirements   (p. 696)

4. Cardiac muscle has abundant mitochondria and depends primarily on aerobic respiration to form ATP.

Heart Physiology   (pp. 696–709)

Electrical Events   (pp. 696–701)

1. Certain noncontractile cardiac muscle cells exhibit automaticity and rhythmicity and can independently initiate action potentials. Such cells have an unstable resting potential called a pacemaker potential that gradually depolarizes, drifting toward threshold for firing. These cells compose the intrinsic conduction system of the heart.

2. The conduction, or nodal, system of the heart consists of the SA and AV nodes, the AV bundle and bundle branches, and the Purkinje fibers. This system coordinates the depolarization of the heart and ensures that the heart beats as a unit. The SA node has the fastest rate of spontaneous depolarization and acts as the heart's pacemaker; it sets the sinus rhythm.

3. Defects in the intrinsic conduction system can cause arrhythmias, fibrillation, and heart block.

4. The heart is innervated by the autonomic nervous system. Autonomic cardiac centers in the medulla include the sympathetic cardioacceleratory center, which projects to the T_1–T_5 region of the spinal cord, which in turn projects to the cervical and upper thoracic chain ganglia. Postganglionic fibers innervate the SA and AV nodes and the cardiac muscle fibers. The parasympathetic cardioinhibitory center exerts its influence via the vagus nerves (X), which project to the heart wall. Most parasympathetic fibers serve the SA and AV nodes.

5. An electrocardiogram is a graphic representation of the cardiac conduction cycle. The P wave reflects atrial depolarization. The QRS complex indicates ventricular depolarization; the T wave represents ventricular repolarization.

SP Exercise: Intrinsic Conduction System of the Heart.

IP Cardiovascular System CD-ROM; Topic: Intrinsic Conduction System, pages 1–7; Topic: Cardiac Action Potential, pages 1–10.

Mechanical Events: The Cardiac Cycle   (pp. 701–703)

6. A cardiac cycle consists of the events occurring during one heartbeat. During mid-to-late diastole, the ventricles fill and the atria contract. Ventricular systole consists of the isovolumetric contraction phase and the ventricular ejection phase. During early diastole, the ventricles are relaxed and are closed chambers until increasing atrial pressure forces the AV valves open and the cycle begins again. At a normal heart rate of 75 beats/min, a cardiac cycle lasts 0.8 s.

7. Pressure changes promote blood flow and valve opening and closing.

IP Cardiovascular System CD-ROM; Topic: Cardiac Cycle, pages 1–20.

Heart Sounds   (pp. 702–704)

8. Normal heart sounds arise chiefly from the closing of heart valves. Abnormal heart sounds, called murmurs, usually reflect valve problems.

Cardiac Output   (pp. 704–709)

9. Cardiac output, typically 5 L/min, is the amount of blood pumped out by each ventricle in 1 minute. Stroke volume is the amount of blood pumped out by a ventricle with each contraction. Cardiac output = heart rate × stroke volume.

10. Stroke volume depends to a large extent on the degree of stretch of cardiac muscle by venous return. Approximately 70 ml, it is the difference between end diastolic volume (EDV) and end systolic volume (ESV). Anything that influences heart rate or blood volume influences venous return, hence stroke volume.

11. Activation of the sympathetic nervous system increases heart rate and contractility; parasympathetic activation decreases heart rate and contractility. Ordinarily, the heart exhibits vagal tone.

12. Chemical regulation of the heart is effected by hormones (epinephrine and thyroxine) and ions (sodium, potassium, and calcium). Imbalances in ions severely impair heart activity.

13. Other factors influencing heart rate are age, sex, exercise, and body temperature.

14. Congestive heart failure occurs when the pumping ability of the heart is inadequate to provide normal circulation to meet body needs. Right heart failure leads to systemic edema; left heart failure results in pulmonary edema.

SP Exercises: Chapter 19, Maintenance of Cardiac Output; Cardiac Output; Case Study: Cardiac arrhythmia.

IP Cardiovascular System CD-ROM; Topic: Cardiac Output, pages 1–10.

Developmental Aspects of the Heart   (pp. 709–711)

1. The heart begins as a simple (mesodermal) tube that is pumping blood by the fourth week of gestation. The fetal heart has two lung bypasses: the foramen ovale and the ductus arteriosus.

2. Congenital heart defects account for more than half of infant deaths. The most common of these disorders lead to inadequate oxygenation of blood or increase the workload of the heart.

3. Age-related changes include sclerosis and thickening of the valve flaps, declines in cardiac reserve, fibrosis of cardiac muscle, and atherosclerosis.

4. Risk factors for cardiac disease include dietary factors, excessive stress, cigarette smoking, and lack of exercise.

REVIEW QUESTIONS

Multiple Choice/Matching

1. When the semilunar valves are open, which of the following are occurring? (a) 2,3,5,6, (b) 1,2,3,7, (c) 1,3,5,6, (d) 2,4,5,7.

(1) coronary arteries fill	(5) blood enters aorta
(2) AV valves are closed	(6) blood enters pulmonary
(3) ventricles are in systole	arteries
(4) ventricles are in diastole	(7) atria contract

2. The portion of the intrinsic conduction system located in the interventricular septum is the (a) AV node, (b) SA node, (c) AV bundle, (d) Purkinje fibers.

3. An ECG provides information about (a) cardiac output, (b) movement of the excitation wave across the heart, (c) coronary circulation, (d) valve impairment.

4. The sequence of contraction of the heart chambers is (a) random, (b) left chambers followed by right chambers, (c) both atria followed by both ventricles, (d) right atrium, right ventricle, left atrium, left ventricle.

5. The fact that the left ventricular wall is thicker than the right reveals that it (a) pumps a greater volume of blood, (b) pumps blood against greater resistance, (c) expands the thoracic cage, (d) pumps blood through a smaller valve.

6. The chordae tendineae (a) close the atrioventricular valves, (b) prevent the AV valve flaps from everting, (c) contract the papillary muscles, (d) open the semilunar valves.

7. In the heart, which of the following apply? (1) Action potentials are conducted from cell to cell across the myocardium via gap junctions, (2) the SA node sets the pace for the heart as a whole, (3) spontaneous depolarization of cardiac cells can occur in the absence of nerve stimulation, (4) cardiac muscle can continue to contract for long periods in the absence of oxygen. (a) all of the above, (b) 1, 3, 4, (c) 1, 2, 3, (d) 2, 3.

8. The activity of the heart depends on intrinsic properties of cardiac muscle and on neural factors. Thus, (a) vagus nerve stimulation of the heart reduces heart rate, (b) sympathetic nerve stimulation of the heart decreases time available for ventricular filling, (c) sympathetic stimulation of the heart increases its force of contraction, (d) all of the above.

9. Freshly oxygenated blood is first received by the (a) right atrium, (b) left atrium, (c) right ventricle, (d) left ventricle.

Short Answer Essay Questions

10. Describe the location and position of the heart in the body.

11. Describe the pericardium and distinguish between the fibrous and the serous pericardia relative to histological structure and location.

12. Trace one drop of blood from the time it enters the right atrium until it enters the left atrium. What is this circuit called?

13. (a) Describe how heart contraction and relaxation influence coronary blood flow. (b) Name the major branches of the coronary arteries, and note the heart regions served by each.

14. The refractory period of cardiac muscle is much longer than that of skeletal muscle. Why is this a desirable functional property?

15. (a) Name the elements of the intrinsic conduction system of the heart in order, beginning with the pacemaker. (b) What is the important function of this conduction system?

16. Draw a normal ECG pattern. Label and explain the significance of its deflection waves.

17. Define cardiac cycle, and follow the events of one cycle.

18. What is cardiac output, and how is it calculated?

19. Discuss how the Frank-Starling law of the heart helps to explain the influence of venous return on stroke volume.

20. (a) Describe the common function of the foramen ovale and the ductus arteriosus in a fetus. (b) What problems result if these shunts remain patent (open) after birth?

Critical Thinking and Clinical Application Questions

1. A gang member was stabbed in the chest during a street fight. He was cyanotic and unconscious from lack of blood delivery to the brain. The diagnosis was cardiac tamponade. What is cardiac tamponade and how does it cause the observed symptoms?

2. You have been called upon to demonstrate the technique for listening to valve sounds. (a) Explain where you would position your stethoscope to auscultate (1) the aortic valve of a patient with severe aortic semilunar valve incompetence and (2) a stenosed mitral valve. (b) During which period(s) would you hear these abnormal valve sounds most clearly? (During atrial diastole, ventricular systole, ventricular diastole, or atrial systole?) (c) What cues would you use to differentiate between an incompetent and a stenosed valve?

3. Florita Santos, a middle-aged woman, is admitted to the coronary care unit with a diagnosis of left ventricular failure resulting from a myocardial infarction. Her history indicated that she was aroused in the middle of the night by severe chest pain. Her skin is pale and cold, and moist sounds are heard over the lower regions of both lungs. Explain how failure of the left ventricle can cause these signs and symptoms.

4. Heather, a newborn baby, needs surgery because she was born with an aorta that arises from the right ventricle and a pulmonary trunk that issues from the left ventricle, a condition called transposition of the great vessels. What are the physiological consequences of this defect?

5. Gabriel, a heroin addict, feels tired, is weak and feverish, and has vague aches and pains. Terrified that he has AIDS, he goes to a doctor and is informed that he is suffering not from AIDS, but from a heart murmur accompanied by endocarditis. What is the most likely way that Gabriel contracted endocarditis?

6. As Cara worked at her dissection, she became frustrated that several of the structures she had to learn about had more than one common name. Provide the second name for each of these structures: (a) atrioventricular groove, (b) tricuspid valve, (c) bicuspid valve (give two synonyms), and (d) atrioventricular bundle.

THE CARDIOVASCULAR SYSTEM: BLOOD VESSELS

20

Overview of Blood Vessel Structure and Function (pp. 718–727)

1. Describe the three layers that typically form the wall of a blood vessel, and state the function of each.

2. Define vasoconstriction and vasodilation.

3. Compare and contrast the structure and function of the three types of arteries.

4. Describe the structure and function of veins, and explain how veins differ from arteries.

5. Describe the structure and function of a capillary bed.

Physiology of Circulation (pp. 727–746)

6. Define blood flow, blood pressure, and resistance, and explain the relationships between these factors.

7. List and explain the factors that influence blood pressure, and describe how blood pressure is regulated.

8. Define hypertension. Describe both its symptoms and consequences.

9. Explain how blood flow is regulated in the body in general and in its specific organs.

10. Outline the factors involved in capillary dynamics, and explain the significance of each.

11. Define circulatory shock. List several possible causes.

Circulatory Pathways: Blood Vessels of the Body (pp. 746, 750–772)

12. Trace the pathway of blood through the pulmonary circuit, and state the importance of this special circulation.

13. Describe the general functions of the systemic circuit. Name and give the location of the major arteries and veins in the systemic circulation.

14. Describe the structure and special function of the hepatic portal system.

Developmental Aspects of Blood Vessels (pp. 746–747)

15. Explain how blood vessels develop in the fetus.

16. Provide examples of changes that often occur in blood vessels as a person ages.

Although blood vessels are sometimes compared to a system of plumbing pipes within which blood circulates, this analogy serves only as a starting point. Unlike rigid pipes, blood vessels are dynamic structures that pulsate, constrict and relax, and even proliferate, as demanded by the changing needs of the body. This chapter examines the structure and function of these important circulatory passageways.

The **blood vessels** of the body form a closed delivery system that begins and ends at the heart. The idea that blood circulates in the body dates back to the 1620s to the inspired experiments of William Harvey, an English physician. Prior to that time, it was thought, as proposed by the ancient Greek physician Galen, that blood moved through the body like an ocean tide, first moving out from the heart and then ebbing back into the heart via the same vessels.

OVERVIEW OF BLOOD VESSEL STRUCTURE AND FUNCTION

The three major types of blood vessels are *arteries*, *capillaries*, and *veins*. As the heart contracts, it forces blood into the large arteries leaving its ventricles. The blood then moves into successively smaller arteries, finally reaching their smallest branches, the *arterioles* (ar-te′re-ōlz; "little arteries"), which feed into the capillary beds of body organs and tissues. Blood draining from the capillaries flows into *venules* (ven′ūlz), and then on into small veins that merge to form larger veins that ultimately empty into the heart. Altogether, the blood vessels in the adult human carry blood on a journey that stretches for about 100,000 km (60,000 miles) through the internal body landscape!

Arteries and veins simply act as conduits for blood. Because arteries carry blood *away from* the heart, they are said to "branch," "diverge," or "fork" as they form smaller and smaller divisions. Veins, by contrast, carry blood *toward* the heart and so are said to "join," "merge," and "converge" into the successively larger vessels approaching the heart. Only the capillaries have intimate contact with tissue cells and directly serve cellular needs. Exchanges between the blood and tissue cells occur primarily through the gossamer-thin capillary walls.

Structure of Blood Vessel Walls

The walls of all blood vessels, except the very smallest, are composed of three distinct layers, or *tunics* ("cloaks"). These tunics surround a central blood-containing space, the vessel **lumen** (Figure 20.1).

The innermost tunic is the **tunica interna,** or *tunica intima* (in′tĭ-mah). The second name is easy to remember once you know that this tunic is in *intimate* contact with the blood in the lumen. This tunic contains the *endothelium,* the simple squamous epithelium that lines the lumen of all vessels. The endothelium is a continuation of the endocardial lining of the heart, and its flat cells fit closely together, forming a slick surface that minimizes friction as blood moves through the vessel lumen. In vessels larger than 1 mm in diameter, a *subendothelial layer* of loose connective tissue (a basement membrane) supports the endothelium.

The middle tunic, the **tunica media** (me′de-ah), is mostly circularly arranged smooth muscle cells and sheets of elastin. The activity of the smooth muscle is regulated by *vasomotor nerve fibers* of the sympathetic division of the autonomic nervous system and a whole battery of chemicals. (The controls are described shortly.) Depending on the needs of the body, either **vasoconstriction** (reduction in lumen diameter due to smooth muscle contraction) or **vasodilation** (widening of the lumen due to smooth muscle relaxation) can be effected. Because small changes in blood vessel diameter greatly influence blood flow and blood pressure, the activities of the tunica media are critical in regulating circulatory dynamics. Generally, the tunica media is the bulkiest layer in arteries, which bear the chief responsibility for maintaining blood pressure and continuous blood circulation.

The outermost layer of a blood vessel wall is the **tunica externa,** or *tunica adventitia* (ad″ven-tish′e-ah; "coming from outside"). This tunic is composed largely of loosely woven collagen fibers that protect and reinforce the blood vessel, and anchor it to surrounding structures. The tunica externa is infiltrated with nerve fibers and lymphatic vessels and, in larger veins, a network of elastin fibers. In larger vessels, a system of tiny blood vessels, the **vasa vasorum** (va′sah va-sor′um)—literally, "vessels of the vessels" which nourish the more external tissues of the blood vessel wall—is found in the tunica externa. The innermost or luminal portion of the vessel obtains its nutrients directly from blood in the lumen.

The three vessel types vary in length, diameter, and the relative thickness and tissue makeup of their walls. These differences are summarized in Table 20.1 and described below. The relationship of the various generations of these blood vascular channels to each other and to vessels of the lymphatic system (which recovers fluids that leak from the circulation) is summarized in Figure 20.2.

What is the functional value of the fact that the capillaries consist of just a single layer of squamous cells underlain by sparse connective tissue?

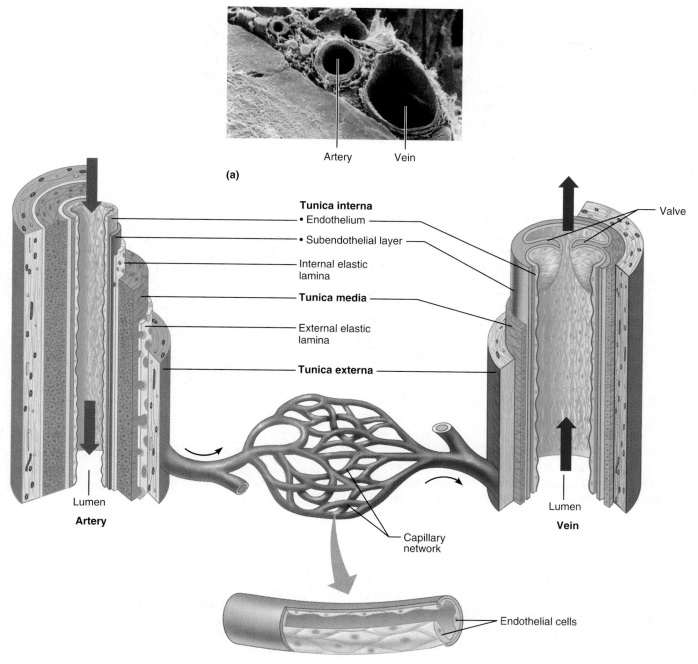

(a)

Artery Vein

Tunica interna
• Endothelium
• Subendothelial layer
Internal elastic lamina
Tunica media
External elastic lamina
Tunica externa

Valve

Lumen
Artery

Capillary network

Lumen
Vein

Endothelial cells

(b)

Capillary

FIGURE 20.1 *Generalized structure of arteries, veins, and capillaries.* **(a)** Scanning electron micrograph of a muscular artery and the corresponding vein in cross section (120×). (© R. G. Kessel and R. H. Kardon, *Tissues and Organs: A Text-Atlas of Scanning Electron Microscopy,* W. H. Freeman and Company, 1979, all rights reserved.) **(b)** The walls of arteries and veins are composed of three tunics: the tunica interna (endothelium underlain by a subendothelial layer of loose connective tissue), tunica media (smooth muscle cells and elastic fibers), and tunica externa (largely collagen fibers). Capillaries, intermediate between arteries and veins in the circulatory pathway, are composed of only the endothelium and a sparse basal lamina. Notice that the tunica media is thick in arteries and thin in veins, while the tunica externa is thin in arteries and relatively thicker in veins.

Capillaries are the exchange sites. The thinner their walls (the barrier to exchanges), the more rapid and efficient the exchange process.

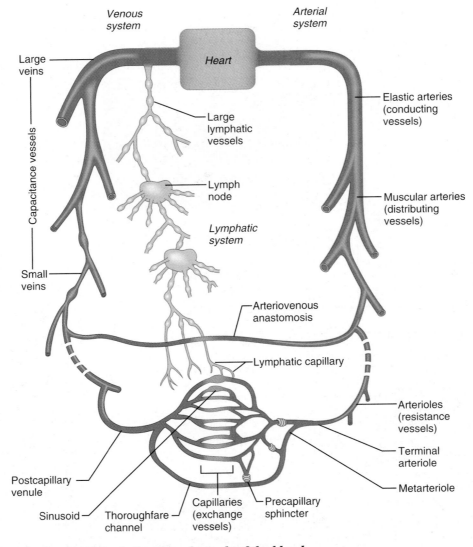

FIGURE 20.2 *Schematic showing the relationship of vessels of the blood vascular system to each other and to vessels of the lymphatic system.*

Arterial System

Arteries are vessels that transport blood *away from* the heart. Although systemic arteries always carry oxygenated blood and systemic veins always carry oxygen-poor blood, this is not true for the vessels of the pulmonary circulation (see Figure 20.17, p. 750) or for the special umbilical vessels of a fetus (see Figure 29.13a, p. 1136).

In terms of relative size and function, arteries can be divided into three groups—elastic arteries, muscular arteries, and arterioles.

Elastic (Conducting) Arteries

Elastic arteries are the thick-walled arteries near the heart—the aorta and its major branches. These arteries are the largest in diameter, ranging from 2.5 cm to 1 cm, and the most elastic (Table 20.1). Their large lumen allows them to serve as low-resistance pathways that conduct blood from the heart to medium-sized arteries. For this reason, they are sometimes referred to as **conducting arteries** (Figure 20.2). The elastic arteries contain more elastin than any other vessel type. It is present in all three tunics, but the tunica media contains the most. There the elastin constructs concentric "holey" sheets (laminae) of elastic connective tissue that look like slices of Swiss cheese interspersed between the layers of smooth muscle cells. The abundant elastin enables these arteries to withstand and smooth out large pressure fluctuations by expanding when the heart forces blood into them, and then recoiling to propel blood onward into the circulation when the heart relaxes. Elastic arteries also contain substantial amounts of smooth muscle, but they are relatively inactive in vasoconstriction. Thus, in terms of their function, the elastic arteries can be visualized as simple elastic tubes.

The elastic arteries expand and recoil passively to accommodate changes in blood volume. Consequently, blood flows fairly continuously rather than starting and stopping with the pulsating rhythm of the heartbeat. If the blood vessels become hard and unyielding, as in arteriosclerosis, blood flows more intermittently, similar to the way water flows through a hard rubber garden hose attached to a faucet. When the faucet (the pump) is on, the high pressure makes the water gush out of the hose. But when the faucet is shut off, the water flow abruptly becomes a trickle and then stops, because the hose walls cannot recoil to keep the water under pressure. Furthermore, without the pressure-smoothing effect of the elastic arteries, the walls of arteries through-out the body experience higher pressures. Battered by high pressures, the arteries eventually weaken and may balloon out or even burst. (These problems are discussed in more detail in *A Closer Look* on pp. 722–723.)

Muscular (Distributing) Arteries

Distally the elastic arteries give way to the **muscular, or distributing, arteries** (see Figure 20.2). The muscular arteries deliver blood to specific body organs and account for most of the named arteries studied in the anatomy laboratory. Their internal diameter ranges from that of a little finger (1 cm) to that of a pencil lead (about 0.3 mm). Proportionately, they have the thickest media of all vessels. Their tunica media contains relatively more smooth muscle and less elastic tissue than that of elastic arteries (Table 20.1); therefore, they are more active in vasoconstriction and are less distensible. However, there *is* an *elastic lamina* on each face of the tunica media in muscular arteries.

Arterioles

Arterioles have a lumen diameter ranging from 0.3 mm down to 10 μm, and are the smallest of the arteries. Larger arterioles exhibit all three tunics, but their tunica media is chiefly smooth muscle with a few scattered elastic fibers. The smaller arterioles, which lead into the capillary beds, are little more than a single layer of smooth muscle cells spiraling around the endothelial lining.

As described shortly, blood flow into the capillary beds from minute to minute is determined by arteriole diameter, which varies in response to changing neural stimuli and local chemical influences. When arterioles constrict, the tissues served are largely bypassed. When arterioles dilate, blood flow into the local capillaries increases dramatically.

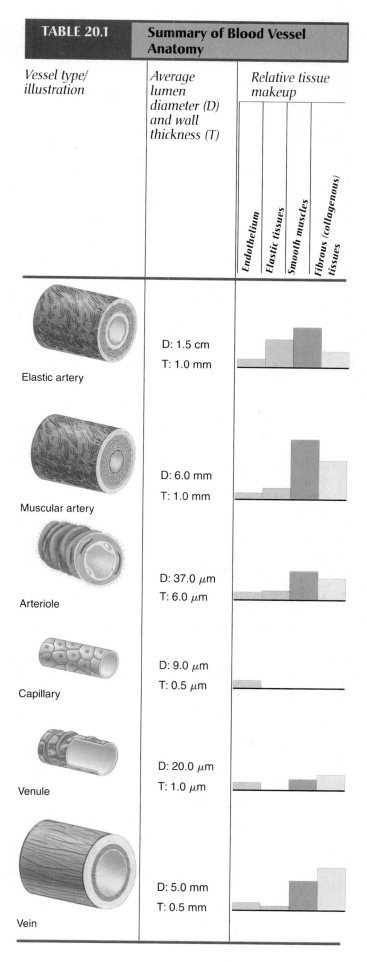

TABLE 20.1	Summary of Blood Vessel Anatomy				
Vessel type/ illustration	*Average lumen diameter (D) and wall thickness (T)*	*Relative tissue makeup*			
		Endothelium	Elastic tissues	Smooth muscles	Fibrous (collagenous) tissues
Elastic artery	D: 1.5 cm T: 1.0 mm				
Muscular artery	D: 6.0 mm T: 1.0 mm				
Arteriole	D: 37.0 μm T: 6.0 μm				
Capillary	D: 9.0 μm T: 0.5 μm				
Venule	D: 20.0 μm T: 1.0 μm				
Vein	D: 5.0 mm T: 0.5 mm				

When pipes get clogged, it is usually because something gets stuck in them—a greasy mass of bone bits or a "hairball." But when arteries are narrowed by **atherosclerosis,** their walls thicken and intrude into the vessel lumen, making it easy to close the vessel completely. A roaming blood clot or arterial spasms can do it.

Although all blood vessels are susceptible to this serious condition, the aorta and the coronary arteries are most often affected. The disease progresses through many stages before the arterial walls actually approach the rigid tube system described in the text, but some of the earlier stages are just as lethal, or more so.

Atherosclerosis: Onset and Stages

What triggers this scourge of blood vessels that indirectly causes half of the deaths in the Western world? According to the *response to injury* hypothesis, the initial event is damage to the tunica interna caused by bloodborne chemicals, by physical factors such as a blow, hypertension, or by viral or bacterial infections. Indeed, the bacterium *Chlamydia pneumoniae,* known primarily as a cause of sexually transmitted disease; *cytomegalovirus (CMV),* a common virus that causes respiratory infections; and even *helicobacter pylorii,* the ulcer-causing bacterium, are suspected of causing the vascular inflammation in some cases and have been found in the vascular lesions.

Irritation to the endothelium sets off the alarm summoning the immune system and the inflammatory process to repair the damage. If it's a one-time injury, when it's over it's over. But most plaques grow slowly, probably through a succession of small injuries that heal, only to be

ruptured again and again. The plaque gradually enlarges as injured endothelial cells release chemotactic agents and growth (mitosis-inducing) factors, and begin to transport and modify greater amounts of lipids picked up from the blood, particularly low-density lipoproteins (LDLs). LDLs are the type that deliver cholesterol to tissue cells via the bloodstream. When the sequestered LDL becomes oxidized (this seems to be a common event), it not only damages neighboring cells but also acts as a chemotactic agent to attract macrophages to the area. The macrophages normally help protect the body by ingesting invading microorganisms and toxic substances including oxidized proteins and lipids. But in a particularly fatty plaque they can become so engorged with the oxidized LDLs that they migrate beneath the interna where they are transformed into lipid-laden *foam cells* and lose their scavenging ability.

The macrophages are joined by smooth muscle cells migrating from the tunica media. These also take up lipids and become foam cells. The accumulating foam cells initiate the **fatty streak stage.** The smooth muscle cells also deposit collagen and elastin fibers, thickening the interna and producing fibrous lesions with a core of dead and dying foam cells called **fibrous** or **atherosclerotic plaques.** At first the vessel walls accommodate the growing plaque by expanding outward, but eventually these fatty mounds of muscle and fibrous tissue begin to protrude into the vessel lumen, producing full-blown atherosclerosis (see photo).

Additionally, impaired endothelial function accounts for at least some cases of coronary artery spasms. Healthy endothelial cells release nitric oxide and prostacyclin—chemicals that promote vasodilation and inhibit platelet aggregation. Atherosclerosis impairs release of these vasodilators and

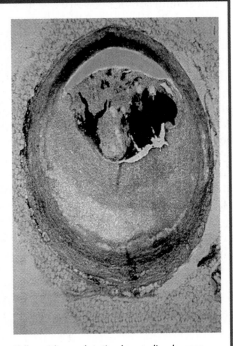

Yellow atherosclerotic plaques line human blood vessel.

antiaggregating factors. Indeed, this may well be one cause of the increased thrombus formation in those with atherosclerosis.

If you add still another ingredient—a chemical one called *lipoprotein (a)*—to the picture, the disease course accelerates. Lipoprotein (a), a special LDL seen only in some people, is most involved in delivering cholesterol to sites where tissue repair is occurring—damaged endothelium for one. Although it is presumed to assist healing, it "backfires" when it is present in excess. Because lipoprotein (a) is similar to a common growth factor, it is thought to promote mitosis of the cells in the vessel wall. Because it resembles plasminogen, it can "stand in" for it at sites where clots have formed. However, it lacks plasminogen's clot-dissolving ability, so by competing successfully with that clot buster, it may *prevent* disposal of unneeded or undesirable clots.

Arteriosclerosis is the end stage of the disease. As enlarging plaques hinder diffusion of nutrients from the blood to the deeper tissues of the artery wall, smooth muscle cells in the tunica media die, and the elastic fibers deteriorate. These elements are replaced by nonelastic scar tissue, and calcium salts are deposited in the lesions, now called *complicated plaques.* These events constrict the vessel and cause the arterial walls to fray and ulcerate, conditions that encourage blood sludging and backup and platelet adhesion and thrombus formation. The increased rigidity of the vessels leads to hypertension. Together, these events increase the risk of myocardial infarcts, strokes, and aneurysms, and are responsible for the pain (angina) that occurs when heart muscle is ischemic.

However, the popular view that most heart attacks are the consequence of severe vessel narrowing and hardening is now being challenged, particularly since some 70% of heart attacks are caused by much smaller obstructions, too small to be seen on an arteriogram or to cause any symptoms in most cases. It now appears that the body's defense system betrays it. The inflammatory process that occurs in the still soft, unstable, cholesterol-rich plaques changes the biology of the vessel wall and makes the plaques susceptible to rupture, exploding off fragments that trigger massive clots that can cause lethal heart attacks. The victim appears perfectly healthy until he drops dead!

Treatment and Prevention

Although this theory of *vulnerable plaque* as the chief culprit has attracted many medical converts, the question of what to do about it remains. Some medical centers are now routinely testing heart patients for elevated levels of cholesterol and C-reactive protein, a well-known marker of inflammation. Additionally, new classes of anti-inflammatory drugs are set to be tested on heart patients, as are some of the tried and true antibiotics like tetracycline.

And what is being done when the heart is at risk because of the mature *arteriosclerotic* plaques in the coronary vessels? Traditionally the choice has been *coronary bypass surgery,* in which veins removed from the legs or small arteries removed from the thoracic cavity are implanted in the heart to restore myocardial circulation. More recently, intravascular devices have been used as ammunition to clear obstructed sites in blood vessels. *Balloon angioplasty* uses a catheter with a balloon tightly packed into its tip. When the catheter reaches the obstruction, the balloon is inflated, and the fatty mass is compressed against the vessel wall. However, this procedure is only useful to clear a few very localized obstructions.

Although angioplasty is faster, cheaper, and much less risky than coronary bypass surgery, it has the same major shortcoming: It does nothing to stop the underlying disease, and in time *restenoses* (new blockages) occur in 30–50% of cases. Insertion of *stents,* short metal-mesh tubes that look much like a ziti noodle, into newly dilated vessels has helped solve the restenosis problem. Bursts of radiation to inhibit reclogging of arteries, a procedure called *brachytherapy,* has reduced restenosis by 50–80% in patients tested.

Even more exciting is a gel that is smeared on the area treated by angioplasty to form a translucent film over the wall. Composed of chemicals called *antisense agents,* the gel blocks the function of a specific gene (c-*myb*) that is essential for smooth muscle proliferation. Newer still is the technique of bathing bypass segments in a solution containing DNA segments targeted to limit smooth muscle proliferation after vascular injury. Another drug ploy in chemical trials is administration of NO donor drugs, which enhance or augment NO activity. NO inhibits the cell proliferation and migration necessary for plaque formation.

It was hoped that cholesterol-lowering drugs such as cholestyramine and lovastatin would act as cardiovascular Drāno and wash the fatty plaques off the walls. They do work to some extent, and a newly discovered side benefit is that they also appear to have some anti-inflammatory activity. However, their side effects—nausea, bloating, and constipation—are so miserable that many people simply stop taking them.

When a clot is trapped by the diseased vessel walls, *thrombolytic (clot-dissolving) agents* are used. These revolutionary drugs include *tissue plasminogen activator (tPA),* a natural substance produced by genetic engineering techniques. Injecting tPA directly into the heart restores blood flow quickly and puts an early end to many heart attacks in progress. Cardiac patients receiving vessel-opening drugs appear to have a lower incidence of ventricular tachycardia and sudden death following a heart attack.

Beyond these clinical measures, the prudent person will also try to prevent the progress of atherosclerosis—by quitting smoking, losing weight to reduce blood lipid (triglyceride) levels, cutting cholesterol and saturated fats from his or her diet, and taking antioxidant vitamins (E and C)—and to "undo" some of the damage caused by free radicals, by exercising to increase blood levels of HDL (the "good" lipoprotein that removes cholesterol from vessel walls and carries it to the liver for elimination). However, getting people to change their lifestyle is not as easy as it seems. Americans like their burgers and butter. But if heart disease can be reversed by reversing atherosclerosis, many people with diseased arteries may be willing to trade lifelong habits for a healthy old age!

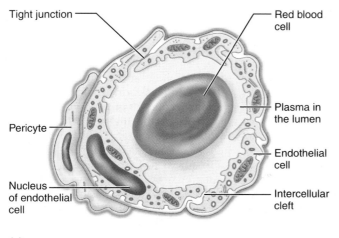

(a)

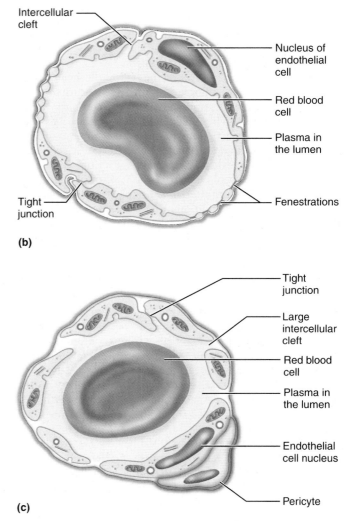

(b)

(c)

FIGURE 20.3   *Capillary structure.* Transverse sections of **(a)** continuous capillary, **(b)** fenestrated capillary, **(c)** discontinuous sinusoidal capillary.

Capillaries

The microscopic **capillaries** are the smallest blood vessels. Their exceedingly thin walls consist of just a thin tunica interna (see Figure 20.1a). In some cases, one endothelial cell forms the entire circumference of the capillary wall. Here and there along the outer surface of the capillaries are spider-shaped *pericytes,* smooth muscle–like cells that stabilize the capillary wall.

Average capillary length is 1 mm and average lumen diameter is 8–10 μm, just large enough for red blood cells to slip through in single file. Most tissues have a rich capillary supply, but there are exceptions. Tendons and ligaments are poorly vascularized. Cartilage and epithelia lack capillaries, but receive nutrients from blood vessels in nearby connective tissues, and the avascular cornea and lens of the eye receive nutrients from the aqueous humor.

If blood vessels are compared to a system of expressways and roads, then the capillaries are the back alleys and driveways that provide direct access to nearly every cell in the body. Given their location

and the thinness of their walls, capillaries are ideally suited for their role—exchange of materials (gases, nutrients, hormones, etc.) between the blood and interstitial fluid (Figure 20.2 and Table 20.1). The mechanisms of these exchanges are described later in this chapter; here we will focus on capillary structure.

Types of Capillaries

Structurally, there are three types of capillaries—*continuous, fenestrated,* and *sinusoidal.* **Continuous capillaries,** abundant in the skin and muscles, are most common. They are continuous in the sense that their endothelial cells provide an uninterrupted lining. Adjacent cells are joined laterally by *tight junctions.* However, these are usually incomplete, and leave gaps of unjoined membrane called **intercellular clefts** (Figure 20.3a) just large enough to allow limited passage of fluids and small solutes. Typically, the endothelial cell cytoplasm exhibits numerous small vesicles believed to ferry fluids across the capillary wall. However, brain capillaries are unique. There the tight junctions of the continuous capillaries are complete around the entire perimeter of the endothelial cells, and constitute the structural basis of the *blood-brain barrier*, described in Chapter 12.

Fenestrated (fen'es-tra-tid) **capillaries** are similar to the continuous variety except that some of their endothelial cells are riddled with oval *pores*, or *fenestrations* (*fenestra* = window) (Figure 20.3b). The fenestrations are usually covered by a very delicate membrane, or diaphragm (probably condensed basal lamina material), but even so this capillary variety has much greater permeability to fluids and small solutes. Fenestrated capillaries are found where active capillary absorption or filtrate formation occurs. For example, fenestrated capillaries in the small intestine receive the digested food nutrients, and those in endocrine organs allow hormones rapid entry into the blood. Fenestrated capillaries with perpetually open pores occur in the kidneys, where rapid filtration of blood plasma is essential.

Sinusoids (si'nŭ-soyds), or **sinusoidal capillaries,** are highly modified, leaky capillaries found only in certain organs (liver, bone marrow, lymphoid tissues, and some endocrine organs). Sinusoids have large, irregularly shaped lumens and are usually fenestrated. Their endothelial lining is modified—it exhibits fewer tight junctions and larger intercellular clefts than ordinary capillaries (Figure 20.3c). These structural adaptations allow large molecules (such as proteins) and even blood cells to pass between the blood and surrounding tissues. In the liver, the endothelium of the sinusoids is *discontinuous* and large macrophages called **Kupffer cells** form part of the lining. In other organs, such as the spleen, phagocytes located just outside the sinusoids stick cytoplasmic extensions through the intercellular clefts into the sinusoidal lumen to get at their "prey." Blood flows sluggishly through the tortuous sinusoid channels, allowing time for it to be modified in various ways. For example, the liver processes nutrient-laden venous blood draining from the digestive organs by absorbing nutrients and removing and destroying any contained bacteria.

Capillary Beds

Capillaries do not function independently. Instead they tend to form interweaving networks called **capillary beds.** The flow of blood from an arteriole to a venule—that is, through a capillary bed—is called the **microcirculation.** In most body regions, a capillary bed consists of two types of vessels: (1) a *vascular shunt* (metarteriole–thoroughfare channel), a short vessel that directly connects the arteriole and venule at opposite ends of the bed, and (2) *true capillaries,* the actual *exchange vessels* (Figure 20.4). The **terminal arteriole** feeding the bed leads into a

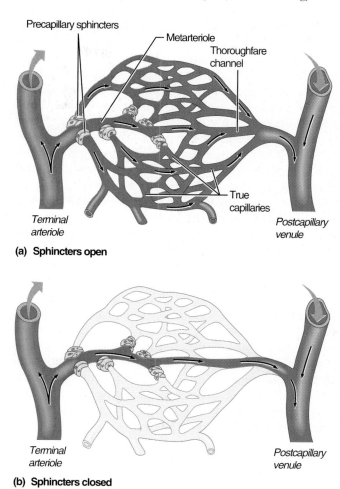

? *Assume the capillary bed depicted is in your calf muscle. Which condition would the capillary bed be in— (a) or (b)—if you were doing calf raises at the gym?*

(a) Sphincters open

(b) Sphincters closed

FIGURE 20.4 *Anatomy of a capillary bed.* The composite metarteriole–thoroughfare channel acts as a shunt to bypass the true capillaries when precapillary sphincters controlling blood entry into the true capillaries are constricted.

metarteriole (a vessel structurally intermediate between an arteriole and a capillary), which is directly continuous with the **thoroughfare channel** (intermediate between a capillary and a venule). The thoroughfare channel, in turn, joins the **postcapillary venule** that drains the bed.

The **true capillaries** number 10 to 100 per capillary bed, depending on the organ or tissues served. They usually branch off the metarteriole (proximal end of the shunt) and return to the thoroughfare channel (the distal end), but occasionally they spring

(a) The true capillaries would be flushed with blood to ensure that the working calf muscles could receive the needed nutrients and dispose of their metabolic wastes.

from the terminal arteriole and empty directly into the venule. A cuff of smooth muscle fibers, called a **precapillary sphincter,** surrounds the root of each true capillary at the metarteriole and acts as a valve to regulate the flow of blood into the capillary. Blood flowing through a terminal arteriole may take one of two routes: through the true capillaries or through the shunt. When the precapillary sphincters are relaxed (open), blood flows through the true capillaries and takes part in exchanges with tissue cells. When the sphincters are contracted (closed), blood flows through the shunts and bypasses the tissue cells.

The relative amount of blood entering a capillary bed is regulated by vasomotor nerve fibers and local chemical conditions. A capillary bed may be flooded with blood or almost completely bypassed, depending on conditions in the body or in that specific organ. For example, suppose you have just eaten and are sitting relaxed, listening to your favorite musical group. Food digestion is going on, and blood is circulating freely through the true capillaries of your gastrointestinal organs to receive the breakdown products of digestion. However, most of these same capillary pathways are closed between meals. Further, when you are exercising vigorously, blood is rerouted from the digestive organs (food or no food) to the capillary beds of your skeletal muscles where it is more immediately needed. This helps explain why vigorous exercise right after a meal can cause indigestion or abdominal cramps.

Venous System

Blood is carried from the capillary beds toward the heart by veins. En route, the venous vessels increase in diameter, and their walls gradually thicken as they progress from venules to the larger and larger veins.

Venules

Venules, ranging from 8 to 100 μm in diameter, are formed when capillaries unite. The smallest venules, the **postcapillary venules,** consist entirely of endothelium around which a few pericytes congregate. They are extremely porous (more like capillaries than veins in this way), and fluid and white blood cells move easily from the bloodstream through their walls. Indeed, a well-recognized sign of an inflamed area is the adhesion of white blood cells to the postcapillary venule endothelium, followed by their migration through the wall into the inflamed tissue. The larger venules have one or two layers of smooth muscle cells (a scanty tunica media) and thin externa as well.

Veins

Venules join to form **veins.** Veins usually have three distinct tunics, but their walls are always thinner and their lumens larger than those of corresponding arteries (see Figure 20.1 and Table 20.1). Consequently, in routine histological preparations, veins are usually collapsed and their lumens appear slit-like. There is relatively little smooth muscle or elastin in the tunica media, which is poorly developed and tends to be thin even in the largest veins. The tunica externa is the heaviest wall layer. Consisting of thick longitudinal bundles of collagen fibers and elastic networks, it is often several times thicker than the tunica media. In the largest veins—the venae cavae, which return blood directly to the heart—the tunica externa is further thickened by longitudinal bands of smooth muscle.

With their large lumens and thin walls, veins can accommodate a fairly large blood volume. Since up to 65% of the body's total blood supply is found in the veins at any time, veins are called **capacitance vessels,** or **blood reservoirs** (see Figures 20.1 and 20.2). Even so, the veins are normally only partially filled with blood.

Because blood pressure within veins is low, they can be much thinner walled than arteries without danger of bursting. However, the low-pressure condition itself demands some special adaptations to help return blood to the heart at the same rate it was pumped into the circulation. The large-diameter lumens of veins (which offer relatively little resistance to blood flow) are one structural adaptation. Another is valves that prevent blood from flowing backward. **Venous valves** are formed from folds of the tunica interna (Figure 20.1), and they resemble the semilunar valves of the heart in both structure and function. Venous valves are most abundant in the veins of the limbs, where the upward flow of blood is opposed by gravity. They are absent in veins of the ventral body cavity.

A simple experiment will demonstrate the effectiveness of your venous valves. Hang one hand by your side until the blood vessels on its dorsal aspect become distended with blood. Next place two fingertips against one of the distended veins, and pressing firmly, move the superior finger proximally along the vein and then release that finger. The vein will remain flattened and collapsed, despite the pull of gravity. Finally, remove the distal fingertip and watch the vein refill rapidly with blood.

◩ *Homeostatic Imbalance*

Varicose veins are veins that have become tortuous and dilated because of incompetent valves. Over 15% of adults suffer from varicose veins, usually in the lower limbs. Several factors contribute, including heredity and conditions that hinder venous return, such as prolonged standing in one position, obesity, or pregnancy. Both the "potbelly" of an overweight person and the enlarged uterus of a pregnant woman exert downward pressure on vessels of the groin, and restrict return of blood to the heart. Consequently, blood tends to pool in the lower limbs and, with time, the valves weaken and the venous walls stretch and become floppy. Superficial veins, which receive little support from surrounding tissues, are especially susceptible. Another cause of varicose veins is elevated venous pressure. For example, straining to have a baby or a bowel movement raises intra-abdominal pressure, preventing blood from draining from the veins of the anal canal. The resulting varicosities in the anal veins are called *hemorrhoids* (hem'ŏ-roidz). ∎

Venous sinuses, such as the *coronary sinus* of the heart and the *dural sinuses* of the brain, are highly specialized, flattened veins with extremely thin walls composed only of endothelium. They are supported by the tissues that surround them, rather than by any additional tunics. The dural sinuses, which receive cerebrospinal fluid and blood draining from the brain, are reinforced by the tough dura mater that covers the brain surface.

Vascular Anastomoses

Where vascular channels unite, they form **vascular anastomoses** (ah-nas"to-mo'sēz; "coming together"). Most organs receive blood from more than one arterial branch, and arteries supplying the same territory often merge, forming **arterial anastomoses.** Arterial anastomoses provide alternate pathways, called **collateral channels,** for blood to reach a given body region. If one branch is cut or blocked by a clot, the collateral channel can often provide the area with an adequate blood supply. Arterial anastomoses occur around joints, where active movement may hinder blood flow through one channel. They are also common in abdominal organs, the brain, and the heart. Arteries that do not anastomose, or which have a poorly developed collateral circulation, supply the retina, kidneys, and spleen. If their blood flow is interrupted, cells supplied by such vessels die.

The metarteriole—thoroughfare channel shunts of capillary beds that connect arterioles and venules are examples of **arteriovenous anastomoses.** Veins interconnect much more freely than arteries, and **venous anastomoses** are common. (You may be able to see venous anastomoses through the skin on the dorsum of your hand.) Because venous anastomoses are abundant, occlusion of a vein rarely blocks blood flow or leads to tissue death.

∎ PHYSIOLOGY ∎ OF CIRCULATION

Have you ever climbed a mountain? Well, get ready to climb a hypothetical mountain as you learn about circulatory dynamics. Like scaling a mountain, tackling blood pressure regulation and other topics of cardiovascular physiology is challenging, but then it's so exhilarating when you succeed. Let's begin the climb.

To sustain life, blood must be kept circulating. It is easy to accept this statement, but to understand how blood is kept moving takes a bit more effort. By now, you are aware that the heart is the pump, the arteries are conduits, the arterioles are resistance vessels, the capillaries are exchange sites, and the veins are conduits and blood reservoirs. Now we need to define three physiologically important terms—blood flow, blood pressure, and resistance—and examine how these factors relate to the physiology of blood circulation.

INTRODUCTION TO BLOOD FLOW, BLOOD PRESSURE, AND RESISTANCE
Definition of Terms

1. **Blood flow** is the actual volume of blood flowing through a vessel, an organ, or the entire circulation in a given period (ml/min). If we consider the entire vascular system, blood flow is equivalent to cardiac output (CO), and under resting conditions, it is relatively constant. At a given time, however, blood flow through *individual* body organs may vary widely and is intimately related to their immediate needs.

2. **Blood pressure (BP)** is the force per unit area exerted on the wall of a blood vessel by its contained blood. Blood pressure is expressed in terms of millimeters of mercury (mm Hg). For example, a blood pressure of 120 mm Hg is equal to the pressure exerted by a column of mercury 120 mm high. Unless stated otherwise, the term *blood pressure* means systemic arterial blood pressure in the largest arteries near the heart. It is the pressure gradient—the *differences* in blood pressure within the vascular system—that provides the driving force that keeps blood moving—always from a higher to a lower pressure area—through the body.

3. **Resistance** is opposition to flow and is a measure of the amount of friction blood encounters as it passes through the vessels. Since most friction is encountered in the peripheral (systemic) circulation, well away from the heart, we generally use the term **peripheral resistance (PR).**

There are three important sources of resistance: blood viscosity, vessel length, and vessel diameter.

- **Blood viscosity.** *Viscosity* (vis-kos'ĭ-te) is the internal resistance to flow and is related to the thickness or "stickiness" of a fluid. The greater the viscosity, the less easily molecules slide past one another and the more difficult it is to get and keep the fluid moving. Blood is much more viscous than water because it contains formed elements and plasma proteins; hence it flows more slowly under the same conditions. Blood viscosity is fairly constant, but infrequent conditions such as polycythemia (excessive numbers of red blood cells) can increase blood viscosity and, hence, resistance. On the other hand, if the RBC count is low, as in some anemias, blood is less viscous, and peripheral resistance declines.

- **Total blood vessel length.** The relationship between total blood vessel length and resistance is straightforward: the longer the total vessel length, the greater the resistance encountered. An extra pound or two of fat requires that miles of small vessels be added to service it. This significantly increases the peripheral resistance.

- **Blood vessel diameter.** Since both blood viscosity and blood vessel length are normally unchanging, the influence of these factors can be considered constant in healthy people. Changes in blood vessel diameter are frequent, however, and significantly alter peripheral resistance. How so? The answer lies in principles of fluid flow. Fluid close to the walls of a tube or channel is slowed by friction as it passes along the wall, whereas fluid in the center of the channel flows more freely and faster. You can verify this by watching the flow of water in a river. Water close to the bank hardly seems to move, while that in the middle of the river flows quite rapidly.

Furthermore, in a tube of a given size, the relative speed and position of fluid in the different regions of the tube's cross section remain constant, a phenomenon called *laminar flow* or *streamlining*. The smaller the tube (or blood vessel), the greater the friction, because relatively more of the fluid contacts the tube walls where its movement is impeded. Resistance varies *inversely* with the *fourth power* of the vessel radius (one-half the diameter). This means, for example, that if the radius of a vessel is doubled, the resistance is one-sixteenth as much ($r^4 = 2 \times 2 \times 2 \times 2 = 16$ and $1/r^4 = 1/16$), and blood will flow much more quickly in the vessel with the larger lumen. Thus, the large arteries close to the heart, which do not change dramatically in diameter, contribute little to peripheral resistance, and the small-diameter arterioles, which can enlarge or constrict in response to neural and chemical controls, are the major determinants of peripheral resistance.

When blood encounters an abrupt change in the tube size or rough or protruding areas of the tube wall (such as the fatty plaques of atherosclerosis), the smooth laminar blood flow is replaced by *turbulent blood flow*, that is, irregular fluid motion where blood from the different laminae mixes. *Turbulence* dramatically increases the resistance.

Relationship Between Blood Flow, Blood Pressure, and Resistance

Now that we have defined these terms, let us summarize briefly the relationships between them. Blood flow (*F*) is *directly* proportional to the difference in blood pressure (ΔP) between two points in the circulation, that is, the blood pressure, or hydrostatic pressure, gradient. Thus, if ΔP increases, blood flow speeds up, and when ΔP decreases, blood flow declines. In addition, blood flow is *inversely* proportional to the peripheral resistance (*R*) in the systemic circulation; if *R* increases, blood flow decreases. These relationships are stated by the formula

$$\text{Blood flow } (F) = \frac{\text{difference in blood pressure } (\Delta P)}{\text{peripheral resistance } (R)}$$

Of these two factors influencing blood flow, *R* is far more important than ΔP in influencing local blood flow. This is because when the arterioles serving a particular tissue dilate (thus decreasing the resistance), blood flow to that tissue increases, even though the systemic pressure is unchanged or may actually be falling.

SYSTEMIC BLOOD PRESSURE

Any fluid driven by a pump through a circuit of closed channels operates under pressure, and the nearer the fluid is to the pump, the greater the pressure it is under. The dynamics of blood flow in blood vessels is no exception, and blood flows through the blood vessels along a pressure gradient, always moving from higher- to lower-pressure areas. Fundamentally, *the pumping action of the heart generates blood flow. Pressure results when flow is opposed by resistance.*

As illustrated in Figure 20.5, systemic blood pressure is highest in the aorta and declines throughout the length of the pathway to finally reach 0 mm Hg in the right atrium. The steepest change (drop) in blood pressure occurs in the arterioles, which offer the greatest resistance to blood flow. However, as long as a pressure gradient exists, blood continues to flow from higher- to lower-pressure areas until it completes the circuit back to the heart.

Arterial Blood Pressure

Essentially, arterial blood pressure reflects two factors: (1) how much the elastic arteries close to the heart can be stretched (their *compliance* or *distensibility*) and (2) the volume of blood forced into them at any time. If the amounts of blood entering and leaving them during any given period were equal, arterial pressure would be constant. Instead, as Figure 20.5 reveals, blood pressure changes (rises and falls) in a regular fashion in the elastic arteries near the heart, that is, it is obviously *pulsatile.*

As the left ventricle contracts and expels blood into the aorta, it imparts kinetic energy to the blood, which in turn stretches the elastic aorta as aortic pressure reaches its peak. Indeed, if the aorta were opened during this period, blood would spurt upward 5 or 6 feet! This pressure peak, called the **systolic** (sis-tah'lik) **pressure,** averages about 120 mm Hg in healthy adults. Blood moves forward into the arterial bed because the pressure in the aorta is higher than the pressure in the more distal vessels. During diastole, the aortic semilunar valve closes, preventing blood from flowing back into the heart, and the walls of the aorta (and other elastic arteries) recoil, maintaining adequate pressure on the reducing blood volume to keep the blood flowing forward into the smaller vessels. During this time, aortic pressure drops to its lowest level (approximately 70 to 80 mm Hg in healthy adults), called the **diastolic** (di-as-tah'lik) **pressure** (see Figure 20.5). Thus, you can picture the elastic arteries as pressure reservoirs that operate as auxiliary pumps to keep blood circulating throughout diastole, when the heart is relaxing. Essentially, the volume and energy of blood stored within them during systole are given back during diastole.

The difference between the systolic and diastolic pressures is called the **pulse pressure.** Pulse pressure is felt as a throbbing pulsation in an artery (a *pulse*) during systole, as the elastic arteries are expanded by the blood being forced into them by ventricular contraction. Increased stroke volume and faster blood ejection from the heart (due to increased contractility) cause *temporary* increases in the pulse pressure. Pulse pressure is chronically increased by arteriosclerosis because the elastic arteries become less stretchy. Because aortic pressure fluctuates up and down with each heartbeat, the important pressure figure to consider is the **mean (average) arterial pressure (MAP),** because it is this pressure that propels the blood to the tissues throughout the cardiac cycle. Because diastole usually lasts longer than systole, the mean pressure is not simply the value halfway between systolic and diastolic pressures. Instead, the MAP is roughly equal to the diastolic pressure plus one-third of the pulse pressure. This can be shown in equation form as:

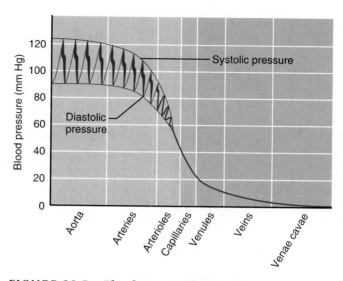

What is it about the anatomy of the largest arteries that leads to the pulsatile pressure changes shown on the left side of the graph?

FIGURE 20.5   *Blood pressure in various blood vessels of the systemic circulation.*

$$MAP = \frac{diastolic}{pressure} + \frac{pulse\ pressure}{3}$$

Thus, the MAP of a person with a systolic blood pressure of 120 mm Hg and a diastolic pressure of 80 mm Hg would be approximately 93 mm Hg:

$$MAP = 80 + (1/3 \times 40)$$

MAP and pulse pressure both decline with increasing distance from the heart. The MAP loses ground to the never-ending friction between the blood and the vessel walls, and the pulse pressure is gradually phased out in the less elastic muscular arteries, where elastic rebound of the vessels ceases to occur. At the end of the arterial tree, blood flow is steady and the pulse pressure has disappeared.

Capillary Blood Pressure

As shown in Figure 20.5, by the time blood reaches the capillaries, blood pressure has dropped to approximately 40 mm Hg and by the end of the

The largest arteries are elastic and they stretch and recoil as blood is pumped into them by the heart and then runs distally into the circulation.

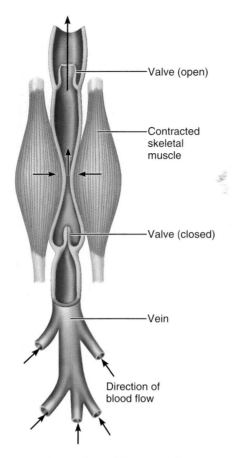

FIGURE 20.6 **Operation of the muscular pump.** As illustrated, when skeletal muscles contract and press against the flexible veins, the valves proximal to the area of contraction are forced open and blood is propelled toward the heart. The valves distal to the point of contraction are closed by the backflowing blood.

capillary beds is further reduced to 20 mm Hg or less. Such low capillary pressures are desirable because (1) capillaries are fragile and high pressures would rupture them, and (2) most capillaries are extremely permeable and thus even the low capillary pressure forces solute-containing fluids (filtrate) out of the bloodstream into the interstitial space. As described later in this chapter, these fluid flows help to distribute nutrients, gases, and hormones between blood and tissue cells and continuously refresh the interstitial fluid.

Venous Blood Pressure

Unlike arterial pressure, which pulsates with each contraction of the left ventricle, venous blood pressure is steady and changes very little during the cardiac cycle. The entire pressure gradient in the veins, from venules to the termini of the venae cavae, is only about 20 mm Hg as opposed to an average pres-

sure gradient of 60 mm Hg from the aorta to the ends of the arterioles. These pressure differences between arteries and veins become very clear when these vessels are cut. If a vein is cut, the blood flows evenly from the wound; a lacerated artery produces rapid spurts of blood. The very low pressure in the venous system reflects the cumulative effects of peripheral resistance, which dissipates most of the energy of blood pressure (as heat) during each circuit.

Factors Aiding Venous Return

Despite the structural modifications of veins (large lumens and valves), venous pressure is normally too low to promote adequate venous return. Hence, functional adaptations, such as the respiratory and skeletal muscle "pumps," are critically important to venous return.

1. Respiratory pump. Pressure changes occurring in the ventral body cavity during breathing create the respiratory pump that sucks blood upward toward the heart. As we inhale, abdominal pressure increases, squeezing the local veins; because venous valves prevent backflow, blood is forced toward the heart. At the same time, the pressure within the chest decreases, allowing thoracic veins to expand and speeding blood entry into the right atrium of the heart.

2. Muscular pump. Skeletal muscle activity, or the so-called muscular pump, is even more important. As the skeletal muscles surrounding the deep veins contract and relax, they "milk" blood toward the heart, and once blood passes each successive valve, it cannot flow back (Figure 20.6). People who earn their living in "standing professions," such as hairdressers and dentists, often have swollen ankles because blood pools in their feet and legs during prolonged periods of skeletal muscle inactivity.

MAINTAINING BLOOD PRESSURE

Maintaining a steady flow of blood from the heart to the toes is vital for proper organ function. But making sure that the energetic person who jumps out of bed in the morning does not keel over from inadequate blood flow to the brain requires the finely tuned cooperation of the heart, blood vessels, and kidneys—all under close supervision of the brain. Central among the homeostatic mechanisms that regulate cardiovascular dynamics are those that maintain blood pressure. So, let's take a quick look back at the variables that affect blood pressure before we examine how they are manipulated to keep blood pressure within the homeostatic range.

The main factors influencing blood pressure are *cardiac output, peripheral resistance,* and *blood vol-*

ume. If we rearrange slightly the formula pertaining to blood flow presented on p. 728, we can see how cardiac output (blood flow of the entire circulation) and peripheral resistance relate to blood pressure.

Blood pressure = cardiac output × peripheral resistance

Since CO, in turn, depends on blood volume (the heart can't pump out what doesn't enter its chambers), it is clear that blood pressure varies *directly* with CO, PR, and blood volume. Thus, in theory, a change (increase or decrease) in any of these variables causes a corresponding change in blood pressure. However, what *really* happens in the body is that changes in one variable that threaten blood pressure homeostasis are quickly compensated for by changes in the other variables.

As described in Chapter 19, CO is equal to *stroke volume* (ml/beat) multiplied by the *heart rate* (beats/min), and normal CO is 5.0 to 5.5 L/min. Figure 20.7 should help you recall the main factors that determine cardiac output—venous return and the neural and hormonal controls. Although the figure doesn't name the cardiac centers in the medulla, remember that the parasympathetic cardioinhibitory center is "in charge" of heart rate most of the time and, via the vagus nerves, it maintains what is referred to as the *resting heart rate*. During such "resting" periods, stroke volume is controlled mainly by venous return (end diastolic volume or EDV). During stressful conditions, the sympathetic cardioacceleratory center takes over and increases both heart rate (by acting on the SA node) and stroke volume (by enhancing cardiac muscle contractility, which decreases end systolic volume [ESV]). The enhanced CO, in turn, results in an increase in the MAP (mean systemic arterial blood pressure).

In the discussion that follows, we will focus on factors that regulate blood pressure by altering peripheral resistance and blood volume, but a flowchart showing the influence of nearly all important factors is provided in Figure 20.10 (p. 736). The *short-term controls* of blood pressure, mediated by the nervous system and bloodborne chemicals, counteract moment-to-moment fluctuations in blood pressure by altering peripheral resistance. By contrast, the slower-acting renal mechanisms that constitute the *long-term controls* of blood pressure regulate blood volume.

Short-Term Mechanisms: Neural Controls

Neural controls of peripheral resistance are directed primarily at two main goals. (1) They alter blood distribution to respond to specific demands. For exam-

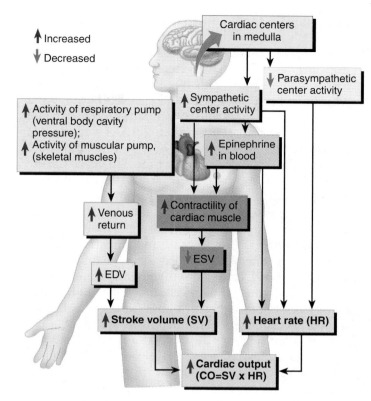

FIGURE 20.7   *Summary of major factors enhancing cardiac output.* (Note: Efferent impulses from cardiac centers travel along autonomic nerves. EDV = end diastolic volume, ESV = end systolic volume.)

ple, during exercise blood is shunted temporarily from the digestive organs to the skeletal muscles, and heat loss from the body is enhanced when blood vessels in the skin dilate and become flushed with warm blood. (2) They maintain adequate MAP by altering blood vessel diameter. (Remember, very small changes in blood vessel diameter cause substantial changes in peripheral resistance, hence in systemic blood pressure.) Under conditions of low blood volume, all vessels except those supplying the heart and brain are constricted to allow as much blood as possible to flow to those vital organs.

Most neural controls operate via reflex arcs chiefly involving the following components: baroreceptors and the associated afferent fibers, the vasomotor center of the medulla, vasomotor fibers, and vascular smooth muscle. Occasionally, inputs from chemoreceptors and higher brain centers also influence the neural control mechanism. Each of these factors is described next.

Role of the Vasomotor Center

The neural center that oversees changes in the diameter of blood vessels is the **vasomotor center,** a cluster of sympathetic neurons in the medulla. This center plus the cardiac centers described earlier make up the **cardiovascular center** that integrates blood pressure control by altering cardiac output and blood vessel diameter. The vasomotor center transmits impulses at a fairly steady rate along sympathetic efferents (motor fibers) called **vasomotor fibers,** which exit from the T_1 through L_2 levels of the spinal cord and run to innervate the smooth muscle of blood vessels, most importantly arterioles. As a result, the arterioles are almost always in a state of moderate constriction, called **vasomotor tone.** However, the degree of tonic vasoconstriction varies from organ to organ. Generally, arterioles of the skin and digestive viscera receive vasomotor impulses more frequently and tend to be more strongly constricted than those of skeletal muscles. Any increase in sympathetic activity produces generalized vasoconstriction and a rise in blood pressure. Decreased sympathetic activity allows the vascular muscle to relax somewhat and causes blood pressure to decline to basal levels. Most vasomotor fibers release norepinephrine, which is a potent vasoconstrictor. In skeletal muscle, however, some vasomotor fibers may release acetylcholine, causing vasodilation. Though important to local blood flow, such vasodilator fibers are *not* important to overall regulation of systemic blood pressure.

Vasomotor activity is modified by inputs from (1) baroreceptors (pressure-sensitive mechanoreceptors that respond to changes in arterial pressure and stretch), (2) chemoreceptors (receptors that respond to changes in blood levels of oxygen, carbon dioxide, and H^+), and (3) higher brain centers (hypothalamus and cerebrum), as well as by certain hormones and other bloodborne chemicals. Let's look at how each of these factors brings about its effect.

Baroreceptor-Initiated Reflexes

Baroreceptors are located not only in the *carotid sinuses* (dilations in the internal carotid arteries, which provide the major blood supply to the brain) and *aortic arch* but also in the wall of nearly every large (elastic) artery of the neck and thorax. When arterial blood pressure rises and stretches these receptors, they send off a rapid stream of impulses to the vasomotor center. This inhibits the vasomotor center, resulting in vasodilation of not only the arterioles but also the veins, and a decline in blood pressure (Figure 20.8). While dilation of the arterioles substantially reduces the peripheral resistance, venodilation shifts blood to the venous reservoirs, causing a decline in venous return and cardiac output. Afferent impulses from the baroreceptors also reach the cardiac centers, where they stimulate parasympathetic activity and inhibit the sympathetic (cardioacceleratory) center, reducing heart rate and contractile force. Conversely, a decline in mean arterial pressure initiates reflex vasoconstriction and increases cardiac output, causing blood pressure to rise. Thus, periph-

eral resistance and cardiac output are regulated in tandem according to baroreceptor inputs so that changes in blood pressure are minimized.

The function of rapidly responding baroreceptors is to protect the circulation against short-term (acute) changes in blood pressure, such as those occurring with changes in posture. For example, blood pressure falls (particularly in the head) when one stands up after reclining. The **carotid sinus reflex** protects the blood supply to the brain, whereas the **aortic reflex** is more concerned with maintaining adequate blood pressure in the systemic circuit as a whole. Baroreceptors are relatively *ineffective* in protecting us against sustained pressure changes, as evidenced by the fact that some people do have chronic hypertension. In such cases, the baroreceptors are apparently "reset" to monitor pressure changes at a higher set point.

Chemoreceptor-Initiated Reflexes

When the oxygen content or pH of the blood drops sharply or carbon dioxide levels rise, chemoreceptors in the aortic arch and large arteries of the neck transmit impulses to the cardioacceleratory center, causing CO to increase, and to the vasomotor center, causing reflex vasoconstriction. The rise in blood pressure that follows speeds the return of blood to the heart and lungs. The most prominent of these receptors are the *carotid* and *aortic bodies* located close by the baroreceptors in the carotid sinus and aortic arch. Because they are more important in regulating respiratory rate and depth than in blood pressure regulation, their function is considered more fully in Chapter 23.

Influence of Higher Brain Centers

Reflexes that regulate blood pressure are integrated at the brain stem (medulla) level. Although the cerebral cortex and hypothalamus are not involved in routine controls of blood pressure, these higher brain centers can modify arterial blood pressure via relays to the medullary centers. For example, the fight-or-flight response mediated by the hypothalamus has profound effects on blood pressure. (Even the simple act of speaking can make your blood pressure jump if the person you are talking to makes you anxious.) The hypothalamus also mediates the redistribution of blood flow and other cardiovascular responses that occur during exercise and changes in body temperature.

Short-Term Mechanisms: Chemical Controls

As we have seen, changing levels of oxygen and carbon dioxide help regulate blood pressure via the

? *Which part of this feedback loop occurs when you rise from a seated position?*

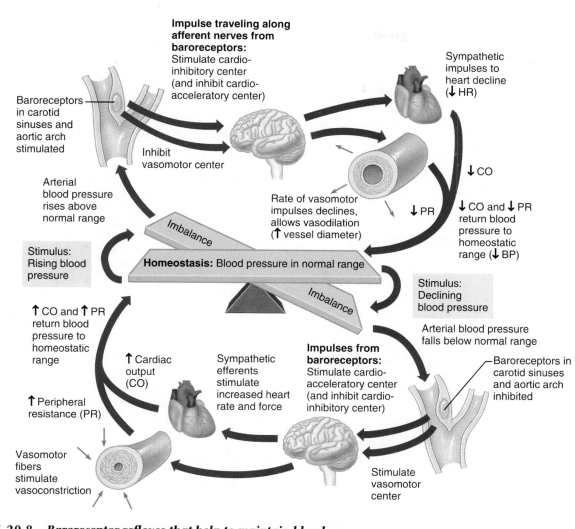

FIGURE 20.8 *Baroreceptor reflexes that help to maintain blood pressure homeostasis.* (CO = cardiac output; PR = peripheral resistance; HR = heart rate; BP = blood pressure.)

chemoreceptor reflexes. But numerous other bloodborne chemicals (described below) influence blood pressure by acting directly on vascular smooth muscle or on the vasomotor center. The most important such agents are hormones (see Table 20.2).

■ **Adrenal medulla hormones.** During periods of stress, the adrenal gland releases norepinephrine (NE) and epinephrine to the blood, and both en-

hance the sympathetic fight-or-flight response (see Figure 20.10). As noted earlier, NE has a vasoconstrictive action. Epinephrine increases cardiac output and promotes generalized vasoconstriction (except in skeletal and cardiac muscle, where it generally causes vasodilation). It is interesting to note that *nicotine*, an important chemical in tobacco and one of the strongest toxins known, mimics the effects of the catecholamines. It causes intense vasoconstriction not only by directly stimulating ganglionic sympathetic neurons but also by prompting release of large amounts of epinephrine and NE.

■ **Atrial natriuretic** (na″tre-u-ret′ik) **peptide (ANP).** The atria of the heart produce a peptide hormone, called atrial natriuretic peptide, which causes blood volume and blood pressure to decline. As

Arterial blood pressure initially drops, resulting in stimulation of the vasomotor and cardioacceleratory centers. The result is increased heart rate and vasoconstriction, which increases arterial blood pressure, restoring homeostasis.

TABLE 20.2		Influence of Selected Hormones on Variables Affecting Blood Pressure		
Variable affecting BP	*Effect on variable*	*Hormone(s)*	*Site of action*	*Result*
Cardiac output	↑ (↑ HR and contractility)	Epinephrine	Heart (β_1 receptors)	↑ BP
		Norepinephrine (NE)	Heart (β_1 receptors)	↑ BP
Peripheral resistance	↑ (via vasoconstriction)	Angiotensin II	Arterioles	↑ BP
		Antidiuretic hormone	Arterioles	↑ BP
		Epinephrine and NE	Arterioles (α receptors)	↑ BP
	↓ (via vasodilation)	Epinephrine and NE	Large veins (β_2 receptors)	↓ BP
		Atrial natriuretic peptide	Arterioles	↓ BP
Blood volume	↓ (salt and water loss)	Atrial natriuretic peptide	Kidney tubule cells	↓ BP
	↑ (salt and water retention)	Aldosterone	Kidney tubule cells	↑ BP
		Cortisol	Kidney tubule cells	↑ BP
	↑ (water retention)	Antidiuretic hormone (ADH)	Kidney tubule cells	↑ BP

noted in Chapter 17, ANP antagonizes aldosterone and prods the kidneys to excrete more sodium and water from the body, causing blood volume to drop. It also causes a generalized vasodilation and reduces cerebrospinal fluid formation in the brain.

- **Antidiuretic hormone (ADH).** Antidiuretic hormone is produced by the hypothalamus and stimulates the kidneys to conserve water. ADH is not usually important in short-term blood pressure regulation. However, when blood pressure falls to dangerously low levels (as during severe hemorrhage), much more ADH is released and helps to restore arterial pressure by causing intense vasoconstriction.

- **Angiotensin II.** Angiotensin (an"je-o-ten'sin) II is generated in response to renin release by the kidneys when renal perfusion is inadequate (Figure 20.9). It causes intense vasoconstriction, promoting a rapid rise in systemic blood pressure. It also stimulates release of aldosterone and ADH, which act in long-term regulation of blood pressure by enhancing blood volume as described shortly.

- **Endothelium-derived factors.** The endothelium is the source of several chemicals that affect vascular smooth muscle (as well as blood clotting). For example, the peptide **endothelin** is one of the most potent vasoconstrictors known. Released in response to low blood flow, it appears to bring about its long-lasting effects by enhancing calcium entry into vascular smooth muscle. Endothelial cells also release **PDGF** (prostaglandin-derived growth factor), another vasoconstrictor chemical.

Nitric oxide (NO), originally called *endothelium-derived relaxing factor (EDRF),* is still another vasoactive substance secreted by endothelial cells. NO is released in direct response to a high blood flow and signaling molecules such as acetylcholine and bradykinin (and the vasodilator nitroglycerine), and its release is a secondary or consequent response to the release of endothelin. NO, which acts via a cyclic GMP second-messenger system, promotes both reflex (systemic) and highly localized vasodilation. However, NO is quickly destroyed and its potent vasodilator effects are very brief. Until recently the sympathetic nervous system was thought to rule blood vessel diameter, but we now know that its primary vascular role is to produce vasoconstriction, and that NO plays the major role in causing vasodilation.

- **Inflammatory chemicals.** Histamine, prostacyclin, kinins, and certain other chemicals released during the inflammatory response and certain allergic responses are potent vasodilators. They also promote fluid loss from the bloodstream by increasing capillary permeability.

- **Alcohol.** Ingestion of alcohol causes blood pressure to drop by inhibiting ADH release, by depressing the vasomotor center, and by promoting vasodilation, especially in the skin. This accounts for the flushed appearance of someone who has drunk a generous amount of an alcoholic beverage.

Long-Term Mechanisms: Renal Regulation

Although the baroreceptors respond to short-term changes in blood pressure, they quickly adapt to prolonged or chronic episodes of high or low pressure. This is where the kidneys step in to restore and maintain blood pressure homeostasis by regulating blood volume. Although blood volume varies with age, body size, and sex, renal mechanisms usually maintain it at close to 5 L.

As noted earlier, blood volume is a major determinant of cardiac output (via its influence on venous pressure, venous return, EDV, and stroke volume). An increase in blood volume is followed by a rise in blood pressure, and anything that increases blood volume, such as excessive salt intake which promotes water retention, raises the MAP because of the greater fluid load in the vascular tree. By the same token, decreased blood volume translates to a fall in blood pressure. Blood loss and the garden-variety dehydration that occurs during vigorous exercise are common causes of reduced blood volume. A sudden drop in BP often signals internal bleeding and blood volume that is becoming too dangerously low to support normal circulation. However, these assertions—increased blood volume leads to rising BP and decreased blood volume results in declining BP—do not tell the whole story because we are dealing with a dynamic system. Increases in blood volume that cause a rise in blood pressure also stimulate the kidneys to eliminate water, which reduces blood volume and, consequently, blood pressure. Likewise, falling blood volume triggers renal mechanisms that increase blood volume and blood pressure. As you can see, blood pressure can be stabilized or maintained within normal limits only when blood volume is stable.

The kidneys act both directly and indirectly to regulate arterial pressure and provide the major long-term mechanism of blood pressure control. The *direct renal mechanism* alters blood volume. When blood volume or blood pressure rises, the rate at which fluid filters from the bloodstream into the kidney tubules is speeded up. In such situations, the kidneys are unable to process the filtrate rapidly enough, and more of this fluid leaves the body in urine. As a result, blood volume and blood pressure fall. Conversely, when blood pressure or blood volume is low, water is conserved and returned to the bloodstream and blood pressure rises (see Figures 20.9 and 20.10). As blood volume goes, so goes the arterial blood pressure.

The *indirect renal mechanism* involves the **renin-angiotensin mechanism.** When arterial blood pressure declines, special cells in the kidneys release the enzyme *renin* (re'nin) into the blood.

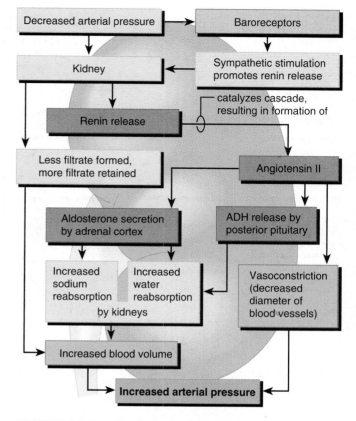

FIGURE 20.9 *Slowly acting renal hormonal mechanisms for blood pressure control.* Notice that sympathetic nervous system stimulation and action, part of the more rapidly acting short-term control pathway, also participate in this system by triggering renin release.

Renin triggers a series of enzymatic reactions that produce **angiotensin II,** as mentioned previously. Angiotensin II is a potent vasoconstrictor which, by promoting an increase in systemic blood pressure, increases the rate of blood delivery to the kidneys and renal perfusion. It also stimulates the adrenal cortex to secrete **aldosterone,** a hormone that enhances renal reabsorption of sodium, and prods the posterior pituitary to release ADH, which promotes more water reabsorption (Figure 20.9). As sodium moves into the bloodstream, water follows; thus, both blood volume and blood pressure rise. The indirect renal mechanism is described in more detail in Chapter 26.

Monitoring Circulatory Efficiency

The efficiency of a person's circulation can be assessed by taking pulse and blood pressure measurements. These measurements, along with those of respiratory rate and body temperature, are referred to collectively as **vital signs** in clinical settings. Let's look at how vital signs are determined or measured.

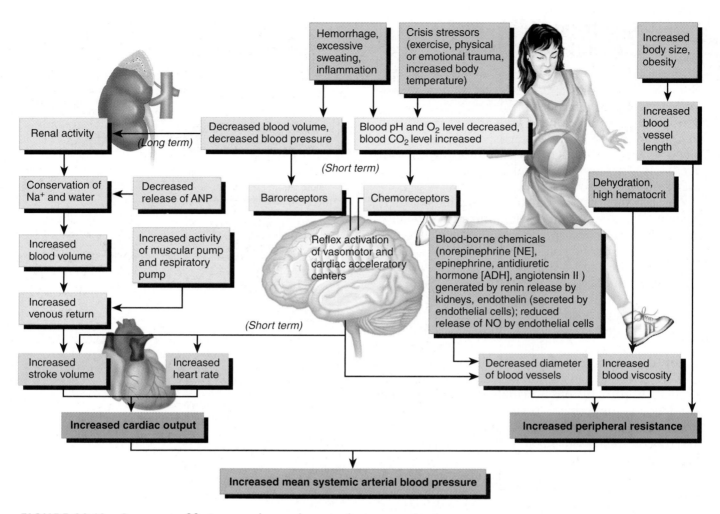

FIGURE 20.10 *Summary of factors causing an increase in mean systemic arterial blood pressure.*

Taking a Pulse

The alternating expansion and recoil of elastic arteries during each cardiac cycle create a pressure wave—a **pulse**—that is transmitted through the arterial tree with each heartbeat. You can feel a pulse in any artery that lies close to the body surface by compressing the artery against firm tissue, and this provides an easy way to count heart rate. Because it is so accessible, the point where the radial artery surfaces at the wrist, the *radial pulse,* is routinely used to take a pulse measurement, but there are several other clinically important arterial pulse points (Figure 20.11). Because these same points are compressed to stop blood flow into distal tissues during hemorrhage, they are also called **pressure points.** For example, if you seriously lacerate your hand, you can slow or stop the bleeding by compressing the radial or the brachial artery.

Monitoring the arterial pulse rate is an easy way to assess the effects of activity, postural changes, and emotions on heart rate. For example, the pulse of a healthy man may be around 66 beats per minute

when he is lying down, rise to 70 when he sits up, and rise to 80 when he suddenly stands. During vigorous exercise or emotional upset, pulse rates between 140 and 180 are not unusual because of sympathetic nervous system effects on the heart.

Measuring Blood Pressure

Most often, systemic arterial blood pressure is measured indirectly by the **auscultatory** (aw-skul′tah-to″re) **method.** This procedure is used to measure blood pressure in the brachial artery of the arm. The *blood pressure cuff,* or *sphygmomanometer* (sfig″mo-mah-nom′ĕ-ter; *sphygmo* = pulse), is wrapped snugly around the arm just superior to the elbow and inflated until the cuff pressure exceeds systolic blood pressure. At this point, blood flow into the arm is stopped and a brachial pulse cannot be felt or heard. As the pressure of the cuff is gradually reduced, the examiner listens (auscultates) carefully for sounds in the brachial artery with a stethoscope. The pressure read as the first soft tapping sounds are heard (the first point at which a small amount of blood is spurting through the constricted artery) is recorded as the systolic pressure. As the cuff pressure is reduced still further, these sounds, called the *sounds of Korotkoff,* become louder and more distinct, but when the artery is no longer constricted and blood flows freely, the sounds can no longer be heard. The pressure at which the sounds disappear is recorded as the diastolic pressure.

In normal adults at rest, systolic blood pressure varies between 110 and 140 mm Hg, and diastolic pressure between 75 and 80 mm Hg, but it is important to recognize that blood pressure varies with age, sex, weight, race, and socioeconomic status. What is "normal" for you may not be normal for your grandfather or your neighbor. Blood pressure also varies with mood, physical activity, and posture. Nearly all of these variations can be explained in terms of the factors affecting blood pressure that have already been discussed.

Alterations in Blood Pressure

Hypotension

Hypotension, or low blood pressure, is a systolic blood pressure below 100 mm Hg. In many cases, hypotension simply reflects individual variations and is no cause for concern. In fact, low blood pressure is often associated with long life and an old age free of illness.

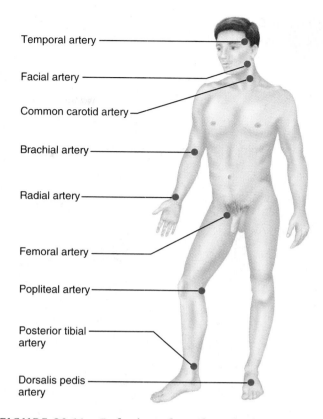

FIGURE 20.11   *Body sites where the pulse is most easily palpated.* (The specific arteries indicated are discussed on pp. 751–763.)

🔼 *Homeostatic Imbalance*

Elderly people are prone to *orthostatic hypotension*— temporary low blood pressure and dizziness when they rise suddenly from a reclining or sitting position. Because the aging sympathetic nervous system does not respond as quickly as it once did to postural changes, blood pools briefly in the lower limbs, reducing blood pressure and consequently blood delivery to the brain. Making postural changes slowly to give the nervous system time to adjust usually prevents this problem.

Chronic hypotension may hint at poor nutrition, because the poorly nourished are often anemic and have inadequate levels of blood proteins. Because blood viscosity is low, blood pressure is also lower than normal. Chronic hypotension also warns of Addison's disease (inadequate adrenal cortex function), hypothyroidism, or severe tissue wasting. *Acute hypotension* is one of the most important signs of circulatory shock (p. 744) and a threat to patients undergoing surgery and those in intensive care units. ▪

Hypertension

Transient elevations in systolic blood pressure occur as normal adaptations during fever, physical exer-

tion, and emotional upset, such as during anger and fear. Persistent **hypertension,** or high blood pressure, is common in obese people because the total length of their blood vessels is relatively greater than that in thinner individuals.

🔼 *Homeostatic Imbalance*

Chronic hypertension is a common and dangerous disease that warns of increased peripheral resistance. An estimated 30% of people over the age of 50 years are hypertensive. Although hypertension is usually asymptomatic for the first 10 to 20 years, it slowly but surely strains the heart and damages the arteries. For this reason, hypertension is often called the "silent killer." Prolonged hypertension is the major cause of heart failure, vascular disease, renal failure, and stroke. Because the heart is forced to pump against greater resistance, it must work harder, and in time the myocardium enlarges. When finally strained beyond its capacity to respond, the heart weakens and its walls become flabby. Hypertension also ravages the blood vessels, and causes small tears in the endothelium that accelerate the progress of

Why is blood delivery to the skin vasculature increased during strenuous exercise?

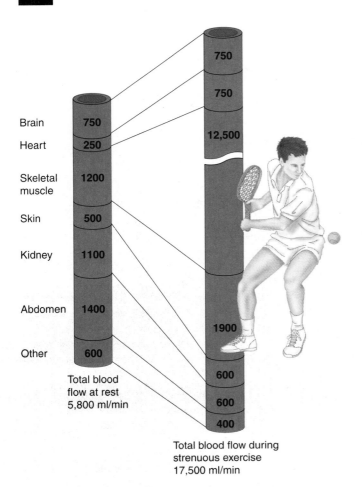

Brain	750
Heart	250
Skeletal muscle	1200
Skin	500
Kidney	1100
Abdomen	1400
Other	600

Total blood flow at rest
5,800 ml/min

750
750
12,500

1900
600
600
400

Total blood flow during strenuous exercise
17,500 ml/min

FIGURE 20.12 *Distribution of blood flow to selected body organs at rest and during strenuous exercise.*

atherosclerosis (see *A Closer Look* on pp. 722–723). As the vessels become increasingly blocked, blood flow to the tissues becomes inadequate, and vascular complications begin to appear in vessels of the brain, heart, kidneys, and retinas of the eyes.

Although hypertension and atherosclerosis are often linked, it is often difficult to blame hypertension on any distinct anatomical pathology. Hypertension is defined physiologically as a condition of sustained elevated arterial pressure of 140/90 or higher, and the higher the blood pressure, the greater the risk for serious cardiovascular problems. As a rule, elevated diastolic pressures are more sig-

Muscle activity generates heat, which must be dissipated if homeostasis is to be maintained. The skin is the body's heat exchange site.

nificant medically, because they always indicate progressive occlusion and/or hardening of the arterial tree.

About 90% of hypertensive people have **primary,** or **essential, hypertension,** in which no underlying cause has been identified. However, the following factors are believed to be involved:

1. Diet. Dietary factors that contribute to hypertension include high sodium, saturated fat, and cholesterol intake and deficiencies in certain metal ions (K^+, Ca^{2+}, and Mg^{2+}).

2. Obesity.

3. Age. Clinical signs of the disease usually appear after age 40.

4. Race. More blacks than whites are hypertensive, and the course of the disease also varies in different population groups.

5. Heredity. Hypertension runs in families. Children of hypertensive parents are twice as likely to develop hypertension as are children of normotensive parents.

6. Stress. Particularly at risk are "hot reactors," people whose blood pressure zooms upward during every stressful event.

7. Smoking. Nicotine enhances the sympathetic nervous system's vasoconstrictor effects.

Primary hypertension cannot be cured, but most cases can be controlled by restricting salt, fat, and cholesterol intake, losing weight, stopping smoking, managing stress, and taking antihypertensive drugs. Drugs commonly used are diuretics, beta-blockers, calcium channel blockers, and angiotensin-converting enzyme (ACE) inhibitors. Inhibiting ACE suppresses the renin-angiotensin mechanism.

Secondary hypertension, which accounts for 10% of cases, is due to identifiable disorders, such as excessive renin secretion by the kidneys, arteriosclerosis, and endocrine disorders such as hyperthyroidism and Cushing's disease. Treatment for secondary hypertension is directed toward correcting the causative problem. ■

BLOOD FLOW THROUGH BODY TISSUES

Blood flow through body tissues, or **tissue perfusion,** is involved in (1) delivery of oxygen and nutrients to, and removal of wastes from, tissue cells, (2) gas exchange in the lungs, (3) absorption of nutrients from the digestive tract, and (4) urine formation by the kidneys. Furthermore, the rate of blood flow to each tissue and organ is almost exactly the right amount to provide for proper function—no more, no less. When the body is at rest, the brain receives about 13% of total blood flow, the heart 4%, kidneys 20%, and the abdominal organs 24%. Skeletal muscles, which make up almost half of body mass, normally receive about 20%. During exercise, however, nearly all of the increased cardiac output flushes into the skeletal muscles and blood flow to the kidneys and digestive organs is actually reduced (Figure 20.12).

Velocity of Blood Flow

As shown in Figure 20.13, the speed or velocity (in cm/s) of blood flow changes as blood travels through the systemic circulation. All other factors being constant, it is fastest in the aorta and other large arteries, slowest in the capillaries, and then picks up speed again in the veins leading back to the heart.

Velocity is *inversely* related to the *cross-sectional area* of the blood vessels to be filled. Blood flows fastest where the total cross-sectional area is least. As the arterial system continues to branch, the total cross-sectional area of the vascular bed increases, and the velocity of blood flow declines proportionately. Even though the individual branches have smaller lumens, their combined cross-sectional area, and thus the volume of blood they can hold, is much greater than that of the aorta. For example, the cross-sectional area of the aorta is 2.5 cm^2, and the average velocity of blood flow in the aorta is 40–50 cm/s. However, the total cross-sectional area of the capillaries is 4500 cm^2, so capillary blood flow is very slow (about 0.03 cm/s). Slow capillary blood flow is beneficial because it allows adequate time for exchanges to be made between the blood and tissue cells.

As capillaries combine to form venules and then veins, total cross-sectional area declines and velocity increases. The cross-sectional area of the venae cavae is 8 cm^2, and the velocity of blood flow varies from 10 to 30 cm/s in those vessels, depending on the activity of the skeletal muscle pump.

Autoregulation: Local Regulation of Blood Flow

Autoregulation is the automatic adjustment of blood flow to each tissue in proportion to its requirements at any point in time. This process is regulated by local conditions and is largely independent of systemic factors. The mean arterial pressure is identical throughout the body and homeostatic mechanisms adjust cardiac output as needed to maintain that pressure constant. Changes in blood flow through individual organs are controlled *intrinsically* by modifying the diameter of local arterioles feeding the capillaries. You can compare blood flow autoregula-

Does the speed of blood flow vary directly or indirectly with the cross-sectional area of the vascular bed?

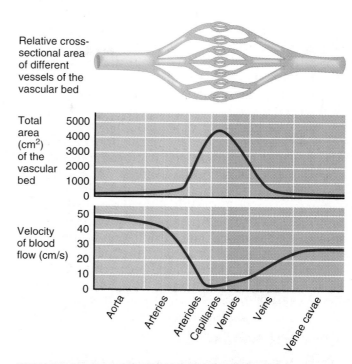

FIGURE 20.13 *Relationship between blood flow velocity and total cross-sectional area in various blood vessels of the systemic circulation.*

tion to water use in your home. To get water in a sink, or the garden hose, you have to turn on a faucet. Whether you have several taps open or none, the pressure in the main water pipe in the street remains relatively constant, as it does in the even larger water lines closer to the pumping station. Similarly, what is happening in the arterioles feeding the capillary beds of a given organ has little effect on pressure in the muscular artery feeding the organ, or on the large elastic arteries. The pumping station is, of course, the heart. The beauty of this system is that as long as the water company (circulatory feedback mechanisms) maintains a relatively constant water pressure (MAP), local demand regulates the actual amount of fluid (blood) delivered to various areas.

Thus, organs regulate their own blood flows by varying the resistance of their arterioles. As described next, these intrinsic control mechanisms may be classed as *metabolic* or *myogenic*.

Blood flow velocity varies indirectly with the cross-sectional area of the vascular bed. As cross-sectional area increases, flow velocity decreases.

Metabolic Controls

In most tissues, declining levels of nutrients, particularly oxygen, are the strongest stimuli for autoregulation. An interesting recent finding is that nitric oxide (NO) induces vasodilation at the capillaries to help get the needed oxygen to the tissue cells. Hemoglobin in the red blood cells delivers the NO to where it is needed: As oxygen is being loaded onto hemoglobin in the lungs, NO latches onto hemoglobin as well. When it reaches the tissue capillaries it is shed, causing vasodilation as oxygen is unloaded. Certain other substances released by metabolically active tissues (such as potassium and hydrogen ions, adenosine, and lactic acid), prostaglandins, and inflammatory chemicals (histamine and kinins) also serve as autoregulation stimuli. Whatever the precise stimulus, the net result of metabolically controlled autoregulation is immediate vasodilation of the arterioles serving the capillary beds of the "needy" tissues, and so blood flow to the local area temporarily increases. This is accompanied by relaxation of the precapillary sphincters (see Figure 20.4), allowing blood to surge through the true capillaries and become available to the tissue cells.

Myogenic Controls

Inadequate blood perfusion through an organ is quickly followed by a decline in the organ's metabolic rate and, if prolonged, organ death. Likewise, excessively high arterial pressure and tissue perfusion can be dangerous because it may rupture the more fragile blood vessels. Such changes in local (arteriolar) blood pressure and volume are important in autoregulation because they directly stimulate vascular smooth muscle, provoking **myogenic** (*myo* = muscle; *gen* = origin) **responses.** Vascular smooth muscle responds directly to passive stretch (increased intravascular pressure) with increased tone, which resists the stretch and causes vasoconstriction. Reduced stretch, on the other hand, promotes vasodilation and increases blood flow into the tissue. Hence, the myogenic mechanism keeps tissue perfusion fairly constant in spite of most variations in systemic pressure.

Generally, both chemical (metabolic) and physical factors determine the final autoregulatory response of a tissue. For example, **reactive hyperemia** (hi"per-e'me-ah) is the dramatic increase in blood flow into a tissue that occurs after the blood supply to an area has been temporarily blocked. It results both from the myogenic response and from an accumulation of metabolic wastes in the area during the time of occlusion.

Long-Term Autoregulation

If the nutrient requirements of a tissue are greater than the short-term autoregulatory mechanism can easily supply, a long-term autoregulation mechanism may evolve over a period of weeks or months to enrich the local blood flow still more. The number of blood vessels in the region increases, and existing vessels enlarge. This phenomenon, called *angiogenesis*, is particularly common in the heart when a coronary vessel is partially occluded. It occurs throughout the body in people who live in high-altitude areas, where the air contains less oxygen.

Blood Flow in Special Areas

Each organ has special requirements and functions that are revealed in its pattern of autoregulation. Autoregulation in the brain, heart, and kidneys is extraordinarily efficient. In those organs, adequate perfusion is maintained even when mean arterial pressure is fluctuating.

Skeletal Muscles

Blood flow in skeletal muscle is extremely changeable and varies with muscle activity. Resting skeletal muscles receive about 1 L of blood per minute, and only about 25% of their capillaries are open. During such periods, myogenic and general neural mechanisms predominate. When muscles become active, blood flow increases *(hyperemia)* in direct proportion to their greater *metabolic* activity, a phenomenon called **active** or **exercise hyperemia.**

The arterioles in skeletal muscle have cholinergic receptors and both alpha and beta (α, β) adrenergic receptors, which bind epinephrine. When epinephrine is present in low levels, it mediates vasodilation by binding chiefly to the β receptors. Likewise, "occupied" cholinergic receptors are believed to promote vasodilation. Consequently, blood flow can increase tenfold or more during physical activity (see Figure 20.12), and virtually all capillaries in the active muscles open to accommodate the increased flow. By contrast, the high levels of epinephrine typical of massive sympathetic nervous system activation (and extremely vigorous exercise involving large numbers of skeletal muscles) cause intense vasoconstriction mediated by the α-adrenergic receptors. This protective response, believed to be initiated by muscle chemoreflexes when muscle oxygen delivery falls below some critical level, ensures that muscle demands for blood flow do not exceed cardiac pumping ability and that vital organs continue to receive an adequate blood supply. Without question, strenuous exercise is one of the most demanding conditions the cardiovascular system faces.

Muscular autoregulation occurs almost entirely in response to the decreased oxygen concentrations (accompanied by local increases in carbon dioxide, lactic acid, and other metabolites) that result from the "revved-up" metabolism of working muscles. However, systemic adjustments, mediated by the vasomotor center, must also occur to ensure that blood delivery to the muscles is both faster and more abundant. Strong vasoconstriction of the vessels of blood reservoirs such as those of the digestive viscera and skin diverts blood away from these regions temporarily, ensuring that more blood reaches the muscles. Ultimately, the major factor determining how long muscles can continue to contract vigorously is the ability of the cardiovascular system to deliver adequate oxygen and nutrients.

The Brain

Total blood flow to the brain averages about 750 ml/min and is maintained at relatively constant levels. The necessity for constant cerebral blood flow becomes crystal clear when one takes into account that neurons are totally intolerant of ischemia.

Although the brain is the most metabolically active organ in the body, it is the least able to store essential nutrients. Cerebral blood flow is regulated by one of the most precise autoregulatory systems in the entire body and is tailored to local neuronal need. Thus, when you make a fist with your right hand, the neurons in the left cerebral motor cortex controlling that movement receive a more abundant blood supply than the adjoining neurons. Brain tissue is exceptionally sensitive to declining pH, and increased blood carbon dioxide levels (resulting in acidic conditions in brain tissue) cause marked vasodilation. Oxygen deficit is a much less potent stimulus for autoregulation. However, greatly excessive carbon dioxide levels abolish autoregulatory mechanisms and severely depress brain activity.

Besides the metabolic controls, the brain also has a myogenic mechanism that protects it from possibly damaging changes in blood pressure. When mean arterial pressure declines, cerebral vessels dilate to ensure adequate brain perfusion. On the other hand, when arterial pressure rises, cerebral vessels constrict, protecting the small, more fragile vessels farther along in the pathway from rupture due to excessive pressure. Under certain circumstances, such as brain ischemia caused by rising intracranial pressure (as with a brain tumor), the brain (via the medullary cardiovascular centers) regulates its own blood flow by triggering a rise in systemic blood pressure. However, when systemic pressure changes are extreme, the brain becomes vulnerable. Fainting, or *syncope* (sin'cuh-pe; "cutting short"), occurs when mean arterial pressure falls below 60 mm Hg. Cerebral edema is the usual result of pressures over 160 mm Hg, which dramatically increase brain capillary permeability.

The Skin

Blood flow through the skin (1) supplies nutrients to the cells, (2) aids in body temperature regulation, and (3) provides a blood reservoir. The first function is served by autoregulation in response to the need for oxygen; the second and third require neural intervention. We will concentrate on the skin's temperature-related function here.

Below the skin surface are extensive venous plexuses, in which the blood flow can change from 50 ml/min to as much as 2500 ml/min, depending on body temperature. This capability reflects neural adjustments of blood flow through unique coiled arteriovenous anastomoses as well as through arterioles. These tiny shunts are located mainly in the fingertips, palms of the hands, toes, soles of the feet, ears, nose, and lips. They are richly supplied with sympathetic nerve endings (a characteristic that sets them apart from the shunts of most other capillary beds), and are controlled by reflexes initiated by temperature receptors or signals from higher CNS centers. The arterioles, by contrast, are also responsive to local and metabolic autoregulatory stimuli.

When the skin surface is exposed to heat, or body temperature rises for other reasons (such as vigorous exercise), the hypothalamic "thermostat" signals for reduced vasomotor stimulation of the skin vessels. As a result, warm blood flushes into the capillary beds and heat radiates from the skin surface. Vasodilation of the arterioles is enhanced even more when we sweat, because an enzyme in perspiration acts on a protein present in tissue fluid to produce *bradykinin*. Bradykinin, in turn, stimulates the vessel's endothelial cells to release NO, the potent vasodilator.

When the ambient temperature is cold and body temperature drops, superficial skin vessels are strongly constricted. Hence, blood almost entirely bypasses the capillaries associated with the arteriovenous anastomoses, diverting the warm blood to the deeper, more vital organs. Paradoxically, the skin may stay quite rosy because some blood gets "trapped" in the superficial capillary loops as the shunts swing into operation.

The Lungs

Blood flow through the pulmonary circuit to and from the lungs is unusual in many ways. The pathway is relatively short, and pulmonary arteries and arterioles are structurally more like veins and venules. That is, they have thin walls and large lumens. Since resistance to blood flow is low, less pressure is needed to propel blood through the pulmonary arterial system. Consequently, arterial pressure in the pulmonary circulation is much lower than in the systemic circulation (24/8 versus 120/80).

In addition, the autoregulatory mechanism in the pulmonary circulation is *exactly opposite* that seen in most tissues. Low oxygen levels cause vasoconstriction, and high levels promote vasodilation.

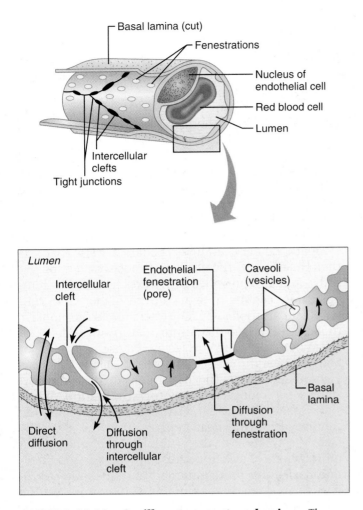

FIGURE 20.14 *Capillary transport mechanisms.* The four possible pathways or routes of transport across the endothelial cell wall of a fenestrated capillary. (The endothelial cell is drawn as if cut in cross section.)

While this may seem odd, it is perfectly consistent with the gas exchange role of this circulation. When the air sacs of the lungs are flooded with oxygen-rich air, the pulmonary capillaries become flushed with blood and ready to receive the oxygen load. If the air sacs are collapsed or blocked with mucus, the oxygen content in those areas will be low, and blood will largely bypass those nonfunctional areas.

The Heart

Movement of blood through the smaller vessels of the coronary circulation of the heart is influenced by aortic pressure and the pumping activity of the ventricles. When the ventricles contract and compress the coronary vessels, blood flow through the myocardium stops. As the heart relaxes, the high aortic pressure forces blood through the coronary circulation. Under normal circumstances, the myoglobin in the cardiac cells stores sufficient oxygen to satisfy their oxygen needs during systole. However, an abnormally rapid heartbeat seriously reduces the ability of the myocardium to receive adequate oxygen and nutrients during diastole.

Under resting conditions, blood flow through the heart is about 250 ml/min and is probably controlled by a myogenic mechanism. During strenuous exercise, the coronary vessels dilate in response to local accumulation of carbon dioxide (leading to acidosis), and blood flow may increase three to four times (see Figure 20.12). Consequently, blood flow remains fairly constant despite wide variation (50 to 140 mm Hg) in coronary perfusion pressure. This enhanced blood flow during increased heart activity is important because cardiac cells use as much as 65% of the oxygen carried to them in blood under resting conditions. (Most other tissue cells use about 25% of the oxygen delivered during each circulatory round.) Thus, increasing the blood flow is the only way to make sufficient additional oxygen available to a more vigorously working heart.

Blood Flow Through Capillaries and Capillary Dynamics

Blood flow through capillary networks is slow and intermittent, rather than steady. This phenomenon, called **vasomotion,** reflects the on/off opening and closing of precapillary sphincters in response to local autoregulatory controls.

Capillary Exchanges of Respiratory Gases and Nutrients

Oxygen, carbon dioxide, most nutrients (amino acids, glucose, lipids), and metabolic wastes pass between the blood and interstitial fluid by **diffusion.** Recall that in diffusion, movement always occurs along a concentration gradient—each substance moving from an area of its higher concentration to an area of its lower concentration. Hence, oxygen and nutrients pass from the blood, where their concentration is fairly high, through the interstitial fluid to the tissue cells. Carbon dioxide and metabolic wastes leave the cells, where their content is higher, and diffuse into the capillary blood. In general, small water-soluble solutes, such as amino acids and sugars, pass through fluid-filled intercellular capillary clefts (and sometimes through fenestrations), while lipid-soluble molecules, such as respiratory gases, diffuse directly through the lipid bilayer of the endothelial cell plasma membranes (see Figure 20.14). Newly forming caveoli (cytoplasmic vesicles) translocate some larger molecules, such as small proteins. As mentioned earlier, capillaries differ in their "leakiness," or permeability. Liver capillaries

? *How would fluid flows change if the OP of the interstitial fluid rose dramatically—say because of a severe bacterial infection in the surrounding tissues?*

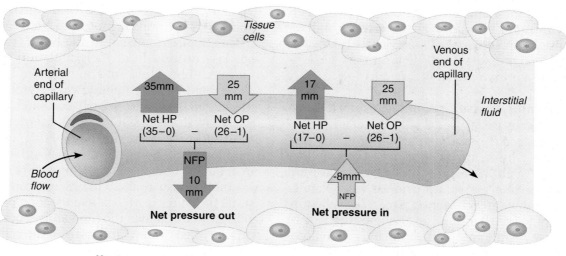

Key to pressure values:
HP_c at arterial end = 35 mm Hg HP_{if} = 0 mm Hg OP_{if} = 1 mm Hg
HP_c at venous end = 17 mm Hg OP_c = 26 mm Hg

FIGURE 20.15 *Forces responsible for fluid flows at capillaries.* Direction of fluid movement depends on the difference between two opposing forces, hydrostatic pressure (HP) and colloid osmotic pressure (OP). The net HP is the force that tends to push fluid out of the capillary. The net HP ($HP_c - HP_{if}$) is equal to the hydrostatic pressure of the blood (HP_c), which varies along the length of the capillary. The net OP is the force that draws fluid back into the capillary. The net OP ($OP_c - OP_{if}$) is a constant value, reflecting the fact that blood has a much higher content of nondiffusible solutes (proteins) than interstitial fluid. At the arterial end, net HP forcing fluid outward exceeds OP drawing water inward, resulting in a NPF (net filtration pressure) that promotes a net fluid loss from the capillary: NFP = 10 mm Hg. At the downstream (venule) end, HP is overpowered by OP (NFP = −8 mm Hg), and fluid returns to the bloodstream.

have large fenestrations that allow even proteins to pass freely, whereas brain capillaries are impermeable to most substances.

Fluid Movements

All the while nutrient and gas exchanges are occurring across the capillary walls by diffusion, bulk fluid flows are also going on. Fluid is forced out of the capillaries through the clefts at the arterial end of the bed, but most of it returns to the bloodstream at the venous end. These fluid flows are relatively unimportant to the capillary exchange process; instead they help determine the relative fluid volumes in the bloodstream and the extracellular space. As described next, the *direction and amount* of fluid that flows across the capillary walls reflect the balance (or imbalance) between two dynamic and opposing forces—hydrostatic and colloid osmotic pressures (Figure 20.15).

Hydrostatic Pressures Hydrostatic pressure is the force exerted by a fluid pressing against a wall. In capillaries, hydrostatic pressure is the same as *capil-*

lary blood pressure—the pressure of blood against the capillary walls. **Capillary hydrostatic pressure (HP$_c$)** tends to force fluids through the capillary walls. Because blood pressure drops as blood flows through the length of a capillary bed, HP_c is higher at the arterial end of the bed (about 35 mm Hg) than at the venous end (approximately 17 mm Hg).

In theory, blood pressure—which forces fluid out of the capillaries—is opposed by the **interstitial fluid hydrostatic pressure (HP$_{if}$)** acting outside the capillaries (back pressure) pushing fluid in. Thus, the *net* hydrostatic pressure acting on the capillaries at any point is the difference between the hydrostatic force that is pushing fluid *out* of the capillary (blood pressure) and the hydrostatic force that is pushing fluid *into* the capillary (interstitial fluid pressure). However, there is usually very little fluid in the interstitial space, because it is constantly withdrawn

Fluid that would ordinarily reenter the circulation at the venous end of the capillary bed will remain in the tissue spaces, held by the elevated OP of the interstitial fluid (and will ultimately be picked up by the lymphatic system vessels).

by the lymphatic vessels (see Chapter 21). Although the hydrostatic pressure in the interstitial space may vary from a slightly negative to slightly positive value, traditionally it is assumed to be zero. For simplicity, that is the value we will use here. The *net effective hydrostatic pressures* at the arterial and venous ends of the capillary bed are essentially equal to the HP_c (blood pressure) at those locations. This information is summarized in Figure 20.15.

Colloid Osmotic Pressures Colloid osmotic pressure is created by the presence in a fluid of large non-diffusible molecules, such as plasma proteins, that are prevented from moving through the capillary membrane. Such molecules draw water toward them; that is, they encourage osmosis (see Chapter 3) whenever the water concentration in their vicinity is relatively lower than it is on the opposite side of the capillary membrane. The abundant plasma proteins in capillary blood (primarily albumin molecules) develop a **capillary colloid osmotic pressure (OP_c)**, also called *oncotic pressure,* of approximately 26 mm Hg. Since interstitial fluid contains few proteins, its colloid osmotic pressure (OP_{if}) is substantially lower—ranging from 0.1 to 5 mm Hg. We use a value of 1 mm Hg for the OP_{if} in Figure 20.15. Unlike hydrostatic pressure, osmotic pressure does not vary significantly from one end of the capillary bed to the other. Thus, in our example, the *net osmotic pressure* that pulls fluid back into the capillary blood is 25 mm Hg.

Hydrostatic-Osmotic Pressure Interactions To determine whether there is a net gain or net loss of fluid from the blood, we have to calculate the **net filtration pressure (NFP)** which considers all of the forces acting at the capillary bed. At any point along a capillary, fluids will leave the capillary if net hydrostatic pressure is greater than the net colloid osmotic pressure, and fluids will enter the capillary if the net colloid osmotic pressure exceeds net hydrostatic pressure. As shown in Figure 20.15, hydrostatic forces dominate at the arterial end:

$$NFP = (HP_c - HP_{if}) - (OP_c - OP_{if})$$
$$NFP = (35 - 0) - (26 - 1)$$
$$NFP = (35 - 25) = 10 \text{ mm Hg}$$

Thus, in this example, 10 mm Hg worth of pressure (net excess of HP) is forcing fluid out of the capillary. On the other hand, osmotic forces dominate at the venous end:

$$NFP = (17 - 0) - (26 - 1)$$
$$NFP = 17 - 25 = 18 \text{ mm Hg}$$

The negative pressure value we see here indicates that the NFP (due to net excess of OP) is driving fluid back *into* the capillary bed. Thus, net fluid flows *out* of the circulation at the arterial ends of capillary beds and *into* the bloodstream at the venous ends. However, more fluid enters the tissue spaces than is returned to the blood, resulting in a net loss of fluid from the circulation of about 1.5 ml/min. This fluid and any leaked proteins are picked up by the lymphatic vessels and returned to the vascular system, which accounts for the relatively low levels of both fluid and proteins in the interstitial space. Were this not so, this "insignificant" fluid loss would empty your blood vessels of plasma in about 24 hours!

Notice also that a change in either the hydrostatic pressures or the colloid osmotic pressures will substantially change the fluid shifts occurring at the capillary beds and to and from tissue cells.

Circulatory Shock

Circulatory shock is any condition in which blood vessels are inadequately filled and blood cannot circulate normally. This results in inadequate blood flow to meet tissue needs. If the condition persists, cells die and organ damage follows.

The most common form of shock is **hypovolemic** (hi"po-vo-le'mik) **shock.** Hypovolemic shock (*hypo* = low, deficient; *volemia* = blood volume) results from large-scale loss of blood, as might follow acute hemorrhage, severe vomiting or diarrhea, or extensive burns. If blood volume drops rapidly, heart rate increases in an attempt to correct the problem. Thus, a weak, "thready" pulse is often the first sign of hypovolemic shock. Intense vasoconstriction also occurs, which shifts blood from the various blood reservoirs into the major circulatory channels and enhances venous return. Blood pressure is stable at first, but eventually drops if blood loss continues. A sharp drop in blood pressure is a serious, and late, sign of hypovolemic shock. The key to managing hypovolemic shock is to replace fluid volume as quickly as possible.

Although many body system responses to hypovolemic shock have yet to be explored, acute bleeding is such a threat to life that it seems important to have a comprehensive flowchart of its recognizable signs and symptoms and an accounting of the body's attempt to restore homeostasis. Figure 20.16 provides such a resource. Study it in part now, and then in more detail later when you have completed your studies of the remaining body systems.

Text continues on page 746.

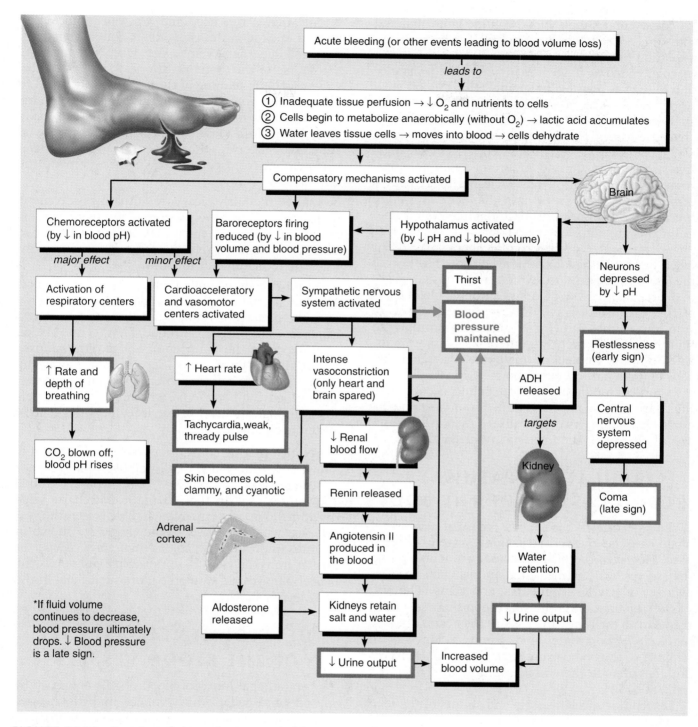

**FIGURE 20.16   *Events and signs of compensated (nonprogressive)
hypovolemic shock.*** (Recognizable clinical signs are shown in *red-bordered boxes*.)

In **vascular shock,** blood volume is normal and constant. Poor circulation is a result of extreme vasodilation that causes an abnormal expansion of the vascular bed. The huge drop in peripheral resistance that follows is revealed by rapidly falling blood pressure. The most common causes of vascular shock are loss of vasomotor tone due to anaphylaxis (anaphylactic shock), a systemic allergic reaction in which bodywide vasodilation is triggered by the massive release of histamine; failure of autonomic nervous system regulation (also referred to as *neurogenic shock*); and septicemia. *Septicemia (septic shock)* is a severe systemic bacterial infection; bacterial toxins are notorious vasodilators.

A transient type of vascular shock may occur when you sunbathe for a prolonged time. The heat of the sun on your skin causes cutaneous blood vessels to dilate. Then, if you stand up abruptly, blood pools briefly (because of gravity) in the dilated vessels of your lower limbs rather than returning promptly to the heart. Consequently, your blood pressure falls. The dizziness you feel at this point is a signal that your brain is not receiving enough oxygen.

Cardiogenic shock, or pump failure, occurs when the heart is so inefficient that it cannot sustain adequate circulation. Its usual cause is myocardial damage, as might follow numerous myocardial infarcts.

CIRCULATORY PATHWAYS: BLOOD VESSELS OF THE BODY

When referring to the body's complex network of blood vessels, the term **vascular system** is often used. However, the heart is actually a double pump that serves two distinct circulations, each with its own set of arteries, capillaries, and veins. The *pulmonary circulation* is the short loop that runs from the heart to the lungs and back to the heart. The *systemic circulation* routes blood through a long loop to all parts of the body before returning it to the heart. Both circuits are shown schematically in Table 20.3 (pp. 750–751).

Except for special vessels and shunts of the fetal circulation (described in Chapter 29), the principal arteries and veins of the systemic circulation are described and illustrated in Tables 20.4 through 20.13. Although there are many similarities between the systemic arteries and veins, there are also important differences:

1. Whereas the heart pumps all of its blood into a single systemic artery—the aorta—blood returning to the heart is delivered largely by two terminal systemic veins, the superior and inferior venae cavae. The single exception to this is the blood draining from the myocardium of the heart, which is collected by the cardiac veins and reenters the right atrium via the coronary sinus (see p. 690).

2. All arteries run deep and are well protected by body tissues along most of their course, but both deep and superficial veins exist. Deep veins parallel the course of the systemic arteries, and with a few exceptions, the naming of these veins is identical to that of their companion arteries. Superficial veins run just beneath the skin and are readily seen, especially in the limbs, face, and neck. Since there are no superficial arteries, the names of the superficial veins do not correspond to the names of any of the arteries.

3. Unlike the fairly clear arterial pathways, venous pathways tend to have numerous interconnections, and many veins are represented by not one but two similarly named vessels. As a result, venous pathways are more difficult to follow.

4. In most body regions, there is a similar and predictable arterial supply and venous drainage. However, the pattern of venous drainage in at least two important body areas is unique. First, venous blood draining from the brain enters large *dural sinuses* rather than typical veins. Second, blood draining from the digestive organs enters a special subcirculation, called the *hepatic portal circulation*, and perfuses through the liver before it reenters the general systemic circulation.

Notice that by convention, oxygen-rich blood is shown red, while blood that is relatively oxygen-poor is depicted blue, regardless of vessel types. A unique convention used in the schematic flowcharts (pipe diagrams) that accompany each table is that the vessels that would be closer to the viewer are shown in brighter, more intense colors than those deeper or farther from the viewer—for example, for veins, the deeper blue vessels would be more superficial in the body region shown.

DEVELOPMENTAL ASPECTS OF THE BLOOD VESSELS

The endothelial lining of blood vessels is formed by mesodermal cells, which collect in little masses called **blood islands** throughout the microscopic embryo. These then form sprouting extensions that reach toward one another and the forming heart to lay down the rudimentary vascular tubes. Meanwhile, adjacent mesenchymal cells surround the endothelial tubes, forming the muscular and fibrous coats of the vessel walls. As noted in Chapter 19, the heart is pumping blood through the rudimentary vascular system by the fourth week of development.

In addition to the fetal shunts that bypass the nonfunctional lungs (the *foramen ovale* and *ductus arteriosus*), other vascular modifications are found in the fetus. A special vessel, the *ductus venosus*, largely bypasses the liver. Also important are the *umbilical vein* and *arteries*, large vessels that circulate blood between the fetal circulation and the placenta where gas and nutrient exchanges occur with the mother's blood (see Chapter 29, Figure 29.13 and p. 1136). Once the fetal circulatory pattern is laid down, few vascular changes occur until birth, when the umbilical vessels and shunts are occluded.

In contrast to congenital heart diseases, congenital vascular problems are rare, and blood vessels are remarkably trouble free during youth. Vessel formation occurs as needed to support body growth, wound healing, and to rebuild vessels lost each month during a woman's menstrual cycle. As we age, signs of vascular disease begin to appear. In some, the venous valves weaken and purple, snakelike varicose veins appear. In others, more insidious signs of inefficient circulation appear: tingling in the fingers and toes and cramping of muscles.

Although the degenerative process of atherosclerosis begins in youth, its consequences are rarely apparent until middle to old age, when it may precipitate a myocardial infarct or stroke. Until puberty, the blood vessels of boys and girls look alike, but from puberty to about the age of 45 years, women have strikingly less atherosclerosis than men because of the protective effects of estrogen. Beneficial effects of estrogen are well documented. By enhancing nitric oxide production, inhibiting endothelin release, and blocking voltage-gated Ca^{2+} channels, its effect on blood vessels is to reduce resistance to blood flow. Estrogen also stimulates the liver to produce enzymes that speed up the catabolism of LDLs and increase the production of HDLs, thus reducing the risk of atherosclerosis (see *A Closer Look*). Between the ages of 45 and 65, when estrogen production wanes in women, this "gap" between the sexes closes and males and females are equally at risk for cardiovascular disease.

Blood pressure changes with age. In a newborn baby, arterial pressure is about 90/55. Blood pressure rises steadily during childhood finally reaching the adult value (120/80). In old age, normal blood pressure averages 150/90, which is a hypertensive pressure in younger people. After the age of 40, the incidence of hypertension increases dramatically. Unlike atherosclerosis, which is increasingly important in old age, hypertension claims many youthful victims and is the single most important cause of sudden cardiovascular death in men in their early 40s and 50s.

At least some vascular disease is a product of our modern technological culture. "Blessed" with high-protein, lipid-rich diets, empty-calorie snacks, energy-saving devices, and high-stress jobs, many of us are struck down prematurely. Cardiovascular disease can be prevented somewhat by diet modifications, regular aerobic exercise, and eliminating cigarette smoking. Poor diet and exercise habits and smoking are probably more detrimental to your blood vessels than aging itself could ever be!

* * *

Now that we have described the structure and function of blood vessels, our survey of the cardiovascular system is complete. The pump, the plumbing, and the circulating fluid form a dynamic organ system that ceaselessly services every other organ system of the body, as summarized by *Making Connections*. However, our study of the so-called *circulatory system* is still unfinished. We have yet to examine the lymphatic system, which acts cooperatively with the cardiovascular system to ensure continuous circulation and to provide sites from which lymphocytes can police the body and provide for immunity. These are the topics of Chapter 21.

RELATED CLINICAL TERMS

Aneurysm (an'u-rizm; *aneurysm* = a widening) A balloon-like outpocketing of an artery wall that places the artery at risk for rupture; most often reflects gradual weakening of the artery by chronic hypertension or arteriosclerosis. The most common sites of aneurysm formation are the abdominal aorta and arteries feeding the brain and kidneys.

Angiogram (an'je-o-gram"; *angio* = a vessel; *gram* = writing) Diagnostic technique involving the infusion of a radiopaque substance into the circulation for X-ray examination of specific blood vessels; the major technique for diagnosing coronary artery occlusion and risk of a heart attack.

Diuretic (*diure* = urinate) A chemical that promotes urine formation, thus reducing blood volume; diuretic drugs are frequently prescribed to manage hypertension.

Phlebitis (flĕ-bi'tis; *phleb* = vein; *itis* = inflammation) Inflammation of a vein accompanied by painful throbbing and redness of the skin over the inflamed vessel; most often caused by bacterial infection or local physical trauma.

Phlebotomy (flĕ-bot'o-me; *tomy* = cut) A venous incision made for the purpose of withdrawing blood or bloodletting.

Sclerotherapy Procedure used for removing varicose or spider veins; tiny needles are used to inject hardening agents into the abnormal vein; the vein scars, closes down, and is absorbed by the body.

Thrombophlebitis Condition of undesirable intravascular clotting initiated by a roughening of a venous lining; often follows severe episodes of phlebitis. An ever-present danger is that the clot may detach and form an embolus.

MAKING CONNECTIONS

SYSTEM CONNECTIONS: Homeostatic Interrelationships Between the Cardiovascular System and Other Body Systems

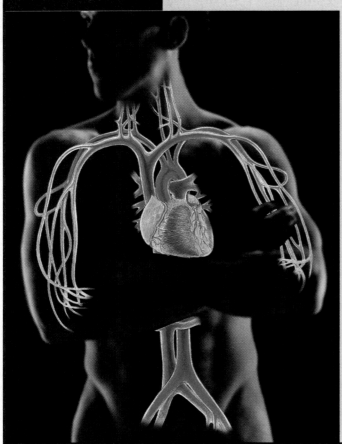

- ANS regulates cardiac rate and force; sympathetic division maintains blood pressure and controls blood distribution according to organ need

Endocrine System
- The cardiovascular system delivers oxygen and nutrients; carries away wastes; blood serves as a transport vehicle for hormones
- Various hormones influence blood pressure (epinephrine, ANP, thyroxine, ADH); estrogen maintains vascular health in women

Lymphatic System/Immunity
- The cardiovascular system delivers oxygen and nutrients to lymphatic organs, which house immune cells; provides transport medium for lymphocytes and antibodies; carries away wastes
- The lymphatic system picks up leaked fluid and plasma proteins and returns them to the cardiovascular system; its immune cells protect cardiovascular organs from specific pathogens

Respiratory System
- The cardiovascular system delivers oxygen and nutrients; carries away wastes
- The respiratory system carries out gas exchange: loads oxygen and unloads carbon dioxide from the blood; respiratory "pump" aids venous return

Digestive System
- The cardiovascular system delivers oxygen and nutrients; carries away wastes
- The digestive system provides nutrients to the blood including iron and B vitamins essential for RBC (and hemoglobin) formation

Urinary System
- The cardiovascular system delivers oxygen and nutrients; carries away wastes; blood pressure maintains kidney function
- The urinary system helps regulate blood volume and pressure by altering urine volume and releasing renin

Reproductive System
- The cardiovascular system delivers oxygen and nutrients; carries away wastes
- Estrogen maintains vascular health in women

Integumentary System
- The cardiovascular system delivers oxygen and nutrients; carries away wastes
- The skin vasculature is an important blood reservoir and provides a site for heat loss from the body

Skeletal System
- The cardiovascular system delivers oxygen and nutrients; carries away wastes
- Bones are the sites of hematopoiesis; protect cardiovascular organs by enclosure: and provide a calcium depot

Muscular System
- The cardiovascular system delivers oxygen and nutrients; carries away wastes
- Aerobic exercise enhances cardiovascular efficiency and helps prevent atherosclerosis; the muscle "pump" aids venous return

Nervous System
- The cardiovascular system delivers oxygen and nutrients; carries away wastes

CLOSER CONNECTIONS

THE CARDIOVASCULAR SYSTEM and Interrelationships with the Muscular, Nervous, and Urinary Systems

The cardiovascular system is the "king of systems." No body system can live without the blood that surges ceaselessly through cardiovascular channels. Likewise, it is nearly impossible to find a system that does not influence the cardiovascular system in some way. The respiratory and digestive systems enrich blood with oxygen and nutrients, respectively, and in turn take a share of the various riches the blood has to offer. By returning leaked plasma fluid to the vascular system, the lymphatic system helps to keep those vessels filled with blood so that circulation is possible. However, the three cardiovascular system partnerships that we will look at more closely here are those it has with the muscular, nervous, and urinary systems.

Muscular System

Talk about "one hand washing the other"—that's a pretty fair analogy for the interaction between the cardiovascular and muscular systems. Muscles cramp and become nonfunctional when deprived of an adequate supply of oxygen-rich blood, and their capillary supply expands (or atrophies) along with changes in muscle mass. The arterioles of skeletal muscle even have special beta-adrenergic and ACh receptors so that they can be dilated by neural reflexes when most other body arterioles are responding to vasoconstrictor "orders." When muscles are active and healthy, so too is the cardiovascular system. Without exercise, the heart weakens and loses mass; but when we exercise aerobically, the heart increases in size and strength. Heart rate goes down as stroke volume rises, and a lower HR means that the heart relaxes more and beats hundreds of thousands times less in a lifetime. Aerobic exercise also enhances HDL blood levels and reduces LDL levels—helping to clear fatty deposits from the vascular walls and deferring atherosclerosis, hypertension, and heart disease. Not a bad trade!

Nervous System

The brain is critically dependent on absolutely continuous oxygen and glucose delivery; thus it should come as no surprise to learn that the brain has the body's most precise autoregulatory circulation mechanism. Acting on acutely sensitive arterioles, this mechanism protects the brain from most deficits. For this we can be profoundly thankful, for when neurons die, an important part of us dies as well. The cardiovascular system is just as dependent on the nervous system. Although an intrinsic conduction system sets sinus rhythm, it does not even begin to adapt to stimuli that mobilize the cardiovascular system to peak efficiency in maintaining blood pressure during position changes and delivering blood faster during times of stress. This is the job of the autonomic nervous system, which initiates reflexes as necessary to increase or decrease cardiac output and peripheral resistance, and to redirect blood from one organ to others to serve specific needs and protect vital organs.

Urinary System

Like the lymphatic system, the urinary system (primarily the kidneys) helps maintain blood volume and circulatory dynamics, but in a much more complicated way. The kidneys use blood pressure (courtesy of the cardiovascular system) to form the filtrate which the kidney tubule cells then process. Essentially this processing involves reclaiming needed nutrients and water while allowing metabolic waste and excess ions (including H^+) to leave the body in urine. As a result, blood is continually refreshed and blood composition and volume are carefully regulated. So dedicated are the kidneys to preserving blood volume that urine output stops entirely when blood volume is severely depressed. On the other hand, the kidneys can increase systemic BP by releasing renin when the blood pressure becomes inadequate.

CLINICAL CONNECTIONS

Cardiovascular System

Case study: Mr. Hutchinson, another middle-aged victim of the collision on Route 91, has a tourniquet around his thigh when admitted in an unconscious state to Noble Hospital. The emergency technician who brings him in states that his right lower limb was pinned beneath the bus for at least 30 minutes. He is immediately scheduled for surgery. Admission notes include the following:

- Multiple contusions of lower limbs
- Compound fracture of the right tibia; bone ends covered with sterile gauze
- Right leg blanched and cold, no pulse

- Blood pressure 90/48; pulse 140/min and thready; patient diaphoretic (sweaty)

1. Relative to what you have learned about tissue requirements for oxygen, what is the condition of the tissues in the right lower limb?

2. Will the fracture be attended to, or will Mr. Hutchinson's other homeostatic needs take precedence? Explain your answer choice and predict his surgical treatment.

3. What do you conclude regarding Mr. Hutchinson's cardiovascular measurements (pulse and BP), and what measures do you expect will be taken to remedy the situation before commencing with surgery?

(Answers in Appendix F)

TABLE 20.3	Pulmonary and Systemic Circulations

Pulmonary Circulation

The pulmonary circulation (Figure 20.17a) functions only to bring blood into close contact with the alveoli (air sacs) of the lungs so that gaseous exchanges can occur. It does not directly serve the metabolic needs of body tissues.

Oxygen-poor, dark red blood enters the pulmonary circulation as it is pumped from the right ventricle into the large **pulmonary trunk** (Figure 20.17b). The pulmonary trunk runs diagonally upward for about 8 cm and then divides abruptly to form the **right** and **left pulmonary arteries.** In the lungs, the pulmonary arteries subdivide into the **lobar** (lo'bar) **arteries** (three in the right lung and two in the left lung), each of which serves one lung lobe. The lobar arteries accompany the main bronchi into the lungs and then branch profusely, forming arterioles and, finally, the dense networks of **pulmonary capillaries** that surround and cling to the delicate air sacs. It is here that oxygen loading and carbon dioxide unloading between the blood and alveolar air occur. As gas exchanges occur and oxygen content of the blood rises, the blood becomes bright red. The pulmonary capillary beds drain into venules, which join to form the two **pulmonary veins** exiting from each lung. The four pulmonary veins complete the circuit by unloading their precious cargo into the left atrium of the heart. Note that any vessel with the term *pulmonary* or *lobar* in its name is part of the pulmonary circulation. All others are part of the systemic circulation.

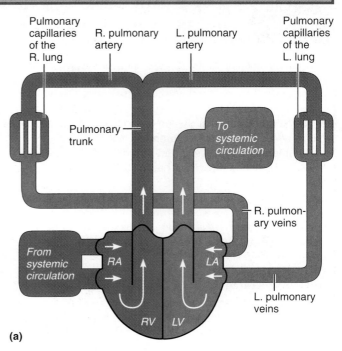

(a)

FIGURE 20.17 *Pulmonary circulation.* (a) Schematic flowchart. (b) Illustration. The arterial system is shown in blue to indicate that the blood carried is oxygen poor; the venous drainage is shown in red to indicate that the blood transported is oxygen rich.

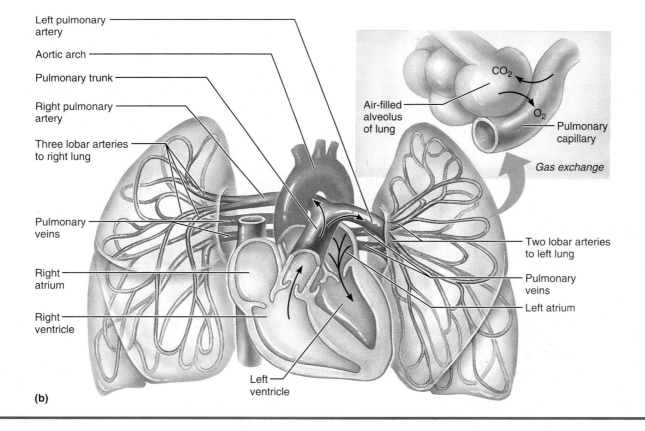

(b)

TABLE 20.3	**Pulmonary and Systemic Circulations** *continued*

Pulmonary arteries carry oxygen-poor, carbon dioxide–rich blood, and pulmonary veins carry oxygen-rich blood.* This is exactly opposite the situation in the systemic circulation, where arteries carry oxygen-rich blood and veins carry carbon dioxide–rich, relatively oxygen-poor blood.

Systemic Circulation

The systemic circulation provides the *functional blood supply* to all body tissues; that is, it delivers oxygen, nutrients, and other needed substances while carrying away carbon dioxide and other metabolic wastes. Freshly oxygenated blood* returning from the pulmonary circuit is pumped out of the left ventricle into the aorta (Figure 20.18). From the aorta, blood can take various routes, since essentially all systemic arteries branch from this single great vessel. The aorta arches upward from the heart and then curves and runs downward along the body midline to its terminus in the pelvis, where it splits to form the two large arteries serving the lower extremities. The branches of the aorta continue to subdivide to produce the arterioles and, finally, the capillaries that ramify through the body organs. Venous blood draining from organs inferior to the diaphragm ultimately enters the inferior vena cava. Except for some thoracic venous drainage (which enters the azygos system of veins), body regions above the diaphragm are drained by the superior vena cava. The venae cavae empty the carbon dioxide–laden blood into the right atrium of the heart.

Two points concerning the two major circulations must be emphasized: (1) Blood passes from systemic veins to systemic arteries only after first moving through the pulmonary circuit (Figure 20.17a), and (2) although the entire cardiac output of the right ventricle passes through the pulmonary circulation, only a small fraction of the output of the left ventricle flows through any single organ (Figure 20.18). The systemic circulation can be viewed as multiple circulatory channels functioning in parallel to distribute blood to all body organs.

As you examine tables that follow and locate the various systemic arteries and veins in the illustrations, be aware of cues that make your memorization task easier. In many cases, the name of a vessel reflects the body region traversed (axillary, brachial, femoral, etc.), the organ served (renal, hepatic, gonadal, etc.), or the bone followed (vertebral, radial, tibial, etc.). Also, notice that arteries and veins tend to run together side by side and, in many places, they also run with nerves. Finally, be alert to the fact that the systemic vessels do not always match on the right and left sides of the body. Some of the large, deep vessels of the trunk region are asymmetrical or unpaired (their initial symmetry is lost during embryonic development), while almost all vessels are bilaterally symmetrical in the head and limbs.

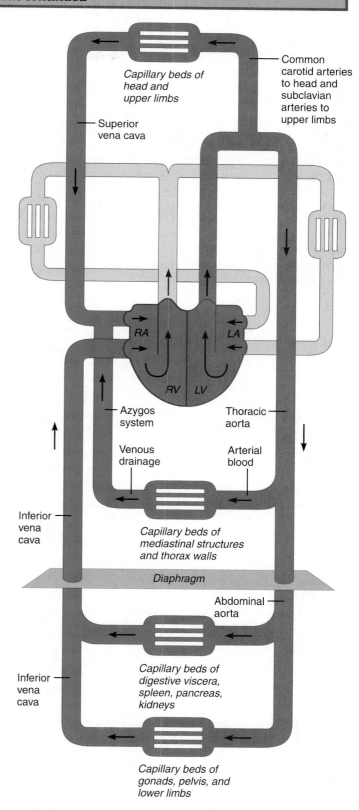

FIGURE 20.18 *Schematic flowchart showing an overview of the systemic circulation.* The pulmonary circulation is shown in gray for comparison.

---

* By convention, oxygen-rich blood is shown *red* and oxygen-poor blood is shown *blue.*

TABLE 20.4   **The Aorta and Major Arteries of the Systemic Circulation**

The distribution of the aorta and major arteries of the systemic circulation is diagrammed in flowchart form in Figure 20.19a and illustrated in Figure 20.19b. Fine points about the various vessels arising from the aorta are provided in Tables 20.5 through 20.8.

Description and Distribution

Aorta.   This vessel is the largest artery in the body. In adults, the aorta (a-or′tah) is approximately the size of a garden hose where it issues from the left ventricle of the heart. Its internal diameter is 2.5 cm, and its wall is about 2 mm thick. It decreases in size slightly as it runs to its terminus. The aortic semilunar valve guards the base of the aorta and prevents backflow of blood during diastole. Opposite each of the semilunar valve cusps is an *aortic sinus,* which contains baroreceptors important in reflex regulation of blood pressure.

Different portions of the aorta are named according to their relative shape or location. The first portion, the **ascending aorta,** runs posteriorly and to the right of the pulmonary trunk. It persists for only about 5 cm

before curving to the left as the aortic arch. The only branches of the ascending aorta are the **right** and **left coronary arteries,** which supply the myocardium (see p. 689). The **aortic arch,** deep to the sternum, begins and ends at the sternal angle (level of T_4). Its three major branches (R to L) are: (1) the **brachiocephalic** (bra′ke-o-sĕ-fal″ik; "armhead") **artery,** which passes superiorly under the right clavicle and branches into the **right common carotid** (kah-rot′id) **artery** and the **right subclavian artery,** (2) the **left common carotid artery,** and (3) the **left subclavian artery.** These three vessels provide the arterial supply of the head, neck, upper limbs, and part of the thorax wall. The **thoracic,** or **descending, aorta** runs along the anterior spine from T_5 to T_{12}, sending off numerous small arteries to the thorax wall and viscera before piercing the diaphragm. As it enters the abdominal cavity, it becomes the **abdominal aorta.** This portion supplies the abdominal walls and viscera and ends at the level of L_4 where it splits into the **right** and **left common iliac arteries,** which supply the pelvis and lower limbs.

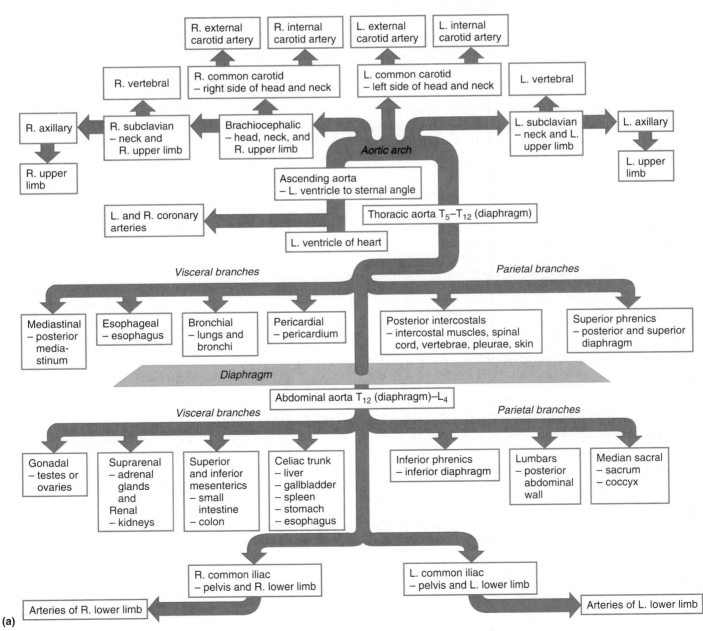

(a)

FIGURE 20.19 *Major arteries of the systemic circulation.* (**a**) Schematic flowchart. (**b**) Illustration, anterior view.

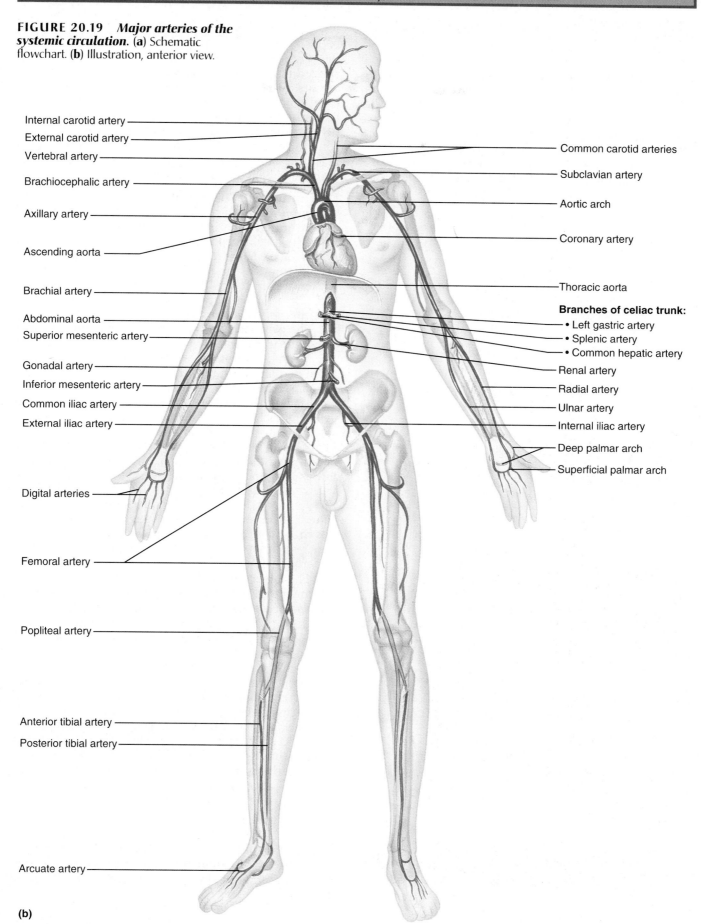

Internal carotid artery

External carotid artery

Vertebral artery

Brachiocephalic artery

Axillary artery

Ascending aorta

Brachial artery

Abdominal aorta

Superior mesenteric artery

Gonadal artery

Inferior mesenteric artery

Common iliac artery

External iliac artery

Digital arteries

Femoral artery

Popliteal artery

Anterior tibial artery

Posterior tibial artery

Arcuate artery

Common carotid arteries

Subclavian artery

Aortic arch

Coronary artery

Thoracic aorta

Branches of celiac trunk:
• Left gastric artery
• Splenic artery
• Common hepatic artery

Renal artery

Radial artery

Ulnar artery

Internal iliac artery

Deep palmar arch

Superficial palmar arch

(b)

TABLE 20.5	Arteries of the Head and Neck

Four paired arteries supply the head and neck. These are the common carotid arteries, plus three branches from each subclavian artery: the vertebral arteries, and the thyrocervical and costocervical trunks illustrated in Figure 20.20b. Of these, the common carotid arteries have the broadest distribution. These arteries, and their branches, are diagrammed in Figure 20.20a.

Each common carotid divides into two major branches (the internal and external carotid arteries). At

the division point, each internal carotid artery has a slight dilation, the **carotid sinus,** that contains baroreceptors that assist in reflex blood pressure control. The **carotid bodies,** chemoreceptors involved in the control of respiratory rate, are located close by. Pressing on the neck in the area of the carotid sinuses can cause unconsciousness (*carot* = stupor) because it mimics high blood pressure, eliciting vasodilation, which interferes with blood delivery to the brain.

Description and Distribution
Common carotid arteries.   The origins of these two arteries differ: The right common carotid artery arises from the brachiocephalic artery; the left is the second branch of the aortic arch. The common carotid arteries ascend through the lateral neck, and at the superior border of the larynx (the level of the "Adam's apple"), each divides into its two major branches, the *external* and *internal carotid arteries.*

The **external carotid arteries** supply most tissues of the head except for the brain and orbit. As each artery runs superiorly, it sends branches to the thyroid gland and larynx (**superior thyroid artery**), the tongue (**lingual artery**), the skin and muscles of the anterior face (**facial artery**), and the posterior scalp (**occipital artery**). Each external carotid artery terminates by splitting into a **superficial temporal artery,** which supplies the parotid salivary gland and most of the scalp, and a **maxillary artery,** which supplies the upper and lower jaws and chewing muscles, the teeth, and the nasal cavity. A clinically important branch of the maxillary artery is the *middle meningeal artery* (not illustrated). It enters the skull through the foramen spinosum and supplies the inner surface of the parietal bone, squamous region of the temporal bone, and the underlying dura mater.

The larger **internal carotid arteries** supply the orbits and most (over 80%) of the cerebrum. They assume a deep course and enter the skull through the carotid canals of the temporal bones. Once inside the cranium, each gives off one main branch, the ophthalmic artery, and then divides into the anterior and middle cerebral arteries. The **ophthalmic** (of-thal'mik) **arteries** supply the eyes, orbits, forehead, and nose. Each **anterior cerebral artery** supplies the medial surface of the frontal and parietal lobes of the cerebral hemisphere on its side and also anastomoses with its partner on the opposite side via a short arterial shunt called the **anterior communicating artery** (Figure 20.20d). The **middle cerebral arteries** run in the lateral fissures of their respective cerebral hemispheres and supply the lateral parts of the temporal and parietal lobes.

Vertebral arteries.   These vessels spring from the subclavian arteries at the root of the neck and ascend through foramina in the transverse processes of the cervical vertebrae to enter the skull through the foramen magnum. En route, they send branches to the vertebrae and cervical spinal cord and some deep structures of the neck. Within the cranium, the right and left vertebral arteries join to form the **basilar** (bas'ĭ-lar) **artery,**

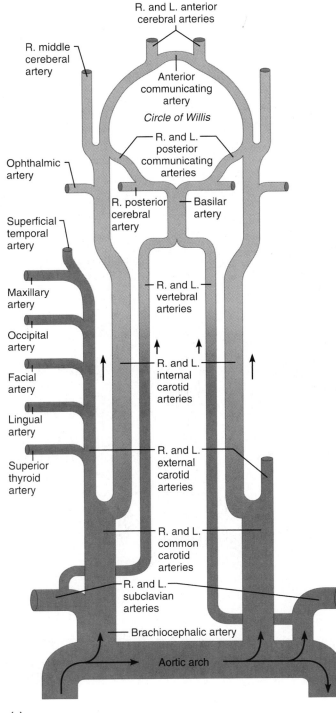

(a)

TABLE 20.5	Arteries of the Head and Neck *continued*

which ascends along the anterior aspect of the brain stem, giving off branches to the cerebellum, pons, and inner ear (see Figure 20.20b and d). At the pons-midbrain border, the basilar artery divides into a pair of **posterior cerebral arteries,** which supply the occipital lobes and the inferior parts of the temporal lobes of the cerebral hemispheres.

Arterial shunts called **posterior communicating arteries** connect the posterior cerebral arteries to the middle cerebral arteries anteriorly. The two posterior and single anterior communicating arteries complete the formation of an arterial anastomosis called the **circle of Willis.** The circle of Willis encircles the pituitary gland and optic chiasma and unites the brain's anterior and posterior blood supplies. It also acts to equalize blood pressure in the two brain areas and provides alternate routes for blood to reach the brain tissue if a carotid or vertebral artery becomes occluded.

Thyrocervical and costocervical trunks. These short vessels arise from the subclavian artery just lateral to the vertebral arteries on each side (Figure 20.20b and Figure 20.21). The thyrocervical trunk mainly supplies the thyroid gland, portions of the cervical vertebrae and spinal cord, and some scapular muscles. The costocervical trunk serves deep neck and superior intercostal muscles.

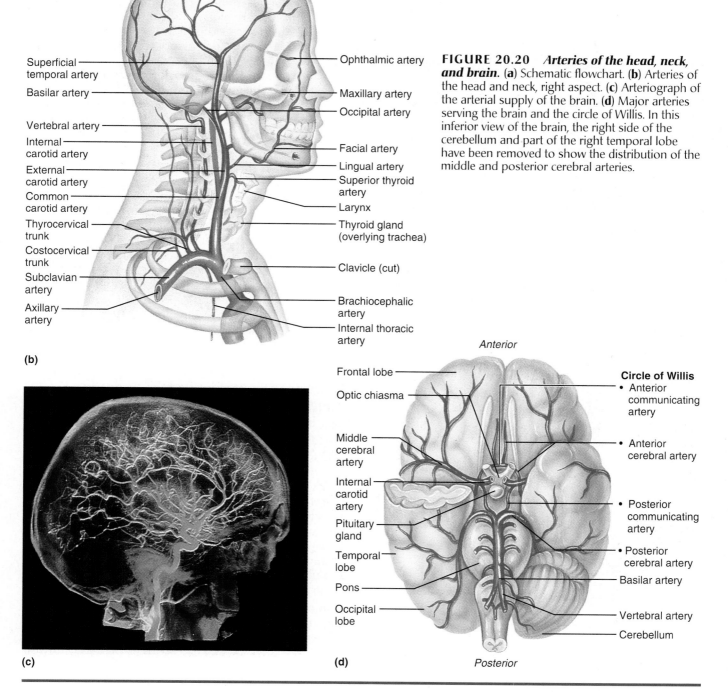

FIGURE 20.20 *Arteries of the head, neck, and brain.* (a) Schematic flowchart. (b) Arteries of the head and neck, right aspect. (c) Arteriograph of the arterial supply of the brain. (d) Major arteries serving the brain and the circle of Willis. In this inferior view of the brain, the right side of the cerebellum and part of the right temporal lobe have been removed to show the distribution of the middle and posterior cerebral arteries.

TABLE 20.6   Arteries of the Upper Limbs and Thorax

The upper limbs are supplied entirely by arteries arising from the **subclavian arteries** (see Figure 20.21a). After giving off branches to the neck, each subclavian artery courses laterally between the clavicle and first rib to enter the axilla, where its name changes to axillary artery. The thorax wall is supplied by an array of vessels that arise either directly from the thoracic aorta or from branches of the subclavian arteries. Most visceral organs of the thorax receive their functional blood supply from small branches issuing from the thoracic aorta. Because these vessels are so small and tend to vary in number (except for the bronchial arteries), they are not illustrated in Figures 20.21a and b, but several of these are listed at the end of this table.

Description and Distribution

Arteries of the Upper Limb

Axillary artery. As it runs through the axilla accompanied by cords of the brachial plexus, each axillary artery gives off branches to the structures of the axilla, chest wall, and shoulder girdle. These include the **thoracoacromial** (tho"rah-ko-ah-kro'me-al) **trunk,** which supplies the superior shoulder (deltoid) and pectoral region; the **lateral thoracic artery,** which serves the lateral chest wall and breast; the **subscapular artery** to the scapula, dorsal thorax wall, and latissimus dorsi muscle; and the **anterior and posterior circumflex arteries,** which wrap around the humoral neck and help supply the shoulder joint and the deltoid muscle. As the axillary artery emerges from the axilla, it becomes the brachial artery.

Brachial artery. The brachial artery runs down the medial aspect of the humerus and supplies the anterior flexor muscles of the arm. One major branch, the **deep brachial artery,** serves the posterior triceps brachii muscle. As the brachial artery nears the elbow, it gives off several small branches that contribute to an anastomosis serving the elbow joint and connecting it to the arteries of the forearm. As the brachial artery crosses the anterior midline aspect of the elbow, it provides an easily palpated pulse point (brachial pulse) (see Figure 20.11). Immediately beyond the elbow, the brachial artery splits to form the radial and ulnar arteries, which more or less follow the course of similarly named bones down the length of the anterior forearm.

Radial artery. The radial artery runs from the median line of the cubital fossa to the styloid process of the radius. It supplies the lateral muscles of the forearm, the wrist, and the thumb and index finger. At the root of the thumb, the radial artery provides a convenient site for taking the radial pulse.

Ulnar artery. The ulnar artery supplies the medial aspect of the forearm, fingers 3–5, and the medial aspect of the index finger. Proximally, the ulnar artery gives off a short branch, the **common interosseous** (in"ter-os'e-us) **artery,** which runs between the radius and ulna to serve the deep flexors and extensors of the forearm.

Palmar arches. In the palm, branches of the radial and ulnar arteries anastomose to form two palmar arches, the **superficial** and **deep palmar arches.** The

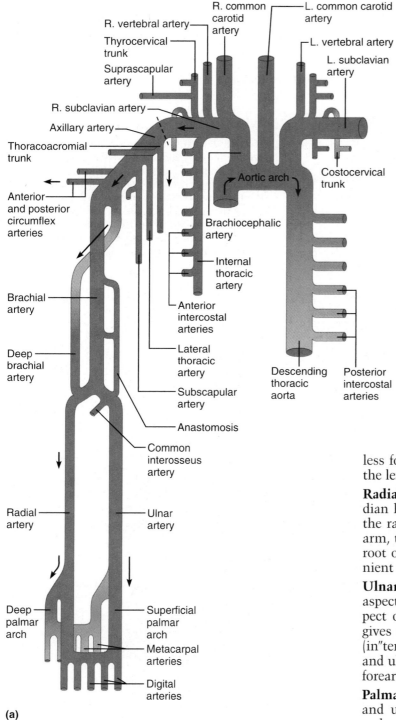

(a)

TABLE 20.6	Arteries of the Upper Limbs and Thorax *continued*

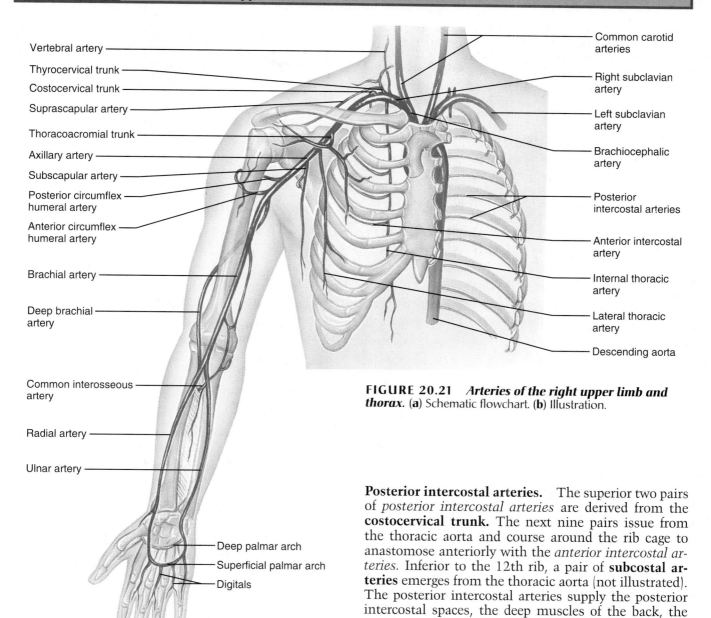

Vertebral artery

Thyrocervical trunk

Costocervical trunk

Suprascapular artery

Thoracoacromial trunk

Axillary artery

Subscapular artery

Posterior circumflex humeral artery

Anterior circumflex humeral artery

Brachial artery

Deep brachial artery

Common interosseous artery

Radial artery

Ulnar artery

Deep palmar arch

Superficial palmar arch

Digitals

Common carotid arteries

Right subclavian artery

Left subclavian artery

Brachiocephalic artery

Posterior intercostal arteries

Anterior intercostal artery

Internal thoracic artery

Lateral thoracic artery

Descending aorta

(b)

FIGURE 20.21 *Arteries of the right upper limb and thorax.* (**a**) Schematic flowchart. (**b**) Illustration.

metacarpal arteries and the **digital arteries** that supply the fingers arise from these palmar arches.

Arteries of the Thorax Wall

Internal thoracic (mammary) arteries. The internal thoracic (mammary) arteries, which arise from the subclavian arteries, supply blood to most of the anterior thorax wall. Each of these arteries descends lateral to the sternum and gives off **anterior intercostal arteries,** which supply the structures of the intercostal spaces anteriorly. The internal thoracic artery also sends superficial branches to the skin and mammary glands and terminates in twiglike branches to the anterior abdominal wall and diaphragm.

Posterior intercostal arteries. The superior two pairs of *posterior intercostal arteries* are derived from the **costocervical trunk.** The next nine pairs issue from the thoracic aorta and course around the rib cage to anastomose anteriorly with the *anterior intercostal arteries.* Inferior to the 12th rib, a pair of **subcostal arteries** emerges from the thoracic aorta (not illustrated). The posterior intercostal arteries supply the posterior intercostal spaces, the deep muscles of the back, the vertebrae, and the spinal cord. Together, the posterior and anterior intercostal arteries supply the intercostal muscles.

Superior phrenic arteries. One or more paired superior phrenic arteries serve the posterior superior aspect of the diaphragm surface.

Arteries of the Thoracic Viscera

Pericardial arteries. Several tiny branches supply the posterior pericardium.

Bronchial arteries. Two left and one right bronchial arteries supply systemic (oxygen-rich) blood to the lungs, bronchi, and pleurae.

Esophageal arteries. Four to five esophageal arteries supply the esophagus.

Mediastinal arteries. Many small mediastinal arteries serve the contents of the posterior mediastinum.

TABLE 20.7   Arteries of the Abdomen

The arterial supply to the abdominal organs arises from the abdominal aorta (Figure 20.22a). Under resting conditions, about half of the entire arterial flow is found in these vessels. Except for the celiac trunk, the superior and inferior mesenteric arteries, and the median sacral artery, all are paired vessels. These arteries supply the abdominal wall, the diaphragm, and visceral organs of the abdominopelvic cavity. The branches are given here in order of their issue.

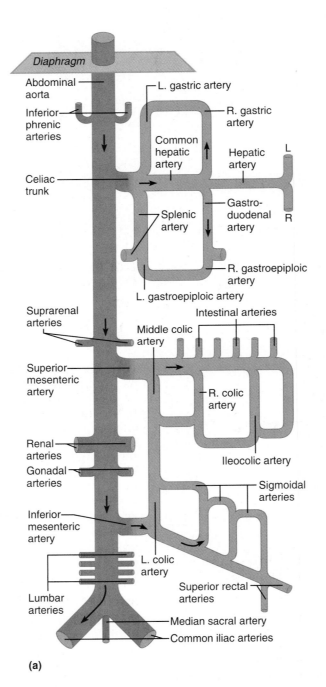

(a)

FIGURE 20.22   Arteries of the abdomen. (a) Schematic flowchart.

TABLE 20.7	Arteries of the Abdomen *continued*

Description and Distribution

Inferior phrenic arteries. The inferior phrenics emerge from the aorta at T_{12}, just inferior to the diaphragm. They serve the inferior diaphragm surface.

Celiac trunk. This very large unpaired branch of the abdominal aorta divides almost immediately into three branches: the common hepatic, splenic, and left gastric arteries (Figure 20.22b). The **common hepatic** (hĕ-pat'ik) **artery** runs superiorly, giving off branches to the stomach, duodenum, and pancreas. Where the **gastroduodenal artery** branches off, it becomes the **hepatic artery proper,** which splits into right and left branches that serve the liver. As the **splenic** (splen'ik) **artery** passes deep to the stomach, it sends branches to the pancreas and stomach and terminates in branches to the spleen. The **left gastric** (*gaster* = stomach) **artery** supplies part of the stomach and the inferior esophagus. The **right** and **left gastroepiploic** (gas"tro-ep"ĭ-plo'ik) **arteries,** branches of the gastroduodenal and splenic arteries, respectively, serve the left (greater) curvature of the stomach. A **right gastric artery,** which supplies the stomach's right (lesser) curvature, may arise from the common hepatic artery or hepatic artery proper.

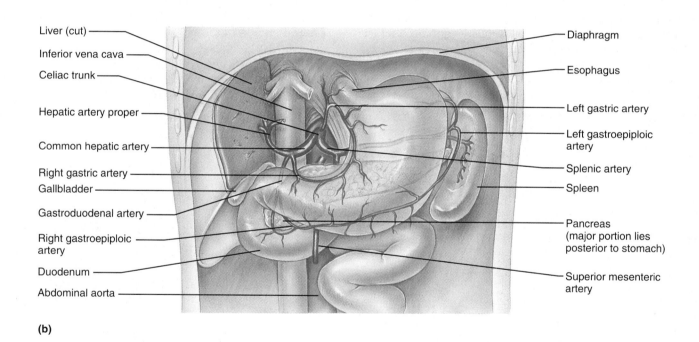

(b)

FIGURE 20.22 *Arteries of the abdomen, continued.* (b) The celiac trunk and its major branches. (See *A Brief Atlas of the Human Body,* Figure 52.)

TABLE 20.7	Arteries of the Abdomen *continued*

Superior mesenteric (mes-en-ter'ik) **artery.** This large, unpaired artery arises from the abdominal aorta at the level of L_1 immediately below the celiac trunk (see Figure 20.22d). It runs deep to the pancreas and then enters the mesentery, where its numerous anastomosing branches serve virtually all of the small intestine via the **intestinal arteries,** most of the large intestine—the appendix, cecum, ascending colon (via the **ileocolic artery**), and part of the transverse colon (via the **right** and **middle colic arteries**).

Suprarenal (soo"prah-re'nal) **arteries.** The suprarenal arteries flank the origin of the superior mesenteric artery as they emerge from the abdominal aorta (Figure 20.22c). They supply blood to the adrenal (suprarenal) glands overlying the kidneys.

Renal arteries. The short, but wide, right and left renal arteries issue from the lateral surfaces of the aorta slightly below the superior mesenteric artery (between L_1 and L_2). Each serves the kidney on its side.

Gonadal (go-nă'dul) **arteries.** The paired gonadal arteries are called the **testicular arteries** in males and the **ovarian arteries** in females. In females, these vessels extend into the pelvis to serve the ovaries and part of the uterine tubes. In males the much longer testicular arteries descend through the pelvis and inguinal canal to enter the scrotal sac, where they serve the testes.

Inferior mesenteric artery. This final major branch of the abdominal aorta is unpaired and arises from the anterior aortic surface at the level of L_3. It serves the distal part of the large intestine—from the midpart of the transverse colon to the midrectum—via its **left colic, sigmoidal,** and **superior rectal branches** (Figure 20.22d). Looping anastomoses between the superior and inferior mesenteric arteries help ensure that blood will continue to reach the various regions of the digestive viscera in cases of trauma to one of these abdominal arteries.

Lumbar arteries. Four pairs of lumbar arteries arise from the posterolateral surface of the aorta in the lumbar region. These segmental arteries supply the posterior abdominal wall.

Median sacral artery. The unpaired median sacral artery issues from the posterior surface of the abdominal aorta at its terminus. This tiny artery supplies the sacrum and coccyx.

Common iliac arteries. At the level of L_4, the aorta splits into the right and left common iliac arteries, which supply blood to the lower abdominal wall, pelvic organs, and lower limbs (see Figure 20.23).

TABLE 20.7   **Arteries of the Abdomen** *continued*

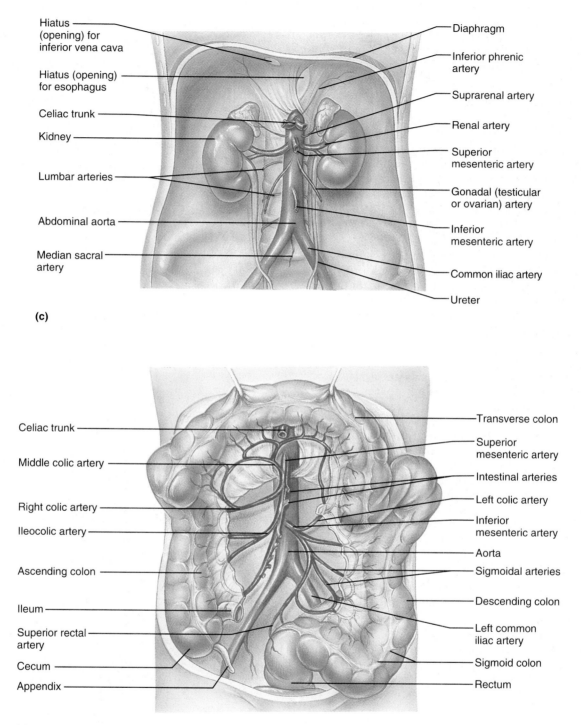

(c)

(d)

FIGURE 20.22   *Arteries of the abdomen, continued.* (**c**) Major branches of the abdominal aorta. (**d**) Distribution of the superior and inferior mesenteric arteries. (The transverse colon has been reflected superiorly to provide a better view of these arteries.)

TABLE 20.8 **Arteries of the Pelvis and Lower Limbs**

At the level of the sacroiliac joints, the **common iliac arteries** divide into two major branches, the internal and external iliac arteries (Figure 20.23a). The internal iliacs distribute blood mainly to the pelvic region. The external iliacs serve the lower limbs; they also send some branches to the abdominal wall.

Description and Distribution

Internal iliac arteries. These paired arteries run into the pelvis and distribute blood to the pelvic walls and viscera (bladder, rectum, uterus, and vagina in the female and prostate gland and ductus deferens in the male). Additionally they serve the gluteal muscles via the **superior** and **inferior gluteal arteries,** adductor muscles of the medial thigh via the **obturator artery,** and external genitalia and perineum via the **internal pudendal artery** (not illustrated).

External iliac arteries. These arteries supply the lower limbs (Figure 20.23b). As they course through the pelvis, they give off branches to the anterior abdominal wall. After passing under the inguinal ligaments to enter the thigh, they become the femoral arteries.

Femoral arteries. As each of these arteries passes down the anteromedial thigh, it gives off several branches to the muscles of the thigh. The largest of the deep branches is the **deep femoral artery** (also more simply called **deep artery of the thigh**) that is the main supply to the thigh muscles (hamstrings, quadriceps, and adductors). Proximal branches of the deep femoral artery, the **lateral** and **medial circumflex femoral arteries,** encircle the neck of the femur. The medial circumflex artery supplies the head and neck of the femur. A long descending branch of the posterior circumflex artery supplies the vastus lateralis muscle. Near the knee the femoral artery passes posteriorly and through a gap in the adductor magnus muscle, the *adductor hiatus,* to enter the popliteal fossa, where its name changes to popliteal artery.

Popliteal artery. This posterior vessel contributes to an arterial anastomosis that supplies the knee region and then splits into the anterior and posterior tibial arteries of the leg.

Anterior tibial artery. The anterior tibial artery runs through the anterior compartment of the leg, supplying the extensor muscles along the way. At the ankle, it becomes the **dorsalis pedis artery,** which supplies the ankle and dorsum of the foot, and gives off a branch, the **arcuate artery,** which issues the metatarsal arteries to the metatarsus of the foot. The superficial dorsalis pedis ends by penetrating into the sole of the foot where it forms the medial part of the **plantar arch.** The dorsalis pedis artery provides a clinically important pulse point, the pedal pulse (see Figure 20.11). If the pedal pulse is easily felt, it is fairly certain that the blood supply to the leg is good.

Posterior tibial artery. This large artery courses through the posteromedial part of the leg and supplies

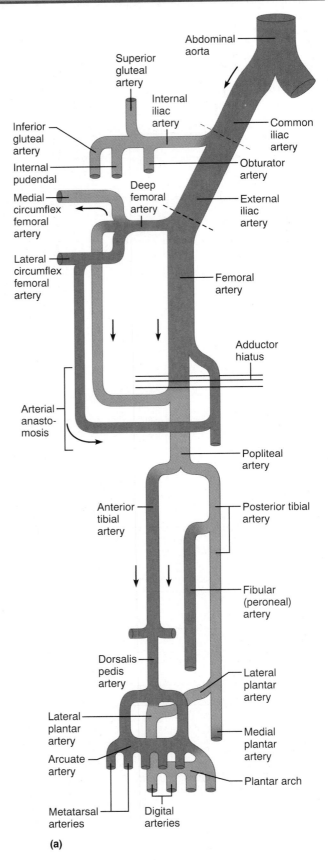

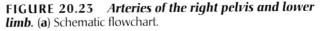

FIGURE 20.23 *Arteries of the right pelvis and lower limb.* **(a)** Schematic flowchart.

TABLE 20.8	Arteries of the Pelvis and Lower Limbs *continued*

the flexor muscles. Proximally, it gives off a large branch, the **fibular (peroneal) artery,** which supplies the lateral fibularis (peroneal) muscles of the leg. At the ankle, the posterior tibial artery divides into **lateral** and **medial plantar arteries** that serve the plantar surface of the foot. The lateral plantar artery forms the lateral end of the plantar arch. The **digital arteries** serving the toes arise from the plantar arch formed by the lateral plantar artery.

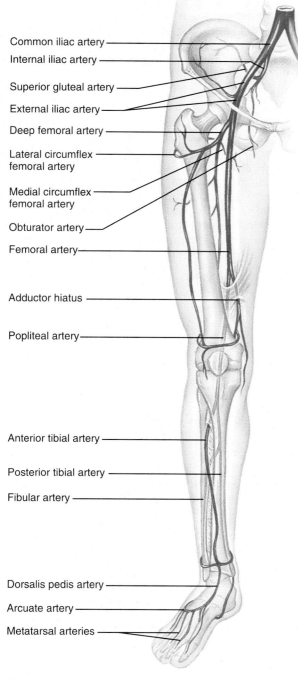

Common iliac artery
Internal iliac artery
Superior gluteal artery
External iliac artery
Deep femoral artery
Lateral circumflex femoral artery
Medial circumflex femoral artery
Obturator artery
Femoral artery
Adductor hiatus
Popliteal artery
Anterior tibial artery
Posterior tibial artery
Fibular artery
Dorsalis pedis artery
Arcuate artery
Metatarsal arteries

(b)

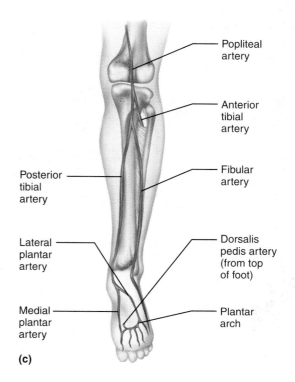

Popliteal artery
Anterior tibial artery
Fibular artery
Posterior tibial artery
Lateral plantar artery
Dorsalis pedis artery (from top of foot)
Medial plantar artery
Plantar arch

(c)

FIGURE 20.23 *Arteries of the right pelvis and lower limb, continued.*
(**b**) Illustration, anterior view. (**c**) Posterior view of the leg and foot.

In our survey of the systemic veins, the major tributaries (branches) of the venae cavae are noted first in Figure 20.24, followed by a description in Tables 20.10 through 20.13 of the venous pattern of the various body regions. Since veins run toward the heart, the most distal veins are named first and those closest to the heart last. Since deep veins generally drain the same areas served by their companion arteries, they are not described in detail.

Description and Areas Drained

Superior vena cava.   This great vein receives systemic blood draining from all areas superior to the diaphragm, except the heart wall. It is formed by the union of the **right** and **left brachiocephalic veins** and empties into the right atrium (see Figure 20.24b). Notice that there are two brachiocephalic veins, but only one brachiocephalic artery. Each brachiocephalic vein is formed by the joining of the **internal jugular** and **subclavian veins** on its side. In most of the flowcharts that follow, only the vessels draining blood from the right side of the body are followed (except for the azygos circulation of the thorax).

Inferior vena cava.   The widest blood vessel in the body, this vein returns blood to the heart from all body regions below the diaphragm. The abdominal aorta lies directly to its left. The distal end of the inferior vena cava is formed by the junction of the paired **common iliac veins** at L_5. From this point, it courses superiorly along the anterior aspect of the spine, receiving venous blood draining from the abdominal walls, gonads, and kidneys. Immediately above the diaphragm, the inferior vena cava ends as it enters the inferior aspect of the right atrium of the heart.

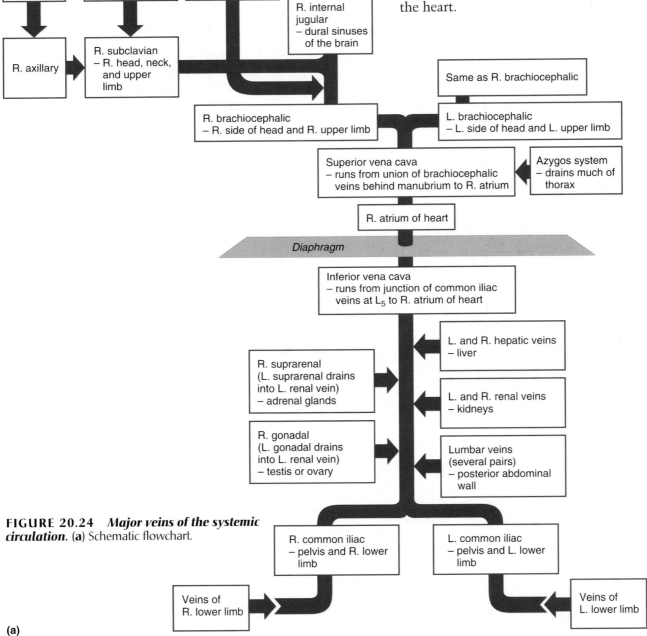

FIGURE 20.24 *Major veins of the systemic circulation.* **(a)** Schematic flowchart.

(a)

TABLE 20.9 **The Venae Cavae and the Major Veins of the Systemic Circulation** *continued*

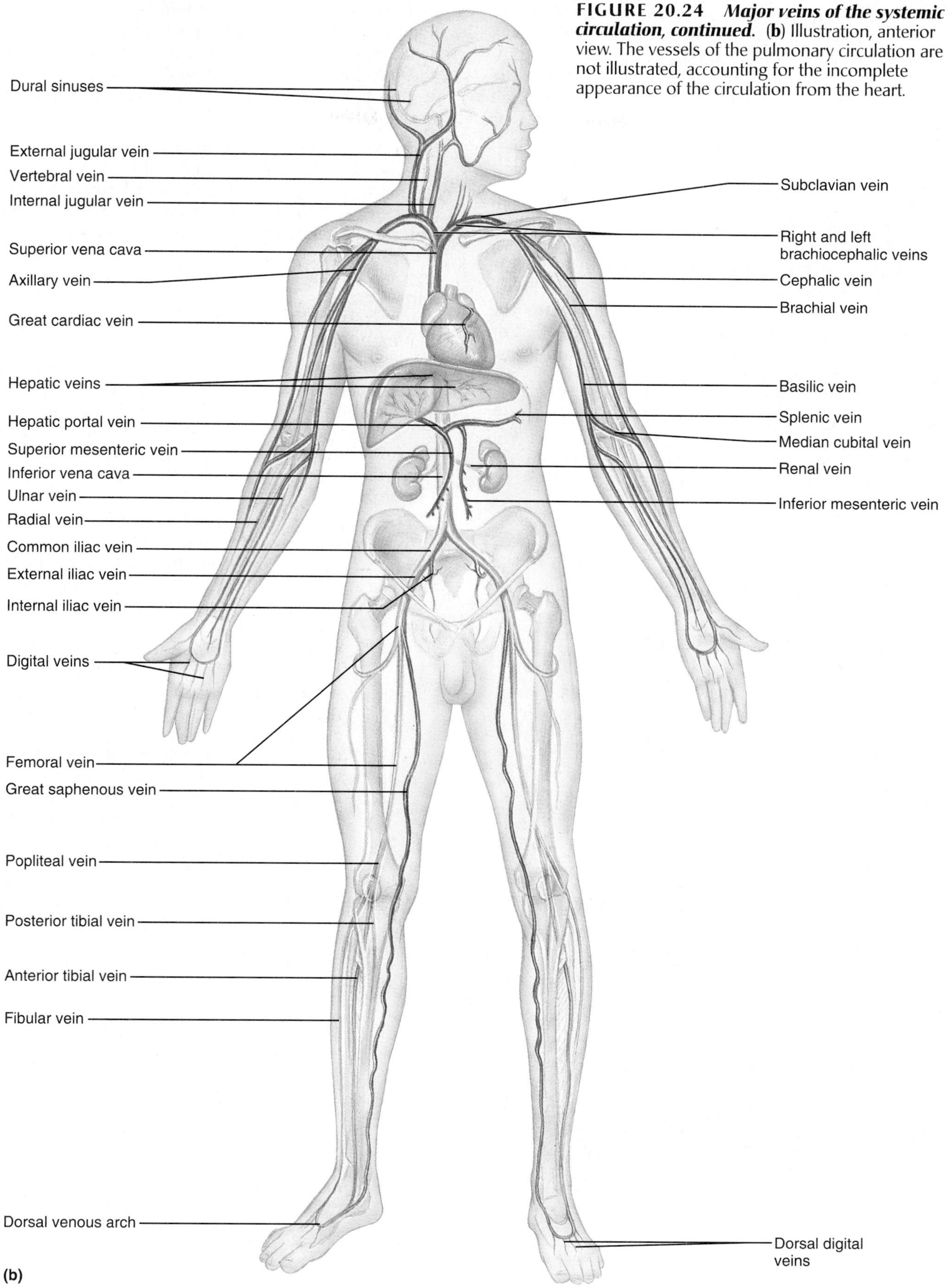

FIGURE 20.24 *Major veins of the systemic circulation, continued.* **(b)** Illustration, anterior view. The vessels of the pulmonary circulation are not illustrated, accounting for the incomplete appearance of the circulation from the heart.

Dural sinuses

External jugular vein
Vertebral vein
Internal jugular vein

Superior vena cava

Axillary vein

Great cardiac vein

Hepatic veins

Hepatic portal vein
Superior mesenteric vein
Inferior vena cava
Ulnar vein
Radial vein
Common iliac vein
External iliac vein
Internal iliac vein

Digital veins

Femoral vein
Great saphenous vein

Popliteal vein

Posterior tibial vein

Anterior tibial vein

Fibular vein

Dorsal venous arch

Subclavian vein

Right and left brachiocephalic veins
Cephalic vein
Brachial vein

Basilic vein
Splenic vein
Median cubital vein
Renal vein
Inferior mesenteric vein

Dorsal digital veins

(b)

TABLE 20.10	Veins of the Head and Neck

Most blood draining from the head and neck is collected by three pairs of veins: the external jugular veins (which empty into the subclavians), the internal jugular veins, and the vertebral veins which drain into the brachiocephalic vein (see Figure 20.25a). Although most of the extracranial veins have the same names as the extracranial arteries (for example, facial, ophthalmic, occipital, and superficial temporal), their courses and interconnections differ substantially.

Most veins of the brain drain into the **dural sinuses,** an interconnected series of enlarged chambers located between the dura mater layers. The **superior** and **inferior sagittal sinuses** are found within the falx cerebri, which dips down between the cerebral hemispheres. The inferior sagittal sinus drains into the **straight sinus** posteriorly (Figures 20.25a and c). The superior sagittal and straight sinuses then empty into the **transverse sinuses,** which run in shallow grooves on the internal surface of the occipital bone. These, in turn, drain into the S-shaped sigmoid sinuses, which become the *internal jugular veins* as they leave the skull through the jugular foramen. The **cavernous si-**

nuses, which flank the sphenoid body, receive venous blood from the **ophthalmic veins** of the orbits and the facial veins, which drain the nose and upper lip area. The internal carotid artery and cranial nerves III, IV, VI, and part of V, all run *through* the cavernous sinus on their way to the orbit and face.

Description and Area Drained

External jugular veins. The right and left external jugular veins drain superficial head (scalp and face) structures served by the external carotid arteries. However, their tributaries anastomose frequently, and some of the superficial drainage from these regions enters the internal jugular veins as well. As the external jugular veins descend through the lateral neck, they pass obliquely over the sternocleidomastoid muscles and then empty into the subclavian veins.

Vertebral veins. Unlike the vertebral arteries, the vertebral veins do not serve much of the brain. Instead they drain only the cervical vertebrae, the spinal cord, and some small neck muscles. The vertebral veins run inferiorly through the transverse foramina of the cervical vertebrae and join the brachiocephalic veins at the root of the neck.

Internal jugular veins. The paired internal jugular veins, which receive the bulk of blood draining from the brain, are the largest of the paired veins draining the head and neck. They arise from the dural venous sinuses, exit from the skull via the *jugular foramina,* and then descend through the neck alongside the internal carotid arteries. As they move inferiorly, they receive blood from some of the deep veins of the face and neck—branches of the **facial** and **superficial temporal veins** (see Figure 20.25b). At the base of the neck, each internal jugular vein joins the subclavian vein on its own side to form a brachiocephalic vein. As already noted, the two brachiocephalic veins unite to form the superior vena cava.

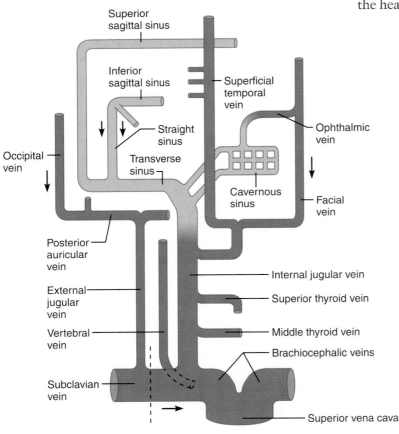

(a)

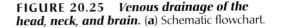

FIGURE 20.25 *Venous drainage of the head, neck, and brain.* **(a)** Schematic flowchart.

TABLE 20.10   **Veins of the Head and Neck** *continued*

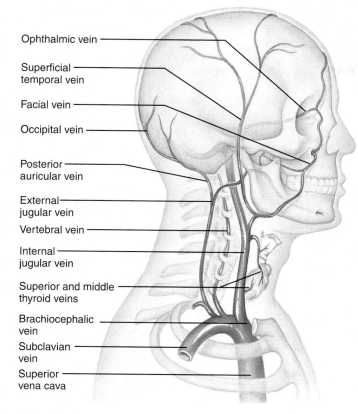

Ophthalmic vein

Superficial temporal vein

Facial vein

Occipital vein

Posterior auricular vein

External jugular vein

Vertebral vein

Internal jugular vein

Superior and middle thyroid veins

Brachiocephalic vein

Subclavian vein

Superior vena cava

(b)

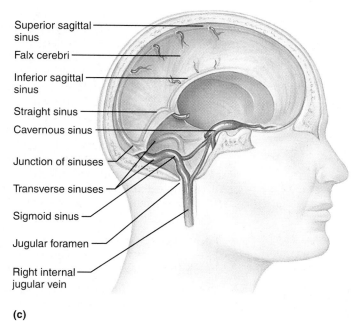

Superior sagittal sinus

Falx cerebri

Inferior sagittal sinus

Straight sinus

Cavernous sinus

Junction of sinuses

Transverse sinuses

Sigmoid sinus

Jugular foramen

Right internal jugular vein

(c)

FIGURE 20.25 *Venous drainage of the head, neck, and brain, continued.* (**b**) Veins of the head and neck, right superficial aspect. (**c**) Dural sinuses of the brain, right aspect.

TABLE 20.11 **Veins of the Upper Limbs and Thorax**

The deep veins of the upper limbs follow the paths of their companion arteries and have the same names (Figure 20.26a). However, except for the largest, most are paired or double veins that flank their artery on both sides. The superficial veins of the upper limbs are larger than the deep veins and are easily seen just beneath the skin. The median cubital vein, crossing the anterior aspect of the elbow, is commonly used to obtain blood samples or administer intravenous medications or blood transfusions.

Blood draining from the mammary glands and the first two to three intercostal spaces enters the **brachiocephalic veins.** However, the vast majority of thoracic tissues and the thorax wall are drained by a rather complex network of veins collectively called the **azygos** (az′ĭ-gos) **system.** The branching nature of the azygos system provides a collateral circulation for draining the abdominal wall and other areas served by the inferior vena cava, and there are numerous anastomoses between the azygos system and the inferior vena cava.

Description and Areas Drained

Deep Veins of the Upper Limb
The most distal deep veins of the upper limb are the radial and ulnar veins. The deep and superficial **palmar venous arches** of the hand empty into the **radial** and **ulnar veins** of the forearm, which then unite to form the **brachial vein** of the arm. As the brachial vein enters the axilla, it becomes the **axillary vein,** which in turn becomes the **subclavian vein** at the level of the first rib.

Superficial Veins of the Upper Limb
The superficial venous system begins with the **dorsal venous arch** (not illustrated), a plexus of superficial veins in the dorsum of the hand. In the distal forearm, this plexus drains into three major superficial veins—the cephalic and basilic veins and the median vein of the forearm—that anastomose frequently as they course upward (see Figure 20.26b). The **cephalic vein** coils around the radius as it travels superiorly and then continues up the lateral superficial aspect of the arm to the shoulder, where it runs in the groove between the deltoid and pectoralis muscles to join the axillary vein. The **basilic vein** courses along the posteromedial aspect of the forearm, crosses the elbow, and then takes a deep course. In the axilla, it joins the brachial vein, forming the axillary vein. At the anterior aspect of the elbow, the **median cubital vein** connects the basilic and cephalic veins. The **median vein of the forearm** lies between the radial and ulnar veins in the forearm and terminates (variably) at the elbow by entering the basilic or the cephalic vein.

The Azygos System
The azygos system consists of the following vessels which flank the vertebral column laterally:

Azygos vein. Found against the right side of the vertebral column, the **azygos** ("unpaired") **vein** originates

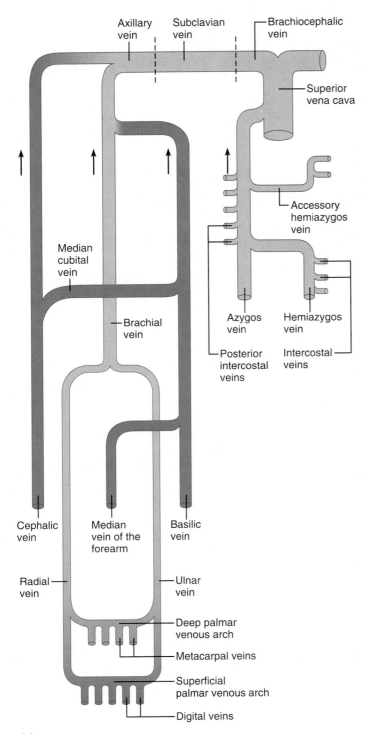

(a)

TABLE 20.11	Veins of the Upper Limbs and Thorax *continued*

in the abdomen, from the **right ascending lumbar vein** that drains most of the right abdominal cavity wall and from the **right posterior intercostal veins** (except the first) that drain the chest muscles. At the level of T₄, it arches over the great vessels that run to the right lung and empties into the superior vena cava.

Hemiazygos (hĕ′me-a-zi′gus; "half the azygos") **vein.** Ascends on the left side of the vertebral column. Its origin, from the **left ascending lumbar vein** and the lower (9th–11th) **posterior intercostal veins**, mirrors that of

the inferior portion of the azygos vein on the right. About midthorax, the hemiazygos vein passes in front of the vertebral column and joins the azygos vein.

Accessory hemiazygos vein. The accessory hemiazygos completes the venous drainage of the left (middle) thorax and can be thought of as a superior continuation of the hemiazygos vein. It receives blood from the 4th–8th posterior intercostal veins and then crosses to the right to empty into the azygos vein. Like the azygos, it receives venous blood from the lungs *(bronchial veins)*.

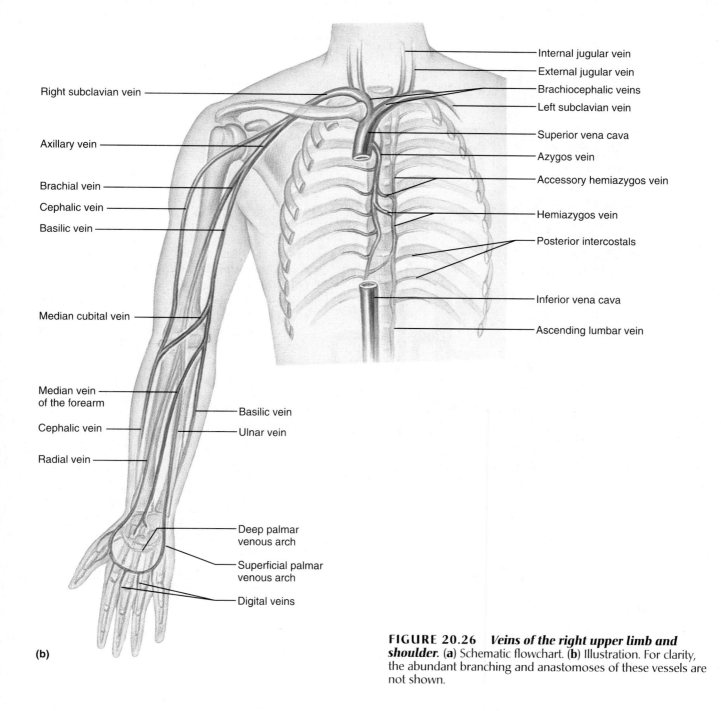

(b)

FIGURE 20.26 *Veins of the right upper limb and shoulder.* **(a)** Schematic flowchart. **(b)** Illustration. For clarity, the abundant branching and anastomoses of these vessels are not shown.

TABLE 20.12   **Veins of the Abdomen**

Blood draining from the abdominopelvic viscera and abdominal walls is returned to the heart by the **inferior vena cava** (Figure 20.27a). Most of its venous tributaries have names that correspond to the arteries serving the abdominal organs.

Veins draining the digestive viscera empty their blood into a common vessel, the **hepatic portal vein,** which transports this venous blood into the liver before it is allowed to enter the major systemic circulation via the hepatic veins (Figure 20.27b). Such a venous system—veins to capillaries (or sinusoids) to veins—is called a *portal system*. Portal systems always serve very

specific regional tissue needs. The hepatic portal system delivers blood laden with nutrients from the digestive organs to the liver. As the blood percolates slowly through the liver sinusoids, the hepatic parenchymal cells remove nutrients required for their various metabolic functions, and phagocytic cells lining the sinusoids rid the blood of bacteria and other foreign matter that has penetrated the digestive mucosa. The role of the liver in nutrient processing and metabolism is described in detail in Chapter 25. The veins of the abdomen are listed below in inferior to superior order.

Description and Areas Drained

Lumbar veins.   Several pairs of lumbar veins drain the posterior abdominal wall. They empty both directly into the inferior vena cava and into the ascending lumbar veins of the azygos system of the thorax.

Gonadal (testicular or ovarian) veins.   The right gonadal vein drains the ovary or testis on the right side of the body and empties into the inferior vena cava. The left member drains into the left renal vein superiorly.

Renal veins.   The right and left renal veins drain the kidneys.

Suprarenal veins.   The right suprarenal vein drains the adrenal gland on the right and empties into the inferior vena cava. The left suprarenal vein drains into the left renal vein.

Hepatic portal system.   The **hepatic portal vein** is a short vessel about 8 cm long; it begins at the level of L_2. Numerous tributaries from the stomach and pancreas contribute to the hepatic portal system (see Figure 20.27c), but the major vessels are as follows:

- **Superior mesenteric vein:** Drains the entire small intestine, part of the large intestine (ascending and transverse regions), and stomach.

- **Splenic vein:** This large vessel collects blood from the spleen, parts of the stomach and pancreas, and then joins the superior mesenteric vein to form the hepatic portal vein.

- **Inferior mesenteric vein:** Drains the distal portions of the large intestine and rectum and joins the splenic vein just before that vessel unites with the superior mesenteric vein to form the hepatic portal vein.

Hepatic veins.   The right and left hepatic veins carry venous blood from the liver to the inferior vena cava.

Cystic veins.   The cystic veins drain the gallbladder and join the hepatic veins.

Inferior phrenic veins.   The inferior phrenic veins drain the inferior surface of the diaphragm.

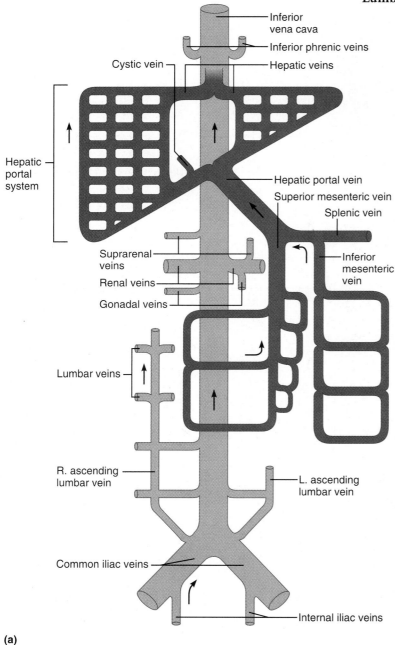

(a)

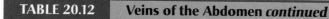

TABLE 20.12   **Veins of the Abdomen** *continued*

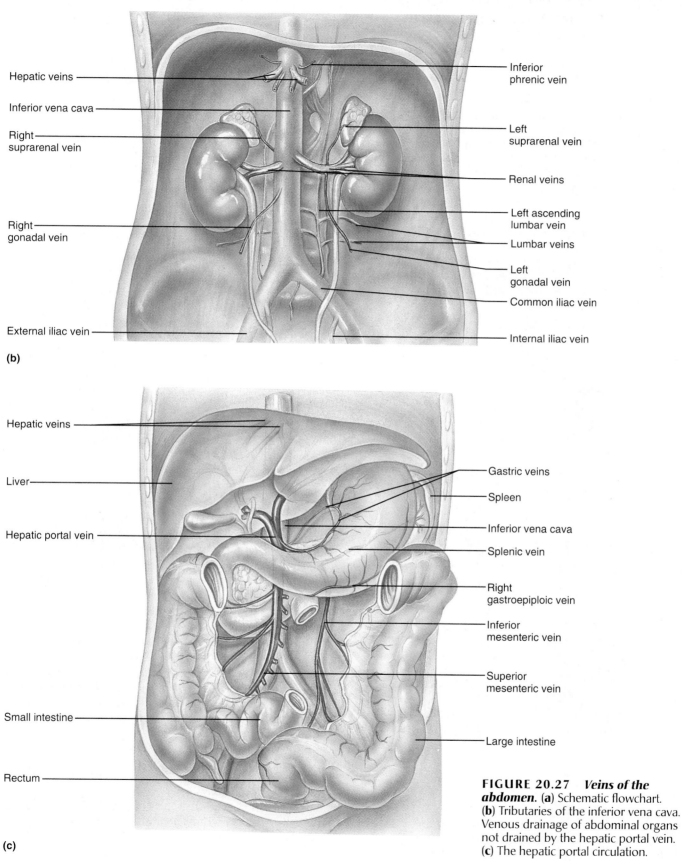

Hepatic veins

Inferior vena cava

Right
suprarenal vein

Right
gonadal vein

External iliac vein

Inferior
phrenic vein

Left
suprarenal vein

Renal veins

Left ascending
lumbar vein

Lumbar veins

Left
gonadal vein

Common iliac vein

Internal iliac vein

(b)

Hepatic veins

Liver

Hepatic portal vein

Small intestine

Rectum

Gastric veins

Spleen

Inferior vena cava

Splenic vein

Right
gastroepiploic vein

Inferior
mesenteric vein

Superior
mesenteric vein

Large intestine

(c)

FIGURE 20.27   *Veins of the abdomen.* (**a**) Schematic flowchart.
(**b**) Tributaries of the inferior vena cava.
Venous drainage of abdominal organs
not drained by the hepatic portal vein.
(**c**) The hepatic portal circulation.

TABLE 20.13   **Veins of the Pelvis and Lower Limbs**

As is the case in the upper limbs, most deep veins of the lower limbs have the same names as the arteries they accompany and many are double. The two superficial saphenous veins (great and small) are poorly supported by surrounding tissues, and are common sites of varicosities. The great saphenous (*saphenous* = obvious) vein is frequently excised and used as a coronary bypass vessel.

Description and Areas Drained

Deep veins.   After being formed by the union of the small **medial** and **lateral plantar veins**, the **posterior tibial vein** ascends deep within the calf muscle (Figure 20.28) and receives the **fibular (peroneal) vein.** The **anterior tibial vein** is the superior continuation of the **dorsalis pedis vein** of the foot. At the knee it unites with the posterior tibial vein to form the **popliteal vein,** which crosses the back of the knee. As the popliteal vein emerges from the knee, it becomes the **femoral vein,** which drains the deep structures of the thigh. The femoral vein becomes the **external iliac vein** as it enters the pelvis. In the pelvis, the external iliac vein unites with the **internal iliac vein** to form the

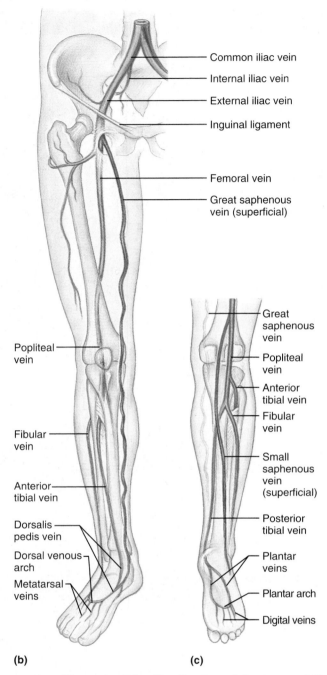

(b)   (c)

common iliac vein. The distribution of the internal iliac veins parallels that of the internal iliac arteries.

Superficial veins.   The **great** and **small saphenous** (sah-fe′nus) **veins** issue from the **dorsal venous arch** of the foot (Figure 20.28b and c). These veins anastomose frequently with each other and with the deep veins along their course. The great saphenous vein is the longest vein in the body. It travels superiorly along the medial aspect of the leg to the thigh, where it empties into the femoral vein just distal to the inguinal ligament. The small saphenous vein runs along the lateral aspect of the foot and then through the deep fascia of the calf muscles, which it drains. At the knee, it empties into the popliteal vein.

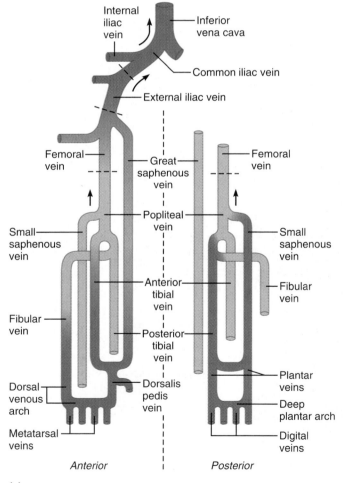

(a)

FIGURE 20.28   Veins of the right lower limb.
(**a**) Schematic flowchart of anterior and posterior vessels.
(**b**) Anterior view of the lower limb. (**c**) Posterior view of the leg and foot.

CHAPTER SUMMARY

Media study tools that could provide you additional help in reviewing specific key topics of Chapter 20 are referenced below. **SP** *= Study Partner;* **IP** *= Interactive Physiology.*

Overview of Blood Vessel Structure and Function (pp. 718–727)

1. Blood is transported throughout the body via a continuous system of blood vessels. Arteries transport blood away from the heart; veins carry blood back to the heart. Capillaries carry blood to tissue cells and are exchange sites.

Structure of Blood Vessel Walls (pp. 718–719)

2. All blood vessels except capillaries have three layers: tunica interna, tunica media, and tunica externa. Capillary walls are composed of the tunica interna only.

Arterial System (pp. 720–723)

3. Elastic (conducting) arteries are the large arteries close to the heart that expand and recoil to accommodate changing blood volume. Muscular (distributing) arteries carry blood to specific organs; they are less stretchy and more active in vasoconstriction. Arterioles regulate blood flow into capillary beds.

4. Arteriosclerosis is a degenerative vascular disease. Initiated by endothelial lesions, it progresses through fatty streak, atherosclerotic, and arteriosclerotic stages.

Capillaries (pp. 724–726)

5. Capillaries are microscopic vessels with very thin walls. Most exhibit clefts, which aid in the exchange between the blood and interstitial fluid. Spider-shaped cells called pericytes help to reinforce the external faces of capillaries.

6. Vascular shunts (metarterioles–thoroughfare channels) connect the terminal arteriole and venule at opposite ends of a capillary bed. Most true capillaries arise from and rejoin the shunt channels. The amount of blood flowing into the true capillaries is regulated by precapillary sphincters.

7. The most permeable capillaries are sinusoids (wide, tortuous channels). Fenestrated capillaries with pores are next most permeable. Least permeable are continuous capillaries, which lack pores.

Venous System (pp. 726–727)

8. Veins have comparatively larger lumens than arteries, and a system of valves prevents backflow of blood. Respiratory and skeletal muscle pumps aid return of venous blood to the heart.

9. Normally most veins are only partially filled with blood; thus, they can serve as blood reservoirs.

Vascular Anastomoses (p. 727)

10. The joining together to provide alternate channels for blood to reach the same organ is called an arterial anastomosis. Vascular anastomoses also form between veins and between arterioles and venules.

IP Cardiovascular System CD-ROM; Topic: Anatomy Review, Blood Vessel Structure and Function, pages 1–28.

PHYSIOLOGY OF CIRCULATION (pp. 727–746)

Introduction to Blood Flow, Blood Pressure, and Resistance (pp. 727–728)

1. Blood flow is the amount of blood flowing through a vessel, an organ, or the entire circulation in a given period of time. Blood pressure is the force per unit area exerted on a vessel wall by the contained blood. Resistance is opposition to blood flow; blood viscosity and blood vessel length and diameter contribute to resistance.

2. Blood flow is directly proportional to blood pressure and inversely proportional to resistance.

IP Cardiovascular System CD-ROM; Topic: Factors that Affect Blood Pressure, pages 1–15.

Systemic Blood Pressure (pp. 728–730)

1. Systemic blood pressure is highest in the aorta and lowest in the venae cavae. The steepest drop in BP occurs in the arterioles, where resistance is greatest.

2. Arterial BP depends on compliance of the elastic arteries and on how much blood is forced into them. Arterial blood pressure is pulsatile, and peaks during systole (systolic pressure). During diastole, as blood is forced distally in the circulation by the rebound of elastic arteries, arterial BP drops to its lowest value, called the diastolic pressure.

3. Pulse pressure is systolic pressure minus diastolic pressure. The mean arterial pressure (MAP) = diastolic pressure plus one-third of pulse pressure and is the pressure that keeps blood moving throughout the cardiac cycle.

4. Low capillary pressure (40 to 15 mm Hg) protects the delicate capillaries from rupture while still allowing adequate exchange across the capillary walls.

5. Venous pressure is nonpulsatile and low (declining to zero) because of the cumulative effects of resistance. Venous valves, large lumens, and functional adaptations (muscular and respiratory pumps) promote venous return.

Maintaining Blood Pressure (pp. 730–732)

1. Blood pressure varies directly with CO, PR, and blood volume. Vessel diameter is the major factor determining PR, and small changes in vessel (chiefly arteriolar) diameter significantly affect blood pressure.

IP Cardiovascular System CD-ROM; Topic: Measuring Blood Pressure, pages 1–13.

Short-Term Mechanisms: Neural Controls (pp. 731–732)

2. BP is regulated by autonomic neural reflexes involving baroreceptors or chemoreceptors, the vasomotor center (a sympathetic center that regulates blood vessel diameter), and vasomotor fibers, which act on vascular smooth muscle.

3. Activation of the receptors by falling BP (and to a lesser extent by a rise in blood CO_2, or falling blood pH or O_2 levels) stimulates the vasomotor center to increase vasoconstriction and the cardioacceleratory center to increase heart rate and contractility. Rising BP inhibits the vasomotor center (permitting vasodilation) and activates the cardioinhibitory center.

4. Higher brain centers (cerebrum and hypothalamus) may modify neural controls of BP via medullary centers.

SP Exercise: Chapter 20, Blood Pressure Homeostasis.

Short-Term Mechanisms: Chemical Controls (pp. 732–734)

5. Bloodborne chemicals that increase BP by promoting vasoconstriction include epinephrine and NE (these also increase heart rate and contractility), ADH, angiotensin II (generated in response to renin release by kidney cells), and PDGF and endothelin released by vascular endothelium cells.

6. Chemicals that reduce BP by promoting vasodilation include atrial natriuretic peptide (also causes a decline in blood volume), nitric oxide released by the vascular endothelium, inflammatory chemicals, and alcohol.

Long-Term Mechanisms: Renal Regulation (p. 735)

7. The kidneys directly regulate blood pressure by regulating blood volume. Rising BP enhances filtrate formation and fluid losses in urine; falling BP causes the kidneys to retain more water, increasing blood volume.

8. Indirect renal regulation of blood volume involves the renin-angiotensin mechanism, a hormonal mechanism. When BP falls, the kidneys release renin, which triggers the formation of angiotensin II (a vasoconstrictor) and release of aldosterone, which causes salt and water to be retained.

IP Cardiovascular System CD-ROM; Topic: Blood Pressure Regulation, pages 1–31.

Monitoring Circulatory Efficiency (pp. 735–737)

9. Pulse and blood pressure measurements are used to assess cardiovascular efficiency.

10. The pulse is the alternating expansion and recoil of arterial walls with each heartbeat. Pulse points are also pressure points.

11. Blood pressure is routinely measured by the auscultatory method. Normal blood pressure in adults is 120/80 (systolic/diastolic). Hypotension is rarely a problem. Hypertension is the major cause of myocardial infarct, stroke, and renal disease.

IP Cardiovascular System CD-ROM; Topic: Measuring Blood Pressure, pages 11, 12.

Alterations in Blood Pressure (pp. 737–738)

12. Hypotension, or low blood pressure (systolic pressure below 100 mm Hg), is a sign of health in the well conditioned. In other individuals it warns of poor nutrition, disease, or circulatory shock.

13. Chronic hypertension (high blood pressure) is persistent BP readings of 140/90 or higher. It indicates increased peripheral resistance, which strains the heart and promotes vascular complications of other organs, particularly the eyes and kidneys. Risk factors are high-fat, high-salt diet, obesity, advanced age, smoking, stress, and being a member of the black race or a family with a history of hypertension.

SP Case Study: Cardiac Arrythmia.

IP Cardiovascular System CD-ROM; Topic: Measuring Blood Pressure, pages 11, 12.

Blood Flow Through Body Tissues (pp. 738–746)

1. Blood flow is involved in delivering nutrients and wastes to and from cells, gas exchange, absorbing nutrients, and forming urine.

Velocity of Blood Flow (p. 739)

2. Blood flows fastest where the cross-sectional area of the vascular bed is least (aorta), and slowest where the cross-sectional area is greatest (capillaries). The slow flow in capillaries allows time for nutrient-waste exchanges.

Autoregulation: Local Regulation of Blood Flow (pp. 739–740)

3. Autoregulation is the local adjustment of blood flow to individual organs based on their immediate requirements. It is largely controlled by local chemical factors that cause vasodilation of arterioles serving the area and open the precapillary sphincters. Myogenic controls respond to changes in blood pressure.

SP Exercise: Chapter 20, Autoregulation.

IP Cardiovascular CD-ROM; Topic: Autoregulation and Capillary Dynamics, pages 1–13.

Blood Flow in Special Areas (pp. 740–742)

4. In most instances, autoregulation is controlled by oxygen deficits and accumulation of local metabolites. However, autoregulation in the brain is controlled primarily by a drop in pH and by myogenic mechanisms; and vasodilation of pulmonary circuit vessels occurs in response to high levels of oxygen.

Blood Flow Through Capillaries and Capillary Dynamics (pp. 742–744)

5. Nutrients, gases, and other solutes smaller than plasma proteins cross the capillary wall by diffusion. Water-soluble substances move through the clefts or fenestrations; fat-soluble substances pass through the lipid portion of the endothelial cell membrane.

6. Fluid flows occurring at capillary beds reflect the relative effect of outward (net hydrostatic pressure) forces minus the effect of inward (net osmotic pressure) forces. In general, fluid flows out of the capillary bed at the arterial end and reenters the capillary blood at the venule end.

IP Cardiovascular CD-ROM; Topic: Autoregulation and Capillary Dynamics, pages 14–38.

7. The small net loss of fluid and protein into the interstitial space is collected by lymphatic vessels and returned to the cardiovascular system.

Circulatory Shock (pp. 744–746)

8. Circulatory shock occurs when blood perfusion of body tissues is inadequate. Most cases of shock reflect low blood volume (hypovolemic shock), abnormal vasodilation (vascular shock), or pump failure (cardiogenic shock).

SP Exercise: Chapter 20, Making Connections: The Cardiovascular System; Case Study: Septic Shock.

Circulatory Pathways: Blood Vessels of the Body (pp. 746, 750–772)

1. The pulmonary circulation transports O_2-poor, CO_2-laden blood to the lungs for oxygenation and carbon dioxide unloading. Blood returning to the right atrium of the heart is pumped via the pulmonary trunk, to the lungs by the right ventricle. Blood issuing from the lungs is returned to the left atrium by the pulmonary veins. (See Table 20.3 and Figure 20.17.)

2. The systemic circulation transports oxygenated blood from the left ventricle to all body tissues via the aorta and its branches. Venous blood returning from the systemic circuit is delivered to the right atrium via the venae cavae.

3. Tables 20.3 to 20.13 and Figures 20.17b to 20.28 illustrate and describe vessels of the systemic circulation.

Developmental Aspects of the Blood Vessels (pp. 746–747)

1. The fetal vasculature develops from embryonic blood islands and mesenchyme and is functioning in blood delivery by the fourth week.

2. Fetal circulation differs from circulation after birth. The pulmonary and hepatic shunts and special umbilical vessels are normally occluded shortly after birth.

3. Blood pressure is low in infants and rises to adult values. Age-related vascular problems include varicose veins, hypertension, and atherosclerosis. Hypertension is the most important cause of sudden cardiovascular death in middle-aged men. Atherosclerosis is the most important cause of cardiovascular disease in the aged.

REVIEW QUESTIONS

Multiple Choice/Matching

1. Which statement does not accurately describe veins? (a) They have less elastic tissue and smooth muscle than arteries, (b) they contain more fibrous tissue than arteries, (c) most veins in the extremities have valves, (d) they always carry deoxygenated blood.

2. Which of the following tissues is mainly responsible for vasoconstriction? (a) elastic tissue, (b) smooth muscle, (c) collagenic tissue, (d) adipose tissue.

3. Peripheral resistance (a) is inversely related to the diameter of the arterioles, (b) tends to increase if blood viscosity increases, (c) is directly proportional to the length of the vascular bed, (d) all of these.

4. Which of the following can lead to decreased venous return of blood to the heart? (a) an increase in blood volume, (b) an increase in venous pressure, (c) damage to the venous valves, (d) increased muscular activity.

5. Arterial blood pressure increases in response to: (a) increasing stroke volume, (b) increasing heart rate, (c) arteriosclerosis, (d) rising blood volume, (e) all of these.

6. Which of the following would *not* result in the dilation of the feeder arterioles and opening of the precapillary sphincters in any capillary bed? (a) a decrease in O_2 content of the blood, (b) an increase in CO_2 content of the blood, (c) a local increase in histamine, (d) a local increase in pH.

7. The structure of a capillary wall differs from that of a vein or an artery because (a) it has two tunics instead of three, (b) there is less smooth muscle, (c) it has a single tonic—only the tunica interna, (d) none of these.

8. The baroreceptors in the carotid and aortic bodies are sensitive to (a) a decrease in carbon dioxide, (b) changes in arterial pressure, (c) a decrease in oxygen, (d) all of these.

9. The myocardium receives its blood supply directly from (a) the aorta, (b) the coronary arteries, (c) the coronary sinus, (d) the pulmonary arteries.

10. Blood flow in the capillaries is steady despite the rhythmic pumping of the heart because of the (a) elasticity of the large arteries, (b) small diameter of capillaries, (c) thin walls of the veins, (d) venous valves.

11. Tracing the blood from the heart to the right hand, we find that blood leaves the heart and passes through the aorta, the right subclavian artery, the axillary and brachial arteries, and through either the radial or ulnar artery to arrive at the hand. Which artery is missing from this sequence? (a) coronary, (b) brachiocephalic, (c) cephalic, (d) right common carotid.

12. Which of the following do not drain directly into the inferior vena cava? (a) lumbar veins, (b) hepatic veins, (c) inferior mesenteric vein, (d) renal veins.

13. In atherosclerosis, which layer of the vessel wall thickens most? (a) tunica media, (b) tunica interna, (c) tunica adventitia, (d) tunica externa.

Short Answer Essay Questions

14. How is the anatomy of capillaries and capillary beds well suited to their function?

15. Distinguish between elastic arteries, muscular arteries, and arterioles relative to location, histology, and functional adaptations.

16. Write an equation showing the relationship between peripheral resistance, blood flow, and blood pressure.

17. (a) Define blood pressure. Differentiate between systolic and diastolic blood pressure. (b) What is the normal blood pressure value for a young adult?

18. Describe the neural mechanisms responsible for controlling blood pressure.

19. Explain the reasons for the observed changes in blood flow velocity in the different regions of the circulation.

20. How does the control of blood flow to the skin for the purpose of regulating body temperature differ from the control of nutrient blood flow to skin cells?

21. Describe neural and chemical (both systemic and local) effects exerted on the blood vessels when one is fleeing from a mugger. (Be careful, this is more involved than it appears at first glance.)

22. How are nutrients, wastes, and respiratory gases transported to and from the blood and tissue spaces?

23. (a) What blood vessels contribute to the formation of the hepatic portal circulation? (b) What is the function of this circulation? (c) Why is a portal circulation a "strange" circulation?

24. Physiologists often consider capillaries and postcapillary venules together. (a) What functions do these vessels share? (b) Structurally, how do they differ?

Critical Thinking and Clinical Application Questions

1. Mrs. Johnson is brought to the emergency room after being involved in an auto accident. She is hemorrhaging and has a rapid, thready pulse, but her blood pressure is still within normal limits. Describe the compensatory mechanisms that are acting to maintain her blood pressure in the face of blood loss.

2. A 60-year-old man is unable to walk more than 100 yards without experiencing severe pain in his left leg; the pain is relieved by resting for 5–10 minutes. He is told that the arteries of his leg are becoming occluded with fatty material and is advised to have the sympathetic nerves serving that body region severed. Explain how such surgery might help to relieve this man's problem.

3. Your friend Joanie, who knows little about science, is reading a magazine article about a patient who had an "aneurysm at the base of his brain that suddenly grew much larger." The surgeons' first goal was to "keep it from rupturing," and the second goal was to "relieve the pressure on the brain stem and cranial nerves." The surgeons were able to "replace the aneurysm with a section of plastic tubing," so the patient recovered. Joanie asks you what all this means. Explain. (Hint: Check this chapter's Related Clinical Terms.)

4. The Agawam High School band is playing some lively marches while the coaches are giving pep talks to their respective football squads. Although it is September, it is unseasonably hot (88°F) and the band uniforms are wool. Suddenly, Harry the tuba player becomes light-headed and faints. Explain his fainting in terms of vascular events.

5. When one is cold or the external temperature is low, most venous blood returning from the distal part of the arm travels in the deep veins where it picks up heat (countercurrent mechanism) from the nearby brachial artery en route. However, when one is hot, and especially during exercise, venous return from the distal arm travels in the superficial veins and those veins tend to bulge superficially in a person who is working out. Explain why venous return takes a different route in the second situation.

THE LYMPHATIC SYSTEM 21

They can't all be superstars! When we mentally tick off the names of the body's organ systems, the lymphatic (lim-fat'ik) system is probably not the first to come to mind. Yet without this quietly working system, our cardiovascular system would stop working and our immune system would be hopelessly impaired. The **lymphatic system** actually consists of two semi-independent parts: (1) a meandering network of *lymphatic vessels* and (2) various *lymphoid tissues* and *organs* scattered throughout the body. The lymphatic vessels transport fluids that have escaped from the blood vascular system back to the blood. The lymphoid organs house phagocytic cells and lymphocytes, which play essential roles in body defense and resistance to disease.

LYMPHATIC VESSELS

As blood circulates through the body, exchanges of nutrients, wastes, and gases occur between the blood and the interstitial fluid. As explained in Chapter 20, the hydrostatic and colloid osmotic pressures operating at capillary beds force fluid out of the blood at the arterial ends of the capillary beds ("upstream") and cause most of it to be reabsorbed at the venous ends ("downstream"). The fluid that remains behind in the tissue spaces, as much as 3 L daily, becomes part of the interstitial fluid. This leaked fluid, plus any plasma proteins that escape from the bloodstream, must be carried back to the blood to ensure that the cardiovascular system has sufficient blood volume to operate properly. This problem of circulatory dynamics is resolved by the **lymphatic vessels,** or **lymphatics,** an elaborate system of drainage vessels that collects the excess protein-containing interstitial fluid and returns it to the bloodstream. Once interstitial fluid enters the lymphatics, it is called **lymph** (*lymph* = clear water).

Distribution and Structure of Lymphatic Vessels

The lymphatic vessels form a one-way system in which lymph flows only toward the heart. This transport system begins in microscopic blind-ended **lymph capillaries** (Figure 21.1a), which weave between the tissue cells and blood capillaries in the loose connective tissues of the body. Lymph capillaries are widespread, occurring almost everywhere blood capillaries occur. However, lymph capillaries are absent from bones and teeth, bone marrow, and the entire central nervous system (where the excess tissue fluid drains into the cerebrospinal fluid).

Although similar to blood capillaries, lymphatic capillaries are so remarkably permeable that they were once thought to be open at one end like a straw. We now know that they owe their permeability to two unique structural modifications:

1. The endothelial cells forming the walls of lymphatic capillaries are not tightly joined; instead, their edges loosely overlap one another, forming easily opened, flaplike *minivalves* (Figure 21.1b).

2. Bundles of fine collagen filaments anchor the endothelial cells to surrounding structures so that any increase in interstitial fluid volume separates the cell flaps, opening up gaps in the wall, rather than causing the lymphatic capillary to collapse.

So, what we have here is a system analogous to one-way swinging doors in the lymphatic capillary wall. The flaps gape open, allowing fluid to enter the lymphatic capillary, when fluid pressure is greater in the interstitial space. However, when the pressure is greater *inside* the lymphatic vessels, the endothelial cell flaps are forced together; this prevents the lymph from leaking back out and forces it along the vessel.

Proteins present in the interstitial space are unable to enter blood capillaries, but they enter lymphatic capillaries easily. In addition, when tissues are inflamed, lymphatic capillaries develop openings that permit uptake of even larger particles such as cell debris, pathogens (disease-causing microorganisms such as bacteria and viruses), and cancer cells. The pathogenic agents and cancer cells can then use the lymphatics to travel throughout the body. This threat to the body is partly resolved by the fact that the lymph takes "detours" through the lymph nodes, where it is cleansed of debris and "examined" by cells of the immune system, as described shortly.

Highly specialized lymphatic capillaries called **lacteals** (lak'te-alz) are present in the fingerlike villi of the intestinal mucosa. The lymph draining from the digestive viscera is milky white (*lacte* = milk) rather than clear because the lacteals play a major role in absorbing digested fats from the intestine. This fatty lymph, called **chyle** ("juice"), is also delivered to the blood via the lymphatic stream.

From the lymphatic capillaries, lymph flows through successively larger and thicker-walled channels—first collecting vessels, then trunks, and finally the largest of all, the ducts (Figure 21.2). The **lymphatic collecting vessels** have the same three tunics as veins, but they are thinner-walled, have more internal valves, and anastomose more. In general the collecting vessels in the skin travel along with superficial *veins* of the blood vascular system, whereas the deep lymphatic vessels of the trunk and digestive viscera travel with the deep *arteries*. Like the larger veins, the larger lymphatics receive their nutrient blood supply from a branching vasa vasorum.

What transport vessels present in the cardiovascular system are absent in the lymphatic system? Why is their absence not a problem?

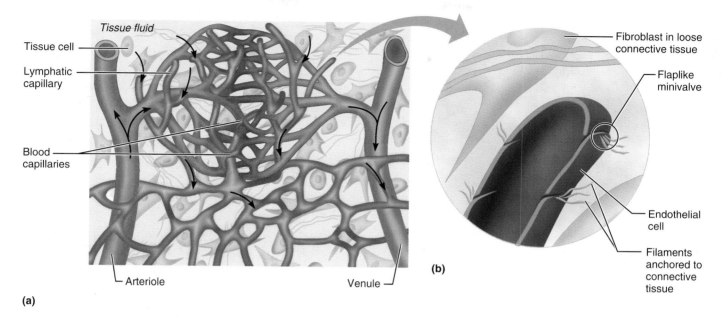

FIGURE 21.1 *Distribution and special structural features of lymphatic capillaries.* (a) Structural relationship between a capillary bed of the blood vascular system and lymphatic capillaries. Arrows indicate direction of fluid movement. **(b)** Lymphatic capillaries begin as blind-ended tubes. The endothelial cells forming their walls overlap one another, forming flaplike minivalves.

⚡ Homeostatic Imbalance

When lymphatic vessels are severely inflamed, the related vessels of the vasa vasorum become congested with blood. As a result, the pathway of the associated superficial lymphatics becomes visible through the skin as red lines that are tender to the touch. This unpleasant condition is called *lymphangitis* (lim″fan-ji′tis; *angi* = vessel). ∎

The **lymphatic trunks** are formed by the union of the largest collecting vessels, and drain fairly large areas of the body. The major trunks, named mostly for the regions from which they collect lymph, are the paired **lumbar, bronchomediastinal, subclavian,** and **jugular trunks,** and the single **intestinal trunk** (Figure 21.2b).

Lymph is eventually delivered to one of two large *ducts* in the thoracic region. The **right lymphatic duct** drains lymph from the right upper arm and the right side of the head and thorax (Figure 21.2a). The much larger **thoracic duct,** which receives lymph from the rest of the body, arises anterior to the first two lumbar vertebrae as an enlarged sac, the **cisterna chyli** (sis-ter′nah ki′li). The cisterna chyli collects lymph from the two large lumbar trunks that drain the lower limbs and from the intestinal trunk that drains the digestive organs. As the thoracic duct runs superiorly, it receives lymphatic drainage from the left side of the thorax and left upper limb and head region. Each terminal duct empties the lymph into the venous circulation at the junction of the internal jugular vein and subclavian vein on its own side of the body (Figure 21.2b).

Lymph Transport

Unlike the cardiovascular system, the lymphatic system lacks an organ that acts as a pump. Under normal conditions, lymphatic vessels are very low pressure conduits, and the same mechanisms that promote venous return in blood vessels act here as well—the milking action of active skeletal muscles, pressure changes within the thorax during breathing, and valves to prevent backflow. Lymphatics are usually bundled together in connective tissue sheaths with blood vessels, and pulsations of the nearby arteries also promote lymph flow. In addition

The arteries. It's not a problem because lymphatic fluid is formed by the blood capillary beds and is only moved in one direction, toward the heart.

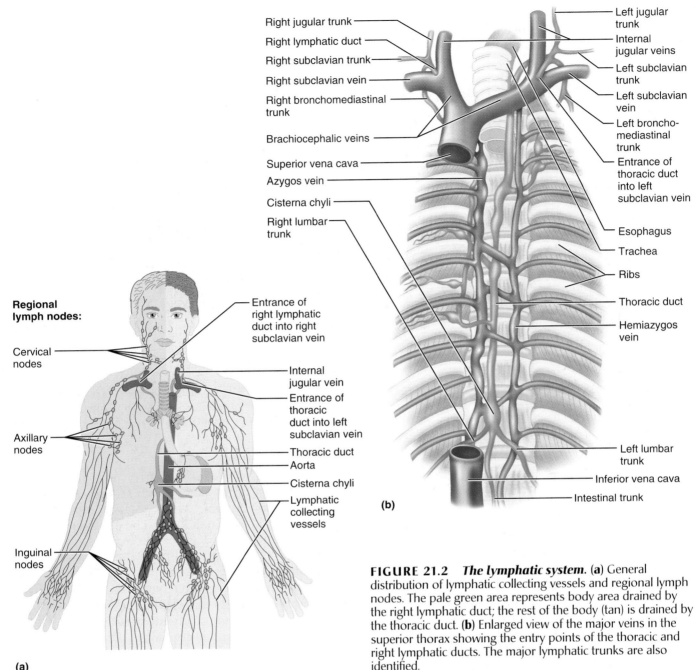

Right jugular trunk
Right lymphatic duct
Right subclavian trunk
Right subclavian vein
Right bronchomediastinal trunk
Brachiocephalic veins
Superior vena cava
Azygos vein
Cisterna chyli
Right lumbar trunk

Left jugular trunk
Internal jugular veins
Left subclavian trunk
Left subclavian vein
Left broncho-mediastinal trunk
Entrance of thoracic duct into left subclavian vein
Esophagus
Trachea
Ribs
Thoracic duct
Hemiazygos vein
Left lumbar trunk
Inferior vena cava
Intestinal trunk

Regional lymph nodes:

Cervical nodes

Axillary nodes

Inguinal nodes

Entrance of right lymphatic duct into right subclavian vein
Internal jugular vein
Entrance of thoracic duct into left subclavian vein
Thoracic duct
Aorta
Cisterna chyli
Lymphatic collecting vessels

(a)

(b)

FIGURE 21.2   *The lymphatic system.* (a) General distribution of lymphatic collecting vessels and regional lymph nodes. The pale green area represents body area drained by the right lymphatic duct; the rest of the body (tan) is drained by the thoracic duct. **(b)** Enlarged view of the major veins in the superior thorax showing the entry points of the thoracic and right lymphatic ducts. The major lymphatic trunks are also identified.

to these mechanisms, smooth muscle in the walls of the lymphatic trunks and thoracic duct contracts rhythmically, actually helping to pump the lymph along. Even so, lymph transport is sporadic and much slower than that occurring in veins. About 3 L of lymph enters the bloodstream every 24 hours, a volume almost exactly equal to the amount of fluid lost to the tissue spaces from the bloodstream in the same time period. Movements of adjacent tissues are extremely important in propelling lymph through the lymphatics. When physical activity or passive movements increase, lymph flows much more rapidly (balancing the greater rate of fluid loss from the bloodstream in such situations). Hence, it

is a good idea to immobilize a badly infected body part to hinder flow of inflammatory material from that region.

Homeostatic Imbalance

Anything that prevents the normal return of lymph to the blood, such as blockage of the lymphatics by tumors or removal of lymphatics during cancer surgery (e.g., mastectomy), results in severe localized edema *(lymphedema)*. However, when lymphatics are removed surgically, lymphatic drainage is eventually reestablished by regrowth from the vessels still remaining in the area. ■

LYMPHOID CELLS, TISSUES, AND ORGANS: AN OVERVIEW

In order to understand some of the basic aspects of the lymphatic system's role in body protection and immunity, we will investigate the components of lymphoid organs—lymphoid cells and lymphoid tissues—before considering the organs themselves.

Lymphoid Cells

Infectious microorganisms, such as bacteria and viruses, that manage to penetrate the body's epithelial barriers quickly proliferate in the underlying loose connective tissues. These invaders are fought off by the inflammatory response, by phagocytes (macrophages), and by lymphocytes (a variety of white blood cell described in Chapter 18).

Lymphocytes, the main warriors of the immune system, arise in red bone marrow (along with other formed elements). They then mature into one of the two main varieties of immunocompetent cells—**T cells (T lymphocytes)** or **B cells (B lymphocytes)**—which act to protect the body against antigens. (Generally speaking, *antigens* are anything the body perceives as foreign, such as bacteria and their toxins, viruses, mismatched RBCs, or cancer cells.) Activated T cells manage the immune response and some of them directly attack and destroy foreign cells. B cells protect the body by producing **plasma cells,** daughter cells that secrete antibodies into the blood (or other body fluids). Antibodies immobilize antigens until they can be destroyed by phagocytes or other means. The precise roles of the lymphocytes and mechanisms of immunity will be explored in more detail in Chapter 22.

Lymphoid **macrophages** play a crucial role in body protection and in the immune response by phagocytizing foreign substances and by helping to activate T cells. So, too, do the spiny-looking **dendritic cells** found in lymphoid tissue. Last but not least are the **reticular cells,** fibroblastlike cells that produce the reticular fiber **stroma** (stro'mah) or network that supports the other cell types in the lymphoid organs (see Figure 21.3).

Lymphoid Tissue

Lymphoid (lymphatic) tissue is an important component of the immune system, mainly because it (1) houses and provides a proliferation site for lymphocytes and (2) furnishes an ideal surveillance vantage point for both lymphocytes and macrophages. Lymphoid tissue, a type of loose connective tissue called **reticular connective tissue,** dominates all the lymphoid organs except the thymus. Macrophages live on the fibers of the network. Within the spaces of

Specifically, how is reticular connective tissue classified?

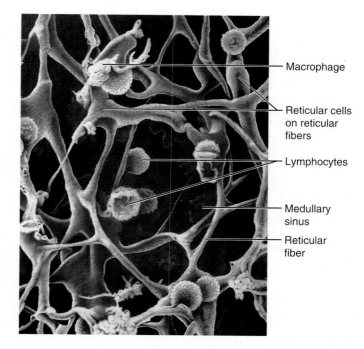

- Macrophage
- Reticular cells on reticular fibers
- Lymphocytes
- Medullary sinus
- Reticular fiber

FIGURE 21.3 *Scanning electron micrograph of reticular tissue in a human lymph node (850×).*

this network are huge numbers of lymphocytes that squeeze through the walls of postcapillary venules coursing through this tissue to reside temporarily in the lymphoid tissue (Figure 21.3), and then leave to patrol the body again. The cycling of lymphocytes between the circulatory vessels, lymphoid tissues, and loose connective tissues of the body ensures that lymphocytes reach infected or damaged sites quickly.

Lymphoid tissue comes in various "packages." **Diffuse lymphatic tissue,** consisting of a few scattered reticular tissue elements, is found in virtually every body organ, but larger collections appear in the lamina propria of mucous membranes and within lymphoid organs. Like diffuse lymphatic tissue, **lymphatic follicles (nodules)** lack a capsule, but follicles tend to be solid, spherical bodies consisting of tightly packed reticular elements and cells. Follicles often exhibit lighter-staining centers, called **germinal centers.** Follicular dendritic cells and B cells predominate in germinal centers, and these centers enlarge dramatically when the B cells are rapidly dividing and producing plasma cells. In many cases, the follicles are found within and forming part of larger lymphoid organs, such as lymph nodes. However, isolated aggregations of lymphatic follicles occur in the

Loose connective tissue proper.

? *What is the benefit of having fewer efferent than afferent lymphatics in lymph nodes?*

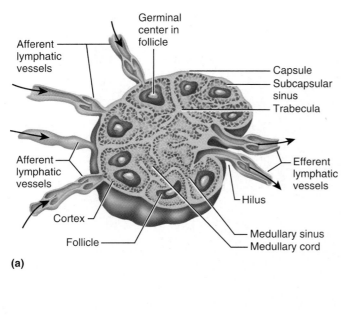

(a)

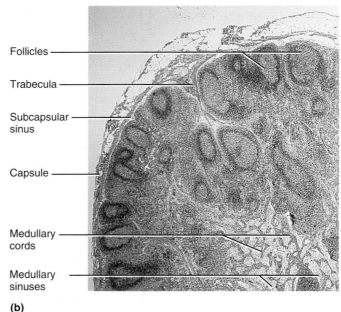

(b)

FIGURE 21.4 *Structure of a lymph node*. (a) Longitudinal view of the internal structure of a lymph node and associated lymphatics. Notice that several afferent lymphatics converge on its convex side, whereas fewer efferent lymphatics exit at its hilus. Arrows indicate the direction of lymph flow into and out of the node.

(b) Photomicrograph of part of a lymph node (28×).

intestinal wall as Peyer's patches (see Figure 21.5) and in the appendix.

Lymphoid Organs

Lymphoid organs, exemplified by lymph nodes, the spleen, and the thymus (Figure 21.5), are discrete, encapsulated collections of diffuse lymphoid tissue and follicles. The exact pattern of the lymphoid tissues differs in the various lymphoid organs. These are described next.

LYMPH NODES

As lymph is transported back to the bloodstream, it is filtered through the **lymph nodes** that cluster along the lymphatic vessels of the body. There are hundreds of these small organs, but because the lymph nodes are usually embedded in connective tissue, they are not ordinarily seen. Large clusters of lymph nodes occur near the body surface in the inguinal, axillary, and cervical regions of the body—

places where the lymphatic vessels converge to form large trunks (see Figure 21.2a).

Lymph nodes have two basic functions, both concerned with body protection. (1) They act as lymph "filters." Phagocytic macrophages in the nodes remove and destroy microorganisms and other debris that enter the lymph from the loose connective tissues, effectively preventing them from being delivered to the blood and spreading further in the body. (2) They play a role in activating the immune system. Lymphocytes, also strategically located in the lymph nodes, monitor the lymphatic stream for the presence of *antigens* and mount an attack against them. Let's look at how the structure of a lymph node supports these defensive functions.

Structure of a Lymph Node

Lymph nodes vary in both shape and size, but most of them are bean shaped and less than 2.5 cm (1 inch) in length. Each node is surrounded by a dense fibrous **capsule** from which connective tissue strands called **trabeculae** extend inward to divide the node into a number of compartments (Figure 21.4). Its internal framework or stroma of reticular fibers physically supports the ever-changing population of lymphocytes.

It causes lymphatic fluid to back up or accumulate somewhat in lymph nodes, allowing more time for its cleansing.

Lymph nodes have two histologically distinct regions, the **cortex** and the **medulla**. The superficial part of the cortex contains densely packed follicles, many with germinal centers heavy with dividing B cells. Dendritic cells nearly encapsulate the follicles and abut the deeper part of the cortex, which primarily houses T cells in transit. The T cells circulate continuously between the blood, lymph nodes, and lymphatic stream, performing their surveillance role.

Medullary cords, thin inward extensions from the cortical lymphoid tissue, contain both types of lymphocytes and plasma cells and define the medulla. Throughout the node are **lymph sinuses,** large lymph capillaries spanned by crisscrossing reticular fibers. Numerous macrophages reside on these reticular fibers and phagocytize foreign matter in the lymph as it flows by in the sinuses. Additionally, some of the lymph-borne antigens in the percolating lymph leak into the surrounding lymphoid tissue, where they activate some of the strategically positioned lymphocytes to mount an immune attack against them.

Circulation in the Lymph Nodes

Lymph enters the convex side of a lymph node through a number of **afferent lymphatic vessels.** It then moves through a large, baglike sinus, the **subcapsular sinus,** into a number of smaller sinuses that cut through the cortex and enter the medulla. The lymph meanders through these sinuses and finally exits the node at its **hilus** (hi'lus), the indented region on the opposite side, via **efferent lymphatic vessels.** Because there are fewer efferent vessels draining the node than afferent vessels feeding it, the flow of lymph through the node stagnates somewhat, allowing time for the lymphocytes and macrophages to carry out their protective functions. Lymph passes through several nodes before its cleansing process is completed.

Homeostatic Imbalance

Sometimes lymph nodes are overwhelmed by the very agents they are trying to destroy. For example, when large numbers of bacteria or virus particles are trapped in the nodes, the nodes become inflamed, swollen, and tender to the touch, a condition often referred to (erroneously) as swollen glands. Such infected lymph nodes are called *buboes* (bu'boz). (Buboes are the most obvious symptom of bubonic plague, the "Black Death" that killed much of Europe's population in the late Middle Ages.) Lymph nodes can also become secondary cancer sites, particularly in metastasizing cancers that enter lymphatic vessels and become trapped there. The fact

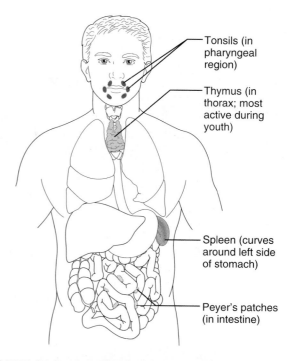

FIGURE 21.5 *Lymphoid organs.* Tonsils, spleen, thymus, and Peyer's patches.

that cancer-infiltrated lymph nodes are swollen but not painful helps to distinguish cancerous lymph nodes from those infected by microorganisms. ■

OTHER LYMPHOID ORGANS

Lymph nodes are just one example of many types of **lymphoid organs** or aggregates of lymphatic tissue in the body. Others are the spleen, thymus gland, tonsils, and Peyer's patches of the intestine (see Figure 21.5), as well as bits of lymphatic tissue scattered in the connective tissues. The common feature of all these organs is their tissue makeup: All are composed of *reticular connective tissue.* Although all lymphoid organs help protect the body, only the lymph nodes filter lymph. The other lymphoid organs and tissues typically have efferent lymphatics draining them, but lack afferent lymphatic vessels. The distinctive features of these other lymphoid organs and aggregates are described next.

Spleen

The soft, blood-rich **spleen** is about the size of a fist and is the largest lymphoid organ. Located in the left side of the abdominal cavity just beneath the diaphragm, it extends to curl around the anterior aspect of the stomach (see Figures 21.5 and 21.6c). It is served by the large *splenic artery* and *vein,* which enter and exit the *hilus* on its slightly concave anterior surface.

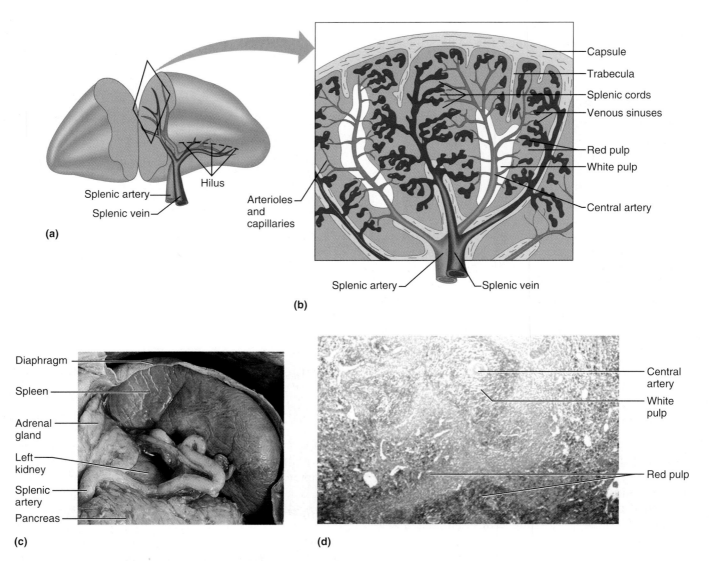

FIGURE 21.6 *Structure of the spleen.* **(a)** Gross structure (diagrammatic view). **(b)** Diagram of the histological structure of part of the spleen. **(c)** Photograph of the spleen in its normal position in the abdominal cavity, anterior view. **(d)** Photomicrograph of spleen tissue showing white and red pulp regions (94×). The white pulp, consisting mainly of lymphocytes cuffing around branches of the splenic artery, is surrounded by red pulp regions containing abundant erythrocytes and venous sinuses.

The spleen provides a site for lymphocyte proliferation and immune surveillance and response. But perhaps even more important are the spleen's blood-cleansing functions. Besides extracting aged and defective blood cells and platelets from the blood, its macrophages remove debris, foreign matter, bacteria, viruses, toxins, and so on from blood flowing through its sinuses. The spleen also performs three additional, and related, functions.

1. It stores some of the breakdown products of red blood cells for later reuse and releases others to the blood for processing by the liver. For example, spleen macrophages salvage and store iron for later use by the bone marrow in making hemoglobin.

2. It is a site of erythrocyte production in the fetus (a capability that normally ceases after birth).

3. It stores blood platelets.

Like lymph nodes, the spleen is surrounded by a fibrous capsule, has trabeculae that extend inward (Figure 21.6), and contains both lymphocytes and macrophages. Consistent with its blood-processing functions, it also contains huge numbers of erythrocytes. Areas composed mostly of lymphocytes suspended on reticular fibers are called **white pulp.** The white pulp clusters or forms "cuffs" around the *central arteries* (small branches of the splenic artery) within the organ and forms what appear to be islands in a sea of red pulp. **Red pulp** is essentially all

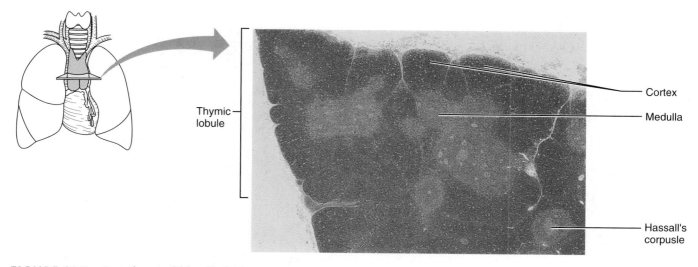

FIGURE 21.7 *Location and histological structure of the thymus.* The photomicrograph of a portion of the thymus shows its lobules with cortical and medullary regions (25×).

remaining splenic tissue, that is, the venous sinuses (blood sinusoids) and the **splenic cords,** regions of reticular connective tissue that are exceptionally rich in macrophages. Red pulp is most concerned with disposing of worn-out red blood cells and bloodborne pathogens, while white pulp is involved with the immune functions of the spleen. Notice that the naming of the pulp regions reflects their appearance in fresh spleen tissue rather than their staining properties. Indeed, as can be seen in the photomicrograph in Figure 21.6d, the white pulp takes on a purplish hue and sometimes appears darker than the red pulp.

⚠ Homeostatic Imbalance

Because the spleen's capsule is relatively thin, a direct blow or severe infection may cause it to rupture, spilling blood into the peritoneal cavity. Under such conditions the spleen must be removed quickly (a procedure called a *splenectomy*) and the splenic artery tied off to prevent life-threatening hemorrhage and shock. Surgical removal of the spleen seems to create few problems because the liver and bone marrow take over most of its functions. However, post-splenectomy patients do have a greater risk of certain bacterial infections. In children younger than 12, the spleen will regenerate if a small part of it is left in the body. ■

Thymus

The bilobed **thymus** (thi'mus) has important functions primarily during the early years of life. In infants, it is found in the inferior neck and extends

into the mediastinum of the superior thorax, where it partially overlies the heart deep to the sternum (see Figures 21.5 and 21.7). By secreting hormones (thymosin and thymopoietin), the thymus causes T lymphocytes to become immunocompetent; that is, it enables them to function against specific pathogens in the immune response (see Chapter 22). The size of the thymus varies with age. Prominent in newborns, it continues to increase in size during childhood, when it is most active. During adolescence its growth stops, and it starts to atrophy gradually. By old age it has been replaced almost entirely by fibrous and fatty tissue and is difficult to distinguish from surrounding connective tissue.

To understand thymic histology, it helps to compare the thymus to a cauliflower head—the flowerets represent *thymic lobules*, each containing an outer cortex and an inner medulla (Figure 21.7). The vast majority of thymic cells are lymphocytes. In the cortical regions the rapidly dividing lymphocytes are densely packed, but a few macrophages are scattered among them. The lighter-staining medullary areas contain fewer lymphocytes; they also contain some rather bizarre structures called **Hassall's** or **thymic corpuscles.** Hassall's corpuscles appear to be areas of degenerating cells, but their significance is unknown. Because the thymus lacks B cells, it has no follicles.

The thymus differs from other lymphoid organs in two other important ways. (1) It functions strictly in T lymphocyte maturation and thus is the only lymphoid organ that does not *directly* fight antigens. In fact, the so-called *blood-thymus barrier* keeps bloodborne antigens from leaking into the cortical regions to prevent premature activation of the immature lymphocytes. (2) The stroma of the thymus

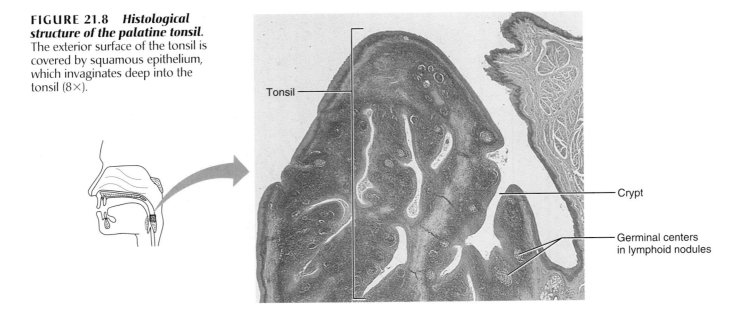

FIGURE 21.8 *Histological structure of the palatine tonsil.* The exterior surface of the tonsil is covered by squamous epithelium, which invaginates deep into the tonsil (8×).

Tonsil

Crypt

Germinal centers in lymphoid nodules

consists of star-shaped epithelial cells rather than reticular fibers. These **thymocytes** secrete the hormones that stimulate the lymphocytes to become immunocompetent.

Tonsils

The **tonsils** are perhaps the simplest lymphoid organs. They form a ring of lymphatic tissue around the entrance to the pharynx (throat), where they appear as swellings of the mucosa (Figure 21.5, and Figure 23.3 on p. 837). The tonsils are named according to location. The **palatine tonsils** are located on either side at the posterior end of the oral cavity. These are the largest of the tonsils and the most often infected. The **lingual tonsil** lies at the base of the tongue. The **pharyngeal tonsil** (referred to as the *adenoids* if enlarged) is in the posterior wall of the nasopharynx. The tiny **tubal tonsils** surround the openings of the auditory tubes into the pharynx. The tonsils gather and remove many of the pathogens entering the pharynx in inhaled air or in food.

The lymphoid tissue of the tonsils contains follicles with obvious germinal centers surrounded by diffusely scattered lymphocytes. The tonsil masses are not fully encapsulated, and the epithelium overlying them invaginates deep into their interior, forming blind-ended structures called **crypts** (Figure 21.8). The crypts trap bacteria and particulate matter, and the bacteria work their way through the mucosal epithelium into the lymphoid tissue, where

most are destroyed. It seems a bit dangerous to "invite" infection this way, but this strategy produces a wide variety of immune cells with a "memory" for the trapped pathogens. Thus, the body takes a calculated risk early on (during childhood) for the benefits of heightened immunity and better health later.

Aggregates of Lymphoid Follicles

Peyer's (pi′erz) **patches,** large isolated clusters of lymph follicles structurally similar to the tonsils, are found in the wall of the distal portion (ileum) of the small intestine (Figures 21.5 and 21.9). Similarly, lymphoid follicles are heavily concentrated in the wall of the **appendix,** a tubular offshoot of the first part (cecum) of the large intestine. Peyer's patches and the appendix are in an ideal position (1) to destroy bacteria (which are present in large numbers in the intestine), thereby preventing these pathogens from breaching the intestinal wall, and (2) to generate many "memory" lymphocytes for long-term immunity. Peyer's patches, the appendix, and the tonsils, all located in the digestive tract, and lymphoid nodules in the walls of the bronchi (organs of the respiratory tract), are part of the collection of small lymphoid tissues referred to as **mucosa-associated lymphatic tissue (MALT).** Collectively, MALT acts to protect the digestive and respiratory tracts from the never-ending onslaughts of foreign matter entering those cavities.

DEVELOPMENTAL ASPECTS OF THE LYMPHATIC SYSTEM

By the fifth week of embryonic development, the beginnings of the lymphatic vessels and the main clusters of lymph nodes are apparent. These arise from the budding of **lymph sacs** from developing veins. The first of these, the *jugular lymph sacs,* arise at the junctions of the internal jugular and subclavian veins and form a branching system of lymphatic vessels throughout the thorax, upper extremities, and head. The two main connections of the jugular lymph sacs to the venous system are retained and become the right lymphatic duct and, on the left, the superior part of the thoracic duct. Caudally the elaborate system of abdominal lymphatics buds largely from the primitive inferior vena cava. The lymphatics of the pelvic region and lower extremities form from sacs on the iliac veins.

Except for the thymus, which is an endodermal derivative, the lymphatic system organs develop from mesoderm. Mesodermal mesenchymal cells migrate to characteristic body sites, where they develop into reticular tissue. The thymus is the first lymphoid organ to appear. It forms as an outgrowth of the lining of the primitive pharynx. It then detaches and migrates caudally into the thorax where it becomes infiltrated with immature lymphocytes derived from the hematopoietic tissues elsewhere in the embryo's body. Except for the spleen and tonsils, the lymphoid organs are poorly developed before birth. Shortly after birth, they become heavily populated by lymphocytes, and their development parallels the maturation of the immune system (see Chapter 22). There is some evidence that the embryonic thymus produces hormones that control the development of the other lymphoid organs.

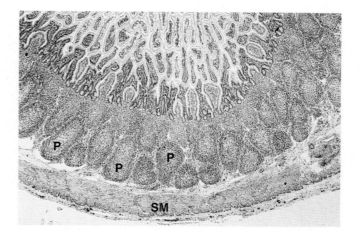

FIGURE 21.9 *Peyer's patches.* Histological structure of aggregated lymphoid follicles—Peyer's patches (P)—in the wall of the ileum of the small intestine (SM = submucosa) (20×).

* * *

Although the functions of the lymphatic vessels and lymphoid organs overlap, each helps to maintain body homeostasis in unique ways, as summarized in *Making Connections.* The lymphatic vessels help to maintain blood volume. The macrophages of lymphoid organs remove and destroy foreign matter in the lymphatic stream and blood. Additionally, lymphoid organs and tissues provide sites from which the immune system can be mobilized. In Chapter 22, we continue this story as we examine the inflammatory and immune responses that allow us to resist the constant barrage of pathogens.

Nervous System

- Lymphatic vessels pick up leaked plasma fluid and proteins in PNS structures; immune cells protect PNS structures from specific pathogens
- The nervous system innervates larger lymphatics; opiate neuropeptides influence immune functions; the brain helps regulate immune response

Endocrine System

- Lymphatic vessels pick up leaked fluids and proteins; lymph distributes hormones; immune cells protect endocrine organs
- The thymus produces hormones that promote development of lymphatic organs and "program" T lymphocytes; stress hormones depress immune activity

Cardiovascular System

- Lymphatic vessels pick up leaked plasma and proteins; spleen destroys aged RBCs, stores iron and platelets, and removes debris from blood; immune cells protect cardiovascular organs from specific pathogens
- Blood is the source of lymph; lymphatics develop from veins; blood provides the route for circulation of immune elements

Respiratory System

- Lymphatic vessels pick up leaked fluids and proteins from respiratory organs; immune cells protect respiratory organs from specific pathogens; the tonsils and plasma cells in the respiratory mucosa (which secrete the antibody IgA) prevent pathogen invasion
- The lungs provide oxygen needed by lymphoid/immune cells and eliminate carbon dioxide; the pharynx houses the tonsils; the respiratory "pump" aids lymph flow

Digestive System

- Lymphatic vessels pick up leaked fluids and proteins from digestive organs; lymph transports some products of fat digestion to the blood; lymphoid follicles in the wall of the intestine prevent invasion of pathogens
- The digestive system digests and absorbs nutrients needed by cells of lymphatic organs; gastric acidity inhibits pathogens' entry into blood

Urinary System

- Lymphatic vessels pick up leaked fluid and proteins from urinary organs; immune cells protect urinary organs from specific pathogens
- Urinary system eliminates wastes and maintains homeostatic water/acid-base/electrolyte balance of the blood for lymphoid/immune cell functioning; urine flushes some pathogens out of the body

Reproductive System

- Lymphatic vessels pick up leaked fluid and proteins from reproductive organs; immune cells protect against pathogens
- Reproductive organs' hormones may influence immune functioning; acidity of vaginal secretions is bacteriostatic

Integumentary System

- Lymphatic vessels pick up leaked plasma fluid and proteins from the dermis; lymphocytes in lymph enhance the skin's protective role by defending against specific pathogens via the immune response
- The skin's keratinized epithelium provides a mechanical barrier to pathogens; Langerhans' cells and dermal macrophages act as antigen presenters in the immune response; acid pH of skin secretions inhibits growth of bacteria on the skin

Skeletal System

- Lymphatic vessels pick up leaked plasma fluid and proteins from the periostea; immune cells protect bones from pathogens
- The bones house hematopoietic tissue which produces the lymphocytes (and macrophages) that populate the lymphoid organs and provide body immunity

Muscular System

- Lymphatic vessels pick up leaked fluids and proteins from skeletal muscle; immune cells protect muscles from pathogens
- The skeletal muscle "pump" aids the flow of lymph; heat produced during muscle activity initiates feverlike effects; muscles protect superficial lymph nodes

All living body cells are bathed by lymph. Nonetheless, the lymphatic system with its ghostly vessels and well-hidden (for the most part) organs and tissues is elusive and hard to pin down. Like a ground mole that tunnels throughout your yard, you know it's there, but you never see it as it quietly goes about its business. Since hormones made by endocrine organs are released into the extracellular space, and lymphatic vessels sequester the fluid containing them, there is little question that lymph is an important medium for delivering hormones throughout the body. Lymph plays a similar delivery role as it delivers fats absorbed by the digestive organs.

Our immune system, charged with protecting the body from specific pathogens, is often considered a separate and independent functional system. However, it is impossible to divorce the immune system from the lymphatic system because lymphoid organs are the programming sites and seedbeds for the immune cells and provide crucial vantage points for monitoring blood and lymph for the presence of foreign substances. Less understood are the intimate interactions between the immune cells that populate the lymphoid tissues and the nervous and endocrine systems, but this interesting topic is deferred to Chapter 22 to give you the opportunity to consider the immune system in detail (see *A Closer Look,* p. 828).

Although the lymphatic system serves the entire body, its chief "master" is the cardiovascular system. It is this relationship that we will explore further here.

Cardiovascular System

By now you are probably fairly comfortable with the fact that the lymphatic system picks up and returns leaked fluid and proteins to the blood vascular system. So,

what's the big deal? We ingest water in virtually everything we eat and drink—lost water is replaceable in a few minutes, right? The answer, of course, is "true." But what would still be missing is the leaked proteins, and these plasma proteins (most made by the liver) take time and energy to make. Since plasma proteins play a *major* role in keeping fluid in blood vessels (or encouraging its return), without them our blood vessels would contain too little fluid to support blood circulation. And without blood circulation, the whole body would die for lack of oxygen and nutrients and drown in its own wastes. Lymphoid organs also help to maintain the health and purity of blood—lymph nodes filter microorganisms and other debris from lymph before it is allowed to reenter the blood, and the spleen performs the same cleansing service for blood. In addition, the spleen disposes of inefficient, deformed, or aged erythrocytes.

However, this relationship is not entirely one-way. Lymphatic vessels spring from veins of the cardiovascular system, and blood delivers oxygen and nutrients to all body organs, including lymphoid ones. Blood also provides a means for (1) rapid transport of lymphocytes (immune cells) that continuously patrol the body for foreign substances and (2) broad distribution of antibodies (made by B cells and plasma cells), which bind to foreign substances and immobilize them until they can be disposed of by phagocytosis or other means. Furthermore, the endothelial cells of capillaries express surface protein "signals" (integrins/adhesion molecules) that lymphocytes can recognize when the surrounding area is injured or infected. Hence, capillaries guide immune cells to the sites where their protective services are needed, and the extremely porous walls of the postcapillary venules allow them to slip through the vessel wall to get there.

CLINICAL CONNECTIONS

Lymphatic System/Immunity

Case study: Back to following the progress of Mr. Hutchinson, we learn that the routine complete blood count (CBC) performed on admission revealed that his leukocyte count is dangerously low and that follow-up lab tests show that his lymphocytes are deficient. One day postsurgery, he complains of pain in his right ring finger (that hand had a crush injury). When examined, it was noted that the affected finger and the dorsum of the right hand were edematous, and red streaks radiated superiorly on his right forearm. Higher than normal doses of antibiotics are prescribed, and a sling is applied to the affected arm. Nurses are instructed to apply gloves and gown when giving Mr. Hutchinson his care.

Relative to these observations:

1. What do the red streaks emanating from the bruised finger indicate? What would you conclude his problem was if there were no red streaks but the right arm was very edematous?

2. Why is it important that Mr. Hutchinson not move the affected arm excessively (i.e., why was the sling ordered)?

3. How might the low lymphocyte count, megadoses of antibiotics, and orders for additional clinical staff protection be related?

4. Do you predict that Mr. Hutchinson's recovery will be uneventful or problematic? Why?

(Answers in Appendix F)

RELATED CLINICAL TERMS

Elephantiasis (el″lĕ-fan-ti′ah-sis) Typically a tropical disease in which the lymphatics (particularly those of the male's lower limbs and scrotum) become clogged with parasitic worms; swelling (due to edema) reaches enormous proportions.

Hodgkin's disease A malignancy of the lymph nodes; symptoms include swollen, nonpainful lymph nodes, fatigue, and often persistent fever and night sweats. Characterized by presence of giant cells of uncertain origin called Reed-Sternberg cells. Etiology unknown; treated with radiation therapy; high cure rate.

Lymphadenopathy (lim-fad″ĕ-nop′ah-the; *adeno* = a gland; *pathy* = disease) Any disease of the lymph nodes.

Lymphangiography (lim-fan″je-og′rah-fe) Diagnostic procedure in which the lymphatic vessels are injected with radiopaque dye and then visualized with X rays.

Lymphoma Any neoplasm (tumor) of the lymphoid tissue, whether benign or malignant.

Mononucleosis A viral disease common in adolescents and young adults; symptoms include fatigue, fever, sore throat, and swollen lymph nodes. Caused by the Epstein-Barr virus, which is transmitted in saliva ("kissing disease") and specifically attacks B lymphocytes. This attack leads to a massive activation of T lymphocytes, which in turn attack the virus-infected B cells. Large numbers of oversized T lymphocytes circulate in the bloodstream.

(These lymphocytes were originally misidentified as monocytes: *mononucleosis* = a condition of monocytes.) Usually lasts 4 to 6 weeks.

Non-Hodgkin's lymphoma Includes all cancers of lymphoid tissues except Hodgkin's disease. Involves uncontrolled multiplication and metastasis of undifferentiated lymphocytes, with swelling of the lymph nodes, spleen, and Peyer's patches; other organs may eventually become involved. The fifth most common cancer. A high-grade type, which primarily affects young people, grows quickly but is responsive to chemotherapy; up to a 50% remission rate. A low-grade type, which affects the elderly, is resistant to chemotherapy and so is often fatal.

Sentinel node The first node that receives lymph drainage from a body area suspected of being cancerous. When examined for presence of cancer cells, this node gives the best indication of whether metastasis through the lymph vessels has occurred.

Splenomegaly (sple″no-meg′ah-le; *mega* = big) Enlargement of the spleen due to accumulation of infectious microorganisms; typically caused by septicemia, mononucleosis, malaria, and leukemia.

Tonsillitis (ton″sĭ-li′tis; *itis* = inflammation) Congestion of the tonsils, typically with infecting bacteria, which causes them to become red, swollen, and sore.

CHAPTER SUMMARY

Media study tools that could provide you additional help in reviewing specific key topics of Chapter 21 are referenced below. **SP** *= Study Partner;* **IP** *= Interactive Physiology.*

1. Lymphatic vessels, lymph nodes, and other lymphoid organs and tissues make up the lymphatic system. This system returns fluids that have leaked from the blood vascular system back to the blood, protects the body by removing foreign material from the lymph stream, and provides a site for immune surveillance.

Lymphatic Vessels (pp. 778–780)

Distribution and Structure of Lymphatic Vessels (pp. 778–779)

1. Lymphatic vessels form a one-way network—lymphatic capillaries, collecting vessels, trunks, and ducts—in which fluid flows only toward the heart. The right lymphatic duct drains lymph from the right arm and right side of the upper body; the thoracic duct receives lymph from the rest of the body. These ducts empty into the blood vascular system at the junction of the internal jugular and subclavian veins in the neck.

Lymph Transport (pp. 779–780)

2. The flow of lymphatic fluid is slow; it is maintained by skeletal muscle contraction, pressure changes in the thorax, and (possibly) contractions of the lymphatic vessels. Backflow is prevented by valves.

3. Lymphatic capillaries are exceptionally permeable, admitting proteins and particulate matter from the interstitial space.

4. Pathogens and cancer cells may spread through the body via the lymphatic stream.

SP Exercise: Chapter 21, The Lymphatic System.

Lymphoid Cells, Tissues, and Organs: An Overview (pp. 781–782)

Lymphoid Cells (p. 781)

1. The cells in lymphoid tissues include lymphocytes (immunocompetent cells called T cells or B cells), plasma cells (antibody-producing offspring of B cells), macrophages (phagocytes that function in the immune response), and reticular cells that form the lymphoid tissue stroma.

Lymphoid Tissue (pp. 781–782)

2. Lymphoid tissue is reticular connective tissue. It houses macrophages and a continuously changing population of lymphocytes. It is an important element of the immune system.

3. Lymphoid tissue may be diffuse or packaged into dense follicles. Follicles often display germinal centers (areas where B cells are proliferating).

Lymphoid Organs (p. 782)

4. Lymphoid organs are discrete encapsulated structures containing both diffusely arranged and dense reticular tissue. The main lymphoid organs are lymph nodes, the spleen, thymus, tonsils, and follicle aggregates.

Lymph Nodes (pp. 782–783)

Structure of a Lymph Node (pp. 782–783)

1. Lymph nodes, clustered along lymphatic vessels, filter lymph. Each lymph node has a fibrous capsule, a cortex, and a medulla. The cortex contains mostly lymphocytes, which act in immune responses; the medulla contains macrophages, which engulf and destroy viruses, bacteria,

and other foreign debris, as well as lymphocytes and plasma cells.

SP Exercise: Chapter 21, Structure of a Lymph Node.

Circulation in the Lymph Nodes (p. 783)

2. Lymph enters the lymph nodes via afferent lymphatic vessels and exits via efferent vessels. There are fewer efferent vessels; therefore, lymph flow stagnates within the lymph node, allowing time for its cleansing.

Other Lymphoid Organs (pp. 783–786)

1. Unlike lymph nodes, the spleen, thymus, tonsils, and Peyer's patches do not filter lymph. However, most lymphoid organs contain both macrophages and lymphocytes.

Spleen (pp. 783–785)

2. The spleen provides a site for lymphocyte proliferation and immune function, and destroys aged or defective red blood cells and bloodborne pathogens. It also stores and releases the breakdown products of hemoglobin as necessary, stores platelets, acts as a hematopoietic site in the fetus.

Thymus (pp. 785–786)

3. The thymus is most functional during youth. Its hormones cause T lymphocytes to become immunocompetent.

Tonsils and Aggregates of Lymphoid Follicles (p. 786)

4. Peyer's patches of the intestinal wall, lymphatic follicles of the appendix, tonsils of the pharynx, and follicles in the bronchial walls of the respiratory tract are known as MALT (mucosa-associated lymphatic tissue). They prevent pathogens in the respiratory and digestive tracts from penetrating the mucous membrane lining.

Developmental Aspects of the Lymphatic System (p. 787)

1. Lymphatics develop as outpocketings of developing veins. The thymus develops from endoderm; the other lymphoid organs derive from mesenchymal cells of mesoderm.

2. The thymus, the first lymphoid organ to appear, plays a role in the development of other lymphoid organs.

3. Lymphoid organs are populated by lymphocytes, which arise from hematopoietic tissue.

SP Exercise: Chapter 21, Making Connections: The Lymphatic System; Case Study: Genetic Immunodeficiency.

REVIEW QUESTIONS

Multiple Choice/Matching

1. Lymphatic vessels (a) serve as sites for immune surveillance, (b) filter lymph, (c) transport leaked plasma proteins and fluids to the cardiovascular system, (d) are represented by vessels that resemble arteries, capillaries, and veins.

2. The saclike initial portion of the thoracic duct is the (a) lacteal, (b) right lymphatic duct, (c) cisterna chyli, (d) lymph sac.

3. Entry of lymph into the lymphatic capillaries is promoted by which of the following? (More than one answer may be appropriate.) (a) one-way minivalves formed by overlapping endothelial cells, (b) the respiratory pump, (c) the skeletal muscle pump, (d) greater fluid pressure in the interstitial space.

4. The structural framework of lymphoid organs is (a) areolar connective tissue, (b) hematopoietic tissue, (c) reticular tissue, (d) adipose tissue.

5. Lymph nodes are densely clustered in all of the following body areas *except* (a) the brain, (b) the axillae, (c) the groin, (d) the cervical region.

6. The germinal centers in lymph nodes are largely sites of (a) macrophages, (b) proliferating B lymphocytes, (c) T lymphocytes, (d) all of these.

7. The red pulp areas of the spleen are sites of (a) venous sinuses, macrophages, and red blood cells, (b) clustered lymphocytes, (c) connective tissue septa.

8. The lymphoid organ that functions primarily during youth and then begins to atrophy is the (a) spleen, (b) thymus, (c) palatine tonsils, (d) bone marrow.

9. Collections of lymphoid tissue (MALT) that guard mucosal surfaces include all of the following *except* (a) appendix nodules, (b) the tonsils, (c) Peyer's patches, (d) the thymus.

Short Answer Essay Questions

10. Compare and contrast blood, interstitial fluid, and lymph.

11. Compare the structure and functions of a lymph node to those of the spleen.

12. (a) What anatomical characteristic ensures that the flow of lymph through a lymph node is slow? (b) Why is this desirable?

Critical Thinking and Clinical Application Questions

1. Mrs. Jackson, a 59-year-old woman, has undergone a left radical mastectomy (removal of the left breast and left axillary lymph nodes and vessels). Her left arm is severely swollen and painful, and she is unable to raise it more than shoulder height. (a) Explain her signs and symptoms. (b) Can she expect to have relief from these symptoms in time? How so?

2. A friend tells you that she has tender, swollen "glands" along the left side of the front of her neck. You notice that she has a bandage on her left cheek that is not fully hiding a large infected cut there. Exactly what are her swollen "glands," and how did they become swollen?

22 THE IMMUNE SYSTEM: INNATE AND ADAPTIVE BODY DEFENSES

Faced by a hostile army of microorganisms swarming on our skin, and invaded by airborne bacteria and ravaging viruses, we stay amazingly healthy most of the time. The body seems to have evolved a single-minded approach toward such foes—if you're not with us, then you're against us. To implement that stance, it relies heavily on two intrinsic defense systems that act both independently and cooperatively to provide resistance to disease, or **immunity** (immun = free).

One of these systems, the **innate** (or **nonspecific**) **system,** like a lowly foot soldier, is always prepared—that is, it responds within minutes to protect the body from all foreign substances. Our innate defenses erect two "barricades." Their *first line of defense* is the external body membranes—intact skin and mucosae—that prevent entry of microorganisms. The *second line of defense,* called into action by chemical signals when the external defenses are penetrated, uses antimicrobial proteins as well as phagocytes and other cells to inhibit the invaders' spread throughout the body. The hallmark and most important mechanism of the second line of defense is inflammation.

The second protective arm, the **adaptive** (or **specific**) **defense system,** * is more like an elite fighting force, equipped with high-tech weapons that can be very specifically aimed. The adaptive immune system mounts the attack against *particular* foreign substances and provides the body's *third line of defense.* This defensive response takes considerably more time to mount than the innate response. Although we consider them separately, the specific and nonspecific defenses always work hand in hand.

Although certain organs of the body (notably lymphatic system organs) are intimately involved with our protective immune response, the **immune system** is a *functional system* rather than an organ system in an anatomical sense. Its "structures" are trillions of individual immune cells (especially lymphocytes), which inhabit lymphatic tissues and circulate in body fluids, and a diverse array of molecules.

When the immune system is operating effectively, it skillfully protects the body from most infectious microorganisms, cancer cells, and transplanted organs or grafts. The immune system does this both directly, by cell attack, and indirectly, by releasing mobilizing chemicals and protective antibody molecules.

■ PART 1 ■
INNATE (NONSPECIFIC) DEFENSES

You could say we come "fully equipped" with innate or nonspecific defenses, because they are part and parcel of our anatomy. The mechanical barriers that cover body surfaces and the cells and chemicals that act on the initial internal battlefronts are in place at birth, ready to ward off invading **pathogens** (harmful or disease-causing microorganisms). In many cases, our nonspecific defenses alone are able to destroy pathogens and ward off infection. In others, the adaptive immune system is called into action to reinforce and enhance the nonspecific mechanisms. In either case, the innate defenses effectively reduce the workload of the adaptive system by preventing entry and spread of microorganisms in the body. Table 22.2 on p. 801 summarizes the most important nonspecific defense mechanisms, which are detailed next.

SURFACE BARRIERS: SKIN AND MUCOSAE

The body's first line of defense—the *skin* and the *mucous membranes,* and the secretions these membranes produce—is highly effective. As long as the epidermis is unbroken, this heavily keratinized epithelial membrane presents a formidable physical barrier to most microorganisms that swarm on the skin. Keratin is also resistant to most weak acids and bases and to bacterial enzymes and toxins. Intact mucosae provide similar mechanical barriers within the body. Recall that mucous membranes line all body cavities that open to the exterior: the digestive, respiratory, urinary, and reproductive tracts. Besides serving as physical barriers, these epithelial membranes produce a variety of protective chemicals:

1. The acidity of skin secretions (pH of 3 to 5) inhibits bacterial growth and sebum contains chemicals that are toxic to bacteria. Vaginal secretions of adult females are also very acidic.

2. The stomach mucosa secretes a concentrated hydrochloric acid solution and protein-digesting enzymes. Both kill microorganisms.

3. Saliva, which cleanses the oral cavity and teeth, and lacrimal fluid, which washes external eye surfaces, contain **lysozyme,** an enzyme that destroys bacteria.

4. Sticky mucus traps many microorganisms that enter the digestive and respiratory passageways.

The respiratory tract mucosae also have structural modifications that counteract potential invaders. The mesh of tiny mucus-coated hairs inside the nose traps inhaled particles, and the mucosa of the upper respiratory tract is ciliated. The cilia sweep dust- and bacteria-laden mucus superiorly toward the mouth, preventing it from entering the lower respiratory passages, where the warm, moist environment provides an ideal site for bacterial growth.

Although the surface barriers are quite effective, they are breached occasionally by small nicks and cuts resulting, for example, from brushing your teeth or shaving. When this happens and microorganisms do invade deeper tissues, the *internal* innate defenses come into play.

INTERNAL DEFENSES: CELLS AND CHEMICALS

The body uses an enormous number of nonspecific cellular and chemical devices to protect itself. Some of these rely on the destructive powers of phagocytes and natural killer cells; others involve antimicrobial proteins that are present in blood or tissue fluid. The inflammatory response enlists the help of macrophages, mast cells, and all types of white blood cells, as well as dozens of chemicals that kill pathogens and help repair tissue. All of these protective ploys identify potentially harmful substances by recognizing surface carbohydrates unique to infectious organisms (bacteria, viruses, and fungi). Fever is also a nonspecific protective response. Although these are only a few of the possible protective mechanisms, they are among the most significant.

Phagocytes

Pathogens that get through the skin and mucosae into the underlying connective tissue are confronted by *phagocytes* (*phago* = eat). The chief phagocytes are **macrophages** ("big eaters"), which derive from circulating white blood cells called **monocytes.** Monocytes that leave the bloodstream and enter the tissues enlarge and develop into macrophages.

Free macrophages, like the alveolar macrophages of the lungs and dendritic cells of the epidermis, wander throughout the region in search of cellular debris or "foreign invaders." Others, like *Kupffer cells* in the liver and microglia of the brain, are *fixed macrophages* (permanent residents of particular or-

gans). Whatever their mobility, all macrophages are similar structurally and functionally.

Neutrophils, the most abundant type of white blood cell, also become phagocytic on encountering infectious material in the tissues. **Eosinophils,** still another type of white blood cell, are only weakly phagocytic. However, they are very important in defending the body against parasitic worms. When they encounter such parasites, they position themselves against the worm, and then discharge the destructive contents of their large cytoplasmic granules all over their prey. Recently **mast cells,** more associated with their role in allergies, have been shown to have a striking ability to bind with and ingest a wide range of bacteria, and then to kill the internalized bacteria. Although mast cells are not normally included in listings of professional phagocytes, they share their capabilities.

Mechanism of Phagocytosis

A phagocyte engulfs particulate matter much the way an amoeba ingests a food particle. Flowing cytoplasmic extensions bind to the particle (Figure 22.1) and then pull it inside, enclosed within a membrane-lined vacuole. The **phagosome** thus formed is then fused with a *lysosome* to form a **phagolysosome** (see steps ① – ③ in Figure 22.1b).

Phagocytic attempts are not always successful. To accomplish ingestion, **adherence** must occur. The phagocytes must first *adhere* or cling to the particle, a feat made possible by recognizing its carbohydrate "signature." This is particularly difficult with microorganisms such as *pneumococcus*, which have external capsules made of complex sugars. These pathogens can sometimes elude capture because phagocytes cannot bind to their capsule polysaccharides. Adherence is both more probable and more efficient when complement proteins and antibodies coat foreign particles, a process called **opsonization** (literally, "to make tasty"), because they provide sites or "handles" to which phagocyte receptors can bind.

Sometimes the way neutrophils and macrophages kill their ingested prey is much more than simple digestion of the engulfed microorganism by lysosomal enzymes. For example, some pathogens such as the tuberculosis bacillus and certain parasites are resistant to lysosomal enzymes and can even multiply within a vacuole inside the macrophage. However, when the macrophage is stimulated by chemicals released by cells of the adaptive immune system, additional enzymes are activated that produce the **respiratory burst,** an event that liberates a deluge of free

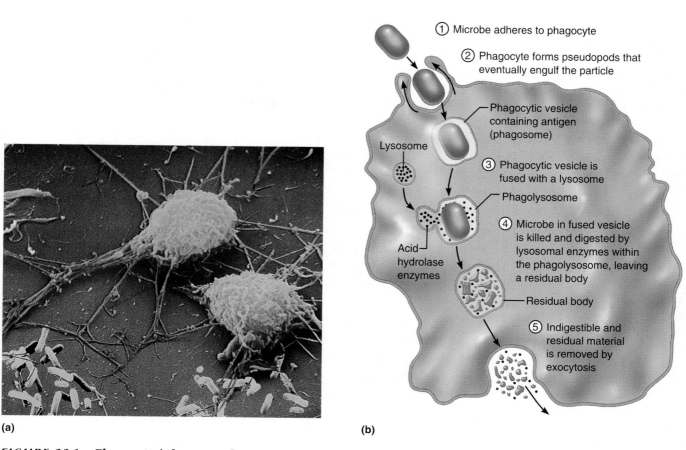

① Microbe adheres to phagocyte

② Phagocyte forms pseudopods that eventually engulf the particle

Phagocytic vesicle containing antigen (phagosome)

Lysosome

③ Phagocytic vesicle is fused with a lysosome

Phagolysosome

④ Microbe in fused vesicle is killed and digested by lysosomal enzymes within the phagolysosome, leaving a residual body

Acid hydrolase enzymes

Residual body

⑤ Indigestible and residual material is removed by exocytosis

(a)　　　　**(b)**

FIGURE 22.1 *Phagocytosis by macrophages.* (a) In this scanning electron micrograph (4300×), two macrophages are shown pulling sausage-shaped *E. coli* bacteria toward them with their long cytoplasmic extensions. Several bacteria on the macrophages' surfaces are being engulfed. (b) Events of phagocytosis.

radicals (including nitric oxide), which have potent cell-killing ability.

Neutrophils also produce antibiotic-like chemicals, called *defensins* (see p. 662), and cause a more widespread and thorough type of cell killing by releasing oxidizing chemicals and a substance identical to household bleach into the extracellular space. Unhappily, the neutrophils also destroy themselves in the process, whereas macrophages, which rely only on intracellular killing, can go on to kill another day. Excessively vigorous and prolonged neutrophil activity may even cause normal human tissues to become cancerous.

Natural Killer Cells

Natural killer (NK) cells, which "police" the body in blood and lymph, are a unique group of defensive cells that can lyse and kill cancer cells and virus-

infected body cells before the immune system is activated and enlisted in the fight. Sometimes called the "pit bulls" of the defense system, NK cells are part of a small but distinct group of *large granular lymphocytes*. Unlike lymphocytes of the adaptive immune system, which can recognize and react only against *specific* virus-infected or tumor cells, NK cells are far less picky. They can eliminate a variety of infected or cancerous cells, apparently by recognizing certain surface sugars. The name "natural" killer cells reflects the nonspecificity of their killing activity.

NK cells are not phagocytic. Their mode of killing involves an attack on the target cell's membrane and release of cytolytic chemicals called *perforins* (see p. 820). Shortly thereafter, channels appear in the target cell's membrane and its nucleus disintegrates rapidly. NK cells also secrete potent chemicals that enhance the inflammatory response.

Inflammation: Tissue Response to Injury

The **inflammatory response** is triggered whenever body tissues are injured. For example, it occurs in response to physical trauma (a blow), intense heat, and irritating chemicals, as well as to infection by viruses, fungi, and bacteria. The inflammatory response

1. Prevents the spread of damaging agents to nearby tissues

2. Disposes of cell debris and pathogens

3. Sets the stage for repair processes

The four *cardinal signs* of short-term, or acute, inflammation are *redness, heat* (*inflam* = set on fire), *swelling,* and *pain.* If the swollen and painful area is a joint, joint movement may be hampered temporarily. This forces the injured part to rest, which aids healing. Some authorities consider *impairment of function* to be the fifth cardinal sign of acute inflammation. We will see how these come about by examining the major events of the inflammatory process, which are depicted in Figure 22.2.

Vasodilation and Increased Vascular Permeability

The inflammatory process begins with a chemical "alarm"—a flood of inflammatory chemicals are released into the extracellular fluid. Injured tissue cells, phagocytes, lymphocytes, mast cells, and blood proteins are all sources of these inflammatory mediators, the most important of which are **histamine** (his′tah-mēn), **kinins** (ki′ninz), **prostaglandins (PGs)** (pros″tah-glan′dinz), **complement,** and **cytokines.** Though some of these mediators have individual inflammatory roles as well (see Table 22.1), they all cause small blood vessels in the vicinity to dilate. As more blood flows into the area, local **hyperemia** (congestion with blood) occurs, accounting for the *redness* and *heat* of an inflamed region.

The liberated chemicals also increase the permeability of local capillaries. Consequently, **exudate**—fluid containing proteins such as clotting factors and antibodies—seeps from the bloodstream into the tissue spaces. This exudate causes the local edema, or *swelling,* that in turn presses on adjacent nerve endings, contributing to a sensation of *pain.* Pain also results from the release of bacterial toxins, lack of nutrition to cells in the area, and the sensitizing effects of released prostaglandins and kinins. Aspirin and some other anti-inflammatory drugs produce their analgesic (pain-reducing) effects by inhibiting prostaglandin synthesis.

Although at first glance edema may seem to be detrimental, it isn't. The surge of protein-rich fluids into the tissue spaces (1) helps to dilute harmful substances that may be present, (2) brings in large quantities of oxygen and nutrients needed for repair, and (3) allows the entry of clotting proteins (Figure 22.2). The clotting proteins form a gel-like fibrin mesh in the tissue space that effectively isolates the

TABLE 22.1	Inflammatory Chemicals	
Chemical	*Source*	*Physiological effects*
Histamine	Granules of basophils and mast cells; released in response to mechanical injury, presence of certain microorganisms, and chemicals released by neutrophils	Promotes vasodilation of local arterioles; increases permeability of local capillaries, promoting exudate formation
Kinins (bradykinin and others)	A plasma protein, kininogen, is cleaved by the enzyme kallikrein found in plasma, urine, saliva, and in lysosomes of neutrophils and other types of cells; cleavage releases active kinin peptides	Same as for histamine; also induce chemotaxis of leukocytes and prompt neutrophils to release lysosomal enzymes, thereby enhancing generation of more kinins; induce pain
Prostaglandins (PGs)	Fatty acid molecules produced from arachidonic acid; found in all cell membranes; generated by lysosomal enzymes of neutrophils and other cell types	Sensitize blood vessels to effects of other inflammatory mediators; one of the intermediate steps of prostaglandin generation produces free radicals, which themselves can cause inflammation; induce pain
Platelet-derived growth factor	Secreted by platelets and endothelial cells	Stimulates fibroblast activity and repair of damaged tissues
Complement	See Table 22.2 (p. 801)	
Cytokines	See Table 22.4 (pp. 818–819)	

? *Why is it important that the capillaries become leaky during the inflammatory response?*

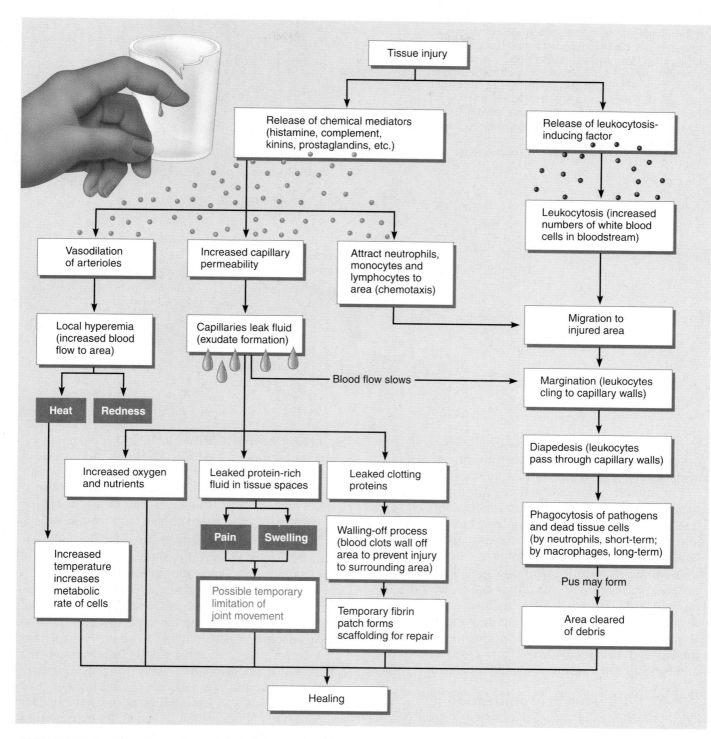

FIGURE 22.2 *Flowchart of events in inflammation.* The four cardinal signs of acute inflammation are shown in red boxes, as is limitation of joint movement, which constitutes a fifth cardinal sign (impairment of function) in some cases.

FIGURE 22.3 *Events of phagocyte mobilization.*
When leukocytosis-inducing factors are released by
injured cells at an inflammatory site, ① in this example
neutrophils are released rapidly from the bone marrow
to the blood. Because fluid is lost from the bloodstream
at the inflammatory site, blood flow slows locally and
neutrophils begin to roll along the vascular endothelium. ② Margination begins to occur as cell
adhesion molecules (CAMs) of the neutrophils cling to
those of the endothelial cells of the capillary walls, and
then ③ diapedesis occurs as they squeeze through
the capillary walls. Continued migration toward the site
of release of the inflammatory chemicals is ④ positive
chemotaxis. Once at the inflamed site, the phagocytes
help rid the area of pathogens and cellular debris.

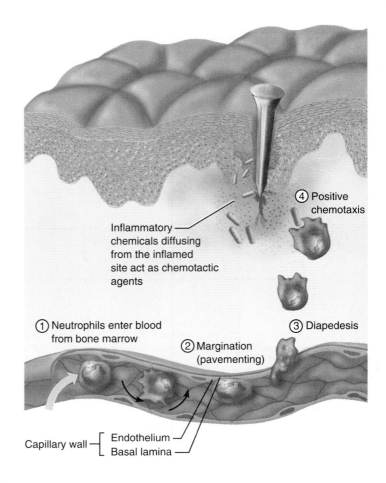

injured area and prevents the spread of bacteria and
other harmful agents into surrounding tissues. It
also forms a scaffolding for permanent repair.

At inflammatory sites where an epithelial barrier
has been breached, an additional chemical player enters the battle—β-defensins. These broad-spectrum
antibiotic-like chemicals are continually present in
epithelial mucosal cells in small amounts and help
to maintain the sterile environment of the body's internal passageways (urinary tract, respiratory
bronchi, etc.). However, when the mucosal surface is
abraded or penetrated and the underlying connective
tissue becomes inflamed, the output of β-defensins
increases dramatically, helping to control bacterial
and fungal colonization in the exposed area.

Phagocyte Mobilization

Soon after inflammation begins, the damaged area is
invaded by more phagocytes—mast cells lead, followed by neutrophils and then macrophages (Figure
22.3). If the inflammation was provoked by
pathogens, a group of plasma proteins known as
complement (discussed shortly) is activated and elements of adaptive immunity (lymphocytes and antibodies) also invade the injured site. The process by
which phagocytes are mobilized to infiltrate the
injured site has three main phases:

① **Leukocytosis.** Chemicals called **leukocytosis-
inducing factors** released by injured cells promote rapid
release of neutrophils from red bone marrow, and within
a few hours the number of neutrophils in the bloodstream may increase as much as four to five fold. This
leukocytosis is a characteristic sign of inflammation.

② **Margination.** Because of the tremendous outpouring of fluid from the blood into the injured area, blood
flow in the region slows and the neutrophils begin to
roll along the vascular endothelium as if "tasting" the
local environment. In inflamed areas, the endothelial
cells sprout specific cell adhesion molecules (CAMs)
called *selectins*. These provide footholds (signal that
"this is the place") for other CAMs (integrins) on the
surfaces of the neutrophils. When the complementary
CAMs connect and bind together, the neutrophils cling
to the inner walls of the capillaries and postcapillary
venules. This phenomenon is known as **margination**
or **pavementing.**

③ **Diapedesis.** The continued chemical signaling
prompts the neutrophils to squeeze through the capillary walls, a process called **diapedesis** or **emigration.**

④ **Chemotaxis.** Neutrophils usually migrate randomly, but inflammatory chemicals act as homing
devices, or more precisely **chemotactic agents** (Figure
22.3), that attract them and other white blood cells to

the site of the injury. Within an hour after the inflammatory response has begun, neutrophils have collected at the injury site and are actively devouring any foreign material present.

As the counterattack continues, monocytes follow neutrophils into the area. Monocytes are fairly poor phagocytes, but within 8–12 hours after leaving the bloodstream and entering the tissues, they swell and develop large numbers of lysosomes, becoming macrophages with insatiable appetites. These late-arriving macrophages continue to wage the battle, replacing the neutrophils on the battlefield. Macrophages are the central actors in the final disposal of cell debris as an acute inflammation subsides, and they predominate at sites of prolonged, or *chronic,* inflammation. The ultimate goal of an inflammatory response is to clear the injured area of pathogens, dead tissue cells, and any other debris so that tissue can be repaired. Once this is accomplished, healing usually occurs quickly.

⚠ *Homeostatic Imbalance*

In severely infected areas, the battle takes a considerable toll on both sides, and creamy, yellow *pus* may accumulate in the wound. **Pus** is a mixture of dead or dying neutrophils, broken-down tissue cells, and living and dead pathogens. If the inflammatory mechanism fails to fully clear the area of debris, the sac of pus may be walled off by collagen fibers, forming an *abscess.* Surgical drainage of abscesses is often necessary before healing can occur.

Bacteria, such as tuberculosis bacilli, that are resistant to digestion by macrophages escape the effects of antibiotics by remaining snugly enclosed within their macrophage hosts. In such cases, *infectious granulomas* form. These tumorlike growths contain a central region of infected macrophages surrounded by uninfected macrophages and an outer fibrous capsule. A person may harbor pathogens walled off in granulomas for years without displaying any symptoms of disease. However, if the person's resistance to infection is compromised or reduced, the bacteria may be activated and break out, leading to clinical disease symptoms. ▪

Antimicrobial Proteins

A variety of **antimicrobial proteins** enhance the innate defenses by attacking microorganisms directly or by hindering their ability to reproduce. The most important of these are interferon and complement proteins (Table 22.2, p. 801).

Interferon

Viruses—essentially nucleic acids surrounded by a protein coat—lack the cellular machinery to generate ATP or synthesize proteins. They do their "dirty

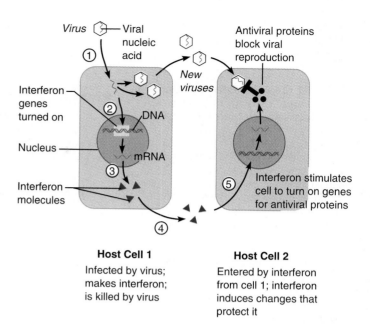

FIGURE 22.4   *The interferon mechanism against viruses.*

work" or damage in the body by invading tissue cells and taking over the cellular metabolic machinery needed to reproduce themselves. Although the virus-infected cells can do little to save themselves, they can help protect cells that have not yet been infected by secreting small proteins called **interferons** (in"ter-fēr'onz), or **IFNs.** The interferon molecules diffuse to nearby cells, where they stimulate synthesis of a protein known as *PKR,* which then "interferes" with viral replication in those cells by blocking protein synthesis at the ribosomes (Figure 22.4). Interferon's protection is not *virus-specific,* so interferon produced against a particular virus protects us against a variety of other viruses.

Interferon is actually a family of related proteins, each with slightly different physiological effects, that is produced by a variety of body cells. Lymphocytes secrete gamma (γ) or immune, interferon, but most other leukocytes secrete alpha (α) interferon. Fibroblasts secrete beta (β) interferon. Besides their antiviral effects, interferons activate macrophages and mobilize natural killer cells. Because both macrophages and natural killer cells can act directly against malignant cells, the interferons play some anticancer role. β interferon, in particular, acts to reduce inflammation in injured areas.

Interferon has also found a niche as an antiviral agent. FDA-approved *alpha IFN* is used to treat genital warts (caused by the herpes virus) and it is the first drug to have some success in combating hepatitis C (spread via blood and sexual intercourse), the most common and most dreaded form of hepatitis. Interferon is also being used against devastating viral infections in organ transplant patients.

? *Which pathway can occur without the need for an immune response?*

FIGURE 22.5 *Events and results of complement activation.* Complement can be activated by either the classical pathway or the alternative pathway. The classical pathway, which is mediated by 11 complement proteins designated C1–C9 (the C1 complex incorporates three of these proteins) requires antigen-antibody interactions. The alternative pathway occurs when plasma proteins called factors B, D, and P interact with the cell wall polysaccharides of certain bacteria and fungi. The two pathways converge to activate C3 (by cleaving it into C3a and C3b), which initiates a common terminal sequence that generates most of the biological activities of complement. Once C3b binds to the target cell's surface, it enzymatically initiates the remaining steps of complement activation. These steps result in the incorporation of MAC, the membrane attack complex (components C5b, and C6 to C9), into the target cell's membrane, creating a funnel-shaped lesion that causes cell lysis. Bound C3b also enhances phagocytosis (a function called opsonization). Released C3a, C5a, and other products of complement activation (C4a) promote the well-known events of inflammation (histamine release, increased vascular permeability, and so on).

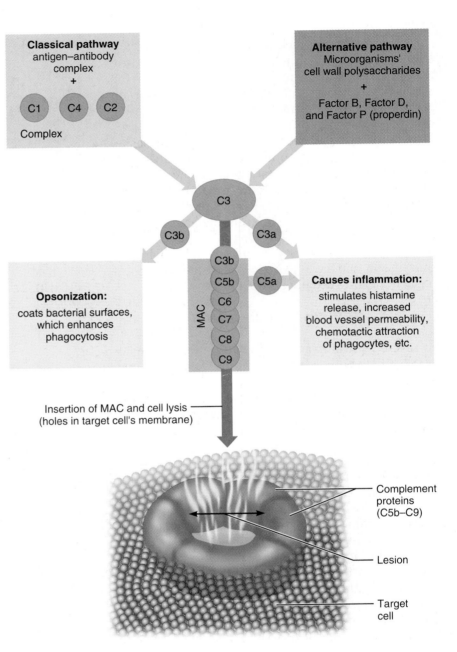

Classical pathway
antigen–antibody
complex
+
C1 C4 C2
Complex

Alternative pathway
Microorganisms'
cell wall polysaccharides
+
Factor B, Factor D,
and Factor P (properdin)

C3

C3b   C3a

Opsonization:
coats bacterial surfaces,
which enhances
phagocytosis

C3b
C5b   C5a
C6
MAC C7
C8
C9

Causes inflammation:
stimulates histamine
release, increased
blood vessel permeability,
chemotactic attraction
of phagocytes, etc.

Insertion of MAC and cell lysis
(holes in target cell's membrane)

Complement
proteins
(C5b–C9)

Lesion

Target
cell

Complement

The term **complement system,** or simply **complement,** refers to a group of at least 20 plasma proteins that normally circulate in the blood in an inactive state. These proteins include C1 through C9 (C1 is actually three proteins), and factors B, D, and P, plus several regulatory proteins. Complement provides a major mechanism for destroying foreign substances in the body. When it is activated, chemical mediators are unleashed that amplify virtually all aspects of the inflammatory process. Another effect is the killing of bacteria and certain other cell types by cell lysis. (Luckily our own cells are equipped with proteins that inactivate complement.) Although complement itself is a nonspecific defensive mechanism, it "complements" or enhances the effectiveness of *both* nonspecific and specific defenses.

Complement can be activated by two pathways as outlined in Figure 22.5: the classical pathway or the alternative pathway. The **classical pathway** is

TABLE 22.2	Summary of Nonspecific Body Defenses
Category/associated elements	**Protective mechanism**

First Line of Defense: Surface Membrane Barriers

Intact skin epidermis	Forms mechanical barrier that prevents entry of pathogens and other harmful substances into body
▪ Acid mantle	Skin secretions (perspiration and sebum) make epidermal surface acidic, which inhibits bacterial growth; sebum also contains bactericidal chemicals
▪ Keratin	Provides resistance against acids, alkalis, and bacterial enzymes
Intact mucous membranes	Form mechanical barrier that prevents entry of pathogens
▪ Mucus	Traps microorganisms in respiratory and digestive tracts
▪ Nasal hairs	Filter and trap microorganisms in nasal passages
▪ Cilia	Propel debris-laden mucus away from lower respiratory passages
▪ Gastric juice	Contains concentrated hydrochloric acid and protein-digesting enzymes that destroy pathogens in stomach
▪ Acid mantle of vagina	Inhibits growth of most bacteria and fungi in female reproductive tract
▪ Lacrimal secretion (tears); saliva	Continuously lubricate and cleanse eyes (tears) and oral cavity (saliva); contains lysozyme, an enzyme that destroys microorganisms
▪ Urine	Normally acid pH inhibits bacterial growth; cleanses the lower urinary tract as it flushes from the body

Second Line of Defense: Nonspecific Cellular and Chemical Defenses

Phagocytes	Engulf and destroy pathogens that breach surface membrane barriers; macrophages also contribute to immune response
Natural killer cells	Promote cell lysis by direct cell attack against virus-infected or cancerous body cells; do not depend on specific antigen recognition
Inflammatory response	Prevents spread of injurious agents to adjacent tissues, disposes of pathogens and dead tissue cells, and promotes tissue repair; chemical mediators released attract phagocytes (and immunocompetent cells) to the area
Antimicrobial proteins	
▪ Interferons (α, β, γ)	Proteins released by virus-infected cells that protect uninfected tissue cells from viral takeover; mobilize immune system
▪ Complement	Lyses microorganisms, enhances phagocytosis by opsonization, and intensifies inflammatory and immune responses
Fever	Systemic response initiated by pyrogens; high body temperature inhibits microbial multiplication and enhances body repair processes

linked to the immune system because it depends on the binding of antibodies to the invading organisms and the subsequent binding of C1 to the antigen-antibody complexes, a step called **complement fixation** (described on page 812). The **alternative pathway** is triggered by interaction between factors B, D, and P and polysaccharide molecules present on the surface of certain microorganisms.

Each pathway involves a cascade in which complement proteins are activated in an orderly sequence—each causing catalysis of the next step. Both the classical and the alternative pathways converge on C3, cleaving it into two fragments (C3a and C3b). This event initiates a common terminal pathway that causes cell lysis, promotes phagocytosis, and enhances inflammation.

The lytic series of events begins when C3b binds to the target cell's surface, triggering the insertion of a group of complement proteins called **MAC (membrane attack complex)** into the cell's membrane. MAC, in turn, forms and stabilizes an open hole in the target cell's membrane that ensures lysis by interfering with the cell's ability to eject Ca^{2+}.

The C3b molecules that coat the microorganism provide "handles" that receptors of macrophages and neutrophils can adhere to, allowing them to engulf the particle more rapidly. As noted earlier, this very important process is called *opsonization*. C3a and other cleavage products produced during complement fixation amplify the inflammatory response by stimulating mast cells and basophils to release histamine and by attracting neutrophils and other inflammatory cells to the area.

Fever

Inflammation represents a localized response to infection, but sometimes the body's response to the invasion of microorganisms is more widespread. **Fever,** or abnormally high body temperature, is a systemic response to invading microorganisms. As described further in Chapter 25, body temperature is regulated by a cluster of neurons in the hypothalamus, commonly referred to as the body's thermostat. Normally, the thermostat is set at approximately 36.2°C (98.2°F). However, the thermostat is reset upward in response to chemicals called **pyrogens** (*pyro* = fire), secreted by leukocytes and macrophages exposed to bacteria and other foreign substances in the body.

High fevers are dangerous because excess heat inactivates (denatures) enzymes. On the other hand, mild or moderate fever is an adaptive response that seems to benefit the body. Bacteria require large amounts of iron and zinc to multiply, but during a fever the liver and spleen sequester these nutrients, making them less available. Fever also increases the metabolic rate of tissue cells in general, speeding up repair processes.

■ PART 2 ■
ADAPTIVE (SPECIFIC)
DEFENSES

Most of us would find it wonderfully convenient if we could walk into a single clothing store and buy a complete wardrobe—hat to shoes—that fit us "to a T" regardless of any special figure problems. We *know* that such a service would be next to impossible to find. And yet we take for granted our **adaptive immune system,** the body's built-in *specific defensive system,* which stalks and eliminates with nearly equal precision almost any type of pathogen that intrudes into the body.

Essentially, the adaptive immune system is a *functional* system that recognizes specific foreign substances and acts to immobilize, neutralize, or destroy them. When it operates effectively, this system protects us from a wide variety of infectious agents, as well as from abnormal body cells. When it fails, malfunctions, or is disabled, some of the most devastating diseases—such as cancer, rheumatoid arthritis, and AIDS—may result. The activity of the adaptive immune system tremendously amplifies the inflammatory response and is responsible for most complement activation. At first glance, the adaptive system seems to have a major shortcoming. Unlike the innate defenses that are always ready and able to react, this arm of the immune system must "meet" or be primed by an initial exposure to a specific foreign substance (antigen) before it can protect the body against that substance, and this priming takes precious time.

The basis of this specific immunity was revealed in the late 1800s, when researchers demonstrated that animals surviving a serious bacterial infection have protective factors in their blood (now known to be proteins called *antibodies*) that defend them against future attacks by the same pathogen. Furthermore, it was shown that if antibody-containing serum from the surviving animals was injected into animals that had *not* been exposed to the pathogen, those animals would also be protected. These landmark experiments were exciting because they revealed three important aspects of the adaptive immune response:

1. It is antigen-specific: It recognizes and is directed against *particular* antigens, that is, against pathogens or foreign substances that initiate the immune response.

2. It is systemic: Immunity is not restricted to the initial infection site.

3. It has "memory": After an initial exposure, it recognizes and mounts even stronger attacks on the previously encountered pathogens.

At first antibodies were thought to be the sole artillery of the adaptive immune system, but then in the mid-1900s it was discovered that injection of antibody-containing serum did *not* always protect the recipient from diseases the donor had survived. In such cases, however, injection of the donor's lymphocytes *did* provide immunity. As the pieces began to fall into place, two separate but overlapping arms of adaptive immunity were recognized, each using a variety of attack mechanisms that vary with the intruder.

Humoral (hu'mor-ul) **immunity,** also called **antibody-mediated immunity,** is provided by antibodies present in the body's "humors," or fluids (blood, lymph, etc.). Though antibodies are produced by lymphocytes (or their offspring), they circulate freely in the blood and lymph, where they bind primarily to bacteria and their toxins and to free viruses, inactivating them temporarily and marking them for destruction by phagocytes or complement.

When lymphocytes themselves defend the body, the immunity is called **cellular** or **cell-mediated immunity** because the protective factor is living cells. The cellular arm also has cellular targets—virus- or parasite-infected tissue cells, cancer cells, and cells of foreign grafts. The lymphocytes act against such targets either *directly,* by lysing the foreign cells, or *indirectly,* by releasing chemical mediators that enhance the inflammatory response or activate other lymphocytes or macrophages. Thus, as you can see, although both arms of the adaptive immune system respond to virtually the same antigens, they do it in very different ways.

Before we describe the humoral and cell-mediated responses individually, we will consider the *antigens* that trigger the activity of the remarkable cells involved in these immune responses.

ANTIGENS

Antigens (an'tĭ-jenz), or **Ags,** are substances that can mobilize the immune system and provoke an immune response. As such, they are the ultimate targets of all immune responses. Most antigens are large, complex molecules (both natural and man-made) that are not normally present in the body. Consequently, as far as our immune system is concerned, they are intruders, or **nonself.**

Complete Antigens and Haptens

Complete antigens exhibit two important functional properties:

1. Immunogenicity is the ability to stimulate proliferation of specific lymphocytes and antibody production. (The word *antigen* is a contraction of "*anti*body *gen*erating," which refers to this particular antigenic property.)

2. Reactivity is the ability to react with the products of these reactions, that is, with the activated lymphocytes and the antibodies released in response to them.

An almost limitless variety of foreign molecules can act as complete antigens, including virtually all foreign proteins, nucleic acids, some lipids, and many large polysaccharides. Of these, proteins are the strongest antigens. Pollen grains and microorganisms—such as bacteria, fungi, and virus particles—are immunogenic because their surface membranes (or coats in the case of viruses) bear many different foreign macromolecules.

As a rule, small molecules—such as peptides, nucleotides, and many hormones—are not immunogenic. But if they link up with the body's own proteins, the adaptive immune system may recognize the combination as foreign and mount an attack that is harmful rather than protective. (These reactions, called *allergies,* are described later in the chapter.) In such cases, the troublesome small molecule is called a **hapten** (hap'ten; *haptein* = grasp) or **incomplete antigen.** Haptens are reactive but not immunogenic unless they are attached to protein carriers. Besides certain drugs (particularly penicillin), chemicals that act as haptens are found in

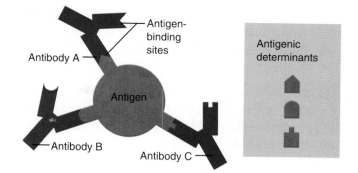

FIGURE 22.6 *Antigenic determinants.* Antibodies bind to antigenic determinants on the surface of an antigen. Most antigens express several different antigenic determinants; thus, different antibodies can bind to the same complex antigen. In this example, three specific types of antibodies react with different antigenic determinants on the same antigen molecule. (Note that each antigenic determinant would actually be represented many times.)

poison ivy, animal dander, some detergents, cosmetics, and a number of common household and industrial products.

Antigenic Determinants

The ability of a molecule to act as an antigen depends both on its size and complexity. Only certain parts of an entire antigen, called **antigenic determinants,** are immunogenic. Free antibodies or activated lymphocytes bind to these sites in much the same manner that an enzyme binds to a substrate.

Most naturally occurring antigens have a variety of antigenic determinants (Figure 22.6) on their surfaces, some more potent than others in provoking an immune response. Because different antigenic determinants are "recognized" by different lymphocytes, a single antigen may mobilize several lymphocyte populations and may stimulate formation of many different kinds of antibodies against it. Large proteins have hundreds of chemically different antigenic determinants, which accounts for their high immunogenicity and reactivity. However, large simple molecules such as plastics, which have many identical, regularly repeating units (and thus are not chemically complex), have little or no immunogenicity. Such substances are used to make artificial implants (hip replacements and others) because they are not seen as foreign and rejected by the body.

Self-Antigens: MHC Proteins

The external surfaces of all our cells are dotted with a huge variety of protein molecules. Assuming our immune system has been properly "programmed," these **self-antigens** are not foreign or antigenic to us, but they are strongly antigenic to other individuals.

(This is the basis of transfusion reactions and graft rejection.) Among these cell surface proteins is a specific group of glycoproteins that marks a cell as *self.* These proteins, called **MHC proteins,** are coded for by genes of the **major histocompatibility complex (MHC).** Because millions of different combinations of these genes are possible, it is unlikely that any two people except identical twins have the same MHC-encoded proteins. Essentially, there are two major groups of MHC proteins, distinguished by their cellular sites. Class I MHC proteins are found on virtually *all* body cells, whereas *class II MHC proteins* are displayed *only by certain cells* that act in the immune response.

Each MHC molecule has a deep groove that typically displays a peptide. In healthy cells, all of the displayed peptides come from the breakdown of cellular proteins during normal protein recycling and tend to be quite diverse. However, in infected cells, the MHC proteins also bind to fragments of (foreign) antigens and, as described shortly, this event plays a crucial role in mobilizing the immune system.

CELLS OF THE ADAPTIVE IMMUNE SYSTEM: AN OVERVIEW

The three crucial cell types of the adaptive immune system are two distinct populations of lymphocytes, and **antigen-presenting cells** or **APCs. B lymphocytes** or **B cells** oversee humoral immunity. **T lymphocytes** or **T cells** are non-antibody-producing lymphocytes that constitute the cell-mediated arm of immunity. Unlike the two types of lymphocytes, APCs do not respond to specific antigens but instead play essential auxiliary roles.

Lymphocytes

Like all blood cells, lymphocytes originate in red bone marrow from hematopoietic stem cells (hemocytoblasts). When released from bone marrow, the immature lymphocytes are essentially identical (Figure 22.8). Whether a given lymphocyte matures into a B cell or a T cell depends on where in the body it becomes **immunocompetent,** that is, able (or competent) to recognize a specific antigen by binding to it. T cells undergo this two- to three-day process in the thymus under the direction of thymic hormones. Within the thymus, the immature lymphocytes divide rapidly and their numbers increase enormously, but only the maturing T cells with the sharpest ability to identify *foreign* antigens survive.

This selection process requires both negative and positive selection (Figure 22.7). The vast majority of lymphocytes, those capable of binding strongly

What cell types act as APCs?

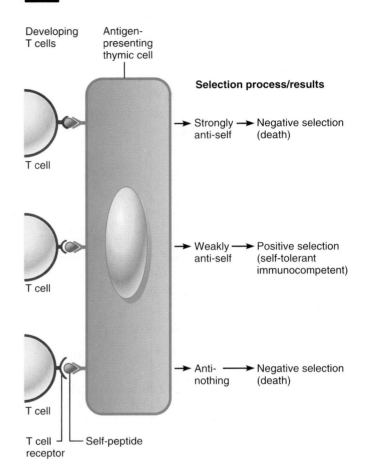

FIGURE 22.7   *T cell selection in the thymus.* T cell receptors with high affinity for self-peptides presented by the thymic cells and those that fail to interact with the presented self-peptide are selected against and die. Only T cells with intermediate affinity for the self-peptides leave the thymus.

to, and mounting a vigorous attack against, *self-antigens* (cell surface proteins of body cells)—are weeded out and destroyed, a process called *negative selection,* which occurs in the thymic medulla. This destruction of T cells that bind too strongly to one's own proteins is crucial because such T cells would attack self-tissues (in autoimmune reactions) if allowed to live. Lymphocytes that interact weakly with self-antigens (can bind without producing a strong activation signal) continue to develop. This latter process, called *positive selection,* occurs in the thymic cortex and selects (identifies) those T cells that will best recognize one's own tissue cells when they are bound to antigen. Thus, accomplishing immunocompetence and developing **self-tolerance,** or

Macrophages and dendritic cells, to name a few.

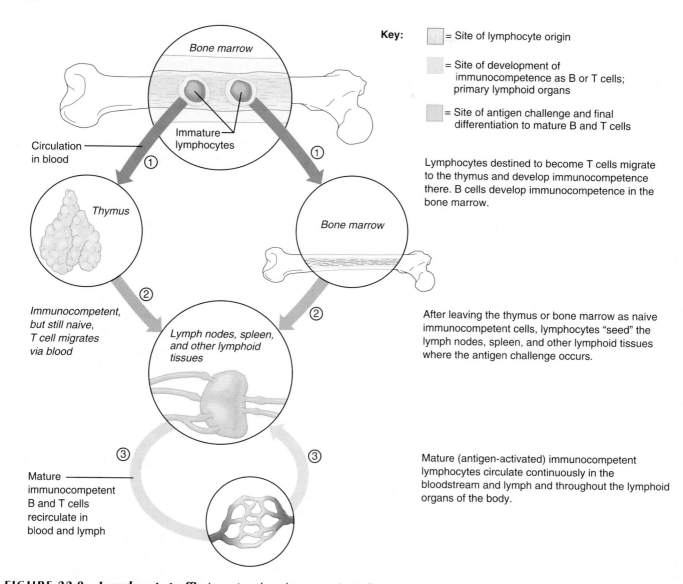

Key:

☐ = Site of lymphocyte origin

☐ = Site of development of immunocompetence as B or T cells; primary lymphoid organs

☐ = Site of antigen challenge and final differentiation to mature B and T cells

Lymphocytes destined to become T cells migrate to the thymus and develop immunocompetence there. B cells develop immunocompetence in the bone marrow.

After leaving the thymus or bone marrow as naive immunocompetent cells, lymphocytes "seed" the lymph nodes, spleen, and other lymphoid tissues where the antigen challenge occurs.

Mature (antigen-activated) immunocompetent lymphocytes circulate continuously in the bloodstream and lymph and throughout the lymphoid organs of the body.

FIGURE 22.8 *Lymphocyte traffic.* Immature lymphocytes arise in bone marrow. (Note that red marrow, from which the lymphocytes originate, is not found in the medullary cavity of the diaphysis of long bones in adults.)

relative unresponsiveness to self-antigens, are both essential parts of a lymphocyte's "education" during fetal life.

B cells become immunocompetent and self-tolerant in bone marrow, but little is known about the factors that control B cell maturation in humans. However, some self-reactive B cells are inactivated (a phenomenon called *anergy*), while others are killed outright. The lymphoid organs where the lymphocytes become immunocompetent (thymus and bone marrow) are called **primary lymphoid organs** (Figure 22.8). All other lymphoid organs are referred to as **secondary lymphoid organs.**

When B or T cells become immunocompetent, they display a unique type of receptor on their surfaces. These receptors (some 10^4 to 10^5 per cell) enable the lymphocyte to recognize and bind to a specific antigen. Once these receptors have appeared, the lymphocyte is committed to react to one distinct antigen, and one only, because *all* of its antigen receptors are the same. For example, receptors of one lymphocyte can recognize only a single antigenic determinant of the hepatitis A virus, those of another

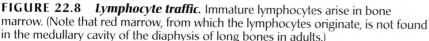

can bind only to a given determinant on pneumococcus bacteria, and so forth. Although the antigen receptors of T cells and B cells differ in overall structure, both T and B lymphocytes are capable of responding to the same antigens.

While many of the details of lymphocyte processing are still unknown, we know that lymphocytes become immunocompetent *before* meeting the antigens they may later attack. Thus, *it is our genes, not antigens, that determine what specific foreign substances our immune system will be able to recognize and resist.* Another way of saying the same thing: It is the immune cell receptors that constitute our genetically acquired knowledge of the microbes that are likely to be in our environment. An antigen determines only which of the existing T or B cells will proliferate and mount the attack against it. Only some of the possible antigens our lymphocytes are programmed to resist will ever invade our bodies. Consequently, only some members of our army of immunocompetent cells are mobilized in our lifetime. The others are forever idle.

After becoming immunocompetent, the naive (still immature) T cells and B cells are exported to the lymph nodes, spleen, and other secondary lymphoid organs, where the encounters with antigens occur (Figure 22.8). Then, when the lymphocytes bind with recognized antigens, they complete their differentiation into fully functional—mature and antigen-activated—T cells and B cells.

Antigen-Presenting Cells

The major role of **antigen-presenting cells (APCs)** in immunity is to engulf foreign particles and present fragments of these antigens, like signal flags, on their own surfaces where they can be recognized by T cells—that is, they act as *antigen presenters.* (This function is described in more detail shortly.) The major types of cells acting as antigen presenters are *dendritic cells* (which include the *interstitial cells* present in connective tissues throughout the body and the *Langerhans' cells* of the skin epidermis), *macrophages*, and *activated B lymphocytes.*

Notice that all of these cell types are in highly convenient sites to encounter and process antigens. Dendritic cells (DCs) are at the body's outposts or frontiers, best situated to act as mobile sentinels. The macrophages are widely distributed throughout the lymphoid organs and connective tissues. Besides their antigen-presenting role, dendritic cells and macrophages also secrete soluble proteins that activate T cells. Activated T cells, in turn, release chemicals that rev up the mobilization and maturation of DCs and prod macrophages to become *activated macrophages*, true "killers" that are insatiable

phagocytes and secrete bactericidal chemicals. As you will see, interactions between lymphocytes, and between lymphocytes and APCs, underlie virtually all phases of the immune response.

Although APCs and lymphocytes are found throughout each lymphoid organ, specific cells tend to be more numerous in certain areas. For example, the paracortical areas of lymph nodes house mostly T cells, and dendritic cells and B cells tend to populate the *germinal centers* (see Figure 21.4, p. 782), whereas relatively more macrophages are clustered around the medullary sinuses. Macrophages tend to remain fixed in the lymphoid organs, as if waiting for antigens to come to them. But lymphocytes, especially the T cells (which account for 65–85% of bloodborne lymphocytes), circulate continuously throughout the body (see Figure 22.8). This greatly increases a lymphocyte's chance of coming into contact with antigens located in different parts of the body, as well as with huge numbers of macrophages and other lymphocytes. Although this recirculation process appears to be random, the emigration of the lymphocytes into the tissues where their protective services are needed is highly specific, regulated by homing signals *(CAMs)* displayed on vascular endothelial cells.

Because lymph capillaries pick up proteins and pathogens from nearly all body tissues, immune cells in lymph nodes are in a strategic position to encounter a large variety of antigens. Lymphocytes and APCs in the tonsils act primarily against microorganisms that invade the oral and nasal cavities, whereas the spleen acts as a filter to trap bloodborne antigens.

The active migration of DCs to the lymph nodes and other secondary lymphoid organs may be an even more important way of ensuring that the immune cells encounter invading antigens, compared to T cell recirculation and passive delivery of antigens to lymphoid organs by lymphatics. With their long veil-like extensions, DCs are very efficient antigen catchers, and once they have internalized antigens by phagocytosis, they enter the nearby lymphatics to get to the lymphoid organs where they will present the antigens to T and B cells. Indeed, recent research has suggested that the DCs are the true initiators of adaptive immunity.

In summary, the adaptive immune system is a two-fisted defensive system that uses lymphocytes, APCs, and specific molecules to identify and destroy all substances—both living and nonliving—that are in the body but are not recognized as being part of the body, or self. Its response to such threats depends on the ability of its cells (1) to recognize foreign substances (antigens) in the body by binding to them, and (2) to communicate with one another so that the whole system mounts a response specific to those antigens.

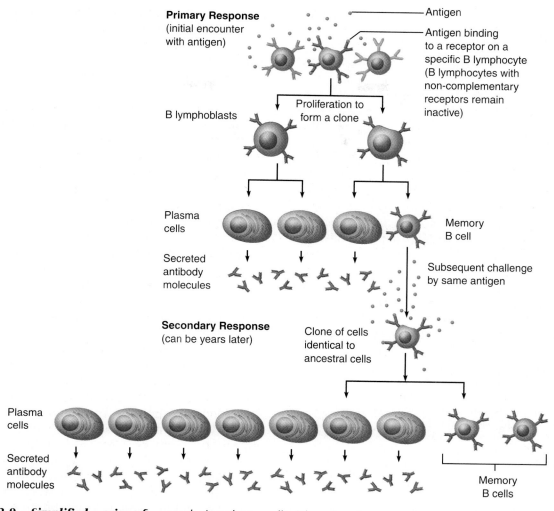

FIGURE 22.9   *Simplified version of clonal selection of a B cell stimulated by antigen binding*. The initial meeting stimulates the primary response in which the B cell proliferates rapidly, forming a clone of like cells, most of which differentiate into antibody-producing plasma cells. (Though not indicated in the figure, the production of mature plasma cells takes about 5 days and eight cell generations.) Cells that do not differentiate into plasma cells become memory cells, which are already primed to respond to subsequent exposures to the same antigen. Should such a meeting occur, the memory cells quickly produce more memory cells and larger numbers of effector plasma cells with the same antigen specificity. Responses generated by memory cells are called secondary responses.

HUMORAL IMMUNE RESPONSE

The **antigen challenge,** the first encounter between a naive immunocompetent lymphocyte and an invading antigen, usually takes place in the spleen or a lymph node, but may happen in any lymphoid tissue. If the lymphocyte is a B cell, the challenging antigen provokes the *humoral immune response,* in which antibodies are produced against the challenger.

Clonal Selection and Differentiation of B Cells

An immunocompetent but naive B lymphocyte is *activated* (stimulated to complete its differentiation) when antigens bind to its surface receptors and cross-link adjacent receptors together. Antigen binding is quickly followed by receptor-mediated endocytosis of the cross-linked antigen-receptor complexes. These activating events (plus some interactions with T cells described shortly) trigger **clonal** (klo'nul) **selection.** That is, they stimulate the B cell to grow and then to multiply rapidly to form an army of cells all exactly like itself and bearing the same antigen-specific receptors (Figure 22.9). The resulting family of identical cells, all descended from the *same* ancestor cell, is called a **clone.** It is the antigen that does the selecting in clonal selection by "choosing" a lymphocyte with complementary receptors.

Most cells of the clone become **plasma cells,** the antibody-secreting effector cells of the humoral response. Although B cells secrete limited amounts of antibodies, plasma cells develop the elaborate internal

? *Why is the secondary response so much faster than the primary response?*

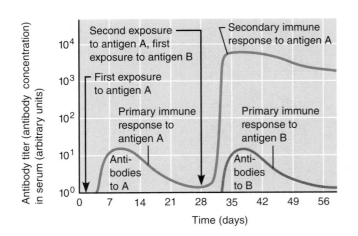

FIGURE 22.10 *Primary and secondary humoral responses.* The blue graph line shows both a primary response and a secondary response to antigen A. In the primary response, there is a time lag and then a gradual rise and fairly rapid decline in the level of antibodies in the blood. A second exposure to antigen A on day 28 elicits a secondary response that is both more rapid and more intense. Additionally, though not shown on the graph, antibody levels remain high for a much longer time. If a second antigen, antigen B, were also encountered at day 28, the reaction to that antigen would be a primary response, not a secondary one, and its graph line (red) would show a pattern similar to that seen during the primary response to antigen A. (Time shown on the horizontal axis is given for example only. The time for response to different antigens varies greatly.)

machinery (largely rough endoplasmic reticulum) needed to secrete antibodies at the unbelievable rate of about 2000 molecules per second. Each plasma cell functions at this breakneck pace for 4 to 5 days and then dies. The secreted antibodies, each with the same antigen-binding properties as the receptor molecules on the surface of the parent B cell, circulate in the blood or lymph, where they bind to free antigens and mark them for destruction by other specific or nonspecific mechanisms. The clone cells that do not differentiate into plasma cells become long-lived **memory cells** that can mount an almost immediate humoral response if they encounter the same antigen again (see Figure 22.9).

Immunological Memory

The cellular proliferation and differentiation just described constitute the **primary immune response,** which occurs on the first exposure to a particular antigen. The primary response typically has a lag period of several (3 to 6) days after the antigen challenge. This lag period mirrors the time required for the few B cells specific for that antigen to proliferate and for their offspring to differentiate into plasma cells. After the mobilization period, plasma antibody levels rise, reaching peak levels in about 10 days, and then begin to decline (Figure 22.10).

If (and when) someone is reexposed to the same antigen, whether it's the second or the twenty-second time, a **secondary immune response** occurs. Secondary immune responses are faster, more prolonged, and more effective, because the immune system has already been primed to the antigen, and sensitized memory cells are already in place and "on alert." These memory cells provide what is commonly called **immunological memory.**

Within hours after recognition of the "old enemy" antigen, a new army of plasma cells is being generated. Within 2 to 3 days the antibody titer in the bloodstream rises steeply to reach much higher levels than were achieved in the primary response. The antibodies of secondary responses not only bind with greater affinity (more tightly), but their levels in the blood remain high for weeks to months. (Hence, when the appropriate chemical signals are present, plasma cells can keep functioning for much longer than the 4 to 5 days seen in primary responses.) Memory cells persist for long periods in humans and many retain their capacity to produce powerful secondary humoral responses for a lifetime.

The same general phenomena occur in the cellular immune response: A primary response sets up a pool of activated lymphocytes (in this case, T cells) and generates memory cells that can then mount secondary responses.

Active and Passive Humoral Immunity

When your B cells encounter antigens and produce antibodies against them, you are exhibiting **active humoral immunity** (Figure 22.11). Active immunity is (1) *naturally acquired* during bacterial and viral infections, during which we may develop the symptoms of the disease and suffer a little (or a lot), and (2) *artificially acquired* when we receive **vaccines.** Indeed, once it was realized that secondary responses are so much more vigorous, the race was on to develop vaccines to "prime" the immune response by providing a first meeting with the antigen.

Most vaccines contain dead or *attenuated* (living, but extremely weakened) pathogens, or their components (parts of the microorganism or genetically engineered versions). Vaccines provide two benefits:

Sensitization and "tooling up" have already occurred. Hence, all that is required is that the memory cells proliferate to increase their number.

1. They spare us most of the symptoms and discomfort of the disease that would otherwise occur during the primary response.

2. Their weakened antigens provide functional antigenic determinants that are both immunogenic and reactive.

So-called *booster shots* that may intensify the immune response at later meetings with the same antigen are also available.

Vaccines have literally wiped out smallpox and have substantially lessened the illness caused by many childhood killers such as whooping cough, polio, and measles. Although vaccines have dramatically reduced hepatitis B, tetanus, and pneumonia in adults, immunization of adults in the U.S. has a much lower priority than that of children and over 50,000 Americans die each year from vaccine-preventable infections (pneumonia, influenza, and hepatitis).

Nonetheless, conventional vaccines have some shortcomings. Although it was originally believed that the response of the immune system was pretty much the same regardless of how the antigen got into the body (under its own power or via a vaccine), that has proved not to be the case. Apparently vaccines mainly target the type of helper T cell (more details shortly) that revs up B cell defenses and antibody formation (the T_H2 cell) as opposed to the type (T_H1) that generates strong cellular responses. As a result, lots of antibodies are formed that provide immediate protection, but cellular immunological memory is only poorly established. (The immune system is deprived of the learning experience that comes with clearing an infection via a T_H1-mediated response.) In rare cases, vaccines cause the very disease they are trying to prevent when the attenuated antigen isn't weakened enough (an infrequent consequence of polio vaccine). In other cases, they may trigger an allergic response. The new "naked DNA" antiviral vaccines, blasted into the skin with a gene gun, and *edible* vaccines taken orally appear to circumvent these problems.

Passive humoral immunity differs from active immunity, both in the antibody source and in the degree of protection it provides (see Figure 22.11). Instead of being made by your plasma cells, the antibodies are harvested (obtained) from the serum of an immune human or animal donor. As a result, your B cells are *not* challenged by antigens, immunological memory does *not* occur, and the protection provided by the "borrowed" antibodies ends when they naturally degrade in the body.

Passive immunity is conferred *naturally* on a fetus when the mother's antibodies cross the placenta and enter the fetal circulation. For several months after birth, the baby is protected from all the antigens to which the mother has been exposed. Passive

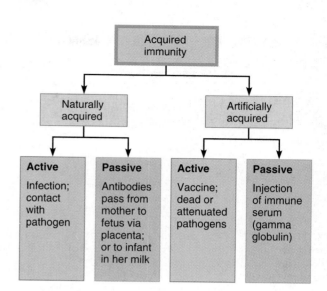

Why is no immunological memory established in passive forms of immunity?

FIGURE 22.11   *Types of acquired immunity*. Green boxes signify active types of immunity in which the immunological memory is established. Orange boxes signify short-lived passive types of immunity; no immunological memory is established.

immunity is *artificially* conferred when serum such as *gamma globulin* is infused. Gamma globulin is commonly administered after exposure to hepatitis. Other immune sera are used to treat poisonous snake bites (an antivenom), botulism, rabies, and tetanus (an antitoxin) because these rapidly fatal diseases would kill a person before active immunity could be established. The donated antibodies provide immediate protection, but their effect is short-lived (two to three weeks).

Antibodies

Antibodies, also called **immunoglobulins** (im″u-no-glob′u-linz), or **Igs**, constitute the **gamma globulin** part of blood proteins. As mentioned, antibodies are soluble proteins secreted by activated B cells or by plasma cells in response to an antigen, and they are capable of binding specifically with that antigen.

Antibodies are formed in response to an incredible number of different antigens. Despite their variety, all antibodies can be grouped into one of five Ig classes, each slightly different in structure and function. We will see how these Ig classes differ, but first let's take a look at how all antibodies are alike.

Because sensitization does not occur and no memory cells are generated. The Ig protection provided by passive immunity is donated and temporary.

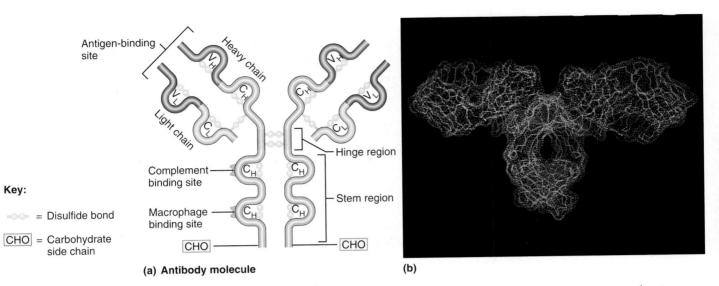

Key:

∽○∽ = Disulfide bond

[CHO] = Carbohydrate side chain

(a) Antibody molecule

(b)

FIGURE 22.12 Basic antibody structure. (a) Basic antibody structure consists of four polypeptide chains that are joined together by disulfide bonds (S—S). Two of the chains are short *light chains*; the other two are long *heavy chains*. Each chain has a V (variable) region (different in different antibodies) and a C (constant) region (essentially identical in different antibodies of the same class). The variable regions are the antigen–binding sites of the antibody; hence, each antibody monomer has two antigen–binding sites. **(b)** Computer-generated image of antibody structure. Each tiny colored dot (sphere) represents an individual amino acid of the polypeptide chains.

Basic Antibody Structure

Regardless of its class, each antibody has a basic structure consisting of four looping polypeptide chains linked together by disulfide (sulfur-to-sulfur) bonds (Figure 22.12). Two of these chains, the **heavy (H) chains,** are identical to each other and contain approximately 400 amino acids each. The other two chains, the **light (L) chains,** are also identical to each other, but are only about half as long. The heavy chains have a flexible *hinge* region at their approximate "middles." Notice that disulfide bonds also bind together amino acids within the same heavy or light chain (see Figure 22.12). Since the amino acids that are bound together are about 110 amino acids apart, the intervening parts of the polypeptide chains loop out. When the four chains are combined, they form a molecule, called an **antibody monomer** (mon'o-mer), with two identical halves, each consisting of a heavy and a light chain. The molecule as a whole is **T** or **Y** shaped.

Each of the four chains forming an antibody has a **variable (V) region** at one end and a much larger **constant (C) region** at the other end. Antibodies responding to different antigens have very different variable regions, but their constant regions are the same (or nearly so) in all antibodies of a given class. The variable regions of the heavy and light chains in each arm combine to form an **antigen-binding site** shaped to "fit" a specific antigenic determinant or an antigen. Hence, each antibody monomer has two such antigen-binding regions.

The constant regions that form the *stem* of the antibody monomer determine the antibody class and serve common functions in all antibodies: These are the *effector regions* of the antibody that dictate (1) the cells and chemicals the antibody can bind to, and (2) how the antibody class will function in antigen elimination. For example, some antibodies can fix complement, some circulate in blood while others are found primarily in body secretions, some cross the placental barrier, and so on.

Antibody Classes

The five major immunoglobulin classes are designated IgM, IgA, IgD, IgG, and IgE, on the basis of the C (constant) regions in their heavy chains. (Remember the name MADGE to recall the five Ig types.) As illustrated in Table 22.3, IgD, IgG, and IgE have the same basic Y-shaped structure and thus are *monomers*. IgA occurs in both monomer and *dimer* (two linked monomers) forms. (Only the dimer is shown in the table.) Compared to the other antibodies, IgM is a huge antibody, a *pentamer* (*penta* = five), constructed from five linked monomers.

The antibodies of each class have different biological roles and locations in the body. For example, IgM is the first antibody class that is *released* to the blood by plasma cells. IgG is the most abundant antibody in plasma and is the only Ig class that crosses the placental barrier. Hence, the passive immunity that a mother transfers to her fetus is "with the compliments" of her IgG antibodies. *Only* IgM and IgG can fix complement. The IgA dimer, also called **secretory IgA,** is found primarily in mucus and other secretions that bathe body surfaces. It plays a major role in preventing pathogens from gaining entry into the body. IgD is always bound to the B cell surface; hence it acts as a B cell receptor. IgE antibodies, almost never found in blood, are the "troublemaker" antibodies involved in some allergies. These and other characteristics unique to each of the immunoglobulin classes are summarized in Table 22.3.

Mechanisms of Antibody Diversity

It is estimated that our various plasma cells can make over a billion different types of antibodies. Since antibodies, like all proteins, are specified by genes, it would seem that an individual must have billions of genes. Not so; each body cell contains perhaps 100,000 genes that code for all the proteins the cells must make.

How can a limited number of genes generate a seemingly limitless number of different antibodies? Recombinant DNA studies have shown that the genes that actually dictate the structure of each antibody are not present as such in embryonic cells. Instead of a complete set of "antibody genes," embryonic cells contain a few hundred genetic bits and pieces that can be thought of as an "erector set for antibody genes." These gene segments are shuffled and combined in different ways (a process called **somatic recombination**) by each B cell as it becomes immunocompetent. The information of these newly assembled genes is then expressed in the surface receptors of B cells and in the antibodies later released by their plasma cell "offspring."

In an immature B cell, the gene segments coding for the light and heavy chains (exons) are physically separated by noncoding DNA segments (introns), but are located on the same chromosome. (See Chapter 3 if you need a refresher on this terminology.) The light and heavy chains are manufactured separately and then joined to form the antibody molecule. The events forming the light and heavy chains are similar, but the process for the heavy chains is more complex because of the larger variety of gene (DNA) segments for the constant regions of the heavy chains (remember, the C regions of the heavy chains determine antibody class). The random mixing of gene segments to make unique antibody genes specifying the H and L chains accounts for only part of the huge variability seen in antibodies. Certain areas of the V gene segments, the so-called *hypervariable regions*, are "hot spots" for somatic mutations that enormously increase antibody variation.

A single plasma cell can "switch" to make various kinds of heavy chains, producing two or more different antibody classes with the same antigen specificity. For example, the first antibody released in the primary response is IgM; then the plasma cell begins to secrete IgG. During secondary responses, almost all of the Ig protein is IgG.

TABLE 22.3	Immunoglobulin Classes
 IgD (monomer)	IgD is virtually always attached to the external surface of a B cell, where it functions as the antigen receptor of the B cell; important in B cell activation.
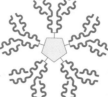 IgM (pentamer)	IgM exists in monomer and pentamer (five united monomers) forms. The monomer, which is attached to the B cell surface, serves as an antigen receptor. The pentamer (illustrated) circulates in blood plasma and is the first Ig class released by plasma cells during the primary response. (This fact is diagnostically useful because presence of IgM in plasma usually indicates current infection by the pathogen eliciting IgM's formation.) Because of its numerous antigen binding sites, IgM is a potent agglutinating agent and readily fixes and activates complement.
 IgG (monomer)	IgG is the most abundant and diverse antibody in plasma, accounting for 75–85% of circulating antibodies. It protects against bacteria, viruses, and toxins circulating in blood and lymph, readily fixes complement, and is the main antibody of both primary and secondary responses. It crosses the placenta and confers passive immunity from the mother to the fetus.
 IgA (dimer)	IgA monomer exists in limited amounts in plasma. The dimer (illustrated), referred to as secretory IgA, is found in body secretions such as saliva, sweat, intestinal juice, and milk, and helps prevent attachment of pathogens to epithelial cell surfaces (including mucous membranes and the epidermis).
 IgE (monomer)	IgE is slightly larger than the IgG antibody. It is secreted by plasma cells in skin, mucosae of the gastrointestinal and respiratory tracts, and tonsils. Its stem region becomes bound to mast cells and basophils, and when its receptor ends are triggered by an antigen, it causes the cells to release histamine and other chemicals that mediate inflammation and an allergic reaction. Typically only traces of IgE are found in plasma, but levels rise during severe allergic attacks or chronic parasitic infections of the gastrointestinal tract.

? *Both complement fixation and agglutination aid the process of phagocytosis, but they do so in different ways. What is different about their mechanisms of action?*

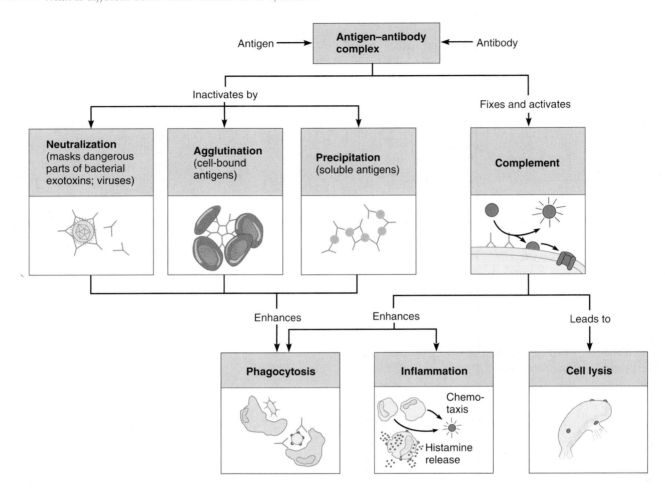

FIGURE 22.13 *Mechanisms of antibody action.* Antibodies act against free viruses, red blood cell antigens, bacterial toxins, and intact bacteria.

Antibody Targets and Functions

Though antibodies cannot themselves destroy antigen-bearing invaders, they can inactivate them and tag them for destruction (Figure 22.13). The common event in all antibody-antigen interactions is formation of **antigen-antibody** (or **immune**) **complexes.** The defensive mechanisms used by antibodies are neutralization, agglutination, precipitation, and complement fixation. Of these, complement fixation and neutralization are the most important.

As a result of complement fixation, C3b molecules are released and coat cellular antigens. These molecules provide roughness, or handles, to which the phagocyte can bind more easily, thus facilitating engulfment. Agglutination forms large clumps of the antigen which then precipitate and can't go anywhere, making them fair game for the phagocytes.

Complement fixation and activation is the chief antibody ammunition used against cellular antigens, such as bacteria or mismatched red blood cells. When antibodies bind to cells, they change shape to expose complement-binding sites on their constant regions (see Figure 22.12a). This triggers complement fixation into the antigenic cell's surface, which is followed by cell lysis. Additionally, as described earlier, molecules released during complement activation tremendously amplify the inflammatory response and promote phagocytosis via opsonization. Hence, a positive feedback cycle that enlists more and more defensive elements is set into motion.

Neutralization, the simplest effector mechanism, occurs when antibodies block specific sites on viruses or bacterial exotoxins (toxic chemicals secreted by bacteria). As a result, the virus or exotoxin loses its toxic effect because it cannot bind to receptors on tissue cells to cause injury. The antigen-antibody complexes are eventually destroyed by phagocytes.

Because antibodies have more than one antigen-binding site, they can bind to the same determinant on more than one antigen at a time; consequently, antigen-antibody complexes can be cross-linked into large lattices. When cell-bound antigens are cross-linked the process causes clumping, or **agglutination,** of the foreign cells. IgM, with ten antigen-binding sites (see Table 22.3), is an especially potent agglutinating agent. Recall that this type of reaction occurs when mismatched blood is transfused (the foreign red blood cells are clumped) and is the basis of tests used for blood typing (see p. 675). **Precipitation** is a similar mechanism, in which soluble molecules (instead of cells) are cross-linked into large complexes that settle out of solution. Like agglutinated bacteria, these immobilized (precipitated) antigen molecules are much more easily captured and engulfed by phagocytes than are freely moving antigens.

A quick and dirty way to remember how antibodies work is to remember they have a PLAN of action—**p**recipitation, **l**ysis (by complement), **a**gglutination, and **n**eutralization.

Monoclonal Antibodies

In addition to their use to provide passive immunity, commercially prepared antibodies are used in research, clinical testing, and treating certain cancers. **Monoclonal antibodies,** used for such purposes, are produced by descendants of a single cell, and are pure antibody preparations that are specific for a single antigenic determinant.

A current technology for making monoclonal antibodies involves fusion of tumor cells and B lymphocytes. The resulting cell hybrids, called **hybridomas** (hi″brĭ-do′mahz), have desirable traits of both parent cells. Like myeloma (tumor) cells, they proliferate indefinitely in culture; like B cells, they produce a single type of antibody.

Monoclonal antibodies are used to diagnose pregnancy, certain sexually transmitted diseases, some types of cancer (e.g., colon cancer), hepatitis, and rabies. Compared with conventional diagnostic tests for these conditions, monoclonal antibody tests are much more specific, sensitive, and rapid. Monoclonal antibodies are also being used to treat leukemia and lymphomas, cancers which are present in the circulation and thus easily accessible to injected antibodies, and as "guided missiles" to deliver anticancer drugs only to cancerous tissue.

CELL-MEDIATED IMMUNE RESPONSE

Despite their immense versatility, antibodies provide only partial immunity. Their prey is the *obvious* pathogen. They are fairly useless against infectious microorganisms such as the tuberculosis bacillus

that quickly slips inside body cells to multiply there. In these cases, the second (cell-mediated) arm of immunity comes into play.

The T cells that mediate cellular immunity are a diverse lot, much more complex than B cells in both classification and function. There are two major populations of effector T cells based on which of a pair of structurally related *cell differentiation glycoproteins*—CD4 or CD8—is displayed by a mature T cell. These cell surface receptors are distinct from the T cell antigen receptors, but they *do* play a role in interactions that occur between a tissue cell and other cells or foreign antigens. **CD4 cells,** also called T4 cells, are primarily **helper T cells** (T_H), whereas most **CD8 cells** (T8 cells) are **cytotoxic T cells** (T_C), whose role is to destroy any cells in the body that harbor anything foreign. In addition to these two major groups of T cells, there are delayed hypersensitivity T cells (T_{DH}) (a special type of T_H cell), suppressor T cells (T_S), and memory T cells. Details of T cells' roles are provided shortly.

Before going into the details of the cell-mediated immune response, we will recap some introductory information and compare the importance of the humoral and cellular responses in the overall scheme of adaptive immunity. The soluble antibodies, produced by plasma cells (B cell progeny), are in many ways the simplest ammunition of the immune response. Antibodies are specialized to latch onto intact bacteria and soluble foreign molecules *in extracellular environments*—that is, free in body secretions and tissue fluid and circulating in blood and lymph. Antibodies never invade the solid tissues unless a lesion is present. At the most basic level, antibody production and pathogen multiplication are in a race against each other and the fate of antibody targets is a case of "the quick or the dead." Remember, however, that forming antibody-antigen complexes does *not* destroy the antigens; instead, it prepares them for disposal by innate defenses and activated T cells.

In contrast to B cells and antibodies, T cells are unable to "see" free antigens or antigens that are in their natural state. T cells can recognize and respond only to *processed* fragments of protein antigens displayed on surfaces of body cells, and then only under specific circumstances. Consequently, T cells are best suited for cell-to-cell interactions, and most of their direct attacks on antigens (mediated by the cytotoxic T cells) target body cells infected by viruses, bacteria, or intracellular parasites; abnormal or cancerous body cells; and cells of infused or transplanted foreign tissues.

Clonal Selection and Differentiation of T Cells

The stimulus for clonal selection and differentiation is the same in B cells and T cells—binding of antigen. However, as described next, the mechanism by which T cells recognize "their" antigen is very different and has some unique restrictions.

Antigen Recognition and MHC Restriction

Like B cells, immunocompetent T cells are activated when the variable regions of their surface receptors bind to a "recognized" antigen. However, T cells must accomplish *double recognition:* They must simultaneously recognize nonself (the antigen) and self (a MHC protein of a body cell).

Two types of MHC proteins are important to T cell activation. **Class I MHC proteins** are displayed by virtually all body cells except red blood cells and are always recognized by CD8 T cells. After being synthesized at the endoplasmic reticulum (Figure 22.14a), class I MHC proteins bind with a peptide fragment 8 or 9 amino acids long ferried into the ER from the cytosol by special transport proteins aptly called TAPs (for **t**ransporter **a**ssociated with **a**ntigen **p**rocessing). The "loaded" class I MHC proteins then migrate to the plasma membrane to display their attached protein fragment. Class I MHC proteins display fragments of proteins synthesized *in the cell*—either bits of cellular (self) proteins or peptides derived from endogenous antigens. **Endogenous antigens** are foreign proteins that are synthesized within a body cell, such as the viral proteins produced by virus-infected cells or alien (mutated) proteins made by a body cell that has become cancerous.

Unlike the widely distributed class I proteins, class II MHC proteins are typically found only on the surfaces of mature B cells, some T cells, and antigen-presenting cells, where they enable the cells of the immune system to recognize one another. Like class I MHC proteins, these are synthesized at the ER and bind to peptide fragments (Figure 22.14b). However, the peptide fragments they bind are longer (14–17 amino acids) and they come from **exogenous antigens** (foreign antigens) that have been phagocytosed and broken down within the phagosome vesicle. While still in the ER, the class II MHC molecules become bound to the so-called *invariant protein,* which prevents them from binding to peptides while in the ER. Class II MHC proteins then move from the ER through the Golgi apparatus and then into the phagosomes. It is there that the invariant chain loosens its hold on the MHC molecule, allowing the "jaws" of MHC to close around the degraded protein fragments. Vesicles are then recycled to the cell surface, where the class II MHC proteins display their booty for recognition by CD4 cells.

The role of the MHC proteins in the immune response is strikingly important—they provide the means for signaling to immune system cells that infectious microorganisms are hiding within body cells. Without such a mechanism, viruses and certain bacteria that thrive in cells could multiply unbothered and unnoticed. When MHC proteins are complexed with fragments of our own (self) proteins, T cells passing by get the signal "Leave this cell alone, it's ours!" and simply ignore them. But when they are complexed to antigenic peptides (either exogenous or endogenous), they betray the invaders and "sound a molecular alarm" that in effect says "I (self) have been invaded" by (1) acting as antigen holders and (2) forming the self part of the self-antiself complexes that T cells must recognize to be activated.

T Cell Activation

T cell activation is actually a two-step process involving antigen binding and costimulation.

Step 1: Antigen Binding The first step mostly entails what has already been described: T cell antigen receptors or **TCRs** bind to an antigen–MHC protein complex on the surface of a body cell. Like B cell receptors, a TCR has variable and constant regions, but it consists of two (usually alpha and beta) rather than four polypeptide chains.

Helper and cytotoxic T cells have different preferences for the class of MHC protein that helps to deliver the activation signal. This constraint, acquired during the "education" process in the thymus, is called **MHC restriction.** Helper T cells (CD4) can bind only to antigens linked to class II MHC proteins, which are typically displayed on the surfaces of APCs. Dendritic cells are the most potent APCs for T_H cells. As described earlier on this page, APCs attach very small parts of the antigens to the MHC II proteins for T cell recognition. In addition, mobile APCs like Langerhans' cells (dendritic cells of the epidermis) actually migrate to the lymph nodes and other lymphoid tissues to do their antigen presentation. As a result of this early alert, the body is spared a good deal of tissue damage that might otherwise occur before the antigens finally entered the blood to be carried to the lymph nodes for recognition.

By contrast, cytotoxic T cells (CD8) are activated by antigen fragments complexed with MHC I proteins (Figure 22.15). Because virtually all body cells exhibit class I MHC proteins, cytotoxic T cells do not require *special* antigen-presenting cells—any body cell (target cell) displaying a recognizable self-antiself (MHC I–antigen fragment) complex will do. Nonetheless, antigen-presenting cells produce costimulatory molecules that *are* required for T_C cell activation. Hence, whether a helper or a cytotoxic T cell is involved, the process of antigen presentation

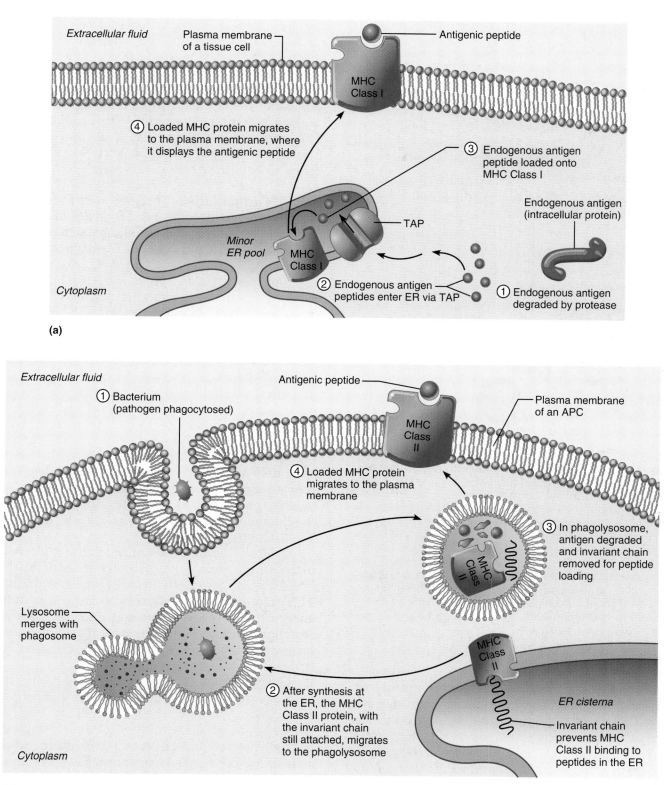

(a)

(b)

FIGURE 22.14   *MHC proteins, and antigen processing and display.*
(a) MHC class I proteins pick up peptide fragments in the ER. These peptides may be derived from self-proteins or foreign (viral or cancer) proteins made within the cell. The latter are called endogenous antigens. **(b)** MHC class II proteins pick up peptide fragments from endocytosed vesicles where phagocytized foreign debris is degraded. Both MHC classes display the peptides at the cell surface.

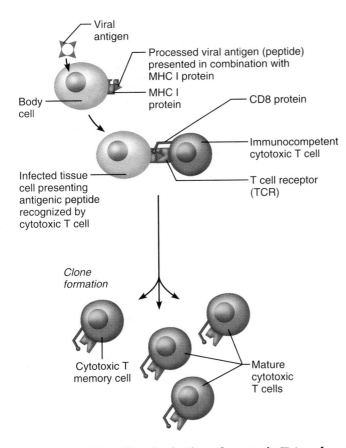

FIGURE 22.15   *Clonal selection of cytotoxic (T$_C$) and helper (T$_H$) T cells involves simultaneous recognition of self and antiself.* Cytotoxic T and helper cells are stimulated to proliferate (clone) and complete their differentiation when they bind to parts of foreign antigens complexed to MHC proteins. Immunocompetent T$_C$ cells are activated when they bind to endogenous antigens (antiself)—part of a virus in this example—complexed to a class I MHC protein. (Class I proteins are displayed on the surfaces of all body cells, enabling cytotoxic T cells to respond to infected or malignant body cells.) Activation of T$_H$ cells is similar, except that the processed antigen is complexed with a class II MHC protein, typically found only on antigen-presenting cells such as macrophages or dendritic cells.

is essentially the same—only the presenting cell type and the MHC class recognized differ. However, very slight variations in the antigenic peptides displayed can ultimately result in very different degrees of T cell activation. The process in which the T cells adhere to and crawl over the surface of other cells examining them for antigens they might recognize is sometimes called **immunologic surveillance.** (Note, however, that this term is commonly used to indicate *only* the NK and dendritic cells' constant monitoring of body cells for foreign antigens.)

The TCR that acts to recognize the antiself-self complex is linked to multiple intracellular signaling pathways. Besides the TCR, other T cell surface proteins are involved in this first step. For example, the CD4 and CD8 proteins used to identify the two ma-

jor T cell groups are adhesion molecules that help maintain the coupling during antigen recognition. Additionally, the CD4 and CD8 proteins are associated with kinase enzymes inside the T cell that phosphorylate cell proteins, activating some and inactivating others when antigen binding occurs. Once antigen binding has occurred, the T cell is stimulated but is still "idling," like a car that has been started but not put into gear. Such T cells are referred to as *naive T cells* at this stage.

Step 2: Costimulation   The story isn't over yet. Before a T cell can proliferate and form a clone, it must recognize one or more **costimulatory signals.** Sometimes the added requirement is T cell binding to still other surface receptors on an APC. For example, macrophages begin to sprout *B7 proteins* on their surfaces when the nonspecific defenses are being mobilized to fend off microorganisms. B7 binding to the CD$_{28}$ receptor on a T cell is a crucial costimulatory signal, particularly to helper T cells. Other costimulators include the cytokine chemicals released by macrophages or T cells; and interleukins 1 and 2 (described shortly) are important examples. As you might guess, there are several types of costimulators, which promote different responses in the activated T cells. The important understanding is that, depending on the receptors to which the costimulators bind, they nudge the T cells either to complete their activation process or to abort activation entirely. Or, to go back to our car analogy, they can (1) put the car in gear and step on the gas, or (2) put the brakes on T cell activity.

A T cell that binds to antigen without receiving the costimulatory signal becomes tolerant to that antigen and is unable to divide or secrete cytokines. This state of unresponsiveness to antigen is called **anergy.** Just why this happens is uncertain, but the current hypothesis is that the two-signal sequence is a safeguard to prevent the immune system from destroying our own healthy cells.

Once activated, a T cell enlarges and proliferates to form a clone of cells that differentiate and perform functions according to their T cell class. This primary response peaks within a week after a single exposure. A period of death then occurs during which activated T cells undergo apoptosis between days 7 and 30; the activated T cells die off and effector activity wanes as the amount of antigen declines. This wholesale disposal of activated effector cells has a critical protective role. Activated T cells are potential hazards. They produce huge amounts of inflammatory cytokines, which contribute to infection-driven hyperplasia of lymph nodes (an important forerunner of lymphoid malignancies). Additionally, once they've done their job the effector T cells are unnecessary and thus disposable. In every case, thousands

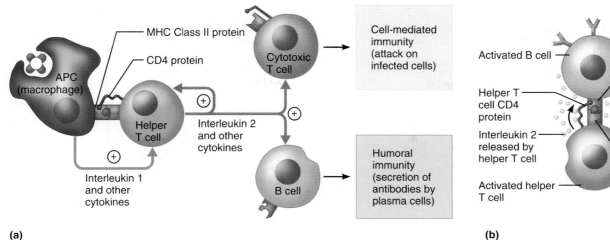

(a)

(b)

FIGURE 22.16 *The central role of helper T cells.* Helper T cells mobilize both arms (cellular and humoral) of the immune response. Once an immuno-competent helper T cell binds to an antigen-presenting cell (APC) such as a macrophage and recognizes the self-antiself complex, it produces a clone of helper T cells (not illustrated), all with identical receptors keyed to MHC II receptors complexed with the same antigen. **(a)** In addition, the macrophage releases interleukin 1, a costimulator that enhances T cell activation. Activated helper T cells release interleukin 2, a costimulator that stimulates proliferation and activity of other helper T cells (specific for the same antigenic determinant). Interleukin 2 also helps activate cytotoxic T cells and B cells. **(b)** T and B cells must sometimes interact directly for full B cell activation to occur. In such cases, helper T cells bind with the self-antiself complexes (processed antigen complexed with MHC proteins) on activated B cell surfaces, and then release interleukin as a costimulatory signal.

of clone members become memory T cells. They persist, perhaps for life, providing a reservoir of T cells that can mediate secondary responses to the same antigen. Because memory cells have more adhesion molecules, they can adhere to APCs more effectively than do the effector T cells of the primary response.

Cytokines

As in the inflammatory response, chemical mediators enhance the immune response. **Cytokines,** the mediators involved in cellular immunity, include hormonelike glycoproteins released by activated T cells and macrophages. As mentioned previously, some cytokines act as costimulators of T cells and T cell proliferation (see Figure 22.16). For example, **interleukin 1 (IL-1),** released by macrophages, costimulates bound T cells to liberate **interleukin 2 (IL-2)** and to synthesize more IL-2 receptors. Interleukin 2 is a key growth factor. Acting as a local hormone, it sets up a positive feedback cycle that encourages activated T cells to divide even more rapidly. (Therapeutic use of genetically engineered IL-2 enhances the activity of the body's own cytotoxic T cells against cancer.)

Additionally, all activated T cells secrete one or more other cytokines that help to amplify and regulate a variety of immune and nonspecific responses. Some cytokines (perforin and lymphotoxin) are cell toxins; others (e.g., gamma interferon) enhance the killing power of macrophages; and still others are inflammatory factors. Cytokines and their effects on target cells are summarized in Table 22.4 on pp. 818–819.

Specific T Cell Roles

Helper T Cells

Helper T cells are regulatory cells that play a central role in the immune response. Once primed by APC presentation of antigen, their major function is to chemically or directly stimulate proliferation of other T cells and of B cells that have already become bound to antigen (Figure 22.16). In fact, without the "director" helper T cells, there is *no* immune response. Their cytokines furnish the chemical help needed to recruit other immune cells to fight off intruders (Figure 22.16a).

Helper T cells also interact directly with B cells in lymph nodes that have mounted antigen fragments on their surfaces bound to MHC II receptors (Figure 22.16b), prodding them into more rapid division and then, like the boss of an assembly line, signaling for antibody formation to begin. Whenever a T_H cell binds to a B cell, the T cell releases interleukin 2 (and other cytokines). Although B cells may be activated solely by binding to certain antigens (**T cell–independent antigens),** most antigens require T cell "help" (costimulation) to activate the B cells to which they bind. This more common variety of antigens is called **T cell–dependent antigens.** In general, T cell–independent antigen responses are weak and short-lived. The B cell division process

TABLE 22.4	Summary of Functions of Cells and Molecules Involved in the Adaptive Immune Response
Element	**Function in immune response**
Cells	
B cell	Lymphocyte that resides in lymph nodes, spleen, or other lymphoid tissues, where it is induced to replicate by antigen binding and helper T cell interactions; its progeny (clone members) form memory cells and plasma cells
Plasma cell	Antibody-producing "machine"; produces huge numbers of antibodies (immunoglobulins) with the same antigen specificity; represents further specialization of B cell clone descendants
Helper T cell (T_H)	A regulatory T cell that binds with a specific antigen presented by an APC; upon circulating into spleen and lymph nodes, it stimulates production of other cells (cytotoxic T cells and B cells) to help fight invader; acts both directly and indirectly by releasing cytokines
Cytotoxic T cell (T_C)	Also called a cytolytic (CTL) or killer T cell; activated by antigen presented by any body cell; recruited and activity enhanced by helper T cells; its specialty is killing virus-invaded body cells and cancer cells; it is involved in rejection of foreign tissue grafts
Suppressor T cell (T_S)	Slows or stops activity of B and T cells once infection (or onslaught by foreign cells) has been conquered
Delayed hypersensitivity T cell (T_{DH})	Presumed to be a specific subgroup of T_H cells that promotes nonspecific cell killing by macrophages; important in delayed hypersensitivity reactions
Memory cell	Descendant of activated B cell or any class of T cell; generated during initial immune response (primary response); may exist in body for years after, enabling it to respond quickly and efficiently to subsequent infections or meetings with same antigen
Antigen-presenting cell (APC)	One of several cell types (dendritic cell, macrophage, activated B cell) that engulfs and digests antigens that it encounters and presents parts of them on its plasma membrane (bound to a MHC protein) for recognition by T cells bearing receptors for same antigen; this function, antigen presentation, is essential for normal cell-mediated responses; macrophages also release chemicals (cytokines) that activate T cells and prevent viral multiplication
Molecules	
Antibody (immunoglobulin)	Protein produced by B cell or by plasma cell; antibodies produced by plasma cells are released into body fluids (blood, lymph, saliva, mucus, etc.), where they attach to antigens, causing complement fixation, neutralization, precipitation, or agglutination, which "mark" the antigens for destruction by complement or phagocytes
Cytokines	
Interferons (IFNs)	
▪ Alpha *(α)*	Secreted by most leukocytes; antiviral effects; activates macrophages and NK cells
▪ Beta *(β)*	Secreted by fibroblasts; effects as above, but also acts to reduce inflammation
▪ Gamma *(γ)*	Secreted by lymphocytes; effects as for alpha IFN; stimulates synthesis and expression of more class I and II MHC proteins; enhances activity of B cells and T_C cells

continues as long as it is stimulated by the T_H cell. Thus, helper T cells function to help unleash the protective potential of the B cells.

Cytokines released by the T_H cells not only mobilize immune cells and macrophages, they also attract other types of white blood cells into the area and tremendously amplify nonspecific defenses. As the released chemicals summon more and more cells into the battle, the immune response gains momentum, and the antigens are overwhelmed by the sheer numbers of immune elements acting against them.

It is interesting, but not surprising, that different cytokine exposure during T_H cell differentiation in-

duces different subsets of helper T cells. For example, IL-12 induces T_H1 differentiation, while IL-4 drives T_H2 differentiation. Generally speaking, the T_H1 cells activate macrophages and promote differentiation of cytotoxic T cells, that is, they mediate classical cell-mediated immunity. By contrast, T_H2 cells promote the migration of eosinophils and basophils to the battlefield and function in parasitic infections of the gut. Both types of T_H cells promote antibody formation, but the classes formed differ.

Cytotoxic T Cells

Cytotoxic T cells, also called **killer T cells,** are the only T cells that can directly attack and kill other

TABLE 22.4	*continued*
Element	**Function in immune response**

Interleukins (ILs)

▪ IL-1	Secreted by activated macrophages; costimulates T and B cells to proliferate; promotes inflammation; causes fever (believed to be the pyrogen that resets the thermostat of the hypothalamus)
▪ IL-2	Secreted by T_H cells; stimulates proliferation of T and B cells; activates NK cells; also called T cell growth factor
▪ IL-3	Stimulates production of leukocytes and mast cells
▪ IL-4	Secreted by T_H cells; costimulates activated B cells; causes plasma cells to secrete more IgE antibodies
▪ IL-5	Secreted by some T_H cells and mast cells; costimulates B cells; causes plasma cells to secrete IgA antibodies; attracts eosinophils
▪ IL-6	Induces differentiation of B cells into plasma cells; enhances proliferation and activity of T cells; stimulates liver to secrete mannose-binding protein, which triggers complement binding to bacteria with mannose sugar in their capsules
▪ IL-7	B cell stimulator; influences macrophage activity
▪ IL-8	Stimulates angiogenesis
▪ IL-9	Stimulates myeloid tissue to produce blood cells
▪ IL-12	Secreted by granulocytes, dendritic cells, and macrophages; stimulates T_C activity and NK cell activity; promotes T_H1 differentiation
Lymphotoxin (LT)	Released by T_C cells; causes DNA fragmentation; promotes inflammation
Macrophage migration inhibitory factor (MIF)	Inhibits macrophage migration and keeps them in the area of antigen deposition
Perforin	Released by T_C cells; causes cell lysis by creating large pores in the target cell's membrane
Suppressor factors	Released by T_S cells; suppresses the immune response
Transforming growth factor beta (TGF-β)	Inhibits activation and proliferation of macrophages, T cells, and B cells; a suppressor of the immune response
Tumor necrosis factors (TNFs)	Produced by lymphocytes and in large amounts by macrophages; enhance nonspecific killing; slow tumor growth; cause selective damage to blood vessels; enhance granulocyte chemotaxis; help activate T cells, phagocytes, and eosinophils
Complement	Group of bloodborne proteins activated after binding to antibody-covered antigens; causes lysis of microorganism and enhances inflammatory response
Antigen	Substance capable of provoking an immune response; typically are large complex molecules (e.g., proteins and modified proteins) that are not normally present in the body

cells. Activated cytotoxic T cells roam the body, circulating in and out of the blood and lymph and through lymphatic organs in search of body cells displaying antigens to which they have been sensitized. Their main targets are virus-infected cells, but they also attack tissue cells infected by certain intracellular bacteria (such as the tubercle bacillus) or parasites, cancer cells, and foreign cells introduced into the body by blood transfusions or organ transplants.

Before the onslaught can begin, the cytotoxic cell must "dock" on the target cell by binding to a self-antiself complex. Remember, all body cells display class I MHC antigens, so all infected or abnormal body cells can be destroyed by these T cells as long as

the appropriate antigen and costimulatory stimuli (typically IL-2 released by helper T cells) are also present. The attack on foreign human cells, such as those of a graft, is more difficult to explain because *all* of the antigens are nonself. However, apparently the cytotoxic cells sometimes "see" the foreign class I MHC antigens as a combination of a class I MHC protein bound to antigen, and NK cells are always on surveillance. By contrast to the T_C cells that recognize antigen-bearing class I MHC molecules, NK cells activate their killing machinery when they bind to a MICA receptor, MHC-related cell surface proteins that are switched on in cancer cells and in cells under stress, such as virus-infected cells and cells of

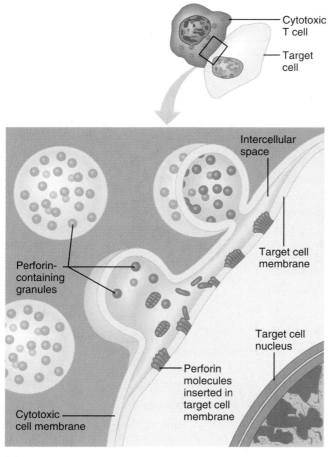

(a)

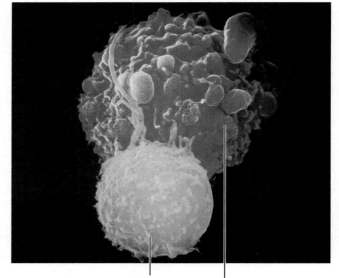

(b)

FIGURE 22.17 *Cytotoxic T cells attack infected and cancerous cells.* **(a)** A proposed mechanism of target cell lysis by cytotoxic T cells. The initial event is the tight binding of the cytotoxic cell to the target cell. During this binding period, granules within the cytotoxic cell fuse with its plasma membrane and release their perforin contents by exocytosis. The perforin molecules insert into the target cell membrane and, in the presence of ionic calcium, they polymerize, forming cylindrical holes that allow free exchange of ions and water. Target cell lysis occurs long after the cytotoxic cell has detached. **(b)** Scanning electron micrograph of a cancer cell that has been lysed by a cytotoxic T cell. (1250×)

transplanted organs and tissues. Thus, NK cells stalk abnormal or foreign cells in the body that T_C cells can't "see."

The mechanism of the cytotoxic cell's **lethal hit** which leads to cell lysis is unclear, but the following events occur in at least some cases:

1. The cytotoxic T cell binds tightly to the target cell and, during this period, a cytotoxic chemical, **perforin,** is released from the T cell's granules and inserted into the plasma membrane of the target cell (Figure 22.17).

2. The cytotoxic cell then detaches and continues the search for another prey.

3. The perforin molecules in the target cell membrane polymerize and cause cell lysis by creating transmembrane pores very much like those produced by complement activation.

Other cytotoxic T cells apparently use different or additional signals to induce target cell death. For example, some secrete **lymphotoxin,** which causes fragmentation of the target cell's DNA; some release **tumor necrosis factor (TNF),** which triggers apop-

tosis, killing the cell slowly over 48 to 72 hours. Still others secrete **gamma interferon,** which stimulates macrophages to killer status and indirectly enhances the killing process. Because of their ability to induce target cell lysis, T_C cells are often called **cytolytic (CTL) T cells.**

Other T Cells

Like the helper T cells, mature **suppressor T (T_S) cells** are regulatory cells. Because they release cytokines that suppress the activity of both T and B cells, T_S cells are thought to be vital for winding down and finally stopping the immune response after an antigen has been successfully inactivated or destroyed. This helps prevent uncontrolled or unnecessary immune system activity. Because of their inhibitory role, suppressors are presumed to be important in preventing autoimmune reactions. However, suppressor T cells are very difficult to study, their activation process is obscure and controversial. Most of what is written about them is still hypothetical.

Delayed-type hypersensitivity cells (T$_{DH}$), generally believed to be a special subclass of T4 cells, now appear to have some members that display CD8 receptors. Functionally, they appear to be most instrumental in promoting allergic reactions called delayed hypersensitivity reactions (see p. 826). By secreting gamma interferon and other cytokines, they mobilize macrophages to killer status. The activated macrophages then have the responsibility of eliminating the antigen.

Gamma delta (γδ) T cells represent about 10% of all T cells. These oddball T cells, which live in the intestine, are triggered into action when their TCRs bind to MICA receptors (or their close relative MICB receptors). Hence, these T cells are more similar to NK cells than to other T cell types.

Summary of T Cell Roles

Each type of T cell has unique roles to play in the immune response, yet at the same time is heavily enmeshed in interactions with other immune cells and elements as summarized in Figure 22.18. Cytotoxic T cells prevent infectious microorganisms, hidden within cells from antibody surveillance, from doing their "dirty work" by killing their cellular hosts. Helper T cells costimulate both other T cells and B cells and recruit nonspecific defenses. Suppressor T cells prevent runaway or undesirable immune reactions. Delayed hypersensitivity T cells (not shown in Figure 22.18) enhance the killing activity of macrophages in certain allergic states. But the lesson to take away with you is that without helper T cells, *there is no adaptive immune response* because the helper cells direct or help complete the activation of *all* other immune cells. Their crucial role in immunity is painfully evident when they are destroyed, as in AIDS (see Immunodeficiencies on p. 823).

Organ Transplants and Prevention of Rejection

Organ transplants, a viable treatment option for many patients with end-stage cardiac or renal disease, have been done with mixed success for over 30 years. Immune rejection presents a particular problem when the goal is to provide such patients with functional organs from a living or recently deceased donor. Essentially, there are four major varieties of grafts:

1. Autografts are tissue grafts transplanted from one body site to another in the *same person*.

2. Isografts are grafts donated by *genetically identical individuals*, the only example being identical twins.

3. Allografts are grafts transplanted from individuals that are *not genetically identical* but belong to the *same species*.

4. Xenografts are grafts taken *from another animal species*, such as transplanting a baboon heart into a human being.

Transplant success depends on the similarity of the tissues because cytotoxic T cells (and antibodies) act vigorously to destroy any foreign tissue in the body. Autografts and isografts are the ideal donor tissues. Given an adequate blood supply and no infection, they are always successful because the MHC-encoded proteins are identical. Xenografts are as yet only temporarily successful.* Consequently, the problematic graft type, and the type that is also most frequently used, is the allograft, which is usually obtained (harvested) from a human donor who has just died.

Before an allograft is even attempted, the ABO and other blood group antigens of the donor and recipient must be determined, because they are also present on most body cells. Next, the recipient and donor tissues are typed to determine their MHC antigen match. MHC variation among human tissues is tremendous, so good tissue matches between unrelated individuals are hard to obtain. However, the donor and recipient tissues are matched as closely as possible (at least a 75% match is essential).

Following surgery the patient is treated with *immunosuppressive therapy* involving one or more of the following: (1) corticosteroid drugs, such as prednisone, to suppress inflammation; (2) cytotoxic drugs; (3) radiation (X ray) therapy; (4) antilymphocyte globulins; and (5) an immunosuppressant drug such as cyclosporine. Many of these drugs kill rapidly dividing cells (such as activated lymphocytes), and all of them have severe side effects.

The major problem with immunosuppressive therapy is that the patient's immune system, while suppressed, cannot protect the body against other foreign agents. Explosive bacterial and viral infection remains the most frequent cause of death in these patients. The key to successful graft survival is to provide enough immunosuppression to prevent graft rejection but not enough to be toxic, and to use antibiotics to keep infection under control.

---

*Though still true at publication of this book, this situation is about to change. By microinjecting human genes into fertilized animal (pig) eggs, researchers have been able to neutralize the attack of complement and inhibit the hyperacute rejection response of a xenograft which can destroy a transplanted organ within minutes by cutting off its blood supply. Genetically altered pigs are now being bred and human trials using these animals' organs are ongoing.

? *Several costimulators exist. What are the possible consequences of costimulation with different costimulatory factors?*

FIGURE 22.18 **Summary of the primary immune response.** In this simple flowchart, the activities of the humoral and cellular arms of the immune system are indicated by different colored backgrounds. Some events known to be costimulated by cytokines are noted. However, other costimulatory signals are also involved in many of these events. Although complement, NK cells, and phagocytes are nonspecific defenses, they are enlisted in the fight by cytokines released by immune cells. (Immune cell receptors are not illustrated here for simplicity.)

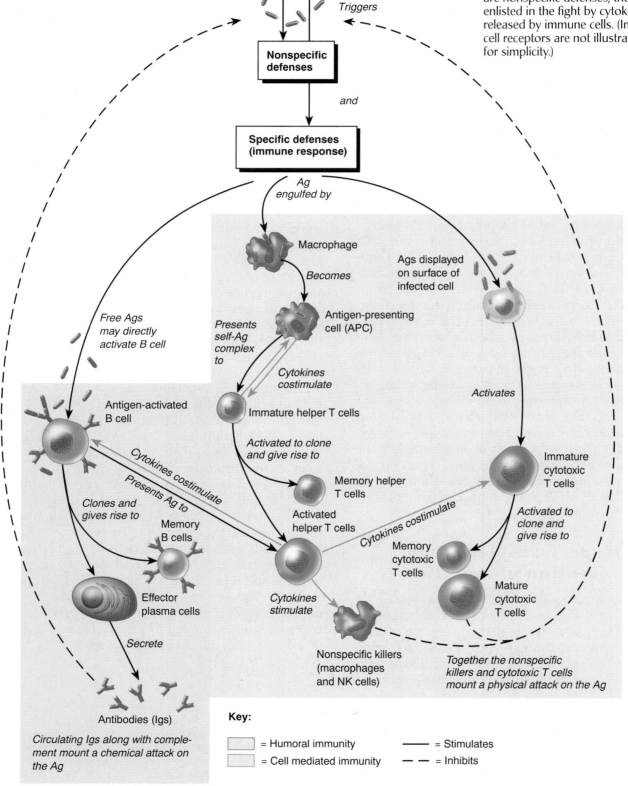

Antigen (Ag) intruder

Triggers

Nonspecific defenses

and

Specific defenses (immune response)

Ag engulfed by

Macrophage

Becomes

Antigen-presenting cell (APC)

Ags displayed on surface of infected cell

Presents self-Ag complex to

Cytokines costimulate

Immature helper T cells

Activates

Activated to clone and give rise to

Memory helper T cells

Immature cytotoxic T cells

Free Ags may directly activate B cell

Antigen-activated B cell

Cytokines costimulate

Presents Ag to

Clones and gives rise to

Memory B cells

Activated helper T cells

Cytokines costimulate

Memory cytotoxic T cells

Activated to clone and give rise to

Effector plasma cells

Cytokines stimulate

Mature cytotoxic T cells

Secrete

Nonspecific killers (macrophages and NK cells)

Together the nonspecific killers and cytotoxic T cells mount a physical attack on the Ag

Antibodies (Igs)

Circulating Igs along with complement mount a chemical attack on the Ag

Key:

☐ = Humoral immunity   —— = Stimulates
☐ = Cell mediated immunity   - - - = Inhibits

Completion or abortion of the T cell activation process.

HOMEOSTATIC IMBALANCES OF IMMUNITY

Under certain circumstances, the immune system is depressed and even fails, or exhibits activities that damage the body itself. Most such problems can be classified as immunodeficiencies, hypersensitivities, or autoimmune diseases.

Immunodeficiencies

Immunodeficiencies include both congenital and acquired conditions in which the production or function of immune cells, phagocytes, or complement is abnormal. The most devastating *congenital condition* is **severe combined immunodeficiency (SCID) syndromes,** which result from various genetic defects that produce a marked deficit of both B and T cells. One such defect causes abnormalities in the receptors for several interleukins. Another results in a defective *adenosine deaminase (ADA)* enzyme. In the absence of normal ADA, metabolites accumulate that are lethal to T cells. Children afflicted with SCID have little or no protection against disease-causing organisms of any type. Interventions must begin in the first months of life because minor infections easily shrugged off by most children cause the victims to become deathly ill and wasted. Untreated, this condition is fatal, but successful bone marrow transplants using bone marrow tissue or cultured stem cells from umbilical cord blood to provide normal stem cells improve the chance of survival of SCID victims. Lacking such a treatment, the only hope for survival has been living behind protective barriers (in a plastic "bubble") that keep out all infectious agents. However, recent genetic engineering techniques using viruses as vectors to transfer ADA minigenes to the victim's hematopoietic stem cells or T cells have shown promise.

There are various *acquired immunodeficiencies.* For example, *Hodgkin's disease,* cancer of the lymph nodes, can lead to immunodeficiency by depressing lymph node cells, and immunosuppression is the goal of certain drugs used to treat cancer. Currently, however, the most devastating of the acquired immunodeficiencies is **acquired immune deficiency syndrome (AIDS),** which cripples the immune system by interfering with the activity of helper T (CD4) cells.

AIDS, first identified in this country in 1981 among homosexual men and intravenous drug users, is characterized by severe weight loss, night sweats, and swollen lymph nodes. Increasingly frequent opportunistic infections occur, including a rare type of pneumonia called *pneumocystis pneumonia* caused by a protozoan, and the bizarre malignancy *Kaposi's sarcoma,* a cancerlike condition of the blood vessels evidenced by purple lesions of the skin. Some AIDS victims develop severe dementia. The course of AIDS is often grim, finally ending in complete debilitation and death from cancer or overwhelming infection.

AIDS is caused by a virus transmitted in body secretions—especially blood, semen, and vaginal secretions. The virus commonly enters the body via blood transfusions or blood-contaminated needles and during intimate sexual contact in which the mucosa is torn or where open lesions caused by sexually transmitted diseases allow the virus access to the blood. The virus is also present in saliva and tears, and there are documented cases of transmission by oral sex.

The virus, called **HIV (human immunodeficiency virus),** destroys helper T cells, thus depressing cell-mediated immunity. Although B cells and T_C cells initially mount a vigorous response to viral exposure, the thymus is depressed and memory T_H cells are preferentially eliminated by the virus. In time a profound deficit of antibodies develops and the cytotoxic cells become nonresponsive to viral cues. The whole immune system is turned topsy-turvy. It is now clear that the virus multiplies steadily in the lymph nodes throughout most of the chronic asymptomatic period. Symptomatic AIDS gradually appears a few months to ten or more years later when the lymph nodes are destroyed and can no longer contain the virus. The virus also invades the brain, which accounts for the dementia of some AIDS patients. Although there are exceptions, most AIDS victims die within a few months to eight years after diagnosis.

The infectious specificity of HIV reflects the fact that the CD4 proteins provide the avenue for HIV's attack. A particular harpoonlike HIV coat glycoprotein (gp120) fits into the CD4 receptor like a plug fits into a socket. However, HIV also needs a nearby protein called gp41 to gain entry; gp41 emerges once gp120 has attached and in split-second action fuses the virus to the target cell. Once inside, HIV "sets up housekeeping," using the enzyme *reverse transcriptase* to produce DNA from the information encoded in its (viral) RNA. This DNA copy, now called a *provirus,* then inserts itself into the host cell's DNA and directs the host cell to crank out new copies of viral RNA (and proteins) so that the virus can multiply and infect other cells. Although the helper T cells are the main HIV targets, other body cells displaying CD4 proteins (macrophages, monocytes, and dendritic cells) are also at risk. The HIV reverse transcriptase enzyme is not very accurate and produces errors rather frequently, causing HIV's relatively high mutation rate and its changing resistance to drugs.

The years since 1981 have witnessed a global AIDS epidemic—worldwide, 16,000 people are infected with the HIV virus every day. Currently more than 30 million people are infected worldwide, and almost 90% of them live in the developing countries of Asia and sub-Saharan Africa. Half of the victims are women and a quarter are children. This case distribution reflects the fact that in the most heavily hit countries, most HIV transmission occurs via heterosexual contacts as opposed to the pattern (homosexual and contaminated needle transmission) seen in western Europe and the United States. The virus can also be transmitted from an infected mother to her fetus. Early in the epidemic, hemophiliacs were also at particular risk because the blood factors they need were isolated from pooled blood donations. Although manufacturers began taking measures to kill the virus in 1984 and new genetically engineered factors became available, an estimated 60% of the hemophiliacs in this country were infected.

By mid-1996, estimates of the number of Americans infected with HIV topped one million. Although anti-HIV antibodies may appear in the blood as early as two weeks after infection, there is a six-month "window" after infection during which antibodies can develop. So for every diagnosed case, there are probably many more asymptomatic carriers of the virus. Moreover, the disease has a long incubation period (from a few months to over ten years) between exposure and the appearance of clinical symptoms. Not only has the number of identified cases in the United States jumped exponentially in the at-risk populations, but the "face of AIDS" is changing too. Though homosexual men still account for the bulk of cases transmitted by sexual contact, more and more heterosexuals are contracting this disease. Particularly disturbing is the near-epidemic increase in diagnosed cases among teenagers and young adults. AIDS is now the leading killer of all Americans ages 25 to 44.

Diagnostic tests to identify HIV carriers have become increasingly sophisticated: For example, besides the simple oral test which scrapes tissue from the oral mucosa, an even easier urine test has been approved as a second painless alternative to the standard HIV blood tests.

No sure cure for AIDS has yet been found. Over 100 drugs are now in the Food and Drug Administration pipeline, and vaccines are undergoing clinical trials, but it is unlikely that an approved vaccine will be available soon. Several antiviral drugs that inhibit the enzymes the HIV virus needs in order to multiply in the body are now clinically available. *Reverse transcriptase inhibitors*, such as AZT, were early on the scene. AZT has been followed by others, including ddI, ddC, d4T, and 3TC. In late 1995 and early 1996, *protease inhibitors* (saquinavir, ritonavir, and others) were approved. Initially, it appeared that combination therapy using drugs from each class delivers a one-two punch to the HIV virus; that is, it postpones drug resistance (a problem with monodrug therapy with AZT alone) and causes the *viral load* (amount of HIV virus per millimeter in the infected person's blood) to plummet while boosting the number of helper T cells. Several patients in the combination drug regimen seemed to revive from near death. Sadly, the treatments are beginning to fail in about half of those treated, as indicated by an ominous rise in viral load. New hope comes from the group of drugs now being developed that block HIV's entry into helper T cells by preventing the fusion promoted by gp41. However, drug research is slow and laborious and the clock is ticking.

Autoimmune Diseases

Occasionally the immune system loses its ability to distinguish friend (self) from foe (foreign antigens). When this happens, the artillery of the immune system, like friendly fire, can turn against itself. That is, the body produces antibodies (*autoantibodies*) and sensitized cytotoxic T cells that destroy its own tissues. This puzzling phenomenon is called **autoimmunity.** If a disease state results, it is referred to as **autoimmune disease.**

Some 5% of adults in North America—two-thirds of them women—are afflicted with autoimmune disease. Most common are:

- *Multiple sclerosis (MS)*, which destroys the white matter of the brain and spinal cord (see p. 407)

- *Myasthenia gravis*, which impairs communication between nerves and skeletal muscles (see p. 289)

- *Graves' disease*, which prompts the thyroid gland to produce excessive amounts of thyroxine

- *Type I (juvenile) diabetes mellitus*, which destroys pancreatic beta cells, resulting in a deficit of insulin and inability to use carbohydrates (see p. 640)

- *Systemic lupus erythematosus (SLE)*, a systemic disease that particularly affects the kidneys, heart, lungs, and skin (see Related Clinical Terms, p. 830)

- *Glomerulonephritis*, a severe impairment of renal function

- *Rheumatoid arthritis (RA)*, which systematically destroys joints (see p. 270)

Current therapies include treatment approaches that depress certain aspects of the immune response. For example, injections of genetically engineered antibodies to the CD4 receptor on helper T cells seem to stabilize the disease of some MS victims. More recent is the discovery that *thalidomide* is a potential boon to those with autoimmune disease. This drug, which was used to alleviate morning sickness in pregnant women in the 1950s until it was found to cause tragic birth defects, inhibits the immune system's production of TNF-α.

How does the normal state of self-tolerance break down? It appears that one or more of the following events may be triggers:

1. Lymphocyte programming is ineffective. Self-reactive T cells or B cells, which should be silenced or eliminated during their programming phase in the thymus and bone marrow, instead escape to the rest of the body. This is believed to occur in multiple sclerosis.

2. New self-antigens appear. Self-proteins that have not been previously exposed to the immune system may appear in the circulation. They may be generated by (1) gene mutations that cause new proteins to appear at the external cell surface and (2) changes in the structure of self-antigens by hapten attachment or as a result of infectious damage. These then become immune system targets.

3. Foreign antigens resemble self-antigens. Antibodies made against a foreign antigen sometimes cross-react with a self-antigen that has very similar determinants. For instance, antibodies produced during a streptococcal infection are known to cross-react with heart antigens, causing lasting damage to the heart muscle and valves, as well as to joints and kidneys. This age-old disease is called *rheumatic fever*.

Hypersensitivities

At first, the immune response was thought to be purely protective. However, it was not long before its dangerous potentials were discovered. **Hypersensitivities** or **allergies** (*allo* = altered; *erg* = reaction) are the result of immune responses in which the immune system causes tissue damage as it fights off a perceived "threat" (such as pollen or animal dander) that would otherwise be harmless to the body. The term **allergen** is used to distinguish this type of antigen from those producing essentially normal responses. People rarely die of allergies; they are just miserable with them.

There are different types of hypersensitivity reactions, which are distinguished by (1) their time course, and (2) whether antibodies or T cells are the principal immune elements involved. Allergies mediated by antibodies are the *immediate* and *subacute hypersensitivities*. The most important cell-mediated allergic state is *delayed hypersensitivity*.

Immediate Hypersensitivities

The **immediate hypersensitivities**, also called **acute** or **type I hypersensitivities**, begin within seconds after contact with the allergen and last about half an hour.

Anaphylaxis The most common type of immediate hypersensitivity is **anaphylaxis** (an"ah-fĭ-lak'sis; "against protection"). The initial meeting with an allergen produces no symptoms but it sensitizes the person. APCs digest the allergen and present its fragments to T cells as usual. The steps that follow are obscure, but appear to be mediated by interleukin 4 (IL-4) secreted by T cells. IL-4 stimulates B cells to mature into IgE-secreting plasma cells, which spew out huge amounts of that antibody. When the IgE molecules attach to **mast cells** and **basophils,** sensitization is complete. Anaphylaxis is triggered at later encounters with the same allergen, which promptly binds and cross-links the IgE antibodies on the surfaces of the mast cells and basophils. This event induces an enzymatic cascade that causes the mast cells and basophils to degranulate, releasing a flood of **histamine** and other inflammatory chemicals (chemotactic factors, prostaglandins, and various cytokines). Together, these chemicals induce the inflammatory response typical of anaphylaxis (Figure 22.19).

Anaphylactic reactions may be local or systemic. Mast cells are particularly abundant in connective tissues of the skin and beneath the mucosa of respiratory passages and the gastrointestinal tract, and these areas are common sites of local allergic reactions. Histamine causes blood vessels to become dilated and leaky, and is largely to blame for the best recognized symptoms of anaphylaxis: a runny nose, itching reddened skin (hives), and watery eyes. When the allergen is inhaled, symptoms of *asthma* appear because smooth muscle in the walls of the bronchioles contracts, constricting those small passages and restricting air flow. When the allergen is ingested (in food or via drugs), gastrointestinal discomfort (cramping, vomiting, or diarrhea) occurs. Over-the-counter (OTC) antiallergy drugs contain antihistamines that counteract these effects.

Fortunately, the bodywide or systemic response, known as **anaphylactic shock,** is fairly rare. It typically occurs when the allergen directly enters the blood and circulates rapidly through the body, as might happen with certain bee stings or spider bites. It may also follow injection of a foreign substance (such as penicillin or other drugs which act as haptens) into susceptible individuals. The mechanism of anaphylactic shock is essentially the same as that of local responses; but when mast cells and basophils are enlisted throughout the entire body, the outcome is life threatening. The bronchioles constrict (and the tongue may swell), making it difficult to breathe, and the sudden vasodilation and fluid loss from the bloodstream may cause circulatory collapse (hypotensive shock) and death within minutes. Epinephrine is the drug of choice to reverse these histamine-mediated effects.

TABLE 23.1	**Principal Organs of the Respiratory System**	
Structure	*Description, general and distinctive features*	*Function*
Nose	Jutting external portion supported by bone and cartilage; internal nasal cavity divided by midline nasal septum and lined with mucosa	Produces mucus; filters, warms, and moistens incoming air; resonance chamber for speech
	Roof of nasal cavity contains olfactory epithelium	Receptors for sense of smell
	Paranasal sinuses around nasal cavity	Same as for nasal cavity; also lighten skull
Pharynx	Passageway connecting nasal cavity to larynx and oral cavity to esophagus; three subdivisions: nasopharynx, oropharynx, and laryngopharynx	Passageway for air and food
	Houses tonsils (lymphoid tissue masses involved in body protection against pathogens)	Facilitates exposure of immune system to inhaled antigens
Larynx	Connects pharynx to trachea; framework of cartilage and dense connective tissue; opening (glottis) can be closed by epiglottis or vocal folds	Air passageway; prevents food from entering lower respiratory tract
	Houses true vocal cords	Voice production
Trachea	Flexible tube running from larynx and dividing inferiorly into two primary bronchi; walls contain C-shaped cartilages that are incomplete posteriorly where connected by trachealis muscle	Air passageway; cleans, warms, and moistens incoming air
Bronchial tree	Consists of right and left primary bronchi, which subdivide within the lungs to form secondary and tertiary bronchi and bronchioles; bronchiolar walls contain complete layer of smooth muscle; constriction of this muscle impedes expiration	Air passageways connecting trachea with alveoli; cleanses, warms, and moistens incoming air
Alveoli	Microscopic chambers at termini of bronchial tree; walls of simple squamous epithelium underlain by thin basement membrane; external surfaces intimately associated with pulmonary capillaries	Main sites of gas exchange
	Special alveolar cells produce surfactant	Reduces surface tension; helps prevent lung collapse
Lungs	Paired composite organs located within pleural cavities of thorax; composed primarily of alveoli and respiratory passageways; stroma is fibrous elastic connective tissue, allowing lungs to recoil passively during expiration	House respiratory passages smaller than the primary bronchi
Pleurae	Serous membranes; parietal pleura lines thoracic cavity; visceral pleura covers external lung surfaces	Produce lubricating fluid and compartmentalize lungs

lower respiratory tract by closing the glottis. The abdominal muscles then contract, and the intra-abdominal pressure rises. These events, collectively known as **Valsalva's maneuver,** help to empty the rectum or bladder and can also splint (stabilize) the body trunk when one lifts a heavy load.

The Trachea

The **trachea** (tra′ke-ah), or *windpipe,* descends from the larynx through the neck and into the mediastinum. It ends by dividing into the two primary bronchi at midthorax (see Figure 23.1). In humans, the trachea is 10–12 cm (about 4 inches) long and 2.5 cm (1 inch) in diameter, and is very flexible and mobile.

The tracheal wall consists of several layers that are common to many tubular body organs (Figure 23.5). From internal to external, these layers are the *mucosa, submucosa,* and *adventitia.* The **mucosa** has the same goblet cell–containing pseudostratified epithelium that occurs throughout most of the respiratory tract. Its cilia (Figure 23.6) continually propel mucus, loaded with dust particles and other debris, toward the pharynx. This epithelium rests on a fairly thick lamina propria that has a rich supply of elastic fibers.

Homeostatic Imbalance

Smoking inhibits and ultimately destroys the cilia. When their function is lost, coughing is the only means of preventing mucus from accumulating in the lungs. For this reason, smokers with respiratory

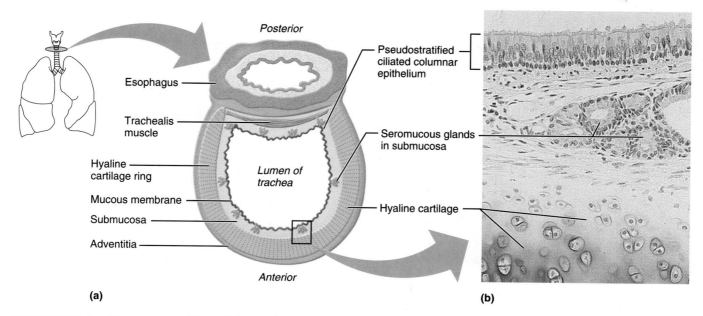

Posterior

Esophagus

Trachealis muscle

Hyaline cartilage ring

Mucous membrane

Submucosa

Adventitia

Lumen of trachea

Anterior

(a)

Pseudostratified ciliated columnar epithelium

Seromucous glands in submucosa

Hyaline cartilage

(b)

FIGURE 23.5 *Tissue composition of the tracheal wall.* (a) Cross-sectional view of the trachea, illustrating its relationship to the esophagus, the position of the supporting hyaline cartilage rings in its wall, and the trachealis muscle connecting the free ends of the cartilage rings. (b) Photomicrograph of a portion of the tracheal wall (cross-sectional view; 212×).

congestion should avoid medications that inhibit the cough reflex. ◼

The **submucosa,** a connective tissue layer deep to the mucosa, contains seromucous glands that help produce the mucus "sheets" within the trachea. The outermost **adventitia** layer is a connective tissue layer that is reinforced internally by 16 to 20 C-shaped rings of hyaline cartilage (Figure 23.5), and is fused to the perichondrium of these cartilages. Because of its elastic elements, the trachea is flexible enough to stretch and move inferiorly during inspiration and recoil during expiration, but the cartilage rings prevent it from collapsing and keep the airway patent despite the pressure changes that occur during breathing. The open posterior parts of the cartilage rings, which abut the esophagus (see Figure 23.5a), are connected by smooth muscle fibers of the **trachealis muscle** and soft connective tissue. Since this portion of the tracheal wall is not rigid, the esophagus can expand anteriorly as swallowed food passes through it. Contraction of the trachealis muscle decreases the trachea's diameter, causing expired air to rush upward from the lungs with greater force. This action helps to expel mucus from the trachea when we cough by accelerating the exhaled air to

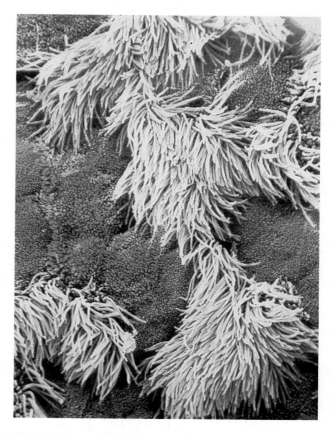

FIGURE 23.6 *Cilia.* Scanning electron micrograph of cilia in the trachea (198,900×). The cilia appear as yellow, grasslike projections. Mucus-secreting goblet cells (orange) with short microvilli are interspersed between the ciliated cells.

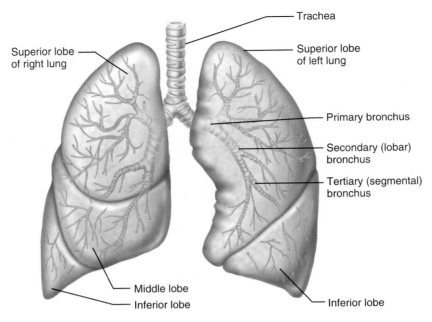

FIGURE 23.7 *Conducting respiratory passages.* The air pathway inferior to the larynx consists of the trachea and primary, secondary, and tertiary bronchi, which in turn branch into the smaller bronchi and bronchioles until the terminal bronchioles of the lungs are reached.

speeds of 100 mph! The last tracheal cartilage is expanded (see Figure 23.1), and a spar of cartilage, called the **carina** (kar-ri'nah; "keel"), projects posteriorly from its inner face, marking the point where the trachea ends by splitting into the two *primary bronchi.* The mucosa of the carina is highly sensitive and violent coughing is triggered when a foreign object makes contact with it.

🪂 *Homeostatic Imbalance*

Tracheal obstruction is life threatening. Many people have suffocated after choking on a piece of food that suddenly closed off their trachea (or the glottis of the larynx). The **Heimlich maneuver,** a procedure in which the air in a person's own lungs is used to "pop out," or expel, an obstructing piece of food, has saved many people from becoming victims of such "café coronaries." The Heimlich maneuver is simple to learn and easy to do. However, it is best learned by demonstration because cracked ribs are a distinct possibility when it is done incorrectly. ■

The Bronchi and Subdivisions: The Bronchial Tree

The Conducting Zone

The **right** and **left primary (principal) bronchi** (brong'ki) are formed by the division of the trachea approximately at the level of T₇ (Figure 23.7) in an erect (standing) person. Each bronchus runs obliquely in the mediastinum before plunging into the medial depression (hilus) of the lung on its own side. The right primary bronchus is wider, shorter, and more vertical than the left and is the more common site for an inhaled foreign object to become lodged. By the time incoming air reaches the bronchi, it is warm, cleansed of most impurities, and saturated with water vapor.

Once inside the lungs, each primary bronchus subdivides into **secondary (lobar) bronchi**—three on the right and two on the left—each of which supplies one lung lobe. The secondary bronchi branch into third-order **tertiary (segmental) bronchi,** which in turn divide repeatedly into smaller and smaller bronchi (fourth-order, fifth-order, etc.). Overall, there are about 23 orders of branching air passageways in the lungs. Air passages under 1 mm in diameter are called **bronchioles** ("little bronchi"), and the tiniest of these, the **terminal bronchioles,** are less than 0.5 mm in diameter. Because of this branching pattern, the conducting network within the lungs is often called the **bronchial** or **respiratory tree.** A resin cast showing the extensive branching of the bronchial tree is illustrated in Figure 23.11.

The tissue composition of the walls of the primary bronchi mimics that of the trachea, but as the conducting tubes become smaller, the following structural changes occur:

1. The cartilage support structures change. The cartilage rings are replaced by irregular *plates* of cartilage, and by the time the bronchioles are reached,

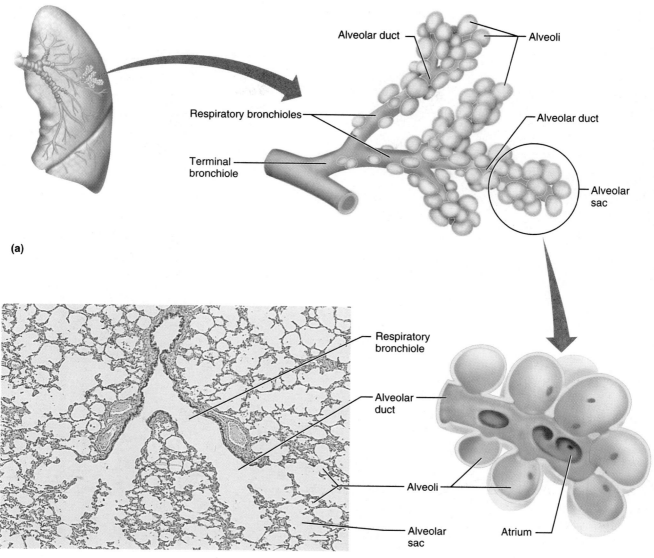

FIGURE 23.8 *Respiratory zone structures.* **(a)** Diagrammatic view of the respiratory unit (respiratory bronchiole, alveolar ducts, alveolar sacs, and alveoli). **(b)** Photomicrograph of a section of the human lung, showing the respiratory structures that form the final divisions of the bronchial tree (30×). Notice the thinness of the alveolar walls.

supportive cartilage is no longer present in the tube walls. However, elastic fibers are found in the tube walls throughout the bronchial tree.

2. The epithelium type changes. The mucosal epithelium thins as it changes from pseudostratified columnar epithelium to columnar epithelium and then cuboidal epithelium in the terminal bronchioles. Cilia are sparse, and mucus-producing cells are absent in the bronchioles. Thus, most airborne debris found at or below the level of the bronchioles is removed by macrophages located in the alveoli.

3. The amount of smooth muscle increases. The relative amount of smooth muscle in the tube walls increases as the passageways become smaller. The complete layer of circular smooth muscle in the bronchioles and the absence of supporting cartilage that would hinder constriction allows them to provide substantial resistance to air passage under certain conditions (as described later).

The Respiratory Zone

The respiratory zone is defined by the presence of thin-walled air sacs called **alveoli** (al-ve′o-li; *alveol* = small cavity). It begins as the terminal bronchioles feed into **respiratory bronchioles** within the lung (Figure 23.8). Protruding from these smallest bronchioles are scattered alveoli. The respiratory bronchioles lead into **alveolar ducts**, winding ducts

? *What is the role of surfactant, produced by the type II cells?*

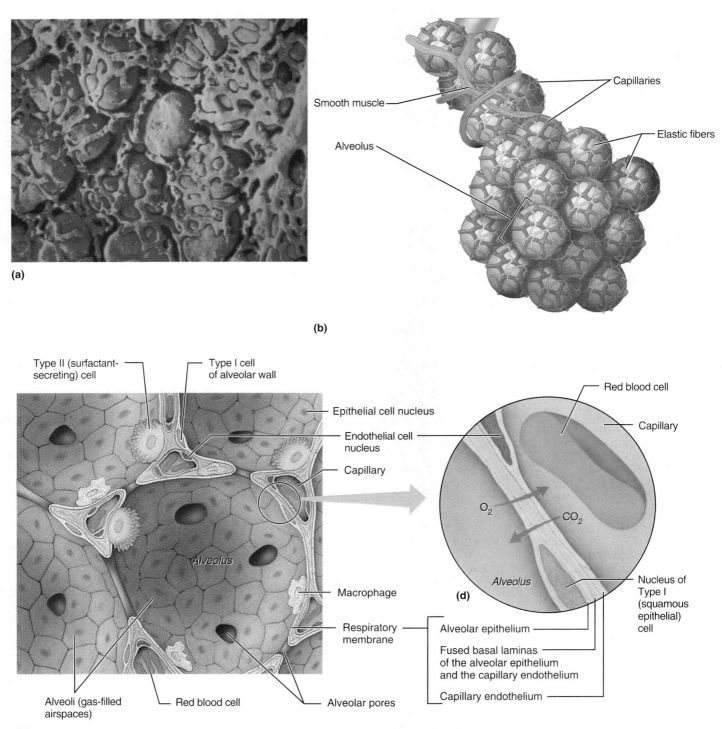

(a)

Capillaries

Smooth muscle

Elastic fibers

Alveolus

(b)

Type II (surfactant-secreting) cell

Type I cell of alveolar wall

Epithelial cell nucleus

Endothelial cell nucleus

Capillary

Red blood cell

Capillary

O_2 CO_2

Alveolus

Macrophage

Respiratory membrane

Alveolus

Alveoli (gas-filled airspaces)

Red blood cell

Alveolar pores

(d)

Nucleus of Type I (squamous epithelial) cell

Alveolar epithelium

Fused basal laminas of the alveolar epithelium and the capillary endothelium

Capillary endothelium

(c)

FIGURE 23.9 *Anatomy of the respiratory (alveolar-capillary) membrane.* (a) Scanning electron micrograph of casts of alveoli and associated pulmonary capillaries (1195×). From *Tissues and Organs* by R. G. Kessel and R. H. Kardon. © 1979 W. H. Freeman.

(b) Diagrammatic view of the pulmonary capillary-alveoli relationship. **(c** and **d)** Detailed anatomy of the respiratory membrane composed of the alveolar squamous epithelial cells (type I cells), the capillary endothelium, and the scant basement membranes (fused basal

laminas) intervening between. Type II (surfactant-secreting) alveolar cells and the alveolar pores that connect adjacent alveoli are also shown. Found within the alveoli are free-roaming alveolar macrophages that phagocytize debris.

whose walls consist of diffusely arranged rings of smooth muscle cells, connective tissue fibers, and outpocketing alveoli. The alveolar ducts then lead into terminal clusters of alveoli called **alveolar sacs.** Many people mistakenly equate alveoli, the actual sites of gas exchange, with alveolar sacs, but they are not the same thing. The alveolar sac is analogous to a bunch of grapes, in which the individual grapes are the alveoli. The 300 million or so gas-filled alveoli in the lungs account for most of the lung volume and provide a tremendous surface area for gas exchange.

The Respiratory Membrane The walls of the alveoli are composed primarily of a single layer of squamous epithelial cells, called **type I cells,** surrounded by a flimsy basal lamina. The thinness of their walls is hard to imagine, but a sheet of tissue paper is much thicker. The external surfaces of the alveoli are densely covered with a "cobweb" of pulmonary capillaries (Figure 23.9a and b). Together, the alveolar and capillary walls and their fused basal laminas form the **respiratory membrane (air-blood barrier),** also called the *alveolar-capillary membrane,* which has gas on one side and blood flowing past on the other (Figure 23.9d). Gas exchanges occur readily by simple diffusion across the respiratory membrane—the oxygen passes from the alveolus into the blood, and carbon dioxide leaves the blood to enter the gas-filled alveolus. The type I cells also are the primary source of *angiotensin converting enzyme (ACE),* an enzyme that plays a role in blood pressure regulation.

Scattered amid the type I squamous cells that form the major part of the alveolar walls are cuboidal **type II cells** (Figure 23.9c). The type II cells secrete a fluid containing surfactant that coats the gas-exposed alveolar surfaces. (Surfactant's role in reducing the surface tension of the alveolar fluid is described later in this chapter.)

The lung alveoli have three other significant features: (1) They are surrounded by fine elastic fibers of the same type that surround the entire respiratory tree (Figure 23.9b). (2) Open pores connect adjacent alveoli (Figure 23.9c). These **alveolar pores** allow air pressure throughout the lung to be equalized and provide alternate air routes to any alveoli whose bronchi have collapsed through disease. (3) **Alveolar macrophages** crawl freely along the internal alveolar surfaces. Also called *dust cells,* these free macrophages are remarkably efficient. Although huge numbers of infectious microorganisms are continually carried into the alveoli, alveolar surfaces are

usually sterile. Because the alveoli are "dead ends," aged and dead macrophages must be prevented from accumulating within them. Most macrophages simply get swept up by the ciliary current of superior regions and are carried passively to the pharynx. In this manner, we clear and swallow over 2 million dust cells per hour!

The Lungs and Pleural Coverings

Gross Anatomy of the Lungs

The paired **lungs** occupy all of the thoracic cavity except the *mediastinum,* which houses the heart, great blood vessels, bronchi, esophagus, and other organs (Figure 23.10). Each cone-shaped lung is suspended in its own pleural cavity and connected to the mediastinum by vascular and bronchial attachments, collectively called its **root.** The anterior, lateral, and posterior lung surfaces lie in close contact with the ribs and form the continuously curving **costal surface.** Just deep to the clavicle is the **apex,** the narrow superior tip of the lung. The concave, inferior surface that rests on the diaphragm is the **base.** On the medial (mediastinal) surface of each lung is an indentation, the **hilus.** Blood vessels of the pulmonary and systemic circulations enter and leave the lungs via the hilus. Each primary bronchus also plunges into the hilus on its own side and begins to branch almost immediately. All conducting and respiratory passageways distal to the primary bronchi are found within the lungs.

Because the apex of the heart is slightly to the left of the median plane, the two lungs differ slightly in shape and size. The left lung is smaller than the right, and the **cardiac notch (impression)**—a concavity in its medial aspect—is molded to and accommodates the heart (see Figure 23.10a). The left lung is subdivided into two (upper and lower) **lobes** by the *oblique fissure,* whereas the right lung is partitioned into three lobes (upper, middle, and lower) by the *oblique* and *horizontal fissures.* Each of the lung lobes, in turn, contains a number of pyramid-shaped **bronchopulmonary segments** separated from one another by connective tissue septa. Each such segment is served by its own artery and vein and receives air from an individual segmental bronchus. Each lung contains 10 bronchopulmonary segments arranged in similar (but not identical) patterns (Figure 23.11). The bronchopulmonary segments are clinically important because pulmonary disease is often confined to one or a few bronchopulmonary segments. Their connective tissue partitions allow diseased segments to be surgically removed without damaging neighboring healthy segments or impairing their blood supply.

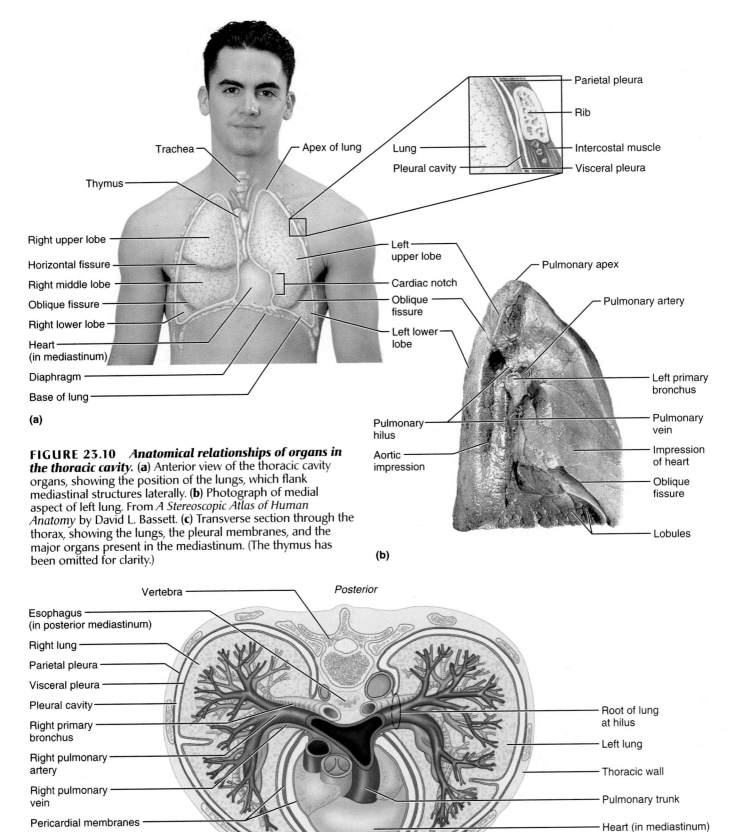

FIGURE 23.10 *Anatomical relationships of organs in the thoracic cavity.* (a) Anterior view of the thoracic cavity organs, showing the position of the lungs, which flank mediastinal structures laterally. (b) Photograph of medial aspect of left lung. From *A Stereoscopic Atlas of Human Anatomy* by David L. Bassett. (c) Transverse section through the thorax, showing the lungs, the pleural membranes, and the major organs present in the mediastinum. (The thymus has been omitted for clarity.)

FIGURE 23.11 *A cast of the bronchial tree in which the individual bronchopulmonary segments have been painted different colors.*

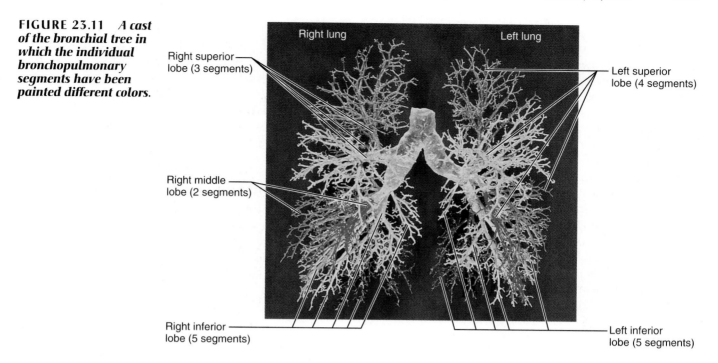

Right lung

Left lung

Right superior lobe (3 segments)

Left superior lobe (4 segments)

Right middle lobe (2 segments)

Right inferior lobe (5 segments)

Left inferior lobe (5 segments)

The smallest subdivision of the lung visible with the naked eye is the **lobule.** Lobules at the lung surface appear as hexagons ranging from the size of a pencil eraser to the size of a penny (Figure 23.10b). Each lobule is served by a large bronchiole and its branches. In most city dwellers and in smokers, the connective tissue that separates the individual lobules is blackened with carbon.

As mentioned earlier, the lungs consist largely of air spaces. The balance of lung tissue, or its **stroma** (literally "mattress" or "bed"), is mostly elastic connective tissue. As a result, the lungs are soft, spongy, elastic organs that together weigh just over 1 kg (2.5 pounds). The elasticity of healthy lungs helps to reduce the work of breathing, as described shortly.

Blood Supply and Innervation of the Lungs

The lungs are perfused by two circulations, the pulmonary and the bronchial, which differ in their size, origin, and function. Systemic venous blood that is to be oxygenated in the lungs is delivered by the **pulmonary arteries,** which lie anterior to the primary bronchi (Figure 23.10c). Within the lungs, the pulmonary arteries branch profusely along with the bronchi, and finally feed into the **pulmonary capillary networks** surrounding the alveoli (see Figure 23.9a). Freshly oxygenated blood is conveyed from the respiratory zones of the lungs to the heart by the **pulmonary veins,** whose tributaries course back to the hilus both with the corresponding bronchi and in the connective tissue septa separating the bronchopulmonary segments.

The large-volume, low-pressure venous blood input through the pulmonary arteries contrasts with the small-volume, high-pressure input via the bronchial arteries. The **bronchial arteries,** which provide systemic blood to the lung tissues, arise from the aorta and enter the lungs at the hilus. Within the lung they run along the branching bronchi, supplying all lung tissues except the alveoli (these are supplied by the pulmonary circulation). Although some systemic venous blood is drained from the lungs by the tiny bronchial veins, there are multiple anastomoses between the two circulations, and most venous blood bypasses the systemic circuit and returns to the heart via the pulmonary veins.

The lungs are innervated by parasympathetic and (rare) sympathetic motor fibers, and visceral sensory fibers. These nerve fibers enter each lung through the **pulmonary plexus** on the lung root and run along the bronchial tubes and blood vessels within the lungs. Parasympathetic fibers constrict the air tubes, and sympathetic fibers dilate them.

The Pleurae

The **pleurae** (ploo're; "the sides") form a thin, double-layered serosa (see Figure 23.10). The **parietal pleura** covers the thoracic wall and superior face of the diaphragm. It continues around the heart and between the lungs, forming the lateral walls of the mediastinal enclosure and snugly enclosing the root of the lung. From here, the pleura extends as the **visceral,** or **pulmonary, pleura** to cover the external lung surface, dipping into and lining its fissures.

The pleurae produce **pleural fluid** that fills the slitlike **pleural cavity** between them. This lubricating secretion allows the lungs to glide easily over the thorax wall during our breathing movements. Although the pleurae slide easily across one another, their separation is strongly resisted by the surface tension of the pleural fluid between them. Consequently, the lungs cling tightly to the thorax wall and are forced to expand and recoil passively as the volume of the thoracic cavity changes (alternately increases and decreases) during breathing.

The pleurae also help divide the thoracic cavity into three separate chambers—the central mediastinum and the two lateral pleural compartments, each containing a lung. This compartmentalization helps prevent one mobile organ (for example, the lung or heart) from interfering with another. It also limits the spread of local infections.

Homeostatic Imbalance

Pleurisy (ploo'rĭ-se), inflammation of the pleurae, often results from pneumonia. Inflamed pleural membranes produce less pleural fluid, and the pleural surfaces become dry and rough, resulting in friction and stabbing pain with each breath. Conversely, the pleurae may produce an excessive amount of fluid, which exerts pressure on the lungs. This type of pleurisy hinders breathing movements, but it is much less painful than the dry rubbing type.

Besides pleural fluid, other fluids that may accumulate in the pleural cavity include blood (leakage from damaged blood vessels) and blood filtrate (the watery fluid that oozes from the lung capillaries when right-sided heart failure occurs). The general term for fluid accumulation in the pleural space is *pleural effusion.* ■

MECHANICS OF BREATHING

Breathing, or **pulmonary ventilation,** consists of two phases: **inspiration,** the period when air flows into the lungs, and **expiration,** the period when gases exit the lungs. The mechanical factors that promote these gas flows are the topics of this section.

Pressure Relationships in the Thoracic Cavity

Before we can begin to describe the breathing process, it is important to understand that *respiratory pressures are always described relative to atmospheric pressure.* **Atmospheric pressure (P_{atm})** is the pressure exerted by the air (gases) surrounding the body. At sea level atmospheric pressure is 760 mm Hg (the pressure exerted by a column of mercury 760 mm high). Thus, a negative respiratory

pressure such as −4 mm Hg indicates that the pressure in that particular area is lower than atmospheric pressure by 4 mm Hg (760 − 4 = 756 mm Hg). By the same token, a positive respiratory pressure is higher than atmospheric pressure and a respiratory pressure of 0 is equal to atmospheric pressure. Now, we are ready to examine the pressure relationships that normally exist in the thoracic cavity.

Intrapulmonary Pressure

The **intrapulmonary** (intra-alveolar) **pressure (P_{alv})** is the pressure within the alveoli of the lungs. Intrapulmonary pressure rises and falls with the phases of breathing, but it *always* eventually equalizes itself with the atmospheric pressure outside the body (Figure 23.12).

Intrapleural Pressure

The pressure within the pleural cavity, the **intrapleural pressure (P_{ip})**, also fluctuates with breathing phases. However, the intrapleural pressure is always about 4 mm Hg less than the pressure in the alveoli. Hence, it is negative relative to *both* the intrapulmonary and atmospheric pressures.

The question often asked is "How is this negative pressure established?" or "What causes it?" Let's examine some of the forces that exist in the thorax to see if we can answer this question. First of all, we know that there are opposing forces acting. Two forces act to pull the lungs (visceral pleura) away from the thorax wall (parietal pleura) and thus to cause lung collapse:

- **The natural tendency of the lungs to recoil.** Because of their elasticity, lungs assume the smallest size possible at any given time.

- **The surface tension of the alveolar fluid.** The surface tension of the fluid film in the alveoli constantly acts to draw the alveoli to their smallest possible dimension.

However, these forces are opposed by:

- **The elasticity of the chest wall.** The natural elasticity of the thoracic cage tends to pull the thorax outward and to enlarge the lungs.

So which force wins? The answer is neither in a healthy person, because of the strong fluid bond (adhesive force) between the parietal and visceral pleura. Pleural fluid secures the pleurae together in the same way that a drop of water holds two glass slides to one another. They slide from side to side easily, but they remain closely apposed, and separating them requires extreme force. The net result of the dynamic interplay between these forces is negative intrapleural pressure.

In addition, the amount of pleural fluid in the pleural cavity must remain minimal for the negative intrapleural pressure to be maintained. Active pumping of the pleural fluid out of the pleural cavity into the lymphatics occurs almost continuously. If it didn't, an excessive amount of fluid would accumulate in the intrapleural space (remember, fluids move from high to low pressure), producing a condition of positive pressure in the pleural cavity.

The importance of the negative pressure in the intrapleural space and the tight coupling of the lungs to the thorax wall cannot be overemphasized. Any condition that equalizes the intrapleural pressure with the intrapulmonary (or atmospheric) pressure causes *immediate lung collapse*. It is the **transpulmonary** (or *transpulmonic*) **pressure**—the difference between the intrapulmonary and intrapleural pressures $(P_{alv} - P_{ip})$—that keeps the airspaces of the lungs open or, phrased another way, keeps the lungs from collapsing.

▲ Homeostatic Imbalance

Atelectasis (at"ĕ-lik'tah-sis), or lung collapse, commonly occurs when air enters the pleural cavity through a chest wound, but it may result from a rupture of the visceral pleura, which allows air to enter the pleural cavity from the respiratory tract. It is a common sequel to pneumonia.

The presence of air in the intrapleural space is referred to as a **pneumothorax** (nu"mo-tho'raks; "air thorax"). A pneumothorax is reversed by closing the "hole" and drawing air out of the intrapleural space with chest tubes, which allows the lung to reinflate and resume its normal function. Note that since the lungs are in completely separate pleural cavities, one lung can be collapsed without interfering with the other. ■

Pulmonary Ventilation: Inspiration and Expiration

Pulmonary ventilation is a mechanical process that depends on volume changes occurring in the thoracic cavity. A rule to keep in mind throughout the following discussion is that *volume changes* lead to *pressure changes*, which lead to the *flow of gases* to equalize the pressure.

$$\Delta V \to \Delta P \to F \text{ (flow of gases)}$$

The relationship between the pressure and volume of gases is given by **Boyle's law**, an *ideal gas law*, which states that when the temperature is constant, the pressure of a gas varies inversely with its volume. That is:

$$P_1V_1 = P_2V_2$$

where P is the pressure of the gas in millimeters of mercury, V is its volume in cubic millimeters, and

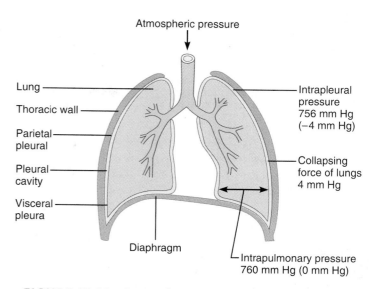

FIGURE 23.12 *Intrapulmonary and intrapleural pressure relationships.* Differences in pressure relative to atmospheric pressure (760 mm Hg) are given in parentheses.

subscripts 1 and 2 represent the initial and resulting conditions respectively.

Gases, like liquids, conform to the shape of their container. However, unlike liquids, gases always *fill* their container. Therefore, in a large volume, the gas molecules will be far apart and the pressure will be low. But if the volume is reduced, the gas molecules will be compressed and the pressure will rise. A good example is an inflated automobile tire. The tire is hard and strong enough to bear the weight of the car because air is compressed to about one-third of its atmospheric volume in the tire, providing the high pressure. Now let us see how this relates to the phases of inspiration and expiration.

Inspiration

The process of **inspiration** (inhalation) is easily understood if you visualize the thoracic cavity as a gas-filled box with a single entrance at the top, the tube-like trachea. The volume of this box is changeable and can be increased by enlarging all of its diameters, thereby decreasing the gas pressure within it. This, in turn, causes air to rush into the box from the atmosphere, because gases always flow down their pressure gradients.

Identical relationships exist during normal quiet inspiration, when the **inspiratory muscles**—the diaphragm and external intercostal muscles—are activated. Here's how quiet inspiration works:

1. Action of the diaphragm. When the dome-shaped diaphragm contracts, it moves inferiorly and flattens out (Figure 23.13, top). As a result, the superior-inferior dimension (height) of the thoracic cavity increases.

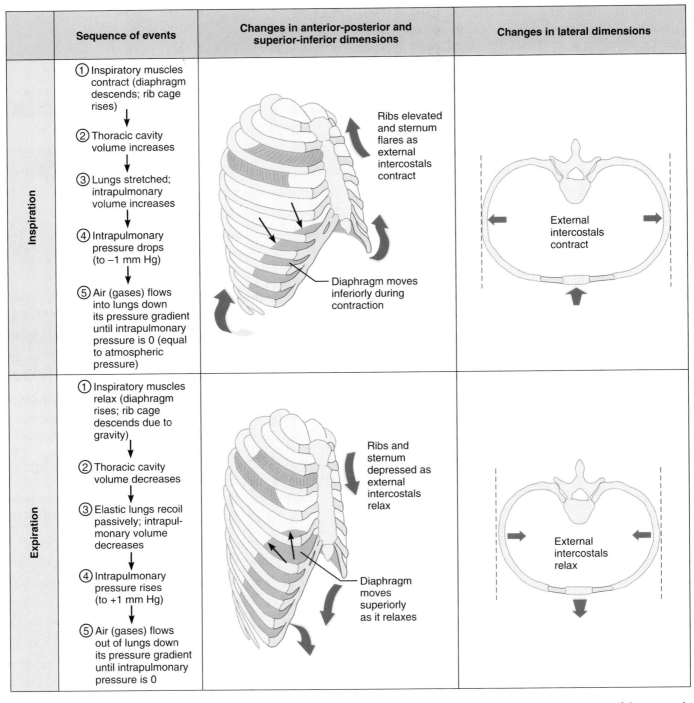

	Sequence of events	Changes in anterior-posterior and superior-inferior dimensions	Changes in lateral dimensions
Inspiration	① Inspiratory muscles contract (diaphragm descends; rib cage rises) ② Thoracic cavity volume increases ③ Lungs stretched; intrapulmonary volume increases ④ Intrapulmonary pressure drops (to –1 mm Hg) ⑤ Air (gases) flows into lungs down its pressure gradient until intrapulmonary pressure is 0 (equal to atmospheric pressure)	Ribs elevated and sternum flares as external intercostals contract Diaphragm moves inferiorly during contraction	External intercostals contract
Expiration	① Inspiratory muscles relax (diaphragm rises; rib cage descends due to gravity) ② Thoracic cavity volume decreases ③ Elastic lungs recoil passively; intrapulmonary volume decreases ④ Intrapulmonary pressure rises (to +1 mm Hg) ⑤ Air (gases) flows out of lungs down its pressure gradient until intrapulmonary pressure is 0	Ribs and sternum depressed as external intercostals relax Diaphragm moves superiorly as it relaxes	External intercostals relax

FIGURE 23.13 *Changes in thoracic volume that lead to the flow of air (gases) during inspiration (top) and expiration (bottom).* In the middle are lateral views of the thorax showing the changes that occur in the superior-inferior dimensions (as the diaphragm alternately contracts and relaxes) and in the anterior-posterior dimensions (as the external intercostal muscles alternately contract and relax). On the right are superior views of transverse thoracic sections showing lateral dimension changes resulting from alternate contraction and relaxation of the external intercostal muscles. Flowcharts of volume changes occurring during inspiration and expiration appear to the left of the diagrams.

2. Action of the intercostal muscles. Contraction of the external intercostal muscles lifts the rib cage and pulls the sternum forward (Figure 23.13, top). Since the ribs curve downward as well as forward around the chest wall, the broadest lateral and anteroposterior dimensions of the rib cage are normally directed obliquely downward. But when the ribs are raised and drawn together, they swing outward, expanding the diameter of the thorax both laterally and in the anteroposterior plane. This is much like the action that occurs when a curved bucket handle is raised.

Although these actions expand the thoracic dimensions by only a few millimeters along each plane, this is enough to increase the volume of the thoracic cavity by almost 500 ml—the usual volume of air that enters the lungs during a normal quiet inspiration. Of the two types of inspiratory muscles, the diaphragm is far more important in producing the volume changes that lead to normal quiet inspiration.

As the thoracic dimensions increase during inspiration, the lungs are stretched and the intrapulmonary volume increases. As a result, intrapulmonary pressure drops about 1 mm Hg relative to atmospheric pressure. Anytime the intrapulmonary pressure is less than the atmospheric pressure ($P_{alv} < P_{atm}$), air rushes into the lungs along the pressure gradient. Inspiration ends when intrapulmonary and atmospheric pressures become equal. During the same period, the intrapleural pressure declines to about −6 mm Hg relative to atmospheric pressure (Figure 23.14).

During the *deep* or *forced inspirations* that occur during vigorous exercise and in some chronic obstructive pulmonary disease victims (see below and 875), the thoracic volume is further increased by activity of accessory muscles. Several muscles, including the scalenes and sternocleidomastoid muscles of the neck and the pectoralis minor of the chest, raise the ribs even more than occurs during quiet inspiration. Additionally, the back extends as the thoracic curvature is straightened by the erector spinae muscles.

Expiration

Quiet **expiration** (exhalation) in healthy individuals is a passive process that depends more on the natural elasticity of the lungs than on muscle contraction. As the inspiratory muscles relax and resume their initial resting length, the rib cage descends and the lungs recoil. Thus, both the thoracic and intrapulmonary volumes decrease. This compresses the alveoli, and the intrapulmonary pressure rises to about 1 mm Hg above atmospheric pressure (see Figure 23.14). When $P_{alv} > P_{atm}$, the pressure gradient forces gases to flow out of the lungs.

In contrast, *forced expiration* is an active process produced by contraction of abdominal wall muscles, primarily the oblique and transversus muscles. These contractions (1) increase the intra-abdominal pressure, which forces the abdominal organs superiorly against the diaphragm, and (2) depress the rib cage. The internal intercostal, latissimus dorsi, and quadratus lumborum muscles may also help to depress the rib cage and decrease thoracic volume.

The ability of a trained vocalist to hold a particular note depends on the coordinated activity of several muscles normally used in forced expiration. Thus, control of accessory muscles of expiration is

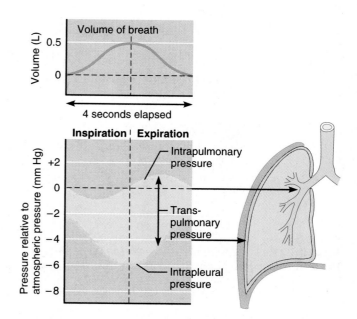

As thoracic volume increases, what happens to the intrapulmonary pressure? To the intrapleural pressure?

FIGURE 23.14 *Changes in intrapulmonary and intrapleural pressures during inspiration and expiration.* Notice that normal atmospheric pressure (760 mm Hg) is given a value of 0 on the scale.

very important when precise regulation of air flow from the lungs is desired.

Physical Factors Influencing Pulmonary Ventilation

As we have seen, the lungs are stretched during inspiration and recoil passively during expiration. The inspiratory muscles consume energy to enlarge the thorax. Energy is also used to overcome various factors that hinder air passage and pulmonary ventilation. These factors—respiratory passageway (airway) resistance, alveolar surface tension forces, and lung compliance—are examined next.

Airway Resistance

The major *nonelastic* source of resistance to gas flow is *friction*, or drag, encountered in the respiratory passageways. The relationship between gas flow *(F)*, pressure *(P)*, and resistance *(R)* is given by the following equation:

$$F = \frac{\Delta P}{R}$$

Notice that the factors determining gas flow in the respiratory passages and blood flow in the cardiovas-

They both decrease.

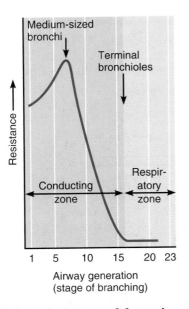

FIGURE 23.15 *Resistance of the various segments of the respiratory passageways.* The airway resistance peaks in the medium-sized bronchi and then declines sharply as the total cross-sectional area of the airway increases rapidly.

cular system are equivalent. The amount of gas flowing in and out of the alveoli is directly proportional to ΔP, the *difference* in pressure, or the pressure gradient, between the external atmosphere and the alveoli, that is, $\Delta (P_{atm} - P_{alv})$. Normally, very small differences in pressure produce large changes in the volume of gas flow. The average pressure gradient during normal quiet breathing is 2 mm Hg or less, and yet it moves 500 ml of air in and out of the lungs with each breath.

But, as the equation also indicates, gas flow changes *inversely* with resistance. That is, gas flow decreases as resistance increases. As with the cardiovascular system, resistance in the respiratory tree is determined mostly by the diameters of the conducting tubes. As a rule, airway resistance is insignificant because

1. The diameters of the airways of the initial part of the conducting zone are huge, relatively speaking.

2. Gas flow stops at the terminal bronchioles (where small airway diameters might start to be a problem) and diffusion takes over as the main force driving gas movements.

Hence, as shown in Figure 23.15, the greatest resistance to gas flow occurs in the medium-sized bronchi.

However, smooth muscle of the bronchiolar walls is exquisitely sensitive to neural controls and certain chemicals. For example, inhaled irritants and inflammatory chemicals such as histamine cause vigorous constriction of the bronchioles and dramat-

ically reduce air passage via reflex activation of the parasympathetic division of the nervous system. Indeed, the strong bronchoconstriction occurring during an *acute asthma attack* can stop pulmonary ventilation almost completely, regardless of the pressure gradient. Conversely, epinephrine released during sympathetic nervous system activation or administered as a drug dilates bronchioles and reduces airway resistance. Local accumulations of mucus, infectious material, or solid tumors in the passageways are important sources of airway resistance in those with respiratory disease.

Whenever airway resistance rises, breathing movements become more strenuous. However, such compensation has its limits. When the bronchioles are severely constricted or obstructed, even the most magnificent respiratory efforts cannot restore ventilation to life-sustaining levels.

Alveolar Surface Tension Forces

At any gas-liquid boundary, the molecules of the liquid are more strongly attracted to each other than to the gas. This unequal attraction produces a state of tension at the liquid surface, called **surface tension,** that (1) draws the liquid molecules ever more closely together and reduces their overall contact with the dissimilar gas molecules, and (2) resists any force that tends to increase the area of the surface.

Water is composed of highly polar molecules and has a very high surface tension. Since water is the major component of the liquid film that coats the alveolar walls, it is always acting to reduce the alveoli to their smallest possible size. If the alveolar film was pure water, the alveoli would actually collapse between breaths. But the alveolar film contains **surfactant** (ser-fak'tant), a detergent-like complex of lipids and proteins produced by the type II alveolar cells. Surfactant interferes with the cohesiveness of water molecules, much the way a laundry detergent reduces the attraction of water for water, allowing water to interact with and pass through fabrics. As a result, the surface tension of alveolar fluid is reduced, and less energy is needed to overcome those forces to expand the lungs and discourage alveolar collapse. According to some authorities, breaths that are deeper than normal stimulate the type II cells to synthesize and secrete more surfactant.

Homeostatic Imbalance

When too little surfactant is present, surface tension forces can collapse the alveoli. This requires that the alveoli be completely reinflated during each inspiration, an effort that uses tremendous amounts of energy. This is the problem faced by newborns with **infant respiratory distress syndrome (IRDS),** a condition peculiar to premature babies. Since inadequate pulmonary surfactant is produced until the last two months of fetal development, babies born prematurely often are unable to keep their alveoli inflated between breaths. IRDS is treated with positive-pressure respirators that force air into the alveoli, keeping them open between breaths. Spraying natural or synthetic surfactant into the newborn's respiratory passageways also helps.

Many survivors of IRDS suffer from *bronchopulmonary dysplasia*, a chronic lung disease, during childhood and beyond. This condition is believed to result from inflammatory injury to respiratory zone structures caused by use of the ventilator on the newborn's delicate lungs. ■

Lung Compliance

Healthy lungs are unbelievably stretchy, or distensible. The ease with which the lungs can be expanded—that is, their distensibility—is referred to as **lung compliance**. Specifically, lung compliance (C_L) is a measure of the change in lung volume (ΔV_L) that occurs with a given change in the transpulmonary pressure $(\Delta[P_{alv} - P_{ip}])$. This is stated as

$$C_L = \frac{\Delta V_L}{\Delta(P_{alv} - P_{ip})}$$

The more the lung expands for a given rise in transpulmonary pressure, the greater its compliance. Said another way, the higher the lung compliance, the easier it is to expand the lungs at any given transpulmonary pressure.

Lung compliance is determined largely by two factors: distensibility of the lung tissue and the surrounding thoracic cage, and surface tension in the alveoli. Because lung (and thoracic) distensibility is generally high and alveolar surface tension is kept low by surfactant, the lungs of healthy people tend to have high compliance, which favors efficient ventilation.

Compliance is diminished by any factor that

1. Reduces the natural resilience of the lungs, such as fibrosis (e.g., scar tissue formed in tuberculosis)

2. Blocks the smaller respiratory passages (e.g., with fluid or thick mucus, as in pneumonia or chronic bronchitis respectively)

3. Reduces the production of surfactant

4. Decreases the flexibility of the thoracic cage or its ability to expand

The lower the lung compliance, the more energy is needed just to breathe.

⬛ *Homeostatic Imbalance*

Deformities of the thorax, ossification of the costal cartilages (common during old age), and paralysis of the intercostal muscles all reduce lung compliance because they hinder thoracic expansion. ■

Respiratory Volumes and Pulmonary Function Tests

Respiratory Volumes and Capacities

The amount of air that is flushed in and out of the lungs varies substantially depending on the conditions of inspiration and expiration. Consequently, several different respiratory volumes can be described. Specific combinations (sums) of these respiratory volumes, called *respiratory capacities*, are measured to gain information about a person's respiratory status. The apparatus that is used to measure respiratory volumes is called a *spirometer*.

Respiratory Volumes The **respiratory,** or **lung, volumes** include tidal, inspiratory reserve, expiratory reserve, and residual volumes. The values recorded in Figure 23.16a and indicated below represent normal values for a healthy 20-year-old male weighing about 70 kg (155 pounds). Figure 23.16b provides average values for both males and females.

During normal quiet breathing, about 500 ml of air moves into (and then out of) the lungs with each breath. This respiratory volume is referred to as the **tidal volume (TV)**. The amount of air that can be inspired forcibly beyond the tidal volume (2100 to 3200 ml) is called the **inspiratory reserve volume (IRV).**

The **expiratory reserve volume (ERV)** is the amount of air—normally 1000 to 1200 ml—that can be evacuated from the lungs after a tidal expiration. Even after the most strenuous expiration, about 1200 ml of air still remains in the lungs; this is the **residual volume (RV)**. Residual volume air helps to keep the alveoli patent (open) and prevent lung collapse.

Respiratory Capacities The **respiratory capacities** include inspiratory capacity, functional residual capacity, vital capacity, and total lung capacity (see Figure 23.16). As noted, the respiratory capacities always consist of two or more lung volumes.

The **inspiratory capacity (IC)** is the total amount of air that can be inspired after a tidal expiration; thus, it is the sum of the tidal and inspiratory reserve volumes. The **functional residual capacity (FRC)** is the combined residual and expiratory reserve volumes and represents the amount of air remaining in the lungs after a tidal expiration.

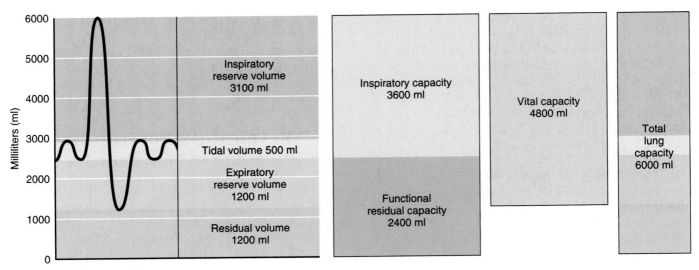

(a) Spirographic record for a male

	Measurement	Adult male average value	Adult female average value	Description
Respiratory volumes	Tidal volume (TV)	500 ml	500 ml	Amount of air inhaled or exhaled with each breath under resting conditions
	Inspiratory reserve volume (IRV)	3100 ml	1900 ml	Amount of air that can be forcefully inhaled after a normal tidal volume inhalation
	Expiratory reserve volume (ERV)	1200 ml	700 ml	Amount of air that can be forcefully exhaled after a normal tidal volume exhalation
	Residual volume (RV)	1200 ml	1100 ml	Amount of air remaining in the lungs after a forced exhalation
Respiratory capacities	Total lung capacity (TLC)	6000 ml	4200 ml	Maximum amount of air contained in lungs after a maximum inspiratory effort: TLC = TV + IRV + ERV + RV
	Vital capacity (VC)	4800 ml	3100 ml	Maximum amount of air that can be expired after a maximum inspiratory effort: VC = TV + IRV + ERV (should be 80% TLC)
	Inspiratory capacity (IC)	3600 ml	2400 ml	Maximum amount of air that can be inspired after a normal expiration: IC = TV + IRV
	Functional residual capacity (FRC)	2400 ml	1800 ml	Volume of air remaining in the lungs after a normal tidal volume expiration: FRC = ERV + RV

(b) Summary of respiratory volumes and capacities for males and females

FIGURE 23.16   *Respiratory volumes and capacities.* (a) Idealized spirographic record of respiratory volumes, in a healthy young adult male weighing approximately 70 kg. **(b)** Summary of respiratory volumes and capacities for males and females.

Vital capacity (VC) is the total amount of exchangeable air. It is the sum of the tidal, inspiratory reserve, and expiratory reserve volumes. In healthy young males, VC is approximately 4800 ml. The **total lung capacity (TLC)** is the sum of all lung volumes and is normally around 6000 ml in males. As indicated in Figure 23.16b, lung volumes and capacities (with the possible exception of tidal volume) tend to be smaller in women than in men because of their smaller size.

Dead Space

Some of the inspired air fills the conducting respiratory passageways and never contributes to gas exchange in the alveoli. The volume of these conducting zone conduits, which make up the **anatomical dead space,** typically amounts to about 150 ml. (The rule of thumb is that the anatomical dead space volume in a healthy young adult is equal to 1 ml per pound of body weight.) This means that if the tidal volume is 500 ml, only 350 ml of this is involved in alveolar ventilation. The remaining 150 ml of the tidal breath is in the anatomical dead space.

TABLE 23.2	Effects of Breathing Rate and Depth on Alveolar Ventilation of Three Hypothetical Patients						

Breathing pattern of hypothetical patient	Dead space volume (DSV)	Tidal volume (TV)	Respiratory rate*	Minute ventilation	Alveolar ventilation	% of (TV) = dead space volume
I—Normal rate and depth	150 ml	500 ml	20/min	10,000 ml/min	7000 ml/min	30%
II—Slow, deep breathing	150 ml	1000 ml	10/min	10,000 ml/min	8500 ml/min	15%
III—Rapid, shallow breathing	150 ml	250 ml	40/min	10,000 ml/min	4000 ml/min	60%

*Respiratory rate values are artificially adjusted to provide equivalent minute respiratory volumes as a baseline for comparison of alveolar ventilation.

If some of the alveoli cease to act in gas exchange (due to alveolar collapse or obstruction by mucus, for example), the **alveolar dead space** is added to the anatomical dead space, and the sum of the nonuseful volumes is referred to as **total dead space.**

Pulmonary Function Tests

Because the various lung volumes and capacities are often abnormal in people with pulmonary disorders, they are routinely measured in such patients. A **spirometer** (spi-rom'ĕ-ter) is a simple instrument utilizing a hollow bell inverted over water. The bell moves as the patient breathes into a connecting mouthpiece, and a graphic recording is made on a rotating drum. Spirometry testing is most useful for evaluating losses in respiratory function and for following the course of certain respiratory diseases. Although it cannot provide a specific diagnosis, spirometry can distinguish between *obstructive pulmonary disease* involving increased airway resistance (such as chronic bronchitis or asthma) and *restrictive disorders* involving a reduction in total lung capacity resulting from structural or functional changes in the lungs (due to diseases such as tuberculosis or polio, or to fibrosis due to exposure to certain environmental agents such as asbestos). Increases in TLC, FRC, and RV may occur as a result of hyperinflation of the lungs in obstructive disease, whereas VC, TLC, FRC, and RV are reduced in restrictive diseases, which limit lung expansion.

More information can be obtained about a patient's ventilation status by assessing the rate at which gas moves into and out of the lungs. The **minute** or **total ventilation** is the total amount of gas that flows into or out of the respiratory tract in 1 minute. During normal quiet breathing, the minute ventilation in healthy people is about 6 L/min (500 ml per breath multiplied by 12 breaths per minute). During vigorous exercise, the minute ventilation may reach 200 L/min, as both the depth and rapidity of breathing increase.

Two other useful tests are the FVC and FEV. **FVC,** or **forced vital capacity,** measures the amount of gas expelled when a subject takes a deep breath and then forcefully exhales maximally and as rapidly as possible. **FEV,** or **forced expiratory volume,** determines the amount of air expelled during specific time intervals of the FVC test. For example, the volume exhaled during the first second is FEV_1. Those with healthy lungs can exhale about 80% of the FVC within 1 second. Those with obstructive pulmonary disease have a low FEV_1, while restrictive disease produces a low FVC.

Alveolar Ventilation

Minute ventilation values provide a rough-and-ready yardstick for assessing respiratory efficiency, but the **alveolar ventilation rate (AVR),** or **alveolar ventilation,** is a better index of effective ventilation. The AVR takes into account the volume of air wasted in dead space areas and measures the flow of fresh gases in and out of the alveoli during a particular time. Alveolar ventilation is computed with the following equation:

$$\underset{\text{(ml/min)}}{\text{AVR}} = \underset{\text{(breaths/min)}}{\text{frequency}} \times \underset{\text{(ml/breath)}}{\text{(TV} - \text{dead space)}}$$

In healthy people, alveolar ventilation is usually about 12 breaths per minute multiplied by the difference of 500 − 150 ml per breath, or 4200 ml/min.

Increasing the volume of each inspiration (the depth of breathing) enhances alveolar ventilation and gas exchange more than raising the respiratory rate because anatomical dead space is constant in a particular individual. Alveolar ventilation drops dramatically during rapid shallow breathing because most of the inspired air never reaches the exchange sites. Furthermore, as tidal volume approaches the dead space value, effective ventilation approaches zero, regardless of rapidity of breathing. The effects of rate and depth of breathing on effective alveolar ventilation are summarized for three hypothetical patients in Table 23.2.

TABLE 23.3	Nonrespiratory Air (Gas) Movements
Movement	Mechanism and result
Cough	Taking a deep breath, closing glottis, and forcing air superiorly from lungs against glottis; glottis opens suddenly and a blast of air rushes upward; can dislodge foreign particles or mucus from lower respiratory tract and propel such substances superiorly
Sneeze	Similar to a cough, except that expelled air is directed through nasal cavities as well as through oral cavity; depressed uvula closes oral cavity off from pharynx and routes air upward through nasal cavities; sneezes clear upper respiratory passages
Crying	Inspiration followed by release of air in a number of short expirations; primarily an emotionally induced mechanism
Laughing	Essentially same as crying in terms of air movements produced; also an emotionally induced response
Hiccups	Sudden inspirations resulting from spasms of diaphragm; believed to be initiated by irritation of diaphragm or phrenic nerves, which serve diaphragm; sound occurs when inspired air hits vocal folds of closed glottis
Yawn	Very deep inspiration, taken with jaws wide open; formerly believed to be triggered by need to increase amount of oxygen in blood but this theory is now being questioned; ventilates all alveoli (not the case in normal quiet breathing)

Nonrespiratory Air Movements

There are many processes other than breathing that move air into or out of the lungs and that may modify the normal respiratory rhythm. Coughs and sneezes clear the air passages of debris or collected mucus. Laughing and crying reflect our emotions. Most of these **nonrespiratory air movements** result from reflex activity, but some are produced voluntarily. Examples of the most common of these movements are given in Table 23.3.

GAS EXCHANGES IN THE BODY
Basic Properties of Gases

Gas exchange in the body occurs by bulk flow of gases (and solutions of gases) and by diffusion of gases through tissues. To fully understand these processes, we must examine some of the physical properties of gases and their behavior in liquids. Two more gas laws—*Dalton's law of partial pressures* and *Henry's law*—provide most of the information we need.

Dalton's Law of Partial Pressures

Dalton's law of partial pressures states that the total pressure exerted by a mixture of gases is the sum of the pressures exerted independently by each gas in the mixture. Further, the pressure exerted by each gas—its **partial pressure**—is directly proportional to its percentage in the total gas mixture.

As noted earlier, atmospheric pressure is approximately 760 mm Hg at sea level. Since air is a mixture of gases, we can examine it more closely in terms of the gases that comprise it. As indicated in Table 23.4, nitrogen makes up about 79% of air, and the partial pressure of nitrogen (written as P_{N_2}) is 78.6% × 760 mm Hg, or 597 mm Hg. Oxygen gas (O_2), which accounts for nearly 21% of the atmosphere, has a partial pressure (P_{O_2}) of 159 mm Hg (20.9% × 760 mm Hg). As you can see, nitrogen and oxygen together contribute about 99% of the total atmospheric pressure. Air also contains 0.04% carbon dioxide (CO_2), up to 0.5% water vapor, and insignificant amounts of inert gases such as argon and helium.

At high altitudes, where the atmosphere is less influenced by gravity, all partial pressure values decline in direct proportion to the decrease in atmospheric pressure. For example, at 10,000 feet above sea level where the atmospheric pressure is 563 mm Hg, the P_{O_2} is 110 mm Hg. Likewise, atmospheric pressure increases by 1 atmosphere (760 mm Hg) for each 33 feet of descent (in water) below sea level. Thus, at 99 feet below sea level, the total pressure on the body is equivalent to 4 atmospheres, or 3040 mm Hg, and the partial pressure exerted by each of the component gases is also quadrupled.

Henry's Law

According to **Henry's law,** when a mixture of gases is in contact with a liquid, each gas will dissolve in the liquid in proportion to its partial pressure. Thus the greater the concentration of a particular gas in the gas phase, the more and the faster the gas will go into solution in the liquid. At equilibrium, the gas partial pressures in both phases are the same. If, however, the partial pressure of a gas later becomes greater in the liquid than in the adjacent gas phase, some of the dissolved gas molecules will reenter the gaseous phase. So the direction (and amount) of gas movements is determined by their partial pressures in the two phases. This flexible situation is exactly

TABLE 23.4	Comparison of Gas Partial Pressures and Approximate Percentages in the Atmosphere and in the Alveoli			
	Atmosphere (sea level)		Alveoli	
Gas	Approximate percentage	Partial pressure (mm Hg)	Approximate percentage	Partial pressure (mm Hg)
N_2	78.6	597	74.9	569
O_2	20.9	159	13.7	104
CO_2	0.04	0.3	5.2	40
H_2O	0.46	3.7	6.2	47
	100.0%	760	100.0%	760

what occurs when gases are exchanged in the lungs and tissues.

The volume of a gas that will dissolve in a liquid at any given partial pressure also depends on the solubility of the gas in the liquid and on the temperature of the liquid. The various gases present in air have very different solubilities in water (or plasma). Carbon dioxide is most soluble. Oxygen is only 1/20 as soluble as CO_2, and nitrogen is nearly insoluble. Thus, at a given partial pressure, much more CO_2 than O_2 will dissolve, and practically no N_2 will go into solution. Beyond this specific condition, the solubility of *any* gas in water decreases with increasing temperature. Think of club soda, which is produced by forcing CO_2 into water under high pressure. If you take the cap off a refrigerated bottle of club soda and allow it to stand at room temperature, within just a few minutes you will have plain water—all the carbon dioxide gas will have escaped from solution.

Hyperbaric oxygen chambers provide clinical applications of Henry's law. These chambers contain oxygen at pressures higher than 1 atmosphere and are used to force greater than normal amounts of oxygen into a patient's blood in cases of carbon monoxide poisoning, circulatory shock, or asphyxiation. Hyperbaric therapy is also used to treat individuals with gas gangrene or tetanus poisoning because the anaerobic bacteria causing these infections cannot live in the presence of high oxygen levels. Scuba diving provides another illustration of Henry's law (see *A Closer Look* on p. 872).

Homeostatic Imbalance

Although breathing oxygen at 2 atmospheres is not a problem for short periods, **oxygen toxicity** develops rapidly when the P_{O_2} is greater than 2.5–3 atmospheres. Excessive oxygen concentrations generate huge amounts of harmful free radicals, resulting in profound central nervous system disturbances, coma, and death. ■

Composition of Alveolar Gas

As shown in Table 23.4, the gaseous makeup of the atmosphere is quite different from that in the alveoli. The atmosphere is almost entirely oxygen and nitrogen; the alveoli contain more carbon dioxide and water vapor and much less oxygen. These differences reflect the effects of the following: (1) gas exchanges occurring in the lungs (oxygen diffuses from the alveoli into the pulmonary blood and carbon dioxide diffuses in the opposite direction), (2) humidification of air by the conducting passages, and (3) the mixing of alveolar gas that occurs with each breath. Because only 500 ml of air is inspired with each tidal inspiration, gas in the alveoli is actually a mixture of newly inspired gases and gases remaining in the respiratory passageways between breaths.

The partial pressures of oxygen and carbon dioxide are easily changed by increasing the depth and rate of breathing. High alveolar ventilation brings more oxygen into the alveoli, increasing the alveolar P_{O_2}, and rapidly eliminates carbon dioxide from the lungs.

Gas Exchanges Between the Blood, Lungs, and Tissues

As described earlier, during *external respiration* oxygen enters and carbon dioxide leaves the blood in the lungs. At the body tissues, where the process is called *internal respiration*, the same gases move in opposite directions by the same mechanism (diffusion). External and internal respiration will be considered consecutively to emphasize their similarities, but keep in mind that gases are transported by the blood between these two exchange sites.

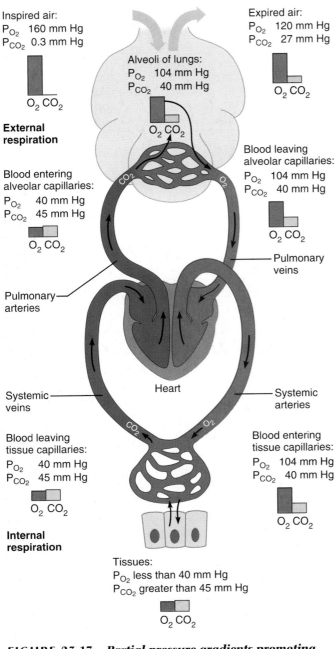

Inspired air:
P_{O_2} 160 mm Hg
P_{CO_2} 0.3 mm Hg

External respiration

Expired air:
P_{O_2} 120 mm Hg
P_{CO_2} 27 mm Hg

Alveoli of lungs:
P_{O_2} 104 mm Hg
P_{CO_2} 40 mm Hg

O_2 CO_2

Blood entering
alveolar capillaries:
P_{O_2} 40 mm Hg
P_{CO_2} 45 mm Hg

O_2 CO_2

Blood leaving
alveolar capillaries:
P_{O_2} 104 mm Hg
P_{CO_2} 40 mm Hg

O_2 CO_2

Pulmonary veins

Pulmonary arteries

Heart

Systemic veins

Systemic arteries

Blood leaving
tissue capillaries:
P_{O_2} 40 mm Hg
P_{CO_2} 45 mm Hg

O_2 CO_2

Internal respiration

Blood entering
tissue capillaries:
P_{O_2} 104 mm Hg
P_{CO_2} 40 mm Hg

O_2 CO_2

Tissues:
P_{O_2} less than 40 mm Hg
P_{CO_2} greater than 45 mm Hg

O_2 CO_2

FIGURE 23.17 *Partial pressure gradients promoting gas movements in the body.* Gradients promoting oxygen and carbon dioxide exchanges across the respiratory membrane in the lungs (external respiration) are shown in the top part of the figure. Gradients promoting gas movements across systemic capillary membranes in the body tissues (internal respiration) are indicated at the bottom. Gradient units are in mm Hg. (Notice that the gaseous composition of alveolar air and expired air are different. This is because there is some mixing of dead space air with the air being expired.)

External Respiration: Pulmonary Gas Exchange

During external respiration, dark red blood flowing through the pulmonary circuit is transformed into the scarlet river that is returned to the heart for distribution by systemic arteries to all body tissues. Although this color change is due to oxygen uptake and

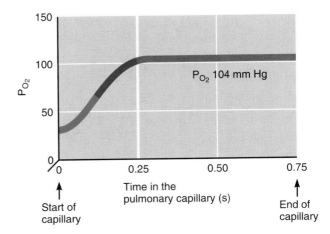

FIGURE 23.18 *Oxygenation of blood in the pulmonary capillaries.* Note that the time expired from the time blood enters the pulmonary capillaries (indicated by 0) until the P_{O_2} is 104 mm Hg is approximately 0.25 second.

its binding to hemoglobin in red blood cells, carbon dioxide exchange (unloading) is occurring equally fast.

Several factors influence the movement of oxygen and carbon dioxide across the respiratory membrane. These include

1. Partial pressure gradients and gas solubilities

2. Functional aspects, such as the matching of alveolar ventilation and pulmonary blood perfusion

3. Structural characteristics of the respiratory membrane (e.g., its thickness and surface area)

Partial Pressure Gradients and Gas Solubilities

Since the P_{O_2} of venous blood in the pulmonary arteries is only 40 mm Hg, as opposed to a P_{O_2} of approximately 104 mm Hg in the alveoli, a steep oxygen partial pressure gradient exists, and oxygen diffuses rapidly from the alveoli into the pulmonary capillary blood (Figure 23.17). Equilibrium—that is, a P_{O_2} of 104 mm Hg on both sides of the respiratory membrane—has usually occurred in 0.25 second, about one-third the time a red blood cell is in a pulmonary capillary (Figure 23.18). The lesson here is that the blood can flow through the pulmonary capillaries three times as quickly and still be adequately oxygenated. Carbon dioxide moves in the opposite direction along a much less steep partial pressure gradient of about 5 mm Hg (45 mm Hg to 40 mm Hg) until equilibrium occurs at 40 mm Hg (see Figure 23.17). Carbon dioxide is then expelled gradually from the alveoli during expiration. Even though the pressure gradient for oxygen diffusion is much steeper than that for carbon dioxide, equal amounts of these gases are exchanged because carbon dioxide is 20 times more soluble in plasma and alveolar fluid than oxygen.

Suppose a patient is receiving oxygen by mask. What will be the condition of the arterioles leading into the oxygen-enriched alveoli? What is the advantage of this response?

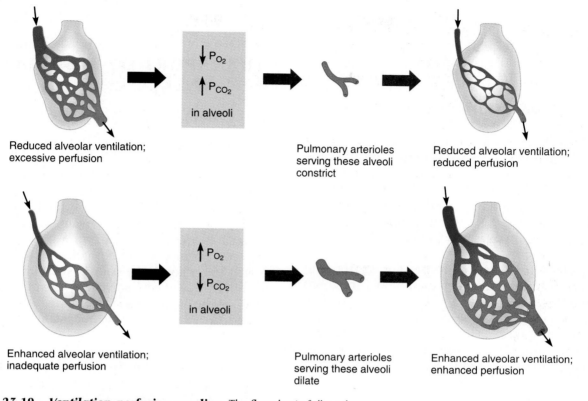

FIGURE 23.19 *Ventilation-perfusion coupling.* The flowcharts follow the autoregulatory events that result in local matching of blood flow (perfusion) through the pulmonary capillaries to current conditions of alveolar ventilation.

In the figure (reading across, top row):
Reduced alveolar ventilation; excessive perfusion → $\downarrow P_{O_2}$ $\uparrow P_{CO_2}$ in alveoli → Pulmonary arterioles serving these alveoli constrict → Reduced alveolar ventilation; reduced perfusion

(bottom row):
Enhanced alveolar ventilation; inadequate perfusion → $\uparrow P_{O_2}$ $\downarrow P_{CO_2}$ in alveoli → Pulmonary arterioles serving these alveoli dilate → Enhanced alveolar ventilation; enhanced perfusion

Ventilation-Perfusion Coupling For gas exchange to be efficient, there must be a close match, or coupling, between *ventilation* (the amount of gas reaching the alveoli) and *perfusion* (the blood flow in pulmonary capillaries). As explained in Chapter 20 and illustrated in Figure 23.19, local autoregulatory mechanisms continuously respond to alveolar conditions.

In alveoli where ventilation is inadequate, the P_{O_2} is low. As a result, the terminal arterioles constrict, and blood is redirected to respiratory areas where P_{O_2} is high and oxygen pickup may be more efficient. On the other hand, in alveoli where ventilation is maximal, pulmonary arterioles dilate, increasing blood flow into the associated pulmonary capillaries. Notice that the autoregulatory mechanism controlling pulmonary vascular muscle is exactly opposite that controlling most arterioles in the systemic circulation.

Changes in the P_{CO_2} in the alveoli cause changes in the diameters of the *bronchioles*. Passageways servicing areas where alveolar CO_2 levels are high dilate, allowing carbon dioxide to be eliminated from the body more rapidly, while those serving areas where the P_{CO_2} is low constrict.

As a result of modifications in these two systems, alveolar ventilation and pulmonary perfusion are synchronized. Poor alveolar ventilation results in low oxygen and high carbon dioxide levels in the alveoli; consequently, the pulmonary capillaries constrict and the airways dilate, bringing air flow and blood flow into closer physiological match. High P_{O_2} and low P_{CO_2} in the alveoli cause respiratory passageways serving the alveoli to constrict, and promote flushing of blood into the pulmonary capillaries. Although these homeostatic mechanisms provide appropriate conditions for efficient gas exchange, they never completely balance ventilation and perfusion in every alveolus because of (1) the shunting of blood from the bronchial and coronary veins, (2) the effects of gravity, and (3) the occasional alveolar duct plugged with mucus. Consequently,

They would be dilated. This response allows matching of blood flow to availability of oxygen.

blood in the pulmonary veins actually has a slightly lower P_{O_2} (100 mm Hg) than alveolar air (104 mm Hg) as opposed to the precise equality indicated in Figure 23.17.

Thickness of the Respiratory Membrane In healthy lungs, the respiratory membrane is only 0.5 to 1 μm thick, and gas exchange is usually very efficient.

Homeostatic Imbalance

If the lungs become waterlogged and edematous (as in pneumonia), the exchange membrane thickens dramatically. Under such conditions, even the total time (0.75 s) that red blood cells are in transit through the pulmonary capillaries may not be enough for adequate gas exchange, and body tissues begin to suffer from oxygen deprivation. ■

Surface Area for Gas Exchange The greater the surface area of the respiratory membrane, the more gas can diffuse across it in a given time period. The alveolar surface area is enormous in healthy lungs. Spread flat, the total gas exchange surface of these tiny sacs in a man's lungs is about 50–70 m^2, or approximately 40 times greater than the surface area of his skin!

Homeostatic Imbalance

In certain pulmonary diseases, the alveolar surface area actually functioning in gas exchange is drastically reduced. This occurs in emphysema, when the walls of adjacent alveoli break through and the alveolar chambers become larger. It also occurs when tumors, mucus, or inflammatory material blocks gas flow into the alveoli. ■

Internal Respiration: Capillary Gas Exchange in the Body Tissues

Although the partial pressure and diffusion gradients are reversed, the factors promoting gas exchanges between the systemic capillaries and the tissue cells are essentially identical to those acting in the lungs (see Figure 23.17). Tissue cells continuously use oxygen for their metabolic activities and produce carbon dioxide. The P_{O_2} in the tissues is always lower than that in the systemic arterial blood (40 mm Hg versus 104 mm Hg). Consequently, oxygen moves rapidly from the blood into the tissues until equilibrium is reached, and carbon dioxide moves quickly along its partial pressure gradient into the blood. As a result, venous blood draining the capillary beds of the tissues and returning to the heart has a P_{O_2} of 40 mm Hg and a P_{CO_2} of 45 mm Hg.

In summary, the gas exchanges that occur between the blood and the alveoli and between the blood and the tissue cells take place by simple diffusion driven by the partial pressure gradients of O_2 and CO_2 that exist on the opposite sides of the exchange membranes.

TRANSPORT OF RESPIRATORY GASES BY BLOOD

Oxygen Transport

Molecular oxygen is carried in blood in two ways, bound to hemoglobin within red blood cells and dissolved in plasma (see Figure 23.22, p. 866). Oxygen is poorly soluble in water, so only about 1.5% of the oxygen transported is carried in the dissolved form. Indeed, if this were the *only* means of oxygen transport, a P_{O_2} of 3 atmospheres or a cardiac output of 15 times normal would be required to provide the oxygen levels needed by body tissues! This problem, of course, has been solved by hemoglobin, and 98.5% of the oxygen ferried from the lungs to the tissues is carried in a loose chemical combination with hemoglobin.

Association and Dissociation of Oxygen and Hemoglobin

As described in Chapter 18, hemoglobin (Hb) is composed of four polypeptide chains, each bound to an iron-containing heme group (see Figure 18.4). Since the iron atoms bind oxygen, each hemoglobin molecule can combine with four molecules of oxygen, and the process of oxygen loading is rapid and reversible.

The hemoglobin-oxygen combination, called **oxyhemoglobin** (ok″sĭ-he″mo-glo′bin), is written **HbO$_2$**. Hemoglobin that has released oxygen is called **reduced hemoglobin**, or **deoxyhemoglobin**, and is written as **HHb**. Loading and unloading of O_2 can be indicated by a single reversible equation:

$$\text{HHb} + O_2 \underset{\text{Tissues}}{\overset{\text{Lungs}}{\rightleftharpoons}} \text{HbO}_2 + \text{H}^+$$

A kind of *cooperation* exists between the four polypeptides of the hemoglobin molecule. After the first O_2 molecule binds to iron, the hemoglobin molecule changes shape. As a result, hemoglobin more readily takes up two more oxygen molecules and uptake of the fourth oxygen molecule is even more facilitated. When all four heme groups are bound to oxygen, a hemoglobin molecule is said to be *fully saturated*. When one, two, or three oxygen molecules are bound, hemoglobin is *partially saturated*. By the same token, unloading of one oxygen molecule enhances the unloading of the next, and so on. Thus, the *affinity* of hemoglobin for oxygen changes with its state of oxygen saturation, and both loading and unloading of oxygen are very efficient.

The rate at which hemoglobin reversibly binds or releases oxygen is regulated by several factors: P_{O_2}, temperature, blood pH, P_{CO_2}, and the concentration of an organic chemical called BPG in the blood. These factors interact to ensure adequate deliveries of oxygen to tissue cells.

Influence of P_{O_2} on Hemoglobin Saturation

Because of the cooperation just described, the relationship between the amount of O_2 bound to hemoglobin (% of Hb saturation) and the P_{O_2} of the blood is not linear. When hemoglobin saturation is plotted against P_{O_2}, the **oxygen-hemoglobin dissociation curve** is produced. This S-shaped curve has a steep slope between 10 and 50 mm Hg P_{O_2} and then flattens out (plateaus) between 70 and 100 mm Hg (Figure 23.20).

Under normal resting conditions (P_{O_2} = 104 mm Hg), arterial blood is 98% saturated, and 100 ml of systemic arterial blood contains about 20 ml of oxygen. The *oxygen content* of arterial blood is written as 20 vol % (volume percent). As arterial blood flows through the systemic capillaries, about 5 ml O_2 per 100 ml of blood is released, yielding a hemoglobin saturation of 75% (and an oxygen content of 15 vol %) in venous blood.

The nearly complete saturation of hemoglobin in arterial blood explains why breathing deeply (which increases both the alveolar and arterial blood P_{O_2} above 104 mm Hg) causes *little* increase in the oxygen saturation of hemoglobin. Remember, P_{O_2} measurements indicate only the amount of oxygen dissolved in plasma, not the amount bound to hemoglobin. However, P_{O_2} values are a good index of lung function, and when arterial P_{O_2} is significantly less than alveolar air P_{O_2}, some degree of respiratory impairment exists.

A hemoglobin saturation curve reveals two other important pieces of information. First, hemoglobin is almost completely saturated at a P_{O_2} of 70 mm Hg, and further increases in P_{O_2} produce only small increases in oxygen binding. The adaptive value of this is that oxygen loading and delivery to the tissues can still be adequate when the P_{O_2} of inspired air is well below its usual levels, a situation common at higher altitudes and in those with cardiopulmonary disease. Moreover, because most oxygen *unloading* occurs on the steep portion of the curve, where the partial pressure changes very little, only 20–25% of bound oxygen is unloaded during one systemic circuit (see Figure 23.20), and substantial amounts of oxygen are still available in venous blood (the *venous reserve*). Thus, if oxygen tension drops to very

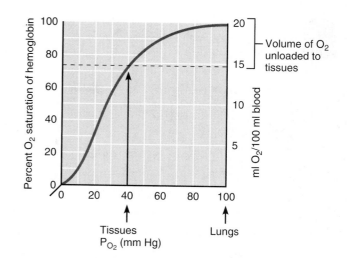

FIGURE 23.20 *Oxygen-hemoglobin dissociation curve.* Percent O_2 saturation of hemoglobin and blood oxygen content are shown at different oxygen partial pressures (P_{O_2}). Notice that hemoglobin is almost completely saturated with oxygen at a P_{O_2} of 70 mm Hg. Rapid loading and unloading of oxygen to and from hemoglobin occur at the oxygen partial pressures corresponding to the steep portion of the curve. During the systemic circuit, approximately 25% of the hemoglobin-bound oxygen is unloaded to the tissues (that is, approximately 5 ml of the 20 ml of O_2 per 100 ml of blood is released). Thus, hemoglobin of venous blood is still about 75% saturated with oxygen after one round through the body.

low levels in the tissues, as might occur during vigorous muscle activity, much more oxygen can dissociate from hemoglobin to be used by the tissue cells without requiring that respiratory rate or cardiac output increase.

Influence of Temperature, pH, P_{CO_2}, and BPG on Hemoglobin Saturation

A number of factors—most importantly temperature, H^+ concentration, P_{CO_2}, and the amount of BPG in the blood—influence hemoglobin saturation at a given P_{O_2}. BPG (2,3-biphosphoglycerate) is a unique compound that binds reversibly with hemoglobin. It is produced by red blood cells (RBCs) as they break down glucose by the anaerobic process called glycolysis.

All of these factors influence hemoglobin saturation by modifying hemoglobin's three-dimensional structure, and thereby its affinity for oxygen. Generally speaking, an *increase* in temperature (Figure 23.21a), P_{CO_2}, H^+ content of the blood (Figure 23.21b), or BPG levels in blood decreases Hb's affinity for O_2 and causes the oxygen-hemoglobin dissociation curve to shift to the right. This enhances oxygen unloading from the blood. Conversely, a *decrease* in any of these factors increases hemoglobin's affinity for oxygen and shifts the dissociation curve to the left. It is interesting to note that BPG levels decrease as RBCs age, providing a useful marker for blood banks to use to determine when fresh blood

Do increasing temperature and P_{CO_2} affect O_2 unloading in the same or opposite directions?

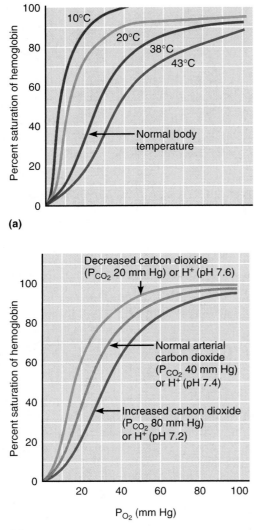

(a)

(b)

FIGURE 23.21 *Effects of temperature, P_{CO_2}, and blood pH on the oxygen-hemoglobin dissociation curve.* Oxygen unloading is accelerated at conditions of (**a**) increased temperature, (**b**) increased P_{CO_2}, and/or hydrogen ion concentration (decreased pH), causing the dissociation curve to shift to the right. This response to decreased pH is called the Bohr effect.

should be discarded. When BPG levels drop too low, O_2 becomes irreversibly bound to hemoglobin, rendering the blood useless for transfusions.

If you give a little thought to how these factors might be related, you'll realize that they all tend to be at their highest levels in the systemic capillaries where oxygen unloading is the goal. As cells metab-

olize glucose and use oxygen, they release carbon dioxide, which in turn increases the P_{CO_2}, and H^+ content of the capillary blood. Since declining blood pH (acidosis) weakens the hemoglobin-oxygen bond, a phenomenon called the **Bohr effect,** oxygen unloading is accelerated where it is most needed. Additionally, heat is a by-product of metabolic activity, and active tissues are usually warmer than less active ones. A rise in temperature affects hemoglobin's affinity for oxygen both directly and indirectly (via its influence on RBC metabolism and BPG synthesis). Collectively, these factors see to it that Hb unloads much more oxygen in the vicinity of hard-working tissue cells.

Certain hormones, such as thyroxine, testosterone, growth hormones, and catecholamines (epinephrine and norepinephrine), increase RBC metabolic rate and BPG formation. As a result, these hormones enhance oxygen delivery to the tissues.

The Hemoglobin–Nitric Oxide Partnership in Gas Exchange

Nitric oxide (NO), secreted by cells of the lungs and vascular endothelial cells, is a well-known vasodilator that plays an important role in blood pressure regulation. Hemoglobin, on the other hand, has a formidable reputation as a vasoconstrictor because it is a NO scavenger—its iron-containing heme group destroys NO. Yet—and here's the paradox—local vessels dilate where gases are being unloaded and loaded.

A second (nonrespiratory) hemoglobin cycle involving NO and actually aiding the loading and unloading of respiratory gases seems to speak to this mystery. In the lungs, as oxygen binds to Hb's heme groups, Hb undergoes shape changes that enable NO to latch onto a dangling cysteine amino acid. The result of this binding is a thiol group in which the NO is protected from being degraded by hemoglobin's iron. The enriched hemoglobin circulates and as it unloads oxygen it also unloads NO, which dilates the local vessels and aids oxygen delivery. Then as deoxygenated hemoglobin picks up carbon dioxide (a process described shortly), it also picks up any circulating NO in the area and carries these gases to the lungs where they are unloaded. What we seem to have here is hemoglobin carrying its own vasodilator!

Impairments of Oxygen Transport

Homeostatic Imbalance

Whatever the cause, inadequate oxygen delivery to body tissues is called **hypoxia** (hi-pok'se-ah). This

The same direction.

condition is easy to recognize in fair-skinned people when hemoglobin saturation falls below 75%, because their skin and mucosae take on a bluish cast (become cyanotic). In dark-skinned individuals, this color change can be observed only in the mucosae and nailbeds. Based on cause, classification of hypoxia is as follows:

1. Anemic hypoxia reflects poor oxygen delivery resulting from too few RBCs or from RBCs that contain abnormal or too little hemoglobin.

2. Ischemic (stagnant) hypoxia results when blood circulation is impaired or blocked. Congestive heart failure may cause body-wide ischemic hypoxia, whereas emboli or thrombi only block oxygen delivery to tissues distal to the obstruction.

3. Histotoxic hypoxia occurs when body cells are unable to use oxygen even though adequate amounts are delivered. This variety of hypoxia is the consequence of metabolic poisons, such as cyanide.

4. Hypoxemic (hypoxic) hypoxia is indicated by reduced arterial P_{O_2}. Possible causes include imbalances in the ventilation-perfusion coupling mechanism, pulmonary diseases that impair ventilation, and breathing air containing scant amounts of oxygen. **Carbon monoxide poisoning** is a unique type of hypoxemic hypoxia, and is the leading cause of death from fire. Carbon monoxide (CO) is an odorless, colorless gas that competes vigorously with oxygen for heme binding sites. Moreover, because hemoglobin's affinity for carbon monoxide is over 200 times greater than that for oxygen, carbon monoxide is a highly successful competitor. Even at minuscule partial pressures, carbon monoxide can displace oxygen.

Carbon monoxide poisoning is particularly dangerous because it does not produce the characteristic signs of hypoxia—cyanosis and respiratory distress. Instead the victim is confused and has a throbbing headache. In rare cases, fair skin becomes cherry red (the color of the hemoglobin–carbon monoxide complex), which the eye of the beholder interprets as a healthy "blush." Those with CO poisoning are given hyperbaric therapy (if available) or 100% oxygen until the carbon monoxide has been cleared from the body. ■

Carbon Dioxide Transport

Normally active body cells produce about 200 ml of carbon dioxide each minute—exactly the amount excreted by the lungs. Carbon dioxide is transported in the blood from the tissue cells to the lungs in three forms: as a gas dissolved in plasma, chemically bound to hemoglobin, and in plasma as the bicarbonate ion (Figure 23.22).

1. Dissolved in plasma. Some 7–10% of the carbon dioxide transported is simply dissolved in plasma.

However, most carbon dioxide molecules entering the plasma quickly enter the red blood cells, where the majority of the chemical reactions that prepare carbon dioxide for transport occur.

2. Chemically bound to hemoglobin in red blood cells. Just over 20% of transported carbon dioxide is carried within RBCs as **carbaminohemoglobin** (kar-bam″ĭ-no-he″mo-glo′bin):

$$CO_2 + \text{hemoglobin} \rightleftharpoons HbCO_2$$
$$\text{(carbaminohemoglobin)}$$

This reaction is rapid and does not require a catalyst. Since carbon dioxide binds directly to the amino acids of *globin* (and not to the heme), carbon dioxide transport in RBCs does not compete with the oxyhemoglobin (or the NO) transport mechanism.

The loading and unloading of carbon dioxide to and from hemoglobin are directly influenced by (1) the P_{CO_2} and (2) the degree of oxygenation of hemoglobin. Carbon dioxide rapidly dissociates from hemoglobin in the lungs, where the P_{CO_2} of alveolar air is lower than that in blood. Carbon dioxide is loaded in the tissues, where the P_{CO_2} is higher than that of the blood. Deoxygenated hemoglobin combines more readily with carbon dioxide than does oxygenated hemoglobin (see the discussion of the Haldane effect on p. 867).

3. As bicarbonate ion in plasma. The largest fraction of carbon dioxide (approximately 70%) is converted to **bicarbonate ions (HCO_3^-)** and transported in plasma. As illustrated in Figure 23.22a, when carbon dioxide diffuses into the RBCs, it combines with water, forming carbonic acid (H_2CO_3). H_2CO_3 is unstable and quickly dissociates into hydrogen ions and bicarbonate ions:

$$CO_2 + H_2O \rightleftharpoons H_2CO_3 \rightleftharpoons H^+ + HCO_3^-$$

Carbon dioxide · Water · Carbonic acid · Hydrogen ion · Bicarbonate ion

Although this reaction also occurs in plasma, it is thousands of times faster in erythrocytes because they (and not plasma) contain **carbonic anhydrase** (kar-bon′ik an-hi′drās), an enzyme that reversibly catalyzes the conversion of carbon dioxide and water to carbonic acid. Hydrogen ions released during the reaction bind to hemoglobin, triggering the Bohr effect; thus, oxygen release is enhanced by carbon dioxide loading (as HCO_3^-). Because of the buffering effect of hemoglobin, the liberated hydrogen ions cause little change in pH under resting conditions. Hence, blood becomes only slightly more acidic (the pH declines from 7.4 to 7.34) as it passes through the tissues.

Once generated, bicarbonate ions diffuse quickly from the RBCs into the plasma, where they are carried to the lungs. To counterbalance the rapid

? *In diagram (a), as O_2 is being unloaded from hemoglobin, the hemoglobin also sheds some nitric oxide. What possible role does the NO play in this gas exchange process?*

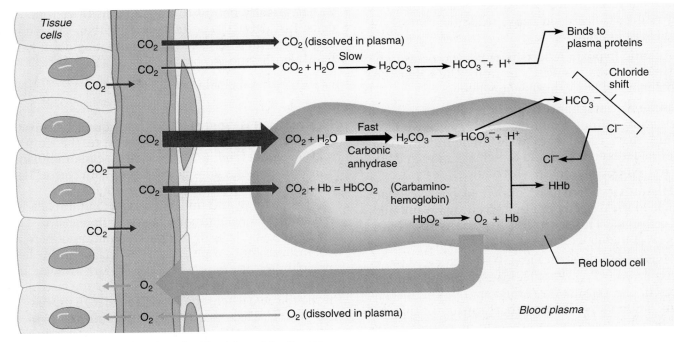

(a) Oxygen release and carbon dioxide pickup at the tissues

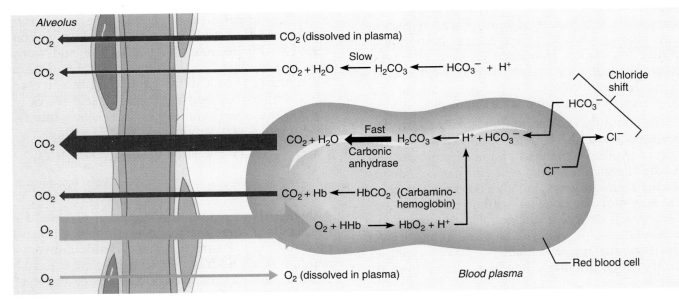

(b) Oxygen pickup and carbon dioxide release in the lungs

FIGURE 23.22 *Transport and exchange of carbon dioxide and oxygen.* Gas exchanges occurring **(a)** at the tissues and **(b)** in the lungs. Carbon dioxide is transported primarily as bicarbonate ion (HCO_3^-) in plasma (70%); smaller amounts are transported bound to hemoglobin (indicated as $HbCO_2$) in the red blood cells (22%) or in physical solution in plasma (7%). Nearly all the oxygen transported in blood is bound to hemoglobin as oxyhemoglobin (HbO_2) in red blood cells. Very small amounts (about 1.5%) are carried dissolved in plasma. (Reduced hemoglobin is indicated as HHb.) Notice that the relative sizes of the transport arrows indicate the proportionate amounts of O_2 and CO_2 being moved by each method.

Causes vasodilation of the local blood vessels, aiding oxygen delivery to the tissue cells and CO_2 loading into the blood.

outrush of negative bicarbonate ions from the RBCs, chloride ions (Cl^-) move from the plasma into the erythrocytes. This ionic exchange process is called the **chloride shift** (see Figure 23.22).

In the lungs, the process is reversed (see Figure 23.22b). As blood moves through the pulmonary capillaries, its P_{CO_2} declines from 45 mm Hg to 40 mm Hg. For this to occur, carbon dioxide must first be freed from its "bicarbonate housing." Bicarbonate ions reenter the RBCs (and chloride ions move into the plasma) and bind with hydrogen ions to form carbonic acid, which is then split by carbonic anhydrase to release carbon dioxide and water. This carbon dioxide, along with that released from hemoglobin and from solution in plasma, then diffuses along its partial pressure gradient from the blood into the alveoli.

The Haldane Effect

The amount of carbon dioxide transported in blood is markedly affected by the degree of oxygenation of the blood. The lower the P_{O_2} and the hemoglobin saturation with oxygen, the more carbon dioxide that can be carried in the blood (Figure 23.23). This phenomenon, called the **Haldane effect,** reflects the greater ability of reduced hemoglobin to form carbaminohemoglobin and to buffer H^+ by combining with it (see Figure 23.22a). As carbon dioxide enters the systemic bloodstream, it causes more oxygen to dissociate from hemoglobin (the Bohr effect), which in turn allows more carbon dioxide to combine with hemoglobin and more bicarbonate ions to be formed (the Haldane effect).

In the pulmonary circulation (Figure 23.22b), the situation is reversed. Uptake of oxygen facilitates the release of carbon dioxide. As hemoglobin becomes saturated with oxygen, the hydrogen ions released combine with HCO_3^-, helping to unload CO_2 from the pulmonary blood. The Haldane effect encourages CO_2 exchange in both the tissues and lungs.

Influence of Carbon Dioxide on Blood pH

Typically, the H^+ released during carbonic acid dissociation is buffered by hemoglobin or other proteins within the red blood cells or in plasma. The bicarbonate ions generated in the red blood cells diffuse into the plasma, where they act as the *alkaline reserve* part of the carbonic acid–bicarbonate buffer system of the blood. The **carbonic acid–bicarbonate buffer system** is very important in resisting shifts in blood pH (see the equation in point 3 on p. 865 concerning CO_2 transport). For example, if the hydrogen ion concentration in blood begins to rise, excess H^+ is removed by combining with HCO_3^- to form carbonic acid (a weak acid that dis-

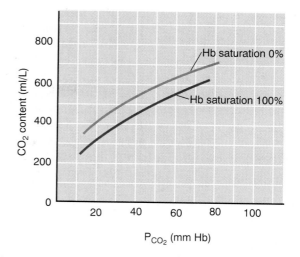

FIGURE 23.23 *The Haldane effect.* Blood equilibrium curves for CO_2 relative to fully oxygenated hemoglobin (Hb) (P_{O_2} = 100%) and deoxygenated Hb show that the lower the degree of Hb saturation, the more CO_2 can be transported by the blood.

sociates very little at either physiological or acidic pH). If H^+ concentration drops below desirable levels in blood, carbonic acid dissociates, releasing hydrogen ions and lowering the pH again.

Changes in respiratory rate or depth can produce dramatic changes in blood pH by altering the amount of carbonic acid in the blood. Slow, shallow breathing allows carbon dioxide to accumulate in the blood. As a result, carbonic acid levels increase and blood pH drops. Conversely, rapid, deep breathing quickly flushes carbon dioxide out of the blood, reducing carbonic acid levels and increasing blood pH. Thus, respiratory ventilation can provide a fast-acting system to adjust blood pH (and P_{CO_2}) when it is disturbed by metabolic factors. Respiratory adjustments play a major role in the acid-base balance of the blood, as discussed in detail in Chapter 27.

CONTROL OF RESPIRATION

Neural Mechanisms and Generation of Breathing Rhythm

Although our tidelike breathing seems so beautifully simple, its control is fairly complex. The basic controls of breathing involve the activity of neurons in the reticular formation of the medulla and pons. Because the medulla sets the respiratory rhythm, we will begin there.

Medullary Respiratory Centers

Clustered neurons in two areas of the medulla oblongata appear to be critically important in respiration. These are (1) the **dorsal respiratory group**

(DRG), located dorsally near the root of cranial nerve IX, and (2) the **ventral respiratory group (VRG)**, a network of neurons that extends within the ventral brain stem from the spinal cord to the pons-medulla junction. The DRG appears to be the pacesetting respiratory center. For this reason it has been tagged the **inspiratory center**. When its neurons fire, a burst of nerve impulses travels along the **phrenic** and **intercostal nerves** to excite the diaphragm and external intercostal muscles, respectively (Figure 23.24). As a result, the thorax expands and air rushes into the lungs. The DRG then becomes dormant, and expiration occurs passively as the inspiratory muscles relax and the lungs recoil. This cyclic on/off activity of the inspiratory neurons repeats continuously and produces a respiratory rate of 12–15 breaths per minute, with inspiratory phases lasting for about 2 seconds followed by expiratory phases lasting about 3 seconds. This normal respiratory rate and rhythm is referred to as **eupnea** (ūp-ne'ah; *eu* = good, *pne* = breath). During severe hypoxia, neurons in the DRG generate gasping (perhaps in a last-ditch effort to restore oxygen to the brain). When the inspiratory center is completely suppressed, as by an overdose of sleeping pills, morphine, or alcohol, respiration stops completely.

Unlike the DRG, which contains neurons mostly involved in initiating inspiration, the VRG contains a more even mix of neurons involved in inspiration and expiration. Both neuronal networks appear to direct the activity of the respiratory muscles, but the contribution of the VRG is murky. It is called into play mainly during forced breathing (especially forced expiration) when more strenuous breathing movements are needed.

Pons Respiratory Centers

Although the medullary inspiratory center generates the basic respiratory rhythm, pons centers influence and modify the activity of medullary neurons. For example, pons centers appear to smooth out the transitions from inspiration to expiration, and vice versa; and when lesions are made in the pneumotaxic center, inspirations become very prolonged.

The **pneumotaxic** (noo"-mo-tak'sik) **center,** the more superior pons center (see Figure 23.24), continuously transmits inhibitory impulses to the inspiratory center of the medulla. When its signals are

Key:

(+) = Positive effect (stimulation)

(−) = Negative effect (inhibition)

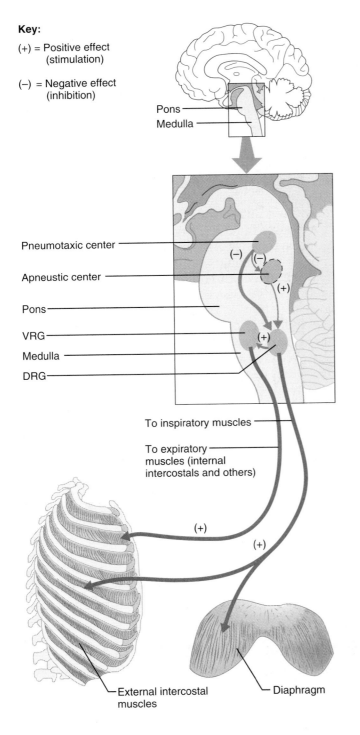

Pons
Medulla

Pneumotaxic center
Apneustic center
Pons
VRG
Medulla
DRG

To inspiratory muscles
To expiratory muscles (internal intercostals and others)

External intercostal muscles
Diaphragm

FIGURE 23.24 *Postulated neural pathways involved in control of respiratory rhythm.* During normal quiet breathing, the neurons of the dorsal respiratory group (DRG), or medullary inspiratory center, set the pace by alternately (1) depolarizing and sending impulses to the inspiratory muscles and then (2) becoming quiescent, allowing passive expiration to occur. Neurons of the ventral respiratory group (VRG) are largely inactive during normal quiet breathing. When forced breathing is required, the inspiratory center enlists the activity of neurons in the VRG which helps activate the muscles of forced expiration. The pons centers interact with the medullary centers in such a way that the breathing pattern is smooth. The pneumotaxic center normally limits the inspiratory phase and smoothes the transitions between inspirations and expirations. The apneustic center in the pons continuously stimulates the DRG unless inhibited by the pneumotaxic center. (Note that the efferent pathway is incomplete. Medullary neurons communicate with lower motor neurons in the spinal cord. The lower motor neurons, which innervate the muscles of respiration, are not illustrated to simplify the figure.)

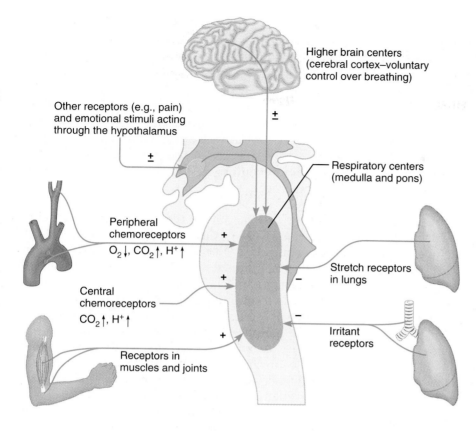

Higher brain centers
(cerebral cortex–voluntary
control over breathing)

Other receptors (e.g., pain)
and emotional stimuli acting
through the hypothalamus

Respiratory centers
(medulla and pons)

Peripheral
chemoreceptors

$O_2\downarrow$, $CO_2\uparrow$, $H^+\uparrow$

Stretch receptors
in lungs

Central
chemoreceptors

$CO_2\uparrow$, $H^+\uparrow$

Irritant
receptors

Receptors in
muscles and joints

FIGURE 23.25 *Neural and chemical influences on the respiratory centers in the medulla.* Excitatory influences (+) increase the frequency of impulses sent to the muscles of respiration and result in deeper, faster breathing. Inhibition of the medullary center (−) has the reverse effect. In some cases, the impulses may be excitatory or inhibitory (indicated by ±), depending on the precise receptors or brain regions activated.

particularly strong, the period of inspiration is shortened. The most important role of the pneumotaxic center is to fine-tune the breathing rhythm and prevent lung overinflation.

The **apneustic** (ap-noo′stik) **center** appears to provide inspiratory drive by continuously stimulating the medullary inspiratory center. Unless inhibited by the pneumotaxic center or by afferent impulses from the lungs, its effect is to prolong inspiration or to cause breath holding in the inspiratory phase. Breathing becomes deep and slow when the pneumotaxic efferents are cut, which indicates that the apneustic center is usually inhibited by pneumotaxic center neurons.

Genesis of the Respiratory Rhythm

Although there is little question that breathing is rhythmic, we still cannot fully explain the origin of its rhythm. One theory is that the inspiratory neurons are *pacemaker neurons,* which have intrinsic automaticity and rhythmicity. Though pacemaker activity has been demonstrated in neurons of the rostral VRG in newborn infants, no such activity has been found in adults.

Another hypothesis is that inputs from stretch receptors in the lungs help to establish the respiratory rhythm (see Inflation Reflex on the next page). However, these influences seem relatively unimportant under resting conditions. The most popular

theory is that normal respiratory rhythm is a result of reciprocal inhibition of interconnected neuronal networks within the medulla. Whatever the influences acting on the medulla, one thing is sure—the medullary centers are themselves capable of maintaining the normal rhythm of breathing.

Factors Influencing the Rate and Depth of Breathing

Depth and rate of breathing can be modified in response to changing body demands. Inspiratory depth is determined by how actively the respiratory center stimulates the motor neurons serving the respiratory muscles. The greater the frequency, the more motor units are excited and the greater the force of respiratory muscle contractions. The rate of respiration is determined by how long the inspiratory center is active or how quickly it is switched off.

The respiratory centers in the medulla and pons are sensitive to both excitatory and inhibitory stimuli. These stimuli are described next and summarized in Figure 23.25.

Pulmonary Irritant Reflexes

The lungs contain receptors that respond to an enormous variety of irritating factors. When activated, these receptors communicate with the respiratory centers via vagal nerve afferents. Accumulated mucus or inhaled debris such as dust, lint, cigarette smoke, or noxious fumes stimulate receptors in the bronchioles that promote reflex constriction of those air passages. The same irritants stimulate a cough when present in the trachea or bronchi, and a sneeze when in the nasal cavity.

The Inflation Reflex

The visceral pleurae and conducting passages in the lungs contain numerous stretch receptors (baroreceptors) that are vigorously stimulated when the lungs are inflated. These receptors, in turn, dispatch inhibitory impulses via afferent fibers of the vagus nerves to the medullary respiratory centers that end inspiration and allow expiration to occur. As the lungs recoil, the stretch receptors become quiet, and so inspiration is initiated once again. This reflex, called the **inflation,** or **Hering-Breuer** (her'ing broy'er), **reflex** is thought to be more a protective response (to prevent excessive stretching of the lungs) than a normal regulatory mechanism because the threshold of these receptors in humans is very high.

Influence of Higher Brain Centers

Hypothalamic Controls Strong emotions and pain acting through the limbic system activate sympathetic centers in the hypothalamus, which can modify respiratory rate and depth by sending signals to the respiratory centers. For example, have you ever touched something cold and clammy and gasped? That response was mediated through the hypothalamus. So too is the breath holding that occurs when we are angry and our increased respiratory rate when we are excited. A rise in body temperature acts to increase the respiratory rate, while a drop in body temperature produces the opposite effect; and sudden chilling of the body (a dip in the North Atlantic Ocean in late October) can cause cessation of breathing (apnea)—or at the very least, gasping.

Cortical Controls Although breathing is normally regulated involuntarily by the brain stem respiratory centers, we can also exert conscious control over the rate and depth of our breathing—choosing to hold our breath or to take an extra deep breath, for example. During voluntary control, the cerebral motor cortex sends direct signals to the motor neurons that stimulate the respiratory muscles, bypassing the medullary centers. Our ability to voluntarily hold our breath is limited, however, because the respiratory centers of the brain stem automatically reinitiate breathing when the concentration of carbon dioxide in the blood reaches critical levels. That explains why drowning victims typically have water in their lungs.

Chemical Factors

Among the factors that influence the rate and depth of breathing set by the medullary inspiratory center, the most important are chemical–changing levels of carbon dioxide, oxygen, and hydrogen ions in arterial blood. Sensors responding to such chemical fluctuations, called **chemoreceptors,** are found in two major body locations. The **central chemoreceptors** are located bilaterally in the medulla. The **peripheral chemoreceptors** are found within the great vessels of the neck.

Influence of P_{CO_2} Of all the chemicals influencing respiration, CO_2 is the most potent and the most closely controlled. Normally, arterial P_{CO_2} is 40 mm Hg and is maintained within ±3 mm Hg of this level by homeostatic mechanisms.

How is respiratory compensation in response to changing P_{CO_2} levels accomplished? Actually, because the peripheral chemoreceptors are only weakly responsive to arterial P_{CO_2} levels, this exquisitely sensitive mechanism is mediated mainly through the influence of rising carbon dioxide levels on the central chemoreceptors of the brain stem (Figure 23.26). Carbon dioxide diffuses easily from the blood into the cerebrospinal fluid, where it is hydrated and forms carbonic acid. As the acid then dissociates, hydrogen ions are liberated. This is the same reaction that occurs when carbon dioxide enters red blood cells (see p. 866). However, unlike RBCs or plasma, cerebrospinal fluid contains virtually no proteins that can buffer these free hydrogen ions. Thus, as P_{CO_2} levels rise, a condition referred to as **hypercapnia** (hi"per-kap'ne-ah), the pH of cerebrospinal fluid drops, exciting the central chemoreceptors, which make abundant synapses with the respiratory regulatory centers. As a result, the depth and perhaps the rate of breathing are increased. This breathing pattern, called **hyperventilation,** enhances alveolar ventilation and quickly flushes CO_2 out of the blood, increasing blood pH. An elevation of only 5 mm Hg in arterial P_{CO_2} results in a doubling of alveolar ventilation, even when arterial oxygen content and pH are unchanged. When P_{O_2} and pH are below normal, the response to elevated P_{CO_2} is even greater. Hyperventilation is normally self-limiting, ending when homeostatic blood P_{CO_2} levels are restored.

Notice that while rising carbon dioxide levels act as the initial stimulus, it is the rising hydrogen ion levels that prod the *central* chemoreceptors into activity. In the final analysis, control of breathing during rest is aimed primarily at *regulating the H^+ concentration in the brain.*

Homeostatic Imbalance

People experiencing anxiety attacks may hyperventilate involuntarily, to the point where they become dizzy and may faint. This reflects the fact that low carbon dioxide levels in the blood (**hypocapnia**) cause cerebral blood vessels to constrict, reducing brain perfusion and producing cerebral ischemia. Such attacks may be averted by breathing into a paper bag. Since the person is rebreathing expired air, rich in carbon dioxide, carbon dioxide is retained in the blood. ■

When P_{CO_2} levels are abnormally low, respiration is inhibited and becomes slow and shallow; that is, **hypoventilation** occurs. In fact, periods of **apnea** (breathing cessation) may occur until the arterial P_{CO_2} rises and again stimulates respiration.

Sometimes competing swimmers voluntarily hyperventilate so they can hold their breath longer during swim meets. This is incredibly dangerous for the following reasons. The oxygen content of blood rarely drops much below 60% of normal during regular breath-holding, because as the P_{O_2} drops, carbon dioxide levels rise enough to make breathing unavoidable. However, strenuous hyperventilation can lower the P_{CO_2} so much that a lag period occurs before it rebounds enough to stimulate respiration again. This lag may allow oxygen levels to fall well below 50 mm Hg, causing the swimmer to black out (and perhaps drown) before he or she has the urge to breathe.

Influence of P_{O_2} Cells sensitive to arterial oxygen levels are found in the peripheral chemoreceptors, that is, in the **aortic bodies** of the aortic arch and in the **carotid bodies** at the bifurcation of the common carotid arteries. Those in the carotid bodies are the main oxygen sensors.

Under normal conditions, the effect of declining P_{O_2} on ventilation is slight and mostly limited to enhancing the sensitivity of central receptors to increased P_{CO_2}. Arterial P_{O_2} must drop *substantially*, to at least 60 mm Hg, before oxygen levels become a major stimulus for increased ventilation. This is not as strange as it may appear. Remember, there is a huge reservoir of oxygen bound to hemoglobin, and hemoglobin remains almost entirely saturated unless or until the P_{O_2} of alveolar gas and arterial blood falls below 60 mm Hg. The central chemoreceptors then begin to suffer from oxygen starvation, and their activity is depressed. At the same time, the peripheral chemoreceptors become excited and stimulate the respiratory centers to increase ventilation, even if P_{CO_2} is normal. Thus, the peripheral chemoreceptor reflex system can maintain ventilation in those whose alveolar oxygen levels are low even though brain stem centers are depressed by hypoxia.

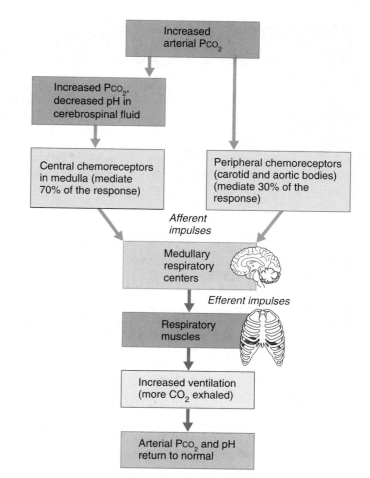

FIGURE 23.26 *Negative feedback mechanism by which changes in P_{CO_2} and H^+ regulate ventilation.*

Homeostatic Imbalance

In people who retain carbon dioxide because of pulmonary disease (e.g., emphysema and chronic bronchitis), arterial P_{CO_2} is chronically elevated and chemoreceptors become unresponsive to this chemical stimulus. In such cases, declining P_{O_2} levels acting on the oxygen-sensitive peripheral chemoreceptors provide the principal respiratory stimulus, or the so-called **hypoxic drive.** Gas mixtures administered to such patients during respiratory distress are only slightly enriched with oxygen. Pure oxygen would stop their breathing, because their respiratory stimulus (low P_{O_2} levels) would be removed. ■

Influence of Arterial pH Changes in arterial pH can modify respiratory rate and rhythm even when carbon dioxide and oxygen levels are normal. Because few hydrogen ions diffuse from the blood into the cerebrospinal fluid, the direct effect of arterial H^+ concentration on central chemoreceptors is insignificant compared to the effect of hydrogen ions generated by elevations in P_{CO_2}. The increased ventilation that occurs in response to falling arterial pH is mediated through the peripheral chemoreceptors.

A CLOSER LOOK   Dangerous Raptures of the Deep

Even the nonswimmers among us are beckoned by photos of scuba divers examining brightly colored coral and other ocean creatures as they glide effortlessly through sparkling blue water. "What a life," we think enviously. But hidden beneath that veneer of breathtaking beauty is life-threatening danger for the inexperienced, the careless, and the unlucky. In addition to the significant risk of drowning, scuba divers are exposed to the omnipresent danger of two other medical emergencies—air embolism and decompression sickness.

As a diver descends to greater and greater depths underwater, the pressure on his or her body rises proportionately (1 atmosphere [760 mm Hg] of pressure for each 10 m [33 ft] of descent) because of the increasing weight of the water. **Scuba** (self-**c**ontained **u**nderwater **b**reathing **a**pparatus) **gear** has freed divers from heavy pressurized suits and air lines to the surface because it permits continual equalization of the air

pressure (provided by the mixture of compressed air in the tank) with the water pressure; that is, air enters the lungs at a higher-than-normal pressure. Thus descent is not usually a problem, unless the diver strives for depths below 100 feet and remains there for an extended time.

Although nitrogen ordinarily has little effect on body functioning, hyperbaric conditions for an extended time force so much nitrogen into solution in the blood that it provokes a narcotic effect called *nitrogen narcosis*. Nitrogen is far more soluble in lipids than in water, so it tends to concentrate in lipid-rich tissues such as the central nervous system, bone marrow, and fat depots. The diver becomes dizzy, giddy, and appears to be intoxicated; for this reason, the condition is sometimes called "rapture of the deep."

Assuming the diver has taken care to avoid these narcotizing effects, and ascends to the surface gradually (see the graph), dissolved nitrogen gas can be driven out of the tissues and then eliminated by the lungs without any problems. But if

the ascent is rapid, the P_{N2} decreases abruptly and the poorly soluble nitrogen gas appears to "boil" from the tissues and out of solution in the body fluids.

Gas bubbles in blood represent potentially lethal emboli, and those formed within joints, bones, and muscles can cause excruciating musculoskeletal pain, commonly called the "bends." Other signs of this *decompression sickness* are mood changes, seizures, nausea, and numbness, as well as "migrating" skin rashes and *chokes*. The syndrome called chokes or cardiovascular bends involves substernal pain, coughing, dyspnea, and, in severe cases, shock due to gas bubbles in the pulmonary capillaries. Although these signs and symptoms usually appear within an hour of surfacing, they can be delayed as long as 36 hours.

Gas emboli may also occur if the diver ascends suddenly and without exhaling. This situation usually occurs when he or she panics after aspiration of seawater, equipment failure, or other hazards, such as extra strong currents or rough waters.

Although changes in P_{CO_2} and H^+ concentrations are interrelated, they are distinct stimuli. A drop in blood pH (acidosis) may reflect carbon dioxide retention, but it may also result from metabolic causes, such as accumulation of lactic acid during exercise or of fatty (or other organic) acids in patients with poorly controlled diabetes mellitus. Regardless of cause, as arterial pH declines, respiratory system controls attempt to compensate and raise blood pH by eliminating carbon dioxide (and carbonic acid) from the blood; thus, respiratory rate and depth increase.

Summary of Interactions of P_{CO_2}, P_{O_2}, and Arterial pH

Although every cell in the body must have oxygen to live, the body's need to rid itself of carbon dioxide is the most important stimulus for breathing in a healthy person. However, as we have seen, carbon dioxide does not act in isolation, and various

chemical factors enforce or inhibit one another's effects. These interactions are summarized here:

1. Rising carbon dioxide levels are the most powerful respiratory stimulant. As carbon dioxide is hydrated in cerebrospinal fluid, liberated H^+ acts directly on the central chemoreceptors, causing a reflexive increase in breathing rate and depth. Low P_{CO_2} levels depress respiration.

2. Under normal conditions, blood P_{O_2} affects breathing only indirectly by influencing chemoreceptor sensitivity to changes in P_{CO_2}. Low oxygen tensions augment P_{CO_2} effects; high P_{O_2} levels diminish the effectiveness of carbon dioxide stimulation.

3. When arterial P_{O_2} falls below 60 mm Hg, it becomes the major stimulus for respiration, and ventilation is increased via reflexes initiated by the peripheral chemoreceptors. This may increase oxygen loading into the blood, but it also causes hypocapnia (low P_{CO_2} blood levels) and an increase in blood pH, both of which inhibit respiration.

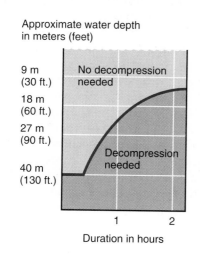

Approximate water depth
in meters (feet)

Even with modern scuba gear, divers need to know how to adjust from high underwater pressures to the lower pressures at the surface.

Under such conditions, the alveoli are likely to rupture. If connections occur between the alveoli and the pulmonary bloodstream, life-threatening gas emboli develop as the diver takes the first breath of air at the surface, and the related problems develop within two minutes. Since the ascent is typically made head-up, the emboli usually invade the cerebral circulation. Seizures, localized motor and sensory deficits, and unconsciousness are all possibilities. Indeed, many scuba-related drownings appear to follow loss of consciousness due to arterial gas embolism.

The usual and most effective treatment for decompression sickness is hyperbaric therapy, where compression is reinstituted and then decompression is redone . . . slowly. When a gas embolus is suspected, oxygen and antiseizure medications are given until hyperbaric therapy can begin. However, the best therapy is to avoid the need for treatment by making sure that divers are well instructed in the use of the equipment and acquainted with the possible risks. Assuming that is done, the vacation should be GREAT!

4. Changes in arterial pH resulting from carbon dioxide retention or metabolic factors act indirectly through the peripheral chemoreceptors to promote changes in ventilation, which in turn modify arterial P_{CO_2} and pH. Arterial pH does not influence the central chemoreceptors directly.

RESPIRATORY ADJUSTMENTS DURING EXERCISE AND AT HIGH ALTITUDES

Effects of Exercise

Respiratory adjustments during exercise are geared to both the intensity and duration of the exercise. Working muscles consume tremendous amounts of oxygen and produce large amounts of carbon dioxide; thus, ventilation can increase 10 to 20 fold during vigorous exercise. Breathing becomes deeper and more vigorous, but the respiratory rate may not be significantly changed. This breathing pattern is called **hyperpnea** (hi"perp-ne′ah) to distinguish it from the deep and often rapid pattern of hyperventilation. Also, the respiratory changes seen in hyperpnea match metabolic demands and so do not lead to significant changes in O_2 or CO_2 levels in the blood. By contrast, hyperventilation may provoke excessive ventilation, resulting in a low P_{CO_2} and alkalosis.

This exercise-enhanced ventilation does *not* appear to be prompted by rising P_{CO_2} and declining P_{O_2} and pH in the blood for two reasons. First, ventilation increases abruptly as exercise begins, followed by a more gradual increase, and finally a steady state of ventilation. Similarly, ventilation declines a bit suddenly when exercise stops, followed by a more gradual decrease to pre-exercise values. Second, while venous levels change, P_{CO_2} and P_{O_2} levels in arterial blood remain surprisingly constant during exercise. In fact, P_{CO_2} levels may decline below normal levels and P_{O_2} levels may rise very slightly because of the efficiency of the respiratory adjustments. Our present understanding of the mechanisms that produce these observations is sketchy, but the most accepted explanation is as follows.

The abrupt increase in ventilation that occurs as exercise begins reflects interaction of the following neural factors:

1. Psychic stimuli (that is, our conscious anticipation of exercise)

2. Simultaneous cortical motor activation of the skeletal muscles and of the respiratory centers

3. Excitatory impulses from proprioceptors in moving muscles, tendons, and joints to the respiratory centers

The subsequent gradual increase and then plateauing of respiration probably reflects the rate of CO_2 delivery to the lungs (the "CO_2 flow").

The small but abrupt decrease in ventilation that occurs as exercise ends reflects the shutting off of the neural control mechanisms. The subsequent gradual decline in ventilation to baseline levels likely reflects a decline in the CO_2 flow. It occurs as the oxygen debt is being repaid. The rise in lactic acid levels that contributes to oxygen debt is *not* a result of inadequate respiratory function, because alveolar ventilation and pulmonary perfusion are as well matched during exercise as during rest. Rather, it reflects cardiac output limitations or inability of the skeletal muscles to further increase their oxygen consumption (see Chapter 9). In light of this fact, the practice of inhaling pure oxygen by mask, used by some football players to replenish their "oxygen-starved" bodies as quickly as possible, is useless. The panting athlete *does* have an oxygen deficit. But extra oxygen will not help, because the shortage is in the muscles—not the lungs.

Effects of High Altitude

Most Americans live between sea level and an altitude of approximately 2400 meters (8000 feet). The differences in barometric pressure within this range are not great enough to cause healthy people any problems when they spend brief periods in the higher-altitude areas. However, when you travel quickly from sea level to elevations above 8000 ft, where air density and P_{O_2} are lower, your body initially responds with symptoms of *acute mountain sickness (AMS)*—headaches, shortness of breath, nausea, and dizziness. This is a common sequel for travelers to ski resorts such as Vail, Colorado (8120 ft), or Brian Head, Utah (a heart-pounding 9600 ft). In severe cases of AMS, lethal pulmonary and cerebral edema may occur.

When you move on a *long-term* basis from sea level to the mountains, your body—bothered to some extent by these symptoms—begins to make several respiratory and hematopoietic adjustments. This multistep adaptive response of the body is called **acclimatization.** As already explained, decreases in arterial P_{O_2} cause the central chemoreceptors to become more responsive to increases in P_{CO_2}, and a substantial decline in P_{O_2} directly stimulates the peripheral chemoreceptors. As a result, increased ventilation occurs as the brain attempts to restore gas exchange to previous levels. Within a few days, the minute respiratory volume stabilizes at a level 2–3 L/min higher than that seen at sea level. Since increased ventilation also reduces arterial carbon dioxide levels, the P_{CO_2} of individuals living at high altitudes is typically below 40 mm Hg (its value at sea level).

Because less oxygen is available to be loaded, high-altitude conditions always result in lower than normal hemoglobin saturation levels. For example, at about 6000 m (19,000 ft) above sea level, the oxygen saturation of arterial blood is only 67% (compared to nearly 98% at sea level). But hemoglobin unloads only 20–25% of its oxygen at sea level pressures; thus, even at these reduced hemoglobin saturations, oxygen needs of the tissues are still met adequately under resting conditions. Additionally, at high altitudes hemoglobin's affinity for oxygen is reduced because of an increased concentration of BPG, and more oxygen is released to the tissues during each circulatory round.

Although the tissues receive adequate oxygen under normal conditions, problems arise when all-out efforts are demanded of the cardiovascular and respiratory systems (as discovered by U.S. athletes competing in 1968 summer Olympic events on the high mesa of Mexico City, at 7370 ft). Unless one is fully acclimatized, such conditions almost guarantee that body tissues will become severely hypoxic.

When blood oxygen tension declines, the kidneys intervene by accelerating the production of erythropoietin, which stimulates bone marrow production of red blood cells (see Chapter 18). This phase of acclimatization, which occurs slowly, provides long-term compensation for living at high altitudes.

HOMEOSTATIC IMBALANCES OF THE RESPIRATORY SYSTEM

The respiratory system is particularly vulnerable to infectious diseases because it is wide open to airborne pathogens. Since many of these inflammatory conditions, such as rhinitis and laryngitis, were considered earlier in the chapter, we will turn our attention to the most disabling respiratory disorders: the diseases collectively called *chronic obstructive pulmonary disease (COPD),* and *asthma, tuberculosis,* and *lung cancer.* COPD and lung cancer are living proof of the devastating effects of cigarette smoking on the body. Long known to promote cardiovascular disease, cigarettes are perhaps even more effective at destroying the lungs.

Chronic Obstructive Pulmonary Disease (COPD)

The **chronic obstructive pulmonary diseases,** exemplified best by chronic bronchitis and obstructive emphysema, are a major cause of death and disability in the United States and are becoming increasingly prevalent. These diseases have certain features in common (Figure 23.27):

1. Patients almost invariably have a history of smoking.

2. Dyspnea (disp-ne'ah), difficult or labored breathing often referred to as "air hunger," occurs and gets progressively more severe.

3. Coughing and frequent pulmonary infections are common.

4. Most COPD victims develop respiratory failure (accompanied by hypoxemia, carbon dioxide retention, and respiratory acidosis).

Obstructive emphysema is distinguished by permanent enlargement of the alveoli, accompanied by deterioration of the alveolar walls. Chronic inflammation leads to lung fibrosis, and invariably the lungs lose their elasticity. Arterial oxygen and carbon dioxide levels remain essentially normal until late in the disease. As the lungs become less elastic, the airways collapse during expiration and obstruct the outflow of air. This has two important consequences: (1) Accessory muscles must be enlisted to breathe, and the victims are perpetually exhausted because they must expend 15–20% of their total body energy supply (as opposed to 5% in healthy individuals) just to breathe. (2) For complex reasons, the bronchioles open during inspiration but collapse during expiration, trapping huge volumes of air in the alveoli. This hyperinflation leads to the development of a permanently expanded "barrel chest" and flattens the diaphragm, thus decreasing ventilation efficiency.

Damage to the pulmonary capillaries increases the resistance in the pulmonary circuit, forcing the right ventricle to overwork, which enlarges it. Emphysema victims are sometimes called "pink puffers" because their breathing is labored, but they do not become cyanotic (blue) because gas exchange remains surprisingly adequate until late in the disease. In addition to cigarette smoking, hereditary

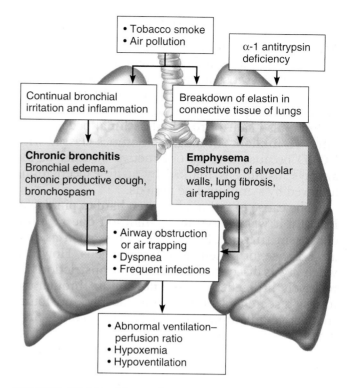

FIGURE 23.27 *The pathogenesis of COPD.*

factors (e.g., α-1 antitrypsin deficiency) may help to cause emphysema in some patients.

In **chronic bronchitis,** inhaled irritants lead to chronic excessive mucus production by the mucosa of the lower respiratory passageways and to inflammation and fibrosis of that mucosa. These responses obstruct the airways and severely impair lung ventilation and gas exchange. Pulmonary infections are frequent because bacteria thrive in the stagnant pools of mucus. Chronic bronchitis patients are sometimes called "blue bloaters" because hypoxia and carbon dioxide retention occur early in the disease and cyanosis is common. However, the degree of dyspnea is usually moderate compared to that of emphysema sufferers. In addition to the major factor of cigarette smoking, environmental pollution may contribute to the development of chronic bronchitis.

COPD is routinely treated with beta agonist drugs (bronchodilators) and corticosteroids in aerosol form (inhalers). Severe dyspnea and hypoxia nearly mandates oxygen use. A new, highly controversial surgical treatment introduced in 1994, called *lung volume reduction surgery,* has been performed on some emphysema patients. In this procedure, part of the grossly enlarged lungs is removed to give the remaining lung tissue room to expand. At least in the short term enough breathing capacity is restored to allow those patients to live something close to a normal life. (The average increase in lung function is 55%.)

endoderm-derived linings and forms the walls of the respiratory passageways and the stroma of the lungs. By 28 weeks, the respiratory system has developed sufficiently to allow a baby born prematurely to breathe on its own. As noted earlier, infants born before this time tend to exhibit infant respiratory distress syndrome resulting from inadequate surfactant production.

During fetal life, the lungs are filled with fluid and all respiratory exchanges are made by the placenta. Vascular shunts (the ductus arteriosus and foramen ovale) cause circulating blood to largely bypass the lungs (see Chapter 29, p. 1136). At birth, the fluid-filled pathway is drained, and the respiratory passageways fill with air. As the P_{CO_2} in the baby's blood rises, the respiratory centers are excited, causing the baby to take its first breath. Consequently, the alveoli inflate and begin to function in gas exchange. However, it is nearly two weeks before the lungs are fully inflated.

Homeostatic Imbalance

Important birth defects of the respiratory system include *cleft palate* and *cystic fibrosis*. **Cystic fibrosis (CF)**, the most common lethal genetic disease in the U.S., strikes one out of every 2400 white children, and every day two children die of it. CF causes oversecretion of a viscous mucus that clogs the respiratory passages, providing a breeding ground for airborne bacteria that predisposes the child to fatal respiratory infections which can only be treated by a lung transplant. It affects secretory processes in other body systems as well. Most importantly it impairs food digestion by clogging ducts that deliver pancreatic enzymes and bile to the small intestine, and sweat glands produce an extremely salty perspiration.

At the root of CF is a faulty gene that codes for the *CFTR* (cystic fibrosis transmembrane conductance regulator) *protein*. The normal CFTR protein works as a chloride channel to control the flow of Cl^- in and out of cells. In those with the mutated genes, CFTR lacks an essential amino acid and so it gets "stuck" in the ER, unable to reach the membrane to perform its normal role. Consequently, less Cl^- is secreted and less water follows, resulting in the thick mucus typical of CF.

Like the gene causing sickle-cell anemia, this mutated gene has beneficial effects in children that carry just *one* of the mutated genes. It seems that it protects against the deadly effects of diarrhea in areas susceptible to cholera.

A new study on CF that dovetails with the earlier findings indicates that in the lungs other forces are at work to initiate the downward spiral typical of this disease. It appears that infection of CF victims' lungs with the bacterium *Pseudomonas aeruginosa* trips another genetic switch that causes the disabled cells to churn out oceans of abnormal mucin (the primary component of mucus). The bacteria then feed on the stagnant pools of mucus and keep sending signals to the cells to make more, thus setting up a positive feedback cycle. Toxins released by the bacteria and the local inflammatory reaction set up by the immune response damage the lungs. Unable to reach the bacteria embedded in the thick mucus, the immune cells begin to attack the lung tissue, turning the air sacs into bloated cysts.

Conventional therapy for CF has been mucus-dissolving drugs, "clapping" the chest to loosen the thick mucus, and antibiotics to prevent infection. Three lines of research being pursued to try to compensate for the deficient CFTR are (1) using disabled cold viruses to carry normal CFTR genes into cells lining the respiratory tract, (2) prodding another channel protein to take over the Cl^- transporting duties, and (3) developing techniques to free the CFTR protein from the ER so that it can assume its normal function at the cell's plasma membrane. Surprisingly, scientists have reversed the signs of CF in mice by feeding them decosahexaenoic acid (DHA), a fatty acid found in fish oils. DHA-based drug testing in humans began in 2000 after it was demonstrated that CF patients also have significant fatty acid imbalances in organs affected by the disease. ■

The respiratory rate is highest in newborn infants (40–80 respirations per minute). At five years of age it is around 25 per minute, and in adults it is between 12 and 18 per minute. In old age, the rate often increases again. At birth, only about one-sixth of the final number of alveoli are present. The lungs continue to mature and more alveoli are formed until young adulthood. However, if smoking begins in the early teens, the lungs never completely mature, and those additional alveoli are lost forever.

In infants, the ribs take a nearly horizontal course. Thus, infants rely almost entirely on descent of the diaphragm to increase thoracic volume for inspiration. By the second year, the ribs are positioned more obliquely, and the adult form of breathing is established.

Homeostatic Imbalance

Most respiratory system problems that occur are the result of external factors—for example, viral or bacterial infections or obstruction of the trachea by a piece of food. For many years, bacterial pneumonia was one of the worst killers in the United States. Antibiotics have greatly decreased its lethality, but it is still a dangerous disease, particularly in the elderly. By far the most problematic diseases *at present* are those described earlier: COPD,

asthma, lung cancer, and newly diagnosed tuberculosis cases in AIDS patients. ■

The maximum amount of oxygen that we can use during aerobic metabolism, the $\dot{V}O_{2max}$, declines about 9% per decade in inactive people beginning in their mid-20s. In those that remain active (work out regularly), $\dot{V}O_{2max}$ still declines but much less—it is comparable to that of inactive folks 20 years younger. As we age, the thoracic wall becomes more rigid and the lungs gradually lose their elasticity, resulting in a decreasing ability to ventilate the lungs. Vital capacity declines by about one-third by the age of 70 years. Blood oxygen levels decline slightly and sensitivity to the stimulating effects of carbon dioxide decreases, particularly in a reclining or supine position. As a result, many old people tend to become hypoxic during sleep and exhibit *sleep apnea* (temporary cessation of breathing during sleep).

The number of glands in the nasal mucosa decreases as does the flow of blood to this mucosa. Thus, the nose dries and produces a thickened mucus that makes us want to clear the throat. Additionally, many of the respiratory system's protective mechanisms become less effective with age. Ciliary activity of its mucosa decreases, and the macrophages in the lungs become sluggish. The net result is that the elderly are more at risk for respiratory tract infections, particularly pneumonia and influenza.

★   ★   ★

Lungs, bronchial tree, heart, and connecting blood vessels—together, these organs fashion a remarkable system that ensures that blood is oxygenated and relieved of carbon dioxide and that all tissue cells of the body have access to these services. Although the cooperation of the respiratory and cardiovascular systems is quite obvious, all organ systems of the body depend on the functioning of the respiratory system, as summarized in *Making Connections* next.

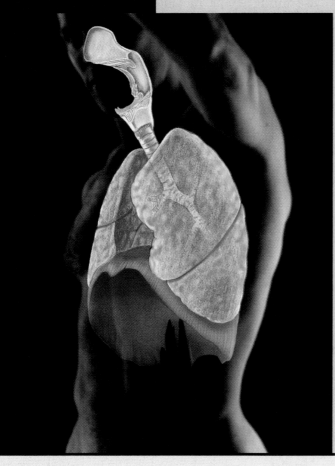

Nervous System
- Respiratory system provides oxygen needed for normal neuronal activity; disposes of carbon dioxide
- Medullary and pons centers regulate respiratory rate and depth; stretch receptors in lungs provide feedback

Endocrine System
- Respiratory system provides oxygen; disposes of carbon dioxide
- Epinephrine dilates the bronchioles; testosterone promotes laryngeal enlargement in pubertal males; glucocorticoids promote surfactant production; catecholamines sensitize peripheral chemoreceptors

Cardiovascular System
- Respiratory system provides oxygen; disposes of carbon dioxide; carbon dioxide present in blood as HCO_3^- and H_2CO_3 contributes to blood buffering
- Blood is the transport medium for respiratory gases

Lymphatic System/Immunity
- Respiratory system provides oxygen; disposes of carbon dioxide; tonsils in pharynx house immune cells
- Lymphatic system helps to maintain blood volume required for respiratory gas transport; immune system protects respiratory organs from bacteria, bacterial toxins, viruses, and cancer

Digestive System
- Respiratory system provides oxygen; disposes of carbon dioxide
- Digestive system provides nutrients needed by respiratory system organs

Urinary System
- Respiratory system provides oxygen; disposes of carbon dioxide
- Kidneys dispose of metabolic wastes of respiratory system organs (other than carbon dioxide)

Reproductive System
- Respiratory system provides oxygen; disposes of carbon dioxide

Integumentary System
- Respiratory system provides oxygen; disposes of carbon dioxide
- Skin protects respiratory system organs by forming surface barriers

Skeletal System
- Respiratory system provides oxygen; disposes of carbon dioxide
- Bones protect lungs and bronchi by enclosure

Muscular System
- Respiratory system provides oxygen needed for muscle activity; disposes of carbon dioxide
- Activity of the diaphragm and intercostal muscles essential for producing volume changes that lead to pulmonary ventilation; regular exercise increases respiratory efficiency

Every day we inhale and exhale nearly 20,000 L of air. This accomplishes two things—it supplies the body with the oxygen it needs to oxidize food and release energy, and it expels carbon dioxide, the major waste product of that process. As crucial as breathing is, most of us don't think very often about the importance of having fresh air on call. But the phenomenon of gulping for air is familiar to every athlete. Indeed, the respiratory rate of a competitive swimmer may jump to over 40 breaths/min and the amount of air inhaled per breath may soar from the usual 500 ml to as much as 6 to 7 L.

The respiratory system is beautifully engineered for its function. Its alveoli are flushed with new air more than 15,000 times each day and the alveolar walls are so indescribably thin that red blood cells moving single-file through the pulmonary capillaries can make the gaseous "swaps" with the air-filled alveoli within a fraction of a second. Although every cell in the body depends on this system for the oxygen it needs to live, the respiratory system interactions that we will consider here are those it has with the cardiovascular, lymphatic, and muscular systems.

Cardiovascular System

For all intents and purposes, the interaction between the respiratory and cardiovascular systems is so intimate that these two systems are inseparable. Respiratory system organs, as important as they are, can make only the external gas exchanges, those that occur in the lungs. Although all body cells depend on the respiratory system to provide needed oxygen, they pick up that oxygen not from the lungs but from the blood. Thus, without blood to act as the intermediary and the heart and blood vessels to act as the hardware to pump the blood around the body, all the efforts of the respiratory system would be useless.

In turn, production of angiotensin converting enzyme by the type II cells of the alveoli plays an important role in blood pressure regulation.

Lymphatic System/Immunity

Of all body systems, only the respiratory system is completely exposed to the external environment (yes, the skin is exposed but its exposed parts are all dead). Because air contains a rich mix of potentially dangerous inhabitants (bacteria, viruses, fungi, asbestos fibers, etc.), the respiratory system is continuously at risk for infection or damage from external agents. Lymphoid outposts help protect the respiratory tract and enhance the defenses (cilia, mucus) that the respiratory system itself erects. Particularly well situated to apprehend intruders at the oral-nasal-pharynx junction are the palatine, pharyngeal, lingual, and tubal tonsils. Their macrophages engulf foreign antigens and they provide sites where lymphocytes are sensitized to mount immune responses. Their effectiveness is revealed by the fact that respiratory tract infections are much more frequent in people who have had their tonsils removed.

Muscular System

Skeletal muscle cells, like other body cells, need oxygen to live. The notable part of this interaction is that most respiratory compensations that occur service increased muscular activity. (It is difficult to think of incidents where this is not the case—the exceptions concern disease conditions.) When we are at rest, the respiratory system operates at basal levels, but any time physical activity becomes more vigorous, the respiratory rhythm picks up the beat to match supply to need and maintain the acid-base balance of the blood.

CLINICAL CONNECTIONS

Respiratory System

Case study: Barbara Joley was in the bus that was hit broadside. When she was freed from the wreckage, she was deeply cyanotic and her respiration had stopped. Her heart was still beating, but her pulse was fast and thready. The emergency medical technician reported that when Barbara was found, her head was cocked at a peculiar angle and it looked like she had a fracture at the level of the C_2 vertebra. The following questions refer to these observations.

1. How might the "peculiar" head position explain Barbara's cessation of breathing?

2. What procedures (do you think) should have been initiated immediately by the emergency personnel?

3. Why is Barbara cyanotic? Explain cyanosis.

4. Assuming that Barbara survives, how will her accident affect her lifestyle in the future?

Barbara survived transport to the hospital and notes recorded at admission included the following observations.

- Right thorax compressed; ribs 7 to 9 fractured
- Right lung atelectasis

Relative to these notes:

5. What is atelectasis and why is only the right lung affected?

6. How do the recorded injuries relate to the atelectasis?

7. What treatment will be done to reverse the atelectasis? What is the rationale for this treatment?

(Answers in Appendix F)

RELATED CLINICAL TERMS

Adenoidectomy (adenotonsillectomy) Surgical removal of an infected pharyngeal tonsil (adenoids).

Adult respiratory distress syndrome (ARDS) A dangerous lung condition that can develop after severe injury to the body. Following the trauma, neutrophils leave the body's capillaries in large numbers and then secrete chemicals that increase capillary permeability. The capillary-rich lungs are heavily affected. As the lungs fill with the fluids of edema, the patient suffocates. Even with mechanical ventilation, ARDS is hard to control and often lethal.

Aspiration (as″pĭ-ra′shun) (1) The act of inhaling or drawing something into the lungs or respiratory passages. Pathological aspiration in which vomit or excessive mucus is drawn into the lungs may occur when a person is unconscious or anesthetized; turning the head to one side helps to prevent this event. (2) Withdrawal of fluid by suction (use of an aspirator); done during surgery to keep an area free of blood or other body fluids; mucus is aspirated from the trachea of tracheotomy patients.

Bronchoscopy (*scopy* = viewing) Use of a viewing tube inserted through the nose or mouth to examine the internal surface of the main bronchi in the lung. Forceps attached to the tip of the tube can remove trapped objects or take samples of mucus for examination.

Cheyne-Stokes (chān′stōks) **breathing** Abnormal breathing pattern sometimes seen just before death (the "death rattle") and in people with combined neurological and cardiac disorders. It consists of bursts of tidal volume breaths (increasing and then decreasing in depth) alternating with periods of apnea. Trauma and hypoxia of the brain stem centers, as well as P_{CO_2} imbalances between arterial blood and cerebrospinal fluid, may be causative factors.

Deviated septum Condition in which the nasal septum takes a more lateral course than usual and may obstruct breathing; often manifests in old age, but may also result from nose trauma.

Endotracheal tube A thin plastic tube threaded into the trachea through the nose or mouth; used to deliver oxygen to patients who are breathing inadequately, in a coma, or under anesthesia.

Epistaxis (ep″ĭ-stak′sis; *epistazo* = to bleed at the nose) Nosebleed; commonly follows trauma to the nose or excessive nose blowing. Most nasal bleeding is from the highly vascularized anterior septum and can be stopped by pinching the nostrils closed or packing them with cotton.

Nasal polyps Mushroomlike benign neoplasms of the nasal mucosa; may occur in response to nasal irritation and may block airflow.

Orthopnea (or″thop-ne′ah; *ortho* = straight, upright) Inability to breathe in the horizontal (lying down) position.

Otorhinolaryngology (o″to-ri″no-lar″in-gol′o-je; *oto* = ear; *rhino* = nose) Branch of medicine that deals with diagnosis and treatment of diseases of the ears, nose, and throat.

Pneumonia Infectious inflammation of the lungs, in which fluid accumulates in the alveoli; the sixth most common cause of death in the United States. Most of the more than 50 different varieties of pneumonia are viral or bacterial.

Pulmonary embolism Obstruction of the pulmonary artery or one of its branches by an embolus (most often a blood clot that has been carried from the lower limbs and through the right side of the heart into the pulmonary circulation). Symptoms are chest pain, productive bloody cough, tachycardia, and rapid, shallow breathing. Can cause sudden death unless treated quickly; usual treatment is oxygen by mask, morphine to relieve pain and anxiety, and anticoagulant drugs to help dissolve the clot.

Stuttering Condition in which the vocal cords of the larynx are out of control and the first syllable of words is repeated in "machine-gun" fashion. The cause is undetermined, but problems with neuromuscular control of the larynx and emotional factors are suspect. Many stutterers become fluent when whispering or singing, both of which involve a change in the manner of vocalization.

Sudden infant death syndrome (SIDS) Unexpected death of an apparently healthy infant during sleep. Commonly called crib death, SIDS is one of the most frequent causes of death in infants under one year old. Believed to be a problem of immaturity of the respiratory control centers. Most cases occur in infants placed in a prone position (on their abdomen) to sleep—a position which results in hypoxia due to rebreathing exhaled (CO_2-rich) air. Since 1992, a campaign recommending placing infants on their back to sleep has led to a decline of 40% or more in the incidence of SIDS in the U.S.

Tracheotomy (tra″ke-ot′o-me) Surgical opening of the trachea; done to provide an alternate route for air to reach the lungs when more superior respiratory passageways are obstructed (as by food or a crushed larynx).

CHAPTER SUMMARY

Media study tools that could provide you additional help in reviewing specific key topics of Chapter 23 are referenced below. **SP** = *Study Partner;* **IP** = *Interactive Physiology.*

1. Respiration involves four processes: ventilation, external respiration, internal respiration, and transport of respiratory gases in the blood. Both the respiratory system and the cardiovascular system are involved in respiration.

Functional Anatomy of the Respiratory System (pp. 835–850)

2. Respiratory system organs are divided functionally into conducting zone structures (nose to bronchioles), which filter, warm, and moisten incoming air; and respiratory zone structures (respiratory bronchioles to alveoli), where gas exchanges occur.

The Nose and the Paranasal Sinuses (pp. 835–839)

3. The nose provides an airway for respiration; warms, moistens, and cleanses incoming air; and houses the olfactory receptors.

4. The external nose is shaped by bone and cartilage plates. The nasal cavity, which opens to the exterior, is divided by the nasal septum. Paranasal sinuses and nasolacrimal ducts drain into the nasal cavities.

The Pharynx   (p. 839)

5. The pharynx extends from the base of the skull to the level of C$_6$. The nasopharynx is an air conduit; the oropharynx and laryngopharynx are common passageways for food and air. Pairs of tonsils are found in the oropharynx and nasopharynx.

The Larynx   (pp. 839–842)

6. The larynx, or voice box, contains the vocal cords. It also provides a patent airway and serves as a switching mechanism to route food and air into the proper channels.

7. The epiglottis prevents food or liquids from entering the respiratory channels during swallowing.

The Trachea   (pp. 842–844)

8. The trachea extends from the larynx to the primary bronchi. The trachea is reinforced by C-shaped cartilage rings, which keep the trachea patent, and its mucosa is ciliated.

The Bronchi and Subdivisions: The Bronchial Tree (pp. 844–846)

9. The right and left main bronchi run into their respective lungs, within which they continue to subdivide into smaller and smaller passageways.

10. The terminal bronchioles lead into respiratory zone structures: alveolar ducts, alveolar sacs, and finally alveoli. Gas exchange occurs in the alveoli, across the respiratory membrane.

11. As the respiratory conduits become smaller, cartilage is reduced in amount and finally lost; the mucosa thins, and smooth muscle in the walls increases.

SP Exercise: Chapter 23, Anatomy of the Respiratory Membrane.

IP Respiratory System CD-ROM; Topic: Anatomy Review, page 6.

The Lungs and Pleural Coverings   (pp. 847–850)

12. The lungs, the paired organs of gas exchange, flank the mediastinum in the thoracic cavity. Each is suspended in its own pleural cavity via its root and has a base, an apex, and medial and costal surfaces. The right lung has three lobes; the left has two.

13. The lungs are primarily air passageways/chambers, supported by an elastic connective tissue stroma.

14. The pulmonary arteries carry blood returned from the systemic circulation to the lungs, where gas exchange occurs. The pulmonary veins return newly oxygenated (and most venous) blood back to the heart to be distributed throughout the body. The bronchial arteries provide the nutrient blood supply of the lungs.

15. The parietal pleura lines the thoracic wall and mediastinum; the pulmonary pleura covers external lung surfaces. Pleural fluid reduces friction during breathing movements.

IP Respiratory System CD-ROM; Topic: Anatomy Review: Respiratory Structure, pages 1–5.

Mechanics of Breathing   (pp. 850–858)

Pressure Relationships in the Thoracic Cavity (pp. 850–851)

1. Intrapulmonary pressure is the pressure within the alveoli. Intrapleural pressure is the pressure within the pleural cavity; it is always negative relative to intrapulmonary and atmospheric pressures.

IP Respiratory System CD-ROM; Topic: Pulmonary Ventilation, pages 7–9.

Pulmonary Ventilation: Inspiration and Expiration (pp. 851–853)

2. Gases travel from an area of higher pressure to an area of lower pressure.

3. Inspiration occurs when the diaphragm and intercostal muscles contract, increasing the dimensions (and volume) of the thorax. As the intrapulmonary pressure drops, air rushes into the lungs until the intrapulmonary and atmospheric pressures are equalized.

4. Expiration is largely passive, occurring as the inspiratory muscles relax and the lungs recoil. When intrapulmonary pressure exceeds atmospheric pressure, gases flow from the lungs.

IP Respiratory System CD-ROM; Topic: Pulmonary Ventilation, pages 3–6, 11–13.

Physical Factors Influencing Pulmonary Ventilation (pp. 853–855)

5. Friction in the air passageways causes resistance, which decreases air passage and causes breathing movements to become more strenuous. The greatest resistance to air flow occurs in the midsize bronchi.

6. Surface tension of alveolar fluid acts to reduce alveolar size and collapse the alveoli. This tendency is resisted in part by surfactant.

7. Lung compliance depends on elasticity of lung tissue and flexibility of the bony thorax. When either is impaired, expiration becomes an active process, requiring energy expenditure.

IP Respiratory System CD-ROM; Topic: Pulmonary Ventilation, pages 14–18.

Respiratory Volumes and Pulmonary Function Tests (pp. 855–857)

8. The four respiratory volumes are tidal, inspiratory reserve, expiratory reserve, and residual. The four respiratory capacities are vital, functional residual, inspiratory, and total lung. Respiratory volumes and capacities may be measured by spirometry.

9. Anatomical dead space is the air-filled volume (about 150 ml) of the conducting passageways. If alveoli become nonfunctional in gas exchange, their volume is added to the anatomical dead space, and the sum is the total dead space.

10. Alveolar ventilation is the best index of ventilation efficiency because it accounts for anatomical dead space.

$$AVR = \left(TV - \text{anatomical dead space}\right) \times \text{respiratory rate}$$
$$\text{(ml/breath)}$$

11. The FVC and FEV tests, which determine the rate at which VC air can be expelled, are particularly valuable in distinguishing between obstructive and restrictive disease.

SP Exercise: Chapter 23, Respiratory Volumes and Capacities.

Nonrespiratory Air Movements   (p. 858)

12. Nonrespiratory air movements are voluntary or reflex actions that clear the respiratory passageways or express emotions.

Gas Exchanges in the Body   (p. 858–862)

Basic Properties of Gases   (pp. 858–859)

1. Gaseous movements in the body occur by bulk flow and by diffusion.

2. Dalton's law states that each gas in a mixture of gases exerts pressure in proportion to its percentage in the total mixture.

3. Henry's law states that the amount of gas that will dissolve in a liquid is proportional to the partial pressure of the gas. Solubility of the gas in the liquid and the temperature are other important factors.

> **IP** Respiratory System CD-ROM; Topic: Gas Exchange, pages 1–6.

Composition of Alveolar Gas   (p. 859)

4. Alveolar gas contains more carbon dioxide and water vapor and considerably less oxygen than atmospheric air.

Gas Exchanges Between the Blood, Lungs, and Tissues   (pp. 859–862)

5. External respiration is the process of gas exchange that occurs in the lungs. Oxygen enters the pulmonary capillaries; carbon dioxide leaves the blood and enters the alveoli. Factors influencing this process include the partial pressure gradients, the thickness of the respiratory membrane, surface area available, and the matching of alveolar ventilation and pulmonary perfusion.

6. Internal respiration is the gas exchange that occurs between the systemic capillaries and the tissues. Carbon dioxide enters the blood, and oxygen leaves the blood and enters the tissues.

> **IP** Respiratory System CD-ROM; Topic: Gas Exchange, pages 6–11, 15–16.

Transport of Respiratory Gases by Blood (pp. 862–867)

Oxygen Transport   (pp. 862–865)

1. Molecular oxygen is carried bound to hemoglobin in the blood cells. The amount of oxygen bound to hemoglobin depends on the P_{O_2} and P_{CO_2} of blood, blood pH, the presence of BPG, and temperature. A small amount of oxygen gas is transported dissolved in plasma. Nitric oxide carried by hemoglobin aids the delivery of O_2 and pickup of CO_2 in the tissues by causing vasodilation.

2. Hypoxia occurs when inadequate amounts of oxygen are delivered to body tissues. When this occurs, the skin and mucosae may become cyanotic.

Carbon Dioxide Transport   (pp. 865–867)

3. CO_2 is transported in the blood dissolved in plasma, chemically bound to hemoglobin, and (primarily) as bicarbonate ion in plasma. Loading and unloading of O_2 and CO_2 are mutually beneficial.

4. Accumulation of CO_2 leads to acidosis; depletion of CO_2 from blood leads to respiratory alkalosis.

> **IP** Respiratory System CD-ROM; Topic: Gas Transport, pages 1–15.

Control of Respiration   (pp. 867–873)

Neural Mechanisms and Generation of Breathing Rhythm   (pp. 867–869)

1. Medullary respiratory centers are the inspiratory center (dorsal respiratory group) and the ventral respiratory group. The inspiratory center is responsible for the rhythmicity of breathing.

2. The pneumotaxic center (and perhaps other respiratory centers in the pons) influence the activity of the medullary inspiratory center.

Factors Influencing the Rate and Depth of Breathing (pp. 869–873)

3. Pulmonary irritant reflexes are initiated by dust, mucus, fumes, and pollutants.

4. The inflation (Hering-Breuer) reflex is a protective reflex initiated by extreme overinflation of the lungs; it acts to initiate expiration.

5. Emotions, pain, body temperature changes, and other stressors can alter respiration by acting through hypothalamic centers. Respiration can also be controlled voluntarily for short periods of time.

6. Important chemical factors modifying baseline respiratory rate and depth are arterial levels of CO_2, H^+, and O_2.

7. An increasing P_{CO_2} is the most powerful respiratory stimulant. It acts (via release of H^+ in the CSF) on the central chemoreceptors to cause a reflexive increase in the rate and depth of breathing. Hypocapnia depresses respiration and results in hypoventilation and, possibly, apnea.

8. Acidosis and a decline in blood P_{O_2} act on peripheral chemoreceptors and enhance the response to CO_2.

9. Arterial P_{O_2} levels below 60 mm Hg constitute the hypoxic drive.

> **IP** Respiratory System CD-ROM; Topic: Control of Respiration, pages 6–14.

Respiratory Adjustments During Exercise and at High Altitudes   (pp. 873–874)

Effects of Exercise   (pp. 873–874)

1. As exercise begins, there is an abrupt increase in ventilation (hyperpnea) followed by a more gradual increase. When exercise stops, there is an abrupt decrease in ventilation followed by a gradual decline to baseline values.

2. P_{O_2}, P_{CO_2}, and blood pH remain quite constant during exercise and hence do not appear to account for changes in ventilation. Psychological factors and proprioceptor inputs may contribute.

Effects of High Altitude   (p. 874)

3. At high altitudes, there is a decrease in arterial P_{O_2} and hemoglobin saturation levels because of the decrease in barometric pressure compared to sea level. Hyperventilation helps restore gas exchange to physiological levels.

4. Long-term acclimatization involves increased erythropoiesis.

Homeostatic Imbalances of the Respiratory System (pp. 874–877)

1. Two major respiratory disorders are COPD (emphysema and chronic bronchitis) and lung cancer; a significant cause is cigarette smoking. A third major disorder is asthma. Tuberculosis is re-emerging as a major health problem.

Chronic Obstructive Pulmonary Disease (COPD) (p. 875)

2. Emphysema is characterized by permanent enlargement and destruction of alveoli. The lungs lose their elasticity, and expiration becomes an active process.

3. Chronic bronchitis is characterized by excessive mucus production in the lower respiratory passageways, which severely impairs ventilation and gas exchange. Patients may become cyanotic as a result of chronic hypoxia.

Asthma (p. 876)

4. An obstructive condition caused by an immune response that causes its victims to wheeze and gasp for air as their inflamed respiratory passages constrict. Marked by exacerbations and periods of relief from symptoms.

Tuberculosis (p. 876)

5. Tuberculosis (TB), an infectious disease caused by an airborne bacterium, mainly affects the lungs. Although most infected individuals remain asymptomatic by walling off the bacteria in tubercles, when immunity is depressed disease symptoms ensue. Recent TB increases in AIDS patients and some patients' failure to complete drug therapy have produced multidrug-resistant TB strains.

Lung Cancer (pp. 876–877)

6. Lung cancer, promoted by free radicals and other carcinogens, is extremely aggressive and metastasizes rapidly.

SP Case Study: Lung Disease.

Developmental Aspects of the Respiratory System (pp. 877–878)

1. The superior respiratory system mucosa develops from the invagination of the ectodermal olfactory placode; that of the inferior passageways develops from an outpocketing of the endodermal foregut lining. Mesoderm forms the walls of the respiratory conduits and the lung stroma.

2. Premature infants have problems keeping their lungs inflated owing to the lack of surfactant in their alveoli, resulting in infant respiratory distress syndrome (IRDS). Surfactant is formed late in fetal development.

3. Cystic fibrosis (CF), the most common fatal hereditary disease, results from an abnormal CFTR protein that fails to form a chloride channel. The result is thick mucus which clogs respiratory passages and invites infection.

4. With age, the thorax becomes more rigid, the lungs become less elastic, and vital capacity declines. In addition, the stimulatory effect of increased arterial levels of carbon dioxide is less apparent, and respiratory system protective mechanisms are less effective.

REVIEW QUESTIONS

Multiple Choice/Matching

1. Cutting the phrenic nerves will result in (a) air entering the pleural cavity, (b) paralysis of the diaphragm, (c) stimulation of the diaphragmatic reflex, (d) paralysis of the epiglottis.

2. Following the removal of his larynx, an individual would (a) be unable to speak, (b) be unable to cough, (c) have difficulty swallowing, (d) be in respiratory difficulty or arrest.

3. Under ordinary circumstances, the inflation reflex is initiated by (a) the inspiratory center, (b) the apneustic center, (c) overinflation of the alveoli and bronchioles, (d) the pneumotaxic center.

4. The detergent-like substance that keeps the alveoli from collapsing between breaths because it reduces the surface tension of the water film in the alveoli is called (a) lecithin, (b) bile, (c) surfactant, (d) reluctant.

5. Which of the following determines the *direction* of gas movement? (a) solubility in water, (b) partial pressure gradient, (c) the temperature, (d) molecular weight and size of the gas molecule.

6. When the inspiratory muscles contract, (a) the size of the thoracic cavity is increased in diameter, (b) the size of the thoracic cavity is increased in length, (c) the volume of the thoracic cavity is decreased, (d) the size of the thoracic cavity is increased in both length and diameter.

7. The nutrient blood supply of the lungs is provided by (a) the pulmonary arteries, (b) the aorta, (c) the pulmonary veins, (d) the bronchial arteries.

8. Oxygen and carbon dioxide are exchanged in the lungs and through all cell membranes by (a) active transport, (b) diffusion, (c) filtration, (d) osmosis.

9. Which of the following would not normally be treated by 100% oxygen therapy? (Choose all that apply.) (a) anoxia, (b) carbon monoxide poisoning, (c) respiratory crisis in an emphysema patient, (d) eupnea.

10. Most oxygen carried in the blood is (a) in solution in the plasma, (b) combined with plasma proteins, (c) chemically combined with the heme in red blood cells, (d) in solution in the red blood cells.

11. Which of the following has the greatest stimulating effect on the respiratory center in the brain? (a) oxygen, (b) carbon dioxide, (c) calcium, (d) willpower.

12. In mouth-to-mouth artificial respiration, the rescuer blows air from his or her own respiratory system into that of the victim. Which of the following statements are correct?

1. Expansion of the victim's lungs is brought about by blowing air in at higher than atmospheric pressure (positive-pressure breathing).
2. During inflation of the lungs, the intrapleural pressure increases.
3. This technique will not work if the victim has a hole in the chest wall, even if the lungs are intact.
4. Expiration during this procedure depends on the elasticity of the alveolar and thoracic walls.

(a) all of these, (b) 1, 2, 4, (c) 1, 2, 3, (d) 1, 4.

13. A baby holding its breath will (a) have brain cells damaged because of low blood oxygen levels, (b) automatically start to breathe again when the carbon dioxide levels in the blood reach a high enough value, (c) suffer heart damage because of increased pressure in the carotid sinus and aortic arch areas, (d) be called a "blue baby."

14. Under ordinary circumstances, which of the following blood components is of no physiological significance? (a) bicarbonate ions, (b) carbaminohemoglobin, (c) nitrogen, (d) chloride.

15. Damage to which of the following would result in cessation of breathing? (a) the pneumotaxic center, (b) the medulla, (c) the stretch receptors in the lungs, (d) the apneustic center.

16. The bulk of carbon dioxide is carried (a) chemically combined with the amino acids of hemoglobin as carbaminohemoglobin in the red blood cells, (b) as the ion HCO_3^- in the plasma after first entering the red blood cell, (c) as carbonic acid in the plasma, (d) chemically combined with the heme portion of Hb.

Short Answer Essay Questions

17. Trace the route of air from the external nares to an alveolus. Name subdivisions of organs where applicable, and differentiate between conducting and respiratory zone structures.

18. (a) Why is it important that the trachea is reinforced with cartilage rings? (b) Of what advantage is it that the rings are incomplete posteriorly?

19. Briefly explain the anatomical "reason" why most men have deeper voices than boys or women.

20. The lungs are mostly passageways and elastic tissue. (a) What is the role of the elastic tissue? (b) Of the passageways?

21. Describe the functional relationships between volume changes and gas flow into and out of the lungs.

22. What is it about the structure of the respiratory membrane that makes the alveoli ideal sites for gas exchange?

23. Discuss how airway resistance, lung compliance, and alveolar surface tension influence pulmonary ventilation.

24. (a) Differentiate clearly between minute respiratory volume and alveolar ventilation rate. (b) Which provides a more accurate measure of ventilatory efficiency, and why?

25. State Dalton's law of partial pressures and Henry's law.

26. (a) Define hyperventilation. (b) If you hyperventilate, do you retain or expel more carbon dioxide? (c) What effect does hyperventilation have on blood pH?

27. Describe age-related changes in respiratory function.

Critical Thinking and Clinical Application Questions

1. Harry, the swimmer with the fastest time on the Springfield College swim team, routinely hyperventilates before a meet, as he says, "to sock some more oxygen into my lungs so I can swim longer without having to breathe." First of all, what basic fact about oxygen loading has Harry forgotten (a lapse leading to false thinking)? Second, how is Harry jeopardizing not only his time but his life?

2. A member of the "Blues" gang was rushed into an emergency room after receiving a knife wound in the left side of his thorax. The diagnosis was pneumothorax and a collapsed lung. Explain exactly (a) why the lung collapsed, and (b) why only one lung (not both) collapsed.

3. A surgeon removed three adjacent bronchopulmonary segments from the left lung of a patient with TB. Almost half of the lung was removed, yet there was no severe bleeding, and relatively few blood vessels had to be cauterized (closed off). Why was the surgery so easy to perform?

THE DIGESTIVE SYSTEM 24

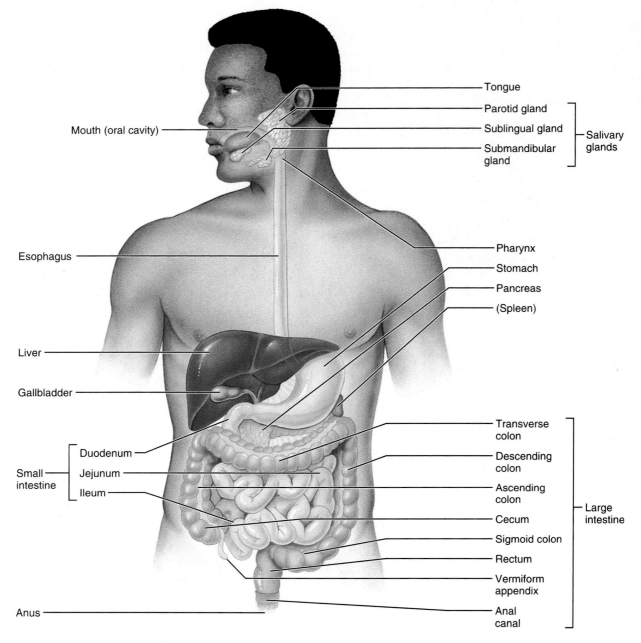

FIGURE 24.1   *Organs of the alimentary canal and related accessory digestive organs.*

Children have a special fascination for the workings of the digestive system. They relish crunching a potato chip, delight in making "mustaches" with milk, and giggle when their stomach "growls." As adults, we know that a healthy digestive system is essential to maintaining life, because it converts foods into the raw materials that build and fuel our body's cells. Specifically, the digestive system takes in food, breaks it down into nutrient molecules, absorbs these molecules into the bloodstream, and then rids the body of the indigestible remains.

OVERVIEW OF THE DIGESTIVE SYSTEM

The organs of the digestive system (Figure 24.1) can be separated into two main groups: (1) those of the *alimentary* (al"ĭ-men'tar-e; *aliment* = nourish) *canal* and (2) *accessory digestive organs.*

The **alimentary canal,** also called the **gastrointestinal (GI) tract,** is the continuous, muscular digestive tube that winds through the body. It **digests** food—breaks it down into smaller fragments (*digest* = dissolved)—and **absorbs** the digested fragments through its lining into the blood. The organs of the alimentary canal are the *mouth, pharynx, esophagus, stomach, small intestine,* and *large intestine.* The large intestine leads to the terminal opening, or *anus.* In a cadaver, the alimentary canal is approximately 9 m (about 30 feet) long, but in a living person, it is considerably shorter because of its muscle tone. Food material within this tube is technically outside the body because the canal is open to the external environment at both ends.

The **accessory digestive** organs are the *teeth, tongue, gallbladder,* and a number of large digestive glands—the *salivary glands, liver,* and *pancreas.* The teeth and tongue are in the mouth, or oral cavity, while the digestive glands and gallbladder lie outside the GI tract and connect to it by ducts. The accessory digestive glands produce a variety of secretions that contribute to the breakdown of foodstuffs.

Digestive Processes

The digestive tract can be viewed as a "disassembly line" in which food becomes less complex at each step of processing and its nutrients become available to the body. The processing of food by the digestive system involves six essential activities: ingestion, propulsion, mechanical digestion, chemical digestion, absorption, and defecation (Figure 24.2).

1. Ingestion is simply taking food into the digestive tract, usually via the mouth.

2. Propulsion moves food through the alimentary canal. It includes *swallowing,* which is initiated voluntarily, and *peristalsis* (per″ĭ-stal′sis), an involuntary process. **Peristalsis** (*peri* = around; *stalsis* = constriction), the major means of propulsion, involves alternate waves of contraction and relaxation of muscles in the organ walls (Figure 24.3a). Its main effect is to squeeze food from one organ to the next, but some mixing occurs as well. In fact, peristaltic waves are so powerful that, once swallowed, food and fluids will reach your stomach even if you stand on your head.

3. Mechanical digestion physically prepares food for chemical digestion by enzymes. Mechanical processes include chewing, mixing of food with saliva by the tongue, churning food in the stomach, and **segmentation,** or rhythmic local constrictions of the intestine (Figure 24.3b). Segmentation mixes food with digestive juices and increases the efficiency of absorption by repeatedly moving different parts of the food mass over the intestinal wall.

4. Chemical digestion is a series of catabolic steps in which complex food molecules are broken down to their chemical building blocks. It is accomplished by enzymes secreted by various glands into the lumen of the alimentary canal. The enzymatic break-

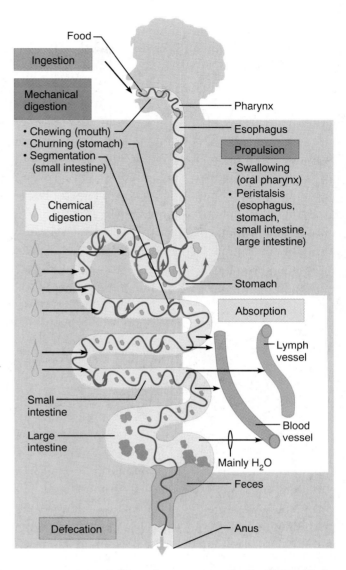

FIGURE 24.2 *Schematic summary of gastrointestinal tract activities.* Gastrointestinal tract activities include ingestion, mechanical digestion, chemical (enzymatic) digestion, propulsion, absorption, and defecation. Sites of chemical digestion are also sites that produce enzymes or that receive enzymes or other secretions made by accessory organs outside the alimentary canal. The mucosa of the GI tract secretes mucus, which protects and lubricates.

down of foodstuffs begins in the mouth and is essentially complete in the small intestine.

5. Absorption is the passage of digested end products (plus vitamins, minerals, and water) from the lumen of the GI tract into the blood or lymph. For absorption to occur, these substances must first enter the mucosal cells by active or passive transport processes. The small intestine is the major absorptive site.

6. Defecation eliminates indigestible substances from the body via the anus in the form of feces.

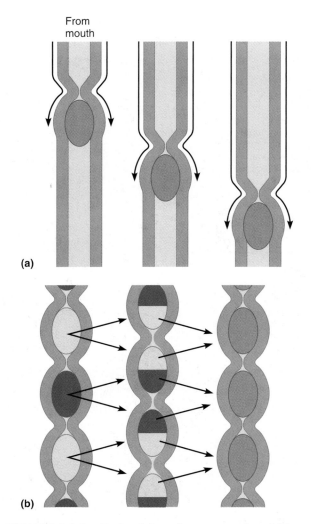

(a)

(b)

FIGURE 24.3 *Peristalsis and segmentation.* **(a)** In peristalsis, adjacent segments of the intestine (or other alimentary tract organs) alternately contract and relax, which results in movement of food along the tract distally. **(b)** In segmentation, nonadjacent segments of the intestine alternately contract and relax. Because the active segments are separated by inactive regions, the food is moved forward and then backward. This results in mixing of the food, rather than food propulsion.

Some of these processes are the job of a single organ. For example, only the mouth ingests and only the large intestine defecates. But most digestive system activities require the cooperation of several organs and occur bit by bit as food moves along the tract. Later, when we discuss the functioning of each GI tract organ, we will consider which of these specific processes it performs (Table 24.1 on p. 908 provides a summary of this information), as well as the neural or hormonal factors that regulate these processes.

Basic Functional Concepts

A theme stressed throughout this book has been the body's efforts to maintain the constancy of its internal environment, particularly of the blood. Most organ systems respond to changes in that environment either by attempting to restore some plasma variable to its former levels or by changing their own function. The digestive system, however, creates an optimal environment for its functioning in the lumen (cavity) of the GI tract, an area that is actually *outside* the body, and essentially all digestive tract regulatory mechanisms act to control luminal conditions so that digestion and absorption can occur there as effectively as possible.

Two facts apply to these regulatory mechanisms:

1. Digestive activity is provoked by a range of mechanical and chemical stimuli. Sensors (mechanoreceptors and chemoreceptors of various types) involved in controls of GI tract activity are located in the walls of the tract organs (Figure 24.4). These sensors respond to several stimuli, the most important being stretching of the organ wall by food in the lumen, osmolarity (solute concentration) and pH of the contents, and the presence of substrates and end products of digestion. When appropriately stimulated, these receptors initiate reflexes that (1) activate or inhibit glands that secrete digestive juices into the lumen or hormones into the blood or (2) mix lumen contents and move them along the length of the tract by stimulating the smooth muscle of the GI tract walls.

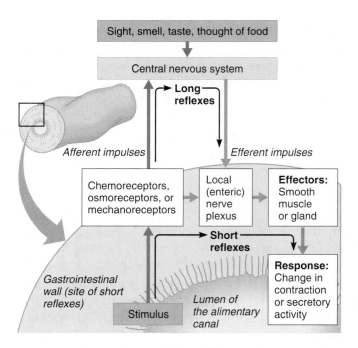

FIGURE 24.4 *Schematic of the long and short neural reflex pathways initiated by stimuli in the gastrointestinal tract.*

2. Controls of digestive activity are both extrinsic and intrinsic. Another novel trait of the digestive tract is that many of its controlling systems are intrinsic—a product of "in-house" nerve plexuses or local hormone-producing cells. As described shortly, the wall of the alimentary canal contains nerve plexuses. These plexuses extend essentially the entire length of the GI tract and influence each other both in the same and in different digestive organs. As a result, two kinds of reflex activity occur, short and long. *Short reflexes* are mediated entirely by the local *(enteric)* plexuses (the so-called gut brain) in response to GI tract stimuli. *Long reflexes* are initiated by stimuli arising within or outside the GI tract and involve CNS centers and extrinsic autonomic nerves (Figure 24.4).

The stomach and small intestine also contain hormone-producing cells that, when appropriately stimulated, release their products to the extracellular space. These hormones are distributed via the blood to their target cells within the same or different digestive tract organs, which they prod into secretory or contractile activity.

Digestive System Organs: Relationship and Structural Plan

Relationship of the Digestive Organs to the Peritoneum

Most digestive system organs reside in the abdominopelvic cavity, the largest of our ventral body cavities. Recall from Chapter 1 that all ventral body cavities contain slippery *serous membranes*. The **peritoneum** of the abdominopelvic cavity is the most extensive of these membranes (Figure 24.5a). The **visceral peritoneum** covers the external surfaces of most digestive organs and is continuous with the **parietal peritoneum** that lines the body wall. Between the two peritoneums is the **peritoneal cavity,** a slitlike potential space containing fluid secreted by the serous membranes. The serous fluid lubricates the mobile digestive organs, allowing them to glide easily across one another and along the body wall as they carry out their digestive activities.

A **mesentery** (mes'en-ter"e) is a double layer of peritoneum—a sheet of two serous membranes fused back to back—that extends to the digestive organs from the body wall. Mesenteries provide routes for blood vessels, lymphatics, and nerves to reach the digestive viscera. They also hold organs in place and store fat. In most places the mesentery is *dorsal* and attaches to the posterior abdominal wall, but there are examples of *ventral* mesenteries as well (Figure 24.5a). Some of the mesenteries, or peritoneal folds, are given specific names (such as the *omenta*). These are described more completely later on.

Not all alimentary canal organs are suspended by a mesentery. For example, some parts of the small intestine originate in the peritoneal cavity but then adhere to the dorsal abdominal wall (Figure 24.5b). In so doing, they lose their mesentery and come to lie posterior to the peritoneum. These organs, which

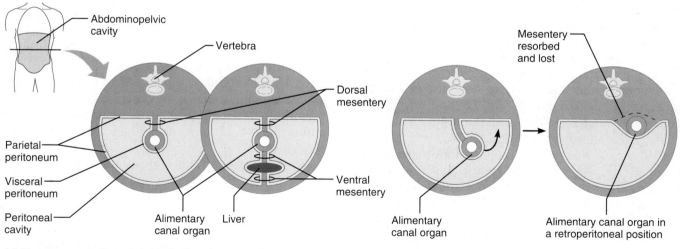

(a) Transverse section of abdominal cavity

(b) Some organs become retroperitoneal

FIGURE 24.5 *The peritoneum and the peritoneal cavity.* **(a)** Simplified cross sections through the abdominal cavity showing the relative locations of the visceral and parietal peritoneums, and dorsal (left) and ventral (right) mesenteries. Note that the peritoneal cavity is much smaller than depicted here, as it is nearly filled by the organs within it. **(b)** Some alimentary canal organs lose their mesentery during development and become retroperitoneal.

include most of the pancreas and parts of the large intestine, are called **retroperitoneal** (*retro* = behind) **organs.** By contrast, digestive organs (like the stomach) that keep their mesentery and remain in the peritoneal cavity are called **intraperitoneal** or **peritoneal organs.** It is not known why some digestive organs end up in the retroperitoneal position, while others are suspended freely. One explanation is that anchoring of various segments of the alimentary canal to the posterior abdominal wall minimizes the chances that this tube will twist or kink during peristaltic movements. (Kinking and obstruction of the tract lead to tissue necrosis and death.)

Homeostatic Imbalance

Peritonitis, or inflammation of the peritoneum, can arise from a piercing wound of the abdomen or from a perforating ulcer that leaks stomach juices into the peritoneal cavity, but most commonly it results from a burst appendix (that sprays bacteria-containing feces all over the peritoneum). In peritonitis, the peritoneal coverings tend to stick together around the infection site. This localizes the infection and provides time for macrophages to mount the attack to prevent the inflammation from spreading. If peritonitis becomes *generalized* (widespread within the peritoneal cavity), it is dangerous and often lethal. Treatment includes removing as much infectious debris as possible from the peritoneal cavity and administering megadoses of antibiotics. ■

Blood Supply: The Splanchnic Circulation

The **splanchnic circulation** includes those arteries that branch off the abdominal aorta to serve the digestive organs and the *hepatic portal circulation*. The arterial supply—the hepatic, splenic, and left gastric branches of the celiac trunk that serve the spleen, liver, and stomach, and the mesenteric arteries (superior and inferior) that serve the small and large intestines (see pp. 759 and 761)—normally receives one-quarter of the cardiac output. This percentage (blood volume) increases after a meal has been eaten. The hepatic portal circulation (described on pp. 770–771) collects nutrient-rich venous blood draining from the digestive viscera and delivers it to the liver. The liver collects the absorbed nutrients for metabolic processing or for storage before releasing them back to the bloodstream for general cellular use.

Histology of the Alimentary Canal

Most digestive organs only have a "share" in the common work of digestion. Consequently, it helps to consider structural characteristics that promote similar functions in virtually all parts of the alimentary canal before we attempt to describe the functional anatomy of the digestive system in a cohesive way.

From the esophagus to the anal canal, the walls of the alimentary canal have the same four basic layers, or *tunics* (Figure 24.6). From the lumen outward, these layers are the *mucosa, submucosa, muscularis externa,* and *serosa.* Each tunic contains a predominant tissue type that plays a specific role in food breakdown.

The Mucosa The **mucosa,** or **mucous membrane**—the innermost layer—is a moist epithelial membrane that lines the lumen of the alimentary canal from the mouth to the anus. Its major functions are (1) *secretion* of mucus, digestive enzymes, and hormones, (2) *absorption* of the end products of digestion into the blood, and (3) *protection* against infectious disease. The mucosa in a particular region of the GI tract may express one or all three of these capabilities.

More complex than most other mucosae in the body, the typical digestive mucosa consists of three sublayers: (1) a lining epithelium, (2) a lamina propria, and (3) a muscularis mucosae. Typically, the **epithelium** of the mucosa is a *simple columnar epithelium* that is rich in mucus-secreting *goblet cells.* The slippery mucus it produces protects certain digestive organs from being digested themselves by enzymes working within their cavities and eases food passage along the tract. In the stomach and small intestine, the mucosa contains both enzyme-secreting and hormone-secreting cells. Thus, in such sites, the mucosa is a diffuse kind of endocrine organ as well as part of the digestive organ.

The **lamina propria** (*proprius* = one's own), which underlies the epithelium, is loose areolar (or reticular) connective tissue. Its capillaries nourish the epithelium and absorb digested nutrients. Its isolated lymph nodules, part of **MALT,** the mucosa-associated lymphatic tissue described on p. 786, are important in the defense against bacteria and other pathogens, which have rather free access to our digestive tract. Particularly large collections of lymph nodules occur at strategic locations, such as within the pharynx (as the tonsils) and in the appendix.

External to the lamina propria is the **muscularis mucosae,** a scant layer of smooth muscle cells that produces local movements of the mucosa. For example, the twitching of this muscle layer dislodges food particles that have adhered to the mucosa. In the small intestine, it throws the mucosa into a series of small folds that immensely increase its surface area.

The Submucosa The **submucosa,** just external to the mucosa, is a moderately dense connective tissue containing blood and lymphatic vessels, lymph nod-

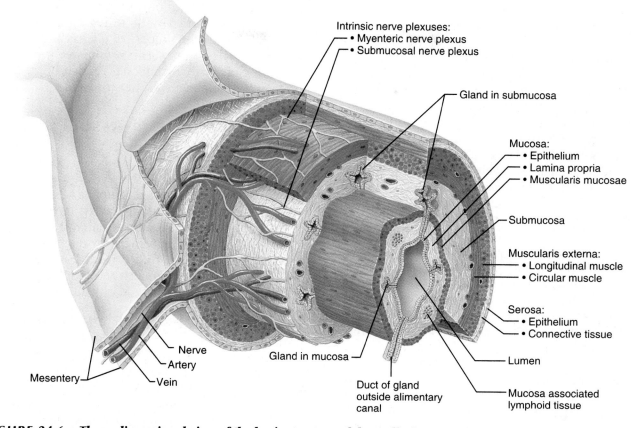

Intrinsic nerve plexuses:
• Myenteric nerve plexus
• Submucosal nerve plexus

Gland in submucosa

Mucosa:
• Epithelium
• Lamina propria
• Muscularis mucosae

Submucosa

Muscularis externa:
• Longitudinal muscle
• Circular muscle

Serosa:
• Epithelium
• Connective tissue

Lumen

Mucosa associated lymphoid tissue

Duct of gland outside alimentary canal

Gland in mucosa

Nerve

Artery

Vein

Mesentery

FIGURE 24.6   *Three-dimensional view of the basic structure of the wall of alimentary canal depicting its four basic layers: the mucosa, submucosa, muscularis externa, and serosa.*

ules, and nerve fibers. Its rich supply of elastic fibers enables the stomach to regain its normal shape after temporarily storing a large meal. Its luxuriant vascular network supplies surrounding tissues of the GI tract wall.

The Muscularis Externa   Just deep to the submucosa is the **muscularis externa,** also simply called the **muscularis.** This layer is responsible for segmentation and peristalsis; that is, it mixes and propels foodstuffs along the digestive tract. This tunic typically has an inner *circular layer* and an outer *longitudinal layer* of smooth muscle cells (Figure 24.6). In several places along the GI tract, the circular layer thickens to form *sphincters* that act as valves to prevent backflow and control food passage from one organ to the next.

The Serosa   The **serosa,** the protective outermost layer of the intraperitoneal organs, is the *visceral*

peritoneum. It is formed of areolar connective tissue covered with *mesothelium,* a single layer of squamous epithelial cells.

In the esophagus, which is located in the thoracic instead of the abdominopelvic cavity, the serosa is replaced by an **adventitia** (ad″ven-tish′e-ah). The adventitia is ordinary fibrous connective tissue that binds the esophagus to surrounding structures. Retroperitoneal organs have *both* a serosa (on the side facing the peritoneal cavity) and an adventitia (on the side abutting the dorsal body wall).

Enteric Nervous System of the Alimentary Canal

The alimentary canal has its own in-house nerve supply, staffed by the so-called **enteric** (*enter* = gut) **neurons,** which communicate widely with each other to regulate digestive system activity. These enteric neurons constitute the bulk of the two major *intrinsic nerve plexuses* found within the walls of the alimentary canal: the submucosal and myenteric nerve plexuses (Figure 24.6).

The **submucosal nerve plexus** occupies the submucosa and chiefly regulates the activity of glands and smooth muscle in the mucosa tunic. The large **myenteric** (mi-en'ter-ik; "intestinal muscle") **nerve plexus** lies between the circular and longitudinal layers of smooth muscle of the muscularis externa. Enteric neurons of the myenteric plexus provide the major nerve supply to the GI tract wall and control GI tract mobility. Control of the patterns of segmentation and peristalsis is largely automatic, involving local reflex arcs between enteric neurons in the same or different plexuses or (even) organs.

The enteric nervous system is also linked to the central nervous system by afferent visceral fibers and by sympathetic and parasympathetic branches (motor fibers) of the autonomic nervous system that enter the intestinal wall and synapse with neurons in the intrinsic plexuses. Thus, as mentioned earlier, digestive activity is also subject to extrinsic controls exerted by autonomic fibers via long reflex arcs. Generally speaking, parasympathetic inputs enhance secretory activity and mobility, whereas sympathetic impulses inhibit digestive activities. But keep in mind that the largely independent enteric ganglia are much more than just way stations for the autonomic nervous system as is the case in other organ systems. Indeed, the enteric nervous system contains some 100 million neurons, as many as the entire spinal cord.

■ PART 1 ■
FUNCTIONAL ANATOMY OF THE DIGESTIVE SYSTEM

Now that we have summarized some points that unify functioning digestive system organs and have surveyed their common "anatomical landscape," we are ready to consider the special structural and functional capabilities of each organ of this system. See Table 24.1 on p. 908 for an overview of organ functions. Most of the digestive organs are shown in their normal body positions in Figure 24.1, so you will probably find it helpful to refer back to that illustration from time to time as you read the following sections.

THE MOUTH, PHARYNX, AND ESOPHAGUS

The mouth is the only part of the alimentary canal involved in ingestion or food entry in the body. However, most digestive functions associated with the mouth reflect the activity of the related accessory organs, such as the teeth, salivary glands, and tongue, because within the mouth food is chewed, and mixed and moistened with saliva containing enzymes that begin the process of chemical digestion. The mouth also begins the propulsive process of swallowing, during which food is carried through the pharynx and esophagus to the stomach.

The Mouth and Associated Organs

The Mouth

The **mouth,** a mucosa-lined cavity, is also called the **oral cavity,** or **buccal** (buk'al) **cavity.** Its boundaries are the lips anteriorly, the cheeks laterally, the palate superiorly, and the tongue inferiorly (Figure 24.7). Its anterior opening is the **oral orifice.** Posteriorly, the oral cavity is continuous with the *oropharynx.* The walls of the mouth withstand considerable friction and are lined with stratified squamous epithelium instead of the more typical simple columnar epithelium. The epithelium on the gums, hard palate, and dorsum of the tongue is slightly keratinized for extra protection against abrasion during eating. The oral mucosa, like all moist surface linings, responds to injury by producing antimicrobial peptides called *defensins.* This may help to explain how the mouth, a site teeming with disease-causing microbes, remains so remarkably healthy.

The Lips and Cheeks The **lips (labia)** and the **cheeks** have a core of skeletal muscle covered externally by skin. The *orbicularis oris muscle* forms the bulk of the fleshy lips. The cheeks are formed largely by the *buccinators.* The lips and cheeks help keep food between the teeth when we chew and play a small role in speech. The recess bounded externally by the lips and cheeks and internally by the gums and teeth is called the **vestibule** ("porch"). The area that lies within the teeth and gums is the **oral cavity proper.**

The lips are much larger than most people think; anatomically they extend from the inferior margin of the nose to the superior boundary of the chin. The reddened area where one applies lipstick or lands a kiss is called the **red margin.** This transitional zone, where keratinized skin meets the oral mucosa, is poorly keratinized and translucent, allowing the red color of blood in the underlying capillaries to show through. There are no sweat or sebaceous glands in the red margin, so it must be moistened with saliva periodically to prevent it from becoming dry and cracked (chapped lips). The **labial frenulum** (fren'u-lum) is a median fold that joins the internal aspect of each lip to the gum (Figure 24.7b).

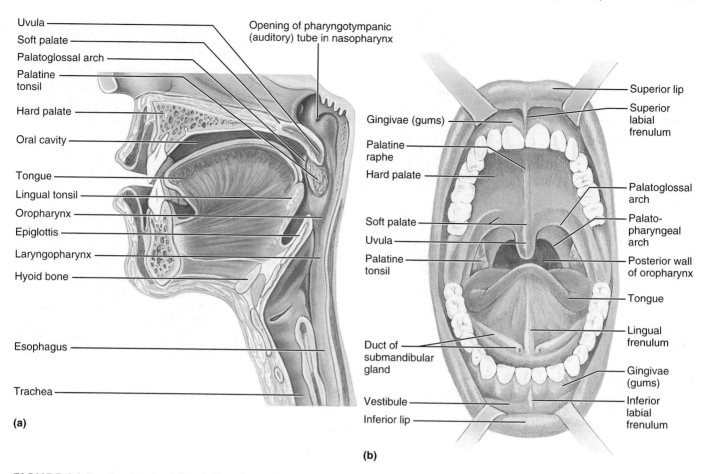

FIGURE 24.7 *Anatomy of the oral cavity (mouth).* **(a)** Sagittal section of the oral cavity and pharynx. **(b)** Anterior view of the oral cavity.

The Palate The **palate,** which forms the roof of the mouth, has two distinct parts: the hard palate anteriorly and the soft palate posteriorly (see Figure 24.7). The **hard palate** is underlain by bone (the palatine bones and the palatine processes of the maxillae), and it forms a rigid surface against which the tongue forces food during chewing. The mucosa on either side of its *raphe* (ra'fe), a midline ridge, is slightly corrugated, which helps to create friction.

The **soft palate** is a mobile fold formed mostly of skeletal muscle. Projecting downward from its free edge is the fingerlike **uvula** (u'vu-lah). The soft palate rises reflexively to close off the nasopharynx when we swallow.

■ To demonstrate this action, try to breathe and swallow at the same time.

Laterally, the soft palate is anchored to the tongue by the **palatoglossal arches** and to the wall of the oropharynx by the more posterior **palatopharyngeal arches.** These two paired folds form the boundaries of the **fauces** (faw'sēz; *fauc* = throat), the arched area of the oropharynx that contains the palatine tonsils.

The Tongue

The **tongue** occupies the floor of the mouth and fills most of the oral cavity when the mouth is closed (see Figure 24.7). The tongue is largely composed of interlacing bundles of skeletal muscle fibers, and during chewing, it grips the food and constantly repositions it between the teeth. Tongue movements also mix food with saliva and form it into a compact mass called a **bolus** (bo'lus; "a lump"), and then initiate swallowing by pushing the bolus posteriorly into the pharynx. The versatile tongue also helps us form consonants (k, d, t, and so on) when we speak.

The tongue has both intrinsic and extrinsic skeletal muscle fibers. The **intrinsic muscles** are confined within the tongue and are not attached to bone. Their muscle fibers, which run in several different planes, allow the tongue to change its shape (but not its position), becoming thicker, thinner, longer, or shorter as needed for speech and swallowing. The **extrinsic muscles** extend to the tongue from their points of origin on bones of the skull or the soft palate, as described in Chapter 10 (see Table 10.2 and Figure 10.7). The extrinsic muscles alter the tongue's position; they protrude it, retract it, and move it from side to side. The tongue is divided by a

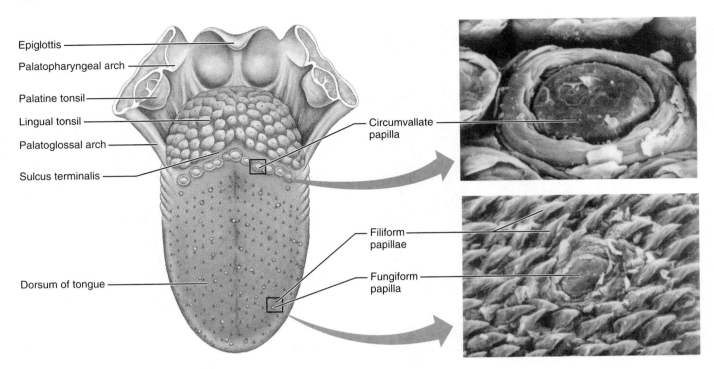

Epiglottis

Palatopharyngeal arch

Palatine tonsil

Lingual tonsil

Palatoglossal arch

Sulcus terminalis

Dorsum of tongue

Circumvallate papilla

Filiform papillae

Fungiform papilla

FIGURE 24.8   *Surface view of the dorsal surface of the tongue.* The tonsils closely associated with the oral cavity are illustrated. Also shown are the locations and detailed structures of the circumvallate, fungiform, and filiform papillae. (Top right, 300×; bottom right, 140×.) © R. G. Kessel and R. H. Kardon, *Tissues and Organs: A Text-Atlas of Scanning Electron Microscopy,* W. H. Freeman and Company, 1979, all rights reserved.

median septum of connective tissue, and each half contains identical groups of muscles.

A fold of mucosa, called the **lingual frenulum,** secures the tongue to the floor of the mouth (see Figure 24.7b). It also limits posterior movements of the tongue.

Homeostatic Imbalance

Children born with an extremely short lingual frenulum are often referred to as "tongue-tied" because of speech distortions that result when movement of the tongue is restricted. This congenital condition, called *ankyloglossia* ("fused tongue"), is corrected surgically by snipping the frenulum. ■

The superior tongue surface bears papillae, peglike projections of the underlying mucosa. The three types of papillae—filiform, fungiform, and circumvallate—are shown in Figure 24.8. The conical **filiform papillae** give the tongue surface a roughness that aids in licking semisolid foods (such as ice cream) and provide friction for manipulating foods in the mouth. These papillae, the smallest and most numerous type, align in parallel rows on the tongue dorsum. They contain keratin, which stiffens them and gives the tongue its whitish appearance.

The mushroom-shaped **fungiform papillae** are scattered widely over the tongue surface. Each has a vascular core that gives it a reddish hue. Ten to twelve large **circumvallate,** or **vallate, papillae** are located in a V-shaped row at the back of the tongue. They resemble the fungiform papillae but have an additional surrounding furrow. As described in Chapter 16, both the fungiform and circumvallate papillae house taste buds.

Immediately posterior to the circumvallate papillae is the **sulcus terminalis,** a groove that distinguishes the anterior two-thirds of the tongue that lies in the oral cavity from its posterior third residing in the oropharynx. The mucosa covering the posteriormost aspect and root of the tongue lacks papillae, but it is still bumpy because of the nodular *lingual tonsil,* which lies just deep to its mucosa (see Figure 24.8).

The Salivary Glands

A number of glands both inside and outside the oral cavity produce and secrete **saliva.** Saliva (1) cleanses the mouth, (2) dissolves food chemicals so that they can be tasted, (3) moistens food and aids in compacting it into a bolus, and (4) contains enzymes that begin the chemical breakdown of starchy foods.

Most saliva is produced by three pairs of **extrinsic salivary glands** that lie outside the oral cavity and empty their secretions into it. Their output is augmented slightly by small **intrinsic salivary glands,** also called **buccal glands,** scattered throughout the oral cavity mucosa.

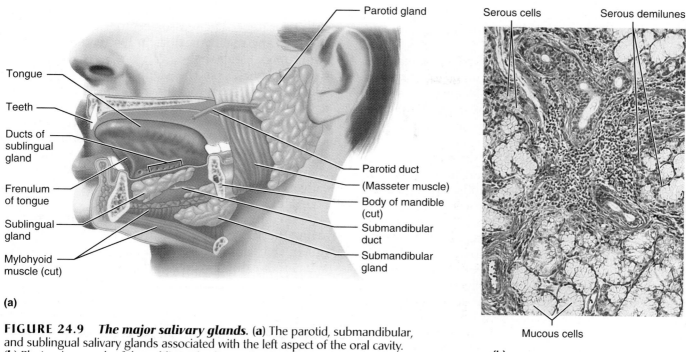

FIGURE 24.9 *The major salivary glands.* **(a)** The parotid, submandibular, and sublingual salivary glands associated with the left aspect of the oral cavity. **(b)** Photomicrograph of the sublingual salivary gland (943×), which is a mixed salivary gland consisting predominantly of mucus-producing cells (light blue) and a few serous-secreting units (purple). The serous cells sometimes form demilunes (caps) around the bases of the mucous cells. Copyright © Science Photo Library/Photo Researchers, Inc.

The extrinsic salivary glands include the parotid, submandibular, and sublingual glands (Figure 24.9), paired compound tubuloalveolar glands that develop from the oral mucosa and remain connected to it by ducts. The large **parotid** (pah-rot'id; *par* = near, *otid* = the ear) **gland** lies anterior to the ear between the masseter muscle and the skin. The prominent **parotid duct** runs parallel to the zygomatic arch, pierces the buccinator muscle, and opens into the vestibule next to the second upper molar. Branches of the facial nerve run through the parotid gland on their way to the muscles of facial expression. For this reason, surgery on this gland can result in facial paralysis.

Homeostatic Imbalance

Mumps, a common children's disease, is an inflammation of the parotid glands caused by the mumps virus *(myxovirus),* which spreads from person to person in saliva. If you check the location of the parotid glands in Figure 24.9a, you can readily understand why people with mumps complain that it hurts to open their mouth or chew. Besides that discomfort, other signs and symptoms of mumps include moderate fever and pain when swallowing acid foods (sour pickles, grapefruit juice, etc.). Mumps viral infections in adult males carry a 25% risk that the testes may become infected as well, leading to sterility. ■

About the size of a walnut, the **submandibular gland** lies along the medial aspect of the mandibular body. Its duct runs beneath the mucosa of the oral cavity floor and opens at the base of the lingual frenulum (see also Figure 24.7). The small **sublingual gland** lies anterior to the submandibular gland under the tongue. It opens via 10 or 12 ducts into the floor of the mouth.

To a greater or lesser degree, the salivary glands are composed of two types of secretory cells: mucous and serous. The **serous cells** produce a watery secretion containing enzymes, the ions of saliva, and a tiny bit of mucin, whereas the **mucous cells** produce **mucus,** a stringy, viscous solution. The parotid glands contain only serous cells. The submandibular and buccal glands contain approximately equal numbers of serous and mucous cells. The sublingual glands (see Figure 24.9b) contain mostly mucous cells.

Several of the substances found in saliva have important physiological functions. For instance, they may enhance the activity of its enzymes or protect the oral cavity. Hence, it is worthwhile to consider the composition of saliva in more detail.

Composition of Saliva Saliva is largely water—97 to 99.5%—and therefore is hypo-osmotic. Its osmolarity depends on the precise glands that are active and the nature of the stimulus for salivation. As a rule, saliva is slightly acidic (pH 6.75 to 7.00), but its pH may vary. Its solutes include electrolytes (sodium, potassium, chloride, phosphate, and bicarbonate ions); the digestive enzyme salivary amylase; the proteins mucin (mu'sin), lysozyme, and IgA; and metabolic wastes (urea and uric acid). When dissolved in water, the glycoprotein *mucin* forms thick mucus that lubricates the oral cavity and hydrates foodstuffs. Protection against microorganisms is provided by (1) *IgA antibodies*; (2) *lysozyme*, a bacteriostatic enzyme that inhibits bacterial growth in the mouth and may help to prevent tooth decay; (3) a cyanide compound; and (4) *defensins* (see discussion on p. 662), which are produced in greater amounts when the oral mucosa is cut or lesions are present. Besides acting as a local antibiotic, defensins function as cytokines to call defensive cells (lymphocytes, neutrophils, etc.) into the mouth for battle. In addition to these three protectors, the friendly bacteria that live on the back of the tongue convert food-derived nitrates in saliva into nitrites which, in turn, are converted into *nitric oxide* in an acid environment. This transformation occurs around the gums, where acid-producing bacteria tend to cluster, and in the hydrochloric acid–rich secretions of the stomach. The highly toxic nitric oxide is believed to act as a bactericidal agent in these locations.

Control of Salivation The intrinsic salivary glands secrete saliva continuously in amounts just sufficient to keep the mouth moist. But when food enters the mouth, the extrinsic glands are activated and copious amounts of saliva pour out. The average output of saliva is 1000–1500 ml per day.

Salivation is controlled primarily by the parasympathetic division of the autonomic nervous system. When we ingest food, chemoreceptors and pressoreceptors in the mouth send signals to the **salivatory nuclei** in the brain stem (pons and medulla). As a result, parasympathetic nervous system activity increases and motor fibers in the *facial (VII)* and *glossopharyngeal (IX) nerves* trigger a dramatically increased output of watery (serous), enzyme-rich saliva. The chemoreceptors are activated most strongly by acidic substances such as vinegar and citrus juice. The pressoreceptors are activated by virtually any mechanical stimulus in the mouth— even rubber bands.

Sometimes just the sight or smell of food is enough to get the juices flowing. The mere thought of hot fudge sauce on peppermint stick ice cream will make many a mouth water! Irritation of the lower regions of the GI tract by bacterial toxins, spicy foods, or hyperacidity—particularly when accompanied by a feeling of nausea—also increases salivation. This response may help wash away or neutralize the irritants.

In contrast to parasympathetic controls, the sympathetic division causes release of a thick mucin-rich saliva. Extremely strong activation of the sympathetic division constricts blood vessels serving the salivary glands and almost completely inhibits the saliva release, causing a dry mouth. Dehydration also inhibits salivation because low blood volume results in reduced filtration pressure at capillary beds.

⚡ *Homeostatic Imbalance*

Any disease process that inhibits saliva secretion causes difficulty in talking, swallowing, and eating. Because decomposing food particles are allowed to accumulate and bacteria flourish, *halitosis* (hal"i-to'sis; "bad breath") can result. ■

The Teeth

The **teeth** lie in sockets (alveoli) in the gum-covered margins of the mandible and maxilla. The role of the teeth in food processing needs little introduction. We **masticate,** or chew, by opening and closing our jaws and moving them from side to side while continually using our tongue to move the food between our teeth. In the process, the teeth tear and grind the food, breaking it down into smaller fragments.

Dentition and the Dental Formula Ordinarily by age 21, two sets of teeth, the **primary** and **permanent dentitions,** have formed (Figure 24.10). The primary dentition consists of the **deciduous teeth** (de-sid'u-us; *decid* = falling off), also called **milk** or **baby teeth.** The first teeth to appear, at about age six months, are the lower central incisors. Additional pairs of teeth erupt at one- to two-month intervals until about 24 months, when all 20 milk teeth have emerged.

As the deep-lying **permanent teeth** enlarge and develop, the roots of the milk teeth are resorbed from below, causing them to loosen and fall out between the ages of 6 and 12 years. Generally, all the teeth of the permanent dentition but the third molars have erupted by the end of adolescence. The third molars, also called the *wisdom teeth*, emerge

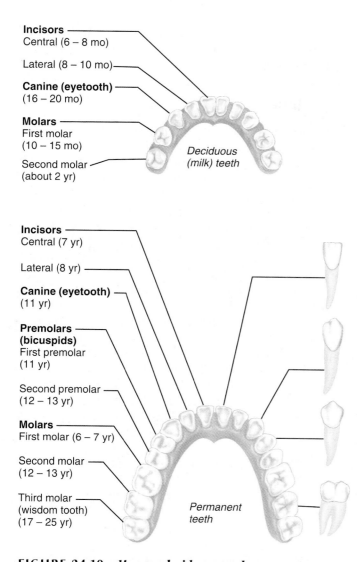

Incisors
Central (6 – 8 mo)

Lateral (8 – 10 mo)

Canine (eyetooth)
(16 – 20 mo)

Molars
First molar
(10 – 15 mo)

Second molar
(about 2 yr)

*Deciduous
(milk) teeth*

Incisors
Central (7 yr)

Lateral (8 yr)

Canine (eyetooth)
(11 yr)

**Premolars
(bicuspids)**
First premolar
(11 yr)

Second premolar
(12 – 13 yr)

Molars
First molar (6 – 7 yr)

Second molar
(12 – 13 yr)

Third molar
(wisdom tooth)
(17 – 25 yr)

*Permanent
teeth*

**FIGURE 24.10   *Human deciduous and permanent
teeth*.** Approximate time of tooth eruption is shown in
parentheses. Since the same number and arrangement of teeth
exist in both upper and lower jaws, only the lower jaw is shown
in each case. The shapes of individual teeth are shown on the
right.

later, between the ages of 17 and 25 years. There are
usually 32 permanent teeth in a full set, but sometimes the wisdom teeth never erupt or are completely absent.

Homeostatic Imbalance

When a tooth remains embedded in the jawbone, it
is said to be *impacted*. Impacted teeth can cause a
good deal of pressure and pain and must be removed
surgically. Wisdom teeth are most commonly impacted. ■

Teeth are classified according to their shape and
function as incisors, canines, premolars, and molars
(see Figure 24.10). The chisel-shaped **incisors** are
adapted for cutting or nipping off pieces of food. The
conical or fanglike **canines** (cuspids or eyeteeth) tear
and pierce. The **premolars** (bicuspids) and **molars**
have broad crowns with rounded cusps (tips) and are
best suited for grinding or crushing. The molars (literally, "millstones"), which have four or five cusps,
are the best grinders. During chewing, the upper and
lower molars repeatedly lock together, the cusps of
the uppers fitting into valleys in the lowers, and vice
versa. This action generates tremendous crushing
forces.

The **dental formula** is a shorthand way of indicating the numbers and relative positions of the different types of teeth in the mouth. This formula is
written as a ratio, uppers over lowers, for *one-half* of
the mouth. Since the other side is a mirror image,
the total dentition is obtained by multiplying the
dental formula by 2. The primary dentition consists
of two incisors (I), one canine (C), and two molars
(M) on each side of each jaw, and its dental formula
is written as

$$\frac{2I, \ 1C, \ 2M \ (upper \ jaw)}{2I, \ 1C, \ 2M \ (lower \ jaw)} \times 2 \ (20 \ teeth)$$

Similarly, the permanent dentition (two incisors,
one canine, two premolars [PM], and three molars)
is

$$\frac{2I, \ 1C, \ 2PM, \ 3M}{2I, \ 1C, \ 2PM, \ 3M} \times 2 \ (32 \ teeth)$$

Tooth Structure   Each tooth has two major regions:
the crown and the root (Figure 24.11). The enamel-
covered **crown** is the exposed part of the tooth above
the **gingiva** (jin' jĭ-vah), or **gum,** which surrounds
the tooth like a tight collar. **Enamel,** an acellular,
brittle material that directly bears the force of chewing, is the hardest substance in the body. It is heavily mineralized with calcium salts, and its densely
packed hydroxyapatite (mineral) crystals are oriented in force-resisting columns perpendicular to
the tooth's surface. The cells that produce the
enamel degenerate when the tooth erupts, leaving
none for repair purposes. Consequently, any
breaches of the enamel caused by decay or cracks
will not heal and must be artificially filled.

The portion of the tooth embedded in the jawbone is the **root.** Canine teeth, incisors, and premolars have one root, although the first upper premolars commonly have two. As for molars, the first two
upper molars have three roots, while the corresponding lower molars have two. The root pattern of the
third molar varies, but a fused single root is most
common.

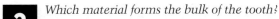

Which material forms the bulk of the tooth?

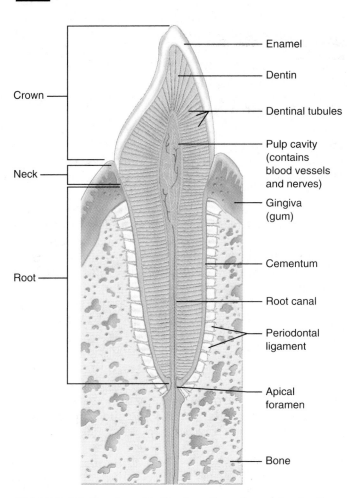

FIGURE 24.11 *Longitudinal section of a canine tooth within its bony alveolus.*

The crown and root are connected by a constricted tooth region called the **neck.** The outer surface of the root is covered by a calcified connective tissue called **cementum,** which attaches the tooth to the thin **periodontal** (per″e-o-don′tal; "around the tooth") **ligament.** This ligament anchors the tooth in the bony alveolus of the jaw, forming a fibrous joint called a *gomphosis* (see p. 250). Where the gingiva borders on a tooth, it dips downward to form a shallow groove called the *gingival sulcus.* In youth, the gingival margin adheres tenaciously to the enamel covering the crown. But as the gums begin to recede with age, the gingiva adheres to the more sensitive cementum covering the superior region of the root. As a result, the teeth *appear* to get longer in old age—hence the expression "long in the tooth" sometimes applied to elderly people.

Dentin, a bonelike material, underlies the enamel cap and forms the bulk of a tooth. It surrounds a central **pulp cavity,** which contains a number of soft tissue structures (connective tissue, blood vessels, and nerve fibers) collectively called **pulp.** Pulp supplies nutrients to the tooth tissues and provides for tooth sensation. Where the pulp cavity extends into the root, it becomes the **root canal.** At the proximal end of each root canal is an **apical foramen** that provides a route for blood vessels, nerves, and other structures to enter the pulp cavity of the tooth. The teeth are served by the superior and inferior alveolar nerves, branches of the trigeminal nerve (see Table 13.2, p. 487) and by the superior and inferior alveolar arteries, branches of the maxillary artery (see Figure 20.20, p. 755).

Dentin contains unique radial striations called *dentinal tubules* (Figure 24.11). Each tubule is filled with an elongated cellular process of an **odontoblast** (o-don′to-blast; "tooth former"), the cell type that secretes and maintains the dentin. The spherical cell bodies of odontoblasts line the underlying pulp cavity just deep to the dentin. Dentin is formed throughout adult life and gradually encroaches on the pulp cavity. New dentin can also be laid down fairly rapidly to compensate for tooth damage or decay.

Homeostatic Imbalance

Death of a tooth's nerve and consequent darkening of the tooth is commonly caused by a blow to the jaw. Swelling in the local area pinches off the blood supply to the tooth and the nerve dies. Typically the pulp becomes infected by bacteria some time later and must be removed by *root canal therapy.* After the cavity is sterilized and filled with an inert material, the tooth is capped. ■

Although enamel, dentin, and cementum are all calcified and resemble bone, they differ from bone in that they are avascular. Enamel also differs from cementum and dentin because it lacks collagen as its main organic component.

Tooth and Gum Disease Dental **caries** (kār′ēz; "rottenness"), or **cavities,** result from a gradual demineralization of enamel and underlying dentin by bacterial action. The decay process begins when **dental plaque** (a film of sugar, bacteria, and other mouth debris) adheres to the teeth. Metabolism of the trapped sugars by the bacteria produces acids, which can dissolve the calcium salts of the teeth. Once the salts are leached out, the remaining organic matrix of the tooth is readily digested by protein-digesting enzymes released by the bacteria. Frequent brushing and flossing daily help prevent damage by removing forming plaque.

Even more serious than tooth decay is the effect of unremoved plaque on the gums. As dental plaque

Dentin.

accumulates, it calcifies, forming **calculus** (kal′ku-lus; "stone"), or tartar. When this occurs in the gingival sulci, it disrupts the seals between the gingivae and the teeth, putting the gums at risk for infection. In the early stages of such an infection, called *gingivitis* (jin″ji-vi′tis), the gums become red, sore, and swollen and may bleed. Gingivitis is reversible if the calculus is removed, but if it is neglected the bacteria eventually invade the bone around the teeth, forming pockets of infection. The immune system attacks not only the intruders but also the body tissues, carving deep pockets around the teeth and dissolving the bone away. This more serious condition is called **periodontal disease, or periodontitis.** Periodontal disease affects up to 95% of all people over the age of 35 and accounts for 80–90% of tooth loss in adults.

Still, tooth loss is not inevitable. Even advanced cases of periodontitis can be treated by scraping the teeth, cleaning the infected pockets, then cutting the gums to shrink the pockets, and following up with antibiotic therapy. Together, these treatments alleviate the bacterial infestations and encourage reattachment of the surrounding tissues to the teeth and bone. Much less painful are (1) a new laser approach for destroying the diseased tissues and (2) a nonsurgical therapy now in clinical trials in which an antibiotic-impregnated film is temporarily glued to the exposed root surface. Of those receiving this topical therapy, 81% were able to avoid surgery or tooth extraction. Clinical treatment is followed up by a home regimen of plaque removal by consistent frequent brushing and flossing and hydrogen peroxide rinses.

Although periodontal disease has traditionally been viewed as a self-limiting low-grade infection, there may be more at risk than teeth. Some contend that it increases the risk of heart disease in at least two ways: (1) the chronic inflammation promotes atherosclerotic plaque formation, and (2) bacteria entering the blood from infected gums stimulate clot formation that helps to clog coronary arteries.

The Pharynx

From the mouth, food passes posteriorly into the **oropharynx** and then the **laryngopharynx** (see Figure 24.7). Both of these are common passageways for food, fluids, and air. (The nasopharynx has no digestive role.)

The histology of the pharyngeal wall resembles that of the oral cavity. The mucosa contains a friction-resistant stratified squamous epithelium well supplied with mucus-producing glands. The external muscle layer consists of two *skeletal muscle* layers. The cells of the inner layer run longitudinally. Those of the outer layer, the *pharyngeal constrictor muscles*, encircle the wall like three stacked fists

(see Figure 10.8, p. 339). Sequential contractions of these muscles propel food into the esophagus below.

The Esophagus

The **esophagus** (ĕ-sof′ah-gus; "carry food"), a muscular tube about 25 cm (10 inches) long, is collapsed when not involved in food propulsion. After food moves through the laryngopharynx, it is routed into the esophagus posteriorly as the epiglottis closes off the larynx to food entry.

As shown in Figure 24.1, the esophagus takes a fairly straight course through the mediastinum of the thorax and then pierces the diaphragm at the **esophageal hiatus** (hi-a′tus; "gap") to enter the abdomen. It joins the stomach at the **cardiac orifice.** The cardiac orifice is surrounded by the **cardiac** or **gastroesophageal** (gas″tro-ĕsof″ah-je′al) **sphincter** (see Figure 24.13), which is a *physiological* sphincter. That is, it acts as a valve, but the only structural evidence of this sphincter is a slight thickening of the circular smooth muscle at that point. The muscular diaphragm, which surrounds this sphincter, helps keep it closed when food is not being swallowed.

◢ *Homeostatic Imbalance*

Heartburn, the first symptom of *gastroesophageal reflux disease (GERD),* is the burning, radiating substernal pain that occurs when the acidic gastric juice regurgitates into the esophagus, causing the pH there to drop below 4. Symptoms are so similar to those of a heart attack that many first-time sufferers of heartburn are rushed to the hospital emergency room. Heartburn is most likely to happen when one has eaten or drunk to excess, and in conditions that force abdominal contents superiorly, such as extreme obesity, pregnancy, and running, which causes stomach contents to splash upward with each step (runner's reflux). It is also common in those with a **hiatus hernia,** a structural abnormality in which the superior part of the stomach protrudes slightly above the diaphragm. Since the diaphragm no longer reinforces the cardiac sphincter, gastric juice may flow into the esophagus, particularly when lying down. If the episodes are frequent and prolonged, *esophagitis* (inflammation of the esophagus) and *esophageal ulcers* may result. An even more threatening sequel is esophageal cancer. However, these consequences can usually be prevented or managed by avoiding late-night snacks and by using antacid preparations. ■

Unlike the mouth and pharynx, the esophagus wall has all four of the basic alimentary canal layers described on pp. 892–893. Some features of interest:

1. The esophageal mucosa contains a nonkeratinized stratified squamous epithelium. At the

? *What is the functional significance of the epithelial change seen in part (b)?*

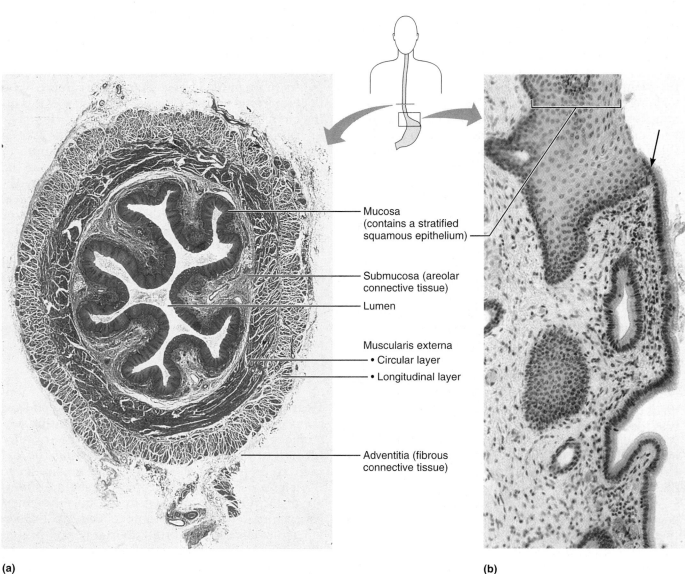

(a)

(b)

FIGURE 24.12   *Microscopic structure of the wall of the esophagus.*
(a) Cross-sectional view of the esophagus showing its four tunics (5×). The section shown is taken from the region close to the stomach junction, so the muscularis is composed of smooth muscle. (Superiorly, at the pharyngeal junction, the esophageal muscularis is composed of skeletal muscle.) **(b)** Longitudinal section through the esophagus-stomach junction (132×). Arrow shows the point of abrupt transition from the stratified squamous epithelium of the esophagus (top) to the simple columnar epithelium of the stomach (bottom).

Labels in figure (a):
Mucosa (contains a stratified squamous epithelium)
Submucosa (areolar connective tissue)
Lumen
Muscularis externa
• Circular layer
• Longitudinal layer
Adventitia (fibrous connective tissue)

esophagus-stomach junction, that abrasion-resistant epithelium changes abruptly to the simple columnar epithelium of the stomach, which is specialized for secretion (Figure 24.12b).

The presence of the stratified squamous epithelium indicates the esophagus is mainly a chute and thus must accommodate high friction, whereas the stomach has a secretory simple columnar epithelium, reflecting its secretory role in chemical digestion.

2. When the esophagus is empty, its mucosa and submucosa are thrown into longitudinal folds (Figure 24.12a). When food is in transit, these folds flatten out.

3. The submucosa contains mucus-secreting *esophageal glands*. As a bolus moves through the esophagus, it compresses these glands, causing them to secrete lubricating mucus that "greases" the esophageal walls and aids food passage.

4. The muscularis externa is skeletal muscle in its superior third, a mixture of skeletal and smooth muscle in its middle third, and entirely smooth muscle in its inferior third.

5. Instead of a serosa, the esophagus has a fibrous adventitia composed entirely of connective tissue, which blends with surrounding structures along its route.

Digestive Processes Occurring in the Mouth, Pharynx, and Esophagus

The mouth and its accessory digestive organs are involved in most digestive processes (see Table 24.1). The mouth (1) ingests and, as described below, (2) begins mechanical digestion by chewing, and (3) initiates propulsion by swallowing. Salivary amylase, the enzyme in saliva, starts the chemical breakdown of polysaccharides (starch and glycogen) into smaller fragments of linked glucose molecules. (If you chew a piece of bread for a few minutes, it will begin to taste sweet as sugars are released.) Except for a few drugs that are absorbed through the oral mucosa (for example, nitroglycerine), essentially no absorption occurs in the mouth.

In contrast to the multifunctional mouth, the pharynx and esophagus merely serve as conduits to pass food from the mouth to the stomach. Their single digestive function is food propulsion, accomplished by the role they play in swallowing.

Since chemical digestion is covered in a special physiology section later in the chapter, only the mechanical processes of chewing and swallowing are discussed here.

Mastication (Chewing)

As food enters the mouth, its mechanical breakdown is begun by **mastication,** or chewing. The cheeks and closed lips hold food between the teeth, the tongue mixes food with saliva to soften it, and the teeth cut and grind solid foods into smaller morsels. Mastication is partly voluntary and partly reflexive. We initiate it by voluntarily putting food into our mouths and contracting the muscles that close our jaws. Continued jaw movements are controlled mainly by stretch reflexes and in response to pressure inputs from receptors in the cheeks, gums, and tongue, but they can also be voluntary as desired.

Deglutition (Swallowing)

To send food on its way from the mouth, it is first compacted by the tongue into a bolus and then swallowed. **Deglutition** (deg″loo-tish′un), or swallowing, is a complicated process that involves coordinated activity of the tongue, soft palate, pharynx, esophagus, and over 22 separate muscle groups. It has two major phases, the buccal and the pharyngeal-esophageal (Figure 24.13).

The **buccal phase** occurs in the mouth and is voluntary. In the buccal phase, we place the tip of the tongue against the hard palate, and then contract the tongue to force the bolus into the oropharynx (Figure 24.13a). As food enters the pharynx and stimulates tactile receptors there, it passes out of our control and into the realm of involuntary reflex activity.

The involuntary **pharyngeal-esophageal phase** of swallowing is controlled by the swallowing center located in the medulla and lower pons. Motor impulses from that center are transmitted via various cranial nerves, most importantly the vagus nerves, to the muscles of the pharynx and esophagus. As illustrated in Figure 24.13b, once food enters the pharynx, all routes except the desired one—into the digestive tract—are blocked off:

- The tongue blocks off the mouth.
- The soft palate rises to close off the nasopharynx.
- The larynx rises so that the epiglottis covers its opening into the respiratory passageways.

Food is squeezed through the pharynx and into the esophagus by wavelike peristaltic contractions (Figure 24.13c–e). Solid foods pass from the oropharynx to the stomach in 4 to 8 seconds; fluids pass within 1 to 2 seconds. Just before the peristaltic wave (and food) reaches the end of the esophagus, the gastroesophageal sphincter relaxes reflexively to allow food to enter the stomach.

◢ *Homeostatic Imbalance*

If we try to talk or inhale while swallowing, the various protective mechanisms may be short-circuited and food may enter the respiratory passageways instead. This event typically triggers the coughing reflex in an attempt to expel the food. ■

THE STOMACH

Below the esophagus, the GI tract expands to form the **stomach** (see Figure 24.1), a temporary "storage tank" where the chemical breakdown of proteins begins and food is converted to a creamy paste called **chyme** (kīm; "juice"). The stomach lies in the upper left quadrant of the peritoneal cavity, nearly hidden by the liver and diaphragm. Specifically, it lies in the left hypochondriac, epigastric, and umbilical

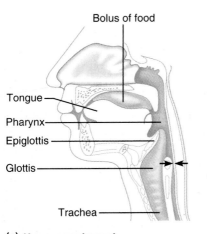

(a) **Upper esophageal sphincter contracted**

(b) **Upper esophageal sphincter relaxed**

(c) **Upper esophageal sphincter contracted**

FIGURE 24.13 *Deglutition (swallowing).* The process of swallowing consists of a voluntary (buccal) phase (a) and involuntary (pharyngeal–esophageal) phases (b–e). **(a)** During the buccal phase, the tongue rises and presses against the hard palate, forcing the food bolus into the oropharynx where the involuntary phase of swallowing begins. **(b)** Food passage into respiratory passageways is prevented by rising of the uvula and larynx; relaxation of the upper esophageal sphincter allows food entry into the esophagus. **(c)** The constrictor muscles of the pharynx contract, forcing food into the esophagus inferiorly, and the upper esophageal sphincter contracts after entry. **(d)** Food is conducted along the length of the esophagus to the stomach by peristaltic waves. **(e)** The gastroesophageal sphincter opens, and food enters the stomach.

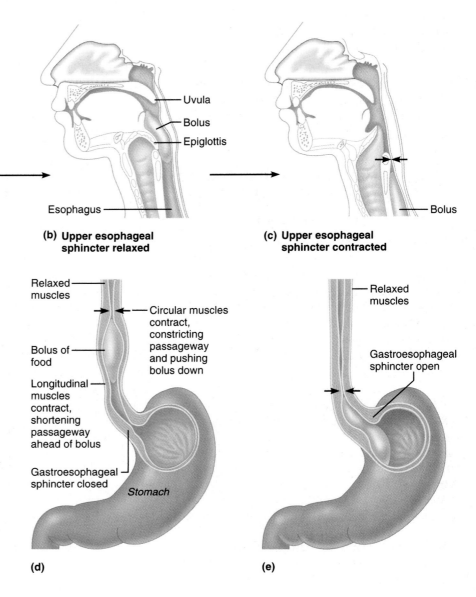

regions of the abdomen. Though relatively fixed at both ends, the stomach is quite movable in between. It tends to lie high and run horizontally in short, stout people (a steer-horn stomach) and is often elongated vertically in tall, thin people (a J-shaped stomach).

Gross Anatomy

The adult stomach varies from 15 to 25 cm (6 to 10 inches) long, but its diameter and volume depend on how much food it contains. An empty stomach has a volume of about 50 ml and a cross-sectional diameter only slightly larger than the large intestine, but when it is really distended it can hold about 4 L (1 gallon) of food and may extend nearly all the way to the pelvis! When empty, the stomach collapses inward, throwing its mucosa (and submucosa) into large, longitudinal folds called **rugae** (roo'ge; *ruga* = wrinkle, fold) (Figure 24.14).

The major regions of the stomach are shown in Figure 24.14a. The small **cardiac region,** or **cardia** ("near the heart"), surrounds the cardiac orifice through which food enters the stomach from the esophagus. The **fundus** is its dome-shaped part, tucked beneath the diaphragm, that bulges superolaterally to the cardia. The **body** is the midportion of the stomach. It is continuous inferiorly with the funnel-shaped **pyloric region.** The wider and more superior part of the pyloric region, the **pyloric antrum** (*antrum* = cave) narrows to form the **pyloric canal,** which terminates at the **pylorus.** The pylorus is continuous with the duodenum (the first part of the small intestine) through the **pyloric sphincter,** which controls stomach emptying (*pylorus* = gatekeeper).

? *What structural modification of a tunic underlies the stomach's ability to mechanically digest food?*

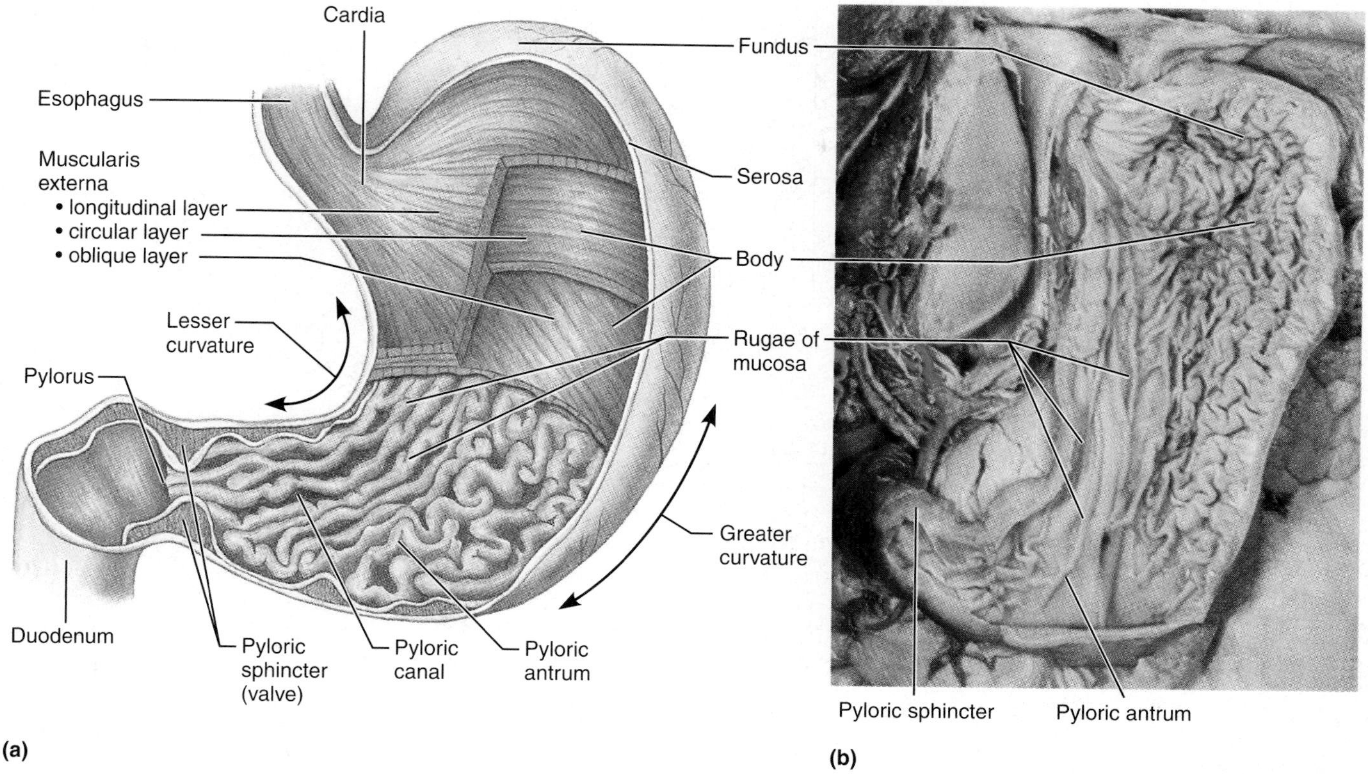

FIGURE 24.14 *Anatomy of the stomach.* **(a)** Gross internal anatomy (frontal section). (See *A Brief Atlas of the Human Body*, Figure 53a.) **(b)** Photograph of internal aspect of stomach; notice the many rugae.

The entire extent of the convex lateral surface of the stomach is its **greater curvature,** and its concave medial surface is the **lesser curvature** (Figure 24.14a). Extending from these curvatures are two mesenteries, called *omenta* (o-men'tah), that help to "tether" the stomach to other digestive organs and the body wall (see Figure 24.30, p. 929). The **lesser omentum** runs from the liver to the lesser curvature of the stomach, where it becomes continuous with the visceral peritoneum covering the stomach. The **greater omentum** drapes inferiorly from the greater curvature of the stomach to cover the coils of the small intestine. It then runs dorsally and superiorly (enclosing the spleen on its way) to wrap the transverse portion of the large intestine before blending with the *mesocolon,* a dorsal mesentery that secures the large intestine to the parietal peritoneum of the posterior abdominal wall. The greater omentum is riddled with fat deposits (*oment* = fatty skin) that give it the appearance of a lacy apron covering and

protecting the abdominal contents. It also contains large collections of lymph nodes. The immune cells and macrophages in these nodes "police" the peritoneal cavity and intraperitoneal organs.

The stomach is served by the autonomic nervous system. Sympathetic fibers from thoracic splanchnic nerves are relayed through the celiac plexus. Parasympathetic fibers are supplied by the vagus nerve. The arterial supply of the stomach is provided by branches (gastric and splenic) of the celiac trunk (see Figure 20.22, p. 758). The corresponding veins are part of the hepatic portal system (see Figure 20.27c, p. 771) and ultimately drain into the hepatic portal vein.

Microscopic Anatomy

The stomach wall exhibits the four tunics typical of most of the alimentary canal, but its muscularis and mucosa are modified for the special roles of the stomach. Besides the usual circular and longitudinal layers of smooth muscle, the muscularis externa has an innermost smooth muscle layer that runs *obliquely* (Figure 24.14a and Figure 24.15). This

It has a third layer of smooth muscle in its muscularis in which the fibers run obliquely, allowing it to exhibit churning movements that physically break down the food.

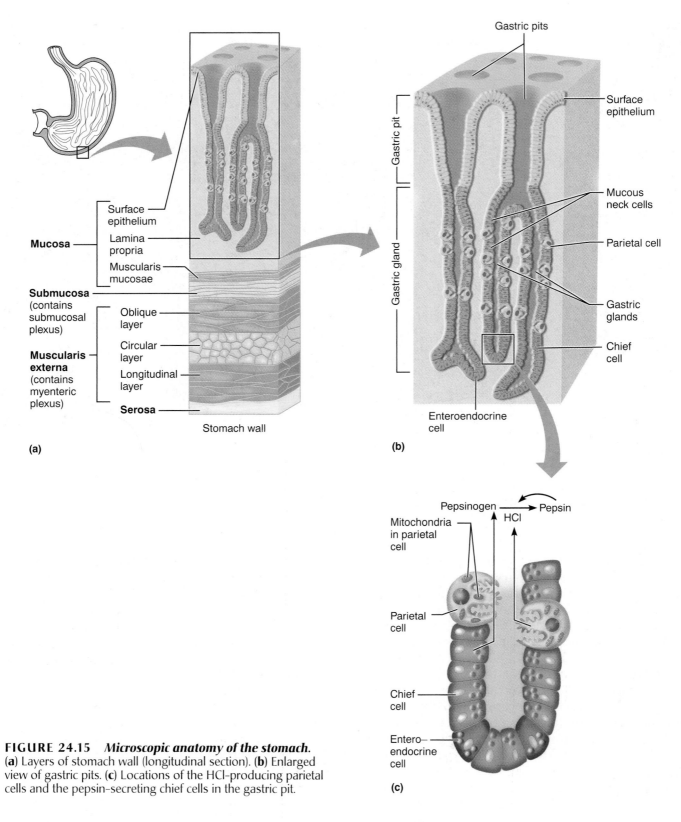

FIGURE 24.15 *Microscopic anatomy of the stomach.*
(a) Layers of stomach wall (longitudinal section). (b) Enlarged view of gastric pits. (c) Locations of the HCl-producing parietal cells and the pepsin-secreting chief cells in the gastric pit.

arrangement allows the stomach not only to move food along the tract, but also to churn, mix, and pummel the food, physically breaking it down into smaller fragments.

The lining epithelium of the stomach mucosa is a simple columnar epithelium composed entirely of goblet cells, which produce a protective coat of alkaline mucus. This otherwise smooth lining is dotted with millions of deep **gastric pits** (Figure 24.15), which lead into the **gastric glands** that collectively produce the stomach secretion called **gastric juice.** The cells forming the walls of the gastric pits are primarily goblet cells, but those composing the gastric glands vary in different regions of the stomach. For example, the cells in the glands of the cardia and pyloris are primarily mucus secreting, whereas cells of the pyloric antrum produce most of the stimulatory hormone called gastrin. The glands of the stomach fundus and body, where most chemical digestion occurs, are substantially larger and produce the majority of the stomach secretions. The glands in these regions contain a variety of secretory cells, including the four types described here:

1. Mucous neck cells, found in the upper, or "neck," regions of the glands, produce a different type of mucus from that secreted by the goblet cells of the surface epithelium. It is not yet understood what special function this *acidic* mucus performs.

2. Parietal (oxyntic) cells, found mainly in the middle region of the glands scattered among the chief cells (described next), secrete *hydrochloric acid (HCl)* and *intrinsic factor.* Although the parietal cells appear spherical when viewed with a light microscope, they actually have three prongs that exhibit dense microvilli (they look like fuzzy pitchforks!). This structure provides a huge surface area for secreting H^+ and Cl^- into the stomach lumen. HCl makes the stomach contents extremely acidic (pH 1.5–3.5), a condition necessary for activation and optimal activity of pepsin, and harsh enough to kill many of the bacteria ingested with foods. Intrinsic factor is a glycoprotein required for absorption of vitamin B_{12} in the small intestine.

3. Chief, or **zymogenic** (zi"mo-jeh'nik), **cells** produce *pepsinogen* (pep-sin'o-jen), the inactive form of the protein-digesting enzyme **pepsin.** The chief cells occur mainly in the basal regions of the gastric glands. When chief cells are stimulated, the first pepsinogen molecules they release are activated by HCl encountered in the apical region of the gland (Figure 24.15c). But once pepsin is present, it also catalyzes the conversion of pepsinogen to pepsin. This positive feedback process is limited only by the amount of pepsinogen present. The activation process involves removal of a small peptide fragment

from the pepsinogen molecule, causing it to change shape and expose its active site. Chief cells also appear to secrete relatively insignificant amounts of lipases (fat-digesting enzymes).

4. Enteroendocrine (en"ter-o-en'do-krin; "gut endocrine") **cells** release a variety of hormones or hormonelike products directly into the lamina propria. These products, including **gastrin, histamine, endorphins** (natural opiates), **serotonin, cholecystokinin,** and **somatostatin,** diffuse into the blood capillaries, and ultimately influence several digestive system target organs (Table 24.2, p. 912). Gastrin, in particular, plays essential roles in regulating stomach secretion and mobility, as described shortly.

The stomach mucosa is exposed to some of the harshest conditions in the entire digestive tract. Gastric juice is corrosively acidic (the H^+ concentration in the stomach can be 100,000 times that found in blood), and its protein-digesting enzymes can digest the stomach itself. However, the stomach is not a passive victim of its formidable environment. It mounts an aggressive counterattack to protect itself, producing what is called the **mucosal barrier.** Four factors create this barrier:

1. A thick coating of bicarbonate-rich mucus is built up on the stomach wall.

2. The epithelial cells of the mucosa are joined together by tight junctions that prevent gastric juice from leaking into the underlying tissue layers.

3. Deep in the gastric glands, where the protective alkaline mucus is absent, the external faces of the plasma membranes of the glandular cells are impermeable to HCl.

4. Damaged epithelial mucosal cells are shed and quickly replaced by division of *undifferentiated stem cells* that reside where the gastric pits join the gastric glands. The stomach surface epithelium is completely renewed every three to six days. (However, glandular cells deep within the gastric glands have a much longer life span.)

Homeostatic Imbalance

Anything that breaches the gel-like mucosal barrier produces inflammation of the underlying layers of the stomach wall, a condition called *gastritis.* Persistent damage to the underlying tissues can promote **gastric ulcers,** erosions of the stomach wall. The most distressing symptom of gastric ulcers is gnawing epigastric pain that seems to bore through to your back. The pain typically occurs 1–3 hours after eating and is often relieved by eating again. The danger posed by ulcers is perforation of the stomach wall followed by peritonitis and, perhaps, massive hemorrhage.

TABLE 24.1	Overview of the Functions of the Gastrointestinal Organs	
Organ	*Major functions**	*Comments/additional functions*
Mouth and associated accessory organs	■ Ingestion: food is voluntarily placed into oral cavity ■ Propulsion: voluntary (buccal) phase of deglutition (swallowing) initiated by tongue; propels food into pharynx ■ Mechanical digestion: mastication (chewing) by teeth and mixing movements by tongue □ Chemical digestion: chemical breakdown of starch is begun by salivary amylase present in saliva produced by salivary glands	Mouth serves as a receptacle; most functions performed by associated accessory organs; mucus present in saliva helps dissolve foods so they can be tasted and moistens food so that tongue can compact it into a bolus that can be swallowed; oral cavity and teeth cleansed and lubricated by saliva
Pharynx and esophagus	■ Propulsion: peristaltic waves move food bolus to stomach, thus accomplishing involuntary (pharyngeal-esophageal) phase of deglutition	Primarily food chutes; mucus produced helps to lubricate food passageways
Stomach	■ Mechanical digestion and propulsion: peristaltic waves mix food with gastric juice and propel it into the duodenum □ Chemical digestion: digestion of proteins begun by pepsin ▨ Absorption: absorbs a few fat-soluble substances (aspirin, alcohol, some drugs)	Also serves as storage site for food until it can be moved into the duodenum; hydrochloric acid produced is a bacteriostatic agent and activates protein-digesting enzymes; mucus produced helps lubricate and protect stomach from self-digestion; intrinsic factor produced is required for intestinal absorption of vitamin B_{12}
Small intestine and associated accessory organs (liver, gallbladder, pancreas)	■ Mechanical digestion and propulsion: segmentation by smooth muscle of the small intestine continually mixes contents with digestive juices and moves food along tract and through ileocecal valve at a slow rate, allowing sufficient time for digestion and absorption □ Chemical digestion: digestive enzymes conveyed in from pancreas and brush border enzymes attached to villi membranes complete digestion of all classes of foods ▨ Absorption: breakdown products of carbohydrate, protein, fat, and nucleic acid digestion; vitamins; electrolytes; and water are absorbed by active and passive mechanisms	Small intestine is highly modified for digestion and absorption (circular folds, villi, and microvilli); alkaline mucus produced by intestinal glands and bicarbonate-rich juice ducted in from pancreas help neutralize acidic chyme and provide proper environment for enzymatic activity; bile produced by liver emulsifies fats and enhances (1) fat digestion and (2) absorption of fatty acids, monoglycerides, cholesterol, phospholipids, and fat-soluble vitamins; gallbladder stores and concentrates bile; bile is released to small intestine in response to hormonal signals
Large intestine	□ Chemical digestion: some remaining food residues are digested by enteric bacteria (which also produce vitamin K and some B vitamins) ▨ Absorption: absorbs most remaining water, electrolytes (largely NaCl), and vitamins produced by bacteria ■ Propulsion: propels feces toward rectum by peristalsis, haustral churning, and mass movements ▨ Defecation: reflex triggered by rectal distension; eliminates feces from body	Temporarily stores and concentrates residues until defecation can occur; copious mucus produced by goblet cells eases passage of feces through colon

* The colored boxes beside the functions correspond to the color coding of digestive functions (gastrointestinal tract activities) illustrated in Figure 24.2.

Common predisposing factors for ulcer formation include hypersecretion of hydrochloric acid and hyposecretion of mucus. For years, factors that favor high HCl or low mucus production such as aspirin and nonsteroidal anti-inflammatory drugs (ibuprofen), smoking, alcohol, coffee, and stress were blamed for causing ulcers. Although acid conditions *are* necessary for ulcer formation, acidity in and of itself is not sufficient to cause ulcer formation. It is now clear that most recurrent ulcers (90%) are the work of acid-resistant, corkscrew-shaped *Helicobacter pylori* bacteria, which burrow beneath the mucus to attach to the epithelium and destroy the protective mucosal layer, leaving denuded areas. These bacteria release several chemicals that help them do their "dirty work" including (1) *urease,* an enzyme that breaks down urea to CO_2 and ammonia (the ammonia then acts as a base to neutralize some of the stomach acid in their locale), (2) a *cytotoxin* that lesions the stomach epithelium, and (3) several proteins that act as chemotactic agents to attract macrophages and other defensive cells into the area. This bacterial-causal theory has been difficult to prove because the bacterium is ubiquitous, found not only in some 70–90% of ulcer and gastritis sufferers but in a substantial number (over 33%) of healthy people as well. Even more troubling are current studies that link this bacterium to some stomach cancers.

In ulcers colonized by *Helicobacter,* the goal is to kill the embedded bacteria. A one- to two-week-long course of antibiotics (preferably two antibiotics with complementary effects, such as metronidazole and tetracycline) in combination with a bismuth-containing compound effectively promotes healing and prevents recurrence. (Bismuth is the active ingredient in Pepto-Bismol.) In cases of active ulcers, a H_2-receptor blocker, which inhibits HCl secretion by blocking histamine's effects, may also be prescribed. In noninfectious cases, the H_2-receptor blockers, drugs such as cimetidine (Tagamet) and ranitidine (Zantac), are the therapy of choice. ■

Digestive Processes Occurring in the Stomach

Except for ingestion and defecation, the stomach is involved in the whole "menu" of digestive activities. Besides serving as a holding area for ingested food, the stomach continues the demolition job begun in the oral cavity by further degrading food both physically and chemically. It then delivers chyme, the product of its activity, into the small intestine at an appropriate rate.

Protein digestion is initiated in the stomach and is essentially the only type of enzymatic digestion that occurs there. The most important protein-digesting enzyme produced by the gastric mucosa is pepsin. In children, however, the stomach glands also secrete **rennin,** an enzyme that acts on milk protein (casein), converting it to a curdy substance that looks like soured milk. Two common lipid-soluble substances—alcohol and aspirin—pass easily through the stomach mucosa into the blood, but some other lipid-soluble drugs also pass. The movement of aspirin and large quantities of alcohol through the gastric mucosa may cause gastric bleeding; thus, these substances should be avoided by those with gastric ulcers.

Despite the obvious benefits of preparing food to enter the intestine, the only stomach function essential to life is secretion of intrinsic factor. **Intrinsic factor** is required for intestinal absorption of vitamin B_{12}, needed to produce mature erythrocytes—in its absence, *pernicious anemia* results. However, if vitamin B_{12} is administered by injection, individuals can survive with minimal digestive problems even after total gastrectomy (stomach removal). (The stomach's activities are summarized in Table 24.1).

Since chemical digestion and absorption are described later, here we will focus on events that (1) control secretory activity of the gastric glands and (2) regulate stomach motility and emptying.

Regulation of Gastric Secretion

Gastric secretion is controlled by both neural and hormonal mechanisms. Under normal conditions the gastric mucosa pours out as much as 3 L of gastric juice—an acid brew so potent it can dissolve nails—every day. Nervous control is provided by long (vagus nerve–mediated) and short (local enteric) nerve reflexes. When the vagus nerves actively stimulate the stomach, secretory activity of virtually all of its glands increases. (By contrast, activation of sympathetic nerves depresses secretory activity.) Hormonal control of gastric secretion is largely the province of gastrin, which stimulates secretion of both enzymes and HCl, and hormones produced by the small intestine, which are mostly gastrin antagonists.

Stimuli acting at three distinct sites—the head, stomach, and small intestine—provoke or inhibit gastric secretory activity and, accordingly, the three phases of gastric secretion are called the *cephalic, gastric,* and *intestinal phases* (Figure 24.16). However, the effector site is the stomach in all cases and, once initiated, one or all three phases may be occurring at the same time.

Phase 1: cephalic (reflex) The **cephalic,** or **reflex, phase** of gastric secretion occurs *before* food enters the stomach. It is triggered by the aroma, taste, sight, or thought of food. During this phase, the brain gets the stomach ready for its upcoming digestive chore. Inputs from activated olfactory receptors and taste buds are relayed to the hypothalamus, which in turn stimulates the vagal nuclei of the medulla oblongata, causing motor impulses to be transmitted via the vagus nerves to the parasympathetic enteric ganglia. Enteric ganglionic neurons in turn stimulate the stomach glands. The enhanced secretory activity that results when we see or think of food is a *conditioned reflex* and occurs only when we like or want the food. If we are depressed or have no appetite, this part of the cephalic reflex is suppressed.

? *Distension of the stomach and duodenal walls have differing effects on stomach secretory activity. What are these effects?*

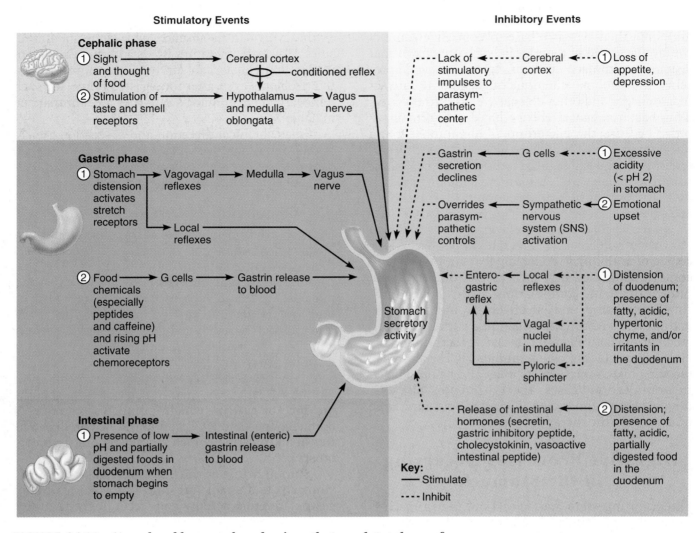

FIGURE 24.16 **Neural and hormonal mechanisms that regulate release of gastric juice.** Stimulatory factors are shown on the left; inhibitory factors are shown on the right.

Phase 2: gastric Once food reaches the stomach, local neural and hormonal mechanisms initiate the **gastric phase.** This phase provides about two-thirds of the gastric juice released. The most important stimuli are distension, peptides, and low acidity. Stomach distension activates stretch receptors and initiates both local (myenteric) reflexes and the long vagovagal reflexes. In the latter type of reflex, impulses travel to the medulla and then back to the stomach via vagal fibers. Both types of reflexes lead to acetylcholine (ACh) release, which in turn stimulates the output of more gastric juice by the secretory cells.

Though neural influences initiated by stomach distension are important, the hormone gastrin probably plays a greater role in stimulating stomach gland secretion during the gastric phase. Chemical stimuli provided by partially digested proteins (peptides), caffeine (present in colas, coffee, and tea), and rising pH directly activate gastrin-secreting entoendocrine cells called **G cells.** Although gastrin also stimulates the release of enzymes, its main target is the HCl-secreting parietal cells, which it prods to spew out even more HCl. Highly acidic (pH below 2) gastric contents *inhibit* gastrin secretion.

Distension of the stomach promotes increased secretory activity whereas stretching the wall of the duodenum causes hormones to be released that inhibit stomach secretory activity.

When protein foods are in the stomach, the pH of the gastric contents generally rises because proteins act as buffers to tie up H$^+$. The rise in pH stimulates gastrin and subsequently HCl release, which in turn provides the acidic conditions needed for protein digestion. The more protein in the meal, the greater the amount of gastrin and HCl released. As proteins are digested, the gastric contents gradually become more acidic, which again inhibits the gastrin-secreting cells. This negative feedback mechanism helps maintain optimal pH and working conditions for the gastric enzymes.

G cells are also activated by the neural reflexes already described. Emotional upsets, fear, anxiety, or anything that triggers the fight-or-flight response inhibits gastric secretion because (during such times) the sympathetic division overrides parasympathetic controls of digestion (Figure 24.16).

The control of the HCl-secreting parietal cells is unique and multifaceted. Basically, HCl secretion is stimulated by three chemicals, all of which work through second-messenger systems (Figure 24.17). *ACh* released by parasympathetic nerve fibers and *gastrin* secreted by G cells bring about their effects by increasing intracellular Ca^{2+} levels. *Histamine*, released by mucosal cells called histaminocytes, acts through cyclic AMP (cAMP). When only one of the three chemicals binds to the parietal cells, HCl secretion is scanty, but when all three bind, volumes of HCl pour forth as if shot out by a high-pressure hose. (As noted earlier, antihistamines, such as cimetidine, which bind to and block the H$_2$ receptors of parietal cells, are used to treat gastric ulcers due to hyperacidity.) The process of HCl formation within the parietal cells is complicated and as yet poorly understood. The consensus is that H$^+$ is actively pumped into the stomach lumen against a tremendous concentration gradient. As hydrogen ions are secreted, chloride ions (Cl$^-$) are also pumped into the lumen to maintain an electrical balance in the stomach. The Cl$^-$ is obtained from blood plasma, while the H$^+$ appears to come from the breakdown of carbonic acid (formed by the combination of carbon dioxide and water) within the parietal cell (Figure 24.17):

$$CO_2 + H_2O \rightarrow H_2CO_3 \rightarrow H^+ + HCO_3^-$$

As H$^+$ is pumped from the cell and HCO$_3^-$ (bicarbonate ion) accumulates within the cell, HCO$_3^-$ is ejected through the basal cell membrane into the capillary blood. As a result, blood draining from the stomach is more alkaline than the blood serving it. This phenomenon is called the **alkaline tide.**

Phase 3: intestinal

The **intestinal phase** of gastric secretion has two components—one excitatory and the other inhibitory (Figure 24.16). The *excitatory* aspect is set into motion as partially digested food

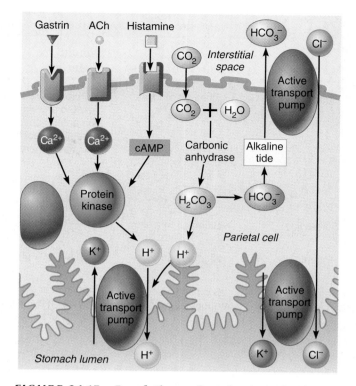

FIGURE 24.17 *Regulation and mechanism of HCl secretion.* Binding of histamine, gastrin, and acetylcholine (ACh) to parietal cell membrane receptors initiates intracellular events (mediated by second-messenger systems) that lead to HCl secretion into the stomach lumen. H$^+$ and HCO$_3^-$ (bicarbonate ions) are generated from the dissociation of carbonic acid (H$_2$CO$_3$). As H$^+$ and Cl$^-$ are pumped into the lumen, HCO$_3^-$ is pumped into the interstitial space in exchange for chloride ions (Cl$^-$).

begins to fill the initial part (duodenum) of the small intestine. This stimulates intestinal mucosal cells to release a hormone that encourages the gastric glands to continue their secretory activity. The effects of this hormone mirror those of gastrin, so it has been named **intestinal (enteric) gastrin.** However, intestinal mechanisms stimulate gastric secretion only briefly. As the intestine distends with chyme containing large amounts of H$^+$, fats, partially digested proteins, and various irritating substances, the *inhibitory* component is triggered in the form of the **enterogastric reflex.**

The enterogastric reflex is actually a trio of reflexes that (1) inhibit the vagal nuclei in the medulla, (2) inhibit local reflexes, and (3) activate sympathetic fibers that cause the pyloric sphincter to tighten and prevent further food entry into the small intestine. As a result, gastric secretory activity declines. As described in more detail in the next section, these "brakes" on gastric activity protect the small intestine from harm due to excessive acidity and match the small intestine's processing abilities to the amount of chyme entering it at a given time.

TABLE 24.2   Hormones and Hormonelike Products That Act in Digestion*

Hormone	Site of production	Stimulus for production	Target organ	Activity
Gastrin	Stomach mucosa	Food (particularly partially digested proteins) in stomach (chemical stimulation); acetylcholine released by nerve fibers	Stomach	■ Causes gastric glands to increase secretory activity; most pronounced effect is on HCl secretion ■ Stimulates gastric emptying
			Small intestine	■ Stimulates contraction of intestinal muscle
			Ileocecal valve	■ Relaxes ileocecal valve
			Large intestine	■ Stimulates mass movements
Serotonin	Stomach mucosa	Food in stomach	Stomach	■ Causes contraction of stomach muscle
Histamine	Stomach mucosa	Food in stomach	Stomach	■ Activates parietal cells to release HCl
Somatostatin	Stomach mucosa; duodenal mucosa	Food in stomach; stimulation by sympathetic nerve fibers	Stomach	■ Inhibits gastric secretion of all products; inhibits gastric motility and emptying
			Pancreas	■ Inhibits secretion
			Small intestine	■ Inhibits GI blood flow; thus inhibits intestinal absorption
			Gallbladder	■ Inhibits contraction and bile release
Intestinal gastrin	Duodenal mucosa	Acidic and partially digested foods in duodenum	Stomach	■ Stimulates gastric glands and motility
Secretin	Duodenal mucosa	Acidic chyme (also partially digested proteins, fats, hypertonic or hypotonic fluids, or irritants in chyme)	Stomach	■ Inhibits gastric gland secretion and gastric motility during gastric phase of secretion
			Pancreas	■ Increases output of pancreatic juice rich in bicarbonate ions; potentiates CCK's action
			Liver	■ Increases bile output
Cholecystokinin (CCK)	Duodenal mucosa	Fatty chyme, in particular, but also partially digested proteins	Liver/pancreas	■ Potentiates secretin's actions on these organs
			Pancreas	■ Increases output of enzyme-rich pancreatic juice
			Gallbladder	■ Stimulates organ to contract and expel stored bile
			Hepatopancreatic sphincter (of Oddi)	■ Relaxes to allow entry of bile and pancreatic juice into duodenum
Gastric inhibitory peptide (GIP)[†]	Duodenal mucosa	Fatty and/or glucose-containing chyme	Stomach	■ Inhibits gastric gland secretion and gastric motility during gastric phase
Vasoactive intestinal peptide (VIP)	Duodenal mucosa	Chyme containing partially digested foods	Duodenum	■ Stimulates buffer secretion; dilates intestinal capillaries
			Stomach	■ Inhibits HCl production ■ Relaxes intestinal smooth muscle

*Except for somatostatin, all of these polypeptides also stimulate the growth (particularly of the mucosa) of the organs they affect.

[†]Also called gastric insulinotropic peptide because it stimulates the release of insulin from the pancreatic islets.

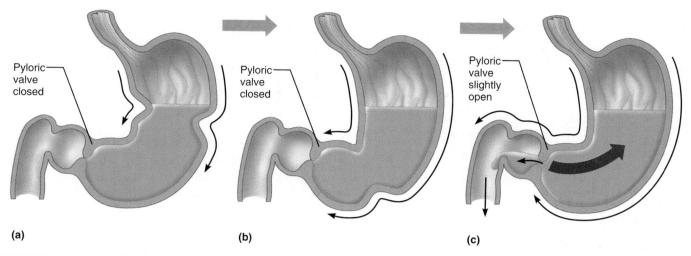

(a)　　　　　　　　　　　**(b)**　　　　　　　　　　　**(c)**

FIGURE 24.18 *Peristaltic waves act primarily in the inferior portion of the stomach to mix and move chyme through the pyloric valve.* (a) Peristaltic waves move toward the pylorus. (b) The most vigorous peristalsis and mixing action occurs close to the pylorus. (c) The pyloric end of the stomach acts as a pump that delivers small amounts of chyme into the duodenum, while simultaneously forcing most of its contained material backward into the stomach, where it undergoes further mixing.

In addition, the factors just named trigger the release of several intestinal hormones, collectively called **enterogastrones,** which include **secretin** (se-kre′tin), **cholecystokinin** (ko″le-sis″to-ki′nin) **(CCK), vasoactive intestinal peptide (VIP),** and **gastric inhibitory peptide (GIP).** All of these hormones inhibit gastric secretion when the stomach is very active. These hormones also play other roles, which are summarized in Table 24.2.

Gastric Motility and Emptying

Stomach contractions, accomplished by the trilayered muscularis, not only cause its emptying but also compress, knead, twist, and continually mix the food with gastric juice to produce chyme. Because the mixing movements are accomplished by a unique type of peristalsis (for example, in the pylorus it is bidirectional rather than unidirectional), the processes of mechanical digestion and propulsion are inseparable in the stomach.

Response of the Stomach to Filling Although the stomach stretches to accommodate incoming food, internal stomach pressure remains constant until about 1 L of food has been ingested. Thereafter, the pressure rises. The relatively unchanging pressure in a filling stomach reflects two phenomena: (1) reflex-mediated relaxation of the stomach muscle and (2) plasticity of visceral smooth muscle.

Reflexive relaxation of stomach smooth muscle in the fundus and body occurs both in anticipation of and in response to food entry into the stomach. As food travels through the esophagus, the stomach muscles relax. This **receptive relaxation** is coordinated by the swallowing center of the brain stem and mediated by the vagus nerves. The stomach also actively dilates in response to gastric filling, which activates stretch receptors in the wall. This phenomenon, called **adaptive relaxation,** appears to depend on local reflexes involving *nitric oxide (NO) releasing neurons.*

Plasticity is the intrinsic ability of visceral smooth muscle to exhibit the stress-relaxation response, that is, to be stretched without greatly increasing its tension and contracting expulsively. As described in Chapter 9 (p. 309), this capability is very important in hollow organs, like the stomach, that must serve as temporary reservoirs.

Gastric Contractile Activity After a meal, peristalsis begins near the cardiac sphincter, where it produces only gentle rippling movements of the stomach wall. But as the contractions approach the pylorus, where the stomach musculature is thicker, they become much more powerful. Consequently, the contents of the fundus remain relatively undisturbed, while foodstuffs close to the pylorus receive a truly lively pummeling and mixing.

The pyloric region of the stomach, which holds about 30 ml of chyme, acts as a "dynamic filter" that allows only liquids and small particles to pass through the barely open pyloric valve during the digestive period. Normally, each peristaltic wave reaching the pyloric muscle "spits" or squirts 3 ml or less of chyme into the small intestine. Because the contraction also *closes* the valve, which is normally partially relaxed, the rest (about 27 ml) is propelled backward into the stomach, where it is mixed further (Figure 24.18). This back-and-forth pumping action effectively breaks up solids in the gastric contents.

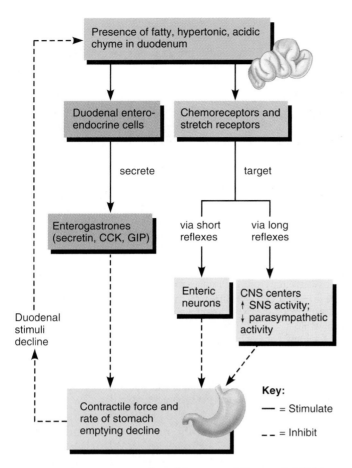

FIGURE 24.19 *Neural and hormonal factors inhibiting gastric emptying.* These controls ensure that the food will be well liquefied in the stomach and prevent the small intestine from being overwhelmed.

Although the intensity of the stomach's peristaltic waves can be modified a good deal, their rate is constant—always around three per minute. This contractile rhythm is set by the spontaneous activity of *pacemaker cells* located at the margins of the longitudinal smooth muscle layer. The pacemaker cells, believed to be muscle-like noncontractile cells called *interstitial cells of Cajal* (ka-hal'), depolarize and repolarize spontaneously three times each minute, establishing the so-called *cyclic slow waves* of the stomach, or its **basic electrical rhythm (BER).** Since the pacemakers are electrically coupled to the rest of the smooth muscle sheet by gap junctions, their "beat" is transmitted efficiently and quickly to the entire muscularis. The pacemakers set the maximum rate of contraction, but they do not initiate the contractions or regulate their force. Instead, they generate subthreshold depolarization waves, which are then "ignited" (enhanced by further depolarization and brought to threshold) by neural and hormonal factors.

Factors that increase the strength of stomach contractions are the same factors that enhance gastric secretory activity. Distension of the stomach wall by food activates stretch receptors and gastrin-secreting cells, which both ultimately stimulate gastric smooth muscle and so increase gastric motility. Thus, the more food there is in the stomach, the more vigorous the stomach mixing and emptying movements will be—within certain limits—as described next.

Regulation of Gastric Emptying The stomach usually empties completely within four hours after a meal. However, the larger the meal (the greater the stomach distension) and the more liquid its contents, the faster the stomach empties. Fluids pass quickly through the stomach. Solids linger, remaining until they are well mixed with gastric juice and converted to the liquid state.

The rate of gastric emptying depends as much—and perhaps more—on the contents of the duodenum as on what is happening in the stomach. The stomach and duodenum act in tandem like a "coupled meter" that functions at less than full capacity. As chyme enters the duodenum, receptors in its wall respond to chemical signals and to stretch, initiating the enterogastric reflex and the hormonal (enterogastrone) mechanisms described earlier. These mechanisms inhibit gastric secretory activity and prevent further duodenal filling by reducing the force of pyloric contractions (Figure 24.19).

A carbohydrate-rich meal moves through the duodenum rapidly, but fats form an oily layer at the top of the chyme and are digested more slowly by enzymes acting in the intestine. Thus, when chyme entering the duodenum is fatty, food may remain in the stomach six hours or more.

Homeostatic Imbalance

Vomiting, or **emesis,** is an unpleasant experience that causes stomach emptying by a different route. Many factors signal the stomach to "launch lunch," but the most common are extreme stretching of the stomach or intestine or the presence of irritants such as bacterial toxins, excessive alcohol, spicy foods, and certain drugs in those organs. Sensory impulses stream from the irritated sites to the **emetic** (e-met'ik) **center** of the medulla and initiate a number of motor responses. The diaphragm and abdominal wall muscles contract, increasing intra-abdominal pressure, the cardiac sphincter relaxes, and the soft palate rises to close off the nasal passages. As a result, the stomach (and perhaps duodenal) contents are forced upward through the esophagus and pharynx and out the mouth. Before vomiting, an individual typically is pale-faced, feels nauseated, and salivates. Excessive vomiting can cause dehydration and may lead to severe disturbances in the electrolyte and acid-base balance of the body. Since large

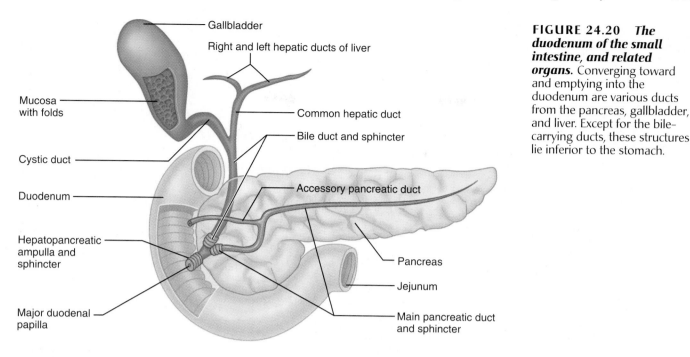

Gallbladder

Right and left hepatic ducts of liver

Mucosa with folds

Cystic duct

Duodenum

Hepatopancreatic ampulla and sphincter

Major duodenal papilla

Common hepatic duct

Bile duct and sphincter

Accessory pancreatic duct

Pancreas

Jejunum

Main pancreatic duct and sphincter

FIGURE 24.20 *The duodenum of the small intestine, and related organs.* Converging toward and emptying into the duodenum are various ducts from the pancreas, gallbladder, and liver. Except for the bile-carrying ducts, these structures lie inferior to the stomach.

amounts of hydrochloric acid are lost in vomitus, the blood becomes alkaline as the stomach attempts to replace its lost acid. ■

THE SMALL INTESTINE AND ASSOCIATED STRUCTURES

In the small intestine, usable food is finally prepared for its journey into the cells of the body. However, this vital function cannot be accomplished without the aid of secretions from the liver (bile) and pancreas (digestive enzymes). Thus, these necessary accessory organs will also be considered in this section.

The Small Intestine

The **small intestine** is the body's major digestive organ. Within its twisted passageways, digestion is completed and virtually all absorption occurs.

Gross Anatomy

The small intestine is a convoluted tube extending from the pyloric sphincter in the epigastric region to the **ileocecal** (il″e-o-se′kal) **valve** in the right iliac region where it joins the large intestine (see Figure 24.1). It is the longest part of the alimentary tube, but its diameter is only about 2.5 cm (1 inch). Although it is 6–7 m long (approximately 20 feet, or as tall as a two-story building) in a cadaver, the small intestine is only about 2–4 m (8–13 feet) long during life because of muscle tone.

The small intestine has three subdivisions: the duodenum, which is mostly retroperitoneal, and the

jejunum and ileum, both intraperitoneal organs. The relatively immovable **duodenum** (du″o-de′num; "twelve finger widths long"), which curves around the head of the pancreas (see Figure 24.20), is about 25 cm (10 inches) long. Although it is the shortest intestinal subdivision, the duodenum has the most features of interest. The bile duct, delivering bile from the liver, and the main pancreatic duct, carrying pancreatic juice from the pancreas, unite in the wall of the duodenum in a bulblike point called the **hepatopancreatic ampulla** (hep″ah-to-pan″kre-at′ik am-pul′ah; *ampulla* = flask). The ampulla opens into the duodenum via the volcano-shaped **major duodenal papilla** (see Figure 24.20). The entry of bile and pancreatic juice is controlled by a muscular valve called the **hepatopancreatic sphincter,** or **sphincter of Oddi.**

The **jejunum** (jĕ-joo′num; "empty") is about 2.5 m (8 feet) long and extends from the duodenum to the ileum. The **ileum** (il′e-um; "twisted"), approximately 3.6 m (12 feet) in length, joins the large intestine at the ileocecal valve. The jejunum and ileum hang in sausagelike coils in the central and lower part of the abdominal cavity, suspended from the posterior abdominal wall by the fan-shaped **mesentery** (see Figure 24.30). These more distal parts of the small intestine are encircled and framed by the large intestine.

Nerve fibers serving the small intestine include parasympathetics from the vagus and sympathetics from the thoracic splanchnic nerves, both relayed through the superior mesenteric (and celiac) plexus. The arterial supply is primarily from the superior mesenteric artery (p. 761). The veins run parallel to the arteries and typically drain into the superior

? *What is a lacteal and what is its function?*

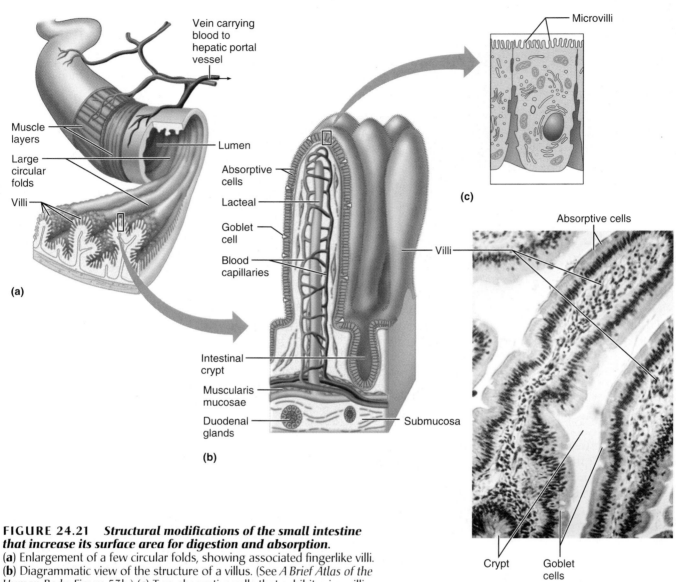

FIGURE 24.21 *Structural modifications of the small intestine that increase its surface area for digestion and absorption.* (**a**) Enlargement of a few circular folds, showing associated fingerlike villi. (**b**) Diagrammatic view of the structure of a villus. (See *A Brief Atlas of the Human Body*, Figure 53b.) (**c**) Two absorptive cells that exhibit microvilli on their free (luminal) surface. (**d**) Photomicrograph of the small intestinal mucosa, showing villi (300×).

mesenteric vein. From there, the nutrient-rich venous blood from the small intestine drains into the hepatic portal vein (p. 771) which carries it to the liver.

Microscopic Anatomy

Modifications for Absorption The small intestine is highly adapted for nutrient absorption. Its length

A lymph capillary that receives some of the fatty materials absorbed from the intestine.

alone provides a huge surface area, and its wall has three structural modifications—plicae circulares, villi, and microvilli—that amplify its absorptive surface enormously. One estimate is that the intestinal surface area is about 200 m² (equal to the floor space of an average two-story house)! Most absorption occurs in the proximal part of the small intestine, so these specializations decrease in number toward its distal end.

The **circular folds,** or **plicae circulares** (pli′ke ser″ku-lar′ēs) are deep, permanent folds of the mucosa and submucosa (Figure 24.21a). Nearly 1 cm

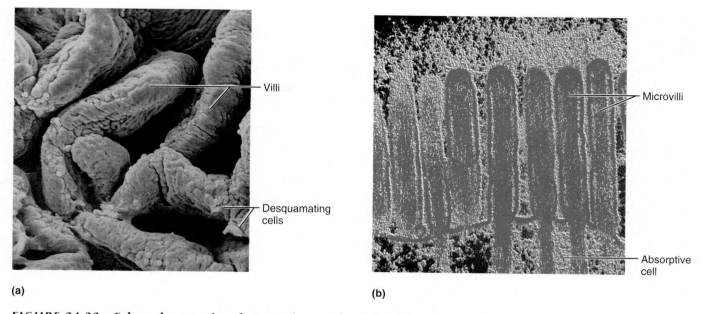

(a)

(b)

FIGURE 24.22   *False-color scanning electron micrographs of the villi and microvilli of the small intestine.* (a) Villi (110×). Dead desquamating cells seen on the ridges of the villi are colored yellow and orange. **(b)** Microvilli of absorptive cells. The microvilli appear as red projections from the surface of the absorptive cell. The yellow granules are mucus granules (26,000×).

tall, these folds force chyme to spiral through the lumen, slowing its movement and allowing time for full nutrient absorption.

Villi (vil′i) are fingerlike projections of the mucosa, over 1 mm high, that give it a velvety texture, much like the soft nap of a towel (Figures 24.21b and 24.22a). The epithelial cells of the villi are chiefly absorptive columnar cells called *enterocytes* ("intestinal cells"). In the core of each villus is a dense capillary bed and a wide lymph capillary called a **lacteal** (lak′te-al). Digested foodstuffs are absorbed through the epithelial cells into both the capillary blood and the lacteal. The villi are large and leaflike in the duodenum (the intestinal site of most active absorption) and gradually become narrower and shorter along the length of the small intestine. A "slip" of smooth muscle in the villus core allows it to alternately shorten and lengthen. These pulsations (1) increase the amount of contact between the villus surface and the contents of the intestinal lumen, making absorption more efficient, and (2) "milk" lymph along through the lacteals.

Microvilli, tiny projections of the plasma membrane of the absorptive cells of the mucosa, give the mucosal surface a fuzzy appearance sometimes called the **brush border** (Figures 24.21c and 24.22b). Besides amplifying the absorptive surface, the plasma membranes of the microvilli bear enzymes, referred to collectively as **brush border enzymes.** These enzymes complete the final stages of digestion of carbohydrates and proteins in the small intestine.

Histology of the Wall Although externally the different subdivisions of the small intestine appear to be nearly identical, their internal and microscopic anatomies reveal some important differences. The four tunics typical of the GI tract are also seen here, but the mucosa and submucosa are modified to reflect the relative location of the intestine in the digestive pathway (see Figure 24.21b).

The epithelium of the mucosa is largely simple columnar *absorptive cells* bound by tight junctions and richly endowed with microvilli. It also has many mucus-secreting *goblet cells;* interspersed T cells called *intraepithelial lymphocytes,* an important immunological compartment; and scattered *enteroendocrine cells* (the source of the enterogastrones—secretin and cholecystokinin to name two).

Between the villi, the mucosa is studded with *pits* (openings) that lead into tubular intestinal glands called **intestinal crypts,** or **crypts of Lieberkühn** (le′ber-kun). The epithelial cells that line these crypts secrete *intestinal juice,* a watery mixture containing mucus that serves as a carrier fluid for absorption of nutrients from chyme. Located deep in the crypts are specialized secretory cells called *Paneth cells.* These cells fortify the small intestine's defenses against certain bacteria by releasing *lysozyme,* an antibacterial enzyme. The crypts decrease in number along the length of the small intestine, but the goblet cells become more abundant.

The various epithelial cells arise from rapidly dividing stem cells at the base of the crypts. The daughter cells gradually migrate up the villi, where they are shed from the villus tips. In this way, the villus epithelium is renewed every three to six days. The rapid replacement of intestinal (and gastric) epithelial cells has clinical as well as physiological significance. Treatments for cancer, such as radiation therapy and chemotherapy, preferentially target the cells in the body that are dividing most quickly. This kills cancer cells, but it also nearly obliterates the GI tract epithelium, causing nausea, vomiting, and diarrhea after each treatment.

The submucosa is typical areolar connective tissue, and it contains both individual and *aggregated lymphoid follicles,* the latter called **Peyer's** (pi'erz) **patches.** Peyer's patches increase in abundance toward the end of the small intestine, reflecting the fact that the large intestine contains huge numbers of bacteria that must be prevented from entering the bloodstream. A set of elaborated mucus-secreting **duodenal** (*Brunner's*) **glands** is found in the submucosa of the duodenum only. These glands (see Figure 24.21b) produce an alkaline (bicarbonate-rich) mucus that helps neutralize the acidic chyme moving in from the stomach. When this protective mucus barrier is inadequate, the intestinal wall is eroded and *duodenal ulcers* result.

The muscularis is typical and bilayered. Except for the bulk of the duodenum, which is retroperitoneal and has an adventitia, the external intestinal surface is covered by visceral peritoneum (serosa).

Intestinal Juice: Composition and Control

The intestinal glands normally secrete 1 to 2 L of **intestinal juice** daily. The major stimulus for its production is distension or irritation of the intestinal mucosa by hypertonic or acidic chyme. Normally, the pH range of intestinal juice is slightly alkaline (7.4–7.8), and it is isotonic with blood plasma. Intestinal juice is largely water but it also contains some mucus, which is secreted both by the duodenal glands and by goblet cells of the mucosa. Intestinal juice is relatively enzyme-poor because intestinal enzymes are largely limited to the bound enzymes of the brush border.

The Liver and Gallbladder

The *liver* and *gallbladder* are accessory organs associated with the small intestine. The liver, one of the body's most important organs, has many metabolic and regulatory roles. However, its *digestive* function is to produce bile for export to the duodenum. Bile is a fat emulsifier; that is, it breaks up fats into tiny particles so that they are more accessible to digestive enzymes. We will describe bile and the emulsification process when we discuss the digestion and absorption of fats later in the chapter. Although the liver also processes nutrient-laden venous blood delivered to it from the digestive organs, this function concerns its metabolic rather than its digestive role. (The metabolic functions of the liver are described in Chapter 25.) The gallbladder is chiefly a storage organ for bile.

Gross Anatomy of the Liver

The ruddy, blood-rich **liver** is the largest gland in the body, weighing about 1.4 kg (3 pounds) in the average adult. Shaped like a wedge, it occupies most of the right hypochondriac and epigastric regions, extending farther to the right of the body midline than to the left. Located under the diaphragm, the liver lies almost entirely within the rib cage, which provides some protection (see Figures 24.1 and 24.23c).

Typically, the liver is said to have four primary lobes. The largest of these, the *right lobe,* is visible on all liver surfaces and separated from the smaller *left lobe* by a deep fissure (Figure 24.23a). The posteriormost *caudate lobe* and the *quadrate lobe,* which lies inferior to the left lobe, are visible in an inferior view of the liver (Figure 24.23b). A mesentery, the **falciform ligament,** separates the right and left lobes anteriorly and suspends the liver from the diaphragm and anterior abdominal wall (see also Figure 24.30a and d, p. 929). Running along the free inferior edge of the falciform ligament is the **round ligament,** or **ligamentum teres** (te'rēz; "round"), a fibrous remnant of the fetal umbilical vein (Figures 24.23 and 24.30a). Except for the superiormost liver area (the *bare area*), which is fused to the diaphragm, the entire liver is enclosed by a serosa (the visceral peritoneum). As mentioned earlier, a dorsal mesentery, the lesser omentum (see Figure 24.30b), anchors the liver to the lesser curvature of the stomach. The **hepatic artery** and the **hepatic portal vein,** which enter the liver at the **porta hepatis** ("gateway to the liver") and the common hepatic duct, which runs inferiorly from the liver, all travel through the lesser omentum to reach their destinations. The gallbladder rests in a recess on the inferior surface of the right liver lobe (Figure 24.23).

The traditional scheme of defining the liver's lobes (outlined above) has been criticized because it is based on superficial features of the liver, such as the ligamentum teres and falciform ligament. Some anatomists emphasize that the primary lobes of the liver should be defined as the territories served by the right and left hepatic ducts. These two territories are delineated by a plane drawn from the indentation (sulcus) of the inferior vena cava to the gallbladder recess (see Figure 24.23b). Those to the right of the plane are the right lobe and those to its left constitute the left lobe. According to this scheme, the small quadrate and caudate lobes are part of the left lobe.

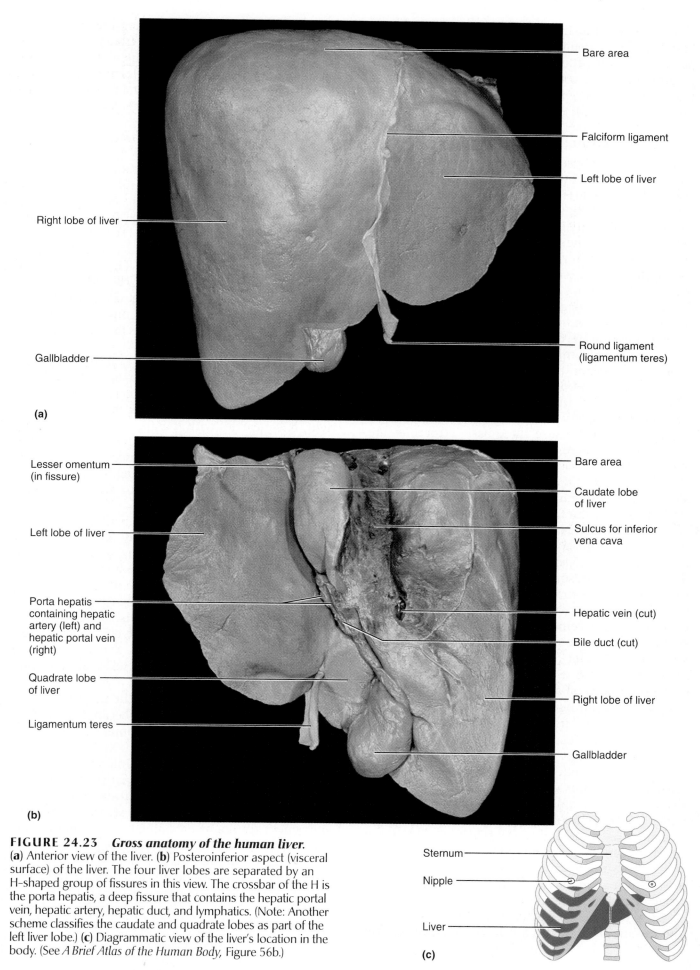

Bare area

Falciform ligament

Left lobe of liver

Right lobe of liver

Round ligament (ligamentum teres)

Gallbladder

(a)

Lesser omentum (in fissure)

Bare area

Caudate lobe of liver

Left lobe of liver

Sulcus for inferior vena cava

Porta hepatis containing hepatic artery (left) and hepatic portal vein (right)

Hepatic vein (cut)

Bile duct (cut)

Quadrate lobe of liver

Right lobe of liver

Ligamentum teres

Gallbladder

(b)

FIGURE 24.23 *Gross anatomy of the human liver.*
(**a**) Anterior view of the liver. (**b**) Posteroinferior aspect (visceral surface) of the liver. The four liver lobes are separated by an H-shaped group of fissures in this view. The crossbar of the H is the porta hepatis, a deep fissure that contains the hepatic portal vein, hepatic artery, hepatic duct, and lymphatics. (Note: Another scheme classifies the caudate and quadrate lobes as part of the left liver lobe.) (**c**) Diagrammatic view of the liver's location in the body. (See *A Brief Atlas of the Human Body*, Figure 56b.)

Sternum

Nipple

Liver

(c)

Bile leaves the liver through several bile ducts that ultimately fuse to form the large **common hepatic duct,** which travels downward toward the duodenum. Along its course, that duct fuses with the **cystic duct** draining the gallbladder to form the **bile duct** (Figure 24.20).

Microscopic Anatomy of the Liver

The liver is composed of sesame seed–sized structural and functional units called **liver lobules** (Figure 24.24). Each lobule is a roughly hexagonal (six-sided) structure consisting of plates of *liver cells,* or **hepatocytes** (hep'ah-to-sīts), organized like bricks in a garden wall. The hepatocyte plates radiate outward from a **central vein** running within the longitudinal axis of the lobule. To make a rough "model" of a liver lobule, open a thick paperback book until its two covers meet: The pages represent the plates of hepatocytes and the hollow cylinder formed by the rolled spine represents the central vein.

If you keep in mind that the liver's main function is to filter and process the nutrient-rich blood delivered to it, the description of its anatomy that follows will make a lot of sense. At each of the six corners of a lobule is a **portal triad** (*portal tract* region), so named because three basic structures are always present there: a branch of the *hepatic artery* (supplying oxygen-rich arterial blood to the liver), a branch of the *hepatic portal vein* (carrying venous blood laden with nutrients from the digestive viscera), and a **bile duct.** Between the hepatocyte plates are enlarged, very leaky capillaries, the **liver sinusoids.** Blood from both the hepatic portal vein and the hepatic artery percolates from the triad regions through these sinusoids and empties into the central vein. From the central veins blood eventually enters the hepatic veins, which drain the liver, and empty into the inferior vena cava. Inside the sinusoids are star-shaped **hepatic macrophages,** also called **Kupffer** (kōōp'fer) **cells** (see Figure 24.24d), which remove debris such as bacteria and worn-out blood cells from the blood as it flows past.

The versatile hepatocytes are virtual organelle storehouses with large amounts of both rough and smooth ER, Golgi apparatuses, peroxisomes, and mitochondria. Thus equipped, the hepatocytes not only produce bile but also (1) process the bloodborne nutrients in various ways (e.g., they store glucose as glycogen and use amino acids to make plasma proteins); (2) store fat-soluble vitamins; and (3) play important roles in detoxification, such as ridding the blood of ammonia by converting it to urea (Chapter 25). Thanks to the various liver cells, the blood leaving the liver contains fewer nutrients and waste materials than the blood that entered it.

Secreted bile flows through tiny canals, called **bile canaliculi** (kan"ah-lik'u-li; "little canals"), that run between adjacent hepatocytes toward the bile duct branches in the portal triads (see Figure 24.24d). Notice that blood and bile flow in opposite directions in the liver lobule. Bile entering the bile ducts eventually leaves the liver via the common hepatic duct to travel toward the duodenum.

⊿ *Homeostatic Imbalance*

Hepatitis (hep"ah-ti'tis), or inflammation of the liver, is most often due to viral infection. Thus far the catalogue of hepatitis-causing viruses numbers six—from the hepatitis A virus (HVA) to the hepatitis F virus (HVF). Of these, two (HVA and HVE) are transmitted enterically and the infections they cause tend to be largely self-limiting. Those transmitted via blood—most importantly HVB and HVC—are linked to chronic hepatitis and cirrhosis of the liver (see discussion below). HVD is a mutated virus that needs HVB to be infectious. Thus far, little is known about HVF.

In the United States, over 40% of hepatitis cases are due to HVB, which is transmitted via blood transfusions, contaminated needles, or sexual contact. Hepatitis B is a serious problem in its own right, but it carries with it a greater menace—an elevated risk of liver cancer. However, childhood immunization using vaccine produced in bacteria is sweeping the feet out from under HVB; the incidence of acute hepatitis from this strain has fallen about 70% since its peak in 1985.

Hepatitis A, accounting for about 32% of cases, is a more benign form frequently observed in day-care centers. It is transmitted via sewage-contaminated food, raw shellfish, water, and by the feces-mouth route, which explains why it is important for restaurant employees to scrub their hands after using the washroom. Hepatitis E, is transmitted in a similar way. It causes waterborne epidemics, largely in developing countries, and is a major cause of death in pregnant women. It is relatively insignificant in the United States.

Hepatitis C has emerged as the most important liver disease in the United States because it produces persistent or chronic liver infections (as opposed to acute infections). More than 10,000 Americans die annually due to sequels of HVC infection. However, the life-threatening C form of hepatitis is now being successfully treated by combination drug therapy (either the immune-suppressing steroid prednisone and genetically engineered alpha interferon or ribaviron, an antiviral drug, and interferon, depending on the hepatitis strain).

Nonviral causes of acute hepatitis include drug toxicity and wild mushroom poisoning.

Cirrhosis (sĭ-ro'sis; "orange colored") is a diffuse and progressive chronic inflammation of the liver

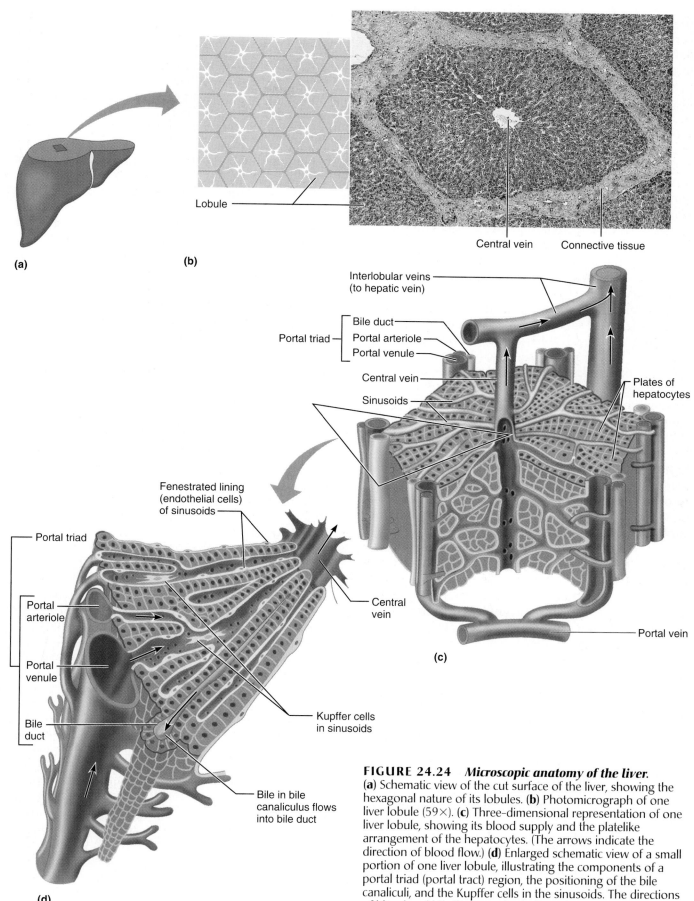

(a)

(b)

Lobule

Central vein   Connective tissue

Interlobular veins
(to hepatic vein)

Bile duct

Portal triad
Portal arteriole

Portal venule

Central vein

Sinusoids

Plates of
hepatocytes

Fenestrated lining
(endothelial cells)
of sinusoids

Portal triad

Portal
arteriole

Portal
venule

Bile
duct

Central
vein

Kupffer cells
in sinusoids

Bile in bile
canaliculus flows
into bile duct

Portal vein

(c)

(d)

FIGURE 24.24 *Microscopic anatomy of the liver.*
(**a**) Schematic view of the cut surface of the liver, showing the
hexagonal nature of its lobules. (**b**) Photomicrograph of one
liver lobule (59×). (**c**) Three-dimensional representation of one
liver lobule, showing its blood supply and the platelike
arrangement of the hepatocytes. (The arrows indicate the
direction of blood flow.) (**d**) Enlarged schematic view of a small
portion of one liver lobule, illustrating the components of a
portal triad (portal tract) region, the positioning of the bile
canaliculi, and the Kupffer cells in the sinusoids. The directions
of blood and bile flow are also indicated.

that typically results from chronic alcoholism or severe chronic hepatitis. The alcohol-poisoned or damaged hepatocytes regenerate (those closer to the blood-supplying portal triads faster than those nearer the central vein), but the liver's connective (scar) tissue regenerates faster. As a result, the liver becomes fatty and fibrous and its activity is depressed.

As the scar tissue eventually shrinks, it obstructs blood flow throughout the hepatic portal system, causing **portal hypertension.** Fortunately, some veins of the portal system anastomose with veins that drain into the venae cavae *(portal-caval anastomoses).* The main ones are (1) veins in the inferior esophagus, (2) hemorrhoidal veins in the anal canal, and (3) superficial veins around the umbilicus. However, these connecting veins are small and tend to burst when forced to carry large volumes of blood. Signs of their failure include vomiting blood, and a snakelike network of distended veins surrounding the navel. This network is called *caput medusae* (kap'ut mě-du'se; "medusa head") after a monster in Greek mythology whose hair was made of writhing snakes. Other complications due to portal hypertension include swollen veins in the esophagus (esophageal varices) and *ascites,* an accumulation of fluid in the peritoneal cavity. ■

Composition of Bile Bile is a yellow-green, alkaline solution containing bile salts, bile pigments, cholesterol, neutral fats, phospholipids (lecithin and others), and a variety of electrolytes. Of these *only* bile salts and phospholipids aid the digestive process.

Bile salts, primarily cholic acid and chenodeoxycholic acids, are cholesterol derivatives. Their role is to *emulsify* fats—that is, to distribute them throughout the watery intestinal contents. (Another example of emulsification is homogenization, which distributes cream throughout the watery phase of milk.) As a result, large fat globules entering the small intestine are physically separated into millions of small fatty droplets that provide large surface areas for the fat-digesting enzymes to work on. Bile salts also facilitate fat and cholesterol absorption (discussed later) and help solubilize cholesterol, both that contained in bile and that entering the small intestine in food. Although many substances secreted in bile leave the body in feces, bile salts are not among them. Instead, bile salts are conserved by means of a recycling mechanism called the **enterohepatic circulation.** In this process, bile salts are (1) reabsorbed into the blood by the distal part of the small intestine (ileum), (2) returned to the liver via the hepatic portal blood, and then (3) resecreted in newly formed bile.

The chief bile pigment is **bilirubin** (bil"ĭ-roo'bin), a waste product of the heme of hemoglobin formed during the breakdown of worn-out erythrocytes. The globin and iron parts of hemoglobin are saved and recycled, but bilirubin is absorbed from the blood by the liver cells and actively excreted into the bile. Most of the bilirubin in bile is metabolized in the small intestine by resident bacteria, and one of its breakdown products, *urobilinogen* (u"ro-bi-lin'o-jen), gives feces a brown color. In the absence of bile, feces are gray-white in color and have fatty streaks (because essentially no fats are digested or absorbed).

The liver produces 500 to 1000 ml of bile daily, and bile production is stepped up when the GI tract contains fatty chyme. Bile salts themselves are the major stimulus for enhanced bile secretion (Figure 24.25), and when the enterohepatic circulation is returning large amounts of bile salts to the liver, its output of bile rises dramatically. *Secretin,* released by intestinal cells exposed to fatty chyme, stimulates the cells lining the bile ducts in the liver to secrete a watery, bicarbonate-rich juice that lacks bile salts.

The Gallbladder

The **gallbladder** is a thin-walled, green muscular sac, roughly the size of a kiwi fruit (about 10 cm [4 inches] long), that snuggles in a shallow fossa on the ventral surface of the liver (see Figures 24.1 and 24.23). Its rounded head protrudes from the inferior margin of the liver. The gallbladder stores bile that is not immediately needed for digestion and concentrates it by absorbing some of its water and ions. (In some cases, bile released from the gallbladder is ten times as concentrated as that entering it.) When empty, or when storing only small amounts of bile, its mucosa is thrown into honeycomb-like folds that, like the rugae of the stomach, allow the organ to expand as it fills. When its muscular wall contracts, bile is expelled into its duct, the **cystic duct,** and then flows into the bile duct. The gallbladder, like most of the liver, is covered by visceral peritoneum.

Regulation of Bile Release into the Small Intestine

When digestion is not occurring, the hepatopancreatic sphincter (guarding the entry of bile and pancreatic juice into the duodenum) is tightly closed and the released bile backs up the cystic duct into the gallbladder, where it is stored until needed. Although the liver makes bile continuously, bile does not usually enter the small intestine until the gallbladder contracts. The major stimulus for gallbladder contraction is *cholecystokinin (CCK),* an intestinal hormone (Figure 24.25). CCK is released to the blood when acidic, fatty chyme enters the duodenum. Besides causing the gallbladder to contract, CCK has

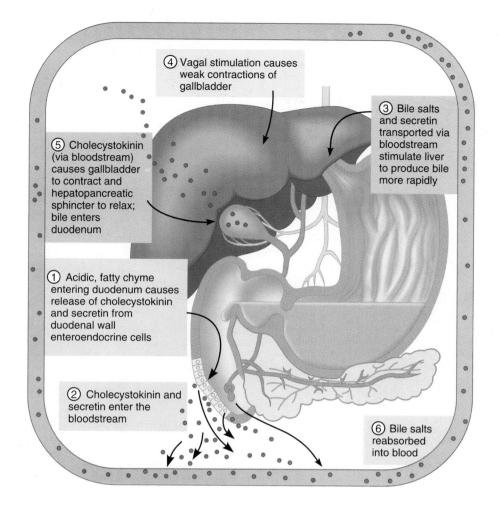

④ Vagal stimulation causes weak contractions of gallbladder

③ Bile salts and secretin transported via bloodstream stimulate liver to produce bile more rapidly

⑤ Cholecystokinin (via bloodstream) causes gallbladder to contract and hepatopancreatic sphincter to relax; bile enters duodenum

① Acidic, fatty chyme entering duodenum causes release of cholecystokinin and secretin from duodenal wall enteroendocrine cells

② Cholecystokinin and secretin enter the bloodstream

⑥ Bile salts reabsorbed into blood

FIGURE 24.25 *Mechanisms promoting the secretion of bile and its entry into the duodenum.* When digestion is not occurring, bile is stored and concentrated in the gallbladder. However, when acidic chyme enters the small intestine, a number of mechanisms are initiated that accelerate the liver's output of bile, cause the gallbladder to contract, and promote the relaxation of the hepatopancreatic sphincter, which allows bile (and pancreatic juice) to enter the small intestine, where it plays a role in fat emulsification. The single most important stimulus of bile secretion by the hepatocytes is an increased level of bile salts in the enterohepatic circulation.

two other important effects: (1) it stimulates secretion of pancreatic juice, and (2) it relaxes the hepatopancreatic sphincter so that bile and pancreatic juice can enter the duodenum. CCK and the other digestive hormones are summarized in Table 24.2, p. 912. Parasympathetic impulses delivered by the vagus nerves have a minor impact on stimulating gallbladder contraction.

◪ *Homeostatic Imbalance*

Bile is the major vehicle for cholesterol excretion from the body, and bile salts keep the cholesterol dissolved within bile. Too much cholesterol or too few bile salts can lead to crystallization of cholesterol, forming **gallstones,** or *biliary calculi* (bil'e-a"re kal'ku-li), which obstruct the flow of bile from the gallbladder. Then, when the gallbladder or its duct contracts, the sharp crystals cause agonizing pain (that radiates to the right thoracic region). Current treatments for gallstones include dissolving the crystals with drugs, pulverizing them with ultrasound vibrations (lithotripsy), vaporizing them with lasers, and the classical treatment, surgically removing the gallbladder. The gallbladder can be removed without

impairing digestive functions because, in such instances, the bile duct simply enlarges to assume the bile-storing role. Gallstones are easy to diagnose because they show up well with ultrasound imaging (see pp. 20–21).

Blockage of the bile duct prevents both bile salts and bile pigments from entering the intestine. As a result, the yellow bile pigments accumulate in the blood and eventually are deposited in the skin, causing it to become yellowish, or *jaundiced.* Jaundice caused by blocked ducts is called *obstructive jaundice,* but jaundice may also reflect liver disease (in which the liver is unable to carry out its normal metabolic duties). ■

The Pancreas

The **pancreas** (pan'kre-as; *pan* = all, *creas* = flesh, meat) is a soft, tadpole-shaped gland that extends across the abdomen from its *tail* (abutting the spleen) to its *head,* which is encircled by the C-shaped duodenum (see Figures 24.1 and 24.20). Most of the pancreas is retroperitoneal and lies deep to the greater curvature of the stomach.

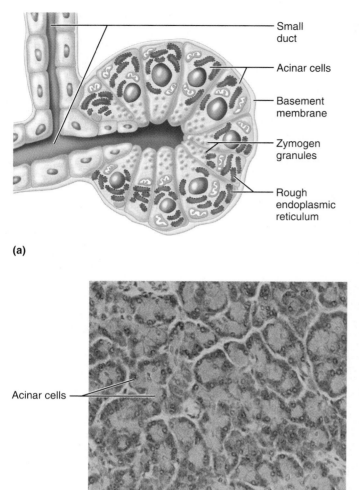

Small duct

Acinar cells

Basement membrane

Zymogen granules

Rough endoplasmic reticulum

(a)

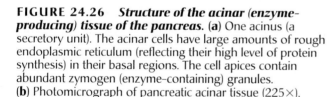

Acinar cells

(b)

FIGURE 24.26 *Structure of the acinar (enzyme-producing) tissue of the pancreas.* **(a)** One acinus (a secretory unit). The acinar cells have large amounts of rough endoplasmic reticulum (reflecting their high level of protein synthesis) in their basal regions. The cell apices contain abundant zymogen (enzyme-containing) granules. **(b)** Photomicrograph of pancreatic acinar tissue (225×).

An accessory digestive organ, the pancreas is important to the digestive process because it produces a broad spectrum of enzymes that break down all categories of foodstuffs, which the pancreas then delivers to the duodenum. This exocrine product, called **pancreatic juice,** drains from the pancreas via the centrally located **main pancreatic duct.** The pancreatic duct generally fuses with the bile duct (conveying bile from the liver and gallbladder) just as it enters the duodenum (at the hepatopancreatic

What is the significance of the fact that pancreatic enzymes are activated in the small intestine?

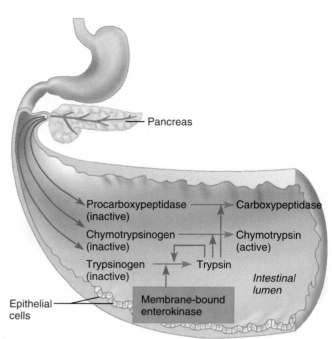

Pancreas

Procarboxypeptidase (inactive)

Carboxypeptidase

Chymotrypsinogen (inactive)

Chymotrypsin (active)

Trypsinogen (inactive)

Trypsin

Intestinal lumen

Epithelial cells

Membrane-bound enterokinase

FIGURE 24.27 *Activation of pancreatic proteases in the small intestine.* Pancreatic proteases are secreted in an inactive form and are activated in the duodenum. Enterokinase, a membrane-bound (brush border) intestinal enzyme, activates trypsinogen to the active trypsin form. Trypsin, itself a proteolytic enzyme, then activates procarboxypeptidase and chymotrypsinogen.

ampulla). A smaller *accessory pancreatic duct* empties directly into the main duct.

Within the pancreas are the **acini** (as′ĭ-ni; singular = acinus), clusters of secretory cells surrounding ducts (Figure 24.26). These cells are full of rough endoplasmic reticulum and exhibit deeply staining **zymogen** (zi′mo-jen; "fermenting") **granules** containing the digestive enzymes they manufacture.

The pancreas also has an endocrine function. Scattered amidst the acini are the more lightly staining *pancreatic islets (islets of Langerhans)*. These mini-endocrine glands release insulin and glucagon, hormones that play an important role in carbohydrate metabolism, as well as several other hormones (see Chapter 17).

They are activated in the small intestine where they work and where the cells exposed to them are protected by mucus. No such protection exists in the pancreas.

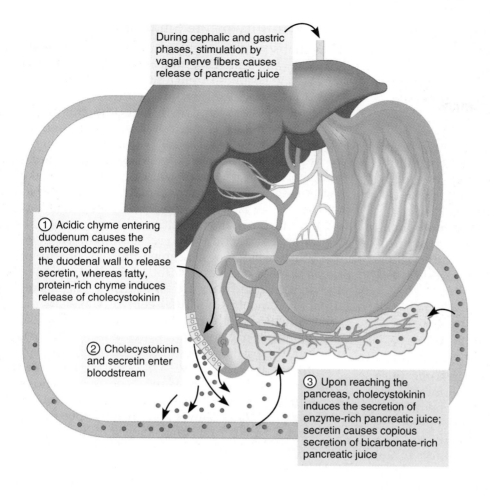

During cephalic and gastric phases, stimulation by vagal nerve fibers causes release of pancreatic juice

① Acidic chyme entering duodenum causes the enteroendocrine cells of the duodenal wall to release secretin, whereas fatty, protein-rich chyme induces release of cholecystokinin

② Cholecystokinin and secretin enter bloodstream

③ Upon reaching the pancreas, cholecystokinin induces the secretion of enzyme-rich pancreatic juice; secretin causes copious secretion of bicarbonate-rich pancreatic juice

FIGURE 24.28 *Regulation of pancreatic juice secretion by hormonal and neural factors.* Hormonal controls, exerted by secretin and cholecystokinin (steps 1–3), are the main regulatory factors. Neural control is mediated by parasympathetic fibers of the vagus nerves, primarily during the cephalic and gastric phases of gastric secretory activity.

Composition of Pancreatic Juice

Approximately 1200 to 1500 ml of clear pancreatic juice is produced daily. It consists mainly of water, and contains enzymes and electrolytes (primarily bicarbonate ions). The acinar cells produce the enzyme-rich component of pancreatic juice. The epithelial cells lining the smallest pancreatic ducts release the bicarbonate ions that make it alkaline (about pH 8). This high pH enables pancreatic fluid to neutralize the acid chyme entering the duodenum and provides the optimal environment for activity of intestinal and pancreatic enzymes. Like pepsin of the stomach, pancreatic proteases (protein-digesting enzymes) are produced and released in inactive (precursor) forms, which are then activated in the duodenum, where they do their work. This prevents the pancreas from self-digestion. For example, within the duodenum, *trypsinogen* is activated to **trypsin** by **enterokinase,** an intestinal brush border enzyme. Trypsin, in turn, activates two other pancreatic proteases (*procarboxypeptidase* and *chymotrypsinogen*) to their active forms, **carboxypeptidase** (kar-bok″se-pep′tĭ-dās) and **chymotrypsin** (ky″mo-trip′sin), respectively (Figure 24.27). Other pancreatic enzymes—**amylase, lipases,** and **nucleases**—are se-

creted in active form, but require that ions or bile be present in the intestinal lumen for optimal activity.

Regulation of Pancreatic Secretion

Secretion of pancreatic juice is regulated both by local hormones and by the parasympathetic nervous system (Figure 24.28). The more important hormonal mechanism is exerted by two intestinal hormones: (1) *secretin*, released in response to the presence of HCl in the intestine, and (2) *cholecystokinin*, released in response to the entry of proteins and fats. Both hormones act on the pancreas, but secretin targets the duct cells, prompting their release of watery *bicarbonate-rich* pancreatic juice, whereas CCK stimulates the acini to release *enzyme-rich* pancreatic juice. Vagal stimulation causes release of pancreatic juice primarily during the cephalic and gastric phases of gastric secretion.

Normally, the amount of HCl produced in the stomach is exactly balanced by the amount of bicarbonate (HCO_3^-) actively secreted by the pancreas, and as HCO_3^- is secreted into the pancreatic juice, H^+ enters the blood. Consequently, the pH of venous blood returning to the heart remains relatively unchanged because alkaline blood draining from the stomach is neutralized by the acidic blood draining the pancreas.

Digestive Processes Occurring in the Small Intestine

Although food reaching the small intestine is unrecognizable, it is far from being digested chemically. Carbohydrates and proteins are partially degraded, but virtually no fat digestion has occurred up to this point. The process of food digestion is accelerated during the chyme's tortuous journey of 3 to 6 hours through the small intestine, and it is here that virtually all nutrient absorption occurs. Like the stomach, the small intestine has no part in ingestion and defecation.

Requirements for Optimal Intestinal Digestive Activity

Although the primary functions of the small intestine are digestion and absorption, intestinal juice provides little of what is needed to perform these functions. Most substances required for chemical digestion—bile, digestive enzymes (except for the brush border enzymes), and bicarbonate ions (to provide the proper pH for enzymatic catalysis)—are *imported* from the liver and pancreas. Hence, anything that impairs liver or pancreatic function or delivery of their juices to the small intestine severely hinders our ability to digest food and absorb nutrients.

Optimal digestive activity in the small intestine also depends on a slow, measured delivery of chyme from the stomach. As indicated earlier, the small intestine can process only small amounts of chyme at one time. Why is this so? Entering chyme is usually hypertonic. Thus, if large amounts of chyme were rushed into the small intestine, the osmotic water loss from the blood into the intestinal lumen would result in dangerously low blood volume. Additionally, the low pH of entering chyme must be adjusted upward and the chyme must be well mixed with bile and pancreatic juice for digestion to continue. These modifications take time. Thus, food movement into the small intestine is carefully controlled by the pumping action of the stomach pylorus (see pp. 913–914), which prevents the duodenum from being overwhelmed. Since the actual chemistry of digestion and absorption is covered in detail later, next we will examine how the small intestine mixes and propels food along its length, and how its motility is regulated.

Motility of the Small Intestine

Intestinal smooth muscle mixes chyme thoroughly with bile and pancreatic and intestinal juices, and moves food residues through the ileocecal valve into the large intestine. In contrast to the peristaltic waves of the stomach, which both mix and propel food, *segmentation* is the most common motion of the small intestine.

If X-ray fluoroscopy is used to examine the small intestine after it is "loaded" with a meal, it looks like the intestinal contents are being "massaged"—that is, the chyme is simply moved backward and forward in the lumen a few centimeters at a time by alternating contraction and relaxation of rings of smooth muscle. These segmenting movements of the intestine (see Figure 24.3b), like the peristalsis of the stomach, are initiated by intrinsic pacemaker cells (presumably interstitial cells of Cajal) in the longitudinal smooth muscle layer. However, unlike the stomach pacemakers, which have only one rhythm, the pacemakers in the duodenum depolarize more frequently (12–14 contractions per minute) than those in the ileum (8 or 9 contractions per minute). As a result, segmentation also moves intestinal contents slowly and steadily toward the ileocecal valve at a rate that allows ample time to complete digestion and absorption. The intensity of segmentation is altered by long and short reflexes (which parasympathetic activity enhances and sympathetic activity decreases) and hormones. The more intense the contractions, the greater the mixing effect; however, the basic contractile rhythms of the various intestinal regions remain unchanged.

True peristalsis occurs only after most nutrients have been absorbed. At this point, segmenting movements wane, and peristaltic waves initiated in the duodenum begin to sweep slowly along the intestine, moving 10–70 cm before dying out. Each successive wave is initiated a bit more distally, and this pattern of peristaltic activity is called the **migrating mobility complex.** A complete "trip" from duodenum to ileum takes about two hours. The process then repeats itself, sweeping out the last remnants of the meal plus bacteria, sloughed-off mucosal cells, and other debris into the large intestine. This "housekeeping" function is critical for preventing the overgrowth of bacteria that migrate from the large intestine into the small intestine. As food again enters the stomach with the next meal, peristalsis is replaced by segmentation.

The local enteric neurons of the GI tract wall coordinate these intestinal mobility patterns. The physiological diversity of the enteric neurons allows a variety of effects to occur depending on which neurons are activated or inhibited. For example, a given ACh-releasing (cholinergic) sensory neuron in the small intestine, once activated, may simultaneously send messages to several different interneurons in the myenteric plexus that regulate peristalsis:

- Impulses sent proximally by cholinergic neurons cause contraction and shortening of the circular muscle layer.

- Impulses sent distally to certain interneurons cause shortening of the longitudinal muscle layer and distension of the intestine, in response to ACh-releasing neurons.

- Other impulses sent distally by activated VIP or NO-releasing enteric neurons cause relaxation of the circular muscle.

As a result, as the proximal area constricts and forces chyme along the tract, the lumen of the distal part of the intestine enlarges to receive it.

Most of the time, the ileocecal sphincter is constricted and closed. However, two mechanisms—one neural and the other hormonal—cause it to relax when ileal mobility increases and allow food residues to enter the cecum. Enhanced activity of the stomach initiates the **gastroileal** (gas"tro-il'e-ul) **reflex,** a long reflex that enhances the force of segmentation in the ileum. In addition, gastrin released by the stomach increases the motility of the ileum and relaxes the ileocecal sphincter. Once the chyme has passed through, it exerts backward pressure that closes the valve's flaps, preventing regurgitation into the ileum.

THE LARGE INTESTINE

The **large intestine** (see Figure 24.1) frames the small intestine on three sides and extends from the ileocecal valve to the anus. Its diameter is greater than that of the small intestine (hence, *large* intestine), but it is less than half as long (1.5 m versus 6 m). Its major function is to absorb water from indigestible food residues (delivered to it in a fluid state) and eliminate them from the body as semisolid **feces** (fe'sēz).

Gross Anatomy

Over most of its length, the large intestine exhibits three features not seen elsewhere—teniae coli, haustra, and epiploic appendages. Except for its terminal end, the longitudinal muscle layer of its muscularis is reduced to three bands of smooth muscle called **teniae coli** (ten'ne-e ko'li; "ribbons of the colon"). Their tone causes the wall of the large intestine to pucker into pocketlike sacs called **haustra** (haw'strah; "to draw up"). The final, and very obvious, unique feature of the large intestine is its **epiploic** (ep"i-plo'ik) **appendages** (Figure 24.29a), small fat-filled pouches of visceral peritoneum that hang from its surface. Their significance is not known.

Subdivisions

The large intestine has the following subdivisions: cecum, appendix, colon, rectum, and anal canal. The saclike **cecum** (se'kum; "blind pouch"), which lies below the ileocecal valve in the right iliac fossa, is the first part of the large intestine (Figure 24.29a). Attached to its posteromedial surface is the blind, wormlike **vermiform appendix.** The appendix contains masses of lymphoid tissue, and as part of MALT (see p. 786), it plays an important role in body immunity. However, it has an important structural shortcoming—its twisted structure provides an ideal location for enteric bacteria to accumulate and multiply.

Homeostatic Imbalance

Acute inflammation of the appendix, or **appendicitis,** results from a blockage (often by feces) that traps infectious bacteria within its lumen. Unable to empty its contents, the appendix swells, squeezing off venous drainage. These events may lead to ischemia and necrosis (death and decay) of the appendix. If the appendix ruptures, feces containing bacteria spray over the abdominal contents, causing *peritonitis.* Although the symptoms of appendicitis are highly variable, the first is usually pain in the umbilical region. Loss of appetite, nausea and vomiting, and relocalization of pain to the lower right abdominal quadrant follow. Immediate surgical removal of the appendix (appendectomy) is the accepted treatment for suspected appendicitis. Appendicitis is most common during adolescence, when the entrance to the appendix is at its widest. ■

The **colon** has several distinct regions. As the **ascending colon,** it travels up the right side of the abdominal cavity to the level of the right kidney. Here it makes a right-angle turn—the **right colic,** or **hepatic, flexure**—and travels across the abdominal cavity as the **transverse colon.** Directly anterior to the spleen, it bends acutely at the **left colic (splenic) flexure** and descends down the left side of the posterior abdominal wall as the **descending colon.** Inferiorly, it enters the pelvis, where it becomes the S-shaped **sigmoid colon.** Except for the transverse and sigmoid parts of the colon, which are anchored to the posterior abdominal wall by mesentery sheets called **mesocolons** (see Figure 24.30c and d), the colon is a retroperitoneal organ.

In the pelvis, at the level of the third sacral vertebra, the sigmoid colon joins the **rectum,** which runs posteroinferiorly just in front of the sacrum.

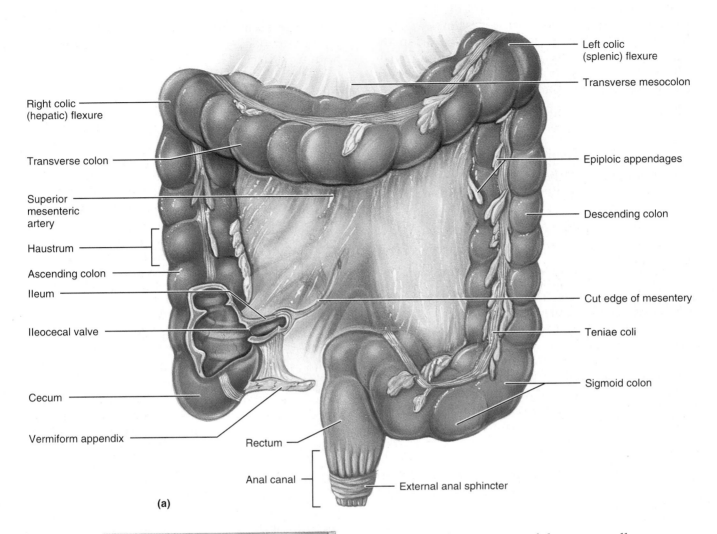

Right colic (hepatic) flexure

Transverse colon

Superior mesenteric artery

Haustrum

Ascending colon

Ileum

Ileocecal valve

Cecum

Vermiform appendix

Left colic (splenic) flexure

Transverse mesocolon

Epiploic appendages

Descending colon

Cut edge of mesentery

Teniae coli

Sigmoid colon

Rectum

Anal canal

External anal sphincter

(a)

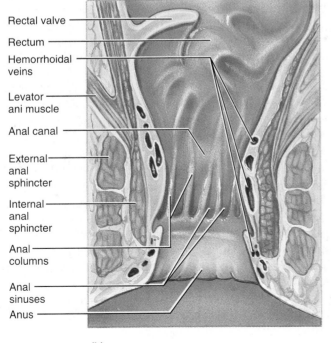

Rectal valve

Rectum

Hemorrhoidal veins

Levator ani muscle

Anal canal

External anal sphincter

Internal anal sphincter

Anal columns

Anal sinuses

Anus

(b)

The natural orientation of the rectum allows a number of pelvic organs to be examined digitally (with a finger) through the anterior rectal wall. This is called a **rectal exam.** Despite its name (*rectum* = straight), the rectum has three lateral curves or bends, represented internally as three transverse folds called **rectal valves** (Figure 24.29b). These valves separate feces from flatus, that is, they stop feces from being passed along with gas.

The **anal canal,** the last segment of the large intestine, lies entirely external to the abdominopelvic cavity. About 3 cm long, the anal canal begins where the rectum penetrates the levator ani muscle of the pelvic floor and it opens to the body exterior at the **anus.** The anal canal has two sphincters (Figure 24.29b), an involuntary **internal anal sphincter** composed of smooth muscle (part of the muscularis), and a voluntary **external anal sphincter** composed of a distinct skeletal muscle. The sphincters, which act rather like purse strings to open and close the anus, are ordinarily closed except during defecation.

FIGURE 24.29 *Gross anatomy of the large intestine.*
(**a**) Diagrammatic view. (**b**) Structure of the anal canal.

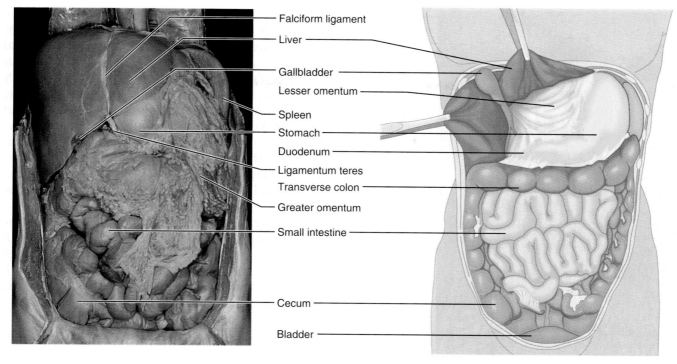

(a)

(b)

Falciform ligament
Liver
Gallbladder
Lesser omentum
Spleen
Stomach
Duodenum
Ligamentum teres
Transverse colon
Greater omentum
Small intestine
Cecum
Bladder

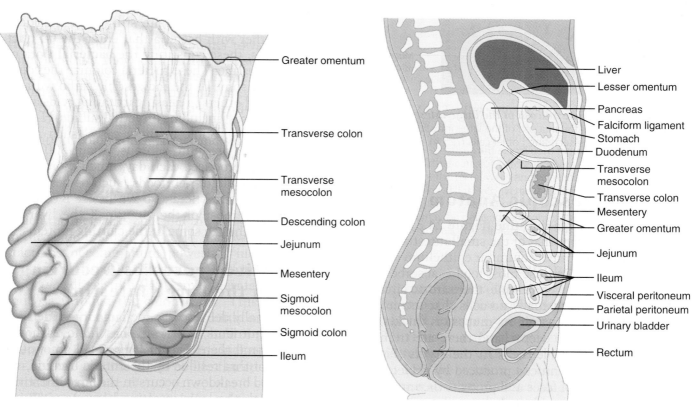

(c)

(d)

Greater omentum
Transverse colon
Transverse mesocolon
Descending colon
Jejunum
Mesentery
Sigmoid mesocolon
Sigmoid colon
Ileum

Liver
Lesser omentum
Pancreas
Falciform ligament
Stomach
Duodenum
Transverse mesocolon
Transverse colon
Mesentery
Greater omentum
Jejunum
Ileum
Visceral peritoneum
Parietal peritoneum
Urinary bladder
Rectum

FIGURE 24.30 *Mesenteries of the abdominal digestive organs.* (a) The greater omentum, a dorsal mesentery connecting the greater curvature of the stomach to the dorsal body wall, is shown in its normal position covering the abdominal viscera. (See *A Brief Atlas of the Human Body,* Figures 56a and b.) **(b)** The lesser omentum, a ventral mesentery attaching the liver to the lesser curvature of the stomach (the liver and gallbladder have been reflected superiorly). **(c)** The greater omentum has been reflected superiorly to reveal the mesentery attachments of the small and large intestine. **(d)** Sagittal section of the abdominopelvic cavity of a male, showing the relationships of the peritoneal attachments.

Chemical Digestion of Specific Food Groups

Carbohydrates

Depending on our food likes and dislikes, most of us ingest between 200 and 600 grams of carbohydrate foods each day. The monomers of carbohydrates are **monosaccharides** (simple sugars), which are absorbed immediately without further ado. Only three of these are common in our diet: *glucose, fructose,* and *galactose.* The more complex carbohydrates that our digestive system is able to break down to monosaccharides are the disaccharides *sucrose* (table sugar), *lactose* (milk sugar), and *maltose* (grain sugar) and the polysaccharides *glycogen* and *starch.* In the average diet, most digestible carbohydrates are in the form of starch, with smaller amounts of disaccharides and monosaccharides. Humans lack enzymes capable of breaking down most other polysaccharides, such as cellulose. Thus, indigestible polysaccharides do not nourish us but they do help move the food along the GI tract by providing bulk, or fiber.

Chemical digestion of starch (and perhaps glycogen) begins in the mouth. **Salivary amylase,** present in saliva, splits starch into *oligosaccharides,* smaller fragments of two to eight linked monosaccharides (in this case glucose molecules). Salivary amylase works best in a slightly acid to neutral environment (pH of 6.75–7.00), which is maintained in the mouth by the buffering effects of bicarbonate and phosphate ions in saliva. Starch digestion continues until amylase is inactivated by stomach acid and broken apart by the stomach's protein-digesting enzymes. Generally speaking, the larger the meal, the longer amylase continues to work in the stomach because foodstuffs in its relatively immobile fundus are poorly mixed with gastric juices.

Starchy foods and other digestible carbohydrates that escape being broken down by salivary amylase are acted on by **pancreatic amylase** in the small intestine (see Figure 24.33). Within about ten minutes of entering the small intestine, starch is entirely converted to various oligosaccharides, mostly maltose. Intestinal brush border enzymes further digest these products to monosaccharides. The most important of these brush border enzymes are **dextrinase** and **glucoamylase,** which act on oligosaccharides composed of more than three simple sugars, and the disaccharidase enzymes **maltase, sucrase,** and **lactase,** which hydrolyze maltose, sucrose, and lactose respectively into their constituent monosaccharides. Because the colon does not secrete digestive enzymes, chemical digestion *officially ends* in the small intestine. As noted earlier, however, resident colon bacteria do break down and metabolize the residual complex carbohydrates further, adding much to their nutrition but essentially nothing to ours.

Homeostatic Imbalance

In some people, intestinal lactase is present at birth but then becomes deficient, probably due to genetic factors. In such cases, the person becomes intolerant of milk products (the source of lactose), and the undigested lactose sugar creates osmotic gradients that not only prevent water from being absorbed in the small and large intestines but also pull water from the interstitial space into the intestines. The result is diarrhea. Bacterial metabolism of the undigested solutes produces large amounts of gas that result in bouts of bloating, flatulence, and cramping pain. The solution to this problem is simple—add lactase enzyme "drops" to your milk or take a lactose tablet before meals containing milk products. ■

Proteins

Proteins digested in the GI tract include not only dietary proteins (typically about 125 g per day), but also 15–25 g of enzyme proteins secreted into the GI tract lumen by its various glands and (probably) an equal amount of protein derived from sloughed and disintegrating mucosal cells. In healthy individuals, virtually all of this protein is digested all the way to its **amino acid** monomers.

Protein digestion begins in the stomach as pepsinogen secreted by the chief cells is quickly activated to **pepsin** (actually a group of proteolytic, or protein-digesting, enzymes). Pepsin functions optimally in the acidic pH range found in the stomach: 1.5–2.5. It preferentially cleaves bonds involving the amino acids tyrosine and phenylalanine so that proteins are broken into polypeptides and small numbers of free amino acids (see Figure 24.34). Pepsin, which hydrolyzes 10–15% of ingested protein, is inactivated by the high pH in the duodenum, so its proteolytic activity is restricted to the stomach. **Rennin** (the enzyme that coagulates milk protein) is apparently not produced in adults.

Protein fragments entering the small intestine from the stomach are greeted by a host of proteolytic enzymes (Figure 24.33). **Trypsin** and **chymotrypsin** secreted by the pancreas cleave the proteins into smaller peptides. These peptides, in turn, become the grist for other enzymes. The pancreatic and brush border enzyme **carboxypeptidase** splits off one amino acid at a time from the end of the polypeptide chain that bears the carboxyl group. Other brush border enzymes such as **aminopeptidase** and **dipeptidase** liberate the final amino acid products (Figure 24.34). Aminopeptidase digests a

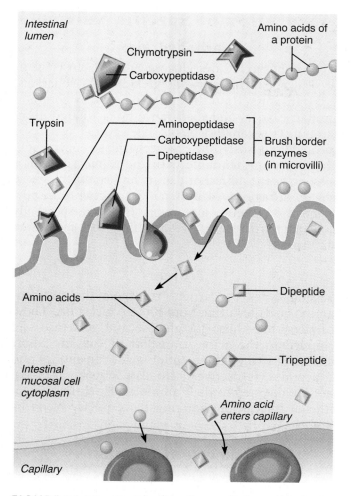

FIGURE 24.34 *Protein digestion and absorption in the small intestine.* Proteins and protein fragments are digested to amino acids by the action of pancreatic proteases (trypsin, chymotrypsin, and carboxypeptidase), and by brush border enzymes of the intestinal mucosal cells. The amino acids are then absorbed by active transport mechanisms into the capillary blood of the villi.

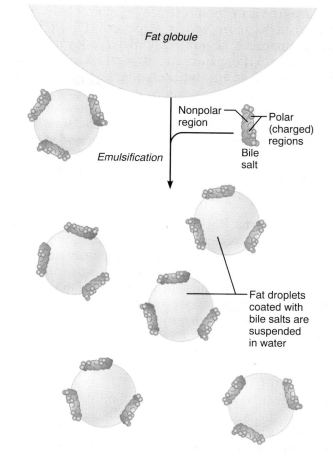

FIGURE 24.35 *Role of bile salts in fat emulsification.* As large aggregates of fats enter the small intestine, bile salts cling via their nonpolar parts to the fat molecules (triglycerides). Their polar parts, facing the aqueous phase, interact with water and repel each other, causing the fatty globule to be physically broken up into smaller fat droplets and forming a stable emulsion.

protein, one amino acid at a time, by working from the amine end. Both carboxypeptidase and aminopeptidase can independently dismantle a protein, but the teamwork between these enzymes and between trypsin and chymotrypsin, which attack the more internal parts of the protein, speeds up the process tremendously.

Lipids

Although the American Heart Association recommends a low-fat diet, the amount of lipids (fats) ingested daily varies tremendously among American adults, ranging from 30 g to 150 g or more. The small intestine is essentially the sole site of lipid digestion because the pancreas is the only significant source of fat-digesting enzymes, or **lipases** (Figure 24.33). Neutral fats (triglycerides or triacylglycerols) are the most abundant fats in the diet.

Since both triglycerides and their breakdown products are insoluble in water, fats need special "pretreatment" with bile salts to be digested and absorbed in the watery environment of the small intestine. In aqueous solutions, triglycerides aggregate to form large fat globules. Consequently, only the triglyceride molecules at the surfaces of such fatty masses are accessible to the water-soluble lipase enzymes. However, this problem is quickly resolved because as the fat globules enter the duodenum, they are coated with detergent-like bile salts (Figure 24.35). Bile salts have both nonpolar and polar regions. Their nonpolar (hydrophobic) parts cling to the fat molecules, and their polar (ionized hydrophilic) parts allow them to repel each other and to interact with water. As a result, fatty droplets are pulled off the large fat globules, and a stable *emulsion*—an aqueous suspension of fatty droplets, each about 1 μm in diameter—is formed. The process of emulsification does *not* break chemical bonds. It just reduces the attraction between fat molecules so that they can be more widely dispersed. This process vastly increases the number of triglyceride molecules exposed to the pancreatic lipases. Without bile, lipids would be incompletely digested in the time food is in the small intestine.

The pancreatic lipases catalyze the breakdown of fats by cleaving off two of the fatty acid chains, thus yielding free **fatty acids** and **monoglycerides** (glycerol with one fatty acid chain attached). Fat-soluble vitamins that ride with fats require no digestion.

Nucleic Acids

Both DNA and RNA, present in small amounts in foods, are hydrolyzed to their **nucleotide** monomers by **pancreatic nucleases** present in pancreatic juice. The nucleotides are then broken apart by intestinal brush border enzymes (**nucleosidases** and **phosphatases**), which release their free bases, pentose sugars, and phosphate ions (see Figure 24.33).

ABSORPTION

Up to 10 L of food, drink, and GI secretions enter the alimentary canal daily, but only 1 L or less reaches the large intestine. Virtually all of the foodstuffs, 80% of the electrolytes, and most of the water are absorbed in the small intestine. Although absorption occurs all along the length of the small intestine, most of it is completed by the time chyme reaches the ileum. Hence, the major absorptive role of the ileum is to reclaim bile salts to be recycled back to the liver for re-secretion. At the end of the ileum, all that remains is some water, indigestible food materials (largely plant fibers such as cellulose), and millions of bacteria. This debris is passed on to the large intestine.

Most nutrients are absorbed through the mucosa of the intestinal villi by *active transport* processes driven directly or indirectly (secondarily) by metabolic energy (ATP). They then enter the capillary blood in the villus to be transported in the hepatic portal vein to the liver. The exception is some of the lipid digestion products, which are absorbed passively by diffusion and then enter the lacteal in the villus to be carried to the blood via lymphatic fluid. Because the epithelial cells of the intestinal mucosa are joined at their luminal (apical) surfaces by tight junctions, substances cannot move *between* the cells. Consequently, materials must pass *through* the epithelial cells and into the interstitial fluid abutting their basal membranes if they are to enter the blood capillaries. This reabsorption mechanism is called **transepithelial transport.** The absorption of each nutrient class is described next.

Absorption of Specific Nutrients

Carbohydrates

The monosaccharides glucose and galactose, liberated by the breakdown of starch and disaccharides, are transported into the epithelial cells by common protein carriers and then move by facilitated diffusion into the capillary blood. The carriers, which are located very close to the disaccharidase enzymes on the microvilli, combine with the monosaccharides as soon as the disaccharides are broken down. These sugars are conveyed by secondary active transport coupled to sodium ion transport (cotransport). By contrast, fructose moves *entirely* by facilitated diffusion.

Proteins

Several types of carriers transport the different amino acids resulting from protein digestion. These carriers, like those for glucose and galactose, are coupled to the active transport of sodium. Short chains of two or three amino acids (dipeptides and tripeptides, respectively) are also actively absorbed, but are digested to their amino acids within the epithelial cells before entering the capillary blood by diffusion.

Homeostatic Imbalance

Whole proteins are not usually absorbed, but in rare cases intact proteins are taken up by endocytosis and released on the opposite side of the epithelial cell by exocytosis. This process, most common in newborn infants, reflects the immaturity of their intestinal mucosa and accounts for many early food allergies. The immune system "sees" the intact proteins as antigenic and mounts an attack. These food allergies usually disappear as the mucosa matures. This mechanism may also provide a route for IgA antibodies present in breast milk to reach an infant's bloodstream. These antibodies confer some passive immunity on the infant (temporary protection against antigens to which the mother has been sensitized). ■

Lipids

Just as bile salts accelerate lipid digestion, they are also essential for the absorption of its end products. As the water-insoluble products of fat digestion—the monoglycerides and free fatty acids—are liberated by lipase activity, they quickly become associated with bile salts and *lecithin* (a phospholipid found in bile) to form micelles. **Micelles** (mi-selz′) are collections of fatty elements clustered together with bile salts in such a way that the polar (hydrophilic) ends of the molecules face the water and the nonpolar portions form the core. Also nestled in the hydrophobic core are cholesterol molecules and fat-soluble vitamins. Although micelles are similar to emulsion droplets, they are much smaller "vehicles" and easily diffuse between microvilli to come into close contact with the mucosal cell surface (Figure 24.36). The various lipid substances then leave the micelles and, because of their high lipid solubility, move through the lipid phase of the plasma membrane by simple diffusion. The micelles provide a continuous source of solubilized fat breakdown products that can be released and absorbed as fat digestion continues. Without the micelles, the lipids simply float on the surface of the chyme (like oil on water), inaccessible to the absorptive surfaces of the epithelial cells. Generally, fat absorption is completed in the ileum, but in the absence of bile (as might occur when a gallstone blocks the cystic duct) it happens so slowly that most of the fat passes into the large intestine and is lost in feces.

Once inside the epithelial cells, the free fatty acids and monoglycerides are resynthesized into triglycerides. The triglycerides are then combined with small amounts of phospholipids and cholesterol, and coated with a "skin" of proteins to form water-soluble lipoprotein droplets called **chylomicrons** (ki"lo-mi'kronz). These are processed by the Golgi apparatus for extrusion from the cell. This series of events is quite different from the absorption of amino acids and simple sugars, which pass through the epithelial cells unchanged.

Although a few free fatty acids enter the capillary blood, the milky-white chylomicrons are too large to pass through the basement membranes of the blood capillaries and instead enter the more permeable lacteals. Thus, most fat enters the lymphatic stream and is eventually emptied into the venous blood in the neck region via the thoracic duct, which drains the digestive viscera. While in the bloodstream, the triglycerides of the chylomicrons are hydrolyzed to free fatty acids and glycerol by **lipoprotein lipase,** an enzyme associated with the capillary endothelium. The fatty acids and glycerol can then pass through the capillary walls to be used by tissue cells for energy or stored as fats in adipose tissue. The residual chylomicron material is combined with proteins by the liver cells, and these "new" lipoproteins are used to transport cholesterol in the blood.

Nucleic Acids

The pentose sugars, nitrogenous bases, and phosphate ions resulting from nucleic acid digestion are transported actively across the epithelium by special carriers in the villus epithelium. They then enter the blood.

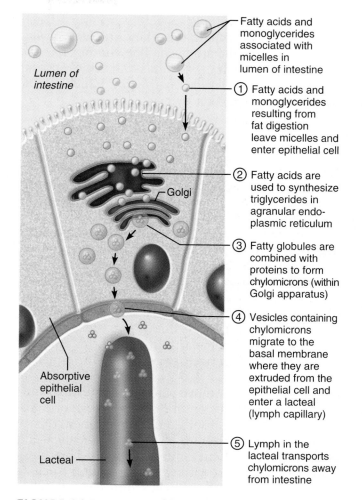

FIGURE 24.36 *Fatty acid absorption.* Digestion products of fat breakdown include glycerol, fatty acids, and monoglycerides. The monoglycerides, free fatty acids, phospholipids, and cholesterol associate with bile salt micelles, which serve to "ferry" them to the intestinal mucosa. They then dissociate and enter the mucosal cells by diffusion. Within the mucosal epithelial cells, they are recombined to lipids and packaged with other lipoid substances (phospholipids and cholesterol) and protein to form chylomicrons. The chylomicrons are extruded from the epithelial cells and enter the lacteal for distribution in the lymph. Free fatty acids and monoglycerides enter the capillary bed (not illustrated).

Labels in figure:
- Fatty acids and monoglycerides associated with micelles in lumen of intestine
- Lumen of intestine
- ① Fatty acids and monoglycerides resulting from fat digestion leave micelles and enter epithelial cell
- ② Fatty acids are used to synthesize triglycerides in agranular endoplasmic reticulum
- Golgi
- ③ Fatty globules are combined with proteins to form chylomicrons (within Golgi apparatus)
- ④ Vesicles containing chylomicrons migrate to the basal membrane where they are extruded from the epithelial cell and enter a lacteal (lymph capillary)
- Absorptive epithelial cell
- ⑤ Lymph in the lacteal transports chylomicrons away from intestine
- Lacteal

Vitamins

Although the small intestine absorbs dietary vitamins, the large intestine absorbs some of the K and B vitamins made by its enteric bacterial "guests." As already noted, fat-soluble vitamins (A, D, E, and K) dissolve in dietary fats, become incorporated into the micelles, and move across the villus epithelium by passive diffusion. It follows that gulping pills containing fat-soluble vitamins without simultaneously eating some fat-containing food results in little or no absorption of these vitamins.

Most water-soluble vitamins (B vitamins and vitamin C) are absorbed easily by diffusion. The exception is vitamin B$_{12}$, which is a very large, charged molecule. Vitamin B$_{12}$ binds to *intrinsic factor*, produced by the stomach. The vitamin B$_{12}$–intrinsic factor complex then binds to specific mucosal receptor sites in the terminal ileum, which trigger its active uptake by endocytosis.

Electrolytes

Absorbed electrolytes come from both ingested foods and gastrointestinal secretions. Most ions are actively absorbed along the entire length of the small intestine. However, iron and calcium absorption is largely limited to the duodenum.

As noted, absorption of sodium ions in the small intestine is coupled to active absorption of glucose and amino acids. For the most part, anions passively follow the electrical potential established by sodium transport. That is, Na$^+$ is actively pumped out of the epithelial cells by a Na$^+$-K$^+$ pump after it enters those cells by diffusion. Chloride ions are also transported actively, and in the terminus of the small intestine HCO$_3^-$ is actively secreted into the lumen in exchange for Cl$^-$.

Potassium ions move across the intestinal mucosa by simple diffusion in response to osmotic gradients. As water is absorbed from the lumen, the resulting rise in potassium levels in chyme creates a concentration gradient for its absorption. Thus, anything that interferes with water absorption (such as diarrhea) not only reduces potassium absorption but also "pulls" K$^+$ from the interstitial space into the intestinal lumen.

For most nutrients, the amount *reaching* the intestine is the amount absorbed, regardless of the nutritional state of the body. In contrast, absorption of iron and calcium is intimately related to the body's need for them at the time.

Ionic iron, essential for hemoglobin production, is actively transported into the mucosal cells, where it binds to the protein **ferritin** (fer′ĭ-tin). This phenomenon is called the *mucosal iron barrier*. The intracellular iron-ferritin complexes then serve as local storehouses for iron. When body reserves of iron are adequate, little is allowed to pass into the portal blood, and most of the stored iron is lost as the epithelial cells later slough off. However, when iron reserves are depleted (as during acute or chronic hemorrhage), iron uptake from the intestine and its release to the blood are accelerated. Menstrual bleeding is a major route of iron loss in females, and the intestinal epithelial cells of women have about four times as many iron transport proteins as do those of men. In the blood, iron binds to **transferrin,** a plasma protein that transports it in the circulation.

Calcium absorption is closely related to blood levels of ionic calcium. It is locally regulated by the active form of **vitamin D,** which acts as a cofactor to facilitate calcium absorption. Decreased blood levels of ionic calcium prompt *parathyroid hormone (PTH)* release from the parathyroid glands. Besides facilitating the release of calcium ions from bone matrix and enhancing the reabsorption of calcium by the kidneys, PTH stimulates activation of vitamin D by the kidneys, which in turn accelerates calcium ion absorption in the small intestine.

Water

Approximately 9 L of water, mostly derived from GI tract secretions, enter the small intestine daily. Water is the most abundant substance in chyme, and 95% of it is absorbed in the small intestine by osmosis. The normal rate of water absorption is 300 to 400 ml per hour. Water moves freely in both directions across the intestinal mucosa, but *net osmosis* occurs whenever a concentration gradient is established by the active transport of solutes (particularly Na$^+$) into the mucosal cells. Thus, water uptake is effectively coupled to solute uptake and, in turn, affects the rate of absorption of substances that normally pass by diffusion. As water moves into the mucosal cells, these substances follow along their concentration gradients.

Malabsorption of Nutrients

Malabsorption, or impaired nutrient absorption, has many and varied causes. It can result from anything that interferes with the delivery of bile or pancreatic juice to the small intestine, as well as factors that damage the intestinal mucosa (severe bacterial infections and antibiotic therapy with neomycin) or reduce its absorptive surface area. A common but poorly understood malabsorption syndrome is *gluten enteropathy,* also called *adult celiac disease.* In this condition, gluten, a protein plentiful in some grains (wheat, rye, barley, oats), damages the intestinal villi and reduces the length of the microvilli of the brush border. The resulting diarrhea and malnutrition are usually controlled by eliminating gluten-containing grains (all grains but rice and corn) from the diet. ■

DEVELOPMENTAL ASPECTS OF THE DIGESTIVE SYSTEM

As described many times before, the very young embryo is flat and consists of three germ layers, which, from top to bottom, are ectoderm, mesoderm, and endoderm. However, this flattened cell mass soon folds to form a cylindrical body, and its internal cavity becomes the cavity of the alimentary tube, which

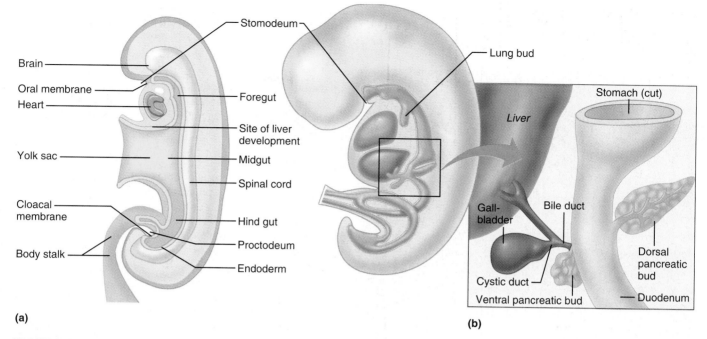

(a)

(b)

FIGURE 24.37 Embryonic development of the digestive system. (a) Three-week-old embryo. The endoderm has folded, and the foregut and hindgut have formed. (The midgut is still open and continuous with the yolk sac.) The anterior and posterior points of ectodermal-endodermal fusion (the oral and cloacal membranes, respectively) will shortly break through to form the mouth and anal openings. **(b)** By eight weeks of development, the accessory organs are budding out from the endodermal layer, as shown in the enlargement.

is initially closed at both ends. The epithelial lining of the developing alimentary tube, or **primitive gut,** forms from endoderm (Figure 24.37a). The rest of the wall arises from mesoderm. The anteriormost endoderm (that of the foregut) touches a depressed area of the surface ectoderm called the **stomodeum** (sto″mo-de′um; "on the way to becoming the mouth"). The two membranes fuse, forming the **oral membrane,** which shortly breaks through to form the opening of the mouth. Similarly, the end of the hindgut fuses with an ectodermal depression, called the **proctodeum** (*procto* = anus), to form the **cloacal** (*cloaca* = sewer) **membrane** and then breaks through to form the anus. By week 8, the alimentary canal is a continuous "tube" extending from the mouth to the anus and is open to the external environment at each end. Shortly after, the glandular organs (salivary glands, liver with gallbladder, and pancreas) bud out from the mucosa at various points along its length (Figure 24.37b). These glands retain their connections, which become ducts leading into the digestive tract.

Homeostatic Imbalance

The digestive system is susceptible to many congenital defects that interfere with feeding. The most common are *cleft palate,* in which the palatine bones or palatine processes of the maxillae (or both) fail to fuse, and *cleft lip,* which often occur together. Of the two, cleft palate is far more serious because the child is unable to suck properly. Another common defect is *tracheoesophageal fistula,* in which there is an opening between the esophagus and the trachea, and the esophagus often lacks a connection to the stomach. The baby chokes and becomes cyanotic during feedings because food enters the respiratory passageways. These defects are usually corrected surgically.

Cystic fibrosis (described in more detail in Chapter 23, pp. 878–879) primarily affects the lungs, but it also significantly impairs the activity of the pancreas. In this genetic disease, the mucous glands produce huge amounts of viscous mucus, which block the ducts or passageways of involved organs. Blockage of the pancreatic duct prevents pancreatic juice from reaching the small intestine. As a result, most fats and fat-soluble vitamins are not digested or absorbed, and the stools are bulky and fat-laden. The pancreatic problems can be handled by administering pancreatic enzymes with meals. ∎

During fetal life, the developing infant receives all of its nutrients through the placenta; hence, obtaining and processing nutrients is no problem if the mother is adequately nourished. Nonetheless, the fetal GI tract is "trained" in utero for future food digestion as the fetus naturally swallows some of the surrounding amniotic fluid. This fluid contains

MAKING CONNECTIONS

SYSTEM CONNECTIONS: Homeostatic Interrelationships Between the Digestive System and Other Body Systems

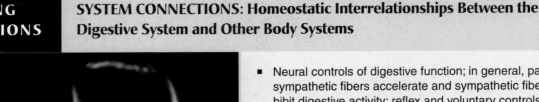

Integumentary System
- Digestive system provides nutrients needed for energy, growth, and repair; supplies fats that provide insulation in the dermis and subcutaneous tissue
- The skin synthesizes vitamin D needed for calcium absorption from the intestine; protects by enclosure

Skeletal System
- Digestive system provides nutrients needed for energy, growth, and repair; absorbs calcium needed for bone salts
- Skeletal system protects some digestive organs by bone; cavities store some nutrients (e.g., calcium, fats)

Muscular System
- Digestive system provides nutrients needed for energy, growth, and repair; liver removes lactic acid, resulting from muscle activity, from the blood
- Skeletal muscle activity increases motility of GI tract

Nervous System
- Digestive system provides nutrients needed for normal neural functioning; nutrient signals influence neural regulation of satiety

- Neural controls of digestive function; in general, parasympathetic fibers accelerate and sympathetic fibers inhibit digestive activity; reflex and voluntary controls of defecation

Endocrine System
- Liver removes hormones from blood, ending their activity; digestive system provides nutrients needed for energy, growth, and repair; pancreas has hormone-producing cells
- Local hormones help regulate digestive function

Cardiovascular System
- Digestive system provides nutrients to heart and blood vessels; absorbs iron needed for hemoglobin synthesis; absorbs water necessary for normal plasma volume; liver secretes bilirubin resulting from hemoglobin breakdown in bile and stores iron for reuse
- Cardiovascular system transports nutrients absorbed by alimentary canal to all tissues of body; distributes hormones of the digestive tract

Lymphatic System/Immunity
- Digestive system provides nutrients for normal functioning; lysozyme of saliva and HCl of stomach provide nonspecific protection against bacteria
- Lacteals drain fatty lymph from digestive tract organs and convey it to blood; Peyer's patches and lymphoid tissue in mesentery house macrophages and immune cells that protect digestive tract organs against infection; plasma cells provide IgA in saliva

Respiratory System
- Digestive system provides nutrients needed for energy metabolism, growth, and repair
- Respiratory system provides oxygen to and carries away carbon dioxide produced by digestive system organs

Urinary System
- Digestive system provides nutrients for energy fuel, growth, and repair; excretes some bilirubin produced by the liver
- Kidneys transform vitamin D to its active form, which is needed for calcium absorption

Reproductive System
- Digestive system provides nutrients for energy, growth, and repair and extra nutrition needed to support fetal growth

CLOSER CONNECTIONS

THE DIGESTIVE SYSTEM and Interrelationships with the Cardiovascular, Lymphatic, and Endocrine Systems

As body systems go, the digestive system is very independent. Although the central nervous system can and does influence its activity, that is not a requirement. Indeed, most digestive system activities can be regulated entirely by local controls. Consider, for example, the following: (1) stomach and intestinal muscle is self-pacing, (2) the enteric neurons of the GI tract's intrinsic nerve plexuses carry out reflex activity that can regulate not only the organ they reside in but also adjacent and more distant GI tract organs, and (3) its enteroendocrine cells are probably even more important than neural influences in controlling digestive activity. Also unique is the fact that the digestive system operates in and regulates an environment that is actually outside the body. This is a very difficult chore indeed—consider how quickly that environment can be changed by incoming foods (pizza and beer) or intestinal outflow (diarrhea).

Every cell in the body needs nutrients to stay healthy and grow, so the importance of the digestive system to virtually all body systems is clear-cut. But what systems are essential to the digestive system? If we ignore universal services, like gas exchange and renal regulation of body fluids, then the most crucial interactions are those with the cardiovascular, lymphatic, and endocrine systems. Let's take a look.

Cardiovascular and Lymphatic Systems

The important elements in this interaction are the pickup vessels. Indeed, all digestive activity would be for naught if the digested end products were not absorbed into the blood capillaries and lacteals, where they are accessible to all body cells. Besides its transport function, the lymphatic system also offers protective services via its tonsils, Peyer's patches, and individual lymphoid follicles dispersed all along the length of the alimentary tube. Again, the fact that the alimentary canal is open at both ends makes it easy prey for both pathogenic and opportunistic bacteria and fungi.

Endocrine System

The digestive tract not only has major interactions with the endocrine system, it is itself the largest and most complex endocrine organ in the body. It makes hormones that regulate GI tract mobility and secretory activity of the stomach, small intestine, liver, and pancreas (e.g., secretin, CCK, GIP, serotonin), as well as cause local blood vessels to dilate to receive the digested food products being absorbed (VIP). Furthermore, some digestive organ hormones (CCK and glucagon, for example) help mediate hunger or satiety, and pancreatic hormones (insulin and glucagon) are involved in bodywide regulation of carbohydrate, fat, and amino acid metabolism.

CLINICAL CONNECTIONS

Digestive System

Case study: Remember Mr. Gutteman, the gentleman who was dehydrating? It seems that his tremendous output of urine was only one of his current problems. Today, he complains of a headache and feeling bloated, and that he has "the runs" (diarrhea). In order to try to pinpoint the problem, he is asked the following questions.

- Have you had these symptoms previously? (Response: "On rare occasions.")
- Are you allergic to any foods? (Response: "Shellfish doesn't like me and milk gives me the runs.")

As a result of his responses, a lactose-free diet is ordered for Mr. Gutteman instead of the regular diet originally prescribed.

1. Why is the new diet prescribed? (What is believed to be his problem?)

Mr. Gutteman's problem continues despite the diet change. In fact, the frequency of diarrhea increases and by the end

of the next day, he is complaining of severe lower right abdominal pain, and his stool is largely bloody mucus. Again, he is asked some questions to probe his condition. One is whether he has traveled outside the country recently. He has not, reducing the possibility of infection with *Shigella* bacteria, which is associated with poor sanitation. Other questions:

- Do you drink alcohol and how much? (Response: "Little or none.")
- Have you recently eaten raw eggs or a salad containing mayonnaise at a gathering? (Response: "No.")
- Have you recently eaten at a "new" restaurant? (Response: "Yes, the day of the accident.")

2. On the basis of these responses, what do you think Mr. Gutteman's diarrhea stems from? How will it be diagnosed and treated?

3. Why do you think such patients are reported to a public health department?

(Answers in Appendix F)

several chemicals that stimulate GI maturation, including gastrin and epidermal growth factor. By contrast, feeding is a newborn baby's most important activity, and several feeding reflexes enhance the infant's ability to obtain food: The *rooting reflex* helps the infant find the nipple, and the *sucking reflex* helps the baby hold onto the nipple and swallow. Newborn babies tend to double their birth weight within six months, and their caloric intake and food processing ability are extraordinary. For example, a six-week-old infant weighing about 4 kg (less than 9 pounds) drinks about 600 ml of milk daily. A 65-kg adult (143 lb) would have to drink 10 L of milk to ingest a corresponding volume of fluid! However, the stomach of a newborn infant is very small, so feeding must be frequent (every 3–4 hours). Peristalsis is inefficient, and vomiting of the feeding is not unusual. As the teeth break through the gums, the infant progresses to solid foods and is usually eating an adult diet by the age of two years.

Unless there are abnormal interferences, the digestive system operates throughout childhood and adulthood with relatively few problems. However, contaminated food or extremely spicy or irritating foods sometimes cause an inflammation of the GI tract, called **gastroenteritis** (gas"tro-en"tĕ-ri'tis). Ulcers and gallbladder problems—inflammation or **cholecystitis** (ko"le-sis-ti'tis), and gallstones—are problems of middle age.

During old age, GI tract activity declines. Fewer digestive juices are produced, absorption is less efficient, and peristalsis slows. The result is less frequent bowel movements and, often, constipation. Taste and smell become less acute, and periodontal disease often develops. Many elderly people live alone or on a reduced income. These factors, along with increasing physical disability, tend to make eating less appealing, and many of our elderly citizens are inadequately nourished.

Diverticulosis, fecal incontinence, and cancer of the GI tract are fairly common problems of the aged.

Cancers of the stomach and colon rarely have early signs, and the disease often has metastasized (making it inoperable) before a person seeks medical attention. Should metastasis occur, secondary cancer of the liver is almost guaranteed because of the "detour" the splanchnic venous blood takes through the liver via the hepatic portal circulation. However, when detected early, these cancers are treatable. Perhaps the best advice is to have regular dental and medical checkups. Most oral cancers are detected during routine dental examinations, 50% of all rectal cancers can be felt digitally, and nearly 80% of colon cancers can be seen and removed during a colonoscopy.

Presently, colon cancer is the second largest cause of cancer deaths (behind lung cancer) in males in the United States. Since most colorectal cancers derive from initially benign mucosal tumors called **polyps,** and the incidence of polyp formation increases with age, a yearly colon examination should be a health priority in everyone over the age of 50. As described in *A Closer Look* on pp. 142–143, development of colon cancer is a gradual process that involves mutations in several regulatory genes. Recent studies have identified the mutant gene responsible for nearly 50% of all colon cancers. Dubbed *p53*, the gene normally acts as a watchdog to make sure that errors in DNA are not passed on before being corrected. When *p53* is damaged or inhibited, its tumor suppressor effect is lost and damaged DNA accumulates, resulting in carcinogenesis.

*   *   *

As summarized in *Making Connections*, the digestive system keeps the blood well supplied with the nutrients needed by all body tissues to fuel their energy needs and to synthesize new proteins for growth and maintenance of health. Now we are ready to examine how these nutrients are used by body cells, the topic of Chapter 25.

RELATED CLINICAL TERMS

Ascites (ah-si'tēz; *asci* = bag, bladder) Abnormal accumulation of fluid within the peritoneal cavity; if excessive, causes visible bloating of the abdomen. May result from portal hypertension caused by liver cirrhosis or by heart or kidney disease.

Barrett's esophagus A pathological change in the epithelium of the lower esophagus from stratified squamous to a metaplastic columnar epithelium. A possible sequel to untreated chronic gastroesophageal reflux due to hiatal hernia which predisposes the individual to aggressive esophageal cancer (adenocarcinoma).

Bruxism (bruck'sizm) Grinding or clenching of teeth, usually at night during sleep in response to stress. Can wear down and crack the teeth.

Bulimia (bu-lim'e-ah; *bous* = ox; *limos* = hunger) Binge-purge behavior in which episodes of massive overeating are followed by some method of purging, that is, self-induced vomiting, taking laxatives or diuretics, excessive exercising. Behavior most common in young women of high school or college age, and in high school males in certain athletic activities (wrestling). Rooted in a pathological fear of being fat, and a need for control; provides a means of handling stress and depression. Consequences include erosion of tooth enamel and stomach trauma or rupture (from vomiting) and severe electrolyte disturbances which impair heart activity. May be linked to deficient secretion of CCK, a gut-brain peptide that plays a role in satiety. Therapy includes hospitalization to control behavior, and nutritional counseling.

Dysphagia (dis-fa'je-ah; *dys* = difficult, abnormal; *phag* = eat) Difficulty swallowing; usually due to obstruction or physical trauma to the esophagus.

Endoscopy (en-dos'ko-pe; *endo* = within, inside; *scopy* = viewing) Visual examination of a ventral body cavity or the interior of a tubular visceral organ with a flexible tubelike device called an endoscope, which contains a light source and a lens. A general term for a colonoscopy (viewing the colon), sigmoidoscopy (viewing the sigmoid colon), etc.

Enteritis (*enteron* = intestine) Inflammation of the intestine, especially the small intestine.

Hemochromatosis (he"mo-kro"mah-to'sis; *hemo* = blood; *chroma* = color; *osis* = condition of) A syndrome caused by a disorder in iron metabolism, owing to excessive/prolonged iron intake or a breakdown of the mucosal iron barrier; excess iron is deposited in the tissues, resulting in increased skin pigmentation and increased incidence of hepatic cancer and liver cirrhosis; also called *bronze diabetes* and *iron storage disease.*

Ileus (il'e-us) A condition in which all GI tract movement stops and the gut appears to be paralyzed. Can result from electrolyte imbalances and blockade of parasympathetic impulses by drugs (such as those commonly used during abdominal surgery); usually reversed when these interferences end. Restoration of mobility is indicated by the reappearance of bowel (intestinal) sounds (gurgling, etc.).

Inflammatory bowel disease A noncontagious, periodic inflammation of the intestinal wall. Affects up to 2 of every 1000 people. Symptoms include cramping, diarrhea, weight loss, and intestinal bleeding. Two subtypes occur: Crohn's disease and ulcerative colitis. Crohn's disease is more serious, with deep ulcers and fissures developing along the whole intestine, but mostly in the terminal ileum. Ulcerative colitis, by contrast, is characterized by a shallow inflammation of the large-intestinal mucosa, mainly in the rectum. Although inflammatory bowel disease was formerly considered a nervous disorder, it is now understood as an abnormal immune and inflammatory response to bacterial antigens that normally occur in the intestine. It is treated by adopting a special diet, reducing stress, using antibiotics, and most effectively, with anti-inflammatory and immunosuppressant drugs.

Laparoscopy (lap"ah-ros'ko-pe) (*lapar* =the flank; *scopy* = observation) Endoscopy of the peritoneal cavity. Examination of the peritoneal cavity and its organs with an endoscope inserted through the anterior abdominal wall. Laparoscopy is often used to assess the condition of the digestive organs and the pelvic reproductive organs of females.

Orthodontics (or"tho-don'tiks; *ortho* = straight) Branch of dentistry that prevents and corrects misaligned teeth.

Pancreatitis (pan"kre-ah-ti'tis) A rare but extremely serious inflammation of the pancreas. May result from excessively high levels of fat in the blood, but most often results from activation of pancreatic enzymes in the pancreatic duct, causing the pancreatic tissue and duct to be digested. This painful condition can lead to nutritional deficiencies because pancreatic enzymes are essential to food digestion in the small intestine.

Peptic ulcers Collective term referring to gastric and duodenal ulcers.

Proctology (prok-tol'o-je; *procto* = rectum, anus; *logy* = study of) Branch of medicine dealing with treatment of diseases of the colon, rectum, and anus.

Pyloric stenosis (pi-lor'ik stĕ-no'sis; *stenosis* = narrowing, constriction) Congenital abnormality in which the pyloric sphincter is abnormally constricted. There is usually no problem until the baby begins to take solid food, and then projectile vomiting begins. Usually corrected surgically.

Xerostomia (ze"ro-sto'me-ah; *zer* = dry, *stom* = mouth) Extreme dryness of the mouth; can be caused by blockage of salivary glands by cysts or autoimmune invasion of the salivary glands or ducts (Sjögren's syndrome).

CHAPTER SUMMARY

Media study tools that could provide you additional help in reviewing specific key topics of Chapter 24 are referenced below. **SP** *= Study Partner;* **IP** *= Interactive Physiology.*

Overview of the Digestive System (pp. 888–894)

1. The digestive system includes organs of the alimentary canal (mouth, pharynx, esophagus, stomach, small and large intestines) and accessory digestive system organs (teeth, tongue, salivary glands, liver, gallbladder, and pancreas).

Digestive Processes (pp. 889–890)

2. Digestive system activities include six functional processes: ingestion, or food intake; propulsion, or movement of food through the tract; mechanical digestion, or processes that physically mix or break foods down into smaller fragments; chemical digestion, or food breakdown by enzymatic action; absorption, or transport of products of digestion through the intestinal mucosa into the blood; and defecation, or elimination of the undigested residues (feces) from the body.

SP Exercise: Chapter 24, Gastrointestinal Tract Activities.

Basic Functional Concepts (pp. 890–891)

3. The digestive system controls the environment within its lumen to ensure optimal conditions for digestion and absorption of foodstuffs.

4. Receptors and hormone-secreting cells in the alimentary canal wall respond to stretch and chemical signals that result in stimulation or inhibition of GI secretory activity or motility. The alimentary canal has a local (intrinsic) nerve supply.

Digestive System Organs: Relationships and Structural Plan (pp. 891–894)

5. The parietal and visceral layers of the peritoneum are continuous with one another via several extensions (mesentery, falciform ligament, lesser and greater omenta), and are separated by a potential space containing serous fluid, which decreases friction during organ activity.

6. The digestive viscera is served by the splanchnic circulation, consisting of arterial branches of the celiac trunk and aorta and the hepatic portal circulation.

7. All organs of the GI tract have the same basic pattern of tissue layers in their walls; that is, all have a mucosa, a submucosa, a muscularis, and a serosa (or adventitia). Intrinsic nerve plexuses (enteric nervous system) are found within the wall.

Part I: Functional Anatomy of the Digestive System (pp. 894–932)

The Mouth, Pharynx, and Esophagus (pp. 894–903)

1. Food enters the GI tract via the mouth, which is continuous with the oropharynx posteriorly. The boundaries of the mouth are the lips and cheeks, palate, and tongue.

2. The oral mucosa is stratified squamous epithelium, an adaptation seen where abrasion occurs.

3. The tongue is mucosa-covered skeletal muscle. Its intrinsic muscles allow it to change shape; its extrinsic muscles allow it to change position.

4. Saliva is produced by many small buccal glands and three pairs of major salivary glands—parotid, submandibular, and sublingual—that secrete their product into the mouth via ducts. Largely water, saliva also contains ions, proteins, metabolic wastes, lysozyme, defensins, IgA, salivary amylase, a cyanide compound, and mucin.

5. Saliva moistens and cleanses the mouth; moistens foods, aiding their compaction; dissolves food chemicals to allow for taste; and begins chemical digestion of starch (salivary amylase). Saliva output is increased by parasympathetic reflexes initiated by activation of chemical and pressure receptors in the mouth and by conditioned reflexes. The sympathetic nervous system depresses salivation.

6. The 20 deciduous teeth begin to be shed at the age of six and are gradually replaced during childhood and adolescence by the 32 permanent teeth.

7. Teeth are classed as incisors, canines, premolars, and molars. Each tooth has an enamel-covered crown and a cementum-covered root. The bulk of the tooth is dentin, which surrounds the central pulp cavity. A periodontal ligament secures the tooth to the bony alveolus.

8. Food propelled from the mouth passes through the oropharynx and laryngopharynx. The mucosa of the pharynx is stratified squamous epithelium; skeletal muscles in its wall (constrictor muscles) move food toward the esophagus.

9. The esophagus extends from the laryngopharynx and joins the stomach at the cardiac orifice, which is surrounded by the gastroesophageal sphincter.

10. The esophageal mucosa is stratified squamous epithelium. Its muscularis is skeletal muscle superiorly and changes to smooth muscle inferiorly. It has an adventitia rather than a serosa.

11. The mouth and associated accessory organs accomplish food ingestion and mechanical digestion (chewing and mixing), initiate the chemical digestion of starch (salivary amylase), and propel food into the pharynx (buccal phase of swallowing).

12. Teeth function to masticate food. Chewing is initiated voluntarily and is then controlled reflexively.

13. The tongue mixes food with saliva, compacts it into a bolus, and initiates swallowing (the voluntary phase). The pharynx and esophagus are primarily food conduits that conduct food to the stomach by peristalsis. The swallowing center in the medulla and pons controls this phase reflexively. When the peristaltic wave approaches the gastroesophageal sphincter, the sphincter relaxes to allow food entry to the stomach.

The Stomach (pp. 903–915)

1. The C-shaped stomach lies in the upper left quadrant of the abdomen. Its major regions are the cardia, fundus, body, and pyloric region. When empty, its internal surface exhibits rugae.

2. The stomach mucosa is simple columnar epithelium dotted with gastric pits that lead into gastric glands. Secretory cells in the gastric glands include pepsinogen-producing chief cells; parietal cells, which secrete hydrochloric acid and intrinsic factor; mucous neck cells, which produce mucus; and enteroendocrine cells, which secrete hormones.

3. The mucosal barrier, which protects the stomach from self-digestion and HCl, reflects the fact that the mucosal cells are connected by tight junctions, secrete a thick mucus, and are quickly replaced when damaged.

4. The stomach muscularis contains a third layer of obliquely oriented smooth muscle that allows it to churn and mix food.

5. Protein digestion is initiated in the stomach by activated pepsin and requires acidic conditions (provided by HCl). Few substances are absorbed.

6. Gastric secretory activity is controlled by both nervous and hormonal factors. The three phases of gastric secretion are the cephalic, gastric, and intestinal. For the most part stimuli acting on the head and stomach (cephalic and gastric respectively) stimulate gastric secretion. Most stimuli acting on the small intestine trigger the enterogastric reflex and release of secretin, CCK, and GIP, all of which inhibit gastric secretory activity. Sympathetic activity also inhibits gastric secretion.

7. Mechanical digestion in the stomach is triggered by stomach distension and coupled to food propulsion and stomach emptying. Food movement into the duodenum is controlled by the pylorus and feedback signals from the small intestine. Pacemaker cells in the smooth muscle sheet set the maximal rate of peristalsis.

The Small Intestine and Associated Structures (pp. 915–927)

1. The small intestine extends from the pyloric sphincter to the ileocecal valve. Its three subdivisions are the duodenum, jejunum, and ileum. The common bile duct and pancreatic duct join to form the hepatopancreatic ampulla and empty their secretions into the duodenum through the hepatopancreatic sphincter (of Oddi).

2. The duodenal submucosa contains elaborate mucus-secreting duodenal glands (Brunner's glands); that of the ileum contains Peyer's patches (lymph follicles). The duodenum is covered with an adventitia rather than a serosa.

3. Circular folds, villi, and microvilli increase the intestinal surface area for digestion and absorption.

4. The small intestine is the major digestive and absorptive organ. Intestinal juice is largely water and is relatively enzyme-poor. The major stimuli for its release are stretch and chemicals.

5. The liver is a lobed organ overlying the stomach. Its digestive role is to produce bile, which it secretes into the common hepatic duct.

6. The structural and functional units of the liver are the hepatic lobules. Blood flowing to the liver via the hepatic artery and hepatic portal vein flows into its sinusoids from which Kupffer cells remove debris and hepatocytes remove nutrients. Hepatocytes store glucose as glycogen, use amino acids to make plasma proteins, and detoxify metabolic wastes and drugs.

SP Case Study: Cirrhosis of the Liver.

7. Bile is made continuously by the hepatocytes. Bile salts, secretin, and vagal stimulation stimulate bile production.

8. The gallbladder, a muscular sac that lies beneath the right liver lobe, stores and concentrates bile.

9. Bile contains electrolytes, a variety of fatty substances, bile salts, and bile pigments in an aqueous medium. Bile salts are emulsifying agents; they disperse fats and form water-soluble micelles, which solubilize the products of fat digestion.

10. Cholecystokinin released by the small intestine stimulates the gallbladder to contract and the hepatopancreatic sphincter to relax, allowing bile (and pancreatic juice) to enter the duodenum.

SP Exercise: Chapter 24, Control of Bile Secretion.

11. The pancreas is retroperitoneal between the spleen and small intestine. Its exocrine product, pancreatic juice, is carried to the duodenum via the pancreatic duct.

12. Pancreatic juice is an HCO_3^--rich fluid containing enzymes that digest all categories of foods. Secretion of pancreatic juice is controlled by intestinal hormones and the vagus nerves.

13. Mechanical digestion and propulsion in the small intestine mix chyme with digestive juices and bile and force the residues, largely by segmentation, through the ileocecal valve. Pacemaker cells set the rate of segmentation. Ileocecal valve opening is controlled by the gastroileal reflex and gastrin.

The Large Intestine (pp. 927–932)

1. The subdivisions of the large intestine are the cecum (and appendix), colon (ascending, transverse, descending, and sigmoid portions), rectum, and anal canal. It opens to the body exterior at the anus.

2. The mucosa of most of the large intestine is simple columnar epithelium containing abundant goblet cells. The longitudinal muscle in the muscularis is reduced to three bands (teniae coli), which pucker its wall, producing haustra.

3. The major functions of the large intestine are absorption of water, and some electrolytes (and vitamins made by enteric bacteria), and defecation (evacuation of food residues from the body).

4. The defecation reflex is triggered when feces enter the rectum. It involves parasympathetic reflexes leading to the contraction of the rectal walls and is aided by Valsalva's maneuver.

SP Exercise: Chapter 24, The Digestive System.

Part II: Physiology of Chemical Digestion and Absorption (pp. 932–938)

Chemical Digestion (pp. 932–936)

1. Chemical digestion is accomplished by hydrolysis, catalyzed by enzymes.

2. Most chemical digestion is done in the small intestine by intestinal (brush border) enzymes and, more importantly, by pancreatic enzymes. Alkaline pancreatic juice neutralizes the acidic chyme and provides the proper environment for operation of the enzymes. Both pancreatic juice (the main source of lipases) and bile are necessary for normal fat breakdown.

Absorption (pp. 936–938)

1. Virtually all of the foodstuffs and most of the water and electrolytes are absorbed in the small intestine. Except for fat digestion products, fat-soluble vitamins, and most water-soluble vitamins (which are absorbed by diffusion), most nutrients are absorbed by active transport processes.

2. Fat breakdown products are solubilized by bile salts (in micelles), resynthesized to triglycerides in the intestinal mucosal cells, and combined with other lipids and protein as chylomicrons that enter the lacteals. Other absorbed substances enter the villus blood capillaries and are transported to the liver via the hepatic portal vein.

Developmental Aspects of the Digestive System (pp. 938–939, 942)

1. The mucosa of the alimentary canal develops from the endoderm, which folds to form a tube. The glandular accessory organs (salivary glands, liver, pancreas, and gallbladder) form from outpocketings of the foregut endoderm. The remaining three tunics of the alimentary canal wall are formed by mesoderm.

2. Important congenital abnormalities of the digestive tract include cleft palate/lip, tracheoesophageal fistula, and cystic fibrosis. All interfere with normal nutrition.

3. Various inflammations plague the digestive system throughout life. Appendicitis is common in adolescents,

gastroenteritis and food poisoning can occur at any time (given the proper irritating factors), ulcers and gallbladder problems increase in middle age.

4. The efficiency of all digestive system processes declines in the elderly, and periodontal disease is common. Diverticulosis, fecal incontinence, and GI tract cancers such as stomach and colon cancer appear with increasing frequency in an aging population.

REVIEW QUESTIONS

Multiple Choice/Matching

1. The peritoneal cavity (a) is the same thing as the abdominopelvic cavity, (b) is filled with air, (c) like the pleural and pericardial cavities, a potential space containing serous fluid, (d) contains the pancreas and all of the duodenum.

2. Obstruction of the hepatopancreatic sphincter impairs digestion by reducing the availability of (a) bile and HCl, (b) HCl and intestinal juice, (c) pancreatic juice and intestinal juice, (d) pancreatic juice and bile.

3. The action of an enzyme is influenced by (a) its chemical surroundings, (b) its specific substrate, (c) the presence of needed cofactors or coenzymes, (d) all of these.

4. Carbohydrates are acted on by (a) peptidases, trypsin, and chymotrypsin, (b) amylase, maltase, and sucrase, (c) lipases, (d) peptidases, lipases, and galactase.

5. The parasympathetic nervous system influences digestion by (a) relaxing smooth muscle, (b) stimulating peristalsis and secretory activity, (c) constricting sphincters, (d) none of these.

6. The digestive juice product containing enzymes capable of digesting all four major foodstuff categories is (a) pancreatic, (b) gastric, (c) salivary, (d) biliary.

7. The vitamin associated with calcium absorption is (a) A, (b) K, (c) C, (d) D.

8. Someone has eaten a meal of buttered toast, cream, and eggs. Which of the following would you expect to happen? (a) Compared to the period shortly after the meal, gastric motility and secretion of HCl decrease when the food reaches the duodenum; (b) gastric motility increases even as the person is chewing the food (before swallowing); (c) fat will be emulsified in the duodenum by the action of bile; (d) all of these.

9. The site of production of GIP and cholecystokinin is (a) the stomach, (b) the small intestine, (c) the pancreas, (d) the large intestine.

10. Which of the following is not characteristic of the large intestine? (a) It is divided into ascending, transverse, and descending portions; (b) it contains abundant bacteria, some of which synthesize certain vitamins; (c) it is the main absorptive site; (d) it absorbs much of the water and salts remaining in the wastes.

11. The gallbladder (a) produces bile, (b) is attached to the pancreas, (c) stores and concentrates bile, (d) produces secretin.

12. The sphincter between the stomach and duodenum is (a) the pyloric sphincter, (b) the cardiac sphincter, (c) the hepatopancreatic sphincter, (d) the ileocecal sphincter.

In items 13–17, trace the path of a single protein molecule that has been ingested.

13. The protein molecule will be digested by enzymes secreted by (a) the mouth, stomach, and colon, (b) the stomach, liver, and small intestine, (c) the small intestine, mouth, and liver, (d) the pancreas, small intestine, and stomach.

14. The protein molecule must be digested before it can be transported to and utilized by the cells because (a) protein is only useful directly, (b) protein has a low pH, (c) proteins in the circulating blood produce an adverse osmotic pressure, (d) the protein is too large to be readily absorbed.

15. The products of protein digestion enter the bloodstream largely through cells lining (a) the stomach, (b) the small intestine, (c) the large intestine, (d) the bile duct.

16. Before the blood carrying the products of protein digestion reaches the heart, it first passes through capillary networks in (a) the spleen, (b) the lungs, (c) the liver, (d) the brain.

17. Having passed through the regulatory organ selected above, the products of protein digestion are circulated throughout the body. They will enter individual body cells as a result of (a) active transport, (b) diffusion, (c) osmosis, (d) phagocytosis.

Short Answer Essay Questions

18. Make a simple line drawing of the organs of the alimentary tube and label each organ. Then add three labels to your drawing—salivary glands, liver, and pancreas—and use arrows to show where each of these organs empties its secretion into the alimentary tube.

19. Sara was on a diet but she could not eat less and kept claiming her stomach had a mind of its own. She was joking, but indeed, there is a "gut brain" called the enteric nervous system. Is it part of the parasympathetic and sympathetic nervous system? Explain.

20. Name the layers of the alimentary tube wall. Note the tissue composition and major function of each layer.

21. What is a mesentery? Mesocolon? Greater omentum?

22. Name the six functional activities of the digestive system.

23. (a) Describe the boundaries of the oral cavity. (b) Why do you suppose its mucosa is stratified squamous epithelium rather than the more typical simple columnar epithelium?

24. (a) What is the normal number of permanent teeth? Of deciduous teeth? (b) What substance covers the tooth crown? Its root? (c) What substance makes up the bulk of a tooth? (d) What and where is pulp?

25. Describe the two phases of swallowing, noting the organs involved and the activities that occur.

26. Describe the role of these cell types found in the gastric glands: parietal, chief, mucous neck, and enteroendocrine.

27. Describe the regulation of the cephalic, gastric, and intestinal phases of gastric secretion.

28. (a) What is the relationship between the cystic, common hepatic, bile, and pancreatic ducts? (b) What name is given to the point of fusion of the bile and pancreatic ducts?

29. Explain why fatty stools result from the absence of bile or pancreatic juice.

30. Note the function of the Kupffer cells and the hepatocytes of the liver.

31. What are (a) brush border enzymes? (b) chylomicrons?

32. Name one inflammatory condition particularly common to adolescents, two common in middle age, and one common in old age.

33. What are the effects of aging on digestive system activity?

Critical Thinking and Clinical Application Questions

1. You are a young research assistant at a pharmaceutical company. Your group has been asked to develop an effective laxative that (1) provides bulk and (2) is nonirritating to the intestinal mucosa. Explain why these requests are important by describing what would happen if the opposite conditions were present.

2. After a heavy meal rich with fried foods, Debby Collins, an overweight 45-year-old woman, was rushed to the emergency room with severe spasmodic pains in her epigastric region that radiated to the right side of her rib cage. She indicated that the attack came on suddenly, and her abdomen was found to be tender to the touch and somewhat rigid. What do you think is this patient's problem and why is her pain discontinuous (colicky)? What are the treatment options and what might happen if the problem is not resolved?

3. A baby is admitted to the hospital with a history of diarrhea and watery feces occurring over the last three days. The baby has sunken fontanels, indicating extreme dehydration. Tests indicate that the baby has a bacterium-induced colitis, and antibiotics are prescribed. Because of the baby's loss of intestinal juices, do you think that his blood pH would indicate acidosis or alkalosis? Explain your reasoning.

4. Gary Francis, a middle-aged salesman, complains of a burning pain in the "pit of his stomach," usually beginning about two hours after eating and abating after drinking a glass of milk. When asked to indicate the site, he points to his epigastric region. The GI tract is examined by X-ray fluoroscopy. A gastric ulcer is visualized, and drug therapy using an antibiotic-bismuth approach is recommended. (a) Why is this treatment suggested? (b) What are the possible consequences of nontreatment?

5. Dr. Dolan used an endoscope to view Mr. Habib's colon. He noted the presence of several polyps and removed them during the same procedure. What is an endoscope? Why did Dr. Dolan opt to remove the polyps immediately?

25 NUTRITION, METABOLISM, AND BODY TEMPERATURE REGULATION

Nutrition (pp. 949–961)

1. Define nutrient, essential nutrient, and calorie.

2. List the six major nutrient categories. Note important sources and their main cellular uses.

3. Distinguish between nutritionally complete and incomplete proteins.

4. Define nitrogen balance and indicate possible causes of positive and negative nitrogen balance.

5. Distinguish between fat- and water-soluble vitamins, and list the vitamins belonging to each group.

6. For each vitamin, list its important sources and functions in the body, and describe consequences of its deficit or excess.

7. List minerals essential for health; note important dietary sources and describe how each is used.

Metabolism (pp. 962–986)

8. Define metabolism. Explain how catabolism and anabolism differ.

9. Define oxidation and reduction and note the importance of these reactions in metabolism. Indicate the role of coenzymes used in cellular oxidation reactions.

10. Explain the difference between substrate-level phosphorylation and oxidative phosphorylation.

11. Follow the oxidation of glucose in body cells. Summarize important events and products of glycolysis, the Krebs cycle, and the electron transport chain.

12. Define glycogenesis, glycogenolysis, and gluconeogenesis.

13. Describe the process by which fatty acids are oxidized for energy.

14. Define ketone bodies, and indicate the stimulus for their formation.

15. Describe how amino acids are metabolized for energy.

16. Describe the need for protein synthesis in body cells.

17. Explain the concept of amino acid or carbohydrate-fat pools and describe pathways by which substances in these pools can be interconverted.

18. List important goals and events of absorptive and postabsorptive states, and explain how these events are regulated.

19. Describe several metabolic functions of the liver.

20. Differentiate between LDLs and HDLs relative to their structures and major roles in the body.

Body Energy Balance (pp. 986–996)

21. Explain what is meant by body energy balance.

22. Describe some current theories of food intake regulation.

23. Define basal metabolic rate and total metabolic rate. Name several factors that influence each.

24. Describe how body temperature is regulated, and indicate the common mechanisms regulating heat production/retention and heat loss from the body.

Developmental Aspects of Nutrition and Metabolism (pp. 996–997)

25. Describe the effects of inadequate protein intake on the fetal nervous system.

26. Describe the cause and consequences of the low metabolic rate typical of the elderly.

27. List ways that medications commonly used by aged people may influence their nutrition and health.

Are you a food lover? People can be divided into two camps according to their reactions to food—those who live to eat and those who eat to live. Still, we all recognize the vital importance of food for life. Some say that "you are what you eat," and this is true in that part of the food we eat is converted to our living flesh. In other words, a certain fraction of nutrients is used to build cell structures, replace worn-out parts, and synthesize functional molecules. However, most foods are used as metabolic fuels. That is, they are oxidized and transformed into **ATP**, the chemical energy form used by cells to drive their many activities. The energy value of foods is measured in units called **kilocalories** (kcal) or "large calories (C)." One kilocalorie is the amount of heat energy needed to raise the temperature of 1 kilogram of water 1°C (1.8°F) and is the unit conscientiously counted by dieters.

In Chapter 24, we considered how foods are digested and absorbed. But what happens to these foods once they have entered the blood? Why do we need bread, meat, and fresh vegetables? Why does everything we eat seem to turn to fat? This chapter will try to answer these questions as it explains both the nature of nutrients and the metabolic roles they play in the body.

NUTRITION

A **nutrient** is a substance in food that is used by the body to promote normal growth, maintenance, and repair. The nutrients needed for health divide neatly into six categories. Three of these—carbohydrates, lipids, and proteins—are collectively called the **major nutrients** and make up the bulk of what we eat.

Vitamins and minerals, though equally crucial for health, are required in minute amounts. In a strict sense, water, which accounts for about 60% of the volume of the food we eat, is also a major nutrient. However, because its importance as a dissolving medium (solvent), as a reactant in many chemical reactions, and in many other aspects of body functioning is described in Chapter 2 (pp. 41–42), only the five nutrient classes listed above will be considered in detail here.

Most foods offer a combination of nutrients. For example, a bowl of cream of mushroom soup contains all the major nutrients plus some vitamins and minerals. A diet consisting of foods from each of the five food groups, that is, grains, fruits, vegetables, meats and fish, and milk products, normally guarantees adequate amounts of all the needed nutrients. A more detailed guide to good nutrition appears in Figure 25.1.

The ability of cells, especially those of the liver, to convert one type of molecule into another is truly remarkable. These interconversions allow the body to use the wide range of chemicals found in different foods and to adjust to varying food intakes. But there are limits to this ability to conjure up new molecules from old. At least 45 and possibly 50 molecules, called **essential nutrients**, cannot be made by such interconversions and must be provided by the diet. As long as all the essential nutrients are ingested, the body can synthesize the hundreds of additional molecules required for life and good health. The use of the word "essential" to describe the chemicals that must be obtained from outside sources is unfortunate and misleading, to say the least, because both the essential and the nonessential nutrients are

FIGURE 25.1 *Official food guide pyramid.* Issued by the U.S. Department of Agriculture in 1992, this food pyramid suggests eating a large variety of foods (with the emphasis on fruits and vegetables and starches) for good health. In this diet, most of the fat comes from the milk and meat-poultry-eggs groups. Caloric intake ranges from 1500 calories to 2800 calories (smallest to largest number of servings) daily.

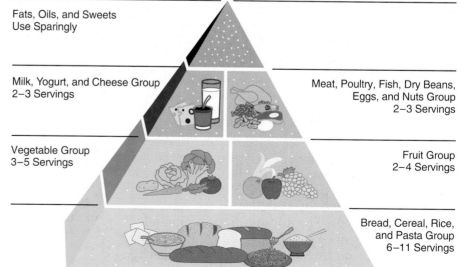

Fats, Oils, and Sweets
Use Sparingly

Milk, Yogurt, and Cheese Group
2–3 Servings

Meat, Poultry, Fish, Dry Beans, Eggs, and Nuts Group
2–3 Servings

Vegetable Group
3–5 Servings

Fruit Group
2–4 Servings

Bread, Cereal, Rice, and Pasta Group
6–11 Servings

equally vital (essential) for normal functioning. In this initial section, we will review the dietary sources and recommended daily requirements of each of the nutrient categories and describe their general importance and uses in the body.

Carbohydrates

Dietary Sources

Except for milk sugar (lactose) and small amounts of glycogen in meats, all the carbohydrates (see Chapter 2) we ingest are derived from plants. Sugars (monosaccharides and disaccharides) come from fruits, sugar cane, sugar beets, honey, and milk. The polysaccharide starch is found in grains, legumes, and root vegetables. Cellulose, a polysaccharide plentiful in most vegetables, is not digested by humans but provides roughage, or fiber, which increases the bulk of the stool and facilitates defecation.

Uses in the Body

The monosaccharide **glucose** is the carbohydrate molecule ultimately used by body cells. Carbohydrate digestion also yields fructose and galactose, but these monosaccharides are converted to glucose by the liver before they enter the general circulation. Glucose is a major body fuel and is readily used to make ATP. Although many body cells use fats as energy sources, neurons and red blood cells rely almost entirely on glucose to supply their energy needs. Because even a temporary shortage of blood glucose can severely depress brain function and lead to death of neurons, the body carefully monitors and regulates blood sugar (glucose) levels.

Other uses of monosaccharides are meager. Small amounts of pentose sugars are used to synthesize nucleic acids, and a variety of sugars are attached to externally facing plasma membrane proteins and lipids (Figure 3.2, p. 68). When glucose is present in excess of what the body needs for ATP synthesis, it is converted to glycogen or fat and stored for later use.

Dietary Requirements

The low-carbohydrate diet of the Inuit (Eskimos) and the high-carbohydrate diet of peoples in the Far East indicate that humans can be healthy even with wide variations in carbohydrate intake. This, no doubt, reflects the body's ability to use fats and amino acids as fuels. The minimum requirement for carbohydrates is not known, but 100 grams per day is presumed to be the smallest amount needed to maintain adequate blood glucose levels. The current recommendation is 125 to 175 grams of carbohy-

drate daily with the emphasis on *complex* carbohydrates. The new battle cry "strive for five" (servings of fruits and vegetables per day) is one of the soundest dietary ideas to come along in a long time. When less than 50 grams per day is consumed, tissue proteins and fats are broken down for energy fuel.

American adults typically consume 200 to 300 grams of carbohydrate each day, accounting for about 46% of dietary food energy. Because starchy foods are less expensive than meat and other high-protein foods, carbohydrates make up an even greater percentage of the diet in low-income groups. Starchy foods and milk have many valuable nutrients, such as vitamins and minerals. By contrast, highly refined carbohydrate foods such as candy and soft drinks provide energy sources only—the term "empty calories" is commonly used to describe such foods. Eating refined, sugary foods instead of more complex carbohydrates (for instance, replacing milk with soft drinks and whole grain breads with cookies) may cause nutritional deficiencies as well as obesity. Other possible consequences of excessive intake of simple carbohydrates are listed in Table 25.1.

Lipids

Dietary Sources

Although we also ingest cholesterol and phospholipids, the most abundant dietary lipids are neutral fats—triglycerides or triacylglycerols (see Chapter 2). We eat saturated fats in animal products such as meat and dairy foods and in a few plant products such as coconut. Unsaturated fats are present in seeds, nuts, and most vegetable oils. Major sources of cholesterol are egg yolk, meats (particularly organ meats like liver), and milk products. Fats (but not cholesterol) are digested to fatty acids and monoglycerides and then reconverted to triglycerides for transport in the lymph (see Chapter 24, p. 937).

Although the liver is adept at converting one fatty acid to another, it cannot synthesize *linoleic* (lin"o-le'ik) *acid*, a fatty acid component of *lecithin* (les'ĭ-thin). Thus, linoleic acid is an *essential fatty acid* that must be ingested. Recent research indicates that linolenic acid may also be essential. Fortunately, these two fatty acids are found in most vegetable oils.

Uses in the Body

The use of triglycerides and cholesterol is controlled mainly by the liver and adipose tissue. Like sugary carbohydrates, fats have fallen into disfavor, particularly in affluent groups where food is plentiful and the "battle of the bulge" is constant. But fats make

TABLE 25.1	Summary of Carbohydrate, Lipid, and Protein Nutrients		
	Recommended daily amounts (RDA) for adults	Problems	
Food sources		Excesses	Deficits
Carbohydrates			
■ *Complex carbohydrates (starches):* bread, cereal, crackers, flour, pasta, nuts, rice, potatoes	125–175 g 55–60% of total caloric intake	Obesity; nutritional deficits; dental caries; gastrointestinal irritation; elevated triglycerides in plasma	Tissue wasting (in extreme deprivation); metabolic acidosis resulting from accelerated fat use for energy
■ *Simple carbohydrates (sugars):* carbonated drinks, candy, fruit, ice cream, pudding, young (immature) vegetables			
■ *Both complex and simple carbohydrates:* pastries (pies, cookies, cakes)			
Lipids			
■ *Animal sources:* lard, meat, poultry, eggs, milk, milk products	80–100 g 30% or less of total caloric intake	Obesity and increased risk of cardiovascular disease (particularly if excesses of saturated fat)	Weight loss; fat stores and tissue proteins catabolized to provide metabolic energy; problems controlling heat loss (due to depletion of subcutaneous fat)
■ *Plant sources:* chocolate; corn, soy, cottonseed, olive oils; coconut; corn; peanuts			
■ *Essential fatty acids:* corn, cottonseed, soy oils; wheat germ; vegetable shortenings	6000 mg	Not known	Poor growth; skin lesions (eczema-like)
■ *Cholesterol:* organ meats (liver, kidneys, brains), egg yolks, fish roe; smaller concentrations in milk products and meat	250 mg or less	Increased levels of plasma cholesterol and LDL lipoproteins; correlated with increased risk of cardiovascular disease	Increased risk of stroke (CVA) in susceptible individuals
Proteins			
■ *Complete proteins:* eggs, milk, milk products, meat (fish, poultry, pork, beef, lamb)	0.8 g/kg body weight	Obesity; possibly aggravation of chronic disease states	Profound weight loss and tissue wasting; growth retardation in children; anemia; edema (due to deficits of plasma proteins) During pregnancy: miscarriage or premature birth
■ *Incomplete proteins:* legumes (soybeans, lima beans, kidney beans, lentils); nuts and seeds; grains and cereals; vegetables			

foods tender, flaky, or creamy, and make us feel full and satisfied. Furthermore, fats *are* essential for several reasons. Dietary fats help the body absorb fat-soluble vitamins, triglycerides are the major energy fuel of hepatocytes and skeletal muscle, and phospholipids are an integral component of myelin sheaths and *all* cellular membranes. Fatty deposits in adipose tissue provide (1) a protective cushion around body organs, such as the kidneys and eyeballs, (2) an insulating layer beneath the skin, and (3) an easy-to-store concentrated source of energy fuel. Regulatory molecules called *prostaglandins* (pros"tah-glan'dinz), formed from linoleic acid via arachidonic (ah"rah-kĭ-don'ik) acid, play a role in smooth muscle contraction, control of blood pressure, and inflammation.

Unlike the neutral fats, cholesterol is not used for energy. It is important as a stabilizing component of plasma membranes and is the precursor from which bile salts, steroid hormones, and other essential functional molecules are formed.

Dietary Requirements

Fats represent over 40% of the calories in American diets. There are no precise recommendations on amount or type of dietary fats, but the American

? *John literally "lives" on baked bean sandwiches. Is he getting all the essential amino acids in this rather restricted diet?*

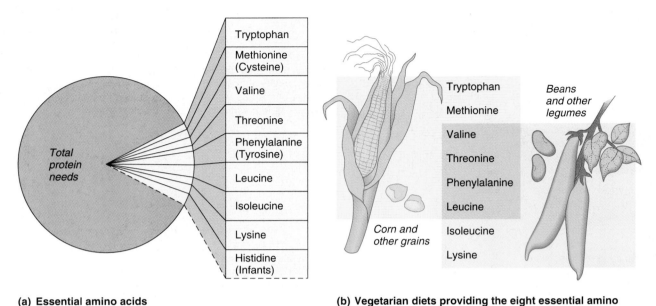

(a) Essential amino acids

(b) Vegetarian diets providing the eight essential amino acids for humans

FIGURE 25.2 *Essential amino acids.* Eight amino acids must be available simultaneously and in the correct relative amounts for protein synthesis to occur. (The ninth, histidine, is an essential amino acid in infants but not in adults.) **(a)** The relative amounts of the essential amino acids and total proteins needed by adults. Notice that the essential amino acids represent only a small percentage of the total recommended amino acid (protein) intake. Histidine is graphed with dashed lines because its requirement in adults has not been established. The amino acids shown in parentheses are *not* essential but can substitute in part for methionine and phenylalanine. **(b)** Vegetarian diets must be carefully constructed to provide all essential amino acids. As shown, corn has little isoleucine and lysine. Beans have ample isoleucine and lysine but little tryptophan and methionine. All essential amino acids can be obtained by consuming a meal of corn and beans.

Heart Association suggests that (1) fats should represent 30% or less of total caloric intake, (2) saturated fats should be limited to 10% or less of total fat intake, and (3) daily cholesterol intake should be no more than 200 mg (the amount in one egg yolk). Because a diet high in saturated fats and cholesterol may contribute to cardiovascular disease, these are wise guidelines. Sources of the various lipid classes and consequences of their deficient or excessive intake are summarized in Table 25.1. However, fat requirements for infants and children are higher than those for adults.

Fat Substitutes

In an attempt to reduce fat intake without losing fat's desirable aspects, many people have turned to fat substitutes or foods prepared with them. Perhaps the oldest fat substitute is air (beaten into a product to make it fluffy). Other fat substitutes are modified starches and gums, and more recently whey protein. Such products are metabolized and provide calories, but the newest substitutes, sucrose polyesters, are not metabolized because they are not absorbed at all.

Most fat substitutes have two easily recognized drawbacks: (1) they don't stand up to the intense heat needed to fry foods, and (2) although the manufacturers claim otherwise, they don't taste nearly as good as the "real thing." The ones that are not absorbed tend to cause flatus (gas) and may interfere with absorption of fat-soluble drugs and vitamins.

Proteins

Dietary Sources

Animal products contain the highest-quality proteins, that is, those with the greatest amount and best ratios of essential amino acids (Figure 25.2). Eggs, milk, and most meat proteins (Table 25.1) are **complete proteins** that meet all the body's amino acid requirements for tissue maintenance and

Yes. He has both grain and beans, which together contain all the essential amino acids.

growth. Legumes (beans and peas), nuts, and cereals are protein-rich but their proteins are nutritionally incomplete because they are low in one or more of the essential amino acids. Leafy green vegetables are well balanced in all essential amino acids except methionine, but contain only small amounts of protein. As you can see, strict vegetarians must carefully plan their diets to obtain all the essential amino acids and prevent protein malnutrition. Cereal grains and legumes when ingested together provide all of the essential amino acids (Figure 25.2b), and some combination of these foods is found in the diets of all cultures (most obviously in the rice and beans seen on nearly every plate in a Mexican restaurant). For nonvegetarians, grains and legumes are useful as partial substitutes for more expensive animal proteins.

Uses in the Body

Proteins are important structural materials of the body, including, for example, keratin of skin, collagen and elastin of connective tissues, and muscle proteins. In addition, functional proteins such as enzymes, hemoglobin, and some hormones regulate an incredible variety of body functions. Even so, whether amino acids are used to synthesize new proteins or are burned for energy depends on a number of factors:

1. The all-or-none rule. All amino acids needed to make a particular protein must be present in a cell at the same time and in sufficient amounts. If one is missing, the protein cannot be made. Because essential amino acids cannot be stored, those not used immediately to build proteins are oxidized for energy or converted to carbohydrates or fats.

2. Adequacy of caloric intake. For optimal protein synthesis, the diet must supply sufficient carbohydrate or fat calories for ATP production. When it doesn't, dietary and tissue proteins are used for energy.

3. Nitrogen balance of the body. In healthy adults the rate of protein synthesis equals the rate of protein breakdown and loss. This homeostatic state is reflected in the **nitrogen balance** of the body, which can be determined by chemical analysis based on the fact that the nitrogen content of protein averages 16%. The body is in nitrogen balance when the amount of nitrogen ingested in proteins equals the amount excreted in urine and feces.

The body is in *positive nitrogen balance* when protein synthesis exceeds protein breakdown and loss—the normal situation in growing children and pregnant women. A positive balance also occurs when tissues are being rebuilt or repaired following illness or injury. Positive nitrogen balance always in-

dicates that the amount of protein being incorporated into tissue is greater than the amount being broken down and used for energy.

In *negative nitrogen balance,* protein breakdown exceeds the use of protein for building structural or functional molecules. This occurs during physical and emotional stress (for example, infection, injury, burns, depression, or anxiety), when the quality of dietary protein is poor, and during starvation. In such cases, the tissues lose protein faster than it is replaced—an undesirable state of affairs.

4. Hormonal controls. Certain hormones, called **anabolic hormones,** accelerate protein synthesis and growth. The effects of these hormones vary continually throughout life. For example, pituitary *growth hormone* stimulates tissue growth during childhood and conserves protein in adults, and the *sex hormones* trigger the growth spurt of adolescence. Other hormones, such as the adrenal *glucocorticoids* released during stress, enhance protein breakdown and conversion of amino acids to glucose.

Dietary Requirements

Besides supplying essential amino acids, dietary proteins furnish the raw materials for making nonessential amino acids and various nonprotein nitrogen-containing substances. The amount of protein a person needs reflects his or her age, size, metabolic rate, and current state of nitrogen balance. However, as a rule of thumb, nutritionists recommend a daily protein intake of 0.8 g per kilogram of body weight—approximately 56 g for a 70-kg (154-pound) man and 48 g for a 58-kg (128-pound) woman. About 30 g of protein is supplied by a small serving of fish and a glass of milk. Most Americans eat far more protein than they need.

Vitamins

Vitamins (*vita* = life) are potent organic compounds needed in minute amounts for growth and good health. Unlike other organic nutrients, vitamins are not used for energy and do not serve as building blocks, but they are crucial in helping the body to use those nutrients that do. Without vitamins, all the carbohydrates, proteins, and fats we eat would be useless.

Most vitamins function as **coenzymes** (or parts of coenzymes); that is, they act with an enzyme to accomplish a particular chemical task. For example, the B vitamins riboflavin and niacin act as coenzymes (FAD and NAD^+, respectively) in the oxidation of glucose for energy. We will describe the roles of some vitamins in the metabolism discussion.

Because most vitamins are not made in the body, they must be taken in via foods or vitamin supplements. The exceptions are vitamin D made in the skin, and vitamins B and K synthesized by intestinal bacteria. In addition, the body can convert β-carotene (kar′o-tēn), the orange pigment in carrots and other foods, to vitamin A. (Thus, β-carotene and substances like it are called *provitamins.*)

Initially vitamins were given a convenient letter designation that indicated the order of their discovery. For example, ascorbic (as-kor′bik) acid is called vitamin C. Although more chemically descriptive terms have been assigned to the vitamins, this earlier terminology is still commonly used.

Vitamins are classed as either **fat soluble** or **water soluble.** The water-soluble vitamins, which include B-complex vitamins and vitamin C, are absorbed along with water from the gastrointestinal tract. (The exception is vitamin B_{12}, which must bind to gastric intrinsic factor to be absorbed.) Insignificant amounts of water-soluble vitamins are stored in the body and, unless used, they are excreted in urine. Consequently, few conditions resulting from excessive levels of these vitamins *(hypervitaminosis)* are known.

Fat-soluble vitamins (vitamins A, D, E, and K) bind to ingested lipids and are absorbed along with their digestion products. Anything that interferes with fat absorption also interferes with the uptake of fat-soluble vitamins. Except for vitamin K, fat-soluble vitamins are stored in the body, and pathologies due to fat-soluble vitamin toxicity, particularly vitamin A hypervitaminosis, are well documented clinically.

Vitamins are found in all major food groups, but no one food contains all the required vitamins. Thus, a balanced diet is the best way to ensure a full vitamin complement. Certain vitamins (A, C, and E) are *antioxidants* that disarm tissue-damaging free radicals and, as such, they have anticancer effects. Diets rich in broccoli, cabbage, and Brussels sprouts (all good sources of vitamins A and C) appear to reduce cancer risk. The whole story of how antioxidants interact in the body is still murky, but chemists propose that they act much like a bucket brigade to pass the dangerous chemical property from one molecule to the next. First, vitamin E reacts with the free radicals, restoring them to their less harmful state. This action converts vitamin E into a free radical which is in turn inactivated by carotenoids (bases of vitamin A). The carotenoid radicals thus formed are inactivated by vitamin C, and the water-soluble radicals formed during this reaction finally flush from the body in urine.

However, controversy abounds concerning the ability of vitamins to work wonders, such as the idea that huge doses (megadoses) of vitamin C will prevent colds. The notion that taking megadoses of vitamin supplements is the road to eternal youth and glowing health is useless at best and, at worst, may cause serious health problems, particularly in the case of fat-soluble vitamins. Table 25.2 lists the vitamins, their sources, their importance in the body, and consequences of their deficiency or excess. (Note that while most vitamins are quantified in mg or μg, some are still measured in the older International Units [IU].)

Minerals

The body requires moderate amounts of seven **minerals** (calcium, phosphorus, potassium, sulfur, sodium, chloride, and magnesium) and trace amounts of about a dozen others (Table 25.3, p. 958). Minerals make up about 4% of the body by weight, with calcium and phosphorus (as bone salts) accounting for about three-quarters of this amount.

Minerals, like vitamins, are not used for fuel but work with other nutrients to ensure a smoothly functioning body. Incorporation of minerals into structures gives added strength. For example, calcium, phosphorus, and magnesium salts harden the teeth and strengthen the skeleton. However, most minerals are ionized in body fluids or bound to organic compounds to form molecules such as phospholipids, hormones, enzymes, and other functional proteins. For example, iron is essential to the oxygen-binding heme of hemoglobin. Sodium and chloride ions, the major electrolytes in blood, help to maintain (1) normal osmolarity and water balance of body fluids and (2) responsiveness of neurons and muscle cells to stimuli. The amount of a particular mineral present in the body gives very few clues to its importance in body function. For example, just a few milligrams of iodine can make a critical difference to health.

A fine balance between uptake and excretion is crucial for retaining needed amounts of minerals while preventing toxic overload. For example, sodium, present in virtually all natural foods (and added in large amounts to processed foods), may contribute to high blood pressure if ingested in excess.

Fats and sugars are practically devoid of minerals, and highly refined cereals and grains are poor sources. The most mineral-rich foods are vegetables, legumes, milk, and some meats. Specific minerals and their roles in the body are summarized in Table 25.3.

Text continues on page 962.

TABLE 25.2	Vitamins				

Vitamin	Description, comments	Sources and recommended daily amounts (RDA) for adults	Importance in body	Problems Excesses	Deficits
Fat-soluble vitamins					
A (retinol)	Group of compounds including retinol and retinal; 90% is stored in liver, which can supply body needs for a year; stable to heat, acids, alkalis; easily oxidized; rapidly destroyed by exposure to light	Formed from provitamin β-carotene in intestine, liver, kidneys; carotene found in deep-yellow and deep-green leafy vegetables; vitamin A found in fish liver oils, egg yolk, liver, fortified foods (milk, margarine) RDA: males 5000 IU (1.5 mg), females 4000 IU	Required for synthesis of photoreceptor pigments, integrity of skin and mucosae, normal tooth and bone development; normal reproductive capabilities; important antioxidant (hence, anticancer and antiathero-sclerosis effects)	Toxic when ingested in excess of 50,000 IU daily for months; symptoms: nausea, vomiting, anorexia, headache, hair loss, bone and joint pain, bone fragility, enlargement of liver and spleen; may increase smokers' risk of lung cancer	Night blindness; epithelial changes; dry skin and hair, skin sores; increased respiratory, digestive, urogenital infections; drying of conjunctiva; clouding of cornea; in pregnant women, developmental defects of the embryo
D (antirachitic factor)	Group of chemically distinct sterols; concentrated in liver and to lesser extent in skin, kidneys, spleen, other tissues; stable to heat, light, acids, alkalis, oxidation	Vitamin D_3 is chief form in body cells; produced in skin by irradiation of 7-dehydrocholesterol by UV light; active form (1,25-dihydroxyvitamin D_3) produced by chemical modification of vitamin D_3 in liver, then kidneys; major food sources: fish liver oils, egg yolk, fortified milk RDA: 400 IU	Functionally a hormone; increases calcium blood levels by enhancing absorption of calcium; with PTH mobilizes calcium from bones; both mechanisms serve calcium home-ostasis of blood (essential for normal neuromuscular function, blood clotting, bone and tooth formation)	1800 IU/day may be toxic to children, massive doses induce toxicity in adults; symptoms: vomiting, diarrhea, fatigue, weight loss, hypercalcemia and calcification of soft tissues, irreversible cardiac and renal damage	Faulty mineralization of bones and teeth; rickets in children, osteomalacia in adults; poor muscle tone, restlessness, irritability
E (antisterility factor)	Group of related compounds called tocopherols, chemically related to sex hormones; stored primarily in muscle and adipose tissues; resistant to heat, light, acids; unstable in presence of oxygen	Wheat germ and vegetable oils, nuts, whole grains, dark-green leafy vegetables RDA: males 15 IU; females 12 IU (actual RDA depends on diet; high-fat diet requires a higher RDA)	An antioxidant that disarms free radicals; helps prevent oxidation of unsaturated fatty acids and cholesterol, thus helps prevent oxidative damage to cell membranes and atherosclerosis	Few side effects even at massive doses; slow wound healing; decreased platelet adhesion and increased clotting time	Extremely rare, precise effects uncertain: possible decreased life span and hemolysis of RBCs, fragile capillaries; spinocerebellar degeneration
K (coagulation vitamin)	Number of related compounds known as quinones; small amounts stored in liver; heat resistant; destroyed by acids, alkalis, light, oxidizing agents; antagonized by certain anticoagulants and antibiotics that interfere with synthetic activity of enteric bacteria	Most synthesized by bacteria in large intestine; food sources: leafy green vegetables, broccoli, cabbage, cauliflower, pork liver RDA: males 70 μg, females 55 μg; obtaining adequate amount is not a problem	Essential for formation of clotting proteins and some other proteins made by liver; as intermediate in electron transport chain, participates in oxidative phosphorylation in all body cells	None known, not stored in appreciable amounts	Easy bruising and bleeding (prolonged clotting time)

►

TABLE 25.2	Vitamins *(continued)*

Vitamin	Description, comments	Sources and recommended daily amounts (RDA) for adults	Importance in body	Problems Excesses	Problems Deficits

Water-soluble vitamins

Vitamin	Description, comments	Sources and recommended daily amounts (RDA) for adults	Importance in body	Excesses	Deficits
C (ascorbic acid)	Simple 6-carbon crystalline compound derived from glucose; rapidly destroyed by heat, light, alkalis; about 1500 mg is stored in body, particularly in adrenal gland, retina, intestine, pituitary; when tissues are saturated, excess excreted by kidneys	Fruits and vegetables, particularly citrus, cantaloupe, strawberries, tomatoes, fresh potatoes, leafy green vegetables RDA: 60 mg; 100 mg for those who smoke (but 200 mg/d appears to be the optimal dose for most males)	Antioxidant; acts in hydroxylation reactions in formation of nearly all connective tissues; in conversion of tryptophan to serotonin (vasoconstrictor) and of cholesterol to bile salts; enhances iron absorption and use; required for activation of folacin (a B vitamin)	Result of megadoses (10 or more times RDA); diarrhea; enhanced mobilization of bone minerals and blood coagulation; high uric acid levels and exacerbation of gout, kidney stone formation	Defective formation of intercellular cement; joint pains, poor tooth and bone growth; poor wound healing, increased susceptibility to infection; extreme deficit causes scurvy (bleeding gums, anemia, degeneration of muscle and cartilage, weight loss)
B$_1$ (thiamine)	Rapidly destroyed by heat; very limited amount stored in body; excess eliminated in urine	Lean meats, liver, fish, eggs, whole grains, leafy green vegetables, legumes RDA: 1.5 mg	Part of coenzyme cocarboxylase, which acts in carbo-hydrate metabolism; required for transformation of pyruvic acid to acetyl CoA, for synthesis of pentose sugars and acetylcholine; for oxidation of alcohol	None known	Beriberi; decreased appetite, vision disturbances; unsteadiness when standing or walking, confusion, loss of memory; profound fatigue; heart enlargement, tachycardia
B$_2$ (riboflavin)	Named for its similarity to ribose sugar: has green-yellow fluorescence; quickly decomposed by UV, visible light, alkalis; body stores are carefully guarded; excess eliminated in urine	Widely varying sources such as liver, yeast, egg white, whole grains, meat, poultry, fish, legumes; major source is milk RDA: 1.7 mg	Present in body as coenzymes FAD and FMN (flavin mononucleotide), both of which act as hydrogen acceptors in body; also is component of amino acid oxidases	None known	Dermatitis; cracking of lips at corners (cheilosis); lips and tongue become purple-red and shiny; ocular problems: light sensitivity, blurred vision; one of the most common vitamin deficiencies
Niacin (nicotinamide)	Simple organic compounds stable to acids, alkalis, heat, light, oxidation (even boiling does not decrease potency); very limited amount stored in body, day-to-day supply is desirable; excess secreted in urine	Diets that provide adequate protein usually provide adequate niacin because amino acid tryptophan is easily converted to niacin; preformed niacin provided by poultry, meat, fish; less important sources: liver, yeast, peanuts, potatoes, leafy green vegetables RDA: 20 mg	Constituent of NAD$^+$ (nicotinamide adenine dinucleotide) coenzyme involved in glycolysis, oxidative phosphorylation, fat breakdown; inhibits cholesterol synthesis; peripheral vascular dilator	Result of megadoses; hyperglycemia; vasodilation leading to flushing of skin, tingling sensations; possible liver damage; gout	Pellagra after months of deprivation (rare in U.S.); early signs are vague: listlessness, headache, weight loss, loss of appetite; pro-gresses to sore, red tongue and lips; nausea, vomiting, diarrhea; photosensitivity: skin becomes rough, cracked, may ulcerate; neurological symptoms also occur: charac-terized by the 4 Ds: dermatitis, diarrhea, dementia, and death (in final stage)

TABLE 25.2	Vitamins *(continued)*				
Vitamin	Description, comments	Sources and recommended daily amounts (RDA) for adults	Importance in body	Problems	
				Excesses	Deficits
B$_6$ (pyridoxine)	Group of three pyridines occurring in both free and phosphorylated forms in body; stable to heat, acids; destroyed by alkalis, light; body stores very limited	Meat, poultry, fish, whole grains, bananas; less important sources: potatoes, sweet potatoes, tomatoes, spinach RDA: 2–3 mg	Active form is coenzyme pyridoxal phosphate, which functions with several enzymes involved in amino acid metabolism; required for conversion of tryptophan to niacin, for glycogenolysis, for formation of antibodies and hemoglobin, and for breakdown of homocysteine	Depressed deep tendon reflexes, numbness and loss of sensation in extremities; difficulty walking; nerve damage	Infants: nervous irritability, convulsions, anemia, vomiting, weakness, abdominal pain; adults: seborrhea lesions around eyes and mouth; increased risk of heart disease
B$_5$ (pantothenic acid)	Quite stable, little loss of activity with cooking except in acidic or alkaline solutions; liver, kidneys, brain, adrenal, heart tissues contain large amounts	Name derived from Greek *panthos* meaning "everywhere"; widely distributed in animal foods, whole grains, legumes; liver, yeast, egg yolk, meat especially good sources; some produced by enteric bacteria RDA: 10 mg	Functions in form of coenzyme A in reactions that remove or transfer acetyl group, e.g., formation of acetyl CoA from pyruvic acid, oxidation and synthesis of fatty acids; also involved in synthesis of steroids and heme of hemoglobin	None known	Symptoms vague: loss of appetite, abdominal pain, mental depression, pains in arms and legs, muscle spasms, neuromuscular degeneration (neuropathy of alcoholics is thought to be related to deficits)
Biotin	Urea derivative containing sulfur; crystalline in its free form; stable to heat, light acids; in tissues, is usually combined with protein; stored in minute amounts, particularly in liver, kidneys, brain, adrenal glands	Liver, egg yolk, legumes, nuts; some synthesized by bacteria in gastrointestinal tract RDA: not stablished, probably about 0.3 mg (assumed that formation of the vitamin by enteric bacteria provides far more than is needed)	Functions as coenzyme for a number of enzymes that catalyze carboxylation, decarboxylation, deamination reactions; essential for reactions of Krebs cycle, for formation of purines and nonessential amino acids, for use of amino acids for energy	None known	Scaly skin, muscle pains, pallor, anorexia, nausea, fatigue; elevated blood cholesterol levels
B$_{12}$ (cyanocobalamin)	Most complex vitamin; contains cobalt; stable to heat; inactivated by light and strongly acidic or basic solutions; intrinsic factor required for transport across intestinal membrane; stored principally in liver; liver stores of 2000–3000 μg sufficient to provide body needs for 3–5 years	Liver, meat, poultry, fish, dairy foods except butter, eggs; not found in plant foods RDA: 3–6 μg	Functions as coenzyme in all cells, particularly in gastrointestinal tract, nervous system, and bone marrow; in bone marrow, acts in synthesis of DNA; when absent, erythrocytes do not divide but continue to enlarge; essential for synthesis of methionine and choline	None known	Pernicious anemia, signified by pallor, anorexia, dyspnea, weight loss, neurological disturbances; most cases reflect impaired absorption rather than actual deficit

➤

TABLE 25.2	Vitamins (continued)				

Vitamin	Description, comments	Sources and recommended daily amounts (RDA) for adults	Importance in body	Problems	
				Excesses	Deficits
Folic acid (folacin)	Pure vitamin is bright-yellow crystalline compound; stable to heat; easily oxidized in acidic solutions and light; stored mainly in liver	Liver, orange juice, deep-green vegetables, yeast, lean beef, eggs, veal, whole grains; synthesized by enteric bacteria RDA: 0.4 μg	Basis of coenzymes that act in synthesis of methionine and certain other amino acids, choline, DNA; essential for formation of red blood cells and for normal neural tube development in the embryo; helps breakdown of homocysteine	None known	Macrocytic or megaloblastic anemia; gastrointestinal disturbances; diarrhea; spina bifida risk in newborn; low birth weight and disabling neurological deficits; increased risk of heart attack and stroke

Notes

1. Each vitamin has specific functions; one vitamin cannot substitute for another. Many reactions in body require several vitamins, and lack of one can interfere with activity of others.

2. Diet that includes recommended amounts of the five food groups will furnish adequate amounts of all vitamins except vitamin D. Each food group makes a special vitamin contribution. For example, fruits and vegetables are principal sources of vitamin C; dark-green leafy vegetables and deep-yellow vegetables and fruits are primary sources of vitamin A (carotene); milk is principal source of riboflavin; meat, poultry, and fish are outstanding sources of niacin, thiamin, and vitamins B_6 and B_{12}. Whole grains are also important sources of niacin and thiamin.

3. Because vitamin D is present in natural foods in only very small amounts, infants, pregnant and lactating women, and people who have little exposure to sunlight should use vitamin D supplements (or supplemented foods, e.g., fortified milk).

4. Vitamins A and D are toxic in excessive amounts and should be used in supplementary form only when prescribed by a physician. Because most water-soluble vitamins are excreted in urine when ingested in excess, the effectiveness of taking supplements is questionable.

5. Many vitamin deficiencies are secondary to disease (including anorexia, vomiting, diarrhea, or malabsorption diseases) or reflect increased metabolic requirements due to fever or stress factors. Specific vitamin deficiencies require therapy with the vitamins that are lacking.

TABLE 25.3	Minerals				

Mineral	Distribution in body, comments	Sources and recommended daily amounts (RDA) for adults	Importance in body	Problems	
				Excesses	Deficits
Calcium (Ca)	Most stored in salt form in bones; the most abundant cation in body; absorbed from intestine in presence of vitamin D; excess excreted in feces; blood levels regulated by PTH and calcitonin	Milk, milk products, leafy green vegetables, egg yolk, shellfish RDA: 1200–1500 mg, dropping to 800–1000 mg after age 25	In salt form, required for hardness of bones, teeth; ionic calcium in blood and cells essential for normal membrane permeability, transmission of nerve impulses, muscle contraction, normal heart rhythm, blood clotting; activates certain enzymes; helps prevent hypertension	Depressed neural function; lethargy and confusion; muscle pain and weakness; calcium salt deposit in soft tissues; kidney stones	Muscle tetany; osteomalacia, osteoporosis; retarded growth and rickets in children

TABLE 25.3	Minerals (continued)

Mineral	Distribution in body, comments	Sources and recommended daily amounts (RDA) for adults	Importance in body	Problems	
				Excesses	Deficits
Chlorine (Cl)	Exists in body almost entirely as chloride ion; principal anion of extracellular fluid; highest concentrations in cerebrospinal fluid and gastric juice; excreted in urine	Table salts (as for sodium); usually ingested in excess RDA: not established; normal diet contains 3–9 g	With sodium, helps maintain osmotic pressure and pH of extracellular fluid; required for HCl formation by stomach glands; activates salivary amylase; aids in transport of CO_2 by blood (chloride shift)	Vomiting	Severe vomiting or diarrhea leads to Cl loss and alkalosis; muscle cramps; apathy
Sulfur (S)	Widely distributed: particularly abundant in hair, skin, nails; excreted in urine	Meat, milk, eggs, legumes (all rich in sulfur-containing amino acids) RDA: not established; diet adequate in proteins meets body's needs	Essential constituent of many proteins (insulin), some vitamins (thiamin and biotin); found in mucopolysaccharides present in cartilage, tendons, bone	Not known	Not known
Potassium (K)	Principal intracellular cation, 97% within cells; fixed proportion of K is bound to proteins, and measurements of body K are used to determine lean body mass; K^+ leaves cells during protein catabolism, dehydration, glycogenolysis; most excreted in urine	Widely distributed in foods; normal diet provides 2–6 g/day; avocados, dried apricots, meat, fish, fowl, cereals RDA: not established; diet adequate in calories provides ample amount, i.e., 2500 mg	Helps maintain intracellular osmotic pressure; needed for normal nerve impulse conduction, muscle contraction, glycogenesis, protein synthesis	Usually a complication of renal failure or severe dehydration, but may result from severe alcoholism; paresthesias, muscular weakness, cardiac abnormalities	Rare but may result from severe diarrhea or vomiting; muscular weakness, paralysis, nausea, vomiting, tachycardia, heart failure
Sodium (Na)	Widely distributed: 50% found in extracellular fluid, 40% in bone salts, 10% within cells; absorption is rapid and almost complete; excretion, chiefly in urine, controlled by aldosterone	Table salt (1 tsp = 2000 mg); cured meats (ham, etc.), sauerkraut, cheese; diet supplies substantial excess RDA: not established; probably about 2500 mg	Most abundant cation in extracellular fluid: principal electrolyte maintaining osmotic pressure of extracellular fluids and water balance; as part of bicarbonate buffer system, aids in acid-base balance of blood; needed for normal neuromuscular function; part of pump for transport of glucose and other nutrients	Hypertension; edema	Rare but can occur with excessive vomiting, diarrhea, sweating, or poor dietary intake; nausea, abdominal and muscle cramping, convulsions

TABLE 25.3	Minerals *(continued)*				

Mineral	Distribution in body, comments	Sources and recommended daily amounts (RDA) for adults	Importance in body	Problems	
				Excesses	Deficits
Magnesium (Mg)	In all cells, particularly abundant in bones; absorption parallels that of calcium; excreted chiefly in urine	Milk, dairy products, whole-grain cereals, nuts, legumes, leafy green vegetables RDA: 350 mg	Constituent of many coenzymes that play a role in conversion of ATP to ADP; required for normal muscle and nerve irritability	Diarrhea	Neuromuscular problems, tremors, muscle weakness, irregular heartbeat; vasospasm and hypertension; sudden cardiac death; seen in alcoholism, severe renal disease, prolonged diarrhea, and with diuretic drugs
Phosphorus (P)	About 80% found in inorganic salts of bones, teeth; remainder in muscle, nervous tissue, blood; absorption aided by vitamin D; about 1/3 dietary intake excreted in feces; metabolic by-products excreted in urine	Diets rich in proteins are usually rich in phosphorus; plentiful in milk, eggs, meat, fish, poultry, legumes, nuts, whole grains RDA: 800 mg	Component of bones and teeth, nucleic acids, proteins, phospholipids, ATP, phosphates (buffers) of body fluids; thus, important for energy storage and transfer, muscle and nerve activity, cell permeability	Not known, but excess in diet may depress absorption of iron and manganese	Rickets, poor growth
Trace Minerals*					
Fluorine (F)	Component of bones, teeth, other body tissues; excreted in urine	Fluoridated water RDA: 1.5–4 mg	Important for tooth structure; may help prevent dental caries (particularly in children) and osteoporosis in adults	Mottling of teeth	Not known
Cobalt (Co)	Found in all cells; larger amounts in bone marrow	Liver, lean meat, poultry, fish, milk RDA: not established	A constituent of vitamin B_{12}, which is needed for normal maturation of red blood cells	Goiter, polycythemia; heart disease	See deficit of vitamin B_{12} (Table 25.2)
Chromium (Cr)	Widely distributed	Liver, meat, cheese, whole grains, brewer's yeast, wine RDA: 0.05–2 mg	Required for synthesis of glucose tolerance factor (GTF) needed for proper glucose metabolism; enhances effectiveness of insulin on carbohydrate metabolism; may increase blood levels of HDLs while decreasing levels of LDLs	Not known	Impairs ability of insulin to work, hence increases insulin secretion and risk of adult onset diabetes mellitus
Copper (Cu)	Concentrated in liver, heart, brain, spleen; excreted in feces	Liver, shellfish, whole grains, legumes, meat; typical diet provides 2–5 mg daily RDA: 2–3 mg	Required for synthesis of hemoglobin; essential for manufacture of melanin, myelin, some intermediates of electron transport chain	Rare; abnormal storage results in Wilson's disease	Rare

*Trace minerals together account for less than 0.005% of body weight.

TABLE 25.3	Minerals (continued)

Mineral	Distribution in body, comments	Sources and recommended daily amounts (RDA for adults	Importance in body	Problems	
				Excesses	Deficits
Iodine (I)	Found in all tissues, but in high concentrations only in thyroid gland; absorption controlled by blood levels of protein-bound iodine; excreted in urine	Cod liver oil, iodized salt, shellfish, vegetables grown in iodine-rich soil RDA: 0.15 mg	Required to form thyroid hormones (T_3 and T_4), which are important in regulating cellular metabolic rate	Depressed synthesis of thyroid hormones	Hypothyroidism: cretinism in infants, myxedema in adults (if less severe, simple goiter); impaired learning and motivation in children
Iron (Fe)	60–70% in hemoglobin; remainder in skeletal muscle, liver, spleen, bone marrow bound to ferritin; only 2–10% of dietary iron is absorbed because of mucosal barrier; lost from body in urine, perspiration, menstrual flow, shed hair; sloughed skin and mucosal cells	Best sources: meat, liver, also in shellfish, egg yolk, dried fruit, nuts, legumes, molasses RDA: males 10 mg, females 15 mg	Part of heme of hemoglobin, which binds bulk of oxygen transported in blood; component of cytochromes, which function in oxidative phosphorylation	Hemochromatosis (inherited condition of iron excess); damage to liver (cirrhosis and liver cancer), heart, pancreas (causing diabetes)	Iron-deficiency anemia; pallor, lethargy, flatulence, anorexia, paresthesias, impaired cognitive performance in children; inability to maintain body temperature; depressed killing by phagocytes (due to inability to produce the respiratory burst)
Manganese (Mn)	Most concentrated in liver, kidneys, spleen; excreted largely in feces	Nuts, legumes, whole grains, leafy green vegetables, fruit RDA: 2.5–5 mg	Acts with enzymes catalyzing synthesis of fatty acids, cholesterol, urea, hemoglobin; needed for normal neural function, lactation, oxidation of carbohydrates, protein hydrolysis	Appears to contribute to obsessive behavior, hallucinations, and violent behavior	Not known
Selenium (Se)	Stored in liver, kidneys	Meat, seafood, cereals RDA: 0.05–2 mg	Antioxidant; constituent of certain enzymes; spares vitamin E	Nausea, vomiting, irritability, fatigue, weight loss, loss of hair	Not known
Zinc (Zn)	Concentrated in liver, kidneys, brain; excreted in feces	Seafood, meat, cereals, legumes, nuts, wheat germ, yeast RDA: 15 mg	Constituent of several enzymes (e.g., carbonic anhydrase); plays a structural role in a variety of proteins (e.g., transcription factors and tumor suppressor protein p53); required for normal growth, wound healing, taste, smell, sperm production	Difficulty in walking; slurred speech; tremors; interferes with body's ability to absorb copper, which can lead to weakened immunity	Loss of taste and smell; growth retardation; learning impairment; depressed immunity

METABOLISM
Overview of Metabolic Processes

Once inside body cells, nutrients become involved in an incredible variety of biochemical reactions. These reactions are known collectively as **metabolism** (*metabol* = change), a broad term referring to all chemical reactions necessary to maintain life. During metabolism, substances are constantly being built up and torn down. Cells use energy to extract more energy from foods, and then use some of this energy to drive their activities. Even at rest, the body uses energy on a grand scale.

Anabolism and Catabolism

Metabolic processes are either *anabolic* (synthetic) or *catabolic* (degradative). **Anabolism** (ah-nab'o-lizm) includes reactions in which larger molecules or structures are built from smaller ones; for example, the bonding of amino acids to make proteins and of proteins and lipids to form cell membranes. **Catabolism** (kah-tab'o-lizm) involves processes that break down complex structures to simpler ones. Hydrolysis of foods in the digestive tract is catabolic. So too is the group of reactions collectively called **cellular respiration,** during which food fuels (particularly glucose) are broken down within cells and some of the energy released is captured to form ATP, the cells' energy currency. ATP then serves as the "chemical drive shaft" that links energy-releasing catabolic reactions to cellular work. Recall from Chapter 2 that reactions driven by ATP are coupled reactions. ATP is never hydrolyzed directly. Instead enzymes shift its high-energy phosphate groups to other molecules, which are then said to be **phosphorylated** (fos"for-ĭ-la'ted). Phosphorylation energizes, or primes, the molecule to change in a way that increases its activity, produces motion, or does work. For example, many regulatory enzymes that catalyze key steps in metabolic pathways are activated by phosphorylation.

Figure 25.3 is a flowchart of the processing of energy-containing nutrients in the body. As you can see, three major stages are involved.

■ Stage 1 is digestion, which occurs in the gastrointestinal tract as described in Chapter 24. The absorbed nutrients are then transported in blood to the tissue cells.

■ In stage 2, which occurs in the cytoplasm of body cells, the newly delivered nutrients are (1) built into cellular molecules (lipids, proteins, and glycogen) by anabolic pathways or (2) broken down by catabolic pathways to *pyruvic* (pi-roo'vik) *acid* and *acetyl CoA* (as'ĕ-til ko-a').

■ Stage 3, which is almost entirely catabolic, occurs in the mitochondria of cells, requires oxygen, and completes the breakdown of newly digested or stored foodstuffs, producing carbon dioxide and water and harvesting large amounts of ATP.

The primary function of *cellular respiration,* consisting of glycolysis of stage 2 and all events of stage 3, is to generate ATP, which traps some of the chemical energy of the original food molecules in its own high-energy bonds. Thus, food fuels, such as glycogen and fats, *store* energy in the body, and these stores are later *mobilized to produce ATP for cellular use.* You do not need to memorize this schematic now, but as you will soon see, it provides a cohesive summary of nutrient processing and metabolism in the body.

Oxidation-Reduction Reactions and the Role of Coenzymes

The energy-yielding (ATP-yielding) reactions within cells are **oxidation reactions.** *Oxidation* was originally defined as the combination of oxygen with other elements. Examples are rusting of iron (the slow formation of iron oxide) and burning of wood and other fuels. In burning, oxygen combines rapidly with carbon, producing (in addition to carbon dioxide and water) an enormous amount of energy, which is liberated as heat and light. But then it was discovered that oxidation *also* occurs when hydrogen atoms are *removed* from compounds, so the definition was expanded to its current form: *Oxidation occurs via the gain of oxygen or the loss of hydrogen.* As explained in Chapter 2, whichever way oxidation occurs, the oxidized substance always *loses* (or nearly loses) electrons as they move to (or toward) a substance that more strongly attracts them.

This process can be explained by reviewing the consequences of different electron-attracting abilities of atoms (see p. 37). Because hydrogen is very electropositive, its lone electron usually spends more time orbiting the other atoms of molecules that hydrogen helps to form. But when a hydrogen *atom* is removed, its electron goes with it, and the molecule as a whole loses that electron. Conversely, oxygen is very electron-hungry (electronegative), so when oxygen binds with other atoms the shared electrons spend more time in oxygen's vicinity. Again, the molecule as a whole loses electrons. As you will soon see, essentially all oxidation of food fuels involves the step-by-step removal of pairs of hydrogen atoms (and thus pairs of electrons) from the substrate molecules, eventually leaving only carbon dioxide (CO_2). Molecular oxygen (O_2) is the *final* electron acceptor. It combines with the removed hydrogen atoms at the very end of the process, to form water (H_2O).

Whenever one substance loses electrons (is oxidized), another substance gains them (is reduced). Thus, oxidation and reduction are coupled reactions

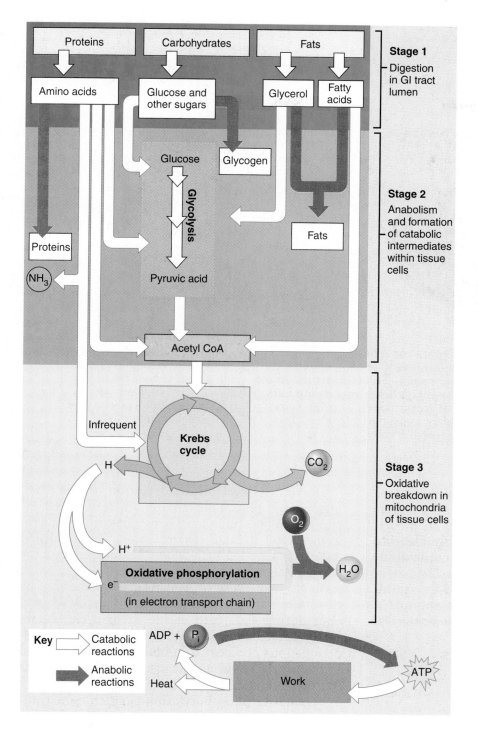

FIGURE 25.3 *The three stages of metabolism of energy-containing nutrients in the body.* In stage 1, foods are degraded to their absorbable forms by digestive enzymes in the gastrointestinal (GI) tract. In stage 2, the absorbed nutrients are transported in blood to the body cells, where they may be incorporated into cellular molecules (anabolism) or broken down via glycolysis and other reactions to pyruvic acid and/or acetyl CoA, and then funneled into the catabolic pathways of stage 3. Stage 3 consists of the pathways of the Krebs cycle and oxidative phosphorylation, which go on within the mitochondria. During the Krebs cycle, acetyl CoA is broken apart: Its carbon atoms are liberated as carbon dioxide (CO_2), and the hydrogen atoms removed are delivered to a chain of receptors (the electron transport chain), which ultimately delivers them (as protons and free electrons) to molecular oxygen so that water is formed. Some of the energy released during the electron transport chain reactions is used to form the bonds of high-energy ATP molecules. Once generated, ATP provides the energy needed by cells to power their many activities. The reactions of glycolysis (stage 2) and the Krebs cycle and oxidative phosphorylation (stage 3) collectively constitute cellular respiration.

and we speak of **oxidation-reduction (O-R)**, or **redox, reactions** in the same breath. The key understanding about redox reactions is that "oxidized" substances *lose* energy and "reduced" substances *gain* energy as energy-rich electrons are transferred from the first to the second. Consequently, as food fuels are oxidized, their energy is transferred to a "bucket brigade" of other molecules and ultimately to ADP to form energy-rich ATP molecules.

Like all other chemical reactions in the body, oxidation-reduction reactions are catalyzed by en-

zymes. Enzymes are discussed in detail in Chapter 2, pp. 55–57). The enzymes that catalyze oxidation-reduction reactions by removing hydrogen are specifically called **dehydrogenases** (de-hi'dro-jen-ās"ez), while those catalyzing the transfer of oxygen are **oxidases.** Most of these enzymes require the help of a specific coenzyme, typically derived from one of the B vitamins. Although the enzymes catalyze the removal of hydrogen atoms to oxidize a substance, they cannot *accept* the hydrogen (hold on or bond to it). Their **coenzymes,** however, act as reversible

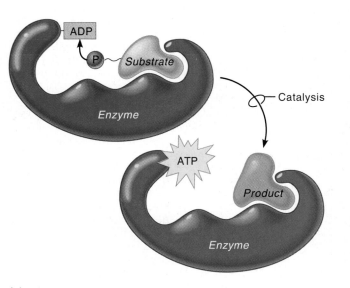

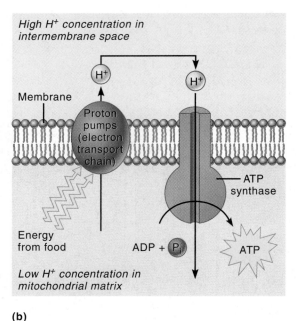

(a)

(b)

FIGURE 25.4 *Mechanisms of phosphorylation.* **(a)** Substrate-level phosphorylation occurs when a high-energy phosphate group is transferred enzymatically directly from a substrate to ADP to form ATP. Substrate phosphorylation occurs both in the cytoplasm and in the mitochondrial matrix. **(b)** Oxidative phosphorylation, which occurs in mitochondria, reflects the activity of electron transport proteins that act as proton "pumps" to create a proton gradient across the cristae membranes. The source of energy for this pumping is energy released during oxidation of food fuels. As the protons flow passively back into the mitochondrial fluid matrix (through ATP synthases), energy of their diffusion gradient is used to bind phosphate groups to ADP.

hydrogen (or electron) acceptors, becoming reduced each time a substrate is oxidized. Two very important coenzymes of the oxidative pathways are **nicotinamide** (nik″o-tin′ah-mīd) **adenine dinucleotide (NAD⁺),** based on *niacin,* and **flavin adenine dinucleotide (FAD),** derived from *riboflavin.* The oxidation of succinic acid to fumaric acid and the simultaneous reduction of FAD to FADH₂, an example of a coupled oxidation-reduction reaction, is as follows:

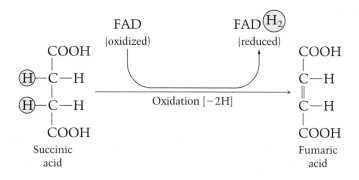

Mechanisms of ATP Synthesis

How do our cells capture some of the energy liberated during cellular respiration to make ATP molecules? There appear to be two mechanisms—substrate-level phosphorylation and oxidative phosphorylation.

Substrate-level phosphorylation occurs when high-energy phosphate groups are transferred directly from phosphorylated substrates (metabolic intermediates) to ADP (Figure 25.4a). Essentially, it occurs because the high-energy bonds attaching the phosphate groups to the substrates are even more unstable than those of ATP. ATP is synthesized by this route during one of the steps of glycolysis, and once during each turn of the Krebs cycle (Figure 25.5). The enzymes catalyzing substrate-level phosphorylations are located in both the cytoplasm and in the watery environment inside the mitochondria (mitochondrial matrix).

Oxidative phosphorylation is much more complicated, but it also releases most of the energy that is eventually captured in ATP bonds during cellular respiration. This process, which is carried out by electron transport proteins forming part of the cristae membranes in the mitochondria, is an example of a **chemiosmotic process.** Chemiosmotic processes couple the movement of substances across membranes to chemical reactions. In this case, some of the energy released during the oxidation of food fuels (the "chemi" part of the term) is used to pump (*osmo* = push) hydrogen ions or protons (H⁺) across the inner (cristae) mitochondrial membrane into the intermembrane space (Figure 25.4b). This creates a steep diffusion gradient for protons across the mem-

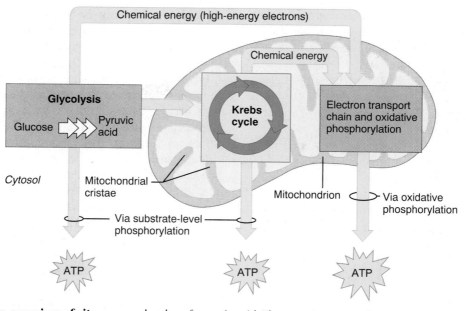

FIGURE 25.5 *An overview of sites of ATP formation during cellular respiration.* Glycolysis occurs outside the mitochondria in the cytosol. The Krebs cycle and the electron transport chain reactions occur within the mitochondria. During glycolysis, each glucose molecule is broken down to two molecules of pyruvic acid. The pyruvic acid enters the mitochondrial matrix, where the Krebs cycle decomposes it to carbon dioxide. During glycolysis and the Krebs cycle, small amounts of ATP are formed by substrate-level phosphorylation. Chemical energy from glycolysis and the Krebs cycle, in the form of energy-rich electrons, is then transferred to the electron transport chain, which is built into the membrane of the cristae. The electron transport chain carries out oxidative phosphorylation, which accounts for most of the ATP generated by cellular respiration.

brane. Then, when protons do flow back across the membrane (through a membrane channel protein called *ATP synthase*), some of this gradient energy is captured and used to attach phosphate groups to ADP.

Carbohydrate Metabolism

Because all food carbohydrates are eventually transformed to glucose, the story of carbohydrate metabolism is really a tale of glucose metabolism. Glucose enters the tissue cells by facilitated diffusion, a process that is greatly enhanced by insulin. Immediately upon entry into the cell, glucose is phosphorylated to form *glucose-6-phosphate* by transfer of a phosphate group (PO_4^{3-}) to its sixth carbon during a coupled reaction with ATP:

Glucose + ATP → glucose-6-PO_4 + ADP

Because most body cells lack the enzymes needed to reverse this reaction, it effectively traps glucose inside the cells. Since glucose-6-phosphate is a *different* molecule from simple glucose, the reaction also keeps intracellular glucose levels low, maintaining a diffusion gradient for glucose entry. Only intestinal mucosa cells, kidney tubule cells, and liver cells have the enzymes needed to reverse this phosphorylation reaction, which reflects their central roles in glucose uptake *and* release. The catabolic and anabolic pathways for carbohydrates all begin with glucose-6-phosphate.

Oxidation of Glucose

Glucose is the pivotal fuel molecule in the oxidative (ATP-producing) pathways. The complete catabolism of glucose is shown by the overall equation

$$C_6H_{12}O_6 + 6O_2 \rightarrow 6H_2O + 6CO_2 + 36\ \text{ATP} + \text{heat}$$
glucose   oxygen   water   carbon
                                      dioxide

This equation gives few hints that glucose breakdown is a very complex process or that it involves three of the pathways included in Figure 25.3 and diagrammed in Figure 25.5:

1. Glycolysis (color-coded pink through the chapter);

2. The Krebs cycle (color-coded pale-green); and

3. The electron transport chain and oxidative phosphorylation (color-coded violet).

These metabolic pathways occur in a definite order, so we will consider them sequentially.

What would happen in this biochemical pathway if for some reason NADH + H⁺ could not transfer its "picked up" hydrogens to pyruvic acid?

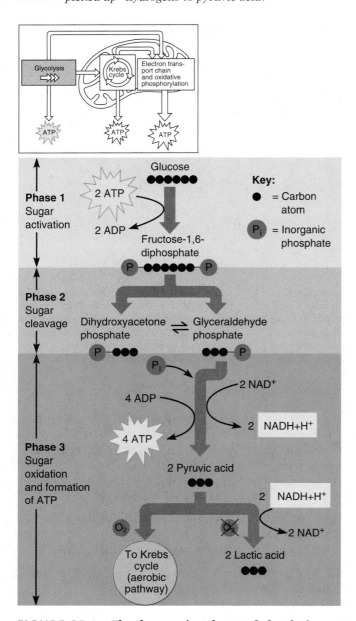

FIGURE 25.6 *The three major phases of glycolysis.* During phase 1, glucose is activated by phosphorylation and converted to fructose-1,6-diphosphate. In phase 2, fructose-1,6-diphosphate is cleaved into two 3-carbon fragments (reversible isomers). In phase 3, the 3-carbon fragments are oxidized (by removal of hydrogen) and 4 ATP molecules are formed. The fate of pyruvic acid depends on whether or not molecular O₂ is available.

Glycolysis would come to a halt because the oxidase enzyme cannot hold onto the H⁺ removed during substrate oxidation. If all the coenzymes are already reduced and can't release hydrogen, the oxidase enzyme couldn't continue to function.

Glycolysis Glycolysis (gli-kol′ĭ-sis; "sugar splitting") is a series of ten chemical steps by which glucose is converted into two *pyruvic acid* molecules, yielding a net gain of 2 ATP per glucose molecule. Glycolysis occurs in the cytosol of cells, where its steps are catalyzed by specific soluble enzymes. Except for the first step during which glucose entering the cell is phosphorylated to glucose-6-phosphate, all of its steps are fully *reversible*. Glycolysis is an *anaerobic* (an-a′er-ōb-ik; *an* = without, *aero* = air) *process*. Although this is sometimes mistakenly interpreted to mean that the pathway occurs only in the absence of oxygen, this is *not* the case. The correct interpretation is that glycolysis *does not use oxygen and occurs whether or not oxygen is present.* Figure 25.6 shows the glycolytic pathway in an abbreviated form that considers its three major phases. The total glycolytic pathway, complete with the names and structural formulas of all intermediates and a description of the events of each step, appears in Appendix C.

1. Sugar activation. In phase 1, glucose is phosphorylated and converted to fructose-6-phosphate, which is then phosphorylated again. These three steps yield fructose-1,6-diphosphate and use two ATP molecules. The two separate reactions of the sugar with ATP provide the *activation energy* needed to prime the later stages of the pathway—hence this phase is sometimes referred to as the *energy investment phase.* (The importance of activation energy is described in Chapter 2.)

2. Sugar cleavage. During phase 2, fructose-1,6-diphosphate is split into two 3-carbon fragments that exist (reversibly) as one of two isomers: glyceraldehyde (glis″er-al′dĕ-hīd) 3-phosphate or dihydroxyacetone (di″hi-drok″se-as′ĕ-tōn) phosphate.

3. Oxidation and ATP formation. In phase 3, consisting of six steps, two major events happen. First, the two 3-carbon fragments are oxidized by the removal of hydrogen, which is picked up by NAD⁺. Hence, some of glucose's energy is transferred to NAD⁺. Second, inorganic phosphate groups (Pᵢ) are attached to each oxidized fragment by high-energy bonds. Later, as these terminal phosphates are cleaved off, enough energy is captured to form a total of four ATP molecules. As noted earlier, formation of ATP by this method is called *substrate-level phosphorylation.*

The final products of glycolysis are two molecules of **pyruvic acid** and two molecules of reduced NAD⁺ (NADH + H⁺),* with a net gain of two ATP molecules per glucose molecule. Four ATPs are produced, but remember that two are used in phase 1 to

---

* NAD carries a positive charge (NAD⁺); thus, when it accepts the hydrogen pair, NADH + H⁺ is the resulting reduced product.

"prime the pump." Each pyruvic acid molecule has the formula $C_3H_4O_3$, and glucose is $C_6H_{12}O_6$. Thus, between them the two pyruvic acid molecules have lost 4 hydrogen atoms, which are now bound to two molecules of NAD^+. Although a small amount of ATP has been harvested, the other two end products (H_2O and CO_2) of glucose oxidation have yet to appear.

The fate of pyruvic acid, which still contains most of glucose's chemical energy, depends on the availability of oxygen at the time it is produced. There is a limited supply of NAD^+, thus glycolysis will continue and pyruvic acid will enter the next processing phase, the Krebs cycle, only if the reduced coenzymes ($NADH + H^+$) formed during glycolysis are relieved of the hydrogen accepted. Only then can they continue to act as hydrogen acceptors. When oxygen is readily available, this is no problem. $NADH + H^+$ simply delivers its burden of hydrogen atoms to the enzymes of the electron transport chain in the mitochondria, which in turn deliver them to molecular oxygen, forming water. However, when oxygen is not present in sufficient amounts, as might occur during strenuous exercise, the $NADH + H^+$ produced during glycolysis unloads its hydrogen baggage *back onto pyruvic acid*, thus reducing it. This addition of two hydrogen atoms to pyruvic acid yields **lactic acid** (see bottom right of Figure 25.6). Some of the lactic acid produced diffuses out of the cells and is transported to the liver. When oxygen is again available, lactic acid is oxidized back to pyruvic acid and enters the **aerobic pathways** (the oxygen-requiring Krebs cycle and electron transport chain within the mitochondria). There it is completely oxidized to water and carbon dioxide. The liver may also convert lactic acid all the way back to glucose-6-phosphate (reverse glycolysis) and then store it as glycogen or free it of its phosphate and release it to the blood if blood sugar levels are low.

Although glycolysis generates ATP very rapidly, glycolysis alone produces only 2 ATP/glucose molecule, as compared to the 36 ATP/glucose harvested when glucose is completely oxidized. Except for red blood cells (which typically carry out *only* glycolysis), prolonged anaerobic metabolism ultimately results in acid-base problems. Consequently, *totally* anaerobic conditions resulting in lactic acid formation provide only a temporary (or emergency) means of rapid ATP production when oxygen is in low supply. It can go on without tissue damage for the longest periods in skeletal muscle, for much shorter periods in cardiac muscle, and almost not at all in the brain.

Krebs Cycle The **Krebs cycle,** named after its discoverer Hans Krebs, is the next stage of glucose oxidation. The Krebs cycle occurs in the aqueous mitochondrial matrix and is fueled largely by pyruvic acid produced during glycolysis and by fatty acids resulting from fat breakdown.

After pyruvic acid enters the mitochondria, the first order of business is to convert it to *acetyl CoA*. This multistep preparation process, which links glycolysis to the Krebs cycle, involves three events:

1. Decarboxylation, in which one of pyruvic acid's carbons is removed and released as carbon dioxide gas (a waste product of metabolism). CO_2 diffuses out of the cells into the blood to be expelled by the lungs.

2. Oxidation by the removal of hydrogen atoms. The removed hydrogens are picked up by NAD^+.

3. Combining the resulting acetic acid with *coenzyme A* to produce the final product, **acetyl coenzyme A,** or simply **acetyl CoA.** Coenzyme A (CoA-SH) is a sulfur-containing coenzyme derived from pantothenic acid, a B vitamin.

Acetyl CoA is now ready to enter the Krebs cycle and be broken down completely by the mitochondrial enzymes. Coenzyme A shuttles the 2-carbon acetic acid to the enzyme that can condense it with a 4-carbon acid called **oxaloacetic** (ok"sah-lo"ah-sēt'ik) **acid** to produce the 6-carbon **citric acid.** Citric acid is the first substrate of the cycle, and biochemists (in particular) prefer to call the Krebs cycle the **citric acid cycle.** Figure 25.7 summarizes the basic events of the Krebs cycle.

As the cycle moves along through its eight successive steps, the atoms of citric acid are rearranged to produce different intermediate molecules, most called **keto acids.** The acetic acid that enters the cycle is broken apart carbon by carbon (decarboxylated) and is oxidized, simultaneously generating $NADH + H^+$ and $FADH_2$. At the end of the cycle, acetic acid has been totally disposed of and oxaloacetic acid, the *pickup molecule,* is regenerated. Because two *decarboxylation* and four *oxidation* events occur, the products of the Krebs cycle are two molecules of carbon dioxide and four molecules of reduced coenzymes (3 $NADH + H^+$ and 1 $FADH_2$). The addition of water at certain steps accounts for some of the released hydrogen. One molecule of ATP is formed (via substrate-level phosphorylation) during each turn of the cycle. The detailed events of each of the eight steps of the Krebs cycle are described in Appendix C.

Now let's account for the pyruvic acid molecules entering the mitochondrion. A total of three carbon dioxides and five molecules of reduced coenzymes—1 $FADH_2$ and 4 $NADH + H^+$ (equal to the removal of 10 hydrogen atoms)—are formed for each pyruvic acid. The products of glucose oxidation in the Krebs cycle are twice that (remember 1 glucose = 2 pyruvic acids): six CO_2, ten molecules of reduced coenzymes, and two ATP molecules. Notice that it is

? *What two major kinds of chemical reactions occur in this cycle, and how are these reactions indicated symbolically?*

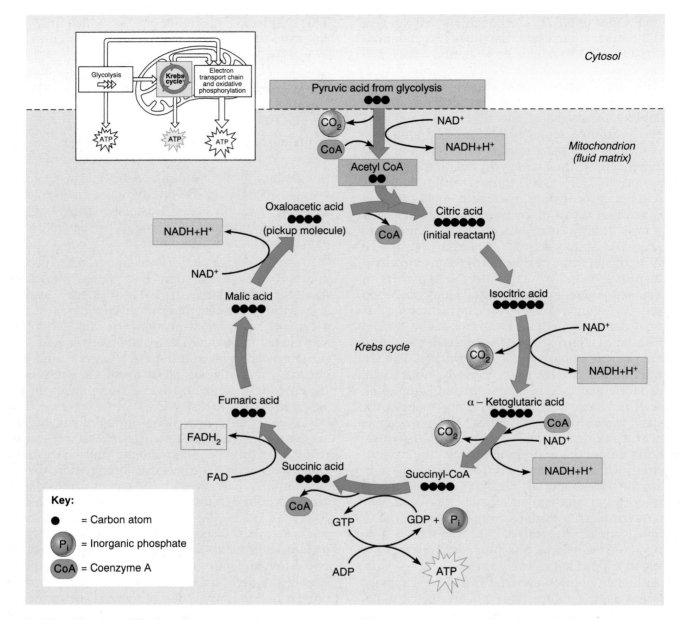

FIGURE 25.7 Simplified version of the Krebs cycle. During each turn of the cycle, two carbon atoms are removed from the substrates as CO_2 (decarboxylation reactions); four oxidations by removal of hydrogen atoms occur, producing four molecules of reduced coenzymes (3 NADH + H$^+$ and 1 FADH$_2$); and one ATP is synthesized by substrate-level phosphorylation. An additional decarboxylation reaction and an oxidation reaction occur to convert pyruvic acid, the product of glycolysis, to acetyl CoA, the molecule that actually enters the Krebs cycle pathway.

these Krebs cycle reactions that produce the CO_2 evolved during glucose oxidation. The reduced coenzymes, which carry their "extra electrons" in high-energy linkages, must now be oxidized if the Krebs cycle (and glycolysis) is to continue.

Although the glycolytic pathway is exclusive to carbohydrate oxidation, breakdown products of carbohydrates, fats, and proteins can feed into the Krebs cycle to be oxidized for energy. Likewise, some Krebs cycle intermediates can be siphoned off to make fatty acids and nonessential amino acids, as described shortly. Thus the Krebs cycle, besides serving as the final common pathway for the oxidation of food fuels, is a source of building materials for anabolic (biosynthetic) reactions.

Decarboxylation, shown as removal of CO_2, and oxidation, shown as the reduction of the coenzymes, e.g., NAD$^+$ → NADH + H$^+$.

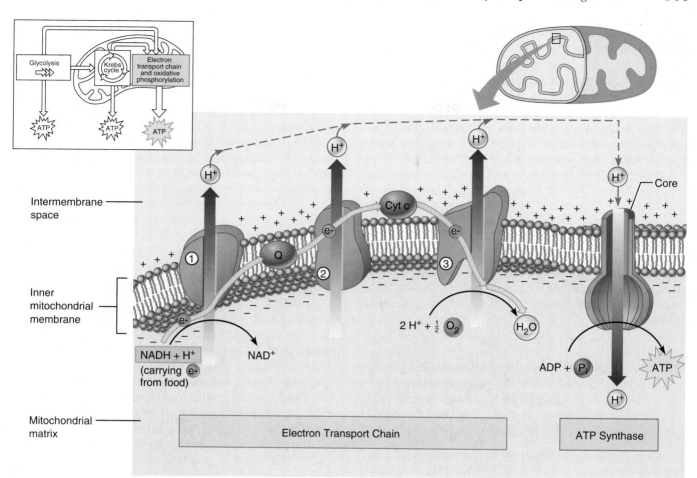

FIGURE 25.8 *One hypothetical mechanism of oxidative phosphorylation.* Schematic diagram showing the flow of electrons through the three major respiratory enzyme complexes—(1) NADH dehydrogenase (FMN, Fe-S), (2) cytochrome b-c$_1$, and (3) cytochrome oxidase (a-a$_3$)—of the electron transport chain during the transfer of two electrons from reduced NAD$^+$ to oxygen. Coenzyme Q and cytochrome c are mobile and act as carriers between the major complexes.

The electron transport chain is an energy converter; it transforms chemical energy into a gradient of H$^+$. As the electrons flow along their energy gradient, some of the energy is used by each complex to pump protons from the mitochondrial matrix into the intermembrane space. These protons create an electrochemical proton gradient that drives them back across the inner membrane through the ATP synthase complex. ATP synthase uses the energy of proton flow (electrical energy) to synthesize ATP from ADP and P$_i$ present in the matrix. Oxidation of each NADH + H$^+$ to NAD$^+$ yields 3 ATP. Because FADH$_2$ unloads its hydrogen atoms beyond the first respiratory complex, less energy (2 ATP) is captured as a result of its oxidation.

Electron Transport Chain and Oxidative Phosphorylation Like glycolysis, none of the reactions of the Krebs cycle directly use oxygen. This is the exclusive function of the **electron transport chain,** which oversees the final catabolic reactions that occur on the inner mitochondrial membranes. However, because the reduced coenzymes produced during the Krebs cycle are the "grist" (substrates) for the electron transport chain "mill," these pathways are coupled, and both phases are considered to be oxygen requiring (aerobic).

In the electron transport chain, the hydrogens removed during the oxidation of food fuels are finally combined with molecular oxygen, and the energy released during those reactions is harnessed to attach inorganic phosphate groups (P$_i$) to ADP. As noted earlier, this type of phosphorylation process is called *oxidative phosphorylation.* Let us peek under the hood of a cell's power plant and look more closely at this rather complicated process.

Most components of the electron transport chain are proteins bound to metal atoms *(cofactors).* These proteins, which form part of the structure of the inner mitochondrial membrane (cristae), are quite varied in composition (Figure 25.8). For example, some of the proteins, the **flavins,** contain flavin mononucleotide (FMN) derived from the vitamin riboflavin. Others contain both sulfur (S) and iron (Fe), but most are brightly colored iron-containing pigments called **cytochromes** (si'to-krōmz; *cyto* = cell, *chrom* = color). Neighboring carriers are clustered together to form three major **respiratory**

enzyme complexes that are alternately reduced and oxidized by picking up electrons and then passing them on to the next complex in the sequence. The first such complex accepts hydrogen atoms from $NADH + H^+$, oxidizing it to NAD^+. $FADH_2$ transfers its hydrogen "baggage" slightly further along into the chain. The hydrogens delivered to the electron transport chain by the reduced coenzymes are quickly split into protons (H^+) and electrons. The electrons are shuttled along the membrane from one acceptor to the next. The protons escape into the watery matrix only to be picked up and deposited ("pumped") across the membrane into the intermembrane space by one of the three respiratory enzyme complexes. Ultimately the electron pairs are delivered to half a molecule of O_2 (i.e., to an oxygen atom), creating oxygen ions (O^-), which strongly attract the hydrogen ions and form water as indicated by the reaction

$$2H (2H^+ + 2e^-) + \frac{1}{2}O_2 \rightarrow H_2O$$

Virtually all the water resulting from glucose oxidation is formed during oxidative phosphorylation. Because $NADH + H^+$ and $FADH_2$ are oxidized as they release their burden of picked-up hydrogen atoms, the net reaction for the electron transport chain is

$$\text{Coenzyme-2H} + \frac{1}{2}O_2 \rightarrow \text{coenzyme} + H_2O$$
$$\underset{\substack{\text{reduced} \\ \text{coenzyme}}}{} \qquad \underset{\substack{\text{oxidized} \\ \text{coenzyme}}}{}$$

The transfer of electrons from $NADH + H^+$ to oxygen releases large amounts of energy (is exergonic). If hydrogen combined directly with molecular oxygen, the energy would be released in one big burst and most of it would be lost to the environment as heat. Instead energy is released in many small steps as the electrons stream from one electron acceptor to the next. Each successive carrier has a greater affinity for electrons than those preceding it. Therefore, the electrons cascade "downhill" from $NADH + H^+$ to progressively lower energy levels until they are finally delivered to oxygen, which has the greatest affinity of all for electrons (Figure 25.9). You could say that oxygen "pulls" the electrons down the chain.

By using the stepwise release of electronic energy to pump protons from the fluid matrix into the intermembrane space, the electron transport chain functions as an energy conversion machine. Because the crista membrane is nearly impermeable to H^+, this chemiosmotic process creates an **electrochemical proton (H^+) gradient** across the inner membrane that temporarily has potential energy (gradient energy). This energy source is referred to as *proton motive force* to emphasize the capacity of the gradient to do work.

The proton gradient has two major aspects: (1) It creates a pH gradient, with the hydrogen ion concentration in the matrix much lower than that in the intermembrane space; and (2) it generates a voltage across the membrane that is negative $(-)$ on the matrix side and positive $(+)$ between the mitochondrial membranes. Both conditions strongly attract the protons back into the matrix. However, the only areas of the membrane that are freely permeable to H^+ are at large enzyme-protein complexes called **ATP synthases.** As the protons take this "route" they create an electrical current, and ATP synthase harnesses this electrical energy to catalyze attachment of a phosphate group to ADP to form ATP (Figure 25.8). The enzyme's several subunits appear to work together like gears. As the ATP synthase core rotates, ADP and inorganic phosphate are pulled in and ATP is churned out, thus completing the process of oxidative phosphorylation.

Notice something here. The ATP synthase works like an ion pump running in reverse. Recall from Chapter 3 that ion (solute) pumps use ATP as their energy source to transport ions against their electrochemical gradients. Here we have ATP synthases using the energy of an existing ion (proton) gradient to power ATP synthesis as the downhill H^+ current flows through the ATP synthase core channels. The proton gradient also supplies energy to pump needed metabolites (ADP, pyruvic acid, inorganic phosphate) and calcium ions across the relatively impermeable inner mitochondrial membrane. The outer membrane is quite freely permeable to these substances, so no "help" is needed there. However, the supply of energy from oxidation is not limitless; so when more of the proton motive force or gradient energy is used to drive these transport processes, less is available to make ATP.

Homeostatic Imbalance

Studies of metabolic poisons, some quite deadly, support this chemiosmotic model of oxidative phosphorylation. For example, cyanide (the gas used in gas chambers) disrupts the process by binding to cytochrome oxidase and blocking electron flow from cytochrome a_3 to oxygen. Poisons commonly called "uncouplers" abolish the proton gradient by making the crista membrane permeable to H^+. Consequently, although the electron transport chain continues to deliver electrons to oxygen at a furious pace, and oxygen consumption rises, no ATP is made. ∎

The stimulus for ATP production is entry of *ADP* into the mitochondrial fluid matrix. As ADP is transported in, ATP is moved out in a coupled transport process.

Summary of ATP Production When oxygen is present, cellular respiration is remarkably efficient. Of the total 686 kilocalories (kcal) of energy present in 1 mole of glucose, as much as 262 kcal can be captured in the bonds of ATP molecules. The rest is liberated as heat. (The mole is explained on p. 32.) This corresponds to an energy capture of about 38%. This is far more efficient than any human-made machines, which use only 10–30% of the energy actually available to them.

During cellular respiration, most energy flows in this sequence: Glucose → NADH + H⁺ → electron transport chain → proton motive force → ATP. Let's do a little bookkeeping to summarize the net energy gain from one glucose molecule. Once we have tallied the net gain of 4 ATP molecules per glucose produced directly by substrate-level phosphorylations (2 during glycolysis and 2 during the Krebs cycle), all that remains to be calculated is the number of ATP molecules produced by oxidative phosphorylation (Figure 25.10).

Each NADH + H⁺ that transfers a pair of high-energy electrons to the electron transport chain contributes enough energy to the proton gradient to generate between 2 and 3 ATP molecules. (We will round it off to 3 to simplify our calculations.) The oxidation of FADH₂ is somewhat less efficient because it doesn't donate electrons to the "top" of the electron transport chain as does NADH + H⁺, but to a lower energy level. So, for each 2 hydrogen atoms delivered by FADH₂, just 2 ATPs are produced (instead of 2+ or 3). Thus, the 8 NADH + H⁺ and the 2 FADH₂ produced during the Krebs cycle are "worth" 24 and 4 ATPs respectively. The 2 NADH + H⁺ generated during glycolysis yields 4 (or 6) ATP molecules. Overall, complete oxidation of 1 glucose molecule to carbon dioxide and water yields 38 or 36 molecules of ATP (Figure 25.10). The alternative figures given represent the present uncertainty about the energy yield of reduced NAD⁺ generated *outside* the mitochondria by glycolysis. The inner mitochondrial membrane is *not* permeable to reduced NAD⁺ generated in the cytosol, so that formed during glycolysis uses a *shuttle molecule* to deliver its extra electron pair to the electron transport chain. As we know, shuttles cost money, and the money used for this shuttle is ATP. At present, the consensus is that the net energy yield for reoxidation of these reduced NAD⁺s is probably the same as for FADH₂, that is, 2 ATP per

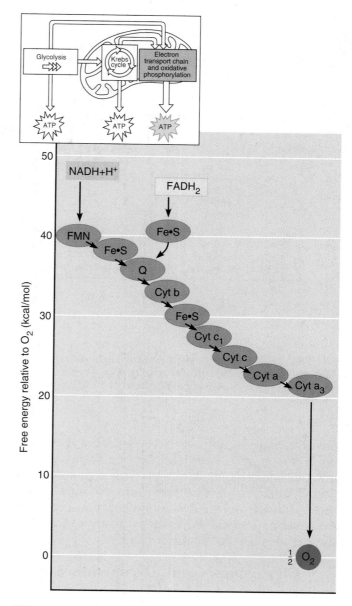

FIGURE 25.9 *Electronic energy gradient in the electron transport chain.* Each member of the chain oscillates between a reduced state and an oxidized state. A component of the chain becomes reduced when it accepts electrons from its "uphill" neighbor (which has a lower affinity for the electrons). Each member of the chain reverts to its oxidized form as it passes electrons to its "downhill" neighbor (which has a greater affinity for the electrons). At the "bottom" of the chain is oxygen, which is *very* electronegative. The overall energy drop for electrons traveling from NADH to oxygen is 53 kcal/mol, but this fall is broken up into a series of smaller steps by the electron transport chain.

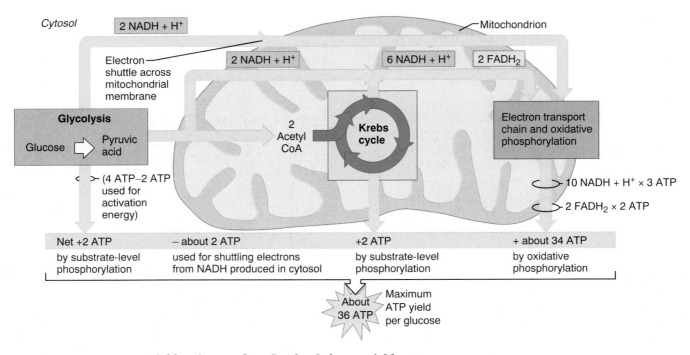

FIGURE 25.10 *Energy yield review: Each molecule of glucose yields many ATP molecules during cellular respiration.* See the text for an explanation of why the figure for the maximum ATP yield of 36 ATP/glucose is an estimate.

electron pair. Thus, we have deducted 2 ATP to cover the "fare" of the shuttle (see Figure 25.10), and our bookkeeping comes up with a grand total of 36 ATP/glucose as the maximum possible energy yield. Actually our figures are probably still too high because, as mentioned earlier, the proton motive force is also used to do other work.

Glycogenesis and Glycogenolysis

Although most glucose is used to generate ATP molecules, unlimited amounts of glucose do *not* result in unlimited ATP synthesis, because cells cannot store large amounts of ATP. When more glucose is available than can be immediately oxidized, rising intracellular ATP concentrations eventually inhibit glucose catabolism and initiate processes that store glucose as glycogen or fat. Because the body can store much more fat than glycogen, fats account for 80–85% of stored energy. (Fat synthesis is considered in the discussion of lipid metabolism.)

When glycolysis is "turned off" by high ATP levels, glucose molecules are combined in long chains to form glycogen, the animal carbohydrate storage product. This process, called **glycogenesis** (*glyco* = sugar; *genesis* = origin), begins when glucose entering cells is phosphorylated to glucose-6-phosphate and then converted into its isomer, *glucose-1-phosphate*. The terminal phosphate group is cleaved off as the enzyme *glycogen synthase* catalyzes the attachment of glucose to the growing glycogen chain

(Figure 25.11). Liver and skeletal muscle cells are most active in glycogen synthesis and storage.

When blood levels of glucose drop, glycogen lysis, or splitting, occurs. This process is known as **glycogenolysis** (gli"ko-jĕ-nol'ĭ-sis). The enzyme *glycogen phosphorylase* oversees phosphorylation and cleavage of glycogen to release glucose-1-phosphate, which is then converted to glucose-6-phosphate, a form that can enter the glycolytic pathway to be oxidized for energy.

In muscle cells and most other cells, the glucose-6-phosphate resulting from glycogenolysis is trapped because it cannot cross the cell membrane. However, hepatocytes (and some kidney and intestinal cells) contain *glucose-6-phosphatase,* a unique enzyme which removes the terminal phosphate, producing free glucose. Because glucose readily diffuses from the cell into the blood, the liver can use its glycogen stores to provide blood sugar for the benefit of other organs when blood glucose levels drop (see Table 25.4 on p. 983). Liver glycogen is also an important energy source for skeletal muscles that have depleted their own glycogen reserves.

■ A common misconception is that athletes need to eat more protein to improve their athletic performance and maintain their greater muscle mass. Actually, a diet rich in complex carbohydrates, which lays in muscle glycogen, is much more effective in sustaining intense muscle activity than are high-protein meals. Notice that the emphasis is on *com-*

plex carbohydrates. Eating a candy bar before an athletic event to provide "quick" energy does more harm than good because it stimulates insulin secretion, which favors glucose use and retards fat use at a time when fat use should be maximal. Building muscle protein or avoiding its loss requires not only extra protein, but also extra (protein-sparing) calories to meet the greater energy needs of the increasingly massive muscles.

Long-distance runners in particular are well aware of the practice of glycogen loading, popularly called "carbo loading," for endurance events. Glycogen loading "tricks" the muscles into storing more glycogen than they normally would. It involves (1) eating meals high in protein and fat while exercising heavily over a period of several days (this depletes muscle glycogen stores) and then (2) two to three days before the event decreasing exercise intensity (to about 40% of the norm) and switching abruptly to a high-carbohydrate diet. This causes a rebound in muscle glycogen stores of two to four times the normal amount. However, glycogen retains water, and the resulting weight increase and fluid retention may hamper the ability of muscle cells to obtain adequate oxygen. Some athletes using the procedure have complained of cardiac and skeletal muscle pain. Consequently, its use should be cautioned against.

Gluconeogenesis

When too little glucose is available to stoke the "metabolic furnace," glycerol and amino acids are converted to glucose. **Gluconeogenesis,** the process of forming new *(neo)* sugar from *noncarbohydrate* molecules, occurs in the liver. It takes place when dietary sources and glucose reserves have been depleted and blood glucose levels are beginning to decline. Gluconeogenesis protects the body, the nervous system in particular, from the damaging effects of low blood sugar *(hypoglycemia)* by ensuring that ATP synthesis can continue. The hormones triggering this process and the way fats and proteins are siphoned into these pathways are considered in the discussions of fat and protein metabolism.

Lipid Metabolism

Fats are the body's most concentrated source of energy. They contain very little water, and the energy yield from fat catabolism is approximately twice that gained from either glucose or protein breakdown; that is, 9 kcal per gram of fat versus 4 kcal per gram of carbohydrate or protein. Most products of fat digestion are transported in lymph in the form of fatty-protein droplets called *chylomicrons* (see Chapter 24). Eventually, the lipids in the chylomicrons are hydrolyzed by plasma enzymes, and the resulting

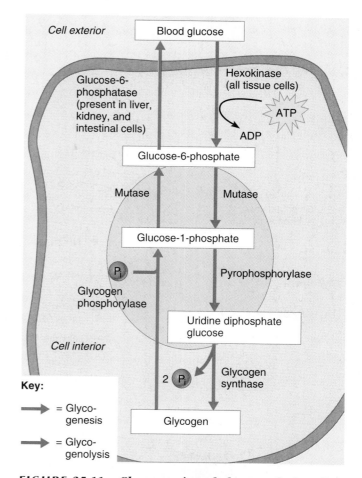

FIGURE 25.11 *Glycogenesis and glycogenolysis.* When glucose supplies exceed cellular need for ATP synthesis, glycogenesis, the conversion of glucose into glycogen for storage, occurs. Glycogenolysis, the breakdown of glycogen to release glucose, is stimulated by falling blood sugar levels. Notice that glycogen is synthesized and degraded by different enzymatic pathways.

fatty acids and glycerol are taken up by body cells where they are processed in various ways. In this section we focus on the oxidation of fats for energy and on some of their anabolic uses.

Oxidation of Glycerol and Fatty Acids

Of the various lipids, only neutral fats are routinely oxidized for energy. Catabolism of neutral fats involves the separate oxidation of their two different building blocks: glycerol and fatty acid chains. Most body cells easily convert glycerol to glyceraldehyde phosphate, a glycolysis intermediate which then flows into the Krebs cycle. Glyceraldehyde is equal to half a glucose molecule, and ATP energy harvest from its complete oxidation is approximately half that of glucose (18 ATP/glycerol).

Beta oxidation, the initial phase of fatty acid oxidation, occurs in the mitochondria. Although many

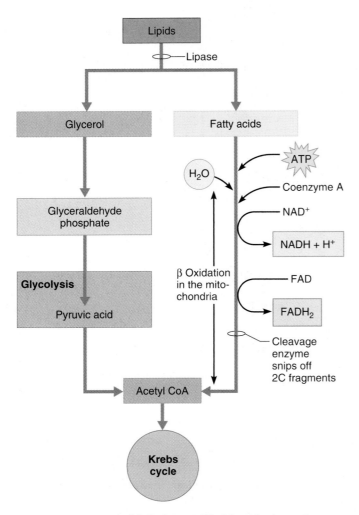

FIGURE 25.12   *Initial phase of lipid oxidation.* The building blocks of lipids are catabolized by different pathways. Glycerol is converted to glyceraldehyde phosphate, a glycolysis intermediate, and completes the glycolytic pathway through pyruvic acid to acetyl CoA. Fatty acids undergo a process called beta oxidation. The fatty acids are first activated by a coupled reaction with ATP and combined with coenzyme A. After being oxidized twice (reducing a NAD$^+$ and FAD), acetyl CoA is cleaved off and the process begins again. Acetyl CoA enters the Krebs cycle as described on p. 967.

different types of reactions (oxidation, dehydration, and others) are involved, the *net result* of beta oxidation is that the fatty acid chains are broken apart into two-carbon *acetic acid* fragments, and reduced coenzymes are produced (Figure 25.12). Each acetic acid molecule is fused to coenzyme A, forming acetyl CoA. The term "beta oxidation" reflects the fact that the carbon in the beta (third) position is oxidized during the process and cleavage of the fatty acid in each case occurs between the alpha and beta carbons. Acetyl CoA is then picked up by oxaloacetic acid and enters the aerobic pathways to be completely oxidized to carbon dioxide and water.

Notice that unlike glycerol, which enters the glycolytic pathway, acetyl CoA resulting from fatty acid

breakdown *cannot* be used for gluconeogenesis because the metabolic pathway is irreversible past pyruvic acid.

Lipogenesis and Lipolysis

There is a continual turnover of neutral fats in adipose tissue. New fats are "put in the larder" for later use, while stored fats are broken down and released to the blood. That bulge of fatty tissue you see today does *not* contain the same fat molecules it did a month ago.

When glycerol and fatty acids from dietary fats are not immediately needed for energy, they are recombined into triglycerides and stored. About 50% ends up in subcutaneous tissue; the balance is stockpiled in other fat depots of the body. Triglyceride synthesis, or **lipogenesis** (see Figure 25.13), occurs when cellular ATP and glucose levels are high. Excess ATP leads also to an accumulation of acetyl CoA and glyceraldehyde-PO$_4$, intermediates of glucose metabolism, which would otherwise feed into the Krebs cycle. But when these two metabolites are present in excess, they are channeled into triglyceride synthesis pathways. Acetyl CoA molecules are condensed together, forming fatty acid chains that grow two carbons at a time. (This accounts for the fact that almost all fatty acids in the body contain an even number of carbon atoms.) Because acetyl CoA, an intermediate in glucose catabolism, is also the *starting point* for synthesis of fatty acids (see Figure 25.13), glucose is easily converted into fat. Glyceraldehyde phosphate is converted to glycerol, which is condensed with fatty acids to form triglycerides. Thus, even if the diet is fat-poor, excessive carbohydrate intake provides *all the raw materials* needed to form neutral fats. When blood sugar is high, lipogenesis is the major activity in adipose tissues and is also an important liver function.

Lipolysis (lĭ-pol'ĭ-sis; "fat splitting"), the breakdown of stored fats into glycerol and fatty acids, is essentially lipogenesis in reverse. The fatty acids and glycerol products are released to the blood, helping to ensure that body organs have continuous access to fat fuels for aerobic respiration. (The liver, cardiac muscle, and resting skeletal muscles actually prefer fatty acids as an energy fuel.) The meaning of the adage "fats burn in the flame of carbohydrates" becomes clear when carbohydrate intake is inadequate. Under such conditions, lipolysis is accelerated as the body attempts to fill the fuel gap with fats. However, the ability of acetyl CoA to enter the Krebs cycle depends on the availability of oxaloacetic acid to act as the pickup molecule. When carbohydrates are deficient, oxaloacetic acid is converted to glucose (to fuel the brain). Without oxaloacetic acid, fat oxidation is incomplete, acetyl CoA accumulates, and the liver converts acetyl CoA molecules to **ketones**, or **ketone bodies,** which are released into the blood. This con-

What is the central molecule in lipid metabolism?

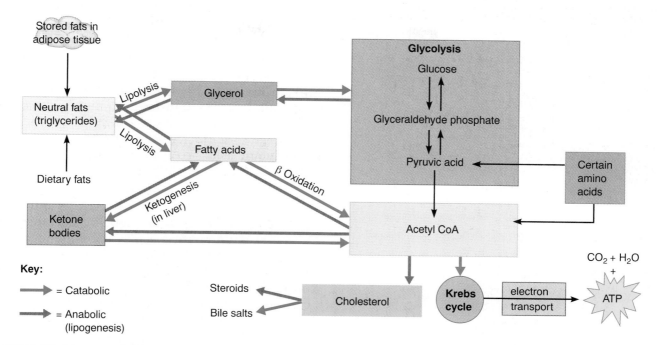

FIGURE 25.13 *Metabolism of triglycerides.* When needed for energy, dietary or stored fats enter the catabolic pathways. Glycerol enters the glycolytic pathway (as glyceraldehyde phosphate) and the fatty acids are broken down by beta oxidation to acetyl CoA, which enters the Krebs cycle. When fats are to be synthesized (lipogenesis) and stored in fat depots, the intermediates are drawn from glycolysis and the Krebs cycle in a reversal of the processes noted above. Likewise, excess dietary fats are stored in adipose tissues. When triglycerides are in excess or are the primary energy source, the liver releases their breakdown products (fatty acids → acetyl CoA) in the form of ketone bodies. Excessive amounts of carbohydrates and amino acids are also converted to triglycerides (lipogenesis).

version process is called **ketogenesis** (Figure 25.13). Ketones include acetoacetic acid, β-hydroxybutyric acid, and acetone, all formed from acetic acid. (The *keto acids* cycling through the Krebs cycle and the *ketone bodies* resulting from fat metabolism are quite different and should not be confused.)

◪ Homeostatic Imbalance

When ketones accumulate in the blood faster than they can be used by tissue cells for fuel, *ketosis* results and large amounts of ketones are excreted from the body in the urine. Ketosis is a common consequence of starvation, unwise dieting (in which inadequate carbohydrates are eaten), and diabetes mellitus. Because most ketone bodies (like fatty acids) are organic acids, the outcome of ketosis is *metabolic acidosis.* The body's buffer systems cannot tie up the acids fast enough, and blood pH drops to dangerously low levels. The person's breath smells fruity as acetone vaporizes from the lungs, and breathing be-

comes more rapid as the respiratory system tries to blow off blood carbonic acid as carbon dioxide to force the blood pH upward. In severe untreated cases, the person may become comatose or even die as the acid pH depresses the nervous system. ∎

Synthesis of Structural Materials

All body cells use phospholipids and cholesterol to build their membranes. Phospholipids are also important components of myelin sheaths of neurons. In addition, the liver (1) synthesizes lipoproteins for transport of cholesterol, fats, and other substances in the blood; (2) makes tissue factor, a clotting factor; (3) synthesizes cholesterol from acetyl CoA; and (4) uses cholesterol for forming bile salts. Certain endocrine organs (ovaries, testes, and adrenal cortex) use cholesterol as the basis for synthesizing their steroid hormones.

Protein Metabolism

Like all biological molecules, proteins have a limited life span and must be broken down to their amino acids and replaced before they begin to deteriorate.

Acetyl CoA.

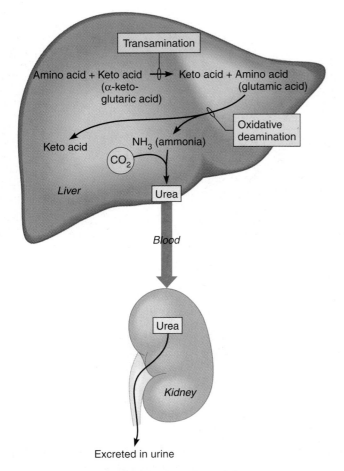

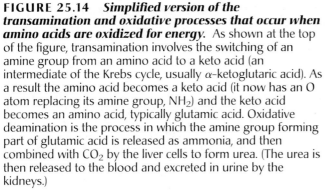

FIGURE 25.14 *Simplified version of the transamination and oxidative processes that occur when amino acids are oxidized for energy.* As shown at the top of the figure, transamination involves the switching of an amine group from an amino acid to a keto acid (an intermediate of the Krebs cycle, usually α–ketoglutaric acid). As a result the amino acid becomes a keto acid (it now has an O atom replacing its amine group, NH_2) and the keto acid becomes an amino acid, typically glutamic acid. Oxidative deamination is the process in which the amine group forming part of glutamic acid is released as ammonia, and then combined with CO_2 by the liver cells to form urea. (The urea is then released to the blood and excreted in urine by the kidneys.)

Newly ingested amino acids transported in the blood are taken up by cells by active transport processes and used to *replace* tissue proteins at the rate of about 100 grams each day. When more protein is ingested than is needed for these constructive (or anabolic) purposes, amino acids are oxidized for energy or converted to fat.

Oxidation of Amino Acids

Before amino acids can be oxidized for energy, they must be *deaminated*, that is, their amine group (NH_2) must be removed. The resulting molecule is then converted to pyruvic acid or one of the keto acid intermediates in the Krebs cycle. The key molecule in these interconversions is *glutamic* (gloo-tam'ik) *acid,* a common nonessential amino acid. The events that occur, shown in simplified form in Figure 25.14, appear to be as follows:

1. Transamination (trans″am-ĭ-na′shun). A number of amino acids can transfer their amine group to α-ketoglutaric acid (a Krebs cycle keto acid) to form glutamic acid. The transferring amino acid becomes a keto acid (that is, it has an oxygen atom where the amine group formerly was), and the keto acid (α-ketoglutaric acid) becomes an amino acid (glutamic acid). This reaction is fully reversible.

2. Oxidative deamination. In the liver, the amine group of glutamic acid is removed as **ammonia** (**NH_3**), and α-ketoglutaric acid is regenerated. The liberated ammonia molecules are combined with carbon dioxide, yielding **urea** and water. The urea is released to the blood and removed from the body in urine. Because ammonia is toxic to body cells, the ease with which glutamic acid funnels amine groups into the **urea cycle** is extremely important. This mechanism rids the body not only of the ammonia produced during oxidative deamination, but also of bloodborne ammonia produced by intestinal bacteria.

3. Keto acid modification. The goal of amino acid degradation is to produce molecules that can be oxidized in the Krebs cycle or converted to glucose. Hence, keto acids resulting from transamination are altered as necessary to produce metabolites that can enter the Krebs cycle. The most important of these metabolites are pyruvic acid (actually an intermediate before Krebs cycle pathways), acetyl CoA, α-ketoglutaric acid, and oxaloacetic acid (see Figure 25.7). Since the reactions of glycolysis are reversible, deaminated amino acids that are converted to pyruvic acid can also be reconverted to glucose and contribute to gluconeogenesis.

Synthesis of Proteins

Amino acids are the most important anabolic nutrients. Not only do they form all protein structures, but they form the bulk of the body's functional molecules as well. As described in Chapter 3, protein synthesis (directed by messenger RNA molecules) occurs on ribosomes, where cytoplasmic enzymes oversee the formation of peptide bonds linking the amino acids together into protein polymers. The amount and type of protein synthesized are precisely controlled by hormones (growth hormone, thyroxine, sex hormones, and others), so protein anabolism reflects the hormonal balance at each stage in the life cycle.

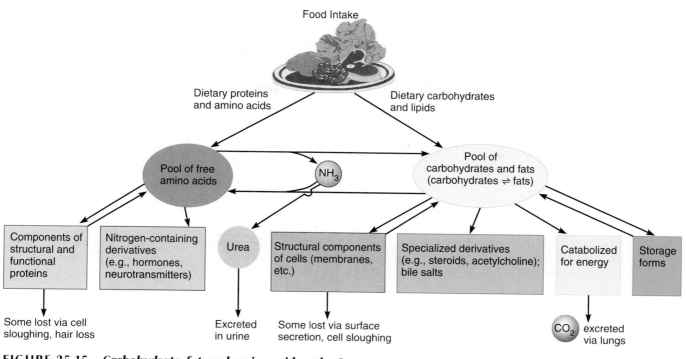

FIGURE 25.15 *Carbohydrate, fat, and amino acid pools.* Sources contributing to the pools and fates of intermediates drawn from the pools are indicated.

During the course of your lifetime, your cells synthesize 225–450 kg (about 500–1000 pounds) of proteins, depending on your size. However, you do not need to consume anywhere near that amount because nonessential amino acids are easily formed by siphoning keto acids from the Krebs cycle and transferring amine groups to them by transamination. Most of these transformations occur in the liver. The liver also provides nearly all the nonessential amino acids needed to produce the relatively small amount of protein that the body synthesizes each day. However, a complete set of amino acids must be present for protein synthesis to take place, so all of the essential amino acids must be provided by the diet. If some are not, the rest are oxidized for energy even though they may be needed for anabolism. In such cases, negative nitrogen balance always results because body protein is broken down to supply the essential amino acids needed.

Catabolic-Anabolic Steady State of the Body

The body exists in a *dynamic catabolic-anabolic state*. With few exceptions (notably DNA), organic molecules are continuously broken down and rebuilt—frequently at a head-spinning rate.

The blood is the common transport pool for all body cells, and it contains many kinds of nutrients—glucose, ketone bodies, fatty acids, glyc-

erol, and lactic acid. Some organs routinely use blood energy sources other than glucose, thus sparing (saving) glucose for tissues with stricter glucose requirements (see Table 25.4 on p. 983).

The body's total supply of nutrients constitutes its **nutrient pools**—amino acid, carbohydrate, and fat stores—which can be drawn on to meet its varying needs (Figure 25.15a). These pools are interconvertible because their pathways are linked by key intermediates (see Figure 25.16). The liver, adipose tissue, and skeletal muscles are the primary effector organs determining the amounts and direction of the conversions shown in the figure.

The **amino acid pool** (Figure 25.15) consists of the body's total supply of free amino acids. A small amount of animo acids and proteins is lost daily in urine and in sloughed hairs or skin cells. Typically, these are replaced via the diet; otherwise, amino acids arising from tissue breakdown return to the pool. This pool is the source of amino acids for resynthesizing body proteins and for forming several amino acid derivatives. In addition, as described above, deaminated amino acids can participate in gluconeogenesis. Not all events of amino acid metabolism occur in all cells. For example, *only* the liver forms urea, and excreting urea from the body is a major role of the kidneys (and to a small extent the skin). Nonetheless, the concept of a common amino acid pool is valid because all cells are connected by the blood.

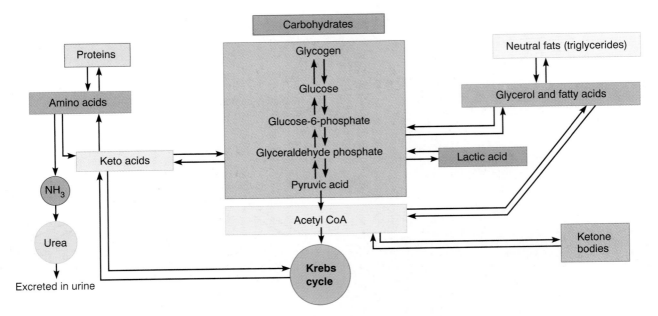

FIGURE 25.16 *Pathways of interconversion of carbohydrates, fats, and proteins.*

Because carbohydrates are easily and frequently converted to fats, the **carbohydrate** and **fat pools** are usually considered together (Figure 25.16). There are two major differences between this pool and the amino acid pool: (1) Fats and carbohydrates are oxidized directly to produce cellular energy, whereas amino acids can be used to supply energy *only after being converted to a carbohydrate intermediate* (a keto acid). (2) Excess carbohydrate and fat can be stored as such, whereas excess amino acids are *not* stored as protein. Instead, they are oxidized for energy or converted to fat or glycogen for storage.

Absorptive and Postabsorptive States: Events and Controls

Metabolic controls act to equalize blood concentrations of nutrients between two nutritional states. Sometimes referred to as the *fed state,* the **absorptive state** is the time during and shortly after eating, when nutrients are flushing into the bloodstream from the gastrointestinal tract. The **postabsorptive (fasting) state** is the period when the GI tract is empty and energy sources are supplied by the breakdown of body reserves. Those who eat "three squares" a day are in the absorptive state for the four hours during and after each meal and in the postabsorptive state in the late morning, late afternoon, and all night. However, postabsorptive mechanisms can sustain the body for much longer intervals if necessary—even to accommodate weeks of fasting—as long as water is taken in.

Absorptive State

During the absorptive state (Figure 25.17) anabolic processes exceed catabolic ones. Glucose is the major energy fuel. Dietary amino acids and fats are used to remake degraded body protein or fat, and small amounts are oxidized to provide ATP. Excess metabolites, regardless of source, are transformed to fat if not used for anabolism. We will consider the fate of each nutrient group during this phase and the hormonal control of its pathways.

Carbohydrates Absorbed monosaccharides are delivered directly to the liver, where fructose and galactose are converted to glucose. Glucose, in turn, is released to the blood or converted to glycogen and fat. Glycogen formed in the liver is stored there, but most of the fat synthesized by the hepatocytes is released to the blood and picked up for storage by adipose tissues. Bloodborne glucose not sequestered by the liver enters body cells to be metabolized for energy; any excess is stored in skeletal muscle cells as glycogen or in adipose cells as fat.

Triglycerides Nearly all products of fat digestion enter the lymph in the form of chylomicrons (see p. 937), which are hydrolyzed to fatty acids and glycerol before they can pass through the capillary walls. *Lipoprotein lipase,* the enzyme that catalyzes this fat hydrolysis, is particularly active in the capillaries of muscle and fat tissues. Adipose cells, skeletal muscle cells, and liver cells use triglycerides (triacylglycerols) as their primary energy source, and when dietary carbohydrates are limited, other cells begin to

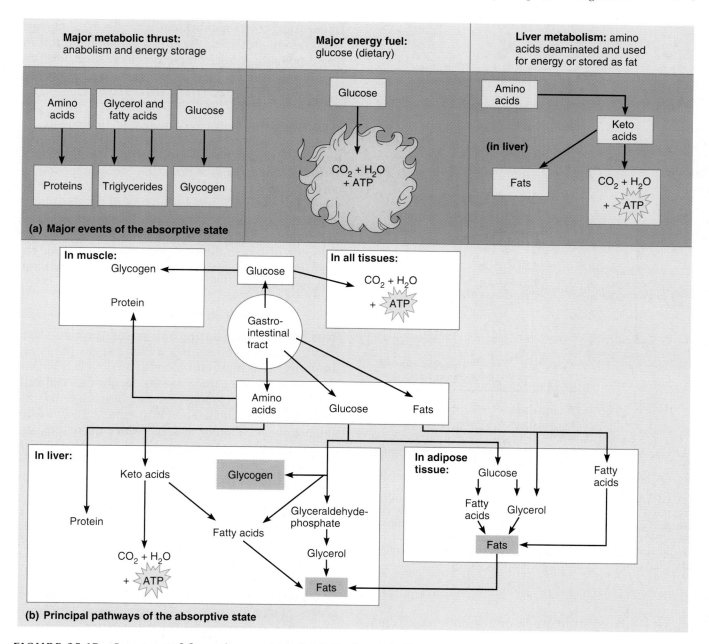

FIGURE 25.17   *Summary of the major events and principal metabolic pathways of the absorptive state.* (Although not indicated in part (b), amino acids are also taken up by tissue cells and used for protein synthesis, and fats are the primary energy fuel of muscle, liver cells, and adipose tissue.)

oxidize more fat for energy. Although some fatty acids and glycerol are used for anabolic purposes by the general population of tissue cells, most enter adipose tissue where they are reconverted to triglycerides and stored.

Amino Acids Absorbed amino acids are delivered to the liver, which deaminates some of them to keto acids. The keto acids may flow into the Krebs cycle pathway to be used for ATP synthesis, or they may be converted to liver fat stores. The liver also uses some of the amino acids to synthesize plasma pro-

teins, including albumin, clotting proteins, and transport proteins. However, most amino acids flushing through the liver sinusoids remain in the blood for uptake by other body cells, where they are used for protein synthesis.

Hormonal Control Insulin directs essentially all the events of the absorptive state (Figure 25.18). Rising blood glucose levels (above 100 mg glucose/100 ml blood) after a carbohydrate-containing meal act as a humoral stimulus that prods the beta cells of the pancreatic islets to secrete more insulin. (This

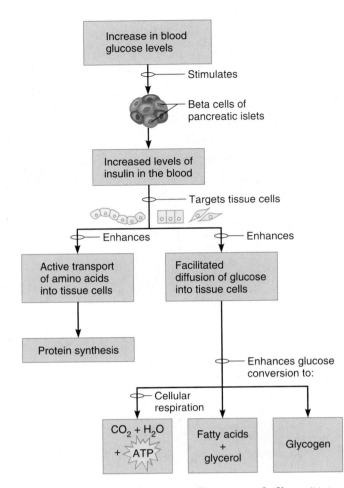

FIGURE 25.18 *Effects of insulin on metabolism.* (Note that not all the effects indicated occur in all tissue cells.) When insulin concentrations are low (i.e., during the postabsorptive period), these normal effects of insulin are inhibited and glycogenolysis and gluconeogenesis occur, that is, the insulin-mediated intracellular reactions are reversed.

glucose-induced stimulation of insulin release is enhanced by several GI tract hormones including gastrin, CCK, and secretin). A second important stimulus for insulin release is elevated amino acid levels in the blood. As insulin binds to membrane receptors of its target cells, it activates carrier-mediated facilitated diffusion of glucose into the cells. Within seconds to minutes, the rate of glucose entry into tissue cells increases 15 to 20 times. (The exception is brain cells, which actively take up glucose whether or not insulin is present.) Once glucose enters the tissue cells, insulin enhances its oxidation for energy and stimulates its conversion to glycogen and, in adipose tissue, to triglycerides. Insulin also "revs up" the active transport of amino acids into cells, promotes protein synthesis, and inhibits virtually all liver enzymes that promote gluconeogenesis.

As you can see, insulin is a **hypoglycemic** (hi"po-gli-se'mik) **hormone.** It sweeps glucose out of the blood into the tissue cells, thereby lowering blood sugar levels. Additionally, it enhances glucose oxidation or storage while simultaneously inhibiting any process that might increase blood sugar levels.

Homeostatic Imbalance

Diabetes mellitus, a consequence of inadequate insulin production or abnormal insulin receptors, is a metabolic disorder with far-reaching consequences. Without insulin or receptors that "recognize" it, glucose becomes unavailable to most body cells. Thus, blood glucose levels remain high, and large amounts of glucose are excreted in urine. Metabolic acidosis, protein wasting, and weight loss occur as large amounts of fats and tissue proteins are used for energy. (Diabetes mellitus is described in more detail in Chapter 17.) ■

Postabsorptive State

The primary goal during the postabsorptive state, that is, between meals when blood sugar levels are dropping, is to maintain blood glucose levels within the homeostatic range (80–100 mg glucose/100 ml). The importance of constant blood glucose has already been explained. The brain almost always uses glucose as its energy source. Most events of the postabsorptive state either make glucose available to the blood or spare glucose for the organs that need it most (Figure 25.19).

Sources of Blood Glucose Glucose can be obtained from stored glycogen, from tissue proteins, and in limited amounts from fats, by several routes (Table 25.4, p. 983).

1. Glycogenolysis in the liver. The liver's stores of glycogen (about 100 g) are the first line of glucose reserves. They are mobilized quickly and efficiently and can maintain blood sugar levels for about four hours during the postabsorptive period.

2. Glycogenolysis in skeletal muscle. Glycogen stores in skeletal muscle are approximately equal to liver reserves. Before liver glycogen is exhausted, glycogenolysis begins in skeletal muscle and to a lesser extent in other tissues. However, the glucose produced is not released to the blood because, unlike the liver, skeletal muscle does not have the enzymes needed to dephosphorylate glucose. Instead, glucose is partly oxidized to pyruvic acid (or, during anaerobic conditions, lactic acid), which enters the blood, is reconverted to glucose by the liver, and is released to the blood again. Thus, skeletal muscle contributes to blood sugar homeostasis indirectly, via liver mechanisms.

3. Lipolysis in adipose tissues and the liver. Adipose and liver cells produce glycerol by lipolysis, and the liver converts it to glucose (gluconeogenesis),

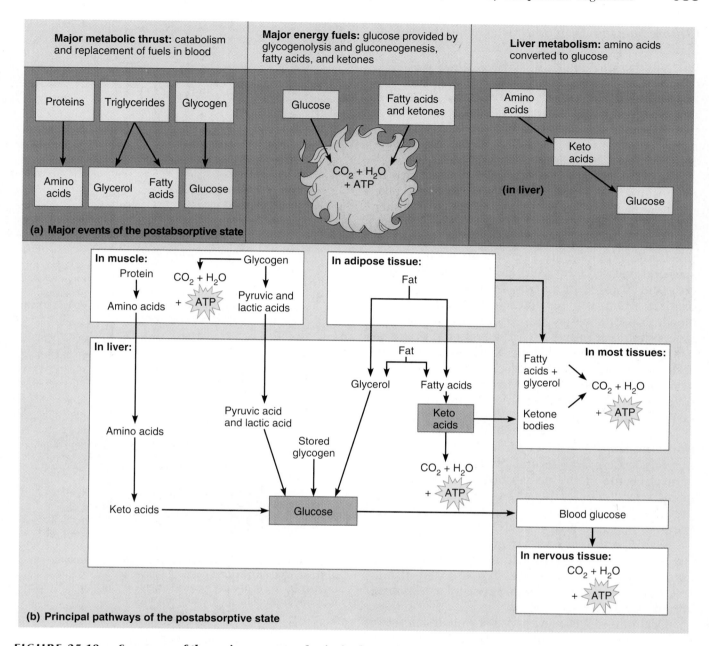

FIGURE 25.19 *Summary of the major events and principal metabolic pathways of the postabsorptive state.*

which is released to the blood. Because acetyl CoA, the beta oxidation product of fatty acids, is produced beyond the *reversible* steps of glycolysis, fatty acids *cannot* be used to bolster blood sugar levels.

4. Catabolism of cellular protein. Tissue proteins become the major source of blood glucose when fasting is prolonged and glycogen and fat stores are nearly exhausted (and during stress accompanied by enhanced glucocorticoid secretion). Cellular amino acids (mostly from muscle) are deaminated and converted to glucose in the liver. During extremely long fasts (several weeks), the kidneys also carry out glu-

coneogenesis and contribute fully as much glucose to the blood as does the liver.

Even during prolonged fasting, the body sets priorities. Muscle proteins are the first to go (to be catabolized). Movement is not nearly as important as maintaining wound healing and the immune response. But as long as life continues, so does the body's production of ATP needed to drive life processes. Obviously, there are limits to the amount of tissue protein breakdown that can occur before the body stops functioning. The heart is almost entirely muscle protein, and when it is severely catabolized, the result is death.

Glucose Sparing Even collectively, all of the manipulations to increase blood glucose are inadequate to provide energy supplies for prolonged fasting periods. Luckily, the body can adapt to burn more fats and proteins, which enter the Krebs cycle along with glucose breakdown products. The increased use of noncarbohydrate fuel molecules (especially triacylglycerols) to conserve glucose is called **glucose sparing.**

As the body progresses from the absorptive to the postabsorptive state, the brain continues to take its share of blood glucose; but virtually every other organ switches to fatty acids as its major energy source, thus sparing glucose for the brain. During this transition phase, lipolysis begins in adipose tissues and released fatty acids are picked up by tissue cells and oxidized for energy. In addition, the liver oxidizes fats to ketones and releases them into the blood for use by tissue cells. If fasting continues for longer than four or five days, the brain too begins to use large quantities of ketone bodies as well as glucose as its energy fuel (Table 25.4). The survival value of the brain's ability to use an alternative fuel source is obvious. Much less tissue protein has to be ravaged to form glucose.

Hormonal and Neural Controls The sympathetic nervous system and several hormones interact to control events of the postabsorptive state. Consequently, regulation of this phase is much more complex than that of the absorptive state when a single hormone, insulin, holds sway.

An important trigger for initiating postabsorptive events is damping of insulin release, which occurs as glucose blood levels begin to drop. As insulin levels decline, all insulin-induced cellular responses are inhibited as well.

Declining glucose levels also stimulate release of the insulin antagonist **glucagon** by the alpha cells of the pancreatic islets. Like other hormones acting during the postabsorptive state, glucagon is a *hyperglycemic hormone*, that is, it promotes a rise in blood glucose. Glucagon targets the liver and adipose tissue (Figure 25.20). The hepatocytes respond by accelerating glycogenolysis and gluconeogenesis. Adipose cells mobilize their fatty stores (lipolysis) and release fatty acids and glycerol to the blood. Thus, glucagon "refurbishes" blood energy fuels by enhancing both glucose and fatty acid levels. Glucagon release is inhibited after the next meal or whenever blood sugar levels rise and insulin secretion is turned on again.

Thus far, the picture is pretty straightforward. Increasing blood sugar levels trigger insulin release, which "pushes" glucose into the cells, and blood sugar levels decline. This stimulates secretion of glucagon, which "pulls" glucose from the cells into

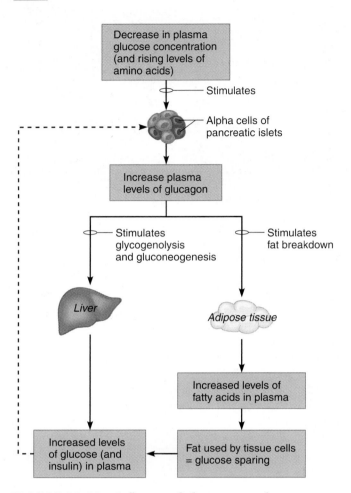

What hormone is glucagon's main antagonist?

FIGURE 25.20 **Influence of glucagon on plasma glucose concentrations.** The negative-feedback control exerted by rising plasma glucose levels on glucagon secretion is also indicated (by the dashed arrow).

the blood. However, there is more than a push-pull mechanism here because *both* insulin and glucagon release are strongly stimulated by rising levels of amino acids in the blood. This effect is insignificant when we eat a normally balanced meal, but it has an important adaptive role when we eat a high-protein, low-carbohydrate meal. In this instance, the stimulus for insulin release is strong, and if it were not counterbalanced, the brain might be damaged by the resulting abrupt onset of hypoglycemia as glucose is rushed out of the blood. Simultaneous release of glucagon modulates the effects of insulin and helps stabilize blood glucose levels.

Insulin.

TABLE 25.4	Profiles of the Major Body Organs in Fuel Metabolism		
Tissue	*Fuel stores*	*Preferred fuel*	*Fuel sources exported*
Brain	None	Glucose (ketone bodies during starvation)	None
Skeletal muscle (resting)	Glycogen	Fatty acids	None
Skeletal muscle (during exertion)	None	Glucose	Lactate
Heart muscle	None	Fatty acids	None
Adipose tissue	Triglycerides	Fatty acids	Fatty acids, glycerol
Liver	Glycogen, triglycerides	Amino acids, glucose, fatty acids	Fatty acids, glucose, ketone bodies

Adapted from Mathews, van Holde, and Ahern, 1999, *Biochemistry* 3/e, San Francisco: Addison Wesley Longman, p. 832.

The sympathetic nervous system plays a crucial role in supplying fuel quickly when blood sugar levels drop suddenly. Adipose tissue is well supplied with sympathetic fibers, and epinephrine released by the adrenal medulla in response to sympathetic activation acts on the liver, skeletal muscle, and adipose tissues. Together, these stimuli mobilize fat and promote glycogenolysis—essentially the same effects prompted by glucagon. Bodily injury, anxiety, anger, or any other stressor that mobilizes the fight-or-flight response, including declining blood glucose levels that act on central glucose receptors in the hypothalamus, will trigger this control pathway.

In addition to glucagon and epinephrine, a number of other hormones—including growth hormone, thyroxine, sex hormones, and corticosteroids—influence metabolism and nutrient flow. Growth hormone secretion is enhanced by prolonged fasting or rapid declines in plasma glucose levels, and it exerts important anti-insulin effects (see Chapter 17, p. 619). However, the release and activity of most of these hormones are not specifically related to absorptive or postabsorptive events of metabolism. A checklist of the typical metabolic effects of various hormones is provided in Table 25.5.

TABLE 25.5	Summary of Normal Hormonal Influences on Metabolism						
Hormone's effects	*Insulin*	*Glucagon*	*Epinephrine*	*Growth hormone*	*Thyroxine*	*Cortisol*	*Testosterone*
Stimulates glucose uptake by cells	✓				✓		
Stimulates amino acid uptake by cells	✓			✓			
Stimulates glucose catabolism for energy	✓				✓		
Stimulates glycogenesis	✓						
Stimulates lipogenesis and fat storage	✓						
Inhibits gluconeogenesis	✓						
Stimulates protein synthesis (anabolic)	✓			✓	✓		✓
Stimulates glycogenolysis		✓	✓				
Stimulates lipolysis and fat mobilization		✓	✓	✓	✓	✓	
Stimulates gluconeogenesis		✓	✓	✓		✓	
Stimulates protein breakdown (catabolic)						✓	

Role of the Liver in Metabolism

The anatomy of the liver and its role in bile formation and digestive activities were described in Chapter 24. Here, we will focus on the liver's metabolic functions. The liver is one of the most biochemically complex organs in the body. It processes nearly every class of nutrients and plays a major role in regulating plasma cholesterol levels. While mechanical contraptions can, in a pinch, stand in for a failed heart, lungs, or kidney, the only thing that can do liver's work is a hepatocyte.

General Metabolic Functions

The hepatocytes carry out some 500 or more intricate metabolic functions. The detailed credits that the liver deserves are well beyond the scope of this text, but a brief summary is provided next.

The liver (1) packages fatty acids into forms that can be stored or transported, (2) synthesizes plasma proteins, (3) forms nonessential amino acids and converts the ammonia resulting from their deamination to urea, a less toxic excretory product, (4) stores glucose as glycogen and regulates blood sugar homeostasis. In addition to these roles in fat, protein, and glucose metabolism, liver cells (5) store certain vitamins, (6) conserve iron salvaged from worn-out red blood cells, (7) degrade hormones, and (8) detoxify substances such as alcohol and drugs. These major metabolic functions are detailed in Table 25.6.

Cholesterol Metabolism and Regulation of Plasma Cholesterol Levels

Cholesterol, though a very important dietary lipid, has received little attention thus far, primarily because it is not used as an energy fuel. It serves instead as the structural basis of bile salts, steroid hormones, and vitamin D and as a major component of plasma membranes. Additionally, cholesterol is part of the structure of a key signaling molecule (the *hedgehog [Hh] protein*) that helps direct embryonic development. About 15% of blood cholesterol comes from the diet. The other 85% is made from acetyl CoA by the liver and to a lesser extent by other body cells, particularly intestinal cells. Cholesterol is lost from the body when it is catabolized and secreted in bile salts, which are eventually excreted in feces.

Lipoproteins and Cholesterol Transport Because triglycerides and cholesterol are completely insoluble in water, they do not circulate free in the bloodstream. Instead, they are transported to and from tissue cells in body fluids bound to small lipid-protein complexes called **lipoproteins.** These complexes solubilize the hydrophobic lipids, and their

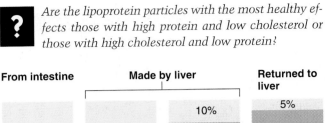

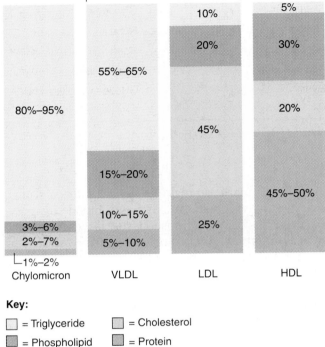

FIGURE 25.21 *Approximate composition of lipoproteins that transport lipids in body fluids.* VLDL = very low density lipoprotein; LDL = low-density lipoprotein; HDL = high-density lipoprotein.

protein parts contain signals that regulate lipid entry and exit at specific target cells.

Lipoproteins vary considerably in their relative fat-protein composition, but they all contain triglycerides, phospholipids, and cholesterol in addition to protein (Figure 25.21). In general, the higher the percentage of lipid in the lipoprotein, the lower its density; and the greater the proportion of protein, the higher its density. On this basis, there are **high-density lipoproteins (HDLs), low-density lipoproteins (LDLs),** and **very low density lipoproteins (VLDLs).** *Chylomicrons,* which transport absorbed lipids from the gastrointestinal tract, are considered to be a separate class and have the lowest density of all.

The liver is the primary source of VLDLs, which transport triglycerides made or processed in the liver to the peripheral tissues, but mostly *to adipose tissues.* Once all the triglycerides are unloaded, the

TABLE 25.6	Summary of Metabolic Functions of the Liver

Metabolic processes targeted	*Functions*
Carbohydrate metabolism Particularly important in maintaining blood glucose homeostasis	**1.** Converts galactose and fructose to glucose **2.** Glucose buffer function: stores glucose as glycogen when blood glucose levels are high; in response to hormonal controls, performs glycogenolysis and releases glucose to blood **3.** Gluconeogenesis; converts amino acids and glycerol to glucose when glycogen stores are exhausted and blood glucose levels are falling **4.** Converts glucose to fats for storage
Fat metabolism Although most cells are capable of some fat metabolism, liver bears the major responsibility	**1.** Primary body site of beta oxidation (breakdown of fatty acids to acetyl CoA) **2.** Converts excess acetyl CoA to ketone bodies for release to tissue cells **3.** Stores fats **4.** Forms lipoproteins for transport of fatty acids, fats, and cholesterol to and from tissues **5.** Synthesizes cholesterol from acetyl CoA; catabolizes cholesterol to bile salts, which are secreted in bile
Protein metabolism Other metabolic functions of the liver could be dispensed with and body could still survive; however, without liver metabolism of proteins, severe survival problems ensue: many essential clotting proteins would not be made, and ammonia would not be disposed of, for example	**1.** Deaminates amino acids (required for their conversion to glucose or use for ATP synthesis); amount of deamination that occurs outside the liver is unimportant **2.** Forms urea for removal of ammonia from body; inability to perform this function (e.g., in cirrhosis or hepatitis) results in accumulation of ammonia in blood **3.** Forms most plasma proteins (exceptions are gamma globulins and some hormones and enzymes); plasma protein depletion causes rapid mitosis of the hepatocytes and actual growth of liver, which is coupled with increase in synthesis of plasma proteins until blood values are again normal **4.** Transamination: intraconversion of nonessential amino acids; amount that occurs outside liver is inconsequential
Vitamin/mineral storage	**1.** Stores vitamin A (1–2 years' supply) **2.** Stores sizable amounts of vitamins D and B_{12} (1–4 months' supply) **3.** Stores iron; other than iron bound to hemoglobin, most of body's supply is stored in liver as ferritin until needed; releases iron to blood as blood levels drop
Biotransformation functions	**1.** Relative to drug metabolism, performs synthetic reactions which yield inactive products that can be secreted by the kidneys and nonsynthetic reactions which may result in products that are more active, changed in activity, or less active **2.** Processes bilirubin resulting from RBC breakdown and excretes bile pigments in bile **3.** Metabolizes bloodborne hormones to forms that can be excreted in urine

residues of these VLDLs are converted into LDLs, which are cholesterol-rich. The role of the LDLs is to transport cholesterol *to the peripheral tissues,* making it available to the tissue cells for membrane or hormone synthesis and for storage for later use. The LDLs also regulate cholesterol synthesis in the tissue cells. Docking of LDL to the LDL receptor triggers endocytosis of the entire particle.

The major function of HDLs, which are particularly rich in phospholipids and cholesterol, is to transport excess cholesterol *from peripheral tissues* (which do not have the ability to degrade or excrete HDL) *to the liver,* where it is broken down and becomes part of bile. The liver makes the protein envelopes of the HDL particles and then ejects them into the bloodstream in collapsed form, rather like

deflated beach balls. Once in the circulation, these still-incomplete HDL particles fill with cholesterol picked up from the tissue cells and "pulled" from the artery walls. HDL also services the needs of steroid-producing organs, like the ovaries and adrenal glands, for their raw material (cholesterol). These organs have the ability to selectively remove cholesterol from the HDL particles without engulfing them.

High plasma levels of cholesterol (above 200 mg cholesterol/100 ml blood) have been linked to risk of atherosclerosis, which clogs the arteries and causes strokes and heart attacks. However, it is not enough to simply measure total cholesterol. The form in which cholesterol is transported in the blood is more important clinically. As a rule, high levels of HDLs are considered *good* because the transported cholesterol is destined for degradation. An HDL level between 35 and 60 is considered okay; levels above 60 are thought to protect against heart disease. High LDL levels (160 or above) are considered *bad* because when LDLs are excessive, potentially lethal cholesterol deposits are laid down in the artery walls.

If LDLs are "bad" cholesterol, then one variety of LDL—lipoprotein (a)—is "really nasty." This lipid appears to promote plaque formation that thickens and stiffens the blood vessel walls. High levels of lipoprotein (a) in the blood can double a man's risk of a heart attack before the age of 55. It is estimated that one in five males has elevated levels of this lipoprotein in the blood. (See *A Closer Look* in Chapters 19 and 20 for more information.)

Factors Regulating Plasma Cholesterol Levels A negative feedback loop partially adjusts the amount of cholesterol produced by the liver according to cholesterol in the diet. A high cholesterol intake inhibits its synthesis by the liver, but it is not a one-to-one relationship because the liver produces a certain basal amount of cholesterol even when dietary intake is excessive. For this reason, severe restriction of dietary cholesterol, although helpful, does not lead to a steep reduction in plasma cholesterol levels.

The relative amounts of saturated and unsaturated fatty acids in the diet have an important effect on plasma cholesterol levels. Saturated fatty acids *stimulate liver synthesis* of cholesterol and *inhibit its excretion* from the body. Thus, moderate decreases in intake of saturated fats (found chiefly in animal fats and coconut oil) can reduce plasma cholesterol levels as much as 15–20%. In contrast, unsaturated fatty acids (found in most vegetable oils) *enhance excretion* of cholesterol and its catabolism to bile salts, thereby reducing total plasma cholesterol levels. The unhappy exception to this good news about unsaturated fats concerns "healthy" oils that have been hardened by hydrogenation to make them more solid. The hydrogenation process changes their fatty acids into the so-called *trans fatty acids*, which cause serum changes worse than those caused by saturated fats. That is, they spark a greater increase in LDLs and a greater reduction in HDLs, thus producing the unhealthiest ratio of total cholesterol to HDL.

In addition, certain unsaturated fatty acids (omega-3 fatty acids) found in especially large amounts in some cold-water fish lower the proportions of both saturated fats and cholesterol. The omega-3 fatty acids have a powerful antiarrhythmic effect on the heart and also make blood platelets less sticky, thus helping to prevent spontaneous clotting

that can block blood vessels and lower blood pressure (even in nonhypertensive people). In those with moderate to high cholesterol levels, replacing half of the lipid and cholesterol rich animal proteins in the diet with soy protein appears to lower cholesterol significantly (10–25%).

Factors other than diet also influence plasma cholesterol levels. For example, cigarette smoking, coffee drinking, and stress appear to increase LDL levels, whereas regular aerobic exercise lowers LDL levels and increases levels of HDLs. Interestingly, body shape, that is, the distribution of body fat, provides clues to risky blood levels of cholesterol and fats. "Apples" (people with upper body and abdominal fat distribution, seen more often in men) tend to have higher levels of cholesterol in general and LDLs in particular than "pears" (fat localized in the hips and thighs, a pattern more common in women).

Most cells other than liver and intestinal cells obtain the bulk of the cholesterol they need for membrane synthesis from the bloodstream. When a cell needs cholesterol, it makes receptor proteins for LDL and inserts them in its plasma membrane. LDL binds to the receptors and is engulfed in coated pits by endocytosis. Within 10 to 15 minutes the endocytotic vesicles fuse with lysosomes, where the cholesterol is freed for use. When excessive cholesterol accumulates within a cell, it inhibits both the cell's own cholesterol synthesis and its synthesis of LDL receptors.

Homeostatic Imbalance

High cholesterol levels undoubtedly put us at risk for cardiovascular disease and heart attack, and cholesterol-lowering drugs such as the *statins* (e.g., lovastatin and pravastatin) are routinely prescribed for cardiac patients with high cholesterol. Now, however, one or more studies have indicated that cholesterol levels below 190 for males and 178 for females may be equally devastating because they may enhance the risk of "bleeding" strokes and death from cerebral hemorrhage. Additionally, almost half of those who get heart disease have normal cholesterol levels, while others with poor lipid profiles remain disease free. Although most Americans would probably benefit from reducing their intake of saturated fats and cholesterol-rich foods, at least for now, the moderate approach to cholesterol control may be the wisest. ■

BODY ENERGY BALANCE

When any fuel is burned, it consumes oxygen and liberates heat. The "burning" of food fuels by our cells is no exception. As described in Chapter 2, energy can be neither created nor destroyed—only converted from one form to another. If we apply this principle (actually the *first law of thermodynamics*) to cell metabolism, it means that bond energy released as foods are catabolized must be precisely balanced by the total energy output of the body. Thus, a dynamic balance exists between the body's energy intake and its energy output:

Energy intake = total energy output
(heat + work + energy storage)

Energy intake is considered equal to the energy liberated during food oxidation. Undigested foods are not part of the equation because they contribute no energy. **Energy output** includes the energy (1) immediately lost as heat (about 60% of the total), (2) used to do work (driven by ATP), and (3) stored in the form of fat or glycogen. (Because losses of organic molecules in urine, feces, and perspiration are very small in healthy people, they are usually ignored.) A close look at this situation reveals that *nearly all the energy derived from foodstuffs is eventually converted to heat.* Heat is lost during every cellular activity—when ATP bonds are formed and when they are cleaved to do work, as muscles contract, and through friction as blood flows through blood vessels. Though the cells cannot use this energy to do work, the heat warms the tissues and blood and helps maintain the homeostatic body temperature that allows metabolic reactions to occur efficiently. Energy storage is an important part of the equation only during periods of growth and net fat deposit.

Regulation of Food Intake

When energy intake and energy outflow are balanced, body weight remains stable. When they are not, weight is either gained or lost. Body weight in most people is surprisingly stable. This fact indicates the existence of physiological mechanisms that control food intake (and hence, the amount of food oxidized) or heat production or both.

Control of food intake poses difficult questions to researchers. For example, what type of receptor could sense the body's total content of calories and alert one to start eating or to put down that fork? Despite heroic research efforts, no such signal or single receptor type has been found.

It has been known for some time that the hypothalamus releases several peptides that influence feeding behavior. For example, a pair of peptides called *orexins* are powerful appetite enhancers, *neuropeptide Y* causes us to crave carbohydrates, and *galanin* produces a yen for fats, while *GLP-1* (glucagon-like peptide) and *serotonin* make us feel full and satisfied. Current theories of how feeding behavior and hunger is regulated focus on one or more of five factors: neural signals from the digestive tract, bloodborne signals related to body energy stores, hormones, body temperature, and psychological factors. All these factors appear to operate through feedback signals to the feeding centers of the brain. Brain receptors include thermoreceptors, chemoreceptors (for glucose, insulin, and others), and receptors that respond to a number of peptides (leptin, neuropeptide Y, and others). Although the hypothalamic nuclei play an essential role in regulating hunger and satiety, brain stem areas are also involved. Sensors in peripheral locations have also been suggested, with the liver and gut itself (alimentary canal) as the prime candidates.

Neural Signals from the Digestive Tract

A novel method by which the brain can evaluate the contents of the gut depends on vagal nerve fibers that carry on a two-way conversation between the gut and brain. Vagal afferents transmit distinctively different patterns of neural impulses in response to the presence of carbohydrates and proteins in the alimentary canal. For example, clinical tests show that two calories' worth of protein produce a 30–40% larger and longer response in vagal afferents than do two calories' worth of glucose. Using these signals, together with others that it receives (see below), the brain can decode what is eaten and how much.

Nutrient Signals Related to Energy Stores

At any time, plasma levels of glucose, amino acids, and fatty acids provide information to the brain that may help adjust energy intake to energy output. For example:

1. When we eat, plasma glucose levels rise and cellular metabolism of glucose increases. Subsequent activation of glucose receptors in the brain ultimately depresses eating. During fasting, this signal would be absent, resulting in hunger and a "turning on" of food-seeking behaviors.

2. Elevated plasma levels of amino acids depress eating, whereas low amino acid levels in blood stimulate it, but the precise mechanism mediating these effects is unknown.

3. Blood concentrations of fatty acids and of *leptin* ("thin"), a peptide released by adipose cells that circulates at levels related to fat reserves, serve as indicators of the body's total energy stores (in adipose tissue) and provide a mechanism for controlling hunger. According to this theory, the greater the fat reserves, the larger the basal amount of fatty acids and leptin released to the blood and the more eating behavior is inhibited.

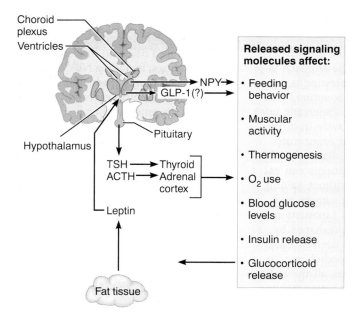

FIGURE 25.22 *Hypothetical scheme of controls of feeding behavior and satiety.*

The overall satiety signal appears to be **leptin,** which is secreted over a period of hours exclusively by fat tissue in response to an increase in fatty mass in the body. Blood levels of glucocorticoids and insulin play a role in regulating leptin release (inhibit it), but the precise mechanism is yet to be worked out.

Leptin binds to receptors in the choroid plexuses of the ventricles, where it gains entry to the brain. There, leptin acts on the hypothalamus to regulate the amount of body fat via controls of appetite and energy output. Its main target of action is the ventromedial hypothalamus, where it suppresses the secretion of neuropeptide Y (NPY), the most potent appetite stimulant known. Hence, leptin decreases food intake and cranks up activity and heat production. Whether the NPY receptors are the final targets of leptin is not known. The hypothalamus also has receptors for GLP-1, a powerful inhibitor of feeding behavior, and some believe that leptin may also act through GLP-1. So far we only have bits and pieces of this story.

Hormones

Insulin, released during food absorption, depresses hunger and is presumed to be an important satiety signal. In contrast, glucagon levels rise during fasting and stimulate hunger. Other hormonal controls include epinephrine (released during fasting), which triggers hunger, and cholecystokinin, an intestinal hormone secreted during food digestion, which depresses it.

Body Temperature

Increased body temperature may inhibit eating behavior. Such a thermal signal would help to explain why people in cold climates normally eat more than those in more temperate or warm regions.

Psychological Factors

Although the mechanisms described thus far are reflexive, their final outcome may be enhanced or inhibited by psychological factors that have little to do with caloric balance, such as just the thought of food. Psychological factors are thought to be very important in the obese. However, even when psychological factors are the underlying cause of obesity, individuals do *not* continue to gain weight endlessly. Their feeding controls still operate, but they have a higher "set point," which maintains total energy content at higher-than-normal levels.

Hypothetical Model: Interrelationship of Factors in Control of Food Intake

Whew! How does all of this information fit together? The best current hypothesis based on animal studies is shown in Figure 25.22.

Metabolic Rate and Body Heat Production

The body's rate of energy output (usually expressed per hour) is called the **metabolic rate.** Metabolic rate is the total heat produced by all the chemical reactions and mechanical work of the body. It can be measured directly or indirectly. In the *direct method,* the person enters a chamber called a **calorimeter** and heat liberated by the body is absorbed by water circulating around the chamber. The rise in the water's temperature is directly related to the heat produced by the person's body. The *indirect method* uses a **respirometer** (Figure 25.23) to measure oxygen consumption, which is directly proportional to heat production. For each liter of oxygen used, the body produces about 4.8 kcal of heat.

Because many factors influence metabolic rate, it is usually measured under standardized conditions. The person is in a postabsorptive state (has not eaten for at least 12 hours), is reclining, and is mentally and physically relaxed. The temperature of the room is a comfortable 20–25°C. The measurement obtained under these circumstances is called the **basal metabolic rate (BMR)** and reflects the energy the body needs to perform only its most essential activities, such as breathing and maintaining resting levels of organ function. The BMR, often referred to as the "energy cost of living," is reported in kilocalories per square meter of body surface per hour (kcal/m^2/h).

An average 70-kg (154-pound) adult has a BMR of approximately 60–72 kcal/h. A quick approximation of one's BMR can be calculated by multiplying weight in kilograms (note that 2.2 pounds = 1 kg) times the factor 1 for males or the factor 0.9 for females. Although named the *basal* metabolic rate, this measurement is *not* the lowest metabolic state of the body. That situation occurs during sleep, when the skeletal muscles are completely relaxed.

The factors that influence BMR include surface area, age, gender, stress, and hormones. Although BMR is related to overall weight and size, the critical factor is surface area rather than weight itself. This reflects the fact that as the ratio of body surface area to body volume increases, heat loss to the environment increases and the metabolic rate must be higher to replace the lost heat. Hence, if two people weigh the same, the taller or thinner person will have a higher BMR than the shorter or fatter person.

In general, the younger a person, the higher the BMR. Children and adolescents require large amounts of energy for growth. In old age, BMR declines dramatically as skeletal muscles begin to atrophy. (This helps explain why those elderly who fail to reduce their caloric intake gain weight.) Gender also plays a role. Metabolic rate is disproportionately higher in males than in females. Males typically have more muscle, which is very active metabolically even during rest. Fatty tissue, present in greater relative amounts in females, is metabolically sluggish compared to muscle.

Body temperature and metabolic rate tend to rise and fall together. Fever (hyperthermia), resulting from infections and other factors, results in a markedly higher metabolic rate. Stress, whether physical or emotional, increases metabolic rate by mobilizing the sympathetic nervous system. Blood-borne norepinephrine and epinephrine (released by adrenal medullary cells) provoke a rise in metabolic rate primarily by stimulating fat catabolism.

The amount of **thyroxine** produced by the thyroid gland is probably the most important hormonal factor in determining BMR; hence, thyroxine has been dubbed the "metabolic hormone." Its direct effect on most body cells (except brain cells) is to increase oxygen consumption, presumably by accelerating the use of ATP to operate the sodium-potassium pump. As ATP reserves decline, cellular respiration accelerates. Thus, the more thyroxine produced, the higher the BMR.

Homeostatic Imbalance

Hyperthyroidism causes a host of problems resulting from the excessive metabolic rate it produces. The body catabolizes stored fats and tissue proteins, and, despite increased hunger and food intake, the person often loses weight. Bones weaken and body muscles, including the heart, begin to atrophy. In contrast, *hypothyroidism* results in slowed metabolism, obesity, and diminished thought processes. ∎

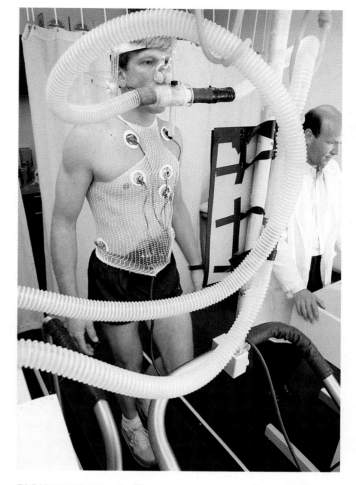

FIGURE 25.23 Indirect measurement of metabolic rate by respirometry. The indirect method for measuring metabolic rate is based on the fact that oxygen use and heat liberation during food oxidation are directly proportional. The person breathes into a respirometer, and the total amount of oxygen consumed during testing (in this case, during walking on a treadmill) is measured. The average amount of oxygen consumed per hour (L/h) is multiplied by 4.83 (the amount of energy liberated when 1 L of oxygen is consumed in the oxidation of protein, carbohydrate, or fat) to calculate the metabolic rate. If, for example, oxygen use is 16 L/h, the metabolic rate is 77.28 kcal/h (16 L/h = 4.83 kcal/L).

The term **total metabolic rate (TMR)** refers to the total rate of kilocalorie consumption to fuel *all* ongoing activities—involuntary and voluntary. The BMR accounts for a surprisingly large part of the TMR. For example, a woman whose total energy needs per day are about 2000 kcal may spend well over half of these calories (1400 kcal or so) supporting vital body activities. Skeletal muscle activity causes the most dramatic short-term changes in TMR, reflecting the fact that skeletal muscles make up nearly half of body mass. Even slight increases in muscular work can cause remarkable leaps in TMR and heat production. When a well-trained athlete exercises vigorously for several minutes, the metabolic

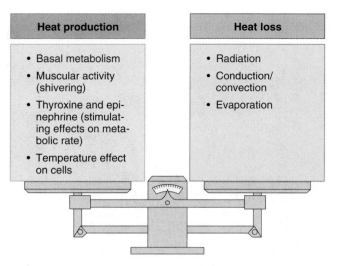

Heat production	Heat loss
• Basal metabolism • Muscular activity (shivering) • Thyroxine and epinephrine (stimulating effects on metabolic rate) • Temperature effect on cells	• Radiation • Conduction/convection • Evaporation

FIGURE 25.24 *As long as heat production and heat loss are properly balanced, body temperature remains constant.* Factors contributing to heat production (and temperature rise) are shown on the left side of the scale; those contributing to heat loss (and temperature fall) are shown on the right side of the scale.

rate may increase to 15 to 20 times normal, and it remains elevated for several hours afterward. Surprisingly, physical training has little effect on BMR. Although it would appear that athletes, especially those with greatly enhanced muscle mass, should have much higher BMRs than nonathletes, there is little difference between those of the same sex and surface area. Food ingestion also induces a rapid increase in the total metabolic rate. This effect, called **dietary,** or **food-induced, thermogenesis,** is greatest when proteins are eaten. The heightened metabolic activity of the liver during such periods probably accounts for the bulk of additional energy use. In contrast, fasting or very low caloric intake depresses metabolic rate and results in a slower breakdown of body reserves.

Regulation of Body Temperature

As shown in Figure 25.24, body temperature represents the balance between heat production and heat loss. Although all body tissues produce heat, those most active metabolically produce the greatest amounts. When the body is at rest, most heat is generated by the liver, heart, brain, and endocrine organs. Inactive skeletal muscles account for 20–30% of body heat. This situation changes dramatically with even slight alterations in muscle tone; and during vigorous exercise, heat production by the skeletal muscles can be 30 to 40 times that of the rest of the body. Clearly, a change in muscle activity is one of the most important means of modifying body temperature.

Body temperature averages 36.2°C (98.2°F) and is usually maintained within a narrow range of 35.6–37.8°C (96–100°F), despite considerable change in external (air) temperature. A healthy individual's body temperature fluctuates approximately 1°C (1.8°F) in 24 hours, with a low occurring in the early morning and a high in the late afternoon or early evening.

The adaptive value of this precise temperature homeostasis becomes obvious when we consider the effect of temperature on the rate of biochemical reactions, specifically enzyme activity. At normal body temperature, conditions are optimal for enzymatic activity. As body temperature rises, catalysis is accelerated: With each rise of 1°C, the rate of chemical reactions is about 10% faster. As the temperature spikes upward beyond the homeostatic range, neurons are depressed and proteins begin to denature (degrade). Most adults go into convulsions when their temperature reaches 41°C (106°F), and 43°C (about 110°F) appears to be the absolute limit for life. In contrast, most body tissues can withstand marked reductions in temperature as long as other conditions are carefully controlled. This fact underlies the use of body cooling during open heart surgery when the heart must be stopped. Low body temperature reduces metabolic rate (and consequently nutrient requirements of body tissues and the heart), allowing more time for surgery without incurring tissue damage.

Core and Shell Temperatures

Different body regions have different temperatures at rest. The body's **core** (that is, organs within the skull and the thoracic and abdominal cavities) has the highest temperature. Its **shell,** or heat loss surface, essentially the skin, has the lowest temperature in most circumstances. Of the two body sites used routinely to obtain body temperature clinically, the rectum typically has a temperature about 0.4°C (0.7°F) higher than the mouth and is a better indicator of core temperature.

It is core temperature that is precisely regulated. Blood serves as the major *agent of heat transfer* or *exchange* between the body core and shell. Whenever the shell is warmer than the external environment, heat is lost from the body. Thus, when heat must be dissipated, warm blood is allowed to flush into the skin capillaries. On the other hand, when it is necessary to conserve heat, blood largely bypasses the skin. This reduces heat loss and allows the shell temperature to fall toward that of the environment. Thus, while the core temperature stays relatively constant, the temperature of the shell may fluctuate substantially (between 20°C [68°F] and 40°C [104°F] for instance) as its "thickness" is adapted to changes in body activity and external temperature. (You really *can* have cold hands and a warm heart.)

Mechanisms of Heat Exchange

The physical mechanisms that govern heat exchange between our skin and the external environment are the same as those controlling heat transfer between inanimate objects. It helps to think of the temperature of an object—whether that object is the skin or a radiator—as a guide to its heat content (*think* "heat concentration"). Then just remember that heat always flows down its concentration gradient from warmer to cooler regions. The body uses four mechanisms of heat transfer—radiation, conduction, convection, and evaporation (Figure 25.25).

Radiation Radiation is the loss of heat in the form of infrared waves (thermal energy). Any dense object that is warmer than objects in its environment—for example, a radiator and (usually) the body—will transfer heat to those objects. Under normal conditions, close to half of body heat loss occurs by radiation.

Because the direction of radiant energy flow is always from warmer to cooler, radiation explains why an initially cold room warms up shortly after it is filled with people (the "body heat furnace," so to speak). The body also gains heat by radiation, as demonstrated by the warming of the skin during sunbathing.

Conduction and Convection Conduction is the transfer of heat between objects that are in direct contact with each other. For example, when we step into a hot tub, some of the heat of the water is transferred to our skin, and warm buttocks transfer heat to the seat of a chair by conduction. Unlike radiation, conduction requires molecule-to-molecule *contact* of objects; that is, the thermal energy must move through a material medium.

When the body shell transfers heat to the surrounding air, convection also comes into play. Because warm air tends to expand and rise and cool air (being denser) falls, the warmed air enveloping the body is continually being replaced by cooler air molecules. This process, called **convection**, substantially enhances heat exchange from the body surface to the air because the cooler air absorbs heat by conduction more rapidly than the already-warmed air. Together, conduction and convection account for 15–20% of heat loss to the environment. These processes are enhanced by anything that moves air more rapidly across the body surface such as wind or a fan, that is, by *forced convection*.

Evaporation Water evaporates because its molecules absorb heat from the environment and become energetic enough (that is, vibrate fast enough) to escape as gas (water vapor). The heat absorbed by water during evaporation is called **heat of vaporiza-**

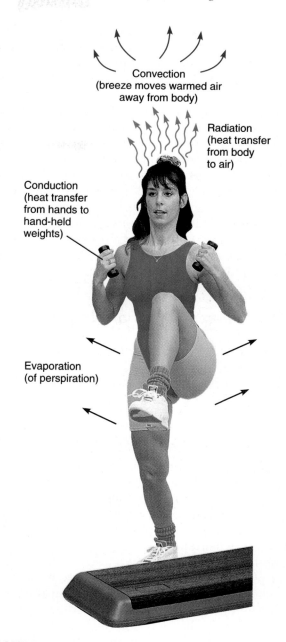

FIGURE 25.25 *Mechanisms of heat exchange between the body and the external environment.*

Convection
(breeze moves warmed air away from body)

Radiation
(heat transfer from body to air)

Conduction
(heat transfer from hands to hand-held weights)

Evaporation
(of perspiration)

tion. Because water absorbs a great deal of heat before vaporizing, its evaporation from body surfaces removes large amounts of body heat. For every gram of water evaporated, about one-half kilocalorie (0.58 kcal) of heat is removed from the body.

There is a basal level of body heat loss due to the continuous evaporation of water from the lungs, from the mucosa of the mouth, and through the skin. The unnoticeable water loss occurring via these routes is called **insensible water loss** and the heat loss that accompanies it is called **insensible heat loss.** Insensible heat loss dissipates about 10%

of the basal heat production of the body and is a constant; that is, it is not subject to body temperature controls. However, the control mechanisms do initiate heat-promoting activities to counterbalance this heat loss when necessary.

Evaporative heat loss becomes an active (sensible) process when body temperature rises and sweating produces increased amounts of water for vaporization. Extreme emotional states activate the sympathetic nervous system, causing body temperature to rise one degree or so, and vigorous exercise can thrust body temperature upward as much as 2–3°C (5–6°F). During vigorous muscular activity, 1–2 L/h of perspiration can be produced and evaporated, removing 600–1200 kcal of heat from the body each hour. This is over 30 times the amount of heat lost via insensible heat loss!

Homeostatic Imbalance

When sweating is heavy and prolonged, especially in untrained individuals, losses of water and salt (NaCl) may cause painful spasms of the skeletal muscles called *heat cramps*. This situation is easily rectified by ingesting fluids. ■

Role of the Hypothalamus

Although other brain regions contribute, the hypothalamus, particularly its *preoptic* region, is the main integrating center for thermoregulation. Together the **heat-loss center** (located more anteriorly) and the **heat-promoting center** comprise the **thermoregulatory centers.**

The hypothalamus receives afferent input from (1) **peripheral thermoreceptors,** which are located in the shell (the skin), and (2) **central thermoreceptors** (receptors sensitive to the temperature of blood), located in the body core including the anterior portion of the hypothalamus itself. Much like a thermostat, the hypothalamus responds to this input by reflexively initiating appropriate heat-promoting or heat-loss activities via autonomic effector pathways. Although the central thermoreceptors are more critically located than the peripheral ones, varying inputs from the shell probably alert the hypothalamus of its need to act to prevent temperature changes in the core, that is, they allow the hypothalamus to anticipate possible changes to be made.

Heat-Promoting Mechanisms

When the external temperature is cold (or the temperature of circulating blood falls), the hypothalamic heat-promoting center is activated. It, in turn, triggers one or more of the following mechanisms to maintain or increase core body temperature (Figure 25.26).

1. Vasoconstriction of cutaneous blood vessels. Activation of the sympathetic vasoconstrictor fibers serving the blood vessels of the skin causes strong vasoconstriction. As a result, blood is restricted to deep body areas and largely bypasses the skin. Because the skin is separated from deeper organs by a layer of insulating subcutaneous (fatty) tissue, heat loss from the shell is dramatically reduced and shell temperature drops toward that of the external environment.

Homeostatic Imbalance

Restricting blood flow to the skin is not a problem for a brief period, but if it is extended (as during prolonged exposure to very cold weather), skin cells deprived of oxygen and nutrients begin to die. This extremely serious condition is called *frostbite*. ■

2. Increase in metabolic rate. Cold stimulates the release of norepinephrine (NE) by sympathetic nerve fibers. NE elevates the metabolic rate, which enhances heat production. This mechanism is called **chemical,** or **nonshivering, thermogenesis.**

3. Shivering. If the mechanisms described above are not enough to handle the situation, shivering is triggered. Brain centers controlling muscle tone are activated and when muscle tone reaches sufficient levels to alternately stimulate stretch receptors in antagonistic muscles, involuntary shuddering contractions of the skeletal muscles begin. Shivering is very effective in increasing body temperature because muscle activity produces large amounts of heat.

4. Enhanced thyroxine release. When environmental temperature decreases gradually, as in the transition from a hot to a cold season, the hypothalamus releases *thyrotropin-releasing hormone.* This activates the anterior pituitary to release *thyroid-stimulating hormone,* which, in turn, induces the thyroid gland to liberate larger amounts of thyroid hormone (T_3 and T_4) to the blood. Because thyroid hormone increases metabolic rate, body heat production increases, allowing us to maintain a constant body temperature in cold environmental conditions.

Beside these involuntary adjustments, we humans make a number of *behavioral modifications* to prevent overcooling of our body core. These include:

■ Putting on more or warmer clothing to restrict heat loss (a hat, gloves, and "insulated" outer garments)

■ Drinking hot fluids

■ Changing posture to reduce exposed body surface area (hunching over or clasping the arms across the chest)

■ Increasing physical activity to generate more heat (jumping up and down, clapping the hands)

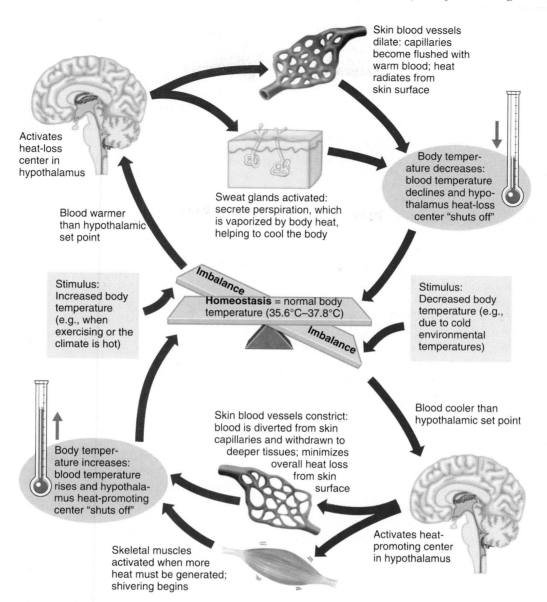

FIGURE 25.26 *Mechanisms of body temperature regulation.*

Heat-Loss Mechanisms

Heat-loss mechanisms protect the body from excessively high temperatures, which can be particularly damaging to the body. Most heat loss occurs through the skin via the physical mechanisms of heat exchange described earlier—radiation, conduction, convection, and evaporation. How do the heat-exchange mechanisms fit into the heat-loss temperature regulation scheme? The answer is quite simple. Whenever core body temperature rises above normal, the hypothalamic heat-promoting center is inhibited. At the same time, the heat-loss center is activated and so triggers one or both of the following (Figure 25.26):

1. Vasodilation of cutaneous blood vessels. Inhibiting the vasomotor fibers serving blood vessels of the skin allows the vessels to dilate. As the blood vessels of the skin swell with warm blood, heat is lost from the shell by radiation, conduction, and convection.

2. Enhanced sweating. If the body is extremely overheated or if the environment is so hot (over 61°C [about 92°F]) that heat cannot be lost by other means, increased heat loss by evaporation becomes necessary. The sweat glands are strongly activated by sympathetic fibers and spew out large amounts of perspiration. Evaporation of perspiration is an efficient means of ridding the body of surplus heat as long as the air is dry. However, when the relative humidity is high, evaporation occurs much more slowly. In such cases, the heat-liberating mechanisms cannot work well, and we feel miserable and irritable.

Behavioral or voluntary measures commonly taken to reduce body heat include:

- Reducing activity ("laying low")

- Seeking a cooler environment (a shady spot) or using a device to increase convection (a fan) or cooling (an air conditioner)

- Wearing light-colored, loosely fitting clothing that reflects radiant energy and reduces heat gain (this is actually cooler than being nude because bare skin absorbs most radiant energy that strikes it)

Homeostatic Imbalance

When normal heat loss processes become ineffective, the **hyperthermia,** or elevated body temperature, that ensues depresses the hypothalamus. As a result, heat-control mechanisms are suspended, creating a vicious positive-feedback cycle. Sharply increasing temperatures increase the metabolic rate, which in turn increases heat production. The skin becomes hot and dry and, as the temperature continues to spiral upward, multiple organ (including brain) damage becomes a distinct possibility. This condition, called **heat stroke,** can be fatal unless corrective measures are initiated immediately (e.g., effective body cooling by immersion in cool water and administration of fluids).

The terms *heat exhaustion* and *exertion-induced heat exhaustion* are often used to describe the heat-associated collapse of an individual during or following vigorous physical activity. This condition, evidenced by elevated body temperature and mental confusion and/or fainting, is due to dehydration and consequent low blood pressure. In contrast to heat stroke, heat-loss mechanisms are still functional in heat exhaustion; indeed, the symptoms are a direct consequence of those mechanisms. However, heat exhaustion can rapidly progress to heat stroke if the body is not cooled and rehydrated promptly. ■

Fever

Fever is *controlled hyperthermia.* Most often, it results from infection somewhere in the body, but it may be caused by other conditions (cancer, allergic reactions, central nervous system injuries). White blood cells, injured tissue cells, and macrophages release *pyrogens* (literally, fire starters), chemical substances now known to be cytokines, that act on the hypothalamus, causing release of prostaglandins.

The prostaglandins, in turn, reset the hypothalamic thermostat to a higher temperature, causing the body to initiate heat-promoting mechanisms. As a result of vasoconstriction, heat loss from the body surface declines, the skin cools, and shivering begins to generate heat. This situation, called "the chills," is a sure sign that body temperature is rising. The temperature rises until it reaches the new setting, and then body temperature is maintained at the "fever setting" until natural body defenses or antibiotics reverse the disease process. The thermostat is then reset to a lower (or normal) level, which causes heat-loss mechanisms to swing into action. Sweating begins and the skin becomes flushed and warm. Physicians have long recognized these signs as signals that body temperature is falling.

As explained in Chapter 22, fever, by increasing the metabolic rate, helps speed the various healing processes, and it also appears to inhibit bacterial growth. The danger of fever is that if the body thermostat is set too high, proteins may be denatured and permanent brain damage may occur.

DEVELOPMENTAL ASPECTS OF NUTRITION AND METABOLISM

Good nutrition is essential *in utero,* as well as throughout life. If the mother is ill nourished, the development of her infant is affected. Most serious is the lack of proteins needed for fetal tissue growth, especially brain growth. Additionally, because brain growth continues for the first three years after birth, inadequate calories and proteins during this time will lead to mental deficits or learning disorders. Proteins are needed for muscle and bone growth, and calcium is required for strong bones. Although anabolic processes are less critical after growth is completed, sufficient nutrients of all categories are still essential to maintain normal tissue replacement and metabolism.

Homeostatic Imbalance

There are many inborn errors of metabolism (or genetic disorders), but perhaps the two most common are *cystic fibrosis* and *phenylketonuria* (fen"il-ke"to-nu're-ah) *(PKU).* Cystic fibrosis is described in Chapter 23. In PKU, the tissue cells are unable to use one particular amino acid, phenylalanine (fen"il-al'ah-nīn), which is present in all protein foods. The defect involves a deficiency of the enzyme that converts phenylalanine to tyrosine. Because phenylalanine cannot be metabolized, it and its deaminated products accumulate in the blood and act as neurotoxins that cause brain damage and retardation within a few months. These consequences are uncommon today because most states require a simple urine or blood screening test to identify affected newborns, and these children are put on a special low-phenylalanine diet. When this special diet should be terminated is controversial. Some believe it should be continued through adolescence; others order it discontinued when the child is 6 years old. Because tyrosine is the basis of melanin, all of these children are very blond and fair-skinned.

There are a number of other enzyme defects, classified according to the impaired process as carbohydrate, lipid, or mineral metabolic disorders. For example, two carbohydrate deficits are galactosemia and glycogen storage disease. *Galactosemia* results from an abnormality or lack of liver enzymes needed to transform galactose to glucose. Galactose accumulates in the blood and leads to mental deficits. In *glycogen storage disease,* glycogen synthesis is normal, but one of the enzymes needed to convert glycogen back to glucose is missing. As excessive amounts of glycogen are stored, its storage organs (liver and skeletal muscles) become glutted with glycogen and enlarge tremendously. ■

With the exception of *insulin-dependent diabetes mellitus,* children free of genetic disorders rarely exhibit metabolic problems. However, by middle age and particularly old age, *non-insulin-dependent diabetes mellitus* becomes a major problem, particularly in the obese. (Diabetes mellitus is described in Chapter 17.)

Metabolic rate progressively declines throughout the life span. In old age, muscle and bone wasting and declining efficiency of the endocrine system take their toll. Because many elderly are also less active, the metabolic rate is sometimes so low that it becomes nearly impossible to obtain adequate nutrition without gaining weight. The elderly also use more medications than any other group, at a time of life when the liver has become less efficient in its detoxifying duties. Consequently, many of the agents prescribed for age-related medical problems influence nutrition. For example:

■ Some diuretics prescribed for congestive heart failure or hypertension (to flush fluids out of the body)

can cause severe hypokalemia by promoting the excessive loss of potassium.

■ Some antibiotics, for example, sulfa drugs, tetracycline, and penicillin, interfere with food digestion and absorption. They may also cause diarrhea, thereby decreasing nutrient absorption further.

■ Mineral oil, a popular laxative with the elderly, interferes with absorption of fat-soluble vitamins, and high-fiber diets inhibit calcium absorption.

■ Alcohol is used by about half the elderly population in this country. When alcohol is substituted for food, nutrient stores can be depleted. Excessive alcohol intake leads to malabsorption problems, certain vitamin and mineral deficiencies, deranged metabolism, and damage to the liver and pancreas.

In short, the elderly are at risk not only from the declining efficiency of the metabolic processes themselves, but also from a huge variety of lifestyle and medication factors that affect their nutrition.

Although malnutrition and a waning metabolic rate present problems to some elderly, certain nutrients—notably glucose—appear to contribute to the aging process in people of all ages. Nonenzymatic reactions (the so-called *browning reactions*) between glucose and proteins, long known to discolor and toughen foods, may have the same effects on body proteins. When enzymes attach sugars to proteins, they do so at specific sites and the glycosylated protein molecules produced play well-defined roles in the body. By contrast, nonenzymatic binding of glucose to proteins (a process that increases with age) is haphazard and eventually causes cross-links to form between proteins. The increase in this type of binding probably contributes to lens clouding, atherosclerotic plaque formation, and the general tissue stiffening and loss of elasticity so common in the aged.

⋆ ⋆ ⋆

Nutrition is one of the most overlooked areas in clinical medicine. Yet, what we eat influences nearly every phase of metabolism and plays a major role in our overall health. Now that we have examined the fates of nutrients within body cells, we are ready to study the urinary system, the organ system that works tirelessly to rid the body of nitrogen wastes resulting from metabolism and to maintain the purity of our internal fluids.

RELATED CLINICAL TERMS

Appetite A desire for food. A psychological phenomenon dependent on memory and associations, as opposed to hunger, which is a physiological need to eat.

Familial hypercholesterolemia (hi″per-ko-les″ter-ol-e′me-ah) An inherited condition in which the LDL receptors are absent or abnormal, the uptake of cholesterol by tissue cells is blocked, and the total concentration of cholesterol (and LDLs) in the blood is enormously elevated (e.g., 680 mg cholesterol/100 ml blood). Affected individuals develop atherosclerosis at an early age, and most die during childhood or adolescence from coronary artery disease. At present the only moderately successful treatment is liver transplantation.

Hypothermia (hi″po-ther′me-ah) Low body temperature resulting from prolonged uncontrolled exposure to cold. Vital signs (respiratory rate, blood pressure, and heart rate) decrease as cellular enzymes become sluggish. Drowsiness sets in and, oddly, the person becomes comfortable even though previously he or she felt extremely cold. Shivering stops at a core temperature of 30–32°C (87–90°F) when the body has exhausted its heat-generating capabilities. Uncorrected, the situation progresses to coma and finally death (by cardiac arrest), when body temperatures approach 21°C (70°F).

Kwashiorkor (kwash″e-or′kor) Severe protein deficiency in children, resulting in mental retardation and failure to grow. A consequence of malnutrition or starvation, it is characterized by a bloated abdomen because the level of plasma proteins is inadequate to keep fluid in the bloodstream.

Marasmus (mah-raz′mus) Protein-calorie malnutrition, accompanied by progressive wasting.

Pica (pi′kah) Craving and eating substances not normally considered nutrients, such as clay.

Skin-fold test Clinical test of body fatness. A skin fold in the back of the arm or below the scapula is measured with a caliper. A fold over 1 inch in thickness indicates excess fat. Also called the fat-fold test.

Suction lipectomy (*lip* = fat; *ectomy* = removal) Surgery in which the fat of the hypodermis is suctioned from places where it is most abundant, such as the thighs, abdomen, and breasts. This is a cosmetic procedure that does not result in a permanent reduction of fat.

CHAPTER SUMMARY

Media study tools that could provide you additional help in reviewing specific key topics of Chapter 25 are referenced below. **SP** *= Study Partner;* **IP** *= Interactive Physiology.*

Nutrition (pp. 949–961)

1. Nutrients include water, carbohydrates, lipids, proteins, vitamins, and minerals. The bulk of the organic nutrients is used as fuel to produce cellular energy (ATP). The energy value of foods is measured in kilocalories (kcal).

2. Essential nutrients are those that cannot be synthesized by body cells and must be ingested in the diet.

Carbohydrates (p. 950)

3. Carbohydrates are obtained primarily from plant products. Absorbed monosaccharides other than glucose are converted to glucose by the liver.

4. Monosaccharides are used primarily for cellular fuel. Small amounts are used for nucleic acid synthesis and to glycosylate plasma membranes.

5. Minimum carbohydrate requirement for adults is 100 g/day.

Lipids (pp. 950–952)

6. Most dietary lipids are triglycerides. The primary sources of saturated fats are animal products; unsaturated fats are present in plant products. The major source of cholesterol is egg yolk.

7. Linoleic and linolenic acids are essential fatty acids.

8. Neutral fats provide reserve energy, cushion body organs, and insulate the body. Phospholipids are used to synthesize plasma membranes and myelin. Cholesterol is used in plasma membranes and is the structural basis of vitamin D, steroid hormones, and bile salts.

9. Fats should represent 30% or less of caloric intake, and saturated fats should be replaced by unsaturated fats if possible. Cholesterol intake should be restricted to 250 mg or less daily.

Proteins (pp. 952–953)

10. Animal products provide high-quality protein containing all (8) essential amino acids. Most plant products lack one or more of the essential amino acids.

11. Amino acids are the structural building blocks of the body and of important regulatory molecules.

12. Protein synthesis can and will occur if all essential amino acids are present and sufficient carbohydrate (or fat) calories are available to produce ATP. Otherwise, amino acids will be burned for energy.

13. Nitrogen balance occurs when protein synthesis equals protein loss.

14. A dietary intake of 0.8 g of protein/kg of body weight is recommended for adults.

SP Exercise: Chapter 25, Nutrient Use by Body Cells.

Vitamins (pp. 953–958)

15. Vitamins are organic compounds needed in minute amounts. Most act as coenzymes.

16. Except for vitamin D and the K and B vitamins made by enteric bacteria, vitamins are not made in the body.

17. Water-soluble vitamins (B and C) are not stored to excess in the body. Fat-soluble vitamins include vitamins A, D, E, and K; all but vitamin K are stored and can accumulate to toxic amounts.

Minerals (pp. 954, 958–961)

18. Besides calcium, phosphorus, potassium, sulfur, sodium, chloride, and magnesium, the body requires trace amounts of at least a dozen other minerals.

19. Minerals are not used for energy. Some are used to mineralize bone; others are bound to organic compounds or exist as ions in body fluids, where they play various roles in cell processes and metabolism.

20. Mineral uptake and excretion are carefully regulated to prevent mineral toxicity. The richest sources of minerals are animal products, vegetables, and legumes.

Metabolism (pp. 962–986)

Overview of Metabolic Processes (pp. 962–965)

1. Metabolism encompasses all chemical reactions necessary to maintain life. Metabolic processes are either anabolic or catabolic.

2. Cellular respiration refers to catabolic processes during which energy is released and some is captured in ATP bonds.

3. Energy is released when organic compounds are oxidized. Cellular oxidation is accomplished primarily by the removal of hydrogen (electrons). When molecules are oxidized, others are simultaneously reduced by accepting hydrogen (or electrons).

4. Most enzymes catalyzing oxidation-reduction reactions require coenzymes as hydrogen acceptors. Two important coenzymes in these reactions are NAD^+ and FAD.

5. In animal cells, the two mechanisms of ATP synthesis are substrate-level phosphorylation and oxidative phosphorylation.

IP Muscular System CD-ROM; Topic: Muscle Metabolism, pages 3–8.

Carbohydrate Metabolism (pp. 965–973)

6. Carbohydrate metabolism is essentially glucose metabolism.

7. Phosphorylation of glucose on entry into cells effectively traps it in most tissue cells.

8. Glucose is oxidized to carbon dioxide and water via three successive pathways: glycolysis, Krebs cycle, and electron transport chain. Some ATP is harvested in each pathway, but the bulk is captured in the electron transport chain.

9. Glycolysis is a reversible pathway in which glucose is converted into two pyruvic acid molecules; two molecules of reduced NAD^+ are formed, and there is a net gain of two ATPs. Under aerobic conditions, pyruvic acid enters the Krebs cycle; under anaerobic conditions, it is reduced to lactic acid.

10. The Krebs cycle is fueled by pyruvic acid (and fatty acids). To enter the cycle, pyruvic acid is converted to acetyl CoA. The acetyl CoA is then oxidized and decarboxylated. Complete oxidation of two pyruvic acid molecules yields 6 CO_2, 8 NADH + H^+, 2 $FADH_2$, and a net gain of 2 ATP. Much of the energy originally present in the bonds of pyruvic acid is now present in the reduced coenzymes.

SP Exercises: Chapter 25, Metabolism; Aerobic Respiration Factory.

11. In the electron transport chain, (a) reduced coenzymes are oxidized by delivering hydrogen to a series of oxidation-reduction acceptors; (b) hydrogen is split into hydrogen ions and electrons (as electrons run downhill from acceptor to acceptor, the energy released is used to pump H^+ into the mitochondrial intermembrane space, which creates an electrochemical proton gradient); (c) the energy stored in the electrochemical proton gradient drives H^+

back through ATP synthase, which uses the energy to form ATP; (d) H^+ and electrons are combined with oxygen to form water.

12. For each glucose molecule oxidized to carbon dioxide and water, there is a net gain of 36 ATP: 4 ATP from substrate-level phosphorylation and 34 ATP from oxidative phosphorylation. The shuttle for reduced NAD^+ produced in the cytosol uses 2 ATP of that amount.

13. When cellular ATP reserves are high, glucose catabolism is inhibited and glucose is converted to glycogen (glycogenesis) or to fat (lipogenesis). Much more fat than glycogen is stored.

14. Gluconeogenesis is the formation of glucose from noncarbohydrate (fat or protein) molecules. It occurs in the liver when blood glucose levels begin to fall.

SP Exercise: Chapter 25, Aerobic Respiration Factory.

IP Muscular System CD-ROM; Topic: Muscle Metabolism, pages 10–22.

Lipid Metabolism (pp. 973–975)

15. End products of lipid digestion (and cholesterol) are transported in blood in the form of chylomicrons.

16. Glycerol is converted to glyceraldehyde-PO_4 and enters the Krebs cycle or is converted to glucose.

17. Fatty acids are oxidized by beta oxidation into acetic acid fragments. These are bound to coenzyme A and enter the Krebs cycle as acetyl CoA. Dietary fats not needed for energy or structural materials are stored in adipose tissue.

18. There is a continual turnover of fats in fatty depots. Breakdown of fats to fatty acids and glycerol is called lipolysis.

19. When excessive amounts of fats are used, the liver converts acetyl CoA to ketone bodies and releases them to the blood. Excessive levels of ketone bodies (ketosis) lead to metabolic acidosis.

20. All cells use phospholipids and cholesterol to build their plasma membranes. The liver forms many functional molecules (lipoproteins, tissue factor, etc.) from lipids.

Protein Metabolism (pp. 975–977)

21. To be oxidized for energy, amino acids are converted to keto acids that can enter the Krebs cycle. This involves transamination, oxidative deamination, and keto acid modification.

22. Amine groups removed during deamination (as ammonia) are combined with carbon dioxide by the liver to form urea. Urea is excreted in urine.

23. Deaminated amino acids may also be converted to fatty acids and glucose.

24. Amino acids are the body's most important building blocks. Nonessential amino acids are made in the liver by transamination.

25. In adults, most protein synthesis serves to replace tissue proteins and to maintain nitrogen balance.

26. Protein synthesis requires the presence of the essential amino acids. If any are lacking, amino acids are used as energy fuels.

Catabolic-Anabolic Steady State of the Body (pp. 977–978)

27. The amino acid pool provides amino acids for synthesis of proteins and amino acid derivatives, ATP synthesis, and energy storage. To be stored, amino acids are first converted to fats or glycogen.

SP Exercise: Chapter 25, Metabolism.

28. The carbohydrate-fat pool primarily provides fuels for ATP synthesis and other molecules that can be stored as energy reserves.

29. The nutrient pools are connected by the bloodstream; fats, carbohydrates, and proteins may be interconverted via common intermediates.

Absorptive and Postabsorptive States: Events and Controls (pp. 978–983)

30. During the absorptive state (during and shortly after a meal), glucose is the major energy source; needed structural and functional molecules are made; excesses of carbohydrates, fats, and amino acids are stored as glycogen and fat.

31. Events of the absorptive state are controlled by insulin, which enhances the entry of glucose (and amino acids) into cells and accelerates its use for ATP synthesis or storage as glycogen or fat.

SP Case Study: Diabetes Mellitus.

32. The postabsorptive state is the period when blood-borne fuels are provided by breakdown of energy reserves. Glucose is made available to the blood by glycogenolysis, lipolysis, and gluconeogenesis. Glucose sparing begins and, if fasting is prolonged (4–5 days), the brain also begins to metabolize ketone bodies.

33. Events of the postabsorptive state are controlled largely by glucagon and the sympathetic nervous system, which mobilize glycogen and fat reserves and trigger gluconeogenesis.

Role of the Liver in Metabolism (pp. 984–986)

34. The liver is the body's main metabolic organ and it plays a crucial role in processing (or storing) virtually every nutrient group. It helps maintain blood energy sources, metabolizes hormones, and detoxifies drugs and other substances.

35. The liver synthesizes cholesterol, catabolizes cholesterol and secretes it in the form of bile salts, and makes lipoproteins.

36. LDLs transport triglycerides and cholesterol from the liver to the tissues, whereas HDLs transport cholesterol from the tissues to the liver (for catabolism and elimination).

37. Excessively high LDL levels are implicated in atherosclerosis, cardiovascular disease, and strokes.

Body Energy Balance (pp. 986–996)

1. Body energy intake (derived from food oxidation) is precisely balanced by energy output (heat, work, and energy storage). Eventually, all of the energy intake is converted to heat.

2. When energy balance is maintained, weight remains stable. When excess amounts of energy are stored, obesity results (condition of excessive fat storage, 20% or more above the norm).

Regulation of Food Intake (pp. 987–990)

3. The hypothalamus and other brain centers are involved in the regulation of eating behavior.

4. Factors thought to be involved in regulating food intake include (a) neural signals from the gut to the brain, (b) nutrient signals related to total energy storage (e.g., large fatty reserves and increased plasma levels of glucose and amino acids depress hunger and eating); (c) plasma concentrations of hormones that control events of the absorptive and postabsorptive states, and provide feedback signals to brain feeding centers; (d) body temperature; and (e) psychological factors.

Metabolic Rate and Body Heat Production (pp. 990–992)

5. Energy used by the body per hour is the metabolic rate.

6. Basal metabolic rate (BMR), reported in kcal/m^2/h, is the measurement obtained under basal conditions: the person is at comfortable room temperature, supine, relaxed, and in the postabsorptive state. BMR indicates energy needed to drive only the resting body processes.

7. Factors influencing metabolic rate include age, sex, size, body surface area, thyroxine levels, specific dynamic action of foods, and muscular activity.

Regulation of Body Temperature (pp. 992–996)

8. Body temperature reflects the balance between heat production and heat loss and is normally 35.6–37.8°C, which is optimal for physiological activities.

9. At rest, most body heat is produced by the liver, heart, brain, kidneys, and endocrine organs. Activation of skeletal muscles causes dramatic increases in body heat production.

10. The body core (organs within the skull and the ventral body cavity) generally has the highest temperature. The shell (the skin) is the heat-exchange surface, and is usually coolest.

11. Blood serves as the major heat-exchange agent between the core and the shell. When skin capillaries are flushed with blood and the skin is warmer than the environment, heat is lost from the body. When blood is withdrawn to deep organs, heat loss from the shell is inhibited.

12. Heat-exchange mechanisms include radiation, conduction, convection, and evaporation. Evaporation, the conversion of water to water vapor, requires the absorption of heat. For each gram of water vaporized, about 0.5 kcal of heat is absorbed.

13. The hypothalamus acts as the body's thermostat. Its heat-promotion and heat-loss centers receive inputs from peripheral and central thermoreceptors, integrate these inputs, and initiate responses leading to heat loss or heat promotion.

14. Heat-promoting mechanisms include vasoconstriction of skin vasculature, increase in metabolic rate (via release of norepinephrine), and shivering. If environmental cold is prolonged, the thyroid gland is stimulated to release more thyroxine.

15. When heat must be removed from the body, dermal blood vessels are dilated, allowing heat loss through radiation, conduction, and convection. When greater heat loss is mandated (or the environmental temperature is so high that radiation and conduction are ineffective), sweating is initiated. Evaporation of perspiration is an efficient means of heat loss as long as the humidity is low.

16. Profuse sweating can lead to heat exhaustion, indicated by a rise in temperature, a drop in blood pressure, and collapse. When the body cannot rid itself of surplus heat, body temperature rises to the point where all thermoregulatory mechanisms become ineffective—a potentially lethal condition called heat stroke.

17. Fever is controlled hyperthermia, which follows thermostat resetting to higher levels by prostaglandins and initiation of heat-promotion mechanisms, evidenced by the chills. When the disease process is reversed, heat-loss mechanisms are initiated.

Developmental Aspects of Nutrition and Metabolism (pp. 996–997)

1. Good nutrition is essential for normal fetal development and normal growth during childhood.

2. Inborn errors of metabolism include cystic fibrosis, PKU, glycogen storage disease, galactosemia, and many others. Hormonal disorders, such as lack of insulin or thyroid hormone, may also lead to metabolic abnormali-

ties. Diabetes mellitus is the most significant metabolic disorder in the young and the old.

SP Exercise: Chapter 25, Diabetes Mellitus.

3. In old age, metabolic rate declines as enzyme and endocrine systems become less efficient and skeletal muscles atrophy. Reduced caloric needs make it difficult to obtain adequate nutrition without becoming overweight.

4. Elderly individuals ingest more medications than any other age group, and many of these drugs negatively affect their nutrition.

REVIEW QUESTIONS

Multiple Choice/Matching

1. Which of the following reactions would liberate the most energy? (a) complete oxidation of a molecule of sucrose to CO_2 and water, (b) conversion of a molecule of ADP to ATP, (c) respiration of a molecule of glucose to lactic acid, (d) conversion of a molecule of glucose to carbon dioxide and water.

2. The formation of glucose from glycogen is (a) gluconeogenesis, (b) glycogenesis, (c) glycogenolysis, (d) glycolysis.

3. The net gain of ATP from the complete metabolism (aerobic) of glucose is closest to (a) 2, (b) 30, (c) 36, (d) 4.

4. Which of the following best defines cellular respiration? (a) intake of carbon dioxide and output of oxygen by cells, (b) excretion of waste products, (c) inhalation of oxygen and exhalation of carbon dioxide, (d) oxidation of substances by which energy is released in usable form to the cells.

5. During aerobic respiration, electrons are passed down the electron transport chain and _____ is formed. (a) oxygen, (b) water, (c) glucose, (d) NADH + H$^+$.

6. Metabolic rate is relatively low in (a) youth, (b) physical exercise, (c) old age, (d) fever.

7. In a temperate climate under ordinary conditions, the greatest loss of body heat occurs through (a) radiation, (b) conduction, (c) evaporation, (d) none of the above.

8. Which of the following is not a function of the liver? (a) glycogenolysis and gluconeogenesis, (b) synthesis of cholesterol, (c) detoxification of alcohol and drugs, (d) synthesis of glucagon, (e) deamination of amino acids.

9. Amino acids are essential (and important) to the body for all the following except (a) production of some hormones, (b) production of antibodies, (c) formation of most structural materials, (d) as a source of quick energy.

10. A person has been on a hunger strike for seven days. Compared to normal, he has (a) increased release of fatty acids from adipose tissue, ketosis, and ketonuria, (b) elevated glucose concentration in the blood, (c) increased plasma insulin concentration, (d) increased glycogen synthetase (enzyme) activity in the liver.

11. Transamination is a chemical process by which (a) protein is synthesized, (b) an amine group is transferred from an amino acid to a keto acid, (c) an amine group is cleaved from the amino acid, (d) amino acids are broken down for energy.

12. Three days after removal of the pancreas from an animal, the researcher finds a persistent increase in (a) acetoacetic acid concentration in the blood, (b) urine volume, (c) blood glucose, (d) all of the above.

13. Hunger, appetite, obesity, and physical activity are interrelated. Thus, (a) hunger sensations arise *primarily* from the stimulation of receptors in the stomach and intestinal tract in response to the absence of food in these organs; (b) obesity, in most cases, is a result of the abnormally high enzymatic activity of the fat-synthesizing enzymes in adipose tissue; (c) in all cases of obesity, the energy content of the ingested food has exceeded the energy expenditure of the body; (d) in a normal individual, increasing blood glucose concentration increases hunger sensations.

14. Body temperature regulation is (a) influenced by temperature receptors in the skin, (b) influenced by the temperature of the blood perfusing the heat regulation centers of the brain, (c) subject to both neural and hormonal control, (d) all of the above.

15. Which of the following yields the greatest caloric value per gram? (a) fats, (b) proteins, (c) carbohydrates, (d) all are equal in caloric value.

Short Answer Essay Questions

16. What is cellular respiration? What is the common role of FAD and NAD$^+$ in cellular respiration?

17. Describe the site, major events, and outcomes of glycolysis.

18. Pyruvic acid is a product of glycolysis, but it is not the substance that enters the Krebs cycle. What is the substance, and what must occur if pyruvic acid is to be transformed into this molecule?

19. Define glycogenesis, glycogenolysis, gluconeogenesis, and lipogenesis. Which is (are) likely to be occurring (a) shortly after a carbohydrate-rich meal, (b) just before waking up in the morning?

20. What is the harmful result when excessive amounts of fats are burned for energy? Name two conditions that might lead to this result.

21. Make a flowchart that indicates the pivotal intermediates through which glucose can be converted to fat.

22. Distinguish between the role of HDLs and that of LDLs.

23. List some factors that influence plasma cholesterol levels. Also list the sources and fates of cholesterol in the body.

24. What is meant by "body energy balance," and what happens if the balance is not precise?

25. Explain the effect of the following on metabolic rate: thyroxine levels, eating, body surface area, muscular exercise, emotional stress, starvation.

26. Explain the terms "core" and "shell" relative to body temperature balance. What serves as the heat-transfer agent from one to the other?

27. Compare and contrast mechanisms of heat loss with mechanisms of heat promotion, and explain how these mechanisms determine body temperature.

 ### Critical Thinking and Clinical Application Questions

1. Calculate the number of ATP molecules that can be harvested during the complete oxidation of an 18-carbon fatty acid. (Take a deep breath and think about it—you *can* do it.)

2. Every year dozens of elderly people are found dead in their unheated apartments and listed as victims of hypo-

thermia. What is hypothermia, and how does it kill? Why are the elderly more susceptible to hypothermia than the young?

3. Frank Moro has been diagnosed as having severe atherosclerosis and high blood cholesterol levels. He is told that he is at risk for a stroke or a heart attack. First, what foods would you suggest that he avoid like the plague? What foods would you suggest he add or substitute? What activities would you recommend?

4. In the 1940s, some physicians prescribed low doses of a chemical called dinitrophenol (DNP) to help patients lose weight. This drug therapy was abandoned after a few patients died. DNP uncouples the chemiosmotic machinery. Explain how this causes weight loss.

5. While attempting to sail solo from Los Angeles to Tahiti, Simon encountered a storm that marooned him on an uninhabited island. He was able, using his ingenuity and a pocket knife, to obtain plenty of fish to eat, and roots were plentiful. However, the island was barren of fruits and soon his gums began to bleed and he started to develop several infections. Analyze his problem. (Some of the material in the tables might help.)

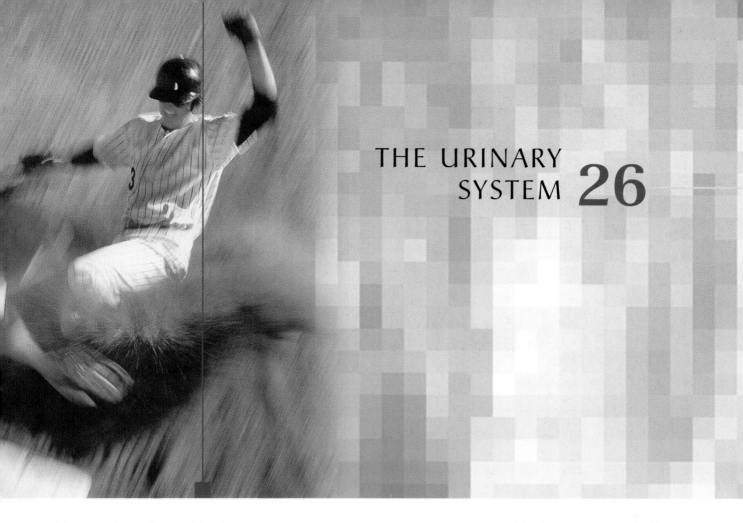

THE URINARY SYSTEM 26

Kidney Anatomy (pp. 1004–1013)

1. Describe the gross anatomy of the kidney and its coverings.
2. Trace the blood supply through the kidney.
3. Describe the anatomy of a nephron.

Kidney Physiology: Mechanisms of Urine Formation (pp. 1013–1029)

4. List several kidney functions that help maintain body homeostasis.
5. Identify the parts of the nephron responsible for filtration, reabsorption, and secretion, and describe the mechanisms underlying each of these functional processes.
6. Explain the role of aldosterone and of atrial natriuretic peptide in sodium and water balance.
7. Describe the mechanism that maintains the medullary osmotic gradient.
8. Explain the formation of dilute versus concentrated urine.
9. Describe the normal physical and chemical properties of urine.
10. List several abnormal urine components, and name the condition when each is present in detectable amounts.

Ureters (pp. 1029–1030)

11. Describe the general location, structure, and function of the ureters.

Urinary Bladder (pp. 1030–1031)

12. Describe the general location, structure, and function of the urinary bladder.

Urethra (pp. 1031–1032)

13. Describe the general location, structure, and function of the urethra.
14. Compare the course, length, and functions of the male urethra with those of the female.

Micturition (pp. 1032–1034)

15. Define micturition and describe its neural control.

Developmental Aspects of the Urinary System (pp. 1034–1036)

16. Trace the embryonic development of the urinary organs.
17. List several changes in urinary system anatomy and physiology that occur with age.

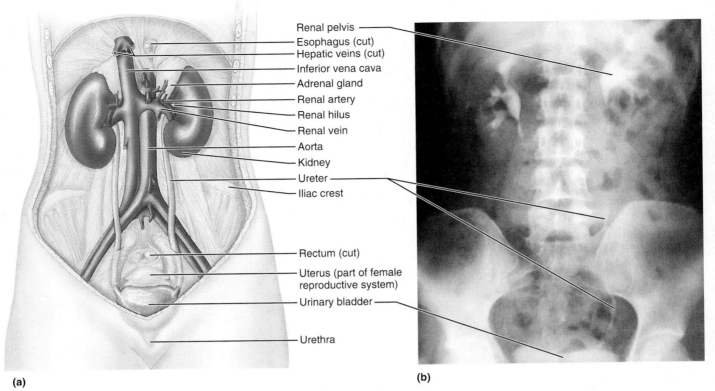

Renal pelvis
Esophagus (cut)
Hepatic veins (cut)
Inferior vena cava
Adrenal gland
Renal artery
Renal hilus
Renal vein
Aorta
Kidney
Ureter
Iliac crest

Rectum (cut)
Uterus (part of female reproductive system)
Urinary bladder
Urethra

(a)   **(b)**

FIGURE 26.1   *Organs of the urinary system.* Anterior view of the female urinary organs. (Most unrelated abdominal organs have been omitted.) **(a)** Diagrammatic view. (See *A Brief Atlas of the Human Body,* Figure 55.) **(b)** X ray of urinary organs.

The kidneys, which maintain the constancy of fluids in our internal environment, are perfect examples of homeostatic organs. Much like a water purification plant that keeps a city's water drinkable and disposes of its wastes, the kidneys are usually unappreciated until they malfunction and body fluids become contaminated with internal "garbage." Every day the kidneys filter nearly 200 liters of fluid from the bloodstream, allowing toxins, metabolic wastes, and excess ions to leave the body in urine while returning needed substances to the blood. Although the lungs and skin also participate in excretion, the kidneys are the major excretory organs.

As the kidneys perform these excretory functions, they simultaneously regulate the volume and chemical makeup of the blood, maintaining the proper balance between water and salts and between acids and bases. Frankly, this would be tricky work for a chemical engineer, but the kidneys do it efficiently most of the time.

Another renal function is gluconeogenesis (see p. 973), a role undertaken only during prolonged fasting. At such times the kidneys can supply approximately one-fifth as much glucose as the liver. The kidneys also act as endocrine glands. They produce the enzyme *renin* (re'nin), which helps regulate blood pressure and kidney function, and the hormone *erythropoietin* (ĕ-rith"ro-poi'ĕ-tin), which stimulates red blood cell production in bone marrow (see Chapter 18). The kidneys also metabolize vitamin D to its active form (see Chapter 17).

Besides the urine-forming kidneys, the **urinary system** includes the *urinary bladder,* which provides a temporary storage reservoir for urine and three tubelike organs—the paired *ureters* (u-re'terz) and the *urethra* (u-re'thrah), which furnish transportation channels for urine (Figure 26.1).

KIDNEY ANATOMY

Location and External Anatomy

The bean-shaped kidneys lie in a retroperitoneal position (between the dorsal body wall and the parietal peritoneum) in the *superior* lumbar region (Figure 26.2). They extend approximately from the level of the twelfth thoracic vertebra to the third lumbar vertebra; thus the kidneys receive some protection from the lower part of the rib cage (see Figure 26.2b). The right kidney is crowded by the liver and lies slightly lower than the left. An adult's kidney weighs about

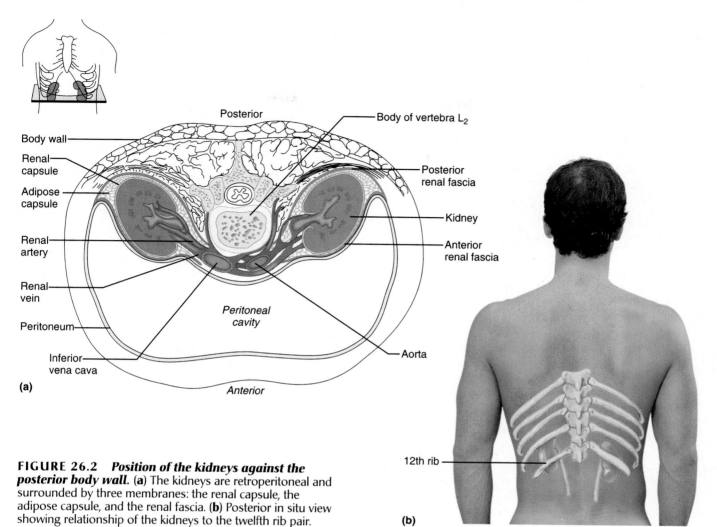

FIGURE 26.2 *Position of the kidneys against the posterior body wall.* **(a)** The kidneys are retroperitoneal and surrounded by three membranes: the renal capsule, the adipose capsule, and the renal fascia. **(b)** Posterior in situ view showing relationship of the kidneys to the twelfth rib pair.

150 g (5 ounces) and its average dimensions are 12 cm long, 6 cm wide, and 3 cm thick, about the size of a large bar of soap. The lateral surface of the kidney is convex. Its medial surface is concave and has a vertical cleft called the **renal hilus** that leads into a space within the kidney called the *renal sinus.* Several structures, including the ureters, the renal blood vessels, lymphatics, and nerves, enter or exit the kidney at the hilus and occupy the sinus. Atop each kidney is an *adrenal* (or *suprarenal*) *gland,* an endocrine gland that is unrelated to the kidney functionally (see Figure 26.1).

Three layers of supportive tissue surround each kidney (Figure 26.2a). The fibrous **renal capsule** (*ren* = kidney) adheres directly to the kidney surface. This transparent capsule provides a strong barrier that prevents infections in surrounding regions from spreading to the kidneys. The middle layer is a fatty mass called the **adipose capsule,** which helps attach the kidney to the posterior body wall and cushions it against blows. The outermost layer, the **renal fascia,** is dense fibrous connective tissue. It surrounds the kidney and its attendant membranes and also the adrenal gland and anchors these organs to surrounding structures.

Homeostatic Imbalance

The fatty encasement of the kidneys is important in holding the kidneys in their normal body position. If the amount of fatty tissue dwindles (as with extreme emaciation or rapid weight loss), the kidneys may drop to a lower position, an event called *ptosis* (to'sis; "a fall"). Ptosis may cause a ureter to become kinked, which creates problems because the urine, unable to drain, backs up into the kidney and exerts pressure on its tissue. This condition, called *hydronephrosis* (hi"dro-ne̯-fro'sis; "water in the kidney"), can severely damage the kidney, leading to necrosis (tissue death) and renal failure. ∎

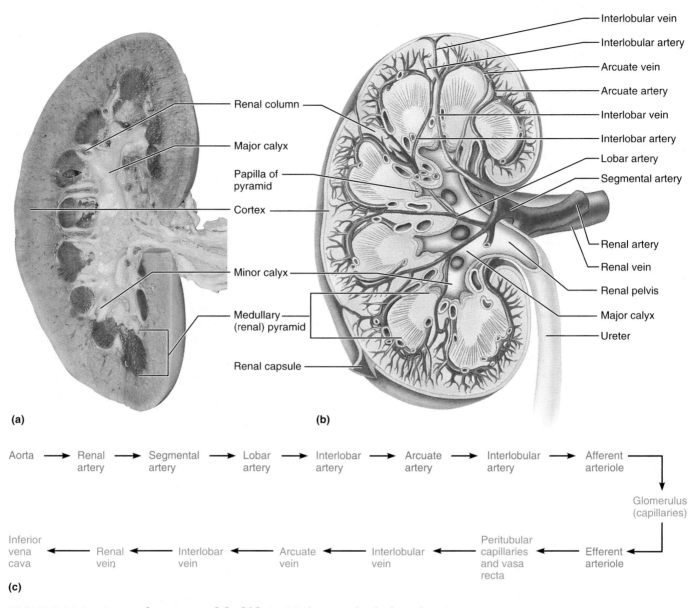

FIGURE 26.3 *Internal anatomy of the kidney.* (a) Photograph of a frontal section of a triple-injected kidney. (Arteries and veins are injected with red and blue latex respectively; pelvis and ureter injected with yellow latex. (b) Diagrammatic view of a coronally sectioned kidney, illustrating major blood vessels. (c) Summary of pathway of renal vasculature.

Internal Anatomy

A frontal section through a kidney reveals three distinct regions (Figure 26.3): the *cortex,* the *medulla,* and the *pelvis.* The most superficial region, the **renal cortex,** is light in color and has a granular appearance. Deep to the cortex is the darker, reddish-brown **renal medulla,** which exhibits cone-shaped tissue masses called **medullary** or **renal pyramids.** The broad *base* of each pyramid faces toward the cortex, and its apex, or *papilla* ("nipple"), points in-

ternally. The pyramids appear striped because they are formed almost entirely of roughly parallel bundles of microscopic urine-collecting tubules. The **renal columns,** inward extensions of cortical tissue, separate the pyramids. Each medullary pyramid and its surrounding capsule of cortical tissue constitutes a **lobe** (one of approximately eight) of a kidney.

Lateral to the hilus within the renal sinus is the **renal pelvis;** this flat, funnel-shaped tube is continuous with the ureter leaving the hilus. Branching extensions of the pelvis form two or three **major ca-**

lyces (ka'lih-sēz; singular, *calyx*), each of which subdivides to form several **minor calyces,** cup-shaped areas that enclose papillae of the pyramids. The calyces collect urine, which drains continuously from the papillae, and empty it into the renal pelvis. Urine then flows through the renal pelvis and into the ureter, which transports it to the bladder to be stored. The walls of the calyces, pelvis, and ureter contain smooth muscle, which contracts rhythmically and propels urine along its course by peristalsis.

⚠ *Homeostatic Imbalance*

Infection of the renal pelvis and calyces produces the condition called *pyelitis* (pi"ĕ-li'tis). Infections or inflammatory conditions that affect the entire kidney are *pyelonephritis* (pi"ĕ-lo-nĕ-fri'tis). Kidney infections in females are usually caused by fecal bacteria *(E. coli)* that spread from the anal region to the urinary tract. Less often they result from bloodborne bacteria (traveling from other infected sites) that lodge and multiply in the kidneys. In severe cases of pyelonephritis, the kidney swells, abscesses form, and the pelvis fills with pus. Untreated, the kidneys may be severely damaged, but antibiotic therapy usually achieves total remission. ∎

Blood and Nerve Supply

The kidneys continuously cleanse the blood and adjust its composition, so it is not surprising that they have a very rich blood supply (Figure 26.3). Under normal resting conditions, the large **renal arteries** deliver approximately one-fourth of the total systemic cardiac output (about 1200 ml) to the kidneys each minute. The renal arteries issue at right angles from the abdominal aorta between the first and second lumbar vertebrae. Because the aorta lies to the left of the midline, the right renal artery is longer than the left. As each renal artery approaches a kidney, it divides into five **segmental arteries** that enter the hilus. Within the renal sinus, each segmental artery branches into several **lobar arteries.** These then divide to form several **interlobar arteries,** which pass between the medullary pyramids toward the cortex.

At the medulla-cortex junction, the interlobar arteries branch into the **arcuate** (ar'ku-āt) **arteries** that arch over the bases of the medullary pyramids. Small **interlobular arteries** radiate outward from the arcuate arteries to supply the cortical tissue (see Figure 26.5a, page 1010). More than 90% of the blood entering the kidney perfuses the cortex, which contains the bulk of the *nephrons* (nef'ronz), the structural and functional units of the kidneys.

Veins of the kidney pretty much trace the pathway of the arterial supply in reverse. Blood leaving the renal cortex drains sequentially into the **interlobular, arcuate, interlobar,** and finally the **renal veins.** (There are no lobar or segmental veins.) The renal veins issue from the kidneys and empty into the inferior vena cava. Since the inferior vena cava lies to the right of the vertebral column, the left renal vein is about twice as long as the right.

The nerve supply of the kidney and its ureter is provided by the **renal plexus,** a variable network of autonomic nerve fibers and ganglia. The renal plexus is an offshoot of the celiac plexus (see Chapter 14). It is largely supplied by sympathetic fibers from the least thoracic and first lumbar splanchnic nerves, which course along with the renal artery to reach the kidney. These sympathetic fibers are vasomotor fibers that regulate renal blood flow by adjusting the diameter of renal arterioles and, as described shortly, they also influence the urine-forming role of the nephrons.

Nephrons

Each kidney contains over 1 million tiny blood-processing units called *nephrons,* which carry out the processes that form urine (Figure 26.4). In addition, there are thousands of *collecting ducts,* each of which collects urine from several nephrons and conveys it to the renal pelvis.

Each **nephron** consists of a **glomerulus** (glomer'u-lus; *glom* = ball of yarn), a tuft of capillaries associated with a **renal tubule.** The end of the renal tubule, called the **glomerular (Bowman's) capsule** is blind, enlarged, and cup-shaped and completely surrounds the glomerulus (much as a well-worn baseball glove encloses a ball). Collectively, the glomerular capsule and the enclosed glomerulus are called the **renal corpuscle.** The glomerular endothelium is *fenestrated* (penetrated by many pores), which makes these capillaries exceptionally porous: They allow large amounts of solute-rich, virtually protein-free fluid to pass from the blood into the glomerular capsule. This plasma-derived fluid or **filtrate** is the raw material that is processed by the renal tubules to form urine.

The external *parietal layer* of the glomerular capsule is simple squamous epithelium. This layer simply contributes to the structure of the capsule and plays no part in forming filtrate. The *visceral layer,* which clings to the glomerulus (Figure 26.4b), consists of highly modified, branching epithelial cells called **podocytes** (pod'o-sīts; "foot cells"). The extensions of the octopus-like podocytes terminate in **foot processes,** which intertwine with one another

? *What would be the path taken by a urea molecule in the glomerular blood to reach the renal pelvis?*

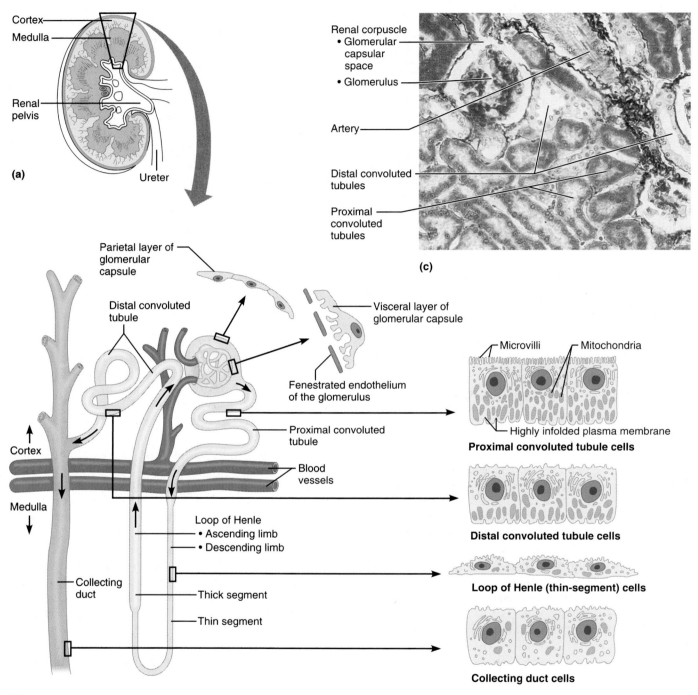

(a)

Cortex
Medulla
Renal pelvis
Ureter

Renal corpuscle
• Glomerular capsular space
• Glomerulus
Artery
Distal convoluted tubules
Proximal convoluted tubules

(c)

Parietal layer of glomerular capsule
Distal convoluted tubule
Visceral layer of glomerular capsule
Fenestrated endothelium of the glomerulus
Proximal convoluted tubule
Cortex
Medulla
Blood vessels
Loop of Henle
• Ascending limb
• Descending limb
Collecting duct
Thick segment
Thin segment

Microvilli Mitochondria
Highly infolded plasma membrane
Proximal convoluted tubule cells
Distal convoluted tubule cells
Loop of Henle (thin-segment) cells
Collecting duct cells

(b)

FIGURE 26.4 *Location and structure of nephrons.* (**a**) Orientation diagram indicating the location of nephrons in the kidney. (**b**) Schematic view of a nephron depicting the structural characteristics of epithelial cells forming its various regions. (**c**) Photomicrograph of renal cortical tissue. Notice the renal corpuscles and the sections through the various tubules. In the proximal convoluted tubules, the lumens appear "fuzzy" because they are filled with the long microvilli of the epithelial cells that form the tubule walls. In the distal tubules, by contrast, the lumens are clear (400×).

It would go from the glomerular blood through the filtration membrane into Bowman's capsule, then through the renal tubule (PCT → loop of Henle ← DCT), then into the collecting duct which carries the urine through the cortex and medulla, and into the minor calyx, into the major calyx, and into the pelvis.

as they cling to the basement membrane of the glomerulus. The clefts or openings between the foot processes, called **filtration slits** or **slit pores** (see Figure 26.8 on p. 1012), allow the filtrate to pass to the interior of the glomerular capsule, a region called the **capsular space.**

The remainder of the renal tubule (Figure 26.4b) is about 3 cm (approximately 1.25 inches) long and has three named parts. It leaves the glomerular capsule as the elaborately coiled **proximal convoluted tubule (PCT)**, makes a hairpin loop called the **loop of Henle,** and then winds and twists again as the **distal convoluted tubule (DCT)** before it empties its product into a collecting duct. The meandering nature of the renal tubule increases its length and enhances its filtrate processing capabilities. (Designating the convoluted parts of the nephron as proximal or distal indicates their relative closeness to the glomerulus—the PCT being closer.)

The **collecting ducts,** each of which receives urine from many nephrons, run through the medullary pyramids and give them their striped appearance. As the collecting ducts approach the renal pelvis, they fuse to form the large **papillary ducts,** which deliver urine into the minor calyces via papillae of the pyramids.

Throughout its length, the renal tubule consists of a single layer of epithelial cells on a basement membrane, but each region of the renal tubule has a unique cellular anatomy, which reflects its specific role in processing filtrate. The walls of the *proximal convoluted tubule* are formed by cuboidal epithelial cells, which actively reabsorb substances from the filtrate and also secrete substances into it. These cells have large mitochondria and their luminal (exposed) surfaces bear dense microvilli (Figure 26.4b and c) that tremendously increase their surface area. This so-called *brush border* dramatically increases their capacity for reabsorbing water and solutes from the filtrate.

The U-shaped *loop of Henle* has *descending* and *ascending limbs.* The proximal part of the descending limb is continuous with the proximal tubule and its cells are similar. The rest of the descending limb, called the **thin segment,** is a simple squamous epithelium that is freely permeable to water. The epithelium becomes cuboidal or even low columnar in the ascending part of the loop of Henle, which therefore becomes the **thick segment.** In some nephrons, the thin segment is found only in the descending limb of the loop of Henle. In others, it extends into the ascending limb as well.

The epithelial cells of the *distal convoluted tubule,* like those of the PCT, are confined to the cortex and are cuboidal, but they are somewhat thinner and almost entirely lack microvilli. These structural features reveal that these tubules may play a greater role in secreting solutes into the filtrate than in re-

absorbing substances from it. Beginning late in the DCT, the tubule cells become heterogeneous, more like those in the collecting ducts beyond than those in most of the DCT. (For this reason, this part of the DCT is sometimes called the *connecting tubule.*) The two most important cell types seen here and in the collecting ducts are *intercalated cells,* cuboidal cells with abundant microvilli, and the more numerous *principal cells,* which lack microvilli. As described shortly, the intercalated cells play a major role in maintaining the acid-base balance of the blood. The principal cells help maintain the body's water and Na^+ (salt) balance.

Cortical nephrons represent 85% of the nephrons in the kidneys. Except for small parts of their loops of Henle that dip into the outer medulla, they are located entirely in the cortex. The remaining nephrons are somewhat different in structure. The renal corpuscles of these **juxtamedullary** (juks"tah-mĕ'dul-ah-re) **nephrons** are located very close to (*juxta* = near to) the cortex-medulla junction. Their loops of Henle deeply invade the medulla, and their thin segments are much more extensive than those of cortical nephrons (Figure 26.5). The juxtamedullary nephrons play an important role in the kidney's ability to produce concentrated urine.

Capillary Beds (Microvasculature) of the Nephron

Every nephron is closely associated with two capillary beds: the *glomerulus* and the *peritubular capillaries* (Figure 26.5). The glomerulus is specialized for filtration. It differs from all other capillary beds in the body in that it is both fed and drained by arterioles—the **afferent arteriole** and the **efferent arteriole,** respectively. The afferent arterioles arise from the *interlobular arteries* that run through the renal cortex (see also Figure 26.3b). Because (1) arterioles are high-resistance vessels and (2) the afferent arteriole has a larger diameter than the efferent, the blood pressure in the glomerulus is extraordinarily high for a capillary bed and easily forces fluid and solutes out of the blood into the glomerular capsule along nearly its entire length. Most of this filtrate (99%) is reabsorbed by the renal tubule cells and returned to the blood in the peritubular capillary beds.

The **peritubular capillaries** arise from the efferent arterioles draining the glomeruli. These capillaries cling closely to adjacent renal tubules and empty into nearby venules of the renal venous system. The peritubular capillaries are adapted for absorption rather than filtrate formation. They are low-pressure, porous capillaries that readily absorb solutes and water from the tubule cells as these substances are reclaimed from the filtrate. Furthermore, all substances secreted by the nephrons are derived from the blood of the peritubular capillaries.

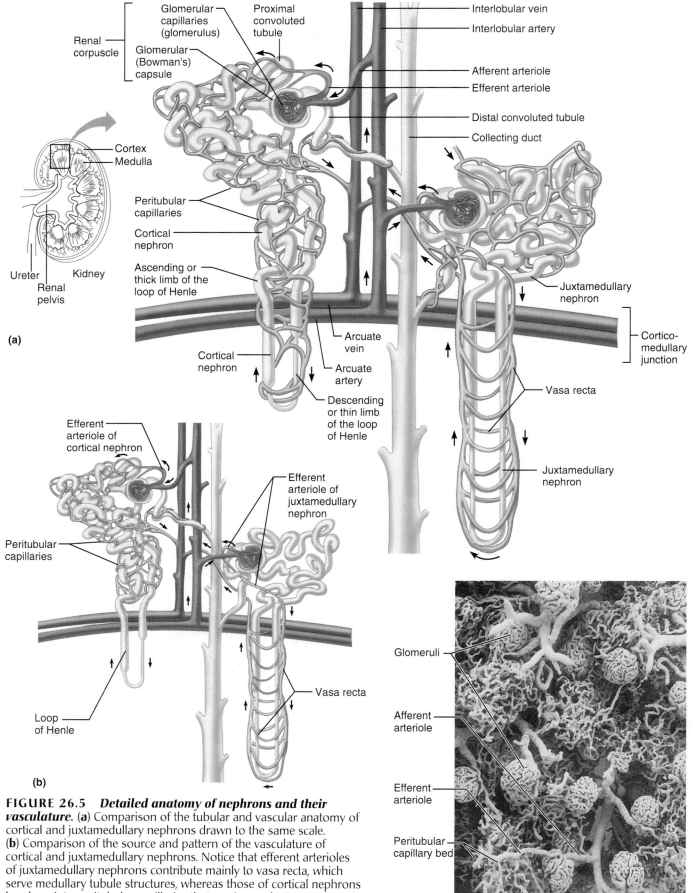

(a)

(b)

(c)

FIGURE 26.5 *Detailed anatomy of nephrons and their vasculature.* (**a**) Comparison of the tubular and vascular anatomy of cortical and juxtamedullary nephrons drawn to the same scale. (**b**) Comparison of the source and pattern of the vasculature of cortical and juxtamedullary nephrons. Notice that efferent arterioles of juxtamedullary nephrons contribute mainly to vasa recta, which serve medullary tubule structures, whereas those of cortical nephrons break up into peritubular capillaries that service nephron structures in the renal cortex. (**c**) Scanning electron micrograph of a cast of blood vessels associated with nephrons (79×). (© R. G. Kessel and R. H. Kardon, *Tissues and Organs: A Text-Atlas of Scanning Electron Microscopy,* W. H. Freeman and Company, 1979, all rights reserved.)

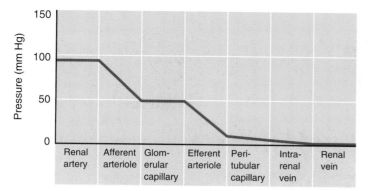

FIGURE 26.6 *Relative blood pressures in the renal circulation.* The two areas of highest resistance (thus, points of greatest decline in blood pressure) are the afferent and efferent arterioles, respectively feeding and draining the glomeruli.

In summary, the microvasculature of the nephrons consists of two capillary beds separated by intervening efferent arterioles. The first capillary bed (glomerulus) produces the filtrate. The second (peritubular capillaries) reclaims most of that filtrate.

Vascular Resistance in the Microcirculation As blood flows through the renal circulation, it encounters high resistance, first in the afferent and then in the efferent arterioles. As a result, renal blood pressure declines from approximately 95 mm Hg in the renal arteries to 8 mm Hg or less in the renal veins (Figure 26.6). The resistance of the afferent arterioles protects the glomeruli from large fluctuations in systemic blood pressure. Resistance encountered in the efferent arterioles reinforces the high glomerular pressure and reduces the hydrostatic pressure in the peritubular capillaries.

Juxtaglomerular Apparatus

Each nephron has a region called a **juxtaglomerular** (juks"tah-glo-mer'u-lar) **apparatus (JGA)**, where the initial portion of its coiling distal tubule lies against the afferent (and sometimes the efferent) arteriole feeding the glomerulus (Figure 26.7). Both of these structures are modified at the point of contact.

Present in the arteriole walls are the granular **juxtaglomerular (JG) cells**—enlarged, smooth muscle cells with prominent secretory granules containing renin. These cells act as mechanoreceptors that directly sense the blood pressure in the afferent arteriole. The **macula densa** (mak'u-lah den'sah; "dense spot") is a group of tall, closely packed distal tubule cells that lies adjacent to the JG cells. The macula

Notice that in the deepest part of the renal cortex, the efferent arterioles serving the juxtamedullary nephrons tend *not* to break up into peritubular capillaries that serve the nearby tubules but instead form bundles of long straight vessels called **vasa recta** (va'sah rek'tah; "straight vessels") that extend deep into the medulla paralleling the courses of the longest loops of Henle (Figure 26.5b). There the vasa recta surround and service the medullary nephron tubular structures. The thin-walled vasa recta play an important role in forming concentrated urine, as described shortly.

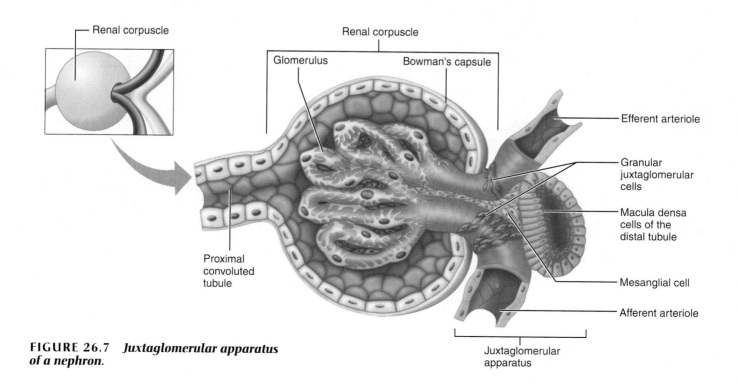

FIGURE 26.7 *Juxtaglomerular apparatus of a nephron.*

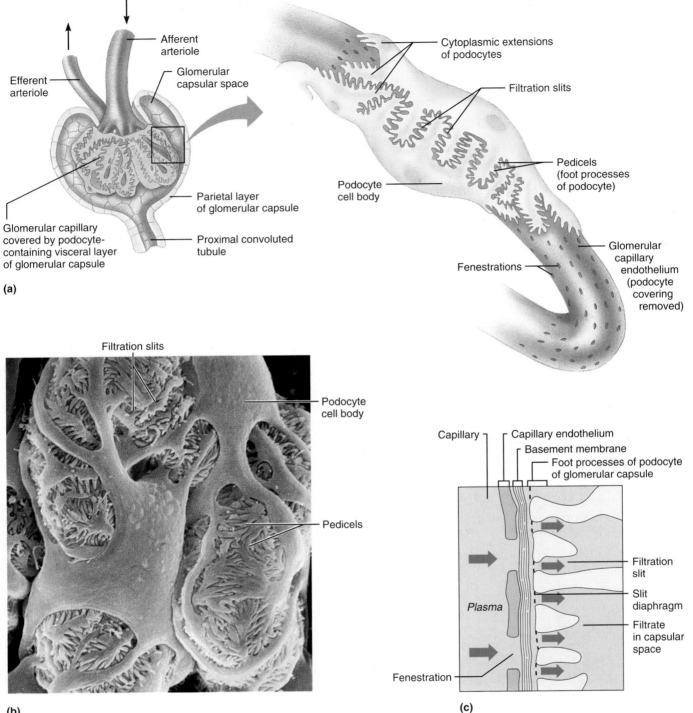

(a)

Afferent
arteriole

Efferent
arteriole

Glomerular
capsular space

Cytoplasmic extensions
of podocytes

Filtration slits

Podocyte
cell body

Pedicels
(foot processes
of podocyte)

Glomerular capillary
covered by podocyte-
containing visceral layer
of glomerular capsule

Parietal layer
of glomerular capsule

Proximal convoluted
tubule

Fenestrations

Glomerular
capillary
endothelium
(podocyte
covering
removed)

(b)

Filtration slits

Podocyte
cell body

Pedicels

(c)

Capillary

Capillary endothelium

Basement membrane

Foot processes of podocyte
of glomerular capsule

Plasma

Fenestration

Filtration
slit

Slit
diaphragm

Filtrate
in capsular
space

FIGURE 26.8   *The filtration membrane.* The filtration membrane is composed of three layers: the fenestrated endothelium of the glomerular capillaries, the podocyte-containing visceral layer of the glomerular capsule, and the intervening basement membrane. (a) Diagrammatic three-dimensional view of the relationship of the visceral layer of the glomerular capsule to the glomerular capillaries. The visceral epithelium is drawn incompletely to show the fenestrations in the underlying capillary wall. (b) Scanning electron micrograph of the visceral layer. Filtration slits between the podocyte foot processes are evident (385,790×). (c) Diagrammatic view of a section through the filtration membrane showing all three structural elements.

densa cells are chemoreceptors (or osmoreceptors) that respond to changes in the solute content of the filtrate in the tubule lumen. These two distinct cell populations play important roles in regulating the rate of filtrate formation and systemic blood pressure as described shortly. The role of the *mesanglial cells,* seemingly part of the JG apparatus, is still unclear but they appear to play a part in controlling the rate of glomerular filtration.

The Filtration Membrane

The **filtration membrane** is the actual filter that lies between the blood and the interior of the glomerular capsule (Figure 26.8). It is a porous membrane that allows free passage of water and solutes smaller than plasma proteins. Its three layers are:

1. The fenestrated endothelium of the glomerular capillaries

2. The visceral membrane of the glomerular capsule made of podocytes

3. The intervening basement membrane composed of the fused basal laminas of the other layers

The capillary pores (fenestrations) prevent passage of blood cells, but all plasma components are allowed to pass. The basement membrane restricts passage of all but the smallest proteins while permitting most other solutes to pass. The structural makeup of the gel-like basement membrane also seems to confer electrical selectivity on the filtration process. Most of its proteins are anionic (negatively charged) glycoproteins that repel other macromolecular anions and hinder their passage into the tubule. Because most plasma proteins also bear a net negative charge, this electrical repulsion reinforces the plasma protein blockage imposed by molecular size. Macromolecules that do manage to make it through the basement membrane may still be denied passage by thin membranes *(slit diaphragms)* that extend across the filtration slits between the podocyte foot processes. Although there is some uncertainty about what happens to macromolecules that get "hung up" in the filtration membrane, the assumption is that they are engulfed by the podocytes and degraded.

KIDNEY PHYSIOLOGY: MECHANISMS OF URINE FORMATION

Of the approximately 1000–1200 ml of blood that passes through the glomeruli each minute, some 650 ml is plasma, and about one-fifth of this (120–125 ml) is forced into the renal tubules. This is equivalent to filtering out your entire plasma vol-

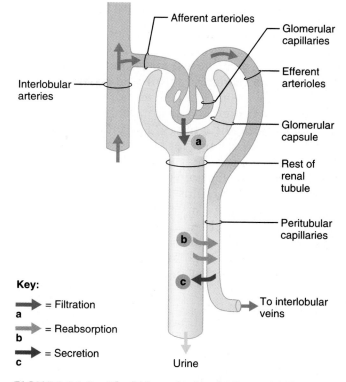

FIGURE 26.9 *The kidney depicted schematically as a single, large, uncoiled nephron.* A kidney actually has millions of nephrons acting in parallel. The three major mechanisms by which the kidneys adjust the composition of plasma are **(a)** glomerular filtration, **(b)** tubular reabsorption, and **(c)** tubular secretion.

ume more than 60 times each day! Considering the magnitude of their task, it is not surprising that the kidneys (which account for only about 1% of body weight) consume 20–25% of all oxygen used by the body at rest.

Filtrate and urine are quite different. Filtrate contains everything that blood plasma does except proteins; but by the time filtrate has percolated into the collecting ducts, it has lost most of its water, nutrients, and essential ions. What remains, now called **urine,** contains mostly metabolic wastes and unneeded substances. The kidneys process about 180 L (47 gallons) of blood-derived fluid daily. Of this amount, only about 1% (1.5 L) actually leaves the body as urine; the rest is returned to the circulation.

Urine formation and the simultaneous adjustment of blood composition involve three major processes (Figure 26.9). *Glomerular filtration* is mainly the job of the glomeruli. The renal tubules carry out *tubular reabsorption* and *secretion,* both carefully regulated by renal and hormonal controls. In addition, the collecting ducts work in concert with the nephrons to concentrate or dilute the urine.

Glomerular Filtration

Urine formation begins with **glomerular filtration.** For the most part, filtration is a passive, nonselective process in which fluids and solutes are forced through a membrane by hydrostatic pressure (see Chapter 3). Forming filtrate does not require the use of metabolic energy, so the glomeruli can be viewed as simple mechanical filters. The principles of fluid dynamics that account for the formation of tissue fluid at all capillary beds (see Chapter 20) apply to the formation of filtrate at the glomerulus as well. However, the glomerulus is a much more efficient filter than other capillary beds because (1) its *filtration membrane* is thousands of times more permeable to water and solutes than are other capillary membranes, and (2) glomerular blood pressure is much higher than that of other capillary beds (approximately 55 mm Hg as opposed to 18 mm Hg or less), resulting in a much higher *net filtration pressure*. As a result of these differences, the kidneys produce about 180 L of filtrate daily, in contrast to the 3 to 4 L formed daily by all other capillary beds of the body combined.

In general, molecules smaller than 3 nm in diameter—such as water, glucose, amino acids, and nitrogenous wastes—pass freely from the blood into the renal tubule. Hence, these substances are usually present in the same concentration in the blood and the glomerular filtrate. Larger molecules pass with greater difficulty, and those larger than 7–9 nm are usually completely barred from entering the tubule. Keeping the plasma proteins *in* the capillaries maintains the colloid osmotic (oncotic) pressure of the glomerular blood, preventing the loss of all of its water to the renal tubules. The presence of proteins or blood cells in the urine usually indicates some problem with the filtration membrane.

Net Filtration Pressure

To determine the **net filtration pressure (NFP)** responsible for filtrate formation, we must examine the forces acting at the glomerular bed (Figure 26.10). **Glomerular hydrostatic pressure (HP_g),** which is essentially glomerular blood pressure, is the chief force pushing water and solutes out of the blood across the filtration membrane. Although theoretically the colloid osmotic pressure in the intracapsular space of the glomerular capsule "pulls" the filtrate into the tubule, this pressure is essentially zero because virtually no proteins enter the capsule. Glomerular hydrostatic pressure (55 mm Hg) is opposed by forces that tend to drive fluids back into glomerular capillaries. These filtration-opposing forces are (1) **colloid osmotic (oncotic) pressure of glomerular blood (OP_g),** which is 28 to 30 mm Hg, and (2) **capsular hydrostatic pressure (HP_c)** exerted by fluids in the glomerular capsule (about 15 mm

Hg). Thus, the NFP responsible for forming renal filtrate from plasma is 10 mm Hg:

$$NFP = HP_g - (OP_g + HP_c)$$

$$NFP = 55 \text{ mm Hg} - (30 \text{ mm Hg} + 15 \text{ mm Hg})$$
$$NFP = 55 \text{ mm Hg} - (45 \text{ mm Hg})$$
$$NFP = 10 \text{ mm Hg}$$

Glomerular Filtration Rate (GFR)

The **glomerular filtration rate** or **GFR** is the total amount of filtrate formed per minute by the kidneys. Factors governing filtration rate at capillary beds are (1) total surface area available for filtration, (2) filtration membrane permeability, and (3) net filtration pressure. The normal GFR in both kidneys in adults is 120–125 ml/min (7.5 L/hour, or 180 L/day). Because glomerular capillaries are exceptionally permeable and have a huge surface area (collectively equal to the surface area of the skin), huge amounts of filtrate can be produced even with the usual modest net filtration pressure of 10 mm Hg. The opposite side of this "coin" is that a drop in glomerular pressure of only about 15% stops filtration altogether.

The GFR is *directly proportional* to the net filtration pressure. Hence, a change in any of the pressures acting at the filtration membrane (see Figure 26.10) changes the NFP and thus the GFR. An increase in arterial (and glomerular) blood pressure in the kidneys increases the GFR, whereas dehydration (which causes an increase in glomerular osmotic pressure) inhibits filtrate formation. Although certain disease states may modify any of these pressures, changes in the GFR *normally* result from changes in glomerular blood pressure, which is subject to both intrinsic and extrinsic controls.

Regulation of Glomerular Filtration

The importance of maintaining a fairly constant GFR is that reabsorption of water and other substances from the filtrate depends partly on the *rate* at which filtrate flows through the renal tubules. When massive amounts of filtrate form and flow is very rapid, needed substances cannot be reabsorbed fast enough and are lost from the body in urine. On the other hand, when filtrate is scanty and flows slowly, nearly all of it is reabsorbed, including most of the wastes that are normally disposed of. Thus, for optimal filtrate processing, the GFR must be precisely regulated.

In humans, the GFR is held relatively constant by at least three important mechanisms that regulate renal blood flow—*renal autoregulation* (the intrinsic system), *neural controls,* and the *renin-angiotensin system* (basically a hormonal mechanism). These plus some less well defined factors are examined next.

Intrinsic Controls: Renal Autoregulation Under ordinary conditions, glomerular blood pressure is regulated by the kidneys' autoregulatory system. By adjusting its own resistance to blood flow, a process called **renal autoregulation,** the kidney can maintain a nearly constant glomerular filtration rate despite fluctuations in systemic arterial blood pressure. It does this by directly regulating the diameter of the afferent and to a lesser extent the efferent arterioles. The mechanism of renal autoregulation entails two types of controls: (1) a *myogenic mechanism* that responds to changes in pressure in the renal blood vessels, and (2) a *tubuloglomerular feedback mechanism* that senses changes in the juxtaglomerular apparatus (Figure 26.11). Additionally, the *renin-angiotensin mechanism*, which mainly functions to maintain homeostasis of systemic blood pressure, acts indirectly to maintain the GFR of the kidneys as discussed later.

The **myogenic** (mi″o-jen′ik) **mechanism** reflects the general tendency of vascular smooth muscle to contract when it is stretched. Increasing systemic blood pressure causes the afferent arterioles to constrict, which restricts blood flow into the glomerulus (and lowers the downstream glomerular pressure) and prevents glomerular blood pressure from rising to damaging levels. On the other hand, a decline in systemic blood pressure causes dilation of afferent arterioles and an increase in glomerular hydrostatic pressure. Both responses help maintain a normal GFR.

Autoregulation by the **tubuloglomerular feedback mechanism** is "directed" by the *macula densa cells* of the *juxtaglomerular apparatus* (see Figure 26.7). These cells, located in the walls of the distal tubules, respond to filtrate flow rate and osmotic signals by causing or not causing the release of a chemical that produces intense vasoconstriction of the afferent arterioles. When the macula densa cells are exposed to slowly flowing filtrate or filtrate with low osmolarity (Figure 26.11), they promote vasodilation of the afferent arterioles. This allows more blood to flow into the glomerulus, thus increasing the NFP and GFR. On the other hand, when the filtrate is flowing rapidly and/or it has a high sodium and chloride content (or high osmolarity in general), the macula densa cells prompt generation of the vasoconstrictor chemical by the JG cells. This hinders blood flow into the glomerulus, which decreases the GFR and allows more time for filtrate processing. The macula densa cells also send messages to the JG cells of the juxtaglomerular apparatus that set the renin-angiotensin mechanism into motion. This mechanism is very important in balancing filtration and tubular reabsorption.

By altering the resistance of the *afferent arterioles* when systemic blood pressure rises and falls, re-

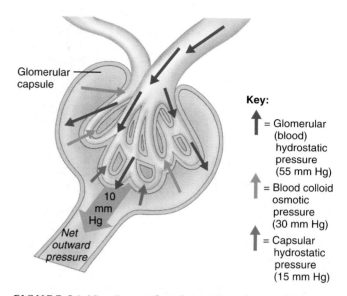

What would be the consequence of liver disease for the process shown here?

Glomerular capsule

10 mm Hg

Net outward pressure

Key:

↑ = Glomerular (blood) hydrostatic pressure (55 mm Hg)

↑ = Blood colloid osmotic pressure (30 mm Hg)

↑ = Capsular hydrostatic pressure (15 mm Hg)

FIGURE 26.10 *Forces that determine glomerular filtration and the effective filtration pressure.* The glomerular hydrostatic (blood) pressure is the major factor forcing fluids and solutes out of the blood. This is opposed by the colloid osmotic pressure of the blood and the hydrostatic pressure that exists within the glomerular capsule. The pressure values cited in the diagram are approximate.

nal autoregulatory mechanisms maintain a relatively constant blood flow through the kidneys over an arterial pressure range from about 80 to 180 mm Hg. Consequently, large changes in water and solute excretion are prevented. However, this intrinsic control system cannot handle extremely low systemic blood pressure, such as might result from serious hemorrhage *(hypovolemic shock)*. Once mean systemic blood pressure drops below 70 mm Hg, autoregulation effectively ceases.

Sympathetic Nervous System Controls Neural controls serve the needs of the body as a whole. When the sympathetic nervous system is at rest, the renal blood vessels are maximally dilated and renal autoregulation mechanisms prevail. With moderate sympathetic activity, the efferent and afferent arterioles are constricted to about the same extent. Because blood flow into and out of the glomerulus is hindered to the same degree, the GFR declines only slightly. However, renal autoregulatory mechanisms may be overcome during periods of extreme stress or

With liver disease you probably wouldn't have a normal supply of plasma proteins. Hence, more filtrate would be formed due to lower osmotic pressure in the blood.

? *Why are the myogenic and tubuloglomerular mechanisms called autoregulatory mechanisms?*

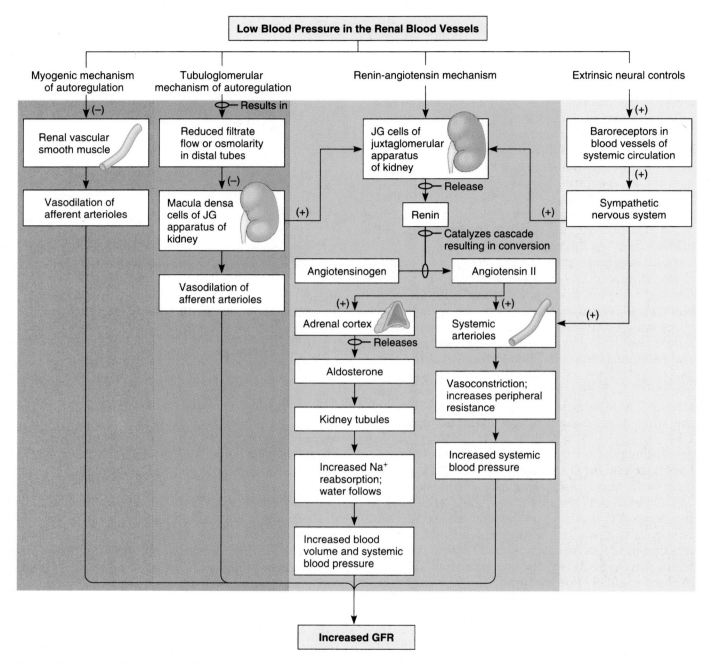

FIGURE 26.11 *Flowchart indicating the mechanisms regulating the glomerular filtration rate (GFR) in the kidneys.* (Note that [+] indicates a stimulatory effect and [−] indicates an inhibitory effect.)

emergency when it is necessary to shunt blood to the heart, brain, and skeletal muscles at the expense of the kidneys. At such times, norepinephrine released

by sympathetic nerve fibers (and epinephrine released by the adrenal medulla) interact with alpha-adrenergic receptors on vascular smooth muscle, causing strong constriction of the afferent arterioles, inhibiting filtrate formation. This, in turn, indirectly trips the renin-angiotensin mechanism by stimulating the macula densa cells. The sympathetic ner-

Because they are initiated by and within the kidney and allow the kidney to control its own (auto) rate of blood flow and GFR.

vous system also directly stimulates the JG cells (via the binding of norepinephrine to beta-adrenergic receptors) to release renin, bringing about an increase in systemic blood pressure via the renin-angiotensin mechanism.

Renin-Angiotensin Mechanism The **renin-angiotensin mechanism** is triggered when the JG cells of the juxtaglomerular apparatus release renin in response to various stimuli. **Renin** acts enzymatically on **angiotensinogen,** a plasma globulin made by the liver, to release **angiotensin I,** which is in turn converted to **angiotensin II** by **angiotensin converting enzyme (ACE)** associated with the capillary endothelium in various body tissues, particularly the lungs. Angiotensin II, a potent vasoconstrictor, activates smooth muscle of arterioles throughout the body, causing mean arterial blood pressure to rise. It also stimulates the adrenal cortex to release aldosterone, which causes the renal tubules to reclaim more sodium ions from the filtrate. Because water follows sodium osmotically, blood volume and blood pressure rise (see Figure 26.11). The afferent arterioles sport fewer angiotensin receptors than do the efferent arterioles, so angiotensin causes the efferent arterioles to constrict to a greater extent, thereby increasing the glomerular hydrostatic (blood) pressure. This defensive mechanism partially restores the GFR to normal levels. Angiotensin also targets the mesanglial cells associated with the glomerulus, causing them to contract and reduce the GFR in that manner.

Renin release is triggered by several factors acting independently or collectively. These include:

1. Reduced stretch of the granular JG cells. A drop in systemic blood pressure below 80 mm Hg (as might be due to hemorrhage, salt depletion, dehydration, etc.) stretches the JG cells to a lesser degree and directly stimulates them to release more renin.

2. Stimulation of the JG cells by input from activated macula densa cells (Figure 26.11). Under conditions in which the macula densa cells prompt a *reduced* release of the vasoconstrictor chemical (promoting vasodilation of the afferent arteriole), they also stimulate the JG cells to release renin.

3. Direct stimulation of the JG cells via β_1-adrenergic receptors by renal sympathetic nerves.

4. Direct stimulation of JG cells by angiotensin II itself.

Although the renin-angiotensin mechanism contributes to renal autoregulation, its main thrust is to stabilize systemic blood pressure and extracellular fluid volume. To that end, angiotensin II, once generated, also stimulates the hypothalamus to release antidiuretic hormone (ADH) and activates the hypothalamic thirst center. These processes are considered in more detail in Chapter 27.

Other Factors Renal cells produce a whole battery of chemicals, many of which act locally as signaling molecules (paracrines). Among these are the following.

1. Prostaglandins (eicosanoids): PGE_2, PGI_2, both vasodilators produced in response to sympathetic stimulation and angiotensin II, are believed to counteract the effect of norepinephrine and angiotensin II on the kidney. The adaptive value of these opposing actions is to prevent renal damage while responding to body demands to increase peripheral resistance.

2. Nitric oxide: A potent vasodilator produced by the vascular endothelium.

3. Adenosine: Although it functions as a vasodilator systemically, it constricts the renal vasculature. Adenosine is the most likely candidate for the vasoconstrictor of the tubuloglomerular mechanism.

4. Endothelin: Secreted by the vascular endothelium and selected tubule cells, endothelin is a powerful vasoconstrictor.

⚠ *Homeostatic Imbalance*

An abnormally low urinary output (less than 50 ml/day) is called *anuria* (ah-nu're-ah). It may indicate that glomerular blood pressure is too low to cause filtration, but renal failure and anuria usually result from situations in which the nephrons cease to function for a variety of reasons (e.g., acute nephritis, transfusion reactions, or crush injuries). ■

Tubular Reabsorption

Our total blood volume filters into the renal tubules about every 45 minutes, so all of our plasma would be drained away as urine within an hour were it not for the fact that most of the tubule contents are quickly reclaimed and returned to the blood. This reclamation process, called **tubular reabsorption,** is a *transepithelial process* that begins as soon as the filtrate enters the proximal tubules. Transported substances move through *three* membrane barriers—the *luminal* and *basolateral membranes* of the tubule cells (which abut the filtrate and interstitial fluid respectively) and the *endothelium* of the peritubular capillaries—to reach the blood. Because the tubule cells are connected by tight junctions, movement of substances between the cells is limited, although some important ions (Ca^{2+}, Mg^{2+}, K^+, and some Na^+) do use the *paracellular pathway* (movement between cells) in some cases.

Given healthy kidneys, virtually all organic nutrients such as glucose and amino acids are completely reabsorbed. The kidneys simply maintain or restore the normal plasma concentrations. On the other hand, the reabsorption of water and many ions is continuously regulated and adjusted in response

? *How are primary and secondary active transport processes (both shown here) different?*

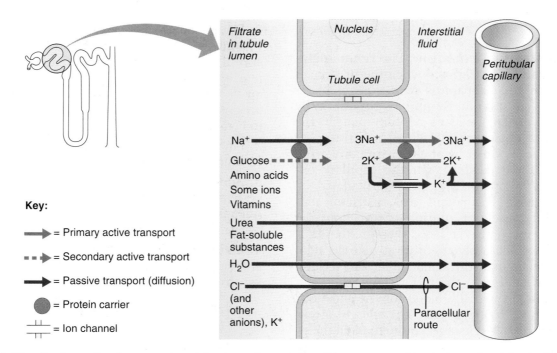

Key:

→ = Primary active transport

- - ▸ = Secondary active transport

▶ = Passive transport (diffusion)

● = Protein carrier

╥ = Ion channel

FIGURE 26.12 Reabsorption by PCT cells. Most often Na⁺ enters at the luminal surface by facilitated diffusion, and its transport is coupled to transport of another solute (a cotransport process referred to as secondary active transport). After Na⁺ enters, it diffuses to the basolateral membrane, where it is pumped into the interstitial space by a sodium pump (a Na^+-K^+ ATPase). From there it diffuses into the peritubular capillary. Active pumping of Na⁺ at the basolateral membrane creates concentration and osmotic gradients that drive reabsorption of water by osmosis, of anions and fat-soluble substances by diffusion, and of organic nutrients and selected cations by secondary active transport (symport with Na⁺ at the luminal membrane). Most of the organic nutrients are reabsorbed in the PCT.

to hormonal signals. Depending on the substances transported, the reabsorption process may be *passive* (none of its membrane transport steps require ATP) or *active* (at least one of its steps is driven by ATP directly or indirectly).

Sodium Reabsorption: Primary Active Transport

Sodium ions are the single most abundant cation in the filtrate, and the bulk of energy (about 80%) used for active transport is devoted to their reabsorption. Sodium reabsorption is almost always active and via the transcellular (transepithelial) route. In general, two basic processes that promote active reabsorption of Na⁺ occur in each tubule segment along the length of the nephron: (1) Sodium enters the tubule cells from the filtrate at the luminal membrane and then (2) sodium is actively transported out of the tubule cell by a Na^+-K^+ ATPase pump present in the basolateral membrane (Figure 26.12). From there, it moves passively by diffusion into the adjacent peritubular capillaries. Movement of Na⁺ and other absorbed substances into the peritubular capillary blood is rapid because of its low hydrostatic pressure and high osmotic pressure (remember, most proteins are not filtered out of the blood into the tubule).

The active pumping of Na⁺ from the tubule cell results in a strong electrochemical gradient that favors the *passive* entry of Na⁺ at the luminal face by carrier-mediated, facilitated diffusion. This occurs because (1) the pump maintains the intracellular Na⁺ concentration at low levels, and (2) the K⁺ that is pumped into the tubule cell almost immediately diffuses out again into the interstitial (extracellular) fluid via leakage channels, leaving the tubule cell interior (and the tubule lumen) with a net negative charge.

Since each of the tubule segments plays a slightly different role in reabsorption, the precise

---

In primary active transport, the energy for the process is provided directly by the cleavage of ATP. In secondary active transport, the substance moves by facilitated diffusion (typically cotransported with Na⁺) and the gradient energy is provided by the active Na⁺ pumping occurring elsewhere in the cell.

mechanism by which Na$^+$ is reabsorbed at the luminal membrane varies (see Figure 26.13, p. 1021). However, the net effect is that Na$^+$ reabsorption by primary active transport provides the energy and the means for reabsorbing most other solutes.

Reabsorption of Water, Ions, and Nutrients: Passive and Secondary Active Transport

In **passive tubular reabsorption,** which encompasses diffusion, facilitated diffusion, and osmosis, substances move along their electrochemical gradient without the use of ATP (see Chapter 3). As positively charged sodium ions move through the tubule cells into the peritubular capillary blood, they establish an electrical gradient. This gradient favors passive reabsorption of anions (Cl$^-$ and HCO$_3^-$ for example) into the peritubular capillaries to restore electrical neutrality in the filtrate and plasma. Just which anion is absorbed depends on the pH of the blood at the time, a topic considered in Chapter 27.

As we have seen, sodium movement also establishes a strong osmotic gradient, and water moves by osmosis (diffuses) into peritubular capillaries. Because water is "obliged" to follow salt, this sodium-linked water flow is referred to as **obligatory water reabsorption.**

As water leaves the tubules, the relative concentration of substances still present in the filtrate increases dramatically and, if able, they too begin to follow their concentration gradients through the tubule cells and into the peritubular capillaries (Figure 26.12). This phenomenon of solutes following the movement of the solvent explains the passive reabsorption of a number of solutes present in the filtrate, such as cations, fatty acids and some urea. It also explains in part why lipid-soluble drugs and environmental toxins are difficult to excrete. Although the gradient provides the driving force, remember that solute size and lipid solubility may be limiting factors.

Substances reabsorbed by **secondary active transport** (the "push" comes from the gradient created by Na$^+$-K$^+$ pumping at the basolateral membrane) include glucose, amino acids, lactate, vitamins, and most cations. In nearly all these cases, transport is linked to that of Na$^+$. A common carrier moves Na$^+$ down its concentration gradient by facilitated diffusion, as it cotransports (symports) another solute (Figure 26.12). Cotransported solutes diffuse through the basolateral membrane and into the peritubular capillaries.

There is some overlap of carriers; for example, fructose and galactose compete with glucose for the same Na$^+$-associated carrier. Even so, the transport systems for the various solutes are quite specific *and limited.*

Excluding Na$^+$, there is a **transport maximum** (**T$_m$**) for nearly every substance that is actively reabsorbed. The T$_m$ (reported in mg/min) reflects the number of carriers in the renal tubules available to ferry each particular substance. In general, there are plentiful carriers and high transport maximums for substances such as glucose that need to be retained and few or no carriers for substances of no use to the body. When the carriers are saturated—that is, all bound to the substance they transport—excesses of that substance are excreted in urine. This is exactly what happens in individuals who become hyperglycemic because of uncontrolled diabetes mellitus. As levels of glucose approach 400 mg/100 ml of plasma, its T$_m$ of 375 mg/min is exceeded and large amounts of glucose are lost in the urine even though the renal tubules are still functioning normally.

Any plasma proteins that squeeze through the filtration membrane are completely removed from the filtrate in the proximal tubule by endocytosis, also an active ATP-requiring process. Once inside the tubule cells, the proteins are digested to their amino acids, which then move into the peritubular blood.

Nonreabsorbed Substances

Some substances are not reabsorbed or are reabsorbed incompletely because they (1) lack carriers, (2) are not lipid soluble, or (3) are too large to pass through plasma membrane pores of the tubular cells. The most important of these substances are the nitrogenous end products of protein and nucleic acid metabolism: **urea, creatinine,** and **uric acid.** Urea is small enough to diffuse through the membrane pores, and 50–60% of the urea present in filtrate is reclaimed. Creatinine, a large, lipid-insoluble molecule, is secreted to a small extent and not reabsorbed at all. Its plasma concentration remains stable as long as muscle mass is stable, which makes it useful for measuring the GFR and assessing glomerular function.

Absorptive Capabilities of Different Regions of the Renal Tubules

Proximal Convoluted Tubule Although the entire length of the renal tubule is involved in reabsorption to some degree (Table 26.1), the proximal tubule cells are by far the most active "reabsorbers" and the events just described (illustrated in Figure 26.12) occur mainly in this tubular segment. Normally, all of the glucose, lactate, and amino acids are reabsorbed in the proximal tubule. Some 65% of the Na$^+$ present in filtrate is absorbed in the proximal tubule

TABLE 26.1	**Reabsorption Capabilities of Different Segments of the Renal Tubules and Collecting Ducts**	
Tubule segment	*Substance reabsorbed*	*Mechanism*
Proximal convoluted tubule	Sodium ions (Na^+)	Primary active transport via ATP-dependent Na^+-K^+ carrier; sets up electrochemical gradient for passive solute diffusion, osmosis, and secondary active transport (cotransport) with Na^+
	Virtually all nutrients (glucose, amino acids, vitamins)	Active transport; cotransport with Na^+
	Cations (K^+, Mg^{2+}, Ca^{2+}, and others)	Active transport; cotransport with Na^+ for most; K^+ mainly by the paracellular route
	Anions (Cl^-, HCO_3^-)	Passive transport; paracellular diffusion driven by electrochemical gradient for Cl^-; active transport (cotransport with Na^+ for HCO_3^-
	Water	Osmosis; driven by solute reabsorption (obligatory)
	Urea and lipid-soluble solutes	Passive diffusion driven by the electrochemical gradient created by osmotic movement of water
	Small proteins	Endocytosed by tubule cells and digested to amino acids within tubule cells
Loop of Henle Descending limb Ascending limb	Water	Osmosis
	Na^+, Cl^-, K^+	Active transport of Cl^-, Na^+, and K^+ via a Na^+-K^+-$2Cl^-$ cotransporter in thick portion; also paracellular transport; Na^+-H^+ antitransport
	Ca^{2+}, Mg^{2+}	Passive transport driven by electrochemical gradient; paracellular route
Distal convoluted tubule	Na^+	Primary active transport; requires aldosterone
	Ca^{2+}	PTH-mediated primary active transport via ATP-dependent Ca^{2+} carrier
	Cl^-	Diffusion; follow electrochemical gradient created by active reabsorption of Na^+; also cotransport with Na^+
	Water	Osmosis; facultative water reabsorption; depends on ADH to increase porosity of tubule epithelium in the most distal portion
Collecting duct	Na^+, H^+, K^+, HCO_3^-, and Cl^-	Aldosterone-mediated primary active transport of Na^+ and the medullary gradient create the conditions for passive transport of some HCO_3^- and Cl^- and cotransport of H^+, K^+, Cl^-, and HCO_3^-
	Water	Osmosis; facultative water reabsorption; depends on ADH to increase porosity of tubule epithelium
	Urea	Facilitated diffusion in response to concentration gradient in the deep medulla region; most remains in medullary interstitial space

by symport with other solutes or by Na^+-H^+ exchange (Figure 26.13a). Because water follows in the same proportion, about 65% of the water is also reabsorbed there. Additionally, 90% of the filtered bicarbonate (HCO_3^-), 50% of the chloride (Cl^-), and about 55% of the potassium (K^+) are reclaimed in the PCT. Likewise, the bulk of the *selective*, or active transport–dependent, reabsorption of electrolytes is accomplished by the time the filtrate reaches the loop of Henle. For the most part, the reabsorption of electrolytes such as calcium, phosphate, and magnesium is hormonally controlled and closely related to

the need to regulate their plasma levels. (These controls are described in Chapter 27.) Nearly all of the uric acid is reabsorbed in the proximal tubule, but it is later secreted back into the filtrate. Of the 125 ml/min of fluid that is filtered into the renal tubules, about 40 ml/min remains to enter the loop of Henle when the reabsorption mechanisms of the PCT have finished their work.

Loop of Henle Beyond the proximal convoluted tubule, the permeability characteristics of the tubule epithelium change dramatically (see Figure 26.15,

p. 1026). Here, for the first time, water reabsorption is not coupled to solute reabsorption. Water can leave the descending limb of the loop of Henle but *not* the ascending limb. For reasons that will be explained shortly, these permeability differences play a vital role in the kidneys' ability to form dilute and concentrated urine. Another 25% of the filtered Na^+, 10% of the water, 35% of the filtered Cl^-, and 30% of the K^+ is reabsorbed in the loop of Henle. However, K^+ recycles—it is reabsorbed in the ascending limb and secreted from the descending limb.

A Na^+-K^+-$2Cl^-$ symporter is the main means of Na^+ entry at the luminal surface in the thick portion of the ascending limb (Figure 26.13b). However, this segment also has Na^+-H^+ antiporters like the PCT, and some 50% of Na^+ passes via the paracellular route in this tubular region. In the thin portion of the ascending limb (found mainly in the inner medulla), Na^+ moves passively but the mechanism is unclear.

Distal Convoluted Tubule and Collecting Duct
By the time the DCT is reached, only about 10% of the originally filtered NaCl and 25% of the water remain in the tubule; fluid flow into the DCT is about 25 ml/min. Na^+-Cl^- symporters absorb Na^+ and Cl^-, but most reabsorption from this point on depends largely on the need of the body at the time and is regulated by hormones. If necessary, nearly all of the water and Na^+ reaching these regions can be reclaimed. In the absence of regulatory hormones, the distal tubule and collecting duct are relatively impermeable to water. Reabsorption of more water depends on the presence of antidiuretic hormone (ADH), which makes the collecting ducts more permeable to water. This mechanism is described shortly.

Reabsorption of the remaining Na^+ is "fine-tuned" by the hormone aldosterone. Several conditions—for example, decreased blood volume or blood pressure, low Na^+ concentration (hyponatremia) or increased K^+ concentration (hyperkalemia) in the extracellular fluid—cause the adrenal cortex to release aldosterone to the blood. Except for hyperkalemia (which directly stimulates the adrenal cortex to secrete aldosterone), just about all of these conditions promote the renin-angiotensin mechanism, which in turn prompts the release of aldosterone (see Figure 26.11). Aldosterone targets the principal cells of the collecting ducts (prodding them to open or synthesize more Na^+ channels—the route by which Na^+ passively enters through the luminal surface—and to synthesize more basolateral Na^+-K^+ transporters and K^+ channels). As a result, little or no Na^+ leaves the body in urine. In the absence of aldosterone, essentially no sodium is reabsorbed by these segments, re-

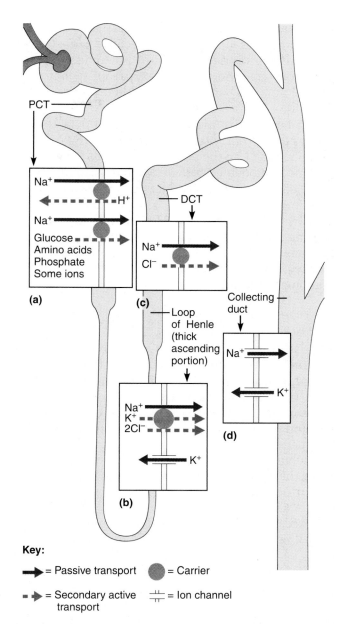

Key:

➤ = Passive transport   ⬤ = Carrier

▪▪➤ = Secondary active transport   ǂ = Ion channel

FIGURE 26.13 *Some of the mechanisms by which Na⁺ enters at the luminal surface of the tubule cells.* In each segment, the gradient for passive Na^+ entry is provided by a Na^+-K^+ ATPase pump at the basolateral membrane (not illustrated). **(a)** In the PCT, Na^+ entry by facilitated diffusion occurs using both symport (glucose and other nutrients cotransported) and antiport (e.g., H^+–Na^+ exchange) carriers. **(b)** Facilitated diffusion of Na^+ entry entails a Na^+-K^+-$2Cl^-$ symport process in the thick portion of the ascending loop of Henle. **(c)** Na^+ entry is via a Na^+-Cl^+ symporter in the DCT. **(d)** Na^+ enters the collecting duct by diffusion through membrane pores. (As Na^+ is reabsorbed, K^+ is secreted.)

sulting in catastrophic Na^+ losses of about 2% of Na^+ filtered daily, an amount that is incompatible with life.

A second (indirect) effect of aldosterone is to promote water absorption because as sodium is reabsorbed, water follows it back into the blood (if it can).

Aldosterone also regulates (reduces) potassium ion concentrations in the blood because aldosterone-induced reabsorption of sodium is coupled to potassium ion secretion in the principal cells, that is, as Na^+ enters, K^+ diffuses into the lumen (Figure 26.13d).

In contrast to aldosterone, which acts to conserve Na^+, *atrial natriuretic peptide (ANP)*, a hormone released by the atrial cardiac cells when blood volume or blood pressure is elevated, acts directly on the medullary collecting ducts to inhibit Na^+ reabsorption. It acts indirectly to inhibit Na^+ reabsorption by counteracting ADH's stimulation of aldosterone release by the adrenal cortex and stimulates an increase in the GFR by targeting renal arterioles. Consequently, ANP also reduces water reabsorption and blood volume.

Tubular Secretion

The failure of tubule cells to reabsorb some or all of certain filtered solutes is an important means of clearing plasma of unwanted substances. A second such mechanism is **tubular secretion**—essentially, reabsorption in reverse. Substances such as hydrogen and potassium ions, creatinine, ammonium ion, and certain organic acids move from blood of the peritubular capillaries through the tubule cells or from the tubule cells themselves into filtrate. Thus, the urine that is eventually excreted is composed of *both filtered and secreted substances*. With one major exception (K^+), the proximal tubule is the main site of secretion (see Figure 26.13), but the cortical parts of the collecting ducts and the late regions of the distal tubules and collecting ducts are also active in secretion (see Figure 26.16, p. 1027).

Tubular secretion is important for:

1. Disposing of substances not already in the filtrate, such as certain drugs (penicillin and phenobarbital, for example)

2. Eliminating undesirable substances or end products that have been reabsorbed by passive processes, such as urea and uric acid

3. Ridding the body of excessive potassium ions

4. Controlling blood pH

Nearly all of the potassium ions in urine are derived from active tubular secretion into the collecting ducts under the influence of aldosterone because virtually all of the K^+ present in the filtrate is reabsorbed in the PCT and ascending loop of Henle. When blood pH begins to drop toward the acid end of its homeostatic range, the renal tubule cells actively secrete hydrogen ions into the filtrate and retain more bicarbonate and potassium ions. As a result, the blood pH rises and the urine drains off the excess acid. Conversely, when blood pH approaches the alkaline end of its range, Cl^- rather than HCO_3^- is reabsorbed and bicarbonate is allowed to leave the body in urine. Renal mechanisms for controlling the acid-base balance of the blood are described in more detail in Chapter 27.

Regulation of Urine Concentration and Volume

Because we will make frequent use of the term *osmolality* (oz"mo-lal'ĭ-te) in this discussion, it is important that this concept be clearly understood. A solution's **osmolality** is the number of solute particles dissolved in one liter (1000 g) of water* and is reflected in the solution's ability to cause osmosis (and to alter certain physical properties of the solvent, such as boiling and freezing points). This capability, called *osmotic activity*, is determined only by the number of nonpenetrating solute particles and is independent of their type and nature. Ten sodium ions have the same osmotic activity as ten glucose molecules or ten amino acids in the same volume of solution. Since the *osmol* (equivalent to 1 mole of a nonionizing substance in 1 L of water) is a fairly large concentration unit, the **milliosmol** (mil"e-oz'mōl) **(mOsm)**, equal to 0.001 osmol, is used to describe the solute concentration of body fluids.

One of the crucial functions of the kidneys is to keep the solute load of body fluids constant at about 300 mOsm, the approximate osmotic concentration of blood plasma, by regulating urine concentration and volume. The kidneys accomplish this feat by the so-called **countercurrent mechanism.** The term *countercurrent* indicates that something is flowing in opposite directions through adjacent channels. In the kidneys the countercurrent mechanism involves the interaction between the flow of filtrate through the long loop of Henle of juxtamedullary nephrons (the *countercurrent multiplier*), and the flow of blood through the limbs of adjacent vasa recta blood vessels, the *countercurrent exchanger.* This relationship establishes and maintains an osmotic gradient extending from the cortex through the depths of the medulla that allows the kidneys to vary urine concentration quite dramatically. Let's look at how this coupled mechanism works.

The Countercurrent Mechanism and the Medullary Osmotic Gradient

The osmolality of the filtrate entering the proximal convoluted tubule is identical to that of plasma, about 300 mOsm. As described earlier, large amounts of water *and* salt are reabsorbed by the

---

*This is as opposed to the term *osmolarity*, which refers to the number of dissolved particles in one liter of *solution*.

A CLOSER LOOK   Renal Failure and the Artificial Kidney

Although renal failure is not common, it occurs when functioning nephrons become too few to carry out normal kidney functions. Possible causes of renal failure are:

1. Repeated damaging infections of the kidneys

2. Physical injury to the kidneys or major physical trauma (crushing injury) to other parts of the body

3. Prolonged pressure on skeletal muscles (causes release of myoglobin, which can clog the renal tubules)

4. Chemical poisoning of the tubule cells by heavy metals (mercury or lead) or organic solvents (dry-cleaning fluid, paint thinner, acetone, and others)

5. Inadequate blood delivery to the tubule cells (as may happen with arteriosclerosis)

In renal failure, filtrate formation decreases or stops completely. Nitrogenous wastes accumulate quickly in the blood when the tubule cells are not working, a condition called *azotemia* (az″o-te′me-ah), and blood pH tumbles to the acidic range. *Uremia* (u-re′me-ah), the bodywide (or systemic) response to azotemia, and electrolyte imbalance set in shortly and totally disrupt life processes. Signs and symptoms of uncontrolled uremia include diarrhea, vomiting, edema (due to Na^+ retention), labored breathing, cardiac irregularities (due to excess K^+), convulsions, coma, and death.

To prevent these consequences of azotemia, the blood must be cleansed of metabolic wastes and its ionic composition adjusted to a normal level by *dialysis* (*dialys* = separate) while the kidneys are shut down. In *hemodialysis*, which uses an "artificial kidney" apparatus, the patient's blood is passed through a membrane tubing

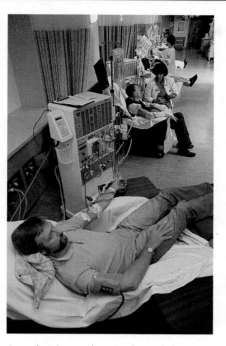

A renal patient undergoing hemodialysis.

(see the diagram) that is permeable only to selected substances, and the tubing is immersed in a bathing solution that differs slightly from normal cleansed plasma. As blood circulates through the tubing, substances such as nitrogenous wastes and K^+ present in the blood (but not in the bath) diffuse out of the blood into the surrounding solution and substances to be added to the blood, mainly buffers for H^+ (and glucose for malnourished patients), move from the bathing solution into the blood. In this way, needed substances are retained in the blood or added to it, while wastes and ion excesses are removed. Hemodialysis is routinely done three times weekly and each session takes 4 to 8 hours. Serious problems occasionally encountered by hemodialysis patients are thrombosis, infection, and ischemia in the

extremity where the shunt is inserted. Also, because the patient's blood must be heparinized to prevent clotting as it travels through the artificial kidney, there is some risk of hemorrhage.

A less efficient but more convenient procedure for patients is *continuous ambulatory peritoneal dialysis (CAPD)*. CAPD uses the patient's own peritoneal membrane as the dialyzing membrane. Isosmotic fluid that is also equal in chemical content to normal plasma and interstitial fluid is introduced into the patient's peritoneal cavity with a catheter and left to equilibrate there for 15 to 60 minutes. Then the dialysate is retrieved from the peritoneal cavity and replaced with fresh dialysis fluid. The procedure is repeated until the patient's blood chemistry reaches normal. Because some ambulatory patients may be inattentive to cloudy or bloody dialysis fluid, infection is more common in CAPD than in hemodialysis.

When renal damage is irreversible, the kidneys become totally incapable of processing plasma or concentrating urine, and a kidney transplant is the only answer. Unhappily, the signs and symptoms of this life-threatening kidney problem become obvious only after about 75% of renal function has been lost. (This large margin of safety reflects the fact that individual nephrons hypertrophy and begin to function even more effectively as others are progressively lost.) At this point, the signs of *renal insufficiency* becomes obvious. Glomerular filtration declines. Azotemia occurs as the kidneys lose their ability to form both concentrated and dilute urine, and the urine becomes isotonic with blood plasma. The end stage of renal failure, uremia, occurs when about 90% of the nephrons have been lost.

proximal tubule cells; and though the amount and flow of filtrate are reduced by 65% by the time the loop of Henle is reached, the filtrate is still isosmotic.

The stage is set for conserving most of this remaining filtrate as the osmolality of the interstitial fluid increases from the cortex to the depths of the medulla. The solute concentration increases from 300 mOsm in the cortex to about 1200 mOsm in the deepest part of the medulla (Figure 26.14). How can this tremendous concentration of the medullary fluids be explained? The answer lies in the unique workings of the loop of Henle of the juxtamedullary nephrons, which penetrates deep into the renal medulla, and their closely associated vasa recta. Notice that in each case the fluids involved—filtrate and blood respectively—first descend and then ascend through parallel limbs. First, we will follow filtrate processing through the loop of Henle, as portrayed in Figure 26.14a, to see how it functions as a **countercurrent multiplier** to establish the osmotic gradient.

1. The descending limb of the loop of Henle is relatively impermeable to solutes and freely permeable to water. Because the osmolality of the medullary interstitial fluid gradually increases along the course of the descending limb (the mechanism of this increase is explained shortly), water passes out of the filtrate osmotically all along its course. Thus, the osmolality of the filtrate (particularly as concerns its salt content) reaches its highest point (1200 mOsm) at the "elbow" of the loop of Henle.

2. The ascending limb of the loop of Henle is impermeable to water and actively transports sodium chloride into the surrounding interstitial fluid. As the filtrate rounds the corner into the ascending limb, the tubule permeability changes, becoming impermeable to water and selectively permeable to salt. The concentration of sodium and chloride ions in the filtrate entering the ascending limb is very high (and interstitial fluid concentrations are lower). Although the thin segment of the ascending limb of the loop of Henle also contributes to Na^+ and Cl^- reabsorption, the thick segment in particular is very active in reabsorbing salt (NaCl) via its Na^+-K^+-$2Cl^-$ cotransporter. As Na^+ and Cl^- are extruded from the filtrate into the medullary interstitial fluid, they contribute to the high osmolality there.

Movement of Na^+ and Cl^- out of the filtrate in the ascending limb establishes a strong osmotic gradient for water reabsorption. However, the ascending limb of the loop of Henle is nearly impermeable to water, which therefore remains in the filtrate. Consequently, the amount of fluid filtrate/min that reaches the DCT is the same as the amount we had moving through the elbow of the loop of Henle, 19 ml/min. Because the filtrate loses salt but not water,

it becomes increasingly dilute until, at 100 mOsm at the DCT, it is hypo-osmotic, or hypotonic, to blood plasma and cortical fluids. Notice in Figure 26.14 that there is a rather constant difference in filtrate concentration between the two limbs of the loop of Henle. Because the ascending limb is relatively more permeable to Na^+ and Cl^-, some salt tends to diffuse back into the filtrate in that portion of the tubule, even as it is being actively reabsorbed. Consequently, the solute content in the filtrate of the ascending limb is always about 200 mOsm lower than in the descending limb and that of the surrounding medullary interstitial fluid. However, because of countercurrent flow, the loop of Henle is able to "multiply" these small changes in solute concentration into a gradient change along the vertical length of the loop (both inside and outside) that is closer to 900 mOsm.

Although the two limbs of the loop of Henle are not in direct contact, they are close enough to influence each other's exchanges with the interstitial fluid they both share. Hence, water diffusing out of the descending limb produces the increasingly "salty" filtrate that the ascending limb uses to raise the osmolality of the medullary interstitial fluid. Furthermore, the more salt the ascending limb extrudes, the more water diffuses out of the descending limb and the saltier the filtrate in the descending limb becomes. Thus, a positive feedback mechanism is established that produces the high osmolality of the fluids in the descending limb and the interstitial fluid. The energy that is used to establish the medullary gradient ultimately comes from active sodium transport in the ascending limb.

3. The collecting ducts in the deep medullary regions are permeable to urea. In a normally hydrated person, the dilute urine passes through the distal convoluted tubule and the superior portions of the collecting ducts without major modification. In the presence of aldosterone some more NaCl is reabsorbed in the DCT and collecting ducts, reducing the filtrate osmolality still more (Figure 26.14a), but the amount of urea in the filtrate remains high because most nephron tubule segments beyond the PCT are impermeable to it. However, as urine passes through the deep medullary regions where the collecting ducts are highly permeable to urea, urea diffuses out of the ducts into the medullary interstitial fluid, where it contributes to the high osmolality in that region. Urea continues to move passively until its concentration inside and outside the duct is equal. Even though the ascending limb of the loop of Henle is poorly permeable to urea, when urea concentration in the medullary interstitial space is very high, some urea does enter the limb. However, it simply gets recycled back to the collecting duct, where it diffuses out again. Unlike the active

? *What portion of the nephron acts as the countercurrent multiplier? As the countercurrent exchanger?*

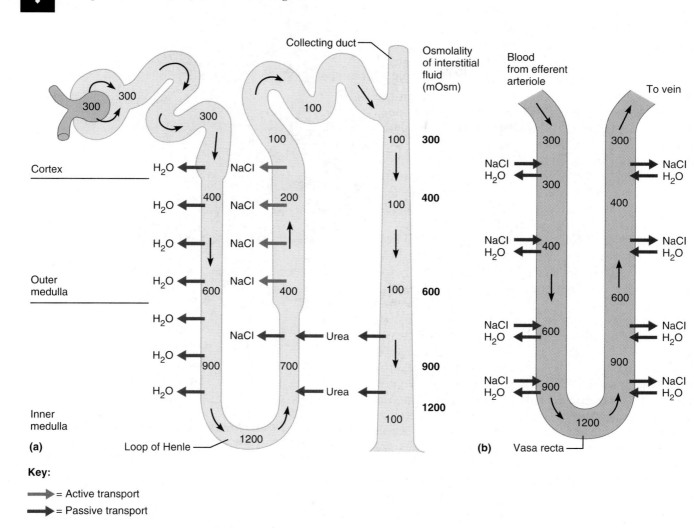

Key:

➤ = Active transport

➤ = Passive transport

FIGURE 26.14 *Countercurrent mechanism for establishing and maintaining the medullary osmotic gradient.* (a) Filtrate entering the descending limb of the loop of Henle is isosmotic (isotonic) to both blood plasma and the surrounding (cortical) interstitial fluid. As the filtrate flows from the cortex to the medulla in the descending limb, water leaves the tubule by osmosis and the relative solute content (osmolality) of the filtrate increases from 300 to 1200 mOsm. As the filtrate rounds the "bend" and flows into the ascending limb of the loop of Henle, the permeability of the tubule epithelium changes from water permeable, largely solute (salt) impermeable to the opposite condition: water impermeable, salt permeable. Consequently, salt leaves the ascending limb, diluting the filtrate as it approaches the cortex. Thus the different permeabilities of the two limbs of the loop of Henle cooperate to establish the osmotic gradient in the medullary interstitial fluid: The descending limb produces increasingly salty filtrate, while the ascending limb uses this high salt concentration to maintain the high osmolality of the interstitial fluid in the medulla. Diffusion of urea from the lower portion of the collecting duct also contributes to the high osmolality in the medulla. (b) Dissipation of the medullary osmotic gradient is prevented because the blood in the vasa recta continuously equilibrates with (becomes isosmotic to) the interstitial fluid—becoming more concentrated as it follows the descending limb of the loop of Henle and less concentrated as it approaches the cortical region. This occurs because of the high porosity and sluggish blood flow in these specialized vessels.

reabsorption of NaCl in the thick part of the ascending limb of the loop of Henle, which is a crucial element of the countercurrent multiplier system, urea's cycling simply serves to equalize its concentration inside and outside the renal tubules.

4. The vasa recta act as a countercurrent exchanger, maintaining the osmotic gradient while delivering the nutrient blood supply to cells in the area. If ordinary capillaries paralleled the deep loops of Henle, the medullary gradient would quickly be destroyed because as massive amounts of Na^+ were absorbed from the medulla, equally large amounts of water would be lost into the medullary interstitial fluid. By contrast, the vasa recta play an important role in maintaining the osmotic gradient established by the cycling of salt. Because these vessels receive only about 10% of the renal blood supply, blood flow through the vasa recta is sluggish. Moreover, the vasa recta are freely permeable to water and salt, and blood in the vasa recta makes passive exchanges with the surrounding interstitial fluid, achieving equilibrium all along its path. Consequently, as the blood flows into the medullary depths, it loses water and gains salts (becomes hypertonic). Then, as it emerges from the medulla into the cortex, the process is reversed. The blood picks up water and loses salt (see Figure 26.14b). Because blood leaving and then reentering the cortex via the vasa recta has the same solute concentration, the vasa recta act as a **countercurrent exchanger.** This system does not create the medullary gradient (that is set up by the loop of Henle), but it protects it by preventing rapid removal of salt from the medullary interstitial space.

Now that the basic mechanisms establishing the medullary gradient have been explained, we can explore how the kidneys excrete dilute and concentrated urine.

Formation of Dilute Urine

Because tubular filtrate is diluted as it travels through the ascending limb of the loop of Henle, all the kidney needs to do to secrete dilute (hypoosmotic) urine is allow the filtrate (described above) to continue on its way into the renal pelvis. Under these conditions, fluid enters the pelvis at a rate of 15–19 ml/min. When antidiuretic hormone is not being released by the posterior pituitary, that is exactly what happens. The collecting ducts remain essentially impermeable to water, and no further water reabsorption occurs. Moreover, as noted, sodium and selected other ions can be removed from the filtrate by active or passive mechanisms by distal and collecting tubule cells so that urine becomes even more dilute before entering the renal pelvis (Figure 26.14a). The osmolality of urine can plunge as low

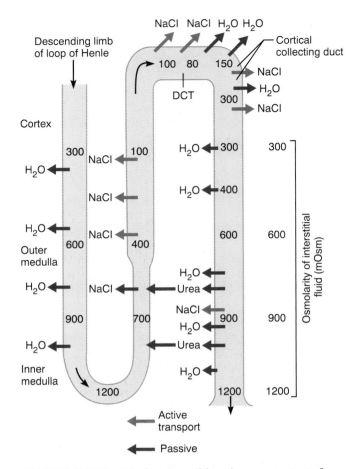

FIGURE 26.15 *Mechanism of forming concentrated urine in the presence of ADH.* ADH activates the Na^+-absorbing principal cells of the late DCT and collecting duct.

as 50 mOsm, about one-sixth the concentration of glomerular filtrate or blood plasma.

Formation of Concentrated Urine

As its name reveals, **antidiuretic hormone (ADH)** inhibits *diuresis* (di"u-re'sis), or urine output. It accomplishes this via a second-messenger system using cyclic AMP that increases the number of water-filled channels in the principal cells of the collecting ducts by stimulating the transport of the protein subunits of the (water) channels to their luminal surfaces where they are inserted. Consequently water passes easily into the cells and then through the water-permeable basolateral membrane into the interstitial space, and the osmolality of the filtrate becomes equal to that of the interstitial fluid.

In the initial parts of the distal tubules in the cortex, the osmolality of the filtrate is approximately 100 mOsm, but as the filtrate flows through the collecting ducts and is subjected to the hyperosmolar conditions in the medulla, water rapidly leaves the filtrate (Figures 26.15 and 26.16e). Depending on

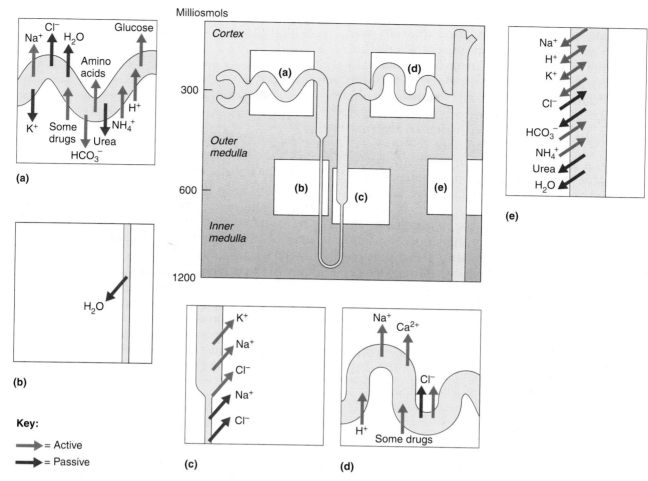

FIGURE 26.16   *Summary of nephron functions.* The glomerulus provides the filtrate that is processed by the renal tubule. The different regions of the renal tubule carry out reabsorption and secretion and maintain a gradient of osmolality within the medullary interstitial fluid. Varying osmolality at different points in the tubule and interstitial fluid is symbolized in the central figure by gradients of color. The inserts that accompany the central figure describe the main transport functions of the four regions of the nephron tubule and of the collecting duct. Blue arrows indicate passive transport processes, red arrows represent primary or secondary active transport. Blood vessels are not shown. **(a) Proximal tubule.** Filtrate that enters the proximal tubule from the glomerular capsule has about the same osmolality as blood plasma. The main activity of the transport epithelium of the proximal tubule is reabsorption of certain solutes from the filtrate back into the blood. Nearly all nutrients, such as glucose and amino acids, and about 65% of Na^+ are actively transported out of the tubule and enter the peritubular capillaries; Cl^- and water follow passively. The proximal tubule cells also secrete specific substances, such as the ammonium ion

and other nitrogenous wastes, into the filtrate and help maintain a constant pH in blood and interstitial fluid by the controlled secretion of H^+ and reabsorption of HCO_3^-. By the end of the proximal tubule, the filtrate volume has been reduced by 65%. **(b) Descending limb.** The descending limb of the loop of Henle is freely permeable to water but not to salt. As filtrate in the limb descends into the medulla, it loses water by osmosis to the interstitial fluid, which is increasingly hypertonic in that direction. Consequently, salt and other solutes become more concentrated in the filtrate. **(c) Ascending limb.** The filtrate enters the ascending limb of the loop of Henle, which is impermeable to water but is permeable to Na^+ and Cl^-. These ions, which have become concentrated within the descending limb, move passively out of the thin portion and are actively pumped out of the thick portion of the ascending limb and contribute to the high osmolality of interstitial fluid in the inner medulla. K^+ is cotransported with Na^+ and Cl^-. Because the tubule epithelium here is not permeable to water, the filtrate becomes more and more dilute as the exodus of salt from the filtrate continues. **(d) Distal tubule.** The distal tubule, like the proximal tubule, is specialized for

selective secretion and reabsorption. Na^+ and Cl^- are cotransported, H^+ may be secreted, and in the presence of aldosterone, more Na^+ is reabsorbed. The water permeability of the distal tubule is extremely low and almost no further H_2O absorption occurs there. **(e) Collecting duct.** The urine, normally quite dilute at this point, begins its journey via the collecting duct back into the medulla, with its increasing osmolality gradient. In the cortical collecting duct, K^+, H^+, and/or HCO_3^- ions may be reabsorbed or secreted depending on what is required to maintain blood pH homeostasis. The wall of the medullary region of the collecting duct is permeable to urea and is made more so by the presence of ADH. A certain amount of urea diffuses out of the collecting duct and contributes to the high osmolality of the inner medulla. In the absence of antidiuretic hormone, the collecting ducts are nearly impermeable to water, and the dilute urine passes out of the kidney. In the presence of ADH, the pores of the collecting duct enlarge, and the filtrate loses water by osmosis as it passes through medullary regions of increasing osmolality. Consequently, water is conserved, and concentrated urine is excreted.

the amount of ADH released (which is keyed to the level of body hydration), urine concentration may rise as high as 1400 mOsm, the concentration of interstitial fluid in the deepest part of the medulla. In the presence of maximal ADH secretion, 99% of the water in filtrate is reabsorbed and returned to the blood, and only a very small amount (about 1 ml/min) of highly concentrated urine is excreted. The ability of our kidneys to produce such concentrated urine is critically tied to our ability to survive without water. Water reabsorption that depends on the presence of ADH is called **facultative water reabsorption.**

ADH is released more or less continuously unless the solute concentration of the blood drops too low. Its release is enhanced by any event that increases plasma osmolality to above 300 mOsm/L, such as excessive water loss through sweating or diarrhea, or reduced blood volume or blood pressure (hemorrhage). These mechanisms are considered in Chapter 27. Although ADH is the "signal" to produce concentrated urine that opens the door (pores) for water reabsorption, the kidneys' ability to respond depends on the high medullary osmotic gradient.

Diuretics

Diuretics are chemicals that enhance urinary output. Any substance that is not reabsorbed or that exceeds the ability of the renal tubules to reabsorb it will act as an *osmotic diuretic* and carry water out with it. For example, the high blood glucose levels of a diabetes mellitus patient act as an osmotic diuretic. Alcohol, essentially a sedative, encourages diuresis by inhibiting release of ADH. Caffeine (found in coffee, tea, and colas) and most diuretic drugs commonly prescribed for hypertension or the edema of congestive heart failure increase urine flow by inhibiting sodium ion reabsorption and thus the obligatory water reabsorption that normally follows. For example, a *loop diuretic*, like furosemide (Lasix), inhibits the Na^+-K^+-$2Cl^-$ symporters in the thick portion of the ascending loop of Henle, and thiazide diuretics (Diuril and others) inhibit Na^+-Cl^- transport in the DCT by competing for the Cl^- site on the symporter.

Renal Clearance

The term **renal clearance** refers to the volume of plasma that is cleared of a particular substance in a given time, usually 1 minute. Renal clearance tests are used to determine the GFR, which provides information about the amount of functioning renal tissue. Renal clearance tests are also done to detect glomerular damage and to follow the progress of diagnosed renal disease.

The renal clearance rate (RC) of any substance, in ml/min, is calculated from the equation

$$RC = UV/P$$

where:

U is the concentration (mg/ml) of the substance in urine

V is the flow rate of urine formation (ml/min)

P is the concentration of the same substance in plasma

Inulin (in'u-lin), a polysaccharide with a molecular weight of approximately 5000, is often used as the standard to determine the GFR by renal clearance because it is not reabsorbed, stored, or secreted by the kidneys. Because infused inulin is eliminated only in the urine, its renal clearance value is equal to the GFR. Measured values for inulin are $U = 125$ mg/ml, $V = 1$ ml/min, and $P = 1$ mg/ml. Therefore, the computation of its renal clearance is RC = (125 × 1)/1 = 125 ml/min, meaning that, in 1 minute, the kidneys have removed (cleared) all the inulin present in 125 ml of plasma.

A clearance value less than that of inulin means that the substance is partially reabsorbed. For example, urea has a clearance value of 70 ml/min, meaning that of the 125 ml of glomerular filtrate formed each minute, approximately 70 ml is completely cleared of urea, while the urea in the remaining 55 ml is recovered and returned to the plasma. If the renal clearance is zero, reabsorption is complete. Glucose has a clearance value of zero in healthy individuals, and HCO_3^-, Na^+, Cl^-, and Ca^{2+} are all close to zero (less than 2 ml/min). If clearance is greater than that of inulin, the tubule cells are secreting the substance into the filtrate. This is the case with creatinine, which has an RC of 140 ml/min, and most drug metabolites. Knowing a drug's renal clearance value is essential because if it is high, then the drug dosage must also be high and administered frequently to maintain a therapeutic dosage.

Characteristics and Composition of Urine

Physical Characteristics

Color and Transparency Freshly voided urine is generally clear and pale to deep yellow in color. Its yellow color is due to **urochrome** (u'ro-krōm), a pigment that results from the body's destruction of hemoglobin (via bilirubin or bile pigments). The more concentrated the urine, the deeper the yellow color. An abnormal color such as pink or brown, or a smoky tinge, may result from eating certain foods (beets, rhubarb) or may be due to the presence of bile pigments or blood in the urine. Additionally, some

TABLE 26.2	Abnormal Urinary Constituents	
Substance	*Name of condition*	*Possible causes*
Glucose	Glycosuria	Nonpathological: excessive intake of sugary foods Pathological: diabetes mellitus
Proteins	Proteinuria, or albuminuria	Nonpathological: excessive physical exertion; pregnancy; high-protein diet Pathological (over 250 mg/day): heart failure, severe hypertension; glomerulonephritis; often initial sign of asymptomatic renal disease
Ketone bodies	Ketonuria	Excessive formation and accumulation of ketone bodies, as in starvation and untreated diabetes mellitus
Hemoglobin	Hemoglobinuria	Various: transfusion reaction, hemolytic anemia, severe burns, etc.
Bile pigments	Bilirubinuria	Liver disease (hepatitis, cirrhosis) or obstruction of bile ducts from liver or gallbladder
Erythrocytes	Hematuria	Bleeding urinary tract (due to trauma, kidney stones, infection, or neoplasm)
Leukocytes (pus)	Pyuria	Urinary tract infection

commonly prescribed drugs and vitamin supplements alter the color of urine. Cloudy urine may indicate an infection in some part of the urinary tract.

Odor Fresh urine is slightly aromatic. But if it is allowed to stand, it develops an ammonia odor due to bacterial metabolism of its urea solutes. Some drugs and vegetables (asparagus) alter the usual odor of urine, as do some diseases. For example, in diabetes mellitus the urine smells fruity because of its acetone content.

pH Urine is usually slightly acidic (around pH 6), but changes in body metabolism or diet may cause it to vary from about 4.5 to 8.0. An *acid-ash* diet that contains large amounts of protein and whole wheat products produces acidic urine. A vegetarian *(alkaline-ash)* diet, prolonged vomiting, and bacterial infection of the urinary tract all cause the urine to become alkaline.

Specific Gravity Because urine is water plus solutes, it weighs more (is more dense) than distilled water. The term used to compare the weight of a substance to the weight of an equal volume of distilled water is **specific gravity.** The specific gravity of distilled water is 1.0. The specific gravity of urine ranges from 1.001 to 1.035, depending on its solute concentration. When urine is extremely concentrated, solutes begin to precipitate out of solution.

Chemical Composition

Water accounts for about 95% of urine volume; the remaining 5% consists of solutes. The largest component of urine by weight, apart from water, is *urea*, which is derived from the normal breakdown of amino acids. Other **nitrogenous wastes** found in urine include *uric acid* (an end product of nucleic acid metabolism) and *creatinine* (a metabolite of creatine phosphate). Creatine phosphate, a molecule that stores energy for the regeneration of ATP, is found in large amounts in skeletal muscle tissue. Normal solute constituents of urine, in order of decreasing concentration, include urea; sodium, potassium, phosphate, and sulfate ions; creatinine; and uric acid. Much smaller but highly variable amounts of calcium, magnesium, and bicarbonate ions are also found in urine. Abnormally high concentrations of any of these urinary constituents may indicate pathology. (Normal urine values are listed in Appendix E.)

With certain diseases urine composition changes dramatically and may include, for example, glucose, blood proteins, red blood cells, hemoglobin, white blood cells (pus), or bile pigments. The presence of such abnormal substances in urine is an important warning of illness and helps in diagnosis (Table 26.2).

URETERS

The **ureters** are slender tubes that convey urine from the kidneys to the bladder (see Figure 26.1). Each ureter begins at the level of L_2 as a continuation of the renal pelvis. From there it descends behind the peritoneum to the base of the bladder, turns medially and runs obliquely through the posterior bladder wall. This arrangement prevents backflow of urine into the ureters during bladder filling because any increase in pressure within the bladder compresses and closes the distal ends of the ureters.

Histologically, the ureter wall is trilayered. The transitional epithelium of its lining *mucosa* is continuous with that of the kidney pelvis superiorly and the bladder medially. Its middle *muscularis* is composed chiefly of two smooth muscle sheets: the inner longitudinal and the outer circular. An additional layer of smooth muscle, an external longitudinal layer, appears in the lower third of the ureter. The *adventitia* covering its external surface is typical fibrous connective tissue (Figure 26.17).

The ureters play an active role in transporting urine. Distension of the ureter by incoming urine stimulates its muscularis to contract, which propels urine into the bladder. (Urine does *not* reach the bladder through gravity alone.) The strength and frequency of the peristaltic waves are adjusted to the rate of urine formation. Although the ureters are innervated by both sympathetic and parasympathetic fibers, neural control of their peristalsis appears to be insignificant compared to the local response of ureteral smooth muscle to stretch.

⚠ Homeostatic Imbalance

On occasion, calcium, magnesium, or uric acid salts in urine may crystallize and precipitate in the renal pelvis, forming **renal calculi** (kal'ku-li; *calculus* = little stone), or kidney stones. Most calculi are under 5 mm in diameter and pass through the urinary tract without causing problems. However, larger calculi can obstruct a ureter and block urine drainage. Increasing pressure within the kidney causes excruciating pain, which radiates from the flank to the anterior abdominal wall on the same side. Pain also occurs when the contracting ureter walls close in on the sharp calculi as they are being eased through a ureter by peristalsis.

Predisposing conditions are frequent bacterial infections of the urinary tract, urinary retention, high concentrations of calcium in the blood, and alkaline urine. Surgical removal of calculi has been nearly replaced by *shock wave lithotripsy*, a newer, noninvasive procedure that uses ultrasonic shock waves to shatter the calculi. The pulverized, sandlike remnants of the calculi are then painlessly eliminated in the urine. People with a history of kidney stones are encouraged to acidify their urine by drinking cranberry juice and to ingest large quantities of water to keep the urine dilute. ■

URINARY BLADDER

The **urinary bladder** is a smooth, collapsible, muscular sac that stores urine temporarily. It is located retroperitoneally on the pelvic floor just posterior to the pubic symphysis. In males, the bladder lies im-

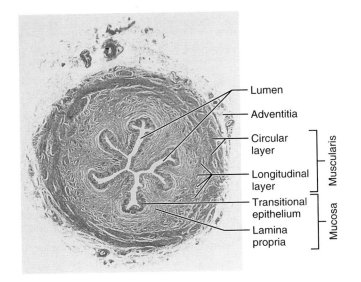

FIGURE 26.17   *Structure of the ureter wall (cross-sectional view, 10×).* The muscularis of the ureter contains two smooth muscle layers: an outer, circular layer and an inner, longitudinal layer. Its lumen is lined by a mucosa containing transitional epithelium. The external surface of the ureter is covered by a loose, connective tissue adventitia.

mediately anterior to the rectum. The prostate gland (part of the male reproductive system) surrounds its neck inferiorly where it empties into the urethra. In females, the bladder is anterior to the vagina and uterus.

The interior of the bladder has openings for both ureters and the urethra (Figure 26.18). The smooth, triangular region of the bladder base outlined by these three openings is the **trigone** (tri'gōn; *trigon* = triangle). The trigon is important clinically because infections tend to persist in this region.

The bladder wall has three layers: a mucosa containing transitional epithelium, a thick muscular layer, and a fibrous adventitia (except on its superior surface, where it is covered by the parietal peritoneum). The muscular layer, called the **detrusor** (de-tru'sor; "to thrust out") **muscle,** consists of intermingled smooth muscle fibers arranged in inner and outer longitudinal layers and a middle circular layer.

The bladder is very distensible and uniquely suited for its function of urine storage. When it contains little or no urine, the bladder collapses into its basic pyramidal shape. Its walls are thick and thrown into folds *(rugae)*. But as urine accumulates, the bladder expands and becomes pear-shaped as it rises superiorly in the abdominal cavity (Figure 26.19). The muscular wall stretches and thins, and rugae disappear. These changes allow the bladder to store more urine without a significant rise in internal pressure (at least until 300 ml has accumulated).

How do the internal and external urethral sphincters differ structurally and functionally?

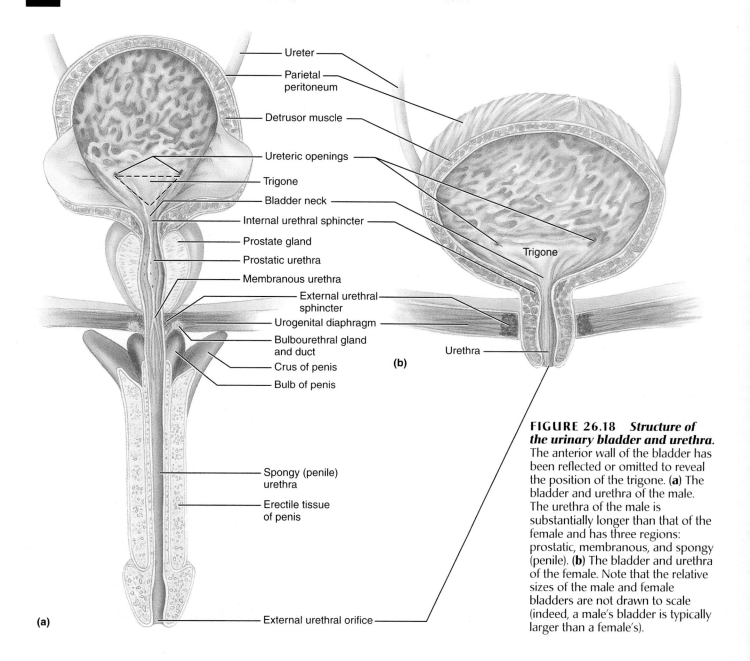

Ureter

Parietal peritoneum

Detrusor muscle

Ureteric openings

Trigone

Bladder neck

Internal urethral sphincter

Prostate gland

Prostatic urethra

Membranous urethra

External urethral sphincter

Urogenital diaphragm

Bulbourethral gland and duct

Crus of penis

Bulb of penis

Spongy (penile) urethra

Erectile tissue of penis

External urethral orifice

Trigone

Urethra

(b)

(a)

FIGURE 26.18 *Structure of the urinary bladder and urethra.* The anterior wall of the bladder has been reflected or omitted to reveal the position of the trigone. **(a)** The bladder and urethra of the male. The urethra of the male is substantially longer than that of the female and has three regions: prostatic, membranous, and spongy (penile). **(b)** The bladder and urethra of the female. Note that the relative sizes of the male and female bladders are not drawn to scale (indeed, a male's bladder is typically larger than a female's).

A moderately full bladder is about 12.5 cm (5 inches) long and holds approximately 500 ml (1 pint) of urine. However, the bladder can hold more than double that amount if necessary. When it is tense with urine, it can be palpated well above the pubic symphysis; when it is overdistended, it may burst. Although urine is formed continuously by the kidneys, it is usually stored in the bladder until its release is convenient.

URETHRA

The **urethra** is a thin-walled muscular tube that drains urine from the bladder and conveys it out of the body. The epithelium of its mucosal lining is mostly pseudostratified columnar epithelium. However, near the bladder it becomes transitional epithelium, and near its external opening it changes to a protective stratified squamous epithelium.

The internal sphincter, composed of smooth muscle, is controlled involuntarily; the external sphincter, of skeletal muscle, is controlled voluntarily.

At the bladder-urethra junction a thickening of the detrusor smooth muscle forms the **internal urethral sphincter.** This involuntary sphincter keeps the urethra closed when urine is not being passed and prevents leaking of urine between voidings. A second sphincter, the **external urethral sphincter (sphincter urethrae),** surrounds the urethra as it passes through the *urogenital diaphragm.* This sphincter is formed of skeletal muscle and is voluntarily controlled. The *levator ani* muscle of the pelvic floor also serves as a voluntary constrictor of the urethra (see Table 10.7, p. 349).

The length and functions of the urethra differ in the two sexes (see Figure 26.18). In females the urethra is 3–4 cm (1.5 inches) long and is tightly bound to the anterior vaginal wall by fibrous connective tissue. Its external opening, the **external urethral orifice (meatus),** lies anterior to the vaginal opening and posterior to the clitoris (see Figure 28.16, p. 1092).

Homeostatic Imbalance

Because the female's urethra is very short and its external orifice is close to the anal opening, improper toilet habits (wiping back to front after defecation) can easily carry fecal bacteria into the urethra. Actually, most urinary tract infections occur in sexually active young women, because intercourse drives bacteria from the vagina and external genital region toward the bladder. The use of spermicides magnifies this problem, because the spermicide kills helpful bacteria, allowing infectious fecal bacteria to colonize the vagina. Overall, 40% of all women get urinary tract infections.

The mucosa of the urethra is continuous with that of the rest of the urinary tract, and an inflammation of the urethra *(urethritis)* can ascend the tract to cause bladder inflammation *(cystitis)* or even renal inflammations *(pyelitis* or *pyelonephritis).* Symptoms of urinary tract infection include *dysuria* (painful urination), urinary *urgency* and *frequency,* fever, and sometimes cloudy or blood-tinged urine. When the kidneys are involved, back pain and a severe headache often occur as well. Most urinary tract infections are easily cured by antibiotics. ■

In males the urethra is approximately 20 cm (8 inches) long and has three named regions. The **prostatic urethra,** about 2.5 cm (1 inch) long, runs within the prostate gland. The **membranous urethra,** which runs through the urogenital diaphragm, extends for about 2 cm from the prostate gland to the beginning of the penis. The **spongy,** or **penile, urethra,** about 15 cm long, passes through the penis and opens at its tip via the **external urethral orifice.** The male urethra has a double function: It carries semen as well as urine out of the body. The repro-

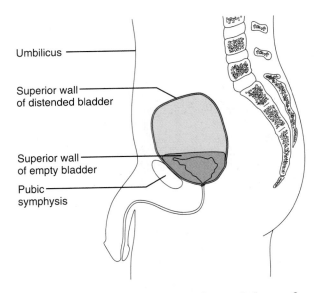

Umbilicus

Superior wall of distended bladder

Superior wall of empty bladder

Pubic symphysis

FIGURE 26.19 *Relative positioning and shape of a distended and an empty urinary bladder in an adult male.*

ductive function of the male urethra is discussed in Chapter 28.

MICTURITION

Micturition (mik″tu-rish′un; *mictur* = urinate), also called **voiding** or **urination,** is the act of emptying the bladder. Ordinarily, as urine accumulates, distension of the bladder walls activates stretch receptors there. Impulses are transmitted via visceral afferent fibers to the sacral region of the spinal cord, setting up spinal reflexes that (1) initiate increased sympathetic outflow to the bladder that inhibits the detrusor muscle and internal sphincter (temporarily) and (2) stimulate contraction of the external urethral sphincter by activating pudendal motor fibers (Figure 26.20a). When about 200 ml of urine has accumulated, afferent impulses are transmitted to the brain; at this point one feels the urge to void. Contractions of the bladder become both more frequent and urgent and if the time is convenient to empty the bladder (a decision made by the inferofrontal region of the cerebral cortex), voiding reflexes are initiated. In this case the visceral afferent impulses activate the *micturition center* of the dorsolateral pons. Acting as an on/off switch for micturition, this center signals the parasympathetic neurons that stimulate contraction of the detrusor muscle and relaxation of the internal and external sphincters, allowing urine to be expelled from the bladder (Figure 26.20.b).

When one chooses not to void, reflex bladder contractions subside within a minute or so and urine continues to accumulate. Because the external sphincter (and the levator ani) is voluntarily

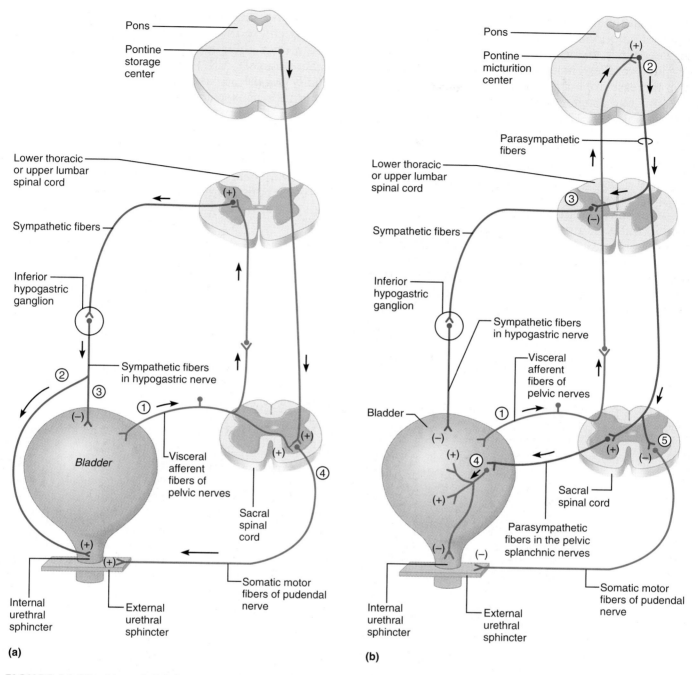

FIGURE 26.20 Neural circuits controlling continence and micturition. (a) Storage reflexes: During bladder filling and storage, distension of the bladder wall stimulates visceral afferent fibers ①. These initiate spinal reflexes that stimulate contraction of the internal and external sphincters and inhibit contraction of the detrusor muscle of the bladder ②–④. (b) Voiding reflexes: upon activation by visceral afferent ①, parasympathetic outflow via the pontine micturition center ② stimulates detrusor muscle contraction, inhibits contraction of the internal sphincter ④, and inhibits sympathetic ③ and somatic (pudendal) outflow to the bladder and external sphincter ⑤, respectively. Consequently, the bladder empties.

controlled, we can choose to keep it closed and postpone bladder emptying temporarily. After 200–300 ml more has collected, the micturition reflex occurs again and, if urination is delayed again, is damped once more. The urge to void eventually becomes irresistible and micturition occurs, whether one wills it or not.

Homeostatic Imbalance

Incontinence, the inability to control micturition voluntarily, is normal in infants, who have yet to learn to control their external urethral sphincter. Reflex voiding occurs each time a baby's bladder fills enough to activate the stretch receptors, but the internal sphincter prevents dribbling of urine between voidings just as it does in adults. After the toddler years (and toilet training), incontinence is usually a result of emotional problems, physical pressure during pregnancy, or nervous system problems (stroke or spinal cord lesions). In *stress incontinence,* a sudden increase in intra-abdominal pressure (during laughing and coughing) forces urine through the external sphincter. This condition is common during pregnancy when the uncommonly heavy uterus stretches the muscles of the pelvic floor and the urogenital diaphragm that support the external sphincter. In *overflow incontinence,* urine dribbles from the urethra whenever the bladder overfills.

In **urinary retention,** the bladder is unable to expel its contained urine. Urinary retention is normal after general anesthesia has been given (it seems that it takes a little time for the detrusor muscle to regain its activity). Urinary retention in men often reflects hypertrophy of the prostate gland, which narrows the urethra, making it difficult to void. When urinary retention is prolonged, a slender rubber drainage tube called a **catheter** (kath'ĕ-ter) must be inserted through the urethra to drain the urine and prevent bladder trauma from excessive stretching. ■

DEVELOPMENTAL ASPECTS OF THE URINARY SYSTEM

As the kidneys develop in a young embryo, it almost seems as if they are unable to "make up their mind" how to go about it. As illustrated in Figure 26.21, three different sets of kidneys develop from the *urogenital ridges,* paired elevations of the intermediate mesoderm that give rise to both the urinary and reproductive organs. Only the last set persists to become the adult kidneys. During the fourth week of development, the first tubule system, the **pronephros** (pro-nef'ros; "prekidney"), forms and then quickly degenerates as a second, lower set appears. Although the pronephros never functions and is gone by the sixth week, the **pronephric duct** that connects it to the cloaca persists and is used by the later-developing kidneys. (The cloaca is the terminal part of the gut that opens to the body exterior.) As the second renal system, the **mesonephros** (mez"o-nef'ros; "middle kidney"), claims the pronephric duct, it comes to be called the **mesonephric duct.** The mesonephric kidneys, in turn, degenerate once the third set, the **metanephros** (met"ah-nef'ros; "after kidney"), makes its appearance.

The metanephros starts to develop at about five weeks as hollow **ureteric buds** that push superiorly from the mesonephric duct into the urogenital ridge, inducing the mesoderm there to form nephrons. The distal ends of the ureteric buds, in turn, form the renal pelves, calyces, and collecting ducts; their unexpanded proximal parts, now called the **ureteric ducts,** become the ureters.

Because the kidneys develop in the pelvis and then ascend to their final position in the abdomen, they receive their blood supply from successively higher sources. Although the lower blood vessels usually degenerate as the upper ones appear, they sometimes persist and multiple renal arteries are common. The metanephric kidneys are excreting urine by the third month of fetal life, and most of the amniotic fluid that surrounds a developing fetus is actually fetal urine. Nonetheless, the fetal kidneys do not work nearly as hard as they will after birth because exchange through the placenta allows the mother's urinary system to clear most of the undesirable substances from the fetal blood.

As the metanephros is developing, the cloaca subdivides to form the **urogenital sinus,** into which the urinary and genital ducts empty, and the future rectum and anal canal. The urinary bladder and the urethra then develop from the urogenital sinus.

◭ *Homeostatic Imbalance*

There are many congenital abnormalities of the urinary system, but three of the most common are horseshoe kidney, hypospadias, and polycystic kidney.

When ascending from the pelvis the kidneys are very close together, and in 1 out of 600 people they fuse across the midline, forming a single, U-shaped *horseshoe kidney.* This condition is usually asymptomatic, but it may be associated with kidney abnormalities, such as obstructed drainage, which place a person at risk for frequent kidney infections.

Hypospadias (hi"po-spa'de-as), found in male infants only, is the most common congenital abnormality of the urethra. It occurs when the urethral orifice is located on the ventral surface of the penis. This problem is corrected surgically when the child is around 12 months old.

Congenital polycystic kidney is an inherited condition in which the baby's kidneys have many urine-filled cysts (sacs). It results from abnormal development of the collecting ducts such that they obstruct urine drainage from some of the nephrons, forcing urine to accumulate within the cysts in the kidney. This disease almost invariably causes renal failure during childhood, but kidney transplants have improved chances of survival for some children.

Adult polycystic kidney disease, another inherited condition, is distinct from congenital polycystic kidney. The cysts, which represent dilations along the nephrons, develop so gradually that they produce no symptoms until about 40 years of age. Then, both kidneys begin to enlarge as blisterlike cysts containing chloride-rich fluid accumulate. The damage caused by these cysts progresses slowly, and many victims live without problems until their 60s. Ulti-

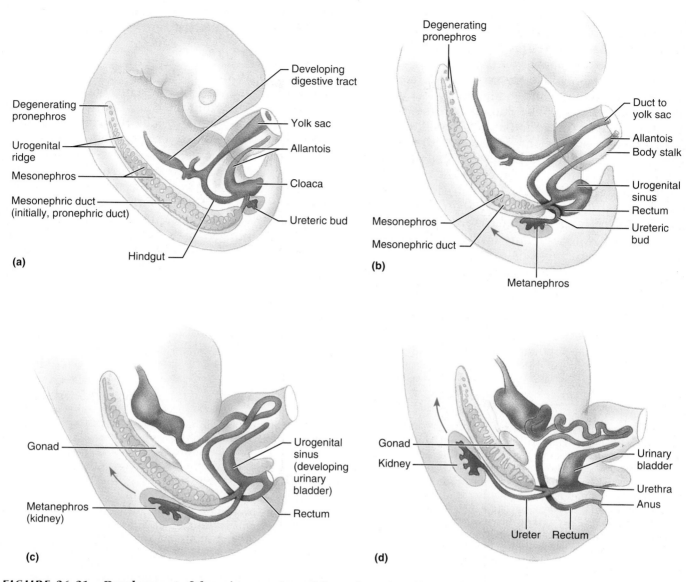

FIGURE 26.21 *Development of the urinary system of the embryo.* (**a**) Fifth week. (**b**) Sixth week. (**c**) Seventh week. (**d**) Eighth week. (Direction of metanephros migration as it develops is indicated by red arrows.)

mately, however, the kidneys become "knobby" and grossly enlarged, reaching weights of up to 14 kg (30 pounds) each. Unless a kidney transplant is done most victims die from kidney failure or hypertension. Adult polycystic disease is relatively common, affecting 1 in 500 people. ∎

Because an infant's bladder is very small and its kidneys are unable to concentrate urine for the first two months, a newborn baby voids 5 to 40 times daily, depending on fluid intake. By 2 months of age, the infant is voiding approximately 400 ml/day, and the amount steadily increases until about adolescence, when adult urine output (about 1500 ml/day) is achieved.

Control of the voluntary urethral sphincter goes hand in hand with nervous system development. By 15 months, most toddlers are aware when they have voided. By 18 months, they can usually hold urine in the bladder for about two hours. This is the first sign that toilet training for micturition can begin. Daytime control usually is achieved well before nighttime control. It is unrealistic to expect complete nighttime control before the age of 4 years.

From childhood through late middle age, most urinary system problems are infectious conditions. *Escherichia coli* (esh″ĕ-rik′e-ah ko′li) are normal residents of the digestive tract and generally cause no problems there, but these bacteria account for 80% of all urinary tract infections. *Sexually transmitted diseases,* which are chiefly reproductive tract infections (discussed in Chapter 28), can also inflame the urinary tract, clogging some of its ducts. Childhood streptococcal infections such as strep throat and scarlet fever, if not treated promptly, may cause long-term inflammatory damage to the kidneys.

Only about 3% of elderly people have histologically normal kidneys, and kidney function declines with advancing age. The kidneys shrink as the nephrons decease in size and number, and the tubule cells become less efficient. By age 70, the GFR is only half that of middle-aged adults. This slowing is believed to result from narrowing of the renal arteries by atherosclerosis. Diabetics are particularly at risk for renal disease, and over 50% of those who have had diabetes mellitus for 20 years (regardless of their age) are in renal failure owing to the vascular ravages of this disease.

The bladder of an aged person is shrunken, with less than half the capacity of a young adult (250 ml versus 600 ml). Loss of bladder tone causes an annoying frequency. *Nocturia* (nok-tu're-ah), the need to get up during the night to urinate, plagues almost two-thirds of this population. Many people eventually experience incontinence, and with it a tremendous blow to their self-esteem.

* * *

The ureters, urinary bladder, and urethra have important roles to play in transporting, storing, and eliminating urine from the body. But when the term "urinary system" is used, it is the kidneys, which produce the urine, that capture center stage. As summarized in *Making Connections* in Chapter 27 (p. 1064), the other organ systems of the body contribute to the well-being of the urinary system in many ways. But without continuous kidney function, internal body fluids quickly become contaminated with nitrogenous wastes, and the electrolyte and fluid balance of the blood is dangerously disturbed. No body cell can escape the harmful effects of such imbalances.

Now that renal mechanisms have been described, we are ready to integrate kidney function into the larger topic of fluid and electrolyte balance of the body as a whole—the focus of Chapter 27.

RELATED CLINICAL TERMS

Acute glomerulonephritis (glo-mer"u-lo-nef-ri'tis) **(GN)** Inflammation of the glomeruli, often arising from an immune response. In some cases, circulating immune complexes (antibodies bound to foreign substances, such as streptococcal bacteria) become trapped in the glomerular basement membranes. In other cases, immune responses are mounted against one's own kidney tissues, leading to glomerular damage. In either case, the inflammatory response that follows damages the filtration membrane, increasing its permeability. Blood proteins and even blood cells begin to pass into the renal tubules and are lost in urine. As the osmotic pressure of blood drops, fluid seeps from the bloodstream into the tissue spaces, causing body-wide edema. Renal shutdown requiring dialysis (see *A Closer Look* on page 1023) may occur temporarily, but normal renal function usually returns within a few months. If permanent glomerular damage occurs, chronic GN and ultimately renal failure result.

Bladder cancer Bladder cancer, five times more common in men than in women, accounts for about 3% of all cancer deaths. It usually involves neoplasms of the bladder's lining epithelium and may be induced by carcinogens from the environment or the workplace that end up in urine. Smoking, exposure to industrial chemicals, and artificial sweeteners also have been linked to bladder cancer. Blood in the urine is a common warning sign.

Cystocele (sis'to-sēl; *cyst* = a sac, the bladder; *cele* = hernia, rupture) Herniation of the urinary bladder into the vagina; a common result of tearing of the pelvic floor muscles during childbirth.

Cystoscopy (sis-tos'ko-pe; *cyst* = bladder; *scopy* = observation) Procedure in which a thin viewing tube is threaded into the bladder through the urethra to examine the bladder's mucosal surface.

Diabetes insipidus (in-sī'pĭ-dus; *insipid* = tasteless, bland) Condition in which large amounts (up to 40 L/day) of dilute urine flush from the body; results from little or no ADH release due to injury to, or a tumor in, the hypothalamus or posterior pituitary. Can lead to severe dehydration and electrolyte imbalances unless the individual drinks large volumes of liquids to maintain normal fluid balance.

Intravenous pyelogram (IVP) (pi'ĕ-lo-gram; *pyelo* = kidney pelvis; *gram* = written) An X-ray of the kidney and ureter obtained after intravenous injection of a contrast medium. Allows assessment of renal blood vessels for obstructions, viewing of renal anatomy (pelvis and calyces), and determination of rate of excretion of the contrast medium.

Nephrotoxin A substance (heavy metal, organic solvent, or bacterial toxin) that is toxic to the kidney.

Nocturnal enuresis (en"u-re'sis) An inability to control urination at night during sleep; bed-wetting. Occurs most often in children who are sound sleepers and in those with a small bladder capacity (children and the aged), but may be due to emotional rather than physical causes.

Renal infarct Area of dead, or necrotic, renal tissue. May result from inflammation, hydronephrosis, or blockage of the vascular supply to the kidney. A common cause of localized renal infarct is obstruction of an interlobar artery. Because interlobar arteries are end arteries (do not anastomose), their obstruction leads to ischemic necrosis of the portions of the kidney they supply.

Urinalysis Analysis of urine as an aid to diagnosing health or disease. The most significant indicators of disease in urine are proteins, glucose, acetone, blood, and pus.

Urologist (u-rol'o-jist) Physician who specializes in diseases of urinary structures in both sexes and in diseases of the reproductive tract of males.

CHAPTER SUMMARY

Media study tools that could provide you additional help in reviewing specific key topics of Chapter 26 are referenced below. **SP** = *Study Partner;* **IP** = *Interactive Physiology.*

Kidney Anatomy (pp. 1004–1013)

Location and External Anatomy (pp. 1004–1005)

1. The paired kidneys are retroperitoneal in the superior lumbar region.

2. A renal capsule, an adipose capsule, and renal fascia surround each kidney. The fatty adipose capsule helps hold the kidneys in position.

Internal Anatomy (pp. 1006–1007)

3. A kidney has a superficial cortex, a deeper medulla consisting mainly of medullary pyramids, and a medial pelvis. Extensions of the pelvis (calyces) surround and collect urine draining from the apices of the medullary pyramids.

Blood and Nerve Supply (p. 1007)

4. The kidneys receive 25% of the total cardiac output/minute.

5. The vascular pathway through a kidney is as follows: renal artery → segmental arteries → lobar arteries → interlobar arteries → arcuate arteries → interlobular arteries → afferent arterioles → glomeruli → efferent arterioles → peritubular capillary beds → interlobular veins → arcuate veins → interlobar veins → renal vein.

6. The nerve supply of the kidneys is derived from the renal plexus.

Nephrons (pp. 1007–1013)

7. Nephrons are the structural and functional units of the kidneys.

8. Each nephron consists of a glomerulus (a high-pressure capillary bed) and a renal tubule. Subdivisions of the renal tubule (from the glomerulus) are the glomerular capsule, proximal convoluted tubule, loop of Henle, and distal convoluted tubule. A second capillary bed, the low-pressure peritubular capillary bed, is closely associated with the renal tubule of each nephron.

9. The more numerous cortical nephrons are located almost entirely in the cortex; only a small part of their loop of Henle penetrates into the medulla. Juxtamedullary nephrons are located at the cortex-medulla junction, and their loop of Henle dips deeply into the medulla. Instead of directly forming peritubular capillaries, the efferent arterioles of many of the juxtamedullary nephrons form unique bundles of straight vessels, called vasa recta, that serve tubule segments in the medulla. Juxtamedullary nephrons and the vasa recta play an important role in establishing the medullary osmotic gradient.

10. Collecting ducts receive urine from many nephrons and help concentrate urine. They form the medullary pyramids.

11. The juxtaglomerular apparatus is at the point of contact between the afferent arteriole and the first part of the distal convoluted tubule. It consists of the juxtaglomerular (JG) cells and the macula densa.

SP Exercise: Chapter 26, Structure of Nephrons.

IP Urinary System CD-ROM; Topics: Anatomy Review, pages 1–20.

12. The filtration membrane consists of the fenestrated glomerular endothelium, the intervening basement membrane, and the podocyte-containing visceral membrane of the glomerular capsule. It permits free passage of substances smaller than (most) plasma proteins.

Kidney Physiology: Mechanisms of Urine Formation (pp. 1013–1029)

1. Functions of the nephrons include filtration, tubular reabsorption, and tubular secretion. Via these functional processes, the kidneys eliminate nitrogenous metabolic wastes, and regulate the volume, composition, and pH of the blood.

SP Exercise: Chapter 26, Nephron Activity.

Glomerular Filtration (pp. 1014–1017)

2. The glomeruli function as filters. High glomerular blood pressure (55 mm Hg) occurs because the glomeruli are fed and drained by arterioles, and the afferent arterioles are larger in diameter than the efferent arterioles.

3. About one-fifth of the plasma flowing through the kidneys is filtered from the glomeruli into the renal tubules.

4. Usually about 10 mm Hg, the net filtration pressure (NFP) is determined by the relationship between forces favoring filtration (glomerular hydrostatic pressure) and forces that oppose it (capsular hydrostatic pressure and blood colloid osmotic pressure).

5. The glomerular filtration rate (GFR) is directly proportional to the net filtration pressure and is about 125 ml/min (180 L/day).

6. Renal autoregulation, which enables the kidneys to maintain a relatively constant renal blood flow and glomerular filtration rate, involves a myogenic mechanism and a tubuloglomerular feedback mechanism mediated by the macula densa.

7. Strong sympathetic nervous system activation causes constriction of the afferent arterioles, which decreases filtrate formation and stimulates renin release by the JG cells.

IP Urinary System CD-ROM; Topic: Glomular Filtration, pages 1–15.

8. The renin-angiotensin mechanism mediated by the JG cells raises systemic blood pressure via generation of angiotensin II, which promotes aldosterone secretion.

Tubular Reabsorption (pp. 1017–1022)

9. During tubular reabsorption, needed substances are removed from the filtrate by the tubule cells and returned to the peritubular capillary blood. The primary active transport of Na^+ by a Na^+-K^+ ATPase pump at the basolateral membrane accounts for Na^+ reabsorption and establishes the electrochemical gradient that drives the reabsorption of most other solutes and H_2O. Na^+ enters at the luminal surface of the tubule cell via facilitated diffusion or via diffusion through a channel.

10. Passive tubular reabsorption is driven by electrochemical gradients established by active reabsorption of sodium ions. Water, many anions, and various other substances (for example, urea) are reabsorbed passively by diffusion via transcellular or paracellular pathways.

11. Secondary active tubular reabsorption occurs by cotransport with Na^+ via protein carriers. Transport of

such substances is limited by the number of carriers available. Substances reabsorbed actively include nutrients and some ions.

12. Certain substances (creatinine, drug metabolites, etc.) are not reabsorbed or are reabsorbed incompletely because of the lack of carriers, their size, or nonlipid solubility.

13. The proximal tubule cells are most active in reabsorption. Most of the nutrients, 65% of the water and sodium ions, and the bulk of actively transported ions are reabsorbed in the proximal convoluted tubules.

14. Reabsorption of additional sodium ions and water occurs in the distal tubules and collecting ducts and is hormonally controlled. Aldosterone increases the reabsorption of sodium (and obligatory water reabsorption); antidiuretic hormone enhances water reabsorption by the collecting ducts.

Tubular Secretion (p. 1022)

15. Tubular secretion is a means of adding substances to the filtrate (from the blood or tubule cells). It is an active process that is important in eliminating drugs, urea, and excess ions and in maintaining the acid-base balance of the blood.

Regulation of Urine Concentration and Volume (pp. 1022–1028)

16. The graduated hyperosmolality of the medullary fluids (largely due to the cycling of NaCl and urea) ensures that the filtrate reaching the distal convoluted tubule is dilute (hypo-osmolar). This allows urine with osmolalities ranging from 65 to 1200 mOsm to be formed.

- The descending limb of the loop of Henle is permeable to water, which leaves the filtrate and enters the medullary interstitium. The filtrate and medullary fluid at the tip of the loop of Henle are hyperosmolar.

- The thick ascending limb is impermeable to water, but Na^+ and Cl^- are actively transported out of the filtrate into the interstitial space. The filtrate becomes more dilute as it continues to lose salt. Na^+ and Cl^- move passively out of the thin portion of the ascending limb.

- As filtrate flows through the collecting ducts in the inner medulla, urea diffuses into the interstitial space. Some urea enters the ascending limb and is recycled.

- The blood flow in the vasa recta is sluggish, and the contained blood equilibrates with the medullary interstitial fluid. Hence, blood entering and exiting the medulla in the vasa recta is isotonic to blood plasma and the high solute concentration of the medulla is maintained.

17. In the absence of antidiuretic hormone, dilute urine is formed because the dilute filtrate reaching the collecting duct is simply allowed to pass from the kidneys.

18. As blood levels of antidiuretic hormone rise, the collecting ducts become more permeable to water, and water moves out of the filtrate as it flows through the hyperosmotic medullary areas. Consequently, more concentrated urine is produced, and in smaller amounts.

SP Exercise: Chapter 26, Early Filtrate Processing.

IP Urinary System CD-ROM; Topics: Early Filtrate Processing, pages 1–22; Late Filtrate Processing, pages 1–13.

Renal Clearance (p. 1028)

19. Renal clearance is the volume flow rate (ml/min) at which the kidneys clear the plasma of a particular solute.

Studies of renal clearance provide information about renal function or the course of renal disease.

Characteristics and Composition of Urine (pp. 1028–1029)

20. Urine is typically clear, yellow, aromatic, and slightly acidic. Its specific gravity ranges from 1.001 to 1.035.

21. Urine is 95% water; solutes include nitrogenous wastes (urea, uric acid, and creatinine) and various ions (always sodium, potassium, sulfate, and phosphate).

22. Substances not normally found in urine include glucose, proteins, erythrocytes, pus, hemoglobin, and bile pigments.

23. Daily urinary volume is typically 1.5–1.8 L, but this depends on the state of hydration of the body.

Ureters (pp. 1029–1030)

1. The ureters are slender tubes running retroperitoneally from each kidney to the bladder. They conduct urine by peristalsis from the renal pelvis to the urinary bladder.

Urinary Bladder (pp. 1030–1031)

1. The urinary bladder, which functions to store urine, is a distensible muscular sac that lies posterior to the pubic symphysis. It has two inlets (ureters) and one outlet (urethra) that outline the trigone. In males, the prostate gland surrounds its outlet.

2. The bladder wall consists of a transitional epithelium-containing mucosa, a three-layered detrusor muscle, and an adventitia.

Urethra (pp. 1031–1032)

1. The urethra is a muscular tube that conveys urine from the bladder to the body exterior.

2. Where the urethra leaves the bladder, it is surrounded by an internal urethral sphincter, an involuntary smooth muscle sphincter. Where it passes through the urogenital diaphragm, the voluntary external urethral sphincter is formed by skeletal muscle.

3. In females the urethra is 3–4 cm long and conducts only urine. In males it is 20 cm long and conducts both urine and semen.

Micturition (pp. 1032–1034)

1. Micturition is emptying of the bladder.

2. Stretching of the bladder wall by accumulating urine initiates the micturition reflex, in which parasympathetic fibers, in response to signals from the micturition center of the pons, cause the detrusor muscle to contract and the internal urethral sphincter to relax (open).

3. Because the external sphincter is voluntarily controlled, micturition can usually be delayed temporarily.

Developmental Aspects of the Urinary System (pp. 1034–1036)

1. Three sets of kidneys (pronephric, mesonephric, and metanephric) develop from the intermediate mesoderm. The metanephros is excreting urine by the third month of development.

2. Common congenital abnormalities are horseshoe kidney, polycystic kidney, and hypospadias.

3. The kidneys of newborns cannot concentrate urine; their bladder is small and voiding is frequent. Neuromuscular maturation generally allows toilet training for micturition to begin by 18 months of age.

4. The most common urinary system problems in children and young to middle-aged adults are bacterial infections.

5. Renal failure is uncommon but serious. The kidneys are unable to concentrate urine, nitrogenous wastes accumulate in the blood, and acid-base and electrolyte imbalances occur.

SP Case Study: Renal Failure.

6. With age, nephrons are lost, the filtration rate decreases, and tubule cells become less efficient at concentrating urine.

7. Bladder capacity and tone decrease with age, leading to frequent micturition and (often) incontinence. Urinary retention is a common problem of elderly men.

REVIEW QUESTIONS

Multiple Choice/Matching

1. The lowest blood concentration of nitrogenous waste occurs in the (a) hepatic vein, (b) inferior vena cava, (c) renal artery, (d) renal vein.

2. The glomerular capillaries differ from other capillary networks in the body because they (a) have a larger area of anastomosis, (b) are derived from and drain into arterioles, (c) are not made of endothelium, (d) are sites of filtrate formation.

3. Damage to the renal medulla would interfere *first* with the functioning of the (a) glomerular capsules, (b) distal convoluted tubules, (c) collecting ducts, (d) proximal convoluted tubules.

4. Which is reabsorbed by the proximal convoluted tubule? (a) Na^+, (b) K^+, (c) amino acids, (d) all of the above.

5. Glucose is not normally found in the urine because it (a) does not pass through the walls of the glomerulus, (b) is kept in the blood by colloid osmotic pressure, (c) is reabsorbed by the tubule cells, (d) is removed by the body cells before the blood reaches the kidney.

6. Filtration at the glomerulus is directly related to (a) water reabsorption, (b) arterial blood pressure, (c) capsular hydrostatic pressure, (d) acidity of the urine.

7. Renal reabsorption (a) of glucose and many other substances is a T_m-limited active transport process, (b) of chloride is always linked to the passive transport of Na^+, (c) is the movement of substances from the blood into the nephron, (d) of sodium occurs only in the proximal tubule.

8. If a freshly voided urine sample contains excessive amounts of urochrome, it has (a) an ammonia-like odor, (b) a pH below normal, (c) a dark yellow color, (d) a pH above normal.

9. Conditions such as diabetes mellitus, starvation, and low-carbohydrate diets are closely linked to (a) ketosis, (b) pyuria, (c) albuminuria, (d) hematuria.

Short Answer Essay Questions

10. What is the importance of the adipose capsule that surrounds the kidney?

11. Trace the pathway a creatinine molecule takes from a glomerulus to the urethra. Name every microscopic or gross structure it passes through on its journey.

12. Explain the important differences between blood plasma and renal filtrate, and relate the differences to the structure of the filtration membrane.

13. Describe the mechanisms that contribute to renal autoregulation.

14. Describe what is involved in active and passive tubular reabsorption.

15. Explain how the peritubular capillaries are adapted for receiving reabsorbed substances.

16. Explain the process and purpose of tubular secretion.

17. How does aldosterone modify the chemical composition of urine?

18. Explain why the filtrate becomes hypotonic as it flows through the ascending limb of the loop of Henle. Also explain why the filtrate at the tip of the loop of Henle (and the interstitial fluid of the deep portions of the medulla) is hypertonic.

19. How does urinary bladder anatomy support its storage function?

20. Define micturition and describe the micturation reflex.

21. Describe the changes that occur in kidney and bladder anatomy and physiology in old age.

Critical Thinking and Clinical Application Questions

1. Mrs. Bigda, a 60-year-old woman, was brought to the hospital by the police after falling to the pavement. She is found to have alcoholic hepatitis. She is put on a salt- and protein-restricted diet and diuretics are prescribed to manage her ascites (accumulated fluid in the peritoneal cavity). How will diuretics reduce this excess fluid? Name and describe the mechanism of action of two types of diuretics. Why is her diet salt-restricted?

2. While repairing a frayed utility wire, Herbert, an experienced lineman, slips and falls to the ground. Medical examination reveals a fracture of his lower spine and transection of the lumbar region of the spinal cord. How will Herbert's micturition be controlled from this point on? Will he ever again feel the need to void? Will there be dribbling of urine between voidings? Explain the reasoning behind all your responses.

3. What is cystitis? Why are women more frequent cystitis sufferers than men?

4. Hattie, aged 55, is awakened by excruciating pain that radiates from her right abdomen to the loin and groin regions on the same side. The pain is not continuous but recurs at intervals of 3 to 4 minutes. Diagnose her problem, and cite factors that might favor its occurrence. Also, explain why Hattie's pain comes in "waves."

5. Why does use of a spermicide increase a woman's risk for urinary tract infection?

27 FLUID, ELECTROLYTE, AND ACID-BASE BALANCE

Body Fluids (pp. 1041–1044)

1. List the factors that determine body water content and describe the effect of each factor.
2. Indicate the relative fluid volume and solute composition of the fluid compartments of the body.
3. Contrast the overall osmotic effects of electrolytes and nonelectrolytes.
4. Describe factors that determine fluid shifts in the body.

Water Balance (pp. 1044–1047)

5. List the routes by which water enters and leaves the body.
6. Describe feedback mechanisms that regulate water intake and hormonal controls of water output in urine.
7. Explain the importance of obligatory water losses.
8. Describe possible causes and consequences of dehydration, hypotonic hydration, and edema.

Electrolyte Balance (pp. 1047–1055)

9. Indicate the routes of electrolyte entry and loss from the body.
10. Describe the importance of ionic sodium in fluid and electrolyte balance of the body, and indicate its relationship to normal cardiovascular system functioning.
11. Briefly describe mechanisms involved in regulating sodium and water balance.
12. Explain how potassium, calcium, magnesium, and anion balance of plasma is regulated.

Acid-Base Balance (pp. 1055–1063)

13. List important sources of acids in the body.
14. List the three major chemical buffer systems of the body and describe how they operate to resist pH changes.

15. Describe the influence of the respiratory system on acid-base balance.
16. Describe how the kidneys regulate hydrogen and bicarbonate ion concentrations in the blood.
17. Distinguish between acidosis and alkalosis resulting from respiratory and metabolic factors. Describe the importance of respiratory and renal compensations to acid-base balance.

Developmental Aspects of Fluid, Electrolyte, and Acid-Base Balance (pp. 1063, 1066)

18. Explain why infants and the aged are at greater risk for fluid and electrolyte imbalances than are young adults.

Have you ever wondered why on certain days you don't urinate for hours at a time, while on others you void large amounts of urine every few minutes? Or why on occasion you cannot seem to quench your thirst? These situations and many others reflect one of the body's most important functions: that of maintaining fluid, electrolyte, and acid-base balance.

Cell function depends not only on a continuous supply of nutrients and removal of metabolic wastes, but also on the physical and chemical homeostasis of the surrounding fluids. This was recognized with great style in 1857 by the French physiologist Claude Bernard, who said, "It is the fixity of the internal environment which is the condition of free and independent life." In this chapter, we will examine the composition and distribution of fluids in the internal environment and then "look backward" to consider the roles of various body organs and functions in establishing, regulating, or altering this balance.

BODY FLUIDS

Body Water Content

If you are a healthy young adult, water probably accounts for about half your body weight (mass). However, not all bodies contain the same amount of water. Total body water is a function not only of weight, age, and sex but also of the relative amount of body fat. Because of their low body fat and low bone mass, infants are 73% or more water; this high level of hydration accounts for their "dewy" skin, like that of a freshly picked peach. But total water content declines throughout life, accounting for only about 45% of body weight in old age. A healthy young man is about 60% water; a healthy young woman about 50%. This difference between the sexes reflects the relatively larger amount of body fat and smaller amount of skeletal muscle in females. Of all body tissues, adipose tissue is *least* hydrated (containing up to 20% water); even bone contains more water than does fat. By contrast, skeletal muscle is about 65% water. Thus, people with greater muscle mass have proportionately more body water.

Fluid Compartments

Water occupies two main **fluid compartments** within the body (Figure 27.1). A little less than two-thirds by volume is in the **intracellular fluid (ICF) compartment,** which actually consists of trillions of tiny individual compartments: the cells. In an adult male of average size (70 kg, or 154 pounds), ICF accounts for about 25 L of the 40 L of body water. The remaining one-third or so of body water is outside

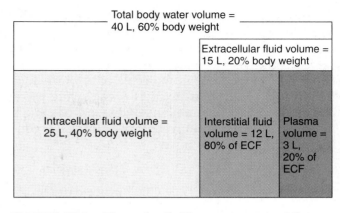

FIGURE 27.1 *The major fluid compartments of the body.* Approximate volume and percentage of body weight values are noted for a 70-kg (154-pound) male.

cells, in the **extracellular fluid (ECF) compartment.** The ECF constitutes the body's "internal environment" referred to by Claude Bernard and the external environment of each body cell. The ECF compartment is, in turn, divisible into two important *subcompartments:* (1) **plasma,** the fluid portion of blood within the blood vessels, and (2) **interstitial fluid (IF),** the fluid in the microscopic spaces between tissue cells. Additionally, there are numerous other examples of ECF that are distinct from both plasma and interstitial fluid, such as lymph, cerebrospinal fluid, humors of the eye, synovial fluid, serous fluid, and secretions of the gastrointestinal tract. However, most of these are similar to interstitial fluid and they are usually considered part of it. In the 70-kg adult male, interstitial fluid accounts for approximately 12 L and plasma about 3 L of the total (15 L) ECF volume (see Figure 27.1).

Composition of Body Fluids

Solutes: Electrolytes and Nonelectrolytes

Water serves as the *universal solvent* in which a variety of solutes are dissolved. Solutes may be classified broadly as *electrolytes* and *nonelectrolytes.* The **nonelectrolytes** have bonds (usually covalent bonds) that prevent them from dissociating in solution; therefore, they have no electrical charge. Most nonelectrolytes are organic molecules—glucose, lipids, creatinine, and urea, for example. In contrast, electrolytes are chemical compounds that *do* dissociate into ions (ionize) in water. (See Chapter 2 if necessary to review these concepts of chemistry.) Because ions are charged particles, they can conduct an electrical current—hence the name **electrolyte.** Typically, electrolytes include inorganic salts, both inorganic and organic acids and bases, and some proteins.

Although all dissolved solutes contribute to the osmotic activity of a fluid, electrolytes have much greater osmotic power than nonelectrolytes because each electrolyte molecule dissociates into at least two ions. For example, a molecule of sodium chloride (NaCl) contributes twice as many solute particles as glucose (which remains undissociated), and a molecule of magnesium chloride ($MgCl_2$) contributes three times as many:

$$NaCl \rightarrow Na^+ + Cl^- \qquad \text{(two particles)}$$

$$MgCl_2 \rightarrow Mg^{2+} + Cl^- + Cl^- \qquad \text{(three particles)}$$

$$glucose \rightarrow glucose \qquad \text{(one particle)}$$

Regardless of the type of solute particle, water moves according to osmotic gradients—from areas of lesser osmolality to those of greater osmolality. Thus, electrolytes have the greatest ability to cause fluid shifts. (To review the difference between the terms *osmolarity* and *osmolality*, see p. 1022.)

Electrolyte concentrations of body fluids are usually expressed in **milliequivalents per liter (mEq/L)**, a measure of the number of electrical charges in 1 liter of solution. Because the total number of negative charges (anions) in a solution is always equal to the number of positive charges (cations), the milliequivalent system of reporting electrolyte concentrations makes it easier to follow electrolyte shifts.

The concentration of any ion in solution in mEq/L can be computed using the equation

$$mEq/L = \frac{\text{concentration of ion (mg/L)}}{\text{atomic weight of ion}} \times \begin{array}{c}\text{no. of}\\\text{electrical}\\\text{charges on}\\\text{one ion}\end{array}$$

Thus, to compute the mEq/L of sodium or calcium ions in solution in plasma, we would determine the normal concentration of these ions in plasma, look up their atomic weights in the periodic table (see Appendix D), and plug these values into the equation:

$$Na^+: \frac{3300 \text{ mg/L}}{23} \times 1 = 143 \text{ mEq/L}$$

$$Ca^{2+}: \frac{100 \text{ mg/L}}{40} \times 2 = 5 \text{ mEq/L}$$

Notice that for ions with a single charge, 1 mEq is equal to 1 mOsm, whereas 1 mEq of bivalent ions (those with a double charge like calcium) is equal to 1/2 mOsm. In either case, 1 mEq provides the same degree of reactivity.

Comparison of Extracellular and Intracellular Fluids

Even a quick glance at the bar graphs in Figure 27.2 reveals that each fluid compartment has a distinctive pattern of electrolytes. But, except for the relatively high protein content in plasma, the extracellular fluids are very similar. Their chief cation is sodium, and their major anion is chloride. However, plasma contains somewhat fewer chloride ions than interstitial fluid, because the nonpenetrating plasma proteins are normally anions and plasma is electrically neutral.

In contrast to extracellular fluids, intracellular fluid contains only small amounts of Na^+ and Cl^-. Its most abundant cation is potassium, and its major anion is phosphate (HPO_4^{2-}). Cells also contain moderate amounts of magnesium ions and substantial quantities of soluble proteins (about three times the amount found in plasma).

Notice that sodium and potassium ion concentrations in extracellular and intracellular fluids are nearly opposite (Figure 27.2). The characteristic distribution of these ions on the two sides of cellular membranes reflects the activity of cellular ATP-dependent sodium-potassium pumps, which keep intracellular Na^+ concentrations low while maintaining high intracellular K^+ concentrations (see Chapter 3). Renal mechanisms can reinforce these ion distributions by secreting potassium as sodium is reabsorbed from the filtrate.

Electrolytes are the most abundant solutes in the various body fluids and determine most of their chemical and physical reactions, but they do not constitute the *bulk* of dissolved solutes in these fluids. Proteins and some of the nonelectrolyte molecules (phospholipids, cholesterol, and neutral fats) found in the ECF are much larger molecules, and they account for about 90% of the mass of dissolved solutes in plasma, 60% in interstitial fluid, and 97% in the intracellular compartment.

Fluid Movement Among Compartments

The continuous exchange and mixing of body fluids are regulated by osmotic and hydrostatic pressures. Although water moves freely between the compartments along osmotic gradients, solutes are unequally distributed because of their molecular size, electrical charge, or dependence on active transport. Anything that changes the solute concentration in any compartment leads to net water flows.

Exchanges between plasma and interstitial fluid occur across capillary membranes. The pressures driving these fluid movements are described in detail in Chapter 20 on pp. 743–744, and the major points are recapped in the discussion of glomerular filtration on p. 1014. Here we will simply review the outcome of these mechanisms. Nearly protein-free plasma is forced out of the bloodstream into the in-

What is the major cation in ECF? In ICF? What is the intra-cellular anion counterpart of ECF's chloride ions?

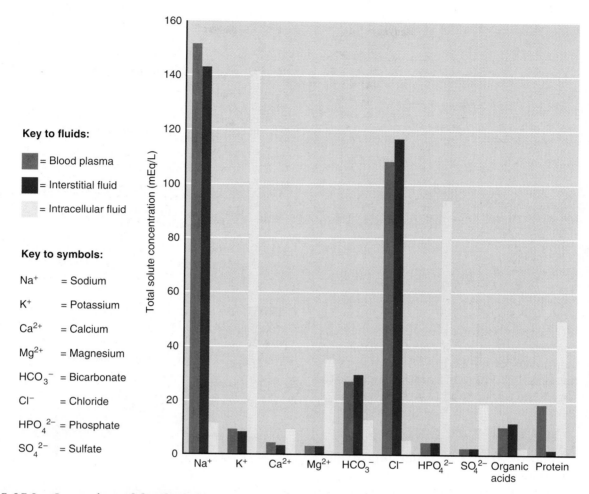

Key to fluids:

■ = Blood plasma

■ = Interstitial fluid

□ = Intracellular fluid

Key to symbols:

Na^+	= Sodium
K^+	= Potassium
Ca^{2+}	= Calcium
Mg^{2+}	= Magnesium
HCO_3^-	= Bicarbonate
Cl^-	= Chloride
HPO_4^{2-}	= Phosphate
SO_4^{2-}	= Sulfate

FIGURE 27.2 *Comparison of the electrolyte composition of blood plasma, interstitial fluid, and intracellular fluid.*

terstitial space by the hydrostatic pressure of blood. This filtered fluid is almost completely reabsorbed into the bloodstream in response to the colloid os-motic (oncotic) pressure of plasma proteins. Under normal circumstances, the small net leakage that re-mains behind in the interstitial space is picked up by lymphatic vessels and returned to the bloodstream.

Exchanges between the interstitial and intracel-lular fluids are more complex because of the selective permeability of cellular membranes. As a general rule, two-way osmotic flows of water are substantial. But ion fluxes are restricted and, in most cases, ions move selectively by active transport. Movements of nutrients, respiratory gases, and wastes are typically unidirectional. For example, glucose and oxygen

move into the cells and metabolic wastes move out of the cells and into the blood.

Of the various body fluids, only plasma circu-lates throughout the body and links the external and internal environments (Figure 27.3). Exchanges oc-cur almost continuously in the lungs, gastrointesti-nal tract, and kidneys. Although these exchanges al-ter plasma composition and volume, compensating adjustments in the other two fluid compartments follow quickly so that balance is restored.

Many factors can cause marked changes in ECF and ICF volumes. However, water moves freely be-tween compartments, so the osmolalities of all body fluids are equal, except for the first few minutes after a change in one of the fluids occurs. Increasing the solute content of the ECF (most importantly, the NaCl concentration) can be expected to cause os-motic changes in the ICF—namely, a shift of water

Na$^+$, K$^+$, Protein anions.

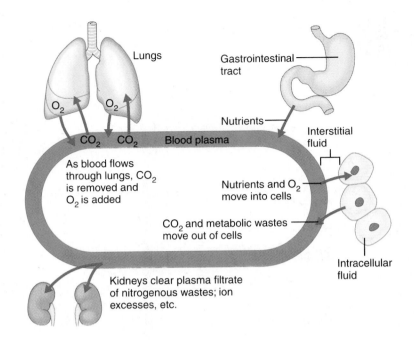

FIGURE 27.3 *The continuous mixing of body fluids.* Blood plasma serves as the communicating medium between the external and internal environments. Exchanges are made between the plasma and tissue cells (intracellular fluid) through the interstitial space.

out of the cells. Conversely, decreasing the osmolality of the ECF causes water to move into the cells. Thus, the volume of the ICF is determined by the solute concentration of the ECF. These concepts underlie all events that control fluid balance in the body and should be understood thoroughly.

WATER BALANCE

To remain properly hydrated, water intake must equal water output. *Water intake* varies widely from person to person and is strongly influenced by habit, but it is typically about 2500 ml a day in adults (Figure 27.4). Most water enters the body through ingested liquids (about 60%) and solid foods (about 30%). About 10% of body water is produced by cellular metabolism; this is called **metabolic water** or **water of oxidation.**

Water output occurs by several routes. Some water (28%) vaporizes out of the lungs in expired air or diffuses directly through the skin; this is called **insensible water loss.** Some is lost in obvious perspiration (8%) and in feces (4%). The balance (60%) is excreted by the kidneys in urine.

Healthy people have a remarkable ability to maintain the tonicity of their body fluids within very narrow limits (285–300 mOsm/L). A rise in plasma osmolality triggers (1) thirst, which prompts us to get a drink of water, and (2) release of antidiuretic hormone (ADH), which causes the kidneys to conserve water and excrete concentrated urine. On the other hand, a decline in osmolality inhibits both thirst and ADH release, the latter followed by output of large volumes of dilute urine.

How would the values shown here be affected by (a) drinking a six-pack of beer? (b) a fast in which only water is ingested?

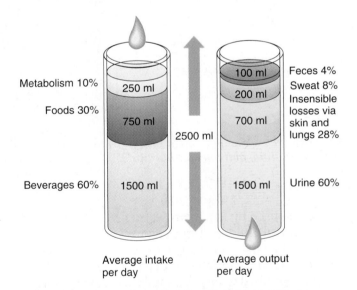

FIGURE 27.4 *Major sources of body water (water intake) and routes of water loss from the body (water output).* When intake and output are in balance, the body is adequately hydrated.

(a) Much more fluid intake from beverages; much greater fluid output in urine (and perhaps if too much alcohol was imbibed, in vomitus, an output route not illustrated). (b) Beverage intake would represent a greater percentage of water intake; fluid intake from foods and metabolic water would drop drastically. Less urine would be made and excreted by the kidneys.

? *What effect would eating pretzels have on this mechanism?*

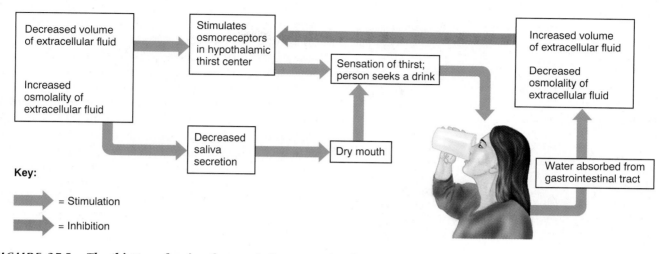

FIGURE 27.5 *The thirst mechanism for regulating water intake.*

Regulation of Water Intake: The Thirst Mechanism

Thirst is the driving force for water intake, but the **thirst mechanism** is poorly understood. It appears that a decrease in plasma volume of 10% (or more, as from hemorrhage) or an increase in plasma osmolality of 1–2% results in a dry mouth and excites the hypothalamic *thirst center*. A dry mouth occurs because the rise in plasma oncotic pressure causes less fluid to leave the bloodstream. Because the salivary glands obtain the water they require from the blood, less saliva is produced.

The hypothalamic thirst center is stimulated when its *osmoreceptors* lose water by osmosis to the hypertonic ECF, an event that causes them to become irritable and depolarize. Collectively, these events cause a subjective sensation of thirst, which motivates us to get a drink (Figure 27.5). This mechanism helps explain the nagging thirst of a hemorrhaging patient who has lost 800 ml or more of blood and, on a lighter note, why it is that some cocktail lounges and bars provide free *salty* snacks to their patrons.

Curiously, thirst is quenched almost as soon as we begin drinking water, even though the water has yet to be absorbed into the blood. The damping of thirst begins as the mucosa of the mouth and throat is moistened and continues as stretch receptors in the stomach and intestine are activated, providing feedback signals that inhibit the thirst center. This premature slaking of thirst prevents us from drinking far more than we need and overdiluting our body fluids, thus allowing time for the osmotic changes to come into play as regulatory factors.

As effective as thirst is, it is not always a reliable indicator of need. This is particularly true during athletic events, when thirst can be satisfied long before sufficient liquids have been drunk to maintain the body in top form. Additionally, elderly or confused people may not recognize or heed thirst signals, and fluid-overloaded renal or cardiac patients may feel thirsty despite their condition.

Regulation of Water Output

Output of certain amounts of water are unavoidable. Such **obligatory water losses** help to explain why we cannot survive for long without drinking. Even the most heroic conservation efforts on the part of the kidneys cannot compensate for zero water intake. Obligatory water losses include *insensible water losses* from the lungs and through the skin, water that accompanies undigested food residues in feces, and a minimum daily **sensible water loss** of 500 ml in urine. Obligatory water loss in urine reflects the facts that (1) when we eat an adequate diet, our kidneys must excrete about 900–1200 mOsm of solutes to maintain blood homeostasis, and (2) human kidneys must flush urine solutes out of the body in water.

Beyond obligatory water loss, the solute concentration and volume of urine excreted depend on fluid intake, diet, and water loss via other avenues. For example, if you jog on a hot day and perspire profusely, much less urine than usual has to be excreted to maintain water balance. Normally, the kidneys

It would increase thirst because body fluids would become more concentrated.

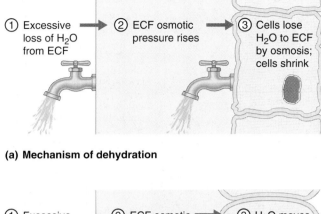

(a) Mechanism of dehydration

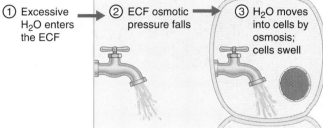

(b) Mechanism of hypotonic hydration (water intoxication)

FIGURE 27.6 *Disturbances in water balance.*

begin to eliminate excess water about 30 minutes after it is ingested. This delay mainly reflects the time required for ADH release to be inhibited. Diuresis reaches a peak in 1 hour and then declines to its lowest level after 3 hours.

The body's water volume is closely tied to a powerful water "magnet," ionic sodium. Moreover, our ability to maintain water balance through urinary output is really a problem of sodium *and* water balance because the two are always regulated in tandem by mechanisms that serve cardiovascular function and blood pressure.

Disorders of Water Balance

Few people really appreciate the importance of water in keeping the body's "machinery" working at peak efficiency. The principal abnormalities of water balance are dehydration, hypotonic hydration, and edema, and each of these conditions offers a special set of problems to its victims.

Dehydration

Dehydration occurs when water loss exceeds water intake over a period of time and the body is in negative fluid balance. Dehydration is a common sequel to hemorrhage, severe burns, prolonged bouts of vomiting or diarrhea, profuse sweating, water deprivation, and diuretic abuse. Dehydration may also be caused by endocrine disturbances, such as diabetes

mellitus or diabetes insipidus (see Chapter 17). Early signs and symptoms of dehydration include a "cottony" or sticky oral mucosa, thirst, dry flushed skin, and decreased urine output *(oliguria)*. If prolonged, dehydration may lead to weight loss, fever, and mental confusion. Another very serious consequence of water loss from the plasma ECF compartment is inadequate blood volume to maintain normal circulation and ensuing *hypovolemic shock* (see p. 744).

In all these situations, water is lost from the ECF (Figure 27.6a). This is followed by the osmotic movement of water from the cells into the ECF, which equalizes the osmolality of the extracellular and intracellular fluids even though the total fluid volume has been reduced. Though the overall effect is called dehydration, it rarely involves only a deficit of water. Most often, as water is lost electrolytes are lost as well.

Hypotonic Hydration

When the osmolality of the ECF starts to drop (usually this reflects a deficit of Na$^+$), several compensatory mechanisms are set into motion. One of these is inhibition of ADH release, and as a result, excess water is quickly flushed from the body in urine. But when there is renal insufficiency or an extraordinary amount of water is drunk very quickly, a type of cellular *overhydration* called **hypotonic hydration** or **water intoxication** may occur. In either case, the ECF is diluted—its sodium content is normal, but excess water is present. Thus, the hallmark of this condition is **hyponatremia** (low ECF Na$^+$ level). This in turn promotes net osmosis into the tissue cells, causing them to swell as they become abnormally hydrated (Figure 27.6b). These events must be reversed quickly—for example, by intravenous administration of hypertonic mannitol, which reverses the osmotic gradient and "pulls" water out of the cells. Otherwise, the resulting electrolyte dilution leads to severe metabolic disturbances evidenced by nausea, vomiting, muscular cramping, and cerebral edema. Water intoxication is particularly damaging to neurons. Uncorrected cerebral edema quickly leads to disorientation, convulsions, coma, and death.

Edema

Edema (ĕ-de'mah; "a swelling") is an atypical accumulation of fluid in the interstitial space, leading to tissue swelling. Edema may be caused by any event that steps up the flow of fluid out of the bloodstream or hinders its return.

Factors that accelerate fluid loss from the bloodstream include increased blood pressure and capillary permeability. Increased blood (capillary hydrostatic) pressure can result from incompetent venous valves, localized blood vessel blockage, congestive heart failure, hypertension, or high blood volume, for instance, during pregnancy or resulting from abnormal retention of sodium (which promotes water retention). Whatever the cause, the abnormally high blood pressure intensifies filtration at the capillary beds.

Increased capillary permeability is usually due to an ongoing inflammatory response. Recall that inflammatory chemicals cause local capillaries to become very porous, allowing large amounts of exudate (containing clotting proteins, nutrients, and immune elements) to form.

Edema caused by hindered fluid return to the bloodstream usually reflects an imbalance in the colloid osmotic pressures on the two sides of the capillary membranes. For example, **hypoproteinemia** (hi"po-pro"te-ĭ-ne′me-ah), a condition of unusually low levels of plasma proteins, results in tissue edema because protein-deficient plasma has an abnormally low colloid osmotic pressure. Fluids are forced out of the capillary beds at the arterial ends by blood pressure as usual, but fail to return to the blood at the venous ends. Thus, the interstitial spaces become congested with fluid. Hypoproteinemia may result from protein malnutrition, liver disease (in which the liver fails to make its normal supply of plasma proteins), or *glomerulonephritis* (in which plasma proteins pass through "leaky" renal filtration membranes and are lost in urine). Although the cause differs, the result is the same when lymphatic vessels are blocked or have been surgically removed. The small amounts of plasma proteins that seep out of the bloodstream are not returned to the blood as usual. As the leaked proteins accumulate, they exert an ever-increasing colloid osmotic pressure, which draws fluid from the blood and holds it in the interstitial space.

Because excess fluid in the interstitial space increases the distance nutrients and oxygen must diffuse between the blood and the cells, edema can impair tissue function. However, the most serious problems resulting from edema affect the cardiovascular system. When fluid leaves the bloodstream and accumulates in the interstitial space, both blood volume and blood pressure decline and the efficiency of the circulation is severely impaired.

ELECTROLYTE BALANCE

Electrolytes include salts, acids, and bases, but the term **electrolyte balance** usually refers to the salt balance in the body. Salts (NaCl and $CaHPO_4$, for example) provide minerals essential for neuromuscular excitability, secretory activity, membrane permeability, and many other cellular functions. Salts are also important in controlling fluid movements. Although many electrolytes are crucial for cellular activity, we will specifically examine the regulation of sodium, potassium, calcium, and magnesium

only. Acids and bases, which are more intimately involved in determining the pH of body fluids, are considered in the next section.

Salts enter the body in foods and fluids, and small amounts are generated during metabolic activity. For example, phosphates are liberated during catabolism of nucleic acids and bone matrix. Obtaining enough electrolytes is usually not a problem. Indeed, most of us have a far greater taste than need for salt (NaCl). We shake table salt on our food even though natural foods contain ample amounts and processed foods contain exorbitant quantities. The taste for very salty foods is learned, but some liking for salt may be innate to ensure adequate intake of these vital ions.

Salts are lost from the body in perspiration, feces, and urine. When we are salt-depleted, our perspiration is more dilute. Even so, on a hot day much more salt than usual can be lost in sweat, and gastrointestinal disorders can lead to large salt losses in feces or vomitus. Thus, the flexibility of renal mechanisms that regulate the electrolyte balance of the blood is a critical asset. Some causes and consequences of electrolyte excesses and deficiencies are summarized in Table 27.1.

Homeostatic Imbalance

In rare instances, severe electrolyte deficiencies prompt a craving for salty foods and an increased willingness to try new substances in the search for salt, such as exotic smoked meats or pickled eggs. This is common in those with *Addison's disease*, a disorder entailing deficient mineralocorticoid hormone production by the adrenal cortex. When electrolytes other than NaCl are deficient, substances not usually considered to be nutrients, like chalk, clay, starch, and burnt match tips, may be eaten. Such an appetite for abnormal substances is called *pica*. ∎

The Central Role of Sodium in Fluid and Electrolyte Balance

Sodium holds a central position in fluid and electrolyte balance and overall body homeostasis. Indeed, regulating the balance between sodium input and output is one of the most important functions of the kidneys. Sodium salts ($NaHCO_3$ and NaCl) account for 90–95% of all solutes in the ECF, and they contribute about 280 mOsm of the total (300 mOsm) ECF solute concentration. At its normal plasma concentration of about 142 mEq/L, sodium is the single most abundant cation in the ECF and is the only one exerting *significant* osmotic pressure. Additionally, cellular plasma membranes are relatively impermeable to Na^+, but some does manage to diffuse in and must be pumped out against its

TABLE 27.1	Causes and Consequences of Electrolyte Imbalances		
Ion	Abnormality/ serum value	Possible causes	Consequences
Sodium	Hypernatremia (Na^+ excess in ECF: >145 mEq/L)	Dehydration; uncommon in healthy individuals; may occur in infants or the confused aged (individuals unable to indicate thirst) or may be a result of excessive intravenous NaCl administration	Thirst: CNS dehydration leads to confusion and lethargy progressing to coma; increased neuromuscular irritability evidenced by twitching and convulsions
	Hyponatremia (Na^+ deficit in ECF: <130 mEq/L)	Solute loss, water retention, or both (e.g., excessive Na^+ loss through burned skin, excessive sweating, vomiting, diarrhea, tubal drainage of stomach, and as a result of excessive use of diuretics); deficiency of aldosterone (Addison's disease); renal disease; excess ADH release	Most common signs are those of neurologic dysfunction due to brain swelling. If sodium amounts are actually normal but water is excessive, the symptoms are the same as those of water excess: mental confusion; giddiness; coma if development occurs slowly; muscular twitching, irritability, and convulsions if the condition develops rapidly; systemic edema; congestive heart failure in cardiac patients. In hyponatremia accompanied by water loss, the main signs are decreased blood volume and blood pressure (circulatory shock)
Potassium	Hyperkalemia (K^+ excess in ECF: >5.5 mEq/L)	Renal failure; deficit of aldosterone; rapid intravenous infusion of KCl; burns or severe tissue injuries which cause K^+ to leave cells	Nausea, vomiting, diarrhea; bradycardia; cardiac arrhythmias, depression, and arrest; skeletal muscle weakness; flaccid paralysis
	Hypokalemia (K^+ deficit in ECF: <3.5 mEq/L)	Gastrointestinal tract disturbances (vomiting, diarrhea), gastrointestinal suction; chronic stress; Cushing's disease; inadequate dietary intake (starvation); hyperaldosteronism; diuretic therapy	Cardiac arrhythmias, flattened T wave; muscular weakness; alkalosis; hypoventilation; mental confusion
Phosphate	Hyperphosphatemia (HPO_4^{2-} excess in ECF: >6 mEq/L)	Increased intestinal absorption; decreased urinary loss due to renal failure; hypoparathyroidism; major tissue trauma	No direct clinical symptoms because an excess or deficit in phosphate is usually accompanied by a decrease or increase in Ca^{2+} levels
	Hypophosphatemia (HPO_4^{2-} deficit in ECF: <1 mEq/L)	Decreased intestinal absorption; increased urinary output; hyperthyroidism	
Chloride	Hyperchloremia (Cl^- excess in ECF: >105 mEq/L)	Dehydration; increased retention or intake; hyperkalemia	Metabolic acidosis due to enhanced loss of bicarbonate; weakness; stupor; rapid, deep breathing; unconsciousness
	Hypochloremia (Cl^- deficit in ECF: <95 mEq/L)	Vomiting; overhydration; hypokalemia; excessive ingestion of alkaline substances; aldosterone deficiency	Metabolic alkalosis due to bicarbonate retention
Calcium	Hypercalcemia (Ca^{2+} excess in ECF: >5.8 mEq/L or 11 mg%)	Hyperparathyroidism; excessive vitamin D; prolonged immobilization; renal disease (decreased excretion); malignancy; Paget's disease; Cushing's disease accompanied by osteoporosis	Bone wasting, pathological fractures; flank and deep thigh pain; kidney stones, nausea and vomiting, cardiac arrhythmias and arrest; depressed respiration, coma
	Hypocalcemia (Ca^{2+} deficit in ECF: <4.5 mEq/L or 9 mg%)	Burns (calcium trapped in damaged tissues); increased renal excretion in response to stress and increased protein intake; diarrhea; vitamin D deficiency; alkalosis	Tingling of fingers, tremors, tetany, convulsions; depressed excitability of the heart, bleeder's disease
Magnesium	Hypermagnesemia (Mg^{2+} excess in ECF: >6 mEq/L)	Rare (occurs when Mg is not excreted normally); deficiency of aldosterone; excessive ingestion of Mg^{2+}-containing antacids	Lethargy; impaired CNS functioning, coma, respiratory depression
	Hypomagnesemia (Mg^{2+} deficit in ECF: <1.4 mEq/L)	Alcoholism; loss of intestinal contents, severe malnutrition; diuretic therapy	Tremors, increased neuromuscular excitability, convulsions

electrochemical gradient. These two qualities give sodium the primary role in controlling ECF volume and water distribution in the body.

It is important to understand that while the sodium content of the body may change, *its concentration in the ECF normally remains stable* because of immediate adjustments in water volume. Remember, *water follows salt.* Furthermore, because all body fluids are in osmotic equilibrium, a change in plasma sodium levels affects not only plasma volume and blood pressure, but also the ICF and interstitial fluid volumes. In addition, sodium ions continually move back and forth between the ECF and body secretions. For example, large volumes (about 8 L) of sodium-containing secretions (gastric, intestinal, and pancreatic juice, saliva, bile) are spewed into the digestive tract daily, only to be almost completely reabsorbed. Finally, renal acid-base control mechanisms (discussed shortly) are coupled to sodium ion transport.

Regulation of Sodium Balance

Despite the crucial importance of sodium, sodium receptors that specifically monitor Na^+ levels in body fluids have yet to be found. Regulation of the sodium-water balance is inseparably linked to blood pressure and blood volume, and involves a variety of neural and hormonal controls. Na^+ reabsorption does *not* exhibit a transport maximum, and in healthy individuals nearly all the Na^+ in the urinary filtrate is reabsorbed. We will begin our coverage of sodium balance by reviewing the regulatory effect of aldosterone. Then we will examine the various feedback loops that interact to regulate sodium and water balance and blood pressure.

Influence and Regulation of Aldosterone

The hormone **aldosterone** "has the most to say" about renal regulation of sodium ion concentrations in the ECF. But whether aldosterone is present or not, some 65% of the sodium in the renal filtrate is reabsorbed in the proximal tubules of the kidneys and another 25% is reclaimed in the loops of Henle (see Chapter 26).

When aldosterone concentrations are high, essentially all the remaining Na^+ (actually NaCl, because Cl^- is cotransported) is actively reabsorbed in the distal convoluted tubules and collecting ducts. Water follows if it can, that is, if the tubule permeability has been increased by ADH. Thus, aldosterone usually promotes both sodium and water retention. However, when aldosterone release is inhibited, virtually no Na^+ reabsorption occurs beyond the distal tubule. So, although urinary excretion of large amounts of sodium *always* results in the excretion of large amounts of water as well, the

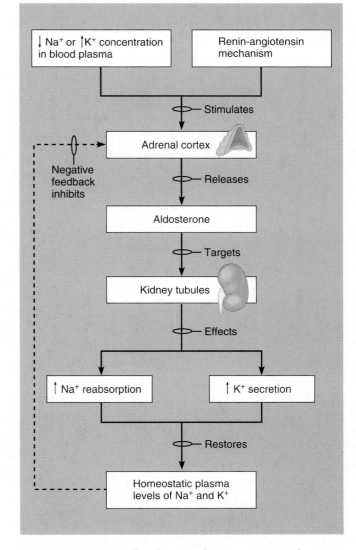

FIGURE 27.7 *Mechanisms and consequences of aldosterone release.*

reverse is *not* true. Substantial amounts of nearly sodium-free urine can be eliminated as needed to achieve water balance.

Recall that aldosterone is produced by adrenal cortical cells. The most important trigger for aldosterone release is the renin-angiotensin mechanism mediated by the juxtaglomerular apparatus of the renal tubules (see Figure 27.9, p. 1051). When the juxtaglomerular (JG) apparatus responds to (1) sympathetic nervous system stimulation, (2) decreased filtrate osmolality, or (3) decreased stretch (due to decreased blood pressure), the JG cells release renin. Renin catalyzes the series of reactions that produce angiotensin II which, in turn, prompts aldosterone release. Conversely, high renal blood pressure and high filtrate osmolality depress renin-angiotensin-aldosterone release. The adrenal cortical cells are also directly stimulated to release aldosterone by elevated K^+ levels in the ECF (Figure 27.7).

At what point in this flowchart would plasma osmolality change dramatically?

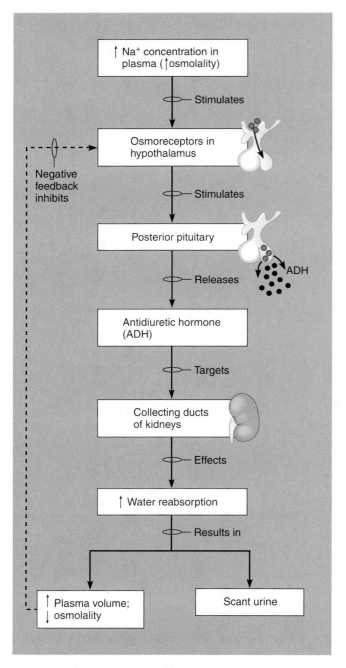

FIGURE 27.8 *Mechanisms and consequences of ADH release.*

Aldosterone brings about its effects slowly, over a period of hours to days, and although it has a dramatic effect on tubular reabsorption of sodium, it has a relatively small impact on blood volume. The principal effects of aldosterone are to diminish uri-

nary output and increase blood volume, but before these can change more than a few percent, basic feedback mechanisms for blood volume control come into play. Even people with Cushing's disease (hyperaldosteronism) rarely have ECF and blood volumes more than 5–10% above normal.

Homeostatic Imbalance

People with Addison's disease (hypoaldosteronism) lose tremendous amounts of salt and water to urine. As long as they ingest adequate amounts of salt and fluids, people with this condition can avoid problems, but they are perpetually teetering on the brink of hypovolemia and dehydration. ■

Cardiovascular System Baroreceptors

Blood volume is carefully monitored and regulated to maintain blood pressure and cardiovascular function. As blood volume (hence pressure) rises, baroreceptors in the heart and in the large vessels of the neck and thorax (carotid arteries and aorta) alert the cardiovascular centers in the brain stem. Shortly after, sympathetic nervous system impulses to the kidneys decline, allowing the afferent arterioles to dilate. As the glomerular filtration rate rises, sodium and water output increase. This phenomenon, called *pressure diuresis*, reduces blood volume and, consequently, blood pressure. In contrast, drops in systemic blood pressure lead to constriction of the afferent arterioles, which reduces filtrate formation and urinary output and increases systemic blood pressure (see Figure 27.9). Thus, the baroreceptors provide information on the "fullness" or volume of the circulation that is critical for maintaining cardiovascular homeostasis. Because sodium ion concentration determines fluid volume, the baroreceptors might be regarded as "sodium receptors." Note however that the direct influence of blood pressure changes on GFR is fairly unimportant because autoregulatory mechanisms kick in quickly.

Influence and Regulation of ADH

The amount of water reabsorbed in the collecting ducts of the kidneys is proportional to ADH release. When ADH levels are low, most of the water reaching the collecting ducts is simply allowed to pass through because the number of water-filled channels in the principal cells is at a minimum level. The result is dilute urine and a reduced volume of body fluids. When ADH levels are high, nearly all of the filtered water is reabsorbed and a small volume of highly concentrated urine is excreted.

Osmoreceptors of the hypothalamus sense the ECF solute concentration and trigger or inhibit ADH release from the posterior pituitary accordingly (Figure 27.8). A decrease in sodium ion concentration (for instance, due to the increased blood volume that

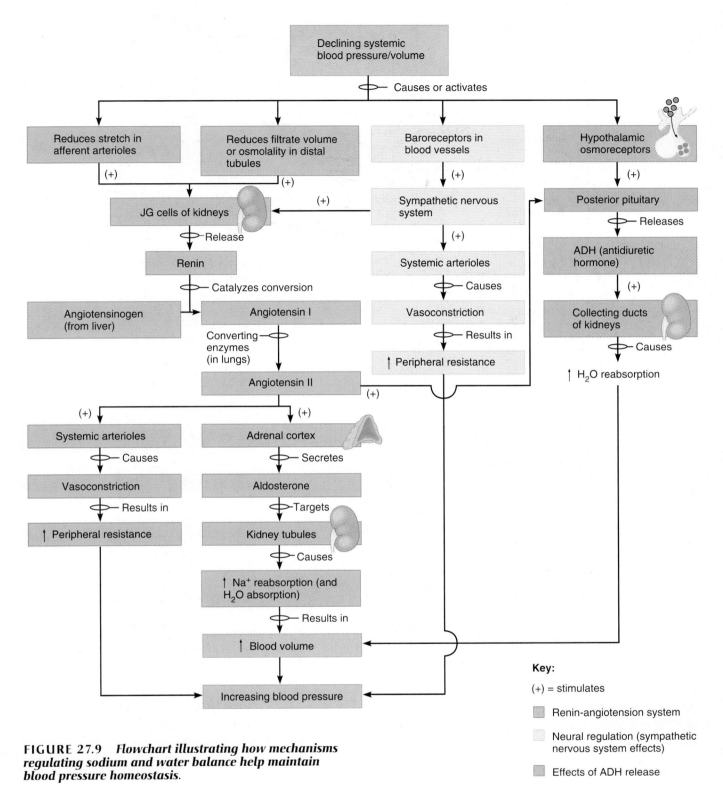

FIGURE 27.9 *Flowchart illustrating how mechanisms regulating sodium and water balance help maintain blood pressure homeostasis.*

results from drinking a lot of beer or punch at the weekend "keg" party) inhibits ADH release and allows more water to be excreted in urine, restoring normal Na$^+$ levels in the blood. An increase in sodium levels (which may be due to decreased blood volume) stimulates ADH release both directly by stimulating the hypothalamic osmoreceptors and in-

directly via the renin-angiotensin mechanism. (The latter stimulus is not illustrated in Figure 27.8 but is indicated in Figure 27.9.) Factors that specifically trigger ADH release by reducing blood volume include prolonged fever; excessive sweating, vomiting, or diarrhea; severe blood loss; and traumatic burns. Figure 27.9 summarizes how renal mechanisms

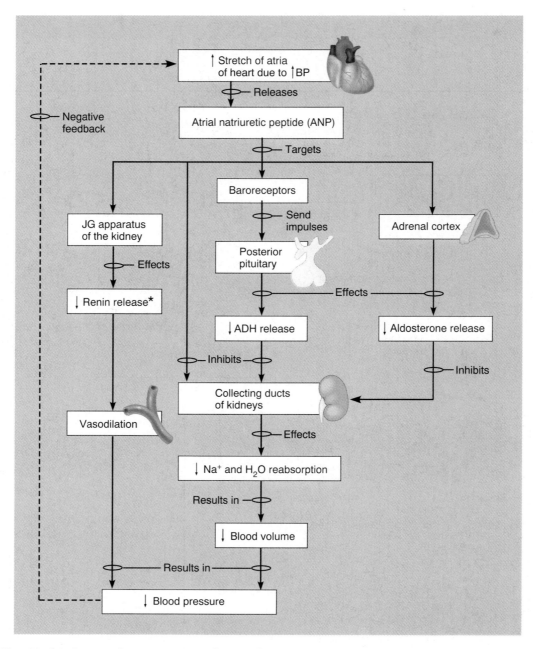

FIGURE 27.10 *Mechanisms and consequences of ANP release.* (*↓ renin release also inhibits ADH and aldosterone release and hence the effects of those hormones.)

involving aldosterone, angiotensin II, and ADH are interrelated and tied into overall controls of blood volume and blood pressure.

Influence and Regulation of Atrial Natriuretic Peptide

The influence of **atrial natriuretic peptide (ANP)** (also called *atrial natriuretic factor, ANF*) can be summarized in one sentence: It reduces blood pressure and blood volume by inhibiting nearly all events that promote vasoconstriction and Na^+ and water retention (Figure 27.10). Atrial natriuretic peptide is a hormone that is released by certain cells of the heart atria when they are stretched by the effects of elevated blood pressure. ANP has potent diuretic

and natriuretic (literally, salt-excreting) effects. It promotes excretion of sodium and water by the kidneys, presumably by directly inhibiting the ability of the collecting duct to reabsorb Na^+ and by suppressing the release of ADH, renin, and aldosterone. Additionally, ANP acts both directly and indirectly (by inhibiting renin-induced generation of angiotensin II) to relax vascular smooth muscle; thus it causes vasodilation. Collectively, these effects reduce blood pressure.

Influence of Other Hormones

Female Sex Hormones The **estrogens** are chemically similar to aldosterone and, like aldosterone, they enhance NaCl reabsorption by the renal tubules. Because water follows, many women retain fluid as their estrogen levels rise during the menstrual cycle. The edema experienced by many pregnant women is also largely due to the effect of estrogens. In contrast, **progesterone** appears to decrease sodium reabsorption by blocking the effect of aldosterone on the renal tubules. Thus, progesterone has a diuretic-like effect and promotes sodium and water loss.

Glucocorticoids The usual effect of **glucocorticoids,** such as cortisol and hydrocortisol, is to enhance tubular reabsorption of sodium, but they also promote an increased glomerular filtration rate which may mask their effects on the tubules. However, when their plasma levels are high, the glucocorticoids exhibit potent aldosterone-like effects and promote edema (see Chapter 17).

Regulation of Potassium Balance

Potassium, the chief intracellular cation, is required for normal neuromuscular functioning as well as for several essential metabolic activities, including protein synthesis. Nevertheless, it can be extremely toxic. Because the relative ICF-ECF potassium ion (K^+) concentration directly affects a cell's resting membrane potential, even a slight change in K^+ concentration on either side of the membrane has profound effects on neurons and muscle fibers. Potassium excesses in the ECF decrease their membrane potential, causing depolarization, which is often followed by reduced excitability. Too little K^+ in the ECF causes hyperpolarization and nonresponsiveness. The heart is particularly sensitive to K^+ levels. Both too much and too little K^+ (hyperkalemia and hypokalemia respectively) can disrupt electrical conduction in the heart, leading to sudden death (see Table 27.1).

Potassium is also part of the body's buffer system, which resists changes in the pH of body fluids. Shifts of hydrogen ions (H^+) into and out of cells induce corresponding shifts of potassium in the opposite direction to maintain cation balance. Thus, ECF potassium levels rise with acidosis, as K^+ leaves and H^+ enters the cells, and fall with alkalosis, as K^+ moves into the cells and H^+ leaves them to enter the ECF. Although these pH-driven shifts do not change the total amount of K^+ in the body, they can seriously interfere with the activity of excitable cells.

Regulatory Site: The Cortical Collecting Duct

Like sodium balance, potassium balance is maintained chiefly by renal mechanisms. However, there are important differences in the way this balance is achieved. The amount of sodium reabsorbed in the tubules is precisely tailored to need, and sodium ions are *never* secreted into the filtrate.

In contrast, the renal tubules predictably reabsorb about 55% of the filtered K^+, and the thick ascending limb of Henle's loop absorbs another 30% or so, leaving less than 15% to be lost in urine regardless of need. The responsibility for K^+ balance falls chiefly on the cortical collecting ducts, and is accomplished mainly by changing the amount of potassium *secreted* into the filtrate.

As a rule, the relative potassium levels in the ECF are excessive, and K^+ secretion by the principal cells of the cortical collecting ducts is accelerated over basal levels. At times, the amount of potassium excreted may actually exceed the amount filtered. However, when ECF potassium concentrations are abnormally low, K^+ moves from the tissue cells into the ECF and the renal principal cells conserve K^+ by reducing its secretion and excretion to the minimal levels possible. Note that these principal cells are the same cells that mediate aldosterone-induced reabsorption of Na^+ and the ADH-stimulated reabsorption of water. Additionally, a unique population of collecting duct cells, called *type A intercalated cells*, can reabsorb some of the K^+ still left in the filtrate (in conjunction with active secretion of H^+), thus helping to reestablish K^+ (and pH) balance. However, keep in mind that the main thrust of renal regulation of potassium is to *excrete* it. Because the kidneys have a very limited ability to retain potassium, it may be lost in urine even in the face of a deficiency. Consequently, failure to ingest potassium-rich substances eventually results in severe potassium deficiency.

Essentially, two factors determine the rate and extent of potassium secretion—the plasma potassium ion concentration and aldosterone levels.

Influence of Plasma Potassium Concentration

The single most important factor influencing potassium secretion is the K^+ concentration in blood plasma. A high-potassium diet increases the K^+ content of the ECF. This favors entry of K^+ into the principal cells of the cortical collecting duct and prompts them to secrete K^+ into the filtrate so that more potassium is excreted. Conversely, a low-potassium diet or accelerated K^+ loss depresses its secretion (and promotes its limited reabsorption) by the collecting ducts.

Influence of Aldosterone

As aldosterone stimulates the principal cells to reabsorb sodium, it simultaneously enhances potassium ion secretion (see Figure 27.7). To maintain electrolyte balance there is a one-for-one exchange of Na^+ and K^+ in the cortical collecting ducts. For each Na^+ reabsorbed, a K^+ is secreted. Thus, as plasma Na^+ levels rise, K^+ levels fall proportionately.

Adrenal cortical cells are *directly* sensitive to the K^+ content of the ECF bathing them. When it increases even slightly, the adrenal cortex is strongly stimulated to release aldosterone, which increases potassium secretion by the exchange process just described. Thus, potassium controls its own concentrations in the ECF via feedback regulation of aldosterone release. Aldosterone is also secreted in response to the renin-angiotensin mechanism previously described (see p. 1051).

⚠ *Homeostatic Imbalance*

In an attempt to reduce NaCl intake, many people have turned to salt substitutes, which are high in potassium. However, heavy consumption of these substitutes is safe only when aldosterone release by the adrenal cortex is normal. In the absence of aldosterone, hyperkalemia is swift and lethal regardless of K^+ intake (see Table 27.1). Conversely, when a person has an adrenocortical tumor that pumps out tremendous amounts of aldosterone, ECF potassium levels fall so low that neurons all over the body hyperpolarize and paralysis occurs. ■

Regulation of Calcium and Phosphate Balance

About 99% of the body's calcium is found in bones in the form of calcium phosphate salts, which provide strength and rigidity to the skeleton. Ionic calcium in the ECF is important for normal blood clotting, cell membrane permeability, and secretory behavior. Like sodium and potassium, ionic calcium has potent effects on neuromuscular excitability. Hypocalcemia increases excitability and causes muscle tetany. Hypercalcemia is equally dangerous because it inhibits neurons and muscle cells and may cause life-threatening cardiac arrhythmias (see Table 27.1).

Calcium is closely regulated, rarely deviating from normal limits. Calcium balance is regulated primarily by the interaction of two hormones—parathyroid hormone and calcitonin—and under normal circumstances about 98% of the filtered Ca^{2+} is reabsorbed. The bony skeleton provides a dynamic reservoir from which calcium and phosphorus can be withdrawn or deposited to maintain the balance of these electrolytes.

Influence of Parathyroid Hormone

The most important controls of Ca^{2+} homeostasis are exerted by **parathyroid hormone (PTH)**, re-leased by the tiny parathyroid glands located on the posterior aspect of the thyroid gland in the pharynx. Declining plasma levels of Ca^{2+} directly stimulate the parathyroid glands to release PTH, which promotes an increase in calcium levels by targeting the following organs (see Figure 17.11, p. 629):

1. Bones. PTH activates osteoclasts (bone-digesting cells), which break down the bone matrix, resulting in the release of Ca^{2+} and PO_4^{2-} to the blood.

2. Small intestine. PTH enhances intestinal absorption of Ca^{2+} indirectly by stimulating the kidneys to transform vitamin D to its active form (1,25-dihydroxycholecalciferol, also called 1,25-dihydroxyvitamin D_3), which is a necessary cofactor for calcium absorption by the small intestine.

3. Kidneys. PTH increases calcium reabsorption by the renal tubules while simultaneously decreasing phosphate ion (PO_4^{2-}) reabsorption. Thus, calcium conservation and phosphate excretion go hand in hand. Hence, the *product* of calcium and phosphate in the ECF remains constant, preventing calcium-salt deposit in bones or soft body tissues.

Some calcium is reabsorbed passively in the proximal convoluted tubule (PCT) via diffusion through the paracellular route (a process driven by the electrochemical gradient). However, PTH-dependent Ca^{2+} reabsorption occurs mainly in the distal convoluted tubule (DCT) and is driven by a Ca^{2+}-ATPase pump.

As a rule, most (75%) filtered phosphate (including $H_2PO_4^-$, HPO_4^{2-}, and PO_4^{3-}) is reabsorbed in the proximal tubules by active transport. In the absence of PTH, phosphate reabsorption is regulated by its transport maximum. Amounts present in excess of that maximum simply flow out in urine. But when the PTH level rises, active transport of phosphate is inhibited.

When calcium levels in the ECF are within normal limits (9–11 mg/100 ml blood) or are high, PTH secretion is inhibited. Consequently, release of calcium from bone is inhibited, larger amounts of calcium are lost in feces and urine, and more phosphate is retained. Hormones other than PTH alter phosphate reabsorption. For example, insulin increases it while glucagon decreases it.

Influence of Calcitonin

Calcitonin, a hormone produced by the parafollicular cells of the thyroid gland, is released in response to rising blood calcium levels. Calcitonin targets bone, where it encourages deposit of calcium salts and inhibits bone reabsorption. Calcitonin is an antagonist of PTH, but its contribution to calcium and phosphate homeostasis is minor to negligible.

Regulation of Magnesium Balance

Magnesium, the second most abundant intracellular cation, activates the coenzymes needed for carbohydrate and protein metabolism and plays an essential role in cardiac function, neurotransmission, and neuromuscular activity. Half of the magnesium in the body is in the skeleton. Most of the remainder is found within cells.

Control of magnesium balance is poorly understood, but a renal transport maximum for magnesium is known to exist. Under ordinary conditions, only 3–5% of the filtered magnesium is excreted; excesses and deficits are rapidly corrected. Both PTH and AMP inhibit Mg^{2+} reabsorption in the PCT.

Regulation of Anions

Chloride is the major anion accompanying sodium in the ECF and, like sodium, it helps maintain the osmotic pressure of the blood. When blood pH is within normal limits or is slightly alkaline, about 99% of filtered chloride ions are reabsorbed. In the PCT, they move passively and simply follow sodium ions out of the filtrate and into the peritubular capillary blood. In most other tubule segments, Na^+ and Cl^- transport are coupled (see Figure 26.13, p. 1021). When acidosis occurs, fewer chloride ions accompany sodium because bicarbonate ion reabsorption is stepped up to restore blood pH to its normal range. Thus, the choice between chloride and bicarbonate ions serves acid-base regulation. Most other anions, such as sulfates and nitrates, have definite transport maximums, and when their concentrations in the filtrate exceed their renal thresholds, excesses spill over into urine.

ACID-BASE BALANCE

Because of their abundant hydrogen bonds, all functional proteins (enzymes, hemoglobin, cytochromes, and others) are influenced by hydrogen ion concentration. It follows then that nearly all biochemical reactions are influenced by the pH of their fluid environment, and the acid-base balance of body fluids is closely regulated. (For a review of the basic principles of acid-base reactions and pH, see Chapter 2, pp. 43–45.)

The optimal pH of various body fluids differs, but not very much. The normal pH of arterial blood is 7.4, that of venous blood and interstitial fluid is 7.35, and that of intracellular fluid averages 7.0. The lower pH in cells and venous blood reflects their greater amounts of acidic metabolites (such as lactic acid) and carbon dioxide, which combines with water to form carbonic acid (H_2CO_3).

Whenever the pH of arterial blood rises above 7.45, a person is said to have **alkalosis** (al″kah-lo′sis) or **alkalemia.** A drop in arterial pH to below 7.35 results in **acidosis** (as″ĭ-do′sis) or **acidemia.** Because a pH of 7.0 is neutral, chemically speaking, 7.35 is not acidic. However, it is a higher-than-optimal H^+ concentration for functioning of most cells, so any arterial pH between 7.35 and 7.0 is called **physiological acidosis.**

Although small amounts of acidic substances enter the body via ingested foods, most hydrogen ions originate as by-products or end products of cellular metabolism. For example, (1) breakdown of phosphorus-containing proteins (and certain other molecules) releases *phosphoric acid* into the ECF, (2) anaerobic respiration of glucose produces *lactic acid*, (3) fat metabolism yields other organic acids, such as fatty acids, and *ketone bodies*, and (4) as described in Chapter 23, the loading and transport of carbon dioxide in the blood as bicarbonate liberates hydrogen ions. Finally, although hydrochloric acid produced by the stomach is not really *in* the body, it is a source of H^+ that must be buffered if digestion is to occur normally in the small intestine.

The concentration of H^+ in blood is regulated sequentially by (1) chemical buffer systems, (2) the respiratory center in the brain stem, and (3) renal mechanisms. Chemical buffers act within a fraction of a second to resist pH changes and are the first line of defense. Within 1–3 minutes, changes in respiratory rate and depth are occurring to compensate for acidosis or alkalosis. The kidneys, although the most potent of the acid-base regulatory systems, ordinarily require hours to a day or more to effect changes in blood pH.

Chemical Buffer Systems

Before examining the body's chemical buffer systems, let's review the definitions of strong and weak acids and bases. Recall that acids are *proton donors*, and that the acidity of a solution reflects only the *free* hydrogen ions, not those bound to anions. *Strong acids*, which dissociate completely and liberate all their H^+ in water (Figure 27.11a), can dramatically change a solution's pH. By contrast, *weak acids*, like carbonic and acetic acids, dissociate only partially (Figure 27.11b), and so have a much slighter effect on pH. However, weak acids are efficient at preventing pH changes, and this feature allows them to play extremely important roles in the chemical buffer systems of the body.

Now, we will attend to the concept of strong and weak bases. Remember that bases are *proton acceptors*. Strong bases are those, like hydroxides, that dissociate easily in water and quickly tie up H^+. Conversely, bicarbonate ion (HCO_3^-) and ammonia

To prevent the pH shift that occurs in situation (a), would it be better to add a strong base or a weak base? Why?

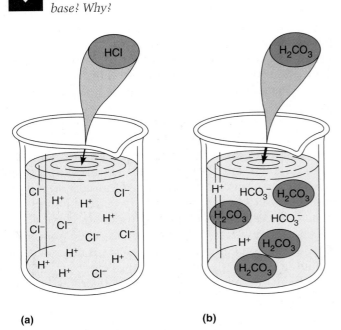

(a)   **(b)**

FIGURE 27.11 *Comparison of dissociation of strong and weak acids.* **(a)** When HCl, a strong acid, is added to water it dissociates completely into its ions (H^+ and Cl^-). **(b)** By contrast, dissociation of H_2CO_3, a weak acid, is very incomplete, and some molecules of H_2CO_3 remain undissociated (symbols shown in green circles) in solution.

(NH_3) are slower to accept protons and are considered weak bases.

Chemical acid-base buffers are systems of one or two molecules that act to resist changes in pH when a strong acid or base is added. They do this by binding to hydrogen ions whenever the pH of body fluids drops and releasing them when pH rises. The three major chemical buffer systems in the body are the *bicarbonate, phosphate,* and *protein buffer systems,* each of which is important in maintaining the pH in one or more fluid compartments. They all work together. Anything that causes a shift in hydrogen ion concentration in one compartment simultaneously causes a change in the others. Thus, the buffer systems actually buffer each other, so that any drifts in pH are resisted by the *entire* chemical buffering system.

A weak base, because the weak base would act as a buffer to tie up the released H^+.

Bicarbonate Buffer System

The **bicarbonate buffer system** is a mixture of carbonic acid (H_2CO_3) and its salt, sodium bicarbonate ($NaHCO_3$), in the same solution. Although it also buffers the ICF, it is the *only* important *ECF* buffer.

Carbonic acid, a weak acid, does not dissociate to any great extent in neutral or acidic solutions. Thus, when a strong acid such as hydrochloric acid (HCl) is added to this system, most existing carbonic acid remains intact. However, the bicarbonate ions of the salt act as weak bases to tie up the H^+ released by the stronger acid, forming *more* carbonic acid:

$$\underset{\text{strong acid}}{HCl} + \underset{\text{weak base}}{NaHCO_3} \rightarrow \underset{\text{weak acid}}{H_2CO_3} + \underset{\text{salt}}{NaCl}$$

Because the strong acid HCl is converted to the weak acid H_2CO_3, it lowers the pH of the solution only slightly.

Similarly, if a strong base such as sodium hydroxide (NaOH) is added to the same buffered solution, a weak base such as sodium bicarbonate ($NaHCO_3$) will not dissociate under such (alkaline) conditions and so will not contribute to the rise in pH. On the other hand, carbonic acid will be forced to dissociate further by the presence of the base (and rising pH), donating more H^+ to tie up the OH^- released by the strong base:

$$\underset{\text{strong base}}{NaOH} + \underset{\text{weak acid}}{H_2CO_3} \rightarrow \underset{\text{weak base}}{NaHCO_3} + \underset{\text{water}}{H_2O}$$

The net result is replacement of a strong base (NaOH) by a weak one ($NaHCO_3$), so that the pH of the solution rises very little.

Although the bicarbonate salt in the example is sodium bicarbonate, the salt is dissociated and the specific cation of the salt is unimportant. Other bicarbonate salts function in the same way. Within cells, where little sodium is present, potassium and magnesium bicarbonates act as part of the bicarbonate buffer system.

The buffering power of this type of system is directly related to the concentrations of the buffering substances. Thus, if acids enter the blood at such a rate that all the available bicarbonate ions, often referred to as the **alkaline reserve,** are tied up, the buffer system becomes ineffective in resisting pH changes. The bicarbonate ion concentration in the ECF is normally around 25 mEq/L and is closely regulated by the kidneys. The concentration of carbonic acid is only about one-twentieth that of bicarbonate. However, the supply of carbonic acid (made available from the CO_2 released during cellular respiration) is almost limitless, so obtaining that member of the buffer pair is usually not a problem. The carbonic acid content of the blood is subject to respiratory controls.

Phosphate Buffer System

The operation of the **phosphate buffer system** is nearly identical to that of the bicarbonate buffer. Its components are the sodium salts of dihydrogen phosphate ($H_2PO_4^-$) and monohydrogen phosphate (HPO_4^{2-}). NaH_2PO_4 acts as a weak acid. Na_2HPO_4, with one less hydrogen atom, acts as a weak base.

Again, hydrogen ions released by strong acids are tied up in weak acids:

$$\underset{\text{strong acid}}{HCl} + \underset{\text{weak base}}{Na_2HPO_4} \rightarrow \underset{\text{weak acid}}{NaH_2PO_4} + \underset{\text{salt}}{NaCl}$$

and strong bases are converted to weak bases:

$$\underset{\text{strong base}}{NaOH} + \underset{\text{weak acid}}{NaH_2PO_4} \rightarrow \underset{\text{weak base}}{Na_2HPO_4} + \underset{\text{water}}{H_2O}$$

Because the phosphate buffer system is present in low concentrations in the ECF (approximately one-sixth that of the bicarbonate buffer system), it is relatively unimportant for buffering blood plasma. However, it is a very effective buffer in urine and in intracellular fluid, where phosphate concentrations are usually higher.

Protein Buffer System

Proteins in plasma and within cells are the body's most plentiful and powerful source of buffers, and constitute the **protein buffer system.** In fact, at least three-quarters of all the buffering power of body fluids resides within the cells, and most of this reflects the buffering activity of intracellular proteins.

As described in Chapter 2, proteins are polymers of amino acids. Some of the linked amino acids have free, or exposed, groups of atoms called *organic acid (carboxyl) groups (—COOH)*, which dissociate to release H^+ when the pH begins to rise:

$$R^\star\text{—COOH} \rightarrow R\text{—COO}^- + H^+$$

Additionally, some amino acids have groups of atoms that can act as bases and accept protons. For example, an exposed —NH_2 group can bind with hydrogen ions, becoming —NH_3^+:

$$R\text{—}NH_2 + H^+ \rightarrow R\text{—}NH_3^+$$

Because this removes free hydrogen ions from the solution, it prevents the solution from becoming too acidic. Consequently, the same protein molecules can function reversibly as either acids or bases depending on the pH of their environment. Molecules with this ability are called **amphoteric** (am"fo-ter'ik) **molecules.**

---

*Note that R indicates the rest of the organic molecule, which contains many atoms.

Hemoglobin of red blood cells is an excellent example of a protein that functions as an intracellular buffer. As explained earlier, carbon dioxide released from the tissues forms carbonic acid, which dissociates to liberate hydrogen and bicarbonate ions in the blood. Meanwhile, hemoglobin is unloading oxygen, becoming reduced hemoglobin, which carries a negative charge. Because hydrogen ions rapidly bind to the hemoglobin anions, pH changes are minimized (see Chapter 23). In this case, carbonic acid, a weak acid, is buffered by an even weaker acid, hemoglobin.

Physiological Buffer Systems

Respiratory System Regulation of Hydrogen Ion Concentration

Respiratory system regulation of acid-base balance provides a *physiological, or functional, buffering system.* Although it acts more slowly than the chemical buffers, it has one to two times the buffering power of all the chemical buffers combined. As described in Chapter 23, the respiratory system eliminates carbon dioxide from the blood while replenishing its supply of oxygen. Carbon dioxide generated by cellular respiration enters erythrocytes in the circulation and is converted to bicarbonate ions for transport in the plasma as shown by the equation

$$CO_2 + H_2O \underset{}{\overset{\text{carbonic anhydrase}}{\rightleftharpoons}} \underset{\substack{\text{carbonic} \\ \text{acid}}}{H_2CO_3} \rightleftharpoons H^+ + \underset{\substack{\text{bicarbonate} \\ \text{ion}}}{HCO_3^-}$$

There is a reversible equilibrium between dissolved carbon dioxide and water on the one hand and carbonic acid on the other, and also between carbonic acid and the hydrogen and bicarbonate ions. Consequently, an increase in any of these chemical species will push the reaction in the opposite direction. Notice also that the right side of the equation is equivalent to the bicarbonate buffer system.

In healthy individuals, carbon dioxide is expelled from the lungs at the same rate it is formed in the tissues. During carbon dioxide unloading, the reaction shifts to the left and H^+ generated from carbonic acid is reincorporated into water. Thus, H^+ produced by CO_2 transport is not allowed to accumulate and has little or no effect on blood pH. However, when hypercapnia occurs, it activates medullary chemoreceptors (via cerebrospinal fluid acidosis promoted by excessive accumulation of CO_2). The medullary chemoreceptors respond by increasing respiratory rate and depth (see Figure 23.26, p. 871). Additionally, a rising plasma H^+ concentration as results from any metabolic process excites the respiratory center indirectly (via peripheral chemoreceptors) to stimulate deeper, more rapid

Integumentary System
- Kidneys dispose of nitrogenous wastes; maintain fluid, electrolyte, and acid-base balance of blood
- Skin provides external protective barrier; serves as site for water loss (via perspiration); vitamin D synthesis site

Skeletal System
- Kidneys dispose of nitrogenous wastes; maintain fluid, electrolyte, and acid-base balance of blood
- Bones of rib cage provide some protection to kidneys

Muscular System
- Kidneys dispose of nitrogenous wastes; maintain fluid, electrolyte, and acid-base balance of blood; renal regulation of Na^+, K^+, and Ca^{2+} content in ECF crucial for muscle excitability and contractility
- Muscles of pelvic diaphragm and external urethral sphincter function in voluntary control of micturition; creatinine is a nitrogenous waste product of muscle metabolism that must be excreted by the kidneys

Nervous System
- Kidneys dispose of nitrogenous wastes; maintain fluid, electrolyte, and acid-base balance of blood; renal control of Na^+, K^+, and Ca^{2+} in ECF essential for normal neural function

- Neural controls involved in micturition; sympathetic nervous system activity triggers the renin-angiotensin mechanism

Endocrine System
- Kidneys dispose of nitrogenous wastes; maintain fluid, electrolyte, and acid-base balance of blood; produce erythropoietin; renal regulation of Na^+ and water balance essential for blood pressure homeostasis and hormone transport in blood
- ADH, aldosterone, ANP, and other hormones help regulate renal reabsorption of water and electrolytes

Cardiovascular System
- Kidneys dispose of nitrogenous wastes; maintain fluid, electrolyte, and acid-base balance of blood; renal regulation of Na^+ / water balance essential for blood pressure homeostasis. Na^+, K^+, and Ca^{2+} regulation maintains cardiac irritability
- Systemic arterial blood pressure is the driving force for glomerular filtration; heart secretes atrial natriuretic peptide; blood transports nutrients, oxygen, etc. to urinary organs

Lymphatic System/Immunity
- Kidneys dispose of nitrogenous wastes; maintain fluid, electrolyte, and acid-base balance of blood
- By returning leaked plasma fluid to cardiovascular system, lymphatic vessels help ensure normal systemic arterial pressure required for kidney function; immune cells protect urinary organs from infection, cancer, and other antigens

Respiratory System
- Kidneys dispose of nitrogenous wastes; maintain fluid, electrolyte, and acid-base balance of blood
- Respiratory system provides oxygen required by kidney cells for their high metabolic activity; disposes of carbon dioxide; cells in the lung convert angiotensin I to angiotensin II

Digestive System
- Kidneys dispose of nitrogenous wastes; maintain fluid, electrolyte, and acid-base balance of blood; also, metabolize vitamin D to the active form needed for calcium absorption
- Digestive organs provide nutrients needed for kidney cell health; liver synthesizes most urea, a nitrogenous waste that must be excreted by the kidneys

Reproductive System
- Kidneys dispose of nitrogenous wastes; maintain fluid, electrolyte, and acid-base balance of blood

A drink of water—such a simple thing. It quenches a powerful thirst, but that's not all. Inadequate hydration is a major cause of poor functioning, both mental and physical, and of fatigue, illness, and even death (if it exists during vigorous activity). The body also needs water to regulate its temperature via perspiration, to wash wastes and toxins out of the body in urine, to maintain proper blood volume and pressure, and to keep the skeletal muscles well hydrated (otherwise they feel weak and tire quickly, and even stop working altogether). A drink provides that needed water, but it takes the kidneys to keep it pure so that all of these vital activities can continue to occur. Seemingly wise, the exceedingly tiny nephrons "know" which blood solutes are valuable and which disposable, and that alcohol and caffeine are diuretics and as such drain water from the body. Make no mistake about it, the kidneys are crucially important to the body—all systems, all cells. The body systems that most influence the kidneys' ability to do their life-sustaining work are the cardio-vascular, endocrine, and nervous systems. Let's take a look.

Cardiovascular System

As important as they are, and as intricate as their functioning is, the kidneys cannot work without blood to process and blood pressure to drive filtrate through the glomerular filters. When we hemorrhage severely, the GFR drops and the kidneys stop functioning altogether. When blood pressure is normal, as both autoregulatory and systemic mechanisms try to keep it, it provides the driving force that allows the nephrons to perform their job. On the other hand, because the kidneys furnish the means for excreting or retaining water, they are indispensable for maintaining the blood volume that furnishes that filtration pressure.

Endocrine System and Nervous System

Although the kidneys have a whole "toolbox" of ways to ensure that their own blood pressure is normal, auto-regulation can be augmented by hormonal mechanisms (the renin-angiotensin mechanism which calls aldosterone and ADH into service, and release of ANP by the heart). In addition, activity of sympathetic inputs makes sure that bodywide blood pressure is maintained within homeostatic limits. Alternatively, the kidneys' needs are overcome as necessary by the same sympathetic fibers to make sure that the heart and brain receive adequate blood perfusion in periods of dramatically falling blood pressure. Now that's flexibility!

CLINICAL CONNECTIONS

Urinary System

Case study: Mr. Heyden, a somewhat stocky 72-year-old man, is brought in to the emergency room (ER). The paramedics report that his left arm and the left side of his body trunk were pinned beneath some of the wreckage, and that when he was freed, his left hypogastric and lumbar areas appeared to be compressed and his left arm was blanched and without sensation. On admission, Mr. Heyden is alert, slightly cyanotic, and complaining of pain in his left side; he loses consciousness shortly thereafter. His vital signs are taken, blood is drawn for laboratory tests, and Mr. Heyden is catheterized and immediately scheduled for a CT scan of his left abdominal region.

Analyze the information that was subsequently recorded on Mr. Heyden's chart:

- Re vital signs: Temperature° 102°F; BP 90/50 mm Hg and falling; heart rate 116 beats/min and thready; 30 respirations/min

1. Given the values above and his attendant cyanosis, what would you guess is Mr. Heyden's immediate problem? Explain your reasoning.

- Re CT scan: CT scan reveals a ruptured spleen and a large hematoma in the upper left abdominal quadrant. Mr. Heyden is scheduled for an immediate splenectomy.

2. Rupture of the spleen results in massive hemorrhage—much more so than when most other organs are severely damaged. Explain this observation. What problems (if any) may ensue from removal of Mr. Heyden's spleen?

- Re hematology data: Most blood tests yield normal results. However, renin, aldosterone, and ADH levels are elevated.

3. Explain the cause and consequence of each of the hematology findings.

- Re urinalysis: Some granular casts (particulate cell debris) are noted, and the urine is brownish-red in color; other values are normal, but urine output is very low. An order is given to force fluids.

4. What might account for the low volume of urine output? (Name at least two possibilities.) What might account for the casts and abnormal color of his urine? Can you see any possible relationship between his crush injury and these findings?

The next day, Mr. Heyden is awake and alert. He says that he now has feeling in his arm, but he is still complaining of pain. However, the pain site appears to have moved from the left upper quadrant to his lumbar region. His urine output is still low. He is scheduled once again for a CT scan, this time of his lumbar region. The order to force fluids is renewed and some additional and more specific blood tests are ordered. We will visit Mr. Heyden again shortly, but in the meantime be thinking about what these new findings may indicate.

(Answers are in Appendix F)

and acids that need to be excreted by the kidneys. This, along with buffer systems that are not yet fully effective, results in a tendency toward acidosis.

4. The high rate of insensible water loss in infants because of their larger surface area relative to body volume (about three times that of adults). Infants lose substantial amounts of water through their skin.

5. The inefficiency of the kidneys in infants. At birth, the kidneys are immature, only about half as proficient as adult kidneys at concentrating urine, and notably inefficient in ridding the body of acids.

All these factors put newborns at risk for dehydration and acidosis, at least until the end of the first month when their kidneys achieve reasonable efficiency. Bouts of vomiting or diarrhea greatly amplify the risk. With such extreme variations in their internal environment, it is little wonder that mortality of premature infants in particular is so high.

In old age, total body water often decreases (the loss is largely from the intracellular compartment) because muscle mass progressively declines and body fat rises. Although few changes in solute concentrations of body fluids occur, the speed with which homeostasis of the internal environment is restored after being disrupted declines with age. Elders may be unresponsive to thirst cues and thus are at risk for dehydration. Additionally, they are the most frequent prey of diseases that lead to severe fluid, electrolyte, or acid-base problems, such as congestive heart failure (and its attendant edema) and diabetes mellitus. Because most fluid, electrolyte, and acid-base imbalances occur when body water content is highest or lowest, the very young and the very old are the most frequent victims.

⋆ ⋆ ⋆

This chapter has examined the chemical and physiological mechanisms that provide the optimal internal environment for survival. Although the kidneys are the "Rolex" among homeostatic organs in regulating water, electrolyte, and acid-base balance, they do not and cannot act alone. Rather, their activity is made possible by a host of hormones and enhanced both by bloodborne buffers, which give the kidneys time to react, and by the respiratory system, which shoulders a substantial responsibility for acid-base balance of the blood. Now that we have discussed the topics relevant to renal functioning, and you have read how the urinary system interacts with other body systems in *Making Connections* (pp. 1064–1065), the topics in Chapters 26 and 27 should be drawn together in an understandable way.

RELATED CLINICAL TERMS

Antacid An agent that counteracts acidity, such as sodium bicarbonate, aluminum hydroxide gel, and magnesium trisilicate. Commonly used in the management of heartburn.

Conn's disease Also called **primary aldosteronism;** a condition of hypersecretion of aldosterone by adrenal cortical cells accompanied by excessive loss of potassium and generalized muscular weakness, hypernatremia, and hypertension. Usual cause is adrenal tumor; usual treatment is adrenal-suppressing drugs prior to tumor removal.

Renal tubular acidosis A metabolic acidosis resulting from impaired renal reabsorption of bicarbonate; the urine is alkaline.

Syndrome of inappropriate ADH secretion (SIADH) A group of disorders associated with excessive ADH secretion in the absence of appropriate (osmotic or nonosmotic) stimuli. Characterized by hyponatremia, concentrated urine, fluid retention, and weight gain. Usual causes are ectopic secretion of ADH by cancer cells (e.g., bronchogenic cancer cells) and brain disorders or trauma affecting the ADH-secreting hypothalamic neurons. Temporary management involves water restriction.

CHAPTER SUMMARY

Media study tools that could provide you additional help in reviewing specific key topics of Chapter 27 are referenced below. **SP** = *Study Partner;* **IP** = *Interactive Physiology.*

Body Fluids (pp. 1041–1044)

Body Water Content (p. 1041)

1. Water accounts for 45–75% of body weight, depending on age, sex, and amount of body fat.

Fluid Compartments (p. 1041)

2. About two-thirds (25 L) of body water is found within cells in the intracellular fluid (ICF) compartment; the balance (15 L) is in the extracellular fluid (ECF) compartment. The ECF includes plasma and interstitial fluid.

Composition of Body Fluids (pp. 1041–1042)

3. Solutes dissolved in body fluids include electrolytes and nonelectrolytes. Electrolyte concentration is expressed in mEq/L.

4. Plasma contains more proteins than does interstitial fluid; otherwise, extracellular fluids are similar. The most abundant ECF electrolytes are sodium, chloride, and bicarbonate ions.

CHAPTER SUMMARY

5. Intracellular fluids contain large amounts of protein anions and potassium, magnesium, and phosphate ions.

IP Fluid, Electrolyte, and Acid/Base Balance CD-ROM; Topic: Introduction to Body Fluids, pages 1–8.

Fluid Movement Among Compartments (pp. 1042–1044)

6. Fluid exchanges between compartments are regulated by osmotic and hydrostatic pressures: (a) Filtrate is forced out of the capillaries by hydrostatic pressure and pulled back in by oncotic pressure. (b) Water moves freely between the ECF and the ICF by osmosis, but solute movements are restricted by size, charge, and dependence on active transport. (c) Water flows always follow changes in ECF osmolality.

7. Plasma links the internal and external environments.

IP Fluid, Electrolyte, and Acid/Base Balance CD-ROM; Topic: Introduction to Body Fluids, pages 19–23.

Water Balance (pp. 1044–1047)

1. Sources of body water are ingested foods and fluids and metabolic water.

2. Water leaves the body via the lungs, skin, gastrointestinal tract, and kidneys.

Regulation of Water Intake: The Thirst Mechanism (p. 1045)

3. Increased plasma osmolality or decreased plasma volume triggers the thirst mechanism, mediated by hypothalamic osmoreceptors. Thirst, inhibited by distension of the gastrointestinal tract by ingested water and then by osmotic signals, may be damped before body needs for water are met.

SP Exercise: Chapter 27, Thirst Mechanism.

Regulation of Water Output (pp. 1045–1046)

4. Obligatory water loss is unavoidable and includes insensible water losses from the lungs, the skin, in feces, and about 500 ml of urine output daily.

5. Beyond obligatory water loss, the volume of urinary output depends on water intake and loss via other routes and reflects the influence of antidiuretic hormone and aldosterone on the renal tubules.

Disorders of Water Balance (pp. 1046–1047)

6. Dehydration occurs when water loss exceeds water intake over time. It is evidenced by thirst, dry skin, and decreased urine output. A serious consequence is circulatory collapse.

7. Hypotonic hydration occurs when body fluids are excessively diluted and cells become swollen by water entry. The most serious consequence is cerebral edema.

8. Edema is an abnormal accumulation of fluid in the interstitial space, which may impair blood circulation.

IP Fluid, Electrolyte, and Acid/Base Balance CD-ROM; Topic: Water Homeostasis, pages 1–27.

Electrolyte Balance (pp. 1047–1055)

1. Most electrolytes (salts) are obtained from ingested foods and fluids. Salts, particularly NaCl, are often ingested in excess of need.

2. Electrolytes are lost in perspiration, feces, and urine. The kidneys are most important in regulating electrolyte balance.

The Central Role of Sodium in Fluid and Electrolyte Balance (pp. 1047–1049)

3. Sodium salts are the most abundant solutes in ECF. They exert the bulk of ECF osmotic pressure and control water volume and distribution in the body.

4. Na^+ transport by the renal tubule cells is coupled to and helps regulate K^+, Cl^-, HCO_3^-, and H^+ concentrations in the ECF.

Regulation of Sodium Balance (pp. 1049–1053)

5. Sodium ion balance is linked to water balance and blood pressure regulation and involves both neural and hormonal controls.

6. Declining blood pressure and falling filtrate osmolality stimulate the juxtaglomerular cells to release renin. Renin, via angiotensin II, enhances systemic blood pressure and aldosterone release.

7. Cardiovascular system baroreceptors sense changing arterial blood pressure, prompting changes in sympathetic vasomotor activity. Rising arterial pressure leads to vasodilation and enhanced Na^+ and water loss in urine. Falling arterial pressure promotes vasoconstriction and conserves Na^+ and water.

8. Atrial natriuretic peptide, released by certain atrial cells in response to rising blood pressure (or blood volume), causes systemic vasodilation and inhibits renin, aldosterone, and ADH release. Hence, it enhances Na^+ and water excretion, reducing blood volume and blood pressure.

SP Exercise: Chapter 27, Blood Pressure Regulatory Mechanism.

9. Estrogens and glucocorticoids increase renal retention of sodium. Progesterone promotes enhanced sodium and water excretion in urine.

Regulation of Potassium Balance (pp. 1053–1054)

10. About 85% of filtered potassium is reabsorbed by the more proximal regions of the nephrons.

11. The main thrust of renal regulation of K^+ is to excrete it. Potassium ion secretion by the principal cells of the cortical collecting ducts is enhanced by increased plasma K^+ content and aldosterone. Type A cells of the collecting duct reabsorb small amounts of K^+ during K^+ deficit.

Regulation of Calcium and Phosphate Balance (p. 1054)

12. Calcium balance is regulated primarily by parathyroid hormone, which targets the bones, intestine, and kidneys, thereby enhancing blood Ca^{2+} levels. Active reabsorption occurs primarily in the DCT.

13. Calcitonin accelerates the deposit of Ca^{2+} in bone and inhibits its release from bone matrix. However, its influence on the kidneys is relatively minor.

Regulation of Magnesium Balance (p. 1055)

14. Magnesium levels are regulated to counteract excesses or deficits in the ECF. The mechanism is unclear.

Regulation of Anions (p. 1055)

15. When blood pH is normal or slightly high, chloride is the major anion accompanying sodium reabsorption. In acidosis, chloride is replaced by bicarbonate.

16. Reabsorption of most other anions appears to be regulated by their T_m.

> **IP** Fluid, Electrolyte, and Acid/Base Balance CD-ROM; Topic: Electrolyte Homeostasis, pages 1–38.

Acid-Base Balance (pp. 1055–1063)

1. Acids are proton (H^+) donors; bases are proton acceptors. Acids that dissociate completely in solution are strong acids; those that dissociate incompletely are weak acids. Strong bases are more effective proton acceptors than are weak bases.

2. The homeostatic pH range of arterial blood is 7.35 to 7.45. A higher pH represents alkalosis; a lower pH reflects acidosis.

3. Some acids enter the body in foods, but most are generated by breakdown of phosphorus-containing proteins, incomplete oxidation of fats or glucose, and the loading and transport of carbon dioxide in the blood.

4. Acid-base balance is achieved by renal regulation of bicarbonate ion (hence, hydrogen ion) concentration of body fluids.

> **IP** Fluid, Electrolyte, and Acid/Base Balance CD-ROM; Topic: Acid/Base Homeostasis, pages 1–15.

Chemical Buffer Systems (pp. 1055–1057)

5. Chemical buffers are single or paired sets (a weak acid and its salt) of molecules that act rapidly to resist excessive shifts in pH by releasing or binding H^+.

6. Chemical buffers of the body include the bicarbonate, phosphate, protein, and ammonium ion buffer systems.

> **IP** Fluid, Electrolyte, and Acid/Base Balance CD-ROM; Topic: Acid/Base Homeostasis, pages 16–26.

Physiological Buffer Systems (pp. 1057–1060)

7. Respiratory regulation of acid-base balance of the blood utilizes the bicarbonate buffer system and the fact that carbon dioxide and water are in reversible equilibrium with H_2CO_3.

8. Acidosis activates the respiratory center to increase respiratory rate and depth, which eliminates more CO_2 and causes blood pH to rise. Alkalosis depresses the respiratory center, resulting in CO_2 retention and a fall in blood pH.

> **IP** Fluid, Electrolyte, and Acid/Base Balance CD-ROM; Topic: Acid/Base Homeostasis, pages 27–28.

9. The kidneys provide the major long-term mechanism for controlling acid-base balance by maintaining stable HCO_3^- levels in the ECF. Metabolic acids (organic acids other than carbonic acid) can be eliminated from the body only by the kidneys.

10. Secreted hydrogen ions come from the dissociation of carbonic acid generated within the tubule cells.

11. Tubule cells are impermeable to bicarbonate in the filtrate, but they can conserve filtered bicarbonate ions indirectly by absorbing HCO_3^- generated within them (by dissociation of carbonic acid to HCO_3^- and H^+). For each HCO_3^- (and Na^+) reabsorbed, one H^+ is secreted into the filtrate where it combines with HCO_3^-.

12. To generate and add new HCO_3^- to plasma to counteract acidosis, either of two mechanisms may be used:

- Secreted H^+, buffered by bases other than HCO_3^-, is excreted from the body in urine (the major urine buffer is the phosphate buffer system).

- NH_4^+ (derived from glutamine catabolism) is excreted in urine.

13. To counteract alkalosis, bicarbonate ion is secreted into the filtrate and H^+ is reabsorbed.

> **IP** Fluid, Electrolyte, and Acid/Base Balance; Topic: Acid/Base Homeostasis, pages 29–37.

Abnormalities of Acid-Base Balance (pp. 1060–1063)

14. Classification of acid-base imbalances as metabolic or respiratory indicates the cause of the acidosis or alkalosis.

15. Respiratory acidosis results from carbon dioxide retention; respiratory alkalosis occurs when carbon dioxide is eliminated faster than it is produced.

> **SP** Case Study: Lung Disease.

16. Metabolic acidosis occurs when fixed acids (lactic acid, ketone bodies, and others) accumulate in the blood or when bicarbonate is lost from the body; metabolic alkalosis occurs when bicarbonate levels are excessive.

> **SP** Case Study: Diabetes Mellitus.

17. Extremes of pH for life are 7.0 and 7.8.

18. Compensations occur when the respiratory system or kidneys act to reverse acid-base imbalances resulting from abnormal or inadequate functioning of the alternate system. Respiratory compensations involve changes in respiratory rate and depth. Renal compensations modify blood levels of HCO_3^-.

> **SP** Case Study: Renal Failure.

> **IP** Fluid, Electrolyte, and Acid/Base Balance CD-ROM; Topic: Acid/Base Homeostasis, pages 38–59.

Developmental Aspects of Fluid, Electrolyte, and Acid-Base Balance (pp. 1063, 1066)

1. Factors that place infants at risk for dehydration and acidosis include low lung residual volume, high rate of fluid intake and output, high metabolic rate, relatively large body surface area, and functional immaturity of the kidneys at birth.

2. The elderly are at risk for dehydration because of their low percentage of body water and insensitivity to thirst cues. Diseases that promote fluid and acid-base imbalances (cardiovascular disease, diabetes mellitus, and others) are most common in the aged.

REVIEW QUESTIONS

Multiple Choice/Matching

1. Body water content is greatest in (a) infants, (b) young adults, (c) elderly adults.

2. Potassium, magnesium, and phosphate ions are the predominant electrolytes in (a) plasma, (b) interstitial fluid, (c) intracellular fluid.

3. Sodium balance is regulated primarily by control of amount(s) (a) ingested, (b) excreted in urine, (c) lost in perspiration, (d) lost in feces.

4. Water balance is regulated primarily by control of amount(s) (use choices in question 3).

5 through 10. Answer questions 5 through 10 by choosing a response from the following:

(**a**) ammonium ions (**e**) hydrogen ions (**h**) potassium
(**b**) bicarbonate   (**f**) magnesium   (**i**) sodium
(**c**) calcium   (**g**) phosphate   (**j**) water
(**d**) chloride

5. Three substances regulated (at least in part) by the influence of aldosterone on the kidney tubules.

6. Two substances regulated by parathyroid hormone.

7. Two substances secreted into the proximal convoluted tubules in exchange for sodium ions.

8. Part of an important chemical buffer system in plasma.

9. Two ions produced during catabolism of glutamine.

10. Substance regulated by ADH's effects on the renal tubules.

11. Which of the following factors will enhance ADH release? (a) increase in ECF volume, (b) decrease in ECF volume, (c) decrease in ECF osmolality, (d) increase in ECF osmolality, (e) increase in blood pressure, (f) decrease in blood pressure.

12. The pH of blood varies directly with (a) HCO_3^-, (b) P_{CO_2}, (c) H^+, (d) none of the above.

13. In an individual with metabolic acidosis, a clue that the respiratory system is compensating is provided by (a) high blood bicarbonate levels, (b) low blood bicarbonate levels, (c) rapid, deep breathing, (d) slow, shallow breathing.

Short Answer Essay Questions

14. Name the body fluid compartments, noting their locations and the approximate fluid volume in each.

15. Describe the thirst mechanism, indicating how it is triggered and terminated.

16. Explain why and how sodium and water balance are jointly regulated.

17. Describe the role of the respiratory system in controlling acid-base balance.

18. Explain how the chemical buffer systems resist changes in pH.

19. Explain the relationship of the following to renal secretion and excretion of hydrogen ions: (a) plasma carbon dioxide levels, (b) phosphate, and (c) sodium bicarbonate reabsorption.

20. List several factors that place newborn babies at risk for acid-base imbalances.

Critical Thinking and Clinical Application Questions

1. Mr. Jessup, a 55-year-old man, is operated on for a cerebral tumor. About a month later, he appears at his physician's office complaining of excessive thirst. He claims to have been drinking about 20 liters of water daily for the past week and that he has been voiding nearly continuously. A urine sample is collected and its specific gravity is reported as 1.001. What is your diagnosis of Mr. Jessup's problem? What connection might exist between his previous surgery and his present problem?

2. For each of the following sets of blood values, name the acid-base imbalance (acidosis or alkalosis), determine its cause (metabolic or respiratory), decide whether the condition is being compensated, and cite at least one possible cause of the imbalance.

Problem 1: pH 7.63; P_{CO_2} 19 mm Hg; HCO_3^- 19.5 mEq/L.

Problem 2: pH 7.22; P_{CO_2} 30 mm Hg; HCO_3^- 12.0 mEq/L.

3. Explain how emphysema and congestive heart failure can lead to acid-base imbalance.

4. Mrs. Bush, a 70-year-old woman, is admitted to the hospital. Her history states that she has been suffering from diarrhea for three weeks. On admission, she complains of severe fatigue and muscle weakness. A blood chemistry study yields the following information: Na^+ 142 mEq/L; K^+ 1.5 mEq/L; Cl^- 92 mEq/L; P_{CO_2} 32 mm Hg. Which electrolytes are within normal limits? Which are so abnormal that the patient has a medical emergency? Which of the following represents the greatest danger to Mrs. Bush? (a) a fall due to her muscular weakness, (b) edema, (c) cardiac arrhythmia and cardiac arrest.

28 THE REPRODUCTIVE SYSTEM

1. Describe the common function of the male and female reproductive systems.

Anatomy of the Male Reproductive System (pp. 1071–1076)

2. Describe the structure and function of the testes, and explain the importance of their location in the scrotum.

3. Describe the location, structure, and function of the accessory organs of the male reproductive system.

4. Describe the structure of the penis, and indicate its role in the reproductive process.

5. Discuss the sources and functions of semen.

Physiology of the Male Reproductive System (pp. 1076–1086)

6. Define meiosis. Compare and contrast it to mitosis.

7. Outline the events of spermatogenesis.

8. Describe the phases of the male sexual response.

9. Discuss hormonal regulation of testicular function and the physiological effects of testosterone on male reproductive anatomy.

Anatomy of the Female Reproductive System (pp. 1086–1094)

10. Describe the location, structure, and function of the ovaries.

11. Describe the location, structure, and function of each of the organs of the female reproductive duct system.

12. Describe the anatomy of the female external genitalia.

13. Discuss the structure and function of the mammary glands.

Physiology of the Female Reproductive System (pp. 1095–1103)

14. Describe the process of oogenesis and compare it to spermatogenesis.

15. Describe the phases of the ovarian cycle, and relate them to events of oogenesis.

16. Describe the regulation of the ovarian and menstrual cycles.

17. Discuss the physiological effects of estrogens and progesterone.

18. Describe the phases of female sexual response.

Sexually Transmitted Diseases (pp. 1103–1104)

19. Indicate the infectious agents and modes of transmission of gonorrhea, syphilis, chlamydia, and genital herpes.

Developmental Aspects of the Reproductive System: Chronology of Sexual Development (pp. 1104–1109)

20. Discuss the determination of genetic sex and prenatal development of male and female structures.

21. Describe the significant events of puberty and menopause.

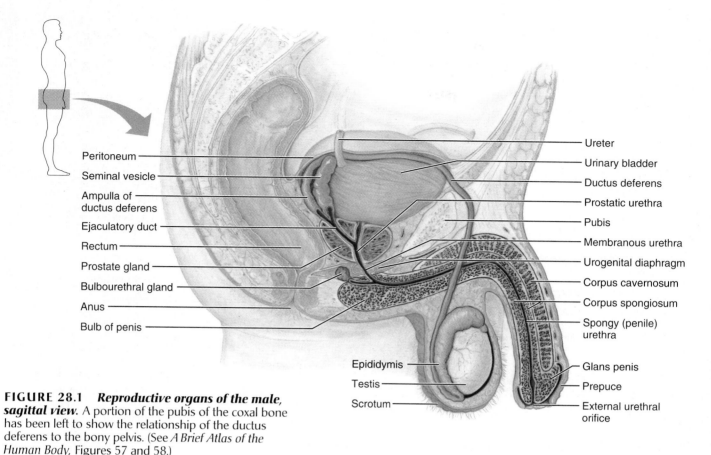

Peritoneum

Seminal vesicle

Ampulla of
ductus deferens

Ejaculatory duct

Rectum

Prostate gland

Bulbourethral gland

Anus

Bulb of penis

Ureter

Urinary bladder

Ductus deferens

Prostatic urethra

Pubis

Membranous urethra

Urogenital diaphragm

Corpus cavernosum

Corpus spongiosum

Spongy (penile)
urethra

Glans penis

Prepuce

External urethral
orifice

Epididymis

Testis

Scrotum

FIGURE 28.1 *Reproductive organs of the male,*
sagittal view. A portion of the pubis of the coxal bone
has been left to show the relationship of the ductus
deferens to the bony pelvis. (See *A Brief Atlas of the*
Human Body, Figures 57 and 58.)

ost organ systems of the body function almost continuously to maintain the well-
being of the individual. The **reproductive**
system, however, appears to "slumber" until puberty. The **primary sex organs,** or **gonads** (go′nadz;
"seeds"), are the *testes* in males and the *ovaries* in females. The gonads produce sex cells, or **gametes**
(gam′ēts; "spouses") and secrete **sex hormones.** The
remaining diverse collection of reproductive structures—ducts, glands, and external genitalia (jen-ĭ-
ta′le-ah)—are referred to as **accessory reproductive**
organs. Although male and female reproductive organs are quite different, their common purpose is to
produce offspring.

The male's reproductive role is to manufacture
male gametes called *sperm* and to deliver them to
the female reproductive tract, where fertilization can
occur. The mutually complementary role of the female is to produce female gametes, called *ova* or
eggs. When these events are appropriately timed, a
sperm and egg fuse to form a fertilized egg, the first
cell of the new individual, from which all body cells
will arise. The male and female reproductive systems are equal partners in events leading up to fertilization, but once fertilization has occurred, the female uterus provides a protective environment in
which the embryo develops until birth.

The sex hormones—androgens in males and estrogens and progesterone in females—play vital
roles both in the development and function of the reproductive organs and in sexual behavior and drives.
These hormones also influence the growth and development of many other organs and tissues of the
body.

ANATOMY OF THE MALE
REPRODUCTIVE SYSTEM

Figure 28.1 presents an overview of the male reproductive system. Males are males because they possess testes. The sperm-producing **testes,** or male gonads, lie within the *scrotum.* From the testes, the
sperm are delivered to the body exterior through a
system of ducts including (in order) the *epididymis,*
the *ductus deferens,* and finally the *urethra,* which
opens to the outside at the tip of the *penis.* The accessory sex glands, which empty their secretions
into the ducts during ejaculation, are the *seminal*
vesicles, prostate gland, and *bulbourethral glands.*
Take a moment to trace the duct system and identify
the accessory glands before continuing.

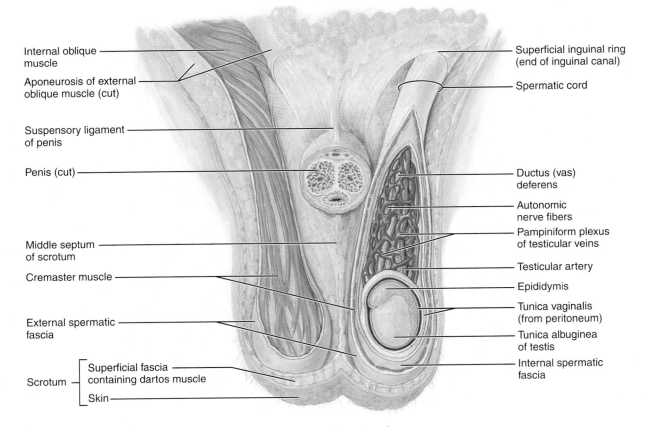

Internal oblique muscle

Aponeurosis of external oblique muscle (cut)

Suspensory ligament of penis

Penis (cut)

Middle septum of scrotum

Cremaster muscle

External spermatic fascia

Scrotum — Superficial fascia containing dartos muscle
— Skin

Superficial inguinal ring (end of inguinal canal)

Spermatic cord

Ductus (vas) deferens

Autonomic nerve fibers

Pampiniform plexus of testicular veins

Testicular artery

Epididymis

Tunica vaginalis (from peritoneum)

Tunica albuginea of testis

Internal spermatic fascia

FIGURE 28.2   *Relationships of the testis to the scrotum and spermatic cord.* The scrotum has been opened and its anterior portion removed. (See *A Brief Atlas of the Human Body,* Figure 58.)

The Scrotum

The **scrotum** (skro′tum; "pouch") is a sac of skin and superficial fascia that hangs outside the abdominopelvic cavity at the root of the penis (Figure 28.2). Covered with sparse hairs, the scrotal skin is more heavily pigmented than that elsewhere on the body. The paired oval testes, or *testicles,* lie suspended in the scrotum. A midline septum divides the scrotum into right and left halves, providing one compartment for each testis. This location seems a rather vulnerable position for a man's testes, which contain his entire genetic heritage. However, viable sperm cannot be produced at core body temperature (36.2°C). Hence the superficial location of the scrotum, which provides a temperature about 3°C lower, is an essential adaption. Furthermore, the scrotum responds to changes in temperature. For example, when it is cold, the testes are pulled closer to the warmth of the body wall, and the scrotum becomes shorter and heavily wrinkled to reduce heat loss. When it is warm, the scrotal skin is flaccid and loose to increase the surface area for cooling, and the testes hang lower. These changes in scrotal surface area help maintain a fairly constant intrascrotal temperature and reflect the activity of two sets of muscles. The **dartos** (dar′tos; "skinned") **muscle,** a layer of smooth muscle in the superficial fascia, wrinkles the scrotal skin. The **cremaster** (kremas′ter; "a suspender") **muscles,** bands of skeletal muscle that arise from the internal oblique muscles of the trunk, elevate the testes.

The Testes

Each plum-sized testis is approximately 4 cm (1.5 inches) long and 2.5 cm (1 inch) in diameter and is surrounded by two tunics. The outer tunic is the two-layered **tunica vaginalis** (vaj″ĭ-nal′is), derived from the peritoneum (see Figures 28.2 and 28.3). Deep to this is the **tunica albuginea** (al″bu-jin′e-ah; "white coat"), the fibrous capsule of the testis. Septa extending from the tunica albuginea divide the testis into 250 to 300 wedge-shaped compartments or *lobules* (Figure 28.3). Each lobule contains one to four tightly coiled **seminiferous** (sem″ĭ-nif′er-us; "sperm-carrying") **tubules,** the actual "sperm factories."

The seminiferous tubules of each lobule converge to form a **tubulus rectus,** a straight tubule that conveys sperm into the **rete** (re′te) **testis,** a tubular network on the posterior side of the testis. From the rete testis, sperm leave the testis through the *efferent ductules* and enter the *epididymis* (ep″ĭ-did′ĭ-mis), which hugs the external surface of the testis.

? *Which of the tubular structures shown here are the "sperm factories"?*

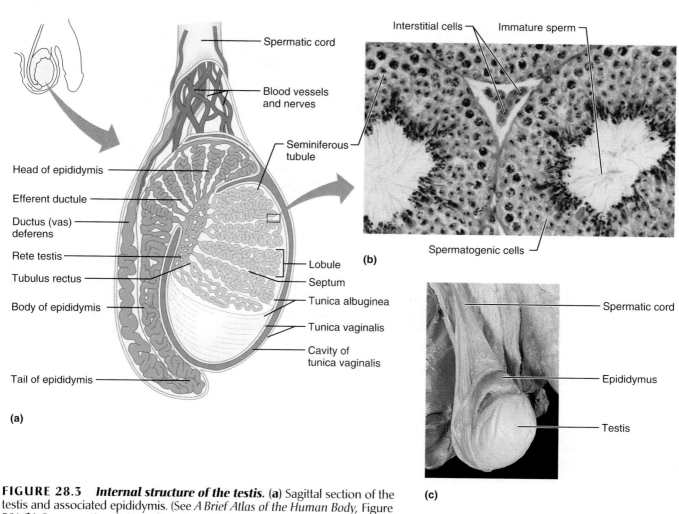

FIGURE 28.3 *Internal structure of the testis.* **(a)** Sagittal section of the testis and associated epididymis. (See *A Brief Atlas of the Human Body,* Figure 58.) **(b)** Cross-sectional view of portions of the seminiferous tubules, showing the spermatogenic (sperm-forming) cells making up the epithelium of the tubule walls and the location of interstitial cells in the loose connective tissue between the seminiferous tubules (1200×). **(c)** External view of a testis from a cadaver; same orientation as in part (a).

Lying in the soft connective tissue surrounding the seminiferous tubules are the **interstitial cells,** or *Leydig* (li'dig) *cells.* These cells produce androgens (most importantly *testosterone*), which they secrete into the surrounding interstitial fluid. Thus, the sperm-producing and hormone-producing functions of the testis are carried out by completely different cell populations.

The long **testicular arteries,** which branch from the abdominal aorta (see *gonadal arteries* in Figure 20.22c, p. 761) supply the testes. The **testicular veins** draining the testes arise from a vinelike network called the **pampiniform** (pam-pin'ĭ-form; "tendril-shaped") **plexus** that surrounds the testicular artery (see Figure 28.2). This plexus absorbs heat from the arterial blood, cooling it before it enters the testes. Thus, it provides an additional avenue for maintaining the testes at their cool homeostatic temperature. The testes are served by both divisions of the autonomic nervous system. Associated sensory nerves transmit impulses that result in agonizing pain and nausea when the testes are hit forcefully. The nerve fibers are enclosed, along with the blood vessels and lymphatics, in a connective tissue sheath called the **spermatic cord** (see Figure 28.2).

Homeostatic Imbalance

Although *testicular cancer* is relatively rare (affecting 1 of every 20,000 males), it is the most common cancer in young men (ages 15 to 35). A history of mumps or orchitis (inflammation of the testis)

Seminiferous tubules.

increases the risk, but the most important risk factor for this cancer is *cryptorchidism* (nondescent of the testes, see p. 1108). Because the most common sign of testicular cancer is a painless solid mass in the testis, self-examination of the testes should be practiced by every male. If detected early, the cure rate is impressive. Over 90% of testicular cancers are cured by surgical removal of the cancerous testis *(orchiectomy)* followed by radiation therapy and chemotherapy. ■

The Penis

The **penis** ("tail") is a copulatory organ, designed to deliver sperm into the female reproductive tract (see Figure 28.4). The penis and scrotum, which hang suspended from the perineum, make up the external reproductive structures, or **external genitalia,** of the male. The **male perineum** (per"ĭ-ne'um; "around the anus") is the diamond-shaped region located between the pubic symphysis anteriorly, the coccyx posteriorly, and the ischial tuberosities laterally. The floor of the perineum is formed by muscles described in Chapter 10 (pp. 348–349).

The penis consists of an attached *root* and a free *shaft* or *body* that ends in an enlarged tip, the **glans penis.** The skin covering the penis is loose, and it slides distally to form a cuff of skin called the **prepuce** (pre'pūs), or **foreskin,** around the glans. Frequently, the foreskin is surgically removed shortly after birth, a procedure called *circumcision* ("cutting around").

Internally, the penis contains the spongy urethra and three long cylindrical bodies *(corpora)* of erectile tissue. *Erectile tissue* is a spongy network of connective tissue and smooth muscle riddled with vascular spaces. During sexual excitement, the vascular spaces fill with blood, causing the penis to enlarge and become rigid. This event, called *erection,* enables the penis to serve as a penetrating organ. The midventral erectile body, the **corpus spongiosum** (spon"je-o'sum; "spongy body") surrounds the urethra. It expands distally to form the glans and proximally to form the part of the root called the **bulb of the penis.** The bulb is covered externally by the sheetlike bulbospongiosus muscle and is secured to the urogenital diaphragm. The paired dorsal erectile bodies, called the **corpora cavernosa** (kor'por-ah kă-ver-no'sah; "cavernous bodies"), make up most of the penis and are bound by fibrous tunica albuginea. Their proximal ends form the **crura** (kroo'rah; "legs") **of the penis** (singular, crus). Each crus is surrounded by an ischiocavernosus muscle and anchors to the pubic arch of the bony pelvis.

The Male Duct System

As mentioned earlier, sperm travel from the testes to the outside of the body through a system of ducts. In order (proximal to distal), the **accessory ducts** are the epididymis, the ductus deferens, and the urethra.

The Epididymis

The comma-shaped **epididymis** (*epi* = beside; *didym* = the testes) is about 3.8 cm (1.5 inches) long (see Figures 28.1 and 28.3a, c). Its *head,* which joins the efferent ductules, caps the superior aspect of the testis. Its *body* and *tail* regions lie on the posterolateral aspects of the testis. The bulk of the epididymis consists of the highly coiled *duct of the epididymis* with a true uncoiled length of about 6 m (20 feet). Some of the cells of the pseudostratified epithelium of the duct mucosa exhibit long, nonmotile microvilli *(stereocilia),* which absorb excess testicular fluid and pass nutrients to the sperm in the lumen.

The immature, nearly nonmotile sperm that leave the testis are moved slowly through the duct of the epididymis. As they move along its tortuous course (a trip that takes about 20 days), the sperm gain the ability to swim. When a male is sexually stimulated and ejaculates, the smooth muscle in the walls of the epididymis contracts, expelling sperm from its tail into the next segment of the duct system, the *ductus deferens.* However, sperm can be stored in the epididymis for several months. If held longer, they are eventually phagocytized by epithelial cells of the epididymis.

The Ductus Deferens

The **ductus deferens** (duk'tus def'er-ens; "carrying away"), or **vas deferens,** is about 45 cm (18 inches) long. It runs upward as part of the spermatic cord from the epididymis through the inguinal canal into the pelvic cavity (see Figure 28.1). It is easily palpated as it passes anterior to the pubic bone. It then loops medially over the ureter and descends along the posterior wall of the bladder. Its terminus expands to form the **ampulla** and then joins with the duct of the seminal vesicle (a gland) to form the short **ejaculatory duct.** Each ejaculatory duct, in turn, passes into the prostate gland where it empties into the urethra.

The ductus deferens propels live sperm from their storage sites, the epididymis and distal part of the ductus deferens, into the urethra. At the moment of ejaculation, the thick layers of smooth muscle in its walls create strong peristaltic waves that rapidly squeeze the sperm forward.

As Figure 28.3 illustrates, part of the ductus deferens lies in the scrotal sac. Some men opt to take full responsibility for birth control by having a **vasectomy** (vah-sek'to-me; "cutting the vas"). In this relatively minor operation, the physician makes a

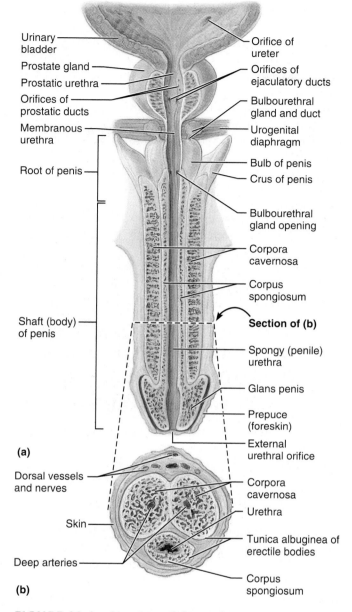

Urinary bladder

Prostate gland

Prostatic urethra

Orifices of prostatic ducts

Membranous urethra

Root of penis

Shaft (body) of penis

(a)

Dorsal vessels and nerves

Skin

Deep arteries

(b)

Orifice of ureter

Orifices of ejaculatory ducts

Bulbourethral gland and duct

Urogenital diaphragm

Bulb of penis

Crus of penis

Bulbourethral gland opening

Corpora cavernosa

Corpus spongiosum

Section of (b)

Spongy (penile) urethra

Glans penis

Prepuce (foreskin)

External urethral orifice

Corpora cavernosa

Urethra

Tunica albuginea of erectile bodies

Corpus spongiosum

FIGURE 28.4 *Structure of the penis.* **(a)** Longitudinal (coronal) section of the penis. **(b)** Transverse section of the penis.

small incision into the scrotum and then cuts through and ligates (ties off) the ductus deferens. Although sperm continue to be produced for the next several years, they can no longer reach the body exterior. Eventually, they deteriorate and are phagocytized. Vasectomy is simple and provides highly effective birth control (close to 100%).

The Urethra

The **urethra** is the terminal portion of the male duct system (see Figures 28.1 and 28.4). It conveys both urine and semen (at different times), so it serves both the urinary and reproductive systems. Its three regions are (1) the *prostatic urethra*, the portion surrounded by the prostate gland; (2) the *membranous urethra* in the urogenital diaphragm; and (3) the *spongy (penile) urethra*, which runs through the penis and opens to the outside at the *external urethral orifice*. The spongy urethra is about 15 cm (6 inches) long and accounts for 75% of urethral length.

Accessory Glands

The **accessory glands** include the paired seminal vesicles and bulbourethral glands and the single prostate gland (Figure 28.1). These glands produce the bulk of *semen* (sperm plus accessory gland secretions).

The Seminal Vesicles

The **seminal** (sem′ĭ-nul) **vesicles** lie on the posterior wall of the bladder. These glands are fairly large, each about the shape and length (5 to 7 cm) of a little finger. Their secretion, which accounts for about 60% of the volume of semen, is a yellowish viscous alkaline fluid containing fructose sugar, ascorbic acid, a coagulating enzyme *(vesiculase)*, and prostaglandins. As noted, the duct of each seminal vesicle joins that of the ductus deferens on the same side to form the ejaculatory duct. Sperm and seminal fluid mix in the ejaculatory duct and enter the prostatic urethra together during ejaculation.

The Prostate Gland

The **prostate** (pros′tāt) **gland** is a single doughnut-shaped gland about the size of a chestnut (see Figures 28.1 and 28.4). It encircles the part of the urethra just inferior to the bladder. Enclosed by a thick connective tissue capsule, it is made up of 20 to 30 compound tubular-alveolar glands embedded in a mass (stroma) of smooth muscle and dense connective tissue. The prostatic gland secretion, accounting for up to one-third of the semen volume, is a milky, slightly acid fluid that contains citrate (a nutrient source) and several enzymes (fibrinolysin, hyaluronidase, acid phosphatase), and prostate-specific antigen (PSA), and it plays a role in activating sperm. It enters the prostatic urethra via several ducts when prostatic smooth muscle contracts during ejaculation.

⚠ *Homeostatic Imbalance*

The prostate gland has a reputation as a health destroyer (perhaps reflected in the common mispronunciation "prostrate"). Hypertrophy of the prostate gland, which affects nearly every elderly male, distorts the urethra. The more the man strains to urinate, the more the valvelike prostatic mass blocks the opening. This troublesome condition also enhances the risk of bladder infections (cystitis) and kidney damage. Traditional treatment has been surgical, but some newer options are gaining popularity, such as using microwaves or drugs (e.g., finasteride) to shrink the prostate; and inserting and inflating a small balloon to compress the prostate tissue away from the prostatic urethra. Another option is TUNA (transurethral needle ablation), a procedure in which a catheter containing a tiny needle is guided into position, and then burstlike releases of radio-frequency radiation are used to incinerate excess prostate tissue.

Prostatitis (pros"tah-ti'tis), inflammation of the prostate, is the single most common reason for a man to consult a urologist, and prostate cancer is the third most prevalent cancer in men. Screening for prostate cancer typically involves digital examination of the prostate through the anterior wall of the rectum and assay of PSA levels in the blood. PSA, although a normal component of semen, is a tumor marker and its rising serum level follows the clinical disease course of prostate cancer. When possible, prostate cancer is treated surgically. Alternative therapies for metastasized cancers involve castration or therapy with drugs (flutamide) that block androgen receptors, or LHRH* analogues (such as leuprolide), which inhibit gonadotropin release. Deprived of the stimulatory effects of androgens, the prostatic tissue regresses and urinary symptoms typically decline. ■

The Bulbourethral Glands

The **bulbourethral** (bul"bo-u-re'thral) **glands,** also called **Cowper's glands,** are pea-sized glands situated inferior to the prostate gland (Figures 28.1 and 28.4). They produce a thick, clear mucus prior to ejaculation that drains into the spongy urethra and neutralizes traces of acidic urine in the urethra.

Semen

Semen (se'men) is a milky white, somewhat sticky mixture of sperm and accessory gland secretions. The liquid provides a transport medium and nutrients and contains chemicals that protect and activate the sperm and facilitate their movement. Mature sperm cells are streamlined cellular "missiles" containing very little cytoplasm or stored nutrients. The fructose in seminal vesicle secretion provides nearly all their fuel. The prostaglandins in semen decrease the viscosity of mucus guarding the entry (cervix) of the uterus and stimulate reverse peristalsis in the uterus and the medial parts of the uterine tubes, facilitating the movement of sperm through

the female reproductive tract. The presence of the hormone *relaxin* and certain enzymes in semen enhance sperm motility. The relative alkalinity of semen as a whole (pH 7.2–7.6) due to bases (spermine and others) helps neutralize the acid environment of the male's urethra and the female's vagina. Thus, it protects the delicate sperm and enhances their motility. Sperm are very sluggish under acidic conditions (below pH 6). Semen also contains an antibiotic chemical called **seminalplasmin,** which destroys certain bacteria. Clotting factors (fibrinogen and others) found in semen coagulate it just after it is ejaculated. Then, its contained fibrinolysin liquefies the sticky mass, enabling the sperm to swim out and begin their journey through the female duct system.

The amount of semen propelled out of the male duct system during ejaculation is relatively small, only 2–5 ml, but there are between 50 and 130 million sperm per milliliter.

PHYSIOLOGY OF THE MALE REPRODUCTIVE SYSTEM

Male Sexual Response

The chief phases of the male sexual response are (1) *erection* of the penis, which allows it to penetrate the female vagina, and (2) *ejaculation,* which expels semen into the vagina.

Erection

Erection, enlargement and stiffening of the penis, results from engorgement of the erectile bodies with blood. When a man is not sexually aroused, the arterioles supplying the erectile tissue are constricted, and the penis is flaccid. However, during sexual excitement a parasympathetic reflex is triggered that promotes release of nitric oxide locally. NO relaxes vascular smooth muscle, causing these arterioles to dilate, which allows the erectile bodies to fill with blood. Expansion of the corpora cavernosa of the penis compresses their drainage veins, retarding blood outflow and maintaining engorgement. The corpus spongiosum expands but not nearly as much as the cavernosa; its main job is to keep the urethra open during ejaculation. Erection of the penis is one of the rare examples of parasympathetic control of arterioles. Another parasympathetic effect is stimulation of the bulbourethral glands, which causes lubrication of the glans penis.

Erection is initiated by a variety of sexual stimuli, such as touching the genital skin, mechanical stimulation of the pressure receptors in the penis, and erotic sights, sounds, and smells. The CNS responds to such stimuli by activating parasympathetic neurons that innervate the internal pudendal arteries serving the penis. Sometimes erection is in-

---

*LHRH (luteinizing hormone–releasing hormone) is analogous to GnRH (gonadotropin-releasing hormone). See discussion of GnRH's effects on p. 620.

duced solely by emotional or higher mental activity (the thought of a sexual encounter). Emotions and thoughts can also inhibit erection, causing vasoconstriction and resumption of the flaccid penile state. Impotence, the inability to attain erection, is considered in *A Closer Look* on p. 1083.

Ejaculation

Ejaculation (*ejac* = to shoot forth) is the propulsion of semen from the male duct system. While erection is under parasympathetic control, *ejaculation* is under sympathetic control. When impulses provoking erection reach a certain critical level, a spinal reflex is initiated, and a massive discharge of nerve impulses occurs over the sympathetic nerves serving the genital organs (largely at the level of L_1 and L_2). As a result,

1. The reproductive ducts and accessory glands contract, emptying their contents into the urethra.

2. The bladder sphincter muscle constricts, preventing expulsion of urine or reflux of semen into the bladder.

3. The bulbospongiosus muscles of the penis undergo a rapid series of contractions, propelling semen at a speed of up to 200 inches/s from the urethra. These rhythmic muscle contractions are accompanied by intense pleasure and many systemic changes, such as generalized muscle contraction, rapid heartbeat, and elevated blood pressure.

The entire ejaculatory event is referred to as **climax** or **orgasm**. Orgasm is quickly followed by muscular and psychological relaxation and vasoconstriction of the penile arterioles, which allows the penis to become flaccid once again. After ejaculation, there is a latent period, ranging in time from minutes to hours, during which a man is unable to achieve another orgasm. The latent period increases in length with aging.

Spermatogenesis

Spermatogenesis (sper″mah-to-jen′ĕ-sis; "sperm formation") is the sequence of events in the seminiferous tubules of the testes that produces male gametes—**sperm** or **spermatozoa.** The process begins during puberty, around the age of 14 years in males, and continues throughout life. Every day, a healthy adult male makes about 400 million sperm. It seems that nature has made sure that the human species will not be endangered for lack of sperm.

Before we lunge into describing the process of spermatogenesis, let's lay some groundwork by defining some terms you need to know. First of all, having two sets of chromosomes, one from each parent, is a key factor in the human life cycle (Figure 28.5). The normal chromosome number in most

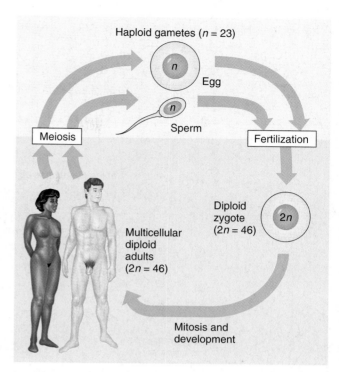

FIGURE 28.5 *The human life cycle.*

body cells is referred to as the **diploid** (dip′loid) or **2n chromosomal number** of the organism. In humans, this number is 46, and such diploid cells contain 23 pairs of similar chromosomes called **homologous** (homol′ŏ-gus) **chromosomes** or **homologues.** One member of each pair is from the male parent (the *paternal chromosome*); the other is from the female parent (the *maternal chromosome*). Generally speaking, the two homologues of each chromosome pair look alike and carry genes that code for the same traits, though not necessarily for identical expression of those traits. (Consider, for example, the homologous genes controlling the expression of freckles; the paternal gene might code for an ample sprinkling of freckles and the maternal gene for no freckles.) In Chapter 30, we will consider how maternal and paternal genes interact to produce our visible traits.

In contrast, the number of chromosomes present in human gametes is 23, referred to as the **haploid** (hap′loid) or **n chromosomal number,** and gametes contain only one member of each homologous pair. When sperm and egg fuse, they form a fertilized egg that reestablishes the typical diploid chromosomal number of human cells.

Gamete formation in both sexes involves **meiosis** (mi-o′sis; "a lessening"), a unique kind of nuclear division that, for the most part, occurs only within the gonads. Recall that *mitosis* (the process by which most body cells divide) distributes replicated chromosomes equally between the two daughter cells. Consequently, each daughter cell receives a set of

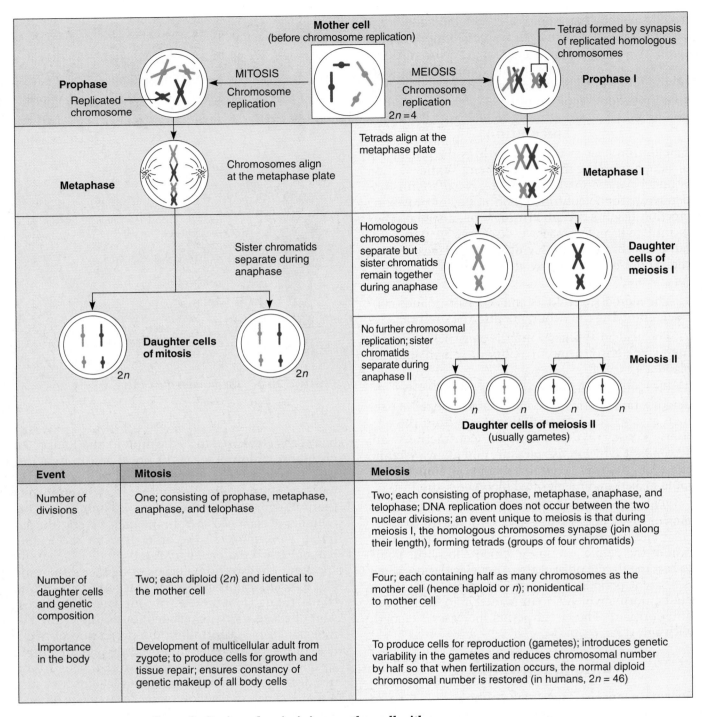

FIGURE 28.6   *Comparison of mitosis and meiosis in a mother cell with a diploid number (2n) of 4.* Mitosis is shown on the left, meiosis on the right. (Not all the phases of mitosis and meiosis are shown.)

chromosomes identical to that of the mother cell. Meiosis, on the other hand, consists of two consecutive nuclear divisions, and its product is four daughter cells instead of two, each with half as many chromosomes as typical body cells. Thus, meiosis provides the *means* for reducing the chromosomal number by half (from 2n to n) in gametes. Mitosis and meiosis are compared in Figure 28.6.

Meiosis

The two nuclear divisions of meiosis, called *meiosis I* and *meiosis II*, are divided into phases for convenience. Although these phases are given the same names as those of mitosis (prophase, metaphase, anaphase, and telophase), events of meiosis I are quite different from those of mitosis, as detailed on page 1080 and in Figure 28.7.

FIGURE 28.7 *Meiotic cell division.* This series of diagrams shows meiotic cell division for an animal cell with a diploid number (2*n*) of 4. The behavior of the chromosomes is emphasized.

Interphase events
As in mitosis, meiosis is preceded by events occurring during interphase that lead to DNA replication and other preparations needed for the cell division process. Just before meiosis begins, the replicated chromatids, held together by centromeres, are ready and waiting.

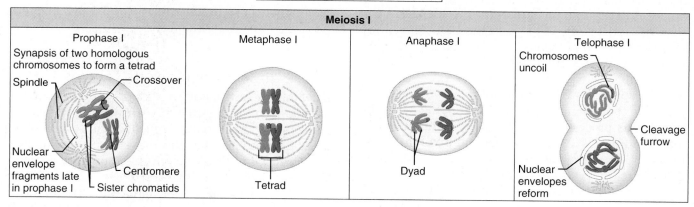

Meiosis I

Prophase I — Synapsis of two homologous chromosomes to form a tetrad; Spindle; Crossover; Nuclear envelope fragments late in prophase I; Centromere; Sister chromatids

Metaphase I — Tetrad

Anaphase I — Dyad

Telophase I — Chromosomes uncoil; Cleavage furrow; Nuclear envelopes reform

Prophase I
As in prophase of mitosis, the chromosomes coil and condense, the nuclear membrane and nucleolus break down and disappear, and the spindle is formed. However, a unique event not seen in mitosis, called synapsis, occurs in prophase I of meiosis. Synapsis involves the coming together of homologous chromosomes to form tetrads, little packets of four chromatids. While in synapsis, the "arms" of adjacent homologous chromatids become wrapped around each other, forming several points of crossover or chiasmata. Generally speaking, the longer the chromatids, the more chiasmata are formed. (This process of crossover is shown in one tetrad of the prophase I view, and the result of that one event of crossing over is followed through meiosis II.) Prophase I is the longest period of meiosis, accounting for about 90% of the total period. By its end, the tetrads have attached to the spindle and are moving toward the spindle equator, and the sister chromatids have exchanged parts at points of crossover.

Metaphase I
During metaphase I, the tetrads align on the spindle equator in preparation for anaphase.

Anaphase I
Unlike the anaphase events of mitosis, the centromeres do not break during anaphase I of meiosis, and so the sister chromatids (dyads) remain firmly attached. However, the homologous chromosomes do separate from each other and the dyads are moved toward opposite poles of the cell.

Telophase I
The nuclear membranes re-form around the chromosomal masses, the spindle breaks down, and the chromatin reappears as telophase and cytokinesis are completed, forming two daughter cells. The daughter cells (now haploid) enter a second interphase-like period, called interkinesis, before meiosis II occurs. There is no second replication of DNA before meiosis II.

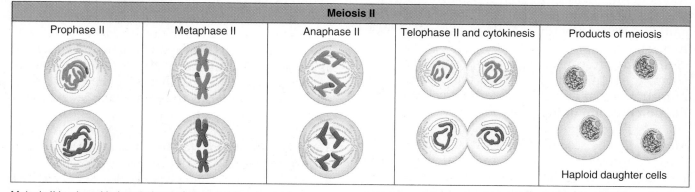

Meiosis II

Prophase II | Metaphase II | Anaphase II | Telophase II and cytokinesis | Products of meiosis — Haploid daughter cells

Meiosis II begins with the products of meiosis I (two haploid daughter cells) and undergoes a mitosis-like nuclear division process referred to as the equational division of meiosis. After progressing through prophase, metaphase, anaphase, and telophase, followed by cytokinesis, the product is four haploid daughter cells each genetically different from the original mother cell. During human spermatogenesis the daughter cells remain interconnected by cytoplasmic extensions.

Recall that prior to mitosis all the chromosomes are replicated. Then the identical copies remain together as *sister chromatids* connected by a centromere throughout prophase and during metaphase. At anaphase, the centromeres split and the sister chromatids separate from each other so that each daughter cell inherits a copy of *every* chromosome possessed by the mother cell (see Figure 28.6). Let's look at how meiosis differs.

Meiosis I As in mitosis, the chromosomes replicate before meiosis begins. But in prophase of meiosis I, an event never seen in mitosis (nor in meiosis II for that matter) occurs. The replicated chromosomes seek out their homologous partners and pair up with them along their entire length (Figure 28.7). This intimate alignment of the homologues takes place at discrete spots along the length of the homologues—more like buttoning together than zipping. As a result of this process, called **synapsis**, little groups of four chromatids called **tetrads**, or **bivalents**, are formed. During synapsis, a second unique event called crossover occurs. **Crossovers** are formed within each tetrad as the free ends of one maternal and one paternal chromatid wrap around each other at one or more points. Crossover allows an exchange of genetic material between the paired maternal and paternal chromosomes (see Figure 28.7).

During metaphase I, the tetrads line up at the spindle equator. This alignment is random; that is, either the paternal or maternal chromosome can be on a given side of the equator. During anaphase I, the sister chromatids representing each homologue behave as a unit—almost as if replication had not occurred—and the *homologous chromosomes* (each still composed of two joined sister chromatids) are distributed to opposite ends of the cell.

Thus, when meiosis I is completed, the following conditions exist: Each daughter cell has (1) *two* copies of one member of each homologous pair (either the paternal or maternal) and none of the other, and (2) a diploid amount of DNA but a *haploid* chromosomal number because the still-united sister chromatids are considered to be a single chromosome. Since meiosis I reduces the chromosome number from *2n* to *n*, it is sometimes called the **reduction division of meiosis.**

Meiosis II The second meiotic division, meiosis II, mirrors mitosis in every way, except that the chromosomes are *not* replicated before it begins. Instead, the chromatids present in the two daughter cells of meiosis I are simply parceled out among four cells. Because the chromatids are distributed equally to the daughter cells (as in mitosis), meiosis II is sometimes referred to as the **equational division of meiosis** (Figure 28.7).

Meiosis accomplishes two important tasks: (1) It reduces the chromosomal number by half and (2) it introduces genetic variability. The random orientation of the homologous pairs during meiosis I provides tremendous variability in the resulting gametes by scrambling genetic characteristics derived from the two parents in different combinations. Variability is increased further by crossover because during late prophase I, the homologues break at points of crossover and exchange chromosomal segments (see Figure 28.7). (This process is described further in Chapter 30.) As a result, it is likely that no two gametes are exactly alike, and all are different from the original mother cells.

Spermatogenesis: Summary of Events in the Seminiferous Tubules

Now that we have described meiosis, let us turn to the specific events of spermatogenesis which take place in the seminiferous tubules. A histological section of an adult testis shows that most of the cells making up the epithelial walls of the seminiferous tubules are in various stages of cell division (Figure 28.8). These cells, collectively called **spermatogenic** ("sperm forming") **cells,** give rise to sperm in the following series of cellular divisions and transformations.

Mitosis of Spermatogonia: Forming Spermatocytes The outermost and least differentiated tubule cells, which are in direct contact with the epithelial basal lamina, are stem cells called **spermatogonia** (sper"mah-to-go'ne-ah; "sperm seed"). The spermatogonia divide more or less continuously by *mitosis* and, until puberty, all their daughter cells become spermatogonia. Spermatogenesis begins during puberty, and from then on, each mitotic division of a spermatogonium results in two distinctive daughter cells (Figure 28.8)—*types A* and *B*. The **type A daughter cell** remains at the basement membrane to maintain the germ cell line. The **type B** cell gets pushed toward the lumen, where it becomes a **primary spermatocyte** destined to produce four sperm. (To keep these cell types straight, remember that just as the letter A is always at the beginning of our alphabet, a type A cell is always at the tubule basement membrane ready to begin a new generation of gametes.)

Meiosis: Spermatocytes to Spermatids Still early in spermatogenesis, each primary spermatocyte generated during the first phase undergoes meiosis I, forming two smaller haploid cells called **secondary spermatocytes.** The secondary spermatocytes continue on rapidly into meiosis II, and their daughter cells, called **spermatids** (sper'mah-tidz), are small round cells with large spherical nuclei seen closer to the lumen of the tubule.

What cell type results from meiosis?

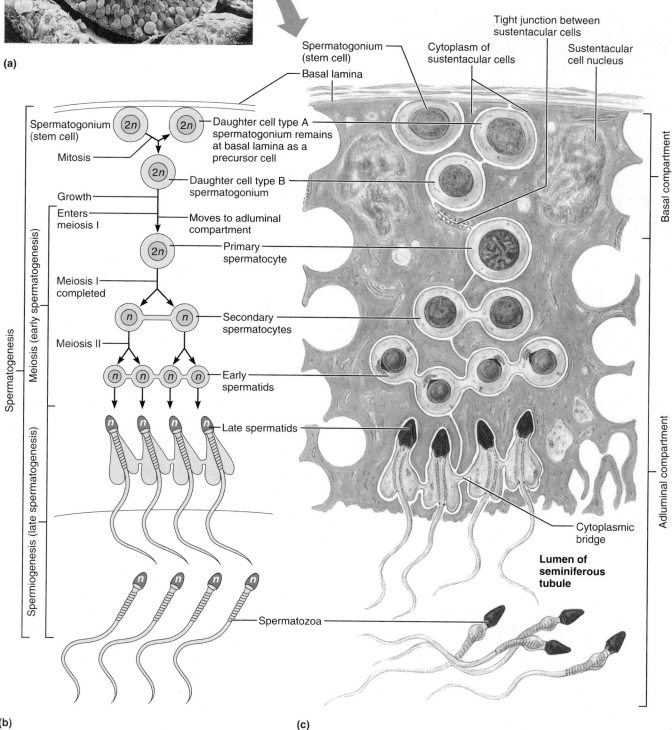

(a)

FIGURE 28.8 *Spermatogenesis.* **(a)** Scanning electron micrograph of a cross-sectional view of a seminiferous tubule (225×). (© R. G. Kessel and R. H. Kardon, *Tissues and Organs: A Text-Atlas of Scanning Electron Microscopy*, W. H. Freeman and Company, 1979, all rights reserved.) **(b)** Flowchart of events of spermatogenesis, showing the relative positioning of the various spermatogenic cells. **(c)** Enlarged view of a portion of the wall of the seminiferous tubule, showing the spermatogenic cells surrounded by sustentacular cells (colored green). (*Note:* The process of spermatogenesis encompasses the phases through the production of sperm. Conversion of the haploid spermatids to sperm cells is specifically called spermiogenesis.)

(b)

(c)

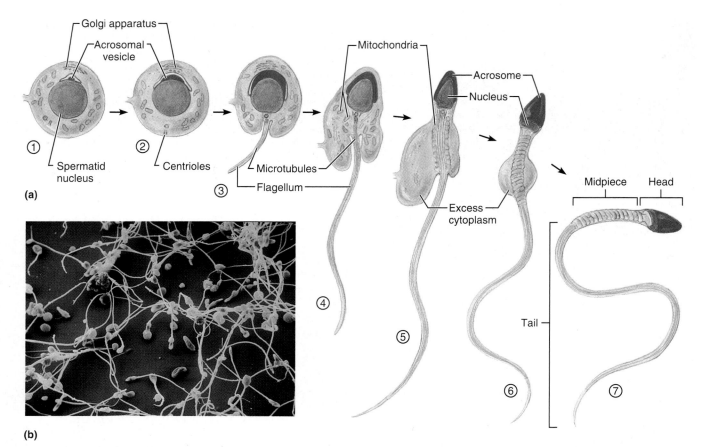

FIGURE 28.9 Spermiogenesis: transformation of a spermatid into a functional sperm. (a) The stepwise process of spermiogenesis consists of ① activity of the Golgi apparatus to package the acrosomal enzymes, ② positioning of the acrosome at the anterior end of the nucleus and of the centrioles at the opposite end of the nucleus, ③ elaboration of microtubules to form the flagellum of the tail, ④ mitochondrial multiplication and their positioning around the proximal portion of the flagellum, and ⑤ sloughing off excess cytoplasm. ⑥ Structure of an immature sperm that has just been released from a sustentacular cell. ⑦ Structure of a fully mature sperm. **(b)** Scanning electron micrograph of mature sperm (430×).

Spermiogenesis: Spermatids to Sperm As late spermatogenesis begins, each spermatid has the correct chromosomal number for fertilization *(n)*, but is nonmotile. It still must undergo a streamlining process called **spermiogenesis**, during which it sheds its excess cytoplasmic baggage and forms a tail. Details of this process appear in Figure 28.9a. The resulting **sperm,** or **spermatozoon** (sper″mah-to-zo′on; "animal seed"), has three major regions: a head, a midpiece, and a tail, which correspond roughly to *genetic, metabolic,* and *locomotor regions,* respectively. The **head** of the sperm consists almost entirely of its flattened nucleus, which contains compacted DNA. Adhering to the top of the nucleus is a helmetlike **acrosome** (ak′ro-sōm; "tip piece"). The lysosomelike acrosome is produced by the Golgi apparatus and contains hydrolytic enzymes (hyaluronidase and others) that enable the sperm to penetrate and enter an egg. The sperm **midpiece** contains mitochondria spiraled tightly around the contractile filaments of the tail. The **tail** is a typical flagellum produced by a centriole. The mitochondria provide the metabolic energy (ATP) needed for the whiplike movements of the tail that will propel the sperm along its way in the female reproductive tract.

Role of the Sustentacular Cells Throughout spermatogenesis, descendants of the same spermatogonium remain closely attached to one another by cytoplasmic bridges (see Figure 28.8). Additionally, they are surrounded by and connected to supporting cells of a special type, called **sustentacular cells** or **Sertoli cells,** which extend from the basal lamina to the lumen of the tubule (see Figure 28.8c). The sustentacular cells, bound to each other by tight junctions, form an unbroken layer within the seminiferous tubule, dividing it into two compartments. The **basal compartment** extends from the basal lamina to their tight junctions and contains spermatogonia and the earliest primary spermatocytes. The **adluminal compartment** lies internal to the tight junc-

A CLOSER LOOK   Erectile Dysfunction and Passion Pills

Erectile dysfunction (ED), more commonly known as **impotence**, is the inability to attain an erection when desired.

So what causes an erection? In two words, nitric oxide. When males become aroused, parasympathetic nerves in the penis release nitric oxide, which activates cyclic GMP, an enzyme that causes smooth muscle in the penile blood vessel walls to relax. As it does so, blood rushes into the spongy cavernous tissue, stiffening the penis. And what blocks nitric oxide release or its effects? Psychological factors, alcohol, or certain drugs can cause temporary impotence in healthy males. When chronic, the condition is largely the result of hormonal, vascular, or nervous system problems.

For a long time, impotence was dismissed as "being in your head," but it now appears that psychological or emotional causes account for less than 20% of cases. A whole bunch of drugs can produce impotence; these include antihypertensives, antihistamines, antidepressants, the antiulcer drug cimetidine (Tagamet), and some appetite suppressants, cold remedies, and antihistamines. Vascular problems that are common causes include arteriosclerosis, injury to the penile blood vessels, or incompetent varicose veins in the penis, which create a "venous leak" that allows blood to leave the penis prematurely. Neural causes typically result from damage to the penile nerves during surgery or radiation therapy, tumors,

or physical trauma. The single most important hormonal cause is diabetes mellitus, which leads not only to arterial blockages that reduce blood flow into the penis, but also to nerve damage that makes erection difficult or impossible.

So, let's assume the cause is determined, but the problem still exists. What's a man to do? Until recently, most remedies were unpleasant and invasive. Using a vacuum pump to suck blood into the penis and thus pump up a flaccid penis often embarrasses its user and does a good deal to take the romance out of the moment. Drugs such as papaverine and prostaglandin E_1 dilate the penile blood vessels but require painful injections into the penis, and alprostadil gel (Muse), a soft pellet that is inserted into the urethra, is said to hurt more than the injections by users. In cases of nonresponsive vascular disease or damage to the penile nerves, the usual approaches are reconstructive surgery to repair damaged arteries or a venous leak, or implanting a device in the penis that will make it erect. However, surgery is expensive and has its shortcomings even in the most capable hands, and the implantable prostheses sometimes break down, requiring still more surgery.

As you might expect, then, the advent of a little blue triangle of a pill called Viagra (sildenafil) has been greeted with great enthusiasm by men with erectile dysfunction that allowed them some erectile ability but not enough to permit penetra-

tion. To say that Viagra has been a big hit among those who still want to have sexual relations is a vast understatement. The drug was approved by the FDA in March of 1998; by September 1998, 4 million prescriptions had been filled. Viagra works, but not for everyone; its success rate is 70%. Among Viagra's advantages are that it (1) works within the hour to produce a sustained erection by increasing the concentration of nitric acid in the penis, (2) is taken orally, and (3) has essentially no significant side effects in healthy males (i.e., those who do not have severe vascular problems or long-standing pyschogenic impotence). Some men experience transient blue-green vision, stuffy nose, headache, or temporary facial flushing, or perhaps mild stomach upset—all trade-offs that they claim they are ready to put up with for sexual potency. The major problem with Viagra use has been among men with preexisting serious health problems (heart disease, hypertension, and diabetes mellitus) who ignore the drug's (or their physician's) warnings that Viagra also increases systemic blood pressure as it works its "miracle." Consequently, fatalities have been reported among men taking Viagra, most involving men taking nitrate drugs to dilate the coronary circulation. Nonetheless, Viagra is seen as a wonder drug. Will it work equally well for women with problems with sexual arousal? The clitoris is physiologically similar to the penis and clinical trials are rolling. More later.

---

tions and includes the meiotically active cells and the tubule lumen.

The tight junctions between the sustentacular cells form the **blood-testis barrier.** This barrier prevents the membrane antigens of differentiating sperm from escaping through the basal lamina into the bloodstream. Because sperm are not formed until puberty, they are absent when the immune system is being programmed to recognize one's own tissues early in life. The spermatogonia, which are recognized as "self," are outside the barrier and can be influenced by bloodborne chemical messengers

that prompt spermatogenesis. Following mitosis of the spermatogonia, the tight junctions of the sustentacular cells open to allow primary spermatocytes to pass into the adluminal compartment—much as locks in a canal open to allow a boat to pass.

In the adluminal compartment, spermatocytes and spermatids are nearly enclosed in recesses in the sustentacular cells (see Figure 28.8). The sustentacular cells deliver nutrients to the dividing cells, move them along to the lumen, secrete **testicular fluid** (rich in androgens and metabolic acids) that provides the transport medium for sperm in the lumen, and dispose of the excess cytoplasm sloughed off the spermatids as they transform into sperm. As described shortly, the sustentacular cells also produce chemical mediators that help regulate spermatogenesis.

Spermatogenesis—from formation of a primary spermatocyte to release of immature sperm into the lumen—takes 64 to 72 days. Sperm in the lumen are unable to "swim" and are incapable of fertilizing an egg. They are pushed by the pressure of the testicular fluid through the tubular system of the testes into the epididymis. There they mature further, gaining increased motility and fertilizing power.

⬛ Homeostatic Imbalance

According to some studies, a gradual decline in male fertility has been occurring in the past 50 years. Not only has the average volume of the ejaculate declined 20%, but average sperm count per milliliter has decreased more than 50%. Sperm quality is dropping as well, with a greater percentage of sluggish and deformed sperm obvious. Any time sperm count is fewer than 20 million/ml, chances of becoming a father plummet. Some believe the main cause is environmental toxins, especially compounds with estrogenic effects. These compounds, which block the action of male sex hormones as they program sexual development, are now found in our meat supply as well as in the air. Common antibiotics such as tetracycline may suppress sperm formation; and radiation, lead, certain components of pesticides, marijuana, and excessive alcohol can cause abnormal (two-headed, multiple-tailed, etc.) sperm to be produced.

Male infertility may also be caused by anatomical obstructions and hormonal imbalances. One of the first test series done when a couple has been unable to conceive is *semen analysis*. Factors analyzed include sperm count, motility, and morphology (shape and maturity), and semen volume, pH, and fructose content. A low sperm count accompanied by a high percentage of immature sperm may hint that a man has a *varicocele* (var'ĭ-ko-sēl"), a condition that hinders drainage of the testicular vein, resulting in an elevated temperature in the scrotum that interferes with normal sperm development. ⬛

Hormonal Regulation of Male Reproductive Function

The Brain-Testicular Axis

Hormonal regulation of spermatogenesis and testicular androgen production involves interactions between the hypothalamus, anterior pituitary gland, and testes, a relationship sometimes called the **brain-testicular axis.** The sequence of regulatory events, shown schematically in Figure 28.10, is as follows:

① The hypothalamus releases **gonadotropin-releasing hormone (GnRH),** which controls the release of the anterior pituitary gonadotropins, **follicle-stimulating hormone (FSH),** and **luteinizing hormone (LH).** (Both FSH and LH were named for their effects on the female gonad.) GnRH reaches the anterior pituitary cells via the blood of the hypophyseal portal system.

② Binding of GnRH to pituitary cells (gonadotrophs) prompts them to secrete FSH and LH into the blood.

③ FSH stimulates spermatogenesis indirectly by stimulating the sustentacular cells to release **androgen-binding protein (ABP).** ABP prompts the spermatogenic cells to bind and concentrate **testosterone,** which in turn stimulates spermatogenesis. Thus, FSH makes the cells receptive to testosterone's stimulatory effects.

④ LH binds to the interstitial cells and stimulates them to secrete testosterone (and a small amount of estrogen). LH is therefore sometimes called **interstitial cell–stimulating hormone (ICSH)** in males. Locally testosterone serves as the final trigger for spermatogenesis. Testosterone entering the bloodstream exerts a number of effects at other body sites.

⑤ Both the hypothalamus and the anterior pituitary are subject to feedback inhibition by bloodborne hormones. Testosterone inhibits hypothalamic release of GnRH and acts directly on the anterior pituitary to inhibit gonadotropin release. **Inhibin** (in-hib'in), a protein hormone produced by the sustentacular cells, serves as a barometer of the normalcy of spermatogenesis. When the sperm count is high, inhibin release increases and it inhibits anterior pituitary release of FSH and GnRH release by the hypothalamus. When the sperm count falls below 20 million/ml, inhibin secretion declines steeply.

As you can see, the amount of testosterone and sperm produced by the testes reflects a balance among three sets of hormones: (1) GnRH, which indirectly stimulates the testes via its effect on FSH and ICSH release; (2) gonadotropins, which directly stimulate the testes; and (3) testicular hormones (testosterone and inhibin), which exert negative feedback controls on the hypothalamus and anterior pituitary. Since the hypothalamus is also influenced by input from other brain areas, the whole axis is under CNS control. In the absence of GnRH and gonadotropins, the testes atrophy, and for all practical purposes, sperm and testosterone production ceases.

Development of male reproductive structures (discussed later in the chapter) depends on prenatal secretion of male hormones, and for a few months after birth, a male infant has plasma gonadotropin and testosterone levels nearly equal to those of a midpubertal boy. Soon thereafter, blood levels of these hormones recede and they remain low throughout childhood. As puberty nears, much higher levels of testosterone are required to suppress hypothalamic release of GnRH. As more GnRH is released, more testosterone is secreted by the testes, but the threshold for hypothalamic inhibition keeps rising until the adult pattern of hormone interaction is achieved. Maturation of the brain-testicular axis takes about three years, and once established, the balance between the interacting hormones remains relatively constant. Consequently, an adult male's sperm and testosterone production also remains fairly stable. (In females, cyclic swings in gonadotropin and female sex hormone levels are the norm.)

Mechanism and Effects of Testosterone Activity

Testosterone, like all steroid hormones, is synthesized from cholesterol. It exerts its effects by activating specific genes to transcribe messenger RNA molecules, which results in enhanced synthesis of certain proteins in the target cells. (See Chapter 17.)

In some target cells, testosterone must be transformed into another steroid to exert its effects. In the prostate gland, for example, testosterone must be converted to *dihydrotestosterone (DHT)* before it can bind within the nucleus. In certain neurons of the brain, testosterone is converted to *estrogen* (es'tro-jen) to bring about its stimulatory effects. Thus, in that case at least, a "male" hormone is transformed into a "female" hormone to exert its masculinizing effects.

As puberty ensues, testosterone not only prompts spermatogenesis but has multiple anabolic effects throughout the body (see Table 28.1). It targets all accessory reproductive organs—ducts, glands, and the penis—causing them to grow and assume adult size and function. In adult males, normal plasma levels of testosterone maintain these organs: When the hormone is deficient or absent, all accessory organs atrophy, semen volume declines markedly, and erection and ejaculation are impaired. Thus, a man becomes both sterile and impotent. This situation is easily remedied by testosterone replacement therapy.

Male secondary sex characteristics—that is, features induced in the *nonreproductive* organs by the male sex hormones (mainly testosterone)—make their appearance at puberty. These include the appearance of pubic, axillary, and facial hair, enhanced hair growth on the chest or other body areas

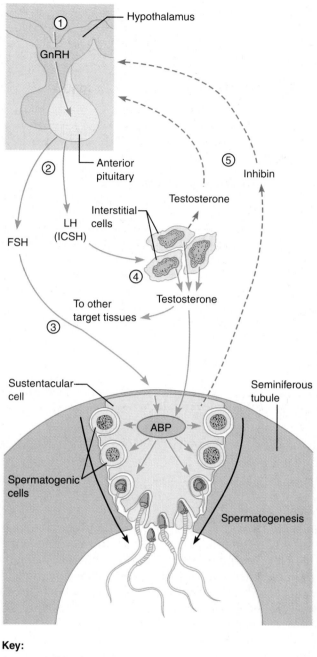

Key:
——→ = Stimulates
- - - → = Inhibits

FIGURE 28.10 *Hormonal regulation of testicular function, the brain-testicular axis.* ① The hypothalamus releases gonadotropin-releasing hormone (GnRH). ② GnRH stimulates the anterior pituitary to release gonadotropins— follicle-stimulating hormone (FSH) and luteinizing hormone (LH), also called interstitial cell–stimulating hormone (ICSH). ③ FSH acts on the sustentacular cells, causing them to release androgen-binding protein (ABP). ④ LH acts on the interstitial cells, stimulating their release of testosterone. ABP binding of testosterone then enhances spermatogenesis. ⑤ Rising levels of testosterone and inhibin (released by the sustentacular cells) exert feedback inhibition on the hypothalamus and pituitary.

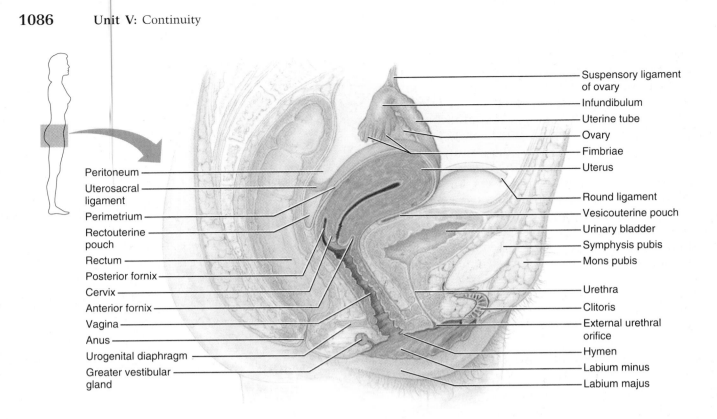

FIGURE 28.11   *Midsagittal section of the female pelvis, showing internal organs of the female reproductive system.* (See *A Brief Atlas of the Human Body,* Figure 59.)

in some men, and a deepening of the voice as the larynx enlarges. The skin thickens and becomes oilier (which predisposes young men to acne), bones grow and increase in density, and skeletal muscles increase in size and mass. The last two effects are often referred to as the *somatic effects* of testosterone (*soma* = body).

Testosterone also boosts basal metabolic rate and influences behavior. It is the basis of the sex drive (libido) in both males and females, be they heterosexual or homosexual. Thus, although testosterone is called a "male" sex hormone, it should not be specifically tagged as a promoter of male sexual activity. In embryos, the presence of testosterone masculinizes the brain.

The testes are not the only source of androgens; the adrenal glands of both sexes also release androgens. However, since the relatively small amounts of adrenal androgens are unable to support normal testosterone-mediated functions when the testes fail to produce androgens, we can assume that it is the testosterone production by the testes that supports male reproductive function.

ANATOMY OF THE FEMALE REPRODUCTIVE SYSTEM

The reproductive role of the female is far more complex than that of a male. Not only must she produce gametes, but her body must prepare to nurture a developing embryo for a period of approximately nine months. **Ovaries,** the **female gonads,** are the primary reproductive organs of a female, and like the male testes, ovaries serve a dual purpose: Besides their gametogenic functions, they produce the female sex hormones, the **estrogens*** and **progesterone** (pro-ges'tĕ-rōn). The accessory ducts (uterine tubes, uterus, and vagina) transport or otherwise serve the needs of the reproductive cells and a developing fetus.

As illustrated in Figure 28.11, the ovaries and duct system, collectively known as the **internal genitalia,** are mostly located in the pelvic cavity. The fe-

---

*The ovaries produce several different estrogens (*estradiol, estrone, and estriol*). Estradiol is the most abundant and is most responsible for estrogenic effects.

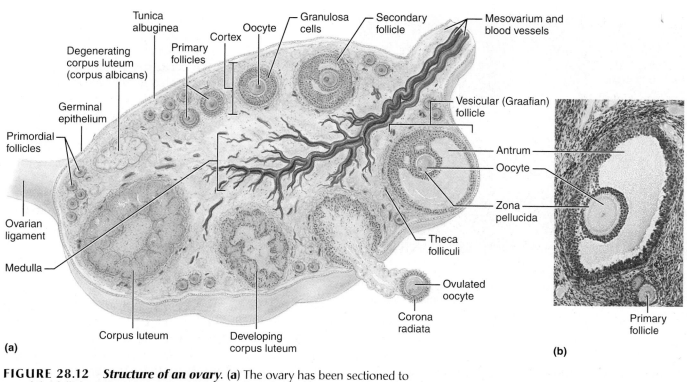

FIGURE 28.12 *Structure of an ovary.* (a) The ovary has been sectioned to reveal the follicles in its interior. Note that not all of these structures would appear in the ovary at the same time. (b) Photomicrograph of a mature vesicular (Graafian) follicle (250×).

male's accessory ducts, from the vicinity of the ovary to the body exterior, are the *uterine tubes*, the *uterus*, and the *vagina*. The external sex organs of females are referred to as the **external genitalia.**

The Ovaries

The paired ovaries, which flank the uterus on each side (see Figure 28.11), are shaped like almonds and about twice as large. Each ovary is held in place within the peritoneal cavity by several ligaments. The **ovarian ligament** anchors the ovary medially to the uterus; the **suspensory ligament** anchors it laterally to the pelvic wall; and the **mesovarium** (mez"o-va're-um) suspends it in between (see Figures 28.11 and 28.14a). Both the suspensory ligament and the mesovarium are part of the **broad ligament,** a peritoneal fold that "tents" over the uterus and supports the uterine tubes, uterus, and vagina. The fibrous ovarian ligaments are enclosed within the broad ligament.

The ovaries are served by the **ovarian arteries,** branches of the abdominal aorta (see Figure 20.22c, p. 761), and by the *ovarian branch of the uterine arteries.* The ovarian blood vessels reach the ovaries by traveling through the suspensory ligaments and mesovaria.

Like a testis, an ovary is surrounded externally by a fibrous **tunica albuginea** (Figure 28.12). The tunica albuginea is in turn covered externally by a layer of cuboidal epithelial cells called the *germinal epithelium,* which is continuous with the peritoneum of the mesovarium. The term *germinal epithelium* is a misnomer because this layer does not give rise to ova. The ovary has an outer cortex, which houses the forming gametes, and an inner medullary region, containing the largest blood vessels and nerves, but the relative extent of each region is poorly defined.

Embedded in the highly vascular connective tissue of the ovary cortex are many tiny saclike structures called **ovarian follicles.** Each follicle consists of an immature egg, called an **oocyte** (o'o-sīt; *oo* = egg), encased by one or more layers of very different cells. The surrounding cells are called **follicle cells** if a single layer is present, and **granulosa cells** when more than one layer is present. Follicles at different stages of maturation are distinguished by their structure. In a **primordial follicle,** one layer of squamouslike follicle cells surrounds the oocyte. A **primary follicle** has two or more layers of cuboidal or columnar-type granulosa cells enclosing the oocyte; it becomes a **secondary follicle** when fluid-filled spaces appear between the granulosa cells and then coalesce to form a central fluid-filled cavity called an *antrum.* At its most mature stage, when it is called a

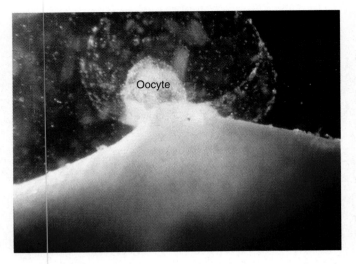

Oocyte

FIGURE 28.13   *Ovulation.* A secondary oocyte is released from a follicle at the surface of the ovary. (The orange mass below the ejected oocyte is part of the ovary. Primary and secondary oocytes are explained in Figure 28.19.)

vesicular, or **Graafian** (graf′e-an) **follicle,** the follicle bulges from the surface of the ovary. The oocyte of the vesicular follicle "sits" on a stalk of granulosa cells at one side of the antrum. Each month in adult women, one of the ripening follicles ejects its oocyte from the ovary, an event called **ovulation** (Figure 28.13). After ovulation, the ruptured follicle is transformed into a very different looking structure called the **corpus luteum** (lu′te-um; plural, corpora lutea), which eventually degenerates. As a rule, most of these structures can be seen within the same ovary. In older women, the surfaces of the ovaries are scarred and pitted, revealing that many oocytes have been released.

The Female Duct System

The Uterine Tubes

The **uterine** (u′ter-in) **tubes,** also called **fallopian tubes** and **oviducts,** form the initial part of the female duct system (Figures 28.11 and 28.14a). They receive the ovulated oocyte and provide a site where fertilization can occur. Each uterine tube is about 10 cm (4 inches) long and extends medially from the region of an ovary to empty into the superolateral region of the uterus via a constricted region called the **isthmus** (is′mus). The distal end of each uterine tube expands as it curves around the ovary, forming the **ampulla;** fertilization usually occurs in this region. The ampulla ends in the **infundibulum** (in″fun-dib′u-lum), an open, funnel-shaped structure bearing ciliated, fingerlike projections called **fimbriae** (fim′bre-e; "fringe") that drape over the ovary. Unlike the male duct system, which is continuous with the tubules of the testes, the uterine

tubes have little or no actual contact with the ovaries. An ovulated oocyte is cast into the peritoneal cavity, and many oocytes are lost there. However, the uterine tube performs a complex sequence of movements to capture oocytes. Specifically, the infundibulum bends to drape over the ovary while the fimbriae stiffen and sweep the ovarian surface. The beating cilia on the fimbriae then create currents in the peritoneal fluid that act to carry the oocyte into the uterine tube, where it begins its journey toward the uterus.

The uterine tube aids the progress of an oocyte. Its wall contains sheets of smooth muscle, and its thick, highly folded mucosa contains both ciliated and nonciliated cells. The oocyte is carried toward the uterus by a combination of muscular peristalsis and the beating of the cilia. Nonciliated cells of the mucosa have dense microvilli and produce a secretion that keeps the oocyte (and sperm, if present) moist and nourished.

Externally, the uterine tubes are covered by visceral peritoneum and supported along their length by a short mesentery (part of the broad ligament) called the **mesosalpinx** (mez″o-sal′pinks; "mesentery of the trumpet"; *salpin* = trumpet), a reference to the trumpet-shaped uterine tube it supports (see Figure 28.14a).

Homeostatic Imbalance

The fact that the uterine tubes are not continuous with the ovaries places women at risk for *ectopic pregnancy* in which an ovum, fertilized in the peritoneal cavity, begins developing there. Such pregnancies naturally abort, often with substantial bleeding. Another potential problem is infection spreading into the peritoneal cavity from other parts of the reproductive tract. Gonorrhea bacteria and other sexually transmitted microorganisms sometimes infect the peritoneal cavity in this way, causing an extremely severe inflammation called **pelvic inflammatory disease (PID).** Unless treated promptly, PID can cause scarring of the narrow uterine tubes and of the ovaries and lead to sterility, and even death. Indeed, scarring and closure of the uterine tubes, which have an internal diameter as small as the width of a human hair in some regions, is one of the major causes of female infertility. ■

The Uterus

The **uterus** (Latin for "womb") is located in the pelvis, anterior to the rectum and posterosuperior to the bladder (see Figures 28.11 and 28.14). It is a hollow, thick-walled organ that functions to receive, retain, and nourish a fertilized ovum. In a premenopausal woman who has never been pregnant, the uterus is about the size and shape of an inverted

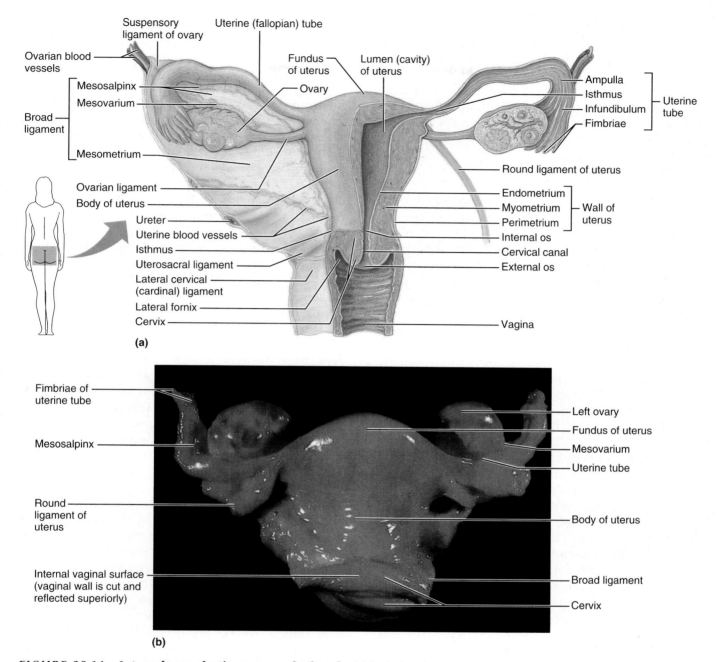

FIGURE 28.14   *Internal reproductive organs of a female.* (a) Posterior view of the female reproductive organs. The posterior walls of the vagina, uterus, and uterine tubes and the broad ligament have been removed on the right side to reveal the shape of the lumen of these organs. **(b)** Anterior view of the reproductive organs of a female cadaver. Compare this to the posterior view of the same organs in (a).

pear, but it is usually somewhat larger in women who have borne children. Normally, the uterus is flexed anteriorly where it joins the vagina, causing the uterus as a whole to be inclined forward or *anteverted*. However, the organ is frequently turned backward, or *retroverted*, in older women.

The major portion of the uterus is referred to as the **body** (see Figure 28.11 and Figure 28.14). The

rounded region superior to the entrance of the uterine tubes is the **fundus,** and the slightly narrowed region between the body and the cervix is the *isthmus*. The **cervix** of the uterus is its narrow neck, or outlet, which projects into the vagina inferiorly. The cavity of the cervix, called the **cervical canal,** communicates with the vagina via the *external os* (*os* = mouth) and with the cavity of the uterine body via the *internal os.* The mucosa of the cervical canal contains *cervical glands* that secrete a mucus that fills the cervical canal and covers the external os, presumably to block the spread of bacteria from the vagina into the uterus. Cervical mucus also blocks the entry of sperm, except at midcycle, when it becomes less viscous and allows sperm to pass through.

⚠ *Homeostatic Imbalance*

Cancer of the cervix is common among women between the ages of 30 and 50. Risk factors include frequent cervical inflammations, sexually transmitted diseases including genital warts, and multiple pregnancies. The cancer cells arise from the epithelium covering the cervical tip. In a *Papanicolaou (Pap) smear,* or cervical smear test, some of these epithelial cells are scraped away and then examined for abnormalities. A Pap smear is the most effective way to detect this slow-growing cancer, and women are advised to have one every year. ■

Supports of the Uterus   The uterus is supported laterally by the **mesometrium** ("mesentery of the uterus") portion of the broad ligament (Figure 28.14a). More inferiorly, the **lateral cervical (cardinal) ligaments** extend from the cervix and superior part of the vagina to the lateral walls of the pelvis, and the paired **uterosacral ligaments** secure the uterus to the sacrum posteriorly. The uterus is bound to the anterior body wall by the fibrous **round ligaments,** which run through the inguinal canals to anchor in the subcutaneous tissue of the labia majora (the outer lips of the vulva). These various ligaments allow the uterus a good deal of mobility, and its position changes as the rectum and bladder fill and empty.

⚠ *Homeostatic Imbalance*

Despite the many anchoring ligaments, the principal support of the uterus is provided by the muscles of the pelvic floor, namely the muscles of the urogenital and pelvic diaphragms (Table 10.7, pp. 349–350). These muscles are sometimes torn during childbirth. Subsequently, the unsupported uterus may sink inferiorly, until the tip of the cervix protrudes through the external vaginal opening. This condition is called **prolapse of the uterus.** ■

The undulating course of the peritoneum around and over the various pelvic structures produces several cul-de-sacs, or blind-ended peritoneal pouches. The most important of these are the *vesicouterine* (ves"i-ko-u'ter-in) *pouch* between the bladder and uterus, and the *rectouterine pouch* between the rectum and uterus (see Figure 28.11).

The Uterine Wall   The wall of the uterus is composed of three layers (Figure 28.14a): the perimetrium, the myometrium, and the endometrium. The **perimetrium,** the outermost serous layer, is the visceral peritoneum. The **myometrium** (mi"o-me'tre-um; "muscle of the uterus") is the bulky middle layer, composed of interlacing bundles of smooth muscle. The myometrium contracts rhythmically during childbirth to expel the baby from the mother's body. The **endometrium** is the mucosal lining of the uterine cavity (Figure 28.15); it is a simple columnar epithelium underlain by a thick lamina propria of highly cellular connective tissue. If fertilization occurs, the young embryo burrows into the endometrium (implants) and resides there for the rest of its development.

The endometrium has two chief *strata* (layers). The **stratum functionalis** (fungk-shun-a'lis), or **functional layer,** undergoes cyclic changes in response to blood levels of ovarian hormones and is shed during menstruation (approximately every 28 days). The thinner, deeper **stratum basalis** (ba-sa'lis), or **basal layer,** forms a new functionalis after menstruation ends. It is unresponsive to ovarian hormones. The endometrium has numerous *uterine glands* that change in length as the endometrial thickness changes.

To understand the cyclic changes of the uterine endometrium (discussed later in the chapter), it is essential to understand the vascular supply of the uterus. The **uterine arteries** arise from the *internal iliacs* in the pelvis, ascend along the sides of the uterus, and send branches into the uterine wall (Figure 28.14a). These branches break up into several **arcuate** (ar'ku-āt) **arteries** within the myometrium. The arcuate arteries send **radial branches** into the endometrium, where they in turn give off **straight arteries** to the stratum basalis and **spiral (coiled) arteries** to the stratum functionalis (see Figure 28.15b). The spiral arteries undergo repeated degeneration and regeneration, and it is their spasms that actually cause the functionalis layer to be shed during menstruation. Veins in the endometrium are thin-walled and form an extensive network with occasional sinusoidal enlargements.

The Vagina

The **vagina** ("sheath") is a thin-walled tube, 8–10 cm (3–4 inches) long. It lies between the bladder and the rectum and extends from the cervix to the body exterior (see Figure 28.11). The urethra is embedded in its anterior wall. Often called the *birth canal,* the

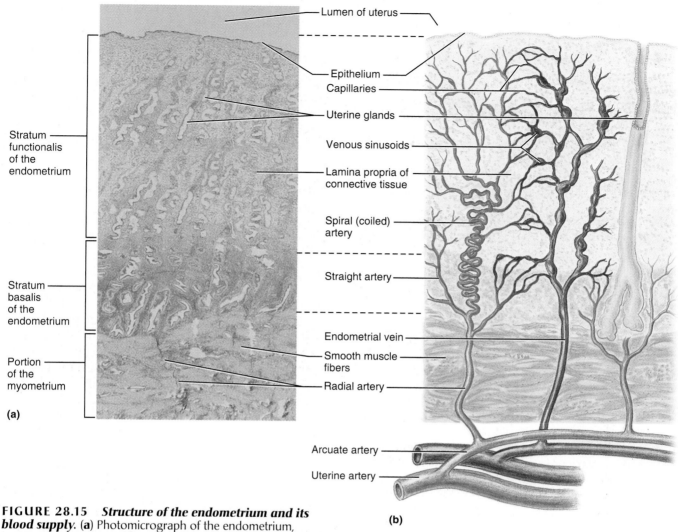

FIGURE 28.15 **Structure of the endometrium and its blood supply.** **(a)** Photomicrograph of the endometrium, longitudinal section, showing its stratum functionalis and basalis regions (20×). **(b)** Diagrammatic view of the endometrium, showing the straight arteries that serve the stratum basalis and the spiral arteries that serve the stratum functionalis. The thin-walled veins and venous sinusoids are also illustrated.

vagina provides a passageway for delivery of an infant and for menstrual flow. Since it receives the penis (and semen) during sexual intercourse, it is the *female organ of copulation.*

The highly distensible wall of the vagina consists of three coats: an outer fibroelastic *adventitia,* a smooth muscle *muscularis,* and a *mucosa* marked by transverse ridges or *rugae,* which stimulate the penis during intercourse. The epithelium of the mucosa is a stratified squamous epithelium adapted to stand up to friction. Certain mucosal cells *(dendritic cells)* act as antigen-presenting cells and are thought to provide the route of HIV transmission from an infected male to the female during sexual intercourse. (AIDS, the immune deficiency disease caused by

HIV, is described in Chapter 22.) The vaginal mucosa has no glands; it is lubricated by the cervical mucous glands. Its epithelial cells release large amounts of glycogen, which is anaerobically metabolized to lactic acid by resident bacteria. Consequently, the pH of a woman's vagina is normally quite acidic. This acidity helps keep the vagina healthy and free of infection, but it is also hostile to sperm. Although vaginal fluid of *adult* females is acidic, it tends to be alkaline in adolescents, predisposing sexually active teenagers to sexually transmitted diseases.

In virgins (females who have never participated in sexual intercourse), the mucosa near the distal **vaginal orifice** forms an incomplete partition called

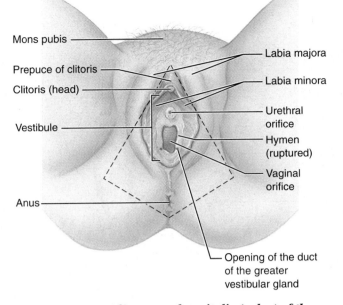

Mons pubis

Prepuce of clitoris

Clitoris (head)

Vestibule

Anus

Labia majora

Labia minora

Urethral orifice

Hymen (ruptured)

Vaginal orifice

Opening of the duct of the greater vestibular gland

FIGURE 28.16 *The external genitalia (vulva) of the female.* The region enclosed by the dashed lines is the perineum.

the **hymen** (hi'men) (Figure 28.16). The hymen is very vascular and tends to bleed when it is ruptured during the first coitus (sexual intercourse). However, its durability varies. In some females, it is ruptured during a sports activity, tampon insertion, or pelvic examination. Occasionally, it is so tough that it must be breached surgically if intercourse is to occur.

The upper end of the vaginal canal loosely surrounds the cervix of the uterus, producing a vaginal recess called the **vaginal fornix.** The posterior part of this recess, the *posterior fornix,* is much deeper than the *lateral* and *anterior fornices* (see Figures 28.11 and 28.14a). Generally, the lumen of the vagina is quite small and, except where it is held open by the cervix, its posterior and anterior walls are in contact with one another. The vagina stretches considerably during copulation and childbirth, but its lateral distension is limited by the ischial spines and the sacrospinous ligaments.

Homeostatic Imbalance

The uterus tilts away from the vagina. Hence, attempts by untrained persons to induce an abortion by entering the uterus with a surgical instrument may result in puncturing of the posterior wall of the vagina, followed by hemorrhage and—if the instrument is unsterile—peritonitis. ■

The External Genitalia

As mentioned earlier, the female reproductive structures that lie external to the vagina are called the *external genitalia* (see Figure 28.16). The external genitalia, also called the **vulva** (vul'vah; "covering") or **pudendum** ("shameful"), include the mons pubis, labia, clitoris, and structures associated with the vestibule.

The **mons pubis** (mons pu'bis; "mountain on the pubis") is a fatty, rounded area overlying the pubic symphysis. After puberty, this area is covered with pubic hair. Running posteriorly from the mons pubis are two elongated, hair-covered fatty skin folds, the **labia majora** (la'be-ah mah-jor'ah; "larger lips"). These are the female counterpart, or *homologue,* of the male scrotum (that is, they derive from the same embryonic tissue). The labia majora enclose the **labia minora** (mi-nor'ah; "smaller"), two thin, hair-free skin folds, homologous to the ventral penis. The labia minora enclose a recess called the **vestibule,** which contains the external opening of the urethra more anteriorly followed by that of the vagina. Flanking the vaginal opening are the pea-size **greater vestibular glands** (see also Figure 28.11), homologous to the bulbourethral glands of males. These glands release mucus into the vestibule and help to keep it moist and lubricated, facilitating intercourse.

Just anterior to the vestibule is the **clitoris** (klit'o-ris; "hill"), a small, protruding structure, composed largely of erectile tissue, that is homologous to the penis of the male. It is hooded by a skin fold called the **prepuce of the clitoris,** formed by the junction of the labia minora folds. The clitoris is richly innervated with sensory nerve endings sensitive to touch, and it becomes swollen with blood and erect during tactile stimulation, contributing to a female's sexual arousal. Like the penis, the clitoris has dorsal erectile columns (corpora cavernosa), but it lacks a corpus spongiosum. In males, the urethra carries both urine and semen and runs through the penis, but the female urinary and reproductive tracts are completely separate, and neither runs through the clitoris.

The female **perineum** is a diamond-shaped region located between the pubic arch anteriorly, the coccyx posteriorly, and the ischial tuberosities laterally. The soft tissues of the perineum overlie the muscles of the pelvic outlet and the posterior ends of the labia majora overlie the *central tendon,* into which most muscles supporting the pelvic floor insert (see Table 10.7).

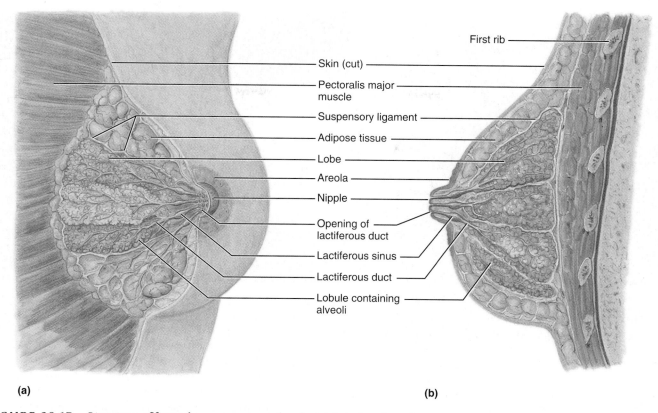

First rib

Skin (cut)

Pectoralis major muscle

Suspensory ligament

Adipose tissue

Lobe

Areola

Nipple

Opening of lactiferous duct

Lactiferous sinus

Lactiferous duct

Lobule containing alveoli

(a)   **(b)**

FIGURE 28.17   *Structure of lactating mammary glands.* (a) Anterior view of a partially dissected breast. **(b)** Sagittal section of a breast.

The Mammary Glands

The **mammary glands** are present in both sexes, but they normally function only in females (Figure 28.17). Since the biological role of the mammary glands is to produce milk to nourish a newborn baby, they are actually important only when reproduction has already been accomplished.

Developmentally, mammary glands are modified sweat glands that are really part of the *skin,* or *integumentary system.* Each mammary gland is contained within a rounded skin-covered breast anterior to the pectoral muscles of the thorax. Slightly below the center of each breast is a ring of pigmented skin, the **areola** (ah-re'o-lah), which surrounds a central protruding **nipple.** Large sebaceous glands in the areola make it slightly bumpy and produce sebum that reduces chapping and cracking of the skin of the nipple. Autonomic nervous system controls of smooth muscle fibers in the areola and nipple cause the nipple to become erect when stimulated by tactile or sexual stimuli and when exposed to cold.

Internally, each mammary gland consists of 15 to 25 **lobes** that radiate around and open at the nipple. The lobes are padded and separated from each other by fibrous connective tissue and fat. The interlobar connective tissue forms **suspensory ligaments** that attach the breast to the underlying muscle fascia and to the overlying dermis. As suggested by their name, the suspensory ligaments provide natural support for the breasts, like a built-in brassiere. Within the lobes are smaller units called **lobules,** which contain glandular **alveoli** that produce milk when a woman is lactating. These compound alveolar glands pass the milk into the **lactiferous** (lak-tif'er-us) **ducts,** which open to the outside at the nipple. Just deep to the areola, each lactiferous duct has a dilated region called a **lactiferous sinus.** Milk accumulates in these sinuses during nursing. The process and regulation of lactation are described in Chapter 29.

The description of mammary glands given here applies only to nursing women or women in the last trimester of pregnancy. In nonpregnant women, the glandular structure of the breast is largely undeveloped and the duct system is rudimentary; hence breast size is largely due to the amount of fat deposits.

Breast Cancer

Invasive breast cancer, the most common malignancy of U.S. women, strikes nearly 200,000 American women each year. One woman in eight will

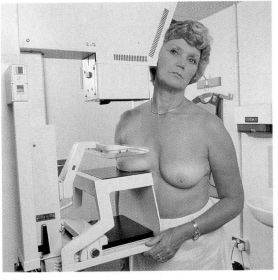

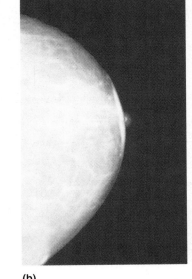

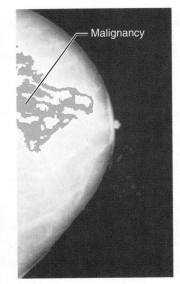

Malignancy

(a)

(b)

(c)

FIGURE 28.18 *Mammograms.* (**a**) Photograph of a woman undergoing mammography. (**b**) Normal breast. (**c**) Breast with tumor.

develop this condition. Breast cancer usually arises from the epithelial cells of the ducts, not from the alveoli. A small cluster of cancer cells grows into a lump in the breast from which cells eventually metastasize.

Known risk factors for developing breast cancer include (1) early onset menses and late menopause; (2) no pregnancies or first pregnancy later in life; (3) previous history of breast cancer; and (4) family history of breast cancer (especially in a sister or mother). Other risk factors proposed but as yet unproven include exposure to high estrogen concentrations while in utero and postmenopause, cigarette smoking, and excessive alcohol intake. Some 10% of breast cancers stem from hereditary defects and half of these can be traced to dangerous mutations in a pair of genes, dubbed *BRCA1* (breast cancer 1) and *BRCA2*, which virtually guarantee that the carriers will develop breast cancer. However, more than 70% of women who develop breast cancer have no known risk factors for the disease.

Breast cancer is often signaled by a change in skin texture, puckering, or leakage from the nipple. Early detection by breast self-examination and mammography is unquestionably the best way to increase one's chances of surviving breast cancer. Since most breast lumps are discovered by women themselves in routine monthly breast exams, this simple examination should be a health maintenance priority in every woman's life. Currently the American Cancer Society recommends scheduling **mammography**— X-ray examination that detects breast cancers too small to feel (less than 1 cm)—every two years for women between 40 and 49 years old and yearly thereafter (Figure 28.18).

Once diagnosed, breast cancer is treated in various ways depending on specific characteristics of the lesion. Current therapies include (1) radiation therapy, (2) chemotherapy, and (3) surgery, often followed by irradiation or chemotherapy to destroy stray cancer cells. Newer on the therapy scene are Herceptin, a drug containing bioengineered antibodies that jam estrogen receptors that control aggressive growth in breast cancer cells; and tamoxifen, an antiestrogen compound that significantly improves the outcome for those with early- or late-stage breast cancer. Until the 1970s, the standard treatment was **radical mastectomy** (mas-tek'to-me; "breast cutting"), removal of the entire affected breast, plus all underlying muscles, fascia, and associated lymph nodes. Medical records reveal that this painful and disfiguring treatment is no more effective at halting the cancer than less extensive surgery. Hence, most physicians now recommend **lumpectomy,** in which only the cancerous part (lump) is excised, or **simple mastectomy,** removal of the breast tissue only (and perhaps some of the axillary lymph nodes).

Many mastectomy patients opt for breast reconstruction to replace the excised tissue. Although silicone gel implants initially appeared to be "just what the doctor ordered," their use was banned by the FDA in 1991 following evidence (still a matter of controversy) that implant ruptures that allowed gel leakage and distribution to other body sites resulted in autoimmune disorders and cancers. Currently tissue "flaps," containing muscle, fat, and skin taken from the patient's abdomen or back, are providing acceptable alternatives for "sculpting" a natural-looking breast.

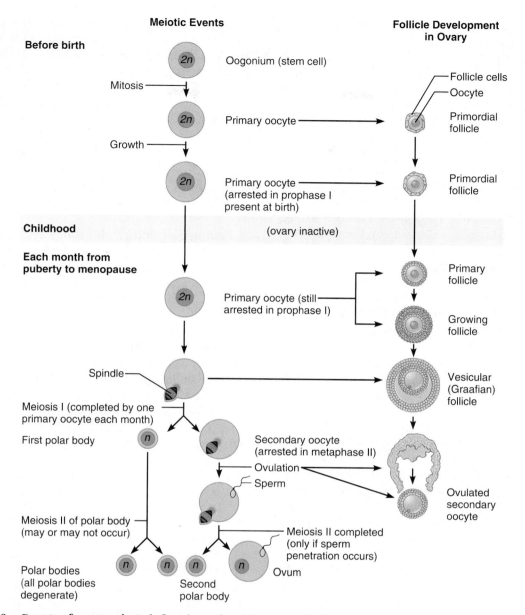

Meiotic Events

Before birth

Mitosis

Growth

2n Oogonium (stem cell)

2n Primary oocyte

2n Primary oocyte (arrested in prophase I present at birth)

Follicle Development in Ovary

Follicle cells
Oocyte
Primordial follicle

Primordial follicle

Childhood (ovary inactive)

Each month from puberty to menopause

2n Primary oocyte (still arrested in prophase I)

Primary follicle

Growing follicle

Spindle

Meiosis I (completed by one primary oocyte each month)

First polar body

Secondary oocyte (arrested in metaphase II)

Ovulation

Sperm

Vesicular (Graafian) follicle

Ovulated secondary oocyte

Meiosis II of polar body (may or may not occur)

Meiosis II completed (only if sperm penetration occurs)

Polar bodies (all polar bodies degenerate)

Second polar body

Ovum

FIGURE 28.19 *Events of oogenesis.* Left, flowchart of meiotic events. Right, correlation with follicle development and ovulation in the ovary.

PHYSIOLOGY OF THE FEMALE REPRODUCTIVE SYSTEM

Oogenesis

Gamete production in males begins at puberty and continues throughout life, but the situation is quite different in females. A female's total supply of eggs is already determined by the time she is born, and the time span during which she releases them extends from puberty to menopause (about the age of 50).

Meiosis, the specialized nuclear division that occurs in the testes to produce sperm, also occurs in the ovaries. In this case, female sex cells are produced, and the process is called **oogenesis** (o"o-gen'e-sis; "the beginning of an egg"). The process of oogenesis (Figure 28.19) takes years to complete. First, in the fetal period the **oogonia**, the diploid stem cells of the ovaries, multiply rapidly by mitosis and then enter a growth phase and lay in nutrient reserves. Gradually, *primordial follicles* (Figure 28.19) begin to appear as the oogonia are transformed into **primary oocytes** and become surrounded by a single layer of flattened follicle cells. The primary oocytes begin the first meiotic division, but become "stalled" late in prophase I and do not complete it. By birth, a female's lifetime supply of primary oocytes, approximately 2 million of them, is already in place in the cortical region of the immature ovary. Since they remain in their state of suspended animation all through childhood, the wait is a long one—10 to 14 years at the very least!

At puberty, perhaps 400,000 oocytes remain and beginning at this time a small number of primary oocytes are activated each month. However, only one is "selected" each time to continue meiosis I, ultimately producing two haploid cells (each with 23 replicated chromosomes) that are quite dissimilar in size. The smaller cell is called the **first polar body.** The larger cell, which contains nearly all the cytoplasm of the primary oocyte, is the **secondary oocyte.** The events of this first maturation division are interesting. A spindle forms at the very edge of the oocyte (see Figure 28.19, left), and a little "nipple," into which the polar body chromosomes will be cast, appears at that edge. This sets up the polarity of the oocyte and ensures that the polar body receives almost no cytoplasm or organelles.

The first polar body may continue its development and undergo meiosis II, producing two even smaller polar bodies. However, in humans, the secondary oocyte arrests in metaphase II and it is this cell (not a functional ovum) that is ovulated. If an ovulated secondary oocyte is not penetrated by a sperm, it simply deteriorates. But, if sperm penetration does occur, it quickly completes meiosis II, yielding one large **ovum** and a tiny **second polar body** (see Figure 28.19). Completion of meiosis II and the union of the egg and sperm nuclei are described in Chapter 29. What you should realize now is that the potential end products of oogenesis are three tiny polar bodies, nearly devoid of cytoplasm, and one large ovum. All of these cells are haploid, but only the ovum is a *functional gamete.* This is quite different from spermatogenesis, where the product is four viable gametes—spermatozoa.

The unequal cytoplasmic divisions that occur during oogenesis ensure that a fertilized egg has ample nutrients for its seven-day journey to the uterus. Without nutrient-containing cytoplasm the polar bodies degenerate and die. Since the reproductive life of a female is at best about 40 years (from the age of 11 to approximately 51) and typically only one ovulation occurs each month, fewer than 500 oocytes out of her potential of 400,000 are released during a woman's lifetime. Again, nature has provided us with a generous oversupply of sex cells.

The Ovarian Cycle

The monthly series of events associated with the maturation of an egg is called the **ovarian cycle.** The ovarian cycle is best described in terms of two consecutive phases. The **follicular phase** is the period of follicle growth, typically indicated as lasting from the first to the fourteenth day of the cycle. The **luteal phase** is the period of corpus luteum activity, days 14–28. The "typical" ovarian cycle repeats at intervals of 28 days, with **ovulation** occurring midcycle. However, cycles as long as 40 days or as short as 21 days are fairly common. In such cases, the length of the follicular phase and timing of ovulation vary, but the luteal phase remains constant: It is 14 days between the time of ovulation and the end of the cycle. Hormonal controls of these events will be described shortly; here we will concentrate on what happens each month within the ovary (Figure 28.20).

The Follicular Phase

Maturation of a primordial follicle to the mature state occupies the first half of the cycle and involves several events as shown in Figure 28.20, steps 1–6.

A Primordial Follicle Becomes a Primary Follicle

When the primordial follicles ① are activated, the squamouslike cells surrounding the primary oocyte grow, becoming cuboidal cells, and the oocyte enlarges. The follicle is now called a primary follicle ②.

A Primary Follicle Becomes a Secondary Follicle

Next the follicular cells proliferate until they form a stratified epithelium around the oocyte ③. As soon as more than one cell layer is present, the follicle cells take on the name *granulosa cells*. The granulosa cells are connected to the developing oocyte by gap junctions, through which ions, metabolites, and signaling molecules can pass. One of the signals passing from the granulosa cells to the oocyte "tells" the oocyte to grow.

In the next stage ④, a layer of connective tissue begins to condense around the follicle, forming the **theca folliculi** (the'kah fah-lik'u-li; "box around the follicle"). As the granulosa cells continue to divide and the follicle grows, the thecal and granulosa cells cooperate to produce estrogens (the inner thecal cells produce androgens, which the granulosa cells convert to estrogens). At the same time, the granulosa cells secrete a glycoprotein-rich substance that forms a thick transparent membrane, called the **zona pellucida** (pě-lu'sid-ah), around the oocyte.

In phase ⑤, clear liquid accumulates between the granulosa cells and eventually coalesces to form a fluid-filled cavity called the **antrum** ("cave"). The presence of an antrum distinguishes the new secondary follicle from the primary follicle.

A Secondary Follicle Becomes a Vesicular Follicle

The antrum continues to expand with fluid until it isolates the oocyte, along with its surrounding capsule of granulosa cells called a **corona radiata** ("radiating crown"), on a stalk on one side of the follicle. When a follicle attains full size (about 2.5 cm, or 1 inch, in diameter), it becomes a vesicular follicle ⑥ and bulges from the external ovarian surface like an "angry boil." This usually occurs by day 14.

As one of the final events of follicle maturation, the primary oocyte completes meiosis I to form the

? *Which of the follicles shown contain primary oocytes? Which contain secondary oocytes?*

FIGURE 28.20 *Schematic view of the ovarian cycle: development and fate of ovarian follicles.* The numbers on the diagram indicate the sequence of events in follicle development, *not* the movement of a developing follicle within the ovary. ① Primordial follicle containing a primary oocyte surrounded by flattened cells. ② A primary follicle containing a primary oocyte surrounded by cuboidal follicle cells. ③, ④ The growing primary follicle. This follicle is secreting estrogen as it continues to mature. ⑤ The secondary follicle with its forming antrum. ⑥ The mature vesicular follicle, ready to be ovulated. Meiosis I, producing the secondary oocyte and first polar body, occurs in the mature follicle. ⑦ The ruptured follicle and ovulated secondary oocyte surrounded by its corona radiata of granulosa cells. ⑧ The corpus luteum, formed from the ruptured follicle under the influence of LH, produces progesterone (and estrogens). ⑨ The scarlike corpus albicans.

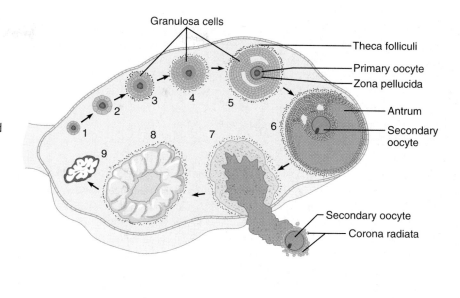

secondary oocyte and first polar body (see Figure 28.19). Once this has occurred (⑥ in Figure 28.20), the stage is set for ovulation. At this point, the granulosa cells send another important signal to the oocyte that says, in effect, "Wait, do not complete meiosis yet!"

Ovulation

Ovulation occurs when the ballooning ovary wall ruptures and expels the secondary oocyte (still surrounded by its *corona radiata*) into the peritoneal cavity ⑦. Some women experience a twinge of pain in the lower abdomen when ovulation occurs. This episode, called *mittelschmerz* (mit'el-shmārts: German for "middle pain"), is caused by the intense stretching of the ovarian wall during ovulation.

In the ovaries of adult females, there are always several follicles at different stages of maturation. As a rule, one follicle outstrips the others to become the *dominant follicle* and is at the peak stage of maturation when the hormonal (LH) stimulus is given for ovulation. How this follicle is selected, or selects itself, is still uncertain, but it is probably the one that attains the greatest FSH sensitivity the quickest. The others degenerate.

In 1–2% of all ovulations, more than one oocyte is ovulated. This phenomenon, which increases with age, can result in multiple births. Since different oocytes are fertilized by different sperm, the sib-

lings are *fraternal*, or nonidentical, twins. Identical twins result from the fertilization of a single oocyte by a single sperm, followed by separation of the fertilized egg's daughter cells, in early development.

The Luteal Phase

After ovulation, the ruptured follicle collapses, and the antrum fills with clotted blood. This *corpus hemorrhagicum* is eventually absorbed. The remaining granulosa cells increase in size and along with the internal thecal cells they form a new, quite different endocrine gland, the *corpus luteum* ("yellow body") (see Figure 28.20, step ⑧). Once formed, the corpus luteum begins to secrete progesterone and some estrogen. If pregnancy does not occur, the corpus luteum starts degenerating in about ten days and its hormonal output ends. In this case, all that ultimately remains is a scar ⑨ called the *corpus albicans* (al'bĭ-kans; "white body"). On the other hand, if the oocyte is fertilized and pregnancy ensues, the corpus luteum persists until the placenta is ready to take over its hormone-producing duties in about three months.

Hormonal Regulation of the Ovarian Cycle

Ovarian events are much more complicated than those occurring in the testes, but the hormonal controls set into motion at puberty are similar in the two sexes. Gonadotropin-releasing hormone (GnRH),

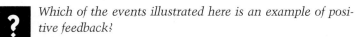

Which of the events illustrated here is an example of positive feedback?

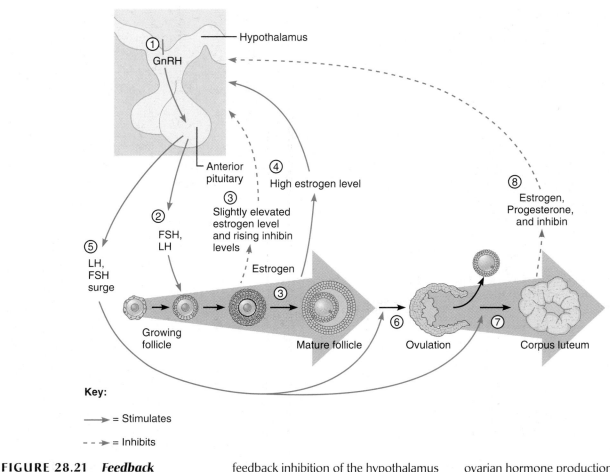

Key:

⟶ = Stimulates

- - -➤ = Inhibits

FIGURE 28.21   *Feedback interactions involved in the regulation of ovarian function.* Numbers refer to events listed in the text. Events that follow step 8 (negative feedback inhibition of the hypothalamus and anterior pituitary by progesterone and estrogens) are not depicted, but involve a gradual deterioration of the corpus luteum and, therefore, a decline in ovarian hormone production. Ovarian hormones reach their lowest levels in the blood around day 28.

the pituitary gonadotropins, and, in this case, ovarian estrogen and progesterone interact to produce the cyclic events occurring in the ovaries.

Establishing the Ovarian Cycle

During childhood, the ovaries grow and continuously secrete small amounts of estrogens which inhibit hypothalamic release of GnRH. But as puberty nears, the hypothalamus becomes less sensitive to estrogen and begins to release GnRH in a rhythmic pulselike manner. GnRH, in turn, stimulates the anterior pituitary to release FSH and LH, which act on the ovaries.

Gonadotropin levels continue to increase for about four years and, during this time, pubertal girls are still not ovulating and thus are incapable of getting pregnant. Eventually, the adult cyclic pattern is achieved, and hormonal interactions stabilize.

These events are heralded by the young woman's first menstrual period, referred to as **menarche** (mĕ-nar′ke; *men* = month, *arche* = first). Usually, it is not until the third year postmenarche that the cycles become regular and all are ovulatory.

Hormonal Interactions During the Ovarian Cycle

The waxing and waning of anterior pituitary gonadotropins (FSH and LH) and ovarian hormones and the positive and negative feedback interactions that regulate ovarian function are as described next. Notice that events 1–8 in the following discussion directly correspond to the same numbered steps in Figure 28.21. A 28-day cycle is assumed.

Stimulation of LH (and FSH) surges by rising estrogen levels, steps 4 and 5.

① On day 1 of the cycle, rising levels of GnRH from the hypothalamus stimulate increased production and release of follicle-stimulating hormone (FSH) and luteinizing hormone (LH) by the anterior pituitary.

② FSH and LH stimulate follicle growth and maturation and estrogen secretion. FSH exerts its main effects on the follicle cells, whereas LH (at least initially) more specifically targets the thecal cells. (Why only *some* follicles respond to these hormonal stimuli is still a mystery. However, there is little doubt that enhanced responsiveness is due to formation of more gonadotropin receptors.) As the follicles enlarge, estrogen secretion begins. LH prods the thecal cells to produce androgens. These diffuse through the basement membrane, where they are converted to estrogens by the FSH-primed granulosa cells. Only tiny amounts of ovarian androgens enter the blood, because they are almost completely converted to estrogens within the ovaries.

③ The rising estrogen levels in the plasma exert *negative feedback* on the anterior pituitary, inhibiting its release of FSH and LH, while simultaneously prodding it to synthesize and accumulate these gonadotropins. Within the ovary, estrogen increases estrogen output by intensifying the effect of FSH on follicle maturation. *Inhibin*, released by the granulosa cells, also exerts negative feedback controls on FSH release during this period.

④ The initial small rise in estrogen blood levels inhibits the hypothalamic-pituitary axis as just described, but high estrogen levels have the opposite effect. Once estrogen reaches a critical concentration in the blood, it exerts *positive feedback* on the brain and anterior pituitary.

⑤ High estrogen levels set a cascade of events into motion. There is a sudden burstlike release of accumulated LH (and, to a lesser extent, FSH) by the anterior pituitary. This occurs about midcycle (also see Figure 28.22a).

⑥ The sudden flush of LH stimulates the primary oocyte of the dominant follicle to complete the first meiotic division. The secondary oocyte that results continues on to metaphase II. LH also triggers ovulation at or around day 14. Perhaps LH induces the synthesis of proteolytic enzymes too, but whatever the mechanism, the blood stops flowing through the protruding part of the follicle wall. Within 5 minutes, that region of the follicle wall bulges out, thins, and then abruptly ruptures. The role (if any) of FSH in this process is unknown. Shortly after ovulation, estrogen levels decline. This probably reflects the damage to the dominant estrogen-secreting follicle during ovulation.

⑦ The LH surge also transforms the ruptured follicle into a corpus luteum (hence the name "luteinizing" hormone), and stimulates the newly formed endocrine gland to produce progesterone and estrogen almost immediately after it is formed.

⑧ As progesterone and estrogen levels rise in the blood, the combination exerts a powerful negative feedback (inhibitory) effect on anterior pituitary release of LH and FSH. Corpus luteum release of inhibin enhances this inhibitory effect. With gonadotropin decline, the development of new follicles is inhibited, and additional LH surges that might cause additional oocytes to be ovulated are prevented.

⑨ As LH blood levels decline, the stimulus for luteal activity ends, and the corpus luteum begins degenerating. As goes the corpus luteum, so go the levels of ovarian hormones, and blood estrogen and progesterone levels drop sharply. (However, if implantation of an embryo has occurred, the activity of the corpus luteum is maintained by an LH-like hormone released by the developing embryo.)

⑩ A marked decline in ovarian hormones at the end of the cycle (days 26–28) ends their blockade of FSH and LH secretion, and the cycle starts anew.

The Uterine (Menstrual) Cycle

Although the uterus is where the young embryo implants and develops, it is receptive to implantation only for a very short period each month. Not surprisingly, this brief interval is exactly the time when a developing embryo would normally begin implanting, about seven days after ovulation. The **uterine,** or **menstrual** (men′stroo-al), **cycle** is a series of cyclic changes that the uterine endometrium goes through each month as it responds to changing levels of ovarian hormones in the blood. These endometrial changes are coordinated with the phases of the ovarian cycle, which in turn are dictated by gonadotropins released by the anterior pituitary.

The events of the three-stage uterine cycle, depicted in Figure 28.22d, are as follows:

1. Days 1–5: Menstrual phase. In this phase, the uterus sheds all but the deepest part of its endometrium. (At the beginning of this stage, gonadotropins are beginning to rise a bit and ovarian hormones are at their lowest normal levels. Then FSH levels begin to rise.) The thick functional layer of the endometrium detaches from the uterine wall, a process that is accompanied by bleeding for 3–5 days. The detached tissue and blood pass out through the vagina as the menstrual flow. By day 5, the growing ovarian follicles are starting to produce more estrogen (see Figure 28.22b).

2. Days 6–14: Proliferative phase. In this phase, the endometrium rebuilds itself: Under the influence of rising blood levels of estrogen, the basal layer of the endometrium generates a new functional layer. As this new layer thickens, its glands enlarge and its spiral arteries increase in number (also see Figure 28.15). Consequently, the endometrium once

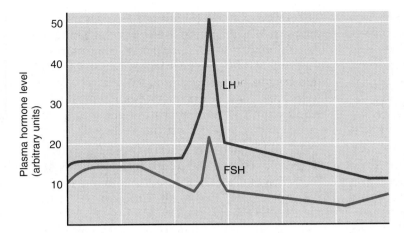

(a) Fluctuation of gonadotropin levels

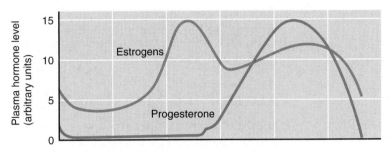

(b) Fluctuation of ovarian hormone levels

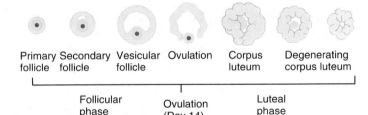

(c) Ovarian cycle

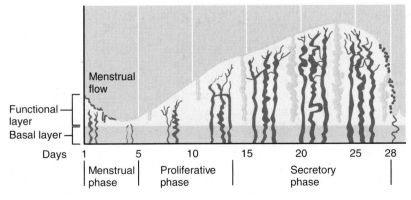

(d) Uterine cycle

FIGURE 28.22 *Correlation of anterior pituitary and ovarian hormones with structural changes of the ovarian and uterine cycles.* The time bar at the bottom of the figure, reading Days 1 to 28, applies to all four parts of this figure. **(a)** The fluctuating levels of pituitary gonadotropins in the blood (FSH = follicle-stimulating hormone; LH = luteinizing hormone). These hormones regulate the events of the ovarian cycle. **(b)** The fluctuating levels of ovarian hormones (estrogens and progesterone) that cause the endometrial changes of the uterine cycle. The high estrogen levels in (b) are also responsible for the LH/FSH surge in (a). **(c)** Structural changes in the ovarian follicles during the 28-day ovarian cycle are correlated with (d) changes in the endometrium of the uterus during the uterine cycle. Though not shown here, some evidence indicates that follicular development may actually *begin* on day 25 of the preceding cycle. **(d)** The three phases of the uterine cycle: menstrual, proliferative, and secretory. Basically, the first phase is a shedding and the second is a rebuilding of the endometrium. Both of these events occur before ovulation. The third phase, which begins immediately after ovulation, enriches the blood supply and provides the nutrients that prepare the endometrium to receive an embryo.

again becomes velvety, thick, and well vascularized. During this phase, estrogens also induce synthesis of progesterone receptors in the endometrial cells, readying them for interaction with progesterone.

Normally, the cervical mucus is thick and sticky, but rising estrogen levels cause it to thin and become crystalline, forming channels that facilitate the passage of sperm into the uterus. **Ovulation** occurs in the ovary at the end of this stage (day 14) in response to the sudden release of LH from the anterior pituitary. LH also converts the ruptured follicle to a corpus luteum.

3. Days 15–28: Secretory phase. In this phase, the endometrium prepares for implantation of an embryo. Rising levels of progesterone from the corpus luteum act on the estrogen-primed endometrium, causing the spiral arteries to elaborate and coil more tightly and converting the functional layer to a secretory mucosa. The uterine glands enlarge, coil, and begin secreting nutritious glycoproteins into the uterine cavity. These nutrients sustain the embryo until it has implanted in the blood-rich endometrial lining. All events of the secretory phase are promoted by progesterone.

Increasing progesterone levels also cause the cervical mucus to become viscous again, forming the *cervical plug,* which blocks sperm entry and plays an important role in keeping the uterus "private" in the event an embryo has begun to implant. Rising progesterone (and estrogen) levels inhibit LH release by the anterior pituitary. If fertilization has not occurred, the corpus luteum begins to degenerate toward the end of the secretory phase as LH blood levels decline. Progesterone levels fall, depriving the endometrium of hormonal support, and the spiral arteries kink and go into spasms. Denied oxygen and nutrients, the endometrial cells begin to die, and as their lysosomes rupture, the functional layer begins to self-digest, setting the stage for menstruation to begin on day 28. The spiral arteries constrict one final time and then suddenly relax and open wide. As blood gushes into the weakened capillary beds, they fragment, causing the functional layer to slough off. The menstrual cycle starts over again on this first day of menstrual flow.

Figure 28.22 also illustrates how the ovarian and uterine cycles fit together. Notice that the menstrual and proliferative phases overlap the follicular stage and ovulation in the ovarian cycle, and that the uterine secretory phase corresponds to the ovarian luteal phase.

Extremely strenuous activity can delay menarche in girls and can disrupt the normal menstrual cycle in adult women, even causing *amenorrhea* (a-men"o-re'ah), cessation of menses. Part of the problem appears to be that female athletes have little body fat, and fat deposits help convert adrenal androgens to estrogens. In addition, hypothalamic controls are blocked in some way by severe physical regimens. These effects are usually totally reversible when the athletic training is discontinued, but a worrisome consequence of amenorrhea in young, healthy adult women is that they suffer dramatic losses in bone mass normally seen only in old age. Once estrogen levels drop and the menstrual cycle stops (regardless of cause), bone loss begins. Currently, female athletes are encouraged to increase their daily calcium intake to 1.5 g, roughly the amount contained in a quart of milk.

Although menses has traditionally been viewed as a somewhat messy way of discarding a uterine lining "fattened" in anticipation of a baby that was never conceived, its adaptive value and expense to the female body in terms of tissue, blood, and nutrient (particularly iron) loss have recently been questioned. Why not just keep the prepared endometrium for the next cycle? A radical and controversial view of menses proposed by Maggie Profit of the University of California at Berkeley may give you something to think about and debate. The uterus is a hospitable receptacle for bacteria and viruses delivered via the male penis and piggy-backing in semen. This being so, she theorizes that menses is an aggressive way of cleaning house. Not only does bleeding rid the body of the uterine lining where pathogens are likely to linger, but menstrual blood is loaded with macrophages which provide active protection. Additionally, since menstrual blood lacks clotting factors, flushing the uterus out appears to be the goal rather than the by-product of bleeding.

Extrauterine Effects of Estrogens and Progesterone

With a name meaning "generators of sexual activity," estrogens are analogous to testosterone, the male steroid. As estrogen levels rise during puberty, they (1) promote oogenesis and follicle growth in the ovary and (2) exert anabolic effects on the female reproductive tract (see Table 28.1). Consequently, the uterine tubes, uterus, and vagina grow larger and become functional—ready to support a pregnancy. The uterine tubes and uterus exhibit enhanced motility; the vaginal mucosa thickens; and the external genitalia mature.

Estrogens also support the growth spurt at puberty that makes girls grow much more quickly than boys during the ages of 12 and 13. But this growth is short-lived because rising estrogen levels also cause earlier closure of the epiphyses of long bones, and females reach their full height between the ages of 15 and 17 years. In contrast, the aggressive growth of males continues until the age of 19 to 21 years.

The estrogen-induced secondary sex characteristics of females include (1) growth of the breasts; (2) increased deposit of subcutaneous fat, especially in the hips and breasts; (3) widening and lightening of the pelvis (adaptations for childbirth); (4) growth of axillary and pubic hair; and (5) several metabolic effects, including maintaining low total blood cholesterol levels (and high HDL levels) and facilitating calcium uptake, which helps sustain the density of the skeleton. (These metabolic effects, though initiated under estrogen's influence during puberty, are not true secondary sex characteristics.)

TABLE 28.1	Summary of Hormonal Effects of Gonadal Estrogens, Progesterone, and Testosterone		
Source, stimulus, effects	*Estrogens*	*Progesterone*	*Testosterone*
Major source	Ovary: developing follicles and corpus luteum	Ovary: mainly the corpus luteum	Testes: interstitial cells
Stimulus for release	FSH (and LH)	LH	ICSH and declining levels of inhibin produced by the sustentacular cells
Feedback effects exerted	Both negative and positive feedback exerted on anterior pituitary release of gonadotropins	Negative feedback exerted on anterior pituitary release of gonadotropins	Negative feedback suppresses release of ICSH by the anterior pituitary (and perhaps release of GnRH by the hypothalamus)
Effects on reproductive organs	Stimulate growth and maturation of reproductive organs and breasts at puberty and maintain their adult size and function. Promote the proliferative phase of the uterine cycle; stimulate production of watery (crystalline) cervical mucus and activity of fimbriae and uterine tube cilia. Promote oogenesis and ovulation by stimulating formation of FSH and LH receptors on follicle cells. Stimulate capacitation of sperm in the female reproductive tract by affecting vaginal and uterine secretions. During pregnancy stimulate growth of the uterus and enlargement of the external genitalia and mammary glands.	Cooperates with estrogen in stimulating growth of breasts and in regulating the uterine cycle (promotes the secretory phase of the uterine cycle); stimulates production of viscous cervical mucus. During pregnancy, quiets the myometrium and acts with estrogen to cause mammary glands to achieve their mature milk-producing state.	Stimulates formation of male reproductive ducts, glands, and external genitalia. Promotes descent of the testes. Stimulates growth and maturation of the internal and external genitalia at puberty; maintains their adult size and function. Required for normal spermatogenesis via effects promoted by its binding to ABP on spermatogenic cells; suppresses mammary gland development.
Promotion of secondary sex characteristics and somatic effects	Promote lengthening of long bones and feminization of the skeleton (particularly the pelvis); inhibit bone reabsorption and then stimulate epiphyseal closure; promote hydration of the skin; stimulate female pattern of fat deposit, and appearance of axillary and pubic hair. During pregnancy act with relaxin (placental hormone) to induce softening and relaxation of the pelvic ligaments and pubic symphysis.		Stimulates the growth spurt at puberty; promotes increased skeletal mass and then epiphyseal closure at the end of adolescence; promotes growth of the larynx and vocal cords and deepening of the voice; enhances sebum secretion and hair growth, especially on the face, axillae, genital region, and chest.
Metabolic effects	Generally anabolic; stimulate Na^+ reabsorption by the renal tubules, hence inhibit diuresis; enhance HDL (and reduce LDL) blood levels (cardiovascular sparing effect)	Promotes diuresis (antiestrogenic effect); increases body temperature	Generally anabolic; stimulates hematopoiesis; enhances the basal metabolic rate
Neural effects	Feminize the brain		Responsible for sex drive (libido) in both sexes; masculinizes the brain; promotes aggressiveness

Progesterone works with estrogen to establish and then help regulate the uterine cycle and promotes changes in cervical mucus (see Figure 28.22d). Its other effects are exhibited largely during pregnancy, when it inhibits the motility of the uterus and takes up where estrogen leaves off in preparing the breasts for lactation. Indeed, progesterone is named for these important roles (*pro* = for; *gestation* = pregnancy). However, the source of progesterone and estrogen during most of pregnancy is the placenta, not the ovaries.

Female Sexual Response

The **female sexual response** is similar to that of males in most respects. During sexual excitement, the clitoris, vaginal mucosa, and breasts engorge with blood; the nipples erect; and increased activity of the vestibular glands lubricates the vestibule and facilitates entry of the penis. These events, though more widespread, are analogous to the *erection* phase in men. Sexual excitement is promoted by touch and psychological stimuli and is mediated along the same autonomic nerve pathways as in males.

The final phase of the female sexual response, *orgasm*, is not accompanied by ejaculation, but muscle tension increases throughout the body, pulse rate and blood pressure rise, and the uterus begins to contract rhythmically. As in males, orgasm is accompanied by a sensation of intense pleasure and followed by relaxation. Orgasm in females is not followed by a refractory period; consequently, females may experience multiple orgasms during a single sexual experience. A man must achieve orgasm and ejaculate if fertilization is to occur, but female orgasm is not required for conception. Indeed, some women never experience orgasm, yet are perfectly able to conceive.

SEXUALLY TRANSMITTED DISEASES

Sexually transmitted diseases (STDs), also called *venereal diseases (VDs)*, are infectious diseases spread through sexual contact. The United States has the highest rates of infection among developed countries. Over 12 million people in the United States, a quarter of them adolescents, get STDs each year.

As a group, STDs are the single most important cause of reproductive disorders. Until recently, the bacterial infections gonorrhea and syphilis were the most common STDs but, within the past decade, viral diseases such as genital herpes and AIDS (mostly transmitted during sexual intercourse) have taken center stage. AIDS, caused by HIV, the virus that cripples the immune system, is described in Chapter 22. The other important bacterial and viral STDs are described here. Condoms are effective in helping to prevent the spread of STDs, and their use is strongly urged particularly since the advent of AIDS.

Gonorrhea

The causative agent of **gonorrhea** (gon"o-re'ah) is *Neisseria gonorrhoeae*, which invades the mucosae of the reproductive and urinary tracts. These bacteria are spread by contact with genital, anal, and pharyngeal mucosal surfaces. Most cases occur in adolescents and young adults.

Commonly called "the clap," the most frequent symptom of gonorrhea in males is *urethritis*, accompanied by painful urination and discharge of pus from the penis (penile "drip"). Symptoms vary in women, ranging from none (about 20% of cases) to abdominal discomfort, vaginal discharge, abnormal uterine bleeding, and occasionally, urethral symptoms similar to those seen in males.

Untreated gonorrhea can cause urethral constriction and inflammation of the entire male duct system. In women, it causes pelvic inflammatory disease and sterility. These consequences declined with the advent in the 1950s of penicillin, tetracycline, and certain other antibiotics. However, strains resistant to those antibiotics are becoming increasingly prevalent.

Syphilis

Syphilis (sif'ĭ-lis), caused by *Treponema pallidum*, a corkscrew-shaped bacterium, is usually transmitted sexually. However, it can be contracted congenitally from an infected mother. Indeed, an increasing number of women trading sex for "crack" (a cocaine derivative) are passing syphilis on to their fetuses. Fetuses infected with syphilis are usually stillborn or die shortly after birth. The bacterium easily penetrates intact mucosae and abraded skin and enters the local lymphatics and the bloodstream. Within a few hours of exposure, an asymptomatic bodywide infection is in progress. The incubation period is typically two to three weeks, at the end of which a red, painless primary lesion called a *chancre* (shang'ker) appears at the site of bacterial invasion. In males, this is typically the penis, but in females the lesion often goes undetected within the vagina or on the cervix. The primary lesion ulcerates and becomes crusty; then it heals spontaneously and disappears after one to a few weeks.

If syphilis is untreated, its secondary signs appear several weeks later. A pink skin rash all over the body is one of the first symptoms. Fever and joint pain are common. Anemia may develop, and the hair may fall out in clumps. These signs and symptoms disappear spontaneously in three to twelve weeks. Then the disease enters the *latent period* and is detectable only by a blood test. The latent stage may last a person's lifetime (or the bacteria may be killed by the immune system), or it may be followed by the signs of *tertiary syphilis*. Tertiary syphilis is characterized by *gummas* (gum'ahs), destructive lesions of the CNS, blood vessels, bones, and skin. Penicillin, which interferes with the ability of dividing bacteria to synthesize new cell walls, is still the treatment of choice for all stages of syphilis.

Chlamydia

Chlamydia (klah-mid'e-ah; *chlamys* = cloak) is a largely unrecognized, silent epidemic that infects perhaps 4 to 5 million people yearly, making it the most common sexually transmitted disease in the United States. Chlamydia is responsible for 25–50% of all diagnosed cases of pelvic inflammatory disease (and, consequently, at least a 1 in 4 chance of ectopic pregnancy), and each year more than 150,000 infants are born to infected mothers. About 20% of men and 30% of women infected with gonorrhea are also infected by *Chlamydia trachomatis*, the causative agent of chlamydia.

Chlamydia is a bacterium with a viruslike dependence on host cells. Its incubation period within the body cells is about one week. Symptoms include urethritis (involving painful, frequent urination and a thick penile discharge); vaginal discharge; abdominal, rectal, or testicular pain; painful intercourse; and irregular menses. In men, it can cause arthritis as well as widespread urogenital tract infection. In women, 80% of whom suffer *no* symptoms from the infection, its most severe consequence is sterility. Newborns infected in the birth canal tend to develop conjunctivitis and respiratory tract inflammations including pneumonia. The disease can be diagnosed by cell culture techniques and is easily treated with tetracycline.

Genital Warts

The *human papillomavirus (HPV)*—actually a group of about 60 viruses—is responsible for the sexual transmission of **genital warts** (Figure 28.23). About a million Americans develop genital warts each year, and it appears that HPV infection increases the risk for certain cancers (penile, vaginal, cervical, and anal). Indeed, the virus is linked to 80% of all cases of invasive cervical cancer.

Treatment is difficult and controversial. Some prefer to leave the warts untreated unless they become very widespread, whereas other clinicians rec-

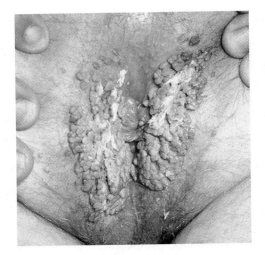

FIGURE 28.23   *Genital warts on the vulva.*

ommend their removal by cryosurgery or laser therapy and/or treatment with alpha interferon.

Genital Herpes

Human herpesviruses (*herpes simplex, Epstein-Barr virus*, and others) are among the most difficult human pathogens to control. They remain silent for weeks or years and then suddenly flare up, causing a burst of blisterlike lesions. Direct transmission of herpes simplex virus type 2, the common cause of **genital herpes** (her'pēz), is via infectious secretions. The resulting painful lesions that appear on the reproductive organs of infected adults are usually more of a nuisance than a threat to life. However, congenital herpes infections can cause severe malformations of a fetus, and herpes infections have been implicated as a cause of cervical cancer. Most people who have genital herpes do not know it, and it has been estimated that one-quarter to one-half of all adult Americans harbor the type 2 herpes simplex virus. The antiviral *acyclovir*, which speeds healing of the lesions and reduces the frequency of flare-ups, is the drug of choice for treatment. Inter Vir-A, an antiviral ointment, provides some relief from the itching and pain that accompany the lesions.

DEVELOPMENTAL ASPECTS OF THE REPRODUCTIVE SYSTEM: CHRONOLOGY OF SEXUAL DEVELOPMENT

Thus far, we have described the reproductive organs as they exist and operate in adults. Now we are ready to look at events that cause us to become reproductive individuals. These events begin long before birth and end, at least in women, in late middle age.

Embryological and Fetal Events

Determination of Genetic Sex

Aristotle believed that the "heat" of lovemaking determined maleness. Not so! Genetic sex is determined at the instant the genes of a sperm combine with those of an ovum, and the determining factor is the **sex chromosomes** each gamete contains. Of the 46 chromosomes in the fertilized egg, two (one pair) are sex chromosomes. The other 44 are called **autosomes.** Two types of sex chromosomes, quite different in size, exist in humans: the large **X chromosome** and the much smaller **Y chromosome.** The body cells of females have two X chromosomes and are designated XX; the ovum resulting from normal meiosis in a female always contains an X chromosome. Males have one X chromosome and one Y in each body cell (XY). Approximately half of the sperm produced by normal meiosis in males contain an X and the other half a Y. If the fertilizing sperm delivers an X chromosome, the fertilized egg and its daughter cells will contain the female (XX) composition, and the embryo will develop ovaries. If the sperm bears the Y, the offspring will be male (XY) and will develop testes. A single gene on the Y chromosome—the *SRY* gene—initiates testes development and hence determines maleness. Thus, the father determines the genetic sex of the offspring, and all subsequent events of sexual differentiation depend on which gonads are formed during embryonic life.

Homeostatic Imbalance

When meiosis distributes the sex chromosomes to the gametes improperly, an event called **nondisjunction,** abnormal combinations of sex chromosomes occur in the zygote and cause striking abnormalities in sexual and reproductive system development. For example, females with a single X chromosome (XO), a condition called *Turner's syndrome,* never develop ovaries. Males with no X chromosome (YO) die during embryonic development. Although most XXX females have normal intelligence, females carrying four or more X chromosomes have underdeveloped ovaries and limited fertility and as a rule are mentally retarded. *Klinefelter's syndrome,* which affects one out of 500 live male births, is the most common sex chromosome abnormality. Affected individuals usually have a single Y chromosome, two or more X chromosomes, and are sterile males. Although XXY males are normal (or only slightly below normal) intellectually, the incidence of mental retardation increases as the number of X sex chromosomes rises. Human males with an extra Y chromosome (XYY) appear to be normal healthy males but may be a little taller than average. ■

Sexual Differentiation of the Reproductive System

The gonads of both males and females begin their development during week 5 of gestation as masses of mesoderm called the **gonadal ridges** (Figure 28.24). The gonadal ridges form bulges on the dorsal abdominal wall just medial to the mesonephros (a transient kidney system, see p. 1034). The **paramesonephric,** or **Müllerian, ducts** (future female ducts) develop lateral to the **mesonephric (Wolffian) ducts** (future male ducts), and both sets of ducts empty into a common chamber called the *cloaca.* At this stage of development, the embryo is said to be in the **sexually indifferent stage,** because the gonadal ridge tissue can develop into either male or female gonads and both duct systems are present.

Shortly after the gonadal ridges appear, **primordial germ cells** migrate to them from an embryonic structure called the *yolk sac* (see Figure 29.7) and seed the developing gonads with stem cells destined to become spermatogonia or oogonia. Once these cells are in residence, the gonadal ridges differentiate into testes or ovaries, depending on the genetic makeup of the embryo. The process begins in week 7 in male (XY) embryos. Seminiferous tubules form in the internal part of the gonadal ridges and join the mesonephric duct via the efferent ductules. Further development of the mesonephric duct produces the duct system of the male. The paramesonephric ducts play no part in male development and soon degenerate unless the tiny testes fail to secrete a hormone called *anti-Müllerian hormone (AMH).* (AMH causes the breakdown of the paramesonephric [Müllerian] ducts, which give rise to the female duct system [oviducts and uterus].)

In female embryos (XX), the process begins slightly later, at about week 8. The outer, or cortical, part of the immature ovaries forms follicles. Shortly thereafter, the paramesonephric ducts differentiate into the structures of the female duct system, and the mesonephric ducts degenerate.

Like the gonads, the external genitalia arise from the same structures in both sexes (Figure 28.25). During the indifferent stage, all embryos exhibit a small projection called the **genital tubercle** on their external body surface. The urogenital sinus, which develops from subdivision of the cloaca (future urethra and bladder), lies deep to the tubercle. The **urethral groove,** which serves as the external opening of the urogenital sinus, is on the tubercle's inferior surface. The urethral groove is flanked laterally by the **urethral folds** and then the **labioscrotal swellings.**

During week 8, the external genitalia begin to develop rapidly. In males, the genital tubercle enlarges, forming the penis. The urethral folds fuse in the midline, forming the spongy urethra in the

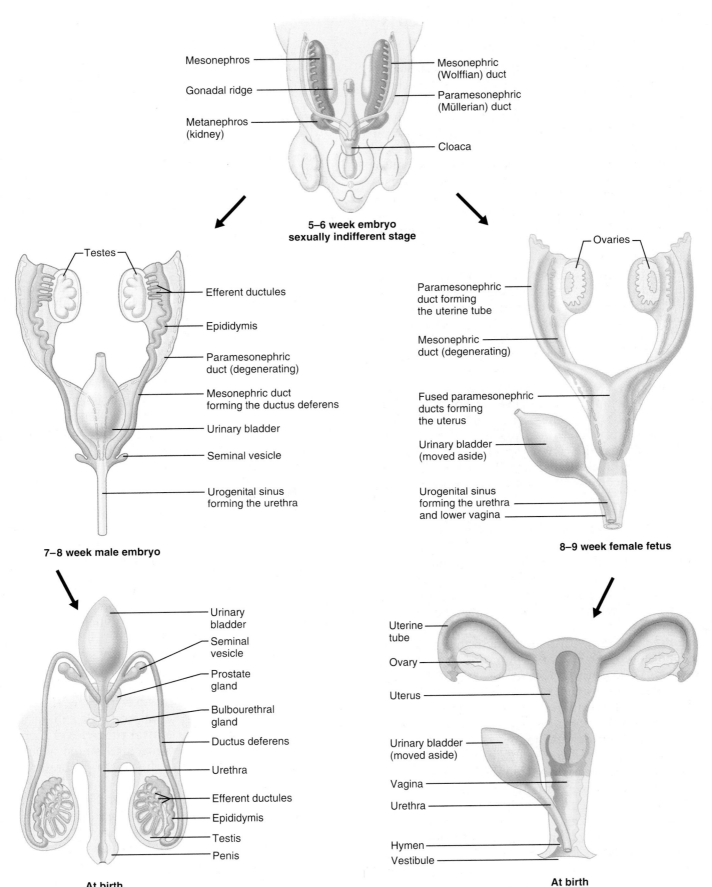

Mesonephros

Gonadal ridge

Metanephros
(kidney)

Mesonephric
(Wolffian) duct

Paramesonephric
(Müllerian) duct

Cloaca

**5–6 week embryo
sexually indifferent stage**

Testes

Efferent ductules

Epididymis

Paramesonephric
duct (degenerating)

Mesonephric duct
forming the ductus deferens

Urinary bladder

Seminal vesicle

Urogenital sinus
forming the urethra

7–8 week male embryo

Ovaries

Paramesonephric
duct forming
the uterine tube

Mesonephric
duct (degenerating)

Fused paramesonephric
ducts forming
the uterus

Urinary bladder
(moved aside)

Urogenital sinus
forming the urethra
and lower vagina

8–9 week female fetus

Urinary
bladder

Seminal
vesicle

Prostate
gland

Bulbourethral
gland

Ductus deferens

Urethra

Efferent ductules

Epididymis

Testis

Penis

**At birth
male development**

Uterine
tube

Ovary

Uterus

Urinary bladder
(moved aside)

Vagina

Urethra

Hymen

Vestibule

**At birth
female development**

FIGURE 28.24   *Development of the internal reproductive organs.*

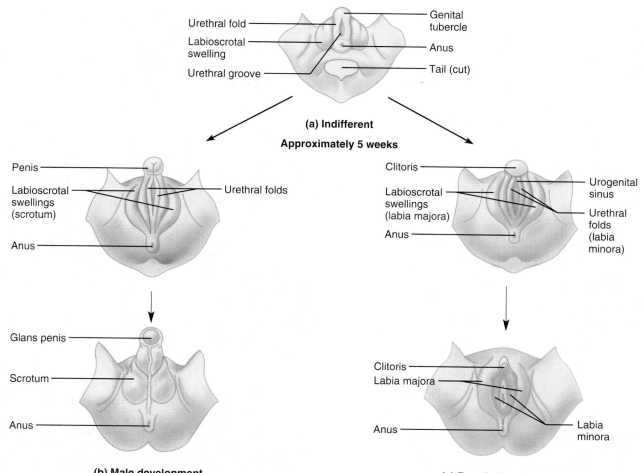

FIGURE 28.25 **Development of external genitalia in humans.** (a) The external reproductive organs are undifferentiated until about the eighth week of gestation. During the indifferent stage, all embryos have a conical elevation called the genital tubercle, which has a ventral opening called the urethral groove. The groove is surrounded by urethral folds, which in turn are bounded by the labioscrotal swellings. (b) In males, under the influence of testosterone the genital tubercle enlarges, forming the shaft of the penis, the urethral groove elongates and closes completely, and the urethral folds give rise to the spongy (penile) urethra. The labioscrotal swellings develop into the scrotum. (c) In females (or in the absence of testosterone), the genital tubercle gives rise to the clitoris, the urethral groove remains open as the vestibule, and the urethral folds become the labia minora. The labioscrotal swellings develop into the labia majora.

penis. Only the tips of the folds remain unfused to form the *urethral orifice* at the tip of the penis. The labioscrotal swellings also fuse in the body midline to form the scrotum. In females, the genital tubercle gives rise to the clitoris and the urethral groove persists as the vestibule. The unfused urethral folds become the labia minora, and the unfused labioscrotal folds become the labia majora.

Differentiation of accessory structures and the external genitalia as male or female structures is dependent on the presence or absence of testosterone. When testes are formed, they quickly begin to release testosterone, which causes the development of male accessory ducts and external genitalia. In the absence of testosterone, the female ducts and external genitalia develop.

Homeostatic Imbalance

Any interference with the normal pattern of sex hormone production in the embryo results in bizarre abnormalities. For example, if the embryonic testes do not produce testosterone, a genetic male develops the female accessory structures and external genitalia. If the testes fail to produce AMH, both the female and male duct systems form, but the external genitalia are those of the male. On the other hand, if a genetic female is exposed to testosterone (as might happen if the mother has an androgen-producing tumor of her adrenal gland), the embryo has ovaries but develops the male ducts and glands, as well as a penis and an empty scrotum. It appears that the female pattern of reproductive structures has an intrinsic ability to develop (is the default condition),

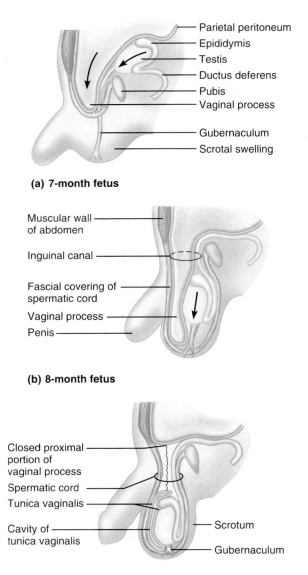

(a) 7-month fetus

- Parietal peritoneum
- Epididymis
- Testis
- Ductus deferens
- Pubis
- Vaginal process
- Gubernaculum
- Scrotal swelling

(b) 8-month fetus

- Muscular wall of abdomen
- Inguinal canal
- Fascial covering of spermatic cord
- Vaginal process
- Penis

(c) 1-month old infant

- Closed proximal portion of vaginal process
- Spermatic cord
- Tunica vaginalis
- Cavity of tunica vaginalis
- Scrotum
- Gubernaculum

FIGURE 28.26   Descent of the testes. (a) A testis begins to descend behind the peritoneum about two months before birth, and the vaginal process (a peritoneal extension) protrudes into the tissue that will become the scrotum (scrotal swellings). **(b)** By 8 months, the testis, accompanied by the vaginal process, has descended into the scrotum via the inguinal canal. **(c)** In the one-month-old infant, descent is complete. The lower part of the vaginal process becomes the tunica vaginalis covering of the testis. The proximal portion of the vaginal process is obliterated.

and in the absence of testosterone it proceeds to do so, regardless of the embryo's genetic makeup.

Individuals with accessory reproductive structures that do not "match" their gonads are called *pseudohermaphrodites* (soo-do-her-maf'ro-dīts). (True *hermaphrodites* are rare and possess both ovarian and testicular tissue.) Many pseudohermaphrodites have sought sex-change operations to match their outer selves (external genitalia) with their inner selves (gonads). ■

Descent of the Gonads

About two months before birth, the testes begin their descent toward the scrotum, dragging their supplying blood vessels and nerves along behind them (Figure 28.26). They finally exit from the pelvic cavity via the inguinal canals, slanting passages through the oblique muscles of the inferior abdominal wall, and enter the scrotum. This migration is stimulated by testosterone made by the male fetus's testes, and is guided mechanically by a strong fibrous cord called the **gubernaculum** ("governor"), which extends from the testis to the floor of the scrotal sac. The *tunica vaginalis* covering of the testis that is pulled into the scrotum with the testis is derived from a fingerlike outpocketing of the parietal peritoneum, the *vaginal process.* The accompanying blood vessels, nerves, and fascial layers form part of the *spermatic cord,* which helps suspend the testis within the scrotum.

Like the testes, the ovaries descend during fetal development, but in this case only to the pelvic brim, where their progress is stopped by the tentlike broad ligament. Each ovary is guided in its descent by a gubernaculum (anchored in the labium majus) that later divides, becoming the ovarian and round ligaments that help support the internal genitalia in the pelvis.

◭ *Homeostatic Imbalance*

Failure of the testes to make their normal descent leads to *cryptorchidism* (*crypt* = hidden, concealed; *orchi* = testicle). Because this condition causes sterility and increases the risk of testicular cancer, surgery is usually performed during early childhood to rectify this problem. ■

Puberty

Puberty is the period of life, generally between the ages of 10 and 15 years, when the reproductive organs grow to their adult size and become functional. These changes occur in response to rising levels of gonadal hormones (testosterone in males and estrogen in females). The secondary sex characteristics and regulatory events of puberty were described earlier. But it is important to remember that puberty represents the earliest time that reproduction is possible.

The events of puberty occur in the same sequence in all individuals, but the age at which they occur varies widely. In males, the event that signals puberty's onset is enlargement of the testes and scrotum between the ages of 8 and 14, followed by the appearance of pubic, axillary, and facial hair. Growth of the penis goes on over the next two years, and sexual maturation is evidenced by the presence of mature sperm in the semen. In the meantime, the young man has unexpected, and often embarrassing, erections and frequent nocturnal emissions ("wet dreams") as his hormones surge and the hormonal control axis struggles to achieve a normal balance.

The first sign of puberty in females is budding breasts, apparent between the ages of 8 and 13 years. Menarche usually occurs about two years later. Dependable ovulation and fertility await the maturation of the hormonal controls, which takes nearly two more years.

Menopause

Most woman reach the peak of their reproductive abilities in their late 20s. After that, ovarian function declines gradually, presumably because the ovaries become less and less responsive to gonadotropin signals. At the age of 30, there are still some 100,000 oocytes in the ovaries; by the age of 50, there are probably 3 eggs left. As estrogen production declines, many ovarian cycles become anovulatory, while in others 2 to 4 oocytes per month are ovulated, a sign of declining control. These increasingly frequent multiple ovulations explain why twins and triplets are more common in women who have deferred childbearing until their 30s. Menstrual periods become erratic and increasingly shorter in length. Eventually, ovulation and menses cease entirely. This normally occurs between the ages of 46 and 54 years, an event called menopause. **Menopause** is considered to have occurred when a whole year has passed without menstruation.

Although ovarian estrogen production continues for a while after menopause, the ovaries finally become nonfunctional as endocrine organs. Without sufficient estrogen (some is made by fat cells after menopause), the reproductive organs and breasts begin to atrophy. The vagina becomes dry; intercourse may become painful (particularly if infrequent); and vaginal infections become increasingly common. Other sequels due to lack of estrogen include irritability and depression (in some); intense vasodilation of the skin's blood vessels, which causes uncomfortable sweat-drenching "hot flashes"; gradual thinning of the skin and loss of bone mass. Slowly rising total blood cholesterol levels and falling HDL levels place postmenopausal women at risk for cardiovascular disorders. Some physicians prescribe low-dose estrogen-progesterone (e.g., Premarin and Provera respectively) preparations to help women through this often difficult period and to prevent the skeletal and cardiovascular complications. However, there is still controversy about whether the estrogen component increases the risk of breast cancer in postmenopausal women. Another "bonus" of postmenopausal estrogen is that it boosts short-term memory and appears to at least partially protect against Alzheimer's disease, a degenerative disorder of the CNS.

Although some women's groups vocally urge rejection of hormone replacement therapy (HRT) because it "interferes with the natural events of a woman's life," it is important to recognize that women are living longer after their estrogen-synthesizing reproductive years than ever before. Consequently, the time span for falling prey to cardiovascular disorders and Alzheimer's disease, and during which osteoporosis can take its toll and cause pathological fractures of the spine and hip, is ever lengthening. Women must make informed choices about their options in this matter.

There is no equivalent of menopause in males. While aging men do exhibit a steady decline in testosterone secretion and a longer latent period after orgasm, a male's reproductive capability seems unending. Healthy men are able to father offspring well into their eighth decade of life. However, there is a noticeable difference in sperm motility with aging. Sperm of a young man can make it up the uterine tubes in 20 to 50 minutes, whereas those of a 75-year-old take 2½ days for the same trip.

* * *

The reproductive system is unique among organ systems in at least two ways: (1) It is nonfunctional during the first 10 to 15 years of life, and (2) it is capable of interacting with the complementary system of another person—indeed, it *must* do so to carry out its biological function. The events of this complex interaction culminate in pregnancy and birth. To be sure, having a baby is not always what the interacting partners have in mind, and we humans have devised a variety of techniques for preventing this outcome (see *A Closer Look* on p. 1110).

The major goal of the reproductive system is ensuring the healthy function of its own organs so that conditions are optimal for producing offspring. However, as illustrated in *Making Connections*, gonadal hormones do influence other body organs, and the reproductive system depends on other body systems for oxygen and nutrients and to carry away and dispose of its wastes.

Now that we know how the reproductive system functions to prepare itself for childbearing, we are ready to consider the events of pregnancy and prenatal development of a new living being, the topics of Chapter 29.

In a society such as ours, where many women opt for professional careers or must work for economic reasons, **contraception** (*contra* = against; *cept* = taking), or birth control, is often seen as a necessity. Although scientists across the country are making some headway in the search for a male contraceptive, and an extremely effective antisperm vaccine exists, thus far the burden for birth control has fallen on women's shoulders, and most birth control products are female directed.

The key to birth control is dependability. As shown by the red arrows in the accompanying flow-chart, the birth control techniques and products currently available have many sites of action for blocking reproduction. Let's examine the relative advantages of a few of these more closely.

The most-used contraceptive product in the United States is the *birth control pill*, or simply "the pill," supplied in 28-tablet packets. The first 20 or 21 tablets contain minute amounts of estrogens and progestins (progesterone-like hormones) taken daily; the last seven tablets are hormone free. The pill tricks the hypothalamic-pituitary axis and "lulls it to sleep," because the relatively constant blood levels of ovarian hormones make it appear that the woman is pregnant (both estrogen and progesterone are produced throughout pregnancy). Ovarian follicles do not develop and ovulation ceases. The endometrium does proliferate slightly and is sloughed off monthly when the pill is discontinued, but menstrual flow is much reduced. However, since hormonal balance in the body is one of the

most precisely controlled of all body functions, some women simply cannot tolerate the changes caused by the pill—they become nauseated and/or hypertensive. For a while, the pill was suspected of increasing the incidence of breast and uterine cancer, but it now appears that the new very low dose preparations may actually help protect against ovarian and endometrial cancer. (The argument is less clear when it comes to breast cancer.) The incidence of serious cardiovascular side effects, such as strokes, heart attacks, and blood clots, that occurred (rarely) with earlier forms of the pill has also been reduced. Presently, well over 50 million women use the pill to prevent pregnancy. Its incidence of failure is less than 6 pregnancies per 100 women per year.

Many combination birth control pills (marketed under various names including Ovral, Tetragynon, and Triphasil) are also being used for *postcoital contraception*, a use now approved by the FDA. In common use in Europe and Canada, widely prescribed on our college campuses, and the therapy of choice for rape victims in U.S. emergency rooms, these *morning-after pills (MAPs)* were one of the best-kept secrets in the U.S.—most teenagers and women in their 30s and 40s had never heard of them. Taken within 72 hours of unprotected intercourse, the somewhat more concentrated estrogen-progesterone combination "messes up" the normal hormonal signals so much that fertilization is prevented altogether or a fertilized egg is prevented from implanting. Although clinical tests in the U.S. for *mifepristone (RU486)*, the so-called *abortion pill* developed in France, did not begin until 1994, it has been tested clinically for over a decade on thou-

sands of women in other parts of the world and has proven to have a 96–98% success rate with virtually no side effects. Mifepristone is an antihormone that, when taken during the first 7 weeks of pregnancy in conjunction with a tiny amount of prostaglandin to induce uterine contradictions, induces miscarriage by blocking progesterone's quieting effect on the uterus.

Other approaches to contraception include a device that is implanted just under the skin and injections of synthetic progesterone. The Norplant implant, six tiny silicone rods that release a progestin hormone over a five-year period, was approved late in 1990 and quickly gained advocates. Now in its second generation, its failure rate (0.05%) is even lower than that of the pill, but there have been some problems with rod removal and recently some hints of possible neurotoxicity. *Depo-Provera,* a synthetic progesterone, has had the longest gestation period of any contraceptive in U.S. history. Developed 35 years before, it was finally approved for use as an injectable contraceptive in October 1992. Administered in a 150-mg dose once every three months. Depo-Provera's failure rate is only 0.4%.

For several years, the second most used contraceptive method was the *intrauterine device (IUD)*, a plastic or metal device inserted into the uterus that prevented implantation of the fertilized egg in the endometrial lining. The failure rate of the IUD was nearly as low as that of the pill; however, IUDs were taken off the market in the United States because of occasional contraceptive failure, uterine perforation, or pelvic inflammatory disease. New IUD products that provide sustained local delivery of synthetic progesterone to the endometrium are particularly recommended for women who have given birth and are in monogamous relationships (i.e., who have a lower risk of developing PID).

Some methods, such as *tubal ligation* or *vasectomy* (cutting or cauterizing the uterine tubes or vas deferens, respectively) are nearly foolproof and are the choice of approximately 33% of couples of childbearing age in the United States. Both procedures can be done in the physician's office. However, these techniques are usually permanent, making them unpopular with individuals who still plan to have children but want to select the time.

Coitus interruptus, or withdrawal of the penis just before ejaculation, is simply against nature, and control of ejaculation is never assured. *Rhythm* or *fertility awareness methods* are based on recognizing the period of ovulation or fertility and avoiding intercourse during those intervals. This may be accomplished by (1) recording daily basal body temperatures (body temperature drops slightly immediately prior to ovulation and then rises slightly after ovulation) or (2) recording changes in the consistency of vaginal mucus (the mucus first become sticky and then clear and stringy, much like egg white, during the fertile period). Both of these rhythm techniques require accurate record keeping for several cycles before they can be used with confidence, but have a high success rate for those willing to take the time necessary. With a failure rate of 10–20%, it is obvious that some people are willing and some are not.

Barrier methods, such as diaphragms, cervical caps, condoms (male and female [vaginal pouch] versions), spermicidal foams, gels, and sponges, are quite effective, especially when used by both partners. But many avoid them because they

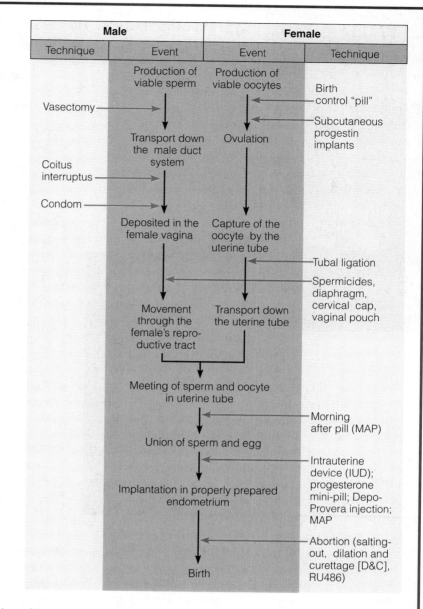

Flowchart of the events that must occur to produce a baby. Techniques or products that interfere with the process are indicated by colored arrows at the site of interference and act to prevent the next step from occurring.

can reduce the spontaneity of sexual encounters.

This summary doesn't even begin to touch on the large number of experimental birth control drugs now awaiting clinical trials, and

other methods are sure to be developed in the near future. In the final analysis, however, the only 100% effective means of birth control is the age-old one—*total abstinence.*

MAKING CONNECTIONS

SYSTEM CONNECTION: Homeostatic Interrelationships Between the Reproductive System and Other Body Systems

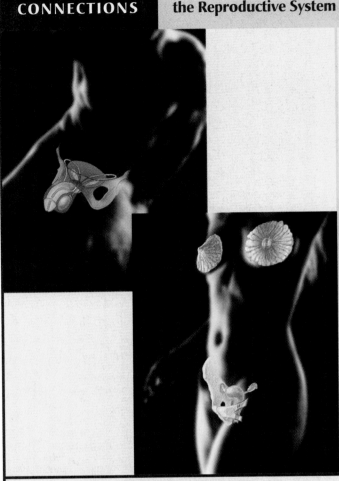

the pelvic floor support reproductive organs and aid erection of penis/clitoris

Nervous System

- Sex hormones masculinize or feminize the brain and influence sex drive
- Hypothalamus regulates timing of puberty; neural reflexes regulate events of sexual response

Endocrine System

- Gonadal hormones exert feedback effects on hypothalamic pituitary axis; placental hormones promote maternal hypermetabolism
- Gonadotropins (and GnRH) help regulate function of gonads

Cardiovascular System

- Estrogens lower blood cholesterol levels and promote cardiovascular health in premenopausal women; pregnancy increases workload of the cardiovascular system
- Cardiovascular system transports needed substances to organs of reproductive system; local vasodilation involved in erection; blood transports sex hormones

Lymphatic System/Immunity

- Developing embryo/fetus escapes immune surveillance (not rejected)
- Lymphatic vessels drain leaked tissue fluids; transport sex hormones; immune cells protect reproductive organs from disease; IgA is present in breast milk

Respiratory System

- Pregnancy impairs descent of the diaphragm, promotes dyspnea
- Respiratory system provides oxygen; disposes of carbon dioxide; vital capacity and respiratory rate increase during pregnancy

Digestive System

- Digestive organs crowded by developing fetus; heartburn, constipation common during pregnancy
- Digestive system provides nutrients needed for health

Urinary System

- Hypertrophy of the prostate gland inhibits urination; compression of bladder during pregnancy leads to urinary frequency and urgency
- Kidneys dispose of nitrogenous wastes and maintain acid-base balance of blood of mother and fetus; semen discharged through the urethra of the male

Integumentary System

- Androgens activate oil glands which lubricate skin and hair; gonadal hormones stimulate characteristic fat distribution and appearance of pubic and axillary hair; estrogen increases skin hydration; enhances facial skin pigmentation during pregnancy
- Skin protects all body organs by external enclosure; mammary gland secretions (milk) nourish the infant

Skeletal System

- Androgens masculinize the skeleton and increase bone density; estrogen feminizes the skeleton and maintains bone mass in females
- The bony pelvis encloses some reproductive organs; if narrow, the bony pelvis may hinder vaginal delivery of an infant

Muscular System

- Androgens promote increase in muscle mass
- Abdominal muscles active during childbirth; muscles of

CLOSER CONNECTIONS	THE REPRODUCTIVE SYSTEM and Interrelationships with the Endocrine, Skeletal, and Muscular Systems

The reproductive system seems to be a particularly "selfish" system—interested only in its own well-being. But when we reflect on the importance of reproduction and continuity in the scheme of things, that selfishness is a benefit, not a drawback. Indeed, the reproductive system ensures (or tries to) that a person's gene pool passes on and, rather than directing its main energy to other body systems, it builds a whole new organism. Quite an incredible feat, don't you think?

Though reproductive system interactions with other body systems are relatively meager, one—that with the endocrine system—stands out and will receive the bulk of our attention. Indeed, these two systems are difficult to divorce from one another, particularly since the gonads themselves function as endocrine organs. Most other reproductive system interactions, such as those with the nervous and cardiovascular systems, are mediated through its hormones. For example, testosterone and estrogen imprint the brain as male or female, respectively, and estrogens help stave off atherosclerosis and help keep a woman's cardiovascular system (particularly her blood vessels) youthful. The gonadal hormones also play major roles in somatic development particularly at puberty, and these are explored below.

Endocrine System

First of all, specific endocrine organs of the body—the hypothalamus via its gonadotropin-releasing hormones and the anterior pituitary gland via FSH and LH—help orchestrate virtually all reproductive functions. Without them, the gonads would never mature to adult functional status during adolescence, gametes would not be pro-

duced, and subsequent effects mediated by gonadal hormones would fail to occur. We would forever remain sexually infantile with a childlike reproductive system. On the other hand, if testosterone were not secreted by the tiny male fetal testes, the male reproductive tract would not be formed to begin with. Furthermore, gonadal hormones produced in response to gonadotropins also help to regulate gonadotropin release (via negative and positive feedback mechanisms), as well as that of GnRH. So what we have here is an interlocking cycle in which one part cannot function without the other.

Skeletal and Muscular Systems

Once the reproductive system gonads are "up and running," they in turn start producing hormones that not only cause maturation of their own system's accessory organs and help maintain gametogenesis, but also are crucial for the incredible growth spurt that occurs during puberty that converts the child's body to an adult's. The most important effects are the anabolic effects exerted on bone and skeletal muscle. The skeleton becomes taller, heavier, and denser, and in females its pelvis is shaped to accommodate birth. All through a woman's reproductive years, estrogen helps maintain a healthy bone mass. (Its lack is quickly obvious after menopause.) The skeletal muscles, too, increase in size and mass, and become capable of great strength. Testosterone in particular is the muscle builder. Although the gonadal hormones do not act alone, and their effects are superimposed on those of growth hormone and thyroid hormone, they certainly promote adult stature and are responsible for the musculoskeletal changes that distinguish the sexes anatomically.

CLINICAL CONNECTIONS

Reproductive System

Case study: We are back to look in on Mr. Heyden today. Since we last saw him (Chapter 27), he has been to radiology to try to determine the cause of his back pain, and his blood test results have come in. The note recorded from X ray indicates:

- X irradiation of skeleton displays numerous carcinomatous metastases in his skull and lumbar vertebrae

The hematology note of interest here reads:

- Serum acid phosphatase levels abnormally high

1. What is carcinoma? What do you suppose is the primary source of the secondary carcinoma lesions in his skull and spine?

2. On what basis did you come to this conclusion?

3. What other tests might be of some diagnostic help here?

4. What type of therapy do you predict Mr. Heyden will be given to treat his carcinoma? Why?

(Answers in Appendix F)

RELATED CLINICAL TERMS

Dysmenorrhea (dis"men-ŏ-re'ah; *dys* = bad; *meno* = menses, a month) Painful menstruation; may reflect abnormally high prostaglandin activity during menses.

Endometriosis (en"do-me"tre-o'sis; *endo* = within; *metrio* = the uterus, womb) An inflammatory condition in which endometrial tissue occurs and grows atypically in the pelvic cavity. Characterized by abnormal uterine or rectal bleeding, dysmenorrhea, and pelvic pain. May cause sterility.

Gynecology (gi"nĕ-kol'o-je; *gyneco* = woman; *ology* = study of) Specialized branch of medicine that deals with the diagnosis and treatment of female reproductive system disorders.

Gynecomastia (gi"nĕ-ko-mas'te-ah) Development of breast tissue in the male; a consequence of adrenal cortex hypersecretion of estrogens, certain drugs (cimetidine, spironolactone, and some chemotherapeutic agents), and marijuana use.

Hysterectomy (his"tĕ-rek'to-me; *hyster* = uterus; *ectomy* = cut out) Surgical removal of the uterus.

Inguinal hernia Protrusion of part of the intestine into the scrotum or through a separation in the abdominal muscles in the groin region. Since the inguinal canals represent weak points in the abdominal wall, inguinal hernia may be caused by heavy lifting or other activities that increase intra-abdominal pressure.

Laparoscopy (lap"ah-ros'ko-pe; *lapar* = the flank, *scopy* = observation) Examination of the abdominopelvic cavity with a laparoscope, a viewing device at the end of a thin tube inserted through the anterior abdominal wall. Laparoscopy is often used to assess the condition of the pelvic reproductive organs of females.

Oophorectomy (o"of-o-rek'to-me; *oophor* = ovary) Surgical removal of the ovary.

Orchitis (or-ki'tis; *orcho* = testis) Inflammation of the testes, sometimes caused by the mumps virus.

Ovarian cancer Malignancy of the ovary; the fourth most common reproductive system cancer; incidence increases with age with the greatest risk between the ages of 50 and 60. Since early symptoms mimic GI disorders (abdominal discomfort, bloating, and flatulence), medical assessment is often delayed until after metastasis has occurred; five-year survival rate is 90% if the condition is diagnosed before metastasis.

Ovarian cysts The most common disorders of the ovary; some are tumors. Types include (1) *simple follicle retention cysts* in which single or clustered follicles become enlarged with a clear fluid; (2) *dermoid cysts*, which are filled with a thick yellow fluid and contain partially developed tissues of hair, teeth, bone, etc.; and (3) *chocolate cysts* filled with dark gelatinous material, which are the result of endometriosis of the ovary. None of these is malignant, but the latter two may become so.

Salpingitis (sal"pin-ji'tis; *salpingo* = uterine tube) Inflammation of the uterine tubes.

CHAPTER SUMMARY

Media study tools that could provide you additional help in reviewing specific key topics of Chapter 28 are referenced below. **SP** = *Study Partner;* **IP** = *Interactive Physiology.*

1. The function of the reproductive system is to produce offspring. The gonads produce gametes (sperm or ova) and sex hormones. All other reproductive organs are accessory organs.

Anatomy of the Male Reproductive System (pp. 1071–1076)

The Scrotum (p. 1072)

1. The scrotum contains the testes. It provides a temperature slightly lower than that of the body, as required for viable sperm production.

The Testes (pp. 1072–1074)

2. Each testis is covered externally by a tunica albuginea that extends internally to divide the testis into many lobules. Each lobule contains sperm-producing seminiferous tubules and interstitial cells that produce androgens.

The Penis (p. 1074)

3. The penis, the male copulatory organ, is largely erectile tissue (corpus spongiosum and corpora cavernosa). Engorgement of the erectile tissue with blood causes the penis to become rigid, an event called erection.

4. The male perineum is the region encompassed by the pubic symphysis, ischial tuberosities, and coccyx.

SP Exercise: Chapter 28, Internal Structure of the Testis.

The Male Duct System (pp. 1074–1075)

5. The epididymis hugs the external surface of the testis and serves as a site for sperm maturation and storage.

6. The ductus (vas) deferens, extending from the epididymis to the urethra, propels sperm into the urethra by peristalsis during ejaculation. Its terminus fuses with the duct of the seminal vesicle, forming the ejaculatory duct.

7. The urethra extends from the urinary bladder to the tip of the penis. It conducts semen and urine to the body exterior.

Accessory Glands (pp. 1075–1076)

8. The accessory glands produce the bulk of the semen, which contains fructose from the seminal vesicles, activating fluid from the prostate gland, and mucus from the bulbourethral glands.

Semen (p. 1076)

9. Semen is an alkaline fluid that dilutes and transports sperm. Important chemicals in semen are nutrients, prostaglandins, and seminalplasmin. An ejaculation contains 2–5 ml of semen, with 50 to 130 million sperm/ml in normal adult males.

Physiology of the Male Reproductive System (pp. 1076–1086)

Male Sexual Response (pp. 1076–1077)

1. Erection is controlled by parasympathetic reflexes.

2. Ejaculation is expulsion of semen from the male duct system, promoted by the sympathetic nervous system. Ejaculation is part of male orgasm, which also includes pleasurable sensations and increased pulse and blood pressure.

Spermatogenesis (pp. 1077–1084)

3. Spermatogenesis, production of male gametes in the seminiferous tubules, begins at puberty.

4. Meiosis, the basis of gamete production, consists of two consecutive nuclear divisions without DNA replication in between. Meiosis reduces the chromosomal number by half and introduces genetic variability. Events unique to meiosis include synapsis and crossover of homologous chromosomes.

5. Spermatogonia divide by mitosis to maintain the germ cell line. Some of their progeny become primary spermatocytes, which undergo meiosis I to produce secondary spermatocytes. Secondary spermatocytes undergo meiosis II, each producing a total of four haploid (n) spermatids.

6. Spermatids are converted to functional sperm by spermiogenesis, during which superfluous cytoplasm is stripped away and an acrosome and a flagellum (tail) are produced.

7. Sustentacular cells form the blood-testis barrier, nourish spermatogenic cells, move them toward the lumen of the tubules, and secrete fluid for sperm transport.

Hormonal Regulation of Male Reproductive Function (pp. 1084–1086)

8. GnRH, produced by the hypothalamus, stimulates the anterior pituitary gland to release FSH and LH (ICSH). FSH causes sustentacular cells to produce androgen-binding protein (ABP). LH stimulates interstitial cells to release testosterone, which binds to ABP, stimulating spermatogenesis. Testosterone and inhibin (produced by sustentacular cells) feed back to inhibit the hypothalamus and anterior pituitary.

9. Maturation of hormonal controls occurs during puberty and takes about three years.

10. Testosterone stimulates maturation of the male reproductive organs and triggers the development of the secondary sex characteristics of the male. It exerts anabolic effects on the skeleton and skeletal muscles, stimulates spermatogenesis, and is responsible for sex drive.

SP Exercise: Chapter 28, Hormonal Regulation of Testicular Function.

Anatomy of the Female Reproductive System (pp. 1086–1094)

1. The female reproductive system produces gametes and sex hormones and houses a developing infant until birth.

The Ovaries   (pp. 1087–1088)

2. The ovaries flank the uterus laterally and are held in position by the ovarian and suspensory ligaments and mesovaria.

3. Within each ovary are oocyte-containing follicles at different stages of development and corpora lutea.

The Female Duct System (pp. 1088–1092)

4. The uterine tube, supported by the mesosalpinx, extends from near the ovary to the uterus. Its fibriated and ciliated distal end creates currents that help move an ovulated oocyte into the uterine tube. Cilia of the uterine tube mucosa help propel the oocyte toward the uterus.

5. The uterus has fundus, body, and cervical regions. It is supported by the broad, lateral cervical, uterosacral, and round ligaments.

6. The uterine wall is composed of the outer perimetrium, the myometrium, and the inner endometrium. The endometrium consists of a functional layer (stratum functionalis), which sloughs off periodically unless an embryo has implanted, and an underlying basal layer (stratum basalis), which rebuilds the functional layer.

7. The vagina extends from the uterus to the exterior. It is the copulatory organ and allows passage of the menstrual flow and a baby.

The External Genitalia   (p. 1092)

8. The female external genitalia (vulva) include the mons pubis, labia majora and minora, clitoris, and the urethral and vaginal orifices. The labia majora house the mucus-secreting vestibular glands.

The Mammary Glands   (pp. 1093–1094)

9. The mammary glands lie over the pectoral muscles of the chest and are surrounded by adipose and fibrous connective tissue. Each mammary gland consists of many lobules, which contain milk-producing alveoli.

Physiology of the Female Reproductive System (pp. 1095–1103)

Oogenesis   (pp. 1095–1096)

1. Oogenesis, the production of eggs, begins in the fetus. Oogonia, the diploid stem cells of female gametes, are converted to primary oocytes before birth. The infant female's ovaries contain about 2 million primary oocytes arrested in prophase of meiosis I.

2. At puberty, meiosis resumes. Each month, one primary oocyte completes meiosis I, producing a large secondary oocyte and a tiny first polar body. Meiosis II of the secondary oocyte produces a functional ovum and a second polar body, but does not occur unless the secondary oocyte is penetrated by a sperm.

3. The ovum contains most of the primary oocyte's cytoplasm. The polar bodies are nonfunctional and degenerate.

The Ovarian Cycle   (pp. 1096–1097)

4. During the follicular phase (days 1–14), several primary follicles begin to mature. The follicle cells proliferate and produce estrogens, and a connective tissue capsule (theca) is formed around the maturing follicle. Generally, only one follicle per month completes the maturation process and protrudes from the ovarian surface. Late in this phase, the oocyte in the dominant follicle completes meiosis I. Ovulation occurs, usually on day 14, releasing the secondary oocyte into the peritoneal cavity. Other developing follicles deteriorate.

5. In the luteal phase (days 15–28), the ruptured follicle is converted to a corpus luteum, which produces progesterone and estrogen for the remainder of the cycle. If fertilization does not occur, the corpus luteum degenerates in about ten days.

Hormonal Regulation of the Ovarian Cycle (pp. 1097–1099)

6. Beginning at puberty, the hormones of the hypothalamus, anterior pituitary, and ovaries interact to establish and regulate the ovarian cycle. Establishment of the mature cyclic pattern, indicated by menarche, takes about four years.

7. The hormonal events of each ovarian cycle are as follows: (1) GnRH stimulates the anterior pituitary to release FSH and LH, which stimulate follicle maturation and estrogen production. (2) When blood estrogen reaches a certain level, positive feedback exerted on the hypothalamic-pituitary axis causes a sudden release of LH that stimulates the primary oocyte to continue meiosis and triggers ovulation. LH then causes conversion of the ruptured follicle to a corpus luteum and stimulates its secretory activity. (3) Rising levels of progesterone and estrogen inhibit the hypothalamic-pituitary axis, the corpus luteum deteriorates, ovarian hormones drop to their lowest levels, and the cycle begins anew.

The Uterine (Menstrual) Cycle   (pp. 1099–1101)

8. Varying levels of ovarian hormones in the blood trigger events of the uterine cycle.

9. During the menstrual phase of the uterine cycle (days 1–5), the functional layer sloughs off in menses. During the proliferative phase (days 6–14), rising estrogen levels stimulate its regeneration, making the uterus receptive to implantation about one week after ovulation. During the secretory phase (days 15–28), the uterine glands secrete glycogen, and endometrial vascularity increases further.

10. Falling levels of ovarian hormones during the last few days of the ovarian cycle cause the spiral arteries to become spastic and cut off the blood supply of the functional layer, and the uterine cycle begins again with menstruation.

SP Exercise: Chapter 28, The Female Menstrual Cycle.

Extrauterine Effects of Estrogens and Progesterone (pp. 1101–1103)

11. Estrogen promotes oogenesis. At puberty, it stimulates the growth of the reproductive organs and the growth spurt and promotes the appearance of the secondary sex characteristics.

12. Progesterone cooperates with estrogen in breast maturation and regulation of the uterine cycle.

Female Sexual Response   (p. 1103)

13. The female sexual response is similar to that of males. Orgasm in females is not accompanied by ejaculation and is not necessary for conception.

Sexually Transmitted Diseases   (pp. 1103–1104)

1. Sexually transmitted diseases (STDs) are infectious diseases spread via sexual contact. Gonorrhea, syphilis, chlamydia, and some forms of vaginitis are bacterial diseases; if untreated, they can cause sterility. Syphilis has broader consequences than most other sexually transmit-

ted bacterial diseases since it can infect organs throughout the body. Genital herpes and genital warts, viral infections, are implicated in cervical cancer. AIDS, a condition of immune suppression, can also be transmitted sexually.

Developmental Aspects of the Reproductive System: Chronology of Sexual Development (pp. 1104–1109)

SP Case Study: Sexually Transmitted Diseases.

Embryological and Fetal Events   (pp. 1105–1108)

1. Genetic sex is determined by the sex chromosomes: an X from the mother, an X or a Y from the father. If the fertilized egg contains XX, it is a female and develops ovaries; if it contains XY, it is a male and develops testes.

2. Gonads of both sexes arise from the mesodermal gonadal ridges. The mesonephric ducts produce the male accessory ducts and glands. The paramesonephric ducts produce the female duct system.

3. The external genitalia arise from the genital tubercle and associated structures. The development of male accessory structures and external genitalia depends on the presence of testosterone produced by the embryonic testes. In its absence, female structures develop.

4. The testes form in the abdominal cavity and descend into the scrotum.

Puberty   (pp. 1108–1109)

5. Puberty is the interval when reproductive organs mature and become functional. It begins in males with penile and scrotal growth and in females with breast development.

Menopause   (p. 1109)

6. During menopause, ovarian function declines, and ovulation and menstruation cease. Hot flashes and mood changes may occur. Postmenopausal events include atrophy of the reproductive organs, bone mass loss, and increasing risk for cardiovascular disease.

REVIEW QUESTIONS

Multiple Choice/Matching

1. The structures that draw an ovulated oocyte into the female duct system are (a) cilia, (b) fimbriae, (c) microvilli, (d) stereocilia.

2. The usual site of embryo implanation is (a) the uterine tube, (b) the peritoneal cavity, (c) the vagina, (d) the uterus.

3. The male homologue of the female clitoris is (a) the penis, (b) the scrotum, (c) the penile urethra, (d) the testis.

4. Which of the following is correct relative to female anatomy? (a) The vaginal orifice is the most dorsal of the three openings in the perineum, (b) the urethra is between the vaginal orifice and the anus, (c) the anus is between the vaginal orifice and the urethra, (d) the urethra is the more ventral of the two orifices in the vulva.

5. Secondary sex characteristics are (a) present in the embryo, (b) a result of male or female sex hormones increasing in amount at puberty, (c) the testis in the male and the ovary in the female, (d) not subject to withdrawal once established.

6. Which of the following produces the male sex hormones? (a) seminal vesicles, (b) corpus luteum, (c) developing follicles of the testes, (d) interstitial cells.

7. Which will occur as a result of nondescent of the testes? (a) male sex hormones will not be circulated in the body, (b) sperm will have no means of exit from the body, (c) inadequate blood supply will retard the development of the testes, (d) viable sperm will not be produced.

8. The normal diploid number of human chromosomes is (a) 48, (b) 47, (c) 46, (d) 23, (e) 24.

9. Relative to differences between mitosis and meiosis, choose the statements that apply *only* to events of meiosis. (a) tetrads present, (b) produces two daughter cells, (c) produces four daughter cells, (d) occurs throughout life, (e) reduces the chromosomal number by half, (f) synapsis and crossover of homologues occur.

10. Match the key choices with the descriptive phrases below.

Key: **(a)** Androgen-binding protein   **(e)** Inhibin
           (b) Estrogens   **(f)** LH
           (c) FSH   **(g)** Progesterone
           (d) GnRh   **(h)** Testosterone

_____, _____ **(1)** Hormones that directly regulate the ovarian cycle

_____, _____ **(2)** Chemicals in males that inhibit the pituitary-testicular axis

_____ **(3)** Hormone that makes the cervical mucus viscous

_____ **(4)** Potentiates the activity of testosterone on spermatogenic cells

_____, _____ **(5)** In females, exerts feedback inhibition on the hypothalamus and anterior pituitary

_____ **(6)** Stimulates the secretion of testosterone

11. The menstrual cycle can be divided into three continuous phases. Starting from the first day of the cycle, their consecutive order is (a) menstrual, proliferative, secretory, (b) menstrual, secretory, proliferative, (c) secretory, menstrual, proliferative, (d) proliferative, menstrual, secretory, (e) secretory, proliferative, menstrual.

12. Spermatozoa are to seminiferous tubules as oocytes are to (a) fimbriae, (b) corpus albicans, (c) ovarian follicles, (d) corpora lutea.

13. Which of the following does not add a secretion that makes a major contribution to semen? (a) prostate, (b) bulbourethral glands, (c) testes, (d) vas deferens.

14. The corpus luteum is formed at the site of (a) fertilization, (b) ovulation, (c) menstruation, (d) implantation.

15. The sex of a child is determined by (a) the sex chromosome contained in the sperm, (b) the sex chromosome contained in the oocyte, (c) the number of sperm fertilizing the oocyte, (d) the position of the fetus in the uterus.

16. FSH is to estrogen as estrogen is to (a) progesterone, (b) LH, (c) FSH, (d) testosterone.

17. A drug that "reminds the pituitary" to produce gonadotropins might be useful as (a) a contraceptive, (b) a diuretic, (c) a fertility drug, (d) an abortion stimulant.

Short Answer Essay Questions

18. Why is the term *urogenital system* more applicable to males than to females?

19. The spermatid is haploid, but it is not a functional gamete. Name and describe the process during which a spermatid is converted to a motile sperm, and describe the major structural (and functional) regions of a sperm.

20. Oogenesis in the female results in one functional gamete—the egg, or ovum. What other cells are produced? What is the significance of this rather wasteful type of gamete production—that is, production of a single functional gamete instead of four, as seen in males?

21. List three secondary sex characteristics of females.

22. Describe the events and possible consequences of menopause.

23. Define menarche. What does it indicate?

24. For three popular types of birth control—the pill, the diaphragm, and coitus interruptus—indicate (a) the point of blockage of the reproductive process, (b) the mode of action, (c) the relative efficiency or effectiveness, and (d) the inherent problems.

25. Trace the pathway of a sperm from the male testes to the uterine tube of a female.

26. In menstruation, the stratum functionalis is shed from the endometrium. Explain the hormonal and physical factors responsible for this shedding. (Hint: See Figure 28.22.)

27. Both the epithelium of the vagina and the cervical glands of the uterus help prevent the invasion and spread of vaginal pathogens. Explain how each of these mechanisms works.

28. Some anatomy students were saying that the bulbourethral glands (and the urethral glands) of males act like city workers who come around and clear parked cars from the street before a parade. What did they mean by this analogy?

29. A man swam in a cold lake for an hour and then noticed that his scrotum was shrunken and wrinkled. His first thought was that he had lost his testicles. What had really happened?

Critical Thinking and Clinical Application Questions

1. Gina Marciano, a 44-year-old mother of eight children, visited her physician complaining of a "bearing down" sensation in her pelvis, low backache, and urinary incontinence. A vaginal examination showed that the external os of her cervix was just inside the vaginal orifice and her perineum exhibited large keloids. Her history revealed that she was a member of a commune located in the nearby mountains that shunned hospital births (if at all possible). What do you think Gina's problem is and what caused it? (Be anatomically specific.)

2. Harry, a sexually active adolescent, appeared in the emergency room complaining of a penile "drip" and pain in urination. An account of his recent sexual behavior was requested and recorded. (a) What do you think Harry's problem is? (b) What is the causative agent of this disorder? (c) How is the condition treated, and what may happen if it isn't treated?

3. A 36-year-old mother of four is considering tubal ligation as a means of ensuring that her family gets no larger. She asks the physician if she will become "menopausal" after the surgery. (a) How would you answer her question and explain away her concerns? (b) Explain what a tubal ligation is.

4. Mr. Scanlon, a 76-year-old gentleman, is interested in a much younger woman. Concerned because of his age, he asks his urologist if he will be able to father a child. What questions would a physician ask of this man, and what diagnostic tests would be ordered?

5. Lucy had both her left ovary and her right uterine tube removed surgically at age 17 because of a cyst and a tumor in these organs. Now, at age 32, she remains healthy and is expecting her second child. How could Lucy conceive a child with just one ovary and one uterine tube, widely separated on opposite sides of the pelvis like this?

29 PREGNANCY AND HUMAN DEVELOPMENT

From Egg to Embryo (pp. 1119–1129)

1. Describe the importance of capacitation to the ability of sperm to penetrate an oocyte.
2. Explain the mechanism of the fast and slow blocks to polyspermy.
3. Define fertilization.
4. Explain the process and product of cleavage.
5. Describe the processes of implantation and placenta formation, and list placental functions.

Events of Embryonic Development (pp. 1129–1137)

6. Describe the process of gastrulation and its consequence.
7. Name and describe the formation, location, and function of the embryonic membranes.
8. Define organogenesis and indicate the important roles of the three primary germ layers in this process.
9. Describe the unique features of the fetal circulation.

Events of Fetal Development (pp. 1137–1139)

10. Indicate the duration of the fetal period, and note the major events of fetal development.

Effects of Pregnancy on the Mother (pp. 1139–1141)

11. Describe changes in maternal reproductive organs and in cardiovascular, respiratory, and urinary system functioning during pregnancy.
12. Indicate the effects of pregnancy on maternal metabolism and posture.

Parturition (Birth) (pp. 1141–1143)

13. Explain how labor is initiated, and describe the three stages of labor.

Adjustments of the Infant to Extrauterine Life (pp. 1143–1144)

14. Outline the events leading to the first breath of a newborn.
15. Describe the changes that occur in the fetal circulation after birth.

Lactation (pp. 1144–1145)

16. Explain how the breasts are prepared for lactation.

ecause the birth of a baby is such a familiar event, we tend to lose sight of the wonder of this accomplishment: How does a single cell, the fertilized egg, grow to become a complex human being consisting of trillions of cells? The details of this process can fill a good-sized book. Our intention here is simply to outline the important events of gestation and to consider briefly the events that occur immediately after birth.

But to get started, we need to define some terms. The term **pregnancy** refers to events that occur from the time of fertilization (conception) until the infant is born. The pregnant woman's developing offspring is called the **conceptus** (kon-sep'tus; "that which is conceived"). The time during which development occurs is referred to as the **gestation period** (*gestare* = to carry) and extends by convention from the last menstrual period until birth, approximately 280 days. Thus, at the moment of fertilization, the mother is officially (but illogically) two weeks pregnant! For two weeks following fertilization, the conceptus undergoes *preembryonic development* and is

known informally as a **preembryo.** From the third through eighth weeks after fertilization, the *embryonic period*, the conceptus is called an **embryo,** and from the ninth week through birth, the *fetal period*, the conceptus is called a **fetus** ("the young in the womb"). At birth, it is an infant. Figure 29.1 shows the changing size and shape of the conceptus as it progresses from fertilization to the early fetal stage.

FROM EGG TO EMBRYO

Accomplishing Fertilization

Before fertilization can occur, sperm must reach the ovulated secondary oocyte. The oocyte is viable for 12 to 24 hours after it is cast out of the ovary, and then the chance of pregnancy drops to almost zero the next day. Although some "super sperm" are viable for five days in the female reproductive tract, most sperm retain their fertilizing power for 24 to 72 hours after ejaculation. Consequently, for successful fertilization to occur, coitus must occur no

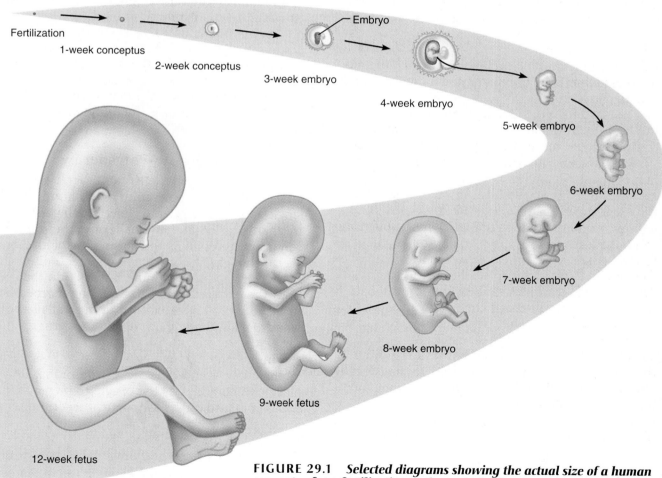

Fertilization

1-week conceptus

2-week conceptus

3-week embryo

Embryo

4-week embryo

5-week embryo

6-week embryo

7-week embryo

8-week embryo

9-week fetus

12-week fetus

FIGURE 29.1 *Selected diagrams showing the actual size of a human conceptus from fertilization to the early fetal stage.* The embryonic stage begins in the third week after fertilization; the fetal stage begins in the ninth week.

more than three days before ovulation and no later than 24 hours after, at which point the oocyte is approximately one-third of the way down the length of the uterine tube. **Fertilization** occurs when a sperm fuses with an egg to form a fertilized egg, or **zygote** (zi′gōt; "yoked together"), the first cell of the new individual. Let's look at the events leading to fertilization.

Sperm Transport and Capacitation

During copulation, a man expels millions of sperm with considerable force into his partner's vaginal canal. Despite this "head start," most sperm don't come within shouting distance of the oocyte just about 5 inches away. Sperm travel a treacherous path as they propel themselves through the female reproductive tract. Millions of them leak from the vagina almost immediately after being deposited there. Of those that remain, millions more are destroyed by the vagina's acidic environment and, unless the thick "curtain" of cervical mucus has been made fluid by estrogens, additional millions fail to make it through the cervix. Those that do manage to enter the uterus are subjected to forceful uterine contractions that act in a washing machine–like manner to disperse them throughout the uterine cavity, where thousands more are destroyed by resident phagocytic leukocytes. Only a few thousand (and sometimes fewer than 200) sperm, out of the millions in the male ejaculate, finally reach the uterine tubes, where the oocyte may be moving leisurely toward the uterus.

These difficulties aside, there is still another hurdle to be overcome. Sperm freshly deposited in the vagina are incapable of penetrating an oocyte. They must first be **capacitated;** that is, their membranes must become fragile so that the hydrolytic enzymes in their acrosomes can be released. While the exact mechanism of capacitation is still a mystery, we do know that as sperm swim through the cervical mucus, uterus, and uterine tubes, the cholesterol that keeps their acrosomal membranes "tough" and stable is depleted, and the sperm undergo a gradual capacitation over the next 6 to 8 hours. Thus, even though the sperm may actually reach the oocyte within a few minutes, they must "wait around" (so to speak) for capacitation to occur. This is a pretty elaborate mechanism for preventing the spilling of acrosomal enzymes, but consider the alternative. If sperm had fragile acrosomal membranes while still within the male reproductive tract, they could rupture prematurely, causing some degree of autolysis (self-digestion) of the male reproductive system.

Acrosomal Reaction and Sperm Penetration

The ovulated oocyte is encapsulated by the corona radiata and, deep to that, the zona pellucida, both of which must be breached before the oocyte itself can be penetrated. Once a sperm binds to the zona pellucida around the oocyte, it undergoes the **acrosomal** (ak′ro-sōm-al) **reaction,** in which acrosomal enzymes (hyaluronidase, acrosin, proteases, and others) are released in the immediate vicinity of the oocyte (Figure 29.2). Hundreds of acrosomes must rupture to break down the intercellular cement, rich in hyaluronic acid, that holds the granulosa cells together and to digest holes in the zona pellucida. This is one case that does not bear out the adage, "The early bird catches the worm." The sperm that arrive first on the scene "sacrifice themselves" for the good of their brothers. A sperm that comes along later, after hundreds of sperm have undergone acrosomal reactions to expose the oocyte membrane, is in the best position to be *the* fertilizing sperm.

Once a path has been cleared and a single sperm makes contact with the oocyte's membrane receptors, its nucleus is pulled into the oocyte cytoplasm (Figure 29.2a). Each sperm carries a special two-part binding apparatus on its surface. The *beta protein* part acts first as it finds and binds to a receptor on the oocyte membrane. This event engages the *alpha protein* part, causing it to insert into the membrane. This somehow causes the egg and sperm membranes to open and then fuse together with such perfect contact that the contents of both cells are combined within a single membrane—all without spilling a drop.

Blocks to Polyspermy

Although **polyspermy** (entry of several sperm into an egg) occurs in some animals, in humans only a single sperm is allowed to penetrate the oocyte. Two mechanisms ensure **monospermy,** the one-sperm condition. As soon as the plasma membrane of one sperm contacts the oocyte membrane, sodium channels open and ionic sodium diffuses into the oocyte from the extracellular space, causing its membrane to depolarize. This electrical event, called the *fast block to polyspermy,* prevents other sperm from fusing with the oocyte membrane. Then, once the sperm has entered, ionic calcium (Ca^{2+}) is released by the oocyte's endoplasmic reticulum into its cytoplasm which activates the oocyte to prepare for cell division. It also causes the **cortical reaction** (see Figure 29.2a), in which the cortical granules located just deep to the plasma membrane spill their enzymatic contents into the extracellular space beneath

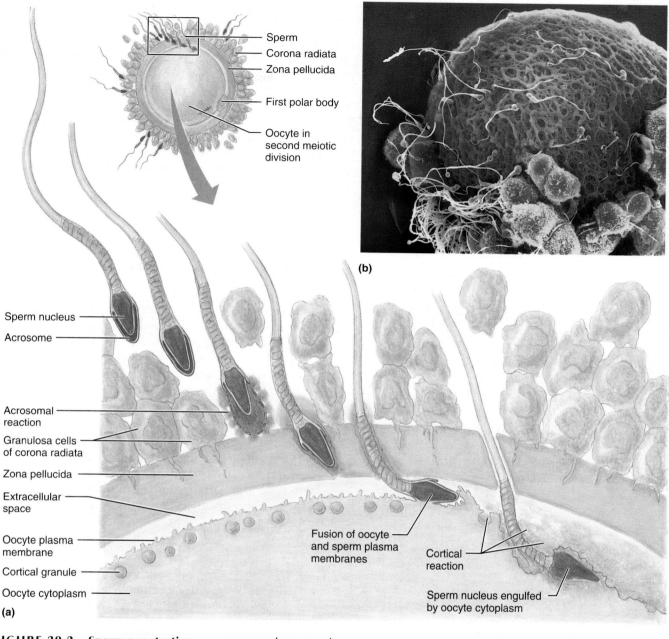

Sperm
Corona radiata
Zona pellucida
First polar body
Oocyte in second meiotic division

(b)

Sperm nucleus
Acrosome

Acrosomal reaction
Granulosa cells of corona radiata
Zona pellucida
Extracellular space
Oocyte plasma membrane
Cortical granule
Oocyte cytoplasm

(a)

Fusion of oocyte and sperm plasma membranes
Cortical reaction
Sperm nucleus engulfed by oocyte cytoplasm

FIGURE 29.2 *Sperm penetration and the cortical reaction (slow block to polyspermy).* (**a**) The sequential steps of oocyte penetration by a sperm are depicted from left to right. Sperm penetration through the corona radiata and zona pellucida surrounding the oocyte is accomplished by release of acrosomal enzymes by many sperm. Fusion of the plasma membrane of a single sperm with the oocyte membrane is followed by engulfment of the sperm by the oocyte cytoplasm and the cortical reaction. The cortical reaction involves the release of the contents of the oocyte's cortical granules into the extracellular space, preventing the entry of more than one sperm. (**b**) Scanning electron micrograph of human sperm surrounding a human oocyte (750×). The smaller spherical cells are granulosa cells of the corona radiata.

the zona pellucida. The enzymes destroy the sperm receptors. Additionally, the spilled material binds water, and as it swells, it detaches all sperm still in contact with receptors on the oocyte membrane, accomplishing the permanent *slow block to polyspermy*. In rare cases of polyspermy that do oc- cur, the embryos contain too much genetic material and are nonviable (die).

Completion of Meiosis II and Fertilization

After a sperm enters the oocyte, it loses its tail and midpiece and then migrates to the center of the

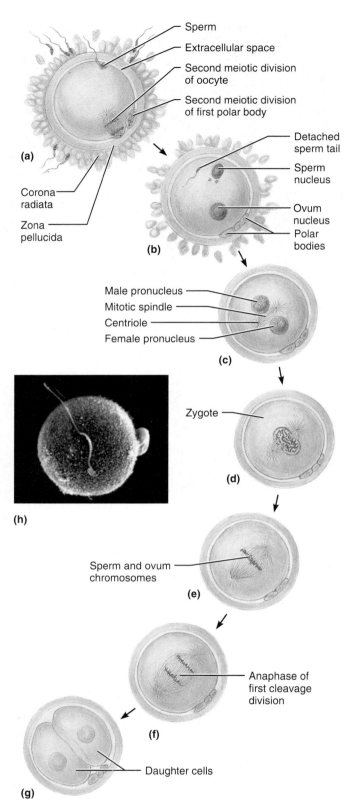

Sperm
Extracellular space
Second meiotic division of oocyte
Second meiotic division of first polar body

(a)

Corona radiata

Zona pellucida

Detached sperm tail
Sperm nucleus
Ovum nucleus
Polar bodies

(b)

Male pronucleus
Mitotic spindle
Centriole
Female pronucleus

(c)

Zygote

(d)

(h)

Sperm and ovum chromosomes

(e)

Anaphase of first cleavage division

(f)

Daughter cells

(g)

oocyte. The secondary oocyte now completes meiosis II to form the ovum nucleus and ejects the second polar body (Figure 29.3a and b). The ovum and sperm nuclei swell, becoming **female** and **male pronuclei** (pro-nu'kle-i; *pro* = before), and approach each other as a mitotic spindle develops between them (Figure 29.3c). The pronuclei membranes then rupture, releasing their chromosomes into the immediate vicinity of the spindle. The true moment of fertilization occurs as the maternal and paternal chromosomes combine and produce the diploid zygote (Figure 29.3d).* Almost as soon as the male and female pronuclei come together, their chromosomes replicate. Then, the first mitotic division of the conceptus begins (Figure 29.3d–f).

Preembryonic Development

Preembryonic development begins with fertilization and continues as the preembryo travels through the uterine tube, floats free in the cavity of the uterus, and finally implants in the uterine wall. Significant events of the preembryonic period are *cleavage*, which produces a structure called a blastocyst, and *implantation* of the blastocyst.

Cleavage and Blastocyst Formation

Cleavage is a period of fairly rapid mitotic divisions of the zygote following fertilization. Since there is little time for *growth* between divisions, the daughter cells become smaller and smaller (Figure 29.4). Cleavage produces cells with a high surface-to-

FIGURE 29.3 *Events immediately following sperm penetration.* (**a**) Once a sperm has entered the secondary oocyte, the oocyte completes meiosis II and casts out the second polar body. (**b–c**) The ovum nucleus formed during meiosis II begins to swell, and the sperm and egg nuclei approach each other. When fully swollen, the nuclei are called pronuclei. (**d–g**) As the male and female pronuclei come together (constituting the act of fertilization), they immediately replicate their DNA, and then form chromosomes which attach to the mitotic spindle and undergo a mitotic division. This division is the first cleavage division and results in two daughter cells. (**h**) Scanning electron micrograph of a fertilized egg (zygote). Notice the second polar body protruding at the right. A single sperm is seen against the zygote's surface (500×).

---

*Some sources define the term *fertilization* simply as the act of oocyte penetration by the sperm. However, unless the chromosomes in the male and female pronuclei are actually joined, the zygote is never formed in humans.

? *Why is the multicellular blastocyst only slightly larger than the single-celled zygote?*

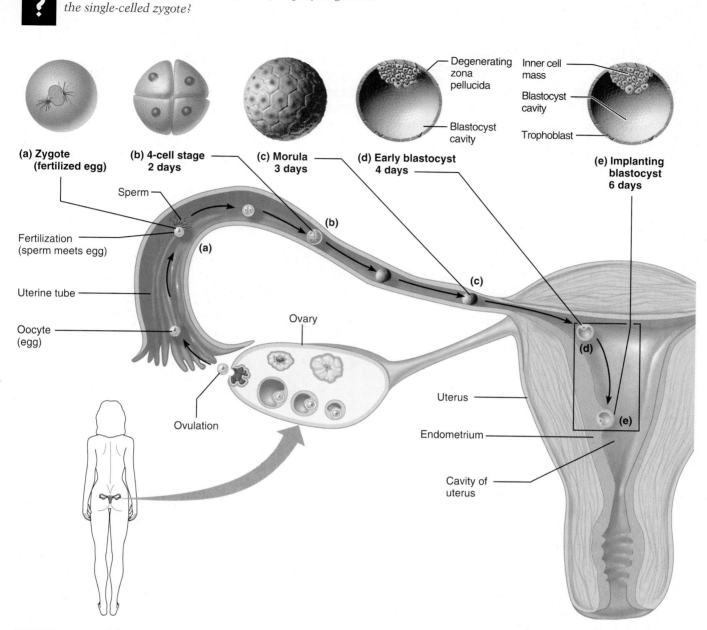

(a) Zygote (fertilized egg)

(b) 4-cell stage 2 days

(c) Morula 3 days

(d) Early blastocyst 4 days

Degenerating zona pellucida

Blastocyst cavity

Inner cell mass

Blastocyst cavity

Trophoblast

(e) Implanting blastocyst 6 days

Sperm

Fertilization (sperm meets egg)

Uterine tube

Oocyte (egg)

Ovary

Ovulation

Uterus

Endometrium

Cavity of uterus

FIGURE 29.4 Cleavage is a rapid series of mitotic divisions that begins with the zygote and ends with the blastocyst. The zygote begins to divide about 24 hours after fertilization, and continues to divide rapidly (undergo cleavage) as it travels down the uterine tube. Three to four days after ovulation, the preembryo reaches the uterus and floats freely for two to three days, nourished by secretions of the endometrial glands. At the late blastocyst stage, the embryo is implanting into the endometrium; this begins at about day 7 after ovulation. (**a**) Zygote. (**b**) Four-cell stage. (**c**) Morula, a solid ball of blastomeres. (**d**) Early blastocyst; the morula hollows out, fills with fluid, and "hatches" from the zona pellucida. (**e**) Late blastocyst, composed of an outer sphere of trophoblast cells and an eccentric cell cluster called the inner cell mass.

volume ratio, which enhances their uptake of nutrients and oxygen and the disposal of wastes. It also provides a large number of cells to serve as building blocks for constructing the embryo. Consider, for a

moment, the difficulty of trying to construct a building from one huge block of granite. If you now consider how much easier it would be if instead you could use hundreds of bricks, you will quickly grasp the importance of cleavage.

By about 36 hours after fertilization, the first cleavage division has produced two identical cells

Because virtually no growth occurs between the successive divisions of cleavage.

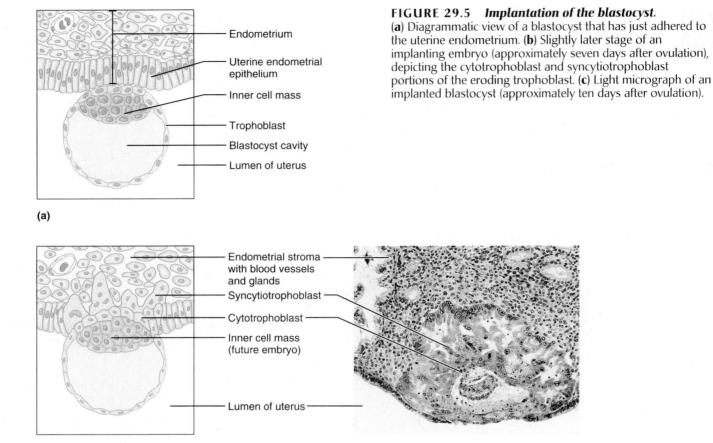

(a)

- Endometrium
- Uterine endometrial epithelium
- Inner cell mass
- Trophoblast
- Blastocyst cavity
- Lumen of uterus

(b)

- Endometrial stroma with blood vessels and glands
- Syncytiotrophoblast
- Cytotrophoblast
- Inner cell mass (future embryo)
- Lumen of uterus

(c)

FIGURE 29.5 *Implantation of the blastocyst.*
(**a**) Diagrammatic view of a blastocyst that has just adhered to the uterine endometrium. (**b**) Slightly later stage of an implanting embryo (approximately seven days after ovulation), depicting the cytotrophoblast and syncytiotrophoblast portions of the eroding trophoblast. (**c**) Light micrograph of an implanted blastocyst (approximately ten days after ovulation).

called *blastomeres.* These then divide to produce four cells, then eight, and so on. By 72 hours after fertilization, a berry-shaped cluster of 16 or more cells called the **morula** (mor'u-lah; "little mulberry") has been formed. All the while, transport of the pre-embryo toward the uterus continues. By the fourth or fifth day after fertilization, the preembryo consists of about 100 cells and is floating freely in the uterus. The zona pellucida now starts to break down and the inner structure, or blastocyst, escapes from it. The **blastocyst** (blas'to-sist) is a fluid-filled hollow sphere composed of a single layer of large, flattened cells called **trophoblast** (trof'o-blast) **cells** and a small cluster of rounded cells, called the **inner cell mass,** at one side. Trophoblast cells take part in placenta formation, a fact hinted at by the literal translation of "trophoblast" (nourishment generator). The inner cell mass becomes the **embryonic disc,** which forms the embryo proper.

Implantation

When the blastocyst reaches the uterus, it floats freely in the uterine cavity for two to three days, receiving nourishment from the uterine secretions. Then, some six to seven days after ovulation,

implantation begins. The blastocyst's outer layer, the trophoblast, tests the readiness of the endometrium for implantation. The receptivity of the endometrium to implantation—the so-called *window of implantation*—is opened by the proper levels of ovarian hormones (estrogens and progesterone) in the blood. If the mucosa is properly prepared, the blastocyst implants high in the uterus. If the endometrium is not yet optimally mature, the blastocyst detaches and floats to a lower level, finally implanting when it finds a site with the proper chemical signals. The trophoblast cells overlying the inner cell mass adhere to the endometrium (Figure 29.5a) and begin to secrete digestive enzymes and growth factors against the endometrial surface. The endometrium quickly thickens at the point of contact and the area takes on the characteristics of an acute inflammatory response—the uterine blood vessels become more permeable and leaky, and inflammatory cells including lymphocytes, natural killer cells, and macrophages invade the area. The trophoblast then proliferates and forms two distinct layers (Figure 29.5b). The cells in the inner layer, called the **cytotrophoblast** (si"to-trof'o-blast), retain their cell boundaries. By contrast, the cells in the outer layer lose their plasma membranes and form a

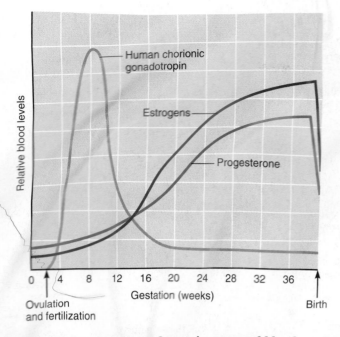

FIGURE 29.6 *Relative change in maternal blood levels of human chorionic gonadotropin, estrogens, and progesterone during pregnancy.* (Actual blood concentrations are not depicted.)

multinuclear cytoplasmic mass called the **syncytiotrophoblast** (sin-sit″e-o-trof′o-blast; *syn* = together, *cyt* = cell), which projects invasively into the endometrium and rapidly digests the uterine cells it cont⸺. As the endometrium is eroded, the blastocyst ⸺ows into this thick, velvety lining and is surro⸺d by a pool of blood leaked from degraded endo⸺l blood vessels. Shortly, the implanted blasto⸺ covered over and sealed off from the uterin⸺y by proliferation of the endometrial cells (F⸺ 9.5c).

Im⸺on takes about a week and is usually comple⸺he fourteenth day after ovulation—the exa⸺hen the endometrium normally begins to ⸺ff in menses. Obviously, menses would f⸺ the embryo as well and must be prevente⸺egnancy is to continue. Viability of the corp⸺is maintained by an LH-like hormone cal⸺**chorionic gonadotropin** (ko″re-on′ik go-⸺(**hCG**), which is secreted by the troph⸺ the blastocyst. hCG bypasses the pituit⸺f the blastocyst. hCG bypasses and prom⸺ontrols at this critical time ing proges⸺ luteum to continue secreting prog⸺trogen. The *chorion*, which develops ⸺⸺oblast after implantation, continues ⸺stimulus; thus, the developing conc⸺stimulus; thus, the developing the uterus ⸺ the hormonal control of hCG is usu⸺y phase of development. ⸺ the mother's blood by

the third week ⸺⸺tation (one week a⸺tion). Blood leve⸺⸺hCG rise until th⸺second month and⸺decline sharply⸺to reach a low valu⸺ich persists for the re⸺⸺der of gestation (F⸺9.6). Between the second and third month, th⸺⸺⸺ta (described next) as⸺sumes the role of pro⸺⸺ and estrogen produc⸺tion for the remainder⸺⸺gnancy. The corpus luteum then degenerate⸺⸺varies remain in⸺active until after birth. ⸺cy tests used to⸺day are antibody tests tha⸺⸺ in a woman's blood or urine.

Initially, the implanted ⸺⸺⸺ its nutri⸺tion by digesting the endom⸺⸺by the second month, the placenta⸺⸺rients and oxygen to the embryo and ⸺⸺bry⸺onic metabolic wastes. Since pla⸺⸺s a continuation of the events of im⸺⸺⸺a consider it next, although we will⸺⸺l ourselves a little as far as embryonⅰ⸺concerned.

Placentation

Placentation (plas″en-ta′shun) refers t⸺tion of a **placenta** ("flat cake"), a temp⸺that originates from both embryonic (tro⸺and maternal (endometrial) tissues. When the ⸺erating trophoblast gives rise to a layer of extrae⸺bryonic mesoderm on its inner surface (Figure 29.7b), it becomes the **chorion**. The chorion develops fingerlike **chorionic villi** (Figure 29.7c), which become especially elaborate where they are in contact with maternal blood. Soon the mesodermal cores of the villi become richly vascularized by newly forming blood vessels, which extend to the embryo as the umbilical arteries and vein. Thus, the embryo's blood vessels are connected to the fetal (chorionic) portion of the placenta. The continuing erosion produces large, blood-filled **lacunae,** or **intervillus spaces,** in the stratum functionalis of the endometrium (see Figure 28.15, p. 1091), and the villi come to lie in these spaces totally immersed in maternal blood (Figure 29.7d). The part of the endometrium that lies between the chorionic villi and the stratum basalis becomes the **decidua basalis** (de-sid′u-ah), whereas that surrounding the uterine cavity face of the implanted embryo forms the **decidua capsularis** (Figure 29.7d and f). Together, the chorionic villi and the decidua basalis form the pancake-shaped placenta.

The fetal side of the placenta is easily recognized because it is slick and smooth, and the umbilical cord projects from its surface. In contrast, the maternal side is "bumpy," revealing the shape of the chorionic villus masses.

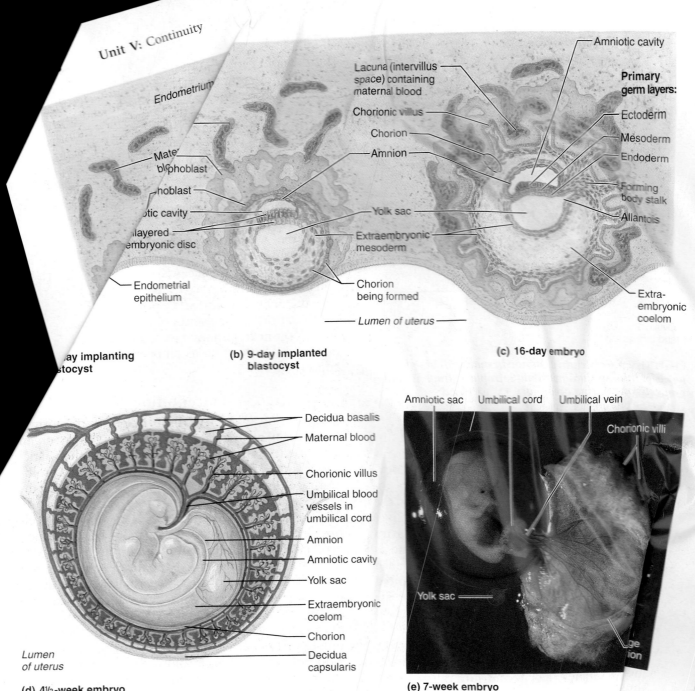

Endometrium

Maternal blood
...bloblast
...hoblast
...tic cavity
...layered embryonic disc

Endometrial epithelium

Lacuna (intervillus space) containing maternal blood

Chorionic villus

Chorion

Amnion

Yolk sac

Extraembryonic mesoderm

Chorion being formed

Lumen of uterus

Amniotic cavity

Primary germ layers:

Ectoderm

Mesoderm

Endoderm

Forming body stalk

Allantois

Extra-embryonic coelom

(a) ...ay implanting ...tocyst

(b) 9-day implanted blastocyst

(c) 16-day embryo

Decidua basalis

Maternal blood

Chorionic villus

Umbilical blood vessels in umbilical cord

Amnion

Amniotic cavity

Yolk sac

Extraembryonic coelom

Chorion

Decidua capsularis

Lumen of uterus

(d) 4½-week embryo

Amniotic sac   Umbilical cord   Umbilical vein

Chorionic villi

Yolk sac

(e) 7-week embryo

FIGURE 29.7 *Events of placentation, early embryonic development, and formation of the embryonic membranes.* Times given refer to the age of the conceptus, which is typically 2 weeks "younger" than the gestation age. (**a**) Implanting blastocyst. The erosion of the endometrium by the syncytiotrophoblast is ongoing, and cells of the embryonic disc are now separated by a fluid-filled space from the amnion. (**b**) Implantation is completed by the ninth day and extraembryonic mesoderm is beginning to form a discrete layer beneath the cytotrophoblast. (**c**) By 16 days, the cytotrophoblast and associated mesoderm have become the chorion, and chorionic villi are elaborating. The embryo now exhibits all three germ layers, a yolk sac (formed by endodermal cells), and an allantois, an outpocketing of the yolk sac that forms the structural basis of the body stalk, or umbilical cord. (**d**) At 4½ weeks, the decidua capsularis (placental portion enclosing the embryo on the uterine lumen face) and decidua basalis (placental portion encompassing the elaborate chorionic villi and maternal endometrium) are well formed. The chorionic villi lie in blood-filled intervillus spaces within the

endometrium. Organogen... ...ring in the embryo, which is n... its nutrition via the umbilical... connect it (via the body s... placenta. The amnion ar... and well developed. (**e**) A 7-... the embryonic membra... 13-contribute to the place... ...e week fetus. (**g**) Detaile... vascular relationships... ment decidua basalis. This... ...f the has been accomplish... third month of deve...

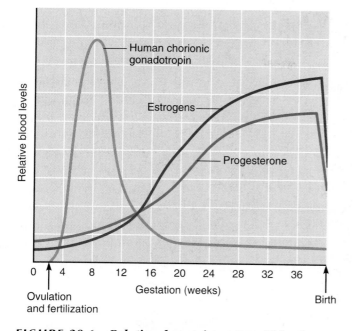

FIGURE 29.6   *Relative change in maternal blood levels of human chorionic gonadotropin, estrogens, and progesterone during pregnancy.* (Actual blood concentrations are not depicted.)

multinuclear cytoplasmic mass called the **syncytiotrophoblast** (sin-sit″e-o-trof′o-blast; *syn* = together, *cyt* = cell), which projects invasively into the endometrium and rapidly digests the uterine cells it contacts. As the endometrium is eroded, the blastocyst burrows into this thick, velvety lining and is surrounded by a pool of blood leaked from degraded endometrial blood vessels. Shortly, the implanted blastocyst is covered over and sealed off from the uterine cavity by proliferation of the endometrial cells (Figure 29.5c).

Implantation takes about a week and is usually completed by the fourteenth day after ovulation—the exact time when the endometrium normally begins to slough off in menses. Obviously, menses would flush away the embryo as well and must be prevented if the pregnancy is to continue. Viability of the corpus luteum is maintained by an LH-like hormone called **human chorionic gonadotropin** (ko″re-on′ik go-nad″o-trōp′in) **(hCG),** which is secreted by the trophoblast cells of the blastocyst. hCG bypasses the pituitary-ovarian controls at this critical time and prompts the corpus luteum to continue secreting progesterone and estrogen. The *chorion,* which develops from the trophoblast after implantation, continues this hormonal stimulus; thus, the developing conceptus takes over the hormonal control of the uterus during this early phase of development. hCG is usually detectable in the mother's blood by

the third week of gestation (one week after fertilization). Blood levels of hCG rise until the end of the second month and then decline sharply by 4 months to reach a low value, which persists for the remainder of gestation (Figure 29.6). Between the second and third month, the placenta (described next) assumes the role of progesterone and estrogen production for the remainder of the pregnancy. The corpus luteum then degenerates and the ovaries remain inactive until after birth. All pregnancy tests used today are antibody tests that detect hCG in a woman's blood or urine.

Initially, the implanted embryo obtains its nutrition by digesting the endometrial cells, but by the second month, the placenta is providing nutrients and oxygen to the embryo and carrying away embryonic metabolic wastes. Since placenta formation is a continuation of the events of implantation, we will consider it next, although we will be getting ahead of ourselves a little as far as embryonic development is concerned.

Placentation

Placentation (plas″en-ta′shun) refers to the formation of a **placenta** ("flat cake"), a temporary organ that originates from both embryonic (trophoblastic) and maternal (endometrial) tissues. When the proliferating trophoblast gives rise to a layer of extraembryonic mesoderm on its inner surface (Figure 29.7b), it becomes the **chorion.** The chorion develops fingerlike **chorionic villi** (Figure 29.7c), which become especially elaborate where they are in contact with maternal blood. Soon the mesodermal cores of the villi become richly vascularized by newly forming blood vessels, which extend to the embryo as the umbilical arteries and vein. Thus, the embryo's blood vessels are connected to the fetal (chorionic) portion of the placenta. The continuing erosion produces large, blood-filled **lacunae,** or **intervillus spaces,** in the stratum functionalis of the endometrium (see Figure 28.15, p. 1091), and the villi come to lie in these spaces totally immersed in maternal blood (Figure 29.7d). The part of the endometrium that lies between the chorionic villi and the stratum basalis becomes the **decidua basalis** (desid′u-ah), whereas that surrounding the uterine cavity face of the implanted embryo forms the **decidua capsularis** (Figure 29.7d and f). Together, the chorionic villi and the decidua basalis form the pancake-shaped placenta.

The fetal side of the placenta is easily recognized because it is slick and smooth, and the umbilical cord projects from its surface. In contrast, the maternal side is "bumpy," revealing the shape of the chorionic villus masses.

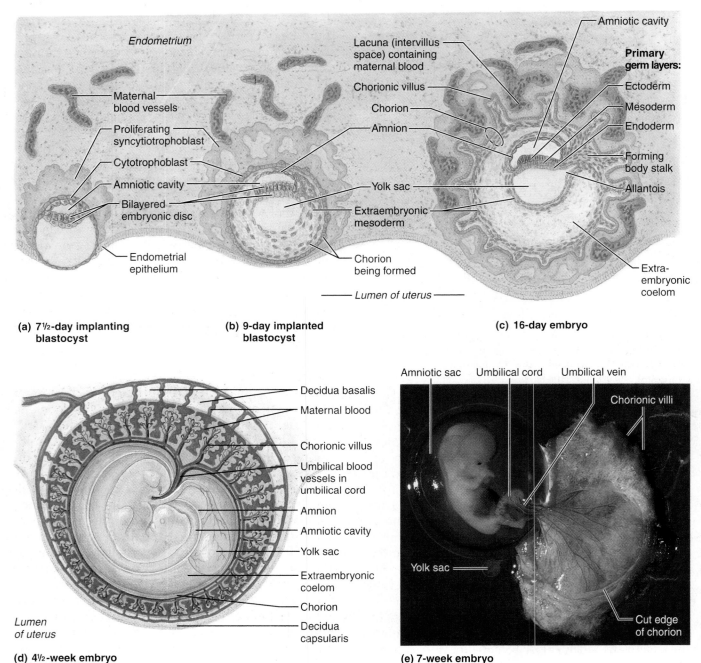

(a) 7½-day implanting blastocyst

Endometrium

Maternal blood vessels

Proliferating syncytiotrophoblast

Cytotrophoblast

Amniotic cavity

Bilayered embryonic disc

Endometrial epithelium

(b) 9-day implanted blastocyst

Lacuna (intervillus space) containing maternal blood

Chorionic villus

Chorion

Amnion

Yolk sac

Extraembryonic mesoderm

Chorion being formed

Lumen of uterus

(c) 16-day embryo

Amniotic cavity

Primary germ layers:
Ectoderm
Mesoderm
Endoderm

Forming body stalk

Allantois

Extra-embryonic coelom

(d) 4½-week embryo

Lumen of uterus

Decidua basalis
Maternal blood
Chorionic villus
Umbilical blood vessels in umbilical cord
Amnion
Amniotic cavity
Yolk sac
Extraembryonic coelom
Chorion
Decidua capsularis

(e) 7-week embryo

Amniotic sac   Umbilical cord   Umbilical vein

Chorionic villi

Yolk sac

Cut edge of chorion

FIGURE 29.7 Events of placentation, early embryonic development, and formation of the embryonic membranes. Times given refer to the age of the conceptus, which is typically 2 weeks "younger" than the gestation age. **(a)** Implanting blastocyst. The erosion of the endometrium by the syncytiotrophoblast is ongoing, and cells of the embryonic disc are now separated by a fluid-filled space from the amnion. **(b)** Implantation is completed by the ninth day and extraembryonic mesoderm is beginning to form a discrete layer beneath the cytotrophoblast. **(c)** By 16 days, the cytotrophoblast and associated mesoderm have become the chorion, and chorionic villi are elaborating. The embryo now exhibits all three germ layers, a yolk sac (formed by endodermal cells), and an allantois, an outpocketing of the yolk sac that forms the structural basis of the body stalk, or umbilical cord. **(d)** At 4½ weeks, the decidua capsularis (placental portion enclosing the embryo on the uterine lumen face) and decidua basalis (placental portion encompassing the elaborate chorionic villi and maternal endometrium) are well formed. The chorionic villi lie in blood-filled intervillus spaces within the endometrium. Organogenesis is occurring in the embryo, which is now receiving its nutrition via the umbilical vessels that connect it (via the body stalk) to the placenta. The amnion and yolk sac are well developed. **(e)** A 7-week embryo and the embryonic membranes that will contribute to the placenta (at right). **(f)** 13-week fetus. **(g)** Detailed anatomy of the vascular relationships in the mature decidua basalis. This state of development has been accomplished by the end of the third month of development.

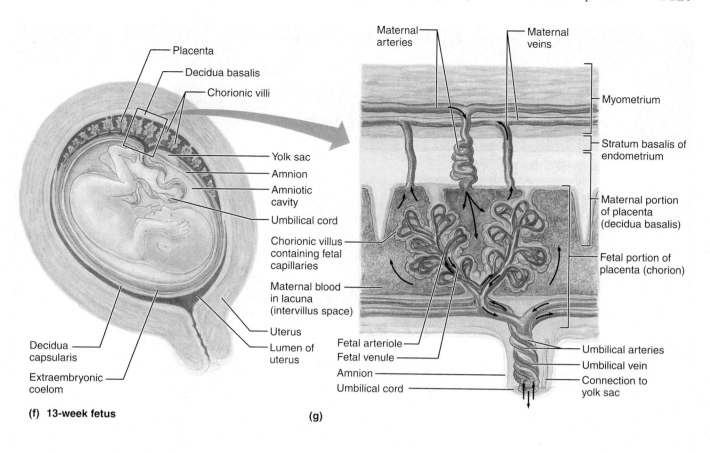

(f) 13-week fetus

(g)

The placenta detaches and sloughs off after the infant is born, so the name of the maternal portion—*decidua* ("that which falls off")—is very appropriate. The decidua capsularis expands to accommodate the growing fetus, which eventually fills and stretches the uterine cavity. As the developing infant grows, the villi in the decidua capsularis are compressed, and they degenerate. As this happens, the villi in the decidua basalis increase in number and branch even more profusely.

The placenta is usually fully formed and functional as a nutritive, respiratory, excretory, and endocrine organ by the end of the third month of pregnancy. However, well before this time, oxygen and nutrients are diffusing from maternal to embryonic blood, and embryonic metabolic wastes are passing in the opposite direction. The barriers to free passage of substances between the two blood supplies are embryonic barriers—the membranes of the chorionic villi and the endothelium of embryonic capillaries. Although the maternal and embryonic blood supplies are very close, they normally do not intermix.

While the placenta secretes hCG from the very beginning, the ability of its syncytiotrophoblast cells (the "hormone manufacturers") to produce the steroid hormones of pregnancy (estrogens and progesterone) matures much more slowly. If, for some reason, the placenta is producing inadequate amounts of these hormones when hCG levels wane, the endometrium degenerates and the pregnancy is aborted. Failures in implantation and placental development are clinically important because they account for spontaneous abortions in fully one-third of human pregnancies.

Throughout pregnancy, blood levels of estrogens and progesterone continue to increase (see Figure 29.6), encouraging growth and further differentiation of the mammary glands and readying them for lactation. The placenta also produces other hormones, such as *human placental lactogen, human chorionic thyrotropin,* and *relaxin.* The effects of these hormones on the mother are described shortly.

A CLOSER LOOK   BABY MAKING: Popsicles, GIFTS, and Surrogate Moms

In the summer of 1978, Louise Joy Brown was born. A cause for her parents to rejoice, for sure, but tiny Louise was unique. She was the first baby born who was conceived in a glass dish, or more precisely, in an in vitro laboratory setting. Most who read of this event were overcome with amazement that a child could be conceived outside the body. For scientists in the field of reproductive physiology, Louise's birth was a stunning breakthrough that fueled assisted reproductive techniques (ART), the subsequent avalanche of new techniques to provide childless couples with offspring.

Making babies outside the privacy of the bedroom is quite a recent phenomenon. In the 1970s, the first and simplest ART became widely available. This technique, *artificial insemination by donor (AID)*, was directed at couples where the male was sterile, or nearly so, and at single women. Almost overnight, sperm banks that collected, froze, and stored sperm from anonymous donors began to spring up. Today, in the United States, there are several sperm banks with frozen ejaculates for sale. The technique is simple by today's standards: All that is required is sperm, a syringe, and a fertile woman. The sperm are deposited in the woman's vagina or cervix at the appropriate time of the month, and nature is allowed to take its course. Presently, AID accounts for over 20,000 American babies each year. It is an "oldie but goodie" that does what it claims to do.

Although the number of *test-tube babies* like Louise is swelling, they are still relatively few because *in vitro fertilization (IVF)* is an elegant, but tricky, procedure. The woman is primed with a flood of hormones (natural or synthetic gonadotropins) to induce superovulation, the development of several oocytes in her ovaries. (The photographs show an unstimulated and stimulated ovary.) The oocytes are then sucked from

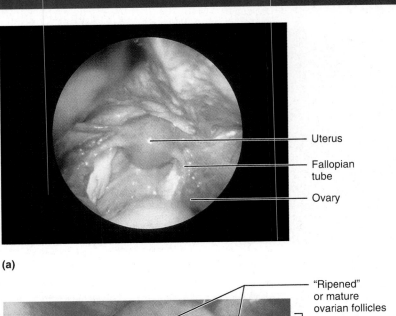

(a)

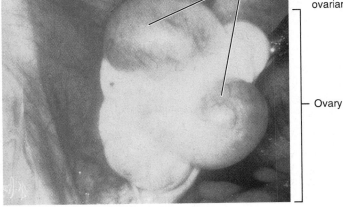

(b)

Photographs of (**a**) an unstimulated ovary and (**b**) an ovary that has been stimulated with hormones to develop multiple mature follicles, which can be seen protruding from the ovary surface.

the ovary through a small opening in the abdomen into which a viewing instrument called a *laparoscope* has been inserted. A few of these oocytes are fertilized with her partner's sperm in a glass receptacle and, after a few days of growth, all (32-cell) preembryos that show normal early development are implanted back into her uterus. An even newer procedure extracts *immature* oocytes from the ovary and pushes them to mature in vitro by bathing the isolated eggs with the gonadotropins. This spares the woman the daily hormone injec-

tions that temporarily turn her life into a "roller coaster" of mood swings, bloating, and nausea, and substantially reduces the time and cost of the IVF procedure.

IVF has an added benefit for couples whose families harbor genes for genetic abnormalities. In a technique called *preimplantation diagnosis*, one or two cells are plucked from the early (8-cell) conceptus and tested for the presence of the deleterious genes. Hence, the decision to implant or not can be made with some sense of security. The other oocytes are also fertilized and frozen "for spares" in case the first implantation is unsuccessful—hence, the catchy term *popsicle babies*. Since it is estimated that only 30% of natural fertilizations result in pregnancy and a healthy baby, the success rate of 15–20% reported by these clinics is actually very good, and at the rate of one IVF birth a day the number in the IVF baby club is beginning to become significant.

Meanwhile, the use of *surrogate* mothers has been growing by leaps and bounds. Surrogates (for a fee) provide ova and substitute uteri for women who have fertile husbands but who have been unsuccessful in becoming pregnant. The surrogate is artificially fertilized with the sponsor husband's sperm (as in AID), and the baby is eventually delivered to the sponsoring parents. Anyone who reads the newspapers is well aware that the use of surrogates is teeming with legal and ethical problems, primarily because some surrogate moth-

ers have refused to relinquish the baby after birth. Whose baby is it?

Surrogates will probably fade from the scene when *embryo transfers* from the womb of one woman into that of another become more common. In this procedure, a conceptus is harvested from a donor, who has been artificially inseminated by the infertile recipient's mate, and then transferred into the infertile recipient's uterus. The harvesting procedure involves flushing the donor's uterus out during the first week of pregnancy through a soft catheter (plastic tube). As you can imagine, this procedure is even more rife with complications than in vitro fertilization, but the first successful transfer occurred in early 1984 with the birth of a healthy baby boy.

Another technique called *gamete intrafallopian transfer (GIFT)* is an attempt to bypass the hazards (high acidity, unsuitable mucus, phagocytes, and others) that sperm encounter in a female's reproductive

tract by bringing oocytes and sperm together in a woman's uterine tubes. As with the in vitro procedure, the woman is given ovulatory stimulant drugs, and oocytes are harvested by vacuum aspiration. The oocytes and the partner's sperm are mixed and then immediately inserted into the woman's uterine tubes via the same laparoscope tubing. GIFT has a cost advantage over in vitro fertilization because everything is done as a single procedure. A new twist on this technique is that of actually injecting the oocytes with sperm or even spermatids—a procedure called *intracytoplasmic sperm injection (ISI)*.

These scientific breakthroughs are good news for childless couples, but for some people they have raised the fear of a "brave new world" where babies will gestate in laboratory bottles and only those with the right gender or selected genetic traits will be chosen to live. This is the nature of every major change—intense joy in a vale of tears.

EVENTS OF EMBRYONIC DEVELOPMENT

Having followed placental development into the fetal stage, we will now backtrack and consider development of the embryo during and after implantation. Even while implantation is occurring, the blastocyst is being converted to the next developmental stage, the **gastrula** (gas'troo-lah), in which the three primary germ layers form, and the embryonic membranes are developing. Before becoming three-layered, the inner cell mass first subdivides into two layers—the upper *epiblast* and the lower *hypoblast*. Extensions of these sheets form two of the four embryonic membranes.

Formation and Roles of the Embryonic Membranes

The **embryonic membranes** that form during the first two to three weeks of development include the amnion, yolk sac, allantois, and chorion (see Figure 29.7c). The **amnion** (am'ne-on) develops when cells of the epiblast fashion themselves into the transparent membranous sac. This sac, the amnion, becomes filled with **amniotic fluid.** Later, as the em-

bryonic disc curves to form the tubular body (a process we will describe shortly), the amnion curves with it. Eventually, the sac extends all the way around the embryo, broken only by the umbilical cord (see Figure 29.7d).

Sometimes called the "bag of waters," the amnion provides a buoyant environment that protects the developing embryo against physical trauma, and helps maintain a constant homeostatic temperature. The fluid also prevents the rapidly growing embryonic parts from adhering and fusing together and allows the embryo considerable freedom of movement, thus aiding musculoskeletal development. Initially, amniotic fluid is derived from the maternal blood, but as the fetal kidneys become functional, fetal urine contributes to amniotic fluid volume.

The **yolk sac** (see Figure 29.7b) forms from cells of the hypoblast. These cells arrange themselves into a sac that hangs from the ventral surface of the embryo (see Figure 29.7b). Thus, the amnion and yolk sac resemble two balloons touching one another with the embryonic disc at the point of contact. In many species, the yolk sac is the main source of nutrition for the embryo, but human eggs contain very little yolk and nutritive functions have been taken over by the placenta. Nevertheless, the yolk sac is important in humans because it (1) forms part of the *gut* (digestive tube), (2) produces the earliest blood cells and blood vessels, and (3) is the source of primordial germ cells that migrate into the embryo's body to seed the gonads.

The **allantois** (ah-lan′to-is) forms as a small outpocketing at the caudal end of the yolk sac (see Figure 29.7c). In animals that develop within shelled eggs, the allantois is a disposal site for solid metabolic wastes (excreta). In humans, the allantois is the structural base for constructing the umbilical cord that links the embryo to the placenta, and becomes part of the urinary bladder. When the umbilical cord is fully formed, it contains a core of embryonic connective tissue (Wharton's jelly), the umbilical arteries and vein, and is covered externally by amniotic membrane.

The *chorion*, which helps to form the placenta, has already been described (see Figure 29.7c). Since it is the outermost membrane, the chorion encloses the embryonic body and all other membranes.

Gastrulation: Germ Layer Formation

During the third week, the two-layered embryonic disc transforms into a three-layered *embryo* in which the **primary germ layers**—*ectoderm, mesoderm,* and *endoderm*—are present. This process, called **gastrulation** (gas″troo-la′shun), involves cellular rearrangements and widespread cell migrations. It begins as a raised groove called the **primitive streak** appears on the dorsal surface of the embryonic disc and establishes the longitudinal axis of the embryo (Figure 29.8e and f).

The cell migrations that follow appear rather frantic when observed microscopically. Surface (epiblast) cells of the embryonic disc migrate medially across other cells and enter the primitive streak. The first cells that enter the groove form the most inferior germ layer, the **endoderm.** Those that follow push laterally between the cells at the upper and lower surfaces, forming the **mesoderm** (see Figure 29.8g and h). The cells that remain on the embryo's dorsal surface form the **ectoderm.** The mesodermal cells immediately beneath the early primitive streak quickly aggregate, forming a rod of mesodermal cells called the **notochord** (no′to-kord), which serves as the first axial support of the embryo (see Figure 29.9a). The conceptus, now called an embryo, is about 2 mm long.

The three primary germ layers differ in the arrangement of their cells and serve as the *primitive tissues* from which all body organs will derive. Ectoderm ("outer skin") fashions structures of the nervous system and the skin epidermis. Endoderm ("inner skin") forms the epithelial linings of the digestive, respiratory, and urogenital systems, and

associated glands. Mesoderm ("middle skin") forms virtually everything else. Both ectoderm and endoderm consist mostly of cells that are securely joined to each other and are considered *epithelia.* Mesoderm, by contrast, is a *mesenchyme* (literally, "poured into the middle [of the embryo]"). Notice, however, that the term **mesenchyme** indicates *any* embryonic tissue with star-shaped cells that are free to migrate widely within the embryo. Table 29.1 (p. 1135) lists the germ layer derivatives; some of the details of the differentiation processes are described next.

Organogenesis: Differentiation of the Germ Layers

Gastrulation lays down the basic structural framework of the embryo and sets the stage for **organogenesis** (or″gah-no-jen′ĕ-sis), formation of body organs and organ systems. During organogenesis, the cells of the embryo continue to rearrange themselves and, by the end of the embryonic period, when the embryo is eight weeks old and about 22 mm (slightly less than 1 inch) long from head to buttocks (referred to as the *crown-rump measurement*), all the adult organ systems are recognizable. It is truly amazing how much organogenesis occurs in such a short time in such a small amount of living matter.

Specialization of the Ectoderm

The first major event in organogenesis is **neurulation,** the differentiation of ectoderm that gives rise to the brain and spinal cord (Figure 29.9a). This process is *induced* (stimulated to happen) by chemical signals from the *notochord,* the rod of mesoderm that defines the body axis, mentioned earlier. The ectoderm overlying the notochord thickens, forming the **neural plate** (Figure 29.9b), and then starts to fold inward as a **neural groove,** which forms prominent **neural folds** as it deepens (Figure 29.9c). By the twenty-second day, the superior margins of the neural folds fuse, forming a **neural tube,** which soon pinches off into the body (Figure 29.9d). As described in Chapter 12, the anterior end of the neural tube becomes the brain and the rest becomes the spinal cord. The associated **neural crest cells** migrate widely and give rise to the cranial, spinal, and sympathetic ganglia (and associated nerves), to the medulla of the adrenal gland, and contribute to some connective tissues.

FIGURE 29.8 *Schematic view of gastrulation: formation of the three primary germ layers.* **(a)—(d)** Orienting diagrams. **(e)** and **(f)** Surface views of an embryonic disc, amnion and yolk sac removed. **(g)** and **(h)** Cross sections of the embryonic disc, showing the germ layers resulting from cell migration. The first epiblast cells that migrate medially into the primitive streak **(g)** become endoderm. Those that follow **(h)** become mesoderm. The epiblast surface is now called ectoderm.

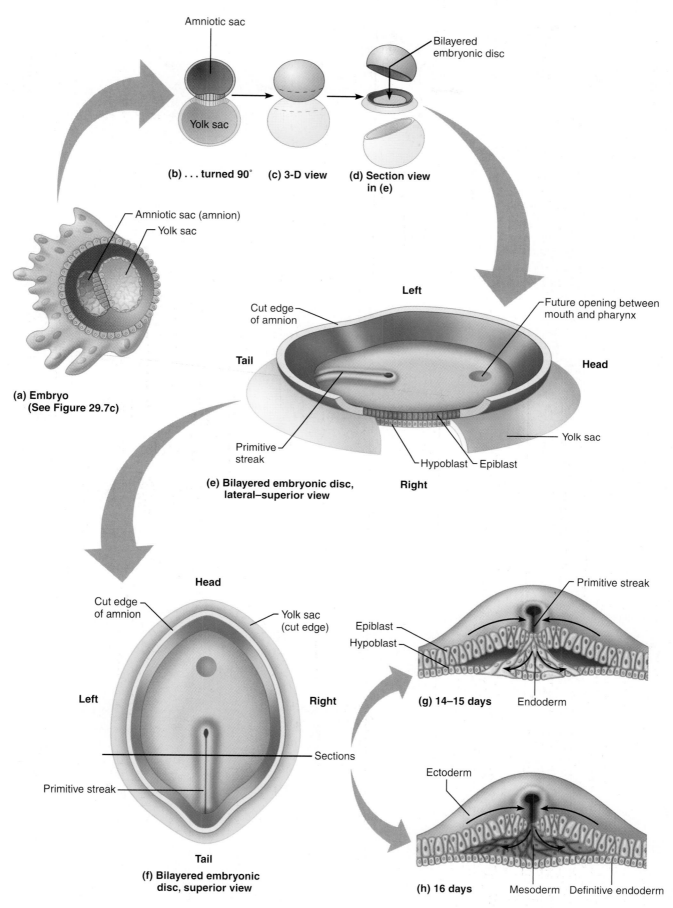

Amniotic sac

Yolk sac

(b) . . . turned 90°

(c) 3-D view

Bilayered embryonic disc

(d) Section view in (e)

Amniotic sac (amnion)

Yolk sac

(a) Embryo (See Figure 29.7c)

Left

Cut edge of amnion

Future opening between mouth and pharynx

Tail

Head

Primitive streak

Yolk sac

Hypoblast — Epiblast

(e) Bilayered embryonic disc, lateral–superior view

Right

Head

Cut edge of amnion

Yolk sac (cut edge)

Left

Right

Sections

Primitive streak

Tail

(f) Bilayered embryonic disc, superior view

Primitive streak

Epiblast

Hypoblast

(g) 14–15 days

Endoderm

Ectoderm

(h) 16 days

Mesoderm   Definitive endoderm

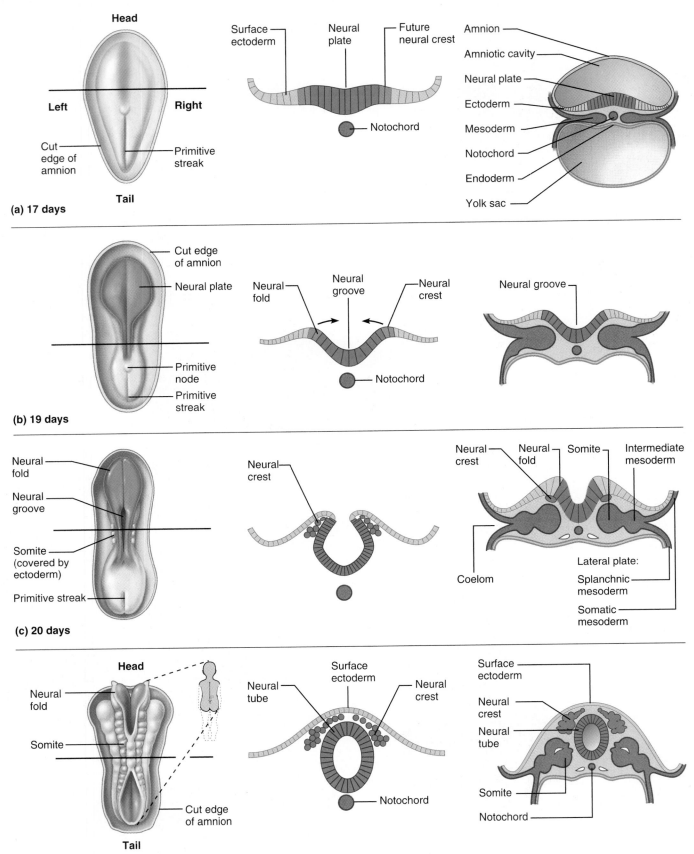

FIGURE 29.9 *Events of neurulation.* Dorsal surface views on the left; frontal sections on the right. **(a)** The flat three-layered embryo, which has accomplished gastrulation. The notochord and neural plate have formed. **(b)** The neural folds form by folding of the neural plate. **(c)** The neural folds begin to close. **(d)** The newly formed neural tube has detached from the surface ectoderm and lies between the surface ectoderm and the notochord; the neural crest is evident and the embryonic body is beginning to fold.

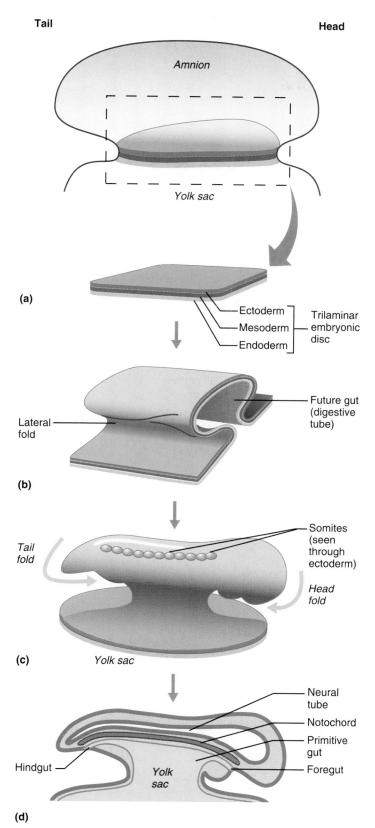

(a)

Ectoderm
Mesoderm
Endoderm
} Trilaminar embryonic disc

(b)

Lateral fold

Future gut (digestive tube)

(c)

Tail fold

Head fold

Somites (seen through ectoderm)

Yolk sac

(d)

Neural tube
Notochord
Primitive gut
Foregut

Hindgut

Yolk sac

FIGURE 29.10 *Folding of the embryonic body, lateral views, head to right.* (**a**) Model of the flat three-layered embryo as three sheets of paper. (**b**)–(**c**) Folding begins with lateral folds, then head and tail folds appear. (**d**) A 24-day embryo in sagittal section. Notice the primitive gut, which derives from the yolk sac, and the notochord and neural tube dorsally.

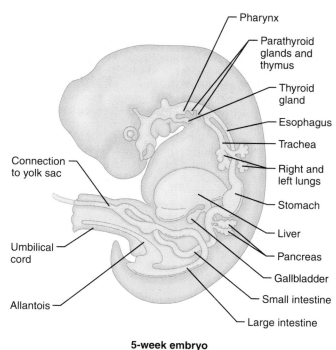

5-week embryo

FIGURE 29.11 *Endodermal differentiation to form the epithelial linings of the digestive and respiratory tracts and associated glands.*

By the end of the first month of development, the three primary brain vesicles (fore-, mid-, and hindbrain) are obvious. By the end of the second month, all brain flexures are evident, the cerebral hemispheres cover the top of the brain stem (see Figure 12.4), and brain waves can be recorded. Most of the remaining ectoderm forming the surface layer of the embryonic body differentiates into the epidermal layer of the skin. Other ectodermal derivatives are indicated in Table 29.1.

Specialization of the Endoderm

As noted, the embryo starts off as a flat embryonic plate, but as it grows rapidly, it folds to achieve a cylindrical body shape (Figure 29.10). In the simplest sense, this process resembles three stacked sheets of paper folding laterally into a tube. At the same time, the head and tail regions fold under as shown in Figure 29.10c. The folding occurs from both ends, and progresses toward the central part of the enbryonic body, where the yolk sac and umbilical vessels protrude. As the endoderm undercuts and its edges come together and fuse, it encloses part of the yolk sac. The tube of endoderm formed, called the **primitive gut,** forms the epithelial lining of the gastrointestinal tract (Figure 29.11). The organs of the GI tract (pharynx, esophagus, etc.) quickly become apparent, and then the oral and anal openings perforate. The mucosal lining of the respiratory tract

FIGURE 29.12   *Early mesodermal differentiation.* (a) Segregation of the mesoderm (with the exception of notochord formation, which occurs earlier) takes place as neurulation is occurring. The notochord becomes flanked bilaterally by a somite, intermediate mesoderm, and lateral mesoderm successively. **(b)** Schematic view of the undercutting of the lateral folds showing how the lateral mesodermal plates are carried ventrally, forming the coelom. **(c)** Mesodermal divisions and relationships in an undercut (folded) embryo. The sclerotome, dermatome, and myotome of each somite contribute to the formation of the vertebrae, skin dermis, and skeletal muscles, respectively. The intermediate mesoderm forms the kidneys, gonads, and associated structures. The somatic mesoderm portion of the lateral mesoderm forms the limb buds, parietal serosa, and contributes to the dermis. The splanchnic mesoderm forms the muscle, connective tissues, and serosae of the walls of the visceral organs, and the heart and blood vessels.

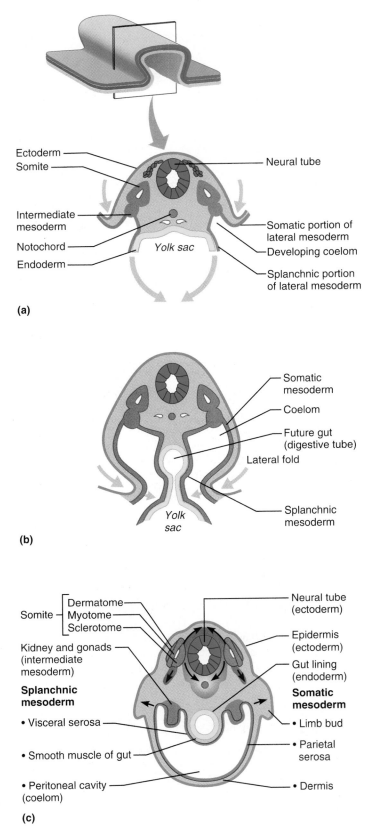

forms as an outpocketing from the *foregut* (pharyngeal endoderm), and the glands that derive from endoderm arise as endodermal outpocketings at various points along the tract. For example, the thyroid, parathyroids, and thymus form from the pharyngeal endoderm, and the liver and pancreas arise from the intestinal mucosa *(midgut)* more inferiorly. Notice that while the glands mentioned arise entirely from endodermal cells, the endoderm forms only the epithelium of the mucosal lining of the hollow organs of the digestive and respiratory tracts. Mesoderm forms the rest of their wall structure.

Specialization of the Mesoderm

Because we have covered embryonic development of the various organ systems in earlier chapters, we will focus only on the very early events of mesodermal segregation and specialization here. The first evidence of mesodermal differentiation is the appearance of the notochord in the embryonic disc (see Figure 29.9c). (Although the notochord is eventually replaced by the vertebral column, its remnants persist in the springy nucleus pulposus of the intervertebral discs until the second decade of life.) Shortly thereafter, three mesodermal aggregates appear on either side of the notochord (Figures 29.9 b–d and 29.12a). The largest of these, the *somites* (so'mīts), are a series of paired mesodermal blocks that hug the notochord on either side. All 40 pairs of somites are present by the end of week 4. Flanking the somites laterally are small clusters of mesoderm called the *intermediate mesoderm* and then double sheets of *lateral mesoderm*.

Each **somite** has three functional parts—*sclerotome*, *dermatome*, and *myotome*. Cells of the **sclerotome** (skle'ro-tōm; "hard piece") on either side migrate medially, gather around the notochord and neural tube, and produce the vertebra and rib at the associated level. The **dermatome** ("skin piece") cells, which form the outer wall of each somite, help form

TABLE 29.1	Derivatives of the Primary Germ Layers	
Ectoderm	*Mesoderm*	*Endoderm*
All nervous tissue	Skeletal, smooth, and cardiac muscle	Epithelium of digestive tract (except that of oral and anal cavities)
Epidermis of skin and epidermal derivatives (hairs, hair follicles, sebaceous and sweat glands, nails)	Cartilage, bone, and other connective tissues	Glandular derivatives of digestive tract (liver, pancreas)
Cornea and lens of eye	Blood, bone marrow, and lymphoid tissues	Epithelium of respiratory tract, auditory tube, and tonsils
Epithelium of oral and nasal cavities, of paranasal sinuses, and of anal canal	Endothelium of blood vessels and lymphatics	Thyroid, parathyroid, and thymus glands
Tooth enamel	Serosae of ventral body cavity	Epithelium of reproductive ducts and glands
Epithelium of pineal and pituitary glands and adrenal medulla	Fibrous and vascular tunics of eyes	Epithelium of urethra and bladder
Melanocytes: some cranial bones and branchial cartilages (derived from neural crest)	Synovial membranes of joint cavities	
	Organs of urogenital system (ureters, kidneys, gonads, and reproductive ducts)	

the dermis of the skin in the dorsal part of the body. The **myotome** (mi′o-tōm; "muscle piece") consists of somite cells that remain behind as the sclerotome and dermatome cells migrate away. The myotomes develop in conjunction with the vertebrae. They form the skeletal muscles of the neck, body trunk, and, via their **limb buds,** the muscles of the limbs.

Cells of the **intermediate mesoderm** form the gonads and kidneys. The **lateral mesoderm** consists of paired mesodermal plates: the *somatic mesoderm* and the *splanchnic mesoderm.* Cells of the **somatic mesoderm** have three major roles: (1) They help to form the dermis of the skin in the ventral body region; (2) they form the parietal serosa that lines the ventral body cavity; and (3) they migrate into the forming limbs and produce the bones, ligaments, and dermis of the limbs (see Figure 29.12c). **Splanchnic mesoderm** provides the mesenchymal cells that form the heart and blood vessels and most connective tissues of the body. Splanchnic mesodermal cells also collect around the endodermal mucosal linings, where they form the smooth muscle, connective tissues, and serosal coverings (in other words, nearly the entire wall) of the digestive and respiratory organs. As you can see in Figure 29.12c, folding of the embryonic body forms the ventral body cavity, called the **coelom** (se′lom), at this time. As already described, the lateral mesodermal layers cooperate to form the serosae of the coelom.

By the end of the embryonic period, the bones have begun to ossify and the skeletal muscles are well formed and contracting spontaneously. The metanephric kidneys are developing, the gonads are formed, and the lungs and digestive organs are attaining their final shape and body position. Blood delivery to and from the placenta via the umbilical vessels is constant and efficient. The heart and the liver are competing with each other for space and form a conspicuous bulge on the ventral surface of the embryo's body. All this by the end of eight weeks in an embryo about 1 inch long from crown to rump!

Development of the Fetal Circulation Embryonic development of the cardiovascular system lays the groundwork for the fetal circulatory pattern, which is converted to the adult pattern at birth. The first blood cells arise in the yolk sac. Before the third week of development, tiny spaces appear within the splanchnic mesoderm. These are quickly lined by endothelial cells, covered with mesenchyme, and linked together into rapidly spreading vascular networks, destined to form the heart, blood vessels, and lymphatics. By the end of the third week, the embryo has a fairly elaborate system of paired blood vessels, and the two vessels forming the heart have fused and bent into an S shape (see Figure 19.24). By 3½ weeks, the miniature heart is pumping blood for an embryo less than a quarter inch long.

Unique vascular modifications seen only during prenatal development include the **umbilical arteries** and **vein** and three *vascular shunts* (Figure 29.13). All of these structures are occluded at birth. The

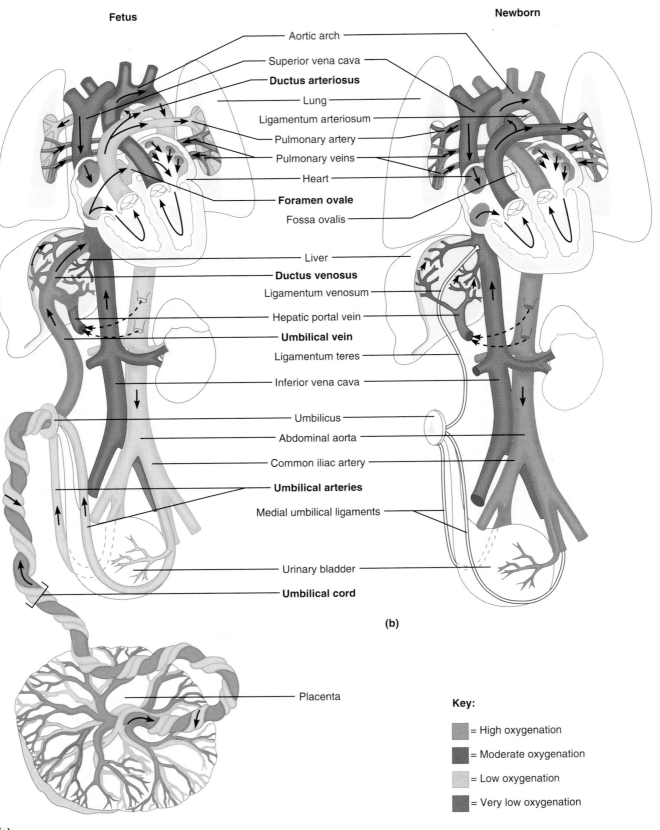

Fetus

Newborn

Aortic arch

Superior vena cava

Ductus arteriosus

Lung

Ligamentum arteriosum

Pulmonary artery

Pulmonary veins

Heart

Foramen ovale

Fossa ovalis

Liver

Ductus venosus

Ligamentum venosum

Hepatic portal vein

Umbilical vein

Ligamentum teres

Inferior vena cava

Umbilicus

Abdominal aorta

Common iliac artery

Umbilical arteries

Medial umbilical ligaments

Urinary bladder

Umbilical cord

(b)

Placenta

Key:

= High oxygenation

= Moderate oxygenation

= Low oxygenation

= Very low oxygenation

(a)

FIGURE 29.13 *Circulation in fetus and newborn.* Arrows indicate direction of blood flow. **(a)** Special adaptations for embryonic and fetal life. The umbilical vein carries oxygen- and nutrient-rich blood from the placenta to the fetus; the umbilical arteries carry waste-laden blood from the fetus to the placenta; the ductus arteriosus and foramen ovale bypass the nonfunctional lungs; and the ductus venosus allows blood to partially bypass the liver. **(b)** Changes in the cardiovascular system at birth. The umbilical vessels are occluded, as are the liver and lung bypasses (ductus venosus and arteriosus, and the foramen ovale).

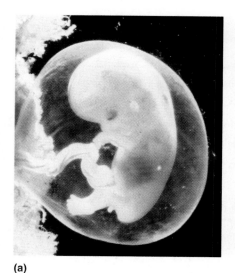

(a)

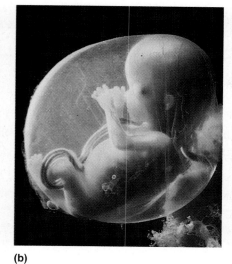

(b)

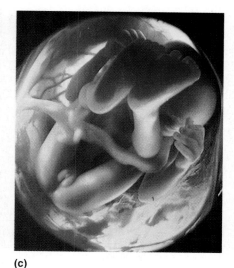

(c)

FIGURE 29.14 *Photographs of developing fetuses.* The major events of fetal development are growth and tissue specialization. All the organ systems are laid down, at least in rudimentary form, during the embryonic period of development. (**a**) Fetus of 9 weeks is about 3 cm long, and every organ system and external feature is present. (**b**) Fetus of 14 weeks, approximately 6 cm long. (**c**) Fetus of 20 weeks, approximately 19 cm long. By birth the fetus is typically 35 cm in length crown to rump.

large umbilical vein carries freshly oxygenated blood returning from the placenta into the embryonic body, where it is conveyed to the liver. There, some of the returning blood percolates through the liver sinusoids and out the hepatic veins. Most of the blood coursing through the umbilical vein, however, enters the **ductus venosus** (duk′tus ve-no′sus), a venous shunt that bypasses the liver entirely. Both the hepatic veins and the ductus venosus empty into the inferior vena cava where the placental blood mixes with deoxygenated blood returning from the lower parts of the fetus's body. The vena cava in turn conveys this "mixed load" of blood directly to the right atrium of the heart. After birth, the liver plays an important role in nutrient processing, but during embryonic life these functions are performed by the mother's liver. Consequently, blood flow through the liver during development is important only to ensure viability of the liver cells.

Blood entering and leaving the heart encounters two more shunt systems, each serving to bypass the nonfunctional lungs. Some of the blood entering the right atrium flows directly into the left side of the heart via the **foramen ovale** ("oval hole"), an opening in the interatrial septum. Blood that enters the right ventricle is pumped out into the pulmonary trunk. However, the second shunt, the **ductus arteriosus**, transfers most of that blood directly into the aorta, again bypassing the pulmonary circuit. (The lungs *do* receive adequate blood to maintain their growth.) Blood enters the two pulmonary bypass shunts because the heart chamber or vessel on the other side

of each shunt is a lower-pressure area, owing to the low volume of venous return from the lungs. Blood flowing distally through the aorta eventually reaches the umbilical arteries, which are branches of the internal iliac arteries serving the pelvis. From here the largely deoxygenated blood, laden with metabolic wastes, is delivered back to the capillaries in the chorionic villi of the placenta. The changes in the circulatory plan that occur at birth are described in Figure 29.13b.

EVENTS OF FETAL DEVELOPMENT

The main events of the fetal period—weeks 9 through 38—are listed chronologically in Table 29.2. The fetal period is a time of rapid growth of the body structures that were established in the embryo. The fetal period is more than just a time of growth, however. During the first half of this period, cells are still differentiating into specific cell types to form the body's distinctive tissues and are completing the fine details of body structure. During the fetal period, the developing fetus grows from a crown-to-rump length of about 30 mm (slightly more than 1 inch) and a weight of approximately 1 g (0.03 ounce) to about 360 mm (14 inches) and 2.7–4.1 kg (6–10 pounds) or more. (Total body length at birth is about 550 mm, or 22 inches.) As you might expect with such tremendous growth, the changes in fetal appearance are quite dramatic (Figure 29.14).

TABLE 29.2	**Developmental Events of the Fetal Period**

Time	Changes and accomplishments
8 weeks (end of embryonic period) 8 weeks	Head nearly as large as body; all major brain regions present; first brain waves in brain stem Liver disproportionately large and begins to form blood cells Limbs present; digits are initially webbed, but fingers and toes are free by the end of this interval Ossification just begun; weak, spontaneous muscle contractions occur Cardiovascular system fully functional (heart has been pumping blood since the fourth week) All body systems present in at least rudimentary form Approximate crown-to-rump length: 30 mm (1.2 inches); weight: 2 grams (0.06 ounce)
9–12 weeks (third month) 12 weeks	Head still dominant, but body elongating; brain continues to enlarge, shows its general structural features; cervical and lumbar enlargements apparent in spinal cord; retina of eye is present Skin epidermis and dermis obvious; facial features present in crude form Liver prominent and bile being secreted; palate is fusing; most glands of endodermal origin are developed; walls of hollow visceral organs gaining smooth muscle Blood cell formation begins in bone marrow Notochord degenerating and ossification accelerating; limbs well molded Sex readily detected from the genitals Approximate crown-to-rump length at end of interval: 90 mm
13–16 weeks (fourth month) 16 weeks	Cerebellum becoming prominent; general sensory organs differentiated; eyes and ears assume characteristic position and shape; blinking of eyes and sucking motions of lips occur Face looks human and body beginning to outgrow head Glands developed in GI tract; meconium is collecting Kidneys attain typical structure Most bones are now distinct and joint cavities are apparent Approximate crown-to-rump length at end of interval: 140 mm
17–20 weeks (fifth month)	Vernix caseosa (fatty secretions of sebaceous glands) covers body; lanugo (silklike hair) covers skin Fetal position (body flexed anteriorly) assumed because of space restrictions Limbs reach near-final proportions Quickening occurs (mother feels spontaneous muscular activity of fetus) Approximate crown-to-rump length at end of interval: 190 mm
21–30 weeks (sixth and seventh months) At birth	Period of substantial increase in weight (may survive if born prematurely at 27–28 weeks, but hypothalamic temperature regulation and lung production of surfactant are still inadequate) Myelination of cord begins; eyes are open Distal limb bones are beginning to ossify Skin is wrinkled and red; fingernails and toenails are present; tooth enamel is forming on deciduous teeth Body is lean and well proportioned Bone marrow becomes sole site of blood cell formation Testes reach scrotum in seventh month (in males) Approximate crown-to-rump length at end of interval: 280 mm
30–40 weeks (term) (eighth and ninth months)	Skin whitish pink; fat laid down in subcutaneous tissue (hypodermis) Approximate crown-to-rump length at end of interval: 360–400 mm (14–16 inches); weight: 2.7–4.1 kg (6–10 pounds)

Homeostatic Imbalance

Since many potentially harmful substances can cross placental barriers and enter the fetal blood, a pregnant woman should be very much aware of what she is taking into her body, particularly during the embryonic period when the body's foundations are laid down. **Teratogens** (ter'ah-to-jenz; *terato = monster*), factors that may cause severe congenital abnormalities or even fetal death, include alcohol, nicotine, many drugs (anticoagulants, sedatives, antihypertensives, and some antibiotics), and maternal infections, particularly German measles. For example, when a woman drinks alcohol, her fetus becomes inebriated as well. However, the fetal consequences may be much more lasting and result in the *fetal alcohol syndrome (FAS)* typified by microcephaly (small head), mental retardation, and abnormal growth. Nicotine hinders oxygen delivery to the fetus, impairing normal growth and development. The sedative **thalidomide** (thah-lid'o-mīd) was used by thousands of pregnant women in the 1960s to alleviate morning sickness. When taken during the period of limb bud differentiation (days 26–56), it sometimes resulted in tragically deformed infants with short flipperlike legs and arms. ■

EFFECTS OF PREGNANCY ON THE MOTHER

Pregnancy can be a difficult time for the mother. Not only are there anatomical changes, but striking changes in her metabolism and physiology occur as well.

Anatomical Changes

As pregnancy progresses, the female reproductive organs become increasingly vascularized and engorged with blood, and the vagina develops a purplish hue *(Chadwick's sign)*. This enhanced vascularity increases vaginal sensitivity and sexual intensity, and some women achieve orgasm for the first time when they are pregnant. The breasts, too, engorge with blood and, prodded by rising levels of estrogen and progesterone, they enlarge and their areolae darken. Some women develop increased pigmentation of facial skin of the nose and cheeks, a condition called *chloasma* (klo-az'mah; "to be green") or the "mask of pregnancy."

The degree of uterine enlargement during pregnancy is remarkable. Starting as a fist-sized organ, the uterus fills most of the pelvic cavity by 16 weeks (Figure 29.15b). Though the fetus is only about 120

mm long at this time, the placenta is fully formed, uterine muscle is hypertrophied, and amniotic fluid volume is increasing. As pregnancy continues, the uterus pushes higher into the abdominal cavity, exerting pressure on both abdominal and pelvic organs (Figure 29.15c). As birth nears, the uterus reaches the level of the xiphoid process and occupies most of the abdominal cavity (Figure 29.15d). The crowded abdominal organs press superiorly against the diaphragm, which intrudes on the thoracic cavity. As a result, the ribs flare, causing the thorax to widen.

The increasing bulkiness of the anterior abdomen changes the woman's center of gravity, and many women develop lordosis (accentuated lumbar curvature) and backaches during the last few months of pregnancy. Placental production of the hormone **relaxin** causes pelvic ligaments and the pubic symphysis to relax, widen, and become more flexible. This increased motility eases birth passage, but it may result in a waddling gait in the meantime.

Considerable weight gain occurs during a normal pregnancy. Because some women are over- or underweight before pregnancy begins, it is almost impossible to state the ideal or desirable weight gain. However, summing up the weight increases resulting from fetal and placental growth, growth of the maternal reproductive organs and breasts, and increased blood volume during pregnancy, a weight gain of approximately 13 kg (about 29 pounds) usually occurs.

Obviously, good nutrition is necessary all through pregnancy if the developing fetus is to have all the building materials (especially proteins, calcium, and iron) needed to form its tissues. Additionally, multivitamins containing folic acid seem to reduce the risk of having babies with neurological problems, including such birth defects as spina bifida and anencephaly. However, a pregnant woman needs only 300 additional calories daily to sustain proper fetal growth. The emphasis should be on eating high-quality food, not just more food.

Metabolic Changes

As the placenta enlarges, it secretes increasing amounts of **human placental lactogen (hPL)**, also called **human chorionic somatomammotropin (hCS)**, which works cooperatively with estrogens and progesterone to stimulate maturation of the breasts for lactation. In addition, hPL promotes growth of the fetus and exerts a glucose-sparing effect in the mother. Consequently, maternal cells metabolize more fatty acids and less glucose than usual,

? *Using this figure as a reference, explain why many pregnant women complain of being "short-winded."*

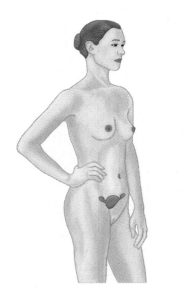 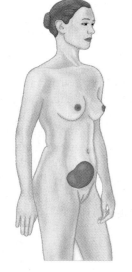 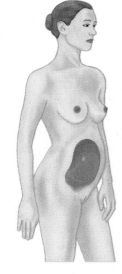 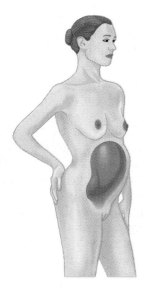

| (a) Before conception | (b) 4 months | (c) 7 months | (d) 9 months |

FIGURE 29.15 *Relative size of the uterus before conception and during pregnancy.* **(a)** Before conception, the uterus is the size of a fist and resides within the pelvis. **(b)** At 4 months, the fundus of the uterus is halfway between the pubic symphysis and the umbilicus. **(c)** At 7 months, the fundus is well above the umbilicus. **(d)** At 9 months, the fundus reaches the xiphoid process.

sparing glucose for use by the fetus. The placenta also releases **human chorionic thyrotropin (hCT),** a glycoprotein hormone similar to thyroid-stimulating hormone of the anterior pituitary. hCT activity increases the rate of maternal metabolism throughout the pregnancy, causing hypermetabolism. Because plasma levels of parathyroid hormone and activated vitamin D rise, pregnant women tend to be in positive calcium balance throughout pregnancy. This ensures that the developing fetus will have adequate calcium to mineralize its bones.

Physiological Changes

Gastrointestinal System

Many women suffer nausea, commonly called *morning sickness,* during the first few months of pregnancy, until their system adjusts to the elevated levels of progesterone and estrogens. (Nausea is also a side effect of many birth control pills.) Reflux of stomach acid into the esophagus, causing *heartburn,*

As the pregnancy advances, the uterus pushes higher in the abdomen, eventually pressing on the diaphragm and inhibiting its downward movement.

is common because the esophagus is displaced and the stomach is crowded by the growing uterus. *Constipation* occurs because motility of the digestive tract declines during pregnancy.

Urinary System

The kidneys produce more urine during pregnancy because they have the additional burden of disposing of fetal metabolic wastes. As the growing uterus compresses the bladder, urination becomes more frequent, more urgent, and sometimes uncontrollable *(stress incontinence).*

Respiratory System

The nasal mucosa responds to estrogens by becoming edematous and congested. Thus, nasal stuffiness and occasional nosebleeds may occur. Vital capacity and respiratory rate increase during pregnancy and hyperventilation is common. But residual volume declines, and many women exhibit *dyspnea* (dispne'ah), or difficult breathing, during the later stages of pregnancy.

Cardiovascular System

The most dramatic physiological changes occur in the cardiovascular system. Total body water rises; blood volume increases 25–40% by the thirty-second week, owing to increases in both formed elements and plasma volume to accommodate the additional needs of the fetus. The rise in blood volume also acts as a safeguard against blood loss during birth. Blood pressure and pulse typically rise and increase cardiac output by 20–40% at various stages of pregnancy. This helps propel the greater circulatory volume around the body. Because the uterus presses on the pelvic blood vessels, venous return from the lower limbs may be impaired, resulting in *varicose veins.*

PARTURITION (BIRTH)

Parturition (par"tu-rish'un; "bringing forth young") is the culmination of pregnancy—giving birth to the baby. It usually occurs within 15 days of the calculated due date (280 days from the last menstrual period). The series of events that expel the infant from the uterus are collectively called **labor.**

Initiation of Labor

While the precise mechanism that triggers labor is not clear, several events and hormones interlock in this process. During the last few weeks of pregnancy, estrogens reach their highest levels in the mother's blood.* This has two important consequences: It stimulates the myometrial cells of the uterus to form abundant oxytocin receptors (Figure 29.16) and it antagonizes progesterone's quieting influence on the uterine muscle. As a result, the myometrium becomes increasingly irritable, and weak, irregular uterine contractions begin to occur. These contractions, called *Braxton Hicks contractions,* have caused many women to go to the hospital, only to be told that they were in **false labor** and sent home.

As birth nears, two more chemical signals cooperate to convert these false labor pains into the real thing. Certain cells of the fetus begin to produce **oxytocin** (ok"sĭ-to'sin), which in turn acts on the placenta (Figure 29.16), causing it to release **prostaglandins** (pros"tah-glan'dinz). Both hormones are powerful uterine muscle stimulants, and since the myometrium is now highly sensitive to oxytocin, contractions become more frequent and more vigorous. At this point, the increasing emotional and physical stresses activate the mother's hypothalamus, which signals for oxytocin release by the posterior pituitary. Together the elevated levels of oxytocin and prostaglandins trigger the rhythmic expulsive contractions of true labor. Once the hypothalamus is involved, a *positive feedback mecha-*

---

*Some studies indicate that the fetus determines its own birth date. Rising levels of fetal adrenocortical hormones (especially cortisol) late in pregnancy have a major effect on stimulating the placenta to release such large amounts of estrogens.

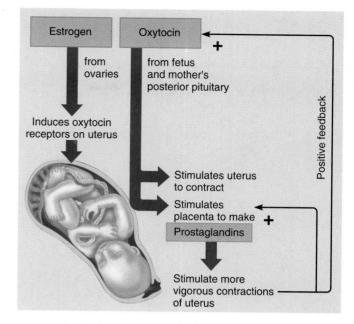

FIGURE 29.16 *Hormonal induction of labor.*

nism is propelled into action—greater contractile force causes the release of more oxytocin, which causes greater contractile force, and so on (Figure 29.16). These expulsive contractions are aided by the fact that *fetal fibronectin,* a natural "stickum" (adhesive protein) that binds the fetal and maternal tissues together throughout pregnancy, changes to a lubricant just before true labor begins.

Both oxytocin and prostaglandins are essential for initiating labor in humans. Untimely spurts of oxytocin may provoke some instances of premature birth, and interfering with production of either of these hormones will hinder onset of labor. For example, antiprostaglandin drugs such as ibuprofen can inhibit the early stages of labor and such drugs are used occasionally to prevent preterm births.

Stages of Labor

Stage 1: Dilation Stage

The **dilation stage** (Figure 29.17a) is the time from labor's onset until the cervix is fully dilated by the baby's head (about 10 cm in diameter). As labor starts, weak but regular contractions (much like peristaltic contractions) begin in the upper part of the uterus and move downward toward the vagina. At first, only the superior uterine muscle is active. These initial contractions are 15–30 minutes apart and last for 10–30 seconds. As labor progresses, the contractions become more vigorous and more rapid, and the lower uterine segment gets involved. As the infant's head is forced against the cervix with each contraction, the cervix softens, becomes thinner (effaces), and dilates. Eventually the amnion ruptures,

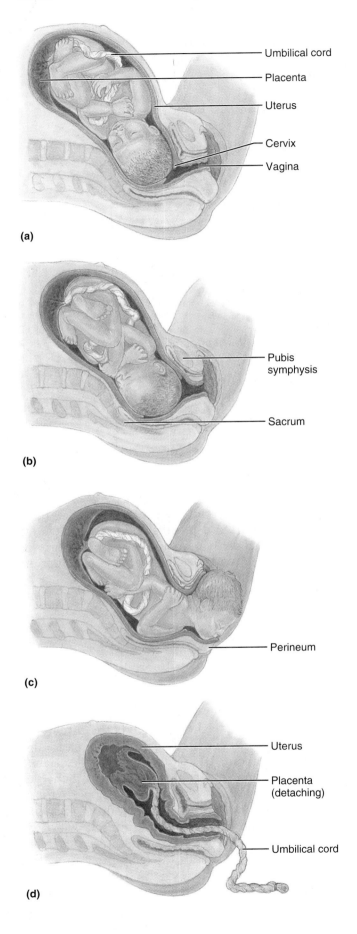

(a)

Umbilical cord
Placenta
Uterus
Cervix
Vagina

(b)

Pubis symphysis
Sacrum

(c)

Perineum

(d)

Uterus
Placenta (detaching)
Umbilical cord

releasing the amniotic fluid, an event commonly called "breaking the water." The dilation stage is the longest part of labor, lasting 6–12 hours (or more). Several events happen during this phase. *Engagement* occurs when the infant's head enters the true pelvis. As descent continues through the birth canal, the baby's head rotates so that its greatest dimension is in the anteroposterior line (Figure 29.17b), which allows it to navigate the narrow dimensions of the pelvic outlet.

Stage 2: Expulsion Stage

The **expulsion stage** (Figure 29.17c) lasts from full dilation to delivery of the infant, or actual childbirth. Ordinarily, by the time the cervix is fully dilated, strong contractions are occurring every 2–3 minutes and lasting about 1 minute, and a mother undergoing labor without local anesthesia has an increasing urge to push or bear down with the abdominal muscles. Although this phase can take as long as 2 hours, it is typically 50 minutes in a first birth and around 20 minutes in subsequent births.

Crowning occurs when the largest dimension of the baby's head is distending the vulva. At this point, an *episiotomy* (e-piz"e-ot'o-me) may be performed to reduce tissue tearing. An episiotomy is an incision made to widen the vaginal orifice. The baby's head extends as it exits from the perineum, and once the head has been delivered, the rest of the baby's body is delivered much more easily. After birth, the umbilical cord is clamped and cut.

When the infant is in the usual *vertex*, or head-first, *presentation*, the skull (its largest diameter) acts as a wedge to dilate the cervix. The head-first presentation also allows the baby to be suctioned free of mucus and to breathe even before it has completely exited from the birth canal (Figure 29.17c). In *breech* (buttock-first) and other nonvertex presentations, these advantages are lost and delivery is much more difficult, often requiring the use of forceps.

Homeostatic Imbalance

If a woman has a deformed or malelike pelvis, labor may be prolonged and difficult. This condition is called *dystocia* (dis-to'se-ah; *dys* = difficult; *toc* = birth). Besides extreme maternal fatigue, another

FIGURE 29.17 *Parturition*. (a) Dilation stage (early). The baby's head has entered the true pelvis and is engaged. The widest head dimension is along the left–right axis. **(b)** Late dilation. The baby's head rotates so that its greatest dimension is in the anteroposterior axis as it moves through the pelvic outlet. Dilation of the cervix is nearly complete. **(c)** Expulsion stage. The baby's head extends as it reaches the perineum and is delivered. **(d)** Placental stage. After the baby has been delivered, the placenta is detached by the continuing uterine contractions and is delivered.

possible consequence of dystocia is fetal brain damage, resulting in cerebral palsy or epilepsy, and decreased viability of the infant. To prevent these outcomes, a *cesarean* (se-sa're-an) *(C-) section* is performed in many such cases. A C-section is delivery of the infant through a surgical incision made through the abdominal and uterine walls. ■

Stage 3: Placental Stage

The **placental stage** (Figure 29.17d), or the delivery of the placenta, is accomplished within 30 minutes after birth of the infant. The strong uterine contractions that continue after birth compress uterine blood vessels, limit bleeding, and cause the placenta to detach from the uterine wall. The placenta and its attached fetal membranes, collectively called the **afterbirth,** are then easily removed by a gentle tug on the umbilical cord. It is very important that all placental fragments be removed to prevent continued uterine bleeding after birth *(postpartum bleeding).* The number of blood vessels in the severed umbilical cord is routinely counted after the placenta has been delivered because absence of one of the umbilical arteries is often associated with cardiovascular disorders in the infant.

ADJUSTMENTS OF THE INFANT TO EXTRAUTERINE LIFE

The **neonatal period** is the four-week period immediately after birth. Here we will be concerned with the events of just the first few hours after birth in a normal infant. As you might suspect, birth represents quite a shock to the infant. Exposed to physical trauma during the birth process, it is suddenly cast out of its watery, warm environment and its placental life supports are severed. Now it must do for itself all that the mother had been doing for it—respire, obtain nutrients, excrete, and maintain its body temperature.

At 1 and 5 minutes after birth, the infant's physical status is assessed based on five signs: heart rate, respiration, color, muscle tone, and reflexes (tested by slaps on the feet). Each observation is given a score of 0 to 2, and the total is called the **Apgar score.** An Apgar score of 8 to 10 indicates a healthy baby. Lower scores reveal problems in one or more of the physiological functions assessed.

Taking the First Breath

The crucial first requirement is to breathe. Once carbon dioxide is no longer removed by the placenta, it accumulates in the baby's blood, causing central acidosis. This excites respiratory control centers in the baby's brain and triggers the first inspiration. The first breath requires a tremendous effort—the air-

ways are tiny, and the lungs are collapsed. However, once the lungs have been inflated in full-term babies, surfactant in alveolar fluid helps to reduce surface tension in the alveoli, and breathing becomes easier. The rate of respiration is rapid (about 45 respirations/min) during the first two weeks and then gradually declines to normal levels.

Keeping the lungs inflated is much more difficult for premature infants (those weighing less than 2500 g, or about 5.5 pounds, at birth) because surfactant production occurs during the last months of prenatal life. Consequently, preemies are usually put on respiratory assistance (a ventilator) until their lungs are mature enough to function on their own.

Occlusion of Special Fetal Blood Vessels and Vascular Shunts

After birth the special umbilical blood vessels and fetal shunts are no longer necessary (see Figure 29.13b). The umbilical arteries and vein constrict and become fibrosed. The proximal parts of the umbilical arteries persist as the *superior vesical arteries* that supply the urinary bladder; their distal parts become the **medial umbilical ligaments.** The remnant of the umbilical vein becomes the **round ligament of the liver,** or **ligamentum teres,** that attaches the umbilicus to the liver. The ductus venosus collapses as blood stops flowing through the umbilical vein and is eventually converted to the **ligamentum venosum** on the liver's undersurface.

As the pulmonary circulation becomes functional, left heart pressure increases and right heart pressure decreases. As a result, the pulmonary shunts close. The flap of the foramen ovale is pushed to the shut position, and its edges fuse to the septal wall. Ultimately, only a slight depression, the **fossa ovalis,** marks its position in adults. The ductus arteriosus constricts and is converted to the cordlike **ligamentum arteriosum,** connecting the aorta and pulmonary trunk.

Except for the foramen ovale, all of the special circulatory adaptations of the fetus are functionally occluded within 30 minutes after birth. Closure of the foramen ovale is usually complete within the year; however, the flap of the foramen ovale *never* fuses in about one-fourth of all people. As a rule, this presents no problem because normal blood pressure in the left atrium causes the "trapdoor" to stay in the closed position. As described in Chapter 19, failure of the ductus arteriosus or foramen ovale to close leads to congenital heart defects.

The Transitional Period

Infants pass through an unstable **transitional period** lasting 6–8 hours after birth, during which they

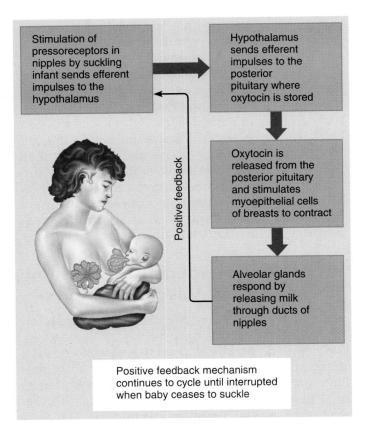

FIGURE 29.18 *Positive feedback mechanism of the milk let-down reflex.*

adjust to extrauterine life. For the first 30 minutes, the baby is awake, alert, and active. Heart rate increases above the normal infant range of 120 to 160 beats per minute, respirations become more rapid and irregular, and body temperature falls. Then activity gradually diminishes, and the baby sleeps for 3 hours or so. A second period of activity then occurs, and the baby gags frequently as it regurgitates mucus and debris. After this, the infant sleeps again and then stabilizes, with waking periods (dictated by hunger) occurring every 3–4 hours.

LACTATION

Lactation is production of milk by the hormone-prepared mammary glands. Rising levels of (placental) estrogens, progesterone, and lactogen toward the end of pregnancy stimulate the hypothalamus to release prolactin-releasing hormone (PRH). The anterior pituitary gland responds by secreting **prolactin.** (This mechanism is described in more detail in Chapter 17, pp. 620–621.) After an initial two- to three-day delay, true milk production begins. During this delay (and during late gestation), a yellowish fluid called **colostrum** (ko-los′trum) is secreted. Colostrum has less lactose than milk and almost no

fat, but it contains more protein, vitamin A, and minerals than true milk. Like milk, colostrum is rich in IgA antibodies. Since these antibodies are resistant to digestion in the stomach, they may help to protect the infant's digestive tract against bacterial infection. Additionally, these IgA antibodies are absorbed by endocytosis and subsequently enter the bloodstream to provide even broader immunity.

After birth, prolactin release gradually wanes, and continual milk production depends on mechanical stimulation of the nipples, normally provided by the sucking infant. Mechanoreceptors in the nipple send afferent nerve impulses to the hypothalamus, stimulating secretion of PRH. This results in a burstlike release of prolactin, which stimulates milk production for the next feeding.

The same afferent impulses also prompt hypothalamic release of oxytocin from the posterior pituitary via a *positive feedback mechanism.* Oxytocin causes the **let-down reflex,** the actual ejection of milk from the alveoli of the mammary glands (Figure 29.18). Let-down occurs when oxytocin binds to myoepithelial cells surrounding the glands, after which milk is ejected from *both* breasts, not just the suckled one. Many obstetricians recommend that mothers nurse because oxytocin also stimulates the recently emptied uterus to contract, helping it to return to (nearly) its prepregnant size.

Breast milk also has advantages for the infant: (1) Its fats and iron are better absorbed and its amino acids are metabolized more efficiently than those of cow's milk. (2) Besides IgA and other immunoglobulins it has a host of other beneficial chemicals, including complement, lysozyme, interferon, and lactoperoxidase, that work to protect infants from life-threatening infections. Mother's milk also contains interleukins and prostaglandins that prevent overzealous inflammatory responses and a glycoprotein that deters the ulcer-causing bacterium *(H. pylori)* from attaching to the stomach mucosa. (3) Its natural laxative effect helps to cleanse the bowels of **meconium** (mĕ-ko′ne-um), a tarry green-black paste containing sloughed-off epithelial cells, bile, and other substances. Since meconium, and later feces, provides the route for eliminating bilirubin from the body, clearing meconium as quickly as possible helps to prevent *physiological jaundice* (see the Related Clinical Terms section). It also encourages bacteria (the source of vitamin K and some B vitamins) to colonize the large intestine.

When nursing is discontinued, the stimulus for prolactin release and milk production ends, and the mammary glands stop producing milk. Women who nurse their infants for six months or more lose a significant amount of calcium from their bones, but those on sound diets usually replace lost bone calcium after weaning the infant. While prolactin levels are high, the normal hypothalamic-pituitary controls of the ovarian cycle are damped, probably because stimulation of the hypothalamus by sucking causes it to release beta endorphin, a peptide hormone that inhibits hypothalamic release of GnRH and hence the release of gonadotropins by the pituitary. Because of this prolactin-induced inhibition of ovarian function, nursing has been called natural birth control. Nonetheless, there is a good deal of "slippage" in these controls, and most women begin to ovulate even while continuing to nurse their infants.

★ ★ ★

In this chapter, we have followed changes that occur during human *in utero* development. But the phenomenon of differentiation has barely been touched upon. How does an unspecialized cell that can become *anything* in the body develop into a specific *something* (a heart cell, for example)? And what paces the developmental sequence, so that if a particular process fails to occur at a precise time, it never occurs at all? Scientists are beginning to believe that there are master switches in the genes. Chapter 30, the final chapter of this book, describes a small part of the "how" as it examines the interaction of genes that determine who we finally become.

RELATED CLINICAL TERMS

Abortion (*abort* = born prematurely) Premature removal of the embryo or fetus from the uterus; may be spontaneous or induced.

Eclampsia (e-klamp'se-ah; *eclamp* = shine) Dangerous condition in which the pregnant woman becomes edematous and hypertensive, and proteinuria and seizures occur; formerly called toxemia of pregnancy.

Ectopic pregnancy (ek-top'ik; *ecto* = outside) A pregnancy in which the embryo implants in any site other than the uterus; most often the site is a uterine tube (tubal pregnancy); since the uterine tube (as well as most other ectopic sites) is unable to establish a placenta or accommodate growth, the uterine tube ruptures unless the condition is diagnosed early, or the pregnancy spontaneously aborts.

Hydatid (hydatidiform) mole (hi'dah-tid; *hydat* = watery) Developmental abnormality of the placenta; the conceptus degenerates and the chorionic villi convert into a mass of vesicles that resemble tapioca. Signs include vaginal bleeding, which contains some of the grapelike vesicles.

Physiological jaundice (jawn'dis) Jaundice sometimes occurring in normal newborns within three to four days after birth. Fetal erythrocytes are short-lived, and they break down rapidly after birth; the infant's liver may be unable to process the bilirubin (breakdown product of hemoglobin pigment) fast enough to prevent its accumulation in blood and subsequent deposit in body tissues.

Placenta abruptio (ah-brup'she-o; *abrupt* = broken away from) Premature separation of the placenta from the uterine wall; if this occurs before labor, it can result in fetal death due to anoxia.

Placenta previa (pre've-ah) Placental formation adjacent to or across the internal os of the uterus; represents a problem because as the uterus and cervix stretch, tearing of the placenta may occur; additionally, the placenta precedes the infant during labor.

Ultrasonography (ul"trah-son-og'rah-fe) Noninvasive technique that uses sound waves to visualize the position and size of the fetus and placenta (see *A Closer Look*, Chapter 1).

CHAPTER SUMMARY

Media study tools that could provide you additional help in reviewing specific key topics of Chapter 29 are referenced below. **SP** = *Study Partner*; **IP** = *Interactive Physiology.*

1. The gestation period of approximately 280 days extends from the woman's last menstrual period to birth. The conceptus undergoes preembryonic development for about two weeks after fertilization, then embryonic development (weeks 3–8) and fetal development (week 9 to birth).

From Egg to Embryo (pp. 1119–1129)

Accomplishing Fertilization (pp. 1119–1122)

1. An oocyte is fertilizable for up to 24 hours; some sperm are viable within the female reproductive tract for up to five days.

2. Sperm must survive the hostile environment of the vagina and become capacitated.

3. Hundreds of sperm must release their acrosomal enzymes to break down the egg's corona radiata and zona pellucida.

4. When one sperm binds to receptors on the egg, it triggers first the fast block to polyspermy (membrane depolarization) and then the slow block (release of cortical granules).

5. Following sperm penetration, the secondary oocyte completes meiosis II. Then the ovum and sperm pronuclei fuse (fertilization), forming a zygote.

Preembryonic Development (pp. 1122–1129)

6. Cleavage, a rapid series of mitotic divisions without intervening growth, begins with the zygote and ends with a blastocyst. The blastocyst consists of the trophoblast and an inner cell mass. Cleavage produces a large number of cells with a favorable surface-to-volume ratio.

7. The trophoblast adheres to, digests, and implants in the endometrium. Implantation is completed when the blastocyst is entirely surrounded by endometrial tissue, about 14 days after ovulation.

8. hCG released by the blastocyst maintains hormone production by the corpus luteum, preventing menses.

hCG levels decline after four months. Typically, the placenta is functional as an endocrine organ by the third month.

9. The placenta acts as the respiratory, nutritive, and excretory organ of the fetus and produces the hormones of pregnancy. It is formed from embryonic (chorionic villi) and maternal (endometrial decidua) tissues. The chorion develops when the trophoblast becomes associated with extraembryonic mesoderm.

Events of Embryonic Development   (pp. 1129–1137)

Formation and Roles of the Embryonic Membranes (pp. 1129–1130)

1. The fluid-filled amnion forms from cells of the superior surface (epiblast) of the embryonic disc. It protects the embryo from physical trauma and adhesion formation, provides a constant temperature, and allows fetal movements.

2. The yolk sac forms from the hypoblast of the embryonic disc; it is the source of primordial germ cells and early blood cells.

3. The allantois, a caudal outpocketing of the yolk sac, forms the structural basis of the umbilical cord.

4. The chorion is the outermost membrane and takes part in placentation.

Gastrulation: Germ Layer Formation (p. 1130)

5. Gastrulation involves cellular migrations that ultimately transform the inner cell mass into a three-layered embryo (gastrula) containing ectoderm, mesoderm, and endoderm. Cells that move through the midline primitive streak become endoderm if they form the most inferior layer of the embryonic disc and mesoderm if they ultimately occupy the middle layer. Cells remaining on the superior surface become ectoderm.

Organogenesis: Differentiation of the Germ Layers (pp. 1130–1137)

6. Ectoderm forms the nervous system and the epidermis of the skin and its derivatives. The first event of organogenesis is neurulation, which produces the brain and spinal cord. By the eighth week, all major brain regions are formed.

SP Case Study: Congenital Defect.

7. Endoderm forms the mucosa of the digestive and respiratory systems, and all associated glands (thyroid, parathyroids, thymus, liver, pancreas). It becomes a continuous tube when the embryonic body undercuts and fuses ventrally.

8. Mesoderm forms all other organ systems and tissues. It segregates early into (1) a dorsal superior notochord, (2) paired somites that form the vertebrae, skeletal trunk muscles, and part of the dermis, and (3) paired masses of intermediate and lateral mesoderm. The intermediate mesoderm forms the kidneys and gonads. The somatic layer of the lateral mesoderm forms the dermis of skin, parietal serosa, and bones and muscles of the limbs; the splanchnic layer forms the cardiovascular system and the visceral serosae.

9. The fetal cardiovascular system is formed in the embryonic period. The umbilical vein delivers nutrient- and oxygen-rich blood to the embryo; the paired umbilical arteries return oxygen-poor, waste-laden blood to the placenta. The ductus venosus allows most of the blood to bypass the liver; the foramen ovale and ductus arteriosus are pulmonary shunts.

SP Exercises: Chapter 29, Derivation of Primary Germ Layers; Circulation of a Fetus.

Events of Fetal Development   (pp. 1137–1139)

1. All organ systems are laid down during the embryonic period; growth and tissue/organ specialization are the major events of the fetal period.

2. During the fetal period, fetal length increases from about 30 mm to 360 mm, and weight increases from less than an ounce to 6–10 pounds.

Effects of Pregnancy on the Mother (pp. 1139–1141)

Anatomical Changes   (p. 1139)

1. Maternal reproductive organs and breasts become increasingly vascularized during pregnancy, and the breasts enlarge.

2. The uterus eventually occupies nearly the entire abdominopelvic cavity. Abdominal organs are pushed superiorly and encroach on the thoracic cavity, causing the ribs to flare.

3. The increased abdominal mass changes the woman's center of gravity; lordosis and backache are common. A waddling gait occurs as pelvic ligaments and joints are loosened by placental relaxin.

4. A typical weight gain during pregnancy in a woman of normal weight is 29 pounds.

Metabolic Changes   (pp. 1139–1140)

5. Human placental lactogen has anabolic effects and promotes glucose sparing in the mother. Human chorionic thyrotropin results in maternal hypermetabolism.

Physiological Changes   (pp. 1140–1141)

6. Many women suffer morning sickness, heartburn, and constipation during pregnancy.

7. The kidneys produce more urine, and pressure on the bladder may cause frequency, urgency, and stress incontinence.

8. Vital capacity and respiratory rate increase, but residual volume decreases. Dyspnea is common.

9. Total body water and blood volume increase dramatically. Heart rate and blood pressure rise, resulting in enhancement of cardiac output in the mother.

Parturition (Birth)   (pp. 1141–1143)

1. Parturition encompasses a series of events called labor.

Initiation of Labor   (p. 1141)

2. When estrogen levels are sufficiently high, they induce oxytocin receptors on the myometrial cells and inhibit progesterone's quieting effect on uterine muscle. Weak, irregular contractions begin.

3. Fetal cells produce oxytocin, which stimulates prostaglandin production by the placenta. Both hormones stimulate contraction of uterine muscle. Increasing stress activates the hypothalamus, causing oxytocin release from the posterior pituitary; this sets up a positive feedback loop resulting in true labor.

SP Exercise: Chapter 29, How Oxytocin Promotes Labor Contractions.

Stages of Labor   (pp. 1141–1143)

4. The dilation stage is from the onset of rhythmic, strong contractions until the cervix is fully dilated. The head of the fetus rotates as it descends through the pelvic outlet.

5. The expulsion stage extends from full cervical dilation until birth of the infant.

6. The placental stage is the delivery of the afterbirth (the placenta and attached fetal membranes).

Adjustments of the Infant to Extrauterine Life (pp. 1143–1144)

1. The infant's Apgar score is recorded immediately after birth.

Taking the First Breath (p. 1143)

2. After the umbilical cord is clamped, carbon dioxide accumulates in the infant's blood, causing respiratory centers in the brain to trigger the first inspiration.

3. Once the lungs are inflated, breathing is eased by the presence of surfactant, which decreases the surface tension of the alveolar fluid.

Occlusion of Special Fetal Blood Vessels and Vascular Shunts (p. 1143)

4. Inflation of the lungs causes pressure changes in the circulation; as a result, the umbilical arteries and vein, ductus venosus, and ductus arteriosus collapse, and the foramen ovale closes. The occluded blood vessels are converted to fibrous cords; the site of the foramen ovale becomes the fossa ovalis.

The Transitional Period (pp. 1143–1144)

5. For the first 8 hours after birth, the infant is physiologically unstable and adjusting. After stabilizing, the infant wakes approximately every 3–4 hours in response to hunger.

Lactation (pp. 1144–1145)

1. The breasts are prepared for lactation during pregnancy by high blood levels of estrogen, progesterone, and placental lactogen.

2. Colostrum, a premilk fluid, is a fat-poor fluid that contains more protein, vitamin A, and minerals than true milk. It is produced toward the end of pregnancy and for the first two to three days after birth.

3. True milk is produced around day 3 in response to suckling, which stimulates the hypothalamus to prompt anterior pituitary release of prolactin and posterior pituitary release of oxytocin. Prolactin stimulates milk production; oxytocin triggers milk let-down. Continued breast-feeding is required for continued milk production.

4. At first, ovulation and menses are absent or irregular during nursing, but in most women the ovarian cycle is eventually reestablished while still nursing.

REVIEW QUESTIONS

Multiple Choice/Matching

1. Indicate whether each of the following statements is describing (a) cleavage or (b) gastrulation.

_____ **(1)** period during which a morula forms
_____ **(2)** period when vast amounts of cell migration occur
_____ **(3)** period when the three embryonic germ layers appear
_____ **(4)** period during which the blastocyst is formed

2. Most systems are operational in the fetus by four to six months. Which system is the exception to this generalization, affecting premature infants? (a) the circulatory system, (b) the respiratory system, (c) the urinary system, (d) the digestive system.

3. The zygote contains chromosomes from (a) the mother only, (b) the father only, (c) both the mother and father, but half from each, (d) each parent and synthesizes others.

4. The outer layer of the blastocyst, which later attaches to the uterus, is the (a) decidua, (b) trophoblast, (c) amnion, (d) inner cell mass.

5. The fetal membrane that forms the basis of the umbilical cord is the (a) allantois, (b) amnion, (c) chorion, (d) yolk sac.

6. In the fetus, the ductus arteriosus carries blood from (a) the pulmonary artery to the pulmonary vein, (b) the liver to the inferior vena cava, (c) the right ventricle to the left ventricle, (d) the pulmonary trunk to the aorta.

7. Which of the following changes occur in the baby's cardiovascular system after birth? (a) blood clots in the umbilical vein, (b) the pulmonary vessels dilate as the lungs expand, (c) the ductus venosus becomes obliterated, as does the ductus arteriosus, (d) all of these.

8. Following delivery of the infant, the delivery of the afterbirth includes the (a) placenta only, (b) placenta and decidua, (c) placenta and attached (torn) fetal membranes, (d) chorionic villi.

9. Identical twins result from the fertilization of (a) one ovum by one sperm, (b) one ovum by two sperm, (c) two ova by two sperm, (d) two ova by one sperm.

10. The umbilical vein carries (a) waste products to the placenta, (b) oxygen and food to the fetus, (c) oxygen and food to the placenta, (d) oxygen and waste products to the fetus.

11. The germ layer from which the skeletal muscles, heart, and skeleton are derived is the (a) ectoderm, (b) endoderm, (c) mesoderm.

12. Which of the following cannot pass through placental barriers? (a) blood cells, (b) glucose, (c) amino acids, (d) gases, (e) antibodies.

13. The most important hormone in initiating and maintaining lactation after birth is (a) estrogen, (b) FSH, (c) prolactin, (d) oxytocin.

14. The initial stage of labor, during which the neck of the uterus is stretched, is the (a) dilation stage, (b) expulsion stage, (c) placental stage.

15. Match each embryonic structure in Column A with its adult derivative in Column B.

Column A	Column B
_____ **(1)** notochord	
_____ **(2)** ectoderm (not neural tube)	**(a)** kidney
_____ **(3)** intermediate mesoderm	**(b)** peritoneal cavity
_____ **(4)** splanchnic mesoderm	**(c)** pancreas, liver
_____ **(5)** sclerotome	**(d)** parietal serosa, dermis
_____ **(6)** coelom	**(e)** nucleus pulposus
_____ **(7)** neural tube	**(f)** visceral serosa
_____ **(8)** somatic mesoderm	**(g)** hair and epidermis
_____ **(9)** endoderm	**(h)** brain
	(i) ribs

Short Answer Essay Questions

16. Fertilization involves much more than a mere restoration of the diploid chromosome number. (a) What does the process of fertilization entail on the part of both the egg and sperm? (b) What are the effects of fertilization?

17. Cleavage is an embryonic event that mainly involves mitotic divisions. How does cleavage differ from mitosis occurring during life after birth, and what are its important functions?

18. The life span of the ovarian corpus luteum is extended for nearly three months after implantation, but otherwise it deteriorates. (a) Explain why this is so. (b) Explain why it is important that the corpus luteum remain functional following implantation.

19. The placenta is a marvelous, but temporary, organ. Starting with a description of its formation, show how it is an intimate part of both fetal and maternal anatomy and physiology during the gestation period.

20. Why is it that only one sperm out of the hundreds (or thousands) available enters the oocyte?

21. What is the function of gastrulation?

22. (a) What is a breech presentation? (b) Cite two problems with this type of presentation.

23. What factors are believed to bring about uterine contractions at the termination of pregnancy?

24. Explain how the flat embryonic disc takes on the cylindrical shape of a tadpole.

Critical Thinking and Clinical Application Questions

1. Jennie, a freshman in your dormitory, tells you she just discovered that she is three months pregnant. She recently bragged that since she came to college she has been drinking alcohol heavily and experimenting with every kind of recreational drug she could find. From the following, select the advice you would give her, and explain why it is the best choice. (a) She must stop taking drugs, but they could not have affected her fetus during these first few months of her pregnancy. (b) Harmful substances usually cannot pass from mother to embryo, so she can keep using drugs. (c) There could be defects in the fetus, so she should stop using drugs and visit a doctor as soon as possible. (d) If she has not taken any drugs in the last week, she is okay.

2. During Mrs. Jones's labor, the obstetrician decided that it was necessary to perform an episiotomy. What is an episiotomy, and why is it done?

3. A woman in substantial pain called her doctor and explained (between sobs) that she was about to have her baby "right here." The doctor calmed her and asked how she had come to that conclusion. She said that her water had broken and that her husband could see the baby's head. (a) Was she right? If so, what stage of labor was she in? (b) Do you think that she had time to make it to the hospital 60 miles away? Why or why not?

4. Mary is a heavy smoker and has ignored a friend's advice to stop smoking during her pregnancy. On the basis of what you know about the effect of smoking on physiology, describe how Mary's smoking might affect her fetus.

5. While Mortimer was cramming for his anatomy test, he read that some parts of the mesoderm become segmented. He suddenly realized that he could not remember what segmentation is. Define segmentation, and give two examples of segmented structures in the embryo.

HEREDITY 30

The Vocabulary of Genetics (pp. 1150–1151)

1. Define allele.
2. Differentiate clearly between genotype and phenotype.

Sexual Sources of Genetic Variation (pp. 1151–1153)

3. Describe events that lead to genetic variability of gametes.

Types of Inheritance (pp. 1153–1157)

4. Compare and contrast dominant-recessive inheritance with incomplete dominance and codominance.

5. Describe the mechanism of sex-linked inheritance.
6. Explain how polygene inheritance differs from that resulting from the action of a single pair of alleles.

The Influence of Environmental Factors on Gene Expression (p. 1157)

7. Provide examples illustrating how gene expression may be modified by environmental factors.

Nontraditional Inheritance (pp. 1159–1160)

8. Describe how genomic imprinting and extrachromosomal (mitochondrial) inheritance differ from classical Mendelian inheritance.

Genetic Screening, Counseling, and Therapy (pp. 1160–1162)

9. List and explain several techniques used to determine or predict genetic diseases.
10. Describe briefly some approaches of gene therapy.

Why are mitotic (metaphase) cells studied instead of those in interphase when a geneticist orders a karyotype?

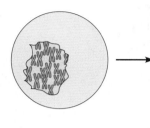

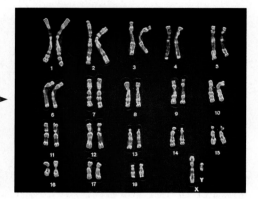

(a) The slide is viewed with a microscope, and the chromosomes are photographed.

(b) The photograph is entered into a computer, and the chromosomes are electronically rearranged into pairs according to size and space.

(c) The resulting display is the karyotype.

FIGURE 30.1   *Preparing a human karyotype.* After lymphocytes are stimulated to divide and grow in culture for several days, they are treated with a drug that arrests mitosis in metaphase, a stage when the chromosomes are easily identified. Then the cells are harvested, treated with a solution that makes their chromosomes spread out, and prepared for microscopic examination. As illustrated in (**a**), the chromosomes are photographed, and the photograph is entered into a computer for analysis and arrangement of the chromosomes into homologous pairs (**b**). The completed karyotype is displayed for examination of chromosomal structure and number (**c**).

The wondrous growth and development of a new individual is guided by the codes of the gene-bearing chromosomes it receives from its parents in the egg and sperm. As described in Chapter 3, individual *genes,* or DNA segments, contain the "recipes" or genetic blueprints for proteins, many of which are enzymes that dictate the synthesis of virtually all of the body's molecules. Consequently, genes are ultimately expressed in your hair color, sex, blood type, and so on. However, genes do not act as free agents. As you will see, the ability of a gene to prompt the development of a specific trait is enhanced or inhibited by interactions with other genes, as well as by environmental factors.

Although the science of **genetics,** which studies the mechanism of heredity (*genes* = give birth to), is still relatively young, our understanding of how human genes act and interact has advanced considerably since the basic principles of heredity were first proposed by the monk Gregor Mendel in the mid-1800s. Part of the problem in unraveling the story of human genetics is that, unlike the pea plants

Mendel studied, humans have long life spans, produce relatively few offspring, and cannot be mated experimentally just to examine the nature of their offspring. Furthermore, Mendel was careful to pick characteristics that vary in an either-or fashion rather than those that vary in a more-or-less way like many human traits. But the urge to understand human inheritance is a powerful one, and recent advances have enabled geneticists to manipulate and engineer human genes to examine their expression and treat or cure disease. Although we touch upon some of these newer topics, we will concentrate on the principles of heredity discovered by Mendel over a century ago.

THE VOCABULARY OF GENETICS

The nuclei of all human cells except gametes contain the diploid number of chromosomes (46), consisting of 23 pairs of homologous chromosomes. Two of this set are **sex chromosomes** (X and Y) that determine genetic sex (that is, male = XY; female = XX). The other 44 are the 22 pairs of **autosomes** that guide the expression of most other traits. The complete human **karyotype** (kar′e-o-tīp), or diploid chromosomal complement displayed in homologous pairs, is illustrated in Figure 30.1. The diploid

Because the genetic material is more dispersed as chromatin during interphase so no individual chromosomes are visible (or identifiable).

genome (je'nōm), or genetic (DNA) makeup, actually represents two sets of genetic instructions—one from the egg and the other from the sperm.

Gene Pairs (Alleles)

Since chromosomes are paired, it follows that the genes within them are paired as well. Consequently, each of us receives two genes, one inherited from each parent (for the most part), that interact to dictate a particular trait. These matched genes, which are at the same *locus* (location) on homologous chromosomes, are called **alleles** (ah-lēlz'). Alleles may code for the same or for alternative forms of a trait. For example, there are two alleles that dictate whether or not you have loose thumb ligaments. One variant codes for tight thumb ligaments; the other codes for loose ligaments (the so-called double-jointed thumb condition). When the two alleles controlling a single trait are the same, a person is said to be **homozygous** (ho-mo-zi'gus) for that particular trait. When the two alleles are different, the individual is **heterozygous** (het"er-o-zi'gus) for the trait.

Sometimes, one allele masks or suppresses the expression of its partner. Such an allele is said to be **dominant,** whereas the allele that is masked is said to be **recessive.** By convention, a dominant allele is represented by a capital letter (for example, *J*), and a recessive allele by a lowercase letter *(j)*. Dominant alleles are expressed, or make themselves "known," when they are present in either single or double dose. But for recessive alleles to be expressed, they must be present in double dose, the homozygous condition. Returning to our thumb example, a person whose genetic makeup includes either the gene pair *JJ* (the homozygous dominant condition) or *Jj* (the heterozygous condition) will have double-jointed thumbs. The combination *jj* (the homozygous recessive condition) is needed to produce tight thumb ligaments.

A common misunderstanding is that dominant traits are automatically seen more often because they are expressed whether the allele is present in single or double dose. However, dominance and recessiveness alone do not determine the frequency of a trait in a population; this also depends on the frequency or relative abundance of the respective dominant or recessive alleles in that population.

Genotype and Phenotype

A person's genetic makeup (that is, whether one is homozygous or heterozygous for the various gene pairs) is referred to as his or her **genotype** (jen'o-tīp).

The way that genotype is expressed in the body is called one's **phenotype** (fe'no-tīp). Thus, the double-jointed condition is the phenotype produced by a genotype of *JJ* or *Jj*.

SEXUAL SOURCES OF GENETIC VARIATION

Before we examine how genes interact, let us consider why each of us is one of a kind, with a unique genotype and phenotype. This variability reflects three events that occur before we are even a twinkle in our parents' eyes; independent assortment of chromosomes, crossover of homologues, and random fertilization of eggs by sperm.

Segregation and Independent Assortment of Chromosomes

As described in Chapter 28 (p. 1080), each pair of homologous chromosomes synapses during meiosis I, forming a tetrad. This happens during both spermatogenesis and oogenesis. Chance determines how the tetrads align on the metaphase I spindle, so maternal and paternal chromosomes are randomly distributed to the daughter nuclei. As illustrated in Figure 30.2, this simple event leads to an amazing amount of variation in the gametes that result. The cell in our example has a diploid chromosome number of 6, so three tetrads are formed. As you can see, the possible combinations of alignments of the three tetrads result in eight different types of gametes. Since the way each tetrad aligns is random, and many mother cells are undergoing meiosis simultaneously, each alignment and each type of gamete occurs with the same frequency as all others. Notice two important points here: (1) The members of the allele pair determining each trait are **segregated,** or distributed to different gametes, during meiosis; and (2) alleles on different pairs of homologous chromosomes are distributed independently of each other. The net result is that each gamete has a single allele for each trait, and that allele represents only one of the four possible parental alleles.

The number of different gamete types resulting from **independent assortment** of the homologues during meiosis I can be quickly calculated for any genome by using the formula 2^n, where n is the number of homologous pairs. In our example, $2^n = 2^3$ (or $2 \times 2 \times 2$), for a total of eight different gamete types. The number of gamete types increases dramatically as chromosome number increases. A cell with six pairs of homologues would produce 2^6, or

? *What event leads to the variability seen here?*

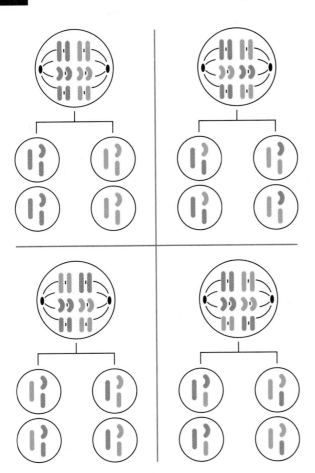

FIGURE 30.2 *Production of variability in gametes by independent assortment of homologous chromosomes during metaphase of meiosis I.* The large circles depict the possible metaphase I alignments of a mother cell with a diploid number of 6. The paternal homologues are depicted as the purple chromosomes; the maternal chromosomes are green. The small circles represent the genetic complements of the daughter cells (gametes) arising from each alignment. Some of the gametes contain all maternal or all paternal chromosomes; others have maternal and paternal chromosomes in varying combinations.

Random alignment of the tetrads on the metaphase I spindle.

64, kinds of gametes. In a man's testes, the number of gamete types that can be produced on the basis of independent assortment alone is 2^{23}, or about 8.5 million possibilities—an incredible variety. Since a human female's ovaries complete at most 500 reduction divisions in her lifetime, the number of different gamete types produced simultaneously is significantly less. Still, each ovulated oocyte will most likely be novel genetically because of the random assortment of chromosomes.

Crossover of Homologues and Gene Recombination

Additional variation results from the crossing over and exchange of chromosomal parts that occur during meiosis I. Genes are arranged linearly along a chromosome's length, and genes on the same chromosome are said to be **linked** because they are physically connected and transmitted as a unit to daughter cells during mitosis. However, during meiosis, paternal chromosomes can precisely exchange gene segments with their homologous maternal counterparts, giving rise to **recombinant chromosomes** with mixed contributions from each parent.

In the hypothetical example shown in Figure 30.3, the genes for hair and eye color are linked. The paternal chromosome contains alleles coding for blond hair and blue eyes, while the maternal alleles code for brown hair and brown eyes. In the **crossover,** or **chiasma,** shown, the break occurs between these linked genes, resulting in some gametes with alleles for blond hair and brown eyes and others with alleles for brown hair and blue eyes. (If the chiasma formed at another position, other gene combinations would occur.) Thus, as a result of crossover, two of the four chromatids present in the tetrad end up with a mixed set of alleles—some maternal and some paternal. This means that when the chromatids segregate, each gamete will receive a unique combination of parental genes.

Only two of the four chromatids in a tetrad appear to take part in crossing over and recombination, but these two may make many crossovers while synapsed. Furthermore, each crossover causes recombination of many genes, not just two as shown in our example. In general, crossovers occur at random points (a sort of molecular lottery), and the longer the chromosome, the greater the number of possible crossovers. Since humans have 23 tetrads, with crossovers going on in most of them during meiosis I, the variability resulting from this factor alone is tremendous.

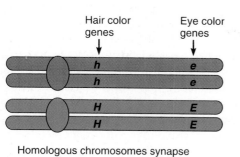

> **?** *What would be the result if the crossover occurred at a different location?*

Hair color genes

Eye color genes

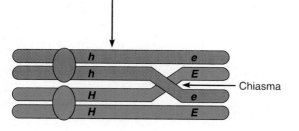

Homologous chromosomes synapse during prophase I of meiosis; each chromosome consists of two sister chromatids.

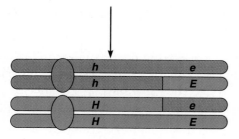

Chiasma

One chromatid segment exchanges positions with a homologous chromatid segment, crossing over occurs, forming a chiasma.

The chromatids forming the chiasma break and the broken-off ends join their corresponding homologues

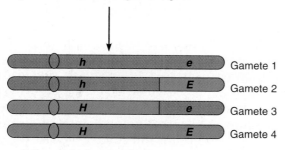

Gamete 1
Gamete 2
Gamete 3
Gamete 4

At the conclusion of meiosis, each haploid gamete has one of the four chromosomes shown; two of the chromosomes are recombinant (they carry new combinations of genes).

Key

H = allele for brown hair   *E* = allele for brown eyes
h = allele for blond hair   *e* = allele for blue eyes

▬ = Paternal chromosome
▬ = Maternal chromosome ⎫ Homologous pair

Random Fertilization

At any point in time, gametogenesis is turning out gametes with all possible variations introduced by independent assortment and random crossovers. Compounding the variety still more is the fact that a single human egg will be fertilized by a single sperm on a totally haphazard basis. If we consider variation resulting only from independent assortment and random fertilization, any resulting offspring represents one out of the close to 72 trillion (8.5 million × 8.5 million) zygotes possible. The additional variation introduced by crossovers increases this number exponentially. Perhaps now you can understand why brothers and sisters are so different, and marvel at how they can also be so alike in many ways.

Once these sources of variability are understood, it is also easy to see that the concept that most people have of their inheritance is inaccurate. We often hear something like "I'm one-half German, one-quarter Scottish, and one-quarter Italian," which reveals a belief that genes from both sides of the family are parceled out in mathematically precise combinations. While it is true that we receive half of our genes from each parent, they may not (indeed, do not) contain one-quarter from each grandparent, or one-eighth from each great grandparent, and there are no such things as German or Italian genes.

TYPES OF INHERITANCE

Although a few phenotypes in humans can be traced to a single gene pair (as described shortly), most such traits are very limited in nature, or reflect variation in a single enzyme. Most human traits are determined by multiple alleles or by the interaction of several gene pairs.

Dominant-Recessive Inheritance

Dominant-recessive inheritance reflects the interaction of dominant and recessive alleles. A simple diagram, called the **Punnett square,** is used to figure out the possible combinations of genes for a single trait that would result from the mating of parents of

FIGURE 30.3 *Crossover and genetic recombination occurring during meiosis I introduce genetic variability in the gametes formed.*

You would get a different selection of linked genes in the resulting gametes.

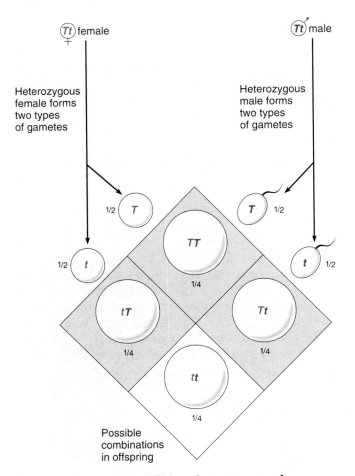

Heterozygous
female forms
two types
of gametes

Heterozygous
male forms
two types
of gametes

Possible
combinations
in offspring

FIGURE 30.4   *Probabilities of genotypes and phenotypes resulting from a mating of two heterozygous parents.* The alleles carried by the mother are shown at left; those carried by the father, along the right side. The Punnett square shows all possible combinations of those alleles in the zygote. In this example, the *T* allele is dominant and determines the tongue-rolling ability; the *t* allele is recessive. Homozygous recessive individuals *(tt)* are unable to exhibit tongue rolling. The *TT* and *Tt* offspring are tongue rollers.

known genotypes (Figure 30.4). In the example shown, both parents can roll their tongues into a U because both are heterozygous for the dominant allele *(T)* that confers this ability. In other words, each parent has the genotype *Tt*. The gamete types (alleles) of one parent are indicated along one side of the Punnett square, and the gamete types for the other parent are shown along an adjacent side. The alleles are combined down and across to determine the possible gene combinations (genotypes) and their expected frequency in their offspring. As you can see from the completed Punnett square, the probability of these parents producing a homozygous dominant child *(TT)* is 25% (1 out of 4); of producing a heterozygous child *(Tt)*, 50% (2 out of 4); and of producing a homozygous recessive child, 25% (1 out

of 4). Since the *T* allele is dominant, the *TT* and *Tt* offspring will be tongue rollers; only the *tt* offspring will be unable to roll their tongues.

Although the Punnett square allows the results of simple genetic crosses to be predicted quickly, it only predicts the *probability* of having a certain percentage of offspring with a particular genotype (and phenotype). The larger the number of offspring, the greater the likelihood that the ratios will conform to the predictions—just as the chances of getting half heads and half tails when we toss a coin increase with the number of tosses. If we toss only twice, we may well get heads both times. Likewise, if the couple in our example has only two children, it would not be surprising if both children had the genotype *Tt*.

What are the chances of having two children of the same genotype? To determine the probability of two events happening in succession, we must multiply the probabilities of the separate events happening. The probability of getting heads for one coin toss is ½, so the probability of getting two heads in a row is ½ × ½ = ¼. (In other words, out of a large number of double coin tosses, we will get two heads one-quarter of the time.) Now consider the chances of our couple having two children, both non–tongue rollers *(tt)*. The probability that one child will be *tt* is ¼, so the probability that both children will be *tt* is ¼ × ¼ = ¹⁄₁₆, or only slightly better than 6%. Remember though, the production of each child, like each coin toss in a series, is an *independent event* that does not influence the others. If you get heads on the first toss, the chance of getting heads the second time is still ½. Likewise, if our couple's first child is a non–tongue roller, they still have a ¼ chance of getting a non–tongue roller the next time.

Dominant Traits

Human traits known to be dictated by dominant alleles include widow's peaks, dimples, freckles, and the condition of detached earlobes, in which the lobe hangs inferior to the point of attachment of the ear.

Genetic disorders caused by dominant genes are fairly uncommon because *lethal dominant genes* are almost always expressed and result in the death of the embryo, fetus, or child. Thus, the deadly genes are not usually passed along to successive generations. However, there are some dominant disorders where the person is less impaired or at least survives long enough to reproduce. One example is *achondroplasia* (a-kon"dro-pla'ze-ah), a rare type of dwarfism resulting from inability of the fetus to form cartilage bone. Another is *Huntington's disease,* a fatal nervous system disease involving degeneration of the basal nuclei. However, in this case, the gene is a *delayed-action gene* that is expressed when

TABLE 30.1	Traits Determined by Simple Dominant-Recessive Inheritance

Phenotype due to expression of:

Dominant genes (ZZ or Zz)	Recessive genes (zz)
Tongue roller	Inability to roll tongue into a U shape
Free (unattached) earlobes	Attached earlobes
Farsightedness	Normal vision
Astigmatism	Normal vision
Freckles	Absence of freckles
Dimples in cheeks	Absence of dimples
Feet with normal arches	Flat feet
PTC* taster	PTC nontaster
Widow's peak	Straight hairline
Double-jointed thumb	Tight thumb ligaments
Broad lips	Thin lips
Polydactyly (extra fingers and toes)	Normal number of fingers and toes
Syndactyly (webbed digits)	Normal digits
Achondroplasia (heterozygous: dwarfism; homozygous: lethal)	Normal cartilage bone formation
Huntington's disease	Absence of Huntington's disease
Normal skin pigmentation	Albinism
Absence of Tay-Sachs disease	Tay-Sachs disease
Absence of cystic fibrosis	Cystic fibrosis
Normal mentation	Schizophrenia

*PTC is phenylthiocarbamide, a harmless bitter chemical.

the affected individual is in his or her late 30s or early 40s. Offspring of a parent with Huntington's disease have a 50% chance of inheriting the lethal gene. (The parent is inevitably heterozygous, since the dominant homozygous condition is lethal to the fetus.) Many informed offspring of such parents are opting not to become parents themselves. These and other dominant gene–determined traits are listed in Table 30.1.

Recessive Traits

Some examples of recessive inheritance are actually the more desirable genetic condition. For example, normal vision is dictated by recessive alleles, whereas farsightedness and astigmatism are prescribed by dominant alleles. However, many if not most genetic disorders are inherited as simple recessive traits. These include conditions as different as *albinism* (lack of skin pigmentation); *cystic fibrosis,* a condition of excessive mucus production which impairs lung and pancreatic functioning; and *Tay-Sachs disease,* a disorder of brain lipid metabolism, caused by an enzyme deficit that shows itself a few months after birth (see p. 88).

Recessive genetic disorders are more frequent than those caused by dominant alleles because those who carry a single recessive allele for a recessive genetic disorder do not themselves express the disease. However, they can pass the gene on to their offspring and, for this reason, such people are called **carriers** of the disorder.

Incomplete Dominance (Intermediate Inheritance)

As opposed to the rules of dominant-recessive inheritance just described, in which one allele variant completely masks the other, some traits exhibit **incomplete dominance,** or **intermediate inheritance.** In such instances, the heterozygote has a phenotype intermediate between those of homozygous dominant and homozygous recessive individuals. Incomplete dominance is common in plants and some animals, but less so in humans.

Perhaps the best human example is inheritance of the *sickling gene (s),* which causes a substitution of one amino acid in the β chain of hemoglobin. Hemoglobin molecules containing the abnormal polypeptide chains crystallize and become sharp when the oxygen tension in blood is low, causing the erythrocytes to assume a sickle shape (see Figure 18.8b, p. 660). The sickling gene is especially widespread among black people (those now living in the malaria belt of Africa or their descendants). Those with a double dose of the sickling allele *(ss)* have **sickle-cell anemia,** and any condition that lowers the oxygen level of their blood, such as respiratory difficulty or excessive exercise, can precipitate a *sickle-cell crisis.* The deformed erythrocytes jam up and fragment in small capillaries, causing intense pain. Vital organs (e.g., the brain or heart) may also suffer ischemic damage. Treatments for sickle-cell anemia include frequent blood transfusions to

TABLE 30.2	ABO Blood Groups			
		Frequency (% of U.S. population)		
Blood group (phenotype)	Genotype	White	Black	Asian
O	ii	45	49	40
A	$I^A I^A$ or $I^A i$	40	27	28
B	$I^B I^B$ or $I^B i$	11	20	27
AB	$I^A I^B$	4	4	5

replace the sickling red blood cells with normal cells or drug therapy with hydroxyurea.

Individuals heterozygous for the sickling gene *(Ss)* make both normal and sickling hemoglobin. As a rule, these individuals are quite healthy, but they can suffer a crisis if there is prolonged reduction in blood oxygen levels, as might happen when traveling in high-altitude areas. It is estimated that 10% of African Americans have the heterozygous condition, known as **sickle-cell trait.** This high frequency is due to the fact that the heterozygous condition confers resistance to malaria—an important adaptation in some parts of Africa. *SS* homozygotes are normal.

Multiple-Allele Inheritance

Although we inherit only two alleles for each gene, some genes actually exhibit more than two alternate forms, leading to a phenomenon called **multiple-allele inheritance.** Inheritance of the ABO blood types is an example. Three alleles determine the ABO blood types in humans: I^A, I^B, and i. Each of us receives two of these. The I^A and I^B alleles are *codominant,* and both are expressed when present. The i allele is recessive to the other two alleles. Genotypes determining the four possible ABO blood groups are indicated in Table 30.2.

Sex-Linked Inheritance

Inherited traits determined by genes on the sex chromosomes are said to be **sex-linked.** The X and Y sex chromosomes are not homologous in the true sense. The Y chromosome, which contains the gene (or genes) that determines maleness, is only about one-third as large as the X chromosome. While the X chromosome bears over 2500 genes, the Y carries only about 15. The Y chromosome lacks many of the genes present on the X that code for nonsexual characteristics. For example, genes coding for certain clotting factors, cone pigments, and even testosterone receptors are present on the X chromosome but not the Y chromosome. A gene found only on the X chromosome is said to be **X-linked.**

When a male inherits an X-linked recessive allele—for example, for hemophilia (bleeder's disease) or for red-green color blindness—its expression is never masked or damped, because there is no corresponding allele on his Y chromosome. Consequently, the recessive gene is always expressed, even when present only in single dose. In contrast, females must have two X-linked recessive alleles to express such a disorder; as a result, very few females exhibit any of these X-linked conditions. X-linked traits are typically passed from mothers to their sons. Since males receive no X chromosome from their fathers, X-linked traits are never passed from father to son. Of course, the mother can also pass the recessive allele to her daughter, but unless the daughter receives another such allele from her father, she will not express the trait.

It should also be noted that some segments of the Y chromosome have no counterpart on the X chromosome. Traits coded for by genes in these segments (including the *SRY* gene necessary for proper male development and that for hairy ear pinnae) appear only in males and are passed from fathers to sons. This type of sex-linked gene transmission is referred to as **Y-linked inheritance.**

Polygene Inheritance

So far, we have considered only those traits inherited by mechanisms of classical Mendelian genetics, which are fairly easily understood. Such traits typically have two, or perhaps three, alternate forms. However, many phenotypes depend on several different gene pairs at different locations acting in tandem. **Polygene inheritance** results in *continuous,* or *quantitative,* phenotypic variation between two extremes and explains many human characteristics. Skin color, for instance, is controlled by three separately inherited genes, each existing in two allelic forms: *A, a; B, b;* and *C, c.* The *A, B,* and *C* alleles confer dark skin pigment, and their effects are additive. The *a, b,* and *c* alleles confer pale skin tone. Hence, an individual with an *AABBCC* genotype would be about as dark-skinned as a human can get, while an *aabbcc* person would be very fair. However, when individuals heterozygous for at least one of these gene pairs mate, a broad range of pigmentation is possible in their offspring. Such polygene inheritance results in a distribution of genotypes and phenotypes that, when plotted, yields a bell-shaped curve (Figure 30.5).

The amount of brown pigment in the iris (which determines eye color) is also regulated by polygenes, as are intelligence and height. Height is controlled by four gene pairs, and different combinations of the tallness or shortness alleles are reflected in differences in stature in much the same way as described for inheritance of skin tone. The phenomenon of polygene inheritance helps explain why parents of average height can produce very tall or very short offspring.

THE INFLUENCE OF ENVIRONMENTAL FACTORS ON GENE EXPRESSION

There are many situations in which environmental factors override or at least influence gene expression. It appears that while our genotype (discounting mutations) is as unchanging as the Rock of Gibraltar, our phenotype is more like clay. If this were not the case, we would never get a tan, women bodybuilders would never develop bulging muscles, and there would be no hope for treating genetic disorders.

Sometimes, maternal factors (drugs, pathogens, and others) alter normal gene expression during embryonic development. Take, for example, the case of the "thalidomide babies" (discussed on p. 1139). As a result of the drug ingestion by their mothers, the embryos developed phenotypes other than that directed by their genes (flipperlike appendages developed to various degrees and lengths). Such environmentally produced phenotypes that mimic conditions that may be caused by genetic mutations (permanent transmissible changes in the DNA) are called **phenocopies.**

Equally significant are environmental factors that may influence genetic expression after birth, such as the effect of poor infant nutrition on subse-

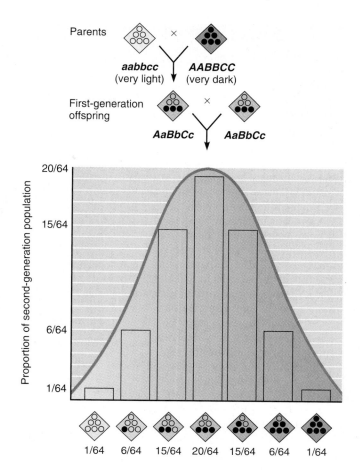

FIGURE 30.5   *Polygene inheritance of skin color based on three separate gene pairs.* Alleles for dark skin are incompletely dominant over those for light skin. Each dominant gene (*A, B,* or *C*) contributes one "unit" of darkness (indicated by a dark dot) to the phenotypes. In the example shown, the parents are homozygotes at opposite ends of the phenotype range. Each of their children inherits 3 "units" of darkness (is a heterozygote) and has intermediate pigmentation. When they mate, their offspring (second generation) may have a wide variation in pigmentation as shown by the histogram (plotted distribution) of skin colors.

quent brain growth, general body development, and height. In this way a person with "tall genes" can be stunted by insufficient nutrition. Furthermore, hormonal deficits during childhood can lead to abnormal skeletal growth and proportions as exemplified by cretinism, a type of dwarfism resulting from hypothyroidism. This is an example of genes (or genetic abnormalities) other than those dictating height having an influence on the height phenotype. Hence, part of a gene's environment consists of the influence of other genes.

A CLOSER LOOK   Stem Cells: Medical Miracle of the Twenty-First Century?

When we are young adults we feel invincible, immortal, and body injuries seem to heal in the blink of an eye, in most cases. But, in time, human tissues and organs can fail, become damaged, or wear out and may be unable to repair themselves. Common examples are the loss of heart muscle following heart attacks, brain degeneration in Parkinson's disease, destruction of pancreatic insulin-producing cells in diabetes, serious burns, spinal cord injuries, and damage to the joints in arthritis.

Some of this damage can be repaired by surgery, or with organ transplants or bionic devices like artificial hips, but these repairs and replacements do not work as well as the original healthy tissues and organs. Regeneration of the damaged tissues would be more desirable. Ways to encourage regeneration within the body itself are currently being investigated, but most progress in "tissue engineering" has been made in culturing human cells and organs in the laboratory for return to the body. For example, cultured skin is now available and a laboratory-grown urinary bladder has been tested successfully in dogs. Tests of lab-grown arteries are also under way.

A problem with culturing adult tissues is that most types of cells from fully differentiated organs are difficult to grow in large quantities in the laboratory. However, immature, **tissue-specific stem cells** can be grown in cell culture for months, and animal studies have shown that when returned to their source they can regenerate all cell types of their parent tissue. Once believed to occur in only a few tissue types, such stem cells are now known to be widespread—perhaps in *every* kind of tissue. Stem cells have been isolated and cultured from nervous tissue of the brain, bone, cartilage, fat, muscle, and blood-forming tissue in bone marrow; and

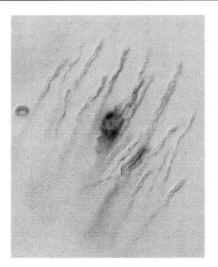

Stem cells

they probably also exist in the liver and pancreas.

Despite their advantages, these tissue-specific stem cells are partially differentiated, meaning they can undergo only a limited number of divisions in culture and when reintroduced into the body. This limits their capacity for regenerating other tissues. Therefore, researchers would prefer to use immortal cells that divide indefinitely and have unlimited potential, such as the earliest versions of the sex cells (primordial germ cells) and cells from early embryos. In 1998, when such **embryonic stem (ES) cells** were isolated from the inner cell mass of a human embryo, the announcement was made with a great deal of excitement and fanfare. These "universal" cells were shown to be capable of long-term culture and able to transform themselves into any type of human tissue when grafted into a developing mouse embryo. (Research on human embryos is always controversial and some say unethical; however, it should be noted that the blastocysts used were left over after fertility treatments and would have been destroyed if not used. Also, individual embryonic stem cells of blastocysts cannot grow into entire

embryos, so their use does not violate current laws that limit research on human embryos.)

The isolation of human ES cells is viewed as a major advance for tissue transplant, perhaps the breakthrough that will one day allow the replacement of any organ. More immediate goals are to use such cells to produce new neurons for victims of Parkinson's disease, new insulin-secreting cells for diabetics, blood-forming cells for victims of bone marrow cancer, and myelinating cells for victims of multiple sclerosis. For example, using ES cells from mice, researchers have produced glial cells that were able to deposit myelin around spinal cord fibers when injected into rats.

Before the long-term goals can become a reality, however, more research must be done. For example, much more must be learned about how to direct cultured ES cells to differentiate into specific, desired cell types. So far, only glia, neurons, cardiac muscle cells, and blood-forming cells have been obtained directly from cultured stem cells—and only from *mouse* stem cells. Medicine is a long way from growing a multi-tissue organ like the brain, the lung, or a bone of the skeleton. Furthermore, before the stem-cell derivatives can be used efficiently, they will probably have to be genetically engineered so the recipient's immune system does not reject them. To avoid this rejection problem, some researchers are seeking ways to reprogram a patient's *own* adult cells back to an undifferentiated state, and use these cells to grow personal organs that will not be rejected. Will ES cells be the miracle scientists and many devastated by illness are hoping for? Who knows.

The paths of new medical technologies often have many blind alleys and twists and turns. Here's hoping this technology takes the throughway.

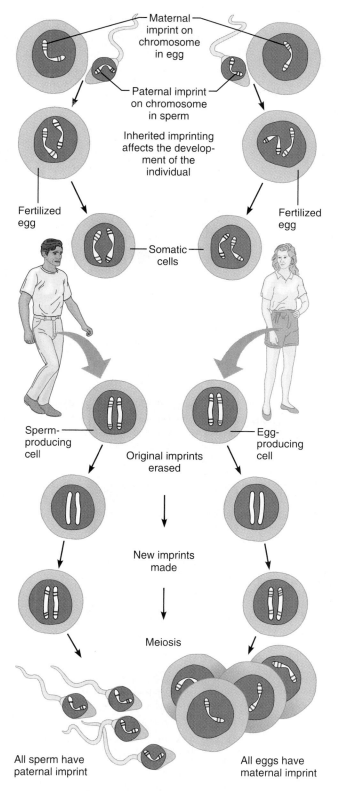

FIGURE 30.6 *Genomic (genetic) imprinting.* Sperm and ova may convey chromosomes that are differently imprinted. The same allele may have a different phenotypic effect in offspring, depending on whether it has the maternal or the paternal imprint. Each generation, the old imprints are "erased" when sperm or ova are produced, and all the chromosomes are newly imprinted according to the sex of the individual.

NONTRADITIONAL INHERITANCE

Mendel's writings underlie mainstream thinking about heredity, but some additional factors fail to fit Mendel's rules. Among these nontraditional types of inheritance are *genomic imprinting* and *extrachromosomal inheritance*, and the story is still unfolding.

Genomic Imprinting

Genomic imprinting accounts for the astonishing fact that the same allele can have a different effect depending on which parent it comes from. For example, victims of Prader-Willi syndrome are mildly to moderately mentally retarded, short, and grossly obese. Children with Angelman syndrome (called *marionette joyeuse* by the French) are severely retarded, unable to speak coherently, laugh uncontrollably, and exhibit jerky, lurching movements as if tied to a puppeteer's strings. The symptoms of these disorders are very different, but the genetic cause is the same—deletion of a particular region of chromosome 15. If the defective chromosome comes from the father, the result is Prader-Willi syndrome; the mother's defective chromosome confers Angelman syndrome. Thus, it seems that the genes of the deleted region normally behave differently depending on whether they are transmitted in the ovum or the sperm.

The explanation for this difference is that during gametogenesis, certain genes in both sperm and eggs are modified by the addition of methyl ($-CH_3$) groups. This process, called **genomic** (or **genetic**) **imprinting,** somehow tags them as paternal or maternal and confers important functional differences in the resulting embryo. The developing embryo "reads" these tags, sometimes expressing the mother's gene while the father's version remains idle (or vice versa). The methylation process is reversible. Each generation, the old imprints are "erased" when new gametes are produced and all the chromosomes are newly imprinted according to the sex of the parents (Figure 30.6).

Extrachromosomal (Mitochondrial) Inheritance

Although we have focused on the chromosomal basis of inheritance, remember that not all genes are in the cell's nucleus. Some are in mitochondria. Since the ovum donates essentially all the cytoplasm in the fertilized egg, mitochondrial genes are transmitted to the offspring almost exclusively by the mother. A growing list of disorders, all rare, is now being linked to errors in mitochondrial genes. Most

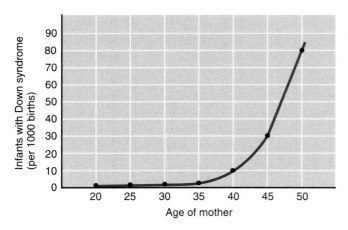

FIGURE 30.7 *Maternal age and Down syndrome.*

involve unusual degenerative muscle disorders or neurological problems. Some researchers suggest that Parkinson's disease may be among them.

GENETIC SCREENING, COUNSELING, AND THERAPY

Genetic screening and *genetic counseling* provide information and options for prospective parents not even dreamed of 100 years ago. Newborn infants are routinely screened for a number of anatomical disorders (congenital hip dysplasia, imperforate anus, and others), and phenylketonuria (PKU) testing is mandated by law in many states. These tests are performed much too late to allow parental choice concerning whether or not to have the child, but they do alert the new parents that treatment is necessary to ensure the well-being of their infant. Anatomical defects are usually treated surgically, and PKU is managed by strict dietary measures that exclude most phenylalanine-containing foods. Adult children of parents with Huntington's disease are obvious candidates for these services, but many other genetic conditions also place babies at risk. For example, a woman pregnant for the first time at the age of 35 may wish to know if her baby has trisomy-21 (Down syndrome), a chromosome abnormality with a high incidence in children of older mothers (Figure 30.7). Depending on the condition to be investigated, the screening process can occur before conception by carrier recognition or during fetal testing.

Carrier Recognition

When a prospective parent displays a recessive genetic disorder, but his or her partner does not, it is important to find out if the partner is heterozygous for the same recessive gene. If not, the offspring will receive only one such gene and will not express the trait. But if the other parent *is* a carrier, then the child's chance of having two harmful genes is 50%.

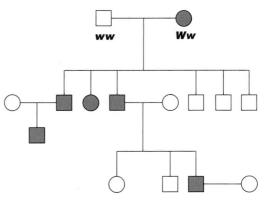

(a) Pedigree of a dominant trait

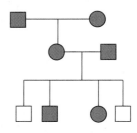

(b) Pedigree of a recessive trait

FIGURE 30.8 *Pedigree analysis is done to detect carriers of particular genes.* Circles represent females; squares represent males. Horizontal lines indicate a mating of the connected individuals; vertical lines show their offspring. (**a**) A pedigree of inheritance of the woolly hair gene in three generations of a family. Colored symbols represent individuals expressing the dominant woolly hair trait. White symbols represent those not expressing the trait; these individuals are homozygous recessive *(ww)*. (**b**) Pedigree tracing the recessive trait albinism (lack of skin pigmentation) through three generations. Both parents in the second generation have pigmented skin (colored symbols), and yet two of their four children have the albino trait (white symbols). We can deduce that both of these parents are heterozygous and have pigmented skin because the allele for that trait is dominant.

There are two major avenues for identifying carriers: pedigrees and blood tests. A **pedigree** traces a particular genetic trait through several generations and helps us predict the future. A genetic counselor collects information on the phenotypes of as many family members as possible and uses it to construct the pedigree. Figure 30.8a uses the rare woolly hair trait of northern Europeans to illustrate how a pedigree is constructed and read. Woolly hair, resulting from the presence of a dominant allele *(W)*, is fuzzy-textured hair that breaks easily. Looking at the pedigree, we know that individuals with the woolly hair phenotype (the colored symbols) have at least one dominant gene *(WW or Ww)*, while those with normal hair must be homozygous recessive *(ww)*. By working backward and applying the rules of dominant-recessive inheritance, we can deduce the genotypes of the parents. Three of their offspring

? *What noninvasive imaging technique is used to determine some aspects of fetal development? (You might want to check A Closer Look in Chapter 1 for this.)*

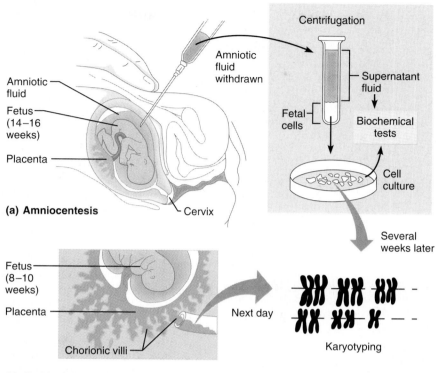

(a) Amniocentesis

(b) Chorionic villi sampling

FIGURE 30.9 *Fetal testing.* **(a)** In amniocentesis, ultrasound is used to locate the fetus, and a small amount of amniotic fluid is extracted for testing. Physicians can diagnose some disorders from chemicals in the fluid itself, while other disorders may show up in tests performed on cells cultured from fetal cells present in the fluid. These tests include biochemical tests to detect the presence of certain enzymes, and karyotyping, to determine whether the chromosomes of the fetal cells are normal in number and microscopic appearance. **(b)** In chorionic villi sampling, a physician inserts a narrow tube through the cervix and suctions out a tiny sample of fetal tissue (chorionic villi) from the placenta. The fetal tissue sample is used immediately for karyotyping.

have normal hair *(ww)*, so we know that each parent must have at least one recessive gene. The mother has woolly hair, determined by the *W* gene, so her genotype must be *Ww*, and the father with normal hair is homozygous recessive. You should be able to take it from here and figure out the most likely genotypes of the offspring. (Try it!) Figure 30.8b illustrates a pedigree analysis for albinism, a recessive trait. See what you can deduce about it. These pedigrees represent two of the simplest cases, because many human traits are controlled by multiple alleles or polygenes and are much more difficult to work out.

Simple blood tests are used to screen for the sickling gene in heterozygotes, and sophisticated *blood chemistry tests* and *DNA probes* can detect the presence of other unexpressed recessive genes. At present, carriers of the Tay-Sachs and cystic fibrosis genes can be identified with such tests.

Fetal Testing

Fetal testing is used when there is a known risk of a genetic disorder. The most common type of fetal testing is **amniocentesis** (am"ne-o-sen-te'sis). In this fairly simple procedure, a wide-bore needle is inserted into the amniotic sac through the mother's abdominal wall (Figure 30.9a), and a small sample (about 10 ml) of fluid is withdrawn. Because there is a chance of injuring the fetus before ample amniotic fluid is present, this procedure is not normally done before the fourteenth week of pregnancy. Using ultrasound to visualize the position of the fetus and the amniotic sac has dramatically reduced the risk of this procedure. Although the fluid is checked for the presence of chemicals (enzymes and others) that serve as markers for specific diseases, most tests are done on the sloughed-off fetal cells in the fluid. These cells are isolated and cultured (grown and proliferated) in laboratory dishes over a period of several weeks. Then the cells are examined for DNA markers of genetic disease and karyotyped (see Figure 30.1) to check for chromosomal abnormalities such as Down syndrome.

Chorionic villi sampling (CVS) suctions off bits of the chorionic villi from the placenta for examination (Figure 30.9b). A small tube is inserted through the vagina and cervical canal and guided by ultrasound to an area where a piece of placental tissue can be removed. CVS allows testing at eight weeks (but waiting until after the tenth week is usually recommended) and is much faster than amniocentesis. Karyotyping can be done almost immediately on the rapidly dividing chorionic cells.

Since both of these procedures are invasive, they carry with them an inherent risk to both fetus and mother. (For example, increased fetal risk of finger and toe defects is linked to CVS.) Although they are routinely ordered for pregnant women over the age of 35 (because of the enhanced risk of Down syndrome), they are performed on younger women only when the probability of finding a severe fetal disorder is greater than the probability of doing harm during the procedure. If a serious genetic or congenital defect is detected in the developing fetus, the parents must decide whether or not to continue the pregnancy.

Human Gene Therapy

As advances occur in diagnosis of human genetic diseases, new applications of "gene therapy" to alleviate or even cure the disorders have been quick to follow, particularly in cases traced to a single defective gene or protein. Indeed, genetic engineering has the potential to replace a defective gene with a normal version. One approach is to "infect" the defective cells with a virus into which a functional gene has been inserted; another is to inject the "corrected" DNA directly into the patient's cells. Such therapies have had mixed success in treating cystic fibrosis and muscular dystrophy. However, genetic engineering processes are prohibitively expensive and raise some thorny ethical, religious, and societal questions: Who will pay? Who determines who will be treated with the new therapies? Are we playing God?

* * *

In this chapter, we have explored some of the most basic principles of genetics, the manner in which genes are expressed, and the means by which gene expression can be modified. Considering the precision required to make perfect copies of genes and chromosomes, and considering the incredible mechanical events of meiotic division, it is amazing that genetic defects are as rare as they are. Perhaps, after reading this chapter, you have a greater sense of wonder that you turned out as well as you did.

RELATED CLINICAL TERMS

Deletion Chromosomal aberration in which part of a chromosome is lost.

Down syndrome Also called **mongolism,** this condition usually reflects the presence of an extra autosome (trisomy of chromosome 21); the child has slightly slanted eyes, flattened facial features, a large tongue, and a tendency toward short stature and stubby fingers. Some, but not all, affected people are mentally retarded. The most important risk factor appears to be advanced maternal (or paternal) age.

Mutation (*mutare* = change) A permanent structural change in a gene; the mutation may or may not affect function, depending on the precise site and nature of the alteration.

Nondisjunction Abnormal segregation of chromosomes during meiosis, resulting in gametes receiving two or no copies of a particular parental chromosome. If the abnormal gamete participates in fertilization, the resulting zygote will have an abnormal chromosomal complement (monosomy or trisomy) for that particular chromosome (as in Down syndrome).

CHAPTER SUMMARY

Media study tools that could provide you additional help in reviewing specific key topics of Chapter 30 are referenced below. **SP** = *Study Partner;* **IP** = *Interactive Physiology.*

1. Genetics is the study of heredity and mechanisms of gene transmission.

The Vocabulary of Genetics   (pp. 1150–1151)

1. A complete diploid set of chromosomes is called the *karyotype* of an organism; the complete genetic comple-

ment is the *genome.* A person's genome consists of two sets of instructions, one from each parent.

Gene Pairs (Alleles)   (p. 1151)

2. Genes coding for the same trait and found at the same locus on homologous chromosomes are called alleles.

3. Alleles may be the same or different in expression. When the allele pair is identical, the person is homozygous for that trait; when the alleles differ, the person is heterozygous.

Genotype and Phenotype (p. 1151)

4. The actual genetic makeup of cells is the genotype; phenotype is the manner in which those genes are expressed.

SP Exercise: Chapter 30, The Vocabulary of Genetics.

Sexual Sources of Genetic Variation (pp. 1151–1153)

Segregation and Independent Assortment of Chromosomes (pp. 1151–1152)

1. During meiosis I of gametogenesis, tetrads align randomly on the metaphase plate, and chromatids are randomly distributed to the daughter cells. This is called independent assortment of the homologues. Each gamete receives only one allele of each gene pair.

2. Each different metaphase I alignment produces a different assortment of parental chromosomes in the gametes, and all combinations of maternal and paternal chromosomes are equally possible.

Crossover of Homologues and Gene Recombination (p. 1152)

3. During meiosis I, two of the four chromatids (one maternal and the other paternal) may cross over at one or more points and exchange corresponding gene segments. The recombinant chromosomes contain new gene combinations, adding to the variability arising from independent assortment.

Random Fertilization (p. 1153)

4. The third source of genetic variation is random fertilization of eggs by sperm.

Types of Inheritance (pp. 1153–1157)

Dominant-Recessive Inheritance (pp. 1153–1155)

1. Dominant genes are expressed when present in single or double dose; recessive genes must be present in double dose to be expressed.

2. For traits following the dominant-recessive pattern, the laws of probability predict the outcome of a large number of matings.

3. Genetic disorders more often reflect the homozygous recessive condition than the homozygous dominant or heterozygous condition because dominant genes are expressed and, if they are lethal genes, the pregnancy is usually aborted. Genetic disorders caused by dominant alleles include achondroplasia and Huntington's disease; recessive disorders include cystic fibrosis and Tay-Sachs disease.

4. Carriers are heterozygotes who carry a deleterious recessive gene (but do not express the trait) and have the potential of passing it on to offspring.

Incomplete Dominance (Intermediate Inheritance) (pp. 1155–1156)

5. In incomplete dominance, the heterozygote exhibits a phenotype intermediate between those of the homozygous dominant and recessive individuals. Inheritance of sickle-cell trait is an example of incomplete dominance.

SP Case Study: Sickle Cell Anemia.

Multiple-Allele Inheritance (p. 1156)

6. Multiple-allele inheritance involves genes that exist in more than two allelic forms in a population. Only two of the alleles are inherited, but on a random basis. Inheritance of ABO blood types is an example of multiple-allele inheritance in which the I^A and I^B alleles are codominant.

SP Exercise: Chapter 30, Punnett Square: ABO Blood Groups.

Sex-Linked Inheritance (p. 1156)

7. Traits determined by genes on the X and Y chromosomes are said to be sex-linked. The small Y chromosome lacks most genes present on the X chromosome. Recessive genes located only on the X chromosome are expressed in single dose in males. Examples of such X-linked conditions, passed from mother to son, include hemophilia and red-green color blindness. There are a few Y-linked genes, passed from father to son only.

Polygene Inheritance (pp. 1156–1157)

8. Polygene inheritance occurs when several gene pairs interact to produce phenotypes that vary quantitatively over a broad range. Height, eye color, and skin pigmentation are examples.

The Influence of Environmental Factors on Gene Expression (p. 1157)

1. Environmental factors may influence the expression of genotype.

2. Maternal factors that cross the placenta may alter expression of fetal genes. Environmentally provoked phenotypes that mimic genetically determined ones are called phenocopies. Nutritional deficits or hormonal imbalances may alter anticipated growth and development during childhood.

SP Case Study: Congenital Defect.

Nontraditional Inheritance (pp. 1159–1160)

Genomic Imprinting (p. 1159)

1. Genomic imprinting, which involves methylation of certain genes during gametogenesis, confers different effects and phenotypes on maternal and paternal genes. It is reversible, and occurs anew each generation.

Extrachromosomal (Mitochondrial) Inheritance (pp. 1159–1160)

2. Cytoplasmic (mitochondrial) genes pass to offspring via the ovum and help to determine certain characteristics. Deletions or mutations in mitochondrial genes are responsible for some rare genetic diseases.

Genetic Screening, Counseling, and Therapy (pp. 1160–1162)

Carrier Recognition (pp. 1160–1161)

1. The likelihood of an individual carrying a deleterious recessive gene may be assessed by constructing a pedigree. Some of these genes can be deduced by various blood tests and DNA probes.

Fetal Testing (pp. 1161–1162)

2. Amniocentesis is fetal testing based on aspirated samples of amniotic fluid. Fetal cells in the fluid are cultured for several weeks, then examined for chromosomal defects (karyotyped) or for DNA markers of genetic disease. Amniocentesis cannot be performed until the fourteenth week of pregnancy.

3. Chorionic villi sampling is fetal testing based on a sample of the chorion. This tissue is rapidly mitotic, so karyotyping can be done almost immediately. Samples may be obtained by the eighth week.

SP Exercise: Chapter 30, Fetal Testing.

Human Gene Therapy (p. 1162)

4. Thus far, gene therapy has been particularly useful for correcting single-gene disorders. The most common approach involves transferring a corrected gene via a virus (or some other vector) to the affected cells to restore normal function.

REVIEW QUESTIONS

Multiple Choice/Matching

1. Match one of the following terms (a–i) with each of the appropriate descriptions below:

(a) alleles   **(d)** genotype   **(g)** phenotype
(b) autosomes   **(e)** heterozygote   **(h)** recessive allele
(c) dominant allele **(f)** homozygote   **(i)** sex chromosomes

_____ **(1)** genetic makeup
_____ **(2)** how genetic makeup is expressed
_____ **(3)** chromosomes that dictate most body characteristics
_____ **(4)** alternate forms of the same gene
_____ **(5)** an individual bearing two alleles that are the same for a particular trait
_____ **(6)** an allele that is expressed, whether in single or double dose
_____ **(7)** an individual bearing two alleles that differ for a particular trait
_____ **(8)** an allele that must be present in double dose to be expressed

2. Match the following types of inheritance (a–e) with the descriptions below:

(a) dominant-recessive   **(d)** polygene
(b) incomplete dominance   **(e)** sex-linked
(c) multiple-allele

_____ **(1)** only sons show the trait
_____ **(2)** homozygotes and heterozygotes have the same phenotype
_____ **(3)** heterozygotes exhibit a phenotype intermediate between those of the homozygotes
_____ **(4)** phenotypes of offspring may be more varied than those of the parents
_____ **(5)** inheritance of ABO blood types
_____ **(6)** inheritance of stature

Short Answer Essay Questions

3. Describe the important mechanisms that lead to genetic variations in gametes.

4. The ability to taste PTC (phenylthiocarbamide) depends on the presence of a dominant gene *T*; nontasters are homozygous for the recessive gene *t*. This is a situation of classical dominant-recessive inheritance. (a) Consider a mating between heterozygous parents producing three offspring. What proportion of the offspring will be tasters? What is the chance that all three offspring will be tasters? Nontasters? What is the chance that two will be tasters and one will be a nontaster? (b) Consider a mating between *Tt* and *tt* parents. What is the anticipated percentage of tasters? Nontasters? What proportion can be expected to be homozygous recessive? Heterozygous? Homozygous dominant?

5. Most albino children are born to normally pigmented parents. Albinos are homozygous for the recessive gene *(aa)*. What can you conclude about the genotypes of the nonalbino parents?

6. A woman with blood type A has two children. One has type O blood and the other has type B blood. What is the genotype of the mother? What are the genotype and phenotype of the father? What is the genotype of each child?

7. In skin color inheritance, what will be the relative range of pigmentation in offspring arising from the following parental matches? (a) *AABBCC × aabbcc*, (b) *AABBCC × AaBbCc*, (c) *AAbbcc × aabbcc*

8. Compare and contrast amniocentesis and chorionic villi sampling as to the time at which they can be performed and the techniques used to obtain information on the fetus's genetic status.

Critical Thinking and Clinical Application Questions

1. A color-blind man marries a woman with normal vision. The woman's father was also color-blind. (a) What is the chance that their first child will be a color-blind son? A color-blind daughter? (b) If they have four children, what is the chance that two will be color-blind sons? (Be careful on this one.)

2. Brian is a college student. His genetics assignment is to do a family pedigree for dimples in the cheeks. Absence of dimples is recessive; presence of dimples reflects a dominant allele. Brian has dimples, as do his three brothers. His mother and maternal grandmother are dimple-free, but his father and all other grandparents have dimples. Construct a pedigree spanning three generations for Brian's family. Show the phenotype and genotype for each person.

3. Mr. and Mrs. Lehman have sought genetic counseling. Mrs. Lehman is concerned because she is unexpectedly pregnant and her husband's brother died of Tay-Sachs disease. She can recall no incidence of Tay-Sachs disease in her own family. Do you think biochemical testing should be recommended to detect the deleterious gene in Mrs. Lehman? Explain your answer.

4. The Browns are both carriers of the recessive allele that causes the metabolic disorder called phenylketonuria. What is the probability of each of the following occurring? (a) All three children will have the disorder. (b) None of their three children will have the disorder. (c) One or more of their children will have the disorder (d) At least one of their children will be phenotypically normal.

APPENDIX A
THE METRIC SYSTEM

Measurement	Unit and abbreviation	Metric equivalent	Metric to English conversion factor	English to metric conversion factor
Length	1 kilometer (km)	= 1000 (10^3) meters	1 km = 0.62 mile	1 mile = 1.61 km
	1 meter (m)	= 100 (10^2) centimeters = 1000 millimeters	1 m = 1.09 yards 1 m = 3.28 feet 1 m = 39.37 inches	1 yard = 0.914 m 1 foot = 0.305 m
	1 centimeter (cm)	= 0.01 (10^{-2}) meter	1 cm = 0.394 inch	1 foot = 30.5 cm 1 inch = 2.54 cm
	1 millimeter (mm)	= 0.001 (10^{-3}) meter	1 mm = 0.039 inch	
	1 micrometer (μm) [formerly micron (μ)]	= 0.000001 (10^{-6}) meter		
	1 nanometer (nm) [formerly millimicron (mμ)]	= 0.000000001 (10^{-9}) meter		
	1 angstrom (Å)	= 0.0000000001 (10^{-10}) meter		
Area	1 square meter (m²)	= 10,000 square centimeters	1 m² = 1.1960 square yards 1 m² = 10.764 square feet	1 square yard = 0.8361 m² 1 square foot = 0.0929 m²
	1 square centimeter (cm²)	= 100 square millimeters	1 cm² = 0.155 square inch	1 square inch = 6.4516 cm²
Mass	1 metric ton (t)	= 1000 kilograms	1 t = 1.103 ton	1 ton = 0.907 t
	1 kilogram (kg)	= 1000 grams	1 kg = 2.205 pounds	1 pound = 0.4536 kg
	1 gram (g)	= 1000 milligrams	1 g = 0.0353 ounce 1 g = 15.432 grains	1 ounce = 28.35 g
	1 milligram (mg)	= 0.001 gram	1 mg = approx. 0.015 grain	
	1 microgram (μg)	= 0.000001 gram		
Volume (solids)	1 cubic meter (m³)	= 1,000,000 cubic centimeters	1 m³ = 1.3080 cubic yards 1 m³ = 35.315 cubic feet	1 cubic yard = 0.7646 m³ 1 cubic foot = 0.0283 m³
	1 cubic centimeter (cm³ or cc)	= 0.000001 cubic meter = 1 milliliter	1 cm³ = 0.0610 cubic inch	1 cubic inch = 16.387 cm³
	1 cubic millimeter (mm³)	= 0.000000001 cubic meter		
Volume (liquids and gases)	1 kiloliter (kl or kL)	= 1000 liters	1 kL = 264.17 gallons	1 gallon = 3.785 L
	1 liter (l or L)	= 1000 milliliters	1 L = 0.264 gallons 1 L = 1.057 quarts	1 quart = 0.946 L
	1 milliliter (ml or mL)	= 0.001 liter = 1 cubic centimeter	1 ml = 0.034 fluid ounce 1 ml = approx. $\frac{1}{4}$ teaspoon 1 ml = approx. 15–16 drops (gtt.)	1 quart = 946 ml 1 pint = 473 ml 1 fluid ounce = 29.57 ml 1 teaspoon = approx. 5 ml
	1 microliter (μl or μL)	= 0.000001 liter		
Time	1 second (s)	= $\frac{1}{60}$ minute		
	1 millisecond (ms)	= 0.001 second		
Temperature	Degrees Celsius (°C)		°F = $\frac{9}{5}$ °C + 32	°C = $\frac{5}{9}$(°F − 32)

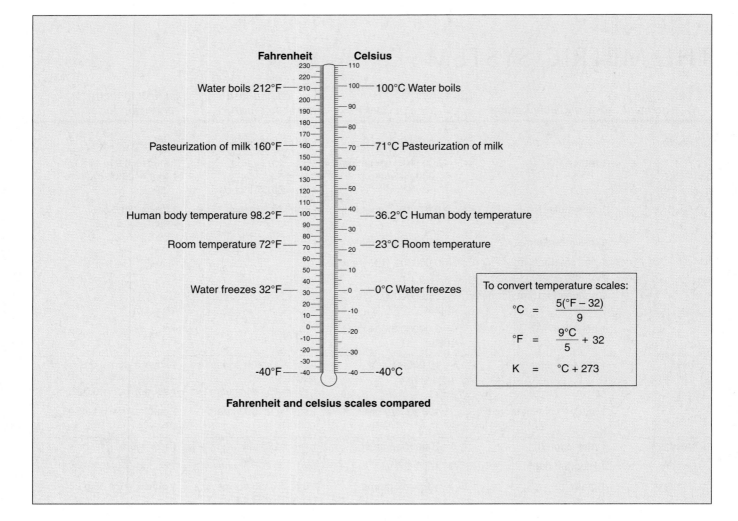

Fahrenheit and celsius scales compared

APPENDIX B
THE AMINO ACIDS

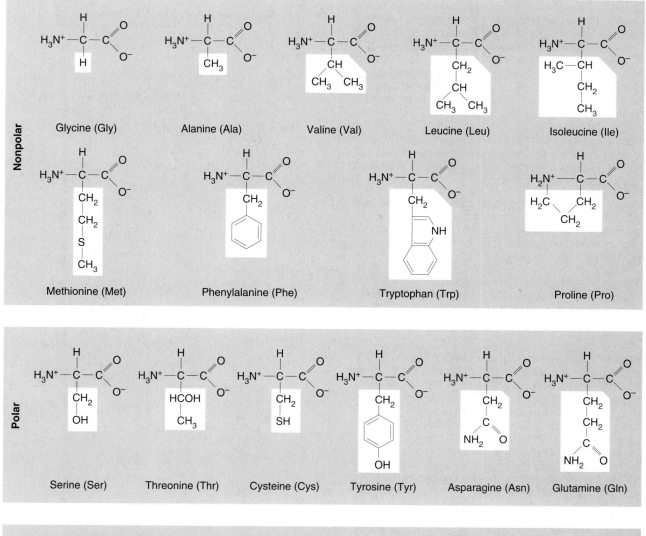

Nonpolar

Glycine (Gly)   Alanine (Ala)   Valine (Val)   Leucine (Leu)   Isoleucine (Ile)

Methionine (Met)   Phenylalanine (Phe)   Tryptophan (Trp)   Proline (Pro)

Polar

Serine (Ser)   Threonine (Thr)   Cysteine (Cys)   Tyrosine (Tyr)   Asparagine (Asn)   Glutamine (Gln)

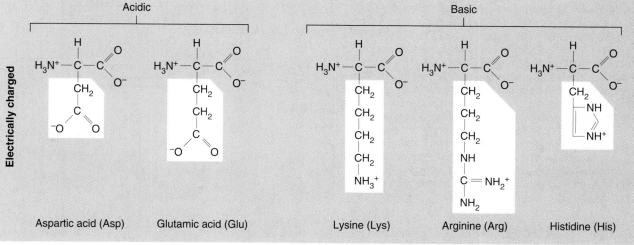

Electrically charged

Acidic

Basic

Aspartic acid (Asp)   Glutamic acid (Glu)   Lysine (Lys)   Arginine (Arg)   Histidine (His)

APPENDIX C
TWO IMPORTANT METABOLIC PATHWAYS

Step 1 Glucose enters the cell and is phosphorylated by the enzyme hexokinase, which catalyzes the transfer of a phospate group, indicated as P_i, from ATP to the number six carbon of the sugar, producing glucose-6-phosphate. The electrical charge of the phosphate group traps the sugar in the cell because of the impermeability of the plasma membrane to ions. Phosphorylation of glucose also makes the molecule more chemically reactive. Although glycolysis is supposed to *produce* ATP, in step 1 ATP is actually consumed—an energy investment that will be repaid with dividends later in glycolysis.

Step 2 Glucose-6-phosphate is rearranged and converted to its isomer, fructose-6-phosphate. Isomers, remember, have the same number and types of atoms but in different structural arrangements.

Step 3 In this step, still another molecule of ATP is used to add a second phosphate group to the sugar, producing fructose-1,6-diphosphate. So far, the ATP ledger shows a debit of -2. With phosphate groups on its opposite ends, the sugar is now ready to be split in half.

Step 4 This is the reaction from which glycolysis gets its name. An enzyme cleaves the sugar molecule into two different 3-carbon sugars: glyceraldehyde 3-phosphate and dihydroxyacetone phosphate. These two sugars are isomers of one another.

Step 5 The isomerase enzyme interconverts the 3-carbon sugars, and if left alone in a test tube, the reaction reaches equilibrium. This does not happen in the cell, however, because the next enzyme in glycolysis uses only glyceraldehyde phosphate as its substrate and is unreceptive to dihydroxyacetone phosphate. This pulls the equilibrium between the two 3-carbon sugars in the direction of glyceraldehyde phosphate, which is removed as fast as it forms. Thus, the net result of steps 4 and 5 is cleavage of a 6-carbon sugar into two molecules of glyceraldehyde phosphate; each will progress through the remaining steps of glycolysis.

Step 6 An enzyme now catalyzes two sequential reactions while it holds glyceraldehyde phosphate in its active site. First, the sugar is oxidized by the transfer of H from the number one carbon of the sugar to NAD, forming NADH+H$^+$. Here we see in metabolic context the oxidation-reduction reaction described in Chapter 25. This reaction releases substantial amounts of energy, and the enzyme capitalizes on this by coupling the reaction to the creation of a high-energy phosphate bond at the number one carbon of the oxidized substrate. The source of the phosphate is inorganic phosphate (P_i) always present in the cytosol. As products, the enzyme releases NADH+H$^+$ and 1, 3-diphosphoglyceric acid. Notice in the figure that the new phosphate bond is symbolized with a squiggle (~), which indicates that the bond is at least as energetic as the high-energy phosphate bonds of ATP.

THE TEN STEPS OF GLYCOLYSIS Each of the ten steps of glycolysis is catalyzed by a specific enzyme found dissolved in the cytoplasm. All steps are reversible. An abbreviated version of the three major phases of glycolysis appears in the lower right-hand corner of the figure.

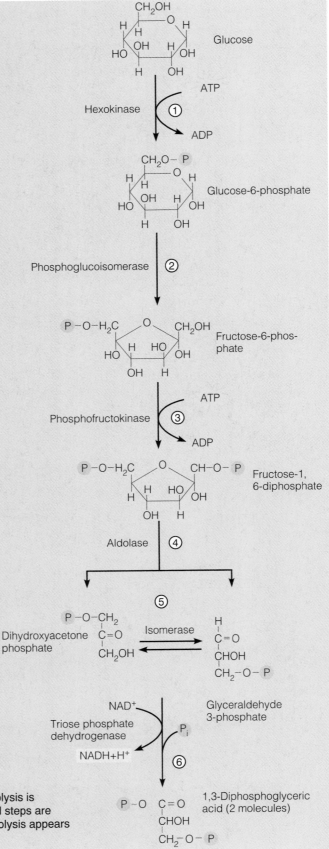

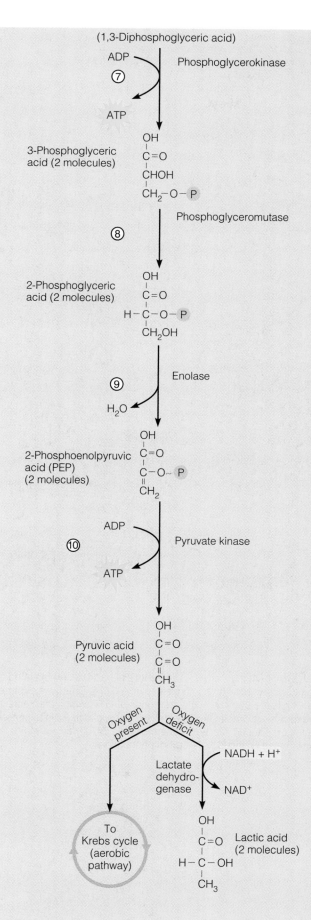

Step 7   Finally, glycolysis produces ATP. The phosphate group, with its high-energy bond, is transferred from 1,3-diphosphoglyceric acid to ADP. For each glucose molecule that began glycolysis, step 7 produces two molecules of ATP, because every product after the sugar-splitting step (step 4) is doubled. Of course, two ATPs were invested to get sugar ready for splitting. The ATP ledger now stands at zero. By the end of step 7, glucose has been converted to two molecules of 3-phosphoglyceric acid. This compound is not a sugar. The sugar was oxidized to an organic acid back in step 6, and now the energy made available by that oxidation has been used to make ATP.

Step 8   Next, an enzyme relocates the remaining phosphate group of 3-phosphoglyceric acid to form 2-phosphoglyceric acid. This prepares the substrate for the next reaction.

Step 9   An enzyme forms a double bond in the substrate by removing a water molecule from 2-phosphoglyceric acid to form phosphoenolpyruvic acid, or PEP. This results in the electrons of the substrate being rearranged in such a way that the remaining phosphate bond becomes very unstable; it has been upgraded to high-energy status.

Step 10   The last reaction of glycolysis produces another molecule of ATP by transferring the phosphate group from PEP to ADP. Because this step occurs twice for each glucose molecule, the ATP ledger now shows a net gain of two ATPs. Steps 7 and 10 each produce two ATPs for a total credit of four, but a debt of two ATPs was incurred from steps 1 and 3. Glycolysis had repaid the ATP investment with 100% interest. In the meantime, glucose has been broken down and oxidized to two molecules of pyruvic acid, the compound produced from PEP in step 10.

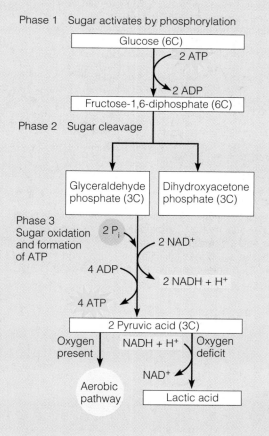

Step 1   Two-carbon acetyl CoA is combined with oxaloacetic acid, a 4-carbon compound. The unstable bond between the acetyl group and CoA is broken as oxaloacetic acid binds and CoA is freed to prime another 2-carbon fragment derived from pyruvic acid. The product is the 6-carbon citric acid, for which the cycle is named.

Step 2   A molecule of water is removed, and another is added back. The net result is the conversion of citric acid to its isomer, isocitric acid.

Step 3   The substrate loses a CO_2 molecule, and the remaining 5-carbon compound is oxidized, forming a α-ketoglutaric acid and reducing NAD^+.

Step 4   This step is catalyzed by a multi-enzyme complex very similar to the one that converts pyruvic acid to acetyl CoA. CO_2 is lost; the remaining 4-carbon compound is oxidized by the transfer of electrons to NAD^+ to form $NADH+H^+$ and is then attached to CoA by an unstable bond. The product is succinyl CoA.

Step 5   Substrate-level phosphorylation occurs in this step. CoA is displaced by a phosphate group, which is then transferred to GDP to form guanosine-triphosphate (GTP). GTP is similar to ATP, which is formed when GTP donates a phosphate group to ADP. The products of this step are succinic acid and ATP.

Step 6   In another oxidative step, two hydrogens are removed from succinic acid, forming fumaric acid, and transferred to FAD to form $FADH_2$. The function of this coenzyme is similar to that of $NADH+H^+$, but $FADH_2$ stores less energy. The enzyme that catalyzes this oxidation-reduction reaction is the only enzyme of the cycle that is embedded in the mitochondrial membrane. All other enzymes of the citric acid cycle are dissolved in the mitochondrial matrix.

Step 7   Bonds in the substrate are rearranged in this step by the addition of a water molecule. The product is malic acid.

Step 8   The last oxidative step reduces another NAD^+ and regenerates oxaloacetic acid, which accepts a 2-carbon fragment from acetyl CoA for another turn of the cycle.

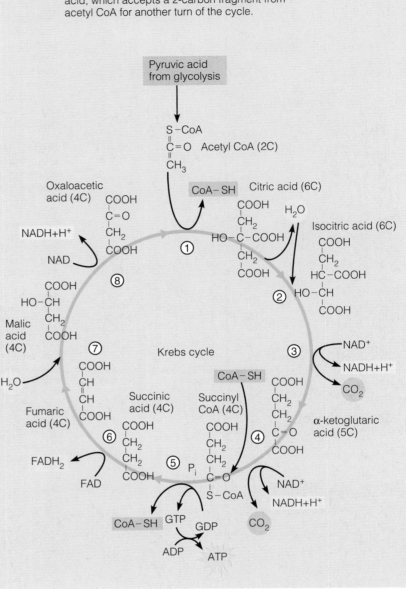

KREBS CYCLE (CITRIC ACID CYCLE)

All but one of the steps (step 6) occur in the mitochondrial matrix. The preparation of pyruvic acid (by oxidation, decarboxylation, and reaction with coenzyme A) to enter the cycle as acetyl CoA is shown above the cycle. Acetyl CoA is picked up by oxaloacetic acid to form citric acid; and as it passes through the cycle, it is oxidized four more times (forming three molecules of reduced NAD [$NADH + H^+$] and one of reduced FAD [$FADH_2$]) and decarboxylated twice (releasing 2 CO_2). Energy is captured in the bonds of GTP, which then acts in a coupled reaction with ADP to generate one molecule of ATP by substrate level phosphorylation.

APPENDIX D
PERIODIC TABLE OF THE ELEMENTS

Periodic Table of the Elements

Representative (main group) elements

Transition metals

Representative (main group) elements

Period	IA	IIA	IIIB	IVB	VB	VIB	VIIB	VIIIB			IB	IIB	IIIA	IVA	VA	VIA	VIIA	VIIIA
1	1 H 1.0079																	2 He 4.003
2	3 Li 6.941	4 Be 9.012											5 B 10.811	6 C 12.011	7 N 14.007	8 O 15.999	9 F 18.998	10 Ne 20.180
3	11 Na 22.990	12 Mg 24.305											13 Al 26.982	14 Si 28.086	15 P 30.974	16 S 32.066	17 Cl 35.453	18 Ar 39.948
4	19 K 39.098	20 Ca 40.078	21 Sc 44.956	22 Ti 47.88	23 V 50.942	24 Cr 51.996	25 Mn 54.938	26 Fe 55.845	27 Co 58.933	28 Ni 58.69	29 Cu 63.546	30 Zn 65.39	31 Ga 69.723	32 Ge 72.61	33 As 74.922	34 Se 78.96	35 Br 79.904	36 Kr 83.8
5	37 Rb 85.468	38 Sr 87.62	39 Y 88.906	40 Zr 91.224	41 Nb 92.906	42 Mo 95.94	43 Tc 98	44 Ru 101.07	45 Rh 102.906	46 Pd 106.42	47 Ag 107.868	48 Cd 112.411	49 In 114.82	50 Sn 118.71	51 Sb 121.76	52 Te 127.60	53 I 126.905	54 Xe 131.29
6	55 Cs 132.905	56 Ba 137.327	57 La 138.906	72 Hf 178.49	73 Ta 180.948	74 W 183.84	75 Re 186.207	76 Os 190.23	77 Ir 192.22	78 Pt 195.08	79 Au 196.967	80 Hg 200.59	81 Tl 204.383	82 Pb 207.2	83 Bi 208.980	84 Po 209	85 At 210	86 Rn 222
7	87 Fr 223	88 Ra 226.025	89 Ac 227.028	104 Rf 261	105 Db 262	106 Sg 263	107 Bh 264	108 Hs 265	109 Mt 266	110 Uun 269	111 Uuu 272	112 Uub 277	114		116			118

Rare earth elements

Lanthanides	58 Ce 140.115	59 Pr 140.908	60 Nd 144.24	61 Pm 145	62 Sm 150.36	63 Eu 151.964	64 Gd 157.25	65 Tb 158.925	66 Dy 162.5	67 Ho 164.93	68 Er 167.26	69 Tm 168.934	70 Yb 173.04	71 Lu 174.967
Actinides	90 Th 232.038	91 Pa 231.036	92 U 238.029	93 Np 237.048	94 Pu 244	95 Am 243	96 Cm 247	97 Bk 247	98 Cf 251	99 Es 252	100 Fm 257	101 Md 258	102 No 259	103 Lr 262

The periodic table arranges elements according to atomic number and atomic weight into horizontal rows called *periods* and 18 vertical columns called *groups* or *families*. The elements in the groups are classified as being in either A or B classes.

Elements of each group of the A series have similar chemical and physical properties. This reflects the fact that members of a particular group have the same number of valence shell electrons, which is indicated by the number of the group. For example, group IA elements have one valence shell electron, group IIA elements have two, and group VA elements have five. In contrast, as you progress across a period from left to right, the properties of the elements change in discrete steps, varying gradually from the very metallic properties of groups IA and IIA elements to the nonmetallic properties seen in group VIIA (chlorine and others), and finally to the inert elements (noble gases) in group VIIIA. This change reflects the continual increase in the number of valence shell electrons seen in elements (from left to right) within a period.

Class B elements are referred to as *transition elements*. All transition elements are metals, and in most cases they have one or two valence shell electrons. (In these elements, some electrons occupy more distant electron shells before the deeper shells are filled.)

In this periodic table, the colors are used to convey information about the phase (solid, liquid, or gas) in which a pure element exists under standard conditions (25 degrees centigrade and 1 atmosphere of pressure). If the element's symbol is solid black, then the element exists as a solid. If its symbol is red, then it exists as a gas. If its symbol is dark blue, then it is a liquid. If the element's symbol is green, the element does not exist in nature and must be created by some type of nuclear reaction.

APPENDIX E
REFERENCE VALUES FOR SELECTED BLOOD AND URINE STUDIES

The reference values listed for the selected blood and urine studies are common ranges for adults, but specific "normals" are established by the laboratory performing the analysis. The values may be affected by a wide range of circumstances, including testing methods and equipment used, client age, body mass, sex, diet, activity level, medications, and extent of disease processes.

Reference values are identified in both standard or conventional units and in the system of international (SI) units. SI units (given in parentheses) are measurements of amount per volume and are used in most countries and scientific journals. SI units are often given as moles or millimoles per liter. Most clinical laboratories and textbooks in the United States use conventional or standard units which measure mass per volume. These values are given as grams, milligrams, or milliequivalents per deciliter or liter. It is anticipated that the United States will eventually use SI units exclusively.

Sample types in column 1 are serum (S), plasma (P), arterial whole blood (A), and whole blood (WB).

Blood Chemistry Studies

Test (sample)	Reference values: Conventional (SI)	Physiologic indication and clinical implications
Aspartate aminotransferase (ALT) or glutamic-oxaloacetic transaminase (SGOT) (S)	10–40 U/ml (5–30 U/L)	Cellular damage. Increased after myocardial infarction, acute liver disease, drug toxicity, muscle trauma. Decreased in ketoacidosis.
Ammonia (P)	15–120 μg/dl (12–65 μmol/L)	Liver and renal function. Increased values in liver disease, renal failure, newborn hemolytic disease, heart failure, cor pulmonale. Decreased values in hypertension.
Amylase (S)	56–190 IU/L (25–125 U/L)	Pancreatic function. Increased values in pancreatitis, mumps, obstruction of pancreatic duct, ketoacidosis. Decreased values in liver disease, perforated bowel, pulmonary infarct, toxemia of pregnancy.
Bilirubin (S)	Total: 0.1–1.0 mg/dl (5.1–17.0 μmol/L) Conjugated: <0.5 (<5.0 μmol/L) Unconjugated: 0.2–1.0 mg/dl (18–20 μmol/L) Newborn: 1.0–12.0 mg/dl (<200 μmol/L)	Liver function and red cell breakdown. Increased levels of conjugated in liver disease and biliary obstruction. Increased levels of unconjugated in hemolysis of red blood cells.
Blood urea nitrogen (S)	7–26 mg/dl (2.5–9.3 mmol/L)	Kidney function. Increased values in renal disease, dehydration, urinary obstruction, congestive heart failure, myocardial infarction, burns. Decreased values in liver failure, overhydration, impaired protein absorption, pregnancy.
Cholesterol	<200 mg/dl (<6.5 mmol/L)	Increased values in diabetes mellitus.
Creatine kinase (CK) (S)	Female: 30–135 U/L (10–55 U/ml) Male: 55–170 U/L (12–70 U/ml)	Cellular damage. Increased values in myocardial infarction, muscular dystrophy, hypothyroidism, pulmonary infarction, cerebral vascular accident (CVA), shock, tissue damage, and trauma.
Creatinine (S)	0.5–1.2 mg/dl (44–97 μmol/L)	Renal function. Increased values in renal disease and acromegaly. Decreased in muscular dystrophy.

Test (sample)	Reference values: Conventional (SI)	Physiologic indication and clinical implications
Gases (A)		Acid-base balance.
Bicarbonate	22–26 mEq/L (22–26 mmol/L)	
Carbon dioxide content	Arterial: 19–24 mEq/L (19–24 mmol/L) Venous: 22–30 mEq/L (22–30 mmol/L)	Increased values in metabolic alkalosis and respiratory acidosis. Decreased values in metabolic acidosis and respiratory alkalosis.
Carbon dioxide partial pressure (P_{CO_2})	Arterial: 35–45 mm Hg (same) Venous: 45 mm Hg (same)	
Oxygen (O_2) saturation	95%–98% (same)	Increased values in polycythemia and decreased in pulmonary disease.
Oxygen partial pressure (P_{O_2})	80–105 mm Hg	Increased values in hyperventilation, polycythemia. Decreased values in anemia, hypoxia.
pH	7.35–7.45 (same)	Increased values in metabolic and respiratory alkalosis. Decreased values in metabolic and respiratory acidosis.
Glucose (S)	Fasting: 70–120 mg/ml (3.9–6.1 mmol/L)	Metabolic function. Increased values in diabetes mellitus, Cushing's syndrome, liver disease, acute stress, and hyperpituitarism. Decreased levels in Addison's disease, insulinomas.
Immunoglobulins (S)		
IgG	560–1800 mg/dl (5.6–18 g/L)	Immune response. Increased levels in chronic infections, rheumatic fever, liver disease, rheumatoid arthritis. Decreased levels in amyloidosis leukemia, and preeclampsia.
IgE	0.01–0.4 mg/dl (0.1–0.4 mg/L)	Allergic responses. Increased values in allergic responses. Decreased values in agammaglobulinemia.
IgA	85–563 mg/dl (0.85–5.6 g/L)	Immune integrity. Increased values in liver disease, rheumatic fever, cancer, autoimmune disorders. Decreased values in immunodeficiency disorders and inflammatory bowel disease.
IgM	55–375 mg/dl (0.5–3.8 g/L)	Immune integrity. Increased in autoimmune disease, rheumatoid arthritis, viral and fungal infections. Decreased in amyloidosis and leukemia.
IgD	0.5–3.0 mg/dl (5–30 mg/L)	Immune integrity. Increased values in myelomas and chronic infection.
Ketone bodies (S or P)	Negative Toxic level >20 mg/dl (0.2 g/L)	Fatty acid catabolism. Values increased in ketoacidosis, starvation, low-carbohydrate diet, dehydration, aspirin ingestion.
Lactic acid (lactate) (WB)	Arterial: 3–7 mg/dl (0.3–0.7 mmol/L) Venous: 5–20 mg/dl (0.5–2.0 μmol/L)	Anaerobic oxygenation of tissues. Increased values in congestive heart failure, shock, hemorrhage, strenuous exercise.
Lactate dehydrogenase (LDH) (S)	115–225 IU/L (0.4–1.7 μmol/L)	Tissue damage of organs or striated muscle. Increased in myocardial infarction, pulmonary infarction, liver disease, cerebral vascular accident, infectious mononucleosis, muscular dystrophy, fractures.
Lipoproteins (S)		
Total	400–800 mg/dl 4.0–8.0 g/L	Metabolism—fat utilization. Increased values in diabetes mellitus, hypothyroidism, hyperlipidemia. Decreased values in hyperthyroidism and fat malabsorption.

➤

Test (sample)	Reference values: Conventional (SI)	Physiologic indication and clinical implications
High-density lipoproteins (HDL)	25% of total Female: >55 mg/dl (>0.1 mmol/L) Male: >45 mg/dl (>0.1 mmol/L)	Increased levels in liver disease. Decreased levels in atherosclerotic heart disease, malnutrition.
Low-density lipoproteins (LDL)	75% of total 60–180 mg/dl (<3.2 mmol/L)	Increased values in hyperlipidemia, atherosclerotic heart disease. Decreased values in fat malabsorption and malnutrition.
Very low density lipoproteins (VLDL)	25%–50% of total (same)	Same
Osmolality (S)	285–295 mOsm/kg H_2O (285–295 mmol/kg H_2O)	Fluid and electrolyte balance. Increased levels in hypernatremia, dehydration, kidney disease, alcohol ingestion. Decreased levels in hyponatremia, overhydration, and syndrome of inappropriate ADH secretion (SIADH).
Phosphate (S) (phosphorus)	2.5–4.5 mg/dl (0.8–1.5 mmol/L)	Parathyroid function; bone disease. Increased levels in hypoparathyroidism, renal failure, bone metastasis, liver disease, hypocalcemia. Decreased values in hyperparathyroidism, hypercalcemia, alcoholism, vitamin D deficiency, ketoacidosis, osteomalacia.
Potassium (S)	3.5–5.1 mEq/L (3.5–5.1 mmol/L)	Fluid and electrolyte balance. Increased levels in renal disease, Addison's disease, ketoacidosis, burns, and crush injuries. Decreased levels in vomiting, diarrhea, Cushing's syndrome, alkalosis.
Protein (S)		
Total	6.0–8.0 g/dl (60–80 g/L)	Osmotic pressure; immune system integrity. Increased values in chronic infections, dehydration, shock, hemoconcentration. Decreased values in protein malnutrition, burns, diarrhea, renal failure.
Albumin	3.2–5.0 g/dl (32–50 g/L)	Osmotic pressure. Increased levels in dehydration. Decreased levels in liver disease, malnutrition, Crohn's disease, nephrotic syndrome, systemic lupus erythematosus.
Sodium (S)	136–145 mEq/L (136–145 mmol/L)	Fluid and electrolyte balance. Increased values in dehydration, diabetes insipidus, Cushing's disease. Decreased values in vomiting, diarrhea, burns, Addison's disease, edema, congestive heart failure, overhydration, syndrome of inappropriate ADH secretion (SIADH).
Triglycerides	10–190 mg/dl (0.1–1.9 g/L)	Increased values in myocardial infarction, liver disease, nephrotic syndrome, pregnancy.
Uric acid (S)	Female: 2.0–7.3 mg/dl (0.09–0.36 nmol/L) Male: 2.1–8.5 mg/dl (0.15–0.48 nmol/L)	Renal function. Increased in lead poisoning, impaired renal function, gout, metastatic cancer, alcoholism, leukemia. Decreased in Wilson's disease.

Hematology Studies

Hemoglobin (S)	Female: 12–16 g/dl (7.4–9.9 mmol/L) Male: 14–18 g/dl (8.7–11.2 mmol/L)	Oxygenation status. Increased values in dehydration, polycythemia, congestive heart failure, chronic obstructive pulmonary disease, high altitudes. Decreased levels in anemia, hemorrhage, cancer, renal disease, systemic lupus erythematosus, nutritional deficiency.
Hematocrit (WB)	Female: 37%–47% (same) Male: 42%–54% (same)	Oxygenation status. Increased levels in polycythemia, dehydration, congestive heart failure, shock, surgery. Decreased levels in anemia, hemorrhage, bone marrow disease, malnutrition, cirrhosis, rheumatoid arthritis.

Test (sample)	Reference values: Standard (SI)	Physiologic indication and clinical implications
Hematology Studies (continued)		
Partial thromboplastin time (activated) (PTT or APTT)	APTT: 30–40 sec (same) PTT: 60–70 sec (same)	Clotting mechanisms. Increased values in clotting factor deficiencies, cirrhosis, vitamin K deficiency, disseminated intravascular coagulation (DIC). Decreased values in early DIC, extensive cancer.
Platelet count (WB)	150,000–400,000/mm^3 (150–400 $\times$ 10^9/L)	Clotting mechanisms. Increased values in polycythemia, cancers, rheumatoid arthritis, cirrhosis, trauma. Decreased values in liver disease, hemolytic and pernicious anemia, disseminated intravascular coagulation (DIC), idiopathic thrombocytopenic purpura (ITP), systemic lupus erythematosus (SLE).
Prothrombin time (PT) (WB)	11–12.5 sec (same) 1.5–2 $\times$ control (evaluating anticoagulant treatment)	Clotting mechanisms. Increased values in liver disease, vitamin K deficiency, salicylate intoxication. Decreased values in disseminated intravascular coagulation (DIC), bile duct obstruction.
Red blood cell count (RBC) (WB)	Female: 4.2–5.4 million/mm^3 (4.2–5.4 $\times$ 10^{12}/L) Male: 4.7–6.1 million/mm^3 (4.7–6.1 $\times$ 10^{12}/L)	Oxygenation status. Increased values in high altitudes, polycythemia, hemoconcentration, cor pulmonale. Decreased values in hemorrhage, hemolysis, anemias, chronic illness, nutritional deficiencies, leukemia, overhydration.
Reticulocyte count	0.5%–2.0% (same) (WB)	Bone marrow function. Increased values in hemolytic anemia, sickle-cell anemia, leukemia, pregnancy. Decreased values in pernicious anemia, folic acid deficiency, cirrhosis, chronic infection, bone marrow depression or failure.
White blood cell count (WBC) (WB)		
Total (males)	5,000–10,000/mm^3 (5–10 $\times$ 10^9/L)	Immune system integrity. Increased values in infection, trauma, stress, tissue necrosis, inflammation. Decreased values in bone marrow depression or failure, drug toxicity, overwhelming infection, malnutrition.
White blood cell count, differential (WB)		
Neutrophils	55%–70% (same)	Immune system integrity. Increased values in stress, Cushing's syndrome, inflammatory disorders, ketoacidosis, gout. Decreased levels in aplastic anemia, bone marrow suppression, viral infections, severe bacterial infections, Addison's disease.
Lymphocytes	20%–40% (same)	Immune system integrity. Increased values in chronic bacterial infections, mumps, rubella, lymphocytic leukemia, infectious mononucleosis, infectious hepatitis. Decreased levels in leukemia, immunodeficiency, lupus erythematosus, bone marrow depressive drugs.
Eosinophils	1%–4% (same)	Immune system integrity. Increased levels in allergic reactions, parasitic infections, leukemia, autoimmune diseases. Decreased levels in excess adrenosteroid production.
Monocytes	2%–8% (same)	Immune system integrity. Increased levels in inflammatory disorders, viral infections, tuberculosis. Decreased levels in steroid therapy.
Basophils	0.5%–1.0% (same)	Immune system integrity. Increased levels in myeloproliferative disorders, leukemia. Decreased levels in allergic reactions, hyperthyroidism, stress.

➤

Test (sample)	Reference values: Conventional (SI)	Physiologic indication and clinical implications
Urine tests		
Amylase (24 hr)	3–35 IU/hr <5000 Somogyi units/24 hr (6.5–48 U/hr)	Pancreatic function. Increased values in pancreatic disease or obstruction, inflammation of the parotid or salivary glands, and cholecystitis.
Bilirubin (random)	Negative (same)	Liver function. Increased values in liver disease, extrahepatic obstruction (gallstones, tumor, inflammation).
Blood (random)	Negative (same)	Urinary system function. Increased values in cystitis, renal disease, hemolytic anemia, transfusion reaction, prostatitis, burns.
Osmolality (random or fasting)	50–1400 mOsm/kg H_2O (50–1400 mmol/kg H_2O) (random) <850 mOsm/kg H_2O (<850 mmol/kg H_2O) fluid restriction	Fluid and electrolyte balance, renal function, and endocrine function. Increased levels in hypernatremia, syndrome of inappropriate ADH secretion (SIADH), congestive heart failure (CHF), cirrhosis, acidosis. Decreased levels in diabetes insipidus, hypercalcemia, fluid intoxication, polynephritis, renal tubular necrosis, aldosteronism.
Phosphate (24 hr)	80%–90% reabsorbed (same)	Parathyroid function. Increased levels in hyperparathyroidism, uremia, osteomalacia, renal disease, autoimmune disease. Decreased levels in hypoparathyroidism.
Potassium (24 hr)	25–120 mEq/24 hr (25–120 mmol/24 hr)	Fluid and electrolyte balance. Increased values in renal disease, metabolic acidosis, dehydraton, aldosteronism, Cushing's syndrome. Decreased values in Addison's disease, malabsorption, acute renal failure, syndrome of inappropriate ADH secretion (SIADH).
Protein (random)	<8 mg/dl (<0.8 mg/L)	Renal function. Increased levels in nephrotic syndrome, trauma, hyperthyroidism, diabetes mellitus, lupus erythematosus.
Sodium (24 hr)	40–220 mEq/24 hr (40–220 mmol/24 hr)	Fluid and electrolyte balance. Increased values in dehydration, ketoacidosis, syndrome of inappropriate ADH secretion (SIADH), adrenocortical insufficiency. Decreased levels in congestive heart failure (CHF), renal failure, diarrhea, aldosteronism.
Uric acid (24 hr)	0.4–1.0 g/24 hr (1.5–4.0 mmol/24 hr)	Renal function and metabolism. Increased in gout, leukemia, liver disease, ulcerative cholitis. Decreased in renal disease, alcoholism, lead toxicity.
Urinalysis (random)		
Color	Straw, yellow, amber	Fluid balance and renal function. Darker in dehydration. Lighter in overhydration, diabetes insipidus. Color varies with disease states, diet, and medications.
Odor	Aromatic	Metabolic function infection. Abnormal odors in infection, ketonuria, rectal fistula, hepatic failure, phenylketonuria.
Specific gravity	1.005–1.030	Kidney function. Increased levels in dehydration, syndrome of inappropriate ADH secretion (SIADH), fever, glycosuria, proteinuria, decreased renal blood flow (stenosis, heart failure). Decreased levels in renal disease.
Urobilinogen (24 hr)	<4 mg/24 hr	Liver function. Increased in hemolytic anemias, hepatitis A, cirrhosis, biliary disease. Decreased in renal disease, common bile duct obstruction.
Volume (24 hr)	1000–2000 ml/24 hr (1.0–2.0 L/24 hr)	Fluid and electrolyte balance, renal function. Increased values in diabetes insipidus, diabetes mellitus, renal disease. Decreased values in dehydration, syndrome of inappropriate ADH secretion (SIADH), renal disease.

APPENDIX F
ANSWERS TO CLINICAL CONNECTIONS, MULTIPLE CHOICE, AND MATCHING QUESTIONS

Chapter 1
1. c; **2.** a; **3.** e; **4.** a, d; **5.** (a) wrist (b) hip bone (c) nose (d) toes (e) scalp; **6.** neither c nor d would be visible in the median sagittal section; **7.** (a) dorsal (b) ventral (c) dorsal (d) ventral; **8.** b; **9.** b

Chapter 2
1. b, d; **2.** d; **3.** b; **4.** a; **5.** b; **6.** a; **7.** c, d; **8.** b; **9.** a; **10.** a; **11.** b; **12.** a, c; **13.** (1)a, (2)c; **14.** c; **15.** d; **16.** e; **17.** d; **18.** d; **19.** a; **20.** b; **21.** b; **22.** c

Chapter 3
1. d; **2.** a, c; **3.** b; **4.** d; **5.** b; **6.** e; **7.** c; **8.** d; **9.** a; **10.** a; **11.** b; **12.** d; **13.** c; **14.** b; **15.** d; **16.** a; **17.** b; **18.** d

Chapter 4
1. a, c, d, b; **2.** c, e; **3.** b, f, a, d, g, d; **4.** b; **5.** c; **6.** b

Chapter 5
Clinical Connections   1. The skin separates and protects the internal environment of the body from potentially dangerous elements in the external environment. Mrs. DeStephano's chart indicates epidermal abrasions, which represent the loss of this barrier. Epidermal loss will also cost Mrs. DeStephano the acid mantle of her skin, protection against UV radiation, and Langerhans' cells, which protect against invasion by microorganisms.   **2.** Dermal macrophages found in the dermis can act as a backup system against bacterial and viral invasion when the epidermis is damaged.   **3.** Suturing brings the edges of wounds close together and promotes faster healing because smaller amounts of granulation tissue need to be formed. This is termed *healing by first intention.*   **4.** Cyanosis signals a decrease in the amount of oxygen carried by hemoglobin in the blood. Respiratory system and/or cardiovascular system impairments can lead to cyanosis.

   Review Questions   1. a; **2.** c; **3.** d; **4.** d; **5.** b; **6.** b; **7.** c; **8.** c; **9.** b; **10.** b; **11.** a; **12.** d; **13.** b

Chapter 6
Clinical Connections   1. Mrs. DeStephano's broken leg has a transverse fracture of the open variety because the broken ends of the bone are protruding through the skin.   **2.** The laceration of the skin caused by the broken end of the bone creates a breach in the protective barrier created by the skin, providing an entry point for bacteria and other microorganisms. In addition, the protruding ends of the bone have now been exposed to the nonsterile external environment. This could result in the development of osteomyelitis, a bacterial infection, which can be treated with antibiotics.   **3.** Reduction of a fracture is the clinical term for "setting the bone." Mrs. DeStephano's physician chose internal reduction, in which surgery is performed and the broken ends of the bone are secured together by pins or wires. A cast was applied to keep the aligned ends of the bone immobile until healing of the fracture has occurred.   **4.** Healing of Mrs. DeStephano's fracture will begin as bony callus formation fills the break in the bone with bony tissue. This process begins 3–4 weeks after the break occurs and is completed within 2–3 months. **5.** Nutrient arteries supply blood to the bone tissue. In order for Mrs. DeStephano's break to heal normally, the bony tissue must be supplied with oxygen (to generate ATP for energy) and nutrients in order to rebuild the bone. Damage to a nutrient artery will decrease the delivery of these building materials and could slow the process of healing.   **6.** For a fracture that is slow to heal, new techniques that promote healing include electrical stimulation, which promotes the deposition of new bone tissue; ultrasound treatments, known to

speed healing; and possibly the addition of bone substitutes to the fractured area.   **7.** At age 45, Mrs. DeStephano will most likely not regenerate her knee cartilage. (Cartilage growth typically ends during adolescence.) Cartilage damage that occurs during adulthood is slow to heal, due to the avascular nature of cartilage, and is usually irreparable. Surgical removal of torn cartilage to allow improved movement of the joint is the usual treatment for this type of damage.

   Review Questions   1. e; **2.** b; **3.** c; **4.** d; **5.** e; **6.** b; **7.** c; **8.** b; **9.** e; **10.** c; **11.** b; **12.** c; **13.** a; **14.** b

Chapter 7
1. (1)b, g; (2)h; (3)d; (4)d, f; (5)e; (6)c; (7)a, b, d, h; (8) i; **2.** (1)g, (2)f, (3)b, (4)a, (5)b, (6)c, (7)d, (8)e; **3.** (1)b, (2)c, (3)e, (4)a, (5)h, (6)e, (7)f

Chapter 8
1. (1)c, (2)a, (3)a, (4)b, (5)c, (6)b, (7)b, (8)a, (9)c; **2.** b; **3.** d; **4.** d; **5.** b; **6.** d; **7.** d

Chapter 9
Clinical Connections   1. The first reaction to tissue injury is the initiation of the inflammatory response. The inflammatory chemicals increase the permeability of the capillaries in the injured area, allowing white blood cells, fluid, and other substances to reach the injured area. The next step in healing involves the formation of granulation tissue, in which the vascular supply for the injured area is regenerated and collagen fibers to knit the torn edges of the tissue together are formed. Skeletal muscle does not regenerate well, so the damaged areas of Mrs. DeStephano's muscle tissue will probably be repaired primarily by the formation of fibrous tissue, creating scar tissue.   **2.** Healing is aided by good circulation of blood within the injured area. Vascular damage compromises healing because the supply of oxygen and nutrients to the tissue is reduced.   **3.** Under normal circumstances, skeletal muscles receive electrical signals from the nervous system continuously. These signals help to maintain muscle tone and readiness. Severing of the sciatic nerve removes this continuous nervous input to the muscles and will lead to muscle atrophy. Immobility of muscles will lead to a replacement of contractile muscle tissue with noncontractile fibrous connective tissue. Distal to the point of transection, the muscle will begin to decrease in size within 3–7 days of becoming immobile. This process can be delayed by electrically stimulating the tissues. Passive range-of-motion exercises also help prevent loss of muscle tone and joint range, and improve circulation in the injured areas.   **4.** Mrs. DeStephano's physician wants to supply her damaged tissues with the necessary building materials to encourage healing. A high-protein diet will provide plenty of amino acids to rebuild or replace damaged proteins, carbohydrates will provide the fuel molecules needed to generate the required ATP, and vitamin C is important for the regeneration of connective tissue.

   Review Questions   1. c; **2.** b; **3.** (1)b, (2)a, (3)b, (4)a, (5)b, (6)a; **4.** c; **5.** a; **6.** a; **7.** d; **8.** a; **9.** (1)a; (2)a, c; (3)b; (4)c; (5)b; (6)b; **10.** a; **11.** c; **12.** c; **13.** c; **14.** b

Chapter 10
1. c; **2.** c; **3.** (1)e, (2)c, (3)g, (4)f, (5)d; **4.** a; **5.** c; **6.** d; **7.** c; **8.** c; **9.** b; **10.** d; **11.** b; **12.** a; **13.** c; **14.** b; **15.** possible answers may be found on pages 323–324; **16.** a

Chapter 11
1. b; **2.** (1)d, (2)b, (3)f, (4)c, (5)a; **3.** b; **4.** c; **5.** a; **6.** c; **7.** b; **8.** d; **9.** c; **10.** c; **11.** a; **12.** (1)d, (2)b, (3)a, (4)c

Chapter 12
1. a; **2.** d; **3.** c; **4.** a; **5.** (1)d, (2)f, (3)e, (4)g, (5)b, (6)f, (7)i, (8)a; **6.** b; **7.** c; **8.** a; **9.** (1)a, (2)b, (3)a, (4)a, (5)b, (6)a, (7)b, (8)b, (9)a; **10.** d

Chapter 13

1. b; **2.** c; **3.** c; **4.** (1)e; (2)d; (3)c; (4)g; (5)b; (6)f; (7)a; **5.** (1)f; (2)i; (3)b; (4)g, h, l; (5)e; (6)i; (7)c; (8)k; (9)l; (10)c, d, f, k; **6.** (1)b 6; (2)d 1, 8; (3)c 2; (4)c 5; (5)a 4; (6)a 3, 9; (7)a 7; (8)a 7; (9)d 1; (10)a 3, 4, 7, 9; **7.** (1)b, 1; (2)a, 3 and 5; (3)a, 4; (4)a, 2; (5)c, 2; **8.** b

Chapter 14

Clinical Connections 1. The location of Jimmy's lacerations and bruises and his inability to rise led the paramedics to suspect a head, neck, or back injury. They immobilized his head and torso to prevent any further damage to the brain and spinal cord. **2.** The worsening neurological signs indicate a probable intracranial hemorrhage. The blood escaping from the ruptured blood vessel(s) will begin to compress Jimmy's brain and increase his intracranial pressure. Jimmy's surgery will involve repair of the damaged vessel(s) and removal of the mass of clotted blood pressing on his brain. **3.** The blow to Jimmy's head caused his brain to swell (cerebral edema) as part of the inflammatory response to tissue injury. Dexamethasone is an anti-inflammatory drug that will decrease swelling and relieve intracranial pressure. **4.** Jimmy is suffering from spinal shock, which occurs as a result of injury to the spinal cord. Spinal shock is a temporary condition in which all reflex and motor activities caudal to the level of spinal cord injury are lost, so Jimmy's muscles are paralyzed, his blood pressure is low due to the loss of sympathetic tone in his vasculature, and his temperature is elevated because he has lost part of his temperature-regulating mechanism (sweating). **5.** This condition is called the *mass reflex reaction* (or *autonomic hyperreflexia*). This is a condition in which a normal stimulus triggers a massive activation of both autonomic and somatic motor neurons. **6.** Extremely high arterial blood pressure can cause a rupture of the cerebral blood vessels (as well as other blood vessels in the body) and put Jimmy's life at risk.

Review Questions 1. d; **2.** (1)S, (2)P, (3)P, (4)S, (5)S, (6)P, (7)P, (8)S, (9)P, (10)S, (11)P, (12)S; **3.** b; **4.** c; **5.** a

Chapter 15

1. c; **2.** d; **3.** c; **4.** e; **5.** b; **6.** (1)d, (2)e, (3)d, (4)a

Chapter 16

1. d; **2.** a; **3.** d; **4.** c; **5.** d; **6.** b; **7.** c; **8.** d; **9.** a; **10.** b; **11.** c; **12.** d; **13.** b; **14.** a; **15.** b; **16.** b; **17.** b; **18.** d; **19.** b; **20.** c; **21.** d; **22.** b; **23.** b; **24.** e; **25.** b; **26.** c; **27.** c; **28.** c; **29.** c

Chapter 17

Clinical Connections 1. Rationale for orders: As Mr. Gutteman is unconscious, the level of damage to his brain is unclear. Monitoring his responses and vital signs every hour will provide information for his care providers about the extent of his injuries. Turning him every 4 hours and providing careful skin care will prevent decubitus ulcers (bedsores) as well as stimulating his proprioceptive pathways. **2.** Mr. Gutteman's condition is termed *diabetes insipidus*, a condition in which insufficient quantities of antidiuretic hormone (ADH) are produced or released. Diabetes insipidus patients excrete large volumes of urine but do not have glucose or ketones present in the urine. The head trauma could have damaged Mr. Gutteman's hypothalamus, which produces the hormone, or injured his posterior pituitary gland, which releases ADH into the bloodstream. **3.** Diabetes insipidus is not life threatening for most individuals with normal thirst mechanisms, as they will be thirsty and drink to replenish the lost fluid. However, Mr. Gutteman is comatose, so his fluid output must be monitored closely so that the volume lost can be replaced by IV line. His subsequent recovery may be complicated if he has suffered damage to his hypothalamus, which houses the thirst center neurons.

Review Questions 1. b; **2.** a; **3.** c; **4.** d; **5.** (1)c, (2)b, (3)f, (4)d, (5)e, (6)g, (7)a, (8)h, (9)b and e, (10)a; **6.** d; **7.** c; **8.** b; **9.** d; **10.** b; **11.** d; **12.** b; **13.** c; **14.** d

Chapter 18

1. c; **2.** c; **3.** d; **4.** b; **5.** d; **6.** a; **7.** a; **8.** b; **9.** c; **10.** d

Chapter 19

1. a; **2.** c; **3.** b; **4.** c; **5.** b; **6.** b; **7.** c; **8.** d; **9.** b

Chapter 20

Clinical Connections 1. The tissues in Mr. Hutchinson's right leg were deprived of oxygen and nutrients for at least one-half hour. When tissues are deprived of oxygen, tissue metabolism decreases and eventually ceases, so these tissues may have died due to anoxia. **2.** Mr. Hutchinson's vital signs (low BP; rapid, thready pulse) indicate that he is facing a life-threatening problem that must be stabilized before other, less vital problems can be addressed. As for surgery, he may be scheduled for open reduction of his crushed bone, depending upon the condition of the tissues in his crushed right leg. If tissue death has occurred in his leg, he may undergo amputation of that limb. **3.** Mr. Hutchinson's rapid, thready pulse and falling blood pressure are indications of hypovolemic shock, a type of shock resulting from decreased blood volume. Because his blood volume is low, his heart rate is elevated to increase cardiac output in an effort to maintain the blood supply to his vital organs. Mr. Hutchinson's blood volume must be increased as quickly as possible with blood transfusions or intravenous saline. This will stabilize his condition and allow his physicians to continue with his surgery.

Review Questions 1. d; **2.** b; **3.** d; **4.** c; **5.** e; **6.** d; **7.** c; **8.** b; **9.** b; **10.** a; **11.** b; **12.** c; **13.** b

Chapter 21

Clinical Connections 1. The red streaks radiating from Mr. Hutchinson's finger indicate that his lymphatic vessels are inflamed. This inflammation may be caused by a bacterial infection. If Mr. Hutchinson's arm had exhibited edema without any accompanying red streaks, the problem would likely have been impaired lymph transport from his arm back to his trunk, due to injury or blockage of his lymphatic vessels. **2.** Mr. Hutchinson's arm was placed in a sling to immobilize it, slowing the drainage of lymph from the infected area in an attempt to limit the spread of the inflammation. **3.** Mr. Hutchinson's low lymphocyte count indicates that his body's ability to fight infection by bacteria or viruses is impaired. The antibiotics and additional staff protection will protect Mr. Hutchinson until his lymphocyte count increases again. Gloving and gowning also protect the staff caring for Mr. Hutchinson from any body infection he might have. **4.** Mr. Hutchinson's recovery may be problematic, as he probably already has an ongoing bacterial infection and his ability to raise a defense to this infection is impaired.

Review Questions 1. c; **2.** c; **3.** a, d; **4.** c; **5.** a; **6.** b; **7.** a; **8.** b; **9.** d

Chapter 22

1. c; **2.** a; **3.** d; **4.** d, e; **5.** a; **6.** d; **7.** b; **8.** c; **9.** d

Chapter 23

Clinical Connections 1. Spinal cord injury from a fracture at the level of the C_2 vertebra would interrupt the normal transmission of signals from the brain stem down the phrenic nerve to the diaphragm, and Barbara would be unable to breathe due to paralysis of the diaphragm. **2.** Barbara's head, neck, and torso should have been immobilized to prevent further damage to the spinal cord. In addition, she required assistance to breathe, so her airway was probably intubated to permit ventilation of her lungs. **3.** Cyanosis is a decrease in the degree of oxygen saturation of hemoglobin. As Barbara's respiratory efforts cease, her alveolar P_{O_2} will fall, so there is less diffusion of oxygen between the lungs and the bloodstream. In addition, her arterial P_{CO_2} will increase. Both of these conditions will promote the unloading of oxygen from hemoglobin. **4.** Injury to the spinal cord at the level of the C_2 vertebra will cause quadriplegia (paralysis of all four limbs). **5.** Atelectasis is the collapse of a lung. Because it is the right thorax that is compressed, only her right lung is affected. Because the lungs are in separate pleural cavities, only the right lung collapsed. **6.** Barbara's fractured ribs probably punctured her lung tissue and allowed air within the lung to enter the pleural cavity. **7.** The atelectasis will be reversed by closing the tear in her lung tissue, then removing the air from the pleural cavity using chest tubes. This will allow her lung to re-inflate.

Review Questions 1. b; 2. a; 3. c; 4. c; 5. b; 6. d; 7. d; 8. b; 9. a, e; 10. c; 11. b; 12. b; 13. b; 14. c; 15. b; 16. b

Chapter 24

Clinical Connections 1. Mr. Gutteman's statement about the effects of milk on his digestive tract suggests that he may be deficient in lactase, a brush border enzyme that breaks down lactose (milk sugar). 2. His responses to the questions reduced the possibility that he has gastric ulcers or *Shigella* poisoning. Mr. Gutteman's diarrhea may be due to a *Salmonella* infection. To verify this diagnosis, Mr. Gutteman's stool will be cultured to check for growth of *Salmonella* bacteria. If he does have *Salmonella* poisoning, he will be treated with painkillers and receive fluid and electrolyte replacement. 3. His name will be given to the health department because individuals infected with *Salmonella* may remain carriers for months after their symptoms subside.

Review Questions 1. c; 2. d; 3. d; 4. b; 5. b; 6. a; 7. d; 8. d; 9. b; 10. c; 11. c; 12. a; 13. d; 14. d; 15. b; 16. c; 17. a

Chapter 25

1. a; 2. c; 3. c; 4. d; 5. b; 6. c; 7. a; 8. d; 9. d; 10. a; 11. b; 12. d; 13. c; 14. d; 15. a

Chapter 26

1. d; 2. b; 3. c; 4. d; 5. c; 6. b; 7. a; 8. c; 9. a

Chapter 27

Clinical Connections 1. Mr. Heyden's vital signs suggest that he is in hypovolemic shock, which is probably due to an internal hemorrhage. 2. The spleen is a highly vascular organ due to its role as a blood filtering organ. The macrophages in Mr. Heyden's liver and bone marrow will help compensate for the loss of his spleen. 3. Elevation of renin, aldosterone, and antidiuretic hormone indicate that Mr. Heyden's body is trying to compensate for his falling blood pressure and blood loss. *Renin:* released when renal blood flow is diminished and blood pressure falls. Renin initiates the angiotensin-aldosterone response. The formation of angiotensin II leads to vasoconstriction, which will increase blood pressure, and to the release of aldosterone. *Aldosterone:* increases Na^+ reabsorption by the kidney. The movement of this reabsorbed Na^+ into the bloodstream will promote the movement of water from the interstitial fluid, resulting in an increase in blood volume. *Antidiuretic hormone (ADH):* released when the hypothalamic osmoreceptors sense an increase in osmolality. ADH has two consequences: it is a potent vasoconstrictor and will increase blood pressure, and it promotes water retention by the kidney, in-

creasing blood volume. 4. Mr. Heyden's urine production may be decreased for several reasons. The severe drop in his blood pressure would reduce renal blood flow, thus reducing glomerular blood pressure and decreasing his glomerular filtration rate. The elevation in his ADH levels can reduce urine output due to increased water reabsorption by the kidney. He may also have damage to the kidney due to his crush injury in the left lumbar region. The presence of casts and a brownish-red color in his urine are probably damaged cells and blood. If he has suffered kidney damage due to being crushed, he may have nephron damage that would include disruption of the filtration membrane, allowing red blood cells to pass into the filtrate and therefore the urine. He may also have damaged renal tubules and peritubular capillaries, allowing entry of blood and damaged renal tubule epithelial cells into the filtrate.

Review Questions 1. a; 2. c; 3. b; 4. b; 5. h, i, j; 6. c, g; 7. e, h; 8. b; 9. a, e; 10. j; 11. b, d, f; 12. a; 13. c

Chapter 28

Clinical Connections 1. *Carcinoma* is the term for cancer originating from epithelial tissue. The primary source of Mr. Heyden's cancer is likely to be the prostate gland. 2. Elevation of serum acid phosphatase levels is diagnostic for carcinoma of the prostate. (In addition, Mr. Heyden's age places him in a group that is at relatively higher risk for this type of cancer.) 3. Digital examination of Mr. Heyden's prostate should detect the presence of carcinoma in this tissue. In addition, serum levels of prostate specific antigen (PSA) should be checked, as an increase in this antigen is specific for prostate cancer. 4. Mr. Heyden's carcinoma has advanced to the point of metastasis. He will probably undergo a treatment that reduces the levels of androgens in his body, as androgens promote growth of the prostate-derived tissue. These treatments could include castration or administration of drugs that block the production and/or effects of androgens.

Review Questions 1. a and b; 2. d; 3. a; 4. d; 5. b; 6. d; 7. d; 8. c; 9. a, c, e, f; 10. (1)c, f; (2)e, h; (3)g; (4)a; (5)b, g (and perhaps e); (6)f; 11. a; 12. c; 13. d; 14. b; 15. a; 16. b; 17. c

Chapter 29

1. (1)a, (2)b, (3)b, (4)a; 2. b; 3. c; 4. b; 5. a; 6. d; 7. d; 8. c; 9. a; 10. b; 11. c; 12. a; 13. c; 14. a; 15. (1)e, (2)g, (3)a, (4)f, (5)i, (6)b, (7)h, (8)d, (9)c

Chapter 30

1. (1)d, (2)g, (3)b, (4)a, (5)f, (6)c, (7)e, (8)h; 2. (1)e, (2)a, (3)b, (4)d, (5)c, (6)d

GLOSSARY

Pronunciation Key:

' = Primary accent
" = Secondary accent
Pronounce:

a, fa, āt	as in	fate	o, no, ōt	as in	note
a, hă, at		hat	ŏ, fr, ŏg		frog
ah		father	oo		soon
ar		tar	or		for
e, stre, ēt		street	ow		plow
ĕ, hĕ, en		hen	oy		boy
er		her	sh		she
ew		new	u, mu, ūt		mute
g		go	ŭ, sŭ, un		sun
i, bi, īt		bite	z		zebra
i, hi, im		him	zh		measure
ng		ring			

Abdomen (ab'do-men) Portion of the body between the diaphragm and the pelvis.

Abduct (ab-dukt') To move away from the midline of the body.

Abscess (ab'ses) Localized accumulation of pus and disintegrating tissue.

Absolute refractory period Period following stimulation during which no additional action potential can be evoked.

Absorption Process by which the products of digestion pass through the alimentary tube mucosa into the blood or lymph.

Accessory digestive organs Organs that contribute to the digestive process but are not part of the alimentary canal; include the tongue, teeth, salivary glands, pancreas, liver.

Accommodation The process of increasing the refractive power of the lens of the eye; focusing.

Acetabulum (as"ĕ-tab'u-lum) Cuplike cavity on lateral surface of the hip bone that receives the femur.

Acetylcholine (ACh) (as"ĕ-til-ko'lēn) Chemical transmitter substance released by some nerve endings.

Acetylcholinesterase (AChE) (as"ĕ-til-ko"lin-es'ter-ās) Enzyme present at the neuromuscular junction that prevents continued muscle contraction in the absence of additional stimulation.

Achilles tendon. *See* Calcaneal tendon.

Acid A substance that releases hydrogen ions when in solution (compare with Base); a proton donor.

Acid-base balance Situation in which the pH of the blood is maintained between 7.35 and 7.45.

Acidosis (as"ĭ-do'sis) State of abnormally high hydrogen ion concentration in the extracellular fluid.

Actin (ak'tin) A contractile protein of muscle.

Action potential A large transient depolarization event, including polarity reversal, that is conducted along the membrane of a muscle cell or a nerve fiber.

Activation energy The amount of energy required to push a reactant to the level necessary for action.

Active immunity Immunity produced by an encounter with an antigen; provides immunologic memory.

Active site Region on the surface of functional (globular) proteins that fit and interact chemically with other molecules of complimentary shape and charge.

Active transport Membrane transport processes for which ATP is provided, e.g., solute pumping and endocytosis.

Adaptation (1) Any change in structure or response to suit a new environment; (2) decline in the transmission of a sensory nerve when a receptor is stimulated continuously and without change in stimulus strength.

Adduct (a-dukt') To move toward the midline of the body.

Adenine (A) (ad'ĕ-nēn) One of the two major purines found in both RNA and DNA; also found in various free nucleotides of importance to the body, such as ATP.

Adenohypophysis (ad"ĕ-no-hi-pof'ĭ-sis) Anterior pituitary; the glandular part of the pituitary gland.

Adenoids (ad'en-noids) Pharyngeal tonsils.

Adenosine triphosphate (ATP) (ah-den'o-sēn tri"fos'fāt) Organic molecule that stores and releases chemical energy for use in body cells.

Adipocyte (ad'ĭ-po-sīt) An adipose, or fat cell.

Adipose tissue (ad'ĭ-pōs) Areolar connective tissue modified to store nutrients; a connective tissue consisting chiefly of fat cells.

Adrenal glands (uh-drē'nul) Hormone-producing glands located superior to the kidneys; each consists of medulla and cortex areas.

Adrenergic fibers (ad"ren-er'jik) Nerve fibers that release norepinephrine.

Adrenocorticotropic hormone (ACTH) (ah-dre'no-kor"tĭ-ko-trō'pik) Anterior pituitary hormone that influences the activity of the adrenal cortex.

Adventitia (ad"ven-tish'e-ah) Outermost layer or covering of an organ.

Aerobic (a'er-ōb"ik) Oxygen-requiring.

Aerobic endurance The length of time a muscle can continue to contract using aerobic pathways.

Aerobic respiration Respiration in which oxygen is consumed and glucose is broken down entirely; water, carbon dioxide, and large amounts of ATP are the final products.

Afferent (af'er-ent) Carrying to or toward a center.

Afferent (sensory) nerve Nerve that contains processes of sensory neurons and carries nerve impulses to the central nervous system.

Afferent (sensory) neuron Nerve cell that carries impulses toward the central nervous system; initiates nerve impulses following receptor stimulation.

Agglutination (ah-gloo"tĭ-na'shun) Clumping of (foreign) cells; induced by cross-linking of antigen-antibody complexes.

Agonist (ag'o-nist) Muscle that bears the major responsibility for effecting a particular movement; a prime mover.

AIDS Acquired immune deficiency syndrome; caused by human immunodeficiency virus (HIV); symptoms include severe weight loss, night sweats, swollen lymph nodes, opportunistic infections.

Albumin (al-bu'min) The most abundant plasma protein.

Aldosterone (al-dos'ter-ōn) Hormone produced by the adrenal cortex that regulates sodium ion reabsorption.

Alimentary canal (al"ĭ-men'tar-e) The continuous hollow tube extending from the mouth to the anus; its walls are constructed by the oral cavity, pharynx, esophagus, stomach, and small and large intestines.

Alkalosis (al"kah-lo'sis) State of abnormally low hydrogen ion concentration in the extracellular fluid.

Allantois (ah"lan'to-is) Embryonic membrane; its blood vessels develop into blood vessels of the umbilical cord.

Alleles Genes coding for the same trait and found at the same locus on homologous chromosomes.

Allergy (hypersensitivity) Overzealous immune response to an otherwise harmless antigen.

Alopecia (al"o-pe'she-ah) Baldness.

Alpha (α) helix The most common type of secondary structure of the amino acid chain in proteins; resembles the coils of a telephone cord.

Alveolar (acinar) gland (al-ve'o-lar) A gland whose secretory cells form small, flasklike sacs.

Alveolar ventilation rate (AVR) An index of respiratory efficiency; measures volume of air wasted and flow of fresh gases in and out of alveoli.

Alveolus (al-ve'o-lus)   One of the microscopic air sacs of the lungs.

Alzheimer's disease (altz'-hi-merz)   Degenerative brain disease resulting in progressive loss of memory and motor control, and increasing dementia.

Amino acid (ah-me'no)   Organic compound containing nitrogen, carbon, hydrogen, and oxygen; building block of protein.

Ammonia (NH₃)   Common waste product of protein breakdown in the body; a colorless volatile gas, very soluble in water and capable of forming a weak base; a proton acceptor.

Amniocentesis   A common form of fetal testing in which a small sample of fluid is removed from the amniotic cavity.

Amnion (am'ne-on)   Fetal membrane that forms a fluid-filled sac around the embryo.

Amoeboid motion (ah-me'boyd)   The flowing movement of the cytoplasm of a phagocyte.

Amphiarthrosis (am"fe-ar-thro'sis)   A slightly movable joint.

Ampulla (am-pul'lah)   A localized dilation of a canal or duct.

Amylase   Digestive system enzyme that breaks down starchy foods.

Anabolism (ah-nab'o-lizm)   Energy-requiring building phase of metabolism in which simpler substances are combined to form more complex substances.

Anaerobic (an-a'er-ōb-ik)   Not requiring oxygen.

Anaerobic glycolysis (gli-kol'ĭ-sis)   Energy-yielding conversion of glucose to lactic acid in various tissues, notably muscle, when sufficient oxygen is not available.

Anaerobic threshold   The point at which muscle metabolism converts to anaerobic glycolysis.

Anaphase   Third stage of mitosis in which a full set of daughter chromosomes move toward the poles of a cell.

Anastomosis (ah-nas"to-mo'sis)   A union or joining of nerves, blood vessels, or lymphatics.

Anatomy   Study of the structure of living organisms.

Androgen (an'dro-jen)   A hormone that controls male secondary sex characteristics, such as testosterone.

Anemia (ah-nē-mē-ah)   Reduced oxygen-carrying ability of blood resulting from too few erythrocytes or abnormal hemoglobin.

Aneurysm (an'u-rizm)   Blood-filled sac in an artery wall caused by dilation or weakening of the wall.

Angina pectoris (an-ji'nah, an'jĭ-nah pek'tor-is)   Severe suffocating chest pain caused by brief lack of oxygen supply to heart muscle.

Angiotensin II (an"je-o-ten'sin)   A potent vasoconstrictor activated by renin; also triggers release of aldosterone.

Anion (an'i-on)   An ion carrying one or more negative charges and therefore attracted to a positive pole.

Anoxia (ah-nŏk'se-ah)   Deficiency of oxygen.

Antagonist (an-tag'o-nist)   Muscle that reverses, or opposes, the action of another muscle.

Anterior pituitary.   *See* Adenohypophysis.

Antibody   A protein molecule that is released by a plasma cell (a daughter cell of an activated B lymphocyte) and that binds specifically to an antigen; an immunoglobulin.

Anticodon (an"ti-ko'don)   The three-base sequence complimentary to the messenger RNA (mRNA) codon.

Antidiuretic hormone (ADH) (an"tĭ-dī"yer-eh'tik)   Hormone produced by the hypothalamus and released by the posterior pituitary; stimulates the kidneys to reabsorb more water, reducing urine volume.

Antigen (Ag) (an'tĭ-jen)   A substance or part of a substance (living or nonliving) that is recognized as foreign by the immune system, activates the immune system, and reacts with immune cells or their products.

Anucleate cells (a-nu'kle-āt)   A cell without a nucleus.

Anus (a'nus)   Distal end of digestive tract; outlet of rectum.

Aorta (a-or'tah)   Major systemic artery; arises from the left ventricle of the heart.

Aortic body   Receptor in the aortic arch sensitive to changing oxygen, carbon dioxide, and pH levels of the blood.

Apgar score   Evaluation of an infant's physical status at 1 and 5 minutes after birth by assessing five criteria: heart rate, respiration, color, muscle tone, and reflexes.

Apnea   Breathing cessation.

Apocrine gland (ap'o-krin)   The less numerous type of sweat gland; produces a secretion containing water, salts, proteins, and fatty acids.

Apoenzyme (ap'ō-en-sī m)   The protein portion of an enzyme.

Aponeurosis (ap"o-nu-ro'sis)   Fibrous or membranous sheet connecting a muscle and the part it moves.

Appendicitis (ap"ĕ n-dĭ-si'tis)   Inflammation of the appendix (wormlike sac attached to the cecum of the large intestine).

Appendicular   Relating to the limbs; one of the two major divisions of the body.

Appositional growth   Growth accomplished by the addition of new layers onto those previously formed.

Aqueous humor (a'kwe-us)   Watery fluid in the anterior chambers of the eye.

Arachnoid (ah-rak-noid)   Weblike; specifically, the weblike middle layer of the three meninges.

Areola (ah-re'o-lah)   Circular, pigmented area surrounding the nipple; any small space in a tissue.

Areolar connective tissue   A type of loose connective tissue.

Arrector pili (ah-rek'tor pi'li)   Tiny, smooth muscles attached to hair follicles; cause the hair to stand upright when activated.

Arrhythmia (a-rith'me-ah)   Irregular heart rhythm caused by defects in the intrinsic conduction system.

Arteries   Blood vessels that conduct blood away from the heart and into the circulation.

Arteriole (ar'tē r'e-ol)   A minute artery.

Arteriosclerosis (ar'tē r'e-o-skler-o'sis)   Any of a number of proliferative and degenerative changes in the arteries leading to their decreased elasticity.

Arthritis   Inflammation of the joints.

Arthroscopic surgery (ar-thro-skop'ik)   Procedure enabling a surgeon to repair the interior of a joint through a small incision.

Articular capsule   Double-layered capsule composed of an outer fibrous capsule lined by synovial membrane; encloses the joint cavity of a synovial joint.

Articular cartilage   Hyaline cartilage covering bone ends at movable joints.

Articulation (joint)   The junction of two or more bones.

Association areas   Functional areas of the cerebral cortex that act mainly to integrate diverse information for purposeful action.

Association neuron (interneuron)   Nerve cell located between motor and sensory neurons that shuttle signals through CNS pathways where integration occurs.

Astigmatism (ah-stig'mah-tizm)   A condition in which unequal curvatures in different parts of the lens (or cornea) of the eye lead to blurred vision.

Astrocyte (as'tro-sī te)   A type of CNS supporting cell; assists in exchanges between blood capillaries and neurons.

Ataxia (ah-tak'se-ah)   Disruption of muscle coordination resulting in inaccurate movements.

Atelectasis (at"ĕ -lik'tah-sis)   Lung collapse.

Atherosclerosis (a"ther-o"skler-o'sis)   Changes in the walls of large arteries consisting of lipid deposits on the artery walls; the early stage of arteriosclerosis.

Atmospheric pressure   Force that air exerts on the surface of the body.

Atom   Smallest particle of an elemental substance that exhibits the properties of that element; composed of protons, neutrons, and electrons.

Atomic mass number   Sum of the number of protons and neutrons in the nucleus of an atom.

Atomic number   The number of protons in an atom.

Atomic symbol   The one- or two-letter symbol used to indicate an element; usually the first letter(s) of the element's name.

Atomic weight The average of the mass numbers of all the isotopes of an element.

ATP (adenosine triphosphate) (ah-den'o-sēn tri"fos'fāt) Organic molecule that stores and releases chemical energy for use in body cells.

Atria (a'tre-ah) The two superior receiving chambers of the heart.

Atrial natriuretic peptide (ANP) (a'tre-al na"tre-u-ret'ik) A hormone released by certain cells of the heart atria that reduces blood pressure and blood volume by inhibiting nearly all events that promote vasoconstriction and Na^1 and water retention.

Atrioventricular (AV) bundle (a"tre-o-ven-trī'kyoo-ler) Bundle of specialized fibers that conduct impulses from the AV node to the right and left ventricles; also called bundle of His.

Atrioventricular (AV) node Specialized mass of conducting cells located at the atrioventricular junction in the heart.

Atrioventricular (AV) valve Valve that prevents backflow into the atria when the ventricles are contracting.

Atrophy (at'ro-fe) Reduction in size or wasting away of an organ or cell resulting from disease or lack of use.

Auditory ossicles (ah'sih-kulz) The three tiny bones serving as transmitters of vibrations and located within the middle ear: the malleus, incus, and stapes.

Auditory tube Tube that connects the middle ear and the pharynx. Also called eustachian tube.

Autoimmune response Production of antibodies or effector T cells that attack a person's own tissue.

Autolysis (aw"-tol'y-sis) Process of autodigestion (self-digestion) of cells, especially dead or degenerate cells.

Autonomic nervous system (ANS) Efferent division of the peripheral nervous system that innervates cardiac and smooth muscles and glands; also called the involuntary or visceral motor system.

Autonomic (visceral) reflexes Reflexes that activate smooth or cardiac muscle and/or glands.

Autoregulation The automatic adjustment of blood flow to a particular body area in response to its current requirements.

Autosomes Chromosomes number 1 to 22; do not include the sex chromosomes.

Avogadro's number (av"o-gad'rōz) The number of molecules in one mole of any substance, 6.02×10^{23}.

Axial Relating to the head, neck, and trunk; one of the two major divisions of the body.

Axolemma (ak"so-lem'ah) The plasma membrane of an axon.

Axon Neuron process that carries impulses away from the nerve cell body; efferent process; the conducting portion of a nerve cell.

B cells Lymphocytes that oversee humoral immunity; their descendants differentiate into antibody-producing plasma cells. Also called B lymphocytes.

Baroreceptor (bayr"o-re-sep-tor) Pressoreceptor; receptor that is stimulated by pressure changes.

Basal body (ba'sal) The elongated part of a cell centriole that forms the bases of cilia and flagella.

Basal lamina (lam'ĭ-nah) Noncellular, adhesive supporting sheet consisting largely of glycoproteins secreted by epithelial cells.

Basal metabolic rate (BMR) Rate at which energy is expended (heat produced) by the body per unit time under controlled (basal) conditions: 12 hours after a meal, at rest.

Basal nuclei (basal ganglia) Gray matter areas located deep within the white matter of the cerebral hemispheres.

Basal surface The surface near the base or interior of a structure; nearest the lower side or bottom of a structure.

Base A substance capable of binding with hydrogen ions; a proton acceptor.

Basement membrane Extracellular material consisting of a basal lamina secreted by epithelial cells and a reticular lamina secreted by underlying connective tissue cells.

Basophil (ba'zo-fil) White blood cell whose granules stain deep blue with basic dye; has a relatively pale nucleus.

Benign (be-nin') Not malignant.

Biceps (bi'seps) Two-headed, especially applied to certain muscles.

Bicuspid (mitral) valve (mi'tral) The left atrioventricular valve.

Bile Greenish-yellow or brownish fluid produced in and secreted by the liver, stored in the gallbladder, and released into the small intestine.

Bilirubin (bil"i-roo'bin) Red pigment of bile.

Biofeedback Training that provides an awareness of visceral activities; enables an element of voluntary control over autonomic body functions.

Biogenic amines (bi"o-jen'ik am'inz, ah'mē nz) Class of neurotransmitters, including catecholamines and indolamines.

Bipolar neuron Neuron with axon and dendrite that extend from opposite sides of the cell body.

Blastocyst (blas'to-sist) Stage of early embryonic development; the product of cleavage.

Blood pressure (BP) Force exerted by blood against a unit area of the blood vessel walls; differences in blood pressure between different areas of the circulation provide the driving force for blood circulation.

Blood-brain barrier Mechanism that inhibits passage of materials from the blood into brain tissues; reflects relative impermeability of brain capillaries.

Bolus (bo'lus) A rounded mass of food prepared by the mouth for swallowing; any soft round mass.

Bone (osseous) tissue (os'e-us) A connective tissue that forms the bony skeleton.

Bone remodeling Process involving bone formation and destruction in response to hormonal and mechanical factors.

Bone resorption The removal of osseous tissue; part of the continuous bone remodeling process.

Bony thorax (thoracic cage) Bones that form the framework of the thorax; includes sternum, ribs, and thoracic vertebrae.

Bowman's capsule (bo-manz). *See* Glomerular capsule.

Boyle's law States that when the temperature is constant, the pressure of a gas varies inversely with its volume.

Bradycardia (brad"e-kar'deah) A heart rate below 60 beats per minute.

Brain death State of irreversible coma, even though life-support measures may have restored other body organs.

Brain stem Collectively the midbrain, pons, and medulla of the brain.

Brain ventricle Fluid-filled cavity of the brain.

Branchial groove (brang'ke-al) An indentation of the surface ectoderm in the embryo; the external auditory canals develop from these.

Bronchioles The branching air passageways inside the lungs.

Bronchus (brong'kus) One of the two large branches of the trachea that leads to the lungs.

Buffer Chemical substance or system that minimizes changes in pH by releasing or binding hydrogen ions.

Bulk (vesicular) transport The movement of large particles and macromolecules across a plasma membrane.

Burn Tissue damage inflicted by intense heat, electricity, radiation, or certain chemicals, all of which denature cell proteins and cause cell death in the affected areas.

Bursa (ber'sa) A fibrous sac lined with synovial membrane and containing synovial fluid; occurs between bones and muscle tendons (or other structures), where it acts to decrease friction during movement.

Bursitis Inflammation of a bursa.

Calcaneal tendon (kal-ka'ne-al) Tendon that attaches the calf muscles to the heelbone (calcaneus); also called the Achilles tendon.

Calcitonin (kal"sih-to'nin) Hormone released by the thyroid that promotes a decrease in calcium levels of the blood; also called thyrocalcitonin.

Calculus (kal'ku-lus) A stone formed within various body parts.

Callus (kal'lus)   (1) Localized thickening of skin epidermis resulting from physical trauma; (2) repair tissue (fibrous or bony) formed at a fracture site.

Calorie (cal)   Amount of energy needed to raise the temperature of 1 gram of water 1° Celsius. Energy exchanges associated with biochemical reactions are usually reported in kilocalories (1 kcal = 1000 cal) or large calories (Cal).

Calyx (ka'liks)   A cuplike extension of the pelvis of the kidney.

Canaliculus (kan"ah-lik'u-lus)   Extremely small tubular passage or channel.

Cancer   A malignant, invasive cellular neoplasm that has the capability of spreading throughout the body or body parts.

Capillaries (kap'il-layr"ē s)   The smallest of the blood vessels and the sites of exchange between the blood and tissue cells.

Capping   Proposed mechanism of B cell activation in which multivalent antigens bind to several adjacent receptors on a B lymphocyte and pull them into a continuous cluster.

Carbohydrate (kar"bo-hi-drāt)   Organic compound composed of carbon, hydrogen, and oxygen; includes starches, sugars, cellulose.

Carbonic acid–bicarbonate system   Chemical buffer system that helps maintain pH homeostasis of the blood.

Carbonic anhydrase (kar-bon'ik an-hi'dras)   Enzyme that facilitates the combination of carbon dioxide with water to form carbonic acid.

Carcinogen (kar'sī'no jin)   Cancer-causing agent.

Cardiac cycle   Sequence of events encompassing one complete contraction and relaxation of the atria and ventricles of the heart.

Cardiac muscle   Specialized muscle of the heart.

Cardiac output (CO)   Amount of blood pumped out of a ventricle in one minute.

Cardiac reserve   The difference between resting and maximal cardiac output.

Cardiogenic shock   Pump failure; the heart is so inefficient that it cannot sustain adequate circulation.

Cardiovascular system   Organ system which distributes the blood to deliver nutrients and remove wastes.

Carotene (kar'o-tē n)   Yellow to orange pigment that accumulates in the stratum corneum epidermal layer and in fatty tissue of the hypodermis.

Carotid body (kar-rot'id)   A receptor in the common carotid artery sensitive to changing oxygen, carbon dioxide, and pH levels of the blood.

Carotid sinus (si-nus)   A dilation of a common carotid artery; involved in regulation of systemic blood pressure.

Cartilage (kar'tĭ-lij)   White, semiopaque connective tissue.

Cartilage bone (endochondral bone)   Bone formed by the calcification of hyaline cartilage structures.

Cartilaginous joints (kar"ti-laj'ĭ-nus)   Bones united by cartilage; no joint cavity is present.

Catabolism (kat-tab'o-lizm)   Process in which living cells break down substances into simpler substances.

Catalyst (kat'ah-list)   Substance that increases the rate of a chemical reaction without itself becoming chemically changed or part of the product.

Cataract   Clouding of the eye's lens; often congenital or age-related.

Catecholamines (kat"ē-kol'ah-menz)   Epinephrine and norepinephrine.

Cation (kat'i-on)   An ion with a positive charge.

Caudal (kaw'dul)   Literally, toward the tail; in humans, the inferior portion of the anatomy.

Cecum (se'kum)   The blind-end pouch at the beginning of the large intestine.

Cell differentiation   The development of specific and distinctive features in cells, from a single cell (the fertilized egg) to all the specialized cells of adulthood.

Cell division period (mitotic (M) phase)   One of two major periods in the cell life cycle; involves the division of the nucleus (mitosis) and the division of the cytoplasm (cytokinesis).

Cell life cycle   Series of changes a cell goes through from the time it is formed until it reproduces itself.

Cell-mediated immune response   Immunity conferred by activated T cells, which directly lyse infected or cancerous body cells or cells of foreign grafts and release chemicals that regulate the immune response.

Cell membrane.   *See* Plasma membrane.

Cellular respiration   Metabolic processes in which ATP is produced.

Cellulose (sel'u-lōs)   A fibrous carbohydrate that is the main structural component of plant tissues.

Central (Haversian) canal (hah-ver'-shan)   The canal in the center of each osteon that contains minute blood vessels and nerve fibers that serve the needs of the osteocytes.

Central nervous system (CNS)   Brain and spinal cord.

Centriole (sen'tre-ol)   Minute body found near the nucleus of the cell; active in cell division.

Centrosome (cell center)   A region near the nucleus which contains paired organelles called centrioles.

Cerebellum (ser"ē-bel'um)   Brain region most involved in producing smooth, coordinated skeletal muscle activity.

Cerebral aqueduct (ser'ē-bral, sē-re'bral)   The slender cavity of the midbrain that connects the third and fourth ventricles; also called the aqueduct of Sylvius.

Cerebral cortex   The outer gray matter region of the cerebral hemispheres.

Cerebral dominance   Designates the hemisphere that is dominant for language.

Cerebral palsy   Neuromuscular disability in which voluntary muscles are poorly controlled or paralyzed as a result of brain damage.

Cerebral white matter   Consists largely of myelinated fibers bundled into large tracts; provides for communication between cerebral areas and lower CNS centers.

Cerebrospinal fluid (CSF) (ser'ē-bro-spi'nal)   Plasmalike fluid that fills the cavities of the CNS and surrounds the CNS externally; protects the brain and spinal cord.

Cerebrovascular accident (CVA) (ser"ē-bro-vas'ku-lar).   *See* Stroke.

Cerebrum (ser'-ē-brum)   The cerebral hemispheres and the structures of the diencephalon.

Cervical vertebrae   The seven vertebrae of the vertebral column located in the neck.

Cervix   Lower outlet of the uterus extending into the vagina.

Chemical bond   An energy relationship holding atoms together; involves the interaction of electrons.

Chemical energy   Energy stored in the bonds of chemical substances.

Chemical equilibrium   A state of apparent repose created by two reactions proceeding in opposite directions at equal speed.

Chemical reaction   Process in which molecules are formed, changed, or broken down.

Chemoreceptor (ke"mo-re-sep'ter)   Receptors sensitive to various chemicals in solution.

Chemotaxis (ke"mo-tak'sis)   Movement of a cell, organism, or part of an organism toward or away from a chemical substance.

Cholecystokinin (CCK) (ko"le-sis"to-ki'nin)   An intestinal hormone that stimulates gallbladder contraction and pancreatic juice release.

Cholesterol (ko-les'ter-ol")   Steroid found in animal fats as well as in most body tissues; made by the liver.

Cholinergic fibers (ko"lin-er'jik)   Nerve endings that, upon stimulation, release acetylcholine.

Chondroblast (kon'dro-blast)   Actively mitotic cell form of cartilage.

Chondrocyte (kon'dro-sīt)   Mature cell form of cartilage.

Chorion (kor'e-on)   Outermost fetal membrane; helps form the placenta.

Chorionic villi sampling (ko"re-on'ik vil'i)   Fetal testing procedure in which bits of the chorionic villi from the placenta are snipped off and the cells karyotyped. This procedure can be done as early as 8 weeks into the pregnancy.

Choroid (ko'roid) The vascular middle tunic of the eye.

Choroid plexus (ko'roid plex'sus) A capillary knot that protrudes into a brain ventricle; involved in forming cerebrospinal fluid.

Chromatin (kro'-mah-tin) Structures in the nucleus that carry the hereditary factors (genes).

Chromosomes (kro'mo-somz) Barlike bodies of tightly coiled chromatin; visible during cell division.

Chronic obstructive pulmonary disease (COPD) Collective term for progressive, obstructive respiratory disorders; includes emphysema, chronic bronchitis.

Chyme (kīm') Semifluid, creamy mass consisting of partially digested food and gastric juice.

Cilia (sil'e-ah) Tiny, hairlike projections on cell surfaces that move in a wavelike manner.

Circle of Willis An arterial anastomosis at the base of the brain.

Circumduction (ser"kum-duk'shun) Movement of a body part so that it outlines a cone in space.

Cirrhosis (sĭ-ro'sis) Chronic disease of the liver, characterized by an overgrowth of connective tissue or fibrosis.

Cisterna chyli (sis-ter'nah ki"ly) An enlarged sac at the base of the thoracic duct; the origin of the thoracic duct.

Cisternae (sis-ter'ne) Any cavity or enclosed space serving as a reservoir.

Cleavage An early embryonic phase consisting of rapid mitotic cell divisions without intervening growth periods; product is a blastocyst.

Clonal selection (klo'nul) Process during which a B cell or T cell becomes sensitized through binding contact with an antigen.

Clone Descendants of a single cell.

Coagulation Process in which blood is transformed from a liquid to a gel; blood clotting.

Cochlea (kok'le-ah) Snail-shaped chamber of the bony labyrinth that houses the receptor for hearing (the organ of Corti).

Codon (ko'don) The three-base sequence on a messenger RNA molecule that provides the genetic information used in protein synthesis.

Coenzyme (ko-en'zim) Nonprotein substance associated with and activating an enzyme, typically a vitamin.

Cofactor Metal ion or organic molecule that is required for enzyme activity.

Collagen fiber The most abundant of the three fibers found in the matrix of connective tissue.

Colloid (kol'oid) A mixture in which the solute particles do not settle out readily and do not pass through natural membranes.

Colloidal osmotic pressure (kol'oid-ahl ahz-ma'tik) Pressure created in a fluid by large nondiffusible molecules, such as plasma proteins that are prevented from moving through a (capillary) membrane. Such substances tend to draw water to them.

Colon Regions of the large intestine; includes ascending, transverse, descending, and sigmoid portions.

Combination (synthesis) reaction Chemical reaction in which larger, more complex atoms or molecules are formed from simpler ones.

Complement A group of bloodborne proteins, which, when activated, enhance the inflammatory and immune responses and may lead to cell lysis.

Complementary base The degree of base pairing between two sequences of DNA and/or RNA molecules.

Complete blood count (CBC) Clinical test that includes a hematocrit, counts of all formed elements and clotting factors, and other indicators of normal blood function.

Complimentarity of structure and function The relationship between a structure and it's function; i.e., structure determines function.

Compound Substance composed of two or more different elements, the atoms of which are chemically united.

Concentration gradient The difference in the concentration of a particular substance between two different areas.

Conducting zone Includes all respiratory passageways that provide conduits for air to reach the sites of gas exchange (the respiratory zone).

Conductivity Ability to transmit an electrical impulse.

Cones One of the two types of photoreceptor cells in the retina of the eye; provide for color vision.

Congenital (kun-jeh'nih-tul) Existing at birth.

Congestive heart failure (CHF) Condition in which the pumping efficiency of the heart is depressed so that circulation is inadequate to meet tissue needs.

Conjunctiva (kon"junk-ti'vah) Thin, protective mucous membrane lining the eyelids and covering the anterior surface of the eye itself.

Connective tissue A primary tissue; form and function vary extensively. Functions include support, storage, and protection.

Connexons (kŏ-nek'sonz) Hollow, cylindrical cells that connect adjacent cells at gap junctions thus allowing chemical substances to pass through.

Consciousness The ability to perceive, communicate, remember, understand, appreciate, and initiate voluntary movements; a functioning sensorium.

Contraception The prevention of conception; birth control.

Contractility Muscle cell's ability to move by shortening.

Contraction To shorten or develop tension, an ability highly developed in muscle cells.

Contralateral Relating to the opposite side.

Control center One of three interdependent components of homeostatic control mechanisms; determines the set point.

Cornea (kor'ne-ah) Transparent anterior portion of the eyeball; part of the fibrous tunic.

Corona radiata (kor-o'nah ra-de-ah'tah) (1) Arrangement of elongated follicle cells around a mature ovum; (2) crownlike arrangement of nerve fibers radiating from the internal capsule of the brain to every part of the cerebral cortex.

Coronary circulation The functional blood supply of the heart; shortest circulation in the body.

Cortex (kor'teks) Outer surface layer of an organ.

Corticosteroids (kor"tĭ-ko-stĕ'roidz) Steroid hormones released by the adrenal cortex.

Cortisol (hydrocortisone) (kor'tih-sol) Glucocorticoid produced by the adrenal cortex.

Covalent bond (ko-va'lent) Chemical bond created by electron sharing between atoms.

Cranial nerves The 12 nerve pairs that arise from the brain.

Craniosacral division Another name for the parasympathetic division of the autonomic nervous system.

Cranium (cranial bones) (kra'ne-um) Bony protective encasement of the brain and organs of hearing and equilibrium; also called the skull.

Creatine kinase (kre'ah-tin) Enzyme that catalyzes the transfer of phosphate from phosphocreatine to ADP, forming creatine and ATP; important in muscle contraction.

Creatine phosphate (CP) (fos-fāt) Compound that serves as an alternative energy source for muscle tissue.

Creatinine (kre-at'ĭ-nin) A nitrogenous waste molecule which is not reabsorbed by the kidney; this characteristic makes it useful for measurement of the GFR and glomerular function.

Crista ampullaris Sensory receptor organ within the ampulla of each semicircular canal of the inner ear; dynamic equilibrium receptor.

Cross section A cut running horizontally from right to left, dividing the body into superior and inferior parts.

Crystals Large arrays of cations and anions held together by ionic bonds; formed when an element or compound solidifies or is in a dry state.

Cutaneous (ku-ta'ne-us) Pertaining to the skin.

Cutaneous sensory receptors Receptors located throughout the skin that respond to stimuli arising outside the body; part of the nervous system.

Cyclic AMP   Important intracellular second messenger that mediates hormonal effects; formed from ATP by the action of adenylate cyclase, an enzyme associated with the plasma membrane.

Cystic fibrosis (CF)   Genetic disorder in which oversecretion of mucus clogs the respiratory passages, predisposes to fatal respiratory infections.

Cytochromes (si"to-krōmz)   Brightly colored iron-containing proteins that form part of the inner mitochondrial membrane and function as electron carriers in oxidative phosphorylation.

Cytokines   Chemical mediators involved in cellular immunity; include lymphokines, and monokines.

Cytokinesis (si"to-ki-ne'sis)   The division of cytoplasm that occurs after the cell nucleus has divided.

Cytoplasm (si'to-plazm)   The cellular material surrounding the nucleus and enclosed by the plasma membrane.

Cytosine (C) (si'to-sēn)   Nitrogen-containing base that is part of a nucleotide structure.

Cytoskeleton   Literally, cell skeleton. An elaborate series of rods running through the cystol, supporting cellular structures and providing the machinery to generate various cell movements.

Cytosol   Viscous, semitransparent fluid substance of cytoplasm in which other elements are suspended.

Cytotoxic T cell   Effector T cell that directly kills (lyses) foreign cells, cancer cells, or virus-infected body cells. Also called a killer T cell.

Deamination (de'am"ih-na'shun)   Removal of an amine group from an organic compound.

Decomposition reaction   Chemical reaction in which a molecule is broken down into smaller molecules or its constituent atoms.

Defecation (def"ih-ka'shun)   Elimination of the contents of the bowels (feces).

Deglutition (deg"loo-tish'un)   Swallowing.

Dehydration (de"hi-dra'shun)   Condition of excessive water loss.

Dehydration synthesis   Process by which a large molecule is synthesized by covalently bonding smaller molecules together.

Dendrite (den'drīt)   Branching neuron process that serves as a receptive, or input, region; transmits the nerve impulse toward the cell body.

Depolarization (de-po"ler-ah-za'shun)   Loss of a state of polarity; loss or reduction of negative membrane potential.

Dermatome (der'mah-tōm)   Portion of somite mesoderm that forms the dermis of the skin; also the area of skin innervated by the cutaneous branches of a single spinal nerve.

Dermis   Layer of skin deep to the epidermis; composed of dense irregular tissue.

Desmosome (dez'muh-sōm)   Cell junction composed of thickened plasma membranes joined by filaments.

Diabetes insipidus (di"ah-be'tez in-sih'pih-dus)   Disease characterized by passage of a large quantity of dilute urine plus intense thirst and dehydration caused by inadequate release of antidiuretic hormone (ADH).

Diabetes mellitus (DM) (meh-li'tus)   Disease caused by deficient insulin release, leading to inability of the body cells to use carbohydrates.

Dialysis (di-al'ah-sis)   Diffusion of solute(s) through a semipermeable membrane.

Diapedesis (di"ah-pĕ-de'sis)   Passage of blood cells through intact vessel walls into tissue.

Diaphragm (di'ah-fram)   (1) Any partition or wall separating one area from another; (2) a muscle that separates the thoracic cavity from the lower abdominopelvic cavity.

Diaphysis (di-af'ĭ-sis)   Elongated shaft of a long bone.

Diarthrosis (di"ar-thro'sis)   Freely movable joint.

Diastole (di-as'to-le)   Period of the cardiac cycle when either the ventricles or the atria are relaxing.

Diastolic pressure (di-as-tah'lik)   Arterial blood pressure reached during or as a result of diastole; lowest level of any given ventricular cycle.

Diencephalon (interbrain) (di"en-seh'fuh-lon)   That part of the forebrain between the cerebral hemispheres and the midbrain including the thalamus, the third ventricle, and the hypothalamus.

Differential white blood cell count   Diagnostic test to determine relative proportion of individual leukocyte types.

Diffusion (dĭ-fu'zhun)   The spreading of particles in a gas or solution with a movement toward uniform distribution of particles.

Digestion   Chemical or mechanical process of breaking down foodstuffs to substances that can be absorbed.

Digestive system   System that processes food into absorbable units and eliminates indigestible wastes.

Dipeptide   A combination of two amino acids united by means of a peptide bond.

Diploë (dip'lo-e)   The internal layer of spongy bone in flat bones.

Diploid chromosomal number   The chromosomal number characteristic of an organism symbolized as $2n$; twice the chromosomal number (n) of the gamete; in humans, $2n = 46$.

Diplopia (dĭ-plo'pe-ah)   Double vision.

Dipole (polar molecule)   Nonsymmetrical molecules that contain electrically unbalanced atoms.

Disaccharide (di-sak'ah-rī d", di-sak'ah-rid)   Literally, double sugar; e.g., sucrose, lactose.

Dislocation (luxation)   Occurs when bones are forced out of their normal alignment at a joint.

Displacement (exchange) reaction   Chemical reaction in which bonds are both made and broken; atoms become combined with different atoms.

Distal (dis'tul)   Away from the attached end of a limb or the origin of a structure.

Diuretics (di"u-ret'iks)   Chemicals that enhance urinary output.

Diverticulum (di"ver-tik'u-lum)   A pouch or sac in the walls of a hollow organ or structure.

DNA (deoxyribonucleic acid) (de-ok"sī-ri"bo-nu-kla'ik)   A nucleic acid found in all living cells; it carries the organism's hereditary information.

DNA replication   Process that occurs before cell division; insures that all daughter cells have identical genes.

Dominant traits   Occurs when one allele masks or suppresses the expression of its partner.

Dominant-recessive inheritance   Reflects the interaction of dominant and recessive alleles.

Dorsal (dor'sul)   Pertaining to the back; posterior.

Double helix   The secondary structure assumed by two strands of DNA, held together throughout their length by hydrogen bonds between bases on opposite strands.

Down-regulation   Decreased response to hormonal stimulation; involves a loss in the number of receptors and effectively prevents the target cells from overreacting to persistently high hormone levels.

Duct (dukt)   A canal or passageway; a tubular structure that provides an exit for the secretions of a gland, or for conducting any fluid.

Ductus (vas) deferens   Extends from the epididymis to the urethra; propels sperm into the urethra by peristalsis during ejaculation.

Duodenum (du"o-de'num)   First part of the small intestine.

Dura mater (du'-rah ma'ter)   Outermost and toughest of the three membranes (meninges) covering the brain and spinal cord.

Dynamic equilibrium   Sense that reports on angular or rotatory movements of the head in space.

Dyskinesia (dis-kĭ ne'ze-ah)   Disorders of muscle tone, posture, or involuntary movements.

Dyspnea (disp-ne'ah)   Difficult or labored breathing; air hunger.

Eccrine glands (ek'-rin)   Sweat glands abundant on the palms, soles of feet, and the forehead.

Ectoderm (ek'to-derm)   Embryonic germ layer; forms the epidermis of the skin and its derivatives, and nervous tissues.

Edema (ĕ de'mah)   Abnormal accumulation of fluid in body parts or tissues; causes swelling.

Effector (ef-ek'ter) Organ, gland, or muscle capable of being activated by nerve endings.

Efferent (ef'er-ent) Carrying away or away from, especially a nerve fiber that carries impulses away from the central nervous system.

Elastic cartilage Cartilage with abundant elastic fibers; more flexible than hyaline cartilage.

Elastic fiber Fiber formed from the protein elastin, which gives a rubbery and resilient quality to the matrix of connective tissue.

Electrical energy Energy formed by the movement of charged particles across cell membranes.

Electrocardiogram (ECG or EKG) (e-lek"tro-car'de-o-gram") Graphic record of the electrical activity of the heart.

Electrochemical gradient The distribution of ions involving both a chemical and an electrical gradient interacting to determine the direction of diffusion.

Electroencephalogram (EEG) (e-lek"tro-en-sef'ah-lo-gram") Graphic record of the electrical activity of nerve cells in the brain.

Electrolyte (e-lek'tro-līt) Chemical substances, such as salts, acids, and bases, that ionize and dissociate in water and are capable of conducting an electrical current.

Electrolyte balance Refers to the balance between input and output of salts (sodium, potassium, calcium, magnesium) in the body.

Electromagnetic (radiant) energy Energy form that travels in waves.

Electron Negatively charged subatomic particle; orbits the atom's nucleus.

Electron shells (energy levels) Regions of space that consecutively surround the nucleus of an atom.

Element One of a limited number of unique varieties of matter that composes substances of all kinds; e.g., carbon, hydrogen, oxygen.

Embolism (em'bo-lizm) Obstruction of a blood vessel by an embolus (blood clot, fatty mass, bubble of air, or other debris) floating in the blood.

Embryo (em'bre-o) Developmental stage extending from gastrulation to the end of the eight week.

Emesis Reflexive emptying of the stomach through the esophagus and pharynx; also known as vomiting.

Encephalitis (en"seh-fuh-lī'tis) Inflammation of the brain.

Endergonic reaction Chemical reaction that absorbs energy, e.g., an anabolic reaction.

Endocardium (en"do-kar'de-um) Endothelial membrane that lines the interior of the heart.

Endochondral ossification (en'do-kon'dral) Embryonic formation of bone by the replacement of calcified cartilage; most skeletal bones formed by this process.

Endocrine glands (en'do-krin) Ductless glands that empty their hormonal products directly into the blood.

Endocrine system Body system that includes internal organs that secrete hormones.

Endocytosis (en"do-si-to'sis) Means by which fairly large extracellular molecules or particles enter cells, e.g., phagocytosis, pinocytosis, receptor-mediated endocytosis.

Endoderm (en'do-derm) Embryonic germ layer; forms the lining of the digestive tube and its associated structures.

Endogenous (en-doj'ĕ-nŭs) Originating or produced within the organism or one of its parts.

Endometrium (en"do-me'tre-um) Mucous membrane lining of the uterus.

Endomysium (en"do-mis'e-um) Thin connective tissue surrounding each muscle cell.

Endoplasmic reticulum (en"do-plaz'mik rĕ-tik'u-lum) Membranous network of tubular or saclike channels in the cytoplasm of a cell.

Endosteum (en-dos'te-um) Connective tissue membrane covering internal bone surfaces.

Endothelium (en"do-the'le-um) Single layer of simple squamous cells that line the walls of the heart, blood vessels, and lymphatic vessels.

Energy The capacity to do work; may be stored (potential energy) or in action (kinetic energy).

Energy intake Energy liberated during food oxidation.

Energy output Sum of energy lost as heat, as work, and as fat or glycogen storage.

Enzyme (en'zīm) A protein that acts as a biological catalyst to speed up a chemical reaction.

Eosinophil (e"o-sin'o-fil) Granular white blood cell whose granules readily take up a stain called eosin.

Ependymal cell (ĕ-pen'dĭ-mul) A type of CNS supporting cell; lines the central cavities of the brain and spinal cord.

Epidermis (ep"ĭ-der'mis) Superficial layer of the skin; composed of keratinized stratified squamous epithelium.

Epididymis (ep"ĭ-dĭ'dĭ-mis) That portion of the male duct system in which sperm mature. Empties into the ductus (or vas) deferens.

Epidural space Area between the bony vertebrae and the dura mater of the spinal cord.

Epiglottis (eh"puh-glah'tis) Elastic cartilage at the back of the throat; covers the opening of the larynx during swallowing.

Epileptic seizures Abnormal electrical discharges of groups of brain neurons, during which no other messages can get through.

Epimysium (ep"ĭ-mis'e-um) Sheath of fibrous connective tissue surrounding a muscle.

Epinephrine (ep"ĭ-nef'rin) Chief hormone produced by the adrenal medulla. Also called adrenalin.

Epiphyseal plate (e"pĭ-fis'e-ul) Plate of hyaline cartilage at the junction of the diaphysis and epiphysis that provides for growth in length of a long bone.

Epiphysis (e-pif'ĭ-sis) The end of a long bone, attached to the shaft.

Epithalamus Most dorsal portion of the diencephalon; forms the roof of the third ventricle with the pineal gland extending from its posterior border.

Epithelium (epithelial tissue) (ep"ĭ-the'le-ul) Pertaining to a primary tissue that covers the body surface, lines its internal cavities, and forms glands.

Erythrocytes (e-rith'ro-sīts) Red blood cells.

Erythropoiesis (ĕ-rith"ro-poi-e'sis) Process of erythrocyte formation.

Erythropoietin (ĕ-rith"ro-poi'ĕ-tin) Hormone that stimulates production of red blood cells.

Esophagus (ĕ-sof'ah-gus) Muscular tube extending from the laryngopharynx through the diaphragm to join the stomach; collapses when not involved in food propulsion.

Estrogens (es'tro-jenz) Hormones that stimulate female secondary sex characteristics; female sex hormones.

Eupnea (ūp-ne'ah) Normal respiratory rate and rhythm.

Eustachian tube (u-sta'shan). *See* Auditory tube.

Exchange (displacement) reaction Chemical reaction in which bonds are both made and broken; atoms become combined with different atoms.

Excitation-contraction coupling Sequence of events by which transmission of an action potential along the sarcolemma leads to the sliding of myofilaments.

Excretion (ek-skre'shun) Elimination of waste products from the body.

Exergonic reaction Chemical reaction that releases energy, e.g., a catabolic or oxidative reaction.

Exocrine glands (ek'so-krin) Glands that have ducts through which their secretions are carried to a particular site.

Exocytosis (ek"so-si-to'sis) Mechanism by which substances are moved from the cell interior to the extracellular space as a secretory vesicle fuses with the plasma membrane.

Exogenous (ek-sah'jeh-nus) Developing or originating outside an organ or part.

Exons Amino acid-specifying informational sequences (separated by introns) in the genes of higher organisms.

Extension Movement that increases the angle of a joint, e.g., straightening a flexed knee.

Exteroceptor (ek"ster-o-sep'tor)   Sensory end organ that responds to stimuli from the external world.

Extracellular fluid   Internal fluid located outside cells; includes plasma and interstitial fluid.

Extracellular materials   Substances found outside the cell; includes interstitial fluid, blood plasma, and cerebrospinal fluid.

Extracellular matrix   Nonliving material that separates the living cells in connective tissue consisting of ground substance and fibers.

Extrasystole (ek"strah-sis'to-le)   Premature heart contraction.

Extrinsic (ek-strin'sik)   Of external origin.

Extrinsic eye muscles   The six skeletal muscles which attach to and move each eye.

Exudate (ek'syoo-dāt)   Material including fluid, pus, or cells that has escaped from blood vessels and has been deposited in tissues.

Facilitated diffusion   Passive transport process used by certain molecules, e.g., glucose and other simple sugars too large to pass through plasma membrane pores.

Fallopian tube (fah-lō'pē-un).   *See* Uterine tube.

Fascia (fash'e-ah)   Layers of fibrous tissue covering and separating muscle.

Fascicle (fas'ĭ-kl)   Bundle of nerve or muscle fibers bound together by connective tissues.

Fatty acids   Linear chains of carbon and hydrogen atoms (hydrocarbon chains) with an organic acid group at one end. A constituent of fat.

Feces (fe'sēz)   Material discharged from the bowel; composed of food residue, secretions, bacteria.

Fenestrated (fen'es-tra-tid)   Pierced with one or more small openings.

Fertilization   Fusion of the sperm and egg nuclei.

Fetus   Developmental stage extending from the ninth week of development to birth.

Fiber   A slender threadlike structure or filament.

Fibrillation   Condition of rapid and irregular or out-of-phase heart contractions.

Fibrin (fi'brin)   Fibrous insoluble protein formed during blood clotting.

Fibrinogen (fi-brin'o-jin)   A blood protein that is converted to fibrin during blood clotting.

Fibrinolysis   Process that removes unneeded blood clots when healing has occurred.

Fibroblast (fi'bro-blast)   Young, actively mitotic cell that forms the fibers of connective tissue.

Fibrocartilage   The most compressible type of cartilage; resistant to stretch. Forms vertebral discs and knee joint cartilages.

Fibrocyte (fi'bro-sīt)   Mature fibroblast; maintains the matrix of fibrous types of connective tissue.

Fibrosis   Proliferation of fibrous connective tissue called scar tissue.

Fibrous joints   Bones joined by fibrous tissue; no joint cavity is present.

Filtrate   A plasma-derived fluid that is processed by the renal tubules to form urine.

Filtration   Passage of a solvent and dissolved substances through a membrane or filter.

First-degree burn   A burn in which only the epidermis is damaged.

Fissure (fih'zher)   (1) A groove or cleft; (2) the deepest depressions or inward folds on the brain.

Fixator (fix'a-ter)   Muscle that immobilizes one or more bones, allowing other muscles to act from a stable base.

Flagellum (flah-jel'lum)   Long, whiplike extension of the plasma membrane of some bacteria and a sperm; propels the cell.

Flexion (flek'shun)   Movement that decreases the angle of the joint, e.g., bending the knee from a straight to an angled position.

Flexor (withdrawal) reflex   Reflex initiated by a painful stimulus (actual or perceived); causes automatic withdrawal of the threatened body part from the stimulus.

Fluid mosaic model   A depiction of the structure of the membranes of a cell as phospholipid bilayers in which proteins are dispersed.

Follicle (fah'lih-kul)   (1) Ovarian structure consisting of a developing egg surrounded by one or more layers of follicle cells; (2) colloid-containing structure of the thyroid gland.

Follicle-stimulating hormone (FSH)   Hormone produced by the anterior pituitary that stimulates ovarian follicle production in females and sperm production in males.

Fontanels (fon'tah-nelz')   Fibrous membranes at the angles of cranial bones that accommodate brain growth in the fetus and infant.

Foramen (fo-ra'men)   Hole or opening in a bone or between body cavities.

Forebrain (prosencephalon)   Anterior portion of the brain consisting of the telencephalon and the diencephalon.

Formed elements   Cellular portion of blood.

Fossa (fos'ah)   A depression, often an articular surface.

Fovea (fo've-ah)   A pit.

Fracture   A break in a bone.

Free radicals   Highly reactive chemicals with unpaired electrons that can scramble the structure of proteins, lipids, and nucleic acids.

Frontal (coronal) plane   Longitudinal (vertical) plane that divides the body into anterior and posterior parts.

Fulcrum   The fixed point on which a lever moves when a force is applied.

Fundus (fun'dus)   Base of an organ; part farthest from the opening of the organ.

Gallbladder   Sac beneath the right lobe of the liver used for bile storage.

Gallstones (biliary calculi)   Crystallized cholesterol that obstructs the flow of bile from the gallbladder.

Gamete (gam'ēt)   Sex or germ cell.

Gametogenesis (gam"eh-to-jen'eh-sis)   Formation of gametes.

Ganglion (gang'gle-on)   Collection of nerve cell bodies outside the CNS.

Gap junction   A passageway between two adjacent cells; formed by transmembrane proteins called connexons.

Gastrin   Hormone secreted in the stomach; regulates gastric juice secretion by stimulating HCl production.

Gastroenteritis   Inflammation of the gastrointestinal tract.

Gastrulation (gas"troo-la'shun)   Developmental process that produces the three primary germ layers (ectoderm, mesoderm, and endoderm).

Gene   One of the biological units of heredity located in chromatin; transmits hereditary information.

Genetic code   Refers to the rules by which the base sequence of a DNA gene is translated into protein structures (amino acid sequences).

Genitalia (jen'i-tā'lē-ă)   The internal and external reproductive organs.

Genome   The complete set of chromosomes derived from one parent (the haploid genome); or, the two sets of chromosomes, i.e., one set from the egg, the other from the sperm (the diploid genome).

Genotype (jen'o-tīp)   One's genetic makeup or genes.

Germ layers   Three cellular layers (ectoderm, mesoderm, and endoderm) that represent the initial specialization of cells in the embryonic body and from which all body tissues arise.

Gestation (jes-ta'shun)   The period of pregnancy; about 280 days for humans.

Gland   Organ specialized to secrete or excrete substances for further use in the body or for elimination.

Glaucoma (glaw-ko'mah)   Condition in which intraocular pressure increases to levels that cause compression of the retina and optic nerve; results in blindness unless detected early.

Glial cells (gle'al)   The supporting cells in the central nervous system.

Glomerular capsule (glō-mer' yoo-ler)   Double-walled cup at end of a renal tubule; encloses a glomerulus. Also called Bowman's capsule.

Glomerular filtration rate (GFR) Rate of filtrate formation by the kidneys.

Glomerulus (glo-mer'u-lus) Cluster of capillaries forming part of the nephron; forms filtrate.

Glottis (glah'tis) Opening between the vocal cords in the larynx.

Glucagon (gloo'kah-gon) Hormone formed by alpha cells of islets of Langerhans in the pancreas; raises the glucose level of blood.

Glucocorticoids (gloo"ko-kor'tĭ-koidz) Adrenal cortex hormones that increase blood glucose levels and aid the body in resisting long-term stressors.

Gluconeogenesis (gloo"ko-ne"o-jen'ĕ-sis) Formation of glucose from noncarbohydrate molecules.

Glucose (gloo'kōs) Principal blood sugar; a hexose.

Glycerol (glis'er-ol) A modified simple sugar (a sugar alcohol).

Glycocalyx (cell coat) (gli"ko-kal'iks) A layer of externally facing glycoproteins on a cell's plasma membrane that determines blood type; involved in the cellular interactions of fertilization, embryonic development, and immunity, and acts as an adhesive between cells.

Glycogen (glī'kō-jin) Main carbohydrate stored in animal cells; a polysaccharide.

Glycogenesis (gli"ko-jen'ĕ-sis) Formation of glycogen from glucose.

Glycogenolysis (gli"ko-jĕ-nol'ĭ-sis) Breakdown of glycogen to glucose.

Glycolipid (gli"ko-lip'id) A lipid with one or more covalently attached sugars.

Glycolysis (gli-kol'ĭ-sis) Breakdown of glucose to pyruvic acid—an anaerobic process.

Goblet cells Individual cells (unicellular glands) that produce mucus.

Golgi apparatus (gol'je) Membranous system close to the cell nucleus that packages protein secretions for export, packages enzymes into lysosomes for cellular use, and modifies proteins destined to become part of cellular membranes.

Golgi tendon organs Proprioceptors located in tendons, close to the point of skeletal muscle insertion; important to smooth onset and termination of muscle contraction.

Gonad (go'nad) Primary reproductive organ; i.e., the testis of the male or the ovary of the female.

Gonadocorticoids (gon"ah-do-kor'tĭ-koidz) Sex hormones, primarily androgens, secreted by the adrenal cortex.

Gonadotropins (go-nad"o-trōp'inz) Gonad-stimulating hormones produced by the anterior pituitary.

Graafian follicle (grā'fe-an). *See* Vesicular follicle.

Graded muscle responses Variations in the degree of muscle contraction by changing either the frequency or strength of the stimulus.

Graded potential A local change in membrane potential that varies directly with the strength of the stimulus, declines with distance.

Graves' disease Disorder resulting from hyperactive thyroid gland.

Gray matter Gray area of the central nervous system; contains cell bodies and unmyelinated fibers of neurons.

Growth hormone (GH) Hormone that stimulates growth in general; produced in the anterior pituitary; also called somatotropin (STH).

Guanine (G) (gwan'ēn) One of two major purines occurring in all nucleic acids.

Gustation (gus-ta'shun) Taste.

Gut-brain peptides Neuropeptides produced by nonneural tissue that are widespread in the gastrointestinal tract.

Gyrus (ji'rus) An outward fold of the surface of the cerebral cortex.

Hair follicle Structure with outer and inner root sheaths extending from the epidermal surface into the dermis and from which new hair develops.

Hapten (hap'ten) An incomplete antigen; has reactivity but not immunogenicity.

Haversian system (hah-ver'zhen). *See* Osteon.

Heart attack (coronary). *See* Myocardial infarction.

Heart block Impaired transmission of impulses from atrium to ventricle resulting in dysrhythmia.

Heart murmur Abnormal heart sound (usually resulting from valve problems).

Heimlich maneuver Procedure in which the air in a person's own lungs is used to expel an obstructing piece of food.

Helper T cell Type of T lymphocyte that orchestrates cellular immunity by direct contact with other immune cells and by releasing chemicals called lymphokines; also helps to mediate the humoral response by interacting with B cells.

Hematocrit (hem"ah'to-krit) The percentage of erythrocytes to total blood volume.

Hematoma (he"mah-to'mah) Mass of clotted blood that forms at an injured site.

Hematopoiesis (hem"ah-tō-poi-ē'sis) Blood cell formation; hemopoiesis.

Heme (hem) Iron-containing pigment that is essential to oxygen transport by hemoglobin.

Hemocytoblast (he"mo-si'to-blast) Bone marrow cell that gives rise to all the formed elements of blood.

Hemoglobin (he'muh-glo-bin) Oxygen-transporting component of erythrocytes.

Hemolysis (he-mah'lĕ-sis) Rupture of erythrocytes.

Hemophilia (he"mo-fil'e-ah) A term loosely applied to several different hereditary bleeding disorders that exhibit similar signs and symptoms.

Hemopoiesis (he"mo-poi-e'sis) Hematopoiesis.

Hemorrhage (hem'or-ij) Loss of blood from the vessels by flow through ruptured walls; bleeding.

Hemostasis (he"mo-sta'sis) Stoppage of bleeding.

Heparin Natural anticoagulant secreted into blood plasma.

Hepatic portal system (hĕ-pat'ik) Circulation in which the hepatic portal vein carries dissolved nutrients to the liver tissues for processing.

Hepatitis (hep"ah-ti'tis) Inflammation of the liver.

Hernia (her'ne-ah) Abnormal protrusion of an organ or a body part through the containing wall of its cavity.

Heterozygous (het"er-o-zi'gus) Having different allelic genes at one locus or (by extension) many loci.

Hilton's law Any nerve serving a muscle producing movement at a joint; also innervates the joint itself and the skin over the joint.

Hilus (hi'lus) The indented region of an organ from which blood and/or lymph vessels and nerves enter and exit.

Hippocampus Limbic system structure that plays a role in converting new information into long-term memories.

Histamine (his'tuh-mēn) Substance that causes vasodilation and increased vascular permeability.

Histology (his-tol'o-je) Branch of anatomy dealing with the microscopic structure of tissues.

HIV (human immunodeficiency virus) Virus that destroys helper T cells, thus depressing cell-mediated immunity; symptomatic AIDS gradually appears when lymph nodes can no longer contain the virus.

Holocrine glands (hol'o-krin) Glands that accumulate their secretions within their cells; secretions are discharged only upon rupture and death of the cell.

Homeostasis (ho'me-o-sta'sis) A state of body equilibrium or stable internal environment of the body.

Homologous (ho-mol'ŏ-gus) Parts or organs corresponding in structure but not necessarily in function.

Homozygous (ho-mo-zi'gus) Having identical genes at one or more loci.

Hormones Steroidal or amino acid-based molecules released to the blood that act as chemical messengers to regulate specific body functions.

Humoral immune response (hu'mer-ul) Immunity conferred by antibodies present in blood plasma and other body fluids.

Huntington's disease Hereditary disorder leading to degeneration of the basal nuclei and the cerebral cortex.

Hyaline cartilage (hi'ah-lĭn)   The most abundant cartilage type in the body; provides firm support with some pliability.

Hydrochloric acid (HCl) (hi"dro-klor'ik)   Acid that aids protein digestion in the stomach; produced by parietal cells.

Hydrogen bond   Weak bond in which a hydrogen atom forms a bridge between two electron-hungry atoms. An important intramolecular bond.

Hydrogen ion (H⁺)   A hydrogen atom minus its electron and therefore carrying a positive charge (i.e., a proton).

Hydrolysis (hi"drah'lă-sis)   Process in which water is used to split a substance into smaller particles.

Hydrophilic (hi"dro-fil'ik)   Refers to molecules, or portions of molecules, that interact with water and charged particles.

Hydrophobic (hi"dro-fo'bik)   Refers to molecules, or portions of molecules, that interact only with nonpolar molecules.

Hydrostatic pressure (hi"dro-stah'tic)   Pressure of fluid in a system.

Hydroxyl ion (OH⁻) (hi-drok'sil)   An ion liberated when a hydroxide (a common inorganic base) is dissolved in water.

Hyperalgesia   Pain amplification.

Hypercapnia (hi"per-kap'ne-ah)   High carbon dioxide levels in the blood.

Hyperemia   An increase in blood flow into a tissue or organ; congested with blood.

Hyperglycemic (hi"per-gli-se'mik)   Term used to describe hormones such as glucagon that elevate blood glucose level.

Hyperopia (hi"per-o'pe-ah)   A condition in which visual images are routinely focused behind the retina; commonly known as farsightedness.

Hyperplasia (hi"per-pla'ze-ah)   Accelerated growth, e.g., in anemia, the bone marrow produces red blood cells at a faster rate.

Hyperpnea (hi"perp-ne'ah)   Deeper and more vigorous breathing, but with unchanged respiratory rate, as during exercise.

Hypersensitivity (allergy)   Overzealous immune response to an otherwise harmless antigen.

Hypertension (hi"per-ten'shun)   High blood pressure.

Hypertonic (hi"per-ton'ik)   Excessive, above normal, tone or tension.

Hypertonic solution   A solution that has a higher concentration of nonpenetrating solutes than the reference cell; having greater osmotic pressure than the reference solution (blood plasma or interstitial fluid).

Hypertrophy (hi-per'trah-fe)   Increase in size of a tissue or organ independent of the body's general growth.

Hyperventilation   Increased depth and rate of breathing.

Hypocapnia   Low carbon dioxide levels in the blood.

Hypodermis (superficial fascia)   Subcutaneous tissue just deep to the skin; consists of adipose plus some areolar connective tissue.

Hypoglycemic (hi"po-gli-se'mik)   Term used to describe hormones such as insulin that decrease blood glucose level.

Hyponatremia   Abnormally low concentrations of sodium ions in extracellular fluid.

Hypoproteinemia (hi"po-pro"te-ĭ-ne'me-ah)   A condition of unusually low levels of plasma proteins causing a reduction in colloid osmotic pressure; results in tissue edema.

Hypotension   Low blood pressure.

Hypothalamic-hypophyseal tract (hi"po-thah-lam'ik–hi"po-fiz'e-al)   Nerve bundles that run through the infundibulum and connect the neurohypophysis and the hypothalamus.

Hypothalamus (hi"po-thal'ah-mus)   Region of the diencephalon forming the floor of the third ventricle of the brain.

Hypotonic (hi"po-ton'ik)   Below normal tone or tension.

Hypotonic solution   A solution that is more dilute (containing fewer nonpenetrating solutes) than the reference cell. Cells placed in hypotonic solutions plump up rapidly as water rushes into them.

Hypoventilation   Decreased depth and rate of breathing.

Hypovolemic shock (hi"po-vo-le'mik)   Most common form of shock; results from extreme blood loss.

Hypoxia (hi-pok'se-ah)   Condition in which inadequate oxygen is available to tissues.

Ileocecal valve (il"e-o-se'kal)   Site where the small intestine joins the large intestine.

Ileum (il'e-um)   Terminal part of the small intestine; between the jejunum and the cecum of the large intestine.

Immune system   A functional system whose components attack foreign substances or prevent their entry into the body.

Immunity (im"un'ĭ-te)   Ability of the body to resist many agents (both living and nonliving) that can cause disease; resistance to disease.

Immunocompetence   Ability of the body's immune cells to recognize (by binding) specific antigens; reflects the presence of plasma membrane-bound receptors.

Immunodeficiency disease   Disease resulting from the deficient production or function of immune cells or certain molecules (complement, antibodies, etc.) required for normal immunity.

In vitro (in ve'tro)   In a test tube, glass, or artificial environment.

In vivo (in ve'vo)   In the living body.

Incompetent valve   Valve which does not close properly.

Incontinence   Inability to control micturition voluntarily.

Indolamines   Biogenic amine neurotransmitters; includes serotonin and histamine.

Infarct (in'farkt)   Region of dead, deteriorating tissue resulting from a lack of blood supply.

Infectious mononucleosis   Highly contagious viral disease; marked by excessive agranulocytes.

Inferior (caudal)   Pertaining to a position near the tail end of the long axis of the body.

Inferior vena cava   Vein that returns blood from body areas below the diaphragm.

Inflammation (in"flah-ma'shun)   A nonspecific defensive response of the body to tissue injury; includes dilation of blood vessels and an increase in vessel permeability; indicated by redness, heat, swelling, and pain.

Infundibulum (in"fun-dib'u-lum)   (1) A stalk of tissue that connects the pituitary gland to the hypothalamus; (2) the distal end of the fallopian (uterine) tube.

Inguinal (ing'wĭ-nal)   Pertaining to the groin region.

Inner cell mass   Accumulation of cells in the blastocyst from which the embryo develops.

Innervation (in"er-vā'shun)   Supply of nerves to a body part.

Inorganic compound   Chemical substances that do not contain carbon, including water, salts, and many acids and bases.

Insertion   Movable attachment of a muscle.

Insulin   A hormone that enhances the carrier-mediated diffusion of glucose into tissue cells, thus lowering blood glucose levels.

Integration   The process by which the nervous system processes and interprets sensory input and makes decisions about what should be done at each moment.

Integumentary system (in-teg"u-men'tare)   Skin and its derivatives; provides the external protective covering of the body.

Intercalated discs (in-ter'kah-la"ted)   Gap junctions connecting muscle cells of the myocardium.

Interferon (IFN) (in-ter-fē'r'on)   Chemical that is able to provide some protection against virus invasion of the body; inhibits viral growth.

Internal capsule   Band of projection fibers that runs between the basal nuclei and the thalamus.

Internal respiration   Exchange of gases between blood and tissue fluid and between tissue fluid and cells.

Interoceptor (in"ter-o-sep'tor)   Sensory receptor nerve ending in the viscera, which is sensitive to changes and stimuli within the body's internal environment; also called visceroceptor.

Interphase   One of two major periods in the cell life cycle; includes the period from cell formation to cell division.

Interstitial cells (Leydig cells)   Cells located in the loose connective tissue surrounding the seminiferous tubules; they produce androgens (most importantly testosterone), which are secreted into the surrounding interstitial fluid.

Interstitial fluid (in"ter-stish'al)   Fluid between the cells.

Interstitial lamellae Incomplete lamellae that lie between intact osteons, filling the gaps between forming osteons, or representing the remnants of an osteon that has been cut through by bone remodeling.

Intervertebral discs (in"ter-ver'teh-brul) Discs of fibrocartilage between vertebrae.

Intracapsular ligament Ligament located within and separate from the articular capsule of a synovial joint.

Intracellular fluid (in"trah-sel'u-ler) Fluid within a cell.

Intrinsic factor Substance produced by the stomach that is required for vitamin B_{12} absorption.

Introns Noncoding segment or portion of DNA that ranges from 60 to 100,000 nucleotides long.

Involuntary muscle Muscles that cannot ordinarily be controlled voluntarily.

Involuntary nervous system The autonomic nervous system.

Ion (ī'on) Atom with a positive or negative electric charge.

Ionic bond (ī-ah'nik) Chemical bond formed by electron transfer between atoms.

Ipsilateral (ip"sih-la'ter-ul) Situated on the same side.

Irritability Ability to respond to stimuli.

Ischemia (is-ke'me-ah) Local decrease in blood supply.

Isograft Tissue graft donated by an identical twin.

Isomer (i'so-mer) One of two or more substances that has the same molecular formula but with its atoms arranged differently.

Isometric contraction (i"so-me'trik) Contraction in which the muscle does not shorten (the load is too heavy) but its internal tension increases.

Isotonic contraction (i"so-tah'nik) Contraction in which muscle tension remains constant at a given joint angle and load, and the muscle shortens.

Isotonic solution A solution with a concentration of nonpenetrating solutes equal to that found in the reference cell.

Isotopes (i'so-tōps) Different atomic forms of the same element, vary only in the number of neutrons they contain; the heavier species tend to be radioactive.

Jejunum (jě-joo'num) The part of the small intestine between the duodenum and the ileum.

Joint (articulation) The junction of two or more bones.

Joint kinesthetic receptor (kin"es-thet'ik) Receptor that provides information on joint position and motion.

Juxtaglomerular apparatus (JGA) (juks"tah-glo-mer'ular) Cells of the distal tubule and afferent arteriole located close to the glomerulus that play a role in blood pressure regulation by releasing the enzyme renin.

Kayrotype (kar'e-o-tīp) The diploid chromosomal complement, typically shown as homologous chromosome pairs arranged from longest to shortest (X and Y are arranged by size rather than paired).

Keratin (ker'ah-tin) Water-soluble protein found in the epidermis, hair, and nails that makes those structures hard and water-repellent; precursor is keratohyalin.

Ketones (ketone bodies) (ke'tōnz) Fatty acid metabolites; strong organic acids.

Ketosis (kē-tō'sis) Abnormal condition during which an excess of ketone bodies is produced.

Killer T cell. *See* Cytotoxic T cell.

Kilocalories (kcal). *See* Calorie.

Kinetic energy (ki-net'ik) The energy of motion or movement, e.g., the constant movement of atoms, or the push given to a swinging door that sets it into motion.

Krebs cycle Aerobic metabolic pathway occurring within mitochondria, in which food metabolites are oxidized and CO_2 is liberated, and coenzymes are reduced.

Labia (la'be-ah) Lips; singular labium.

Labor Collective term for the series of events that expel the infant from the uterus.

Labyrinth (lab'ī-rinth") Bony cavities and membranes of the inner ear.

Lacrimal (lak'ri-mal) Pertaining to tears.

Lactation (lak-ta'shun) Production and secretion of milk.

Lacteal (lak'te-al) Special lymphatic capillaries of the small intestine that take up lipids.

Lactic acid (lak'tik) Product of anaerobic metabolism, especially in muscle.

Lacunae (lah-ku'ne) A small space, cavity, or depression; lacunae in bone or cartilage are occupied by cells.

Lamella (lah-mel'ah) A layer, such as of bone matrix in Haversian systems of compact bone.

Lamina (lam'ī-nah) (1) A thin layer or flat plate; (2) the portion of a vertebra between the transverse process and the spinous process.

Large intestine Portion of the digestive tract extending from the ileocecal valve to the anus; includes the cecum, appendix, colon, rectum, and anal canal.

Larynx (lar'ingks) Cartilaginous organ located between the trachea and the pharynx; voice box.

Latent period Period of time between stimulation and the onset of muscle contraction.

Lateral Away from the midline of the body.

Leukemia Refers to a group of cancerous conditions of white blood cells.

Leukocytes (loo'ko-sīts) White blood cells; formed elements involved in body protection that take part in inflammatory and immune responses.

Leukocytosis An increase in the number of leukocytes (white blood cells); usually the result of a microbiological attack on the body.

Leukopenia (loo"ko-pe'ne-ah) Abnormally low white blood cell count.

Leukopoiesis The production of white blood cells.

Lever system Consists of a lever (bone), effort (muscle action), resistance (weight of object to be moved), and fulcrum (joint).

Ligament (lig'ah-ment) Band of regular fibrous tissue that connects bones.

Ligands Signaling chemicals that bind specifically to membrane receptors.

Limbic system (lim'bik) Functional brain system involved in emotional response.

Lipid (lih'pid) Organic compound formed of carbon, hydrogen, and oxygen; examples are fats and cholesterol.

Lipolysis (lī-pol'ī-sis) The breakdown of stored fats into glycerol and fatty acids.

Liver Lobed accessory organ that overlies the stomach; produces bile to help digest fat, and serves other metabolic and regulatory functions.

Lumbar (lum'bar) Portion of the back between the thorax and the pelvis.

Lumbar vertebrae The five vertebrae of the lumbar region of the vertebral column, commonly called the small of the back.

Lumen (loo'min) Cavity inside a tube, blood vessel, or hollow organ.

Luteinizing hormone (LH) (lu'te-in-ī z"ing) Anterior pituitary hormone that aids maturation of cells in the ovary and triggers ovulation in females. In males, causes the interstitial cells of the testis to produce testosterone.

Lymph (limf') Protein-containing fluid transported by lymphatic vessels.

Lymph node Small lymphatic organ that filters lymph; contains macrophages and lymphocytes.

Lymphatic system (lim'fa-tik) System consisting of lymphatic vessels, lymph nodes, and other lymphoid organs and tissues; drains excess tissue fluid from the extracellular space and provides a site for immune surveillance.

Lymphatics General term used to designate the lymphatic vessels that collect and transport lymph.

Lymphocyte Agranular white blood cell that arises from bone marrow and becomes functionally mature in the lymphoid organs of the body.

Lymphokines (lim'fo-kīnz) Proteins involved in cell-mediated immune responses that enhance immune and inflammatory responses.

Lysosomes (li'so-sōmz) Organelles that originate from the Golgi apparatus and contain strong digestive enzymes.

Lysozyme (li'so-zī m) Enzyme in sweat, saliva, and tears that is capable of destroying certain kinds of bacteria.

Macromolecules Large, complex molecules containing from 100 to over 10,000 amino acids.

Macrophage (mak'ro-fā j") Protective cell type common in connective tissue, lymphatic tissue, and certain body organs that phagocytizes tissue cells, bacteria, and other foreign debris; important as an antigen-presenter to T cells and B cells in the immune response.

Macula (mak'u-lah) (1) Static equilibrium receptor within the vestibule of the inner ear; (2) a colored area or spot.

Malignant (muh-lig'nent) Life threatening; pertains to neoplasms that spread and lead to death, such as cancer.

Malignant melanoma (mel"ah-no'mah) Cancer of the melanocytes; can begin wherever there is pigment.

Mammary glands (mam'mer-e) Milk-producing glands of the breast.

Mandible (man'dĭ-bl) Lower jawbone; U-shaped, largest bone of the face.

Mast cells Immune cells that function to detect foreign substances in the tissue spaces and initiate local inflammatory responses against them; typically found clustered deep to an epithelium or along blood vessels.

Mastication (mas"tĭ-ka'shun) Chewing.

Meatus (me-a'tus) External opening of a canal.

Mechanical advantage (power lever) Condition that occurs when the load is close to the fulcrum and the effort is applied far from the fulcrum; allows a small effort exerted over a relatively large distance to move a large load over a small distance.

Mechanical disadvantage (speed lever) Condition that occurs when the load is far from the fulcrum and the effort is applied near the fulcrum; the effort applied must be greater than the load to be moved.

Mechanical energy The energy directly involved in moving matter; e.g., in bicycle riding, the legs provide the mechanical energy that moves the pedals.

Mechanoreceptor (meh"kĕ-no-rē-sep'-ter) Receptor sensitive to mechanical pressure such as touch, sound, or exerted by muscle contraction.

Medial (me'-de-ahl) Toward the midline of the body.

Medial lemniscal system (lem-nis'kul) The pathway to the cerebral cortex for discriminative touch, pressure, vibration, and conscious proprioception.

Median (midsagittal) plane Specific sagittal plane that lies exactly in the midline.

Mediastinum (me"de-ah-sti'-num) A subdivision of the thoracic cavity containing the pericardial cavity.

Medulla (mĕ-dul'ah) Central portion of certain organs.

Medulla oblongata (mĕ-dul'ah ob"long-gah'tah) Inferior most part of the brain stem.

Meiosis (mi-o'sis) Nuclear division process that reduces the chromosomal number by half and results in the formation of four haploid (*n*) cells; occurs only in certain reproductive organs.

Melanin (mel'ah-nin) Dark pigment formed by cells called melanocytes; imparts color to skin and hair.

Melatonin (mel'ah-to'nin) A hormone secreted by the pineal gland that inhibits secretion of GnRH by the hypothalamus; secretion peaks at night.

Membrane potential Voltage across the plasma membrane.

Membrane receptors A large, diverse group of integral proteins and glycoproteins that serve as binding sites.

Memory cells Members of T cell and B cell clones that provide for immunologic memory.

Menarche (mĕ-nar'ke) Establishment of menstrual function; the first menstrual period.

Meninges (mĕ-nin'-jē z) Protective coverings of the central nervous system; from the most external to the most internal, the dura mater, arachnoid, and pia mater.

Meningitis (mĕ-nin-ji'tis) Inflammation of the meninges.

Menopause Period of life when, prompted by hormonal changes, ovulation and menstruation cease.

Menstruation (men"stroo-a'shun) The periodic, cyclic discharge of blood, secretions, tissue, and mucus from the mature female uterus in the absence of pregnancy.

Merocrine glands (mer'o-krin) Glands that produce secretions intermittently; secretions do not accumulate in the gland.

Mesencephalon (mes"en-sef'ah-lon) Midbrain.

Mesenchyme (meh'zin-kī m) Common embryonic tissue from which all connective tissues arise.

Mesenteries (mes"en-ter'ē z) Double-layered extensions of the peritoneum that support most organs in the abdominal cavity.

Mesoderm (mez'o-derm) Primary germ layer that forms the skeleton and muscles of the body.

Mesothelium (mez"o-the'le-um) The epithelium found in serous membranes lining the ventral body cavity and covering its organs.

Messenger RNA (mRNA) Long nucleotide strands that reflect the exact nucleotide sequences of the genetically active DNA and carry the message of the latter.

Metabolic (fixed) acid Acid generated by cellular metabolism that must be eliminated by the kidneys.

Metabolic rate (mĕt"ah-bol'ik) Energy expended by the body per unit time.

Metabolism (mĕ-tab'o-lizm) Sum total of the chemical reactions occuring in the body cells.

Metaphase Second stage of mitosis.

Metastasis (mĕ-tas'tah-sis) The spread of cancer from one body part or organ into another not directly connected to it.

Metencephalon (afterbrain) Anterior portion of the rhombencephalon; composed of the pons and the cerebellum.

Microfilaments (mi"kro-fil'ah-ments) Thin strands of the contractile protein actin.

Microglia (mi-krog'le-ah) A type of CNS supporting cell; can transform into phagocytes in areas of neural damage or inflammation.

Microtubules (mi"kro-tu'būlz) One of three types of rods in the cytoskeleton of a cell; hollow tubes made of spherical protein that determine the cell shape as well as the distribution of cellular organelles.

Microvilli (mi"kro-vil'i) Tiny projections on the free surfaces of some epithelial cells; increase surface area for absorption.

Micturition (mik"tu-rish'un) Urination, or voiding; emptying the bladder.

Midbrain (mesencephalon) Region of the brain stem between the diencephalon and the pons.

Midsagittal (median) plane Specific sagittal plane that lies exactly in the midline.

Milliequivalents per liter (mEq/L) The units used to measure electrolyte concentrations of body fluids; a measure of the number of electrical charges in 1 liter of solution.

Mineralocorticoid (min"er-al'ō-kor'tih-koyd) Steroid hormone of the adrenal cortex that regulates mineral metabolism and fluid balance.

Minerals Inorganic chemical compounds found in nature; salts.

Mitochondria (mi"to-kon'dre-ah) Cytoplasmic organelles responsible for ATP generation for cellular activities.

Mitosis Process during which the chromosomes are redistributed to two daughter nuclei; nuclear division. Consists of prophase, metaphase, anaphase, and telophase.

Mixed nerves Nerves containing the processes of motor and sensory neurons; their impulses travel to and from the central nervous system.

Molar (mo'lar) A solution concentration determined by mass of solute—one liter of solution contains an amount of solute equal to its molecular weight in grams.

Molarity (mo-lar'ĭ-te) A way to express the concentration of a solution; moles per liter of solution.

Mole (mōl) A mole of any element or compound is equal to its atomic weight or its molecular weight (sum of atomic weights) measured in grams.

Molecule Particle consisting of two or more atoms joined together by chemical bonds.

Monoclonal antibodies (mon"o-klo'nal) Pure preparations of identical antibodies that exhibit specificity for a single antigen.

Monocyte (mon'o-sīt) Large single-nucleus white blood cell; agranular leukocyte.

Monokines Chemical mediators that enhance the immune response; secreted by macrophages.

Monosaccharide (mon"o-sak'ah-rīd) Literally, one sugar; building block of carbohydrates; e.g., glucose.

Morula (mor'u-lah) The mulberry-like solid mass of blastomeres resulting from cleavage in the early conceptus.

Motor areas Functional areas in the cerebral cortex that control voluntary motor functions.

Motor end plates Troughlike part of a muscle fiber's sarcolemma that helps form the neuromuscular junction.

Motor nerves Nerves that carry impulses leaving the brain and spinal cord, and destined for effectors.

Motor unit A motor neuron and all the muscle cells it stimulates.

Mucous membranes Membranes that form the linings of body cavities open to the exterior (digestive, respiratory, urinary, and reproductive tracts).

Mucus (myoo'kus) A sticky, thick fluid secreted by mucous glands and mucous membranes; keeps the free surface of membranes moist.

Multinucleate cell (mul"tĭ-nu'kle-āt) Cell with more than one nucleus, e.g., skeletal muscle cells, liver cells.

Multiple sclerosis (MS) Demyelinating disorder of the CNS; causes patches of sclerosis in the brain and spinal cord.

Multipolar neurons Neurons with three or more processes; most common neuron type in the CNS.

Muscarinic receptors (mus"kah-rin'ik) Acetylcholine-binding receptors of the autonomic nervous system's target organs; named for activation by the mushroom poison muscarine.

Muscle fiber A muscle cell.

Muscle spindle (neuromuscular spindle) Encapsulated receptor found in skeletal muscle that is sensitive to stretch.

Muscle tension The force exerted by a contracting muscle on some object.

Muscle tone Sustained partial contraction of a muscle in response to stretch receptor inputs; keeps the muscle healthy and ready to act.

Muscle twitch The response of a muscle to a single brief threshold stimulus.

Muscular dystrophy A group of inherited muscle-destroying diseases.

Muscular system The organ system consisting of the skeletal muscles of the body and their connective tissue attachments.

Myelencephalon (spinal brain) Lower part of the hindbrain, especially the medulla oblongata.

Myelin sheath (mi'ĕ-lin) Fatty insulating sheath that surrounds all but the smallest nerve fibers.

Myoblasts Embryonic mesoderm cells from which all muscle fiber develops.

Myocardial infarction (MI) (mi"o-kar'de-al in-fark'shun) Condition characterized by dead tissue areas in the myocardium; caused by interruption of blood supply to the area.

Myocardium (mi"o-kar'de-um) Layer of the heart wall composed of cardiac muscle.

Myofibril (mi"o-fi'bril) Rodlike bundle of contractile filaments (myofilaments) found in muscle cells.

Myofilament (mi"o-fil'ah-ment) Filament that constitutes myofibrils. Of two types: actin and myosin.

Myoglobin (mi"o-glo'bin) Oxygen-binding pigment in muscle.

Myogram A graphic recording of mechanical contractile activity produced by an apparatus that measures muscle contraction.

Myometrium (mi"o-me'tre-um) Thick uterine musculature.

Myopia (mi-o'pe-ah) A condition in which visual images are focused in front of rather than on the retina; nearsightedness.

Myosin (mi'o-sin) One of the principal contractile proteins found in muscle.

Myxedema (mik"sĕ-de'mah) Condition resulting from underactive thyroid gland.

Nares (na'rez) The nostrils.

Natural killer (NK) cells Defensive cells that can lyse and kill cancer cells and virus-infected body cells before the immune system is activated.

Necrosis (nĕ-kro'sis) Death or disintegration of a cell or tissues caused by disease or injury.

Negative feedback mechanisms The most common homeostatic control mechanism. The net effect is that the output of the system shuts off the original stimulus or reduces its intensity.

Neonatal period The four-week period immediately after birth.

Neoplasm (ne'o-plazm) An abnormal mass of proliferating cells; benign neoplasms remain localized; malignant neoplasms are cancers, which can spread to other organs.

Nephron (nef'ron) Structural and functional unit of the kidney; consists of the glomerulus and renal tubule.

Nerve cell A neuron.

Nerve fiber Axon of a neuron.

Nerve growth factor (NGF) Protein that controls the development of sympathetic postganglionic neurons; secreted by the target cells of postganglionic axons.

Nerve impulse A self-propagating wave of depolarization; also called an action potential.

Nerve plexuses Interlacing nerve networks that occur in the cervical, brachial, lumbar, and sacral regions and primarily serve the limbs.

Nervous system Fast-acting control system that triggers muscle contraction or gland secretion.

Neural tube Fetal tissue which gives rise to the brain, spinal cord, and associated neural structures; formed from ectoderm by day 23 of embryonic development.

Neuroglia (nu-rog'le-ah) Nonexcitable cells of neural tissue that support, protect, and insulate the neurons.

Neurohypophysis (nu"ro-hi-pof'ĭ-sis) Posterior pituitary; portion of the pituitary gland derived from the brain.

Neuromuscular junction Region where a motor neuron comes into close contact with a skeletal muscle cell.

Neuron (nu-ron') Cell of the nervous system specialized to generate and transmit nerve impulses.

Neuron cell body The biosynthetic center of a neuron; also called the perikaryon, or soma.

Neuronal pools Functional groups of neurons that process and integrate information.

Neuropeptides (nu"ro-pep'tĭds) A class of neurotransmitters including beta-endorphins and enkephalins (which act as euphorics and reduce perception of pain) and gut-brain peptides.

Neurotransmitter Chemical released by neurons that may, upon binding to receptors of neurons or effector cells, stimulate or inhibit them.

Neutral fats Consist of fatty acid chains and glycerol; also called triglycerides or triacylglycerols. Commonly known as oils when liquid.

Neutralization reaction Displacement reaction in which mixing an acid and a base forms water and a salt.

Neutron (nu'tron) Uncharged subatomic particle; found in the atomic nucleus.

Neutrophil (nu'tro-fil) Most abundant type of white blood cell.

Nicotinic receptors (nik"o-tin'ik) Acetylcholine-binding receptors of all autonomic postganglionic neurons and skeletal muscle neuromuscular junctions; named for activation by nicotine.

Nociceptor (no"se-sep'tor) Receptor sensitive to potentially damaging stimuli that result in pain.

Nondisjunction Failure of sister chromatids to separate during mitosis or failure of homologous pairs to separate during meiosis; results in abnormal numbers of chromosomes in the resulting daughter cells.

Nonpolar molecules Electrically balanced molecules.

Norepinephrine (nor"ep-ĭ-nef'rin) A catecholamine (biogenic amine) neurotransmitter and adrenal medullary hormone, associated with sympathetic nervous system activation.

Nuclear envelope The double membrane barrier of a cell nucleus.

Nucleic acid (nuc-kla'ik) Class of organic molecules that includes DNA and RNA.

Nucleoli (nu-kle'o-li) Dense spherical bodies in the cell nucleus involved with ribosomal subunit synthesis and storage.

Nucleoplasm (nu'kle-o-plazm") A colloidal fluid, enclosed by the nuclear membrane.

Nucleosome (nu'kle-o-sōm) Fundamental unit of chromatin; consists of a strand of DNA wound around a cluster of eight histone proteins.

Nucleotide (nu'kle-o-tīd) Building block of nucleic acids; consists of a sugar, a nitrogen-containing base, and a phosphate group.

Nucleus (nu'kle-is) Control center of a cell; contains genetic material.

Nutrients Chemical substances taken in via the diet that are used for energy and cell building.

Oblique section A cut made diagonally between the horizontal and vertical plane of the body or an organ.

Occlusion (ah-kloo'zhun) Closure or obstruction.

Octet rule (rule of eights) (ok-tet') The tendency of atoms to interact in such a way that they have eight electrons in their valence shell.

Olfaction (ol-fak'shun) Smell.

Oligodendrocyte (ol'ĭ-go-den'dro-sīt) A type of CNS supporting cell that composes myelin sheaths.

Oocyte (o'o-sīt) Immature female gamete.

Oogenesis (o"o-jen'ĕ-sis) Process of ovum (female gamete) formation.

Ophthalmic (op-thal'mik) Pertaining to the eye.

Optic (op'tik) Pertaining to the eye or vision.

Optic chiasma (op'tik ki-az'muh) The partial crossover of fibers of the optic nerves.

Organ A part of the body formed of two or more tissues and adapted to carry out a specific function; e.g., the stomach.

Organ system A group of organs that work together to perform a vital body function; e.g., the nervous system.

Organelles (or"gah-nelz') Small cellular structures (ribosomes, mitochondria, and others) that perform specific metabolic functions for the cell as a whole.

Organic compound Any compound composed of atoms (some of which are carbon) held together by covalent (shared electron) bonds.

Organic Pertaining to carbon-containing molecules, such as proteins, fats, and carbohydrates.

Organism The living animal (or plant), which represents the sum total of all its organ systems working together to maintain life.

Origin Attachment of a muscle that remains relatively fixed during muscular contraction.

Osmolality The number of solute particles dissolved in one liter (1000 g) of water; reflects the solution's ability to cause osmosis.

Osmolarity (oz"mo-lar'ĭ-te) The total concentration of all solute particles in a solution.

Osmoreceptor (oz"mo-re-sep'tor) Structure sensitive to osmotic pressure or concentration of a solution.

Osmosis (oz-mo'sis) Diffusion of a solvent through a membrane from a dilute solution into a more concentrated one.

Osmotic pressure The pressure applied to a solution to prevent the passage of solvent into it; the tendency to resist further (net) water entry.

Ossicles (os'sik-kilz). *See* Auditory ossicles.

Ossification (os"ĭ-fi-ka'shun). *See* Osteogenesis.

Osteoblasts (os-'te-o-blasts) Bone-forming cells.

Osteoclasts (os'te-o-klasts) Large cells that resorb or break down bone matrix.

Osteocyte (os'te-o-sīt) Mature bone cell.

Osteogenesis (os"te-o-jen'e-sis) The process of bone formation; also called ossification.

Osteoid (os'te-oid) Unmineralized bone matrix.

Osteomalacia (os"te-o-mah-la'she-ah) Disorder in which bones are inadequately mineralized; soft bones.

Osteon (os'te-on) System of interconnecting canals in the microscopic structure of adult compact bone; unit of bone; also called Haversian system.

Osteoporosis (os"te-o-po-ro'sis) Increased softening of the bone resulting from a gradual decrease in rate of bone formation.

Ovarian cycle (o-vayr'e-an) Monthly cycle of follicle development, ovulation, and corpus luteum formation in an ovary.

Ovary (o'var-e) Female sex organ in which ova (eggs) are produced; female gonad.

Ovulation (ov"u-la'shun) Ejection of an immature egg (oocyte) from the ovary.

Ovum (o'vum) Female gamete; egg.

Oxidases Enzymes that catalyze the transfer of oxygen in oxidation-reduction reactions.

Oxidation (oks'i-da-shun) Process of substances combining with oxygen or the removal of hydrogen.

Oxidation-reduction (redox) reaction A reaction that couples the oxidation (loss of electrons) of one substance with the reduction (gain of electrons) of another substance.

Oxidative phosphorylation (ok'sĭ-da'tiv fos"for-ĭ-la-shun) Process of ATP synthesis during which an inorganic phosphate group is attached to ADP; occurs via the electron transport chain within the mitochondria.

Oxygen debt The volume of oxygen required after exercise to oxidize the lactic acid formed during exercise.

Oxyhemoglobin (ok"sĭ-he"mo-glo'bin) Oxygen-bound form of hemoglobin.

Oxytocin (ok"sĭ-to'sin) Hormone secreted by hypothalamic cells that stimulates contraction of the uterus during childbirth and the ejection of milk during nursing.

Paget's disease (paj'ets) Disorder characterized by excessive bone breakdown and abnormal bone formation.

Palate (pal'at) Roof of the mouth.

Pancreas (pan'kre-us) Gland located behind the stomach, between the spleen and the duodenum; produces both endocrine and exocrine secretions.

Pancreatic juice (pan"kre-at'ik) Bicarbonate-rich secretion of the pancreas containing enzymes for digestion of all food categories.

Papilla (pah-pil'ah) Small, nipplelike projection; e.g., dermal papillae are projections of dermal tissue into the epidermis.

Parasagittal planes All sagittal planes offset from the midline.

Parasympathetic nervous system The division of the autonomic nervous system that oversees digestion, elimination, and glandular function; the resting and digesting subdivision.

Parasympathetic tone State of parasympathetic effects; e.g., unnecessary heart accelerations; normal activity levels of digestive and urinary tracts.

Parathyroid glands (par"ah-thi'roid) Small endocrine glands located on the posterior aspect of the thyroid gland.

Parathyroid hormone (PTH) Hormone released by the parathyroid glands that regulates blood calcium level.

Parietal (pah-ri'ĕ-tal) Pertaining to the walls of a cavity.

Parietal serosa The part of the double-layered membrane that lines the walls of the ventral body cavity.

Parkinson's disease Neurodegenerative disorder of the basal nuclei involving abnormalities of the neurotransmitter dopamine; symptoms include persistent tremor and rigid movement.

Partial pressure The pressure exerted by a single component of a mixture of gases.

Parturition (par"tu-rish'un) Culmination of pregnancy; giving birth.

Passive immunity Short-lived immunity resulting from the introduction of "borrowed antibodies" obtained from an immune animal or human donor; immunological memory is not established.

Passive transport Membrane transport processes that do not require cellular energy (ATP), e.g., diffusion, which is driven by kinetic energy.

Pathogen (path'o-jen) Disease-causing microorganism.

Pectoral (pek'tor-al) Pertaining to the chest.

Pectoral (shoulder) girdle Bones that attach the upper limbs to the axial skeleton; includes the clavicle and scapula.

Pedigree Traces a particular genetic trait through several generations and helps predict the genotype of future offspring.

Pelvic girdle (hip girdle) Consists of the paired coxal bones that attach the lower limbs to the axial skeleton.

Pelvis (pel'vis) (1) Basin-shaped bony structure composed of the pelvic girdle, sacrum, and coccyx; (2) expanded proximal portion of the ureter within the kidney.

Penis (pe'nis) Male organ of copulation and urination.

Pepsin Enzyme capable of digesting proteins in an acid pH.

Peptide bond (pep'tid) Bond joining the amine group of one amino acid to the acid carboxyl group of a second amino acid with the loss of a water molecule.

Perforating canals Canals that run at right angles to the long axis of the bone, connecting the vascular and nerve supplies of the periosteum to those of the central canals and medullary cavity; also called Volkmann's canals.

Pericardium (per"ĭ-kar'de-um) Double-layered serosa enclosing the heart and forming its superficial layer.

Perichondrium (per"ĭ-kon'dre-um) Fibrous, connective-tissue membrane covering the external surface of cartilaginous structures.

Perimysium (per"ĭ-mis'e-um) Connective tissue enveloping bundles of muscle fibers.

Perineum (per"ĭ-ne'um) That region of the body spanning the region between the ischial tuberosities and extending from the anus to the scrotum in males and from the anus to the vulva in females.

Periosteum (per"e-os'te-um) Double-layered connective tissue that covers and nourishes the bone.

Peripheral congestion Condition caused by failure of the right side of the heart; results in edema in the extremities.

Peripheral nervous system (PNS) Portion of the nervous system consisting of nerves and ganglia that lie outside of the brain and spinal cord.

Peripheral resistance (PR) A measure of the amount of friction encountered by blood as it flows through the blood vessels.

Peristalsis (per"i-stal'sis) Progressive, wavelike contractions that move foodstuffs through the alimentary tube organs (or that move other substances through other hollow body organs).

Peritoneum (per"ĭ-to-ne'um) Serous membrane lining the interior of the abdominal cavity and covering the surfaces of abdominal organs.

Peritonitis (per"ĭ-to-ni'tis) Inflammation of the peritoneum.

Permeability That property of membranes that permits passage of molecules and ions.

Peroxisomes (pĕ-roks'ĭ-sōmz) Membranous sacs in cytoplasm containing powerful oxidase enzymes that use molecular oxygen to detoxify harmful or toxic substances, such as free radicals.

Petechiae (pe-te'ke-ah) Small, purplish skin blotches caused by widespread hemorrhage due to thrombocytopenia.

Peyer's patches (pi'erz) Lymphoid organs located in the small intestine.

pH unit (pe-āch) The measure of the relative acidity or alkalinity of a solution.

Phagocytosis (fag"o-si-to'sis) Engulfing of foreign solids by (phagocytic) cells.

Phagosome (fag'o-sōm) Vesicle formed as a result of phagocytosis; formed when an endocytic vesicle fuses with a lysosome.

Pharynx (fayr'inks) Muscular tube extending from the region posterior to the nasal cavities to the esophagus.

Phenotype (fe-no'tīp) Observable expression of the genotype.

Phospholipid (fos"fo-lip'id) Modified lipid, contains phosphorus.

Phosphorylation The mechanism of ATP synthesis; includes substrate-level and oxidative phosphorylation.

Photoreceptor (fo"to-re-sep'tor) Specialized receptor cells that respond to light energy.

Physiological acidosis (as"ĭ-do'sis) Arterial pH lower than 7.35 resulting from any cause.

Physiology (fiz"e-ol'o-je) Study of the function of living organisms.

Pineal gland (body) (pin'e-al) A hormone-secreting part of the diencephalon of the brain thought to be involved in setting the biological clock and influencing reproductive function.

Pinocytosis (pe"no-si-to'sis) Engulfing of extracellular fluid by cells.

Pituitary gland (pĭ-tu'ih-tayr"e) Neuroendocrine gland located beneath the brain that serves a variety of functions including regulation of gonads, thyroid, adrenal cortex, lactation, and water balance.

Placenta (plah-sen'tah) Temporary organ formed from both fetal and maternal tissues that provides nutrients and oxygen to the developing fetus, carries away fetal metabolic wastes, and produces the hormones of pregnancy.

Plasma (plaz'mah) The nonliving fluid component of blood within which formed elements and various solutes are suspended and circulated.

Plasma cells Members of a B cell clone; specialized to produce and release antibodies.

Plasma membrane Membrane, composed of three lamina layers, that encloses cell contents; outer limiting cell membrane.

Platelet (plāt'let) Cell fragment found in blood; involved in clotting.

Pleura (ploo'rah) Two-layered serous membrane that lines the thoracic cavity and covers the external surface of the lung.

Pleural cavities (ploo'ral) A subdivision of the thoracic cavity; each houses a lung.

Plexus (plek'sus) A network of converging and diverging nerve fibers, blood vessels, or lymphatics.

Polar molecules Nonsymmetrical molecules that contain electrically unbalanced atoms.

Polarized State of a plasma membrane of an unstimulated neuron or muscle cell in which the inside of the cell is relatively negative in comparison to the outside; the resting state.

Polycythemia (pol"e-si'the'me-ah) An excessive or abnormal increase in the number of erythrocytes.

Polymer A substance of high molecular weight with long, chainlike molecules consisting of many similar (repeated) units.

Polypeptide (pol"e-pep'tīd) A chain of amino acids.

Polyps Benign mucosal tumors.

Polysaccharide (pol"e-sak'ah-rīd) Literally, many sugars, a polymer of linked monosaccharides; e.g., starch, glycogen.

Pons (1) Any bridgelike structure or part; (2) the part of the brain stem connecting the medulla with the midbrain, providing linkage between upper and lower levels of the central nervous system.

Pore The surface opening of the duct of a sweat gland.

Positive feedback mechanisms Feedback that tends to cause the level of a variable to change in the same direction as an initial change.

Posterior pituitary. *See* Neurohypophysis.

Postganglionic neuron (post"gan-gle-ah'nik) Autonomic motor neuron that has its cell body in a peripheral ganglion and projects its axon to an effector.

Potential energy Stored or inactive energy.

Preganglionic neuron Autonomic motor neuron that has its cell body in the central nervous system and projects its axon to a peripheral ganglion.

Presbyopia (pres"be-o'pe-ah) A condition that results in the loss of near focusing ability; typical onset is around age 40.

Pressoreceptor A nerve ending in the wall of the carotid sinus and aortic arch sensitive to vessel stretching.

Pressure gradient Difference in hydrostatic pressure that drives filtration.

Primary active transport A type of active transport in which the energy needed to drive the transport process is provided directly by hydrolysis of ATP.

Prime mover Muscle that bears the major responsibility for effecting a particular movement; an agonist.

Process (1) Prominence or projection; (2) series of actions for a specific purpose.

Progesterone (pro-jes'ter-ōn) Hormone partly responsible for preparing the uterus for the fertilized ovum.

Prolactin (PRL) (pro-lak'tin) Adenohypophyseal hormone that stimulates the breasts to produce milk.

Pronation (pro-na'shun) Inward rotation of the forearm causing the radius to cross diagonally over the ulna—palms face posteriorly.

Prophase The first stage of mitosis, consisting of linear contraction and increase in thickness of the chromosomes accompanied by migration of the two daughter centrioles toward the poles of the cell.

Proprioceptor (pro"pre-o-sep'tor) Receptor located in a joint, muscle, or tendon; concerned with locomotion, posture and muscle tone.

Prostaglandin (PG) (pros"tah-glan'din) A lipid-based membrane-associated chemical messenger synthesized by most tissue cells that acts locally as a hormonelike substance.

Prostate gland Accessory reproductive gland; produces one-third of semen volume, including fluids that activate sperm.

Protein (pro'tēn) Complex substance containing carbon, oxygen, hydrogen, and nitrogen; composes 10% to 30% of cell mass.

Prothrombin time Diagnostic test to determine status of hemostasis system.

Proton (pro'ton) Subatomic particle that bears a positive charge; located in the atomic nucleus.

Proton acceptor A substance that takes up hydrogen ions in detectable amounts. Commonly referred to as a base.

Proton donor A substance that releases hydrogen ions in detectable amounts; an acid.

Proximal (prok'si-mul) Toward the attached end of a limb or the origin of a structure.

Pseudounipolar neuron (soo"do-u"nĭ-po'lar) Another term for unipolar neuron.

Psychosomatic illnesses Emotion-induced illnesses.

Puberty Period of life when reproductive maturity is achieved.

Pulmonary (pul'muh-nayr-e) Pertaining to the lungs.

Pulmonary arteries Vessels that deliver blood to the lungs to be oxygenated.

Pulmonary circuit System of blood vessels that serves gas exchange in the lungs; i.e., pulmonary arteries, capillaries, and veins.

Pulmonary congestion Condition in which vascular pressure increases, resulting in tissue edema; caused by failure of the left side of the heart.

Pulmonary edema (ĕ-de'muh) Leakage of fluid into the air sacs and tissue of the lungs.

Pulmonary veins Vessels that deliver freshly oxygenated blood from the respiratory zones of the lungs to the heart.

Pulmonary ventilation Breathing; consists of inspiration and expiration.

Pulse Rhythmic expansion and recoil of arteries resulting from heart contraction; can be felt from outside the body.

Pupil Opening in the center of the iris through which light enters the eye.

Purkinje fibers (pur-kin'je) Modified cardiac muscle fibers of the conduction system of the heart.

Pus Fluid product of inflammation composed of white blood cells, the debris of dead cells, and a thin fluid.

Pyloric sphincter (pi-lor'ik sfink'ter) Valve of the distal end of the stomach that controls food entry into the duodenum.

Pyramidal (corticospinal) tracts Major motor pathways concerned with voluntary movement; descend from the frontal lobes of each cerebral hemisphere.

Pyruvic acid An intermediate compound in the metabolism of carbohydrates.

Radioactivity The process of spontaneous decay seen in some of the heavier isotopes, during which particles or energy is emitted from the atomic nucleus; results in the atom becoming more stable.

Radioisotope (ra"de-o-i'so-tōp) Isotope that exhibits radioactive behavior.

Ramus (ra'mus) Branch of a nerve, artery, vein, or bone.

Rapid eye movement (REM) sleep Stage of sleep in which rapid eye movements, an alert EEG pattern, and dreaming occur; also called paradoxical sleep.

Reactant A substance taking part in a chemical reaction.

Receptor (re-sep'tor) (1) A cell or nerve ending of a sensory neuron specialized to respond to particular types of stimuli; (2) molecule that binds specifically with other molecules, e.g., neurotransmitters, hormones, antigens.

Receptor-mediated endocytosis One of three types of endocytosis in which engulfed particles attach to receptors before endocytosis occurs.

Receptor potential A graded potential that occurs at a sensory receptor membrane.

Recessive traits A trait due to a particular allele that does not manifest itself in the presence of other alleles that generate traits dominant to it; must be present in double dose in order to be expressed.

Reduction Chemical reaction in which electrons and energy are gained by a molecule (often accompanied by gain of hydrogen ions) or oxygen is lost.

Referred pain Pain felt at a site other than the area of origin.

Reflex Automatic reaction to stimuli.

Refraction The bending of a light ray when it meets a different surface at an oblique rather than right angle.

Refractory period Period during which an excitable cell is not responsive to a threshold stimulus.

Regeneration Replacement of destroyed tissue with the same kind of tissue.

Relative refractory period Follows the absolute refractory period; interval when a threshold stimulus is unable to trigger an action potential.

Renal (re'nal) Pertaining to the kidney.

Renal autoregulation Process the kidney uses to maintain a nearly constant glomerular filtration rate despite fluctuations in systemic blood pressure.

Renal clearance The volume flow rate (ml/min) at which the kidneys clear the plasma of a particular solute; provides information about renal function.

Renin (re'nin) Substance released by the kidneys that is involved with raising blood pressure.

Rennin Stomach-secreted enzyme that acts on milk protein; not produced in adults.

Repolarization Movement of the membrane potential to the initial resting (polarized) state.

Reproductive system Organ system that functions to produce offspring.

Resistance exercise High-intensity exercise in which the muscles are pitted against high resistance or immovable forces and, as a result, muscle cells increase in size.

Respiration The processes involved in supplying the body with oxygen and disposing of carbon dioxide.

Respiratory system Organ system that carries out gas exchange; includes the nose, pharynx, larynx, trachea, bronchi, lungs.

Resting membrane potential The voltage that exists across the plasma membrane during the resting state of an excitable cell; ranges from -50 to -200 millivolts depending on cell type.

Reticular fibers (ret-tik'u-lar) Fine network of connective tissue fibers that form the internal supporting framework of lymphoid organs.

Reticular formation Functional system that spans the brain stem; involved in regulating sensory input to the cerebral cortex, cortical arousal, and control of motor behavior.

Reticular lamina A layer of extracellular material containing a fine network of collagen protein fibers; together with the basal lamina it is a major component of the basement membrane.

Reticulocyte (rĕ-tik'u-lo-sīt) Immature erythrocyte.

Retina (ret'ĭ-nah) Neural tunic of the eyeball; contains photoreceptors (rods, cones).

Rhombencephalon (hindbrain) (romb"en-sef'ah-lon) Caudal portion of the developing brain; constricts to form the metencephalon and myelencephalon; includes the pons, cerebellum, and medulla oblongata.

Ribosomal RNA (rRNA) A constituent of ribosome; exists within the ribosomes of cytoplasm and assists in protein synthesis.

Ribosomes (ri'bo-sōmz) Cytoplasmic organelles at which proteins are synthesized.

RNA (ribonucleic acid) (ri'bo-nu-kle'ik) Nucleic acid that contains ribose and the bases A, G, C, and U. Carries out DNA's instructions for protein synthesis.

Rods One of the two types of photosensitive cells in the retina.

Rotation The turning of a bone around its own long axis.

Rugae (ru'ge) Elevations or ridges, as in stomach mucosa.

Rule of nines Method of computing the extent of burns by dividing the body into a number of areas, each accounting for 9% of the total body area.

S (synthetic) phase The part of the interphase period of the cell life cycle in which DNA replicates itself, ensuring that the two future cells will receive identical copies of genetic material.

Sagittal plane (sa'jih-tul) A longitudinal (vertical) plane that divides the body or any of its parts into right and left portions.

Saliva Secretion of the salivary glands; cleanses and moistens the mouth and begins chemical digestion of starchy foods.

Saltatory conduction Transmission of an action potential along a myelinated fiber in which the nerve impulse appears to leap from node to node.

Sarcolemma The plasma membrane surface of a muscle fiber.

Sarcomere (sar'ko-mēr) The smallest contractile unit of muscle; extends from one Z disc to the next.

Sarcoplasm The nonfibrillar cytoplasm of a muscle fiber.

Sarcoplasmic reticulum (SR) (sar"ko-plaz'mik rĕ-tik'u-lum) Specialized endoplasmic reticulum of muscle cells.

Satellite cell A type of supporting cell in the PNS; surrounds neuron cell bodies within ganglia.

Schwann cell A type of supporting cell in the PNS; forms myelin sheaths and is vital to peripheral nerve fiber regeneration.

Sclera (skle'rah) White opaque portion of the fibrous tunic of the eyeball.

Scrotum (skro'tum) External sac enclosing the testes.

Sebaceous glands (oil glands) (se-ba'shus) Epidermal glands that produce an oily secretion called sebum.

Sebum (se'bum) Oily secretion of sebaceous glands.

Second-degree burn A burn in which the epidermis and the upper region of the dermis are damaged.

Second messenger Intracellular molecule generated by the binding of a chemical (hormone or neurotransmitter) to a plasma membrane receptor; mediates intracellular responses to the chemical messenger.

Secondary sex characteristics Anatomic features, not directly involved in the reproductive process, that develop under the influence of sex hormones, e.g., male or female pattern of muscle development, bone growth, body hair distribution, etc.

Secretion (se-kre'shun) (1) The passage of material formed by a cell to its exterior; (2) cell product that is transported to the exterior of a cell.

Secretory vesicles (granules) Vesicles containing proteins that migrate to the plasma membrane of a cell and discharge their contents from the cell by exocytosis.

Section A cut through the body (or an organ) that is made along a particular plane; a thin slice of tissue prepared for microscopic study.

Segregation During meiosis, the distribution of the members of the allele pair to different gametes.

Selectively permeable membrane A membrane that allows certain substances to pass while restricting the movement of others; also called differentially permeable membrane.

Semen (se'men) Fluid mixture containing sperm and secretions of the male accessory reproductive glands.

Semilunar valves (sĕ"me-loo'ner) Valves that prevent blood return to the ventricles after contraction.

Seminiferous tubules (sem"ĭ-nif'er-us) Highly convoluted tubes within the testes; form sperm.

Sense organs Localized collections of many types of cells working together to accomplish a specific receptive process.

Sensory (afferent) nerves Nerves that contain processes of sensory neurons and carry impulses to the central nervous system.

Sensory areas Functional areas of the cerebral cortex that provide for conscious awareness of sensation.

Sensory receptor Dendritic end organs, or parts of other cell types, specialized to respond to a stimulus.

Serosa (serous membrane) (se-ro'sah) The moist membrane found in closed ventral body cavities.

Serous fluid (sēr'us) Clear, watery fluid secreted by cells of a serous membrane.

Serum (sēr'um) Amber-colored fluid that exudes from clotted blood as the clot shrinks and then no longer contains fibrinogen.

Sesamoid bones (ses'ah-moid) Short bones embedded in tendons, variable in size and number, many of which influence the action of muscles; largest is the patella (kneecap).

Severe combined immunodeficiency disease (SCID) Congenital condition resulting in little or no protection against disease-causing organisms of any type.

Sex chromosomes The chromosomes, X and Y, that determine genetic sex (XX = female; XY = male); the 23rd pair of chromosomes.

Sex-linked inheritance Inherited traits determined by genes on the sex chromosomes, e.g., X-linked genes are passed from mother to son, Y-linked genes are passed from father to son.

Sexually transmitted disease (STD) Infectious disease spread through sexual contact.

Signal sequence A short peptide segment present in a protein being synthesized that causes the associated ribosome to attach to the membrane of rough ER.

Simple diffusion The unassisted transport across a plasma membrane of a lipid-soluble or very small particle.

Sinoatrial (SA) node (si"no-a'tre-al) Specialized myocardial cells in the wall of the right atrium; pacemaker of the heart.

Sinus (si'nus) (1) Mucous-membrane-lined, air-filled cavity in certain cranial bones; (2) dilated channel for the passage of blood or lymph.

Skeletal cartilage Comprises most of the skeleton in early fetal life; articular cartilage, nasal cartilage in the adult skeleton.

Skeletal muscle Muscle composed of cylindrical multinucleate cells with obvious striations; the muscle(s) attached to the body's skeleton; voluntary muscle.

Skeletal system   System of protection and support composed primarily of bone and cartilage.

Skull   Bony protective encasement of the brain and the organs of hearing and equilibrium; includes the facial bones. Also called the cranium.

Small intestine   Convoluted tube extending from the pyloric sphincter to the ileocecal valve where it joins the large intestine; the site where digestion is completed and virtually all absorption occurs.

Smooth muscle   Spindle-shaped cells with one centrally located nucleus and no externally visible striations (bands). Found mainly in the walls of hollow organs.

Sodium-potassium (Na^+-K^+) pump   A primary active transport system that simultaneously drives Na^1 out of the cell against a steep gradient and pumps K^1 back in.

Sol-gel transformation   Reversible change of a colloid from a fluid (sol) to a more solid (gel) state.

Solute (sol'yoot)   The substance that is dissolved in a solution.

Solute pump   Enzyme-like protein carrier that mediates active transport of solutes such as amino acids and ions uphill against their concentration gradients.

Somatic nervous system (so-ma'tik)   Division of the peripheral nervous system that provides the motor innervation of skeletal muscles; also called the voluntary nervous system.

Somatic reflexes   Reflexes that activate skeletal muscle.

Somatosensory system   That part of the sensory system dealing with reception in the body wall and limbs; receives inputs from exteroceptors, proprioceptors, and interoceptors.

Somite (so'mīt)   A mesodermal segment of the body of an embryo that contributes to the formation of skeletal muscles, vertebrae, and dermis of skin.

Spatial discrimination   The ability of neurons to identify the site or pattern of stimulation.

Special senses   The senses of taste, smell, vision, hearing, and equilibrium.

Specific gravity   Term used to compare the weight of a substance to the weight of an equal volume of distilled water.

Sperm (spermatozoa)   Male gamete.

Spermatogenesis (sper"mah-to-jen'ĕ-sis)   The process of sperm (male gamete) formation; involves meiosis.

Sphincter (sfink'ter)   A circular muscle surrounding an opening; acts as a valve.

Spinal cord   The bundle of nervous tissue that runs from the brain to the first to third lumbar vertebrae and provides a conduction pathway to and from the brain.

Spinal nerves   The 31 nerve pairs that arise from the spinal cord.

Splanchnic circulation (splangk'nik)   The blood vessels serving the digestive system.

Spleen   Largest lymphoid organ; provides for lymphocyte proliferation, immune surveillance and response, and blood-cleansing functions.

Spongy bone   Internal layer of skeletal bone. Also called cancellous bone.

Sprain   Ligaments reinforcing a joint are stretched or torn.

Static equilibrium   Sense of head position in space with respect to gravity.

Stenosis (stĕ-no'sis)   Abnormal constriction or narrowing.

Steroids (stĕ'roidz)   Group of chemical substances including certain hormones and cholesterol; they are fat soluble and contain little oxygen.

Stimulus (stim'u-lus)   An excitant or irritant; a change in the environment that evokes a response.

Stomach   Temporary reservoir in the gastrointestinal tract where chemical breakdown of proteins begins and food is converted into chyme.

Stressor   Any stimulus that directly or indirectly causes the hypothalamus to initiate stress-reducing responses, such as the fight or flight response.

Stroke   Condition in which brain tissue is deprived of a blood supply, as in blockage of a cerebral blood vessel; a cerebrovascular accident (CVA).

Stroke volume (SV)   Amount of blood pumped out of a ventricle during one contraction.

Stroma (stro'mah)   The basic internal structural framework of an organ.

Structural (fibrous) proteins   Consist of extended, strandlike polypeptide chains forming a strong, ropelike structure that is linear, insoluble in water, and very stable; e.g., collagen.

Subcutaneous (sub"kyu-ta'ne-us)   Beneath the skin.

Substrate   A reactant on which an enzyme acts to cause a chemical action to proceed.

Sudoriferous gland (su"do-rif'er-us)   Epidermal gland that produces sweat.

Sulcus (sul'kus)   A furrow on the brain, less deep than a fissure.

Summation   Accumulation of effects, especially those of muscular, sensory, or mental stimuli.

Superficial   Located close to or on the body surface.

Superior   Refers to the head or upper body regions.

Superior vena cava   Vein that returns blood from body regions superior to the diaphragm.

Supination (soo"pĭ-na'shun)   The outward rotation of the forearm causing palms to face anteriorly.

Suppressor T cells   Regulatory T lymphocytes that suppress the immune response.

Surfactant (ser-fak'tant)   Secretion produced by certain cells of the alveoli that reduces the surface tension of water molecules, thus preventing the collapse of the alveoli after each expiration.

Suspension   Heterogeneous mixtures with large, often visible solutes that tend to settle out.

Suture (soo'cher)   An immovable fibrous joint; with one exception, all bones of the skull are united by sutures.

Sweat gland.   *See* Sudoriferous gland.

Sympathetic nervous system   The division of the autonomic nervous system that activates the body to cope with some stressor (danger, excitement, etc.); the fight, fright, and flight subdivision.

Sympathetic (vasomotor) tone   State of partial vasoconstriction of the blood vessels maintained by sympathetic fibers.

Symphysis (sim'fih-sis)   A joint in which the bones are connected by fibrocartilage.

Synapse (sin'aps)   Functional junction or point of close contact between two neurons or between a neuron and an effector cell.

Synapsis (sĭ-nap'sis)   Pairing of homologous chromosomes during the first meiotic division.

Synaptic cleft (si-nap'tik)   Fluid-filled space at a synapse.

Synaptic delay   Time required for an impulse to cross a synapse between two neurons.

Synaptic knobs (boutons)   The bulbous distal endings of the telodendria.

Synaptic vesicles   Small membranous sacs containing the neurotransmitter acetylcholine.

Synarthrosis (sin"ar-thro'sis)   Immovable joint.

Synchondrosis (sin"kon-dro'sis)   A joint in which the bones are united by hyaline cartilage.

Syndesmosis (sin"des-mos'sis)   A joint in which the bones are united by a ligament or a sheet of fibrous tissue.

Synergist (sin'er-jist)   Muscle that aids the action of a prime mover by effecting the same movement or by stabilizing joints across which the prime mover acts to prevent undesirable movements.

Synostosis (sin"os-to'sis)   A completely ossified joint; a fused joint.

Synovial fluid   Fluid secreted by the synovial membrane; lubricates joint surfaces and nourishes articular cartilages.

Synovial joint   Freely movable joint exhibiting a joint cavity; also called a diarthrosis.

Synthesis (combination) reaction   A chemical reaction in which larger, more complex atoms or molecules are formed from simpler ones.

Systemic (sis-tem'ik)   Pertaining to the whole body.

Systemic circuit   System of blood vessels that serves gas exchange in the body tissues.

Systole (sis'to-le)   Period when either the ventricles or the atria are contracting.

Systolic pressure (sis-tah'lik)   Pressure exerted by blood on the blood vessel walls during ventricular contractions.

T cells   Lymphocytes that mediate cellular immunity; include helper, killer, suppressor, and memory cells. Also called T lymphocytes.

T tubule (transverse tubule)   Extension of the muscle cell plasma membrane (sarcolemma) that protrudes deeply into the muscle cell.

Tachycardia (tak"e-kar'de-ah)   A heart rate over 100 beats per minute.

Target cell   A cell that is capable of responding to a hormone because it bears receptors to which the hormone can bind.

Taste buds   Sensory receptor organs that house gustatory cells, which respond to dissolved food chemicals.

Telencephalon (endbrain) (tel"en-seh'fuh-lon)   Anterior subdivision of the primary forebrain that develops into olfactory lobes, cerebral cortex, and corporal striata.

Telodendria   The terminal branches of an axon.

Telophase   The final phase of mitosis; begins when migration of chromosomes to the poles of the cell has been completed and ends with the complete separation of the two daughter cells.

Tendon (ten'dun)   Cord of dense fibrous tissue attaching muscle to bone.

Tendonitis   Inflammation of tendon sheaths, typically caused by overuse.

Testis (tes'tēz)   Male primary sex organ that produces sperm; male gonad.

Testosterone (tes-tos'tĕ-rōn)   Male sex hormone produced by the testes; during puberty promotes virilization, and is necessary for normal sperm production.

Tetanus (tet'ah-nus)   (1) A smooth, sustained muscle contraction resulting from high-frequency stimulation; (2) an infectious disease caused by an anaerobic bacterium.

Thalamus (tha'luh-mis)   A mass of gray matter in the diencephalon of the brain.

Thermogenesis (ther"mo-jen'ĕ-sis)   Heat production.

Thermoreceptor (ther"mo-re-sep'ter)   Receptor sensitive to temperature changes.

Third-degree burn   A burn that involves the entire thickness of the skin; also called a full-thickness burn. Usually requires skin grafting.

Thoracic duct   Large duct that receives lymph drained from the entire lower body, the left upper extremity, and the left side of the head and thorax.

Thorax (tho'raks)   That portion of the body trunk above the diaphragm and below the neck.

Threshold stimulus   Weakest stimulus capable of producing a response in an irritable tissue.

Thrombin (throm'bin)   Enzyme that induces clotting by converting fibrinogen to fibrin.

Thrombocyte (throm'bo-sīt)   Platelets; cell fragments that participate in blood coagulation.

Thrombocytopenia (throm"bo-si"to-pe'ne-ah)   A reduction in the number of platelets circulating in the blood.

Thrombus (throm'bus)   A clot that develops and persists in an unbroken blood vessel.

Thymine (T) (thi'mēn)   Single-ring base (a pyrimidine) in DNA.

Thymus gland (thi'mus)   Endocrine gland active in immune response.

Thyrocalcitonin (thi"ro-kal"sī-to'nin).   *See* Calcitonin.

Thyroid gland (thi'roid)   One of the largest of the body's endocrine glands; straddles the anterior trachea.

Thyroid hormone (TH)   The major hormone secreted by thyroid follicles; stimulates enzymes concerned with glucose oxidation.

Thyroid-stimulating hormone (TSH)   Adenohypophyseal hormone that regulates secretion of thyroid hormones.

Thyroxine (T₄) (thi-rok'sin)   Iodine-containing hormone secreted by the thyroid gland; accelerates cellular metabolic rate in most body tissues.

Tight junction   Area where plasma membranes of adjacent cells are fused.

Tissue   A group of similar cells and their intercellular substance specialized to perform a specific function; primary tissue types of the body are epithelial, connective, muscle, and nervous tissue.

Tissue perfusion   Blood flow through body tissues or organs.

Tonicity (to-nis'ĭ-te)   A measure of the ability of a solution to cause a change in cell shape or tone by promoting osmotic flows of water.

Tonsils   A ring of lymphocyte tissue around the entrance to the pharynx; there are three pairs of tonsils, each named for its location.

Trabecula (trah-bek'u-lah)   (1) Any of the fibrous bands extending from the capsule into the interior of an organ; (2) struts or thin plates of bone in spongy bone.

Trachea (tra'ke-ah)   Windpipe; cartilage-reinforced tube extending from larynx to bronchi.

Tract   A collection of nerve fibers in the central nervous system having the same origin, termination, and function.

Transcription   One of the two major steps in the transfer of genetic code information involving the transfer of information from a DNA gene's base sequence to the complimentary base sequence of an mRNA molecule.

Transduction (trans-duk'shun)   The conversion of the energy of a stimulus into an electrical event.

Transepithelial transport (trans-ep"ĭ-the'le-al)   Movement of substances through, rather than between, adjacent epithelial cells connected by tight junctions, such as absorption of nutrients in the small intestine.

Transfer RNA (tRNA)   Short-chain RNA molecules that transfer amino acids to the ribosome.

Transfusion reaction   Occurs when the donor's red blood cells are attacked by the recipient's plasma agglutinins.

Translation   One of the two major steps in the transfer of genetic code information in which the information carried by mRNA is decoded and used to assemble polypeptides.

Transverse (horizontal) plane   A plane running from right to left, dividing the body into superior and inferior parts.

Tricuspid valve (tri-kus'pid)   The right atrioventricular valve.

Triglycerides (tri-glis'er-īdz)   Fats and oils composed of fatty acids and glycerol; are the body's most concentrated source of energy fuel; also known as neutral fats.

Triiodothyronine (T₃) (tri"i-o"do-thi'ro-nēn)   Secretion and function similar to those of thyroxine.

Tripeptide   A combination of three amino acids united by means of a peptide bond.

Trophoblast (tro'fo-blast)   Outer sphere of cells of the blastocyst.

Tropic hormone (trōp'ik)   A hormone that regulates the function of another endocrine organ.

Trypsin   Proteolytic enzyme secreted by the pancreas.

Tubular reabsorption   The movement of filtrate components from the renal tubules into the blood.

Tubular secretion   The movement of undesirable substances (such as drugs, urea, excess ions) from blood into filtrate.

Tumor   An abnormal growth of cells; a swelling; cancerous at times.

Tunica (too'nĭ-kah)   A covering or tissue coat; membrane layer.

Tympanic membrane (tim-pan'ik)   Eardrum.

Ulcer (ul'ser)   Lesion or erosion of the mucous membrane, such as gastric ulcer of stomach.

Umbilical cord (um-bĭ'lĭ-kul)   Structure bearing arteries and veins connecting the placenta and the fetus.

Umbilicus (um-bĭ'lĭ-kus)   Navel; marks site where umbilical cord was attached in fetal stage.

Unipolar neuron   Neuron in which embryological fusion of the two processes leaves only one process extending from the cell body.

Unmyelinated fibers (un-mi'ĕ-lĭ-nāt"ed)   Axons lacking a myelin sheath and therefore conducting impulses quite slowly.

Up-regulation   Increased target cell formation of receptors in response to increasingly higher levels of the hormones to which they respond.

Uracil (U) (u'rah-sil)   A smaller, single-ring base (a pyrimidine) found in RNA.

Urea (u-re'ah)   Main nitrogen-containing waste excreted in urine.

Ureter (u'-re-ter)   Tube that carries urine from kidney to bladder.

Urethra (u-re'thrah)   Canal through which urine passes from the bladder to outside the body.

Uric acid   The nitrogenous waste product of nucleic acid metabolism; component of urine.

Urinary bladder   A smooth, collapsible, muscular sac that stores urine temporarily.

Urinary system   System primarily responsible for water, electrolyte, and acid-base balance and removal of nitrogenous wastes.

Uterine tube (u'-ter-in)   Tube through which the ovum is transported to the uterus. Also called fallopian tube.

Uterus (u'ter-us)   Hollow, thick-walled organ that receives, retains, and nourishes fertilized egg; site where embryo/fetus develops.

Uvula (u'vu-lah)   Tissue tag hanging from soft palate.

Vaccine   Preparation that provides artificially acquired active immunity.

Vagina   Thin-walled tube extending from the cervix to the body exterior; often called the birth canal.

Valence shell (va'lens)   Outermost electron shell (energy level) of an atom that contains electrons.

Varicosities   Knoblike swellings containing mitochondria and synaptic vesicles.

Vas (vaz')   A duct; vessel.

Vasa recta (va'sah rek'tah)   Capillary branches that supply loops of Henle in the medulla region.

Vascular   Pertaining to blood vessels or richly supplied with blood vessels.

Vascular spasm   Immediate response to blood vessel injury; results in constriction.

Vasoconstriction (vas"o-kon-strik'shun)   Narrowing of blood vessels.

Vasodilation (vas"o-di-la'shun)   Relaxation of the smooth muscles of the blood vessels producing dilation.

Vasomotion (vas"o-mo'shun)   Intermittent contraction or relaxation of the precapillary sphincter beds resulting in a staggered blood flow when tissue needs are not extreme.

Vasomotor center (vas"o-mo'ter)   Brain area concerned with regulation of blood vessel resistance.

Vasomotor fibers   Sympathetic nerve fibers that regulate the contraction of smooth muscle in the walls of blood vessels, thereby regulating blood vessel diameter.

Veins (vānz")   Blood vessels that return blood toward the heart from the circulation.

Ventral   Pertaining to the front; anterior.

Ventricle   (1) Paired, inferiorly located heart chambers that function as the major blood pumps; (2) cavities in the brain.

Venule (ven'ūl)   A small vein.

Vertebral column (spine) (ver'ti-brul)   Formed of a number of individual bones called vertebrae and two composite bones (sacrum and coccyx).

Vesicle (vĕ'sĭ-kul)   A small liquid-filled sac or bladder.

Vesicular follicle   Mature ovarian follicle.

Vestibule   An enlarged area at the beginning of a canal, i.e., inner ear, nose, larynx.

Villus (vil'us)   Fingerlike projections of the small intestinal mucosa that tremendously increase its surface area for absorption.

Visceral (vis'er-al)   Pertaining to an internal organ of the body or the inner part of a structure.

Visceral muscle   Type of smooth muscle; its cells contract as a unit and rhythmically, are electrically coupled by gap junctions, and often exhibit spontaneous action potentials.

Visceral organs (visera)   A group of internal organs housed in the ventral body cavity.

Visceral serosa (se-ro'sah)   The part of the double-layered membrane that lines the outer surfaces of organs within the ventral body cavity.

Viscosity (vis'kos'ĭ-te)   State of being sticky or thick.

Visual field   The field of view seen when the head is still.

Vital capacity (VC)   The volume of air that can be expelled from the lungs by forcible expiration after the deepest inspiration; total exchangeable air.

Vital signs   Includes pulse, blood pressure, respiratory rate, and body temperature measurements.

Vitamins   Organic compounds required by the body in minute amounts.

Vocal folds   Mucosal folds that function in voice production (speech); also called the true vocal cords.

Volatile acid   An acid that can be eliminated by the lungs; carbonic acid is converted to CO_2, which diffuses into the alveoli.

Volkmann's canals.   *See* Perforating canals.

Voluntary muscle   Muscle under strict nervous control; skeletal muscle.

Voluntary nervous system   The somatic nervous system.

Vulva (vul'vuh)   Female external genitalia.

Wallerian degeneration (wal-er"ĕ-an)   A process of disintegration of an axon that occurs when it is crushed or severed and cannot receive nutrients from the cell body.

Water of oxidation (metabolic water)   Water produced from cellular metabolism (about 10% of our body's water).

White matter   White substance of the central nervous system; myelinated nerve fibers.

Xenograft   Tissue graft taken from another animal species.

Yolk sac (yōk)   Endodermal sac that serves as the source for primordial germ cells.

Zygote (zi'gōt)   Fertilized egg.

Text, Photography, and Illustration Credits

of Human Anatomy by David L. Bassett.
29.14a–c: © Lennart Nilsson. From A Child is Born. © 1990 Dell Publishing, New York, New York.
A Closer Look, a and b: Courtesy of the IVF Clinic, Stanford Medical Center.
Chapter 30 Chapter Opener: © Pictor.
30.1b: © SIU/Visuals Unlimited.
30.1c: © CNRI/SPL/Photo Researchers, Inc.
A Closer Look: © Oliver Brüstle, University of Bonn Medical Center.

Illustration Credits

Chapter 1 1.1: Jeanne Koelling/Kristin Mount, 1.2a–l: Wendy Hiller Gee, 1.3: Adapted from Campbell, Reece, and Mitchell, Biology, 5e, F40.10a, p788 (San Francisco: Benjamin Cummings, 1999). © 1999 Addison Wesley Longman, Inc., 1.4: Imagineering, 1.5: Imagineering, 1.6: Imagineering, 1.7a and b: Imagineering, 1.9a and b: Imagineering, 1.10a: Adapted from Seeley, Stephens, Tate, Anatomy & Physiology, 4e, F1.15a, p20 (New York: WCB/McGraw-Hill, 1998).© 1998 The McGraw-Hill Companies, Inc., 1.10b: Imagineering, 1.11a–b: Imagineering, 1.12: Imagineering, Table 1.1: Imagineering.

Chapter 2 2.1: Imagineering, 2.2: Imagineering, 2.3: Imagineering, 2.4a and b: Imagineering, 2.5a Imagineering, 2.5b: Adapted from Campbell, Reece, and Mitchell, Biology, 5e, F2.13, p31 (San Francisco: Benjamin Cummings, 1999). © 1999 Addison Wesley Longman, Inc., 2.6a–c: Imagineering, 2.7a and b: Imagineering, 2.8: Imagineering, 2.9: Imagineering, 2.10a–c: Imagineering, 2.11: Imagineering, 2.12: Imagineering, 2.13a–c: Imagineering, 2.14a–c: Imagineering, 2.15a–e: Imagineering, 2.16: Imagineering, 2.17a–e: Imagineering, 2.18: Adapted from Mathews, Van Holde, Ahern, Biochemistry, 3e, F6.13c, d, f, p175 (San Francisco: Benjamin Cummings, 2000) © 2000 Addison Wesley Longman, Inc., 2.19a and b: Imagineering, 2.20a and b: Imagineering, 2.21: Imagineering, 2.22a and b: Imagineering, 2.23: Imagineering, 2.24: Imagineering.

Chapter 3 3.1: Tomo Narashima, 3.2: Carla Simmons/Kristin Mount, 3.3: Imagineering, 3.4: Tomo Narashima, 3.4b, c: Adapted from Lodish, et al., Molecular Cell Biology, 3e,

F24–35, p1156 and F24–38, p1159 (New York: W.H. Freeman and Company, 1995) © 1995 Scientific American Books. 3.5a and b: Imagineering, 3.6a and b: Imagineering, 3.7a and b: Imagineering, 3.9: Imagineering, 3.10: Imagineering, 3.11a: Imagineering, 3.12a–c: Imagineering, 3.13: Imagineering, 3.14: Imagineering, 3.15a: Tomo Narashima, 3.16a and c: Tomo Narashima, 3.17: Imagineering, 3.18: Tomo Narashima, 3.19: Tomo Narashima, 3.21: Adapted from Campbell, Mitchell, Reece, Biology, 5e, F7.16, p116 (San Francisco: Benjamin Cummings, 1999) © 1999 Addison Wesley Longman, Inc., 3.22: Tomo Narashima, 3.23: From Campbell, Biology, 4e, F7.21, p128 (Menlo Park: Benjamin Cummings, 1996) © 1996 Addison Wesley Longman, Inc., 3.24a: Tomo Narashima, 3.25a–c: Imagineering, 3.26a: Tomo Narashima, 3.28: Imagineering, 3.29: Imagineering, 3.30: Adapted from Campbell, Mitchell, Reece, Biology, 5e, F12.5, p210 (San Francisco: Benjamin Cummings, 1999) © 1999 Addison Wesley Longman, Inc., 3.31: Adapted from Campbell, Mitchell, Reece, Biology, 5e, F12.14, p218 (San Francisco: Benjamin Cummings, 1999) © 1999 Addison Wesley Longman, Inc., 3.32: Imagineering, 3.33: Imagineering, 3.34a: Imagineering, 3.35: Imagineering, 3.36: Adapted from Campbell, Mitchell, Reece, Biology, 5e, F19.11, p357 (San Francisco: Benjamin Cummings, 1999) © 1999 Addison Wesley Longman, Inc.

Chapter 4 4.1a and b: Imagineering, 4.3a and b: Imagineering, 4.4: Imagineering, 4.5a and b: Imagineering, 4.6: Imagineering, 4.7: Imagineering, 4.9a–c: Imagineering, 4.12a–c: Imagineering, 4.13: Imagineering, A Closer Look: Imagineering.

Chapter 5 5.2: Imagineering, 5.3: Tomo Narashima, 5.4: Imagineering, 5.7a and b: Imagineering, 5.9a: Imagineering, Making Connections: Vincent Perez/Wendy Hiller Gee.

Chapter 6 6.1: Precision Graphics/Kristin Mount, 6.2: Adapted from Tortora and Grabowski, Principles of Anatomy and Physiology, 9e, F7.2, p186 (New York: John Wiley & Sons, Inc.) © 2000 Biological Sciences Textbooks, Inc. and Sandra Reynolds Grabowski, 6.3a–c: Barbara Cousins/Kristin Mount, 6.4: Tomo Narashima, 6.5a and b: Barbara Cousins/Kristin Mount, 6.6: Imagineering, 6.7:Imagineering, 6.8: Barbara

Cousins/Kristin Mount, 6.10: Imagineering, 6.11: Imagineering, 6.13: Barbara Cousins/Kristin Mount, Making Connections: Vincent Perez/Wendy Hiller Gee.

Chapter 7 7.1a and b: Nadine Sokol, 7.2a Nadine Sokol, 7.2b: Kristin Mount, 7.3a and b: Nadine Sokol, 7.4a–c: Nadine Sokol, 7.5: Nadine Sokol, 7.6 and b: Nadine Sokol, 7.7: Nadine Sokol, 7.8a–c: Nadine Sokol, 7.9b: Nadine Sokol, 7.10a and b: Nadine Sokol, 7.11a and b: Vincent Perez/Kristin Mount, 7.12: Nadine Sokol, 7.13: Kristin Otwell/Kristin Mount, 7.14a and b: Laurie O'Keefe/Kristin Mount, 7.15: Laurie O'Keefe/Kristin Mount, 7.16a–c: Laurie O'Keefe/Kristin Mount, 7.17a–c: Laurie O'Keefe/Kristin Mount, 7.18a and b: Laurie O'Keefe/Kristin Mount, 7.19a and b: Imagineering, 7.20a and b: Laurie O'Keefe/Kristin Mount, 7.21: Nadine Sokol, 7.22a–f: Laurie O'Keefe/Kristin Mount, 7.23a and b: Laurie O'Keefe/Kristin Mount, 7.24a and b: Laurie O'Keefe/Kristin Mount, 7.26a: Laurie O'Keefe/Kristin Mount, 7.27a–c: Laurie O'Keefe/Kristin Mount, 7.28b: Laurie O'Keefe/Kristin Mount, 7.29a: Laurie O'Keefe/Kristin Mount, 7.31a–c: Kristin Otwell/Kristin Mount, 7.32: Imagineering, 7.33a and b: Nadine Sokol, 7.34a and b: Imagineering, Table 7.2: Kristin Otwell/Kristin Mount, Mount, Table 7.3: Kristin Otwell/Kristin Mount, Table 7.4: Laurie O'Keefe/Kristin Mount.

Chapter 8 8.1a and b: Precision Graphics/Kristin Mount, 8.2a–c: Barbara Cousins/Kristin Mount, 8.3a: Precision Graphics/Kristin Mount, 8.4a and b: Barbara Cousins/Kristin Mount, 8.7: Imagineering, 8.8a and b: Barbara Cousins/Kristin Mount, 8.9a, c and d: Barbara Cousins/Kristin Mount, 8.10a, b and d: Barbara Cousins/Kristin Mount, 8.11a–c and e: Barbara Cousins/Kristin Mount, 8.12: Imagineering, Table 8.2: Imagineering.

Chapter 9 9.1a: Raychel Ciemma/Kristin Mount, 9.2b–d: Imagineering, 9.3a–d: Imagineering, 9.4: Imagineering, 9.6a–d: Imagineering, 9.7: Imagineering, 9.8a–c: Imagineering, 9.9a–d: Imagineering, 9.10: Imagineering, 9.11a: Imagineering, 9.12a and b: Imagineering, 9.13: Imagineering, 9.14: Imagineering, 9.15: Adapted from Martini, Fundamentals of Anatomy & Physiology, 4e, F10.14, p296 (Upper Saddle River, NJ: Prenctice-Hall, Inc., 1998) © 1998 Frederic H. Martini, Inc.,

9.16a–c: Imagineering, 9.17: Imagineering, 9.18a and b: Imagineering, 9.19a–c: Imagineering, 9.20a and b: Imagineering, 9.21a–c: Imagineering, 9.22a and b: Imagineering, 9.23: Imagineering, 9.24: Imagineering, 9.25a and b: Imagineering, Table 9.1: Imagineering, Table 9.4: Imagineering, Making Connections: Vincent Perez/Wendy Hiller Gee.

Chapter 10 10.1: Adapted from Martini, Fundamentals of Anatomy & Physiology, 4e, F11.1, p317 (Upper Saddle River, NJ: Prentice-Hall, Inc., 1998) © 1998 Frederic H. Martini, Inc., 10.2a and b: Imagineering, 10.3a–c: Imagineering, 10.4b: Raychel Ciemma/Kristin Mount, 10.5b: Raychel Ciemma/Kristin Mount, 10.6: Raychel Ciemma/Kristin Mount, 10.7a–c: Raychel Ciemma/Kristin Mount, 10.8a and b: Raychel Ciemma/Kristin Mount, 10.9s, b, d and e: Raychel Ciemma/Kristin Mount, 10.10a and b: Raychel Ciemma/Kristin Mount, 10.11a–c: Raychel Ciemma/Kristin Mount, 10.12a–c: Raychel Ciemma/Kristin Mount, 10.13a and b: Raychel Ciemma/Kristin Mount, 10.14a–d: Raychel Ciemma/Kristin Mount, 10.15a–c: Raychel Ciemma/Kristin Mount, 10.16a and b: Raychel Ciemma/Kristin Mount, 10.17a and b: Raychel Ciemma/Kristin Mount, 10.18a–d: Laurie O'Keefe/Kristin Mount, 10.19a–c: Raychel Ciemma/Kristin Mount, 10.20a–c: Wendy Hiller Gee/Kristin Mount, 10.21a–d: Raychel Ciemma/Kristin Mount, 10.22a–c: Raychel Ciemma/Kristin Mount, 10.23a–f: Raychel Ciemma/Kristin Mount, 10.24: Imagineering, 10.25a–f: Laurie O'Keefe/Kristin Mount.

Chapter 11 11.1: Imagineering, 11.2a and b: Imagineering, 11.3a–c: Imagineering, 11.4b: Charles W. Hoffman/Kristin Mount, 11.5a–d: Imagineering, 11.6a and b: Imagineering, 11.7: Imagineering, 11.8: Imagineering, 11.9a and b: Imagineering, 11.10a and b: Imagineering, 11.11: Imagineering, 11.12: Imagineering, 11.13a–c: Imagineering, 11.14: Imagineering, 11.15: Imagineering, 11.16: Imagineering, 11.17a: Imagineering, 11.18: Imagineering, 11.19a and b: Imagineering, 11.20a–d: Imagineering, 11.21: Imagineering, 11.22a and b: Imagineering, 11.23: Imagineering, 11.24a–f: Imagineering, 11.25: Imagineering, Table 11.1: Imagineering, Table 11.3: Imagineering.

Chapter 12 12.2a–d: Imagineering, 12.3a–e: Imagineering, 12.4a–d: Imagineering, 12.5a–d: Imagineering, 12.6a and b: Imagineering, 12.8a and b: Imagineering, 12.9: Imagineering, 12.10a–b: Imagineering, 12.11a–b: Imagineering, 12.12: Imagineering, 12.13a and b: Imagineering, 12.15: Imagineering, 12.16a–c: Imagineering, 12.18: Imagineering, 12.19: Imagineering, 12.20a: Imagineering, 12.21: Imagineering, 12.22a and b: Imagineering, 12.24a: Imagineering, 12.26: Imagineering, 12.27a and b: Stephanie McCann/Kristin Mount, 12.28: Imagineering, 12.29: Imagineering, Table 12.1: Imagineering.

Chapter 13 13.1: Imagineering, 13.2b: Charles W. Hoffman/Kristin Mount, 13.3a–d: Imagineering, 13.4a and b: Imagineering, 13.5: Imagineering, 13.6a and b: Stephanie McCann/Kristin Mount, 13.7: Imagineering, 13.8a–c: Imagineering, 13.9a and b: Imagineering, 13.10a and b: Imagineering, 13.11a and b: Imagineering, 13.12: Imagineering, 13.13: Imagineering, 13.14a and b: Imagineering, 13.15a and b: Imagineering, 13.16: Imagineering, 13.17: Imagineering, Table 13.1: Imagineering, Table 13.2: Imagineering, A Closer Look: Imagineering.

Chapter 14 14.1: Imagineering, 14.2: Imagineering, 14.3: Imagineering, 14.4: Imagineering, 14.5: Imagineering, 14.6a and b: Imagineering, 14.7: Imagineering, 14.8: Imagineering, 14.9: Imagineering, Making Connections: Vincent Perez/Wendy Hiller Gee.

Chapter 15 15.1: Imagineering, 15.2a and b: Imagineering, 15.3: Imagineering, 15.4a and b: Imagineering, 15.5a: © Alexander Tsiaras/Science Source/Photo Researchers, Inc. 15.5b: Imagineering, 15.6: Imagineering, 15.7: Imagineering, 15.8a and b: Imagineering, Table 15.3: Imagineering.

Chapter 16 16.1a–c: Imagineering, 16.2a and b: Imagineering, 16.3: Imagineering, 16.5a and b: Charles W. Hoffman/Kristin Mount, 16.6a–c: Imagineering, 16.7a: Imagineering, 16.8: Imagineering, 16.9a and b: Imagineering, 16.11: Imagineering, 16.13: Imagineering, 16.15a and b: Imagineering, 16.16a and b: Imagineering, 16.17a–c:

Imagineering, 16.18a and b: Imagineering, 16.19: Imagineering, 16.20: Imagineering, 16.21: Imagineering, 16.23: Imagineering, 16.24a and b: Charles W. Hoffman/Kristin Mount, 16.25: Precision Graphics/Kristin Mount, 16.26: Precision Graphics/Kristin Mount, 16.27a–c: Imagineering, 16.28a–c: Imagineering, 16.29a and b: Imagineering, 16.30: Imagineering, 16.31a–c: Imagineering, 16.33: Imagineering, 16.34a and b: Imagineering, 16.35: Imagineering, 16.36a: Kristin Mount, 16.36b: Charles W. Hoffman/Kristin Mount, 16.37: Imagineering.

Chapter 17 17.1a and b: Imagineering, 17.2: Imagineering, 17.3a–c: Imagineering, 17.4: Imagineering, 17.5: Imagineering, 17.6: Imagineering, 17.8a: Imagineering, 17.9: Imagineering, 17.10a: Imagineering, 17.11: Imagineering, 17.12: 1: Imagineering, 17.13: Imagineering, 17.15: Imagineering, 17.17: Imagineering, 17.18: Imagineering, 17.19: Imagineering, Table 17.1: Imagineering, Table 17.3: Imagineering, Making Connections: Vincent Perez/Wendy Hiller Gee.

Chapter 18 18.1: Imagineering, 18.3: Imagineering, 18.4a and b: Nadine Sokol, 18.5: Imagineering, 18.6: Imagineering, 18.7: Imagineering, 18.9: Imagineering, 18.11: Imagineering, 18.12: Imagineering, 18.13a and b: Imagineering, Table 18.2: Imagineering, Table 18.4: Imagineering.

Chapter 19 19.1a–c: Wendy Hiller Gee/Kristin Mount, 19.2: Barbara Cousins, 19.3: Barbara Cousins/Kristin Mount, 19.4b, d and e: Barbara Cousins, 19.5: Imagineering, 19.6: Barbara Cousins/Kristin Mount, 19.7a and b: Imagineering, 19.8a: Barbara Cousins, 19.9a and b: Barbara Cousins, 19.10a and b: Barbara Cousins, 19.11b: Imagineering, 19.12a and b: Imagineering, 19.13: Imagineering, 19.14a: Barbara Cousins/Kristin Mount, 19.14b: Kristin Mount, 19.15: Imagineering, 19.16: Imagineering, 19.17: Imagineering, 19.18a–d: GTS Graphics, 19.19a and b: Imagineering, 19.20: Imagineering, 19.21a and b: GTS Graphics, 19.22: Imagineering, 19.23: Imagineering, 19.24a–e: Imagineering, 19.25a–c: Imagineering, A Closer Look: Imagineering.

Chapter 20 20.1: Adapted from Tortora and Grabowski, Principles of Anatomy and Physiology, 9e, F21.1, p671 (New York: John Wiley & Sons, Inc., 2000) © 2000 Biological Sciences Textbooks, Inc. and Sandra Reynolds Grabowski., 20.2: Imagineering, 20.3–c: Imagineering, 20.4a and b: Imagineering, 20.5: Imagineering, 20.6: Imagineering, 20.7: Imagineering, 20.8: Imagineering, 20.9: Imagineering, 20.10: Imagineering, 20.11: Imagineering, 20.12: Imagineering, 20.13: Imagineering, 20.14: Imagineering, 20.15: Imagineering, 20.16: Imagineering, 20.17a: Kristin Mount, 20.17b: Barbara Cousins/Kristin Mount, 20.18: Imagineering, 20.19a: Imagineering, 20.19b: Barbara Cousins, 20.20a: Imagineering, 20.20b and d: Barbara Cousins, 20.21a: Imagineering, 20.21b: Barbara Cousins, 20.22a: Imagineering, 20.22b–d: Barbara Cousins, 20.23a: Imagineering, 20.23b–d: Barbara Cousins, 20.24a: Imagineering, 20.24b: Barbara Cousins, 20.25a: Imagineering, 20.25b and c: Barbara Cousins, 20.26a: Imagineering, 20.26b: Barbara Cousins, 20.27a: Imagineering, 20.27b and c: Barbara Cousins, 20.28a: Imagineering, 20.28b and c: Barbara Cousins, Table 20.1: Imagineering, Making Connections: Vincent Perez/Wendy Hiller Gee.

Chapter 21 21.1a and b: Imagineering, 21.2a and b: Imagineering, 21.4a: Imagineering, 21.5: Imagineering, 21.6a and b: Imagineering, 21.7: Imagineering, 21.8: Imagineering, Making Connections: Vincent Perez/Wendy Hiller Gee.

Chapter 22 22.1b: Imagineering, 22.2: Imagineering, 22.3: Imagineering, 22.4: Imagineering, 22.5: Imagineering, 22.6: Imagineering, 22.7: Imagineering, 22.8: Imagineering, 22.9: Imagineering, 22.10: Imagineering, 22.11: Imagineering, 22.12a: Imagineering, 22.13: Imagineering, 22.14a and b: Imagineering, 22.15: Imagineering, 22.16: Imagineering, 22.17a: Imagineering, 22.18: Imagineering, 22.19: Imagineering, Table 22.3: Imagineering, A Closer Look: Imagineering.

Chapter 23 23.1: Imagineering, 23.2b: Imagineering, 23.3b: Imagineering, 23.4a and b: Imagineering, 23.5a: Imagineering, 23.7: Imagineering,

23.8a and b: Imagineering, 23.9b–d: Imagineering, 23.10a and c: Imagineering, 23.12: Imagineering, 23.13: Imagineering, 23.14: Imagineering, 23.15: Imagineering, 23.16a and b: Imagineering, 23.17: Imagineering, 23.18: Imagineering, 23.19: Imagineering, 23.20: Imagineering, 23.21a and b: Imagineering, 23.22a and b: Imagineering, 23.23: Adapted from Berne and Levy, Physiology, 3e, F36.9, p598 (St. Louis, MO: Mosby-Year Book, Inc., 1993) © 1993 Mosby-Year Book Europe Limited, 23.24: Imagineering, 23.25: Imagineering, 23.26: Imagineering, 23.27: Imagineering, 23.28a and b: Imagineering, A Closer Look: Imagineering, Making Connections: Vincent Perez/Wendy Hiller Gee.

Chapter 24 24.1: Kristin Mount, 24.2: Imagineering, 24.3a and b: Precision Graphics/Karl Miyajima, 24.4: Imagineering, 24.5a and b: Precision Graphics/Karl Miyajima, 24.6: Adapted from Seeley, Stephens, Tate, Anatomy & Physiology, 4e, F24.2, p778 (New York: WCB/McGraw Hill, 1998) © 1998 The McGraw-Hill Companies, Inc., 24.7a and b: Cyndie Wooley/Karl Miyajima, 24.8: Imagineering, 24.9a: Imagineering, 24.10: Imagineering, 24.11: Imagineering, 24.13a–e: Imagineering, 24.14a: Kristin Otwell/Karl Miyajima, 24.15a–c: Imagineering, 24.16: Imagineering, 24.17: Imagineering, 24.18a–c: Imagineering, 24.19: Imagineering, 24.20: Imagineering, 24.21a and b: Imagineering, 24.23c: Karl Miyajima, 24.24c and d: Imagineering, 24.25: Imagineering, 24.26a: Imagineering, 24.27: Imagineering, 24.28: Imagineering, 24.29a: Karl Miyajima, 24.29b: Precision Graphics/Karl Miyajima, 24.30b–d: Imagineering, 24.32: Imagineering, 24.33: Imagineering, 24.34: Imagineering, 24.35: Imagineering, 24.36: Imagineering, 24.37a and b: Imagineering, Making Connections: Vincent Perez/Wendy Hiller Gee.

Chapter 25 25.1: Imagineering, 25.2a and b: Precision Graphics/Karl Miyajima, 25.3: Imagineering, 25.4a and b: Imagineering, 25.6: Precision Graphics/Karl Miyajima, 25.7: Imagineering, 25.8: Imagineering, 25.9: Precision Graphics/Karl Miyajima, 25.10: Imagineering,

Imagineering, 25.6: Precision Graphics/Karl Miyajima, 25.7: Imagineering, 25.8: Imagineering, 25.9: Precision Graphics/Karl Miyajima, 25.10: Imagineering, 25.11: Imagineering, 25.12: Precision Graphics/Karl Miyajima, 25.13: Precision Graphics/Karl Miyajima, 25.14: Precision Graphics/Karl Miyajima, 25.15: Imagineering, 25.16: Precision Graphics/Karl Miyajima, 25.17: Precision Graphics/Karl Miyajima, 25.18: Precision Graphics/Karl Miyajima, 25.19a and b: Precision Graphics/Karl Miyajima, 25.20: Precision Graphics/Karl Miyajima, 25.21: Precision Graphics/Karl Miyajima, 25.22: Kristin Mount/Karl Miyajima, 25.24: Precision Graphics/Karl Miyajima, 25.26: Imagineering.

Chapter 26 26.1a: Linda McVay/Karl Miyajima, 26.2a: Imagineering, 26.3b and c: Imagineering, 26.4a and b: Imagineering, 26.5a and b: Imagineering, 26.6: Precision Graphics/Karl Miyajima, 26.7: Imagineering, 26.8a and c: Imagineering, 26.9: Imagineering,

26.10: Imagineering, 26.11: Imagineering, 26.12: Imagineering, 26.13: Precision Graphics/Karl Miyajima, 26.14: Imagineering, 26.15: Imagineering, 26.16a–e: Precision Graphics/Karl Miyajima, 26.18a and b: Linda McVay/Karl Miyajima, 26.19: Precision Graphics/Karl Miyajima, 26.20a and b: Imagineering, 26.21a–d: Linda McVay/Kristin Mount.

Chapter 27 27.1: Precision Graphics/Karl Miyajima, 27.2: Precision Graphics/Karl Miyajima, 27.3: Precision Graphics/Karl Miyajima, 27.4: Precision Graphics/Karl Miyajima, 27.5: Imagineering, 27.6a and b: Precision Graphics/Karl Miyajima, 27.7: Imagineering, 27.8: Precision Graphics/Karl Miyajima, 27.9: Imagineering, 27.10: Imagineering, 27.11a and b: Precision Graphics/Karl Miyajima, 27.12: Imagineering, 27.13: Imagineering, 27.14: Imagineering, Making Connections: Vincent Perez/Wendy Hiller Gee.

Chapter 28 28.1: Martha Blake/Kristin Mount, 28.2: Martha Blake/Karl Miyajima, 28.4a and b: Martha Blake/Kristin Mount, 28.5: Imagineering, 28.6: Precision Graphics/Karl Miyajima, 28.7: Imagineering, 28.8c: Unknown/Kristin Mount, 28.9a: Martha Blake/Kristin Mount, 28.10: Precision Graphics/Karl Miyajima, 28.11: Martha Blake/Kristin Mount, 28.12a: Martha Blake/Karl Miyajima, 28.14a: Martha Blake/Kristin Mount, 28.15b: Martha Blake/Kristin Mount, 28.16: Carla Simmons/Karl Miyajima, 28.17a and b: Martha Blake/Karl Miyajima, 28.19: Imagineering, 28.20: Imagineering, 28.21: Imagineering, 28.22a–d: Imagineering, 28.24: Imagineering, 28.25a–c: Imagineering, 28.26a–c: Imagineering, A Closer Look: Precision Graphics/Karl Miyajima, Making Connections: Vincent Perez/Wendy Hiller Gee.

Chapter 29 29.1: Imagineering, 29.2a: Martha Blake/Kristin Mount, 29.3a–g: Martha

Blake/Kristin Mount, 29.4a–e: Imagineering, 29.5a: Imagineering, 29.6: Precision Graphics/Karl Miyajima, 29.7a–g: Martha Blake/Karl Miyajima, 29.8: Imagineering, 29.9: Imagineering, 29.10: Imagineering, 29.11: Precision Graphics/Karl Miyajima, 29.12a–c: Imagineering, 29.13a and b: Imagineering, 29.15a–d: Imagineering, 29.16: Imagineering, 29.17a–d: Martha Blake/Kristin Mount, 29.18: Imagineering, Table 29.2: Precision Graphics/Karl Miyajima.

Chapter 30 30.2: Precision Graphics/Karl Miyajima, 30.3: Precision Graphics/Karl Miyajima, 30.4: Precision Graphics/Karl Miyajima, 30.5: Imagineering, 30.6: Imagineering, 30.7: Precision Graphics/Karl Miyajima, 30.8: Precision Graphics/Karl Miyajima, 30.9a and b: Imagineering.

Appendices Appendix B: Karl Miyajima, Appendix C: Karl Miyajima, Appendix D: GTS Graphics/Karl Miyajima.

INDEX

Note: Page numbers in **boldface** indicate a definition. A *t* following a page number indicates tabular material, an *f* indicates an illustration, and a *b* indicates boxed material.

Dear Student:

My goal for revising **Human Anatomy & Physiology** was to not only help you succeed in the course, but also to make the learning process exciting and rewarding. I would appreciate hearing about your experiences with this textbook and its multimedia support, and I invite your suggestions for improvements!

Many thanks,

Elaine N. Marieb

Elaine N. Marieb

1. Did you use the **Study Partner CD-ROM**?   ☐ Yes ☐ No

If so, which features were most useful to you? Did you use it on your own, or did your instructor require you to use it? _____

_____

_____

_____

Did the **Study Partner CD-ROM** help you to study? Was it easy to use, or did you experience problems? How can we improve the **Study Partner CD-ROM**? _____

_____

_____

_____

2. Did you use **The A&P Place**?   ☐ Yes ☐ No

If so, which features were most useful to you? Did you use it on your own, or did your instructor require it? _____

_____

_____

_____

Did **The A&P Place** help you to study? Was it easy to use, or did you experience problems? How can we improve **The A&P Place**? _____

_____

_____

_____

3. Which study tool(s) - e.g. Chapter Summary, Review Questions, "Closer Connections" feature - did you find most useful? _____

_____

_____

_____

_____

4. What did you like most about *Human Anatomy & Physiology,* **Fifth Edition**? Please
provide three examples, in order of priority. _____

_____

_____

_____

_____

5. Do you have any ideas about how this book could be improved? If so, please write your specific
suggestions, citing page numbers if appropriate. _____

_____

_____

_____

_____

‖‖ ‖ ‖ ‖

BUSINESS REPLY MAIL

FIRST-CLASS MAIL   PERMIT NO. 275   SAN FRANCISCO CA

POSTAGE WILL BE PAID BY ADDRESSEE

BENJAMIN/CUMMINGS SCIENCE
PEARSON EDUCATION
1301 SANSOME STREET
SAN FRANCISCO CA 94111-9328

‖‖‖‖‖‖‖‖‖‖‖‖‖‖‖‖‖‖‖‖‖‖‖‖‖‖‖‖‖‖‖

School: _____

Instructor's Name: _____

Optional:

Your name: _____

Email: _____

Date: _____

May Benjamin Cummings have permission
to quote your comments in promotions for
Human Anatomy & Physiology? ☐ Yes ☐ No

oxy- *oxygen* oxygenation, the saturation of a substance with oxygen

pan- *all, universal* panacea, a cure-all

papill- *nipple* dermal papillae, projections of the dermis into the epidermal area

para- *beside, near* paraphrenitis, inflammation of tissues adjacent to the diaphragm

pect-, pectus *breast* pectoralis major, a large chest muscle

pelv- *a basin* pelvic girdle, which cradles the pelvic organs

peni- *a tail* penis; penile urethra

penna- *a wing* unipennate, bipennate muscles, whose fascicles have a feathered appearance

pent- *five* pentose, a 5-carbon sugar

pep-, peps-, pept- *digest* pepsin, a digestive enzyme of the stomach; peptic ulcer

per-, permea- *through* permeate; permeable

peri- *around* perianal, situated around the anus

phago- *eat* phagocyte, a cell that engulfs and digests particles or cells

pheno- *show, appear* phenotype, the physical appearance of an individual

phleb- *vein* phlebitis, inflammation of the veins

pia *tender* pia mater, delicate inner membrane around the brain and spinal cord

pili *hair* arrector pili muscles of the skin, which make the hairs stand erect

pin-, pino- *drink* pinocytosis, the process of a cell in small particles

platy- *flat, broad* platysma, broad, flat muscle of the neck

pleur- *side, rib* pleural serosa, the membrane that lines the thoracic cavity and covers the lungs

plex-, plexus *net, network* brachial plexus, the network of nerves that supplies the arm

pneumo *air, wind* pneumothorax, air in the thoracic cavity

pod- *foot* podiatry, the treatment of foot disorders

poly- *multiple* polymorphism, multiple forms

post- *after, behind* posterior, places behind (a specific) part

pre-, pro- *before, ahead of* prenatal, before birth

procto- *rectum, anus* proctoscope, an instrument for examining the rectum

pron- *bent forward* prone; pronate

propri- *one's own* proprioception, awareness of body parts and movement

pseudo- *false* pseudotumor, a false tumor

psycho- *mind, psyche* psychogram, a chart of personality traits

ptos- *fall* renal ptosis, a condition in which the kidneys drift below their normal position

pub- *of the pubis* puberty

pulmo- *lung* pulmonary artery, which brings blood to the lungs

pyo- *pus* pyocyst, a cyst that contains pus

pyro- *fire* pyrogen, a substance that induces fever

quad-, quadr- *four-sided* quadratus lumborum, a muscle with a square shape

re- *back, again* reinfect

rect- *straight* rectus abdominis, rectum

ren- *kidney* renal, renin, an enzyme secreted by the kidney

retin, retic- *net, network* endoplasmic reticulum, a network of membranous sacs within a cell

retro- *backward, behind* retrogression, to move backward in development

rheum- *watery flow, change, or flux* rheumatoid arthritis, rheumatic fever

rhin-, rhino- *nose* rhinitis, inflammation of the nose

ruga- *fold, wrinkle* rugae, the folds of the stomach, gallbladder, and urinary bladder

sagitt- *arrow* sagittal (directional term)

salta- *leap* saltatory conduction, the rapid conduction of impulses along myelinated neurons

sanguin- *blood* consanguineous, indicative of a genetic relationship between individuals

sarco- *flesh* sarcomere, unit of contraction in skeletal muscle

saphen- *visible, clear* great saphenous vein, superficial vein of the thigh and leg

sclero- *hard* sclerodermatitis, inflammatory thickening and hardening of the skin

seb- *grease* sebum, the oil of the skin

semen *seed, sperm* semen, the discharge of the male reproductive system

semi- *half* semicircular, having the form of half a circle

sens- *feeling* sensation; sensory

septi- *rotten* sepsis, infection; antiseptic

septum *fence* nasal septum

sero- *serum* serological tests, which assess blood conditions

serrat- *saw* serratus anterior, a muscle of the chest wall that has a jagged edge

sin-, sino- *a hollow* sinuses of the skull

soma- *body* somatic nervous system

somnus *sleep* insomnia, inability to sleep

sphin- *squeeze* sphincter

splanchn- *organ* splanchnic nerve, autonomic supply to abdominal viscera

spondyl- *vertebra* ankylosing spondylitis, rheumatoid arthritis affecting the spine

squam- *scale, flat* squamous epithelium, squamous suture of the skull

steno- *narrow* stenocoriasis, narrowing of the pupil

strat- *layer* strata of the epidermis, stratified epithelium

stria- *furrow, streak* striations of skeletal and cardiac muscle tissue

stroma *spread out* strome, the connective tissue framework of some organs

sub- *beneath, under* sublingual, beneath the tongue

sucr- *sweet* sucrose, table sugar

sudor- *sweat* sudoriferous glands, the sweat glands

super- *above, upon* superior, quality or state of being above others or a part

supra- *above, upon* supracondylar, above a condyle

sym-, syn- *together, with* synapse, the region of communication between two neurons

synerg- *work together* synergism

systol- *contraction* systole, contraction of the heart

tachy- *rapid* tachycardia, abnormally rapid heartbeat

tact- *touch* tactile sense

telo- *the end* telophase, the end of mitosis

templ-, tempo- *time* temporal summation of nerve impulses

tens- *stretched* muscle tension

tertius *third* peroneus tertius, one of three peroneus muscles

tetan- *rigid, tense* tetanus of muscles

therm- *heat* thermometer, an instrument used to measure heat

thromb- *clot* thrombocyte; thrombus

thyro- *a shield* thyroid gland

tissu- *woven* tissue

tono- *tension* tonicity; hypertonic

tox- *poison* antitoxic, effective against poison

trab- *beam, timber* trabeculae, spicules of bone in spongy bone tissue

trans- *across, through* transpleural, through the pleura

trapez- *table* trapezius, the four-sided muscle of the upper back

tri- *three* trifurcation, division into three branches

trop- *turn, change* tropic hormones, whose targets are endocrine glands

troph- *nourish* trophoblast, from which develops the fetal portion of the placenta

tuber- *swelling* tuberosity, a bump on a bone

tunic- *covering* tunica albuginea, the covering of the testis

tympan- *drum* tympanic membrane, the eardrum

ultra- *beyond* ultraviolet radiation, beyond the band of visible light

vacc- *cow* vaccine

vagin- *a sheath* vagina

vagus *wanderer* the vagus nerve, which starts at the brain and travels into the abdominopelvic cavity

valen- *strength* valence shells of atoms
venter, ventr- *hollow cavity, belly* ventral (directional term); ventricle
ventus *the wind* pulmonary ventilation
vert- *turn* vertebral column
vestibul- *a porch* vestibule, the anterior entryway to the mouth and nose
vibr- *shake, quiver* vibrissae, hairs of the nasal vestibule
villus *shaggy hair* microvilli, which have the appearance of hair in light microscopy
viscero- *organ, viscera* visceroinhibitory, inhibiting the movements of the viscera
viscos- *sticky* viscosity, resistance to flow
vita- *life* vitamin
vitre- *glass* vitreous humor, the clear jelly of the eye
viv- *live* in vivo
vulv- *a covering* vulva, the female external genitalia
zyg- *a yoke, twin* zygote

Suffixes

-able *able to, capable of* viable, ability to live or exist
-ac *referring to* cardiac, referring to the heart
-algia *pain in a certain part* neuralgia, pain along the course of a nerve
-apsi *juncture* synapse, where two neurons connect
-ary *associated with, relating to* coronary, associated with the heart
-asthen *weakness* myasthenia gravis, a disease involving paralysis
-atomos *indivisible* anatomy, which involves dissection
-bryo *swollen* embryo
-cide *destroy or kill* germicide, an agent that kills germs
-cipit *head* occipital
-clast *break* osteoclast, a cell that dissolves bone matrix
-crine *separate* endocrine organs, which secrete hormones into the blood
-dips *thirst, dry* polydipsia, excessive thirst associated with diabetes
-ectomy *cutting out, surgical removal* appendectomy, cutting out of the appendix
-ell, -elle *small* organelle
-emia *condition of the blood* anemia, deficiency of red blood cells
-esthesi *sensation* anesthesia, lack of sensation
-ferent *carry* efferent nerves, nerves carrying impulses away from the CNS
-form, -forma *shape* cribriform plate of the ethmoid bone
-fuge *driving out* vermifuge, a substance that expels worms of the intestine
-gen *an agent that initiates* pathogen, any agent that produces disease
-glea, -glia *glue* neuroglia, the connective tissue of the nervous system
-gram *data that are systematically recorded, a record* electrocardiogram, a recording showing action of the heart
-graph *an instrument used for recording data or writing* electrocardiograph, an instrument used to make an electrocardiogram
-ia *condition* insomnia, condition of not being able to sleep
-iatrics *medical specialty* geriatrics, the branch of medicine dealing with disease associated with old age
-ism *condition* hyperthyroidism
-itis *inflammation* gastritis, inflammation of the stomach
-lemma *sheath, husk* sarcolemma, the plasma membrane of a muscle cell
-logy *the study of* pathology, the study of changes in structure and function brought on by disease

-lysis *loosening or breaking down* hydrolysis, chemical decomposition of a compound into other compounds as a result of taking up water
-malacia *soft* osteomalacia, a process leading to bone softening
-mania *obsession, compulsion* erotomania, exaggeration of the sexual passions
-nata *birth* prenatal development
-nom *govern* autonomic nervous system
-odyn *pain* coccygodynia, pain in the region of the coccyx
-oid *like, resembling* cuboid, shaped as a cube
-oma *tumor* lymphoma, a tumor of the lymphatic tissues
-opia *defect of the eye* myopia, nearsightedness
-ory *referring to, of* auditory, referring to hearing
-pathy *disease* osteopathy, any disease of the bone
-phasia *speech* aphasia, lack of ability to speak
-phil, -philo *like, love* hydrophilic, water-attracting molecules
-phobia *fear* acrophobia, fear of heights
-phragm *partition* diaphragm, which separates the thoracic and abdominal cavities
-phylax *guard, preserve* anaphylaxis, prophylactic
-plas *grow* neoplasia, an abnormal growth
-plasm *form, shape* cytoplasm
-plasty *reconstruction of a part, plastic surgery* rhinoplasty, reconstruction of the nose through surgery
-plegia *paralysis* paraplegia, paralysis of the lower half of the body or limbs
-rrhagia *abnormal or excessive discharge* metrorrhagia, uterine hemorrhage
-rrhea *flow or discharge* diarrhea, abnormal emptying of the bowels
-scope *instrument used for examination* stethoscope, instrument used to listen to sounds of parts of the body
-some *body* chromosome
-sorb *suck in* absorb
-stalsis *compression* peristalsis, muscular contractions that propel food along the digestive tract
-stasis *arrest, fixation* hemostasis, arrest of bleeding
-stitia *come to stand* interstitial fluid, between the cells
-stomy *establishment of an artificial opening* enterostomy, the formation of an artificial opening into the intestine through the abdominal wall
-tegm *cover* integument
-tomy *to cut* appendectomy, surgical removal of the appendix
-trud *thrust* protrude, detrusor muscle
-ty *condition of, state* immunity, condition of being resistant to infection or disease
-uria *urine* polyuria, passage of an excessive amount of urine
-zyme *ferment* enzyme